HANDBOOK OF APPLIED OPTIMIZATION

Handbook of Applied Optimization

EDITED BY

Panos M. Pardalos and Mauricio G. C. Resende

OXFORD
UNIVERSITY PRESS

2002

OXFORD
UNIVERSITY PRESS

Oxford New York
Athens Auckland Bangkok Bogotá Buenos Aires Calcutta
Cape Town Chennai Dar es Salaam Delhi Florence Hong Kong Istanbul
Karachi Kuala Lumpur Madrid Melbourne Mexico City Mumbai
Nairobi Paris São Paulo Shanghai Singapore Taipei Tokyo Toronto Warsaw
and associated companies in
Berlin Ibadan

Published by Oxford University Press, Inc.
198 Madison Avenue, New York, New York 10016
www.oup.com

Library of Congress Cataloging-in-Publication Data

Handbook of applied optimization / edited by Panos Pardalos and Mauricio Resende.
p. cm.
Includes bibliographical references and index.
ISBN-13 978-0-19-512594-8
ISBN 0-19-512594-0 (hardcover : alk. paper)
1. Mathematical optimization. 2. Programming (Mathematics)—Industrial applications.
I. Pardalos, P. M. (Panos M.), 1954- II. Resende, Mauricio G. C., 1955-

QA402.5 .H334 2001

519.3–dc21 00-068146

EDITORIAL AND PRODUCTION STAFF
Acquiring Editor: Sean Pidgeon
Development Editor: Sean Pidgeon
Assistant Project Editor: Merilee Johnson
Manufacturing Controller: Genieve Shaw
Book Designer: Joan Greenfield
Managing Editor: Matthew Giarratano
Publisher: Karen Day

To our parents,
Kalypso and Miltiades,
Renalva and Roberto (in memorium)

CONTENTS

PART TWO: APPLICATIONS

PREFACE

Optimization is a basic tool in applied research, engineering, business, and the sciences. In the last decade, algorithmic advances as well as hardware and software improvements have provided an excellent framework on which to build optimization-based decision support systems, as well as multidiscipline design tools. Despite the dynamic state of applied optimization, we feel that the time is ripe to bring together in one volume the major algorithmic and software advances in this field. Leading experts in applied optimization have contributed to this volume.

The *Handbook of Applied Optimization* is aimed at engineers, scientists, operations researchers, and other applications specialists who are looking for the most appropriate and recent optimization tools to solve particular problems. The handbook provides a broad spectrum of advances in applied optimization with a focus on the algorithmic and computational aspects of this field.

The handbook consists of three main parts: algorithms, applications, and software. In the algorithms section of the handbook, the important algorithms in the major fields of optimization are described. The purpose of the section on applications is to provide the practitioner with a description of the relevant optimization issues in a number of specific application areas. We include overview articles discussing broad problem types, such as scheduling, vehicle routing, network design, bin packing, inventory management, travelling salesman, satisfiability, location, and assignment problems, as well as some examples of applied optimization in specific areas. These applications cover a broad spectrum of fields, such as transportation, agriculture, manufacturing, aerospace, telecommunications, energy, biology, finance, and the environment. In the software section of the handbook, the tools of applied optimization are described, with an emphasis on practical details: how to implement and test optimization algorithms, use existing optimization packages and the Internet, and use modelling languages to build optimization systems.

We would like to take the opportunity to thank Oxford University Press for inviting us to edit this handbook, the authors for their contributions, and the anonymous referees for their valuable comments and suggestions.

—Panos M. Pardalos and Mauricio G. C. Resende

INTRODUCTION

The field of optimization is concerned with the characterization of solutions and development of algorithmic techniques for solving mathematical problems of the form

$$\max(\min) f(x)$$

such that

$$x \in X \subseteq \mathbb{R}^n$$

where x denotes the decision variables in the mathematical model, $f(x)$ is the objective function that measures the quality of the solution, and X is the set of feasible solutions.

Many problems in decisionmaking, engineering, and the sciences concern the efficient allocation of scarce resources, and can be modelled and formulated as optimization problems. Optimization technology has become a key component of modern-day decisionmaking in activities ranging from aerospace, biotechnology, finance, agriculture, energy, telecommunication, and the Internet. A recent survey of the five hundred largest companies in the United States showed that nearly 85 percent of them used linear programming (Fang and Puthenpura, 1993).

Optimization, as a scientific discipline, can be traced to the birth of the digital computer. During World War II, optimization was applied to model and solve logistics and operational military problems. George B. Dantzig was involved in these activities and during that time he developed linear programming models and the simplex method for solving linear programs. At the same time, Leonid V. Kantorovich (1975 Nobel Laureate in Economics) developed independently the method of resolving multipliers for linear programming and identified its application to problems in equipment work distribution, the best use of sowing area, rational material cutting, use of complex resources, and distribution of transport flows. Since the 1950s we have witnessed an accelerated growth of optimization techniques. For example, in linear programming, Leonid Khachian in 1979 published the ellipsoid algorithm, the first polynomial-time algorithm for linear programming, and in 1984 Narendra Karmarkar published a polynomial-time interior-point algorithm for linear programming. Interior-point methods pushed the limits of the dimension of problems that could be solved. Variants of interior-point methods were also shown to perform well in practice and are today common in modern optimization packages.

Linear programming deals with optimization problems in which $f(x)$ is a linear function of n variables, and X can be represented by a set of linear inequalities. Linear programming is an example of a problem belonging to a much broader class of problems known as *combinatorial optimization*, where the feasible domain is finite but often of exponential size. A large number of practical problems can be modelled as combinatorial optimization problems. *Integer programming* imposes additional integrality constraints on some subsets of the decision variables of a linear program. Almost all combinatorial optimization problems can be modelled as integer programs. Advances in computational algorithms, such as cutting-plane methods, branch and bound, branch and cut, column generation, decomposition techniques, and polyhedral combinatorics have led to powerful commercial integer programming systems (e.g., CPLEX [ILOG 2001] and Xpress [Dash Optimization 2001]). Today, we can routinely solve to optimality large integer programs that occur in practice. However, because of the hardness of these problems, many combinatorial optimization instances still cannot be solved exactly. Developments in methods for finding approximate (i.e., suboptimal) solutions to these instances have taken off in the last decade. *Metaheuristics*, such as tabu search (Glover and Laguna, 1997), simulated annealing (Kirkpatrick et al., 1983), GRASP (Feo and Resende, 1995), and genetic algorithms (Holland, 1975) enable practitioners to find good-quality solutions to difficult problems in reasonable computational time. Combinatorial optimization techniques are today applied to solve problems in transportation and logistics, forestry and agriculture, manufacturing and production, telecommunications, energy, architecture, chemistry, molecular biology, and so on. This handbook illustrates the wide applicability of combinatorial optimization.

Another important area in combinatorial optimization is *network optimization*. After the publication of the first graph theory book by D. König (König, 1936) and Hitchcock's first algorithm for the transportation problem in 1931, research in this field took off with important algorithms for minimum spanning tree (Kruskal, 1956; Prim, 1957) and shortest-path problems (Dijkstra, 1959). Many of these early network algorithms are still widely used today. In 1962, the book *Flows in Networks* by Ford and Fulkerson appeared (Ford and Fulkerson, 1962). With the advent of computers and the development of data structures (Tarjan, 1983), network algorithms are used today to solve large real-world optimization problems in fields as diverse as telecommunications, the Internet, transportation, and logistics.

In *nonlinear optimization*, the objective $f(x)$ and constraints that define the feasible domain X can be, in general, nonlinear functions. In the late 1950s, we witnessed the first attempts to solve nonlinear optimization problems. In the early 1960s, J. B. Rosen published the gradient projection method for nonlinear programming (Rosen, 1960; Rosen, 1961). This work triggered much research in the 1960s for developing computational algorithms for nonlinear optimization. The interplay between continuous nonlinear optimization and combinatorial optimization has motivated the development of new algorithmic techniques for large-scale problems. As an example, the field of interior-point methods uses nonlinear programming approaches to solve linear programming, a combinatorial optimization problem. Seeds to these ideas can be found in the book by Fiacco and McCormick (1968) and the work of Dikin (1967). Similarly, the ellipsoid method for linear programming (Khachian, 1979) comes from nonlinear programming (Shor, 1970). In general, a combinatorial constraint of the form $x \in \{0, 1\}$ is equivalent to the nonconvex constraint $x^2 - x = 0$. This simple observation indicates that the presence of nonconvexity can reflect the difficulty of an optimization problem (Horst et al., 2000).

Since the 1970s, progress on algorithmic aspects of nonlinear programming as well as implementation techniques has led to the development of powerful software systems (e.g., MINOS [Murtagh and Saunders, 1983] and Lancelot [Conn et al., 1992]). Nonlinear optimization is used in many fields of engineering, such as aerospace, chemical, nuclear, mechanical, electrical, and computer engineering. This handbook illustrates a wide range of applications of nonlinear optimization. In addition, it is a key component of decision support systems in finance, logistics, supply-chain, and electronic commerce.

In many real situations, such as water resource management, energy, inventory control, and finance, some or all of the data defining an optimization problem are stochastic in nature. Such problems are solved via stochastic programming. *Stochastic programming* (Prékopa, 1995) lies on the borderline of probability theory and optimization. Stochastic programming techniques provide us with tools to design and control stochastic systems with the help of optimization algorithms.

In many practical problems, since convexity of the objective (or the feasible domain) cannot be easily verified, it is reasonable to assume that the problems have multiple local optimal solutions. *Global optimization* deals with these multiextremal problems. Global optimization can be continuous or discrete. Only in the last couple of decades have techniques to solve practical global optimization problems been developed (Horst and Pardalos, 1995). Recent developments in this field include algorithms for DC (difference of convex functions) programs, monotonic optimization, metaheuristics, multilevel optimization, multiobjective optimization, and stochastic methods. Global optimization applications arise in many situations, including allocation and location problems, structural engineering, chemical engineering design and control, bioengineering, and finance.

Many large applied optimization problems can be solved by using *decomposition techniques* (Lasdon, 1970). These techniques break up a large problem into subproblems of lower dimension that can be solved. From the optimal solutions of the subproblems, an optimal solution of the large problem can be computed. Decomposition techniques apply to problems having special structural characteristics. For example, in linear programming, when the constraint matrix is block diagonal or block angular, the Dantzig–Wolfe decomposition method (Dantzig and Wolfe, 1960) applies. Benders decomposition (Benders, 1962) is another well-known technique that is used to solve large mixed-integer programming problems. Both of these methods have been extended to solve more general structured optimization problems. Decomposition can easily take advantage of parallel and distributed computing environments.

Parallelism in hardware, software, and algorithms enables today's practitioner to solve large-scale optimization problems more efficiently (Migdalas et al., 1997). In some situations, parallel processing can reduce the run time of an algorithm to solve a problem from some days or months to a few minutes or even seconds. For example, CPLEX's interior-point algorithm for linear programming is available for parallel computers. In this version, the solution of linear systems at each iteration of the algorithm is done in parallel, speeding up the solution for many problems. Another example is the use of distributed computing environments to solve hard problems, such as a 13,509-city traveling salesman problem (Applegate et al., 1998) and quadratic assignment problems of dimension $n = 30$ (Anstreicher et al., 2001).

Practitioners of applied optimization can utilize advanced *optimization modelling languages* that facilitate the decisionmaking process based on the optimization paradigm. These modelling languages support many steps in the modelling process, such as problem definition, data collection, mathematical programming formulation, translation into a form that can be handled by a optimization solver, solution, and analysis of the results. There is a great variety of modelling languages. Two well-known and widely used languages are GAMS (Brooke et al., 1992) and AMPL (Fourer et al., 1993).

In the last few years, the Internet has created enormous opportunities for users of applied optimization. Such opportunities include increased information transfer, alternative channels for software and technical document distribution, and online optimization.

Overall, the field of applied optimization has reached maturity and will continue to have a great impact in all aspects of decisionmaking. Applied optimization has grown, together with computer and communication technology, and the continued growth of these technologies will contribute to a more widespread use of applied optimizations.

References

Anstreicher, K., N. Brixius, J.-P. Goux, and J. Linderoth. 2001. "Solving large quadratic assignment problems on computational grids." *Mathematical Programming*. Forthcoming.

Applegate, D., R. Bixby, V. Chvátal, and W. Cook. 1998. "On the solution of traveling salesman problems." *Doc. Math. J. DMV, Extra ICM* **III**: 645–656.

Benders, J. F. 1962. "Partitioning procedures for solving mixed variables programming problems." *Numerische Mathematik* **4**: 238–252.

Brooke A., D. Kendrick, and A. Meeraus. 1992. *GAMS: A User's Guide*. Redwood City, CA: The Scientific Press.

Conn A. R., N. I. M. Gould, and Ph. L. Toint. 1992. *LANCELOT: A Fortran Package for Large-Scale Nonlinear Optimization (Release A)*. Springer-Verlag.

Dantzig, D. B., and P. Wolfe. 1960. "The decomposition principle for linear programs." *Operations Research* **8**: 101–111.

Dash Optimization. Xpress, 2001. http://www.dashoptimization.com

Dijkstra, E. W. 1959. "A note on two problems in connection with graphs." *Numerische Mathematik* **1**: 269–271.

Dikin, I. I. 1967. "Iterative solution of problems of linear and quadratic programming." *Soviet Mathematics Doklady* **8**: 674–675.

Fang, S.-C., and S. Puthenpura. 1993. *Linear Optimization and Extensions: Theory and Algorithms*. Prentice-Hall.

Feo, T. A., and M. G. C. Resende. 1995. "Greedy randomised adaptive search procedures." *Journal of Global Optimization* **6**: 109–133.

Fiacco, A. V., and G. P. McCormick. 1968. *Nonlinear Programming: Sequential Unconstrained Minimization Techniques*. John Wiley and Sons. Reprinted as volume 4 of the SIAM Classics in Applied Mathematics Series, 1990.

Ford, L. R., and D. R. Fulkerson. 1962. *Flows in Networks*. Princeton University Press.

Fourer, R., D. Gay, and B. W. Kernighan. 1993. *AMPL: A Modeling Language for Mathematical Programming AMPL*. San Francisco: The Scientific Press.

Glover, F., and M. Laguna. 1997. *Tabu Search*. Kluwer Academic Publishers.

Holland, J. H. 1975. *Adaptation in Natural and Artificial Systems*. University of Michigan Press.

Horst, R., and P. M. Pardalos. 1995. *Handbook of Global Optimization*. Kluwer Academic Publishers.

Horst R., P. M. Pardalos, and N. V. Thoai. 2000. *Introduction to Global Optimization*. 2nd ed. Kluwer Academic Publishers.

ILOG. ILOG CPLEX, 2001. http://www.ilog.com/products/cplex.

Khachian, L. G. 1979. "A polynomial algorithm in linear programming." *Soviet Mathematics Doklady* **20**: 1093–1096.

Kirkpatrick, S., C. D. Gellat, Jr., and M. P. Vecchi. 1983. "Optimization by simulated annealing." *Science* **220**: 671–680.

König, D. 1936. *Theorie der endlichen und unendlichen Graphen*. Reprinted Chelsea, New York, 1950.

Kruskal, J. B. 1956. "On the shortest spanning subtree of a graph and the travelling salesman problem." *Proceedings of the American Mathematical Society* **7**: 48–50.

Lasdon, L. S. 1970. *Optimization Theory for Large Systems*. MacMillan.

Migdalas, A., P. M. Pardalos, and S. Storøy, eds. 1997. *Parallel Computing in Optimization*. Kluwer Academic Publishers.

Murtagh, B. A., and M. A. Saunders. 1983. "MINOS 5.0 user's guide." Technical Report 83-20R, Systems Optimization Laboratory, Stanford University, Stanford, CA.

Prékopa, A. 1995. *Stochastic Programming*. Kluwer Academic Publishers.

Prim, R. C. 1957. "Shortest connection networks and some generalizations." *Bell System Technological Journal* **36**: 1389–1401.

Rosen, J. B. 1960. "The gradient projection method for nonlinear programming. Part I. Linear constraints." *J. Soc. Ind. Appl. Math.* **8**: 181–217.

Rosen, J. B. 1961. "The gradient projection method for nonlinear programming. Part II. Non-linear constraints." *J. Soc. Ind. Appl. Math.* **9**: 514–553.

Shor, N. Z. 1970. "Utilization of the operation of space dilatation in the minimization of convex functions." *Cybernetics* **6**: 7–15.

Tarjan, R. E. 1983. *Data Structures and Network Algorithms*. Philadelphia: SIAM.

—Panos M. Pardalos and Mauricio G. C. Resende

PART ONE

ALGORITHMS

1 LINEAR PROGRAMMING

1.1 Introduction

Tamás Terlaky

ABSTRACT Linear programming (LP) is the fundamental problem in mathematical optimization. It has an enormous number of applications in economics, engineering, basic sciences, and many other fields. A large number of optimization algorithms make use of LP-based algorithms, thus the need to develop efficient algorithms for solving LP problems is evident. Various algorithms have been developed in the past fifty years. There are two computationally highly efficient and successful cases of algorithms. The first class contains simplex-type algorithms, originally developed by Dantzig. The second is the class of interior-point methods (IPMs). After a brief introduction both classes will be discussed.

1. The LP Problem

In plain words we can say that linear programming (LP) consists of optimizing, that is, minimizing or maximizing, a linear function—for example, $c^T x$ where $c \in \mathbb{R}^n$ is given and $x \in \mathbb{R}^n$ is the vector of unknowns—on a certain domain. The domain is given by a set of linear constraints. The constraints can be either equalities or inequalities, for example, we have

$$a_i^T x - b_i \gtreqqless 0, \quad i = 1, \ldots, n,$$

where $a_i \in \mathbb{R}^n$ and $b_i \in \mathbb{R}^n$ are given. At first sight, LP problems have a quite simple structure. Only linear functions are involved, but not only a set of linear equations has to be solved; therefore our task is made more difficult in two ways. Inequalities are also involved, and an optimization component—"find a solution that has the best possible value of the function $c^T x$"—is present. Before discussing mathematical properties of LP problems and presenting algorithmic issues, let us first devote some paragraphs to the history of LP.

Although its roots go back to the previous centuries, LP—as an important discipline with ever growing numbers of applications—was born in the middle of the twentieth century. To keep the story short, just some of the major milestones are briefly mentioned here. The two major components, optimizing a function and solving a linear system, can be traced back for centuries.

Optimization is a natural activity. Optimization elements can be found in ancient Greek mathematics too. For our discussions, Lagrange's (Prékopa, 1980) work is a major milestone. He considered the minimization of certain functions while a set of (possibly nonlinear) equality constraints are satisfied. Although Lagrange did not consider inequalities, he did lay the foundations of duality theory.

On the side of solving linear systems, an epoch-making step was made by Gauss by establishing his celebrated Gauss elimination algorithm (Strang, 1988), which is still one of the main techniques used to solve systems of linear equations. As already mentioned, besides equations, LP involves inequalities as well. An intellectually appealing, but computationally inefficient algorithm to solve systems of linear inequalities was designed by Fourier (Prékopa, 1980) and was later rediscovered by Motzkin (Prékopa, 1980). Systems of linear inequalities were thoroughly studied by Farkas (1902). He developed the theory of alternative systems: one, and only one, of two related inequality systems has a solution (see Corollary 3 below for a rigorous mathematical formulation). Farkas' fundamental result is still one of the most frequently cited theorems in LP literature. It is essentially equivalent to the Strong Duality Theorem (Theorem 2 below) of LP.

The significance of Farkas' results was not realized until the middle of the twentieth century when LP was born as a new, turbulent research field. The founding father of LP was George Dantzig (1963). Although some specially structured LP problems were formulated and systematically studied earlier by Kantorovich (Prékopa, 1980), Dantzig was the first to develop a concise theory: geometrical analysis; duality theory; a library of typical practical models; and a computationally efficient algorithm, the simplex method. Illuminating true stories about the birth of LP and the simplex method can be found in Lenstra et al. (1991), including Dantzig's recollection of those historic years (Dantzig, 1991). Because solving large-scale LP problems is computationally a highly intense activity, from the 1950s the algorithmic developments in the area of LP are strongly related to advances in computer technology. As the speed and capacity of computers increases, new avenues open to solve larger and larger LP problems. In the 1950s, problems with hundreds of variables were considered large-scale problems. Today everyone can solve problems of tens of thousands of variables on their own desktop PCs in minutes and LP problems with millions of variables have already been solved. The reader is referred to Bixby (1994) and Andersen et al. (1996) for surveys of computational state of the art.

The simplex algorithm and some of its variants are presented in Chapter 1.2. The simplex method makes use of the algebraic, geometric, and combinatorial properties of LP problems. It is in fact a family of algorithms, allowing a lot of flexibility for exploration in actual implementations. Although, even today, implementations of the simplex method are among the most efficient, it has become clear that some variants of the simplex method require a prohibitively long time to solve particular LP problems. This was made explicit in the early seventies by Klee and Minty (1972), who showed that a variant of the simplex method requires exponentially many steps to solve an LP problem where the set of feasible solutions is a slightly perturbed hypercube. On the other hand, under a probabilistic model, it has been proved that the average case behavior of some simplex methods is polynomial (Borgwardt, 1987).

The Klee–Minty example stimulated research for alternative algorithms for LP. It was an open problem for a long time whether LP problems were polynomially solvable. The solution was presented by Khacijan (1979). Khacijan's ellipsoid method was able to solve LP problems in polynomial time. This major result reached the front page of the *New York Times*, but it soon turned out that in practice the ellipsoid method is not computationally competitive with the simplex method. In practice the simplex method performs much better than its theoretical worst-case bound suggests, while the practical behavior of the ellipsoid method is close to its worst-case bound. Therefore, it has been decided not to discuss the ellipsoid method in more detail here. Nevertheless, the ellipsoid method made it possible to prove polynomial solvability of large classes of optimization problems (Grötschel et al., 1988) and opened the way to search further for more practical polynomial algorithms.

The next breakthrough was reached by Karmarkar in 1984 (Karmarkar, 1984). He presented an interior-point method (IPM) whose worst-case complexity bound was better than that for the ellipsoid method. Karmarkar also claimed superior computational performance. Karmarkar's results and claims started the interior-point revolution (Wright, 1998; Roos et al., 1997; Wright, 1996; Ye, 1997). Nowadays, the superior theoretical properties of IPMs are obvious; highly efficient implementations of IPMs are available; the implementations are at least competitive with the implementations of the simplex method; and they typically have superior performance on large-scale problems (Andersen et al., 1996).

2. A Practical Model

Let us first try to understand via a practical example what we are talking about, what an LP problem actually is. The mixing problem naturally leads to an LP problem. Mixing liquids, or blending different materials, in order to obtain a certain material with desired properties is a typical problem that arises in various areas of engineering and economics. Let us consider a variant of the problem from a civil-engineering perspective.

For a certain construction, high-quality concrete is needed. Engineering theory tells us the ideal granule composition of sand and gravel. Let us assume for simplicity that we distinguish $1, \ldots, 6$ different granule size classes. The ideal sand and gravel mixture contains $g_1, \ldots, g_6$ percent granule from those classes, respectively. There are three sand and gravel mines within reasonable distance, where the granule composition of the sand and gravel is $a_{1i}, \ldots, a_{6i}$,

$i = 1, 2, 3$, respectively. The civil engineer has to make a mixture of the material obtained from the three mines to approximate the ideal sand and gravel mixture as closely as possible. Thus, in a unit amount mixture, he takes x_1, x_2, and x_3 percent of the sand and gravel from the three respective mines. Mathematically, the granule composition $r_1, ..., r_6$ of the resulting mixture can be calculated as follows: Because x_1, x_2, and x_3 are percentages in a mixture, we have

$$x_1 + x_2 + x_3 = 1, \quad x_1, x_2, x_3 \geqslant 0.$$

Then the mixture has

$$r_j = a_{j1}x_1 + a_{j2}x_2 + a_{j3}x_3$$

percent of granules from size class j. The civil engineer faces the problem of how to find the optimal mixing percentages. He realizes that the number of granule classes is larger than the number of mines, hence his limited freedom does not allow him in general to obtain the ideal mixture. Thus his objective is to minimize the deviation $D(r\|g)$ between the obtained, $r = (r_1, ..., r_6)^T$, and the ideal, $g = (g_1, ..., g_6)^T$, mixture. (Here the superscript ^{T} denotes the transpose of the corresponding vector.) The deviation can, for example, be measured as the largest absolute value deviation at different granule classes. This largest deviation should be as small as possible. Thus we want to

$$\text{minimize } D(r\|g) = \max_{1 \leqslant j \leqslant 6} | r_j - g_j | .$$

The minimization of the largest absolute value deviation (the ℓ_∞-norm) can be modeled by introducing a new variable t and using only linear inequalities. The resulting optimization problem reads as

$$\begin{aligned} &\text{minimize} && t \\ &\text{subject to} && -t \leqslant a_{j1}x_1 + a_{j2}x_2 + a_{j3}x_3 - g_j \leqslant t \quad j = 1, ..., 6, \\ &&& x_1 + x_2 + x_3 = 1, \\ &&& x_1, x_2, x_3 \geqslant 0. \end{aligned}$$

Observe that in this problem we want to find a specific solution to a system of linear equations/inequalities. The set of feasible solutions is determined by a number of linear inequalities and an equation, called constraints. The specific solution we seek is a vector (x_1, x_2, x_3, t) that respects all the prescribed constraints while allowing the smallest possible value of t among all possibilities. Such a solution is called an optimal solution of this optimization problem, which is a specific LP problem.

This mixing problem illustrates how LP problems arise naturally in practice. The books by Williams (1990) and Murty (1995) contain several practical models leading to LP problems. After this real-life example, let us define the general LP problem in standard form.

3. The LP Problem and Its Dual, Duality Theorems

The LP problem in standard form can be defined as

$$\begin{aligned} &\text{minimize} && c^T x \\ &\text{subject to} && Ax = b, \\ &&& x \geqslant 0, \end{aligned} \qquad \text{(LP)}$$

where c, $x \in \mathbb{R}^n$, $b \in \mathbb{R}^n$ are vectors, and $A \in \mathbb{R}^{m \times n}$ is a matrix. The following terminology is commonly used. The coordinates x_i of the vector x are unknown and they are considered as variables. The linear function $c^T x$ is called the objective function, A is called the coefficient matrix, vector b is the right-hand-side vector. Furthermore, we have equality constraints $Ax = b$ and inequality constraints $x \geqslant 0$. The latter are referred to as nonnegativity constraints.

A vector x is called a solution if it satisfies $Ax = b$; it is called a feasible solution, or in short, feasible if the solution x is nonnegative and it is called infeasible if it contains at least one negative coordinate. Finally, a vector $\bar{x}$ is an optimal solution of the LP problem if it is feasible with the smallest possible objective values $c^T\bar{x}$, that is, for all feasible solutions the relation

$$c^T\bar{x} \leqslant c^T x$$

holds.

Each LP problem, usually referred to as the primal problem, is associated with another LP problem, called the dual problem. The dual problem can be derived, for example, as follows. Let us try to design a function whose maximum is a guaranteed lower bound for the optimal value of the primal problem (LP).

Let $y \in \mathbb{R}^m$. Then the value

$$y^T(Ax - b)$$

is identically zero for any primal solution x, thus the function

$$c^T x + y^T(b - Ax) = b^T y + x^T(c - A^T y)$$

clearly has the same value as $c^T x$. Now, because $x \geqslant 0$, if we require that $c - A^T y \geqslant 0$, then $x^T(c - A^T y) \geqslant 0$. Hence for any primal feasible x we have

$$c^T x = b^T y + x^T(c - A^T y) \geqslant b^T y \quad \text{whenever} \quad A^T y \leqslant c,$$

that is, a lower bound for the optimal value of the primal problem is derived. For obvious reasons we are looking for the best possible lower bound. To find the best (largest) possible lower bound results in the optimization problem:

$$\begin{aligned} &\text{maximize} \quad b^T y \\ &\text{subject to} \quad A^T y \leqslant c. \end{aligned}$$

This problem is the dual LP problem. Analogous terminology is used to the case of the primal problem so, for example, a vector $y \in \mathbb{R}^m$ is a dual feasible solution if it satisfies all the inequality constraints of the dual LP, a vector $\bar{y}$ is an optimal solution of the dual problem if for all dual feasible y the inequality $b^T\bar{y} \geqslant b^T y$ holds.

Observe that due to the way the dual problem was derived, we have actually given a proof for the Weak Duality Theorem of LP.

Theorem 1 (Weak Duality Theorem). *Let the vectors* x *and* y *be primal and dual feasible solutions. Then*

$$c^T x \geqslant b^T y$$

with equality if and only if $x^T(c - A^T y) = 0$ *or, equivalently, if*

$$x_j(c - A^T y)_j = 0 \quad \forall j = 1, \ldots, n.$$

The last conditions are called complementarity conditions. One can easily verify the following result.

Corollary 1. *If the complementarity conditions hold for a pair of feasible primal and dual solutions* $\bar{x}$ *and* $\bar{y}$ *then one has* $c^T\bar{x} = b^T\bar{y}$ *and then* $\bar{x}$ *is a primal, and* $\bar{y}$ *is a dual optimal solution.*

Observe, that this corollary contains only an implication: "if the objective values are equal then the feasible solutions are optimal." The reverse statement is a deep, fundamental result, the so-called Strong Duality Theorem of LP.

Theorem 2 (Strong Duality Theorem). *If both the primal and dual problems admit a feasible solution, then for any optimal solutions* $\bar{x}$ *and* $\bar{y}$ *we have* $c^T\bar{x} = b^T\bar{y}$, *that is, for optimal solutions the primal and dual objective values are equal. Moreover, if either the primal or the dual problem has no feasible solution and the other has a feasible solution, then the feasible problem is unbounded.*

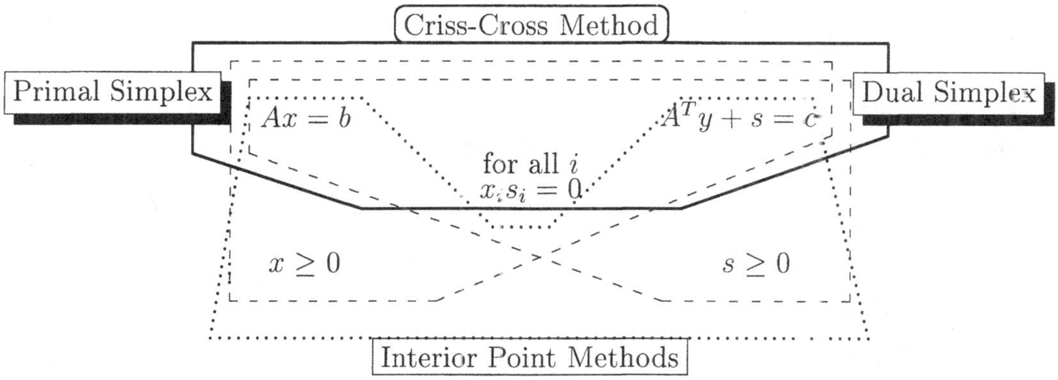

Figure 1. All algorithms for linear programming keep a part of the optimality criteria valid while working toward the others

The theorem of alternatives, or, as most frequently cited, the Farkas Lemma (Farkas, 1902; Roos et al., 1997) is equivalent to the Strong Duality Theorem (Theorem 2). Here we present just two of its variants as a corollary because they follow easily from this theorem.

Corollary 2 (Farkas Lemma, primal form). *One and only one of the following two systems of linear inequalities has a feasible solution:*

$$(I_p)\quad \begin{array}{l} Ax = b \\ x \geqslant 0 \end{array} \qquad (II_p)\quad \begin{array}{l} A^Ty \leqslant 0 \\ b^Ty > 0. \end{array}$$

Corollary 3 (Farkas Lemma, dual form). *One and only one of the following two systems of linear inequalities has a feasible solution:*

$$(I_d)\quad \begin{array}{l} c^Tx < 0 \\ Ax = 0 \\ x \geqslant 0 \end{array} \qquad (II_d)\quad A^Ty \leqslant c.$$

The duality theorems and the variants of the Farkas Lemma have many important implications. Starting backward, one sees that the second variant of the Farkas Lemma gives an easy-to-check certificate for the infeasibility and unsolvability of the dual problem. If we obtain a solution to (I_d), then the dual problem is certainly infeasible. The corollary also says that if the dual problem is infeasible, then such a certificate must exist. Similarly, from the first variant we learn that if we have a solution to the system (II_p) then this solution is a certificate for the infeasibility of the primal problem. Moreover, the infeasibility of the primal problem implies that such a certificate must exist.

Duality theorems can be used in many different ways. They can be used to check optimality of solutions. If two solutions $\bar{x}$ and $\bar{y}$ are given, then we need to check if $\bar{x}$ is primal feasible, that is, if

$$A\bar{x} = b, \quad \bar{x} \geqslant 0$$

holds; if $\bar{y}$ is dual feasible, that is, if

$$A^T\bar{y} \leqslant c$$

holds; and finally if the two objective values are equal, that is, if

$$c^T\bar{x} = b^T\bar{y} \text{ or equivalently, if } \bar{x}_j(c - A^T\bar{y})_j = 0 \quad \forall j = 1, \ldots, n.$$

These relations provide the fundamentals of different algorithmic concepts that are expanded briefly in the next section.

4. Fundamentals of Algorithms

The results of the duality theorems show that in order to find optimal solutions for both the primal and dual problems, we need to solve the following system:

$$\begin{aligned} Ax &= b, \qquad & x &\geqslant 0, \\ A^T y + s &= c, \qquad & s &\geqslant 0, \\ x_j s_j &= 0, \qquad & j &= 1, \dots, n. \end{aligned}$$

Here the first line represents primal feasibility, the second line dual feasibility, and the last is the complementarity condition that guarantees by Corollary 1 that the primal and dual objective values are equal:

$$c^T x = b^T y.$$

The specific algorithms for LP approach the solution of the system of optimality conditions differently. The basic structures are illustrated in Figure 1.

Simplex-type or "pivoting" algorithms can be found on the upper part of Figure 1. All of them, primal and dual simplex methods and criss-cross type algorithms, use basic solutions (see Chapter 1.2 for a detailed discussion) and hence they always keep the primal and dual equality and the complementarity conditions satisfied. Criss-cross methods stay with these minimal requirements and work toward both primal and dual feasibility. Apart from these, primal (dual) simplex methods require and preserve primal (dual) feasibility while working toward optimality.

Interior-point methods (IPMs) seek to approach the optimal solution through a sequence of points that are always strictly feasible. Such methods have been known for a long time (Fiacco and McCormick, 1968), but because they are more demanding for storage and reliable floating-point arithmetic than simplex methods, were not considered to be effective. Contrary to simplex-type algorithms, IPMs—at least the most popular primal-dual ones—require and preserve both primal and dual feasibility while the complementarity condition is relaxed. In fact, the requirement is even stronger: not only x, $s \geqslant 0$ is required, but x, $s > 0$; that is, both the primal and dual solutions are strictly positive. Finally, note that IPMs are discussed in detail in Chapter 1.3.

ACKNOWLEDGMENTS

This paper benefited from several suggestions from Tibor Illés, who read an earlier version of this paper. The author gratefully acknowledges his contribution. A part of this work was done while the author was working at the Department CROSS, Faculty ITS, Delft University of Technology and is partially supported by grant number OPG 0048923 of the National Sciences and Engineering Research Council of Canada.

References

Andersen, E. D., J. Gondzio, Cs. Mészáros, and X. Xu. 1996. "Implementation of interior point methods for large scale linear programming." In *Interior Point Methods of Mathematical Programming*, edited by T. Terlaky, pp. 189–252. Dordrecht, The Netherlands: Kluwer Academic Publishers.

Bixby, R. E. 1994. "Progress in linear programming." *ORSA J. Comput.* **6(1)**: 15–22.

Borgwardt, K. H. 1987. *The Simplex Method: a Probabilistic Analysis*. Berlin: Springer Verlag.

Dantzig, G. B. 1963. *Linear Programming and Extensions*. Princeton: Princeton University Press.

Dantzig, G. B. 1991. "Linear programming." In *History of Mathematical Programming. A Collection of Personal Reminiscences*, edited by J. K. Lenstra, A. H. G. Rinnooy-Kan, and A. Schrijver. The Netherlands: CWI, North-Holland.

Farkas, J. 1902. "Theorie der einfachen Ungleichungen." *J. Reine Angewandte Math.* **124**: 1–27.

Fiacco, A. V. and G. P. McCormick. 1968. *Nonlinear Programming: Sequential Unconstrained Minimization Techniques*. New York: John Wiley & Sons. Reprinted from *SIAM Classics in Applied Mathematics*, Vol. 4. Philadelphia: SIAM Publications.

Grötschel, M., L. A. Lovász, and A. Schrijver. 1988. *Geometric Algorithms and Combinatorial Optimization*. Berlin: Springer Verlag.

Karmarkar, N. K. 1984. "A new polynomial-time algorithm for linear programming." *Combinatorica* 4: 373–395.

Khacijan, L. G. 1979. "A polynomial time algorithm in linear programming." *Sov. Math. Doklady* **20**: 191–194.

Klee, V. and G. J. Minty. 1972. "How good is the simplex algorithm?" In *Inequalities III*, edited by O. Shisha. New York: Academic Press.

Lenstra, J. K., A. H. G. Rinnooy Kan, and A. Schrijver, eds. 1991. *History of Mathematical Programming. A Collection of Personal Reminiscences*, The Netherlands: CWI, North-Holland.

Murty, K. G. 1995. *Operations Research: Deterministic Optimization Models*. Englewood Cliffs, NJ: Prentice-Hall Inc.

Prékopa, A. 1980. "On the development of optimization." *Am. Math. Monthly* **87**: 527–542.

Roos, C., T. Terlaky, and J.-Ph. Vial. 1997. *Theory and Algorithms for Linear Optimization: an Interior Point Approach*. New York: John Wiley & Sons.

Strang, G. 1988. *Linear Algebra and Its Applications*. Orlando, FL: Harcourt Brace Jovanovich.

Williams, H. P. 1990. *Model Building in Mathematical Programming*. 3d ed. New York: John Wiley & Sons.

Wright, M. H. 1998. "The interior-point revolution in constrained optimization." Numerical Analysis Manuscript 98-4-09. Murray Hill: AT & T Bell Labs.

Wright, S. 1996. *Primal–Dual Interior-Point Methods*. Philadelphia: SIAM.

Ye, Y. 1997. *Interior Point Algorithms*. New York: John Wiley & Sons.

1.2 Simplex-Type Algorithms

Tamás Terlaky

ABSTRACT Although interior-point methods (IPMs) have polynomial complexity, simplex-type methods—originally developed by Dantzig—are still one of the most efficient algorithms to solve linear programming problems. In many situations, simplex methods compete favorably with IPMs, especially when they are embedded in branch-and-cut schemes to solve combinatorial and integer optimization problems. In this paper a short introduction to the theory of simplex algorithms is given. Some implementation issues, first-phase strategies, and a brief overview of state-of-the-art implementations is given as well.

1. Introduction

Simplex methods (Dantzig, 1963) proceed from basic solution to basic solution toward an optimal basic solution. Geometrically the path of basic solutions is a path on the boundary of the set of feasible solutions. It is actually a path from vertex to vertex, because the set of feasible solutions is a polyhedron and basic solutions are associated with vertices of this polyhedron. This process could require many steps to go around a multifaceted feasible region. Intuitively, a more direct path through the interior of the region is appealing since the possibility exists of moving through the polytope in very few steps. This is what interior-point methods (IPMs) do (see Chapter 1.3). Despite the contention that simplex methods could require a large number of steps, they still compete, sometimes favorably, with polynomial IPMs (Bixby, 1994; Andersen et al., 1996; Lustig et al., 1994; Maros and Mitra, 1996).

2. The LP Problem: Basic Solutions

In this paper the LP problem and its dual is considered in standard form:

$$\begin{array}{llll} & \text{minimize} & c^Tx & & \text{maximize} & b^Ty \\ \text{(LP)} & \text{subject to} & Ax = b, & \text{(LD)} & \text{subject to} & A^Ty \leqslant c, \\ & & x \geqslant 0, & & & \end{array}$$

where $c, x \in \mathbb{R}^n$, $b, y \in \mathbb{R}^m$ are vectors, and $A \in \mathbb{R}^{m \times n}$ is a matrix. In the sequel, without loss of generality, we may assume that the matrix A has full row rank. By introducing slack variables, the dual constraints can be rewritten as $A^Ty + s = c$, $s \geqslant 0$. We have seen in Section 4 of Chapter 1.1 that to solve the primal and dual LP problems one needs to solve the following system:

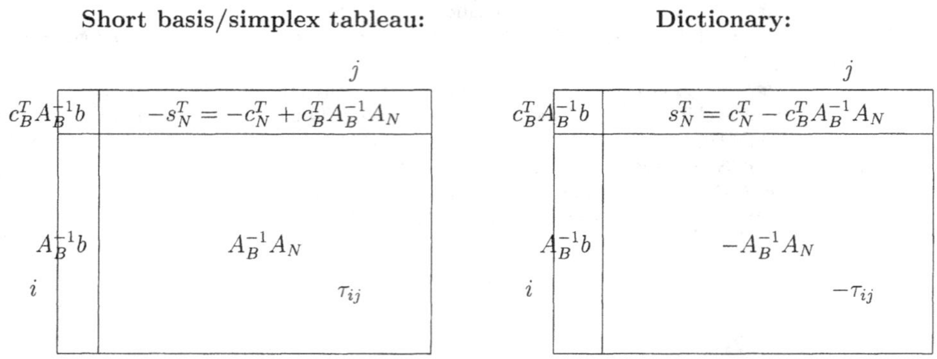

Figure 1 The structure of a basis tableau and a dictionary

$$\begin{aligned} Ax = b, &\quad x \geqslant 0, \\ A^Ty + s = c, &\quad s \geqslant 0, \\ x_js_j = 0, &\quad i = 1, ..., n. \end{aligned}$$

As illustrated in Figure 1 in Chapter 1.1 the candidate solutions in simplex-type algorithms always satisfy the equality and complementarity constraints. This property follows from the fact that they use basic solutions and in the subsequent iterations they go from basis to basis while approaching optimality. Let us see first what can we gain by using basic solutions.

Definition 1. *Let* $\mathrm{A_B} \in \mathbb{R}^{\mathrm{m \times m}}$ *be a nonsingular submatrix of* A, *that is,* $\mathrm{A_B}$ *contains* m *linearly independent columns of the matrix* $\mathrm{A} = [\mathrm{a}_1, ..., \mathrm{a}_j, ..., \mathrm{a}_n]$. *Then* $\mathrm{A_B}$ *is called a basis of the LP problem.*

Let $I_B \subset \{1, ..., n\}$, with $j \in I_B$ if, and only if, a_j is a column of A_B, be the set of basis indices. Furthermore, let A_N denote the complement matrix, the nonbasic part of the matrix A. Also let the vectors c, x, and s be partioned analogously, that is, c_B, x_B, and s_B denote the subvectors containing the coordinates whose index is in I_B while the complement vectors are denoted by c_N, x_N and s_N.

Clearly, by setting all the coordinates of the vector x_N equal to zero a unique solution of the linear equation system

$$Ax = A_Bx_B = b$$

can be obtained. The unique solution is given by $x_B = A_B^{-1}b$. The solution x obtained this way is called a primal basic solution and can equivalently be defined as $x_j = 0$ when $j \in I_N = I \backslash I_B$ and $x_j = (A_B^{-1}b)_i$ if a_j is the i-th column of the basis matrix A_B.

Analogously, a dual basic solution can be constructed by taking the dual constraints associated with the basis indices I_B and setting all the coordinates of s_E equal to zero. Then the dual basis solution y is the unique solution of the linear equation system.

$$A_B^T y = c_B.$$

Thus $y = A_B^{-T} c_B$, where the superscript $^{-T}$ denotes the inverse of the transposed matrix A_B^T. Then the vector s_N can easily be computed as $s_N = c_N - A_N^T y = c_N - A_N^T A_B^{-T} c_B$.

If we take a pair, x and (y, s), of primal and dual basic solutions associated with the same basis A_B, then by definition not only the primal equality constraints, $Ax = b$, and the dual equality constraints, $A^T y + s = c$, are satisfied but the complementary conditions, $x_j s_j = 0$, $\forall 1 \leqslant j \leqslant n$, are satisfied as well. This property is obvious because $x_j = 0$ if $j \notin I_B$ while $s_j = 0$ if $j \in I_B$.

The discussions so far demonstrate that basic solutions satisfy all the equality and complementarity constraints. When using basic solutions we have to work to satisfy the inequality constraints. To make our discussions more plausible we need some further terminology. A basis, A_B, and the corresponding basic solutions, x and (y, s), are called primal feasible if $x_B \geqslant 0$, it is dual feasible if $s_N \geqslant 0$, and it is called optimal if it is both primal and dual feasible.

All simplex-type algorithms proceed from basic solution to basic solution toward an optimal basic solution. Geometrically the path of basic solutions is a path on the boundary of the set of feasible solutions. It is actually a path from vertex to vertex, because the set of feasible solutions is a polyhedron and basic solutions are associated with vertices of this polyhedron. At each step a basis is at hand, optimality, that is, primal and/or dual feasibility is checked and a new basis is selected. We consider three classes of simplex-type algorithms.

1. *Criss-cross methods.* Criss-cross methods can be initialized by an arbitrary basic solution. Then optimality is checked. If the basic solution is optimal then we are done, else a primal or dual infeasible variable (coordinate) is chosen and we go over to a new basis in which this specific coordinate becomes both primal and dual feasible. The process is repeated as long as optimality is reached or evidence of primal or dual infeasibility is established (see e.g., Terlaky, 1985, and Fukuda and Terlaky, 1997).

2. *Primal simplex methods.* Primal simplex algorithms require that at each stage the basic solution is primal feasible. Having a primal feasible basic solution at hand, primal simplex methods check if the basis is dual feasible. If so, it is optimal and we are done. If it is not dual feasible, a dual infeasible coordinate is chosen and we pass over to a new basis that is again primal feasible and the previously dual infeasible variable enters the basis and hence becomes both primal and dual feasible. The process is repeated as long as optimality is reached or evidence of dual infeasibility is established (see e.g., Dantzig, 1963, and Murty, 1995).

   To obtain a primal feasible basis is not a trivial task; a so-called first-phase problem needs to be solved to get a primal feasible basic solution. The optimal solution of the first-phase problem either provides a primal feasible basic solution or gives evidence of infeasibility of the problem (LP). Later on, some first-phase strategies will be discussed.

3. *Dual simplex methods.* Dual simplex algorithms require that at each stage the basic solution is dual feasible. Having a dual feasible basic solution at hand, dual simplex methods check if the basis is primal feasible. If so, it is optimal and we are done. If it is not primal feasible, a primal infeasible coordinate is chosen and we pass over to a new basis that is again dual feasible and the previously primal infeasible variable enters the basis and hence becomes both primal and dual feasible. The process is repeated as long as optimality is reached or evidence of primal infeasibility is established (see e.g., Dantzig, 1963, and Murty, 1995).

   To obtain a dual feasible basis is not a trivial task; a so-called first-phase problem needs to be solved to get a dual feasible basic solution. The optimal solution of the first phase problem either provides a dual feasible basic solution or gives evidence of infeasibility of the problem (LD).

After this verbal discussion let us give a rigorous description of all these methods. For this purpose we need to introduce the notion of a simplex or basis tableau. In simplex-type methods, basic tableaus, or in a different but equivalent setting, dictionaries are used to encode the current basis representation of LP problems with respect to a given basis. (For completeness, the frequently used dictionary notation [Fukuda and Terlaky, 1997] and its relaxation to the simplex tableau is presented as well.) Both tableaus and dictionaries are actually matrices where the same data is presented in a slightly different form.

2.1. *Basic Tableaus and Dictionaries*

In a basis tableau, the basis representation of the system of equations

$$x_0 = c^T x \tag{1}$$

$$b = Ax$$

is encoded. When a basis A_B is given, then we have

$$A_B^{-1}b = x_B + A_B^{-1}A_N x_N. \tag{2}$$

From here we obtain a basic solution when we set $x_N = 0$ and then $x_B = A_B^{-1}b$. Recall that to have the basis primal feasible we need x_B to be nonnegative.

By using $x_B = A_B^{-1}b - A_B^{-1}A_N x_N$ we can reformulate the first equation in (1) as

$$x_0 = c^T x = c_B^T x_B + c_N^T x_N = c_B^T(A_B^{-1}b - A_B^{-1}A_N x_N) + c_N^T x_N = c_B^T A_B^{-1}b + (c_N^T - c_B^T A_B^{-1}A_N)x_N. \tag{3}$$

Choosing $x_N = 0$ we obtain the current objective value

$$x_0 = c_B^T A_B^{-1}b = c_B^T x_B = y^T b,$$

where we used the definition of the dual basis vector, $y = A_B^{-T}c_B$. Recall that the nonzero part of the dual vector s_N equals $c_N - A_N^T A_B^{-T}c_B$, hence to have dual feasibility we need the coefficient vector of x_N in Equation (3) to be nonnegative.

As we discussed, if we have both primal and dual feasibility of the basic solution, then an optimal pair of solutions is obtained. If at least one primal or dual infeasible variable exists, then, based on the current basis representation of the problem data, a new basis is selected. In order to ease the description of the basis exchange procedure, we introduce the notion of simplex/basis tableaus and the equivalent dictionary.

Basis/Simplex Tableaus and Dictionaries

The problem data are the current objective value, $c_B^T A_B^{-1}b$; the dual solution, $s_N^T = c_N^T - c_B^T A_B^{-1}A_N$ (see Equation (3)); the current form of the right-hand side, which is the actual primal basic solution, $x_B = A_B^{-1}b$; and finally the basis representation of the nonbasis part of the coefficient matrix, $A_B^{-1}A_N$ (see Equation (2)). These data are arranged in a matrix in the respective notation of the simplex tableau and dictionary. The (short) simplex tableau is based on the representation $c_B^T A_B^{-1}b = x_0 - s_N^T x_N = x_0 - (c_N^T - c_B^T A_B^{-1}A_N)x_N$ and $A_B^{-1}b = x_B + A_B^{-1}A_N x_N$ while the dictionary is based on $x_0 = c_B^T A_B^{-1}b + s_N^T x_N = c_B^T A_B^{-1}b + (c_N^T - c_B^T A_B^{-1}A_N)x_N$ and $x_B = A_B^{-1}b - A_B^{-1}A_N x_N$. The data is then given, in the respective forms, in Figure 1. In these tableaus we will use the convention that row i will belong to basic variable x_i and column j to nonbasic variable x_j, thus the basis tableau element τ_{ij} is the coefficient of the basis vector a_i in the current basis representation of the nonbasic vector a_j. Let us stress again that the index pair "ij" does not indicate the geographical coordinate of the element τ_{ij} in the tableau, but it refers to the basis–nonbasis pair (a_i, a_j). To keep our discussions short, in the sequel only the simplex tableau will be used. If the reader wants to translate the algorithms to the dictionary notation, the only difference is that the right blocks of the basis tableau must be multiplied by minus one to obtain the corresponding dictionary.

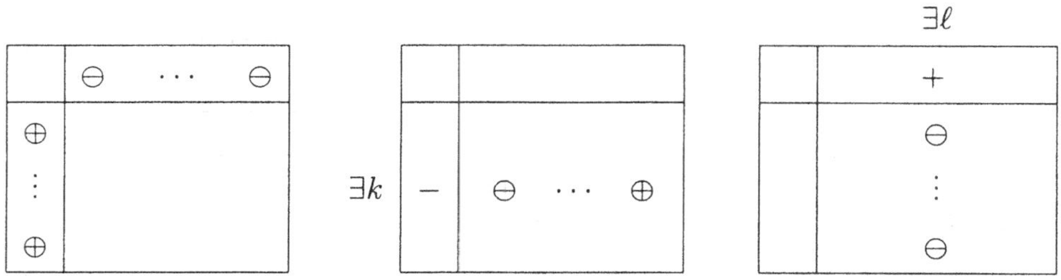

(I) optimal (II) primal inconsistent (III) dual inconsistent

Figure 2 Terminal basis tableau structures

Terminal Tableaus

As we discussed earlier, an optimal pair of solutions is obtained when both x_B and s_N are nonnegative. This situation—that is, the structure of an optimal tableau—is represented by part (I) of Figure 2. Furthermore, we can conclude that the primal, respectively dual, problem is inconsistent when the tableau structures are as indicated by parts (II) and (III) of Figure 2. These three tableaus are called terminal; when either of these tableaus is reached the solution process terminates.

Optimality of the basic solutions related to tableau (I) has already been discussed. It is easy to see that tableau (II) gives evidence of the inconsistency of the primal problem. In that case the equality constraint represented by the k-constraint is given as

$$(A_B^{-1}b)_i = x_k + (A_B^{-1}A_N)_i x_N = x_k + \sum_{j \in I_N} \tau_{kj} x_j,$$

if a_k is the i-th element in the basis. From part (II) of the figure we have $(A_B^{-1}b)_k < 0$, $(A_B^{-1}A_N)_k \geqslant 0$ and in any primal feasible solution all the variables must be nonnegative. Clearly the left-hand side of this equation is negative, while the right-hand side is nonnegative for all nonnegative vector x, thus this equation can never be satisfied by a primal feasible solution. The proof of the dual inconsistent case is analogous, it easily follows by dualization as well

We will see how these terminal tableaus are reached by simplex-type methods and how these tableaus serve as stopping criteria in the respective algorithms.

2.2. *Pivoting*

Pivoting is the basic operation in all simplex-type methods. Pivoting essentially means moving from one basis to another one, where the two bases differ exactly by two variables. There is precisely one variable that is not in the second basis but it is in the first one—this variable is referred to as the leaving variable—and there is exactly one variable that is not in the first but is in the second basis; this variable is called the entering variable. There is a simple, necessary, and sufficient condition to test if the selection of a leaving-entering pair, say (k, ℓ), will result in a new basis in a possible pivot. By classical linear algebra we know that the pivot (k, ℓ) is responsible if, and only if, a_k has a nonzero coefficient in the basis representation of the nonbasic vector a_ℓ, that is, if $\tau_{k\ell} \neq 0$. This criteria guarantees that the new set of basic variables (their index set I_B' is given by $I_B' := I_B \cup \{\ell\} \setminus \{k\}$) not only has cardinality m but it is linearly independent as well.

When the new basis is known, one can easily recompute the new tableau by computing the inverse of the basis matrix. However, this does not have to be done from scratch; it is more efficient to update the basis tableau as indicated in Figure 3, and exposed by the following formulas where the notation given in Figure 3 is used. We have $I_B' := I_B \cup \{\ell\} \setminus \{k\}$:

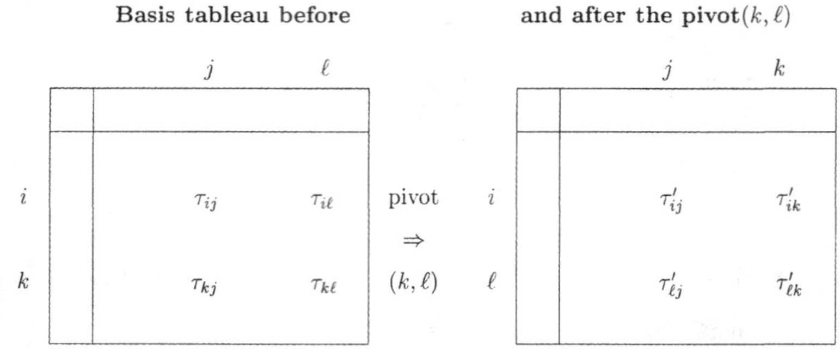

Figure 3 Pivot operation

$$\tau'_{ij} = \tau_{ij} - \frac{\tau_{i\ell}\tau_{kj}}{\tau_{k\ell}}, \quad \forall i \in I'_B\ \{\ell\};\ j \in I'_N\ \{k\} \quad \tau'_{ik} = -\frac{\tau_{i\ell}}{\tau_{k\ell}}, \quad \forall i \in I'_B \setminus \{\ell\}$$

$$\tau'_{\ell j} = \frac{\tau_{kj}}{\tau_{k\ell}} \qquad \forall j \in I_N\ \{\ell\} \qquad \tau'_{\ell k} = -\frac{1}{\tau_{k\ell}}.$$

Such pivot operations need to be performed repeatedly, thus it is extremely important to develop efficient computational procedures. In practice, advanced numerical techniques, updates of *LU*-factorization, and sparse matrix techniques are implemented. The reader can get advice about implementation issues by consulting the literature (Nazareth, 1987; Forrest and Tomlin, 1992; Maros and Mitra, 1996).

3. Simplex-Type Algorithms

After this preparation, we are ready to present our three classes of simplex-type algorithms. Before doing so, we make a general observation. All simplex-type algorithms—primal, dual, or criss-cross—use only so-called admissible pivots. There are two types of admissible pivots as illustrated in Figure 4. A pivot is allowed either in a row showing infeasibility of a basic variable or in a column that demonstrates violation of dual feasibility of a nonbasic variable. When the pivot row or column is chosen, then admissible pivots allow natural pivots only, namely pivots indicating that the problem is not primal or dual inconsistent. An admissible pivot exists if, and only if, the current tableau is not terminal. Moreover, admissible pivots have the nice property that after the pivot, both pivot variables became primal (dual) feasible at Type I (Type II) admissible pivots.

The reader may verify that all the algorithms considered in the sequel adapt either or both types of admissible pivots.

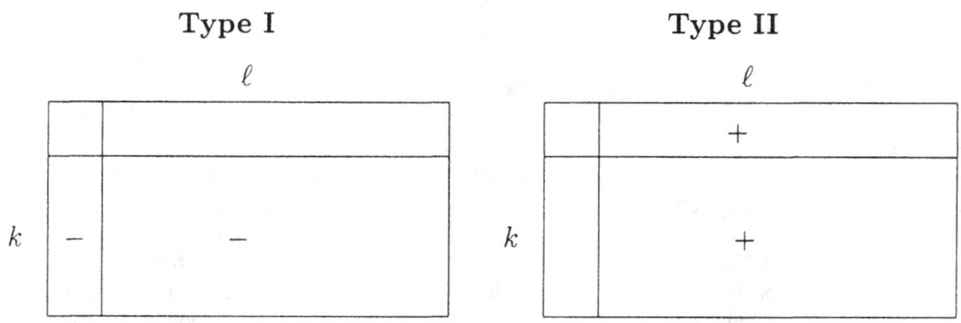

Figure 4 Admissible pivots

3.1. Criss-Cross Methods

Here we just present the finite least-index criss-cross method (Terlaky, 1985; Fukuda and Terlaky, 1997) that is extremely simple and solves LP problems in a finite number of steps. Let us first mention some nice characteristics of criss-cross methods:

- They can start with any basis and thus no first-phase procedure is needed; they consist of one phase only.
- They do not require feasibility to be preserved, thus no work is invested in preserving feasibility.
- Only the signs, not the magnitude of the components in a basis tableau and a prefixed ordering of variables determine the pivot selection.

LEAST-INDEX CRISS-CROSS METHOD
INITIALIZATION

```
  let A_B be an arbitrary initial basis;
  I_B, respectively, I_N is the index set of the basis and nonbasis variables;
while true do
  if x_B ≥ 0 and s_N ≥ 0 then
(I)  stop: the current solution solves the LP problem;
  else
    p := min{i ∈ I_B : x_i < 0};
    q := min{j ∈ I_N : s_j < 0};
    r : min{p, q};
    if r = p then
      if the p-row of the tableau is nonnegative then
(II)   stop: (LP) is inconsistent;
      else
        let q := min{j ∈ I_N : τ_pj < 0};
      endif
    else (i.e., r = q)
      if the q-column of the tableau is nonpositive then
(III)   stop: (LD) is inconsistent;
      else
        let p := min{i ∈ I_B : τ_iq > 0};
      endif
    endif
    perform a pivot: I_B := I_B ∪ {q} \ {p};
  endwhile
end.
```

To ease understanding, a flowchart of the least-index criss-cross method is given in Figure 5. Observe that the criss-cross method terminates at one of the three terminal tableaus demonstrated in Figure 3.

The following striking result can be proved:

Theorem 1. *The least-index criss-cross method is finite.*

The least-index criss-cross rule described here seems to be completely rigid, there is no flexibility in pivot selection. It is worth noting that the finiteness proofs allow significant flexibility in pivot selection (Fukuda and Matsui, 1991). While exploring this flexibility, several variants, for example, the first-in-first-out and the most-often-selected variable criss-cross methods, were developed by Zhang (1999; see also Fukuda and Terlaky, 1997). If we alleviate all these consistent selections and apply the criss-cross scheme by making arbitrary pivot choices that respect only the sign requirement at the respective steps, then the general criss-cross scheme is obtained. In this generality nothing can be said about finiteness of criss-cross rules.

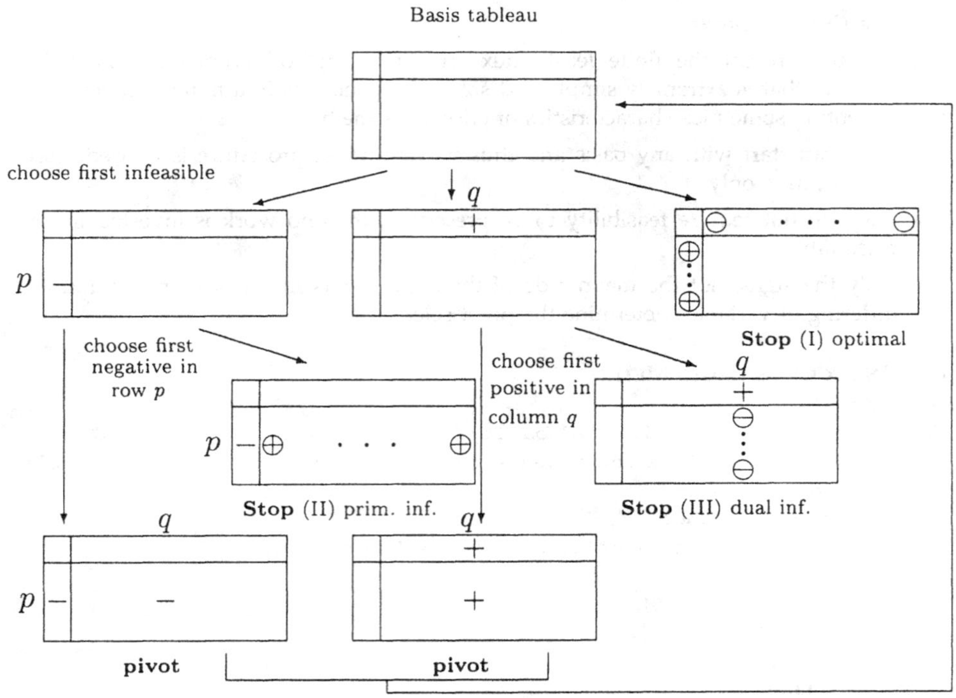

Figure 5 Scheme of the least-index criss-cross method

Finally, note that in spite of their elegance and simplicity, to date, criss-cross methods are not efficient in practice.

3.2. Primal Simplex Methods

Primal simplex methods, as mentioned earlier, need a primal feasible basis to start with. Such a basic solution can be obtained by solving a so-called first-phase problem, presented later in this section.

The pivot selection is designed in such a way that primal feasibility is preserved while traversing through the different basic solutions. The main characteristics of primal simplex methods are as follows.

- A primal feasible basis is needed to start with.
- A first-phase procedure is needed to obtain a primal feasible basis.
- Primal feasibility of the basis is preserved by implementing a so-called ratio test.
- The primal objective value is monotonically decreasing in the subsequent iterations.
- Additional precautions, for example, least-index resolution (Bland, 1977; Terlaky and Zhang, 1993) or lexicographic refinement (Charnes, 1952; Terlaky and Zhang, 1993) are needed to guarantee finite termination.
- Allow extremely efficient implementations.

Primal simplex methods work as follows:

```
Initialization
  let A_B be a primary feasible basis, that is, x_B⩾0;
  I_B, respectively, I_N is the index set of the basis and nonbasis variables;
while true do
  if s_N⩾0 then
```

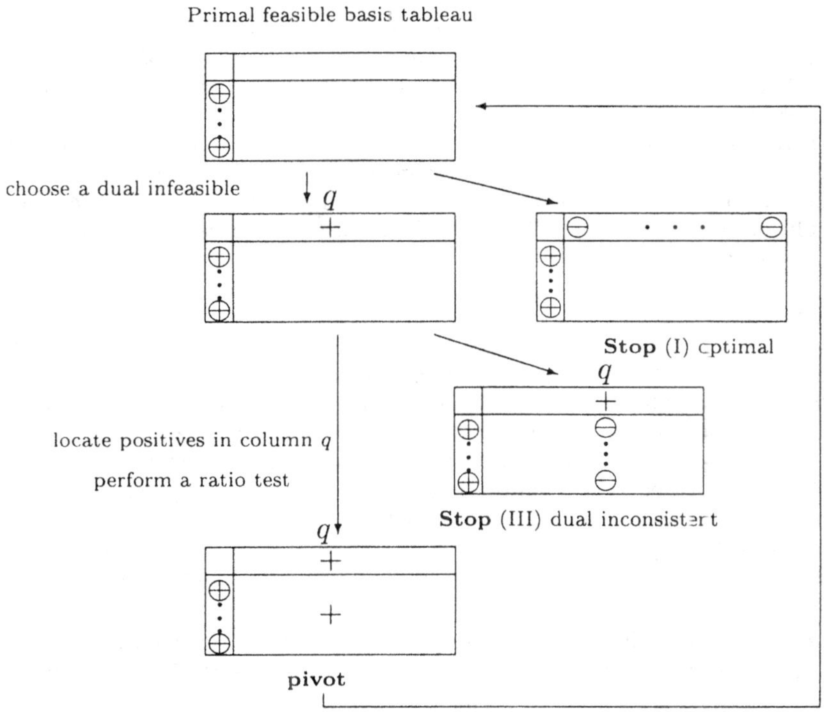

Figure 6 Scheme of the primal simplex method

(I) **stop**: the current solution solves the LP problem;
 else
  let $q \in I_N$ be an index with $s_q < 0$;
  (in the least-index version let $q := \min\{j \in I_N : s_j < 0\}$).
   if the q-column of the tableau is nonpositive **then**
(III) **stop**: (LD) is inconsistent;
   else
    let $\vartheta := \min\{\frac{x_i}{\tau_{iq}} : i \in I_B \text{ and } \tau_{iq} > 0$;
    let $p \in I_B$ such that $\frac{x_p}{\tau_{pq}} = \vartheta$; (ratio test)
    (in the least-index version let $p := \min\{i \in I_B : \frac{x_i}{\tau_{iq}} = \vartheta\}$);
   endif
  endif
  perform a pivot: $I_B := I_B \cup \{q\} \setminus \{p\}$;
 endwhile
end.

Observe that when dual inconsistency is detected, one may also conclude that the primal problem is unbounded. This is due to the extra information that a primal feasible solution is at hand.

To ease understanding, a flowchart of the primal simplex method is given in Figure 6. The following result can be proved:

Theorem 2. *The primal simplex method produces a feasible basic solution with monotonically decreasing objective values. Moreover, it is finite when implemented with the least-index resolution.*

The vast literature on LP contains an extensive family of primal simplex methods. The inherent flexibility of the entering variable allows development of various pivot selection criteria. Here only the recently most popular steepest edge criteria is mentioned (Goldfarb and Forrest, 1991).

There are numerous finite variants like the lexicographic and double lexicographic method, variants that buildup the solution in some specific way, and so on. The reader can find most classic simplex method variants in Dantzig (1963) and Murty (1995) and is advised to consult the recent survey by Terlaky and Zhang (1993) to gain an impression of the latest results about simplex methods.

INITIALIZATION, FIRST-PHASE

Finding a primal feasible solution needs extra work. Actually the solution of another LP problem is needed. The theoretically simple solution is just to discard the objective function (consider the vector c to be zero) and then solve the resulting problem by the least-index criss-cross method. One can easily verify that the result is either a certificate of infeasibility or a primal feasible basis that can be used to initialize the primal simplex method.

This latter method is very simple, but not suitable for use in practice, because no efficient implementation of the criss-cross method is known yet.

A practically more efficient solution is to form and solve an artificial first-phase problem. The first-phase problem is constructed by applying the following steps when considering the primal LP problem.

1. Let us multiply all the equations by -1 where the original right-hand-side element is negative. Thus we obtain a new equivalent equation system $Ax = b$ where we have $b \geqslant 0$.
2. Let us introduce an artificial vector $z \geqslant 0$ and add this to the equation system. Then we get the inequality system

$$Ax + z = b, \quad x \geqslant 0,\ z \geqslant 0. \tag{4}$$

   This system is not equivalent to the original one, but a solution (x, z) of this with $z = 0$ always gives a solution to the original system.
3. Inequality system (4) has a trivial feasible basis, namely the unit matrix as the coefficient matrix of the vector z. The associated basis solution $x = 0$, $z = b$ is feasible because $b \geqslant 0$.
4. To force equivalence to the original system we need to drive all the coordinates of the vector z to zero. This can be achieved by defining the next LP problem where $e^T = (1, \ldots, 1)$.

$$\begin{aligned} \min \quad & e^T z \\ & Ax + z = b \\ & x, z \geqslant 0. \end{aligned}$$

   This is the first-phase problem. Clearly, its objective function value is nonnegative for any feasible solution because it is just the sum of the values of the artificial variables.
5. This artificially formulated LP problem can now be solved by the primal simplex method because a primal feasible initial basis by construction is available.
6. After solving this first-phase problem the following cases might occur.
   - The optimal objective value is positive. In this case the original primal problem is clearly infeasible, because any primal feasible solution would give a solution with objective value zero for the first-phase problem.
   - The optimal objective value is zero and the optimal basis contains columns only from the matrix A. Then a primal feasible basis for the original problem is obtained, we can discard the artificial variables and, as the second phase of the solution process, return to the solution of the original problem. The required feasible basis is now available.

- The optimal objective value is zero and the optimal basis contains some unit vectors corresponding to some artificial variables. Then the artificial variables in the basis have the value zero. Although we do not have a primal feasible basis for the original problem, which contains only original vectors, we go on with solving the original problem. Let us discard the artificial but not the basic variables, fix the artificial basic variables at the zero level, and return to the solution of the original problem. The required feasible basis is now available. When pivoting in the simplex method, we monitor the artificial variables. If any of them leaves the basis, then we discard that variable immediately.

Now it is clear that the primal simplex method essentially requires the solution of two problems. This seems to double the computational cost. Theoretically this is certainly true; however, in practice, the first phase requires less effort than the second phase. Moreover, combined phase one–phase two strategies were developed as well. For details the reader is referred to literature that describes the computational state of the art (Bixby, 1994; Maros and Mitra, 1996; Nazareth, 1987).

3.3. *Dual Simplex Methods*

Dual simplex methods can straightforwardly be derived by dualizing primal simplex methods. Thus, they have analogous features to the primal ones. Their significance lies in the fact that there are many situations, especially in solving integer programming problems in branch-and-bound schemes, when dual methods are more powerful than primal ones. Due to space limitations, only the algorithmic frame of dual simplex methods is presented here.

Initialization

```
  let A_B be a dual feasible basis, that is, s_N ≥ 0;
  I_B, respectively, I_N is the index set of the basis and nonbasis variables;
while true do
  if x_B ≥ 0 then
(I)  stop: the current solution solves the LP problem;
  else
    let p ∈ I_B be an index with x_p < 0;
    (in the least-index version let p := min{i ∈ I_B : x_i < 0});
     if the p-row of the tableau is nonnegative then
(II)  stop: (LP) is inconsistent;
     else
       let ϑ := min{s_j/(−τ_pj) : j ∈ I_N and τ_pj < 0};
       let q ∈ I_N be such that s_q/(−τ_pq) = ϑ; (ratio test)
       (in the least-index version let q := min{j ∈ I_N : s_j/(−τ_pj) = ϑ});
     endif
   endif
   perform a pivot: I_B := I_B ∪ {q}\{p};
  endwhile
end
```

The following result can be proved:

> **Theorem 3**. *The dual simplex method produces dual feasible basic solutions with monotonically increasing objective values. Moreover, when it is implemented with the least-index resolution then it is finite.*

Some variants of the dual simplex method can be found in Dantzig (1963) and Terlaky and Zhang (1993) as well.

4. Complexity

Simplex methods (Dantzig, 1963) proceed from basic solution to basic solution toward an optimal basic solution. The number of basic solutions might be an exponential function of the problem dimension and, in the worst case, simplex methods might visit all those basic solutions. This was made explicit by the famous results of Klee and Minty (1972). They showed that if the set of feasible solutions is a slightly out-of-kilter n-dimensional hypercube then all the feasible basic solutions can be visited by a simplex method. The actual path starts at the origin and visits all 2^n vertices so that the objective function is strictly improved at each step. This result proves that the worst-case complexity of some variants of the simplex method is exponential. An interesting result is that the dual simplex method solves the Klee–Minty problem in a polynomial number of iterations (Murty, 1983). A more challenging exponential example is given by Clausen (1979; see also Murty, 1983). The main feature of Clausen's example is that the primal simplex method is exponential on the primal problem while the dual simplex method is exponential on the dual problem.

Since Klee and Minty's work has been published, the exponential behavior of several simplex method variants has been demonstrated. However, there are still some variants, like Zadeh's rule (see Terlaky and Zhang, 1993) whose exponential worst-case complexity is not established yet.

An interesting result was proved recently by Fukuda and Terlaky (1998). They proved that starting from any basis, a short admissible pivot sequence exists, containing at most n-pivot steps, leading to an optimal basis. This result indicates that polynomial pivot algorithms might exist, at least, there always exists a short admissible pivot path leading to an optimal basis.

There was a huge gap between the theoretical worst-case complexity and practical performance of simplex methods. It was obvious that no practically implemented simplex method would use the exponentially long Klee–Minty path. Actually, commercial software solves the Klee–Minty problem in one step. This is primarily due to the quite sophisticated multiple pricing pivot selection methods implemented. The need to explain the efficiency of practical simplex methods became transparent. It was soon shown that, under a probabilistic model, the expected (average) number of steps required by the shadow-vertex simplex algorithm is linear (Borgwardt, 1987).

5. Degeneracy, Cycling, and Finiteness

In the previous sections, variants of simplex methods were given that solve linear programs in a finite number of steps. For this purpose the least-index resolution was always applied. Without this precaution, simplex methods may fail to solve the LP problem. This phenomenon is caused by degeneracy, the possibility of cycling.

Let us first discuss what we mean by degeneracy. A basis (basic solution) is called primal degenerate if x_B contains a zero coordinate, that is, if there is an $i \in I_B$ with $x_i = 0$. Analogously, a basis (basic solution) is called dual degenerate if s_N contains a zero coordinate, that is, if there is a $j \in I_N$ with $s_j = 0$. Finally, the basis is called degenerate is it is either primal or dual degenerate.

A severe consequence of degeneracy is that in primal or dual simplex methods the primal, respectively, dual objective value remains the same in subsequent iterations. This property opens the possibility for cycling, that is, starting from a certain basic solution, the same basic solutions are revisited again and again. This problem was recognized and demonstrated early, for example, by Beale (1955). Various tools were developed to avoid cycling. The two most famous ones are the lexicographic simplex method (Charnes, 1952), which is equivalent to a perturbation technique (Dantzig et al., 1955), and the utmost simple least-index rule (Bland, 1977), which has been discussed in this article.

Another implication of degeneracy is that multiple optimal solutions occur. This might sound appealing; after all, we will have the freedom to choose from the multiple optimal solutions. However, it turned out that having multiple solutions causes more problems than benefits. Typically, we do not know all the multiple optimal solutions and have limited or no control at all over which optimal solutions will be produced by simplex methods.

6. Sensitivity Analysis

Sensitivity analysis has a great impact when LP problems are used in practice. Sometimes sensitivity information is more important for practical use than the solution itself. Traditionally, sensitivity information is provided by commercial packages. Unfortunately, even today most packages neglect the traps arising from degeneracy and provide sensitivity information based only on the generated optimal basic solution. In the case of degeneracy, such sensitivity information is incomplete and incorrect and, thus, might lead to erroneous decisions with severe consequences (Ward and Wendell, 1990). Recently, easily computable methods were developed to provide correct and complete sensitivity information (Adler and Monteiro, 1992; Jansen et al., 1997; Roos et al., 1997; Koltai and Terlaky, 2000). The key observation is that for correct and complete sensitivity information, not only a single optimal solution but a representation of the set of optimal solutions is needed. Which representation of the optimal set is available might depend on the algorithm used.

To give an impression of how to calculate correct sensitivity information we briefly discuss here how to calculate correct shadow prices and the intervals where these shadow prices are valid. Let us assume that the vector b varies, so the actual right-hand side is $b(\beta) = b + \beta \Delta b$. We are interested in how the optimal value function $f(\beta)$ varies, where

$$f(\beta) := \min_x \{c^T x \mid Ax = b + \beta \Delta b,\ x \geqslant 0\}.$$

It is well known that $f(\beta)$ is a piecewise linear function, thus it is not differentiable everywhere. At a given β value, its right and left derivative might be different. The left $f'_-(\beta)$ [right $f'_+(\beta)$] derivative of $f(\beta)$ is called the left (right) shadow price. We are interested in calculating these shadow prices, for example, at $\beta = 0$, and the intervals (left and right again) where these derivatives are constant.

Let P^*_β and D^*_β denote the set of optimal solution for the primal and dual problems, respectively, that is,

$$P^*_\beta := \{x \mid c^T x = f(\beta),\ Ax = b + \beta \Delta b,\ x \geqslant 0\},$$

$$D^*_\beta := \{(y, s) \mid (b + \beta \Delta b)^T y = f(\beta),\ A^T y + s = c,\ s \geqslant 0\}.$$

Thus, the left and right shadow prices can be calculated by solving some subsidiary LP problems defined on the dual optimal set.

Theorem 4. *Let* $f'_-(\beta)$ *and* $f'_+(\beta)$ *be the left and right derivative of* $f(\cdot)$ *in* β. *Then*

$$f'_-(\beta) = \min\{\Delta b^T y : y \in D^*_\beta\}$$

$$f'_+(\beta) = \max\{\Delta b^T y : y \in D^*_\beta\}.$$

The respective linearity intervals (ranges) can similarly be calculated by solving subsidiary LP problems defined on the primal optimal set.

Theorem 5. *Let* $[\beta_1, \beta_2]$ *be a linearity interval of the optimal value function* $f(\beta)$. *Let* $\bar\beta \in (\beta_1, \beta_2)$ *and define* $\bar D^* := D^*_{\bar\beta}$. *Then*

$$\beta_1 = \min_{\beta,x}\{\beta : Ax - \beta \Delta b = b, x \geqslant 0, x^T s = 0, \text{ for an } s \in \bar D^*\},$$

$$\beta_2 = \max_{\beta,x}\{\beta : Ax - \beta \Delta b = b, x \geqslant 0, x^T s = 0, \text{ for an } s \in \bar D^*\}.$$

The analogous results for reduced costs and their validity ranges can be found in the literature already cited; however, by dualization the reader might easily derive them.

It seems to be quite costly to solve a linear program for each shadow price and range, but one must realize that in case of degenerate LP there is no cheaper method to obtain correct sensitivity information. It is worth mentioning that these subsidiary problems can be solved quite efficiently by appropriately choosing primal or dual methods for which the previous optimal solution still provides a feasible solution to start with.

7. Computational State of the Art

Implementations of simplex methods are still the core of LP packages. High-performance packages like XPRESSMP, CPLEX, and OSL offer both simplex and IPM solvers. Major developments in the last few years are the implementation of automatized dualization and dual simplex methods. These allow flexibility in choosing which of the primal or the dual problems allows a more efficient solution. Dual simplex methods are very efficient at solving relaxations of discrete optimization problems and are particularly powerful in branch-and-cut methods. Another significant development, again especially efficient when solving combinatorial optimization problems, is the implementation of the by now classical steepest-edge simplex algorithm (Goldfarb and Forrest, 1991; Bixby, 1994; Forrest and Tomlin, 1992). Besides these papers, the computational state of the art of simplex-type algorithms can be reviewed by studying the following literature: Bixby (1994), Nazareth (1987), and Maros and Mitra (1996).

The reader can gain updated information at the following Internet sites about availability, capability and performance of state-of-the-art LP packages.

CPLEX `http://www.cplex.com`

XPRESSMP `http://www.dash.co.uk`

OSL `http://www.research.ibm.com/osl/`

At the following sites extensive information about various optimization software and some comparative performance results can be found:

`NEOS: Optimization Software Guide http://www-fp.mcs.anl.gov/otc/Guide/`

`Benchmark for optimization software http://plato.la.asu.edu/bench`^{m}`.html`

`Decision tree for optimization software http://plato.la.asu.edu/guide.html`

ACKNOWLEDGMENTS

This paper benefited from several suggestions from Tibor Illés, who read an earlier version of this paper. The author gratefully acknowledges his contribution. A part of this work was done while the author was working at the Department CROSS, Faculty ITS, Delft University of Technology, and was partially supported by grant number OPG 0048923 of the National Sciences and Engineering Research Council of Canada.

References

Adler, I. and R. D. C. Monteiro. 1992. "A geometric view of parametric linear programming." *Algorithmica* **8**: 161–176.

Andersen, E. D., J. Gondzio, Cs. Mészáros, and X. Xu. 1996. "Implementation of interior point methods for large scale linear programming." In *Interior Point Methods of Mathematical Programming*, edited by T. Terlaky. Dordrecht, The Netherlands: Kluwer Academic Publishers.

Beale, E. M. L. 1955. "Cycling in the dual simplex algorithm." *Naval Res. Logistics Q.* **2**: 269–275.

Bixby, R. E. 1994. "Progress in linear programming." *ORSA J. Comput.* **6(1)**: 15–22.

Bland, R. G. 1977. "New finite pivoting rules for the simplex method." *Math. Op. Res.* **2**: 103–107.

Borgwardt, K. H. 1987. *The Simplex Method: a Probabilistic Analysis*. Berlin: Springer Verlag.

Charnes, A. 1952. "Optimality and degeneracy in linear programming." *Econometrica* **20(2)**: 160–170.

Clausen, J. 1979. "A tutorial note on the complexity of the simplex algorithm" Technical Report NR79/16, DIKU. Copenhagen, Denmark: Copenhagen University.

Dantzig, G. B. 1963. *Linear Programming and Extensions.* Princeton: Princeton University Press.

Dantzig, G. B., A. Orden, and Ph. Wolfe. 1955. "Notes on linear programming: Part I—the generalized simplex method for minimizing a linear form under linear inequality restrictions." *Pacific J. Math.* **5(2)**: 183–195.

Forrest, J. J. H. and J. A. Tomlin. 1992. "Implementing the simplex method for the optimization subroutine library." *IBM Syst. J.* **31**: 11–25.

Fukuda, K. and T. Matsui. 1991. "On the finiteness of the criss-cross method." *Eur. J. Op. Res.* **52**: 119–124.

Fukuda, K. and T. Terlaky. 1997. "Criss-cross methods; a fresh view on pivot algorithms." *Math. Program. Series B* **79**: 369–395.

Fukuda K. and T. Terlaky. 2000. "On the existence of short admissible pivot sequences." PUMA: Mathematics of Optimization **10/4**: 431–488.

Goldfarb, D. and J. J. H. Forrest. 1991. Steepest edge simplex algorithm for linear programming IBM Research Report June. Yorktown Heights: T. J. Watson Research Center.

Jansen, B., J. J. de Jong, C. Roos, and T. Terlaky. 1997. "Sensitivity analysis in linear programming: just be careful!" *Eur J. Op. Res.* **101**: 15–27.

Klee, V. and G. J. Minty. 1972. "How good is the simplex algorithm?" In *Inequalities III*, edited by O. Shisha. New York: Academic Press.

Koltai, T. and T. Terlaky. 2000. "The difference between the managerial and mathematical interpretation of sensitivity analysis results in linear programming." *Int. J. Prod. Econ.* **65/3**: 257–274.

Lustig, I. J., R. E. Marsten, and D. F. Shanno. 1994. "Interior point methods: computational state of the art." *ORSA J. Comput.* **6**: 1–15.

Maros, I. and G. Mitra. 1996. "Simplex algorithms." In *Advances in Linear and Integer Programming*, edited by J. Beasley. Oxford: Oxford University Press.

Murty, K. G. 1983. *Linear Programming.* New York: John Wiley & Sons.

Murty, K. G. 1995. *Operations Research: Deterministic Optimization Models.* Englewood Cliffs, NJ: Prentice-Hall Inc.

Nazareth, J. L. 1987. *Computer Solution of Linear Programs.* Oxford, U.K: Oxford University Press.

Roos, C., T. Terlaky, and J.-Ph. Vial. 1997. In *Theory and Algorithms for Linear Optimization: an Interior Point Approach.* New York: John Wiley & Sons.

Terlaky, T. 1985. "A convergent criss-cross method." *Math. Op. und Stat. Ser. Optimization* **16**: 683–690.

Terlaky, T. and S. Zhang. 1993. "Pivot rules for linear programming—a survey." *Ann. Op. Res.* **46**: 203–233.

Ward, J. E. and R. E. Wendell. 1990. "Approaches to sensitivity analysis in linear programming." *Ann. Op. Res.* **27**: 3–38.

Zhang, S. 1999. "A new variant of criss-cross pivot algorithm for linear programming." *Eur. J. Op. Res.* **116(3)**: 607–614.

1.3 Interior-Point Methods for Linear Optimization

Kees Roos

ABSTRACT The aim of this article is to give an introduction to the recently developed interior-point methods. We restrict ourselves to the case of linear optimization. Much emphasis is put on the role of the central path of a problem, which plays a crucial role both in the theory and in the design of algorithms. The subject being extremely rich, some relevant references are given for topics that are not covered in this article.

1. Introduction

Everyone with some background in mathematics knows how to solve a system of linear equalities, since it is the basic subject in linear algebra. For many practical problems, however, inequalities also play a role. For example, usually a budget may not be larger than some specified amount. In such situations one may end up with a system of linear relations that not only contains equalities but also inequalities. Solving such a system requires methods and theory that go beyond standard mathematical knowledge. Nevertheless, the topic has a rich history and is closely related to the important topic of linear optimization, where the object is to find the optimal (minimal or maximal) value of a linear function subject to linear constraints on the variables; the constraints may be either equality or inequality constraints. Both from a theoretical and computational point of view these topics are equivalent. In this article we describe the ideas underlying a new class of solution methods, that has become known as interior-point methods (IPMs).

2. Short History of Linear Optimization

Linear optimization (LO) is one of the most widely applied mathematical techniques. The last ten years have given rise to revolutionary developments both in computer technology and algorithms for LO. As a consequence, LO problems that ten years ago required a computational time of one year, say, can now be solved within a few minutes. The acceleration achieved is due partly to advances in computer technology and for a significant part also to the new IPMs for LO. The field of linear optimization has been given the name linear programming in the past. The origin of this name goes back to the Dutch Nobel prize winner Koopmans (see Dantzig, 1991). Nowadays the word "programming" usually refers to the activity of writing computer programs, and its use instead of the more natural word optimization gives rise to confusion.

Until very recently, the standard method for solving LO problems was the simplex method, first proposed by Dantzig in 1947 (Dantzig, 1963). Since then, it has been routinely used to solve problems in business, logistics, economics and engineering. In an effort to explain the remarkable efficiency of the simplex method, using the theory of complexity, it has been tried to prove that the computational effort to solve an LO-problem via the simplex method is polynomially bounded in terms of the size of a problem instance. This question is still unsettled today, but it stimulated two important proposals of new algorithms for LO. The first polynomial method is the so-called ellipsoid method, due to Khachiyan (1979). It is based on the ellipsoid technique for nonlinear optimization of Shor (1987). With this technique, Khachiyan proved that LO belongs to the class of polynomially solvable problems. Although this result had a great theoretical impact, it failed to deliver its promises in actual computational efficiency. The second proposal was made in 1984 by Karmarkar (1984). Karmarkar's algorithm is also polynomial, with a better complexity bound than Khachiyan, but it has the further advantage of being highly efficient in practice. After an initial controversy it has been established that for very large, sparse problems, subsequent variants of Karmarkar's method often outperform the simplex method.

Though the field of LO was at that time considered more or less mature, after Karmarkar's paper it suddenly surfaced as one of the most active areas of research in optimization. In the period 1984–1989 more than thirteen hundred papers were published on the subject. A recent (annotated) bibliography was given by Roos and Terlaky (1997). A valuable source of information is the World Wide Web interior-point archive: `http://www-unix.mcs.anl.gov/otc/InteriorPoint/`. Originally the aim of the research was to get a better understanding of the so-called projective method of Karmarkar. Soon it became apparent that this method was related to classical methods like the affine scaling method of Dikin (Dikin, 1967, 1974, 1988), the logarithmic barrier method of Frisch (Frisch, 1955, 1956), and the center method of Huard (1967), and that the last two methods, when tuned properly, could also be proved to be polynomial. Moreover, it turned out that the IPM approach to LO has a natural generalization to the related field of convex nonlinear optimization, which resulted in a new stream of research and an excellent monograph by Nesterov and Nemirovski (1993). The monograph of Nesterov and Nemirovski opened the way into other new subfields of optimization. Two major new subfields are semidefinite optimization and second-order cone optimization, with important applications in system theory, discrete optimization, and many other areas. For a survey of these developments the reader may consult Boyd and Vandenberghe (1996), and the book by Ben-Tal and Nemirovski (2001).

This article offers a brief but rather complete introduction to IPMs for LO. Most of the material is treated in more detail in the book by Roos et al. (1997). We deal with the duality theory for LO, putting much emphasis on the central path, which is at the heart of all current methods. We present a polynomial method for solving an LO problem, including an elegant way to initialize the method. Although the method presented has the best theoretical iteration bound up till now, in practice it is outperformed by so-called barrier methods, a topic that we do not cover in this article. For this and other topics not covered the reader is referred to the aforementioned book or to the book by Wright (1996).

We mention here one other topic that we do not touch in this article, infeasible start methods. These methods have drawn a lot of attention in the literature. They deal with the situation when no feasible starting point is available. See, for example Potra (1994), Bonnans and Potra (1997), and Wright (Wright 1993, 1994, 1996). We want to point out that the embedding method described extensively in Sections 3–6 provides a much more elegant and reliable solution to this problem: any given LO problem can be embedded in a self-dual problem for which a feasible interior starting point is known. It has been implemented by the Andersen brothers, in their code MOSEK (Andersen and Andersen, 1999).

We have strived to give a self-contained treatment of the subject. It is hoped that this article answers the needs of professors who want to teach their students the basic principles of IPMs, of users of LO who want to understand what is behind the new IPM solvers in commercial codes (CPLEX, OSL, XPRESS, MOSEK, etc.) and how to interpret results from those codes, and of other users who want to exploit the new algorithms as part of a more general software toolbox in optimization.

3. Reduction to a Self-Dual LO Problem

3.1. *Canonical LO Problem and Its Dual*

We say that a linear optimization problem is in canonical form if it is written in the following way:

$$(P)\quad \min\{c^T x : Ax \geqslant b, x \geqslant 0\}, \tag{1}$$

where the matrix A is of size $m \times n$, the vectors c and x are in $\mathbb{R}^n$ and b in $\mathbb{R}^m$. Note that all the constraints in (P) are inequality constraints and the variables are nonnegative. Each LO problem can be transformed to an equivalent canonical problem. For this we refer to any text book on LO. Roos et al. (1997) is also a suitable reference. Given the above canonical problem (P), we consider a second problem, denoted by (D) and called the dual problem of (P), given by

$$(D)\quad \max\{b^T y : A^T y \leqslant c, y \geqslant 0\}. \tag{2}$$

The two problems (P) and (D) share the matrix A and the vectors b and c in their description. But the role of b and c has been interchanged: the objective vector c of (P) is the right-hand side vector of (D), and, similarly, the right-hand side vector b of (P) is the objective vector of (D). Moreover, the constraint matrix in (D) is the transposed matrix A^T, where A is the constraint matrix in (P). In both problems the variables are nonnegative. The problems differ in that (P) is minimization problem whereas (D) is a maximization problem, and, moreover, the inequality symbols in the constraints have opposite direction.

Now let x be feasible for (P) and y for (D). Then $x \geqslant 0$, $y \geqslant 0$, $Ax \geqslant b$, and $A^T y \leqslant c$. As a consequence we have

$$b^T y \leqslant (Ax)^T y = x^T(A^T y) \leqslant c^T x. \tag{3}$$

Hence, any y that is feasible for (D) provides a lower bound $b^T y$ for the value of $c^T x$, whenever x is feasible for (P). Conversely, any x that is feasible for (P) provides an upper bound $c^T x$ for the value of $b^T y$, whenever y is feasible for (D). This phenomenon is known as the weak duality property. An immediate consequence is that if $c^T x = b^T y$ for a feasible pair (x, y) then x is optimal for (P) and y is optimal for (D).

The (nonnegative) difference $c^Tx - b^Ty$ between the primal objective value at a primal feasible x and the dual objective value at a dual feasible y is called duality gap for the pair (x, y). We have just established that x is optimal for (P) and y is optimal for (D) if the duality gap vanishes. The most important result in LO states that the converse statement is also true: if x is an optimal solution of (P) and y is an optimal solution of (D) then the duality gap vanishes at the pair (x, y). This result is known as the strong duality property. A proof of this result will be given in Section 4.4.

3.2. Reduction to Inequality System

The strong duality property implies that whenever both (P) and (D) have an optimal solution then the system

$$\begin{aligned} &Ax \geqslant b, && x \geqslant 0, \\ &-A^Ty \geqslant -c, && y \geqslant 0, \\ &b^Ty - c^Tx \geqslant 0 \end{aligned} \tag{4}$$

has a solution, and any such solution provides optimal solutions for (P) and (D). If $\kappa = 1$, the following inequality system is equivalent to (4), as can easily be verified.

$$\begin{pmatrix} 0_{m\times m} & A & b \\ -A^T & 0_{n\times n} & c \\ b^T & -c^T & 0 \end{pmatrix} \begin{pmatrix} y \\ x \\ \kappa \end{pmatrix} \geqslant \begin{pmatrix} 0_m \\ 0_n \\ 0 \end{pmatrix}, \quad x \geqslant 0, y \geqslant 0, \kappa \geqslant 0. \tag{5}$$

Here 0_k denotes the zero vector of length k and $0_{k\times \ell}$ the zero matrix of size $k \times \ell$. Since the right-hand side in (5) is the zero vector, this system is homogeneous: whenever (y, x, κ) solves the system then $\lambda(y, x, \kappa)$ also solves the system, for any positive λ. The new variable κ is called the homogenizing variable. Now, given any solution (x, y, κ) of (5) with $\kappa > 0, (x/\kappa, y/\kappa, 1)$ yields a solution of (4). This makes clear that, in fact, the two systems are completely equivalent unless every solution of (5) has $\kappa = 0$. But if $\kappa = 0$ for every solution of (5), it is obvious that the first system cannot have a solution. Evidently, we can work with the second system without loss of information about the solution set of the first system.

Hence, defining the matrix $\bar{M}$ and the vector $\bar{z}$ by

$$\bar{M} := \begin{pmatrix} 0 & A & -b \\ -A^T & 0 & c \\ b^T & -c^T & 0 \end{pmatrix}, \quad \bar{z} := \begin{pmatrix} y \\ x \\ \kappa \end{pmatrix},$$

where we omitted the size indices of the zero blocks, we have reduced the problem of solving (P) and (D) to finding a solution of the inequality system

$$\bar{M}z \geqslant 0, \quad \bar{z} \geqslant 0. \tag{6}$$

If this system has a nonzero solution with $\kappa > 0$ then it gives optimal solutions for (P) and (D); otherwise these problems have no optimal solutions. Thus our task has been reduced to finding a solution of (6) with $\kappa > 0$, or to prove that such a solution does not exist. In the following we will deal with this problem. In doing so, we will strongly use the fact that the matrix $\bar{M}$ is skew-symmetric, that is, $\bar{M}^T = -\bar{M}$.

3.3. Interior-Point Condition

The method we are going to use for solving (6) is an IPM, and for this we need the system to satisfy the interior-point condition (IPC). We say that any linear system of equalities and inequalities satisfies the IPC if a feasible solution exists that strictly satisfies all inequality constraints. Unfortunately system (6) does not satisfy the IPC. Because if $z = (x, y, \kappa)$ satisfies all (inequality) constraints strictly then $\kappa > 0$, and hence x/κ is feasible for (P) and y/κ is feasible for (D). But then $(c^Tx - b^Ty)/\kappa \geqslant 0$, by weak duality. On the other hand, the last constraint in (6) requires $b^T y - c^T x \geqslant 0$. It follows that $b^T y - c^T x = 0$ and hence no feasible solution of (6) strictly satisfies the last inequality in (6).

To overcome this shortcoming of system (6) we increase the dimension by adding one more nonnegative variable ϑ to the vector $\bar{z}$, and by extending $\bar{M}$ with one extra column and row, according to

$$M := \begin{pmatrix} \bar{M} & r \\ -r^T & 0 \end{pmatrix}, \quad z := \begin{pmatrix} \bar{z} \\ \vartheta \end{pmatrix},$$

where

$$r = e - \bar{M}e,$$

with e denoting an all-one vector of length $m + n + 1$. Letting q be the vector of length $m + n + 2$ given by

$$q := \begin{pmatrix} 0 \\ m + n + 1 \end{pmatrix},$$

we consider the system

$$Mz \geqslant -q, \quad z \geqslant 0. \tag{7}$$

First we observe that this system satisfies the IPC. The all-one vector does the work because, taking $\bar{z} = e$ and $\vartheta = 1$, we have

$$Mz + q = \begin{pmatrix} \bar{M} & r \\ -r^T & 0 \end{pmatrix}\begin{pmatrix} e \\ 1 \end{pmatrix} + \begin{pmatrix} 0 \\ m + n + 2 \end{pmatrix} = \begin{pmatrix} \bar{M}e + r \\ -r^T e + m + n + 2 \end{pmatrix} = \begin{pmatrix} e \\ 1 \end{pmatrix}, \tag{8}$$

since

$$-r^T e + m + n + 2 = -(e - \bar{M}e)^T e + m + n + 2 = 1.$$

Here we used $e^T \bar{M} e = 0$, since $\bar{M}$ is skew-symmetric. The usefulness of the new system stems not only from the fact that it satisfies the IPC. Another crucial property is that there is a correspondence between the solutions of (6) and the solutions of (7) with $\vartheta = 0$. To see this it is useful to write (7) in terms of $\bar{z}$, and ϑ:

$$\begin{pmatrix} \bar{M} & r \\ -r^T & 0 \end{pmatrix}\begin{pmatrix} \bar{z} \\ \vartheta \end{pmatrix} + \begin{pmatrix} 0 \\ m + n + 2 \end{pmatrix} \geqslant 0. \quad \bar{z} \geqslant 0, \quad \vartheta \geqslant 0.$$

Obviously, if $z = (\bar{z}, 0)$ satisfies (7), this implies $\bar{M}\bar{z} \geqslant 0$ and $\bar{z} \geqslant 0$, and hence $\bar{z}$ satisfies (6). On the other hand, if $\bar{z}$ satisfies (6) then $\bar{M}\bar{z} \geqslant 0$ and $\bar{z} \geqslant 0$, and $z = (\bar{z}, 0)$ satisfies (7) if and only if

$$r^T \bar{z} \leqslant m + n + 2.$$

If $r^T \bar{z} \leqslant 0$ this certainly holds. Otherwise, if $r^T \bar{z} \geqslant 0$, a (small enough) positive multiple of $\bar{z}$ will certainly satisfy. Since a positive multiple preserves signs, this is sufficient for our goal.

3.4. Embedding into a Self-Dual LO Problem

In the previous section we reduced the problem of solving (P) and (D) to finding a solution of (7) with $\vartheta = 0$ and $\kappa > 0$, or to prove that such a solution does not exist. For that purpose we consider the LO problem

$$(\mathrm{SP}) \quad \min \{q^T z : Mz \geqslant -q, z \geqslant 0\}. \tag{9}$$

Since $q^T z = (m + n + 2)\vartheta \geqslant 0$, the optimal value is nonnegative. On the other hand, since $q \geqslant 0$, the zero vector $(z = 0)$ is feasible and yields zero as objective value. We conclude that the optimal solutions of (9) are precisely the vectors z satisfying (7) with $\vartheta = 0$. Thus our task becomes to find an optimal solution of (9) with $\kappa > 0$, or to establish that such an optimal solution does not exist.

An interesting feature of the LO problem (9) is that it is self-dual: the dual problem is

$$\max \{-q^T u : M^T u \leqslant q, u \geqslant 0\};$$

since M is skew-symmetric, $M^T u \leqslant q$ is equivalent to $-Mu \leqslant q$, or $Mu \geqslant -q$, and maximizing $-q^T u$ is equivalent to minimizing $q^T u$, and thus the dual problem is the same problem as (9).

The rest of the article is devoted to our main task, namely to find an optimal solution of (9) with $z_{m+n+1} = \kappa > 0$ or to establish that such a solution does not exist.

To simplify notation we let $\bar{n}$ denote the order of the matrix M. So, $\bar{n} = m + n + 2$. Furthermore, we introduce index sets B and N according to

$$B := \{i : z_i > 0 \text{ for some optimal } z\}$$

$$N := \{i : s_i(z) > 0 \text{ for some optimal } z\}.$$

So, B contains all indices i for which an optimal solution z with positive z_i exists. We also write $z_i \in B$ if $i \in B$. Note that we certainly have $\vartheta \notin B$. The main question we have to answer is whether $\kappa \in B$ holds or not. Because if $\kappa \in B$ then an optimal solution z exists with $\kappa > 0$, in which case (P) and (D) have optimal solutions, and otherwise not.

The next lemma implies that the sets B and N are disjoint. In this lemma, and further on, we use some new notations. First we associate to any nonnegative vector z its slack vector $s(z)$ as follows.

$$s(z) := Mz + q.$$

Then z is a feasible vector if and only if $z \geqslant 0$ and $s(z) \geqslant 0$. Furthermore, to any vector $u \in \mathbb{R}^k$, we associate the diagonal matrix U whose diagonal entries are the elements of u, in the same order. Finally, if $v \in \mathbb{R}^k$, then Uv will be denoted shortly as uv. Thus uv is a vector whose entries are obtained by multiplying u and v componentwise.

Lemma 1. *Let* z^1 *and* z^2 *be feasible for (SP). Then* z^1 *and* z^2 *are optimal solutions of (SP) if and only if* $z^1 s(z^2) = z^2 s(z^1) = 0$.

Proof: *Recall that, since* M *is skew-symmetric,*

$$u^T M u = 0, \quad \forall u \in \mathbb{R}^{\bar{n}}. \tag{10}$$

Using this we may write

$$q^T z = z^T(s(z) - Mz) = z^T s(z) - z^T Mz = z^T s(z). \tag{11}$$

As a consequence, $z \geqslant 0$ *is optimal if and only if* $s(z) \geqslant 0$ *and* $z^T s(z) = 0$. *Since, by* (10),

$$(z^1 - z^2)^T M (z^1 - z^2) = 0,$$

we have

$$(z^1 - z^2)^T (s(z^1) - s(z^2)) = 0.$$

Expanding the product on the left and rearranging the terms we get

$$(z^1)^T s(z^2) + (z^2)^T s(z^1) = (z^1)^T s(z^1) + (z^2)^T s(z^2).$$

Now z^1 *is optimal if and only if* $(z^1)^T s(z^1) = 0$, *by* (11), *and similarly for* z^2. *Hence, since* z^1, z^2, $s(z^1)$ *and* $s(z^2)$ *are all nonnegative,* z^1 *and* z^2 *are optimal if and only if*

$$(z^1)^T s(z^2) + (z^2)^T s(z^1) = 0,$$

which is equivalent to

$$z^1 s(z^2) = z^2 s(z^1) = 0,$$

proving the lemma.

Later, in Section 4.3, we prove that the every index belongs to either B or N. Thus B and N form a partition of the whole index set. This partition is the so-called the optimal partition of (SP).

4. The Central Path

4.1. *Definition of the Central Path*

Recall from (8) that $s(e) = e$, where e (as always) denotes the all-one vector of appropriate length (in this case, $\bar{n}$). As a consequence, we have a vector z such that $z_i s_i(z) = 1 (1 \leqslant i \leqslant \bar{n})$, which, using our shorthand notation can also be expressed as

$$zs(z) = e. \tag{12}$$

Now we come to a very fundamental notion, both from a theoretical and algorithmic point of view, namely the central path of LO problem at hand. The underlying theoretical property is that for every positive number μ there exist a unique nonnegative vector z, such that

$$zs(z) = \mu e, \quad z \geqslant 0, s(z) \geqslant 0. \tag{13}$$

If $\mu = 1$, the existence of such a vector is guaranteed by (12). Also note that if we put $\mu = 0$ in (13) then the solutions are just the optimal solutions of (SP), and then the system may have multiple solutions. Therefore, the following lemma is of interest.

Lemma 2. *If* $\mu > 0$, *then there exists at most one nonnegative vector* z *such that* (13) *holds.*

Proof: *Let* z^1 *and* z^2 *be nonnegative vectors statisfying* (13), *and let* $\mathrm{s}^1 = \mathrm{s}(\mathrm{z}^1)$ *and* $\mathrm{s}^2 = \mathrm{s}(\mathrm{z}^2)$. *Since* $\mu > 0, \mathrm{z}^1, \mathrm{z}^2, \mathrm{s}^1, \mathrm{s}^2$ *are all positive. Define* $\Delta\mathrm{z} := \mathrm{z}^2 - \mathrm{z}^1$, *and similarly* $\Delta\mathrm{s} := \mathrm{s}^2 - \mathrm{s}^1$. *Then we may easily verify that*

$$M\Delta\mathrm{z} = \Delta\mathrm{s} \tag{14}$$

$$\mathrm{z}^1\Delta\mathrm{s} + \mathrm{s}^1\Delta\mathrm{z} + \Delta\mathrm{s}\Delta\mathrm{z} = 0. \tag{15}$$

Using the fact that M *is skew-symmetric,* (14) *implies that* $\Delta\mathrm{z}^T\Delta\mathrm{s} = 0$, *or, equivalently,*

$$e^T(\Delta\mathrm{z}\Delta\mathrm{s}) = 0. \tag{16}$$

Rewriting (15) *gives*

$$(\mathrm{z}^1 + \Delta\mathrm{z})\Delta\mathrm{s} + \mathrm{s}^1\Delta\mathrm{z} = 0.$$

Since $\mathrm{z}^1 + \Delta\mathrm{z} = \mathrm{z}^2 > 0$ *and* $\mathrm{s}^1 > 0$, *this implies that no two corresponding entries in* $\Delta\mathrm{z}$ *and* $\Delta\mathrm{s}$ *have the same sign. So it follows that*

$$\Delta\mathrm{z}\Delta\mathrm{s} \leqslant 0. \tag{17}$$

Combining (16) *and* (17), *we obtain* $\Delta\mathrm{z}\Delta\mathrm{s} = 0$. *Hence either* $(\Delta\mathrm{z})_\mathrm{i} = 0$ *or* $(\Delta\mathrm{s})_\mathrm{i} = 0$, *for each* i. *Using* (15) *once more, we conclude that* $(\Delta\mathrm{z})_\mathrm{i} = 0$ *and* $(\Delta\mathrm{z})_\mathrm{i} = 0$, *for each* i. *Hence* $\Delta\mathrm{z} = \Delta\mathrm{s} = 0$, *whence* $\mathrm{z}^1 = \mathrm{z}^2$ *and* $\mathrm{s}^1 = \mathrm{s}^2$. *This proves the lemma.*

To prove the existence of a solution to (13) requires more effort. We postpone this to the next section. For the moment, let us take the existence of a solution to (13) for granted and denote it as $z(\mu)$. We call it the μ-center of (SP). The set

$$\{z(\mu) : \mu > 0\}$$

of all μ-centers represents a parametric curve in the feasible region of (SP) and is called the central path of (SP). Note that

$$q^T z(\mu) = s(\mu)^T z(\mu) = \bar{n}\mu. \tag{18}$$

This proves that along the central path, when μ approaches zero, the objective value $q^T z(\mu)$ monotonically decreases to zero, at a linear rate.

In the next section we will use an other important fact, namely

$$e^T z + e^T s(z) = \bar{n} + q^T z. \tag{19}$$

This identity follows by applying the orthogonality property (10) to the vector $e - z$, which gives

$$(z - e)^T (s(z) - s(e)) = 0.$$

Since $s(e) = e, e^T e = \bar{n}$ and $z^T s(z) = q^T z$, the above relation follows.

4.2. *Existence of the Central Path*

In this section we give an algorithmic proof of the existence of a solution to (13). Starting at $z = e$ we construct the μ-center for any $\mu > 0$. This is done by using the so-called Newton direction as a search direction. The results in this section will also be used later when dealing with a polynomial time method for solving (SP).

Newton Direction

Assume that z is a positive solution of (SP) such that its slack vector $s = s(z)$ is positive, and let Δz denote a displacement in the z-space. Our aim is to find Δz such that $z + \Delta z$ is the μ-center. We denote

$$z^+ := z + \Delta z,$$

and the new slack vector as s^+:

$$s^+ := s(z^+) = M(z + \Delta z) + q = s + M\Delta z.$$

Thus, the displacement Δs in the s-space is simply given by

$$\Delta s = s^+ - s = M\Delta z.$$

Observe that Δz and Δs are orthogonal, by using (10):

$$(\Delta z)^T \Delta s = (\Delta z)^T M \Delta z = 0. \tag{20}$$

We want z^+ to be the μ-center, which means $(z + \Delta z)(s + \Delta s) = \mu e$, or $zs + z\Delta s + s\Delta z + \Delta z \Delta s = \mu e$. The last equation is nonlinear. Applying Newton's method, we omit the nonlinear term $\Delta z \Delta s$, leaving us with the following linear system in the unknown vectors Δz and Δs:

$$M\Delta z - \Delta s = 0$$

$$z\Delta s + s\Delta z = \mu e - zs. \tag{21}$$

This system has a unique solution, as may easily be verified, by using the fact that M is skew-symmetric, and $z > 0$ and $s > 0$. The solution Δz is called the Newton direction. Since we omitted the quadratic term $\Delta z \Delta s$ in our calculation of the Newton direction, $z + \Delta z$ will (in general) not be the μ-center, but hopefully it will be a good approximation. In fact, after Newton step, one has

$$z^+ s(z^+) = (z + \Delta z)(s + \Delta s) = zs + z\Delta s + s\Delta z + \Delta z \Delta s = \mu e + \Delta z \Delta s. \tag{22}$$

Comparing this with our desired result, namely $z^+ s(z^+) = \mu e$, we see that the error is precisely the quadratic term $\Delta z \Delta s$. From (22) and (20) we deduce

$$(z^+)^T s(z^+) = \mu e^T e = \mu \bar{n}, \tag{23}$$

showing that after the Newton step the duality gap already has its desired value

Proximity Measure

To measure the quality of the approximation we introduce a proximity measure $\delta(z, \mu)$ that vanishes if $z = z(\mu)$ and is positive otherwise. It is convenient to introduce the vector

$$v := \sqrt{\frac{zs(z)}{\mu}}, \tag{24}$$

where all operations are componentwise. The proximity measure $\delta(z, \mu)$ is then defined by

$$\delta(z, \mu) := \tfrac{1}{2}\|v - v^{-1}\|. \tag{25}$$

Note that $z = z(\mu)$ if and only if $v = e$ and then $\delta(z, \mu) = 0$, and otherwise $\delta(z, \mu) > 0$. We proceed with some lemmas. The first lemma is a simple technical result on the componentwise product of two arbitrary orthogonal vectors. In this lemma $\|\cdot\|$ denotes the Eucledian norm (or 2-norm) and $\|\cdot\|_\infty$ the Chebychev norm (or infinity norm).

Lemma 3. *If* u *and* v *are orthogonal in* $\mathbb{R}^n$, *then*

$$\|uv\|_\infty \leqslant \tfrac{1}{4}\|u + v\|^2, \quad \|uv\| \leqslant \frac{1}{2\sqrt{2}}\|u + v\|^2.$$

Proof: *We may write*

$$uv = \tfrac{1}{4}[(u + v)^2 - (u - v)^2]. \tag{26}$$

From this we derive the componentwise inequality

$$-\tfrac{1}{4}(u - v)^2 \leqslant uv \leqslant \tfrac{1}{4}(u + v)^2.$$

This implies

$$-\tfrac{1}{4}\|u - v\|^2 e \leqslant uv \leqslant \tfrac{1}{4}\|u + v\|^2 e.$$

Since u *and* v *are othogonal, the vectors* $u - v$ *and* $u + v$ *have the same 2-norm, and hence the first inequality in the lemma follows. For the second inequality we derive from (26) that*

$$\|uv\|^2 = e^T(uv)^2 = \tfrac{1}{16}e^T((u + v)^2 - (u - v)^2)^2 \leqslant \tfrac{1}{16}e^T((u + v)^4 + (u - v)^4).$$

Since $e^T z^4 \leqslant \|z\|^4$ *for any* $z \in \mathbb{R}^n$, *we obtain*

$$\|uv\|^2 \leqslant \tfrac{1}{16}(\|u + v\|^4 + \|u - v\|^4).$$

Again using that $\|u - v\| = \|u + v\|$, *the second inequality follows.*

The next lemma estimates the error term in terms of the proximity measure.

Lemma 4. *If* $\delta := \delta(z, \mu)$, *then* $\|\Delta z \Delta s\|_\infty \leqslant \mu\delta^2$ *and* $\|\Delta z \Delta s\| \leqslant \mu\delta^2\sqrt{2}$.

Proof: *Componentwise division of (21) by* $\sqrt{\mu}v = \sqrt{zs}$ *yields*

$$\sqrt{\frac{z}{s}}\Delta s + \sqrt{\frac{s}{z}}\Delta z = \sqrt{\mu}(v^{-1} - v).$$

The terms at the left represent orthogonal vectors whose componentwise product is $\Delta z \Delta s$. *Applying Lemma 3 to these vectors, and using that* $\|v^{-1} - v\| = 2\delta$, *the result immediately follows.*

Quadratic Convergence of the Newton Process

We are now ready for the main result on the Newton direction.

Lemma 5. *If* $\delta := \delta(z, \mu) < 1$, *then the Newton step is strictly feasible, that is,* $z^+ > 0$ *and* $s^+ > 0$. *Moreover,*

$$\delta(z^+, \mu) \leqslant \frac{\delta^2}{\sqrt{2(1 - \delta^2)}}.$$

Proof: *Let* $0 \leqslant \alpha \leqslant 1, z^\alpha = z + \alpha \Delta z$ *and* $s^\alpha = s + \alpha \Delta s$. *We then have, using* (21),

$$z^\alpha s^\alpha = (z + \alpha \Delta z)(s + \alpha \Delta s) = zs + \alpha(z\Delta s + s\Delta z) + \alpha^2 \Delta z \Delta s.$$

$$= zs + \alpha(\mu e - zs) + \alpha^2 \Delta z \Delta s = (1-\alpha)zs + \alpha(\mu e - \alpha \Delta z \Delta s)$$

By Lemma 4, $\mu e - \alpha \Delta z \Delta s \geqslant \mu(1-\alpha\delta^2)e > 0$. *Hence, since* $(1-\alpha)zs \geqslant 0$, *we have* $z^\alpha s^\alpha > 0$, *for all* $\alpha \in [0,1]$. *Therefore, the components of* z^α *and* s^α *cannot vanish for any* $\alpha \in [0,1]$. *Hence, since* $z > 0$ *and* $s > 0$, *by continuity,* z^α *and* s^α *must be positive for any such* α, *especially for* $\alpha = 1$. *This proves the first statement in the lemma.*

Now let us turn to the proof of the second statement. Let $\delta^+ := \delta(z^+, \mu)$ *and*

$$v^+ := \sqrt{\frac{z^+ s^+}{\mu}}.$$

Then, by definition,

$$2\delta^+ = \|(v^+)^{-1} - v^+\| = \|(v^+)^{-1}[e - (v^+)^2]\|.$$

Recall from (22) *that* $z^+s^+ = \mu e + \Delta z \Delta s$. *In other words,* $\mu(v^+)^2 = \mu e + \Delta z \Delta s$. *Substitution gives*

$$2\delta^+ = \left\| \frac{\frac{\Delta z \Delta s}{\mu}}{\sqrt{e + \frac{\Delta z \Delta s}{\mu}}} \right\| \leqslant \frac{\left\|\frac{\Delta z \Delta s}{\mu}\right\|}{\sqrt{1 - \left\|\frac{\Delta z \Delta s}{\mu}\right\|_\infty}} \leqslant \frac{\delta^2\sqrt{2}}{\sqrt{1-\delta^2}};$$

in the last inequality we used Lemma 4 twice. Thus the proof is complete.

Lemma 5 implies that when $\delta \leqslant 1/\sqrt{2}$, then after a Newton step the proximity to the μ-center satisfies $\delta(z^+, \mu) \leqslant \delta^2$. In other words, Newton's method is quadratically convergent.

EXISTENCE PROOF

Now suppose that we know $z := z(\mu^0)$ for some $\mu^0 > 0$. Note that this is true for $\mu^0 = 1$, since $s(e) = e$. Let $\mu > 0$ be such that $\delta(z, \mu) \leqslant 1/\sqrt{2}$. Since $zs(z) = \mu^0 e$, this holds if and only if

$$\left| \sqrt{\frac{\mu^0}{\mu}} - \sqrt{\frac{\mu}{\mu^0}} \right| \|e\| \leqslant \frac{1}{\sqrt{2}}.$$

Using $\|e\| = \sqrt{\bar{n}}$, one may easily verify that this certainly holds if

$$\frac{1}{1 + \frac{1}{2\sqrt{2\bar{n}}}} \leqslant \sqrt{\frac{\mu}{\mu^0}} \leqslant 1 + \frac{1}{2\sqrt{2\bar{n}}}. \tag{27}$$

Now starting the Newton process at $z = z^0$, with μ fixed, we can generate an infinite sequence $z^0, z^1, \ldots z^k, \ldots$ such that

$$\delta(z^k, \mu) \leqslant \frac{1}{2^{2^k - 1}}.$$

Hence

$$\lim_{k \to \infty} \delta(z^k, \mu) = 0.$$

The generated sequence has at least one accumulation point z^*, since the iterates $z^1, \ldots z^k, \ldots$ lie in the compact set

$$e^T z + e^T s(z) = \bar{n}(1 + \mu), \quad z \geqslant 0, s(z) \geqslant 0,$$

due to (19) and (23). Since $\delta(z^*, \mu) = 0$, we obtain $z^* s(z^*) = \mu e$. This proves that the μ-center exists if μ satisfies (27) with $\mu^0 = 1$, that is if

$$\frac{1}{1+\frac{1}{2\sqrt{2\bar{n}}}} \leqslant \sqrt{\mu} \leqslant 1+\frac{1}{2\sqrt{2\bar{n}}}.$$

By redefining μ^0 as one of the endpoints of the above interval we can repeat the above procedure, and extend the interval where the μ-center exists to

$$\frac{1}{\left(1+\frac{1}{2\sqrt{2\bar{n}}}\right)^2} \leqslant \sqrt{\mu} \leqslant \left(1+\frac{1}{2\sqrt{2\bar{n}}}\right)^2,$$

and so on. After applying the procedure k times the interval where the μ-center certainly exists is given by

$$\frac{1}{\left(1+\frac{1}{2\sqrt{2\bar{n}}}\right)^k} \leqslant \sqrt{\mu} \leqslant \left(1+\frac{1}{2\sqrt{2\bar{n}}}\right)^k.$$

For arbitrary $\mu > 0$, we have to apply the above procedure at most

$$\frac{|\log \mu|}{2\log\left(1+\frac{1}{2\sqrt{2\bar{n}}}\right)}$$

times, to obtain the μ-center. This completes the proof of the existence of the central path.

4.3. Existence of a Strictly Complementary Solution

Now that we have proven the existence of the central path we can use it as a guide to the optimal set, by letting the parameter μ approach to zero. As we show in this section, in this way we obtain an optimal solution z such that $z+s(z)>0$. Recall that the optimality of z means that $zs(z)=0$, which is usually expressed by saying that z and $s(z)$ are complementary vectors. If moreover $z+s(z)>0$ we say that z and $s(z)$ are strictly complementary vectors. Then for every index i, either $z_i>0$ or $s_i(z)>0$. This implies that the index sets B and N, introduced in Section 3.4 form a partition of the index set, the so-called optimal of (SP).

It is convenient to introduce some more notation. It z is a nonnegative vector, we define its support, denoted by $\sigma(z)$, as the set of coordinate indices i for which $z_i>0$. Note that if $s=s(z)$, then $\sigma(z)\cap\sigma(s)=\varnothing$ if and only if z is an optimal solution. Furthermore, z is a strictly complementary optimal solution if and only if it is optimal and $\sigma(z)\cup\sigma(s)=\{1,2,\ldots,\bar{n}\}$.

Theorem 1. *There an optimal solution* z^* *exists such that* $z^*+s(z^*)>0$.

Proof: *Let* $\{\mu_k\}_{k=1}^{\infty}$ *be a positive sequence such that* $\mu_k\to 0$ *if* $k\to\infty$, *and let* $s(\mu_k):=s[z(\mu_k)]$. *Due to (19) the set* $\{[z(\mu_k), s(\mu_k)]\}$ *lies in compact set, and hence it contains a subsequence converging to a point* (z^*, s^*), *with* $s^*=s(z^*)$. *Since* $z(\mu_k)^T s(\mu_k) = \bar{n}\mu_k\to 0$, *we have* $(z^*)^T s^*=0$. *Hence, from (11),* $q^T z^*=0$. *So* z^* *is an optimal solution. We claim that* (z^*, s^*) *is a strictly complementary pair. To prove this, we apply the orthogonality property (10) to the points* z^* *and* $z(\mu_k)$, *which gives*

$$[z(\mu_k)-z^*]^T[s(\mu_k)-s^*]=0.$$

Rearranging the terms, and using $z(\mu_k)^T s(\mu_k)=\bar{n}\mu_k$ *and* $(z^*)^T s^*=0$, *we arrive at*

$$\sum_{j\in\sigma(z^*)} z_j^* s_j(\mu_k) + \sum_{j\in\sigma(s^*)} z_j(\mu_k) s_j^* = \bar{n}\mu_k.$$

Dividing both sides by μ_k *and recalling that* $z_j(\mu_k)s_j(\mu_k)=\mu_k$, *we obtain*

$$\sum_{j\in\sigma(z^*)} \frac{z_j^*}{z_j(\mu_k)} + \sum_{j\in\sigma(s^*)} \frac{s_j^*}{s_j(\mu_k)} = \bar{n}.$$

Letting $k \to \infty$, *the first sum on the left becomes equal to the number of nonzero coordinates in* z^*. *Similarly, the second sum becomes equal to the number of nonzero coordinates in* s^*. *The sum of these numbers being* $\bar{n}$, *we conclude that the optimal pair* (z^*, s^*) *is strictly complementary.*

By using a similar proof technique it can be shown that the limit of $z(\mu)$ exists if μ goes to zero. In other words, the central path converges. The limit point is (of course) a uniquely determined optimal solution of (SP), which can further be characterized as the so-called analytic center of the set optimal solutions (Roos et al., 1997).

Theorem 1 shows that a strictly complementary solution z of (SP) exists. Having such a solution, the classes B and N simply follow from

$$B = \{i : z_i > 0\}, \quad N = \{i : s_i(z) > 0\}.$$

Let us finally mention that Theorem 1 is equivalent to an old result of Goldman and Tucker, which states that every feasible linear system of equalities and inequalities has a strictly feasible solution (Goldman and Tucker, 1956). We want to make explicit that Theorem 1 also implies the strong duality theorem for LO. We devote a separate section to this.

4.4. Proof of the Strong Duality Theorem

The strong duality theorem for LO states that if one of the problems (P) and (D) is feasible, then so is the other problem, and then both problems have optimal solutions and their optimal objective values coincide. As we will show, this result is tightly related to the fact (proved below) that (P) and (D) have optimal solutions if and only $\kappa \in B$. This goes as follows.

In Section 3.2 we established that (P) and (D) have optimal solutions with zero duality gap if and only if a solution of (SP) with $\kappa > 0$ exists. This happens exactly when $\kappa \in B$. For the proof of the strong duality theorem it remains to show that if $\kappa \notin B$, then neither (P) nor (D) has an optimal solution. Thus we consider the case where $\kappa \notin B$. Then, by Theorem 1, we may conclude that $\kappa \in N$ and hence a solution $(\tilde{x}, \tilde{y}, k)$ to (5) exists with a positive slack value for the third constraint. In other words, $\tilde{x}$ and $\tilde{y}$ satisfy

$$\tilde{x} \geqslant 0, \quad \tilde{y} \geqslant 0, \quad A\tilde{x} \geqslant 0, \quad A^T\tilde{y} \leqslant 0, \quad b^T\tilde{y} - c^T\tilde{x} > 0.$$

By the last inequality we cannot have $b^T\tilde{y} \leqslant 0$ and $c^T\tilde{x} \geqslant 0$. Hence, either $b^T\tilde{y} > 0$ or $c^T\tilde{x} < 0$. If $b^T\tilde{y} > 0$ then (P) is infeasible, since otherwise $x \geqslant 0$ such that $Ax \geqslant b$ and we have

$$b^T\tilde{y} \leqslant (Ax)^T\tilde{y} = x^T A^T\tilde{y} \leqslant 0,$$

a contradiction. Moreover, if (D) is not infeasible, then $y \geqslant 0$ such that $A^T y \leqslant c$. Then $y + \alpha\tilde{y}$ is feasible for (D), for any $\alpha > 0$. The dual objective value $b^T(y + \alpha\tilde{y}) = b^T y + \alpha b^T\tilde{y}$ goes to infinity if α grows. This shows that if $b^T\tilde{y} > 0$, then (P) is infeasible and (D) is infeasible or unbounded.

The case $c^T\tilde{x} < 0$ can be handled in the same way. Then (D) is infeasible, since otherwise there exists $y \geqslant 0$ such that $A^T y \leqslant c$, which would imply the contradiction

$$c^T\tilde{x} \geqslant (A^T y)^T\tilde{x} = y^T A\tilde{x} \geqslant 0.$$

If (P) is not infeasible, then we have $x \geqslant 0$ such that $Ax \geqslant b$ and $x + \alpha\tilde{x}$ is feasible for (P), for any $\alpha > 0$. If α grows then the objective value $c^T(x + \alpha\tilde{x}) = c^T x + \alpha c^T\tilde{x}$ goes to minus infinity, showing that (P) is unbounded in that case. We conclude that if $c^T\tilde{x} < 0$ then (D) is infeasible and (P) is infeasible or unbounded. This completes the proof of the strong duality theorem for LO.

5. A Polynomial Solution Method for LO

After all the theoretical results of the previous sections we are now ready to present an algorithm that finds a strictly complementary solution of (SP) in polynomial time.

5.1. *Newton Step Algorithm*

The idea of the algorithm is quite simple. Starting at $z = e$, we choose $\mu < 1$ such that

$$\delta(z, \mu) \leqslant \tau := \frac{1}{\sqrt{2}}, \tag{28}$$

and perform a Newton step targeting at $z(\mu)$. After the step the new iterate z satisfies $\delta(z, \mu) \leqslant \tau^2 = \frac{1}{2}$. Then we decrease μ such that (28) holds for the new values of z and μ and repeat the procedure. Note that after each Newton step we have $q^T z = \bar{n}\mu$. Thus, if μ approaches zero, then z will approach the set of optimal solutions. Formally the algorithm can be stated as follows.

Full Newton step algorithm

```
Input:
  An accuracy parameter ε > 0;
  an update parameter θ, 0 < θ < 1.
begin
  z = e; μ := 1;
  while nμ ≥ ε do
  begin
    μ := (1 − θ)μ;
    z := z + Δz;
  end
end
```

Note that the reduction of the barrier parameter μ is realized by the multiplication with the factor $1-\theta$. In the next section we discuss how an appropriate value of the update parameter θ can be obtained, so that during the course of the algorithm the iterates are kept within the region where Newton's method is quadratically convergent.

5.2. *Complexity Analysis*

At the start of the algorithm we have $\mu = 1$ and $z = z(1) = e$, hence $q^T z = \bar{n}$ and $\delta(z, \mu) = 0$. In each iteration μ is first reduced with the factor $1-\theta$ and then the Newton step is made to the new μ-center. It will be clear that the reduction of μ has effect on the value of the proximity measure. This effect is fully described by the following lemma.

Lemma 6. *Let* $z > 0$ *and* $\mu > 0$ *be such that* $s = s(z) > 0$ *and* $q^T z = \bar{n}\mu$. *Moreover, let* $\delta := \delta(z, \mu)$ *and* $\mu' = (1-\theta)\mu$. *Then*

$$\delta(z, \mu')^2 = (1-\theta)\delta^2 + \frac{\theta^2 n}{4(1-\theta)}.$$

Proof: *Let* $\delta^+ := \delta(z, \mu')$ *and* $v = \sqrt{zs/\mu}$, *as defined in (24). Then, by definition,*

$$4(\delta^+)^2 = \left\| \sqrt{1-\theta}\, v^{-1} - \frac{v}{\sqrt{1-\theta}} \right\|^2 = \left\| \sqrt{1-\theta}\, (v^{-1} - v) + \frac{\theta v}{\sqrt{1-\theta}} \right\|^2.$$

From $z^T s = \bar{n}\mu$ *it follows that* $\|v\|^2 = \bar{n}$. *Hence,* v *is orthogonal to* $v^{-1} - v$:

$$v^T(v^{-1} - v) = \bar{n} - \|v\|^2 = 0.$$

Therefore,

$$4(\delta^+)^2 = (1-\theta)\|v^{-1} - v\|^2 + \frac{\theta^2 \|v\|^2}{1-\theta}.$$

Finally, since $\|v^{-1} - v\| = 2\delta$ *and* $\|v\|^2 = n$ *the result follows.*

Now let us have

$$q^T z = \bar{n}\mu, \quad \text{and} \quad \delta(z,\mu) \leqslant \tfrac{1}{2} \tag{29}$$

at the start of some iteration. It certainly holds in the first iteration. We claim that these properties are maintained during the course of the algorithm if $\theta = 1/(2\sqrt{\bar{n}})$.

When the barrier parameter is updated to $\mu' = (1-\theta)\mu$, the lemma gives

$$\delta(z,\mu')^2 \leqslant \frac{1-\theta}{4} + \frac{\theta^2 n}{4(1-\theta)}.$$

Due to the choice of θ, we get

$$\delta(z,\mu')^2 < \frac{1}{4} + \frac{1}{16(1-\theta)} \leqslant \frac{1}{4} + \frac{1}{8} < \frac{1}{2}.$$

Therefore, after the Newton step we obtain z' such that $\delta(z',\mu') \leqslant 1/2$. Also, $q^T z' = n\mu'$. This proves the claim.

How many iterations are needed by the algorithm? The answer is provided by the following lemma.

Lemma 7. *After at most*

$$\left\lceil \frac{1}{\theta} \log \frac{\bar{n}}{\varepsilon} \right\rceil$$

iterations we have $\bar{n}\mu \leqslant \varepsilon$.

Proof: *Initially, the objective value is* $\bar{n}$ *and in each iteration it is reduced by the factor* $1-\theta$. *Hence, after* k *iterations the objective value, given by* $q^T x(\mu) = \bar{n}\mu$, *is smaller than, or equal to* ε *if*

$$(1-\theta)^k \bar{n} \leqslant \varepsilon.$$

Taking logarithms, this becomes

$$k \log(1-\theta) + \log \bar{n} \leqslant \log \varepsilon.$$

Since $-\log(1-\theta) \geqslant \theta$, *this certainly holds if*

$$k\theta \geqslant \log \bar{n} - \log \varepsilon = \log \frac{\bar{n}}{\varepsilon}.$$

This implies the lemma.

The above results are summarized in the next theorem.

Theorem 2. *If* $\theta = 1/(2\sqrt{\bar{n}})$ *then the algorithm with full Newton steps requires at most*

$$\left\lceil 2\sqrt{\bar{n}} \log \frac{\bar{n}}{\varepsilon} \right\rceil$$

iterations. The output is a feasible $z > 0$ *such that* $q^T z = \bar{n}\mu \leqslant \varepsilon$ *and* $\delta(z,\mu) \leqslant \frac{1}{2}$.

This theorem shows that we can get an ε-solution z of our self-dual model with ε as small as is desirable. But note that z will always be an interior solution, so $z > 0$ and $s = s(z) > 0$.

A crucial question for us is whether the variable $\kappa = z_{\bar{n}-1}$ is positive or zero in the limit, when μ goes to zero. In practice, for a small enough ε it is usually no serious problem to decide which of the two cases occurs. But the theory can help us in an elegant way, as we explain in the next section, where we show that the optimal partition can be found in polynomial time. This requires some further analysis of the central path.

Before proceeding, we want to agree upon the following. Because of (29), the iterates move in a narrow neighborhood around the central path. In the next sections we will assume that the iterates are on the central path, because this simplifies the analysis significantly. At the cost of some additional technicalities in the analysis we can obtain similar results for the iterates of the practical algorithm. For the technical details the reader is referred to Roos et al. (1997).

5.3. *Finding the Optimal Partition*

Let S denote the set of optimal solutions of (SP). If $z \in S$, then, for each index i, either $z_i = 0$ or $s_i(z) = 0$, due to the complementarity property. On the other hand, given i, due to the existence of strictly complementary solutions, there exists an optimal z such that $z_i + s_i(z) > 0$. Hence the number

$$\sigma := \min_i \max_{z \in S} (z_i + s_i(z))$$

is positive. One also has $\sigma \leqslant \bar{n}$, due to (19). We call σ the condition number of (SP). The calculation of this number is more cumbersome than solving (SP). For our purpose, however, it is sufficient to know the following lower bound

$$\sigma \geqslant \frac{1}{\prod_{j=1}^{n} \|M_j\|}, \tag{30}$$

where M_j denotes the jth column of M (Theorem I.33 in Roos et al., 1997).

For this bound we need to make the assumption that the entries of M and q are integral. In the following, in all the bounds where the condition number occurs, it can be safely replaced by this easily computable lower bound.

Lemma 8. *For any positive* μ *one has*

$$z_i(\mu) \geqslant \frac{\sigma}{\bar{n}}, \quad i \in B, \quad z_i(\mu) \leqslant \frac{\bar{n}\mu}{\sigma}, \quad i \in N.$$

Proof: *Let* $i \in N$ *and* $\tilde{z} \in S$ *be such that* $\tilde{s}_i := s_i(\tilde{z})$ *is maximal. Then the definition of the condition number* σ *implies that* $\tilde{s}_i \geqslant \sigma > 0$. *Using (10) once more, we get*

$$(z(\mu) - \tilde{z})^T (s(\mu) - \tilde{s}) = 0.$$

Since $q^T \tilde{z} = \tilde{z}^T \tilde{s} = 0$, *it follows that*

$$z(\mu)^T \tilde{s} + s(\mu)^T \tilde{z} = \bar{n}\mu.$$

This implies

$$z_i(\mu)\tilde{s}_i \leqslant z(\mu)^T \tilde{s} \leqslant \bar{n}\mu.$$

Dividing by $\tilde{s}_i$ *and using that* $\tilde{s}_i \geqslant \sigma$ *we obtain the second inequality of the lemma:*

$$z_i(\mu) \leqslant \frac{\bar{n}\mu}{\tilde{s}_i} \leqslant \frac{\bar{n}\mu}{\sigma}.$$

If $i \in B$, *then it can be derived in a similar way that*

$$s_i(\mu) \leqslant \frac{\bar{n}\mu}{\sigma}.$$

Since $z_i(\mu) s_i(\mu) = \mu$, *this implies the first inequality.*

Lemma 8 gives a complete separation between variables in B and variables in N, provided that μ is so small that

$$\frac{\sigma}{\bar{n}} > \frac{\bar{n}\mu}{\sigma}.$$

or, equivalently,

$$\bar{n}\mu < \frac{\sigma^2}{\bar{n}}.$$

Thus, applying the algorithm with

$$\varepsilon = \frac{\sigma^2}{\bar{n}},$$

we obtain a solution z, which reveals the optimal partition according to

$$B = \{i : z_i > s_i(z)\}, \quad N = \{i : z_i < s_i(z)\}.$$

By substitution of the above value of ε in Theorem 7, the required number of iterations follows:

$$\left\lceil 2\sqrt{\bar{n}} \log \frac{\bar{n}^2}{\sigma^2} \right\rceil.$$

After this number of iterations we know the optimal partition, and hence if κ belongs to B or not. In other words, we then know whether the original problem has an optimal solution or not. In the last case we are done, otherwise we are usually interested in finding a solution. If we are interested in an ε-solution we can proceed with the algorithm until such a solution is obtained. An alternative approach may be to use the rounding procedure described in the next section, which yields an exact solution.

5.4. Rounding to an Exact Solution

Knowing the optimal partition, we aim for finding a strictly complementary solution of (SP). We first establish that the set B in the optimal partition cannot be empty, because then $z = 0$ is a strictly complementary solution, which implies that $s(0)$ is positive. Since $s(0) = q$, and q has zero entries, this cannot happen. Therefore, the set B is not empty. Knowing this we describe a rounding procedure that can be applied to any z generated by the algorithm to yield a vector $\bar{z}$ such that $\bar{z}$ and its slack vector $\bar{s} = s(\bar{z})$ are complementary (in the sense that $\bar{z}_N = \bar{s}_B = 0$). In general, however, these $\bar{z}$ and $\bar{s}$ are not necessarily nonnegative. But, as we will see, after sufficiently many additional iterations of the algorithm the separation between the small and large variables is strong enough to get a strictly complementary solution. All of this can be done in polynomial time.

In the following we refer to the variables $z_i, i \in B$ and $s_i, i \in N$ as large variables and to the remaining variables as small variables. Recall that the large variables converge to positive values if μ approaches zero, and the small variables to zero. Partitioning the matrix M and the vectors z and s according to the optimal partition, the relation $s = Mz + q$ can be rewritten as

$$\begin{pmatrix} s_B \\ s_N \end{pmatrix} = \begin{pmatrix} M_{BB} M_{BN} \\ M_{NB} M_{NN} \end{pmatrix} \begin{pmatrix} z_B \\ z_N \end{pmatrix} + \begin{pmatrix} q_B \\ q_N \end{pmatrix}. \tag{31}$$

Now let $\tilde{z}$ be a strictly complementary solution. Then $q_B^T \tilde{z}_B = 0$ and $\tilde{z}_B > 0$, which implies $q_B = 0$. Using this we derive form (31) that

$$s_B = M_{BB} z_B + M_{BN} z_N. \tag{32}$$

Now let z be generated by the algorithm. We consider the system of equations in the unknown vector ξ given by

$$M_{BB} \xi = s_B - M_{BN} z_N. \tag{33}$$

Note that $\xi = z_B$ is a large solution of (33); its entries are large variables. We can easily see that (33) has more solutions. This follows by using the strictly complementary solution $\tilde{z}$ once more. Since $\tilde{z}_N = 0$ and $s_B(\tilde{z}) = 0$, it follows from (32) that $M_{BB}\tilde{z}_B = 0$. Since $z_B > 0$, we conclude that the matrix M_{BB} must be singular, and hence (33) has multiple solutions.

Now let ξ be any solution of (33) and consider the vector $\bar{z}$ defined by

$$\bar{z}_B = z_B - \xi, \quad \bar{z}_N = 0.$$

Define $\bar{s} = s(\bar{z})$. Since $\bar{z}_N = 0$, we have

$$\bar{s}_B = M_{BB} \bar{z}_B = M_{BB}(z_B - \xi) = 0.$$

Therefore, $\bar{z}_N = \bar{s}_B = 0$, showing that the vectors $\bar{z}$ and $\bar{s}$ are complementary. It will be clear, however, that the vectors $\bar{z}$ and $\bar{s}$ are not necessarily nonnegative, let alone strictly complementary. This only holds if

$$\bar{z}_B = z_B - \xi > 0, \tag{34}$$

and

$$\bar{s}_N = M_{NB}\bar{z}_B + M_{NN}\bar{z}_N + q_N = M_{NB}(z_B - \xi) + q_N = s_N - M_{NN}z_N - M_{NB}\xi > 0. \tag{35}$$

Note that if we run the algorithm long enough, z_B and s_N converge to positive vectors, whereas z_N and s_B (and hence also ξ) converge to zero. This makes it plausible that if ε is made small enough in the algorithm, then we will get a solution ξ of (33) that satisfies (34) and (35), hence giving rise to a strictly complementary solution $\bar{z}$ of (33).

Omitting further details, we conclude this section by stating the main complexity result: If we execute $7\sqrt{\bar{n}}L$ iterations, and apply the above rounding procedure to the point z then generated by the algorithm, we obtain a strictly complementary solution $\bar{z}$ of (SP). Here L denotes the binary input size of the problem (SP).

6. Other Interior-Point Methods

In the preceding sections we have shown that the LO problem can be solved in polynomial time. If the problem is infeasible or unbounded then this is detected by the method, and otherwise it generates an exact (strictly complementary) solution. The number of iterations required by the algorithm will be about the same as predicted by the theory. In practice this means that the method is much slower than the simplex method. There are several ways to make the method more efficient and competitive with the simplex method. We briefly discuss some of them. For a more extensive discussion we refer the reader to Roos et al. (1997) and to some relevant references given below. We restrict ourselves below to strategies that may decrease the number of iterations. The computational effort per iteration, needed to compute the Newton step, is another issue that is critical for the performance of the algorithm, but we do not discuss that here.

6.1. *Adaptive-Update Methods*

The method we considered generates iterates in a narrow neighborhood of the central path, since after each Newton step we have $\delta(x, \mu) \leqslant \frac{1}{2}$. In the theoretical analysis this property is maintained by taking small updates of the barrier parameter ($\theta = 1/(2\sqrt{\bar{n}})$). In practice, Newton's method behaves much better than predicted by Lemma 5. As a consequence, after the Newton step $\delta(x, \mu)$ will usually be much smaller than $\frac{1}{2}$. It means that we can take larger values of θ without loosing the above property. The adaptive-update method seeks to make θ as large as possible so that after each Newton step we still have $\delta(x, \mu) \leqslant \frac{1}{2}$. This strategy accelerates the method significantly and does not deteriorate the theoretical iteration bound (Jarre and Saunders, 1991; Mizuno and Todd, 1991; Mizuno et al., 1993).

6.2. *Large-Update Methods*

A popular strategy is to use larger values of θ, say $\theta = 0.5$, or even $\theta = 0.99$. In this case, after the barrier update, $\delta(x, \mu)$ will be much larger than 1, and Lemma 5 can no longer be used to measure the performance of the Newton process. Even worse, the (full) Newton step may be no longer feasible. The remedy is to take a damped Newton step $z' = z + \alpha\Delta z$, with damping factor α; this factor can be chosen so that z' is feasible and at the same time the proximity decreases sufficiently to get a provable polynomial method. In practice this approach yields very efficient methods (Anstreicher and Bosch, 1992; Gonzaga, 1991a,b; den Hertog et al., 1992; Jansen et al., 1996; Kojima et al., 1993a,b; Roos and Vial, 1989; Todd and Ye, 1996).

6.3. *Predictor-Corrector Method*

This is the most popular method. We describe a simple variant. It is based on a very greedy strategy that uses the Newton step targeting at the zero vector. So, in the definition of the

Newton step one takes $\mu = 0$. The resulting direction is called the affine scaling direction. To stay feasible, this step is damped again, but with a greedy damping factor. As a result of such a step we may end up far from the central path. To restore the proximity a so-called centering step is taken. This is a (damped) Newton step with $\mu = q^T z/\bar{n}$, where z is the current iterate. This method has not only turned out to be extremely efficient, but also has the nice property that asymptotically the objective value converges quadratically to zero (Lustig et al., 1992; Mehrotra, 1992; Potra, 1994; Zhang and Zhang, 1995).

References

Andersen, E. D. and K. D. Andersen. 1999. "The MOSEK interior-point optimizer for linear programming: an implementation of the homogeneous algorithm." In *High Performance Optimization Techniques,* edited by S. Zhang, H. Frenk, C. Roos, and T. Terlaky, pp. 197–232. Dordrecht, The Netherlands: Kluwer Academic Publishers.

Anstreicher, K. M. and R. A. Bosch. 1992. "Long steps in a $O(n^3L)$ algorithm for linear programming." *Math. Programming* **54**: 251–265.

Ben-Tal, A. and A. S. Nemirovskii. 2001. In *Lectures on Modern Convex Optimization. Analysis, Algorithms, Engineering Applications.* Philadelphia: SIAM Publications.

Bonnans, J. F. and F. A. Potra. 1997. "On the convergence of the iteration sequence of infeasible path following algorithms for linear complementarity problems." *Math. Oper. Res.* **22(2)**: 378–407.

Boyd, S. E. and L. Vandenberghe. 1996. "Semidefinite programming." *SIAM Rev.* **38(1)**: 49–96.

Dantzig, G. B. 1963. In *Linear Programming and Extensions.* Princeton: Princeton University Press.

Dantzig, G. B. 1991. "Linear programming." In *History of Mathematical Programming. A Collection of Personal Reminiscences,* edited by J. K. Lenstra, A. H. G. Rinnooy Kan, and A. Schrijver. The Netherlands: CWI, Amsterdam.

Dikin, I. I. 1967. "Iterative solution of problems of linear and quadratic programming." *Dokl. Akad. Nauk SSSR* **174**: 747–748. Translated in: *Sov. Math. Dokl.* **8**: 674–675.

Dikin, I. I. 1974. "On the convergence of an iterative process." *Upravlyaemye Sistemi* **12**: 54–60 (in Russian).

Dikin, I. I. 1988. "Letter to the editor." *Math. Programming* **41**: 393–394.

Frisch, K. R. 1956. "La resolution des problemes de programme lineaire par la methode du potential logarithmique." *Cahiers du Seminaire D'Econometrie* **4**: 7–20.

Frisch, R. 1955. "The logarithmic potential method for solving linear programming problems." Memorandum. Oslo, Norway: University Institute of Economics.

Goldman, A. J. and A. W. Tucker. 1956. "Theory of linear programming." In *Linear Inequalities and Related Systems, Annals of Mathematical Studies, No. 38,* edited by H. W. Kuhn and A. W. Tucker, pp. 53–97. Princeton: Princeton University Press.

Gonzaga, C. C. 1991a. "Large step path-following methods for linear programming. Part I: barrier function method." *SIAM J. Optim.* **1**: 268–279.

Gonzaga, C. C. 1991b. "Large step path-following methods for linear programming. Part II: potential reduction method." *SIAM J. Optimization* **1**: 280–292.

den Hertog, D., C. Roos, and J.-Ph. Vial. 1992. "A complexity reduction for the long-step path-following algorithm for linear programming." *SIAM J. Optim.* **2**: 71–87.

Huard, P. 1967. "Resolution of mathematical programming with nonlinear constraints by the method of centers." In *Nonlinear Programming,* edited by J. Abadie, pp. 207–219. Amsterdam: North-Holland.

Jansen, B., C. Roos, T. Terlaky, and J.-Ph. Vial. 1996. "Long-step primal-dual target following algorithms for linear programming." *Math. Methods Oper. Res.* **44**: 11–30.

Jarre, F. and M. A. Saunders. 1991. "An adaptive primal–dual method for linear programming." *COAL Newsletter* **19**: 7–16.

Karmarkar, N. K. 1984. "A new polynomial-time algorithm for linear programming." *Combinatorica* **4**: 373–395.

Khachiyan, L. G. 1979. "A polynomial algorithm in linear programming." *Dokl. Akad. Nauk SSSR* **244**: 1093–1096. Translated into English in *Sov. Math. Dokl.* **20**: 191–194.

Kojima, M., Y. Kurita, and S. Mizuno. 1993a. "Large-step interior point algorithms for linear complementarity problems." *SIAM J. Optim.* **3(2)**: 398–412.

Kojima, M., N. Megiddo, and S. Mizuno. 1993b. "Theoretical convergence of large-step primal–dual interior point algorithms for linear programming." *Math. Programming* **59**: 1–22.

Lustig, I. J., R. E. Marsten, and D. F. Shanno. 1992. "On implementing Mehrotra's predictor-corrector interior point method for linear programming." *SIAM J. Optim.* **2**: 435–449.

Mehrotra, S. 1992. "On the implementation of a primal–dual interior method." *SIAM J. Optim.* **2(4)**: 575–601.

Mizuno, S. and M. J. Todd. 1991. "An $O(n^3L)$ adaptive path following algorithm for a linear complementarity problem." *Math. Programming* **52**: 587–595.

Mizuno, S., M. J. Todd, and Y. Ye. 1993. "On adaptive-step primal-dual interior-point algorithms for linear programming." *Math. Oper. Res.* **18**: 964–981.

Nesterov, Y. E. and A. S. Nemirovskii. 1993. In *Interior Point Polynomial Methods in Convex Programming: Theory and Algorithms*. Philadephia: SIAM Publications.

Potra, F. A. 1994. "A quadratically convergent predictor-corrector method for solving linear programs from infeasible starting points." *Math. Programming* **67(3)**: 383–406.

Roos, C. and T. Terlaky. 1997. "Advances in linear optimization." In *Annotated Bibliography in Combinatorial Optimization*, edited by M. Dell'Amico, F. Maffioli, and S. Martello, Chapter 7. New York: John Wiley & Sons.

Roos, C., T. Terlaky, and J.-Ph. Vial. 1997. *Theory and Algorithms for Linear Optimization. An Interior Approach.* Chichester, UK: John Wiley & Sons.

Roos, C. and J.-Ph. Vial. 1989. "Long steps with the logarithmic penalty barrier function in linear programming." In *Economic Decision-making: Games, Economics and Optimization*, edited by J. Gabszevwicz, J. F. Richard, and L. Wolsey, pp. 433–441. Amsterdam: Elsevier Science B.V.

Shor, N. Z. 1987. "Quadratic optimization problems." *Sov. J. Comput. Syst. Sci.* **25**: 1–11.

Todd, M. J. and Y. Ye. 1996. "A lower bound on the number of iterations of long-step and polynomial interior-point linear programming algorithms." *Ann. Oper. Res.* **62**: 233–252.

Wright, S. J. 1993. "A path-following infeasible-interior-point algorithm for linear complementarity problems." *Optim. Methods Softw.* **2**: 79–106.

Wright, S. J. 1994. "An infeasible-interior-point algorithm for linear complementarity problems." *Math. Programming* **67(1)**: 29–52.

Wright, S. J. 1996. In *Primal–dual Interior-point Methods*. Philadephia: SIAM Publications.

Zhang, Y. and D. Zhang. 1995. "On polynomiality of the Mehrotra-type predictor-corrector interior-point algorithms." *Math. Programming* **68**: 303–318.

2 SEMIDEFINITE PROGRAMMING

Henry Wolkowicz

ABSTRACT Semidefinite programming is an extension of linear programming where (some of) the vector variables are replaced by matrix variables and (some of) the nonnegativity elementwise constraints are replaced by positive semidefiniteness constraints. These are convex problems that can be solved efficiently by interior-point methods. They arise in many applications, for example, in combinatorial optimizations, matrix completion problems, stability of differential systems, and, more generally, as the dual of Lagrangian relaxations of quadratic models of numerically hard problems.

1. Introduction

1.1. Background

Semidefinite programming (denoted SDP) is an extension of linear programming (LP), where (some of) the vector variables are replaced by matrix variables and (some of) the nonnegativity elementwise constraints are replaced by positive semidefiniteness constraints. (The semidefiniteness constraints are also referred to as linear matrix inequalities.) We can express a linear (primal) SDP as

$$(PSDP) \qquad \begin{aligned} \mu^* := \min \quad & C \bullet X \\ \text{s.t.} \quad & \mathcal{A}X = b \\ & X \succeq 0, \end{aligned}$$

where C and X are in $\mathcal{S}^n$, the space of symmetric, real, $n \times n$ matrices; the inner product $C \bullet X = \operatorname{trace} CX$; $\succeq$ denotes nonnegativity in the Löwner partial order, in other words, $A \succeq B$ if $A - B \succeq 0$, that is, $A - B \in \mathcal{S}^n_+$, the cone of positive semidefinite matrices; and

$$\mathcal{A}: \mathcal{S}^n \longrightarrow \mathcal{R}^m$$

is a linear operator, that is, $(\mathcal{A}X)_i = \operatorname{trace}(A_i X)$, for the given $A_i \in \mathcal{S}^n$, $i = 1, \ldots m$. In fact, SDP is a special case of the cone programming problem

$$\begin{aligned} \min \quad & f(x) \\ \text{s.t.} \quad & g(x) \succeq_K 0, \end{aligned}$$

where: f and g are appropriate functions, K is a convex cone, and $g(x) \succeq_K 0$ denotes the cone partial order, that is, $g(x) \in K$. This is a very general mathematical program that can include standard equality and inequality constraints, since K can consist of the cross product of many cones, for example the nonnegative orthant, the semidefinite cone, and the origin $\{0\}$, $K = \mathcal{R}^n_+ \otimes \mathcal{S}^n_+ \otimes \{0\}$.

Thus, SDP includes LP as a special case; and so it has all the elegant duality and geometric results that LP does. In addition, it is a nonlinear problem and so has extra subtleties that create many interesting difficulties; the geometry of the cone $\mathcal{P}$ is particularly elegant and well studied. Linear SDPs are convex programs that can be solved very efficiently using interior point methods (IPMs). And, SDP also has many very important applications. It arises as the dual of Lagrangian relaxations for quadratic models of numerically hard problems. Thus, we have an efficient tool for finding bounds and estimates for many hard problems in a variety of different areas, for example in engineering, combinatorics, statistics.

SDP Relaxation for Max-Cut Problem

The success and popularity of SDP is exemplified by the results of the relaxation for the max-cut problem. Let $G = (V, E)$ be an undirected graph with the edge set at $V = \{v_i\}_{i=1}^n$ and weights w_{ij} on the edges $(v_i, v_j) \in E$. We want to find the index set $\mathcal{I} \subset \{1, 2, \ldots, n\}$, to maximize the weight of the edges with one end point with the index in $\mathcal{I}$ and the other in the complement. This is equivalent to

$$(MC) \quad mc^* := \max \tfrac{1}{2}\sum_{i<j} w_{ij}(1 - x_i x_j), \quad x \in \mathcal{F},$$

where $\mathcal{F} := \{\pm 1\}^n$, and $x_i = 1$ if $i \in \mathcal{I}$ and -1 otherwise. The objective function is a (homogeneous) quadratic form, $x^T Qx$. Therefore, we get the equivalent quadratic model

$$(QMC) \quad mc^* = \max x^T Qx, \quad x_i^2 = 1, \quad \forall i.$$

The LP relaxation is

$$(LPMC) \quad \max x^T Qx, \quad -1 \leqslant x_i \leqslant 1, \quad \forall i,$$

which yields very little useful information, for example the maximum is $x = 0$ when $Q \preceq 0$ and is intractable otherwise.

The SDP relaxation can be obtained easily from (QMC) by noting $x^T Qx = \operatorname{trace} x^T Qx = \operatorname{trace} Qxx^T$. We then substitute $X = xx^T$ and replace/relax the rank one condition on X to $X \succeq 0$. This yields

$$(SDPMC) \quad pmc^* := \max \operatorname{trace} QX, \quad \operatorname{diag}(X) = e, \quad X \succeq 0,$$

where diag(X) denotes the vector formed from the diagonal of X and e is the vector of ones.

There are several other equivalent forms. Another approach is to use the Lagrangian relaxation of the quadratic model (QMC).

$$mc^* \leqslant dmc^* := \min_{\lambda} \max_{x} x^T Qx - \sum_{i=1}^{n} \lambda_i (x_i^2 - 1) = \min_{Q - \operatorname{Diag}(\lambda) \succeq 0} e^T \lambda,$$

where the semidefinite constraint arises from assuming boundedness of the inner unconstrained quadratic maximization problem, which then attains a maximum at $x = 0$. Since the Slater constraint qualification holds, $Q - \operatorname{Diag}(\lambda) \succ 0$, we conclude that strong duality holds with (SDPMC), that is $mc^* \leqslant dmc^* = pmc^*$ and (SDPMC) is attained.

An alternative approach to this relaxation uses the fact that, for any fixed normalized vector $u \in \mathcal{R}^n$, we can reformulate (MC) as

$$(MC) \quad mc^* = \max \tfrac{1}{2}\sum_{i<j} w_{ij}(1 - v_i^T v_j), \quad v_i \in \{-u, u\}, \quad \forall i. \tag{1}$$

This leads to the relaxation used in Goemans and Williamson (1995) which, again, is equivalent to the SDP relaxation. (Surprisingly, see Poljak and Wolkowicz, 1995 and Poljak et al., 1995, this relaxation is equivalent to several others such as the LP relaxation (box) with perturbations on the objective function). Goemans and Williamson (1995) use (1) and a randomized algorithm, which provides an approximate solution to (MC) whose objective value is within 1.14 of the optimal value. Previous results were within 1.5 of the optimal value. In practice, much better results (e.g., 1.06 on average in Helmberg et al., 1996) are obtained empirically.

In summary, the SDP relaxation provides an efficient way of solving the Lagrangian relaxation which has been shown to be very strong both theoretically (Goemans and Williamson, 1995) and in practice (Helmberg et al., 1996; Helmberg and Rendl, 2000; Benson et al., 2000; Burer and Monteiro, 1999).

1.2. Historical Notes

It is only recently that SDP has become a mathematical discipline in its own right. This is due to the many important applications that have arisen, as well as to the development of efficient primal–dual interior point (denoted PDIP) algorithms that can solve SDPs.

Up until this time, LP models were, and still are, the principal tool used in approximations of numerically hard problems. Dantzig (1991) discusses his well-known tale about his confrontation with Hotelling at his first presentation of the simplex method for LP. After his presentation Hotelling stated: "But we all know that the world is nonlinear" and Von Neumann replied for him: "The speaker titled his talk 'linear programming' and carefully stated his axioms. If you have an application that satisfies the axioms, well use it. If it does not, then don't." And, this provides a very good reason for the success of LP: there are many applications and the software that has been developed over the years provides us with very efficient solution techniques, for example Nemhauser et al. (1989). The main algorithms were originally based on Dantzig's simplex method. Then, begun by the work of Karmarkar (1984), a revolution started in the way we solve LPs and led to the PDIP methods that have resulted in a much greater increase in size and speed, for example Lustig et al. (1994).

However, in his article, Dantzig himself states: "In the final analysis, of course, Hotelling was right. The world is highly nonlinear. Fortunately, systems of linear inequalities (as opposed to equalities) permit us to approximate most of the kinds of nonlinear relations encountered in practical planning." Thus we can see the reason for the great success of LP: good approximations of many problems and efficient algorithms. However, what if we could efficiently solve quadratic approximations? That is where SDP has stepped in, in other words it can efficiently solve approximations of quadratic models of many applied problems and, thus, obtain improved results compared to LP models.

As mentioned above, SDP is a special case of cone programming, which is not a new area; nor is SDP itself. Much of the theory of duality extends directly from LP to SDP. Indeed, this was recognized for cone programs as early as 1963 by Bellman and Fan (1963). Since that time many researchers in such diverse areas as engineering, matrix theory, statistics, and optimization have contributed to results on SDP.

2. Theory

2.1. *Derivation of SDP Relaxation for a General QQP*

The SDP relaxation for general quadratically-constrained quadratic programs, denoted QQP, can be derived from the Lagrangian relaxation. If the quadratic function is defined as

$$q_i(y) = \tfrac{1}{2}y^t Q_i y + y^t b_i + c_i, \quad y \in \mathcal{R}^n$$

and the QQP as

$$(\text{QQP}) \qquad \begin{array}{lll} q^* := \min & & q_0(y) \\ & \text{s.t.} & q_i(y) \leqslant 0, \quad i = 1, \dots m, \end{array} \tag{2}$$

then the Lagrangian, with Lagrange multipliers x_i, is

$$L(y, x) = q_0(y) + \sum_{i=1}^{m} x_i q_i(y), \tag{3}$$

or equivalently,

$$L(y, x) = \tfrac{1}{2}y^t\left(Q_0 + \sum_{i=1}^{m} x_i Q_i\right)y + y^t\left(b_0 + \sum_{i=1}^{m} x_i b_i\right) + \left(c_0 + \sum_{i=1}^{m} x_i c_i\right). \tag{4}$$

Weak duality follows from the hidden constraint, in other words the primal constraints arise from the maximization over $x \geqslant 0$ in

$$d^* := \max_{x \geqslant 0} \min_{y} L(y, x) \leqslant q^* = \min_{y} \max_{x \geqslant 0} L(y, x).$$

We now homogenize, that is we multiply the linear term by a new variable y_0 to get

$$y_0 y^t\left(b_0 + \sum_{i=1}^{m} x_i b_i\right), \quad y_0^2 = 1.$$

Therefore the dual value

$$
\begin{aligned}
d^* &= \max_{x \geqslant 0} \min_{y} L(y,x) \\
&= \max_{x \geqslant 0} \min_{y_0^2=1} \tfrac{1}{2} y^t \left(Q_0 + \sum_{i=1}^m x_i Q_i \right) y + t y_0^2 + y_0 y^t \left(b_0 + \sum_{i=1}^m x_i b_i \right) + \left(c_0 + \sum_{i=1}^m x_i c_i \right) - t \\
&= \max_{x \geqslant 0, t} \min_{y} \tfrac{1}{2} y^t \left(Q_0 + \sum_{i=1}^m x_i Q_i \right) y + t y_0^2 + y_0 y^t \left(b_0 + \sum_{i=1}^m x_i b_i \right) + \left(c_0 + \sum_{i=1}^m x_i c_i \right) - t.
\end{aligned}
$$

We now have a new hidden constraint, that is, a quadratic is only bounded below if the Hessian is positive semidefinite, therefore we get

$$
B := \begin{pmatrix} 0 & b_0^t \\ b_0 & Q_0 \end{pmatrix},
$$

$$
\mathscr{A} : \mathscr{R}^{m+1} \to \mathscr{S}_{n+1}
$$

$$
\mathscr{A}\begin{pmatrix} t \\ x \end{pmatrix} := - \begin{bmatrix} t & \sum_{i=1}^m x_i b_i^t \\ \sum_{i=1}^m x_i b_i & \sum_{i=1}^m x_i Q_i \end{bmatrix}
$$

and the SDP constraint

$$
B - \mathscr{A}\begin{pmatrix} t \\ x \end{pmatrix} \succeq 0.
$$

The dual program is equivalent to the SDP (with $c_0 = 0$)

$$
\begin{aligned}
& d^* = \sup \quad -t + \sum_{i=1}^m x_i c_i \\
\text{(D)} \qquad & \text{s.t.} \quad \mathscr{A}\begin{pmatrix} t \\ x \end{pmatrix} \preceq B \\
& x \in \mathscr{R}^m, \quad t \in \mathscr{R}.
\end{aligned}
$$

We can now form the Lagrangian dual again and obtain the dual of this SDP to get the SDP relaxation of the QQP. (More details on duality are given below in Section 2.3.)

$$
\begin{aligned}
& d^* = \inf \quad \operatorname{trace} BU \\
\text{(DD)} \qquad & \text{s.t.} \quad \mathscr{A}^* U = \begin{pmatrix} -1 \\ c \end{pmatrix} \\
& U \succeq 0.
\end{aligned}
$$

(See Shor, 1987; Boyd et al., 1994; and Poljak et al., 1995.)

Remark 1. *There is no hidden constraint on the* Q_i *if all* q_i *are convex. There is no need for SDP relaxations in the convex case and better algorithms do exist.*

2.2. *Geometry*

Much of the elegant geometry of polyhedral sets developed for LP can be extended to SDP, although the results are much more difficult, and, if anything, more interesting (see

Bohnenblust, 1948; Taussky, 1968; and Barker and Carlson, 1975; and more recently Lewis, 1998). The semidefinite cone is self-polar

$$\mathcal{P} = \mathcal{P}^+ := \{Y : X \cdot Y \geqslant 0, \quad \forall X \in \mathcal{P}\};$$

and it is homogeneous, that is, for any $X, Y \in \text{int}(\mathcal{P})$, an invertible linear operator $\mathcal{A}$ exists that leaves $\mathcal{P}$ invariant and $\mathcal{A}(X) = Y$. The faces of $\mathcal{P}$ can be characterized using the null space, $\mathcal{N}$, (or range space, $\mathcal{R}$) of any matrix in the relative interior, in other words, if $\mathcal{F}$ is a face and X is in its relative interior then

$$\mathcal{F} = \{Y : \mathcal{N}(Y) \supset \mathcal{N}(X)\}.$$

The faces $\mathcal{P}$ are exposed, in other words, $\mathcal{F} = \mathcal{P} \cap \phi^\perp$, where ϕ is any matrix in the conjugate face, such as in $\mathcal{P} \cap \mathcal{F}^\perp$. (Here, $\perp$ denotes the orthogonal complement.) Also the faces are projectionally exposed (Borwein and Wolkowicz, 1981), for example $\mathcal{F} = P\mathcal{P}P$, where P is a projection matrix, and $\mathcal{R}(P) = \mathcal{R}(X)$. In addition, for any face $\mathcal{F}$, $\mathcal{P} + \mathcal{F}^\perp$ is always closed while $\mathcal{P} + \text{span}(\mathcal{F})$ is never closed (Ramana et al., 1997). This latter condition can cause problems in optimality characterizations. Further results on the facial structure and multiplicity of eigenvalues in optimal faces can be found in Pataki (1996, 1998).

One reason for the special nature of SDP is that the cone $\mathcal{P}$ is both homogeneous and self dual. Such cones have been completely classified and only include the Lorentz cone (used for convex quadratic programming with convex quadratic constraints), the semidefinite cone $\mathcal{P}$ and its extensions (of which the most important is the cone of positive semidefinite complex Hermitian matrices), and products of these (including $\mathcal{R}^n_+$, a product of n trivial semidefinite cones). For more details, see Güler and Tuncel (1998).

2.3. *Duality Theory and Optimality Conditions*

Much of the theory of duality extends directly from LP to SDP. As mentioned above, this was recognized as early as 1963 by Bellman and Fan (1963). They also observed that Strong Duality Theorems require a Slater-type constraint qualification (strict feasibility). Note that we get weak duality from

$$\mu^* = \min_{X \succeq 0} \max_y C \cdot X + y^T(b - \mathcal{A}X) \leqslant \max_y \min_{X \succeq 0} y^T b + (C - \mathcal{A}^* y) \cdot X = v^*,$$

where $\mathcal{A}^*$ denotes the adjoint operator of $\mathcal{A}$, $\mathcal{A}^* y = \sum_i y_i \mathcal{A}_i$, and

$$\text{(DSDP)} \qquad \begin{array}{ll} v^* := \max & b^T y \\ \text{s.t.} & \mathcal{A}^* y \preceq C \end{array}$$

defines the dual program. If Slater's constraint qualification, a strictly feasible solution, exists for the primal (dual, respectively) then there is a zero duality gap, $\mu^* = v^*$, and the dual (primal, respectively) is attained.

If Slater's constraint qualification fails, then one can work with the minimal cones and obtain characterizations of optimality and Strong Duality Theorems (see Borwein and Wolkowicz, 1980). (For PSDP this is the minimal face of $\mathcal{P}$ containing the feasible set. For DSDP this is the minimal face of $\mathcal{P}$, which contains $C - \mathcal{A}^*\mathcal{F}$, where $\mathcal{F}$ denotes the feasible set.) One can also obtain a strong dual program whose size is polynomial in the data of the original problem (Ramana, 1997). The relation between these duals is given in Ramana et al. (1997).

If a constraint qualification holds, then the duality theory gives rise to the characterization of optimality

$$\begin{aligned} \mathcal{A}^*y + Z - C = 0 &\quad \text{dual feasibility} \\ b - \mathcal{A}(X) = 0 &\quad \text{primal feasibility} \\ ZX = 0 &\quad \text{complementary slackness} \\ Z, X \succeq 0. & \end{aligned} \tag{5}$$

We call X, (y, Z) a primal-dual optimal pair. Z is the (dual) slack variable.

First- and second-order optimality conditions for general cone programming are given in Shapiro (1997, 2000).

2.4. *Degeneracy and Strict Complementarity*

The notions of degeneracy and strict complementarity do not follow through directly from LP to SDP. Primal (dual) nondegeneracy implies a unique dual (primal) solution. Moreover, the nondegeneracy does not imply strict complementary slackness. A theorem of Goldman and Tucker shows that every (finite optimal valued) LP has an optimal primal–dual pair that satisfies strict complementarity (for SDP this translates to $Z + X \succ 0$). However, this result fails to hold for SDP. Furthermore, the formulation of nondegeneracy conditions requires study of the geometry of the semidefinite cone $\mathcal{P}$. The boundary of $\mathcal{P}$, like that of $\mathcal{R}^n_+$, is composed of smooth pieces, but the pieces are nonlinear. Degeneracy and strict complementary slackness is studied in Alizadeh et al. (1997) and Bonnans and Shapiro (1998); generic results are given in Pataki and Tuncel (1997).

2.5. *Complexity*

SDP are convex programs and fall into the class of problems that can be approximately solved in polynomial time. The complexity is based on self-concordant barriers for the cone $\mathcal{P}$. The smallest (best) barrier parameter is given in Güler and Tuncel (1998).

The complexity of determining the (exact) feasibility of a system of linear matrix inequalities is discussed in Porkolab and Khachiyan (1997) and Ramana (1993).

3. Algorithms

3.1. *Interior-Point Algorithms*

SDP research is highlighted by the elegant PDIP algorithms. The IPMs are extensions of polynomial time algorithms developed originally for LP. The seminal work by Nesterov and Nemirovski (1994) shows that $\log\det(X)$ is a self-concordant barrier function for SDP. Thus, SDP can be solved in polynomial time using a sequence of barrier subproblems. Alizadeh (1995) presented a transparent approach to extend potential function methods from LP to SDP. Simultaneously, strong numerical results for SDP and the special case of the min–max eigenvalue problem appeared in Helmberg et al. (1996) and Jarre (1993). This started a flood of results in SDP from researchers in LP.

The most popular methods are the PDIP methods. These are based on applying Newton's method to solving the optimality conditions from the barrier problems, in other words, the perturbed optimality conditions

$$\begin{aligned} &Z + C - \mathcal{A}^*y = 0, \\ &b - \mathcal{A}(X) = 0, \\ &X - \mu Z^{-1} = 0. \end{aligned} \tag{6}$$

As in LP, the last equation can be written in many ways, such as

$$ZX = \mu I. \tag{7}$$

Different choices give rise to different search directions. However, unlike LP, choices like (7) result in (6) being an overdetermined system of nonlinear equations and so may not have a Newton direction. There are many ways to overcome this problem, for example, by various symmetrization techniques (Monteiro and Zhang, 1998; Todd, 1999), or by using the Gauss–Newton approach (Kruk et al., 1998).

This is still an active area of research, in particular for large sparse problems.

3.2. *Bundle Trust Algorithms*

As seen above, many SDP problems arise from eigenvalue problems. Conversely, many SDP's can be reformulated as eigenvalue optimization problems. For example, in Helmberg and Rendl (2000) the dual SDP is transformed into an eigenvalue optimization problem by reformulating the SDP constraint as a nonnegativity constraint on the minimal eigenvalue of the slack matrix variable and lifting this constraint into the cost function by means of a Lagrange multiplier. The correct value of the Lagrange multiplier is known in advance if the primal feasible matrices have constant trace, which is the case for many combinatorial applications.

The result is the problem of maximizing the minimal eigenvalue of an affine matrix function with an additional linear objective term. This is a nonsmooth convex optimization problem. At each step of a descent method one needs a subgradient and a function value. The subgradients are collected into a bundle to form a cutting-plane model. This model is then used to get the next iterate. More details can be found in Helmberg and Rendl (2000) and Oustry (1998).

3.3. *Large Sparse Case*

The bundle trust method is motivated by the many large SDP problems with special structure that arise from combinatorial optimization. It is particularly well suited to exploit special structure. Other approaches for exploiting special structure in SDPs are given in Benson et al. (2000), Fujisawa et al. (1995), and Kojima et al. (1997).

4. Applications

What has made this area so interesting and exciting is that the theory and algorithms are motivated by many applications, for example, in statistics, combinatorial problems, engineering (mentioned above), and matrix completion problems.

4.1. *Combinatorial Optimization*

Semidefinite programming plays an important role in combinatorial optimization where it is used to solve nonlinear convex relaxations of NP-hard problems. The interest in SDP techniques for combinatorial optimization was motivated by remarkable theoretical results that characterize the quality of the bounds obtained via SDP. We have already seen some details of the result of Goemans–Williamson on the complexity and strength of the SDP relaxations for the max-cut problem. In general, one can model 0, 1 discrete optimization problems using $x_i^2 - x_i = 0$ though this often leads to subproblems that are too hard to solve. One can then take the Lagrangian relaxation and its dual to obtain the tractable SDP relaxation. Thus, SDP provides a means of efficiently solving relaxations of quadratic approximations, although the SDPs themselves are often very large problems. Early examples of applying SDP are the stable set problem and the max-cut problem (see Grötschel et al., 1981 and Lovász and Schrijver, 1991). Surveys in this area are Rendl (1999) and Goemans (1997).

4.2. *Eigenvalue Problems*

SDP is an eigenvalue optimization problem in the sense that the semidefinite constraint is equivalent to nonnegative bounds on eigenvalues. Solutions to SDP generally have multiple zero eigenvalues (just as solutions to LP have many nonbasic, i.e., zero, variables). However, it is not possible to formulate separate differentiable bounds on individual eigenvalues.

Min–max eigenvalue problems can be formulated as an SDP, that is, one can maximize a variable λ subject to $\mathscr{A} - \lambda I \preceq 0$. An early example of applications of min–max eigenvalues is given in Cullum et al. (1975). The converse is not true in general, but is true when the trace is a constant function on the feasible set (see Helmberg and Rendl, 2000). This can be used to exploit structure and solve large-scale problems.

For more on eigenvalue optimization see Lewis and Overton (1996).

4.3. Engineering

Many early applications of SDP appeared in engineering literature, for example, Boyd et al. (1993) and Yakubovich (1967a,b,c, 1973). This was particularly evident in control theory, where it emerged as an important practical tool for the analysis of nonlinear and time-varying systems, and for controller synthesis. This is of great interest in computer-aided control system design. SDPs also arise in other fields of engineering and mathematics, such as structural optimization, circuit design, signal processing, and statistics (see Boyd et al., 1993; Vandenberghe and Balakrishnan, 1997; Vandenberghe and Boyd, 1994; Fletcher, 1981, 1985; Pukelsheim, 1993; and Wolkowicz et al., 2000). For other applications such as solving Ricatti equations see Boyd et al. (1994) and Vandenberghe and Boyd (1994) (In this arena SDP is usually known as LMI: Linear Matrix Inequalities.) A typical example of an SDP application in systems and control is the numerical search for Lyapunov inequalities and also for Lyapunov functions for stability.

4.4. Matrix Completion Problems

The matrix completion problem consists of completing the missing elements of a matrix, subject to certain conditions being met, for example, positive definiteness. This is an extensively studied area in its own right (see e.g., Grone et al., 1984, Laurent, 1998; Johnson, 1990; Bakonyi and Johnson, 1995).

4.5. Nonlinear Programming

The trust region subproblem is a basic tool in unconstrained nonlinear programming, providing a local quadratic model of the objective function that is to be "trusted" only in a certain region. In this case, the SDP relaxation of the quadratic problem provides an exact application to large scale minimization (Rendl and Wolkowicz, 1997). In addition, this provides an SDP framework that allows one to efficiently exploit sparsity when solving unconstrained nonlinear programming problems. In the constrained case, one obtains a quadratically constrained quadratic program as the local model. Again, an approximation of this model can be solved efficiently using SDP (Kruk and Wolkowicz, 1998).

5. Conclusions

The World Wide Web is a tremendous resource for information on SDP. This includes research papers, public domain software, and bibliography files. Several relevant Web sites are listed below. These contain further pointers to other Web sites with more information.

1. http://www.zib.de/helmberg/semidef.html contains a special home page for SDP organized by C. Helmberg, at Berlin Center for Scientific Computing, Konrad Zuse Zentrum fur Informationstechnik, Berlin.
2. ftp://orion.uwaterloo.ca/pub/henry/reports/psd.bib.gz contains a bibfile with papers related to SDP.
3. http://www.mcs.anl.gov:80/home/otc/InteriorPoint is an archive of technical reports and papers on IPMs maintained by Nathan Brixius and S. J. Wright at Argonne National Laboratory. Many of the recent reports discuss PDIP methods for SDP, a subject not discussed in this volume.

SDP is an exciting subject that remains very active, and will no doubt be so for some time to come. Here is an area where nonlinear convex programming has shown its strength, beauty, and versatility. One finds elegant theoretical results, provably efficient algorithms, and a wide variety of applications.

ACKNOWLEDGMENT
This research was supported by the Natural Sciences Research Council, Canada.

References

Alizadeh, F. 1995. "Interior point methods in semidefinite programming with applications to combinatorial optimization." *SIAM J. Optim.* **5**: 13–51.

Alizadeh, F., J.-P. A. Haeberly, and M. L. Overton. 1997. "Complementarity and nondegeneracy in semidefinite programming." *Math. Programming* **77**: 111–128.

Bakonÿi, M. and C. R. Johnson. 1995. "The Euclidean distance matrix completion problem." *SIAM J. Matrix Anal. Appl.* **16(2)**: 646–654.

Barker, G. P. and D. Carlson. 1975. "Cones of diagonally dominant matrices." *Pacific J. Math.* **57**: 15–32.

Bellman, R. and K. Fan. 1963. "On systems of linear inequalities in Hermitian matrix variables". In *Proceedings of Symposia in Pure Mathematics*, Vol 7. AMS.

Benson, S. J., Y. Ye, and X. Zhang. "Solving large-scale sparse semidefinite programs for combinatorial optimization." To appear in *SIAM J. Optim.*

Bohnenblust, F. 1948. "Joint positiveness of matrices". Manuscript.

Bonnans, J. F. and A. Shapiro. 1998. "Nondegeneracy and quantitative stability of parameterized optimization problems with multiple solutions." *SIAM J. Optim.* **8(4)**: 940–946.

Borwein, J. M. and H. Wolkowicz. 1980. "Characterization of optimality for the abstract convex program with finite-dimensional range." *J. Austral. Math. Soc. Ser. A* **30(4)**: 390–411.

Borwein, J. M. and H. Wolkowicz. 1981. "Regularizing the abstract convex program." *J. Math. Anal. Appl.* **83(2)**: 495–530.

Boyd, S., V. Balakrishnan, E. Feron, and L. El Ghaoui. 1993. "Control system analysis and synthesis via linear matrix inequalities." *Proc. ACC*, pp. 2147–2154.

Boyd, S., L. El Ghaoui, E. Feron, and V. Balakrishnan. 1994. Linear matrix inequalities in system and control theory. In *Studies in Applied Mathematics*. Philadelphia, PA: SIAM.

Burer, S. and R. D. C. Monteiro. 1999. An efficient algorithm for solving the MAXCUT SDP relaxation. Technical report. Georgia Tech, Atlanta, GA.

Cullum, J., W. E. Donath, and P. Wolfe. 1975. "The minimization of certain nondifferentiable sums of eigenvalues of symmetric matrices." *Math. Programming Study* **3**: 35–55.

Dantzig, G. 1991. "Linear programming." In *History of Mathematical Programming: A Collection of Personal Reminiscences*. Amsterdam, The Netherlands: CWI North-Holland.

Fletcher, R. 1981. "A nonlinear programming problem in statistics (educational testing)." *SIAM J. Sci. Stat. Comput.* **2**: 257–267.

Fletcher, R. 1985. "Semi-definite matrix constraints in optimization." *SIAM J. Control Optim.* **23**: 493–513.

Fujisawa, K., M. Kojima, and K. Nakata. 1995. "SDPA semidefinite programming algorithm". Technical report. Dept. of Information Sciences, Tokyo Institute of Technology, Tokyo, Japan.

Goemans, M. X. 1997. "Semidefinite programming in combinatorial optimization." *Math. Programming* **79**: 143–162.

Goemans, M. X. and D. P. Williamson. 1995. "Improved approximation algorithms for maximum cut and satisfiability problems using semidefinite programming." *J. Assoc. Comput. Mach.* **42(6)**: 1115–1145.

Grone, B., C. R. Johnson, E. Marques de Sa, and H. Wolkowicz. 1984. "Positive definite completions of partial Hermitian matrices." *Linear Algebra Appl.* **58**: 109–124.

Grötschel, M., L. Lovász, and A. Schrijver. 1981. "The ellipsoid method and its consequences in combinatorial optimization." *Combinatorica* **1**: 169–197.

Güler, O. and L. Tuncel. 1998. "Characterization of the barrier parameter of homogeneous convex cones." *Math. Programming* **81**: 55–76.

Helmberg, C. and F. Rendl. 2000. "A spectral bundle method for semidefinite programming." *SIAM J. Optim.* **10(3)**: 673–696.

Helmberg, C., F. Rendl, R. J. Vanderbei, and H. Wolkowicz. 1996. "An interior-point method for semidefinite programming." *SIAM J. Optim.* **6(2)**: 342–361.

Jarre, F. 1993. "An interior-point method for minimizing the maximum eigenvalue of a linear combination of matrices." *SIAM J. Control Optim.* **31**: 1360–1377.

Johnson, C. R. 1990. "Matrix completion problems: a survey." *Proc. Sym. Appl. Math.* **40**: 171–198.

Karmarkar, N. K. 1984. "A new polynomial-time algorithm for linear programming." *Combinatorica* **4**: 373–395.

Kojima, M., S. Kojima, and S. Hara. 1997. "Linear algebra for semidefinite programming". Technical Report 1004. Dept. of Information Sciences, Tokyo Institute of Technology, Tokyo, Japan. Linear matrix inequalities and positive semidefinite programming (Japanese) (Kyoto, 1996).

Kruk, S. and H. Wolkowicz. 1998. "SQ^2P, sequential quadratic constrained quadratic programming." *Appl. Optim.* pp. 177–204.

Kruk, S., M., Muramatsu, F. Rendl, R. J. Vanderbei, and H. Wolkowicz. 1998. "The Gauss–Newton direction in linear and semidefinite programming". Technical Report CORR 98-16. University of Waterloo, Waterloo, Canada.

Laurent, M. 1998. "A tour d'horizon on positive semidefinite and Euclidean distance matrix completion problems". In *Topics in Semidefinite and Interior-Point Methods, Volume 18 of The Fields Institute for Research in Mathematical Sciences, Communications Series*, pp. 51–76. Providence, Rhode Island: American Mathematical Society.

Lewis, A. S. 1998. "Eigenvalue-constrained faces." *Linear Algebra Appl.* **269**: 159–181.

Lewis, A. S. and M. L. Overton. 1996. "Eigenvalue optimization." *Acta Numerica* **5**: 149–190.

Lovász, L. and A. Schrijver. 1991. "Cones of matrices and set-functions and 0–1 optimization." *SIAM J. Optim.* **1(2)**: 166–190.

Lustig, I. J., R. E. Marsten, and D. F. Shanno. 1994. "Interior point methods for linear programming: computational state of the art." *ORSA J. Comput.* **6**: 1–14.

Monteiro, R. D. C. and Y. Zhang. 1998. "A unified analysis for a class of long-step primal–dual path-following interior-point algorithms for semidefinite programming." *Math. Programming* **81(3, Ser. A)**: 281–299.

Nemhauser, G. L., A. H. G. Rinnooy Kan, and M. J. Todd, eds. 1989. *Optimization*, Volume 1 of *Handbooks in Operations Research and Management Science*. Amsterdam, The Netherlands: North-Holland Publishing Co.

Nesterov, Y. E. and A. S. Nemirovski. 1994. *Interior Point Polynomial Algorithms in Convex Programming*. SIAM, Philadelphia: SIAM Publications.

Oustry, F. 1998. "A second-order bundle method to minimize the maximum eigenvalue function". Submitted to *Math. Programming*, February 1998.

Pataki, G. 1996. "Cone programming and eigenvalue optimization: geometry and algorithms". PhD thesis. Carnegie Mellon University, Pittsburgh, PA.

Pataki, G. 1998. "On the rank of extreme matrices in semidefinite programs and the multiplicity of optimal eigenvalues." *Math. Oper. Res.* **23(2)**: 339–358.

Pataki, G. and L. Tuncel. 1997. "On the generic properties of convex optimization problems in conic form". Technical Report CORR 97-16. Department of Combinatorics and Optimization, University of Waterloo, Waterloo, Canada.

Poljak, S. and H. Wolkowicz. 1995. "Convex relaxations of (0, 1)-quadratic programming." *Math. Oper. Res.* **20(3)**: 550–561.

Poljak, S., F. Rendl, and H. Wolkowicz. 1995. "A recipe for semidefinite relaxation for (0, 1)-quadratic programming." *J. Global Optim.* **7(1)**: 51–73.

Porkolab, L. and L. Khachiyan. 1997. "On the complexity of semidefinite programs." *J. Global Optim.* **10(4)**: 351–365.

Pukelsheim, F. 1993. *Optimal Design of Experiments*. New York: John Wiley and Sons Inc.

Ramana, M. V. 1993. "An algorithmic analysis of multiquadratic and semidefinite programming problems". PhD thesis. Johns Hopkins University, Baltimore, MD.

Ramana, M. V. 1997. "An exact duality theory for semidefinite programming and its complexity implications." *Math. Programming* **77**: 129–162.

Ramana, M., L. Tuncel, and H. Wolkowicz. 1997. "Strong duality for semidefinite programming." *SIAM J. Optim.* **7(3)**: 641–662.

Rendl, F. 1999. "Semidefinite programming and combinatorial optimization." *Appl. Numer. Math.* **29**: 255–281.

Rendl, F. and H. Wolkowicz. 1997. "A semidefinite framework for trust region subproblems with applications to large scale minimization." *Math. Programming* **77(2, Ser. B)**: 273–299.

Shapiro, A. 1997. "First and second order analysis of nonlinear semidefinite programs." *Math. Programming* **77**: 301–320.

Shapiro, A. 2000. "Duality and optimality conditions." *Handbook of Semidefinite Programming: Theory, Algorithms, and Applications*, edited by H. Wolkowicz, R. Saigal, and L. Vandenberghe. Boston MA: Kluwer Academic Publishers.

Shor, N. Z. 1987. "Quadratic optimization problems." *Izv. Akad. Nauk SSSR Tekhn. Kibernet.* **222(1)**: 128–139.

Taussky, O. 1968. "Positive-definite matrices and their role in the study of the characteristic roots of general matrices." *Adv. Math.* **2**: 175–186.

Todd, M. J. 1999. "A study of search directions in primal-dual interior-point methods for semidefinite programming." *Optim. Methods Softw.* **11&12**: 1–46.

Vandenberghe, L. and V. Balakrishnan. October 1997. "Algorithms and software for LMI problems in control." *IEEE Control Syst. Mag.* 89–95.

Vandenberghe, L. and Boyd, S. 1994. "Positive definite programming". In *Mathematical Programming: State of the Art, 1994*, pp. 276–308. The University of Michigan, Michigan.

Vandenberghe, L. and S. Boyd. 1996. "Semidefinite programming." *SIAM Rev.* **38(1)**: 49–95.

Wolkowicz, H., R. Saigal, and L. Vandenberghe (eds). 2000. *Handbook of Semidefinite Programming: Theory, Algorithms, and Applications*, Boston, MA: Kluwer Academic Publishers.

Yakubovich, V. A. 1967a. "The method of matrix inequalities in the stability theory of nonlinear control systems I." *Automat. Remote Control* **25**: 905–917.

Yakubovich, V. A. 1967b. "The method of matrix inequalities in the stability theory of nonlinear control systems II." *Automat. Remote Control* **26**: 577–592.

Yakubovich, V. A. 1967c. "The method of matrix inequalities in the stability theory of nonlinear control systems III." *Automat. Remote Control* **26**: 753–763.

Yakubovich, V. A. 1973. "Minimization of quadratic functionals under the quadratic constraints and the necessity of a frequency condition in the quadratic criterion for absolute stability of nonlinear control systems." *Dokl. Akad. Nauk SSSR* **209(14)**: 1039–1042.

3 COMBINATORIAL OPTIMIZATION

3.1 Introduction

Panos M. Pardalos and Mauricio G. C. Resende

ABSTRACT Combinatorial optimization has gone through many phases during the last few decades, prompted by the recognition that its domain encompasses a wide range of important and challenging practical applications. We provide pointers to books on combinatorial optimization and introduce topics in the field that are covered in more detail in this handbook.

Combinatorial optimization is a very active area of research in applied optimization, combining techniques for combinatorics, linear and nonlinear programming, and the theory of algorithms and data structures, to solve problems over discrete structures. Recent advances in combinatorial optimization, including polyhedral combinatorics, cutting-plane methods, branch-and-cut, branch-and-bound, local search, and metaheuristics, together with advances in computer technology, have made it possible to apply combinatorial optimization to a wide range of practical problems.

A number of classical books in the field cover a variety of techniques and applications (e.g., Lawler, 1976; Papadimitiou and Steiglitz, 1982; Schrijver, 1986 and Nemhauser and Wolsey, 1988). More recent surveys include the *Handbook of Combinatorial Optimization* (Du and Pardalos 1998, 1999, 2001) and many books dedicated to specific combinatorial optimization problems (e.g., Lawler et al., 1985) on the traveling salesman problem; Mirchandani and Francis, 1990 and Drezner, 1995, on location theory; Martello and Toth 1990 on the knapsack problem, Pardalos and Wolkowicz, 1994 and Çela, 1998 on the quadratic assignment problem; Jensen and Toft, 1995, on graph coloring; Pinedo, 1995, on scheduling; Du et al., 1997, on the satisfiability problem; and Pardalos and Pitsoulis, 2000, on nonlinear assignment problems).

In this handbook, several algorithmic aspects of combinatorial optimization are covered. Integer programming is treated separately from combinatorial optimization later in the handbook. However, branch-and-bound and branch-and-cut methods, which can also be applied to integer programming and many nonconvex optimization problems, are treated in this chapter. Following that, dynamic programming is discussed. The above techniques are exact methods, that is, they compute the exact optimal solution. Because of the complexity of combinatorial optimization problems, heuristics are often employed. Local search and metaheuristics are covered next. The metaheuristics described in this handbook include ant systems, population heuristics, memetic algorithms, GRASP, scatter search, tabu search, simulated annealing, and variable neighborhood search.

Branch-and-bound (Nemhauser and Wolsey, 1988) is the most developed technique for exact solution of combinatorial optimization problems. In a branch-and-bound scheme, the problem is decomposed into subproblems of a smaller size. Each of the subproblems is decomposed until each decomposed problem is either solved or proved not to yield an optimal solution of the original problem. Several search strategies, such as depth first search, best-bound search, breadth-first search, and so on, define the problem decomposition. In addition, the efficiency of a branch-and-bound algorithm depends strongly on the quality of the bounds.

Another exact method for combinatorial optimization is the cutting-plane method, introduced by Gomory (1963). A linear constraint that does not exclude any integer-feasible solution is called a cutting plane, or simply a cut. Cutting-plane methods generate a sequence of cutting planes until the optimal solution of the resulting linear program is integer. Advances in polyhedral theory and the development of strong, problem-specific cutting planes make cutting-plane methods today the method of choice for a variety of combinatorial optimization problems. Since it is usually not possible to solve a combinatorial optimization problem efficiently relying on cutting-plane methods alone, branching is usually applied, resulting in a branch-and-cut method.

Dynamic programming, introduced by Bellman (1957), is related to branch-and-bound, in the sense that it performs an intelligent enumeration of all feasible points, but does so in a different way. A problem is decomposed into a nested family of subproblems. Recursion is applied to solve the original problem from the solution of the subproblems. Although dynamic programming is efficient for many classes of problems, on others it can suffer from the exponential growth of the state space, and in those cases requires a large amount of memory.

Though exact methods have matured greatly, in many instances they are still unable to efficiently produce optimal solutions to combinatorial optimization problems. In addition, as the size of the instances increases, these algorithms tend to degenerate to complete enumeration. Local search methods provide a framework for searching the solution space by focusing on local neighborhoods. Given a solution, the method searches the neighborhood of the solution for another solution with a better objective function value and moves to it if one exists. The local search proceeds until it encounters a solution for which no better solution exists in its neighborhood. This is called a locally optimal solution. An example of local search is the Lin–Kernighan local search for the traveling salesman problem (Lin and Kernighan, 1973). Local search plays a key role in connection with metaheuristics.

Combinatorial optimization has gone through many phases during the last few decades, spurred by the recognition that its domain encompasses a wide range of important and challenging practical applications (Glover, 1986).

References

Bellman, R. 1957. *Dynamic Programming*. Princeton, NJ: Princeton University Press.

Çela, E. 1998. *The Quadratic Assignment Problem: Theory and Algorithms*. Norwell, MA: Kluwer Academic Publishers.

Drezner, Z. (ed). 1995. In *Facility Location: a Survey of Applications and Methods*. New York, NY: Springer-Verlag.

Du, D.-Z., J. Gu, and P. M. Pardalos. (eds). 1997. In *Satisfiability Problem: Theory and Applications*. DIMACS Series on Discrete Mathematics and Theoretical Computer Science, Volume 35. Providence, RI: American Mathematical Society.

Du, D.-Z. and P. M. Pardalos. (eds). 1998. In *Handbook of Combinatorial Optimization*, Volumes 1–3. Norwell, MA: Kluwer Academic Publishers.

Du, D.-Z. and P. M. Pardalos. (eds). 1999. In *Handbook of Combinatorial Optimization* Supplement volume A. Norwell, MA: Kluwer Academic Publishers.

Du, D.-Z. and P. M. Pardalos. (eds). 2001. In *Handbook of Combinatorial Optimization* Supplement volume B. Norwell, MA: Kluwer Academic Publishers.

Glover, F. 1986. "Future paths for integer programming and links to artificial intelligence." *Comput. Oper. Res.* **13**: 533–549.

Gomory, R. E. 1963. "An algorithm for integer solutions to linear programs." In *Recent Advances in Mathematical Programming*, edited by R. L. Graves and P. Wolfe, pp. 269–302. New York, NY: McGraw-Hill.

Jensen, T. R. and B. Toft. 1995. In *Graph Coloring Problems*. New York, NY: Wiley-Interscience.

Lawler, E. L. 1976. *Combinatorial Optimization: Networks and Matroids*. Austin, TX: Holt, Reinhart, & Winston.

Lawler, E. L., J. K. Lentra, A. H. G. Rinnooy-Kan, and D. B. Shmoys. 1985. *The Traveling Salesman Problem*. New York, NY: John Wiley and Sons.

Lin, S. and B. W. Kernighan. 1973. "An effective heuristic algorithm for the traveling-salesman problem." *Oper. Res.* **21**: 498–516.

Martello, S. and P. Toth. 1990. *Knapsack Problems: Algorithms and Computer Implementations*. Chichester, UK: John Wiley and Sons.

Mirchandani, P. B. and R. L. Francis. (eds). 1990. In *Discrete Location Theory*. New York, NY: Wiley-Interscience.

Nemhauser, G. L. and L. A. Wolsey. 1988. *Integer and Combinatorial Optimization*. New York, NY: John Wiley and Sons.

Papadimitou, C. H. and K. Steiglitz. 1982. *Combinatorial Optimization: Algorithms and Complexity*. Engelwood Cliffs, NJ: Prentice-Hall.

Pardalos, P. M. and L. S. Pitsoulis. (eds). 2000. In *Nonlinear Assignment Problems: Algorithms and Application.* Norwell, MA: Kluwer Academic Publishers.
Pardalos, P. M. and H. Wolkowicz. (eds). 1994. In *Quadratic Assignment and Related Problems.* DIMACS Series on Discrete Mathematics and Theoretical Computer Science, Volume 16. Providence, RI: American Mathematical Society.
Pinedo, M. 1995. *Scheduling: Theory, Algorithms, and Systems.* Englewood Cliffs, NY: Prentice-Hall.
Schrijver, A. 1986. *Theory of Linear and Integer Programming.* New York, NY: Wiley-Interseiences.

3.2 Branch-and-Bound Methods

Eva K. Lee

ABSTRACT In this article the classical branch-and-bound approach for solving integer programming problems is described. Various branching variable selection and node selection schemes are discussed; and the techniques of problem preprocessing and reformulation, heuristics, and continuous reduced-cost fixing are outlined. These latter techniques have been shown to be very effective when embedded within a branch-and-bound algorithm. The use of an interior-point method as a subproblem solver is also described. Finally, Lagrangian relaxation is analyzed as an alternative relaxation for use within the branch-and-bound tree search environment.

1. Introduction

An integer programming problem (IP) is an optimization problem in which some or all of the variables are restricted to take on only integer values. The exposition presented here will focus on the case in which the objective and constraints of the optimization problem are defined via linear functions. In addition, for simplicity, it will be assumed that all of the variables are restricted to be nonnegative integer values. Thus, the mathematical formulation of the problem under consideration can be stated as:

$$\begin{aligned} &\text{maximize} \quad c^T x \\ &\text{subject to} \quad Ax \leqslant b \\ &\qquad\qquad\quad x \in Z^n_+ \end{aligned} \tag{1}$$

where $A \in \mathscr{R}^{m \times n}$, $b \in \mathscr{R}^m$, and $c \in \mathscr{R}^n$. For notational convenience, let S denote the constraint set of the Equation (1); that is,

$$S := \{x \in Z^n_+ : Ax \leqslant b\}.$$

The classical approach to solving integer programs is branch and bound (Land and Doig, 1960). The branch-and-bound method is based on the idea of iteratively partitioning the set S (branching) to form subproblems of the original integer program. Each subproblem is solved—either exactly or approximately—to obtain an upper bound on the subproblem objective value. The driving force behind the branch-and-bound approach lies in the fact that if an upper bound for the objective value of a given subproblem is less than the objective value of a known integer-feasible solution (e.g., obtained by solving some other subproblem) then the optimal solution of the original integer program cannot lie in the subset of S associated with the given subproblem. Hence, the upper bounds on subproblem objective values are, in essence, used to construct a proof of optimality without exhaustive search.

One concept that is fundamental to obtaining upper bounds on subproblem objective values is that of problem relaxation. A relaxation of the optimization problem

$$\max\{c^T x : x \in S\}$$

is an optimization problem

$$\max\{r(x) : x \in S_R\},$$

where $S \subseteq S_R$ and $c^T x \leqslant r(x)$ for all $x \in S$. Clearly, solving a problem relaxation provides an upper bound on the objective value of the underlying problem. Perhaps the most common relaxation of Equation (1) is the linear programming (LP) relaxation formed by relaxing the integer restrictions and enforcing appropriate bound conditions on the variables; that is, $r(x) = c^T x$ and $S_R = \{x \in \mathscr{R}_+^n : Ax \leqslant b,\ l \leqslant x \leqslant u\}$, where l and u are the lower and upper integer bounds for x.

A formal statement of a general branch-and-bound algorithm (Nemhauser and Wolsey, 1988) is presented below. The notation L is used to denote the list of active subproblems $\{\mathrm{IP}^i\}$, where $\mathrm{IP}^0 = \mathrm{IP}$ denotes the original integer program. The notation $\bar{z}_i$ denotes an upper bound on the optimal objective value of IP^i, and $\underline{z}_{ip}$ denotes the incumbent objective value (i.e., the objective value corresponding to the current best integral-feasible solution to IP; see Equation 1).

1.1. General Branch-and-Bound Algorithm

1. (Initialization): Set $L = \{\mathrm{IP}^0\}$, $\bar{z}_0 = +\infty$, and $\underline{z}_{ip} = -\infty$.
2. (Termination): If $L = \varnothing$, then the solution x^*, which yielded the incumbent objective value $\underline{z}_{ip}$, is optimal. If no such x^* exists (i.e., $\underline{z}_{ip} = -\infty$) then IP is infeasible.
3. (Problem selection and relaxation): Select and delete a problem IP^i from L. Solve a relaxation of IP^i. Let z_i^R denote the optimal objective value of the relaxation, and let x^{iR} be an optimal solution if one exists. (Thus, $z_i^R = c^T x^{iR}$, or $z_i^R = -\infty$.)
4. (Fathoming and Pruning):
   (a) If $z_i^R \leqslant \underline{z}_{ip}$ go to Step 2.
   (b) If $z_i^R > \underline{z}_{ip}$ and x^{iR} is integral feasible, update $\underline{z}_{ip} = z_i^R$. Delete from L all problems with $\bar{z}_i \leqslant \underline{z}_{ip}$. Go to Step 2.
5. (Partitioning): Let $\{S^{ij}\}_{j=1}^{j=k}$ be a partition of the constraint set S^i of problem IP^i. Add problems $\{\mathrm{IP}^{ij}\}_{j=1}^{j=k}$ to L, where IP^{ij} is IP^i with feasible region restricted to S^{ij} and $\bar{z}_{ij} = z_i^R$ for $j = 1, \ldots, k$. Go to Step 2.

The actual implementation of a branch-and-bound algorithm is typically viewed as a tree search, where the problem at the root node of the tree is the original IP. The tree is constructed in an iterative fashion with new nodes formed by branching on an existing node for which the optimal solution of the relaxation is fractional (i.e., some of the integer restricted variables have fractional values). Typically, two child nodes are formed by selecting a fractional valued variable and adding appropriate constraints in each child subproblem to ensure that the associated constraint sets do not include solutions for which this chosen branching variable assumes the same fractional value.

The phrase "fathoming a node" is used in reference to criteria that imply that a node need not be explored further. As indicated in Step 4, these criteria include:

(a) the objective value of the subproblem relaxation at the node is less than or equal to the incumbent objective value; and

(b) the solution for the subproblem relaxation is integer valued.

Note that (a) includes the case when the relaxation is infeasible, since in that case its objective value is $-\infty$. Condition (b) provides an opportunity to prune the tree, effectively fathoming nodes for which the objective value of the relaxation is less than or equal to the updated incumbent objective value. The tree search ends when all nodes are fathomed.

Since the branch-and-bound algorithm is an enumeration procedure, the algorithm terminates with a proven optimal solution for IP (Equation 1), if one exists, or concludes that the integer programming problem is infeasible. A variety of strategies have been proposed for intelligently selecting branching variables, for problem partitioning, and for selecting nodes to process. However, no single collection of strategies stands out as being best in all cases. In the remainder of this article, some of the strategies that have been implemented or proposed are summarized. An illustrative example is also presented. Some of the related computational strategies—preprocessing and reformulation, heuristic procedures, and the concept of reduced-cost fixing—which have proven to be highly effective in branch-and-bound implementations are considered. In addition, there is a discussion of recent LP-based branch-and-bound algorithms that have employed interior-point methods (IPMs) for the subproblem relaxation solver, which is in contrast to using the more traditional simplex-based solvers. Besides the LP relaxation, the notion of Lagrangian relaxation is also presented.

Though branch and bound is a classic approach for solving integer programs, there are practical limitations to its success in applications. Often integer-feasible solutions are not readily available, and node pruning becomes impossible. In this case, branch and bound fails to find an optimal solution due to memory explosion as a result of excessive accumulation of active nodes. In fact, general integer programs are NP-hard; and consequently, at the time of this writing, no known polynomial-time algorithm for solving general integer programs exists (Garey and Johnson, 1979).

In 1983, a breakthrough in the computational possibilities of branch-and-bound came as a result of the research by Crowder, Johnson, and Padberg. In their paper (Crowder et al., 1983), cutting planes were added at the root node to strengthen the LP formulation before branch and bound was called. In addition, features such as reduced-cost fixing, heuristics, and preprocessing were added within the tree search algorithm to facilitate the solution process.

Most commercial integer programming solvers use a branch-and-bound algorithm with LP relaxations. Unless otherwise mentioned, the descriptions of the strategies discussed herein are based on using the LP relaxation.

Readers are referred to the books by Schrijver (1986) and Nemhauser and Wolsey (1988) for detailed theoretical and algorithmic expositions on integer programming and the branch-and-bound algorithms, and for further bibliographic references. The text by Parker and Rardin (1988) also includes useful material about branch and bound.

2. Partitioning Strategies

When LP relaxation is employed, partitioning is done via the addition of linear constraints. Typically, two new nodes are formed on each division. Suppose x^R is an optimal solution to the relaxation of a branch-and-bound node. Common partitioning strategies include:

- *Variable dichotomy (Dakin, 1965).* If x_j^R is fractional, then two new nodes are created, one with the simple bound $x_j \leqslant \lfloor x_j^R \rfloor$ and the other with $x_j \geqslant \lceil x_j^R \rceil$; where $\lfloor \cdot \rfloor$ and $\lceil \cdot \rceil$ denote the floor and the ceiling of a real number. In particular, if x_j is restricted to be binary, then the branching reduces to fixing $x_j = 0$ and $x_j = 1$, respectively. One advantage of simple bounds is that they maintain the size of the basis among branch-and-bound nodes, since the simplex method can be implemented to handle both upper and lower bounds on variables without explicitly increasing the dimensions of the basis.
- *Generalized-upper-bound (GUB) dichotomy (Beale and Tomlin, 1970).* If the constraint $\sum_{j \in Q} x_j = 1$ is present in the original integer program, and x_i^R, $i \in Q$ are fractional, one can partition $Q = Q_1 \cup Q_2$ such that $\sum_{j \in Q_1} x_j^R$ and $\sum_{j \in Q_2} x_j^R$ are approximately of equal value. Then two branches can be formed by setting $\sum_{j \in Q} x_j = 0$ and $\sum_{j \in Q_2} x_j = 0$, respectively.
- *Multiple branches for bounded integer variable.* If x_j^R is fractional, and $x_j \in \{0, ..., l\}$, then one can create $l + 1$ new nodes, with $x_j = k$ for node k, $k = 0, ..., l$. This idea was proposed in the first branch-and-bound algorithm by Land and Doig (1960), but currently is not commonly used.

3. Branching Variable Selection

During the partitioning process, branching variables must be selected to help create the children nodes. Clearly the choice of a branching variable affects the running time of the algorithm. Many different approaches have been developed and tested on different types of integer programs. Some common approaches are listed below:

- *Most/least infeasible integer variable.* In this approach, the integer variable whose fractional value is farthest from (closest to) an integral value is chosen as the branching variable.
- *Driebeck–Tomlin penalties* (*Driebeck, 1966; Tomlin, 1971*). Penalties give a lower bound on the degradation of the objective value for branching in each direction from a given variable. The penalties are the cost of the dual pivot needed to remove the fractional variable from the basis. If many pivots are required to restore primal feasibility, these penalties are not very informative. The up penalty, when forcing the value of the kth basic variable up, is

$$u_k = \min_{j:a_{kj}<0} \frac{(1-f_k)\bar{c}_j}{-a_{kj}}$$

where f_k is the fractional part of x_k, $\bar{c}_j$ is the reduced cost of variable, x_j, and the a_{kj} are the transformed matrix coefficients from the kth row of the optimal dictionary for the LP relaxation. The down penalty d_k is calculated as

$$d_k = \min_{j:a_{kj}>0} \frac{f_k\bar{c}_j}{a_{kj}}.$$

Once the penalties have been computed, a variety of rules can be used to select the branching variable [e.g., $\max_k \max(u_k, d_k)$, or $\max_k \min(u_k, d_k)$]. A penalty can be used to eliminate a branch if the LP objective value for the parent node minus the penalty is worse than the incumbent integer solution. Penalties are out of favor because their cost is considered too high in relation to the benefits they give.

- *Pseudocost estimate.* Pseudocosts provide a way to estimate the degradation to the objective value by forcing a fractional variable to an integral value. The technique was introduced in 1970 by Benichou et al. (1971). Pseudocosts, attempt to reflect the total cost, not just the cost of the first pivot, as with penalties. Once a variable x_k is labeled as a candidate branching variable, the pseudocosts are computed as

$$U_k = \frac{\bar{z}_k - z_k^u}{1-f_k} \quad \text{and} \quad D_k = \frac{\bar{z}_k - z_k^d}{f_k},$$

where $\bar{z}_k$ is the objective value of the parent, z_k^u is the objective value resulting from forcing up, and z_k^d is the objective value from forcing down. (If the subproblem is feasible, the associated pseudocost is not calculated.) If a variable has been branched upon repeatedly, an average may be used.

The branching variable is chosen as that with the maximum degradation, where the degradation is computed as $D_k f_k + U_k(1-f_k)$. Pseudocosts are not considered to be beneficial on problems where there is a large percentage of integer variables.

- *Pseudoshadow prices* (*Fenelon, 1991; Land and Powell, 1979*). Similar to pseudocosts, pseudo-shadow prices estimate the total cost to force a variable to an integral value. Given the up and down pseudoshadow prices (π_i^+ and π_i^-) for each constraint and pseudo-shadow prices (q_j^+ and q_j^-) for each integer variable (these values can be specified by the user), the degradation in the objective function for forcing an integer variable x_k up to an integral value is

$$S_k^+ = (1-f_k)\left[q_k^+ + \sum_i h(\pi_i^+, \pi_i^-) \right]$$

and for forcing down, it is

$$S_k^- = f_k\left[q_k^- - \sum_i h(\pi_i^+, \pi_i^-)\right]$$

where

$$h(\pi_i^+, \pi_i^-) = \max(a_{ik}\pi_i^+, -a_{ik}\pi_i^-, a_{ik}r_i).$$

Here, a_{ik} is the matrix coefficient in row i and column k, and r_i is the dual value for row i. The branching variable is chosen using criteria similar to penalties and pseudocosts.

- *Strong branching.* This branching strategy arose in connection with research on solving difficult instances of the traveling salesman problem and general mixed 0/1 integer programming problems (Applegate et al., 1995; Bixby et al., 1995). Applied to 0/1 integer programs within a simplex-based branch-and-cut setting, strong branching works as follows. Let N and K be positive integers. Given the solution of some LP relaxation, make a list of N binary variables that are fractional and closest to 0.5 (if there are fewer than N fractional variables, take all fractional variables). Suppose that I is the index set of this list. Then, for each $i \in I$, fix x_i first to 0 and then to 1 and perform K iterations (starting with the optimal basis for the LP relaxation of the current node) of the dual simplex method with steepest-edge pricing. Let L_i, U_i, $i \in I$, be the objective values that result from these simplex runs, where L_i corresponds to fixing x_i to 0 and U_i to fixing x_i to 1. A branching variable can be selected based on the best weighted-sum of these two values.
- *Priorities selection.* Variables are selected based on their priorities. Priorities can be user-assigned, or based on objective function coefficients, or on pseudocosts.

4. Node Selection

Given a list of active problems, one has to decide which subproblem should be selected to be examined next. This in turn will affect the possibilities of improving the incumbent, the chance of node fathoming, and the total number of problems needed to be solved before optimality is achieved. Below, various strategies given in Beale (1979), Benichou et al. (1971, 1977), Breu and Burdet (1974), Fenelon (1991), Forrest et al. (1974), Gauthier and Ribiere (1977), and Mitra (1973) are presented.

- *Depth first search with backtracking.* Choose a child of the previous node as the next node; if it is pruned, choose the other child. If this node is also pruned, choose the most recently created unexplored node, which will be the other child node of the last successful node.
- *Best bound.* Among all unexplored nodes, choose the one that has the best LP objective value. In the case of maximization, the node with the largest LP objective value will be chosen. The rationale is that since nodes can only be pruned when the relaxation objective value is less than the current incumbent objective value, the node with largest LP objective value cannot be pruned, since the best objective value corresponding to an integer feasible solution cannot exceed this largest value.
- *Sum of integer infeasibilities.* The sum of infeasibilities at a node is calculated as

  $$s = \sum_j \min(f_j, 1 - f_j).$$

  Choose the node with either the maximum or minimum sum of integer infeasibilities.
- *Best estimate using pseudocosts.* This technique was introduced (Benichou et al., 1971) along with the idea of using pseudocosts to select a branching variable. The individual pseudocosts can be used to estimate the resulting integer objective value attainable from node k:

$$\varepsilon_k = \bar{z}_k - \sum_i \min(D_i f_i, U_i(1 - f_i))$$

where $\bar{z}_k$ is the value of the LP relaxation at node k. The node with the best estimate is chosen.

- *Best estimate using pseudoshadow prices.* Pseudoshadow prices can also be used to provide an estimate of the resulting integer objective value attainable from the node, and the node with the best estimate can then be chosen.
- *Best projection (Forrest et al., 1974; Mitra, 1973).* Choose the node among all unexplored nodes that has the best projection. The projection is an estimate of the objective function value associated with an integer solution obtained by following the subtree starting at this node. It takes into account both the current objective function value and a measure of the integer infeasibility. In particular, the projection p_k associated with node k is defined as

$$p_k = \bar{z}_k - \frac{s_k(\bar{z}_0 - z_{ip})}{s_0},$$

where $\bar{z}_0$ denotes the objective value of the LP at the root node, z_{ip} denotes an estimate of the optimal integer solution, and s_k denotes the sum of the integer infeasibilities at node k. The projection is a weighting between the objective function and the sum of infeasibilities. The weight $(\bar{z}_0 - z_{ip})/s_0$ corresponds to the slope of the line between node 0 and the node producing the optimal integer solution. It can be thought of as the cost to remove one unit of infeasibility. Let n_k be the number of integer infeasibilities at node k. A more general projection is to let $w_k = \mu n_k + (1 - \mu)s_k$, where $\mu \in [0, 1]$, and define

$$p_k = \bar{z}_k - \frac{w_k(\bar{z}_0 - z_{ip})}{w_0}.$$

5. Illustrative Example

In this section, a two-variable integer program is solved using branch and bound. The most infeasible integer variable is used as the branching variable, and best bound is used for node selection. Consider the problem

$$\begin{aligned} \text{maximize} \quad & 13x_1 + 8x_2 \\ \text{subject to} \quad & x_1 + 2x_2 \leqslant 10 \\ & 5x_1 + 2x_2 \leqslant 20 \\ & x_1 \geqslant 0, x_2 \geqslant 0 \\ & x_1, x_2 \ \text{integer} \end{aligned} \qquad (2)$$

Initially, L consists of just this problem IP^0 (Equation 1). The solution to the LP relaxation is $x_1^0 = 2.5$, $x_2^0 = 3.75$, with value $z_0^R = 59.5$. The most infeasible integer variable is x_1, so two new subproblems are created, IP^1 where $x_1 \geqslant 3$ and IP^2 where $x_1 \leqslant 2$, and $L = \{\text{IP}^1, \text{IP}^2\}$.

Both problems in L have the same bound 59.5, so assume the algorithm arbitrarily selects IP^1. The optimal solution to the LP relaxation of IP^1 is $x_1^1 = 3$, $x_2^1 = 2.5$, with value $z_1^R = 59$. The most infeasible integer variable is x_2, so two new subproblems of IP^1 are created, IP^3 where $x_2 \geqslant 3$ and IP^4, where $x_2 \leqslant 2$, and now $L = \{\text{IP}^2, \text{IP}^3, \text{IP}^4\}$.

The algorithm next examines IP^2, since this is the problem with the best bound. The optimal solution to the LP relaxation is $x_1^2 = 2$, $x_2^2 = 4$, with value $z_2^R = 58$. Since x^2 is integral feasible, $\underline{z}_{ip}$ is then updated to 58 and IP^2 is fathomed.

Both of the two problems remaining in L have a best bound greater than 58, so neither can yet be fathomed. Since these two subproblems have the same bound 59, assume the algorithm arbitrarily selects IP^3 to examine next. The LP relaxation to this problem is infeasible, since it requires that x satisfy $x_1 \geqslant 3$, $x_2 \geqslant 3$, and $5x_1 + 2x_2 \leqslant 20$ simultaneously. Therefore, $z_3^R = -\infty$ and this node can be fathomed by bounds since $z_3^R \leqslant \underline{z}_{ip}$.

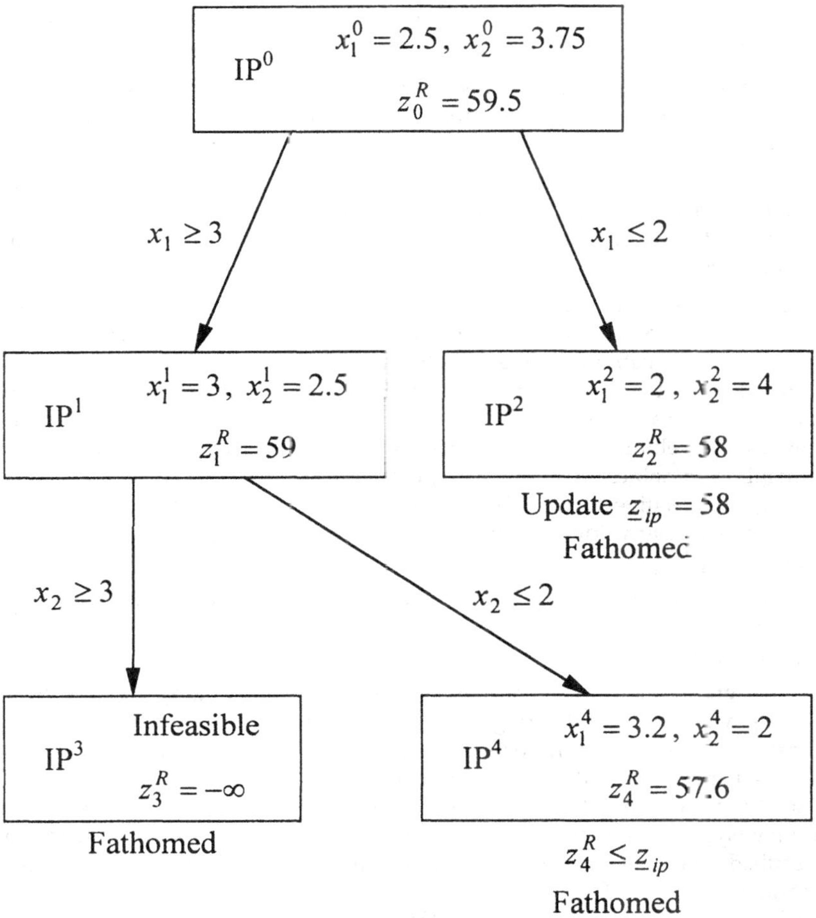

Figure 1 Branch-and-bound example

That leaves the single problem IP^4 in L. The solution to the LP relaxation of this problem is $x_1^4 = 3.2$, $x_2^4 = 2$, with value $z_4^R = 57.6$. Since $z_4^R \leqslant \underline{z}_{ip}$, this subproblem can also be fathomed by bounds. The set L is now empty, hence x^2 is an optimal solution for the integer programming problem (Equation 2).

The progress of the algorithm is indicated in Figure 1. Each box contains the name of the subproblem, the solution to the LP relaxation, and its associated objective value.

6. Preprocessing and Reformulation

Problem preprocessing and reformulation has been shown to be a very effective way of improving integer programming formulations prior to, and during, branch and bound (Bixby and Lee, 1993; Bixby and Wagner, 1987; Bordoefer, 1997; Bradley et al., 1975; Brearley et al., 1975; Crowder et al., 1983; Dietrich and Escudero, 1990; Guignard and Spielberg, 1981; Hoffman and Padberg, 1991; Savelsbergh, 1994). Below, some commonly employed preprocessing techniques are listed. For more details on these procedures, see the references

1. Removal of empty (all zeros) rows and columns. Detection of implicit bounds and implicit slack variables.

2. Removal of rows dominated by multiples of other rows, including pairs of rows for which the support of one is a subset of the support of the other.
3. Strengthen the bounds within rows by comparing individual variables and coefficients to the right-hand side. Additional strengthening may be possible for integral variables using integer rounding.
4. Use variable bounds to determine upper and lower bounds for the left-hand side of a constraint, and compare these bounds to the right-hand side. Where possible, conclude that a constraint is inconsistent, redundant, or forces the fixing of some or all variables in the support. Several of these row-driven operations can be dualized to columns.
5. Aggregation. Given an equality constraint where the bound on some variable is implied by the satisfaction of the bounds on the other variables, this variable can be substituted out, and the constraint deleted. Note that free variables always satisfy this condition. Note also that in order to control fill-in (and coefficient growth), not all such substitutions may be desirable. For integral variables, there is the added restriction that they can be eliminated only if their integrality is implied by the integrality of the remaining variables. For integer programming problems, an added advantage of aggregation relative to LP's, is that the reduction in the number of equality constraints increases the relative dimension of the underlying polytope.
6. Coefficient reduction. Consider a constraint $\sum_{j\in K} a_j x_j \geqslant b$ in which all $a_j \geqslant 0$ and all $x_j \geqslant 0$. If x_j is a 0/1 variable and $a_j > b$, for some $j \in K$, replace a_j by b. A stronger version of this procedure is possible when the problem formulation involves other constraints of appropriate structure.
7. Conflict graph, logical implications and probing. Conflict graph, logical implications and probing work by investigating the implications among other variables when a certain variable is fixed to 0 or 1. For example, this analysis may yield logical implications such as $x_k = 1$ implies $x_j = 0$, or $x_k = 1$ implies $x_j = 1$, for some other variable x_j. The implied equality is then added as an explicit constraint. Logical implications can be performed recursively, as well as in conjunction with some of the preprocessing techniques described above. Computational issues and numerical experiments regarding these techniques can be found in Bixby and Lee (1993), Bordoefer (1997), Guignard and Spielberg (1981), and Savelsbergh (1994).

7. Heuristics

Heuristic procedures provide a means for obtaining integer-feasible solutions quickly, and can be used repeatedly within the branch-and-bound search tree. A good heuristic—one that produces good integer-feasible solutions—is a crucial component in the branch-and-bound algorithm since it provides an upper bound for reduced-cost fixing (discussed later) at the root, and thus allows reduction in the size of the linear program that must be solved. This in turn may reduce the time required to solve subsequent linear programs at nodes within the search tree. In addition, a good upper bound increases the likelihood of being able to fathom active nodes, which is extremely important when solving large-scale integer programs as they tend to create many active nodes leading to memory explosion.

Broadly speaking, five ideas are commonly used in developing heuristics. The first idea is that of greediness. Greedy algorithms work by successively choosing variables based on best improvement in the objective value. Kruskal's algorithm, which is an exact algorithm for finding the minimum-weight spanning tree in a graph, is one of the most well-known greedy algorithms. Greedy algorithms have been applied to a variety of problems, including 0/1 knapsack problems, uncapacitated facility location problems, set-covering problems, and the traveling salesman problem.

A second idea is that of local search, which involves searching in a local neighborhood of a given integer-feasible solution for a feasible solution with a better objective value. The k-interchange heuristic is a classic example of a local search heuristic. Simulated annealing is another example, but with a variation. It allows, with a certain probability, updated solutions with less favorable objective values in order to increase the likelihood of escaping from a local optimum. The readers should refer to Chapters 3.5 and 3.6 of this handbook for detailed descriptions of local search and metaheuristics.

Randomized enumeration is a third idea that is used to obtain integer-feasible solutions. One such method is that of genetic algorithms, where the randomness is modeled on the biological mechanisms of evolution and natural selection (Goldberg, 1989). Recent work on applying a genetic algorithm to the set-covering problem can be found in Beasley and Chu (1996).

The term primal heuristics refers to certain LP-based procedures for constructing integral-feasible solutions from points that are in some sense good, but fail to satisfy integrality. Typically, these nonintegral points are obtained as optimal solutions of LP relaxations. Primal heuristic procedures involve successive variable fixing and rounding (according to rules usually governed by problem structure) and subsequent re-solves of the modified primal LP (Baldick, 1992; Bixby et al., 1995; Bixby and Lee, 1993; Hoffman and Padberg, 1991).

The fifth general principle is that of exploiting the interplay between primal and dual solutions. For example, an optimal or heuristic solution to the dual of an LP relaxation may be used to construct a heuristic solution for the primal IP. Problem-dependent criteria based on the generated primal-dual pair may suggest seeking an alternative heuristic solution to the dual, which would then be used to construct a new heuristic solution to the primal. Iterating back and forth between primal and dual heuristic solutions would continue until an appropriate termination condition is satisfied.

It is not uncommon that a heuristic involves more than one of these ideas. For example, pivot-and-complement is a simplex-based heuristic in which binary variables in the basis are pivoted out and replaced by slack variables. When a feasible integer solution is obtained, the algorithm performs a local search in an attempt to obtain a better integer-feasible solution (Balas and Martin, 1980). Obviously, for a specific branch-and-bound implementation, the structure of the problems at which the implementation is targeted influences the design of an effective heuristic (Applegate et al., 1995; Bixby et al., 1995; Bixby and Lee, 1993; Crowder et al., 1983; Erlenkotter, 1978; Hoffman and Padberg, 1991; Lee and Mitchell, 1997).

8. Continuous Reduced-Cost Implications

Reduced-cost fixing is a well-known and important idea in the literature of integer programming (Crowder et al., 1983). Given an optimal solution to an LP relaxation, the reduced costs $\bar{c}_j$ are nonpositive for all nonbasic variables x_j at the lower bound, and nonnegative for all nonbasic variables at their upper bounds. Let x_j be a nonbasic variable in a continuous optimal solution having objective value z_{LP}, and let z_{ip} be the objective value associated with an integer-feasible solution to IP Equation 1. The following facts specify circumstances when integer variables can be fixed without affecting the optimal objective value:

(a) If x_j is at its lower bound in the continuous solution and $z_{LP} - z_{ip} \leqslant -\bar{c}_j$, then an optimal solution to the integer program with x_j at its lower bound exists.

(b) If x_j is at its upper bound in the continuous solution and $z_{LP} - z_{ip} \leqslant \bar{c}_j$, then an optimal solution to the integer program with x_j at its upper bound exists.

When reduced-cost fixing is applied to the root node of a branch-and-bound tree, variables that are fixed can be removed from the problem, resulting in a reduction in the size of the integer program. A variety of studies have examined the effectiveness of reduced-cost fixing within the branch-and-bound tree search (Bixby et al., 1995; Bixby and Lee, 1993; Crowder et al., 1983; Hoffman and Padberg, 1991; Padberg and Rinaldi, 1989).

9. Subproblem Solver

When linear programs are employed as the relaxations within a branch-and-bound algorithm, it is common to use a simplex-based algorithm to solve each subproblem, using dual simplex to reoptimize from the optimal basis of the parent node. This technique of advanced basis has been shown to reduce the number of simplex iterations required to solve the child node to optimality, and thus speed up the overall computational effort. Recently, with the advancement of computational technology, the increase in the size of integer programs, and the success of IPMs in solving large-scale linear programs (Andersen et al., 1996; Lustig et al., 1994; Zhang et al., 1992), there have been some research efforts devoted to exploring branch-and-bound implementations that employ interior-point algorithms as the LP solver (Borchers and Mitchell, 1991; Lee and Mitchell, 1996, 1997; de Silva and Abramson, 1998). In such an implementation, the notion of advanced basis is no longer applicable, and care is needed to take advantage of warmstart vectors for the interior-point solver so as to facilitate effective computational results. In Lee and Mitchell (1996, 1997), the notion of advanced warmstart is discussed and computational results are presented.

10. Lagrangian Relaxation

LP relaxation has been widely used within branch-and-bound implementations. However, other relaxations can also be used, some of which may lead to an improvement (compared to LP relaxation) to the upper bound of the optimal objective to Equation 1. One such relaxation is known as Lagrangian relaxation. There were a number of forays prior to 1970 into the use of Lagrangian methods in discrete optimization (see Fisher, 1981, and Nemhauser and Wolsey, 1988) for an extensive list of references. However, the spark that ignited a significant amount of research into the computational possibilities of the Lagrangian approach occurred when Held and Karp (1970) used a Lagrangian problem based on minimum spanning trees to devise a remarkably successful algorithm for the traveling salesman problem. Encouraged by this success, Lagrangian methods were applied in the early 1970s to scheduling problems, and the general integer programming problem. Geoffrion (1974) coined the term "Lagrangian Relaxation" for this approach. Lagrangrian relaxation works by partitioning the constraints into two sets. The relaxation is then obtained by "dualizing" one set of constraints. Consider an integer programming problem of the form

$$
\begin{aligned}
z_{ip} = \text{maximize} \quad & c^T x \\
\text{subject to} \quad & Ax = b \\
& Dx \geqslant f \\
& x \in Z_+^n
\end{aligned}
\tag{3}
$$

We consider two possible Lagrangian problems associated with (Equation 3). The first is obtained by dualizing the constraints $Ax = b$, and the second by dualizing the constraints $Dx \geqslant f$. Let

$$
\begin{aligned}
z_{\text{LR1}}(u) = \text{maximize} \quad & c^T x + u^T(Ax - b) \\
\text{subject to} \quad & Dx \geqslant f \\
& x \in Z_+^n
\end{aligned}
\tag{4}
$$

and

$$
\begin{aligned}
z_{\text{LR2}}(v) = \text{maximize} \quad & c^T x + v^T(Dx - f) \\
\text{subject to} \quad & Ax = b \\
& x \in Z_+^n
\end{aligned}
\tag{5}
$$

In (4), u is free, and in (5), $v \geqslant 0$. The following properties of Equations (4) and (5) establish that they are indeed relaxations of Equation (3).

1. For all u, $z_{\text{LR1}}(u) \geqslant z_{ip}$.
2. For all $v \geqslant 0$, $z_{\text{LR2}}(v) \geqslant z_{ip}$.

Either (4) or (5) can be used in place of LP relaxation to derive upper bounds in a branch-and-bound algorithm. Note that in contrast to LP relaxation, integrality restrictions are imposed in these Lagrangian relaxations. Hence, the Lagrangian relaxations (Equations 4 and 5) have the potential to provide better upper bounds to the original integer program than the corresponding LP relaxation. Besides providing bounds, a Lagrangian relaxation can also be used as a medium for selecting branching variables and selecting nodes. In addition, good feasible integral solutions to (IP) may be obtained by perturbing nearly feasible solutions to (4) and (5) (Erlenkotter, 1978). Subgradient algorithms have been used to solve Lagrangian relaxations for various classes of combinatorial problems (Corneujols et al., 1977; Fisher, 1981; Geoffrion and McBride, 1978; Goffin, 1977; Held et al., 1974; Neebe and Rao, 1983; Shapiro, 1979). Although Lagrangian relaxation is not appropriate for all cases of integer programming problems, it is particularly suited for some; for example, the generalized assignment problem.

References

Andersen, E. D., J. Gondzio, C. Mészáros, and X. Xu. 1996. "Implementation of interior point methods for large scale linear programming." In *Interior Point Methods in Mathematical Programming*, edited by T. Terlakychapter 6, pp. 189–252. Boston: Kluwer Academic Publishers.

Applegate, D., R. E. Bixby, V. Chvátal, and W. Cook. 1995. "Finding cuts in the tsp (a preliminary report)." Technical Report 95-05. DIMACS, Rutgers University, New Brunswick, NJ.

Balas, E. and C. H. Martin. 1980. "Pivot and complement-a heuristic for 0/1 programming." *Mgmnt Sci.* **26**: 86–96.

Baldick, R. 1992. "A randomised heuristic for inequality-constrained mixed-integer programming." Technical Report. Department of Electrical and Computer Engineering, Worcester Polytechnic Institute, Worcester, MA.

Beale, E. M. L. 1979. "Branch and bound methods for mathematical programming systems." *Ann. Discrete Math.* **5**: 201–219.

Beale, E. M. L. and J. A. Tomlin. 1970. "Special facilities in a general mathematical programming system for non-convex problems using ordered sets of variables." In *Proceedings of the Fifth International Conference on Operations Research*, edited by J. Lawrence, pp. 447–454. Tavistock Publications.

Beasley, J. E. and P. C. Chu. 1996. "A genetic algorithm for the set covering problem." *Eur. J. Oper. Res.* **194**: 392–404.

Benichou, M., J. M. Gauthier, P. Girodet, G. Hehntges, G. Ribiere, and O. Vincent. 1971. "Experiments in mixed integer linear programming." *Math. Programming* **1**: 76–94.

Benichou, M., J. M. Gauthier, G. Hehntges, and G. Ribiere. 1977. "The efficient solution of large-scale linear programming problems—some algorithmic techniques and computational results." *Math. Programming* **13**: 280–322.

Bixby, R. E. and E. K. Lee. 1993. "Solving a truck dispatching scheduling problem using branch-and-cut." Technical Report TR93-37. Department of Computational and Applied Mathematics, Rice University, Houston. Companion paper appeared in *Oper. Res.*, **43(3)**, 355–367.

Bixby, R. E. and D. K. Wagner. 1987. "A note on detecting simple redundancies in linear systems." *Oper. Res. Lett.* **6**: 15–18.

Bixby, R. E., W. Cook, A. Cox, and E. K. Lee. 1995. "Parallel mixed integer programming." Technical Report CRPC-TR95554. Center for Research on Parallel Computation, Rice University, Houston, TX. Revised paper appeared in *Ann. Oper. Res., Special Issue on Parallel Optimization*, **90**, 19–43.

Borchers, B. and J. E. Mitchell. 1991. "Using an interior point method in a branch and bound algorithm for integer programming." Technical Report 195. Mathematical Sciences, Rensselaer Polytechnic Institute, Troy, NY. Revised July 7, 1992.

Bordoefer, R. 1997. "Aspects of set packing, partitioning, and covering." Technical Report. Technischen Universitäte, Berlin.

Bradley, G. H., P. L. Hammer, and L. Woley. 1975. "Coefficient reduction in 0–1 variables." *Math. Programming* **7**: 263–282.

Brearley, A. L., G. Mitre, and H. P. Williams. 1975. "Analysis of mathematical programming problems prior to applying the simplex method." *Math. Programming* **5**: 54–83.

Breu, R. and C. A. Burdet. 1974. "Branch and bound experiments in zero-one programming." *Math. Programming* **2**: 1–50.

Corneujols, G., M. L. Fisher, and G. L. Nemhauser. 1977. "Location of bank accounts to optimize float: an analytic study of exact and approximate algorithms." *Mgmnt Sci.* **23**: 789–810.

Crowder, H., E. L. Johnson, and M. Padberg. 1983. "Solving large-scale zero-one linear programming problem." *Oper. Res.* **31**: 803–834.

Dakin, R. J. 1965. "A tree search algorithm for mixed integer programming problems." *Comp. J.* **8**: 250–255.

de Silva, A. and D. Abramson. 1998. "A parallel interior point method and its application to facility location problems." *Comput. Optim. Appl.* **9**: 249–273.

Dietrich, B. and L. Escudero. 1990. "Coefficient reduction for knapsack-like constraints in 0/1 programs with variable upper bounds." *Oper. Res. Lett.* **9**: 9–14.

Driebeek, N. J. 1966. "An algorithm for the solution of mixed integer programming problems." *Mgmnt Sci.* **21**: 576–587.

Erlenkotter, D. 1978. "A dual-based procedure for uncapacitated facility location." *Oper. Res.* **26**: 992–1009.

Fenelon, M. 1991. "Branching strategies for mip." *CPLEX.*

Fisher, M. L. 1981. "The Lagrangian relaxation method for solving integer programming problems." *Mgmnt Sci.* **27(1)**: 1–18.

Forrest, J. J., J. P. H. Hirst, and J. A. Tomlin. 1974. "Practical solution of large mixed integer programming problems with umpire." *Mgmnt Sci.* **20**: 736–773.

Garey, M. R. and D. S. Johnson. 1979. *Computers and Intractability—a Guide to the Theory of NP-completeness.* Oxford: W. H. Freeman and Company.

Gauthier, J. M. and G. Ribiere. 1977. "Experiments in mixed integer programming using pseudocosts." *Math. Programming* **12**: 26–47.

Geoffrion, A. M. 1974. "Lagrangian relaxation and its uses in integer programming." *Math. Programming* **2**: 82–114.

Geoffrion, A. M. and R. McBride. 1978. "Lagrangian relaxation applied to capacitated facility location problems." *AIIE Trans.* **10**: 40–47.

Goffin, J. L. 1977. "On the convergence rates of subgradient optimization methods." *Math. Programming* **13**: 329–347.

Goldberg, D. E. 1989. *Genetic Algorithms in Search, Optimization, and Machine Learning.* Addison-Wesley.

Guignard, M. and K. Spielberg. 1981. "Logical reduction methods in zero-one programming." *Oper. Res.* **29**: 49–74.

Held, M. and R. M. Karp. 1970. "The traveling salesman problem and minimum spanning trees." *Oper. Res.* **18**: 1138–1162.

Held, M., P. Wolfe, and H. D. Crowder. 1974. "Validation of subgradient optimization." *Math. Programming* **6**: 61–88.

Hoffman, K. L. and M. Padberg. 1991. "Improving LP-representations of zero-one linear programs for branch-and-cut." *ORSA J. Comput.* **3**: 121–134.

Land, A. H. and A. G. Doig. 1960. "An automatic method for solving discrete programming problems." *Econometrica* **28**: 497–520.

Land, A. H. and S. Powell. 1979. "Computer codes for problems of integer programming." *Ann. Discrete Math.* **5**: 221–269.

Lee, E. K. and J. E. Mitchell. 1996. "Computational experience in nonlinear mixed integer programming." In *The Operations Research Proceedings, 1996*, pp. 95–100. Springer-Verlag.

Lee, E. K. and J. E. Mitchell. 1997. "Computational experience of an interior-point sqp algorithm in a parallel branch-and-bound framework." In *Proceedings of High Performance Optimization Techniques 1997*, pp. 329–347. Springer-Verlag.

Lustig, I. J., R. E. Marsten, and D. F. Shanno. 1994. "Interior point methods for linear programming: computational state of the art." *ORSA J. Comput.* **6(1)**: 1–14.

Mitra, G. 1973. "Investigations of some branch and bound strategies for the solution of mixed integer linear programs." *Math. Programming* **4**: 155–170.

Neebe, A. W. and M. R. Rao. 1983. "An algorithm for the fixed charge assignment of users to sources problem." *J. Oper. Res. Soc.* **34**: 1107–1115.

Nemhauser, G. L. and L. A. Wolsey. 1988. In *Integer and Combinatorial Optimization.* New York: Wiley.

Padberg, M. and G. Rinaldi. 1989. "A branch-and-cut approach to a traveling salesman problem with side constraints." *Mgmnt Sci.* **35**: 1393–1412.

Parker, R. G. and R. L. Rardin. 1988. *Discrete Optimization.* San Diego: Academic Press.

Savelsbergh, M. W. P. 1994. "Preprocessing and probing for mixed integer programming problems." *ORSA J. Comput.* **6**: 445–454.

Schrijver, A. 1986. *Theory of Linear and Integer Programming.* New York: Wiley.

Shapiro, J. F. 1979. "A survey of Lagrangian techniques for discrete optimization." *Ann. Oper. Res.* **5**: 113–138.

Tomlin, J. A. 1971. "An improved branch and bound method for integer programming." *Oper. Res.* **19**: 1070–1075.
Zhang, Y., R. Tapia, and J. Dennis, Jr. 1992. "On the superlinear and quadratic convergence of primal–dual interior-point linear programming algorithms." *SIAM J. Optim.* **2**: 304–324.

3.3 Branch-and-Cut Algorithms for Combinatorial Optimization Problems

John E. Mitchell

ABSTRACT Branch-and-cut methods are very successful techniques for solving a wide variety of integer programming problems, and they can provide a guarantee of optimality. We describe how a branch-and-cut method can be tailored to a specific integer programming problem, and how families of general cutting planes can be used to solve a wide variety of problems. Other important aspects of successful implementations are discussed in this article. The area of branch-and-cut algorithms is constantly evolving, and it promises to become even more important with the exploitation of faster computers and parallel computing.

1. Introduction

Many combinatorial optimization problems can be formulated as mixed-integer linear programming (LP) problems. They can then be solved by branch-and-cut methods, which are exact algorithms consisting of a combination of a cutting-plane method with a branch-and-bound algorithm. These methods work by solving a sequence of LP relaxations of the integer programming problem. Cutting-plane methods improve the relaxation of the problem to more closely approximate the integer programming problem, and branch-and-bound algorithms proceed, by a sophisticated divide-and-conquer approach, to solve problems. Branch-and-bound methods are discussed elsewhere in this book. In this article we discuss cutting-plane methods and their integration with branch-and-bound into branch-and-cut methods. Many problems in the various application areas discussed in this book have been attacked using these methods.

Cutting-plane algorithms for general integer programming problems were first proposed by Gomory (1963; see Section 4.1). Unfortunately, the cutting planes proposed by Gomory did not appear to be very strong, leading to slow convergence of these algorithms, so the algorithms were neglected for many years. The development of polyhedral theory and the consequent introduction of strong, problem-specific cutting planes led to a resurgence of cutting-plane methods in the 1980s, and cutting-plane methods are now the method of choice for a wide variety of problems. Perhaps the best known branch-and-cut algorithms are those that have been used to solve the traveling salesman problem (TSP; see Section 4.2). This approach is able to solve and prove optimality of far larger instances than other methods. Two papers that describe some of this research and also serve as good introductions to the area of branch-and-cut algorithms are Grötschel and Holland (1991) and Padberg and Rinaldi (1991). A more recent work on the branch-and-cut approach to the TSP is Applegate et al. (1994). The successful research on the TSP is based upon the use of polyhedral theory to find strong cutting planes.

Branch-and-cut methods have also been used to solve other combinatorial optimization problems, again through the exploitation of strong cutting planes arising from polyhedral theory. Problems attacked recently with cutting-plane or branch-and-cut methods include the linear ordering problem, maximum cut problems, scheduling problems, network design problems, packing problems, the maximum satisfiability problem, biological and medical applications, and finding maximum planar subgraphs. Recent surveys include Caprara and Fischetti (1997) and Jünger et al. (1995).

Branch-and-cut methods for general integer programming problems are also of great interest (see, for example, Balas et al., 1996a; Ceria et al., 1998; Cordier et al., 1999; Crowder et al., 1983; Jünger and Thienel, 1998; Nemhauser et al., 1994). It is usually not possible to efficiently solve a general integer programming problem using just a cutting-plane approach, and it is therefore necessary to also branch, resulting in a branch-and-cut approach. A pure branch-and-bound approach can be sped up considerably by the employment of a cutting-plane scheme, either just at the top of the tree, or at every node of the tree, because the cutting planes lead to a considerable reduction in the size of the tree.

For general problems, the specialized facets used when solving a specific combinatorial optimization problem are not available. Useful families of general inequalities include cuts based on knapsack problems (Crowder et al., 1983), Gomory cutting planes (Gomory, 1963; Balas et al., 1996b), and lift and project cutting planes (Balas et al., 1996a). Cutting planes and polyhedral theory are discussed in more detail in Section 4. Before that, an example of a branch-and-cut algorithm is given in Section 2, and an algorithm is outlined in Section 3.

The software packages MINTO (Nemhauser et al., 1994) and ABACUS (Jünger and Thienel, 1998) implement branch-and-cut algorithms to solve integer programming problems. The packages use standard LP solvers to solve the relaxations and they have a default implementation available. They also offer the user many options, including how to add cutting planes and how to branch. The commercial packages CPLEX and XPRESS-MP have also incorporated cutting-plane generation into their branch-and-bound algorithms. Many refinements are required for an efficient implementation of a branch-and-cut algorithm and some of these are detailed in Section 5.

Nemhauser and Wolsey (1988) and Wolsey (1998) provide excellent and detailed descriptions of cutting-plane algorithms and the other material in this entry, as well as other aspects of integer programming. Schrijver (1986) is an excellent source of additional material.

One aspect of a branch-and-cut approach that should not be overlooked is that it can be used to provide bounds. In particular, if we are minimizing but we are unable to prove optimality, a lower bound on the optimal value can be deduced from the algorithm, which can be used to provide a guarantee on the distance from optimality. Therefore, for large and/or hard problems, branch and cut can be used in conjunction with heuristics or metaheuristics to obtain a good (possibly optimal) solution and also to indicate how far from optimality this solution may be.

2. A Simple Example

The integer programming problem

$$\begin{array}{llcrcl} \min & z := -6x_1 & - & 5x_2 & & \\ \text{subject to} & 3x_1 & + & x_2 & \leqslant & 11 \\ & -x_1 & + & 2x_2 & \leqslant & 5 \\ & & & x_1, x_2 & \geqslant & 0, \quad \text{integer}. \end{array} \tag{1}$$

is illustrated in Figure 1. The feasible integer points are marked. The linear programming relaxation (or LP relaxation) is obtained by ignoring the integrality restrictions and is indicated by the polyhedron contained in the solid lines.

A branch-and-cut approach first solves the LP relaxation, giving the point $\left(2\frac{3}{7}, 3\frac{5}{7}\right)$, with value $-33\frac{1}{7}$. There is now a choice: should the LP relaxation be improved by adding a cutting plane, for example, $x_1 + x_2 \leqslant 5$, or should the problem be divided into two by splitting on a variable?

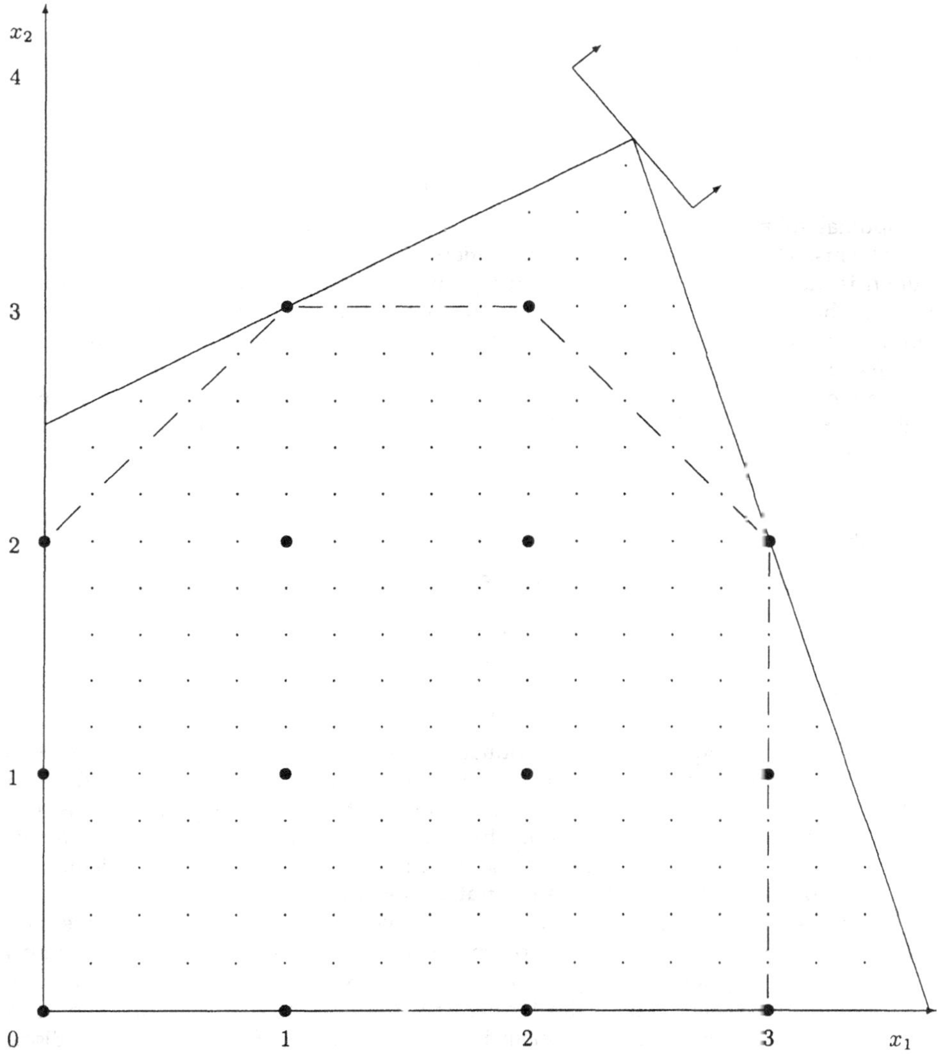

Figure 1 A two-dimensional integer programming problem

If the algorithm splits on x_1, two new problems are obtained:

$$
\begin{array}{llcccl}
\min\ z := & -6x_1 & - & 5x_2 & & \\
\text{subject to} & 3x_1 & + & x_2 & \leqslant & 11 \\
 & -x_1 & + & 2x_2 & \leqslant & 5 \\
 & \mathbf{x_1} & & & \geqslant & \mathbf{3} \\
 & & & x_1, x_2 & \geqslant & 0, \quad \text{integer.}
\end{array}
\tag{2}
$$

and

$$\begin{array}{llcll} \min\ z := & -6x_1 & - & 5x_2 & \\ \text{subject to} & 3x_1 & + & x_2 & \leqslant 11 \\ & -x_1 & + & 2x_2 & \leqslant 5 \\ & \mathbf{x_1} & & & \leqslant \mathbf{2} \\ & & & x_1, x_2 & \geqslant 0, \quad \text{integer.} \end{array} \tag{3}$$

The optimal solution to the original problem will be the better of the solutions to these two subproblems. The solution to the LP relaxation of Equation (2) is (3, 2) with value −28. This solution is integral, so it solves Equation (2), and becomes the incumbent best-known feasible solution. The LP relaxation of Equation (3) has optimal solution (2, 3.5), with value −29.5. This point is nonintegral, so it does not solve Equation (3), and it must be attacked further.

Assume the algorithm uses a cutting-plane approach and adds the inequality $2x_1 + x_2 \leqslant 7$ to Equation (3). This is a valid inequality, in that it is satisfied by every integral point that is feasible in Equation (3). Further, this inequality is violated by (2, 3.5), so it is a cutting plane. The resulting subproblem is

$$\begin{array}{llcll} \min\ z := & -6x_1 & - & 5x_2 & \\ \text{subject to} & 3x_1 & + & x_2 & \leqslant 11 \\ & -x_1 & + & 2x_2 & \leqslant 5 \\ & x_1 & & & \leqslant 2 \\ & \mathbf{2x_1} & + & \mathbf{x_2} & \leqslant \mathbf{7} \\ & & & x_1, x_2 & \geqslant 0, \quad \text{integer.} \end{array} \tag{4}$$

The LP relaxation of Equation (4) has optimal solution (1.8, 3.4), with value −27.8. Notice that the optimal value for this modified relaxation is larger than the value of the incumbent solution. The value of the optimal integral solution to the second subproblem must be at least as large as the value of the relaxation. Therefore, the incumbent solution is better than any feasible integral solution for Equation (4), so it actually solves the original problem.

The progress of the algorithm is illustrated in Figure 2.

Of course, there are several issues to be resolved with this algorithm. These include the major questions of deciding whether to branch or to cut (Section 5.5) and deciding how to branch and how to generate cutting planes (Section 4). Notice that the cutting plane introduced in the second subproblem is not valid for the first subproblem. This inequality can be modified to make it valid for the first subproblem by using a lifting technique, which is discussed in Section 5.6.

3. Outline of an Algorithm

We regard the mixed-integer LP problem

$$\begin{array}{lll} \min & c^T x & \\ \text{subject to} & Ax & \leqslant b \\ & x & \geqslant 0 \\ & x_i & \quad \text{integer}, \quad i = 1, \ldots, p. \end{array} \tag{5}$$

as our standard form, where x and c are n-vectors, b is an m-vector, and A is an $m \times n$ matrix. The first p variables are restricted to be integer, and the remainder may be fractional. If $p = n$ then this is an integer programming problem. If a variable is restricted to take the values 0 or 1 then it is a binary variable. If all variables are binary then the problem is a binary program. There is no loss of generality with restricting attention to such a format.

A branch-and-cut algorithm is outlined in Figure 3. Notice that L is the set of active nodes in the branch-and-cut tree. The value of the best-known feasible point for (ILP) is $\bar{z}$,

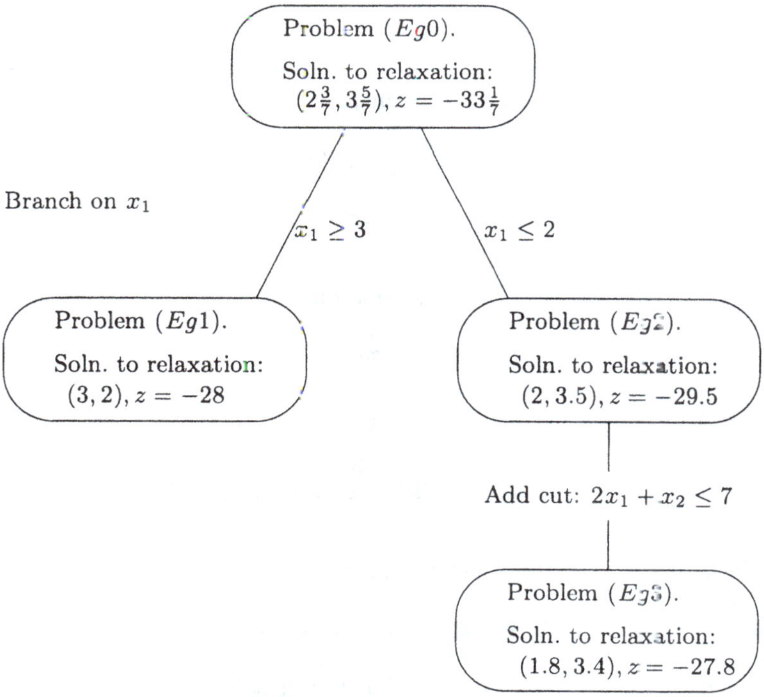

Figure 2 Progress of branch-and-cut on the two-dimensional integer programming problem

which provides an upper bound on the optimal value of (5). Further, $\underline{z}_l$ is a lower bound on the optimal value of the current subproblem under consideration. The value of the LP relaxation of the subproblem can be used to update $\underline{z}_l$. In some situations, a very large number of violated cutting planes are found in Step 5, in which case it is common to sort the cutting planes somehow (perhaps by violation), and add just a subset. The subproblems formed in Step 7 are called child subproblems, with the previous problem S^l being the parent subproblem. Usually, the partitioning takes the form of a variable disjunction: $x_i \leqslant a$ versus $x_i \geqslant a + 1$ for some variable x_i and integer a, as in the example. Other choices are possible, and they are discussed more in Chapter 3.2.

The relaxations can be solved using any method for LP problems. Typically, the initial relaxation is solved using the simplex method. Subsequent relaxations are solved using the dual simplex method, since the dual solution for the relaxation of the parent subproblem is still feasible in the relaxation of the child subproblem. Further, when cutting planes are added in Step 5, the current iterate is still dual feasible, so again the modified relaxation can be solved using the dual simplex method. It is also possible to use an interior-point method (IPM), and this can be a good choice if the LP relaxations are large—see Section 5.8 for more discussion of this option.

If the objective function and/or the constraints in (5) are nonlinear, the problem can still be attacked with a branch-and-cut approach. For more information about such problems, see Chapter 9.4 on mixed-integer nonlinear programming problems.

4. Polyhedral Theory and Cutting Planes

We say that any inequality $\pi^T x \leqslant \pi_0$ that is satisfied by all the feasible points of (5) is a valid inequality. The convex hull of the set of feasible solutions to (5) is a polyhedron. Every valid inequality defines a face of this polyhedron, namely the set of all the points in the polyhedron that satisfy $\pi^T x = \pi_0$. A facet is a face of a polyhedron that has dimension one less than the

1. *Initialization*: Denote the initial integer programming problem by ILP^0 and set the active nodes to be $L = \{ILP^0\}$. Set the upper bound to be $\bar{z} = +\infty$. Set $\underline{z}_l = -\infty$ for the one problem $l \in L$.

2. *Termination*: If $L = \emptyset$, then the solution x^*, which yielded the incumbent objective value $\bar{z}$, is optimal. If no such x^* exists (i.e., $\bar{z} = +\infty$) then ILP is infeasible.

3. *Problem selection*: Select and delete a problem ILP^l from L.

4. *Relaxation:* Solve the linear programming relaxation of ILP^l. If the relaxation is infeasible, set $\underline{z}_l = +\infty$ and go to Step 6. Let $\underline{z}_l$ denote the optimal objective value of the relaxation if it is finite and let x^{lR} be an optimal solution; otherwise set $\underline{z}_l = -\infty$.

5. *Add cutting planes*: If desired, search for cutting planes that are violated by x^{lR}; if any are found, add them to the relaxation and return to Step 4.

6. *Fathoming and Pruning*:

   (a) If $\underline{z}_l \geq \bar{z}$ go to Step 2.

   (b) If $\underline{z}_l < \bar{z}$ and x^{lR} is integral feasible, update $\bar{z} = \underline{z}_l$, delete from L all problems with $\underline{z}_l \geq \bar{z}$, and go to Step 2.

7. *Partitioning*: Let $\{S^{lj}\}_{j=1}^{j=k}$ be a partition of the constraint set S^l of problem ILP^l. Add problems $\{\text{ILP}^{lj}\}_{j=1}^{j=k}$ to L, where ILP^{lj} is ILP^l with feasible region restricted to S^{lj} and $\underline{z}_{lj}$ for $j = 1, \ldots, k$ is set to the value of $\underline{z}_l$ for the parent problem l. Go to Step 2.

Figure 3 A general branch-and-cut algorithm

dimension of the polyhedron, and it is necessary to have an inequality that represents each facet in order to have a complete linear inequality description of the polyhedron. If all the facets of the convex hull of the set of integer-feasible points are known, then the integer problem can be solved as an LP problem by minimizing the objective function over this convex hull. Unfortunately, it is not easy to obtain such a description. In fact, for an NP-Complete problem, such a description must contain an exponential number of facets, unless P = NP

In the example above, the convex hull of the set of feasible integer points has dimension 2, and all of the dashed lines represent facets. The valid inequality $3x_1 + x_2 \leqslant 11$ represents a face of the convex hull of dimension 0, namely the point (3, 2).

Chvátal–Gomory, strong, and general cutting planes are described in Sections 4.1, 4.2, and 4.3, respectively.

4.1. Chvátal–Gomory Cutting Planes

Cutting planes can be obtained by first combining together inequalities from the current LP relaxation and then exploiting the fact that the variables must be integral. This process is known as integer rounding, and the cutting planes generated are known as Chvátal–Gomory cutting planes. Integer rounding was described implicitly by Gomory (1963) and described explicitly by Chvátal (1973).

Consider again the example given earlier. The first step is to take a weighted combination of the inequalities. For example,

$$\tfrac{1}{6}(3x_1 + x_2 \leqslant 11) + \tfrac{5}{12}(-x_1 + 2x_2 \leqslant 5)$$

gives the valid inequality for the relaxation:

$$\tfrac{1}{12}x_1 + x_2 \leqslant 3\tfrac{11}{12}$$

Since all the variables are constrained to be nonnegative, rounding down the left-hand side of this inequality will only weaken it, giving $x_2 \leqslant 3\frac{11}{12}$, also valid for the LP relaxation. In any feasible solution to the integer programming problem, the left-hand side of this inequality must take an integer value. Therefore, the right-hand side can be rounded down to give the following valid inequality for the integer programming problem:

$$x_2 \leqslant 3.$$

Gomory originally derived constraints of this form directly from the optimal simplex tableau. These constraints are then added to the tableau and the modified relaxation is solved. For example, after introducing slack variables x_3 and x_4, the optimal simplex tableau for the example problem can be written

$$\begin{array}{llcll}
\min & & & 2\tfrac{3}{7}x_3 + 1\tfrac{2}{7}x_4 & =: z + 33\tfrac{1}{7} \\
\text{subject to} & x_1 & + & \tfrac{2}{7}x_3 - \tfrac{1}{7}x_4 & = 2\tfrac{3}{7} \\
 & x_2 & + & \tfrac{1}{7}x_3 + \tfrac{3}{7}x_4 & = 3\tfrac{5}{7} \\
 & & & x_i \geqslant 0, \quad i = 1, \ldots, 4 &
\end{array}$$

Since x_1 and x_2 are constrained to be integer, it follows that x, x_3, and x_4 must also all be integer. For each row, the fractional parts of the left-hand side and the right-hand side must be equal. Thus, the objective function row of this tableau indicates that the constraint

$$\tfrac{3}{7}x_3 + \tfrac{2}{7}x_4 \geqslant \tfrac{1}{7} \tag{6}$$

must be satisfied. This constraint can be added to the tableau and the modified relaxation solved using the dual simplex method. The constraint can be written in terms of the original variables:

$$\tfrac{1}{7} \leqslant \tfrac{3}{7}x_3 + \tfrac{2}{7}x_4 = \tfrac{3}{7}(11 - 3x_1 - x_2) + \tfrac{2}{7}(5 + x_1 - 2x_2) = 6\tfrac{1}{7} - x_1 - x_2,$$

or, rearranging,

$$x_1 + x_2 \leqslant 6.$$

If this constraint is added to the relaxation, the optimal value is $z = -32.5$, achieved at $x = (2.5, 3.5)$. The Gomory cutting plane generated at this new point can be written as $2x_1 + x_2 \leqslant 8$. Adding this constraint gives an optimal point of $x = (2.2, 3.6)$ with value $z = -31.2$, and the process can be iterated further.

Cutting planes can be generated from any constraint where the corresponding basic variable is fractional. The two constraint rows of the tableau given above imply the constraints

$$\tfrac{2}{7}x_3 + \tfrac{6}{7}x_4 \geqslant \tfrac{3}{7} \tag{7}$$

$$\tfrac{1}{7}x_3 + \tfrac{3}{7}x_4 \geqslant \tfrac{5}{7} \tag{8}$$

which are equivalent to

$$2x_2 \leqslant 7$$

and

$$x_2 \leqslant 3,$$

respectively.

Every valid inequality for the convex hull of the set of feasible points for (5) can be derived by repeatedly applying the Chvátal–Gomory rounding procedure (Chvátal, 1973). If a cutting plane is always generated from the first possible row then Gomory's cutting plane algorithm will solve an integer program in a finite number of iterations (Gomory, 1963). Balas et al. (1996b) have shown that the algorithm can be made competitive with other methods if certain techniques are used, such as adding many Chvátal–Gomory cuts at once. In the example, adding all three of the cuts (1)–(3) would give an optimal solution of $x = \left(2\tfrac{2}{3}, 3\right)$ with value $z = -31$; the Gomory cuts that would be generated next are equivalent to $x_1 + x_2 \leqslant 5$ and $2x_1 + x_2 \leqslant 8$, and adding these constraints would lead to the optimal solution $x = (3, 2)$, $z = -28$. Gomory cuts can contain a large number of nonzeros, so care is required to ensure that the LP relaxation does not become very hard with large memory requirements. The cuts are generated directly from the basis inverse, so care must also be taken to avoid numerical difficulties.

Gomory cutting planes can also be derived for mixed-integer LP problems. See, for example, Nemhauser and Wolsey (1988) and Wolsey (1998) for more details.

4.2. Strong Cutting Planes from Polyhedral Theory

The resurgence of interest in cutting-plane algorithms in the 1980s was due to the development of polyhedral combinatorics and the consequent implementation of cutting-plane algorithms that used facets of the convex hull of integral feasible points as cuts. Typically, a partial polyhedral description of the convex hull of the set of integer-feasible points is determined theoretically. This description will usually contain families of facets of certain types. Separation routines for those families can often be developed; such a routine will take as input a point (for example, the optimal solution to the LP relaxation), and return violated constraints from the family as output, if any exist.

In the remainder of this subsection, we illustrate the use of this approach through consideration of the TSP. In this problem, a set of cities is provided along with distances between the cities. A route that visits each city exactly once and returns to the original city is called a tour. It is desired to choose the shortest tour. One application of the TSP is in printed circuit board (PCB) production: a PCB needs holes drilled in certain places to hold electronic components such as resistors, diodes, and integrated circuits. These holes can be regarded as the cities, and the objective is to minimize the total distance traveled by the drill.

The TSP can be represented on a graph, $G = (V, E)$, where V is the set of vertices (or cities) and E is the set of edges (or links between the cities). Each edge $e \in E$ has an associated cost (or length), c_e. If the incidence vector x is defined by

$$x_e = \begin{cases} 1 & \text{if edge } e \text{ is used} \\ 0 & \text{otherwise} \end{cases}$$

then the TSP can be formulated as

$$\begin{array}{ll} \min & \sum c_e x_e \\ \text{subject to} & x \text{ is the incidence vector of a tour.} \end{array}$$

The incidence vector of a tour must satisfy the degree constraints that the sum of the edge variables must be two at each vertex, giving the following relaxation:

$$\begin{array}{lll} \min & \sum c_e x_e & \\ \text{s.t.} & \sum_{e \in \delta(v)} x_e = 2 & \text{for all vertices } v \\ & x_e = 0 \text{ or } 1 & \text{for all edges } e. \end{array} \qquad (9)$$

where $\delta(v)$ denotes the set of all edges incident to vertex v. All tours are feasible in this formulation, but it also allows infeasible solutions corresponding to subtours, consisting of several distinct unconnected loops. Consider, for example, the graph shown in Figure 4, where the labeled edge lengths are $a = 1$, $b = 2$, and unshown edges have length 10. The point $x_{12} = x_{23} = x_{13} = x_{45} = x_{46} = x_{56} = 1$, $x_{ij} = 0$ for all other edges, solves (9).

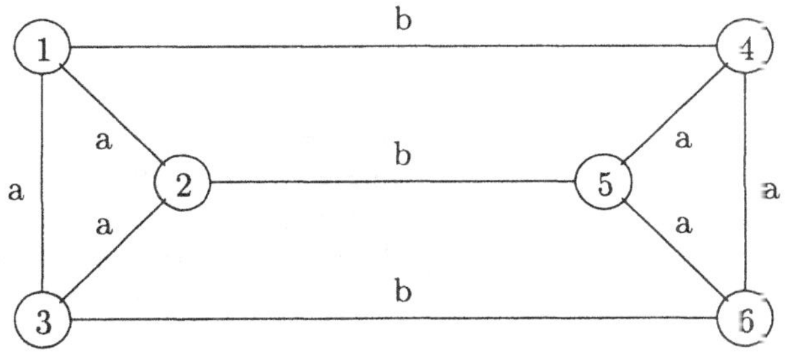

Figure 4 A traveling salesman example

Any tour must use two of the edges between the set of vertices {1, 2, 3} and the set of vertices {4, 5, 6}. In general, subtour elimination constraints can be added to the relaxation. These take the form

$$\sum_{e \in \delta(U)} x_e \geqslant 2 \qquad (10)$$

for every subset $U \subseteq V$ with cardinality $3 \leqslant |U| \leqslant \frac{|V|}{2}$, where $\delta(U)$ denotes the set of edges with exactly one endpoint in U. Any feasible solution to the relaxation given above, which also satisfies the subtour elimination constraints, must be the incidence vector of a tour. All of these inequalities are facets of the convex hull of the set of incidence vectors of tours. Now, the number of subtour elimination constraints is exponential in the number of cities, so the subtour elimination constraints are added as cutting planes as needed.

The degree constraints and the subtour elimination constraints, together with the simple bounds $0 \leqslant x_e \leqslant 1$, are still not sufficient to describe the convex hull of the set of incidence vectors of tours. Consider again the graph shown in Figure 4, where now the edge lengths are $a = 2$ and $b = 1$. All edges that are not shown have length 10. The point $x_{12} = x_{23} = x_{13} = x_{45} = x_{46} = x_{56} = 0.5$, $x_{14} = x_{25} = x_{36} = 1$ is feasible in (9) and satisfies (10). However, this point is not in the convex hull of the set of incidence vectors of tours. It violates the 2-matching inequality

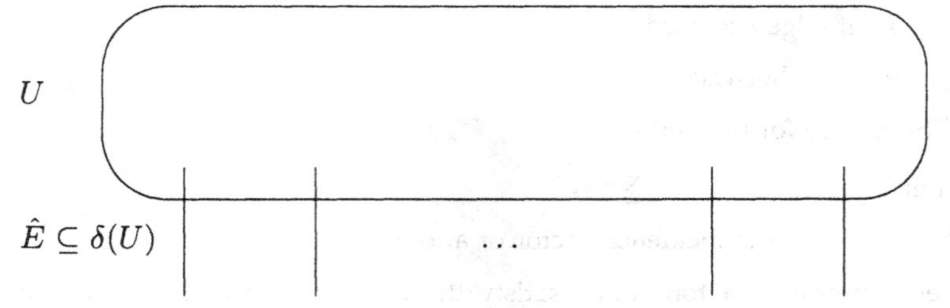

Figure 5 Graph for a 2-matching inequality

$$x_{12} + x_{23} + x_{13} + x_{14} + x_{25} + x_{36} \leqslant 4 \tag{11}$$

To obtain the general form of this inequality, consider Figure 5. Here, U is a subset of the vertices, and $\hat{E}$ is an odd set of disjoint edges, each with exactly one endpoint in U. Let $E(U)$ be the set of edges that have both endpoints in U. A tour can use at most $|U| - 1$ of the edges in $E(U)$, in which case it can use only two of the edges from $\hat{E}$. If more edges are used from $\hat{E}$, then fewer edges must be used from $E(U)$. The general form of constraint (11) is

$$\sum_{e \in E(U)} x_e + \sum_{e \in \hat{E}} x_e \leqslant |U| + \frac{|\hat{E}| - 1}{2} \tag{12}$$

This is a facet defining inequality, provided $\hat{E}$ contains at least three edges. It can be generalized further to give the family of generalized comb inequalities (Nemhauser and Wolsey, 1988), which are also facet defining.

Many other families of valid inequalities have been discovered. Branch-and-cut methods search for violated inequalities from these families. These are the most successful methods for solving large instances of the TSP (Applegate et al., 1994; Grötschel and Holland, 1991; Padberg and Rinaldi, 1991) and similar ideas have been used to solve a variety of problems (Caprara and Fischetti, 1997; Jünger et al., 1995; Nemhauser and Wolsey, 1988).

4.3. Alternative General Cutting Planes

A knapsack problem is an integer programming problem with just one linear inequality constraint. A general integer programming problem can be regarded as the intersection of several knapsack problems, one for each constraint. This observation was used in the breakthrough paper by Crowder et al. (1983) to solve general integer programming problems. The approach consists of finding facets and strong cutting planes for the knapsack problem and adding these constraints to the LP relaxation of the integer program as cutting planes. These inequalities have been extended to knapsacks with general integer variables and one continuous variable (Ceria et al., 1998) and to binary problems with generalized upper bounds.

Another family of useful inequalities are lift-and-project or disjunctive inequalities. These were originally introduced by Balas, and it is only in the last few years that the value of these cuts has become apparent for general integer programming problems (Balas et al., 1996a). Given the feasible region for a binary programming problem $S := \{x : Ax \leqslant b, x_i = 0, 1 \ \forall i\}$, each variable can be used to generate a set of disjunctive inequalities. Let $S_j^0 := \{x : Ax \leqslant b, 0 \leqslant x_i \leqslant 1 \ \forall i, x_j = 0\}$ and $S_j^1 := \{x : Ax \leqslant b, 0 \leqslant x_i \leqslant 1 \ \forall i, x_j = 1\}$. Then $S \subseteq S_j^0 \cup S_j^1$, so valid inequalities for S can be generated by finding valid inequalities for the convex hull of $S_j^0 \cup S_j^1$. These inequalities are generated by solving LP problems. Because of the expense, the cuts are usually only generated at the root node. Nonetheless, they can be very effective computationally.

Other general cutting planes have been developed. Several families and routines for identifying violated inequalities are described and investigated computationally by Cordier et al. (1999).

These alternative general cutting planes are not usually strong enough on their own to solve an integer programming problem, and they are most successfully employed in branch-and-cut algorithms for integer programming.

5. Techniques for a Good Implementation

Many refinements to the basic algorithm are necessary to get the best possible performance out of a branch-and-cut code. These include using reduced costs to eliminate variables (Section 5.1), working with a subset of the variables and then adding in the omitted variables later if necessary (Section 5.2), using primal heuristics to generate good solutions so that nodes can be pruned by bounds (Section 5.3), preprocessing the problem (Section 5.4), maintaining an appropriate balance between cutting and branching (Section 5.5), and strengthening cuts through lifting (Section 5.6). Further details are discussed in Section 5.7 and the possibility of using an IPM to solve the LP relaxations is discussed in Section 5.8.

5.1. *Fixing Variables*

Nonbasic variables can be guaranteed to take their current values in an optimal solution to the integer programming problem if their reduced costs are sufficiently large. For example, if the binary variable x_j equals zero in the optimal solution to an LP relaxation and the reduced cost of this variable is r_j, then any feasible point in the relaxation with $x_j = 1$ must have value at least $\underline{z}+r_j$, where $\underline{z}$ is the optimal value of the relaxation. The value $\bar{z}$ of the best-known feasible integral solution provides an upper bound on the optimal value of the integer program, so we must have $x_j = 0$ if $r_j > \bar{z} - \underline{z}$. Similar tests can be derived for nonbasic variables at their upper bounds. It is also possible to fix variables when an IPM is used to solve the relaxations (Mitchell et al., 1998).

Once some variables have been fixed in this manner, it is often possible to fix further variables using logical implications. For example, in a TSP, if x_e has been set equal to one for two edges incident to a particular vertex, then all other edges incident to that vertex can have their values fixed to zero.

In order to fully exploit the fixing of variables within the branching process, parent node reconstruction (Padberg and Rinaldi, 1991) is performed as follows. Once a parent node has been selected, it is not immediately divided into two children, but is solved again using the cutting-plane algorithm. When the cutting-plane procedure terminates, the optimal reduced cost vector has been reconstructed and this is used to perform variable fixing.

5.2. *Column Generation*

For many problems, including the TSP, it is impractical to work explicitly with all the variables. Thus, the typical approach for the TSP is to find, for example, the ten closest neighbors for each vertex and work only with the corresponding edges. In order to verify optimality, it is necessary to check at the end that none of the omitted edges are actually necessary. In LP terms, this requires checking whether a column should be added to the relaxation corresponding to the omitted variable. The reduced cost of this column can be found from the values of the dual variables: the column should be added if the extra dual constraint would be violated by the optimal dual solution to the current relaxation. If a large number of columns need to be added, it is usual to just add a subset of them, resolve, and check whether more need to be added.

5.3. *Primal Heuristics*

In the example problem, the existence of a good incumbent solution made it possible to prune (4) by bounds. In many cases, it takes many stages until the solution to a relaxation is integral. Therefore, it is often useful to have good heuristics for converting the fractional solution of a relaxation into a good integral solution that can be used to prune other subproblems. Primal heuristics that use the solution to the LP relaxation are discussed further in Chapter 3.2 on branch-and-bound algorithms. Branch-and-cut can also be combined with other heuristics such as local search, and also with metaheuristics.

5.4. *Preprocessing*

A very important component of a practical branch-and-cut algorithm is preprocessing to eliminate unnecessary constraints, determine any fixed variables, and simplify the problem in other ways. Preprocessing techniques are discussed in Chapter 3.2 on branch-and-bound algorithms.

5.5. *When to Add Cutting Planes*

Usually, there comes a point at which the cutting-plane loop in Steps 4 and 5 tails off, that is, the solution to one relaxation is not much better than the solutions to recent relaxations. The algorithm should then proceed to Step 6. It is believed that tailing off is a function of lack of knowledge about the polyhedral structure of the relaxation, rather than a fundamental weakness of the cutting-plane approach (Padberg and Rinaldi, 1991).

The computational overhead of searching for cutting planes can be prohibitive. Therefore, it is common to not search at some nodes of the tree. Alternatives include searching at every eighth node, say, or at every node at a depth of a multiple of eight in the tree. In some implementations, a fixed number of rounds of cutting-plane searching is performed at a node, with perhaps several rounds performed at the root node, and fewer rounds performed lower in the tree.

The cut-and-branch variant adds cutting planes only at the root node of the tree. Usually, an implementation of such a method will expend a great deal of effort on generating cutting planes, requiring time far greater than just solving the relaxation at the root. With cut-and-branch, all generated cuts are valid throughout the tree. Cut-and-branch is an excellent technique for many general integer programs, but it lacks the power of branch-and-cut for some hard problems. See Cordier et al. (1999) for more discussion of the relative computational performance of cut-and-branch and branch-and-cut.

5.6. *Lifting Cuts*

A cut added at one node of the branch-and-cut tree may well not be valid for another subproblem. Of course, it is not necessary to add the cut at any other node, in which case the cut is called a local cut. This cut will then only affect the current subproblem and its descendants. The drawback to such an approach is in the potential memory requirement of needing to store a different version of the problem for each node of the tree. In order to make a cut valid throughout the tree (or global), it is necessary to lift it. Lifting can be regarded as a method of rotating a constraint.

For binary problems, the inequality generated at a node in the tree will generally only use the variables that are not fixed at that node. If the inequality

$$\sum_{j\in J} a_j x_j \leqslant h \quad \text{for some subset } J \subseteq \{1, \ldots, n\}$$

is valid at a node where x_i has been fixed to zero, the lifted inequality takes the form

$$\sum_{j\in J} a_j x_j + \alpha_i x_i \leqslant h$$

for some scalar α_i. This scalar should be maximized in order to make the inequality as strong as possible. Now, maximizing α_i requires solving another integer program where x_i is fixed at one, so it may be necessary to make an approximation and underestimate α_i. This process has to be applied successively to each variable that has been fixed at the node. The order in which the variables are examined may well affect the final inequality, and other valid inequalities can be obtained by lifting more than one variable at a time. See Ceria et al. (1998) for discussion of lifting in the case of general mixed-integer LP problems.

5.7. *Implementation Details*

Many details of tree management can be found in Chapter 3.2 on branch-and-bound algorithms. These include node selection, branching variable selection, and storage requirements, among other issues. Typically, a branch-and-bound algorithm stores the solution to a node as a list of the indices of the basic variables. If cuts are added locally then it is necessary to store a representation of the constraint set for each active node.

Many branch-and-cut implementations use a pool of cuts (Padberg and Rinaldi, 1991), which is a set of constraints that have been generated earlier and either not included in the relaxation or subsequently dropped because they no longer appeared to be active. These cuts can be checked quickly for violation before more involved separation routines are invoked. The pool of cuts also makes it possible to reconstruct the parent node more efficiently.

5.8. *Solving Large Problems*

The difficulty of a particular integer programming problem is not purely a function of the size of the problem. For example, there are problems in the standard MIPLIB test set (Bixby et al., 1998) with just a few hundred variables that prove resistant to standard solution approaches, because of an explosion in the size of the tree.

For some problems, difficulties are caused by the size of the LP relaxation, and IPMs may be useful in such cases. IPMs are superior to simplex methods for many LP problems with thousands of variables. For large integer programs, the first relaxation at the top node of the tree can be solved using an IPM, and subsequent relaxations can be solved using the (dual) simplex method. For some problems, the relaxations are just too large to be handled with a simplex method, so IPMs are used throughout the algorithm. IPMs handle degeneracy better than the simplex method. Therefore, for example, the branch-and-cut solver described by Applegate et al. (1994) occasionally uses an IPM to handle some subproblems. Restarting is harder with an IPM than with a simplex method when the relaxation is only slightly changed, so various techniques are needed for a successful interior-point cutting-plane algorithm. The use of IPMs in branch-and-cut algorithms is surveyed by Mitchell et al. (1998). Interior-point branch-and-bound methods are discussed in Chapter 3.2.

One way to enable the solution of far larger problems is to use a parallel computer. The nature of branch-and-cut and cut-and-branch algorithms makes it possible for them to exploit coarse grain parallel computers efficiently: typically, an LP relaxation is solved on a node of the computer. It is possible to use one node to manage the distribution of linear programs to nodes. Alternatively, methods have been developed where a common data structure is maintained, and all nodes across this data structure, to obtain a relaxation that requires solution. It is also possible to generate cutting planes in parallel.

6. Conclusions

Branch-and-cut methods have proven to be a very successful approach for solving a wide variety of integer programming problems. In contrast with metaheuristics, they can guarantee optimality. For a structured integer programming problem, the branch-and-cut method should be tailored to the problem with an appropriate selection of cutting planes. These methods are also useful for general integer programming problems, where families of general cutting planes are used. Other important aspects of successful implementations have also been discussed in this article. The area of branch-and-cut algorithms is constantly evolving, and it promises to become even more important with the exploitation of faster computers and parallel computing.

ACKNOWLEDGMENT
This research was supported in part by NSF Grant CCR = 9901822.

References

Applegate, D., R. Bixby, V. Chvátal, and W. Cook. 1994. "The traveling salesman problem." Technical report. DIMACS, New Brunswick, NJ: Rutgers University.

Balas, E., S. Ceria, and G. Cornuéjols. 1996a. "Mixed 0–1 programming by lift-and-project in a branch-and-cut framework." *Mgmnt Sci.* **42**: 1229–1246.

Balas, E., S. Ceria, G. Cornuéjols, and N. Natraj. 1996b. "Gomory cuts revisited." *Oper. Res. Lett.* **19**: 1–9.

Bixby, R. E., S. Ceria, C. M. McZeal, and M. W. P. Savelsbergh. 1998. "An updated mixed integer programming library: MIPLIB 3.0." *Optima* **58**: 12–15. Problems available at http://www.caam.rice.edu/bixby/miplib/miplib.html.

Caprara, A. and A. Fischetti. 1997. "Branch and cut algorithms." In *Annotated Bibliographies in Combinatorial Optimization*, edited by M. Dell'Amico, F. Maffioli, and S. Martellochapter 4. John Wiley.

Ceria, S., C. Cordier, H. Marchand, and L. A. Wolsey. 1998. "Cutting plane algorithms for integer programs with general integer variables." *Math. Programming* **81**: 201–214.

Chvátal, V. 1973. "Edmonds polytopes and a hierarchy of combinatorial problems." *Discrete Math.* **4**: 305–337.

Cordier, C., H. Marchand, R. Laundy, and L. A. Wolsey. 1999. "bc-opt: a branch-and-cut code for mixed integer programs." *Math. Programming* **86(2)**: 335–353.

Crowder, H. P., E. L. Johnson, and M. Padberg. 1983. "Solving large-scale zero-one linear programming problems." *Oper. Res.* **31**: 803–834.

Gomory, R. E. 1963. "An algorithm for integer solutions to linear programs." In *Recent Advances in Mathematical Programming*, edited by R. L. Graves and P. Wolfe, pp. 269–302. New York: McGraw-Hill.

Grötschel, M. and O. Holland. 1991. "Solution of large-scale travelling salesman problems." *Math. Programming* **51(2)**: 141–202.

Jünger, M. and S. Thienel. 1998. "Introduction to ABACUS—A Branch-And-Cut System." *Oper. Res. Lett.* **22**: 83–95.

Jünger, M., G. Reinelt, and S. Thienel. 1995. "Practical problem solving with cutting plane algorithms in combinatorial optimization." In *Combinatorial Optimization: DIMACS Series in Discrete Mathematics and Theoretical Computer Science*, **20**: 111–152: AMS.

Mitchell, J. E., P. P. Pardalos, and M. G. C. Resende. 1998. "Interior point methods for combinatorial optimization." In *Handbook of Combinatorial Optimization*, edited by D.-Z. Du, and P. Pardalos, **1**: 189–297: Kluwer Academic Publishers.

Nemhauser, G. L. and L. A. Wolsey. 1988. *Integer and Combinatorial Optimization*. New York: John Wiley.

Nemhauser, G. L., M. W. P. Savelsbergh, and G. C. Sigismondi. 1994. "MINTO, a Mixed INTeger Optimizer." *Oper. Res. Lett.* **15**: 47–58.

Padberg, M. and G. Rinaldi. 1991. "A branch-and-cut algorithm for the resolution of large-scale symmetric traveling salesman problems." *SIAM Rev.* **33(1)**: 60–100.

Schrijver, A. 1986. *Theory of Linear and Integer Programming*. Chichester, England: John Wiley.

Wolsey, L. A. 1998. *Integer Programming*. New York: John Wiley.

3.4 Dynamic Programming Approaches

Augustine O. Esogbue

ABSTRACT This article presents an overview of the various aspects of dynamic programming from its inception to modern-day developments. Using a state-space approach, both the fundamentals and computational aspects are presented. In addition to nonserial dynamic programming, a subject not traditionally treated in basic operations research programs or literature, it includes aspects of dynamic programming models of the novel type such as fuzzy dynamic programming and neurodynamic programming that have only appeared in the literature in the past few years. An extensive bibliography is also presented.

1. Introduction

In this article, we present a subject of considerable importance in applied mathematics with many ramifications, applications, and impacts in diverse fields. Dynamic programming or DP has been introduced in operations research (OR) by all sorts of appellation leading to various interpretations and consequently various representations and classifications of the subject. What approach should we choose to demystify the subject?

Consider a few of these descriptions by different authors in the OR literature. Dynamic programming is

1. "a mathematical procedure designed primarily to improve the computational efficiency of select mathematical programming problems...,"
2. "not a mathematical model with which we can associate an algorithm..., rather it is a computational method that allows us to break up a complex problem into a sequence of easier subproblems...,"
3. "a technique that can be used to solve many optimization problems,"
4. "an approach to optimization,"
5. "an optimization procedure,"
6. "a slightly misleading name for what is really a mathematical theory of multistage decision processes."

Descriptions of the subject in different disciplines to which it has been applied are even more mind boggling. Which of the foregoing catalog of representation best approximates DP? Although the last is closest to the heart of Bellman, we add to it "a philosophy of problem solving." Clearly, optimization is of interest with optimal decisionmaking of central importance. We adopt a systems theoretic, and particularly the state-space, formalism as instructive. This approach, employed in our monogram (Esogbue, 1972) aids better problem modeling via dynamic programming by centering attention on the concept and role of state in dynamic programming. Appropriate choice of the state variable helps to reduce dimensionality, the bane of otherwise widespread use of dynamic programming, *ab initio*. Resultant model types and, *ipso facto*, solvability issues, depend on the structure of state transformations as the process evolves. This is where the philosophy and art dimension enters the picture. This critical component is usually not articulated well or often enough in presentations and discussions of dynamic programming but significant applications of DP to large-scale real-world problems have exploited this property with beneficial results (Hall, 1989).

2. Types of Decision Problem Situations

In view of the centrality of decisionmaking to the subject, we initiate our presentation with an overview of its classification. Matrix formulation of decision problems, as discussed in the classical literature, permits recognition of four distinct types of decision problems. These four types differ mainly in the extent and kind of information that is known or assumed about the states of nature.

1. *Decisionmaking under certainty.* In this situation, it is assumed that the decisionmakers have complete information about the states of nature and therefore know exactly which of the possibilities is the true state of nature. This is the simplest decision problem situation. The most notable example is the Wilson, or economic order quantity model, for deterministic demand and allowing no stock outs. This scenario leads to deterministic optimization models.
2. *Decisionmaking under risk.* Here, the decisionmakers acknowledge that, due to uncertainties of a random type, they do not know the true states of nature. In other words, they have partial information that can be expressed in terms of probabilities applicable to all of the possible states. To determine the best course of action they explore the action that maximizes the expectation or expected value of their returns. These considerations lead to stochastic optimization models.
3. *Decisionmaking under complete ignorance.* This is again decisionmaking under uncertainty in which we assume complete ignorance about the likelihood of the occurrence of the various possible states of nature. Each course of action will lead to a series of possible specified outcomes; but which it will be is unknown. Indeed, even the probability of the

occurrence of these is unknown. Determination of the best course of action then depends on the stipulation of criterion such as minimax, maximax, Laplace, Hurwitz, and so on. This line of pursuit again generates stochastic optimization models.

4. *Decisionmaking under conflict.* When the states of nature are subject to the control of an adverse intellect, as in the courses of action of an opponent or adversary, then each of the opposing decisionmakers will try to optimize at the other's expense. Thus, neither of them knows exactly which states of nature will eventually materialize but each has some basis for analysis. Some examples include business competition, bargaining for spare parts, ordering decisions in crisis or war situations, and so on. The set of techniques, developed for dealing with this type of problem situation, is discussed under the methodology of game theory. Their inventory theoretic applications are given in our book in the chapters on minimax solutions of economic criteria models and multistage game with probability models for this type. Again, the resultant optimization models are primarily of the stochastic type.
5. *Decisionmaking in a fuzzy environment.* This new class of decisionmaking is an attempt to recognize a type of situation possessing a special kind of uncertainty other than those due to randomness. By decisionmaking in a fuzzy environment, we focus on decision processes in which the goals and/or the constraints, but not necessarily the system under control, are fuzzy, ambiguous, vague, or imprecise in nature. This means that the goals and/or the constraints constitute classes of alternatives whose boundaries are not sharply defined. Under this relatively new category, we consider fuzzy optimization models.

An example of a fuzzy constraint is: "the cost of any spare parts in inventory class A should not be *substantially* higher than α," where α is some specified constant value. We note here that α can be a fuzzy number. Similarly, an example of a fuzzy goal is: "the number of spare parts in inventory class A, x should be in the *vicinity* of x_0," where x_0 is again some specified constant and again, as remarked before, a fuzzy number. The words in italics are the sources of fuzziness in these examples.

Fuzzy goals, and fuzzy constraints can be defined precisely as fuzzy sets in the space of alternatives. Usually, they are represented via their respective membership functions $\mu_G(x)$ and $\mu_c(x)$. These play a similar role for fuzzy sets as does the characteristic function of crisp sets. A fuzzy decision, then, may be viewed as the confluence of the given goals and constraints. In other words, the fuzzy decision D, represented by its membership function $\mu_D(x)$, is the intersection of $= \min\{\mu_G(x), \mu_c(x)\}$. A maximizing decision is defined as a point in the space of alternatives at which the membership function of a fuzzy decision attains its maximum value. Decisionmaking in a fuzzy environment is discussed in detail in this handbook in the chapter dealing with fuzzy optimization.

3. Decision-Oriented Processes

Because of the key role of process decomposition and decisionmaking in DP, we begin by focusing our attention on processes in which the element of decisionmaking is of central importance. We discuss two types of such processes: single-stage and multistage decision processes.

3.1. *A Single-Stage Decision Process*

Consider a process or a mechanism for transforming an input into associated outputs. Bellman aptly refers to a process as a system in a flux and abstractly, as a study of transformations. Adding the element of decisionmaking to a formally defined process when it is in a given state and stage results in a quadruple consisting of inputs, decisions, returns, and outputs. We call this a decision process. If the process has only one stage, then we have a single-stage decision process. An appropriate diagrammatic representation of such a process is given in Figure 1. For such a system, $\vec{p}_n = \tau_n(\vec{p}_0, \vec{q}_n)$. In the foregoing, when the system is in state $\vec{p}_0$, and action $\vec{q}$ is taken, there are two immediate consequences, namely, an immediate return $g(\vec{p}_0, \vec{q})$, and a final output state $\tau_n(\vec{p}_0, \vec{q}_n)$.

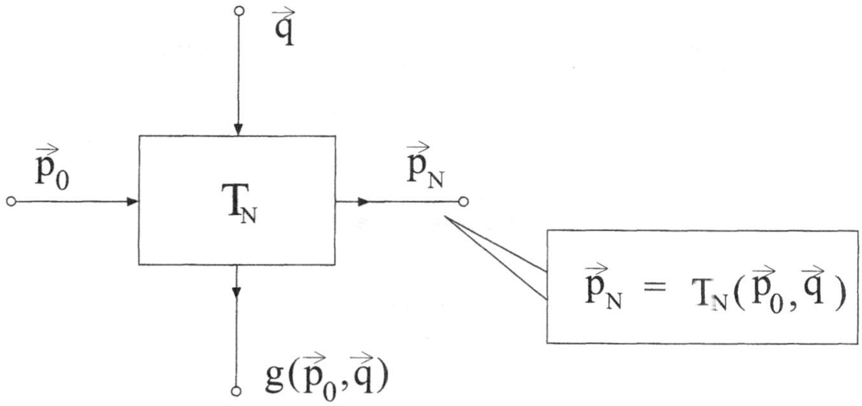

Figure 1 Single-stage process

3.2. Multistage Decision Process or a Dynamic Programming Process

Decomposing the single-stage process into a sequence of its constituent and related subparts through appropriate mappings results in a multistage process. This is the so-called separability property. In practice, this can often be achieved with some creativity and some ingenuity even when it is not initially readily transparent. If in a multistage process, the element of decisionmaking is incorporated, the resulting process is a quadruple $DP = [P, Q, T(\vec{p}, \vec{q})$ and $G(\vec{p}, \vec{q})]$, called a multistage decision processor simply a dynamic programming process. In Figure 2, we give a diagrammatic representation of this basic model, the serial multistage process or processes whose inputs to any stage are outputs from only one stage, usually the preceding one. Similarly, the output of any stage is the input to only one stage, the next one.

We note that although the foregoing process is almost exclusively the one known to operations researchers, it is certainly not the only type known or of interest to the literature of dynamic programming. A nonserial structure, on the other hand, represents systems where at least one stage in the system either receives inputs from more than one stage or sends outputs to more than one stage (Esogbue and Marks, 1972; Meier and Beightler, 1967). We shall introduce the reader to this system in a later section.

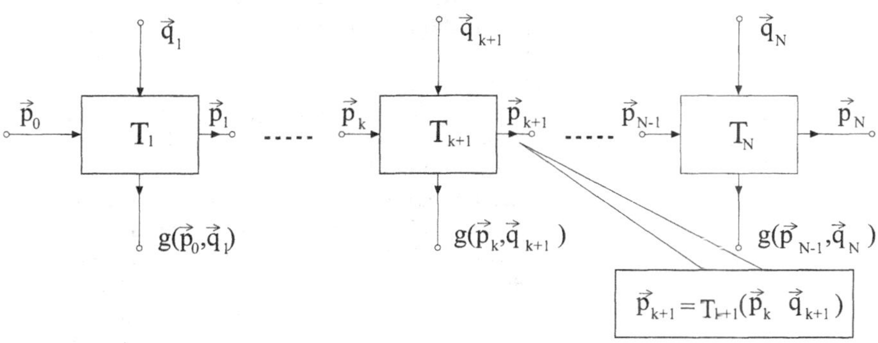

Figure 2 Multistage process

4. Discrete Dynamic Programming (DDP; Esogbue, 1972)

Let us now briefly formalize our description of a discrete deterministic dynamic programming or sequential decision process (DDP). Define a variable point p, the state vector, in a space $P, (p \in P)$ called the state space, and another point q, the decision vector, in a space Q called the policy space. We note that each policy in this space is a concatenation of elementary decisions. Let the transition function or transformation $T(p, q)$ possessing the closure property be such that whenever the decision variable q belongs to Q, then $p_1 = T(p_0, q_0)$ belongs to P. Consider a sequence $\{q_i\}_{i=0}^{N}$, which generates a corresponding sequence of points $\{p_i\}_{i=0}^{N}$ in P:

$$p_i = T(p_{i-1}, q_i), \quad i = 1, 2, ..., N. \tag{1}$$

Suppose that the aim is to find the sequence of decisions that makes extreme some prescribed scalar return function G_N of the state and decision variables defined as

$$G_N = G(p_0, p_1, ..., p_N; q_1, q_2, ..., q_N). \tag{2}$$

For this rather general formulation to be meaningfully posed, we must impose some structure upon G. Such a superimposed structure, however, is not intrinsic to the basic character of the process. Some typical functions derived from experience with frequently occurring problems in real-life processes, which G can assume, are:

$$\text{Allocation problems} \rightarrow G_N = \sum_{k=0}^{N} G(p_k) \tag{3}$$

$$\text{Reliability problems} \rightarrow G_N = \prod_{k=0}^{N} G(p_k) \tag{4}$$

$$\text{Terminal control problems} \rightarrow G_N = \max_{0 \leqslant k \leqslant N} G(p_k) \tag{5}$$

$$\text{Game-oriented problems} \rightarrow G_N = \min \max G(p_k) \tag{6}$$

$$\text{Infinite-stage Markov process} \rightarrow G_N = \max \sum_{k=0}^{N-1} G(p_k, p_{k+1}). \tag{7}$$

All these types of return functions possess something in common, namely: separability and monotonicity properties. We avoid the problem of proving that the maximum exists by assuming either that all the variables are discrete or that appropriate continuity conditions have been imposed. We next introduce a sequence of functions $f_N(p), p \in P, N = 1, 2 ...$ defined by the relation

$$f_N(p) = \max_{\{q_i\}} G_N, \quad \forall i. \tag{8}$$

The introduction of the function is motivated by the recognition that the maximum value depends on the initial state p and on the number of transformations or stages N. What we have done is to imbed a particular optimization problem within a family of related problems. This is generally referred to in dynamic programming as invariant imbedding, a process with the property that the solution to the family contains the solution of the originally posed optimization problem. We can then develop a functional equation that relates the cost of members of this family of varying duration and varying initial condition or initial state to the cost of nearby family members (in the nearest-neighbor sense).

We must state, at this point, that although the above classical form of imbedding, in terms of the initial state, is the most prevalent, it is not the only one possible. For example, in terminal control problems, a different form of imbedding in which the functional equation is defined in terms of the effect (or contribution) of the initial state upon the terminal state rather than in terms of the initial state itself, is not only possible but most instructive with resultant computational benefits. This flexibility in formulating the functional equation is one of the principal attributes of dynamic programming.

Suppose we take a return function that has the separability property, that is,

$$G_N = \sum_{i=0}^{N-1} G(p_i, q_{i+1}) + \phi(p_N), \tag{9}$$

where $\phi(p_N)$ is the reward function of the terminal state p_N. We see that for $N = 1$

$$f_1(p) = \max_{q_1} G(p_0, q_1) + \phi(p_1) \tag{10}$$

and for $N \geqslant 2$

$$f_N(p) = \max_{\{q_1, q_2, \ldots, q_N\}} G(p_0, p_1, \ldots, p_N; q_1, q_2, \ldots, q_N). \tag{11}$$

Using the obvious properties of the maximum operator and exploiting the separable structure of the return function, we can write

$$f_N(p) = \max_{q_1}\left\{ \max_{q_2, q_3, \ldots, q_N} \left\{ \sum_{i=0}^{N-1} G(p_i, q_{i-1}) + \phi(p_N) \right\}\right\}$$

$$= \max_{q_1}\left\{ G(p_0, q_1) + \max_{q_2, \ldots, q_N} \left\{ \sum_{i=2}^{N-1} G(p_i, q_{i+1}) + \phi(p_N) \right\}\right\}$$

$$= \max_{q_1}\{G(p_0, q_1) + f_{N-1}[T(p_0, q_1)]\} \tag{12}$$

We thus have an inductive technique for deducing the properties of $f_N(p)$ from that of the simpler function $f_1(p)$. For the case where $N = \infty$, an unbounded process, a novel nonlinear function equation results:

$$f(p) = \max_{q_1}\{G(p_0, q_1) + f[T(p_0, q_1)]\}. \tag{13}$$

What we have above is, in essence, the mathematical transliteration of a major theorem in dynamic programming which we now verbalize in its traditional form.

4.1. Bellman's Principle of Optimality

Let us first state this principle as initially stated by Bellman, the founder of dynamic programming. It is also referred to in control theory as the Bellman theorem or the Bellman equation.

"An optimal policy has the property that whatever the initial state and initial decision are, the remaining decisions must constitute an optimal policy with regard to the state resulting from the first decision."

Bellman calls this principle intuitive. This statement may be somewhat misleading because it has caused many zealous neophytes in dynamic programming, using a rather brute force approach, to misapply the method.

Let us interpret this principle in the context of our preceding formulation. We begin by assuming that the N-stage process that we are controlling satisfies the condition of separating the past and future. In other words, that it possesses the Markovian property. Let us assume further that we are starting the process in state p_0 and proceed to derive an optimal sequence of decisions $q_0, q_1, \ldots, q_{N-1}$ in such a way that the new stage $p_1 = T(p_0, q_0)$. The next state $p_2 = T(p_1, q_1)$, and so on. Then, regardless of what policy obtained the new state p_1, the sequence $q_1, \ldots, q_N$ must be the optimal sequence of decisions that we must follow from now on to the end of the remaining $(N-1)$ stage process. We note that this principle is not only mathematically correct but it has philosophical underpinnings, which have been exploited by expert dynamic programming practitioners in decomposing and reducing otherwise complex and intricate large-scale problems into simpler and yet optimal subdivisions for solvability (Hall, 1989).

4.2. Functional Equations in Dynamic Programming

THEORY

The mathematical transliteration of the principle of optimality generates the functional equations or the recurrence relations of dynamic programming. Once a problem has been given a structure, a functional equation that expresses the optimal returns of the multistage process as a function of the stage and state variables must next be written. We note that the type of functional equation associated with the theory of dynamic programming is nonlinear in character, thus leading to a nonlinear functional equation.

The optimal policy is determined by the solution of the functional equation, which, in turn, determines the solution. This duality and equivalence between policies and functions is important in many ways. In particular, the method of approximation in policy space provides a straightforward way of obtaining bounds on the solution of the functional equation, upper bounds if a minimization process, and lower bounds if a maximization process. The technique of quasi-linearization, for example, possesses this basic property.

SOME FUNCTIONAL EQUATIONS FOR TYPICAL RETURN FUNCTIONS DISCUSSED EARLIER

Let us now exhibit the functional equations associated with the return functions discussed earlier.

1. For allocation processes:

$$G = \sum_{k=0}^{n} G(p_k),$$

$$f_N(p) = \text{opt}_{q_0}\{g(p,q) + f_{N-1}[T(p,q_0)]\} \quad \forall N \geqslant 1, \tag{14}$$

with the boundary condition

$$f_0(p) = \text{opt}_{q_0} g(p,q_0).$$

2. For reliability problems:

$$G = \prod_{k=0}^{N} G(p_k)$$

$$f_N(p) = \text{opt}_{q_0}\{g_N(p,q_0) \cdot f_{N-1}[T(p,q_0)]\}. \tag{15}$$

3. For terminal control problems:

$$G = G(p_N)$$

$$f_N(p) = \text{opt}\{f_{N-1}[T(p,q_0)]\}, \tag{16}$$

with

$$f_0(p) = q_0 g(p). \tag{17}$$

4. For stationary problems with finite Markov chains:

$$G = \sum_{i=0}^{N-1} G(p_i, p_{i+1})$$

$$f_N(p) = \text{opt}\{g[p, T(p)] + f_{N-1}[T(p)]\}. \tag{18}$$

5. For game theoretic problems:

$$G = \max_{k \geqslant 0} G(p_k, q_k)$$

$$f(p) = \max_{\{q\}} \max_{k \geqslant 0} G(p_k, q_k)$$

$$= \max_{q_0}\left\{G(p), \max_{q_0} f[T(p, q_0)]\right\}. \quad (19)$$

VARIANTS OF FUNCTIONAL EQUATIONS

Various types of functional equations can arise depending on the nature of the processes being studied. For example, it depends on whether our formulation is deterministic or stochastic, discrete or continuous, Markov transition type, serial or nonserial, finite or infinite staged, feedback or adaptive, fuzzy or nonfuzzy, and so on. Another factor that affects the type of the functional equation is whether the solution procedure to be used is a backward or a forward type. By and large, the type of data we are given, or can obtain, is an influential factor in the formulation of the functional equation.

5. Nonserial Dynamic Programming

While considerable attention has been devoted in the literature to the analysis of serial systems, a natural extension is the study of nonserial systems. These are usually treated by employing the tools developed for serial systems. Meier and Beightler (1967) define a nonserial structure as a structure where at least one stage in the system receives inputs from more than one stage or sends outputs to more than one stage. This situation exists whenever a system is a combination of serial and parallel processes. Such systems are found in the study of chemical processing systems, natural gas transmission pipelines, water resource systems, and various other processes.

The four basic nonserial systems usually discussed in the literature are:

1. *A diverging branch system* (Figure 3). A simple example of this is a linear allocation problem occurring when a parent organization breaks up into several small firms as may result from antitrust sanctions (Meier and Beightler, 1967). Another example is found in water resource management systems when a river basin may consist of a main stream branching off into several tributaries (Esogbue and Warsi, 1986).
2. *A converging branch system* (Figure 4). This system is illustrated by a linear allocation problem that arises when a number of companies merge into a larger one. Another example from water resources systems analysis involves the analysis of reservoir or treatment plants located in various separate streams flowing into a main one in a delta (Esogbue and Warsi, 1986).
3. *A feedforward loop system* (Figure 5). Such a system may arise in gas transmission pipelines and networks where some nodes in adjoining areas are fed by a bypass line that later on joins the parent transmission line. Many examples of this structure can be

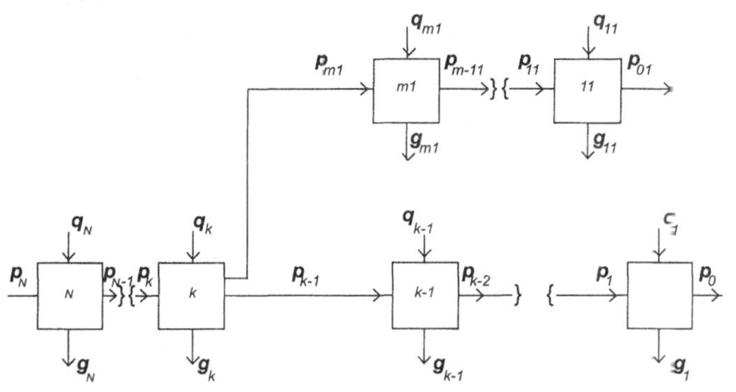

Figure 3 A nonserial system with a diverging branch structure

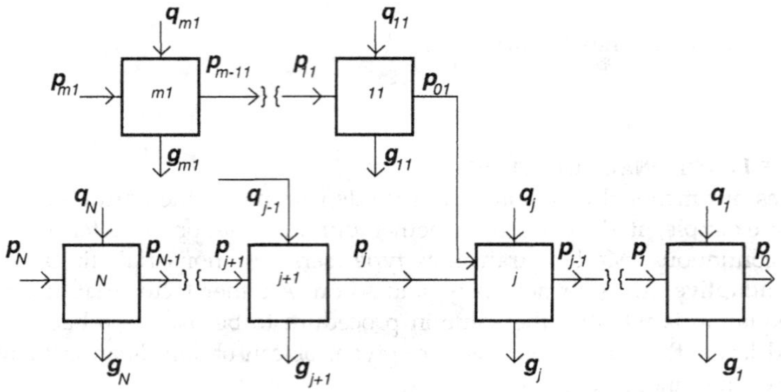

Figure 4 A nonserial system with a converging branch structure

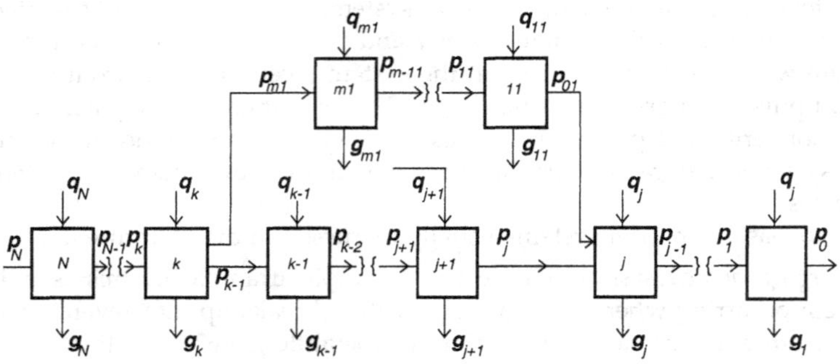

Figure 5 A nonserial system with a feedforward loop structure

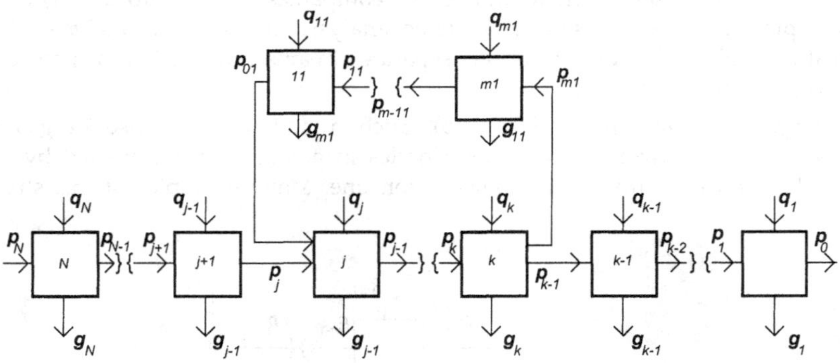

Figure 6 A nonserial system with a feedback loop structure

found in sewer or water distribution systems in the urban environment (Esogbue and Lee, 1988).

4. *A feedback loop system* (Figure 6). For example, in chemical engineering processes with product recycle characteristics, fresh feed entering a stage with a certain quality level and flow rate is mixed with part of the product stream leaving another stage at a different quality level and recycle rate (Esogbue and Warsi, 1986).

More complex nonserial processes can always be constructed by various combinations of these four basic nonserial systems. Unlike serial systems, the literature is devoid of any set of sufficiency or other similar conditions that are developed primarily for nonserial systems. Nemhauser (1966), employing the concept of state and decision inversion, extended the principle of optimality to the analysis of nonserial systems. For diverging branch systems, however, the principle of optimality can be directly applied without the use of state or decision inversion. Converging branches that have their initial states free can be optimized using state inversion. The mechanics of inversion is an approach employed to transform a system output variable into the inverse of the input variable, and vice versa. In the converging branch system of Figure 4, if the state variable p_{m1} is free, then the state inversion is performed on stages $m1$ to 11. However, converging branches that have their initial states fixed must use decision inversion instead on the stage just before the junction stage. In our example, decision inversion would be performed at stage 11. Both feedforward loop systems and feedback loop systems require decision inversion. Such an inversion can take place at either stage just before the junction with the main serial system or the stage just after leaving the main serial system. These stages are 11 and $m1$, respectively, both in the feedback and feedforward loop systems. The mathematical models for these systems can be found in Esogbue and Marks (1974).

In the absence of a general theory for the analysis of complex systems, it is always helpful to be provided with a set of usable rules that will guide the optimal analysis of such systems. To do this, Meier and Beightler (1967) presented the following six rules:

1. If any stage has no return function and if all its outputs are system outputs, then that stage and its decision variables may be eliminated without loss of structure.
2. Optimization via dynamic programming should proceed in the direction opposite to the flow of the system.
3. Elimination of outputs by stage combination should be considered for any stage having as many or more output than input variables.
4. Any fixed output should be transformed into an input by either decision or state inversion.
5. Any loop with less than four decision variables should be optimized with respect to all decisions simultaneously.
6. A feedforward or feedback loop should be cut at a state that is an input to a stage with multiple inputs. Decision inversion is then performed on the stage whose output is the cut state.

It must be emphasized, however, that these rules serve only as a guide for the analysis of nonserial systems. Nothing can be substituted for the usual degree of ingenuity called for in identifying the proper variables to be assigned the roles of state and decision variables. This is the art portion of dynamic programming.

In the first part of this section, the traditional theory of nonserial dynamic programming was briefly presented. Bertele and Brioschi (1973) introduced a different version of nonserial dynamic programming, which is concerned with optimizing a nonlinear function of the following type:

$$\text{Optimize}: \ \Phi(X) = \text{opt.} \sum_{k \in K} \xi_k(X^k),$$

where:

$$X = \{X_1, X_2, ..., X_n\} \tag{20}$$

is a set of discrete variables and

$$K = 1, 2, ..., n-1.$$

If $X^k = \{X_k, X_{k+1}\}$ for all k then the problem is a serial optimization problem. However, if the set X^i does not have this form, but is specified by $X^k \subset X$, then the problem is termed a nonserial optimization problem. In the nonserial dynamic programming formulation of the nonserial optimization problem, each variable X_k represents a stage. Therefore, the order in which the variables X_k are optimized affects the number of computations and amount of fast storage needed in the dynamic programming procedure. To determine the best order of elimination of the variables, Bertele and Brioschi (1973) construct a secondary optimization problem.

We note that none of the foregoing advances presented any efficient computational algorithm for the numerical solution of complex real world nonserial problems until a systematic attack on the problem was embarked upon by Esogbue and Marks (1972, 1974), Esogbue and Warsi (1986), and later Lee and Esogbue (1986). Additionally, these studies presented computational complexity analysis of the resultant algorithms. To reduce computational problems in complex nonserial systems, it is always useful to first determine the main chain. Such a determination may be quite difficult in many systems. However, Lee and Esogbue (1993) present a theorem that shows how to, almost routinely, determine the optimal main chain in any complex nonserial systems. Additionally, they discuss the computational complexities of the resultant models, and then apply the algorithms developed to complex water resources conveyance networks. These later studies opened the door for the use of nonserial dynamic programming in the direct analysis and efficient solution of systems with complex nonserial structures without the necessity of resorting to sometimes incorrect decompositions via serial systems.

6. Stochastic Multistage Decision Processes

As pointed out earlier, inclusion of uncertainty in our models leads to different functional equations. It is important to remark here that this does not pose any problems in dynamic programming. For example, whereas in classical optimization methods this generally introduces added complexity and results in models with completely different structures from the deterministic versions, in dynamic programming, on the other hand, both the deterministic and stochastic versions result in structures with invariant properties that have certain computational niceties.

To sketch some of the issues here, let us focus on multistage decision processes in which the uncertainty is of the random type but the stochastic effects introduced are restricted only to the discrete type of time. Bellman and Kalaba (1965) consider the case in which the transformation T is of the following form,

$$p' = T(p, q, r). \tag{21}$$

In the foregoing, r is a stochastic variable with probability distribution Φ. In this case, let us agree to employ the criterion of maximizing the expected value as the yardstick with which to measure the goodness of our optimal policy. We must point out that this expected value is not necessarily the criterion function itself, but can be the average of the criterion function. For example, we wish to maximize the expected value of

$$\sum_{n=1}^{N} g_n(p_n, q_n),$$

where $p_1 = p_0$ and $p_{n+1} = T(p_n, q_n, r_n)$, $n = 1, 2, \cdots, N-1$, and r_n are stochastic variables with distributions Φ_n, $n = 1, 2, ..., N-1$, respectively. Invoking Bellman's principle of optimality, we obtain the following associated functional equations of dynamic programming,

$$\begin{cases} f_N(p) = \max\limits_q g_N(p, q) \\ f_n(p) = \max\limits_q \left\{ g_n(p, q) + \int_{\mathscr{R}} f_{n+1}(T(p, q, r))\, d\Phi_n(r) \right\} \\ n \leqslant N-1. \end{cases} \tag{22}$$

The invariant property of the dynamic programming functional equation implies that both the deterministic and stochastic decision processes can be treated by the same formalism.

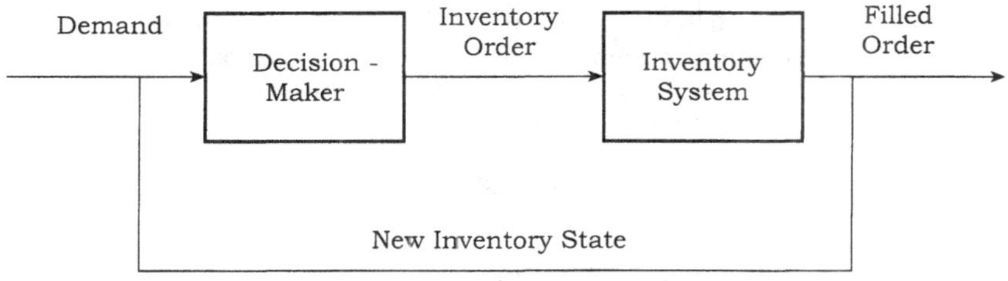

Figure 7 Simple feedback inventory system

Let us recap what takes place in the stochastic multistage decision process under discussion. Starting with an initial state $p_1 = p_0$, we choose a decision q_1. The new state $p_2 = T(p_1, q_1, r_1)$ is generated by realizing r_1 from its distribution $\Phi_1(r)$. This implies that p_2 has to be observed after the realization of stochastic variable r_1. Through a feedback mechanism we can compensate for the uncertainty introduced. This type of control, called feedback control, is illustrated in Figure 7 for a simple inventory management system.

Notice that in the deterministic case the maximum returns and optimal policies are the same whether the decisions q_n are chosen sequentially as in Figure 7 or all at the same time. This means that the feedback aspects can be ignored. In the stochastic case, however, we can employ the feedback control, which is based on the assumption that we can observe the states of the system at each stage. As configured in Figure 7, again using an inventory system as a motivating example, a transformation of the inventory system's output (filled order/sales) is fed back to the decisionmaker (manager) who combines it with the received demand to generate a system order. We see that the feedback control involves a choice of policy functions $\{q_n(p)\}$. Hence, it is on a more sophisticated level than the usual maximization over the vector $(q_1, q_2, ..., q_N)$.

Another interesting extension of multistage decision processes is the adaptive control process, as shown in Figure 8, again for an inventory model. As before, the stochastic multistage decision process assumes that the transformation,

$$p' = T(p, q, r), \tag{23}$$

involves a stochastic parameter r with known distribution. Here, however, we suppose that r is a stochastic variable with a fixed, but unknown, probability distribution function. Usually, the unknown distribution functions at stage n are assumed to have the following respective distributions,

$$\Phi_n(r \mid \bar{r}_n), \qquad n = 1, 2, ..., N, \tag{24}$$

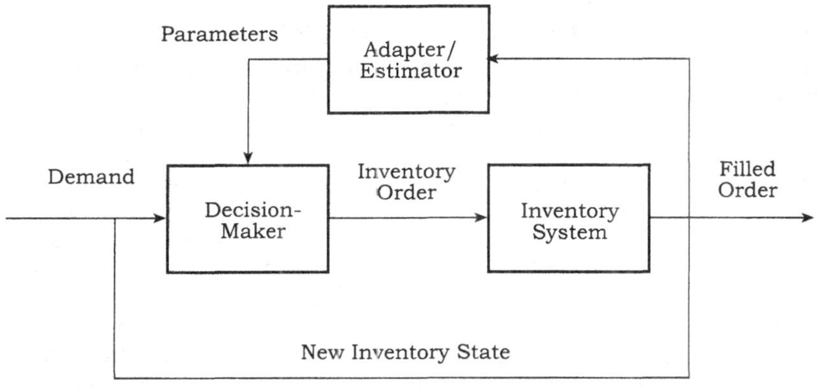

Figure 8 Adaptive feedback inventory system

where $\bar{r}_n = (r_1, r_2, \ldots, r_{n-1})$ and r_i are obtained by observation. This means that the distribution function at the current stage depends upon a prior estimation and the past history of realizations of stochastic variables. Under this assumption, the associated dynamic programming model has the following form,

$$\begin{cases} f_N(p \mid \bar{r}_N) = \max_q g_N(p, q) \\ f_n(p \mid \bar{r}_n) = \max_q \left[g_n(p, q) + \int_{\mathcal{R}} f_{n+1}(T(p, q, r) \mid \bar{r}_{n+1})\, d\Phi_n(r \mid \bar{r}_n) \right] \\ n \leqslant N - 1. \end{cases}$$

Notice that by employing our approach, we have bypassed one of the fundamental difficulties associated with adaptive control processes, that is, the estimation problem. An essential part of the adaptive control process is the learning process. What is the optimal use of the acquired information? How should one modify the prior estimation by using the new data? We can employ a Bayes estimation procedure because of its simplicity and intuitive character. But there is no reason to believe that this is an optimal estimation procedure for the general case. We have exploited this property in the development of our powerful new neuro-fuzzy adaptive controller, an aspect of which is treated in a later section.

Let us finally mention another and more recent variation, that of fuzzy adaptive control processes. The injection of fuzziness as a vehicle for capturing imprecise uncertainty has proven very helpful in developing more realistic control systems and algorithms of general type. Effective, intelligent automated systems, such as fuzzy adaptive controllers capable of learning from process data as well as incorporating linguistic data, possess significant advantages that make them more attractive candidates for the much-needed technology.

An adaptive fuzzy controller could automatically generate a set of fuzzy control rules and improve on them as the control process evolves. A neuro-fuzzy adaptive controller, using dynamic programming for feedback and optimality, was initially developed by Esogbue and Murrell (1993) and subsequently refined and extended with vastly improved capabilities. Such advances in this area of fuzzy adaptive control have also reported. See, for example, Esogbue and Murrell (1994) and more comprehensively, Esogbue (2000). These approaches have great potential for developing cost-effective control policies for an array of complex real-world problem situations.

Let us finally remark that other forms of stochastic dynamic programming, notably those with Markov transition matrices, exist. In fact, this form is usually the one focused on by most operations researchers using Howard's algorithm (Howard, 1960) and, under simplifying assumptions, some analytic solutions especially via methods inspired by linear programming. For an introduction to this algorithm, see section on computational procedures for stochastic systems.

7. Computational Procedures for Dynamic Programming

Numerical solution of the functional equation, usually via the predictor-corrector algorithm, is normally routine for problems involving one or two state variables. For high-dimensional problems, however, the computational requirements (space and time complexity) can be quite taxing. The principal problem, usually referred to as the curse of dimensionality, has been responsible for limited use of dynamic programming in many large-scale systems such as water resources systems. To circumvent or mitigate these problems, numerous computational procedures or algorithms have emerged while others are being developed. Although time complexity is an important consideration, the critical constraint revolves around the space complexity or the usually exponential growth in high-speed memory requirement with the number of variables. The thrust of various algorithms is therefore to minimize this requirement.

Before discussing these algorithms, it is constructive to provide the following remarks. We cannot overemphasize the critical role played by adroit problem formulation, especially with an optimal choice of the state variables, as well as a compact representation of the state space, in ameliorating the curse of dimensionality. Another issue, always proposed by Bellman, is the availability of improved computing devices. Vast strides have been made in recent times by the phenomenal advances in high-powered computers and peripherals including hand-held computers. Yet another issue is the technology of parallel computing, a field that was greatly influenced by the early writings of Bellman especially in the context of dynamic programming and reliability theory where parallel redundancy was discussed (Bellman and Dreyfus, 1962). With his technology, certain large-scale problems that had previously defied solution via dynamic programming are now routine.

7.1. Overview of Some Dimensionality Reduction Algorithms

A discussion of the comparative effectiveness of these methods in solving problems of different structures and sizes is beyond the scope of this article. However, a general discussion of the computational aspects is provided by Morin in a volume edited by Puterman (1978) as well as by other researchers. In general, however, the solution procedures available can be grouped into several categories, based on their efficiency in handling the problems in the realm of each of the following types: (1) deterministic, (2) stochastic, (3) nonserial. Note that because of the extra difficulties (as mentioned earlier) caused by continuous systems, they are most often approximated by discrete systems using piecewise linearization. Hence, we concentrate on methods for discrete systems.

Some of the classical approximation techniques reported in the literature for resolving aspects of the storage problems are: approximation in policy space, approximation in state space, and approximation in function space. All of these are based on the method of successive approximation. Other approaches include Durling's method of the tube, Larson's state increment dynamic programming, differential dynamic programming, Morin and Esogbue's imbedded state space approach, and Collins' diagonal decomposition method. The concepts behind these approaches will be briefly reviewed in the following sections.

COMPUTATIONAL PROCEDURES FOR DETERMINISTIC PROBLEMS

1. *Conventional DP (CDP; Bellman and Kalaba, 1965).* At any stage, $k \in N$, given the quantized values of $f_{k-1}(.)$, p_k, q_k at a finite number of grid points, it is required to compute $f_k(p)$ from Equation (6). This can be done in two steps using the predictor-corrector algorithm. First, the optimum $f_k(p)$ is determined by backward solution of the functional equation and by straightforward comparison over discretized points of q_k and stored on each of the predetermined grid points. This is the predictor move. Then, in the second step, the corrector move or reconstruction phase, the state transition equation is solved and any other values $f_{k-1}(.)$ not precomputed at previously identified grid points are obtained by an appropriate interpolation or extrapolation scheme, to improve $f_k(p)$. Realization of this algorithm via the digital computer is usually accomplished via a computer program consisting of three nested DO loops.
2. *State increment DP (SIDP; Morin and Esogbue, 1974).* This method is sometimes more aptly referred to as dynamic programming with controlled time increments. SIDP, like the CDP, is based on iterative application of Bellman's principle of optimality. However, instead of using a fixed time interval (stage) as in CDP, SIDP determines the time interval as the minimum time required for at least one of the state variables to change by one increment. This choice of interval ensures that the next state for any given control lies within a small neighborhood of the point at which control is applied. Another important aspect of this approach relates to the transfer of large amounts of data from low-speed to high-speed memory (10^{-3} seconds).

   The computations are carried out in blocks. Blocks are processed in an order which exploits knowledge of the preferred direction of motion. In general, the characteristics of a given problem situation will give a clear indication of this direction. Lack of this knowledge usually results in increased computational effort. In general, however, it can be demonstrated that the computational burden and memory requirement of this algorithm does not depend on the state dimension, thus making it a competitive method for reducing the so-called curse of dimensionality.

3. *Discrete differential DP (DDDP; Jacobson and Mayne, 1996).* This is basically a successive approximation method with a computational orientation to nonlinear dynamical systems, hence its appeal to water resources systems. The method starts with an initial guess of the optimal values of the state variables. Then a small band of values on either side of this initial guess is used as the state space of the problem. Applying the recursive equation over this reduced state space, if a locally improved solution is obtained, the procedure is repeated with the new solution replacing the initial solution. This iterative method stops when the new solution does not improve the objective function, hence a local optimum is usually reached. The method therefore guarantees a global solution only when a problem's structure is such that any local solution is also a global solution. The utility is thus dependent on our ability to determine this structure.
4. *Diagonal decomposition method (DDM; Collins, 1970a).* This is another successive approximation technique in which the problem of solving a functional equation with a domain space of dimension N is converted to the iterative solution of N-functional equations with a domain space of dimension one. The underlying idea behind this method is the fact that a matrix A can be decomposed into its diagonal and off-diagonal parts. This decoupling procedure, applicable to state and decision equations, transforms the N-dimensional optimization problem into N one-dimensional optimization problems.
5. *Functional decomposition method (Rose, 1969).* The commutative product of a contraction mapping leads to a nested sequence of multistage processes. This process can be considered as dynamic programming processes within dynamic programming processes. This forms the basis for decomposing the functional equation into a series of one-dimensional functional equations, which can then be solved routinely. The objective of this approach is therefore akin to that of the diagonal decomposition approach described earlier.
6. *Durling's concept of a tube (COT; Durling, 1964).* The basic concept of storing the optimal value function in fast-access memory is replaced by representing a function of several variables by a series of polynomials. The function at grid points is reconstructed from the stored values of the coefficients of the approximating series orthogonal function. The trajectory approximation procedure is initiated by constructing an approximate feasible trajectory, over coarse grid points.

   A tube is constructed enclosing the neighborhood region of this trajectory, and using CPD, the new trajectory is evaluated over the region with finer grid points bounded by the tube. If calculations show that an optimal point lies outside the tube, that point within the objective function is chosen instead as the best approximation to the optimal path. The procedure stops when an approximate trajectory is found which does not violate boundaries.
7. *Imbedded state space dynamic programming (ISSDP; Morin and Esogbue, 1974).* By exploiting the discontinuity properties of the maximum convolution, the usual M-dimensional search over the entire state space can be reduced to a one-dimensional search over an imbedded state space. This approach suggested by Morin and Esogbue successfully mitigates the curse of dimensionality in many cases. In particular, when the return functions are discontinuous or can be transformed into step functions, then the return $g(p, q)$ obtained by starting in p and using policy q as well as the optimal return $f_k(p)$ can also be shown to be step functions of their arguments. This implies that calculation of the resultant functional equation reduces to a process of simply updating a list of efficient points and a list-merging operation. This algorithm, developed in connection with a capacity expansion problem involving water resources projects, has been applied to a wide variety of problems such as the separable nonlinear multidimensional knapsack problem and generalized combinatorial optimization problems. The method not only performs very efficiently with regard to core storage and computational time

requirements but also obtains correct solutions to problems where brute force application of conventional dynamic programming will yield incorrect results.

8. *Other techniques with reduced computational requirements (Bellman and Dreyfus, 1962).* There are a number of other solution procedures for dynamic programming formulations, but because of space constraints we will not review them here. These include quasi-linearization (Bellman and Kalaba, 1965), Lagrange multipliers (Bellman and Dreyfus, 1962), approximation in function space and in policy space (Bellman and Dreyfus, 1962), terminal state dynamic programming (Collins, 1970b), and the continuous version of differential dynamic programming approach (Jacobson and Mayne, 1996). Although the latter is known to be very efficient as well as to possess quadratic convergence properties including the capability of circumventing the limitations due to the curse of dimensionality, it is, in general, very difficult to implement. Further, it requires certain differentiability properties, which may not be guaranteed in certain situations.

COMPUTATIONAL PROCEDURES FOR STOCHASTIC SYSTEMS

One of the many attractions of dynamic programming is that the functional equation formulations for both deterministic and stochastic processes are very similar in structure. Bellman's "approximation in function space" and "approximation in policy space techniques" can be used for solving most stochastic dynamic programming problems. To discuss solution of stochastic dynamic programs it is fruitful to consider specific structures.

A method developed particularly for stochastic dynamic programming with finite Markov chain structure and which has been used in analyzing many systems including water resources systems is Howard's algorithm. A popular version is known as the policy iteration algorithm or the iteration in policy space procedure (IPS; Howard, 1960). We may consider the discounted and nondiscounted cases, but for illustrative purposes let us consider the problem of maximizing the discounted total returns over a finite time or infinite time horizon. This procedure, which is related to linear programming, is composed of two parts. The first is a value determination operation, where for a given policy k, the value of the gain is determined. Using this gain a new optimal control function is found in the second step, called the policy improvement routine. This new policy is sent to the first step and so on. The procedure stops when there is no strict improvement in value (gain). The algorithm is quite simple and elegant with convergence to the optimal strategy being fairly rapid. A schematic representation of this algorithm is given in Figure 9. As can be seen, the solution reduces to solving sets of linear simultaneous equations as well as subsequent comparisons, with each succeeding policy possessing a higher gain than the previous one, hence the implications for linear programming solutions.

To summarize, in the above, we see that (a) the value determination operation yields g and v_i corresponding to a given choice of q_i and p_{ij}; in other words, it yields values as a function of policy; and (b) the policy improvement routine yields p_i and q_i, increase the gain for a given set of v_1; in other words, it yields the policy as a function of values. We note that in general one may enter the iteration in any box. If, for example, one chooses (a), one needs the initial policy while if one chooses (b), then one needs the set of returns.

COMPUTATIONAL PROCEDURES FOR NONSERIAL SYSTEMS

Although the conventional dynamic programming algorithm can be used to solve nonserial systems from a computational feasibility standpoint, only four of the various schemes that were discussed for serial deterministic systems seem to be applicable. The most promising for large systems are: Collins' diagonal decomposition method (DDM), Rose's functional decomposition method (FDM), discrete differential dynamic programming (DDDP), and the imbedded state space approach (ISS).

The decomposition method (DM) is developed specifically for nonserial dynamic programming problems, especially the type referred to as secondary optimization problems (Bertele and Brioschi, 1973). Basically, dynamic programming tries to eliminate variables in a nonserial dynamic programming problem in such a manner as to minimize the amount of

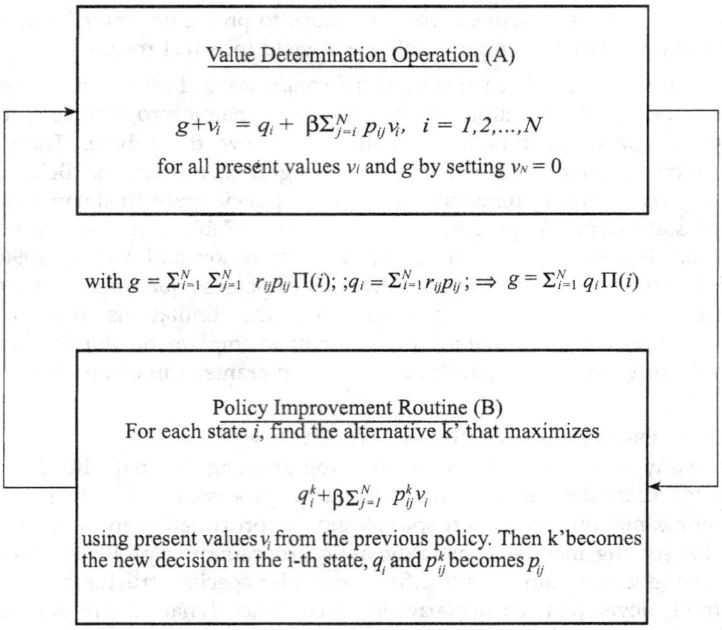

Figure 9 Schematic representation of Howard's algorithm

storage and the number of computations required. This idea is accomplished by choosing an order of elimination of variables, by a secondary optimization problem, for which the largest degree of the eliminated variables is minimal.

The ISSDP method can be used to reduce a multidimensional state variable problem to a sequence of one-dimensional problems. To minimize the storage problems, which would be normally involved if the conventional algorithm were employed to process data, Esogbue and Warsi (1986) report success with a special high-level computing algorithm developed specifically for the four basic nonserial structures. Of particular interest are the computational complexity analyses provided for the algorithms developed.

8. Recent Advances in Dynamic Programming of Novel Types

There are a number of important advances not just in computational dynamic programming but in modeling and solution philosophy, which are reported in a somewhat diffuse literature. In the following, we consider three of these significant breakthroughs, namely fuzzy dynamic programming, fuzzy criterion dynamic programming, and neuro-dynamic programming.

8.1. Essentials of Multistage Decisionmaking (Control) Under Fuzziness—Bellman and Zadeh's Approach

We present a general framework for dealing with multistage decisionmaking (control) under fuzziness as given by Bellman and Zadeh in their 1970 seminal work. This has become the basis for other extensions and versions of what is now known as fuzzy dynamic programming. Several expository articles on this subject exist but the reader is recommended to consult Esogbue and Bellman (1984), and Esogbue and Kacprzyk (1998). First, we assume that the dynamics of the problem considered can be equated with a deterministic dynamic system described by a state transition equation

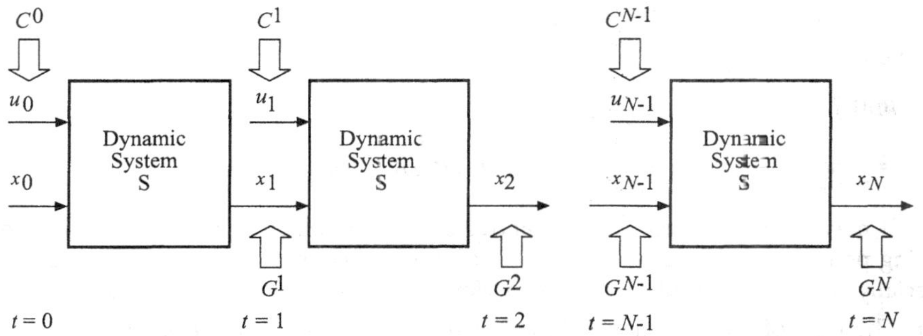

Figure 10 A general framework of the multistage decisionmaking under fuzziness

$$x_{t+1} = f(x_t, u_t), \quad t = 0, 1, ..., \tag{25}$$

where $x_t, x_{t+1} \in X = \{x\} = \{s_1, ..., s_n\}$ are the states (equated here, for simplicity but without loss of generality, with outputs) at time (stage) t and $t+1$, respectively, and $u \in U = \{u\} = \{c_1, ..., c_m\}$ is the decision (control or input) at t; the state and decision spaces, X and U, are assumed finite throughout this article. We note that a stage need not be a time instant though this is assumed here for a better intuitive appeal.

The essence of multistage decisionmaking under fuzziness may then be depicted as in Figure 10. Basically, at $t = 0$, u_0, attain a state at time $t = 1$, x_1, make a decision $u_1, ...$ Finally, being at stage $t = N-1$ in state x_{N-1} we make decision u_{N-1} and attain the final state x_N.

The state transitions from state x_t to x_{t+1} under decision u_t are given by (25), the consecutive decisions applied u_t are subjected to fuzzy constraints C^t, and on the states attained x_{t+1} fuzzy goals G^{t+1} are imposed, $t = 0, 1, ..., N-1$.

The performance of the multistage decisionmaking process is evaluated by the fuzzy decision, which, assumed to be a decomposable fuzzy set in $U \times X \times ... \times U \times X$, is

$$\begin{aligned} &\mu_D(u_0, ..., u_{N-1}|x_0) \\ &= \mu_C^0(u_0) \wedge \mu_{G^1}(x_1) \wedge \mu_{C^{N-1}}(u_{N-1}) \wedge \mu_{G^N}(x_N) \\ &= \bigwedge_{t=0}^{N-1} [\mu_{C^t}(u_t) \wedge \mu_{G^{t+1}}(x_{t+1})], \end{aligned} \tag{26}$$

where $x_0 \in X$ is an initial state, the x_t's are given by (25), and N is some termination time that is fixed and specified in the basic case.

The problem is to find the optimal sequence of decisions $u_0^*, ..., u_{N-1}^*$ such that

$$\begin{aligned} &\mu_D(u_0^*, ..., u_{N-1}^*|x_0) \\ &= \max_{u_0,...,u_{N-1}} [\mu_C^0(u_0) \wedge \mu_{G^1}(x_1) \wedge \mu_{C^{N-1}}(u_{N-1}) \wedge \mu_{G^N}(x_N)] \\ &= \max_{u_0,...,u_{N-1}} \bigwedge_{t=0}^{N-1} [\mu_{C^t}(u_t) \wedge \mu_{G^{t+1}}(x_{t+1})]. \end{aligned}$$

For simplicity, we assume, as is normally the case (unless otherwise specified), that at each stage t fuzzy constraints are given, $\mu_{C^0}(u_0), ..., \mu_{C^{N-1}}(u_{N-1})$, and a fuzzy goal, $\mu_{G^N}(x_N)$, is imposed only on the final state x_N.

Then, the fuzzy decision is

$$\mu_D(u_0, ..., u_{N-1}|x_0)$$

$$= \mu_C^0(u_0) \wedge \ldots \wedge \mu_{C^{N-1}}(u_{N-1}) \wedge \mu_{G^N}(x_N) \tag{28}$$

and the problem is to find $u_0^*, \ldots, u_{N-1}^*$ such that

$$\mu_D(u_0^*, \ldots, u_{N-1}^* | x_0)$$

$$= \max_{u_0, \ldots, u_{N-1}} [\mu_C^0(u_0) \wedge \ldots \wedge \mu_{C^{N-1}}(u_{N-1}) \wedge \mu_{G^N}(x_N)]. \tag{29}$$

It may readily be seen that the above general problem formulation can be viewed as the starting point for numerous extensions. A convenient classification of resultant problem classes, especially the discrete models, may be performed using the following basis:

- The type of the termination time: fixed and specified in advance, fuzzy, implicitly given (by entering a termination set of states), or infinite.
- The type of the dynamic system: deterministic, stochastic, fuzzy, or fuzzy-stochastic.
- The type of objective function: cost minimization, profit/benefit maximization, or fuzzy criterion set based on a satisfactory degree of maximization.

8.2. Fuzzy Dynamic Programming for Multistage Decisionmaking with a Fixed and Specified Termination Time

To illustrate the derivation of a fuzzy dynamic programming model, let us consider one of the above-mentioned cases. The case of a fixed and specified (in advance) termination time is fundamental and so we should employ it as our basic fuzzy DP model. In fact, it provides a suitable point of departure for virtually all fuzzy dynamic programming model extensions. We will therefore discuss it now in some detail with special focus on the deterministic version while referring the reader to Esogbue and Kacprzyk (1998) for others such as the stochastic and fuzzy dynamic systems.

8.3. The Case of a Deterministic Dynamic System

Recall that a deterministic system is described by its state transition equation (25), that is, $x_{t+1} = f(x_t, u_t); x_t, x_{t+1} \in X = \{s_1, \ldots, s_n\}, u_t \in U = \{c_1, \ldots, c_m\}; t = 0, 1, \ldots, N-1; x_0 \in X$ is the initial state, and $N < \infty$ is a fixed and specified termination time (planning horizon).

The fuzzy constraints are $\mu_{C^0}(u_0), \ldots, \mu_{C^{N-1}}(u_{N-1})$ and the fuzzy goal is $\mu_{G^N}(x_N)$. The fuzzy decision is

$$\mu_D(u_0, \ldots, u_{N-1} | x_0)$$

$$= \mu_{C^0}(u_0) \wedge \ldots \wedge \mu_{C^{N-1}}(u_{N-1}) \wedge \mu_{G^N}(x_N), \tag{30}$$

and the problem is to find $u_0^*, \ldots, u_{N-1}^*$ such that

$$\mu_D(u_0^*, \ldots, u_{N-1}^* | x_0)$$

$$= \max_{u_0, \ldots, u_{N-1}} [\mu_{C^0}(u_0) \wedge \ldots \wedge \mu_{C^{N-1}}(u_{N-1}) \wedge \mu_{G^N}(x_N)]. \tag{31}$$

It is easy to see that only the two last terms in this problem, that is,

$$\mu_{C^{N-1}}(u_{N-1}) \wedge \mu_{G^N}(f(x_{N-1}, u_{N-1}))$$

depend on u_{N-1} (and not on other decisions), hence (31) can be rewritten as

$$\max_{u_0, \ldots, u_{N-1}} [\mu_{C^0}(u_0) \wedge \ldots \wedge \mu_{C^{N-1}}(u_{N-1}) \wedge \mu_{G^N}(x_N)]$$

$$= \max_{u_0, \ldots, u_{N-2}} [\mu_{C^0}(u_0) \wedge \ldots \wedge \mu_{C^{N-2}}(u_{N-2})$$

$$\wedge \max_{u_{N-1}}(\mu_{C^{N-1}}(u_{N-1}) \wedge \mu_{G^N}(f(x_{N-1}, u_{N-1})))] \tag{32}$$

On repeating this backward iteration for $u_{N-2}, ..., u_0$, we obtain the set of dynamic programming recurrence equations

$$\begin{cases} \mu_{G^{N-i}}(x_{N-i}) = \max\limits_{u_{N-i}}[\mu_{C^{N-i}}(u_{N-i}) \wedge \mu_{G^{N-i+1}}(x_{N-i+1})] \\ x_{N-i+1} = f(x_{N-i}, u_{N-i});\ i = 1, ..., N, \end{cases} \tag{33}$$

where $\mu_{G^{N-i}}(.)$ is a fuzzy goal at $t = N - i$ induced by a fuzzy goal at $t = N - i + 1$.

An optimal sequence of decisions sought, $u_0^*, ..., u_{N-1}^*$, is given by the successive maximizing values of u_{N-i} in (33). It is convenient to represent the solution, u_t^*, by an optimal policy $a_t^* : X \rightarrow U$, such that $u_t^* = a_t^*(x_t), t = 0, 1, ..., N-1$, that is, relating an optimal decision to the current state.

Finally, let us remark that in virtually all of the foregoing cases, a dynamic programming type algorithm can be devised. These models and algorithms give us more flexibility in modeling and solution of complex real-world decisions and control problems thus enhancing the utility of dynamic programming. For more discussions about these algorithms and sample applications see Esogbue and Kacprzyk (1998).

9. Fuzzy Criterion Dynamic Programming

An essentially different approach to optimization is via the introduction of a noncost objective function (economic criteria). This approach has led to the development of a new and powerful type of dynamic programming, which we discuss in this section. We motivate our presentation by introducing the concept of decision criteria.

9.1. The Concept of Decision Criteria

In any decision problem, it is necessary to possess some performance measure that we can use to evaluate the effectiveness of our actions or decisions. We call such a measure a decision criterion. In the usual mathematical programming model, for example, it is called the objective function. If we wish to make the value of the accepted measure of performance as high as possible, then we have a maximization problem. Analogously, if we wish for this value of measure to be as low as possible, then we have a minimization problem. In the theory of mathematical programming, the objective function is usually assumed to be known perfectly. This is not usually the case in practice. It is, however, an important aspect of a decision problem because it guides the decisionmaking process. This fact provides the basic motivation for us to discuss the decision criterion problem particularly as it appears in multistage decision processes.

To discuss the foregoing criterion model more formally, let us assume that the planning horizon has N stages, labeled by $1, 2, ..., N$ from beginning to end. At each stage i, let x_i be the initial inventory level, d_i the order quantity at the beginning of the stage i, and ξ the stochastic demand characterized by a probability density function $\phi(\xi)$. Then, the economic criterion is to minimize the following function,

$$J_n(\mathbf{x},) = \sum_{i=n}^{N} \theta^{i-n}[c(d_i) + L(x_i + d_i)], \tag{34}$$

which is called economic criterion model, where $c(d)$ is the ordering cost function, $L(y)$ is the expected holding and shortage cost function, and θ is the discount rate, $0 \leqslant \theta \leqslant 1$. Economic criterion models have been discussed extensively, especially by mathematicians and operations researchers during the past half century. Researchers tended to favor or utilize this type of model by arguing that it is reasonable to assume

$$\mu_A(\mathbf{x}) = \rho_1\mu_{A_1}(x_1) + \rho_2\mu_{A_2}(x_2) + \cdots + \rho_n\mu_{A_n}(x_n), \tag{35}$$

where $\mathbf{x} = (x_1, x_2, ..., x_n), x_i \in X_i$ and $\rho_1, \rho_2, ..., \rho_n$ are coefficients of convex combination. If there is a stochastic perturbation ξ in fuzzy criterion set A, then the expected fuzzy criterion function is defined by

$$\int_{\mathscr{R}} \mu(x(\xi))\, d\Phi(\xi), \tag{36}$$

where Φ is the probability distribution of ξ, in which $\mu(x(\xi))$ may be $\mu(x(\xi)) = \mu(x+\xi)$, $\mu(x(\xi)) = \mu(x-\xi)$ and so on.

9.2. *Dynamic Programming and Fuzzy Criterion Models*

In the real world, it is not always possible or in fact necessary to precisely define the criteria. For example, in an inventory system, if the demand must be satisfied regardless of the physical constraints on the warehouse, then the state of the system can be described by an imaginary inventory level, which is the inventory level after a decision and a stochastic demand. When this imaginary inventory level is less than the dead inventory level, then the demand is not fulfilled. In this case, the difference represents the shortage quantity. When the imaginary inventory level is larger than the largest physical storage, then the amount exceeds the capacity of the warehouse, and the difference represents the degree that the capacity level has been exceeded. Usually, there exists a best state at which we may define the value of the fuzzy criterion as 1. When the inventory level deviates from this best state, then the fuzzy criterion value decreases. In other words, one seeks to attain some satisfactory state or level of performance. Thus, the set of all satisfactory states is a fuzzy criterion set whose fuzzy criterion function (Figure 11) is the satisfactory degree of elements.

For a given N-stage decision process, inventory control or reservoir operation, let $A_1, A_2, \ldots, A_N$ be fuzzy criterion sets of satisfactory states with fuzzy criterion functions $\mu_1, \mu_2, \ldots, \mu_N$ at the stages $1, 2, \ldots, N$, respectively, on the real line $\mathscr{R}$. Assume that $\rho_1, \rho_2, \ldots, \rho_N$ are coefficients of convex combination representing the relative importance among $A_1, A_2, \ldots, A_N$. Let x_i, d_i and ξ_i be the state, decision, and stochastic variables at stage i, respectively. Then, using the mass balance formula of inventory theory, the state transition equationcan be written as follows:

$$x_{i+1} = x_i + d_i + \xi_i, \quad i = 1, 2, \ldots, N-1. \tag{37}$$

Our problem then translates to the control of this system in such a manner that the states over all stages are satisfactory. In other words, at stage n, the objective is to maximize the expected fuzzy criterion function of the product $A_n \otimes A_{n+1} \otimes \cdots \otimes A_N$. We call this decision criterion the fuzzy criterion, which is to maximize the weighted sum of all membership functions such that the states over their respective stages are satisfactory.

Based on the fuzzy criterion set operations, the expected fuzzy criterion function J_n of the product $A_n \otimes A_{n+1} \otimes \cdots \otimes A_N$ is

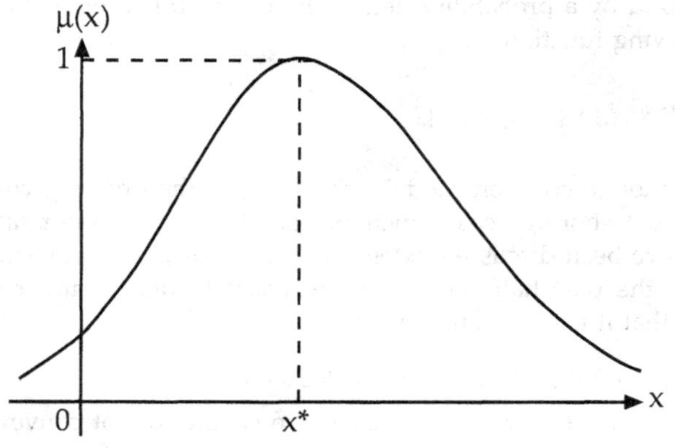

Figure 11 Fuzzy criterion function $\mu(x)$

$$J_n(\mathbf{x};) = \sum_{i=n}^{N} \gamma_i^{(n)} \int_{\mathscr{R}} \mu_i(x_i + d_i + \xi_i)\, d\Phi_i(\xi_i), \tag{38}$$

where $\mathbf{x} = (x_1, x_2, ..., x_N)$ is a state vector, $= (d_1, d_2, ..., d_N)$ is a decision vector, and

$$\gamma_i^{(n)} = \frac{\rho_i}{\rho_n + \rho_{n+1} + \cdots + \rho_N}. \tag{39}$$

For the quantities d_i and ξ_i, positivity implies that it is an input, while negativity implies that it is an output. We note that $\gamma_n^{(n)}, \gamma_{n+1}^{(n)}, ..., \gamma_N^{(n)}$ are simply coefficients of convex combination and

$$\gamma_n^{(n)} : \gamma_{n+1}^{(n)} : \cdots : \gamma_N^{(n)} = \rho_n : \rho_{n+1} : \cdots : \rho_N. \tag{40}$$

9.3. A Fuzzy Criterion Dynamic Programming Model

Let us now formally introduce a fuzzy criterion dynamic programming model, which is associated with problem (38) as follows:

$$\begin{cases} f_N(x) = \sup\limits_{d \in D_N} L_N(x + d) \\ f_n(x) = \sup\limits_{d \in D_n} \left\{ \theta_n L_n(x + d) + (1 - \theta_n) \int_{\mathscr{R}} f_{n+1}(x + d + \xi)\, d\Phi_n(\xi) \right\} \\ \quad n \leqslant N - 1, \end{cases} \tag{41}$$

where

$$L_n(y) = \int_{\mathscr{R}} \mu_n(y + \xi)\, d\Phi_n(\xi), \tag{42}$$

and $\theta_n = \gamma_n^{(n)}, D_n = [q_n, Q_n]$ is a set of feasible policies q_n and Q_n are not necessarily finite and positive. The foregoing fuzzy criterion dynamic programming-based model is called a fuzzy criterion model.

We have applied the foregoing optimization model and its variants to address a number of complex problems especially those with noncost or profit functions. An array of examples can be drawn from inventory theory, water resources systems, and cluster analysis. For an extensive development of fuzzy criterion dynamic programming see Liu and Esogbue (1996).

10. Neuro-Dynamic Programming

Let us now turn our attention to a recent approach to computational dynamic programming with many ramifications and application possibilities. The term neuro-dynamic programming (or the acronym NDP) was used by Bertsekas and Tsitsikilis (1996) to refer to the relatively new body of work involving the hybrid algorithms of dynamic programming and neuron-like learning methods of modern control theory. Although these methods were initially extensively developed within the artificial intelligence community, they are increasingly popularized in intelligent control and recently in the new domains of computational intelligence, where the hybridization of such tools as fuzzy logic, neural networks, genetic algorithms, and evolutionary computation are espoused. The specific class of mathematical learning theory most favored in these techniques is reinforcement learning.

One way to motivate these methods is to consider them as being able to learn to make good decisions through observations of their own behavior, and adaptively correcting their mistakes via a mechanism involving a series of carefully designed reinforcements. These processes are tied to the methods of simulations and successive approximations of computational dynamic programming. Both reinforcement learning and approximate dynamic programming have now become the foundations of several intelligent online control models, which have proven effective in systematic searches of the control/action spaces in a possibly dynamic environment. The two most popular and favored reinforcement learning approaches are Sutton's method of temporal differences and Watkin's Q-learning.

The temporal differences method may be viewed as a form of policy iteration method, akin to that employed in Howard's (1960) algorithm. Specifically, starting with an initial policy, the prediction problem posed is that of learning the expected discounted rewards for each state in the state space using this initial policy. Using the learned expected discounted rewards, a new policy can be determined, which improves upon the initial policy. The process continues until the algorithm converges to the optimal policy in the sense of Howard's algorithm.

The other competitive and sometimes superior learning method is Q-learning algorithm, a form of successive approximation technique of dynamic programming. Unlike policy iteration, Q-learning fixes a Q-value to each state-action pair. This initial value represents the estimate of the optimal value function after k updates of starting in state i, taking action a, and proceeding in an optimal manner as estimated by the current Q-values. These Q-values are then updated and the process continues until convergence is obtained. If each state-action pair is visited infinitely often, then Q-learning can be shown to converge to the optimal Q-values for the problem.

While the theory behind NDP is still embryonic, new results about their successful application to erstwhile intractable large-scale dynamic programming problems are encouraging. For example, Esogbue and Hearnes (1999) report the development of an approximate dynamic programming approach to reinforcement learning for continuous action set-point regulator problems, which learns near-optimal control policies based on scalar performance measures. A continuous action space (CAS) algorithm that uses derivative free line search methods to obtain the optimal action in continuous space has been developed to enhance the performance of their new neuro-fuzzy adaptive controller. They discuss the theoretical convergence properties of the algorithm and investigate several heuristic stopping criteria. A practical application of the algorithm to the inverted pendulum problem and a power system stabilization problem is presented.

The foregoing methods and others in their family are the subject of ongoing research, including their application in conjuction with neuro-fuzzy adaptive controllers, for complex problems where there is no explicit model of the plant and the cost (Esogbue, 2000). This is predicated on our ability to calculate, usually via simulations, the cost corresponding to a given policy. The use of these methods in operations research, especially in scheduling, online communication network routing, and intelligent manufacturing are bound to become more frequent as operations research breaks into new terrains.

11. Overview of Applications

Dynamic programming is undeniably a powerful apparatus for addressing an array of decision- and control-oriented problems, both of a theoretical and applied nature, occurring in a diverse array of fields. Consequently, its applications are numerous. It is clearly beyond the scope of our current effort to satisfactorily address the question of applications. We will therefore only give an outline of these, highlighting in the process the key areas in operations research and management science. We will note that its major area of application is in control where it is credited with being the most powerful vehicle for treating numerous problems especially those of a nonlinear dynamic, and even those of a stochastic, nature.

Although Bellman's book on dynamic programming (Bellman, 1957) contained exercises illustrating applications of DP to diverse areas, perhaps the first major volume addressing computational issues and DP applications to problems typically encountered in operations research is Bellman and Dreyfus (1962). Among the areas are knapsack (cargo loading), reliability of multicomponent devices, scheduling, equipment replacement, inventory, production scheduling and manufacturing, multidimensional resource allocation problems, location-allocation problems, optimal search problems, Markovian decision processes, and various control models. Both deterministic and stochastic versions are treated. We reiterate the contributions of concepts proposed in this volume to various areas of operations research and the new areas of parallel and distributed computing (Bertsekas, 1982).

A detailed discussion of the important problem of scheduling via dynamic programming and hybrids of DP and integer programming may be found in Bellman et al. (1982). Here, problems of routing, network flow, traveling salesman, sequencing, flow-shop scheduling, and job shop scheduling are extensively treated and computational issues addressed. A number of these tools are fundamental to the efficient treatment of manufacturing sytems and production. For example, in inventory processes, another classic prototypical OR problem area with important ramifications for manufacturing, is fuzzy criterion dynamic programming (Liu and Esogbue, 1996, 1998). Additionally, Esogbue and Hearnes (1998a) extend some of the analysis presented by Bellman and Dreyfus (1962) in their treatment of equipment replacement problems to more realistic scenarios involving fuzzy and imprecise variables usually encountered in replacement analysis.

Let us briefly cite some applications to various real world systems. Water resources is an important area for containing many problems whose optimal analysis is best effected via dynamic programming modeling or philosophies (Hall, 1989). It is also important for influencing the development of significant advances in computational dynamic programming. For example, reservoir output optimization and design, optimal ground water pumping strategies and irrigation systems, optimal sequencing and scheduling of capacity expansion water supply projects, and so on, are influenced by dynamic programming (Esogbue and Morin, 1974). Similarly, optimal operations planning and analysis of water resources located in river streams and expansive basins led to the development of new and powerful algorithms for nonserial dynamic programming. See, for example Esogbue and Warsi (1986, 1988) and Esogbue and Lee (1988). For classical dynamic programming, its influences in the development of algorithms such as the imbedded states space method for reducing dimensionality in dynamic programming are best illustrated by Morin and Esogbue (1974). For the development of new and powerful more realistic models of dynamic programming, see Esogbue and Liu (1996), who use the practical problems encountered in reservoir operations and inventory systems to develop fuzzy criterion dynamic programming. For a detailed account of these applications, see the book by Esogbue (1989).

We conclude this sketch of the applications of dynamic programming to real-world systems by considering the field of health care delivery systems. Here again, there are numerous applications. Let us mention a few, primarily ours. With Bellman, for example, we had begun a yet unfinished manuscript on dynamic programming of health care. Applications range from surgical scheduling (Esogbue, 1969), to anesthesia administration (Esogbue, 1985; Esogbue et al., 1976), medical diagnosis (Esogbue, 1999; Esogbue and Elder, 1983), nuclear medicine and tumor detection (Bellman et al., 1973a,b), to treatment (Esogbue and Elder, 1983), and to program planning including resource allocation processes in cancer research (Esogbue, 1983).

12. Conclusions

We have attempted a presentation of dynamic programming that touches its various facets, from theory to computational aspects to novel dynamic programming models and applications. It is known that most students of dynamic programming are frequently taught that the major problem, which had hitherto thwarted routine application of dynamic programming is the frequently mentioned curse of dimensionality. Consequently, considerable attention was paid to the re-creation of tools and various algorithms developed for the amelioration of this sometimes exaggerated problem.

A unique feature of this article on dynamic programming is the discussion, in one article, of deterministic and stochastic versions for both serial and nonserial systems while introducing the reader to novel facets such as fuzzy dynamic programming, fuzzy criterion dynamic programming, and neuro-dynamic programming. A limitation is the lack of coverage of the interesting area of hybrid systems involving the issue of adroit wedding of techniques in solving large-dimensional problems. For example, in some instances, it is instructive to combine dynamic programming and linear programming, dynamic programming and neural networks, or dynamic programming and genetic algorithms, using one tool's powers to reinforce or compensate for the limitations of the other. This is in contrast to dependence on one tool. Although space constraints prevent an exhaustive treatment of the numerous applications, we have attempted to expose the serious reader to some of the possibilities and provide an extensive bibliography for additional inquiry into the subject.

References

Barnard, E. 1993. "Temporal difference methods and markov models." *IEEE Trans. Sys., Man Cybernet.* **23**: 357–365.

Barto, A. G., R. S. Sutton, and C. W. Anderson. 1983. "Neuron-like adaptive elements that can solve difficult learning control problems." *IEEE Trans. Sys., Man Cybernet.* **13**: 835–846.

Barto, A. G., S. J. Bradtke, and S. P. Singh. 1995. "Learning to act using real-time dynamic programming." *Artificial Intell.* **72**: 81–138.

Bellman, R. E. 1957. *Dynamic Programming*. Princeton, NJ: Princeton University Press.

Bellman, R. E. 1961. *Adaptive Control Processes: A Guided Tour*. Princeton, NJ: Princeton University Press.

Bellman, R. E. 1971. *Introduction to the Mathematical Theory of Control Processes*, Volume 2. New York, NY: Academic Press.

Bellman, R. E. 1984. *The Eye of the Hurricane: An Autobiography*. Singapore: World Scientific.

Bellman, R. E. and S. E. Dreyfus. 1959. "Functional approximation and dynamic programming." *Math. Tables Other Aids to Computation* **13**: 247–251.

Bellman, R. E. and S. E. Dreyfus. 1962. *Applied Dynamic Programming*. Princeton, NJ: Princeton University Press.

Bellman, R. E., A. O. Esogbue, and I. Nabeshima. 1982. *Mathematical Aspects of Scheduling and Applications*. Oxford, England: Pergamon Press.

Bellman, R. E. and R. E. Kalaba. 1965. *Dynamic Programming and Modern Control Theory*. Academic Press.

Bellman, R. E., R. Kalaba, and B. Kotkin. 1962. Polynomial approximation—a new computational technique in dynamic programming—I: allocation process. RM-3086-PR. The Rand Corporation.

Bellman, R., R. Kalaba, and B. Kotkin. 1973a. "Polynomial approximation—a new computational technique in dynamic programming: allocation processes." *Math. Comput.* **17**: 155–161.

Bellman, R., B. Kashef, H. Sugiyama, and S. Friedland. 1973b. "Scan–rescan processes and dynamic programming: nuclear medicine and tumor detection with additional applications to the other related areas." In *Proceedings of the 6th International Conference on System Sciences*, p. 322.

Berenji, H. R. 1992. "A reinforcement learning-based architecture for fuzzy logic control." *Int. J. Approximate Reasoning* **6(2)**: 267–292.

Bertele, D. and F. Brioschi. 1973. *Nonserial Dynamic Programming*. New York: Academic Press.

Bertsekas, D. P. 1982. "Distributed dynamic programming." *IEEE Trans. Automatic Control* **AC-27**: 610–616.

Bertsekas, D. P. and J. N. Tsitsiklis. 1996. *Neuro-Dynamic Programming*. Belmont, MA: Athena Scientific.

Collins, D. C. 1970a. "Reduction of dimensionality in dynamic programming via the method of diagonal decomposition." *J. Math. Anal. Appl.* **30**: 223–234.

Collins, D. C. 1970b. "Terminal state dynamic programming for differential equations." *J. Math. Anal. Appl.* **31**: 487–503.

Durling, A. E. 1964. "Computational aspects of dynamic programming in higher dimensions." Technical Report TR-64-3, May 1964. Electrical Engineering Department, Syracuse University, Syracuse, NY.

Esogbue, A. O. 1969. "Dynamic programming and optimal control of variable multichannel stochastic service systems with applications." *Math. Biosci.* **5(l/2).**

Esogbue, A. O. 1972. "Fundamentals of modern dynamic programming." Mimeographed Lecture Notes. Department of Operations Research, Case Western Reserve University.

Esogbue, A. O. 1983. "Dynamic programming, fuzzy sets and the modeling of R&D management control systems." *IEEE Trans. Sys., Man Cybernet.* **13(1).**

Esogbue, A. O. 1985. "A fuzzy dynamic programming model of intra-operative anesthesia administration." In *Management Decision Support Systems Using Fuzzy Sets and Possibility Theory*, edited by J. Kacprzyk and R. Yager, ISR Series, pp. 155–161. Cologne: Verlag TUV Rheinland.

Esogbue, A. O. 1989. *Dynamic Programming for Optimal Water Resources Systems Analysis*. Englewood Cliffs, NJ: Prentice Hall.

Esogbue, A. O. 1991. "Computational aspects and applications of a branch and bound algorithm for fuzzy multistage decision processes." *J. Comput. Math. Appl.* **21(11/12)**: 117–128.

Esogbue, A. O. August 1998. "Neuro fuzzy adaptive control of complex processes." Completion Report Grant N0. ECS 91216004. National Science Foundation.

Esogbue, A. O. 1999. "Fuzzy dynamic programming, fuzzy adaptive neuro control, and the general medical diagnosis problem." *J. Comput. Math. Appl.* **37(11-12)**: 37–45.

Esogbue, A. O. 2000. "Neuro-fuzzy adaptive control: structure, algorithms and performance." *Appl. Math. Rev.* **1.**

Esogbue, A. O., V. Aggarwal, and D. Kaujalgi. 1976. "Computer aided anesthesia administration." *Int J. Biomed. Comput.* **7(4).**

Esogbue, A. O. and C. W. Ahn. 1989. "Computational experiments with a class of dynamic programming algorithms of higher dimensions." *J. Comput. Math. Appl.* **19(11).**

Esogbue, A. O. and R. E. Bellman. 1984. "Fuzzy dynamic programming and its extensions." *Fuzzy Sets and Decision Analysis, TIMS/Studies Mgmnt Sci.* **20**: 147–167.

Esogbue, A. O. and R. C. Elder. 1983. "Measurement and valuation of a fuzzy mathematical model for medical diagnosis." *Int. J. Fuzzy Sets Sys.* **10(3).**

Esogbue, A. O. and W. E. Hearnes. 1998a. "On replacement models via a fuzzy set theoretic framework." *IEEE Trans., Sys., Man, Cybernet., Part C* **87(4)**: 549–560.

Esogbue, A. O. and W. E. Hearnes. 1998b. "Approximate policy improvement for continuous action set-point regulation problems." In *Proceedings of the 1998 IEEE World Congress on Computational Intelligence*, Vol. II, Anchorage, Alaska, May 4–9.

Esogbue, A. O. and W. E. Hearnes. 1999. "A learning algorithm for the control of continuous action set-point regulator systems." *J. Comput. Anal. Appl.* **1(2)**: 121–145.

Esogbue, A. O. and J. Kacprzyk. 1998. "Fuzzy dynamic programming." In *Fuzzy Sets in Decision Analysis, Operations Research and Statistics*, edited by R. Slowinski, pp. 281–307. Boston: Kluwer Academic Publishers.

Esogbue, A. O. and C. Y. Lee. 1988. "Optimal analysis of large complex water resources conveyance systems via nonserial dynamic programming." *J. Comput. Math. Appl.* **15(10/11).**

Esogbue, A. O. and C. Y. Lee. 1990. "Determination of the optimal main chain in nonserial dynamic programming networks." *Eur. J. Oper. Res.*

Esogbue, A. O. and B. R. Marks. 1972. "The status of nonserial dynamic programming." *Mgmnt Sci (Theory).*

Esogbue, A. O. and B. R. Marks. 1974. "Nonserial dynamic programming: a survey." *Oper. Res. Q.* **25(2).**

Esogbue, A. O. and B. R. Marks. 1977. "Dynamic programming models of the nonserial critical path-cost problem." *Mgmnt Sci.* **24(2).**

Esogbue, A. O. and J. A. Murrell. 1993. "A fuzzy adaptive controller using reinforcement learning networks." In *Proceedings of the 2nd IEEE International Conference on Fuzzy Systems, San Francisco, CA,* pp. 178–183.

Esogbue, A. O. and J. A. Murrell. 1994. "Advances in fuzzy adaptive control." *J Comput. Math. Appl.* **27(9/10)**: 29–35.

Esogbue, A. O. and B. Liu. 1996a. "On fuzzy criterion dynamic programming." In *New Frontiers in Fuzzy Logic and Soft Computing Biennial Conference of the North American Fuzzy Information Processing Society—NAFIPS*, pp. 70–74. Piscataway, NJ: IEEE.

Esogbue, A. O. and B. Liu. November 1996b. "Modeling and control of open inventory networks via fuzzy criterion dynamic programming." In *Proceedings of the ISAI/IFIS 1996, USA Collaboration in Intelligent Systems Technologies, Cancun, Mexico*, pp. 413–417.

Esogbue, A. O. and B. Liu. 1996c. "Optimal fuzzy criterion clustering based on fuzzy Prototypes." In *Proceedings of the Fourth International Fuzzy Systems and Intelligent Controls Conference, Maui, Hawaii,* pp. 401–410.

Esogbue, A. O. and B. Liu. 1997a. "Stochastic multijoint reservoir operation via fuzzy criterion dynamic programming." In *Proceedings of the Seventh International Fuzzy Systems Association World Congress, Prague, Czech Republic*, pp. 177–182.

Esogbue, A. O. and B. Liu. 1997b. "On stochastic multistage decisionmaking under fuzzy criteria." In *Proceedings of FUZZY'97: The International Conference on Fuzzy Logic and Applications, Zichron Yachov, Israel*, pp. 307–315.

Esogbue, A. O. and B. Liu. 1997c. "Fuzzy criterion dynamic programming and optimal control of multidimensional open inventory networks." In *Proceedings of FUZZ-IEEE'97: The Sixth IEEE International Conference on Fuzzy Systems, Barcelona, Spain*, pp. 755–760.

Esogbue, A. O. and T. L. Morin. 1974. "A useful theorem in the dynamic programming solution of sequencing and scheduling problems occurring in capital expenditure planning." *Water Resources Res.* **10(1).**

Esogbue, A. O., M. Theologidu, and K. Guo. 1992. "On the application of fuzzy sets theory to the optimal flood control problem arising in water resources systems." *Fuzzy Sets Sys.* **48**: 155–172.

Esogbue, A. O. and N. A. Warsi. 1986. "A high level computing algorithm for diverging and converging branch nonserial dynamic programming systems." *J. Comput. Math. Appl.* **12A(6).**

Esogbue, A. O. and N. A. Warsi. 1988. "A high level dynamic programming algorithm for processing nonserial looped systems." *J. Comput. Math. Appl.* **16(10/11).**

Hall, W. A. 1989. "Dynamic programming for practical water resources systems engineering." In *Dynamic Programming for Optimal Water Resources Systems Analysis*, edited by A. Esogbue. NJ: Prentice-Hall.

Howard, R. 1960. *Dynamic Programming and Markov Chains*. New York: Wiley and Sons.

Ignizio, J. P. 1973. "Validating claims for algorithms proposed for publication." *Oper. Res.* **21(3)**: 852–854.

Jacobson, D. and D. N. Mayne. 1996. In *Neuro-dynamic Programming*. Belmont, MA: Athena Scientific.

Kacprzyk, J. 1983. *Multistage Decision-making Under Fuzziness*. Cologne: Verlag TÜV Rheinland.

Larson, R. E. and J. L. Casti. 1973. In *Principles of Dynamic Programming*. New York: Marcel Dekker, Inc.

Lee, C. and A. O. Esogbue. 1986. "Optimal procedures for dynamic programs with complex loop structures." *J. Math. Anal. Appl.* **119(1/2).**

Lee, C. Y. and A. O. Esogbue. 1993. "Determination of the optimal main chain in nonserial dynamic programming networks." *Eur. J. Oper. Res.* **67**: 248–258.

Liu, B. and A. O. Esogbue. 1996. "Fuzzy criterion set and fuzzy criterion dynamic programming." *J. Math. Anal. Appl.* **199(1)**: 293–311.

Liu, B. and A. O. Esogbue. 1998. In *Decision Criteria and Optimal Inventory Processes*. Norwell, MA: Kluwer Academic Publishers.

Meier, W. L. Jr. and C. J. Beightler. 1967. "An optimization method for branching multistage water resources systems." *J. Water Resources Res.* **8(6)**: 645–652.

Meier, W. L. Jr. and C. J. Beightler. 1968. "Branch compression and in nonserial multistaged systems." *J. Math. Anal. Appl.* **21**: 426–430.

Morin, T. L. and A. O. Esogbue. 1974. "Imbedded state space approach to reduction of dimensionality in dynamic programming of higher dimensions." *J. Math. Anal. Appl.* **48(3).**

Nemhauser, G. L. 1978. 1966. *Introduction to Dynamic Programming*. New York: Wiley.

Puterman, M. L. (ed). 1978. *Dynamic Programming and Its Applications*. New York: Academic Press.

Puterman, M. L. 1994. *Markovian Decision Problems*. New York: Wiley.

Rose, C. J. 1969. "Dynamic programming processes within dynamic programming processes." *J. Math. Anal. Appl.* **26**: 121–145.

Sniedovich, M. 1992. *Dynamic Programming*. New York: Marcel Dekker, Inc.

Tesauro, G. J. 1992. "Practical issues in temporal difference learning." *Machine Learning* **8**: 25–277.

Tsitsiklis, J. N. and Van Roy. 1996. "Feature-based methods for large-scale dynamic programming." *Machine learning* **22**: 59–94.

Wong, P. J. and D. G. Luenberger. 1968. "Reducing the memory requirement of dynamic programming." *Oper. Res.* **18(1)**: 119–131.

3.5 Local Search

Mutsunori Yagiura and Toshihide Ibaraki

ABSTRACT Local search is one of the basic methods used to find approximate solutions for hard combinatorial optimization problems, in particular those known as NP-hard. In this article, we describe a general framework of local search, and then give some numerical results showing how locally optimal solutions are distributed. These provide intuitive support to the successful application of local search algorithms to many optimization problems. Various techniques to enhance the ability of local search and some theoretical results are also mentioned.

1. Introduction

The optimization problems we consider in this article are generally defined as follows:

$$\begin{array}{ll} \text{minimize} & f(x) \\ \text{subject to} & x \in F \end{array} \quad (1)$$

F is the set of solutions x that satisfy all the given constraints. F is called the feasible region and each $x \in F$ (respectively, $x \notin F$) is called a feasible (respectively, an infeasible) solution. A feasible solution $x^* \in F$ is called optimal if $f(x^*) \leqslant f(x)$ holds for all $x \in F$. When F is combinatorial in some sense, we call Equation (1) a combinatorial (or discrete) optimization problem.

There are numerous combinatorial optimization problems, for which computing exact optimal solutions is considered to be computationally intractable, for example, those known as NP-hard (Garey and Johnson, 1979). However, in practice, we are usually satisfied with good solutions, and in this respect, approximate (or heuristic) algorithms are very important. There are several useful tools used to design approximate algorithms. The most common one is perhaps the greedy method, which directly constructs approximate solutions by successively determining the values of variables on the basis of some local information. Another important tool is the local search (LS), which starts from an initial solution x and repeatedly replaces it with a better solution in its neighborhood $N(x)$ until no better solution is found in $N(x)$; where $N(x)$ is a set of solutions obtainable from x by slight perturbations. The resulting solution x, which cannot be improved by a solution in $N(x)$, is called locally optimal.

The use of LS in combinatorial optimization has a long history. Two papers appeared in the late 1950s, in which edge-exchange LS algorithms were applied to the traveling salesman problem (TSP; Bock, 1958; Croes, 1958). It might be interesting to mention that Croes solved a fortytwo-city TSP instance by hand in about seventy hours. Then Lin (1965) and Reiter and Sherman (1965) examined various neighborhoods for TSP and introduced the multistart strategy. In subsequent years, LS has been applied to a variety of problems and its concept has been generalized in various ways. For example, a sophisticated neighborhood construction mechanism was successfully applied to the graph partitioning and TSPs by Kernighan and Lin (1970) and Lin and Kernighan (1973), respectively. The idea in these papers is called the variable depth search procedure. See for example Chapter 19 of Papadimitriou and Steiglitz (1982) and the book edited by Aarts and Lenstra (1997) for more about LS.

In the 1980s, more generalized approaches, called simulated annealing (Černý, 1985; Kirkpatrick et al., 1983) and tabu search (Glover, 1986, 1989, 1990) were proposed, in which moves to nonimproved solutions are allowed during the search. LS has also been combined with other heuristic algorithms, such as the greedy method, genetic algorithm (or evolutionary computation), ant system, and so on. These approaches have recently been called metaheuristics, and have been successfully applied to many combinatorial optimization problems. Metaheuristics cover a wide range of heuristic approaches, but LS is one of the central ingredients in most cases. Recent surveys on metaheuristics are found, for example, in Chapter 3.6 of this handbook.

In this article we describe a general framework of LS, and show some numerical results in which locally optimal solutions are distributed. These results provide intuitive support to the successful applications of LS algorithms to many optimization problems.

We then introduce basic components of LS, such as neighborhoods and move strategies. For each of these components, we introduce various techniques to enhance the ability of LS, such as the variable depth search mentioned above, and sophisticated data structures to speed up the search. Some ideas from metaheuristics, such as simulated annealing and tabu search, are also illustrated.

One of the important questions about LS is how many steps are needed to reach a locally optimal solution. To analyze this from the viewpoint of complexity theory, Johnson et al. (1988) defined a complexity class called PLS (polynomial-time local search). We briefly summarize known results about this class and related topics.

Regarding complexity classes, we note here that LS is inherently related to class NP. Let the decision problem version of optimization problem (1) be defined by introducing a constant K and by asking for a solution $x \in F$ satisfying $f(x) \leqslant K$. Such a decision problem belongs to NP if enumeration of all solutions x can be done in nondeterministic polynomial time and if the conditions $x \in F$ and $f(x) \leqslant K$ can be checked in polynomial time for each x. It answers "yes" if, and only if, there is at least one solution x that satisfies the two conditions. This NP computation may also become tractable for deterministic computation if enumeration of all solutions can be replaced by generation of good solutions, which constitute only a small subset of the solution space. We emphasize that LS appears to be quite an effective tool to achieve this goal; that is, locally optimal solutions can be regarded as the good solutions in the above argument if the neighborhood and other components of LS are carefully designed. In this article, we would like to show that this is in fact the case.

To illustrate the above concepts in the following discussion, we use some combinatorial optimization problems as concrete examples, which include the maximum satisfiability (MAXSAT), generalized assignment (GAP), graph coloring (GCP), and traveling salesman (TSP). All of these problems are known to be NP-hard.

In MAXSAT, we are given n 0–1 variables, y_j, $j = 1, ..., n$, m clauses C_i and weights w_i ($\geqslant 0$), $i = 1, ..., m$, and are asked to determine a 0–1 assignment that maximizes the sum of the weights of satisfied classes. Denoting $\bar{y}_j = 1 - y_j$ and $L = \cup_{j=1}^{n} \{y_j, \bar{y}_j\}$ (the set of literals), clauses are defined by $C_i \subseteq L$ (e.g., $C_i = \{y_1, \bar{y}_3, y_8\}$). For a $v \in \{0,1\}^n$, let

$$P_i(v) = \{j \mid y_j \in C_i \text{ and } v_j = 1, \text{ or } \bar{y}_j \in C_i \text{ and } v_j = 0\}$$

and

$$\theta_i(v) = \begin{cases} 1, & \text{if } \mid P_i(v) \mid \geqslant 1 \\ 0, & \text{otherwise.} \end{cases}$$

Then the problem is formally described as follows.

Maximum Satisfiability Problem (MAXSAT)

Input: Clauses $C_1, ..., C_m \subseteq L$ consisting of n 0–1 variables, and weights, $w_1, ..., w_m$.

Output: A 0–1 assignment $v \in \{0,1\}^n$ that maximizes $\sum_{i=1}^{m} w_i \theta_i(v)$.

The satisfiability problem (SAT) is a decision problem asking whether a $v \in \{0,1\}^n$ exists that attains $\sum_{i=1}^{m} \theta_i(v) = m$; such a v is called a satisfying assignment.

The GAP is defined for n jobs $J = \{1, ..., n\}$ and m agents $I = \{1, ..., m\}$, and we are asked to determine a minimum-cost assignment that assigns each job to exactly one agent under the resource constraint of each agent. Assigning job j to agent i incurs a cost of c_{ij}, and consumes an amount a_{ij} of resource, where the total amount of the resource available at agent i is b_i. An assignment is a mapping $\sigma : J \rightarrow I$, where $\sigma(j) = i$ means that job j is assigned to agent i.

Generalized Assignment Problem (GAP)

Input: Costs c_{ij} and resource requirements a_{ij} for $i \in I$ and $j \in J$, and the available amount of resources b_i for $i \in I$.

Output: An assignment $\sigma : J \rightarrow I$ that minimizes $\sum_{j \in J} c_{\sigma(j),j}$ subject to $\sum_{j \in J, \sigma(j)=i} a_{ij} \leqslant b_i$ for all $i \in I$.

The following two problems are also representative combinatorial problems frequently discussed in the literature.

Graph Coloring Problem (GCP)

Input: An undirected graph with n vertices.

Output: A coloring of vertices with the minimum number of colors, in which the end vertices of each edge have different colors.

Traveling Salesman Problem (TSP)

Input: n cities and distances d_{ij} between cities i and j, $1 \leqslant i, j \leqslant n$.

Output: A tour (i.e., Hamiltonian cycle), of the minimum length, that visits every city exactly once and returns to the initial city.

In TSP, a tour is represented by a permutation of n cities, and the pairs of neighboring cities in the tour are called edges.

2. Framework of Local Search

In this section, we first explain the search space, and then give a general framework of LS.

2.1. *Search Space*

The set of solutions of an optimization problem (1), which may potentially be visited in a LS algorithm, is called the search space. If generating feasible solutions is easy, then it is reasonable to define the search space to be the feasible region F. On the other hand, if generating feasible solutions is not easy (i.e., searching only within F is difficult), we may define an appropriate search space $\tilde{F}$ by taking the problem structures into account. If we adopt a search space different from F, we need to modify the objective function of (1) so that we can evaluate the amount of infeasibility of given solutions as well. Let us denote such a function by $\tilde{f}$. It is preferable to define $\tilde{F}$ and $\tilde{f}$ in such a way that an optimal solution in $\tilde{F}$ with respect to $\tilde{f}$ is also an optimal feasible solution of the original problem (1).

For example, generating feasible solutions is easy for TSP and MAXSAT, since any permutation of n cities is a feasible tour for TSP, and any n-dimensional 0–1 vector is a feasible assignment for MAXSAT. On the other hand, the problems of finding a feasible coloring with a given number of colors for GCP and finding a feasible solution for GAP are known to be NP-complete. In these cases, we have to consider how to define $\tilde{F}$ and $\tilde{f}$.

For GCP, the following definitions are possible among others.

GCP-1: Fix the number of colors to k. Then $\tilde{F}$ is the set of solutions with k colors. Define $\tilde{f}$ to be the number of edges having the same colors on their end vertices. Then a solution, x, is a feasible k-coloring if and only if $\tilde{f}(x) = 0$. In this case, various numbers of colors, k, should be tried.

GCP-2: Consider the number of colors, k, as a variable. Given a solution, x, let $\tilde{f}(x)$ be the weighted sum of the number of edges $s(x)$ having the same colors on their end vertices and the number of the colors used, $k(x)$. The search space, $\tilde{F}$, is the set of all color assignments to vertices.

GCP-3: Assume that the colors are indexed with $1, 2, \ldots$, and let the search space, $\tilde{F}$, be the set of all permutations of n vertices. A feasible coloring is generated from each permutation by assigning colors to the vertices in the order of the permutation, where each vertex receives the minimum color among those feasibly assignable to it. Then $\tilde{f}$ is the number of colors used in the obtained feasible solution.

In GCP-2, the weight of $s(x)$ should be sufficiently large to maintain feasibility.

For GAP, most natural search spaces may be defined as follows.

GAP-1: Let the search space, $\tilde{F}$, be the set of all assignments, $\sigma : J \rightarrow I$, and let $\tilde{f} = \sum_{j\in J} c_{\sigma(j),j} + \sum_{i\in I} \alpha_i \max\{(\sum_{j\in J,\sigma(j)=i} a_{ij}) - b_i, 0\}$, where the values of α_i (>0) are parameters. The second part represents the penalty to the infeasibility of resource constraints at $i \in I$. Thus, a solution, $\sigma \in \tilde{F}$, is feasible if, and only if, this penalty part is zero.

As seen in these examples, there are usually many possible definitions of the search space and the evaluation function. To design efficient LS algorithms, it is crucial to use appropriate $\tilde{F}$ and $\tilde{f}$. However, the appropriateness depends on the structure of the given problem, and general discussion appears to be difficult. See for example the report on extensive computational experiments for GCP by Johnson et al. (1991).

2.2. *Local Search*

An intuitive explanation of LS has already been given in Section 1. Here we define LS in the following more general framework.

LS ALGORITHM

Start from the given initial solution, x, and repeat replacing x with the solution in its neighborhood, $N(x)$ (i.e., a move), determined by the specified move strategy, until some stopping criterion is satisfied.

Therefore, to implement the LS algorithm we must specify:

1. the neighborhood, $N(x)$,
2. the evaluation function, $\tilde{f}$, of solutions,
3. the move strategy, and
4. the stopping criterion.

Rule (1) specifies how to generate candidate solutions from the current solution. Rule (2) specifies the goodness criterion for the solutions in the neighborhood. Rule (3) specifies the order of solutions to be searched in the neighborhood, and the rule to choose the next solution. Rule (4) specifies when the LS algorithm terminates.

LS is usually used under the following multistart framework called multistart local search (MLS).

MLS ALGORITHM

Repeat Steps I and II until some stopping criterion is satisfied.

Step I. Generate an initial solution x.

Step II. Improve x by using the LS algorithm.

To design an efficient LS algorithm, rules (1) to (4), as well as the rule to generate initial solutions in Step I of MLS, are crucial. The neighborhood of rule (1) and the evaluation function of rule (2) should be carefully designed so that they can exploit the structure of the given problem as much as possible. Traditionally, LS has been understood as the following simple algorithm.

BASIC LS [LS with rules (3) and (4) defined as follows]:

Rule (3): Choose any solution, $x' \in N(x)$, satisfying $\tilde{f}(x') < \tilde{f}(x)$ and set $x := x'$.

Rule (4): Stop the search if the current solution, x, is locally optimal, that is, $\tilde{f}(x) \leqslant \tilde{f}(x')$ holds for all $x' \in N(x)$.

However, the recent trend has been to use the terminology LS in the more general framework described at the beginning of this section. Many ideas of metaheuristics can be explained within this framework. More discussion of the design of these rules will be given in Section 4.

3. Neighborhood Structure: Experimental Results

The effectiveness of LS is sometimes explained by the proximate optimality principle (POP) (Glover and Laguna, 1997). The meaning of the POP is that good solutions often have similar structures. This concept is vague, but it is a key to the successful design of LS to find such "hidden" structures common to good solutions. It is important to define the neighborhood so that such structures are preserved after the neighborhood operations.

For TSP, for example, the set of edges in a tour (i.e., the pairs of neighboring cities in the tour) are known to be important. It was reported by Lin and Kernighan (1973) that the number of edges common to any two locally optimal solutions with respect to the Lin–Kernighan neighborhood is about 85 percent on average.

To clarify these points, we present some numerical results to see how locally optimal solutions of MAXSAT and GAP are distributed under various definitions of neighborhood. Similar experiments were also conducted for TSP by Boese et al. (1994). For MAXSAT, we consider the k-flip neighborhood, N_k, for $k = 1$, 2 and 3, where

$$N_k(v) = \{v' \in \{0,1\}^n \mid \text{Hamming distance between } v \text{ and } v' \text{ is at most } k\}.$$

Clearly, $N_k(v) \subseteq N_{k+1}(v)$ holds for all $k \geqslant 1$. Solutions are evaluated in this experiment by the function

$$\tilde{f}_{\text{MSAT}}(v) = \sum_{i=1}^{m} w_i \theta_i(v)$$

where $\tilde{f}_{\mathrm{MSAT}}(v)$ is maximized. For GAP, we consider the shift neighborhood, N_{shift}, and the swap neighborhood, N_{swap}, where

$$N_{\mathrm{shift}}(\sigma) = \{\sigma' \mid \sigma' \text{ is obtained from } \sigma \text{ by changing the assignment of one job}\},$$

$$N_{\mathrm{swap}}(\sigma) = \{\sigma' \mid \sigma' \text{ is obtained from } \sigma \text{ by exchanging the assignments of two jobs}\}.$$

Solutions are evaluated by the function

$$\tilde{f}_{\mathrm{GAP}}(\sigma) = \sum_{j\in J} c_{\sigma(j),j} + \sum_{i\in I} \alpha_i \max\left\{\left(\sum_{j\in J,\sigma(j)=i} a_{ij}\right) - b_i, 0\right\} \quad (2)$$

where $\alpha_i = 2$ are used for all $i \in I$ in our experiment. For both problems, the initial solutions in Step I of MLS are generated randomly, and BASIC LS described in the previous section is used to find locally optimal solutions.

The MAXSAT instances were taken from Web site `http://www.research.att.com/~mgcr/data/maxsat.tar.gz`, which are randomly generated instances of size $n = 100$ and $m = 850$. They are generated by assigning random weights to the DIMACS benchmark instances called `jnh`, where the weights are integers randomly chosen from [1, 1000]. The GAP instances were taken from Web site `http://msemga.ms.ic.ac.uk/job/crlib/gapinfo.html`, which are randomly generated instances of sizes up to $n = 200$ and $m = 20$. These instances are classified into four types, A, B, C, and D, according to their generation rules. Although we tested many instances, we show the results for one MAXSAT instance called `jnh01` and one GAP instance of type D with $n = 200$ and $m = 10$, as the results for other instances are similar. The algorithms were coded in C and run on a workstation Sun Ultra 2 Model 2300 (two UltraSPARC-II 300 MHz processors with 1 GB memory), where the computation was executed on a single processor.

Let x^* be the best solution obtained after repeating Steps I and II of MLS one hundred times, where neighborhood N_3 is used for MAXSAT and neighborhood $N_{\mathrm{shift}} \cup N_{\mathrm{swap}}$ is used for GAP. Figures 1–8 show, for the various solutions, x, to be explained later, the deviation of $\tilde{f}(x)$ in percent from $\tilde{f}(x^*)$ against the distance between x and x^*. The distance between two solutions is defined as the Hamming distance for MAXSAT, and the number of jobs assigned to different agents for GAP. Figures 1–4 are the results for MAXSAT, and Figures 5–8 are those for GAP. In Figures 1 and 5, ten solutions, x, are randomly generated for each distance from 1 to n, and their percent deviation of $\tilde{f}(x)$ are shown. Figures 2–4 (respectively, 6–8) are the results for the locally optimal solutions after the solutions in Figure 1 (respectively, five) are improved by BASIC LS with given neighborhoods. Figures 2–4 are the results for N_1, N_2 and N_3, respectively, and Figures 6–8 are the results for N_{shift}, N_{swap} and $N_{\mathrm{shift}} \cup N_{\mathrm{swap}}$, respectively.

From these figures, we may make the following observations.

- From Figures 1 and 5, we can observe that the distance and quality of random solutions have a strong positive correlation; apart from the x^*, the quality degrades rapidly and linearly.
- In Figures 2–4 and 6–8, the strong correlation between distance and quality of locally optimal solutions can still be observed. An important difference from the case of random selections is that the quality of solutions is much improved. This implies that LS is very effective for finding good solutions even if it starts from random solutions.
- If we use a larger neighborhood, better locally optimal solutions are obtained. With larger neighborhoods (e.g., N_2 and N_3 for MAXSAT, and N_{swap} and $N_{\mathrm{shift}} \cup N_{\mathrm{swap}}$ for GAP), even better solutions than x^* can frequently be found.
- A large number of locally optimal solutions exist in the search space.
- In Figure 7, a strange distribution is observed. This is partially because the number of jobs assigned to each agent does not change by swap moves. This indicates that LS with

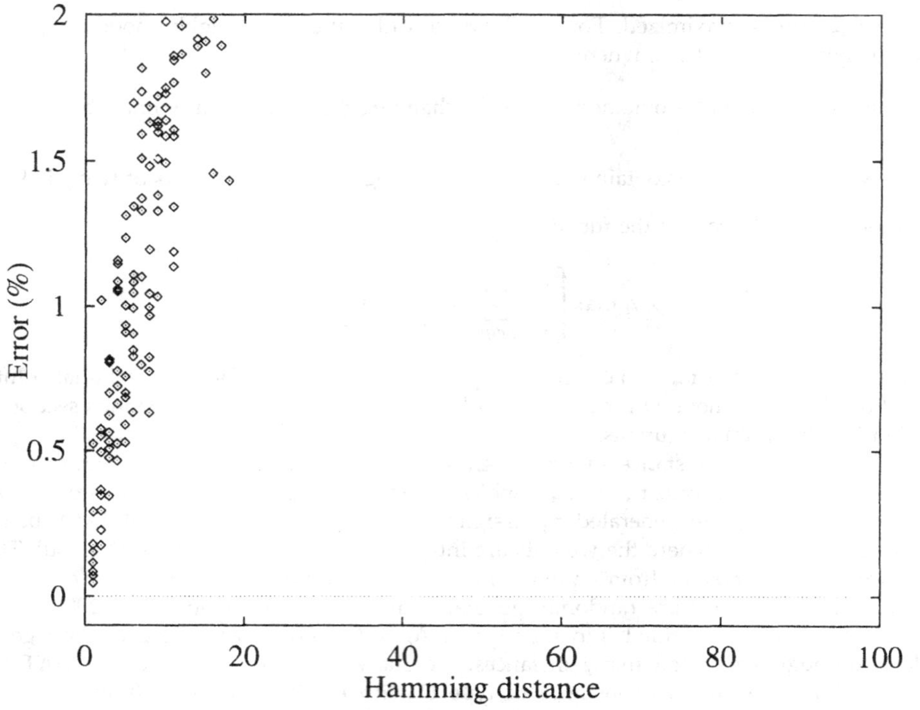

Figure 1 Distribution of random MAXSAT solutions around a good solution x^*

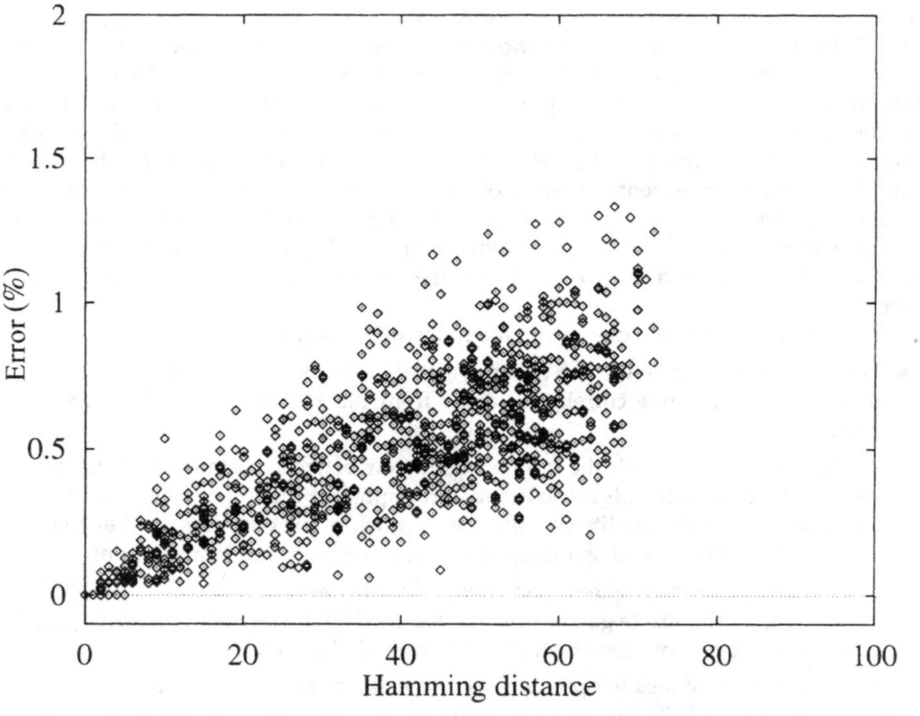

Figure 2 Distribution of locally optimal MAXSAT solutions with respect to N_1 around a good solution x^*

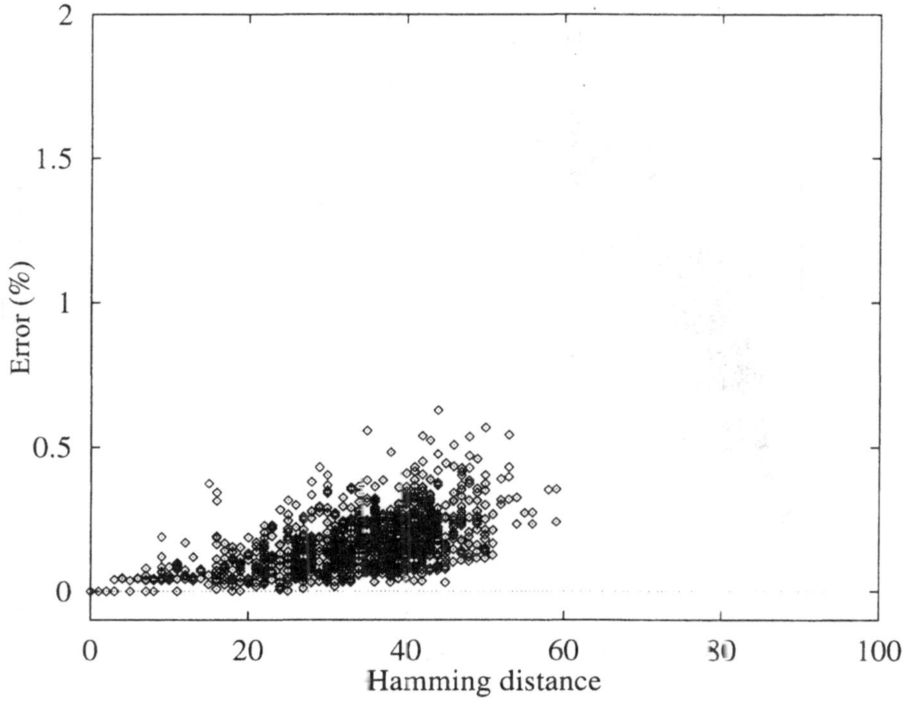

Figure 3 Distribution of locally optimal MAXSAT solutions with respect to N_2 around a good solution x^*

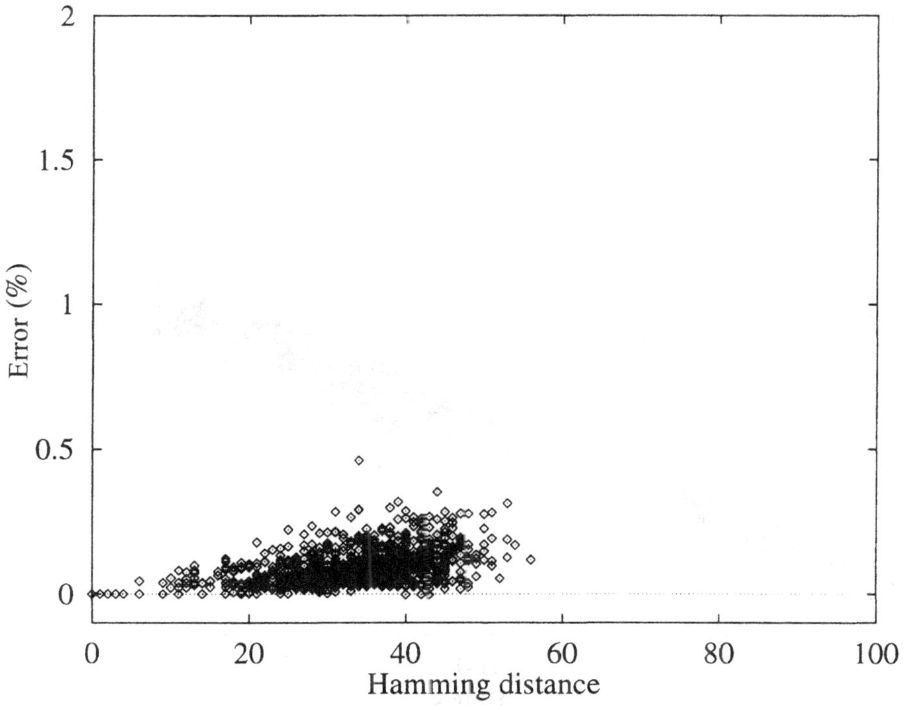

Figure 4 Distribution of locally optimal MAXSAT solutions with respect to N_3 around a good solution x^*

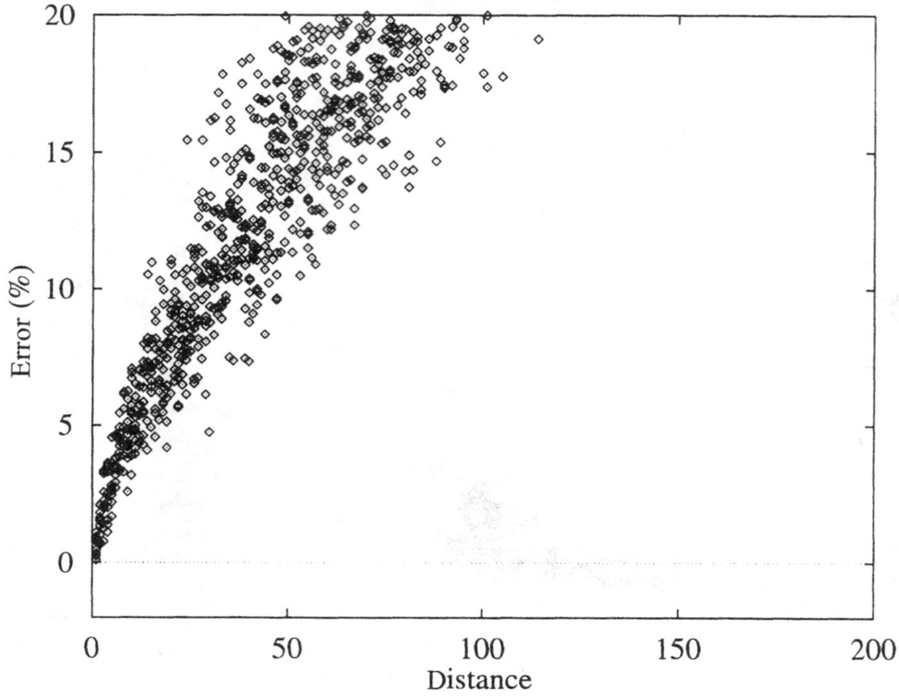

Figure 5 Distribution of random GAP solutions around a good solution x^*

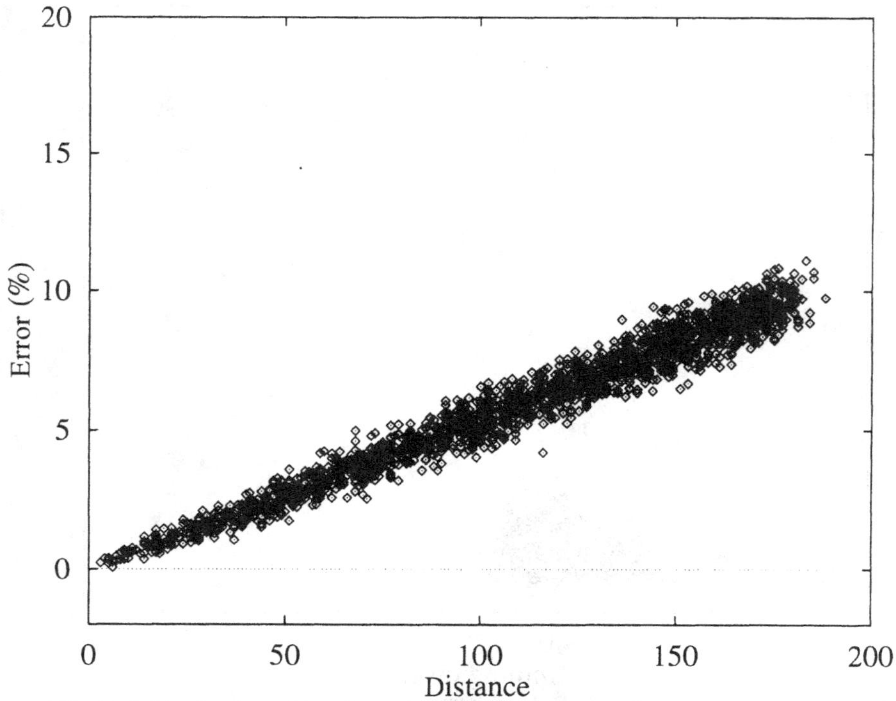

Figure 6 Distribution of locally optimal GAP solutions with respect to N_{shift} around a good solution x^*

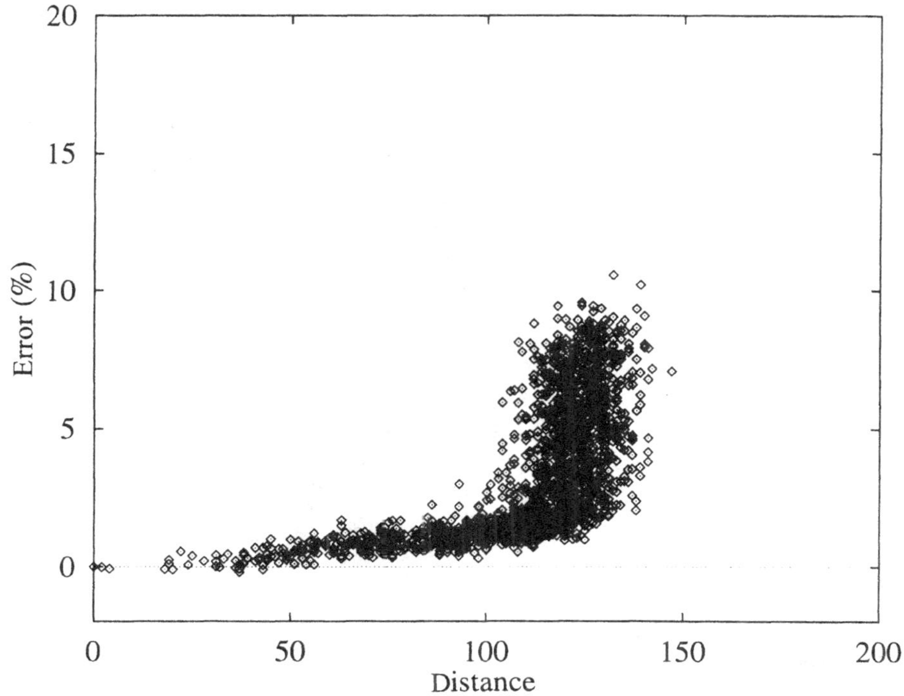

Figure 7 Distribution of locally optimal GAP solutions with respect to N_{swap} around a good solution x^*

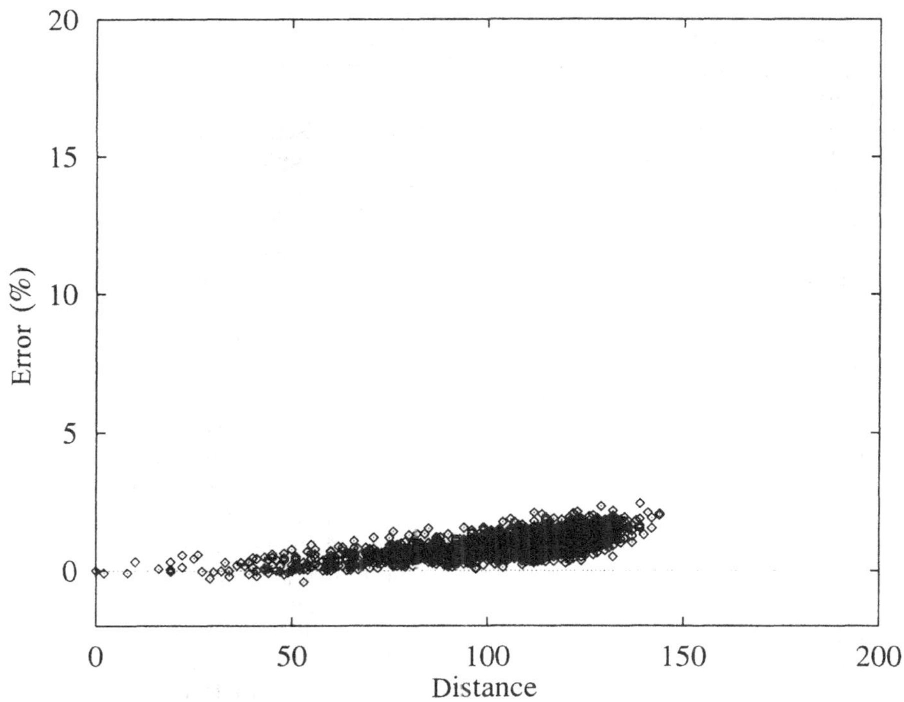

Figure 8 Distribution of locally optimal GAP solutions with respect to $N_{shift} \cup N_{swap}$ around a good solution x^*

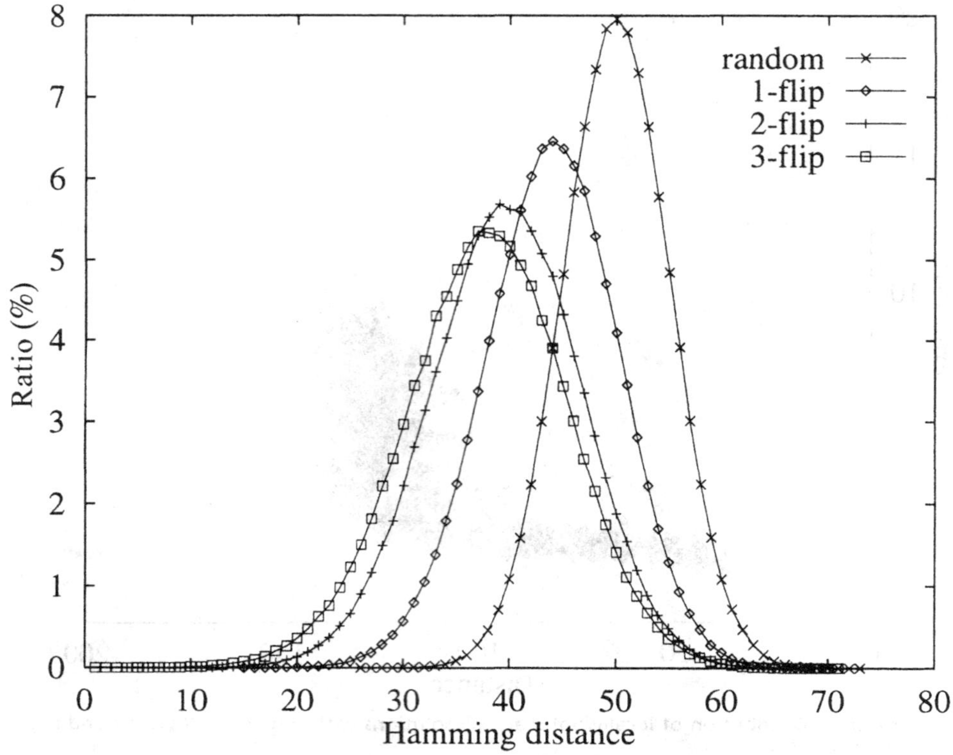

Figure 9 Distances between locally optimal solutions (MAXSAT)

a swap neighborhood only may be awkward, and should be combined with other neighborhoods such as N_{shift}.

Next, Figures 9 and 10 show the distribution of distances between locally optimal solutions of MAXSAT and GAP, respectively, where one thousand solutions are generated and distances between all pairs are considered. For comparison purposes, we show the results for different neighborhoods, and also for randomly generated solutions. The expected distance between two random solutions is $n/2 = 50$ for MAXSAT, and $n(1-1/m) = 180$ for GAP. From these figures, we can observe that the average distance between locally optimal solutions is smaller than that of randomly generated solutions, and the average distance becomes closer if we use larger neighborhoods. A similar tendency can also be observed in Figures 2–4 and 6–8; the solutions are more shifted to left as we use a larger neighborhood.

In the third experiment, we consider the number of moves needed to find a better solution from the initial locally optimal solution. The following two strategies, called best uphill and random uphill, are tested to escape from the current locally optimal solutions. The best uphill strategy moves to the best solution in $N(x)\setminus\{x\}$, while the random uphill strategy chooses a solution randomly from $N(x)\setminus\{x\}$. In each case, BASIC LS is then applied to the new solution to obtain a different locally optimal solution, where neighborhood N_1 is used for MAXSAT and N_{shift} is used for GAP. To avoid cycling in this process, we prohibit the reverse moves that are defined as follows. For MAXSAT, if the assignment of a variable v_j is changed after leaving the initial locally optimal solution, then the move that changes the assignment of v_j again is specified as the reverse move. Therefore, the assignment of one variable is fixed in each move. For GAP, if the assignment of job j is changed from i to i', then the move that assigns job j to agent i again is defined as the reverse move. Therefore, the assignment of a job, i, is uniquely fixed after its assignment has been changed $m-1$ times. Consequently, at most n

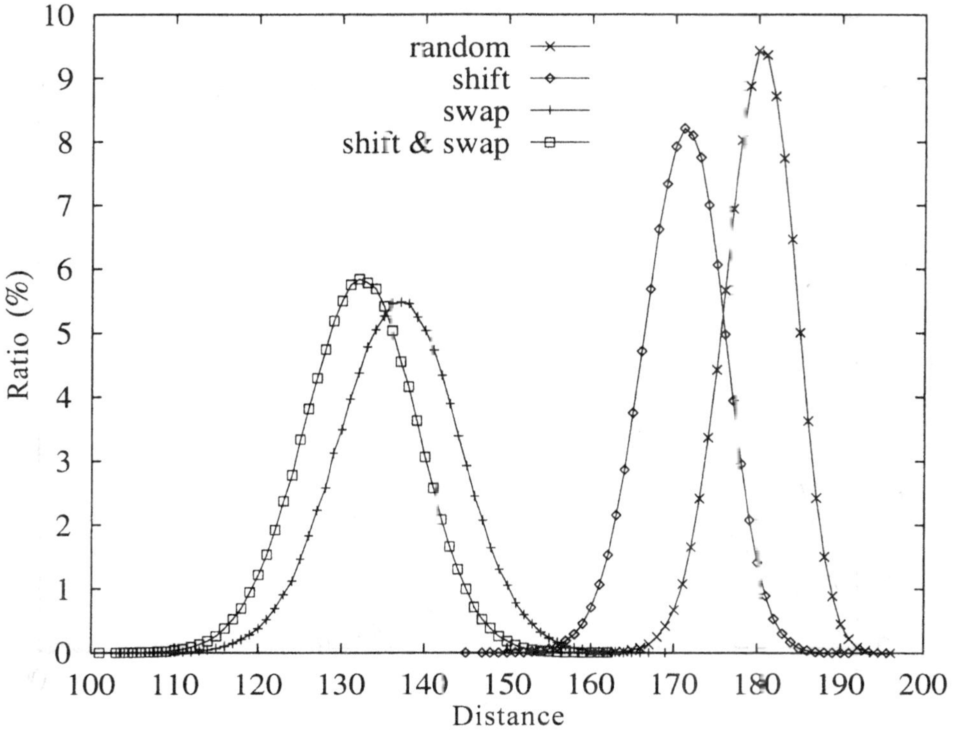

Figure 10 Distances between locally optimal solutions (GAP)

(respectively, $n(m-1)$), moves are possible for MAXSAT (respectively, GAP) after leaving the initial locally optimal solution. If the locally optimal solution obtained is not better than the initial locally optimal solution, then the above procedure is repeated from the new locally optimal solution, without resetting the set of reverse moves. Whenever this procedure reaches the stage in which no further move is possible (since all moves are considered to be reverse moves), then we conclude a failure of the search and terminate the computation.

Figures 11 and 12 show the distribution of the number of moves needed to find a better solution from the initial locally optimal solution, for MAXSAT and GAP, respectively, where ten thousand independent runs are tested. From the figures, we can observe the following.

- A better solution is very often found within a few moves in both best uphill and random uphill strategies.
- The best uphill strategy is better than the random uphill strategy (at the cost of testing all solutions in the neighborhood of each locally optimal solution).

All the results in this section support the concept of POP, and they give intuitive support to the successful application of LS algorithms. The results also tell that searching around locally optimal solutions is worth trying to get solutions of better quality.

4. Designing Efficient Local Search Algorithms

To design effective LS algorithms, it is important to carefully define each component of LS, such as the neighborhood, the evaluation function, the move strategy and the construction of initial solutions. In this section, we explain various techniques that are useful to enhance the ability of LS.

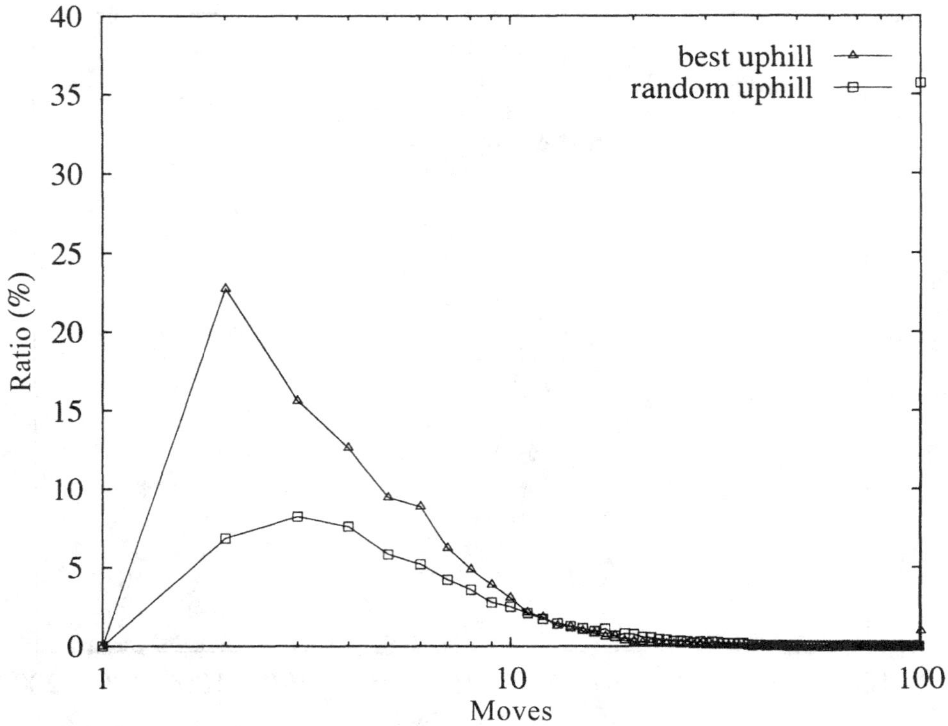

Figure 11 Number of moves needed to find a better solution (MAXSAT)

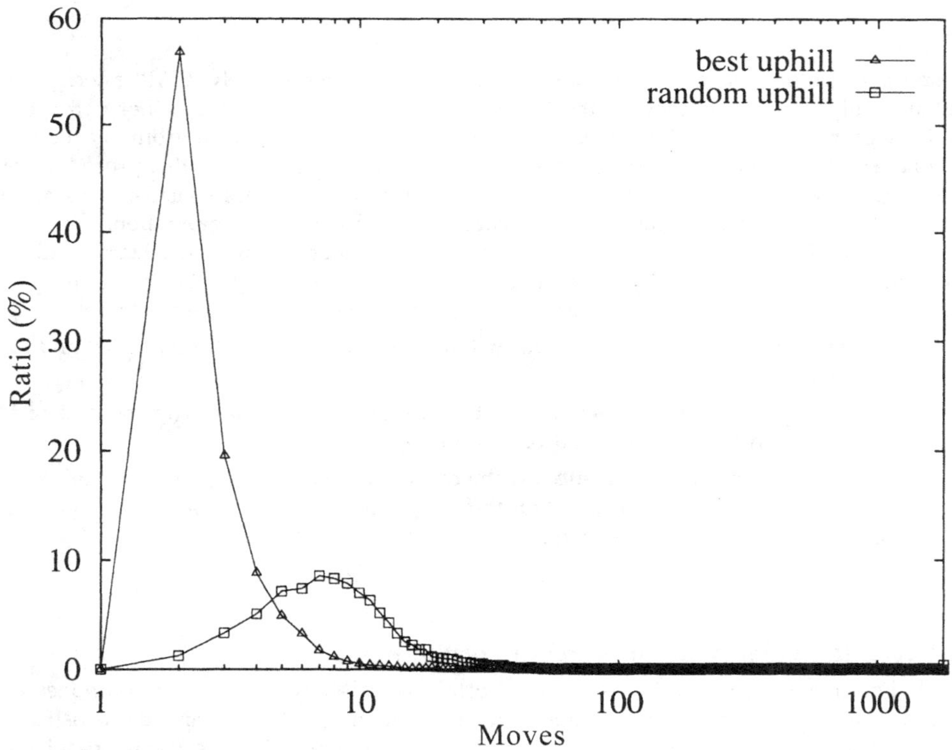

Figure 12 Number of moves needed to find a better solution (GAP)

4.1. Neighborhood

In defining the neighborhood in LS, it is important to preserve the structure common to good solutions. To achieve this, we may consider more than one neighborhood of varied sizes. We give some examples designed for TSP.

City-swap neighborhood. The set of solutions obtainable by exchanging the positions of two cities in the permutation.

City-insert neighborhood. The set of solutions obtainable by deleting a city from the current position in the permutation and inserting it into another position.

Or-opt neighborhood. The set of solutions obtainable by deleting a subpath of length at most three, and inserting it into another position.

λ-opt neighborhood. The set of solutions obtainable by changing at most λ edges of the tour, where λ ($\geqslant 2$) is a constant.

The city-insert neighborhood is a subset of the Or-opt neighborhood, and the Or-opt neighborhood is a subset of the 3-opt neighborhood. Among these, the λ-opt neighborhood with $\lambda = 2, 3$ and the Or-opt neighborhood are known to be effective, and the city-swap neighborhood is not competitive with the others. This is explained by POP as follows As the cost of TSP is the sum of distances of tour edges, the number of different edges between two tours might be the most appropriate distance measure. The distance of the current tour and a tour in the neighborhood is four for the city-swap neighborhood, three for the city-insert neighborhood, three for the Or-opt neighborhood, and λ for the λ-opt neighborhood. This means that the solutions in the city-swap neighborhood are not close enough to the current solution compared to other neighborhoods. Neighborhood size is also important. As observed in the experiments in Section 3, the quality of locally optimal solutions usually improves if a larger neighborhood is used; however, the computational time to search the neighborhood also increases. Therefore, the neighborhood size is usually kept within an appropriate bound. For example, the size of the λ-opt neighborhood is $O(n^{\lambda})$, and parameter values such as $\lambda = 2$, 3 are considered to be appropriate. For more details, see, for example, Johnson and McGeoch (1997) and Lawler et al. (1985).

There is a sophisticated mechanism called variable depth search, which is an embedded neighborhood construction that compounds simple moves to create more complex and powerful moves. This approach was first successfully applied to the graph partitioning problem by Kernighan and Lin (1970), and then to TSP by Lin and Kernighan (1973). A similar but slightly more general idea called the ejection chain is applied to various problems under the framework of tabu search (Glover and Laguna, 1997; Yagiura et al., 1999). The idea is general, but the neighborhood size becomes exponentially large if we use this idea without restriction. Therefore, it is necessary to carefully choose simple moves, which are then used to create complex moves, so that the problem structure is exploited as much as possible. This idea is explained, for example, in Chapter 19 of Papadimitriou and Steiglitz (1982) and in Johnson and McGeoch (1997).

It is also important to consider how to search the neighborhood efficiently. For this purpose, it is often effective to eliminate from consideration those solutions having no possibility of improvement. Here we show a simple example for the 2-opt neighborhood of TSP. Assume that a better solution is found by deleting two edges $\{i, j\}$ and $\{k, l\}$ from the current tour and by adding $\{i, k\}$ and $\{j, l\}$; that is, $d_{ij} + d_{kl} > d_{ik} + d_{jl}$. This implies that at least one of the conditions $d_{ij} > d_{jl}$ or $d_{kl} > d_{ik}$ holds. Hence, while considering the 2-opt moves in which edge $\{i, j\}$ is deleted, we can limit the candidate edges to add to those $\{j, l\}$ with $d_{jl} < d_{ij}$ or those $\{i, k\}$ with $d_{ik} < d_{ij}$. Note that if we decide to delete edge $\{i, j\}$ and to add edge $\{j, l\}$, then the remaining edges $\{k, l\}$ to be deleted and $\{i, k\}$ to be added are uniquely determined to maintain feasibility. A similar reduction is also possible for the 3-opt neighborhood. The Lin–Kernighan neighborhood can be considered as a generalization of this idea to the λ-opt neighborhood with general λ, although it is not guaranteed to reach locally optimal solutions. For the Euclidean TSP, in which the coordinates of the cities are given and the distances are the Euclidean distances between them, a special data structured called the K-d tree can be used to enumerate all cities, l, satisfying $d_{jl} < d_{ij}$ for a given edge $\{i, j\}$ efficiently (Bentley, 1992). Data structures for efficiently executing various basic neighborhood operations are also extensively studied, for example, Fredman et al. (1995). Other techniques to improve the efficiency of LS for TSP can be found in Johnson and McGeoch (1997).

Similar reductions of neighborhoods are also possible for MAXSAT. Yagiura and Ibaraki (1999) proposed an efficient data structure to reduce the sizes of the 2- and 3-flip neighborhoods. They show that the expected size of the reduced 2-flip neighborhood is $O(n + m)$, assuming randomly generated solutions, while the original size of N_2 is $O(n^2)$. Similar results are also reported for the 3-flip neighborhood.

A general idea to improve the efficiency of searching the neighborhood, called the candidate list strategy, is often used under the framework of tabu search (Glover and Laguna, 1997). If the problem can be naturally formulated as an integer programming problem, a reduction of the neighborhood size using the information of its Lagrangian or linear programming (LP) relaxation problem is also possible (e.g., Yagiura et al., 1999).

4.2. Evaluation Function

A simple evaluation function, which is defined over the search space, $\tilde{F}$ ($\subseteq F$), may be

$$\tilde{f}(x) = \begin{cases} f(x), & \text{if } x \in F \\ f(x) + M, & \text{otherwise} \end{cases} \tag{3}$$

where M is a sufficiently large constant. A more sophisticated evaluation function such as (2) is usually more effective than (3), since the amount of infeasibility is also taken into account. These are also called penalty functions and have been applied to a wide range of optimization problems. In penalty function (2), penalty weights α_i are crucially important. They can be given as fixed constants or can be adaptively controlled during the search. An example of such an automatic controlling mechanism can be found in Yagiura et al. (1999).

Evaluation functions need not be directly related to the objective function f. For example Johnson et al. (1991) used the following function for GCP, where the number of colors, k, is considered to be a variable and the search space is the set of all color assignments to vertices. For a given solution, x, let V_i be the set of vertices colored with i and E_i be the set of edges both of whose end vertices are in V_i, $i = 1, \ldots, k$. Then the evaluation function is defined as

$$\tilde{f}(x) = -\sum_{i=1}^{k} |V_i|^2 + \sum_{i=1}^{k} 2|V_i| \cdot |E_i|, \tag{4}$$

which is to be minimized. By the first term of (4), a color set V_i with a larger size tends to get more vertices than smaller ones, and as a side effect of this, the number of colors, k, is minimized. If the neighborhood of x is defined to be the set of solutions obtainable by changing the color assignment of one vertex, then locally optimal solutions are always feasible (i.e., $E_i = \emptyset$ for all i). To see this, suppose that E_i is nonempty, and let v be an end vertex of an edge in E_i. If the color of v is changed to $k + 1$ (i.e., a new color), then the second term of (4) is reduced by at least $2|V_i|$ while the first term is increased by at most

$$|V_i|^2 - ((|V_i| - 1)^2 + 1^2) = 2|V_i| - 2.$$

Compared to the straightforward evaluation function, such as the weighted sum of k and $\sum_{i=1}^{k} |E_i|$, a merit of using (4) is that we can evaluate the gain of a move in which the number of colors is not changed but the size of a small V_i is decreased.

It is also effective to dynamically change the function $\tilde{f}$ during the search, in order to take into account the previous search history. Here we show a simple example for SAT, which is called the weighting strategy (Selman and Kautz, 1993). The evaluation function is defined by

$$\tilde{f}_{SAT}(v) = \sum_{i=1}^{m} w_i'(1-\theta_i(v))$$

where $w_1', \ldots, w_m'$ are parameters satisfying $w_i' > 0$ for all i. Problem SAT asks to find a 0-1 assignment, v, satisfying $\tilde{f}_{SAT}(v) = 0$. The values of w_i' are initially set to one, and increased by one for those i with $\theta_i(v) = 0$ whenever a locally optimal solution v is reached. By this rule, the weights for those clauses that are hard to satisfy will become larger as the search proceeds, and the probability that such clauses are left unsatisfied will decrease. A similar idea applicable to various problems including TSP is called the guided local search (Voudouris and Tsang, 1999). The long-term memory strategy (Glover and Laguna, 1997) used under the framework of tabu search is also similar, but is more general in that it allows more flexible use of search history, such as the frequency that a variable takes a specified value and the frequency that the value of a variable has been changed.

4.3. Move Strategy

As discussed in Section 2.2, the basic move strategy in LS is to choose an improved solution in the neighborhood. In this case, the two strategies, called the first admissible move strategy and the best admissible move strategy, are usually used. In the first admissible move strategy, solutions in $N(x)$ are scanned according to a prespecified order, and the first improved solution found is immediately accepted as the next solution. In this strategy, the order of searching the neighborhood may affect the computational time and/or solution quality. In the best admissible move strategy, the best solution in $N(x)$ is chosen as the next solution.

Move strategies in which moves to nonimproved solutions are also allowed, are used in the metaheuristic framework such as simulated annealing and tabu search. By allowing such moves, the search can escape from a locally optimal solution and can examine a wider area of solutions around the locally optimal solution. As discussed in Section 3, this will increase the probability of finding better solutions. The basic rule of the simulated annealing method is to repeat the following steps, where t is a parameter called temperature.

1. Choose a solution $x' \in N(x)$ randomly.
2. Set $\Delta := \tilde{f}(x') - \tilde{f}(x)$.
3. Set $x := x'$ with probability $\min\{1, e^{-\Delta/t}\}$.

The value of t is initially set large, and is gradually decreased as the search proceeds. The behavior of the simulated annealing algorithm becomes essentially the same as that of the first admissible strategy if $t \to 0$. There are variants of this, called the threshold accepting and great deluge algorithms, in which the rule of Step 3 is slightly modified. Simulated annealing was proposed independently by Kirkpatrick et al. (1983) and Černý (1985). Reports on extensive computational results are found in Johnson et al. (1989, 1991), and a recent survey is in Chapter 3.6.8 of this handbook.

In the tabu search, the best solution in $N(x)\backslash(\{x\} \cup T)$ is chosen as the next solution, where the set T, called tabu list (or short-term memory), is a set of solutions that includes those solutions recently visited. Within this restricted neighborhood, a move to a new solution is executed even if the current solution is already locally optimal; but cycling of a short period can be avoided as a result of introducing T. Another type of memory, called long-term memory, is often employed in tabu search algorithms, as already discussed in Section 4.2. Tabu search was proposed by Glover (1986, 1989, 1990). Details of tabu search can be found in the book by Glover and Laguna (1997), and in Chapter 3.6.7 of this handbook.

The idea called strategic oscillation is often used in the framework of tabu search and is effective in dealing with the search space that contains infeasible solutions. In this method, more than one neighborhood operator is adaptively used so that the search visits the feasible and infeasible regions alternately. This strategy is useful for problems, such as the knapsack problem and the maximum clique problem, to which an LS within the feasible region is not very effective. For these problems, the behavior of LS would be similar to that of a greedy algorithm if it is restricted to the feasible region. Other useful ideas are also proposed for tabu search. For details see Glover and Laguna (1997).

Selman et al. (1994) proposed an LS algorithm called WALKSAT for problem SAT. This algorithm is specialized to problem SAT, but its idea is similar to the candidate list strategy of tabu search and is applicable to other problems. In this method, neighborhood N_i is used and the next solution is chosen by the following simple rule. First a clause i with $\theta_i(v) = 0$ (i.e., an unsatisfied clause) is randomly chosen, and then the variable to be flipped is chosen from $\{j \mid y_j \in C_i \text{ or } \bar{y}_j \in C_i\}$ by (a) choosing the best flip or by (b) choosing a random flip, where rule (a) or (b) is chosen randomly for each move. By using this rule, at most $| C_i |$ candidates are searched for each move, which is usually much smaller than $| N_1 |= n$. Now suppose the instance is satisfiable and let v^* be a satisfying assignment. One of the important features of this method is that, if rule (b) is used, the probability of getting closer to v^* is at least $1/| C_i |$, since at least one variable in C_i must have different assignment in v^*. Because of this property, it is known that, if $| C_i | \leqslant 2$ holds for all i and the instance is satisfiable, then algorithm WALKSAT with rule (b) will find a satisfying assignment in $O(n^2)$ time with probability arbitrarily close to one (Papadimitriou, 1991).

4.4. Initial Solutions

Initial solutions in LS may be generated randomly. The use of greedy methods to construct initial solutions is also possible. There is a variant of these, called greedy randomized adaptive search procedure (GRASP), in which initial solutions are generated by a randomized greedy method. GRASP was proposed by Feo and Resende (1989). See for example Feo and Resende (1995) and Chapter 3.6.5 of this handbook.

Making use of the good solutions found during the previous search is also possible. There are many ideas for this purpose. The simplest one may be the iterated local search proposed by Johnson (1990). In this approach, initial solutions are generated by randomly perturbing the best solution found in the search by then. A variant of this, in which the amount of perturbation is dynamically controlled during the entire search, is called the variable neighborhood search (Mladenović and Hänsen, 1997). See Chapter 3.6.9 of this handbook. Another variant of this, in which the solution maintained for the generation of initial solutions is not necessarily the best solution, is called the chained local optimization or the large-step Markov chain approach (Martin et al., 1992). A similar approach, in which more than one solution is maintained, is the genetic local search. This is a hybrid approach of the genetic algorithm and LS. The genetic algorithm is a probabilistic algorithm, whose idea comes from evolution. The initial solutions are generated by combining more than one candidate solution by the crossover operations. For more details, see for example Ulder et al. (1990). There is a more general approach, called the scatter search, in which initial solutions are generated by using the information of many reference solutions (Glover and Laguna, 1997). (Historically, Johnson's iterated LS was inspired by the prior work of Martin et al.; see Johnson and McGeoch, 1997), and the proposal of genetic LS is much earlier (e.g., Brady, 1985) than chained local optimization.)

Boese et al. (1994) proposed to generate initial solutions for TSP on the basis of the statistical data of the previous search. They maintain the frequency of appearances of each edge in locally optimal solutions, and generate initial solutions randomly according to a biased distribution so that the edges frequently used in good locally optimal solutions are more likely to be chosen. This idea is motivated by the results of experiments analogous to those in Section 3. There is another, but similar, approach called the ant system, which is inspired by the behavior of ants. Ants are able to find good paths between a food source and their home colony by communicating via pheromones. This mechanism stores the information of past good solutions in the form of pheromone trails, which bias the later search to promising directions. In the ant system approach, initial solutions are generated by using a greedy method, in which the pheromone trails are used. See, for example, Dorigo and Gambardella (1997) and Chapter 3.6.2 of this handbook.

5. Theoretical Results

Here we consider theoretical aspects of LS in the framework of BASIC LS. Computational time to search the neighborhood of the current solution is usually polynomially bounded; however, the number of moves to reach a locally optimal solution from an initial solution may not be polynomial. (However, we can bound the number of such moves by the number of values $\tilde{f}$ can take, which is usually of pseudo-polynomial order.) In fact, instances and starting solutions of TSP exist such that the local search with 2-opt neighborhood will make $\Theta(2^{n/2})$ moves before halting. A similar result is also shown for the 3-opt by Chandra et al. (1999), and this type of result for TSP is summarized in Johnson and McGeoch (1997). To analyze the complexity of finding a locally optimal solution, Johnson et al. (1988) introduced a complexity class called PLS (polynomial-time local search). Each LS problem in PLS is defined as a pair of optimization problems and the definition of neighborhood. Many important LS problems are known to be complete for class PLS, which means that the time to find a locally optimal solution by any method cannot be polynomially bounded unless all PLS problems can be solved in polynomial time. For example, the LS problem for TSP under the λ-opt neighborhood for a sufficiently large constant λ, and that for MAXSAT under the 1-flip neighborhood are known to be PLS-complete. See an extensive survey of Yannakakis (1997) for the theory of PLS-completeness.

Contrary to these theoretical results, the time taken to reach a locally optimal solution is usually fast in practice. To investigate this direction, there are attempts to obtain theoretical bounds on the average time to find a locally optimal solution under appropriate probabilistic models. For example, the average number of moves that the LS algorithm with 2-opt neighborhood will make is bounded by $O(n^{10} \log n)$ for randomly generated two-dimensional Euclidean TSP instances (Chandra et al., 1999). However, it is reported in the experimental results by Johnson and McGeoch (1997) that the average number of moves is approximately $\Theta(n \log n)$, which is much smaller than the theoretical bound above

There is another attempt to represent the neighborhood structure using a graph whose set of vertices is equal to $\tilde{F}$ and the set of edges is given by $\{\{x, x'\} \mid x' \in N(x)\}$, and to analyze the time to find a locally optimal solution on general classes of such graphs. For details see Tovey (1997).

6. Conclusions

In this article, we surveyed local search algorithms for combinatorial optimization problems. We also presented some experimental results that examine the distribution of locally optimal solutions, the distances between them, and so on. These results may intuitively explain why LS and metaheuristic algorithms are successful for many optimization problems. We also summarized additional techniques that have been proposed to improve the performance of LS. Some theoretical results on the performance of LS are also mentioned.

LS is one of the oldest methods for combinatorial optimization problems; however, it is still developing, inspired by the recent metaheuristic algorithms, and many of its theoretical aspects have yet to be understood. We hope this article will be of some help for the future research of LS.

ACKNOWLEDGMENTS

The authors are grateful to Wolfgang Slany and Shunji Umetani for their valuable comments. This research was partially supported by a Scientific Grant-in-Aid, by the Ministry of Education, Culture, Sports and Science and Technology of Japan.

References

Aarts, E. H. L. and J. K. Lenstra (eds). 1997. *Local Search in Combinatorial Optimization*. Chichester, England: John Wiley & Sons.

Bentley, J. J. 1992. "Fast algorithms for geometric traveling salesman problems." *ORSA J. Comput.* 4: 387–410.

Bock, F. 1958. "An algorithm for solving 'traveling-salesman' and related network optimization problems." Manuscript associated with talk presented at the Fourteenth National Meeting of the Operations Research Society of America, St Louis, Missouri, 1958. Abstract in *Bulletin Fourteenth National Meeting of the Operations Research Society of America*, p. 897.

Boese, K. D., A. B. Kahng, and S. Muddu. 1994. "A new adaptive multistart technique for combinatorial global optimisations." *Oper. Res. Lett.* **16**: 101–113.

Brady, R. M. 1985. "Optimization strategies gleaned from biological evolution." *Nature* **317**: 804–806.

Černý, V. 1985. "Thermodynamical approach to the traveling salesman problem: an efficient simulation algorithm." *J. Optim. Theory Appl.* **45**: 41–51.

Chandra, B., H. Karloff, and C. A. Tovey. 1999. "New results on the old *k*-opt algorithm for the traveling salesman problem." *SIAM J. Comput* **28**: 1998–2029.

Croes, G. A. 1958. "A method for solving traveling-salesman problems." *Oper. Res.* **6**: 791–812.

Dorigo, M. and L. M. Gambardella. 1997. "Ant colony system: a cooperative learning approach to the traveling salesman problem." *IEEE Trans. Evolut. Comput.* **1**: 53–66.

Feo, T. A. and M. G. C. Resende. 1989. "A probabilistic heuristic for a computationally difficult set covering problem." *Oper. Res. Lett.* **8**: 67–71.

Feo, T. A. and M. G. C. Resende. 1995. "Greedy randomised adaptive search procedures." *J. Global Optim.* **6**: 109–133.

Fredman, M. L., D. S. Johnson, L. A. McGeoch, and G. Ostheimer. 1995. "Data structures for traveling salesmen." *J. Algorithms* **18**: 432–479.

Garey, M. R. and D. S. Johnson. 1979. *Computers and Intractability: A Guide to the Theory of NP-Completeness.* New York: W.H. Freeman.

Glover, F. 1986. "Future paths for integer programming and links to artificial intelligence." *Comput. Oper. Res.* **13**: 533–549.

Glover, F. 1989. "Tabu search—Part I." *ORSA J. Comput.* **1**: 190–206.

Glover, F. 1990. "Tabu search—Part II." *ORSA J. Comput.* **2**: 4–32.

Glover, F. and M. Laguna. 1997. *Tabu Search.* Boston: Kluwer Academic Publishers.

Johnson, D. S. 1990. "Local optimization and the traveling salesman problem." In *Proceedings of the 17th Colloquium on Automata, Languages and Programming*, pp. 446–461.

Johnson, D. S. and L. A. McGeoch. 1997. "The traveling salesman problem: a case study." In *Local Search in Combinatorial Optimization*, edited by E. H. L. Aarts and J. K. Lenstra, pp. 215–310. Chichester, England: John Wiley & Sons.

Johnson, D. S., C. R. Aragon, L. A. McGeoch, and C. Schevon. 1989. "Optimization by simulated annealing: an experimental evaluation; Part I, Graph partitioning." *Oper. Res.* **37**: 865–892.

Johnson, D. S., C. R. Aragon, L. A. McGeoch, and C. Schevon. 1991. "Optimization by simulated annealing: an experimental evaluation; Part II, Graph coloring and number partitioning." *Oper. Res.* **39**: 378–406.

Johnson, D. S., C. H. Papadimitriou, and M. Yannakakis. 1988. "How easy is local search?" *J. Comput. Sys. Sci.* **37**: 79–100.

Kernighan, B. W. and S. Lin. 1970. "An efficient heuristic procedure for partitioning graphs." *Bell Sys. Tech. J.* **49**: 291–307.

Kirkpatrick, S., C. D. Gelatt, Jr., and M. P. Vecchi. 1983. "Optimization by simulated annealing." *Science* **220**: 671–680.

Lawler, E. L., J. K. Lenstra, A. H. G. Rinnooy Kan, and D. B. Shmoys (eds). 1985. *The Traveling Salesman Problem: a Guided Tour of Combinatorial Optimization.* Chichester, England: John Wiley & Sons.

Lin, S. 1965. "Computer solutions of the traveling salesman problem." *Bell Sys. Tech. J.* **44**: 2245–2269.

Lin, S. and B. W. Kernighan. 1973. "An effective heuristic algorithm for the traveling salesman problem." *Oper. Res.* **21**: 498–516.

Martin, O., S. W. Otto, and E. W. Felten. 1992. "Large-step Markov chains for the TSP incorporating local search heuristic." *Oper. Res. Lett.* **11**: 219–224.

Mladenović, N. and P. Hansen. 1997. "Variable neighborhood search." *Comput. Oper. Res.* **24**: 1097–1100.

Papadimitriou, C. H. 1991. "On selecting a satisfying truth assignment." Extended Abstract. In *Proceedings of the 32nd Annual Symposium on Foundations of Computer Science*, pp. 163–169.

Papadimitriou, C. H. and K. Steiglitz. 1982. *Combinatorial Optimization: Algorithms and Complexity.* Prentice-Hall. Re-issued by Dover Publications, New York, NJ, 1998.

Reiter, S. and G. Sherman. 1965. "Discrete optimizing." *J. Soc. Ind. Appl. Math.* **13**: 864–889.

Selman, B. and H. Kautz. A. 1993. "Domain-independent extensions to GSAT: solving large structured satisfiability problems." In *Proceedings of the 13th International Joint Conference on Artificial Intelligence*, pp. 290–295.

Selman, B., H. A. Kautz, and B. Cohen. 1994. "Noise strategies for improving local search." In *Proceedings of the Twelfth National Conference on Artificial Intelligence*, pp. 337–343.

Tovey, C. A. 1997. "Local improvement on discrete structures." In *Local Search in Combinatorial Optimization*, edited by E. H. L. Aarts and J. K. Lenstra, pp. 57–89. Chichester, England: John Wiley & Sons.

Ulder, N. L. J., E. H. L. Aarts, H.-J. Bandelt, P. J. M. Van Laarhouven, and E. Pesch. 1990." Genetic local search algorithms for the traveling salesman problem." In *Proceedings of the 1st International Workshop on Parallel Problem Solving from Nature*, pp. 109–116.

Voudouris, C. and E. Tsang. 1999. "Guided local search and its application to the traveling salesman problem." *Eur. J. Oper. Res.* **113**: 469–499.

Yagiura, M. and T. Ibaraki. 1999. "Analyses on the 2 and 3-flip neighborhoods for the MAX SAT." *J. Combinatorial Optim.* **3**: 95–114.

Yagiura, M. T. Ibaraki, and F. Glover. 1999. "An ejection chain approach for the generalized assignment problem." Technical Report #99013. Department of Applied Mathematics and Physics, Graduate School of Informatics, Kyoto University, Kyoto, Japan.

Yannakakis, M. 1997. "Computational complexity." In *Local Search in Combinatorial Optimization*, edited by E. H. L. Aarts and J. K. Lenstra, pp. 19–55. Chichester, England: John Wiley & Sons.

3.6 Metaheuristics

3.6.1 Introduction

Bruce L. Golden and Edward A. Wasil

ABSTRACT When applied to minimization problems, traditional local improvement methods explore the neighborhood of a current solution and only select new solutions that strictly decrease the total objective function value (e.g., length or distance). Simulated annealing, deterministic annealing, smoothing algorithms, tabu search, genetic algorithms, and other related techniques can select new solutions that increase the total length and, in this way, they can avoid becoming trapped in poor local minima. We refer to these new techniques as metaheuristics and we review them in this article.

1. Introduction

Metaheuristics first appeared in the early 1980s (the term was coined by Glover, 1986) and, from the beginning, they have been iterative solution procedures designed for solving difficult optimization problems (typically, combinatorial optimization problems). Metaheuristics often have more traditional heuristics (e.g., improvement heuristics) as subprocedures. Another feature that sets metaheuristics apart from more traditional heuristics is that uphill as well as downhill intermediate moves are allowed (for minimization problems). In addition, infeasible intermediate moves may also be allowed. For the purposes of this article, we will focus on the solution of minimization problems.

There is a vast open literature on the theory and practice of metaheuristics that includes bibliographies, edited volumes, books, and articles. For information on metaheuristics at the overview level, we refer the reader to the article by Osman and Kelly (1996) and the edited volumes by Aarts and Lenstra (1997), Osman and Laporte (1996), and Reeves (1993).

2. Simulated Annealing

In the early 1980s, Kirkpatrick et al. (1983), using an analogy between combinatorial optimization problems such as the traveling salesman problem (TSP) and large physical systems of the kind studied in statistical mechanics, proposed a novel solution procedure in which uphill moves could be accepted probabilistically. Before this time, only downhill moves were considered in solving TSPs and related problems. We now briefly describe the simulated annealing procedure.

Let S be the current solution and $N(S)$ be a neighborhood of S that contains alternative solutions that are in the "vicinity" of S. We randomly select $S' \in N(S)$ and compute the difference $D = f(S') - f(S)$, where $f(S)$ is the objective function value of S. If $D < 0$, then S' is selected as the new solution (that is, downhill moves are always accepted). If $D > 0$ and $e^{-D/T} > q$ (where $0 < q < 1$ is randomly generated from a uniform distribution), then S' is selected as the new solution (that is, an uphill move may be accepted). T is simply a control parameter known as the temperature. Typically, the temperature is gradually lowered during the search process so that the probability of accepting an uphill move steadily decreases. This is handled via an annealing schedule of the form $A = \{t_1, t_2, \ldots, t_n\}$, where $t_1 > t_2 > \ldots > t_{n-1} > t_n$. At each temperature, the procedure continues until a stopping condition is satisfied.

3. Deterministic Annealing

In allowing the possibility of nonimproving moves, simulated annealing opened the doors for new ways of thinking about heuristic search. As an alternative to simulated annealing, Dueck and Scheuer (1990) and Dueck (1993) developed several deterministic variants, including threshold accepting, record-to-record travel, and the great deluge algorithm.

In threshold accepting, a threshold value, T, is specified as the upper bound on the amount of objective function increase allowed from one move to the next. If $f(S') < f(S) + T$, then S' is selected as the new solution. Thus, uphill moves can be made. The threshold value is decreased according to a predetermined schedule, until it equals zero. At each threshold, the procedure continues until a stopping criterion is satisfied.

In record-to-record travel, Record is defined as the total distance of the best solution observed so far. Deviation is defined as $k\% \times$ Record. If $f(S') <$ Record + Deviation, then S' is selected as the new solution. Again, uphill moves can be made. We point out that record-to-record travel has been used extensively in the solution of vehicle routing problems (for example, see Chao et al., 1995).

The great deluge algorithm is similar in structure to threshold accepting. Toussaint and Golden (1994) described the procedure in a minimization context and interpreted it from a scuba diver's point of view.

Wood and Downs (1998) have recently proposed a demon algorithm for handling minimization problems, for example, TSPs. The demon algorithm is based on the concept of a creditor, known as the demon. When the algorithm accepts a solution of shorter length, the decrease in length is credited to the demon. Similarly, when the algorithm accepts a solution of increased length, the increase in length is debited from the demon. An uphill move will be accepted only if the demon has enough credit to "pay" for the increase in length. Minimization is encouraged by either imposing an upper bound on the demon value or annealing the demon value as the algorithm progresses. This leads to two separate implementations of the demon algorithm.

The upper bound prevents the demon from receiving all of the credit from accepting an improved solution. If the demon value is annealed, the value of the demon is decreased regularly. In both approaches, the credit given to the demon is reduced. The algorithm will have more difficulty accepting uphill moves after accepting downhill moves, forcing the algorithm to seek out solutions of shorter length.

Numerous variants of the demon algorithm have been proposed by Wood and Downs (1998) and Pepper (1999). Some of these are deterministic and some are probabilistic. Extensive computational tests on the TSP have been conducted by Pepper (1999).

4. Smoothing Algorithms

Gu and Huang (1994) introduced smoothing algorithms and applied the idea to the TSP. Smoothing algorithms (also called search space smoothing and fine-tuned learning in the literature) work in the following way. Start with intercity distances that have been smoothed by a specified function. Gu and Huang used the following function:

$$d_{ij}(\alpha) = \begin{cases} \bar{d} + (d_{ij} - \bar{d})^{\alpha}, & d_{ij} \geqslant \bar{d} \\ \bar{d} - (\bar{d} - d_{ij})^{\alpha}, & d_{ij} < \bar{d} \end{cases}$$

where d_{ij} is the distance from city i to city j, $\bar{d}$ is the average intercity distance, and α is the smoothing factor. Apply a local search (LS) heuristic (e.g., 2-opt) and generate a locally optimal TSP solution (called the current tour). Smooth the distances again to a lesser extent (Gu and Huang specified the schedule for the smoothing factor as α = 5 (begin), 4, 3, 2, 1), apply the LS heuristic using the current tour as the starting tour, and generate a new current tour. Continue until the LS heuristic is applied to the TSP with the original distances (that is, α = 1). We point out that smoothing can be applied to a wide variety of combinatorial optimization problems.

Gu and Huang's heuristic avoids becoming trapped in a poor local minimum by making some moves that increase total length and not making some moves that slightly decrease total length. Coy et al. (1998) provided insight that revealed how smoothing worked and why it worked so well. Numerous variants of Gu and Huang's original smoothing function have been proposed and tested by Schneider et al. (1997) and Coy et al. (2000). Coy et al. applied eight smoothing heuristics to 70 TSPs that ranged in size from 100 to 2392 cities and compared the results produced by the smoothing heuristics to standard versions of 2-opt and 3-opt. Overall, a version of smoothing known as sequential smoothing generated the least-cost tours.

5. Tabu Search

Tabu search was developed by Glover (1986) as a deterministic alternative to simulated annealing, in which memory (rather than probability) guides intelligent search. To prevent cycling, intermediate solutions are recorded (implicitly) and the search procedure is prohibited from revisiting these "tabu" solutions for a prespecified number of iterations. This is accomplished via a tabu list of moves. Not only is cycling prevented, but search diversification is also encouraged. The key idea is to explore diverse regions of the search space. A number of advanced concepts such as long-term memory, intensification, and strategic oscillation (see Glover and Laguna, 1997) have contributed to the widespread success of this metaheuristic.

As an illustration of the effectiveness of tabu search, we focus on the well-known vehicle routing problem (VRP). In Golden et al. (1998), VRP heuristics using simulated annealing, neural networks, and tabu search are compared with respect to 14 benchmark problems from the literature. Typically, the tabu search heuristics generate solution values that are within 1 percent of the best-known solution values. The simulated annealing heuristics outperform neural network heuristics, but they are not competitive with the tabu search heuristics. We point out that the tabu search heuristics have not been very successful at solving TSPs effectively.

6. Genetic Algorithms

Genetic algorithms were developed by Holland (1975) in the 1960s and 1970s. Applications to combinatorial optimization problems only date back to the 1980s, however. The basic idea is to evolve a population of candidate solutions to the given problem, using operators inspired by natural genetic variation and natural selection. The basic genetic algorithm may be summarized as follows:

Step 1. *Initialization*. Construct an initial population of solutions.

Step 2. *Crossover*. Augment the population by adding offspring solutions.

Step 3. *Mutation*. Randomly perform small modifications to the offspring.

Step 4. *Evaluation*. Obtain fitness values for the offspring.

Step 5. *Selection*. Reduce the population size by selecting the appropriate number of survivors (with the largest fitness values) from the current population.

Step 6. *Evolution*. Repeat Steps 2 to 5 until a stopping criterion is satisfied.

Genetic algorithms have been applied to a wide variety of combinatorial optimization problems as well as nonlinear optimization and experimental design problems (see Reeves, 1993 for details). In general, they are outperformed by simulated annealing and (sometimes) tabu search heuristics, but the field is developing rapidly and this situation may change.

7. Guided Local Search

Guided local search (GLS) was developed by Voudouris and Tsang (1996) as a general method for solving combinatorial optimization problems. In GLS, a problem's minimization objective function is augmented with a set of penalty terms and this augmented function is passed to an LS procedure. The LS is restricted by the penalty terms and is focused on promising regions of the search space. When the search settles into a local minimum, the penalty terms are altered, and LS is used to minimize the augmented cost function and escape from the local minimum. As pointed out by Voudouris and Tsang (1999), GLS is closely related to the frequency-based memory approaches used in tabu search. GLS tries to perform functions similar to short-term memory and long-term memory via a single penalty modification mechanism. Unlike simulated annealing, GLS is a deterministic procedure.

GLS has been used to solve a wide range of problems including workforce scheduling (Tsang and Voudouris, 1997), the VRP (DeBacker et al. 2000), and the TSP (Voudouris and Tsang, 1999). In solving the TSP, Voudouris and Tsang (1999) combined GLS with a neighborhood reduction scheme called Fast Local Search (FLS) and tested several variants of GLS–FLS that used 2-opt, 3-opt, and the Lin–Kernighan heuristic as the LS procedures. The variants were tested on problems from TSPLIB that ranged from 48 to 2392 cities. The computational results were impressive. For example, on 28 medium-size problems (48 to 318 cities) with 10 runs for each problem, GLS–FLS with 2-opt found the optimal solution in 269 of the 280 runs, usually in less than 10 seconds of CPU time on a DEC Alpha 3000/600. In further testing, GLS–FLS with 2-opt was found to be very competitive with iterated Lin–Kernighan and genetic LS (two of the best algorithms for the TSP).

8. Greedy Randomized Adaptive Search Procedure

The greedy randomized adaptive search procedure (GRASP) is a two-phase method that consists of a construction phase in which a feasible solution is constructed iteratively followed by a local search phase that tries to improve each constructed solution. GRASP was developed in the late 1980s and was first applied to a difficult set covering problem (Feo and Resende, 1989). Feo and Resende (1995) give one of the first tutorials on GRASP, while Resende (1998) provides a recent survey of the basic method, enhancements, and applications.

In the first phase of GRASP, a feasible solution is constructed one element at a time. All of the elements that can be added to a solution are ordered in a candidate list using an adaptive greedy function that measures the benefit of selecting a particular candidate element. The function is adaptive because a candidate's benefit is updated at each iteration of the construction phase to take into account the previously chosen elements. A restricted candidate list of the best candidates is formed and the next element to enter the constructed solution is randomly chosen from this list. The process continues until an initial feasible solution is constructed. GRASP's construction phase is similar to the semi-greedy heuristic that was developed independently by Hart and Shogan (1987).

In the second phase of GRASP, an LS is applied to improve the constructed solution. In typical implementations, the entire two-phase method is repeated several times.

As pointed out by Kontoravdis and Bard (1995), for GRASP to work well, it is important that high-quality solutions (feasible solutions that are close to the optimum) be constructed during the first phase, with the most promising solutions passed to the second phase. This is different from tabu search and simulated annealing, which do not need high-quality, initial feasible solutions and expend much effort in trying to improve the current solution.

GRASP has few parameters whose values need to be set, so it is easy to implement. Several enhancements to GRASP have been proposed in the literature including the use of path relinking and long-term memory (see Resende, 1998, for more details).

GRASP has been used to successfully solve a wide variety of problems. Resende (1998) lists over 50 applications of GRASP to operations research problems (including production scheduling, vehicle routing, and network flow) and industrial problems (including manufacturing equipment selection and electric transmission expansion planning).

9. Scatter Search

Scatter search is an evolutionary procedure (like genetic algorithms) that combines existing solutions to a problem in order to construct new, better solutions. The origins of scatter search go back to the job shop scheduling literature of the 1960s, in which researchers tried to obtain improved decision rules for scheduling by forming weighted combinations of existing decision rules. The notion of combining solutions (or constraints) is also found in the integer and nonlinear programming literature. Glover et al. (2000a) provide a tutorial of scatter search that gives background information, a detailed outline of the process, and implications for future research.

Basically, there are four steps in the scatter search process:

Step 1. Generate a starting set of diverse solutions. Apply a heuristic procedure to improve the starting solutions. Extract the best solutions and designate them as reference solutions.

Step 2. Construct new solutions by combining subsets of the current reference solutions.

Step 3. Apply a heuristic procedure to improve the new solutions.

Step 4. Extract the best solutions from the improved, new solutions and add them to the set of reference solutions.

Steps 2, 3, and 4 are repeated until the set of reference solutions does not change. The process stops when a specified iteration limit is reached. The set of reference solutions can be diversified by restarting from Step 1. Typically, the reference set has 20 solutions or less in order to keep the number of possible combinations in the reference set at a tractable level.

Extensions to the basic scatter search process like the use of intensification and diversification and a discussion of the more general path relinking methods are given in Glover (1997, 1998). Applications of scatter search to assignment problems and mixed integer programming problems are presented in Martí et al. (2000) and Glover et al. (2000b).

10. Ant Colony Optimization

In the early 1990s, Marco Dorigo used an ant colony metaphor to construct an ant algorithm to solve discrete optimization problems. Real ants establish shortest routes between feeding sources and their colony by laying down a chemical substance known as pheromone along a trail. When an ant encounters a previously pheromone-marked trail, the ant decides with high probability to follow it and further marks the trail with its own pheromone. Thus, there is positive feedback in the foraging process—the probability of selecting a particular trail increases with the number of ants that have previously selected the same trail. Thus, ants engage in cooperative learning every time they search for food and they seem to be quite efficient. The goal of ant colony optimization is to mimic this natural process in order to search a complex solution space.

Dorigo and his colleagues (see Dorigo et al., 1996) developed an ant system (AS) for solving the TSP (and other combinatorial optimization problems in general). Artificial ants built solutions to a problem and a common memory (corresponding to a pheromone trail) was updated each time a new solution was constructed. The artificial ants cooperated to construct a solution to a TSP by exchanging information via pheromone that had been laid down on a graph's edges.

Dorigo and Gambardella (1997) extended the AS concept to include, for example, a local updating rule that shuffles TSP tours and named their approach ant colony system (ACS). Furthermore, the authors combined ACS with an LS procedure (3-opt) and computationally tested ACS–3-opt on asymmetric and symmetric TSPs. They found that ACS–3-opt produced very good (sometimes optimal) solutions that were competitive with a genetic algorithm-based approach.

Extensions to the basic AS (such as the use of a queen process) and hybrid ASs (such as the fast AS) are presented by Taillard (2000). In addition to the TSP, ASs have been applied to the quadratic assignment problem (Gambardella et al., 1999), vehicle routing (Bullnheimer et al., 1999), and routing on communications networks (Bonabeau et al., 1999).

11. Variable Neighborhood Search

Mladenovic and Hansen (1997) used a simple idea—systematically change the neighborhhood in an LS—to produce an effective metaheuristic known as variable neighborhood search (VNS). The basic version of VNS has two steps:

Step 1. *Initialization.* Let $N_k(x)$ be the set of solutions in the kth neighborhood of x. Select a finite set of neighborhood structures N_k, $k = 1, 2, ..., k_{\max}$ to use in the search. Find an initial solution x.

Step 2. *Main step.* Set $k = 1$. Repeat the following steps until $k = k_{\max}$. Randomly generate x' from $N_k(x)$. Apply LS using x' as the initial solution. If the local optimum obtained (say x'') is better, move from x to x'' and continue the search with N_1. Otherwise, $k = k + 1$.

VNS stops when a criterion like maximum number of iterations or maximum computation time has been reached. VNS is easy to implement and program (there is only parameter $k_{\max}$) and can be extended to include ideas from other metaheuristics [like intensification and diversification from tabu search; Mladenovic and Hansen (1997) proposed several variants of the basic VNS].

As pointed out by Mladenovic and Hansen, VNS "explores increasingly distant neighborhoods" of the current solution and "jumps" to a new solution when an improvement can be made. The notion is that most variables are already at their optimal values and these "favorable characteristics" are kept and then used to find "promising neighboring solutions."

Mladenovic and Hansen combined basic VNS with an LS procedure (the GENIUS algorithm of Gendreau et al., 1992) to solve randomly generated Euclidean TSPs with and without backhauls for 100 to 1000 cities. VNS + GENIUS was slightly more accurate than GENIUS (by an average of 0.75 percent on the TSPs without backhauls and 0.40 percent on TSPs with backhauls). VNS has also been applied to the p-median problem (Hansen and Mladenovic, 1997) and the problem of finding an extremal graph (Caprossi and Hansen, 2000).

12. Conclusions

Since the early 1980s, researchers have developed and applied a wide variety of metaheuristics to solve difficult optimization problems and have achieved some successes (e.g., tabu search has generated best-known solutions to many benchmark vehicle routing problems). There are drawbacks—procedures can have many parameters whose values need to be set and fine tuned, procedures can be computationally expensive (e.g., running times can be hours or even days), and, for the most part, little computational testing has been carried out on large-size problems. However, the outlook is very bright for the continued use and growth of metaheuristics. Over the next decade, we expect researchers to greatly improve the power and speed of metaheuristics and expand their applicability to important, real-world problems.

References

Aarts, E. and J. K. Lenstra. (eds). 1997. *Local Search in Combinatorial Optimization.* Chichester, England: Wiley.

Bonabeau, E., M. Dorigo, and G. Theraulaz. 1999. *Swarm Intelligence: From Natural to Artificial Systems*. New York: Oxford University Press.

Bullnheimer, B., R. Hartl, and C. Strauss. 1999. "An improved ant system for the vehicle routing problem." *Ann. Oper. Res.* **89**: 319–328.

Caprossi, G. and P. Hansen. 2000. "Variable neighborhood search for extremal graphs: 1 The AutoGraphiX System." *Discrete Math.* **212(1–2):** 29–44.

Chao, I-M., B. Golden, and E. Wasil. 1995. "An improved heuristic for the period vehicle routing problem." *Networks* **26**: 25–44.

Coy, S., B. Golden, G. Runger, and E. Wasil. 1998. "See the forest before the trees: fine-tuned learning and its application to the traveling salesman problem." *IEEE Trans. Sys. Man, Cybernetics* **28(4)**: 454–464.

Coy, S., B. Golden, and E. Wasil. 2000. "A computational study of smoothing heuristics for the traveling salesman problem." *Eur. J. Oper. Res.* **124(1):** 15–27.

DeBacker, B., V. Furnon, P. Shaw, P. Kilby, and P. Prosser. 2000. "Solving vehicle routing problems using constraint programming and metaheuristics." *J. Heuristics* **6(4):** 501–523.

Dorigo, M. and L. Gambardella. 1997. "Ant colony system: a cooperative learning approach to the traveling salesman problem." *IEEE Trans. Evolut. Comput.*, **1(1)**, 53–66 (available at http://iridia.ulb.ac.be/~mdorigo/ACO/ACO.html).

Dorigo, M., V. Maniezzo, and A. Colorni. 1996. "The ant system: optimization by a colony of cooperating agents." *IEEE Trans. Sys., Man, Cybernetics*, **26(1)**, 29–41 (available at http://iridia.ulb.ac.be/~mdorigo/ACO/ACO.html).

Dueck, G. 1993. "New optimization heuristics: the great deluge algorithm and the record-to-record travel." *J. Comput. Phys.* **104**: 86–92.

Dueck, G. and T. Scheuer. 1990. "Threshold accepting: a general purpose optimization algorithm appearing superior to simulated annealing." *J. Comput. Phys.* **90**: 161–175.

Feo, T. and M. Resende. 1989. "A probabilistic heuristic for a computationally difficult set covering problem." *Oper. Res. Lett.* **8(2)**: 67–71.

Feo, T. and M. Resende. 1995. "Greedy randomized adaptive search procedures." *J. Global Optim.* **6**: 109–133.

Gambardella, L., É Taillard, and M. Dorigo. 1999. "Ant colonies for the quadratic assignment problem." *J. Oper. Res. Soc.*, **50**, 167–176 (available at www.idsia.ch/~luca).

Gendreau, M., A. Hertz, and G. Laporte. 1992. "New insertion and postoptimization procedures for the traveling salesman problem." *Operations Research* **40**: 1086–1094.

Glover, F. 1986. "Future paths for integer programming and links to artificial intelligence." *Comput. Oper. Res.* **13**: 533–549.

Glover, F. 1997. "Tabu search and adaptive memory programming—advances, applications, and challenges." In *Interfaces in Computer Science and Operations Research: Advances in Metaheuristics, Optimization, and Stochastic Modeling Techniques*, edited by R. Barr, R. Helgason, and J. Kennington, pp. 1–75. Boston: Kluwer Academic Publishers.

Glover, F. 1998. "A template for scatter search and path relinking." In *Artificial Evolution, Lecture Notes in Computer Science*, 1363, edited by J-K. Hao, E. Lutton, E. Ronald, M. Schoenauer, and D. Snyers, pp. 13–54. Berlin: Springer-Verlag.

Glover, F. and M. Laguna. 1997. *Tabu Search*. Boston: Kluwer Academic Publishers.

Glover, F., M. Laguna, and R. Martí. 2000a. "Scatter search." Forthcoming in *Theory and Applications of Evolutionary Computation: Recent Trends*, edited by A. Ghosh and S. Tsutsui. Springer-Verlag, Berlin (available at http://bus.colorado.edu/faculty/laguna).

Glover, F., A. Løkketangen, and D. Woodruff. 2000b. "Scatter search to generate diverse MIP solutions." In *Computing Tools for Modeling, Optimization, and Simulation: Interfaces in Computer Science and Operations Research*, edited by M. Laguna and J. L. González-Velarde, pp. 299–317. Boston: Kluwer Academic Publishers.

Golden, B., E. Wasil, J. Kelly, and I-M. Chao. 1998. "The impact of metaheuristics on solving the vehicle routing problem: algorithms, problem sets, and computational results." In *Fleet Management and Logistics*, edited by T. Crainic and G. Laporte, pp. 33–56. Boston: Kluwer Academic Publishers.

Gu, J. and X. Huang. 1994. "Efficient local search with search space smoothing: a case study of the traveling salesman problem." *IEEE Trans. Sys., Man, Cybernetics* **24(5)**: 728–735.

Hart, J. and A. Shogan. 1987. "Semi-greedy heuristics: an empirical study." *Oper. Res. Lett.* **6(3)**: 107–114.

Hansen, P. and N. Mladenovic. 1997. "Variable neighborhood search for the p-median." *Location Sci.* **5(4)**: 207–226.

Holland, J. 1975. *Adaptation in Natural and Artificial Systems*. Ann Arbor: University of Michigan Press.

Kirkpatrick, S., C. Gelatt, and M. Vecchi. 1983. "Optimization by simulated annealing." *Science* **220**: 671–680.

Kontoravdis, G. and J. Bard. 1995. "A GRASP for the vehicle routing problem with time windows." *ORSA J. Comput.* **7(1)**: 10–23.

Martí, R., H. Lourenço, and M. Laguna. 2000. "Assigning proctors to exams with scatter search." In *Computing Tools for Modeling, Optimization, and Simulation: Interfaces in Computer Science and Operations Research*, edited by M. Laguna and J. L. González-Velarde, pp. 215–227. Boston: Kluwer Academic Publishers.

Mladenovic, N. and P. Hansen. 1997. "Variable neighborhood search." *Comput. Oper. Res.* **24(11)**: 1097–1100.

Osman, I. and J. Kelly. 1996. "Meta-heuristics: an overview." In *Meta-heuristics: Theory & Applications*, edited by I. Osman and J. Kelly, pp. 1–21. Boston: Kluwer Academic Publishers.

Osman, I. and G. Laporte (eds) 1996. *Metaheuristics in Combinatorial Optimization*, Annals of Operations Research, Volume 63, The Netherlands: Baltzer.

Pepper, J. 1999. "A computational study of annealing-based heuristics for the traveling salesman problem." Master's thesis. University of Maryland, College Park.

Reeves, C. (ed.) 1993. *Modern Heuristic Techniques for Combinatorial Optimization*, New York: Halsted Press.

Resende, M. 1998. "Greedy randomized adaptive search procedures (GRASP)." Technical Report 98.41.1. AT&T Labs Research, Florham Park, NJ (available at www.research.att.com).

Schneider, J., M. Dankesreiter, W. Fettes, I. Morgenstern, M. Schmid, and J. Singer. 1997. "Search-space smoothing for combinatorial optimization problems." *Physica A* **243**: 77–112.

Taillard, É. 2000. "An introduction to ant systems." In *Computing Tools for Modeling, Optimization, and Simulation: Interfaces in Computer Science and Operations Research*, edited by M. Laguna and J. L. González-Velarde, pp. 131–144. Boston: Kluwer Academic Publishers.

Toussaint, K. and B. Golden. 1994. "Exchange heuristics to improve the clarity of base/time plots." *Comput. Oper. Res.* **21**: 573–586.

Tsang, E. and C. Voudouris. 1997. "Fast local search and guided local search and their application to British Telecom's workforce scheduling problem." *Oper. Res. Lett.* **20(3)**: 119–127.

Voudouris, C. and E. Tsang. 1999. "Guided local search and its application to the traveling salesman problem." *Eur. J. Oper. Res.* **113**: 469–499.

Voudouris, C. and E. Tsang. 1996. "Partial constraint satisfaction problems and guided local search." In *Proceedings of the Second International Conference on the Practical Application of Constraint Technology*, pp. 337–356.

Wood, I. and T. Downs. 1998. "Demon algorithms and their application to optimization problems." In *IEEE World Congress on Computational Intelligence*, pp. 1661–1666, New York: IEEE.

3.6.2 Ant Systems

Éric D. Taillard

ABSTRACT This article describes ant systems, a metaheuristic based on an ant-foraging metaphor. The presentation of ant systems has been somewhat generalized by adding a "queen" process in charge of coordinating classical "ant" processes, so that recent ant systems can be naturally included while remaining close to the metaphor. To illustrate how ant systems are practically implemented, a number of applications to the quadratic assignment problem are reviewed.

1. A Model of Real Ants

The metaphor on which ant systems are based can be illustrated by observation of ants of the species *Linepithema humile* by Deneubourg et al. (1990) and illustrated in Figure 1. An ant colony nest is isolated, and a food source is provided that is accessible by a bridge composed of two branches of the same length. Although the ants are totally free to choose the left or the right branch of the bridge, it is rapidly observed that almost all ants use a given branch, even if there is no reason to prefer the left or the right one. This phenomenon is explained by the fact that ants deposit a chemical substance while traveling. They are able to detect this substance with their antennae. This substance carries information and is called a pheromone.

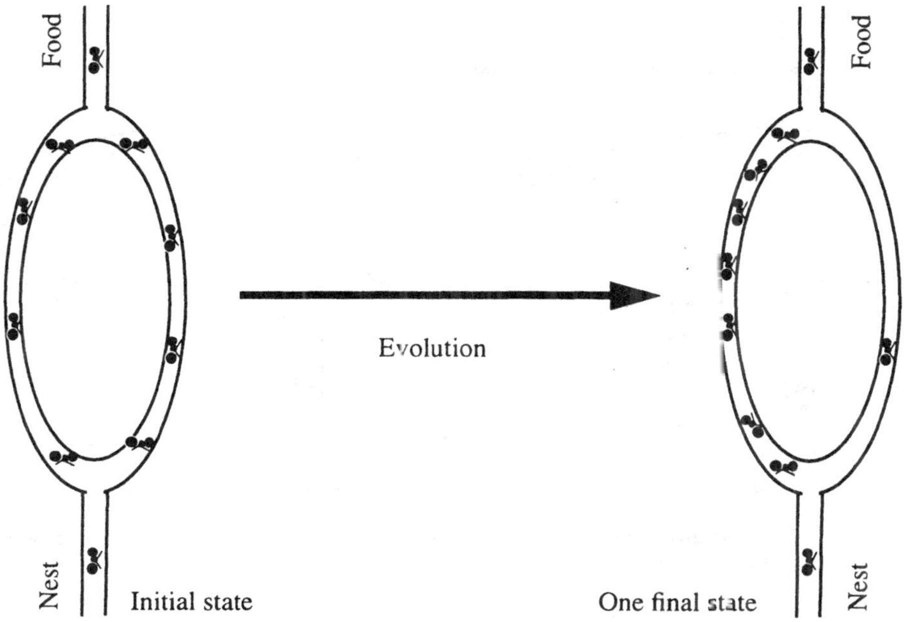

Figure 1 Double-bridge experiment. After a while almost all ants choose the same branch

A real ant is modeled as a probabilistic process: in the absence of pheromone, the ant explores the surrounding area in a totally random manner. If a pheromone trail is present, the ant follows the pheromone trail with a high probability. If two pheromone trails cross each other, the ant follows the trail with the larger amount of pheromone with a higher probability. Since the ant deposits additional pheromone when traveling, the foraging process evolves with positive feedback. Moreover, pheromones evaporate, meaning that a trail that is not used gradually disappears, amplifying the positive feedback effect.

The observation has been repeated by Goss et al. (1989) with a bridge whose branches have different lengths. The experiment is illustrated in Figure 2. Applying the same probabilistic behavior model, it can be deduced that the ants will rapidly choose the shorter branch and this is in fact observed. The phenomenon is explained as follows: at the beginning, when no pheromone is present, the ants choose the left or the right branch with equal probability. Since the ants that chose the shorter branch arrive earlier at the food (and come back to the nest earlier), the pheromone quantity on the shorter branch grows faster. So, the ants are able to find the shortest path between nest and food, even if each ant has an extremely limited view of the surrounding area.

Inspired by this natural optimization process, Colorni et al. (1992) designed a metaheuristic that can be applied to combinatorial optimization problems. The basic idea is to consider artificial ant processes that repeatedly build solutions to the optimization problem. Each process builds a solution with the help of a common memory that can be read and modified by any artificial ant. The common memory plays the role of trails in the real model.

However, when translated into a combinatorial optimization process, the simplified foraging process presented above is generally not able to create efficient heuristic methods and this (historical) model must be extended to cover recent and efficient applications inspired by ant colonies. Indeed, a simple positive feedback mechanism is difficult to tune correctly and often leads to degenerate situations: when the feedback is too strong, the only existing trail is that of the first ant that built a solution and when it is too weak, the trails are uniformly distributed in the whole search space. In the first situation, all ants use the same path and in the second, the ants just perform a random walk.

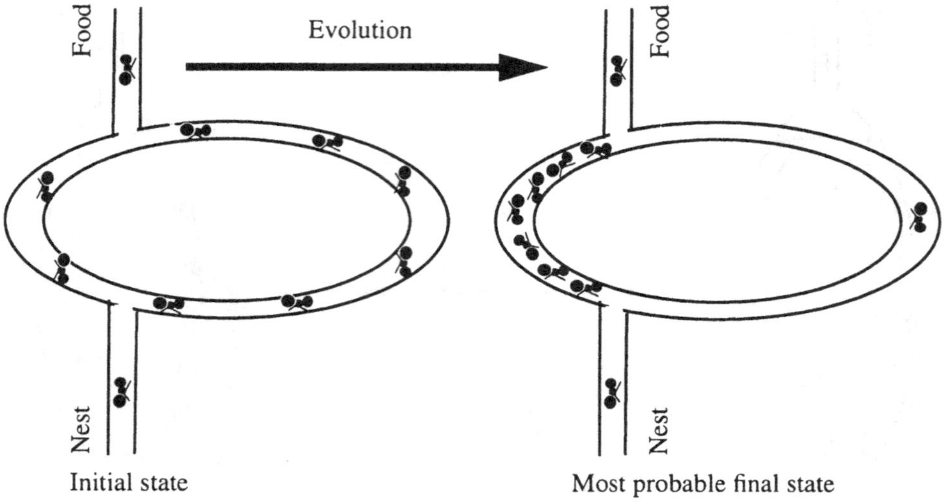

Figure 2 Asymmetric double-bridge experiment. After a while almost all ants choose the shortest branch

The model is extended by considering an ant colony composed of differentiated ants such as workers, soldiers, or sexual individuals. Each of these specialized individuals is assigned a different task. The most important individual of the colony is the queen, which has the ability to create new ants and to choose the type of the ants created. Therefore, the queen coordinates the whole colony and also has a strategic role. Inclusion of an "intelligent" queen helps in the design of efficient algorithms, since search strategies can be included within the artificial ant system, while remaining close to the ant metaphor.

2. Artificial Ant Systems

Since the introduction of the ant system by Colorni et al. (1992), algorithms based on the ant metaphor have evolved. Instead of reviewing successive evolutions of these algorithms, a general ant system model, developed from the real ant model, is presented. This general model is properly called the ant system metaheuristic, and is defined by a set of principles or a methodology that allows the development of heuristics for a wide range of combinatorial optimization problems.

Artificial ant systems are derived from the real ant model by making three analogies:

1. Real ants correspond to the processes in charge of building solutions to the problem considered; these processes are often referred to as artificial ants, or simply ant processes.
2. The pheromone trails correspond to a common memory that is updated each time a new solution is built; more precisely, the memory records a value associated with each component of a solution.
3. The queen corresponds to a central process in charge of activating and coordinating artificial ants and of managing the common memory.

A metaheuristic based on ant behavior can be described by a set of processes that collaborate through a common memory. The first set of processes, corresponding to the ants, built solutions in a probabilistic way, with probabilities depending on information stored in memory. All these artificial ant processes are activated and coordinated by a queen process that also manages the common memory. Very schematically, an ant system can be specified by two different processes:

- *Ant process*
  1. Receive problem data, memory state, and, perhaps, other parameters from the queen process.
  2. With the help of the memory, build a new solution probabilistically.
  3. Send the new solution to queen process.
- *Queen process*
  1. Initialize the memory.
  2. Repeat (in parallel) until a stopping criterion is met:
     (a) Choose parameters for a new ant process and activate the ant process
     (b) Receive a solution from an ant process and update the memory,
  3. Return the best solution produced by the system.

This formulation of ant systems extends and generalizes previous formulations. Indeed, the very first systems did not mention a queen process and the memory consisted of "trails," that is, quantities associated with elements constituting a solution (for example, these elements could be edges for problems in which a path in a graph is searched). In the first ant system applications, there was a fixed number of ant processes that were activated simultaneously. These processes were executed in parallel and directly updated the trail values. Therefore, there was no explicit coordination of ant processes, but rather an implicit one, realized through the trails. Unfortunately, such low-level interactions between simple artificial ants that work asynchronously and that do not know explicitly what the other ants are doing is not really efficient for designing practical applications. Recent developments of ant systems have introduced a kind of synchronization by waiting for the completion of all ant processes before updating the trails. Updates are then performed in a more intelligent way, following search strategies. For example, only an elite solution may be used to update the trails, thus implementing an intensification strategy (Glover and Laguna, 1997). Another example is to decrease the value of the trails corresponding to the elements chosen in the solution that has been built (negative feedback), thus implementing a diversification strategy. More recently, Taillard et al. (1998) have interpreted the concept of artificial trails as a memory storing information about the use of particular components in the solutions previously built. Finally, in order to integrate a stronger, more intelligent form of cooperation between ant processes, the queen process has been introduced by Taillard (1998). A similar process is called Ant Colony Optimization (ACO) as described by Dorigo and Di Caro (1999).

The art of designing a heuristic based on ant systems resides principally in the choice of the information contained in the memory (and their update) and in the way this information is used for constructing a new solution In order to give a flavor of how a practical ant systems-based algorithm may be designed, a few implementations that discuss the design choices are briefly reviewed here. These implementations illustrate very different ways of designing an ant systems-based heuristic. The first implementation discussed is the "historical" one that first introduced the ant system metaphor. Then we review implementations for the quadratic assignment problem for which ant systems have proven to be among the best heuristic methods for some problem-instance classes.

2.1. Historical Application to the Traveling Salesman Problem (TSP)

The first ant system application to a combinatorial optimization problem was programmed for the traveling salesman problem (TSP) by Colorni et al. (1992). In this application, the memory is a vector, **m**, of real numbers (m_e) associated with each edge $e \in E$ of the graph $G(V, E)$ in which a shortest tour passing exactly once by each vertex of V is sought. The values contained in **m** are interpreted as the pheromone quantity left on a trail. Initially all the entries of **m** are set to the same small value. The constructive procedure simulates the behavior of an ant that, starting from a vertex, can only move to a vertex it has not already visited (unless all vertices have been visited; in this case the ant completes its tour by returning to the first vertex). At each step, the edge, e, that is used to move to another vertex is randomly chosen with a probability proportional to $(m_e)^\alpha(d_e)^\beta$, where e is an edge with one end being the vertex on which the ant stands and the other being a vertex not yet visited, $0 \leqslant \alpha$, $0 \geqslant \beta$ are two parameters, and d_e is the length of edge e. This probabilistic rule favors the choice of short edges that frequently appear in the solution generated by the system. A very first implementation of the algorithm had no ant process coordination: a fixed number, k, of totally asynchronous ant processes repeatedly constructed new solutions. The memory updates were also performed asynchronously by the ant processes during the solution construction. Unfortunately, this system was found to be less efficient than another implementation with synchronized ant processes that can be described as follows.

In the best implementation of the algorithm of Colorni et al. (1992), the queen process activates k parallel ant processes at each loop. Each ant process builds a solution with the probabilistic rule described above, sends the solution to the queen process, and dies. The queen therefore receives k solutions $s_1, \ldots, s_k$ of length $L_1, \ldots, L_k$ at each loop. The memory is updated as follows: first $\mathbf{m} \leftarrow \rho\mathbf{m}$ where $0 \leqslant \rho \leqslant 1$ is a parameter that simulates the evaporation of the trails. Then, $m_e \leftarrow m_e + Q/L_i$, $\forall s_i$ and $\forall e \in s_i$, where $0 \leqslant Q$ is another parameter controlling the amount of positive feedback. The value m_e associated with each edge e gives, therefore, a stronger weight to edges that appear in good solutions and to edges that are often chosen.

This approach was able to produce, for small problem instances, solutions of slightly better quality than the heuristic of Lin and Kernighan (1973). This approach gives a very limited role to the queen process, which is why it was not initially integrated into ant systems explicitly. Recently, the system has been greatly improved by:

1. The use of a more elaborated construction procedure, including candidate lists and additional parameters.
2. The improvement of each constructed solution with a local search (LS).
3. The addition of intensification and diversification strategies in memory management, that is, the introduction of a stronger coordination between the ant processes.

The new method has been named ACS by Dorigo and Gambardella (1997).

2.2. *Applications to the Quadratic Assignment Problem*

It is possible to design completely different ant system-based heuristics by:

1. Modifying the type of information memorized.
2. Changing the memory update rules (or learning mechanism).
3. Modifying the constructive procedure.
4. Including alternative search strategies.

The quadratic assignment problem (QAP) is perhaps the problem for which the largest range of ant system-based heuristics have now been designed. This section reviews a few of these methods, focusing on various choices made by their designers. In essence the QAP can be described by the search of a permutation $\pi \in \Pi$ minimizing $\sum_{i=1}^{n}\sum_{j=1}^{n} a_{ij}b_{\pi_i\pi_j}$, where Π is the set of all the permutations of n elements, and $A = (a_{ij})$ and $B = (b_{ij})$ are two matrices of size $n \times n$. A typical QAP application is the placement of interacting facilities in spatial locations. The interaction (or flow) between facilities i and j is given by a_{ij} and the distance between location r and s is given by b_{rs}. The aim is to find locations for the facilities such that the sum of the distance $\times$ flow products is minimized.

For all applications reviewed in this section, the memory is implemented by a matrix M of size $n \times n$. Matrix entry m_{ir} records that facility i was placed on location r in solutions previously generated by the algorithm. The higher the quality and the larger the number of solutions generated with facility i placed on location r, the higher the m_{ir} value is.

MAX-MIN ANT SYSTEM (MMAS)

The main idea of the max-min ant system of Stützle and Hoos (1999) is to bound the values contained in the memory in an interval: $m_{\min} \leqslant m_{ir} \leqslant m_{\max}$, $\forall m_{ir}$, where $m_{\min}$ and $m_{\max}$ are two parameters. By choosing appropriate $m_{\min}$ and $m_{\max}$ values, the search is prevented from stagnating, that is, repeatedly producing the same solution.

Ant process. The constructive procedure works as follows: at each step, a facility i that has not yet been assigned is randomly chosen. This facility is assigned to an unoccupied location r using a "pseudorandom-proportional" rule: either location r, which maximizes m_{ir}, is chosen or the location is chosen from several alternatives, which associated probabilities are proportional to m_{ir}. The former choice has a probability q, the latter $(1-q)$. This step is repeated until all facilities are allocated. Then an LS is applied to improve the solution that is finally sent to the queen process. There are two versions of MMAS: the first one uses a descent procedure stopping at the first local optimum as LS, and the second one uses the tabu search of Taillard (1991).

Queen process. Initially, the queen process sets all the m_{ir} values to $m_{\max}$. Then, the process enters the main iteration that consists of activating k ant processes, where k is a parameter, and of waiting for k solutions $\pi^1, \ldots, \pi^k$ of cost $c^1, \ldots, c^k$ before updating the memory. The memory is updated as follows. First, all the entries of M are decreased by setting $M \leftarrow \rho M$, where $0 \leqslant \rho \leqslant 1$ is a parameter. Let π^b be the best solution among $\pi^1, \ldots, \pi^k$ and let π^* be the best solution produced by the search so far (c^b and c^* are their corresponding costs). Every odd iteration, the queen process updates the memory by setting $m_{i\pi_i^b} \leftarrow m_{i\pi_i^b} + 1/c^b$, $\forall i$, while every even iteration, the memory is updated with π^* and c^* instead of π^b and c^b. Finally, the entries of M that are lower than $m_{\min}$ are set to $m_{\min}$ and those larger than $m_{\max}$ are set to $m_{\max}$.

The max-min ant system has also been applied to the TSP and to the flow shop sequencing problem.

APPROXIMATE NONDETERMINISTIC TREE SEARCH (ANTS)

The main innovation of the approximate nondeterministic tree search (ANTS) designed by Maniezzo (1998) lies in the construction procedure. At each step, the procedure randomly places a facility on a location with a probability depending on the memory, as is usual for ant systems, but also depending on the computation of a lower bound, similar to what is done in implicit enumeration methods, such as branch and bound.

Ant process. First, the ant process receives, from the queen process, an order in which the locations must be filled by the facilities. At each step, the constructive procedure assigns a facility to the next empty location r as follows: first a lower bound LB_{ir} of the cost of assigning facility i to location r is computed, taking into account the assignments already performed in previous steps. Then, the facility to assign is chosen with a probability proportional to $\alpha m_{ir} + (1-\alpha)LB_{ir}$, where $0 \leqslant \alpha \leqslant 1$ is a parameter. Once all facilities are allocated, the solution is improved with an LS that stops at the first local optimum.

Queen process. First, the queen process computes a lower bound LB on the solution cost by solving a linear relaxation of the problem. The dual variables obtained during the computation are used on the one hand for initializing the m_{ir} values of the memory (noted m_{ir}^0) and, on the other hand, these variables are used to compute the order in which the ant process constructive procedure must fill the location. The order depends only on the problem data and is computed only once. The rationale behind this ordering is that the higher the value of a dual variable associated with a location is, the higher the impact of the assignment of a facility to that location should be on the solution cost, and therefore the earlier the location must be filled.

The queen process then activates k ant processes and receives k solutions $\pi^1, \ldots, \pi^k$ of cost $c^1, \ldots, c^k$. A moving average $\bar{c}$ of the values of the last m solutions received is computed and the memory entries are updated as follows:

$$m_{i\pi_i^b} \leftarrow m_{i\pi_i^b} + m_{i\pi_i^b}^0 \frac{\bar{c} - c^b}{\bar{c} - LB}, \quad \forall 1 \leqslant i \leqslant n, \quad \forall 1 \leqslant b \leqslant k$$

HYBRID ANT SYSTEM (HAS)

A first principle of the hybrid ant system (HAS) of Gambardella et al. (1999) is to replace the ant process constructive procedure by a probabilistic LS. Instead of building a brand-new solution each time an ant process is activated, the ant process receives a starting solution and applies a given number of LS moves to that solution. The moves are randomly chosen with a probability that depends on the values in memory. The second basic idea of the HAS is to introduce search strategies in the queen process.

Ant process. First, the ant process receives a solution π from the queen, in addition to the problem data. The process then repeats m local search steps as follows. A facility i is randomly chosen in 1, ..., n. A second facility j is chosen with the pseudorandom-proportional rule and facilities i and j exchange their location. The pseudorandom-proportional rule chooses either the facility j, maximizing $m_{i\pi_j} + m_{j\pi_i}$ (with probability q), or chooses j randomly, with a probability proportional to $m_{i\pi_j} + m_{j\pi_i}$. After m probabilistic LS steps, the solution is improved with an LS (driven by the objective function) that ends at the first local optimum before being sent to the queen process.

Queen process. In addition to the memory matrix M, the queen process manages a pool of k solutions π^1, ..., π^k. Initially, the first k solutions are randomly chosen and improved with an LS. In addition to the pool of solutions, the queen manages π^*, the best solution found by the system (c^* denotes its value). The memory matrix entries are initially set to the same small value. The queen repeats the following cycle.

First, k ant processes are activated, each of them receiving a different solution from the pool. Then, the queen process waits for k solutions μ^1, ..., μ^k from the ant processes and updates π^*. Then, the memory is updated as follows. First, all entries are weakened by setting $M \leftarrow \rho M$, where $0 \leqslant \rho \leqslant 1$ is a parameter. Then, the entries corresponding to the best solution are reinforced by setting $m_{i\pi_i^*} \leftarrow m_{i\pi_i^*} + (1-\rho)/c^*, \ \forall i$. Finally, the pool of solutions is updated using diversification and intensification strategies: normally, the queen process is in the intensification phase and replaces π^b by the best solution among π^b and μ^b, $\forall 1 \leqslant b \leqslant k$. The queen process remains in intensification phase while at least one μ^b is better than π^b. But if no μ^b is better than π^b, the queen process performs a diversification step that consists of replacing $k-1$ solutions in the pool by random ones and by replacing the last one with π^*.

FAST ANT SYSTEM (FANT)

The basic idea of the fast ant system (FANT) of Taillard (1998) is to design a method that is as simple as possible while incorporating diversification and intensification strategies. This is realized, on the one hand, by systematically reinforcing the attractiveness of the m_{ir} values corresponding to the best solution found so far by the search, and on the other hand, by clearing the memory while giving less weight to the best solution if the process appears to be stagnating.

Ant process. The ant process constructs a new solution by randomly choosing the location r of facility i with a probability proportional to m_{ir}. Then the solution is improved with an LS and sent to the queen process.

Queen process. While implementing an automatic intensification and diversification, the queen process requires only one parameter R and manages, in addition to the memory matrix M, a variable v and the best solution π^* found by the system so far. Initially $v = 1$ and $m_{ir} = v, \ \forall i, r$. The queen then repeats the following cycle:

1. Activate an ant process.
2. Wait for a solution π from an ant process.
3. If $\pi = \pi^*$ then set $v \leftarrow v + 1$ and $m_{ir} \leftarrow v, \ \forall i, r$.
4. If π is better than π^* then set $\pi^* \leftarrow \pi$, $v \leftarrow 1$ and $m_{ir} \leftarrow v, \ \forall i, r$.
5. Set $m_{i\pi_i} \leftarrow m_{i\pi_i} + v, \ \forall i$.
6. Set $m_{i\pi_i^*} \leftarrow m_{i\pi_i^*} + R \ \forall i$.

To intensify the search in the neighborhood of π^*, the memory entries corresponding to π^* are reinforced at each iteration by R (Step 6). In order to "memorize" the solution π

generated by the ant process, the memory entries corresponding to π are reinforced by v (Step 5). When π^* is improved, variable v is set to 1 (to give a relatively higher weight to R, and the memory is re-initialized (Step 4). When the solution constructed by the ant process is equal to π^*, meaning that the memory entries that correspond to π^* have been reinforced too much, variable v is incremented (to give a relatively lower weight to R in subsequent queen iterations) and the memory is re-initialized (Step 3). This simple ant system has been applied to other combinatorial optimization problems with success, such as the p-median, the balancing of turbine runners, and the biquadratic assignment problems. It has been implemented on a sequential computer and the activation of only one ant process at a time allows relatively good solutions to be obtained with a very low computational effort. However, the method can be easily extended to the case of an arbitrary number of ant processes.

3. General Considerations

Ant systems belong to the wide class of metaheuristics that work with a memory or Adaptive Memory Programming (Taillard et al., 1998). Ant systems share common features with genetic algorithms, greedy randomized adaptive search procedures (Feo and Resende, 1995), tabu search, and scatter search. Indeed, ant systems are based on a population of artificial agents, the ant processes, but the interaction between individuals, through a common memory, is quite unlike classical genetic operators. The constructive procedure embedded in the ant processes looks a bit like those of a greedy randomized adaptive search procedure and can also be compared to a probabilistic scatter search, since the solutions generated are a kind of combination of solutions previously built. Finally, the ant processes are coordinated by a process that can incorporate various search strategies proposed in tabu search.

Ant systems have now matured and successful applications are more and more numerous. Efficient ant systems very often embed elements that are unrelated to the metaphor that inspired the first ant systems. Typical foreign elements are the use of an LS or memory central management. This is perhaps a natural development for metaheuristics; it has been noted by Taillard et al. (1998) that the best heuristic methods are often inspired by many basic metaheuristics. For example, most efficient genetic algorithms include an LS. Therefore, it is not surprising that ant systems are evolving in a similar way and incorporate components of other metaheuristics; conversely, components of ant systems will certainly be combined with methods based on other metaheuristics. Future developments of ant systems will include the treatment of dynamic and stochastic problems and the design of new distributed systems.

References

Colorni, A., M. Dorigo, and V. Maniezzo. 1992. "Distributed optimization by ant colonies." In *Toward a Practice of Autonomous Systems: Proceedings of the First European Conference on Artificial Life (ECAL 91)*, edited by F. Varela and P. Bourgine, pp. 134–142. Cambridge, England: MIT Press.

Deneubourg, J.-L., S. Aron, S. Goss, and J. Pasteels. 1990. "The self-organizing exploratory pattern of the argentine ant." *J. Insect Behav.* **3**: 159–168.

Dorigo, M. and G. Di Caro. 1999. "The ant colony optimization metaheuristic." In *New Ideas in Optimization*, edited by D. Corne, M. Dorigo, and F. Glover. McGraw-Hill.

Dorigo, M. and L.-M. Gambardella. 1997. "Ant colony system: a cooperative learning approach to the traveling salesman problem." *IEEE Trans. Evolut. Comput.* **1**: 53–66.

Feo, T. and M. Resende. 1995. "Greedy randomised adaptive search procedures." *J. Global Optim.* **16** 109–133.

Gambardella, L. M., E. D. Taillard, and M. Dorigo. 1999. "Ant colonies for the quadratic assignment problem." *J. Oper. Res. Soc.* **50**: 167–176.

Glover, F. and M. Laguna. 1997. *Tabu Search.* Boston: Kluwer Academic Publishers.

Goss, S., S. Aron, J.-L. Deneubourg, and J. Pasteels. 1989. "Self-organized shortcuts in the argentine ant." *Naturwissenschaften* **76**: 579–581.

Lin, S. and B. W. Kernighan. 1973. "An effective heuristic algorithm for the traveling-salesman problem." *Oper. Res.* **21**: 498–516.

Maniezzo, V. 1998. "Exact and approximate nondeterministic tree-search procedures for the quadratic assignment problem." *INFORMS J. Comput.* **11(4)**: 359–369.

Stützle, T. and H. H. Hoos. 1999. "The max–min ant system and local search for combinatorial optimization problems." In *Meta-Heuristics: Advances and Trends in Local Search Paradigms for Optimization*, edited by S. Voss and S. Martello, pp. 313–329. Boston: Kluwer Academic Publishers.
Taillard, E. D. 1991. "Robust taboo search for the quadratic assignment problem." *Parallel Comput.* **17**: 443–455.
Taillard, E. D. 1998. "FANT: fast ant system." Technical report IDSIA-46-98. Lugano, Switzerland: IDSIA.
Taillard, E. D., L.-M. Gambardella, M. Gendreau, and J. Potvin. 1998. "Adaptive memory programming: a unified view of meta-heuristics." To appear in *European J. of Oper. Res.*

3.6.3 Population Heuristics

John E. Beasley

ABSTRACT Standard heuristics in operations research (such as greedy, tabu search and simulated annealing) work on improving a single current solution. Population heuristics use a number of current solutions and combine them together to generate new solutions. Heuristic algorithms encountered in the literature that can generically be classified as population heuristics include genetic algorithms, hybrid genetic algorithms, memetic algorithms, scatter-search algorithms, and bionomic algorithms. In this article we discuss the basic features of population heuristics and provide practical advice about their effective use for solving operations research problems.

1. Introduction

In operations research (OR) over the years, a number of standard heuristic algorithms have appeared. In the early years of OR, heuristics were especially tailored to the problem under consideration. As OR advanced, a number of basic heuristic ideas began to be formalized. These included ideas such as greedy (seek the best possible improvement) and local interchange (e.g., swap two parts of a solution). Further advances came with schemes such as tabu search (accept a change in the solution, for better or worse, unless it is tabu) and simulated annealing (accept a change in the solution, for better or worse, according to some probabilistic criterion). These latter schemes are sometimes called metaheuristics as they provide general guidelines about the approach to be followed, yet need tailoring for each specific problem under consideration. Most readers will probably have encountered these ideas during their career, but for those who have not, some appropriate references are Aarts and Lenstra (1997), Glover and Laguna (1997), and Reeves (1993a).

All of the above have a common characteristic, namely that they have a single solution to a problem and continually seek to improve this single solution in some way.

Population heuristics, by contrast, explicitly work with a population of solutions and combine them together in some way to generate new solutions.

In recent years, the author has been working with such heuristics, both in the commercial world and in the academic world (Beasley and Chu, 1996; Chu and Beasley, 1997, 1998a,b; Chang et al., 2000; Beasley et al., 1999). As a result of this he has, through personal experience, come to believe that:

There is something about a population heuristic that means it is often capable of producing better-quality solutions than single-solution heuristics.

Note here, however, as will be discussed below, population heuristics, by their very nature, may often be slower than single-solution heuristics. Hence we encounter the trade-off we often see in OR between quality of solution and computer time consumed. However, with continuing advances in the computational power available at the desktop, as well as the acknowledged value of even small percentage improvements in solution quality for many decision problems, we believe that population heuristics are coming of age.

Populations heuristics are not particularly demanding intellectually (especially in contrast to a number of areas of OR) and are easy to understand. However their effective use can demand skills in creative algorithm design.

The objectives of this article are:

1. to introduce population heuristics to a wider audience than they have currently reached within the OR community;
2. to clearly explain such heuristics through a number of simple examples; and
3. to provide practical advice as to their effective use for solving OR problems.

2. Structure

In order to structure this article we are going to consider three problems:

- the multidimensional knapsack problem (MKP).
- the equality constrained multidimensional knapsack problem (ECMKP).
- a linear program related to MKP.

and illustrate how a population heuristic can be used to solve them.

In this section we first discuss terminology, then we introduce the MKP and the basic features of population heuristics.

2.1. Terminology

Population heuristics started with the development of genetic algorithms by Holland (1975). As the field advanced (evolved), new terms began to appear in the literature: hybrid genetic algorithms, memetic algorithms, scatter-search algorithms, bionomic algorithms. While there are differences between these approaches, and these will be discussed below, we believe that this profusion of names is obscuring the field and a single unifying name is preferable. Hence throughout this article we refer just to population heuristics.

2.2. Multidimensional Knapsack Problem (MKP)

The multidimensional knapsack problem (MKP) is:

$$\text{maximize} \quad \sum_{j=1}^{n} p_j x_j \tag{1}$$

$$\text{subject to} \quad \sum_{j=1}^{n} r_{ij} x_j \leqslant b_i \quad i = 1, \ldots, m \tag{2}$$

$$x_j \in (0, 1) \quad j = 1, \ldots, n \tag{3}$$

where p_j, r_{ij} and $b_i \geqslant 0$. Each of the m constraints in equation (2) is called a knapsack constraint, which is how the problem derives its name. More about this problem can be found in Chu and Beasley (1998a). Throughout this article we shall use the notation (x_j) to signify variables and (X_j) to signify a set of values for these variables that may, or may not, be feasible.

To make the MKP clearer we present a small example with $n = 7$ variables (columns) and $m = 2$ constraints (rows) in Figure 1, which we shall use throughout this article. Each of the seven columns shown in Figure 1 has an associated profit, and choosing a column consumes a certain amount of the resource available in each of the two rows shown (e.g., if we choose column 6 we get a profit of five but consume three of the four units of resource available in row 1 and 1 of the three units of resource available in row 2). The objective is to maximize the profit of the columns chosen subject to the resource constraints.

column j	*1 2 3 4 5 6 7*
profit p_j	2 2 3 4 5 5 5
row 1 r_{1j}	1 0 2 0 1 3 1 $\leq$ 4 (b_1)
row 2 r_{2j}	1 1 1 2 2 1 3 $\leq$ 3 (b_2)

Figure 1 An example of the multidimensional knapsack problem (MKP)

2.3. *Population Heuristics*

A population heuristic (henceforth abbreviated to PH) can be described as an "intelligent" probabilistic search algorithm and is based on the evolutionary process of biological organisms in nature. During the course of evolution, natural populations evolve according to the principles of natural selection and "survival of the fittest." Individuals who are more successful in adapting to their environment will have a better chance of surviving and reproducing, while individuals who are less fit will be eliminated. This means that the genes from highly fit individuals will spread to an increasing number of individuals in each successive generation. The combination of good characteristics from highly adapted parents may produce even more fit offspring. In this way, species evolve to become increasingly better adapted to their environment.

A PH simulates these processes by taking an initial population of individuals and applying genetic operators in each reproduction. In optimization terms, each individual in the population is encoded into a string or chromosome that represents a possible solution to a given problem. The fitness of an individual is evaluated with respect to a given objective function. Highly fit individuals or solutions are given opportunities to reproduce by exchanging pieces of their genetic information in a crossover procedure with other highly fit individuals. This produces new "offspring" solutions (i.e., children), who share some characteristics taken from both parents. Mutation is often applied after crossover by altering some genes in the strings. The offspring can either replace the whole population (generational approach) or replace less fit individuals (steady-state approach). This evaluation-selection-reproduction cycle is repeated until a satisfactory solution is found. The basic steps of a simple PH are shown below:

```
Generate an initial population
Evaluate fitness of individuals in the population
repeat
    Select individuals from the population to be parents
    Recombine (mate) parents to produce children
    Mutate the children
    Evaluate fitness of the children
    Replace some or all of the population by the children
until
    you decide to stop whereupon report the best solution encountered
```

3. A Population Heuristic for the MKP

So far we have seen what the MKP problem is and have introduced PHs. How then can we bring the two together so as to produce a PH for the MKP?

3.1. *What Does an Individual in our PH World Look Like?*

In the real world we know what individuals look like. In the PH world, what individuals look like (their representation or chromosome) is our choice.

In our PH world for the MKP we shall choose individuals to be n bit binary strings (recall here that n is the number of columns in the MKP). Referring to our example with $n = 7$ shown in Figure 1:

bit (column)	1 2 3 4 5 6 7
individual	0 1 0 0 0 1 0

individual		fitness
1	1 1 0 0 0 0 0	4
2	1 0 0 1 0 0 0	6
3	0 0 0 0 0 0 1	5
4	0 0 1 0 1 0 0	8
5	0 1 1 0 0 0 0	5
6	0 1 0 0 0 1 0	7

Figure 2 An initial population containing six individuals for the multidimensional knapsack problem (MKP)

has an interpretation in terms of the MKP of $X_2 = X_6 = 1$ and $X_1 = X_3 = X_4 = X_5 = X_7 = 0$. Note especially here how many of the 0–1 integer decision problems we deal with in OR have a natural representation as binary bit strings in a PH.

3.2. How Shall We Choose to Evaluate the Fitness (Value, Worth) of Individuals?

In terms of the MKP this choice is clear. The objective function value ($\sum_{j=1}^{n} p_j X_j$) equates to how good a solution is, that is, its fitness. For example, consider the individual 0100010 mentioned above. We have:

column j	*1 2 3 4 5 6 7*
profit p_j	2 2 3 4 5 5 5
individual	0 1 0 0 0 1 0

Hence the fitness value for this individual is $2 + 5 = 7$ (and recall here that we are trying to maximize this fitness, see Equation (1)).

Figure 2 shows an initial population containing six distinct individuals and their associated fitnesses. In general, an initial population is randomly generated in some way. Note here that all the individuals shown in Figure 2 are feasible. As will become apparent below, generating feasible solutions for the MKP is a trivial task.

3.3. How Shall We Select Individuals from the Population to be Parents?

In the real world, individuals are independent beings who for their own reasons decide to become parents. But in the PH world we have to make a choice as to who will become a parent.

In our PH world for the MKP we shall choose to select parents by binary tournament selection. In binary tournament selection we first randomly select two individuals from the population (e.g., from Figure 2 individuals 1 and 5, with fitnesses 4 and 5 respectively). We then select from these two the individual with the best fitness to be the first parent (individual 5 in this case).

Repeating this procedure we might randomly select individuals 3 and 6 with fitnesses 5 and 7, respectively, so our second parent would be individual 6.

3.4. How Shall We Recombine (Mate) Parents to Produce Children?

In the real world we know how parents combine to have children. But how shall we choose to recombine (mate) parents to produce children in the PH world?

In our PH world for the MKP we shall have a single child from two parents by uniform crossover. In uniform crossover each bit in the child solution is created by:

```
repeat for each bit in turn:
    choose one of the two parents at random
    set the child bit equal to the bit in the chosen parent
```

For example, taking the two parents we generated previously (individuals 5 and 6 in Figure 2) we have:

parent 1 0 1 1 0 0 0 0

parent 2 0 1 0 0 0 1 0

child 0 1 ? 0 0 ? 0

In the above we have already filled in part of the child solution (since if the two bits are the same in both parents they are automatically the same in the child). However, the remaining bits (bits 3 and 6 above) must be filled in by random selection of a parent (where both parents have an equal probability of being chosen). Suppose we randomly select parent 1 for bit 3 and parent 2 for bit 6, then the child solution is 0110010.

3.5. Other Ways to Have Children

In the real world there are many different ways to have children. Currently there are fourteen different ways to conceive children (Aldridge, 1998). Similarly uniform crossover is not the only way to have children in the PH world. Other ways are (briefly) outlined below.

One-Point Crossover

In one-point crossover we randomly select a point between two adjacent bits, "cut" the parents into two segments and create two children by rejoining the segments. For example, cutting the parents we had before between bits 3 and 4:

parent	1 0 1 1 0 0 0 0	produces segments	0 1 1	and	0 0 0 0
parent	2 0 1 0 0 0 1 0	produces segments	0 1 0	and	0 0 1 0
to give		child 1	0 1 1		0 0 1 0
		child 2	0 1 0		0 0 0 0

where child 1 (0110010) is composed of the first segment of parent 1 and the second segment of parent 2; child 2 (0100000) is composed of the first segment of parent 2 and the second segment of parent 1.

Restricted One-Point Crossover

Observe that in the one-point crossover example presented above we would have produced children who were identical to the parents (duplicates, clones) if we had chosen to cut the parents between bits 1 and 2, bits 2 and 3: or bits 6 and 7. Restricted one-point crossover restricts the cut point to ensure that the children are different from the parents. This is easily done, simply restrict the cut point to be between the first bit where the two parents differ (bit 3 above) and the last bit where the two parents differ (bit 6 above).

Other Schemes

Other crossover schemes exist, for example:

- fusion, as uniform crossover except that bits are taken from the parents with probabilities proportional to their fitnesses;
- two-point crossover, as one-point crossover (where each parent was cut into two segments) except that each parent is cut into three segments and two children produced by taking alternate segments from each parent.

Indeed, any way of combining two bit strings together could be used to produce children from two parents. Note here however, that one property that crossover schemes typically have in common is that bits which are the same in the parents are the same in the children (i.e., characteristics common to both parents are passed to the children).

3.6. Mutation

Mutation corresponds to small changes that are stochastically applied to the children. Taking our child 0110010 produced by uniform crossover we could decide to make a small change, typically to randomly select one bit and to change its value ("flip it"). For example we might randomly select bit 2 and flip it to give 0010010. Alternatively, we might decide (according to some probabilistic criterion) to make no mutation changes to the child.

Mutation can be applied with a constant probability or with an adaptive probability that changes over the course of the algorithm (perhaps in response to the number of iterations that have passed or in response to population characteristics).

3.7. *Infeasibility*

One problem that must be addressed is that (most likely) not every individual (binary bit string) represents a feasible solution in terms of the underlying problem that is being solved, for example, for our example an individual may violate the constraints of the MKP.

Consider the child 0010010 that we produced by uniform crossover and mutation as discussed above. Hence we have:

column		j 1 2 3 4 5 6 7
child		0 0 1 0 0 1 0
row 1	r_{1j}	1 0 2 0 1 3 1 $\leqslant$ 4 (b_1)
row 2	r_{2j}	1 1 1 2 2 1 3 $\leqslant$ 3 (b_2)

It is clear that this child is infeasible, since the total resource consumed in row 1 is $2+3=5$, exceeding the four units of resource available.

There are a number of strategies for dealing with constraints and infeasible solutions in PHs and these are detailed in Table 1. Note here that the order in which the strategies are presented in Table 1 is the order that we would recommend to others (first try Strategy 1 on that list, if it fails try Strategy 2, etc.). In particular, we would regard use of a penalty function as a method of last resort.

The first strategy in Table 1 is to use a representation that automatically ensures that all solutions are feasible. For some problems such representations exist, for example, the set covering problem (Beasley and Chu, 1996), but for the majority of constrained problems this strategy is not possible. For the MKP this strategy is not possible.

The second strategy in Table 1 is to design a heuristic operator (often called in the literature a repair operator) that guarantees to quickly transform any infeasible solution into a feasible solution. Such a strategy is possible for the MKP and we illustrate this below.

We would note here, however, that our experience has been that dealing with constraints and infeasible solutions in PHs in an effective way can often demand skills in creative algorithm design.

3.8. *Heuristic Operator*

For the MKP, designing a heuristic operator that guarantees to quickly transform any infeasible solution into a feasible solution is trivial, for example,

```
repeat until solution feasible:
    take a randomly chosen column j with X_j = 1
    set X_j = 0
```

Table 1 Strategies for Dealing with Constraints and Infeasible Solutions in Population Heuristics

Strategy	*Description*
1	To use a representation that automatically ensures that all solutions are feasible.
2	To design a heuristic operator (often called in the literature a repair operator) that guarantees to quickly transform any infeasible solution into a feasible solution.
3	To separate the evaluation of fitness and infeasibility.
4	To apply a penalty function to penalize the fitness of any infeasible solution.

To illustrate that this operator works let us take the infeasible child 0010010 we had before and apply it. Since the child is infeasible we randomly select a column from {3, 6}. Suppose we select column 3, then the child becomes 0000010, which from Figure 1 is feasible and of fitness 5. It is clear that whatever solution we start with, the above procedure always guarantees a feasible solution (and is self-evidently a computationally quick procedure).

Note here that this heuristic operator can also be used to generate an initial population of feasible solutions (apply it to any set of random binary bit strings).

3.9. Population Replacement

Our feasible child 0000010 now needs to enter the population. We will use a steady-state population replacement strategy. With this strategy each new child is placed in the population as soon as it is ready (after mutation and application of the heuristic operator in this case). It is common in PHs to keep the population size constant. Hence placing the child in the population means selecting a member of the population to delete ("kill").

A logical approach is to kill the member of the population with the worst fitness. If we were to follow this approach then, from Figure 2, we would choose to kill individual 1 of fitness 4.

3.10. Population Heuristic Summary

Our PH can be summarized as:

```
Generate an initial population
Evaluate fitness of individuals in the population
repeat
    Select individuals from the population to be parents
    Recombine (mate) parents to produce children
    Mutate the children
    Make the children feasible using the heuristic operator
    Evaluate fitness of the children
    Replace some or all of the population by the children
until
    you decide to stop whereupon report the best solution encountered
```

where the difference between this description and the algorithmic description given previously (Section 2.3) is the insertion of a step making the (mutated) children feasible. The operation of this algorithm for one iteration, using the MKP example given above, can be seen in Figure 3.

3.11. Computational Considerations

The PH presented above (with a few modifications) produces results that are superior in quality to other leading heuristics (which are mostly based on tabu search) for the MKP (Chu and Beasley, 1998a). However, as already mentioned, that PH is much slower than other heuristics. Hence, we have the trade-off we often see in OR between quality of solution and computer time consumed.

As a result of our accumulated experience (Beasley and Chu, 1996; Chu and Beasley, 1997, 1998a,b; Chang et al., 2000; Beasley et al., 1999) we have come to believe that PHs, by their very nature, will never be the fastest heuristics in the world. Simply put, they often require the production and evaluation of many different children. However, our experience has been that PHs are capable of generating high-quality solutions to many problems within reasonable computation times.

3.12. Comments

One of the difficulties of working with PHs is that there are many schemes for doing the basic operations that you need:

Generate an initial population
Evaluate fitness of individuals in the population

individual		fitness
1	1 1 0 0 0 0 0	4
2	1 0 0 1 0 0 0	6
3	0 0 0 0 0 0 1	5
4	0 0 1 0 1 0 0	8
5	0 1 1 0 0 0 0	5
6	0 1 0 0 0 1 0	7

Select individuals from the population to be parents

Parents are individuals 5 and 6 after binary tournament selection

Recombine (mate) parents to produce children

Child after uniform crossover is 0110010

Mutate the children

Randomly select bit 2 to flip so child becomes 0010010

Make the children feasible using the heuristic operator

Child changes to 0000010

Evaluate fitness of the children

Child has fitness 5

Replace some or all of the population by the children

Individual 1 chosen for replacement so the population becomes

individual		fitness
1	0 0 0 0 0 1 0	5
2	1 0 0 1 0 0 0	6
3	0 0 0 0 0 0 1	5
4	0 0 1 0 1 0 0	8
5	0 1 1 0 0 0 0	5
6	0 1 0 0 0 1 0	7

Figure 3 Operation of the PH for the multidimensional knapsack problem (MKP) for one iteration

- *Parent selection*: binary tournament, K tournament, ranking, fitness proportionate, and so on.
- *Having children*: uniform, one-point, restricted one-point, fusion, two-point, and so on.
- *Mutation*: constant, adaptive.
- *Population replacement*: steady-state, generational (with/without elitism), island models.

Our experience over a number of different problems has been that sometimes the choice of scheme makes a significant difference, sometimes not.

For example, with respect to crossover, Table 2 gives the average percentage deviation from the best-known solution for different crossover operators (when used in the context of a complete algorithm) for different problems. The key point to note here is that although there are differences between each crossover column in Table 2 they are not extreme differences. If we were simply to choose one of the crossover operators at random then the results would be reasonable across all problems.

Table 2 Performance of Different Crossover Operators

	Crossover Operator *One-Point*	*Two-Point*	*Uniform*
65 set covering problems	0.02	0.03	0.06
9 hard set partitioning problems	2.97	2.51	2.07
10 hard generalized assignment problems	0.37	0.39	0.37

4. Equality Constrained MKP (ECMKP)

What happens if, instead of an inequality in the MKP with respect to the resource constraints (equation (2)), we insist on an equality, that is, that the resource consumed by the columns chosen must equal the resource available. In terms of our previous example given in Figure 1 this implies that we need to choose columns so as to consume exactly 4 units of resource in row 1 and 3 units of resource in row 2.

Reflecting on what we did for the MKP above it is clear that some steps in designing a PH for the ECMKP are "obvious."

1. We can use the same representation and fitness:

   bit (column) 1 2 3 4 5 6 7

   individual 0 1 0 0 0 1 0

   has an interpretation in terms of the ECMKP of $X_2 = X_6 = 1$, $X_1 = X_3 = X_4 = X_5 = X_7 = 0$ and has fitness 7.

2. Parent selection can be carried out as before.
3. Children can be produced using any of the standard schemes (uniform, one-point, etc.) as before.
4. Mutation can be carried out as before.

The key difficulty with designing a PH for the ECMKP lies with the constraints.

Recall Table 1, which shows the strategies that we recommend for dealing with constraints and infeasible solutions. For the MKP considered above we were able to use Strategy 2 in that table. For the ECMKP, however, unlike the MKP, it is impossible to design a heuristic operator that guarantees to quickly transform any infeasible solution into a feasible solution. Hence we need to resort to Strategy 3, to separate the evaluation of fitness and infeasibility. We illustrate this below.

4.1. *Unfitness*

We associate two numbers (unfitness, fitness) with each individual in our PH for the ECMKP where fitness is (as usual) the objective function value and unfitness is a measure of constraint violation. Unfitness is simply any convenient numeric measure (of our own devising) that tells us the degree of infeasibility of a solution. Typically, however, we define this unfitness measure to have some convenient properties:

1. Unfitness $\geqslant 0$ and unfitness $= 0$ if and only if an individual is feasible.
2. The higher the value of unfitness the "more infeasible" a solution is.

For the ECMKP a convenient unfitness measure that has these properties is:

$$\sum_{i=1}^{m} \left| \sum_{j=1}^{n} r_{ij} X_j - b_i \right|, \tag{4}$$

which is simply the sum of the absolute differences between the left-hand side and the right-hand side of the equality constraints.

Consider the individuals in the initial population shown in Figure 2. The first such individual is 1100000 and hence we have:

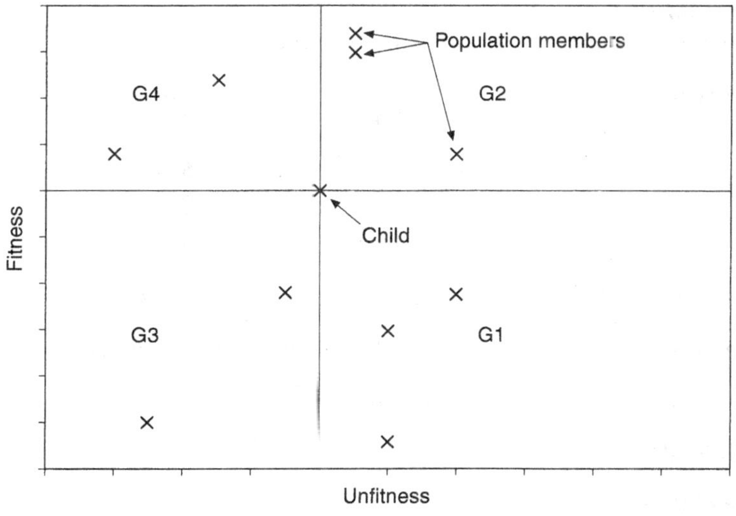

Figure 4 Population replacement

individual	1 1 0 0 0 0 0
profit p_j	2 2 3 4 5 5 5
row 1 r_{1j}	1 0 2 0 1 3 1 = 4 (b_1)
row 2 r_{2j}	1 1 1 2 2 1 3 = 3 (b_2)

so fitness = 4 (as before) and using equation (4) unfitness = $|1-4| + |2-3| = 4$. Since unfitness is not zero we automatically know that this individual is not feasible.

This use of two numbers (unfitness, fitness) for each individual is the key to dealing with the constraints here. Their use is clear when we consider placing a newly generated child in the population (steady-state approach).

Figure 4 shows the situation conceptually. In that figure we have plotted a newly generated child (which will have fitness and unfitness values). However, members of the population will also have fitness and unfitness values and these are also shown in Figure 4. We need to choose one member of the population to kill (replace by the child), but which one?

Suppose we were to draw, as in Figure 4, horizontal and vertical lines through the child. These divide the population into four groups (labelled G1 to G4 in Figure 4). Recall now that we wish to obtain solutions that have unfitness zero (i.e., are feasible) and which maximize fitness; colloquially, we wish to obtain solutions that are as high up the y-axis in Figure 4 as possible. Hence:

1. Any members of the population that lie in group G1 are worse than the child with respect to both measures (fitness and unfitness) and so choosing to kill a member from this group would be ideal.
2. It may be, however, that there are no members of the population in G1. Where should we look next? Plainly we would look in group G4 last, since members of this group are better than the child with respect to both measures. Whilst there are differences relating to whether we look in G2 or G3 first (e.g., see Chu and Beasley, 1998b, for a more detailed discussion of this) here we will look first in G2; then if that is empty, in G3; and finally if that is empty, in G4.

Note here that if we do have to look in G4 (because G1, G2, and G3 are empty) we will be replacing a member of the population by a child that is worse with respect to both measures. While this might seem perverse, our reasoning is that unless this is done the PH may fail to make progress.

Generate an initial population
Evaluate fitness and unfitness of individuals in the population

individual		fitness	unfitness
1	1 1 0 0 0 0 0	4	4
2	1 0 0 1 0 0 0	6	3
3	0 0 0 0 0 0 1	5	3
4	0 0 1 0 1 0 0	8	1
5	0 1 1 0 0 0 0	5	3
6	0 1 0 0 0 1 0	7	2

Select individuals from the population to be parents

Parents are individuals 5 and 6 after binary tournament selection

Recombine (mate) parents to produce children

Child after uniform crossover is 0110010

Mutate the children

Randomly select bit 7 to flip so child becomes 0110011

Evaluate fitness and unfitness of the children

Child has fitness 15 and unfitness 5

Replace some or all of the population by the children

Individual 1 chosen for replacement so the population becomes

individual		fitness	unfitness
1	0 1 1 0 0 1 1	15	5
2	1 0 0 1 0 0 0	6	3
3	0 0 0 0 0 0 1	5	3
4	0 0 1 0 1 0 0	8	1
5	0 1 1 0 0 0 0	5	3
6	0 1 0 0 0 1 0	7	2

Figure 5 Operation of the PH for the equality constrained multidimensional knapsack problem (ECMKP) for one iteration

Once the group in which to look has been decided as above, a logical approach is to kill the member of that group with the worst unfitness (i.e., most infeasible).

4.2. *Example*

The operation of our PH for the ECMKP for one iteration can be seen in Figure 5. Note that Figure 5, in algorithmic terms, is very similar to Figure 3. The differences are that we need to calculate unfitness (as well as fitness) when evaluating the initial population and each child; and that the step relating to making children feasible using the heuristic operator is absent.

Figure 6 shows the situation prevailing in Figure 5 when we come to choose a member of the population to kill. Note that groups G1 and G2 are empty so we choose an individual to kill from group G3. In Figure 5, we have chosen the member of G3 with the worst unfitness, individual 1, in accordance with the approach given above.

4.3. *Penalty Function*

The use in the literature of a penalty function to deal with constrained problems is quite widespread (Strategy 4 in Table 1). A penalty function requires a definition of unfitness

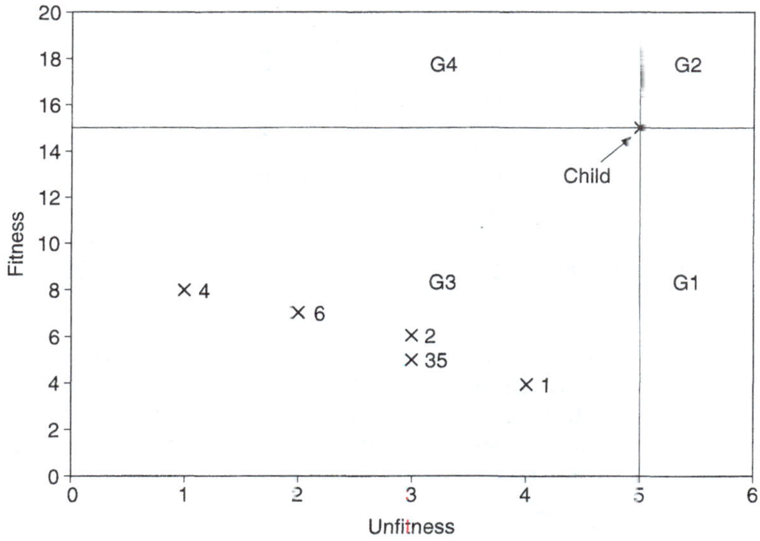

Figure 6 Population replacement—equality constrained multidimensional knapsack problem (ECMKP)

(constraint violation) and a modification of the fitness value by using a weighting parameter λ ($\geqslant 0$) so that penalized fitness value = original fitness value − λ(unfitness). These penalized fitness values are then used in place of the original fitness values (e.g., in binary tournament selection for parent selection).

We said before that we would regard use of a penalty function as a method of last resort. This is because it is dominated intellectually by Strategy 3 in conjunction with the population replacement scheme outlined above. Simply put, the penalty function approach requires:

1. A definition of unfitness, whereupon Strategy 3 is immediately applicable.
2. The setting of a parameter value λ, the appropriate value for which is inevitably data dependent.

In essence then the penalty function (Strategy 4) approach requires us to make more decisions than a Strategy 3 approach.

5. A Linear Program Related to MKP

Some OR problems contain continuous variables and to illustrate how a PH can be developed for such problems consider the linear program (LP) given below. For ease of illustration this LP is related to the MKP example considered previously.

$$\text{maximize} \quad \sum_{j=1}^{n} p_j x_j \tag{5}$$

$$\text{subject to} \quad \sum_{j=1}^{n} r_{ij} x_j \leqslant b_i \quad i = 1, \ldots, m \tag{6}$$

$$L_j \leqslant x_j \leqslant U_j \quad j = 1, \ldots, n \tag{7}$$

In this LP each variable j is a continuous variable restricted to lie in the range $[L_j, U_j]$. To make this LP clearer we present a small example with $n = 7$ variables (columns) and $m = 2$

column j	1	2	3	4	5	6	7
profit p_j	2	2	3	4	5	5	5
row 1 r_{1j}	1	0	2	0	1	3	1 $\leq$ 4 (b_1)
row 2 r_{2j}	1	1	1	2	2	1	3 $\leq$ 3 (b_2)
lower bound L_j	-1	0	1	-1	0	0	0
upper bound U_j	3	3	2	3	1	2	1

Figure 7 An example linear program (LP)

constraints (rows) in Figure 7. The key to solving this LP using a PH is the representation of the continuous variables and we consider this below.

Note here however that we are not advocating solving LPs via a PH. Simplex and interior point solvers are well established and obviously superior for that. Rather we are illustrating how PHs can deal with continuous variables.

5.1. Representation

In order to represent each continuous variable we use a real-numbered representation comprising n real numbers between 0 and 1 (inclusive). Letting Y_j represent the value for variable j then Y_j $(0 \leqslant Y_j \leqslant 1)$ represents the proportion of the interval $[L_j, U_j]$ that is less than X_j, that is, $Y_j = (X_j - L_j)/(U_j - L_j)$ so that $X_j = L_j + Y_j(U_j - L_j)$. For example, referring to Figure 7, if $Y_1 = 0.12$ then 0.12 (12 percent) of the interval from -1 to +3 is less than X_1 and so $X_1 = -1 + 0.12[3 - (-1)] = -0.52$.

This representation automatically ensures that all solutions are feasible with respect to the constraints relating to bounds on the variables (Equation (7), cf. Strategy 1 in Table 1). Typically OR problems formulated with continuous variables contain (explicitly or implicitly) variable bounds, so for such problems this representation is immediately applicable.

5.2. Other Components of a PH for our Example LP

It is clear that some steps in designing a PH for our example LP are "obvious".

1. Fitness can be computed as the objective function value, for example:

variable (column j)	1	2	3	4	5	6	7
individual (Y_j)	0.12	0.61	0.90	0.25	0.77	0.72	0.91
variable value (X_j)	−0.52	1.83	1.90	0	0.77	1.44	0.91

   has fitness 23.92

2. Parent selection can be carried out as before.
3. Children can be produced using any of the standard schemes (uniform, one-point, etc.) as before.
4. Mutation can be carried out as before. In fact mutation plays more of a key role here than it might appear at first sight. Consider the example population for this problem shown at the start of Figure 8. Under crossover the set of distinct real numbers shown there can never be changed, that is, no new real numbers can be introduced into the population. This plainly may be unsatisfactory. Hence the role of mutation with this representation is to introduce diversity (e.g., by making a small change—say ± 5 percent—to a real number we may introduce a new real number into the population).
5. The nature of this LP is that Strategy 2 in Table 1 can be applied. However, for the purposes of illustration, we will deal with the constraints (other than the bound constraint in Equation (7), which is automatically satisfied) using Strategy 3 via unfitness. For this LP [equations (5)–(7)] a convenient measure of unfitness is:

Generate an initial population
Evaluate fitness and unfitness of individuals in the population

individual								fitness	unfitness
1	0.12	0.61	0.90	0.25	0.77	0.72	0.91	23.92	11.20
2	0.05	0.94	0.13	0.44	0.23	0.01	0.93	16.37	4.94
3	0.80	0.80	0.45	0.60	0.70	0.02	0.02	22.95	9.29
4	0.01	0.23	0.45	0.94	0.76	0.73	0.88	30.35	13.28
5	0.25	0.44	0.68	0.69	0.36	0.84	0.27	26.27	11.76
6	0	0	0	0	0	0.20	0.40	1	0

Select individuals from the population to be parents

Parents are individuals 5 and 3 after binary tournament selection

Recombine (mate) parents to produce children

Child after uniform crossover is

0.25 0.80 0.68 0.60 0.36 0.84 0.27

Mutate the children

Randomly select variable 4 and an increase in value, increase it by 5% to 1.05(0.60) so child becomes

0.25 0.80 0.68 0.63 0.36 0.84 0.27

Evaluate fitness and unfitness of the children

Child has fitness 27.47 and unfitness 12.35

Replace some or all of the population by the children

Individual 4 chosen for replacement so the population becomes

individual								fitness	unfitness
1	0.12	0.61	0.90	0.25	0.77	0.72	0.91	23.92	11.20
2	0.05	0.94	0.13	0.44	0.23	0.01	0.93	16.37	4.94
3	0.80	0.80	0.45	0.60	0.70	0.02	0.02	22.95	9.29
4	0.25	0.80	0.68	0.63	0.36	0.84	0.27	27.47	12.36
5	0.25	0.44	0.68	0.69	0.36	0.84	0.27	26.27	11.76
6	0	0	0	0	0	0.20	0.40	1	0

Figure 8 Operation of the PH for our example linear program (LP) for one iteration.

$$\sum_{i=1}^{m} \max\left(0, \sum_{j=1}^{n} r_{ij}X_j - b_i\right) \quad (8)$$

so that, for example, the individual given in Step (1) above has unfitness 11.20

5.3. Example

The operation of our PH for this LP example can be seen in Figure 8. Figure 9 shows the situation prevailing in Figure 8 when we come to choose a member of the population to kill. Note that Figure 8, in algorithmic terms, is identical to Figure 5.

6. Advice

We commented above that one of the difficulties of working with PHs is that there are many schemes for doing the basic operations that you need. Our experience has been that unless one is careful, it is easy to "drown in possibilities" when trying to develop an effective PH. In this section we offer advice based on our experience.

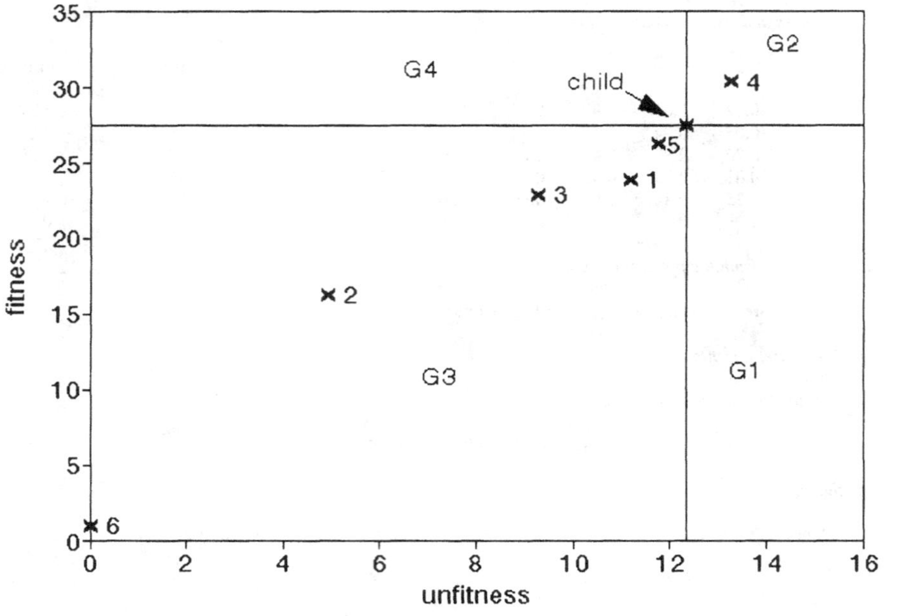

Figure 9 Population replacement—linear program (LP)

Our guiding principle in designing a PH is to make standard choices for the various components that you need, and only to try varying those choices if the computational results are not satisfactory. Below we indicate what, in our view, are good initial choices:

- population size: 100
- initial population: randomly generated
- parent selection: binary tournament
- crossover: uniform
- mutation: constant probability
- duplicates: rejected
- population replacement: steady state
- stopping criterion: based on the number of nonduplicate children generated
- replication: desirable

Note especially here that we advocate the rejection of duplicate children (i.e., they are not allowed to enter the population) and a stopping criterion based on the number of nonduplicate children generated (this equates to the number of children that have entered the population). We regard children as duplicates when they are identical to *any* member of the population.

Precise numeric values for parameters such as the probability of mutation or the number of nonduplicate children to generate before stopping are inevitably problem dependent and must be decided as a result of numeric experimentation.

However, we would say that in the PH world our advice is that having many thousands of children is perfectly acceptable (unlike the real world!).

Also note here that we believe that replicating (repeating) the PH is desirable. The easiest way of doing this is simply to rerun the PH using a different random number stream. As with all heuristic algorithms that contain stochastic elements improvement in the best solution found may be obtained when this is done.

However, by far the most important advice we can offer relates to Table 1. Recall that we said before that our experience has been that dealing with constraints and infeasible solutions in PHs in an effective way can often demand skills in creative algorithm design. To illustrate this we must now regard the strategies presented in Table 1 as strategies on top of which we:

always look for opportunities to improve the child solution in terms of its (unfitness, fitness) values by applying a suitable heuristic operator

As an illustration of this for the MKP the heuristic operator

```
for each column j in turn:
    set X_j = 1 - X_j (i.e., flip the bit) if by so doing we increase fitness
```

would be a suitable operator to improve the child solution (after it has been made feasible). Plainly it has no guarantee of success (compare Strategy 2 in Table 1) but it may improve the child solution. In standard OR terms such a heuristic operator is a local search (LS) heuristic. For the ECMKP an equivalent heuristic operator is:

```
for each column j in turn:
    set X_j = 1 - X_j (i.e., flip the bit) if by so doing we:
    (1)  reduce unfitness; or
    (2)  leave unfitness unchanged but increase fitness
```

For the LP example above an equivalent heuristic operator is:

```
for each column j in turn:
    set a step size γ_j ≥ 0
    examine both X_j = max(L_j, X_j - γ_j) and X_j = min(U_j, X_j + γ_j) and change X_j to one
    of these values if by so doing we:
    (1)  reduce unfitness; or
    (2)  leave unfitness unchanged but increase fitness
```

This heuristic corresponds to making a step change in a continuous variable (while keeping it within its bounds) if it is worthwhile to do so.

Our experience has been that simple, computationally cheap LS heuristics are indispensable in terms of developing computationally effective PHs.

Note too that the strategies presented in Table 1 may be mixed together. For example, in dealing with a commercial scheduling problem we used both Strategies 1 and 3. Strategy 1 was used to ensure that one set of constraints were automatically satisfied and Strategy 3 was used to deal with the other set via unfitness (the observant reader might well have noted that we did precisely this with our LP example above).

7. Common Population Heuristics

As mentioned above, there are a number of heuristic algorithms encountered in the literature that can generically be classified as PHs and in this section we discuss these and highlight their essential characteristics.

7.1. Genetic and Memetic Algorithms

Classical genetic algorithms, as developed by Holland (1975), did not incorporate any problem-specific knowledge (such as we need, for example, to develop a heuristic operator if we were to apply Strategy 2 from Table 1 to a problem). As the field advanced, however, it became clear that incorporating such knowledge was necessary for computational effectiveness and so the term hybrid genetic algorithm was born to distinguish the approach from a classical genetic algorithm. An essentially equivalent term is memetic algorithm (see Chapter 3.6.4 on memetic algorithms in this handbook). A hybrid genetic algorithm is also sometimes called a genetic local search algorithm.

7.2. *Scatter Search*

The scatter search approach of Glover (Glover, 1977, 1994a,b, 1995, 1997; Glover et al., 1995) has a number of distinct characteristics, specifically:

1. Each child is produced as a weighted linear combination of members of a parent set. Weights may typically be related to the relative fitness of parents (so that, for example, more fit parents have a higher weight).
2. Each child is subjected to a heuristic procedure:
    1. to restore a valid representation; and
    2. to improve fitness.

The step relating to restoring a valid representation is a crucial one. For example, with a binary representation, a weighted linear combination of the parents is unlikely to produce binary values for the child.

Scatter search can also be used with tabu search (Fleurent et al., 1996). In their approach there are essentially three sets of individuals:

1. An elite set E from which parents are chosen.
2. A tabu set T containing individuals who have been parents recently and who cannot currently participate in the process.
3. A remaining set R of individuals.

Individuals "circulate" between these sets, from R some are chosen to join E (which is of fixed cardinality). After they have been selected from E to be parents they join T. After an appropriate number of iterations have passed they rejoin R. Individuals disappear from the process only if they are in R and have been selected for replacement by a child.

We note in passing that this circulation system for population individuals can be used with any PH, not just with scatter search.

7.3. *Bionomic Algorithm*

The bionomic algorithm of Christofides (1994, 1996) is a PH based on a generational approach. It involves:

1. an explicit step "maturing/improving/growing" children, which can be done using a problem-specific heuristic or any of the generic LS approaches (greedy, tabu search, simulated annealing);
2. a structured approach to parent selection based upon population structure and the construction of parent sets; and
3. the generation of children from parent sets.

Parent Selection

Parent selection in a bionomic algorithm is as follows:

1. Choose an integer Δ and set $Q = \emptyset$.
2. Allocate an integer "frequency of inclusion" F_i ($\geqslant 0$) to each individual i in the population based upon their fitness.
3. Generate a graph G. G contains one node for each individual i with $F_i \geqslant 1$ and an edge between nodes i and j if the Hamming (bit) difference between individuals i and j is $\leqslant \Delta$ (i.e., if they have similar solution structures).
4. Generate a parent set P as a maximal independent set of G (an independent set in G is a set of nodes, none of which are adjacent; a maximal independent set is an independent set that cannot be enlarged); set $Q = Q \cup P$.
5. Set $F_i = F_i - 1$ for all individuals $i \in P$; remove from G any nodes i with $F_i = 0$ and go to (4) if G is not null.

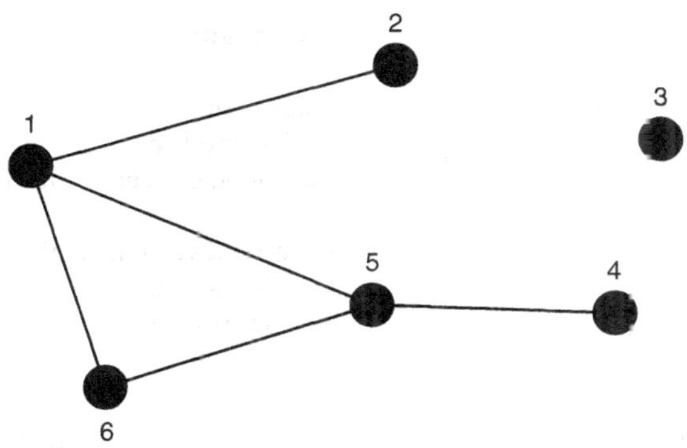

Figure 10 Graph for bionomic algorithm

6. We now have a set Q containing a number of parent sets.

For example, taking the population shown in Figure 2, the corresponding Hamming distances are:

	1	2	3	4	5	6
1	–	2	3	4	2	2
2		–	3	4	4	4
3			–	3	3	3
4				–	2	4
5					–	2

if $\Delta = 2$ G is as shown in Figure 10 (assuming $F_i \geqslant 1\ \forall i$) and there are three maximal independent sets: {1, 3, 4}, {2, 3, 5}, and {2, 3, 4, 6}. Note, however, that we need only to identify a single maximal independent set to form a parent set and this can be easily accomplished, see Bron and Kerbosch (1973) and Christofides (1975).

The logic behind this approach is that if the parents are similar (have a low Hamming distance) then after crossover we will effectively be exploring regions of the search space local to the parents. These "local regions" (solutions that lie within a Hamming distance Δ of one or other parent) will most likely have already been explored during the process of previously maturing the children that led to the current parents. Hence, we need to diversify the search. The parameter Δ controls the degree of diversification.

CHILDREN

Child generation in a bionomic algorithm is as follows:

1. Define an operator that, given a parent set P, generates a single child. This operator may include a stochastic element so that, for example, applying it more than once to the same parent set can lead to different children.
2. For each of the parents sets $P \in Q$ found previously:
   (a) decide how many children to have from P
   (b) apply the operator to P to generate the required number of children.

The bionomic algorithm is a generational approach and a subset of the generated children are used to replace the current population.

8. Conclusions

Clearly it has been impossible within the confines of this article to present everything there is to know about PHs. However, we hope that we have conveyed to the reader sufficient information for them to make a good start in applying such heuristics to their decision problems.

With regard to using PHs for OR problems, our conclusion is that population heuristics work. That is, population heuristics, suitably designed, are capable of generating high-quality solutions to OR problems. Certainly our experience, over a number of different problems, supports this view.

Finally, we would just like to say that we believe that in the next few years we will see more and more applications of PHs to OR problems and we hope that this article contributes to their understanding and use among the OR community.

ACKNOWLEDGMENTS

The author would like to acknowledge comments made by Peter Ross and Mhand Hifi on an earlier version of this article.

FURTHER READING

Useful references for those wanting to learn more are:

- for classical genetic algorithms: Mitchell (1996); Bäck et al. (1997)
- for genetic algorithms applied to OR problems: Reeves (1993b, 1997)
- for memetic algorithms: http://www.densis.fee.unicamp.br/~moscato/memetic_home.html
- for scatter search algorithms: Glover (1977, 1994a,b, 1995, 1997); Glover et al. (1995); Fleurent et al. (1996)
- for bionomic algorithms: Christofides (1994, 1996); Elia et al. (1997); Maniezzo et al. (1998).

References

Aarts, E. H. L. and J. K. Lenstra. (eds). 1997. *Local Search in Combinatorial Optimization*. Chichester, England: Wiley.

Aldridge, S. 1998. Ethical dilemmas. Inside Science Number 114. *New Scientist*, 17th October 1998.

Bäck, T., D. B. Fogel, and Z. Michalewicz. (eds). 1997. *Handbook of Evolutionary Computation*. Oxford: Oxford University Press.

Beasley, J. E. and P. C. Chu. 1996. "A genetic algorithm for the set covering problem." *Eur. J. Oper. Res.* **94**: 392–404.

Beasley, J. E., N. Meade, and T.-J. Chang. 1999. "Index tracking." Working paper (1999) available from the first author at The Management School, Imperial College, London SW7 2AZ, England.

Bron, C. and J. Kerbosch. 1973. "Algorithm 457—finding all cliques of an undirected graph." *Commun. ACM* **16**: 575–577.

Chang, T.-J., N. Meade, J. E. Beasley, and Y. M. Sharaiha. 2000. "Heuristics for cardinality constrained portfolio optimization." *Comp. Oper. Res.* **27**: 1271–1302.

Christofides, N. 1975. In *Graph Theory: an Algorithmic Approach*. London, England: Academic Press, pp. 33–35.

Christofides, N. 1994. "The bionomic algorithm." Presented at *AIRO'94*, Savona, Italy.

Christofides, N. 1996. "Optimal design of telecommunication networks." Presented at *CO96*, London, England.

Chu, P. C. and J. E. Beasley. 1997. "A genetic algorithm for the generalised assignment problem." *Comput. Oper. Res.* **24**: 17–23.

Chu, P. C. and J. E. Beasley. 1998a. "A genetic algorithm for the multidimensional knapsack problem." *J. Heuristics* **4**: 63–86.

Chu, P. C. and J. E. Beasley. 1998b. "Constraint handling in genetic algorithms: the set partitioning problem." *J. Heuristics* **4**: 323–357.

Elia, E., G. Salkin, and N. Christofides. 1997. "The bionomic algorithm for the tax structuring problem." Working paper (1997) available from the Centre for Quantitative Finance, Imperial College, London SW7 2AZ, England.

Fleurent, C., F. Glover, P. Michelon, and Z. Valli. 1996. "A scatter search approach for unconstrained continuous optimization." In *Proc. 1996 IEEE Int. Conf. Evolut. Comput.* New York: IEEE, pp. 643–648.

Glover, F. 1977. "Heuristics for integer programming using surrogate constraints." *Decision Sci.* **8**: 156–166.

Glover, F. 1994a. "Genetic algorithms and scatter search: unsuspected potentials." *Statistics and Computing* **4**: 131–140.

Glover, F. 1994b. "Tabu search for nonlinear and parametric optimization (with links to genetic algorithms)." *Discrete Appl. Math* **49**: 231–255.

Glover, F. 1995. "Scatter search and star-paths: beyond the genetic metaphor." *OR Spektrum* **17**: 125–137.

Glover, F. 1997. "A template for scatter search and path relinking." In *Artificial Evolution. Lecture Notes in Computer Science*, edited by J. K. Hao, E. Lutton, E. Ronald, M. Schoenauer and D. Synders, Volume 1363, pp. 3–51.

Glover, F., J. P. Kelly, and M. Laguna. 1995. "Genetic algorithms and tabu search: hybrids for optimization." *Comput. Oper. Res.* **22**: 111–134.

Glover, F. W. and M. Laguna. 1997. In *Tabu Search*. Dordrecht, The Netherlands: Kluwer Academic Publishers.

Holland, J. H. 1975. In *Adaptation in Natural and Artificial Systems: an Introductory Analysis with Applications to Biology, Control, and Artificial Intelligence*. Michigan: University of Michigan Press.

Maniezzo, V., A. Mingozzi, and R. Baldacci. 1998. "A bionomic approach to the capacitated p-median problem." *J. Heuristics* **4**: 263–280.

Mitchell, M. 1996. In *An Introduction to Genetic Algorithms*. Cambridge, Massachusetts: MIT Press.

Reeves, C. R. (ed.) 1993. *Modern Heuristic Techniques for Combinatorial Problems*. Oxford: Blackwell Scientific Publications.

Reeves, C. R. 1993. "Genetic algorithms." In *Modern Heuristic Techniques for Combinatorial Problems*, edited by C. R. Reeves, pp. 151–196. Oxford: Blackwell Scientific Publications.

Reeves, C. R. 1997. "Genetic algorithms for the Operations Researcher." *INFORMS J. Comp.* **9**: 231–250.

3.6.4 Memetic Algorithms

Pablo Moscato

ABSTRACT A memetic algorithm is a population-based approach for optimization that tries to synergetically use all possible information about the problem being addressed. This knowledge is generally, but not restricted to, some sort of previously-known heuristics or local search algorithms. We only describe here some of the most important applications of incomplete and stochastic memetic algorithm implementations when used as a metaheuristic. We particularly highlight their performance on "classical" combinatorial optimization problems. These results can be regarded as benchmarks of the proposed methodology in comparison with other metaheuristics. We also give pointers to related work in the areas of scheduling, timetabling, computational biology, and molecular design.

1. Introduction

For more than a decade now, the denomination memetic algorithms (MAs; Moscato, 1989) has been used to encompass a class of metaheuristics of proven practical success in a variety of optimization problems. In the last few years, many researchers have become more attracted to this methodology as a way out of some limitations present in other approaches. This is particularly the case with genetic algorithms (GAs) since MAs are orders of magnitude faster than GAs in some problem domains, a conspicuous case being the traveling salesman problem (TSP; Radcliffe and Surry, 1994b). For this reason, MAs are sometimes called hybrid genetic algorithms, Lamarkian evolutionary algorithms, or knowledge-augmented GAs. Other denominations include genetic local search, parallel genetic algorithms, or genetic search agents (Slootmaekers et al., 1998). These denominations are misleading but the MAs are easily recognizable from the so-called "standard GAs" since in MAs the new configurations/

solutions (or "children") are created from highly evolved "parents" of previous generations. The independent evolution of parents is done using problem-domain knowledge. In nonlinear settings this is generally implemented by using gradient information. In combinatorial optimization problems, MAs generally use some local search (LS) or constructive algorithm, but truncated exact procedures (like truncated branch-and-bound or branch-and-cut techniques) can also be used. In all cases, the existence of theorems and properties related to the structure of the optimum solution greatly help to reduce the whole search process.

The term "memetic" comes after Richard Dawkins' introduction of the term "meme" to denote an analogy to the gene in the field of cultural evolution. In a Caltech report written in 1989, we discussed several MAs for combinatorial optimization and we conjectured on the reasons that would explain the success over standard GA approaches. One of them was the working hypothesis of correlation of local optima (Moscato, 1989), a feature that can be efficiently exploited by the recombination operators. A later analysis of this hypothesis appeared in Moscato (1993), and more recently Merz and Freisleben 1998a; 1999b have been also studying these issues.

Currently, at `http://www.densis.fee.unicamp.br/~moscato/memetic_home.html` we constantly update a home page that contains links to all available papers and relevant bibliography for MAs. In addition, we provide information about several recent books that discuss MAs (Laguna and Moscato, 1996; Gen and Cheng, 1997; Corne et al., 1999).

As we said before, MAs and GAs are both population-based approaches to optimization (Moscato, 1993). In particular, the strategy adopted by MAs is inherently parallel, asynchronous, and demands little communication between processes running in a parallel computer. This results in a highly suitable approach to optimization in distributed computing systems composed of heterogeneous machines (e.g., networks of workstations).

1.1. Classification

We can classify MAs as a metaheuristic since we can view them as general-purpose strategies that guide other basic single-agent heuristic or truncated exact methods. Other metaheuristics, like simulated annealing (SA) or guided local search (GLS) (which guide some underlying "hill-climbing" procedures) are applied to a single optimizer and thus they cannot be categorized as population-based. Basic tabu search (TS) techniques also use a single optimizer, but more evolved metaheuristics like scatter search (SS), also use procedures based on "multiple-point" optimization. GAs also try to evolve solutions in configuration space by using a "population" of individuals which represent alternative solutions to the problems. Good reviews stressing the similarities and differences of these approaches can be found in Glover (1994, 1995) and Corne et al. (1999).

One of the main differences between GAs and MAs is that the latter approach tries to use all possible knowledge available to solve the problem. Most SS procedures in the literature can be classified as MAs. The multipoint approach of SS does not make much sense without local optimization. A more formal discussion of MAs and computational complexity can be found in Moscato (1999).

2. Application Areas

2.1. "Classical" NP Optimization Problems

MAs, although under different names, have a remarkable record of successes in a variety of classical NP optimization problems, such as graph partitioning (Bui and Moon, 1998; Merz and Freisleben, 1998a; 1999b), min number partitioning (Berretta and Moscato, 1999), max independent set (Aggarwal et al., 1997; Hifi, 1997; Sakamoto et al., 1997), bin packing (Reeves, 1996), min graph coloring (Costa et al., 1995; Fleurent and Ferland, 1997; Dorne and Rao, 1998; Galinier and Hao, 1998; Coll et al., 1999), set covering (Beasley and Chu, 1996), single machine scheduling with setup times and due dates (França et al., 1999), parallel machine scheduling (Cheng and Gen, 1997; Mendes et al., 1999), min generalized assignment (Chu and Beasley, 1997), multidimensional knapsack (Beasley and Chu, 1998), nonlinear integer programming

(Tagachi et al., 1998), quadratic assignment (Carrizo et al., 1992; Merz and Freisleben, 1997a, 1999a,b), set partitioning (Levine, 1996), and particularly on the min traveling salesman problem (Gorges-Shleuter, 1989, 1991b, 1997; Moscato and Norman, 1992; Radcliffe and Surry, 1994b; Freisleben and Merz, 1996a,b; Merz and Freisleben, 1997b; 1999b; Holstein and Moscato, 1999). Regarding the theory of NP-completeness, most of them can be catalogued as "classical" as they appeared in Karp's notorious 1972 paper on the reducibility of combinatorial problems. Remarkably, in most of them the authors claim that they have developed the best heuristic for the problem at hand. This is important since these problems have a long-standing tradition in combinatorial optimization and almost all general-purpose algorithmic techniques have been tested on them.

2.2. *Timetabling, Scheduling, and Other Applications*

Other applications of MAs include partial shape matching (Ozcan and Mohan, 1998), the binary perceptron learning problem (Moscato, 1993), Kauffman NK landscapes (Merz and Freisleben, 1998b), maintenance scheduling (Burke and Smith, 1997, 1999), timetabling (Burke et al., 1995; 1997, 1998; Paechter et al., 1996; Burke and Newall, 1997), university timetabling (Burke et al., 1996; Paechter et al., 1998), nurse rostering (DeCausmaecker et al., 1999), and sport games scheduling (Costa, 1995).

This list is very limited and its purpose is to document the wide applicability of the approach for combinatorial optimization.

2.3. *Molecular Optimization Problems*

We have selected a particular class of computational problems, involving nonlinear optimization issues, to help the reader to identify a common trend in the literature. Unfortunately, authors continue referring to their technique as genetic, although they are closer in spirit to MAs.

The Caltech report, which gave its name to the then incipient field of MAs (Moscato, 1989), discussed a metaheuristic that can be viewed as a hybrid of GAs and SA developed with M. G. Norman in 1988. In recent years, several papers have applied hybrids of GAs with SA or other methods to a variety of molecular optimization problems (Zacharias et al., 1998; Shankland et al., 1997, 1998; Lorber and Shoichet, 1998; Bayley et al., 1998; De Souza et al., 1998; White et al., 1998; Kariuki et al., 1997; Li et al., 1997; Gunn, 1997; Landree et al., 1997; Jones et al., 1997; Zwick et al., 1996; Dandekar and Argos, 1996; Miller et al., 1996; Doll and VanHove, 1996; MacKay, 1995; Tam and Compton, 1995). In all of these papers the authors refer to their metaheuristic as "genetic," probably being unaware of the existence of MAs. For instance, Fu et al. (1997) say that GAs "combined with conjugated gradient or molecular dynamics has been demonstrated to be efficient for the ground-state configuration search in materials research, for example fullerene formation, in this paper, based on the generalized tight-binding molecular dynamics, we apply the GA to study the surface reconstruction of Silicon (001) for the first time." Hybrid population approaches like this can hardly be catalogued as being genetic, but this denomination has appeared in previous work by Deaven and Ho (1995), which got extra recognition due to an article by J. Maddox in *Nature* (Maddox, 1995).

Other fields of application include cluster physics. Niesse and Mayne (1996) said, "In a recent paper, Gregurick, Alexander and Hartke (S. K. Gregurick, M. H. Alexander, and B. Hartke, *J. Chem. Phys.* **104**, 2684 [1996]) proposed a global geometry optimization technique using a modified Genetic Algorithm approach for clusters. They refer to their technique as a deterministic/stochastic genetic algorithm (DS-GA). In this technique, the stochastic part is a traditional GA, with the manipulations being carried out on binary-coded internal coordinates (atom-atom distances). The deterministic aspect of their method is the inclusion of a coarse gradient descent calculation on each geometry. This step avoids spending a large amount of computer time searching parts of the configuration space which correspond to high-energy geometries. Their tests of the technique show it is vastly more efficient than searches without this local minimization." Other papers in the area of cluster physics include Pucello et al.

(1997), Hobday and Smith (1997), Pullan (1997), White et al. (1998), Ho et al. (1998), Deaven et al. (1996).

Other evolutionary approaches to a variety of molecular problems can be found in vanKampen et al. (1996), Raymer et al. (1997), Doll and VanHove (1996), Meza et al. (1996), May and Johnson (1994), Hirsch and Mullergoymann (1995), and Hartke (1993). Their use for design problems is particularly appealing (see Kariuki et al., 1997; Willett, 1995; Clark and Westhead, 1996). They have also found a way into protein design (Desjarlais and Handel, 1995; Lazar et al., 1997), probably due to their landscape structure (Moscato, 1993).

3. Structure

To give an intuitive idea of what can be considered a genetic MA, we will introduce a pseudocode and describe it in some detail. A more formal introduction can be found in Moscato (1999).

The pseudocode in Figure 1 is helpful in understanding the general structure of an LS-based MA. The use of some reserved words is in order to take into account many MAs used in practice. Starting with `initializePopulation` these reserved words can be executed in different ways. Some authors initialize the population using some kind of constructive or randomized algorithms (or a combination of both).

In the pseudocode, we have also used functions, like `Local-Search-Engine()` for instance, which receives as input an individual and tries to improve it by iteratively using some local-search-based algorithm. The `Local-Search-Engine()` continues to return a better individual until it can no longer improve it or some other termination condition has

```
begin
  initializePopulation Pop using FirstPop();
  foreach individual i ∈ Pop do i := Local-Search-Engine(i);
  foreach individual i ∈ Pop do Evaluate(i);
  repeat /* generations loop */
    parfor j := 1 to #recombinations do
      selectToMerge a set S_par ⊆ Pop;
      offspring := Recombine(S_par, x);
      offspring := Local-Search-Engine(offspring);
      Evaluate(offspring);
      addInPopulation individual offspring to Pop;
    endparfor;
    parfor j := 1 to #mutations do
      selectToMutate an individual i ∈ Pop;
      i_m := Mutate(i);
      i_m := Local-Search-Engine(i_m);
      Evaluate(i_m);
      addInPopulation individual i_m to Pop;
    endparfor;
    Pop := SelectPop(Pop);
    if Pop meetsPopConvCriteria then Pop := RestartPop(Pop);
  until termination-condition=True;
end;
```

Figure 1 The local-search-based memetic algorithm

been reached. After the initial population has been created it is subsequently optimized and at least one recombination process involving more than one individual is performed. Some authors prefer all individuals to be involved in recombinations at each generation loop, so in their MAs `#recombinations` does not need to be specified as a numerical parameter. Other MAs also recombine the solutions with previous features of past populations, as with blackboard systems (Hogg and Williams, 1993), and the pocket-current methodology used in Moscato and Tinetti (1992), Paechter et al. (1996), and Berretta and Moscato (1999).

The reserved word parfor indicates operations that can be executed in parallel if possible. The `selectToMerge` is executed by selecting a subset of individuals (called S_{par}) to be used as input of the `Recombine()` algorithm. Some MAs use a random function for selecting parent configurations, while others use a more complex approach. For instance, some authors have been advocating for the benefits of using a population structure for interaction of individuals (Gorges-Schleuter, 1989, 1991a; Mühlenbein, 1991; Moscato and Norman, 1992; Moscato, 1993). More specifically, Moscato and Tinetti (1992), Paechter et al. (1996), Berretta and Moscato (1999), Mendes et al. (1999), and França et al. (1999) gave supporting evidence of the benefits of using a hierarchical population structure.

The `Recombine()` function is responsible for the creation of new individuals which are identified by `selectToMerge`. This can be implemented in a wide variety of ways, ranging from a single application of any available efficient (i.e., polynomial-time) algorithm up to more complex recombination operators (which means that they need not be efficient) such as the more systematic searches (Aggarwal et al., 1997). The symbol x, passed as input for the `Recombine()` function stands for the instance of the problem which is being addressed. For instance, if the problem is the TSP, then x represents the distance matrix between cities. Afterward, the new individual is optimized and added, or not, to the population according to some criteria (decided by `addInPopulation`). A typical criterion generally used is to avoid introducing a new individual that is a duplicate so as to maintain the diversity of the population.

Analogously, a number of selected individuals (identified by `selectToMutate`) are subjected to some mutations. It is easy to implement the `Mutate()` function as a random (biased or unbiased) process based on the same neighborhood definition implicit by the `Local-Search-Engine` function. However, this is not a strict rule and some other more clever forms of mutation are possible (such as those based on other neighborhood definitions; Berretta and Moscato, 1999). Again, the individual is optimized at the end of the mutation process and added, or not, to the population.

The `SelectPop()` function acts on the population, having the net effect of reducing its size. The selection of the subset to be preserved is not always determined by the objective function or the fitness function. Other features of the individuals are relevant to bias the selective pressure, for instance the interest in maximizing some measure of diversity of the selected set. Also in this case, the `SelectPop()` function can be implemented in a variety of ways. The convergence of the whole population is tested with `meetsPopConvCriteria`. This criteria might include the lack of diversity of a certain subpopulation (appropriately defined when using a population structure). This situation is generally referred in the literature as a diversity crisis. It is a measure that indicates, when it goes under a certain threshold, that the whole population (or subpopulation) has very similar configurations. In this case a `RestartPop()` function is used. In general, it seems reasonable to preserve the best individual (or incumbent solution) and a new population is created using a randomized biased procedure. All the individuals of the population are optimized and evaluated afterward and the whole process is repeated.

The termination-condition for an MA (as is the case of most metaheuristics) can also be implemented in many ways. It can be a time-expiration or generation-expiration criteria as well as more adaptive procedures, like a dynamic measure of lack of improvement.

4. Implementation

In this section, we would like to give several guidelines on how to build up an MA from several software components. Since MAs can eventually be composed of highly evolved interacting codes, we believe that a few general recommendations can help the practitioner in the development process.

We have remarked on several occasions that MAs constitute an "exercise of humbleness and common sense." If we have several alternative algorithms, heuristics, and metaheuristics for the problem, it seems to be a wise idea to use them together. MAs offer the practitioner the possibility of integrating existing codes they have been relying on, but that only enables locally optimal solutions and requires some extra global optimization (for instance, by adding stochasticity and some sort of controlled or biased randomness to the search process).

4.1. General Issues

In the process of developing an MA for a problem, we generally go through the following steps:

- *Approximate solutions.* The first step generally is to write a program that allows generation of approximate solutions (feasible, if possible) for the problem. For instance, this program can be implementing a constructive heuristic. We also check if there is some possibility of randomizing the procedure so as to create an initial population of solutions (so as to be the base to build a `FirstPop()` procedure).
- *Local optimality.* The second task is to write a program that, given as input an approximate solution, can evolve it toward feasibility and (if applicable) improve it according to the objective function of the associated optimization problem. This will be the core function for the `Local-Search-Engine()` procedure. Again, it is important to check if there is some possibility of randomizing the procedure. It may be useful that different individuals run different types of search algorithms. Truncated exact procedures, like branch-and-bound or some other kind of systematic search method, can also be used.
- *Test multistart optimization.* Before adding new features, like recombination of solutions, the next step is to test the average performance of what has been developed so far. It is essential to develop some kind of tests that will help to show the benefits statistically of the addition of future software components.
- *Test an MA without recombination.* Based on the pseudocode of Figure 1, it might be convenient to implement an MA that only differs from the pseudocode in that the `Recombine()` function is never used. At this step it is necessary to establish how several things are going to be implemented, for instance, if a population structure for the interaction of individuals will be used, and so on. This means that all the reserved words as well as the functions (like `Mutate()` and `RestartPop()`) must be implemented. Several different mutation and starting procedures can be compared against the multi-start optimization tests.
- *Recombination.* Considering the results so far, and especially the type of `Mutate()` and `Local-Search-Engine()` functions that have been implemented, the last step is to finally find a recombination procedure [for the `Recombine()` function] which gives significant improvement over the previous results.

4.2. Further Remarks

After these issues have been addressed, it is possible to identify other features present in the implementation. The following remarks might help to address additional problems most commonly found.

- *Control loss of diversity.* This is an important issue in MAs. In general, when the power of the independent search performed by the agents is increased, we expect that diversity (regarding some *ad hoc* measure of similarity between solutions) should decrease. This can be addressed in several ways: for instance, by increasing the role of the `Mutate()` function (Merz and Freisleben, 1999b; 2000); by using a recombination mechanism that takes this into account (such as the distance preserving crossover proposed in Merz and Freisleben, 1997b); or by letting diversity drop but using an efficient scheme for `RestartPop()` (such as the heavy mutation alternative proposed in Burke et al., 1996). Other possibilities are offered by choosing some kind of population structure allowing isolation by distance (Norman and Moscato, 1989; Gorges-Schleuter, 1989, 1991a; Moscato and Norman, 1992; Berretta and Moscato, 1999) or avoiding recombining solutions if they are too similar (Eshelman, 1991). Another possibility is to extend the basic recombination mechanism with conciliator, rebel, and obsequent behaviors as in Berretta and Moscato (1999).

- *Excessive concern for diversity*. An MA that does not evolve in time allowing a clear diversity loss might be a very inefficient stochastic search procedure. A good rule of thumb is first to design an MA with diversity loss before becoming concerned about how to increase/preserve it. Otherwise, a too-dispersed population might not find a way to exploit useful information via recombination.
- *Transient period*. During the first iterations of the so-called generation loop of an MA, it may be the case that the population is not able to generate new incumbent solutions. This transitory period where little evolution is observed can be shortened by selecting a different `FirstPop()` function. The reason is that sometimes a good (regarding the objective function) set of initial solutions conforms the initial population, but it is too diverse. Subsequent recombinations and reoptimizations do little more than resampling (in a random uniform way) the configuration space between configurations. This initial population is then not gaining any advantage from the correlation of local optima (Moscato, 1993; Radcliffe and Surry, 1994a; Boese, 1995; Merz and Freisleben, 1998a). Generating a good initial solution and perturbing it to a certain extent as in Holsten and Moscato (1999) seems to be a good way of handling this phenomenon.
- *A properly diversified restart*. Some authors have implemented MAs that are based on other metaheuristics, which guide the agents (Moscato and Norman, 1992; Costa, 1995; DeCausmaecker et al., 1999). The most commonly used are SA and TS. The latter has been proposed in Moscato (1993) as a good mechanism for increasing the diversity of the population. However, it should be taken into account that the population should diversify but not too much. The reasoning behind this relies on the same argument that we used to explain the initial transient period. When a restart mechanism creates a too diversified population, it shows lack of improvement since useful correlations of local optima are not exploited by the recombination and individual search processes (Moscato, 1989; Radcliffe and Surry, 1994a; Merz and Freisleben, 1999b).
- *Keep a good generic code*. Most implementations of MAs can be easily written in such a way that allows software reuse. A good practice is to keep the general rules that govern the metaheuristic (and the code associated to them) the same across different problems. This helps to identify potential pitfalls of your methodology and can provide insights for new (generic) heuristic problem-solving strategies.

ACKNOWLEDGMENTS

I would like to acknowledge useful discussions with other members of the MemePool Project, and in particular with R. Berretta, L. Buriol, P. M. França, D. Holstein, A. S. Mendes, P. Merz, M. G. Norman, and A. Schaerf. Many thanks to all the authors who are helping to make the Memetic Algorithms Home Page a useful resource. Although this article does not attempt to be a survey, I apologize to anyone inadvertently left out. This work is supported by FAPESP, Brazil.

References

Aggarwal, C. C., J. B. Orlin, and R. P. Tai. 1997. "Optimized crossover for the independent set problem." *Oper. Res.* **45(2)**: 226–234.

Bayley, M. J., G. Jones, P. Willett, and M. P. Williamson. 1998. "Genfold: a genetic algorithm for folding protein structures using nuclear magnetic resonance restraints." *Protein Sci.* **7(2)**: 491–499.

Beasley, J. and P. C. Chu. 1996. "A genetic algorithm for the set covering problem." *Eur. J. Oper. Res.* **94(2)**: 393–404.

Beasley, J. and P. C. Chu. 1998. "A genetic algorithm for the multidimensional knapsack problem." *J. Heuristics* **4**: 63–86.

Berretta, R. and P. Moscato. 1999. "The number partitioning problem: an open challenge for evolutionary computation?" In *New Ideas in Optimization,* edited by D. Corne, M. Dorigo, and F. Glover. Maidenhead, Berkshire, England: McGraw-Hill.

Boese, K. D. 1995. "Cost versus distance in the traveling salesman problem." Technical Report TR-950018. UCLA Computer Science Department, Los Angeles, CA: University of California at Los Angeles.

Bui, T. N. and B. R. Moon. 1998. "GRCA: a hybrid genetic algorithm for circuit ratio-cut partitioning." *IEEE Trans. Comput.-Aided Des. Integrat. Circuits Sys.* **17(3)**: 193–204.

Burke, E. K. and J. P. Newall. 1997. "A phased evolutionary approach for the timetable problem: an initial study." In *Proceedings of the ICONIP/ANZIIS/ANNES'97 Conference, 24–28 November 1997, Dunedin, New Zealand,* pp. 1038–1041. Springer.

Burke, E. K. and A. J. Smith. 1997. "A memetic algorithm for the maintenance scheduling problem." In *Proceedings of the ICONIP/ANZISS/ANNES'97 Conference, 24–28 November 1997, Dunedin, New Zealand.* pp. 469–472. Springer.

Burke, E. K. and A. J. Smith. 1999. "A memetic algorithm to schedule grid maintenance." In *Proceedings of the International Conference on Computational Intelligence for Modelling Control and Automation, Vienna: Evolutionary Computation and Fuzzy Logic for Intelligent Control, Knowledge Acquisition and Information Retrieval, 17–19 February 1999,* pp. 122–127. IOS Press.

Burke, E. K., D. G. Elliman, and R. F. Weare. 1995. "A hybrid genetic algorithm for highly constrained timetabling problems." In *6th International Conference on Genetic Algorithms (ICGA'95, Pittsburgh, 15–19 July 1995),* pp. 605–610. San Francisco: Morgan Kaufmann.

Burke, E. K., J. P. Newall, and R. F. Weare. 1996. "A memetic algorithm for university exam timetabling." In *The Practice and Theory of Automated Timetabling. Lecture Notes in Computer Science,* Volume 1153, edited by E. K. Burke and P. Ross, pp. 241–250. Springer Verlag.

Burke, E. K., K. S. Jackson, J. H. Kingston, and R. F. Weare. 1997. "Automated timetabling: the state of the art." *Commut. J.* **40(9)**: 565–571.

Burke, E. K., J. P. Newall, and R. F. Weare. 1998. "Initialisation strategies and diversity in evolutionary timetabling." *Evolut. Comput.* **6(1)**: 81–103.

Carrizo, J., F. G. Tinetti, and P. Moscato. 1992. "A computational ecology for the quadratic assignment problem." In *Proceedings of the 21st Meeting on Informatics and Operations Research, Buenos Aires, 1992.* Buenos Aires, Argentina: SADIO.

Cheng, R. W. and M. Gen. 1997. "Parallel machine scheduling problems using memetic algorithms." *Comput. Ind. Eng.* **33(3-4)**: 761–764.

Chu, P. C. and J. Beasley. 1997. "A genetic algorithm for the generalised assignment problem." *Comput. Oper. Res.* **24**: 17–23.

Clark, D. E. and D. R. Westhead. 1996. "Evolutionary algorithms in computer-aided molecular design." *J. Comput-aided Molec. Des.* **10(4)**: 337–358.

Coll, P. E., G. A. Durán, and P. Moscato. 1999. "On worst-case and comparative analysis as design principles for efficient recombination operators: a graph coloring case study." In *New Ideas in Optimization,* edited by D. Corne, M. Dorigo, and F. Glover. Maidenhead, Berkshire, England: McGraw-Hill.

Corne, D., M. Dorigo, and F. Glover. 1999. In *New Ideas in Optimization.* Maidenhead, Berkshire, England: McGraw-Hill.

Costa, D. 1995. "An evolutionary tabu search algorithm and the NHL scheduling problem." *INFOR* **33(3)**: 161–178.

Costa, D., N. Dubuis, and A. Hertz. 1995. "Embedding of a sequential procedure within an evolutionary algorithm for coloring problems in graphs." *J. Heuristics* **1(1)**: 105–128.

Dandekar, T. and P. Argos. 1996. "Identifying the tertiary fold of small proteins with different topologies from sequence and secondary structure using the genetic algorithm and extended criteria specific for strand regions." *J. Molec. Biol.* **256(3)**: 645–660.

De Souza, P. A., R. Garg, and V. K. Garg. 1998. "Automation of the analysis of Mossbauer spectra." *Hyperfine Interact.* **112(1-4)**: 275–278.

Deaven, D. M. and K. O. Ho. 1995. "Molecular-geometry optimization with a genetic algorithm." *Phys. Rev. Lett.* **75(2)**: 288–291.

Deaven, D. M., N. Tit, J. R. Morris, and K. M. Ho. 1996. "Structural optimization of Lennard-Jones clusters by a genetic algorithm." *Chem. Phys. Lett.* **256(1-2)**: 195–200.

DeCausmaecker, P., G. VandenBerghe, and E. K. Burke. 1999. "Using tabu search as a local heuristic in a memetic algorithm for the nurse rostering problem." In *Proceedings of the Thirteenth Conference on Quantitative Methods for Decision Making, ORBEL 13, 21–22 January 1999*, Brussels. Abstract only, poster presentation.

Desjarlais, J. R. and T. M. Handel. 1995. "New strategies in protein design." *Curr. Op. Biotechnol.* **6(4)**: 460–466.

Doll, R. and M. A. VanHove. 1996. "Global optimization in LEED structure determination using genetic algorithms." *Surface Sci.* **355(1-3)**: L383–L398.

Dorne, R. and J. K. Hao. 1998. "A new genetic local search algorithm for graph coloring." In *Parallel Problem Solving From Nature V*, Volume 1498, pp. 745–754. Berlin, Germany.

Eshelman, L. J. 1991. "The CHC adaptive search algorithm: how to have safe search when engaging in non-traditional genetic recombination." In *Foundations of Genetic Algorithms*, edited by G. J. E. Rawlings, pp. 265–283. San Mateo, CA: Morgan Kaufmann.

Fleurent, C. and J. A. Ferland. 1997. "Genetic and hybrid algorithms for graph coloring." *Ann. Oper. Res.* **63**: 437–461.

França, P. M., A. S. Mendes, and P. Moscato. 1999. "Memetic algorithms to minimize tardiness on a single machine with sequence-dependent setup times." In *Proceedings of the 5th International Conference of the Decision Sciences Institute*, Atlanta, CA: Decision Sciences Institute.

Freisleben, B. and P. Merz. 1996a. "A genetic local search algorithm for solving symmetric and asymmetric traveling salesman problems." In *Proceedings of the 1996 IEEE International Conference on Evolutionary Compuation*, pp. 616-621. Piscataway, NJ: IEEE Press.

Freisleben, B. and P. Merz. 1996b. "New genetic local search operators for the traveling salesman problem." In *Proceedings of the 4th Conference on Parallel Problem Solving from Nature—PPSN IV. Lecture Notes in Computer Science*, Volume 1141. Edited by H.-M. Voigt, W. Ebeling, I. Rechenberg, and H.-P. Schwefel, pp. 890–900. Springer.

Fu, R. T., K. Esfarjani, Y. Hashi, J. Wu, X. Sun, and Y. Kawazoe. 1997. "Surface reconstruction of Si (001) by genetic algorithm and simulated annealing method." *Sci. Rep. Res. Inst Tohoku Univ. Series A-Phys. Chem. Metall.* **44(1)**: 77–81.

Galinier, P. and J. K. Hao. 1999. "Hybrid evolutionary algorithms for graph coloring." *J. Combinat. Optim.* **3(4)**: 379–397.

Gen, M. and R. Cheng. 1997. In *Genetic Algorithms and Engineering Design*. Wiley Series in Engineering Design and Automation: John Wiley & Sons (Sd).

Glover, F. 1994. "Genetic algorithms and scatter search—unsuspected potentials." *Stat. Comput.* **4(2)**: 131–140.

Glover, F. 1995. "Scatter search and star-paths—beyond the genetic metaphor." *OR Spektrum* **17(2-3)**: 125–137.

Gorges-Schleuter, M. 1989. ASPARAGOS: an asynchronous parallel genetic optimization strategy. In *Proceedings of the Third International Conference on Genetic Algorithms*, edited by J. David Schaffer, pp. 422–427. San Mateo, CA: Morgan Kaufmann Publishers.

Gorges-Schleuter, M. 1991a. "Explicit parallelism of genetic algorithms through population structures." In *Parallel Problem Solving From Nature*, edited by H. Schwefel and R. Manner, pp. 150–159. Berlin, Germany: Springer-Verlag.

Gorges-Schleuter, M. 1991b. "Genetic algorithms and population structures—a massively parallel algorithm." Ph.D. diss. Dortmund, Germany: University of Dortmund.

Gorges-Schleuter, M. 1997. "Asparagos 96 and the traveling salesman problem." In *Proceedings of the 1997 IEEE International Conference on Evolutionary Computation*. Piscataway, NJ: IEEE Press.

Gunn, J. R. 1997. "Sampling protein conformations using segment libraries and agenetic algorithm." *J. Chem. Phys.* **106(10)**: 4270–4281.

Hartke, B. 1993. "Global geometry optimization of clusters using genetic algorithms." *J. Phys. Chem.* **97(39)**: 9973–9976.

Hifi, M. 1997. "A genetic algorithm-based heuristic for solving the weighted maximum independent set and some equivalent problems." *J. Oper. Res. Soc.* **48(6)**: 612–622.

Hirsch, R. and C. C. Mullergoymann. 1995. "Fitting of diffusion-coefficients in a 3-compartment sustained-release drug formulation using a genetic algorithm." *Int. J. Pharma.* **120(2)**: 229–234.

Ho, K. M., A. A. Schvartsburg, B. C. Pan, Z. Y. Lu, C. Z. Wang, J. G. Wacker, J. L. Fye, and M. F. Jarrold. 1998. "Structures of medium-sized silicon clusters." *Nature* **392(6676)**: 582–585.

Hobday, S. and R. Smith. 1997. "Optimisation of carbon cluster geometry using a genetic algorithm." *J. Chem. Soc. Faraday Trans.* **93(22)**: 3919–3926.

Hogg, T. and C. P. Williams. 1993. "Solving the really hard problems with cooperative search." In *Proceedings of AAAI93*, pp. 231–236. Menlo Park, CA: AAAI Press.

Holstein, D. and P. Moscato. 1999. "Memetic algorithms using guided local search: a case study." In *New Ideas in Optimization*, edited by D. Corne, M. Dorigo, and F. Glover. Maidenhead, Berkshire, England: McGraw-Hill.

Jones, G., P. Willett, R. C. Glen, A. R. Leach, and R. Taylor. 1997. "Development and validation of a genetic algorithm for flexible docking." *J. Molec. Biol.* **267(3)**: 727–748.

Kariuki, B. M., H. Serrano-Gonzales, R. L. Johnston, and K. D. M. Harris. 1997. "The application of a genetic algorithm for solving crystal structures from powder diffraction data." *Chem. Phys. Lett.* **280(3-4)**: 189–195.

Laguna, M. and P. Moscato. 1996. "Capitulo 3: algoritmos geneticos." In *Las Nuevas Técnicas Heurísticas y las Redes Neuronales*, edited by B. A. Diaz, pp. 67–103. Madrid: Ed. Paraninfo.

Landree, E., C. Collazo-Davila, and L. D. Marks. 1997. "Multi-solution genetic algorithm approach to surface structure determination using direct methods." *Acta Crystallogr. Sec. B* **53**: 916–922.

Lazar, G. A., J. R. Desjarlais, and T. M. Handel. 1997. "De novo design of the hydrophobic core of ubiquitin." *Protein Sci.* **6(6)**: 1167–1178.

Levine, D. 1996. "A parallel genetic algorithm for the set partitioning problem." In *Meta-Heuristics: Theory & Applications*, edited by I. H. Osman and J. P. Kelly, pp. 23–35. Boston, MA: Kluwer Academic Publishers.

Li, L. P., T. A. Darden, S. J. Freedman, B. C. Furie, B. Furie, J. D. Baleja, H. Smith, R. G. Hiskey, and L. G. Pedersen. 1997. "Refinement of the nuclear magnetic resonance solution structure of the gamma-carboxyglutamic acid domain of coagulation factor IX using molecular dynamics simulation with initial Ca^{2+} positions determined by a genetic algorithm." *Biochemistry* **36(8)**: 2132–2138.

Lorber, D. M. and B. K. Shoichet. 1998. "Flexible ligand docking using conformational ensembles." *Protein Sci.* **7(4)**: 938–950.

MacKay, A. L. 1995. "Generalized crystallography." *THEOCHEM-J. Molec. Struct.* **336(2-3)**: 293–303.

Maddox, J. 1995. "Genetics helping molecular-dynamics." *Nature* **376(6537)**: 209.

May, A. C. W. and M. S. Johnson. 1994. "Protein-structure comparisons using a combination of a genetic algorithm, dynamic-programming and least-squares minimization." *Protein Eng.* **7(4)**: 475–485.

Mendes, A. S., F. M. Muller, P. M. França, and P. Moscato. 1999. "Comparing meta-heuristic approaches for parallel machine scheduling problems with sequence-dependent setup times." In *Proceedings of the 15th International Conference on CAD/CAM Robotics & Factories of the Future, 18–20 August 1999, Aguas de Lindoia, Brazil.* Technological Center for Informatics Foundation, Campinas, SP, Brazil.

Merz, P. and B. Freisleben. 1997a. "A genetic local search approach to the quadratic assignment problem." In *Proceedings of the 7th International Conference on Genetic Algorithms*, edited by T. Bäck, pp. 465–472. San Francisco, CA: Morgan Kaufmann.

Merz, P. and B. Freisleben. 1997b. "Genetic local search for the TSP: new results." In *Proceedings of the 1997 IEEE International Conference on Evolutionary Computation*, pp. 159–164. Piscataway, NJ: IEEE Press.

Merz, P. and B. Freisleben. 1998a. "Memetic algorithms and the fitness landscape of the graph bi-partitioning system." In *Proceedings of the 5th Conference on Parallel Problem Solving from Nature—PPSN V. Lecture Notes in Computer Science*, Volume 1498, edited by A. E. Eiben, T. Back, M. Schoenauer and H.-P. Schwefel, pp. 765–774. Berlin, Germany: Springer.

Merz, P. and B. Freisleben. 1998b. "On the effectiveness of evolutionary search in high-dimensional NK-landscapes." In *Proceedings of the 1998 IEEE International Conference on Evolutionary Computation*, pp. 741–745. Piscataway, NJ: IEEE Press.

Merz, P. and B. Freisleben. 1999a. "A comparison of memetic algorithms, tabu search, and ant colonies for the quadratic assignment problem." In *Proceedings of the 1999 International Congress of Evolutionary Computation (CEC'99), Washington DC*, pp. 2063–2070. Piscataway, NJ: IEEE Press.

Merz, P. and B. Freisleben. 1999b. "Fitness landscapes and memetic algorithm design." In *New Ideas in Optimization*, edited by D. Corne, M. Dorigo, and F. Glover. Maidenhead, Berkshire, England: McGraw-Hill.

Merz, P. and B. Freisleben. 2000. "Fitness landscapes: memetic algorithms and greedy operators for graph bi-partitioning." Technical Report 98-01. Siegen, Germany: University of Siegen. *Evolut. Comput.* **8(1)**: 61–91.

Meza, J. C., R. S. Judson, R. T. Faulkner, and A. M. Treasurywala. 1996. "A comparison of a direct search method and a genetic algorithm for conformational searching." *J. Comput. Chem.* **17(9)**: 1142–1151.

Miller, S. T., J. M. Hogle, and D. J. Filman. 1996. "A genetic algorithm for the ab initio phasing of icosahedral viruses." *Acta Crystallogr. Sect. D* **52**: 235–251.

Moscato, P. 1989. "On evolution, search, optimization, genetic algorithms and martial arts: towards memetic algorithms." Technical Report Caltech Concurrent Computation Program, Report. 826. Pasadena, CA: California Institute of Technology.

Moscato, P. 1993. "An introduction to population approaches for optimization and hierarchical objective functions: the role of tabu search." *Ann. Oper. Res.* **41(1-4)**: 85–121.

Moscato, P. 1999. "Memetic algorithms: a short introduction." In *New Ideas on Optimization*, edited by D. Corne, M. Dorigo, and F. Glover. Maidenhead, Berkshire, England McGraw-Hill.

Moscato, P. and M. G. Norman. 1992. "A memetic approach for the traveling salesman problem implementation of a computational ecology for combinatorial optimization on message-passing systems." In *Parallel Computing and Transputer Applications*, edited by M Valero, E. Onate, M. Jane, J. L. Larriba, and B. Suarez, pp. 177–186. Amsterdam: IOS Press.

Moscato, P. and F. Tinetti. 1992. "Blending heuristics with a population-based approach: a memetic algorithm for the traveling salesman problem." Report 92-12. Universidad Nacional de La Plata, La Plata, Argentina.

Mühlenbein, H. 1991. "Evolution in time and space—the parallel genetic algorithm." In *Foundations of Genetic Algorithms*, edited by J. E. Rawlins, pp. 316–337. San Mateo, CA: Morgan Kaufmann Publishers.

Niesse, J. A. and H. R. Mayne. 1996. "Global geometry optimization of atomic clusters using a modified genetic algorithm in space-fixed coordinates." *J. Chem. Phys.* **105(11)**: 4700–4706.

Norman, M. G. and P. Moscato. 1989. "A competitive and cooperative approach to complex combinatorial search." Technical Report Caltech Concurrent Computation Program, Report 790. California Institute of Technology, Pasadena, CA. Expanded version published at the Proceedings of the 20th Informatics and Operations Research Meeting, Buenos Aires (20th JAIIO), August 1991, pp. 3.15–3.29.

Ozcan, E. and C. K. Mohan. 1998. "Steady state memetic algorithm for partial shape matching." In *Evolutionary Programming VII—7th International Conference, (EP98, San Diego, CA, 25–27 March 1998) Lecture Notes in Computer Science*, Volume 1447, edited by V. W. Porto, N. Saravanan and D. Waagen, pp. 527–536. Berlin: Springer-Verlag.

Paechter, B., A. Cumming, M. G. Norman, and H. Luchian, 1996. "Extensions to a memetic timetabling system." In *The Practice and Theory of Automated Timetabling. Lecture Notes in Computer Science*, Volume 1153, edited by E. K. Burke and P. Ross, pp. 251–265. Berlin: Springer-Verlag.

Paechter, B., R. C. Rankin, and A. Cumming. 1998. "Improving a lecture timetabling system for university wide use." In *The Practice and Theory of Automated Timetabling II. Lecture Notes in Computer Science*, Volume 1408, edited by E. K. Burke and M. Carter, pp. 156–165. Berlin: Springer-Verlag.

Pucello, N., M. Rosati, G. D'Agostino, F. Pisacane, V. Rosato, and M. Celino. 1997. "Search of molecular ground state via genetic algorithm: implementation on a hybrid SIMD-MIMD platform." *Int. J. Modern Phys. C* **8(2)**: 239–252.

Pullan, W. J. 1997. "Structure prediction of benzene clusters using a genetic algorithm." *J. Chem. Inf. Comput. Sci.* **37(6)**: 1189–1193.

Radcliffe, N. J. and P. D. Surry. 1994a. "Fitness variance of formae and performance prediction." In *Proceedings of the Third Workshop on Foundations of Genetic Algorithms*, edited by L. D. Whitley and M. D. Vose, pp. 51–72. San Francisco: Morgan Kaufmann.

Radcliffe, N. J. and P. D. Surry. 1994b. "Formal memetic algorithms." In *Evolutionary Computing: AISB Workshop. Lecture Notes in Computer Science*, ed. T. Fogarty, pp. 1–16. Berlin: Springer-Verlag.

Raymer, M. L., P. C. Sanschagrin, W. F. Punch, S. Venkataraman, E. D. Goodman, and L. A. Kuhn. 1997. "Predicting conserved water-mediated and polar ligand interactions in proteins using a k-nearest neighbors genetic algorithm." *J. Molec. Biol.* **265(4)**: 445–464.

Reeves, C. 1996. "Hybrid genetic algorithms for bin-packing and related problems." *Ann. Oper. Res.* **63**: 371–396.

Sakamoto, A., X. Z. Liu, and T. Shimamoto. 1997. "A genetic approach for maximum independent set problems." *IEICE Trans. Fund. Electron. Commun. Comput. Sci.* **E80A(3)**: 551–556.

Shankland, K., W. I. F. David, and T. Csoka. 1997. "Crystal structure determination from powder diffraction data by the application of a genetic algorithm." *Z. Kristallogri.* **212(8)**: 550–552.

Shankland, K., W. I. F. David, T. Csoka, and L. McBride. 1998. "Structure solution of ibuprofen from powder diffraction data by the application of a genetic algorithm combined with prior conformational analysis." *Int. J. Pharm.* **165(1)**: 117–126.

Slootmaekers, R., H. Van Wulpen, and W. Joosen. 1998. "Modelling genetic search agents with a concurrent object-oriented language." In *Proceedings of High-Performance Computing and Networking: International*

Conference and Exhibition, Amsterdam, The Netherlands, 21–23 April 1998. Lecture Notes in Computer Science. Volume 1401, edited by P. Sloot, M. Bubak, and B. Hertzberger, pp. 843–853. Berlin: Springer-Verlag.

Taguchi, T., T. Yokota, and M. Gen. 1998. "Reliability optimal design problem with interval coefficients using hybrid genetic algorithms." *Comput. Ind. Eng.* **35(1-2)**: 373–376.

Tam, K. Y. and R. G. Compton. 1995. "GAMATCH—a genetic algorithm-based program for indexing crystal faces." *J. Appl. Crystallogr.* **28**: 640–645.

VanKampen, A. H. C., C. S. Strom, and L. M. C. Buydens. 1996. "Lethalization, penalty and repair functions for constraint handling in the genetic algorithm methodology." *Chemom. Intell. Lab. Syst.* **34(1)**: 55–68.

White, R. P., J. A. Niesse, and H. R. Mayne. 1998. "A study of genetic algorithm approaches to global geometry optimization of aromatic hydrocarbon microclusters." *J. Chem. Phys.* **108(5)**: 2208–2218.

Willett, P. 1995. "Genetic algorithms in molecular recognition and design." *Trends Biotechnol.* **13(12)**: 516–521.

Zacharias, C. R., M. R. Lemes, and A. D. Pino. 1998. "Combining genetic algorithm and simulated annealing: a molecular geometry optimization study." *'THEOCHEM-J. Molec. Struct.*, **430**(29–39).

Zwick, M., B. Lovell, and J. Marsh. 1996. "Global optimization studies on the 1-d phase problem." *Int. J. General Syst.* **25(1)**: 47–59.

3.6.5 Greedy Randomized Adaptive Search Procedures

Leonidas S. Pitsoulis and Mauricio G. C. Resende

ABSTRACT GRASP (greedy randomized adaptive search procedure) is a metaheuristic for combinatorial optimization. GRASP is usually implemented as a multistart procedure, where each iteration is made up of a construction phase, where a randomized greedy solution is constructed, and a local search phase, which starts at the constructed solution and applies iterative improvement until a locally optimal solution is found. This article gives an overview of GRASP. Besides describing the basic building blocks of a GRASP, the article covers enhancements to the basic procedure, including reactive GRASP, hybrid GRASP, and intensification strategies.

1. Introduction

Consider a combinatorial optimization problem where one is given a discrete set X of solutions and an objective function $f(x) : x \in X \rightarrow \mathbb{R}$ to be minimized and seeks a solution $x^* \in X$ such that $f(x^*) \leqslant f(x)$, for all $x \in X$. Problems of this type are sometimes easy to solve, that is, they can be solved in polynomial time, but more often polynomial time algorithms (Garey and Johnson, 1979) to solve them are not known and one usually resorts to heuristics that are not guaranteed to find an optimal solution in polynomial time.

Local search (see Section 3.5 in this handbook) is based on searching a local neighborhood for an improving solution. Given a solution $x \in X$, the elements of the neighborhood $N(x)$ of x are those solutions that can be obtained by applying an elementary modification (often called a move) to x. For example, consider $x = (0,1,0)$ and the 1-flip neighborhood of a 0/1 array, where neighbors are all 0/1 arrays that differ from x by exactly one element. For this example $N(x) = \{(0,0,0),(1,1,0),(0,1,1)\}$. Now consider $x = (2,1,3)$ and 2-swap neighborhood of a permutation array. We have $N(x) = \{(3,1,2),(2,3,1),(1,2,3)\}$. Local search starts from an initial solution $x^0 \in X$ and iteratively generates a series of improving solutions $x^1, x^2, ..., x^k$. At the kth iteration, $N(x^k)$ is searched for an improving solution x^{k+1} such that $f(x^{k+1}) < f(x^k)$. If such a solution is found, it is made the current solution. Otherwise, the search ends with x^k as a local optimum.

The effectiveness of local search depends on several factors, such as the neighborhood structure, the function to be minimized, and the starting solution. A solution x is said to be in the basin of attraction of the global optimum if local search starting from x leads to the global optimum. Once the neighborhood and objective function are determined, different starting solutions can be used to start the local search in a multistart procedure. If the starting solution is in the basin of attraction of the global optimum, local search finds the global optimum. Otherwise, a nonglobal local minimum is found. By repeatedly using a randomly generated solution as the starting solution for local search, one can expect to eventually produce a starting solution in the basin of attraction of the global optimum.

Randomly generated solutions are on average of poor quality. Even if a randomly generated solution is in the basin of attraction of a good quality local optimal solution, the number of moves needed to reach the local optimum can be large, even exponential in the problem size (Johnson et al., 1988). The greedy algorithm usually produces solutions of better quality than those of randomly generated solutions. A greedy algorithm constructs a solution, one element at a time. At each step in the construction, a set C of candidate elements that can be added to the solution at that step is constructed. A greedy function is applied to each candidate element and the elements are ranked according to their greedy function values. A best ranked element is added to the solution. Taking this into account, the set C is updated, causing the greedy function to change. The process is repeated until $C = \emptyset$.

With randomly generated initial solutions, a multistart method will eventually find a global optimum. However, using greedy solutions as starting points for local search in a multistart procedure will usually lead to good, though most often suboptimal, solutions. This is because the amount of variability in greedy solutions is small and it is less likely that a greedy starting solution will be in the basin of attraction of a global optimum. If there are no ties in the greedy function values, or if a deterministic rule is used to break ties, there is no variability and a multistart procedure would produce the same solution at each iteration. A semi-greedy heuristic (Hart and Shogan, 1987; Feo and Resende, 1989) adds variability to the greedy algorithm. After the candidate elements are ranked according to their greedy function values, well-ranked candidate elements are placed in a restricted candidate list (RCL) and an element from the RCL is selected at random and is added to the solution.

Hart and Shogan (1987) and Feo and Resende (1989) proposed two schemes to build an RCL. In the cardinality-based scheme, an integer k is fixed and the k top-ranked candidates are placed in the RCL. In the value-based scheme, all candidate elements with greedy function values within 100 α percent of the greedy value are placed in the RCL, where $\alpha \in [0, 1]$. A scheme that is less prone to the scaling problem was proposed by Resende et al. (2000), where the RCL elements are those that are within $\alpha(\bar{g} - \underline{g})$ of the greedy function value, where $\bar{g} = \max\{g(c) \mid c \in C\}$ and $\underline{g} = \min\{g(c) \mid c \in C\}$. For a minimization problem, $\text{RCL} = \{c \in C \mid g(c) \leqslant \underline{g} + \alpha(\bar{g} - \underline{g})$. Note that if $\alpha = 0$, then the semi-greedy construction reduces to a greedy algorithm, and if $\alpha = 1$, it is a random construction.

We can now describe a basic greedy randomized adaptive search procedure (GRASP) (Feo and Resende, 1989, 1995) for a minimization problem. Let x^* be the best solution found and $f^* = f(x^*)$.

1. $f^* = \infty$;
2. Repeat until a stopping criterion is satisfied:
   (a) Generate a greedy randomized solution x;
   (b) Find local optimum x_l with local search starting from x;
   (c) If $f(x_1) < f^*$ then
      (i) $f^* = f(x_l)$;
      (ii) $x^* = x_l$;

2. The Restricted Candidate List

The cardinality-based RCL construction mechanism has not been frequently used in the literature. Most GRASP implementations have used some type of value-based RCL construction scheme. In such a scheme, the RCL parameter α determines the level of greediness or randomness in the construction. Early implementations of GRASP used a fixed value for α, usually determined through experimentation. In some cases, however, simply changing the problem class, or the instance of a particular problem class, required a different parameter value. Mockus et al. (1997) pointed out that, excluding the case of random construction, a GRASP with a fixed RCL parameter value may not converge (asymptotically) to a global minimum because the construction mechanism may exclude all solutions in the basin of attraction of a global minimizer.

A few remedies for this have been proposed. Resende et al. (2000) proposed using a different randomly generated $\alpha \in \text{UNIF}[0,1]$ in each GRASP iteration. A mechanism for a self-adjusting α was proposed by Prais and Ribeiro (2000) in a strategy they named Reactive GRASP. As in Resende et al. (2000), a different randomly generated α is used in each GRASP iteration. However, instead of being chosen from a uniform distribution, α is chosen from a discrete set of values $\{\alpha_1, \alpha_2, ..., \alpha_m\}$. The probability that α_k is selected is $p(\alpha_k)$. Reactive GRASP adaptively changes the probabilities $\{p(\alpha_1), p(\alpha_2), ..., p(\alpha_m)\}$ to favor values of α that produce good solutions. Let f^* be the value of the best solution found so far during a GRASP run and let α_i be the average value of the solutions found using α_i in the construction phase. Periodically (e.g., every N_α GRASP iterations), the values $q_i = f^*/a_i$ and $p(\alpha_i) = q_i / \sum_{j=1}^{m} q_j$ are computed for $i = 1, ..., m$. Note that the more suitable a value of α_i is, the larger the value of q_i is, and consequently, the higher the value of $p(\alpha_i)$, making α_i more likely to be selected.

An alternative to using a RCL was proposed by Bresina (1996). In this scheme, candidates are not excluded from selection during construction, but are assigned probabilities of being selected that are determined by their greedy function values. The greedy function $g(c)$ is applied to all candidate elements $c \in C$ and a bias b_r is assigned to the rth ranked element. Several biases are proposed. In logarithmic bias, $b_r = 1/\log(r+1)$; in linear bias, $b_r = 1/r$; in polynomial (n) bias, $b_r = 1/r^n$; in exponential bias, $b_r = 1/e^r$; and in random bias, $b_r = 1$. The probability that the rth ranked candidate element is selected is $p(r) = b_r / \sum_{j=1}^{|C|} b_j$. Binato et al. (2001) used bias function to select an element from the RCL.

Glover and Laguna (1997) referred to the observation that "good solutions at one level are likely to be found 'close to' good solutions at an adjacent level" as the proximate optimality principle (POP). Fleurent and Glover (1999) pointed out that POP can be interpreted in relation to GRASP. Imperfections introduced during the steps of GRASP construction can be ironed out by applying local search during (and not only at the end of) construction. POP local search is not applied after each construction iteration, but rather two or three times during construction (Fleurent and Glover, 1999; Binato et al., 2001). This avoids the reduction in solution variability that would result if it is applied after each construction step.

3. Long-Term Memory

One possible shortcoming of the standard GRASP framework is the independence of the GRASP iterations, that is, the fact that GRASP does not learn from the history of solutions found in previous iterations. This is so because the standard GRASP discards information about any solution encountered that does not improve upon the incumbent. Besides reactive GRASP, which can be viewed as a long-term memory strategy, other strategies that make use of information gathered throughout GRASP iterations have been proposed, either to speed up the solution process, avoiding unnecessary searches, or to intensify the search around good quality solutions.

A hash table (Aho et al., 1974; Cormen et al., 1990) is a data structure for implementing dictionaries (dynamic sets with the operations of insert, delete, and search). The expected time to search an element in a hash table is $O(1)$, which makes hash tables a computationally effective data structure. A hash table is used in GRASP to keep track of all solutions used as initial solutions for local search. After construction of the greedy randomized solution, the hash table is consulted to verify if the solution is new. If it is, it is added to the hash table, and local search is applied starting from that solution. Otherwise, local search is skipped, and a new GRASP iteration is started.

Path relinking was first introduced in the context of tabu search (Glover and Laguna, 1997) as an approach to integrate intensification and diversification strategies in the search. See Glover et al. (2000) for a survey of path relinking. It explores trajectories that connect high quality solutions by starting from an initial solution and generating a path in the neighborhood of this solution toward another solution, called the guiding solution. This path is generated by selecting moves that introduce attributes of the guiding solution in the initial solution. At each step, all moves that incorporate attributes of the guiding solution are analyzed and the move that best improves (or least deteriorates) the initial solution is chosen. Path relinking in the context of GRASP was first introduced by Laguna and Martí (1999).

A refinement of GRASP with path relinking was made by Aiex et al. (2000a). A pool P of elite solutions is formed with the solutions found in the first $|P|$ GRASP iterations. After this initial phase, each solution s_g produced by the GRASP local search phase is relinked with one or more elite solutions. Given s_g and elite solution s_e, two paths are generated, one from s_g to s_e, and another from s_e to s_g. Two paths are used because they often visit different intermediate solutions. Two strategies for selecting s_e are used. The first is the one proposed by Laguna and Marti (1999), where s_e is selected at random from the pool. The second relinks s_g with all elite solutions in P.

Laguna and Martí (1999) update their pool by maintaining in it three best quality solutions. Aiex et al. (2000a) used an approach proposed by Fleurent and Glover (1999) for using elite solutions within the GRASP framework. The elite set is made up of $p = |P|$ elements. Let c_{best} and c_{worst} be the objective function values of the best and the worst solutions in P, respectively. Given two solutions s and t, let $\Delta(s,t)$ be a measure of dissimilarity of solutions s and t.

Solution s_{gmin} output from the path relinking procedure is a candidate for insertion into the pool and is accepted if it satisfies one of the following acceptance criteria:

1. $c_{\text{gmin}} < c_{\text{best}}$, that is s_{gmin} is the best solution found so far;
2. $c_{\text{best}} < c_{\text{gmin}} < c_{\text{worst}}$ and for all elite solutions $s_p \in P$, $\Delta(s_{\text{gmin}}, s_p) > \delta$, where δ is a cutoff parameter, that is, s_{gmin} is better than the worst elite solution and differs significantly from all elite solutions.

Once accepted for insertion into P, s_{gmin} replaces the worst elite solution, which is discarded from P.

Path relinking can also be used as an intensification phase for the elite set. This is accomplished by applying path relinking to each pair of elite solutions in the pool and updating the pool if necessary. The procedure is repeated until no further change in the pool occurs. This type of intensification can be done in a postoptimization phase (using the final pool of elite solutions), or periodically during the optimization (using the current set of elite solutions).

When applying path relinking as a postoptimization step, after no further change in the elite set occurs, the local search procedure is applied to each elite solution, as the solutions produced by path relinking are not always local optima. The local optima found are candidates for insertion into the elite set. If a change in the elite set occurs, the entire postprocessing step is repeated.

Another use of the information obtained from the good solutions is to implement a memory-based procedure to influence the construction phase by modifying the probabilities assigned to each RCL element. Fleurent and Glover (1999) introduced a memory-based scheme that uses long-term memory in multistart heuristics such as GRASP. Their scheme maintains a set P of elite solutions to be used in the construction phase. As in the description of the elite set for path relinking, for a solution s to become elite, it must be either better than the best member of P, or better than the worst member of P, and sufficiently different from the other elite solutions. For example, one can count identical solution vector components and set a threshold for rejection. A strongly determined variable is one that cannot be changed without eroding the objective or significantly changing other variables. A consistent variable is one that receives a particular value in a large portion of the elite solution set. Let $I(c)$ be a measure of the strongly determined and consistent features of choice c, that is, $I(c)$ becomes larger as e resembles solutions in elite set P. The intensity function $I(c)$ is used in the construction phase as follows. Recall that $g(c)$ is the greedy function. Let $E(c) = F[g(c), I(c)]$ be a function of the greedy and the intensification functions. For example, $E(c) = \lambda g(c) + I(c)$. The intensification scheme biases selection from the RCL to those elements c with a high value of $E(c)$ by setting the probability of selecting c to be $p(c) = E(c)/\sum_{s\in RCL} E(s)$. The function $E(c)$ can vary with time by changing the value of λ, for example, initially λ is set to a large value and, when diversification is called for, λ is decreased. A procedure for changing the value of λ is given by Fleurent and Glover (1999). See also Binato et al. (2001) for an application of this long-term memory strategy.

4. Grasp in Hybrid Metaheuristics

It is easy to incorporate GRASP in a hybrid metaheuristic. Since there is freedom to choose the local search algorithm, a number of metaheuristics can be chosen for this task. See Laguna and González-Velarde (1991), Colomé and Serra (1998), Delmaire et al. (1999), and Liu et al. (2000) for examples of the use of tabu search and simulated annealing as a GRASP local search procedure. Ahuja et al. (1998) use local search with large neighborhoods in a GRASP framework. Variable neighborhood search (Hansen and Mladenović, 1998) is another candidate for hybridization with GRASP.

Path relinking, which was described in the previous section, can be incorporated into GRASP to form a powerful metaheuristic (Laguna and Martí, 1999; Aiex et al., 2000a).

GRASP has also been used in conjunction with genetic algorithms as a mechanism to generate initial solutions (Ahuja et al., 2000), as well as in cross-over (Lourenço et al., 1998).

5. Parallel GRASP

As with any multistart heuristic for combinatorial optimization, a GRASP can be implemented in parallel by dividing the k independent iterations among ρ processors. Another approach is to give a target value τ of the objective function to each processor and execute the algorithm until the first processor finds a solution with a value at least as good as τ at which point all processors halt. Some care is needed to assure that no two iterations start with identical random number generator seeds (Pardalos et al., 1996). These are examples of multiple independent walk parallelism (Verhoeven and Aarts, 1995).

Many parallel implementations of GRASP using the above strategies have been reported in the literature (e.g., Martins et al., 1998, 2000; Murphey et al., 1998b; Pardalos et al., 1995, 1996). In most of these papers, a common observation was made. The speedups in the measured running times were proportional to the number of processors. A typical example can be seen in Pardalos et al. (1996) where, for a PVM-based implementation of a parallel GRASP for the MAXSAT, average speedups that were almost identical to the number of processors were measured (see Table 1).

This observation can be explained if the random variable solution time to target is exponentially distributed, as indicated by the following proposition (Verhoeven and Aarts, 1995).

Proposition 1. *Let* $P_\rho(t)$ *be the probability of not having found a given (target) solution in* t *time units with* ρ *independent processes. If* $P_1(t) = e^{-t/\lambda}$ *with* $\lambda \in \mathbb{R}^+$, *that is* P_1 *corresponds to an exponential distribution, then* $P_\rho(t) = e^{-\rho t/\lambda}$.

The above proposition follows from the definition of the exponential distribution. It implies that the probability of finding a solution of a given value in time ρt with a sequential process is equal to the probability of finding a solution at least as good as that given value in

Table 1 CPU Time (in seconds) and Speedup on MAXSAT Problems. Average Speedup is Shown for Five Ten, and Fifteen Processors

	1 Processor	*5 Processors*		*10 Processors*		*15 Processors*	
Problem	*Time*	*Time*	*Speedup*	*Time*	*Speedup*	*Time*	*Speedup*
jnh201	310.4	62.8	4.9	30.5	10.2	22.2	14.0
jnh202	312.2	59.8	5.2	31.2	10.0	23.4	13.3
jnh203	351.2	72.3	4.9	35.2	10.0	23.2	15.1
jnh205	327.8	63.4	5.2	32.1	10.2	22.5	14.6
jnh207	304.7	56.7	5.4	29.6	10.3	19.8	15.4
jnh208	355.2	65.6	5.4	33.2	10.7	21.0	16.9
jnh209	339.0	60.5	5.6	33.6	10.1	21.6	15.7
jnh210	318.5	57.6	5.5	32.5	9.8	20.8	15.3
jnh301	414.5	85.3	4.9	45.2	9.2	28.3	14.5
jnh302	398.7	88.6	4.5	48.2	8.3	27.0	14.7
Average Speedup			5.2		9.9		15.0

time t with ρ independent parallel processes. Hence, it is possible to achieve linear speedup in solution time to target solution by multiple independent processes.

An analogous proposition can be stated for a two parameter (shifted) exponential distribution.

Proposition 2. *Let* $P_\rho(t)$ *be the probability of not having found a given (target) solution in t time units with* ρ *independent processes. If* $P_1(t) = e^{-(t-\mu)/\lambda}$ *with* $\lambda \in \mathbb{R}^+$ *and* $\mu \in \mathbb{R}$, *that is* P_1 *corresponds to a two parameter exponential distribution, then* $P_\rho(t) = e^{-\rho(t-\mu)/\lambda}$.

Analogously, this proposition follows from the definition of the two parameter exponential distribution. It implies that the probability of finding a solution of a given value in time ρt with a sequential process is equal to $1-e^{-(\rho t-\mu)/\lambda}$, while the probability of finding a solution at least as good as that given value in time t with ρ independent parallel processes is $1-e^{-\rho(t-\mu)/\lambda}$. Note that if $\mu = 0$, then both probabilities are equal and correspond to the nonshifted exponential distribution. Furthermore, if $\rho\mu \ll \lambda$, then the two probabilities are approximately equal and it is possible to achieve approximately linear speedup in solution time to target solution by multiple independent processes.

Aiex et al. (2000b) studied the empirical probability distributions of the random variable time to target solution in five GRASP implementations. They showed that, given a target solution value, the time it takes GRASP to find a solution at least as good as the target fits a two-parameter exponential distribution. As an example, Figure 1 shows a plot of the empirical and estimated theoretical distributions for GRASP on a particular instance. This plot was generated using 200 independent runs of GRASP and the methodology described in Aiex et al. (2000b).

6. Applications of GRASP

Since the late 1980s, GRASP has been applied to a wide range of operations research and industrial optimization problems. These include problems in scheduling, routing, logic, partitioning, location and layout, graph theory, assignment, manufacturing, transportation, telecommunications, automatic drawing, electrical power systems, and VLSI design. Festa and Resende (2001) presented an extensive annotated bibliography of GRASP literature. In this section, we will review a small part of these applications, limiting ourselves to applications in logic, assignment, and location.

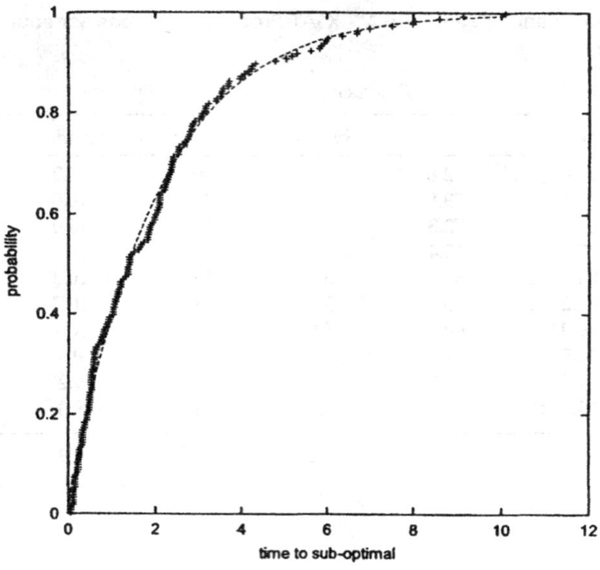

Figure 1 Superimposed empirical and theoretical distributions

6.1. Logic

THE SATISFIABILITY PROBLEM

Let $\mathscr{C}_1, \mathscr{C}_2, ..., \mathscr{C}_m$ be m clauses, where each clause $\mathscr{C}_i$ contains n_i boolean variables $x_1, x_2, ..., x_{n_i}$, which can take on only the values true or false (1 or 0). Each clause $\mathscr{C}_i$ contains a set of literals connected by the **or** ($\vee$) operator,

$$\mathscr{C}_i = \bigvee_{j=1}^{n_i} l_{ij},$$

where the literals $l_{ij} \in \{x_i, \bar{x}_i \mid i = 1, ..., n\}$, and $\bar{x}_i$ is the negation of x_i. A formula is a conjunction of clauses, that is, a set of clauses connected by the **and** ($\wedge$) operator,

$$\mathscr{C}_1 \wedge \mathscr{C}_2 \wedge \cdots \wedge \mathscr{C}_m.$$

In the satisfiability problem (SAT) we are given m clauses involving n Boolean variables, and we are to determine whether there exists an assignment of truth values to the Boolean variables, such that the formula evaluates to true. SAT was the first problem shown to be NP-complete (Cook, 1971; Garey and Johnson, 1979). The satisfiability problem is a central problem in artificial intelligence, mathematical logic, and combinatorial optimization. Problems in computer vision, VLSI design, databases, automated reasoning, and computer-aided design and manufacturing involve the solution of instances of the satisfiability problem. Furthermore, SAT is the basic problem in computational complexity (Cook, 1971; Garey and Johnson, 1979). Developing efficient exact algorithms and heuristics for satisfiability problems can lead to general approaches for solving combinatorial optimization problems. Resende and Feo (1996) presented four GRASP implementations for solving the SAT problem, with extensive computational results. In their paper they demonstrate how different GRASP implementations can be achieved by changing the greedy adaptive function of the construction phase, and the local search that GRASP performs. Their computational experiments were conducted on most of the problems from the second DIMACS challenge benchmark problems for SAT (Johnson and Trick, 1996), and comparisons were made with GSAT (Selman et al., 1992).

INFERRING LOGICAL CLAUSES FROM EXAMPLES

The problem of inferring logical clauses from examples is also related to the satisfiability problem. An example with complete data is a vector $e \in \{0, 1\}^n$, while an example with incomplete data is a vector $e \in \{0, 1, *\}^n$. An example can be thought of as an assignment of truth values to n Boolean variables. The $*$ symbol indicates that there is no assignment (it can take the values of either true or false). Given sets of examples E^+ and E^-, the problem is to find a formula (preferably with a small number of clauses) such that it will take the value of true for all examples in E^+, and the value of false for all examples in E^-. Deshpande and Triantaphyllou (1998) presented two GRASPs for solving the above problem, one for solving the problem with complete data, and the other with incomplete data. Computational experiments were conducted on problems involving eighteen thousand examples. Previous best results involved only up to one thousand examples.

Maximum Satisfiability

The weighted maximum satisfiability problem (MAXSAT) is a generalization of the satisfiability problem. In the MAXSAT, for each clause $\mathscr{C}_i$ we are given a weight w_i, and the problem is to find an assignment of truth values to the boolean variables $x_1, ..., x_n$ such that the sum of the weights of the satisfied clauses is maximized. Resende, Pitsoulis, and Pardalos (Resende et al., 1997, 2000; Pardalos et al., 1996) presented a series of papers where GRASP is implemented for solving the MAXSAT problem.

In Resende et al. (1997) the following mixed integer programming formulation of the MAXSAT is used for GRASP implementation. Let $y_j = 1$ if Boolean variable x_j is true and $y_j = 0$ otherwise. Furthermore, the continuous variable $z_i = 1$ if clause $\mathscr{C}_i$ is satisfied and $z_i = 0$ otherwise. Consider the mixed-integer linear program

$$\max \; F(y, z) = \sum_{i=1}^{m} w_i z_i$$

Subject to

$$\sum_{j \in I_i^+} y_j + \sum_{j \in I_i^-} (1 - y_j) \geqslant z_i, \quad i = 1, ..., m,$$

$$y_j \in \{0, 1\}, \quad j = 1, ..., n,$$

$$0 \leqslant z_i \leqslant 1, \quad i = 1, ..., m,$$

where I_i^+ (respectively, I_i^-) denotes the set of variables appearing unnegated (respectively, negated) in clause $\mathscr{C}_i$. Resende et al. (1997) described the GRASP implementation for MAXSAT in detail and conducted extensive computational experiments from modified SAT instances taken from the second DIMACS implementation challenge (Johnson and Trick, 1996), where the optimal solution was also computed using the CPLEX solver for comparison. The design and implementation of the actual fortran subroutines is described in Resende et al. (2000). Moreover a parallel implementation of GRASP for solving the MAXSAT is given in Pardalos et al. (1996), where the computational results indicate average linear speedup.

6.2. *Assignment and Location Problems*

GRASP has been successfully applied to many combinatorial optimization problems arising in the context of assignment and location problems. Among others we mention intermodal trailer assignment (Feo and González-Velarde, 1995), the generalized assignment problem (Lourenço and Serra, 1998), satellite traffic assignment (Prais and Ribeiro, 2000), aerial photographic map mosaicing (Fernández and Martí, 1999), and location of concentrators (Resende and Ulular, 1997). In the following, examples of GRASP implementations for assignment and location problems that have appeared in the literature are given.

Quadratic Assignment Problem

The quadratic assignment problem (QAP) was introduced by Koopmans and Beckmann (1957) as a mathematical model for the location of a set of indivisible economical activities. Consider the problem of allocating a set of facilities to a set of locations, with the cost being a function of the distance and flow between the facilities, plus costs associated with a facility being placed at a certain location. The objective is to assign each facility to a location such that the total cost is minimized. Specifically, we are given three $n \times n$ input matrices with real elements $F = (f_{ij})$, $D = (d_{kl})$ and $B = (b_{ik})$, where f_{ij} is the flow between facility i and facility j, d_{kl} is the distance between location k and location l, and b_{ik} is the cost of placing facility i at location k. The Koopmans–Beckmann version of the QAP can be formulated as follows. Let n be the number of facilities and locations and denote by N the set $N = \{1, 2, ..., n\}$, then the QAP is formulated as

$$\min_{p \in \Pi_N} \sum_{i=1}^{n} \sum_{j=1}^{n} f_{ij} d_{p(i)p(j)} + \sum_{i=1}^{n} b_{ip(i)},$$

where Π_N is the set of all permutations $p : N \rightarrow N$. Each individual product $f_{ij}d_{p(i)p(j)}$ is the cost of assigning facility i to location $p(i)$ and facility j to location $p(j)$. In the context of facility location, matrices F and D are symmetric with zeros in the diagonal, and all the elements of the matrices are nonnegative. The QAP is one of the hardest optimization problems and, until now, no exact algorithm has been able to solve problems of size $n > 30$ in reasonable computational time. In fact, Sahni and Gonzalez (1976) have shown that the QAP is NP-hard and that even finding an approximate solution within some constant factor from the optimal solution cannot be done in polynomial time unless $P = NP$. There are numerous applications of the QAP in different fields, for example, hospital layout (Elshafei, 1977), ranking of archaeological data (Krarup and Pruzan, 1978), ranking of a team in a relay race (Heffley, 1977), scheduling parallel production lines (Geoffrion and Graves, 1976), and analyzing chemical reactions for organic compounds (Ugi et al., 1979).

Li et al. (1994) presented a GRASP for the QAP and reported computational results on problems from the QAP test problem set QAPLIB (Burkard et al., 1997). The authors describe the construction stage and present different local search strategies based on different neighborhood structures for the QAP. Computational results showed that GRASP finds the best-known solution for almost all instances in QAPLIB, and in some cases it produces a better solution than previously reported. Resende et al. (1996) presented in detail the design, implementation, and usage of the code from Li et al. (1994), and performed extensive computational results where they demonstrate the effect of running time on the solution quality of GRASP for solving the QAP.

Define the sparsity of a matrix to be the ratio of the number of zero entries to the number of all the entries in the matrix. Given a QAP instance with input matrices F and D, define the sparsity of the instance to be the maximum sparsity of F and D. Pardalos et al. (1997) modified the GRASP for QAP presented in Li et al. (1994) to exploit sparsity. In Pardalos et al. (1997) the Fortran subroutines are described, and computational results are presented where the dense and sparse versions of GRASP for the QAP are compared in terms of solution time. The results show that, on average, the sparse version is 32 percent faster than the dense version without affecting the solution quality, while when sparsity is large ($\geqslant 0.8$), the sparse version can be up to 300 percent faster. They take advantage of the inherent parallel nature of the GRASP for the QAP, to implement a parallel version of the heuristic. Pardalos et al. (1997) presented computational results on a parallel GRASP implementation for solving large scale QAP instances, using a Kendall Square Research KSR-1 parallel computer, using up to sixtyfour processors in parallel. The results indicate an average speedup that is almost linear with respect to the number of processors.

GRASP can be thought of as a constructive multistart procedure, that is, a procedure that performs local search each time using a different initial solution obtained from a construction procedure. An immediate observation is that GRASP is memoryless, or it does not use information that could be obtained from previously computed solutions for the construction of new solutions. Fleurent and Glover (1999) suggest modifications in which information from previous solutions can be incorporated into GRASP for solving the QAP, and numerical results indicate consistent improvement in the solution quality for all the QAP instances tested.

Numerical results in Fleurent and Glover (1999) indicate that memory-based modifications improve the solution quality of GRASP for solving the QAP.

Another modification of GRASP for the QAP (Li et al., 1994) can be found in Rangel et al. (1998), where a different local search scheme has been incorporated that results in a smaller number of iterations.

The Biquadratic Assignment Problem

A generalization of the QAP is the biquadratic assignment problem (BiQAP), which is essentially a quartic assignment problem with cost coefficients formed by the products of two four-dimensional arrays. More specifically, consider two $n \times n \times n \times n$ arrays, $F = (f_{ijkl})$ and $D = (d_{mpst})$. The BiQAP can then be stated as:

$$\min_{p\in\Pi_N} \sum_{i=1}^{n}\sum_{j=1}^{n}\sum_{k=1}^{n}\sum_{l=1}^{n} f_{ijkl} d_{p(i)p(j)p(k)p(l)},$$

where Π_N denotes the set of all permutations of $N = \{1, 2, ..., n\}$ as usual. The major application of the BiQAP arises in VLSI circuit design. The majority of VLSI circuits are sequential. Their design process is composed of first translating the circuit specifications into a state transition table, then modeling the system using finite state machines, and finally trying to find an encoding of the states such that the actual implementation is of minimum size. A detailed description of the mathematical modeling of the VLSI problem to a BiQAP can be found in Burkard et al. (1994). Several heuristics have been developed for the BiQAP by Burkard and Çela (1995). In particular, deterministic improvement methods and variants of simulated annealing and tabu search are proposed. Computational experiments on test problems of size up to $n = 32$, with known optimal solutions (a test problem generator is presented in Burkard et al., 1994), suggested that one version of simulated annealing is the best among those tested. GRASP has also been applied to the BiQAP by Mavridou et al. (1998), and produced optimal solutions for all the test problems generated in Burkard et al. (1994)

Multidimensional Assignment Problems

In the multidimensional assignment problem (MAP), the input data consists of one cost array, $C \in \mathbb{R}^{n_1 \times n_2 \times \cdots \times n_M}$, where, without loss of generality, we assume that $n_1 \leqslant n_2 \leqslant \cdots \leqslant n_M$. An integer programming formulation of the M-dimensional assignment problem is

$$\min \quad \sum_{i_1=1}^{n_1} \cdots \sum_{i_M=1}^{n_M} C_{i_1 \cdots i_M} x_{i_1 \cdots i_M}$$

$$\text{subject to} \quad \sum_{i_2=1}^{n_2} \cdots \sum_{i_M=1}^{n_N} x_{i_1 \cdots i_M} = 1, \quad i_1 = 1, \cdots, n_1,$$

$$\sum_{i_1=1}^{n_1} \cdots \sum_{i_{k-1}=1}^{n_{k-1}} \sum_{i_{k+1}=1}^{n_{k+1}} \cdots \sum_{i_M=1}^{n_M} x_{i_1 \cdots i_M} = 1,$$

$$\text{for } i_k = 1, ..., n_k \text{ and } k = 2, ..., M-1$$

$$\sum_{i_1=1}^{n_1} \cdots \sum_{i_{M-1}=1}^{n_{M-1}} x_{i_1 \cdots i_M} = 1, \quad i_M = 1, ..., n_M,$$

$$x_{i_1 \cdots i_M} \in \{0, 1\} \text{ for all } i_i, ..., i_M.$$

Multidimensional assignment problems, in their general form, have found many applications as a means of modeling data association problems. More specifically, the central problem in any multitarget tracking task and multisensor surveillance is the data association problem of partitioning the observations into tracks and false alarms. General classes of these problems can be formulated as multidimensional assignment problems. For a detailed description on the application of MAPs for multiple target tracking applications, see Murphey et al. (1998a). Various applications are also contributed to special cases of the MAP, where the dimension of the problem M is constant. Specifically, the five-dimensional assignment problem has been successfully used for tracking elementary particles. By solving a five-dimensional assignment problem, physicists reconstruct tracks generated by charged elementary particles produced by the large electron–positron colider (LEP) at CERN (Pusztaszeri et al., 1995). Another well known special case is the three-index assignment problem, where $M = 3$ (see Aiex et al., 2000a for a GRASP for the three-index assignment problem).

Murphey et al. (1998a,b) present a GRASP for solving the MAP for arbitrary M. In Murphey et al. (1998a), the authors focus on the mathematical modeling of the multitarget multisensor problem into an M-dimensional assignment problem, and present a preliminary GRASP for solving the resulting MAP, coded in MATLAB. In their subsequent paper (Murphey et al., 1998b) they present a parallel GRASP implementation for solving the MAP, where specialized data structures used for the construction and local search phases are described in detail. In (Murphey et al., 1998b), the MAP is shown to be equivalent to

$$\min \quad \sum_{i=1}^{n_1} C_{ip_2(i)\cdots p_{M-1}(i)p_M(i)}$$

$$\text{subject to} \quad p_i \in \Pi_{N_i}, \quad i = 2, \ldots, M,$$

where Π_{N_i} is the set of all permutations of the sets of integers $N_i = \{1, 2, \ldots, n_i\}$ for $i = 2, \ldots, M$. Experimental results in Murphey et al. (1998b) for a set of realistic test problems indicate that in real-time GRASP produced either optimal solutions or solutions very close to the optimal partitions of observations to targets, for problems with closely spaced targets.

Robertson (2001) presents four GRASP implementations for solving the MAP, also in the context of the multitarget multisensor problem. Computational results are reported for two randomly generated problem sets, and a third problem set that resulted from track initiation scenarios generated from a tracking simulation engine.

PLANT LOCATION

In the capacitated plant location problem (CPLP), we have a set of alternative plants and a set of clients to be assigned to the plants. Each plant has a given capacity and a certain fixed opening cost, and each client has a given demand and a cost associated with the assignment to a plant. The objective is to find which plants to open and assign clients to those opened plants at a minimum cost, such that the demands of the clients are satisfied and the capacities of the plants are not exceeded. In the case where we are constrained to assign each client to only one plant, the problem becomes the pure integer capacitated plant location problem (PI-CPLP). The PI-CPLP is also referred to as the single source CPLP in the literature. The PI-CPLP can be formulated as

$$\min \quad \sum_{i\in I}\sum_{j\in J} c_{ij}x_{ij} + \sum_{j\in J} f_j y_j$$

$$\text{subject to} \quad \sum_{j\in J} x_{ij} = 1, \quad \forall i \in I,$$

$$\sum_{i\in I} d_i x_{ij} \leqslant b_j y_j \quad \forall j \in J,$$

$$x_{ij}, y_j \in \{0, 1\} \quad \forall i \in I, j \in J,$$

where sets I and J are the sets of clients and plants, respectively; f_i is the fixed opening cost for plant i; c_{ij} is the cost of assigning client i to plant j; d_i is the demand of client i; and b_j is the capacity of plant j. The decision variables are x_{ij} and y_j. x_{ij} is equal to 1 if client i is assigned to plant j, and 0 otherwise, and y_j is equal to 1 if plant j is open, and 0 otherwise.

Delmaire et al. (1999) develop four types of heuristics for solving the PI-CPLP, based on four different heuristic approaches: evolution algorithms (EA), GRASP, simulated annealing

(SA), and tabu search (TS). In an experimental study, they compare the performance of the algorithms for solving the PI-CPLP and conclude that a hybrid approach where the speed of GRASP and the solution quality of TS are combined would be very effective. It is noted that, in Delmaire et al. (1999), the maximum number of iterations that GRASP was allowed to run was fifty, which resulted in running times that were, on average, an order of magnitude smaller than the other heuristic procedures, while overall GRASP was second in terms of solution quality.

As a sequel to Delmaire et al. (1999), Díaz and Fernández (1998) present a hybrid reactive GRASP and tabu search for solving the PI-CPLP. Díaz and Fernández (1998) perform computational runs on the same set of problems as in Delmaire et al. (1999), where the hybrid GRASP–tabu search performs well. Specifically, the average deviation of the new approach, from the best known solutions never exceeded 0.3 percent.

Another important problem in the context of plant location is the one that involves economies of scale. Let I and J be the sets of clients and plants, respectively, as defined previously in this subsection. Each client i has a demand b_i, and the cost of shipping one unit of product from plant i to client j is c_{ij}. The location problem with economies of scale is formulated as

$$\min \quad \sum_{i\in I}\sum_{j\in J} c_{ij}x_{ij} + \sum_{j\in J} g_j(y_j)$$

$$\text{subject to} \quad \sum_{i\in I} x_{ij} = y_j, \quad \forall j \in J,$$

$$\sum_{j\in J} x_{ij} = b_i, \quad \forall i \in I,$$

$$y_j \geqslant 0, \quad \forall j \in J,$$

$$x_{ij} \geqslant 0, \quad \forall i \in I, j \in J,$$

where x_{ij} is the amount of products shipped from plant j to client i, and y_j is the total amount of products shipped from plant j. The function $g(\cdot)$ is a cost function that can have several forms depending on the application. For example, in the CPLP, g is constant and represents a fixed opening cost. Holmqvist et al. (1998) present a GRASP for solving the above problem where the function g has the form of warehousing costs, that is,

$$g_j(y_j) = \begin{cases} 0 & \text{if } y_j = 0, \\ a_{1j} + a_{2j}y_j + a_{3j}\sqrt{y_j} & \text{if } y_j > 0, \end{cases}$$

where a_{1j}, a_{2j}, and a_{3j} are nonnegative parameters. They exploit properties of the given problem in the GRASP implementation, and perform computational experiments for a set of problems from the OR-library (Beasley, 1990). Compared to the solutions obtained with Lagrangian relaxation, GRASP found a better solution for all but two (out of fifteen) problem instances.

FREQUENCY ASSIGNMENT PROBLEM

The frequency assignment problem (FAP) arises in the context of resource management in cellular networks, and refers to the problem of assigning a limited number of frequencies to transmitters in the network, such that the electromagnetic interference is minimized. Consider that we have N transmitters in the network, and let the $N \times N$ matrix C represent the minimal frequency separation values for each pair of transmitters. Then, given two transmitters i and j and assigned frequencies f_i and f_j, respectively, the following constraint should be satisfied:

$$| f_i - f_j | > c_{ij}.$$

Given the minimal frequency separation matrix C in the FAP we have the following objective function:

$$\min \sum_{i=1}^{N} \sum_{j=1}^{N} \delta(|f_i - f_j| > c_{ij}),$$

where

$$\delta(|f_i - f_j| > c_{ij}) = \begin{cases} c_{ij} - |f_i - f_j|, & \text{if } |f_i - f_j| \leqslant c_{ij}, \\ 0 & \text{otherwise.} \end{cases}$$

Liu et al. (2000) present a GRASP for solving the FAP in mobile radio networks. The construction phase is a randomized version of a sequential coloring heuristic based on saturation degrees, and the local search phase is a fast simulated annealing procedure. Computational experiments were performed on two data sets, one randomly generated and one with instances from the European EUCLID project CALMA (Combinatorial Algorithms for Military Applications). In the first data set, GRASP compares favorably with pure saturation with the simulated annealing heuristic, while in the second set, GRASP improves the best-known solutions for all instances.

THE *P*-HUB LOCATION PROBLEM

In the p-hub location problem we are given a set of nodes where there is a certain level of traffic between each pair of nodes (traffic can be data, products, etc.), and a cost per unit transfered between any pair of nodes. The objective is to determine p nodes, which will serve as hubs; that is, a set of nodes that is fully interconnected, and every other node is connected to at least one hub, such that the total cost is minimized. Given such a configuration, any two nodes in the network can communicate through the hubs (in the worst case two hubs will be used as intermediate nodes). Specifically, consider that we have n nodes, and w_{ij} is the number of units transfered from node i to node j, and c_{ij} is the cost per unit transferred between i and j. Let y_j be equal to 1 if node j is a hub and 0 otherwise, and x_{ij} be equal to 1 if node i is connected to hub j and 0 otherwise. The p-hub location problem can then be stated as

$$\begin{aligned}
\min \quad & \sum_{ij} w_{ij} \left[\sum_{k} x_{ik} c_{ik} + \sum_{m} x_{jm} c_{jm} + a \sum_{k,m} x_{ik} x_{im} c_{km} \right] \\
\text{subject to} \quad & \sum_{j} x_{ij} = 1, \quad i = 1, \dots, n, \\
& \sum_{j} y_j = p, \\
& x_{ij} \leqslant y_j, \quad i,j = 1, \dots, n, \\
& x_{ij}, y_j \in \{0,1\}, \quad i,j = 1, \dots, n,
\end{aligned}$$

where a is discount parameter. Klincewicz (1992) presents a GRASP and a tabu search for solving the p-hub location problem. The author considers two classes of test problems. The first consists of instances with up to twenty-five nodes where $p = 3, 4$ and the optimal solutions, under the assumption that nodes are to be assigned to the nearest hub, are known. The second problem class consists of instances with fifty-two nodes, where $p = 4, 10$. For the first class of problems, GRASP found the optimal solution in over 90 percent of the instances, while for the second problem class it generated the best-known solutions.

7. Concluding Remarks

GRASP has found its way into many applications in academic as well as industrial settings. Besides producing good quality solutions to hard combinatorial optimization problems, it has several advantages over other metaheuristics. It can usually be implemented quickly, since construction and local search algorithms are readily available. Parameter turning is minimal.

Finally, because of the probability distribution of its running time to find a suboptimal target solution, GRASP can be implemented in parallel with optimal speedup.

Many problems can be approached with GRASP. However, since GRASP is a constructive method, it is difficult to apply it to problems for which feasible solutions are difficult to construct.

References

Aho, A. V., J. E. Hopcroft, and J. D. Ullman. 1974. *The Design and Analysis of Computer Algorithms*. Reading, MA: Addison-Wesley.

Ahuja, R. K., J. B. Orlin, and D. Sharma. 1998. "New neighborhood search structures for the capacitated minimum spanning tree problem." Technical report. Department of ISE, Gainesville, FL 32611-6595: University of Florida.

Ahuja, R. K., J. B. Orlin, and A. Tiwari. 2000. "A greedy genetic algorithm for the quadratic assignment problem." *Comput. Oper. Res.* **27**: 917–934.

Aiex, R. M, M. G. C. Resende, P. M. Pardalos, and G. Toraldo. 2000a. "GRASP with path relinking for the three-index assignment problem." Technical report. NJ 07932: AT&T Labs Research, Florham Park.

Aiex, R. M., M. G. C. Resende, and C. C. Ribeiro. 2000b. "Probability distribution of solution time in GRASP: an experimental investigation." To appear in *J. Heuristics*.

Beasley, J. E. 1990. "OR-Library: distributing test problems by electronic mail." *J. Oper. Res. Soc* **41**: 1069–1072.

Binato, S., W. J. Hery, D. Loewenstern, and M. G. C. Resende. 2001. "A greedy randomized adaptive search procedure for job shop scheduling." In *Essays and Surveys on Metaheuristics*, edited by P. Hansen and C.C. Ribeiro. Norwell, MA: Kluwer Academic Publishers.

Bresina, J. L. 1996. "Heuristic-biased stochastic sampling." In *Proceedings of the AAAI-96*, pp. 271–278. Portland, OR: American Association of Artificial Intelligence.

Burkard, R. E. and E. Çela. 1995. "Heuristics for biquadratic assignment problems and their computational comparison." *Eur. J. Oper. Res.* **83**: 283–300.

Burkard, R. E., E. Çela, and B. Klinz. 1994. "On the biquadratic assignment problem." In *Quadratic Assignment and Related Problems*, edited by P. M. Pardalos and H. Wolkowicz. DIMACS Series on Discrete Mathematics and Theoretical Computer Science, pp. 117–146. Providence, RI: American Mathematical Society.

Burkard, R. E., S. Karisch, and F. Rendl. 1997. "QAPLIB—a quadratic assignment problem library." *J. Global Optim.* **10**: 391–403. An on-line version is available via World Wide Web at the following URL: http://www.opt.math.tu-graz.ac.at/gaplib/.

Colomé, R. and D. Serra. 1998. "Consumer choice in competitive location models: formulations and heuristics." To appear in *Papers in Regional Sci.*

Cook, S. A. 1971. "The complexity of theorem-proving procedures." In *Proceedings of the Third Annual ACM Symposium on Theory of Computing*, pp. 151–158. Shaker Heights, OH: Association of Computing Machinery.

Cormen, T. H., C. E. Leiserson, and R. L. Rivest. 1990. In *Introduction to Algorithm*. Cambridge, MA: The MIT Press.

Delmaire, H., J. A. Díaz, E. Fernández, and M. Ortega. 1999. "Reactive GRASP and tabu search based heuristics for the single source capacitated plant location problem." *INFOR* **37**: 194–225.

Deshpande, A. S. and E. Triantaphyllou. 1998. "A greedy randomized adaptive search procedure (GRASP) for inferring logical clauses from examples in polynomial time and some extensions." *Math. Comput. Modelling* **27**: 75–99.

Díaz, J. A. and E. Fernández. 1998. A hybrid GRASP–tabu search algorithm for the single source capacitated plant location problem. Technical report. Department of Statistics and Operations Research, Universitat Politecnica de Catalunya, Barcelona, Spain.

Elshafei, A. N. 1977. "Hospital layout as a quadratic assignment problem." *Oper. Res. Q.* **28**: 167–179.

Feo, T. A. and J. L. González-Velarde. 1995. "The intermodal trailer assignment problem: models, algorithms, and heuristics." *Transport. Sci.* **29**: 330–341.

Feo, T. A. and M. G. C. Resende. 1989. "A probabilistic heuristic for a computationally difficult set covering problem." *Oper. Res. Lett.* **8**: 67–71.

Feo, T. A. and M. G. C. Resende. 1995. "Greedy randomized adaptive search procedures." *J. Global Optim.* **6**: 109–133.

Fernández, E. and R. Martí. 1999. "GRASP for seam drawing in mosaicking of aerial photographic maps." *J. Heuristics* **5**: 181–197.

Festa, P. and M. G. C. Resende. 2001. "GRASP: an annotated bibliography." In *Essays and Surveys on Metaheuristics*, edited by P. Hansen, and C. C. Ribeiro. Norwell, MA: Kluwer Academic Publishers.

Fleurent, C. and F. Glover. 1999. "Improved constructive multistart strategies for the quadratic assignment problem using adaptive memory." *INFORMS J. Comput.* **11**: 198–204.

Garey, M. R. and D. S. Johnson. 1979. Computers and Intractability: A Guide to the Theory of NP-completeness. New York: W.H. Freeman and Company.

Geoffrion, A. M. and G. W. Graves. 1976. "Scheduling parallel production lines with changeover costs: practical applications of a quadratic assignment/LP approach." *Oper. Res.* **24**: 595–610.

Glover, F. and M. Laguna. 1997. "Tabu Search". Norwell, MA: Kluwer Academic Publishers.

Glover, F., M. Laguna, and R. Martí. 2000. "Fundamentals of scatter search and path relinking." Technical report. Graduate School of Business and Administration, University of Colorado, Boulder, CO 80309-0419.

Hansen, P. and N. Mladenović. 1998. "An introduction to variable neighborhood search." In *Meta-heuristics, Advances and Trends in Local Search Paradigms for Optimization*, edited by S. Voss, S. Martello, I. H. Osman, and C. Roucairol, pp. 433–458. Norwell, MA: Kluwer Academic Publishers.

Hart, J. P. and A. W. Shogan. 1987. "Semi-greedy heuristics: an empirical study." *Oper. Res. Lett.* **6**: 107–114.

Heffley, D. R. 1977. "Assigning runners to a relay team." In *Optimal Strategies in Sports*, edited by S. P. Ladany and R. E. Machol, pp. 169–171. Amsterdam: North-Holland.

Holmqvist, K., A. Migdalas, and P. M. Pardalos. 1998. "A GRASP algorithm for the single source uncapacitated minimum concave-cost network flow problem." In *Network Design: Connectivity and Facilities Location*, edited by P. M. Pardalos and D.-Z. Du. DIMACS Series on Discrete Mathematics and Theoretical Computer Science, **40**: 131–142. Providence, RI: American Mathematical Society.

Johnson, D. S., C. H. Papadimitrion, and M. Yannakakis. 1988. "How easy is local search?" *J. Comput. Sys. Sci.* **17**: 79–100.

Johnson, D. S. and M. A. Trick. (eds). 1996. In *Cliques, Coloring, and Satisfiability: Second DIMACS Implementation Challenge*. DIMACS Series in Discrete Mathematics and Theoretical Computer Science. Providence, RI: American Mathematical Society.

Klincewicz, J. G. 1992. "Avoiding local optima in the P-hub location problem using tabu search and GRASP." *Ann. Oper. Res.* **40**: 283–302.

Koopmans, T. C. and M. J. Beckman. 1957. "Assignment problems and the location of economic activities." *Econometrica* **25**: 53–76.

Krarup, J. and P. M. Pruzan. 1978. "Computer-aided layout design." *Math. Programming Study* **9**: 75–94.

Laguna, M. and J. L. González-Velarde. 1991. "A search heuristic for just-in-time scheduling in parallel machines." *J. Intell. Manufacturing* **2**: 253–260.

Laguna, M. and R. Martí. 1999. "GRASP and path relinking for 2-layer straight line crossing minimization." *INFORMS J. Comput.* **11**: 44–52.

Li, Y., P. M. Pardalos, and M. G. C. Resende. 1994. "A greedy randomized adaptive search procedure for the quadratic assignment problem." In *Quadratic Assignment and Related Problem*, edited by P. M. Pardalos and H. Wolkowicz. DIMACS Series on Discrete Mathematics and Theoretical Computer Science, **16**: 237–261. Providence, RI: American Mathematical Society.

Liu, X., P. M. Pardalos, S. Rajasekaran, and M. G. C. Resende. 2000. "A GRASP for frequency assignment in mobile radio networks." In *Mobile Networks and Computing*, edited by S. Rajasekaran, P. M. Pardalos, and F. Hsu. DIMACS Series on Discrete Mathematics and Theoretical Computer Science, **52**: 195–201. Providence, RI: American Mathematical Society.

Lourenço, H. R. and D. Serra. 1998. "Adaptive approach heuristics for the generalized assignment problem." Technical report. Department of Economics and Management, Universitat Pompeu Fabra, Barcelona, Spain.

Lourenço, H. R., J. P. Paixão, and R. Potugal. 1998. "Metaheuristics for the bus-driver scheduling problem." Technical report. Department of Economics and Management, Universitat Pompeu Fabra, Barcelona, Spain.

Martins, S. L., C. C. Ribeiro, and M. C. Souza. 1998. "A parallel GRASP for the Steiner problem in graphs." In *Proceedings of IRREGULAR'98—5th International Symposium on Solving Irregulary Structured Problems in Parallel*, edited by A. Ferreira and J. Rolim. Lecture Notes in Computer Science, **1457**: 285–297. Berlin: Springer-Verlag.

Martins, S. L., M. G. C. Resende, C. C. Ribeiro, and P. M. Pardalos. 2000. "A parallel GRASP for the Steiner tree problem in graphs using a hybrid local search strategy." *J. Global Optim.* **17**: 267–283.

Mavridou, T., P. M. Pardalos, L. S. Pitsoulis, and M. G. C. Resende. 1998. "A GRASP for the biquadratic assignment problem." *Eur. J. Oper. Res.* **105**: 613–621.

Mockus, J., E. Eddy, A. Mockus, L. Mockus, and G. V. Reklaitis. 1997. In *Bayesian Discrete and Global Optimization*. Norwell, MA: Kluwer Academic Publishers.

Murphey, R. A., P. M. Pardalos, and L. S. Pitsoulis. 1998a. "A greedy randomized adaptive search procedure for the multitarget multisensor tracking problem." In *Network Design: Connectivity and Facilities Location*, edited by P. M. Pardalos and D.-Z. Du. DIMACS Series on Discrete Mathematics and Theoretical Computer Science, **40**: 277–301. Providence, RI: American Mathematical Society.

Murphey, R. A., P. M. Pardalos, and L. S. Pitsoulis. 1998b. "A parallel GRASP for the data association multidimensional assignment problem." In *Parallel Processing of Discrete Problems*, edited by P. M. Pardalos. The IMA Volumes in Mathematics and Its Applications, **106**: 159–180. New York: Springer-Verlag.

Pardalos, P. M., L. S. Pitsoulis, and M. G. C. Resende. 1995. "A parallel GRASP implementation for the quadratic assignment problem." In *Parallel Algorithms for Irregular Structured Problems—IRREGULAR'94*, edited by A. Ferreira and J. Rolim, pp. 111–130. Norwell, MA: Kluwer Academic Publishers.

Pardalos, P. M., L. S. Pitsoulis, and M. G. C. Resende. 1996. "A parallel GRASP for MAX-SAT problems." *Lecture Notes Comput. Sci.* **1184**: 575–585.

Pardalos, P. M., L. S. Pitsoulis, and M. G. C. Resende. 1997. "Algorithm 769: Fortran subroutines for approximate solution of sparse quadratic assignment problems using GRASP." *ACM Trans Math. Softw.* **23**: 196–208.

Prais, M. and C. C. Ribeiro. 2000. "Reactive GRASP: an application to a matrix decomposition problem in TDMA traffic assignment." *INFORMS J. Comput.* **12**: 164–176.

Pusztaszeri, J., P. E. Rensing, and T. M. Liebling. 1995. "Tracking elementary particles near their primary vertex: a combinatorial approach." *J. Global Optim.* **16**: 422–431.

Rangle, M. C., N. M. M. de Abreu, P. O. Boaventura Netto, and M. C. S. Boeres. 1998. "A modified local search for GRASP in the quadratic assignment problem." Technical report. Rio de Janeiro, Brazil: Production Engineering Program, COPPE, Federal University of Rio de Janeiro.

Resende, M. G. C. and T. A. Feo. 1996. "A GRASP for satisfiability." In *Cliques, Coloring, and Satisfiability The Second DIMACS Implementation Challenge*, edited by D. S. Johnson and M. A. Trick. DIMACS Series on Discrete Mathematics and Theoretical Computer Science, **26**: 499–520. Providence, RI: American Mathematical Society.

Resende, M. G. C., P. M. Pardalos, and Y. Li. 1996. "Algorithm 754: Fortran subroutines for approximate solution of dense quadratic assignment problems using GRASP." *ACM Trans. Math. Softw.* **22**: 104–118.

Resende, M. G. C., L. S. Pitsoulis, and P. M. Pardalos. 1997. "Approximate solution of weighted MAX-SAT problems using GRASP." In *Satisfiability Problems*, edited by J. Gu and P. M. Pardalos. DIMACS Series on Discrete Mathematics and Theoretical Computer Science, **35**: 393–405. Providence, RI: American Mathematical Society.

Resende, M. G. C., L. S. Pitsoulis, and P. M. Pardalos. 2000. "Fortran subroutines for computing approximate solutions of MAX-SAT problems using GRASP." *Discrete Appl. Math.* **100**: 95–113.

Resende, M. G. C. and O. Ulular. 1997. "SMART: a tool for AT&T Worldnet access design—location of Cascade 9000 concentrators." Technical report. AT&T Labs Research, Florham Park, NJ 07932.

Robertson, A. J. 2001. "A set of greedy randomized adaptive local search procedure (GRASP) implementations for the multidimensional assignment problem." To appear in *Comput. Optim. Appl.*

Sahni, S. and T. Gonzalez. 1976. "P-complete approximation problems." *J. Assoc. Comput. Machinery* **23**: 555–565.

Selman, B., H. Levesque, and D. Mitchell. 1992. "A new Method for solving hard satisfiability problems." In *Proceedings of the Tenth National Conference on Artificial Intelligence (AAAI92)*, pp. 440–446. San Jose, CA: American Association of Artificial Intelligence.

Ugi, I., J. Bauer, J. Friedrich, J. Gasteiger, C. Jochum, and W. Schubert. 1979. "Neue anwendungsgebiete fur computer in der chemie." *Angew. Chemie* **91**: 99–184.

Verhoeven, M. G. A. and E. H. L. Aarts. 1995. "Parallel local search." *J. Heuristics* **1**: 43–66.

3.6.6 Scatter Search

Manuel Laguna

ABSTRACT This article explores the metaheuristic approach called scatter search, which is an evolutionary method that has recently been shown to yield promising outcomes for solving combinatorial and nonlinear optimization problems. Based on formulations originally proposed in the 1960s for combining decision rules and problem constraints, this method uses strategies for combining solution vectors that have proved effective in a variety of problem areas. Scatter search can be implemented in multiple ways, and offers numerous alternatives for exploiting its fundamental ideas. We identify a general design and illustrate the main features in the context of a classical integer programming problem.

1. Scatter Search Background

Scatter search, from the standpoint of metaheuristic classification, may be viewed as an evolutionary (or also called population-based) algorithm that constructs solutions by combining others. It derives its foundations from strategies originally proposed for combining decision rules and constraints (in the context of integer programming). The goal of this methodology is to enable the implementation of solution procedures that can derive new solutions from combined elements. The way scatter search combines solutions and updates the set of reference solutions used for combination sets this methodology apart from other population-based approaches.

As discussed in Glover (1998, 1999), the approach of combining existing solutions or rules to create new solutions originated in the 1960s. In the area of scheduling, researchers introduced the notion of combining rules to obtain improved local decisions. Numerically weighted combinations of existing rules, suitably restructured so that their evaluations embodied a common metric, generated new rules. The conjecture that information about the relative desirability of alternative choices is captured in different forms by different rules motivated this approach. The combination strategy was devised with the belief that this information could be exploited more effectively when integrated than when treated in isolation (i.e., when existing selection rules are selected one at a time). In general, the decision rules created from such combination strategies produced better empirical outcomes than standard applications of local decision rules. They also proved superior to a "probabilistic learning approach" that used stochastic selection of rules at different junctures, but without the integration effect provided by generating combined rules.

In integer and nonlinear programming, associated procedures for combining constraints were developed, which likewise employed a mechanism for creating weighted combinations. In this case, nonnegative weights were introduced to create new constraint inequalities, called surrogate constraints. The approach isolated subsets of constraints that were gauged to be most critical, relative to trial solutions based on the surrogate constraints. This critical subset was used to produce new weights that reflected the degree to which the component constraints were satisfied or violated Glover (1977).

The main function of surrogate constraints was to provide ways to evaluate choices that could be used to create and modify trial solutions. A variety of heuristic processes that employed surrogate constraints and their evaluations evolved from this foundation. As a natural extension, these processes led to the related strategy of combining solutions. Combining solutions, as manifested in scatter search, can be interpreted as the primal counterpart to the dual strategy of combining constraints.

Scatter search operates on a set of solutions, the reference set, by combining these solutions to create new ones. When the main mechanism for combining solutions is such that a new solution is created from the linear combination of two other solutions, the reference set may evolve as illustrated in Figure 1. This figure assumes that the original reference set of solutions consists of the circles labeled A, B and C. After a nonconvex combination of reference solutions A and B, Solution 1 is created. More precisely, a number of solutions in the line segment defined by A and B are created; however, only Solution 1 is introduced into the reference set. (The criteria used to select solutions for membership in the reference set are discussed later.) In a similar way, convex and nonconvex combinations of original and newly created reference solutions create points 2, 3 and 4. The complete reference set shown in Figure 1 consists of 7 solutions (or elements).

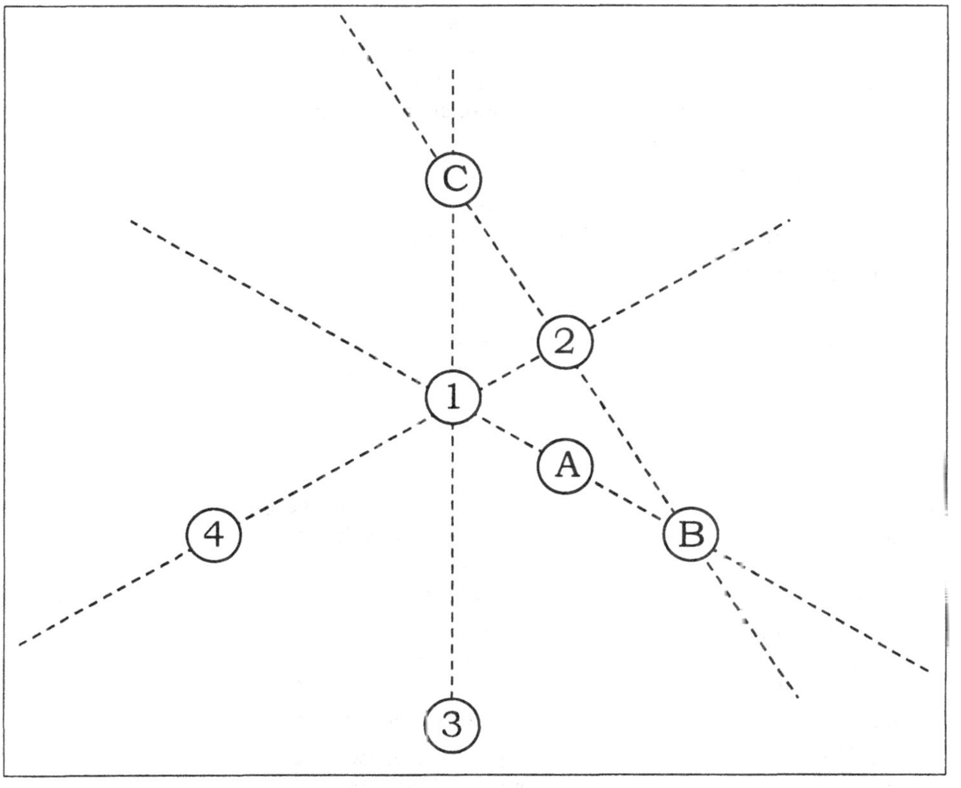

Figure 1 Two-dimensional reference set

Unlike a population in genetic algorithms, the reference set of solutions in scatter search tends to be small. In genetic algorithms, two solutions are randomly chosen from the population and a crossover or combination mechanism is applied to generate one or more offspring. A typical population size in a genetic algorithm consists of one hundred elements, which are randomly sampled to create combinations. In contrast, scatter search chooses two or more elements of the reference set in a systematic way with the purpose of creating new solutions. Since the combination process considers, at a minimum, all pairs of solutions in the reference set, there is a practical need for keeping the cardinality of the set small. Typically, the reference set in scatter search has twenty solutions or less. In general, if the reference set consists of b solutions, the procedure examines approximately $(3b - 7)b/2$ combinations of four different types (Glover, 1998). The basic type consists of combining two solutions; the next type combines three solutions, and so on and so forth. Limiting the scope of the search to a selective group of combination types can be used as a mechanism for controlling the number of possible combinations in a given reference set.

2. Scatter Search Template

The scatter search process, building on the principles that underlie the surrogate constraint design, is organized to (1) capture information not contained separately in the original vectors, and (2) take advantage of auxiliary heuristic solution methods to evaluate the combinations produced and to generate new vectors (Glover, 1998). Specifically, the scatter search approach may be sketched as follows:

1. Generate a starting set of solution vectors to guarantee a critical level of diversity and apply heuristic processes designed for the problem considered, as an attempt to improve these solutions. Designate a subset of the best vectors to be reference solutions.

(Subsequent iterations of this step, transferring from Step 4 below, incorporate advanced starting solutions and best solutions from previous history as candidates for the reference solutions.) The notion of best in this step is not limited to a measure given exclusively by the evaluation of the objective function. In particular, a solution may be added to the reference set if the diversity of the set improves even when the objective value of such a solution is inferior to other solutions competing for admission in the reference set.

2. Create new solutions consisting of structured combinations of subsets of the current reference solutions. The structured combinations are:
   (a) chosen to produce points both inside and outside the convex regions spanned by the reference solutions;
   (b) modified to yield acceptable solutions. (For example, if a solution is obtained by a linear combination of two or more solutions, a generalized rounding process that yields integer values for integer-constrained vector components may be applied. Note that an acceptable solution may or may not be feasible with respect to other constraints in the problem.)
3. Apply the heuristic processes used in Step 1 to improve the solutions created in Step 2. (Note that these heuristic processes must be able to operate on infeasible solutions and may or may not yield feasible solutions.)
4. Extract a collection of the best improved solutions from Step 3 and add them to the reference set. The notion of best is once again broad; making the objective value one among several criteria for evaluating the merit of newly created points. Repeat Steps 2, 3, and 4 until the reference set does not change. Diversify the reference set, by restarting from Step 1. Stop when reaching a specified iteration limit.

The first notable feature in scatter search is that its structured combinations are designed with the goal of creating weighted centers of selected subregions. This adds nonconvex combinations that project new centers into regions that are external to the original reference solutions (see, e.g., Solution 3 in Figure 1). The dispersion patterns created by such centers and their external projections have been found useful in several application areas (Glover, 1994).

Another important feature relates to the strategies for selecting particular subsets of solutions to combine in Step 2. These strategies are typically designed to make use of a type of clustering to allow new solutions to be constructed within clusters and across clusters. Finally, the method is organized to use ancillary improving mechanisms that are able to operate on infeasible solutions, removing the restriction that solutions must be feasible in order to be included in the reference set.

The following principles summarize the foundations of the scatter search methodology (Glover, 1998, 1999):

- Useful information about the form (or location) of optimal solutions is typically contained in a suitably diverse collection of elite solutions.
- When solutions are combined as a strategy for exploiting such information, it is important to provide mechanisms capable of constructing combinations that extrapolate beyond the regions spanned by the solutions considered. Similarly, it is also important to incorporate heuristic processes to map combined solutions into new solutions. The purpose of these combination mechanisms is to incorporate both diversity and quality.
- Taking account of multiple solutions simultaneously, as a foundation for creating combinations, enhances the opportunity to exploit information contained in the union of elite solutions.

The fact that the mechanisms within scatter search are not restricted to a single uniform design allows the exploration of strategic possibilities that may prove effective in a particular implementation. These observations and principles lead to the following template for implementing scatter search.

1. A diversification generation method to generate a collection of diverse trial solutions, using an arbitrary trial solution (or seed solution) as an input.
2. An improvement method to transform a trial solution into one or more enhanced trial solutions. (Neither the input nor the output solutions are required to be feasible, though the output solutions will usually be expected to be so. If no improvement of the input trial solution results, the enhanced solution is considered to be the same as the input solution.)
3. A reference set update method to build and maintain a reference set consisting of the b best solutions found (where the value of b is typically small, e.g., no more than 20), organized to provide efficient accessing by other parts of the method. Solutions gain membership to the reference set according to their quality or their diversity.
4. A subset generation method to operate on the reference set, to produce a subset of its solutions as a basis for creating combined solutions.
5. A solution combination method to transform a given subset of solutions produced by the subset generation method into one or more combined solution vectors.

In the next section, we employ this template to illustrate the design of a simple scatter search procedure for an instance of the well-known 0–1 knapsack problem.

3. A Scatter Search Illustration

Consider the following 0–1 knapsack problem, where the coefficients in the objective function are considered profit values and the coefficients in the constraint are considered weights:

$$\text{Maximize} \quad 11x_1 + 10x_2 + 9x_3 + 12x_4 + 10x_5 + 6x_6 + 7x_7 + 5x_8 + 3x_9 + 8x_{10}$$

$$\text{Subject to} \quad 33x_1 + 27x_2 + 16x_3 + 14x_4 + 29x_5 + 30x_6 + 31x_7 + 33x_8 + 14x_9 + 18x_{10} \leqslant 100$$

$$x_i = \{0, 1\} \quad \text{for } i = 1, \ldots, 10.$$

For the purpose of this illustration, we first describe each of the methods that are needed in the overall procedure. We then discuss the iterative nature of the resulting solution method.

3.1. *Diversification Generation Method*

An appropriate diversification generator for this problem is described in Glover (1998) This generator operates as follows. Choose a value for the parameter $h \leq n - 1$, where n is the number of variables in the problem. Let us choose h = 5. We also choose the initial seed to be x = (0, 0, 0, 0). For each value of h, two solutions are generated. Type 1 solutions are denoted by x' and are initialized as $x_i' = 0$ for all i Then some of the individual elements are modified as follows:

$$x_1' = 1 - x_1$$

$$x_{1+hk}' = 1 - x_{1+hk} \quad \text{for } k = 1, \ldots, n/h.$$

Type 2 solutions are denoted by x'' and obtained as the complement of Type 1 solutions, that is $x_i'' = 1 - x_i'$. Table 1 shows the ten solutions generated with this method using the parameters given above.

The ten solutions in Table 1 can now be used to create the reference set by applying the following improvement method.

3.2. *Improvement Method*

Since the solutions generated with the diversification generation method are not guaranteed to be feasible, the improvement method should be capable of handling starting solutions that are either feasible or infeasible. To illustrate, we consider a simple procedure that operates as follows:

Table 1 Diversified Solutions

h	x'	x''
1	(1,1,1,1,1,1,1,1,1,1)	(0,0,0,0,0,0,0,0,0,0)
2	(1,0,1,0,1,0,1,0,1,0)	(0,1,0,1,0,1,0,1,0,1)
3	(1,0,0,1,0,0,1,0,0,1)	(0,1,1,0,1,1,0,1,1,0)
4	(1,0,0,0,1,0,0,0,1,0)	(0,1,1,1,0,1,1,1,0,1)
5	(1,0,0,0,0,1,0,0,0,0)	(0,1,1,1,1,0,1,1,1,1)

1. If the trial solution is infeasible, then the method attempts to improve it by first finding a feasible solution. A feasible solution is found by changing variable values from one to zero until the constraint is no longer violated. The variables are considered in increasing order of their profit-to-weight ratio, starting with the one with the smallest ratio. Once the solution becomes feasible, the following procedure is applied.
2. If the trial solution is feasible, then the method attempts to improve it by changing variable values from zero to one. This is done in a greedy fashion, so that the first variable to be considered is the one with the largest profit-to-weight ratio. (For example, x_4 has the largest ratio value of $12/14 = 0.857$ in our problem instance.) The procedure stops when no more variables can be given a value of one without violating the constraint.

Table 2 shows the solutions obtained by the application of the improvement method to the trial solutions generated by the diversification generation method. Note that trial Solutions 1, 9 and 10 in Table 2 became the same improved solution with an objective value of 42. In these cases, which are not atypical, the diversification generation method can be applied until a desired number of different improved solutions are found. Also note that the improvement method was able to find the optimal solution starting from the infeasible trial Solution 8. Table 3 shows the moves that the improvement method performed to obtain the optimal solution starting from the corresponding infeasible trial solution.

The iterations in Table 3 show that while the knapsack constraint is violated (i.e., the total weight of the current solution is larger than 100), the candidate moves consist of changing variable values from one to zero. In the first iteration, x_8 is selected because its profit-to-weight ratio is the minimum in the candidate list. After x_6 is also set to zero in iteration 2 using the same criteria as before, the current solution becomes feasible in iteration 3. The procedure now looks for opportunities to improve the objective function value. The candidates are those variables that are currently set to zero (excluding x_6 and x_8) and whose weight is less than or equal to the current slack (i.e., $100 - 86 = 14$). The only variable that meets such criteria is x_4,

Table 2 Initial and Improved Solutions

Solution	*Trial Solution*	*Objective Value*	*Improved Solution*	*Objective Value*
1	(1,1,1,1,1,1,1,1,1,1)	81†	(0,1,1,1,0,0,0,0,1,1)	42
2	(1,0,1,0,1,0,1,0,1,0)	40†	(1,0,1,1,1,0,0,0,0,0)	42
3	(1,0,0,1,0,0,1,0,0,1)	38	(1,0,0,1,0,0,1,0,0,1)	38
4	(1,0,0,0,1,0,0,0,1,0)	24	(1,0,0,1,1,0,0,0,1,0)	36
5	(1,0,0,0,0,1,0,0,0,0)	17	(1,0,1,1,0,1,0,0,0,0)	38
6	(0,0,0,0,0,0,0,0,0,0)	0	(0,1,1,1,0,0,0,0,0,1)	39
7	(0,1,0,1,0,1,0,1,0,1)	41†	(0,1,0,1,0,1,0,0,0,1)	36
8	(0,1,1,0,1,1,0,1,1,0)	43†	(0,1,1,1,1,0,0,0,1,0)	44*
9	(0,1,1,1,0,1,1,1,0,1)	57†	(0,1,1,1,0,0,0,0,1,1)	42
10	(0,1,1,1,1,0,1,1,1,1)	64†	(0,1,1,1,0,0,0,0,1,1)	42

† Denotes an infeasible solution.
* Denotes the optimal solution.

Table 3 Iterations of the Improvement Method

Iteration	Current Solution	Objective Value	Total Weight	Candidate Moves (profit/weight)	Selected Move
1	(0,1,1,0,1,1,0,1,1,0)	43	149	$x_2 = 0$ (0.370)	$x_8 = 0$
				$x_3 = 0$ (0.563)	
				$x_5 = 0$ (0.345)	
				$x_6 = 0$ (0.200)	
				$x_8 = 0$ (0.152)	
				$x_9 = 0$ (0.214)	
2	(0,1,1,0,1,1,0,0,1,0)	38	116	$x_2 = 0$ (0.370)	$x_6 = 0$
				$x_3 = 0$ (0.563)	
				$x_5 = 0$ (0.345)	
				$x_6 = 0$ (0.200)	
				$x_9 = 0$ (0.214)	
3	(0,1,1,0,1,0,0,0,1,0)	32	86	$x_4 = 1$ (0.857)	$x_4 = 1$
4	(0,1,1,1,1,0,0,0,1,0)	44	100	None	

which also happens to have the largest profit-to-weight ratio of 0.857. When the value of x_4 is changed to 1, the optimal solution is found. No more moves are available in iteration 4 and the improvement process stops.

3.3. Reference Set Update Method

This method is used to create and maintain a set of reference solutions. We divide this set of b solutions into two subsets: the subset of b_1 high-quality solutions and the subset of b_2 diverse solutions. Let us consider a reference set of size $b = 5$, where $b_1 = 3$ and $b_2 = 2$. According to these definitions, the subset of high-quality solutions would consist of the improved Solutions 1, 2 and 8.

In order to find a diverse set to complement the subset of high-quality solutions, it is necessary to define a diversity measure. We define the distance between two solutions as the sum of the absolute difference between its corresponding variable values. For example, the distance between the improved Solution 1 and the improved Solution 8 is calculated as follows:

```
(0,1,1,1,0,0,0,0,1,1)
(0,1,1,1,1,0,0,0,1,0)
(0+0+0+0+1+0+0+0+0+1)=2
```

We then use this distance measure to select the two solutions for the subset of diverse solution. In particular, we look for the solution that is not currently in the reference set and that maximizes the minimum distance to all the solutions currently in the reference set. Table 4 shows the distance values from each candidate solution to each solution currently in the reference set.

The list of candidate solutions in Table 4 includes neither Solution 9 nor Solution 10 because these solutions are the same as Solution 1, which is already in the reference set. The distance values show that in order to maximize the minimum distance, improved Solution 3 should be added to the reference set. The final member of the reference set becomes Solution 7,

Table 4 Distance Measures

Candidate Solution	Distance to Solution 1	2	8	Minimum Distance
3 (1,0,0,1,0,0,1,0,0,1)	5	4	7	4
4 (1,0,0,1,1,0,0,0,1,0)	5	2	3	2
5 (1,0,1,1,0,1,0,0,0,0)	5	2	5	2
6 (0,1,1,1,0,0,0,0,0,1)	1	4	3	1
7 (0,1,0,1,0,1,0,0,0,1)	3	6	5	3

Table 5 Subset Generation

Type	*Subsets*
1	(1, 2) (1, 3) (1, 7) (1, 8)
	(2, 3) (2, 7) (2, 8)
	(3, 7) (3, 8)
	(7, 8)
2	(1, 2, 8) (1, 3, 8) (1, 7, 8)
	(2, 3, 8) (2, 7, 8)
	(3, 7, 8)
3	(1, 2, 3, 8) (1, 2, 7, 8)
	(1, 3, 7, 8)
4	(1, 2, 3, 7, 8)

because Solutions 4, 5, 6, and 7 have the same distance (equal to 4) to Solution 3. So, Solution 7 achieves the maximum–minimum distance from the set of candidate solutions.

It is important to note that the reference set does not consist of all the "best" solutions as measured by the objective function value only. Because we are interested in a balance between high quality solutions and diverse solutions, improved Solution 6 with an objective value of thirtynine is bypassed in favor of other solutions that will add more diversity to the set. As indicated in Table 4, the distance between Solution 1 and Solution 6 is only one unit, making this solution unattractive from the standpoint of diversification.

3.4. Subset Generation Method

This method consists of generating the subsets that will be used for creating structured combinations in the next step. The method is typically designed to generate the following types of subsets:

1. All two-element subsets.
2. Three-element subsets derived from the two-element subsets by augmenting each two-element subset to include the best solution (as measured by the objective value) not in this subset.
3. Four-element subsets derived from the three-element subsets by augmenting each three-element subset to include the best solution (as measured by the objective value) not in this subset.
4. The subsets consisting of the best i elements (as measured by the objective value), for $i = 5$ to b.

Table 5 shows the subsets generated using our current reference set. As expected, there are ten possible subsets of type 1 in Table 5. Type 2 subsets are formed by augmenting the two-element subsets. The first two-element subset is formed by Solutions 1 and 2, and we add Solution 8 to form the three-element subset (1, 2, 8). This is done starting from each of the two-element subsets. After eliminating repetitions, there are six subsets of type 2. Subsets of type 3 are formed in a similar way, starting this time from the three-element subsets.

3.5. Solution Combination Method

This method uses the subsets generated in the previous step to combine the elements in each subset with the purpose of creating new trial solutions. The combination method is typically problem-specific since it is directly related to the solution representation. Depending on the specific form of the solution combination method, each subset can create one or more new solutions. For illustration purposes, we use a combination method that creates only one solution from each subset. Specifically, our Solution Combination Method calculates a score for each variable, based on the solutions in the subset and their corresponding objective values. The score for variable i that corresponds to the solutions in subset S is calculated with the

Table 6 Score Calculation for Subset Combination (3, 7, 8)

Solution	x_1	x_2	x_3	x_4	x_5	x_6	x_7	x_8	x_9	x_{10}
3	0.322	0.000	0.000	0.322	0.000	0.000	0.322	0.000	0.000	0.322
7	0.000	0.305	0.000	0.305	0.000	0.305	0.000	0.000	0.000	0.305
8	0.000	0.373	0.373	0.373	0.373	0.000	0.000	0.000	0.373	0.000
Total	0.322	0.678	0.373	1.000	0.373	0.305	0.322	0.000	0.373	0.627

following formula:

$$\text{score}\ (i) = \frac{\sum_{j \in S} OV(j)^* x_i^j}{\sum_{j \in S} OV(j)}$$

where $OV(j)$ is the objective value of solution j, and x_i^j is the value of the ith variable in solution j. Then, the trial solution is constructed by rounding the score for each variable to the nearest integer, that is,

$$x_i' = \begin{cases} 1 & \text{if score } (i) > 0.5 \\ 0 & \text{if score } (i) \leqslant 0.5 \end{cases}.$$

Note that the combination mechanism constructs solutions that may be infeasible. This does not represent a problem in general, since the improvement method is always applied to each trial solution created after the application of the solution combination method. Recall that the improvement method must be designed to deal with either feasible or infeasible starting solutions.

To illustrate the use of the solution combination method, consider the subset type 2 given by solutions 3, 7 and 8. The objective value of solution 3 is 38, the objective value of solution 7 is 36 and the objective value of solution 8 is 44, that is, $OV(3) = 38$, $OV(7) = 36$, and $OV(8) = 44$. The score for each variable can be calculated as shown in Table 6.

After rounding the scores in Table 6, the new solution becomes `(0,1,0,1,0,0,0,0,0,1)` with an objective value of 30 and a total weight of 59. This feasible solution can be improved with the improvement method. In this case, the value of x_5 is changed from zero to one, because x_5 has the highest profit-to-weight ratio of the variables that are currently set to zero. The improved solution has an objective value of 40 and a total weight of 88. Since no more variable values can be switched from zero to one without violating the constraint, the improvement process stops.

The new solutions constructed in this step of the procedure are considered for membership in the reference set. A solution may become a member if its objective value is better than the objective value of any of the solutions in the high-quality subset. Alternatively, if a new solution improves the diversity of the reference set, this solution can replace one that is currently in the diverse set. If the reference set is modified, then the subset generation method, the solution combination method, and the improvement method are applied in sequence. The application of these methods continues until the reference set converges (i.e., no elements in the set are replaced). At this point, the diversification method can be applied again from a different seed and the search continues until termination criteria are satisfied.

4. A Summary of the Procedure

In the previous section, we illustrated the operations that take place within each of the methods that shape a scatter search implementation. We now finish our introduction to scatter search with an overall view of the procedure. This outline (or pseudocode) uses the following parameters:

PSize = the size of the set of diverse solutions generated by the diversification generation method

1. Start with $P = \emptyset$. Use the Diversification Generation Method to construct a solution x. Apply the Improvement Method to x to obtain the improved solution x^*. If $x^* \notin P$ then, add x^* to P (i.e., $P = P \cup x^*$), otherwise, discard x^*. Repeat this step until $|P| = PSize$.
2. Order the solutions in P according to their objective function value (where the best overall solution is first on the list).

For (*Iter* = 1 **to** *MaxIter*)

3. Build $RefSet = RefSet_1 \cup RefSet_2$ from P, with $|RefSet| = b$, $|RefSet_1| = b_1$ and $|RefSet_2| = b_2$. Take the first b_1 solutions in P and add them to $RefSet_1$. For each solution x in P-$RefSet$ and y in $RefSet$, calculate a measure of distance or dissimilarity $d(x,y)$. Select the solution x' that maximizes $d_{\min}(x)$, where $d_{\min}(x) = \min_{y \in RefSet} \{d(x,y)\}$. Add x' to $RefSet_2$, until $|RefSet_2| = b_2$. Make *NewElements* = TRUE.

While (*NewElements*) **do**

4. Calculate the number of subsets (*MaxSubset*) that include at least one new element. Make *NewElements* = FALSE.

For (*SubsetCounter* = 1, …, *MaxSubset*) **do**

5. Generate the next subset s from *RefSet* with the Subset Generation Method. This method generates one of four types of subsets with number of elements ranging from 2 to $|RefSet|$. Let subset $s = \{ s_1, \ldots, s_k \}$, for $2 \le k \le |RefSet|$. (We consider that the Subset Generation Method skips subsets for which the elements considered have not changed from previous iterations.)
6. Apply the Solution Combination Method to s to obtain one or more new solutions x_s.
7. Apply the Improvement Method to x_s, to obtain the improved solution x_s^*.

If (x_s^* is not in *RefSet* and the objective function value of x_s^* is better than the objective function value of the worst element in $RefSet_1$) **then**

8. Add x_s^* to $RefSet_1$ and delete the worst element currently in $RefSet_1$. (The worst element is the solution with worst objective value.)
9. Make *NewElements* = TRUE.

Else

If (x_s^* is not in $RefSet_2$ and $d_{\min}(x_s^*)$ is larger than $d_{\min}(x)$ for a solution x in $RefSet_2$) **then**

10. Add x_s^* to $RefSet_2$ and delete the worst element currently in $RefSet_2$. (The worst element is the solution x with the smallest $d_{\min}(x)$ value.)
11. Make *NewElements* = TRUE.

End if

End if

End for

End while

If (*Iter* < *MaxIter*) **then**

12. Build a new set P using the Diversification Generation Method. Initialize the generation process with the solutions currently in $RefSet_1$. That is, the first b_1 solutions in the new P are the best b_1 solutions in the current *RefSet*.

End if

End for

Figure 2 Scatter search outline

b = the size of the reference set.

b_1 = the size of the high-quality subset.

b_2 = the size of the diverse subset.

MaxIter = maximum number of iterations.

The procedure consists of the steps in the outline of Figure 2, where P denotes the set of solutions generated with the diversification generation method and *RefSet* is the set of solution in the reference set.

The procedure starts with the generation of *PSize* distinct solutions. These solutions are originally generated to be diverse and subsequently improved by the application of the improvement method (Step 1). The set *P* of *PSize* solutions is ordered in Step 2, in order to facilitate the task of creating the reference set in Step 3. The reference set (*RefSet*) is constructed with the first b_1 solutions in *P* and b_2 solutions that are diverse with respect to the members in *RefSet*.

The search consists of three main loops: (1) a for-loop that controls the maximum number of iterations; (2) a while-loop that monitors the presence of new elements in the reference set; and (3) a for-loop, that controls the examination of all the subsets with at least one new element. In Step 4, the number of subsets with at least one new element is counted and this value is assigned to *MaxSubset*. Also, the Boolean variable *NewElements* is made FALSE before the subsets are examined, since it is not known whether a new solution will enter the reference set in the current examination of the subsets. The actual generation of the subsets occurs in Step 5. Note that only subsets with at least one new element are generated in this step. A solution is generated in Step 6 by applying the combination method. Step 7 attempts to improve this solution with the application of the improvement method. If the improved solution from Step 7 is better (in terms of the objective function value) than the worst solution in *RefSet*$_1$, then the improved solution becomes a new element of *RefSet*. Note that *RefSet*$_1$ is the subset of the reference set that contains the best solutions as measured by the objective function value. The solution is added in Step 8 and the *NewElements* indicator is switched to TRUE in Step 9.

If a solution is not admitted to the *RefSet* due to its quality, the solution is tested for its diversity merits. If a solution adds diversity to *RefSet*, then the solution is added to the reference set and the less diverse solution is deleted (as indicated in Steps 10 and 11). Finally, Step 12 is performed if additional iterations are still available. This step provides a seed for set *P* by adding the solutions in *RefSet*$_1$ before a new application of the diversification generation method.

Note that the general procedure outlined in Figure 2 can be modified in different ways. One possibility is to eliminate the outer for-loop along with Step 12. In this case, set *P* is generated only one time and the process stops after no new elements are admitted in the reference set. That is the search is abandoned when the while-loop that contains Steps 4 to 11 becomes false. This variation is useful for problems in which the search has to be performed within a relatively short computer time. Also, there are some settings in which a large percentage of the best solutions are found during the first iteration (i.e., when *Iter* = 1). In this case, bypassing additional iterations has a small effect on the average performance (Campos et al., 1999).

A second variation consists of eliminating Steps 10 and 11 and the if-statement associated with these steps. This variation considers that the reference set will be initially constructed with both high-quality solutions and diverse solutions. However, after Step 3, membership to the reference set is obtained only due to quality. Therefore, in addition to eliminating Steps 10 and 11, Step 8 and its associated if-statement are modified as follows:

If (x_s^* is not in *RefSet* and the objective function value of x_s^* is better than the objective function value of the worst element in *RefSet*) **then**
8. Add x_s^* to *RefSet* and delete the worst element currently in *RefSet*. (The worst element is the solution with worst objective value.)

That is, all references to *RefSet*$_1$ are substituted with references to *RefSet*. In other words, after Step 3, the elements of *RefSet* are always ordered according to the objective function value, making element 1 the best and element *b* the worst.

Implementing both of these variations at the same time results in a very aggressive search method that attempts to find high quality solutions fast. While this may be desirable in some settings, there are also settings in which a more extensive search can be afforded, and therefore making the full procedure outlined in Figure 2 more attractive.

References

Campos, V., M. Laguna, and R. Martí. 1999. "Scatter search for the linear ordering problem." Chapter 21, In *New Methods in Optimisation,* edited by D. Corne, M. Dorigo, and F. Glover. London: McGraw-Hill.

Glover, F. 1977. "Heuristics for integer programming using surrogate constraints." *Decision Sci.* **8(1)**: 156–166.

Glover, F. 1994. "Tabu search for nonlinear and parametric optimization (with links to genetic algorithms)." *Discrete Appl. Math.* **49**: 231–255.

Glover, F. 1998. "A template for scatter search and path relinking." In *Artificial Evolution. Lecture Notes in Computer Science,* Volume 1363, edited by J.-K. Hao, E. Lutton, E. Ronald, M. Schoenauer, and D. Snyers. Berlin: Springer Verlag.

Glover, F. 1999. "Scatter search and path relinking." Chapter 19, In *New Methods in Optimisation,* edited by D. Corne, M. Dorigo, and F. Glover. London: McGraw-Hill.

3.6.7 Tabu Search

Fred Glover and Manuel Laguna

ABSTRACT This article explores the metaheuristic approach called tabu search, which is dramatically changing our ability to solve a host of problems in applied science, business, and engineering. Tabu search has important links to evolutionary and genetic methods, often overlooked, through its intimate connection with scatter search and path relinking, evolutionary procedures that have recently attracted attention for their ability to facilitate the solution of complex problems. The adaptive memory designs of tabu search have also provided useful alternatives and supplements to the types of memory embodied in neural networks, allowing enhancements of neural network processes in practical settings.

1. Tabu Search Background and Relevance

Faced with the challenge of solving hard optimization problems that abound in the real world, classical methods often encounter great difficulty. Vitally important applications in business, engineering, economics, and science cannot be tackled with any reasonable hope of success, within practical time horizons, by solution methods that have been the predominant focus of academic research throughout the past three decades (and which are still the focus of many textbooks).

The metaheuristic approach called tabu search (TS) is dramatically changing our ability to solve problems of practical significance. Current applications of TS span the realms of resource planning, telecommunications, VLSI design, financial analysis, scheduling, space planning, energy distribution, molecular engineering, logistics, pattern classification, flexible manufacturing, waste management, mineral exploration, biomedical analysis, environmental conservation, and scores of others. In recent years, journals in a wide variety of fields have published tutorial articles and computational studies documenting successes by TS in extending the frontier of problems that can be handled effectively—yielding solutions whose quality often significantly surpasses that obtained by methods previously applied. Table 1 gives a partial catalog of example applications. A more comprehensive list, including summary descriptions of gains achieved from practical implementations, can be found in Glover and Laguna (1997). Reports of recent TS implementations can also be found on the Web page `http://www.upt.pt/tabusearch`.

A distinguishing feature of TS is embodied in its exploitation of adaptive forms of memory, which equips it to penetrate complexities that often confound alternative approaches. Yet we are only beginning to tap the rich potential of adaptive memory strategies, and the discoveries that lie ahead promise to be as important and exciting as those made to date. The knowledge and principles that have emerged from the TS framework give a foundation to create practical systems whose capabilities markedly exceed those available earlier. At the same

Table 1 Illustrative Tabu Search Applications

Scheduling	Flow-Time Cell Manufacturing Heterogeneous Processor Scheduling Workforce Planning Classroom Scheduling Machine Scheduling Flow Shop Scheduling Job Shop Scheduling Sequencing and Batching
Design	Computer-Aided Design Fault Tolerant Networks Transport Network Design Architectural Space Planning Diagram Coherency Fixed Charge Network Design Irregular Cutting Problems
Location and Allocation	Multicommodity Location/Allocation Quadratic Assignment Quadratic Semiassignment Multilevel Generalized Assignment Layout Planning Offshore Oil Exploration
Logic and Artificial Intelligence	Maximum Satisfiability Probabilistic Logic Clustering Pattern Recognition/Classification Data Integrity Neural Network/Training and Design
Technology	Seismic Inversion Electrical Power Distribution Engineering Structural Design Minimum Volume Ellipsoids Space Station Construction Circuit Cell Placement
Telecommunications	Call Routing Bandwidth Packing Hub Facility Location Path Assignment Network Design for Services Customer Discount Planning Failure Immune Architecture Synchronous Optical Networks
Production, Inventory, and Investment	Flexible Manufacturing Just-In-Time Production Capacitated MRP Part Selection Multi-Item Inventory Planning Volume Discount Acquisition Fixed Mix Investment
Routing	Vehicle Routing Capacitated Routing Time Window Routing

	Multimode Routing Mixed Fleet Routing Traveling Salesman Traveling Purchaser
Graph Optimization	Graph Partitioning Graph Coloring Clique Partitioning Maximum Clique Problems Maximum Planner Graphs p-Median Problems
General Combinational Optimization	Zero–One Programming Fixed Charge Optimization Nonconvex Nonlinear Programming All-or-None Networks Bilevel Programming General Mixed-Integer Optimization

time, there are many untried variations that may lead to further advances. A conspicuous feature of TS is that it is dynamically growing and evolving, drawing on important contributions by many researchers. (The Web site previously cited provides links to the work of these researchers.)

1.1. General Tenets

The word tabu (or taboo) comes from Tongan, a language of Polynesia, where it was used by the aborigines of Tonga island to indicate things that cannot be touched because they are sacred. According to *Webster's Dictionary*, the word now also means "a prohibition imposed by social custom as a protective measure" or of something "banned as constituting a risk." These current more pragmatic senses of the word accord well with the theme of TS. The risk to be avoided in this case is that of following a counter-productive course, including one which may lead to entrapment without hope of escape. On the other hand, as in the broader social context where "protective prohibitions" are capable of being superseded when the occasion demands, the "tabus" of TS are to be overruled when evidence of a preferred alternative becomes compelling.

The most important association with traditional usage, however, stems from the fact that tabus as normally conceived are transmitted by means of a social memory, which is subject to modification over time. This creates the fundamental link to the meaning of tabu in TS. The forbidden elements of TS receive their status by reliance on an evolving memory, which allows this status to shift according to time and circumstance.

More particularly, TS is based on the premise that problem solving, in order to qualify as intelligent, must incorporate adaptive memory and responsive exploration. The adaptive memory feature of TS allows the implementation of procedures that are capable of searching the solution space economically and effectively. Since local choices are guided by information collected during the search, TS contrasts with memoryless designs that heavily rely on semirandom processes that implement a form of sampling. Examples of memoryless methods include semigreedy heuristics and the prominent genetic and annealing approaches inspired by metaphors of physics and biology. Adaptive memory also contrasts with rigid memory designs typical of branch-and-bound strategies. (It can be argued that some types of evolutionary procedures that operate by combining solutions, such as genetic algorithms, embody a form of implicit memory. Special links with evolutionary methods, and implications for establishing more effective variants of them, are discussed in Glover and Laguna, 1997.)

The emphasis on responsive exploration in TS, whether in a deterministic or probabilistic implementation, derives from the supposition that a bad strategic choice can yield more

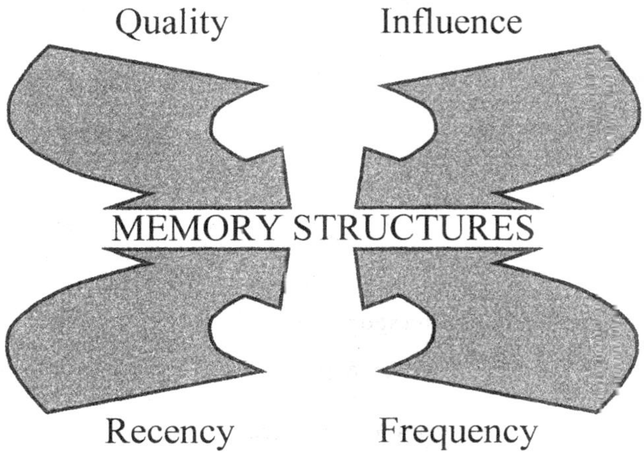

Figure 1 Four tabu search dimensions

information than a good random choice. In a system that uses memory, a bad choice based on strategy can provide useful clues about how the strategy may profitably be changed. (Even in a space with significant randomness a purposeful design can be more adept at uncovering the imprint of structure.)

Responsive exploration integrates the basic principles of intelligent search, that is, exploiting good solution features while exploring new promising regions. TS is concerned with finding new and more effective ways of taking advantage of the mechanisms associated with both adaptive memory and responsive exploration. The development of new designs and strategic mixes makes TS a fertile area for research and empirical study.

1.2. Use of Memory

The memory structures in TS operate by reference to four principal dimensions, consisting of recency, frequency, quality, and influence (Figure 1). Recency-based and frequency-based memory complement each other, and have important characteristics we amplify in later sections. The quality dimension refers to the ability to differentiate the merit of solutions visited during the search. In this context, memory can be used to identify elements that are common to good solutions or to paths that lead to such solutions. Operationally, quality becomes a foundation for incentive-based learning, where inducements are provided to reinforce actions that lead to good solutions and penalties are provided to discourage actions that lead to poor solutions. The flexibility of these memory structures allows the search to be guided in a multiobjective environment, where the goodness of a particular search direction may be determined by more than one function. The TS concept of quality is broader than the one implicitly used by standard optimization methods.

The fourth dimension, influence, considers the impact of the choices made during the search, not only on quality but also on structure. (In a sense, quality may be regarded as a special form of influence.) Recording information about the influence of choices on particular solution elements incorporates an additional level of learning. By contrast, in branch-and-

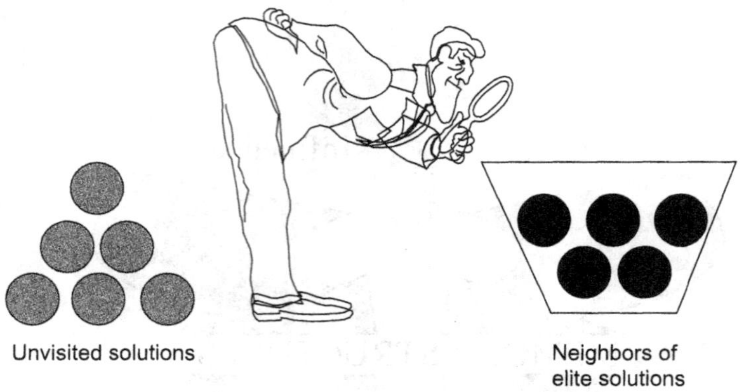

Figure 2 Intensification and diversification

bound, for example, the separation rules are prespecified and the branching directions remain fixed, once selected, at a given node of a decision tree. It is clear, however, that certain decisions have more influence than others as a function of the neighborhood of moves employed, and the way that this neighborhood is negotiated (e.g., choices near the root of a branch-and-bound tree are quite influential when using a depth-first strategy). The assessment and exploitation of influence by a memory more flexible than embodied in such tree searches is an important feature of the TS framework.

The memory used in TS is both explicit and attributive. Explicit memory records complete solutions, typically consisting of elite solutions visited during the search. An extension of this memory records highly attractive but unexplored neighbors of elite solutions. The memorized elite solutions (or their attractive neighbors) are used to expand the local search.

Alternatively, TS uses attributive memory for guiding purposes. This type of memory records information about solution attributes that change in moving from one solution to another. For example, in a graph or network setting, attributes can consist of nodes or arcs that are added, dropped or repositioned by the moving mechanism. In production scheduling, the index of jobs may be used as attributes to inhibit or encourage the method to follow certain search directions.

1.3. *Intensification and Diversification*

Two highly important components of TS are intensification and diversification strategies. Intensification strategies are based on modifying choice rules to encourage move combinations and solution features historically found to be good. They may also initiate a return to attractive regions to search them more thoroughly. Since elite solutions must be recorded in order to examine their immediate neighborhoods, explicit memory is closely related to the implementation of intensification strategies. As Figure 2 illustrates, the main difference between intensification and diversification is that during an intensification stage the search focuses on examining neighbors of elite solutions.

Here the term "neighbors" has a broader meaning than in the usual context of neighborhood search. That is, in addition to considering solutions that are adjacent or close to elite solutions by means of standard move mechanisms, intensification strategies generate neighbors by either grafting together components of good solutions or by using modified evaluation strategies that favor the introduction of such components into a current (evolving) solution. The diversification stage, on the other hand, encourages the search process to examine unvisited regions and to generate solutions that differ in various significant ways from those seen before. Again, such an approach can be based on generating subassemblies of solution components that are then "fleshed out" to produce full solutions, or can rely on modified evaluations as embodied, for example, in the use of penalty/incentive functions.

Intensification strategies require a means for identifying a set of elite solutions as a basis for incorporating good attributes into newly created solutions. Membership in the elite set is often determined by setting a threshold that is connected to the objective function value of the best solution found during the search. However, considerations of clustering and "anticlustering" are also relevant for generating such a set, and more particularly for generating subsets of solutions that may be used for specific phases of intensification and diversification. In the following sections, we show how the treatment of such concerns can be enhanced by making use of special memory structures. The TS notions of intensification and diversification are beginning to find their way into other metaheuristics. It is important to keep in mind that these ideas are somewhat different than the old control theory concepts of exploitation and exploration, especially in their implications for developing effective problem solving strategies.

2. Tabu Search Foundations and Short Term Memory

TS can be applied directly to verbal or symbolic statements of many kinds of decision problems, without the need to transform them into mathematical formulations. Nevertheless, it is useful to introduce mathematical notation to express a broad class of these problems, as a basis for describing certain features of TS. We characterize this class of problems as that of optimizing (minimizing or maximizing) a function $f(x)$ subject to $x \in \mathbf{X}$, where $f(x)$ may be linear or nonlinear, and the set $\mathbf{X}$ summarizes constraints on the vector of decision variables x. The constraints may include linear or nonlinear inequalities, and may compel all or some components of x to receive discrete values. While this representation is useful for discussing a number of problem solving considerations, we emphasize again that in many applications of combinatorial optimization, the problem of interest may not be easily formulated as an objective function subject to a set of constraints. The requirement $x \in \mathbf{X}$, for example, may specify logical conditions or interconnections that would be cumbersome to formulate mathematically, but may be better left as verbal stipulations that can then be coded as rules.

TS begins in the same way as an ordinary local or neighborhood search, proceeding iteratively from one point (solution) to another until a chosen termination criterion is satisfied. Each $x \in \mathbf{X}$ has an associated neighborhood $\mathbf{N}(x) \subset \mathbf{X}$, and each solution $x' \in \mathbf{N}(x)$ is reached from x by an operation called a move.

As an initial point of departure, we may contrast TS with a simple descent method where the goal is to minimize $f(x)$ (or a corresponding ascent method where the goal is to maximize $f(x)$). Such a method only permits moves to neighbor solutions that improve the current objective function value and ends when no improving solutions can be found. A pseudocode of a generic descent method is presented in Figure 3. The final x obtained by a descent method is called a local optimum, since it is at least as good or better than all solutions in its neighborhood. The evident shortcoming of a descent method is that such a local optimum in most cases will not be a global optimum, that is, it usually will not minimize $f(x)$ over all $x \in \mathbf{X}$.

The version of a descent method called steepest descent scans the entire neighborhood of x in search of a neighbor solution x' that gives a smallest $f(x')$ value over $x' \in \mathbf{N}(x)$. Steepest descent implementations of some types of solution approaches (such as certain path

1. Choose $x \in \mathbf{X}$ to start the process.
2. Find $x' \in \mathbf{N}(x)$ such that $f(x') < f(x)$.
3. If no such x' can be found, x is the local optimum and the method stops.
4. Otherwise, designate x' to be the new x and go to 2 .

Figure 3 Descent method

augmentation algorithms in networks and matroids) are guaranteed to yield globally optimal solutions for the problems they are designed to handle, while other forms of descent may terminate with local optima that are not global optima. In spite of this attractive feature, in certain settings steepest descent is sometimes impractical because it is computationally too expensive, as where $\mathbf{N}(x)$ contains many elements or each element is costly to retrieve or evaluate. Still, it is often valuable to choose an x' at each iteration that yields a good if not smallest $f(x')$ value.

2.1. *Candidate List Strategies*

The relevance of choosing good solutions from current neighborhoods is magnified when the guidance mechanisms of TS are introduced to go beyond the locally optimal termination point of a descent method. Thus, an important first level consideration for TS is to determine an appropriate candidate list strategy for narrowing the examination of elements of $\mathbf{N}(x)$, in order to achieve an effective trade-off between the quality of x' and the effort expended to find it. Here quality may involve considerations beyond those narrowly reflected by the value of $f(x')$. If a neighborhood space is totally random, then of course nothing will work better than a totally random choice. (In such a case there is no merit in trying to devise an effective solution procedure.) Assuming that neighborhoods can be identified that are reasonably meaningful for a given class of problems, the challenge is to define solution quality appropriately so that evaluations, likewise, will have meaning. By the TS orientation, the ability to use history in creating such evaluations then becomes important for devising effective methods. This is also true in the use of candidate list strategies. Several key forms of such strategies are highlighted in Glover and Laguna (1997).

3. Memory and Tabu Classifications

An important distinction in TS arises by differentiating between short-term memory and longer-term memory. Each type of memory is accompanied by its own special strategies. However, the effect of both types of memory may be viewed as modifying the neighborhood $\mathbf{N}(x)$ of the current solution x. The modified neighborhood, which we denote by $\mathbf{N}^*(x)$, is the result of maintaining a selective history of the states encountered during the search.

In the TS strategies based on short-term considerations, $\mathbf{N}^*(x)$ is characteristically a subset of $\mathbf{N}(x)$, and the tabu classification serves to identify elements of $\mathbf{N}(x)$ excluded from $\mathbf{N}^*(x)$. In TS strategies that include longer-term considerations, $\mathbf{N}^*(x)$ may also be expanded to include solutions not ordinarily found in $\mathbf{N}(x)$. Characterized in this way, TS may be viewed as a dynamic neighborhood method. This means that the neighborhood of x is not a static set, but rather a set that can change according to the history of the search. This feature of a dynamically changing neighborhood also applies to the consideration of selecting different component neighborhoods from a compound neighborhood that encompasses multiple types or levels of moves, and provides an important basis for parallel processing. Characteristically, a TS process based strictly on short-term strategies may allow a solution x to be visited more than once, but it is likely that the corresponding reduced neighborhood $\mathbf{N}^*(x)$ will be different each time. With the inclusion of longer-term considerations, the likelihood of duplicating a previous neighborhood upon revisiting a solution, and more generally of making choices that repeatedly visit only a limited subset of $\mathbf{X}$, is all but nonexistent. From a practical standpoint, the method will characteristically identify an optimal or near optimal solution long before a substantial portion of $\mathbf{X}$ is examined.

A crucial aspect of TS involves the choice of an appropriate definition of $\mathbf{N}^*(x)$. Due to the exploitation of memory, $\mathbf{N}^*(x)$ depends upon the trajectory followed in moving from one solution to the next (or upon a collection of such trajectories in a parallel processing environment).

The approach of storing complete solutions (explicit memory) is generally used in a highly selective manner, because it can consume an enormous amount of space and time when applied to each solution generated. A scheme that emulates this approach with limited memory requirements is given by the use of hash functions. (Also, explicit memory has a valuable role when selectively applied in strategies that record and analyze certain "special" solutions.) Regardless of the implementation details, short-term and long-term memory functions provide important cornerstones of the TS methodology. These functions characteristically make use of recency-based memory and frequency-based memory. For the purposes of illustration, due to the limited space available in this article, we will restrict our discussion to short-term recency-based memory.

3.1. *Recency-Based Memory*

Recency-based memory, as its name suggests, keeps track of solutions attributes that have changed during the recent past. (A liberal definition of recent also allows such memory to have a role in longer-term strategies.)

To exploit this memory, selected attributes that occur in solutions recently visited are labeled tabu-active, and solutions that contain tabu-active elements, or particular combinations of these attributes, are those that become tabu. This prevents certain solutions from the recent past from belonging to $\mathbf{N}^*(x)$ and hence from being revisited. Other solutions that share such tabu-active attributes are also similarly prevented from being visited. Note that while the tabu classification strictly refers to solutions that are forbidden to be visited, by virtue of containing tabu-active attributes (or more generally by violating certain restrictions based on these attributes), we also often refer to moves that lead to such solutions as being tabu. We illustrate these points with the following example.

MINIMUM *K*-TREE PROBLEM EXAMPLE

The Minimum k-Tree problem seeks a tree consisting of k edges in a graph so that the sum of the weights of these edges is minimum (Lokketangen et al., 1994). An instance of this problem is given in Figure 4, where nodes are shown as numbered circles, and edges are shown as lines that join pairs of nodes (the two endpoint nodes that determine the edge). Edge weights are shown as the numbers attached to these lines. A tree is a set of edges that contains no cycles, that is, that contains no paths that start and end at the same node (without retracing any edges).

Assume that the move mechanism is defined by edge-swapping, as subsequently described, and that a greedy procedure is used to find an initial solution. The greedy construction starts by choosing the edge (i, j) with the smallest weight in the graph, where i and j are the indexes of the nodes that are the endpoints of the edge. The remaining $k-1$ edges are chosen successively to minimize the increase in total weight at each step, where the

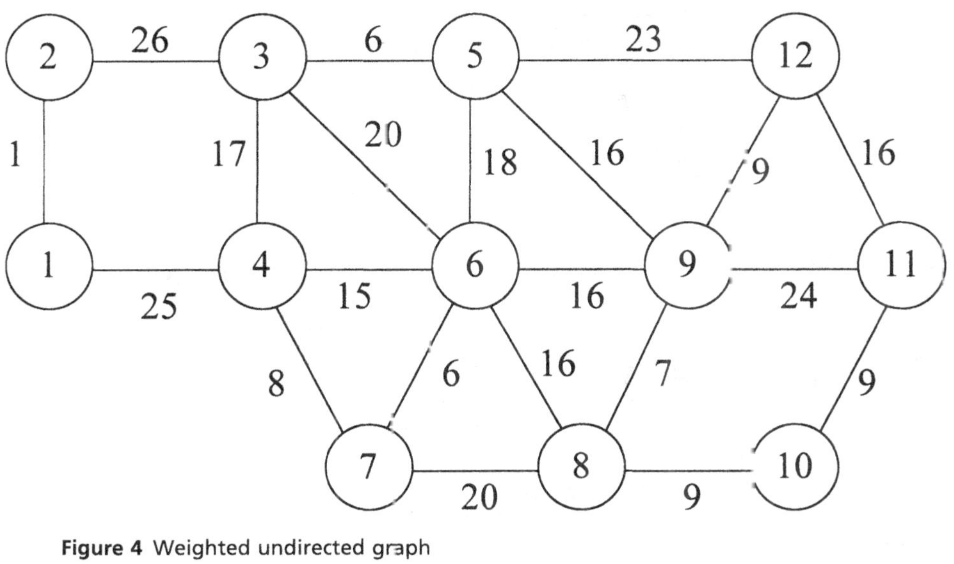

Figure 4 Weighted undirected graph

Table 2 Greedy Construction

Step	*Candidates*	*Selection*	*Total Weight*
1	(1, 2)	(1, 2)	1
2	(1, 4), (2, 3)	(1, 4)	26
3	(2, 3), (3, 4), (4, 6), (4, 7)	(4, 7)	34
4	(2, 3), (3, 4), (4, 6), (6, 7), (7, 8)	(6, 7)	40

edges considered meet exactly one node from those that are endpoints of edges previously chosen. For $k = 4$, the greedy construction performs the steps in Table 2.

The construction starts by choosing edge (1, 2) with a weight of 1 (the smallest weight of any edge in the graph). After this selection, the candidate edges are those that connect the nodes in the current partial tree with those nodes not in the tree [i.e., edges (1, 4) and (2, 3)]. Since edge (1, 4) minimizes the weight increase, it is chosen to be part of the partial solution. The rest of the selections follow the same logic, and the construction ends when the tree consists of 4 edges (i.e., the value of k). The initial solution in this particular case has a total weight of 40.

The swap move mechanism, which is used from this point onward, replaces a selected edge in the tree by another selected edge outside the tree, subject to requiring that the resulting subgraph is also a tree. There are actually two types of such edge swaps, one that maintains the current nodes of the tree unchanged (static) and one that results in replacing a node of the tree by a new node (dynamic). Figure 5 illustrates the best swap of each type that can be made starting from the greedy solution. The added edge in each case is shown by a heavy line and the dropped edge is shown by a dotted line.

The best move of both types is the static swap of Figure 6, where for our present illustration we are defining best solely in terms of the change on the objective function value.

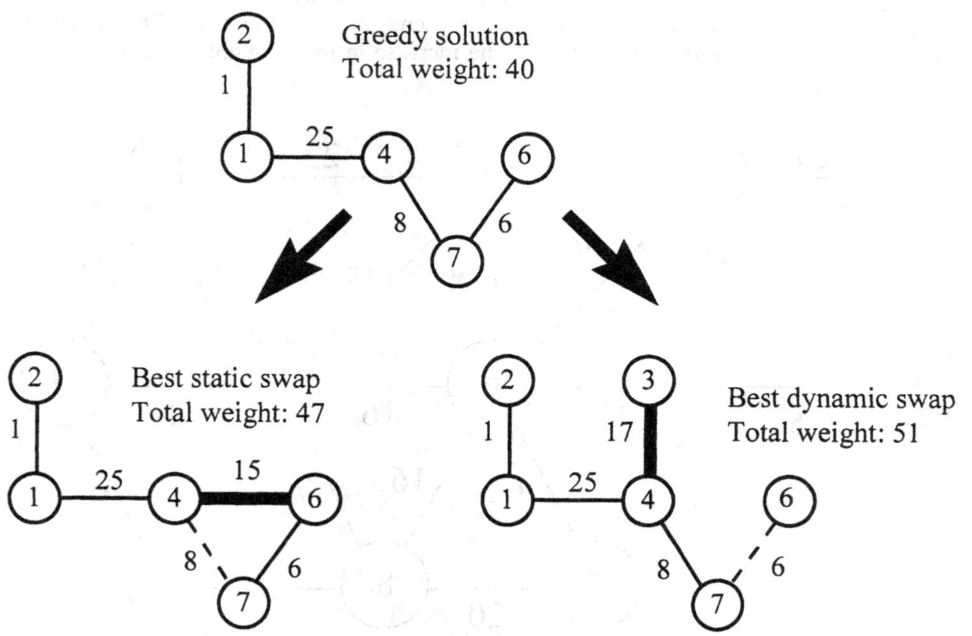

Figure 5 Swap move types

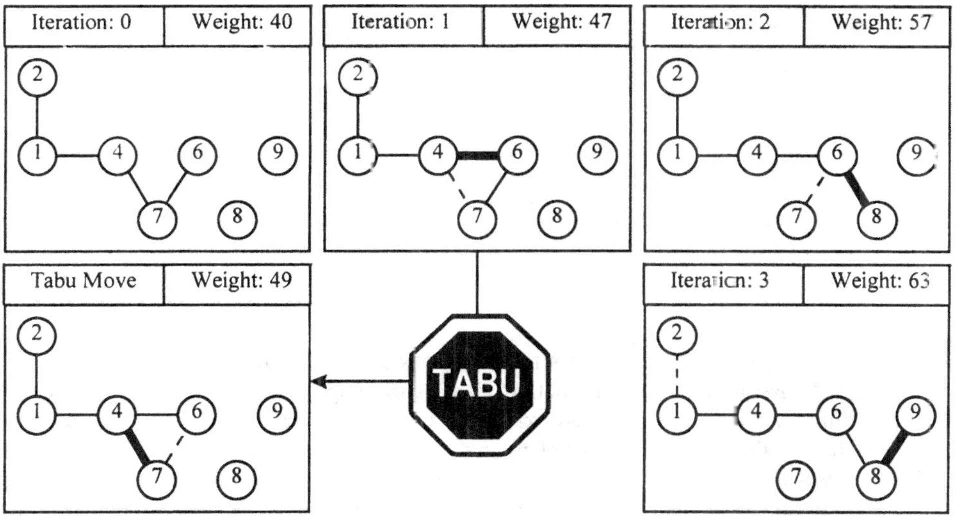

Figure 6 Effects of attributive short term memory

Since this best move results in an increase of the total weight of the current solution, the execution of such a move abandons the rules of a descent approach and sets the stage for a TS process. (The feasibility restriction that requires a tree to be produced at each step is particular to this illustration, since in general the TS methodology may include search trajectories that violate various types of feasibility conditions.)

Given a move mechanism, such as the swap mechanism we have selected for our example, the next step is to choose the key attributes that will be used for the tabu classification. TS is very flexible at this stage of the design. Problem-specific knowledge can be used as guidance to settle on a particular design. In problems where the moves are defined by adding and deleting elements, the labels of these elements can be used as the attributes for enforcing tabu status. Here, in the present example, we can simply refer to the edges as attributes of the move, since the condition of being in or out of the tree (which is a distinguishing property of the current solution) may be assumed to always be automatically known by a reasonable solution representation.

Choosing Tabu Classifications

Tabu classifications do not have to be symmetric, that is, the tabu structure can be designed to treat added and dropped elements differently. Suppose for example that after choosing the static swap of Figure 5, which adds edge (4, 6) and drops edge (4, 7), a tabu status is assigned to both of these edges. Then one possibility is to classify both of these edges tabu-active for the same number of iterations. The tabu-active status has different meanings depending on whether the edge is added or dropped. For an added edge, tabu-active means that this edge is not allowed to be dropped from the current tree for the number of iterations that defines its tabu tenure. For a dropped edge, on the other hand, tabu-active means the edge is not allowed to be included in the current solution during its tabu tenure. Since there are many more edges outside the tree than in the tree, it seems reasonable to implement a tabu structure that keeps a recently dropped edge tabu-active for a longer period of time than a recently added edge. Notice also that for this problem the tabu-active period for added edges is bounded by k, since if no added edge is allowed to be dropped for k iterations, then within k steps all available moves will be classified tabu.

The concept of creating asymmetric tabu classifications can be readily applied to settings where add/drop moves are not used.

Table 3 Tabu Search Iterations

Iteration	*Tabu-Active Net Tenure* 1	2	*Add*	*Drop*	*Weight*
1			(4, 6)	(4, 7)	47
2	(4, 6)	(4, 7)	(6, 8)	(6, 7)	57
3	(6, 8), (4, 7)	(6, 7)	(8, 9)	(1, 2)	63

Illustrative Tabu Classifications for the Min *k*-Tree Problem

As previously remarked, the tabu-active classification may in fact prevent the search from visiting solutions that have not been examined yet. We illustrate this phenomenon as follows. Suppose that in the Min *k*-Tree problem instance of Figure 4, dropped edges are kept tabu-active for two iterations, while added edges are kept tabu-active for only one iteration. (The number of iterations an edge is kept tabu-active is called the tabu tenure of the edge.) Also assume that we define a swap move to be tabu if either its added or dropped edge is tabu-active. If we examine the full neighborhood of available edge swaps at each iteration, and always choose the best that is not tabu, then the first three moves are as shown in Table 3 (starting from the initial solution found by the greedy construction heuristic). The move of iteration 1 is the static swap move previously identified in Figure 5. Diagrams showing the successive trees generated by these moves, starting with the initial greedy solution, are given in Figure 6.

The net tenure values of 1 and 2 in Table 3 for the currently tabu-active edges indicate the number of iterations that these edges will remain tabu-active (including the current iteration).

At iteration 2, the reversal of the move of iteration 1 [that is, the move that now adds (4, 7) and drops (4, 6)] is clearly tabu, since both of its edges are tabu-active at iteration 2. In addition, the move that adds (4, 7) and drops (6, 7) is also classified tabu, because it contains the tabu-active edge (4, 7) (with a net tenure of 2). This move leads to a solution with a total weight of 49, a solution that clearly has not been visited before (see Figure 6). The tabu-active classification of (4, 7) has modified the original neighborhood of the solution at iteration 2, and has forced the search to choose a move with an inferior objective function value (i.e., the one with a total weight of 57). In this case, excluding the solution with a total weight of 49 has little effect on the quality of the best solution found (since we have already obtained one with a weight of 40).

In other situations, however, additional precautions must be taken to avoid missing good solutions. These strategies are known as aspiration criteria. Alternative forms of aspiration criteria are very important in TS. For the moment we observe simply that if the tabu solution encountered at the current step had a weight of 39 instead, which is better than the best weight of 40 so far seen, then we would allow the tabu classification of this solution to be overridden and consider the solution admissible to be visited. The aspiration criterion that applies in this case is called the improved-best aspiration criterion. (It is important to keep in mind that aspiration criteria do not compel particular moves to be selected, but simply make them available, or alternately rescind evaluation penalties attached to certain tabu classifications.)

One other comment about tabu classification deserves to be made at this point. In our preceding discussion of the Min *k*-Tree problem we consider a swap move tabu if either its added edge or its dropped edge is tabu-active. However, we could instead stipulate that a swap move is tabu only if both its added and dropped edges are tabu-active. In general, the tabu status of a move is a function of the tabu-active attributes of the move (i.e., of the new solution produced by the move).

Table 4 Iterations of a First Level Tabu Search Procedure

Iteration	*Tabu-Active Net Tenure* 1	2	*Add*	*Drop*	*Move Value*	*Weight*
3	(6, 8), (4, 7)	(6, 7)	(8, 9)	(1, 2)	6	63
4	(6, 7), (8, 9)	(1, 2)	(4, 7)	(1, 4)	−17	46
5	(1, 2), (4, 7)	(1, 4)	(6, 7)	(4, 6)	−9	37*
6	(1, 4), (6, 7)	(4, 6)	(6, 9)	(6, 8)	0	37
7	(4, 6), (6, 9)	(6, 8)	(8, 10)	(4, 7)	1	38
8	(6, 8), (8, 10)	(4, 7)	(9, 12)	(6, 7)	3	41
9	(4, 7), (9, 12)	(6, 7)	(10, 11)	(6, 9)	−7	34*
10	(6, 7), (10, 11)	(6, 9)	(5, 9)	(9, 12)	7	41

3.2. A First-Level Tabu Search Approach

We now have on hand enough ingredients for a first-level TS procedure. Such a procedure is sometimes implemented in an initial phase of a TS development to obtain a preliminary idea of performance and calibration features, or simply to provide a convenient staged approach for the purpose of debugging solution software. While this naive form of a TS method omits a number of important short term memory considerations, and does not yet incorporate longer term concerns, it nevertheless gives a useful starting point for demonstrating several basic aspects of TS.

We start from the solution with a weight of 63 as shown previously in Figure 6, which was obtained at iteration 3. At each step we select the least weight nontabu move from those available, and use the improved-best aspiration criterion to allow a move to be considered admissible in spite of leading to a tabu solution. The reader may verify that the outcome leads to the series of solutions shown in Table 4, which continues from iteration 3 (just executed). For simplicity, we select an arbitrary stopping rule that ends the search at iteration 10.

The successive solutions identified in Table 4 are shown graphically in Figure 7. In addition to identifying the dropped edge at each step as a dotted line, we also identify the dropped edge from the immediately preceding step as a dotted line, which is labeled 2*, to indicate its current net tabu tenure of two. Similarly, we identify the dropped edge from one further step back by a dotted line, which is labeled 1*, to indicate its current net tabu tenure of one. Finally, the edge that was added on the immediately preceding step is also labeled 1* to indicate that it likewise has a current net tabu tenure of one. Thus, the edges that are labeled with tabu tenures are those that are currently tabu-active, and which are excluded from being chosen by a move of the current iteration (unless permitted to be chosen by the aspiration criterion).

As illustrated in Table 4 and Figure 7, the method continues to generate different solutions, and over time the best-known solution (denoted by an asterisk) progressively improves. In fact, it can be verified for this simple example that the solution obtained at iteration 9 is optimal. (In general, of course, there is no known way to verify optimality in polynomial time for difficult discrete optimization problems, i.e., those that fall in the class called NP-hard. The Min *k*-Tree problem is one of these.)

It may be noted that at iteration 6 the method selected a move with a move value of zero. Nevertheless, the configuration of the current solution changes after the execution of this move, as illustrated in Figure 7.

The selection of moves with certain move values, such as zero move values, may be strategically controlled, to limit their selection as added insurance against cycling in special

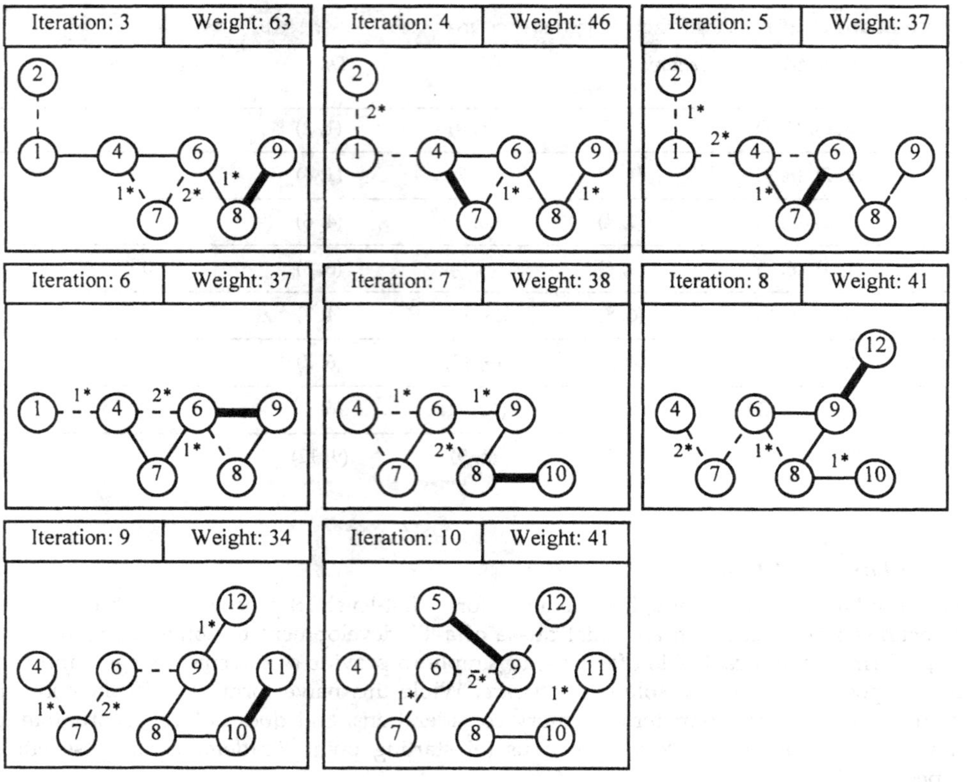

Figure 7 Graphical representation of tabu search iterations

settings. We will soon see how considerations beyond this first level implementation can lead to an improved search trajectory, but the nonmonotonic, gradually improving, behavior is characteristic of TS in general. Figure 8 provides a graphic illustration of this behavior for the current example.

We have purposely chosen the stopping iteration to be small to illustrate an additional relevant feature, and to give a foundation for considering certain types of longer term considerations. One natural way to apply TS is to periodically discontinue its progress, particularly if its rate of finding new best solutions falls below a preferred level, and to restart the method by a process designated to generate a new sequence of solutions.

Classical restarting procedures based on randomization can evidently be used for this purpose, but TS often derives an advantage by employing more strategic forms of restarting. We illustrate a simple instance of such a restarting procedure, which also serves to introduce a useful memory concept.

Critical Event Memory

Critical event memory in TS, as its name implies, monitors the occurrence of certain critical events during the search, and establishes a memory that constitutes an aggregate summary of these events. For our current example, where we seek to generate a new starting solution, a critical event that is clearly relevant is the generation of the previous starting solution. Correspondingly, if we apply a restarting procedure multiple times, the steps of generating all preceding starting solutions naturally qualify as critical events. That is, we would prefer to depart from these solutions in some significant manner as we generate other starting solutions.

Different degrees of departure, representing different levels of diversification, can be achieved by defining solutions that correspond to critical events in different ways (and by

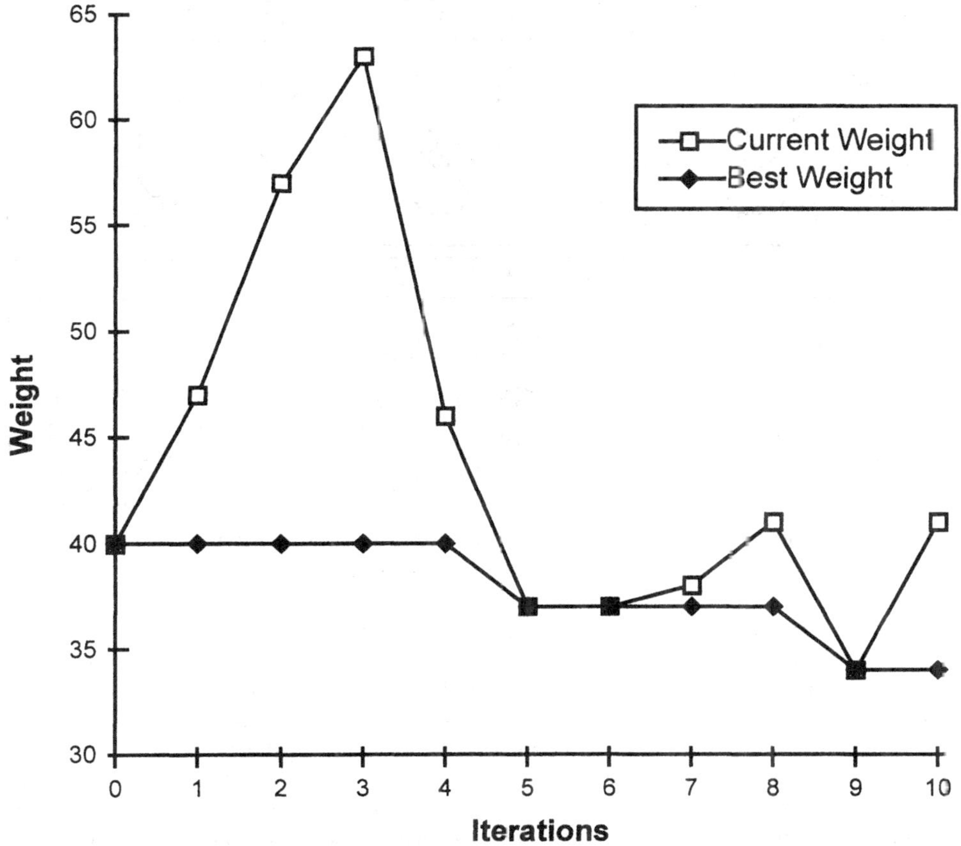

Figure 8 Tabu search trajectory

activating critical event memory by different rules). In the present setting we consider it important that new starting solutions not only differ from preceding starting solutions, but that they also differ from other solutions generated during previous passes. One possibility is to use a blanket approach that considers each complete solution previously generated to represent a critical event. The aggregation of such events by means of critical event memory makes this entirely practicable, but often it is quite sufficient (and sometimes, preferable) to isolate a smaller set of solutions.

For the current example, therefore, we will specify that the critical events of interest consist of generating not only the starting solution of the previous pass(es), but also each subsequent solution that represents a local TS optimum, that is, whose objective function value is better (or no worse) than that of the solution immediately before and after it. Using this simple definition we see that four solutions qualify as critical (i.e., are generated by the indicated critical events) in the first solution pass of our example: the initial solution and the solutions found at iterations 5, 6, and 9 (with weights of 40, 37, 37, and 34, respectively).

Since the solution at iteration 9 happens to be optimal, we are interested in the effect of restarting before this solution is found. Assume we had chosen to restart after iteration 7, without yet reaching an optimal solution. Then the solutions that correspond to critical events are the initial solution and the solutions of iterations 5 and 6. We treat these three solutions in aggregate by combining their edges, to create a subgraph that consists of the edges (1, 2), (1, 4), (4, 7), (6, 7), (6, 8), (8, 9) and (6, 9). (Frequency-based memory refines this representation by accounting for the number of times each edge appears in the critical solutions, and allows the inclusion of additional weighting factors.)

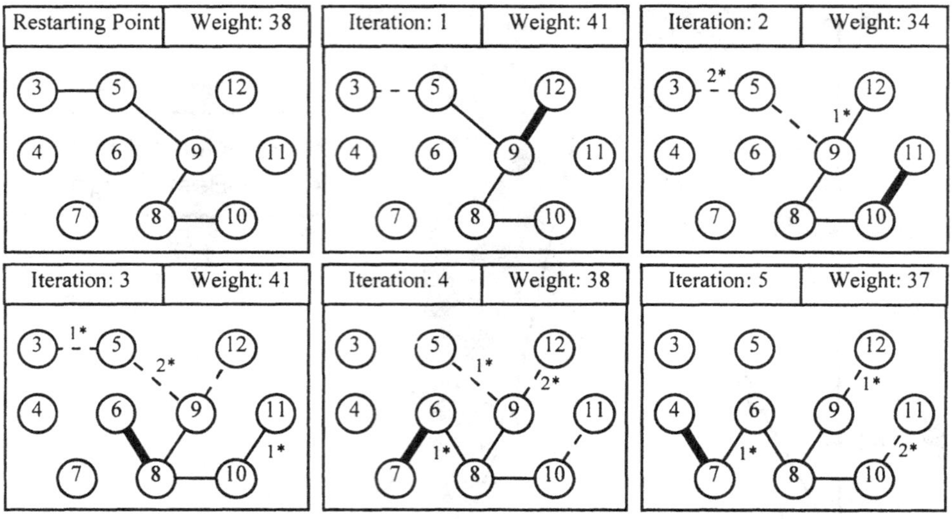

Figure 9 Graphical representation of tabu search iterations after restarting

To execute a restarting procedure, we penalize the inclusion of the edges of this subgraph at various steps of constructing the new solution. It is usually preferable to apply this penalty process during early steps, implicitly allowing the penalty function to decay rapidly as the number of steps increases. It is also sometimes useful to allow one or more intervening steps after applying such penalties before applying them again.

For our illustration, we will use the memory embodied in the subgraph of penalized edges by introducing a large penalty that effectively excludes all these edges from consideration on the first two steps of constructing the new solution. Then, because the construction involves four steps in total, we will not activate the critical event memory on subsequent construction steps, but will allow the method to proceed in its initial form.

Applying this approach, we restart the method by first choosing edge (3, 5), which is the minimum weight edge not in the penalized subgraph. This choice and the remaining choices that generate the new starting solution are shown in Table 5.

Beginning from the solution constructed in Table 5, and applying the first level TS procedure exactly as it was applied on the first pass, generates the sequence of solutions shown in Table 6 and depicted in Figure 9. (Again, we have arbitrarily limited the total number of iterations, in this case to five.)

It is interesting to note that the restarting procedure generates a better solution (with a total weight of 38) than the initial solution generated during the first construction (with a total weight of 40). Also, the restarting solution contains two optimal edges (i.e., edges that appear in the optimal tree). This starting solution allows the search trajectory to find the optimal

Table 5 Restarting Procedure

Step	*Candidates*	*Selection*	*Total Weight*
1	(3, 5)	(3, 5)	6
2	(2, 3), (3, 4), (3, 6), (5, 6), (5, 9), (5, 12)	(5, 9)	22
3	(2, 3), (3, 4), (3, 6), (5, 6), (5, 12), (6, 9), (8, 9), (9, 12)	(8, 9)	29
4	(2, 3), (3, 4), (3, 6), (5, 6), (5, 12), (6, 8), (6, 9), (7, 8), (8, 10), (9, 12)	(8, 10)	38

Table 6 Tabu Search Iterations Following Restarting

Iteration	*Tabu-Active Net Tenure* 1	2	*Add*	*Drop*	*Move Value*	*Weight*
1			(9, 12)	(3, 5)	3	41
2	(9, 12)	(3, 5)	(10, 11)	(5, 9)	−7	34*
3	(3, 5), (10, 11)	(5, 9)	(6, 8)	(9, 12)	7	41
4	(5, 9), (6, 8)	(9, 12)	(6, 7)	(10, 11)	−3	38
5	(9, 12), (6, 7)	(10, 11)	(4, 7)	(8, 10)	−1	37

solution in only two iterations, illustrating the benefits of applying a critical event memory within a restarting strategy. Related memory structures can also be valuable for strategies that drive the search into new regions by partial restarting or by directly continuing a current trajectory (with modified decision rules).

In its complete form, TS contains a variety of elements that go beyond these first level concerns, and that open up possibilities for creating more powerful solution approaches. An associated collection of problem-solving principles is emerging that invites fuller exploration as a basis for exploiting structure in diverse contexts. The reader may consult Glover and Laguna (1997) for further information.

Portions of this manuscript, including the tables and figures, previously appeared in F. Glover and M. Laguna, _Tabu Search_ , Norwell, MA: Kluwer Academic Publishers, 1997. Reprinted with permission.

References

Glover, F. and M. Laguna. 1997. *Tabu Search*. Boston: Kluwer Academic Publishers.
Lokketangen, A., K. Jornsten, and S. Storoy. 1994. "Tabu search within a pivot and complement framework." *Int. Trans. Oper. Res.* **1(3)** 305–316.

3.6.8 Simulated Annealing

E. H. L. Aarts and H. M. M. Ten Eikelder

ABSTRACT Simulated annealing is a powerful randomized local search technique for finding optimal or nearly optimal solutions of combinatorial optimization problems. The method has its origin in the annealing process in condensed matter physics. Simulated annealing presupposes the definition of a neighborhood. The algorithm simulates a walk through the solution space that is obtained by repeatedly moving from a solution to a neighboring one. Steps that give rise to improvements are always accepted, while deteriorating steps are only accepted with a certain probability. This acceptance probability is controlled by a parameter and usually tends to zero during the search process. In this article we describe theoretical convergence results as well as some practical aspects of simulated annealing.

1. Introduction

Many problems in engineering, planning, and manufacturing can be modeled as the search for a solution that minimizes a cost function over a finite set of alternative solutions. This class of combinatorial optimization problems has received much attention over the last two decades and major achievements have been made in its analysis (Papadimitriou and Steiglitz, 1982, Cook et al., 1998). One of these achievements is the distinction of a subclass of combinatorial optimization problems that are notoriously hard—often referred to as NP-hard—and for which it is generally believed that no algorithms can exist that solve each instance in polynomial time. Consequently, there are instances that require superpolynomial or exponential time to be solved to optimality.

To handle NP-hard problems one often resorts to the use of heuristics that find approximative solutions in acceptable amounts of computation time. Over the years a class of heuristics called local search algorithms has been developed, which can be viewed as slow approximation algorithms that provide high-quality solutions. Simulated annealing is among the best-known examples in this class, since it has been proven to find good solutions to a wide range of problems.

2. Threshold Algorithms

Threshold algorithms constitute a general class of local search algorithms that are based on stepwise comparison of the difference in value of the cost function of a solution and one of its neighbors with a threshold.

Let $(\mathcal{S}, f)$ be an instance of a combinatorial minimization problem with solution set $\mathcal{S}$ and cost function $f : \mathcal{S} \to \mathbb{R}$. Furthermore, let $\mathcal{N} : \mathcal{S} \to \mathcal{P}(\mathcal{S})$ be a neighborhood function, which defines, for each $i \in \mathcal{S}$, a set $\mathcal{N}(i) \subseteq \mathcal{S}$ of neighboring solutions. The question is to find a solution $i^* \in \mathcal{S}$ that minimizes the cost over all solutions in $\mathcal{S}$. A threshold algorithm starts off with an initial solution and then continually selects a neighbor of a current solution and compares the difference in cost between these solutions to a threshold t. If the cost difference is smaller than the threshold, the neighbor replaces the current solution. Otherwise, the search continues with the current solution. The algorithm terminates when some stop criterion is satisfied. The value of the threshold may be adjusted at each iteration. Let $(t_k \mid k = 0, 1, 2, ...)$ denote the sequence of thresholds, where t_k is used at iteration k of the algorithm.

The above definitions can be used to build a graph model of threshold algorithms in the following way. We construct a so-called neighborhood graph in which each node corresponds to a solution, and an arc is directed from each solution to each of its neighbors. A threshold algorithm then defines a walk in the neighborhood graph starting at an initial node.

We distinguish between the following three types of threshold algorithms depending on the nature of the thresholds.

- *Iterative improvement:*

  $t_k = 0, \quad k = 0, 1, 2, ...$

  Clearly, this is a variant of the classical greedy local search in which only cost-reducing neighbors are accepted.

- *Threshold accepting:*

  $t_k = c_k, \quad k = 0, 1, 2, ...,$

  where

$$c_k \geqslant 0, \quad c_{k+1} \leqslant c_k, \quad \text{and} \quad \lim_{k \to \infty} c_k = 0. \tag{1}$$

  Threshold-accepting algorithms use a nonincreasing sequence of deterministic thresholds. Due to the use of positive thresholds, neighboring solutions with larger costs are accepted in a limited way. In the course of the algorithm's execution, the threshold values are gradually lowered to eventually become close to 0, in which case almost no deteriorations are accepted.

- *Simulated annealing:*

  $t_k =$ a random variable with probability distribution F_{c_k}, $k = 0, 1, 2, ...$

  This means that the probability of a threshold t_k being at most $y \in \mathbb{R}^+$ is given by $\mathbb{P}_{c_k}\{t_k \leqslant y\} = F_{c_k}(y)$. It is assumed that all t_k are nonnegative and have an expected value $\mathrm{E}(t_k) = c_k$, where c_k satisfies (1).

This implies that each neighboring solution can be chosen with a finite probability to replace the current solution. In practice the function F is chosen such that solutions corresponding to large increases in cost have a small probability of being accepted, whereas solutions corresponding to small increases in cost have a larger probability of being accepted.

In the original simulated annealing version, Kirkpatrick et al. (1983) and Černy (1985) take the negative exponential distribution with parameter $1/c_k$ for F_{c_k}. This choice leads to the following acceptance criterion. For any two solutions with $i, j \in \mathcal{S}$ the probability of accepting j from i at the kth iteration is given by

$$\mathbb{P}_{c_k}\{\text{accept } j \text{ from } i\} = \begin{cases} 1 & \text{if } f(j) \leqslant f(i) \\ \exp\left(\dfrac{f(i) - f(j)}{c_k}\right) & \text{if } f(j) > f(i). \end{cases} \quad (2)$$

The parameter c_k is used in the simulated annealing algorithm as a control parameter, and it plays an important role in the convergence of the algorithm.

2.1. Some Preliminary Convergence Results

For iterative improvement no nontrivial convergence properties are known. An example of a trivial result is the fact that for an infinite number of restarts from randomly chosen start solutions an optimal solution is obtained with probability 1. This variant is often referred to as multistart iterative improvement.

Threshold accepting was introduced by Dueck and Scheuer (1990) as a deterministic version of simulated annealing. A major unresolved problem is the determination of appropriate values for the thresholds. Furthermore, as in the case of iterative improvement, no general convergence results can be proved, but one can do slightly better. Althöfer and Koschnick (1991) have related some convergence properties of threshold accepting to those of simulated annealing. The proof of the convergence results are not constructive. They make use of the fact that in some sense simulated annealing generalizes threshold accepting. Furthermore, they give a simple example for which suboptimal solutions can be reached with threshold sequences of any length, demonstrating that even asymptotically the algorithm can get trapped in local minima.

For simulated annealing there exists general convergence results, which state that under certain mild conditions an optimal solution is found with probability 1. These results are obtained by analyzing the algorithm in terms of Markov chains. This central issue is the subject of Section 4 of this article. Before we present these quantitative performance results, we first discuss some qualitative results.

3. The Physics Analogy

The origin of simulated annealing and the choice of the acceptance criterion lie in the physical annealing process (Kirkpatrick et al., 1983; Černy, 1985). In condensed matter physics, annealing is a thermal process for obtaining low energy states of a solid in a heat bath. It consists of the following two steps: first, the temperature of the heat bath is increased to a maximum value at which the solid melts; second, the temperature is carefully decreased until the particles of the melted solid arrange themselves in the ground state of the solid. In the liquid phase all particles of the solid arrange themselves randomly. In the ground state the particles are arranged in a highly structured lattice and the energy of the system is minimal.

The physical annealing process can be modeled successfully by computer simulation methods based on the early Monte Carlo techniques proposed by Metropolis et al. (1953). Their algorithm simulates the evolution of a solid in a heat bath to thermal equilibrium and is based on Monte Carlo techniques. It generates a sequence of states of the solid in the following way. Given a current state i of the solid with energy E_i, a subsequent state j is generated by applying a perturbation mechanism, which transforms the current state into the next state by a small distortion, for instance by displacement of a single particle. The energy of the next state is E_j. If the energy difference, $E_j - E_i$, is less than or equal to 0, the state j is accepted as the current state. If the energy difference is greater than 0, the state j is accepted with a probability that is given by

$$\exp\left(\frac{E_i - E_j}{k_B T}\right),$$

where T denotes the temperature of the heat bath and k_B is a physical constant known as the Boltzmann constant. The acceptance rule described above is known as the Metropolis criterion and the corresponding algorithm as the Metropolis algorithm.

The simulation is based on the assumption that if the temperature is lowered sufficiently slowly, the solid can reach thermal equilibrium at each temperature. Furthermore, it is assumed that thermal equilibrium is characterized by the Boltzmann distribution, which relates the probability of the solid being in a state i with energy E_i at temperature T, and is given by

$$\mathbb{P}_T\{\mathbf{X} = i\} = \frac{\exp(-E_i/k_B T)}{\sum_j \exp(-E_j/k_B T)},$$

where $\mathbf{X}$ is a random variable denoting the current state of the solid and the summation extends over all possible states. As we show in Section 4, the Boltzmann distribution plays an essential role in analysis of the simulated annealing algorithm.

Returning to simulated annealing, we can apply the Metropolis criterion to generate a sequence of solutions to a combinatorial optimization problem. For this purpose we assume an analogy between a physical many-particle system and a combinatorial optimization problem based on the following equivalences.

- Solutions in a combinatorial optimization problem are equivalent to states of a physical system.
- The cost of a solution is equivalent to the energy of a state.
- Neighboring solutions correspond to subsequent states obtained by the perturbation mechanism.

Next, we introduce a parameter c that plays the role of the temperature. This parameter is the same as the control parameter used in (2).

We can now let go of the physical analogy and formulate simulated annealing in terms of a local search algorithm. Usually in the simulated annealing algorithm the control parameter is only decreased after a number of iterations, say L, with a fixed control parameter. So, given an instance $(\mathcal{S}, f)$ of a combinatorial optimization problem and a neighborhood structure $\mathcal{N}$, the simulated annealing algorithm in the case of minimization can be described in pseudo-PASCAL as shown in Figure 1.

The meaning of the four functions in the program is obvious: INITIALIZE computes a start solution and initial values of the parameters c and L; GENERATE selects a solution from the neighborhood of the current solution; CALCULATE_LENGTH and CALCULATE_CONTROL compute new values of the parameters L and c, respectively. Note that the probability of accepting deteriorations is implemented by comparing the value of $\exp((f(i) - f(j))/c)$ with a random number generated from a uniform distribution on the interval [0, 1). Furthermore, it is obvious that the speed of convergence of the algorithm is determined by the parameters L_k and c_k, $k = 0, 1, \ldots$, and by the stop criterion.

As already mentioned, a typical feature of simulated annealing is that, besides accepting improvements in cost, it also, to a limited extent, accepts deteriorations in cost. Initially, at large values of c, large deteriorations are accepted; as c decreases, only small deteriorations are accepted, and finally, as the value of c approaches 0, almost no deteriorations are accepted at all. Furthermore, there is no limitation on the size of a deterioration with respect to its acceptance, as there is for instance in threshold accepting. In simulated annealing, arbitrarily large deteriorations are accepted with positive probability; for these deteriorations the acceptance probability is small, however.

```
procedure SIMULATED_ANNEALING;
begin
  INITIALIZE (i_start, c_0, L_0);
  k := 0;
  i := i_start;
  repeat
    for l := 1 to L_k do
    begin
      GENERATE (j from N_i);
      if f(j) ≤ f(i) then i := j
      else
      if exp((f(i) - f(j))/c_k) > random[0, 1) then i := j
    end;
    k := k + 1;
    CALCULATE_LENGTH (L_k);
    CALCULATE_CONTROL (c_k);
  until stopcriterion
end;
```

Figure 1 A pseudo-PASCAL program of the simulated annealing algorithm

4. Markov Models

The simulated annealing algorithm can be mathematically modeled using the theory of finite Markov chains (Seneta, 1981).

Definition 1 (Markov chain) *Let $\mathcal{O}$ denote a set of possible outcomes of a sampling process. A Markov chain is a sequence of trials, where the probability of the outcome of a given trial depends only on the outcome of the previous trial. Let* $\mathbf{X}(\mathrm{k})$ *be a random variable denoting the outcome of the* k^{th} *trial. Then the transition probability at the kth trial for each pair of outcomes* $\mathrm{i}, \mathrm{j} \in \mathcal{O}$ *is defined as*

$$P_{ij}(k) = \mathbb{P}\{\mathbf{X}(k) = j \mid \mathbf{X}(k-1) = i\}. \tag{3}$$

The matrix $P(k)$, whose elements are given by (3), is called the transition matrix. A Markov chain is called finite if the set of outcomes is finite. It is called inhomogeneous if the transition probabilities depend on the trial number k. If they do not depend on the trial number, the Markov chain is called homogeneous.

In the case of simulated annealing, a trial corresponds to a transition, and the set of outcomes is given by the finite set of solutions. Furthermore, the outcome of a trial only depends on the outcome of the previous trial. In this section we assume that c_k is the control parameter at the kth iteration. Since successive values of c_k may be equal, this includes the situation in Figure 1 where the control parameter is piecewise constant

Hence simulated annealing can indeed be modeled by a finite Markov chain.

Definition 2 (transition probability). *Let ($\mathcal{S}$, f) be an instance of a combinatorial optimization problem and $\mathcal{N}$ a neighborhood function. Then the transition probabilities for the simulated annealing algorithm with control parameter values* $\mathrm{c_k}$ $\mathrm{k} = 0, 1, 2, \ldots$, *are defined as*

$$\forall i, j \in \mathcal{S}: \quad P_{ij}(k) = \begin{cases} G_{ij}(c_k)A_{ij}(c_k) & \text{if } i \neq j \\ 1 - \sum\limits_{l \in \mathcal{S}, l \neq i} G_{il}(c_k)A_{il}(c_k) & \text{if } i = j, \end{cases} \tag{4}$$

where $G_{ij}(c_k)$ *denotes the generation probability, that is, the probability of generating a solution* j *from a solution* i, *and* $A_{ij}(c_k)$ *denotes the acceptance probability, that is, the probability of accepting a solution* j *that is generated from solution* i.

$G_{ij}(c_k)$ and $A_{ij}(c_k)$ of (4) are conditional probabilities, that is, $G_{ij}(c_k) = \mathbb{P}_{c_k}\{\text{generate } j \mid \mathbf{X}(k-1) = i\}$ and $A_{ij}(c_k) = \mathbb{P}_{c_k}\{\text{accept } j \mid \mathbf{X}(k-1) = i\}$. The corresponding matrices $G(c_k)$ and $A(c_k)$ are the generation matrix and acceptance matrix, respectively.

For simplicity we assume that the generation probabilities are chosen to be independent of the control parameter and spread uniformly over the neighborhoods $\mathcal{N}(i)$, where it is assumed that all neighborhoods are of equal size, that is, $|\mathcal{N}(i)| = \Theta$, for all $i \in \mathcal{S}$. It is also assumed that the neighborhood relation is symmetric, that is, $j \in \mathcal{N}(i) \Leftrightarrow i \in \mathcal{N}(j)$. This means that the neighborhood graph is undirected. Furthermore, we assume that the acceptance probability equals the Metropolis criterion of Section 3. This leads to the following definitions.

Definition 3. (generation probability)

$$\forall i, j \in \mathcal{S}: \quad G_{ij}(c_k) = \begin{cases} \dfrac{1}{\Theta} & \text{if} \quad j \in \mathcal{N}(i) \\ 0 & \text{otherwise,} \end{cases} \tag{5}$$

(acceptance probability)

$$\forall i, j \in \mathcal{S}: \quad A_{ij}(c_k) = \exp\left(-\frac{(f(j) - f(i))^+}{c_k}\right), \tag{6}$$

where $a^+ = max(a, 0)$ *for all* $a \in \mathbb{R}$.

For many combinatorial optimization problems, neighborhoods can be defined along the lines of the definitions given above, and close examination of the literature reveals that in many practical applications the original form of simulated annealing introduced by Kirkpatrick et al. (1983)—or minor variations of it—is indeed used. It is however also possible to formulate a set of conditions guaranteeing asymptotic convergence for a more general class of acceptance and generation probabilities. We return to this subject below. We now concentrate on the asymptotic convergence of simulated annealing. Simulated annealing algorithm finds an optimal solution with probability 1 if, after a possibly large number of trials, say k, we have

$$\mathbb{P}\{\mathbf{X}(k) \in \mathcal{S}^*\} = 1,$$

where $\mathcal{S}^*$ denotes the set of optimal solutions. Below, we show that under certain conditions the simulated annealing converges asymptotically to the set of optimal solutions, that is,

$$\lim_{k \to \infty} \mathbb{P}\{\mathbf{X}(k) \in \mathcal{S}^*\} = 1.$$

An essential property in the study of Markov chains is that of stationarity. Under certain conditions, on the transition probabilities associated with a Markov chain a unique stationary distribution exists, which can be defined as follows.

Definition 4 (stationarity). *Consider a finite homogeneous Markov chain with transition matrix* P *on a set of outcomes* $\mathcal{O}$. *If for all states* $i \in \mathcal{O}$ *the limit*

$$q_i = \lim_{k \to \infty} \mathbb{P}\{\mathbf{X}(k) = i \mid \mathbf{X}(0) = j\}$$

exists and does not depend on the initial state j, *then the stochastic* $|\mathcal{O}|$*-vector* q *with components* q_i *is the stationary distribution of the Markov chain.*

More specifically, q is the probability distribution of the outcomes after an infinite number of trials, and it is the left eigenvector of P with eigenvalue 1.

Next, we discuss two related Markov models for simulated annealing. The first one uses homogeneous Markov chains, and the second one uses inhomogeneous Markov chains. The two models yield related but different convergence properties, see for instance Aarts and Korst (1989).

4.1. A Homogeneous Model

We can now prove asymptotic convergence of simulated annealing based on a model in which the algorithm is viewed as a sequence of homogeneous Markov chains of infinite length. In such a homogeneous Markov chain the value of the control parameter and the transition matrix are independent of k, that is, $c_k = c$ and $P(k) = P$ for all k. This leads to the following result.

> **Theorem 1.** *Let* ($\mathcal{S}$, f) *be an instance of a combinatorial optimization problem,* $\mathcal{N}$ *a neighborhood function, and* P *the transition matrix of the homogeneous Markov chain associated with the simulated annealing algorithm defined by* (4), (5), *and* (6), *with* $c_k = c$, *for all* k. *If the neighborhood graph is connected, the associated homogeneous Markov chain has a stationary distribution* q(c), *whose components are given by*
>
> $$q_i(c) = \frac{\exp(-f(i)/c)}{\sum_{j \in \mathcal{S}} \exp(-f(j)/c)} \quad \textit{for all} \quad i \in \mathcal{S}. \tag{7}$$

As a consequence of Theorem 1 we have

$$q_i^* \stackrel{\text{def}}{=} \lim_{c \downarrow 0} q_i(c) = \begin{cases} \dfrac{1}{|\mathcal{S}^*|} & \text{if } i \in \mathcal{S}^* \\ 0 & \text{otherwise,} \end{cases} \tag{8}$$

where $\mathcal{S}^*$ denotes the set of optimal solutions. This implies that

$$\lim_{c \downarrow 0} \lim_{k \to \infty} \mathbb{P}\{\mathbf{X}(k) \in \mathcal{S}^*\} = 1. \tag{9}$$

This result reflects the basic property of the simulated annealing, that is, the guarantee that the algorithm asymptotically finds an optimal solution. Furthermore, (9) expresses the characteristic of the homogeneous model for simulated annealing, namely, first take the limit of the homogeneous Markov chain for an infinite number of trials, and next take the limit for the control parameter to zero. In the inhomogeneous model these two limits are combined to obtain a single one.

4.2. An Inhomogeneous Model

We now discuss conditions to guarantee asymptotic convergence for the case where each Markov chain is of length 1, that is, after each transition the value of the control parameter is changed. Thus, the simulated annealing algorithm is modeled as an inhomogeneous Markov chain with transition probabilities given by (4), (5), and (6). We assume that

$$c_{k+1} \leqslant c_k \quad \text{for all } k \geqslant 0 \tag{10}$$

and that

$$\lim_{k \to \infty} c_k = 0. \tag{11}$$

> **Theorem 2.** *Let* ($\mathcal{S}$, f) *be an instance of a combinatorial optimization problem,* $\mathcal{N}$ *a neighborhood function, and* P(k) *the transition matrix of the inhomogeneous Markov chain associated with the simulated annealing algorithm defined by* (4), (5), *and* (6). *Furthermore suppose that the neighborhood graph is connected and that the control parameter satisfies* (10) *and* (11). *Then, a constant* $\Gamma > 0$ *exists, such that for each series* c_k *with*
>
> $$\sum_{k=0}^{\infty} \exp\left(\frac{-\Gamma}{c_k}\right) = \infty \tag{12}$$
>
> *the Markov chain converges in distribution to the vector* q*, *with components given by* (8), *or in other words*

$$\lim_{k \to \infty} \mathbb{P}\{\mathbf{X}(k) \in \mathcal{S}^*\} = 1. \tag{13}$$

Elementary mathematics show that (13) holds if for instance

$$c_k \geqslant \frac{\Gamma}{\log(k + k_0)}$$

for any $k_0 \geqslant 1$. Several authors have addressed the minimal value of Γ. Aarts et al. (1997) provide an overview of various results. One of the first is due to Mitra et al. (1986) who found that Γ can be chosen as

$$\Gamma = r\Delta \tag{14}$$

with

$$\Delta = \max_{i \in \mathcal{S}} \max_{j \in \mathcal{N}(i)} \{| f(j) - f(i) |\} \quad \text{and} \quad r = \min_{i \in \mathcal{S} \setminus \tilde{\mathcal{S}}} \max_{j \in \mathcal{S}} d(i, j), \tag{15}$$

where $\tilde{\mathcal{S}}$ denotes the set of all locally maximal solutions, and the distance $d(i, j)$ is the distance between two solutions i, j in the neighborhood graph. Theorem 2 with Γ given by (14) gives sufficient but not necessary conditions for convergence of the Markov chain. Necessary and sufficient conditions for convergence have been given by Hajek (1988). Before discussing this result we need some definitions. A solution j is reachable at height h from a solution i if there is a path in the neighborhood graph from i to j such that all solutions on that path have costs at most h. The depth $D(i)$ of a local but nonglobal minimum i is the smallest number h such that a solution j with $f(j) < f(i)$ is reachable at height $f(i) + h$. Hajek showed that (12) is necessary and sufficient for convergence if

$$\Gamma = D \stackrel{\text{def}}{=} \max_{i \in \hat{\mathcal{S}} \setminus \mathcal{S}^*} D(i), \tag{16}$$

where $\hat{\mathcal{S}}$ is the set of local minima. Hence the smallest value for Γ that yields convergence is D, the maximum of the depth of the local but nonglobal minima.

Here we assume that the neighborhood is symmetric, that is, $j \in \mathcal{N}(i) \Leftrightarrow i \in \mathcal{N}(j)$. Hajek's result even holds if this condition is relaxed to weak reversibility, defined as follows. The neighborhood is called weak reversible if, for any two solutions i and j and any h, the reachability of j at height h from i implies that i is also reachable at height h from j.

Kern (1993) has addressed the problem of calculating the value of D. In particular, he showed for a number of combinatorial optimization problems that it is unlikely that D can be calculated in polynomial time. In addition, the paper presents bounds on the value of D for several combinatorial optimization problems.

The necessity of Hajek's conditions for convergence of the inhomogeneous model is illustrated by the following example. Consider a local but nonglobal minimum i that has the maximum depth D, and suppose that this local minimum has only one neighbor, say j, with cost $f(j) = f(i) + D$. The limit behavior (13) for the inhomogeneous case requires that the search process escapes from i with probability 1. We now compute this escape probability. Let $p_k = \exp(-D/c_k)$ be the probability of a transition of i to j with control parameter c_k. Now the probability that the kth attempt to escape from i is successful equals $p_k \prod_{m=0}^{k-1}(1 - p_m)$. Hence the probability of escaping from i in at most N steps equals

$$\sum_{k=0}^{N} p_k \prod_{m=0}^{k-1} (1 - p_m) = 1 - \prod_{k=0}^{N} (1 - p_k),$$

where the last equality is easily proved by induction. Taking the limit $N \to \infty$ we conclude that escaping from i happens eventually with probability 1 if $\prod_{k=0}^{\infty}(1 - p_k) = 0$. By taking the logarithm this condition becomes

$$\sum_{k=0}^{\infty} \log(1 - p_k) = -\infty.$$

Since $0 < \dots p_2 \leqslant p_1 \leqslant p_0 < 1$ there is a constant $\delta > 1$ such that for all k,

$$-\delta p_k \leqslant \log(1-p_k) \leqslant -p_k.$$

Hence the condition above is equivalent to

$$\sum_{k=0}^{\infty} p_k = \sum_{k=0}^{\infty} \exp(-D/c_k) = \infty,$$

which exactly matches (12) with for Γ the optimal value (16) as given by Hajek (1988).

We end this section with some remarks; details can be found in Aarts and Korst (1989) and Aarts et al. (1997).

- It is possible to formulate a more general class of acceptance and generation probabilities than those we considered above, and prove asymptotic convergence to optimality in that case. However, the probabilities we used above are imposed by this more general class in a natural way. Moreover, as already mentioned, these probabilities are used in practically all applications reported in the literature.
- Asymptotical estimates of the rate of convergence show that the stationary distribution of the simulated annealing algorithm can only be approximated arbitrarily closely if the number of transitions is proportional to $|\mathcal{S}|^2$. For hard problems $|\mathcal{S}|$ is necessarily exponential in the size of the problem instance, thus implying that approximating the asymptotic behavior arbitrarily close results in an exponential-time execution of the simulated annealing algorithm. Similar results have been derived for the asymptotic convergence of the inhomogeneous algorithm.

In summary, simulated annealing behaves as an optimization algorithm if it is allowed an infinite number of transitions—or this behavior can be approximated arbitrarily close, if at least an exponential amount of transitions is allowed. In the next section we show how a more efficient finite time implementation of the simulated annealing algorithm can be obtained. Evidently, this will be at the cost of the guarantee of obtaining optimal solutions. Nevertheless, practice shows that high-quality solutions can be obtained in this way.

5. Cooling Schedules

We now leave the issue of optimal annealing and turn to finite-time implementations of the algorithm, which, as a consequence of the above, can no longer guarantee finding an optimal solution, but may result in much faster executions of the algorithm without losing too much of the quality of the solution. A finite-time implementation of simulated annealing is obtained by generating a sequence of homogeneous Markov chains of finite length at descending values of the control parameter. For this, a set of parameters must be specified that govern the convergence of the algorithm. These parameters are combined in what is called a cooling schedule.

Definition 5 (cooling schedule). *A cooling schedule specifies a finite sequence of values of the control parameter, and a finite number of transitions at each value of the control parameter. More precisely, it is specified by*

- *an initial value of the control parameter* c_0,
- *a decrement function for lowering the value of the control parameter,*
- *a final value of the control parameter specified by a stop criterion, and*
- *a finite length of each homogeneous Markov chain.*

The search for adequate cooling schedules has been the subject of many studies over the past ten years. Reviews are given by Van Laarhoven and Aarts (1987), Collins et al. (1988), and Romeo and Sangiovani-Vincentelli (1991). Below we discuss some of the results.

5.1. Optimal Schedules

Recently, researchers have been investigating optimal finite-time schedules, where optimal refers to the best average cost obtained in finite time. Strenski and Kirkpatrick (1991) analyzed a small instance of a graph partitioning problem and used an approach based on exactly evaluating the probability distributions of outcomes of the Markov chain associated with the simulated annealing algorithm. They found that different schedules, including iterative improvement, may be optimal depending on the schedule length employed. When a sufficiently long schedule is employed, annealing replaces iterative improvement as the optimal schedule. Furthermore, they observed that optimal schedules may be nonmonotone. This result was rather unexpected since the convergence proofs of simulated annealing suggest a monotone lowering of the control parameter value; see for instance Aarts and Korst (1989). Nevertheless, it was in accordance with earlier theoretical results obtained by Hajek and Sasaki (1989), who found for a small artificial problem that the control parameter values of an optimal annealing schedule are all either 0 or ∞. The approach of Strenski and Kirkpatrick (1991) has been further pursued by Boese and Kahng (1994). They introduce the concept of best-so-far versus where-you-are. Specifically, they use an acceptance criterion based on the cost of the best solution found so far, instead of the cost of the current solution. They determine optimal cooling schedules for two small instances of the traveling salesman and the graph partitioning problem and found that optimal sequences of control parameter values may not be monotone. The analysis of optimal finite-time schedules is interesting, but the results obtained so far are only proved to hold for extremely small instances. At present it is not clear what impact they have on larger instances. One might argue that the whimsical structure of small instances may introduce artifacts that are absent from the more regularly structured large instances. In that case the nonmonotonicity results would only hold for a specific class of small problem instances.

5.2. *Heuristic Schedules*

Most of the existing work on cooling schedules presented in the literature deals with heuristic schedules. We distinguish between two broad classes: static and dynamic schedules. In a static cooling schedule the parameters are fixed; they cannot be changed during execution of the algorithm. In a dynamic cooling schedule the parameters are adaptively changed during execution of the algorithm. Below we present some examples.

STATIC COOLING SCHEDULES

The following simple schedule is known as the geometric schedule. It originates from the early work on cooling schedules by Kirkpatrick et al. (1983) and is still used in many practical situations.

INITIAL VALUE OF THE CONTROL PARAMETER

To ensure a sufficiently large value of c_0, one may choose $c_0 = \Delta$, where Δ is the maximal difference in cost between any two neighboring solutions as defined in (15). Exact calculation of Δ is quite time consuming in many cases. However, one often can give simple estimates of its value.

LOWERING THE CONTROL PARAMETER VALUE

A frequently used decrement function is given by

$$c_{k+1} = \alpha \cdot c_k, \quad k = 0, 1, \ldots,$$

where α is a positive constant smaller than but close to 1. Typical values lie between 0.8 and 0.99.

FINAL VALUE OF THE CONTROL PARAMETER

The final value is fixed at some small value, which may be related to the smallest possible difference in cost between two neighboring solutions.

MARKOV CHAIN LENGTH

The length of Markov chains is fixed by some number that may be related to the size of the neighborhoods in the problem instance at hand.

Dynamic Cooling Schedules

There are many extensions of the simple static schedule presented above that lead to a dynamic schedule. For instance, a sufficiently large value of c_0 may be obtained by requiring that the initial acceptance ratio χ_0—defined as the number of accepted transitions at c_0—is close to 1. This can be achieved by starting off at a small positive value of c_0 and multiplying it with a constant factor, larger than 1, until the corresponding value of χ_0, which is calculated from a number of generated transitions, is close to 1. Typical values of χ_0 lie between 0.9 and 0.99. An adaptive calculation of the final value of the control parameter may be obtained by terminating the execution of the algorithm at a c_k-value for which the value of the cost function of the solution obtained in the last trial of a Markov chain remains unchanged for a number of consecutive chains. Clearly such a value exists for each local minimum that is found. The length of Markov chains may be determined by requiring that at each value c_k a minimum number of transitions is accepted. However, since transitions are accepted with decreasing probability, one would obtain $L_k \rightarrow \infty$ for $c_k \downarrow 0$. Therefore, L_k is usually bounded by a constant L_{max} to avoid extremely long Markov chains for small values of c_k.

In addition to this basic dynamic schedule, the literature presents a number of more elaborate schedules. Most of these schedules are based on a statistical analysis of the simulated annealing process, thus allowing a more theoretical estimation of the parameters. An example is the schedule proposed by Huang et al. (1986), which is quoted in the literature as the most efficient one among those whose implementation require only a modicum of sophistication.

6. Practice

To apply simulated annealing in practice, four basic ingredients are needed: a concise solution representation, a neighborhood function, a transition mechanism, and a cooling schedule. The algorithm is usually implemented as a sequence of homogeneous Markov chains of finite length, generated at descending values of the control parameter specified by a cooling schedule. As for the choice of the cooling schedule, we have seen in the previous section that there are some general guidelines. However, for the other ingredients no general rules are known that guide their choice. The way they are handled is still a matter of experience, taste, and skill left to the annealing practitioner, and we expect that this will not change in the near future.

Simulated annealing has been applied to a large variety of problems ranging from practical real-life problems to theoretical test problems. Two appealing examples of real-life applications are the scheduling of the Australian state cricket season by Willis and Terrill (1994) and the design of keyboards for typewriters by Light and Anderson (1993). VLSI design, atomic and molecular physics, and picture processing are the three problem areas in which simulated annealing is most frequently applied. The set of theoretical test problems includes almost all of the well-known problems in discrete mathematics and operations research such as coding, graph coloring, graph partitioning, sequencing, and scheduling problems; see Aarts and Lenstra (1977).

General overviews of applications of simulated annealing are given by Aarts and Korst (1989), Collins et al. (1988), Dowsland (1993), van Laarhoven and Aarts (1987), and Vidal (1993). Overviews of applications in operations research are given by Eglese (1990) and Koulamas et al. (1994). Studies emphasizing performance issues for theoretical test problems are given by several authors. One of the most elaborate studies is presented by Johnson et al. (1989, 1991), who report on an extensive numerical study for several combinatorial optimization problems including graph partitioning, graph coloring, and number partitioning problems. This work provides many practical findings that, in our opinion, reflect the general experience of annealing practitioners. Perhaps the most striking element is the performance ambivalence observed. For the graph partitioning problem, simulated annealing seems to outperform all existing approximation algorithms, whereas for the number partitioning problem the performance is hopelessly poor. Although this bad performance for the number partitioning problem can be understood from analytical arguments, there seems to be no way to adapt the algorithm in order to improve it.

Furthermore, the literature presents results of studies in which the performance of simulated annealing is compared with that of other local search algorithms. Results for the job shop scheduling problem are presented by Vaessens et al. (1996) and for the traveling salesman problem by Ulder et al. (1991). With some restraint one may conclude from these studies that simulated annealing, if long running times are allowed, can outperform about all other algorithms with respect to effectiveness. More general conclusions cannot be drawn due to the many different quality measures that can be applied. To illustrate this we mention the performance studies for quadratic assignment. Pardalos et al. (1993) report that simulated annealing can find acceptable solutions with fewer iterations than tabu search. Batitti and Tecchiolli (1994) question this conclusion and argue that it no longer holds for difficult problem instances if high-quality solutions are required.

Broadly speaking, simulated annealing can find good solutions for a wide variety of problems, but often at the cost of substantial running times. As a result of this, the true merits of the algorithm become obvious in industrial problem settings, where running times are of little or no concern. As an example we mention design problems, since in such cases one is primarily interested in finding high-quality solutions, whereas design time often only plays a minor role. A well-known successful simulated annealing area in this respect is VLSI design (Shahookar and Mazumder, 1991; Wong et al., 1988).

The success of simulated annealing can be explained by the fact that the algorithm is easy to implement and capable of handling almost any optimization problem and any constraint, either by appropriate neighborhoods or by relaxation through the use of penalty functions. These properties are, however, not unique to simulated annealing. They also hold for simple local search algorithms. The main advantage of simulated annealing is that it is able to improve upon the relatively poor performance of local search by simply replacing the deterministic (strict improvement) acceptance criterion by a stochastic criterion, thus circumventing the need for an in-depth study of the problem structure in order to construct more effective neighborhoods, or to design more tailored algorithms. It goes without saying that in an industrial environment this is a great advantage, since often the required expertise is unavailable and more importantly, it saves design time.

References

Aarts, E. H. L. and J. K. Lenstra. 1997. *Local Search in Combinatorial Optimization*. Chichester, England: Wiley.

Aarts, E. H. L. and J. H. M. Korst. 1989. *Simulated Annealing and Boltzmann Machines*. Chichester, England: Wiley.

Aarts, E. H. L., J. H. M. Korst, and P. J. M. van Laarhoven. 1997. "Simulated annealing." In *Local Search in Combinatorial Optimization*, edited by E. H. L. Aarts and J. K. Lenstra, pp. 91–120. Chichester, England: Wiley.

Althöfer, I. and K. U. Koschnick. 1991. "On the convergence of threshold accepting." *Appl. Math. Optim.* **24**: 183–195.

Batitti, R. and G. Tecchiolli. 1994. "Simulated annealing and tabu search in the long run: a comparison on QAP tasks." *Comput. Math. Appl.* **28**: 1–8.

Boese, K. D. and A. B. Kahng. 1994. "Best-so-far vs. where-you-are: implications for optimal finite-time annealing." *Syst. Control Lett.* **22**: 71–78.

Černy, V. 1985. "Thermodynamical approach to the traveling salesman problem: an efficient simulation algorithm." *J. Optim. Theory Appl.* **45**: 41–51.

Collins, N. E., R. W. Eglese, and B. L. Golden. 1988. "Simulated annealing—an annotated bibliography." *Am. J. Math. Mgmnt Sci.* **8**: 209–307.

Cook, W. J., W. H. Cunningham, W. R. Pulleyblank, and A. Schrijver. 1998. *Combinatorial Optimization*. Chichester, England: Wiley.

Dowsland, K. A. 1993. "Simulated annealing." *Modern Heuristic Techniques for Combinatorial Problems*, edited by C. R. Reeves. Oxford: Blackwell Scientific Publications.

Dueck, G. and T. Scheuer. 1990. "Threshold accepting: a general purpose optimization algorithm." *J. Comput. Phys.* **90**: 161–175.

Eglese, R. W. 1990. "Simulated annealing: a tool for operational research." *Eur. J. Oper. Res.* **46**: 271–281.

Hajek, B. 1988. "Cooling schedules for optimal annealing." *Math. Oper. Res.* **13**: 311–329.

Hajek, B. and G. Sasaki. 1989. "Simulated annealing—to cool it or not." *Syst. Control Lett.* **12**: 443–447.

Huang, M. D., F. Romeo, and A. Sangiovanni-Vincentelli. 1986. "An efficient general cooling schedule for simulated annealing." *Proc. IEEE Int Conf. Comput.-Aided Des.*, pp. 381–384.

Johnson, D. S., C. R. Aragon, L. A. McGeoch, and C. Schevon. 1989. "Optimization by simulated annealing: an experimental evaluation, Part I graph partitioning." *Oper. Res.* **37** 865–892.

Johnson, D. S., C. R. Aragon, L. A. McGeoch, and C. Schevon. 1991. "Optimization by simulated annealing: an experimental evaluation, Part II graph colouring and number partitioning." *Oper. Res.* **39**: 378–406.

Kern, W. 1993. "On the depth of combinatorial optimization problems." *Discrete Appl. Math.* **43**: 115–129.

Kirkpatrick, S., C. D. Gelatt, and M. P. Vecchi. 1983. "Optimization by simulated annealing." *Science* **220**: 671–680.

Koulamas, C., S. R. Antony, and R. Jaen. 1994. "A survey of simulated annealing applications to operations research problems." *Omega* **22**: 41–56.

van Laarhoven, P. J. M. and E. H. L. Aarts. 1987. *Simulated Annealing: Theory and Applications.* Dordrecht, The Netherlands: Reidel.

Light, L. W. and P. Anderson. 1993. "Designing better keyboards via simulated annealing." *AI Expert* **9**: 20–27.

Metropolis, M., A. Rosenbluth, M. Rosenbluth, A. Teller, and E. Teller. 1953 "Equation of state calculations by fast computing machines." *J. Chem. Phys.* **21**: 1087–1092.

Mitra, D., F. Romeo, and A. L. Sangiovanni-Vincentelli. 1986. "Convergence and finite-time behavior of simulated annealing." *Adv. Appl. Probab.* **18**: 747–771.

Papadimitriou, C. H. and K. Steiglitz. 1982. *Combinatorial Optimization: Algorithms and Complexity.* New York: Prentice-Hall.

Pardalos, P. M., K. A. Murty, and T. P. Harrison. 1993. "A computational comparison of local search heuristics for solving quadratic assignment problems." *Informatica* **4**: 172–187.

Romeo, F. and A. Sangiovanni-Vincentelli. 1991. "A theoretical framework for simulated annealing." *Algorithmica* **6**: 302–345.

Seneta, E. 1981. *Non-negative Matrices and Markov Chains.* 2d ed. New York: Springer.

Shahookar, K. and P. Mazumder. 1991 "VLSI cell placement techniques." *Comput. Surveys* **23**: 143–220.

Strenski, P. N. and S. Kirkpatrick. 1991. "Analysis of finite length annealing schedules." *Algorithmica* **6**: 346–366.

Ulder, N. L. J., E. H. L. Aarts, H. J. Bandelt, P. J. M. van Laarhoven, and E. Pesch. 1991. "Genetic local search for the traveling salesman problem." *Lecture Notes in Computer Science*, Vol. 496, pp. 109–116. New York: Springer.

Vaessens, R. J., E. H. L. Aarts, and J. K. Lenstra. 1996. "Job shop scheduling by local search." *INFORMS J. Comput.* **8**: 302–317.

Vidal, R. V. V. (Ed.) 1993. "Applied simulated annealing." In *Lecture Notes in Economics and Mathematical Systems*, Vol. 396. New York: Springer.

Willis, R. J., and B. J. Terrill. 1994. "Scheduling the Australian state cricket season using simulated annealing." *J. Oper. Res. Soc.* **45**: 276–280.

Wong, D. F., H. W. Leong, and C. L. Liu. 1988. *Simulated Annealing for VLSI Design.* Boston: Kluwer.

3.6.9 Variable Neighborhood Search

Pierre Hansen and Nenad Mladenović

ABSTRACT Variable neighborhood search is a recent metaheuristic based on the idea of systematic change of neighborhood during both a descent phase and an exploration phase, allowing to get out of local optima. A basic scheme and extensions allowing solution of large problems are presented. Applications to well-known combinatorial optimization problems as well as to stabilization of column-generation algorithms and computer-aided discovery of conjectures in graph theory are outlined.

1. Introduction

Heuristics are an essential tool in applied optimization, and for many large and often messy problems encountered in practice they are the only applicable one. Their main aim is to provide, in reasonable time, near-optimal solutions to constrained or unconstrained optimization problems. Moreover, they may also be very useful within complex exact algorithms, to accelerate many steps. Finally, they are an important building block in optimization-based knowledge discovery systems. Optimization problems are ubiquitous in operations research and its applications to numerous fields. Artificial intelligence tends to focus more on constraint satisfaction problems. Heuristics then seek a feasible solution. Both types of problems are less different than might appear at first glance as constraint satisfaction problems can be expressed as optimization problems in a variety of ways, for example as minimization of a sum of artificial variables representing constraint violations.

General frameworks for building heuristics, known as metaheuristics, have been extensively studied in the last two decades. They include adaptive multistart (Boese et al., 1994), variable depth search (Lin and Kernighan, 1973; Papadimitriou and Steiglitz, 1982), genetic search (Holland, 1999), simulated annealing (Kirkpatrick et al., 1983), tabu search (Glover, 1989, 1990, 1995; Glover and Laguna, 1993, 1997; Hansen and Jaumard, 1990), GRASP (Feo and Resende, 1995), and variable neighborhood search (Mladenović and Hansen, 1997; Hansen and Mladenović, 1998), the topic of the present article, as well as several others (see Reeves, 1993 for a multiauthored book-length survey; Osman and Laporte, 1996, for an extensive bibliography; and other chapters of this handbook). Hybrids have also been explored at length.

While traditional heuristics, such as, for example simple descent methods, are blocked in the first local optimum found, this is not the case for heuristics built within metaheuristic paradigms. All of them provide methods, deterministic or stochastic, for getting out of poor local optima. As such local optima often differ considerably in value from the global optimum, particularly if there are many, the practical impact of metaheuristics has been immense. In contrast to this success, the theory of metaheuristics is lagging. While good heuristics are often obtained, with some ingenuity and a lot of parameter setting, the reason(s) why they work as well as they do are largely unknown. In this respect, the situation is even worse for hybrids. So, some reflection on desirable properties of metaheuristics, which would guarantee both their practical and theoretical interest, may be in order. A tentative list of such properties is the following:

1. *Simplicity.* The metaheuristic should be based on a simple and clear principle, which should be largely applicable.
2. *Coherence.* All steps of heuristics for particular problems should follow naturally from the metaheuristic's principle.
3. *Efficiency.* Heuristics for particular problems should provide optimal or near-optimal solutions for all, or at least most, realistic instances. Preferably, they should find optimal solutions for most problems of benchmarks, when available.
4. *Effectiveness.* Heuristics for particular problems should take moderate computing time to provide optimal or near-optimal solutions.
5. *Robustness.* Performance of heuristics should be consistent over a variety of instances, that is, not just fine-tuned to a particular training set and less good elsewhere.
6. *User friendliness.* Heuristics should be well-defined, easy to understand and, most important, easy to use. This implies they should have as few parameters as possible and ideally none.
7. *Innovation.* Preferably, the metaheuristic's principle and/or the efficiency and effectiveness of the heuristics derived from it should lead to new types of applications.

Variable neighborhood search (VNS) is a recent metaheuristic that strives to obtain the qualities listed above. It is based on a simple and relatively unexplored principle: systematic change of neighborhood during the search. (Note that precise rules for such a change are crucial: several authors have proposed combining different types of moves, or neighborhoods, within the same heuristic—without, however, doing so systematically, nor with that being the main idea of their heuristics.)

This article is organized as follows: the basic VNS scheme is presented and discussed in the next section. Extensions are reviewed in Section 3. Solutions to various combinatorial optimization problems by VNS and comparison with results of other heuristics are illustrated in these two sections. Other types of applications are discussed in Section 4. They include use of VNS within an exact algorithm for stabilized column generation and the knowledge discovery system AutoGraphiX for graph theory. Brief conclusions are drawn in Section 5.

2. Basic VNS

A basic scheme for local search is presented in Figure 1. As discussed above, this heuristic stops as soon as a local optimum is reached. It uses one type of move or its equivalent, one neighborhood structure.

A first improvement of this scheme is multistart: local search is iterated many times from randomly generated initial solutions and the best solution obtained is kept. However, if there are many local optima (i.e., a much larger number than the maximum number of descents that can be performed in reasonable time) even this best local optimum may be very far in value from the global optimum (a phenomenon known as the Tchebycheff catastrophe; Boese et al., 1994).

Clearly, multistart disperses its efforts, exploring many valleys but without focusing on promising regions. This appears to be all one could do if both local minima and their values were randomly distributed in solution space. However, this is often not the case. Indeed, it has been observed many times that local minima tend to conglomerate in one or a few small regions of this solution space (as can be checked for problems where all local optima may be obtained from first-order conditions, e.g., unconstrained quadratic optimization in 0–1 variables). So once a local optimum has been found, it is worth exploiting the information so-obtained, that is, that many variables may have the same or a close value to the global optimum (but which ones remains unknown). Therefore, multistart can be modified to explore close, and then increasingly far, neighborhoods of the incumbent (or best-known) solution in a probabilistic way. This is precisely what is done by the basic version of VNS, when one uses the simplest set of neighborhood structures, that is, nested ones. Next, we present a general formulation of the basic VNS scheme and then discuss this particular case.

$\mathcal{N}_k (k = 1, 2, ..., k_{max})$ denotes a finite set of preselected neighborhood structures, and $\mathcal{N}_k(x)$ the set of solutions in the kth neighborhood of x. Steps of the basic VNS scheme are presented in Figure 2.

One can choose, among others, maximum CPU time allowed, maximum number of iterations, or maximum number of iterations since the last improvement as the stopping condition.

An easy way to obtain a set of neighborhood structures is to consider an elementary type of move and define the neighborhood $\mathcal{N}_k(x)$ as the set of solutions obtained by applying

Initialization. Select a neighborhood structure $\mathcal{N}$ that will be used in the search; find an initial solution x;
Repeat the following until the stopping condition (i.e, finding a local optimum) is met:
(a) Find the best neighbor $x' \in \mathcal{N}(x)$ of x;
(b) If x' is not better than x, stop. Otherwise, set $x = x'$ and return to (a)

Figure 1 Steps of a local search heuristic

Initialization. Select the set of neighborhood structures $\mathcal{N}_k$, $k = 1, \ldots, k_{max}$, that will be used in the search; find an initial solution x; choose a stopping condition;

Repeat the following until the stopping condition is met:

(1) Set $k \leftarrow 1$; (2) Until $k = k_{max}$, repeat the following steps:

(a) *Shaking*. Generate a point x' at random from the kth neighborhood of x ($x' \in \mathcal{N}_k(x)$);

(b) *Local search*. Apply some local search method with x' as initial solution; denote with x'' the so-obtained local optimum;

(c) *Move or not*. If this local optimum is better than the incumbent, move there ($x \leftarrow x''$) and continue the search with $\mathcal{N}_1$ ($k \leftarrow 1$); otherwise, set $k \leftarrow k+1$

Figure 2 Steps of the basic VNS

k such moves to x (without reversal). For instance, if x is a 0–1 vector, the elementary move could complement a component. Then the neighborhoods $\mathcal{N}_k(x)$ are the sets of vectors at Hamming distance k from x. Another example arises for the traveling salesman problem (TSP) and other related routing problems. The elementary type of move could then be as in the 2–opt heuristic (Croes, 1958), that is, delete a pair of edges from the current tour and reconnect the endpoints in the only way that maintains a connected tour.

The scheme of Figure 2 relies upon a local search method. Any one could, in principle, be used and changes in a code to transform a local descent algorithm into a VNS heuristic are often minimal. However, one can do much better by also using several neighborhoods in the descent phase. Indeed, local minima for one neighborhood are not necessarily also local minima for another. Moreover, exploration of neighborhoods can be fast or time-consuming according to the simplicity of the move upon which they are built and their size. This suggests a basic variable neighborhood descent (VND) scheme, which is given in Figure 3.

A VND heuristic can easily be constructed for a given problem by putting together available heuristics from the literature. The order in which they are applied is of great importance and a good rule of thumb is to rank them by nondecreasing size of the neighborhoods they use. One can go further in exploiting this idea: neighborhoods used in a heuristic may be split into a set of nested neighborhoods of increasing size. For instance, when applying 2–opt moves to the TSP, one may consider the ℓ shortest edges from each vertex, then the 2ℓ shortest, and so on for some parameter ℓ (Hansen et al., 2000a). Also, if some neighborhoods are much easier to explore than others, it may be worthwhile to return to (1), that is, to the first neighborhood as soon as an improved local optimum is found with one of the subsequent neighborhoods, instead of repeating all steps in the sequence.

2.1. *Example 1: Traveling Salesman Problem*

Given n cities and all distances between pairs of them, the TSP is to find a minimum length tour x (i.e., a permutation of the cities, which minimizes the sum of the n distances between adjacent cities in the tour). This classical NP-hard problem is perhaps the most studied one in combinatorial optimization. A detailed survey and comparison of heuristics for the TSP is

Initialization. Select the set of neighborhood structures $\mathcal{N}_k'$, $k = 1, \ldots, k'_{max}$, that will be used in the descent; find an initial solution x;

Repeat the following until no improvement is obtained:

(1) Set $k \leftarrow 1$; (2) Until $k = k'_{max}$, repeat the following steps:

(a) *Exploration of neighborhood*. Find the best neighbor x' of x ($x' \in \mathcal{N}_k'(x)$);

(b) *Move or not*. If the solution thus obtained x' is better than x, set $x \leftarrow x'$; otherwise, set $k \leftarrow k+1$.

Figure 3 Steps of the basic VND

Table 1 TSP: Average Results for Random Euclidean Problems over 100 Trials for $n = 100, ..., 500$ and 10 Trials for $n = 600, ..., 1000$. Computing Times in Seconds CPU on a Sun Sparc 10, 135.5 Mips (as all other results in this paper)

	Best value found			Percent improvement		CPU times		
n	*2–OPT*	*VNS-1*	*VNS-2*	*VNS-1*	*VNS-2*	*2–OPT*	*VNS-1*	*VNS-2*
100	825.69	817.55	811.95	0.99	1.66	0.25	0.18	0.17
200	1156.98	1143.19	1132.63	1.19	2.10	3.88	3.21	2.82
300	1409.24	1398.16	1376.76	0.79	2.30	12.12	10.29	9.35
400	1623.60	1602.59	1577.42	1.29	2.84	46.13	40.03	34.37
500	1812.08	1794.59	1756.26	0.96	3.07	110.64	99.57	91.00
600	1991.56	1959.76	1925.51	0.97	3.32	204.60	191.85	173.07
700	2134.86	2120.59	2089.33	0.67	2.13	347.77	307.93	259.06
800	2279.18	2242.11	2190.83	1.63	3.88	539.94	480.50	462.23
900	2547.43	2399.52	2342.01	5.81	8.06	699.33	656.96	624.74
1000	2918.10	2555.56	2483.95	12.42	14.88	891.61	844.88	792.88
Av.	1869.87	1803.36	1768.67	2.73	4.43	285.63	263.54	244.97

given in Johnson and McGeoch (1997). An early, and still much studied, heuristic for the TSP is 2–opt, which has been described above.

Embedding 2–opt in the basic VNS scheme gives substantial improvements, as shown in Table 1, within smaller computing times. Using the full neighborhood (VNS-1) gives an average reduction in length of 2.73 percent and neglecting 40–80 percent of the largest edges gives an improvement of 4.43 percent. In comparison, a 0.5 percent improvement for the 2.5–opt heuristic (Bentley, 1992) over the 2–opt heuristic, at a cost of a 30–40 percent increase in running time is reported in Johnson and McGeoch (1997). It is well known that TSP solutions obtained by 2–opt alone tend to have errors that rapidly increase with problem size. A full application of VNS, incorporatives 1–opt, 2–opt, 2.5–opt and Lin–Kernighan (1973) heuristics, plus others, is currently under study (Hansen et al., 2000a).

2.2. *Example 2: p-Median Problem*

Given a set L of m potential locations for p facilities and a set U of given locations for n users, the p-Median problem (PM) is to simultaneously locate the p facilities in order to minimize the total transportation distance (or costs) from users to facilities, each user being served by the closest facility. Solutions of PM are thus characterized by 0–1 vectors x with p components among $|L|$ equal to 1 indicating where facilities are located. Neighborhood structures are then defined by moving $1, 2, ..., k_{max}$ facilities and correspond to sets of 0–1 vectors at Hamming distance $2, 4, ..., 2k_{max}$ from x. The descent heuristic used is 1–interchange, with the efficient fast interchange (FI) computational scheme of Whitaker (1983). Results of a comparison of heuristics for difficult PM problems, with randomly generated data (not satisfying the triangle inequality) are given in Table 2 (Hansen and Mladenović, 1998).

VNS is compared with results of

1. The FI local search;
2. A sophisticated tabu search (TS) heuristic (Rolland et al., 1996), which includes strategic oscillation (Glover, 1989, 1990, 1995; Glover and Laguna, 1993);
3. The recent heuristic concentration method (Rosing and ReVelle, 1997; Rosing et al., 1998); and
4. Chain-substitute tabu search (Mladenović et al., 1996).

Table 2 PM: Results for Test Problems from Rolland et al. (1996); Maximum Time Allowed for CSTS (Mladenović et al., 1996) and VNS is Set to be 30 Times those of FI; the Best Solution Found in 50 Trials of FI, CSTS, and VNS are Reported. CPU Times for HC Method were not Available to Us

			Percent error					*CPU time*			
n	*p*	*Best known*	*FI*	*VNS*	*HC*	*CSTS*	*TS*	*FI*	*VNS*	*CSTS*	*TS*
200	10	48912.00	1.21	0.00	0.00	0.02	0.68	5.2	80.3	59.1	381.9
	15	31153.00	1.29	0.00	0.00	0.00	2.80	7.5	121.2	118.9	401.1
	20	23323.00	1.96	0.00	0.65	0.00	0.74	9.6	161.6	163.2	416.6
300	10	82664.00	3.58	0.00	0.00	1.08	0.47	12.3	248.2	179.2	1241.0
	15	52685.00	4.47	0.00	0.17	0.50	1.98	19.4	373.3	311.8	1321.6
	20	38244.00	3.34	0.00	1.35	0.72	2.49	24.4	475.8	467.0	1378.3
400	10	123464.00	0.19	0.00	0.00	0.12	3.79	24.5	463.4	361.3	2910.6
	15	79872.00	5.20	0.00	0.75	2.50	5.15	32.0	631.5	463.1	3096.8
	20	58459.00	7.08	0.46	0.00	0.92	1.17	45.4	958.6	653.3	3218.3
500	10	150112.00	2.03	0.00	0.00	0.14	1.52	40.6	864.9	474.3	9732.2
	15	97624.00	3.54	0.00	0.00	1.55	0.79	54.7	1164.2	887.2	9731.1
	20	72856.00	4.86	0.00	0.41	1.46	0.41	74.5	1593.4	1350.0	9748.4
Average			3.23	0.04	0.27	0.75	1.83	29.2	594.7	457.4	3631.5

It appears that VNS and HC outperform the two TS heuristics. The best-known solution is reached eleven times by VNS, with an average error of 0.04 percent, and seven times by HC with an average error of 0.27 percent. FI is faster than the other heuristics but gives substantially worse solutions.

2.3. *Example 3: Minimum Sum-of-Squares Clustering*

Given a set of n points in Euclidean q-dimensional space, the minimum sum-of-squares clustering problem (MSSC) is to partition this set into classes, or clusters, in such a way that the sum of squared distances between entities and the centroids of their clusters is the minimum possible.

Among many heuristics for MSSC, the K-MEANS local search heuristic (Jancey, 1966; MacQueen, 1967) is the most popular. It is an interchange heuristic, where points are reassigned to a cluster other than their own, one at a time, until a local optimum is reached. Another popular heuristic, called H-MEANS (Anderberg, 1973), selects an initial partition, computes the centroids of its clusters, then reassigns entities to the closest centroid and iterates until stability. A new local search heuristic, called J-MEANS is proposed in Hansen and Mladenović (2000b); centroids are relocated at some of the given points, which do not yet coincide with one of them. Results of a comparison of K-MEANS, H-MEANS, $H+K$-MEANS (where H-MEANS and K-MEANS are applied in sequence), and two versions of VNS are given in Table 3. VNS−1 is an extension of K-MEANS and gives slightly better results than $H+K$-MEANS; VNS−2, which extends J-MEANS, proves to be by far the best heuristic.

3. Extensions

Studies of the distribution of local optima for a variety of combinatorial optimization problems show that they are often concentrated in the lower parts of a few large valleys. In other words, there are one or a few large close-to-convex regions, with some noise. The basic VNS scheme is efficient in finding the best local minima within one such region (the first explored) but less good at finding other large regions, possibly with better local optima. Indeed, the information contained in the incumbent solution decreases in importance when one goes far from it, and when going very far, the basic VNS tends to degenerate into multistart.

A first attempt to address this difficulty (Hansen et al., 2000b) was to choose low points within the neighborhoods of x. One way to do so is to proceed from x to one among m (a

Table 3 MSSC: Results for a 1060-entity Problem; Average and Best Results in 10 Trials; Stopping Rule: 150 s for Each Trial

		Percent deviation from best found									
		K-MEANS		*H-MEANS*		*H+K-MEANS*		*VNS-1*		*VNS-2*	
m	*Best found*	*Av.*	*Best*	*Av.*	*Best*	*Av.*	*Best*	*Av.*	*Best*	*Av.*	*Best*
10	1754840264.9	0.19	0.19	0.01	0.00	0.01	0.00	0.14	0.00	0.14	0.04
20	791925963.7	2.89	0.00	1.87	1.29	1.31	0.46	3.50	1.84	0.74	0.01
30	482302357.1	9.34	6.68	11.77	9.01	8.20	5.79	10.64	5.16	0.85	0.00
40	342844809.0	15.50	12.04	20.01	15.28	16.86	11.53	15.95	9.64	0.81	0.00
50	256892529.0	27.16	24.57	35.60	30.59	28.66	17.14	29.95	13.50	1.42	0.00
60	199151542.6	35.79	32.53	44.59	33.56	36.21	30.00	35.19	20.50	0.59	0.00
70	159781533.1	44.28	33.08	56.63	47.90	47.39	38.89	43.85	25.59	0.73	0.00
80	130038918.6	53.15	46.64	62.34	50.77	56.69	43.98	51.08	35.07	0.75	0.00
90	111322621.7	56.12	48.94	63.94	51.38	54.40	48.38	44.94	28.17	0.73	0.00
100	97352045.7	60.41	54.74	46.21	46.21	35.95	35.95	42.99	28.00	1.16	0.00
110	86287804.2	60.69	52.78	59.92	49.73	41.44	40.79	43.97	23.76	1.34	0.00
120	76380389.5	62.90	54.00	62.32	52.66	48.96	41.28	38.58	33.56	1.02	0.00
130	68417681.6	65.91	50.73	54.66	38.95	42.34	24.64	38.46	26.30	0.67	0.00
140	61727504.5	62.16	49.82	53.05	45.51	36.00	36.00	30.85	22.04	1.43	0.00
150	56679822.6	66.06	55.05	47.82	40.74	33.43	26.88	25.41	20.05	1.34	0.00
160	52210995.2	59.37	53.16	41.74	34.88	30.85	25.61	25.83	19.32	0.70	0.00
Av. Error		42.62	35.93	41.40	34.28	32.42	26.71	30.08	19.53	0.90	0.00

parameter) of its neighbors with the smallest objective value, which gives a point in $\mathcal{N}_1(x)$, and iterate to get points in $\mathcal{N}_2(x), \mathcal{N}_3(x), \ldots$. This is reminiscent of GRASP (Feo and Resende, 1995) where, however, such moves are performed at the outset, within a multistart-like framework, and not after a first or an improved local optimum has been found. The resulting low-neighborhood VNS was tested on the weighted MAXSAT problem, discussed below. On average, performance was not better, and sometimes worse than that of the basic VNS scheme.

A second modification addressed the problem of reaching valleys far from the incumbent more directly. As in the basic scheme, or the low-neighborhood scheme, a point x' is drawn in $\mathcal{N}_k(x)$ and a local search applied, yielding a local optimum x''. A correction proportional to the distance between x and x'' is subtracted from the value $f(x'')$ in the test for relocating the search (which is now distinct from the test for improvement of the incumbent). The resulting *skewed VNS scheme* (SVNS) is presented in Figure 4.

Initialization. Select the set of neighborhood structures $\mathcal{N}_k$, $k = 1, \ldots, k_{max}$, that will be used in the search; find an initial solution x and its value $f(x)$; set $x_{opt} \leftarrow x$, $f_{opt} \leftarrow f(x)$: choose a stopping condition, a parameter value α, and a measure ρ of the distance between vectors;

Repeat the following until the stopping condition is met:

(1) Set $k \leftarrow 1$; (2) Until $k = k_{max}$, repeat the following steps:

(a) *Shaking*. Generate a point x' at random from the kth neighborhood of x ($x' \in \mathcal{N}_k(x)$);

(b) *Local search*. Apply some local search method with x' as initial solution; denote with x'' the so-obtained local optimum;

(c) *Improve or not*. If $f(x'') < f_{opt}$ set $f_{opt} \leftarrow f(x)$ and $x_{opt} \leftarrow x''$;

(d) *Move or not*. If $f(x'') - \alpha\rho(x, x'') < f(x)$ set $x \leftarrow x''$ and $k \leftarrow 1$; otherwise set $k \leftarrow k + 1$.

Figure 4 Steps of the Skewed VNS

Table 4 Results for *GERAD* Test Problems for WMAXSAT ($n = 500$)

	VNS	*VNS-low*	*SVNS-low*	*TS*
Number of instances where best solution is found	6	4	23	5
Percent average error in 10 trials	0.2390	0.2702	0.0404	0.0630
Percent best error in 10 trials	0.0969	0.1077	0.0001	0.0457
Total number of instances	25	25	25	25

3.1. *Example 4: WMAXSAT*

The satisfiability problem, in clausal form, consists of determining if a given set of m clauses (all in disjunctive or all in conjunctive form) built upon n logical variables has a solution or not (Garey and Johnson, 1978). The maximum satisfiability problem consists of finding a solution satisfying the largest possible number of clauses. In the weighted maximum satisfiability problem (WMAXSAT; Resende et al., 1997) positive weights are assigned to the clauses and a solution maximizing the sum of weights of satisfied clauses is sought.

Results of experiments with instances having 500 variables, 4,500 clauses and 3 variables, in direct or complemented form per clause are given in Table 4.

It appears that SVNS improves notably upon results of basic VNS and also upon those of a TS heuristic.

When the size of problem instances considered grows, time required by VNS may become large and even prohibitive. Moreover, finding very good (or improved optimal) solutions becomes more difficult. Yet, with the advent of data warehousing, demand for solution of very large problems is important and increasing.

Again, two ways to address this problem have been considered. The first aim is to reduce the amount of computation of the basic scheme. This can be done by an elementary change: eliminate completely the local descent phase (which is usually by far the most time consuming). The resulting scheme is called reduced VNS (RVNS; Hansen et al., 2000c). Surprisingly it provides good results, in very moderate computing time.

3.2. *Example 5. p-Median (Again)*

Results of RVNS for large p-median problems (derived from the 3,038 users example of the ORLIB database, with various values of p) are given in Table 5.

Table 5 Results for PM Problem with 3,038 Users; Abbreviations: VNS – Basic VNS; FI – Fast Interchange; RVNS – Reduced VNS; VNDS – VN Decomposition Search

	Objective values				*CPU time*			*Percent Error*		
p	*VNS*	*FI*	*RVNS*	*VNDS*	*FI*	*RVNS*	*VNDS*	*FI*	*RVNS*	*VNDS*
50	507809.5	510330.2	510216.4	507655.2	612.9	60.7	311.1	0.50	0.47	-0.03
100	354488.7	356005.1	356666.3	353255.2	1040.7	132.4	885.8	0.43	0.61	-0.35
150	281911.9	284159.0	283024.6	281772.1	1459.2	128.5	1432.4	0.80	0.39	-0.05
200	239086.4	240646.2	241355.6	238623.0	1943.6	107.6	1796.8	0.65	0.95	-0.19
250	209718.0	210612.9	210727.7	209343.3	2395.6	150.3	2189.7	0.43	0.48	-0.18
300	188142.3	189467.5	188709.3	187807.1	2583.4	130.6	1471.7	0.70	0.30	-0.18
350	171726.8	172668.5	172388.5	171009.3	2804.3	153.1	2270.3	0.55	0.39	-0.42
400	157910.1	158549.5	158805.0	157079.7	4083.3	158.7	3670.9	0.40	0.57	-0.53
450	146087.8	146727.2	147062.0	145449.0	4223.8	179.5	1652.7	0.44	0.67	-0.44
500	136081.7	136680.5	136665.0	135468.0	4649.3	209.7	2599.8	0.44	0.43	-0.45
Average					2579.6	141.1	1828.12	0.53	0.53	-0.28

It appears that RVNS gives solutions with the same average error level as FI in eighteen times less computing time.

The second aim is to get better solutions (while still keeping computing time moderate). This is done by combining VNS with decomposition. More precisely, we build a two-level VNS scheme by embedding the basic scheme within a successive approximation method. Then for a given solution x, all but k attributes (or variables) are fixed in the local search phase. All such fixations define a neighborhood $\mathcal{N}_k(x)$. As in VNS a random solution x' is drawn at random from $\mathcal{N}_1(x)$. Then the local search is performed from x' but only on the free variable. A new value for this variable is computed, as well as the corresponding value of the objective function. The other steps are similar to those of VNS: if the new solution is not better than the incumbent, we set $k = k + 1$, that is, we look for improvements in the space where all but two variables are fixed, and so on. Otherwise, a move to the best solution found is made and k is reset to 1. The steps of the basic VNDS heuristic just described are given in Figure 5.

Note that VNDS differs from VNS only in Steps (2a) and (2b). The local search in Step (2b) can be done by the basic VNS instead of a cruder descent heuristic. This gives a two-level VNS scheme.

3.3. *Example 6: p-Median (Continued)*

Table 5 also presents results of VNDS for the test problems described above. It appears that VNDS outperforms FI within a similar computing time and that VNDS gives better solutions than VNS in all cases, in five times less time.

4. New Types of Applications

As mentioned in the introduction a desirable property of metaheuristics is innovation, that is, that the new way to look at problems they embed suggests new types of applications. We now review three cases in which the VNS paradigm led to such results.

The first case is the discovery of a strange effect of move selection in heuristics. As mentioned above, various VNS heuristics have been applied to the TSP. The most extensive study made up to now was on the 2-opt heuristic, also described above. This led to the discovery of an unexpected phenomenon (Hansen and Mladenović, 1999a): choosing the best improvement at each iteration, that is, the exchange of 2 edges that reduce the objective function most, tends to be worse on average than choosing the first improvement. This appears to hold both if the first improvement is defined by examining the edges from a chosen origin along the tour, or by ranking edges to be inserted in order of increasing length from each

Initialization. Select the set of neighborhood structures $\mathcal{N}_k$, $k = 1, \ldots, k_{max}$, that will be used in the search; find initial solution x; choose stopping condition;

Repeat the following until the stopping condition is met:

(1) Set $k \leftarrow 1$; (2) Until $k = k_{max}$, repeat the following steps:

(a) *Shaking*. Generate a point x' at random from the kth neighborhood of x ($x' \in \mathcal{N}_k(x)$); in other words, let y be a set of k solution attributes present in x' but not in x ($y = x'\ x$).

(b) *Local search*. Find the local optimum in the space of y either by inspection or by some heuristic; denote the best solution found with y' and with x'' the corresponding solution in the whole space S ($x'' = (x'\ y) \cup y'$);

(c) *Move or not*. If the solution thus obtained is better than the incumbent, move there ($x \leftarrow x''$), and continue the search with $\mathcal{N}_1$ ($k \leftarrow 1$); otherwise, set $k \leftarrow k + 1$;

Figure 5 Steps of the basic VNDS

vertex. An explanation, corroborated by further detailed analysis, is easily obtained in the latter case: if the best exchange is chosen, two edges of small but not necessarily very small length are introduced in the tour and are difficult to remove at later iterations; if a first improvement is chosen, a very small edge and an average length edge are introduced in the tour and the latter is easy to remove at a later iteration. Systematic study of such phenomena in a variety of heuristics could lead to many improvements.

The second case pertains to the use of heuristics within exact algorithms. It is well known that many combinatorial optimization problems such as the p-median or the multisource Weber problem (du Merle et al., 1999), the clustering or partitioning problem with various objectives (Hansen and Jaumard, 1997; du Merle et al., 2000), the aircrew scheduling problem (Desrosiers et al., 1995), and so on, can be solved by combining column generation with integer programming. As the optimal solution of the linear relaxation is often highly degenerate, convergence may be slow, and even when it is found, many iterations may be needed to prove that it is indeed optimal. The dual of the column-generation procedure is Kelley's (1960) outer approximation method. Slow convergence comes from the generation of many hyperplanes far from the cone vertexed at the optimal dual solution. One would therefore want to estimate this last solution and focus the generation procedure by penalizing hyperplanes far from it.

Consider a primal problem P and its dual D, solved by column generation:

$$(P)\quad \begin{array}{rl} & \min c^T x \\ \text{subject to} & Ax = b \\ & x \geqslant 0 \end{array} \qquad (D)\quad \begin{array}{rl} & \min b^T \pi \\ \text{subject to} & A^T \pi \leqslant c \\ & \pi \geqslant 0. \end{array}$$

In du Merle et al. (1999) a linear stabilization scheme is proposed that remains entirely within the column-generation framework. It merges a perturbation and an exact penalty method.

A primal problem $\tilde{P}$ and its dual $\tilde{D}$ are defined as follows:

$$(\tilde{P})\quad \begin{array}{rl} & \min c^T \tilde{x} - \delta_- y_- + \delta_+ y_+ \\ \text{subject to} & A\tilde{x} - y_- + y_+ = b \\ & y_- \leqslant \varepsilon_- \\ & y_+ \leqslant \varepsilon_+ \\ & \tilde{x}, y_-, y_+ \geqslant 0 \end{array} \qquad (\tilde{D})\quad \begin{array}{rl} & \max b^T \tilde{\pi} - \varepsilon_- w_- - \varepsilon_+ w_+ \\ \text{subject to} & A^T \tilde{\pi} \leqslant c \\ & -\tilde{\pi} - w_- \leqslant -\delta_- \\ & \tilde{\pi} - w_+ \leqslant \delta_+ \\ & \tilde{\pi}, w_-, w_+ \geqslant 0. \end{array}$$

In the primal problem y_- and y_+ are vectors of slack and surplus variables with upper bounds ε_- and ε_+, respectively. These variables are penalized in the objective function by vectors $[-\delta_-, \delta_+]$. When applying column generation, after finding no more columns with negative reduced cost, the parameters δ and ε are updated following problem-specific rules.

For this stabilized column generation scheme to work well, good estimates of the optimal dual variables should be obtained at the outset. Ways to find ranges heuristically for these dual variables by sensitivity analysis are sketched for a few problems in du Merle et al. (1999). For the p-median problem, this led to the exact solution of instances with over 3,000 users versus 900 in the literature; for its counterpart in the plane, that is, the multisource Weber problem (Krau, 1997; Hansen et al., 2000a) it led to exact solution of instances with over 1,000 users versus 30 in the literature. For the MSSC problem combining stabilized column generation with the ACCPM interior-point method, hyperbolic 0–1 programming quadratic 0–1 programming, and VNS in several steps (du Merle et al., 2000) led to exact solution of problems with up to 150 entities, including the famous 150 iris example of Fisher (1936).

The third case belongs to the field of computer-aided scientific discovery. A system for study of extremal graphs based upon VNS and called AutoGraphiX (AGX) has recently been developed (Caporossi and Hansen, 1999, 2000; Caporossi et al., 1999a,b,c; Cvetković et al., 2000; Caporossi, 1999; Gutman et al., 1999). It comprises a VND component with a series of simple graph transformations: deletion or addition of an edge; move of an edge; detour, that is, removal of an edge and addition of two edges joining endpoints of this deleted edge to a third vertex adjacent to neither of them; shortcut, that is, the reverse of the previous operation; 2–opt as in the TSP, *insertion* or *deletion* of a vertex, and so on. The neighborhood structure of VNS itself is defined by the Hamming distance between graphs (or, in other words, by successive edge moves).

AGX addresses the following problems: (a) find a graph satisfying given constraints; (b) find optimal or near optimal values for an invariant, subject to constraints (with possibly some parameters such as numbers of vertices and edges); (c) refute a conjecture; (d) suggest a conjecture, or sharpen one; (e) suggest a proof.

Up to now, AGX has found over forty novel conjectures, more than a dozen of which have been proved and one refuted. Some of them are much simpler than related results found before. For instance a study with AGX of the energy E of a chemical graph (Caporossi et al., 1999b) led to the novel formulas

$$E \geqslant 2\sqrt{m}$$

and

$$E \geqslant \frac{4m}{n}.$$

where m and n are the numbers of edges and vertices of a graph, both of which are proved (as well as others).

Another surprising result was obtained when studying the Hosoya or distance polynomial of a tree, defined by

$$H(T) = a_0 + a_1 x + a_2 x^2 + \ldots + a_d x^d.$$

where $a_0 = n$, $a_1 = m$, a_k, is the number of pairs of vertices of T at distance k, and d is the diameter of the tree. Gutman (1992, 1993) conjectured that there were no trees where $H(T)$ was palindromic, that is, $a_0 = a_d$, $a_1 = a_{d-1}$, and so on. AGX found a counter-example with twenty-one vertices and several slightly larger ones (Caporossi et al., 1999c). The case of an odd diameter is more difficult than that of an even one, which leaves one coefficient free in the middle. Consider then the distance to the palindrome condition defined by

$$dp(T) = \sum_{i=0}^{\lfloor d/2 \rfloor} | a_{d-i} - a_i | .$$

For all n, it was found that

$$dp(T) \geqslant \lceil \frac{n}{2} \rceil$$

and for $n \geqslant 10$ the bound always appears to be sharp. Further results obtained by AGX for chemical and general graphs are described in Cvetković et al. (2000), Caporossi et al. (1999a), and Caporossi and Hansen (1999), this last reference also contains an $O(n^3)$ algorithm for finding a basis of affine relations between invariants on a sample of extremal graphs. Applied to the set of trees with minimum index (or largest eigenvalue of the adjacency matrix) studied with AGX in Cvetković et al. (2000) it gave the unexpected relation (or conjecture)

$$2\alpha - m - n_1 + 2r - d = 0$$

where α is the independence number, n_1 the number of pending vertices, and r the radius of the tree. It is doubtful that this relation could have been found without computer assistance, due to its large number of invariants.

5. Conclusions

Reviewing the seven desirable properties of metaheuristics mentioned in the introduction, it appears that VNS possesses them all to a large degree. Indeed, its principle is simple and all steps of the basic and extended schemes rely upon it. VNS has proved efficient in solving the problems of several benchmarks with optimal or very close to optimal results, and within moderate (or at least reasonable) computing times. Moreover, its performance appears to be robust, its basic principles are easy to apply and very easy to use, parameters being kept to a minimum and sometimes absent. Simplicity is probably its main feature. Indeed, VNS gets as good or better results on many problems than most other metaheuristics and in a much simpler way. This explains its potential for innovation already illustrated by several new types of applications: analysis of moves and their selection for the TSP, stabilized column generation and the computer-aided scientific discovery program AGX. All this suggests VNS may play an important role in the design of new heuristics for numerous problems as well as in leading to a better understanding of the reasons for their success.

ACKNOWLEDGMENTS

The authors research was supported by NSERC grant OGPOO 39682 and FCAR grant 32EQ 1048. We thank Jack Brimberg, Gilles Caporossi, Michel Gendreau, Fred Glover, Brigitte Jaumard, Anderson Parreira, and Dionisio Perez-Brito for discussions of VNS; Dragoš Cvetković, Ivan Gutman, Paul Seymour, and Bjarne Toft for discussions of AGX; and David Schilling for providing test problems for the *p*-median problem.

References

Anderberg, M. R. 1973. In *Cluster Analysis for Applications*. New York: Academic Press.

Bentley, J. L. 1992. "Fast algorithms for geometric traveling salesman problem." *ORSA J. Comput.* **4**: 387–411.

Boese, K. D., A. B. Kahng, and S. Muddu. 1994. "A new adaptive multi-start technique for combinatorial global optimizations." *Oper. Res. Lett.* **16**: 101–113.

Caporossi, G. and P. Hansen. 1999. "Finding relations in polynomial time." In *Proceedings of the XVIth International Joint Conference on Artificial Intelligence, Stockholm.* Vol. 2, pp. 780–785.

Caporossi, G. and P. Hansen. 2000. "Variable neighborhood search for extremal graphs. 1. The AutoGraphiX system." *Discrete Math.* **212**: 29–44.

Caporossi, G., I. Gutman, and P. Hansen. 1999a. "Variable Neighborhood Search for Extremal Graphs. 4. Chemical trees with extremal connectivity index." *Comput. Chem.* **23**: 469–477.

Caporossi, G., D. Cvetković, I. Gutman, and P. Hansen. 1999b. "Variable neighborhood search for extremal graphs. 2. Finding graphs with extremal energy." *J. Chem. Inf. Comput. Sci.* **39**: 984–996.

Caporossi, G., A. A. Dobrynin, I. Gutman, and P. Hansen. 1999c. "Trees with palindromic Hosoya polynomials." In *Graph Theory Notes of New York XXXVII*, pp. 10–16.

Croes, G. A. 1958. "A method for solving traveling salesman problems." *Oper. Res.* **6**: 791–812.

Cvetković, D., S. Simic, G. Caporossi, and P. Hansen. 2000. "Variable neighborhood search for extremal graphs. 3. On the largest eigenvalue of color-constrained trees." *Linear Multilinear Algebra.*

Desrosiers, J., Y. Dumas, M. M. Solomon, and F. Soumis. 1995. "Time constrained routing and scheduling." In *Network Routing, Handbooks in Operations Research and Management Science*, Volume 8, edited by M. O. Ball *et al.*, pp. 35–139. Amsterdam: Elsevier Science.

Feo, T. and M. Resende. 1995. "Greedy randomized adaptive search." *J. Global Optim.* **6**: 109–133.

Fisher, R. A. 1936. "The use of multiple measurements in taxonomic problems." *Ann. Eugenics VII, part II* 179–188.

Garey, M. R. and D. S. Johnson. 1978. In *Computers and Intractability: A Guide to the Theory of NP-Completeness.* New York: W.H. Freeman and Co.

Glover, F. 1989. "Tabu search—Part I." *ORSA J. Comput.* **1**: 190–206.

Glover, F. 1990. "Tabu search—Part II." *ORSA J. Comput.* **2**: 4–32.

Glover, F. 1995. "Tabu thresholding: improved search by nonmonotonic trajectories." *ORSA J. Comput.* **7**: 426–442.

Glover, F. and M. Laguna. 1993. "Tabu Search." In *Modern Heuristic Techniques for Combinatorial Optimization*, edited by C. Reeves, pp. 70–150. Oxford: Blackwell.

Glover, F. and M. Laguna. 1997. In *Tabu Search.* Boston: Kluwer.

Gutman, I. 1992. "Independent vertex palindromic graphs." In *Graph Theory Notes of New York XXIII: 3*, pp. 21–24. New York: New York Academy of Sciences.

Gutman, I. 1993. "A contribution to the study of palindromic graphs." In *Graph Theory Notes of New York XXIV: 6*, pp. 51–56. New York: New York Academy of Sciences.

Gutman, I., O. Miljković, G. Caporossi, and P. Hansen. 1999. "Alkanes with small and large Randić connectivity indices." *Chem. Phys. Lett.* **306**: 366–372.

Hansen, P. and B. Jaumard. 1990. "Algorithms for the maximum satisfiability problem." *Computing* **44**: 279–303.

Hansen, P. and B. Jaumard. 1997. "Cluster analysis and mathematical programming." *Math. Programming* **79**: 191–215.

Hansen, P. and N. Mladenović. 1998. "Variable neighborhood search for the p-median." *Location Sci.* **5**: 207–226.

Hansen, P. and N. Mladenović. 1999a. "First improvement may be better than best improvement: an empirical study for the TSP." *Les Cahiers du GERAD, G–99–40.*

Hansen, P. and N. Mladenović. 1999c "An Introduction to Variable Neighbourhood Search." In *Metaheuristics, Advances and Trends in Local Search Paradigms for Optimization*, edited by S. Voss *et al.*, pp. 433–458. Dordrecht, The Netherlands: Kluwer.

Hansen, P. and N. Mladenović. 2000a. "Variable neighborhood search: principles and applications." *Eur. J. Oper. Res.*

Hansen, P. and N. Mladenović. 2000b. "J-MEANS, a new local search heuristic for minimum sum-of-squares clustering." *Pattern Recognition.*

Hansen, P., N. Mladenović, M. Labbé, and M. Gendreau. 2000a. "Variable neighborhood search for the traveling salesman problem."

Hansen, P., B. Jaumard, N. Mladenović, and A. Parreira. 2000b. "Variable neighborhood search for weighted maximum satisfiability."

Hansen, P., N. Mladenović, and D. Perez-Brito. 2000c. "Variable neighborhood decomposition search." *J. Heuristics.*

Hansen, P., B. Jaumard, S. Krau, and O. du Merle. 2000d. "A stabilized Column Generation Algorithm for the Multisource Weber Problem."

Holland, J. H. 1992. In *Adaptation in Natural and Artificial Systems*. Cambridge MA: MIT Press.

Jancey, R. C. 1966. "Multidimensional Group Analysis." *Aust. J. Botany* **14**: 127–130.

Johnson, D. S. and L. A. McGeoch. 1997. "The traveling salesman problem: a case study in local optimization." In *Local Search in Combinatorial Optimization*, edited by E. H. L. Aarts and J. K. Lenstra, pp. 215–310. London: Wiley.

Krau, S. 1997. "Extensions du problème de Weber." Ph.D. diss, École Polytechnique de Montréal (under direction of P. Hansen, B. Jaumard).

Kelley, J. E. 1960. "The cutting-plane method for solving convex programs." *J. SIAM* **8**: 703–712.

Kirkpatrick, S., C. D. Gelatt, and M. Vecchi 1983. "Optimization by simulated annealing." *Science* **220**: 671–680.

Lin, S. and B. W. Kernighan. 1973. "An effective heuristic algorithm for the traveling salesman problem." *Oper. Res.* **21**: 498–516.

MacQueen, J. B. 1967. "Some methods for classification and analysis of multivariate observations." *Proc. 5th Berkeley Symp. Math. Stat. Probab.* **1**: 281–297.

du Merle, O., D. Villeneuve, J. Desrosiers, and P. Hansen. 1999. "Stabilized column generation." *Discrete Math.* **194**: 229–237.

du Merle, O., P. Hansen, B. Jaumard, and N. Mladenović. 2000. "An Interior Point Algorithm for Minimum Sum of Squares Clustering." *SIAM J. Sci. Comput.*

Mladenović, N., J. P. Moreno, and J. Moreno-Vega. 1996. "A chain-interchange heuristic method." *Yugoslav J. Oper. Res.* **6(1)**: 41–54.

Mladenović, N. and P. Hansen. 1997. "Variable neighborhood search." *Comput. Oper. Res.* **24**: 1097–1100.

Osman, I. H. and G. Laporte. 1996. "Metaheuristics: a bibliography." *Ann. Oper. Res.* **63**: 513–628.

Papadimitriou, C. H. and K. Steiglitz. 1982. In *Combinatorial Optimization, Algorithms and Complexity*. NJ: Prentice Hall.

Reeves, C. (ed). 1993. In *Modern Heuristic Techniques for Combinatorial Problems* Oxford: Blackwell.

Resende, M. G. C., L. S. Pitsoulis, and P. M. Pardalos. 1997. "Approximate solution of weighted MAX-SAT problems using GRASP." In *Satisfiability Problem: Theory and Applications*, edited by D. Du, J. Gu, and P. M. Pardalos. *DIMACS Series in Discrete Mathematics and Theoretical Computer Science* 35, Providence: American Mathematical Society.

Rosing, K. E. and C. S. ReVelle. 1997. "Heuristic concentration: two stage solution construction." *Eur. J. Oper. Res.* **97**: 75–86.

Rosing, K. E., C. S. ReVelle, E. Rolland, D. A. Schilling, and J. R. Current. 1998. "Heuristic concentration and tabu search: a head to head comparison." *Eur. J. Oper. Res.* **104**: 93–99.

Rolland, E., D. A. Schilling, and J. R. Current. 1996. "An efficient tabu search procedure for the p-median problem." *Eur. J. Oper. Res.* **96**: 329–342.

Whitaker, R. 1983. "A fast algorithm for the greedy interchange for large-scale clustering and median location problems." *INFOR* **21**: 95–108.

4 QUADRATIC PROGRAMMING

Yinyu Ye

ABSTRACT Quadratic programming (QP) plays an important role in optimization theory. In one sense it is a continuous optimization and a fundamental subroutine for general nonlinear programming, but it is also considered one of the most challenging combinatorial optimization problems. The aim of this chapter is to describe several of the most recent results on solving quadratic programs.

1. Introduction

Consider the following quadratic optimization problem:

$$\text{(QP)} \quad \text{minimize} \quad q(x) := \tfrac{1}{2}x^T Q x + c^T x$$
$$\text{subject to} \quad x \in \mathscr{F},$$

where column vector $c \in \mathscr{R}^n$, $Q \in \mathscr{M}^n$ (the set of all symmetric matrices in $\mathscr{R}^{n \times n}$), superscript ^{T} denotes the transpose operation, and $\mathscr{F}$ is a compact set in $\mathscr{R}^n$. Quadratic programming (QP) plays an important role in optimization theory. In one sense it is a continuous optimization and a fundamental subroutine for general nonlinear programming, but it is also considered one of the most challenging combinatorial optimization problems. This chapter describes several of the most recent results on solving this problem, especially for the nonconvex case, that is, when Q is indefinite. For detailed descriptions of the problems presented here, the reader is referred to the books by Horst et al. (1995), Roos et al. (1997), Vanderbei (1997), Wright (1996), Vavasis (1991), and Ye (1997).

2. Convex Quadratic Programming

Given $A \in \mathscr{R}^{m \times n}$ with rank m and $b \in \mathscr{R}^n$, consider the QP problem, where

$$\mathscr{F} = \{x : Ax = b, x \geqslant 0\}.$$

A feasible point x^* is called a KKT point, where KKT stands for Karush–Kuhn–Tucker, if the following KKT conditions hold: there exists $(y^* \in \mathscr{R}^m, s^* \in \mathscr{R}^n)$ such that (x^*, y^*, s^*) is feasible for the following dual problem:

$$\text{(QD)} \quad \text{maximize} \quad d(x,y) := b^T y - (1/2)x^T Q x$$
$$\text{subject to} \quad A^T y + s - Qx = c, \quad x, s \geqslant 0,$$

and satisfies the complementarity condition

$$(x^*)^T s^* = (1/2)(x^*)^T Q x^* + c^T x^* - (b^T y^* - (1/2)(x^*)^T Q x^*) = 0.$$

We may write the KKT condition as:

$$(x, y, s) \in (\mathscr{R}^n_+, \mathscr{R}^m, \mathscr{R}^n_+)$$

and

$$\begin{aligned} Xs &= 0 \\ Ax &= b \\ -A^T y + Qx - s &= -c, \end{aligned} \tag{1}$$

where X is the diagonal matrix whose diagonal components are those of x, and $\mathscr{R}^n_+ = \{x \in \mathscr{R}^n : x \geqslant 0\}$. Note that this system has a total of $2n + m$ unknowns and $2n + m$ equations including n nonlinear equations.

The above condition is also called the first-order necessary condition. If Q is positive semidefinite, then QP is a convex quadratic program and x^* is an optimal solution for QP if, and only if, x^* is a KKT point for QP. In this case, the solution set for QP is characterized by a system of linear inequalities and equations of (1).

Many polynomial interior-point algorithms have been developed to solve convex quadratic programs. One fundamental theory of interior-point algorithms is the central path, which can be expressed as

$$\mathscr{C} = \{(x, y, s) : Xs = \mu e, A^T y + s - Qx = c, x, s > 0, 0 < \mu < \infty\},$$

where e is the vector of all ones.

We have the following theorem:

Theorem 1. *Let the convex* QP *have interior (strictly positive) feasible points. Then for any* $0 < \mu < \infty$, *the central path point* $[\mathrm{x}(\mu), \mathrm{y}(\mu), \mathrm{s}(\mu)]$ *exists and is unique. Moreover,*

1. *The central pair* $[\mathrm{x}(\mu), \mathrm{y}(\mu), \mathrm{s}(\mu)]$ *is bounded where* $0 < \mu \leqslant \mu^0$ *for any given* $0 < \mu^0 < \infty$.
2. $[\mathrm{x}(\mu), \mathrm{y}(\mu), \mathrm{s}(\mu)]$ *converges to an optimal solution pair of* QP *and* QD.

Thus, optimal solutions of QP and QD are on the path corresponding to $\mu = 0$. The path-following algorithm, described next, is in fact a homotopy algorithm, which first reduces μ, then uses Newton's method to solve the above system, until μ is reduced to zero. The sequences of (x^k, y^k, s^k), therefore, are always close to the central path from μ^0 down to 0.

2.1. Path-Following Algorithm

Given x^k and y^k with

$$0 < x^k \in \mathscr{F}, s^k = Qx^k + c - A^T y^k > 0, \text{ and } \|X^k s^k - \mu^k e\| \leqslant \alpha \mu^k, \tag{2}$$

where $\mu^k = (x^k)^T s^k / n$. Let

$$\gamma = \left(1 - \frac{\alpha}{\sqrt{n}}\right)$$

for some constant α (for example, $\alpha = 1/3$). Then, Newton's method is used to solve the system of linear equations

$$X^k d_s + S^k d_x = \gamma \mu^k e - X^k s^k$$

$$Ad_x = 0 \text{ and } d_s = Qd_x - A^T d_y.$$

Note that the system of linear equations can be also written as

$$\begin{pmatrix} Q + (X^k)^{-1} S^k & -A^T \\ A & 0 \end{pmatrix} \begin{pmatrix} d_x \\ d_y \end{pmatrix} = \begin{pmatrix} \gamma \mu^k (X^k)^{-1} e - s^k \\ 0 \end{pmatrix}, \tag{3}$$

It has been shown that

$$\|(X^k)^{-1} d_x\|^2 + \|(S^k)^{-1} d_s\|^2 \leqslant 2\left(\frac{\alpha}{1-\alpha}\right)^2,$$

$$\left(1 - \frac{\alpha}{\sqrt{n}}\right)\mu^k \leqslant \mu^{k+1} \leqslant \left(1 - \frac{\alpha}{\sqrt{n}} + \frac{\alpha^2}{(1-\alpha)n}\right)\mu^k,$$

and

$$\|X^{k+1} s^{k+1} - \mu^{k+1} e\| \leqslant \frac{\alpha^2 \mu^k}{1-\alpha}.$$

The first ensures that

$$x^{k+1} = x^k + d_x \text{ and } s^{k+1} = s^k + d_s$$

are interior feasible points, the second indicates that μ^k is reduced at a fixed ratio, and the third guarantees that x^{k+1} and s^{k+1} still satisfy the so-called centering condition (2) for a suitable α.

The algorithm can be described as follows:

PATH-FOLLOWING ALGORITHM: CONVEX QP

Given x^0 and y^0 (s^0) satisfy the centering condition (2);
set $k = 0$;

while $\mu^k \geqslant \epsilon$ **do**

begin

let $\gamma = (1 - \alpha/\sqrt{n})$;

let d_x and d_y solve the linear equations (3):

$x^{k+1} = x^k + d_x$;

$y^{k+1} = y^k + d_y$ and $s^{k+1} = Qx^{k+1} + c - A^T y^{k+1}$;

$\mu^{k+1} = (x^{k+1})^T s^{k+1}/n$;

$k = k+1$;

end.

The following theorem guarantees that the algorithm will terminate in $O(\sqrt{n}\ln(\mu^0/\epsilon))$ iterations.

Theorem 2. *The sequences* $\{x^k, s^k, y^k\}$ *generated by the algorithm satisfy*

$$Ax^k = b, x^k > 0, \quad s^k = Qx^k + c - A^T y^k > 0,$$

$$\|X^k s^k - \mu^k e\| \leqslant \alpha \mu^k,$$

and

$$\mu^{k+1} \leqslant \left(1 - \frac{\alpha}{4\sqrt{n}}\right)\mu^k.$$

In the context of scaling, one can verify that the above algorithm uses the scaling matrix of $D = (X^k)^{1/2}(S^k)^{-1/2}$. Since both primal and dual variables are used in the scaling, this algorithm is called the primal-dual scaling algorithm. There are two other scalings that possess the same computation complexity.

PRIMAL-SCALING ALGORITHM

This algorithm uses $D = X^k$. Obviously, if the solutions are perfectly centered, the two scalings are identical. The system of linear equations being solved is

$$X^k d_s + \gamma\mu^k (X^k)^{-1} d_x = \gamma\mu^k e - X^k s^k$$

$$Ad_x = 0 \text{ and } d_s = Qd_x - A^T d_y.$$

DUAL-SCALING ALGORITHM

Dual scaling, that is, $D = (S^k)^{-1}$, is used in this algorithm. Essentially the following system is solved:

$$\gamma\mu^k (S^k)^{-1} d_s + S^k d_x = \gamma\mu^k e - X^k s^k$$

$$Ad_x = 0 \text{ and } d_s = Qd_x - A^T d_y.$$

2.2. *Potential-Reduction Algorithm*

An adaptive potential-reduction algorithm, which uses long steps without jeopardizing the complexity bound above, was developed soon after. It was first shown that the path-following algorithm in fact reduces the Tanabe–Todd–Ye primal-dual potential function

$$\psi(x,s) = \rho \ln(x^T s) - \sum_{j=1}^{n} \ln(x_j s_j), \tag{4}$$

where $\rho = n + \sqrt{n}$, by a constant at each iteration. This potential function has an ideal property. It can be written as

$$\psi(x,s) = (\rho - n)\ln(x^T s) - \sum_{j=1}^{n} \ln\left(\frac{x_j s_j}{x^T s}\right).$$

The inequality of the geometric and arithmetic means yields

$$-\sum_{j=1}^{n} \ln\left(\frac{x_j s_j}{x^T s}\right) \geqslant n \ln n.$$

Hence,

$$(\rho - n)\ln(x^T s) \leqslant \psi(x,s) - n \ln n.$$

This gives the exact amount, $-(\rho - n)\ln(1/\epsilon)$, by which ψ should be reduced to reach $x^T s \leqslant \epsilon$.

The algorithm can be described as follows. Given $0 < x^k \in \mathcal{F}$ and $s^k = Qx^k + c - A^T y^k > 0$, solve the following system of linear equations for (d_x, d_y, d_s) :

$$\begin{pmatrix} Q + (X^k)^{-1} S^k & -A^T \\ A & 0 \end{pmatrix} \begin{pmatrix} d_x \\ d_y \end{pmatrix} = \begin{pmatrix} \dfrac{(x^k)^T s^k}{\rho} (X^k)^{-1} e - s^k \\ 0 \end{pmatrix}, \tag{5}$$

where $\rho \geqslant n + \sqrt{n}$. Let

$$x^{k+1} = x^k + \bar{\theta} d_x, \quad s^{k+1} = s^k + \bar{\theta} d_s, \quad \text{and } y^{k+1} = y^k + \bar{\theta} d_y,$$

where

$$\bar{\theta} = \frac{\sqrt{3}/4}{\left\| (X^k S^k)^{-0.5} \| \, \| (X^k S^k)^{-0.5} \left(\dfrac{(x^k)^T s^k}{\rho} e - X^k S^k e \right) \right\|}. \tag{6}$$

Then, we have

$$\psi(x^{k+1}, s^{k+1}) \leqslant \psi(x^k, s^k) - 3/16.$$

Note that

$$\frac{1}{\|(X^k S^k)^{-0.5}\|} = \min_{1 \leqslant j \leqslant n} \left(\sqrt{x_j^k s_j^k}\right).$$

The algorithm can be stated as follows.

POTENTIAL-REDUCTION ALGORITHM: CONVEX QP

Given $Ax^0 = b, x^0 > 0$ and $s^0 = Qx^0 + c - A^T y^0 > 0$;

set $k = 0$;

while $(x^k)^T s^k \geqslant 2^{-L}$ **do**

begin

compute d_x and d_y of (5);

let $\bar{\theta}$ be given by (6);

$x^{k+1} = x^k + \bar{\theta} d_x$;

$y^{k+1} = y^k + \bar{\theta} d_y$;

$s^{k+1} = Qx^{k+1} + c - A^T y^{k+1};$

$\delta^{k+1} = (x^{k+1})^T s^{k+1};$

$k = k + 1;$

end.

Theorem 3. *Let* $\psi(x^0, s^0) \leqslant O((\rho - n)\ln(\mu^0/\epsilon))$ *and* $\rho \geqslant n + \sqrt{n}$ *Then, the algorithm terminates in* $O((\rho - n)\ln(\mu^0/\epsilon))$ *iterations.*

In practice, $\bar{\theta}$ can be chosen either by line search as

$$\bar{\theta} = \arg\min_{\theta \geqslant 0} \psi(x^k + \theta d_x, s^k + \theta d_s)$$

or according to a heuristic based upon the problem,

$$\bar{\theta} = \beta \max \theta, \text{ subject to } x^k + \theta d_x \geqslant 0 \text{ and } s^k + \theta d_s \geqslant 0,$$

for some $0 < \beta < 1$. The algorithm can start at any interior points x^0 and y^0 without jeopardizing its convergence.

This algorithm is used to solve quadratic knapsack problems and very encouraging results have been reported. By using long steps and $\rho = n + n^{1.5}$, the number of iterations increases at most ln n as n increases.

2.3. Remarks and Extensions

The primal-dual path-following algorithm described here is from Kojima et al. (1989) and Monteiro and Adler (1989), which is based on the central path theory of Megiddo (1986) and Sonnevend (1985); also see Güler (1993). The algorithm was proposed earlier by Tanabe (1988). Also see McShane (1994), Monteiro et al. (1990), Nazareth (1991), Shanno and Simantiraki (1995), Tseng (1992), Vanderbei (1997), Wright (1996), Zhang (1994), and Zhao (1998). The symmetric primal-dual potential algorithm was developed by Kojima et al. (1991). Other potential reduction algorithms are by Todd (1996), Huang and Kortanek (1991), Sturm and Zhang (1995), and Tunçel (1994).

For the primal-scaling algorithm, see Anstreicher et al. (1993), Ben-Daya and Shetty (1990), Bonnans and Bouhtou (1995), Ding and Li (1991), Freund (1991), Gonzaga (1989), Goldfarb and Liu (1990), Güler (1993), Ji et al. (1995), Jarre (1990), Kapoor and Vaidya (1986), Kortanek and Zhu (1993), Mehrotra and Sun (1990), Pardalos et al. (1991), Ponceleon (1990), Todd and Ye (1990), and Ye and Tse (1989), Ye (1987); for the dual-scaling algorithm, see Jarre (1990), Mehrotra and Sun (1990), Nesterov and Nemirovskii (1993), and Renegar (1988).

There are polynomial interior-point algorithms for convex quadratic programming where $\mathscr{F}$ is the intersection of convex quadratic inequalities; see Andersen and Ye (1999), Anstreicher and Vial (1994), Ben-Tal and Nemirovskii (1994), Coleman and Li (1992), Hertog (1992), Faybusovich (1996), Jarre (1992), Luo and Sun (1999), Mehrotra and Sun (1991), Nash and Sofer (1993), Nesterov and Nemirovskii (1993), and Sun and Qi (1992).

3. Nonconvex Quadratic Programming over a Ball

One QP problem is to minimize a quadratic function over an ellipsoid constraint. Since any ellipsoid can be transformed to a ball by an affine transformation, without loss of generality, we consider the following ball-constrained QP problem:

$$\text{(BQP)} \quad \begin{array}{ll} \text{Minimize} & q(x) \\ \text{subject to} & x \in \{x \in \mathscr{R}^n : \|x\|^2 \leqslant 1\}, \end{array} \tag{7}$$

where the radius of the ball is 1. One of the main recent results is that this problem is an easy problem, even when the objective function is nonconvex.

We begin with a brief history of this problem. There is a class of nonlinear programming algorithms called model trust region methods. In these algorithms, a quadratic function is used as an approximate model of the true objective function around the current iterate. Then the main step is to minimize the model function. In general, however, the model is expected to be accurate or trusted only in a neighborhood of the current iterate. Accordingly, the quadratic model is minimized in an L_2-norm neighborhood, which is a ball, around the current iterate.

It is well known that the solution x of problem BQP satisfies the following necessary and sufficient conditions:

$$\begin{aligned} &(Q+\mu I)x = -c \\ &\qquad \mu \geqslant \max\{0, -\underline{\lambda}\} \\ &\text{and } \|x\| = 1, \end{aligned} \tag{8}$$

where $\underline{\lambda}$ denotes the least eigenvalue of matrix Q. Since Q is not positive semidefinite, we must have $\underline{\lambda} < 0$.

Let μ^* and x^* satisfy conditions (8). It has been shown that μ^* is unique and

$$\mu^* \leqslant |\underline{\lambda}| + \|c\|. \tag{9}$$

It is also known that

$$|\underline{\lambda}| \leqslant n \max\{|q_{ij}|\},$$

where q_{ij} is the (i, j)th component of matrix Q. Thus, we have

$$0 \leqslant \mu^* \leqslant \mu^0 := n \max\{|q_{ij}|\} + \|c\|, \tag{10}$$

where μ^0 is a computable upper bound. It is further proved that

$$\tfrac{1}{2}|\underline{\lambda}| \leqslant \tfrac{1}{2}\mu^* \leqslant q(0) - q(x^*) \leqslant \tfrac{1}{2}|\underline{\lambda}| + \|c\|. \tag{11}$$

We now analyze the complexity of solving BQP. A simple bisection method was proposed. For any given μ, denote solutions of the top linear equations by x_μ in conditions (8), that is,

$$x_\mu := -(Q+\mu I)^{-1}c, \quad \forall \mu > |\underline{\lambda}|. \tag{12}$$

For any given μ we can check to see if $\mu \geqslant |\underline{\lambda}|$ by checking the positive definiteness of matrix $Q + \mu I$, which can be solved as a LDL^T decomposition. These facts lead to a bisection method to search for the root of $\|x_\mu\| = 1$ over the interval $\mu \in [|\underline{\lambda}|, \mu^0] \subset [0, \mu^0]$. Obviously, for a given $\epsilon' \in (0, 1)$, a μ such that, for example, $0 \leqslant \mu - \mu^* \leqslant \epsilon' \mu^*/8$, can be obtained in $O(\ln(\mu^0/\mu^*) + \ln(1/\epsilon'))$ bisection steps, and the cost of each step is $O(n^3)$ arithmetic operations (for performing LDL^T decomposition).

The remaining question is what ϵ' would be sufficient to generate an ϵ-minimizer of $q(x)$ over the ball, that is, an x satisfying

$$\frac{q(x) - q(x^*)}{q(0) - q(x^*)} \leqslant \epsilon.$$

Let μ denote the right endpoint of the interval generated by the bisection search. Then, $\mu \geqslant \mu^*$. If $\mu = \mu^*$, then we get an exact solution $x^* = x_{\mu^*}$. Thus, we assume $\mu > \mu^* \geqslant |\underline{\lambda}|$. By the positive semidefiniteness of $Q + \mu^* I$, we have

$$\|x_\mu\| < \|x^*\| = 1.$$

We consider two cases.

CASE I
In the first case we assume

$$\left(1-\frac{\epsilon}{8\sqrt{n}}\right)\mu^* \geqslant |\underline{\lambda}| \text{ or } \mu^* \geqslant |\underline{\lambda}| + \frac{\epsilon}{8\sqrt{n}}\mu^*.$$

Using relation (12) and simplifying, we obtain

$$\begin{aligned}\|x^*\|^2 - \|x_\mu\|^2 &= (x^*)^T(I-(Q+\mu^* I)(Q+\mu I)^{-2}(Q+\mu^* I))x^* \\ &= (x^*)^T(2(\mu-\mu^*)(Q+\mu I)^{-1} - (\mu-\mu^*)^2(Q+\mu I)^{-2})x^*.\end{aligned}$$

Next we bound the above expression by using the smallest eigenvalue $\underline{\lambda}$ of Q. This gives

$$\begin{aligned}\|x^*\|^2 - \|x_\mu\|^2 &\leqslant \left(\frac{2(\mu-\mu^*)}{(\mu-|\underline{\lambda}|)} - \frac{(\mu-\mu^*)^2}{((\mu-|\underline{\lambda}|))^2}\right)\|x^*\|^2 \\ &= \left(\frac{2(\mu-\mu^*)}{(\mu-\mu^*)+(\mu^*-|\underline{\lambda}|)} - \frac{(\mu-\mu^*)^2}{((\mu-\mu^*)+(\mu^*-|\underline{\lambda}|))^2}\right)\|x^*\|^2 = \frac{(\mu-\mu^*)^2+2(\mu-\mu^*)(\mu^*-|\underline{\lambda}|)}{((\mu-\mu^*)+(\mu^*-|\underline{\lambda}|))^2} \\ &= \left(1-\frac{(\mu^*-|\underline{\lambda}|)^2}{((\mu-\mu^*)+(\mu^*-|\underline{\lambda}|))^2}\right) \leqslant \left(1-\frac{(\epsilon\mu^*/8\sqrt{n})^2}{((\mu-\mu^*)+\epsilon\mu^*/8\sqrt{n})^2}\right),\end{aligned}$$

where in the last step we used the assumption $\mu^* \geqslant |\underline{\lambda}| + \frac{\epsilon}{8\sqrt{n}}\mu^*$. Therefore, if we have $\mu-\mu^* \leqslant \epsilon'\mu^*/8$, then

$$\|x^*\|^2 - \|x_\mu\|^2 \leqslant \frac{(2\sqrt{n}\epsilon'/\epsilon)+(\sqrt{n}\epsilon'/\epsilon)^2}{(1+(\sqrt{n}\epsilon'/\epsilon))^2} \leqslant \frac{2\sqrt{n}\epsilon'}{\epsilon}. \tag{13}$$

On the other hand, note that

$$\begin{aligned}q(x_\mu)-q(x^*) &= \tfrac{1}{2}x_\mu^T Q x_\mu + c^T x_\mu - \tfrac{1}{2}(x^*)^T Q x^* - c^T x^* \\ &= \tfrac{1}{2}(Qx_\mu + c)^T(x_\mu - x^*) + \tfrac{1}{2}(Qx^*+c)^T(x_\mu - x^*) \\ &= -\tfrac{1}{2}\mu x_\mu^T(x_\mu - x^*) - \tfrac{1}{2}\mu^*(x^*)^T(x_\mu - x^*) \\ &= -\tfrac{1}{2}(\mu-\mu^*)x_\mu^T(x_\mu - x^*) - \tfrac{1}{2}\mu^*(\|x_\mu\|^2 - \|x^*\|^2).\end{aligned} \tag{14}$$

Now we use the bound (13), the assumption $\mu-\mu^* \leqslant \epsilon'\mu^*/8$, and the fact $\|x_\mu\| \leqslant \|x^*\| = 1$ to obtain:

$$q(x_\mu)-q(x^*) \leqslant \mu^*\epsilon'/8 + \mu^*\frac{\sqrt{n}\epsilon'}{\epsilon} = \left(\epsilon'/4 + \frac{2\sqrt{n}\epsilon'}{\epsilon}\right)\mu^*/2 \leqslant \left(\epsilon'/4 + \frac{2\sqrt{n}\epsilon'}{\epsilon}\right)(q(0)-q(x^*)),$$

where the last step is due to (11). Thus, if we select

$$\epsilon' \leqslant \frac{\epsilon^2}{2\sqrt{n}+1/4},$$

then x_μ is feasible for BQP and

$$q(x_\mu)-q(x^*) \leqslant \epsilon(q(0)-q(x^*)),$$

that is, x_μ is an ϵ-minimizer to x^*.

CASE II
In this case, we have

$$\left(1-\frac{\epsilon}{8\sqrt{n}}\right)\mu^* < |\underline{\lambda}|| \text{ or } \mu^* < |\underline{\lambda}| + \mu^*\frac{\epsilon}{8\sqrt{n}}.$$

Again, if we have $\mu-\mu^* < \epsilon'\mu^*/8$, then $\mu - |\underline{\lambda}| < \epsilon'\mu^*/8 + \mu^*\epsilon/8\sqrt{n}$. However, unlike Case I, we find $\|x_\mu\|$ is not sufficiently close to 1. When we observe this fact, we do the following computation to enhance x_μ.

Let q, $\|q\|=1$, be an eigenvector associated with the eigenvalue $\underline{\lambda}$. Then, one of the unit vector $e_j, j=1,\ldots,n-m$, must have $|e_j^T q| \geqslant 1/\sqrt{n}$. (In fact, we can use any unit vector q to replace e_j as long as $q^T \underline{q} \geqslant 1/\sqrt{n}$. A randomly generated q will do it with high probability.) Now we solve for y from

$$(Q+\mu I)y = e_j$$

and let

$$x = x_\mu + \alpha y,$$

where α is chosen such that $\|x\|=1$. Note we have

$$(Q+\mu I)x = -c + \alpha e_j,$$

and in the computation of x_μ and y, matrix $Q+\mu I$ needs to be factorized only once.

It is easy to show that

$$\|y\| \geqslant \frac{1}{\sqrt{n}(\mu - |\underline{\lambda}|)}$$

and

$$|\alpha| \leqslant 2(\mu - |\underline{\lambda}|)\sqrt{n} \leqslant 2(\epsilon' \mu^*/8 + \epsilon\mu^*/8\sqrt{n})\sqrt{n}.$$

Then, we have from (14)

$$\begin{aligned}
q(x) - q(x^*) &= \tfrac{1}{2}(Qx+c)^T(x-x^*) + \tfrac{1}{2}(Qx^*+c)^T(x-x^*) \\
&= \tfrac{1}{2}(Qx + c - \alpha e_j)^T(x-x^*) + \tfrac{1}{2}\alpha e_j^T(x-x^*) - \tfrac{1}{2}\mu^*(x^*)^T(x-x^*) \\
&= -\tfrac{1}{2}\mu x^T(x-x^*) + \tfrac{1}{2}\alpha e_j^T(x-x^*) - \tfrac{1}{2}\mu^*(x^*)^T(x-x^*) \\
&= -\tfrac{1}{2}(\mu x + \mu^* x^*)^T(x-x^*) + \tfrac{1}{2}\alpha e_j^T(x-x^*) = -\tfrac{1}{2}(\mu - \mu^*)x^T(x-x^*) + \tfrac{1}{2}\alpha e_j^T(x-x^*),
\end{aligned}$$

where the last step follows from $\|x\| = \|x^*\| = 1$. Now we use $\mu - \mu^* < \epsilon'\mu^*/8$ and the preceding upper bound on α to estimate the right-hand side:

$$\begin{aligned}
q(x) - q(x^*) &\leqslant \mu^*\epsilon'/8 + 2(\epsilon'\mu^*/8 + \epsilon\mu^*/8\sqrt{n})\sqrt{n} = (\epsilon'/4 + \sqrt{n}\epsilon'/2 + \epsilon/2)\mu^*/2 \\
&\leqslant (\epsilon'/4 + \sqrt{n}\epsilon'/2 + \epsilon/2)(q(0) - q(x^*)),
\end{aligned}$$

where the last step is due to (11). Thus, if we choose

$$\epsilon' \leqslant \frac{\epsilon}{\sqrt{n} + 1/2},$$

then x is feasible for BQP and

$$q(x) - q(x^*) \leqslant \epsilon(q(0) - q(x^*)) \leqslant \epsilon(\bar{z} - \underline{z}),$$

that is, x is an ϵ-minimizer of $q(x)$ over the ball.

Hence, the bisection method will terminate with an ϵ-minimizer of (BQP) in at most

$$O(\ln(\mu^0/\mu^*) + \ln(1/\epsilon) + \ln n)$$

steps, or in a total of $O(n^3(\ln(\mu^0/\mu^*) + \ln(1/\epsilon) + \ln n))$ arithmetic operations.

Theorem 4. *The total running time of the bisection algorithm for generating an ϵ-minimal solution to the ball-constrained QP is bounded by $O(n^3(\ln(\mu^0/\mu^*) + \ln(1/\epsilon) + \ln n))$ arithmetic operations.*

3.1. *Remarks and Extensions*

The model trust region method is due to Levenberg (1963) and Marquardt (1963). These authors only considered the case where Q is positive definite. Moré (1977) proposed an algorithm with a convergence proof for this case. Gay (1981) and Sorenson (1982) proposed algorithms for the general case, also see Celis et al. (1984), Dennis Jr., and Schnable (1983), Martinez (1994), and Yuan (1990). These algorithms work very well in practice, but no theoretical complexity result was established for this problem then. The simple polynomial bisection method was proposed by Vavasis and Zippel (1990) and Ye (1992). This result was used to develop an approximation algorithm for general quadratic optimization, see Luo et al. (1998). On the applications of ball-constrained nonconvex QP, Gibbons et al. (1996) have demonstrated that a class of combinatorial optimization problems can be solved by solving a sequence of ball-constrained QP problems.

Recently, Rendl and Wolkowicz (1997) showed that BQP can be reformulated as a positive semidefinite problem, which is a convex nonlinear problem.

The polynomial complexity in Theorem 4 can be further improved. In particular, Ye (1994) developed a mixed bisection and Newton's method for solving BQP and established an arithmetic operation bound $O(n^3\ln(\ln(\mu^0/\mu^*)+\ln(1/\epsilon')))$ to yield a μ such that $0\leqslant\mu-\mu^*\leqslant\epsilon'$. Briefly, the idea of the method is to first find an approximate $\underline{\mu}$ to the absolute value of the least eigenvalue $|\underline{\lambda}|$ and an approximate eigenvector q to the true $\underline{q}$, such that $0\leqslant\underline{\mu}-|\underline{\lambda}|\leqslant\epsilon'$ and $q^Tq^k\geqslant 1-\epsilon'$. This approximation can be done in $O(n^3\ln(\ln(1/\epsilon')))$ arithmetic operations. Then, we will use q to replace e_j in Case II (i.e., $\|x_{\underline{\mu}}\| < r$) to enhance $x(\underline{\mu})$ and generate a desired approximation. Otherwise, we know $\mu^*>\underline{\mu}$ and, using the mixed method in Ye (1994), we will generate a $\mu\in(\underline{\mu},\mu^0)$ such that $|\mu-\mu^*|\leqslant\epsilon'\mu^*/8$ in $O(n^3\ln(\ln(\mu^0/\mu^*)+\ln(1/\epsilon')))$ arithmetic operations.

Finally, let Q and c have integer data. Consider the decision problem: is there an $x\in\mathscr{R}^n$ satisfying $\|x\|\leqslant 1$, and $q(x)<0$? Under the Turing machine computational model, Vavasis and Zippel (1990) showed that this problem can be answered in polynomial time.

4. Nonconvex Quadratic Programming over a Cube

In this section we consider approximating the global minimizer of the following QP problem over a cube constraint:

$$\text{(CQP)}\qquad \begin{array}{ll} \text{minimize} & x^TQx \\ \text{subject to} & x\in\{x\in\mathscr{R}^n:\ -e\leqslant x\leqslant e\}. \end{array}$$

We denote by $\underline{x}(Q)$ its global minimizer.

Initially, there is a linear term, c^Tx, in the objective function. However, the problem can be homogenized as

$$\begin{array}{ll} \text{minimize} & x^TQx+t\cdot 2c^Tx \\ \text{subject to} & -e\leqslant x\leqslant e,\ -1\leqslant t\leqslant 1 \end{array}$$

by adding a scalar variable t. There is always an optimal solution $(\underline{x}, \underline{t})$ for this problem in which $\underline{t}=1$ or $\underline{t}=-1$. If $\underline{t}=1$, then $\underline{x}$ is also optimal for the original problem; if $\underline{t}=-1$, then $-\underline{x}$ is optimal for the original problem. Thus, without loss of generality, we can assume $q(x)=x^TQx$ throughout this section.

The function $q(x)$ has a global maximizer over the bounded feasible set as well. Let $\bar{q}:=-\underline{q}(-Q)$ and $\underline{q}:=\underline{q}(Q)$ denote their maximal and minimal objective values, respectively. We now present a polynomial algorithm to compute a 4/7-approximate minimizer, that is, to compute a feasible $\hat{x}$, such that

$$\frac{q(\hat{x})-\underline{q}}{\bar{q}-\underline{q}} \leqslant 4/7.$$

4.1. Positive Semidefinite Relaxation

The algorithm for approximating the QP minimizer is to solve a positive semidefinite programming relaxation problem:

$$\begin{aligned} \underline{p}(Q) := \quad & \text{minimize} \quad & Q \bullet X \\ & \text{subject to} \quad & d(X) \leqslant e, X \succeq 0. \end{aligned} \tag{15}$$

Here, unknown $X \in \mathcal{M}^n$, $\bullet$ is the matrix inner product $Q \bullet X = \text{trace}(Q^T X)$, and $X \succeq Z$ means that $X - Z$ is positive semidefinite. Furthermore, $d(X)$ is a vector containing the diagonal components of X. Note that $d(X) \leqslant e$ can be written as $I_j \bullet X \leqslant 1, j = 1, \ldots, n$, where I_j is the all-zero matrix except the jth diagonal component equal to 1.

The dual of the relaxation is

$$\begin{aligned} \underline{p}(Q) = \quad & \text{maximize} \quad & e^T z \\ & \text{subject to} \quad & Q \succeq D(z), z \leqslant 0, \end{aligned} \tag{16}$$

where $D(z)$ is the diagonal matrix such that $d(D(z)) = z \in \mathcal{R}^n$. Note that the relaxation is feasible and its dual has an interior feasible point so that there is no duality gap between the primal and dual. $\underline{X}(Q)$ and $(\underline{y}(Q), \underline{z}(Q))$ denote an optimal solution pair for the primal (15) and dual (16). For simplicity, we let $\underline{x} = \underline{x}(Q)$ and $\underline{X} = \underline{X}(Q)$ in what follows.

We have the following relations between the CQP problem and its relaxation:

Proposition 1. *Let* $\underline{q} = \underline{q}(Q), \bar{q} = -\underline{q}(-Q), \underline{q} = \underline{p}(Q), \bar{p} = -\underline{p}(-Q), (\bar{y}, \bar{z}) = (-\underline{y}(-Q), -\underline{z}(-Q))$. *Then,*

1. $\bar{q}$ *is the maximal objective value of* $x^T Q x$ *in the feasible set of the* CQP *problem;*
2. $\bar{p} = e^T \bar{z}$ *and is the maximal objective value of* $Q \bullet X$ *in the feasible set of the relaxation, and* $D(\bar{z}) - Q \succeq 0$;
3. $\underline{p} \leqslant \underline{q} \leqslant \bar{q} \leqslant \bar{p}$.

Since $\underline{X}$ is positive semidefinite, there is a matrix factorization $\underline{V} = (\underline{v}_1, \ldots, \underline{v}_n) \in \mathcal{R}^{n \times n}$, that is, $\underline{v}_j$ is the jth column of $\underline{V}$, such that $\underline{X} = \underline{V}^T \underline{V}$. Then, after obtaining $\underline{X}$ and $\underline{V}$, we generate a random vector u uniformly distributed on the unit sphere in $\mathcal{R}^n$ and assign

$$\hat{x} = \underline{D} \sigma(\underline{V}^T u), \tag{17}$$

where

$$\underline{D} = \text{diag}(\|\underline{v}_1\|, \ldots, \|\underline{v}_n\|) = \text{diag}(\sqrt{\underline{x}_{11}}, \ldots, \sqrt{\underline{x}_{nn}}),$$

and, for any $x \in \mathcal{R}^n, \sigma(x) \in \mathcal{R}^n$ is the vector whose jth component is $\text{sign}(x_j)$:

$$\text{sign}(x_j) = \begin{cases} 1 & \text{if } x_j \geqslant 0 \\ -1 & \text{otherwise.} \end{cases}$$

It is easy to see that $\hat{x}$ is a feasible point for the (CQP) problem and we will show later that its expected objective value, $E[q(\hat{x})]$, satisfies

$$\frac{E[q(\hat{x})] - \underline{q}}{\bar{q} - \underline{q}} \leqslant \frac{\pi}{2} - 1 \leqslant \frac{4}{7}.$$

That is, $\hat{x}$ is expected to be a 4/7-approximate minimizer for the CQP problem. One can generate u repeatedly and choose the best $\hat{x}$ in the process. Thus, we will almost surely generate a $\hat{x}$ that is a 4/7-approximate minimizer.

4.2. Approximation Analysis

For any function of one variable $f(t)$ and $X \in \mathscr{R}^{n\times n}$, let $f[X] \in \mathscr{R}^{n\times n}$ be the matrix with the components $f(x_{ij})$. Then we have the following technical lemma.

Lemma 1. *Let* $X \succeq 0$ *and* $d(X) \leqslant 1$. *Then* $\arcsin[X] \succeq X$.

For any $X = V^T V \succ 0, d(X) \leqslant e$, we have

$$E(\sigma(v_i^T u)\sigma(v_j^T u)) = 1 - 2\Pr\{\sigma(v_i^T u) \neq \sigma(v_j^T u)\} = 1 - 2\Pr\left\{\sigma\left(\frac{v_i^T u}{\|v_i\|}\right) \neq \sigma\left(\frac{v_j^T u}{\|v_j\|}\right)\right\}.$$

We have the following relation:

$$\Pr\left\{\sigma\left(\frac{v_i^T u}{\|v_i\|}\right) \neq \sigma\left(\frac{v_j^T u}{\|v_j\|}\right)\right\} = \frac{1}{\pi}\arccos\left(\frac{v_i^T v_j}{\|v_i\|\,\|v_j\|}\right).$$

Then, noting $\arcsin(t) + \arccos(t) = \frac{\pi}{2}$ we have

$$E(\sigma(v_i^T u)\sigma(v_j^T u)) = \frac{2}{\pi}\arcsin\left(\frac{v_i^T v_j}{\|v_i\|\,\|v_j\|}\right),$$

and

$$E(\sigma(V^T u)^T DQD\sigma(V^T u)) = \sum_{i=1}^{n}\sum_{j=1}^{n} q_{ij}\|v_i\|\,\|v_j\|\frac{2}{\pi}\arcsin\left(\frac{v_i^T v_j}{\|v_i\|\,\|v_j\|}\right)$$

$$= \frac{2}{\pi} Q \bullet (D \arcsin[D^{-1}XD^{-1}]D).$$

Theorem 5. *We have*

1. $\bar{p} - \underline{q} \geqslant \frac{2}{\pi}(\bar{p} - \underline{p})$.
2. $\bar{q} - \underline{p} \geqslant \frac{2}{\pi}(\bar{p} - \underline{p})$.
3. $\bar{p} - \underline{p} \geqslant \bar{q} - \underline{q} \geqslant \frac{4-\pi}{\pi}(\bar{p} - \underline{p})$.

The result indicates that the positive semidefinite relaxation value $\bar{p} - \underline{p}$ is a constant approximation of $\bar{q} - \underline{q}$. Similarly, the following corollary can be devised:

Corollary 1. *Let* $X = V^T V \succ 0, d(X) \leqslant e, D = \operatorname{diag}(\sqrt{x_{11}}, \ldots, \sqrt{x_{nn}})$, *and* $\hat{x} = D\sigma(V^T u)$ *where* u *with* $\|u\| = 1$ *is a random vector uniformly distributed on the unit sphere. Moreover, let* $X \succ 0 \rightarrow \underline{X}$. *Then,*

$$\lim_{X\rightarrow \underline{X}} E[q(\hat{x})] = \lim_{X\rightarrow \underline{X}} \frac{2}{\pi} Q \bullet (D \arcsin[D^{-1}XD^{-1}]D) \leqslant \frac{2}{\pi}\underline{p} + \left(1 - \frac{2}{\pi}\right)\bar{p}.$$

Finally, we have the following theorem:

Theorem 6. *Let* $\hat{x}$ *be randomly generated from* $\underline{X}$. *Then*

$$\frac{E(q(\hat{x})) - \underline{q}}{\bar{q} - \underline{q}} \leqslant \frac{\pi}{2} - 1 < 4/7.$$

4.3. Remarks and Extensions

Use of the positive semidefinite relaxation was first proposed by Lovász and Shrijver (1990); also see the recent papers by Alizadeh (1991), Anstreicher and Wolkowicz (1998), Fujie and Kojima (1997), and Polijak et al. (1995). This relaxation problem can be solved in polynomial time, for example, see Nesterov and Nemirovskii (1993).

The approximation result described in this section is from Ye (1999b), based on the work of Goemans and Williamson (1995) and Nesterov (1998) for approximating the max-cut problem. In fact, the result can be extended to a more general $\mathcal{F}$ such that

$$\mathcal{F} = \left\{ x : \sum_{j=1}^{n} a_{ij} x_j^2 = b_i, \quad i = 1, \ldots, m, \quad -e \leqslant x \leqslant e \right\}.$$

Note that $\mathcal{F}$ is not even a convex set for some given a_{ij} and b_i.

Other continuous optimization algorithms for solving cube-constrained quadratic programs were proposed by Bonnans and Bouhtou (1995), Coleman and Li (1992), Dang (1998), Kamath et al. (1992), Pardalos et al. (1991), Ye (1992), among others.

5. Nonconvex Quadratic Programming over a Convex Set

Consider the QP problem with general convex quadratic inequality constraints

$$\begin{aligned} \bar{q}(Q) := \quad & \text{maximize} & & x^T Q x \\ & \text{subject to} & & \|A_i x + b_i\|^2 \leqslant 1, \quad i = 1, \ldots, m, \end{aligned}$$

where $A_i \in \mathcal{R}^{r_i \times n} (r_i \leqslant n)$ and $b_i \in \mathcal{R}^{r_i}$ are given. Note that if $r_i = 1$, then it becomes a linear constraint with (nonsymmetric) upper and lower bounds. Denote by $\bar{q} = \bar{q}(Q)$ the maximum value of the problem.

We assume that the QP problem is feasible, bounded, and has nonempty interior; and thereby, assume that $x = 0$ is a center of the convex feasible region, which implies that $1 - \|b_i\|^2 > 0, i = 1, \ldots, m$. For simplicity, we also assume that Q is positive semidefinite. Thus, the global minimum value of $q(x)$ in the feasible region, $\underline{q} = -\bar{q}(-Q)$, is 0.

5.1. Positive Semidefinite Relaxation

Let

$$Q_0 := \begin{pmatrix} 0 & 0 \\ 0 & Q \end{pmatrix} \succeq 0 \quad \text{and} \quad Q_i := \begin{pmatrix} \|b_i\|^2 & b^T A_i \\ A_i^T b & A_i^T A_i \end{pmatrix} \succeq 0.$$

Then, the problem is equivalent to

$$\begin{aligned} \bar{q} := \quad & \text{maximize} & & (1;x)^T Q_0 (1;x) \\ & \text{subject to} & & (1;x)^T Q_i (1;x) \leqslant 1, \quad i = 1, \ldots, m, \end{aligned}$$

where column vector $(1;x) \in \mathcal{R}^{n+1}$ with 1 at its top position. Our approximation algorithm is to solve the positive semidefinite programming (SDP) relaxation problem

$$\begin{aligned} \text{(SDP)} \quad \bar{p}(Q) := \quad & \text{maximize} & & Q_0 \bullet X \\ & \text{subject to} & & I_1 \bullet X = 1, \\ & & & Q_i \bullet X \leqslant 1, \quad i = 1, \ldots, m, \\ & & & X \succeq 0, \end{aligned} \tag{18}$$

where I_1 is the diagonal matrix with 1 in its first diagonal position and 0 everywhere else. Here, unknown $X \in \mathcal{R}^{(n+1) \times (n+1)}$ is a symmetric matrix. Since the original QP problem is feasible, so is the SDP relaxation.

The dual of SDP is

$$\begin{aligned} \bar{p}(Q) = \quad & \text{minimize} & & y_0 + e^T y \\ & \text{subject to} & & y_0 I_1 + \sum_{i=1}^{m} y_i Q_i \succeq Q_0, \quad y \geqslant 0 \in \mathcal{R}^m. \end{aligned} \tag{19}$$

Note that the primal is feasible and the dual has an interior so that there is no duality gap between the primal and dual. Denote by $\bar{X}(Q)$ and $(\bar{y}_0(Q), \bar{y}(Q))$ an optimal solution pair for the primal (18) and dual (19).

We have the following relations between the QP problem and its SDP relaxation.

Proposition 2. *Let* $\bar{q} = \bar{q}(Q), \underline{q} = -\bar{q}(-Q), \bar{p} = \bar{p}(Q)$, *and* $\underline{p} = -\bar{p}(-Q)$. *Then,*

$$\underline{p} \leqslant \underline{q} \leqslant \bar{q} \leqslant \bar{p}.$$

5.2. Rank Reduction

Let $\bar{X} = \bar{X}(Q) \succeq 0$ be an optimizer of the SDP relaxation, and let the lower triangle factorization matrix of $\bar{X}$ be $\bar{L} = (\bar{l}_1; ...; \bar{l}_n) \in \mathscr{R}^{n \times 1}$, that is, $\bar{l}_j$ is the jth row of $\bar{L}$, such that $\bar{X} = \bar{L}\bar{L}^T$. Then, we have

$$(\bar{L}^T Q_0 \bar{L}) \bullet I = Q_0 \bullet (\bar{L}\bar{L}^T) = Q_0 \bullet \bar{X} = \bar{p},$$

and for $i = 1, ..., m$,

$$(\bar{L}^T Q_i \bar{L}) \bullet I = Q_i \bullet (\bar{L}\bar{L}^T) = Q_i \bullet \bar{X} \leqslant 1.$$

Also, since $\bar{X}_{11} = I_1 \bullet X = 1, \bar{l}_1$ has 1 in its first position and 0 everywhere else.

We generate a random vector $\hat{u} \in \mathscr{R}^{n+1}$ with independent components taking the value 1 or -1 with equal probability, then assign

$$\hat{v} = \bar{L}\hat{u} \in \mathscr{R}^{n+1}. \tag{20}$$

Note that with probability one

$$\hat{v}_1^2 = (\bar{l}_1 \hat{u})^2 = (\hat{u}_1)^2 = 1,$$

$$\hat{v}^T Q_0 \hat{v} = (\bar{L}^T Q_0 \bar{L}) \bullet (\hat{u}\hat{u}^T) \leqslant \bar{p} I \bullet (\hat{u}\hat{u}^T) = n\bar{p},$$

and, for $i = 1, ..., m$,

$$\hat{v}^T Q_i \hat{v} = (\bar{L}^T Q_i \bar{L}) \bullet (\hat{u}\hat{u}^T) \leqslant I \bullet (\hat{u}\hat{u}^T) = n.$$

Now, we generate a vector solution $\hat{x} \in \mathscr{R}^n$ from

$$\hat{x} = (\hat{v}_2; ..., \hat{v}_{n+1})/\hat{v}_1. \tag{21}$$

Then, the expected value

$$E(\hat{x}^T Q \hat{x}) = E(\hat{v}^T Q_0 \hat{v}) = E((\bar{L}^T Q_0 \bar{L}) \bullet (\hat{u}\hat{u}^T)) = (\bar{L}^T Q_0 \bar{L}) \bullet E(\hat{u}\hat{u}^T) = (\bar{L}^T Q_0 \bar{L}) \bullet I = \bar{p},$$

and

$$E(\|A_i \hat{x} + b_i\|^2) = E(\hat{v}^T Q_i \hat{v}) = E((\bar{L}^T Q_i \bar{L}) \bullet \hat{u}\hat{u}^T)) = (\bar{L}^T Q_i \bar{L}) \bullet E(\hat{u}\hat{u}^T) = (\bar{L}^T Q_i \bar{L}) \bullet I \leqslant 1.$$

To summarize, we have

Proposition 3. *With probability one*

$$\hat{v}_1^2 = 1,$$

$$q(\hat{x}) = \hat{v}^T Q_0 \hat{v} \leqslant n\bar{p},$$

$$\|A_i \hat{x} + b_i\|^2 = \hat{v}^T Q_i \hat{v} \leqslant n, \quad i = 1, ..., m.$$

Their expected values

$$E(q(\hat{x})) = E(\hat{x}^T Q x\hat{)} = E(\hat{v}^T Q_0 \hat{v}) = \bar{p}$$

$$E(\|A_i \hat{x} + b_i\|^2) = E(\hat{v}^T Q_i \hat{v}) \leqslant 1, \quad i = 1, ..., m.$$

Although $\hat{x}$ is expected to meet the constraints of the QP problem, we have no guarantee that it will do so for sure. If not, we need to shrink it from

$$\tilde{x} = \hat{\alpha}\hat{x}, \tag{22}$$

where

$$\hat{\alpha} = \min_i\{\hat{\alpha}_i\};$$

and $\hat{\alpha}_i = 1$ if $\|A_i\hat{x} + b_i\|^2 \leqslant 1$, else $\hat{\alpha}_i < 1$ is the positive root of the quadratic equation of α:

$$q_i(\alpha) := \|\alpha(A_i\hat{x}) + b_i\|^2 = 1.$$

Next, we will prove that $\tilde{x}$ is feasible for the QP problem.

5.3. *Approximation Analysis*

We first present a technical lemma.

Lemma 2. *Let* $\beta_i \geqslant 0$ *and*

$$\beta_i := \|A_i\hat{x} + b_i\|.$$

Then, for any $0 \leqslant \alpha \leqslant 1$,

$$\|A_i(\alpha\hat{x}) + b_i\|^2 \leqslant \alpha(\beta_i^2 - \|b_i\|^2) + \|b_i\|^2 \leqslant \max\{\beta_i^2, \|b_i\|^2\}.$$

If $\beta_i > 1$, *then the root of the quadratic equation of* $q_i(\alpha) = 1$

$$\hat{\alpha}_i \geqslant \frac{1 - \|b_i\|}{\sqrt{\beta_i^2 - \|b_i\|^2}} \geqslant \frac{1 - \|b_i\|}{\beta_i}.$$

These lead to the following theorem.

Theorem 7. *With probability one,* $\tilde{x}$ *generated in the above rank reduction procedure is a feasible solution for the QP problem.*

Now we analyze the objective function value

$$q(\tilde{x}) = \tilde{x}^T Q \tilde{x} = \hat{\alpha}^2 \hat{x}^T Q \hat{x} = \hat{\alpha}^2 \hat{v}^T Q_0 \hat{v}.$$

Note that $\hat{\alpha}$ is also a random number here. Recall

$$\hat{\alpha} = \min_i\{\hat{\alpha}_i\} \geqslant \min_{\{i, \|A_i\hat{x}+b_i\|^2 > 1\}} \left(\frac{1 - \|b_i\|}{\sqrt{\|A_i\hat{x} + b_i\|^2 - \|b_i\|^2}} \right).$$

Since

$$\beta_i^2 = \|A_i\hat{x} + b_i\|^2 = \hat{v}^T Q_i \hat{v},$$

the following lemma is a straightforward extension of the result proved by Nemirovskii et al. (1998).

Lemma 3. *For any* $\beta \geqslant 1$, *the probability*

$$\Pr(\max_i\{\|A_i\hat{x} + b_i\|^2\} > \beta^2) < 2m\bar{r}\exp(-\beta^2/2),$$

where $\bar{r} = \max_i\{r_i\}$, *the row dimension of* A_i.

Hence, we have, in particular,

$$\Pr[\max_i\{\beta_i^2 = \|A_i\hat{x} + b_i\|^2\} > 2\ln(4mn\bar{r})] < 2m\bar{r}\exp(-\ln(4mn\bar{r})) = \frac{1}{2n}.$$

Now we have the following theorem.

Theorem 8. *The expected value*

$$E(q(\tilde{x})) = E(\tilde{x}^T Q\tilde{x}) \geqslant \frac{(1-\max_i\{\|b_i\|\})^2}{4\ln(4mn\bar{r})} \cdot \bar{p} \geqslant \frac{(1-\max_i\{\|b_i\|\})^2}{4\ln(4mn\bar{r})} \cdot \bar{q},$$

The quality of the approximation bound is dependent on the value $(1-\max_i\{\|b_i\|\})$, which is related to the condition number of the convex feasible set measuring how round the interior of the set is. This condition number also dictates the computational complexity of linear optimization over this convex set.

If we relax the feasibility by a constant fraction ϵ, then we have

Corollary 2. *In the rank reduction procedure, let $\hat{\alpha}_i$ be the root of*

$$q_i(\alpha) = \|\alpha(A_i\hat{x}) + b_i\|^2 = 1 + \epsilon.$$

Then, with probability one

$$\|A_i\tilde{x} + b_i\|^2 \leqslant 1 + \epsilon, \quad i = 1, \ldots, m,$$

and

$$E(\tilde{x}^T Q\tilde{x}) \geqslant \frac{(\epsilon + 1 - \max_i\{\|b_i\|\})^2}{4\ln(4mn\bar{r})} \cdot \bar{q} \geqslant \frac{\epsilon^2}{4\ln(4mn\bar{r})} \cdot \bar{q}.$$

5.4. Linear Inequality Constraints

Consider the problem

$$\text{(LQP)} \qquad \begin{aligned} \bar{q} := \quad & \text{maximize} \quad y^T Q y \\ & \text{subject to} \quad -l \leqslant A^T y \leqslant u, \end{aligned}$$

where $A \in \mathcal{R}^{n\times m}$ and $y = 0$ is the analytic center of the polytope $\{y : -l \leqslant A^T y \leqslant u\}$ (hence the given bounds $l, u > 0$). Then, the problem can be written as

$$\begin{aligned} &\text{maximize} && y^T Q y \\ &\text{subject to} && (a_i^T y + l_i)(a_i^T y - u_i) \leqslant 0, \quad i = 1, \ldots, m, \end{aligned}$$

where a_i is the ith column of A.

The quadratic constraints can be further written as

$$\left(a_i^T y + \frac{l_i - u_i}{2}\right)^2 \leqslant \frac{(l_i - u_i)^2}{4}$$

or

$$\left(\frac{2}{l_i + u_i} a_i^T y + \frac{l_i - u_i}{l_i + u_i}\right)^2 \leqslant 1.$$

Thus, they are in the quadratic constraint form. Furthermore, since y is the analytic center (see, e.g., Nesterov and Nemirovskii, 1993),

$$\frac{1}{m} \leqslant \frac{l_i}{u_i} \leqslant m.$$

Therefore, from Theorem 8 and noting $\bar{r} = 1$, we have

Corollary 3. *The approximation algorithm presented in this chapter will generate a feasible point $\tilde{x}$ for the problem* LQP *and*

$$E(\tilde{x}^T Q\tilde{x}) \geqslant \frac{1}{m^2 \ln(4mn)} \cdot \bar{q}.$$

Note that this bound is worse than the bound of $\frac{1}{m^2}$ of Bellare and Rogaway (1995), Luo et al. (1998), Vavasis (1991), and Ye (1992). However, if $\frac{l_i}{u_i}$ is close to 1, then the SDP relaxation and the rank reduction could yield a better approximation.

Consider the assignment polytope

$$\sum_{j=1}^{n} x_{ij} = 1, \quad i = 1, ..., n,$$

$$\sum_{i=1}^{n} x_{ij} = 1, \quad j = 1, ..., n,$$

$$x_{ij} \geqslant 0, \quad i = 1, ..., n, \quad j = 1, ..., n.$$

Let us transform x_{ij} to

$$y_{ij} = x_{ij} - \frac{1}{n}.$$

Then, the system becomes

$$\sum_{j=1}^{n} y_{ij} = 0, \quad i = 1, ..., n,$$

$$\sum_{i=1}^{n} y_{ij} = 0, \quad j = 1, ..., n,$$

$$y_{ij} \geqslant -\frac{1}{n}, \quad i = 1, ..., n, \quad j = 1, ..., n.$$

Adding seemly redundant upper bounds and eliminating $2n$ variables from $2n$ homogeneous equations, the system can be written as

$$-\frac{1}{n} \leqslant a_{ij}^T y \leqslant 1 - \frac{1}{n}, \quad i = 1, ..., n, \quad j = 1, ..., n,$$

for given vectors $a_{ij} \in \mathscr{R}^{n^2-2n}$ and unknown $y \in \mathscr{R}^{n^2-2n}$. Note that $y = 0$ is the analytic center of the polytope and

$$1 - \frac{l_i - u_i}{l_i + u_i} = 1 - \left(1 - \frac{2}{n}\right) = \frac{2}{n}.$$

Let the objective be maximizing $y^T Q y$ and C to be positive semidefinite, and the global maximum value be $\bar{q}$. Then, we have

Corollary 4. *The approximation algorithm presented in this chapter will generate a feasible point $\tilde{y}$ for the problem* LQP *over the assignment polytope, and*

$$\mathrm{E}(\tilde{y}^T Q \tilde{y}) \geqslant \frac{1}{n^2 \ln(4n^4)} \cdot \bar{q}.$$

This bound is better than the bound of $\frac{1}{n^3}$ of Luo et al. (1998).

5.5. Remarks and Extensions

The approximation algorithm described here is based on Nemirovskii et al. (1998) and Ye (1999a).

The following is a summary of results on approximating the quadratic optimization over a convex set. Bellare and Rogaway (1995) established several negative results on approximating this problem; Goemans and Williamson (1995) proved an approximation result for the max-cut problem; Nesterov (1998) generalized their result to approximating the max-cut problem with a more general objective matrix; Ye (1999b) extended their result to solving the continuous QP with simple bound constraints; Ye (1999b) and Nesterov (1998) proved constant approximation quality for problem QP such that $A_i^T A_i$ is diagonal and $b_i = 0$; Nesterov (1998) and Nemirovskii et al. (1998) established a quality bound when $b_i = 0$; Luo et al. (1998) constructed a quality bound for approximating QP without those assumptions. Their results can be summarized as follows.

- Bellare and Rogaway (1995) show that if the feasible region is a polytope, then there is a constant δ such that the problem has no polynomial algorithm to yield a solution x with

  $$q(x) \geqslant 2^{-\ln^{\delta} n} \bar{q},$$

  unless $NP \subset \tilde{P}$, where $\tilde{P}$ denotes the class of languages recognizable in quasi-polynomial time; there is a constant $\epsilon \leqslant 1/3$ such that it has no polynomial algorithm to yield a solution x with

  $$q(x) \geqslant (1-\epsilon)\bar{q},$$

  unless P = NP.
- If $i = 1, A_1^T A_1$ is positive definite and $b_i = 0$, then Vavasis (1991) and Ye (1992) have developed a procedure to obtain a feasible x with

  $$q(x) \geqslant (1-\epsilon)\bar{q},$$

  whose running time is polynomial in n and $| \ln \epsilon |$ for any $1 > \epsilon > 0$.
- If all $A_i^T A_i$ are diagonal (or mutually commute) and $b_i = 0$; then Nesterov (1998) and Ye (1999b) are able to generate a feasible (randomized) solution $\hat{x}$ such that

  $$\mathrm{E}(q(\hat{x})) \geqslant \frac{2}{\pi} \bar{q}.$$

- If all $b_i = 0$, then Nesterov (1998) and Nemirovskii et al. (1998) produce a feasible (randomized) solution $\hat{x}$ such that, with a constant probability,

  $$q(\hat{x}) \geqslant \frac{1}{2 \ln(2m \min(m, n))} \bar{q}.$$

- If the feasible region is bounded and has a nonempty interior, Luo et al. (1998) (also see Bellare and Rogaway, 1995; Vavasis, 1991; and Ye, 1992) compute a feasible solution x such that

  $$q(x) \geqslant \frac{1}{m^2} \bar{q}.$$

6. Nonconvex Quadratic Programming over a Nonconvex Set

Here we consider QP with general linear and quadratic terms both in its objective function and in its inequality constraints. One can always reduce such a problem to a QP with a linear objective function and quadratic terms only in its inequality constraints, so that we only deal with QP of the form

$$\text{(QOP)} \quad \text{maximize} \quad c^T x$$
$$\text{subject to} \quad x \in \mathscr{F},$$

where $F \equiv \{x \in C_0 : p(x) \leqslant 0 (\forall p(\cdot) \in \mathscr{P}_F)\}$, $\mathscr{P}_F$ denotes a set of quadratic functions on $\mathscr{R}^n$, and C_0 denotes a nonempty compact convex subset of $\mathscr{R}^n$.

There were two types of conceptual successive convex relaxation methods for approximating the convex hull of F, c.hull(F), which are extensions of the Lovász–Schrijver lift-and-project procedure (Balas et al., 1993; Lovász and Shrijver, 1990), with the use of SDP and the semiinfinite LP relaxation (the reformulation-convexification techniques in Sherali and Tuncbilek, 1995) originally developed for combinatorial optimization problems to general nonconvex quadratic optimization problems. We will call the first method the SSDP (successive semidefinite programming) relaxation method, and the second the SSILP (successive semi-infinite linear programming) relaxation method. Kojima and Tunçel established that both of the methods generate a sequence of compact convex sets C_k of $\mathscr{R}^n (k = 0, 1, ...)$ satisfying

1. $\text{c.hull}(F) \subseteq C_{k+1} \subseteq C_k$ (monotonicity),
2. $\bigcap_{k=1}^{+\infty} C_k = \text{c.hull}(F)$ (asymptotic convergence).

6.1. The SSDP Method

Here we present an implementable variant of the SSDP relaxation method according to the paper by Kojima and Tunçel (1998b). For our theoretical analyses and results, we allow cases where $\mathscr{P}_F$ consists of infinitely many quadratic inequalities. To develop implementable variants of the SSDP relaxation method, however, we need to assume that $\mathscr{P}_F$ is finite.

Let $\mathscr{S}^n$ and $\mathscr{S}^n_+$ denote the set of $n \times n$ symmetric matrices and the set of $n \times n$ symmetric positive semidefinite matrices, respectively. We use the notation $qf(\cdot\,; \gamma, q, Q)$ for a quadratic function having the constant term γ, the linear term $2q^Tx$, and the quadratic term x^TQx;

$$qf(x; \gamma, q, Q) \equiv \gamma + 2q^Tx + x^TQx \ (\forall x \in \mathscr{R}^n),$$

where $\gamma \in \mathscr{R}, q \in \mathscr{R}^n$ and $Q \in \mathscr{S}^n$. Let

$$Q \bullet X \equiv \text{the (trace) inner product of two symmetric matrices } Q \text{ and } X,$$

$$\bar{D} \equiv \{d \in \mathscr{R}^n : \|d\| = 1\}.$$

Let C be a subset of C_0 such that $F \subset C$. For $\forall d \in \bar{D}$, let $\alpha(C, d) \equiv \max\{d^Tx : x \in C\}$. For $\forall D \subset \bar{D}$ and $\forall D' \subset \bar{D}$, define sets $\mathscr{P}^L(C, D)$ and $\mathscr{P}^2(C, D, D')$ of quadratic functions by

$$\mathscr{P}^L(C, D) \equiv \{d^Tx - \alpha(C, d) : d \in D\},$$

$$\mathscr{P}^2(C, D, D') \equiv \{-(d_1^Tx - \alpha(C, d_1))(d_2^Tx - \alpha(C, d_2)) : d_1 \in D, d_2 \in D'\}.$$

Then each $p(\cdot) \in \mathscr{P}^L(C, D) \cup \mathscr{P}^2(C, D, D')$ induces a valid inequality $p(x) \leqslant 0$ for F.

Algorithm 1.

Step 0: Let $C_1 = C_0$. *Choose a* $D_0 \subset \bar{D}$. *Compute* $\alpha(C_0, d) = \max\{d^Tx : x \in C_0\} (\forall d \in D_0)$. *Let* $k = 1$.

Step 1: If $C_k = \emptyset$ *then* $\zeta_k = -\infty$ *and stop; otherwise, compute* $\zeta_k = \max\{c^Tx : x \in C_k\}$.

Step 2: Choose a $D_k \subseteq \bar{D}$. *Compute* $\alpha(C_k, d) = \max\{d^Tx : x \in C_k\} (\forall d \in D_k)$.

Step 3: Let $\mathscr{P}_k = \mathscr{P}^L(C_k, D_0) \cup \mathscr{P}^2(C_k, D_0, D_k)$, *and let*

$$C_{k+1} \equiv \left\{ x \in C_0 : \begin{array}{c} \exists X \in \mathscr{S}^n \text{ such that } \begin{pmatrix} 1 & x^T \\ x & X \end{pmatrix} \in \mathscr{S}^{1+n}_+ \text{ and} \\ \gamma + 2q^Tx + Q \bullet X \leqslant 0 \ [\forall qf(\cdot\,; \gamma, q, Q) \in \mathscr{P}_F \cup \mathscr{P}_k] \end{array} \right\}.$$

Step 4: Let $k = k + 1$, *and go to Step 1.*

To implement Algorithm 1, we need to choose a finite $D_0 \subset \bar{D}$ in Step 0, and a finite $D_k \subset \bar{D}$ in Step 2. Let $D \subseteq \bar{D}$. For $\forall \delta \geqslant 0$, a subset D' of $\bar{D}$ is a δ-net of D if for $\forall d \in D$ a $d' \in D'$ exists such that $\|d' - d\| \leqslant \delta$. It should be noted that we can take a finite δ-net D' of D whenever $\delta > 0$. Thus a δ-net D' of D with $\delta > 0$ is a finite discretization of D.

Theorem 9. *(Theorem 3.3 of Kojima and Tunçel, 1998b). Let* $D_V \equiv \{v_1, v_2, ..., v_n, -v_1, -v_2, ..., -v_n\}$, *where the set of vectors* $v_1, v_2, ..., v_n$ *denotes a basis of* $\mathscr{R}^n$. *Let* $U \subset \mathscr{R}^n$ *be an arbitrary open convex set containing* F. *Then there exists a*

$\delta > 0$ such that if we take a set $D_0 \supseteq D_V$ *and a* $\delta-$net D_k *of* $\bar{D}$, *then* $\{C_k(k=1,2,...)\}$ *satisfies property (1) above and*

(2)′ c.hull(F) $\subseteq C_k \subset U$ *for* $\exists$ *finite number* k.

Usually property (2)′ is unnecessary in practice but property (2)″ below is enough to know upper bounds on the maximum objective value of the QOP.

(2)″ $\zeta^* \equiv \max\{c^T x : x \in F\} \leqslant \zeta_k \equiv \max\{c^T x : x \in C_k\} < \zeta^* + \epsilon$ for $\exists$ finite number k.

For such cases, we can use a localization technique, which avoids some redundant work to make the convex relaxations accurate in unnecessary directions.

Theorem 10. *(Theorem 3.5 of Kojima and Tunçel, 1998b). Let $\kappa > 0$ and $\epsilon > 0$. There exists a $\delta > 0$ such that if we take* $D_0 \supseteq D_V$ *and a δ-net* D_k *of* $D(c,\kappa) \equiv \{d \in \bar{D} : \|d-c\| \leqslant \kappa\}$, *then* $\{C_k(k=1,2,...)\}$ *and* $\{\zeta_k(k=1,2,...)\}$ *satisfy the properties (1) and (2)″.*

6.2. *Remarks and Extensions*

The two successive convex relaxation methods presented in this section are based on Kojima and Tunçel (1998a). These two methods are conceptual (or purely theoretical) in the sense that they require the solutions of an infinite number of semi-infinite SDPs or (semi-infinite LPs) to generate a new C_{k+1} at each iteration. See also a more recent paper (Kojima and Takeda, 1999) for computational complexity analyses of the conceptual successive convex relaxation methods.

Algorithm 1 might be still impractical because as a higher accuracy ϵ is required, not only the number of the semidefinite programming problems (or linear programming problems) to be solved at each iteration but also their sizes explode quite rapidly. More recently, Takeda et al. (1999) presented practical versions of successive convex relaxation methods by further slimming down Algorithm 1. Although those versions no longer enjoy property (2)″, numerical results reported in that paper look promising for some problems.

7. Finding a Local Solution of Nonconvex Quadratic Programming

Again, given $A \in \mathscr{R}^{m\times n}$ with rank m and $b \in \mathscr{R}^m$, consider the QP problem where

$$\mathscr{F} = \{x : Ax = b, x \geqslant 0.\}$$

In this section, unlike in Section 2, we assume that QP is a nonconvex problem, for example, Q has at least one negative eigenvalue in $\mathscr{N}(A)$, the null space of A.

No time complexity bounds have been developed for various QP methods. (Of course, an enumerative search approach will solve QP but it possesses an exponential time bound.) These algorithms generally generate a sequence of points that converge to a stationary or KKT point associated with QP, as discussed in Section 2.

Here we assume that the feasible set, $\mathscr{F}$, of QP has a strictly positive feasible solution. To find out if $\mathscr{F}$ possesses a strictly positive feasible solution, for any given (A, b), a single linear program in polynomial time can be solved, so that the above assumption is without any loss of generality. We make an additional assumption that the feasible region is bounded. With this assumption QP has a minimizer and a maximizer. Let $\underline{q}$ and $\bar{q}$ be their minimal and maximal objective values, respectively.

Again, an ϵ-minimal solution or ϵ-minimizer, $\epsilon \in (0,1)$, for QP is defined as an $x \in \mathscr{F}$ such that:

$$\frac{q(x)-\underline{q}}{\bar{q}-\underline{q}} \leqslant \epsilon.$$

Similarly, we define an ϵ-KKT solution for QP as an (x,y,s) such that $x \in \mathscr{F}$, $s = Qx + c - A^T y \geqslant 0$, and

$$\frac{x^T s}{\bar{q}-\underline{q}} = \frac{q(x)-d(x,y)}{\bar{q}-\underline{q}} \leqslant \epsilon.$$

Note that the minimizer of QP is a special KKT point such that $q(x) = d(x,y) = \underline{q}$.

In this section we present a potential reduction method to compute an ϵ-KKT point in $O\left(\left(\frac{n^6}{\epsilon}\ln\frac{1}{\epsilon} + n^4 \ln n\right)\left(\ln\frac{1}{\epsilon} + \ln n\right)\right)$ arithmetic operations. We also show that Q is positive semi-definite in the null space of the active set at the limit of this point, indicating that the limit satisfies the second-order necessary condition to be a local minimal solution. The result is the first approximation algorithm whose running time is almost linear in $\frac{1}{\epsilon}$, which was an open question in the area of nonlinear optimization complexity.

7.1. Potential Reduction Algorithm

The potential function used to solve QP is

$$\mathscr{P}(x) = \mathscr{P}_{n+\rho}(x,z) := (n+\rho)\ln(q(x)-z) - \sum_{j=1}^{n} \ln(x_j),$$

where $0 < x \in \mathscr{F}$, parameter $\rho > 0$, and $z \leqslant \underline{q}$.

Starting from a point $0 < x^0 \in \mathscr{F}$, the potential reduction algorithm will generate a sequence of $\{x^k\} \in \mathscr{F}$ such that $\mathscr{P}(x^{k+1}) < \mathscr{P}(x^k)$. For simplicity and convenience, we assume $x^0 = e$, and x^0 is the analytic center of $\mathscr{F}$. Our results hold even if x^0 is replaced by an approximate center. Therefore, this assumption is also without loss of generality.

To determine how much the potential function must be reduced to generate an ϵ-KKT point, we note that x^0 is the analytic center of $\mathscr{F}$,

$$\begin{aligned} \mathscr{F} \supset \mathscr{E}_{\text{in}} &:= \{x \in \mathscr{F} : \|(X^0)^{-1}(x-x^0)\| \leqslant 1\} \\ \mathscr{F} \subset \mathscr{E}_{\text{out}} &:= \{x \in \mathscr{F} : \|(X^0)^{-1}(x-x^0)\| \leqslant n\}. \end{aligned} \tag{23}$$

In other words, $\mathscr{F}$ is inscribed by ellipsoid $\mathscr{E}_{\text{in}}$ and outscribed by ellipsoid $\mathscr{E}_{\text{out}}$. The two ellipsoids are concentric, where the ratio of their radii is n.

Thus, for any $x \in \mathscr{F}$

$$\sum_{j=1}^{n} \ln(x_j^0/x_j) \geqslant -n\ln(n+1). \tag{24}$$

Note that if

$$(n+\rho)\ln(q(x^k)-z) - (n+\rho)\ln(q(x^0)-z) \leqslant (n+\rho)\ln\epsilon$$

or

$$\mathscr{P}(x^k) - \mathscr{P}(x^0) \leqslant (n+\rho)\ln\epsilon + \sum_{j=1}^{n}\ln(x_j^0/x_j^k) \tag{25}$$

we must have

$$\frac{q(x^k)-\underline{q}}{\bar{q}-\underline{q}} \leqslant \frac{q(x^k)-\underline{q}}{q(x^0)-\underline{q}} \leqslant \frac{q(x^k)-z}{q(x^0)-z} \leqslant \epsilon,$$

which implies that x^k is an ϵ-minimizer and, thereby, an ϵ-KKT point. From relations (24) and (25), x^k becomes an ϵ-minimizer as soon as

$$\mathscr{P}(x^k) - \mathscr{P}(x^0) \leqslant (n+\rho)\ln\epsilon - n\ln(n+1). \tag{26}$$

Given $0 < x \in \mathscr{F}$, let $\Delta = q(x) - z$ and let $d_x \in \mathscr{N}(A)$, be a vector such that $x^+ := x + d_x > 0$. Then

$$(n+\rho)\ln(q(x^+)-z) - (n+\rho)\ln(q(x)-z)$$

$$= (n+\rho)\ln\left(\Delta + \frac{1}{2}d_x^TQd_x + (Qx+c)^Td_x\right) - (n+\rho)\ln\Delta$$

$$= (n+\rho)\ln\left(1 + \left(\frac{1}{2}d_x^TQd_x + (Qx+c)^Td_x\right)/\Delta\right)$$

$$\leqslant \frac{n+\rho}{\Delta}\left(\frac{1}{2}d_x^TQd_x + (Qx+c)^Td_x\right).$$

On the other hand, if $\|X^{-1}d_x\| \leqslant \alpha < 1$ then

$$-\sum_{j=1}^{n}\ln(x_j^+) + \sum_{j=1}^{n}\ln(x_j) \leqslant -e^TX^{-1}d_x + \frac{\alpha^2}{2(1-\alpha)}.$$

Thus, if $\|X^{-1}d_x\| \leqslant \alpha < 1$ then $x^+ = x + d_x > 0$ and

$$\mathscr{P}(x^+) - \mathscr{P}(x) \leqslant \frac{n+\rho}{\Delta}\left(\frac{1}{2}d_x^TQd_x + \left(Qx + c - \frac{\Delta}{n+\rho}X^{-1}e\right)^Td_x\right) + \frac{\alpha^2}{2(1-\alpha)}. \tag{27}$$

To achieve a potential reduction, we minimize a quadratic function subject to an ellipsoid constraint. We solve the following problem at the kth iteration:

$$\begin{aligned} &\text{minimize} && \frac{1}{2}d_x^TQd_x + \left(Qx^k + c - \frac{\Delta^k}{n+\rho}(X^k)^{-1}e\right)^Td_x \\ &\text{subject to} && Ad_x = 0, \\ & && \|(X^k)^{-1}d_x\|^2 \leqslant \alpha^2. \end{aligned}$$

Let

$$Q^k = X^kQX^k, c^k = X^kQx^k + X^kc - \frac{\Delta^k}{n+\rho}e, \quad A^k = AX^k, \text{ and } \quad d_x' = (X^k)^{-1}d_x.$$

Then the above problem becomes

$$\begin{aligned} &\text{(BQP)} \quad \text{minimize} && q'(d_x') := \tfrac{1}{2}(d_x')^TQ^kd_x' + (c^k)^Td_x' \\ & \quad\quad\quad \text{subject to} && A^kd_x' = 0, \\ & && \|d_x'\|^2 \leqslant \alpha^2. \end{aligned}$$

Let $N^k \in \mathscr{R}^{n\times(n-m)}$ be an orthonormal basis spanning the null space of A^k, where $(N^k)^TN^k = I$, and let $H^k = (N^k)^TQ^kN^k \in \mathscr{R}^{(n-m)\times(n-m)}$ and $g^k = (N^k)^Tc^k \in \mathscr{R}^{n-m}$. Then $d_x' = N^kv$ for some $v \in \mathscr{R}^{n-m}$ and problem BQP can be rewritten as

$$\begin{aligned} &\text{(BHP)} \quad \text{minimize} && \tfrac{1}{2}v^TH^kv + (g^k)^Tv \\ & \quad\quad\quad \text{subject to} && \|v\| \leqslant \alpha^2. \end{aligned}$$

Again, this is the so-called ball-constrained quadratic problem, which can be solved efficiently.

The solution d_x' of problem BQP satisfies the following necessary and sufficient conditions:

$$\begin{aligned} (Q^k + \mu^kI)d_x' - (A^k)^Ty(\mu^k) &= -c^k \text{ for some } y(\mu^k), \\ A^kd_x' &= 0, \\ \mu^k &\geqslant \max\{0, -\lambda^k\}, \\ \text{and} \quad \|d_x'\| &= \alpha, \end{aligned} \tag{28}$$

or, equivalently, the solution v of BHP satisfies the following necessary and sufficient conditions:

$$
\begin{aligned}
&(H^k + \mu^k I)v = -g^k, \\
&\qquad \mu^k \geqslant \max\{0, -\lambda^k\}, \\
&\quad \text{and} \quad \|v\| = \alpha,
\end{aligned} \tag{29}
$$

where $\lambda^k = \underline{\lambda}(H^k)$, and $\underline{\lambda}(H)$ denotes the least eigenvalue of matrix H. Since Q is not positive semidefinite in the null space of A and x^k is strictly positive, we must have $\lambda^k < 0$.

Let

$$
\begin{aligned}
&s(\mu^k) = Q(x^k + d_x) + c - A^T y(\mu^k) \text{ and} \\
&p^k = Q^k d_x' + c^k - (A^k)^T y(\mu^k) = X^k s(\mu^k) - \frac{\Delta^k}{n+\rho} e.
\end{aligned} \tag{30}
$$

Then,

$$
\mu^k = \|p^k\|/\alpha, \quad d_x' = -\frac{\alpha p^k}{\|p^k\|} \tag{31}
$$

and

$$
\begin{aligned}
q'(d_x') &= \tfrac{1}{2}(d_x')^T Q^k d_x' + (c^k)^T d_x' = (d_x')^T (Q^k d_x' + c^k) - \tfrac{1}{2}(d_x')^T Q^k d_x' \\
&= (d_x')^T (Q^k d_x' + c^k - (A^k)^T y(\mu^k)) - \tfrac{1}{2}(d_x')^T Q^k d_x' = -\alpha^2 \mu^k - \tfrac{1}{2}(d_x')^T Q^k d_x' \\
&= -\alpha^2 \mu^k - \tfrac{1}{2} v^T (N^k)^T Q^k N^k (N^k)^T v \leqslant -\alpha^2 \mu^k - \tfrac{1}{2}\underline{\lambda}((N^k)^T Q^k N^k)\|v\|^2 = -\alpha^2 \mu^k + \frac{\alpha^2}{2} \mid \lambda^k \mid \\
&\leqslant -\frac{\alpha^2 \mu^k}{2} = -\frac{\alpha \|p^k\|}{2}.
\end{aligned}
$$

This implies that

$$
\frac{n+\rho}{\Delta}\left(\frac{1}{2} d_x^T Q d_x + \left(Q x^k + c - \frac{\Delta^k}{n+\rho} X^{-1} e\right)^T d_x\right) = \frac{\Delta^k}{n+\rho} q'(d_x') \leqslant -\frac{\alpha}{2}\frac{n+\rho}{\Delta^k}\|p^k\|.
$$

Thus, not only do we have $x^{k+1} := x^k + d_x > 0$ but also, from (27),

$$
\mathscr{P}(x^{k+1}) - \mathscr{P}(x^k) \leqslant -\frac{\alpha}{2}\frac{n+\rho}{\Delta^k}\|p^k\| + \frac{\alpha^2}{2(1-\alpha)}. \tag{33}
$$

Here we see that if

$$
\frac{n+\rho}{\Delta^k}\alpha\mu^k = \frac{n+\rho}{\Delta^k}\|p^k\| \geqslant \frac{3}{4}
$$

and if α is chosen around $1/4$, then

$$
\mathscr{P}(x^{k+1}) - \mathscr{P}(x^k) \leqslant -\frac{5}{96}.
$$

Therefore, according to the implication of (26) we have

> **Theorem 11.** *Let* μ^k *and* $\alpha = 1/4$ *be in condition (28). Then, if* $\frac{n+\rho}{\Delta^k}\alpha\mu^k = \frac{n+\rho}{\Delta^k}\|p^k\| \geqslant \frac{3}{4}$ *for all* k, *the algorithm returns an* ϵ*-minimizer of* QP *in* $O((n+\rho)\ln\frac{1}{\epsilon} + n \ln n)$ *iterations.*

The question is what can we say if $\frac{n+\rho}{\Delta^k}\|p^k\| < \frac{3}{4}$ at some k. The next section will show how to calculate a suitable z such that using $\rho = \frac{2n^2(n+\sqrt{n})}{\epsilon} - n$ guarantees that x^{k+1} must be an ϵ-KKT point whenever $\frac{n+\rho}{\Delta^k}\|p^k\| < 1$. Thus, the number of total iterations to generate an ϵ-minimal or ϵ-KKT solution is bounded by $O(\frac{n^3}{\epsilon}\ln\frac{1}{\epsilon} + n \ln n)$. The algorithm can be simply stated as follows:

> **Algorithm 2.** *Let* $n+\rho = 2n^2(n+\sqrt{n})/\epsilon, \alpha = 1/4, z \leqslant \underline{q}$, *and* x^0 *be an approximate analytic center of* $\mathscr{F}$. Set $k := 0$.

While $\frac{n+\rho}{\Delta^k}\alpha\mu^k = \frac{n+\rho}{\Delta^k}\|p^k\| < \frac{3}{4}$ *in* (28) *or* $\frac{q(x^k)-z}{q(x^0)-z} < \epsilon$ **do**

1. *Solve* BQP.
2. *Let* $\mathrm{x}^{\mathrm{k+1}} = \mathrm{x}^{\mathrm{k}} + \mathrm{X}^{\mathrm{k}}\mathrm{d}'_{\mathrm{x}}$.
3. *Let* $\mathrm{k} := \mathrm{k}+1$ *and return to Step 1.*

7.2. *Generating an* ϵ*-KKT point*

We first illustrate how to compute a suitable lower bound, z, for the QP minimal value $\underline{q}$. A z can be generated by solving, again, a ball-constrained problem:

$$\begin{aligned} &\text{minimize} \quad q(x)-q(x^0) = \tfrac{1}{2}(x-x^0)^T Q(x-x^0) + (Qx^0+c)^T(x-x^0)\\ &\text{subject to} \quad A(x-x^0)=0,\\ &\qquad\qquad x \in \mathscr{E}_{in}. \end{aligned}$$

Let $\hat{x}$ be the minimizer. Then, from relation (23), that is, the ratio of the radii of the inscribing and circumscribing ellipsoids to $\mathscr{F}$ is $1/n$, we have

$$q(x^0)-q(\hat{x}) \geqslant \frac{1}{n^2}(q(x^0)-\underline{q}).$$

Thus, we can assign

$$z := q(x^0) - n^2(q(x^0)-q(\hat{x})). \tag{34}$$

Note that an approximate x, say $q(x^0)-q(x) \geqslant (q(x^0)-q(\hat{x}))/1.1$, would establish a lower bound $z := q(x^0) - 1.1n^2(q(x^0)-q(x))$. This bound is perfectly acceptable for establishing our complexity result as well.

We now refer back to the case where $\frac{n+\rho}{\Delta^k}\|p^k\| < \frac{3}{4}$. Actually we shall address a weaker case where $\frac{n+\rho}{\Delta^k}\|p^k\| < 1$, that is,

$$\left\|\frac{n+\rho}{\Delta^k}X^k s(\mu^k) - e\right\| < 1.$$

First, we must have

$$s(\mu^k) = Q(x^k+d_x) + c - A^T y(\mu^k) = Qx^{k+1} + c - A^T y(\mu^k) > 0.$$

Furthermore,

$$\begin{aligned} &\left\|\frac{n+\rho}{\Delta^k}X^k s(\mu^k) - e\right\|^2\\ &= \left(\frac{n+\rho}{\Delta^k}\right)^2 \left\|X^k s(\mu^k) - \frac{(x^k)^T s(\mu^k)}{n}e\right\|^2 + \left\|\frac{(n+\rho)(x^k)^T s(\mu^k)}{n\Delta^k}e - e\right\|^2\\ &\geqslant \left\|\frac{(n+\rho)(x^k)^T s(\mu^k)}{n\Delta^k}e - e\right\|^2\\ &\geqslant \left(\frac{(n+\rho)(x^k)^T s(\mu^k)}{n\Delta^k} - 1\right)^2 n. \end{aligned}$$

Hence, $\frac{n+\rho}{\Delta^k}\|p^k\| < 1$ implies

$$\frac{n-\sqrt{n}}{n+\rho} \leqslant \frac{(x^k)^T s(\mu^k)}{\Delta^k} \leqslant \frac{n+\sqrt{n}}{n+\rho}.$$

Moreover,

$$\begin{aligned} (x^{k+1})^T s(\mu^k) &= (x^k)^T (X^k)^{-1} X^{k+1} s(\mu^k) \leqslant \|(X^k)^{-1}X^{k+1}\|(x^k)^T s(\mu^k)\\ &\leqslant (1+\alpha)(x^k)^T s(\mu^k) \leqslant 2(x^k)^T s(\mu^k). \end{aligned}$$

Therefore, we have

$$\frac{(x^{k+1})^T s(\mu^k)}{q(x^k)-z} = \frac{(x^{k+1})^T s(\mu^k)}{\Delta^k} \leqslant \frac{2(n+\sqrt{n})}{n+\rho} \leqslant \frac{\epsilon}{n^2}$$

or

$$\frac{(x^{k+1})^T s(\mu^k)}{q(x^k)-q(x^0)+n^2(q(x^0)-q(\hat{x}))} \leqslant \frac{\epsilon}{n^2}.$$

Consequently, if $q(x^k) \geqslant q(x^0)$ then

$$\frac{(x^{k+1})^T s(\mu^k)}{n^2(q(x^k)-q(\hat{x}))} \leqslant \frac{\epsilon}{n^2};$$

otherwise

$$\frac{(x^{k+1})^T s(\mu^k)}{n^2[q(x^0)-q(\hat{x})]} \leqslant \frac{\epsilon}{n^2}.$$

Both of them imply that

$$\frac{(x^{k+1})^T s(\mu^k)}{\bar{q}-q(\hat{x})} \leqslant \epsilon$$

since $q(x^0) \leqslant \bar{q}$ and $q(x^k) \leqslant \bar{q}$, which further implies that

$$\frac{(x^{k+1})^T s(\mu^k)}{\bar{q}-\underline{q}} \leqslant \epsilon,$$

since $q(\hat{x}) \geqslant \underline{q}$. That is, x^{k+1} is an ϵ-KKT point for QP. To summarize, we have

Theorem 12. *Let* z *and* ρ *be chosen as above and let* $\alpha < 1$ *in condition (28). Then,* $\mu^k \|d'_x\| = \|p^k\| < \frac{\Delta^k}{n+\rho}$ *implies that* $x^{k+1} := x^k + X^k d'_x$ *is an* ϵ-KKT *point for* QP.

7.3. *Remarks and Extensions*

Many other algorithms were developed for solving certain linear complementarity problems, which include several subclasses of quadratic programs. One popular algorithm is the smoothing method, developed by Chen and Mangasarian (1996), Chen (1996), Gabriel and Moré (1995), and Kanzow (1997), among others.

ACKNOWLEDGMENT

This work was done while the author was visiting the Department of Mathematics and supported by the Liku Bie Ju Center for Mathematical Sciences, City University of Hong Kong, 83 Tat Chee Avenue, Kowloon, Hong Kong. The author would like to thank Professor Kojima for supplying most materials in section 6.

References

Alizadeh, F. 1991. "Combinatorial optimization with interior point methods and semi-definite matrices." Ph.D. diss. University of Minnesota.

Andersen, E. D. and Y. Ye. 1999. "On a homogeneous algorithm for the monotone complementarity problem." *Math. Programming* **84**: 375–400.

Anstreicher, K. and H. Wolkowicz. 1998. "On Lagrangian relaxation of quadratic matrix constraints." Technical Report CORR 98-24. University of Waterloo.

Anstreicher, K. M., D. den Hertog, C. Roos, and T. Terlaky. 1993. "A long step barrier method for convex quadratic programming." *Algorithmica* **10**: 365–382.

Anstreicher, K. M. and J. P. Vial. 1994. "On the convergence of an infeasible primal-dual interior-point method for convex programming." *Optim. Methods Softw.* **3**: 273–283.

Balas, E., S. Ceria, and G. Cornuéjols. 1993. "A lift-and-project cutting plane algorithm for mixed 0–1 programs." *Math. Programming* **58**: 295–323.

Bellare, M. and P. Rogaway. 1995. "The complexity of approximating a nonlinear program." *Math. Programming* **69**: 429–442.

Ben-Daya, M. and C. M. Shetty. 1990. "Polynomial barrier function algorithm for convex quadratic programming." *Arabian J. Sci. Engng* **15**: 657–670.

Ben-Tal, A. and A. S. Nemirovskii. 1994. "Interior point polynomial time method for truss topology design." *SIAM J. Optim.* **4**: 596–612.

Bonnans, F. and M. Bouhtou. 1995. "The trust region affine interior point algorithm for convex and nonconvex quadratic programming." *RAIRO—Recherche Opérationnelle* **29**: 195–217.

Celis, M. R., J. E. Dennis, and R. A. Tapia. 1984. "A trust region strategy for nonlinear equality constrained optimization." In *Numerical Optimization*, edited by P. T. Boggs, R. Byrd, and R. Schnable, pp. 71–82. Philadelphia: SIAM Publication.

Chen, C. and O. L. Mangasarian. 1996. "A class of smoothing functions for nonlinear and mixed complementarity problems." *Comput. Optim. Appl.* **5**: 97–138.

Chen, X. 1996. "Superlinear convergence of smoothing quasi-Newton methods for nonsmooth equations." To appear in *J. Comp. Appl. Math.*

Coleman, T. F. and Y. Li. 1992. "A globally and quadratically convergent affine scaling method for linear L_1 problems." *Math. Programming* **52**: 189–222.

Dang, C. 1998. "A continuous deformation descent direction method for approximating a solution of the box constrained concave minimization problem." To appear in *Journal of Optimization Theory and Applications.*

Dennis Jr., J. E. and R. E. Schnable. 1983. In *Numerical Methods for Unconstrained Optimization and Nonlinear Equations*. New York: Prentice-Hall.

Ding, J. and T. Y. Li. 1991. "A polynominal-time predictor–corrector algorithm for a class of linear complementarity problems." *SIAM J. Comput.* **1**: 83–92.

Faybusovich, L. 1996. "Semidefinite programming: a path-following algorithm for a linear-quadratic functional." *SIAM J. Optim.* **6**: 1007–1024.

Freund, R. M. 1991. "Polynomial–time algorithms for linear programming based only on primal scaling and projected gradients of a potential function." *Math. Programming* **51**: 203–222.

Fujie, T. and M. Kojima. 1997. "Semidefinite programming relaxation for nonconvex quadratic programs." *J. Global Optim.* **10**: 367–380.

Gabriel, S. A. and J. J. Moré. 1995. "Smoothing of mixed complementarity problems." Technical Report MCS-P541-0995. Mathematics and Computer Science Division, Argonne National Laboratory, Chicago.

Gay, D. M. 1981. "Computing optimal locally constrained steps." *SIAM J. Sci. Stat. Comput.* **2**: 186–197.

Gibbons, L. E., D. W. Hearn, and P. M. Pardalos. 1996. "A continuous based heuristic for the maximum clique problem." In *Cliques, Coloring, and Satisfiability*, edited by D. S. Johnson and M. A. Trick. DIMACS Series in Discrete Mathematics and Theoretical Computer Science, **26**: 103–124. New Jersey: American Mathematical Society.

Goemans, M. X. and D. P. Williamson. 1995. "Improved approximation algorithms for maximum cut and satisfiability problems using semidefinite programming." *J. Assoc. Comput. Mach.* **42**: 1115–1145.

Goldfarb, D. and S. Liu. 1990. "An $O(n^3L)$ primal interior point algorithm for convex quadratic programming." *Math. Programming* **49**: 325–340.

Gonzaga, C. C. 1989. "An algorithm for solving linear programming problems in $O(n^3L)$ operations." In *Progress in Mathematical Programming: Interior Point and Related Methods*, edited by N. Megiddo, pp. 1–28. New York: Springer Verlag.

Güler, O. 1993. "Existence of interior points and interior paths in nonlinear monotone complementarity problems." *Math. Oper. Res.* **18**: 128–147.

den Hertog, D. 1992. "Interior point approach to linear, quadratic and convex programming, algorithms and complexity." Ph.D. diss. Faculty of Mathematics and Informatics, TU Delft, NL-2628 BL Delft, The Netherlands.

Horst, R., P. M. Pardalos, and N. V. Thoai. 1995. In *Introduction to Global Optimization*. Boston: Kluwer Academic Publishers.

Huang, S. and K. O. Kortanek. 1991. "A note on a potential reduction algorithm for LP with simultaneous primal-dual updating." *Oper. Res. Lett.* **10**: 501–507.

Jarre, F. 1990. "On the convergence of the method of analytic centers when applied to convex quadratic programs." *Math. Programming* **49**: 341–358.

Jarre, F. 1992. "Interior-point methods for convex programming." *Appl. Math. Optim.* **26**: 287–311.

Ji, J., F. A. Potra, and S. Huang. 1995. "Predictor–corrector method for linear complementarity problems with polynomial complexity and superlinear convergence." *J. Optim. Theory Appl.* **84**: 187–199.

Kamath, A. P., N. K. Karmarkar, K. G. Ramakrishnan, and M. G. C. Resende. 1992. "A continuous approach to inductive inference." *Math. Programming* **57**: 215–238.

Kanzow, C. 1997. "Some noninterior continuation methods for linear complementarity problems." *SIAM J. Matrix Anal. Appl.*

Kapoor, S. and P. M. Vaidya. 1986. "Fast algorithms for convex quadratic programming and multicommodity flows." In *Proceedings of the 18th Annual ACM Symposium on Theory of Computing*, pp. 147–159. New York: ACM Press.

Kojima, M., S. Mizuno, and A. Yoshise. 1989. "A polynomial-time algorithm for a class of linear complementarity problems." *Math. Programming* **44**: 1–26.

Kojima, M., S. Mizuno, and A. Yoshise. 1991. "An $O(\sqrt{n}L)$ iteration potential reduction algorithm for linear complementarity problems." *Math. Programming* **50**: 331–342.

Kojima, M. and A. Takeda. 1999. "Complexity analysis of conceptual successive convex relaxation of nonconvex sets." Technical Report B-350. Department of Mathematical and Computing Sciences, Tokyo Institute of Technology, Tokyo, Japan.

Kojima, M. and L. Tunçel. 1998a. "Cones of matrices and successive convex relaxations of nonconvex sets." Technical Report B-338. Department of Mathematical and Computing Sciences, Tokyo Institute of Technology, Tokyo, Japan.

Kojima, M. and L. Tunçel. 1998b. "Discretization and localization in successive convex relaxation methods for nonconvex quadratic optimization problems." Technical Report B-341. Department of Mathematical and Computing Sciences, Tokyo Institute of Technology, Tokyo, Japan.

Kortanek, K. O. and J. Zhu. 1993. "A polynomial barrier algorithm for linearly constrained convex programming problems." *Math. Oper. Res.* **18**: 116–127.

Levenberg, K. 1963. "A method for the solution of certain non-linear problems in least squares." *Q. Appl. Math.* **2**: 164–168.

Lovász, L. and A. Shrijver. 1990. "Cones of matrices and setfunctions, and 0–1 optimization." *SIAM J. Optim.* **1**: 166–190.

Luo, Z. and J. Sun. 1999. "An analytic center based column generation algorithm for convex quadratic feasibility problems." *SIAM J. Optim.* **9.**

Luo, Z.-Q., M. Fu, and Y. Ye. 1998. "Approximation algorithms for quadratic programming." *J. Combinat. Optim.* **2**: 29–50.

Marquardt, D. W. 1963. "An algorithm for least-squares estimation of nonlinear parameters." *J. SIAM* **11**: 431–441.

Martinez, J. M. 1994. "Local minimizers of quadratic functions on euclidean balls and spheres." *SIAM J. Optim.* **4**: 159–176.

McShane, K. A. 1994. "Superlinearly convergent $O(\sqrt{n}L)$-iteration interior-point algorithms for linear programming and the monotone linear complementarity problem." *SIAM J. Optim.* **4**: 247–261.

Megiddo, N. 1986. "Pathways to the optimal set in linear programming." In *Progress in Mathematical Programming: Interior Point and Related Methods*, edited by N. Megiddo, pp. 1–35. New York: Springer Verlag.

Mehrotra, S. and J. Sun. 1990. "An algorithm for convex quadratic programming that requires $O(n^{3.5}L)$ arithmetic operations." *Math. Oper. Res.* **15**: 342–363.

Mehrotra, S. and J. Sun. 1991. "A method of analytic centers for quadratically constrained convex quadratic programs." *SIAM J. Numerical Anal.* **28**: 529–544.

Monteiro, R. D. C. and I. Adler. 1989. "Interior path following primal-dual algorithms: Part II: Convex quadratic programming." *Math. Programming* **44**: 43–66.

Monteiro, R. D. C., I. Adler, and M. G. C. Resende. 1990. "A polynominal–time primal-dual affine scaling algorithm for linear and convex quadratic programming and its power series extension." *Math. Oper. Res.* **15**: 191–214.

Moré, J. J. 1977. "The Levenberg–Marquardt algorithm: implementation and theory." In *Numerical Analysis*, edited by G. A. Watson. New York: Springer-Verlag.

Nash, S. G. and A. Sofer. 1993. "A barrier method for large-scale constrained optimization." *ORSA J. Comput.* **5**: 40–53.

Nazareth, J. L. 1991. "The homotopy principle and algorithms for linear programming." *SIAM J. Optim.* **1**: 316–332.

Nemirovskii, A., C. Roos, and T. Terlaky. 1998. "On maximization of quadratic form over intersection of ellipsoids with common center." Technical report. Delft University of Technology, Faculty of Technical Mathematics and Informatics, Delft, The Netherlands.

Nesterov, Yu. E. 1998. "Semidefinite relaxation and nonconvex quadratic optimization." *Optim. Methods Softw.* **9**: 141–160.

Nesterov, Yu. E. and A. S. Nemirovskii. 1993. In *Interior Point Polynomial Methods in Convex Programming: Theory and Algorithms*. Philadelphia: SIAM Publications.

Pardalos, P. M., Y. Ye, and C. G. Han. 1991. "Algorithms for the solution of quadratic knapsack problems." *Linear Algebra Appl.* **152**: 69–91.

Polijak, S., F. Rendl, and H. Wolkowicz. 1995. "A receipe for semidefinite relaxation for 0-1 quadratic programming." *J. Global Optim.* **7**: 51–73.

Ponceleon, D. B. 1990. "Barrier methods for large-scale quadratic programming." Ph.D. diss. California: Standford University.

Rendl, F. and H. Wolkowicz. 1997. "A semidefinite framework for trust region subproblems with applications to large scale minimization." *Math. Programming* **77**: 273–300.

Renegar, J. 1988. "A polynominal-time algorithm, based on Newton's method, for linear programming." *Math. Programming* **40**: 59–93.

Roos, C., T. Terlaky, and J.-Ph. Vial. 1997. In *Theory and Algorithms for Linear Optimization: an Interior Point Approach*. New York: John Wiley & Sons.

Shanno, D. F. and E. Simantiraki. 1995. "An infeasible interior-point method for linear complementarity problems." Technical Report 7-95. RUTCOR, Rutgers University, New Jersey.

Sherali, H. D. and C. H. Tuncbilek. 1995. "A reformulation–convexification approach for solving nonconvex quadratic programming problems." *J. Global Optim.* **7**: 1–31.

Sonnevend, G. 1985. "An analytic center for polyhedrons and new classes of global algorithms for linear (smooth, convex) programming." In *System Modelling and Optimization: Proceedings of the 12th IFIP-Conference*, edited by A. Prekopa, J. Szelezsan, and B. Strazicky, Lecture Notes in Control and Information Sciences, **84**: 866–875. New York: Springer Verlag.

Sorenson, D. C. 1982. "Newton's method with a model trust region modification." *SIAM J. Numer Anal.* **19**: 409–426.

Sturm, J. F. and S. Zhang. 1995. "On a wide region of centers and primal-dual interior point algorithms for linear programming." Technical Report. Tinbergen Institute Rotterdam, Erasmus University, Netherlands.

Sun, J. and L. Qi. 1992. "An interior point algorithm of $O(\sqrt{n} \mid \ln \varepsilon \mid)$ iterations for C^1-convex programming." *Math. Programming* **57**: 239–257.

Takeda, A., Y. Dai, M. Fukuda, and M. Kojima. 1999. "Towards the implementation of successive convex relaxation method for nonconvex quadratic optimization problem." Technical Report B-347. Tokyo, Japan: Department of Mathematical and Computing Sciences, Tokyo Institute of Technology.

Tanabe, K. 1988. "Centered Newton method for mathematical programming." In *System Modeling and Optimization, Proceedings of the 13th IFIP Conference*, edited by M. Iri and K. Yajima, pp. 197–206. New York: Springer-Verlag.

Todd, M. J. 1996. "Potential-reduction methods in mathematical programming." *Math. Programming* **76**: 3–45.

Todd, M. J. and Y. Ye. 1990. "A centered projective algorithm for linear programming." *Math. Oper Res.* **15**: 508–529.

Tseng, P. 1992. "Complexity analysis of a linear complementarity algorithm based on a Lyapunov function." *Math. Programming* **53**: 297–306.

Tunçel, L. 1994. "Constant potential primal-dual algorithms: a framework." *Math. Programming* **66**: 145–159.

Vanderbei, R. J. 1997. *Linear Programming: Foundations and Extensions*. Boston: Kluwer Academic Publishers.

Vavasis, S. A. 1991. *Nonlinear Optimization: Complexity Issues*. Oxford: Oxford University Press.

Vavasis, S. A. and R. Zippel. 1990. "Proving polynomial time for sphere-constrained quadratic programming." Technical Report 90-1182. New York: Department of Computer Science, Cornell University.

Wright, S. J. 1996. *Primal–dual Interior-point Methods*. Philadelphia: SIAM.

Ye, Y. 1987. "Further development on the interior algorithm for convex quadratic programming." Technical report. California: Department of Engineering Economic Systems, Stanford University.

Ye, Y. 1992. "On an affine scaling algorithm for nonconvex quadratic programming." *Math. Programming* **52**: 285–300.

Ye, Y. 1994. "Combining binary search and Newton's method to compute real roots for a class of real functions." *J. Complexity* **10**: 271–280.

Ye, Y. 1997. *Interior-Point Algorithm: Theory and Analysis*. Wiley-Interscience Series in Discrete Mathematics and Optimization. New York: John Wiley & Sons.

Ye, Y. 1999a. "Approximating global quadratic optimization with convex quadratic constraints." *J Global Optim.* **15**: 1–17.

Ye, Y. 1999b. "Approximating quadratic programming with bound and quadratic constraints." *Math. Programming* **84**: 219–226.

Ye, Y. and E. Tse. 1989. "An extension of Karmarkar's projective algorithm for convex quadratic programming." *Math. Programming* **44**: 157–179.

Yuan, Y. 1990. "On a subproblem of trust region algorithms for constrained optimization." *Math. Programming* **47**: 53–63.

Zhang, Y. 1994. "On the convergence of a class of infeasible interior-point methods for the horizontal linear complementarity problem." *SIAM J. Optim.* **4**: 208–227.

Zhao, G. 1998. "Interior point algorithms for linear complementarity problems based on large neighborhoods of the central path." *SIAM J. Optim* **8**: 397–413.

5 NONLINEAR PROGRAMMING

5.1 Introduction

Gianni Di Pillo and Laura Palagi

ABSTRACT Nonlinear programming (NLP) is the broad area of applied mathematics that addresses optimization problems when nonlinearity in the functions is involved. In this article we introduce the problem, and making reference to the smooth case, we review the standard optimality conditions that are at the basis of most algorithms for its solution. Then, we give basic notions concerning the performance of algorithms, in terms of convergence, rate of convergence, and numerical behavior.

1. Problem Definition

We consider the problem of determining the value of a vector of decision variables $x \in \mathbb{R}^n$ that minimizes an objective function $f: \mathbb{R}^n \to \mathbb{R}$, when x is required to belong to a feasible set $\mathcal{F} \subseteq \mathbb{R}^n$; that is we consider the problem:

$$\min_{x \in \mathcal{F}} f(x). \tag{1}$$

Two cases are of main interest:

- the feasible set $\mathcal{F}$ is $\mathbb{R}^n$, so that problem (1) becomes:

$$\min_{x \in \mathbb{R}^n} f(x); \tag{2}$$

  in this case we say that problem (1) is unconstrained. More generally, problem (1) is unconstrained if $\mathcal{F}$ is an open set. The optimality conditions for unconstrained problems stated in Section 2 also hold in the general case. Here for simplicity we assume that $\mathcal{F} = \mathbb{R}^n$.

- the feasible set is described by inequality and/or equality constraints on the decision variables:

$$\mathcal{F} = \{x \in \mathbb{R}^n : g_i(x) \leqslant 0, \quad i = 1, \dots, p; \quad h_j(x) = 0, \quad j = 1, \dots, m\};$$

  then problem (1) becomes:

$$\begin{aligned} \min_{x \in \mathbb{R}^n} \quad & f(x) \\ & g(x) \leqslant 0 \\ & h(x) = 0, \end{aligned} \tag{3}$$

  with $g: \mathbb{R}^n \to \mathbb{R}^p$ and $h: \mathbb{R}^n \to \mathbb{R}^m$. In this case we say that problem (1) is constrained.

Problem (1) is a nonlinear programming (NLP) problem when at least one of the problem functions $f, g_i, i = 1, \dots, p,\ h_j, j = 1, \dots, m$ is nonlinear in its argument x.

Usually it is assumed that in problem (3) the number m of equality constraints is not larger than the number n of decision variables. Otherwise the feasible set could be empty, unless there is some dependency among the constraints. If only equality (inequality) constraints are present, problem (3) is called an equality (inequality) constrained NLP problem.

In the following we assume that the problem functions f, g, h are at least continuously differentiable in $\mathbb{R}^n$. For nonsmooth NLP problems we refer the reader to Chapter 5.4 of this handbook.

When f is a convex function and $\mathcal{F}$ is a convex set, problem (1) is a convex NLP problem. In particular, $\mathcal{F}$ is convex if inequality constraint functions g_i are convex and the equality constraint functions h_j are affine. Convexity adds a lot of structure to the NLP problem, and can be exploited widely, both from the theoretical and the computational point of view. If f is convex quadratic and g, h are affine, we have a quadratic programming problem. This is a case of special interest and is treated extensively in Chapter 4 of this volume. Here we will confine ourselves to general NLP problems, without convexity assumptions.

A point $x^* \in \mathcal{F}$ is a global solution of problem (1) if $f(x^*) \leqslant f(x)$, for all $x \in \mathcal{F}$; it is a strict global solution if $f(x^*) < f(x)$, for all $x \in \mathcal{F}, x \neq x^*$. A main existence result for a constrained problem states that a global solution exists if $\mathcal{F}$ is compact (Weierstrass Theorem). An easy consequence for unconstrained problems is that a global solution exists if the level set $\mathcal{L}_\alpha = \{x \in \mathbb{R}^n : f(x) \leqslant \alpha\}$ is compact for some finite α.

A point $x^* \in \mathcal{F}$ is a local solution of problem (1) if an open neighborhood $\mathcal{B}_{x^*}$ of x^* exists, such that $f(x^*) \leqslant f(x)$, for all $x \in \mathcal{F} \cap \mathcal{B}_{x^*}$; it is a strict local solution if $f(x^*) < f(x)$, for all $x \in \mathcal{F} \cap \mathcal{B}_{x^*}, x \neq x^*$. Of course, a global solution is also a local solution.

To determine a global solution of a NLP problem is in general a very difficult task. Usually, NLP algorithms are only able to determine local solutions. Nevertheless, in practical applications, to also get a local solution can be of great worth.

We introduce some notation. We denote by the apex $'$, the transpose of a vector or a matrix. Given a function $v : \mathbb{R}^n \to \mathbb{R}$, we denote the gradient vector by $\nabla v(x)$ and the Hessian matrix by $\nabla^2 v(x)$. Given a vector function $w : \mathbb{R}^n \to \mathbb{R}^q$, we denote the $n \times q$ matrix whose columns are $\nabla w_j(x), j = 1, \ldots, q$ by $\nabla w(x)$. Given a vector $y \in \mathbb{R}^q$, we denote by $\|y\|$ its Euclidean norm. Let y be a vector with components $y_i, i = 1, \ldots, q$, and A be a matrix with columns $a_j, j = 1, \ldots, q$. Let $\mathcal{K} \subset \{1, \ldots, q\}$ be an index subset, we denote by $y_{\mathcal{K}}$ the subvector of y with components y_i such that $i \in \mathcal{K}$ and by $A_{\mathcal{K}}$ the submatrix of A made up of the columns a_j with $j \in \mathcal{K}$.

2. Optimality Conditions

Local solutions must satisfy necessary optimality conditions (NOCs).

For the unconstrained problem (2) we have the well-known result of classical calculus:

Proposition 1. *Let* x* *be a local solution of problem (2), then*

$$\nabla f(x^*) = 0; \tag{4}$$

moreover, if f *is twice continuously differentiable, then*

$$y'\nabla^2 f(x^*) y \geqslant 0, \quad \forall y \in \mathbb{R}^n. \tag{5}$$

For the constrained problem (3), most of the NOCs commonly used in the development of algorithms assume that at a local solution the constraints satisfy some qualification condition to prevent the occurrence of degenerate cases. These conditions are usually called constraint qualifications and among them the linear independence constraints qualification (LICQ) is the simplest and by far the most invoked.

Let $\hat{x} \in \mathcal{F}$. We say that the inequality constraint g_i is active at $\hat{x}$ if $g_i(\hat{x}) = 0$. We denote by $\mathcal{I}_a(\hat{x})$ the index set of inequality constraints active at $\hat{x}$:

$$\mathcal{I}_a(\hat{x}) = \{i \in \{1, \ldots, p\} : g_i(\hat{x}) = 0\}. \tag{6}$$

Of course, any equality constraint h_j, is active at $\hat{x}$. LICQ is satisfied at $\hat{x}$ if the gradients of the active constraints $\nabla g_{\mathcal{I}_a}(\hat{x})$, $\nabla h(\hat{x})$, are linearly independent.

Under LICQ, the NOCs for problem (3) are stated making use of the (generalized) Lagrangian function:

$$L(x, \lambda, \mu) = f(x) + \lambda' g(x) + \mu' h(x), \tag{7}$$

where $\lambda \in \mathbb{R}^p$, $\mu \in \mathbb{R}^m$ are called (generalized) Lagrange multipliers, or dual variables.

The so called Karush–Kuhn–Tucker (KKT) NOCs are stated as follows:

Proposition 2. *Assume that* x* *is a local solution of problem (3) and that the LICQ holds at* x*; *then multipliers* $\lambda^* \geqslant 0, \mu^*$ *exist such that:*

$$\nabla_x L(x^*, \lambda^*, \mu^*) = 0, \tag{8}$$

$$\lambda^{*\prime} g(x^*) = 0;$$

moreover, if f, g, h *are twice continuously differentiable, then:*

$$y' \nabla_x^2 L(x^*, \lambda^*, \mu^*) y \geqslant 0, \quad \forall y \in \mathcal{N}(x^*),$$

where:

$$\mathcal{N}(x^*) = \{y \in \mathbb{R}^n : \nabla g_{\mathcal{I}_c}(x^*)' y = 0; \nabla h(x^*)' y = 0\}.$$

A point x^* satisfying the NOCs (8) together with some multipliers λ^*, μ^* is called a KKT point.

If a point $x^* \in \mathcal{F}$ satisfies a sufficient optimality condition, then it is a local solution of problem (1). For general NLP problems, sufficient optimality conditions can be stated under the assumption that the problem functions are twice continuously differentiable, so that we have second order sufficiency conditions (SOSCs).

For the unconstrained problem (2) we have the SOSC:

Proposition 3. *Assume that* x^* *satisfies the NOC (4) of Proposition 1. Assume further that*

$$y' \nabla^2 f(x^*) y > 0, \quad \forall y \in \mathbb{R}^n, y \neq 0,$$

that is assume that $\nabla^2 f(x^*)$ *is positive definite. Then* x^* *is a strict local solution of problem (2).*

For the constrained problem (3) we have the SOSC:

Proposition 4. *Assume that* $x^* \in \mathcal{F}$ *and* λ^*, μ^* *satisfy the* NOCs *(8) of Proposition 2. Assume further that*

$$y' \nabla_x^2 L(x^*, \lambda^*, \mu^*) y > 0, \quad \forall y \in \mathcal{P}(x^*), y \neq 0, \tag{9}$$

where:

$$\mathcal{P}(x^*) = \{y \in \mathbb{R}^n : \nabla g_{\mathcal{I}_a}(x^*)' y \leqslant 0, \ \ \nabla h(x^*)' y = 0; \ \ \nabla g_i(x^*)' y = 0, i \in \mathcal{I}_a(x^*) \text{ with } \lambda_i^* > 0\};$$

then x^* *is a strict local solution of problem (3).*

Note that $\mathcal{P}(x^*)$ is a polyhedral set, while $\mathcal{N}(x^*)$ is the null space of a linear operator, with $\mathcal{N}(x^*) \subseteq \mathcal{P}(x^*)$. However, if $\lambda_i^* > 0$ for all $i \in \mathcal{I}_a(x^*)$, then $\mathcal{N}(x^*) = \mathcal{P}(x^*)$. When this happens, we say that the strict complementarity assumption is satisfied by $g(x^*)$ and λ^*. Strict complementarity is a very favorable circumstance, because it is much simpler to deal with the null space of a linear operator than with a general polyhedron. In particular, there is a simple algebraic criterion to test whether the quadratic form $y' \nabla_x^2 L(x^*, \lambda^*, \mu^*) y$ is positive definite on $\mathcal{N}(x^*)$.

Note also that, if in Proposition 4 we substitute the set $\mathcal{P}(x^*)$ with the set

$$\mathcal{N}^+(x^*) = \{y \in \mathbb{R}^n : \nabla h(x^*)' y = 0, \quad \nabla g_i(x^*)' y = 0, \quad i \in \mathcal{I}_a(x^*) \ \text{ with } \ \lambda_i^* > 0\},$$

we still have a sufficient condition, since $\mathcal{P}(x^*) \subseteq \mathcal{N}^+(x^*)$, and again $\mathcal{N}^+(x^*)$ is the null space of a linear operator. However, the condition obtained is stronger then the original one, and, indeed, it is known as the strong second order sufficient condition (SSOSC).

Finally, we point out that, under the strict complementarity assumption, $\mathcal{N}(x^*) = \mathcal{P}(x^*) = \mathcal{N}^+(x^*)$, so that all the SOSCs for problem (3) considered above reduce to the same condition.

A main feature of convex problems is that a (strict) local solution is also a (strict) global solution. Moreover, when f is (strictly) convex and, if present, g_i are convex and h_j are affine, the NOCs given in terms of first order derivatives are also sufficient for a point x^* to be a (strict) global solution.

Optimality conditions are fundamental in the solution of NLP problems. If it is known that a global solution exists, the most straightforward method to employ them is as follows:

find all points satisfying the first order necessary conditions, and declare as a global solution the point with the smallest value of the objective function. If the problem functions are twice differentiable, we can also check the second order necessary condition, filtering out those points that do not satisfy it; for the remaining candidates, we can check a second order sufficient condition to find local minima.

It is important to realize, however, that except for very simple cases, using optimality conditions as described above does not work. The reason is that, even for an unconstrained problem, to find a solution of the system of equations $\nabla f(x) = 0$ is nontrivial; algorithmically, it is usually as difficult as solving the original minimization problem.

The principal context in which optimality conditions become useful is the development and analysis of algorithms. An algorithm for the solution of problem (1) produces a sequence $\{x^k\}, k = 0, 1, ...,$ of tentative solutions, and terminates when a stopping criterion is satisfied. Usually the stopping criterion is based on satisfaction of necessary optimality conditions within a prefixed tolerance; moreover, necessary optimality conditions often suggest how to improve the current tentative solution x^k in order to get the next one x^{k+1} closer to the optimal solution. Thus, necessary optimality conditions provide the basis for the convergence analysis of algorithms. On the other hand, sufficient optimality conditions play a key role in the analysis of the rate of convergence.

3. Performance of Algorithms

3.1. *Convergence and Rate of Convergence*

Let $\Omega \subset \mathscr{F}$ be the subset of points that satisfy the first order NOCs for problem (1). From a theoretical point of view, an optimization algorithm stops when a point $x^* \in \Omega$ is reached. From this point of view, the set Ω is called the target set. Convergence properties are stated with reference to the target set Ω. In the unconstrained case, a possible target set is $\Omega = \{x \in \mathbb{R}^n : \nabla f(x) = 0\}$ whereas in the constrained case Ω may be the set of KKT points satisfying (8).

Let $\{x^k\}, k = 0, 1, ...$ be the sequence of points produced by an algorithm. Then, the algorithm is globally convergent if a limit point x^* of $\{x^k\}$ exists such that $x^* \in \Omega$ for any starting point $x^0 \in \mathbb{R}^n$; it is locally convergent if the existence of the limit point $x^* \in \Omega$ can be established only if the starting point x^0 belongs to some neighborhood of Ω.

The notion of convergence stated before is the weakest that ensures that a point x^k arbitrarily close to Ω can be obtained for a large enough k. In the unconstrained case this implies that

$$\liminf_{k \to \infty} \|\nabla f(x^k)\| = 0.$$

Nevertheless, stronger convergence properties can often be established. For instance, that any subsequence of $\{x^k\}$ possesses a limit point, and any limit point of $\{x^k\}$ belongs to Ω; for the unconstrained case, this means that

$$\lim_{k \to \infty} \|\nabla f(x^k)\| = 0.$$

The strongest convergence requirement is that the whole sequence $\{x^k\}$ converges to a point $x^* \in \Omega$.

In order to define the rate of convergence of an algorithm, it is assumed for simplicity that the algorithm produces a sequence $\{x^k\}$ converging to a point $x^* \in \Omega$. The most widely employed notion of rate of convergence is the Q-rate of convergence, which considers the quotient between two successive iterates given by $\|x^{k+1} - x^*\| / \|x^k - x^*\|$. Then we say that the rate of convergence is Q-linear if there is a constant $r \in (0, 1)$ such that:

$$\frac{\|x^{k+1} - x^*\|}{\|x^k - x^*\|} \leqslant r, \quad \text{for all sufficiently large } k;$$

we say that the rate of convergence is Q-superlinear if:

$$\lim_{k \to \infty} \frac{\|x^{k+1} - x^*\|}{\|x^k - x^*\|} = 0;$$

and we say that the rate of convergence is Q-quadratic if:

$$\frac{\|x^{k+1} - x^*\|}{\|x^k - x^*\|^2} \leqslant R, \quad \text{for all sufficiently large } k,$$

where R is a positive constant. More generally, the rate of convergence is of Q-order p if there is a positive constant R such that:

$$\frac{\|x^{k+1} - x^*\|}{\|x^k - x^*\|^p} \leqslant R, \quad \text{for all sufficiently large } k;$$

however, a Q-order larger than 2 is very seldom achieved. Algorithms in common use are either superlinearly or quadratically convergent.

3.2. *Numerical Behavior*

Besides the theoretical performance, an important aspect of a numerical algorithm is its practical performance. Indeed, a fast convergence rate can be overcome by a great amount of algebraic operations needed at each iteration. Different measures of numerical performance exist, even if direct comparison of CPU time remains the best way of verifying the efficiency of a given method on a specific problem. However, when the computational burden in terms of algebraic operations per iteration is comparable, possible measures of performance can be given in terms of number of iterations, number of objective function/constraint evaluations, and number of gradient evaluations.

Measures of performance are mainly relevant for large-scale NLP problems. The notion of large-scale is machine dependent so that it could be difficult to state *a priori* when a problem is large. Usually, an unconstrained problem with $n \geqslant 1000$ is considered large, whereas a constrained problem without particular structure is considered large when $n \geqslant 100$ and $p + m \geqslant 100$. A basic feature of an algorithm for large-scale problems is that a low storage overhead is needed to make its implementation practicable, and a measure of performance is usually the number of matrix–vector products required. The main difficulty in dealing with large-scale problems is that effective algorithms for small-scale problems do not necessarily translate into efficient ones when applied to solve large problems.

References

Several textbooks on applied optimization devote chapters to NLP theory and practice. Among them the books by Gill et al. (1981), Minoux (1983), Luenberger (1994), Fletcher (1987), Nash and Sofer (1996), Bonnans et al. (1997), and Nocedal and Wright (1999).

More specialized textbooks, strictly devoted to NLP, are the ones by Zangwill (1969), Avriel (1976), McCormick (1983), Evtushenko (1985), Bazaraa et al. (1993), Polak (1997), and Bertsekas (1999). A collection of survey papers on algorithmic aspects of NLP is given in Spedicato (1994).

A particular emphasis on the mathematical foundations of NLP is given in Hestenes (1975), Giannessi (1982), Polyak (1987), Peressini et al. (1988), Jahn (1994), and Rapsáck (1997). A classical reference on optimality conditions and constraint qualifications is the book by Mangasarian (1969).

The theoretical analysis of algorithms for NLP is developed in Zangwill (1969), Orthega and Rheinboldt (1970), Bazaraa et al. (1993), Luenberger (1994), and Polak (1997). For practical aspects of the implementation of algorithms we refer again to Gill et al. (1981).

The search for global optima is treated, for instance, in Törn and Žilinskas (1989), Horst and Pardalos (1995), and Pinter (1996).

This article does not mention multiobjective optimization. Multiobjective NLP problems are considered in Rustem (1998) and Miettinen (1999).

Avriel, M. 1976. *Nonlinear Programming Analysis and Methods*. Englewood Cliffs, New York: Prentice-Hall.

Bazaraa, M. S., H. D. Sherali, and C. M. Shetty. 1993. *Nonlinear Programming—Theory and Algorithms*. (2d ed.). New York: John Wiley & Sons.

Bertsekas, D. P. 1999. *Nonlinear Programming*. (2d ed.). Belmont, Massachusetts: Athena Scientific.
Bonnans, J. F., J. C. Gilbert, C. Lemarèchal, and C. Sagastizàbal. 1997. *Optimisation Numérique: Aspects théoriques et pratiques*. Paris: Springer-Verlag.
Evtushenko, Y. G. 1985. *Numerical Optimization Techniques*. New York: Optimization Software, Inc.
Fletcher, R. E. 1987. *Practical Methods of Optimization*. (2d ed.). Chichester: John Wiley & Sons.
Giannessi, F. 1982. *Metodi Matematici della Programmazione. Problemi lineari e non lineari*. Bologna: Pitagora Editrice.
Gill, P. E., W. Murray, and M. H. Wright. 1981. *Practical Optimization*. New York: Academic Press.
Hestenes, M. R. 1975. *Optimization Theory. The Finite Dimensional Case*. New York: John Wiley & Sons.
Horst, R. and P. Pardalos. 1995. *Handbook of Global Optimization*. Dordrecht: Kluwer Academic Publishers.
Jahn, K. 1994. *Introduction to the Theory of Nonlinear Optimization*. Berlin: Springer-Verlag.
Luenberger, D. G. 1973. *Linear and Nonlinear Programming*. (2d ed.). Menlo Park, California: Addison-Wesley.
Mangasarian, O. L. 1969. *Nonlinear Programming*. New York: McGraw-Hill.
McCormick, G. P. 1983. *Nonlinear Programming—Theory, Algorithms, and Applications*. New York: John Wiley & Sons.
Miettinen, K. 1999. *Nonlinear Multi-objective Optimization*. Boston: Kluwer Academic Publishers.
Minoux, M. 1986. *Mathematical Programming. Theory and Algorithms*. New York: John Wiley.
Nash, S. G. and A. Sofer. 1996. *Linear and Nonlinear Programming*. New York: McGraw-Hill.
Nocedal, J. and S. J. Wright. 1999. *Numerical Optimization*. New York: Springer-Verlag.
Orthega, J. M. and W. C. Rheinboldt. 1970. *Iterative Solution of Nonlinear Equations in Several Variables*. New York: Academic Press.
Peressini, A. L., F. E. Sullivan, and J. J. Uhl, Jr. 1988. *The Mathematics of Nonlinear Programming*. New York: Springer-Verlag.
Pinter, J. D. 1996. *Global Optimization in Action: Continuous and Lipschitz Optimization—Algorithms, Implementations and Applications*. Dordrecht: Kluwer Academic Publishers.
Polak, E. 1997. *Optimization: Algorithms and Consistent Approximations*. New York: Springer-Verlag.
Polyak, B. T. 1987. *Introduction to Optimization*. New York: Optimization Software, Inc.
Rapsáck, T. 1997. *Smooth Nonlinear Optimization in* $\mathbb{R}^n$. Dordrecht: Kluwer Academic Publishers.
Rustem, B. 1998. *Algorithms for Nonlinear Programming and Multiple-objective Decisions*. Chichester: John Wiley & Sons.
Spedicato, E. 1994. *Algorithms for Continuous Optimization: The State of the Art*. Dordrecht: Kluwer Academic Publishers.
Törn, A. and A. Žilinskas. 1989. *Global Optimization*. Berlin: Springer-Verlag.
Zangwill, W. I. 1969. *Nonlinear Programming—a Unified Approach*. Englewood Cliffs, New York: Prentice Hall.

5.2 Unconstrained Nonlinear Programming

Gianni Di Pillo and Laura Palagi

ABSTRACT In this article we consider the main classes of algorithms for the solution of unconstrained NLP problems. As a first step, we describe line-search techniques that are at the basis of most unconstrained algorithms. Then methods that use first- and second-order derivatives are reviewed. The last section is devoted to derivative-free methods. Far from being exhaustive, this article aims to give the main ideas and tools at the basis of unconstrained methods, and to suggest how to go more deeply into more technical and/or recent results.

1. Introduction

In this article we consider algorithms for solving the unconstrained nonlinear programming problem

$$\min_{x\in\mathbb{R}^n} f(x), \qquad (1)$$

where $x \in \mathbb{R}^n$ is the vector of decision variables and $f: \mathbb{R}^n \to \mathbb{R}$ is the objective function. The algorithms described in this section can be easily adapted to solve the constrained problem $\min_{x \in \mathscr{F}} f(x)$ when $\mathscr{F}$ is an open set, provided that a feasible point $x^0 \in \mathscr{F}$ is available.

In the following we assume for simplicity that the problem function f is twice continuously differentiable in $\mathbb{R}^n$ even if in many cases continuous differentiability is only needed once. Nonsmooth NLP algorithms are treated in Chapter 5.4 of this handbook.

We also assume that the standard assumption ensuring the existence of a solution to Equation (1) holds, namely that:

Assumption 1. *The level set* $\mathscr{L}^0 = \{x \in \mathbb{R}^n : f(x) \leqslant f(x^0)\}$ *is compact, for some* $x^0 \in \mathbb{R}^n$

The reader should be familiar with the first- and second-order necessary optimality conditions (NOC) for problem (1), as well as with the basic notions on the performance of algorithms, given in Chapter 5.1 of this handbook.

We employ the same notation as adopted in Chapter 5.1.

2. Unconstrained Optimization Algorithms

The algorithm models treated in this section generate a sequence $\{x^k\}$, starting from x^0, by the following iteration

$$x^{k+1} = x^k + \alpha^k d^k, \tag{2}$$

where d^k is a search direction and α^k is a step size along d^k.

Methods differ in the way the direction and the step size are chosen. Of course different choices of d^k and α^k yield different convergence properties. Roughly speaking, the search direction affects the local behavior of an algorithm and its rate of convergence, whereas global convergence is often tied to the choice of the step size α^k. The following proposition states a rather general convergence result (Orthega, 1988).

Proposition 1. *Let* $\{x^k\}$ *be the sequence generated by iteration* (2). *Assume that: (i)* $d^k \neq 0$ *if* $\nabla f(x^k) \neq 0$; *(ii) for all* k *we have* $f(x^{k+1}) \leqslant f(x^k)$; *(iii) we have*

$$\lim_{k \to \infty} \frac{\nabla f(x^k)' d^k}{\|d^k\|} = 0; \tag{3}$$

(iv) for all k *with* $d^k \neq 0$ *we have*

$$\frac{|\nabla f(x^k)' d^k|}{\|d^k\|} \geqslant c \|\nabla f(x^k)\|, \text{ with } c > 0.$$

Then, either $\bar{k}$ *exists such that* $x^{\bar{k}} \in \mathscr{L}^0$ *and* $\nabla f(x^{\bar{k}}) = 0$, *or an infinite sequence is produced such that:*

a. $\{x^k\} \in \mathscr{L}^0$;

b. $\{f(x^k)\}$ *converges;*

c. $\lim_{k \to \infty} \|\nabla f(x^k)\| = 0. \quad (4)$

Condition (3) of Proposition 1 corresponds to saying that any subsequence of $\{x^k\}$ possess a limit point, and any limit point of $\{x^k\}$ belongs to the target set Ω introduced in Chapter 5.1. However, weaker convergence results can be proved (see Chapter 5.1 of this handbook). In particular, in some cases, only condition

$$\lim_{k \to \infty} \inf \|\nabla f(x^k)\| = 0 \tag{5}$$

holds.

Many methods are classified according to the information they use regarding the smoothness of the function. In particular we will briefly describe gradient methods, conjugate gradient methods, Newton's and truncated Newton's methods, quasi-Newton methods, and derivative-free methods.

Most methods determine α^k by means of a line-search technique. Hence we start with a short review of the most-used techniques and more recent developments in line-search algorithms.

2.1. Line-Search Algorithms

Line-search algorithms (LSA) determine the step size α^k along direction d^k. The aim of different choices of α^k is mainly to ensure that the algorithm defined by (2) is globally convergent, possibly without deteriorating the rate of convergence. A first possibility is to set $\alpha^k = \alpha^*$ with

$$\alpha^* = \arg \min_{\alpha} f(x^k + \alpha d^k),$$

that is α^k is the value that minimizes function f along direction d^k. However, unless function f has some special structure (being quadratic), an exact line search is usually quite computationally costly, and its use may not be worthwhile. Hence approximate methods are used. In cases when ∇f can be computed, we assume that the condition

$$\nabla f(x^k)' d^k < 0, \quad \text{for all } k \tag{6}$$

holds, namely we assume that d^k is a descent direction for f at x^k.

Among the simplest LSAs is Armijo's method, reported in Algorithm 1.

Algorithm 1. Armijo's line-search algorithm.

Data: $\delta \in (0,1)$, $\gamma \in (0,1/2)$, $c \in (0,1)$.

Step 1. Choose Δ^k such that

$$\Delta^k \geqslant c \frac{|\nabla f(x^k)' d^k|}{\|d^k\|^2}$$

and set $\alpha = \Delta^k$.

Step 2. If

$$f(x^k + \alpha d^k) \leqslant f(x^k) + \gamma \alpha \nabla f(x^k)' d^k$$

then $\alpha^k = \alpha$ and stop.

Step 3. Otherwise set $\alpha = \delta\alpha$ and go to Step 2.

The choice of the initial step Δ^k is tied to the direction d^k used. By choosing Δ^k as in Step 1, it is possible to prove that Armijo's LSA finds, in a finite number of steps, a value α^k such that

$$f(x^{k+1}) < f(x^k), \tag{7}$$

and that condition (3) holds.

Armijo's LSA finds a step size α^k that satisfies the condition of sufficient decrease of the objective function and implicitly the condition of sufficient displacement from the current point x^k.

In many cases it can be necessary to impose stronger conditions on the step size α^k. In particular, Wolfe's conditions are widely employed. In this case, Armijo's condition

$$f(x^k + \alpha d^k) \leqslant f(x^k) + \gamma \alpha \nabla f(x^k)' d^k \tag{8}$$

is coupled either with the condition

$$\nabla f(x^k + \alpha^k d^k)' d^k \geqslant \beta \nabla f(x^k)' d^k, \tag{9}$$

or with the stronger one

$$|\nabla f(x^k + \alpha^k d^k)' d^k| \leqslant \beta |\nabla f(x^k)' d^k|, \tag{10}$$

where $\beta \in (\gamma, 1)$, γ being the value used in (8). It is possible to prove that a finite interval of values of α exists where conditions (8) and (10) are satisfied (and hence also (8) and (9)). Of course, an LSA satisfying (8) and (9) enforces conditions (3) and (7). LSAs for finding a step size α^k that satisfies Wolfe's conditions are more complicated than Armijo's LSA, and we do not go into detail here.

In some cases, such as for derivative-free methods, it can be necessary to consider LSAs that do not require the use of derivatives. In this case, condition (6) cannot be verified and a direction d^k cannot be ensured to be a descent one. Hence line-search techniques must also take into account the possibility of using a negative step size, that is, to move along the opposite direction $-d^k$. Moreover, suitable conditions for terminating the line search with $\alpha^k = 0$ must be imposed when it is likely that $f(x^k + \alpha d^k)$ has a minimizer at $\alpha = 0$. A possible LSA is obtained by substituting the Armijo acceptance rule (8) by a condition of the type

$$f(x^k + \alpha d^k) \leqslant f(x^k) - \gamma\alpha^2 \|d^k\|^2, \tag{11}$$

which does not require gradient information. We point out that derivative-free LSAs usually guarantee satisfaction of the condition $\lim_{k\to\infty} \|x^{k+1} - x^k\| = 0$, which can be useful to prove convergence whenever it is not possible to control the angle between the gradient $\nabla f(x^k)$ and the direction d^k. See De Leone et al. (1984) and Grippo et al. (1988) for more sophisticated derivative-free line-search algorithms.

We observe that all the line-search algorithms described so far ensure that condition (7) is satisfied, namely they ensure a monotonic decrease in the objective function values. Actually, enforcing monotonicity may not be necessary to prove convergence results and, in the case of highly nonlinear functions, may cause minimization algorithms to be trapped within some narrow region. A very simple nonmonotone line-search scheme (Grippo et al., 1986) consists of relaxing Armijo's condition (8), requiring that the function value in the new iterate satisfies an Armijo-type rule with respect to a prefixed number of previous iterates, namely that

$$f(x^k + \alpha d^k) \leqslant \max_{0\leqslant j\leqslant J} \{f(x^{k-j})\} + \gamma\alpha\nabla f(x^k)' d^k, \tag{12}$$

where J is the number of previous iterates that are taken into account. Numerical experiences have proved the efficiency of the use of nonmonotone schemes (see Ferris et al., 1996; Raydan, 1997; Lucidi et al., 1998b).

2.2. *Gradient Methods*

The gradient method, also known as the steepest-descent method, is considered a basic method in unconstrained optimization and is obtained by setting $d^k = -\nabla f(x^k)$ in Algorithm (2). It requires only information about first-order derivatives and hence is very attractive because of its limited computational cost and storage requirement. Moreover, it is considered the most significant example of a globally convergent algorithm, in the sense that, using a suitable LSA, it is possible to establish a global convergence result. However, its rate of convergence is usually very poor and hence it is not widely used on its own, even if recent developments have produced more efficient versions of the method. A possible algorithmic scheme is reported in Algorithm 2.

Algorithm 2. The gradient method.

Step 1. Choose $x^0 \in R^n$ and set $k = 0$.
Step 2. If $\nabla f(x^k) = 0$ stop.
Step 3. Set $d^k = -\nabla f(x^k)$ and find α^k by using the Armijo's LSA of Algorithm 1.
Step 4. Set

$$x^{k+1} = x^k - \alpha^k \nabla f(x^k)$$

$k = k + 1$ and go to Step 2.

The choice of the initial step size Δ^k in Armijo's LSA is critical and can significantly affect the behavior of the gradient method. As regards the convergence of the method, Proposition 1 can easily be adapted; nonetheless we also have the following result that holds even without requiring the compactness of the level set $\mathscr{L}^0$.

Proposition 2. *If ∇f is Lipschitz continuous and f is bounded from below, then the sequence generated by the gradient method of Algorithm 2 satisfies condition* (4).

Hence the gradient method is satisfactory with regard to global convergence. However, only a linear convergence rate can be proved. Indeed, from a practical point of view, the rate of convergence is usually very poor and it strongly depends on the condition number of the Hessian matrix $\nabla^2 f$.

Barzilai and Borwein (1988) proposed a particular rule for computing the step size in gradient methods that leads to a significant improvement in numerical performance. In particular the step size is chosen as

$$\alpha^k = \frac{\|x^k - x^{k-1}\|^2}{(x^k - x^{k-1})'(\nabla f(x^k) - \nabla f(x^{k-1}))}. \tag{13}$$

In Raydan (1997) a globalization strategy has been proposed that tries to accept the step size given by (13) as frequently as possible. The globalization strategy is based on Armijo's nonmonotone line search of type (12). Condition (4) can still be proved. Moreover, numerical experience shows that this method is also competitive with some versions of conjugate gradient methods.

2.3. *Conjugate Gradient Methods*

The conjugate gradient method is very popular due to its simplicity and low computational requirements. The method was originally introduced for solving the minimization of a strictly convex quadratic function with the aim of accelerating the gradient method without requiring the overhead needed by Newton's method (see Section 2.4). Indeed it only requires first-order derivatives and no matrix storage and operations are needed.

The basic idea is that the minimization on $\mathbb{R}^n$ of the strictly convex quadratic function

$$f(x) = \tfrac{1}{2}x'Qx + a'x$$

with Q being symmetric positive definite, can be split into n minimizations over $\mathbb{R}$. This is done by means of n directions $d^0, \ldots, d^{n-1}$ conjugate with respect to the Hessian Q, that is directions such that $(d^j)'Qd^i = 0$ for $j \neq i$. Along each direction d^j an exact line search is performed in closed form. This corresponds to the conjugate directions algorithm (CDA) that finds the global minimizer of a strictly convex quadratic function in at most n iterations and that is reported in Algorithm 3.

Algorithm 3. Conjugate directions algorithm for quadratic functions (CDA).

Data. Q-conjugate directions $d^0, \ldots, d^{n-1}$.

Step 1. Choose $x^0 \in R^n$ and set $k = 0$.

Step 2. If $\nabla f(x^k) = 0$ stop.

Step 3. Set

$$\alpha^k = \frac{\nabla f(x^k)'d^k}{(d^k)'Qd^k}.$$

Step 4. Set

$$x^{k+1} = x^k + \alpha^k d^k,$$

$k = k + 1$ and go to Step 2.

The value of α^k at Step 3 is the exact minimizer of the quadratic function $f(x^k + \alpha d^k)$. Note that CDA can be seen equivalently as an algorithm for solving $\nabla f(x) = 0$, that is for solving the linear system $Qx = -a$.

In the CDA, the conjugate directions are given, whereas in the conjugate gradient algorithm (CGA) the n conjugate directions are generated iteratively according to the rule

$$d^k = \begin{cases} -\nabla f(x^k) & k = 0 \\ -\nabla f(x^k) + \beta^{k-1} d^{k-1} & k \geq 1. \end{cases}$$

The scalar β^k is chosen in order to enforce conjugacy among the directions. The most common choices for β^k are either the Fletcher–Reeves formula (Fletcher and Reeves, 1964):

$$\beta^k_{FR} = \frac{\|\nabla f(x^{k+1})\|^2}{\|\nabla f(x^k)\|^2}, \tag{14}$$

or the Polak–Ribiére formula (Polak and Ribiére, 1969):

$$\beta^k_{PR} = \frac{\nabla f(x^{k+1})'(\nabla f(x^{k+1}) - \nabla f(x^k))}{\|\nabla f(x^k)\|^2}. \tag{15}$$

Formulas (14) and (15) are equivalent in the case of quadratic objective function.

When f is not a quadratic function the two formulas (14) and (15) give different values for β and an inexact line search is performed to determine the step size α^k. Usually an LSA that enforces the strong Wolfe's conditions (8) and (10) is used. A scheme of CGA is reported in Algorithm 4.

Algorithm 4. Conjugate gradient algorithm for nonlinear functions (CGA).

Step 1. Choose $x^0 \in R^n$ and set $k = 0$.

Step 2. If $\nabla f(x^k) = 0$ stop.

Step 3. Compute β^{k-1} by either (14) or (15) and set the direction

$$d^k = \begin{cases} -\nabla f(x^k) & k = 0 \\ -\nabla f(x^k) + \beta^{k-1} d^{k-1} & k \geq 1; \end{cases}$$

Step 4. Find α^k by means of an LSA that satisfies the strong Wolfe's conditions (8) and (10).

Step 5. Set

$$x^{k+1} = x^k + \alpha^k d^k,$$

$k = k + 1$ and go to Step 2.

The simplest device used to guarantee global convergence properties of CG methods is a periodic restart along the steepest-descent direction. Restarting can also be used to deal with the loss of conjugacy due to the presence of nonquadratic terms in the objective function that may cause the method to generate inefficient or even meaningless directions. Usually the restart procedure occurs every n steps or if a test on the loss of conjugacy such as

$$|\nabla f(x^k)'\nabla f(x^{k-1})| > \delta \|\nabla f(x^{k-1})\|^2,$$

with $0 < \delta < 1$, is satisfied.

However, in practice, the use of a regular restart may not be the most convenient technique for enforcing global convergence. Indeed CGAs are used for large-scale problems when n is very large and the expectation is to solve the problem in less than n iterations, that is, before restart occurs. Without restart, the CGA with $\beta^k = \beta^k_{FR}$ and an LSA that guarantees the satisfaction of the strong Wolfe's conditions is globally convergent in the sense that (5) holds (Al-Baali, 1985). Actually, in Gilbert and Nocedal (1992), the same type of convergence has been proved for any value of β^k such that $0 < |\beta^k| \leq \beta^k_{FR}$. Analogous convergence results do not hold for the Polak–Ribiére formula even if the numerical performance of this formula is usually superior to the Fletcher–Reeves one. This motivated a big effort to find a globally convergent version of the Polak–Ribiére method. Indeed, condition (5) can be established for a modification of the Polak–Ribiére CG method that uses only positive values of $\beta^k = \max\{\beta^k_{PR}, 0\}$ and a LSnnA that ensures satisfaction of the strong Wolfe's conditions and of the additional condition

$$\nabla f(x^k)' d^k \leqslant -c\|\nabla f(x^k)\|^2$$

for some constant $c > 0$ (Powell, 1986).

Recently it has been proved that the Polak–Ribiére CG method satisfies the stronger convergence condition (4) if a more sophisticated line-search technique is used (Grippo and Lucidi, 1997). Indeed an LSA that enforces condition (11) coupled with a sufficient descent condition of the type

$$-\delta_2\|\nabla f(x^{k+1})\|^2 \leqslant \nabla f(x^{k+1})' d^{k+1} \leqslant -\delta_1\|\nabla f(x^{k+1})\|^2$$

with $0 < \delta_1 < 1 < \delta_2$ must be used.

With respect to the numerical performance, the CGA is affected by the condition number of the Hessian $\nabla^2 f$. This leads to the use of preconditioned conjugate gradient methods and many papers have been devoted to the study of efficient preconditioners.

2.4. Newton's Methods

The Newton's method is considered one of the most powerful algorithms for solving unconstrained optimization problems (see Moré and Sorensen, 1984, for a review). It relies on the quadratic approximation

$$q^k(s) = \tfrac{1}{2} s' \nabla^2 f(x^k) s + \nabla f(x^k)' s + f(x^k)$$

of the objective function f in the neighborhood of the current iterate x^k and requires the computation of ∇f and $\nabla^2 f$ at each iteration k. Newton's direction d_N^k is obtained as a stationary point of the quadratic approximation q^k, that is, as a solution of the system:

$$\nabla^2 f(x^k) d_N = -\nabla f(x^k). \tag{16}$$

Provided that $\nabla^2 f(x^k)$ is nonsingular, Newton's direction is given by $d_N^k = -[\nabla^2 f(x^k)]^{-1}\nabla f(x^k)$ and the basic algorithmic scheme is defined by the iteration

$$x^{k+1} = x^k - [\nabla^2 f(x^k)]^{-1}\nabla f(x^k); \tag{17}$$

we refer to (17) as the pure Newton's method. The reputation of Newton's method is due to the fact that if the starting point x^0 is close enough to a solution x^*, then the sequence generated by iteration (17) converges to x^* superlinearly (quadratically if the Hessian $\nabla^2 f$ is Lipschitz continuous in a neighborhood of the solution).

However, the pure Newton's method presents some drawbacks. Indeed $\nabla^2 f(x^k)$ may be singular, and hence Newton's direction cannot be defined; the starting point x^0 can be such that the sequence $\{x^k\}$ generated by iteration (17) does not converge and convergence to a maximum point may even occur. Therefore, the pure Newton's method requires some modification that enforces global convergence to a solution, preserving the rate of convergence. A globally convergent Newton's method must produce a sequence $\{x^k\}$ such that: (i) it admits a limit point; (ii) any limit point belongs to the level set $\mathcal{L}^0$ and is a stationary point of f; (iii) no limit point is a maximum point of f; and (iv) if x^* is a limit point of $\{x^k\}$ and $\nabla^2 f(x^*)$ is positive definite, then the convergence rate is at least superlinear.

There are two main approaches for designing a globally convergent Newton's method, namely the line-search approach and the trust-region approach. We briefly describe them in the next two sections.

Line-Search Modifications of Newton's Method

In the line-search approach, the pure Newton's method is modified by controlling the magnitude of the step along the Newton's direction d_N^k; the basic iteration (17) becomes

$$x^{k+1} = x^k - \alpha^k[\nabla^2 f(x^k)]^{-1}\nabla f(x^k), \tag{18}$$

where α^k is chosen by a suitable LSA, for example, the Armijo's LSA with initial estimate $\Delta^k = 1$. Moreover, the Newton's direction d_N^k must be perturbed in order to ensure that a globally convergent algorithm is obtained. The simplest way of modifying the Newton's method consists of using the steepest-descent direction (possibly after positive diagonal scaling) whenever d_N^k does not satisfy some convergence conditions.

A possible scheme of a globally convergent Newton's algorithm is shown below.

Algorithm 5. Line search–based Newton's method.

Data. $c_1 > 0$, $c_2 > 0$, $p \geqslant 2$, $q \geqslant 3$.

Step 1. Choose $x_0 \in \mathbb{R}^n$ and set $k = 0$.

Step 2. If $\nabla f(x^k) = 0$ stop.

Step 3. If a solution d_N^k of the system

$$\nabla^2 f(x^k) d_N = -\nabla f(x^k)$$

exists and satisfies the conditions

$$\nabla f(x^k)' d_N^k \leqslant -c_1 \|\nabla f(x^k)\|^q, \quad \|d_N^k\|^p \leqslant c_2 \|\nabla f(x^k)\|$$

then set the direction $d^k = d_N^k$; otherwise set $d^k = -\nabla f(x^k)$

Step 4. Find α^k by using the Armijo's LSA of Algorithm 1.

Step 5. Set

$$x^{k+1} = x^k + \alpha^k d^k,$$

$k = k + 1$ and go to Step 2.

We observe that at Step 3, the steepest-descent direction is taken whenever $\nabla f(x^k)' d_N^k \geqslant 0$. Another modification of the Newton's methods consists of taking $d^k = -d_N^k$ if it satisfies $|\nabla f(x^k)' d^k| \geqslant c_1 \|\nabla f(x^k)\|^q$ and $\|d^k\|^p \leqslant c_2 \|\nabla f(x^k)\|$. This corresponds to the use of a negative curvature direction d^k that may result in practical advantages.

The use of the steepest-descent direction is not the only possibility for constructing globally convergent modifications of Newton's method. Another way is that of perturbing the Hessian matrix $\nabla^2 f(x^k)$ by a diagonal matrix Y^k so that the matrix $\nabla^2 f(x^k) + Y^k$ is positive definite and a solution d of the system

$$(\nabla^2 f(x^k) + Y^k) d = -\nabla f(x^k) \tag{19}$$

exists and provides a descent direction. A way to construct such a perturbation is to use a modified Cholesky factorization (see Gill et al., 1981, and Lin and Moré, 1999).

The most common versions of globally convergent Newton's methods are based on the enforcement of a monotonic decrease of the objective function values, although neither strong theoretical reasons nor computational evidence support this requirement. Indeed, the convergence region of Newton's method is often much larger than one would expect, but a convergent sequence in this region does not necessarily correspond to a monotonically decreasing sequence of function values. It is possible to relax the standard line-search conditions in a way that preserves a global convergence property. More specifically, in Grippo et al. (1986) a modification of the Newton's method that uses the Armijo-type nonmonotone rule (12) has been proposed. The algorithmic scheme remains essentially the same as that of Algorithm 5 except for Step 4. It is reported in Algorithm 6 below.

Algorithm 6. Nonmonotone Newton's method

Data. $c_1 > 0$, $c_2 > 0$, $p \geqslant 2$, $q \geqslant 3$ and an integer M.

Step 1. Choose $x_0 \in \mathbb{R}^n$. Set $k = 0$, $m(k) = 0$.

Step 2. If $\nabla f(x^k) = 0$ stop.

Step 3. If a solution d_N^k of the system

$$\nabla^2 f(x^k) d_N = -\nabla f(x^k)$$

exists and satisfies the conditions

$$\nabla f(x^k)' d_N^k \leqslant -c_1 \|\nabla f(x^k)\|^q, \quad \|d_N^k\|^p \leqslant c_2 \|\nabla f(x^k)\|$$

then set the direction $d^k = d_N^k$; otherwise set $d^k = -\nabla f(x^k)$.

Step 4. Find α^k by means of an LSA that enforces the nonmonotone Armijo's rule (12) with $J = m(k)$.
Step 5. Set

$$x^{k+1} = x^k + \alpha^k d^k,$$

$k = k + 1$, $m(k) = \min\{m(k-1) + 1, M\}$ and go to Step 2.

Actually, much more sophisticated nonmonotone stabilization schemes have been proposed (Grippo et al., 1991). We do not go into detail about these schemes. However, it is worthwhile to remark that only the magnitude of $\|d^k\|$ is checked at each iteration of these schemes and if the test is satisfied, the step size one is accepted without evaluating the objective function. Nevertheless, a test on the decrease of the values of f is performed at least every M steps.

The line search-based Newton's methods described so far converge in general to points satisfying only the first-order NOC for Equation (1). This is essentially due to the fact that Newton's method does not exploit all the information contained in the second derivatives. Indeed stronger convergence results can be obtained by using the pair of directions (d^k, s^k) and using a curvilinear line search, that is, a line search along the curvilinear path

$$x^{k+1} = x^k + \alpha^k d^k + (\alpha^k)^{1/2} s^k, \tag{20}$$

where d^k is a Newton-type direction and s^k is a particular direction that includes negative curvature information with respect to $\nabla^2 f(x^k)$. By using a suitable modification of the Armijo's rule for the step size α^k, algorithms that are globally convergent toward points that satisfy the second-order NOC for Equation (1) can be defined both for the monotone (Moré and Sorensen, 1979) and the nonmonotone versions (Ferris et al., 1996).

TRUST-REGION MODIFICATIONS OF NEWTON'S METHOD

In trust-region modifications of the Newton's method, the iteration becomes

$$x^{k+1} = x^k + s^k,$$

where the step s^k is obtained by minimizing the quadratic model q^k of the objective function not on the whole space $\mathbb{R}^n$ but on a suitable trust region where the model is supposed to be reliable. The trust region is usually defined as an ℓ_p-norm of the step s. The most common choice is the Euclidean norm, so that at each iteration k, the step s^k is obtained by solving

$$\begin{aligned} \min_{s\in\mathbb{R}^n} \quad & \tfrac{1}{2} s' \nabla^2 f(x^k) s + \nabla f(x^k)' s \\ & \|s\|^2 \leqslant (a^k)^2, \end{aligned} \tag{21}$$

where a^k is the trust-region radius. Often a scaling matrix D^k is used and the spherical constraint becomes $\|D^k s\|^2 \leqslant (a^k)^2$; this can be seen as a kind of preconditioning. However, by rescaling, and without loss of generality, we can assume for the sake of simplicity that $D^k = I$, where by I we denote the identity matrix of suitable dimension.

In trust-region schemes, the radius a^k is iteratively updated. The idea is that when $\nabla^2 f(x^k)$ is positive definite then the radius a^k should be large enough that the minimizer of Equation (21) is unconstrained and the full Newton step is taken. The updating rule of a^k depends on the ratio ρ^k between the actual reduction $f(x^k) - f(x^{k+1})$ and the expected reduction $f(x^k) - q^k(s^k)$. A possible scheme is described in Algorithm 7 below where the condition at Step 3 is equivalent to the satisfaction of the NOC for Equation (1).

Algorithm 7. Trust region–based Newton's method.

Data. $0 < \gamma_1 \leqslant \gamma_2 < 1$, $0 < \delta_1 < 1 \leqslant \delta_2$.
Step 1. Choose $x_0 \in R^n$ and a radius $a^0 > 0$. Set $k = 0$.
Step 2. Find $s^k = \arg\min_{\|s\| \leqslant a^k} q^k(s)$
Step 3. If $f(x^k) = q^k(s^k)$ stop.

Step 4. Compute the ratio

$$\rho^k = \frac{f(x^k) - f(x^k + s^k)}{f(x^k) - q^k(s^k)}$$

If $\rho^k \geqslant \gamma_1$ set $x^{k+1} = x^k + s^k$, otherwise set $x^{k+1} = x^k$.

Step 5. Update the radius a^k

$$a^{k+1} = \begin{cases} \delta_1 a^k & \text{if } \rho^k < \gamma_1 \\ a^k & \text{if } \rho^k \in [\gamma_1, \gamma_2] \\ \delta_2 a^k & \text{if } \rho^k > \gamma_2 \end{cases}$$

Set $k = k + 1$ and go to Step 2.

If f is twice continously differentiable, there is a limit point of the sequence $\{x^k\}$ produced by Algorithm 7, that satisfies the first- and second-order NOC for Equation (1). Furthermore, if x^k converges to a point where the Hessian $\nabla^2 f$ is positive definite, then the rate of convergence is superlinear.

This approach was first proposed in Sorensen (1982) and Moré and Sorensen (1983) and it relies on the fact that Equation (21) always admits a global solution and that a characterization of a global minimizer exists even in the case where the Hessian $\nabla^2 f(x^k)$ is not positive definite. Indeed, the optimal solution s^k satisfies a system of the form

$$[\nabla^2 f(x^k) + \lambda^k I]s^k = -\nabla f(x^k), \tag{22}$$

where λ^k is a KKT multiplier for Equation (21) such that $\nabla^2 f(x^k) + \lambda^k I$ is positive semidefinite. Hence, the trust-region modification of Newton's method also fits in the framework of using a correction of the Hessian matrix (19).

The main computational effort in the scheme of Algorithm 7 is the solution of the trust-region subproblem at Step 2. The peculiarities of Equation (21) led to the development of *ad hoc* algorithms for finding a global solution. The first ones proposed in literature (Gay, 1981; Sorensen, 1982; Moré and Sorensen, 1983) were essentially based on the solution of a sequence of linear systems of the type (22) for a sequence $\{\lambda^k\}$. These algorithms produce an approximate global minimizer of the trust-region subproblem, but rely on the ability to compute a Cholesky factorization of the matrix $(\nabla^2 f(x^k) + \lambda^k I)$ at each iteration k. Hence they are appropriate when forming a factorization for different values of λ^k is realistic in terms of both computer storage and time requirements. However, for large-scale problems without special structure, one cannot rely on factorizations of the matrices involved, even if, recently, incomplete Cholesky factorizations for large-scale problems have been proposed (Lin and Moré, 1999). Recent papers concentrate on iterative methods of conjugate gradient type that require only matrix-vector products. Different approaches have been proposed (Pham Dinh and Hoai An, 1998; Lucidi et al., 1998a; Rendl and Wolkowicz, 1997; Sorensen, 1997; Gould et al., 1999).

Actually, the exact solution of Equation (21) is not required. Indeed, in order to prove a global convergence result of a trust-region scheme, it suffices that the model value $q^k(s^k)$ is not larger than the model value at the Cauchy point, which is the minimizer of the quadratic model within the trust region along the (preconditioned) steepest-descent direction (Powell, 1975). Hence any descent method that moves from the Cauchy point ensures convergence (Toint, 1981; Steihaug, 1983).

Truncated Newton's Methods

Newton's method requires at each iteration k the solution of a system of linear equations. In the large-scale setting, exactly solving the Newton's system can be too burdensome and storing or factoring the full Hessian matrix can be difficult if not impossible. Moreover, the exact solution when x^k is far from a solution and $\|\nabla f(x^k)\|$ is large may be unnecessary. On the basis of these remarks, inexact Newton's methods have been proposed (Dembo et al., 1982) that approximately solve the Newton system (16) and still retain a good convergence rate. If $\tilde{d}_N^k$ is any approximate solution of the Newton system (16), the measure of accuracy is given by the residual r^k of the Newton equation, that is

$$r^k = \nabla^2 f(x^k)\tilde{d}_N^k + \nabla f(x^k).$$

By controlling the magnitude of r^k, superlinear convergence can be proved: if $\{x^k\}$ converges to a solution and if

$$\lim_{k\to\infty} \frac{\|r^k\|}{\|\nabla f(x^k)\|} = 0, \tag{23}$$

then $\{x^k\}$ converges superlinearly. This result is at the basis of truncated Newton's methods (see Nash, 2000, for a recent survey).

These methods require only matrix-vector products and hence they are suitable for large-scale problems. The classical truncated Newton's method (TNA; Dembo and Steihaug, 1983) was thought of for the case of positive definite Hessian $\nabla^2 f(x^k)$. It is obtained by applying a CG scheme to find an approximate solution $\tilde{d}_N^k$ to the linear system (16). We refer to outer iterations k for the sequence of points $\{x^k\}$ and to inner iterations i for the iterations of the CG scheme applied to the solution of system (16).

The inner CG scheme generates vectors d_N^i that iteratively approximate the Newton direction d_N^k. It stops either if the corresponding residual $r^i = \nabla^2 f(x^k)d_N^i + \nabla f(x^k)$ satisfies $\|r^i\| \leqslant \varepsilon^k$, where $\varepsilon^k > 0$ is such to enforce (23), or if a nonpositive curvature direction is found, that is, if a conjugate direction s^i is generated such that

$$(s^i)'\nabla^2 f(x^k)s^i \leqslant 0. \tag{24}$$

The same convergence results of Newton's method hold, and if the smallest eigenvalue of $\nabla^2 f(x^k)$ is sufficiently bounded away from zero, the rate of convergence is still superlinear. We show the inner CG method for finding an approximate solution of system (16) in Algorithm 8 below.

For sake of simplicity, we eliminate the dependencies on iteration k and we set $H = \nabla^2 f(x^k)$, $g = \nabla f(x^k)$. The method generates conjugate directions s^i and vectors p^i that approximate the solution $\tilde{d}_N^k$ of the Newton's system.

Algorithm 8. Truncated Newton's algorithm (TNA).

Data. k, H, g and a scalar $\eta > 0$.

Step 1. Set

$$\varepsilon = \eta\|g\|\min\left\{\frac{1}{k+1}, \|g\|\right\},$$

$i = 0$.

$$p^0 = 0,\ r^0 = -g,\ s^0 = r^0,$$

Step 2. If $(s^i)'Hs^i \leqslant 0$, set

$$\tilde{d}_N = \begin{cases} -g & \text{if } i = 0 \\ p^i & \text{if } i > 0 \end{cases} \quad \text{and stop.}$$

Step 3. Compute

$$\alpha^i = \frac{(s^i)'r^i}{(s^i)'Hs^i}$$

and set

$$p^{i+1} = p^i + \alpha^i s^i,$$

$$r^{i+1} = r^i - \alpha^i Hs^i,$$

Step 4. If $\|r^{i+1}\| > \varepsilon$, compute

$$\beta^i = \frac{\|r^{i+1}\|^2}{\|r^i\|^2},$$

set

$$s^{i+1} = r^{i+1} + \beta^i s^i,$$

$i = i + 1$ and go to Step 2;

Step 5. Set $\tilde{d}_N = p^{i+1}$ and stop.

By applying TNA to find an approximate solution of the linear system at Step 3 of Algorithm 5 (Algorithm 6) we obtain the monotone (nonmonotone) truncated version of the Newton's method.

In Grippo et al. (1989) a more general truncated Newton's algorithm based on a CG method has been proposed. The algorithm attempts to find a good approximation of the Newton direction even if $\nabla^2 f(x^k)$ is not positive definite. Indeed it does not stop when a negative curvature direction is encountered and this direction is used to construct additional vectors that allow us to proceed in the CG scheme. Of course this negative curvature truncated Newton's algorithm (NCTNA) encompasses the standard TNA.

For the sake of completeness, we remark that, also in the truncated setting, it is possible to define a class of line search-based algorithms that are globally convergent toward points that satisfy second-order NOCs for Equation (1). This is done by using a curvilinear line search (20) with d^k and s^k obtained by a Lanczos-based iterative scheme. These algorithms were shown to be very efficient in solving large-scale unconstrained problems (Lucidi and Roma, 1997; Lucidi et al., 1998).

2.5. *Quasi-Newton Methods*

Quasi-Newton methods were introduced with the aim of defining efficient methods that do not require the evaluation of second-order derivatives. They are obtained by setting d^k as the solution of

$$B^k d = -\nabla f(x^k), \tag{25}$$

where B^k is a $n \times n$ symmetric and positive definite matrix that is adjusted iteratively in such a way that the direction d^k tends to approximate the Newton direction. Formula (25) is referred to as the direct quasi-Newton formula; in turn the inverse quasi-Newton formula is

$$d^k = -H^k \nabla f(x^k).$$

The idea at the basis of quasi-Newton methods is that of obtaining the curvature information not from the Hessian but only from the values of the function and its gradient. The matrix B^{k+1} (H^{k+1}) is obtained as a correction of B^k (H^k), namely as $B^{k+1} = B^k + \Delta B^k$ ($H^{k+1} = H^k + \Delta H^k$). Let us define the quantities:

$$\delta^k = x^{k+1} - x^k, \quad \gamma^k = \nabla f(x^{k+1}) - \nabla f(x^k).$$

Taking into account that, in the case of a quadratic function f, the so-called quasi-Newton equation

$$\nabla^2 f(x^k)\delta^k = \gamma^k \tag{26}$$

holds, the correction ΔB^k (ΔH^k) is chosen such that

$$(B^k + \Delta B^k)\delta^k = \gamma^k, \quad (\delta^k = (H^k + \Delta H^k)\gamma^k).$$

Methods differ in the way they update the matrix B^k (H^k). Essentially they are classified according to a rank one or a rank two updating formula. The most-used class of quasi-Newton methods is the rank two Broyden class. The updating rule of the matrix H^k is given by the following correction term

$$\Delta H^k = \frac{\delta^k(\delta^k)'}{(\delta^k)'\gamma^k} - \frac{H^k\gamma^k(H^k\gamma^k)'}{(\gamma^k)'H^k\gamma^k} + c(\gamma^k)'H^k\gamma^k v^k(v^k)', \tag{27}$$

where

$$v^k = \frac{\delta^k}{(\delta^k)'\gamma^k} - \frac{H^k\gamma^k}{(\gamma^k)'H^k\gamma^k},$$

and c is a nonnegative scalar. For $c = 0$ we have the Davidon–Fletcher–Powell (DFP) formula (Davidon, 1959; Fletcher and Powell, 1963), whereas for $c = 1$ we have the Broyden–Fletcher–Goldfarb–Shanno (BFGS) formula (Broyden, 1970a,b; Fletcher, 1970; Goldfarb, 1970; Shanno, 1970). An important property of methods in the Broyden class is that if H^0 is positive definite and if, for each k, it results $(\delta^k)'\gamma^k > 0$, then positive definiteness is maintained for all the matrices H^k.

The BFGS method has been generally considered very effective. The BFGS method is a line-search method, where the step size α^k is obtained by an LSA that enforces the strong Wolfe conditions (8) and (10). We describe a possible scheme of the inverse BFGS method in Algorithm 9 below.

Algorithm 9. Quasi-Newton inverse BFGS algorithm.

Step 1. Choose $x_0 \in \mathbb{R}^n$. Set $H^0 = I$ and $k = 0$.
Step 2. If $\nabla f(x^k) = 0$ stop.
Step 3. Set the direction

$$d^k = -H^k\nabla f(x^k);$$

Step 4. Find α^k by means of an LSA that satisfies the strong Wolfe's conditions (8) and (10).
Step 5. Set

$$x^{k+1} = x^k + \alpha^k d^k,$$

$$H^{k+1} = H^k + \Delta H^k$$

with ΔH^k given by (27) with $c = 1$.
Set $k = k+1$ and go to Step 2.

As regards convergence properties of quasi-Newton methods, satisfactory results exist for the convex case (Powell, 1971, 1976; Byrd et al., 1987). In the nonconvex case, only partial results exist. In particular, if a constant ρ exists such that for every k

$$\frac{\|\gamma^k\|^2}{(\gamma^k)'\delta^k} \leqslant \rho, \tag{28}$$

then the sequence $\{x^k\}$ generated by the BFGS algorithm satisfies condition (5) (Powell, 1976). Condition (28) holds in the convex case.

A main result on the rate of convergence of quasi-Newton methods is given in the following proposition (Dennis and Moré, 1974).

Proposition 3. *Let* $\{B^k\}$ *be a sequence of nonsingular matrices and let* $\{x^k\}$ *be given by*

$$x^{k+1} = x^k - (B^k)^{-1}\nabla f(x^k).$$

Assume that $\{x^k\}$ *converges to a point* x^* *where* $\nabla^2 f(x^*)$ *is nonsingular. Then, the sequence* $\{x^k\}$ *converges superlinearly to* x^* *if and only if*

$$\lim_{k\to\infty} \frac{\|[B^k - \nabla^2 f(x^*)](x^{k+1} - x^k)\|}{\|x^{k+1} - x^k\|} = 0.$$

For large-scale problems forming and storing the matrix B^k (H^k) may be impracticable and limited-memory quasi-Newton methods have been proposed. These methods use the information of the last few iterations for defining an approximation of the Hessian avoiding the storage of matrices. Among limited-memory quasi-Newton methods, by far the most used is the limited-memory BFGS method (L-BFGS). In the L-BFGS method the matrix H^k is not formed explicitly and only pairs of vectors (δ^k, γ^k) are stored that define H^k implicitly. The product $H^k \nabla f(x^k)$ is obtained by performing a sequence of inner products involving $\nabla f(x^k)$ and the M most recent pairs (δ^k, γ^k). Usually small values of M $(3 \leqslant M \leqslant 7)$ give satisfactory results, hence the method is very efficient for large-scale problems (Liu and Nocedal, 1989; Nash and Nocedal, 1991).

2.6. *Derivative-Free Methods*

Derivative-free methods neither compute nor explicitly approximate derivatives of f. They are useful when ∇f is either unavailable, or unreliable (for example, due to noise) or computationally too expensive. Nevertheless, for convergence analysis purposes, these methods assume that f is continuously differentiable.

Recently, there has been a growing interest in derivative-free methods and many new results have been developed. We mention specifically only direct search methods, which aim to find the minimum point by using the value of the objective function on a suitable set of trial points. Direct search methods try to overcome the lack of information on the derivatives by investigating the behavior of the objective function in a neighborhood of the current iterate by means of samples of the objective function along a set of directions. Algorithms in this class present features that depend on the particular choice of the directions and of the sampling rule. Direct search methods are not the only derivative-free methods proposed in the literature, but they include pattern-search algorithms (PSA) (Torczon, 1995, 1997; Lewis et al., 1998) and derivative-free line-search algorithms (DFLSA) (De Leone et al., 1984; Lucidi and Sciandrone, 1996) that nowadays enjoy great favor.

The algorithmic schemes for PSA and DFLSA are quite similar and differ in the assumptions on the set of directions that are used and in the rule for finding the step size along these directions. Both for PSA and DFLSA, we refer to a set of directions $\mathscr{D}^k = \{d^1, \ldots d^r\}$ with $r \geqslant n+1$, which positively span $\mathbb{R}^n$. For the sake of simplicity we assume that $\|d^j\| = 1$. Moreover, for PSA we assume that

Assumption 2. *The directions* $d^j \in \mathscr{D}^k$ *are the jth column of the matrix* $(B\Gamma^k)$ *with* $B \in \mathbb{R}^{n \times n}$, *a nonsingular matrix, and* $\Gamma^k \in \mathscr{M} \subseteq Z^{n \times r}$, *where* $\mathscr{M}$ *is a finite set of full row-rank integral matrices.*

This means that in PSA the iterates x^{k+1} lie on a rational lattice centered in x^k. Indeed, pattern-search methods are restricted in the nature of the steps s^k they take. Let $\mathscr{P}^k$ be the set of candidates for the next iteration, namely

$$\mathscr{P}^k = \{x^{k+1} : x^k + \alpha^k d^j, \quad \text{with } d^j \in \mathscr{D}^k\}.$$

The set $\mathscr{P}^k$ is called the pattern. The step $s^k = \alpha^k d^j$ is restricted in the values it assumes for preserving the algebraic structure of the iterates and it is chosen so as to satisfy a simple reduction of the objective function $f(x^k + s^k) < f(x^k)$.

As regards DFLSA we assume:

Assumption 3. *The directions* $d^j \in \mathscr{D}^k$ *are the jth column of a full row-rank matrix* $B^k \in \mathbb{R}^{n \times r}$.

Hence, in DFLSA there is no underlying algebraic structure and the step s^k is a real vector, but it must satisfy a rule of sufficient reduction of the objective function.

For both schemes a possible choice of $\mathscr{D}^k$ is $\{\pm e_j, \ j = 1, \ldots, n\}$ where e_j is the jth column of the identity matrix.

We report simplified versions of PSA and DFLSA, respectively, in Algorithm 10 and in Algorithm 11.

Algorithm 10. Pattern-search algorithm (PSA).

Data. A set $\mathcal{D}^k$ satisfying Assumption (2), $\tau \in \{1,2\}$, $\theta = 1/2$.
Step 1. Choose $x^0 \in \mathbb{R}^n$, set $\Delta^0 > 0$ and $k = 0$;
Step 2. Check for convergence.
Step 3. If an index $j \in \{1, \ldots, r\}$ exists such that

$$f(x^k + \alpha^k d^j) < f(x^k), \quad \alpha^k = \Delta^k$$

set $x^{k+1} = x^k + \alpha^k d^j$, $\Delta^{k+1} = \tau\Delta^k$, $k = k+1$ and go to Step 2.
Step 4. Set $x^{k+1} = x^k$, $\Delta^{k+1} = \theta\Delta^k$, $k = k+1$ and go to Step 2.

Algorithm 11. Derivative-free line-search algorithm (DFLSA).

Data. A set $\mathcal{D}^k$ satisfying Assumption (3), $\gamma > 0$, $\theta \in (0,1)$.
Step 1. Choose $x^0 \in \mathbb{R}^n$, set $\Delta^0 > 0$ and $k = 0$;
Step 2. Check for convergence.
Step 3. If an index $j \in \{1, \ldots, r\}$ exists such that

$$f(x^k + \alpha^k d^j) \leqslant f(x^k), -\gamma(\alpha^k)^2, \quad \alpha^k \geqslant \Delta^k$$

set $x^{k+1} = x^k + \alpha^k d^j$, $\Delta^{k+1} = \alpha^k$, $k = k+1$ and go to Step 2.
Step 4. Set $x^{k+1} = x^k$, $\Delta^{k+1} = \theta\Delta^k$, $k = k+1$ and go to Step 2.

Both schemes produce a sequence that satisfies condition (5). We remark that by choosing at Step 3 of the PSA of Algorithm 10 the index j such that

$$f(x^k + \alpha^k d^j) = \min_{i:d^i \in \mathcal{D}^k} f(x^k + \alpha^k d^i) < f(x^k),$$

the stronger convergence result (4) can be obtained. In DFLSA, condition (4) can be enforced by choosing the step size α^k at Step 3 of Algorithm 11 according to a derivative-free LSA that enforces condition (11) together with a condition of the type

$$f\left(x^k + \frac{\alpha^k}{\delta} d^k\right) \geqslant \max\left\{f(x^k + \alpha^k d^k),\ f(x^k) - \gamma\left(\frac{\alpha^k}{\delta}\right)^2\right\}, \qquad \delta \in (0,1).$$

In both schemes, at Step 2 a check for convergence is needed. Of course, ∇f cannot be used to this aim. The standard test is given by

$$\left(\frac{\sum_{i=1}^{n+1}(f(x^i) - \bar{f})^2}{n+1}\right)^{1/2} \leqslant \text{tolerance},$$

where $\bar{f} = \frac{1}{n+1}\sum_{i=1}^{n+1} f(x^i)$, and $\{x^i,\ i = 1, \ldots n+1\}$ includes the current point and the last n points produced along n directions.

3. Selected Bibliography

We refer to the selected bibliography in Chapter 5.1 for a list of textbooks of general reference that develop the theory and practice on NLP. In the present article, several references have been given in correspondence to particular subjects. Here we add a list of some books and surveys of specific concern to unconstrained NLP. Of course the list is far from exhaustive; it includes only texts widely employed and currently found.

An easy introduction to the practice of unconstrained minimization is given in McKeown et al. (1990); more technical books are the ones by Wolfe (1978) and by Kelley (1999). A basic treatise on unconstrained minimization is the book by Dennis and Schnabel (1983); to be mentioned also is the survey due to Nocedal (1992). With particular concern for conjugate gradient methods, a basic reference is Hestenes (1975). A comprehensive treatment of trust-region methods is given in Conn et al. (2000). A survey paper on direct search methods is by Powell (1998) and a survey paper on large-scale unconstrained methods is due to Nocedal (1996).

References

Al-Baali, M. 1985. "Descent properties and global convergence of the Fletcher–Reeves method with inexact line search." *IMA J. Numerical Anal.* **5**: 121–124.

Barzilai, J. and J. M. Borwein. 1988. "Two point step size gradient method" *IMA J. Numerical Anal.* **8**: 141–148.

Broyden, C. G. 1970a. "The convergence of a class of double-rank minimization algorithms, part I." *J. Inst. Math. Appl.* **6**: 76–90.

Broyden, C. G. 1970b. "The convergence of a class of double-rank minimization algorithms, part II." *J. Inst. Math. Appl.* **6**: 222–231.

Byrd, R. H., J. Nocedal, and Y. Yuan. 1987. "Global convergence of a class of quasi-Newton methods on convex problems." *SIAM J. Numerical Anal.* **24**: 1171–1190.

Conn, A. R., N. I. M. Gould, and Ph. L. Toint. 2000. In *Trust Region Methods*. Philadelphia: SIAM.

Davidon, W. C. 1959. Variable metric method for minimization. Technical Report ANL 5990 (revised), Argonne National Laboratory. Published in *SIAM Journal On Optimization* **1**: 1–17, (1991).

De Leone, R., M. Gaudioso, and L. Grippo. 1984. "Stopping criteria for linesearch methods without derivatives." *Math. Programming* **30**: 285–300.

Dembo, R. S., S. Eisenstat, and T. Steihaug. 1982. "Inexact Newton methods." *SIAM J. Numerical Anal.* **19**: 400–408.

Dembo, R. S. and T. Steihaug. 1983. "Truncated-Newton methods algorithms for large-scale unconstrained optimization." *Math. Programming* **26**: 190–212.

Dennis, J. E. and J. Moré. 1974. "A characterization of superlinear convergence and its application to quasi-Newton methods." *Math. Comput.* **28**: 549–560.

Dennis, J. E. and R. B. Schnabel. 1983. In *Numerical Methods for Unconstrained Optimization and Nonlinear Equations*. Englewood Cliffs: Prentice-Hall.

Ferris, M. C., S. Lucidi, and M. Roma. 1996. "Nonmonotone curvilinear stabilization techniques for unconstrained optimization." *Comput. Optim. Appl.* **6**: 117–136.

Fletcher, R. E. 1970. "A new approach to variable metric algorithms." *Comput. J.* **13**: 317–322.

Fletcher, R. E. and M. J. D. Powell. 1963. "A rapidly convergent descent algorithm for minimization." *Comput. J.* **6**: 163–168.

Fletcher, R. E. and C. M. Reeves. 1964. "Function minimization by conjugate gradients." *Comput. J* **7**: 149–154.

Gay, D. M. 1981. "Computing optimal locally constrained steps." *SIAM J. Sci. Stat. Comput.* **2**: 186–197.

Gilbert, J. C. and J. Nocedal. 1992. "Global convergence properties of conjugate gradient methods for optimization." *SIAM J. Optim.* **2**: 21–42.

Gill, P. E., W. Murray, and M. H. Wright. 1981. In *Practical Optimization*. New York: Academic Press.

Goldfarb, A. 1970. "A family of variable metric methods derived by variational means." *Math. Comput.* **24**: 23–26.

Gould, N. I. M., S. Lucidi, M. Roma, and P. Toint. 1999. "Solving the trust-region subproblems using the Lanczos methods." *SIAM J. Optim.* **9**: 504–525.

Grippo, L., F. Lampariello, and S. Lucidi. 1986. "A nonmonotone line search technique for Newton's method." *SIAM J. Numerical Anal.* **23**: 707–716.

Grippo, L., F. Lampariello, and S. Lucidi. 1988. "Global convergence and stabilization of unconstrained minimization methods without derivatives." *J. Optim. Theory Appl.* **56**: 385–406.

Grippo, L., F. Lampariello, and S. Lucidi. 1989. "A truncated Newton method with nonmonotone linesearch for unconstrained optimization." *J. Optim. Theory Appl.* **60**: 401–419.

Grippo, L., F. Lampariello, and S. Lucidi. 1991. "A class of nonmonotone stabilization methods in unconstrained optimization." *Numerische Math.* **59**: 779–805.

Grippo, L. and S. Lucidi. 1997. "A globally convergent version of the Polak–Ribiere conjugate gradient method." *Math. Programming* **78**: 375–391.

Hestenes, M. R. 1975. *Optimization Theory. The Finite Dimensional Case*. New York: John Wiley & Sons.

Kelley, C. T. 1999. *Iterative Methods for Optimization*. Philadelphia: SIAM.

Lewis, R. M., V. Torczon, and M. W. Trosset. 1998. "Why pattern search works." *Optima: Mathematica Programming Newsletter*, (59)

Lin, C.-J. and J. J. Moré. 1999. "Incomplete Cholesky factorizations with limited memory." *SIAM J. Sci Comput.* **21**: 24–45.

Liu, D. C. and J. Nocedal. 1989. "On the limited memory BFGS method for large scale optimization." *Math. Programming* **45**: 503–528.

Lucidi, S., L. Palagi, and M. Roma. 1998a. "On some properties of quadratic programs with a convex quadratic constraint." *SIAM J. Optim.* **8**: 105–122.

Lucidi, S., F. Rochetich, and M. Roma. 1998b. "Curvilinear stabilization techniques for truncated Newton methods in large scale unconstrained optimization." *SIAM J. Optim.* **8**: 916–939.

Lucidi, S. and M. Roma. 1997. "Numerical experience with new truncated Newton methods in large scale unconstrained optimization." *Comput. Optim. Appl.* **7**: 71–87.

Lucidi S. and M. Sciandrone. 1996. "On the global convergence of derivative-free methods for unconstrained optimization without derivatives (revised version)." Technical Report 18-96. DIS, Università di Roma La Sapienza, Italy.

McKeown, J. J., D. Meegan, and D. Sprevak. 1990. In *An Introduction to Unconstrained Optimization*. Bristol: IOP Publishing Ltd.

Moré, J. J. and D. C. Sorensen. 1979. "On the use of direction of negative curvature in a modified Newton direction." *Math. Programming* **16**: 1–20.

Moré, J. J. and D. C. Sorensen. 1983. "Computing a trust region step." *SIAM J. Sci. Stat. Comput.* **4**: 553–572.

Moré, J. J. and D. C. Sorensen. 1984. "Newton's method." In *Studies in Numerical Analysis*, edited by G. H. Golub, MAA Studies in Mathematics, **24**: 29–82. Washington: The Mathematical Association of America.

Nash, S. G. 2000. "A survey of truncated-Newton methods." *J. Comput. Appl. Math* **124**: 45–59.

Nash, S. G. and J. Nocedal. 1991. "A numerical study of the limited memory BFGS method and the truncated-Newton method for large scale optimization." *SIAM J. Optim.* **1**: 358–372.

Nocedal, J. 1992. "Theory of algorithms for unconstrained optimization." *Acta Numerica* **2**: 199–242.

Nocedal, J. 1997. "Large scale unconstrained optimization." In *The State of the Art in Numerical Analysis*, edited by A. Watson and I. Duff, pp. 311–338. Oxford: Oxford University Press.

Nocedal, J. and S. J. Wright. 1999. In *Numerical Optimization*: Springer-Verlag.

Orthega, J. M. 1988. *Introduction to Parallel and Vector Solution of Linear Systems*. New York: Plenum Press.

Orthega, J. M. and W. C. Rheinboldt. 1970. In *Iterative Solution of Nonlinear Equations in Several Variables*: Academic Press.

Pham Dinh, T. and L. T. Hoai An. 1998. "D.C. optimization algorithm for solving the trust region subproblem." *SIAM J. Optim.* **8**: 476–505.

Polak, E. 1997. *Optimization: Algorithms and Consistent Approximations*: Springer-Verlag.

Polak, E. and G. Ribiére. 1969. "Note sur la convergence de méthodes de directions conjuguées." *Revue Francaise d'Informatique et de Recherche Opérationnelle* **16**: 35–43.

Powell, M. J. D. 1971. "On the convergence of the variable metric algorithm." *J. Inst. Math. Appl.* **7**: 21–36.

Powell, M. J. D. 1975. "Convergence properties of a class of minimization algorithms." In *Nonlinear Programming 2*, edited by O. L. Mangasarian, R. Meyer, and S. M. Robinson. New York: Academic Press.

Powell, M. J. D. 1976. "Some global convergence properties of a variable metric algorithm for minimization without exact line searches." In *Nonlinear Programming, SIAM-AMS Proceedings*, edited by R. W. Cottle and C. E. Lemke. Philadelphia: SIAM Publications.

Powell, M. J. D. 1986. "Convergence properties of algorithms for nonlinear optimization." *SIAM Rev.* **28**: 487–500.

Powell, M. J. D. 1998. "Direct search algorithms for optimization calculations." *Acta Numerica* **7**: 287–336.

Raydan, M. 1997. "The Barzilai and Borwein gradient method for the large scale unconstrained minimization problems." *SIAM J. Optim.* **7**: 26–33.

Rendl, F. and H. Wolkowicz. 1997. "A semidefinite framework to trust region subproblems with applications to large scale minimization." *Math. Programming* **77**: 273–299.

Shanno, D. F. 1970. "Conditioning of quasi-Newton methods for function minimization." *Math. Comput.* **27**: 647–656.

Sorensen, D. C. 1982. "Newton's method with a model trust region modification." *SIAM J. Sci. Stat. Comput.* **19**: 409–427.

Sorensen, D. C. 1997. "Minimization of a large-scale quadratic function subject to an ellipsoidal constraint." *SIAM J. Optim.* **7**: 141–161.

Steihaug, T. 1983. "The conjugate gradient method and trust regions in large scale optimization." *SIAM J. Numerical Anal.* **20**: 626–637.

Toint, P. 1981. "Towards an efficient sparsity exploiting Newton methods for minimization." In *Sparse Matrices and Their Uses*, edited by I. S. Duff, pp. 57–87. London: Academic Press.

Torczon, V. 1995. Pattern search methods for nonlinear optimization. *SIAG/OPT Views-and-News*, (6).

Torczon, V. 1997. "On the convergence of pattern search methods." *SIAM J. Optim.* **7**: 1–25.

Wolfe, M. A. 1978. *Numerical Methods for Unconstrained Optimization. An Introduction*. Workingham: Van Nostrand, Reinhold Company.

5.3 Constrained Nonlinear Programming

Gianni Di Pillo and Laura Palagi

ABSTRACT We consider the basic approaches for the solution of constrained nonlinear programming problems. The first approach is based on the use of unconstrained minimization techniques applied to some merit function, either in sequential penalty methods or in exact penalty methods. The second approach is based on the sequential solution of quadratic programming subproblems. The third approach tries to maintain feasibility of tentative solutions, by employing feasible descent directions. Far from being exhaustive, this article aims to give the main ideas and tools of practical use, and to suggest how to go deeper into more technical and/or more recent results.

1. Introduction

We consider methods for the solution of the nonlinear programming (NLP) problem:

$$\min_{x \in \mathcal{F}} f(x), \tag{1}$$

where $x \in \mathbb{R}^n$ is a vector of decision variables, $f : \mathbb{R}^n \rightarrow \mathbb{R}$ is an objective function, and the feasible set $\mathcal{F}$ is described by inequality and/or equality constraints on the decision variables:

$$\mathcal{F} = \{x \in \mathbb{R}^n : g_i(x) \leqslant 0, \quad i = 1 \ldots, p; \ h_j(x) = 0, j = 1, \ldots, m, m \leqslant n\};$$

that is, we consider methods for the solution of the constrained NLP problem:

$$\begin{aligned} \min_{x \in \mathbb{R}^n} \quad & f(x) \\ & g(x) \leqslant 0 \\ & h(x) = 0, \end{aligned} \tag{2}$$

where $g : \mathbb{R}^n \rightarrow \mathbb{R}^p$, and $h: \mathbb{R}^n \rightarrow \mathbb{R}^m$, with $m \leqslant n$.

Here, for simplicity, the problem functions f, g, h are twice continuously differentiable in $\mathbb{R}^n$, even if in many cases only they need to be once continuously differentiable.

We assume that the reader is familiar with the Karush–Kuhn–Tucker (KKT) necessary optimality conditions (NOC), and with the second order sufficiency condition (SOSC) for Equation (2), as well as with the basic notions on the performances of algorithms, given in Chapter 5.1 of this handbook.

For the solution of Equation (2), two main approaches have been developed. Since, in principle, an unconstrained minimization is simpler than a constrained minimization, the first approach is based on the transformation of the constrained problem into a sequence of unconstrained problems, or even into a single unconstrained problem. Basic tools for this approach are the unconstrained minimization methods of Chapter 5.2 of this handbook. The second approach is based on the transformation of the constrained problem into a sequence of simpler constrained problems, that is, either linear or quadratic programming problems; thus the basic tools become the algorithms considered in this handbook for linear or quadratic programming. Due to space limitations, we will confine ourselves to methods based on quadratic programming, which are by far the most widely employed

We do not treat problems with special structure, such as bound and/or linearly constrained problems, for which highly specialized algorithms exist.

We employ the same notation as that adopted in Chapter 5.1. The additional notation $\max\{y, z\}$, where $y, z \in \mathbb{R}^p$, denotes the vector with components $\max\{y_i, z_i\}$ for $i = 1, \ldots, p$.

2. Unconstrained Sequential Methods

In this section we consider methods that are very basic in NLP. These are: the quadratic penalty method, which belongs to the class of exterior penalty functions methods, the logarithmic barrier method, which belongs to the class of interior penalty functions or barrier functions methods; and the augmented Lagrangian method, which can be considered as an improved exterior penalty method.

2.1. *The Quadratic Penalty Method*

The quadratic penalty method dates back to an idea of Courant (1943), which has been fully developed in the book by Fiacco and McCormick (1968). Let us consider Equation (2) and let $p(x)$, $p:\mathbb{R}^n \to \mathbb{R}$ be a continuous function such that $p(x)=0$ for all $x \in \mathcal{F}$, $p(x)>0$ for all $x \notin \mathcal{F}$. Then, we can associate to Equation (2) the unconstrained problem:

$$\min_{x\in\mathbb{R}^n} \left[f(x)+\frac{1}{\epsilon}p(x)\right],$$

where $\epsilon>0$. The function

$$P(x;\epsilon)=f(x)+\frac{1}{\epsilon}p(x),$$

parameterized by the penalty parameter ϵ, is called a penalty function for Equation (2), since it is obtained by adding a term that penalizes the constraint violations to the objective function of the original problem. It is called an exterior penalty function because its minimizers are usually exterior to the feasible set $\mathcal{F}$. The penalization of the constraint violations becomes more severe as the penalty parameter ϵ is smaller. Given a sequence of positive numbers $\{\epsilon^k\}$, $k=0,1,\ldots$ such that $\epsilon^{k+1}<\epsilon^k$, $\lim_{k\to\infty}\epsilon^k=0$, the exterior penalty method for solving Equation (2) is reported in Algorithm 1 below.

Algorithm 1. The exterior penalty function algorithm.

Data. $\{\epsilon^k\}$ such that $\epsilon^{k+1}<\epsilon^k$ and $\lim_{k\to\infty}\epsilon^k=0$.

Step 1. Choose $x^s \in \mathbb{R}^n$ and set $k=0$.

Step 2. Starting from x^s, find an unconstrained local minimizer

$$x^k = \arg\min_{x\in\mathbb{R}^n} P(x;\epsilon^k).$$

Step 3. If x^k is a KKT point, then stop.

Step 4. Set $x^s=x^k$, $k=k+1$ and go to Step 2.

Under weak assumptions, it can be proved that limit points of the sequence $\{x^k\}$ produced by the method of Algorithm 1 are local solutions of the constrained problem (2).

The most widely used penalty function is the quadratic penalty function and the corresponding method is called the quadratic penalty method. The function P is obtained by using penalty terms that are the squares of the constraint violations and is given by:

$$P(x;\epsilon)=f(x)+\frac{1}{\epsilon}\{\|\max\{g(x),0\}\|^2+\|h(x)\|^2\}.$$

The quadratic penalty method is very simple; however, it suffers from a major disadvantage: as ϵ becomes smaller, the Hessian matrix $\nabla^2 P(x;\epsilon)$ becomes more ill conditioned, so that the unconstrained minimization of $P(x;\epsilon)$ becomes more difficult and slow. To alleviate the effects of ill conditioning, the scheme of Algorithm 1 can be modified so that at Step 2 a starting point x^s closer to x^k than x^{k-1} can be given by extrapolating the trajectory of minimizers $x^0, x^1, \ldots x^{k-1}$. In any case, at present, the method has been overcome by methods based on the augmented Lagrangian function.

As a final remark, the function $P(x;\epsilon)$ is not twice continuously differentiable at points where some constraint g_i is active, due to the presence in P of the term $\max\{g_i, 0\}$, and this must be taken into account in the minimization of P.

2.2. *The Logarithmic Barrier Method*

The original idea of the logarithmic barrier method is due to Frisch (1955); again, a comprehensive treatment is given in the book by Fiacco and McCormick (1968).

This method can be employed for NLP problems with inequality constraints only, that is:

$$\min_{x \in \mathbb{R}^n} \quad f(x) \qquad g(x) \leqslant 0. \tag{3}$$

It is assumed, moreover, that the strictly feasible region

$$\mathcal{F}^o = \{x \in \mathbb{R}^n : g_i(x) < 0, \quad i = 1, ..., p\} \tag{4}$$

is nonempty.

A barrier term for the feasible set $\mathcal{F}$ of Equation (3) is a continuous function $v(x)$ defined in $\mathcal{F}^o$, such that $v(x) \to \infty$ as $x \in \mathcal{F}^o$ approaches the boundary of $\mathcal{F}$. Then, associated to Equation (3), we consider the unconstrained problem

$$\min_{x \in \mathcal{F}^o} [f(x) + \epsilon v(x)],$$

where $\epsilon > 0$. The function

$$V(x; \epsilon) = f(x) + \epsilon v(x),$$

parameterized by ϵ, is called a barrier function for Equation (3), since it is obtained by adding a term that establishes a barrier on the boundary of the feasible set to the objective function of the problem, thus preventing a descent path for V starting in the interior of $\mathcal{F}$ crossing the boundary. The barrier gets sharper as the barrier parameter ϵ gets larger. Algorithms based on barrier functions are also called interior-point algorithms because the sequences that are produced are interior to the strictly feasible region.

Given a sequence of positive numbers $\{\epsilon^k\}$, $k = 0, 1, ...$ such that $\epsilon^{k+1} < \epsilon^k$, and $\lim_{k \to \infty} \epsilon^k = 0$, the barrier function method for solving Equation (3) is shown in Algorithm 2 below.

Algorithm 2. The barrier function algorithm.

Data. $\{\epsilon^k\}$ such that $\epsilon^{k+1} < \epsilon^k$ and $\lim_{k \to \infty} \epsilon^k = 0$.

Step 1. Choose $x^s \in \mathcal{F}^o$ and set $k = 0$.

Step 2. Starting from x^s, find an unconstrained local minimizer

$$x^k = \arg \min_{x \in \mathcal{F}^o} V(x; \epsilon^k).$$

Step 3. If x^k is a KKT point, then stop.

Step 4. Set $x^s = x^k$, $k = k + 1$ and go to Step 2.

As for the quadratic penalty method, the convergence analysis of the method shows that, under weak assumptions, limit points of the sequence $\{x^k\}$ produced by the method in Algorithm 2 are local solutions of the constrained Equation (3).

The most important barrier function for Equation (3) is the logarithmic barrier function, and the corresponding method is called the logarithmic barrier method. The function V is obtained by using as barrier terms the natural logarithm of the constraints:

$$V(x; \epsilon) = f(x) - \epsilon \sum_{i=1}^{p} \log(-g_i(x)).$$

The logarithmic barrier method is also very simple; however, the same remarks made for the quadratic penalty method hold. Namely, the method suffers from ill conditioning as ϵ becomes smaller, and also in this case a better starting point x^s can be found by extrapolating the trajectory of minimizers $x^0, x^1, ..., x^{k-1}$. An additional disadvantage is that x^s must be strictly feasible, so that the feasibility subproblem must be solved at the beginning. At present, the approach based on logarithmic barrier terms is not widely used for general NLP problems. However, in linearly constrained optimization it is of major importance for treating bound constraints, and indeed it is at the basis of the interior-points methods for linear and quadratic programming (see Hertog, 1994).

Finally, we point out that the logarithmic barrier method can be combined with the quadratic penalty method to deal with problems that also have equality constraints: in this case a penalty term $\frac{1}{\epsilon}\|h(x)\|^2$ can be added to the function V, retaining the behavior of both approaches.

2.3. The Augmented Lagrangian Method

The augmented Lagrangian method has been proposed independently by Hestenes (1969) and Powell (1969) for the equality constrained problem; the extension to the inequality constrained problem is due to Rockafellar (1973). Miele et al. (1971a,b, 1972) contributed to publicizing the method; and, in a series of papers that are at the core of his book, Bertsekas (1982) established the superiority of this method over the quadratic penalty method.

Let us consider first an NLP problem with only equality constraints:

$$\begin{aligned} \min_{x\in\mathbb{R}^n} \quad & f(x) \\ & h(x) = 0. \end{aligned} \tag{5}$$

The Lagrangian function for Equation (5) is

$$L(x, \mu) = f(x) + \mu' h(x). \tag{6}$$

The augmented Lagrangian function for Equation (5) is the function

$$L_a(x, \mu; \epsilon) = L(x, \mu) + \frac{1}{\epsilon}\|h(x)\|^2, \tag{7}$$

obtained by adding the quadratic penalty term for the constraint violations to L.

The main property of function L_a is the following:

Proposition 1. *Assume that* x^* *is a local solution of Equation (5) that, together with the multiplier* μ^**, satisfies the SOSC (restricted to equality constraints) given by Proposition 4 of Chapter 5.1; then a threshold value* ϵ^* *exists such that, for all* $\epsilon \in (0, \epsilon^*]$, x^* *is a local solution of the unconstrained problem*

$$\min_{x\in\mathbb{R}^n} L_a(x, \mu^*; \epsilon). \tag{8}$$

Conversely, if for some μ^* *and* ϵ, x^* *satisfies the SOSC to be a local solution of Equation (8) given by Proposition 3 of Chapter 5.1, and* $h(x^*) = 0$, *then* x^* *is a local solution of Equation (5), and* μ^* *is the corresponding multiplier.*

By Proposition 1, $L_a(x, \mu^*; \epsilon^k)$ can be employed in the quadratic penalty method of Algorithm 1 in place of $P(x; \epsilon^k)$, without requiring that ϵ^k decreases to zero. It is only required that the condition $\epsilon^k \leqslant \epsilon^*$ becomes fulfilled, thereby mitigating the ill conditioning associated with the use of P. However, since μ^* is not known in advance, a procedure that iteratively updates an estimate μ^k of μ^* must be included in the method.

A generic algorithmic scheme of an augmented Lagrangian method for solving Equation (5) is given in Algorithm 3.

Algorithm 3. An augmented Lagrangian algorithm.

Step 1. Choose ϵ^0, μ^0 and $x^s \in \mathbb{R}^n$; set $k = 0$.

Step 2. Starting from x^s, find an unconstrained local minimizer

$$x^k = \arg\min_{x\in\mathbb{R}^n} L_a(x, \mu^k; \epsilon^k)$$

Step 3. If (x^k, μ^k) is a KKT pair, then stop.

Step 4. Choose a penalty parameter $\epsilon^{k+1} \in (0, \epsilon^k]$.
Step 5. Update the multiplier estimate to μ^{k+1}.
Step 6. Set $x^s = x^k$, $k = k+1$ and go to Step 2.

The penalty parameter ϵ^{k+1} is usually set to $\rho\epsilon^k$, with $\rho = 1$ if $\|h(x^k)\|^2$ is sufficiently smaller than $\|h(x^{k-1})\|^2$, and $\rho < 1$ otherwise.

Updating formulas for the multiplier estimate are based on the fact that, under the assumptions of Proposition 1, the pair (x^*, μ^*) is a saddle point of the function L_a, so that an ascent strategy with respect to the dual function

$$D_a(\mu;\epsilon) = \min_{x \in \mathscr{B}_x*} L_a(x,\mu;\epsilon),$$

is effective. The simplest updating formula is the first-order formula $\mu^{k+1} = \mu^k + \frac{2}{\epsilon^k} h(x^k)$, which corresponds to a gradient ascent step for D_a; second-order formulas are based on a Newton's step.

Using the augmented Lagrangian method of Algorithm 3 we get a sequence $\{x^k\}$ and a sequence $\{\mu^k\}$. If the sequence $\{\mu^k\}$ is bounded, the sequence $\{x^k\}$ has the same convergence properties as the sequence produced by the quadratic penalty method of Algorithm 1 The rate of convergence depends on the updating formula employed for the multiplier.

Let us now consider an NLP problem with only inequality constraints, that is, Equation (3). An inequality constraint $g_i(x) \leqslant 0$ can be transformed into the equality constraint $g_i(x) + y_i^2 = 0$ by using an additional squared slack variable y_i^2. Equation (3) can be rewritten as the equality constrained problem:

$$\begin{aligned} &\min_{x \in \mathbb{R}^n, y \in \mathbb{R}^p} \quad f(x) \\ &g_i(x) + y_i^2 = 0, \quad i = 1, \ldots, p. \end{aligned} \tag{9}$$

This transformation is only instrumental; indeed, for algorithmic purposes, it has the disadvantage of increasing the number of variables from n to $n+p$.

Let λ_i be the multiplier associated with the ith constraint; then the augmented Lagrangian function for the equality constrained Equation (9) turns out to be:

$$L_a(x,y,\lambda;\epsilon) = f(x) + \sum_{i=1}^{p} \lambda_i[g_i(x) + y_i^2] + \frac{1}{\epsilon}\sum_{i=1}^{p}(g_i(x) + y_i^2)^2. \tag{10}$$

The function $L_a(x, y, \lambda; \epsilon)$ could be employed in the algorithmic scheme described in Algorithm 3, by minimizing with respect to x, y for a given λ and ϵ. However, the structure of L_a in (10) reveals that minimization with respect to each y_i^2 can be easily performed in closed form. Given x, λ, and ϵ, the minimizer y_i^2 of L_a is given by $y_i^2(x,\lambda;\epsilon) = -\min\left\{0, g_i(x) + \frac{\epsilon}{2}\lambda_i\right\}$. By substituting in (10) we get the augmented Lagrangian function in the original variables of Equation (3):

$$L_a(x,\lambda;\epsilon) = f(x) + \lambda' \max\left\{g(x) \ -\frac{\epsilon}{2}\lambda\right\} + \frac{1}{\epsilon}\left\|\max\left\{g(x), -\frac{\epsilon}{2}\lambda\right\}\right\|^2. \tag{11}$$

Accordingly, the first-order formula for updating the multiplier λ is $\lambda^{k+1} = \max\left\{0, \lambda^k + \frac{2}{\epsilon^k} g(x^k)\right\}$. Note that function (11) is not twice continuously differentiable at points where $g_i(x) = -\frac{\epsilon}{2}\lambda_i$ for some i; however, the lack of second-order differentiability does not occur locally in a neighborhood of a KKT pair (x^*, λ^*) if $(g(x^*), \lambda^*)$ satisfies the strict complementarity condition.

Problems with both equality and inequality constraints can be handled by combining the two cases.

The augmented Lagrangian method with first-order multiplier updating is also referred to as the multiplier method. A less commonly used term is the shifted barrier method.

Vast computational experience shows that the augmented Lagrangian function approach is reasonably effective, and many commercial software packages for the solution of NLP problems are based on it.

3. Unconstrained Exact Penalty Methods

Exact penalty methods attempt to solve NLP problems by means of a single minimization of an unconstrained function, rather then by means of a sequence of unconstrained minimizations. Exact penalty methods are based on the construction of a function depending on a penalty parameter $\epsilon > 0$, such that the unconstrained minimum points of this function are also solutions of the constrained problem for all values of ϵ in the interval $(0, \epsilon^*]$, where ϵ^* is a threshold value.

We can subdivide exact penalty methods into two classes: methods based on exact penalty functions and methods based on exact augmented Lagrangian functions. The term exact penalty function is used when the variables of the unconstrained problem are the same as that of the original constrained problem, whereas the term "exact augmented Lagrangian function" is used when the unconstrained problem is defined in the product space of the problem variables and of the multipliers (primal and dual variables).

An exact penalty function, or an augmented Lagrangian function, possesses different exactness properties, depending on which kind of correspondence can be established, under suitable assumptions, between the set of the local (global) solutions of the constrained problem and the set of local (global) minimizers of the unconstrained function. Different definitions of exactness can be found in Di Pillo and Grippo (1989).

The simplest function that possesses exactness properties is obtained by adding the ℓ_1 norm of the constraint violations to the objective function, weighted by a penalty parameter ϵ. The so-called ℓ_1 exact penalty function for Equation (2), introduced by Zangwill (1967), is given by:

$$L_1(x;\epsilon) = f(x) + \frac{1}{\epsilon}\left\{\sum_{i=0}^{p} \max\{g_i(x), 0\} + \sum_{j=1}^{m} | \, h_j(x) \, |\right\}. \tag{12}$$

In spite of its simplicity, function (12) is not continuously differentiable, so that its unconstrained minimization cannot be performed using the techniques seen in Chapter 5.2 of this handbook; instead one should resort to the less efficient nonsmooth optimization techniques of Chapter 5.4 in this handbook.

Continuously differentiable exact penalty functions can be obtained from the augmented Lagrangian function of Section 2.3 by substituting the multiplier vectors λ, μ with the continuously differentiable multiplier functions $\lambda(x)$, $\mu(x)$, which have the property that $\lambda(x^*) = \lambda^*$, $\mu(x^*) = \mu^*$ whenever the triplet (x^*, λ^*, μ^*) satisfies the KKT necessary conditions.

For the equality constrained Equation (5) a multiplier function is

$$\mu(x) = -[\nabla h(x)'\nabla h(x)]^{-1}\nabla h(x)'\nabla f(x);$$

then Fletcher's exact penalty function, introduced in Fletcher (1970), is:

$$U_F(x;\epsilon) = f(x) + \mu(x)'h(x) + \frac{1}{\epsilon}\|h(x)\|^2. \tag{13}$$

Similarly, for the inequality constrained Equation (3) a multiplier function is:

$$\lambda(x) = -[\nabla g(x)'\nabla g(x) + \theta G^2(x)]^{-1}\nabla g(x)'\nabla f(x),$$

where $\theta > 0$ and $G(x)$ is the diagonal matrix with entries $g_i(x)$; an exact penalty function is (Glad and Polak, 1979, Di Pillo and Grippo, 1985):

$$U(x;\epsilon) = f(x) + \lambda'(x)\max\left\{g(x), -\frac{\epsilon}{2}\lambda(x)\right\} + \left\|\max\left\{g(x), -\frac{\epsilon}{2}\lambda(x)\right\}\right\|^2. \tag{14}$$

Exact augmented Lagrangian functions are obtained by incorporating in the augmented Lagrangian function L_a terms that penalize the violation of necessary optimality conditions. For instance, for the equality constrained problem (Equation (5)) an exact augmented Lagrangian function, studied by Di Pillo and Grippo (1979) is:

$$S_a(x, \mu; \epsilon) = f(x) + \mu' h(x) + \frac{1}{\epsilon} \|h(x)\|^2 + \theta \|\nabla h(x)' \nabla_x L(x, \mu)\|^2, \tag{15}$$

where $\theta > 0$. It is evident that the last term in the right-hand side of (15) penalizes in some way the violation of the necessary optimality condition $\nabla_x L(x, \mu) = 0$.

Exact penalty functions and exact augmented Lagrangian functions can also be endowed with barrier terms that act on the boundary of a set that strictly contains the feasible set. In this way it is possible to build functions that not only have the solution of the constrained problem as unconstrained minimizers, but have also compact level sets. Both these properties are quite important for the design of globally convergent algorithms. Indeed, making use of these functions it is possible to develop algorithms for constrained problems that behave like the algorithms for unconstrained problems in terms of convergence and convergence rate (see for instance Di Pillo et al., 1980, 1982, 1986).

Algorithms based on continuously differentiable exact penalty functions have not been widely implemented. This is mainly due to the fact that the evaluation of the multiplier function requires the inversion of a matrix of the dimension of the constraints vector. More promising is the use of exact augmented Lagrangian functions, even if their minimization must be performed on the enlarged space of the decision variables and of the multipliers.

Functions that possess exactness properties are also referred to as exact merit functions when they are used in sequential algorithms to measure the merit of subsequent iterates in moving toward the solution of the constrained problem; in fact their minimization combines the often conflicting requirements of reducing both the objective function and constraint violations (and the violation of other NOCs).

Exact penalty methods, in any case, require some strategy to select a penalty parameter value in the interval $(0, \epsilon^*]$ where exactness holds.

4. Sequential Quadratic Programming Methods

The sequential quadratic programming (SQP) approach can be viewed as a generalization for constrained optimization of Newton's method for unconstrained optimization: in fact it goes a step away from the current point by minimizing a quadratic model of the problem. The SQP approach was introduced by Wilson (1963); the main developments are due to Garcia-Palomares and Mangasarian (1976), Han (1976; 1977), Powell (1978a,b,c).

For simplicity, let us consider the equality constrained problem (Equation (5)), with Lagrangian function (Equation (6)). In principle, a solution of this problem could be found by solving, with respect to (x, μ), the KKT first-order NOC:

$$\nabla L(x, \mu) = \begin{pmatrix} \nabla f(x) + \nabla h(x)\mu \\ h(x) \end{pmatrix} = 0. \tag{16}$$

The Newton iteration for the solution of (16) is given by

$$\begin{pmatrix} x^{k+1} \\ \mu^{k+1} \end{pmatrix} = \begin{pmatrix} x^k \\ \mu^k \end{pmatrix} + \begin{pmatrix} d_x^k \\ d_\mu^k \end{pmatrix},$$

where the Newton's step (d_x^k, d_μ^k) solves the so-called KKT system:

$$\begin{pmatrix} \nabla_x^2 L(x^k, \mu^k) & \nabla h(x^k) \\ \nabla h(x^k)' & 0 \end{pmatrix} \begin{pmatrix} d_x \\ d_\mu \end{pmatrix} = -\begin{pmatrix} \nabla f(x^k) + \nabla h(x^k)\mu^k \\ h(x^k) \end{pmatrix}.$$

By adding the term $\nabla h(x^k)\mu^k$ to both sides of the first equation of the KKT system, the Newton's iteration for (16) is determined equivalently by means of the equations:

$$x^{k+1} = x^k + d_x^k, \tag{17}$$

$$\begin{pmatrix} \nabla_x^2 L(x^k, \mu^k) & \nabla h(x^k) \\ \nabla h(x^k)' & 0 \end{pmatrix} \begin{pmatrix} d_x^k \\ \mu^{k+1} \end{pmatrix} = - \begin{pmatrix} \nabla f(x^k) \\ h(x^k) \end{pmatrix}. \tag{18}$$

The iteration defined by (17)–(18) constitutes the so-called Newton–Lagrange method for solving Equation (5).

Consider now the quadratic programming (QP) problem:

$$\begin{aligned} \min_{s \in \mathbb{R}^n} \quad & \tfrac{1}{2} s' \nabla_x^2 L(x^k, \mu^k) s + \nabla f(x^k)' s \\ & \nabla h(x^k)' s + h(x^k) = 0. \end{aligned} \tag{19}$$

The first-order NOC for (19) can be written as:

$$\begin{pmatrix} \nabla_x^2 L(x^k, \mu^k) & \nabla h(x^k) \\ \nabla h(x^k)' & 0 \end{pmatrix} \begin{pmatrix} s \\ \eta \end{pmatrix} = - \begin{pmatrix} \nabla f(x^k) \\ h(x^k) \end{pmatrix}, \tag{20}$$

where η is a multiplier vector for the linear equality constraints of Equation (19). By comparing (18) with (20) we see that (d_x^k, μ^{k+1}) and a solution (s^k, η^k) of (20) satisfy the same system. Therefore, in order to solve Equation (5) we can employ an iteration based on the solution of the QP Equation (19) rather than on the solution of the linear system (18).

The SQP method for Equation (5) in its simplest form is described in Algorithm 4.

Algorithm 4. The sequential quadratic programming (SQP) method for equality constrained problems.

Step 1. Choose an initial pair (x^0, μ^0); set $k = 0$.
Step 2. If (x^k, μ^k) is a KKT pair of problem (5), then stop.
Step 3. Find (s^k, η^k) as a KKT pair of problem (19).
Step 4. Set $x^{k+1} = x^k + s^k$; $\mu^{k+1} = \eta^k$, $k = k+1$ and go to Step 2.

The SQP method is preferable to the Newton–Lagrange method for the following reason. Assume that the sequence $\{(x^k, \mu^k)\}$ produced by the Newton–Lagrange method converges to some (x^*, μ^*); then, in the limit, $d_x = 0$, and from (18) it results, as expected, that (x^*, μ^*) satisfies the first-order NOC for Equation (5). Assume now that the sequence $\{(x^k, \mu^k)\}$ produced by the SQP method converges to some (x^*, μ^*); then, in the limit, $s = 0$, and from (20) it results again that (x^*, μ^*) satisfies the first-order NOC. However, from (19) and (20) we also have, in the limit, that $s' \nabla^2 L_x(x^*, \mu^*) s \geqslant 0$ for all $s \in \mathbb{R}^n$ such that $\nabla h(x^*)' s = 0$. Therefore, in this case, (x^*, μ^*) also satisfies a second-order NOC. In essence, the minimization process of SQP drives the sequence toward beneficial KKT solutions.

Note that, thanks to the constraint, the objective function of the quadratic Equation (19) could also be written as:

$$\tfrac{1}{2} s' \nabla_x^2 L(x^k, \mu^k) s + \nabla_x L(x^k, \mu^k)' s + L(x^k, \mu^k).$$

This leads to an alternative motivation of the SQP method. Indeed, Equation (5) can be written equivalently as

$$\begin{aligned} \min_{x \in \mathbb{R}^n} \quad & L(x, \mu^*) \\ & h(x) = 0, \end{aligned}$$

where μ^* is the KKT multiplier associated with the solution x^* of (5). Then solving Equation (19) corresponds to finding (x^*, μ^*) by iteratively solving the problem of minimizing a quadratic approximation of the Lagrangian $L(x, \mu^k)$ subject to a linear approximation of the constraint $h(x) = 0$.

The last observation suggests that the SQP framework can be extended easily to the general NLP Equation (2). Indeed, by also linearizing the inequality constraints we get the quadratic programming problem:

$$\min_{s \in \mathbb{R}^n} \ \tfrac{1}{2} s' \nabla_x^2 L(x^k, \lambda^k, \mu^k) s + \nabla f(x^k)' s$$
$$\nabla g(x^k)' s + g(x^k) \leqslant 0 \qquad (21)$$
$$\nabla h(x^k)' s + h(x^k) = 0.$$

The inequality constrained case puts forward as evidence an additional advantage of the SQP approach over the Newton–Lagrange method: indeed it is much simpler to solve the QP subproblem (21) than to solve the set of equalities and inequalities obtained by linearizing the KKT conditions for Equation (2).

The SQP algorithm for Equation (2) is given in Algorithm 5.

Algorithm 5. The sequential quadratic programming (SQP) method for general constrained problems.

Step 1. Choose an initial triplet (x^0, λ^0, μ^0); set $k = 0$.
Step 2. If (x^k, λ^k, μ^k) is a KKT triplet of problem (2), then stop.
Step 3. Find (s^k, ζ^k, η^k) as a KKT triplet of problem (21).
Step 4. Set $x^{k+1} = x^k + s^k$; $\lambda^{k+1} = \zeta^k$; $\mu^{k+1} = \eta^k$, $k = k + 1$ and go to Step 2.

The equivalence between the SQP method and the Newton–Lagrange method provides the basis for the main convergence and rate of convergence results of the former.

Proposition 2. *Assume that* f, g, h *are twice differentiable with Lipschitz continuous second derivatives; assume that* (x^*, λ^*, μ^*) *satisfy the SOSC of Proposition 4 of Chapter 5.1 under strict complementarity between* $g(x^*)$ *and* λ^**. Then, an open neighborhood* $\mathcal{B}_{(x^*,\lambda^*,\mu^*)}$ *exists such that, if* $(x^0, \lambda^0, \mu^0) \in \mathcal{B}_{(x^*,\lambda^*,\mu^*)}$*, the sequence* $\{(x^k, \lambda^k, \mu^k)\}$ *produced by the SQP method is well defined and converges to* (x^*, λ^*, μ^*) *with quadratic rate.*

The SQP method outlined above requires the computation of the second-order derivatives that appear in $\nabla_x^2 L(x^k, \lambda^k, \mu^k)$. As in the Newton's method in unconstrained optimization, this can be avoided by substituting the Hessian matrix $\nabla_x^2 L(x^k, \lambda^k, \mu^k)$ with a quasi-Newton approximation B^k, which is updated at each iteration. An obvious update strategy would be to define:

$$\delta^k = x^{k+1} - x^k, \quad \gamma^k = \nabla L(x^{k+1}, \lambda^k, \mu^k) - \nabla L(x^k, \lambda^k, \mu^k),$$

and to update the matrix B^k by using the direct BFGS formula

$$B^{k+1} = B^k + \frac{\gamma^k (\gamma^k)'}{(\gamma^k)' \delta^k} - \frac{B^k \delta^k (\delta^k)' B^k}{(\delta^k)' B^k \delta^k} \qquad (22)$$

for unconstrained optimization. However, one of the properties that make the BFGS formula appealing for unconstrained problems, the maintenance of positive definiteness of B^k is no longer assured since $\nabla_x^2 L(x^*, \lambda^*, \mu^*)$ is usually positive definite only on a subset of $\mathbb{R}^n$. This difficulty may be overcome by modifying γ^k as proposed by Powell (1978b). Let:

$$\tilde{\gamma}^k = \tau \gamma^k + (1 - \tau) B^k \delta^k,$$

where the scalar τ is defined as

$$\tau = \begin{cases} 1 & \text{if } (\delta^k)' \gamma^k \geqslant 0.2 (\delta^k)' B^k \delta^k \\ 0.8 \dfrac{(\delta^k)' B^k \delta^k}{(\delta^k)' B^k \delta^k - (\delta^k)' \gamma^k} & \text{if } (\delta^k)' \gamma^k < 0.2 (\delta^k)' B^k \delta^k, \end{cases}$$

and update B^k by the damped BFGS formula, obtained by substituting $\tilde{\gamma}^k$ for γ^k in (22); then, if B^k is positive definite the same holds for B^{k+1}. This also ensures that if the quadratic programming problem is feasible, it admits a solution. Of course, by employing the quasi-Newton approximation, the rate of convergence becomes superlinear.

Proposition 2 states a local convergence result. If, as often happens, a point $(x^0, \lambda^0, \mu^0) \in \mathcal{B}_{(x^*,\lambda^*,\mu^*)}$ is not known, the following difficulties have to be tackled:

- at iteration k the QP subproblem may have no solution;
- the sequence $\{(x^k, \lambda^k, \mu^k)\}$ may not be convergent.

The first difficulty arises either because the quadratic objective function is unbounded from below on the linearized feasible set or because the linearized feasible set is empty. The first occurrence can be avoided by substituting $\nabla_x^2 L(x^k, \lambda^k, \mu^k)$ with a positive definite approximation B^k, for instance, as described before; or by employing the Hessian of the augmented Lagrangian $\nabla_x^2 L_a(x^k, \lambda^k, \mu^k; \epsilon)$ in place of $\nabla_x^2 L(x^k, \lambda^k, \mu^k)$. The second occurrence can be overcome by defining a suitable relaxation of the QP subproblem so as to obtain a problem that is always feasible, even if of larger dimension. As an example, the following subproblem has been proposed (Pantoja and Mayne, 1991):

$$\begin{aligned} \min_{s\in\mathbb{R}^n, w\in\mathbb{R}} \quad & \tfrac{1}{2} s' B^k s + \nabla f(x^k)' s + \frac{1}{\varepsilon} w \\ & \nabla g_i(x^k)' s + g_i(x^k) \leqslant w, \quad i = 1, \ldots, p \\ & | \nabla h_j(x^k)' s + h_j(x^k) | \leqslant w, \quad j = 1, \ldots, m \\ & w \geqslant 0, \end{aligned} \tag{23}$$

where $\varepsilon > 0$ is a suitable penalty parameter. It is easy to verify that the feasible set of Equation (23) is not empty, so that if B^k is positive definite, the problem admits a solution.

With respect to the convergence of the sequence $\{(x^k, \lambda^k, \mu^k)\}$, we observe preliminarily that if $\{x^k\}$ converges to some x^* where the linear independence constraints qualification (LICQ) holds, then $\{(\lambda^k, \mu^k)\}$ also converges to some (λ^*, μ^*) such that (x^*, λ^*, μ^*) satisfies the KKT NOCs for Equation (2). In order to enforce the convergence of the sequence $\{x^k\}$, in the same line of the Newton's method for unconstrained optimization, two strategies have been devised with reference to a suitable merit function associated with the NLP problem: the line-search approach and the trust-region approach. As said before, a merit function aims to reduce both the objective function and the constraint violations (as well as the violation of other necessary optimality conditions); in this way it provides a measure of the progress toward a solution of the constrained problem.

4.1. The Line-Search Approach

In the line-search approach the sequence $\{x^k\}$ is given by the iteration:

$$x^{k+1} = x^k + \alpha^k s^k,$$

where the direction s^k is obtained by solving the kth QP subproblem, and the step size α^k is evaluated so as to get a sufficient decrease in the merit function. A merit function widely employed in SQP methods is the ℓ_1 exact penalty function $L_1(x; \epsilon)$ given by (12). Although L_1 is not differentiable everywhere, it always has a directional derivative. It can be shown that, if the quadratic form in the QP subproblem is positive definite, and if the penalty parameter ϵ in L_1 is sufficiently small, then s^k is a descent direction at x^k for L_1, so that an Armijo-like line search along s^k can be performed (Han, 1977). However, due to its nondifferentiability, the merit function L_1 suffers from a major drawback known as the Maratos effect (Maratos, 1978): even close to a solution, the step size $\alpha^k = 1$ may not be accepted, thus deteriorating the convergence rate of the pure SQP method. To avoid the Maratos effect, different techniques have been proposed. Among them a curvilinear line search (Mayne and Polak, 1982), based on the formula $x^{k+1} = x^k + \alpha^k s^k + (\alpha^k)^2 \tilde{s}^k$, where the direction $\tilde{s}^k$, provides a second-order correction, is evaluated so as to further reduce the constraint violations; and the watchdog technique (Chamberlain et al., 1982), based on nonmonotone strategies, by which we allow the merit function L_1 to increase on certain iterations.

The Maratos effect can also be overcome by employing merit functions that are continuously differentiable. For equality constraints, Fletcher's exact penalty function U_F given by (13) is usually used (Powell and Yuan, 1986); the extension to inequality constraints is based on the exact penalty function U given in (14) (Di Pillo et al., 1992; Facchinei, 1997). It can be shown that for a sufficiently small value of the penalty parameter ϵ, s^k gives a descent direction for the continuously differentiable exact penalty functions at x^k, and eventually the step size $\alpha = 1$ is accepted, so that it is possible to conciliate global convergence and superlinear convergence rate. Continuously differentiable exact augmented Lagrangian functions have also been studied as merit functions in SQP methods, with similar results (Dixon, 1979; Lucidi, 1990). For instance, when ϵ is small, the direction given by $(s^k, (\eta^k - \mu^k))$ is a descent direction at (x^k, μ^k) for the function $S_a(x, \mu; \epsilon)$ given by (15). Continuously differentiable exact merit functions are very attractive for their theoretical properties, but have not yet been widely used, mainly because of the fact that the computations needed are rather cumbersome.

We mention that the simple augmented Lagrangian function has also been used as a merit function; however, as is shown, for instance, from Proposition 1, the augmented Lagrangian function is an exact merit function only if proper values of the multipliers are selected. This fact makes arguments on global convergence questionable.

4.2. The Trust-Region Approach

As in the unconstrained case, the trust-region approach consists of introducing a trust-region constraint in the quadratic subproblem that bounds the region where the subproblem is trusted to represent well the original constrained problem:

$$\begin{aligned} \min_{s \in \mathbb{R}^n} \quad & \tfrac{1}{2} s' \nabla_x^2 L(x^k, \lambda^k, \mu^k) s + \nabla f(x^k)' s \\ & \nabla g(x^k)' s + g(x^k) \leqslant 0 \\ & \nabla h(x^k)' s + h(x^k) = 0 \\ & \|s\|^2 \leqslant (a^k)^2. \end{aligned} \tag{24}$$

However, due to the presence of the linear constraints, the solution of the trust-region subproblem (Equation (24)) is much more difficult than in the unconstrained case, and may not even exist at all. A possibility, proposed by Fletcher (1981), is to take into account the linear constraints by ℓ_1 penalty terms, and define the trust region by means of the ℓ_∞ norm of s; thus the problem becomes:

$$\begin{aligned} \min_{s \in \mathbb{R}^n} \quad & \tfrac{1}{2} s' \nabla_x^2 L(x^k, \lambda^k, \mu^k) s + \nabla f(x^k)' s \\ & + \frac{1}{\epsilon^k} \sum_{i=1}^{p} \max\{\nabla g_i(x^k)' s + g_i(x^k), 0\} \\ & + \frac{1}{\epsilon^k} \sum_{j=1}^{m} | \nabla h_j(x^k)' s + h_j(x^k) | \\ & -a^k \leqslant s \leqslant a^k. \end{aligned} \tag{25}$$

It is easily seen that Equation (25) always admits a solution s^k, even when $\nabla_x^2 L(x^k, \lambda^k, \mu^k)$ is substituted by some approximation to avoid the computation of second-order derivatives. Moreover, by suitable manipulations, Equation (25) can be put in the form of a QP problem that can be solved by standard algorithms.

The $L_1(x; \epsilon)$ exact penalty function is used as a merit function. The objective function of Equation (25) is used as an approximation of the L_1 function in the trust region. Similarly to the unconstrained case, the step s^k is accepted if the ratio between the actual and the predicted reduction in the L_1 function is not too small; otherwise the trust-region radius a^k is decreased and the problem is re-solved to obtain a new candidate step. Once $x^{k+1} = x^k + s^k$ has been determined, some multiplier updating formula is employed to get the new multiplier estimates $(\lambda^{k+1}, \mu^{k+1})$. Moreover some updating rule for the penalty parameter is needed to define the next subproblem in (25). The resulting algorithm is known as the SL_1QP algorithm. The SL_1QP algorithm has good global convergence properties; however, it also suffers from the Maratos effect, which can be overcome by techniques similar to the ones described for the L_1 merit function in the line-search approach.

Finally, we mention feasible sequential quadratic programming algorithms, which force all iterates to be feasible. Feasibility may be required when the objective function f is not defined outside the feasible set, or when termination at an infeasible point (which may happen in practice with most algorithms) is undesirable. To maintain feasibility while retaining fast local convergence properties, the step is obtained along a direction that combines the QP direction, a direction that points to the interior of the feasible set, and possibly a second-order correction direction (Paniers and Tits, 1993). Of course feasible SQP algorithms are more computationally expensive than standard SQP algorithms, but they have the additional advantage that the objective function f itself can be used as a merit function.

5. Feasible Direction Methods

For NLP problems with only linear constraints, several methods can be devised based on search directions d^k that are both of descent for the objective function and feasible for the constraints, that is, directions such that, if $x^k \in \mathcal{F}$ is not a minimizer, then not only $f(x^k + \alpha d^k) < f(x^k)$, but also $x^k + \alpha d^k \in \mathcal{F}$ for small-enough values of the step size α. For instance, the feasible direction method of Zoutendijk (1960), the gradient projection method of Rosen (1960, 1961), the reduced gradient method of Wolfe (1963), and the more recent interior-points methods (see Ye, 1997).

The topic of linearly constrained NLP problems is highly specialized and strictly related to quadratic programming, which just represents a particular case. Even a short review would exceed the space allowed for this article. For a comprehensive treatment we refer the reader to Chapter 10 of Bazaraa et al. (1993).

Here we mention that many attempts have been made to generalize feasible direction methods to problems with general constraints, usually by linearizing the constraints in the neighborhood of the current point x^k. However, in doing so, the difficulty arises that, if x^{k+1} is feasible for the linearized constraints, it may not be feasible for the original constraints, especially when nonlinear equality constraints are present. In this case the direction d^k may be tangent to some constraints and an additional correction step must be evaluated so as to restore feasibility while retaining a decrease in the objective function; how to perform this restoration step is the critical issue in feasible direction methods for general NLP problems.

We shall briefly describe the generalized reduced gradient method only (Abadie and Carpentier, 1969), since it is implemented in some currently available codes for NLP.

In this context, it is usual to consider an NLP problem in the form

$$\begin{aligned} \min_{x \in \mathbb{R}^n} \quad & f(x) \\ & h(x) = 0 \\ & l \leqslant x, \end{aligned} \tag{26}$$

where some components of the n-vector l can be $-\infty$. Equation (2) can be put in the form (26) by introducing linear slack variables y_i, so that the inequality constraint $g_i(x) \leqslant 0$ is substituted by the equality constraint $g_i(x) + y_i = 0$ plus the bound constraint $y_i \geqslant 0$.

Assume that x^k is a given feasible point. Under the LICQ assumption and after some reordering, we can partition x^k as $[(x_B^k)' \mid (x_N^k)']'$, where $x_B^k \in \mathbb{R}^m$ is a subset of basic variables with the property that the gradient matrix $\nabla_{x_B} h(x^k)$ is nonsingular, and $x_N^k \in \mathbb{R}^{n-m}$ is a subset of nonbasic variables. Accordingly l is partitioned into $[(l_B)' \mid (l_N)']'$ and for simplicity, the nondegeneracy assumption that $l_B < x_B^k$ is made. By the implicit function theorem, there is a neighborhood $\mathcal{B}_{x_N^k} \subseteq \mathbb{R}^{n-m}$ of x_N^k and a continuously differentiable function $z^k : \mathbb{R}^{n-m} \to \mathbb{R}^m$ such that $x_B^k = z^k(x_N^k)$ and $h(z^k(x_N), x_N) = 0$ for all $x_N \in \mathcal{B}_{x_N^k}$. Then, locally in a neighborhood of x^k, Equation (26) can be rewritten as

$$\begin{aligned} \min_{x_N \in \mathbb{R}^{n-m}} \quad & f(z^k(x_N), x_N) \\ & l_N \leqslant x_N \\ & l_B \leqslant z^k(x_N). \end{aligned} \tag{27}$$

By the nondegeneracy assumption, the last bounds are nonactive at x_N^k. Hence a search direction $d_N^k \in \mathbb{R}^{n-m}$ at x_N^k can be determined by using algorithms for simple bound constrained problems, making use of the reduced gradient formula:

$$\nabla_{x_N} f[z^k(x_N), x_N] = \nabla_{x_N} f(x) - (\nabla_{x_N} h(x) \nabla_{x_B} h(x)^{-1}) \nabla_{x_B} f(x).$$

Algorithms for bound constrained problems are highly specialized and can be obtained by adapting the unconstrained methods of Chapter 5.2 of this handbook.

Given x_N^k and d_N^k, a line-search procedure that uses the objective function f as a merit function is adopted. Letting $x_N(\alpha) = x_N^k + \alpha d_N^k$ in the line search f needs to be evaluated at $(z^k(x_N(\alpha)), x_N(\alpha))$. To this aim the value of $z^k(x_N(\alpha))$ is obtained by applying Newton's method to solve, with respect to x_B, the constraint $h(x_B, x_N) = 0$; in doing this, care must also be taken that the bound constraint on x_B is not violated. In this way, feasibility of the successive iterates is maintained.

It is easily understood that the convergence arguments of feasible direction methods for general NLP problems are quite involved.

Finally, we point out that feasible direction methods require that a feasible starting point is available, and this is not always easy for general nonlinear constraints; on the other hand, since the feasibility subproblem is considered solved, these methods do not rely on merit functions depending on penalty parameters.

6. Selected Bibliography

We refer to the Selected Bibliography in Chapter 5.1 for a list of textbooks of general reference that develop the theory and practice of NLP. In the present article, several references have been given in correspondence to particular subjects. Here we add a list of some books and surveys of specific concern to constrained NLP.

The sequential penalty approach, both exterior and interior, is fully developed in the classical book by Fiacco and McCormick (1968) that has been recently revisited by Nash (1993). For interior-point methods related to the logarithmic barrier approach, and specialized for linear and quadratic programming problems, we refer the reader to Hertog (1994), Roos et al. (1997), Ye (1997), Wright (1997), and the recent survey by Freund and Mizuno (2000); a particular emphasis on theoretical aspects is given in Nesterov and Nemirovskii (1994). For the augmented Lagrangian method the basic reference is the book by Bertsekas (1982). Exact penalty methods have been surveyed by Di Pillo (1994). A survey on penalty, exact penalty, and augmented Lagrangian methods is given in Boukary and Fiacco (1995). Sequential quadratic programming has been surveyed by Boggs and Tolle (1996) and is extensively treated in Conn et al. (2000). Particular attention to feasible direction methods is given in the book by Bazaraa et al. (1993). Algorithms for large-scale problems are surveyed in Conn et al. (1994).

References

Abadie, J. and J. Carpentier. 1969. "Generalization of the Wolfe reduced gradient method to the case of nonlinear constraints." In *Optimization*, edited by R. E. Fletcher, pp. 37–47. London: Academic Press.

Bazaraa, M. S., H. D. Sherali, and C. M. Shetty. 1993. In *Nonlinear Programming—Theory and Algorithms*. 2d ed. New York: John Wiley & Sons.

Bertsekas, D. P. 1982. *Constrained Optimization and Lagrange Multiplier Methods*. New York: Academic Press

Boggs, P. T. and J. W. Tolle. 1996. "Sequential quadratic programming." *Acta Numerica* 4: 1–51.

Boukary, D. and A. V. Fiacco. 1995. "Survey of penalty, exact-penalty and multiplier methods from 1968 to 1993." *Optimization* **32**: 301–334.

Chamberlain, R., C. Lemarechal, H. C. Pedersen, and M. J. D. Powell. 1982. "The watchdog technique for forcing convergence in algorithms for constrained optimization." *Math. Programming Study* **16**: 1–17.

Conn, A. R., N. Gould, and P. L. Toint. 1994. "Large-scale nonlinear constrained optimization: a current survey." In *Algorithms for Continuous Optimization: The State of the Art*, edited by E. Spedicato, pp. 287–332. Dordrecht: Kluwer Academic Publishers.

Conn, A. R., N. I. M. Gould, and Ph. L. Toint. 2000. *Trust Region Methods*. Philadelphia: SIAM.

Courant, R. 1943. "Variational methods for the solution of problems with equilibrium and vibration." *Bull. Amer. Math. Soc.* **49**: 1–23.

Di Pillo, G. 1994. "Exact penalty methods." In *Algorithms for Continuous Optimization: The State of the Art*, edited by E. Spedicato, pp. 1–45. Dordrecht: Kluwer Academic Publishers.

Di Pillo, G., F. Facchinei, and L. Grippo. 1992. "An RQP algorithm using a differentiable exact penalty function for inequality constrained problems." *Math. Programming* **55**: 49–68.

Di Pillo, G. and L. Grippo. 1979. "A new class of augmented Lagrangians in nonlinear programming." *SIAM J. Control Optim.* **17**: 618–628.

Di Pillo, G. and L. Grippo. 1985. "A continuously differentiable exact penalty function for nonlinear programming problems with inequality constraints." *SIAM J. Control Optim.* **23**: 72–84.

Di Pillo, G. and L. Grippo. 1989. "Exact penalty functions in constrained optimization." *SIAM J. Control Optim.* **27**: 1333–1360.

Di Pillo, G., L. Grippo, and F. Lampariello. 1980. "A method for solving equality constrained optimization problems by unconstrained minimization." In *Optimization Techniques*, edited by K. Iracki, K. Malanowski, and S. Walukiewicz. pp. 96–105. Berlin: Springer-Verlag.

Di Pillo, G., L. Grippo, and F. Lampariello. 1982. "A class of methods for the solution of optimization problems with inequalities." In *System Modelling and Optimization*, edited by R. F. Drenick and F. Kozin. pp. 508–519. Berlin: Springer-Verlag.

Di Pillo, G., L. Grippo, and S. Lucidi. 1986. "Globally convergent exact penalty algorithms for constrained optimization." In *System Modelling and Optimization*, edited by A. Prekopa, J. Szelezsan, and B. Strazicky. pp. 694–703. Berlin: Springer-Verlag.

Dixon, L. 1979. "Exact penalty function methods in nonlinear programming." Technical Report 103. Numerical Optimization Centre, Hatfield Polytechnic.

Facchinei, F. 1997. "Robust quadratic programming algorithm model with global and superlinear convergence properties." *J Optim. Theory Appl.* **92**: 543–579.

Fiacco, A. V. and G. P. McCormick. 1968. *Nonlinear Programming: Sequential Unconstrained Minimization Techniques*. New York: John Wiley & Sons.

Fletcher, R. E. 1970. "A class of methods for nonlinear programming with termination and convergence properties." In *Integer and Nonlinear Programming*, edited by J. Abadie, pp. 157–173. Amsterdam: North-Holland.

Fletcher, R. E. 1981. "Numerical experiments with an exact l_1 penalty function method." In *Nonlinear Programming*, edited by O. L. Mangasarian, R. Meyer, and S. M. Robinson **4**: 99–129. New York: Academic Press.

Freund, R. M. and S. Mizuno. 2000. "Interior point methods: current status and future directions." In *High Performance Optimization*, edited by H. Frenk, K. Roos, T. Terlaky, and S. Zhang, pp. 441–466. Dordrecht: Kluwer Academic Publishers.

Frisch, K. R. 1955. "The logarithmic potential method of convex programming." Memorandum of May 13, 1955. Oslo: University Institute of Economics.

Garcia-Palomares, U. M. and O. L. Mangasarian. 1976. "Superlinearly convergent quasi-Newton methods for nonlinearly constrained optimization problems." *Math. Programming* **11**: 11–13.

Glad, T. and E. Polak. 1979. "A multiplier method with automatic limitation of penalty growth." *Math. Programming* **17**: 140–155.

Han, S. P. 1976. "Superlinearly convergent variable metric algorithms for general nonlinear programming problems." *Math. Programming* **11**: 263–282.

Han, S. P. 1977. "A globally convergent method for nonlinear programming." *J. Optim. Theory Appl.* **22**: 297–309.

Hertog, D. den 1994. *Interior Point Approach to Linear, Quadratic and Convex Programming*. Dordrecht: Kluwer Academic Publishers.

Hestenes, M. R. 1969. "Multiplier and gradient methods." *J. Optim. Theory Appl.* **4**: 303–320.

Lucidi, S. 1990. "Recursive quadratic programming algorithm that uses an exact augmented Lagrangian function." *J. Optim. Theory Appl.* **67**: 227–245.

Maratos, N. 1978. "Exact penalty function algorithms for finite dimensional and control optimization problems." Ph.D. diss. London: University of London.

Mayne, D. Q. and E. Polak. 1982. "A superlinearly convergent algorithm for constrained optimization problems." *Math. Programming Study* **16**: 45–61.

McCormick, G. P. 1983. In *Nonlinear Programming—Theory, Algorithms, and Applications*: John Wiley & Sons.

Miele, A., E. E. Cragg, R. R. Iyer, and A. V. Levy. 1971. "Use of the augmented penalty function in mathematical programming, Part I and II." *J. Optim. Theory Appl.* **8**: 115–130 and 131–153.

Miele, A., P. Moseley, A. V. Levy, and G. H. Coggins. 1972. "On the method of multipliers for mathematical programming problems." *J. Optim. Theory Appl.* **10**: 1–33.

Nash, S. G. 1998. "SUMT (revisited)." *Oper. Res.* **46**: 763–775.

Nesterov, Y. and A. Nemirovskii. 1994. In *Interior-point Polynomial Algorithms in Convex Programming*. Philadelphia: SIAM.

Paniers, E. R. and A. L. Tits. 1993. "On combining feasibility, descent and superlinear convergence in inequality constrained optimization." *Math. Programming* **59**: 261–276

Pantoja De O., J. F. A. and D. Q. Mayne. 1991. "Exact penalty function algorithm with simple updating of the penalty parameter." *J. Optim. Theory Appl.* **69**: 441–467.

Powell, M. J. D. 1969. "A method for nonlinear constraints in minimization problems." In *Optimization*, edited by R. E. Fletcher, pp. 283–298. New York: Academic Press.

Powell, M. J. D. 1978a. "Algorithms for nonlinear constraints that use Lagrangian functions." *Math. Programming* **14**: 224–248.

Powell, M. J. D. 1978b. "The convergence of variable metric methods for nonlinear constrained optimization calculation." In *Nonlinear Programming 3*, edited by O. L. Mangasarian, R. Meyer, and S. M. Robinson. New York: Academic Press.

Powell, M. J. D. 1978c. "A fast algorithm for nonlinearly constrained optimization calculations." In *Numerical Analysis, Dundee 1977*, edited by G. A. Watson, pp. 144–157. Berlin: Springer-Verlag.

Powell, M. J. D. and Y. Yuan. 1986. "A recursive quadratic programming algorithm that uses differentiable exact penalty functions." *Math. Programming* **35**: 265–278.

Rockafellar, R. T. 1973. "The multiplier method of Hestenes and Powell applied to convex programming." *J. Optim. Theory Appl.* **12**: 555–562.

Roos, C., T. Terlaky, and J.-Ph. Vial 1997. *Theory and Algorithms for Linear Optimization: An Interior Point Approach*. New York: John Wiley & Sons.

Rosen, J. B. 1960. "The gradient projection method for nonlinear programming, Part I: linear constraints." *SIAM J. Appl. Math.* **8**: 181–217.

Rosen, J. B. 1961. "The gradient projection method for nonlinear programming, Part II: nonlinear constraints." *SIAM J. Appl. Math.* **9**: 514–553.

Wilson, R. B. 1963. A simplicial algorithm for concave programming, Ph.D. diss. Harvard University.

Wolfe, P. 1963. "Methods of nonlinear programming." In *Recent Advances in Mathematical Programming*, edited by R. L. Graves and P. Wolfe, pp. 67–86. New York: McGraw-Hill.

Wright, S. J. 1997. *Primal–Dual Interior-Point Methods*. Philadelphia: SIAM.

Ye, Y. 1997. *Interior Point Algorithms. Theory and Analysis*. New York: Wiley-Interscience.

Zangwill, W. I. 1967. "Nonlinear programming via penalty functions." *Mgmnt Sci.* **13**: 344–358.

Zoutendijk, G. 1960. *Methods of Feasible Directions*. Amsterdam: Elsevier.

5.4 Nonsmooth Optimization

Manlio Gaudioso

ABSTRACT We survey some of the most recognized proposals in the area of numerical methods for minimizing convex functions that are not necessarily differentiable. The approach is aimed at exposing the ideas at the basis of the methods, without discussing the complex technicalities that the interested reader may find in the extremely rich literature of the sector. We focus on the algorithms connected to the cutting-plane idea (bundle methods, methods of centers and the proximal point algorithm), on the finite min-max problem, and on some auxiliary problems relevant from a practical point of view.

1. Introduction

Nondifferentiable or, equivalently, nonsmooth optimization deals with the problem of finding the minima (or the maxima) of functions that exhibit "kinks," that is, points at which the derivatives do not exist.

In practical situations it often happens that a minimum (or a maximum) coincides with a kink and thus the ordinary optimality conditions of differential calculus do not hold. Moreover, even when the minimum (or maximum) is located in a region where the function is differentiable, the presence of kinks can prevent standard optimization methods from convergence.

In general, a considerably greater computational effort is needed for minimizing non-differentiable functions than for tackling smooth problems.

This is mainly due to the search direction finding, a step that is more complex (either in the presence or absence of line search) in nonsmooth optimization since, in general, it requires the solution of an auxiliary optimization problem, often of the quadratic programming type.

The area has grown in connection with the development of convex analysis (Rockafellar, 1970). Most of the research has focused on the numerical algorithms for the unconstrained minimization of convex functions. On the other hand, several methods can be generalized to approach constrained minimization problems as well as to deal with nonconvex objective functions.

The early methods, mainly proposed in the 1960s by scientists coming from the Soviet Union, are known as subgradient-type and space dilatation methods, the latter being strongly connected to the ellipsoid method. The interested reader can find detailed descriptions in Shor (1998) and Polyak (1987).

As for the methods developed mainly in the West, from the 1970s, a really exhaustive review is provided by Hiriart-Urruty and Lemaréchal (1993). A shorter but stimulating survey is in Lemaréchal (1989).

In this article, we present some of the main research streams in the area of numerical methods for minimizing nondifferentiable functions. The article is organized as follows. In Sections 2 and 3 we discuss the basic cutting-plane approach and the methods of centers for dealing with convex functions, respectively. In Section 4 we discuss finite min-max problems (not necessarily convex) in a context derived from the exact penalty function approach for nonlinear programming. In Section 5 we describe proximal-point methods and its variants, which are strictly connected to bundle methods. Finally in Section 6 we briefly discuss some auxiliary problems of the quadratic programming type, which arise as subproblems in many nondifferentiable optimization algorithms.

Throughout the paper we adopt the following notation. We denote by x^* and f^*, respectively, an optimal point and the optimal value of a function $f : \mathbb{R}^n \mapsto \mathbb{R}$, by $\|\cdot\|$ the Euclidean norm in $\mathbb{R}^n$ and by $a^T b$ the inner product of the vectors a and b. By $x_i \in \mathbb{R}^n$ we indicate the estimate of the minimizer available at iteration i of any optimization algorithm. The subdifferential of a convex function f at any point x is denoted by $\partial f(x)$. We recall that the subdifferential of a convex function f at the point x is the set of subgradients of f at x, that is, the set of vectors $g \in \mathbb{R}^n$ satisfying the subgradient inequality:

$$f(y) \geqslant f(x) + g^T(y - x) \quad \forall y \in \mathbb{R}^n.$$

At any point where the function is differentiable, the subdifferential reduces to a singleton and its unique element is the ordinary gradient.

The epigraph of a convex function $f : \mathbb{R}^n \mapsto \mathbb{R}$, indicated by epi$f$, is the set of $\mathbb{R}^{n+1}$ defined as follows:

$$\text{epi}f = \{(x, v) \mid x \in \mathbb{R}^n, \quad v \in \mathbb{R}, \quad v \geqslant f(x)\}.$$

By bundle we indicate the information available at a certain iteration (for example, k) of any algorithm. In particular, the bundle is the set of couples (x_i, g_i), $g_i \in \partial f(x_i)$ for $i = 1, \ldots, k$ or a subset of such couples indexed by $I_k \subset \{1, \ldots, k\}$ in case deletion is allowed. Often, to make it possible to implement the algorithms with limited storage availability, the bundle is made up of couples obtained via appropriate aggregation procedures (Kiwiel, 1983) of the couples (x_i, g_i), for $i = 1, \ldots, k$. Also in this case we refer to I_k as to the bundle index set.

2. The Cutting Plane

The cutting-plane approach may be considered as the prototype algorithm among those working on the basis of a model not simply local to the objective function (e.g., the subgradient method).

The basic idea is to construct a sequence of piecewise affine functions, each of them minorizing the original convex function to be minimized. The sequence is such that the approximation provided by the affine functions becomes better and better. Thus, from the sequence of the minimizers of the approximating functions, it is possible to extract a minimizing sequence for the original function.

Consider the problem of minimizing a convex function

$$\min_{x \in Q} f(x)$$

where $f : \mathbb{R}^n \to \mathbb{R}$ and Q is a nonempty compact convex subset of $\mathbb{R}^n$. Assume that f^* is the optimal value of the above problem; assume also that L_Q is the Lipschitz constant of f on Q and that for each $x \in Q$ we can compute the objective function $f(x)$ and a subgradient $g \in \partial f(x)$. The cutting-plane model (Cheney and Goldstein, 1959; Kelley, 1960; Polyak, 1987) works as follows: at the k-th iteration f is approximated from below by the piecewise affine model:

$$f_k(x) = \max_{1 \leqslant i \leqslant k} \{f(x_i) + g_i^T(x - x_i)\},$$

where x_i, $i = 1, \ldots, k$ are the points previously obtained and $g_i \in \partial f(x_i)$, $i = 1, \ldots, k$. We observe in passing that the construction of a piecewise affine model is not the only possibility (see Gaudioso and Monaco, 1991, for the use of a piecewise quadratic model).

Of course, to define the model it is necessary to have a point $x_1 \in Q$ together with a subgradient $g_1 \in \partial f(x_1)$.

The point x_{k+1} is obtained by minimizing $f_k(x)$ over Q, that is,

$$x_{k+1} = \arg\min_{x \in Q} f_k(x) = \arg\min_{x \in Q} \max_{1 \leqslant i \leqslant k} \{f(x_i) + g_i^T(x - x_i)\}.$$

In the sequel we will refer to f_k as to the cutting-plane function at iteration k. We observe that the calculation of x_{k+1} requires one to solve a problem that is still nondifferentiable. Of course, it is expected to be easier than the original problem and in fact it reduces to the equivalent linear program:

$$\begin{aligned} &\min_{(x,v)} \quad v \\ &v \geqslant f(x_i) + g_i^T(x - x_i) \quad i = 1, \ldots, k. \end{aligned} \tag{1}$$

Observe that the transformation has required the introduction of the additional scalar variable v and consequently (1) is an optimization problem defined in $\mathbb{R}^{n+1}$ and no longer in $\mathbb{R}^n$. It will be apparent in the next section that reasoning in terms of sets in $\mathbb{R}^{n+1}$ instead of in terms of functions defined in $\mathbb{R}^n$ is particularly profitable in devising effective algorithms.

We note that, since by convexity $f_k(x) \leqslant f(x)$ $\forall x \in Q$, the optimal value f_k^* of $f_k(x)$ provides a lower bound on f^*. Moreover, since

$$f_{k+1}(x) = \max\{f_k(x), f(x_{k+1}) + g_{k-1}^T(x - x_{k+1})\} \geqslant f_k(x), \quad \forall x \in Q,$$

the sequence $\{f_k^*\}$ is monotonically nondecreasing and thus the quality of the lower bound becomes increasingly better.

We now state the cutting plane algorithm in a slightly more general setting than the usual one (see also Polyak, 1987).

The following input parameters are given:

- x_1: starting point, $g_1 \in \partial f(x_1)$;
- $\epsilon > 0$: stopping tolerance.

Cutting-Plane Algorithm

Step 1. Initialize $k = 1$.

Step 2. Calculate $x_{k+1} \in B_k$, where B_k is the nonempty set $\{x \mid x \in Q, f_k(x) \leqslant f^*\}$.

Step 3. If $f(x_{k+1}) - f_k(x_{k+1}) < \epsilon$ then stop. Else set $k = k + 1$ and return to Step 2.

The proof of the convergence of the algorithm is rather simple and relies on convexity of f and on compactness of Q.

Theorem 1. *The cutting-plane algorithm terminates at an ϵ-optimal point.*

Proof. *We observe first that, since $f_k(x_{k+1}) \leqslant f^*$, satisfaction of the stopping condition at Step 3 implies that*

$$f(x_{k+1}) - f^* < \epsilon$$

that is, that the point x_{k+1} is ϵ-optimal.

Now suppose that the stopping condition is never satisfied and consequently that

$$f(x_{k+1}) - f_k(x_{k+1}) \geqslant \epsilon$$

holds $\forall k$. Then, from convexity of f and from the definition of f_k, the following inequalities hold for $i = 1, \ldots, k$

$$f(x_i) \geqslant f(x_{k+1}) + g_{k+1}^T(x_i - x_{k+1}) \geqslant f_k(x_{k+1}) + \epsilon + g_{k+1}^T(x_i - x_{k+1}) \geqslant f(x_i) + g_i^T(x_{k+1} - x_i)$$
$$+ \epsilon + g_{k+1}^T(x_i - x_{k+1}),$$

which imply

$$0 \geqslant \epsilon - 2L_Q \|x_{k+1} - x_i\| \quad i = 1, \ldots, k. \tag{2}$$

A consequence of (2) is that all points generated by the algorithm are at a distance bounded away from zero, which is in contradiction with compactness of Q.

We remark that the standard cutting plane, where

$$x_{k+1} = \arg\min_{x \in Q} f_k(x),$$

is a possible implementation of Step 2 of the algorithm. However, it is important to note that this is not the only possible choice and that convergence is guaranteed, provided that x_{k+1} is sufficiently inside the level set of f_k.

The cutting-plane method, in its basic version, suffers from some relevant problems. Firstly we observe that the dimension of the linear program to be solved increases at each iteration as a consequence of the introduction of a new constraint. The convergence proof given above is in fact based on the hypothesis of infinite dimension storage, which is of course not possible to implement.

Another drawback comes from the intrinsic instability of the method. In fact, not only is monotonicity of the values $f(x_k)$ not ensured (this, however, would not in principle be unacceptable), but it may happen that, while points very close to the minimum have been attained at a certain stage of the algorithm, nevertheless points successively generated may be located very far from the minimum point. A description via some interesting numerical examples of such poor behavior may be found in Hiriart-Urruty and Lemaréchal (1993, XV.1).

The above observations have stimulated the search for effective developments stemming from the cutting-plane approach. In particular the whole family of bundle methods (Hiriart-Urruty and Lemaréchal, 1993) can be considered as an effective answer to the questions raised above. The evolution of bundle methods will be outlined in Section 5.

We note that a nested sets structure can be associated to the algorithm. In fact the epigraphs of the approximating functions are nested sets, that is,

$$\text{epi} f_{k+1} \subset \text{epi} f_k \quad \forall k.$$

Moreover

$$\text{epi} f_k \supseteq \text{epi} f \quad \forall k.$$

Thus the term "cutting plane" derives from the fact that, at each step, whenever a new affine piece is added to the previous ones in the definition of the new cutting-plane function, part of the epigraph of the previous approximation of f is cut. It is apparent that to get rapidly good approximations to the epigraph of f it is reasonable to proceed by introducing cuts that substantially reduce the epigraph. Such an observation is the basis of the approach described in the next section.

Finally we mention the possibility of conceiving a version of the cutting-plane approach suitable for parallel computation (Capalbo et al., 1998).

3. Methods of Centers

To describe some of the developments stemming from the cutting-plane idea, it is useful to assume a set-oriented point of view instead of a function-oriented one. In fact it is useful to introduce at iteration k the localization set $F_k(z_k) \subset R^{n+1}$:

$$F_k(z_k) = \{(x, v) \mid x \in Q, f_k(x) \leqslant v \leqslant z_k\},$$

where z_k is any upper bound on the optimal value of f_k, for example, the value of f calculated at any feasible point. The set $F_k(z_k)$ is contained in epif_k (it is obtained by horizontally cutting epif_k) and the point (x^*, f^*) belongs to $F_k(z_k)$.

Thus, a possible way to minimize f is to construct a nested sequence of sets $F_k(z_k)$ shrinking as fast as possible around the point (x^*, f^*).

Of course, aiming at obtaining substantial cuts in passing from $F_k(z_k)$ to $F_{k+1}(z_{k+1})$, it seems natural to look for a central cut, that is, a cut generated on the basis of some appropriate notion of the center of $F_k(z_k)$. Several proposals in this context can be found in the literature stemming from Levin's Center of Gravity method (Levin, 1965).

The approach known as the analytic center cutting-plane method (ACCPM; Goffin et al., 1992) has been shown to be particularly effective, based on the notion of an analytic center introduced in Sonnevend (1985) in connection with the success of the interior-point methods for linear programming. In the ACCPM the required central point of the localization set is calculated as the unique maximizer of the function

$$\psi_k(x, v) = \log(z_k - v) + \sum_{i=1}^{k} \log[v - f(x_i) - g_i^T(x - x_i)].$$

The techniques devised to solve linear programs by means of interior-point methods apply to such maximization (the objective function is a type of potential function).

Once the central point (x_{k+1}, v_{k-1}) has been obtained as an (approximate) maximizer of $\psi_k(x, v)$, a new cut is generated rooted at x_{k+1}, and the upper and lower bounds on the optimal objective function value are possibly updated. The stopping test is based on the gap between the upper and lower bound.

4. Finite Min–Max Problems

Many, and maybe most, of the nondifferentiable optimization problems of practical interest can be put in the finite min-max form:

$$\min_{x \in \mathbb{R}^n} f(x) = \min_{x \in \mathbb{R}^n} \max_{i=1,\ldots,m} \phi_i(x) \tag{3}$$

where the functions $\phi_i : \mathbb{R}^n \to \mathbb{R}$, $i = 1, \ldots, m$ are smooth and not necessarily convex.

Many problems of mathematical interest can be put in the above form. Moreover, many applications (e.g., engineering design problems based on a worst-case-analysis philosophy) give rise to finite min-max problems. From the historical point of view, they have represented a strong motivation for developing the theory of convex and, more generally, nonsmooth analysis.

Of course, the min-max problem can be approached by general-purpose nondifferentiable optimization methods. Nevertheless, its special structure has suggested the development of *ad hoc* algorithms. We cite here Demyanov and Malozemov (1974), Hald and Madsen (1981), and Polak et al. (1991).

The nonsmooth problem (Equation (3)) can be put in a smooth equivalent form. A well known transformation allows us to rewrite it as

$$\begin{aligned} &\min_{(x,v)\in\mathbb{R}^{n+1}} \quad v \\ &\phi_i(x) - v \leqslant 0 \quad i = 1, \dots, m. \end{aligned} \tag{4}$$

The above problem has been approached via smooth nonlinear programming techniques by several authors. In particular, in Polak et al. (1991) a barrier function approach has been adopted.

Here we focus on the method described in Di Pillo et al. (1993) and generalized in Di Pillo et al. (1997), which is based on the construction of a continuously differentiable exact penalty function, whose minimizers are solutions to Equation (4), for finite values of the penalty coefficient.

To this aim, it is necessary to define a continuously differentiable multiplier function that provides an estimate of the multiplier vector associated with Equation (4) as a function of the variable (x, v).

The Karush–Kuhn–Tucker conditions for Equation (4) are the following:

$$\begin{aligned} &\sum_{i=1}^{m} \lambda_i \nabla\phi_i(x) = 0 \\ &\sum_{i=1}^{m} \lambda_i = 1 \\ &(\phi_i(x) - v)\lambda_i = 0 \quad i = 1, \dots, m \\ &\lambda_i \geqslant 0 \quad i = 1, \dots, m \\ &v - \phi_i(x) \geqslant 0 \quad i = 1, \dots, m. \end{aligned} \tag{5}$$

Satisfaction of the Karush–Kuhn–Tucker conditions (5) at the point $(x, f(x))$ does not ensure, unless the functions $\phi_i(x)$ are convex, that the point x is a minimizer of f. In fact it only implies that x is a critical point for f, that is, a point at which the directional derivative of f is nonnegative in all directions. For example, for f defined in $\mathbb{R}$ as $f(x) = \max\{x, -x^2\}$ the point $x = 0$ is a critical point but not a minimizer, whereas for f defined as $f(x) = \max\{x, x^2\}$, the point $x = 0$ is both a critical point and a minimizer.

A point x is a critical point for f if, and only if, $\lambda \in \mathbb{R}^m$ exists such that:

$$\begin{aligned} &\sum_{i=1}^{m} \lambda_i \nabla\phi_i(x) = 0 \\ &\sum_{i=1}^{m} \lambda_i = 1 \\ &(\phi_i(x) - f(x))\lambda_i = 0 \quad i = 1, \dots, m \end{aligned} \tag{6}$$

$\lambda_i \geqslant 0 \quad i = 1, ..., m.$

The multiplier function associates to any point (x, v) a multiplier vector $\lambda(x, v) \in \mathbb{R}^m$ obtained as the unique minimizer of the quadratic function

$$\chi(\lambda) = \left\|\sum_{i=1}^{m} \lambda_i \nabla\phi_i(x)\right\|^2 + \sum_{i=1}^{m}((\phi_i(x) - v)\lambda_i)^2 + \left(\sum_{i=1}^{m} \lambda_i - 1\right)^2 + \gamma(x, v)\|\lambda\|^2,$$

where the term $\gamma(x, v)$, defined as

$$\gamma(x, v) = \sum_{i=1}^{m} [\max(0, \phi_i(x) - v)]^2,$$

is introduced to ensure positive definiteness of the Hessian of $\chi(\lambda)$ also at points (x, v) where some of the inequalities $\phi_i(x) - v \leqslant 0$ are violated. In fact, apart from the term depending on $\gamma(x, v)$, the function $\chi(\lambda)$ penalizes violation at the point (x, v) of those conditions among the Karush–Kuhn–Tucker necessary conditions that are in the form of equality.

The main property of the multiplier function is summarized in the following proposition (Di Pillo et al., 1993, Proposition 2.4).

Proposition 1. *Let* x^* *be a critical point for problem* f *and let* $v^* = f(x^*)$. *Assume that* λ^* *is the associated multiplier vector satisfying* (6). *Then the following holds:*

$$\lambda(x^*, v^*) = \lambda^*.$$

Once the multiplier function is defined, it is possible to introduce an augmented Lagrangian function where the role of the multiplier vector is played by the multiplier function $\lambda(x, v)$. Some assumptions about the structure of the problem are needed.

Assumption 1. A point x_a exists such that the set $\mathscr{C} = \{x \mid f(x) \leqslant f(x_a)\}$ is compact and there is also a point x_b such that $f(x_b) < f(x_a)$.

Assumption 2. For each $x \in \mathscr{C}$, the vectors

$$\begin{pmatrix} \nabla\phi_i(x) \\ -1 \end{pmatrix}, \quad i \in \mathscr{I}_A(x)$$

are linearly independent where $\mathscr{I}_A(x) = \{i \mid f(x) = \phi_i(x)\}$.

The augmented Lagrangian function $P(x, v; \epsilon)$ is defined as follows:

$$P(x, v; \epsilon) = v + \lambda(x, v)^T c(x, v; \epsilon) + \frac{1}{\epsilon} c(x, v; \epsilon)^T B(x)^{-1} c(x, v; \epsilon)$$

where $\epsilon > 0$, and

$$B(x) = \text{Diag}(b_i(x)), \; b_i(x) = f(x_a) - \phi_i(x)$$

$$c_i(x, v; \epsilon) = \phi_i(x) - v + y_i^2(x, v; \epsilon)$$

$$y_i(x, v; \epsilon) = \left\{-\min\left[0, \phi_i(x) - v + \tfrac{1}{2}\epsilon b_i(x)\lambda_i(x, v)\right]\right\}^{\frac{1}{2}}.$$

The penalty term contains weights, defined as shifted barrier terms, on the values of the functions $\phi_i(x)$.

The basic result described in Di Pillo et al. (1993) is the correspondence between local or global minimizers of the max function f and those of the continuously differentiable penalty function.

The following theorem holds

Theorem 2. *A positive threshold* ϵ^* *exists such that for* $\epsilon \in (0, \epsilon^*]$, *if* $(x_\epsilon, v_\epsilon) \in \mathscr{L}_\epsilon$ *is a local minimum point of* $P(x, v; \epsilon)$, *then* x_ϵ *is a local minimum of* f *and* $v_\epsilon = f(x_\epsilon)$ *where*

$$\mathscr{L}_\epsilon = \{(x, v) \mid P(x, v; \epsilon) \leqslant f(x_b)\}.$$

It is worth noting that, under some additional hypotheses, the opposite of the above theorem can be stated (Di Pillo et al., 1993, Theorem 4.4), which ensures correspondence between local minima of the min-max problem and those of the penalty function. On the basis of the approach described above, it is possible to construct an implementable algorithm whose global convergence toward critical points of the min-max problem is proved. We remark that standard library software for differentiable unconstrained minimization can be fruitfully used within the algorithm.

5. Proximal Methods and Variants

The ideas stemming from the cutting-plane approach have been successfully associated to those derived from the general proximal point or Moreau–Yosida regularization approach, giving rise to some of the most effective methods for convex minimization.

Given the problem

$$\min_{x \in \mathbb{R}^n} f(x),$$

where f is a closed proper convex function, the proximal-point algorithm (Rockafellar, 1976) generates a sequence of points

$$x_{k+1} = \arg \min_{x \in \mathbb{R}^n} f(x) + \tfrac{1}{2} u_k \|x - x_k\|^2 \tag{7}$$

starting from any $x_1 \in \mathbb{R}^n$ and by fixing the scalar weighting parameter $u_k > 0 \ \forall k$.

Of course, the calculation of x_{k+1} in the above schema is intolerably expensive. Thus the basic idea for constructing a method that can be implemented is to replace in (7) the original function f with the cutting-plane function f_k defined on the basis of the bundle, the currently available information on the behavior of f:

$$f_k(x) = \max_{i \in I_k} \ \{f(x_i) + g_i^T (x - x_i)\}.$$

More generally, a commonly adopted scheme is the following

$$x_{k+1} = \arg \min_{x \in \mathbb{R}^n} f_k(x) + \tfrac{1}{2} u_k \|x - y_k\|^2, \tag{8}$$

where the point y_k (the current stability center) has substituted x_k in the proximity term of formula (7). In particular, whenever y_k coincides with x_k, the stability center is selected as the last point obtained in the iterative procedure. On the other hand, different possibilities are viable, for example, to select y_k as the best point (in terms of objective function) obtained so far.

Variants of this approach have beeen successfully adopted by Kiwiel (1990) and Schramm and Zowe (1992). In particular in Kiwiel (1990), much attention is paid to devising heuristic techniques to update the weighting parameter u_k.

The introduction of the proximal-point approach has generated some interesting perspectives, mainly connected to the possibility of replacing the Euclidean norm in the proximity term of (8) by any ellyptic norm, giving rise to methods resembling the variable metric ones of differentiable optimization. An interesting analysis may be found in Bonnans et al. (1995) and in Lemaréchal and Sagastizábal (1997), where the authors start from Moreau–Yosida regularization f^p of function f defined as follows:

$$f^p(x) = \min_{y \in \mathbb{R}^n} f(y) + \tfrac{1}{2}(y - x)^T M (y - x) \tag{9}$$

where M is any symmetric positive definite matrix.

The main properties of f^p are differentiability and the fact that f^p retains the same set of minima as f.

The basic idea is to apply a variable metric approach to minimize f^p instead of dealing directly with the minimization of the nondifferentiable function f.

To describe the approach, let us introduce the following notation for the proximal point, that is, the minimizer of the function at the right-hand side of (9):

$$p_M(x) = \arg\min_{y \in \mathbb{R}^n} f(y) + \tfrac{1}{2}(y-x)^T M(y-x). \tag{10}$$

It can be proved (Hiriart-Urruty and Lemaréchal, 1993, XV.4) that the gradient of $f^p(x)$ has the following expression:

$$\nabla f^p(x) = M[x - p_M(x)].$$

Thus, the possibility of adopting a quasi-Newton algorithm to minimize f^p appears viable, at least theoretically.

A possible scheme for a generic iteration of a quasi-Newton algorithm is the following, where the point x_k is assumed to be the current estimate of the minimum and M_k is any symmetric positive definite matrix intended as an approximation to the Hessian of $f^p(x_k)$.

Quasi-Newton algorithm

Step 1. Calculate $p_M(x_k)$ and $\nabla f^p(x_k) = M(x_k - p_M(x_k))$.

Step 2. Perform a quasi-Newton step, letting

$$x_{k+1} = x_k - M_k^{-1}\nabla f^p(x_k) = x_k - M_k^{-1}M(x_k - p_M(x_k)).$$

Step 3. Compute $p_M(x_{k+1})$ and obtain $\nabla f^p(x_{k+1})$.

Step 4. Update M_k, getting M_{k+1}.

We remark that the update at Step 4 may be performed by means of any quasi-Newton formula based on calculation of the two vectors

$$\delta_k = x_{k+1} - x_k, \quad \gamma_k = \nabla f^p(x_{k+1}) - \nabla f^p(x_k).$$

(In fact M_{k+1} satisfies the quasi-Newton equation $M_{k+1}\delta_k = \gamma_k$.)

We observe that the algorithm with $M_k = M\ \forall k$ reduces to the proximal-point algorithm.

It must be noted that the computation of the proximal point is particularly expensive. Consequently the approach pursued in Lemaréchal and Sagastizábal (1997) is to replace function f by the current cutting-plane function f_k, with the aim at merging the implementable bundling mechanism with the quasi-Newton philosophy. Technical variants of this approach are described in Fukushima and Qi (1996), Chen and Fukushima (1999), and Mifflin (1996).

In adopting the proximal-point or the Moreau–Yosida regularization approach, one might be moved by several motivations, mainly that of keeping some kind of proximity control in order to prevent instability and that of hopefully dealing with a function that is more regular than the original one.

If we focus on proximity control, it appears clear that the presence of a quadratic proximal term is not the only way to ensure proximity. In fact several authors have introduced proximity control based on different distance-like functions. Here we mention Kiwiel (1999), where a detailed bibliography is also available. Let h be any strictly convex differentiable function defined in $\mathbb{R}^n$. Then for any couple of points (x, y) of $\mathbb{R}^n$ it is possible to define the D-function of h:

$$D_h(x,y) = h(x) - h(y) - \nabla h(y)^T(x-y), \tag{11}$$

representing the difference (nonnegative for convexity) between the value of h at x and the linearization of h "rooted" at y and evaluated at x.

With slight modifications that account for possible nondifferentiability of h (in such a case the gradient of h in (11) is replaced by any subgradient), the concept of D-function may be applied to strictly convex functions belonging to a class defined as B-functions, that is, generalized Bregman functions (Kiwiel, 1999; in particular, the Euclidean norm belongs to such a class).

The fundamental property of a D-type function (i.e., $D_h(x,y) = 0 \Leftrightarrow x = y$) is that it allows such a function to be embedded in a proximal-point scheme to act as a nonquadratic proximity measure.

In particular the results described in Kiwiel (1999; the bundle Bregman proximal minimization) are based on the use of a polyhedral model involving accumulated subgradients and of a D-function of an appropriate generalized Bregman function to account for the proximity. The subproblem that needs to be solved at each step is of the type:

$$x_{k+1} = \arg\min \min_{x \in \mathbb{R}^n} f_k(x) + \frac{1}{u_k} D_h(x, x_k). \tag{12}$$

In particular, the convergence of the method does not require the exact solution to the subproblem (Equation (12)). Different variants of the very basic approach above described may be found in Kiwiel (1999).

6. Auxiliary Problems

With many methods it is necessary to repeatedly solve problems of the type:

$$\min_{x \in \mathbb{R}^n} f_k(x) + \tfrac{1}{2} u_k \|x - x_k\|^2, \tag{13}$$

where f_k is the cutting-plane function at iteration k, x_k is the current stability center and u_k is the scalar proximity parameter.

Equation (13) can be put in the form of a semidefinite quadratic program. Alternatively its dual can be solved:

$$\min \frac{1}{2u_k} \left\| \sum_{i \in I_k} \lambda_i g_i \right\|^2 - \sum_{i \in I_k} \lambda_i [f(x_i) + g_i^T (x_k - x_i)]$$

$$\sum_{i \in I_k} \lambda_i = 1 \tag{14}$$

$$\lambda_i \geqslant 0 \quad i \in I_k,$$

where I_k is the bundle index set at iteration k.

In fact, once an optimal dual solution λ_i^* $i \in I_k$ has been obtained, the optimal primal solution x_{k+1} is expressed as

$$x_{k+1} = x_k - \frac{1}{u_k} \sum_{i=1}^{k} \lambda_i^* g_i.$$

The literature on this class of problems is extremely rich. Among the others, we cite here the algorithms described in Kiwiel (1994) and Frangioni (1996) for which efficient implementations have been provided. In general the authors prefer to tackle the dual problem (Equation (14)) instead of the primal (13), since in many large-scale nondifferentiable optimization applications n, the dimension of the space is much larger than cardinality of I_k, the current bundle size.

Kiwiel (1994) has presented a method based on an active set strategy coupled with the Cholesky factorization of the active constraint matrix. It is worth noting that the method also allows management of constraints, such as box constraints, on the variables in the primal space.

Similar characteristics, together with a careful exploitation of the problem structure, are found in the very effective algorithm described in Frangioni (1996).

7. Some Notes on the Bibliography

Throughout the paper we have listed a number of references. We emphasize that such references are by no means exhaustive.

We wish to call attention to some papers and books whose reading is strongly suggested to people who wish to obtain a good knowledge of the field.

The mathematical basis is contained in Rockafellar's book (Rockafellar, 1970). A complete treatment of both the theoretical and practical aspects is in the book by Hiriart-Urruty and Lemaréchal (1993), where an extensive bibliography is also given

A short and clear presentation of the basic numerical aspects of nondifferentiable optimization is in Chapter 6 of the book by Bertsekas (1995). Shor's book (1998) is strongly suggested for the methods (e.g., subgradient and space dilatation) not covered by our article.

Finally, we wish to direct the reader's attention to the very seminal paper (Kelley, 1960), where it is possible to retrace *in nuce* many ideas developed during the successive decades.

References

Bertsekas, D. P. 1995. *Nonlinear Programming*. MA: Athena Scientific, Belmont.

Bonnans, J. F., J. Ch. Gilbert, C. Lemaréchal, and C. A. Sagastizábal. 1995. "A family of variable metric proximal methods." *Math. Programming* **68**: 15–47.

Capalbo, V., R. De Leone, and M. Gaudioso. 1998. "The cobweb method for minimizing convex functions." In *High Performance Algorithms and Software in Nonlinear Optimization*, edited by R. De Leone, A. Murli, P. M. Pardalos, and G. Toraldo. Boston: Kluwer Academic Publishers.

Chen, X. and M. Fukushima. 1999. "Proximal quasi-Newton methods for nondifferentiable convex optimization." *Math. Programming* **85**: 313–334.

Cheney, E. W. and A. A. Goldstein. 1959. "Newton's method for convex programming and Tchebycheff approximation." *Numerische Mathematik* **1**: 253–268.

Demyanov, V. F. and V. N. Malozemov. 1974. In *Introduction to Minimax*. New York: John Wiley & Sons.

Di Pillo, G., L. Grippo, and S. Lucidi. 1993. "A smooth method for the finite minimax problem." *Math. Programming* **60**: 187–214.

Di Pillo, G., L. Grippo, and S. Lucidi. 1997. "Smooth transformation of the generalized minimax problem." *J. Optim. Theory Appl.* **95**: 1–24.

Frangioni, A. 1996. "Solving semidefinite quadratic problems within nonsmooth optimization algorithms." *Comput. Oper. Res.* **23**: 1099–1118.

Fukushima, M. and L. Qi. 1996. "A globally and superlinearly convergent algorithm for nonsmooth convex minimization." *SIAM J. Optim.* **6**: 1106–1120.

Gaudioso, M. and M. F. Monaco. 1991. "Quadratic approximations in convex nondifferentiable optimization." *SIAM J. Control Optim.* **29**: 1–10.

Goffin, J.-L., A. Haurie, and J.-P. Vial. 1992. "Decomposition and nondifferentiable optimization with the projective algorithm." *Mgmnt Sci.* **38**: 284–302.

Hald, J. and K. Madsen. 1981. "Combined LP and Quasi-Newton methods for minimax optimization." *Math. Programming* **20**: 49–62.

Hiriart-Urruty, J. B. and C. Lemaréchal. 1993. In *Convex Analysis and Minimization Algorithms* Vol. I and II. Berlin: Springer-Verlag.

Kelley, J. E. 1960. "The cutting plane method for solving convex programs." *J. SIAM* **8**: 703–712.

Kiwiel, K. C. 1983. "An aggregate subgradient method for nonsmooth convex minimization." *Math. Programming* **27**: 320–341.

Kiwiel, K. C. 1990. "Proximity control in bundle methods for convex nondifferentiable minimization." *Math. Programming* **46**: 105–122.

Kiwiel, K. C. 1994. "A Cholesky dual method for proximal piecewise linear programming." *Numerische Mathematik* **68**: 325–340.

Kiwiel, K. C. 1999. "A bundle Bregman proximal method for convex nondifferentiable minimization." *Math. Programming* **85**: 241–258.

Lemaréchal, C. 1989. "Nondifferentiable Optimization." In *Handbooks in OR & MS*, edited by G. L. Nemhauser*et al.* Vol 1. Amsterdam: North-Holland.

Lemaréchal, C. and C. Sagastizábal. 1997. "Variable metrics bundle methods: from conceptual to implementable forms." *Math. Programming* **76**: 393–410.

Levin, A. Yu. 1965. "On an algorithm for minimization of convex functions." *Sov. Math. Doklady* **6**: 286–290.

Mifflin, R. 1996. "A quasi-second order proximal bundle algorithm." *Math. Programming* **73**: 51–72.

Polak, E., D. Q. Mayne, and J. E. Higgins. 1991. "Superlinearly convergent algorithm for min–max problems." *J. Optim. Theory Appl.* **69**: 407–439.

Polyak, B. T. 1987. In *Introduction to Optimization*. New York: Optimization Software Inc.

Rockafellar, R. T. 1970. *Convex Analysis*. Princeton: Princeton University Press.

Rockafellar, R. T. 1976. "Monotone operators and the proximal point algorithm." *SIAM J. Control Optim.* **14**: 877–898.

Schramm, H. and J. Zowe. 1992. "A version of the bundle idea for minimizing a nonsmooth function: conceptual idea, convergence analysis, numerical results." *SIAM J. Optim.* **2**: 121–152.

Shor, N. Z. 1998. *Nondifferentiable Optimization and Polynomial Problems.* Boston: Kluwer Academic Publishers.

Sonnevend, G. 1985. "An analytic center for polyhedrons and new classes of global algorithms for linear (smooth, convex) programming." In *Lecture notes in Control and Information Sciences* **84**: 866–876. New York: Springer.

6 DETERMINISTIC GLOBAL OPTIMIZATION AND ITS APPLICATIONS

Christodoulos A. Floudas

ABSTRACT This article provides an introduction and overview of the activities in the area of deterministic global optimization. Section 1 introduces the key issues, definitions, and motivation of global optimization problems and contrasts them with local optimization problems. Section 2 presents the structure of the three classes of mathematical models that we focus on in this handbook. Section 3 presents a number of illustrative examples of important global optimization applications. Section 4 discusses the approaches that have been proposed in deterministic global optimization and the classes of problems that are addressed. Section 5 provides a large number of research areas that cross several disciplines and a number of important applications. Section 6 lists books published, conferences held, and the introduction of a journal and a book series in the area of global optimization. Section 7 presents an overview of the contributions that have come from the field of chemical engineering.

1. Motivation

Global optimization addresses the computation and characterization of global optima (i.e., minima and maxima) of nonconvex functions constrained in a specified domain. Given an objective function f that is to be minimized and a set of equality and inequality constraints S, the area of deterministic global optimization focuses on the following important issues:

1. Determine a global minimum of the objective function f (i.e., f has the lowest possible value in S) subject to the set of constraints S.
2. Determine lower and upper bounds on the global minimum of the objective function f on S that are valid for the whole feasible region S.
3. Enclose all solutions of the set of equality and inequality constraints S.

In contrast to the aforementioned objectives of the area of deterministic global optimization, the area of local optimization aims at determining a feasible solution that is a local minimum of the objective function f on S (i.e., it is a minimum in its neighborhood but not necessarily of the lowest value for the objective function f). As a result, in general, for nonlinear optimization problems where multiple local minima exist, a local minimum (and also any function evaluation point of f) represents only an upper bound on the global minimum of the objective function f on S.

It is remarkable that, despite the fact that the vast majority of the challenging and important problems that arise in science, applied science, and engineering exhibit nonconvexities and hence multiple minima, there has been relatively little effort devoted to the area of global optimization in comparison to the developments in the area of local optimization. This is partly attributed to the use of local optimization techniques as components of global optimization approaches and also due to the difficulties that arise in the development of global optimization methods.

The theoretical and algorithmic advances that have taken place in the area of deterministic global optimization during the last three decades, and the explosive growth of computing capabilities, show great promise towards addressing these issues. Theoretical and algorithmic advances along with a variety of applications of global optimization are discussed in the recent book by Floudas (2000).

2. Structure of Global Optimization Models

Nonconvex mathematical models from the deterministic global optimization point of view belong to three classes. The first class takes the form:

$$\begin{aligned} \min_{x} \quad & f(x) \\ \text{subject to} \quad & h(x) = 0 \\ & g(x) \leqslant 0 \\ & x \in X \subseteq \mathcal{R}^n, \end{aligned} \tag{1}$$

where f, h, and g are the objective function, and the vectors of the equality and inequality constraints, respectively, and x is a vector of continuous variables of size n.

Within this first class of global optimization models (1) we have (i) the biconvex and/or bilinear mathematical structure; (ii) the signomial mathematical structure that arises in generalized geometric programming; and (iii) the general mathematical structure of f, h, and g belonging to $\mathcal{C}^2$, the set of twice continuously differentiable functions.

The second class of global optimization models (2) corresponds to mixed integer nonlinear MINLP optimization problems that take the form:

$$\begin{aligned} \min_{x,y} \quad & f(x, y) \\ \text{subject to} \quad & h(x, y) = 0 \\ & g(x, y) \leqslant 0 \\ & x \in X \subseteq \mathcal{R}^n \\ & y \in Y \text{ integer}, \end{aligned} \tag{2}$$

where x represents a vector of n continuous variables and y is a vector of integer variables.

The third class of mathematical models (3) focuses on enclosing all solutions of a nonlinear system of equations subject to inequality constraints and variable bounds. This class of models, can be formulated as:

$$\begin{aligned} & h_j(x) = 0, \quad j \in \mathcal{N}_E \\ & g_k(x) \leqslant 0, \quad k \in \mathcal{N}_I \\ & x^L \leqslant x \leqslant x^U, \end{aligned} \tag{3}$$

where $\mathcal{N}_E$ is the set of equalities, $\mathcal{N}_I$ the set of inequality constraints, and x the vector of n continuous variables.

3. Illustrative Applications of Global Optimization

In this section, we introduce a number of important applications of global optimization in the form of illustrative examples. These applications, along with global optimization methods, are discussed in detail in the recent book by Floudas (2000). The applications introduced in this section and their respective mathematical classification are as follows:

- Pooling and blending—bilinear
- Minimization of Gibbs free energy—biconvex
- Tangent plane stability analysis—biconvex
- Robust stability analysis—signomial
- Structure prediction in clusters—box constrained twice differentiable NLP
- Protein folding—box constrained twice differentiable NLP
- Parameter estimation—general constrained twice differentiable NLP
- Heat exhanger network synthesis—separable MINLP
- Pump network synthesis—general MINLP
- Multiple steady states—all solutions of nonlinear constrained equations

- Homogeneous azeotropes—all solutions of nonlinear constrained equations

3.1. *Pooling Problems*

Pooling problems represent an important component in refinery operations. In petrochemical processes it is necessary to model not only the product flows but also the properties of intermediate streams. These streams are usually combined in one or several tanks or pools. The pools are subsequently used in downstream processing or blending. The presence of these streams in the model introduces nonconvexities, which are due to the interactions between the qualities of the input streams and the blended products.

The general formulation of pooling problems, as studied by Ben-Tal et al. (1994), Floudas and Aggarwal (1990), Visweswaran and Floudas (1990, 1993, 1996b), Adjiman et al. (1998b) (see also references mentioned in these publications), take the form:

$$\max \quad -\sum_i \sum_l c_i x_{il} + \sum_l \sum_j d_j y_{lj} + \sum_i \sum_j (d_j - c_i) z_{ij}$$

$$\text{subject to} \qquad \sum_l x_{il} + \sum_j z_{ij} \leqslant A_i \tag{4}$$

$$\sum_i x_{il} + \sum_j y_{lj} = 0$$

$$\sum_i x_{il} \leqslant S_l$$

$$-\sum_i q_{ik} x_{il} + p_{lk} \sum_j y_{lj} = 0$$

$$\sum_l y_{lj} + \sum_i z_{ij} \leqslant D_j$$

$$\sum_l (p_{lk} - Q_{jk}) y_{lj} + \sum_i (q_{ik} - Q_{jk}) z_{ij} \leqslant 0,$$

where the sets I, J, K, and L represent the components i, the products j, the qualities k, and the pools l, respectively. The variables are:

x_{il} amount of component i allocated to pool l

y_{lj} amount going from pool l to product j

z_{ij} amount of component i going to product j

p_{lk} level of quality k in pool l.

The parameters are:

A_i upper bounds for component availabilities

D_j upper bounds for product demands

S_l upper bounds for pool sizes

Q_{jk} upper bounds for product qualities

q_{ik} level of quality k in component i

c_i unit price of component i

d_j unit price of product j.

This mathematical model features nonconvexities in the form of bilinear terms in the constraints. These bilinear terms involve the products of the p_{lk} and y_{lj} variables.

3.2. Minimization of Gibbs Free Energy

The minimization of the Gibbs free energy problem is a fundamental problem in thermodynamics and more specifically in phase and chemical reaction equilibrium. It can be stated as:

Given a set of components $C = \{i\}$; the elements that constitute these components $E = \{e\}$; the set of phases $P = \{k\}$ consisting of vapor and liquid phases, denoted as P_V and P_L, respectively; the liquid phase mol fractions $\boldsymbol{x} \equiv \{x_i\}$; and the vapor phase mol fractions $\boldsymbol{y} \equiv \{y_i\}$, we would like to determine the mol vector $\boldsymbol{n}$ that minimizes the value of the Gibbs free energy while also satisfying the appropriate material balance constraints.

For simultaneous phase and chemical equilibrium where reaction does occur, conservation of the constituent atoms must be satisfied:

$$\sum_{i \in C} \sum_{k \in P} a_{ei} n_i^k = b_e \quad \forall e \in E, \tag{5}$$

where a_{ei} represents the number of gram-atoms of element e in component i, and b_e the total number of gram-atoms of element e in the system.

For phase equilibrium without chemical reaction, conservation over the components need only hold:

$$\sum_{k \in P} n_i^k = n_i^T \quad \forall i \in C, \tag{6}$$

where n_i^T is the total number of moles of component i in the initial mixture.

The material balance constraints for any system, reacting or nonreacting, can be written in the following general form:

$$\mathbf{A} \cdot \boldsymbol{n} - \mathbf{b} = 0, \tag{7}$$

where $\boldsymbol{n}$ represents the column vector of the component mol numbers, $\mathbf{A}$ is the appropriate matrix, and $\mathbf{b}$ is the column vector of the total amounts of elements or compounds in the system.

A physically realizable solution requires that

$$0 \leqslant n_i^k \leqslant n^T \quad \forall i \in C,\ k \in P, \tag{8}$$

where n^T is the total number of mols in the system.

The Gibbs free energy expression for the case of an ideal vapor phase and liquid phases modeled using the NRTL activity coefficient expression (Renon and Prausnitz, 1968) and after application of the theoretical properties developed by McDonald and Floudas (1994) can be written as a combination of a convex part and a nonconvex part:

$$\min \ \hat{G}(\boldsymbol{n}) = \sum_{k \in P} \mathscr{C}_N^k + \sum_{i \in C} \sum_{k \in P_L} n_i^k \left\{ \sum_{j \in C} \frac{\mathscr{G}_{ij} \tau_{ij} n_j^k}{\sum_{l \in C} \mathscr{G}_{lj} n_l^k} \right\}, \tag{9}$$

where $\mathscr{C}_N^k$ is defined as follows:

$$\mathscr{C}_N^k = \sum_{i \in C} n_i^k \left\{ \frac{\Delta G_i^{k,f}}{RT} + \ln \frac{n_i^k}{\sum_{j \in C} n_j^k} \right\} \quad \forall k \in P$$

and is a convex function.

If the following new variables are introduced:

$$\Psi_i^k = \frac{n_i^k}{\sum_{j \in C} \mathscr{G}_{ji} n_j^k} \quad \forall i \in C,\ k \in P_L \tag{10}$$

then the transformed objective function becomes:

$$\min \ \hat{G}(\boldsymbol{\Psi}, \boldsymbol{n}) = \sum_{k \in P} \mathscr{C}^k + \sum_{i \in C} \sum_{k \in P_L} + n_i^k \left\{ \sum_{j \in C} \mathscr{G}_{ij} \tau_{ij} \Psi_i^k \right\}. \tag{11}$$

Hence, the minimization of Gibbs free energy problems using the NRTL equation takes the form:

$$\begin{aligned} &\min && \hat{G}(\boldsymbol{\Psi}, \boldsymbol{n}) \\ &\text{subject to} && \Psi_i^k \left\{ \sum_{j \in C} \mathscr{G}_{ji} n_j^k \right\} = n_i^k \quad \forall i \in C,\ k \in P_L \\ &&& \mathbf{A}\boldsymbol{n} - \mathbf{b} = \mathbf{0} \\ &&& 0 \leqslant n_i^k \leqslant n^T \quad \forall i \in C,\ k \in P. \end{aligned}$$

This mathematical formulation features nonconvexities in the form of a biconvex objective function and a set of bilinear constraints.

3.3. *Tangent Plane Stability Analysis*

The tangent plane stability criterion is also a fundamental problem in thermodynamics and, in particular, in the phase and chemical reaction equilibrium. It focuses on establishing whether a postulated solution is thermodynamically stable with respect to perturbations in any or all of the phases. A necessary and sufficient condition for a postulated solution to be an equilibrium is that the tangent plane distance function, denoted $\mathscr{F}(x)$, be nonnegative for all possible phases represented in the system, with $\mathscr{F}(x)$ defined as follows:

$$\mathscr{F}(x) = \sum_{i \in C} x_i \{\mu_i(x) - \mu_i^0(z)\}, \tag{12}$$

where $\mu_i(x)$ and $\mu_i^0(z)$ represent the chemical potentials calculated at x and z, respectively. The tangent plane distance function is defined as the distance between the Gibbs surface and the tangent plane constructed to this surface at z. If we define

$$g(x) = \sum_{i \in C} x_i \mu_i(x) \quad \text{Molar Gibbs surface (for a single phase)}$$

$$\mathscr{T}(x) = \sum_{i \in C} x_i \mu_i^0(z) \quad \text{Tangent plane to this surface at } z$$

$$\mathscr{F}(x) = g(x) - \mathscr{T}(x) \quad \text{Tangent plane function}$$

then the postulated solution corresponds to a global minimum of the Gibbs free energy as long as the tangent plane lies completely below the Gibbs surface.

The formulation for the tangent plane stability problem is:

$$\begin{aligned} &\min && \mathscr{F}(x) = g(x) - \sum_{i \in C} x_i \mu_i^0(z) \\ &\text{subject to} && \sum_{i \in C} x_i = 1 \\ &&& 0 \leqslant x_i \leqslant 1. \end{aligned}$$

Note that the tangent plane distance function is explicitly minimized. If a nonnegative global optimum solution is obtained, then the stability of the postulated solution is guaranteed.

Based on the analysis and properties proposed by McDonald and Floudas (1995b) for the NRTL equation, the objective function can be written as:

$$\mathscr{F}(x) = \mathscr{C}(x) + \sum_{i \in C} x_i \frac{\sum_{j \in C} \tau_{ij} \mathscr{G}_{ij} x_j}{\sum_{l \in C} \mathscr{G}_{lj} x_l}, \tag{13}$$

where

$$\mathscr{C}(x) = \sum_{i \in C} x_i \{\Delta G_i^f + \ln x_i - \mu_i^0(z)\} \tag{14}$$

is a convex function.

By introducing new variables labeled χ_i,

$$\chi_i = \frac{x_i}{\sum_{j \in C} \mathscr{G}_{ji} x_j} \quad \forall i \in C \tag{15}$$

results in the following tangent plane stability analysis problem for the NRTL activity coefficient equation:

$$\begin{aligned}
\min \quad & \mathscr{F}(x) = \mathscr{C}(x) + \sum_{i \in C} x_i \cdot \sum_{j \in C} \mathscr{G}_{ij} \tau_{ij} \chi_j \\
\text{subject to} \quad & \chi_i \cdot \sum_{j \in C} \mathscr{G}_{ji} x_j = x_i \quad \forall i \in C \\
& \sum_{i \in C} x_i = 1 \\
& 0 \leqslant x_i \leqslant 1 \quad \forall i \in C.
\end{aligned}$$

This mathematical model has nonconvexities in the form of a biconvex objective function and a set of bilinear constraints.

3.4. *Robust Stability Analysis*

Robust stability analysis of linear systems involves the identification of the largest possible region in the uncertain model parameter space for which the controller manages to attenuate any disturbances in the system. The stability of a feedback structure is determined by the roots of the closed loop characteristic equation:

$$\det(I + P(s, q)C(s, q)) = 0,$$

where q is the vector of the uncertain model parameters, and $P(s)$, $C(s)$ are the transfer functions of the plant and controller, respectively. After expanding the determinant we have the characteristic polynomial:

$$P(s, q) = a_n(q)s^n + a_{n-1}(q)s^{n-1} + \cdots + a_1(q)s + a_0(q) = 0,$$

where the coefficients $a_i(q)$, $i = 0, \ldots, n$ are typically multivariable polynomial functions. Real parameter uncertainty is typically expressed as positive and negative deviations of the real parameters from some nominal values.

Checking the stability of a particular system with characteristic equation $P(j\omega, q)$ involves the solution of the following nonconvex optimization problem (Psarris and Floudas, 1995):

$$\min_{q_i, k \geqslant 0, \omega \geqslant 0} \quad k \tag{16}$$

$$\begin{aligned}
\text{subject to} \quad & \mathrm{Re}[P(j\omega, q)] = 0 \\
& \mathrm{Im}[P(j\omega, q)] = 0
\end{aligned}$$

$$q_i^N - \Delta q_i^- k \leqslant q_i \leqslant q_i^N + \Delta q_i^+ k, \quad i = 1, \ldots, n,$$

where q^N is a stable nominal point for the uncertain parameters and Δq^+, Δq^- are the estimated bounds. Stability is ensured if, and only if, the optimal solution k is greater or equal to one.

This mathematical model features nonconvexities in the form of multivariable polynomial functions, known as signomials, that appear in the set of constraints.

3.5. *Structure Prediction in Clusters*

Clusters are aggregates of atoms, ions, or molecules, sufficiently small that a significant proportion of these units is present on their surfaces. The structure prediction of microclusters is of primary importance in a variety of disciplines that range from thermodynamics to material science, catalysis, and astrophysics. The cluster structure prediction problem can be stated as:

Given N particles interacting with two-body central forces, find their configuration(s) in three-dimensional Euclidean space involving the global minimum total potential energy.

Assuming that: (i) many-body and angle dependent interactions need not be taken into account, (ii) conformational entropy is not considered, and (iii) the particles are spherical and are of the same size, the structure prediction problem of a N-particle cluster can be written as:

$$\min_{x,y,z} \; V = \sum_{i=1}^{N-1} \sum_{j=i+1}^{N} \upsilon(r_{ij}), \tag{17}$$

where

$$r_{ij} = \sqrt{(x_i - x_j)^2 + (y_i - y_j)^2 + (z_i - z_j)^2}$$

and

$$x_1 = y_1 = z_1 = y_2 = z_2 = z_3 = 0,$$

where V is the total potential energy of the microcluster as the summation of all two-body interaction terms, $\upsilon(r_{ij})$ is the potential energy term corresponding to the interaction of particle i with particle j, and r_{ij} is the Euclidean distance between i and j.

Pair potentials that have been used in cluster studies include the following (Hoare, 1979):

Mie: $\upsilon(r) = (n-m)^{-1}[nr^{-m} - mr^{-n}]$,

Lennard–Jones: $\upsilon(r) = 4\epsilon\left\{\left(\frac{\sigma}{r}\right)^{12} - \left(\frac{\sigma}{r}\right)^{6}\right\}$,

Morse: $\upsilon(r) = [1 - e^{a(1-r)}]^2 - 1$,

Gaussian: $\upsilon(r) = Ae^{-ar^2} - Be^{-br^2}$,

Born–Meyer: $\upsilon(r) = z^{\alpha}z^{\beta}/r + Ae^{-r/\rho}$.

Due to the nonlinear form of the pairwise potentials, the resulting mathematical model features a twice diffentiable nonlinear and nonconvex objective function subject to box constraints. This problem exhibits a very large number of local minima that grows significantly as the number of particles increases.

3.6. *Structure Prediction in Protein Folding*

Proteins are very complex molecules and this complexity is due to the intra- and inter-molecular interactions that define the native three-dimensional structure of the system and subsequently its biological functionality. The structure prediction of proteins is a fundamental problem of utmost importance in computational chemistry and molecular biology (Leach, 1991; Scheraga, 1992; Pardalos et al., 1994; Vásquez et al., 1994; Pardalos et al., 1996; Dill and Chan, 1997; Neumaier, 1997; Chan and Dill, 1998; Floudas et al., 1999b).

In the internal coordinate space and employing an atomistic level mathematical model that establishes all the interactions (e.g., AMBER (Weiner et al., 1984, 1986), CHARMM (Brooks et al., 1983), DISCOVER (Dauber-Osguthorpe et al., 1988), ECEPP/3 (Némethy et al., 1992)), the structure prediction problem in protein folding can be stated as:

$$\min \quad E(\phi_i, \psi_i, \omega_i, \chi_i^k, \theta_j^N, \theta_j^C) \tag{18}$$

$$\begin{aligned}
\text{subject to} \quad & -\pi \leqslant \phi_i \leqslant \pi, \quad i = 1, \dots, N_{RES} \\
& -\pi \leqslant \psi_i \leqslant \pi, \quad i = 1, \dots, N_{RES} \\
& -\pi \leqslant \omega_i \leqslant \pi, \quad i = 1, \dots, N_{RES} \\
& -\pi \leqslant \chi_i^k \leqslant \pi, \quad i = 1, \dots, N_{RES}, \quad k = 1, \dots, K^i \\
& -\pi \leqslant \theta_j^N \leqslant \pi, \quad j = 1, \dots, J^N, \\
& -\pi \leqslant \theta_j^C \leqslant \pi, \quad j = 1, \dots, J^C,
\end{aligned}$$

where: E represents the total potential energy function (using for instance ECEPP/3) and the free energy of solvation, which can be calculated via methods based on the area accessible to the solvent or methods based on the volume accessible to the solvent; $i = 1, \dots, N_{RES}$ is an indexed set describing the sequence of amino acid residues in the peptide chain; ϕ_i, ψ_i, ω_i, $i = 1, \dots, N_{RES}$ are the dihedral angles along the backbone of this peptide; χ_i^k, $i = 1, \dots, N_{RES}$, $k = 1, \dots, K^i$ denote the dihedral angles of the side chains for the ith residue; and θ_j^N, $j = 1, \dots, J^N$ and θ_j^C, $j = 1, \dots, J^C$ are the amino and carboxyl end group dihedral angles, respectively.

The resulting optimization problem for structure prediction in protein folding is a box constrained nonlinear twice differentiable optimization problem where the nonconvexities are in the highly nonlinear form of the objective function.

3.7. *Parameter Estimation and Data Reconciliation*

Nonlinear mathematical models that accurately predict physical phenomena are essential in engineering and applied science. The parameter estimation and data reconciliation problem is important in the design, synthesis, control, and operation of chemical processes (Ricker, 1984; Stewart et al., 1992; Kim et al., 1990).

The general model consists of an implicit algebraic system of equations

$$\mathbf{f}(\theta, \mathbf{z}) = \mathbf{0}, \tag{19}$$

where θ is a vector of p unknown parameters, $\mathbf{z}$ is a vector of n experimentally measured variables, and $\mathbf{f}$ represents the system of l algebraic functions. All experimentally measured variables are affected to some extent by error. The measurements are related to the true values through:

$$\mathbf{z}_\mu = \zeta_\mu + \mathbf{e}_\mu \quad \mu = 1 \dots m, \tag{20}$$

where ζ_μ is a vector of unknown true values of the experimentally measured variables, $\mathbf{z}_\mu$, at the μ data point, and $\mathbf{e}_\mu$ is the vector of additive error.

Based on the maxiumum likelihood approach and assuming that the errors in each experiment are known, equal, and independent, the parameter estimation and data reconciliation problem known as the error-in-variables formulation takes the form:

$$\min_{\hat{\mathbf{z}}_\mu \theta} \sum_{\mu=1}^{r} \sum_{i=1}^{m} \frac{(\hat{z}_{\mu,i} - z_{\mu,i})^2}{\sigma_i^2} \tag{21}$$

subject to

$$\mathbf{f}(\hat{\mathbf{z}}_\mu, \theta) = \mathbf{0} \quad \mu = 1 \dots m,$$

where the optimization variables, $\hat{\mathbf{z}}_\mu$, are the fitted data variables that are approximations of the true values of the experimental data, ζ_μ, and σ_i is the standard deviation assigned to the ith variable in all experiments.

The resulting mathematical model of the parameter estimation and data reconciliation problem features a convex objective function while the nonconvexities are in the set of constraints that depend on the form of the nonlinear model.

3.8. Heat Exhanger Network Synthesis

The optimization-based approach in the area of process synthesis and, more specifically, for the synthesis of heat exchanger networks consists of (i) generating a superstructure of all alternatives of interest via graph theory, (ii) modeling the superstructure as a mixed integer nonlinear optimization problem, and (iii) developing efficient algorithmic and solution strategies that extract the optimal structure from the set of alternatives embedded in the postulated superstructure (Floudas, 1995).

Given the set ST of temperature locations, the number of stages NS, the set HP of hot process streams, and the set CP of cold process streams, employing the superstructure and formulation of Yee and Grossmann (1991), and using the Chen approximation for the logarithmic mean of the temperature difference (Chen, 1987), the general problem formulation takes the form:

$$\min \sum_{i \in HP} C_{CU} Q_{CU,i} + \sum_{j \in CP} C_{HU} Q_{HU,j}$$

$$+ \sum_{i \in HP} \sum_{j \in CP} \sum_{k \in ST} CF_{ij} z_{ijk} + \sum_{i \in HP} CF_{i,CU} z_{CU,i} + \sum_{j \in CP} CF_{j,HU} z_{HU,j}$$

$$+ \sum_{i \in HP} \sum_{j \in CP} \sum_{k \in ST} \frac{C_{ij} Q_{ijk}}{U_{ij}[\Delta T_{ijk} \Delta T_{ijk+1} (\Delta T_{ijk} + \Delta T_{ijk+1})/2]^{1/3}}$$

$$+ \sum_{i \in HP} \frac{C_{A,i,CU} Q_{CU,i}}{U_{CU,i}[\Delta T_{CU,i}(T_{\text{out},i} - T_{\text{in},CU})(\Delta T_{CU,i} + T_{\text{out},i} - T_{\text{in},CU})/2]^{1/3}}$$

$$+ \sum_{j \in CP} \frac{C_{A,j,HU} Q_{HU,j}}{U_{HU,j}[\Delta T_{HU,j}(T_{\text{in},HU} - T_{\text{out},j})(\Delta T_{HU,j} + T_{\text{in},HU} - T_{\text{out},j})/2]^{1/3}}$$

$$\text{subject to } (T_{\text{in},i} - T_{\text{out},i}) Fcp_i = \sum_{k \in ST} \sum_{j \in CP} Q_{ijk} + Q_{CU,i}, \quad \forall i \in HP$$

$$(T_{\text{out},j} - T_{\text{in},j}) Fcp_j = \sum_{k \in ST} \sum_{i \in HP} Q_{ijk} + Q_{HU,j}, \quad \forall j \in CP$$

$$(T_{i,k} - T_{i,k+1}) Fcp_i = \sum_{j \in CP} Q_{ijk}, \quad \forall k \in ST, \quad \forall i \in HP$$

$$(T_{j,k} - T_{j,k+1}) Fcp_j = \sum_{i \in HP} Q_{ijk}, \quad \forall k \in ST, \quad \forall j \in CP$$

$$T_{\text{in},i} = T_{i,1}, \quad \forall i \in HP$$

$$T_{\text{in},j} = T_{j,NS+1}, \quad \forall j \in CP$$

$$T_{i,k} \geqslant T_{i,k+1}, \quad \forall k \in ST, \quad \forall i \in HP$$

$$T_{j,k} \geqslant T_{j,k+1}, \quad \forall k \in ST, \quad \forall j \in CP$$

$$T_{\text{out},i} \leqslant T_{i,NS+1}, \quad \forall i \in HP$$

$$T_{\text{out},j} \geqslant T_{j,1}, \quad \forall j \in CP$$

$$(T_{i,NS+1} - T_{\text{out},i})Fcp_i = Q_{CU,i}, \quad \forall i \in HP$$

$$(T_{\text{out},j} - T_{j,1})Fcp_j = Q_{HU,j}, \quad \forall j \in CP$$

$$Q_{ijk} - \Omega z_{ijk} \leqslant 0, \quad \forall k \in ST, \quad \forall i \in HP, \quad \forall j \in CP$$

$$Q_{CU,i} - \Omega z_{CU,i} \leqslant 0, \quad \forall i \in HP$$

$$Q_{HU,j} - \Omega z_{HU,j} \leqslant 0, \quad \forall j \in CP$$

$$T_{i,k} - T_{j,k} + \Gamma(1 - z_{ijk}) \geqslant \Delta T_{ijk}, \quad \forall k \in ST, \quad \forall i \in HP, \quad \forall j \in CP$$

$$T_{i,k+1} - T_{j,k+1} + \Gamma(1 - z_{ijk}) \geqslant \Delta T_{ijk+1}, \quad \forall k \in ST, \quad \forall i \in HP, \quad \forall j \in CP$$

$$T_{i,NS+1} - T_{\text{out},CU} + \Gamma(1 - z_{CU,i}) \geqslant \Delta T_{CU,i}, \quad \forall i \in HP$$

$$T_{\text{out},HU} - T_{j,1} + \Gamma(1 - z_{HU,j}) \geqslant \Delta T_{HU,j}, \quad \forall j \in CP$$

$$\Delta T_{ijk} \geqslant 10, \quad \forall k \in ST, \quad \forall i \in HP, \quad \forall j \in CP$$

$$z_{ijk}, z_{CU,i}, z_{HU,j} \in \{0, 1\}, \quad \forall k \in ST, \quad \forall i \in HP, \quad \forall j \in CP,$$

where the continuous variables are: T_{ik}, the temperature of hot stream i at the hot end of stage k; T_{jk}, the temperature of cold stream j at the cold end of stage k; Q_{ijk}, the heat exchanged between hot stream i and cold stream j at temperature location k; $Q_{CU,i}$, the heat exchanged between hot stream i and the cold utility; $Q_{HU,j}$, the heat exchanged between cold stream j and the hot utility; ΔT_{ijk}, the temperature approach for the match of hot stream i and cold stream j at temperature location k; $\Delta T_{CU,i}$, the temperature approach for the match of hot stream i and the cold utility; $\Delta T_{HU,j}$, the temperature approach for the match of cold stream j and the hot utility.

The binary variables are: z_{ijk}, for the existence of a match between hot stream i and cold stream j at temperature location k; $z_{CU,i}$, for the existence of a match between hot stream i and the cold utility; $z_{HU,j}$, for the existence of a match between cold stream j and the hot utility.

The parameters are: T_{in}, the inlet temperature of a stream; T_{out}, the outlet temperature; Fcp, the heat capacity flowrate of a stream; Ω, the upper bound on heat exchange; Γ, the upper bound on the temperature difference.

The resulting mathematical model corresponds to a mixed integer nonlinear optimization problem where the binary variables appear linearly and separably from the continuous variables in both the objective function and constraints. All the constraints are linear while the nonconvexities appear in the objective function.

3.9. Pump Network Synthesis

The synthesis of pump networks involves the postulation of a multilevel superstructure and its formulation as a general MINLP model (Westerlund et al., 1994). The objective is to identify the least costly configuration of centrifugal pumps that achieves a prespecified pressure rise based on a given total flow rate.

The structural decisions for an L level superstructure are represented by a number of discrete variables. The binary variables z_i, $i = 1, \ldots, L$, denote the existence of level i. The integer variables Np_i, $i = 1, \ldots, L$, denote the number of parallel lines at level i ($Np_i \in \{0, \ldots, NP\}$, where NP is the maximum number of parallel lines). The integer variables Ns_i, $i = 1, \ldots, L$, denote the number of pumps in series at level i ($Ns_i \in \{0, \ldots, NS\}$, where NS is the maximum number of pumps in series). The key continuous variables are the fraction of total flow going to level i, x_i; the flow rate on each line at level i, $\dot{v}_i$, the rotation speed of all pumps on level i, ω_i, the power requirements at level i, P_i, and the pressure rise at level i, Δp_i.

The optimization model for a three-level (i.e., $L = 3$) superstructure, that minimizes the annualized cost takes the form:

$$\min \sum_{i=1}^{3}(C_i + C_i' P_i)Np_i Ns_i z_i$$

$$\begin{aligned}
\text{subject to}\quad & P_1 - 19.9\left(\frac{\omega_1}{\omega_{\max}}\right)^3 - 0.1610\left(\frac{\omega_1}{\omega_{\max}}\right)^2 \dot{v}_1 + 0.000561\left(\frac{\omega_1}{\omega_{\max}}\right)\dot{v}_1^2 = 0\\
& P_2 - 1.21\left(\frac{\omega_2}{\omega_{\max}}\right)^3 - 0.0644\left(\frac{\omega_2}{\omega_{\max}}\right)^2 \dot{v}_2 + 0.000564\left(\frac{\omega_2}{\omega_{\max}}\right)\dot{v}_2^2 = 0\\
& P_3 - 6.52\left(\frac{\omega_3}{\omega_{\max}}\right)^3 - 0.1020\left(\frac{\omega_3}{\omega_{\max}}\right)^2 \dot{v}_3 + 0.000232\left(\frac{\omega_3}{\omega_{\max}}\right)\dot{v}_3^2 = 0\\
& \Delta p_1 - 629\left(\frac{\omega_1}{\omega_{\max}}\right)^2 - 0.696\left(\frac{\omega_1}{\omega_{\max}}\right)\dot{v}_1 + 0.0116\dot{v}_1^2 = 0\\
& \Delta p_2 - 215\left(\frac{\omega_2}{\omega_{\max}}\right)^2 - 2.950\left(\frac{\omega_2}{\omega_{\max}}\right)\dot{v}_2 + 0.0115\dot{v}_2^2 = 0 \qquad (22)\\
& \Delta p_3 - 361\left(\frac{\omega_3}{\omega_{\max}}\right)^2 - 0.530\left(\frac{\omega_3}{\omega_{\max}}\right)\dot{v}_3 + 0.00946\dot{v}_3^2 = 0\\
& x_1 + x_2 + x_3 = 1\\
& \omega_i - \omega_{\max} \leqslant 0, \quad i = 1,2,3\\
& \dot{v}_i - \frac{x_i}{Np_i} V_{\text{tot}} = 0, \quad i = 1,2,3\\
& \Delta P_{\text{tot}} - \Delta p_i Ns_i = 0, \quad i = 1,2,3\\
& Np_i = \{0,1,2,3\}, \quad i = 1,2,3\\
& Ns_i = \{0,1,2,3\}, \quad i = 1,2,3\\
& z_i = \{0,1\}, \quad i = 1,2,3.
\end{aligned}$$

This mathematical formulation is mixed integer nonlinear and does not exhibit any particular structure. It is nonconvex in both the continuous and integer variables. The nonconvexities appear in the objective function and constraints, and integer and binary variables exist.

3.10. Multiple Steady States in Reactors

The enclosure of all solutions of nonlinear constrained systems of equations is regarded as a very challenging task in applied mathematics, engineering, and sciences. A class of such problems is the location of all steady states in chemical reactor systems.

As an illustration, we consider here a model of two continuous, nonadiabatic, stirred tank reactors in series at steady state with recycling and an exothermic, first order, irreversible reaction (Kubicek et al., 1980). The modelling equations of this system involve products between linear and exponential terms and are:

$$(1-R)(1-y_1)\exp\left(\frac{10\phi_1}{1+10\phi_1/\gamma}\right) - y_1 = 0$$

$$(1-R)D(1-y_1)\exp\left(\frac{10\phi_1}{1+10\phi_1/\gamma}\right) - 10(1+\beta_1)\phi_1 + \beta_1\phi_{c1} = 0$$

$$y_{11} - y_2 + (1-R)(1-y_2)\exp\left(\frac{10\phi_2}{1+10\phi_2/\gamma}\right) = 0$$

$$10\phi_1 - 10(1+\beta_2)\phi_2 + (1-R)D(1-y_2)\exp\left(\frac{10\phi_2}{1+10\phi_2/\gamma}\right) + \beta_2\phi_{c2} = 0,$$

where the variables are:

y_i = reactant conversion in reactor $i = 1, 2$

ϕ_i = dimensionless temperature in reactor $i = 1, 2$.

and the parameters are:

$\gamma = 1000$ (dimensionless activation energy)

$D = 22$ (dimensionless adiabatic temperature rise)

$\phi_{c1} = \phi_{c2} = 0$ (dimensionless coolant temperatures)

$\beta_1 = \beta_2 = 2$ (dimensionless heat transfer coefficient in reactors 1, 2)

$0 \leqslant R \leqslant 1$ (recycle ratio).

For different values of the recycle ratio R, starting with $R = 0.935$ and ending at $R = 0.995$ with 0.005 increments, the above system of nonlinear constrained equations exhibits one, three, five, or seven solutions. This mathematical model corresponds to a nonlinear constrained system of equations, which is nonconvex.

3.11. Enclosure of All Homogeneous Azeotropes

The ability to predict whether a given mixture will form one or more homogeneous, heterogeneous and reactive azeotropes and to calculate the conditions and compositions of each azeotrope is essential in modeling separation processes.

Considering here the case of homogeneous azeotropes (i.e., one liquid phase), there are three thermodynamic conditions which a system must meet in order for an homogeneous azeotrope to exist. These conditions are: (i) equilibrium, (ii) the composition of vapor phase must be identical to the overall composition of the liquid phase, and (iii) the mole fractions of the components in each phase must sum to unity and must be nonnegative.

The equilibrium condition requires that the chemical potential of each component must be the same in all phases. Since an homogeneous azeotropic system contains a vapor phase (V), and only one liquid phase (L), this condition can be written as:

$$\mu_i^V = \mu_i^L \quad \forall i \in N, \tag{23}$$

where μ_i^V and μ_i^L represent the chemical potential of component i in the vapor and liquid phases. From the definition of the fugacity of component i in a mixture, $\hat{f}_i$,

$$\hat{f}_i^V = \hat{f}_i^L \quad \forall i \in N \tag{24}$$

we get,

$$y_i \hat{\phi}_i^V P = x_i \gamma_i^L f_i^L. \tag{25}$$

The symbol $\hat{\phi}_i^V$ represents the mixture fugacity coefficient of component i in the vapor phase. For the liquid phase, γ_i^L is the activity coefficient, and f_i^L is the fugacity of component i in the liquid phase. Assuming that the vapor phase can be modeled as an ideal gas, we have:

$$\frac{y_i}{x_i} = \frac{\gamma_i^L P_i^{\text{sat}}}{P}. \tag{26}$$

Note that all of the nonidealities of the system are confined to the activity coefficient term. For instance, employing the Wilson activity coefficient (Wilson, 1964) equation we have:

$$\ln \gamma_i = 1 - \ln\left(\sum_{j \in N} x_j \Lambda_{ij} \right) - \sum_{j \in N} \frac{x_j \Lambda_{ji}}{\sum_{k \in N} x_k \Lambda_{jk}}, \tag{27}$$

where Λ_{ij} is the nonsymmetric binary interaction parameter between components i and j with $\Lambda_{ii} = 1$.

The saturated vapor pressure is calculated using the Antoine equation:

$$\ln P_i^{\text{sat}} = a_i - \frac{b_i}{T + c_i}, \tag{28}$$

where a_i, b_i, and c_i are constant parameters with a_i and b_i being positive, while c_i may be positive or negative but $| c_i | < T$.

Condition (ii) takes the form:

$$y_i = x_i \quad \forall i \in N. \tag{29}$$

Condition (iii) requires that the mole fractions in each phase sum to unity and have values between 0 and 1.

$$\begin{aligned} &\sum_{i \in N} y_i = \sum_{i \in N} x_i = 1 \\ &0 \leqslant y_i, x_i \leqslant 1 \quad \forall i \in N. \end{aligned} \tag{30}$$

This mathematical model corresponds to a nonlinear constrained system of equations that exhibits multiplicity of solutions and each solution is an homogeneous azeotrope.

4. Deterministic Global Optimization: Approaches and Classes of Problems

In the last three decades, and to a larger extent in the last fifteen years, there has been a lot of interest in theoretical and computational investigations of challenging global optimization problems. This has resulted in the development and application of novel deterministic global optimization methods to important problems in science, applied science, and engineering. Exciting and intriguing theoretical findings accompanied by algorithmic developments have taken place not only for special classes of nonconvex problems but also for general constrained nonlinear problems, as well as for mixed integer nonlinear problems

Deterministic global optimization approaches and the classes of global optimization problems can be identified as follows.

4.1. Deterministic Global Optimization Approaches

Branch-and-bound methods

Cutting plane methods

Primal-dual decomposition methods

Outer approximation methods

Inner approximation methods

Difference of convex (D.C.) and reverse convex methods

Reformulation-linearization methods

Lipschitzian methods

Trajectory and homotopy methods

Interval analysis methods

4.2. Classes of Global Optimization Problems

Concave minimization problems

D.C. and reverse convex problems

Lipschitzian optimization problems
Quadratic optimization problems
Biconvex optimization problems
Complementarity problems
Multiplicative programming problems
Bilevel and multilevel optimization problems
Fractional programming problems
Generalized geometric programming problems
Twice continuously differentiable, C^2, NLP problems
All solutions of nonlinear constrained systems of equations problems
Mixed integer nonlinear optimization problems
Max-min optimization problems
Integer programming problems
Network optimization problems.

5. Research Areas and Important Applications

Global optimization applications are widespread in all disciplines of science, applied science, and engineering. Research areas and important applications include:

5.1. Research Areas

Computational chemistry
Molecular biology
Economics
Thermodynamics
Chemical engineering process operations
Chemical engineering process design
Chemical engineering process synthesis
VLSI design
Structural optimization
Engineering design
Nuclear and mechanical design
Production management
Transportation and communications network design
Logistics
Finance
Facility location
Water resources management
Design under uncertainty
Packing
Statistics
Parameter estimation
Data reconciliation

5.2. *Important Applications*

Fixed charge linear and nonlinear problems in plant location
Fixed charge linear and nonlinear problems in production lot-sizing
Economies of scale problems
Portfolio selection
Multicommodity network flow problems
Optimal design of trusses
Robust stability of control systems under parametric uncertainty
Location-allocation of wells and platforms
Air traffic control
Economic equilibria and competition
Phase and chemical equilibrium
Structure prediction of atomic and molecular clusters
Structure prediction of molecules
Protein folding
Peptide docking
Structure refinement based on nuclear magnetic resonance and crystallographic data
Location of first and higher order transition states
Feasibility test and flexibility index problems
Design under uncertainty of batch processes
Parameter estimation of nonlinear algebraic and dynamic models
Optimal control of nonlinear processes
Pooling and blending problems
Design of nonideal separations
Location of homogeneous, heterogeneous, and reactive azeotropes
Synthesis of heat exchanger networks
Synthesis of complex reactor networks
Synthesis of distillation systems
Synthesis of pump network configurations
Chip layout and compaction problems
Process layout problems
Trim loss problems
Fabric and metal nesting problems.

6. Publications and Conferences

The significant growth in the area of global optimization, especially during the last fifteen years, is shown not only by the remarkable increase in the number of journal publications across all disciplines, but also by the emergence of (i) the *Journal of Global Optimization*, which started in 1990, (ii) the book series on *Nonconvex Optimization and Its Applications*, which started in 1994, (iii) the impressive list of textbooks and monographs that have been published, and (iv) five international conferences held in 1990, 1991, 1995, and 1999.

A representative list of textbooks and monographs focusing on global optimization includes the following:

6.1. Textbooks and Monographs

Interval Analysis, Moore (1966)

Towards Global Optimization I, Dixon and Szego (1975)

Towards Global Optimization II, Dixon and Szego (1978)

Methods and Applications of Interval Analysis, Moore (1979)

Computer Methods for the Range of Functions, Ratschek and Rokne (1984)

Constrained Global Optimization: Algorithms and Applications, Pardalos and Rosen (1987)

New Computer Methods for Global Optimization, Ratschek and Rokne (1988)

Global Optimization, Torn and Zilinskas (1989)

Global Optimization: Deterministic Approaches, Horst and Tuy (1996, 1993, 1990)

A Collection of Test Problems in Constrained Global Optimization Algorithms, Floudas and Pardalos (1990)

Interval Methods for Systems of Equations, Neumaier (1990)

Global Optimization Using Interval Analysis, Hansen (1992)

Recent Advances In Global Optimization, Floudas and Pardalos (1992)

Handbook of Global Optimization, Horst and Pardalos (1994)

Introduction to Global Optimization, Horst et al. (1995)

Minimax and Applications, Du and Pardalos (1995)

Advances in Multicriteria Analysis, Pardalos et al. (1995)

Global Optimization in Action—Continuous and Lipschitz Optimization, Pinter (1995)

State of the Art in Global Optimization: Computational Methods and Applications, Floudas and Pardalos (1996)

Global Optimization In Engineering Design, Grossmann (1996)

Rigorous Global Search: Continuous Problems, Kearfott (1996)

Optimization on Low Rank Nonconvex Structures, Konno et al. (1997)

Bayesian Heuristic Approach to Discrete and Global Optimization, Mockus et al. (1996)

Developments in Global Optimization, Bomze et al. (1997)

Multilevel Optimization: Algorithms and Applications, Migdalas et al. (1997)

Nonconvex Optimization in Mechanics, Mistakidis and Stavroulakis (1997)

Convex Analysis and Global Optimization, Tuy (1998)

Nondifferentiable Optimization and Polynomial Problems, Shor (1998)

Generalized Convexity, Generalized Monotonicity: Recent Results, Crouzeix et al. (1998)

Nonsmooth Approach to Optimization Problems with Equilibrium Constraints, Outrata et al. (1998)

A Reformulation Linearization Technique for Solving Discrete and Continuous Nonconvex Problems, Sherali and Adams (1999),

Handbook of Test Problems in Local and Global Optimization, Floudas et al. (1999a),

Deterministic Global Optimization: Theory, Algorithms and Applications, Floudas (2000)

The aforementioned textbooks and monographs were complemented with five international conferences with focal themes in the area of global optimization. These took place in the 1990s and are as listed in the following section.

6.2. Global Optimization Conferences

Workshop on Global Optimization, Sopron, Hungary, December 1990, Organizers: IIASA.

Recent Advances in Global Optimization, Princeton University, Princeton, NJ, May 1991. Organizers: C. A. Floudas and P. M. Pardalos.

State of the Art in Global Optimization: Computational Methods and Applications, Princeton University, Princeton, NJ, May 1995. Organizers: C. A. Floudas and P. M. Pardalos.

Workshop on Global Optimization, Szeged, Hungary, December 1995. Organizers: I. M. Bomze, T. Csendes, R. Horst, and P. M. Pardalos.

Optimization in Computational Chemistry and Molecular Biology: Local and Global Approaches, Princeton University, Princeton, NJ, May 1999. Organizers: C. A. Floudas and P. M. Pardalos.

7. Chemical Engineering Contributions

Significant effort has been expended in the last four decades toward theoretical and algorithmic studies of applications that arise in chemical engineering process design, process synthesis, process control, as well as in computational chemistry and molecular biology. In the last decade we have experienced a dramatic growth of interest in chemical engineering for new methods of global optimization and their application to important engineering, as well as computational chemistry and molecular biology problems.

Contributions from the chemical engineering community to the area of global optimization can be traced to the work of Stephanopoulos and Westerberg (1975), and Westerberg and Shah (1978). Renewed interest in seeking global solutions was motivated by the work of Floudas et al. (1989). The first exact primal-dual global optimization approach was proposed by Floudas and Visweswaran (1990, 1993) and its features were explored for quadratically constrained and polynomial problems in the work of Visweswaran and Floudas (1990, 1992, 1993). Swaney (1990) proposed a branch-and-bound global optimization approach and more recently Quesada and Grossmann (1993, 1995) combined convex underestimators in a branch-and-bound framework for linear fractional and bilinear programs. Manousiouthakis and Sourlas (1992) proposed a reformulation to a series of reverse convex problems, and Tsirukis and Reklaitis (1993) proposed a feature extraction algorithm for constrained global optimization. Maranas and Floudas (1992, 1993, 1994a,b) proposed a novel branch-and-bound method combined with a difference of convex functions transformation for the global optimization of molecular conformation problems that arise in computational chemistry. A very important theoretical advance was made by Liu and Floudas (1993, 1995, 1996) who showed that the GOP can be applied to very general classes of NLPs. This result is very significant because it extends the classes of mathematical problems on which the GOP can be applied to nonlinear objective function and constraints that are twice continuously differentiable. Vaidyanathan and El-Halwagi (1994) proposed an interval analysis based global optimization method. In a series of papers, McDonald and Floudas (1994, 1995a,b,c), addressed the fundamental problems of (i) minimization of the Gibbs free energy and (ii) the tangent plane stability criterion that arise in phase and chemical reaction equilibrium, as global optimization problems for the first time. They proposed decomposition based approaches for biconvex problems that result from the use of the NRTL equation, and branch-and-bound approaches for the UNIQUAC, UNIFAC, ASOG, and TK-Wilson equations. Psarris and Floudas (1995) proposed a global optimization approach based on the GOP for the robust stability of control systems under real parametric uncertainty. Ryoo and Sahinidis (1995) suggested the application of reduction tests within the framework of branch-and-bound methods and they proposed a branch-and-reduce global optimization approach (Ryoo and Sahinidis, 1996). Dorneigh and Sahinidis (1995) proposed a global optimization approach for the chip layout and compaction problem. Androulakis et al. (1995) proposed the global optimization method αBB which addresses general continuous optimization problems with nonconvexities in the objective function and/or constraints. This approach classifies the nonconvexities as special structure (e.g., bilinear, signomial, univariate) or generic structure and is based on convex relaxations and a branch-and-bound framework. Maranas and Floudas

(1995) proposed a new approach for enclosing all ϵ-feasible solutions of nonlinearly constrained systems of equations. This approach transforms the problem into a min-max form and corresponds to enclosing all multiple global optima via the αBB global optimization approach.

The books of Floudas and Pardalos (1996) and Grossmann (1996) contain a number of recent chemical engineering contributions, and the review paper by Floudas (1997) presents an overview of recent chemical engineering contributions and advances in global optimization based on the decomposition approach GOP, generalized geometric programming, and the αBB approach.

Some of the aforementioned chemical engineering contributions in 1996 are briefly discussed in the following. Adjiman and Floudas (1996) proposed novel approaches for the rigorous determination of the α parameters that are employed in the αBB global optimization approach. These methods are based on a modified Kharitonov theorem for interval polynomials, interval analysis of the Hessian matrices and on calculations of rigorous bounds on the minimum eigenvalue for general twice differentiable problems. Staus et al. (1996) formulated the combined adaptive controller design and estimation problem as a nonconvex problem with convex objective and bilinear constraints, and proposed a branch-and-bound global optimization method, which is based on the McCormick underestimators. Visweswaran et al. (1996) addressed bilevel linear and quadratic programming problems to global optimality by employing the basic principles of the GOP and developing additional theoretical properties that exploit the underlying mathematical structure of such problems. Androulakis et al. (1996) developed a distributed version of the GOP, and discussed the key theoretical and implementation issues along with extensive computational results on large-scale indefinite quadratic and pooling/blending problems. Shectman and Sahinidis (1996) proposed a finite algorithm for separable concave programs, discussed the design of such branch-and-bound approaches, and presented computational results employing domain reduction tests. McKinnon et al. (1996) addressed the global optimization in phase and chemical reaction equilibrium using interval analysis coupled with the tangent plane stability criterion of Gibbs. Hua et al. (1996) applied an interval analysis method for the phase stability computations of binary and ternary mixtures using activity coefficient models. Mockus and Reklaitis (1996) proposed a continuous formulation for the short term batch scheduling problem, and developed a global optimization approach based on the Bayesian heuristic. Lucia and Xu (1996) studied nonconvexity issues in sparse successive quadratic programming, and Banga and Seider (1996) introduced a stochastic global optimization approach which they applied to the optimal design of a fermentation process, phase equilibrium problems, and optimal control problems. Epperly and Swaney (1996a,b) proposed a branch-and-bound method with a new linear programming underestimating problem, and provided extensive computational studies. This approach is applicable to NLPs in factorable form which includes quadratic objective and constraints, and twice differentiable transcendental functions. Visweswaran and Floudas (1996a) proposed new formulations for the GOP algorithm which are based on a branch-and-bound framework. These allow the implicit solution of relaxed dual problems, which are formulated in a single MILP model, and feature a linear branching scheme. Visweswaran and Floudas (1996b) discussed the implementation issues of the GOP and provide extensive computational experience on a variety of chemical engineering problems. Byrne and Bogle (1996a,b) and Vaidyanathan and El-Halwagi (1996) proposed global optimization methods that are based on interval analysis for constrained NLPs and MINLPs, respectively. Liu et al. (1996) proposed an approach for planning of chemical process networks which is based on global concave minimization. The approach is based on their earlier work on finite global optimization approaches and their computational studies revealed the efficiency of the proposed approach. Ierapetritou and Pistikopoulos (1996) studied the global optimization of stochastic planning, scheduling and design problems, applied the decomposition-based approach, GOP, and demonstrated that significant reductions in the number of relaxed dual problems can be achieved by further exploiting the mathematical structure. Iyer and Grossmann (1996) extended the global optimization approach of Quesada and Grossmann (1993) to the multi-period heat exchanger networks that feature fixed configuration, linear cost functions, arithmetic mean driving forces, and isothermal mixing. Quesada and Grossmann (1996) studied the use of alternative bounding approximations further and applied them to a variety of engineering design problems that include structural design, batch processes, layout design, and portfolio problems. Smith and Pantelides (1996) proposed a symbolic manipulation algorithm for the automatic reformulation of algebraic constraints and introduced a spatial type branch-and-bound approach within the gPROMS framework.

Maranas and Floudas (1997) proposed a global optimization approach for generalized geometric programming problems that have many applications in robust control and engineering design problems. McDonald and Floudas (1997) proposed the combination of the Gibbs free energy minimization problem with the stability criterion problem, developed a special purpose program GLOPEQ, and performed an extensive computational study on difficult phase equilibrium problems. Harding et al. (1997) proposed novel formulations and global optimization methods for the enclosure of all homogeneous azeotropes. This approach offers a theoretical guarantee of enclosing all homogeneous azeotropes and it was applied to a variety of activity coefficient models. Androulakis et al. (1997a) proposed a novel global optimization approach based on atomistic level modeling and the αBB principles for the structure prediction of oligopeptides. Androulakis et al. (1997b) proposed a decomposition based global optimization approach for the peptide docking problem of amino acids into the binding sites of the HLA-DR1 protein. Harding and Floudas (1997) proposed novel global optimization approaches for batch design under uncertainty of multiproduct and multipurpose plants. These approaches are based on the exact calculation of the α parameters within the αBB global optimization framework and they result in the tightest possible underestimators. Epperly et al. (1997) proposed a global optimization approach for stochastic batch design problems that is based on a reduction of the number of variables needed for branching, as suggested in their reduced space branch-and-bound approach (Epperly and Pistikopoulos, 1997). Sriniwas and Arkun (1997) introduced a global optimization approach based on the GOP principles for nonlinear model predictive control problems with polynomial ARX models. VanAntwerp et al. (1997) introduced a global optimization approach based on branch-and-bound for the design of a robust controller for time-varying nonlinear plant perturbations. Smith and Pantelides (1997) further studied their reformulation/spatial branch-and-bound approach for the global optimization of MINLP problems.

Klepeis et al. (1998) proposed new global optimization approaches for structure prediction of solvated peptides using the area and volume accessible to the solvent models. A review of the global optimization activities in the areas of protein folding and peptide docking can be found in Floudas et al. (1999b). Esposito and Floudas (1998) proposed the first global optimization method for parameter estimation and data reconcilliation of nonlinear algebraic models using the principles of the αBB approach. McKinnon and Mongeau (1998) proposed a generic global optimization approach for the phase and chemical reaction equilibrium problem that is based on interval analysis and combines the stability criterion with minimization of the Gibbs free energy. Hua et al. (1998) applied an interval analysis method for phase stability computations of binary and ternary mixtures using equation of state models. Maier et al. (1998) applied an interval analysis based approach for the enclosure of homogeneous azeotropes. They employed the formulations proposed by Harding et al. (1997) and studied systems with activity coefficient and equation of state models. Meyer and Swartz (1998) proposed a new approach for testing convexity for phase equilibrium problems that can be coupled with the global optimization methods of McDonald and Floudas (1994, 1995a,b,c, 1997). Shectman and Sahinidis (1998) proposed a finite global optimization method for separable concave problems. Yamada and Hara (1998) proposed a global optimization approach based on the triangle covering for H-infinity control with constant diagonal scaling. Androulakis and Floudas (1998) studied the parallel computation issues that arise using the αBB global optimization approach. Adjiman et al. (1998a) proposed several rigorous new methods for the calculation of the α parameters for (i) uniform diagonal shift of the Hessian matrix and (ii) nonuniform diagonal shift of the Hessian matrix, and established their potential trade-offs. Adjiman et al. (1998b) presented the detailed implementation of the αBB approach and computational studies in process design problems such as heat exchanger networks, reactor-separator networks, and batch design under uncertainty. Westerlund et al. (1998) proposed an extended cutting plane approach for the global optimization of pseudoconvex MINLP problems. Zamora and Grossmann (1998a) proposed a hybrid branch-and-bound and outer approximation method for the global optimization of heat exchanger networks with no stream splits. Zamora and Grossmann (1998b) introduced a deterministic branch-and-bound approach for structured process systems that have univariate concave, bilinear, and linear fractional terms. They proposed several properties of the contraction operation, embedded them in the global optimization algorithm, and studied the contraction effects on several applications. Adjiman et al. (2000) proposed two novel global optimization approaches for nonconvex mixed integer nonlinear programming problems. The first approach is for separable continuous and integer domains, while the second approach addresses general mixed integer nonlinear problems. Klepeis and Floudas (1999a) presented a detailed comparative study of the putative global minimum structures of solvated oligopeptides where the solvation models employed were based on the area and volume accessible to the solvent. Klepeis and Floudas (1999b) proposed a novel deterministic global optimization approach for free energy calculations of peptides and presented extensive computational studies on solvated and unsolvated pentapeptides. Westerberg and Floudas (1999a,b) introduced a global optimization framework for the enclosure of all transition states of potential energy hypersurfaces, and studied the reaction pathways, and the dynamics of folding solvated and unsolvated oligopeptides. Smith and Pantelides (1999) presented a symbolic reformulation/spatial branch-and-bound approach for the global optimization of nonconvex MINLPs. Adhya et al. (1999) introduced a Lagrangian approach to address small and medium size pooling problems. Floudas (1999a) and Floudas and Pardalos (1999) reviewed the recent global optimization developments in process design and control. Floudas (1999b) reviewed the advances in enclosing all solutions of nonlinear constrained systems of equations via deterministic global optimization approaches. Klepeis et al. (1999) introduced deterministic global optimization approaches to NMR structure refinement and studied the structure prediction of Compstatin. Harding and Floudas (2000) developed rigorous global optimization approaches for the phase stability problem when cubic equations of state are used. Harding and Floudas (2000b) introduced deterministic global optimization approaches for enclosing all heterogeneous and reactive azeotropes. Esposito and Floudas (2000a) proposed two novel global optimization approaches, based on collocation and integration, for the parameter estimation and optimal control of differential-algebraic models. Floudas et al. (1999a,b) introduced a handbook of global optimization test problems that cover a broad spectrum of engineering and applied science applications.

References

Adhya, N., M. Tawarmalani, and N. V. Sahinidis. 1999. "A Lagrangian approach to the pooling problem." *I&EC Res.* **38**: 1956–1972.

Adjiman, C. S. and C. A. Floudas. 1996. "Rigorous Convex Underestimators for General Twice-Differentiable Problems." *J. Global Optim* **9**: 23–40.

Adjiman, C. S., S. Dallwig, C. A. Floudas, and A. Neumaier. 1998a. "A global optimization method, αBB, for general twice-differentiable NLPs—I. theoretical advances." *Comput. Chem. Engng.* **22(9)**: 1137–1158.

Adjiman, C. S., I. P. Androulakis, and C. A. Floudas. 1998b. "A global optimization method, αBB, for general twice-differentiable NLPs—II. Implementation and computational results." *Comput. Chem. Engng.* **22(9)**: 1159–1179.

Adjiman, C. S., I. P. Androulakis, and C. A. Floudas. 2000. "Global optimization of mixed-integer nonlinear problems." *AIChE J.* **46**: 1769–1797.

Androulakis, I. P. and C. A. Floudas. 1998. "Distributed branch and bound algorithms in global optimization." In *Parallel Processing of Discrete Problems*, edited by P. M. Pardalos, IMA Volumes in Mathematics and Its Applications, **106**: 1–36. Berlin: Springer-Verlag.

Androulakis, I. P., C. D. Maranas, and C. A. Floudas. 1995. "αBB: a global optimization method for general constrained nonconvex problems." *J. Global Optim.* **7**: 337–363.

Androulakis, I. P., V. Visweswaran, and C. A. Floudas. 1996. "Distributed decomposition-based approaches in global optimization." In *Proceedings of the State of the Art in Global Optimization: Computational Methods and Applications*, edited by C. A. Floudas and P. M. Pardalos, Kluwer Book Series on Nonconvex Optimization and Its Applications, pp. 285–302. Kluwer Academic Publishers.

Androulakis, I. P., C. D. Maranas, and C. A. Floudas. 1997a. "Global minimum potential energy conformations of oligopeptides." *J. Global Optim.* **11**: 1–34.

Androulakis, I. P., N. N. Nayak, M. G. Ierapetritou, D. S. Monos, and C. A. Floudas. 1997b. "A predictive method for the evaluation of peptide binding in pocket 1 of HLA-DRB1 via global minimization of energy interactions." *Proteins* **29**: 87–102.

Banga, J. R. and W. D. Seider. 1996. "Global optimization of chemical processes using stochastic algorithms." In *Proceedings of the State of the Art in Global Optimization: Computational Methods and Applications*, edited by C. A. Floudas and P. M. Pardalos, Kluwer Book Series on Nonconvex Optimization and Its Applications. Kluwer Academic Publishers.

Ben-Tal, A., G. Eiger, and V. Gershovitz. 1994. "Global Minimization by Reducing the Duality Gap." *Math. Programming* **63**: 193–212.

Bomze, I. M., T. Csendes, R. Horst, and P. M. Pardalos. 1997. *Developments in Global Optimization* Kluwer Academic Publishers.

Brooks, B., R. Bruccoleri, B. Olafson, D. States, S. Swaminathan, and M. Karplus. 1983. "CHARMM: A Program for Macromolecular Energy Minimization and Dynamics Calculations." *J. Comp. Chem.* **4**: 187–217.

Byrne, R. P. and I. D. L. Bogle. 1996a. "An accelerated interval method for global optimisation." *Comput. Chem. Engng. Suppl.* **20**: S49–S54.

Byrne, R. P. and I. D. L. Bogle. 1996b. "Solving nonconvex process optimization problems using interval subdivision algorithms." In *Global Optimization in Engineering Design*, edited by I. E. Grossmann, Kluwer Book Series in Nonconvex optimization and its Applications, pp. 155–174. Kluwer Academic Publishers.

Chan, H. S. and K. A. Dill. 1998. "Protein folding in the landscape perspective: chevron plots and non-Arrhenius kinetics." *Proteins* **30**: 2–33.

Chen, J. J. J. 1987. "Comments on improvements on a replacement for the logarithmic mean." *Chem. Engng. Sci.* **42**: 2488–2489.

Crouzeix, J. P., J. E. Martinez-Legaz, and M. Volle. 1998. *Generalized Convexity, Generalized Monotonicity: Recent Results.* Kluwer Academic Publishers.

Dauber-Osguthorpe, P., V. A. Roberts, D. J. Osguthorpe, J. Wolff, M. Genest, and A. T. Hagler. 1988. "Structure and energetics of ligand-binding to proteins—*Escherichia-coli* dihydrofolate reductase trimethoprim, a drug-receptor system." *Proteins* **4**: 31–47.

Dill, K. A. and H. S. Chan. 1997. "From levinthal to pathways funnels." *Nature Struc. Biol.* **4**: 10–19.

Dixon, L. C. W. and G. P. Szego. 1975. *Towards Global Optimisation.* Amsterdam: North-Holland.

Dixon, L. C. W. and G. P. Szego. 1978. *Towards Global Optimisation 2.* Amsterdam: North-Holland.

Dorneigh, M. C. and N. V. Sahinidis. 1995. "Global Optimization Algorithms for Chip Layout and Compaction." *Engng. Optim.* **25(2)**: 131–154.

Du, D. Z. and P. M. Pardalos. 1995. *Minimax and Applications.* Kluwer Academic Publishers.

Epperly, T. G. W., M. G. Ierapetritou, and E. N. Pistikopoulos. 1997. "On the global and efficient solution of stochastic batch plant design problems." *Comput. Chem. Engng.* **21**: 1411–1431.

Epperly, T. G. W. and E. N. Pistikopoulos. 1997. "A reduced space branch-and-bound algorithm for global optimization." *J. Global Optim.* **11**: 287–311.

Epperly, T. G. W. and R. Swaney. 1996a. "Branch and bound for global NLP: new bounding LP." In *Global Optimization in Engineering Design*, edited by I. E. Grossmann, Kluwer Book Series in Nonconvex Optimization and its Applications, pp. 1–36. Kluwer Academic Press.

Epperly, T. G. W. and R. Swaney. 1996b. "Branch and bound for global NLP: iterative LP algorithm and results." In *Global Optimization in Engineering Design*, edited by I. E. Grossmann, Kluwer Book Series in Nonconvex Optimization and its Applications, pp. 37–74. Kluwer Academic Press.

Esposito, W. R. and C. A. Floudas. 1998. "Global optimization in parameter estimation of nonlinear algebraic models via the error-in-variables approach." *I&EC Res.* **35(5)**: 1841–1858.

Esposito, W. R. and C. A. Floudas. 2000. "Global optimization in parameter estimation of differential algebraic systems." *I&EC Res.* **39(5)**: 1291–1310.

Floudas, C. A. 1995. *Nonlinear and Mixed Integer Optimization: Fundamentals and Applications.* New York: Oxford University Press.

Floudas, C. A. 1997. "Deterministic global optimization in design, control, and computational chemistry." In *IMA Volumes in Mathematics and its Applications: Large Scale Optimization with Applications, Part II*

Volume 93, edited by L. T. Biegler, T. F. Coleman, A. R. Conn, and F. N. Santosa, pp. 129–184. New York: Springer-Verlag.

Floudas, C. A. 1999a. "Global optimization in design and control of chemical process sytems." *J. Process Control*.

Floudas, C. A. 1999b. "Recent advances in global optimization for process synthesis, design and control: enclosure of all solutions." *Comput. Chem. Engng.* **23**: S963–S974.

Floudas, C. A. 2000. In *Deterministic Global Optimization: Theory, Methods and Applications*. Kluwer Academic Publishers.

Floudas, C. A. and A. Aggarwal. 1990. "A decomposition strategy for global optimum search in the pooling problem." *ORSA J. Comput.* **2(3)**.

Floudas, C. A. and P. M. Pardalos. 1990. In *A Collection of Test Problems for Constrained Global Optimization Algorithms*, Volume 455, Lecture Notes in Computer Science. Berlin: Springer-Verlag.

Floudas, C. A. and P. M. Pardalos. 1992. In *Recent Advances in Global Optimization*. Princeton: Princeton University Press.

Floudas, C. A. and P. M. Pardalos. 1996. In *State of the Art in Global Optimization: Computational Methods and Applications*. Kluwer Academic Publishers.

Floudas, C. A. and P. M. Pardalos. 1999. "Recent developments in global optimization and their relevance to process design." In *FOCAPD'99*. AICHE Symposium Series.

Floudas, C. A. and V. Visweswaran. 1990. "A global optimization algorithm (GOP) for certain classes of nonconvex NLPs: I. Theory." *Comput. Chem. Engng.* **14**: 1397–1417.

Floudas, C. A. and V. Visweswaran. 1993. "A primal-relaxed dual global optimization approach." *J. Optim. Theory Appl.* **78(2)**: 187–225.

Floudas, C. A., A. Aggarwal, and A. R. Ciric. 1989. "Global optimum search for nonconvex NLP and MINLP problems." *Comput. Chem. Engng.* **13(10)**: 1117–1132.

Floudas, C. A., P. M. Pardalos, C. S. Adjiman, W. R. Esposito, Z. H. Gümüş, S. T. Harding, J. L. Klepeis, C. Meyer, and C. A. Schweiger. 1999a. In *Handbook of Test Problems in Local and Global Optimization*. Kluwer Academic Publishers.

Floudas, C. A., J. L. Klepeis, and P. M. Pardalos. 1999b. "Global optimization approaches in protein folding and peptide docking." In *DIMACS Series In Discrete Mathematics and Theoretical Computer Science*, Volume 47, edited by M. Farach-Colton, F. S. Roberts, M. Vingron, and M. Waterman, pp. 141–171.

Grossmann, I. E. 1996. *Global Optimization in Engineering Design*. Dordrecht, The Netherlands: Kluwer Academic Publishers.

Hansen, E. 1992. *Global Optimization using Interval Analysis*. Marcel Dekker, Inc.

Harding, S. T. and C. A. Floudas. 1997. "Global optimization in multiproduct and multipurpose batch design under uncertainty." *Ind. Engng. Chem. Res.* **36(5)**: 1644–1664.

Harding, S. T. and C. A. Floudas. 2000a. "Locating all heterogeneous and reactive azeotropes in multicomponent mixtures." *Ind. Engng. Chem. Res.* **39(6)**: 1576–1595.

Harding, S. T. and C. A. Floudas. 2000b. "Phase stability for equations of state: a global optimization approach." *AICHE J.* **46(7)**: 1422–1440.

Harding, S. T., C. D. Maranas, C. M. McDonald, and C. A. Floudas. 1997. "Locating all homogeneous azeotropes in multicomponent mixtures." *Ind. Engng. Chem. Res.* **36(1)**: 160–178.

Hoare, M. R. 1979. "Structure and dynamics of simple microclusters." *Adv. Chem. Phys.* **40**: 49.

Horst, R. and P. M. Pardalos. 1994. In *Handbook of Global Optimization*. Kluwer Academic Publishers.

Horst, R. and H. Tuy. 1990. *Global Optimization: Deterministic Approaches*. 1st ed. Berlin: Springer-Verlag.

Horst, R. and H. Tuy. 1993. *Global Optimization: Deterministic Approaches*. 2d ed. Berlin: Springer-Verlag.

Horst, R. and H. Tuy. 1996. *Global Optimization: Deterministic Approaches*. 3d ed. Berlin: Springer-Verlag.

Horst, R., P. M. Pardalos, and N. V. Thoai. 1995. *Introduction to Global Optimization*. Kluwer Academic Publishers.

Hua, Z., J. F. Brennecke, and M. A. Stadtherr. 1996. "Reliable prediction of phase stability using an interval-newton method." *Fluid Phase Equilibria* **116**: 52–59.

Hua, Z., J. F. Brennecke, and M. A. Stadtherr. 1998. "Enhanced interval analysis for phase stability: cubic equation of state models." *Ind. Engng. Chem. Res.* **37**: 1519–1527.

Ierapetritou, M. G. and E. N. Pistikopoulos. 1996. "Batch plant design and operations under uncertainty." *Ind. Engng. Chem. Res.* **35(2)**: 772–787.

Iyer, R. R. and I. E. Grossmann. 1996. "Global optimization of heat exchanger networks with fixed configuration for multiperiod design." In *Global Optimization in Engineering Design*, edited by I. E. Grossmann, Kluwer Book Series in Nonconvex Optimization and its Applications, pp. 289–308.

Kearfott, R. B. 1996. *Rigorous Global Search: Continuous Problems*. Kluwer Academic Publishers.

Kim, I., M. J. Liebman, and T. F. Edgar. 1990. "Robust error-in-variables estimation using nonlinear programming techniques." *AIChE J.* **36(7)**: 985.

Klepeis, J. L. and C. A. Floudas. 1999a. "A comparative study of global minimum energy conformations of hydrated peptides." *J. Comput. Chem.* **20(6)**: 636–654.

Klepeis, J. L. and C. A. Floudas. 1999b. "Free energy calculations for peptides via deterministic global optimization." *J. Chem. Phys.* **110(15)**: 7491–7512.

Klepeis, J. L., I. P. Androulakis, M. G. Ierapetritou, and C. A. Floudas. 1998. "Predicting solvated peptide conformations via global minimization of energetic atom to atom interactions." *Comput. Chem. Engng.* **22(6)**: 765–788.

Klepeis, J. L., C. A. Floudas, D. Morikis, and J. D. Lambris. 1999. "Predicting peptide structures using NMR data and deterministic global optimization." *J. Comput. Chem.* **20(16)**.

Konno, H., P. T. Thach, and H. Tuy. 1997. In *Optimization on Low Rank Nonconvex Structures*. Kluwer Academic Publishers.

Kubicek, M., H. Hofmann, V. Hlavacek, and J. Sinkule. 1980. "Multiplicity and stability in a sequence of two nonadiabatic nonisothermal CSTRs." *Chem. Engng. Sci.* **35**: 987.

Leach, A. R. 1991. "A survey of methods for searching the conformational space of small and medium-sized molecules." In *Reviews in Computational Chemistry*, volume 2, pp. 1–55. VCH Publishers.

Liu, M. L., N. V. Sahinidis, and J. P. Shectman. 1996. "Planning of chemical process networks via global concave minimization." In *Global Optimization in Engineering Design*, edited by I. E. Grossmann, Kluwer Book Series in Nonconvex Optimization and its Applications, pp. 195–230. Kluwer Academic Publishers.

Liu, W. B. and C. A. Floudas. 1993. "A remark on the GOP algorithm for global optimization." *J. Global Optim.* **3(3)**: 519–521.

Liu, W. B. and C. A. Floudas. 1995. "Convergence of the GOP algorithm for a large class of smooth optimization problems." *J. Global Optim.* **6**: 607–611.

Liu, W. B. and C. A. Floudas. 1996. "Generalized primal-relaxed dual approach for global optimization." *J. Optim. Theory Appl.* **90(2)**: 417–434.

Lucia, A. and J. Xu. 1996. "Nonconvexity and descent in nonlinear programming." In *Proceedings of the State of the Art in Global Optimization: Computational Methods and Applications* edited by C. A. Floudas and P. M. Pardalos, Kluwer Book Series on Nonconvex Optimization and Its Applications. Kluwer Academic Publishers.

Maier, R. W., J. F. Brennecke, and M. A. Stadtherr. 1998. "Reliable computation of homogeneous azeotropes." *AIChE J.* **44**: 1745–1755.

Manousiouthakis, V. and D. Sourlas. 1992. "A global optimization approach to rationally constrained rational programming." *Chem. Engng. Commun.* **115**: 127.

Maranas, C. D. and C. A. Floudas. 1992. "A global optimization approach for Lennard–Jones microclusters." *J. Chem. Phys.* **97(10)**: 7667–7677.

Maranas, C. D. and C. A. Floudas. 1993 "Global optimization for molecular conformation problems." *Ann. Oper. Res.* **42**: 85–117.

Maranas, C. D. and C. A. Floudas. 1994a. "A deterministic global optimization approach for molecular structure determination." *J. Chem. Phys.* **100(2)**: 1247–1261.

Maranas, C. D. and C. A. Floudas. 1994b. "Global minimum potential energy conformations of small molecules." *J. Global Optim.* **4**: 135–170.

Maranas, C. D. and C. A. Floudas. 1995. "Finding all solutions of nonlinearly constrained systems of equations." *J. Global Optim.* **7(2)**: 143–182.

Maranas, C. D. and C. A. Floudas. 1997. "Global optimization in generalized geometric programming." *Comput. Chem. Engng.* **21(4)**: 351–370.

McDonald, C. M. and C. A. Floudas 1994. "Decomposition based and branch and bound global optimization approaches for the phase equilibrium problem." *J. Global Optim.* **5**: 205–251.

McDonald, C. M. and C. A. Floudas. 1995a. "Global optimization for the phase and chemical equilibrium problem: application to the NRTL equation." *Comput. Chem. Engng.* **19(11)**: 1111–1141.

McDonald, C. M. and C. A. Floudas. 1995b. "Global optimization for the phase stability problem." *AIChE J* **41(7)**: 1798–1814.

McDonald, C. M. and C. A. Floudas. 1997 "GLOPEQ: a new computational tool for the phase and chemical equilibrium problem." *Comput. Chem. Engng.* **21**: 1–23.

McDonald, C. M. and C. A. Floudas. 1995c. 'Global optimization and analysis for the Gibbs free energy function using the UNIFAC, Wilson and ASOG equations." *IE&C Res.* **34**: 1674.

McKinnon, K. and M. Mongeau. 1998. "A generic global optimization algorithm for the chemical and phase equilibrium problem." *J. Global Optim.* **12**: 325–351.

McKinnon, K. I. M., C. Millar, and M. Mongeau. 1996. "Global optimization for the chemical and phase equilibrium problem using interval analysis." In *Proceedings of the State of the Art in Global*

Optimization: Computational Methods and Applications, edited by C. A. Floudas and P. M. Pardalos, Kluwer Book Series on Nonconvex Optimization and Its Applications. Kluwer Academic Publishers.

Meyer, C. A. and C. L. E. Swartz. 1998. "A regional convexity test for global optimization: application to the phase equilibrium problem." *Comput. Chem. Engng.* **22**: 1407–1418.

Migdalas, A., P. M. Pardalos, and P. Varbrand. 1997. *Multilevel Optimization: Algorithms and Applications*. Kluwer Academic Publishers.

Mistakidis, E. S. and G. E. Stavroulakis. 1997. *Nonconvex Optimization in Mechanics: Algorithms, Heuristics and Engineering Applications by the F.E.M.* Kluwer Academic Publishers.

Mockus, L. and G. V. Reklaitis. 1996. "A new global optimization algorithm for batch process scheduling." In *State of the Art In Global Optimization*, edited by C. A. Floudas and P. M. Pardalos, pp. 521–538. Kluwer Academic Publishers.

Mockus, J., W. Eddy, A. Mockus, L. Mockus, and G. V. Reklaitis. 1996. *Bayesian Heuristic Approach to Discrete and Global Optimization: Algorithms, Visualization, Software and Applications*. Kluwer Academic Publishers.

Moore, R. E. 1966. *Interval Analysis*. NJ: Prentice-Hall, Englewood Cliffs.

Moore, R. E. 1979. *Methods and Applications of Interval Analysis*. Philadelphia: SIAM.

Némethy, G., K. D. Gibson, K. A. Palmer, C. N. Yoon, G. Paterlini, A. Zagari, S. Rumsey, and H. A. Scheraga. 1992. "Energy parameters in polypeptides. 10. Improved geometrical parameters and nonbonded interactions for use in the ECEPP/3 algorithm with application to proline-containing peptides." *J. Phys. Chem.* **96**: 6472–6484.

Neumaier, A. 1990. In *Interval Methods for Systems of Equations*. Encyclopedia of Mathematics and its Applications. Cambridge, England: Cambridge University Press.

Neumaier, A. 1997. "Molecular modeling of proteins and mathematical prediction of protein structure." *SIAM Rev.* **39(3)**: 407–460.

Outrata, J., M. Kocvara, and J. Zowe. 1998. *Nonsmooth Approach to Optimization Problems with Equilibrium Constraints: Theory, Applications and Numerical Results*. Kluwer Academic Publishers.

Pardalos, P. M. and J. B. Rosen. 1987. *Constrained Global Optimization: Algorithms and Applications*, Volume 268, Lecture Notes in Computer Science. Berlin: Springer-Verlag.

Pardalos, P. M., D. Shalloway, and G. Xue. 1994. "Optimization methods for computing global minima of nonconvex potential energy functions." *J. Global Optim.* **4**: 117–133.

Pardalos, P. M., Y. Siskos, and C. Zopounidis. 1995. In *Advances in Multicriteria Analysis*. Kluwer Academic Publishers.

Pardalos, P. M., D. Shalloway, and G. Xue. 1996. *Global Minimization of Nonconvex Energy Functions: Molecular Conformation and Protein Folding*, Volume 23, DIMACS Series in Discrete Mathematics and Theoretical Computer Science. American Mathematical Society.

Pinter, J. D. 1995. In *Global Optimization In Action—Continuous and Lipschitz Optimization*. Kluwer Academic Publishers.

Psarris, P. and C. A. Floudas. 1995. "Robust stability analysis of systems with real parameter uncertainty: a global optimization approach." *Int. J. Robust Nonlinear Control* **6**: 699–717.

Quesada, I. and I. E. Grossmann. 1993. "Global optimization algorithm for heat exchanger networks." *I&EC Res.* **32**: 487.

Quesada, I. and I. E. Grossmann. 1995. "A global optimization algorithm linear fractional and bilinear programs." *J. Global Optim.* **6**: 39–76.

Quesada, I. and I. E. Grossmann. 1996. "Alternative bounding approximations for the global optimization of various engineering design problems." In *Global Optimization in Engineering Design*, edited by I. E. Grossmann, Kluwer Book Series in Nonconvex Optimization and its Applications, pp. 309–332.

Ratschek, H. and J. Rokne. 1984. In *Computer Methods for the Range of Functions*. Ellis Horwood Series in Mathematics and its Applications. Chichester, England: Halsted Press.

Ratschek, H. and J. Rokne. 1988. In *New Computer Methods for Global Optimization*. Chichester, England: Halsted Press.

Renon, H. and J. M. Prausnitz. 1968. "Local compostions in thermodynamic excess functions for liquid mixtures." *AIChE J.* **14(1)**: 135.

Ricker, N. L. 1984. "Comparison of methods for nonlinear parameters estimation." *Ind. Engng. Chem. Process Design Dev.* **23(2)**: 283–286.

Ryoo, H. S. and N. V. Sahinidis. 1995. "Global optimization of nonconvex NLPs and MINLPs with applications in process design." *Comput. Chem. Engng* **19(5)**: 551–566.

Ryoo, H. S. and N. V. Sahinidis. 1996. "A branch and reduce approach to global optimization." *J. Global Optim.* **8(2)**: 107–139.

Scheraga, H. A. 1992. "Predicting three-dimensional structures of oligopeptides." In *Reviews in Computational Chemistry*, Volume 3, edited by K. B. Lipkowitz and D. B. Boyd, pp. 73–142. VCH Publishers.

Shectman, J. P. and N. V. Sahinidis. 1996. "A finite algorithm for global minimization of separable concave programs." In *Proceedings of the State of the Art in Global Optimization: Computational Methods and Applications*, edited by C. A. Floudas and P. M. Pardalos, Kluwer Academic Series on Nonconvex Optimization and Its Applications. Kluwer Academic Publishers.

Shectman, J. P. and N. V. Sahinidis. 1998. "A finite algorithm for global optimization of separable concave functions." *J. Global Optim.* **12**: 1–36.

Sherali, H. D. and W. P. Adams. 1999. *A Reformulation Linearization Technique for Solving Discrete and Continuous Nonconvex Problems*. Kluwer Academic Publishers.

Shor, N. Z. 1998. *Nondifferentiable Optimization and Polynomial Problems*. Kluwer Academic Publishers.

Smith, E. M. B. and C. C. Pantelides. 1996. "Global optimisation of general process models." In *Global Optimization in Engineering Design*, edited by I. E. Grossmann, pp. 355–386. Kluwer Academic Publishers.

Smith, E. M. B. and C. C. Pantelides. 1997. "Global optimisation of nonconvex MINLPs." *Comput. Chem. Engng Suppl.* **21**: S791–S796.

Smith, E. M. B. and C. C. Pantelides. 1999. 'A symbolic reformulation/spatial branch and bound algorithm for the global optimization of nonconvex MINLPs." *Comput. Chem. Engng* **23**: 457–478.

Sriniwas, G. R. and Y. Arkun. 1997. "A global solution to the nonlinear model predictive control algorithms using polynomial ARX models." *Comput. Chem. Engng* **21**: 431–439.

Staus, G. H., L. T. Biegler, and B. E. Ydstie. 1996. "Adaptive control via nonconvex optimization." In *Proceedings of the State of the Art in Global Optimization: Computational Methods and Applications*, edited by C. A. Floudas and P. M. Pardalos, Kluwer Academic Series on Nonconvex Optimization and Its Applications. Kluwer Academic Publishers.

Stephanopoulos, G. and A. W. Westerberg. 1975. "The use of Hestenes method of multipliers to resolve dual gaps in engineering system optimization." *JOTA* **15**: 285.

Stewart, W. E., M. Caracotsios, and J. P. Sorensen. 1992. "Parameter estimation from multiresponse data." *AIChE J.* **38(5)**: 641.

Swaney, R. E. 1990. Global solution of algebraic nonlinear programs. *Presented at Annual AIChE Meeting*. Chicago.

Torn, A. and A. Zilinskas. 1989. *Global Optimization*, Volume 350, Lecture Notes in Computer Science. Berlin: Springer-Verlag.

Tsirukis, A. G. and G. V. Reklaitis. 1993. "Feature extraction algorithms for constrained global optimization: Parts I and II." *Ann. Oper. Res.* **42**: 229–275.

Tuy, H. 1998. *Convex Analysis and Global Optimization*. Kluwer Academic Publishers.

Vaidyanathan, R. and M. El-Halwagi. 1994. "Global optimization of nonconvex nonlinear programs via interval analysis." *Comput. Chem. Engng*. **18**: 889–897.

Vaidyanathan, R. and M. El-Halwagi. 1996. "Global optimization of nonconvex MINLP's by interval analysis." In *Global Optimization in Engineering Design*, edited by I. E Grossmann, pp. 175–193. Kluwer Academic Publishers.

VanAntwerp, J. G., R. D. Braatz, and N. V. Sahinidis. 1997. "Globally optimal robust control for systems with time-varying nonlinear perturbations." *Comput. Chem. Engng*. **21**: S125–S130.

Vásquez, M., G. Némethy, and H. A. Scheraga. 1994. "Conformational energy calculations on polypeptides and proteins." *Chem. Rev.* **94**: 2183–2239.

Visweswaran, V. and C. A. Floudas. 1990. "A global optimization algorithm GOP for certain classes of nonconvex NLPs: II. Application of theory and test problems." *Comput. Chem. Engng*. **14**: 1419–1434.

Visweswaran, V. and C. A. Floudas. 1992. "Unconstrained and constrained global optimization of polynomial functions in one variable." *J. Global Optim.* **2**: 73.

Visweswaran, V. and C. A. Floudas. 1993. "New properties and computational improvement of the GOP algorithm for problems with quadratic objective function and constraints." *J. Global Optim.* **3(3)**: 439–462.

Visweswaran, V. and C. A. Floudas. 1996a. "New formulations and branching strategies for the GOP algorithm." In *Global Optimization in Engineering Design*, edited by I. E. Grossmann, Kluwer Book Series in Nonconvex Optimization and its Applications, pp. 75–110. Kluwer Academic Publishers.

Visweswaran, V. and C. A. Floudas. 1996b. "Computational results for an efficient implementation of the GOP algorithm and its variants." In *Global Optimization in Engineering Design*, edited by I. E. Grossmann, Kluwer Book Series in Nonconvex Optimization and its Applications, pp. 110–154. Kluwer Academic Publishers.

Visweswaran, V., C. A. Floudas, M. G. Ierapetritou, and E. N. Pistikopoulos. 1996. "A decomposition-based global optimization approach for solving bilevel linear and quadratic programs." In *State of the Art In Global Optimization: Computational Methods and Applications*, edited by C. A. Floudas and P.

M. Pardalos, Kluwer Book Series in Nonconvex Optimization and its Applications, pp. 139–162. Kluwer Academic Publishers.

Weiner, S. J., P. A. Kollman, D. A. Case, U. C. Singh, C. Ghio, G. Alagona, S. Profeta, and P. Weiner. 1984. "A new force field for molecular mechanical simulation of nucleic acids and proteins." *J. Am. Chem. Soc.* **106**: 765–784.

Weiner, S. J., P. A. Kollman, D. T. Nguyen, and D. A. Case. 1986. "An all atom force field for simulations of proteins and nucleic acids." *J. Comput. Chem.* **7(2)**: 230–252.

Westerberg, A. W. and J. V. Shah. 1978. "Assuring a global optimum by the use of an upper bound on the lower (dual) bound." *Comput. Chem. Engng.* **2**: 83.

Westerberg, K. M. and C. A. Floudas. 1999a. "Locating all transition states and studying the reaction pathways of potential energy surfaces." *J. Chem. Phys* **110(18)**: 9259–9295.

Westerberg, K. M. and C. A. Floudas. 1999b. "Dynamics of peptide folding: transition states and reaction pathways of solvated and unsolvated tetraalanine." *J. Global Optim.*

Westerlund, T., F. Pettersson, and I. E. Grossmann. 1994. "Optimization of pump configuration problems as a MINLP problem." *Comput. Chem. Engng* **18(9)**: 845–858.

Westerlund, T., H. Skrifvars, I. Harjunkoski, and R. Pörn. 1998. "An extended cutting plane method for a class of non-convex MINLP problems." *Comput. Chem. Engng* **22(3)**: 357–365.

Wilson, G. M. 1964. "Vapor liquid equilibrium XI: a new expression for the excess Gibbs energy of mixing." *J. Am. Chem. Soc.* **86**: 127.

Yamada, Y. and S. Hara. 1998. "Global optimization for H-infinity control with constant diagonal scaling." *IEEE Trans. Automatic Control* **43**: 191–203.

Yee, T. F. and I. E. Grossmann. 1991. "Simultaneous optimization model for heat exchanger network synthesis." In *Chemical Engineering Optimization Models with GAMS*, Volume 6, edited by I. E. Grossmann, CACHE CACHE Design Case Studies Series.

Zamora, J. M. and I. E. Grossmann. 1998a. "A global MINLP optimization algorithm for the synthesis of heat exchanger networks with no stream splits." *Comput. Chem. Engng* **22(3)**: 367–384.

Zamora, J. M. and I. E. Grossmann. 1998b. "Continuous global optimization of structured process systems models." *Comput. Chem. Engng* **22(12)**: 1749–1770.

7 DECOMPOSITION METHODS FOR MATHEMATICAL PROGRAMMING

Philippe Mahey

ABSTRACT Decomposition methods are mainly used to solve iteratively large-scale optimization problems that present a structure of interrelated subsystems. Reducing dimension is not the only motivation; other important aspects are partitioning heterogeneous problems, decentralizing local decisions, and parallelizing computations among different processors. We review here the classical decomposition techniques in mathematical programming among which are price- and resource-directive schemes. Duality as well as column or row generation play a fundamental role, and we complete the exposition with regularization techniques based on symmetric duality, which leads to primal-dual schemes acting like separable augmented Lagrangian algorithms.

1. Introduction

Decomposition methods for mathematical programming have mostly been motivated by the necessity to solve large-scale problems, and they have naturally been designed initially for structured linear programs. Early in the sixties, at the time LP software was still in its infancy, the most significant contributions to the subject were published by Dantzig and Wolfe (1961) for block-angular linear programs and by Benders (1962) for LP with coupling integer variables, and they remain today the most cited algorithms for decomposition purposes. Besides these pioneering papers, two references are also fundamental because they were the first to attempt to unify both theory and applications, we refer to Lasdon's textbook (Lasdon, 1970) and Geoffrion's seminal paper on large-scale mathematical programming (Geoffrion, 1970a). It is worth mentioning that, in the 1970s, a disproportionate effort was made to develop high-performance codes based on the simplex method to solve larger and larger linear programs, acting coherently with the interests of the computer industry, which gained arguments to sell powerful centralized computers able to run these sophisticated codes on very large data sets. As a consequence of that unequal competition between centralized and decentralized computing tools, we can observe that, until recently, the general feeling that underlies many research reports on compared behavior of these algorithms is of low-performance of decomposition techniques. But other motivations for decomposing can be added to the simple objective of reducing the dimension, giving in the last few decades a new impulse to the subject as summarized below:

- Partitioning heterogeneous problems. As in Benders decomposition (Benders, 1962), we can decompose large-scale NP-hard combinatorial optimization problems by isolating easier-to-solve, eventually-polynomial, subproblems.
- Decentralizing complex multilevel decision models. By this we mean the ability to make the subsystems autonomous enough so that they can take their own local decisions, and, eventually with the help of some simple aggregate indicators, to drive them toward the globally optimal solution. Arrow and Hurwicz (1960) have analyzed the economic implications of decentralization on general resource allocation problems. In Arrow et al. (1958), they also gave the corresponding interpretation of Lagrangian relaxation (or price-coordination, see Section 3 below), which was shown to be completely decentralized at least when some strict convexity assumptions are satisfied.
- Parallelizing local computations on parallel or distributed computing units. The recent textbook by Bertsekas and Tsitsiklis (1989) is a very comprehensive exposition of the benefits and limits of parallel and distributed computation on optimization problems.

These motivations are often cumulated in real-life models and we will illustrate the different aspects of decomposition on the network design problem, which appears in many different applications like transportation, communication, and distribution systems. These problems often present the following characteristics, each of which leads to potential decomposition strategies:

- Multicommodity flow models: the necessity to distinguish between each origin-destination pair over the same network makes the flow models much harder to solve than single-commodity flow problems. Observe that the difficulty is not only linked with the huge number of variables that these models induce. Indeed, multicommodity flows do not satisfy the basic properties of minimum-cost single flows like integrality of the basic solutions, max flow-min cut theorem or the negative cycle optimality condition.
- Complicating Boolean variables associated with the design of the topology and the dimensioning of the link capacities induce the combinatorial complexity of the combined dimensioning and routing problem. Decoupling design decisions from routing strategies is the aim of decomposition here.
- Decomposition in subnetworks: modern communication networks often present a natural spatial structure, either from a geographical point of view (regional subnetworks), or from a technological point of view (Local Access Networks, ring networks or backbone networks), or even from an economic point of view (when different operators search for a spatial economic equilibrium, see Patriksson, 1994, for instance).
- Monotropic programming is an interesting modeling issue for network optimization (see Rockafellar, 1984) which can lead to the distribution of computation among arcs, nodes, and paths in the network.

Other classical problems that are commonly treated by decomposition techniques are multidivisional firm planning, multistage stochastic programs, and optimal control of differential systems with multiple timescales.

The present exposition is not meant to be exhaustive and decomposition is as versatile as the structures of large-scale real-life problems are varied. To keep things clear and to avoid repetitions, we have chosen not to include algorithmic issues linked to sparse matrix computations and basis partitioning in linear programming, to block-iterative algorithms (block Gauss–Seidel or Jacobi algorithms), or to partially separable functions. Moreover, we refer to Chapter 13 for the treatment of general multilevel optimization problems.

In Section 2, we begin by an exposition of the fundamental role of separability in decomposable models. We follow this in Section 3 by the presentation of price-directive decomposition schemes that are based on Lagrangian relaxation and duality. Section 4 is devoted to partitioning of heterogeneous problems and primal resource-directive decomposition schemes. Finally, in Section 5, we expose some mixed-decomposition schemes that combine primal and dual information to achieve decentralization or to regularize the coordination task in a distributed way.

2. Fundamentals: The Role of Separability

Many complex decision problems are made of different local subsystems the decision spaces of which are constrained by coupling relations or external variables. The aim of decomposition in optimization is to substitute a complex mathematical program by iterative solutions of easier subproblems, which are generally coordinated by a master program, leading to a multilevel hierarchical optimization scheme. Thus, a natural cooperation is required between the subsystems to combine their local criteria into a global objective function. The theory of cooperative equilibrium shows that the best collective choice is associated with the optimization of a linear combination of the local criteria. As first-order optimality conditions also lead to linear combinations of active constraints, we can see the role of linearity in the decomposition of mathematical programs. To fix the basic ideas, we will first focus on the properties of separable programs.

Separable functions will be defined here as simple extensions of linear functions when a partition of the variable set is at hand.

Definition 1 (Separable functions). *Suppose the working space* X *is partitioned in a product space* $X_1 \times \ldots \times X_p$ *such that* $x \in X$ *will be read as* $x = (x_1, \ldots, x_p)$, $x_j \in X_j$, $j = 1, \ldots, p$. *A function f defined on* X *is separable with respect to the partition if there exist* p *functions* f_j, *defined on* X_j, $j = 1, \ldots, p$, *and such that:*

$$f(x) = \sum_{j=1}^{p} f_j(x_j)$$

We recall briefly the basic property of separable functions for optimization purposes:

Proposition 1. *Suppose each* f_j *is continuous and bounded below on a subset* S_j *of* X_j; *then:*

$$\inf\left\{\sum_{j=1}^{p} f_j(x_j),\ x_j \in S_j,\ j = 1, \ldots, p\right\} = \sum_{j=1}^{p} \inf\{f_j(x_j),\ x_j \in S_j\}$$

In practice, that property which may be called natural decomposition will be affected by coupling relations and/or variables, or by coupled submodels.

At the opposite of that ideal situation, suppose that we want to minimize a nonseparable function, assumed to be convex and proper, on the product set $S_1 \times \cdots \times S_p$ where each local set S_j is closed and convex, we can choose to iteratively solve some local subproblems:

$$\underset{x_j \in S_j}{\text{Minimize}}\ f_j^t(x_j) + \frac{1}{2\lambda}\|x_j - x_j^t\|^2,$$

where t is the iteration count and f_j^t is an approximation of $f(x_1^t, \ldots, x_{j-1}^t, x_j, x_{j+1}^t, \ldots, x_p^t)$ at $x = x^t$ (see Martinet and Auslender, 1974). Observe that the role of the quadratic term in the subproblems is to regularize the subproblems and to define a step size (the λ parameter) in a valid trust region. Without convexity, but with a smooth f, we can take for f_j^t the block-partitioned first-order approximation of f at x^t:

$$f_j^t(x_j) = \langle \nabla_j f(x^t),\ x_j - x_j^t \rangle.$$

To analyze another situation where separability is not perfect, let us now consider a general optimization problem with coupling submodels acting in the same decision set. In the basic problem (P0) below, let $f_1, \ldots, f_p$ be proper closed convex functions from $\mathbb{R}^n$ into $\mathbb{R}$:

$$\textbf{(P0)}\ \text{Minimize} \sum_{j=1}^{p} f_j(x).$$

We assume here that $\cap_{j=1}^{p} \text{ri}\ (\text{Dom}\ f_j) \neq \emptyset$ where ri$(\cdot)$ denotes the relative interior set. To decouple the submodels, we can substitute (P0) by an equivalent problem, called hereafter the master program, in the product space $(\mathbb{R}^n)^p$, that is, we create p copies of the original variables x:

$$(\text{MP0})\begin{cases} \text{Minimize} \displaystyle\sum_{j=1}^{p} f_j(x^j) \\ (x^1, \ldots, x^p) \in A = \{(x^1, \cdots, x^p) \in (\mathbb{R}^n)^p \mid x^1 = \cdots = x^p\}. \end{cases}$$

Then we can apply the following splitting algorithm:

Algorithm 1. *At iteration* t, *let* $(x)^t$ *be the current estimate of the solution and* $(x^j)^t$ *be the unique minimizer of:*

$$(\text{SP0})\ \text{Minimize}\ f_j(x^j) + \frac{1}{2\lambda}\|x^j - (x)^t\|^2$$

for each $j = 1, \ldots, p$. *Then,*

$$(x)^{t+1} = \frac{1}{p}\sum_{j=1}^{p}(x^j)^t$$

In many applications f_j is the indicator function of a closed convex set C_j of $\mathbb{R}^n$. In that case, the above decomposition strategy reduces to successive projections on each set (eventually in parallel) with averaging steps. This is why it is sometimes referred to as constraint splitting (see Lions and Temam, 1970). But Algorithm 1 should be understood as a theoretical scheme and, in general, additional hypotheses are needed to derive efficient methods.

In some models, a smooth, strongly convex functional f_0 is added to f in the original problem:

$$\text{(P0*)} \quad \text{Minimize } f_0(x) + \sum_{j=1}^{p} f_j(x).$$

Fenchel duality can be directly applied to build the dual (master) problem [recall that the conjugate function of f is defined by $f^*(y) = \sup_x\{\langle y, x\rangle - f(x)\}$]:

$$\text{(D0*)} \quad \text{Minimize } f_0^*(y^1 + \cdots + y^p) + \sum_j f_j^*(-y^j).$$

As f_0 is smooth and strongly convex, so is f_0^* and the decomposition results in the following steps:

Algorithm 2.

1. *Let* x^t *be the unique minimizer of:*

$$\text{Minimize } f_0(x) - \langle x, (y^1)^t + \cdots + (y^p)^t\rangle$$

2. *Let, for* $j = 1, \ldots, p$, $(z^j)^t$ *be the unique minimizer of:*

$$(SP0^*) \quad \text{Minimize } f_j(z^j) + \frac{1}{2\lambda} \;\|(y^j)^t + \lambda(z^j - x^t)\|^2$$

3. $(y^j)^{t+1} = (y^j)^t + \lambda((z^j)^t - x^t)$

The resulting algorithm is highly parallelizable and decentralized, as the strongly convex subproblems will exhibit unique solutions. Convergence has been proved even if the subproblems (SP0*) are approximately solved (see Mouallif et al., 1991). A nice formalization of a large family of such decomposition algorithms, which exploit the role of strong convexity, has been given by Cohen under the denomination of the auxiliary problem principle (see Cohen, 1980).

In the next two sections, we will assume that all functions are naturally separable, and consider the structural coupling introduced by the constraints or by some additional common variables.

3. Coupling Constraints and Dual Decomposition Methods

Large-scale problems with a block-separable structure and a reasonably small number of coupling constraints can be decomposed by relaxing these constraints. To simplify, we will consider only the case of equality constraints as inequalities introduce additional difficulties due to the identification of the set of active constraints.

3.1. *Lagrangian Relaxation and Price-Directive Schemes*

Let us consider a separable problem with coupling equality constraints:

$$(\mathbf{P1})\begin{cases} \text{Minimize} & \sum_{j=1}^{p} f_{0j}(x_j) \\ & \sum_{j=1}^{p} f_{ij}(x_j) = 0,\ i = 1, \ldots, m \\ & x_j \in S_j\ , j = 1, \ldots, p. \end{cases} \tag{1}$$

Here, each function f_{0j} is a proper closed convex function from $\mathbb{R}^{nj}$ into $\mathbb{R}$ and each function f_{ij}, $j = 1, \ldots, p$ is an affine function. Each S_j is a compact convex set in $\mathbb{R}^{n_j}$. We assume that (**P1**) has an optimal solution. Let us denote by f_j the m-vector with components f_{ij}, $i = 1, \ldots, m$.

The Lagrangian relaxation of constraints (1) consists of assigning a vector of prices to the coupling resource and building the corresponding Lagrangian function, which is naturally separable with respect to the partition of the variables among the p subsystems. Minimizing the Lagrangian for fixed prices on the product of the local feasibility sets S_j decomposes into p subproblems. Indeed, let $u \in \mathbb{R}^m$ be the vector of dual variables associated with the coupling constraints, then the jth subproblem is as follows:

$$\text{(SP1-u)}\ \ h_j(u) = \inf_{x_j \in S_j} \{f_{0j}(x_j) + \langle u, f_j(x_j)\rangle\}.$$

The assumption on the boundedness of set S_j is useful here to guarantee the existence of local solutions, but it is not necessary. The Lagrangian dual master problem is thus:

$$\text{(DMP1)}\ \ \underset{u \in \mathbb{R}^m}{\text{Maximize}}\ H(u) = \sum_{j=1}^{p} h_j(u),$$

which is a concave maximization problem in $\mathbb{R}^m$ (indeed this is true even if ($P1$) is not a convex program, see Chapter 9.5 for the application of Lagrangian relaxation to integer programming).

Resuming the price-coordination principle, the coordinator (or master problem) must update the dual variables in order to solve the dual problem (DMP1) at the upper level. The subproblems (SP1-u) act as an oracle that can predict the behavior of the local subsystems for a given master decision u. They may be solved in parallel but the speedup will be limited by the bottleneck induced by the master.

The questions that arise to get an efficient scheme are:

- How should the coordinator update the dual multipliers to converge to the solution of the dual problem?
- Are the subproblems sufficiently decentralized so that their solutions can converge to the local part of the globally optimal decisions?

As each subproblem computes a solution $x_j(u)$, the coordinator can collect all these proposals to build a subgradient of the dual function H. Indeed, $g = \sum_{j=1}^{p} f_j(x_j(u))$ is one of these subgradients and optimality is attained if $g = 0$, that is, if the local proposals satisfy the coupling constraints. Unfortunately, this is unlikely to happen, even if we know the optimal multipliers, because H is in general nonsmooth. Thus, g is in general different from zero and is even not always an ascent direction for H. In this case, a subgradient algorithm (see Chapter 5.4) could be used to update the dual multipliers, but it will be unable to recover either a feasible or an optimal primal solution. To cope with that difficulty, past subgradients should be stored to build ascent directions and converge toward feasible solutions. Algorithms to achieve these goals are bundle methods and cutting plane algorithms.

- *Bundle methods.* Small steps are performed locally to gather a bundle of ϵ-subgradients until an ascent direction, defined by the projection of the origin on the convex hull of the bundle, is obtained, allowing a serious step in the dual space. First applied to decomposition by Medhi (1987), bundle methods converge significantly faster than classical subgradient methods. Regularized versions called proximal subgradients or ballstep subgradients have been proposed recently and have shown quite robust performance in solving large-scale nonlinear programs (see Kiwiel et al., 1999).
- *Cutting planes.* Successive subgradients are stored to build concave piecewise affine approximations of the dual function; maximizing (by linear programming) these functions yields not only monotonically decreasing upper bounds of the dual maximal value, but also allows primal feasible solutions to be built. This is the way the Dantzig–Wolfe's decomposition principle works (as detailed below).

In the differentiable case (i.e., when strict convexity of each f_{0j} imply that the dual function is smooth), the coordinator can update the dual variables by a single gradient step:

$$u^{t+1} = u^t + \lambda_t g^t,$$

where λ_t is a positive step and g^t is the current gradient of the dual function at u^t, that is, $g^t = \sum_{j=1}^{p} f_j(x_j(u^t))$. The resulting method is sometimes referred as Uzawa's algorithm (Arrow et al., 1958, Chapter 10). Curiously, Uzawa considered it as a modification of a general gradient method for saddle-point computation with strict convexity with respect to one variable and linearity with respect to the other variable. In that latter method, known as the Arrow–Hurwicz gradient method (Arrow et al., 1958, Chapter 6), two gradient steps with fixed length are performed in parallel in the primal and dual spaces. More recently, nongradient ascent steps based on successive overrelaxation schemes have been analyzed with success under the denomination of dual-ascent algorithms (Tseng, 1990).

Finally, observe that second-order methods are very unlikely to be of any help when dealing with the lack of smoothness of the Lagrangian dual function. On the other hand, modified Lagrangian functions combining penalty terms to force the coupling constraints will generally destroy the separability of the objective function. For instance, augmented Lagrangian functions will introduce the additional term $\|\sum_j f_j(x_j)\|^2$, which is no longer separable. We will see in the last section how to build separable augmented Lagrangian methods.

3.2. Column-Generation and Dantzig–Wolfe's Decomposition Principle

The Dantzig–Wolfe's decomposition method (Dantzig and Wolfe, 1961) was built to solve block-angular linear programs, that is, problems of the following form:

$$(BLP)\begin{cases} \text{Minimize} & \displaystyle\sum_{j=1}^{p} c_j x_j \\ & \displaystyle\sum_{j=1}^{p} A_j x_j = b_0 \\ & \left.\begin{array}{r} B_j x_j = b_j \\ x_j \geqslant 0 \end{array}\right\} j = 1, \ldots, p \end{cases} \tag{2}$$

Each local variable x_j belongs to $\mathbb{R}^{n_j}$, the number of coupling constraints (2) is m. Assume that each local polyhedron $S_j = \{x_j \in \mathbb{R}^{n_j} \mid B_j x_j = b_j,\ x_j \geqslant 0\}$ is bounded and that (BLP) has an optimal solution $(x_1^*, \ldots, x_p^*)$. Let $u^* \in \mathbb{R}^m$ be the optimal multipliers associated with the coupling constraints.

The key strategy to solve (BLP) by decomposition-coordination is column-generation. That classical technique of solving linear programs with a huge number of columns will be introduced here in connection with the Lagrangian relaxation strategy described before.

Recall that the main difficulty in generating feasible solutions of the coupled subsystems is that these are convex combinations of the subproblem proposals. In the linear case, these proposals correspond to vertices of the local polyhedra S_j. An equivalent formulation of (BLP) where the variables are the coefficients λ_{jt} of convex combinations of the vertices x_j^t is obtained as

$$x_j \in S_j \Leftrightarrow x_j = \sum_t \lambda_{jt} x_j^t, \quad \sum_t \lambda_{jt} = 1, \ \lambda_{jt} \geqslant 0, \ \forall t.$$

The resulting linear program, called the master program, has $m + p$ constraints (the m coupling constraints and the p convexity constraints) and a very large number of columns. Moreover, its columns are not known *a priori*, which makes column generation necessary. At each iteration t of the revised simplex method (see Chapter 1.2), one needs to price out a new column, given the current multipliers $(u^t, \sigma_1^t, \ldots, \sigma_p^t)$, where σ_j is the dual multiplier associated with the jth convexity constraint $\sum_t \lambda_{jt} = 1$. Look at the general form of the column associated with the vertice τ of polyhedron S_j:

$$\begin{bmatrix} l_{j\tau} \\ 0 \\ \vdots \\ 1 \\ \vdots \\ 0 \end{bmatrix},$$

where $l_{j\tau} = A_j x_j^\tau$. Let the cost of that column in the master program be denoted by $c_{j\tau} = c_j x_j^\tau$; then, its reduced cost is simply:

$$\bar{c}_{j\tau} = c_{j\tau} - (u^t) l_{j\tau} - \sigma_j^t.$$

Column generation is used here to compute the smallest reduced cost by solving, for each j, the following subproblem:

$$\text{(BLSP)} \quad \underset{x_j \in S_j}{\text{Minimize}} \ c_j x_j - (u^t) A_j x_j.$$

Indeed, the subproblem solution is a vertex of S_j and substracting σ_j from the subproblem optimal value yields its reduced cost. Forcing the new columns in the basis (multiple pricing strategy) will allow the multipliers to be updated until no new column is generated and an optimal solution is attained. We can verify that the pricing subproblem is exactly the subproblem induced by the Lagrangian relaxation of the coupling constraints (2). We will now show that the restricted master program associated with the generated columns is the dual of the maximization of the current approximation of the Lagrangian dual function in the cutting-plane method. The restricted master at iteration t is as follows, denoting by T_{jt} the set of indexes of the already generated vertices for each j:

$$\text{Minimize} \quad \sum_j \sum_{\tau \in T_{jt}} c_{j\tau} \lambda_{j\tau}$$

$$\text{(BLMP)} \quad \sum_j \sum_{\tau \in T_{jt}} l_{j\tau} \lambda_{j\tau} = b_0$$

$$\sum_{\tau \in T_{jt}} \lambda_{j\tau} = 1, \ \lambda_{j\tau} \geqslant 0, \ j = 1, \ldots, p.$$

Keeping coherent notations for the dual variables, we can write the corresponding dual problem:

$$\text{Maximize} \qquad b_0 u + \sum_j \sigma_j$$

$$(l_{j\tau})u + \sigma_j \leqslant c_{j\tau}, \quad \forall \tau \in T_{jt}.$$

This is easily seen to be equivalent to:

$$\underset{u}{\text{Maximize}} \left\{ b_0 u + \min_{\tau \in T_{jt}} \sum_j (c_{j\tau} - l_{j\tau} u) \right\}.$$

Just noting that $H(u^\tau) = b_0 u^\tau + \sum_j (c_{j\tau} - l_{j\tau} u^\tau)$ and that $g^\tau = b_0 - \sum_j l_{j\tau}$ is the corresponding subgradient, we can finally rewrite the dual as:

$$\underset{u}{\text{Maximize}} \min_{\tau \in T_{jt}} \{H(u^\tau) + g^\tau (u - u^\tau)\},$$

which is the required restricted dual of the cutting plane method.

A nice feature of the decomposition algorithm is inherited from its interpretation by the cutting plane method: the dual approximate value $H(u^\tau)$ is a lower bound of the optimal value of (BLP); moreover, the primal solution computed by (BLMP) is globally feasible and its value z^t (the optimal value of the restricted dual) is a monotonically decreasing upper bound of the optimal value of (BLP).

The algorithm has been applied to many structured LPs and one of the most successful applications is the decomposition of multicommodity flow problems (see Chapter 8.5). In this case, the column generation is very natural as it amounts to solving shortest-path subproblems, which could be performed very quickly by grouping destinations for a given origin. For general block-angular LPs, however, the algorithm presents the typical slow convergence of the cutting plane method. This is mainly due to numerical instabilities of the dual sequence $\{u^t\}$. Trust regions can be added to the restricted dual to stabilize the sequence as in the box-step algorithm (Marsten et al., 1975) but the choice of the dimensions of the box is a difficult problem. A better idea that relates the decomposition principle to the interior-point method has been given by Goffin et al. (1992), who propose to substitute the maximization of the restricted dual by the computation of the analytic center of a polyhedral approximation of the dual epigraph.

Finally, we see that the nonsmoothness of the dual function in the general case is also a limitation to achieving complete decentralization. This is even clearer in the linear case where the subproblems (BLSP) can have non-unique solutions when the allocated prices are optimal. It can easily be seen that degeneracy will occur in at least one subproblem (and at most m), so that only the coordinator is able to build the optimal solution.

4. Projection and Primal Decomposition

4.1. The Projection Strategy

Many problems can be modeled by representing the interactions between subsystems with the aid of coupling variables. In the first class of problems, these variables may be structural variables, which act as external decisions that may affect any subsystem. Such typical situations are listed below:

1. *Stochastic programming with recourse.* In these problems (see Chapter 11), the decision variables can act at different levels with increasing dependencies on random events. For example, in two-stage stochastic linear programs, the variables are partitioned in two groups: deterministic variables y at the first stage and random variables x at the lower level, which are linked to the first-level data by randomly distributed constraints. The two-stage decision problem can be written in the following way:

$$\underset{y \in Y}{\text{Minimize}} \; \{cy + E[\min\{dx \mid Bx = b - Ay, \; x \geqslant 0\}]\},$$

where b is a random recourse vector.

Planning models can often be viewed in this way, letting y be the first period decisions, which are to be defined, satisfying some deterministic relations, but taking into account the operational decisions, which depend on uncertain data associated with a large number of scenarios.

2. *Process control with multiple subsystems.* Many complex planning problems are made of various subsystems that are interrelated by control or decision variables. The control variables may affect the total cost function, the constraints' functional data, and the right-hand sides.
3. *Mixed-integer design problems.* Network design problems arise in many distribution, transportation, and communication problems. These problems induce combinatorial decisions about the topology and the dimensioning of a graph where some flows circulate, respecting some flow constraints. Thus, a compromise should be aimed at between low congestion networks and low-cost topologies. The simplest example is the uncapacitated plant-location problem. In this problem, n clients with known demands b_j, $j = 1, ..., n$ must be delivered by a single plant for which p location sites are available at different installation costs. A demand satisfaction problem that appears to be a simple routing problem corresponds to each location. A basic model associates flow variables x_{ij} between each site i and each client j, and Boolean location decision variables y_i:

$$\text{Minimize} \quad \sum_{i=1}^{p} c_i y_i + \sum_{i=1}^{p} \sum_{j=1}^{n} d_{ij} x_{ij}$$

$$\sum_{i=1}^{p} x_{ij} \geqslant b_j, \; j = 1, ..., n$$

$$x_{ij} \leqslant b_j y_i, \; x_{ij} \geqslant 0, \; \forall i, j$$
$$y_i \in \{0, 1\}, \; i = 1, ..., p.$$

For all these formulations, the projection strategy, as it was initially called by Geoffrion, (1970a), consists of solving, at an upper level, an implicit problem in the coupling variables y for which the subproblems in the x variables will compute subgradients or marginal costs.

Observe that the price-directive procedure described previously for the decomposable problem with coupling constraints is equivalent to a projection on the space of the dual multipliers associated with the coupling constraints. Indeed, the coordination level just fixes these dual variables and transmits them to the subproblems. Another way to say the same thing in the linear case is to observe that the dual problem of (BLP) is a block-angular problem with coupling variables.

The iterative solution of the implicit upper-level problem is very well illustrated by Benders decomposition method, which its author (Benders, 1962) mainly designed to address mixed-integer linear programs. We will detail these computations below, but we must stress that the main advantage it bears with respect to price-directive approaches is that it may solve NP-hard problems to global optimality, whereas Lagrangian duality is limited by the duality gaps it carries with it.

In the second class of problems, artificial coupling variables may be introduced to induce decomposition. The most distinctive example is the resource-directive decomposition first introduced by Geoffrion (1970b). It is a primal alternative to decompose the former model (*P*1), which was treated by price-directive schemes in Section 3. To decouple the subsystems, resource variables y_j are allocated to each subsystem j such that:

$$\begin{cases} f_j(x_j) + y_j = 0, \; j = 1, ..., p \\ \displaystyle\sum_{j=1}^{p} y_j = 0 \end{cases}$$

The role of the coordinator is now to solve the implicit primal master program:

$$\text{(PMP1)}\quad \text{Minimize} \left\{ v(y) \mid \sum_j y_j = 0,\ y \in Y \right\},$$

where $v(y) = \inf\left\{\sum_j v_j(y_j)\right\}$ and $v_j(y)$ is the optimal value of the resource-directive subproblem for a fixed allocation vector y_j:

$$\text{(SP1-y)}\begin{cases} \text{Minimize} & f_{0j}(x_j) \\ & f_j(x_j) + y_j = 0 \\ & x_j \in S_j. \end{cases} \tag{3}$$

The implicit set Y is the set of feasible allocations y for which activity levels $x_j \in S_j$ exist such that $f_j(x_j) + y_j = 0$ for each j. In most practical situations, the subproblems are able to compute the optimal multipliers u_j associated with each local resource constraint (3). It is easily seen that the vector $(u_1, \ldots, u_p)$ is a subgradient of v and that the optimality condition is simply that $u_1 = \cdots = u_p$, that is, that the dual local proposals $(u_1, \ldots, u_p)$ lie in the orthogonal subspace to the resource allocation feasibility subspace.

The same arguments as in the previous section can be used to justify iterative methods to solve the master program, at least in the convex case (the implicit function v is then an extended convex function): projected subgradient algorithms or even cutting planes, and we refer the reader to Chapter 8.5 for a nice application of that scheme.

Again, complete decentralization is not guaranteed even in the linear case. Indeed, when solving the primal subproblem (SP1-y) for $y = y^*$, the optimal resource allocation vector, the jth subsystem manager will indeed obtain his optimal solution x_j^*, but he will be unable to recognize its optimality without the help of the coordinator. Indeed, the local dual multipliers associated with the resource constraints in (SP1-y) are unlikely to be unique and dual degeneracy will occur in most subproblems.

4.2. Benders Decomposition for Mixed-Integer Programs

Benders (1962) has considered linear programs with complicating discrete variables, called mixed-integer programs (see Chapter 9.4), of the following form:

$$\text{(P2)}\begin{cases} \text{Minimize} & cx + dy \\ \text{subject to} & Ax + Dy \geqslant b \\ & x \geqslant 0,\ \ y \in S, \end{cases}$$

where x is a continuous n-vector, y is a p-vector, and S is a subset of Z^p. Suppose also that the right-hand side b is in $\mathbb{R}^m$.

The key idea lies in the fact that, for fixed feasible y, the linear program in x

$$\text{(SP2-y)}\begin{cases} \text{Minimize} & cx \\ \text{subject to} & Ax \geqslant b - Dy \\ & x \geqslant 0 \end{cases}$$

is easy to solve. Moreover, its dual (DP2-y) is a parametric linear program defined by:

$$\text{(DP2-y)}\begin{cases} \text{Maximize} & (b - Dy)u \\ \text{subject to} & A^T u \leqslant c \\ & u \geqslant 0. \end{cases}$$

Observe that the polyhedron Π of feasible dual solutions does not depend on y and that it is nonempty if (P2) has an optimal solution. On the other hand, (DP2-y) allows us to describe the set Y of the y such that there exist $x \geqslant 0$ satisfying $Ax \geqslant b - Dy$:

$$Y = \{y \in \mathbb{R}^p \mid (b - Dy)v \leqslant 0,\ \forall v \in K = \{u \in \mathbb{R}^m \mid A^T u \leqslant 0,\ u \geqslant 0\}\}.$$

Indeed, the convex polyhedral cone K is the set of infinite rays of Π and the primal subproblem (SP2-y) is empty if, and only if, (DP2-y) is unbounded. Denoting by v^j, $j = 1, \ldots, J$ the extreme rays generating K, the set Y is defined by a finite (but probably very large) set of linear inequalities:

$$Y = \{y \in \mathbb{R}^p \mid v^j D y \geqslant v^j b,\ j = 1, \ldots, J\}.$$

To update the complicating variables y, the coordinator must aim at solving the following implicit problem:

$$\begin{cases} \text{Minimize} & v(y) + dy \\ \text{subject to} & y \in S \cap Y, \end{cases}$$

where v is the optimal value of (SP2-y), or equivalently the optimal value of (DP2-y), adding the value $+\infty$ when it is unbounded. v is a piecewise affine convex function on Y and, consequently, it can be represented by the supremum of a finite (but again probably very large) set of affine functions defining the supporting hyperplanes of its epigraph. Indeed, these are given by the extreme points u^i, $i = 1, \ldots, I$ of the dual polyhedron Π as, for any $y \in Y$,

$$v(y) = \max\{(b - Dy)u^i,\ i = 1, \ldots, I\}.$$

Finally, the master program can be seen as an integer program with a large number of implicit constraints:

$$\begin{aligned} &\text{Minimize} && z \\ & && z - dy + u^i D y \geqslant u^i b,\ i = 1, \ldots, I && (4) \\ &\text{(MP2)} && v^j D y \geqslant v^j b,\ j = 1, \ldots, J && (5) \\ & && z \in \mathbb{R}\quad y \in S. \end{aligned}$$

At iteration t of the procedure, the coordinator solves a relaxed master program with only a small number of constraints. Clearly, if it is infeasible, the original problem (P2) is infeasible. Otherwise, an optimal (z^t, y^t) is computed. We now show how violated constraints (sometimes called Benders cuts) can be identified by solving the subproblem (DP2-y) for $y = y^t$. Two situations can arise:

1. (DP2-y) has an optimal solution denoted by u^t, which is indeed an extreme point of Π; then,
   - either its optimal value is such that $(b - Dy^t)u^t = z^t - dy$ and the current solution is optimal for (P2); indeed, all relaxed constraints in (4) are satisfied as (SP2-y) is feasible and $z^t = dy^t + \max\{(b - Dy^t)u^i,\ \forall i\}$;
   - or $z^t - dy^t < (b - Dy^t)u^t$ and the corresponding violated constraint in (4) can be added to the restricted master.
2. (DP2-y) is unbounded and an extreme ray v^t can be computed to add the corresponding constraint of type (5) to the restricted master.

Observe that the sequence $\{z^t\}$ is a monotonically increasing sequence of lower bounds for the optimal value because the optimal solution y^* is feasible for each relaxed master, and the values $(b - Dy^t)u^t + dy^t$ are upper bounds of the optimal value as y^t turns the primal subproblem (SP2-y) feasible.

It is straightforward to verify that Benders decomposition is again a special implementation of the cutting-plane method and, in that sense, it is dual to Dantzig–Wolfe's algorithm when applied to the dual of the block-angular linear program described in Section 3. Thus, we may expect again that the number of generated cuts is much less than the total number of constraints necessary to completely describe the implicit feasible and optimal sets. The main difference here is that it can cope with complicating integer variables. But we must stress that the method is only attractive if the restricted master programs are relatively easy to solve. Another important point is our ability to avoid infeasible subproblems as they jeopardize the progress of the algorithm toward optimality. For some examples like the uncapacitated plant-location problem described above, it is an easy task, as forcing at least one Boolean variable to be nonzero suffices to satisfy the demand. In more general capacitated network design problems, the search for efficient metric inequalities (Pareto-optimal cuts) to force feasibility is a crucial issue to move Benders decomposition competitive (see Magnanti and Wong, 1981, for more details).

A generalized version of Benders decomposition has been proposed by Geoffrion (1972) to take in consideration convex subproblems. More recently, Higle and Sen (1991) proposed a randomized version of Benders decomposition to solve the two-stage stochastic linear program with recourse.

5. Regularized and Decentralized Mixed Decomposition Schemes

We will now give additional insight into decomposition methods by bringing together the advantages of primal and dual decomposition schemes.

5.1. *Symmetric Duality*

We have seen in the previous sections that price- and resource-directive decomposition schemes have a strong duality relationship. Going back to the decomposable model with coupling constraints (**P1**), we can represent the primal and dual coupling in the product space $\mathbb{R}^{m \times p}$ by the following variables and orthogonal coupling subspaces:

$$\mathcal{Y} = (y_1, \ldots, y_p) \quad \mathcal{Y} \in A = \left\{ \sum_{j=1}^{p} y_j = 0 \right\}$$

$$\mathcal{U} = (u_1, \ldots, u_p) \quad \mathcal{U} \in A^{\perp} = \{u_1 = \cdots = u_p\}.$$

The implicit master problem can now be derived in its two symmetric dual forms:

Minimize $v(\mathcal{Y})$ subject to $\mathcal{Y} \in A$

and

Maximize $h(\mathcal{U})$ subject to $\mathcal{U} \in A^{\perp}$,

where v and h are separable functions on the product space as defined in the previous sections:

$$v(\mathcal{Y}) = \sum_j v_j(y_j) \text{ with } v_j(y_j) = \min\{f_{0j}(x_j) \mid f_j(x_j) + y_j = 0,\ x_j \in S_j\}$$

$$h(\mathcal{U}) = \sum_j h_j(u_j) \text{ with } h_j(u_j) = \max_{u_j} \ \min\{f_{0j}(x_j) + u_j f_j(x_j) \mid x_j \in S_j\}.$$

5.2. *Proximal Decomposition*

The proximal decomposition method is an adaptation of the alternating direction method of multipliers originally proposed by Glowinski and Marocco (1975). It is known to be equivalent to Douglas–Rachford splitting for finding a zero of the sum of two maximal monotone operators (see Lions and Mercier, 1979). When applied to any of the symmetric master problems defined above, it results in the following steps:

Algorithm 3.

1. *Regularized subproblem: it corresponds to applying an augmented Lagrangian iteration (cf. Chapter 5.3) to the primal subproblem (SP1-y), yielding, if* y_j^t *and* u_j^t *are the current primal and dual estimates, the following subproblem :*

$$\underset{x_j \in S_j}{\text{Minimize}}\, f_{0j}(x_j) + u_j^t(f_j(x_j) + y_j^t) + \frac{\lambda}{2}\|f_j(x_j) + y_j^t\|^2$$

Let x_j^{t+1} *be its unique minimizer and* r^{t+1} *be the residual of the coupling constraints, that is* $r^{t+1} = \sum_j f_j(x_j^{t+1})$.

2. *Primal and dual updates: the algorithm alternates updating each* y_j *and* u_j *and projecting them on the two feasibility subspaces. As these projections are easy averaging calculations, we can integrate them in the individual updates (one single formula for u as each subsystem still receives the same prices and one distinct formula for each resource allocation* y_j*)* :

$$u^{t+1} = u^t + \frac{\lambda}{p} r^{t+1}$$

$$y_j^{t+1} = -f_j(x_j^{t+1}) + \frac{1}{p} r^{t+1},\ j = 1, ..., p$$

Observe that, even if the method may be interpreted as a separable augmented Lagrangian technique, the parameter λ is a scaling parameter and not a penalty one, which must be estimated to drive the two primal and dual sequences at the same pace toward the optimal fixed point (see details in Mahey et al., 1995). Eckstein (1989) has shown the potential of that algorithm for massively distributed computing and different versions have been shown to be very efficient in solving multicommodity flow problems with linear or convex costs (Eckstein and Fukushima, 1993; Ouorou et al., 2000).

The Progressive Hedging method is a particular adaptation of the proximal decomposition algorithm designed by Rockafellar and Wets (1991) to solve multistage stochastic optimization problems.

The model is a generalization of the two-stage stochastic model considered in Section 4. An eventually very large set of independent scenarios Ω defines the probabilistic events that affect the decision variables, which are grouped in different stages. The dynamics linking these stages can be written, denoting by $x(\omega) = (x_1(\omega), ..., x_T(\omega))$ the multistage decision variables for a given scenario $\omega \in \Omega$, as:

$$B_t x_{t-1}(\omega) + W_t x_t(\omega) = b_t(\omega),\ t = 2, ..., T.$$

For each stage, the set Ω_t of observable events is known and increasing, which means that $\Omega_t \subseteq \Omega_{t+1}$. The key fact in that model is that the decisions at stage t may depend only on the information available at that stage. Thus, the decision variables at stage t satisfy the following nonanticipativity conditions :

$$x_t(\omega) = x_t(\omega'),\ \forall \omega, \omega' \in \Omega_t.$$

The multistage stochastic optimization problem consists of determining feasible decisions satisfying the dynamics equations and the nonanticipativity conditions that minimize the total expectation cost $E\left[\sum_{t=1}^{T} c_t(x_t(\omega), \omega)\right]$, where the cost functions c_t are supposed to be convex and measurable with respect to the set Ω_t.

A decomposition with respect to stages would lead to nested computations where the techniques of Sections 3 and 4 could be applied (see the L-shaped method of Van Slyke and Wets, 1969, in the linear case). However, we focus here on the coupling induced by the nonanticipativity constraints, which are seen to form a subspace of the multistage decision space. The application of proximal decomposition to the former model induces multistage decision subproblems for each scenario, which take the following simple form at iteration k:

$$\text{Minimize}\ \ p(\omega)c(x(\omega), \omega) + \langle u^k(\omega), x(\omega)\rangle + \frac{\lambda}{2}\|x(\omega) - x^k(\omega)\|^2,$$

where $p(\omega)$ is the probability of event ω.

Denoting the unique minimizer by $\tilde{x}^{k+1}$, the primal and dual updates can thus be explicitly computed:

$$x_t^{k+1}(\omega) = \frac{1}{|\Omega_t|}\sum_{\sigma\in\Omega_t}\tilde{x}_t^{k+1}, \ \forall t \text{ and } \forall\omega\in\Omega_t$$

$$u^{k+1} = u^k + \lambda(\tilde{x}^{k+1} - x^k)$$

Other regularization techniques leading to separable augmented Lagrangian algorithms for stochastic optimization problems can be found in the survey by Ruszczynski (1997).

6. Concluding Remarks

The decomposition of block-angular linear programs have given rise to the first, and still very popular, decomposition algorithms by Dantzig and Wolfe, and Benders. But they have also shown some limitations concerning slow convergence (due to nonsmoothness of the primal or dual master functions), the lack of complete decentralization (due to subproblems degeneracy), and disappointing speedup by parallelization (due to the master bottleneck). Some authors have attempted to combine primal and dual allocations to take care of both coupling variables and constraints (like in Van Roy's cross-decomposition, 1983), to achieve decentralization (see Burton and Obel, 1980) or to distribute computation among small regularized subproblems as with the proximal decomposition presented in the last section. These latter algorithms seem very efficient for certain nonlinear monotropic programs like convex-cost multicommodity flow problems, but one must remain aware that all decomposition algorithms rely heavily on the structure of the problems and are, thus, dramatically dependent on their formulations.

ACKNOWLEDGMENTS

The interest of the author for decomposition methods has been gained from many discussions with Guy Cohen, Henrique Pacca Luna, and Nguyen Van Hien, specially remembered and acknowledged here among many others.

References

Arrow, K. and L. Hurwicz. 1960. "Decentralization and computation in resource allocation." In *Essays in Economics and Econometrics*, edited by R. Phouts. Chapel Hill: University of North Carolina Press.

Arrow, K., L. Hurwicz, and H. Uzawa. 1958. In *Studies in Linear and Nonlinear Programming*. Stanford, California: Stanford University Press.

Benders, J. F. 1962. "Partitioning procedures for solving mixed variables programming problems." *Numerische Mathematik* **4**: 238–252.

Bertsekas, D. P. and J. N. Tsitsiklis. 1989. In *Parallel and Distributed Computation*. Englewood Cliffs, New Jersey: Prentice-Hall.

Burton, R. and B. Obel. 1980. "The efficiency of the price, budget and mixed approaches under varying a priori information levels for decentralized planning." *Mgmnt Sci.* **26**: 401–417.

Cohen, G. 1980. "Optimization by decomposition and coordination: a unified approach." *IEEE Trans. Automatic Control* **AC-23**: 222–232.

Dantzig, G. B. and P. Wolfe. 1961. "The decomposition algorithm for linear programming." *Econometrica* **29(4)**: 767–778.

Eckstein, J. 1989. "Splitting methods for monotone operators with applications to parallel optimisation." Ph.D. diss. MIT.

Eckstein, J. and M. Fukushima. 1993. "Some reformulations and applications of the alternating direction method of multiplier." In *Large-scale Optimization: State of the Art*, edited by D. W. Hearn, W. W. Hager, and P. M. Pardalos. Dordrecht, The Netherlands: Kluwer Academic Publishers.

Geoffrion, A. M. 1970a. "Elements of large-scale mathematical programming." *Mgmnt Sci.* **16(11)**: 25–64.

Geoffrion, A. M. 1970b. "Primal resource-directive approaches for optimising nonlinear decomposable systems." *Oper. Res.* **18**: 375–403.

Geoffrion, A. M. 1972. "Generalized benders decomposition." *J. Optim. Theory Appl.* **10**: 237–260.

Glowinski, R. and A. Marocco. 1975. "Sur l'approximation par éléments finis d'ordre 1 et la résolution par pénalisation-dualité d'une classe de problèmes de dirichlet." *RAIRO* **2**: 41–76.

Goffin, J. L., A. Haurie, and J. P. Vial. 1992. "Decomposition and nondifferentiable optimisation with the projective algorithm." *Mgmnt Sci.* **38**: 284–302.

Higle, J. L. and S. Sen. 1991. "Stochastic decomposition: an algorithm for stage linear programs with recourse." *Math. Oper. Res.* **16**: 650–669.

Kiwiel, K. C., T. Larsson, and P. O. Lindberg. 1999. "Dual properties of ballstep subgradient methods with applications to Lagrangian relaxation." Technical Report LiTH-MAT-R-1999. Sweden: Department of Mathematics, University of Linköping

Lasdon, L. S. 1970. In *Optimization for Large Systems*. New York: Macmillan.

Lions, J. L. and R. Temam. 1970. "Eclatement et décentralisation en calcul des variations." *Lecture Notes Math.* **32**: 196–217.

Lions, P. L. and B. Mercier. 1979. "Splitting algorithms for the sum of two nonlinear operators." *SIAM J. Numer. Anal.* **16**: 964–979.

Magnanti, T. L. and R. T. Wong. 1981. "Accelerating benders decomposition algorithmic enhancement and model selection criteria." *Oper. Res.* **29**: 464–484.

Mahey, P., S. Oualibouch, and D. T. Pham. 1995. "Proximal decomposition on the graph of a maximal monotone operator." *SIAM J. Optim.* **5(2)**: 225–236.

Marsten, R. E., W. Hogan, and J. W. Blankenship. 1975. "The boxstep method for large-scale optimisation." *Oper. Res.* **23**: 389–405.

Martinet, B. and A. Auslender. 1974. "Méthodes de décomposition pour la minimisation d'une fonction sur un espace produit." *SIAM J. Control* **12**: 635–642.

Medhi, D. 1987. "Decomposition of structured large scale optimisation problems and parallel optimisation." Ph.D. diss. University of Wisconsin at Madison.

Mouallif, K., V. H. Nguyen, and J. J. Strodiot. 1991. "A perturbed parallel decomposition method for a class of nonsmooth convex minimization problems." *SIAM J. Control Optim.* **29**: 829–847.

Ouorou, A., P. Mahey, and J. P. Vial. 2000. "A survey of algorithms for convex multicommodity flow problems." *Mgmnt Sci.* **46**: 126–147.

Patriksson, M. 1994. In *The Traffic Assignment Problem: Models and Methods* Topics in Transportation. The Netherlands: VSP, Utrecht.

Rockafellar, R. T. 1984. In *Monotropic Programming*. New York: Wiley & Sons.

Rockafellar, R. T. and R. J. B. Wets. 1991. "Scenarios and policy aggregation in optimization under uncertainty." *Math. Oper. Res.* **16**: 1–23.

Van Roy, T. J. 1983. "Cross decomposition for mixed-integer programming." *Math. Programming* **25**: 46–63.

Ruszczynski, A. 1997. "Decomposition methods in stochastic programming." *Math. Programming* **79**: 333–353.

van Slyke, R. and R. J. B. Wets. 1969. "L-shaped linear programs with applications to optimal control and stochastic linear programs." *SIAM J. Appl. Math.* **17**: 638–663.

Tseng, P. 1990. "Dual ascent methods for problems with strictly convex costs and linear constraints: a unified approach." *SIAM J. Control Optim.* **28**: 214–242.

8 NETWORK OPTIMIZATION

8.1 Introduction

Ravindra K. Ahuja, Thomas L. Magnanti, and James B. Orlin

ABSTRACT Due to its widespread applications, network optimization is an important subfield within the broad field of optimization. This article introduces some core network optimization problems that are studied in greater depth in subsequent sections. The article introduces the following fundamental network optimization problems: maximum flow problem, shortest-path problem, minimum-cost single-commodity flow problem, minimum-cost multicommodity flow problem, and the minimum spanning tree problem. The focus of the article is to describe several sample applications of each of the fundamental network optimization problems. These applications are intended to illustrate a range of problem contexts and to be suggestive of how network optimization problems arise in practice.

1. Introduction

Networks are all-pervasive. Electrical and power networks bring lighting and entertainment into our homes. Telephone networks permit us to communicate with each other almost effortlessly within our local communities and across regional and international borders. National highway systems, rail networks, and airline service networks provide us with the means to cross great geographical distances to accomplish our work, to see our loved ones, and to visit new places and enjoy new experiences. Manufacturing and distribution networks give us access to life's essential foodstock and to consumer products. In each of these settings, we often wish to send some goods (vehicles, messages, electricity, or water) from one point to another, typically as efficiently as possible. The field of study concerning the optimal flow of goods on networks is known as network optimization. Network optimization has its roots in the 1940s with the development of linear programming. Since then the field of network optimization has grown at an almost dizzying pace with thousands of scientific papers and multitudes of applications modeling a remarkably wide range of practical situations.

In the following sections, we study the following fundamental network optimization problems: maximum flow problem, shortest-path problem, minimum-cost single-commodity flow problem, minimum-cost multicommodity flow problem, and the minimum spanning tree problem. These problems form the core of network optimization and arise both as stand-alone models and as subproblems in more complex problem settings. Network optimization problems also arise in surprising ways for problems that on the surface might not involve networks at all. Sometimes these applications are linked to a physical entity, and at other times they are not. Indeed, these various applications of network optimization problems seem to be more widespread than the applications of physical networks. In the following sections, we describe several sample applications of each of the fundamental network optimization problems. These applications are intended to illustrate a range of problem contexts and to be suggestive of how network optimization problems arise in practice. We also give a brief description of each generic network optimization problem before presenting its applications. The book by Ahuja et al. (1993) presents additional material and applications of these problems.

2. Preliminaries

In this section, we introduce some basic notation and definitions from graph theory as well as a mathematical programming formulation of the minimum-cost flow problem, which is the core network flow problem that lies at the heart of network optimization.

Let $G = (N, A)$ be a directed network defined by a set N of n nodes and a set A of m directed arcs. Each arc $(i, j) \in A$ has an associated cost c_{ij} that denotes the cost per unit flow on that arc. We assume that the flow cost varies linearly with the amount of flow on the arc. We also associate with each arc $(i,j) \in A$ a capacity u_{ij} that denotes the maximum amount that can flow on the arc and a lower bound l_{ij} that denotes the minimum amount that must flow on the arc. We associate with each node $i \in N$ an integer number $b(i)$ representing its supply/demand. If $b(i) > 0$, then node i is a supply node; if $b(i) < 0$, then node i is a demand node; and if $b(i) = 0$, then node i is a transshipment node. The decision variables in the minimum cost flow problem are arc flows. We represent the flow on an arc $(i,j) \in A$ by x_{ij}. The minimum cost flow problem is an optimization model formulated as follows:

$$\text{Minimize} \quad \sum_{(i,j)\in A} c_{ij}x_{ij} \tag{1a}$$

$$\text{Subject to} \quad \sum_{\{j:(i,j)\in A\}} x_{ij} - \sum_{\{j:(j,i)\in A\}} x_{ji} = b(i), \quad \text{for all } i \in N, \tag{1b}$$

$$l_{ij} \leqslant x_{ij} \leqslant u_{ij}, \quad \text{for all } (i,j) \in A \tag{1c}$$

The data for this model satisfies the feasibility condition $\sum_{i\in N} b(i) = 0$ (that is, total supply must equal the total demand). We refer to the constraints in (1b) as the mass balance constraints. The first term in this constraint represents the total outflow of a node (i.e., the flow emanating from the node) and the second term represents the total inflow of that node (i.e., the flow entering the node). The mass balance constraint states that the outflow minus inflow must equal the supply/demand of the node. The flow must also satisfy the lower bound and capacity constraints (1c), which we refer to as the flow bound constraints. The flow bounds typically model physical capacities or restrictions imposed upon the flows' operating ranges. In most applications, the lower bounds on arc flows are zero; therefore, if we do not state lower bounds for any problem, we assume they have value zero.

We now collect together several basic definitions and describe some notation. A walk in $G = (N,A)$ is a sequence of nodes and arcs $i_1, (i_1, i_2), i_2, (i_2, i_3), i_3, \ldots, (i_{r-1}, i_r), i_r$ satisfying the property that either $(i_k, i_{k+1}) \in A$ or $(i_{k+1}, i_k) \in A$. A walk might revisit nodes. A path is a walk whose nodes (and, hence, arcs) are distinct. For simplicity, we often refer to a path as a sequence of nodes $i_1, i_2, i_3, \ldots, i_r$ when its arcs are apparent from the problem context. A directed path is similarly defined. In this case, for any two consecutive nodes i_k and i_{k+1} on the path, the path must contain the arc (i_k, i_{k+1}). A directed cycle is a directed path together with the arc (i_r, i_1), and a cycle is a path together with the arc (i_r, i_1) or (i_1, i_r).

A graph $G' = (N', A')$ is a subgraph of $G = (N,A)$ if $N' \subseteq N$ and $A' \subseteq A$. A graph G' is a spanning subgraph of $G = (N,A)$ if $N' = N$ and $A' \subseteq A$. Two nodes i and j are said to be connected if the graph contains at least one undirected path between these nodes; otherwise it is disconnected. The connected subgraphs of a graph are called components. A tree is a connected graph that contains no cycle. A subgraph T is a spanning tree of G if T is a tree of G containing all its nodes.

3. Maximum Flow Problem

In the maximum flow problem, we wish to send the maximum amount of flow from a specified source node s to another specified sink node t in a network with arc capacities u_{ij}'s. If we interpret u_{ij} as the maximum flow rate of arc (i, j), then the maximum flow problem identifies the maximum steady-state flow that the network can send from node s to node t per unit time. The maximum flow problem is a special case of the minimum cost flow problem. In the formulation (1), if we set $b(i) = 0$ for all $i \in N$, $c_{ij} = 0$ for all $(i,j) \in A$, and introduce an additional arc (t, s) with cost $c_{ts} = -1$ and flow bound $u_{ts} = 8$, then the minimum cost flow solution maximizes the flow on arc (t, s) (because any flow on arc (t, s) must travel from node s to node t through the arcs in A). With these choices for the data, the solution to the minimum cost flow problem will maximize the flow from node s to node t in the original network.

The maximum flow problem arises in a wide variety of situations and in several forms. Sometimes, it arises directly in combinatorial applications that on the surface might not appear to be maximum flow problems at all; at other times, it occurs as a subproblem in the solution of more difficult network optimization problems. In this section, we describe three applications of the maximum flow problem.

3.1. Application 1: Capacity of Physical Networks

An oil company needs to ship oil from a refinery to a storage facility using the pipelines of its underlying distribution network. In this problem context, the refinery corresponds to a particular node s in the distribution network and the storage facility corresponds to another node t. The capacity of each arc is the maximum amount of oil per unit time that can flow along it. The value of a maximum $s-t$ flow determines the maximum flow rate from the source node s to the sink node t. Similar applications arise in other settings, for example, determining the transmission capacity between two nodes of a telecommunications network.

3.2. Application 2: Feasible Flow Problem

The feasible flow problem consists of finding a feasible flow satisfying the following constraints:

$$\sum_{\{j:(i,j)\in A\}} x_{ij} - \sum_{\{j:(j,i)\in A\}} x_{ji} = b(i), \quad \text{for all } i \in N, \tag{2a}$$

$$0 \leqslant x_{ij} \leqslant u_{ij}, \quad \text{for all } (i,j) \in A \tag{2b}$$

We assume that $\sum_{i\in N} b(i) = 0$. The following distribution scenario illustrates how the feasible flow problem arises in practice. Suppose that merchandise is available at some seaports and desired by other ports. We know the stock of merchandise available at the supply ports, the amount required at the other ports, and the maximum quantity of merchandise that can be shipped on a particular sea route. We wish to know whether we can satisfy all of the demands by using the available supplies.

We can solve the feasible flow problem by solving a maximum flow problem defined on an augmented network as follows. We introduce two new nodes, a source node s and a sink node t. For each node i with supply (that is, with $b(i) > 0$), we add an arc (s, i) with capacity $b(i)$, and for each node i with demand (that is, with $b(i) < 0$), we add an arc (i, t) with capacity $-b(i)$. We refer to the new network as the transformed network. We then solve a maximum flow problem from node s to node t in the transformed network. It is easy to show that model (1) has a feasible solution if, and only if, the maximum flow saturates all the arcs emanating from the source node, that is, $x_{sj} = u_{sj}$ for all arcs $(s,j) \in A(s)$. Moreover, if each $b(i)$ and u_{ij} is integer, then model (2) always has an integer-feasible solution whenever it has a feasible solution (see Theorem 2 in Chapter 8.2).

Sometimes in a feasible flow problem arcs have nonzero lower bounds, that is, the flow bound constraints are $l_{ij} \leqslant x_{ij} \leqslant u_{ij}$ instead of $0 \leqslant x_{ij} \leqslant u_{ij}$, for some constants l_{ij} for each $(i,j) \in A$. By substituting $y_{ij} = x_{ij} - l_{ij}$ for x_{ij}, we can transform this problem to the formulation (2). Then (2b) reduces to $0 \leqslant y_{ij} \leqslant (u_{ij} - l_{ij})$ and (2a) reduces to the same set of equations, but with a different right-hand side vector b'.

3.3. Application 3: The Tournament Problem (Adapted from Ford and Johnson, 1959)

Consider a round-robin tournament between n teams, assuming each team plays against every other team c times. Assume that no game ends in a draw. Suppose that someone claims that α_i for $1 \leqslant i \leqslant n$ denotes the number of victories accrued by the ith team at the end of the tournament. How can we determine whether the given set of nonnegative integers $\alpha_1, \alpha_2, \ldots, \alpha_n$ represents a possible winning record for the n teams?

Define a directed network $G = (N, A)$ with node set $N = \{1, 2, \ldots, n\}$ and arc set $A = \{(i,j) \in N \times N : i < j\}$. Therefore, each node i is connected to the nodes $i+1, i+2, \ldots, n$. Let x_{ij} represent the number of times that team i defeats team j for $i < j$. Observe that the total number of times team i beats team $i+1, i+2, \ldots, n$ is $\sum_{\{j:(i,j)\in A\}} x_{ij}$. Also observe that the number of times that team i beats a team $j < i$ is $c - x_{ji}$. Consequently, the total number of times that team i beats teams $1, 2, ..., i-1$ is $(i-1)c - \sum_{\{j:(j,i)\in A\}} x_{ij}$. The total number of wins of team i must equal the total number of times it beats the teams $1, 2, ..., n$. The preceding observations show that

$$\sum_{\{j:(i,j)\in A\}} x_{ij} - \sum_{\{j:(j,i)\in A\}} x_{ji} = \alpha_i - (i-1)c, \quad \text{for all } i \in N, \tag{3}$$

In addition, a possible winning record must also satisfy the following lower and upper bound conditions:

$$0 \leqslant x_{ij} \leqslant c, \quad \text{for all } (i,j) \in A \tag{4}$$

This discussion shows that the record α_i is a possible winning record if the constraints defined by (3) and (4) have a feasible solution x. Let $b(i) = \alpha_i - (i-1)c$. Observe that both the expressions $\sum_{i\in N} \alpha_i$ and $\sum_{i\in N}(i-1)c$ are equal to $cn(n-1)/2$, which is the total number of games played. Consequently, $\sum_{i\in N} b(i) = 0$. The problem of finding a feasible solution for a network flow system like (3) and (4) is called a feasible flow problem and can be solved by solving a maximum flow problem (see Application 2).

4. Shortest-Path Problem

The shortest-path problem is among the simplest network flow problems. For this problem, we wish to find a directed path of minimum cost (length) from a specified source node s to another specified sink node t in a directed network in which each arc (i, j) has an associated cost (or length) c_{ij}. The shortest-path problem is a special case of the minimum-cost flow problem. In the minimum-cost flow formulation (1), if we set $b(s) = 1$, $b(t) = -1$, and $b(i) = 0$ for all other nodes, then the optimal solution to the problem will send one unit of flow from node s to node t along the shortest path.

Shortest-path problems are alluring to both researchers and practitioners for several reasons: (i) they arise frequently in practice in a wide variety of application settings when we wish to send some material (for example, a computer data packet, a telephone call or a vehicle) between two specified points in a network as quickly, as cheaply, or as reliably as possible; (ii) they are easy to solve efficiently; (iii) as the simplest network models, they capture many of the most salient core ingredients of network flows and so they provide both a benchmark and a point of departure for studying more complex network models; and (iv) they arise frequently as subproblems when solving many combinatorial and network optimization problems. In this section, we describe two problems that do not seem to be related to shortest-paths but can be transformed to shortest-path problems.

4.1. Application 4: Assortment of Structural Steel Beams (Adapted from Frank, 1965)

In its various construction projects, a construction company needs structural steel beams of a uniform cross section but of varying lengths. For each $i = 1, ..., n$, let $D_i > 0$ denote the demand of the steel beam of length L_i, and assume that $L_1 < L_2 < ... < L_n$. The company could meet its needs by maintaining and drawing upon an inventory of stock containing exactly D_i units of the steal beam of length L_i. It might not be economical to carry all the demanded lengths in inventory, however, because of the high cost of setting up the inventory facility to handle each length in inventory. In that case, if the company needs a beam of length L_i that is not carried in inventory, it can cut a beam of longer length down to the desired length. The cutting operation will typically produce unusable steel as scrap. Let K_i denote the cost for setting up the inventory facility to handle beams of length L_i, and let C_i denote the cost of a beam of length L_i. The company wants to determine the lengths of beams to be carried in inventory so that it will minimize the total cost of (i) setting up the inventory facility, and (ii) the cost of usable steel lost as scrap.

We formulate this problem as a shortest-path problem as follows. We construct a directed network G on $(n+1)$ nodes numbered $0, 1, 2, ..., n$; the nodes in this network correspond to various beam lengths. Node 0 corresponds to a beam of length zero and node n

corresponds to the longest beam. For each node i, the network contains a directed arc to every node $j = i+1, i+2, \ldots, n$. We interpret the arc (i, j) as representing a storage strategy in which we hold beams of length L_j in inventory and use them to satisfy the demand of all the beams of lengths $L_{i+1}, L_{i+2}, \ldots, L_j$. The cost c_{ij} of the arc (i, j) is

$$c_{ij} = K_j + C_j \sum_{k=i+1}^{j} D_k.$$

The cost of arc (i, j) has two components: (i) the fixed cost K_j of setting up the inventory facility to handle beams of length L_j, and (ii) the cost of using beams of length L_j to meet the demands of beams of lengths $L_{i+1}, \ldots, L_j$. A directed path from node 0 to node n specifies an assortment of beams to carry in inventory and the cost of the path equals the cost associated with this inventory scheme. For example, the path 0–4–6–9 corresponds to the situation in which we set up the inventory facility for handling beams of lengths L_4, L_6, and L_9. Consequently, the shortest path from node 0 to node n would prescribe the least cost assortment of structural steel beams.

4.2. *Application 5: Tramp Steamer Problem (Lawler, 1966)*

A tramp steamer travels from port to port, carrying cargo and passengers. A voyage of the steamer from port i to port j earns p_{ij} units of profit and requires τ_{ij} units of time. The captain of the steamer would like to know which tour W of the steamer (i.e., a directed cycle) achieves the largest possible mean daily profit when we define the daily profit for any tour W by the expression

$$\mu(W) = \frac{\sum_{(i,j)\in W} p_{ij}}{\sum_{(i,j)\in W} \tau_{ij}}.$$

We assume that $\tau_{ij} \geqslant 0$ for every arc$(i, j) \in A$, and that $\sum_{(i,j)\in W} \tau_{ij} > 0$ for every directed cycle W in the network. Now consider a more restricted version of the tramp steamer problem: the captain of the steamer wants to know whether some tour W will be able to achieve a mean daily profit greater than a specified threshold μ_0. In this restricted version of the tramp steamer problem, we wish to determine whether the underlying network G contains a directed cycle W satisfying the following condition:

$$\frac{\sum_{(i,j)\in W} p_{ij}}{\sum_{(i,j)\in W} \tau_{ij}} > \mu_0.$$

Alternatively, we wish to determine whether there exists a directed cycle W for which

$$\sum_{(i,j)\in W} (\mu_0 \tau_{ij} - p_{ij}) < 0.$$

Hence the restricted version of the tramp steamer problem transforms to identifying whether the network G contains a directed cycle when the cost of each arc (i, j) is $(\mu_0 \tau_{ij} - p_{ij})$. This problem is known as the negative cycle detection problem. Shortest-path algorithms are capable of detecting the presence of negative cycles. The general tramp steamer problem can be solved by solving a sequence of restricted tramp steamer problems.

5. Minimum-Cost Single-Commodity Flow Problem

The minimum cost flow model is the most fundamental of all network flow problems. In this problem, as we described in Section 2, we wish to determine a least cost shipment of a commodity through a network that will satisfy demands at certain nodes from available supplies at other nodes. This model has a number of familiar applications: the distribution of a product from manufacturing plants to warehouses, or from warehouses to retailers; the flow of raw material and intermediate goods through the various machining stations in a production line; the routing of automobiles through an urban street network; and the routing of calls through the telephone system. Minimum-cost flow problems arise in almost all industries, including agriculture, communications, defense, education, energy, health care, manufacturing, medicine, retailing, and transportation. Indeed, minimum-cost flow problems are pervasive in practice. In this section, by considering two selected applications that arise in distribution systems planning and capacity planning, we give a passing glimpse of these applications.

5.1. *Application 6: Distribution Problems (Adapted from Glover and Klingman, 1976)*

A large class of network flow problems center around distribution applications. One core model is often described in terms of shipments from plants to warehouses (or, alternatively, from warehouses to retailers). Suppose a firm has p plants with known supplies and q warehouses with known demands. It wishes to identify a flow that satisfies the demands at the warehouses from the available supplies at the plants and that minimizes its shipping costs. This problem is a well-known special case of the minimum-cost flow problem, known as the transportation problem. We next describe in more detail a slight generalization of this model that also incorporates manufacturing costs at the plants.

A car manufacturer has several manufacturing plants and produces several car models at each plant that it then ships to geographically dispersed retail centers throughout the country. Each retail center requests a specific number of cars of each model. The firm must determine the production plan of each model at each plant and a shipping pattern that satisfies the demand of each retail center while minimizing the overall cost of production and transportation.

We describe this formulation through an example. Figure 1 illustrates a situation with two manufacturing plants, two retailers, and three car models. This model has four types of nodes: (i) plant nodes representing various plants; (ii) plant/model nodes corresponding to

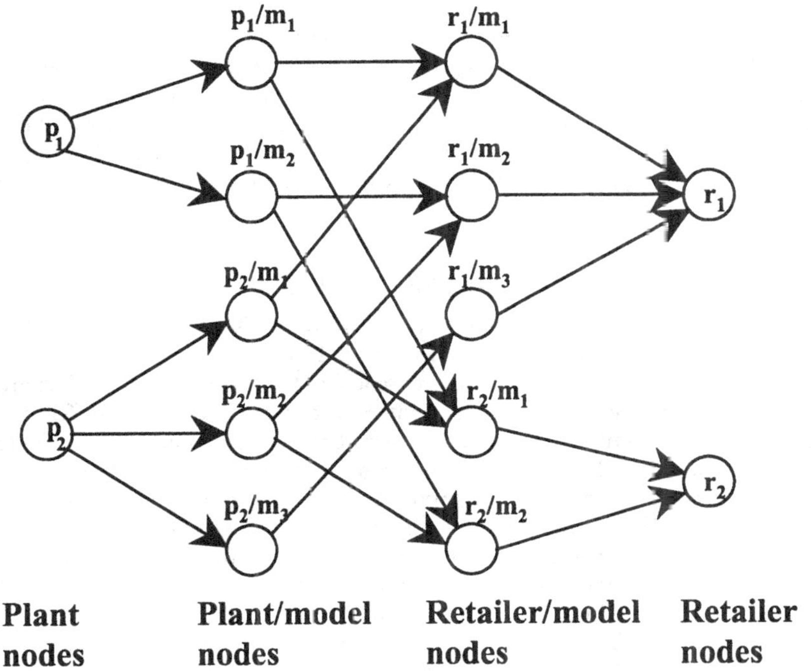

Figure 1 Formulating the production-distribution problem

each model made at each plant; (iii) retailer/model nodes, corresponding to the models required by each retailer; and (iv) retailer nodes corresponding to each retailer. The network contains three types of arcs: (i) production arcs; (ii) transportation arcs; and (iii) demand arcs. The production arcs connect a plant node to a plant/model node; the cost of this arc is the cost of producing the model at that plant. We might place lower and upper bounds on production arcs to control for the minimum and maximum production of each car model at the plants. Transportation arcs connect plant/model nodes to retailer/model nodes; the cost of any such arc is the total cost of shipping one car from the manufacturing plant to the retail center. The transportation arcs might have lower or upper bounds imposed upon their flows to model contractual agreements with shippers or capacities imposed upon any distribution channel. Finally, demand arcs connect retailer/model nodes to the retailer nodes. These arcs have zero costs and positive lower bounds that equal the demand of that model at that retail center.

The production and shipping schedules for the automobile company correspond in a one-to-one fashion with the feasible flows in this network model. Consequently, a minimum-cost flow provides an optimal production and shipping schedule.

5.2. *Application 7: Airplane Hopping Problem*

A small commuter airline uses a plane, with a capacity to carry at most p passengers, on a hopping flight as shown in Figure 2(a). The hopping flight visits the cities $1, 2, 3, \ldots, n$, in a fixed sequence. The plane can pick up passengers at any node and drop them off at any other node. Let b_{ij} denote the number of passengers available at node i who want to go to node j, and let f_{ij} denote the fare per passenger from node i to node j. The airline would like to determine the number of passengers that the plane should carry between the various origins to destinations to maximize the total fare per trip while never exceeding the plane's capacity.

Figure 2(b) shows a minimum-cost flow formulation of this hopping plane flight problem. The network contains data for only those arcs with nonzero costs and with finite capacities: any arc listed without an associated cost has a zero cost; any arc listed without an associated capacity has an infinite capacity. Consider, for example, node 1. Three types of passengers are available at node 1, those whose destination is node 2, node 3, or node 4. We represent these three types of passengers in a new derived network by the nodes 1–2, 1–3, and 1–4 with supplies b_{12}, b_{13}, and b_{14}. A passenger available at any such node, say 1–3, could board the plane at its origin node represented by flowing through the arc (1–3, 1) and incurring a cost of $-f_{13}$ units (or profit of f_{13} units). Or, the passenger might never board the plane, which we represent by the flow through the arc (1–3, 3). It is easy to establish a one-to-one correspondence between feasible flows in Figure 2(b) and feasible loading of the plane with passengers. Consequently, a minimum cost flow in Figure 2(b) will prescribe the most profitable loading of the plane.

6. Minimum-Cost Multicommodity Flow Problem

The minimum-cost flow problem models the flow of a single commodity over a network. Multicommodity flow problems arise when several commodities use the same underlying network. The commodities might be differentiated either by their physical characteristics or simply by their origin-destination pairs. Different commodities have different origins and destinations, and commodities have separate mass balance constraints at each node. Sharing of the common arc capacities binds the different commodities together. In fact, the essential issue addressed by the multicommodity flow problem is the allocation of the capacity of each arc to the individual commodities in a way that minimizes overall flow costs. Multicommodity flow problems arise in many practical situations including: (i) the transportation of passengers from different origins to different destinations within a city; (ii) the routing of nonhomogeneous tankers (nonhomogeneous in terms of speed, carrying capability, and operating costs); (iii) the worldwide shipment of different varieties of grains (such as corn, wheat, rice, and soybeans) from countries that produce grain to those that consume it; and (iv) the transmission of messages in a communication network between different origin-destination pairs. In this

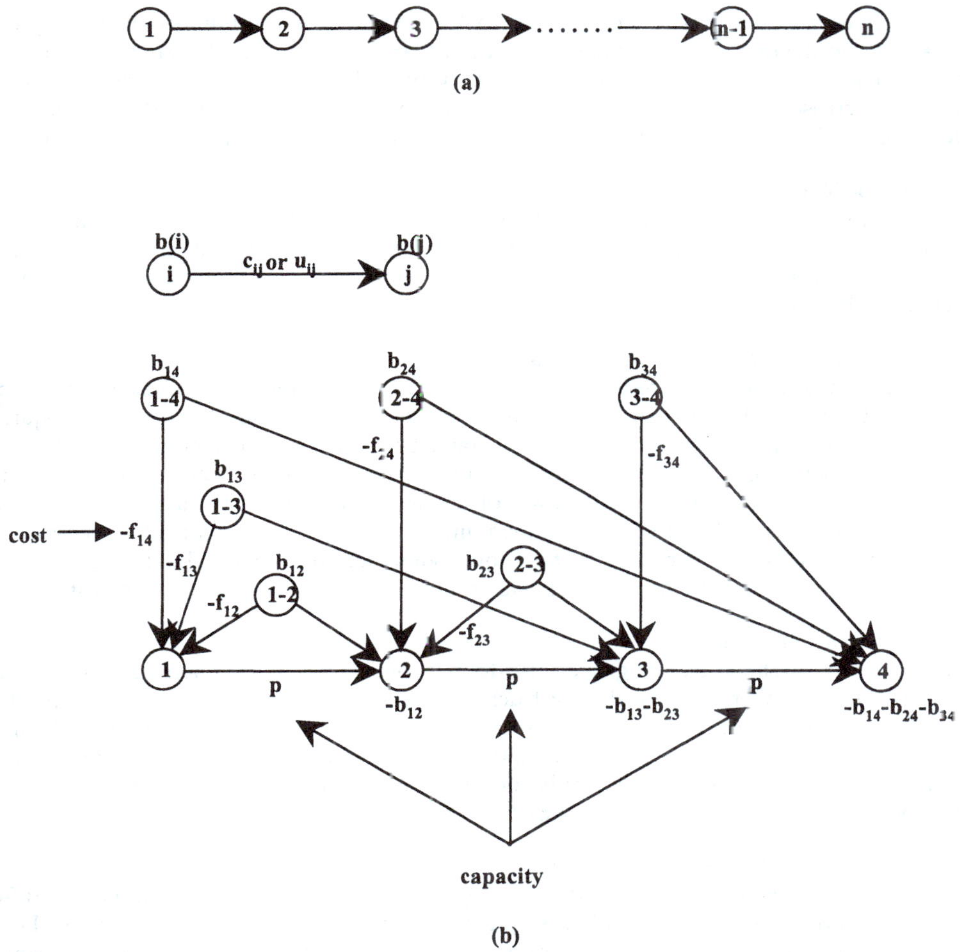

Figure 2 Formulating the hopping plane flight problem as a minimum-cost flow problem

section, we consider several instances of one very general type of application as well as an application arising in social sciences.

6.1. Application 8: Routing of Multiple Commodities

In many applications of the multicommodity flow problem, we distinguish commodities because they are different physical goods, and/or because they have different points of origin and destination; that is, either (i) several physically distinct commodities, for example different manufactured goods, share a common network, or (ii) a single physical good (e.g., messages or products) flows on a network, but the good has multiple points of origin and destination defined by different pairs of nodes in the network that need to send the good to each other. This second type of application arises frequently in problem contexts such as communication systems or distribution/transportation systems. In this section, we briefly introduce several application domains of both types.

COMMUNICATION NETWORKS

In a communication network, nodes represent origin and destination stations for messages, and arcs represent transmission lines. Messages between different pairs of nodes define distinct commodities; the supply and demand for each commodity is the number of messages to be sent between the origin and destination nodes of that commodity. Each transmission line has a fixed capacity (in some applications the capacity of each arc is fixed, in others, we might be able to increase the capacity at a certain cost per unit). In this network, the problem of determining the minimum cost routing of messages is a multicommodity flow problem.

Computer Networks

In a computer communication network, the nodes represent storage devices, terminals, or computer systems. The supplies and demands correspond to the data transmission rates between the computer, terminals, and storage devices, and the transmission line capacities define the bundle constraints.

Railroad Transportation Networks

In a rail network, nodes represent yard and junction points, and arcs represent track sections between the yards. The demand is measured by the number of cars (or, any other equivalent measure of tonnage) to be loaded on any train. Since the system incurs different costs for different goods, we divide traffic demand into different classes. Each commodity in this network corresponds to a particular class of demand between a particular origin-destination pair. The bundle capacity of each arc is the number of cars that we can load on the trains that are scheduled to be dispatched on that arc (over some period of time). The decision problem in this network is to meet the demands of cars at the minimum possible operating cost.

Distribution Networks

In distribution systems planning, we wish to distribute multiple (nonhomogeneous) products from plants to retailers using a fleet of trucks or railcars, and using a variety of railheads and warehouses. The products define the commodities of the multicommodity flow problem, and the joint capacities of the plants, warehouses, railyards, and the shipping lanes define the bundle constraints. Note that this application has important bundle constraints imposed upon the nodes (plants, warehouses) as well as the arcs.

Foodgrain Export-Import Network

The nodes in this network correspond to geographically dispersed locations in different countries, and the arcs correspond to shipments by rail, truck, and ocean freighter. Between these locations, the commodities are various foodgrains such as corn, wheat, rice, and soybean. The capacities at the ports define the bundle constraints.

6.2. Application 9: Racial Balancing of Schools (Adapted from Clarke and Surkis, 1968)

In 1968, the Supreme Court of the United States ruled that all school systems in the country should begin admitting students to schools on a nondiscriminatory basis, and should employ faster techniques to promote desegregated schools across the nation. This decision made it necessary for many school systems to develop radically different procedures for assigning students to schools. Since the Supreme Court did not specify what constitutes an acceptable racial balance, the individual school boards used their own best judgments to arrive at acceptable criteria upon which to base their desegregation plans. This application describes a multicommodity flow model for determining an optimal assignment of students to schools that minimizes the total distance traveled by the students, given a specification of lower and upper limits on the required racial balance in each school.

Suppose that a school district has S schools and the capacity of the jth school is u_j. For the purpose of this formulation, we divide the school district into L population centers. These locations might, for example, be census tracts, bus stops, or city blocks. The only restriction on the population centers is that they be finite in number, and that a single distance measure reasonably approximates the distance any student at center i must travel if he or she is assigned to school j. Let S_{ik} denote the available number of students of the kth ethnic group at the ith population center. The objective is to assign students to schools in a way that achieves the desired ethnic composition for each school and minimizes the total distance traveled by the

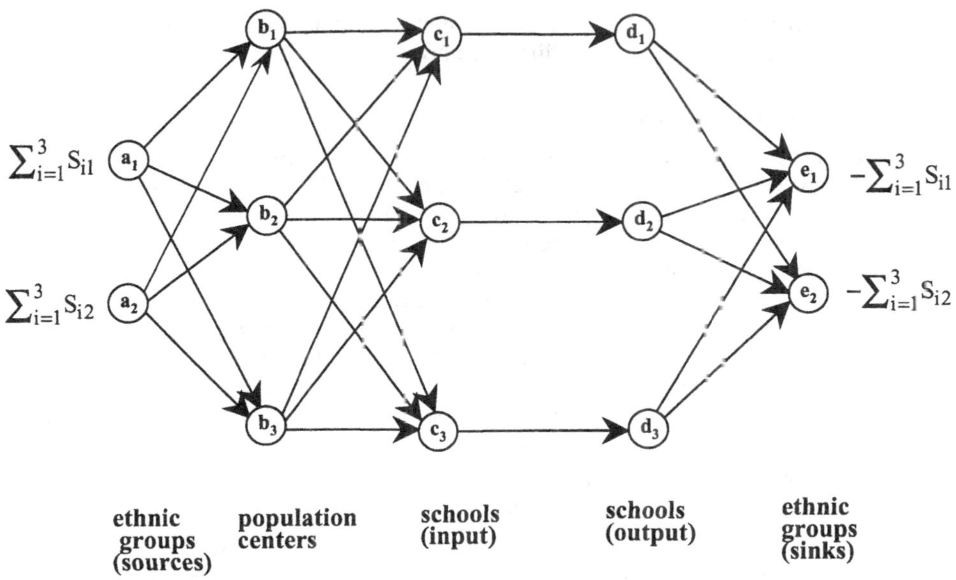

Figure 3 Formulating the racial balancing problem as a multicommodity flow problem

students. Each school j has the ethnic requirement that it must enroll at least l_{jk} and no more than u_{jk} students from the kth ethnic group.

We can model this problem as a multicommodity flow problem on an appropriately defined network. Figure 3 shows this network representation for a problem with three population centers and three schools. This network has one node for each population center and for each school as well as a source and a sink node for each ethnic group. The commodities represent the students of different ethnic groups. The students of the kth ethnic group flow from the source a_k to the sink e_k via population center and school nodes. We set the upper bound on arc (a_k, b_i) connecting the kth ethnic group source node and the ith population center equal to s_{ik} and the cost of the arc (b_i, c_j) connecting the ith population center and jth school equal to f_{ij}, the distance between that population center and that school. By setting the capacity of the arc (c_j, d_j) equal to u_j, we ensure that the total number of students (of all ethnic groups) allocated to the jth school do not exceed the maximum student population for this school. The students of all ethnic groups must share the capacity of each school. Finally, we incorporate constraints on the ethnic compositions of the schools by setting the lower and upper bounds on the arc (d_j, e_k) equal to l_{jk} and u_{jk}. It is fairly easy to verify that the multicommodity flow problem models the racial balancing problem and so a minimum multicommodity flow will specify an optimal assignment of students to the schools.

7. Minimum Spanning Tree Problem

As we noted previously, a spanning tree is a tree (i.e., a connected acyclic graph) that spans (touches) all the nodes of an undirected network. The cost of a spanning tree is the sum of the costs (or lengths) of its arcs. In the minimum spanning tree problem, we wish to identify a spanning tree of minimum cost (or length). Minimum spanning trees generally arise in one of two ways, directly or indirectly. In direct applications, we wish to connect a set of points using the least cost or least length collection of arcs. In indirect applications, we either (i) wish to connect some set of points using a measure of performance that on the surface bears little resemblance to the minimum spanning tree objective (sum of arc costs), or (ii) the problem itself bears little resemblance to an optimal tree problem—these instances often require creativity in modeling so that they become a minimum spanning tree problem. In this section, we discuss several direct as well as indirect applications.

7.1. Application 10: Designing Physical Systems

The design of physical systems can be a complex task involving an interplay between performance objectives (such as throughput and reliability), design costs and operating economics, and available technologies. In many settings, the major criterion is fairly simple: we need to design a network that will connect geographically dispersed system components or that will provide the infrastructure needed so that users can communicate with each other. In many of these settings, the system need not possess any redundancy, so we are interested in the simplest possible connection, namely, a spanning tree. This type of application arises in numerous problem settings including the building of certain types of highways, computer networks, leased-line telephone networks, or railroads, or the installation of cable television lines or high voltage electrical power transmission lines. The following few application settings are representative:

- Connecting terminals in cabling the panels of electrical equipment. How should we wire the terminals to use the least possible length of the wire?
- Constructing a pipeline network to connect a number of towns using the smallest possible total length of pipeline.
- Linking isolated villages in a remote region that are connected by roads, but not yet by telephone service. In this instance, we wish to determine along which stretches of roads we should place telephone lines, using the minimum possible total miles of the lines, to link every pair of villages.
- Constructing a digital computer system composed of high frequency circuitry when it is important to minimize the lengths of wires between different components to reduce both capacitance and delay line effects. Since the system must connect all the components, we obtain a spanning tree problem.
- Connecting a number of computer sites by high-speed lines. Each line is available for leasing at a certain monthly cost, and we wish to determine a configuration that connects all the sites at the least possible cost.

7.2. Application 11: Measuring Homogeneity of Bimetallic Objects (Adapted from Shier, 1982)

This application shows how a minimum spanning tree problem can be used to determine the degree to which a bimetallic object is homogeneous in composition. To use this approach, we measure the composition of the bimetallic object at a set of sample points. We then construct a network with nodes corresponding to the sample points and with an arc connecting physically adjacent sample points. We assign a cost with each arc (i, j) equal to the product of the physical (Euclidean) distance between the sample points i and j and a homogeneity factor between 0 and 1. This homogeneity factor is 0 if the composition of the corresponding samples is exactly alike, and is 1 if the composition is very different; otherwise, it is a number between 0 and 1. Note that this measure gives greater weight to two points if they are different and are far apart. The cost of the minimum spanning tree is a measure of the homogeneity of the bimetallic object. The cost of the tree is 0 if all the sample points are exactly alike, and high cost values imply that the material is quite nonhomogeneous.

7.3. Application 12: Reducing Data Storage (Adapted from Kang et al., 1977)

In several different application contexts, to save memory space, we wish to store data specified in the form of a two-dimensional array more efficiently than storing all the elements of the array. We assume that the rows of the array have many similar entries and differ at only a few places. One such situation arises in the sequence of amino acids in a protein found in the mitochondria of different animals and higher plants.

Since the entities in the rows are similar, one approach for saving memory is to store one row, called the reference row, completely, and to store only the differences between some of the rows so that we can derive each row from these differences and the reference row. Let c_{ij} denote the number of different entries in rows i and j; that is, if we are given row i, then by making c_{ij} changes to the entries in this row we can obtain row j, and vice versa. As an example, suppose that the array contains four rows, represented by R_1, R_2, R_3, and R_4, and we decide to treat R_1 as a reference row. Then one plausible solution is to store the differences between R_1 and R_2, R_2 and R_4, and R_1 and R_3. From this solution, we can obtain rows R_2 and R_3 by making c_{12} and c_{13} changes to the elements in row R_1. Having obtained row R_2, we can make c_{24} changes to the elements of this row to obtain R_4.

It is easy to see that it is sufficient to store differences between those rows that correspond to arcs of a spanning tree. These differences permit us to obtain each row from the reference row. The total storage requirement for a particular storage scheme will be the length of the reference row (which we can take as the row with the least amount of data) plus the sum of the differences between the rows. Therefore, a minimum spanning tree would provide the least cost storage scheme.

ACKNOWLEDGMENTS

The first author gratefully acknowledges the support of the National Science Foundation through grant award number DMI-9900087. The third author acknowledges support from the Office of Naval Research under contract N00014-98-1-0317 and the National Science Foundation through grant award number DMI-9820998.

References

Ahuja, R. K., T. L. Magnanti, and J. B. Orlin. 1993. In *Network Flows: Theory, Algorithms, and Applications*. NJ: Prentice Hall.

Clarke, S. and J. Surkis. 1968. "An operations research approach to racial desegregation of school systems." *Socio-Econ. Planning Sci.* **1**: 259–272.

Ford, L. R. and S. M. Johnson. 1959. "A tournament problem." *Am. Math. Monthly* **66**: 387–389.

Frank, C. R. 1965. "A note on the assortment problem." *Mgmnt Sci.* **11**: 724–726.

Glover, F. and D. Klingman. 1976. "Network applications in industry and government." *AIIE Trans.* **9**: 363–376.

Kang, A. N. C., R. C. T. Lee, C. L. Chang, and S. K. Chang. 1977. "Storage reduction through minimal spanning trees and spanning forests." *IEEE Trans. Comput.* **C-26**: 425–434.

Lawler, E. L. 1966. "Optimal cycles in doubly weighted linear graphs." In *Theory of Graphs: International Symposium, Dunod, Paris*, pp. 209–213. New York: Gordon and Breach.

Shier, D. R. 1982. "Testing for homogeneity using minimum spanning trees." *UMAP J.* **3**: 273–283.

8.2 Maximum Flow Problem

Ravindra K. Ahuja, Thomas L. Magnanti, and James B. Orlin

ABSTRACT The maximum flow problem seeks the maximum possible flow in a capacitated network from a specified source node s to a specified sink node t without exceeding the capacity of any arc. A closely related problem is the minimum cut problem, which is to find a set of arcs with the smallest total capacity whose removal separates node s and node t. We described several applications of the maximum flow problem in Chapter 8.1. In this article we describe algorithmic methods for solving these problems, beginning by considering a generic augmenting path algorithm for solving the maximum flow problem. A byproduct of this algorithm is the well-known max-flow min-cut theorem stating that the value of the maximum flow in a network equals the capacity of the minimum cut. The generic version of the augmenting path algorithm does not run in polynomial time; we describe two specific implementations that run in polynomial time. We next study another class of algorithms, known as preflow-push algorithms, that have recently emerged as the most powerful techniques, both theoretically and computationally, for solving maximum flow problems.

1. Introduction

The maximum flow problem seeks the maximum possible flow in a capacitated network from a specified source node s to a specified sink node t without exceeding the capacity of any arc. A closely related problem is the minimum cut problem, which is to find a set of arcs with the smallest total capacity whose removal separates node s and node t. The maximum flow and minimum cut problems arise in a variety of application settings as diverse as manufacturing, communication systems, distribution planning, matrix rounding, and scheduling. These problems also arise as subproblems in the solution of more difficult network optimization problems. In this article, we study the maximum flow and minimum cut problems, briefly introducing the underlying theory and algorithms. The book by Ahuja et al. (1993) contains a wealth of additional material that expands on this discussion.

Let $G=(N,A)$ be a directed network defined by a set N of n nodes and a set A of m directed arcs. We refer to nodes i and j as endpoints of arc (i, j). A directed path $i_1 - i_2 - i_3 - \ldots - i_k$ is a set of arcs $(i_1, i_2), (i_2, i_3), \ldots, (i_{k-1}, i_k)$. Each arc (i, j) has an associated capacity u_{ij} denoting the maximum amount of flow on this arc. We assume that each arc capacity u_{ij} is an integer, and let $U = \max\{u_{ij} : (i,j) \in A\}$. The network has two distinguished nodes, a source node s and a sink node t. To help in representing a network, we use the arc adjacency list $A(i)$ of node i, which is the set of arcs emanating from it, that is, $A(i) = \{(i,j) \in A : j \in N\}$.

The maximum flow problem is to find the maximum flow from the source node s to the sink node t that satisfies the arc capacities and mass balance constraints at all nodes. We can state the problem formally as follows:

$$\text{Maximize } v \tag{1a}$$

subject to

$$\sum_{\{j:(i,j)\in A\}} x_{ij} - \sum_{\{j:(j,i)\in A\}} x_{ji} = \begin{cases} v & \text{for } i = s, \\ 0 & \text{for all } i \notin \{s,t\}, \\ -v & \text{for } i = t, \end{cases} \tag{1b}$$

$$0 \leqslant x_{ij} \leqslant u_{ij}, \quad \text{for all } (i,j) \in A. \tag{1c}$$

We refer to a vector $x = \{x_{ij}\}$ satisfying (1b) and (1c) as a flow and the corresponding value of the scalar variable v as the value of the flow. We refer to the constraints (1b) as the mass balance constraints, and refer to the constraints (1c) as the flow bound constraints.

In examining the maximum flow problem, we impose two assumptions: (i) all arc capacities are integer; and (ii) whenever the network contains arc (i, j), then it also contains arc (j, i). The second assumption is nonrestrictive since we allow arcs with zero capacity.

Sometimes the flow vector x might be required to satisfy lower bound constraints imposed upon the arc flows; that is, if l_{ij} specifies the lower bound on the flow on arc $(i,j) \in A$, we impose the condition $x_{ij} = l_{ij}$. We refer to this problem as the maximum flow problem with nonnegative lower bounds. It is possible to transform a maximum flow problem with nonnegative lower bounds into a maximum flow problem with zero lower bounds.

The minimum cut problem is a close relative of the maximum flow problem. A cut $[S, \bar{S}]$ partitions the node set N into two subsets S and $\bar{S} = N - S$. It consists of all arcs with one endpoint in S and the other in $\bar{S}$. We refer to the arcs directed from S to $\bar{S}$, denoted by $(S, \bar{S})$, as forward arcs in the cut and the arcs directed from $\bar{S}$ to S, denoted by $(\bar{S}, S)$, as backward arcs in the cut. The cut $[S, \bar{S}]$, is called an $s-t$ cut if $s \in S$ and $t \in \bar{S}$. We define the capacity of the cut $[S, \bar{S}]$, denoted as $u[S, \bar{S}]$, as $\sum_{(i,j)\in(S,\bar{S})} u_{ij}$. A minimum cut in G is an $s-t$ cut of minimum capacity. We will show that any algorithm that determines a maximum flow in the network also determines a minimum cut in the network.

The remainder of this article is organized as follows. In Section 2, we present some preliminary results concerning flows and cuts. We next discuss two important classes of algorithms for solving the maximum flow problem: (i) augmenting path algorithms, and (ii) preflow-push algorithms. Section 3 describes augmenting path algorithms, which augment flow along directed paths from the source node to the sink node. The proof of the validity of the augmenting path algorithm yields the well-known max-flow min-cut theorem, which states that the value of a maximum flow in a network equals the capacity of a minimum cut in the network. In Section 4, we describe two polynomial-time implementations of the generic augmenting path algorithm. In Section 5, we study preflow-push algorithms that flood the network so that some nodes have excesses and then incrementally relieve the flow from nodes with excesses by sending flow from excess nodes forward toward the sink node or backward toward the source node.

We would like to design maximum flow algorithms that are guaranteed to be efficient in the sense that their worst-case running times, that is, the total number of multiplications, divisions, additions, subtractions, and comparisons in the worst case, grow slowly in some measure of the problem's size. We say that a maximum flow algorithm is an $O(n^3)$ algorithm, or has a worst-case complexity of $O(n^3)$, if it is possible to solve any maximum flow problem using a number of computations that is asymptotically bounded by some constant times the term n^3. We say that an algorithm is a polynomial time algorithm if its worst-case running time is bounded by a polynomial function of the input size parameters. For a maximum flow problem, the input size parameters are n, m, and $\log U$ (the number of bits needed to specify the largest arc capacity). We refer to a maximum flow algorithm as a pseudopolynomial time algorithm if its worst-case running time is bounded by a polynomial function of n, m, and U. For example, an algorithm with worst-case complexity of $O(nm \log U)$ is a polynomial time algorithm, but an algorithm with worst-case complexity of $O(nmU)$ is a pseudopolynomial time algorithm.

2. Preliminaries

In this section, we discuss some elementary properties of flows and cuts. We will use these properties to prove the celebrated max-flow min-cut theorem and to establish the correctness of the augmenting path algorithm described in Section 3.

2.1. *Residual Network*

The concept of residual network plays a central role in the development of maximum flow algorithms. Given a flow x, the residual capacity r_{ij} of any arc $(i, j) \in A$ is the maximum additional flow that can be sent from node i to node j using the arcs (i, j) and (j, i). (Recall our assumption from Section 1 that whenever the network contains arc (i, j), it also contains the arc (j, i).) The residual capacity r_{ij} has two components: (i) $u_{ij} - x_{ij}$, the unused capacity of arc (i, j), and (ii) the current flow x_{ji} on arc (j, i), which we can cancel to increase the flow from node i to node j. Consequently, $r_{ij} = u_{ij} - x_{ij} + x_{ji}$. We refer to the network $G(x)$ consisting of the arcs with positive residual capacities as the residual network (with respect to the flow x). Figure 1 gives an example of a residual network.

2.2. *Flow Across an $s-t$ Cut*

Let x be a flow in the network. Adding the mass balance constraint (1b) for the nodes in S, we obtain the equation

$$v = \sum_{i\in S}\left[\sum_{\{j:(i,j)\in A\}} x_{ij} - \sum_{\{j:(j,i)\in A\}} x_{ji}\right] = \sum_{\{(i,j)\in(S,\bar{S})\}} x_{ij} - \sum_{\{(i,j)\in(\bar{S},S)\}} x_{ij}. \tag{2}$$

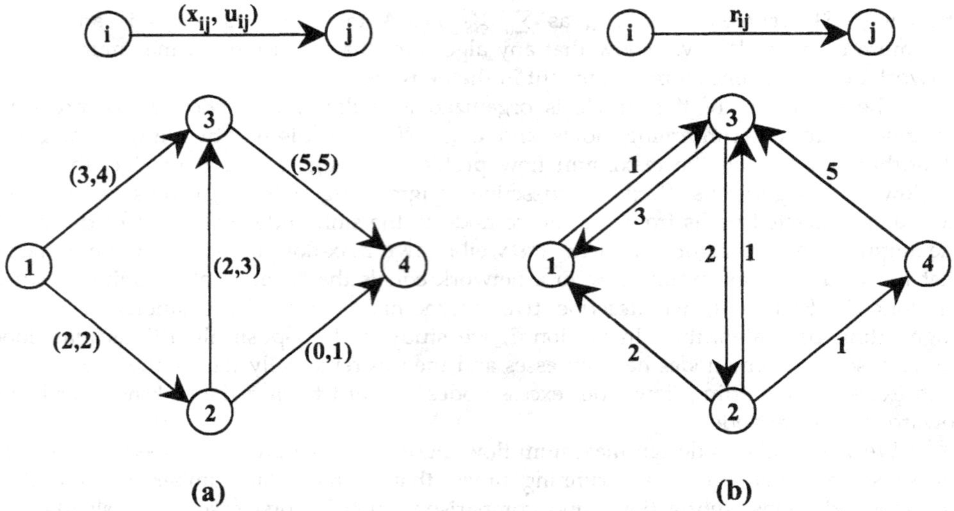

Figure 1 Illustrating the construction of residual network. (a) The original network with arc capacities and a flow x. (b) The residual network $G(x)$

The second equality uses the fact that whenever both the nodes p and q belong to the node set S and $(p,q) \in A$, the variable x_{pq} in the first term within the bracket (for node $i = p$) cancels the variable $-x_{pq}$ in the second term within the bracket (for node $j = q$). The first expression in the right-hand side of equation (2) denotes the amount of flow from the nodes in S to nodes in $\bar{S}$, and the second expression denotes the amount of flow returning from the nodes in $\bar{S}$ to the nodes in S. Therefore, the right-hand side denotes the total (net) flow across the cut, and (2) implies that the flow across any $s-t$ cut $[S, \bar{S}]$ equals v. Substituting $x_{ij} \leqslant u_{ij}$ in the first expression of (2) and $x_{ij} \geqslant 0$ in the second expression yields: $v \leqslant \sum_{(i,j) \in (S,\bar{S})} u_{ij} = u[S, \bar{S}]$ implying that the value of any flow can never exceed the capacity of any cut in the network. We record this result formally for future reference.

Lemma 1. *The value of any flow can never exceed the capacity of any cut in the network. Consequently, if the value of some flow* x *equals the capacity of some cut* [S, S̄], *then* x *is a maximum flow and the cut* [S, S] *is a minimum cut.*

The max-flow min-cut theorem, to be proved in the next section, states that the value of some flow always equals the capacity of some cut.

3. Generic Augmenting Path Algorithm

In this section, we describe one of the simplest and most intuitive algorithms for solving the maximum flow problem, an algorithm known as the augmenting path algorithm. Ford and Fulkerson (1956) and Elias et al. (1956) independently developed the basic augmenting path algorithm.

Let x be a feasible flow in the network G, and let $G(x)$ denote the residual network corresponding to the flow x. We refer to a directed path from the source to the sink in the residual network $G(x)$ as an augmenting path. We define the residual capacity $\delta(P)$ of an augmenting path P as the maximum amount of flow that can be sent along it, that is, $\delta(P) = \min\{r_{ij} : (i,j) \in P\}$. Since the residual capacity of each arc in the residual network is strictly positive, the residual capacity of an augmenting path is strictly positive. Therefore, we can always send a positive flow of δ units along it. Consequently, whenever the network contains an augmenting path, we can send additional flow from the source to the sink. (Sending an additional δ units of flow along an augmenting path decreases the residual capacity of each arc

```
algorithm augmenting path;
begin
    x : = 0;
    while G(x) contains a directed path from node s to node t do
    begin
        identify an augmenting path P from node s to node t;
        set δ : = min {r_ij: (i, j) ∈ P};
        augment δ units of flow along path P and update G(x);
    end;
end;
```

Figure 2 The generic augmenting path algorithm

(i, j) in the path by δ units.) The generic augmenting path algorithm is essentially based upon this simple observation. The algorithm identifies augmenting paths in $G(x)$ and augments flow on these paths until the network contains no such path. Figure 2 describes the generic augmenting path algorithm.

We can identify an augmenting path P in $G(x)$ by using a graph search algorithm. A graph search algorithm starts at node s and progressively finds all nodes that are reachable from the source node using directed paths. Most search algorithms run in time proportional to the number of arcs in the network, that is, $O(m)$ time, and either identify an augmenting path or conclude that $G(x)$ contains no augmenting path; the latter happens when the sink node is not reachable from the source node.

For each arc $(i, j) \in P$, augmenting δ units of flow along P decreases r_{ij} by δ units and increases r_{ji} by δ units. The final residual capacities r_{ij} when the algorithm terminates specifies a maximum (arc) flow in the following manner. Since $r_{ij} = u_{ij} - x_{ij} + x_{ji}$, the arc flows satisfy the equality $x_{ij} - x_{ji} = u_{ij} - r_{ij}$. If $u_{ij} > r_{ij}$, we can set $x_{ij} = u_{ij} - r_{ij}$ and $x_{ji} = 0$; otherwise, we set $x_{ij} = 0$ and $x_{ji} = r_{ij} - u_{ij}$.

We use the maximum flow problem given in Figure 3 to illustrate the algorithm. Figure 3(a) shows the residual network corresponding to the starting flow $x = 0$, which is identical to the original network. The residual network contains three augmenting paths: 1–3–4, 1–2–4, and 1–2–3–4. Suppose the algorithm selects the path 1–3–4 for augmentation. The residual capacity of this path is $\delta = \min\{r_{13}, r_{34}\} = \min\{4, 5\} = 4$. This augmentation reduces the residual capacity of arc (1, 3) to zero (thus we delete it from the residual network) and increases the residual capacity of arc (3, 1) to 4 (so we add this arc to the residual network). The augmentation also decreases the residual capacity of arc (3, 4) from 5 to 1, and increases the residual capacity of arc (4, 3) from 0 to 4. Figure 3(b) shows the residual network at this stage. In the second iteration, the algorithm selects the path 1–2–3–4 and augments 1 unit of flow; Figure 3(c) shows the residual network after the augmentation. In the third iteration, the algorithm augments one unit of flow along the path 1–2–4. Figure 3(d) shows the corresponding residual network. Now the residual network contains no augmenting path and so the algorithm terminates.

Does the augmenting path algorithm always find a maximum flow? The algorithm terminates when the search algorithm fails to identify a directed path in $G(x)$ from node s to node t, indicating that no such path exists (we prove later that the algorithm would terminate finitely). At this stage, let S denote the set of nodes in N that are reachable in $G(x)$ from the source node using directed paths, and $\bar{S} = N - S$. Clearly, $s \in S$ and $t \notin \bar{S}$. Since the search algorithm cannot reach any node in $\bar{S}$ and it can reach each node in S, we know that $r_{ij} = 0$ for each $(i, j) \in (S, \bar{S})$. Recall that $r_{ij} = (u_{ij} - x_{ij}) + x_{ji}, x_{ij} \geqslant u_{ij}$, and $x_{ji} > 0$. If $r_{ij} = 0$, then $x_{ij} = u_{ij}$ and $x_{ji} = 0$. Since $r_{ij} = 0$ for each $(i, j) \in (S, \bar{S})$, by substituting these flow values in expression (3), we find that $v = u[S, \bar{S}]$. Therefore, the value of the current flow x equals the capacity of the cut

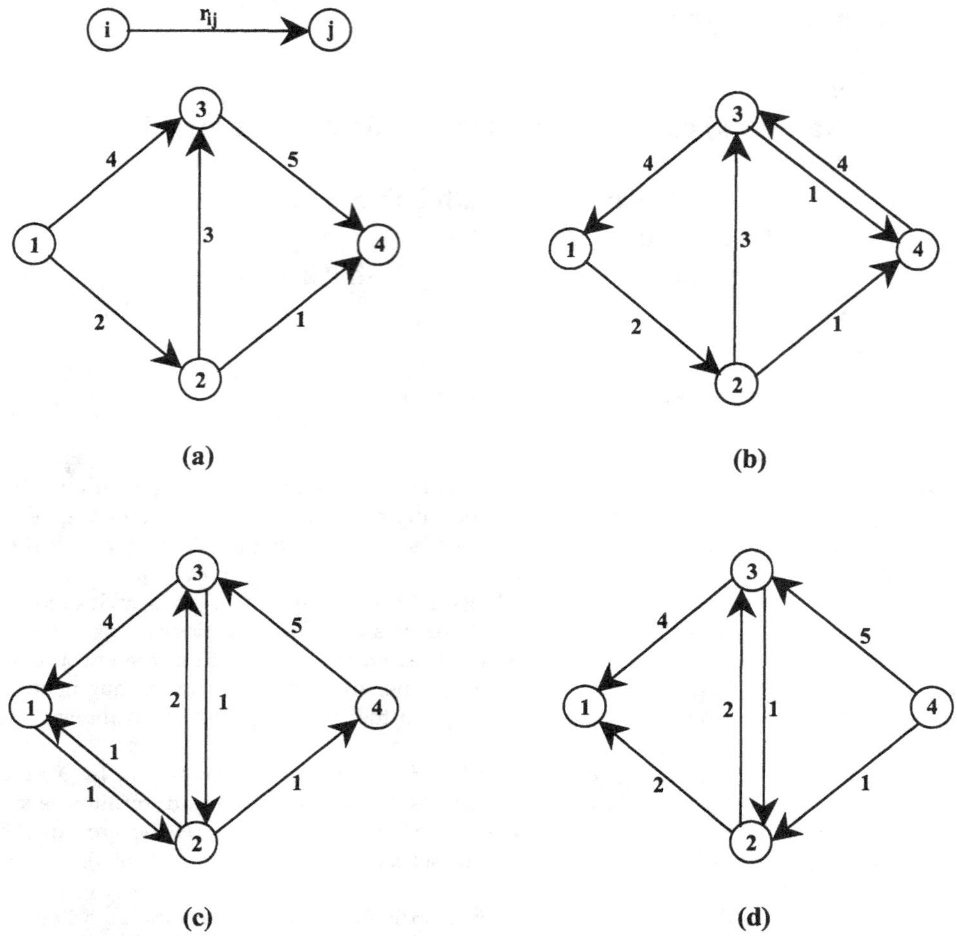

Figure 3 Illustrating the augmenting path algorithm. (a) The residual network $G(x)$ for $x = 0$. (b) The residual network after augmenting four units along the path 1–3–4. (c) The residual network after augmenting one unit along the path 1–2–3–4. (d) The residual network after augmenting one unit along the path 1–2–4

$[S, \bar{S}]$. Lemma 1 implies that x is a maximum flow and $[S, \bar{S}]$ is a minimum cut. This conclusion establishes the correctness of the generic augmenting path algorithm and, as a byproduct, proves the following max-flow min-cut theorem.

Theorem 1 (Max-flow min-cut theorem). *The maximum value of the flow from a source node* s *to a sink node* t *in a capacitated network equals the minimum capacity among all* s – t *cuts.*

The proof of the max-flow min-cut theorem shows that when the augmenting path algorithm terminates, it also discovers a minimum cut $[S, \bar{S}]$, with S defined as the set of all nodes reachable from the source node in the residual network corresponding to the maximum flow. For our previous numerical example, the algorithm finds the minimum cut in the network, which is $[S, \bar{S}]$ with $S = \{1\}$.

The augmenting path algorithm also establishes another important result:

Theorem 2 (Integrality theorem). *If all arc capacities are integer, then the maximum flow problem always has an integer maximum flow.*

This result follows from the facts that the initial (zero) flow is integer and all arc capacities are integer; consequently, all initial residual capacities will be integer. Since subsequently all arc flows change by integer amounts (because residual capacities are integer), the residual capacities remain integer throughout the algorithm. Further, the final integer residual capacities determine an integer maximum flow. The integrality theorem does not imply that every optimal solution of the maximum flow problem is integer. The maximum flow problem might have noninteger solutions and, most often, it has such solutions. The integrality theorem shows that the problem always has at least one integer optimal solution.

What is the worst-case running time of the algorithm? An augmenting path is a directed path in $G(x)$ from node s to node t. We have seen earlier that each iteration of the algorithm requires $O(m)$ time. In each iteration, the algorithm augments a positive integer amount of flow from the source node to the sink node. To bound the number of iterations, we will determine a bound on the maximum flow value. By definition, U denotes the largest arc capacity, and so the capacity of the cut $(\{s\}, S-\{s\})$ is at most nU. Since the value of any flow can never exceed the capacity of any cut in the network, we obtain a bound of nU on the maximum flow value and also on the number of iterations performed by the algorithm. Consequently, the running time of the algorithm is $O(nmU)$, which is a pseudopolynomial time bound. We summarize the preceding discussion with the following theorem.

Theorem 3. *The generic augmenting path algorithm solves the maximum flow problem in* O(nmU) *time.*

The augmenting path algorithm is possibly the simplest algorithm for solving the maximum flow problem. Empirically, the algorithm performs reasonably well. However, the worst-case bound on the number of iterations is poor for large values of U. For example, if $U = 2^n$, the bound is exponential in the number of nodes. Moreover, as shown by known examples, the algorithm can indeed perform this many iterations. A second drawback of the augmenting path algorithm is that if the capacities are irrational, the algorithm might not terminate. For some pathological instances of the maximum flow problem, the augmenting path algorithm does not terminate in a finite number of iterations and although the successive flow values converge to some value, they might converge to a value strictly less than the maximum flow value. (Note, however, that the max-flow min-cut theorem is valid even if arc capacities are irrational.) Therefore, if the augmenting path algorithm is to be guaranteed to be effective in all situations, it must select augmenting paths carefully.

Researchers have developed specific implementations of the generic augmenting path algorithms that overcome these drawbacks. Of these, the following three implementations are particularly noteworthy: (i) the maximum capacity augmenting path algorithm, which always augments flow along an augmenting path with the maximum residual capacity and can be implemented to run in $O(m^2 \log U)$ time; (ii) the capacity scaling algorithm, which uses a scaling technique on arc capacities and can be implemented to run in $O(nm \log U)$ time; and (iii) the shortest augmenting path algorithm, which augments flow along a shortest-path (as measured by the number of arcs) in the residual network and runs in $O(n^2 m)$ time. We describe the first two implementations in the next section; the third implementation is described in Edmonds and Karp (1972).

4. Specific Implementations of the Generic Augmenting Path Algorithm

In this section, we describe two specific implementations of the generic augmenting path algorithm: the maximum capacity augmenting path algorithm, and the capacity scaling algorithm.

The maximum capacity augmenting path algorithm always augments flow along a path with the maximum residual capacity. Let x be any flow and let v be its flow value. As before, let v^* be the maximum flow value. It can be shown using flow decomposition theory that in the residual network $G(x)$ we can find m or fewer directed paths from the source to the sink whose residual capacities sum to $(v^* - v)$. Thus the maximum capacity augmenting path has residual capacity of at least $v^* - v/m$. Now consider a sequence of $2m$ consecutive maximum capacity augmentations starting with the flow x. If in each of these augmentations we augment at least $(v^* - v)/2m$ units of flow, then within $2m$ or fewer iterations we will establish a maximum flow. Note, however, that if one of these augmentations carries less than $(v^* - v)/2m$ units of flow, then from the initial flow vector x, we have reduced the residual capacity of the maximum capacity augmenting path by a factor of at least 2. This argument shows that within $2m$ consecutive iterations, the algorithm either establishes a maximum flow or reduces the residual capacity of the maximum capacity augmenting path by a factor of at least 2. Since the residual capacity of any augmenting path is at most $2U$ and is at least 1, after $O(m \log U)$ iterations, the flow must be maximum.

As we have seen, the maximum capacity augmentation algorithm reduces the number of augmentations in the generic augmenting path algorithm from $O(nU)$ to $O(m \log U)$. However, the algorithm performs more computations per iteration, since it needs to identify an augmenting path with the maximum residual capacity and not just any augmenting path. We shall now suggest a variation of the maximum capacity augmentation algorithm that does not perform more computations per iteration and yet establishes a maximum flow within $O(m \log U)$. Since this algorithm scales the arc capacities implicitly, we refer to it as the capacity scaling algorithm.

The essential idea underlying the capacity scaling algorithm is conceptually quite simple: we augment flow along a path with a sufficiently large residual capacity, instead of a path with the maximum augmenting capacity because we can obtain a path with a sufficiently large residual capacity fairly easily in $O(m)$ time. To define the capacity scaling algorithm, let us introduce a parameter Δ and, with respect to a given flow x, define the Δ-residual network as a network containing arcs whose residual capacity is at least Δ. Let $G(x, \Delta)$ denote the Δ-residual network. Note that $G(x, 1) = G(x)$ and $G(x, \Delta)$ is a subgraph of $G(x)$. The capacity scaling algorithm works as shown in Figure 4.

Let us refer to a phase of the algorithm during which Δ remains constant as a scaling phase and a scaling phase with a specific value of Δ as a Δ-scaling phase. Observe that in a Δ-scaling phase, each augmentation carries at least Δ units of flow. The algorithm starts with $\Delta = 2^{\lfloor \log U \rfloor}$ and halves its value in every scaling phase until $\Delta = 1$. Consequently, the algorithm performs $1 + \lfloor \log U \rfloor = O(\log U)$ scaling phases. In the last scaling phase, $\Delta = 1$ and so $G(x, \Delta) = G(x)$. This result shows that the algorithm terminates with a maximum flow.

```
algorithm capacity scaling;
begin
    x := 0;
    Δ := 2^⌊log U⌋;
    while Δ ≥ 1 do
    begin
        while G(x, Δ) contains a path from node s to node t do
        begin
            identify a path P in G(x, Δ);
            δ : = min {r_ij : (i, j) ∈ P };
            augment δ units of flow along P and update G(x, Δ);
        end;
        Δ := Δ/2 ;
    end;
end ;
```

Figure 4 The capacity scaling algorithm

The efficiency of the algorithm depends upon the fact that it performs at most $2m$ augmentations per scaling phase. To establish this result, consider the flow at the end of the Δ-scaling phase. Let x' be this flow and let v' denote its flow value. Furthermore, let S be the set of nodes reachable from s in $G(x', \Delta)$. Since $G(x', \Delta)$ contains no augmenting path from the sources to the sink, $t \notin S$. Therefore, $[S, \bar{S}]$ forms an $s-t$ cut. The definition of S implies that the residual capacity of every arc in $(S, \bar{S})$ is strictly less than Δ, and so the residual capacity of the cut $(S, \bar{S})$ is at most $m\Delta$. Consequently, $v^* - v' \leqslant m\Delta$. In the next scaling phase, each augmentation carries at least $\Delta/2$ units of flow and so this scaling phase can perform at most $2m$ such augmentations. The algorithm described earlier requires $O(m)$ time to identify an augmenting path and updating the Δ-residual network also requires $O(m)$ time. These arguments establish the following.

Theorem 4. *The capacity scaling algorithm solves the maximum flow problem within* O(m *log* U) *augmentations and runs in* O(m^2 *log* U) *time.*

It is possible to reduce the complexity of the capacity scaling algorithm even further to $O(nm \log U)$. This improvement is described in Gabow (1985) and Ahuja and Orlin (1991).

5. Generic Preflow-Push Algorithm

Another class of algorithms for solving the maximum flow problem, known as preflow-push algorithms, is more decentralized than augmenting path algorithms. Augmenting path algorithms send flow by augmenting along a path. This basic operation further decomposes into the more elementary operation of sending flow along individual arcs. Sending a flow of δ units along a path of k arcs decomposes into k basic operations of sending a flow of δ units along each of the arcs of the path. We shall refer to each of these basic operations as a push. The preflow-push algorithms push flows on individual arcs instead of on augmenting paths.

A path augmentation has one advantage over a single push: it maintains conservation of flow at all nodes. The preflow-push algorithms violate conservation of flow at all steps except at the very end, and instead maintain a preflow at each iteration. A preflow is a vector x satisfying the flow bound constraints and the following relaxation of the mass balance constraints (1b):

$$\sum_{\{j:(i,j)\in A\}} x_{ij} - \sum_{\{j:(j,i)\in A\}} x_{ji} \geqslant 0, \quad \text{for all } i \in N - \{s, t\}. \tag{3}$$

Each element of a preflow vector is either a real number or equals $+\infty$. The preflow-push algorithms maintain a preflow at each intermediate stage. For a given preflow x, we define the excess for each node $i \in N - \{s, t\}$ as

$$e(i) = \sum_{\{j:(j,i)\in A\}} x_{ji} - \sum_{\{j:(i,j)\in A\}} x_{ij}. \tag{4}$$

We refer to a node with positive excess as an active node. We adopt the convention that the source and sink nodes are never active. In a preflow-push algorithm, the presence of an active node indicates that the solution is infeasible. Consequently, the basic operation in this algorithm is to select an active node i and try to remove the excess by pushing flow out of it. When we push flow out of an active node, we need to do it carefully. If we just push flow to an adjacent node in an arbitrary manner and the other nodes do the same, then it is conceivable that some nodes keep pushing flow among themselves resulting in an infinite loop, which is not a desirable situation. Since ultimately we want to send the flow to the sink node, it seems reasonable for an active node to push flow to another node that is closer to the sink. If all nodes maintain this rule, then the algorithm could never encounter an infinite loop. The concept of distance labels defined next allows us to implement this algorithmic strategy.

The preflow-push algorithms maintain a distance label $d(i)$ with each node in the network. The distance labels are nonnegative (finite) integers defined with respect to the residual network $G(x)$. We say that distance labels are valid with respect to a flow x if they satisfy the following two conditions:

$$d(t) = 0, \quad \text{and} \tag{5a}$$

$$d(i) = d(j) + 1 \text{ for every arc } (i, j) \text{ in the residual network } G(x). \tag{5b}$$

We refer to the conditions (5a) and (5b) as the validity conditions. It is easy to demonstrate that $d(i)$ is a lower bound on the length of any directed path (as measured by number of arcs) from node i to node t in the residual network, and thus is a lower bound on the length of the shortest path between nodes i and j. Let $i = i_1 - i_2 - i_3 - \cdots - i_k - t$ be any path of length k in the residual network from node i to node t. The validity conditions (5) imply that $d(i) = d(i_1) \leqslant d(i_2) + 1$, $d(i_2) \leqslant d(i_3) + 1, \ldots, d(i_k) \leqslant d(t) + 1 = 1$. Adding these inequalities shows that $d(i) \leqslant k$ for any path of length k in the residual network, and therefore any (shortest) path from node i to node t contains at least $d(i)$ arcs. We say that an arc (i, j) in the residual network is admissible if it satisfies the condition $d(i) = d(j) + 1$; we refer to all other arcs as inadmissible.

The basic operation in the preflow-push algorithm is to select an active node i and try to remove the excess by pushing flow to a node with smaller distance label. (We will use the distance labels as estimates of the length of the shortest path to the sink node.) If node i has an admissible arc (i, j), then $d(j) = d(i) - 1$ and the algorithm sends flow on admissible arcs to relieve the node's excess. If node i has no admissible arc, then the algorithm increases the distance label of node i so that node i has an admissible arc. The algorithm terminates when the network contains no active nodes, that is, excess resides only at the source and sink nodes. Figure 5 describes the generic preflow-push algorithm.

The algorithm first saturates all arcs emanating from the source node; then each node adjacent to node s has a positive excess, so that the algorithm can begin pushing flow from active nodes. Since the preprocessing operation saturates all the arcs incident to node s, none of these arcs is admissible and setting $d(s) = n$ will satisfy the validity condition (6). But then, since $d(s) = n$, and a distance label is a lower bound on the length of the shortest path from that node to node t, the residual network contains no directed path from s to t. The subsequent

(a)

```
algorithm preflow-push;
begin
    set x := 0 and d(j) := 0 for all j ∈ N;
    set x_sj = u_sj for each arc (s, j) ∈ A(s);
    d(s) : = n;
    while the residual network G(x) contains an active node do
    begin
        select an active node i;
        push/relabel(i);
    end;
end;
```

(b)

```
procedure push/relabel(i);
begin
        if the network contains an admissible arc (i, j) then
            push d :=min{e(i), r_ij } units of flow from node i to node j
        else replace d(i) by min {d(j) + 1: (i, j) ∈ A(i) and r_ij > 0};
end;
```

Figure 5 The generic preflow-push algorithm

pushes maintain this property and drive the solution toward feasibility. Consequently, when there are no active nodes, the flow is a maximum flow.

A push of δ units from node i to node j decreases both the excess $e(i)$ of node i and the residual r_{ij} of arc (i, j) by δ units and increases both $e(j)$ and r_{ji} by δ units. We say that a push of δ units of flow on an arc (i, j) is saturating if $\delta = r_{ij}$ and is nonsaturating otherwise. A nonsaturating push at node i reduces $e(i)$ to zero. We refer to the process of increasing the distance label of a node as a relabel operation. The purpose of the relabel operation is to create at least one admissible arc on which the algorithm can perform further pushes.

It is instructive to visualize the generic preflow-push algorithm in terms of a physical network: arcs represent flexible water pipes, nodes represent joints, and the distance function measures how far nodes are above the ground. In this network, we wish to send water from the source to the sink. We visualize flow in an admissible arc as water flowing downhill. Initially, we move the source node upward, and water flows to its neighbors. Although we would like water to flow downhill toward the sink, occasionally flow becomes trapped locally at a node that has no downhill neighbors. At this point, we move the node upward, and again water flows downhill toward the sink. Eventually, no more flow can reach the sink. As we continue to move nodes upward, the remaining excess flow eventually flows back toward the source. The algorithm terminates when all the water flows either into the sink or flows back to the source.

To illustrate the generic preflow-push algorithm, we use the example given in Figure 6. Figure 6(a) specifies the initial residual network. We first saturate the arcs emanating from the source node, node 1, and set $d(1) = n = 4$. Figure 6(b) shows the residual graph at this stage. At this point, the network has two active nodes, nodes 2 and 3. Suppose that the algorithm selects node 2 for the push/relabel operation. Arc (2, 4) is the only admissible arc and the algorithm performs a saturating push of value $\delta = \min\{e(2), r_{24}\} = \min\{2, 1\} = 1$. Figure 6(c) gives the residual network at this stage. Suppose the algorithm again selects node 2. Since no admissible arc emanates from node 2, the algorithm performs a relabel operation and gives node 2 a new distance label $d(2) = \min\{d(3) + 1, d(1) + 1\} = \min\{2, 5\} = 2$. The new residual network is the same as the one shown in Figure 6(c) except that $d(2) = 2$ instead of 1. Suppose this time the algorithm selects node 3. Arc (3, 4) is the only admissible arc emanating from node 3, and so the algorithm performs a nonsaturating push of value $\delta = \min\{e(3), r_{34}\} = \min\{4, 5\} = 4$. Figure 6(d) specifies the residual network at the end of this iteration. Using this process for a few more iterations, the algorithm will determine a maximum flow.

The analysis of the computational (worst-case) complexity of the generic preflow-push algorithm is somewhat complicated. Without examining the details, we might summarize the analysis as follows. It is possible to show that the preflow-push algorithm maintains valid distance labels at all steps of the algorithm and increases the distance label of any node at most $2n$ times. The algorithm performs $O(nm)$ saturating pushes and $O(n^2m)$ nonsaturating pushes. The nonsaturating pushes are the limiting computational operation of the algorithm and so it runs in $O(n^2m)$ time.

The preflow-push algorithm has several attractive features, particularly its flexibility and its potential for further improvements. Different rules for selecting active nodes for the push/relabel operations create many different versions of the generic algorithm, each with different worst-case complexity. As we have noted, the bottleneck operation in the generic preflow-push algorithm is the number of nonsaturating pushes and many specific rules for examining active nodes can produce substantial reductions in the number of nonsaturating pushes. The following specific implementations of the generic preflow-push algorithms are noteworthy: (i) the FIFO preflow-push algorithm examines the active nodes in the first-in, first-out (FIFO) order and runs in $O(n^3)$ time; (ii) the highest label preflow-push algorithm pushes flow from an active node with the highest value of a distance label and runs in $O(n^2m^{1/2})$ time; and (iii) the excess-scaling algorithm uses the scaling of arc capacities to attain a time bound of $O(nm + n^2 \log U)$. These algorithms are due to Goldberg and Tarjan (1988), Cheriyan and Maheshwari (1989), and Ahuja and Orlin (1989), respectively. These preflow-push algorithms are more general, more powerful, and more flexible than augmenting path algorithms. The best reflow-

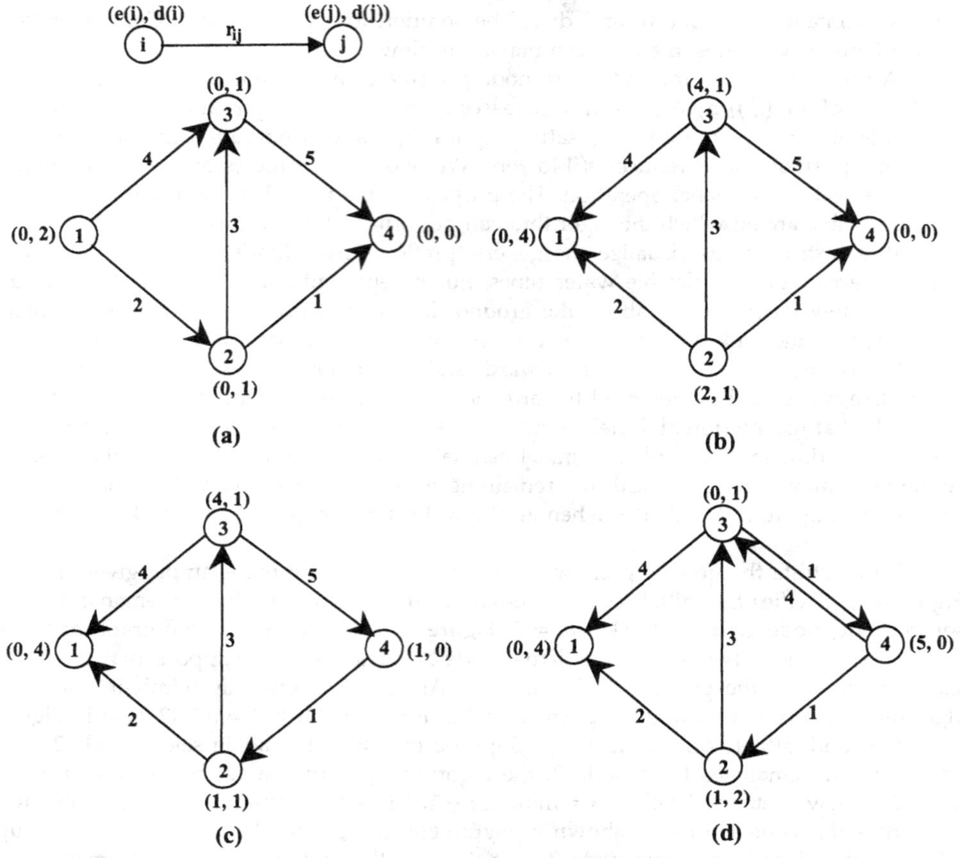

Figure 6 Illustrating the preflow-push algorithm. (a) The residual network $G(x)$ for $x = 0$. (b) The residual network after saturating arcs emanating from the source. (c) The residual network after pushing flow on arc (2, 4). (d) The residual network after pushing flow on arc (3, 4)

push algorithms currently outperform the best augmenting path algorithms in theory as well as in practice [see, for example, Ahuja et al. (1997)].

ACKNOWLEDGMENTS

The first author gratefully acknowledges the support of the National Science Foundation through grant award number DMI-9900087. The third author acknowledges support from the Office of Naval Research under contract N00014-98-1-0317 and the National Science Foundation through grant award number DMI-9820998.

References

Ahuja, R. K. and J. B. Orlin. 1989. "A fast and simple algorithm for the maximum flow problem." *Oper. Res.* **37**: 748–759.

Ahuja, R. K. and J. B. Orlin. 1991. "Distance-directed augmenting path algorithms for maximum flow and parametric maximum flow problems." *Oper. Res.* **37**: 748–759.

Ahuja, R. K., T. L. Magnanti, and J. B. Orlin. 1993. In *Network Flows: Theory, Algorithms, and Applications.* NJ: Prentice Hall.

Ahuja, R. K., M. Kodialam, A. K. Mishra, and J. B. Orlin. 1997. "Computational investigations of maximum flow algorithms." *Eur. J. Oper. Res.* **97**: 509–542.

Cheriyan, J. and S. N. Maheshwari. 1989. "Analysis of preflow-push algorithms for maximum network flow." *SIAM J. Comput.* **18**: 1057–1086.

Edmonds, J. and R. M. Karp. 1972. "Theoretical improvements in algorithmic efficiency for network flow problems." *J. ACM* **19**: 248–264.

Elias, P., A. Feinstein, and C. E. Shannon. 1956. "Note on maximum flow through a network." *IRE Trans. Inf. Theory* **IT-2**: 117–119.

Ford, L. R. and D. R. Fulkerson. 1956. "Maximal flow through a network." *Canadian J. Math.* **8**: 399–404.

Gabow, H. N. 1985. "Scaling algorithms for network problems." *J. Comput. Syst. Sci.* **31**: 148–168.

Goldberg, A. V. and R. E. Tarjan. 1988. "A new approach to the maximum flow problem." *J. ACM* **35**: 921–940.

8.3 Shortest-Path Algorithms

Edith Cohen

ABSTRACT Finding the shortest path to a destination is a fundamental computational problem. Applications range from ones predating the computer age (geographic route selection) to widely-deployed routing algorithms in modern computer networks (e.g., OSPF routing). Shortest-path computations often arise as subroutines when solving a more complex problem. We survey some classical and more recent work on efficiently computing shortest paths and distances. We list several interesting variants of the problem, sketch algorithms when possible, and refer the reader to the rich literature on the subject.

1. Introduction

Computing shortest paths in a network is one of the most fundamental optimization problems, both from theoretical and practical perspectives. Shortest-path instances often arise either as stand-alone problems or as a step in the solution of a more complex problem. Instances of the shortest-paths problem typically constitute from a network and interpoint distances. The network is typically represented as a graph $G = (V, E)$, where V denotes the set of nodes and $E \subset V \times V$ denotes the set of edges. Each edge $(i, j) \in E$ has an associated numeric weight $w(i, j)$. The weight of a path $p = (v_1, v_2, ..., v_k)$ is defined as the sum of the weights of edges $w(p) = \sum_{i=1}^{k-1} w(v_i, v_{i+1})$. A shortest path connecting v_1 and v_k is a path of minimum weight ($+\infty$ if there is no path from v_1 to v_k).

Different applications and problem instances have varying properties, which may set a preference for a particular algorithm. Some of the important distinguishing properties are whether the network is weighted or unweighted (has uniform weights); edges and paths are directed or undirected; or whether instances have some special structure such as planar. The objectives could also differ: a faster approximate solution may be preferred over an exact slower-to-obtain one; the goal could be to find single-source or all-pairs distances, explicit shortest paths, or neighborhood sizes (number of nodes within a certain distance of a given node). The problem instance may also be either static, requiring a one-time computation, or dynamically changing, requiring frequent recomputation after small modifications. Finally, according to the size of the instances and desired performance, we may pursue sequential, parallel, or external-memory algorithms.

In terms of simplicity and speed, the unweighted version seems easier than the weighted version (known algorithms are faster) and nonnegative edge-weights are easier than general weights. The undirected variant is reducible to, and often easier than, the directed one. The single-pair problem seems asymptotically as hard as single-source, but single-source is easier than all-pairs. Most algorithms we consider here obtain the corresponding shortest-paths along with distances with little extra effort.

Below we survey some classic and some newer results. For more detail on the classic algorithms see (Aho et al., 1974; Ahuja et al., 1993; Cormen et al., 1990; Lawler, 1976; Sedgewick, 1988). We focus on theoretical work and asymptotic running times. Unless stated otherwise, we assume networks are represented by adjacency lists (for each node there is a list of adjacent nodes), and we measure performance asymptotically using the RAM (Random Access Memory) model of computation.

2. Single-Source Shortest-Paths

The single-source shortest-paths (SSSP) problem amounts to computing shortest paths from a designated source node s to all other nodes in the graph.

2.1. *Breadth-First Search*

We start by describing a simple algorithm, breadth-first search (BFS), for computing SSSP in directed networks with uniform edge weights. BFS also applies to undirected networks by representing each undirected edge (u, v) as two directed edges $u \rightarrow v$ and $v \rightarrow u$. The output of the BFS algorithm is a spanning tree $T \subset E$ rooted at s with the following properties: for each node v, there is a unique path from s to v on the tree. This tree path to v is such that it has a minimum number of edges among all paths from s to v in the original network G. The BFS algorithm runs in linear time to the number of edges reachable from s, hence, in $O(m + n)$ time. BFS generates the tree T layer by layer, where the ith layer contains all nodes of distance i from s. BFS starts by placing s in the root, and scans nodes in phases such that in phase i, BFS finds all nodes of layer i. Initially, BFS looks at all outgoing edges from s and uncovers all nodes of distance 1. To find nodes of distance $i + 1$, the algorithm looks at all nodes of distance i. For each node v of distance i, it looks at all outgoing edges. For each such edge $v \rightarrow u$, if the other end-point u is not already in T, then the node u must be of distance $i + 1$. Consequently, the edge (v, u) is added to T. Correctness can be established by induction on i, using the fact that every node of distance $i + 1$ must have a neighbor of distance i. Each edge is looked at once (once from each end-point in the undirected case) and each node is scanned once. Hence, the running time is linear to the number of edges and nodes reachable from s.

The BFS algorithm also solves two other fundamental problems, namely, connectivity and reachability. These problems amount to computing all nodes such that there is a path from s to v. When the network is undirected, BFS finds the connected component of s. When the network is directed, it finds the reachability set of s. Since any correct algorithm has to look at all reachable edges from s, BFS has optimal asymptotic running time.

When the edges of the network have varying numeric weights, the shortest-paths problem amounts to finding paths such that the sum of the weights of its edges is minimized, and the BFS algorithm is not applicable. As we shall soon see, nonnegative edge-weights do seem to make the problem more involved, but it can still be solved by an algorithm that in its essence generalizes BFS. Allowing negative edge weights, on the other hand, seems to fundamentally change the algorithmic approach needed and known algorithms are significantly slower. Before proceeding to discuss algorithms for SSSP in weighted graphs, we introduce some important concepts.

2.2. *Difference Constraints*

Each instance of the SSSP problem has a corresponding system of difference constraints, that are linear inequalities of specific form. For each edge $u \rightarrow v$ form u to v with weight w there is a corresponding inequality $d(v) - d(u) \leqslant w$. This constraint implies that the shortest path from s to v cannot be longer than the shortest path to u plus the weight of the edge (u, v). These constraints can be used to verify that a given set of paths from s to all other nodes is indeed a set of shortest paths. Consider a set of values $d(u)$ $(u \in V)$ such that $d(u)$ is the weight of some path from s to u. We argue that $d(u)$ $(u \in V)$ are the corresponding distances from s if, and only if, the difference constraints are satisfied. Obviously, if a condition corresponding to an edge (u, v) is not satisfied, we can obtain a shorter path to v than the one we currently have by using the path to u and the edge (u, v). Conversely, it is also not hard to see that a set of paths that satisfy the difference constraints must be shortest paths.

The system of different constraints, combined with the equality $d(s) = 0$, can also be used to compute distances. It is not hard to show that for every $v \in V$, the maximum value of $d(v)$ in a feasible solution is in fact the distance of v from s. Moreover, a feasible solution exists that simultaneously maximizes $d(v)$ for all v. Therefore, SSSP distances can be obtained by solving the system of constraints subjected to the objective function:

$$\text{Maximize} \sum_{v \in V} d(v).$$

Note that such systems constitute a special-structure subclass of linear programs. Shortest paths can be retrieved from the optimal solution of the LP by considering tight constraints (constraints where equality holds). A constraint is tight if, and only if, the corresponding edge participates in some shortest path from s.

2.3. *Shortest-Paths Tree*

The set of all (directed) edges for which the corresponding difference constraint is tight forms an acyclic spanning subgraph of G. The subgraph contains all shortest paths from s to every other node in G. Moreover, every path originating in s is a shortest path. This subgraph must contain at least one spanning tree of G rooted at s, with directed paths to all other nodes.

Each such tree is a shortest-paths tree, and for each node v, the unique tree-path from s to u is a shortest-path in G. The shortest-paths tree is a compact way to represent $n-1$ shortest-paths from s to each of the $n-1$ other nodes in the network. Note that T contains exactly $n-1$ edges, whereas an explicit representation of a set of shortest paths could be of size $\Theta(n^2)$. Hence, for an SSSP algorithm to run in worst-case subquadratic time, it must represent the output compactly.

2.4. *Relaxing Edges*

A basic step in some shortest-path algorithms is that of updating distances by edge relaxations. The general framework is that this algorithm maintains a tentative distance label $d(u)$ for each node u. The distance label may initially overestimate the actual distance. The labels are decreased in the course of the algorithm until they equal the actual distances. The basic update operations is that of relaxing an edge. Relaxing (u, v) means updating $d(v)$ to be $\min\{d(v), d(u) + w(u,v)\}$. It is clear that edge relaxation preserves the property that labels $d(u)$ $(u \in V)$ upper-bound with the corresponding distances. It is not hard to show that when relaxations are no longer possible, the labels are such that all difference constraints are satisfied and hence, labels equal the distances.

2.5. *Dijkstra's Algorithm*

A classic algorithm for computing single-source shortest-paths (SSSP) in graphs with nonnegative edge weights w: E is Dijkstra's algorithm. The algorithm maintains a distance label $d(u)$ for every node $v \in V$, which always upper bounds the actual distance. When the algorithm terminates, distance labels equal the corresponding distance.

Like BFS, the algorithm grows a set of nodes X. Initially, X includes only the source s, and when the algorithm terminates, X includes all nodes. Nodes are appended to X in the order of their increasing distance from s. Once a node is placed in X, its distance label remains unchanged, and as we shall see, corresponds to the true distance.

Algorithm: Dijkstra(G, s)

$X = s$

$d(s) = 0$

for each $u \in V \backslash \{s\}$, $d(u) = \begin{cases} w(s,u), & \text{if } (s,u) \in E \\ +\infty & \text{if } (s,u) \notin E \end{cases}$

while $X \neq V$

let $u \in V\backslash X$ be such that $d(u)$ is minimum
$X = X \cup \{u\}$
for each $v \in V\backslash X$ such that $(u, v) \in E$
$d(v) = \min(d(v), d(u) + w(u, v))$

CORRECTNESS

Correctness can be established by induction, by showing that the following two invariants hold through the execution of the algorithm:

1. For any $u \in V$, $d(u)$ is the length of the shortest path from s to u among all paths that only go through nodes in X.
2. For any $u \in X$, $v \notin X$, $d(u) \leqslant d(v)$

When the algorithm terminates, $X = V$ and thus, by the first invariant, $d(u)$ $(u \in V)$ are the actual distances.

RUNNING TIME

We outline a simple implementation of Dijkstra's algorithm. First, at each node the list of incident outgoing edges is sorted by increasing weight. A pointer is used with each list, initially pointing at the first (smallest-weight) edge. Pointers can only be advanced forward, according to increasing weight. We use them to find the smallest-weight incident edge that is not directed to a node already in X. To do so, it suffices to advance the pointer forward until it points to an edge that is not directed to X. Hence, the algorithm advances pointers $O(m)$ times in total. Dijkstra's algorithm iteratively selects the next node to add to X. The next node added is always the closest node to s not already in X. It is easy to see that it suffices to consider, for each node in X, the smallest-weight incident edge not directed to X. It takes $O(m \log m)$ time to sort the edges and $O(n^2)$ time to do the rest [X is augmented n times; each time you look at $O(n)$ edges]. Hence the running time of this simple implementation is $O(m \log m + n^2)$. We note that more efficient (and involved) data structures yield asymptoticly-faster implementations: a simple priority queue gives a running time of $O(m \log n)$ whereas Fibonacci heaps implementation gives a running time of $O(m + n \log n)$.

2.6. *Allowing for Negative-Weight Edges*

The incremental approach of Dijkstra's algorithm is not applicable when negative-weight edges are present. Dijkstra's algorithm selects, at each point, the node $u \notin X$ that is closest to s. To do so it considers only paths with prefix in X and one additional edge from a node already in X. The nonnegativity of edge weights guarantees that a shortest path from s to the closest non-X node must be of that form. This may not be true when negative-weight edges are present, all shortest paths to the non-X node closest to s may traverse $\Theta(n)$ non-X nodes.

Presence of negative edge weights seems to make the problem considerably more complex. If the graph contains negative-weight cycles (that is cycles where the sum of weights is negative), shortest paths are not well-defined, since we can always find a path of smaller weight by traversing the cycle enough times (in this case we say that the distance is $-\infty$). If all cycles are positive, then shortest paths are always simple (each edge is traversed at most once), and if all cycles are nonnegative then there is always at least one simple shortest path. If problem instances can have negative cycles and the goal is to compute minimum-weight simple paths, then the problem becomes NP-complete (Garey and Johnson, 1979). Here we focus on polynomial time algorithms that allow inputs with negative-edge weights. The algorithms detect the presence of negative-weight cycles and compute shortest paths when they are well defined.

THE BELLMAN–FORD ALGORITHM

The Bellman–Ford algorithm is a classic algorithm. Similarly to Dijkstra's algorithm, the Bellman–Ford algorithm maintains a distance label $d(u)$ for each node that upper bounds the actual distance. Initially, $d(s) = 0$ and $d(u) = w(s, u)$ when there is an edge (s, u), and $+\infty$

otherwise. The Bellman–Ford algorithm works in iterations, where in each iteration, all edges in the graph are relaxed. Each edge relaxation can be performed in constant time. Hence, each iteration requires $O(m)$ time.

A fairly simple induction argument establishes that after i iterations, $d(u)$ $(u \in V)$ is the weight of the shortest paths from s to u containing i or fewer edges. Thus, when the distance to u is well defined (that is, there is no path from s to u that intercepts a negative cycle), then the label $d(u)$ after n iterations is the distance, and would remain unchanged under further edge relaxations. It is also not hard to see that a negative-weight cycle exists if, and only if, there is a node u such that $d(u)$ after $2n$ iterations is strictly smaller than $d(u)$ after n iterations.

Hence, if we run the Bellman–Ford algorithm for $2n$ iterations, we can either detect the presence of a negative cycle or obtain distances. The running time of Bellman–Ford is $O(mn)$. With appropriate bookkeeping, the Bellman–Ford algorithm can also provide the corresponding paths or negative cycle. It can also be enhanced to identify all nodes of distance $-\infty$ from s (and find actual distances and corresponding paths for others).

JOHNSON'S ALGORITHM

Johnson's algorithm is applicable to graphs with no negative-weight cycles, and it converts all edge weights to nonnegative quantities while preserving the identity of the shortest paths.

Johnson's algorithm first applies the Bellman–Ford algorithm (or any other SSSP algorithm that allows for negative-weight edges) to compute distances from the source s These distances, $d(u)$ $u \in V$, are used to transform edge weights such that the weight of the directed edge (u, v) becomes $w(u, v) + d(u) - d(v)$. The difference constraints imply that the transformed weights are nonnegative, and an edge has zero transformed weight if, and only if, it belongs to a shortest-path. Moreover, a path from s is shortest under the original weights if, and only if, it has zero weight under the transformed weights.

The transformation changes the weight of each path from u to v by an additive term of $d(v) - d(u)$. Thus, the transformation preserves the difference of the weights of any two paths between the same pair of nodes, and therefore, the set of shortest paths for any pair is preserved.

After transforming the weights, any shortest-paths algorithm designed for nonnegative edge weights is applicable. In particular, we may apply Dijkstra's algorithm to compute SSSP for different source nodes. We refer the reader to Cormen et al. (1990) for further details on the Bellman–Ford and Johnson algorithms.

2.7. *Selected Recent Results*

We briefly mention some recent work on computing SSSP.

Thorup recently presented an optimal linear time $O(m)$ algorithm for SSSP with nonnegative edge weights on undirected graphs (Thorup, 1997). His algorithm makes use of bit operations in the RAM model. This result is of great theoretical interest, breaking the $O(m + n \log n)$ bound provided by Fibonacci heaps implementation of Dijkstra's algorithm, and matching the lower bound.

The intriguing gap between the best-known asymptotic running times for SSSP with and without negative-weight edges is addressed by a scaling algorithm due to Goldberg (Goldberg, 1995). His algorithm is suited for instances with integral edge weights when weights of negative-weight edges have small absolute values. Goldberg's algorithm has a running time of $O(m\sqrt{n} \log M)$, where M is the largest absolute value of a negative-weight edge in the graph.

Our primary focus here is on theoretical results. We refer the reader to Cherkassky et al. (1996) and Goldberg and Silverstein (1997) for some interesting experimental work on implementing and evaluating different SSSP algorithms.

3. All-Pairs Shortest-Paths

All-pairs shortest paths (APSP) is the problem of computing shortest paths, or distances, between all pairs of nodes. APSP can be solved by applying a SSSP algorithm separately from every node in the graph. We first discuss the performance of this simple approach, and then continue to discuss algorithms that compute APSP directly.

A related problem to APSP is that of computing the transitive closure of a graph, that is, all pairs (u, v) of nodes such that v is reachable from u via a directed path. Transitive closure can be computed in $O(mn)$ time by applying a reachability (e.g., BFS) algorithm from every node. The undirected version of the TC problem is computing the connected components of the input graph, that is, all pairs of nodes such that there is an undirected path between them. Since undirected reachability is symmetric (u is reachable from v if, and only if, v is reachable from u), the output is simply a partition of the nodes, and thus has $O(n)$ representation. Furthermore, connected components can be computed in linear $O(m)$ time using BFS. (We pick an arbitrary unscanned node and run BFS; everything BFS finds forms a single component and is marked scanned; we iterate until there are no unscanned nodes.)

The APSP problem for directed graphs with nonnegative edge weights can be solved in $O(mn + n^2 \log n)$ time by applying Dijkstra's algorithm n times, once from each node. Note that the same running time can be obtained when negative edge weights are present by first applying Johnson's algorithm (see Section 2.6) to convert all edge weights to be nonnegative. When edge weights are uniform, BFS can be applied from every node yielding a $O(mn)$ time algorithm.

When the graph is dense, that is, the number of edges is $\Omega(n^2)$, then the $\tilde{O}(mn)$ algorithms for transitive closure and APSP computations exhibit a cubic $O(n^3)$ dependence on the number of nodes. As we shall soon see, the transitive closure and APSP problems are closely related to matrix multiplications (MM). The straightforward numeric MM algorithm performs $\Theta(n^3)$ operations to multiply two $n \times n$ square matrices. Faster matrix multiplication algorithms were developed in the last few decades, starting with Strassen (1969). Generally, the notation ω is used for the exponent of the asymptotically best known square MM algorithm. The tightest known bounds on the value of ω are that $2 \leqslant \omega \leqslant 2.38$, stemming from a trivial lower bound of 2 and the currently best known $O(n^{2.38})$ fast MM algorithm (Coppersmith and Winograd, 1990).

A natural question that arises is whether fast MM algorithms can be applied to obtain faster APSP algorithms. We first show how shortest-path computations can be reduced to min plus matrix multiplications, and then mention results that apply to standard numeric MM to solve APSP.

3.1. Distance Products

Weighted directed graphs have a natural matrix representation. Let $G = (V, E)$ with weights $\omega{:}E$, and $V = \{v_1, v_2, ..., v_n\}$ be a weighted graph. The corresponding $n \times n$ matrix A is defined as follows. The entry $[A]_{ij}$ is the weight of the edge (v_i, v_j) if it exists and $+\infty$ otherwise. Diagonal entries are set to zero ($\forall i$, $[A]_{ii} = 0$). When graphs are undirected, their matrix representation is symmetric ($[A]_{ij} = [A]_{ji}$ for all i, j).

The distance product C of two matrices A and B is defined such that $[C]_{ij} = \min_{k=1}^{n}\{[A]_{ik} + [B]_{kj}\}$. This definition resembles the one for numeric matrix products, but the min operation replaces addition and addition replaces multiplication. It is fairly easy to see that $[A^k]_{ij}$ provides the length of the shortest paths from v_i to v_j constituting of at most k edges. (A^k is A raised to the kth power using distance products.) When distances are well-defined (there are no negative-weight cycles), there are always shortest-paths that are simple and thus of size at most $n-1$. Hence, A^k for any $k \geqslant n$ provides the distances. Moreover, negative cycles exist in G if, and only if, A^{k+n} differs from A^k for any $k \geqslant n$ (in particular, if $A^{2n} \neq A^n$).

Similarly, a transitive closure computation can be performed using boolean matrix multiplication, where the boolean operations *or* and *and* replace addition and multiplication. Each graph G is represented by its adjacency matrix A, which has entry of *true* (1) when there is an edge between the corresponding nodes and *false* (0) otherwise. Boolean matrix multiplication can be performed via a numeric multiplication, since an entry in the boolean product is true if, and only if, the corresponding entry in the numeric product is nonnegative. Thus, TC can be computed using numeric MM by considering the numeric ℓth power of the adjacency matrix. In this case, there is a directed path between two nodes if, and only if, the corresponding entry in A^ℓ (any $\ell \geqslant n$) is nonzero. Although boolean matrix multiplication is a special case of numeric multiplication, it is not known if it accommodates algorithms with a better exponent.

The standard matrix multiplication procedure carries through for distance products. Hence, using $O(\log n)$ repeated squarings, the matrix can be raised to a power between n and $2n$ in time $O(n^3 \log n)$. Unfortunately, there is no straightforward way to apply fast MM techniques to distance products. The known fast MM algorithms are applicable over any ring, but the set of real numbers (or integers) does not form a ring with respect to the operations min and plus.

3.2. Dynamic Programming: The Floyd–Warshall Algorithm

The Floyd–Warshall algorithm takes a dynamic programming approach for APSP. It is applicable to directed graphs with real-valued edge weights and extends to computations over semirings.

The FW algorithm maintains tentative distances that are updated iteratively. The iterations are guided by some (arbitrary) ordering $(v_1, \ldots, v_n)$ of the nodes of the graph. After the kth iteration, the tentative distances are the weights of respective shortest-paths among all paths that traverse through the nodes $(v_1, \ldots, v_k)$. The tentative distances are initialized such that $\forall i,j\; d_{ij}^{(0)} = w(i,j)$ if $(i,j) \in E$ and $d_{ii}^{(0)} = +\infty$ otherwise. The kth iteration distances, denoted $d_{ij}^{(k)}$ $(\forall i \forall j)$, are computed from the $k-1$ iteration distances using the relation

$$d_{ij}^{(k)} = \min\{d_{ij}^{(k-1)}, d_{ik}^{(k-1)} + d_{kj}^{(k-1)}\}.$$

If the graph contains no negative cycles, then the shortest distance through $v_1, \ldots, v_k$ is the minimum of the distance through $v_1, \ldots, v_{k-1}$ and the sum of the distances through $v_1, \ldots, v_{k-1}$ from v_i to v_k and from v_k to v_j. Thus, an induction argument on k establishes that after the kth iteration, for each pair (v_i, v_j) of nodes, the tentative distance label is the shortest path between them traversing through a subset of the nodes $(v_1, \ldots, v_k)$.

The tentative distances after the nth iteration correspond to actual distances if the graph contains no negative-weight cycles. Each iteration requires $O(n^2)$ operations (one min and one addition for each pair of nodes), and, hence, n iterations take $O(n^3)$ time. The algorithm can be adapted to produce the corresponding paths within the same time bounds.

3.3. Some Recent Results on APSP

More recent techniques yield algorithms with improved asymptotic running times for the APSP problem. Fredman (1976) proved that APSP can be solved in $O(n^{2.5})$ time nonuniformly. Nonuniformity means that the existence of such an algorithm for any given value of n is proven, but different algorithms may be needed for different values and the description of the algorithms may be large. Fredman's techniques also provide a slightly subcubic uniform bound of $O(n^3(\log \log n/\log n)^{1/3})$, which was later improved to $O(n^3(\log \log n/\log n)^{1/2})$ by Takaoka (1992).

In recent years, techniques were developed that reduce shortest-path computations to numeric matrix multiplication. The first to use fast matrix multiplication algorithms for APSP were Alon et al. (1997). Their algorithm is applicable to directed graphs and allows for negative edge-weights. Their algorithm runs in time $\tilde{O}(n^{(3+\omega)/2})$ for edge weights in $\{-1, +1\}$. Zwick presented an algorithm with an improved exponent of $\tilde{O}(n^{2+\mu})$, where μ solves the equation $1 + 2\mu = \omega(\mu)$, and where $\omega(\mu)$ and is the exponent of the rectangular MM of an $n \times n^\mu$ matrix by an $n^\mu \times n$ matrix (Zwick, 1998). Substituting the best-known bounds for rectangular MM, Zwick obtained an $O(n^{2.575})$ APSP algorithm. The dependence of the running time of both Zwick's and Alon et al.'s algorithms is polynomial in the maximum magnitude (absolute value) of the edge weights, hence, these algorithms are only effective for small weights.

A better exponent is known for undirected graphs. Galil and Margalit (1997) and, independently, Seidel (1995) obtained a $\tilde{O}(n^\omega)$ algorithm for APSP on unweighted undirected graphs. Seidel's algorithm is considerably simpler but the algorithm by Galil and Margalit generalizes to input graphs with small nonnegative integral weights. The best (to date) dependence on the magnitude of the weights is achieved by an algorithm of Shoshan and Zwick (1999) and is $\tilde{O}(Mn^{\omega})$, where M is largest edge weight, assuming that edge weights are (scaled to be) integral.

Thus far we have bounded the running time of the shortest-path algorithm in terms of n and m. An algorithm due to Karger et al. (1991) solves undirected APSP with nonnegative edge weights in $O(m^*n + n^2 \log n)$ time, where m^* is the number of edges participating in shortest-paths. In some cases, it is possible that $m^* = m$, obtaining no improvement over n executions of Dijkstra's algorithm. Often, however, m^* is significantly smaller than m.

4. Variants of the Shortest-Path Problem

4.1. *Trading Precision and Running Time*

The algorithms we had seen so far output exact distances. Significantly faster algorithms, however, are often possible when approximate answers are acceptable. Approximate shortest-path algorithms return paths that may not be the shortest, but their weight is guaranteed to be within some multiplicative factor or additive term from the true distance. Since the output is actual paths, approximate distances are equal or exceed the true values. The stretch of an approximate SP algorithm is a bound on its multiplicative error. For example, stretch 2 means that the approximate distance are guaranteed to be at most twice the corresponding actual distances.

Zwick presented a $\tilde{O}(n^{\omega} \log M)$ time approximation scheme for APSP in directed graphs (Zwick, 1998) with nonnegative edge weights, that is, for any fixed $\epsilon > 0$ there is an algorithm with stretch $1 + \varepsilon$. Note that the exponent value (ω) is significantly below what is known for the exact problem. For undirected graphs, Cohen and Zwick presented a $\tilde{O}(mn^{1/2})$ algorithm with an additive error of $2M$ and a $\tilde{O}(n^2)$ algorithm with an additive error of at most $O(M \log n)$ (Cohen and Zwick, 1997), where M is the maximum edge weight. The two algorithms have stretch 2 and $\log n$, respectively, and do not use fast MM.

An interesting related problem is computing graph spanners. A spanner $G' = (V, E')$ of $G = (V, E)$ is a sparse graph such that distances on the spanner G' are within some multiplicative factor (stretch) of corresponding distances in G. Some tradeoffs between stretch and size (number of edges in the spanner) are provided in Cohen (1999) for undirected graphs. A long line of research has focused on constructing spanners, often considering a restricted family of graphs (such as geometric graphs), and sometimes, motivated by practical applications, imposing additional restrictions on the spanner such as bounded degree (Peleg and Schäffer, 1989; Althöfer et al., 1993).

Below we shall see that approximation allows not only for faster sequential shortest-path algorithms but also for more efficient parallel algorithms.

4.2. *Parallel Algorithms for Shortest Paths*

In the PRAM (Parallel Random Access Machine) computational model multiple processors operate simultaneously over a shared memory. PRAM algorithms are measured by two parameters: their running time and the number of processors used. The product of these two quantities is termed work. The work performed by a parallel algorithm can never exceed the running time of the best-known sequential algorithm (otherwise, a sequential implementation of the parallel algorithm would yield a faster sequential algorithm). The goal when designing a PRAM algorithm is to obtain a good tradeoff between work and time: speed up computation while incurring work that is as close as possible to that of the best-known sequential algorithm. The PRAM model is a clean and simple platform that allows us to study and analyze the inherent parallelizeability of problems and algorithms. We refer the reader to JáJá (1992) and Leighton (1992) for an introduction to parallel computation and the PRAM model.

Some sequential algorithms can be naturally partitioned to a small number of phases, such that each phase can be performed quickly in parallel. One such example, is the standard

MM procedure, which applies to both numeric and distance products. Square $n \times n$ MM is performed in $O(n^3)$ time sequentially and constitutes of n^2 dot products that can each be computed independently. In turn, each dot product is computed using n independent scalar multiplications and a summation of the resulting n products (the respective operations for distance products are addition and min). Summation (or min operation) of n scalars can be performed in logarithmic time using $n/\log n$ processors by repeatedly adding (or comparing) pairs of values. Hence, MM can be sped up to logarithmic in n time while using $O(n^3)$ work.

The Bellman–Ford algorithm, on the other hand, does not parallelize as well. Each phase can be performed in logarithmic time without inflating work, but the number of phases is $O(n)$, and each phase builds on the results of the previous phase. A similar barrier applies to Dijkstra's algorithm. Thus, parallel implementation of these algorithms seems to require linear time in n.

APSP for dense graphs parallelizes efficiently whereas highly-parallel SSSP seems to require inflated work over that of a sequential algorithm. The best-known polylog time parallel SSSP algorithms to date perform the same work as the APSP algorithms. This remains true even for two special cases of BFS, directed reachability and undirected SSSP with uniform weights. (We note that connected components and undirected reachability in particular can be computed efficiently in parallel: in logarithmic time with optimal linear work.) This intriguing gap imposes a long-standing open problem and is referred to in the literature as the transitive closure bottleneck. Algorithms with a range of time-work tradeoffs were proposed (Ullman and Yannakakis, 1991; Spencer, 1997; Klein and Sairam, 1992), some obtaining efficient parallelizations for special classes of graphs (Kao and Klein, 1993; Cohen, 1996; Klein and Sairam, 1993). The bottleneck gap is partially addressed by a construction (Cohen, 1994) that, for any fixed $\varepsilon > 0$ and $\varepsilon_0 > 0$, gives a stretch $(1+\varepsilon)$ algorithm for undirected SSSP from s distinct sources that runs in polylogarithmic time using $O[(m+sn)n^{\varepsilon_0}]$ work.

4.3. *Fast Estimation of Neighborhood Sizes*

A related problem to computing shortest paths is computing neighborhood sizes: For a given node v and radius r, determine how many nodes are within distance r of a node v.

Problems related to TC computation are:

Determine the size of the TC, that is, the number of pairs of nodes are such that u can reach v.

Determine the sizes of reachability sets, that is, for each node v, the number of nodes reachable from v.

Surprisingly perhaps, these size estimation questions can be answered considerably faster than the corresponding explicit computations of neighborhoods and reachability sets. Explicitly, computation of all neighborhoods amounts to an APSP computation and explicit computation of all reachability sets is a TC computation. A neighborhood-size estimation algorithm due to Cohen (1997) inputs directed graphs with nonnegative edge weights and constructs a data structure in $\tilde{O}(m)$ time such that each query in the form of a node v and a radius r can be answered in $\log \log n$ time with high confidence and accuracy. The algorithm is applicable to directed graphs with nonegative edge weights. The size of the transitive closure along with the sizes of all reachability sets can be estimated in $O(m)$ time, with any desired fixed confidence and relative error.

4.4. *Dynamic Networks*

Fully dynamic graph algorithms allow for a quick recomputation of a desired graph property after a graph is modified (edges are inserted, deleted, or labels are changed). Designing dynamic algorithms is important, since, in many applications, networks are subject to small frequent changes, and the cost of recomputation from scratch after each change may be prohibitive.

The goal of a fully dynamic shortest-path algorithm is to quickly recompute distances and shortest paths after edge insertions, deletions, and modifications of edge weights. Such an algorithm maintains a data structure that is updated when the graph changes. With this data structure, distance queries can be processed quickly. A dynamic transitive closure algorithm responds to queries about the reachability of a given node from another. A dynamic shortest-path algorithm returns the current distance from one node to another. Such algorithms may also be required to provide the corresponding paths.

Fully dynamic transitive closure and directed nonnegative all-pairs shortest-path algorithms were recently presented by King (1999). The dynamic transitive closure algorithm has an update time of $O(n^2 \log n)$ and constant query time. King presented several algorithms for dynamic shortest-paths with trade-offs between accuracy and update time. All variants feature constant query time. In particular: (i) estimates within $(2+\varepsilon)$ times the actual distance (for any positive constant ε) can be computed with $\tilde{O}(n^2)$ (amortized) time per update; (ii) estimates within $(1+\varepsilon)$ of actual distances can be obtained with $\tilde{O}(n^2 \log^3 M)$, where M is the largest edge weight and edge weights are integral; (iii) exact shortest paths can be generated with an update time of $\tilde{O}(n^{2.5} M^{0.5})$. These algorithms can also return a corresponding shortest path, but then the query time may increase to linear in the path length. The algorithms are simple and natural and are not based on fast MM. Note that the exact algorithm performs better than recomputation from scratch only when edge weights are small.

4.5. *Parametric Shortest-Paths*

Edge weights can sometimes be presented as parametric expressions rather than simple scalars. The simplest such form is a linear form with one parameter (the weight of $e \in E$ is expressed as $a(e)\delta + b(e)$, where $a(e)$ and $b(e)$ are scalars). With such parametric weights, the shortest path is a piecewise linear concave function of the parameter. Questions of interest are to find the maximum point on the curve (the value of λ for which the shortest-path is maximized), bound the number of breakpoints on the curve, or compute the complete curve efficiently (Gusfield, 1983; Gallo et al., 1989; Cohen and Megiddo, 1993; Carstensen, 1983).

References

Aho, A. V., J. E. Hopcroft, and J. D. Ullman. 1974. In *The Design and Analysis of Computer Algorithms*. Reading, MA: Addison-Wesley Publishing Co.

Ahuja, R. K., T. L. Magnanti, and J. B. Orlin. 1993. In *Network Flows*. Englewood Cliffs, NJ: Prentice-Hall.

Alon, N., Z. Galil, and O. Margalit. 1997. "On the exponent of the all pairs shortest path problem." *J. Comput. System Sci.* **54**: 255–262.

Althöfer, I., G. Das, D. Dobkin, D. Joseph, and J. Soares. 1993. "On sparse spanners of weighted graphs." *Discrete Comput. Geometry* **9**: 81–100.

Carstensen, P. J. 1983. "The complexity of some problems in parametric, linear, and combinatorial programming." Ph.D. diss. Ann Arbor: Department of Mathematics, University of Michigan.

Cherkassky, B. V., A. V. Goldberg, and T. Radzik. 1996. "Shortest paths algorithms: theory and experimental evaluation." *Math. Programming* **73**: 129–174.

Cohen, E. 1994. "Polylog-time and near-linear work approximation scheme for undirected shortest-paths." In *Proceedings of the 26th Annual ACM Symposium on Theory of Computing* 16–26. Montreal, Canada: ACM.

Cohen, E. 1996. "Efficient parallel shortest-paths in diagraphs with a separator decomposition." *J. Algorithms* **21**: 331–357.

Cohen, E. 1997. "Size-estimation framework with applications to transitive closure and reachability." *J. Comput. System Sci.* **55**: 441–453.

Cohen, E. 1999. "Fast algorithms for *t*-spanners and stretch-*t* paths." *SIAM J. Comput.* **28**: 210–236.

Cohen, E. and N. Megiddo. 1993. "Maximizing concave functions in fixed dimension." In *Complexity in Numerical Optimization*, edited by P. M. Pardalos pp. 74–87. London: World Scientific.

Cohen, E. and U. Zwick. 1997. "All pairs small stretch paths." In *Proceedings of the 8th ACM-SIAM Symposium on Discrete Algorithms*, pp. 93–102. New Orleans, Louisiana: ACM-SIAM.

Coppersmith, D. and S. Winograd. 1990. "Matrix multiplication via arithmetic progressions." *J. Symb. Comput.* **9**: 251–280.

Cormen, T., C. Leiserson, and R. Rivest. 1990. In *Introduction to Algorithms*. New York: McGraw-Hill Book Co.

Fredman, M. L. 1976. "New bounds on the complexity of the shortest path problem." *SIAM J. Comput.* **5**: 49–60.

Galil, Z. and O. Margalit. 1997. "All pairs shortest paths for graphs with small integer length edges." *J. Comput. System Sci.* **54**: 243–254.

Gallo, G., M. D. Grigoriadis, and R. E. Tarjan. 1989. "A fast parametric maximum flow algorithm and applications." *SIAM J. Comput.* **18**: 30–55.

Garey, M. R. and D. S. Johnson. 1979. In *Computers and Intractability.* San Francisco, CA: W. H. Freeman and Co.

Goldberg, A. V. 1995. "Scaling algorithms for the shortest-paths problem." *SIAM J. Comput.* **24**: 494–504.

Goldberg, A. V. and C. Silverstein. 1997. "Implementations of Dijkstra's algorithm based on multilevel buckets." In *Network Optimizations, Springer Lecture Notes in Economics and Mathematical Systems,* Volume 450, edited by P. M. Pardalos, D. W. Hearn, and W. W. Hager pp. 292–327. Berlin: Springer-Verlag.

Gusfield, D. 1983. "Parametric combinatorial computing and a problem of program module distribution." *J. Assoc. Comput. Mach.* **30**: 551–563.

JáJá, J. 1992. In *Parallel Algorithms.* Reading, MA: Addition-Wesley Publishing Co.

Kao, M-Y. and P. N. Klein. 1993. "Towards overcoming the transitive-closure bottleneck: efficient parallel algorithms for planar digraphs." *J. Comput. System Sci.* **47**: 459–500.

Karger, D. R., D. Koller, and S. Phillips. 1991. "Finding the hidden path: time bounds for all-pairs shortest paths." In *Proceedings of the 32nd IEEE Annual Symposium on Foundations of Computer Science,* pp. 560–568. San Juan, Puerto Rico: IEEE.

King, V. 1999. "Fully dynamic algorithms for maintaining all-pairs shortest-paths and transitive closure." In *Proceedings of the 40th IEEE Annual Symposium on Foundations of Computer Science.* New York, NY: IEEE.

Klein, P. N. and S. Sairam. 1992. "A parallel randomized approximation scheme for shortest paths." In *Proceedings of the 24th Annual ACM Symposium on Theory of Computing* 750–758. Victoria, British Columbia: ACM.

Klein, P. N. and S. Sairam. 1993. "A linear-processor polylog-time algorithm for shortest paths in planar graphs." In *Proceedings of the 34th IEEE Annual Symposium on Foundations of Computer Science,* pp. 259–270. Palo Alto, CA: IEEE.

Lawler, E. L. 1976. In *Combinatorial Optimization: Network and Matroids.* New York: Holt, Reinhart and Winston.

Leighton, F. T. 1992. In *Parallel Algorithms and Architectures: Arrays, Trees, Hypercubes.* CA: Morgan Kaufmann, San Mateo.

Peleg, D. and A. A. Schäffer. 1989. "Graph spanners." *J. Graph Theory* **13**: 99–116.

Sedgewick, R. 1988. In *Algorithms.* Reading, MA: Addison-Wesley Publishing Co.

Seidel, R. 1995. "On the all-pairs shortest-path problem in unweighted undirected graphs." *J. Comput. System Sci.* **51**: 400–403.

Shoshan, A. and U. Zwick. 1999. "All pairs shortest paths in undirected graphs with integer weights." In *Proceedings of the 40th IEEE Annual Symposium on Foundations of Computer Science.* New York, New York: IEEE.

Spencer, T. H. 1997. "Time-work tradeoffs for parallel algorithms." *J. Assoc. Comput. Mach.* **44**: 742–778.

Strassen, V. 1969. "Gaussian elimination is not optimal." *Numerische Mathematik* **14(3)**: 345–356.

Takaoka, T. 1992. "A new upper bound on the complexity of the all pairs shortest path problem." *Information Processing Let.* **43**: 195–199.

Thorup, M. 1997. "Undirected single-source shortest paths in linear time." In *Proceedings of the 38th IEEE Annual Symposium on Foundations of Computer Science.* Miami Beach, Florida: IEEE.

Ullman, J. D. and M. Yannakakis. 1991. "High-probability parallel transitive closure algorithms." *SIAM J. Comput.* **20**: 100–125.

Zwick, U. 1998. "All pairs shortest paths in weighted directed graphs—exact and almost exact algorithms." In *Proceedings of the 39th IEEE Annual Symposium on Foundations of Computer Science.* Palo Alto, CA: IEEE.

8.4 Minimum-Cost Single-Commodity Flow

S. Thomas McCormick

ABSTRACT This article defines minimum-cost single-commodity flow, or just min cost flow (MCF) for short, and surveys what is known about algorithms for solving it. It shows that many network optimization problems can be modeled using MCF. Concepts of duality and complementary slackness that are crucial for constructing algorithms and recognizing optimality are covered. We show how to compute feasible primal and dual solutions, and fundamental operations for finding an improvement to a primal or dual solution. Classic algorithms, such as primal-dual and out-of-kilter, and the use of scaling to make these run in polynomial time, are briefly covered. We cover the practically important successive approximation algorithm in more detail, including the technique used to make it strongly polynomial. Finally, the most widely used MCF algorithm in practice, network simplex, is covered in some detail.

1. Introduction

The minimum-cost single-commodity flow (or min cost flow (MCF) for short) problem is the following. We are given a directed graph G with nodes N and arcs A. An arc $a \in A$ is an ordered pair of nodes (i, j) with $i, j \in N$. We call i the tail of a, and j the head of a, and denote the arc as $i \to j$ to emphasize this. For each ordered pair $i \to j$ of nodes we usually assume that there is at most one arc with tail i and head j, so that we can write $i \to j \in A$ without ambiguity.

We want to find a flow x on the arcs of this graph, so x is indexed as x_a for $a \in A$. Flow x_a should be between lower bound l_a (we allow $l_a = -\infty$) and upper bound u_a (we allow $u_a = +\infty$), that is, we should have $l_a \leqslant x_a \leqslant u_a$ for all $a \in A$. We call this condition boundedness. In most applications it is reasonable to assume that all $l_a = 0$ (and it is possible to transform any instance into this form). Some applications have all $u_a = +\infty$, in which case they are called uncapacitated.

Many applications have supplies and/or demands at the nodes. We unify these into a vector d of demands at the nodes, so that for $i \in N$, we interpret $d_i > 0$ as an external demand for flow at node i, and $d_i < 0$ as an external supply of $|d_i|$ units of flow at node i. It will be necessary to assume that $\sum_{i \in N} d_i = 0$, that is, that the sum of the supplies equals the sum of the demands. In applications where this assumption is not naturally satisfied, it is straightforward to add a dummy node to soak up any excess supply or demand.

Given the external supplies and demand, we require that our flow x satisfies conservation of flow. For $i \in N$, define δ_i^+ as the subset of arcs with tail i, and δ_i^- as the subset of arcs with head i. Then the flow out of node i is $d_i + \sum_{a \in \delta_i^+} x_a$, and the flow into node i is $\sum_{a \in \delta_i^-} x_a$. Requiring flow in equals flow out at every node is then $\sum_{a \in \delta_i^-} x_a = d_i + \sum_{a \in \delta_i^+} x_a$ for all $i \in N$.

Finally, we have a cost of c_a per unit of flow on arc a, and we want to find a flow satisfying boundedness and conservation at minimum total cost. This gives us our MCF problem:

$$\text{(MCF)} \qquad \text{Min} \qquad \sum_{a \in A} c_a x_a$$

subject to :

$$\sum_{a \in \delta_i^-} x_a = d_i + \sum_{a \in \delta_i^+} x_a \quad \text{for all } i \in N \quad \text{(conservation)}$$

$$l_a \leqslant x_a \leqslant u_a \quad \text{for all } a \in A \quad \text{(boundedness).}$$

Let M denote the coefficient matrix of this linear program, so its constraints are $Mx = d$. Note that M is a 0, ± 1 matrix whose rows are indexed by the nodes and whose columns are indexed by the arcs, the node-arc incidence matrix of the network. Also define C as the maximum absolute value of any cost, and U as the maximum absolute value of any finite u_{ij} or l_{ij}.

This is a linear program, and so could be solved by general linear programming methods. However, MCF has many nice properties that make problems of this form orders of magnitude easier to solve in practice than general linear programs of the same size. Also, MCF is a model that has a huge number of applications, both directly and as a subproblem arising from other models, such as many from Chapter 2.

1.1. Network Problems that can be Modeled as MCF

The maximum flow problem of Chapter 8.2 in this handbook is a special case of MCF: given a maximum flow instance with source s and sink t, add a return arc (t, s) with $c_{ts} = -1$, and with all other costs zero. Put the lower bounds equal to zero, the upper bounds equal to the capacities, and the demands equal to zero.

The shortest-path (SP) problem of Chapter 8.3 is also a special case of MCF: given an SP instance with source s and sink t, make $d_s = -1$, $d_t = +1$, and $d_i = 0$ otherwise. Put all lower bounds equal to zero, and all upper bounds equal to one. Make the cost of arc a equal to its length. Then an MCF picks out the shortest path from s to t.

In the transportation problem we are given sources i in S each with supply s_i, and sinks j in T, each with demand d_j. The cost of supplying one unit of j's demand from source i is c_{ij}. The problem can be capacitated or uncapacitated. This is an MCF problem with $d_i = -s_i$ for $i \in S$.

The assignment problem is a transportation problem with $| S | = | T |$, where all supplies and demands are one. Here we can, for example, think of the sources as workers and the sinks as tasks, and c_{ij} is the cost of assigning worker i to task j.

1.2. Duality in MCF

Understanding how to solve MCF problems requires understanding something about the dual linear program for MCF. Since there is a primal conservation constraint at each node i, there is a dual variable π_i called a node potential on each node i. There are also dual variables λ_{ij} and μ_{ij} corresponding to the lower and upper bounds $l_{ij} \leqslant x_{ij}$ and $x_{ij} \leqslant u_{ij}$. Then the dual linear program is:

$$\text{(DMCF)} \qquad \text{Max} \qquad \sum_i \pi_i d_i + \sum_{i \to j} (\lambda_{ij} l_{ij} - \mu_{ij} u_{ij})$$

$$\text{subject to:}$$

$$\pi_j - \pi_i + \lambda_{ij} - \mu_{ij} = c_{ij} \qquad \forall i \to j \in A$$

$$\pi \text{ free}$$

$$\lambda, \mu \geqslant 0.$$

If $\varepsilon_{ij} = \min(\lambda_{ij}, \mu_{ij}) > 0$, then define

$$\lambda'_{ij} = \lambda_{ij} - \varepsilon_{ij} \geqslant 0 \quad \text{and}$$

$$\mu'_{ij} = \mu_{ij} - \varepsilon_{ij} \geqslant 0.$$

This is still dual feasible, since $\lambda_{ij} - \mu_{ij} = \lambda'_{ij} - \mu'_{ij}$. The change in objective value is $\varepsilon_{ij}(u_{ij} - l_{ij}) \geqslant 0$. Therefore, we can assume without loss of generality that $\lambda_{ij} \cdot \mu_{ij} = 0 \;\; \forall i \to j$, and we will henceforth assume this.

Define the reduced cost of $i \to j$ with respect to π as $c^{\pi}_{ij} = \pi_i + c_{ij} - \pi_j$. Then our dual constraint can be rewritten as $c^{\pi}_{ij} = \lambda_{ij} - \mu_{ij}$. Since we have $\lambda_{ij}, \mu_{ij} \geqslant 0, c^{\pi}_{ij}$ free, and $\lambda_{ij} \cdot \mu_{ij} = 0$, this constraint is equivalent to

$$\lambda_{ij} = \max(c^{\pi}_{ij}, 0) = (c^{\pi}_{ij})^{+}$$

$$\mu_{ij} = \max(-c^{\pi}_{ij}, 0) = (c^{\pi}_{ij})^{-}$$

(where $z^{+} = \max(z, 0)$ is the positive part of z, and $z^{-} = \max(-z, 0)$ is the negative part of z). Using this we can write our dual problem in a much prettier way:

$$\text{(DMCF)} \qquad \text{Max} \quad \sum_i \pi_i d_i + \sum_{a \in A} [(c^{\pi}_a)^{+} l_a - (c^{\pi}_a)^{-} u_a]$$

subject to

$$c^{\pi}_{ij} = \pi_i + c_{ij} - \pi_j \qquad \forall i \rightarrow j$$

$$\pi \text{ free.}$$

Unfortunately, this version of (DMCF) is no longer an LP, since the objective function is nonlinear. However, it is piecewise linear and concave, which is almost as easy as LP.

Note that since the sum of the rows of M is zero, we have a degree of freedom in any dual solution. If we define $\pi_i' = \pi_i + \alpha \ \forall i$, this doesn't affect any of the constraints or the objective value of DMCF, since $\sum_i d_i = 0$. Therefore, we can select some root node r and set $\pi_r = 0$ without loss of generality.

1.3. Optimality Conditions for MCF: Complementary Slackness

To check a feasible flow x for optimality, we look for node potentials π that satisfy complementary slackness with x. It is useful to think of π_i as the cost of sending one unit of flow from r to i. Then for each arc $i \rightarrow j$, if $x_{ij} < u_{ij}$, then we should have that $\pi_i + c_{ij} \geqslant \pi_j$ (otherwise, we could send flow from r to j more cheaply than π_j by sending flow up to i at cost π_i, then over $i \rightarrow j$ at cost c_{ij}). Similarly if $x_{ij} > l_{ij}$, then we should have that $\pi_j - c_{ij} \geqslant \pi_i$ (otherwise, we could effectively send a unit of flow from r to i more cheaply than π_i by sending it to j at cost π_j, then decreasing the flow on $i \rightarrow j$ thereby saving cost c_{ij}). We can summarize by saying that for flow x to be optimal, there must be a set of node potentials π with associated reduced costs c^{π} such that whenever $x_a < u_a$ we have $c^{\pi}_a \geqslant 0$, and whenever $x_a > l_a$ we have $c^{\pi}_a \leqslant 0$.

1.4. Integrality of Optimal Solutions

MCF problems are LPs, so they have basic optimal solutions x^*, where $Bx^* = b$ for optimal basis B, so $x^* = B^{-1}b$. Applying Cramer's Rule, we get $x_j^* = \det\ B^j / \det\ B$, where B^j is matrix B with its jth column replaced by vector b. If B and b are integral, then $\det B^j$ and $\det B$ are integral, so x_j^* is rational, and the only way for x_y^* to be fractional is if $|\det\ B| > 1$. We call an integral matrix T totally unimodular if for all square submatrices B of T, $|\det\ B| = 0, +1$, or -1. This implies T is a $0, \pm 1$ matrix, since the 1×1 case must also hold.

Theorem 1. *Any LP with a totally unimodular constraint matrix* T *and integer right-hand-side has integer optimal solutions.*

Proof: *Let* B *be a basis of* T. *Then* B *nonsingular, so* det B$\neq$0. T *is totally unimodular, so* det B = +1 *or* −1. *By Cramer's Rule,* x_j^* *is an integer for each* j.

Theorem 2. *The constraint matrix for MCF is totally unimodular.*

Proof: *Let* M′ *be a square submatrix of* M; *we need to show that det* (M′) *is unimodular. We have three cases to consider:*

Case 1: Every column of M′ *has exactly* 2 *nonzero entries* $\Rightarrow 1^{T}M' = 0 \Rightarrow M'$ singular $\Rightarrow \det M' = 0$.

Case 2: $\exists$ *column of* M′ *with exactly one nonzero entry. Since the determinant is invariant up to change of sign under column and row permutations, move that column to the first column, and the nonzero to the first row. Now expand the determinant of this matrix along the first column. The only nonzero term comes from the submatrix with the first row and column deleted, which is again a network matrix, and the result follows by induction.*

Case 3: $\exists$ *an all-zero column. Then* $\det M' = 0$.

Corollary 1. *If a MCF instance has integral* d, l, *and* u, *then it has an integral optimal solution.*

Corollary 2. *If a dual MCF problem has integral* c, *then it has an integral optimal solution.*

These corollaries are quite important in many applications. For example, in the assignment problem it is important to know that there exists an integral and hence a 0–1 optimal solution that corresponds to an assignment of workers to tasks.

1.5. *Finding an Initial Feasible Flow*

Not all networks possess a feasible flow. Consider for example a network with nodes 1 and 2, arc $1 \rightarrow 2$, lower bounds zero, upper bound 2, $d_1 = -3$, and $d_2 = +3$. Clearly there is no way to send the supply of 3 to the demand of 3 while staying within the capacity of 2.

We can use max flow to help us find a feasible flow. We start by choosing any initial flow x^0 that satisfies the bounds. Typically x^0 will violate conservation. We then compute the excesses at the nodes, where e_i is defined as $\sum_{k:k \rightarrow i \in A} x_{ki} - \sum_{j:i \rightarrow j \in A} x_{ij} - d_i$, the net flow into i. Note that

$$\sum_{i \in N} e_i = \sum_{i \in N} \left(\sum_{k:k \rightarrow i \in A} x_{ki} - \sum_{j:i \rightarrow j \in A} x_{ij} - d_i \right) = \sum_{i \in N} d_i = 0,$$

because each x_{ij} appears exactly twice in this sum, once positively, once negatively. Define subsets of nodes with respect to the excesses as

$E = \{i \mid e_i > 0\}$ (excess nodes)

$D = \{i \mid e_i < 0\}$ (deficit nodes)

$Z = \{i \mid e_i = 0\}$ (zero nodes).

Now we want to send flow along paths from positive excess nodes to negative excess nodes. To do this we define a new network $G'(x^0)$ based on the residual network $G(x^0)$. We start with $G(x^0)$ and add two new nodes, a source s and a sink t. For $i \in E$ we add an arc $s \rightarrow i$ with capacity $r_{si} = e_i$; and for $i \in D$ we add an arc $j \rightarrow t$ with capacity $r_{jt} = | e_j |$ $(= -e_j)$. Note that $G'(x^0)$ is now a max flow network with capacities r_{ij}.

Now compute a max flow y in $G'(x^0)$. There are two cases, depending on whether flow y saturates all arcs at s or not. (Since $\sum_i e_i = 0$, y saturates all arcs at s iff it also saturates all arcs at t.)

Case 1 (flow y saturates all arcs at s): Let $x = x^0 + y$ mean that $x_{ij} = x^0_{ij} + y_{ij} - y_{ji}$ for all $i \rightarrow j \in A$. Then it is easy to see that x is a feasible flow. Thus this procedure has computed a feasible flow x in this case, at the cost of just one max flow computation.

Case 2 (flow y does not saturate all arcs at s): This implies a min cut S has $\text{cap}(S) < e(E)$. Now

$$e(E) > \text{cap}(S) = e(E - S) - e(D \cap S) + r(\delta^+(S))$$

$$\Rightarrow r(\delta^+(S)) < e(S \cap E) + e(D \cap S) + e(Z \cap S) \text{ (since } e(Z \cap S) = 0)$$

$$= e(S) = x^0(\delta^-(S)) - (x^0(\delta^+(S)) + d(S)).$$

Now by the definition of residual capacity we get

$$r(\delta^+(S)) = (u - x^0)(\delta^+(S)) + (x^0 - l)(\delta^-(S)),$$

implying that $u(\delta^+(S)) - l(\delta^-(S)) < -d(S)$, or $V(S) > 0$, where $V(S)$ is defined as $l(\delta^-(S)) - (u(\delta^+(S) + d(S))$. Note that $l(\delta^-(S))$ is the min possible feasible flow into S, and $u(\delta^+(S)) + d(S)$ is the max possible feasible flow out of S. Thus $V(S) > 0 \Rightarrow$ min flow into $S >$ max flow out of S, so that S clearly proves that no feasible flow exists. Thus we call a node subset S with $V(S) > 0$ an obvious infeasibility, or a positive cut. Intuitively, we can think of $V(S)$ as being the amount of infeasibility caused by S. Thus this case proves that no feasible flow exists, also at the cost of one max flow computation.

Thus this procedure, which we call Phase I max flow (Phase I MF), either computes a max flow or produces a positive cut proving that there is no feasible flow. The following theorem summarizes this:

Theorem 3. *[Hoffman's Circulation Theorem (HCT)—see Hoffman (1960) and Gale (1957)] System* $Mx = d, l \leqslant x \leqslant u$ *is feasible iff there are no positive cuts.*

1.6. *Finding an Initial Feasible* π

If all l, u are finite, then any π is dual feasible. If $u_a = +\infty$, then μ_a doesn't exist $\Rightarrow$ a dual feasible π must satisfy $c_a^\pi \geqslant 0$. Similarly, if $l_a = -\infty$ then λ_a doesn't exist $\Rightarrow$ a dual feasible π must satisfy $c_a^\pi \leqslant 0$. Thus infinite bounds imply sign constraints on the reduced costs.

This is reminiscent of SP feasibility conditions: $\pi_i + c_{ij} \geqslant \pi_j$ or $c_{ij}^\pi \geqslant 0$. The only problem is that we have some arcs where we want c_a^π nonpositive instead of nonnegative. We deal with this by constructing an auxiliary graph G'. When we have an arc $a = i \to j$ in the original where we want $c_a^\pi \leqslant 0$, we put an arc $j \to i$ in the auxiliary with $\tilde{c}_{ji} = -c_{ij}$; we leave $\tilde{c}_{ij} = c_{ij}$ for arcs where we want $c_a^\pi \geqslant 0$. Thus, if both $u_{ij} = +\infty$ and $l_{ij} = -\infty$, the auxiliary graph has both $i \to j$, and $j \to i$; if both u_{ij}, l_{ij} are finite then the auxiliary graph has neither $i \to j$, nor $j \to i$.

Add an artificial source node s to G', with arcs $s \to i$ for all $i \in N$ with $\tilde{c}_{si} = 0$. We now use the SP methods of Chapter 8.3 to look for SP distances in this graph. As in Section 1.5 we get two possible cases.

Case 1: We are successful in finding SP distances π in $G' + s$. If $l_{ij} = -\infty$ we have $j \to i$ in the auxiliary graph and the SP optimality condition says that $\pi_j + \tilde{c}_{ji} \geqslant \pi_i$, or $\pi_i + c_{ij} \leqslant \pi_j$ as desired. If $u_{ij} = +\infty$ we have $i \to j$ in the auxiliary graph and the SP optimality condition says that $\pi_i + \tilde{c}_{ij} \geqslant \pi_j$, or $\pi_i + c_{ij} \geqslant \pi_j$ as desired. Thus π is feasible for DMCF.

Case 2: We are blocked from finding SP distances in $G' + s$ by a negative cycle Q. Then Q proves that there is no feasible π for DMCF.

This discussion establishes:

Theorem 4. *(DMCF feasibility) DMCF is feasible iff an undirected cycle* Q *does not exist in the original graph such that* $c(Q) < 0$ *and all forward* $a \in Q$ *have* $u_a = +\infty$*, and all backward* $a \in Q$ *have* $l_a = -\infty$.

We call this procedure Phase I shortest-path (Phase I SP). Networks where DMCF is infeasible are very rare in practice. Note the primal/dual parallelism:

P feasible $\Longleftrightarrow$ $\nexists$ positive cut Phase I MF

D feasible $\Longleftrightarrow$ $\nexists$ negative cycle Phase I SP.

1.7. *Changing Flows via the Residual Network*

Given any flow x, we define a residual network $G(x)$ as follows. If $x_{ij} < u_{ij}$ in the original network, then we put forward residual arc $i \to j$ in $G(x)$ with residual capacity $r_{ij} = u_{ij} - x_{ij}$ and cost $\tilde{c}_{ij} = c_{ij}$; if $x_{ij} > l_{ij}$ in the original network, then we put backward residual arc $j \to i$ in $G(x)$

with residual capacity $r_{ji} = x_{ij} - l_{ij}$ and cost $\tilde{c}_{ji} = -c_{ij}$. Note that we can define $G(x)$ for any flow x, even x's that do not satisfy conservation and/or boundedness.

Suppose that $\tilde{Q}$ is a directed cycle in $G(x)$. Note that some arcs of $\tilde{Q}$ will be forward in the original network, and some will be backward in the original network, so that $\tilde{Q}$ corresponds to an undirected cycle Q in the original network. Define $r(\tilde{Q}) = \min_{i \to j \in \tilde{Q}} r_{ij}$, the residual capacity of $\tilde{Q}$. Note that $r(\tilde{Q})$ is always strictly positive. Also define $\tilde{c}(\tilde{Q}) = \sum_{i \to j \in \tilde{Q}} \tilde{c}_{ij}$, the cost of $\tilde{Q}$. We have defined $\tilde{c}$ so that $\tilde{c}(\tilde{Q}) = \sum_{i \to j \text{ forward in } Q} c_{ij} - \sum_{i \to j \text{ backward in } Q} c_{ij}$.

If we have node potentials π, then we consider them to apply as well to the nodes of $G(x)$. Note that if both $i \to j$ and $j \to i$ are arcs of $G(x)$, then $\tilde{c}^{\pi}_{ji} = \pi_j + \tilde{c}_{ji} - \pi_i = -(\pi_i + \tilde{c}_{ij} - \pi_j) = -\tilde{c}^{\pi}_{ij}$, so reduced costs are negated on backward arcs as well. Because of this the optimality conditions translate simply into wanting that $\tilde{c}^{\pi}_{ij} \geqslant 0$ for all arcs of $G(x)$.

Also, it is often useful to know that $\tilde{c}(\tilde{Q}) = \tilde{c}^{\pi}(\tilde{Q})$ (since the potentials cancel out around the cycle), that is, that the cost of a cycle is the same whether we measure it in terms of its original or its reduced cost.

If we have a directed cycle $\tilde{Q}$ in $G(x)$, we can change flows around $\tilde{Q}$ by setting

$$x'_{ij} = \begin{cases} x_{ij} + r(\tilde{Q}) & \text{if } i \to j \text{ is forward residual} \\ x_{ij} - r(\tilde{Q}) & \text{if } i \to j \text{ is backward residual} \\ x_{ij} & \text{otherwise.} \end{cases}$$

We call this procedure canceling the cycle $\tilde{Q}$. Because of the way we defined the residual capacities, if x satisfied boundedness, then x' will also satisfy boundedness. If x violated boundedness on $i \to j$, then x''s violation of boundedness on $i \to j$ will be no worse than x's violation. Since we changed flow on a cycle, if x satisfied conservation, then x' will also satisfy conservation. Note that

$$c^T x' = c^T x + r(\tilde{Q}) \left(\sum_{i \to j \text{ forward in } Q} c_{ij} - \sum_{i \to j \text{ backward in } Q} c_{ij} \right) = c^T x + r(\tilde{Q}) \tilde{c}(\tilde{Q}).$$

Thus canceling $\tilde{Q}$ improves the primal objective value if, and only if, $\tilde{c}(\tilde{Q}) < 0$. When $\tilde{c}(\tilde{Q}) < 0$ we call $\tilde{Q}$ a negative cycle.

Note that we can check $G(x)$ for negative cycles by running Phase I SP on it. If this finds a negative cycle then we can cancel it, re-form $G(x)$ for the new x, and repeat. If no negative cycle is found, then we have optimal SP distances π_i in $G(x)$ satisfying $c^{\pi}_{ij} \geqslant 0$ for all arcs of $G(x)$. We saw in Section 1.3 that this proves that x is optimal. Thus we have proved:

Theorem 5. *(Busacker and Gowen, 1961) Flow* x *is optimal if there are no negative directed cycles in* G(x) *with respect to* $\tilde{c}$.

This theorem suggests a rudimentary algorithm for MCF, called negative cycle canceling: find and cancel negative cycles until none remain. Various negative cycle canceling algorithms exist; we can cancel min mean cycles (see Section on Min Mean Cycles) (Goldberg and Tarjan, 1989), most helpful cycles (Weintraub, 1974, Barahona and Tardos, 1989), min ratio cycles (Wallacher, 1991) or relaxed most negative cycles (Shigeno et al., 1997). There are dual cut canceling versions of all of these as well: max mean cuts (Ervolina and McCormick, 1993a), most helpful cuts (Ervolina and McCormick, 1993b), max ratio cuts (McCormick and Shioura, 1999), and relaxed most positive cuts (Shigeno et al., 1997). All of these algorithms are based in polynomial time, but none of them appear to be practical (see McCormick and Liu, 1993); see Shigeno et al. (1997) for a more detailed survey of these algorithms.

2. Min-Cost Flow Algorithms

Many, many MCF algorithms exist. Some classic algorithms are the out-of-kilter algorithm (Fulkerson, 1961; Minty, 1960) and the successive shortest-path algorithm (Jewell, 1958; Iri, 1960; Busacker and Gowen, 1961). Both of these algorithms are only pseudopolynomial, that is, their running time can depend on the size of the data. Furthermore they perform poorly in practice.

When the data is integral (as is most common in practice), these algorithms can be improved in both theory and in practice by applying scaling techniques, introduced by Edmonds and Karp (1972) and later refined by Gabow (1985) and Gabow and Tarjan (1989). The idea is to start out with only a rough approximation of the data, and to successively refine the approximation until we are computing with the original data. What makes this worthwhile is that an optimal solution to the previous scaled data makes a very good initial solution to the current scaled data.

Here is a typical way to use scaling. Suppose that we have a positive integer datum k. Define $b = \lfloor \log_2 k \rfloor = O(\log k)$, one less than the number of bits in k. Then for $j = 0, 1, \ldots, b$, define $k_j = \lfloor \frac{k}{2^j} \rfloor$, the j-th scale of k. Note that k_j is really deleting the last j bits from the binary representation of k.

Example 1. *Consider* $\mathrm{k} = 45 \equiv 101101_2$. *Then* $\mathrm{b} = 5$, $\mathrm{k}_5 = 1$, $\mathrm{k}_4 = 10_2 = 2$, $\mathrm{k}_3 = 101_2 = 5$, $\mathrm{k}_2 = 1011_2 = 11$, $\mathrm{k}_1 = 10110_2 = 22$, $\mathrm{k}_0 = 101101_2 = 45$.

The fact that there is a difference of at most one between k_j and $2k_{j+1}$ is what makes the optimal solution for scale $j+1$ a good initial solution for scale j.

This technique can be applied to the costs c in the out-of-kilter algorithm to get a cost scaling version whose running time is $O(\log C \cdot n \cdot \mathrm{MF})$, where "MF" is the time to solve a max flow problem (Röck, 1980; Bland and Jensen, 1992). Alternatively, it can be applied to the capacities l and u in the successive shortest-path algorithm to get a capacity scaling version whose running time is $O(m \cdot DSP \cdot \log U)$, where "DSP" is the time to solve a SP problem with nonnegative costs via some version of Dijkstra's algorithm (Edmonds and Karp, 1972). There is some empirical evidence that the cost scaling algorithm, when properly implemented, is a good algorithm in practice, see Bland et al. (1993). Empirical testing indicates that choosing a scaling factor of 4 or 8 in place of 2 often works better in practice. Note that these algorithms are weakly polynomial, in that their running times depend on the logarithms of the size of the data.

There is a further theoretical improvement possible for these algorithms that works even when the data are not integral. This technique leads to strongly polynomial algorithms, that is, algorithms whose running times are independent of the size of the data. Although this improvement does not seem to work well in practice, it depends on a concept that will lead to an algorithm that does perform well in practice in Section 2.1.

Recall that flow x is optimal if, and only if, there are node potentials π such that $c_{ij}^{\pi} \geqslant 0$ for all arcs of $G(x)$. Given a scalar $\varepsilon > 0$, a useful relaxation of this condition is to say that x and π are ε-optimal if $c_{ij}^{\pi} \geqslant -\varepsilon$ for all arcs of $G(x)$. Thus 0-optimality is the same as optimality.

Complementary slackness says that if $c_{ij}^{\pi} > 0$ then any optimal x^* has $x_{ij}^* = l_{ij}$, and if $c_{ij}^{\pi} < 0$ then any optimal x^* has $x_{ij}^* = u_{ij}$. It is possible to get a relaxed version of this with ε-optimality:

Lemma 1. *(Primal MCF proximity lemma (Tardos, 1985)) Suppose that* (x, π) *is* ϵ-*optimal. If* $\mathrm{c}_{\mathrm{a}}^{\pi} \geqslant n\epsilon$ *then* $\mathrm{x}_{\mathrm{a}}^* = \mathrm{l}_{\mathrm{a}}$ *for all optimal* x^*. *If* $\mathrm{c}_{\mathrm{a}}^{\pi} \leqslant -n\epsilon$ *then* $\mathrm{x}_{\mathrm{a}}^* = \mathrm{u}_{\mathrm{a}}$ *for all optimal* x^*.

This lemma can be combined with cost scaling techniques to get an $O(mn \log n \cdot \mathrm{MF})$ algorithm for MCF (Tardos, 1985). There is also a dual version of the proximity lemma that can be combined with capacity scaling techniques to get an $O(m^2 \log n \cdot \mathrm{DSP})$ algorithm for MCF (Fujishige, 1986).

2.1. Successive Approximation

This algorithm is due to Goldberg and Tarjan (1990). In addition to being practical, it also has one of the best weakly polynomial running times. Successive approximation is also the name of a general technique. It uses the ε-optimality and scaling ideas of the previous section, but instead of scaling the data, it scales ε.

How large does ε need to be at the beginning? We assume all l, u are finite so that $\pi = 0$ is feasible. Find some feasible x. Then $(x, 0)$ is C-optimal, since $|c_a^\pi| = |c_a| < C$.

We have a black box subroutine called REFINE into which we feed a 2ε-optimal (x, π), and out of which comes an ε-optimal (x', π'). When l and u are integral we will always have that x and x' are integral. However, even when c is integral we might have that π and π' are fractional when $\varepsilon < 1$.

How low do we need to drive ε in order to allow us to conclude we are exactly optimal?

Lemma 2. *(Bertsekas, 1986) If* (x, π) *is* ε*-optimal with* $\varepsilon < \frac{1}{n}$ *then* x *is optimal.*

Proof: *The* ε*-optimality of* (x, π) *implies* $x_a^\pi \geqslant -\varepsilon > -\frac{1}{n}$ *for all* a*. So if* Q *is any cycle of* $G(x)$*, then* $c(Q) = c^\pi(Q) > |Q|\left(-\frac{1}{n}\right) > -1$*. But* $c(Q)$ *is an integer, so* $c(Q) \geqslant 0$*. This is true for all cycles* Q*, so* x *is optimal.*

Since there is a gap of nC between the initial value of C for ε and its final value of $1/n$, and since REFINE halves ε at each iteration, we get:

Corollary 3. *Successive approximation needs* $O[\log(nC)]$ *iterations of* REFINE.

Generic Outline of REFINE

$\varepsilon \leftarrow \frac{\varepsilon}{2}$

If $c_a^\pi > 0$, set $x_a = l_a$

If $c_a^\pi < 0$, set $x_a = u_a$.

[At this point x is 0-optimal, but we have violated conservation, so we have excesses.]

[Now do something to push positive excesses to negative excesses, preserving ε-optimality.]

We could use successive shortest-path to push positive excesses to negative excesses at min cost (this would preserve 0-optimality). But this approach is too slow because we are trying too hard; we are not taking advantage of the flexibility that ε-optimality gives us. To do this, we make use of the push-relabel ideas from max flow. First we redefine what is meant by an active node and admissible arc in the context of ε-optimality.

Definition 1. *We call node* i *active if* $e_i > 0$*, and we call arc* $i \rightarrow j$ *admissible if* $r_{ij} > 0$ *and* $-\varepsilon \leqslant c_{ij}^\pi < 0$.

Note that the $-\varepsilon$ lower bound in admissibility is necessary to preserve ε-optimality, and the upper bound of zero means that we only push flow on negative reduced cost arcs. We also redefine the PUSH and RELABEL subroutines in the context of ε-optimality.

PUSH $(i \rightarrow j)$

Input: i active, $i \rightarrow j$ admissible

Action:

$\theta = \min(e_i, r_{ij})$

$x_{ij} \leftarrow x_{ij} + \theta$

$e_i \leftarrow e_i - \theta$

$e_j \leftarrow e_j + \theta$

Definition 2. *A saturating* PUSH *is one where* $\theta = r_{ij}$*, while an unsaturating* PUSH *is one where* $\theta = e_i < r_{ij}$*. Note that a* PUSH *can cause a node with negative excess to become positive, but can never cause a positive excess node to become negative*

RELABEL (i)

Input: An active node i such that there are no admissible arcs out of i

Action: $\pi_i \leftarrow \pi_i - \varepsilon$

Lemma 3. *Both* PUSH *and* RELABEL *preserve* ε*-optimality.*

Proof: *Suppose we perform* PUSH *on arc* $i \to j$. *Clearly,* $c_{ij}^{\pi} \geqslant -\varepsilon$ *since there is no change in the reduced cost for arc* $i \to j$. *However, the process of pushing excess flow from* i *to* j *could create the new arc* $j \to i \in G(x)$. *Hence, also we need to check that* $c_{ji}^{\pi} \geqslant -\varepsilon$. *To do this, note that* $c_{ij}^{\pi} < 0$ *since* $i \to j$ *is admissible, and so, we have* $c_{ji}^{\pi} = -c_{ij}^{\pi} > 0 > -\varepsilon$. *This shows that* PUSH *preserves* ε*-optimality.*

Next, suppose that a RELABEL *on the active node* i *changes* π_i *to* $\pi_i' = \pi_i - \varepsilon$. *Note that* RELABEL(i) *changes the reduced costs on the arcs incident with* i *so we need to check that these arcs still satisfy* $c_a^{\pi'} \geqslant -\varepsilon$ *after* RELABEL. *Now if* $k \to i$ *is an arc into* i, *then* $c_{ki}^{\pi} \geqslant -\varepsilon$ *by* ε*-optimality. Since* $\pi_i' = \pi_i - \varepsilon$, *it follows that* $c_{ki}^{\pi'} = c_{ki}^{\pi} + \varepsilon \geqslant 0$ *and we are fine. Also, if* $i \to j$ *is an arc out of* i, *then* $i \to j$ *is not admissible, and so,* $c_{ij}^{\pi} \geqslant 0$. *Hence,* $c_{ij}^{\pi'} = c_{ij}^{\pi} - \varepsilon \geqslant -\varepsilon$, *and again, we are fine. This shows that* RELABEL *preserves* ε*-optimality.*

In the MF versions of push-relabel the key to the complexity analysis was showing that there are only $O(n)$ RELABELs per node. Here we can get a similar lemma:

Lemma 4. *The number of MCF* RELABELs *per node is* $\leqslant 3n = O(n)$.

Proof: *Let* x *and* π *be the current solutions in* REFINE *and let* x' *and* π' *be* 2ε*-optimal solutions at the end of the previous phase of successive approximation. We superscript excesses by* x *or* x' *to indicate which flow they are relative to.*

Claim 1. *For all active* i *(with respect to* x*), there exists* j *such that* $e_j^x < 0$ *and a path* P *from* i *to* j *such that* P *is a directed path in* $G(x)$, *and* $\bar{P}$ = *reverse of* P *(from* j *to* i*) is a directed path in* $G(x')$.

Proof of Claim: *Let* i *be an active node and define*

$$S = \{j : \exists \text{path } P \text{ from } i \text{ to } j \text{ subject to } P \in G(x), \bar{P} \in G(x')\}.$$

If $\exists j \in S$ *with* $e_j < 0$, *then we are done. To get a contradiction, assume that* $e_j \geqslant 0 \ \forall j \in S$. *Since* e *is modular,* $i \in S$ *and* x' *is a feasible flow, we get*

$$0 < \sum_{j \in S} e_j^x = e^x(S) = x(\delta^-(S)) - x(\delta^+(S)) - d(S),$$

and

$$0 = \sum_{j \in S} e_j^{x'} = e^{x'}(S) = x'(\delta^-(S)) - x'(\delta^+(S)) - d(S).$$

Subtracting the second equation from the first, we obtain

$$0 < (x - x')[\delta^-(S)] - (x - x')[\delta^+(S)].$$

Next, note that every $a \in \delta^+(S)$ *must have that either* $r_{vw}^x = 0$, *which implies that* $x_{vw} = u_{vw}$, *or* $r_{wv}^{x'} = 0$, *which implies that* $x_{vw}' = l_{vw}$. *In either case, we have* $x_{vw} \geqslant x_{vw}'$. *Hence,* $(x - x')[\delta^+(S)] \geqslant 0$. *Similarly we get that* $(x - x')[\delta^-(S)] \leqslant 0$. *Thus we get*

$$(x - x')[\delta^-(S)] - (x - x')[\delta^+(S)] \leqslant 0$$

which is a contradiction.

Let i be an active node and let j, P and $\bar{P}$ be as in the above claim. Since $P \in G(x)$ and x is ε-optimal, it follows by definition that $c_a^{\pi} \geqslant -\varepsilon \ \forall a \in P$. Adding these inequalities gives us $\pi_i + c(P) - \pi_j \geqslant -\varepsilon \mid P \mid$. Similarly, since $\bar{P} \in G(x')$ and x' is 2ε-optimal, it follows that $c_a^{\pi'} \geqslant -2\varepsilon \ \forall a \in \bar{P}$. Again, adding these inequalities gives us $\pi_j' + c(\bar{P}) - \pi_i' \geqslant -2\varepsilon \mid \bar{P} \mid$. Clearly, $\mid P \mid = \mid \bar{P} \mid$ and $c(P) = -c(\bar{P})$. Furthermore, PUSH and RELABEL never create new deficit nodes and RELABEL changes π only on active nodes, and so we must have $\pi_j = \pi_j'$. Adding the previous two inequalities gives us $\pi_i - \pi_i' \geqslant -3\varepsilon \mid P \mid > -3\varepsilon n$. If we do k RELABELs to i in this phase, we get $\pi_i = \pi_i' - k\varepsilon$, and so

$$(\pi_i' - k\varepsilon) - \pi_i' \geqslant -3\varepsilon n \Rightarrow k \leqslant 3n.$$

Since there are $O(n)$ RELABELs per node in this context, just like in the max flow context, we hope that all push-relabel complexity statements from max flow will still hold. It is easy to check that the total number of RELABELs performed in each execution of REFINE is still $O(n^2)$, the total time for all the RELABELs is still $O(mn)$, the number of saturating PUSHes is still $O(mn)$, and the total time for all saturating PUSHes is still $O(mn)$. It is possible to prove that the number of unsaturating PUSHes is still $O(mn^2)$, but the argument from the max flow case does not work here. Roughly speaking, the reason is that max flow has cost 0 on every arc while MCF has general costs. With a technical argument it is still possible to prove a $O(mn^2)$ bound on the number of unsaturating PUSHes, and hence show that generic push–relabel REFINE is $O(mn^2)$. We note that the max flow analyses of both FIFO push-relabel and max distance push-relabel also do not work in this case.

Goldberg and Tarjan (1990) describe how to use data structures called dynamic trees (Sleator and Tarjan, 1983) and finger search trees to get successive approximation to run in $O(mn \cdot \log(n^2/m)) \log(nC)$ time. Note that this bound on successive approximation is the same as the dynamic trees bound for max flow with the extra factor of $O(\log(nC))$ coming from the scaling.

2.2. *A Strongly Polynomial Bound on Successive Approximation*

MIN MEAN CYCLES

The $\log(nC)$ factor incurred by the successive approximation algorithm is really $\Theta(\log(nC))$ in our current implementation, since we only decrease ε by a factor of 2 in each iteration of REFINE. To break out of this we need some method to dramatically lower ε all at once. This observation then raises the question: how far down can we push ε in a single iteration?

More formally, suppose that x is a nonoptimal primal feasible flow. The problem of finding a feasible set of node potentials π and (simultaneously) the smallest ε such that (x, π) are ε-optimal can be formalized as the following linear program:

$$\text{Min} \quad \varepsilon$$

subject to :

$$c_a^\pi \geqslant -\varepsilon \quad \forall a \in G(x)$$

Letting $M(x)$ denote the node-arc incidence matrix of $G(x)$ and noting that $c^\pi = c - \pi^T M$ we can rewrite this linear program in the following equivalent form:

$$\text{Min} \quad \varepsilon$$

subject to :

$$\varepsilon \mathbb{1}^T - \pi^T M(x) \geqslant -c^T.$$

After some manipulation, the dual of this linear program can be written as

$$\text{(MMC)} \quad \text{Min} \quad \frac{c^T y}{\mathbb{1}^T y}$$

subject to :

$$M(x)y = 0$$

$$y \geqslant 0.$$

(We have eliminated a constraint by moving it into the objective function as a normalization factor to get a fractional problem.)

For a cycle Q, define $\mu(Q) = \frac{c(Q)}{|Q|}$, the mean value of Q. Now it is easy to prove that an optimal solution y to the min mean cycle problem exists that is the incidence vector of a cycle Q. Thus the optimal objective value for MMC is given by $\min_{Q \in G(x)} \mu(Q)$—that is, the optimal objective value is achieved by a min mean cycle of $G(x)$. For a feasible flow x, let $\mu(x)$ denote the mean value of min mean cycle of $G(x)$, and let $\varepsilon(x)$ denote the minimum ε for which there is feasible π with (x, π) ε-optimal. The following lemma follows from our discussion so far.

Lemma 5. *Let* x *be a feasible flow, and suppose* $\hat{Q}$ *is a cycle in* G(x).

1. $\hat{Q}$ *is a min mean cycle if, and only if, for* $\varepsilon = -\mu(\hat{Q})$ *a feasible* π *exists with* $c_a^\pi \geqslant -\varepsilon$ *for all* $a \in G(x)$ *and* $c_a^\pi = -\varepsilon$ *for all* $a \in \hat{Q}$.
2. $\varepsilon(x) = -\mu(x)$.

There are several known algorithms for computing min mean cycles (and the associated π). The fastest strongly polynomial algorithm is a dynamic programming-based algorithm by Karp (1978) that runs in $O(mn)$ time. The fastest weakly polynomial algorithms are based on approximate binary search, and they run in $O(\sqrt{n}m \log(nC))$ time, as described by Orlin and Ahuja (1992), Iwano et al. (1994), and McCormick (1993). But for all three of these algorithms their worst-case performance is equal to their average-case performance, so the best algorithm in practice is probably a parametric SP algorithm by Young et al. (1991) that is $O(mn + n^2 \log n)$.

MIN MEAN CYCLES MAKE SUCCESSIVE APPROXIMATION STRONGLY POLYNOMIAL

Call $\lceil \log n \rceil$ iterations of REFINE a block. Suppose at the beginning of each block we compute a min mean cycle $\hat{Q}$, update π accordingly, and set $\varepsilon_0 \leftarrow \varepsilon(x) = -\mu(\hat{Q})$. Since $c_a^\pi = -\varepsilon_0$ for all $a \in \hat{Q}$ (by Lemma 5) we have $c(\hat{Q}) = c^\pi(\hat{Q}) = -|\hat{Q}|\varepsilon_0$. Now, at the end of the block, since ε is cut in half in each of the $\lceil \log n \rceil$ phases, $\varepsilon \leqslant \frac{\varepsilon_0}{n}$. Thus there must be some arc $a \in \hat{Q}$ with $c_a^\pi \leqslant -n\varepsilon$. Then the MCF proximity lemma applies to a and we can fix a to either its upper or lower bound (depending on whether a is forward or backward in the original graph, respectively) for the remainder of the algorithm.

So, if we update ε with a min mean cycle computation at the beginning of each block, we have a bound of $O(m \log n)$ on the number of calls to REFINE needed. If we use, say, Karp's algorithm for computing min mean cycles we perform

$$O(mn) + O\left(mn \log n \log\left(\frac{n^2}{m}\right)\right) = O\left(mn \log n \log\left(\frac{n^2}{m}\right)\right)$$

work per block and get an overall (strongly polynomial) time bound of

$$O\left(m^2 n \log n \log\left(\frac{n^2}{m}\right)\right).$$

More generally, running our two implementations of the successive approximation algorithm in parallel (stopping when one of them halts) gives a MCF algorithm with running time

$$O\left(mn \cdot \min(\log(nC), m \log n) \cdot \log\left(\frac{n^2}{m}\right)\right).$$

Remark 1. *Notes on the successive approximation algorithm for MCF:*

1. *This is the best-known weakly polynomial time bound for certain values of* m, n, C *and* U.
2. *When* $\log U = o(n^2/m)$ *the double scaling algorithm of Ahuja et al. (1992) is faster with its* $O(mn \log\log U \log(nC))$ *time bound, and other algorithms exist with better weakly polynomial time bounds for some parameter values.*
3. *There is a faster strongly polynomial algorithm, due to Orlin (1993), that runs in time* $O(m \log n \cdot \mathrm{DSP})$.
4. *Cherkassky and Goldberg (1997) have done extensive algorithm engineering and empirical testing on the Successive Approximation algorithm as compared to other MCF algorithms. They found that it seems to work well in practice, but this statement needs some qualifications.*
   (a) *No algorithm has demonstrated consistent superiority on all classes of instances.*

(b) *All published testing of MCF algorithms has occurred on instances coming from programs that randomly generate instances with various properties. Thus there is no guarantee at all that existing empirical results will be useful in predicting which algorithms might work well for your real world instance.*

A C program of their implementation, called CS2, is available for research purposes from URL www.intertrust.com/star/goldberg/soft.html; *information on a commercial version is available from* igsys@eclipse.net. *A web page listing other network optimization codes is at* www-fp.mcs.anl.gov/otc/Guide/SoftwareGuide/Categories/netwrkopt.html

5. *Preliminary investigations suggest that strongly polynomial algorithms are not of much practical import; however, not much work has been done in this area.*
6. *As a rule of thumb, nearly every weakly polynomial MCF algorithm based on the Successive approximation framework can be made strongly polynomial by periodically computing either min mean cycles (cost scaling and its relatives), max mean cuts (capacity scaling and its relatives), or approximations of these* (Shigeno et al., 1997).

2.3. *Network Simplex*

The network simplex algorithm is the most popular and computationally one of the best algorithms for MCF. The idea in this algorithm is to specialize simplex (with bounds) to MCF. In ordinary simplex we assume that all lower bounds are zero. In simplex with bounds we have general lower and upper bounds. Ordinary simplex sets nonbasic variables to zero, whereas simplex with bounds allows each nonbasic variable to be fixed at either its lower bound or its upper bound.

A Basis is a Spanning Tree

First, we define what is meant by a basis in the context of MCF. In the MCF LP, a basis is a set of variables x_a corresponding to a maximal linearly independent subset of columns (arcs) of the node-arc incidence matrix M of the given network. We will use the word basis to mean the set of arcs corresponding to the basic variables. We claim that the set of arcs of a cycle Q can never be part of a linearly independent arc subset. This is because if we take the sum of the columns of M of the forward arcs of Q, minus the sum of the columns of M of the backward arcs of Q, we get zero, and no linearly independent set of columns can have a nontrivial linear combination of zero. Thus a basis in this context is a maximal undirected acyclic subgraph of the network.

Now assume that G is connected, so that it has a spanning tree T. Then it is easy to check that the submatrix whose columns are the arcs of T has full column rank, implying that rank(M) $= n-1$. Thus a basis is an arc subset of size $n-1$, which forms an undirected acyclic graph. A classic theorem in graph theory says that such a graph must be a spanning tree. Hence, we conclude that a basis for the MCF LP is a spanning tree of the graph G.

Now suppose we are given a basis represented by a spanning tree T. In order to completely represent the basis we need to know not only which arcs belong to T, but also, for each nonbasic (non-T) arc a, if the flow on a is currently fixed at its lower bound l_a or its upper bound u_a. To accomplish this we represent the basis by a STATUS vector indexed by arcs a: (1) $\text{status}_a = \text{lb}$ if a is nonbasic and $x_a = l_a$; (2) $\text{status}_a = B$ if a is basic (i.e., $a \in T$); (3) $\text{status}_a = \text{ub}$ if a is nonbasic and $x_a = u_a$.

The STATUS vector determines the flow on the nonbasic (NB) arcs. We can compute a unique flow x on T that satisfies conservation as follows. Recall from graph theory that every tree has at least two leaf nodes (a node i incident to exactly one arc of the tree). Choose a leaf node i. The flows on all arcs incident to i are completely determined by STATUS except for the unique T-arc a incident to i. Therefore conservation at i uniquely determines the flow x_a. Now mark this leaf as processed, delete it from T, and recursively repeat. This will uniquely determine all T-arc flows in $O(m)$ time.

This unique flow will satisfy conservation, but not necessarily $l \leqslant x \leqslant u$. Thus some spanning trees correspond to feasible flows while others do not. Now if a feasible flow exists in the network, then we can find it with Phase I MF. But it might not be a basic feasible flow. In particular, we could have a cycle Q with $l_a < x_a < u_a \;\; \forall a \in Q$. In this case, either $c(Q) \leqslant 0$ or $c(\text{reverse of } Q) \leqslant 0$. Hence we can cancel Q forward (if $c(Q) \leqslant 0$) or backward (if $c(Q) > 0$) without hurting the objective function value. Naively, it takes $O(m)$ time to find and cancel such a cycle, and we might have to cancel up to $m - n + 1$ such cycles to arrive at a basic feasible flow, for a total time bound of $O(m^2)$. A more clever approach uses DFS and gives a running time of $O(mn)$. An even more clever approach uses dynamic trees and takes $O(m \log n)$ time.

However, in practice nobody uses Phase I MF to initialize network simplex. People use Phase I network simplex instead, which works as follows. Add a dummy root node r to the network. Make a Phase I cost $c'_a = 0$, and set $x_a = l_a$ (and so $\text{status}_a = \text{lb}$) for all $a \in A$, thereby generating excesses e_i. For each node $i \neq r$ with $e_i \leqslant 0$ make an artificial arc $r \rightarrow i$ with $l_{ri} = 0$, $u_{ri} = x_{ri} = | e_i |$, and $c'_{ri} = 1$; for each node $j \neq r$ with $e_j > 0$ make an artificial arc $j \rightarrow r$ with $l_{jr} = 0$, $u_{jr} = x_{jr} = e_j$, and $c'_{jr} = 1$. Let T be the union of these artificial arcs.

Note that T is an initial (all-artificial) feasible basis. Now use network simplex as a subroutine to minimize the Phase I objective $\sum_a c'_a x_a$. If the network is feasible, the optimal Phase I solution will have flow zero in all artificial arcs. Then the upper bounds on the artificial arcs can be dropped to zero, the Phase I objective c' is replaced with the real objective c, and Phase II of the algorithm can proceed.

If the network is infeasible, then the optimal Phase I solution will have positive flow in at least one artificial arc. In this case it can be shown that the set of nodes reachable from r via forward augmenting paths with regard to the optimal Phase I solution is a positive cut proving infeasibility via Theorem 3.

In practice it is often faster to use a hybrid Phase I/Phase II strategy where we use an objective of the form $\alpha c' + c$ for some scalar α. We solve for a fairly small value of α, and if there is still a positive artificial flow we double the value of α and repeat. It can be shown that if α reaches the value nC and there is still a positive artificial flow, then the network is infeasible.

PRICING OUT A BASIC TREE

Next, we discuss how to price out T, that is, compute dual variables π with respect to T. We will use the fact that the dual prices are determined by wanting to have zero reduced cost on the basis, or $c^\pi_a = 0 \;\; \forall a \in T$. This gives us $n - 1$ equations (one for each arc of T), in n variables (one for each node of T). We fix some node r as the root, set $\pi_r = 0$, and then compute π for all $i \in N$ via $c^\pi_a = 0 \;\; \forall a \in T$ (see Figure 1). That is, we start with $\pi_r = 0$ and fan-out the equation $c^\pi_a = 0$ for $a \in T$ from the root outwards, which uniquely determines π_i for every $i \in N$. Note that π_i equals the length of the unique path in T from r to i for every $i \in N$. This takes $O(n)$ time.

PERFORMING A PIVOT

Now that we have π, we can now compute the reduced cost c^π_a for a NB arc a. To check optimality, we check the complementary slackness conditions on NB arcs:

1. If $\text{STATUS}_a = \text{lb}$, we want $c^\pi_a \geqslant 0$.
2. If $\text{STATUS}_a = \text{ub}$, we want $c^\pi_a \leqslant 0$.

If all these complementary slackness conditions are satisfied, then we have achieved optimality and we can stop. Else, we select some bad arc that violates complementary slackness:

1. $\text{STATUS}_a = \text{lb}$ but $c^\pi_a < 0$, or
2. $\text{STATUS}_a = \text{ub}$ but $c^\pi_a > 0$.

We refer to the arc selected in (1) or (2) to be the INARC (incoming arc, corresponding to the incoming variable of ordinary simplex). Typically there will be a large number of possible INARCs to choose from; later we will discuss pivot rules for choosing one.

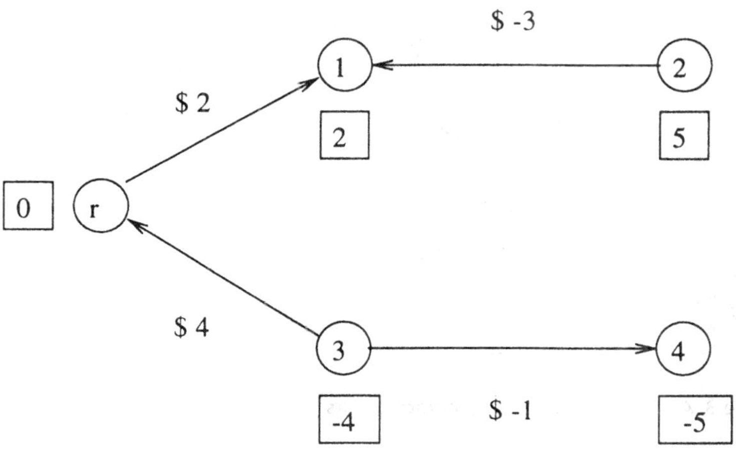

Figure 1 Illustration of how to price out T; the computed prices are boxed

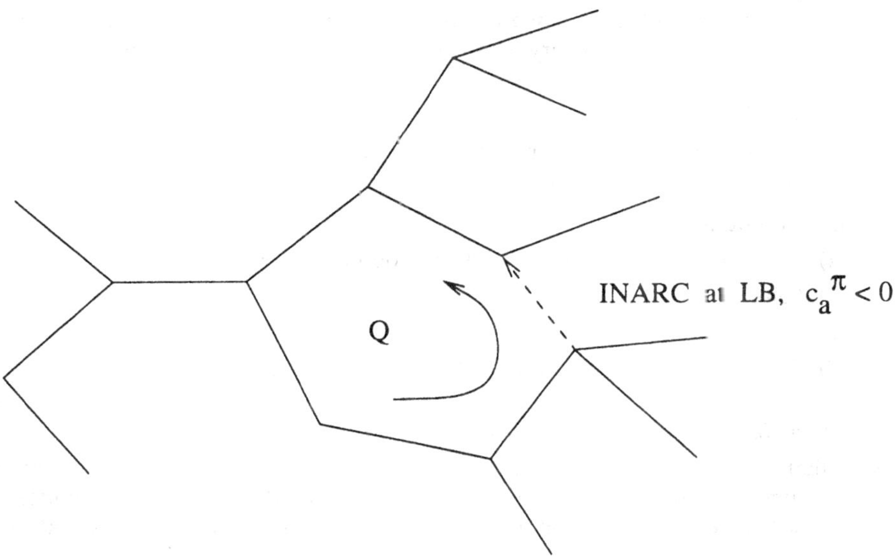

Figure 2 Selecting an INARC results in a fundamental cycle

When an INARC enters the basis (spanning tree T), graph theory tells us that this creates a unique cycle Q in $T + \text{inarc}$ (see Figure 2), called the fundamental cycle of INARC in T, denoted $C(T + \text{inarc})$. We orient Q in the direction that we want to push flow, that is so that: (1) INARC is forward in Q if $\text{status}_{\text{inarc}} = \text{lb}$ (we want to push flow forward in INARC), or (2) INARC is backward in Q if $\text{status}_{\text{inarc}} = \text{ub}$ (we want to push flow backward in INARC). Since $c_a^\pi = 0 \;\; \forall a \in T$, we get

$$c(Q) = c^\pi(Q) = \chi(Q)_{\text{inarc}}\, c^\pi_{\text{inarc}} + \sum_{b \neq \text{inarc}} c^\pi_b = \chi(Q)_{\text{inarc}}\, c^\pi_{\text{inarc}} < 0,$$

showing that Q is a negative cycle. Next, we cancel Q. Note that this is just like ordinary negative cycle canceling except that we could have $r(Q) = 0$ (degeneracy), as in Figure 3. Computing $r(Q)$ is the network simplex version of the min ratio test of ordinary simplex We

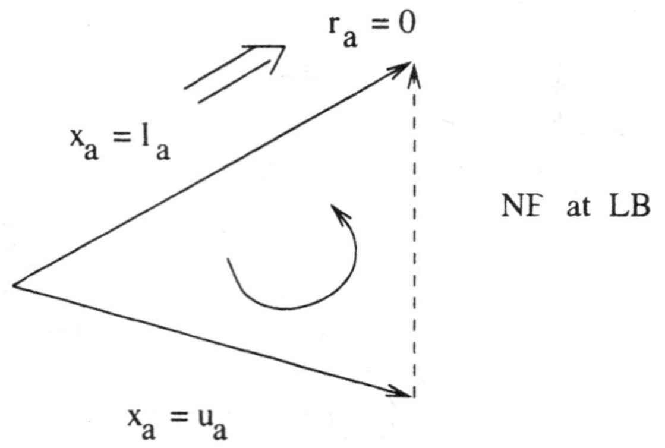

Figure 3 A situation where degeneracy occurs

then pick some b with $r_b = r(Q)$. We call this arc b the OUTARC (outgoing arc, which corresponds to the outgoing variable of ordinary simplex).

Note that we can have inarc = outarc as shown in Figure 4. Otherwise, we set $T \leftarrow T +$ inarc − outarc, update STATUS, and we are ready for the next iteration. One such iteration is called a pivot, just like in ordinary simplex. We keep pivoting until we are optimal. We summarize as follows:

Network Simplex

(1) Use Phase I network simplex to get an initial basic feasible T.
(2) Price out T (compute π).
(3) Compute $c_a^\pi \quad \forall a \notin T$.
(4) If $\nexists$ violating arc, then the solution is optimal. Stop.
(5) Select a violating INARC.
(6) Set $Q = C(T + \text{inarc})$ and cancel Q.
(7) $T \leftarrow T +$ inarc − outarc. Update STATUS and go to (2).

2.4. *Speeding Up Network Simplex in Practice*

Suppose that Step 2 of network simplex transforms the old tree T into a new tree $T' = T +$ inarc − outarc. If inarc = outarc, then $T' = T$ and we do not need to reprice the tree. Otherwise, note that $T -$ outarc consists of exactly two connected components that form a cut,

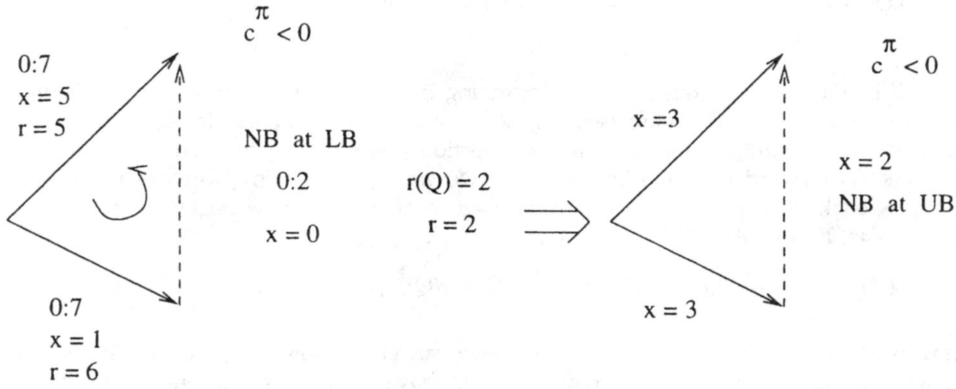

Figure 4 A situation where inarc = outarc

the fundamental cut of OUTARC in T, denoted $C(T-\text{outarc})$. Instead of computing prices in T' from scratch, we could instead note that the prices π' in T' can be obtained from the prices π in T by changing π_i by a constant on one side of $C(T-\text{outarc})$, and leaving π_j alone on the other side. This constant is determined by the requirement that $c^{\pi'}_{\text{inarc}}=0$, that is, it is $\pm\, c^{\pi}_{\text{inarc}}$.

Note that the side of $C(T-\text{outarc})$ not containing r will always be a subtree of T. There are data structures, which allow traversal of any subtree S of T in $O(|S|)$ time, see, for example, the Bazaraa et al. (1990) text. Using such a data structure turns Step 2 into an $O(|S|)$ operation instead of an $O(n)$ operation, a substantial savings.

A naive way to implement Step 6 is to represent T using predecessors that point toward the root. Then when we look for the fundamental cycle $C(T+i\to j)$ $(i\to j=\text{inarc})$, we would use the predecessors to trace the path from i to the root node r and also the path from j to r. Then we determine the JOIN, which is the first node where these paths meet, and from there, we can obtain $C(T+i\to j)$. Note that this process unnecessarily traverses the path from the JOIN to r at least once. There are data structures (Bazaraa et al., 1990) that allow us to avoid having to traverse this path, allowing us to implement Step 6 in $O(|C(T+\text{inarc})|)$ time instead of $O(n)$ time, again a substantial savings.

What we have done so far allows us to reduce the $O(n)$ time for Steps 2 and 6 to something that is heuristically smaller. But there is a much larger opportunity for improvement in the process of selecting an INARC in Steps 3–5, which naively take $O(m)$ time. Empirical tests on network simplex codes reveal that more than 95 percent of the running time is taken up by computing reduced costs.

A good strategy for improvement here is to use partial pricing. Periodically we sample some subset of the nonbasic arcs, either a fixed fraction or a fixed number of them, and we compute reduced costs just for that subset. It is worthwhile to arrange (either via randomization or by stepping the start point of our subset through the arc list) that successive subsets come from different parts of the network in order to ensure that we do not expend too much effort in optimizing one region of the network at the expense of other regions.

We keep a list of all nonbasic arcs whose reduced costs were small enough (a tuning parameter). Then in subsequent pivots we just take INARCs off this list instead of having to compute any further reduced costs. Once the list is exhausted we compute a new list from a new subset of arcs. With a good set of tuning parameters this can reduce the time for Steps 3–5 to something close to constant, at least until we are close to optimality (where we are forced to do full pricing to verify optimality). More sophisticated strategies are used in practice as well, where the number of NB arcs priced, and the frequency of updating the list of potential INARCs, are changed dynamically as the algorithm proceeds.

Finally we address the question of the number of iterations of network simplex. If the problem is nondegenerate, then $c^Tx-c^Tx^*$ decreases by at least one in each iteration, so there can be at most $O(mCU)$ iterations. There are families of nondegenerate instances (with exponentially large data) where network simplex (using Dantzig's pivot rule, see below) takes an exponential number of iterations.

But degeneracy is very common in real world instances. With degeneracy, there are instances where network simplex cycles (iterates through a degenerate cycle of pivots; see Gassner, 1964). There is a theoretical way to avoid degeneracy called Cunningham's Rule (Cunningham, 1979). This further restricts the set of trees we work with to the so-called strongly feasible trees, which imply that we can perform only $O(nC)$ consecutive degenerate pivots. However, Cunningham's Rule tends not be used in practice as it slows down the algorithm too much. The practical pivot rules described below very rarely lead to cycling in the real world.

The speed of any simplex algorithm depends on the pivot rule used to choose the incoming variable, here the INARC. One extreme is Dantzig's rule, which is to choose a most negative reduced cost arc as INARC. This leads to high quality pivots that make a lot of progress toward optimality, but it forces us to look at every nonbasic arc at every iteration, which is too expensive. At the other extreme, we can take the first negative cost nonbasic arc we find. This is very fast per iteration, but usually leads to a low quality of progress towards optimality, and hence to a lot of iterations.

Thus the real trick to get a good real world implementation of network simplex is to find a heuristic rule that will allow us to find a nonbasic arc with a quite small reduced cost, but without having to look at very many arcs. Typically this is done in combination with partial pricing (Mulvey, 1978): we keep the partial pricing list of nonbasic arcs (nearly) sorted by reduced costs, and take the more negative reduced cost arcs first as a sort of local approximation to Dantzig's rule. Empirically this sort of strategy has been found to lead to a reasonably small number of iterations (growing roughly linearly with m), with a reasonably small amount of work in computing reduced costs.

We also note that there are network simplex algorithms with polynomial bounds, even strongly polynomial bounds. In such algorithms, the idea is basically to force network simplex to simulate other algorithms (Shigeno et al., 1997 surveys these algorithms). It is often also useful to know that there is a dual version of the network simplex which is useful in sensitivity analysis.

Remark 2. *Notes on how network simplex performs in practice:*

1. *It is widely used in practice. There are stand-alone network simplex codes available, and there is a network simplex option in many LP codes, for example CPlex* `www.cplex.com`, *and OSL* `www6.software.ibm.com/es/oslv2/features/welcome.htm`.
2. *It is attractive since it is easy to add side constraints.*
3. *It is highly tuned and works very well in practice. Modern codes have many heuristics that make the empirical number of iterations quite small, almost linear in m.*

ACKNOWLEDGMENTS

This chapter was significantly helped by the students in my OR&IE 633 class at Cornell University, Spring 1999. This research was supported by an NSERC Operating Grant. Dedicated to the memory of Mort Klein, one of the pioneers of this area, (to recognize Mort, who passed away a few days ago, and who was one of the early giants of min cost flow, as well as my first department chair).

References

Ahuja, R. K., A. V. Goldberg, J. B. Orlin, and R. E. Tarjan. 1992. "Finding minimum-cost flows by double scaling." *Math. Programming* **53**: 243–266.

Ahuja, R. K., T. L. Magnanti, and J. B. Orlin. 1993. In *Network Flows: Theory, Algorithms, and Applications.* Englewood Cliffs: Prentice-Hall.

Armstrong, R. D. and Z. Jin. 1997. "A new strongly polynomial dual network simplex algorithm." *Math. Programming* **78**: 131–148.

Barahona, F. and É. Tardos. 1989. "Note on Weintraub's minimum-cost circulation algorithm." *SIAM J. Comput.* **18**: 579–583.

Bazaraa, M. S., J. J. Jarvis, and H. D. Sherali. 1990.. 2d ed. In *Linear Programming and Network Flows.* New York: Wiley.

Bertsekas, D. P. 1986. "Distributed asynchronous relaxation methods for linear network flow problems." Working Paper. Cambridge, MA: Laboratory for Information and Decision Systems, MIT.

Bland, R. G. and D. L. Jensen. 1992. "On the computational behavior of a polynomial-time network flow algorithm." *Math. Programming* **54**: 1–39.

Bland, R. G., J. Cheriyan, D. L. Jensen, and L. Ladányi. 1993. "An empirical study of recent min cost flow algorithms." In *Network Flows and Matching*, edited by D. S. Johnson and C. S. McGeoch, American Mathematical Society DIMACS Series in Discrete Mathematics and Theoretical Computer Science **12**: 119–156.

Busaker, R. G. and P. J. Gowen. 1961. "A procedure for determining minimal-cost network flow patterns." ORO Technical Report 15. Baltimore: OR Office, Johns Hopkins.

Cherkassky, B. and A. V. Goldberg. 1997. "An efficient implementation of a scaling minimum-cost flow algorithm." *J. Algorithms* **22**: 1–29.

Cunningham, W. H. 1979. "Theoretical properties of the network simplex method." *Math. Oper. Res.* **4**: 196–208.

Edmonds, J. and R. M. Karp. 1972. "Theoretical improvements in algorithmic efficiency for network flow problems." *J. ACM* **19**: 248–264.

Ervolina, T. R. and S. T. McCormick. 1993a. "Two strongly polynomial cut canceling algorithms for minimum cost network flow." *Discrete Appl. Math.* **46**: 133–165.

Ervolina, T. R. and S. T. McCormick. 1993b. "Canceling most helpful total cuts for minimum cost network flow." *Networks* **23**: 41–52.

Fujishige, S. 1986. "Capacity-rounding algorithm for the minimum cost circulation problem: a dual framework of tardos algorithm." *Math. Programming* **35**: 298–309.

Fulkerson, D. R. 1961. "An out-of-kilter method for minimum cost flow problems." *SIAM J. Appl. Math* **9**: 18–27.

Gabow, H. N. 1985. "Scaling algorithms for network problems." *J. Comput. Syst. Sci.* **31**: 148–168.

Gabow, H. N. and R. E. Tarjan. 1989. "Faster scaling algorithms for network problems." *SIAM J. Comput.* **18**: 1013–1036.

Gale, D. 1957. "A theorem on flows in networks." *Pacific J. Math.* **7**: 1073–1082.

Gassner, B. J. 1964. "Cycling in the transportation problem." *Naval Res. Logistics Q.* **11**: 43–58.

Goldberg, A. V. and R. E. Tarjan. 1989. "Finding minimum-cost circulations by canceling negative cycles." *J. ACM* **36**: 873–886.

Goldberg, A. V. and R. E. Tarjan. 1990. "Finding minimum-cost circulations by successive approximation." *Math. Oper. Res.* **15**: 430–466.

Hadjiat, M. 1994. "Un algorithme fortement polynomial pour la tension de coût minimum basé sur les cocycles de coûts moyens minimums." Technical Report. Marseille Groupe Intelligence Artificielle, Faculté des Sciences de Luminy.

Hassin, R. 1983. "The minimum cost flow problem: a unifying approach to dual algorithms and a new tree-search algorithm." *Math. Programming* **25**: 228–239.

Hoffman, A. J. 1960. "Some recent applications of the theory of linear inequalities to extremal combinatorial analysis." In *Proceedings of Symposia in Applied Mathematics Vol. X, Combinatorial Analysis*, edited by R. Bellman and M. Hall Jr. pp. 113–127. Providence: American Mathematical Society.

Iri, M. 1960. "A new method of solving transportation-network problems." *J. Oper. Soc. Res. Japan* **3**: 27–87.

Iwano, K., S. Misono, S. Tezuka, and S. Fujishige. 1994. "A new scaling algorithm for the maximum mean cut problem." *Algorithmica* **11**: 243–255.

Jewell, W. S. 1958. "Optimal flow through networks." Interim Technical Report 8. Cambridge, MA: MIT OR Center.

Karp, R. M. 1978. "A characterization of the minimum cycle mean in a digraph." *Discrete Math.* **23** 309–311.

Klein, M. 1967. "A primal method for minimal cost flows with applications to the assignment and transportation problems." *Mgmnt. Sci.* **14**: 205–220.

McCormick, S. T. 1993. "Approximate binary search algorithms for mean cuts and cycles." *Oper. Res. Lett.* **14**: 129–132.

McCormick, S. T. and L. Liu. 1993. "An experimental implementation of the dual cancel and tighten algorithm for minimum cost network flow." In *Network Flows and Matching*, edited by D. S. Johnson and G. S. McGeoch, American Mathematical Society DIMACS Series in Discrete Mathematics and Theoretical Computer Science **12**: 247–266.

McCormick, S. T. and A. Shioura. 1999. "Minimum ratio canceling is oracle polynomial for linear programming, but not strongly polynomial, even for networks." Submitted to *Oper. Res. Lett.*

Megiddo, N. 1983. "Applying parallel computation algorithms in the design of serial algorithms." *J. ACM* **30**: 852–865.

Minty, G. J. 1960. "Monotone networks." *Proc. Royal Soc. Lond.* **257A**: 194–212.

Mulvey, J. 1978. "Pivot Strategies for Primal-Simplex Codes." *J. ACM* **25**: 266–270.

Orlin, J. B. 1993. "A faster strongly polynomial minimum cost flow algorithm." *Oper. Res.* **41**: 338–350.

Orlin, J. B. 1997. "A polynominal time primal network simplex algorithm for minimum cost flows." *Math. Programming* **78**: 109–129.

Orlin, J. B. and R. K. Ahuja. 1992. "New scaling algorithms for the assignment and minimum cycle mean problems." *Math. Programming* **54**: 41–56.

Orlin, J. B., S. Plotkin, and É. Tardos. 1993. "Polynomial dual network simplex algorithms." *Math. Programming* **60**: 255–276.

Queyranne, M. N. 1980. "Theoretical efficiency of the algorithm capacity for the maximum flow problem." *Math. Oper. Res.* **5**: 258–266.

Shigeno, M., S. Iwata, and S. T. McCormick. 1997. "Relaxed most negative cycle and most positive cut canceling algorithms for minimum cost flow." Submitted to *Math. Oper. Res.*

Röck, H. 1980. "Scaling Techniques for Minimum Cost Network Flows." In *Discrete Structures and Algorithms,* edited by U. Pape pp. 181–191: Hanser.

Sleator, D. D. and R. E. Tarjan. 1983. "A data structure for dynamic trees." *J. Comput. Syst. Sci.* **26**: 362–391.

Tardos, É. 1985. "A strongly polynomial minimum cost circulation algorithm." *Combinatorica* **5**: 247–256.

Wallacher, C. 1991. "A generalization of the minimum-mean cycle selection rule in cycle canceling algorithms." Report. Inst. Für Ang. Math., Braunschweig.

Weintraub, A. 1974. "A primal algorithm to solve network flow problems with convex costs." *Mgmnt. Sci.* **21**: 87–97.

Young, N. E., R. E. Tarjan, and J. B. Orlin. 1991. "Faster parametric shortest path and minimum balance algorithms." *Networks* **21**: 205–221.

Zadeh, N. 1973. "A bad network problem for the simplex method and other minimum cost flow algorithms." *Math. Programming* **5**: 255–266.

8.5 Minimum-Cost Multicommodity Flow

Pierre Chardaire and Abdel Lisser

ABSTRACT We present a number of specializations of simplex and interior-point-based techniques for the solution of minimum-cost multicommodity flow models. These specializations exploit price-directive decomposition, resource-directive decomposition, and partitioning of linear systems. They are illustrated by considering a nonoriented minimum-cost multicommodity flow model of interest to telecommunication network design.

1. Introduction

Multicommodity network flow models arise when several commodities, each subjected to its own flow constraints, share the same network facilities. These models are composed of one group of flow constraints per commodity and a group of coupling constraints, which ties together the various commodities. Solution methods for multicommodity network flow problems usually exploit both the block structure of the problem and the structure of each block of flow constraints. Multicommodity network flow models are indeed examples of block diagonal problems with complicating constraints for which the Dantzig–Wolfe decomposition technique was invented. Moreover, the network structure of each block of flow constraint is amenable to efficient combinatorial techniques when the coupling constraints are relaxed. It is therefore not surprising that such models have been widely used as test models in the design of solution methods for large-size linear and nonlinear problems. Another reason for the importance of multicommodity network flow models is the large variety of applications that can thus be modeled. Recent applications published since 1990 are mainly in the following areas.

1. Transportation and logistics. Routing of hazardous materials in marine waters (Iakovou et al., 1999), management of disaster relief operations (Haghani and Oh, 1996; Oh and Haghani, 1997), routing of packages among vehicles in transportation networks (Greenwald and Dean, 1996), prediction of traffic congestion in highway networks (Tsujino et al., 1994), optimization of interdependent flows in regional transportation networks (Moiseenko, 1990), dynamic fleet management (Powell and Carvalho, 1997),

aircraft capacity assignment (Berge and Hopperstad, 1993), and airline fleet assignment (Gu et al., 1994; Rushmeier and Kontogiorgis, 1997).

2. Design of integrated circuits. Cell layout (Hong et al., 1997), routing (Shiraishi et al., 1993; Carden et al., 1996), circuit clustering (Yeh et al., 1995; see Raghavan, 1992, for a survey of recent developments in the application of combinatorial optimization to VLSI design).
3. Scheduling, manufacturing and planning. Scheduling of multiple-depot vehicle (Lobel, 1998), scheduling of freight cars on rail networks (Holmberg et al., 1998; Kwon et al., 1998), design of transportation strategies in manufacturing industries (Popken, 1996), aircraft routing and scheduling (Desaulniers et al., 1997), scheduling of hydrothermal power systems (Nabona, 1993), multi-echelon production planning (Chopra and Tsai, 1997), two-stage distribution planning (Hindi and Basta, 1994), effect of lock delay on grain marketing costs (Fuller and Grant, 1997), design of mail processing facilities (Bard et al., 1993), and planning and scheduling of production batches in flexible manufacturing systems (Ghosh and Gaimon, 1993).
4. Telecommunication network design and management. Traffic management of Asynchronous Transfer Mode (ATM) networks (Medova, 1998), virtual path routing for survivable ATM networks (Murakami and Kim, 1996), design of centralized telecommunication networks (Gouveia and Janssen, 1998), design of telecommunication loss networks (Medhi and Guptan, 1997; Girard and Sanso, 1998), design of fiber transport networks (Yoon et al., 1998), routing and wavelength assignment in optical networks (Banerjee and Mukherjee, 1996), terminal layout with hop constraints (Gouveia, 1996), design of token ring LANs (LeBlanc et al., 1996), routing in virtual circuit data networks with unreliable links (Sung and Lee, 1995), telephone network traffic management (Chang et al., 1992; Chang, 1994), ring network design (Arbib et al., 1992), backbone network design in communications network (Bogdanowicz, 1993), routing in packed-switched networks (Ribeiro and Elbaz, 1992), virtual circuit routing in computer networks (Chang et al., 1992; Yee and Lin, 1992; Zhang and Hartmann, 1992), fair integration of routing and flow control in communication networks (Chang, 1992), design of lightwave networks (Labourdette and Acampora, 1991), and design of survivable telecommunication networks (Lisser et al., 1995a,b).

In this article we present a number of general techniques that have been applied to various multicommodity flow problems. These techniques exploit the aforementioned properties of multicommodity flow models to reduce the problem to a sequence of solutions of smaller problems or to speed up direct linear programming solution methods. We use a nonoriented multicommodity flow model, which arises in the routing of circuits in telecommunication transmission networks, as an illustration of the application of these various techniques to multicommodity flow problems.

The remainder of this paper is organized as follows. In Section 2 we introduce and formulate the nonoriented minimum cost multicommodity flow problem. In Section 3 we present a number of price-directive decomposition techniques: Lagrangian relaxation (Section 3.1), Dantzig–Wolfe's decomposition (Section 3.2), and the analytic center cutting-plane method (Section 3.3). In Section 4 we discuss resource-directive decomposition. Besides decomposition approaches, there are other types of specialization. They are based on the partitioning of the linear systems that are to be solved at each step of a linear programming algorithm. Those techniques are presented in Section 5 where we give details on the primal simplex partitioning algorithm (Section 5.1) and on the specialization of the dual affine scaling interior-point solution method (Section 5.2).

2. The Problem to be Solved

A transmission network can be viewed as an undirected graph. An edge (i.e., a link of the network) is characterized by a pair of nodes, a transmission cost per circuit, and a transmission capacity in the number of circuits. The problem to be solved is the determination of the most economical way of using the available transmission capacities to route a traffic matrix (a number of circuits between various pairs of nodes) through the network. This problem can be expressed as follows:

$$\min \sum_{k,j} c_j^k \mid x_j^k \mid$$

$$\text{subject to} \quad Ax^k = r^k, \quad k = 1, \ldots, \bar{K},$$

$$\sum_k \mid x_j^k \mid \leqslant b_j, \quad j = 1, \ldots, \bar{J},$$

where:

- A is the node-arc incidence matrix corresponding to an arbitrary orientation of the network graph (i.e., $A_{ij} = +1$ if arc j is directed away from node i, $A_{ij} = -1$ if arc j is directed towards node i, $A_{ij} = 0$ otherwise),
- k is a commodity characterized by a number of circuits, d^k, to be routed through the network between a given pair of nodes, s^k and t^k,
- r^k is the requirement vector for commodity k (i.e., $r_{s^k}^k = d^k$, $r_{t^k}^k = -d^k$ and $r_l^k = 0$ when $l \notin \{s^k, t^k\}$),
- x^k is the flow vector for a given commodity k (i.e., x_j^k is the flow on arc $j = (s, t)$ with $x_j^k > 0$ if the flow goes from s to t and $x_j^k < 0$ if the flow goes from t to s),
- b_j is the capacity of link j,
- c_j^k is the cost per unit of flow on link j for commodity k.

Assuming that $c_j^k > 0$, one can introduce nonnegative variables, x_j^{k+} and x_j^{k-}, such that $x_j^k = x_j^{k+} - x_j^{k-}$ and $\mid x_j^k \mid = x_j^{k+} + x_j^{k-}$ to obtain an equivalent linear formulation (Minoux, 1986):

$$\min \quad \textstyle\sum_{k,j} c_j^k (x_j^{k+} + x_j^{k-})$$

$$\text{subject to} \quad A(x^{k+} - x^{k-}) = r^k, \quad k = 1, \ldots, \bar{K},$$

$$\sum_k (x_j^{k+} + x_j^{k-}) \leqslant b_j, \quad j = 1, \ldots, \bar{J},$$

$$x_j^{k+} \geqslant 0, \quad k = 1, \ldots, \bar{K}; \quad j = 1, \ldots, \bar{J},$$

$$x_j^{k-} \geqslant 0, \quad k = 1, \ldots, \bar{K}; \quad j = 1, \ldots, \bar{J}.$$

As the objective coefficients are nonnegative the supports of the flows corresponding to an optimal solution to this problem do not contain any circuits. Hence $x_j^{k+} x_j^{k-} = 0$ for all commodity k and all links j at optimality, which guarantees that $\mid x_j^{k+} - x_j^{k-} \mid = x_j^{k+} + x_j^{k-}$. The problem can be rewritten in a more compact form as

$$\text{NMP} \begin{cases} \min & \sum_k [C^k]^{\mathrm{T}} x^k \\ \text{subject to} & Bx^k = r^k, \quad k = 1, \ldots, \bar{K}, \qquad (1) \\ & \sum_k D x^k \leqslant b, \qquad (2) \\ & x^k \geqslant 0, \quad k = 1, \ldots, \bar{K}, \end{cases}$$

with

$$[C^k]^{\mathrm{T}} = [\, [c^k]^{\mathrm{T}} \;\; [c^k]^{\mathrm{T}} \,], \quad [x^k]^{\mathrm{T}} = [\, [x^{k+}]^{\mathrm{T}} \;\; [x^{k-}]^{\mathrm{T}} \,],$$

$$B = [A \quad -A], \quad D = [I \quad I]$$

and $b = (b_1, ..., b_{\bar{J}})^{\mathrm{T}}$.

If we denote by $\bar{I}$ the number of nodes of the network, the problem to be solved has $M = \bar{K}\bar{I} + \bar{J}$ constraints and $(2\bar{K}+1)\bar{J}$ variables in its standard form. Contrary to many transportation applications where the number of nodes and edges may be large but the number of commodities is quite small (for example the commodities may be different manufactured goods), applications in network planning are characterized by a rather small number of nodes and edges and a large number of commodities (often one commodity for each pair of nodes in the network). Note, however, that in the case of particular instances in which the routing costs do not depend on the commodities (i.e., $c_j^k = c_j^1$ for any k) the problem can be reformulated with a drastic reduction in the number of commodities (see Chardaire and Lisser, 1996, for more details).

3. Price Decomposition Methods

If the coupling (also called bundle) constraints (2) of NMP were ignored, the problem would decompose in one shortest-path problem for each commodity. Assume then, that these shortest-path problems are solved. If none of the bundle constraints is violated by the solution obtained we are done. Otherwise, we may modify the routing costs by introducing prices for the use of the links corresponding to the violated bundle constraints and we may solve new shortest-path problems associated with the modified routing costs. This suggests an iterative process that alternates between price computation and shortest-path computation. The shortest-path information found at some iteration may be used alone or combined with shortest-path information found at previous iterations to determine new prices (possibly from previous prices) and to try to find feasible solutions to NMP. We present three solution methods that formalize the idea with varying degrees of sophistication.

3.1. Lagrangian Relaxation

Lagrangian relaxation is an attractive technique for finding tight lower bounds on the optimal value of large linear programs in a fraction of the computing time required for solving those programs. The method places prices (Lagrangian multipliers) on the coupling constraints and includes them in the objective to obtain a problem that decomposes into separate minimum-cost flow problems for each commodity. Appropriate prices may be found by using subgradient techniques (see Minoux, 1986, for more details). In our case, a vector of Lagrangian multipliers, $\lambda \in \mathbb{R}^{\bar{J}+}$, is associated with constraint (2) of NMP to create the following Lagrangian sub-
problem:

$$\begin{cases} L(\lambda) = \min \sum_k [C^k]^{\mathrm{T}} x^k + \lambda^{\mathrm{T}} \left(\sum_k D x^k - b \right) \\ \\ \text{subject to} \quad B x^k = r^k, \quad k = 1, ..., \bar{K}, \\ \qquad\qquad\quad x^k \geqslant 0, \quad k = 1, ..., \bar{K}. \end{cases}$$

For any fixed value of the multipliers, the term $\lambda^{\mathrm{T}} b$ is a constant that can be ignored. The Lagrangian subproblem reduces to $\bar{K}$ subproblems of the form

$$SP_k(\lambda) : \pi_k(\lambda) = \min\{([C^k]^{\mathrm{T}} + \lambda^{\mathrm{T}} D) x^k : x^k \in P_k\} \text{ where } P_k = \{x^k : Bx^k = r^k, x^k \geqslant 0\}.$$

$SP_k(\lambda)$ may be solved by using a shortest-path algorithm between node s_k and t_k in a directed graph where the links j and $j + \bar{J}$ have length $c_j + \lambda_j$. However, as the optimal solution to $SP_k(\lambda)$ is such that the support of the flow does not contain any circuit, we can simply solve a shortest-path problem between s_k and t_k in an undirected graph. The resulting shortest-path corresponds to a directed-path in the graph associated with problem $SP_k(\lambda)$. The corresponding solution, x^k, obtained by routing the overall traffic requirement, d^k, on that path is an optimal solution to $SP_k(\lambda)$ and will be called an optimal shortest-path solution. The dual function $L(\lambda)$ is a lower bound on the optimal value of NMP for any choice of the vector λ. This function is concave and continuous but not differentiable. To maximize it one may rely on subgradient optimization procedures (see Minoux, 1986, or Ahuja et al., 1992, for example). Such procedures are easy to implement and quite robust. They usually converge provided that sufficiently small step sizes are taken in the subgradient update formula. However, even if the Lagrangian multiplier vector obtained at the end of the procedure is optimal (i.e., equal to a vector, λ^*, of optimal values of the dual variables associated with the coupling constraints of NMP) the solution, x, obtained when solving the $\bar{K}$ shortest-path, problems is not feasible for NMP. It is indeed clear that for nontrivial problems some of the commodities should be routed on multiple paths. For a given commodity, k, the optimal solution, x^k, to NMP corresponds to an optimal solution to $SP_k(\lambda^*)$ obtained as some convex combination of optimal shortest-path solutions to $SP_k(\lambda^*)$. The results of the Lagrangian relaxation may be used to try to determine an initial starting basis for a solution approach based on the simplex algorithm. They may also be used to fix some variables to zero.

The optimal value of subproblem $SP_k(\lambda)$ is reached at some extreme point of the polyhedron P_k. (We do not need to worry about extreme rays as the objective coefficients in SP_k are nonnegative.) In addition, as B is a network matrix each basic solution corresponds to a spanning tree (see Section 5.1) and each spanning tree corresponds to a shortest-path solution that defines an extreme point of the polyhedron. Note that each extreme point corresponds to several bases (as many as there are ways of supplementing a path connecting s_k and t_k with extra arcs to form a spanning tree). As all extreme points of P_k are vectors with components in $\{0, d^k\}$, we will consider the normalized extreme points, $y_1^k, \ldots, y_{n_k}^k$, of P_k corresponding to the characteristic vectors of the shortest-paths that support the shortest-path solutions.

The Lagrangian dual problem, $\max\{L(\lambda) : \lambda \geqslant 0\}$, can be restated in terms of these extreme points as the following linear program

$$\text{DLP}\begin{cases} \max \sum_k d_k \pi_k - \lambda^{\mathrm{T}} b \\ \text{subject to} \quad \pi_k - \lambda^{\mathrm{T}} D y_l^k \leqslant [C^k]^{\mathrm{T}} y_l^k \quad k = 1, \ldots, \bar{K}, \quad l = 1, \ldots, n_k, \\ \lambda \geqslant 0. \end{cases}$$

The normalization results in a problem where all constraints coefficients are in $\{-1, 0, 1\}$. The point of interest of the formulation DLP is that the set of constraints saturated at optimality gives us all the extreme point solutions to the subproblems $SP_k(\lambda^*)$ but more importantly that the corresponding values of the dual variables give us information to build an optimal solution to NMP (see Section 3.2). Problem DLP contains an exponential number of constraints. Fortunately, there is no need to make explicit all the constraints as we can use constraint generation techniques to solve DLP or column generation techniques to solve its dual.

3.2. Dantzig–Wolfe Decomposition

The Dantzig–Wolfe decomposition method (Dantzig and Wolfe, 1961) was developed for block-angular systems of which the node-arc formulation of multicommodity flow problems is an important subset. Associating a dual variable α_l^k with each constraint (k, l) of DLP, we obtain the dual problem

$$\min \sum_{k,l} [C^k]^{\mathrm{T}} y_l^k \alpha_l^k$$

$$\text{subject to} \sum_l \alpha_l^k = d^k, \quad k = 1, \ldots, \bar{K}, \tag{3}$$

$$\sum_{k,l} \alpha_l^k D y_l^k \leqslant b, \tag{4}$$

$$\alpha_l^k \geqslant 0, \quad l = 1, \ldots, n_k, k = 1, \ldots, \bar{K}.$$

This formulation could have been derived from NMP directly by using the property that x^k must be a convex combination of extreme points of P_k and by replacing x^k with the resulting expression in the coupling constraints of NMP. Except for a renormalization, the constraints (3) in the formulation above state that x^k is a convex combination of extreme points of P_k for all k, whereas the constraints (4) are the transformed coupling constraints obtained when the convex combinations have been substituted for the original variables.

Introducing vectors v_l^k defined as

$$v_{l_j}^k = y_{l_j}^k + y_{l_{j+J}}^k, \qquad j = 1, \ldots, \bar{J}, \tag{5}$$

we obtain

$$\text{NMP}\begin{cases} \min \sum_{k,l} [c^k]^{\mathrm{T}} v_l^k \alpha_l^k \\ \text{subject to } \sum_l \alpha_l^k = d^k, \quad k = 1, \ldots, \bar{K}, & (6) \\ \sum_{k,l} \alpha_l^k v_l^k \leqslant b, & (7) \\ \alpha^k \geqslant 0, \quad k = 1, \ldots, \bar{K}. \end{cases}$$

v_l^k is the characteristic vector of a path, p_l^k, that connects s_k to t_k in the undirected network graph, that is, the component j of v_l^k is 1 if link j belongs to p_l^k and is 0 otherwise. $[c^k]^{\mathrm{T}} v_l^k$ is the cost of p_l^k per unit of flow routed on that path (i.e., the sum of the costs c_j^k of its links), and α_l^k is the amount of traffic routed on path p_l^k. Constraints (6) express that d_k units of traffic have to be routed between s_k and t_k whereas constraints (7) express the limitation on the capacity available on the links to route the traffic. The formulation above is known as the node-path formulation of the multicommodity flow problem.

NMP has an exponential number of columns but we will not have to express all the columns of this linear program. Indeed, the reduced cost of variable α_l^k is $w_l^k = [c^k + \lambda]^{\mathrm{T}} v_l^k - \pi_k$, where $-\lambda$ is the vector of simplex multipliers associated with the capacity constraints (7) and π_k is the vector of simplex multipliers associated with the convexity constraint corresponding to commodity k. Therefore $\min_l w_l^k$ can be obtained by computing a shortest-path between s_k and t_k in the undirected network graph where the edge lengths are given by the vector $c^k + \lambda$.

In the decomposition algorithm the formulation of NMP given above is called the (full) master program. The $\bar{K}$ shortest-path problems that are solved to generate columns of the master program are called the satellite problems. At each stage of the algorithm a reduced master program, that is, a program that contains a subset of the columns of the master program, is solved. The first reduced master program is initialized with a set of columns corresponding to a basic matrix and the columns corresponding to the slack variables. If no such initial set of columns is available, artificial variables are added to the formulation and a first phase or a big-M method have to be used to drive the program to a feasible basis. After solving a reduced master program to optimality, the $\bar{K}$ satellite problems are solved to check whether the solution obtained is optimal for the full master program. If the reduced costs corresponding to the solution of the satellite problems are nonnegative we are done, otherwise the master program is increased with the new columns candidate for basis entry and the simplex algorithm goes on. After the solution of a reduced master program the reduced cost of all variables in that program are nonnegative. Hence, as λ is the reduced cost vector associated with the slack variables of the coupling constraints we deduce that $\lambda \geqslant 0$ and that in turn the edge lengths in the satellite problems are nonnegative. The satellite problems can therefore be solved using Dijkstra's algorithm (see Johnson, 1972, and Williams, 1964, for details of efficient implementations).

It is easily seen that after solving the satellite problems, the necessary information is available to compute the value of the Lagrangian dual function $L(\lambda)$ at little cost. $L(\lambda)$ is a lower bound on the optimal value of NMP that may be used to speed up termination of the algorithm when only a good approximate solution is required. Some variations of the algorithm may be applied. For example, some columns generated may be discarded from the reduced master program at later stages, the generation of new columns may be carried out before optimality is reached provided that the simplex multipliers λ are nonnegative.

3.3. *Analytic Center Cutting-Plane Method*

Several cutting-plane algorithms have been designed to solve the convex programming problem, $P : \min(f(y) : y \in Y)$, where $Y \subseteq \mathbb{R}^n$ is a convex set and $f : \mathbb{R}^n \mapsto \mathbb{R}$ is a convex function. These algorithms solve P by generating a series of linear approximations to P and by exploiting those linear approximations to find improving feasible solutions as well as improving bounds on the optimal value of P. The Dantzig–Wolfe decomposition algorithm described above may be viewed as a specialization of Kelley's cutting plane algorithm to the convex programming problem $\min(f(\lambda) : \lambda \geqslant 0)$ where $f = -L$. The function $f(\lambda)$ can be written as $f(\lambda) = \sum_k f_k(\lambda) + \lambda^{\mathrm{T}} b$ where the functions f_k are convex functions. f_k is of the form

$$f_k(\lambda) = \max\{-([C^k]^{\mathrm{T}} + \lambda^{\mathrm{T}} D) d_k x^k : x^k \in \mathrm{ext}(Q_k)\}$$

where $\mathrm{ext}(Q_k)$ is the set of extreme points of $Q_k = \{x^k : Bx^k = r^k/d_k, x^k \geqslant 0\}$. Problem P can also be reformulated as

$$P : \min(z : (z, \lambda) \in \mathrm{epi}\, f) \tag{8}$$

where $\mathrm{epi}\, f = \{(z, \lambda) : \lambda \geqslant 0, z \geqslant f(\lambda)\}$ is the epigraph of f. The epigraph of f can also be expressed as

$$\mathrm{epi}\, f = \left\{(z, \lambda) : z = \sum_{k=1}^{\bar{K}} z_k + \lambda^{\mathrm{T}} b, \quad (z_k, \lambda) \in \mathrm{epi}\, f_k\right\}.$$

As the functions f_k are convex their supporting hyperplanes are below their epigraph. Mathematically, for any fixed nonnegative value $\hat{\lambda}$ of λ and for all $\lambda \geqslant 0$

$$f_k(\lambda) \geqslant f_k(\hat{\lambda}) + (\lambda - \hat{\lambda})^{\mathrm{T}} \xi_k,$$

where ξ_k is any element of the subdifferential, $\partial f_k(\hat{\lambda})$, of f_k at $\hat{\lambda}$. Now, it is clear that

$$\begin{aligned} f_k(\hat{\lambda}) &= \max\{-([C^k]^{\mathrm{T}} + \hat{\lambda}^{\mathrm{T}} D) d_k x^k : x^k \in \mathrm{ext}(Q_k)\} \\ &= -([C^k]^{\mathrm{T}} + \lambda^{\mathrm{T}} D) d_k y_l^k + (\lambda - \hat{\lambda})^{\mathrm{T}} D d_k y_l^k \leqslant f_k(\lambda) + (\lambda - \hat{\lambda})^{\mathrm{T}} D d_k y_l^k, \end{aligned}$$

where $y_l^k = \mathrm{argmax}\{-([C^k]^{\mathrm{T}} + \hat{\lambda}^{\mathrm{T}} D) d_k x^k : x^k \in \mathrm{ext}(Q_k)\}$. Hence, $-Dd_k y_l^k$ is a subgradient of f_k at $\hat{\lambda}$ and the constraint

$$z_k \geqslant f_k(\hat{\lambda}) - (\lambda - \hat{\lambda})^{\mathrm{T}} D d_k y_l^k \tag{9}$$

is satisfied by all points (z_k, λ) of $\mathrm{epi}\, f_k$. As $f_k(\hat{\lambda}) = -([C^k]^{\mathrm{T}} + \hat{\lambda}^{\mathrm{T}} D) d_k y_l^k$, constraint (9) can be rewritten as

$$z_k + \lambda^{\mathrm{T}} D d_k y_l^k \geqslant -[C^k]^{\mathrm{T}} d_k y_l^k. \tag{10}$$

Assume that a number of constraints of that form have been generated. We obtain the following linear approximation to P

$$\mathrm{LAP}\left\{ \quad \min \sum_{k=1}^{\bar{K}} d_k z_k + \lambda^{\mathrm{T}} b \text{subject to} z_k + \lambda^T D y_l^k \geqslant -[C^k]^T y_l^k \quad y_l^k \in S_k \subseteq \mathrm{Ext}(Q_k), k = 1, \ldots, \bar{K}, \right. \quad \lambda \geqslant$$

where the sets S_k specify the subsets of constraints generated. Observe that LAP is equivalent to DLP when $S_k = \mathrm{Ext}(Q_k)$ for all k. Kelley's algorithm solves LAP to optimality. The optimal value of LAP is a lower bound on the optimal value of P. Let $x=(\hat{z}_1,\ldots,\hat{z}_k,\hat{\lambda})$ be the optimal solution to LAP. Evaluating f at $\hat{\lambda}$ we obtain an upper bound on the optimal value of P and we generate new constraints violated at x by using the process described above. It is easily seen that this specialization of Kelley's algorithm is simply the dual presentation of the Dantzig–Wolfe algorithm which in turn may be viewed as an application of Benders decomposition technique (see Bazaraa and Jarvis, 1977, for example). In general the stopping criterion of Kelley's algorithm is based on the value of the relative gap between the upper and lower bound (the upper bound corresponds to the value of $L(\lambda)$ that may be used in the Dantzig–Wolfe decomposition algorithm to speed up termination). As we are in the particular situation where epi f is a polyhedron, an alternative stopping criterion is used. The algorithm is stopped when no violated constraints are generated (which corresponds to the termination criterion in the Dantzig–Wolfe decomposition algorithm presented above). The optimal solution to the dual of the last linear approximation solved by the cutting-plane algorithm is an optimal solution to NMP.

Other cutting-plane algorithms differ from Kelley's in the way the linear approximation is refined. They include Bundle methods (Lemaréchal, 1974, 1975; Wolfe, 1975) and central cutting-plane algorithms (Elzinga and Moore, 1973; Levin, 1965; Vaidya, 1996; Goffin et al., 1992). We will focus on the analytic center cutting-plane method (ACCPM). This method is based on the concept of an analytic center. Instead of solving LAP to optimality, a so-called localization set is built from the approximation of epi f defined by the constraints of LAP. Let z_u and z_l be respective upper and lower bounds to the optimal value, z^*, of problem (8). Adding so-called box constraints, $0\leqslant\lambda\leqslant h$, we can construct a compact set containing the point, (z^*, λ^*), associated with the optimal solution to problem (8). This set, known as the localization set and denoted by $\mathscr{L}(z_u)$, is defined by the inequalities

$$\sum_{k=1}^{\bar{K}} d_k z_k + \lambda^{\mathrm{T}} b \leqslant z_u$$

$$z_k + \lambda^{\mathrm{T}} D y_l^k \geqslant -[C^k]^{\mathrm{T}} y_l^k, \quad y_l^k \in S_k \subseteq \mathrm{Ext}(Q_k), k=1,\ldots,\bar{K},$$

$$0\leqslant\lambda\leqslant h.$$

Introducing vectors v_k^l, as in Section 3.2 (see equation 5), $\mathscr{L}(z_u)$ can be expressed in the form

$$\sum_{k=1}^{\bar{K}} d_k z_k + \lambda^{\mathrm{T}} b \leqslant z_u,$$

$$z_k + \lambda^{\mathrm{T}} v_l^k \geqslant -[c^k]^{\mathrm{T}} v_l^k, \quad v_l^k \in L_k, \quad k=1,\ldots,\bar{K},$$

$$0\leqslant\lambda\leqslant h,$$

where L_k are induced from S_k. Let $|L_k|=l_k$ and let the $\bar{J}\times l_k$ matrix $G_k=[v_1^k \ldots v_{l_k}^k]$ for $k=1,\ldots,\bar{K}$. Moreover, let $\hat{c}_k=(-[c^k]^{\mathrm{T}}v_1^k \ldots -[c^k]^{\mathrm{T}}v_{l_k}^k)$ for $k=1,\ldots,\bar{K}$. The localization set is $\mathscr{L}(z_u)=\{\hat{u}:\hat{A}^{\mathrm{T}}\hat{u}\geqslant\hat{c}, \hat{b}^{\mathrm{T}}\hat{u}\leqslant z_u\}$, where

$$\hat{A}=\begin{bmatrix} e^{\mathrm{T}} & & & & & \\ & e^{\mathrm{T}} & & & & \\ & & \ddots & & & \\ & & & e^{\mathrm{T}} & & \\ G_1 & G_2 & \ldots & G_{\bar{K}} & I & -I \end{bmatrix}, \tag{11}$$

$$\hat{c}=(\hat{c}_k,\ldots,\hat{c}_{\bar{K}},0,-h),$$

$$\hat{b} = (d_1, ..., d_k, b)$$

and the sizes of vectors $e^{\mathrm{T}} = (1, ..., 1)$ are defined accordingly.

Introducing slack variables s_0 and $s = (s_1, ..., s_n)$, where n is the dimension of $\hat{c}$, the localization set can be expressed as $\mathscr{L}(z_u) = \{u : \hat{A}^{\mathrm{T}}u - s = \hat{c}, \hat{b}^{\mathrm{T}}u + s_0 = z_u, s \geqslant 0, s_0 \geqslant 0\}$. The analytic center of $\mathscr{L}(z_u)$ is the only solution u that maximizes the product $\prod_{i=0}^{n} s_i$ over the relative interior of $\mathscr{L}(z_u)$. The analytic center could be found by solving the minimization problem

$$\min \quad \psi(s, u) = -\ln(z_u - \hat{b}^{\mathrm{T}}u) - \sum_{j=1}^{n} \ln s_j \text{subject to} \hat{A}^{\mathrm{T}}u - s = \hat{c}, s > 0, \quad z_u - \hat{b}^{\mathrm{T}}u > 0. \tag{12}$$

However, consider the following problem

$$\min_{x} \quad \varphi(x; z_u) = (n+1)\ln(z_u - \hat{c}^{\mathrm{T}}x) - \sum_{j=1}^{n} \ln x_j \text{subject to} \hat{A}x = \hat{b}, x > 0, \quad z_u - \hat{c}^{\mathrm{T}}x > 0,. \tag{13}$$

where $\varphi(x;\, z_u)$ is Karmarkar's potential function. It can be shown (Goffin et al., 1992) that problems (12) and (13) have the same optimality conditions.

First, ACCPM solves problem (13) using the version of the projective algorithm of Karmarkar (1984) described in De Ghellinck and Vial (1986). As a by-product an optimal solution, $(s^*,\, u^*)$ to problem (12) is obtained where $u^* = (z_1^c, ..., z_{\bar{K}}, \lambda^c)$ is the analytic center of $\mathscr{L}(z_u)$ (in practice, (12) is solved to near optimality and the solution $(s^*,\, u^*)$ gives an approximate analytic center). Let x^* denote the optimal solution obtained for problem (13). $z = c^{\mathrm{T}}x^*$ provides a new lower bound for problem (8) and z_l is updated by $z_l \leftarrow \max\{z_l, z\}$.

Secondly, new hyperplanes supporting the epigraph of f are determined by solving

$$f_k(\lambda^c) = \max\{-([C^k]^{\mathrm{T}} + [\lambda^c]^{\mathrm{T}}D)d_k x^k : x^k \in \mathrm{ext}(Q_k)\}$$

for all $k \in \{1, ..., \bar{K}\}$. This can be done by using a shortest-path algorithm. $f(\lambda^c) = \sum_{k=1}^{\bar{K}} f_k(\lambda^c) + [\lambda^c]^{\mathrm{T}}b$ provides an upper bound for problem (8) and $z_u \leftarrow \min\{f(\lambda^c), z_u\}$.

The process is repeated until $(z_u - z_l)/z_l \leqslant \epsilon$ where ϵ is a tolerance value.

4. Resource-Directive Decomposition Methods

Resource decomposition methods have been studied in Sakarovitch (1967), Lasdon (1970), Geoffrion (1970), Grinold (1972), and Kennington and Shalaby (1977). Instead of using prices to decompose the problem, the resource-directive decomposition (right-hand-side allocation) allocates the joint capacity of each link to the commodities. Let us denote by y^k the vector of capacities allocated to commodity k. NMP can be reformulated as

$$\text{NMP}\begin{cases} \min \sum_k [C^k]^{\mathrm{T}} x^k & \\ \text{subject to} \quad Bx^k = r^k, & k = 1, ..., \bar{K}, \\ \qquad Dx^k \leqslant y^k, & k = 1, ..., \bar{K}, \\ \qquad \sum_k y_j^k \leqslant b_j, & j = 1, ..., \bar{J}, \\ \qquad x^k \geqslant 0, & k = 1, ..., \bar{K}, \\ \qquad y^k \geqslant 0 & k = 1, ..., \bar{K}. \end{cases}$$

For any fixed value of the vectors, $y^k, k = 1, ..., \bar{K}$, the problem reduces to $\bar{K}$ subproblems of the form

$$SP_k(y^k)\begin{cases} z^k(y^k) = \min[C^k]^{\mathrm{T}} x^k \\ \text{subject to} \quad Bx^k = r^k, & (14) \\ x_j^k \leqslant y_j^k, \quad j = 1, \ldots, \bar{J} & (15) \\ x_{j+\bar{J}}^k \leqslant y_j^k, \quad j = 1, \ldots, \bar{J} & (16) \end{cases}$$

Note that each constraint, $x_j + x_{j+\bar{J}} \leqslant y_j^k$, of the system, $Dx^k \leqslant y^k$, has been replaced with the two constraints, $x_j \leqslant y_j^k$ and $x_{j+\bar{J}} \leqslant y_j^k$, since the support of the flow at the optimum of subproblem k does not contain any directed cycle. Hence, $SP_k(y^k)$ reduces to a standard minimum-cost flow problem. It is easy to show that NMP is equivalent to the resource-allocation problem

$$\text{RAP}\begin{cases} z(y) = \min \sum_k z_k(y^k) \\ \text{subject to} \quad \sum_k y_j^k \leqslant b_j, \quad j = 1, \ldots, \bar{J}, \\ y^k \geqslant 0 \quad k = 1, \ldots, \bar{K}. \end{cases}$$

We assume that the subproblems, $SP_k(y^k)$, are always feasible as artificial arcs may be added and therefore that the functions, $z_k(y^k)$, are defined for all allocations, y^k. The function, $z(y)$, is a nondifferentiable piecewise linear convex function of y, which can be minimized using subgradient techniques.

First we need to determine a subgradient of $z(y)$ at any given point, y. For a given value of allocation of capacity, y^k, denote by ρ^k the reduced-cost vector associated with an optimal solution, x^{k^*}, to $SP_k(y^k)$ and by π^k the corresponding vector of simplex multipliers. Let θ^k be the vector of dimension $\bar{J}$ defined by

$$\theta_j^k = 0 \qquad \text{if } x_j^{k^*} < y_j^k \text{ and } x_{j+\bar{J}}^{k^*} < y_j^k,$$

$$\theta_j^k = \rho_j^k \qquad \text{if } x_j^{k^*} = y_j^k,$$

$$\theta_j^k = \rho_{j+\bar{J}}^k \quad \text{otherwise.}$$

For any feasible solution, x^k, to the flow constraints of $SP_k(y^k)$, $[C^k]^{\mathrm{T}} x^k = \rho^k x^k + \pi^k r^k$. Therefore,

$$z_k(y^k) = \rho^k x^{k^*} + \pi^k r^k = \theta^k y^k + \pi^k r^k. \tag{17}$$

The last equality comes from the first of the following optimality conditions (complementary slackness conditions):

$$\forall j \in \{1, \ldots, 2\bar{J}\} \quad (0 < x_j^{k^*} < u_j \Rightarrow \rho_j^k = 0),$$

$$\forall j \in \{1, \ldots, 2\bar{J}\} \quad (x_j^{k^*} = 0 \Rightarrow \rho_j^k \geqslant 0),$$

$$\forall j \in \{1, \ldots, 2\bar{J}\} \quad (x_j^{k^*} = u_j \Rightarrow \rho_j^k \leqslant 0),$$

where u_j is either y_j or $y_{j-\bar{J}}$ depending on whether $j \leqslant \bar{J}$ or $j > \bar{J}$ and from the condition

$$\forall j \in \{1, \ldots, \bar{J}\} \quad x_j^{k^*} x_{j+\bar{J}}^{k^*} = 0$$

(the support of the optimal flow does not contain directed cycles). For any nonnegative vector, w, of dimension $\bar{J}$, the optimal solution, s, to $SP_k(w)$ satisfies

$$z_k(w) = \rho^k s^{k^*} + \pi^k r^k \geqslant \theta^k w + \pi^k r^k. \tag{18}$$

Indeed, as $s_j^{k^*} s_{j+\bar{J}}^{k^*} = 0$ for all j we will assume with no loss of generality that $0 \leqslant s_j^{k^*} \leqslant w_j$ and $s_{j+\bar{J}}^{k^*} = 0$. If $\rho_j^k > 0$, we deduce from $\theta_j^k \leqslant 0$ and $s_j^{k^*} \geqslant 0$ that $\rho_j^k s_j^{k^*} \geqslant \theta_j^k w_j$. If $\rho_j^k \leqslant 0$ then $\theta_j^k \leqslant \rho_j^k \leqslant 0$ and $0 \leqslant s_j^{k^*} \leqslant w_j$, implying that $\rho_j^k s_j^{k^*} \geqslant \theta_j^k w_j$.

By combining (17) and (18) we deduce that

$$z_k(w) \geqslant z_k(y^k) + \theta^k (w - y^k).$$

Hence, θ^k is a subgradient of z_k at point y^k. The vector $\theta = (\theta^1, \ldots, \theta^{\bar{K}})$ is therefore a subgradient of z at $y = (y^1, \ldots, y^{\bar{K}})$.

At each step of the subgradient procedure a new allocation is determined by a two-stage process that comprises a move in the direction of the subgradient and a projection onto the set of feasible allocations. First, the vector $\hat{y}$ is computed by a formula of the form $\hat{y} \leftarrow y - t\theta$. Then, $\hat{y}$ is projected onto the set of feasible allocations to obtain an updated vector, y. y is obtained by solving

$$y = \operatorname{argmin}\{\|z_j - \hat{y}_j\|_2 : \sum_k z_j^k \leqslant b_j, z_j \geqslant 0\}$$

for each $j \in \bar{J}$ using a simple and efficient algorithm that can be derived from the KKT optimality conditions (see Held et al., 1974, for details).

5. Partitioning Techniques for the Solution of Linear Systems

An alternative to the decomposition techniques presented above are lower levels of decomposition of the problem, which are used in direct applications of linear programming solution methods to the initial node-arc formulation of NMP.

In the simplex algorithm this amounts to exploiting the structure of the basic matrices. Updating techniques, which are used to implement this exploitation efficiently, are quite involved and are therefore not described in this section. The reader interested in such aspects is referred to Chardaire and Lisser (1996). In the dual affine scaling algorithm the main specialization is in the exploitation of the sparse block structure of the matrix of the linear system to solve the computation of a movement direction. In the next two sections we present such specializations. Another specialization of interior-point techniques to the solution of multicommodity flow problems is presented in Castro (1999). Kamath and Palmon (1995) also propose a fast approximation algorithm based on interior-point techniques. A still better deterministic approximation algorithm is, however, proposed in Kamath et al. (1995).

5.1. *Primal Simplex Specialization*

It is well known that the simplex method can be specialized very efficiently to network flow problems. This specialization, known as the network simplex algorithm, can be transposed to some extent to multicommodity network flow problems (see Kennington and Helgason, 1980, and Chardaire and Lisser, 1996).

Let A be the node-arc incidence matrix of a connected graph with $\bar{I}$ nodes and consider the matrix $[A \mid e^{\bar{I}}]$ where the unit vector $e^{\bar{I}}$ is introduced to avoid row-rank deficiency. The graph induced by the columns of a basis is a spanning tree. Conversely, any spanning tree corresponds to a basis. It follows that the basic matrices are triangular and that solutions to linear systems involving those matrices can be obtained by forward or backward substitution. In fact, as the basic matrices are node-arc incidence matrices of trees and as the right-hand side of one of the two systems to solve at each iteration of the simplex algorithm is a node-arc incidence vector, these substitutions can be implemented very efficiently using a graph algorithm without need for reference to a matrix representation.

In order to exploit such properties the method used for multicommodity network flow problems requires the maintenance of a partitioning of the basic matrix. Consider the matrix, E, of the form,

$$E = \begin{bmatrix} L_1 & R_1 & 0 \\ L_2 & R_2 & 0 \\ L_3 & R_3 & I \end{bmatrix}, \qquad (19)$$

where L_1, R_2 and I are square matrices. It can be shown (Kennington and Helgason, 1980) that if $\det(L_1) \neq 0$ and $\det(E) \neq 0$, the matrix, $Q = R_2 - L_2 L_1^{-1} R_1$, is invertible. Therefore, we have the following propositions, which can easily be proved by block multiplications.

Proposition 1. *The solution to* $\mathrm{E}\hat{\mathrm{x}} = \hat{\mathrm{u}}$, *where* $\hat{\mathrm{x}} = (\mathrm{x}, \mathrm{y}, \mathrm{z})$ *and* $\hat{\mathrm{u}} = (\mathrm{u}, \mathrm{v}, \mathrm{w})$ *are vectors partitioned to be compatible with the partitioning of* E, *is given by*

$$\begin{aligned} y &= Q^{-1}(v - L_2 L_1^{-1} u), \\ x &= L_1^{-1} u - L_1^{-1} R_1 y, \\ z &= w - L_3 L_1^{-1} u - (R_3 - L_3 L_1^{-1} R_1) y. \end{aligned}$$

Proposition 2. *The solution to* $\mathrm{E}^{\mathrm{T}}\hat{\mathrm{x}} = \hat{\mathrm{u}}$, *where* $\hat{\mathrm{x}} = (\mathrm{x}, \mathrm{y}, \mathrm{z})$ *and* $\hat{\mathrm{u}} = (\mathrm{u}, \mathrm{r}, 0)$ *are vectors partitioned to be compatible with the partitioning of* E^{T}, *is given by*

$$\begin{aligned} z &= 0, \\ y &= [Q^{-1}]^{\mathrm{T}}(r - (L_1^{-1} R_1)^{\mathrm{T}} u), \\ x &= [L_1^{-1}]^{\mathrm{T}}(u - L_2^{\mathrm{T}} y). \end{aligned}$$

The constraint matrix associated with the standard form of NMP may be displayed as follows:

$$\hat{A} = \begin{bmatrix} [A \mid -A \mid e^{\bar{I}}] & & & & \\ & [A \mid -A \mid e^{\bar{I}}] & & & \\ & & \ddots & & \\ & & & [A \mid -A \mid e^{\bar{I}}] & \\ [I \mid I \mid 0] & [I \mid I \mid 0] & \cdots & [I \mid I \mid 0] & I \end{bmatrix},$$

where the vectors, $e^{\bar{I}}$, correspond to artificial variables that are introduced to avoid row deficiency and the last column partition corresponds to the slack variables associated with the bundle constraints.

It can easily be shown that every basis, $\hat{B}$, for $\hat{A}$ may be expressed in the form

$$\hat{B} = \begin{bmatrix} B^1 & & & R^1 & & & \\ & \ddots & & & \ddots & & \\ & & B^{\bar{K}} & & & R^{\bar{K}} & \\ P^1 & \cdots & P^{\bar{K}} & T^1 & \cdots & T^{\bar{K}} & \\ S^1 & \cdots & S^{\bar{K}} & U^1 & \cdots & U^{\bar{K}} & I \end{bmatrix},$$

where $B^1, \ldots, B^{\bar{K}}$ are basis matrices for $[A \mid -A \mid e^{\bar{I}}]$. Note that any of the partitions, $R^1, \ldots, R^{\bar{K}}$, may be empty. The structure of the basis, $\hat{B}$, can be used to efficiently calculate the simplex multipliers and a direction corresponding to an entering column. Indeed, $\hat{B}$ can be partitioned in the form of the matrix E given in equation (19) by putting

$$L_1 = \begin{bmatrix} B^1 & & \\ & \ddots & \\ & & B^{\bar{K}} \end{bmatrix}, \quad R_1 = \begin{bmatrix} R^1 & & \\ & \ddots & \\ & & R^{\bar{K}} \end{bmatrix},$$

$$L_2 = \begin{bmatrix} P^1 & \dots & P^{\bar{K}} \end{bmatrix}, \quad R_2 = \begin{bmatrix} T^1 & \dots & T^{\bar{K}} \end{bmatrix},$$

$$L_3 = \begin{bmatrix} S^1 & \dots & S^{\bar{K}} \end{bmatrix}, \quad R_3 = \begin{bmatrix} U^1 & \dots & U^{\bar{K}} \end{bmatrix}.$$

The most important feature of this decomposition is that the solution of linear systems involving the matrix L_1 (see Propositions 1 and 2) decomposes into the solution of linear systems involving the matrices $B^1, \dots, B^{\bar{K}}$. These systems can be solved by the graph techniques used in the network simplex algorithm.

Implementation details and updating techniques, in particular of the inverse matrix Q^{-1}, are detailed in Chardaire and Lisse (1996).

5.2. *Dual Affine Scaling Specialization Algorithm*

The dual affine scaling (DAS) algorithm, discussed in Jansen et al. (1993a,b) and Ling (1992), is used to solve the dual of linear programs of the form

$$\begin{aligned} &\text{minimize} \quad c^{\mathrm{T}}x, \\ &\text{subject to} \quad Ax = b, \quad x \geqslant 0, \end{aligned} \tag{20}$$

where $A \in \mathbb{R}^{m \times n}$, $b \in \mathbb{R}^m$, and $c, x \in \mathbb{R}^n$, and the matrix, A, has full row rank. The dual of problem (20) can be written as

$$\begin{aligned} &\text{maximize} \quad b^{\mathrm{T}}y, \\ &\text{subject to} \quad A^{\mathrm{T}}y + s = c, \quad s \geqslant 0. \end{aligned} \tag{21}$$

Problem (21) contains both unrestricted variables $y \in \mathbb{R}^m$ and nonnegative variables $s \in \mathbb{R}^n$. In this case, $(y; s)$ is defined to be an interior feasible solution if $A^{\mathrm{T}}y + s = c$ and $s > 0$. Given such a solution, $(\hat{y}; \hat{s})$, we want to determine a direction of improvement, $(d_{\hat{y}}; d_{\hat{s}})$. To preserve feasibility this direction must belong to the null space of the system of equality constraints. The two conditions of feasibility and improvement can be expressed as follows

$$A^{\mathrm{T}}d_{\hat{y}} + d_{\hat{s}} = 0, \tag{22}$$

$$b^{\mathrm{T}}d_{\hat{y}} > 0. \tag{23}$$

Now, an affine scaling transformation is applied to re-center the current slack vector at point $e = (1, \dots, 1)^{\mathrm{T}}$. The ball of radius 1 and center e is included in the nonnegative orthant of $\mathbb{R}^n$. Therefore, any movement within distance 1 will preserve the positivity constraints. The affine scaling operator is the inverse of the matrix, $S = \mathrm{diag}(\hat{s})$. This operator transforms the variables, s, into the variables, $u = S^{-1}s$. Any direction of improvement, d_u, in the transformed space corresponds to a direction of improvement,

$$d_{\hat{s}} = Sd_u, \tag{24}$$

in the original space. By exploiting equations (22), (23), and (24), one can show that the direction, $d_u = -B^{\mathrm{T}}b$ with $B = -(AS^2A^{\mathrm{T}})^{-1}AS^{-1}$, is a direction of improvement in the transformed space. This translates into the directions,

$$d_{\hat{y}} = (AS^{-1}A^{\mathrm{T}})^{-1}b, \tag{25}$$

$$d_{\hat{s}} = -A^{\mathrm{T}}d_{\hat{y}}, \tag{26}$$

in the original space. The algorithm starts with an initial interior-point solution, moves to a better interior solution, and iterates the process until an optimal dual solution is obtained. At each iteration of the algorithm a direction of improvement in the original space is obtained from a direction of improvement in a transformed space using formulas (25) and (26)

The DAS algorithm has been implemented by a number of researchers because of its good numerical behavior. An estimate of the primal solution can be obtained during the iterations. However, primal feasibility is usually only reached at the end of the algorithm (see Fang and Puthenpura, 1993, for a more detailed presentation). Implementation of interior-point methods are discussed in Adler et al. (1989a,b) and Resende and Veiga (1993).

The specialization of the DAS algorithm to NMP lies in the specialization of the solution of the linear system (25), which is the most time-consuming part of the DAS algorithm. The matrix of NMP may be displayed as follows:

$$\hat{A} = \begin{bmatrix} [A\vdots -A] & & & & \\ & [A\vdots -A] & & & \\ & & \ddots & & \\ & & & [A\vdots -A] & \\ [I\vdots I] & [I\vdots I] & \dots & [I\vdots I] & I \end{bmatrix},$$

where each block A corresponds to the submatrix obtained from the incidence matrix of the underlying graph after removing the last row of that matrix to avoid row rank deficiency. The identity matrices correspond to the bundle constraints. The search direction, $X = d_{\hat{y}}$, is obtained by solving the following linear system

$$\hat{A}\hat{D}^2\hat{A}^{\mathrm{T}}X = b, \tag{27}$$

where $\hat{D}$ is the diagonal matrix

$$\hat{D} = \begin{bmatrix} D_1 & & & & \\ & D_2 & & & \\ & & \ddots & & \\ & & & D_{2\bar{K}} & \\ & & & & D_{2\bar{K}+1} \end{bmatrix}$$

with diagonal block matrices, D_i, of size $\bar{J}$. Therefore, by block multiplication we obtain

$$\hat{A}\hat{D}^2\hat{A}^{\mathrm{T}} = \begin{bmatrix} A(D_1^2 + D_2^2)A^{\mathrm{T}} & & & A(D_1^2 - D_2^2) \\ & \ddots & & \vdots \\ & & A(D_{2\bar{K}-1}^2 + D_{2\bar{K}}^2)A^{\mathrm{T}} & A(D_{2\bar{K}-1}^2 - D_{2\bar{K}}^2) \\ (D_1^2 - D_2^2)A^{\mathrm{T}} & \cdots & (D_{2\bar{K}-1}^2 - D_{2\bar{K}}^2)A^{\mathrm{T}} & \sum_{i=1}^{2\bar{K}+1} D_i^2 \end{bmatrix}.$$

Let $F = \sum_{i=1}^{2\bar{K}+1} D_i^2$, $\bar{B}_j = D_{2j-1}^2 - D_{2j}^2$ and $B_j = D_{2j-1}^2 + D_{2j}^2$ for $1 \leqslant j \leqslant \bar{K}$. Then, $\hat{A}\hat{D}^2\hat{A}^T$ becomes

$$\hat{A}\hat{D}^2\hat{A}^T = \begin{bmatrix} AB_1A^T & & & A\bar{B}_1 \\ & \ddots & & \vdots \\ & & AB_{\bar{K}}A^T & A\bar{B}_{\bar{K}} \\ \bar{B}_1A^T & \cdots & \bar{B}_{\bar{K}}A^T & F \end{bmatrix}.$$

Let G be the $\bar{K}\bar{I} \times \bar{J}-$matrix such that $G^T = (\bar{B}_1A^T, \cdots, \bar{B}_{\bar{K}}A^T)$ and let N be the $\bar{K}\bar{I}$-diagonal matrix with diagonal blocks $N_i = AB_iA^T$, for $i = 1, \cdots, \bar{K}$. Then,

$$\hat{A}\hat{D}^2\hat{A}^T = \begin{bmatrix} N_1 & & & G_1 \\ & \ddots & & \vdots \\ & & N_{\bar{K}} & G_{\bar{K}} \\ G_1^T & \cdots & G_{\bar{K}}^T & F \end{bmatrix}.$$

The computation of the search direction can be performed by applying block Gaussian elimination to system (27) where the matrix is partitioned in the form given above. The last block of rows is transformed by subtracting from it each of the other row blocks, $i = 1, ..., \bar{K}$, after multiplication by $G_i^T \cdot N_i^{-1}$. The transformed matrix is

$$\begin{bmatrix} N_1 & & G_1 & \\ & \ddots & \vdots & \\ & N_{\bar{K}} & G_{\bar{K}} & G_i^T N_i^{-1} G_i \\ & & F - \sum_{i=1}^{\bar{K}} & \end{bmatrix}.$$

The right-hand side obtained by performing the same operations is

$$\begin{bmatrix} b_{11} & \\ \vdots & \\ b_{1\bar{K}} & G_i^T N_i^{-1} b_{1i} \\ b_2 - \sum_{i=1}^{\bar{K}} & \end{bmatrix}.$$

The linear system is reduced to a block triangular system, which may be solved by backward substitution. In doing so, we obtain

$$X_2 = \left(F - \sum_{i=1}^{\bar{K}} G_i^T N_i^{-1} G_i\right)^{-1} \left(b_2 - \sum_{i=1}^{\bar{K}} G_i^T N_i^{-1} b_{1i}\right) \tag{28}$$

$$X_{1i} = N_i^{-1}(b_{1i} - G_i X_2), \qquad i = 1, \cdots, \bar{K}, \tag{29}$$

where $(X_{1i}, ..., X_{1\bar{K}}, X2)$ is a partitioning of the solution vector compatible with the partitioning of the linear system.

Therefore, the solution of equation (27) reduces to the solution of $\bar{K}(1 + \bar{J})$ linear systems involving the $\bar{I} \times \bar{I}$ matrices, N_i, and of one linear system involving the dense $\bar{J} \times \bar{J}$ matrix, $F - \sum_{i=1}^{\bar{K}} G_i^T N_i^{-1} G_i$.

Chardaire and Lisser (1999) use a Choleski factorization to solve these systems, whereas Castro (1999) uses a mixed strategy where the larger system is solved using a preconditioned conjugate gradient method.

6. Conclusion

Minimum-cost multicommodity flow models arise whenever several commodities, each governed by its own flow requirements, share the same underlying network. As shown by the survey of the recent literature given in the introduction of this paper, such models are used in a wide variety of applications. This abundance of applications has motivated a lot of research in the design of specialized solution methods for multicommodity flow models. We have presented various solution methods that have been considered in the literature. They range from the well-established Dantzig–Wolfe decomposition technique of the sixties to the promising ACCPM technique of the nineties. Our presentation of the solution methods has been illustrated with a nonoriented multicommodity flow model rather than with the more familiar oriented model. By considering this model of significant interest to telecommunication network planning, we have illustrated the versatility of the specialization techniques. We recommend the paper from Chardaire and Lisser (1999) for further details and information on numerical experiments. We also recommend the following list of books on network programming and modeling for further reading: Ahuja et al. (1992), Bazaraa and Jarvis (1977), Glover et al. (1992), and Kennington and Helgason (1980).

References

Adler, I., N. Karmarkar, M. Resende, and G. Veiga. 1989a. "Data structures and programming techniques for the implementation of Karmarkar's algorithm." *ORSA J. Comput.* **1**: 84–106.

Adler, I., M. Resende, G. Veiga, and N. Karmarkar. 1989b. "An implementation of Karmarkar's algorithm for linear programming." *Math. Programming* **44**: 297–335.

Ahuja, R., T. Magnanti, and J. Orlin. 1992. In *Network Flows*. NJ: Prentice-Hall.

Arbib, C., U. Mocci, and C. Scoglio. 1992. "Methodological aspects of ring network design." *Lecture Notes Control Info. Sci.* **180**: 135–144.

Banerjee, D. and B. Mukherjee. 1996. "A practical approach for routing and wavelength assignment in large wavelength-routed optical networks." *IEEE J. Selected Areas Telecommun.* **14(5)**: 903–908.

Bard, J., A. Desilva, T. Feo, and S. Wert. 1993. "Design of semiautomated mail processing facilities." *IEEE Trans.* **25(4)**: 88–101.

Bazaraa, M. S. and J. J. Jarvis. 1977. In *Linear Programming and Network Flows*. New York: Wiley.

Berge, M. and C. Hopperstad. 1993. "Demand driven dispach—a method for dynamic aircraft capacity assignment, models and algorithms." *Oper. Res.* **41(1)**: 153–168.

Bogdanowicz, Z. 1993. "A new optimal algorithm for backbone topology design in communications networks." *Math. Comput. Modelling* **17(8)**: 49–61.

Carden, R., J. Li, and C. Cheng. 1996. "A global router with a theoretical bound on the optimal solution." *IEEE Trans. Comput. Aided Des. Integrated Circuits Syst.* **15(2)**: 208–216.

Castro, J. 2000. "A specialized interior point algorithm for multicommodity network flows." *SIAM J. Optim.* **10(3)**: 852–877.

Chang, C. 1994. "An efficient approach to real-time traffic routing for telephone network management." *J. Oper. Res. Soc.* **45(2)**: 187–201.

Chang, S. 1992. "Fair integration of routing and flow control in communication-networks." *IEEE Trans Commun.* **40(4)**: 821–834.

Chang, Y., J. Wu, and H. Ho. 1992. "Optimal virtual circuit routing in computer networks." *IEEE Proc.-I Commun. Speech Vision* **139(6)**: 625–632.

Chardaire, P. and A. Lisser. 1996. "A primal simplex specialization for solving non-oriented multicommodity flow problems." *Investigation Operativa* **5(2–3)**: 117–152.

Chardaire, P. and A. Lisser. 1999. "Simplex and interior point specialized algorithms for solving non-oriented multicommodity flow problems." Forthcoming paper in *Operations Res.*

Chopra, S., M. Rao, and C. Tsai. 1997. "Computational study of the multiechelon production planning problem." *Naval Research Logistics* **44(1)**: 1–19.

Dantzig, G. B. and P. Wolfe. 1961. "The decomposition algorithm for linear programming." *Econometrica* **29**: 767–778.

De Ghellinck, G. and J.-P. Vial. 1986. "A polynomial Newton method for linear programming." *Algorithmica* **1**: 425–453.

Desaulniers, G., J. Desrosiers, Y. Dumas, M. Solomon, and F. Soumis. 1997. "Daily aircraft routing and scheduling." *Mgmnt Sci.* **43(6)**: 841–855.

Elzinga, J. and T. Moore. 1973. "A central cutting plane algorithm for the convex programming problem." *Math. Programming* **8**: 134–145.

Fang, S. C. and S. Puthenpura. 1993. In *Linear Optimization and Extensions*. Englewood Cliffs, New Jersey: Prentice-Hall.

Fuller, S. and W. Grant. 1997. "Effect of lock delay on grain marketing costs—an examination of the Upper Mississippi and Illinois waterways." *Logistics Transport. Rev.* **29(1)**: 81–95.

Geoffrion, A. 1970. "Elements of large scale mathematical programming I and II." *Mgmnt Sci.* (16) 652–691.

Ghosh, S. and C. Gaimon. 1993. "Production scheduling in flexible manufacturing system with setups." *IIE Trans.* **25(5)**: 21–35.

Girard, A. and B. Sanso. 1998. "Multicommodity flow models, failure propagation, and reliable loss network design." *IEEE-ACM Trans. Networking* **6(1)**: 82–93.

Glover, F., D. Klingman, and N. V. Phillips. 1992. In *Network Models in Optimization and Their Applications in Practice*. New York: Wiley-Interscience.

Goffin, J.-L., A. Haurie, and J.-P. Vial. 1992. "Decomposition and nondifferentiable optimization with the projective algorithm." *Management Science* **37(12)**: 393–409.

Gouveia, L. 1996. "Multicommodity flow models for spanning trees with hop constraints." *Eur. J. Oper. Res.* **95(1)**: 178–190.

Gouveia, L. and E. Janssen. 1998. "Designing reliable tree networks with two cable technologies." *Eur. J. Oper. Res.* **105(3)**: 552–568.

Greenwald, L. and T. Dean. 1996. "Package routing in transportation networks with fixed vehicle schedules." *Networks* **27(1)**: 81–93.

Grinold, R. 1972. "Steepest ascent for large scale linear programs." *J. SIAM* **10(2)**: 260–283.

Gu, Z., E. Johnson, G. Nemhauser, and Y. Wang. 1994. "Some properties of the fleet assignment problem." *Oper. Res. Lett.* **15(2)**: 59–71.

Haghani, A. and S. Oh. 1996. "Formulation and solution of a multi-commodity, multi-modal network flow model for disaster relief operations." *Transport. Res. Part A* **30(3)**: 231–250.

Held, M., P. Wolfe, and H. Crowder. 1974. "Validation of subgradient optimization." *Math. Programming* **6**: 62–88.

Hindi, K. and T. Basta. 1994. "Computationally efficient solution of multiproduct, 2-stage distribution location problem." *J. Oper. Res. Soc.* **45(11)**: 1316–1323.

Holmberg, K., M. Joborn, and J. Lundgren. 1998. "Improved empty freight car distribution." *Transport. Sci.* **32(2)**: 163–173.

Hong, X., T. Xue, J. Huang, C. Cheng, and E. Kuh. 1997. "TIGER: an efficient timing-driven global router for gate array and standard cell layout design." *IEEE Trans. Comput.-Aided des. Integrated Circuits Syst.* **16(11)**: 1323–1331.

Iakovou, E., C. Douligeris, H. Li, C. Ip, and L. Yudhbir. 1999. "A maritime global route planning model for hazardous materials transportation." *Transport. Sci.* **33(1)**: 34–48.

Jansen, B., C. Roos, and T. Terlaky. 1993a. "A family of polynomial affine scaling algorithms for positive semi-definite linear complementarity problems." Technical Report 93-112. Delft, The Netherlands: Delft University of Technology, Faculty of Technical Mathematics and Informatics.

Jansen, B., C. Roos, and T. Terlaky. 1993b. "A polynomial primal–dual Dikin-type algorithm for linear programming." Technical Report 93-36. Delft, The Netherlands: Delft University of Technology, Faculty of Technical Mathematics and Informatics.

Johnson, E. 1972. "On shortest path and sorting." In *Proceedings of the Annual Conference of the ACM, Boston*, p. 510. New York: ACM Press.

Kamath, A. and O. Palmon. 1995. "Improved interior-point algorithms for exact and approximate solutions of multicommodity flow problems." In *Proceedings of the 16th ACM-SIAM Symposium on Discrete Algorithms*, pp. 147–159. New York: ACM Press.

Kamath, A., O. Palmon, and S. Plotkin. 1995. "Fast approximation algorithm for minimum cost multicommodity flow." Technical Report STAN-CS-TN-95-19. Stanford University, Stanford, England. Preliminary version in proceedings of SODA 95, pp. 493–501.

Karmarkar, N. 1984. "A new polynomial time algorithm for linear programming." *Combinatorica* **4**: 373–395.

Kennington, J. and M. Shalaby. 1977. "An effective subgradient procedure for minimal cost multicommodity flow problems." *Mgmnt Sci.* **23(9)**: 994–1004.

Kennington, J. L. and R. V. Helgason. 1980. In *Algorithms for Network Programming*. New York: Wiley-Interscience.

Kwon, O., C. Martland, and J. Sussman. 1998. "Routing and scheduling temporal and heterogeneous freight car traffic on rail networks." *Transport. Res. Part E* **34(2)**: 101–115.

Labourdette, J. and A. Acampora. 1991. "Logically rearrangable multihop lightwave networks." *IEEE Trans. Commun.* **39(8)**: 1223–1230.

Lasdon, L. 1970. In *Optimization theory for large systems*. London: Macmillan.

LeBlanc, L., J. Park, V. Sridhar, and J. Kalvenes. 1996. "Topology design and bridge-capacity assignment for interconnecting token ring lans: a simulated annealing approach." *Telecommun. Syst.* **6(1)**: 21–43.

Lemaréchal, C. 1974. "An algorithm for minimizing convex functions." In *Proceedings of the IFIP'74 Congress*, pp. 552–556. Amsterdam: North-Holland.

Lemaréchal, C. 1975. "An extension of davidson methods to nondifferentiable problems." *Math. Programming* **3**: 95–109.

Levin, A. 1965. "On an algorithm for the minimization of convex functions over convex sets." *Soviet Math. Doklady* **6**: 286–290.

Ling, P. D. 1992. "The theory of interior point algorithms for linear programming." Working Paper. Norwich, England: University of East Anglia.

Lisser, A., R. Sarkissian, and J.-P. Vial. 1995a. "Joint optimal synthesis of base and reserve telecommunication networks." Technical report NT/PAA/ATR/ORI/4491. Moulineaux, France: France Telecom/CNET.

Lisser, A., R. Sarkissian, and J.-P. Vial. 1995b. "Survivability in telecommunication networks." Technical Report NT/PAA/ATR/ORI/4230. Moulineaux, France: France Telecom/CNET.

Lobel, A. 1998. "Vehicle scheduling in public transit and Lagrangean pricing." *Mgmnt Sci.* **44(12:1)**: 1637–1649.

Medhi, D. and S. Guptan. 1997. "Network dimensioning and performance of multiservice, multirate loss networks using dynamic routing." *IEEE-ACM Trans. Networking* **5(6)**: 944–957.

Medova, E. 1998. "Chance constrained stochastic programming for integrated services network management." *Ann. Oper. Res.* **81**: 213–229.

Minoux, M. 1986. In *Mathematical Programming: Theory and Algorithms*. New York: Wiley-Interscience.

Moiseenko, G. 1990. "Vehicle flow optimization in regional transportation networks." *Automat. Remote Control* **51(9:2)**: 1277–1285.

Murakami, K. and H. Kim. 1996. "Virtual path routing for survivable ATM networks." *IEEE-AM Trans. Networking* **4(1)**: 22–39.

Nabona, N. 1993. "Multicommodity network flow model for long term hydro-generation optimization." *IEEE Trans. Power Syst.* **8(2)**: 395–404

Oh, S. and A. Haghani. 1997. "Testing and evaluation of a multi-commodity multi-modal network flow model for disaster relief management." *J. Adv. Transport.* **31(3)**: 249–282.

Popken, D. 1996. "An analytical framework for routing multiattribute multicommodity freight." *Transport. Res. Part B* **30(2)**: 133–145.

Powell, W. and T. Carvalho. 1997. "Dynamic control of multicommodity fleet management problems." *Eur. J. Oper. Res.* **98(3)**: 522–541.

Raghavan, P. 1992. "Integer programming in VLSI design." *Discrete Appl. Math* **40(1)**: 29–43.

Resende, M. and G. Veiga. 1993. "An efficient implementation of a network interior point method." In *Network Flows and Matching: First DIMACS Implementation Challenge*, edited by D. Johnson and C. McGeoch. DIMACS Series on Discrete Mathematics and Theoretical Computer Science, **12**: 299–348: American Mathematical Society.

Ribeiro, C. and D. Elbaz. 1992. "A parallel optimal routing algorithm." *Parallel Comput.* **18(12)**: 1393–1402.

Rushmeier, R. and S. Kontogiorgis. 1997. "Advances in the optimization of airline fleet assignment." *Transport. Sci.* **31(2)**: 159–169.

Sakarovitch, M. 1967. "An extension of generalized upper bounding techniques for structured linear programs." *SIAM J.* (15) 906–914.

Shiraishi, Y., J. Sakemi, and K. Fukuda. 1993. "A global routing algorithm-based on the multicommodity network flow method." *IEICE Trans. Fund. Electronics Commun. Comput. Sci.* **E76A(10)**: 1746–1754.

Sung, C. and K. Lee. 1995. "A routing algorithm for virtual circuit data-networking with unreliable links" *Engng Optim.* **24(2)**: 119–136.

Tsujino, M., M. Fukushima, and T. Ibaraki. 1994. "A model of time-varying flows on a congested multidestination network." *Asia-Pacific J. Oper. Res.* **11(2)**: 141–153.

Vaidya, P. M. 1996. "A new algorithm for minimizing convex functions over convex sets." *Math. Programming* **73(3)**: 291–341.

Williams, J. W. J. 1964. "Algorithm 232, heapsort." *Commun. ACM* **7(6)**: 347–348.

Wolfe, P. 1975. "A method of conjugate subgradients for minimizing nondifferentiable convex functions." *Math. Programming* **3**: 145–173.

Yee, G. and F. Lin. 1992. "A routing algorithm for virtual circuit data-networks with multiple sessions per o-d pair." *Networks* **22(2):** 185–208.

Yeh, C., C. Cheng, and T. Lin. 1995. "Circuit clustering using a stochastic flow-injection method." *IEEE Trans. Comput. Aided Des. Integrated Circuits Syst.* **14(2)**: 154–162.

Yoon, M., Y. Baek, and D. Tcha. 1998. "Design of distributed fiber transport network with hubbing topology." *Eur. J. Oper. Res.* **104(3)**: 510–520.

Zhang, Z. and H. Hartmann. 1992. "On the non-bifurcated routing in virtual circuit communication networks." *Eur. Trans. Telecommun.* **3(1)**: 45–53.

8.6 Minimum Spanning Tree Problem

Ravindra K. Ahuja, Thomas L. Magnanti, and James B. Orlin

ABSTRACT The minimum spanning tree problem is perhaps the simplest, and certainly one of the most central, models in the field of network optimization. The minimum spanning tree problem arises in a number of applications, both as a stand-alone problem and as a sub-problem in more complex problem settings. We described several applications of the minimum spanning tree problem in Chapter 8.1. In this article, we first consider combinatorial-based optimality conditions for assessing whether a given spanning tree is a minimum spanning tree. We consider two such optimality conditions—path and cut optimality conditions. We then describe two algorithms, Kruskal's algorithm and Prim's algorithm, based upon these optimality conditions. We illustrate these algorithms using numerical examples and also analyze their worst-case complexity.

1. Introduction

Spanning trees play a central role within the field of network optimization. As shown in Chapter 8.3, algorithms for solving shortest-path problems typically construct spanning trees rooted at a source node (shortest-path trees). The simplex method for solving minimum-cost flow problems provides another example. As discussed in Chapter 8.6, it is a spanning tree manipulation algorithm that iteratively moves from one spanning tree to another, at each step introducing one arc into the spanning tree in place of another. In this article, we consider yet another instance where spanning trees arise in network optimization by examining a spanning tree model known as the minimum spanning tree problem. A spanning tree T of G is a connected acyclic subgraph that spans all the nodes, that is, a subgraph containing a unique path between every pair of nodes. Given an undirected graph $G = (N, A)$ with $n = |N|$ nodes and $m = |A|$ arcs and with a length or cost c_{ij} associated with each arc (i, j), we wish to find a spanning tree, called a minimum spanning tree, that has the smallest total cost (or length) of its constituent arcs, as measured by the sum of the costs of the arcs in the spanning tree. Two elementary properties of spanning trees are worth noting: (i) every spanning tree of G contains $n-1$ arcs, and (ii) adding an arc to a spanning tree creates a graph with exactly one cycle.

The minimum spanning tree problem arises in a number of applications, both as a stand-alone problem and as a subproblem in more complex problem settings. We begin this section by considering two combinatorial-based optimality conditions for assessing whether a given spanning tree is a minimum spanning tree. The first condition, based upon paths, compares the cost of a nontree arc with the tree arcs in the path that connects the endpoints of the nontree arc. The second condition, based upon cuts, compares the cost of any tree arc with the other arcs contained in the cut defined by removing that arc from the tree. These two path and cut optimality conditions are easy to state and to develop, and yet they quite naturally motivate several algorithms for solving the minimum spanning tree problem. We then describe two algorithms, Kruskal's algorithm and Prim's algorithm, based upon these optimality conditions. Our treatment of the minimum spanning trees is fairly introductory. The book by Ahuja et al. (1993) contains a more in-depth account of this topic.

In solving minimum spanning tree problems, we wish to use algorithms that are as efficient as possible, as measured by the number of computational operations they require. In referring to computational operations, we use a shorthand "big C" notation, saying that an algorithm requires $O(n^2)$ running time, if the number of computations it requires is bounded by cn^2 for some constant c.

2. Optimality Conditions

Optimality conditions for the minimum spanning tree problem play a central role in developing algorithms and establishing their validity. For the minimum spanning tree problem, we can formulate the optimality conditions in two important ways: path optimality conditions and cut optimality conditions. In this section, we develop these conditions.

Let T^* be a spanning tree in G. We refer to the arcs in T^* as tree arcs and arcs not in T^* as nontree arcs. For every nontree arc (k, l) in T^*, let $P(k, l)$ denote a unique path in T^* from node k to node l. Consider, for example, the graph G shown in Figure 1(a). Figure 1(b) shows a spanning tree in G. With respect to this spanning tree, the path $P(3, 9)$ contains the arcs (3, 4), (4, 5), and (5, 9). It is easy to see that the cost of the arc (k, l) must be greater than or equal to the cost of every arc in $P(k, l)$; for otherwise swapping the nontree arc (k, l) with the lower-cost tree arc gives another spanning tree with a cost lower than that of T^*, contradicting the optimality of the spanning tree T^*. It is also possible to establish the converse statement, namely, if every nontree arc (k, l) with respect to T^* has a cost greater than or equal to the cost of every arc in $P(k, l)$, then T^* is a minimum spanning tree. We thus obtain the following path optimality conditions for minimum spanning trees.

Theorem 1. *(Path Optimality Conditions). A spanning tree* T^* *is a minimum spanning tree if and only if it satisfies the following path optimality condition: for every nontree arc* (k, l) *of* G, $c_{ij} \leqslant c_{kl}$ *for every arc* (i, j) *contained in the path in* T^* *connecting nodes* k *and* l.

To develop another characterization of minimum spanning trees, let us examine cuts instead of paths. Consider any minimum spanning tree T^* of G. If we delete any tree arc (i, j) from the spanning tree T, then the resulting subgraph partitions the node set N into two connected subsets S and $\bar{S}$. The arcs from the underlying graph G whose two endpoints belong to different subsets constitute a cut. We can denote this cut either by specifying the node partition $[S, \bar{S}]$ or by specifying the set of arcs $Q(i, j)$ in the cut. For the minimum spanning tree shown in Figure 1(c), by deleting the arc (4, 5) from the spanning tree, we partition the node set into two subsets S and $\bar{S}$ with $S = \{1, 2, 3, 4\}$ and $\bar{S} = \{5, 6, 7, 8, 9, 10\}$. Figure 1(c) shows the resulting cut $Q(4, 5)$; the cut $Q(4, 5)$ contains the arcs (2, 6), (3, 9), (4, 5), (4, 6), and (4, 9). It is easy to see that the cost of any tree arc (i, j) in the spanning tree T^* must be less than or equal to the cost of every nontree arc in the cut $Q(i, j)$. This result is easy to prove. For example, in a minimum spanning tree T^* if some arc (k, l) in the cut $Q(i, j)$ has a cost lower than (i, j), then replacing the arc (i, j) by the arc (k, l) gives us a spanning tree of G with a lower cost than T^* contradicting the optimality of T^*. It is possible to show that these conditions are also sufficient; that is, if every arc in a spanning tree T^* satisfies the cut optimality conditions, then T^* is a minimum spanning tree. These results yield the following cut optimality conditions.

Theorem 2. *(Cut Optimality Conditions). A spanning tree* T^* *is a minimum spanning tree if and only if it satisfies the following cut optimality condition: for every tree arc* $(i, j) \in T^*$, $c_{ij} \leqslant c_{kl}$ *for every arc* (k, l) *contained in the cut formed by deleting arc* (i, j) *from* T^*.

The cut optimality conditions imply that we can always include any minimum cost arc in any cut in some minimum spanning tree, which we state in the following somewhat stronger form.

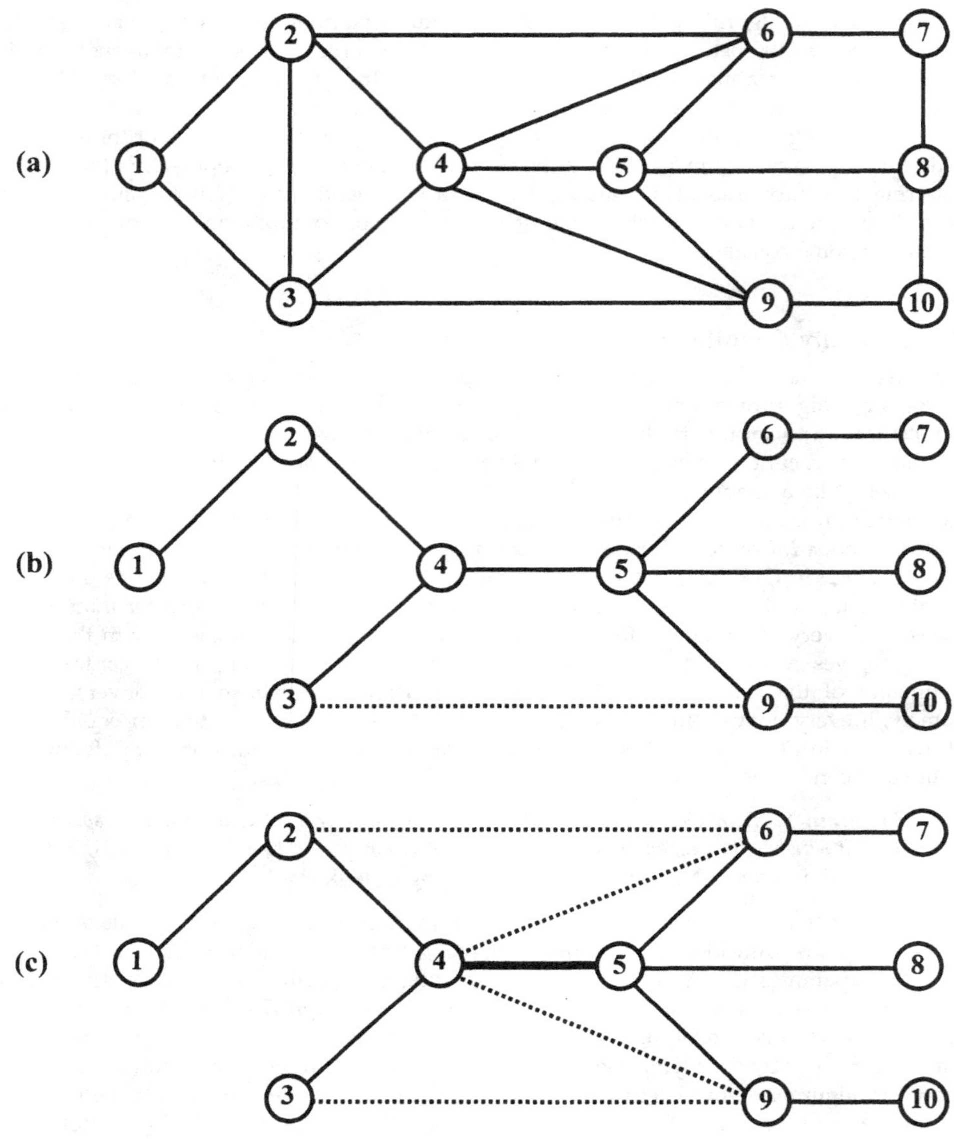

Figure 1 The path-and-cut optimality conditions

Property 1. *Let* F *be a subset of arcs in some minimum-cost spanning tree and let* S *be the set of nodes in some connected component of* F. *Suppose that* (i, j) *is a minimum-cost arc in the cut* $[S, \bar{S}]$. *Then some minimum spanning tree contains all the arcs in* F *as well as the arc* (i, j).

We have established two optimality conditions for the minimum spanning tree problem. In the next section, we will use these conditions to derive algorithms to solve the problem.

3. Kruskal's Algorithm

The path optimality conditions immediately suggest the following straightforward algorithm for solving the minimum spanning tree problem. We start with any arbitrary spanning tree *T* and test the path optimality conditions. If T satisfies these conditions, then it is an optimal tree; otherwise, $c_{ij} > c_{kl}$ for some nontree arc (k, l) and some tree arc (i, j) contained in the unique path in T connecting nodes k and l. In this case, adding arc (k, l) to T in place of arc (i, j) gives us a spanning tree with a lower cost. Repeating this step will give us a minimum spanning tree within a finite number of iterations. Though this algorithm is strikingly simple, in any conceivable implementation, its running time is not very attractive.

To derive an alternate and more efficient algorithm, known as Kruskal's algorithm, from the path optimality conditions, we consider an algorithm that builds an optimal spanning tree from scratch by adding one arc at a time. The algorithm, due to Kruskal (1956), first sorts all the arcs in the nondecreasing order of their costs and defines a set CHOSEN, which is the set of arcs we have chosen as part of a minimum spanning tree. Initially, the set CHOSEN is empty. The algorithm examines the arcs in the sorted order, one by one, and checks whether adding the arc currently being examined to CHOSEN creates a cycle with the arcs already contained in CHOSEN. If it doesn't, then it adds the arc to CHOSEN; otherwise, it discards the arc. We terminate when $|\text{CHOSEN}| = n - 1$. At termination, the arcs in CHOSEN constitute a minimum spanning tree. Let T^* represent this minimum spanning tree.

To illustrate Kruskal's algorithm in a numerical example, consider the network shown in Figure 2(a). Sorted in the order of their costs, the arcs are (2, 4), (3, 5), (3, 4), (2, 3), (4, 5), (2, 1), and (3, 1). In the first three iterations, the algorithm adds the arcs (2, 4), (3, 5), and (3, 4) to CHOSEN (see Figure 2(b)–(d)). In the next two iterations, the algorithm examines arcs (2, 3) and (4, 5) and discards them because the addition of each arc to CHOSEN creates a cycle (see Figure 2(e)–(f)). Then, the algorithm adds arc (2, 1) to CHOSEN and terminates. Figure 2(g) shows the minimum spanning tree.

The correctness of Kruskal's algorithm follows from the fact that we discarded each nontree arc (k, l) with respect to T^* at some stage because it created a cycle with the arcs already in CHOSEN. Consequently, the cost of arc (k, l) is greater than or equal to the cost of every arc in that cycle because the algorithm has examined the arcs in nondecreasing order of their costs. Therefore, the spanning tree T^* satisfies the path optimality conditions and so it is an optimal tree.

We might view the running time of Kruskal's algorithm as being composed of the time for sorting the arcs and the time for detecting cycles. For a network with arbitrarily large arc costs, sorting requires $O(m \log m) = O(m \log n^2) = O(m \log n)$ time. The time to detect a cycle depends upon the method we use for this step. One naive method would work as follows. The set CHOSEN at any stage of the algorithm defines a forest, that is, a collection of subtrees. For example, the set CHOSEN corresponding to Figure 2(c) consists of three subtrees containing the nodes {1}, {2, 4}, and {3, 5}, respectively. We denote these sets of nodes for a collection of subtrees by $S[1]$, $S[2]$, $S[3]$, ... We can store these sets as different singly linked lists (that is, as a list of elements with a pointer to the next element on the list). While examining an arc (k, l), we scan through these linked lists and check whether both the nodes k and l belong to the same list. If so, then adding arc (k, l) to CHOSEN creates a cycle and we discard this arc. If nodes k and l belong to different lists, then we add arc (k, l) to CHOSEN, which requires merging the linked lists containing nodes k and l into a single list. If implemented naively, this data structure requires $O(n)$ time for each arc that we examine and so if we use this data structure, Kruskal's algorithm runs in $O(nm)$ time.

To more efficiently implement Kruskal's algorithm, we can use a special data structure known as the union-find data structure. The union-find data structure maintains a set of (mutually exclusive and exhaustive) node subsets, $S[1]$, $S[2]$, $S[3]$, .., each containing the nodes in the subtree in the forest maintained by Kruskal's algorithm. The union-find data structure performs two operations:

union($S[p]$, $S[q]$): This operation performs a union of the two subsets $S[p]$ and $S[q]$ (that is, merges the two subsets into a single subset).

find(p, $S[p]$): This operation determines the subset $S[p]$ containing element p.

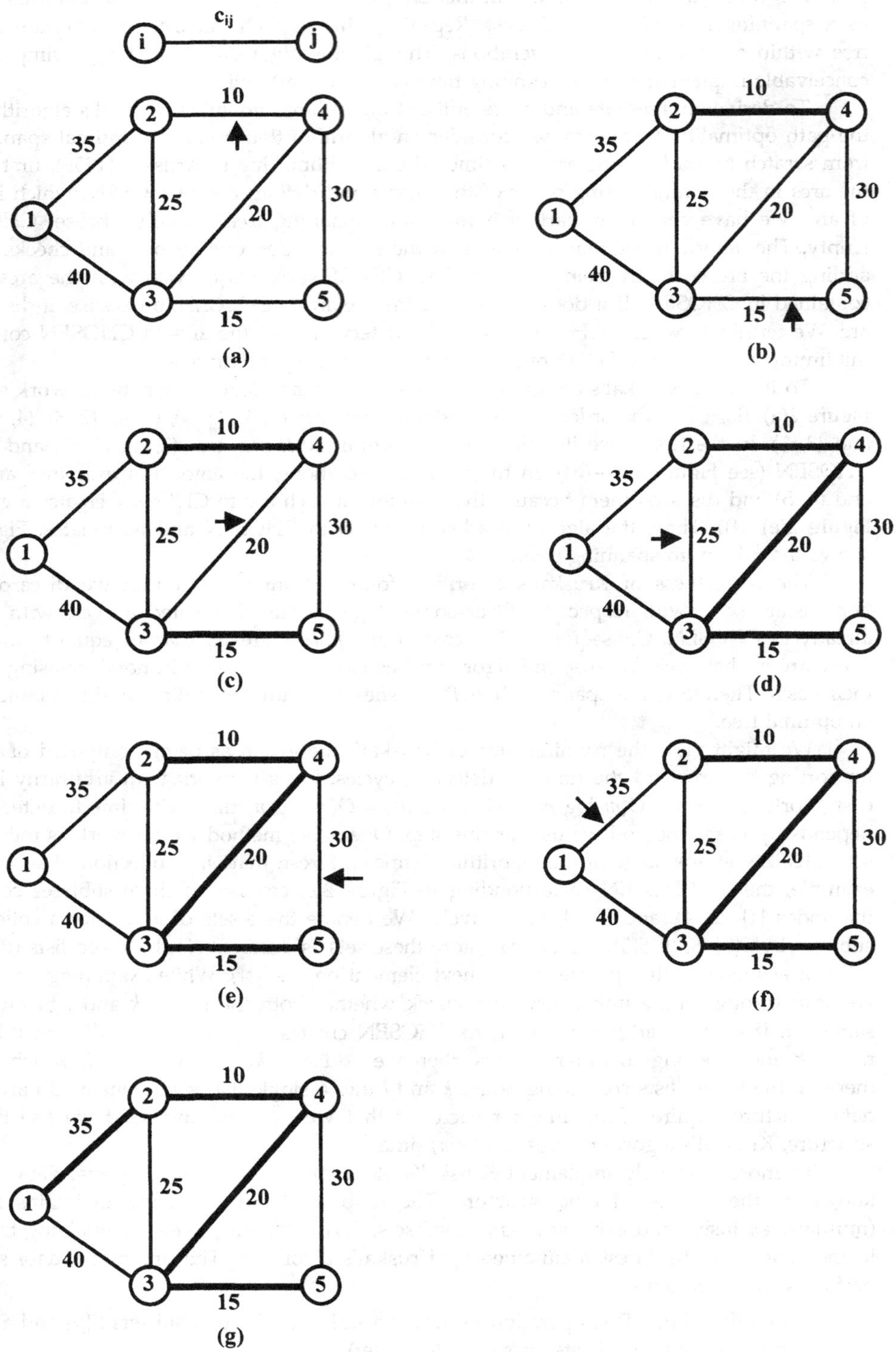

Figure 2 Kruskal's algorithm

Using the union-find data structures, we can implement Kruskal's algorithm in the following manner:

```
algorithm union-find-Kruskal;
begin
  sort the arcs in nonincreasing order of their costs;
  let (i1, j1), (i2, j2), . . ., (im, jm) denote the arcs in this order;
  CHOSEN: = φ;
  S[i]: = {i} for all i∈N;
  for all k: = 1 to m do
  begin
    find(ik, S[ik]);
    find(jk, S[jk]);
    if S[ik] ≠ S[jk] then
    begin
      CHOSEN: = CHOSEN∪{(ik, jk)};
      union(S[ik], S[jk]);
    end;
  end;
    CHOSEN is a minimum spanning tree;
end;
```

Kruskal's algorithm performs $O(m)$ find and $O(m)$ union operations. The book by Cormen et al. (1990) discusses several data structures for implementing union-find operations. One data structure, due to Tarjan (1983), performs each union-find operation in $O(\alpha(n, m))$ time, where $\alpha(n, m)$ is the Ackermann function that, for all practical purposes, is smaller than six. A version of Kruskal's algorithm using this data structure runs in $O(m\ \alpha(n, m))$ time on networks whose arcs are given in the sorted order.

4. Prim's Algorithm

Just as the path optimality conditions allowed us to develop Kruskal's algorithm, the cut optimality conditions permit us to develop another simple algorithm for the minimum spanning tree problem, known as Prim's algorithm. This algorithm, due to Prim (1957), builds a spanning tree from scratch by fanning out from a single node, adding arcs one at a time. It maintains a tree that spans a subset S of nodes and at each step adds a nearest neighbor to S. The algorithm does so by identifying an arc (i, j) of minimum cost in the cut $[S, \bar{S}]$. It adds arc (i, j) to the tree and node j to S, and repeats this basic step until $S = N$. The correctness of the algorithm follows directly from Property 1, which implies that each arc that it adds to the tree is contained in some minimum spanning tree with the arcs that it has selected in the previous steps.

To illustrate Prim's algorithm, we consider the same example, shown in Figure 3(a), that we used earlier to illustrate Kruskal's algorithm. Suppose, initially, $S = \{1\}$. The cut $[S, \bar{S}]$ contains two arcs, (1, 2) and (1, 3), and the algorithm selects the arc (1, 2) (see Figure 3(b)). At this point, $S = \{1, 2\}$ and the cut $[S, \bar{S}]$ contains the arcs (1, 3), (2, 3), and (2, 4). The algorithm selects arc (2, 4) since it has the minimum cost among these three arcs (see Figure 3(c)). In the next two iterations, the algorithm adds arc (4, 3) and then arc (3, 5); Figure 3(d) and (e) show the details of these iterations. Figure 3(f) shows the minimum spanning tree produced by the algorithm.

To analyze the running time of Prim's algorithm, we consider each of the $n-1$ iterations that the algorithm performs as it adds one arc at a time to the tree until it has become a spanning tree containing $n-1$ arcs. In each iteration, the algorithm selects the minimum-cost arc in the cut $[S, \bar{S}]$. If we scan the entire arc list to identify the minimum-cost arc, this operation requires $O(m)$ time, resulting in an $O(nm)$ time bound for the algorithm. Therefore, this naive implementation of Prim's algorithm runs in $O(nm)$ time. However, as we show next, we can improve substantially on this implementation.

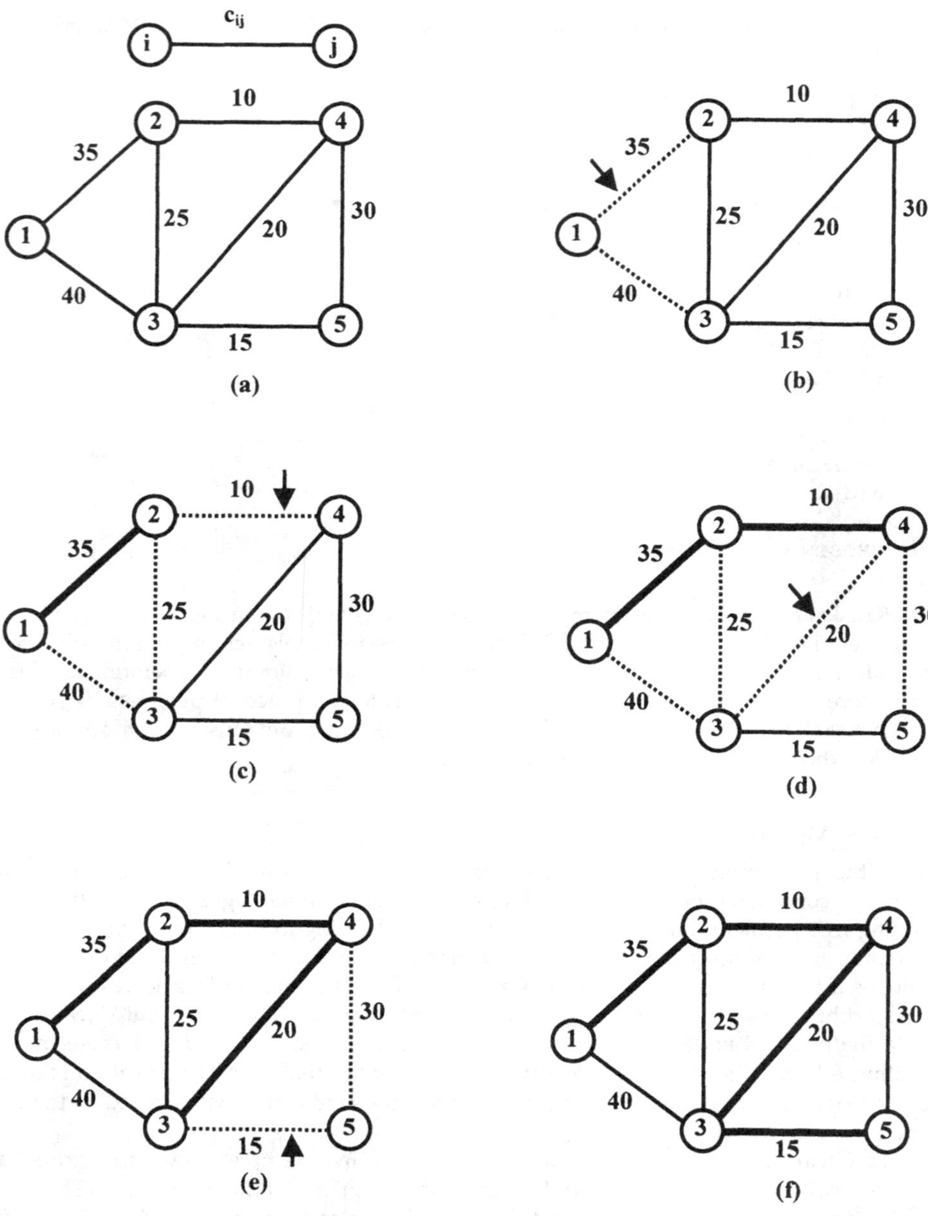

Figure 3 Prim's algorithm

The bottleneck step in the $O(nm)$ implementation of Prim's algorithm is the identification of a minimum-cost arc in the cut $[S, \bar{S}]$. We can improve the efficiency of this step by maintaining two indices for each node j in $\bar{S}$: (i) a distance label $d(j)$, which represents the minimum cost of arcs in the cut incident to a node j in $\bar{S}$, that is, $d(j) = \min\{c_{ij}: (i, j) \in [S, \bar{S}]\}$, and (ii) a predecessor label $pred(j)$, which represents the other endpoint of the minimum-cost arc in the cut incident to node j. For example, in Figure 3(d), three arcs (1, 3), (2, 3), and (4, 3) in the cut are incident to node 3. Among these arcs, the arc (4, 3) has the minimum cost of 20. Therefore, $d(3) = 20$ and $pred(3) = 4$. For the same figure, $d(5) = 30$ and $pred(5) = 4$. If we maintain these indices, then we can easily find the minimum cost of an arc in the cut: we simply compute $\min\{d(j): j \in \bar{S}\}$. If node i achieves this minimum, then $(pred(i), i)$ is a minimum cost arc in the cut. Observe that if we move node i from S to $\bar{S}$, then we need to update the distance and predecessor labels only for the nodes adjacent to node i. For any node $j \neq i$ and $j \in \bar{S}$, $d(j)$ will become the minimum of its previous value and c_{ij}.

We can implement Prim's algorithm using various types of heaps (or priority queues). A heap or a priority queue is a data structure that permits us to perform the following operations on a collection H of objects, each having an associated real number called its key.

create-heap(*H*): Create an empty heap H.

find-min(*H*): Find and return an object with the minimum key.

insert(*i*, *H*): Insert a new object i with a predefined key into a collection of objects.

decrease-key(*i*, value, *H*): Reduce the key of an object i to value, which must be smaller than the key it is replacing.

delete-min(*i*, *H*): Delete an object i with the minimum key from the collection H of objects.

To implement Prim's algorithm using a heap, we could choose H as the collection of nodes in $\bar{S}$ and the key of a node as its distance label. We would then implement Prim's algorithm as shown in Figure 4. (In this description, C denotes the maximum arc cost in the graph G.)

As is clear from its description, Prim's algorithm performs the operations find-min, delete-min, and insert at most n times and the operation decrease-key at most m times. When implemented with different heaps, the algorithm would have different running times. The book by Cormen et al. (1990) describes several heap data structures. By using the so-called the d-heap data structure with $d = \lceil m/n \rceil$, we could perform each find-min, decrease-key, and insert operation in $O(\log_d n)$ time and each delete-min operation in $O(d \log_d n)$ time; the overall algorithm would, therefore, have a total running time of $O(m \log_d n + nd \log_d n) = O(m \log_d n)$.

```
algorithm heap-Prim;
begin
  create-heap(H)
  for each node j∈N - {1} do d(j):=C+1;
  set d(1):=0; and pred(1): = 0;
  for each node j∈N do insert(j, H);
  CHOSEN: =φ;
  while |CHOSEN|<(n - 1) do
  begin
    find-min(i, H);
    delete-min(i, H);
    CHOSEN: = CHOSEN ∪(pred(i), i);
    for each arc (i, j)∈A(i) with j∈H do
      if d(j) > cij then
      begin
        d(j): = cij;
        pred(j): = i;
        decrease-key(j, cij, H);
      end;
  end;
  CHOSEN is a minimum spanning tree;
```

Figure 4 Heap implementation of Prim's algorithm

By using another data structure known as Fibonacci heaps, due to Fredman and Tarjan (1984), we could perform each find-min, decrease-key, and insert operation in $O(1)$ time and each delete-min operation in $O(\log n)$ time; this implementation has a total running time of $O(m + n \log n)$.

5. Summary

The minimum spanning tree problem is perhaps the simplest and one of the most important models in the field of combinatorial optimization. In this article, we have proved two (equivalent) necessary and sufficient conditions—the path and cut optimality conditions—for characterizing the optimality of minimum spanning trees. The path optimality conditions state that a spanning tree T^* is a minimum spanning tree if, and only if, the cost of every nontree arc (k, l) is greater than or equal to the cost of every tree arc in the path in T^* between the nodes k and l. The cut optimality conditions (in a sense, they are the dual of the path optimality conditions) state that a spanning tree T^* is a minimum spanning tree if and only if the cost of the tree arc (i, j) is less than or equal to the cost of every nontree arc in the cut formed by deleting arc (i, j) from T^*. We also described two well-known algorithms for solving the minimum spanning tree problem: Kruskal's and Prim's algorithms. We showed how to use data structures effectively to improve upon naïve implementation of these solution procedures. These algorithms are easy to implement, have excellent running times, and are very efficient in practice.

ACKNOWLEDGMENTS

The first author gratefully acknowledges the support of the National Science Foundation through grant award number DMI-9900087. The third author acknowledges the support from the Office of Naval Research under contract N00014-98-1-0317 and the National Science Foundation through grant award number DMI-9820998.

References

Ahuja, R. K., T. L. Magnanti, and J. B. Orlin. 1993. In *Network Flows: Theory, Algorithms, and Applications*. NJ: Prentice Hall.

Cormen, T. H., C. L. Leiserson, and R. L. Rivest. 1990. In *Introduction to Algorithms*. New York: MIT Press and McGraw Hill.

Fredman, M. L. and R. E. Tarjan. 1984. "Fibonacci heaps and their uses in improved network optimization algorithms." *Proceedings of the 25th Annual Symposium on Foundations of Computer Science*, pp. 338–346. Full paper in Journal of *ACM* **34(1987)**: 596–615.

Kruskal, J. B. 1956. "On the shortest spanning tree of graph and the traveling salesman problem." *Proc. Am. Math. Soc.* **7**: 48–50.

Prim, R. C. 1957. "Shortest connection networks and some generalizations." *Bell Syst. Tech. J.* **36**: 1389–1401.

Tarjan, R. E. 1983. In *Data Structures and Network Algorithms*. Philadelphia: SIAM.

9 INTEGER PROGRAMMING

9.1 Introduction

Nelson Maculan

ABSTRACT A brief introduction is given to the general linear integer programming problem. Cutting planes and enumeration methods are discussed.

1. Introduction

If a linear programming problem has some or all its variables constrained to be integer, we have an integer linear programming problem (IP) or a discrete linear programming problem. In the case where some, but not all, of the variables are integer, we call this problem a mixed integer (linear) programming problem (MIP). Consider the following MIP as an example.

$$(P): \text{maximize } \; x_0 = x_1 - 3x_2 - 4x_3$$

subject to:

$$\begin{aligned} 2x_1 + x_2 - x_3 &\leqslant 4 \\ 4x_1 - 3x_2 &\leqslant 2 \\ 3x_1 + 2x_2 + x_3 &\leqslant 3 \\ x_1 &\geqslant 0 \\ x_2 &\geqslant 0 \\ x_3 &\geqslant 0 \end{aligned}$$

x_2 and x_3 integer.

The topic of integer programming is highly developed and a number of books on the subject have been published. (For example, Korbut and Finkelstein, 1969; Hu, 1969; Greenberg, 1971; Garfinkel and Nemhauser, 1972; Simonnard, 1973; Zionts, 1974; Salkin, 1975; Taha, 1975; Murty, 1976; Maculan, 1978; Schijver, 1986; Nemhauser and Wolsey, 1988; Parker and Rardin, 1988; Wolsey, 1998; Kovács, 1980.) When we do not consider the integrality constraints, we obtain the linear relaxation problem $(\bar{P})$ of (P):

$$(\bar{P}): \text{maximize } \; x_0 = x_1 - 3x_2 - 4x_3$$

subject to:

$$\begin{aligned} 2x_1 + x_2 - x_3 &\leqslant 4 \\ 4x_1 - 3x_2 &\leqslant 2 \\ 3x_1 + 2x_2 + x_3 &\leqslant 3 \\ x_1 &\geqslant 0 \\ x_2 &\geqslant 0 \\ x_3 &\geqslant 0. \end{aligned}$$

We can observe that $\text{val}(P) \leqslant \text{val}(\bar{P})$, where $\text{val}(\cdot)$ is the value of the objective function at the optimal of $(\cdot)$. The set of feasible solutions of $(\bar{P})$ contains the set of feasible solutions of (P), that is, $F(P) \subseteq F(\bar{P})$, where $F(\cdot)$ is the set of feasible solutions of $(\cdot)$.

To solve $(\bar{P})$ we can use the simplex method (Dantzig, 1963; Chvátal, 1983) or interior point algorithms (Karmarkar, 1984; Gonzaga, 1989). One could solve (P) by computing an optimal solution for $(\bar{P})$ and rounding the fractional solutions associated with the variables constrained to be integer. For some linear programming problems we know, *a priori*, that all basic solutions are integer, for example, see Garfinkel and Nemhauser (1972). To illustrate an example of rounding, consider the following problem (P).

$$(P): \text{maximize } \ x_0 = x_1 + 19x_2$$

subject to:

$$x_1 + 20x_2 \leqslant 50$$

$$x_1 + \quad x_2 \leqslant 20$$

$$x_1 \qquad \geqslant 0$$

$$x_2 \geqslant 0$$

$$x_1 \text{ and } x_2 \text{ integer.} \tag{1}$$

We solve $(\bar{P})$, the linear relaxation of (P), that is, ignoring the integrality constraints (1) in (P). Using a primal simplex method, we obtain an optimal solution of $(\bar{P}): \bar{x}_1 = 18\frac{8}{19}$, $\bar{x}_2 = 1\frac{11}{19}$, and $\text{val}(\bar{P}) = 48\frac{8}{19}$. We know that $\text{val}(P) \leqslant \text{val}(\bar{P}) = 48\frac{8}{19}$, but since the coefficients of the objective function in (P) are integer and all variables are constrained to be integer, then we can write $\text{val}(P) \leqslant \lfloor 48\frac{8}{19} \rfloor = 48$, where $\lfloor \alpha \rfloor$ is the largest integer less than or equal to α. Next, we try to round the solutions $\bar{x}_1$ and $\bar{x}_2$ of $(\bar{P})$ in order to obtain feasible solutions of (P):

- $x_1 = 19,\ x_2 = 2$, infeasible solution of (P);
- $x_1 = 18,\ x_2 = 1,\ x_0 = 37$;
- $x_1 = 18,\ x_2 = 2$, infeasible solution of (P);
- $x_1 = 19,\ x_2 = 1,\ x_0 = 38$.

We may consider other feasible solutions of (P), which are not obtained by rounding an optimal solution of $(\bar{P})$. For example,

- $x_1 = 0,\ x_2 = 2,\ x_0 = 38$;
- $x_1 = 10,\ x_2 = 2,\ x_0 = 48$, an optimal solution of (P).

If we had taken the best rounding solution of $(\bar{P})$ we would obtain $x_0 = 38$, which has a relative gap of $\frac{48-38}{48} = \frac{10}{48} \cong 0.2083$. This means that there is a relative gap of more than 20 percent between the optimal solution of (P) and the best rounding solution of $(\bar{P})$.

In this last example we obtained from $\text{val}(\bar{P})$ a very good upper bound for $\text{val}(P)$, since we have no gap between $\lfloor \text{val}(\bar{P}) \rfloor$ and $\text{val}(P) : \text{val}(P) = \lfloor \text{val}(\bar{P}) \rfloor$. Of course, we can have cases where $\text{val}(P) < \text{val}(\bar{P})$, and the relative gap between these quantities can be large.

Consider, without loss of generality, the integer linear programming problem:

$$(P): \text{maximize } \ x_0 = c^T x$$

subject to:

$$Ax = b$$

$$x \geqslant 0$$

$$x \text{ integer.} \tag{2}$$

where $c \in \mathcal{R}^n$, $b \in \mathcal{R}^m$, and $A \in \mathcal{R}^{m \times n}$ are given and $x \in \mathcal{R}^n$ is the vector of the decision variables. If we do not consider constraint (2) in (P), we have its LP relaxation $(\bar{P})$.

We assume that the rank of A is equal to m. Thus we can choose an $m \times m$ submatrix B of A, such that $det(B) \neq 0$, and write $A = (B \ \ N)$. We consider partitions of c and x associated with $(B \ \ N) : x^T = (x_B^T x_N^T)$ and $c^T = (c_B^T c_N^T)$, and $(\bar{P})$ can be written as follows:

$$(\bar{P}): \text{maximize} \;\; x_0 = c_B^T x_B + c_N^T x_N$$

subject to:

$$Bx_B + Nx_N = b \tag{3}$$

$$x_B \geqslant 0, \;\; x_N \geqslant 0.$$

We express x_B as a function of x_N in (3):

$$x_B = B^{-1}b - B^{-1}Nx_N. \tag{4}$$

Using (4), $(\bar{P})$ can also be written as follows:

$$(\bar{P}): \text{maximize} \;\; x_0 = c_B^T B^{-1}b - (c_B^T B^{-1}N - c_N^T)x_N$$

subject to:

$$x_B = B^{-1}b - B^{-1}Nx_N$$

$$x_B \geqslant 0, \;\; x_N \geqslant 0.$$

Let $u^T = c_B^T B^{-1}$, $\bar{x}_B = B^{-1}b$, $z_j = u^T a_j$ (a_j is a column of A, $j = 1, 2, ..., n; a_j \in \mathscr{R}^m$), $y_j = B^{-1}a_j (j = 1, 2, ..., n)$, and $\bar{x}_0 = c_B^T B^{-1}b = u^T b = c_B^T \bar{x}_B$. Also, let I_N be the set of indices of the columns in N and I_B be the set of indices of the columns in B. Thus, $(\bar{P})$ can be written as:

$$(\bar{P}): \text{maximize} \;\; x_0 = \bar{x}_0 - \sum_{j \in I_N} (z_j - c_j)x_j$$

subject to:

$$x_B = \bar{x}_B - \sum_{j \in I_N} y_j x_j \tag{5}$$

$$x_B \geqslant 0, \;\; x_j \geqslant 0, \;\; j \in I_N.$$

If $y_j^T = (y_{1j}, y_{2j}, ..., y_{mj}), x_B^T = (x_{B(1)}, x_{B(2)}, ..., x_{B(m)})$, and $\bar{x}_B^T = (\bar{x}_{B(1)}, \bar{x}_{B(2)}, ..., \bar{x}_{B(m)})$, then (5) can be written as

$$x_{B(i)} = \bar{x}_{B(i)} - \sum_{j \in I_N} y_{ij} x_j, \quad i = 1, 2, ..., m.$$

We now present two different techniques to solve integer linear programming problems. These techniques are known as the cutting plane and implicit enumeration methods. They have been combined to define the branch-and-cut methods, see Chapter 3.3 of this handbook.

2. Cutting-Plane Methods

Cutting-plane methods are based on the computation of the convex hull of the integer-feasible solutions of (P). Once the convex hull is obtained, we apply the simplex method considering the maximization of the objective function of (P) subject to the convex hull of $F(P)$. It is very hard to obtain the convex hull of the integer points that satisfy a set of linear constraints. The polyhedron, which defines the convex hull, can often only be described by an exponentially large number of linear constraints.

Cutting plane methods introduce at each iteration one or more valid linear inequalities to the LP relaxation of (P) that eliminate a fractional solution of $(\bar{P})$ without eliminating any feasible solution of (P). We define a valid linear inequality for (P) to be a linear inequality $d^T x \leqslant \alpha$, such that for all $\bar{x} \in F(P)$ we have $d^T \bar{x} \leqslant \alpha$, where $d \in \mathscr{R}^n$ and $\alpha \in \mathscr{R}$.

How can we find interesting valid linear inequalities for (P)? We consider an interesting valid linear inequality for (P) to be an inequality $E = \{x \in \mathscr{R}^n \mid d^T x \leqslant \alpha\}$ such that $E \cap F(\bar{P}) \neq \varnothing$. E is a valid inequality for (P) because $F(P) \subseteq E$.

A classical cutting plane method starts by solving $(\bar{P})$ using the simplex algorithm. If the optimal solution of $(\bar{P})$ is integer, then we have also solved (P). Otherwise, we introduce a valid linear inequality for (P) that eliminates this $(\bar{P})$-optimal solution and a new linear programming problem is recursively optimized. We now present a Gomory cutting plane algorithm (see Gomory, 1963).

Suppose we have solved $(\bar{P})$ using the simplex algorithm and B is an optimal basis: $\bar{x}_B = B^{-1}b \geqslant 0$, and $z_j - c_j \geqslant 0, j \in I_N$. Then, for a $k \in \{1, 2, ..., m\}$ we have $\bar{x}_{B(k)} \notin \mathscr{Z}$, and therefore this optimal solution of $(\bar{P})$ does not solve (P). We consider

$$x_{B(k)} = \bar{x}_{B(k)} - \sum_{j \in I_N} y_{kj} x_j. \tag{6}$$

This can be written as

$$x_{B(k)} + \sum_{j \in I_N} y_{kj} x_j = \bar{x}_{B(k)},$$

where $x_{B(k)} \geqslant 0, \bar{x}_{B(k)} \geqslant 0, x_j \geqslant 0, j \in I_N$, and therefore

$$x_{B(k)} + \sum_{j \in I_N} \lfloor y_{kj} \rfloor x_j \leqslant \bar{x}_{B(k)}.$$

In addition, $x_{B(k)}$ and x_j, $j \in I_N$ have to be integer in (P). Thus

$$x_{B(k)} + \sum_{j \in I_N} \lfloor y_{kj} \rfloor x_j \leqslant \lfloor \bar{x}_{B(k)} \rfloor. \tag{7}$$

It is easy to see that all solutions of $F(P)$ are feasible for (7). In order to transform (7) into an equality, we introduce a slack variable $s \geqslant 0$ and obtain

$$x_{B(k)} + \sum_{j \in I_N} \lfloor y_{kj} \rfloor x_j + s = \lfloor \bar{x}_{B(k)} \rfloor. \tag{8}$$

Taking into consideration equations (6) and (8), we can have

$$\bar{x}_{B(k)} - \sum_{j \in I_N} y_{kj} x_j + \sum_{j \in I_N} \lfloor y_{kj} \rfloor x_j + s = \lfloor \bar{x}_{B(k)} \rfloor$$

or

$$s = \lfloor \bar{x}_{B(k)} \rfloor - \bar{x}_{B(k)} + \sum_{j \in I_N} (y_{kj} - \lfloor y_{kj} \rfloor) x_j.$$

Define $f_{kj} = y_{kj} - \lfloor y_{kj} \rfloor$, $j \in I_N$ and $f_{k0} = \bar{x}_{B(k)} - \lfloor \bar{x}_{B(k)} \rfloor$. It is easy to verify that $0 \leqslant f_{kj} < 1$ and $0 < f_{k0} < 1$. Thus

$$s = -f_{k0} + \sum_{j \in I_N} f_{kj} x_j,$$

and $s \geqslant 0$ give us

$$\sum_{j \in I_N} f_{kj} x_j \geqslant f_{k0}. \tag{9}$$

Inequality (9) is a valid inequality for (P). This inequality is known as a Gomory cut (Gomory, 1963).

To illustrate this method consider the example:

$$(P): \text{maximize } z = 2x_1 + x_2 + 0x_3 + 0x_4 + 0x_5$$

subject to:

$$\begin{aligned} x_1 + x_2 + x_3 \qquad &= 5 \\ -x_1 + x_2 \qquad + x_4 \quad &= 0 \\ 6x_1 + 2x_2 \qquad\qquad + x_5 &= 21 \\ x_j \geqslant 0, \quad j = 1, ..., 5 & \end{aligned}$$

x_j integer $j = 1, \ldots, 5.$ (10)

To solve $(\bar{P})$ we will use the simplex algorithm. We have

$$A = (a_1 \quad a_2 \quad a_3 \quad a_4 \quad a_5) = \begin{pmatrix} 1 & 1 & 1 & 0 & 0 \\ -1 & 1 & 0 & 1 & 0 \\ 6 & 2 & 0 & 0 & 1 \end{pmatrix}, \quad b = \begin{pmatrix} 5 \\ 0 \\ 21 \end{pmatrix},$$

$$c^T = (2 \quad 1 \quad 0 \quad 0 \quad 0).$$

We will consider

$$B = (a_1 \quad a_2 \quad a_4) = \begin{pmatrix} 1 & 1 & 0 \\ -1 & 1 & 1 \\ 6 & 2 & 0 \end{pmatrix}, \quad \text{then } B^{-1} = \begin{pmatrix} -\frac{1}{2} & 0 & \frac{1}{4} \\ \frac{3}{2} & 0 & -\frac{1}{4} \\ -2 & 1 & \frac{1}{2} \end{pmatrix},$$

$$\bar{x}_B = B^{-1}b = \begin{pmatrix} -\frac{1}{2} & 0 & \frac{1}{4} \\ \frac{3}{2} & 0 & -\frac{1}{4} \\ -2 & 1 & \frac{1}{2} \end{pmatrix} \begin{pmatrix} 5 \\ 0 \\ 21 \end{pmatrix} = \begin{pmatrix} \frac{11}{4} \\ \frac{9}{4} \\ \frac{1}{2} \end{pmatrix} = \begin{pmatrix} \bar{x}_{B(1)} \\ \bar{x}_{B(2)} \\ \bar{x}_{B(3)} \end{pmatrix}.$$

We have $\bar{x}_{B(1)} = \bar{x}_1, \bar{x}_{B(2)} = \bar{x}_2$, and $\bar{x}_{B(3)} = \bar{x}_4$. It is easy to see that basis B is associated with an optimal solution of $(\bar{P})$, that is, $\bar{x}_1 = \frac{11}{4}$, $\bar{x}_2 = \frac{9}{4}, \bar{x}_4 = \frac{1}{2}$, and $\bar{x}_3 = \bar{x}_5 = 0$, which is not an integer solution. We select the row associated with $x_{B(3)}$ to develop a Gomory cut: $x_{B(3)} = x_4 = \frac{1}{2} + 2x_3 - \frac{1}{2}x_5$. From this constraint, we obtain $f_{30} = \frac{1}{2}$, $f_{43} = 0$, and $f_{45} = \frac{1}{2}$. The generated Gomory cut is $-\frac{1}{2}x_5 + s_1 = -\frac{1}{2}$, and $s_1 \geqslant 0$. We join a new row and a new column to A.

We know that the new basis formed by the columns associated with x_1, x_2, x_4, and s_1 is dual feasible, and hence we solve the resulting optimization problem using the dual-simplex algorithm. The pivot rule for this new problem forces the column associated with s_1 to leave the basis and the column associated with x_5 to enter the basis. Let B_1 be the new basis,

$$B_1 = (a_1, a_2, a_4, a_5) = \begin{pmatrix} 1 & 1 & 0 & 0 \\ -1 & 1 & 1 & 0 \\ 6 & 2 & 0 & 1 \\ 0 & 0 & 0 & -\frac{1}{2} \end{pmatrix}, \quad \text{then } B_1^{-1} = \begin{pmatrix} -\frac{1}{2} & 0 & \frac{1}{4} & \frac{1}{2} \\ \frac{3}{2} & 0 & -\frac{1}{4} & -\frac{1}{2} \\ -2 & 1 & \frac{1}{2} & 1 \\ 0 & 0 & 0 & -2 \end{pmatrix},$$

$$\bar{x}_{B_1} = B_1^{-1}b = \begin{pmatrix} -\frac{1}{2} & 0 & \frac{1}{4} & \frac{1}{2} \\ \frac{3}{2} & 0 & -\frac{1}{4} & -\frac{1}{2} \\ -2 & 1 & \frac{1}{2} & 1 \\ 0 & 0 & 0 & -2 \end{pmatrix} \begin{pmatrix} 5 \\ 0 \\ 21 \\ -\frac{1}{2} \end{pmatrix} = \begin{pmatrix} \frac{5}{2} \\ \frac{5}{2} \\ 0 \\ 1 \end{pmatrix} = \begin{pmatrix} \bar{x}_1 \\ \bar{x}_2 \\ \bar{x}_4 \\ \bar{x}_5 \end{pmatrix}.$$

B_1 is associated with an optimal solution, but it is still not integer. If we take the row associated with $\bar{x}_1$ to generate the new Gomory cut, then we obtain $s_2 = -\frac{1}{2} - \frac{1}{2}x_3 + \frac{1}{2}s_1$, $s_2 \geqslant 0$. The simplex pivot rule tells us that the column associated with s_2 leaves the basis while the column associated with x_3 enters. Let B_2 be the new basis,

$$B_2 = (a_1 \quad a_2 \quad a_3 \quad a_4 \quad a_5) = \begin{pmatrix} 1 & 1 & 1 & 0 & 0 \\ -1 & 1 & 0 & 1 & 0 \\ 6 & 2 & 0 & 0 & 1 \\ 0 & 0 & 0 & 0 & -\frac{1}{2} \\ 0 & 0 & -\frac{1}{2} & 0 & 0 \end{pmatrix},$$

$$\text{then } B_2^{-1} = \begin{pmatrix} -\frac{1}{2} & 0 & \frac{1}{4} & \frac{1}{2} & -1 \\ \frac{3}{2} & 0 & -\frac{1}{4} & -\frac{1}{2} & 3 \\ 0 & 0 & 0 & 0 & -2 \\ -2 & 1 & \frac{1}{2} & 1 & -4 \\ 0 & 0 & 0 & -2 & 0 \end{pmatrix},$$

$$\bar{x}_{B_2} = B_2^{-1}b = \begin{pmatrix} -\frac{1}{2} & 0 & \frac{1}{4} & \frac{1}{2} & -1 \\ \frac{3}{2} & 0 & -\frac{1}{4} & -\frac{1}{2} & 3 \\ 0 & 0 & 0 & 0 & -2 \\ -2 & 1 & \frac{1}{2} & 1 & -4 \\ 0 & 0 & 0 & -2 & 0 \end{pmatrix} \begin{pmatrix} 5 \\ 0 \\ 21 \\ -\frac{1}{2} \\ -\frac{1}{2} \end{pmatrix} = \begin{pmatrix} 3 \\ 1 \\ 1 \\ 2 \\ 1 \end{pmatrix} = \begin{pmatrix} \bar{x}_1 \\ \bar{x}_2 \\ \bar{x}_3 \\ \bar{x}_4 \\ \bar{x}_5 \end{pmatrix}.$$

We observe that this solution is now an optimal integer solution which solves (P). With just two cuts we solved this small example.

The Gomory cutting plane method has finite convergence under certain conditions (see Gomory, 1963; Salkin, 1975; and Maculan 1978). The cuts generated in the last example are not facets of the convex hull of $F(P)$. Cutting plane methods have been used for solving large 0–1 linear programming problems for which we can characterize a family of facets of the convex hull of $F(P)$, (see Lucena and Beasley, 1996; Nemhauser and Wolsey, 1988; Nemhauser et al., 1989; Wolsey, 1998).

3. Implicit Enumeration Methods

Implicit enumeration methods have been shown to be a practical way to solve IP and MIP problems. Land and Doig (1960) presented the first general enumeration method for IP and MIP. These implicit enumeration methods are known as branch-and-bound schemes (Little et al., 1963). See Chapter 3.2 of the handbook. We present a branch-and-bound method developed by Dakin (1965). We consider again

$$(P): \text{maximize } x_0 = c^T x$$

subject to:

$$Ax = b$$

$$x \geqslant 0$$

$$x_j \text{ integer}, \quad j \in S \subseteq \{1, 2, ..., n\}. \tag{11}$$

As before, let $(\bar{P})$ be the LP relaxation of (P) and let $(P_0) \equiv (\bar{P})$. That is, we do not consider the constraints defined by (11). Suppose $F(P_0)$ is bounded. Then, we solve (P_0) to obtain an optimal solution $x_j^0, j = 1, 2, ..., n$. If $x_j^0 \in \mathscr{Z}, j \in S$, then we have solved (P). Otherwise, we have $k \in S$ such that $x_k^0, \notin \mathscr{Z}$, and we define two linear programming problems (we branch using x_k):

- (P_1), for which $F(P_1) = F(P_0) \cap \{x \in \mathscr{R}^n \mid x_k \leqslant \lfloor x_k^0 \rfloor\}$,
- (P_2), for which $F(P_2) = F(P_0) \cap \{x \in \mathscr{R}^n \mid x_k \geqslant \lfloor x_k^0 \rfloor + 1\}$.

The solution of (P_1) or (P_2) is easy to compute using an appropriate dual simplex algorithm (Délorme, 1974; Chvátal, 1983; Maculan et al., 1998). Define $x_j^i, j = 1, 2, ..., n$ to be an optimal solution for (P_i). An optimal may not exist since sometimes $F(P_i) = \varnothing$.

We associate this enumeration scheme with a binary tree. For each (P_i) we can have two descendants: (P_q) and (P_{q+1}). These two descendants are built joining a bounding variable constraint to (P_i). We know that $\text{val}(P_q) \leqslant \text{val}(P_i)$, and $\text{val}(P_{q+1}) \leqslant \text{val}(P_i)$.

Define open and closed nodes in this enumeration scheme. A node associated with (P_i) is closed when it has two descendants, or $F(P_i) = \varnothing$, or $x_j^i, \in \mathscr{Z}, j \in S$, or $\text{val}(P_i) \leqslant \text{val}(H)$, where $\text{val}(H)$ is the objective function value of (P) for a known solution of $F(P)$ ($\text{val}(H)$ is a lower bound for $\text{val}(P)$).

A good $\text{val}(H)$ can, *a priori*, be obtained with heuristics or metaheuristics. It is important to also get a good upper bound with, for example, dual or cutting plane procedures for each $\text{val}(P_i)$. Sometimes we obtain an upper bound for all feasible solutions of (P), which are also descendants of (P_i). This upper bound can be less than $\text{val}(P_l)$, when (P_i) is a direct descendant of (P_l), and can be used to prune an open node (P_i). These upper and lower bounds are updated while the enumeration scheme is running. The procedure stops when all nodes are closed. This method converges because we assumed that $F(P_0)$ is bounded.

At each open node we choose a variable to branch on, and evaluate and update bounds. One needs to determine how to choose an open node to be branched on and how to choose a variable for this branching. In Forrest et al. (1974) ideas and computational results to answer this question are discussed. See also Fox et al. (1978) and Ibaraki (1977, 1982). Artificial intelligence principles are also important for branch-and-bound methods (Nilsson, 1980)

To illustrate the Dakin algorithm (Dakin, 1965), we consider the following knapsack problem.

$$(KP): \text{maximize } x_0 = 16x_1 + 24x_2 + 7x_3 + 2x_4$$

subject to:

$$4x_1 + 7x_2 + 3x_3 + x_4 \leqslant 27$$

$$x_j \geqslant 0, \; j = 1, 2, 3, 4$$

$$x_j \text{ integer}, \quad j = 1, 2, 3, 4.$$

We start solving the relaxed problem (KP_0) derived from (KP). As the coefficients of the objective function of (KP) are integer, we can take $\lfloor \text{val}(KP_0) \rfloor$ as the first upper bound for $\text{val}(KP)$. A classical greedy (heuristic) algorithm for (KP) will give us a feasible solution $x_1^H = 6, x_2^H = 0, x_3^H = 1, x_4^H = 0 \rightarrow x_0^H = 103$. This shows that $\text{val}(H) = 103$ is a lower bound of $\text{val}(KP)$. The optimal solution found for (KP_0) is $x_1^0 = \frac{27}{4} = 6.75, x_2^0 = x_3^0 = x_4^0 = 0, \rightarrow x_0^0 = 108 = \text{val}(KP_0)$. Since $103 \leqslant \text{val}(KP) \leqslant 108$, if we take 103 as the value of the optimal solution of (KP), then the relative gap is $\frac{108-103}{108} \cong 0.046$ (4.6 percent). For the knapsack problem, we have at most one variable which is not integer in (KP_0), namely x_1.

We build two new linear programming problems: (KP_1), where $F(KP_1) = F(KP_0) \cap \{x_1 \in \mathcal{R} \mid x_1 \leqslant 6\}$; and (KP_2), where $F(KP_2) = F(KP_0) \cap \{x_1 \in \mathcal{R} \mid x_1 \geqslant 7\}$. Since $F(KP_2) = \emptyset$ then (KP_2) is closed. The optimal solution for (KP_1) is given by $x_1^1 = 6, x_2^1 = \frac{3}{7}, x_3^1 = x_4^1 = 0, \rightarrow x_0^1 = \frac{744}{7} \cong 106.286$. We update the upper bound, which is $\lfloor \frac{744}{7} \rfloor = 106$. Now, if we took 103 as the value of an optimal solution of (KP), the relative gap would be $\frac{106-103}{106} \cong 0.028$ (2.8 percent).

We close (KP_1) again, creating two new linear programming problems: (KP_3), where $F(KP_3) = F(KP_1) \cap \{x_2 \in R \mid x_2 \leqslant 0\}$; and (KP_4), where $F(KP_4) = F(KP_1) \cap \{x_2 \in R \mid x_2 \geqslant 1\}$. Solving (KP_3), we obtain an integer solution identical to the one given by the greedy algorithm. Since no additional information has been obtained solving (KP_3), this node is closed. Solving (KP_4), we find $x_1^4 = 5, x_2^4 = 1, x_3^4 = x_4^4 = 0, \rightarrow x_0^4 = 104$. Since in (KP_4) we obtained a new upper bound of val(KP) associated with an integer solution, then we close (KP_4).

All developed nodes are closed and the best solution found is associated with (KP_4). Thus, val$(KP) = 104$. Once an optimal solution for (KP) is known, the relative gap between the value given by the greedy algorithm and this optimal solution is $\frac{104-103}{104} \cong 0.0096$ (0.96 percent). In this case, the heuristic procedure supplies a good feasible solution. The worst case relative gap for this heuristic is asymptotically equal to one (100 percent) (see Campello and Maculan, 1994).

From the constraints of (KP), we easily determine the feasible range for each variable independently: $x_1 \in \{0, 1, 2, ..., 6\}, x_2 \in \{0, 1, 2, 3\}, x_3 \in \{0, 1, 2, ..., 9\}$, and $x_4 \in \{0, 1, 2, ..., 27\}$. If we try to solve (KP) by explicit enumeration we will have to verify 7840 possible solutions, and keep the solutions which satisfy $4x_1 + 7x_2 + 3x_3 + x_4 \leqslant 27$. Among all solutions satisfying $4x_1 + 7x_2 + 3x_3 + x_4 \leqslant 27$, we take that for which the objective function value is the largest. A (KP) with as few as 100 integer variables cannot be solved by such an enumeration procedure, since the solution time needed would be prohibitive. It is impossible to determine, *a priori*, whether a given instance of integer programming can be solved in a suitable amount of time. However, large instances of (KP) have been solved using branch-and-bound algorithms (see Martello and Toth, 1990).

To improve a branch-and-bound procedure we can introduce cutting planes to the relaxation (P_i). Such a method is called branch-and-cut (see Chapter 3.3 of this handbook).

The branch-and-bound procedures can also be developed using Lagrangian relaxation or Lagrangian decomposition techniques that are presented in this handbook (see Chapter 9.5 of this handbook).

Polyhedral combinatorics have been used in connection with cutting plane algorithms (Nemhauser and Wolsey, 1988; Balas et al., 1993; Lovász and Schrijver, 1991).

Balas established the foundations of disjunctive programming (see Balas, 1998). Many combinatorial optimization problems can be modeled as disjunctive programming and solved as an integer programming problem (see also Balas et al., 1993).

Another powerful technique, used in integer programming is column generation. Such techniques have been presented and analyzed in, for example, Barnhart et al., 1998; Desrochers and Soumis, 1989; Hansen et al., 1987, 1991; Jardim Campos, 1987; Jardim Campos and Maculan, 1988, 1995; Maculan et al., 1992; Vance et al., 1994; Vance, 1998; Vanderbeck, 1994; Dantzig and Wolfe, 1960; Ribeiro et al., 1989).

References

Balas, E. 1998. "Disjunctive programming: properties of the convex hull of feasible points." *Discrete Appl. Math.* **89**: 3–44.

Balas, E., S. Ceria, and G. Cornuéjols. 1993. "A lift-and-project cutting plane algorithm for mixed 0–1 programs." *Math Programming* **58**: 295–324.

Barnhart, C., E. L. Johnson, G. L. Nemhauser, M. W. P. Savelsbergh, and P. H. Vance. 1998. "Branch-and-price: column generation for solving huge integer programs." *Oper. Res.* **46**: 316–329.

Campello, R. E. and N. Maculan. 1994. In *Algoritmos e Heurísticas: Desenvolvimento e Avaliação de Performance*. Editora da Universidade Federal Fluminense (EDUFF).

Chvátal, V. 1983. In *Linear programming*. W. H. Freeman and Company.

Dakin, R. 1965. "A tree search algorithm for mixed integer programming problems." *Comput. J.* **8**: 250–255.

Dantzig, G. B. 1963. In *Linear programming and extensions*. Princeton University Press.

Dantzig, G. B. and P. Wolfe. 1960. "Decomposition principle for linear programming." *Oper. Res.* **8**: 101–111.

Délorme, J. 1974. Contributions à la résolution du problème de recouvrement: méthodes de troncature. Docteur-ingénieur dissertation. Paris: Université Paris VI.

Desrochers, J. and F. Soumis. 1989. "A column-generation approach to the urban transit crew scheduling problem." *Transport. Sci.* **23**: 1–13.

Forrest, J. J. H., J. P. H. Hirst, and J. A. Tomlin. 1974. "Practical solution of large mixed integer programming problems with UMPIRE." *Mgmnt Sci.* **20**: 739–773.

Fox, B. L., J. K. Lenstra, A. H. G. Rinnooy Kan, and L. E. Schrage. 1978. "Branching from the largest upper bound—folklore and facts." *Eur. J. Oper. Res.* **2**: 191–194.

Garfinkel, R. S. and G. L. Nemhauser. 1972. In *Integer Programming*. John Wiley & Sons.

Gomory, R. E. 1963. "An algorithm for integer solutions to linear programs." In *Recents Advances in Mathematical Programming*, edited by R. L. Graves, and P. Wolfe, pp. 269–302. McGraw-Hill.

Gonzaga, C. C. 1989. Algoritmos de pontos interiores em programação linear. Rio de Janeiro: Primeira Escola Brasileira de Otimização, UFRJ.

Greenberg, H. 1971. In *Integer Programming*. Academic Press.

Hansen, P., M. Minoux, and M. Labbé. 1987. "Extension de la programmation linéaire généralisée au cas des programmes mixtes." *C.R. Acad. Sci. Paris* **305(I)**: 569–572.

Hansen, P., B. Jaumard, and M. Poggi de Aragão. 1991. "Un algorithme primal de programmation linéaire généralisée pour les programmes mixtes." *C.R. Acad. Sci. Paris* **313(I)**: 557–560.

Hu, T. C. 1969. In *Integer Programming and Network Flows*. Addison Wesley.

Ibaraki, T. 1977. "The power of dominance relations in branch-and-bound algorithms." *J. ACM* **24**: 264–279.

Ibaraki, T. 1982. "The power of upper and lower bounding functions in branch-and-bound algorithms." *J. Oper. Res. Soc. Jpn* **25**: 292–320.

Jardim Campos, M. H. 1987. "Solução por métodos de subgradientes de primeira ordem dos problemas de relaxação lagrangeana em programação combinatória." Ph.D. thesis. Rio de Janeiro: COPPE, Federal University of Rio de Janeiro.

Jardim Campos, M. H. and N. Maculan. 1988. "Problemas de otimização relacionados ao corte de bobinas de papel." In *Proceedings of the XVI Colóquio Brasileiro de Matemática*, pp. 159–182.

Jardim Campos, M. H. and N. Maculan. 1995. "Optimization problems related to the cut of paper reels: a dual approach." *Investigación Operativa* **5**: 45–53.

Karmarkar, N. 1984. "A new polynomial-time algorithm for linear-programming." *Combinatorica* **4**: 373–395.

Korbut, A. and Y. Finkelstein. 1969. In *Diskretnoje Programmirovanie*. Nauka.

Kovács, L. B. 1980. In *Combinatorial Methods of Discrete Programming*. Akadémia Kiadó.

Land, A. and A. Doig. 1960. "An automatic method for solving discrete programming problems." *Econometrica* **28**: 497–520.

Little, J., K. Murty, D. Sweeney, and C. Karel. 1963. "An algorithm for the traveling salesman problem." *Oper. Res.* **11**: 972–989.

Lovász, L. and A. Schrijver. 1991. "Cones of matrices and set-functions and 0–1 optimization." *SIAM J. Optim.* **1**: 166–190.

Lucena, A. and J. E. Beasley. 1996. "Branch and cut algorithms." In *Advances in Linear and Integer Programming*, edited by J. E. Beasley, pp. 187–221. Oxford Science Publications.

Maculan, N. 1978. In *Programação Linear Inteira*. Editora COPPE-UFRJ.

Maculan, N., P. Michelon, and G. Plateau. 1992. "Column-generation in linear programming with bounding variable constraints and its applications in integer programming." *Pesquisa Operacional* **12**: 45–57.

Maculan, N., M. H. C. Fampa, and P. Michelon. 1998. "Programação linear e inteira." Notes. Rio de Janeiro: Systems Engineering and Computer Science, COPPE, Federal University of Rio de Janeiro.

Martello, S. and P. Toth. 1990. In *Knapsak Problems: Algorithms and Computer Implementations*. John Wiley & Sons.

Murty, K. G. 1976. In *Linear and Combinatorial Programming*. John Wiley & Sons.

Nemhauser, G. L., A. H. G. Rinnooy Kan, and M. J. Todd. 1989. In *Optimization*. Volume 1, Handbooks in Operations Research and Management Science. Amsterdam: North-Holland.

Nemhauser, G. L. and L. A. Wolsey. 1988. In *Integer and Combinatorial Optimization*. John Wiley & Sons.

Nilsson, N. J. 1980. In *Principles of Artificial Intelligence*. Tioga Publishing Company.

Parker, R. G. and R. L. Rardin. 1988. In *Discrete Optimization*. Academic Press.

Ribeiro, C. C., M. Minoux, and M. C. Penna. 1989. "An optimal column-generation-with-ranking algorithm for very large scale partitioning problems in traffic assignment." *Eur. J. Oper. Res.* **41**: 232–239.

Salkin, H. M. 1975. In *Integer Programming*. Addison Wesley.

Schijver, A. 1986. In *Theory of Linear and Integer Programming*. John Wiley & Sons.

Simonnard, M. 1973. In *Programmation Linéaire—Technique du Calcul Economique*. Dunod.
Taha, H. A. 1975. In *Integer Programming—Applications, and Computations*. Academic Press.
Vance, P. H. 1988. "Branch-and-price algorithms for the one-dimensional cutting stock problem." *Comput. Optim. Appl.* **9**: 211–228.
Vance, P. H., C. Barnhart, E. L. Johnson, and G. L. Nemhauser. 1994. "Solving binary cutting stock problems by column generation and branch-and-bound." *Comput. Optim. Appl.* **3**: 111–130.
Vanderbeck. F. 1994. "Decomposition and column generation for integer programming." Ph.D. thesis. Belgium: Universté Catholique de Louvain.
Wolsey, L. A. 1998. In *Integer Programming*. John Wiley & Sons.
Zionts, S. 1974. In *Linear and Integer Programming*. Prentice-Hall.

9.2 Linear 0–1 Programming

Nelson Maculan

ABSTRACT We discuss topics related to linear 0–1 programming, including problem formulation, the algorithm of Balas for linear 0–1 programming, and the use of linear relaxation in cutting plane and branch-and-bound procedures.

1. Introduction

When all variables in a linear programming problem are constrained to be 0 or 1, we have a linear 0–1 programming problem, also known as a binary linear, or a pseudo-Boolean linear program. An example of a linear 0–1 programming problem is

$$(P): \text{maximize } x_0 = x_1 - 3x_2 - 4x_3$$

subject to:

$$2x_1 + x_2 - x_3 \leqslant 4$$

$$4x_1 - 3x_2 \leqslant 2$$

$$3x_1 + 2x_2 + x_3 \leqslant 3$$

$$x_1 \in \{0,1\}$$

$$x_2 \in \{0,1\}$$

$$x_3 \in \{0,1\}.$$

A large number of combinatorial optimization problems are modeled as linear 0–1 programs. Examples include the traveling salesman problem, the set partitioning problem, the set covering problem, the set packing problem, and the Steiner tree problem in graphs (see Lawler et al., 1985; Garfinkel and Nemhauser, 1972; Salkin, 1975; Taha, 1975; Maculan, 1987; Winter; 1987, and other sections of this handbook).

Several industrial applications, such as facility location, fixed charge cost, scheduling, and network design, have been formulated and solved by linear 0–1 programming techniques (see Nemhauser and Wolsey, 1988). Some of these problems are described in this handbook.

Given a combinatorial optimization problem, one can formulate many different models for it. Computationally, a good 0–1 formulation is of crucial importance (Nemhauser and Wolsey, 1988). If one assumes that the integer variables in a general integer linear programming problem are bounded, then one can transform this general integer linear program into a linear 0–1 program. In the knapsack example (KP) of Chapter 10, variables x_1, x_2, x_3, x_4 that are nonnegative and integer. The knapsack constraint implies that $0 \leqslant x_1 \leqslant 6$, $0 \leqslant x_2 \leqslant 3$, $0 \leqslant x_3 \leqslant 9$, and $0 \leqslant x_4 \leqslant 27$. A simple way to make the transformation considers, for example, that $x_1 = \sum_{i=1}^{6}\lambda_i, \lambda_i \in \{0,1\}$, for $i = 1, 2, ..., 6$. That is, x_1 is replaced by $\sum_{i=1}^{6}\lambda_i, \lambda_i \in \{0,1\}$, $i = 1, 2, ..., 6$ in (KP). The same idea is applied to x_2, x_3, and x_4, resulting in $6 + 3 + 9 + 27 = 45$ binary variables for this equivalent linear 0–1 knapsack problem.

Suppose $x \in \mathbb{Z}$, $p \in \mathbb{Z}_+$, and $0 \leqslant x \leqslant p$. Then, a better idea consists of finding $q \in \mathbb{Z}_+$, such that $2^{q+1} \leqslant p < 2^{q+2}$, and

$$x = \sum_{i=0}^{q} 2^i \lambda_i + [p - (2^{q+1} - 1)]\lambda_{q+1}, \quad \lambda_i \in \{0,1\}, \quad i = 1, 2, ..., q, q+1.$$

For (KP), we have

$$x_1 = \lambda_1 + 2\lambda_2 + 3\lambda_3$$

$$x_2 = \lambda_4 + 2\lambda_5$$

$$x_3 = \lambda_6 + 2\lambda_7 + 4\lambda_8 + 2\lambda_9$$

$$x_4 = \lambda_{10} + 2\lambda_{11} + 4\lambda_{12} + 8\lambda_{13} + 12\lambda_{14}$$

$$\lambda_i \in \{0,1\}, \quad i = 1, 2, ..., 14,$$

resulting in a new equivalent linear 0–1 knapsack problem with only 14 binary variables, that is,

$$\text{(KP)} : \text{maximize } x_0 = 16\lambda_1 + 32\lambda_2 + 48\lambda_3 + 24\lambda_4 + 48\lambda_5 + 7\lambda_6 + 14\lambda_7 + 28\lambda_8 + 14\lambda_9 + 2\lambda_{10} + 4\lambda_{11} + 8\lambda_{12} + 16\lambda_{13} + 24\lambda_{14},$$

subject to:

$$4\lambda_1 + 8\lambda_2 + 12\lambda_3 + 7\lambda_4 + 14\lambda_5 + 3\lambda_6 + 6\lambda_7 + 12\lambda_8 + 6\lambda_9 + \lambda_{10} + 2\lambda_{11} + 4\lambda_{12} + 8\lambda_{13} + 12\lambda_{14} \leqslant 27,$$

$$\lambda_i \in \{0,1\}, \quad i = 1, 2, ..., 14.$$

In some applications, a variable x can be assumed to have discrete or continuous values. For example, $x \in X = \{2\} \cup [5,7] \cup \{10\} \cup [15, +\infty)$. Suppose that x has an upper bound M. Then, we can write x as a function of binary variables λ_1, λ_2, λ_3, and λ_4:

$$2\lambda_1 + 5\lambda_2 + 10\lambda_3 + 15\lambda_4 \leqslant x \leqslant 2\lambda_1 + 7\lambda_2 + 10\lambda_3 + M\lambda_4,$$

$$\sum_{i=1}^{4}\lambda_i = 1, \quad \lambda_i \in \{0,1\}, \quad i = 1, 2, 3, 4.$$

If $\lambda_1 = 1$, then $2 \leqslant x \leqslant 2$; if $\lambda_2 = 1$, then $5 \leqslant x \leqslant 7$; if $\lambda_3 = 1$, then $10 \leqslant x \leqslant 10$; and if $\lambda_4 = 1$, then $15 \leqslant x \leqslant M$.

The formulation using binary variables is sometimes a useful tool for optimization problems. We are able to formulate a large number of optimization problems using binary variables, but we cannot guarantee to solve all of them in a reasonable amount of time since the number of binary solutions (feasible of infeasible) can be enormous. Suppose we have 100 binary variables in a linear 0–1 programming problem, the number of feasible solutions can be $O(2^{100})$. We invite the reader to evaluate the amount of time using a large computer, just to inspect the feasibility of these 2^{100} solutions in a linear 0–1 program. Despite this, we can routinely solve large linear 0–1 programming problems using classical branch-and-bound schemes implemented in modern commercial codes for mixed-integer programming (see e.g., Lisser et al., 1998 for the solution of mixed-integer programming problems associated with network design problems with 150,000 variables, where 2,000 are binary, and 45,000 linear constraints).

General cutting plane and implicit enumeration methods can be used for solving linear 0–1 programming problems, but specialized methods for this are available. As an example, consider the Balas algorithm for pure linear 0–1 programming (Balas, 1963, 1965), which is an interesting method that uses additions and comparisons. Special ordered sets, as the multiple choice constraints, give us strategies to branch in an enumeration procedure or in a cutting plane method (see Balas, 1979 and Parker and Rardin, 1988).

2. Balas Algorithm

Balas (1963, 1965) presented an algorithm to solve

$$(P): \text{minimize } \ x_0 = \sum_{j=1}^{n} c_j x_j$$

subject to:

$$\sum_{j=1}^{n} a_{ij} x_j \leqslant b_i, \quad i = 1, 2, ..., m$$

$$x_j \in \{0, 1\}, \quad j = 1, 2, ..., n,$$

where c_j, a_{ij}, and b_i are given real numbers and x_j, $j = 1, 2, ..., n$, are binary variables. Without loss of generality, we can suppose $c_j \geqslant 0$, $j = 1, 2, ..., n$, since if we had a $c_k < 0$, we would take $x_k = 1 - t_k$, $t_k \in \{0, 1\}$ and replace x_k by $1 - t_k$, $t_k \in \{0, 1\}$ in (P). If $I = \{1, 2, ..., n\}$, then we have $x_0 = \sum_{j \in I - \{k\}} c_j x_j - c_k t_k + c_k$ and minimize

$$x_0 - c_k = \sum_{j \in I - \{k\}} c_j x_j - c_k t_k.$$

Define $c^T = (c_1 \ c_2 \ ... \ c_n)$, $b^T = (b_1 \ b_2 \ ... \ b_m), A = (a_{ij})_{m \times n,}$ to be a matrix with m rows and n columns, and $x^T = (x_1 \ x_2 \ ... \ x_n)$. Then, (P) can be written as

$$(P): \text{minimize } \ x_0 = c^T x \tag{1}$$

subject to:

$$Ax \leqslant b \tag{2}$$

$$x \in \{0, 1\}^n. \tag{3}$$

A solution of (1)–(3) is represented by a vector $x^p = (x_1^p \ x_2^p \ ... \ x_n^p)^T$, or by the set $J_p = \{j \mid x_j = 1\}$. For example, if $x^7 = (0\ 1\ 1\ 1\ 0\ 0\ 1\ 1)^T \in \{0, 1\}^8$, then $J_7 = \{2, 3, 4, 7, 8\}$.

A solution x^q of (1)–(3) is a descendant of another solution x^p of (1)–(3) if $J_p \subset J_q$. For example, $x^q = (1\ 1\ 1\ 0\ 1\ 1)^T$ is a descendant of $x^p = (0\ 0\ 1\ 0\ 1\ 1)^T$.

Sometimes we can set, *a priori*, one or more binary variables as follows:

$$3x_1 + 7x_2 + 2x_3 + x_4 \leqslant 5 \Rightarrow x_2 = 0,$$

$$2x_1 + 8x_2 + x_3 + x_4 \geqslant 9 \Rightarrow x_2 = 1.$$

In this example, if both constraints were in (2), then $F(P) = \varnothing$, that is, (P) is infeasible.

We can easily verify that:

- If x^k is a descendant of x^p, then $\sum_{j \in J_k} \geqslant \sum_{j \in J_p} c_j x_j$.
- If $x^0 = (0\ 0 \ ... \ 0)^T$ also satisfies (2), then x^0 is an optimal solution of (P).

We use these two assertions to develop the implicit enumeration scheme for the Balas algorithm. This enumeration scheme starts from node zero associated with $x^0 = (0\ 0 \ ... \ 0)T$, or $J_0 = \varnothing$. If x_0 satisfies (2), then x^0 is an optimal solution, as we verified above.

Suppose we are in a node associated with x^p, and we would like to find a descendant x^q of x^p, such that, $J_q = J_p \cup \{l\}$, $l \notin J_p$. To choose $l \notin J_p$, consider three stopping rules:

1. If x^p is a feasible solution (P), in this case, then all descendants of x^p will not give a better value for x^0, as was pointed out before. Therefore, the node associated with x^p will be closed.

2. Suppose $\bar{x}^0$ is the best known value for x_0, which is a feasible solution for (P). If no feasible solution is known, define $\bar{x}^0 = +\infty$. If

$$\sum_{j \in J_p} c_j + c_l \geqslant \bar{x}_0, \quad \forall l \notin J_p,$$

then all descendants of x^p will not give better values for x_0.

3. If an index i exists, such that

$$b_i - \sum_{j \in J_p} a_{ij} - \sum_{j \notin J_p} \min\{0, a_{ij}\} < 0,$$

then all descendants of x^p will be infeasible for (P).

If a node associated with x^{P} verifies one of these three rules, then this node will be closed.

Now, suppose that x^{P} is associated with an open node. Let $C_p = \{1, 2, ..., n\} \backslash J_p$ be the set of indices of binary variables to be chosen to form J_q, that is, $J_q = J_p \cup \{l\}$, $l \in C_p$. Which index l is to be selected?

Define $s_i = b_i - \sum_{j=1}^{n} a_{ij} x_j$ and $s_i \geqslant 0$, $i = 1, 2, ..., m$, as a slack variable associated with the ith constraint in (P). Let $s_i^p = b_i - \sum_{j \in J_p} a_{ij}$. Slack variable s_i^p gives the value s_i when $x = x^p$. Let

$$d_j^p = \sum_{i=1}^{m} \min\{0, s_i^p - a_{ij}\}, \quad j \in C_p \text{ and } d_l^p = \max_{j \in C_p}\{d_j^p\},$$

where $d_j^p \leqslant 0$ is an empirical measure of infeasibility resulting from setting $x_j = 1$. If $d_l^p = 0$, then $J_q = J_p \cup \{l\}$ is associated with a feasible solution for (P). Suppose $L_p = \{j \in C_p \mid d_j^p = 0\}$. Index $l \in L_p$ is associated with c_l such that $c_l = \min_{j \in L_p}\{c_j\}$. Once x^q is chosen, we update $C_p := C_p \backslash \{l\}$. Now, from x^q we continue the enumeration.

When we close a node, backtracking is provided, that is, if x^q is associated with a closed node we return to its parent x^p, for which we have a new C_p. The enumeration scheme is complete when $C_0 = \varnothing$.

The convergence of the algorithm needs some additional attention to avoid enumerating the same solution twice. To guarantee convergence, we can use a simple scheme as presented in Glover (1965), Geoffrion (1967), and Taha (1975).

3. LP Relaxation

LP relaxations of linear 0–1 programming problems are used in cutting plane methods or in branch-and-bound procedures. We will give some ideas concerning their use.

We consider the linear 0–1 program (P), defined by (1)–(3) and replace $x \in \{0, 1\}^n$ by $x \in [0, 1]^n$. Thus, we obtain the linear relaxation of (P)

$$(\bar{P}) : \text{minimize } x_0 = c^T x \tag{4}$$

subject to:

$$Ax \leqslant b \tag{5}$$

$$x \in [0, 1]^n. \tag{6}$$

We assume $F(\bar{P}) = \{x \in [0, 1]^n \mid Ax \leqslant b\} \neq \varnothing$ is a bounded polyhedron and $F(P) = \{x \in \{0, 1\}^n \mid Ax \leqslant b\} \neq \varnothing$.

Proposition 1. *All* $x \in F(P)$ *is an extreme point of* $F(\bar{P})$.

For a proof of Proposition 1 see Nemhauser and Wolsey (1988) and Taha (1975). Noting that extreme point of $F(\bar{P})$ may not be in $F(P)$, we can develop heuristics to obtain a good feasible solution for (P) from a noninteger optimal solution for $F(\bar{P})$ (see Nemhauser and Wolsey, 1988). In a branch-and-bound scheme to solve a linear 0–1 programming problem we can also use Proposition 1 and such heuristics to prune branches in the tree procedure.

Sometimes we wish to eliminate a solution $\bar{x} \in \{0, 1\}^n$ from further consideration. This can be done with a column generation procedure during a branch-and-bound method. We would like to avoid the generation of a column associated with $\bar{x}$. How can one find a valid inequality for $\mathscr{B} = \{0,1\}^n - \{\bar{x}\}$? We propose to consider all points of a sphere centered at $\bar{x}$ with radius equal to 1, that is, satisfying $\sum_{j=1}^{n}(x_j - \bar{x}_j)^2 = 1$. All the neighbors of $\bar{x}$ in the cube whose vertices are all points of {0, 1}^{n} belong to this sphere. If we want to consider all points of $\mathscr{B}$ we can write

$$\sum_{j=1}^{n}(x_j - \bar{x}_j)^2 \geqslant 1.$$

Since we just consider 0–1 points, that is, $x_j \in \{0,1\}$, then $x_j^2 = x_j$. Thus, $\sum_{j=1}^{n}(x_j - \bar{x}_j)^2 \geqslant 1 \Rightarrow \sum_{j=1}^{n}(x_j^2 - 2\bar{x}_j x_j + \bar{x}_j^2) \geqslant 1$ or $\sum_{j=1}^{n}(x_j - 2\bar{x}_j x_j + \bar{x}_j) \geqslant 1$, which implies

$$\sum_{j=1}^{n}(1 - 2\bar{x}_j)x_j \geqslant 1 - \sum_{j=1}^{n}\bar{x}_j. \tag{7}$$

It is easy to verify that (7) is facet of the convex hull of $\mathscr{B}$.

As an example, suppose $\bar{x} = (0\ 1\ 1\ 0)^T \in \{0,1\}^4$. Then (7) is $x_1 - x_2 - x_3 + x_4 \geqslant -1$.

The Problem

$$(P_1): \text{minimize}\ \ x_0 = c^T x$$

subject to

$$Ax \leqslant b$$

$$x \in \mathscr{B},$$

can also be written as

$$(P_1): \text{minimize}\ \ x_0 = c^T x$$

subject to

$$Ax \leqslant b$$

$$\sum_{j=1}^{n}(1 - 2\bar{x}_j)x_j \geqslant 1 - \sum_{j=1}^{n}\bar{x}_j$$

$$x \in \{0,1\}^n.$$

As an example, suppose we want to solve

$$(P): \text{minimize}\ \ x_0 = -2x_1 - 3x_2$$

subject to

$$9x_1 + 15x_2 \leqslant 20,$$

$$30x_1 + 12x_2 \leqslant 35$$

$$x_j \in \{0,1\},\quad j = 1,2.$$

Solving the linear relaxation $(\bar{P})$ of (P), we obtain $\bar{x}_1 = \bar{x}_2 = \frac{5}{6}$, and $\bar{x}_0 = -\frac{25}{6}$, which is infeasible for (P). Since the coefficients of x_1 and x_2 in the objective function are both negative, we take $x_1 = x_2 = 1$, which is infeasible for (P), resulting in $x_0 = -5$. We will include in $(\bar{P})$ a facet (7) for the convex hull of $\{0,1\}^2 \backslash \{(1\ 1)^T\}$, that is $x_1 + x_2 \leqslant 1$. This new problem has the optimal solution $x_1 = 0$, $x_2 = 1$, and $x_0 = -3$, which is also the optimal of (P).

In linear 0–1 programming problems, we can have special constraints of the type

$$\sum_{j \in Q} x_j = 1.$$

If we choose x_k, $k \in Q$ to branch on in a branch-and-bound procedure, in conventional branching we can add $x_k = 0$ or $\sum_{j \in Q \backslash \{k\}} x_j = 0$ to the linear relaxation problem considered. But, in this case, we could consider another branching scheme, namely $\sum_{j \in Q_1} x_j = 0$ or $\sum_{j \in Q_2} x_j = 0$, where $Q_1 \cup Q_2 = Q$, and $Q_1 \cap Q_2 = \emptyset$.

4. Remarks

In this handbook, the reader can find applications of linear 0–1 programming for which special techniques, based on the ideas developed here, are described.

References

Balas, E. 1963. "Programare liniara cu variabili bivalente." In *Proceedings of the Third Scientific Session on Statistics.*

Balas, E. 1965. "An additive algorithm for solving linear programs with zero–one variables." *Oper. Res.* **13**: 517–526.

Balas, E. 1979. "Disjunctive programming.' *Ann. Discrete Math.* **5**: 5–51.

Garfinkel, R. S. and G. L. Nemhauser. 1972. In *Integer Programming.* John Wiley & Sons.

Geoffrion, A. M. 1967. "Integer programming by implicit enumeration and Balas's method." *SIAM Rev.* **7**: 178–190.

Glover, F. 1965. "A multiphase-dual algorithm for the zero–one integer programming problem." *Oper. Res.* **13**: 879–919.

Lawler, E. L., J. K. Lenstra, A. H. G. Rinnooy Kan, and D. B. Schmoys 1985. In *The Traveling Salesman Problem: A Guide Tour of Combinatorial Optimization.* John Wiley & Sons.

Lisser, A., N. Maculan, M. Passini, and G. Weintraub. 1998. "A new model for 2-architecture problem in telecommunication network design." Presented at IX Congreso Latino-Ibero-Americano de Investigación Operativa (IX CLAIO). Buneos Aires, Argentina.

Maculan, N. 1987. "The Steiner problem in graphs." *Ann. Discrete Math.* **31**: 185–212.

Nemhauser, G. L. and L. A. Wolsey. 1988. In *Integer and Combinatorial Optimization.* John Wiley & Sons.

Parker, R. G. and R. L. Rardin. 1988. In *Discrete Optimization.* Academic Press.

Salkin, H. M. 1975. In *Integer Programming.* Addison-Wesley.

Taha, H. A. 1975. In *Integer Programming—Theory, Applications, and Computations.* Academic Press

Winter, P. 1987. "Steiner problems in networks: a survey." *Networks* **17**: 129–167.

9.3 Pseudo-Boolean Optimization

Yves Crama and Peter L. Hammer

ABSTRACT This article briefly surveys some fundamental results concerning pseudo-Boolean functions, that is, real-valued functions of 0–1 variables. Hundreds of papers have been devoted to the investigation of pseudo-Boolean functions, and a rich and diversified theory has now emerged from this literature. The article states local optimality conditions, outlines basic techniques for global pseudo-Boolean optimization, and shows connections between best linear approximations of pseudo-Boolean functions and game theory. It also describes models and applications arising in various fields, ranging from combinatorics to management science, and points to several classes of functions of special interest.

1. Pseudo-Boolean Functions

1.1. Definitions and Representations

A pseudo-Boolean function is a mapping from $\{0, 1\}^n$ to $\mathcal{R}$, that is, a real-valued function of a finite number of 0–1 variables. Pseudo-Boolean functions have been introduced in Hammer et al. (1963), and extensively studied in Hammer and Rudeanu (1968) and in numerous subsequent publications; a detailed survey appears in Boros and Hammer (2001).

Pseudo-Boolean functions generalize Boolean functions, which are exactly those pseudo-Boolean functions whose values are in $\{0, 1\}$, that is, those $f(X)$ for which $f^2(X) - f(X) \equiv 0$ in $\{0,$

$1\}^n$. Since the elements of $\{0, 1\}^n$ are in one-to-one correspondence with the subsets of $N = \{1, 2, ..., n\}$, every pseudo-Boolean function can be interpreted as a real-valued set function defined on $\mathcal{P}(N)$, the power set of $N = \{1, 2, ..., n\}$. Viewing pseudo-Boolean functions as defined on $\{0, 1\}^n$, rather than on $\mathcal{P}(N)$, provides an algebraic viewpoint, which sometimes carries clear advantages. It is easy to see for instance that the set of all pseudo-Boolean functions in n variables forms a vector space over $\mathcal{R}$, and that the elementary monomials $\prod_{i \in A} x_i$ $(A \in \mathcal{P}(N))$ define a basis of this space. In particular, every pseudo-Boolean function $f(x_1, x_2, ..., x_n)$ can be uniquely represented as a multilinear polynomial of the form

$$f(x_1, x_2, ..., x_n) = c_0 + \sum_{k=1}^{m} c_k \prod_{i \in A_k} x_i, \tag{1}$$

where $c_0, c_1, ..., c_m$ are real coefficients, and $A_1, A_2, ..., A_m$ are nonempty subsets of N. When viewed as a function on $[0, 1]^n$, the right-hand side of (1) defines a continuous extension of the pseudo-Boolean function f, to be denoted f^c.

Note that every pseudo-Boolean function also admits (many) representations of the form

$$f(x_1, x_2, ..., x_n) = b_0 + \sum_{k=1}^{m} b_k \left(\prod_{i \in A_k} x_i \prod_{j \in B_k} \bar{x}_j \right), \tag{2}$$

where $b_0, b_1, ..., b_m$ are real coefficients, and $\bar{x}_j = 1 - x_j$ for $j = 1, 2, ..., n$. If $b_k \geqslant 0$ for all $k = 1, 2, ..., m$, then we say that expression (2) is a posiform of f. It is easy to see that every pseudo-Boolean function can be expressed as a posiform. Other representations of pseudo-Boolean functions have recently been investigated in Foldes and Hammer (2000).

1.2. *Representative Models*

Besides nonlinear binary optimization, pseudo-Boolean functions can also be used to model a wide variety of problems in different fields of application.

Maximum Satisfiability

Consider a collection of Boolean clauses

$$\left(\bigvee_{i \in A_k} \bar{x}_i \vee \bigvee_{j \in B_k} x_j : k = 1, 2, ..., m \right).$$

The maximum satisfiability problem is to find a vector $(x_1, x_2, ..., x_n)$ in $\{0, 1\}^n$ that satisfies the largest possible number of clauses in the collection. This problem, which generalizes the NP-complete satisfiability problem, is equivalent to that of minimizing the posiform (2) with $b_k = 1$ for $k = 1, 2, ..., m$; see Hansen and Jaumard (1990).

Graph Theory

Consider a graph $G = (N, E)$ with positive weights $w : N \rightarrow \mathcal{R}$ on its vertices, and capacities $c : E \rightarrow \mathcal{R}$ on its (undirected) edges. For every $S \subseteq N$, the cut $(S, N \backslash S)$ is the set of edges having exactly one endpoint in S; the capacity of this cut is $\sum_{\{i,j\} \in (S, N \backslash S)} c(i, j)$. The max-cut problem is to find a cut of maximum capacity in G. This problem is eqivalent to maximizing the quadratic pseudo-Boolean function

$$f(x_1, x_2, ..., x_n) = \sum_{1 \leqslant i < j \leqslant n} c(i, j)(x_i \bar{x}_j + \bar{x}_i x_j), \tag{3}$$

under the interpretation that $(x_1, x_2, ..., x_n)$ is the characteristic vector of S.

A stable set in G is a set $S \subseteq N$ such that no edge has both of its endpoints in S; the weight of S is $\sum_{i \in S} w(i)$. The weighted stability problem is to find a stable set of maximum weight in G. This can be seen to be equivalent to maximizing the quadratic pseudo-Boolean function

$$f(x_1, x_2, ..., x_n) = \sum_{i=1}^{n} w(i)x_i - \left(1 + \max_{1 \leqslant i \leqslant n} w(i)\right) \sum_{1 \leqslant i<j \leqslant n} x_i x_j, \tag{4}$$

where $(x_1, x_2, ..., x_n)$ is the characteristic vector of S. Other connections between the weighted stability problem and pseudo-Boolean optimization (in particular, posiforms) have been exploited in Ebenegger et al. (1984).

LINEAR 0–1 PROGRAMMING
Consider the linear 0–1 program

$$\text{maximize} \quad z(x_1, x_2, ..., x_n) = \sum_{j=1}^{n} c_j x_j \tag{5}$$

$$\text{subject to} \quad \sum_{j=1}^{n} a_{ij} x_j = b_i, \quad i = 1, 2, ..., m \tag{6}$$

$$(x_1, x_2, ..., x_n) \in \{0, 1\}^n. \tag{7}$$

This problem is equivalent to the quadratic pseudo-Boolean optimization problem

$$\text{maximize} \quad f(x_1, x_2, ..., x_n) = \sum_{j=1}^{n} c_j x_j - M \sum_{i=1}^{m} \left(\sum_{j=1}^{n} a_{ij} x_j - b_i \right)^2 \tag{8}$$

$$\text{subject to} \quad (x_1, x_2, ..., x_n) \in \{0, 1\}^n, \tag{9}$$

for a sufficiently large M.

MANAGEMENT APPLICATIONS
Constraints of the form:

$$x_i = 1 \text{ if and only if } y = 1, \quad i \in A$$

are encountered in many management applications (e.g., capital budgeting, plant location, tool management problems, and so on; see Crama, 1997; Hammer and Rudeanu, 1968; Stecke, 1983). Since such constraints simply express that $y = \prod_{i \in A} x_i$, pseudo-Boolean formulations of such problems often arise quite naturally by elimination of the y-variables.

GAME THEORY
A game in characteristic function form is nothing but a pseudo-Boolean function f. If $(x_1, x_2, ..., x_n)$ is the characteristic vector of the set S, then $f(x_1, x_2, ..., x_n)$ is interpreted as the payoff that the players indexed in S can secure by acting together. The multilinear representation of f and its continuous extension f^c play an interesting role in this context; see Bilbao (2000) and Section 3 below.

1.3. Special Classes of Pseudo-Boolean Functions

Several authors have investigated special classes of functions with nice properties. Let us simply mention here monotonic functions (whose first derivatives, see below, have constant sign on $\{0, 1\}^n$), supermodular functions (whose second derivatives are nonnegative on $\{0, 1\}^n$), polar functions (which have a posiform (2) such that, for every k, either $A_k = \emptyset$ or $B_k = \emptyset$), unimodular functions (which are polar up to a switch $x_i \leftrightarrow \bar{x}_i$ on a subset of variables), (completely) unimodal functions (which have a unique local maximizer in (every face of) $\{0, 1\}^n$), and so on. Strongly polynomial combinatorial algorithms for the maximization of supermodular functions have recently been proposed in Iwata et al (2000), and Schrijver (2000). Unimodular functions can be maximized by max-flow algorithms (Hansen and Simeone, 1986). Unimodal and related classes of functions have been introduced in de Werra et al. (1988), where the applicability of greedy algorithms for their optimization has also been investigated. The recognition problem for all these classes of functions is examined in Crama (1989).

2. Optimization

Optimization of pseudo-Boolean functions over subsets of $\{0, 1\}^n$ is also known as nonlinear 0–1 optimization. A survey of this field is presented in Hansen et al. (1993). We shall only mention a few fundamental facts, restricting ourselves mostly to the unconstrained case.

2.1. Local Optima

If $f(x_1, x_2, ..., x_n)$ is a pseudo-Boolean function, let us define its ith derivative Δ_i to be the pseudo-Boolean function

$$\Delta_i = f(x_1, ..., x_{i-1}, 1, x_{i+1}, ..., x_n) - f(x_1, ..., x_{i-1}, 0, x_{i+1}, ..., x_n).$$

It was shown in Hammer and Rudeanu (1968) that all the local maxima of the function f are characterized by the system of implications:

$$\text{if } \Delta_i > 0 \text{ then } x_i = 1, \quad \text{if } \Delta_i < 0 \text{ then } x_i = 0, \quad \text{for } i = 1, 2, ..., n.$$

Now let m_i and M_i be arbitrary lower and upper bounds of Δ_i, for example, the sums of the negative and, respectively, the positive coefficients in the polynomial representation of Δ_i. Then it is clear that an equivalent characterization of the local maxima of the pseudo-Boolean function f is given by the system of inequalities

$$\Delta_i - M_i x_i \leqslant 0, \quad \Delta_i - m_i \bar{x}_i \geqslant 0, \quad \text{for } i = 1, 2, ..., n.$$

2.2. Global Optima

The continuous extension f^c has the attractive feature that its global maximizers are at the vertices of the hypercube $[0, 1]^n$ and hence, coincide with those of f (this is because (1) is linear in every variable). This implies that continuous global optimization techniques can be applied to f^c in order to compute the maximum of f. This approach did not prove computationally efficient in past experiments, but remains conceptually valuable. As an amusing corollary, one may note for instance that the optimum of the max-cut function (3) is at least $f^c(\frac{1}{2}) = \frac{1}{2}\sum_{1 \leqslant i<j \leqslant n} c(i, j)$. Thus, we find that every graph contains a cut of capacity at least equal to one-half the total edge capacity, a well-known result of graph theory. Moreover, such a cut can be found efficiently.

A combinatorial variable elimination algorithm for pseudo-Boolean optimization was proposed by Hammer et al. (1963) and Hammer and Rudeanu (1968). The following streamlined version and an efficient implementation of this algorithm are described in Crama et al. (1990). Let $f(x_1, x_2, ..., x_n)$ be the function to be maximized. We can write

$$f(x_1, x_2, ..., x_n) = x_1 \Delta_1(x_2, x_3, ..., x_n) + h(x_2, x_3, ..., x_n),$$

where h and the first derivative Δ_1, do not depend on x_1. Clearly, there exists a maximizer of f, say $(x_1^*, x_2^*, ..., x_n^*)$, with the property that $x_1^* = 1$ if and only if $\Delta_1(x_2^*, x_3^*, ..., x_n^*) > 0$. This suggests the introduction of a function $t(x_2, x_3, ..., x_n)$ such that $t(x_2, x_3, ..., x_n) = \Delta_1(x_2, x_3, ..., x_n)$ if $\Delta_1(x_2, x_3, ..., x_n) > 0$ and $t(x_2, x_3, ..., x_n) = 0$ otherwise. Letting $f_1 = t + h$, we have reduced the maximization of the original function f in n variables to the maximization of f_1, which only depends on $n-1$ variables. Repeating this elimination process n times eventually allows us to determine a maximizer of f. An efficient implementation of this algorithm is proposed in Crama et al. (1990), where it is also proved that the algorithm runs in polynomial time on pseudo-Boolean functions with bounded tree-width.

Another classical approach consists of transforming the problem $\max\{f(X) : X \in \{0, 1\}^n\}$ into an equivalent linear 0–1 programming problem by substituting a variable y_k for the kth monomial T_k of (1) (or (2)) and setting up a collection of linear constraints that enforce the equality $y_k = T_k$. The continuous relaxation of this linear formulation yields an easily computable upper-bound on the maximum of f. Properties of this upper bound and of related formulations of f have been investigated in Balas and Mazzola (1984) and Hammer et al. (1984) and in a series of subsequent papers; see Crama (1993) for a brief account and Section 2.3 for related considerations.

2.3. *Quadratic 0–1 Optimization*

Quadratic 0–1 optimization is an important special case of pseudo-Boolean optimization, both because numerous applications appear in this form, and because the more general case is easily reduced to it. Indeed, consider a function f of the form (1), assume that $|A_1| > 2$ and select $j, l \in A_1$. Then, the function

$$g(x_1, x_2, ..., x_n, y) = c_0 + c_1\left(\prod_{i \in A_1 \setminus \{j,l\}} x_i\right)y + \sum_{k=2}^{m} c_k \prod_{i \in A_k} x_i - M(x_j x_l - 2x_j y - 2x_l y + 3y),$$

where y is a new variable, and M is large enough, has the same maximum value as f ($y = x_j x_l$ in every maximizer of g). Applying this procedure recursively eventually yields a function of degree 2.

Best linear majorants and minorants of pseudo-Boolean functions can provide important information on the function. It was shown in Hammer et al. (1984) that for any quadratic pseudo-Boolean function f, one can construct a linear function

$$l(x_1, x_2, ..., x_n) = l_0 + \sum_{j=1}^{n} l_j x_j,$$

called the roof dual of f, majorizing $f(x_1, x_2, ..., x_n)$ in every binary point $(x_1, x_2, ..., x_n)$ and having the following property of strong persistency: If l_j is strictly positive (respectively, negative), then x_j must be equal to 1 (respectively, 0) in every maximizer of f. In other words, roof duality allows the determination of the optimal values of a subset of variables. Moreover, $\max\{f(X) : X \in \{0,1\}^n\} = \max\{l(X) : X \in \{0,1\}^n\}$ if and only if an associated 2-SAT problem is satisfiable. While the determination of the roof dual in Hammer et al. (1984) was accomplished via linear programming, it was later shown (Boros and Hammer, 1989) that this problem can be reduced to a max-flow problem in an associated network.

3. Linear Approximations

In order to find the best linear approximation $L(f)$ of a pseudo-Boolean function f in the norm L_2, it is sufficient to know how to determine the best linear L_2-approximation of a monomial. Indeed, considering the polynomial representation (1), it is clear that

$$L(f) = c_0 + \sum_{k=1}^{m} c_k L\left(\prod_{i \in A_k} x_i\right).$$

On the other hand, it was shown in Hammer and Holzman (1992) that

$$L\left(\prod_{i \in A} x_i\right) = \frac{1}{2^{|A|}}\left(1 - |A| + 2\sum_{i \in A} x_i\right) \quad \text{for all } A \subseteq N.$$

It was shown in the same paper that the best quadratic, cubic,... L_2-approximations can also be obtained by similar simple closed formulas.

Important game-theoretical applications of best L_2-approximations consist of finding the Banzhaf indices of the players of a simple game, or the Shapley values of the players of an n-person characteristic function game. Indeed, as shown in Hammer and Holzman (1992), these indices are simply the coefficients of best (weighted) linear L_2-approximations of pseudo-Boolean functions describing these games.

Another important application of these results allows the efficient determination of excellent heuristic solutions of unconstrained nonlinear binary optimization problems (Davoine et al., 2000).

ACKNOWLEDGMENTS

This work was partially supported by ONR grants N00014-92-J-1375 and N00014-92-J-4083, by NSF grant DMS-9806389, and by research grants from the Natural Sciences and Engineering Research Council of Canada (NSERC).

References

Balas, E. and J. B. Mazzola. 1984. "Nonlinear 0–1 programming. I. Linearization techniques." *Math. Programming* **30**: 1–22.

Bilbao, J. M. 2000. *Cooperative Games on Combinatorial Structures.* Boston-Dordrecht-London: Kluwer Academic Publishers.

Boros, E. and P. L. Hammer. 1989. A max-flow approach to improved roof duality in quadratic 0–1 minimization. Technical Report RRR 15-1989. Piscataway, NJ: RUTCOR—Rutgers University Center for Operations Research.

Boros, E. and P. L. Hammer. 2001. "Pseudo-Boolean optimization." To appear in *Discrete Appl. Math.*

Crama, Y. 1989. "Recognition problems for special classes of polynomials in 0–1 variables." *Math. Programming* **44**: 139–155.

Crama, Y. 1993. "Concave extensions for nonlinear 0–1 maximization problems." *Math. Programming* **61**: 53–60.

Crama, Y. 1997. "Combinatorial optimization models for production scheduling in automated manufacturing systems." *Eur. J. Oper. Res.* **99**: 136–153.

Crama, Y., P. Hansen, and B. Jaumard. 1990. "The basic algorithm for pseudo-Boolean programming revisited." *Discrete Appl. Math.* **29**: 171–185.

Davoine, T., P. L. Hammer, and B. Vizvari. 2000. "A heuristic for Boolean optimization problems." Technical Report RR 42-2000. Piscataway, NJ: RUTCOR—Rutgers University Center for Operations Research.

Ebenegger, Ch., P. L. Hammer, and D. de Werra. 1984. "Pseudo-boolean functions and stability of graphs." *Ann. Discrete Math.* **19**: 83–97.

Foldes, S. and P. L. Hammer. 2000. Disjunctive and conjunctive normal forms of pseudo-Boolean functions. Technical Report RRR 1-2000. Piscataway, NJ: RUTCOR—Rutgers University Center for Operations Research.

Hammer, P. L. and R. Holzman. 1992. "Approximations of pseudo-Boolean functions: applications to game theory." *ZOR—Methods and Models Oper. Res.* **36**: 3–21.

Hammer, P. L. and S. Rudeanu. 1968. *Boolean Methods in Operations Research and Related Areas.* Berlin/Heidelberg/New York: Springer-Verlag. Edition francaise: Dunod, Paris, 1970.

Hammer, P. L., I. Rosenberg, and S. Rudeanu. 1963. "On the determination of the minima of pseudo-boolean functions." *Stud. Cerc. Mat.* **14**: 359–364 (in Romanian).

Hammer, P. L., P. Hansen, and B. Simeone. 1984. "Roof duality, complementation and persistency in quadratic 0–1 optimization." *Math. Programming* **28**: 121–155.

Hammer, P. L., T. Liebling, B. Simeone, and D. de Werra. 1988. "From linear separability to unimodality: a hierarchy of pseudo-Boolean functions." *SIAM J. Discrete Math.* **1**: 174–184.

Hansen, P. and B. Jaumard. 1990. "Algorithms for the maximum satisfiability problem." *Computing* **44**: 279–303.

Hansen, P. and B. Simeone. 1986. "Unimodular functions." *Discrete Appl. Math.* **14**: 269–281.

Hansen, P., B. Jaumard, and V. Mathon. 1993. "Constrained nonlinear 0–1 programming." *ORSA J. Comput.* **5**: 97–119.

Iwata, S., L. Fleischer, and S. Fujishige. 2000. "A combinatorial, strongly polynomial-time algorithm for minimizing submodular functions." In *Proceedings of the 32nd ACM Symposium on Theory of Computing*, pp. 97–106. New York: Association for Computing Machinery.

Schrijver, A. 2000. "A combinatorial algorithm minimizing submodular functions in strongly polynomial time." *J. Combinat. Theory, Series B* **80**: 346–355.

Stecke, K. E. 1983. "Formulation and solution of nonlinear integer production planning problems for flexible manufacturing systems." *Mgmnt Sci.* **29**: 273–288.

9.4 Mixed-Integer Nonlinear Optimization

Christodoulos A. Floudas

ABSTRACT This article presents an introduction to the class of mixed-integer nonlinear (MINLP) optimization problems. Section 1 presents the motivation for studying MINLP problems Section 2 discusses the mathematical formulation and outlines the theoretical and algorithmic challenges. Section 3 provides an overview of the local MINLP optimization algorithms, which can address convex MINLP problems rigorously. Finally, Section 4 outlines global MINLP optimization approaches for nonconvex MINLP optimization problems.

1. Motivation

A wide range of nonlinear optimization problems involve integer or discrete variables in addition to continuous variables. These classes of optimization problems arise from a variety of applications and are denoted as mixed-integer nonlinear programming (MINLP) problems.

The integer variables can be used to model, for instance, sequences of events, alternative candidates, existence or nonexistence of units (in their zero-one representation), while discrete variables can model, for instance, different equipment sizes. The continuous variables are used to model the input-output and interaction relationships among individual units/operations and different interconnected systems.

The nonlinear nature of these mixed integer optimization problems may arise from (i) nonlinear relations in the integer domain exclusively (e.g., products of binary variables in the quadratic assignment model), (ii) nonlinear relations in the continuous domain only (e.g., complex nonlinear input-output model in a distillation column or reactor unit), (iii) nonlinear relations in the joint integer-continuous domain (e.g., products of continuous and binary variables in the scheduling/planning of batch processes and retrofit of heat recovery systems). The book by Nemhauser and Wolsey (1988) studies mixed-integer linear optimization and combinatorial optimization, while the book by Floudas (1995) studies mixed-integer nonlinear optimization problems.

The coupling of the integer domain and the continuous domain with their associated nonlinearities make the class of MINLP problems very challenging from the theoretical, algorithmic, and computational point of view. Mixed-integer nonlinear optimization problems are encountered in a variety of applications in all branches of engineering and applied science, applied mathematics, and operations research. These represent very important and active research areas that include:

Process Synthesis

- Heat exchanger networks
- Retrofit of heat recovery systems
- Distillation sequencing
- Mass exchange networks
- Reactor-based systems
- Reactor-separator-recycle systems
- Utility systems
- Total process systems
- Metabolic engineering

Process Design

- Reactive distillation
- Design of dynamic systems

Plant layout

Environmental design

Process Synthesis and Design Under Uncertainty

Uncertainty analysis

Dynamic systems

Batch plant design

Molecular Design

Solvent selection

Design of polymers and refrigerants

Property prediction under uncertainty

Interaction of Design, Synthesis and Control

Steady state operation

Dynamic operation

Process Operations

Scheduling of multiproduct plants

Design and retrofit of multiproduct plants

Synthesis, design, and scheduling of multipurpose plants

Planning under uncertainty

Facility Location and Allocation

Facility Planning and Scheduling

Topology of Transportation Networks

The applications in the area of process synthesis in chemical engineering include: (i) the synthesis of grassroot heat recovery networks (Floudas and Ciric, 1989; Ciric and Floudas, 1989, 1990; Yee et al., 1990a,b; Yee and Grossmann, 1990); (ii) the retrofit of heat exchanger systems (Ciric and Floudas, 1990; Papalexandri and Pistikopoulos, 1993); (iii) the synthesis of distillation-based separation systems (Paules and Floudas, 1988, 1992; Viswanathan and Grossmann, 1990; Aggarwal and Floudas, 1990, 1992; Novak et al., 1996; (iv) the synthesis of mass exchange networks (Gupta and Manousiouthakis, 1993; Papalexandri et al., 1994); (v) the synthesis of complex reactor networks (Kokossis and Floudas, 1990, 1994a,b; Schweiger and Floudas, 1998b); (vi) the synthesis of reactor-separator-recycle systems (Kokossis Floudas, 1991); (vii) the synthesis of utility systems (Kalitventzeff and Marechal, 1988); (viii) the synthesis of total process systems (Kocis and Grossmann, 1988, 1989a; Kravanja and Grossmann, 1990, 1996; Daichendt and Grossmann, 1994a,b; Papalexandri and Pistikopoulos, 1996); and (ix) the analysis and synthesis of metabolic pathways (Dean and Dervakos, 1996; Hatzimanikatis et al., 1996a,b; Petkov and Maranas, 1997a). Reviews of the mixed-integer nonlinear optimization frameworks and applications in process synthesis are provided in Grossmann (1990, 1996), Floudas (1995), and Adjiman et al. (1998c), while algorithmic advances for logic and global optimization in process synthesis are reviewed in Floudas and Grossmann (1995).

The MINLP applications in the area of process design include: (i) reactive distillation processes (Ciric and Gu, 1994); (ii) design of dynamic systems (Barton et al., 1998; Allgor and Barton, 1997; Schweiger and Floudas, 1997, 1998a); (iii) plant layout systems (Penteado and Ciric, 1996; Georgiadis et al., 1997); and (iv) environmentally benign systems (Ciric and Huchette, 1993; Stefanis et al., 1997).

The MINLP applications in the area of process synthesis and design under uncertainty include: (i) deterministic and stochastic uncertainty analysis (Grossmann and Floudas, 1987; Acevedo and Pistikopoulos, 1996; Dua and Pistikopoulos, 1998); (ii) design of dynamic systems under uncertainty (Dimitriadis and Pistikopoulos, 1995; Mohideen et al., 1996); and (iii) design of batch processes under uncertainty (Pistikopoulos and Ierapetritou, 1995; Ierapetritou and Pistikopoulos, 1996; Harding and Floudas, 1997; Petkov and Maranas, 1998).

In the area of molecular design, the MINLP applications include: (i) the computer-aided molecular design aspects of selecting the best solvents (Odele and Macchietto, 1993); (ii) design of polymers and refrigerants (Maranas, 1996; Churi and Achenie, 1996, 1997; Duvedi and Achenie, 1997; Raman and Maranas, 1999; Camarda and Maranas, 1999; Vaidyaraman and Maranas, 1999); and (iii) property prediction under uncertainty (Maranas, 1997).

The MINLP applications in the area of interaction of synthesis, design and control include: (i) studies under steady state operation of chemical processes (Luyben and Floudas, 1994a,b; Papalexandri and Pistikopoulos, 1994a,b); and (ii) studies under dynamic operation (Mohideen et al., 1996, 1997; Schweiger and Floudas, 1998a).

Applications of MINLP approaches have also emerged in the area of process operations and include: (i) short-term scheduling of batch and semicontinuous processes (Zhang and Sargent, 1994; Mockus and Reklaitis, 1996), (ii) the design of multiproduct plants (Grossmann and Sargent, 1979; Birewar and Grossmann, 1989, 1990); (iii) the synthesis, design and scheduling of multipurpose plants (Vaselenak et al., 1987a,b; Faqir and Karimi, 1990; Papageorgaki and Reklaitis, 1990a,b; Wellons and Reklaitis, 1991a,b; Sahinidis and Grossmann, 1991; Fletcher et al., 1991; Barbosa-Povoa and Macchietto, 1994; Xia and Macchietto, 1997); and (v) planning under uncertainty (Ierapetritou et al., 1996; Ierapetritou and Pistikopoulos, 1994, 1996; Liu and Sahinidis, 1997; Petkov and Maranas, 1997b). Reviews of the advances in the design, scheduling and planning of batch plants can be found in Reklaitis (1991) and Grossmann et al. (1992), while a collection of recent contributions can be found in the proceedings of the 1998 FOCAPO meeting.

MINLP applications have received significant attention in other engineering disciplines. These include (i) facility location in a multiattribute space (Ganish et al., 1983); (ii) optimal unit allocation in an electric power system (Bertsekas et al., 1983); (iii) facility planning of an electric power generation plant (Bloom, 1983; Rouhani et al., 1985); (iv) chip layout and compaction (Dorneigh and Sahinidis, 1995); (v) topology optimization of transportation networks (Hoang, 1982); and (vi) optimal scheduling of thermal generating units (Geromel and Belloni, 1986).

2. Mathematical Description

The general algebraic MINLP formulation can be stated as:

$$\min_{x,y} f(x,y) \tag{1}$$

$$\begin{aligned} \text{subject to} \quad & h(x,y)=0 \\ & g(x,y)\leqslant 0 \\ & x\in X\subseteq \mathcal{R}^n \\ & y\in Y \ \text{integer}. \end{aligned}$$

Here x represents a vector of n continuous variables (e.g., flows, pressures, compositions, temperatures, sizes of units), and y is a vector of integer variables (e.g., alternative solvents or materials); $h(x,y)=\mathbf{0}$ denote the m equality constraints (e.g., mass, energy balances, equilibrium relationships); $g(x,y)\leqslant\mathbf{0}$ are the p inequality constraints (e.g., specifications on purity of distillation products, environmental regulations, feasibility constraints in heat recovery systems, logical constraints); and $f(x, y)$ is the objective function (e.g., annualized total cost, profit, thermodynamic criteria).

Remark 1. *The integer variables* y *with given lower and upper bounds*

$$y^L\leqslant y\leqslant y^U$$

can be expressed through 0–1 variables (i.e., binary) denoted as, z, by the following formula:

$$y = y^L + z_1 + 2z_2 + 4z_3 + \ldots + 2^{N-1}z_N,$$

where N *is the minimum number of 0–1 variables needed. This minimum number is given by:*

$$N = 1 + INT\left\{\frac{\log(y^U - y^L)}{\log\ 2}\right\},$$

where the INT *function truncates its real argument to an integer value.*

Then, formulation (1) can be written in terms of 0–1 variables:

$$\min_{x,y}\ f(x,y) \tag{2}$$

$$\begin{aligned} \text{subject to}\quad & h(x,y) = 0 \\ & g(x,y) \leqslant 0 \\ & \quad x \in X \subseteq \mathscr{R}^n \\ & \quad y \in \{0,1\}^q, \end{aligned}$$

where y now is a vector of q 0–1 variables (e.g., existence of a process unit ($y_i = 1$) or nonexistence ($y_i = 0$)).

2.1. *Challenges in MINLP*

Dealing with mixed-integer nonlinear optimization models of the form (1) or (2) presents two major challenges. These difficulties are associated with the nature of the problem, namely, the combinatorial domain (y-domain) and the continuous domain (x-domain).

As the number of binary variables y in (2) increase, one is faced with a large combinatorial problem, and the complexity analysis results characterize MINLP problems as NP-complete (Nemhauser and Wolsey, 1988). At the same time, due to the nonlinearities, the MINLP problems are in general nonconvex, which implies the potential existence of multiple local solutions. The determination of a global solution of the nonconvex MINLP problems is also NP-hard, since even the global optimization of constrained nonlinear programming problems (Pardalos and Schnitger, 1988) and quadratic problems with one negative eigenvalue (Pardalos and Vavasis, 1991) are NP-hard. An excellent book on complexity issues for nonlinear optimization is the one by Vavasis (1991).

Despite the aforementioned discouraging results from complexity analysis, which are worst-case results, significant progress has been achieved in the MINLP area from the theoretical, algorithmic, and computational perspective. As a result, several algorithms have been proposed for convex and nonconvex MINLP models, their convergence properties have been investigated, and a large number of applications now exist that cross the boundaries of several disciplines. In the following, we will discuss these developments.

3. Overview of Local Optimization Approaches for Convex MINLP Models

A representative collection of local MINLP algorithms developed for solving convex MINLP models of the form (1) or restricted classes of (2) includes the following:

1. Generalized benders decomposition, GBD (Geoffrion, 1972; Paules and Floudas, 1989; Floudas et al., 1989);
2. Outer approximation, OA (Duran and Grossmann, 1986);
3. Outer approximation with equality relaxation, OA/ER (Kocis and Grossmann, 1987);

4. Outer approximation with equality relaxation and augmented penalty, OA/ER/AP (Viswanathan and Grossmann, 1990);
5. Generalized outer approximation, GOA (Fletcher and Leyffer, 1994);
6. Generalized cross decomposition, GCD (Holmberg, 1990);
7. Branch-and-bound, BB (Beale, 1977; Gupta, 1980; Ostrovsky et al., 1990; Borchers and Mitchell, 1991; Quesada and Grossmann, 1992; Fletcher and Leyffer, 1998);
8. Feasibility approach, FA (Mawengkang and Murtagh, 1986);
9. Extended cutting plane, ECP (Westerlund et al., 1994; Westerlund and Pettersson, 1995);
10. Logic-based approaches (Turkay and Grossmann, 1996; Vecchietti and Grossmann, 1999).

In the pioneering work of Geoffrion (1972) on generalized Benders decomposition (GBD) two sequences of updated upper (nonincreasing) and lower (nondecreasing) bounds are created that converge within ϵ in a finite number of iterations. The upper bounds correspond to solving subproblems in the x variables by fixing the y variables, while the lower bounds are based on duality theory.

The outer approximation (OA) addresses problems with nonlinear inequalities, and creates sequences of upper and lower bounds like the GBD, but it has the distinct feature of using primal information, that is the solution of the upper bound problems, so as to linearize the objective and constraints around that point. The lower bounds in OA are based upon the accumulation of the linearized objective function and constraints, around the generated primal solution points.

The OA/ER algorithm, extends the OA to handle nonlinear equality constraints by relaxing them into inequalities according to the sign of their associated multipliers.

The OA/ER/AP algorithm introduces an augmented penalty function in the lower bound subproblems of the OA/ER approach.

The generalized outer approximation (GOA) extends the OA to the MINLP problems that the GBD addresses and introduces exact penalty functions.

Generalized cross decomposition (GCD) simultaneously utilizes primal and dual information by exploiting the advantages of Dantzig–Wolfe and generalized Benders decomposition.

An overview of these local MINLP algorithms and extensive theoretical, algorithmic, and applications of GBD, OA, OA/ER, OA/ER/AP, GOA, and GCD algorithms can be found in the book by Floudas (1995).

The branch-and-bound (BB) approaches start by solving the continuous relaxation of the MINLP and subsequently perform an implicit enumeration where a subset of the 0–1 variables is fixed at each node. The lower bound corresponds to the NLP solution at each node and it is used to expand on the node with the lowest lower bound or it is used to eliminate nodes if the lower bound exceeds the current upper bound. If the continuous relaxation, NLP in most cases with the exception of the algorithm of Quesada and Grossmann (1992) where an LP problem is obtained, of the MINLP has a 0–1 solution for the y variables, then the BB algorithm will terminate at that node. With a similar argument, if a tight NLP relaxation results in the first node of the tree, then the number of nodes that would need to be eliminated can be low. However, loose NLP relaxations may result in having a large number of NLP subproblems to be solved. The algorithm terminates when the lowest lower bound is within a pre-specified tolerance of the best upper bound.

The feasibility approach, FA, rounds the relaxed NLP solution to an integer solution with the least local degradation by successively forcing the superbasic variables to become nonbasic based on the reduced cost information. The premise of this approach is that the problems to be treated are sufficiently large so that techniques requiring the solution of several NLP relaxations, such as the branch-and-bound approach, have prohibitively large costs. They therefore wish to account for the presence of the integer variables in the formulation and solve the mixed integer problem directly. This is achieved by fixing most of the integer variables to one of their bounds (the nonbasic variables) and allowing the remaining small subset (the basic variables) to take discrete values in order to identify feasible solutions. After each iteration, the reduced costs of the variables in the nonbasic set are computed to measure their effect on the objective function. If a change causes the objective function to decrease, the appropriate variables are removed from the nonbasic set and allowed to vary for the next iteration. When no more improvement in the objective function is possible, the algorithm is terminated. This strategy leads to the identification of a local solution.

The cutting plane algorithm proposed by Kelley (1960) for NLP problems has been extended to MINLPs (Westerlund et al., 1994; Westerlund and Pettersson, 1995). The ECP algorithm relies on the linearization of one of the nonlinear constraints at each iteration and the solution of the increasingly tight MILP made up of these linearizations. The solution of the MILP problem provides a new point on which to base the choice of the constraint to be linearized for the next iteration of the algorithm. The ECP does not require the solution of any NLP problems for the generation of an upper bound. As a result, a large number of linearizations are required for the approximation of highly nonlinear problems and the algorithm does not perform well in such cases. Due to the use of linearizations, convergence to the global optimum solution is guaranteed only for problems involving inequality constraints, which are convex in the x and relaxed y-space.

An alternative to the direct solution of the MINLP problem was proposed by Turkay and Grossmann (1996). Their approach stems from the work of Kocis and Grossmann (1989b) on a modeling/decomposition strategy, which avoids the zero flows generated by the nonexistence of a unit in a process network. The first stage of the algorithm is the reformulation of the MINLP into a generalized disjunctive program. A vector of boolean variables indicate the status of a disjunction (*true* or *false*) and are associated with the alternatives. The set of disjunctions allows the representation of several alternatives. A set of logical relationships between the boolean variables is introduced. Instead of resorting to binary variables within a single model, the disjunctions are used to generate a different model for each alternative. Since all continuous variables associated with the nonexisting alternatives are set to zero, this representation helps to reduce the size of the problems to be solved. Two algorithms are suggested by Turkay and Grossmann (1996). They are logic-based variants of the outer approximation and generalized Benders decomposition. Vecchietti and Grossmann (1999) introduced LOGMIP, a computer code for disjunctive programming and MINLP problems, and studied modeling alternatives and process synthesis applications.

4. Overview of Global Optimization Approaches for Nonconvex MINLP Models

In the previous section we discussed local MINLP algorithms, which are applicable to convex MINLP models. While identification of the global solution for convex problems can be guaranteed, a local solution is often obtained for nonconvex problems. The recent book by Floudas (2000) discusses the theoretical, algorithmic, and application oriented advances in the global optimization of mixed integer nonlinear models. A number of global MINLP algorithms that have been developed to address different types of nonconvex MINLPs are presented in this section. These include:

1. Branch-and-reduce approach (Ryoo and Sahinidis, 1995);
2. interval analysis based approach (Vaidyanathan and El-Halwagi, 1996);
3. extended cutting plane approach (Westerlund et al., 1994, 1998);
4. reformulation/spatial branch-and-bound approach (Smith and Pantelides, 1996, 1999);
5. hybrid branch-and-bound and outer approximation approach (Zamora and Grossmann, 1998a,b);
6. the SMIN-αBB approach (Adjiman et al., 1997, 2000);
7. the GMIN-αBB approach (Adjiman et al., 1997, 2000).

In the following, we will briefly discuss approaches 1–7.

4.1. *Branch-and-Reduce Algorithm*

Ryoo and Sahinidis (1995) extended the scope of branch-and-bound algorithms to problems for which valid convex underestimating NLPs can be constructed for the nonconvex relaxations. The range of application of the proposed algorithm encompasses bilinear problems and separable problems involving functions for which convex underestimators can be built (McCormick, 1976; Al-Khayyal, 1990). Because the nonconvex NLPs must be underestimated at each node, convergence can only be achieved if the continuous variables are branched on. A number of tests are suggested to accelerate the reduction of the solution space. They are summarized in the following.

Optimality-Based Range Reduction Tests

For the first set of tests, an upper bound U on the nonconvex MINLP must be computed and a convex lower bounding NLP must be solved to obtain a lower bound L. If a bound constraint for variable x_i, with $x_i^L \leqslant x_i \leqslant x_i^U$, is active at the solution of the convex NLP and has multiplier $\lambda_i^* > 0$, the bounds on x_i can be updated as follows:

1. If $x_i - x_i^U = 0$ at the solution of the convex NLP and $\kappa_i = x_i^U - \frac{U-L}{\lambda_i^*}$ is such that $\kappa_i > x_i^L$, then $x_i^L = \kappa_i$.
2. If $x_i - x_i^L = 0$ at the solution of the convex NLP and $\kappa_i = x_i^L + \frac{U-L}{\lambda_i^*}$ is such that $\kappa_i < x_i^U$, then $x_i^U = \kappa_i$.

If neither bound constraint is active at the solution of the convex NLP for some variable x_j, the problem can be solved by setting $x_j = x_j^U$ or $x_j = x_j^L$. Tests similar to those presented above are then used to update the bounds on x_j.

Feasibility Based Range Reduction Tests

In addition to ensuring that tight bounds are available for the variables, the underestimators of the constraints are used to generate new constraints for the problem. Consider the constraint $g_i(x, y) \leqslant 0$. If its underestimating function $g_i(x, y) = 0$ at the solution of the convex NLP and its multiplier is $\mu_i^* > 0$, the constraint

$$g_i(x, y) \geqslant -\frac{U-L}{\mu_i^*}$$

can be included in subsequent problems.

The branch-and-reduce algorithm has been tested on a set of small problems.

4.2. *Interval Analysis-Based Approach*

An approach based on interval analysis was proposed by Vaidyanathan and El-Halwagi (1996) to solve to global optimality problems with a twice-differentiable objective function and once-differentiable constraints. Interval arithmetic allows the computation of guaranteed ranges for these functions (Moore, 1979; Ratschek and Rokne, 1984; Neumaier, 1990). The approach relies on the same concepts of successive partitioning of the domain and bounding of the objective function, while the branching takes place on the discrete and continuous variables. The main difference with the branch-and-bound algorithms is that bounds on the problem solution in a given domain are not obtained through optimization. Instead, they are based on the range of the objective function in the domain under consideration, as computed with interval arithmetic. As a consequence, these bounds may be quite loose and efficient fathoming techniques are required in order to enhance convergence. Vaidyanathan and El-Halwagi (1996) suggested node fathoming tests and branching strategies which are outlined in the following. Convergence is declared when the best upper and lower bounds are within a prespecified tolerance and when the width of the corresponding region is below a prespecified tolerance.

Node Fathoming Tests

The upper bound test is a classical criterion used in all branch-and-bound schemes: if the lower bound for a node is greater than the best upper bound for the MINLP, the node can be fathomed.

The infeasibility test is also used by all branch-and-bound algorithms. However, the identification of infeasibility using interval arithmetic differs from its identification using optimization schemes. An inequality constraint $g_i(x, y) \leqslant 0$ is declared infeasible if its interval inclusion over the current domain, is positive. If a constraint is found to be infeasible, the current node is fathomed.

The monotonicity test is used in interval-based approaches. If a region is feasible, the monotonicity properties of the objective function can be tested. For this purpose, the inclusion of the gradients of the objective with respect to each variable are evaluated. If all the gradients have a constant sign for the current region, the objective function is monotonic and only one point needs to be retained from the current node.

The nonconvexity test is used to test the existence of a solution (local or global) within a region. If such a point exists, the Hessian matrix of the objective function at this point must be positive semidefinite. A sufficient condition is the nonnegativity of at least one of the diagonal elements of its interval Hessian matrix.

Vaidyanathan and El-Halwagi (1996) suggested two additional tests to accelerate the fathoming process. The first is denoted as the lower bound test. It requires the computation of a valid lower bound on the objective function through a method other than interval arithmetic. If the upper bound at a node is less than this lower bound, the region can be eliminated. The second test, the distrust region method, aims to help the algorithm identify infeasible regions so that they can be removed from consideration. Based on the knowledge of an infeasible point, interval arithmetic is used to identify an infeasible hypercube centered on that point.

BRANCHING STRATEGIES

The variable with the widest range is selected for branching. It can be a continuous or a discrete variable. In order to determine where to split the chosen variable, a relaxation of the MINLP is solved locally.

Case 1—Continuous branching variable: If the optimal value of the continuous branching variable, x^*, is equal to one of the variable bounds, branch at the midpoint of the interval. Otherwise, branch at $x^* - \beta$, where β is a very small scalar.

Case 2—Discrete branching variable: If the optimal value of the discrete branching variable, y^*, is equal to the upper bound on the variable, define a region with $y = y^*$ and one with $y^L \leqslant y \leqslant y^* - 1$, where y^L is the lower bound on y. Otherwise, create two regions $y^L \leqslant y \leqslant \text{int}(y^*)$ and $\text{int}(y^*) + 1 \leqslant y \leqslant y^U$, where y^U is the upper bound on y.

This algorithm has been tested on a small example problem and a molecular design problem (Vaidyanathan and El-Halwagi, 1996).

4.3. Extended Cutting Plane for Pseudoconvex MINLPs

The use of the ECP algorithm for nonconvex MINLP problems was suggested in Westerlund et al. (1994), using a modified algorithmic procedure as described in Westerlund et al. (1998). The main changes occur in the generation of new constraints for the MILP at each iteration (Step 4). In addition to the construction of the linear function $l_k(x, y)$ at iteration k, the following steps are taken:

1. Remove all constraints for which $l_i(x^k, y^k) > g_{j_i}(x^k, y^k)$. These correspond to linearizations that did not underestimate the corresponding nonlinear constraint at all points due to the presence of nonconvexities.
2. Replace all constraints for which $l_i(x^k, y^k) = g_{j_i}(x^k, y^k) = 0$ by their linearization around (x^k, y^k).
3. If constraint i is such that $g_i(x^k, y^k) > 0$, add its linearization around (x^k, y^k).

The convergence criterion is also modified. In addition to the test used in Step 3, the following two conditions must be met:

1. $(x^k - x^{k-1})^T(x^k - x^{k-1}) \leqslant \delta$, a pre-specified tolerance.
2. $y^k - y^{k-1} = 0$.

The ECP algorithm for pseudoconvex MINLPs has been used to address a trim loss problem arising in the paper industry (Westerlund et al., 1998). A comparative study between outer approximation, generalized Benders decomposition and the extended cutting plane algorithm for convex MINLPs was presented in Skrifvars et al. (1998).

4.4. Reformulation/Spatial Branch-and-Bound Algorithm

A global optimization algorithm of the branch-and-bound type was proposed in Smith and Pantelides (1996, 1997, 1999). It can be applied to problems in which the objective and constraints are functions involving any combination of binary arithmetic operations (addition, subtraction, multiplication, and division) and functions that are either concave over the entire solution space (such as ln) or convex over this domain (such as exp).

The algorithm starts with an automatic reformulation of the original nonlinear problem into a problem that involves only linear, bilinear, linear fractional, simple exponentiation, univariate concave, and univariate convex terms. This is achieved through the introduction of new constraints and variables. The reformulated problem is then solved to global optimality using a branch-and-bound approach. Its special structure allows the construction of a convex relaxation at each node of the tree. It should be noted that due to the introduction of many new constraints and variables the size of the convex relaxation of the reformulated problem increases substantially even for modest size problems. The integer variables can be handled in two ways during the generation of the convex lower bounding problem. The integrality condition on the variables can be relaxed to yield a convex NLP which can then be solved globally. Alternatively, the integer variables can be treated directly and the convex lower bounding MINLP can be solved using a branch-and-bound algorithm. This second approach is more computationally intensive but is likely to result in tighter lower bounds on the global optimum solution.

In order to obtain an upper bound on the optimum solution, a local MINLP algorithm can be used. Alternatively, the MINLP can be transformed to an equivalent nonconvex NLP by relaxing the integer variables. For example, a variable $y \in \{0, 1\}$ can be replaced by a continuous variable $z \in [0, 1]$ by including the constraint $z - z \cdot z = 0$.

This algorithm has been applied to reactor selection, distillation column design, nuclear waste blending, heat exchanger network design, and multilevel pump configuration problems.

4.5. Hybrid Branch-and-Bound and Outer Approximation

Zamora and Grossmann (1998a) proposed a global optimization MINLP approach for the synthesis of heat exchanger networks without stream splitting. This approach is a hybrid branch-and-bound with outer approximation. It is based on two alternative convex underestimators for the heat transfer area. The first type of these convex underestimators along with the variable bounds and techniques for the bound contraction are based on a thermodynamic analysis. The second type is based on a relaxation and transformation so as to employ specific underestimation schemes. These convex underestimators result in a convex MINLP that is solved using the outer approximation approach and which provides valid lower bounds on the global solution. This approach has been applied to five heat exchanger network examples that employ the MINLP model of Yee and Grossmann (1990) that contains linear constraints and nonconvex objective function.

Zamora and Grossmann (1998b) introduced a deterministic branch-and-contract approach for structured process systems that have univariate concave, bilinear, and linear fractional terms. They proposed properties of the contraction operation and studied their effect on several applications.

4.6. The SMIN-αBB Algorithm

The SMIN-αBB global optimization algorithm, proposed by Adjiman et al. (1997, 2000), is designed to solve mathematical models to global optimality where the binary/integer variables appear linearly and hence separably from the continuous variables and/or appear in at most bilinear terms, while nonlinear terms in the continuous variables appear separably from the binary/integer variables. These mathematical models take the form:

$$\begin{aligned} \min_{x,y} \quad & f(x) + x^T A_0 y + c_0^T y \\ \text{subject to} \quad & h(x) + x^T A_1 y + c_1^T y = 0 \\ & g(x) + x^T A_2 y + c_2^T y \leqslant 0 \\ & x \in \mathbf{X} \subseteq \mathscr{R}^n \\ & y \in Y \text{ integer,} \end{aligned}$$

where c_0^T, c_1^T, and c_2^T are constant vectors, A_0, A_1, and A_2 are constant matrices, and $f(x)$, $h(x)$, and $g(x)$ are functions with continuous second-order derivatives.

The theoretical, algorithmic, and computational studies of the SMIN-αBB algorithm are presented in detail in the recent book by Floudas (2000).

4.7. The GMIN-αBB Algorithm

The GMIN-αBB global optimization algorithm proposed by Adjiman et al. (1997, 2000) operates within a branch-and-bound framework. The main difference between it and the algorithms of Gupta and Ravindran (1985), Ostrovsky et al. (1990), and Borchers and Mitchell (1991) is its ability to identify the global optimum solution of a much larger class of problems of the form

$$\begin{aligned} \min_{x,y} \quad & f(x,y) \\ \text{subject to} \quad & h(x,y) = \mathbf{0} \\ & g(x,y) \leqslant \mathbf{0} \\ & x \in \mathbf{X} \subseteq \mathscr{R}^n \\ & y \in \mathscr{N}^q, \end{aligned}$$

where $\mathscr{N}$ is the set of nonnegative integers and the only condition imposed on the functions $f(x, y)$, $g(x, y)$, and $h(x, y)$ is that their continuous relaxations possess continuous second-order derivatives. This increased applicability results from the use of the αBB global optimization algorithm for continuous twice-differentiable NLPs (Androulakis et al., 1995; Adjiman and Floudas, 1996; Adjiman et al., 1998a,b).

The theoretical, algorithmic, and computational studies of the GMIN-αBB algorithm are presented in detail in the recent book by Floudas (2000).

References

Acevedo, J. and E. N. Pistikopoulos. 1996. "A parametric MINLP algorithm for process synthesis problems under uncertainty." *Ind. Engng Chem. Res.* **35(1)**: 147–158.

Adjiman, C. S. and C. A. Floudas. 1996. "Rigorous convex underestimators for general twice-differentiable problems." *J. Global Optim.* **9**: 23–40.

Adjiman, C. S., I. P. Androulakis, and C. A. Floudas. 1997. "Global optimization of MINLP problems in process synthesis and design." *Comput. Chem. Engng Suppl.* **21**: S445–S450.

Adjiman, C. S., S. Dallwig, C. A. Floudas, and A. Neumaier. 1998a. "A global optimization method, αBB, for general twice-differentiable NLPs—I. Theoretical advances." *Comput. Chem. Engng* **22(9)**: 1137–1158.

Adjiman, C. S., I. P. Androulakis, and C. A. Floudas. 1998b. "A global optimization method, α BB, for general twice-differentiable NLPs—II. Implementation and computational results." *Comput. Chem. Engng* **22(9)**: 1159–1179.

Adjiman, C. S., C. A. Schweiger, and C. A. Floudas. 1998c. "Mixed-integer nonlinear optimization in process synthesis." In *Handbook of Combinatorial Optimization*, edited by D.-Z. Du and P. M. Pardalos. Kluwer Academic Publishers.

Adjiman, C. S., I. P. Androulakis, and C. A. Floudas. 2000. "Global optimization of mixed-integer nonlinear problems." *AIChE J.* **46**: 1769–1797.

Aggarwal, A. and C. A. Floudas. 1990. "Synthesis of general distillation sequences-nonsharp separations." *Comput. Chem. Engng* **14(6)**: 631.

Aggarwal, A. and C. A. Floudas. 1992. "Synthesis of heat integrated nonsharp distillation sequences." *Comput. Chem. Engng* **16**: 89.

Al-Khayyal, F. A. 1990. "Jointly constrained bilinear programs and related problems: an overview." *Comput. Math. Appl.* **19**: 53.

Allgor, R. I. and P. I. Barton. 1997. "Mixed integer dynamic optimization." *Comput. Chem. Engng* **21**: S451–S456.

Androulakis, I. P., C. D. Maranas, and C. A. Floudas. 1995. "αBB: a global optimization for general constrained nonconvex problems." *J. Global Optim.* **7**: 337–363.

Barbosa-Povoa, A. P. and S. Macchietto. 1994. "Detailed design of multipurpose batch plants." *Comput. Chem. Engng* **18(11/12)**: 1013–1042.

Barton, P. I., R. J. Allgor, and W. F. Feehery. 1998. "Dynamic optimization in a discontinuous world." *Ind. Engng Chem. Res.* **37(3)**: 966–981.

Beale, E. M. L. 1977. "Integer programming." In *The State of the Art in Numerical Analysis*, pp 409–448. Academic Press.

Bertsekas, D. L., G. S. Lower, N. R. Sandell, and T. A. Posbergh. 1983. "Optimal short term scheduling of large-scale power systems." *IEEE Trans. Automatic Control* **AC-28**: 1.

Birewar, D. B. and I. E. Grossmann. 1989. "Incorporating scheduling in the optimal design of multiproduct batch plants." *Comput. Chem. Engng* **13**: 141–161.

Birewar, D. B. and I. E. Grossmann. 1990. "Simultaneous synthesis, sizing and scheduling of multiproduct batch plants." *Comput. Chem. Engng* **29(11)**: 2242.

Bloom, J. A. 1983. "Solving an electricity generating capacity expansion planning problem by generalized benders' decomposition." *Oper. Res.* **31(5)**: 84.

Borchers, B. and J. E. Mitchell. 1991. "An improved branch-and-bound algorithm for mixed integer nonlinear programs." Technical Report 200. Renssellaer Polytechnic Institute.

Camarda, K. V. and C. D. Maranas. 1999. "Optimization in polymer design using connectivity indices." *Ind. Engng Chem. Res.* **38**: 1884–1892.

Churi, N. and L. E. K. Achenie. 1996. "Novel mathematical programming model for computer aided molecular design." *Ind. Engng Chem. Res.* **35(10)**: 3788–3794.

Churi, N. and L. E. K. Achenie. 1997. "The optimal design of refrigerant mixtures for a two-evaporator refrigeration system." *Comput. Chem. Engng* **21(13)**: 349–354.

Ciric, A. R. and C. A. Floudas. 1989. "A retrofit approach of heat exchanger networks." *Comput. Chem. Engng* **13(6)**: 703.

Ciric, A. R. and C. A. Floudas. 1990. "A mixed-integer nonlinear programming model for retrofitting heat exchanger networks." *Ind. Engng Chem. Res.* **29**: 239.

Ciric, A. R. and D. Y. Gu. 1994. "Synthesis of nonequilibrium reactive distillation processes by MINLP optimization." *AIChE J.* **40(9)**: 1479–1487.

Ciric, A. R. and S. G. Huchette. 1993. "Multiobjective optimization approach to sensitivity analysis waste treatment costs in discrete process synthesis and optimization problems." *Ind. Engng Chem. Res.* **32(11)**: 2636–2646.

Daichendt, M. M. and I. E. Grossmann. 1994a. "Preliminary screening for the MINLP synthesis of process systems: I. Aggregation techniques." *Comput. Chem. Engng* **18**: 663.

Daichendt, M. M. and I. E. Grossmann. 1994b. "Preliminary screening for the MINLP synthesis of process systems: II. Heat exchanger networks." *Comput. Chem. Engng* **18**: 679.

Dean, J. P. and G. A. Dervakos. 1996. "Design of process-compatible biological agents." *Comput. Chem. Engng, Suppl. A* **20**: S67–S72.

Dimitriadis, V. D. and E. N. Pistikopoulos. 1995. "Flexibility analysis of dynamic systems." *Ind. Engng Chem. Res.* **34(12)**: 4451–4462.

Dorneigh, M. C. and N. V. Sahinidis. 1995. "Global optimization algorithms for chip layout and compaction." *Engng Optim.* **25(2)**: 131–154.

Dua, V. and E. N. Pistikopoulos. 1998. "Optimization techniques for process synthesis and material design under uncertainty." *Chem. Engng Res. Des.* **76(A3)**: 408–416.

Duran, M. A. and I. E. Grossmann. 1986. "An outer approximation algorithm for a class of mixed-integer nonlinear programs." *Math. Programming* **36**: 307.

Duvedi, A. and L. E. K. Achenie. 1997. "On the design of environmentally benign refrigerant mixtures: a mathematical programming approach." *Comput. Chem. Engng* **21(8)**: 915–923.

Faqir, N. M. and I. A. Karimi. 1990. "Design of multipurpose batch plants with multiple production routes." In *Proceedings FOCAPD'89 Snowmass, Colorado*, p. 451.

Fletcher, R. and S. Leyffer. 1994. "Solving mixed integer nonlinear programs by outer approximation." *Math. Programming* **66(3)**: 327–349.

Fletcher, R. and S. Leyffer. 1998. "Numerical experience with lower bounds for MIQP branch and bound." *SIAM J. Optim.* **8(2)**: 604–616.

Fletcher, R., J. A. J. Hall, and W. R. Johns. 1991. "Flexible retrofit design of multiproduct batch plants." *Comput. Chem. Engng* **15(12)**: 843.

Floudas, C. A. 1995. In *Nonlinear and Mixed Integer Optimization: Fundamentals and Applications.* New York: Oxford University Press.

Floudas, C. A. 2000. In *Deterministic Global Optimization: Theory, Methods and Applications.* Kluwer Academic Publishers.

Floudas, C. A. and A. R. Ciric. 1989. "Strategies for overcoming uncertainties in heat exchanger network synthesis." *Comput. Chem. Engng* **13(10)**: 1133.

Floudas, C. A. and I. E. Grossmann. 1995. "Algorithmic approaches to process synthesis: logic and global optimization." In *FOCAPD'94*, Volume 91 of *AICHE Symposium Series*, pp. 198–221

Floudas, C. A., A. Aggarwal, and A. R. Ciric. 1989. "Global optimum search for nonconvex NLP and MINLP problems." *Comput. Chem. Engng* **13(10)**: 1117–1132.

Ganish, B., D. Horsky, and K. Srikanth. 1983. "An approach to optimal positioning of a new product." *Mgmnt Sci.* **29**: 1277.

Geoffrion, A. M. 1972. "Generalized Benders decomposition." *J. Optim. Theory Appl.* **10**: 237–260.

Georgiadis, M. C., G. E. Rotstein, and S. Macchietto. 1997. "Optimal layout design in multipurpose batch plants." *Ind. Engng Chem. Res.* **36(11)**: 4852–4863.

Geromel, J. C. and M. R. Belloni. 1986. "Nonlinear progress with complicating variables: theoretical analysis and numerical experience." *IEEE Trans. Syst. Man. Cybernetics* **SMC-16**: 231.

Grossmann, I. E. 1990. "MINLP optimization strategies and algorithms for process synthesis." In *Proceedings of the 3rd. International Conference on Foundations of Computer-aided Process Design*, p. 105.

Grossmann, I. E. 1996. "Mixed integer optimization techniques for algorithmic process synthesis." In *Advances in Chemical Engineering, Process Synthesis*, Volume 23, edited by J. L. Anderson, pp. 171–246. Academic Press.

Grossmann, I. E. and C. A. Floudas. 1987. "Active constraint strategy for flexibility analysis in chemical processes." *Comput. Chem. Engng* **11(6)**: 675.

Grossmann, I. E. and R. W. H. Sargent. 1979. "Optimal design of multipurpose chemical plants." *Ind. Engng Chem. Proc. Des. Dev.* **18**: 343.

Grossmann, I. E., I. Quesada, R. Ramon, and V. T. Voudouris. 1992. "Mixed-integer optimization techniques for the design and scheduling of batch processes." In *Proceedings of NATO Advanced Study Institute on Batch Process Systems Engineering.*

Gupta, A. and V. Manousiouthakis. 1993. "Minimum utility cost of mass exchange networks with variable single component supplies and targets." *Ind. Engng Chem. Res.* **32(9)**: 1937–1950.

Gupta, O. K. 1980. "Branch and bound experiments in nonlinear integer programming." Ph.D. diss. Purdue University.

Gupta, O. K. and R. Ravindran. 1985. "Branch and bound experiments in convex nonlinear integer programming." *Mgmnt Sci.* **31(12)**: 1533–1546.

Harding, S. T. and C. A. Floudas. 1997. "Global optimization in multiproduct and multipurpose batch design under uncertainty." *Ind. Engng Chem. Res.* **36(5)**: 1644–1664.

Hatzimanikatis, V., C. A. Floudas, and J. E. Bailey. 1996a. "Analysis and design of metabolic reaction networks via mixed-integer linear optimization." *AIChE J.* **42(5)**: 1277–1292.

Hatzimanikatis, V., C. A. Floudas, and J. E. Bailey. 1996b. "Optimization of regulatory architectures in metabolic reaction networks." *Biotechnol. Bioengng* **52**: 485–500.

Hoang, H. H. 1982. "Topological optimization of networks: a nonlinear mixed integer model employing generalized Benders decomposition." *IEEE Trans. Aut. Contr.* **AC-27**: 164.

Holmberg, K. 1990. "On the convergence of the cross decomposition." *Math. Programming* **47**: 269.

Ierapetritou, M. G. and E. N. Pistikopoulos. 1994. "Simultaneous incorporation of flexibility and economic risk in operational planning under uncertainty." *Comput. Chem. Engng* **18(3)**: 163–189.

Ierapetritou, M. G. and E. N. Pistikopoulos. 1996. "Batch plant design and operations under uncertainty." *Ind. Engng Chem. Res.* **35(2)**: 772–787.

Ierapetritou, M. G., E. N. Pistikopoulos, and C. A. Floudas. 1996. "Operational planning under uncertainty." *Comput. Chem. Engng* **20(12)**: 1499–1516.

Kalitventzeff, F. and F. Marechal. 1988. "The management of a utility network." In *Process Systems Engineering. PSE'88, Sydney, Australia*, p. 223.

Kelley, J. E. 1960. "The cutting plane method for solving convex programs." *J. SIAM* **8(4)**: 703–712.

Kocis, G. R. and I. E. Grossman. 1987. "Relaxation strategy for the structural optimization of process flow sheets." *Ind. Engng Chem. Res.* **26(9)**: 1869.

Kocis, G. R. and I. E. Grossmann. 1988. "Global optimization of nonconvex MINLP problems in process synthesis." *Ind. Engng Chem. Res.* **27(8)**: 1407.

Kocis, G. R. and I. E. Grossmann. 1989a. "Computational experience with DICOPT solving MINLP problems in process systems engineering." *Comput. Chem. Engng* **13**: 307.

Kocis, G. R. and I. E. Grossmann. 1989b. "A modelling and decomposition strategy for the MINLP optimization of process flowsheets." *Comput. Chem. Engng* **13(7)**: 797–819.

Kokossis, A. C. and C. A. Floudas. 1990. "Optimization of complex reactor networks—I. Isothermal operation." *Chem. Engng Sci.* **45(3)**: 595.

Kokossis, A. C. and C. A. Floudas. 1991. "Optimal synthesis of isothermal reactor–separator–recycle systems." *Chem. Engng Sci.* **46**: 1361.

Kokossis, A. C. and C. A. Floudas. 1994a. 'Optimization of complex reactor networks—II. Nonisothermal operation." *Chem. Engng Sci.* **49(7)**: 1037.

Kokossis, A. C. and C. A. Floudas. 1994b. "Stability in optimal design: synthesis of complex reactor networks." *AIChE* **40(5)**: 849–861.

Kravanja, Z. and I. E. Grossmann. 1990. "PROSYN—an MINLP process synthesizer." *Comput. Chem. Engng* **14**: 1363.

Kravanja, Z. and I. E. Grossmann. 1996. "A computational approach for the modeling/decomposition strategy in the MINLP optimization of process flowsheets with implicit models." *Ind. Engng Chem. Res.* **35(6)**: 2065–2070.

Liu, M. L. and N. V. Sahinidis. 1997. "Process planning in a fuzzy environment." *Eur. J. Oper. Res.* **100(1)**: 142–169.

Luyben, M. L. and C. A. Floudas. 1994a. "Analyzing the interaction of design and control. Part 1: A multiobjective framework and application to binary distillation synthesis." *Comput. Chem. Engng* **18(10)**: 933–969.

Luyben, M. L. and C. A. Floudas. 1994b. "Analyzing the interaction of design and control. Part 2: Reactor–separator–recycle system." *Comput. Chem. Engng* **18(10)**: 971–994.

Maranas, C. D. 1996. "Optimal computer-aided molecular design: a polymer design case study." *Ind. Engng Chem. Res.* **35(10)**: 3403–3414.

Maranas, C. D. 1997. "Optimal molecular design under property prediction uncertainty." *AIChE J.* **43(5)**: 1250–1264.

Mawengkang, H. and B. A. Murtagh. 1986. "Solving nonlinear integer programs with large scale optimization software." *Ann. Oper. Res.* **5**: 425.

McCormick, G. P. 1976. "Computability of global solutions to factorable nonconvex programs: Part I—Convex underestimating problems." *Math. Programming* **10**: 147–175.

Mockus, L. and G. V. Reklaitis. 1996. "A new global optimization algorithm for batch process scheduling." In *State of the Art In Global Optimization*, edited by C. A. Floudas and P. M. Pardalos, pp. 521–533. Kluwer Academic Publishers.

Mohideen, M. J., J. D. Perkins, and E. N. Pistikopoulos. 1996. "Optimal design of dynamic systems under uncertainty." *AIChE J.* **42(8)**: 2251–2272.

Mohideen, M. J., J. D. Perkins, and E. N. Pistikopoulos. 1997. "Robust stability considerations in optimal design of dynamic systems under uncertainty." *J. Process Control* **7(5)**: 371–385.

Moore, R. E. 1979. In *Methods and Applications of Interval Analysis*. Philadelphia: SIAM.

Nemhauser, G. L. and L. A. Wolsey. 1988. In *Integer and Combinatorial Optimization*. John Wiley & Sons.

Neumaier, A. 1990. In *Interval Methods for Systems of Equations*. Encyclopedia of Mathematics and its Applications. Cambridge, England: Cambridge University Press.

Novak, Z., Z. Kravanja, and I. E. Grossmann. 1996. "Simultaneous synthesis of distillation sequences in overall process schemes using an improved MINLP approach." *Comput. Chem. Engng* **20(12)**: 1425–1440.

Odele, O. and S. Macchietto. 1993. "Computer aided molecular design: a novel method for optimal solvent selection." *Fluid Phase Equilibria* **82**: 47.

Ostrovsky, G. M., M. G. Ostrovsky, and G. W. Mikhailow. 1990. "Discrete optimization of chemical processes." *Comput. Chem. Engng* **14**: 111–117.

Papageorgaki, S. and G. V. Reklaitis. 1990a. "Optimal design of multipurpose batch plants: 1. Problem formulation." *Ind. Engng Chem. Res.* **29(10)**: 2054.

Papageorgaki, S. and G. V. Reklaitis. 1990b. "Optimal design of multipurpose batch plants: 2. A decomposition solution strategy." *Ind. Engng Chem. Res.* **29(10)**: 2062.

Papalexandri, K. P. and E. N. Pistikopoulos. 1993. "An MINLP retrofit approach for improving the flexibility of heat exchanger networks." *Ann. Oper. Res.* **42**: 119.

Papalexandri, K. P. and E. N. Pistikopoulos. 1994a. "Synthesis and retrofit design of operable heat exchanger networks: 1. Flexibility and structural controllability aspects." *Ind. Engng Chem. Res.* **33**: 1718.

Papalexandri, K. P. and E. N. Pistikopoulos. 1994b. "Synthesis and retrofit design of operable heat exchanger networks: 2. Dynamics and control structure considerations." *Ind. Engng Chem. Res.* **33**: 1738.

Papalexandri, K. P. and E. N. Pistikopoulos. 1996. "Generalized modular representation framework for process synthesis." *AIChE J.* **42**: 1010.

Papalexandri, K. P., E. N. Pistikopoulos, and C. A. Floudas. 1994. "Mass exchange networks for waste minimization: a simultaneous approach." *Chem. Engng Res. Dev.* **72**: 279.

Pardalos, P. M. and G. Schnitger. 1988. "Checking local optimality in constrained quadratic programming is NP-hard." *Oper. Res. Lett.* **7(1)**: 33.

Pardalos, P. M. and S. A. Vavasis. 1991. "Quadratic programming with one negative eigenvalue is NP-hard." *J. Global Optim.* **1**: 15.

Paules, IV, G. E. and C. A. Floudas. 1988. "Synthesis of flexible distillation sequences for multiperiod operation." *Comput. Chem. Engng* **12(4)**: 267.

Paules, IV, G. E. and C. A. Floudas. 1989. "APROS: algorithmic development methodology for discrete-continuous optimization problems." *Oper. Res.* **37(6)**: 902.

Paules, IV, G. E. and C. A. Floudas. 1992. "Stochastic programming in process synthesis: a two-stage model with MINLP recourse for multiperiod heat-integrated distiallation sequences." *Comput. Chem. Engng* **16(3)**: 189.

Penteado, F. D. and A. R. Ciric. 1996. "An MINLP approach for safe process plant layout." *Ind. Engng Chem. Res.* **35(4)**: 1354–1361.

Petkov, S. B. and C. D. Maranas. 1997a. "Quantitative assesment of uncertainty in the optimization of metabolic pathways." *Biotechnol. Bioengng* **56(2)**: 145–161.

Petkov, S. B. and C. D. Maranas. 1997b. "Multiperiod planning and scheduling of multiproduct batch plants under demand uncertainty." *Ind. Engng Chem. Res.* **36(11)**: 4864–4881.

Petkov, S. B. and C. D. Maranas. 1998. "Design of single product campaign batch plants under demand uncertainty." *AIChE J.* **44(4)**: 896–911.

Pistikopoulos, E. N. and M. G. Ierapetritou. 1995. "Novel approach for optimal process design under uncertainty." *Comput. Chem. Engng* **19(10)**: 1089–1110.

Quesada, I. and I. E. Grossmann. 1992. "An LP/NLP based branch and bound algorithm for convex MINLP optimization problems." *Comput. Chem. Engng* **16**: 937–947.

Raman, V. S. and C. D. Maranas. 1999. "Optimization in product design with properties correlated with topological indices." *Comput. Chem. Engng* **45**: 997–1017.

Ratschek, H. and J. Rokne. 1984. In *Computer Methods for the Range of Functions*. Halsted Press.

Reklaitis, G. V. 1991. "Perspectives on scheduling and planning process operations." In *Proceedings of the 4th International Symposium on Process Systems Engineering, Montreal, Canada.*

Rouhani, R., W. Lasdon, L. Lebow, and A. D. Warren. 1985. "A generalized Benders decomposition approach to reactive source planning in power systems." *Math. Programming Study* **25**: 62.

Ryoo, H. S. and N. V. Sahinidis. 1995. "Global optimization of nonconvex NLPs and MINLPs with applications in process design." *Comput. Chem. Engng* **19(5)**: 551–556.

Sahindis, N. V. and I. E. Grossmann. 1991. "Convergence properties of generalized Benders decomposition." *Comput. Chem. Engng* **15(7)**: 481.

Schweiger, C. A. and C. A. Floudas. 1998a. "Interaction of design and control: optimization with dynamic models." In *Optimal Control: Theory, Algorithms, and Applications*, edited by W. W. Hager and P. M. Pardalos, pp. 388–435. Kluwer Academic Publishers.

Schweiger, C. A. and C. A. Floudas. 1999. "Optimization framework for the synthesis of complex reactor networks." *Ind. Engng Chem. Res* **38**: 744–766.

Schweiger, C. A. and C. A. Floudas. 1997. In *MINOPT: A Software Package for Mixed-Integer Nonlinear Optimization, User's Guide.* Princeton, NJ: Computer-Aided Systems Laboratory, Department of Chemical Engineering, Princeton University.

Skrifvars, H., I. Harjunkoski, T. Westerlund, Z. Kravanja, and R. Pörn. 1996. "Comparison of different MINLP methods applied on certain chemical engineering problems." *Comput. Chem. Engng Suppl.* **20**: S333–S338.

Smith, E. M. B. and C. C. Pantelides. 1996. "Global optimisation of general process models." In *Global Optimization in Engineering Design*, edited by I. E. Grossmann, pp. 355–386. Kluwer Academic Publishers.

Smith, E. M. B. and C. C. Pantelides. 1997. "Global optimisation of nonconvex MINLPs." *Comput. Chem. Engng Suppl.* **21**: S791–S796.

Smith, E. M. B. and C. C. Pantelides. 1999. "A symbolic reformulation/spatial branch and bound algorithm for the global optimization of nonconvex MINLPs." *Comput. Chem. Engng* **23**: 457–478.

Stefanis, S. K., A. G. Livingston, and E. N. Pistikopoulos. 1997. "Environmental impact considerations in the optimal design and scheduling of batch processes." *Comput. Chem. Engng* **21(10)**: 1073–1094.

Turkay, M. and I. E. Grossman. 1996. "Logic-based MINLP algorithms for the optimal synthesis of process networks." *Comput. Chem. Engng* **20(8)**: 959–978.

Vaidyanathan, R. and M. El-Halwagi. 1996. "Global optimization of nonconvex MINLP's by interval analysis." In *Global Optimization in Engineering Design*, edited by I. E. Grossman, pp. 175–193. Kluwer Academic Publishers.

Vaidyaraman, S. and C. D. Maranas. 1999. "Optimal synthesis of refrigation cycles and selection of refrigerants." *AIChE J.* **45**: 997–1017

Vaselenak, J., I. E. Grossmann, and A. W. Westerberg. 1987a. "An embedding formulation for the optimal scheduling and design of multiproduct batch plants." *Ind. Engng Chem. Res.* **26(1)**: 139.

Vaselenak, J., I. E. Grossmann, and A. W. Westerberg. 1987b. "Optimal retrofit design of multipurpose batch plants." *Ind. Engng Chem. Res.* **26(4)**: 718.

Vavasis, S. 1991. In *Nonlinear Optimization: Complexity Issues*. New York, NY: Oxford University Press.

Vecchietti, A. and I. E. Grossmann. 1999. "LOGMIP : a disjunctive 0–1 nonlinear optimizer for process system models." *Comput. Chem. Engng* **23**: 555–565.

Viswanathan, J. and I. E. Grossman. 1990. "A combined penalty function and outer-approximation for MINLP optimization." *Comput. Chem. Engng* **14(7)**: 769–782.

Wellons, M. C. and G. V. Reklaitis. 1991a. "Scheduling of multipurpose batch chemical plants: 1. Formulation of single-product campaigns." *Ind. Engng Chem.* **30(4)**: 671.

Wellons, M. C. and G. V. Reklaitis. 1991b. "Scheduling of multipurpose batch chemical plants: 2. Multiple product campaign formulation and production planning." *Ind. Engng Chem. Res.* **30(4)**: 688.

Westerlund, T. and F. Pettersson. 1995. "An extended cutting plane method for solving convex MINLP problems." *Comput. Chem. Engng* **19**: 131–136.

Westerlund, T., F. Pettersson, and I. E. Grossmann. 1994. "Optimization of pump configuration problems as a MINLP problem." *Comput. Chem. Engng* **18(9)**: 845–858.

Westerlund, T., H. Skrifvars, I. Harjunkoski, and R. Pörn. 1998. "An extended cutting plane method for a class of non-convex MINLP problems." *Comput. Chem. Engng* **22(3)**: 357–365.

Xia, Q. and S. Macchietto. 1997. "Design and synthesis of batch plants: MINLP solution based on a stochastic method." *Comput. Chem. Engng* **21**: S697–S702.

Yee, T. F. and I. E. Grossmann. 1990. "Simultaneous optimization models for heat integration—II. Heat exchanger network synthesis." *Comput. Chem. Engng* **14(10)**: 1165–1184.

Yee, T. F., I. E. Grossmann, and Z. Kravanja. 1990a. "Simultaneous optimization models for heat integration—I. Area and energy targeting and modeling of multi-stream exchangers." *Comput. Chem. Engng* **14(10)**: 1151–1164.

Yee, T. F., I. E. Grossmann, and Z. Kravanja. 1990b. "Simultaneous optimization models for heat integration—III. Area and energy targeting and modeling of multi-stream exchangers." *Comput. Chem. Engng* **14(11)**: 1185–1200.

Zamora, J. M. and I. E. Grossmann. 1998a. "A global MINLP optimization algorithm for the synthesis of heat exchanger networks with no stream splits." *Comput. Chem. Engng* **22(3)**: 367–384.

Zamora, J. M. and I. E. Grossmann. 1998b. "Continuous global optimization of structured process systems models." *Comput. Chem. Engng* **22(12)**: 1749–1770.

Zhang, X. and R. W. H. Sargent. 1994. "The optimal operation of mixed production facilities: general formulation and some solution approaches." In *Proceedings of 5th International Symposium on Process Systems Engineering*, pp. 171–177.

9.5 Lagrangian Relaxation

Monique Guignard

ABSTRACT This article presents material necessary for understanding and designing efficient Lagrangian relaxations for integer programming problems. It suggests various ways in which Lagrangian relaxation can be applied and presents possible extensions. It reviews essential properties of the Lagrangian function, and describes several algorithms to solve the

Lagrangian dual problem. Finally it mentions Lagrangian heuristics, *ad hoc* or generic, as these are an integral part of any Lagrangian approximation scheme.

1. Introduction

The article starts with a description of relaxations, in particular Lagrangian relaxation (LR). It continues with the geometric interpretation of LR, and shows how the effectiveness of a particular LR scheme depends on how it exploits problem structure. Column generation, a dual way of obtaining the same bounds, is briefly discussed. Extensions of LR are also reviewed: Lagrangian decomposition and substitution. The integer linearization property is then described, as its detection may considerably reduce the computational burden.

The next section concentrates on solution methods for the dual problem: subgradient optimization and constraint generation methods, bundle methods, and a more recent, hybrid approach. It starts with the review of some characteristics of the Lagrangian function, important for the design of efficient optimization methods.

The last section is devoted to Lagrangian heuristics, which complement Lagrangian bounding by making an attempt at transforming infeasible Lagrangian solutions into good feasible solutions.

2. Notation

If (P) is an optimization problem, the following notation is used:

FS(P) the set of feasible solutions of (P)

OS(P) the set of optimal solutions of (P)

v(P) the optimal value of (P).

3. Definition of Relaxations

Geoffrion (1974) formally defines a relaxation of an optimization problem (P): $\text{Max}\{f(x) \mid x \in X\}$ as a problem (RP) over the same decision variable x: $\text{Max}\{g(x) \mid x \in Y\}$ such that (i) the feasible set of (RP) contains that of (P), that is, $Y \supseteq X$, and (ii) over the feasible set of (P), the objective function of (RP) dominates (is better than) that of (P), that is, $\forall x \in X, g(x) \geqslant f(x)$. It clearly follows that $v(\text{RP}) \geqslant v(\text{P})$, in other words (RP) is an optimistic version of (P): it has more feasible solutions than (P), and for feasible solutions of (P), its own objective function is better.

The role of relaxations is twofold: they provide bounds on the optimal value of difficult problems, and their solutions, while usually infeasible for the original problem, can often be used as a starting point (a guide) for specialized heuristics.

We concentrate here on linear integer programming problems, in which the constraint set X is defined by polyhedral constraints, plus integrality conditions on at least a subset of the components of x. The most widely used relaxation is the continuous relaxation, in which $g(x) = f(x) = fx$, and the integrality conditions on x are ignored, that is, Y is defined only by the polyhedral constraints of X.

4. Lagrangian Relaxation

We now introduce LR. Without loss of generality, we assume that (P) is of the form $\text{Max}_x\{fx \mid Ax \leqslant b, Cx \leqslant d, x \in X\}$, where X now only contains the integrality restrictions. The constraints $Ax \leqslant b$ are assumed complicating, in the sense that problem (P) without them would be much simpler to solve. The constraints $Cx \leqslant d$ (possibly empty) will be kept together with X to form the Lagrangian relaxation of (P) as follows.

Let λ be a nonnegative vector of weights, called Lagrangian multipliers. The Lagrangian relaxation (LR_λ) of (P) (also called Lagrangian subproblem) is the problem $\text{Max}_x\{fx + \lambda(b - Ax) \mid Cx \leqslant d, x \in X\}$, in which the slacks of the complicating constraints $Ax \leqslant b$ have been added to the objective function with weights λ, and the constraints $Ax \leqslant b$ have been dropped. One says that the constraints $Ax \leqslant b$ have been dualized. (LR_λ) is a relaxation of (P), since (i) FS(LR_λ) contains FS(P), and (ii) for any x feasible for (P), $fx + \lambda(b - Ax)$ is greater than or equal to fx. It follows that $v(\mathrm{LR}_\lambda) \geqslant v(\mathrm{P})$, for all $\lambda \geqslant 0$, that is, the optimal value $v(\mathrm{LR}_\lambda)$, which depends on λ, is an upper bound on the optimal value of (P).

The problem (LR): $\mathrm{Min}_{\lambda \geqslant 0} v(\mathrm{LR}_\lambda)$ of finding the tightest upper bound is called the Lagrangian dual of (P) relative to the complicating constraints $Ax \leqslant b$.

5. Feasible Lagrangian Solution

An optimal solution of (LR_λ) is called a Lagrangian solution. Suppose that x^*, an optimal solution of $(\mathrm{LR}_{\lambda^*})$ for $\lambda^* \geqslant 0$, is a feasible solution of (P), that is, it also satisfies the relaxed constraint $Ax^* \leqslant b$. In general, such a solution is not optimal for (P), and the optimal value of (P), $v(\mathrm{P})$, lies in the bracket $[fx^*, fx^* + \lambda^*(b - Ax^*)]$. If, however, complementary slackness holds, that is, if $\lambda^*(b - Ax^*)$ is 0, then obviously, x^* is an optimal solution for (P).

6. Geometric Interpretation

Theorem *(Geoffrion 1974). The Lagrangian dual (LR) is equivalent to the primal relaxation (PR):* $\mathrm{Max}_x\{fx \mid Ax \leqslant b, x \in Co\{x \in X \mid Cx \leqslant d\}\}$, *in the sense that* $v(LR) = v(PR)$.

This result is based on LP duality and properties of optimal solutions of linear programs. It follows that

- If $Co\{x \in X \mid Cx \leqslant d\} = \{x \mid Cx \leqslant d\}$, then $v(\mathrm{P}) \leqslant v(\mathrm{PR}) = v(\mathrm{LR}) = v(\mathrm{LP})$. In that case, one says that (LR) has the integrality property, and the LR bound is equal to (no better than) the LP bound.
- If $Co\{x \in X \mid Cx \leqslant d\} \subset \{x \mid Cx \leqslant d\}$, then $v(\mathrm{P}) \leqslant v(\mathrm{PR}) = v(\mathrm{LR}) \leqslant v(\mathrm{LP})$, and it may happen that the Lagrangian bound is strictly better than the LP bound.

Column generation (CG) has been used extensively, in particular for solving very large scheduling problems (airline, buses, etc.). One can show that the process is dual to LR. The separation of a problem into a master- and a subproblem is equivalent to the separation of the constraints into kept and dualized constraints. The columns generated are solutions of integer subproblems that have the same constraints as the Lagrangian subproblems. The value of the LP relaxation of the master problem is equal to the optimal Lagrangian bound. The strength of a CG or LR scheme would then seem to be based on the fact that the subproblems do not have the integrality property. It may happen, however, that such a scheme can be successful at solving problems with the integrality property because it permits the indirect computation of $v(\mathrm{LP})$ when this value could not be computed directly, for example, because of an exponential number of constraints (Held and Karp, 1970, 1971).

7. Constructing a Lagrangian Relaxation

There are often many ways in which a given problem can be relaxed in a Lagrangian fashion. We will list a few here, mostly to point out that often some reformulation prior to relaxation can help, and that for many complex models, intuition and some understanding of the constraint interactions may suggest ingenious and efficient relaxation schemes.

1. Isolate an interesting subproblem and dualize the other constraints. This is the most commonly used approach.

2. If there are two (or more) interesting subproblems with common variables, split these variables first, then dualize the copy constraint. This is called Lagrangian decomposition (LD) or variable splitting (Soenen, 1977). Problem (P): $\mathrm{Max}_x\{fx \mid Ax \leqslant b, Cx \leqslant d, x \in X\}$ is equivalent to problem (P′): $\mathrm{Max}_{x,y}\{fx \mid Ax \leqslant b, x \in X, Cy \leqslant d, y \in X, x = y\}$. Dualizing the copy constraint $x = y$ separates the problem into an x-problem and a y-problem. The LD bound can dominate the LR bounds obtained by dualizing either set of constraints:

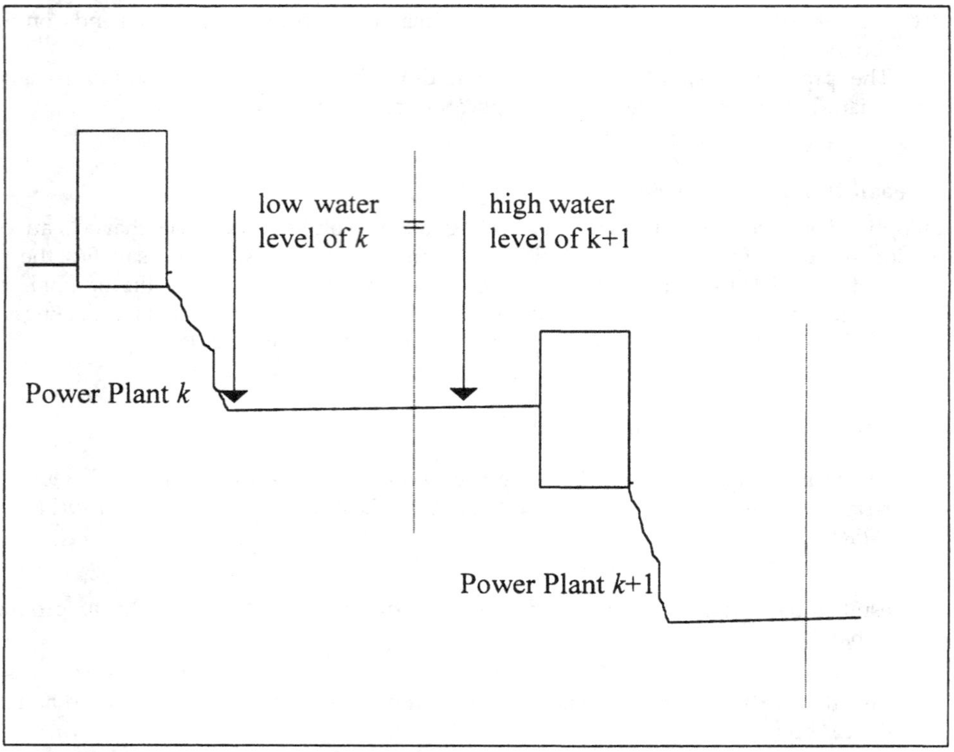

Figure 1 Example of variable splitting: split water level according to power plant

Theorem *(Guignard and Kim, 1987)*

$$v(LD) = Max\{fx \mid x \in Co\{x \in X \mid Ax \leqslant b\} \cap Co\{x \in X \mid Cx \leqslant d\}\}.$$

If one of the subproblems has the integrality property, v(LD) is equal to the better LR bound. If both have the integrality property, $v(\text{LD}) = v(\text{LP})$.

Example 1. *Hydroelectric power management (Guignard and Yan, 1993; see Figure 1). A series of power plants are located along a river, separated by reservoirs and falls. Managing the power generation requires decisions concerning water releases at each plant k in each time period. This results in a large complex mixed-integer program. A possible decomposition of the problem consists in cutting each reservoir in half, that is, splitting the water level variable in each reservoir, and dualizing the copy constraint: high water level in* k + 1 = low *water level in* k. *This Lagrangian decomposition produces one power management problem per power plant* k. *This subproblem does not have a special structure, but it is much simpler and smaller than the original problem, it is readily solvable by commercial software, and it does not have the integrality property. The resulting bound is much stronger than the LP bound.*

3. Dualize linking constraints. Sometimes naturally, sometimes after reformulation, problems may contain independent structures linked by some constraints: $\text{Max}_{x,y}\{fx + gy \mid Ax \leqslant b, x \in X, Cy \leqslant d, y \in Y, Ex + Fy \leqslant h\}$. Dualizing the linking constraints $Ex + Fy \leqslant h$ splits the problem into an x-problem and a y-problem. For instance, a production problem over multiple facilities may contain constraints related to individual facilities, and the demand constraints link all plant productions. Another example is that of a multiperiod model in which facilities built in one period can be used in that, or a later, period. One may be able to use action

(building) variables (for example, x) in the design part of the model and state (existence) variables (for example, y) in the rest of the model. Dualizing the linking relationship between built in period t and built by period t may split the model into a facility-building problem and a facility-using problem (Guignard et al., 1994). This is a special case of Lagrangian substitution (LS), where $Ex + Fy \leqslant h$ is the copy constraint introduced in the reformulation.

One can sometimes dualize aggregate rather than individual copies of variables (Guignard, 1989).

Example 2. *In a capacitated facility location problem, the constraint* $\sum_j d_j x_{ij} \leqslant a_i y_i$ *imposes a capacity restriction on the total amount that can be shipped from potential facility* i. *Creating new variables* x'_{ij} *and* y'_i *in this constraint, one can create an LS by dualizing the aggregate (linking) copy* $\sum_j d_j x_{ij} = \sum_j d_j x'_{ij}$ *and* $a_i y_i = a_i y'_i$. *The LS bound for this problem is as strong as the LD bound obtained by dualizing individual copies* $x_{ij} = x'_{ij}$ *and* $y_i = y'_i$ *(Chen and Guignard, 1998).*

8. Integer Linearization Property

Geoffrion (1974) and Geoffrion and McBride (1978) described and used the following important property of some Lagrangian subproblems. Suppose (LR_λ) is of the form $\text{Max}_x\{fx + gy \mid A_i x_i \leqslant p_i y_i, \text{ all } i, x \in X, By \leqslant b, y_i = 0 \text{ or } 1, \text{ all } i\}$, where there is one set of constraints $A_i x_i \leqslant p_i y_i$ for each i, the constraints $By \leqslant b$ are over y alone, and $X = \Pi_i(X_i)$ may contain some integrality requirement. x_i may be a vector. To solve (LR_λ), one can proceed as follows:

1. Ignore at first the constraints over integer variables only, $By \leqslant b$.
2. The problem separates into one problem for each i:

   $$(LR^i_\lambda) \quad \text{Max}\{f_i x_i + g_i y_i \mid A_i x_i \leqslant p_i y_i, x_i \in X_i, y_i = 0 \text{ or } 1, \text{ all } i\}$$

   where y_i plays the role of a 0–1 parameter:

   for $y_i = 0$, $x_i = 0$, and $f_i x_i + g_i y_i = 0$.

   for $y_i = 1$, solve $(LR^i_\lambda \mid y_i = 1)$: $v_i = \text{Max}\{f_i x_i + g_i \mid A_i x_i \leqslant p_i, x_i \in X_i\}$.

   v_i is the contribution of $y_i = 1$ in the objective function.

3. Replace $v(LR_\lambda)$ by $v(PL_\lambda)$ where problem (PL_λ), usually much simpler to solve than (LR_λ), is

   $$(PL_\lambda) \quad \text{Max}\left\{\sum_i v_i y_i \mid By \leqslant b, y_i = 0 \text{ or } 1, \text{ all } i\right\}.$$

   This process makes use of the integrality constraint on variable y_i and therefore even in cases where both (PL_λ) and $(LR^i_\lambda \mid y_i = 1)$ have the integrality property, it is possible to have

   $$v(LR) = \underset{\lambda}{\text{Min}}\ v(PL_\lambda) = \underset{\lambda}{\text{Min}}\ v(LR_\lambda) < v(LP).$$

9. Characteristics of the Lagrangian Function

The Lagrangian function $z(\lambda) = v(LR_\lambda)$ is an implicit function of λ. Suppose that the set $\{x \in X \mid Cx \leqslant d\} = \{x^1, x^2, \ldots, x^K\}$ is finite, then

$$\underset{x}{\text{Max}}\ \{fx + \lambda(b - Ax) \mid Cx \leqslant d, x \in X\}$$

$$= \underset{k=1,\ldots,K}{\text{Max}}\ \{fx^k + \lambda(b - Ax^k)\},$$

and $z(\lambda)$, the upper envelope of a family of linear functions of λ, is a convex function of λ, with breakpoints where it is not differentiable (see Figure 2). The dual problem (LR) is

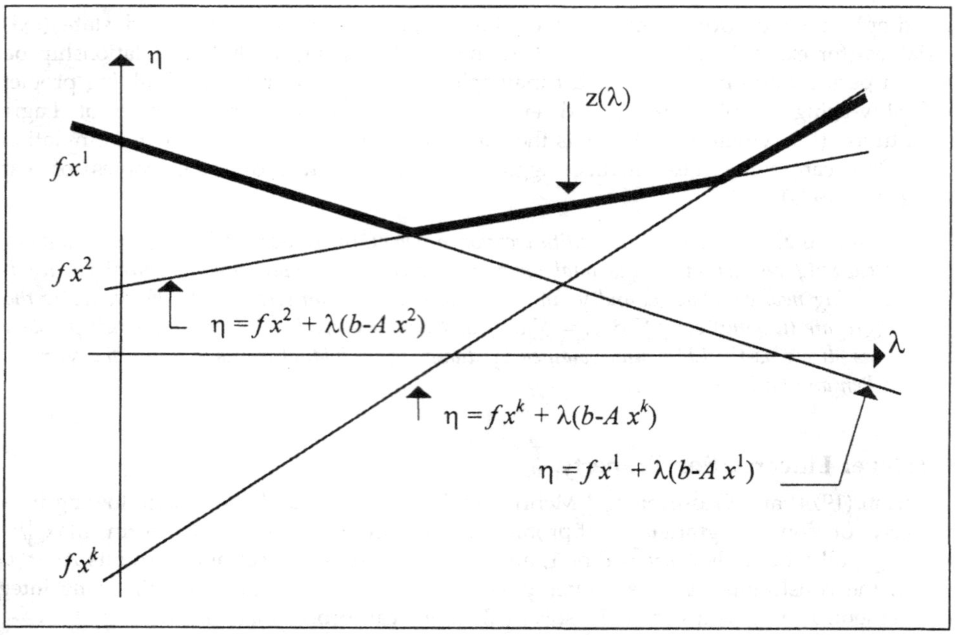

Figure 2 A 2-dimensional representation of z(λ) as the upper envelope of a family of planes

$$\text{(LR)} \quad \operatorname*{Min}_{\lambda \geqslant 0} v\,(\text{LR}_\lambda) = \operatorname*{Min}_{\lambda \geqslant 0} z(\lambda)$$

$$= \operatorname*{Min}_{\lambda \geqslant 0} \operatorname*{Max}_{k=1,\dots,K} \{fx^k + \lambda(b - Ax^k)\}$$

$$= \operatorname*{Min}_{\lambda \geqslant 0, \eta} \{\eta \mid \eta \geqslant fx^k + \lambda(b - Ax^k), k = 1, \dots, K\}.$$

Let λ^* be a minimizer of $z(\lambda)$, and let $\eta^* = z(\lambda^*)$. Let λ^k be a current guess at λ^*, let $\eta^k = z(\lambda^k)$, and let $H_k = \{\lambda \mid fx^k + \lambda(b - Ax^k) = \eta^k\}$ be a level hyperplane passing through λ^k.

- If $z(\lambda)$ is differentiable at λ^k, it has a gradient $\nabla z(\lambda^k)$ at λ^k:

$$\nabla z(\lambda^k) = (b - Ax^k) \perp H_k.$$

- If $z(\lambda)$ is nondifferentiable at λ^k, it has a subgradient s^k at λ^k:

$$s^k = (b - Ax^k) \perp H_k.$$

Remark 1. *A subgradient is not necessarily a direction of increase for the function, even locally, as seen in Figure 3.*

10. Primal and Dual Methods to Solve Relaxation Duals

10.1. Subgradient Method (Held and Karp, 1970; Held et al., 1974)

It is an iterative method in which steps are taken along the negative of a subgradient of $z(\lambda)$. Let λ^k be the current iterate, and x^k be an optimal solution of (LR^k_λ). Then $s^k = b - Ax^k$ is a subgradient of $z(\lambda)$ at λ^k.

If λ^* is an optimal solution of (LR), with $\eta^* = z(\lambda^*)$, let λ^{k+1} be the projection of λ^k on the hyperplane H^* parallel to H_k:

$$H^* = \{\lambda \mid fx^k + \lambda(b - Ax^k) = \eta^*\}.$$

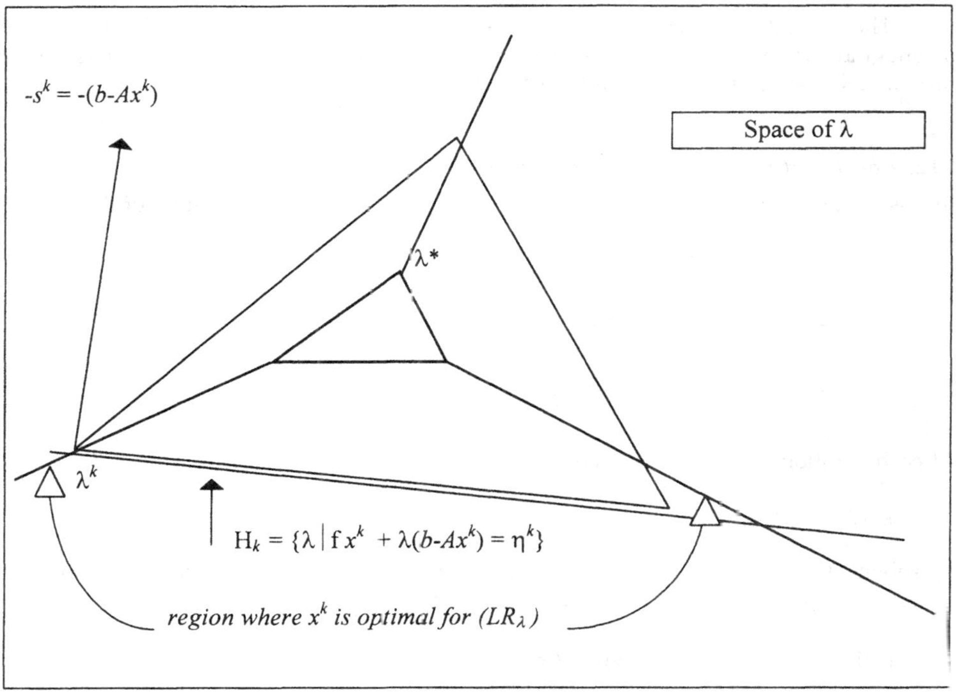

Figure 3 Contour of $z(\lambda)$ through λ^k, subgradient s^k of $z(\lambda)$ at λ^k and move from λ^k to λ'^{k+1}

The vector s^k is perpendicular to both H_k and H^*, therefore $\lambda'^{k+1} - \lambda^k$ is a negative multiple of s^k:

$$\lambda'^{k+1} - \lambda^k = -\mu s^k, \quad \mu \geqslant 0.$$

Also, λ'^{k+1} belongs to H^*:

$$fx^k + \lambda'^{k+1}(b - Ax^k) = \eta^*$$

therefore

$$fx^k(s^k)^t - \mu s^k(b - Ax^k) = \eta^k - \mu s^k \,.\, (s^k)^t = \eta^*$$

and

$$\mu = (\eta^k - \eta^*)/\|s^k\|^2,$$

so that

$$\lambda'^{k+1} = \lambda^k + s^k \,.\, (\eta^* - \eta^k)/\|s^k\|^2.$$

Finally, define $\lambda^{k+1} = [\lambda'^{k+1}]^+$, that is define the next iterate as the projection of λ'^{k+1} onto the nonnegative orthant, as λ must be nonnegative. λ^{k+1} is closer to λ^* than λ^k.

> **Remark 2**. *This formula unfortunately uses the unknown optimal value η^* of (LR). One can try to use an estimate for that value, but then one may be using either too small or too large a multiple of $-s^k$. If one sees that the objective function values do not improve for too many iterations, one should suspect that η^* has been underestimated (for a minimization problem) and one should reduce the difference $\eta^k - \eta^*$:*
>
> $$\lambda^{k+1} = \lambda^k + s^k \,.\, \varepsilon_k(\eta^* - \eta^k)/\|s^k\|^2,$$
>
> *where the scalar ε_k is reduced when there is no improvement for too long.*

The interested reader is referred to Lemaréchal (1974) and Zowe (1985) for an extension of subgradient methods, called bundle methods, in which past information is collected to provide a better approximation of the Lagrangian function.

10.2. Constraint Generation Method (Fisher et al., 1975)

In this method, one uses the fact that $z(\lambda)$ is the upper envelope of a family of linear functions:

$$\text{(LR)} \quad \underset{\lambda \geqslant 0}{\text{Min}}\ v(\text{LR}_\lambda) = \underset{\lambda \geqslant 0}{\text{Min}}\ z(\lambda)$$

$$= \underset{\lambda \geqslant 0}{\text{Min}}\ \underset{k=1,\dots,K}{\text{Max}}\ \{fx^k + \lambda(b - Ax^k)\}$$

$$= \underset{\lambda \geqslant 0,\eta}{\text{Min}}\ \{\eta \mid \eta \geqslant fx^k + \lambda(b - Ax^k), \quad k = 1, \dots, K\}.$$

At each iteration, one generates one or more cuts of the form

$$\eta \geqslant fx^k + \lambda(b - Ax^k),$$

by solving the Lagrangian subproblem (LR_λ^k) with solution x^k. These cuts are added to those generated in previous iterations to form the current LP master problem:

$$(\text{MP}^k) \quad \underset{\lambda \geqslant 0,\eta}{\text{Min}}\ \{\eta \mid \eta \geqslant fx^h + \lambda(b - Ax^h), \quad h = 1, \dots, k\},$$

whose solution is the next iterate λ^{k+1}. The process terminates when $v(\text{MP}^k) = z(\lambda^{k+1})$. This value is the optimal value of (LR).

10.3. Two-Phase Hybrid Method (Guignard and Zhu, 1994)

This combines the subgradient method in a first phase, and constraint generation in a second phase. The multipliers are first adjusted according to the subgradient formula, and at the same time, constraints corresponding to all known solutions of the Lagrangian subproblems are added to the LP master problem. The value of the LP master problem is taken as the current estimate of the optimum of the Lagrangian dual. The estimate gets more and more accurate as iterations go by, so there is no need for any adjustment: $\varepsilon_k = 1\ \forall k$. The Lagrangian bound and the value of the master problem provide a bracket on the dual optimum, and this yields a convergence test, as for the pure constraint generation method.

One must make sure that the process does not cycle. If constraints get repeated, the master problem cannot improve. After the same cut has been generated a given number of times (e.g., 5 times), one can switch to a pure constraint generation phase.

11. Lagrangian Heuristics

Lagrangian relaxation provides bounds, but it often generates nearly feasible integer solutions. One may try to get feasible solutions in different ways:

1. By modifying the solution to correct its infeasibilities while keeping the objective function deterioration small.
   For example: in production scheduling, if one relaxes the demand constraints, one may try to change production (down or up) so as to meet the demand (de Matta and Guignard, 1994).
2. By fixing (at 1 or 0) some of the meaningful decision variables according to their value in the current Lagrangian solution and solving optimally the remaining problem. We call this the lazy heuristic. One guiding principle may be to fix variables that satisfy relaxed constraints.

12. Conclusion

- Lagrangian relaxation is a powerful family of tools for solving approximately integer programming problems. It provides

  stronger bounds than LP relaxation when the problem(s) don't have the integrality property;

  good starting points for heuristic search.

- The availability of powerful interfaces (GAMS, AMPL, etc.) and of flexible IP packages makes it possible for the user to try various schemes and to implement and test them.
- It is not necessary to have special structures embedded in a problem to try to use Lagrangian schemes. If it is possible to decompose the problem structurally into meaningful components and to split them through constraint dualization, possibly after having introduced new variable expressions, it is probably worth trying.
- Solutions to one or more of the Lagrangian subproblems might lend themselves to Lagrangian heuristics, possibly followed by interchange heuristics, to obtain good feasible solutions.
- Lagrangian bounds coupled with Lagrangian heuristics provide the analyst with brackets around the optimal integer value. These are usually much tighter than with LP based bounds and heuristics.

References

Chen, B. and M. Guignard. 1991. "Polyhedral analysis and decompositions for capacitated plant location-type problems." *Discrete Appl. Math,* 1998.

de Matta, R. and M. Guignard. 1994. "Dynamic production scheduling for a process industry." *Oper. Res.* **42**: 492–503.

Fisher, M. L., W. D. Northup, and J. F. Shapiro. 1975. "Using duality to solve discrete optimization problems: theory and computational experience." *Math. Programming Study* **3**: 56–94.

Geoffrion, A. M. 1974. "Lagrangian relaxation for integer programming." *Math. Programming Study* **2**: 82–114.

Geoffrion, A. M. and R. McBride. 1978. "Lagrangian relaxation applied to capacitated facility location problems." *AIIE Trans.* **10**: 40–47.

Guignard, M. 1989. "General aggregation schemes in Lagrangian decomposition: theory and potential applications." Working Paper 89-12-07. PA: Department of Decision Sciences, University of Pennsylvania.

Guignard, M., E. Chajakis and C. Ryu. 1994. "Lagrangean bounds and heuristics for integrated resource planning in forestry." *Proceedings, Symposium on Systems Analysis and Forest Management Problems.* Valdivia, Chile, March 1993, pp. 350–363.

Guignard, M. and S. Kim. 1987. "Lagrangian decomposition: a model yielding stronger Lagrangian bounds." *Math. Programming* **39**: 215–228.

Guignard, M. and H. Yan. 1993. "Structural decomposition methods for dynamic multi-hydropower plant optimization." Working paper 93-12-01. PA: Operations and Information Management Department, University of Pennsylvania.

Guignard, M. and S. Zhu. 1994. "A two-phase dual algorithm for solving Lagrangian duals in mixed integer programming." Presented at Proceedings of the VII CLAIO, Santiago de Chile, July 1994.

Held, M. and R. M. Karp. 1970. "The travelling salesman problem and minimum spanning trees: part I." *Oper. Res.* **18**: 1138–1162.

Held, M. and R. M. Karp. 1971. "The travelling salesman problem and minimum spanning trees: part II." *Math. Programming* **1**: 6–25.

Held, M., P. Wolfe, and H. Crowder. 1974. "Validation of subgradient optimization." *Math. Programming* **6**: 62–88.

Lemaréchal, C. 1974. "An algorithm for minimizing convex functions." *Proceedings IFIP'74 Congress,* pp. 552–556. Amsterdam: North-Holland.

Soenen, R. 1977. "Contribution l'étude des systèmes de conduite en temps réel en vue de la commande d'unités de fabrication." Thèse de Doctorat d'Etat. Lille, France.

Yan, H. 1996. "Solving some difficult mixed-integer programming problems in production and forest management." Ph.D. diss. PA: University of Pennsylvania.
Zowe, J. 1985. "Nondifferentiable optimaization." In *Computational Mathematical Programming*, pp. 323–356. Berlin: Springer-Verlag.

9.6 Heuristics for 0–1 Mixed-Integer Programming

Arne Løkketangen

ABSTRACT Most combinatorial optimization problems of general interest can be modeled as 0–1 mixed-integer programming problems, and as is shown in this handbook, there are multitudes of ways to try to solve them. This article describes ideas and solution methods, mainly based on the metaheuristics of tabu and scatter search that have been tried with good results. The notion of chunking is used as a basis for a statistically based distance measure that can be used for diversification purposes for these kinds of problems. Specialized, domain-specific heuristics and general branch-and-bound techniques are not treated here, and neither are general mixed-integer programming problems.

1. Introduction

When describing heuristics to be used for mixed-integer programming (MIP), it is necessary to first define what is meant by the term heuristic. We will rely on the definition found in the excellent Mathematical Programming Glossary (see Greenberg, 1999), which defines a heuristic to usually mean a procedure that seeks a solution but does not guarantee it will find one. This is often contrasted with an algorithm, so branch and bound would not be considered a heuristic in this sense. On the other hand, in artificial intelligence a heuristic is an algorithm (with some guarantees) that uses a heuristic function to estimate the cost of branching from a given node to a leaf of the search tree.

In this article we will look at heuristic methods that apply to both 0–1 IP and 0–1 MIP problems, with focus on the latter. It is generally held that applying problem or application domain knowledge can enhance most optimization methods. This use of domain knowledge is treated in the applications section of this handbook (Part 2), while we look at more general methods. The formulation of the problems treated here is thus

$$\text{Min } c^T x$$

subject to

$$Ax \geqslant b \tag{1}$$

$$x_i \in \{0,1\} \quad i \in I \tag{2}$$

$$l \leqslant x \leqslant u, \tag{3}$$

where the index set I gives the variables that must take on integer values. We refer to solutions that satisfy all constraints, except perhaps constraint (2), as LP feasible, and to solutions that satisfy all the constraints, including constraint (2), as MIP feasible, or MIP solutions.

After looking at the pivot and complement heuristic, we will look at pivot-based local search methods, designed to focus on obtaining integer feasibility, with associated diversification and speedup techniques, while maintaining a good objective function value. It is clear, however, that the general practitioner might not necessarily be interested in the optimum solution, as defined by the model, but rather a set of good solutions to choose from, as there will always be considerations that are not modeled. We show that scatter search can be used to generate diverse set of good quality solutions, and that the concepts of chunking and Mahalanobis distances can be used to identify this diversity.

As the descriptions below can be rather brief, the reader is referred to the bibliography for additional details.

1.1. Pivot and Complement

One early well-known heuristic is the pivot and complement (P&C) heuristic by Balas and Martin (1980). The P&C heuristic is invoked after a relaxed LP solution to the original MIP has been found. It consists of two main phases: the search phase and the improvement phase. In the search phase, one tries to perform pivot moves that reduce the total amount of integer infeasibility. Priority is given to pivots that exchange a noninteger or slack variable with a basic integer variable. These are called Type 1 moves, and the first one found in an iteration is used. Type 2 moves are any moves that exchange a nonbasic integer variable with a basic integer variable. Among the possible Type 2 moves, the one chosen is the one resulting in the smallest amount of integer infeasibility. If there are no Type 1 or 2 moves, the heuristic tries to find a Type 3 move. In this we try to find the pivot giving the smallest amount of integer infeasibility when using a noninteger variable as the entering variable, and possibly sacrificing primal feasibility. If primal feasibility is lost, then one tries to recover it by first searching for a feasible basis by forcing every basic variable to its upper or lower bound. If still infeasible, perform single integer variable complements that reduce the total amount of primal infeasibility, and, as a last resort, try to find double complements. The search phase ends either as a result of finding an integer-feasible solution, in which case the heuristic goes to the improvement phase, or when primal feasibility is irrecoverably lost. The improvement phase is started when the search phase has found a feasible integer solution. It works by complementing integer variables, and choosing the complement that gives the best improvement in the original objective function value. It first tries single complements. If there are no improving single complements, it tries to find an improving double complement, using the first one found.

1.2. Tabu Search in Pivot and Complement

The P&C heuristic is inherently greedy. In both phases of the search either the first improving or best improving move is taken. The search is also layered, in that in the search phase the best move types are considered first, while in the improvement phase the neighborhood is expanded (to double complements) when no improving moves are found with single complements.

An appealing idea is to replace this greedy search with the metaheuristic tabu search (TS; see Glover and Laguna, 1997. This is an excellent book on TS, and also contains a large bibliography). This is the approach taken by Løkketangen, Jörnsten, and Storøy (1994). In the search phase, Type 1 and 2 moves are considered together, while Type 3 moves are never considered as the search relies on the tabu restrictions to escape the local optima. The search also continues past the first integer-feasible solution found, trying to find improved solutions and remembering the best found. In the improvement phase, only single flips are considered, also relying on the tabu restrictions for escaping from local minima The resulting method gives significant improvements over the original P&C heuristic.

1.3. TS Guided Pivoting Heuristics

It is well known that the optimal solution for a 0–1 MIP problem may be found at an extreme point of the LP feasible set. In Løkketangen and Glover (1998) this property is exploited in a TS guided extreme point pivoting heuristic, using a bounded variable simplex method as a subroutine.

In broad outline, the MIP solution approach may be expressed in the following form. Let x^* denote the best MIP feasible solution found and let z^* denote its objective function value. (To begin, when x^* is unknown, z^* may be assigned a value of infinity.)

TS–MIP in Overview

Step 0. Begin by solving the LP relaxation of the 0–1 MIP problem to obtain an optimal LP basic (extreme point) solution.

Step 1. From a current LP feasible basic solution, consider the feasible pivot moves that lead to adjacent basic feasible solutions.

(a) Isolate and examine a candidate subset of these feasible pivot moves.

(b) If a candidate move creates an MIP feasible solution x whose associated z value yields $z > z^*$, record x as the new x^* and update z^*.

Step 2. Select a pivot move that has the highest evaluation from those in the candidate set, applying tabu search rules to exclude or penalize moves based on their tabu status.

Step 3. Execute the selected pivot, updating the associated tabu search memory and guidance structures, and return to Step 1.

The neighborhood solutions consist of the set of adjacent extreme points on the polytope; those that can be reached by a pivot. The full neighborhood size is thus equal to the number of structural variables. This neighborhood is very costly to evaluate in full for large problems, and in Løkketangen and Glover (1999) various speedup techniques related to different candidate list strategies are described.

The move evaluation function is based on a surrogate constraint-based combined measure of the resulting change in objective function value and amount of integer infeasibility. A variable that has left the basis is kept tabu for the duration of its tabu tenure, while the standard aspiration criterion of accepting a move if it leads to a new best solution is used together with one based on integer infeasibility levels.

Target Analysis (see Glover and Laguna, 1997) was used in Løkketangen and Glover (1998) to discover the reasons (and remedies) for lack of search convergence for some test cases, as well as effective diversification techniques.

Løkketangen and Glover (1998) also describe how the use of controlled randomization (which uses the biases of probability) can be used as a substitute for memory. In particular, the concept of tabu tenure was replaced by the concept of probabilistic move acceptance, where the moves gathered on a sorted candidate list have a finite probability of being accepted, with this probability decreasing as one moves down the candidate list.

The results obtained with the pivot-based methods are good, but rather time consuming due to the expensive neighborhood evaluation.

1.4. Other Similar Approaches

In Blue and Bennett (1998), a hybrid of extreme-point TS and ordinary descent algorithm based on linear programming is described. The descent algorithm is used to find a local minimum, while the TS is then applied to escape the local minimum and diversify the search. These types of approaches can be applied to many important classes of optimization problems, for example, bilinear programming. Similarly, Sun et al. (1998) applies extreme-point TS to a network-based simplex method.

2. Generating Diversified Solutions

There are a number of reasons why decisionmakers are interested in seeing a variety of solutions to the model problem, and not just those that achieve the optimum.

MIP solutions are sometimes used as starting points for more detailed models (e.g., simulations). Such automated, but stochastic, optimization systems often benefit from a variety of starting points. Moreover, in practice MIP models typically leave out a lot of details and make use of very approximate data. The user of a decision support system (DSS) would often like to look at a variety of solutions and mull over the so-called intangibles omitted from the model.

It is highly preferable to identify a set of solutions that are decently good and, especially, diverse. We can reasonably rely on the objective function to quantify the notion of decently good. Methods for quantifying the notion of mutually diverse are given below, as well as illustrations of the use of scatter search (Glover and Laguna, 1997) to generate the required diverse solutions.

2.1. *Scatter Search for 0–1 MIP Problems*

The ideas outlined here can be found in great detail in Glover and Laguna (1997), with a practical implementation in Glover et al. (2000), where the solutions generated by the scatter search are subjected to chunking and testing for diversity. The main idea behind the scatter search approach for 0–1 MIP problems is to generate a set of solution points that span (parts of) the solution space by using a carefully designed diversification generator. These points are then combined using directional rounding and starpaths, which in a sense is a projection onto the constraint polytope of the path between two of the solutions as seen from a third.

2.2. *Chunking and Measures of Diversity*

When some methods have produced a set of different, good solutions, we would like to have a measure of which solutions are most different and which are similar. As is shown in Glover et al. (2000), ordinary metrics like hamming or Euclidian distance does not work properly for 0–1 MIP problems. The measure called Mahalanobis distance, based on the covariance matrix of the solutions found, overcomes the inherent weaknesses of the other measures mentioned and gives a reasonable measure of solution distance. There is, however, one major difficulty. To generate the covariance matrix of a vector of n variables, one needs (at least) $n+1$ vectors. This might entail generating thousands of different solutions to the problem, which might be highly impractical.

To reduce the dimension of the solution vectors, and thus the number of required solutions, one can resort to chunking (see Woodruff, 1998). This works by grouping the variables of the original 0–1 MIP problem into a small number of units, or new variables, called chunks. The Mahalanobis distance is then calculated using the chunked variables instead of the original ones. Glover et al. (2000) show how to select good chunking functions. The paper also shows how the application of scatter search described above produces much more diversified solutions than ordinary branch and bound.

References

Balas, E. and C. Martin. 1980. "Pivot and complement—a heuristic for 0–1 programming." *Mgmnt Sci.* **26(1)**: 86–96.

Blue, J. A. and K. P. Bennett. 1998. "Hybrid extreme point tabu search." *Eur. J. Oper. Res.* **106(2-3)**: 680.

Glover, F. and M. Laguna. 1997. *Tabu Search.* Boston: Kluwer Academic Publishers.

Glover, F., A. Løkketangen, and D. Woodruff. L. 2000. "Scatter search to generate diverse MIP solutions." In *OR Computing Tools for the New Millennium*, Conference Volume from 6th Informs Computing Society Conference, Cancun, Mexico, 2000. Boston: Kluwer Academic Publishers.

Greenberg, H. 1999. "Mathematical programming glossary." URL: http://www.cudenver.edu/~hgreenbe/glossary.

Løkketangen, A. and F. Glover. 1998. "Solving zero-one mixed integer programming problems using tabu search." *Eur. J. Oper. Res.* **106(2-3)**: 624–658.

Løkketangen A. and F. Glover. 1999. "Candidate List and Exploration Strategies for Solving 0/1 MIP Problems using a Pivot Neighborhood." In *MIC, Metaheuristic International Conference, '97*, pp. 141–154. Boston: Kluwer Academic Publishers.

Løkketangen, A., K. Jörnsten, and S. Storøy. 1994. "Tabu Search within a pivot and complement framework." *Int. Trans. Oper. Res.* **1(3)**: 305–317.

Sun, M., J. E. Aronson, P. G. McKeown, and D. Drinka. 1998. "A tabu search heuristic for the fixed charge transportation problem." *Eur. J. Oper. Res.* **106(2-3)**: 441–456.

Woodruff, D. L. 1998. "Proposals for chunking and tabu search." *Eur. J. Oper. Res.* **106(2-3)**: 585–598.

10 ARTIFICIAL NEURAL NETWORKS IN OPTIMIZATION AND APPLICATIONS

Theodore B. Trafalis and Suat Kasap

ABSTRACT This chapter reviews the use of artificial neural networks (ANNs) for optimization problems. Connections with statistical physics are examined with an extensive description of mean field equations. Several energy functions are investigated. Applications to image processing and learning are also presented.

1. Introduction

In the last two decades, there has been significant interest in using artificial neural networks (ANNs) in optimization, machine learning, image processing, pattern recognition (classification/prediction), and engineering. The use of ANNs for classification and prediction problems can be considered as an extension of traditional function approximation techniques. On the other hand, ANNs bring new ideas to the solution of combinatorial and nonlinear optimization problems (Haykin, 1994).

Over the last three decades, the combinatorial optimization problems are some of the most challenging problems that have received considerable attention. The traditional solution methodologies for combinatorial optimization problems can be categorized into three groups as exact, heuristic, and approximation methods, respectively. Exact methods such as simplex and branch and bound guarantee a global optimum, but the time needed to find it grows exponentially with the size of the problem. Heuristic methods such as simulated annealing, tabu search, and genetic algorithms are based on full or partial exploration of different configurations. On the other hand, approximation methods generate (near) global optimal solutions in a reasonable amount of time without a full or partial exploration. ANN-based methods are perceived as approximation methods.

ANNs have constituted a new direction in solving combinatorial optimization problems since the pioneering work of Hopfield and Tank (1985). The basic idea behind using ANNs to solve an optimization problem is to map the optimization problem onto a highly interconnected network of neurons. A particular configuration of neurons being on or off yields an associated value of the network energy, which is the objective function of the optimization problem. The new problem is to find a configuration of neurons that minimizes the network energy that corresponds to an optimal solution for the optimization problem.

This chapter reviews the use of ANNs for applied optimization problems and provides a survey of most of the ANN approaches that have been applied to optimization and image processing. Issues of learning are also examined. In Section 2, basic definitions, classifications and applications of ANNs are introduced. In Section 3, the mathematical basis of problem formulation for ANNs based on the Hopfield energy function and the mean field energy function are explained. In Section 4, various optimization problems formulated as ANNs are presented. In Sections 5 and 6, learning and image processing in ANNs are presented, respectively. Finally, Section 7 gives the conclusion.

2. Artificial Neural Networks (ANNs)

The modern era of ANNs is said to have begun with the introductory work of McCullough and Pitts (1943). They proposed a general theory of information processing based on networks of neurons. Each one of these neurons can only take the output values 1 or 0 by representing the active and resting states of neurons, respectively. In general, ANNs are dynamic systems built from a large number of interconnected neurons that interact with each other in some prescribed way. ANNs have come a long way from the early days of McCullough and Pitts and established themselves as an interdisciplinary subject with rich connections in the fields of neuroscience, psychology, physical sciences, mathematics, and engineering.

ANN models are algorithms for intellectual tasks such as learning and optimization that are based on the concept of how the human brain works. An ANN model is composed of a large number of processing elements called neurons. Each neuron is connected to other neurons by links, each with an associated weight. Neurons without links toward them are called input neurons and those with no link leaving away from them are called output neurons. The neurons are represented by state variables. State variables are functions of the weighted-sum of input variables and other state variables. Each neuron performs a simple transformation at the same time in a parallel-distributed manner. The input-output relation of the transformation in a neuron is characterized by an activation function. The following sigmoid function is the most-used activation function in the ANN literature

$$f(x) = \frac{1}{1 + e^{-\alpha x}}, \tag{1}$$

where α is a slope parameter. The combination of input neurons, output neurons, and links between neurons with associated weights constitutes the architecture of the ANN. Mathematically, an ANN model is a directed graph with the following properties (Haykin, 1994):

1. Each neuron (node) is associated with a state variable S and an activation threshold (bias) v.
2. Each link (edge) between neurons is associated with a weight w.
3. State of the neuron is determined by an activation function $f(S, w, v)$.

The classification of ANNs can be done in different ways. Depending on the nature of the activation function, they can be categorized either as deterministic or stochastic ANNs. Deterministic ANNs activate their states by using deterministic activation functions such as the threshold, linear, or sigmoidal functions. Most of the existing ANN models are deterministic. Stochastic ANNs activate their states according to a probability distribution. A typical example of stochastic ANNs is the Boltzmann Machine (BM). According to the nature of the connectivity, ANNs can be categorized either as feedforward or recurrent ANNs. If a directed graph has no closed paths, then it is called a feedforward ANN. A typical example of the feedforward ANN is the popular multilayer perceptron. Conversely, if a directed graph has closed paths, then it is called a recurrent ANN. A typical example of the recurrent ANN is the Hopfield neural network (NN).

ANNs have very close ties with optimization, and the ties are manifested mainly in two aspects. One aspect, learning algorithms, has been developed based on optimization techniques to train an ANN to perform modeling tasks (Ackley et al., 1985; Couellan, 1997; Peterson and Anderson, 1987; Rumelhart et al., 1986; Trafalis and Tutunji, 1995, 1997; Tutunji, 1996; Xu and Yuille, 1995). On the other hand, ANNs have been developed to solve difficult optimization problems like combinatorial optimization problems that are discussed in Section 4. Many difficult optimization problems can be formulated as the minimization of a quadratic form and thus can be mapped onto ANNs. One of the main advantages of ANN approaches to classical optimization approaches is the inherently parallel and distributed nature of the dynamic solution procedure. Therefore, ANNs are capable of solving large-scale optimization problems in real time (Ansari et al., 1995; Bilbro et al., 1988; Lagerholm et al., 1997; Lee and Takefuji, 1996; van DenBout and Miller, 1990).

ANNs have been used to solve a lot of optimization problems since the pioneering work of Hopfield and Tank (1985). To solve optimization problems using ANNs requires a mapping of the problem onto an ANN in such a way that one can identify a solution from the outputs of neurons. In the next section, this mapping is explained.

3. Energy Functions and ANNs

The procedure of ANN approaches to optimization generally begins with the formulation of an energy function. The first step in designing ANNs for optimization problems is to formulate an energy function that maps the optimization problem onto ANNs. Indeed, the minimum of the energy function corresponds to the optimal solution of the optimization problem. Most ANN approaches formulate the energy function by incorporating the objective function and constraints of combinatorial optimization problems through functional transformation and numerical weighting. A functional transformation is usually used to convert constraints to a penalty function that penalizes the violations. Numerical weighting is often used to balance constraint satisfaction and objective minimization. In general, such an energy function will have the following form:

$$E = \text{cost} + \sum_i A_i \, (\text{constraint violation})_i, \tag{2}$$

where A_i is a weight parameter and cost is the objective function that is independent from the constraint violations. By minimizing the energy function E, one attempts to minimize the cost while at the same time minimizing the constraint violations.

The second step in designing ANNs for combinatorial optimization problems is to derive a dynamic equation (also known as state equation or motion equation). The dynamic equation prescribes the motion of the activation states of the ANN. A properly derived dynamic equation can ensure that the state of the ANN reaches equilibrium. The equilibrium state of the ANN satisfies the constraints and optimizes the objective function of the problem.

The last step is to determine the architecture of the ANN in terms of neurons and links between neurons with associated weights based on the dynamic equation in such a way that one can identify a solution from the outputs of the neurons. The success of optimization using ANNs lies in the formulation of the energy function and the derivation of the dynamic equation. ANNs do not guarantee globally optimal solutions but they compute locally optimal solutions.

3.1. Hopfield Energy Function

ANNs have been used to solve many combinatorial optimization problems since the pioneering work of Hopfield and Tank (1985). They pioneered an approach for solving minimization problems by utilizing the collective computational capabilities of ANNs. The objective function and constraints of the problem are mapped onto a quadratic energy function of neuron states. The energy function has the following form:

$$E(S) = -\frac{1}{2}\sum_{i=1}^{N}\sum_{j=1}^{N} C_{ij} S_i S_j - \sum_{i=1}^{N} S_i I_i, \tag{3}$$

where $S_i = \{0, 1\}$ represents the output state of neurons, C_{ij} represents the weight of the link between neurons i and j, and I_i represents the input bias; the input of neurons is denoted as x_i. The dynamic equations of neurons are defined by

$$\frac{dx_i}{dt} = -x_i + \sum_{j \neq i}^{N} C_{ij} S_i + I_i, \quad \text{where } S_i = f_i(x_i) \text{ and } f_i \text{ is the activation function.} \tag{4}$$

Hopfield and Tank showed that an ANN with symmetric connections and nonnegative elements on the diagonal of the C matrix will always converge by performing gradient descent to a stable state, which is the local minimum of $E(S)$. The local minimum of $E(S)$ is obtained when the network is iterated from an initial state by updating each neuron asynchronously in accordance with the following updating rule:

$$S_i = \operatorname{sgn}\left(\sum_{j=1}^{N} C_{ij} S_j + I_i\right). \tag{5}$$

Hopfield (1984) extended the discrete model to an analog model where $0 \leqslant S_i \leqslant 1$. The modified energy function is:

$$E(S) = -\frac{1}{2}\sum_{i=1}^{N}\sum_{j=1}^{N}C_{ij}S_iS_j - \sum_{i=1}^{N}S_iI_i + \sum_{i=1}^{N}\frac{1}{R_i}\int_0^{S_i} g_i^{-1}(S)dS, \tag{6}$$

where R_i is the input resistance to unit i, and g_i is the activation function.

ANNs to minimize the energy function are composed of either discrete or continuous neurons. The possibilities of using different types of ANNs such as discrete-state and discrete-time, continuous-state and discrete-time, and continuous-state and continuous-time networks to solve combinatorial optimization problems were investigated in Melamed (1994). One usually prefers to use continuous neurons rather than discrete ones for two reasons. First, a continuous network tends to avoid oscillations between stable states. Secondly, solutions are much better than those provided by discrete ANNs because the valley of energy landscapes are wider and neuron outputs are not restricted to corners of the hypercube during convergence.

3.2. *Mean Field Energy Function*

ANNs are similar to some models of magnetic materials namely the spin-glass models encountered in statistical physics (Reichl, 1980). In the last two decades, spin-glass theories of statistical physics have been extensively studied because of their application to other areas like optimization and ANNs (Mezard et al., 1987; Müller and Reinhardt, 1990). Little (1974) pointed out a similarity between an ANN and a system of elementary magnetic spins. Later, Hopfield (1982) studied how such an ANN or a spin system can store and retrieve information. He pointed out that the energy function describing a recurrent network with symmetric connectivity is identical to the energy function describing the dynamics of the spin-glass model.

A spin-glass is a set of spins (i.e., magnetic moments) whose spins S_i and S_j interact through random couplings J_{ij}. In a magnetic model, if all J_{ij} are positive, this is called a ferromagnetic model. Conversely, if all J_{ij} are negative, this is an antiferromagnetic model, and if all J_{ij} are different in sign, this is a disordered magnetic model. A disordered magnetic model is called a spin-glass model. In order to produce such a model, two ingredients are necessary: frustration and disorder. Competing interactions between spins are in conflict with each other leading to frustration. Positions of the spins are random leading to disorder. Spins in a system interact with each other according to the following Hamiltonian:

$$H(S) = -\sum_{i=1}^{N}\sum_{j=1}^{N}J_{ij}S_iS_i. \tag{7}$$

There is a number of problems outside physics that share some of the essential features—frustration and disorder—that characterize spinglasses. Insights and techniques can be borrowed from the spin-glass theory and brought usefully to other problems such as optimization and ANNs. The analogy to ANNs was realized by identifying each spin with a neuron and associating +1 or −1 values of spins with the acting or resting state of neurons. The ground state configuration of spins corresponds to the stable state of ANNs that minimizes the energy function of the network. The ANN input represents the field acting on a spin, the threshold acts as the external field, and the connection or weight matrix corresponds to the random couplings for interaction between spins. Minimization of network energy function describes the dynamics of the spins aligning with their local fields.

There are two models of spin glasses: Ising-glass models and Potts-glass models. In Ising-glass models, the spin S_i can only take two values, $S_i = \pm 1$. A Potts-glass model is a generalization of the two-state Ising-glass model to N-states. Spin glasses were studied extensively in Fischer and Hertz (1991) and Mezard et al. (1987).

The theory of spin glasses relies strongly on the mean field theory (MFT). The MFT assumes that the value of S_i, the state of a spin at a particular spin site i, depends on its local environment because of interactions with other spins. Therefore, the value of S_i is approximated by its mean value, $\langle S_i \rangle$, as the name of the theory implies. The $\langle S \rangle$ over a distribution of given parameters are expressed in terms of derivatives of an effective energy with respect to auxiliary fields according to the MFT approximation, as explained later in this section. The MFT approximation is an analytic manipulation that simplifies the study of a physical system at the equilibrium. Conceptually, the MFT approximation replaces the value of a spin state that occurs in an energy field by its mean value when evaluating the probability distribution of any other spins. The most probable state is the one with the lowest energy. Consequently, finding the most probable state is equivalent to minimizing the energy of the system.

The probability of the system existing in any specific state $S = (S_1, S_2, ..., S_N)$ is given by the following Boltzmann distribution

$$P(S) = \frac{e^{-E(S)/T}}{\sum_S e^{-E(S)T}}, \tag{8}$$

where the term in the denominator is known as the partition function, Z. The partition function requires summing over all possible state configurations of the spin glass $S = (S_1, S_2, ..., S_N)$. Specifically,

$$Z = \sum_S e^{-E(S)/T} = \sum_{S_1=\pm 1} \sum_{S_2=\pm 1} \cdots \sum_{S_N=\pm 1} e^{-E(S)/T}. \tag{9}$$

Note that this partition function refers to the Ising model. Generally, computing the value of Z is not easy because of the combinatorial structure. If summations in equation (9) are given as integrals, the value of Z can be approximated by using saddle point approximation.

Each discrete sum over $S_i = \pm 1$ in equation (9) can be replaced with an integral over a continuous variable V. Then, by introducing the variable U through the inverse Laplace transform representation of the delta function and by making a simple shifting manipulation of the delta function we obtain

$$Z = \sum_S e^{-E(S)/T} = \frac{1}{\pi i} \prod_{i=1}^{N} \int_{-\infty}^{+\infty} \int_{-i\infty}^{+i\infty} e^{-E_{eff}(V,U,T)/T} dU_i dV_i, \tag{10}$$

where the multiplication symbol denotes a multiple integral corresponding to multiple summations, and E_{eff} is the effective energy or free energy of the system, and it can be defined as follows,

$$E_{eff}(V, U, T) = E(V) + T\left(\sum_{j=1}^{n} U_j V_j - \ln(\cosh(U_j))\right). \tag{11}$$

Notice that the evaluation of multiple integrals in equation (10) is not easy. The saddle point approximation is used to evaluate these multiple integrals. With the saddle point approximation, the partition function is dominated by the value of $E_{eff}(V, U, T)$ at a saddle point. Therefore, we seek to determine the values of the saddle points of $E_{eff}(V, U, T)$. They are determined by using a gradient descent approach. The saddle point approximation to the partition function results into a set of equations, called MFT equations, whose solutions are the desired equilibrium states of the system. The general form of the MFT equations for the Ising model is

$$V_i = \tanh U_i \tag{12}$$

$$U_i = -\frac{\partial E(V)}{T \partial V_i}. \tag{13}$$

Note that the discrete variables S_i have been replaced by continuous variables V_i and U_i, and V_i is the mean value of S_i. For the ANN with the energy function given by equation (3), the mean field equations can be reduced to

$$V_i = \tanh\left(\frac{1}{T}\sum_{j=1}^{N} C_{ij}V_j\right), \tag{14}$$

where for simplicity, we have set $I_i = 0$.

Each spin of the Ising model has only two states, either −1 or +1. This makes the Ising model inconvenient to implement for optimization problems with multistate variables. An alternative approach, proposed by Peterson and Soderberg (1989), is to use the Potts model. Instead of allowing each spin to be −1 or +1 independently, spins are arranged into sites, and only one spin within each site is allowed to be +1. Introducing a second index for the spin S_{ij}, the first index denotes the site i and the second index denotes the spin j within each site such that

$$\sum_j S_{ij} = 1. \tag{15}$$

In the Potts model, two possible values of each spin are changed to 0 or 1 rather than −1 or +1. Thus, for every i, S_{ij} has only one value for j and zero for the remaining values of j. Equation (15) guarantees that each site parameter takes only one value. Note that, by using the Potts model, some of the soft constraints of the problem are treated as hard constraints. This makes the solution space smaller than the Ising model.

The effective energy for the Potts model will be as follows

$$E_{eff} = E(V) + T\left(\sum_{j=1}^{N} U_j V_j - \ln\sum_{S} e^{US}\right). \tag{16}$$

The saddle points of $E_{eff}(V, U, T)$ are determined again by using a gradient descent approach. The general form of MFT equations for the Potts model is

$$V_j = \frac{\sum_S S e^{US}}{\sum_S e^{US}} \tag{17}$$

$$U_j = -\frac{\partial E(V)}{T\partial V_j}. \tag{18}$$

It should be noted that the saddle point applies to any energy function whether it is quadratic or not, and whether its arguments are discrete or continuous. Note that the effective energy function can be given without reference to an underlying Boltzmann distribution and to the saddle point approximation by using the Lagrangian method (Elfadel, 1995; Yuille and Kosowsky, 1994). The cost function is replaced with the Lagrangian, and the constraints are imposed using the Lagrange multipliers. Arguments of the effective energy function $E_{eff}(V, U, T)$ require the following two remarks. First, the new variables V are auxiliary variables to play, in the context of constrained mean field optimization, a role similar to that of the Lagrange multipliers in classical optimization where they are associated with the constraint set. The second remark is that the saddle point method introduces a new parameter into the optimization problem: the temperature T. In classical optimization algorithms, one can think of the temperature as the barrier or penalty parameter in constrained optimization (Elfadel, 1995; Yuille and Kosowsky, 1994). The effective energy provides a unified framework for mapping constrained optimization problems, combinatorial and continuous, onto unconstrained optimization problems.

4. ANNs and Optimization

ANNs have been proposed as a model of computation to solve optimization problems. To solve optimization problems using ANNs requires a mapping of the problem onto the ANNs in such a way that one can identify a solution from outputs of the neurons. Optimization problems can be solved using ANNs by the following two approaches.

1. By formulating an optimization problem in terms of minimizing an energy function of either discrete or continuous variables,
2. By designing competition-based ANNs in which neurons are allowed to compete to become active under certain conditions.

These two approaches suggest that ANN approaches are an alternative for solving optimization problems as compared to other optimization techniques.

The use of MFT and ANNs to derive optimization algorithms with examples of both combinatorial and continuous optimization, as well as an example with mixed (binary and continuous) variables has been demonstrated in Bilbro et al. (1992). The MFT approximation is combined with notions from the area of constrained optimization, specifically Lagrange multipliers and adaptive simulated annealing (Stolorz, 1991). The connections between MFT approaches and other optimization techniques like barrier function technique and interior-point methods were demonstrated by Yuille and Kosowsky (1994). A generalization of the effective energy function is investigated by using the notion of a conjugate function (Elfadel, 1995).

ANNs and MFT approaches are extended to optimization problems with inequality constraints and studied in Abe et al. (1992), Fang and Li (1990), Ohlsson et al. (1993), Ohlsson and Pi (1997). Inequality constraints can be taken into account by adding an appropriate penalty term to the energy function. As one can see, equality constraints are taken into account by adding a quadratic penalty term to the energy function. In the case of inequality constraints, for example, $f(x) \leqslant 0$, a penalty function is used that penalizes points for which $f(x) > 0$.

ANNs have constituted a new direction in solving combinatorial optimization problems since the pioneering work of Hopfield and Tank (1985). They showed that an energy function can be defined for an analog ANN, and the ANN always converges by performing the gradient descent to a stable state, which is the local minimum of the energy function. To find the (near) global minimum of the energy function, the Boltzmann Machine (BM) was proposed (Ackley et al., 1985). The BM was the first attempt to combine SA and ANNs to solve combinatorial optimization problems. However, BM is designed for discrete variables with the disadvantage of being slow. Another approach to combining SA and ANNs was proposed in Levy and Adams (1987) by adding a noise term to neural dynamics.

A new method to solve combinatorial optimization problems using the MFT and ANNs was introduced by Peterson and Anderson (1988). That was the first detailed attempt to use the MFT and ANNs for solving combinatorial optimization problems after a brief description by Hopfield and Tank. A problem was mapped onto an ANN by using the Ising model and then the MFT approximation was used in order to escape from local minima. As an extension of this study, a new method based on the Potts model was presented by Peterson and Soderberg (1989). They reduced the dimension of the solution space by one dimension using Potts's model instead of Ising's. The solution quality and parameter insensitivity were advantages of the new method.

The mathematical basis of the behavior of the Hopfield NN and the relationship between the Hopfield NN and combinatorial optimization problems were presented in Uesaka (1993). The relationship between MFT models and the Hopfield model was analyzed mathematically by using the theory of dynamical systems (Kurita and Funahashi, 1996). A systematic approach to design competition-based ANNs for combinatorial optimization problems was presented by Fang and Li (1990). Competition-based ANNs were studied in detail in Fausett (1994). ANNs for combinatorial optimization problems were reviewed in Fang and Li (1990), Herault and Niez (1991), Looi (1992), Melamed (1994), Ramanujam and Sadayappan (1995), Uesaka (1993), and Wang (1994). The following combinatorial optimization problems have been studied using MFT and ANN approaches.

4.1. The Graph Bipartitioning Problem

This problem has been studied in Bilbro et al. (1988), Peterson (1991), Peterson and Anderson (1988), Peterson and Soderberg (1989), and Ramanujam and Sadayappan (1995). The graph bipartitioning problem was mapped onto an ANN such that a neuron being on corresponded to a certain decision by using the Ising model, and then the MFT approximation was used in order to escape from local minima (Peterson and Anderson, 1988). As an extension of this study, a new method that maps the graph bipartitioning problem onto ANNs by using the Potts model was presented in Peterson and Soderberg (1989).

4.2. The Graph Partitioning (GP) Problem

This problem has been studied in Herault and Niez (1991), Melamed (1994), Peterson (1991), Peterson and Soderberg (1989), Ramanujam and Sadayappan (1995), and van DenBout and Miller (1990). A new method that maps the GP problem onto ANNs by using the Potts model by allowing only one state at every node to be on was presented in Peterson and Soderberg (1989). The concept of MFT in the context of the graph partitioning problem was presented in van DenBout and Miller (1990).

4.3. The Traveling Salesman Problem (TSP)

This is the most studied problem of combinatorial optimization problems using ANN approaches. The TSP has been studied in Hopfield and Tank (1985), Looi (1992), Melamed (1994), Peterson (1990, 1991), Peterson and Soderberg (1989), Stolorz (1991), Wang (1994), and Yuille (1990). Hopfield and Tank (1985) formulated the traveling salesman problem (TSP) on a highly interconnected ANN and made exploratory numerical studies. They showed that an energy function can be defined for an analog ANN and the ANN always converges by performing the gradient descent to a stable state, which is the local minimum of the energy function. A new method that maps the TSP problem onto ANNs by using the Potts model by allowing only one state at every node to be on was presented in Peterson and Soderberg (1989). An alternative approach based on the elastic net to solve TSP was developed in Durbin et al. (1989).

4.4. The Quadratic Assignment Problem (QAP)

The generalized QAP represents a large class of combinatorial optimization problems arising in a variety of planning and designing concepts. It has been studied in Fang and Li (1990), Fang et al. (1990), Melamed (1994), and Wang (1994).

4.5. The Knapsack Problem

This has been studied in Abe et al. (1992), Fang and Li (1990), Melamed (1994), Ohlsson et al. (1993), and Ohlsson and Pi (1997). In Ohlsson et al. (1993), a methodology for finding good solutions to the knapsack problem using ANNs and MFT was presented. Inequality constraints were taken into account by adding an appropriate penalty term to the energy function. In Ohlsson and Pi (1997), the MFT approach to the knapsack problem was extended to multiple knapsacks and generalized assignment problems. A hybrid algorithm that combines the LP with the MFT approximation was proposed to increase the performance of the proposed approach.

4.6. The Assignment Problem

A novel method, named as the invisible hand algorithm, for solving the assignment problem using MFT based on the Potts model is proposed in Kosowsky and Yuille (1994). The proposed MFT method and the assignment problem were used to demonstrate connections between MFT and other optimization techniques, in particular, barrier functions and interior-point methods (Yuille and Kosowsky, 1994). A method using a Hopfield NN to determine the optimum frequency assignment so that the signal interference is minimized in the satellite communication was proposed in Kurokawa and Kozuka (1994). The MFT approach was extended to the generalized assignment problem (Ohlsson and Pi, 1997). In another study (Fang and Li, 1990), a competition-based ANN was proposed to solve the generalized assignment problem.

4.7. The Circuit Partitioning

An MFT algorithm based on the Potts glass model for the circuit partitioning problem using the net-cut model was proposed in Bultan and Aykanat (1995).

4.8. The Clustering Problem

A Potts spin-glass model to solve the standard clustering problem with unlabeled data is proposed in Bengtsson and Roivainen (1995). In another study (Hofmann and Buhmann, 1997), a maximum entropy framework, namely deterministic annealing for pairwise data clustering, was developed, and mean field approximation of pairwise clustering was proposed to calculate expectation values for the data assignments. Two other ANN formulations for the clustering problem were given in Melamed (1994) and Urahama (1995).

4.9. The Maximum Clique Problem

Mathematical models combining the Ising spin-glass model and ANNs to solve the maximum clique problem are proposed in Lee and Takefuji (1996). ANN approaches using SA and MFT to the maximum clique problem are presented in Herault and Niez (1991). Another ANN formulation for the maximum clique problem was presented in Ramanujam and Sadayappan (1995).

4.10. The Scheduling Problems

A simplified scheduling problem where P teachers lecture Q classes in X classrooms at T time slots was mapped onto a Potts model by using ANNs (Gislen et al., 1989). The problem was to find a solution where all teachers give a lecture to each of the classes using available classrooms and time slots. In another study (Elmohamed et al., 1997), a similar problem, namely an academic class scheduling problem at the university level, was investigated. A new method to solve the satellite broadcasting scheduling problem was proposed in Ansari et al. (1995). A mean field approach based on the Potts model for solving resource allocation problems with nontrivial topology was developed and applied to airline crew scheduling problems (Lagerholm et al., 1997). A competition-based ANN was proposed in Fang and Li (1990) to solve job scheduling with deadlines.

4.11. Other Problems

A number of interesting combinatorial optimization problems such as vertex cover, maximum independent set, number partitioning, maximum matching, set cover, graph morphism, and graph coloring were studied in Herault and Niez (1991), Melamed (1994), and Ramanujam and Sadayappan (1995).

5. ANNS and Learning

One of the fascinating properties of ANNs is the ability to learn from its environment and to improve ANNs performance through learning. Learning is defined in Haykin (1994) as a procedure by which the free parameters of ANNs are suited through a continuing procedure of stimulation by the environment in which the network is embedded. The type of learning is determined by the manner in which the parameter changes take place. The teaching of ANNs is a difficult task. The backpropagation algorithm (BP) is the most fundamental technique for teaching feedforward ANNs. It was first developed by Werbos (1974) and later studied by Rumelhart et al. (1986). The idea is to backpropagate the output errors to the hidden layers by using what is known as the generalized delta rule. Since the backpropagation algorithm, a great deal of effort has been given to increase the speed of learning. Successful algorithms based on the BP framework were proposed. One of the new directions is to use interior-point methods to solve the learning error minimization problem. Some preliminary work has already given successful results (Couellan, 1997; Trafalis and Tutunji, 1995, 1997; Tutunji, 1996).

The Boltzmann Machine (BM), which was proposed in Ackley et al. (1985), is a stochastic learning algorithm. The BM operates by choosing a neuron and converts the state of this neuron at some temperature with some probability. This conversion changes the energy function. If this conversion is performed repeatedly in a random way, the BM will reach a stable state where no more changes in the energy are possible. This state corresponds to the global minimum of the energy function. In order to reach the global minimum of the energy, the BM uses simulated annealing (SA). The stochastic SA process in the BM was replaced by a set of deterministic equations in the so-called MFT approximation (Peterson and Anderson, 1987) to provide a substantial speedup over the BM. Using MFT in the context of the BM results in a fast deterministic learning algorithm. That was the first attempt to use the MFT in ANN training. In this study, an MFT learning algorithm for ANN training is developed. Results were very encouraging on the XOR, encoder, and line symmetry problems. In another study (Peterson and Hartman, 1989), the MFT learning was elaborated and explored with respect to a variety of tasks. The MFT learning was benchmarked against the BP procedure (Rumelhart et al., 1986) on two different feature recognition problems: two-dimensional mirror symmetry and multidimensional statistical pattern classification. In addition to feature recognition problems, a content addressable memory (CAM) with the MFT learning was investigated. Another CAM study using statistical physics techniques is proposed in Yuille (1990). Various applications of the MFT-like learning, feature recognition, and content addressable memory were reviewed in Peterson (1991).

Principal Component Analysis (PCA) is an important technique for data compression and feature extraction, and has been widely used in statistical data analysis, communication theory, pattern recognition, and image processing. Statistical physics approaches have been adapted to tackle the problem of robust PCA (Xu and Yuille, 1995). This study attempted to develop self-organizing learning rules for robust PCA. The commonly used PCA learning rules are connected to energy functions. These energy functions are generalized by adding a binary decision field with a given prior distribution, so that outliers in the data are dealt with explicitly in order to make PCA robust. Each of the generalized energy functions is then used to define a Boltzmann distribution from which a marginal distribution is obtained by summing over the binary decision field. The marginal distributions define an energy function, from which self-organizing rules have been developed for robust PCA. The energy function is used to form the corresponding Boltzmann distribution, and to determine the related energy function by statistical physics approaches like mean field approximation and the saddle point method. Thereafter, robust self-organizing rules for PCA are derived from the effective energy function.

6. ANNs and Image Processing

Recently, there have been a number of successes in applying ANN and MFT approaches to a variety of image processing problems. The MFT has been used in computer vision and image processing problems. Some researchers (Bilbro et al., 1992; Geiger and Girosi, 1989; Geman and Geman, 1984; Yuille, 1990; Zerubia and Chellappa, 1993; Zhang, 1993) used the MFT for noise removal, image reconstruction, edge detection, and image estimation. A number of problems in image processing can be formulated in terms of minimizing cost function, usually involving both binary fields representing edges in the image and continuous fields representing the smoothed image. A problem formulated in terms of minimizing an energy function can be given a probabilistic interpretation by use of the Boltzmann distribution. This formulation connects to ideas and techniques in statistical physics.

An analogy between images and statistical physics is presented in Geman and Geman (1984). Pixel gray levels and the presence and orientation of edges are considered as states of atoms in a latticelike physical system. The assignment of an energy function in the physical system determines its Boltzmann distribution. Because of the Boltzmann distribution this assignment also determines a Markov random field (MRF) model. In the last decade researchers (Geiger and Girosi, 1989; Geman and Geman, 1984; Zerubia and Chellappa, 1993; Zhang, 1993) have studied the use of MRFs for image processing. One of the main attractions of MRF models in image processing is that they can deal directly with discontinuities. Two fields are usually required in the MRF formulation of a problem: one represents the function that has to be reconstructed, and the other is associated with its discontinuities (edges). The essence of the MRF model is that the probability distribution of the configuration of the fields, for a given set of data, has a Boltzmann distribution for some cost functional dependent upon a small neighborhood. Since the fields have a discrete range, finding the solution becomes a combinatorial optimization problem.

Geman and Geman (1984) proposed a method for solving the image restoration and edge detection problem in noisy data using line processes. In Geiger and Girosi (1989), it is shown that by applying MFT to those MRF models ANNs are obtained. An MRF model-based EM (expectation-maximization) procedure for simultaneously estimating the degradation model and restoring the image, also known as blind image restoration, is described in Zhang (1993). The problem of edge detection and image estimation from a nonstationary image corrupted by additive Gaussian white noise is considered in Zerubia and Chellappa (1993). Another application of MFT (Bilbro et al., 1992), resulted in algorithms that for some problems are qualitatively identical to Graduated NonConvexity (GNC) algorithms. GNC algorithms are considered as image restoration techniques. A number of problems in image processing involve matching features between image frames. A general method for modeling matching problems in terms of elastic network and deformable templates with binary matching fields is described in Yuille (1990). Deformable templates are simple parameterized shapes that interact with an image to obtain best fit of their parameters. The ideas of elastic nets and deformable templates are generalized to include binary matching fields. The resulting models, generalized deformable models, are applied to image processing for both the vision problems of stereo and long-range motion.

7. Conclusions

The use of ANN approaches as applied to optimization, learning, and image processing problems were reviewed. These approaches suggest that ANNs are an alternative for solving optimization problems as compared to other optimization techniques. Motivations for using ANNs include the improvement in the speed of the operation through massively parallel computation, and possible hardware design advantages. One of the main advantages of ANN approaches to classical optimization approaches is the inherently parallel and distributed nature of the dynamic solution procedure. Therefore, ANNs are capable of solving large-scale optimization problems in real time. This is essential in many engineering design, control, and optimization problems.

References

Abe, S., J. Kawakami, and K. Hirasawa. 1992. "Solving inequality constrained combinatorial optimization problems by the Hopfield neural networks." *Neural Networks* **5**: 663–670.

Ackley, D. H., G. E. Hinton, and T. J. Sejnowski. 1985. "A learning algorithm for Boltzman machines." *Cognitive Sci.* **9**: 147–169.

Ansari, N., E. S. H. Hou, and Y. Yu. 1995. "A new method to optimize the satellite broadcasting schedules using the mean field annealing of a Hopfield neural network." *IEEE Trans. Neural Networks* **6(2)**: 470–482.

Bengtsson, M. and P. Roivainen. 1995. "Using the Potts glass for solving the clustering problem." *Int. J. Neural Syst.* **6(2)**: 119–132.

Bilbro, G., R. Mann, T. K. Miller, W. E. Snyder, D. E. VanDenBout, and M. White. 1988. "Optimization by mean field annealing." In *Proceedings of the Annual Conferences on Advances in Neural Information Processing Systems,* Volume 1, edited by D. S. Touretzky, pp. 91–98. San Francisco, CA: Morgan Kaufmann Publishers.

Bilbro, G. L., W. E. Snyder, S. J. Garnier, and J. W. Gault. 1992. "Mean field annealing: a formalism for constructing GNC-like algorithms." *IEEE Trans. Neural Networks* **3(1)**: 131–138.

Bultan, T. and C. Aykanat. 1995. "Circuit partitioning using mean field annealing." *Neurocomputing* **8**: 171–194.

Couellan, N. 1997. Primal–dual techniques for nonlinear programming and applications to artificial neural network training. Ph.D. diss. Norman, OK: School of Industrial Engineering, University of Oklahoma.

DenBout, D. E. van and T. K. Miller. 1990. "Graph partitioning using annealed networks." *IEEE Trans. Neural Networks* **1(2)**: 192–203.

Durbin, R., R. Szeliski, and A. L. Yuille. 1989. "An analysis of the elastic net approach to the travelling salesman problem." *Neural Comput.* **1**: 348–358.

Elfadel, I. M. 1995. "Convex potential and their conjugates in analog mean-field optimization." *Neural Comput.* **7**: 1079–1104.

Elmohamed, S., P. Coddington, and G. Fox. 1997. "A comparison of annealing techniques for academic course scheduling." Technical Report SCCS-777, January 25, 1997. Northeast Parallel Architectures Center (NPAC). NY: Syracuse University.

Fang, L. and T. Li. 1990. "Design of competition-based neural networks for combinatorial optimization." *Int. J. Neural Syst.* **1(3)**: 221–235.

Fang, L., W. H. Wilson, and T. Li. 1990. "Mean field annealing neural net for quadratic assignment." In *International Neural Network Conference, July 9–13, Paris, France*, pp. 282–286. Dordrecht, The Netherlands: Kluwer Academic Publishers.

Fausett, L. 1994. In *Fundamentals of Neural Networks: Architectures, Algorithms, and Applications*. New Jersey: Prentice-Hall.

Fischer, K. H. and J. A. Hertz. 1991. In *Spin Glasses*. Cambridge, England: Cambridge University Press.

Geiger, D. and F. Girosi. 1989. "Coupled Markov random fields and mean field theory." In *Proceedings of the Annual Conferences on Advances in Neural Information Processing Systems*, Volume 2, edited by D. S. Touretsky, pp. 660–667. San Francisco, CA: Morgan Kaufmann Publishers.

Geman, S. and D. Geman. 1984. "Stochastic relaxation, Gibbs distributions, and the Bayesian restoration of images." *IEEE Trans. Pattern Anal. Machine Intelligence* **PAMI-6(6)**: 721–741.

Gislen, L., C. Peterson, and B. Soderberg. 1989. "Teachers and classes with neural networks." *Int. J. Neural Syst.* **1(1)**: 167–183.

Haykin, S. 1994. In *Neural Networks: A Comprehensive Foundation*. New Jersey: Macmillan College Publishing Company.

Herault, L. and J.-J. Niez. 1991. "Neural networks and combinatorial optimization: a study of NP-complete graph problems." In *Neural Networks: Advances and Applications*, edited by E. Gelenbe, pp. 165–213. Amsterdam: Elsevier Science B.V.

Hofmann, T. and J. M. Buhmann. 1997. "Pairwise data clustering by deterministic annealing." *IEEE Trans. Pattern Anal. Machine Intelligence* **19(1)**: 1–14.

Hopfield, J. J. 1982. "Neural networks and physical systems with emergent collective computational abilities." In *Proceedings of the National Academy of Sciences of the U.S.A., Biophysics*, pp. 2554–2558. Washington, D.C.

Hopfield, J. J. 1984. "Neurons with graded response have collective computational properties like those of two-state neurons." In *Proceedings of the National Academy of Sciences of the U.S.A., Biophysics*, pp. 3088–3092. Washington, D.C.

Hopfield, J. J. and D. W. Tank. 1985. "Neural computation of decisions in optimization problems." *Biol. Cybernet.* **52**: 141–152.

Kosowsky, J. J. and A. L. Yuille. 1994. "The invisible hand algorithm: solving the assignment problem with statistical physics." *Neural Networks* **7(3)**: 477–490.

Kurita, N. and K.-I. Funahashi. 1996. "On the Hopfield neural networks and mean field theory." *Neural Networks* **9(9)**: 1531–1540.

Kurokawa, T. and S. Kozuka. 1994. "Use of neural networks for the optimum frequency assignment problem." *Elect. Commun, Japan. Part 1*, **77(11)**: 106–116.

Lagerholm, M., C. Peterson, and B. Soderberg. 1997. "Airline crew scheduling with Potts neurons." *Neural Comput.* **9**: 1589–1599.

Lee, K.-C. and Y. Takefuji. 1996. "Maximum clique problems: part 1 and 2." In *Neural Computing for Optimization and Combinatorics*, edited by Y. Takefuji and J. Wang, pp. 31–77. Singapore: World Scientific.

Levy, B. C. and M. C. Adams. 1987. "Global optimization with stochastic neural networks." In *Neural Networks for Optimization and Signal Processing, Proceedings of the First International Conference on Neural Networks, San Diego*, pp. 681–690. IEEE, NJ.

Little, W. A. 1974. "The existence of persistent states in the brain." *Math. Biosci.* **19**: 101–120.

Looi, C.-K. 1992. "Neural network methods in combinatorial optimization." *Comput. Oper. Res.* **19(3/4)**: 191–208.

McCullough, W. S. and W. Pitts. 1943. "A logical calculus of the ideas imminent in nervous activity." *Bull. Math. Biophys.* **5**: 115–133.

Melamed, I. I. 1994. "Neural networks and combinatorial optimization." *Automat. Remote Control* **55(11)**: 1553–1584.

Mezard, M., G. Parisi, and M. A. Virasoro. 1987. In *Spin Glass Theory and Beyond*. Singapore: World Scientific.

Müller, B. and J. Reinhardt. 1990. In *Neural Networks: an Introduction*. Berlin: Springer-Verlag.

Ohlsson, M. and H. Pi. 1997. "A study of the mean field approach to knapsack problems." *Neural Networks* **10(2)**: 263–271.

Ohlsson, M., C. Peterson, and B. Soderberg. 1993. "Neural networks for optimization problems with inequality constraints: the knapsack problem." *Neural Comput.* **5**: 331–339.

Peterson, C. 1990. "Parallel distributed approaches to combinatorial optimization: benchmark studies on traveling salesman problem." *Neural Comput.* **2**: 261–269.

Peterson, C. 1991. "Mean field theory neural networks for feature recognition, content addressable memory and optimization." *Connection Sci.* **3(1)**: 3–33.

Peterson, C. and J. R. Anderson. 1987. "A mean field theory learning algorithm for neural networks." *Complex Syst.* **1**: 995–1019.

Peterson, C. and J. R. Anderson. 1988. "Neural networks and NP-complete optimization problems: a performance study on the graph bisection problem." *Complex Syst.* **2**: 59–89.

Peterson, C. and E. Hartman. 1989. "Explorations of the mean field theory learning algorithm." *Neural Networks* **2**: 475–494.

Peterson, C. and B. Soderberg. 1989. "A new method for mapping optimization problems onto neural networks." *Int. J. Neural Syst.* **1(1)**: 3–22.

Ramanujam, J. and P. Sadayappan. 1995. "Mapping combinatorial optimization problems onto neural networks." *Inf. Sci.* **82**: 239–255.

Reichl, L. E. 1980. *A Modern Course in Statistical Physics*. TX: University of Texas Press.

Rumelhart, D. E., G. E. Hinton, and R. J. Williams. 1986. "Learning representations by backpropagating errors." *Nature (London)* **323**: 533–536.

Stolorz, P. 1991. "Merging constrained optimisation with deterministic annealing to solve combinatorially hard problems." In *Advances in Neural Information Processing Systems,* edited by J. E. Moody, S. J. Hanson, and R. P. Lippmann, pp. 1025–1032. San Francisco, CA: Morgan Kaufmann, Publishers.

Trafalis, T. B. and T. A. Tutunji. 1995. "A stochastic logarithmic barrier function method for neural network training with bounded weights." In *Intelligent Engineering Systems Through Artificial Neural Networks,* Volume 5, edited by C. H. Dagli, M. Akay, C. L. P. Chen, B. Fernandez, and J. Ghosh, pp. 167–172. New York: ASME Press.

Trafalis, T. B. and T. A. Tutunji. 1997. "Deterministic and stochastic logarithmic barrier function methods for neural network training." In *Parallel Computing in Optimization*, edited by A. Migdalas et al., pp. 529–574. Dordrecht, The Netherlands: Kluwer Academic Publishers.

Tutunji, T. A. 1996. "Logarithmic barrier functions and Newton-type methods with applications to neural network training." Ph.D. diss. Norman, OK: School of Industrial Engineering, University of Oklahoma.

Uesaka, Y. 1993. "Mathematical basis of neural networks for combinatorial optimization problems." *Opto-electron.—Devices Technol.* **8(1)**: 1–9.

Urahama, K. 1995. "Mathematical programming formulation for neural combinatorial optimization algorithms." *Electron. Commun. Japan, Part 3* **78(9)**: 67–75.

Wang, J. 1994. "Deterministic neural networks for combinatorial optimization." In *Progress in Neural Networks*, edited by O. M. Omidvar, pp. 319–340. Westport, CT: Ablex.

Werbos, P. 1974. "Beyond regression: new tools for prediction and analysis in the behavioural sciences." Ph.D. diss. Cambridge, MA: Applied Mathematics, Harvard University.

Xu, L. and A. L. Yuille. 1995. "Robust principal component analysis by self-organizing rules based on statistical physics approach." *IEEE Trans. Neural Networks* **6(1)**: 131–143.

Yuille, A. L. 1990. "Generalized deformable models, statistical physics, and matching problems." *Neural Comput.* **2**: 1–24.

Yuille, A. L. and J. J. Kosowsky. 1994. "Statistical physics algorithms that converge." *Neural Comput.* **6**: 341–356.

Zerubia, J. and R. Chellappa. 1993. "Mean field annealing using compound Gauss–Markov random fields for edge detection and image estimation." *IEEE Trans. Neural Networks* **4(4)**: 703–709.

Zhang, J. 1993. "The mean field theory in EM procedures for blind Markov random field image restoration." *IEEE Trans. Image Process.* **2(1)**: 27–40.

11 STOCHASTIC PROGRAMMING

John R. Birge

ABSTRACT Stochastic programming considers models for optimal decision making where certain parameters are not known with certainty. The emphasis in the field is generally on the construction of the models and on the solution of these models to obtain an optimal value, given assumed parameter distributions, or bounds (perhaps with some confidence level) on an optimal value. Due to recent advances in computational capabilities, many applications appear now in finance, transportation, manufacturing, electric power, telecommunications, and other areas. This chapter describes some samples of these applications. It also provides an overview of stochastic programming model formulations and basic computational techniques.

1. Introduction

Stochastic programming models focus on optimal decisionmaking while certain parameters are stochastic. The first models involve an action and reaction (or recourse) as given in Beale (1955) and Dantzig (1955). An alternative model (called chance or probabilistically constrained programming) was also developed quite early, principally by Charnes and Cooper (1959). Both forms have their origin in statistical decision theory (see, e.g., Wald 1950) but, in contrast to decision theory, stochastic programming has emphasized methods of solution and analytical solution properties over procedures for constructing objectives and updating probabilities.

Since the size of a stochastic program can grow rapidly as additional uncertainties are modeled, many of the algorithmic developments in the area have concentrated on methods to exploit stochastic program structure. For example, Dantzig and Madansky (1961) introduced Dantzig–Wolfe decomposition as a possible method, while Van Slyke and Wets (1969) developed a form of Benders (1962) decomposition for stochastic programs. In general, these methods concentrate on linear models but nonlinearities can apply as well. Current methodologies now allow the solution of very large problems.

The range of stochastic programming applications is as broad as any in optimization. Early applications include an airline fleet assignment model by Ferguson and Dantzig (1956). In that model, planes are assigned to routes to maximize expected revenue given a distribution on passenger demand on each route. Their solution also yielded a net gain of \$142,000 or 9 percent over a solution that ignores uncertainty.

Typically, practical examples display similar or greater improvements. Many applications appear now in finance, transportation, manufacturing, electric power, telecommunications, and other areas. Birge (1997) and Birge and Louveaux (1997) provide several such examples. The sections below provide an overview of stochastic programming formulations, computational techniques, and sample applications.

2. Formulating Stochastic Programs

The simplest form of a stochastic program may be the news vendor (also known as the "newsboy") problem. This familiar example illustrates some of the fundamental characteristics of a stochastic program and broader uses.

The news vendor must determine how many papers (x) to buy now at cost c without knowing the demand represented by random variable ξ, with distribution function F, and given selling price p. Assuming no salvage value, any papers bought in excess of demand are simply discarded with no return.

If the vendor observes the same distribution on demand every day, a reasonable objective is to maximize expected daily profit:

$$\max_{x\geqslant 0} \phi(x) = -cx + \int_0^x p\xi dF(\xi) + \int_x^\infty pxdF(\xi), \tag{1}$$

which has the basic form of the stochastic linear program with recourse action x, followed by observation (ξ), and recourse (sales at price p).

Program 1 has the added advantage of an analytical solution. In this case, assuming F is continuous with density f, differentiation in (1) yields

$$-c + px^*f(x^*) + \int_{x^*}^{\infty} p dF(\xi) - px^*f(x^*) = 0,$$

or

$$F(x^*) = \frac{p-c}{p}.$$

The formula above illustrates the problem in simply substituting expected values for random quantities, as is often done in practice and widely thought to be of little consequence. If this is done in (1), the obvious solution is $x^{\text{det}} = \bar{\xi} = \int \xi f(\xi) d\xi$, the mean demand. This value only matches x^* if $F(x^{\text{det}}) = \frac{p-c}{p}$. For a symmetric distribution where the mean equals the median, that result only occurs if $p = 2c$.

A more likely case is when margins $(p-c)$ are low. Here, the results are quite different. Suppose, for example, that $p = \$0.25$, while $c = \$0.20$. In this case, $F(x^*) = 0.2$. Suppose a uniform demand on [50, 150] with mean of 100. In this case, $x^* = 70$ while $x^{\text{det}} = 100$. The result is that $\phi(x^*) = -20(70) + 25(12) + 25(70)(4/5) = \$3.00$ while $\phi(x^{\text{det}}) = -20(100) + 25(75/2) + 25(100)(1/2) = \$1.875$. Thus, the stochastic model yields an increase of \$1.125 or 60 percent in profits over the deterministic formulation. This difference is called the value of the stochastic solution (VSS see Birge, 1982).

The VSS is similar to the expected value of perfect information (EVPI, see, e.g., Raiffa and Schlaifer, 1961), but the VSS measures a different aspect of uncertainty. For example, EVPI compares the stochastic program solution value with the value of a situation in which all uncertainty is revealed before a decision is made (i.e., with perfect information). In the news vendor case, we have a perfect information solution, $x^{PI} = \xi$, with expected value, $\int_{\Xi}(p-c)\xi f(\xi)d\xi = (p-c)\bar{\xi}$. For the example given, we would have an expected value with perfect information (EVWPI) of \$5.00 and $EVPI = EVWPI - \phi(z^*) = \$2.00$.

In this case, $EVPI > VSS$ but this is not always true. In fact, either value can be zero while the other is positive (see Birge, 1982, and Birge and Louveaux, 1997). The difference is that the value of resolving uncertainty can be quite different from the cost of ignoring uncertainy in making a decision. Stochastic programming often emphasizes this latter form where uncertainty cannot be resolved further (in some economical way) but computational short cuts, such as substituting mean values for random parameters, might still be used.

This simple example illustrates another difficulty for modelers attempting to use mean values alone. If the deterministic model is followed, then one might expect a profit of $(p-c)\bar{\xi}$ or \$5.00 in this case. The vendor would, of course, be disappointed to find only \$1.875, or \$3.125 (62.5 percent) less than the value using means alone. In fact, this loss compared to the mean value problem always occurs when mean values for random constraint parameters are substituted in convex stochastic programs (see, e.g., Madansky, 1960, or Birge and Wets, 1986). This result may explain some losses of confidence in optimization modelers when results do not turn out as well as the model predicts.

General forms of the news vendor problem to other models preserve the characteristic of action, observation, and reaction. This general form is called a two-stage stochastic linear program with (fixed) recourse (2S-SLPR), which is written:

$$\begin{aligned} &\min_{x \in \mathscr{R}^{n_1}} \quad c^T x + \mathscr{Q}(x) \\ &\text{subject to} \quad Ax = b, \qquad (2S-\text{SLPR}) \\ &\qquad\qquad\qquad x \geqslant 0, \end{aligned}$$

where $\mathscr{Q}(x) = \int_{\Xi} Q(x, \xi) P(d\xi)$, with $Q(x, \xi) =$

$$\min_{y \in \mathscr{R}^{n_2}} \quad q^T y$$

$$\text{subject to} \quad Wy = h - Tx, \quad \text{(LPR)}$$

$$y \geqslant 0,$$

and $b \in \mathscr{R}^{m_1}$, $W \in \mathscr{R}^{m_2 \times m_2}$, and $\xi = (q, h, T_{\cdot 1}, ..., T_{\cdot N_2})$ is an underlying random vector defined on a probability space $(\Xi, \mathscr{A}, P)$. The resource is called fixed because W is deterministic. While W could also be random, it is usually assumed to be fixed to obtain clear definitions of feasible solutions (see, for example, Kall, 1976).

To see how the news vendor problem fits the 2S-SLPR framework, note that in (1), we have

$$Q^{\text{news}}(x, \xi) = \{\min - py \mid 0 \leqslant y \leqslant x, y \leqslant \xi\}. \tag{2}$$

The news vendor model can also be written as a special form of (2S-SLPR) called simple recourse in which $W = [I \mid -I]$, forcing the y values to equal the components of $h - Tx$ or $Tx - h$. This corresponds to under- or overselling. While several methods are quite efficient for this problem (see, Wets, 1983, and Nazareth and Wets, 1986), the simple nature of the recourse makes $Q(x)$ straightforward to evaluate so that the simple recourse model can be solved as a convex (with finite random variables, also piecewise linear) program with linear constraints. (Note that $Q(x, \xi)$ is convex in x, yielding $\mathscr{Q}(x)$ convex.)

An additional common characteristic of stochastic programs is to have a constraint that holds with a certain probability. For example, the news vendor may want to have papers for at least a fraction $\alpha < 1$ of the potential customers to maintain good public relations. The addition of these probabilistic constraints or chance constraints, generally of the form,

$$P\{T'x \geqslant h'\} \geqslant \alpha, \tag{3}$$

creates a chance or probabilistically constrained stochastic program. These constraints are often simply replaced by a deterministic equivalent form, $\bar{T}'x \geqslant \bar{h}'$, and added to the deterministic constraints. Since this action does not generally change problem structure, the remainder of the development here only considers recourse problems with all probabilistic constraints implicitly represented in the deterministic constraints. Details on general chance-constrained methods appear in Prékopa (1995) and elsewhere.

The choice of a recourse model, a chance-constrained model, or some combination of these depends on the decisionmakers' attitute toward risk. While (2S-SLPR) appears to imply risk neutrality with its linear objective, we can model many kinds of risk attitudes through appropriate discount factors and constraint modification without creating nonlinearties (see Birge, 2000). In some cases, the decision makers may have multiple objectives or other utility forms. The (2S-SLPR) model can be easily updated to replace $c^T x$ by $c(x)$ and $q^T y$ by, for example, $U(q, y)$.

Many problems involve not just a single observation and reaction but a sequence of these events. In this case, with linear objectives, we have the multistage stochastic linear program with recourse (MS-SLPR):

$$\begin{array}{llllll}
\min\ z & = c_1 x_1 & + E_\Omega[c_2(\omega^1) x_2(\omega^1) + & & \cdots + c_H(\omega^{H-1}) x_H(\omega^{H-1})] & \\
\text{subject to} & W_1 x_1 & & & & = h_1, \\
& T_1(\omega^1) x_1 & + W_2 x_2(\omega^1) & & & = h_2(\omega^1), \\
& & \cdots & & \cdots & \vdots, \\
& & & T_{H-1}(\omega^{H-1}) x_{H-1}(\omega^{H-1}) & + W_H x_H(\omega^{H-1}) & = h_H(\omega^{H-1}), \\
& x^1 \geqslant 0; & & x_t(\omega^{t-1}) \geqslant 0, \text{a.s.}; & t = 2, .., H; & \text{(MS-SLPR)}
\end{array}$$

where c_1 is a known vector in $\mathscr{R}^{n_1}$, h_1 is a known vector in $\mathscr{R}^{m_1}$, each W_t is a known $m_t \times n_t$ matrix (for fixed recourse), and random observations at time $t + 1$ are represented through a random element, ω_t, and a history process, $\omega^t = \{\omega_1, ..., \omega_t\}$, defined on a probability space, $(\Omega_1 \times \cdots \times \Omega_t, \mathscr{A}^t, P^t)$. We then have the subscript notation for a single period and the superscript to denote all previous periods (and an iteration counter in the algorithm discussion below). The complete data process, $\{\omega = (\omega_1, ..., \omega_{H-1})\}$, is defined on $(\Omega, \mathscr{A}, P)$. The random observations at $t+1$ are N_t-vectors, $\xi_{t+1}(\omega^t) = (c_{t+1}(\omega^t), h_{t+1}(\omega^t), (T_t)_1 \cdot (\omega^t), ..., (T_t)_{m_{t+1}} \cdot (\omega^t))$.

This type of formulation becomes somewhat cumbersome since we have implicitly restricted the decisions to depend only on past data. The more general formulation includes explicit constraints to maintain this property.

More generally, we may have nonlinear functions in the objectives of constraints as well as integer variable restrictions. In this case, we can formulate a general multistage stochastic program with recourse:

$$\inf_{x \in \mathscr{N}} \quad \mathrm{E}\left[\sum_{t=0}^{H} f_{t+1}(\omega, x^t(\omega), x_{t+1}(\omega))\right] \tag{4}$$

$$\text{subject to} \quad x_t(\omega) \in X_t(\omega) \text{ a.s.},$$

where E is expectation with respect to the overall sigma field of events, $\mathscr{A}$, and $\mathscr{N}$ denotes the closed linear subspace of nonanticipative processes. Nonanticipativity of the decision process at time t implies that decisions must only depend on the data up to time t, that is, $x^t(\omega)$ must be $\mathscr{A}^t$-measurable, where $\mathscr{A}^t$ is the sigma-field of events up to time t. An alternative characterization of this nonanticipative property is that $x^t(\omega) = \mathrm{E}\{x^t(\omega) \mid \mathscr{A}^t\}$ a.s., $t = 0, ...,$ where $\mathrm{E}\{\cdot \mid \mathscr{A}^t\}$ is conditional expectation with respect to the σ-field $\mathscr{A}^t$. Using the projection operator $\Pi^t : z \to \Pi^t z := \mathrm{E}\{z \mid \mathscr{A}^t\}, t = 0, ...,$ this is equivalent to

$$(I - \Pi^t)x^t = 0, \quad t = 0, ... \tag{5}$$

In our previous examples, the constraints in (2S-SLPR) were implicit by replacing $x(\omega)$ with a single value, x, wherever $x(\omega)$ might appear.

The overwhelming difficulty in stochastic programming stems from the evaluation of random functions and their expectations, such as $\mathscr{Q}(x)$ in (2S-SLPR) or an analogously defined value function in (MS-SLPR), $\mathscr{Q}_t(x_t)$, to reflect the expectation of all future objective terms given decision x_t at time t. Except in special cases, such as the simple recourse case mentioned above, practical solution requires some form of approximation. In some (perhaps most practical) situations, a distribution may not be known. In these cases, an approximating distribution may be chosen to yield results that can be far superior to a single best estimate or mean value substitution for all random parameters.

The majority of approximation methods involve some discretization of an underlying probability measure. The discretization may be chosen to provide upper or lower bounds on values with a known probability measure or unknown distribution with certain known characteristics (such as mean and variance ranges; see Birge and Wets, 1986; Kall et al., 1988, and Birge and Louveaux, 1997). Similar continuous bounding approximations replace the original high-dimensional integral with a combination of low-dimensional integrals (see Birge and Wallace, 1988; Birge and Wets, 1989; Wallace, 1986; Birge and Qi, 1995).

The basic approach then in stochastic programming is to obtain an approximation of the distribution (that may have statistical or absolute error bounds relative to some unknown true distribution) and to formulate a problem (MS-SLPR). The structure of (MS-SLPR) is then used to compute solutions through a variety of optimization techniques.

3. Solution Techniques

Stochastic programming methods rely on a variety of optimization approaches and often statistical analysis. One optimization procedure is to take advantage of the stochastic program structure with special purpose factorizations. These have particular advantages for interior-point methods for stochastic programs. The availability of parallel processing and capabilities for performing necessary routines in parallel may determine which form that factorization should take. Other methods, which seem to offer even greater efficiencies, are based on decomposing a problem or relaxing some constraints. The next subsections consider three of these methods: a standard decomposition approach, decomposition with sampling, and an augmented Lagrangian approach.

3.1. Decomposition Approaches

The general idea behind decomposition methods for (2S-SLPR) and (MS-SLPR) is that a smaller problem in x (or x_1) alone can be solved with some approximation of the recourse function, $\mathcal{Q}(x)$. This idea started with a method called *L*-shaped by Van Slyke and Wets (1969) that is a form of Benders (1962) decomposition. From that, various extensions have been suggested.

Consider the general formulation in (2S-SLPR). The basic idea of the L-shaped method is to approximate the nonlinear term in the objective function of this problem. A general principle behind this approach is that, since the nonlinear objective function term (the recourse function) involves a solution of all second-stage recourse linear programs, we would like to avoid numerous function evaluations for it. We therefore use that term to build a master problem in x but we only evaluate the recourse function exactly as a subproblem.

To make this approach possible, we assume again that the random vector ξ has finite support. The (2S-SLPR) problem has a dual-block-angular structure that leads to the name, *L*-shaped method, for the decomposition method we describe here. The method is essentially a Dantzig–Wolfe decomposition (inner linearization; Dantzig and Wolfe, 1960) of the dual or a Benders decomposition (outer linearization; Benders, 1962) of the primal. This method, due to Van Slyke and Wets (1969) for stochastic programming, also considers feasibility questions of particular relevance for stochastic programs. In the following, we suppose the first stage decision, x, is bounded, that (2S-SLPR) has a finite optimal value, and that the random vector ξ takes on K values, $\xi_1, \ldots, \xi_K$, each with probability p_k. Procedures to remove such assumptions are straightforward.

L-shaped Method

Step 0. Set $s = t = v = 0$.

Step 1. Set $v = v + 1$. Solve the linear program (6)–(8).

$$\min \; z = c^T x + \theta \tag{6}$$

$$\text{subject to } \; Ax = b,$$

$$D_\ell x \geqslant d_\ell, \quad \ell = 1, \ldots, r; \tag{7}$$

$$E_\ell x + \theta \geqslant e_\ell, \quad \ell = 1, \ldots, s; \tag{8}$$

$$x \geqslant 0, \quad \theta \in \mathcal{R}.$$

Let (x^v, θ^v) be an optimal solution. If no constraint (8) is present, θ^v is set equal to $-\infty$ and not considered in the computation of x^v.

Step 2. For $k = 1, \ldots, K$ solve the linear program

$$\min \; w' = e^T v^+ + e^T v^- \tag{9}$$

$$\text{subject to } \; Wy + Iv^+ - Iv^- = h_k - T_k x^v, \tag{10}$$

$$y \geqslant 0, \quad v^+ \geqslant 0, \quad v^- \geqslant 0,$$

where $e^T = (1, \ldots, 1)$, until, for some k, the optimal value $w' > 0$. In this case, let σ^v be the associated dual multipliers and define

$$D_{r+1} = \sigma^v T_k \tag{11}$$

and

$$d_{r+1} = \sigma^{\nu} h_k \tag{12}$$

to generate a constraint (called a feasibility cut) of type (7). Set $r = r+1$, add to the constraint set (7), and return to Step 1. If, for all k, $w' = 0$, go to Step 3.

Step 3. For $k = 1, ..., K$ solve the linear program

$$\begin{aligned} \min \quad & w = q_k^T y \\ \text{subject to} \quad & Wy = h_k - T_k x^{\nu} \\ & y \geqslant 0. \end{aligned} \tag{13}$$

Let π_k^{ν} be the simplex multipliers associated with the optimal solution of Problem k of type (13). Define

$$E_{s+1} = \sum_{k=1}^{K} p_k \pi_k^{\nu} T_k \tag{14}$$

and

$$e_{s+1} = \sum_{k=1}^{K} p_k \pi_k^{\nu} h_k. \tag{15}$$

Let $w^{\nu} = e_{s+1} - E_{s+1} x^{\nu}$.

If $\theta^{\nu} \geqslant w^{\nu}$, stop. The solution x^{ν} is optimal. Otherwise, set $s = s+1$, add to the constraint set (8), and return to Step 1.

The method solves an approximation of (2S-SLPR) using an outer linearization of $\mathcal{Q}$. Two types of constraints are sequentially added: (i) feasibility cuts (7) determining $\{x \mid \mathcal{Q}(x) < +\infty\}$, and (ii) optimality cuts (8) which are linear supports of $\mathcal{Q}$ on its domain of finiteness.

This method has also been generalized to multiple stages (see Birge, 1985) as a nested decomposition method. The basic idea of the nested L-shaped or Benders decomposition method is to place cuts on the period t value function, $\mathcal{Q}_{t+1}(x_t)$, and to add other cuts to achieve an x_t that has a feasible completion in the future. The cuts represent successive linear approximations of $\mathcal{Q}_{t+1}$. Due to the polyhedral structure of $\mathcal{Q}_{t+1}$, this process converges to an optimal solution in a finite number of steps.

3.2. *Decomposition Methods with Sampling*

As mentioned earlier, stochastic programming formulations often include some approximation of an underlying probability distribution. When that distribution is known, it is common to use a sample approximation. A potential disadvantage of sampling, however, is that some effort might be wasted on an inaccurate approximation. An approach to avoid this problem is to use sampling within another algorithm without complete optimization. A natural candidate is to embed sampling into the L-shaped method described in the previous section. Dantzig and Glynn (1990) and Dantzig and Infanger (1991) follow this approach directly by computing sample cuts in the L-shaped method (and extensions) and then using importance sampling to reduce the error in these cuts. Higle and Sen (1991) present another approach based on a sample that grows as the algorithm progresses.

This method, called stochastic decomposition, generates many cuts with increasingly large samples built on previous samples and adjusts these cuts to drop as the algorithm continues processing. The basic method assumes that $Q(x, \xi)$ is feasible for all x and that there is a known (probability one) lower bound on $Q(x, \xi)$ (e.g., 0), where the dual solutions to the recourse problem (13) are bounded, and the sets $\{x \mid Ax = b, x \geqslant 0\}$ and Ξ are also compact.

With these assumptions, the basic stochastic decomposition method generates iterates, x^k, and observations, ξ^k, in the following way.

Stochastic Decomposition Method

Step 1. Set $k = 0$, $\xi^0 = \bar{\xi}$, and let x^1 solve

$$\min_{Ax=b, x \geqslant 0} \{c^T x + Q(x, \xi^0)\}. \tag{16}$$

Step 2. Let $k = k + 1$ and ξ^k be an independent sample generated from ξ. Find $\mathcal{Q}^k(x^k) = \frac{1}{k}\sum_{t=1}^{k} Q(x^k, \xi^t) = \frac{1}{k}\sum_{t=1}^{k} \pi_t^k(\xi^t - Tx^k)$. Let $E_k = \frac{1}{k}\sum_{t=1}^{k} \pi^t T$ and $e_k = \frac{1}{k}\sum_{i=1}^{k} \pi_t^k \xi^t$.

Step 3. Update all previous cuts by $E_t \leftarrow \frac{k-1}{k} E_t$ and $e_t \leftarrow \frac{k-1}{k} e_t$ for $t = 1, ..., k-1$.

Step 4. Solve the L-shaped master problem as in (6)–(8) to obtain x^{k+1}. Go to Step 1.

This method differs slightly from the basic method in Higle and Sen (1991) in that we assume π_t^k to be optimal dual solutions in each iteration. Higle and Sen allow a restricted set of dual optima that may decrease the solution effort (with perhaps fewer effective cuts). Various approaches may be used for practical stopping conditions. Their results show good convergence behavior on a variety of problems, especially those arising in network design (see Sen et al., 1994).

3.3. *Augmented Lagrangian Methods*

Lagrangian methods apply to stochastic programs as in other optimization problems and ensure strictly convex objective terms. These methods follow the general augmented Lagrangian approach (see Bertsekas, 1982), where the constraint on nonanticipativity is relaxed and placed into the objective. In this case, for a constraint that $\mathbf{x}^0 = E[\mathbf{x}^0]$ a.e., for a sample of K realizations, $x_k^0 - \sum_{k=1}^{K} p_k x_k^0 = 0$ for $k = 1, ..., K$. The result is a penalty $\frac{r}{2}\left\|x_k^0 - \sum_{k=1}^{K} p_k x_k^0\right\|^2$ so that a problem of the general form in (4) for a two-stage ($H = 1$) problem becomes:

$$\inf_{x \in \mathcal{N}} \sum_{k=1}^{k} f_1(\omega_k, x_k^0, x_k^1) + \pi_k^\nu \cdot (x_k^0 - \sum_{l=1}^{k} p_l x_l^0) + \frac{r}{2}\left\|x_k^0 - \sum_{l=1}^{K} p_l x_l^0\right\|^2$$

$$\text{subject to } x_t(\omega^k) \in X_t(\omega^k), \quad t = 0, 1; \; k = 1, ..., K, \tag{17}$$

where ω_k denotes all realizations for the kth event and ν is an iteration number. The algorithm performs iterations with a fixed step size such that $\pi_k^{\nu+1} = \pi_k^\nu + rp_k\left(x_k^0 - \sum_{k=1}^{K} p_k x_k^0\right)$. The advantage (see, e.g., the discussion in Dempster, 1988) is that Newton type steps can be applied since the method maintains a nonsingular Hessian.

This method is similar to the progressive hedging algorithm (PHA) of Rockafellar and Wets (1991), which separates individual scenario problems for each iteration by replacing the $\sum_{k=1}^{K} p_k x_k^0$ terms with the expected value result from the previous iteration, $\hat{x}^{0,\nu} = \sum_{k=1}^{K} p_k x_k^{0,\nu-1}$ where $x^{0,\nu-1}$ was optimal on iteration $\nu - 1$. This substitution means that (17) splits into separate problems for each k.

PHA offers computational advantages at least for structured problems as reported by Mulvey and Vladimirou (1991). A related approach is the extension of the row action algorithm of Censor and Lent (1981) by Nielsen and Zenios (1993a,b), which has particular efficiencies for network constraints. The key to these methods' success is that individual subproblem structure is maintained throughout the algorithm. The convergence of this method is based on Rockafellar's (1976) proximal point method. The basis for this approach is not dual ascent but the contraction of the pair $(\hat{x}^{\nu+1}, \pi^{\nu+1})$ about a saddle point.

4. Applications

Stochastic programming has seen great gains since its original applications in airline fleet assignment by Ferguson and Dantzig (1956). This section presents the advantages obtained in solving stochastic problems in place of deterministic models. Examples of these values appear in all sectors of the economy as the following examples demonstrate (see King, 1988 for other areas).

4.1. *Finance*

Financial problems inherently involve uncertainty. These problems are especially attractive for stochastic formulations because deterministic approximations are quite obviously flawed. For example, using a deterministic expected return results in investments only in the highest risk (and also highest expected return) asset, but few people are willing to follow such a strategy. The essence of a stochastic program is to develop a hedging strategy that mixes investments in a way that a strictly deterministic approximation cannot reproduce.

As an example, we consider the Russell–Yasuda Kasai Model (Cariño et al., 1994), which won the second prize in the 1993 Franz Edelman Awards for Management Science Achievement. In that model, decisions are made for a Japanese insurance company on how best to invest in assets to meet a random liability stream over time. The investment returns are also random. Additional complications include legal restrictions about the use of income to meet liabilities.

The overall Russell–Yasuda Kasai model has the structure in (MS-SLPR) with a piecewise linear convex objective. A decomposition method was implemented to solve realistically sized models.

For such financial models, deterministic solutions clearly provide inadequate solutions since they ignore risk. Static portfolio models, based on Markowitz's (1952) model, have, however, been used in many cases. Since these models consider the asset side without liabilities, their implementations may result in significant transaction costs and inopportune liquidation of assets.

The solution of the Russell–Yasuda Kasai model improved significantly over the static portfolio procedures that had been used previously. The authors report that the Russell–Yasuda Kasai model yielded additional income of $79 million in its first two years of use.

4.2. *Transportation*

Transportation models remain a prominent area for application of stochastic programming models. From Dantzig and Ferguson's model, several other areas have developed. An area of particular focus has been in dynamic vehicle allocation (DVA, see Powell, 1988, 1990; Frantzeskakis and Powell, 1990; Powell and Chung, 1993).

The basic problem in DVA is to route a set of vehicles (notably trucks or freight cars) to meet demands along routes and to position them for anticipated future loads. The overall goal is to maximize returns over given horizons. The model again fits the form of (MS-SLPR) but the wide distribution on load demands makes approximation schemes critically important.

Powell and his colleagues have developed a range of approximations that resemble the value function, $\mathcal{Q}_t(x)$, at each time period and yield a form of dynamic programming approximation method. Their results with realistic vehicle fleet sizes of up to one thousand vehicles indicate approximation errors that are often below 10 percent. Other experiments (Powell, 1990) show that a combination of their methods with a dispatcher's decisionmaking yields 10 to 15 percent gains over purely manual dispatches.

4.3. *Manufacturing*

The complexity and inherent uncertainty in both demand and supply aspects of manufacturing make this area also particularly fruitful for stochastic programming models. Eppen et al. (1989) provide a particularly interesting example for a capacity planning model at General Motors. The goal was to determine capacity for various products at a number of plants to maximize an expected profit objective with a downside risk constraint. The model fits the form in (2S-SLPR) with the addition of the risk constraint imposing an additional linking constraint across scenarios.

This model was to aid a high-level planning decision (in contrast to the operational-level vehicle allocation model). A limited number of scenarios was sufficient to represent the distribution on demands that were critical for the decision. The result of their stochastic model was an increase of over 6 percent in expected profit over two potential deterministic scenarios and complete avoidance of downside risk that occurred in another potential scenario. In this case, no individual deterministic scenario could provide the optimal stochastic programming (hedging) solution.

4.4. *Power Systems*

Power system models have been a common area of application and source of developments in stochastic programming methods. Pereira and Pinto (1985, 1991) use decomposition procedures for models of the Brazilian power system, which includes a number of reservoirs that must be controlled with uncertain inputs from precipitation as well as uncertain loads and other sources. A similar model, SOCRATES (Jacobs et al., 1993), is used at Pacific Gas and Electric for coordination of their wide range of hydrological power sources.

As an example of the potential of these models, Takriti et al. (1996) consider a model of the Michigan power system. This model is designed for daily scheduling and, therefore, explicitly includes binary variables for an individual unit's state of operation (on or off). Their procedure uses the progressive hedging method to achieve convergence to near-optimal solutions quickly. The results for a sample week showed potential savings over a deterministic procedure of almost $150,000 in generation costs for that week alone.

4.5. *Economic Policy*

Stochastic programming models have often been used in economic policy considerations as well. For example, Manne (1974) considered the U.S. decision of whether to begin or delay a breeder reactor program. The result showed a low cost for waiting for more information that may have influenced the decision to delay.

As another example, Birge and Rosa (1995) consider a model of the U.S. economy with random returns on investment in new technologies with reduced carbon emissions. They consider various methods of limiting carbon emissions in the future due to environmental concerns. The stochastic nature of investment in this situation can again have a large effect on the preferred policy. Deterministic models all yield solutions that concentrate investment on a given winner technology. The stochastic program solution hedges and produces acceptable results in a variety of circumstances The overall value of the stochastic solution is as high in this case as 2 percent of the total economic activity in the United States.

References

Beale, E. M. L. 1955. "On minimizing a convex function subject to linear inequalities." *J. R. Stat. Soc., Series B* **17**: 173–184.

Benders, J. F. 1962. "Partitioning procedures for solving mixed-variables programming problems." *Numer Math.* **4**: 238–252.

Bertsekas, D. P. 1982. In *Constrained Optimization and Lagrange Multiplier Methods*. New York: Academic Press.

Birge, J. R. 1982. "The value of the stochastic solution in stochastic linear programs with fixed recourse." *Math. Programming* **24**: 314–325.

Birge, J. R. 1985. "Decomposition and partitioning methods for multi-stage stochastic linear programs." *Oper. Res.* **33**: 989–1007.

Birge, J. R. 1997. "Stochastic programming computation and applications state-of-the-art survery." *INFORMS J. Comput.* **9**: 111–133.

Birge, J. R. 2000. "Option methods for incorporating risk into linear capacity planning models." *Manufact. Service Oper. Mgmnt* **2**: 19–31.

Birge, J. R. and F. V. Louveaux. 1997. In *Introduction to Stochastic Programming*. New York: Springer-Verlag.

Birge, J. R. and L. Qi. 1995. "Continuous approximation schemes for stochastic programs." *Ann. Oper Res.* **56**: 15–38.

Birge, J. R. and C. H. Rosa. 1995. "Modelling investment uncertainty in the costs of global CO_2 emission policy." *Eur. J. Oper. Res.* **83**: 466–488.

Birge, J. R. and S. W. Wallace. 1988. "A separable piecewise linear upper bound for stochastic linear programs." *SIAM J. Control Optim.* **26**: 725–739.

Birge, J. R. and R. J-B. Wets. 1986. "Designing approximation schemes for stochastic optimization problems, in particular, for stochastic programs with recourse." *Math. Programming Study* **27**: 54–102.

Birge, J. R. and R. J-B. Wets. 1989. "Sublinear upper bounds for stochastic programs with recourse." *Math. Programming* **43**: 131–149.

Cariño, D. R., T. Kent, D. H. Meyers, C. Stacy, M. Sylvanus, A. L. Turner, K. Watanabe, and W. T. Ziemba. 1994. "The Russel–Yasuda Kasai model: an asset/liability model for a Japanese company using multistage stochastic programming." *Interfaces* **24(1)**: 29–49.

Censor, Y. and A. Lent. 1981. "An iterative row-action method for interval convex programming." *J. Optim. Theory Appl.* **343**: 321–353.

Charnes, A. and W. W. Cooper. 1959. "Chance-constrained programming." *Mgmnt Sci.* **5**: 73–79.

Dantzig, G. B. 1955. "Linear programming under uncertainty." *Mgmnt Sci.* **1**: 197–206.

Dantzig, G. B. and P. Glynn. 1990. "Parallel processors for planning under uncertainty." *Ann. Oper. Res.* **22**: 1–21.

Dantzig, G. B. and G. Infanger. 1991. "Large-scale stochastic linear programs—importance sampling and Benders decomposition." In *Computational and Applied Mathematics, I (Dublin, 1991)*, edited by C. Brezinski and V. Kulisch, pp. 111–120. Amsterdam: North-Holland.

Dantzig, G. B. and A. Madansky. 1961. "On the solution of two-stage linear programs under uncertainty." In *Proceedings of the Fourth Berkeley Symposium on Mathematical Statistics and Probability*, edited by J. Neyman, pp. 165–176. Berkeley: University of California Press.

Dantzig, G. B. and P. Wolfe. 1960. "The decomposition principle for linear programs." *Oper. Res.* **8**: 101–111.

Dempster, M. A. H. 1988. "On stochastic programming II: dynamic problems under risk." *Stochastics* **25**: 15–42.

Eppen, G. D., R. K. Martin, and L. Schrage. 1989. "A scenario approach to capacity planning." *Oper. Res.* **37**: 517–527.

Ferguson, A. and G. B. Dantzig. 1956. "The allocation of aircraft to routes: an example of linear programming under uncertain demands." *Mgmnt Sci.* **3**: 45–73.

Frantzeskakis, L. and W. B. Powell. 1990. "A successive linear approximation procedure for stochastic, dynamic vehicle allocation problems." *Transport Sci.* **24**: 40–57.

Higle, J. and S. Sen. 1991. "Stochastic decomposition: an algorithm for two stage linear programs with recourse." *Math. Oper. Res.* **16**: 650–669.

Jacobs, J., G. Freeman, J. Grygier, D. Morton, G. Schultz, K. Staschus, J. Stedinger, and B. Zhang. 1993. "Stochastic optimal coordination of river-basin and thermal electric systems (SOCRATES): a system for scheduling hydroelectric generation under uncertainty." Technical Report. San Francisco, CA: Pacific Gas and Electric Company.

Kall, P. 1976. In *Stochastic Linear Programming*. Berlin: Springer-Verlag.

Kall, P., A. Ruszczyński, and K. Frauendorfer. 1988. "Approximations in stochastic programming." In *Numerical Techniques for Stochastic Optimization*, edited by Y. Ermoliev and R. Wets, pp. 33–64. Berlin: Springer-Verlag.

King, A. 1988. "Stochastic programming problems: examples from the literature." In *Numerical Techniques for Stochastic Optimization*, edited by Y. Ermoliev and R. Wets, pp. 543–567. Berlin: Springer-Verlag.

Madansky, A. 1960. "Inequalities for stochastic linear programming problems." *Mgmnt Sci.* **6**: 197–204.

Manne, A. S. 1974. "Waiting for the breeder." In *Review of Economic Studies Symposium on the Economics of Exhaustible Resources* **41**: 47–65.

Markowitz, H. M. 1952. "Portfolio selection." *J. Finance* **7**: 77–91.

Mulvey, J. M. and H. Vladimirou. 1991. "Applying the progressive hedging algorithm to stochastic generalized networks." *Ann. Oper. Res.* **31**: 399–424.

Nazareth, J. L. and R. J-B. Wets. 1986. "Algorithms for stochastic programs: the case of nonstochastic tenders." *Math. Programming Study* **28**: 1–28.

Nielsen, S. S. and S. A. Zenios. 1993a. "Proximal minimizations with *D*-functions and the massively parallel solution of linear stochastic network programs." *Int. J. Supercomput. Appl.* **7**: 349–364.

Nielsen, S. S. and S. A. Zenios. 1993b. "A massively parallel algorithm for non-linear stochastic network problems." *Oper. Res.* **41**: 319–337.

Pereira, M. V. F. and L. M. V. G. Pinto. 1985. "Stochastic optimization of a multireservoir hydroelectric system—a decomposition approach." *Water Resources. Res.* **21**: 779–792.

Pereira, M. V. F. and L. M. V. G. Pinto. 1991. "Multi-stage stochastic optimization applied to energy planning." *Math. Programming* **52**: 359–375.

Powell, W. B. 1988. "A comparative review of alternative algorithms for the dynamic vehicle allocation program." In *Vehicle Routing: Methods and Studies*, edited by B. Golden and A. Assad. Amsterdam: North-Holland.

Powell, W. B. 1990. "Real-time optimization for truckload motor carriers." *OR/MS Today* **17(2)**: 28–33.

Powell, W. B. and R. K.-M. Cheung. 1993. "A network recourse decomposition method for dynamic networks with random arc capacities." Technical Report. Princeton: Department of Civil Engineering and Operations Research, Princeton University.

Prékopa, A. 1995. In *Stochastic Programming*. Dordrecht, The Netherlands: Kluwer.

Raiffa, H. and R. Schlaifer. 1961. In *Applied Statistical Decision Theory*. Boston: Harvard University.

Rockafellar, R. T. 1976. "Monotone operators and the proximal point algorithm." *SIAM J. Control Optim.* **14**: 877–898.

Rockafellar, R. T. and R. J-B. Wets. 1991. "Scenarios and policy aggregation in optimization under uncertainty." *Math. Oper. Res.* **16**: 119–147.

Sen, S., R. D. Doverspike, and S. Cosares. 1994. "Network planning with random demand." *Telecommun. Sys.* **3**: 11–30.

Takriti, S., J. R. Birge, and E. Long. 1996. "A stochastic model for the unit commitment problem." *IEEE Trans. Power Sys.* **11**: 1497–1508.

Van Slyke, R. and R. J-B. Wets. 1969. "L-shaped linear programs with application to optimal control and stochastic programming." *SIAM J. Appl. Math.* **17**: 638–663.

Wald, A. 1950. In *Statistical Decision Functions*. New York: Wiley.

Wallace, S. W. 1986. "Solving stochastic programs with network recourse." *Networks* **16**: 295–317.

Wets, R. J-B. 1983. "Solving stochastic programs with simple recourse." *Stochastics* **10**: 219–242.

12 HIERARCHICAL OPTIMIZATION

Hoang Tuy

ABSTRACT Hierarchical optimization deals with mathematical programming problems whose feasible set is implicitly determined by a sequence of nested optimization problems. The most important case is the bilevel linear programming problem (BLP), which has attracted much attention from researchers due to numerous applications in various fields. Basic properties of BLP, together with solution methods proposed so far in the literature, are discussed, based on different reformulations of BLP as a single mathematical program. In particular, a branch-and-bound method is described in detail. The relation of BLP with multicriteria optimization is clarified, and the current research situation in nonlinear bilevel programming is briefly reviewed.

1. Introduction

In hierarchial systems involving two levels of decisionmaking, with different, sometimes conflicting objectives, the following optimization problem may arise. The higher level (leader) controls the decision variables $x \in X$ (for instance, transfer prices, resource allocation, capital investment channelling, etc.), while the lower level (follower) controls the decision variables $y \in Y$ (activity level, production volume, technology alternatives, etc.). The higher level does not exercise a direct control on the operations of the lower level, but its decision influences the decision of the lower level through constraints imposed on the operations of the latter. To each decision x made by the higher level, the lower level responds by a $y = y(x)$ optimizing its objective function $f(y)$ over the set of feasible operations determined by the decision of the higher level. Assuming that the higher level knows perfectly the reaction of the lower level to each of its decisions, what is the optimal decision of the higher level, that is, the vector x that ensures the best values of its objective function $F(x, y)$ under these conditions? Mathematically, this problem, known as the bilevel programming problem can be formulated as

$$\text{(GBP)} \quad \underset{x,y}{\text{minimize}} \quad F(x,y)$$

$$\text{subject to} \quad G(x,y) \leqslant 0, \quad x \in X$$

$$y = y(x) \text{ solves}$$

$$\min_y \{f(y) \mid g(x,y) \leqslant 0, \quad y \in Y\}$$

where $X \subset R^{n_1}$, $Y \subset R^{n_2}$, $F: X \times Y \rightarrow R$, $G: X \times Y \rightarrow R^{m_1}$, $f: Y \rightarrow R$, $g: X \times Y \rightarrow R$, and $G: X \times Y \rightarrow R^{m_2}$. A more general optimization problem, called the multilevel programming problem, is the extension of the above model to hierarchical systems composed of several (more than two) levels of decisionmaking.

In this chapter we will focus on the bilevel linear program, which is by far the most extensively studied problem of multilevel optimization. A bilevel linear program is a GBP where all the functions involved are linear, that is, a problem of the form

$$\text{(BLP)} \quad \underset{x,y}{\text{minimize}} \quad L(x,y)$$

$$\text{subject to} \quad A_1 x + B_1 y \geqslant b^1, \quad x \geqslant 0 \tag{1}$$

$$y = y(x) \text{ solves}$$

$$\min_y \{\langle d, y\rangle \mid A_2 x + B_2 y \geqslant b^2, \quad y \geqslant 0\} \tag{2}$$

where $x \in R^{n_1}$, $y \in R^{n_2}$, $d \in R^{n_2}$, $A_i = R^{m_i \times n_1}$, $B_i = R^{m_i \times n_2}$, and $L: R^{n_1} \times R^{n_2} \rightarrow R$ is a real-valued linear function.

The leader/follower strategy was first proposed for a duopoly by von Stackelberg (1952). In the absence of the constraint (1) and under the assumption that the lower-level problem $R(x)$ always has a unique optimal solution, the above model is actually a static Stackelberg game in which the higher level is the first player (leader) and the lower level is the second player (follower).

Originally formulated and studied as a mathematical program by Bracken and McGill (1973, 1974), multilevel optimization has risen to become a subject of extensive research during the last two decades. It has met with numerous applications in diverse fields: economic development policy, agriculture economics, road network design, oil industry regulation, international water systems and flood control, energy policy, traffic assignment, structural analysis in mechanics (e.g., cable structures, elastoplastic structures, structures with fractal geometries). For a review of some of these applications, see for example, Ben-Ayed (1993) and Vicente and Calamai (1994). Multilevel optimization can also be useful for the study of hierarchical designs in many complex systems, and, in particular, in biological systems (see Vincent, 1990 and Baer et al., 1992).

Hierarchial optimization problems are known to be computationally difficult. In fact BLP has been shown to be NP-hard by Jeroslow (1985) and strongly NP-hard by Hansen et al. (1992). A more detailed discussion of recent results on the complexity of multilevel problems can be found in Pardalos and Deng (1997). Despite the notorious difficulty of this class of problems, significant progress has been achieved during the last two decades in the understanding of their mathematical structure and the development of reasonably efficient techniques for solving them.

In the special case when the follower's objective function is the negative of the leader's, BLP reduces to the linear max-min problem, which was first studied by Falk (1973). It is easy to show that the linear max-min problem is actually equivalent to a convex maximization linear constraints, hence can be approached by methods of convex maximization. The general bilevel linear programming problem is more complicated and fraught with pitfalls. Actually, several methods proposed in the literature for its solution turned out to be nonconvergent, or incorrect or convergent to a local nonglobal optimum (see Ben Ayed, 1993, where some of these errors have been reported). Furthermore, suboptimal solutions may lead to both theoretical and real-world paradoxes, as for instance in network design problems, where increase in link capacities may deteriorate the performance of the network (Berechman, 1984).

Among the best computational results reported earlier in the literature, one should mention those of the branch-and-bound method of Bard and Moore (1990), the algorithm using complementary pivoting of Judice and Faustino (1988) and the branch-and-bound algorithm of Hansen et al. (1992). More recently global optimization methods have been developed to take advantage of the reverse convexities and monotonicity properties present in BLP. In particular, due to these properties the nonconvexity rank of this problem is at most equal to rankA_2. This allows a significant simplification of the search procedure when rankA_2 is small compared to the overall dimension, as in many cases of practical interest.

2. General Properties of BLP

Denote by $R(x)$ the follower's subproblem (2). As it stands, the above formulation of BLP implies that any optimal solution $(\bar{x}, \bar{y})$ of BLP must satisfy $A_1\bar{x} + B_1\bar{y} \geqslant b^1$ and $L(\bar{x}, \bar{y}) \leqslant L(\bar{x}, y')$ for every optimal solution y' of $R(\bar{x})$ such that $A_1\bar{x} + B_1y' \geqslant b^1$. This condition, sometimes referred to as the tie cooperation assumption, means that when the lower-level problem $R(x)$ has more than one optimal solution, the follower is supposed to choose among these solutions the best one according to the leader's constraints and preferences. Of course, other models can be considered that correspond to different assumptions on the follower's behavior when $R(x)$ has more than one optimal solution (then an extra constraint has to be added to $R(x)$).

For the study of BLP it is convenient to convert it into an ordinary mathematical program. Define

$$D := \{(x, y) \mid Ax + By \geqslant b, \quad x \geqslant 0, y \geqslant 0\}. \tag{3}$$

where

$$A = \begin{bmatrix} A_1 \\ A_2 \end{bmatrix}, \quad B = \begin{bmatrix} B_1 \\ B_2 \end{bmatrix}, \quad b = \begin{bmatrix} b^1 \\ b^2 \end{bmatrix}.$$

For simplicity we assume that D has a nonempty interior and that the polyhedron

$$A_2x + B_2y \geqslant b^2, \quad x \geqslant 0, y \geqslant 0$$

is bounded (so both polyhedrons are bounded and have a nonempty interior). Then for every $x \geqslant 0$ either the problem $R(x)$ is infeasible or it has a finite optimal value.

Theorem 1. *(Bard and Falk, 1982) BLP is equivalent to the following single level mathematical programming problem:*

$$\text{(MP)} \qquad \min_{x,y} L(x,y) \tag{4}$$

$$\text{subject to } Ax + By \geqslant b \tag{5}$$

$$B_2^T\lambda \leqslant d \tag{6}$$

$$\langle \lambda, A_2x - b^2 \rangle + \langle d, y \rangle = 0 \tag{7}$$

$$x \geqslant 0, \quad y \geqslant 0, \quad \lambda \geqslant 0. \tag{8}$$

Proof: *From linear programming theory,* y *solves* R(x), *if, and only if,* $\lambda \in \mathrm{R}_+^{m_2}$ *exists, satisfying (6) and (7), hence the result.*

Note that the objective and all constraints in (MP) are linear, except the complementarity constraint (7).

The above mathematical program involves extra variables λ. To reformulate BLP as a single level mathematical program with no extra variables, we define

$$\varphi(u) := \min\{\langle d, y \rangle \mid u + B_2y \geqslant b^2, \quad y \geqslant 0\}, \quad E = \{u \mid u + B_2y \geqslant b^2, \quad y \geqslant 0\}. \tag{9}$$

By the duality theory of linear programming, we also have

$$\varphi(u) = \sup\{\langle \lambda, b^2 - u \rangle \mid B_2^T\lambda \leqslant d, \quad \lambda \geqslant 0\} \tag{10}$$

Lemma 1. *The function* $\varphi(\mathrm{u})$ *is convex, finite, lower semicontinuous on the polyhedron* E, *and satisfies:*

$$\varphi(u') \leqslant \varphi(u) \quad \text{whenever } u' \geqslant u. \tag{11}$$

Proof: *Clearly* E *is a polyhedron and* $\varphi(\mathrm{u}) < +\infty$ *for* $\mathrm{u} \in \mathrm{E}$. *Since by* (10), $\varphi(\mathrm{u})$ *is the pointwise supremum of a family of affine functions* $\mathrm{u} \mapsto \langle \lambda, \mathrm{b}^2 - \mathrm{u} \rangle$, *it follows that* $\varphi(\mathrm{u})$ *is convex and lower semicontinuous on* E. *To prove the monotonicity property (11), observe that if* $\mathrm{u}' \geqslant \mathrm{u}$ *then* $\mathrm{u} + \mathrm{B}_2\mathrm{y} \geqslant \mathrm{b}^2$ *implies* $\mathrm{u}' + \mathrm{B}_2\mathrm{y} \geqslant 2$, *hence,* $\varphi(\mathrm{u}') \leqslant \varphi(\mathrm{u})$.

Theorem 2. *(Tuy et al., 1993) BLP is equivalent to the following linear program with an additional reverse convex constraint:*

$$\text{(RC)} \qquad \min_{x,y} L(x,y) \tag{12}$$

$$\text{subject to } (x,y) \in D \tag{13}$$

$$\varphi(A_2x) - \langle d, y \rangle \geqslant 0. \tag{14}$$

Proof: *Clearly,* y *solves* R(x) *if, and only if,* $\langle d, y\rangle = \varphi(A_2x)$, *that is if, and only if,* (14) *holds because* $\varphi(A_2x) - \langle d, y\rangle \leqslant 0\ \forall(x, y) \in D$. *By Lemma 1,* $\varphi(A_2x)$ *is convex in* x, *so (14) is a reverse convex constraint.*

The reverse convex structure implies a number of important properties, which have been proved in various ways by Candler and Townsley (1982), Bialas and Karwan (1984), Bard (1983) (see also Benson, 1989 for multilevel linear programs):

Proposition 1. *The feasible set of BLP is a union of faces of* D.

Proof: *By definition* $\varphi(A_2x) - \langle d, y\rangle \leqslant 0\ \forall(x, y) \in D$, *therefore, (14) implies that the feasible set of (MP) consists of all global maximizers of the convex function* $\varphi(A_2x) - \langle d, y\rangle$ *on the polytope* D. *It is well known that this set is a union of faces of* D *(see Rockafellar, 1970).*

In earlier models of bilevel linear programming the constraint (1) is absent. In that case, it can be shown that the feasible set of BLP is connected (Paraga, 1981), a property which is no longer true in the general case (see Konno et al., 1997).

Proposition 2. *If a relative interior point of a face of* D *is feasible then the whole face is feasible. In particular, if BLP is solvable then at least one optimal solution is an extreme point of* D.

Proof: *The first assertion is simply a consequence of Proposition 1. The second assertion also follows, because the optimal value in BLP must then be the minimum of the linear function* L(x, y) *over a certain face of* D *and hence must be achieved at an extreme point of this face, that is also an extreme point of* D.

Theorem 3. *(Anandalingam and White, 1990) There exists a positive constant* M_0 *such that for any* $M \geqslant M_0$ *every optimal solution of the following linearly constrained concave minimization problem will solve BLP*

$$\text{(CM)}\quad \min\{g_M(\lambda) \mid B_2^T\lambda \leqslant d, \quad \lambda \geqslant 0\} \tag{15}$$

where

$$g_M(\lambda) = \min\{L(x, y) + M(\langle\lambda, A_2x - b^2\rangle + \langle d, y\rangle) \mid (x, y) \in D\} \tag{16}$$

Proof: *Denote by* $V_0(D)$ *the set of extreme points* (x, y) *of* D *that are feasible, that is that satisfy* $\langle\lambda, A_2x - b^2\rangle + \langle d, y\rangle = 0$ *for some* $\lambda \geqslant 0$ *satisfying* $B_2^T\lambda \leqslant d$, *and by* $V_+(D)$ *the set of extreme points* (x, y) *such that* $\langle\lambda, A_2x - b^2\rangle + \langle d, y\rangle > 0$ *for some* $\lambda \geqslant 0$ *satisfying* $B_2^T\lambda \leqslant d$, *Since both these sets are finite, there exists* M_0 *such that for all* $M \geqslant M_0$:

$$\min_{(x,y)\in V_+(D)} [L(x, y) + M(\langle\lambda, A_2x - b^2\rangle + \langle d, y\rangle)] > \min_{(x,y)\in V_0(D)} L(x, y)$$

Then clearly every optimal solution of the problem

$$\min\{L(x, y) + M(\langle\lambda, A_2x - b^2\rangle + \langle d, y\rangle) \mid (x, y) \in D, \quad B_2^T\lambda \leqslant d, \lambda \geqslant 0\} \tag{17}$$

will minimize L(x, y) *over the set* $V_0(D)$, *hence will solve BLP by Proposition 2.*

Note that the term $\langle\lambda, A_2x - b_2\rangle + \langle d, y\rangle$ expresses the gap between the objective function values of the lower level problem and its dual, so $M\langle\lambda, A_2x - b_2\rangle + \langle d, y\rangle$ can be viewed as a penalty for the gap.

3. Some Basic Solution Approaches

Most solution approaches proposed to date for the bilevel programming problem use, directly or indirectly, one of the above reformulations of BLP as a single-level mathematical program. Since (CM) is a linearly constrained concave minimization problem, while (MP) and (RC) are linear programs with an additional complementarity or reverse convex constraint, these approaches can actually be interpreted as specialized methods for concave minimization or reverse convex programming problems discussed in deterministic global optimization (see Horst and Tuy, 1996). A major cause of difficulty of all these problems, however, is that their feasible set is not robust (i.e., is not equal to the closure of its interior), as is usually assumed in continuous optimization.

One of the earliest algorithms for BLP using the formulation (MP) is the parametric complementary pivot algorithm proposed by Bialas and Karwan (1982). However, it was later shown by Judice and Faustino (1988) and Ben-Ayed and Blair (1990) that this algorithm may not find an optimal solution. Judice and Faustino (1988) gave a modified parametric complementary pivot method that provides an ε-optimal solution for any given $\varepsilon > 0$.

Bard and Falk (1982) converted the complementary slackness condition (7) into a piecewise separable form and solved the problem by a branch-and-bound technique. Candler and Townsley (1982) proposed an algorithm based on an implicit enumeration of vertices of the feasible polytope of the lower level problem, such that at each iteration, global information is generated and used to avoid returning to previously explored vertices.

Bialas and Karwan (1984) developed the kth best algorithm, which can be considered a variant of Thuong–Tuy's method for linear programs with an additional reverse convex constraint. Among the methods using (MP) the best known are the branch-and-bound method of Bard and Moore (1990) in which branching is done on the parameters λ, and that of Hansen et al. (1992), in which branching is based on conditions expressed in terms of tightness of the follower's constraints (i.e., in terms of positiveness of the corresponding dual variables). Another method by Visweswaran and Floudas (1996) decomposes the problem (MP) into a series of subproblems yielding lower and upper bounds to the global optimum.

Tuy et al. (1992) proposed a polyhedral annexation method applied to the reverse convex program (RC), and later, a conical branch-and-bound method applied to the quasiconcave minimization problem dual to (RC) (Tuy et al., 1994). Both these methods work on a space of reduced dimension, namely of dimension at most $1 + \text{rank}A_2$. In a more recent approach by Tuy and Ghannadan (1998), the reverse convex (RC) program was shown to belong to the class of nonconvex problems with separated nonconvexity (see Tuy, 1992) and solved by a simplicial branch-and-bound procedure exploiting this property.

Using the formulation (CM), White and Anandalingam (1993) developed a penalty approach to BLP which consists in solving the concave minimization problem (CM), for a sufficiently large M, by a method borrowing some essential ideas from the earlier cone splitting algorithm of Tuy (1964). Another penalty method for BLP, proposed by Amouzegar and Moshirvaziri (1998), solves BLP via the bilinear program (17), equivalent to (CM), by using parts of Thuong–Tuy's (1984) method for moving from one local minimizer to a better one.

Finally, since most exact methods can handle only moderate size problems, Gendreau et al. (1996) have proposed an adaptive search method relative to the tabu metaheuristic, which is reported to have been able to solve large problem instances with good accuracy.

4. A Branch-and-Bound Algorithm

The computational burden of a branch-and-bound algorithm usually increases exponentially with the dimension of the space in which branching is performed. In this respect, it is important to note that (BLP) can easily be transformed into an equivalent problem with p nonconvex variables only, where $p = \text{rank}A_2$. This allows the problem to be solved by a branch-and-bound procedure operating mainly in a p-dimensional space, with p at most equal to, and often much smaller than, $\min\{m_2, n_2\}$. First observe the following

Proposition 3. *There exists* K_0 *such that for any* $K \geqslant K_0$ *the function*

$$\varphi_K(u) = \min\left\{\langle d, y\rangle + K\sum_{i=1}^{m_2} z_i \mid B_2 y + z \geqslant b^2 - u, \quad y \geqslant 0, z \geqslant 0\right\} \tag{18}$$

$$= \max\{\langle \lambda, b^2 - u\rangle \mid B_2^T\lambda \leqslant d, \quad 0 \leqslant \lambda \leqslant K\} \tag{19}$$

is an extension of $\varphi(u)$ *to the whole space* R^{m_2}. *This function is convex and satisfies the monotonicity condition:* $\varphi_K(u') \leqslant \varphi_K(u)$ *whenever* $u' \geqslant u \in R^{m_2}$.

Proof: *Let* K_0 *be so large that* $0 \leqslant \lambda \leqslant K_0$ *for every vertex* λ *of the polyhedron* $B_2^T\lambda, \lambda \geqslant 0$. *For every* $u \in E = \text{dom}\varphi$ *[see (9)] we have (10), hence for* $u \in E$ *and* $K \geqslant K_0$:

$$\varphi(u) = \max\{\langle \lambda, b^2 - u\rangle \mid B_2^T\lambda \leqslant d, \quad 0 \leqslant \lambda \leqslant K\}$$

which implies (18) by the duality theorem of linear programming. The function φ_K *is convex as the pointwise supremum of a family of affine functions in* u. *Finally, if* $u' \geqslant u$ *then* $\langle \lambda, b^2 - u'\rangle \leqslant \langle \lambda, b^2 - u\rangle$ *for every* $\lambda \geqslant 0$, *hence* $\varphi_K(u') \leqslant \varphi_K(u)$.

Note that the lower level subproblem $R(x)$ is infeasible if $A_2x \notin E$, so Kz can be interpreted as a penalty to be incurred if the constraint $(u + B_2y - b^2)_i \geqslant 0$ is violated, with $z = (b_2 - u - B_2y)_i > 0$.

The problem BLP in which the lower level subproblem $R(x)$ is replaced by

$$\min\left\{\langle d, y\rangle + K\sum_{i=1}^{m_2} z_i \mid A_2x + B_2y + z \geqslant b^2, \quad y \geqslant 0, z \geqslant 0\right\} \tag{20}$$

will be referred to as the K-regularization of (BLP). This modification amounts to inflicting a penalty of $K\sum_{i=1}^{m_2} z_i$ to the follower for any violation of the constraint $A_2x + B_2y \geqslant b^2$. Any optimal solution of (BLP/K) is an approximate optimal solution of (BLP), and by Proposition 3, the two problems are equivalent when $K \geqslant K_0$.

Denote the problem (RC) in which φ is replaced with φ_K by (RC/K). We next describe a branch-and-bound algorithm proposed by Tuy and Ghannadan (1998) for solving (BLP/K) via (RC/K).

Let Q be a matrix formed by p linearly independent rows of A_2. Since $p = \text{rank}A_2$, we have $A_2x' = A_2x$ if and only if $Qx' = Qx$. Then, setting $h(t) = \varphi_K(A_2x)$ whenever $Qx = t$, we have a well defined function $h: R^p \to R$ because for any $t \in R^p$: $Qx' = Qx = t$ implies $A_2x' = A_2x$, hence $\varphi_K(A_2x') = \varphi_K(A_2x)$. Let

$$Q = [Q_M, Q_N], \quad x = \begin{bmatrix} x_M \\ x_N \end{bmatrix}$$

where Q_M is a nonsingular $p \times p$ submatrix of Q. From the relation $Qx = t$ we have

$$x = \begin{bmatrix} Q_M^{-1} \\ 0 \end{bmatrix} t + \begin{bmatrix} -Q_M^{-1}Q_Nx_N \\ x_N \end{bmatrix}.$$

Therefore, setting

$$Z = \begin{bmatrix} Q_N^{-1} \\ 0 \end{bmatrix}, \quad w = \begin{bmatrix} -Q_M^{-1}Q_Nx_N \\ x_N \end{bmatrix}$$

yields

$$x = Zt + w \text{ with } Qw = -Q_Nx_N + Q_Nx_N = 0. \tag{21}$$

Since $Qw = 0$ implies $A_2w = 0$ it follows that $h(t) = \varphi_K(A_2Zt)$, hence $h(t)$ is a convex function. We have thus proved that:

Theorem 4. *(BLP/K) is equivalent to the reverse convex program*

$$\min\{L(x, y) \mid (x, y) \in D, Qx = t, h(t) - \langle d, y\rangle \geqslant 0\}. \tag{22}$$

where $h(t) = \varphi_K(A_2Zt)$ *is a convex function on* R^p, *with* $p = \text{rank}A_2$.

This new formulation shows that the problem becomes linear when $t = (t_1, ..., t_p)$ is fixed. Therefore, in the following branch-and-bound method, we will branch upon t.

We start with a simplex S_1 in the t-space R^p, large enough to contain the image of the polyhedron D under the mapping $(x,y)\mapsto Qx$. This simplex can be constructed by computing

$$\tau_1 = \min\{(Qx)_i \mid (x,y) \in D\}, \quad i = 1, \ldots, p,$$

$$\rho = \max\left\{\sum_{i=1}^{p} t_i \mid Qx = t, (x,y) \in D\right\},$$

and letting

$$S_1 = \left\{t \mid \tau_i \leqslant t_i > (i = 1, \ldots, p); \quad \sum_{i=1}^{p} t_i \leqslant \rho\right\}.$$

Note that the $(p+1)$ vertices of this simplex are $v^0 = (\tau_1, \ldots, \tau_p)$, and $v^j = v^0 + (\rho - \sum_{i=1}^{p} \tau_i)e^j$ for $j = 1, \ldots, p$. The algorithm then proceeds by simplicial partition of S_1 according to a general branch-and-bound scheme characterized by two basic operations: bounding and simplicial subdivision.

4.1. Bounding

Given a simplex $S = [v^0, v^1, \ldots, v^p) \subset S_1$ denote by $l_S(t)$ the affine function that agrees with $h(t)$ at the vertices $v^0, v^1, \ldots, v^p$ of S. Then $l_S(t) \geqslant h(t)$ for all $t \in S$, so the constraint

$$h(t) - \langle d, y \rangle \geqslant 0, \quad t \in S$$

will be relaxed if $h(t)$ is replaced by $l_S(t)$. It follows that a lower bound for the minimum of $L(x,y)$ over the set

$$\{(x,y) \in D \mid Qx = t \in S, h(t) - \langle d, y \rangle = 0\} \tag{23}$$

is provided by the optimal value $\beta(S)$ of the linear program

$$\min\{L(x,y) \mid (x,y) \in D, Qx = t \in S, l_S(t) - \langle d, y \rangle = 0\}. \tag{24}$$

Since any point $t \in S$ can be represented as $t = \sum_{i=0}^{p} \alpha_i v^i$ with $\sum_{i=0}^{p} \alpha_i = 1$, $\alpha_i \geqslant 0$, the linear program (24) can be solved by solving

$$LP(S)| \quad \begin{aligned} &\min && L(x,y) \\ &\text{subject to} && Ax + By \geqslant b, \quad Qx = \sum_{i=0}^{p} \alpha_i v^i \\ & && \sum_{i=0}^{p} \alpha_i \varphi_K(A_2 Z v^i) - \langle d, y \rangle \geqslant 0 \\ & && \sum_{i=0}^{p} \alpha_i = 1, \quad \alpha_i \geqslant 0 \;\; i = 0, 1, \ldots, p \end{aligned} \tag{25}$$

If $[x(S), y(S), \alpha(S)]$ is a basic optimal solution of this linear program then $[x(S), y(S), t(S)]$ with $t(S) = \sum_{i=0}^{p} \alpha_i(S) v^i$ is a basic optimal solution of (24).

Remark 1. *If a basic optimal solution* $[x(S), y(S), \alpha(S)]$ *of* LP(S) *satisfies* $\alpha_i(S) = 1$ *for some* i *(so that* $\alpha_j = 0\ \forall_j \neq i$) *then* $t(S)$ *is a vertex of* S, *hence* $l_S[t(S)] = h[t(S)]$, *so that* $[x(S), y(S), t(S)]$ *is feasible to (23). In that case,* $\beta(S)$ *yields the exact minimum of* $L(x, y)$ *over the set (23).*

4.2. Simplicial Subdivision

Recall first some definitions. Given a simplex $S = [v^0, v^1, \ldots, v^p]$ and a point $t \in S$ that is not a vertex of S, that is, such that $t = \sum_{i \in I} \alpha_i v^i$ with $\sum_{i \in I} \alpha_i = 1$, $0 < \alpha_i < 1$, let, for each $i \in I, S_i$ denote the simplex whose vertex set is obtained from that of S by replacing v^i with t. Then the simplices $S_i, i \in I$, form a partition of S via t. This partition is called a bisection if t is the midpoint of a longest edge of S.

Now to describe our subdivision rule, select a natural number N and let $\tau(S)$ be the generation index of S, that is, a number defined such that $\tau(S_i) = 1$ and $\tau(S') = \tau(S) + 1$ whenever S' is a child of S via some partition. Then if $\tau(S)$ is a multiple of N apply a bisection of S: otherwise, partition S via $t(S) = \sum_{i=0}^{p} \alpha_i(S) v^i$, where $(x(S), y(S), \alpha(S))$ is a basic optimal solution of LP(M).

ALGORITHM

Initialization. Solve the linear program $\min\{L(x,y) \mid Ax + By \geqslant b, x \geqslant 0, y \geqslant 0\}$. If an optimal solution of this linear program satisfies $\varphi_K(A_2x) - \langle d, y\rangle = 0$ then terminate: this is an optimal solution of (BLP/K). Otherwise construct S_1 as above indicated. Set $\tau(S_1) = 1$, $\mathscr{P}_1 = \mathscr{R}_1 = \{S_1\}$, CBS = best feasible solution available of (BLP/K), $CBV = L(x,y)$ for $CBS = (x,y)$. (If no feasible solution is readily available, let $CBV = +\infty$.) Set $k = 1$.

Step 1. For every simplex $S = [v^0, v^1, \ldots, v^p] \in \mathscr{P}_k$ compute the values $h(v^i) = \varphi_K(A_2 Z v^i)$, $i = 0, 1, \ldots, p$, and solve the linear program $LP(S)$ for a basic optimal solution $(x(S), y(S), \alpha(S))$ and the optimal value $\beta(S)$. Let $t(S) = \sum_{i=0}^{p} \alpha_1(S) v^i$.

Step 2. Update CBS and CBV from the newly obtained feasible solutions (including those $[x(S), y(S)]$ such that $t(S)$ is a vertex of S; see Remark 1).

Step 3. Delete every simplex $S \in R_k$ such that $\beta(S) \geqslant CBV$ and let S_k be the collection of remaining simplices. If $S_k = \varnothing$, then terminate: CBS solves (BLP/K) if CBV $< +\infty$; otherwise (BLP) is infeasible.

Step 4. Select $S_k \in \operatorname{argmin}\{\beta(S) \mid S \in S_k\}$, $t^k = t(S_k)$. Subdivide M_k following the specified rule and let $\mathscr{P}_{k+1}$ be the partition of $S_k, \tau(S) = \tau(S_k) + 1$ for every $S \in \mathscr{P}_{k+1}$. Let $\mathscr{R}_{k+1} = (S_k \setminus \{S_k\}) \cup \mathscr{P}_{k+1}$. Increment k and return to Step 1.

Theorem 5. *Let $CBV = \gamma_k$ and $CBS = (\bar{x}^k, \bar{y}^k)$ at iteration k. If Algorithm 1 is infinite, then $\gamma_k \searrow \gamma$ = optimal value of (BLP/K) and any cluster point of the sequence $(\bar{x}^k, \bar{y}^k)$ solves (BLP/K).*

Proof: *Let (x^k, y^k, t^k) be an optimal solution of $LP(S_k)$. By a known result from the theory of simplicial subdivision (see Horst and Tuy, 1996 or Tuy, 1998), the subdivision rule used for the algorithm ensures that if the procedure is infinite, an infinite nested sequence $\{S_k, k \in \Delta\}$, $\Delta \subset \{1, 2, \ldots\}$ exists, such that $t^k := t(S_k) \to \bar{t}$ as $k \to \infty (k \in \Delta)$, and $\bar{t}$ is a vertex of the limit simplex $S_\infty = \cap_{k \in \Delta} S_k$. Then $\lim l_{S_k}(t^k) = h(\bar{t})$ and since $l_{S_k}(t^k) - \langle d, y^k\rangle \geqslant 0$ for every k, it follows that $h(\bar{t}) - \langle d, \bar{y}\rangle \geqslant 0$, and consequently, $h(\bar{t}) - \langle d, \bar{y}\rangle = 0$. Thus, $(\bar{x}, \bar{y}, \bar{t})$ is feasible to (23) and hence is optimal to (23).*

Remark 2. *Let (x^k, y^k, t^k) be a basic optimal solution of $LP(S_k)$, with $t^k = t(S_k) = \sum_{i=0}^{p} \alpha_i^k v^{ki}$, where v^{ki} denotes the ith vertex of S_k. Then (x^k, y^k, t^k) satisfies every linear constraint of (22), and will become an optimal solution of (BLP/K) as soon as it satisfies the nonlinear constraint as well. The infeasibility of (x^k, y^k, t^k) is due to the fact that $l_{S_k}(t^k) > h(t^k)$ and can be measured by the quantity*

$$l_{S_k}(t^k) - h(t^k) = \sum_{i=0}^{p} \alpha_i^k [h(v^{ki}) - h(t^k)]$$

Thus, given a tolerance $\varepsilon > 0$ we can terminate when the above quantity becomes less than ε. The solution obtained is then an ε-approximate optimal solution of (BLP) (assuming $K \geqslant K_0$) in the sense that it solves the approximate problem which differs from (BLP) only in that the constraint y solves $R(x)$ is relaxed to y is an ε-optimal solution to $R(x)$

Remark 3. *In practical implementation of the method it will generally not be too difficult to estimate a correct value of K_0. If for some value of K the solution $(\bar{x}, \bar{y})$ of the problem (BLP/K) satisfies (at least approximately) the constraint y solves R(x), then K is correct* *and* $(\bar{x}, \bar{y})$ *can be accepted as an optimal solution to (BLP). Otherwise, one must increase K. Since* $\varphi_K(u) \leqslant \varphi(u)$, *the optimal value found for (BLP/K) gives an upper bound for the optimal value of (BLP) and can be used to accelerate the convergence of the branch-and-bound procedure in the next round of computation.*

5. Related Problems

Several authors have attempted to establish a link between (BLP) and linear multiobjective programming. Bard (1983) proved that an optimal solution of (BLP) must be an efficient (Pareto-optimal) point for the associated bicriteria programming problem

$$\min_{x,y}\{(L(x,y), \langle d, y\rangle) \mid (x,y) \in D\}. \tag{26}$$

Unfortunately, counter examples have been found to this result and hence, to two algorithms of Bard based on it (see, Ben-Ayed and Blair, 1990). Actually, an optimal solution of (BLP) is an efficient point for a more complicated multiobjective program, namely:

$$\min_{x,y}\{(A_2x, -A_2x, \langle d, y\rangle) \mid (x,y) \in D\} \tag{27}$$

with criteria matrix

$$\begin{bmatrix} A_2 & 0 \\ -A_2 & 0 \\ 0 & d^T \end{bmatrix}.$$

This can be seen by noting that if $(\bar{x}, \bar{y})$ solves (BLP) then for any $(x,y) \in D$, the relations $A_2x \leqslant A_2\bar{x}$, $-A_2x \leqslant -A_2\bar{x}$, $\langle d, y\rangle \leqslant \langle d, \bar{y}\rangle$ imply $A_2x = A_2\bar{x}$, $\langle d, y\rangle = \langle d, \bar{y}\rangle$, which is exactly the definition of an efficient solution for (27). Therefore, if D_e denotes the set of efficients of (27) then (BLP) is equivalent to the problem

$$\min\{L(x,y) \mid (x,y) \in D_e\}$$

which is an instance of the general problem of optimization over the efficient set (OES). The latter has been studied by various authors during the last decade (see, Konno et al., 1997, Chapter 15, and references therein). It consists of finding the minimum of a linear function $L(z)$ over the set Z_e of all efficient solutions of a multiobjective program

$$\min\{Cz \mid z \in Z\}, \tag{28}$$

where Z is a polyhedron in R^n, and C is a given $m \times n$ matrix of rows $c^1, \ldots, c^m$. Conversely, the problem $\min\{L(z) \mid z \in Z_e\}$ can be rewritten as the following special (BLP):

$$\text{minimize } L(z) \text{ subject to } z \text{ solves } \min\left\{\sum_{i=1}^{m} \langle c^i, v\rangle \mid Cv \leqslant Cz, z \in Z\right\}, \tag{29}$$

which is actually a bilevel linear program if we define $L(z,v) := L(z)$, $D = \{(z,v) \mid z \in Z, Cv - Cz \leqslant 0\}$, $\langle d, v\rangle = \sum_{i=1}^{m} \langle c^i, v\rangle$.

Thus any method for solving (BLP) can be adapted to solve (OES) and conversely. This equivalence between (BLP) and (OES) was first established by Fülöp (1993).

6. Nonlinear Bilevel Programming

A bilevel program (GBP) is said to be nonlinear when some of the functions $F(x,y)$, $G(x,y)$, $f(x,y)$, $g(x,y)$ are nonlinear. To date the convex bilevel programming problem (CBP) has attracted most of the attention in the literature. This problem corresponds to the case when all the mentioned functions are convex (in particular, quadratic convex). Just as with (BLP), a natural method for converting (CBP) into a single-level mathematical program is to replace the lower-level problem by Karush–Kuhn–Tucker optimality conditions. Based on this reformulation branch-and-bound methods have been proposed by Bard (1988), Bard and

Moore (1990), and Al-Khayyal et al. (1992). Descent methods and gradient information based methods have been developed by Vicente et al. (1994), and by Kolstad and Lasdon (1990), respectively. Other approaches using a penalty function (Aiyoshi and Shimizu, 1984), or a double-penalty function (Ishizuka and Aiyoshi, 1992) or an implicit function combined with a bundle algorithm (Dempe, 1998), are rather confined to computing stationary points and local minima. On the other hand, theoretical questions such as stability properties, existence of solutions, and so on, which may arise in the convergence theory of the above algorithms, have received little attention (Loridan and Morgan, 1989; Lignola and Morgan, 1989).

Global optimization methods can be applied to (CBP) by reformulating it as a convex program with an additional reverse convex constraint (Tuy, 1998):

$$\min\{F(x,y) \mid G(x,y) \leqslant 0,\ g(x,y) \leqslant 0,\ f(y) - \phi(x) \leqslant 0\} \tag{30}$$

where $\phi(x)$ is the optimal value of the lower level problem $R(x)$ for a given x (as can easily be verified, $\phi(x)$ is a convex function, so $f(y) - \phi(x)$ is actually a reverse convex constraint). However, a difficulty of this approach is that the effective domain of the convex function $\phi(x)$ may not contain an open neighbourhood of D. To overcome this difficulty the lower level constraint is relaxed to $g(x,y) \leqslant \varepsilon$, with $\varepsilon > 0$, so that the function $\phi(x)$ is replaced by

$$\phi_\varepsilon(x) = \min\{f(y) \mid g(x,y) \leqslant \varepsilon, y \geqslant 0\}.$$

If $(x^\varepsilon, y^\varepsilon, z^\varepsilon)$ is an optimal solution of the resulting problem, then as $\varepsilon \to 0$, any accumulation point of the sequence $\{(x^\varepsilon, y^\varepsilon)\}$ will solve (CBP).

For general nonlinear bilevel programs very few exact methods exist. An approach using a bundle method in the upper-level problem with subgradient information from the lower level problem has been developed by Falk and Liu (1995), but no computational experience has been available yet. So far most studies on nonlinear bilevel programs deal with theoretical properties. An excellent review of the current state of research and development in nonlinear bilevel programming, especially in problems arising in traffic planning (such as the origin-destination matrix adjustment problem), has been given by Migdalas (1995). More recent existence results and necessary optimality conditions for this class of problems can be found in Chen and Florian (1998). These results also indicate that a sensitivity-based approach is rather promising for bilevel problems involving variational inequalities in the lower level such as those encountered in various applications in mechanics (e.g., shape and structure optimization) and engineering.

References

Aiyoshi, E. and K. Shimizu. 1984. "A solution method for the static constrained Stackelberg problem via penalty method." *IEEE Trans. Automat. Control* **29**: 1111–1114.

Anandalingam, G. and D. J. White. 1990. "A solution method for the linear static Stackelberg problem using penalty functions." *IEEE Trans. Automat. Control* **35**: 1170–1173.

Baer, E., A. Hiltner, and R. Morgan. 1992. "Biological and synthetic hierarchical composites." *Phys. Today* **46**: 60–67.

Bard, J. F. 1983. "An efficient point algorithm for a linear two-stage optimization problem." *Oper. Res.* **31**: 670–684.

Bard, J. F. 1998. In *Practical Bilevel Optimization Algorithms and Optimization*. Kluwer Academic Publishers.

Bard, J. F. and J. E. Falk. 1982. "An explicit solution to the multilevel programming problem." *Comput. Oper. Res.* **9**: 77–100.

Bard, J. F. and J. T. Moore. 1990. "A branch and bound algorithm for the bilevel programming problem." *SIAM J. Sci. Stat. Comput.* **11**: 281–292.

Bialas, W. F. and M. H. Karwan. 1982. "On two-level linear optimization." *IEEE Trans. Automat. Control* **AC 27**: 151–155.

Bialas, W. F. and M. H. Karwan. 1984. "Two-level linear programming." *Mgmt. Sci.* **30**: 1004–1016.

Ben-Ayed, O. 1993. "Bilevel linear programming." *Comput. Oper. Res.* **16**: 485–501.

Ben-Ayed, O. and C. E. Blair. 1990. "Computational difficulties of bilevel linear programming." *Oper. Res.* **38**: 556–560.

Benson, H. P. 1989. "On the structure and properties of a linear multilevel programming problem." *J. Optim. Theory Appl.* **60**: 353–373.

Berechman, J. 1984. "Highway-capacity utilization and investment in transportation corridors." *Environ. Planning* **16A**: 1475–1488.

Bracken, J. and J. M. McGill. 1973. "Mathematical programs with optimization problems in the constraints." *Oper. Res.* **21**: 37–44.

Bracken, J. and J. M. McGill. 1974. "A method for solving mathematical programs with non-linear programs in the constraints." *Oper. Res.* **22**: 1097–1101.

Candler, W. and R. Townsley. 1982. "A linear two-level programming problem." *Comput. Oper. Res.* **9**: 59–76.

Chen, Y. and M. Florian. 1998. "Congested O-D trip demand adjustment problem: bilevel programming formulation and optimality conditions." In *Multilevel Optimization: Algorithms and Applications*, edited by A. Migdalas, P. Pardalos, and P. Värbrand, pp. 1–22. Kluwer Academic Publishers.

Chen, P. C., P. Hansen, and B. Jaumard. 1991. "On-line and off-line vertex enumeration by adjacent lists." *Oper. Res. Lett.* **10**: 403–409.

Dempe, S. 1992. "A necessary and sufficient optimality condition for bilevel programming problems." *Optimization* **25**: 341–354.

Dempe, S. 1998. "An implicit function approach to bilevel programming problems." In *Multilevel Optimization: Algorithms and Applications*, edited by A. Migdalas, P. Pardalos, and P. Värbrand, pp. 273–289. Kluwer Academic Publishers.

Edmunds, T. and J. Bard. 1991. "Algorithms for nonlinear bilevel mathematical programming." *Trans. Syst., Man., Cynbernet.* **21**: 83–89.

Falk, J. E. 1973. "A linear max–min problem." *Math. Programming* **5**: 169–188.

Falk, J. E. and J. Liu. 1995. "On bilevel programming, Part I: general nonlinear cases." *Math. Programming* **70**: 47–72.

Fülöp, J. 1993. "On the equivalence between a linear bilevel programming problem and linear optimization over the efficient set." Working Paper WP 93-1. Budapest, Hungary: LORDS, Computer and Automation Institute.

Gendreau, M., P. Marcotte, and G. Savard. 1996. "A hybrid tabu-ascent algorithm for the linear bilevel programming problem." *J. Global Optim.* **8**: 217–233.

Hansen, P., B. Jaumard, and G. Savard. 1992. "New branch and bound rules for linear bilevel programming." *SIAM J. Sci. Stat. Comput.* **13**: 1194–1157.

Horst, R. and H. Tuy. 1996. In *Global Optimization.* 3d ed. Berlin: Springer-Verlag.

Judice, J. J. and A. M. Faustino. 1988. "The solution of the linear bilevel programming problem by using the linear complementarity problem." *Investigacao Oper.* **8**: 77–95.

Kolstad, C. and L. Lasdon. 1990. "Derivative evaluation and computational experience with large bilevel mathematical programs." *J. Optim. Theory Appl.* **65**: 485–499.

Konno, H., P. T. Thach, and H. Tuy. 1997. In *Optimization on Low Rank Nonconvex Structures.* Kluwer Academic Publishers.

Lignola, M. B. and J. Morgan. 1998. "Existence of solutions to generalized bilevel programming problem." In *Multilevel Optimization: Algorithms and Applications*, edited by A. Migdalas, P. Pardalos, and P. Värbrand, pp. 315–330. Kluwer Academic Publishers.

Loridan, P. and J. Morgan. 1989. "A theoretical approximation scheme for Stackelberg problems." *J. Optim. Theory Appl.* **61**: 95–110.

Loridan, P. and J. Morgan. 1990. In *Quasiconvex lower level problem and applications to two level optimization.* Lecture Notes in Economics and Mathematical Systems, **345**: 325–341. Berlin: Springer-Verlag.

Migdalas, A. 1995. "Bilevel programming in traffic planning: models, methods and challenge." *J. Global Optim.* **7**: 381–405.

Migdalas, A., P. M. Pardalos, and P. Värbrand. (eds). 1998. In *Multilevel Optimization: Algorithms and Applications.* Kluwer Academic Publishers.

Pardalos, P. M. and X. Deng. 1997. In *Complexity issues in hierarical optimization*, DIMACS Series, **37**: 219–224. American Mathematical Society.

Parraga, F. A. 1981. "Hierarchy programming and applications to economic policy." Ph.D. diss. Tucson: Department of Systems and Industrial Engineering, University of Arizona.

Shimizu, K., Y. Ishizuka, and J. F. Bard. 1997. In *Nondifferentiable and Two-Level Mathematical Programming.* Kluwer Academic Publishers.

Von Stackelberg, H. 1952. In *The Theory of the Market Economy.* London, England: William Hodge and Company.

Tuy, H. 1992. "On nonconvex optimization problems with separated nonconvex variables." *J. Global Optim.* **2**: 133–144.

Tuy, H. 1998. In *Convex Analysis and Global Optimization.* Kluwer Academic Publishers.

Tuy, H., A. Migdalas, and P. Värbrand. 1992. "A global optimization approach for the linear two-level program." *J. Global Optim.* **3**: 1–23.

Tuy, H., A. Migdalas, and P. Värbrand. 1994. "A quasiconcave minimization method for solving linear two-level programs." *J. Global Optim.* **4**: 243–263.

Tuy, H. and S. Ghannadan. 1998. "A new branch and bound method for bilevel linear programs." In *Multilevel Optimization: Algorithms and Applications*, edited by A. Migdalas, P. Pardalos, and P. Värbrand, pp. 231–241. Kluwer Academic Publishers.

Vicente, L. N. and P. H. Calamai. 1994. "Bilevel and multilevel programming: a bibliography review." *J. Global Optim.* **5**: 291–306.

Vicente, L., G. Savard, and J. Judice. 1994. "Descent approaches for quadratic bilevel programming." *J. Optim. Theory Appl.* **81**: 379.

Vincent, J. F. V. 1990. In *Structural biomaterials*. Princeton: Princeton University Press.

White, D. J. and G. Anandalingam. 1993. "A penalty function approach for solving bilevel linear programs." *J. Global Optim.* **3**: 397–419.

13 COMPLEMENTARITY AND RELATED PROBLEMS

Michael C. Ferris and Christian Kanzow

ABSTRACT This survey gives an introduction to some of the recent developments in the field of complementarity and related problems. After presenting two typical examples and the basic existence and uniqueness results, we focus on some new trends for solving nonlinear complementarity problems. Extensions to mixed complementarity problems, variational inequalities, and mathematical programs with equilibrium constraints are also discussed.

1. Introduction

In this chapter, we give a survey of complementarity problems that covers some standard results and presents a number of new developments in the field. These developments are first described for the most basic problem, the nonlinear complementarity problem (NCP), which is to find a vector $x \in \mathbb{R}^n$ satisfying the system of equations and inequalities

$$x \geqslant 0, \quad F(x) \geqslant 0, \quad x^T F(x) = 0$$

or, equivalently,

$$x_i \geqslant 0, \quad F_i(x) \geqslant 0, \quad x_i F_i(x) = 0 \quad \text{for } i = 1, \ldots, n.$$

Here $F : X \to \mathbb{R}^n$ is a given function defined on a subset $X \subseteq \mathbb{R}^n$ containing at least the nonnegative orthant.

Due to expository and space considerations, we do not always present a result in its most general form; for example, we will introduce the class of monotone functions in Section 3 and state a couple of important results for this class even though some of them actually hold for the somewhat broader classes of pseudomonotone or P_0-functions.

Many of the basic theoretical results in complementarity problems have been known for a long time; an excellent survey of much of the research in this area prior to 1990 can be found in Harker and Pang (1990). Further references and more recent work can also be found in Ferris and Pang (1997a), Harker (1993), and Pang (1994). Here, we give just a flavor of the known existence and uniqueness results and concentrate more on the algorithmic point of view.

The organization of this paper is as follows. Section 2 gives two typical applications of complementarity problems as they arise in economics and engineering sciences. In Section 3 we then state a few existence and uniqueness results for the standard NCP, and Section 4 describes a couple of different solution techniques for the NCP. Section 5 deals with some more general equilibrium-type problems, namely mixed complementarity problems, variational inequalities, and mathematical programs with equilibrium constraints.

The notation used in this is assumed to be rather standard: the n-dimensional Euclidean space is denoted by $\mathbb{R}^n$ with $\|x\|$ being the Euclidean norm of a vector $x \in \mathbb{R}^n$. The scalar product of two vectors $x, y \in \mathbb{R}^n$ is written as $x^T y$. An inequality $x \geqslant 0$ means that all components of the vector $x \in \mathbb{R}^n$ are nonnegative. If $F : \mathbb{R}^n \to \mathbb{R}^n$ is a continuously differentiable function, we denote its Jacobian at a point $x \in \mathbb{R}^n$ by $F'(x)$. Furthermore, if $M \in \mathbb{R}^{n \times n}$ denotes an arbitrary matrix with elements m_{ij} and $I, J \subseteq \{1, \ldots, n\}$ are two given subsets, then M_{IJ} is an $|I| \times |J|$-dimensional submatrix with entries m_{ij} for $i \in I$ and $j \in J$.

2. Examples

Complementarity problems arise frequently in the general equilibrium theory (Arrow and Debreu, 1954) from economics. Many complex computable general equilibrium models are used for various aspects of policy design and analysis, including carbon abatement (Manne and Rutherford, 1993) and trade reform (Harrison et al., 1997). Other economic applications use game theory (Nash, 1950); new examples of this are becoming popular due to deregularization of electricity markets (Smeers, 1997).

Our particular example is taken from Scarf (1973) and is a schematic Arrow–Debreu equilibrium problem (Arrow and Debreu, 1954). The economy contains ℓ commodities, m utility-maximizing consumers, and n profit-maximizing producers. The key variables of the model are the prices p of the commodities. Given these prices, it is assumed that the agents demand certain amounts of the commodities, which are supplied so as to maximize profit in each production sector. Complementarity is derived from Walras's law, namely that supply will equal demand in the equilibrium state.

The production side of the economy depends on a technology matrix

$$A = [a^1 \quad a^2 \quad \cdots \quad a^n] \in \mathbb{R}^{\ell \times n},$$

that determines the unit output level of each sector. Generally, the entries of A are nonlinear functions of the price vector p and can be calculated using production functions (see Ferris and Pang, 1997b). For simplicity, we will assume here a constant technology matrix, and convert the activity level z_j of sector j into commodities using $a^j z_j$. Under the aforementioned assumption of price taking, each sector maximizes profits for given prices p:

$$\max\{p^T a^j z_j : z_j \geqslant 0\}.$$

The necessary and sufficient optimality conditions for production are therefore:

$$z \geqslant 0, \quad A^T p \leqslant 0, \quad p^T A z = 0. \tag{1}$$

For consumption, we derive a demand function for each of the consumers in the economy. The ith consumer is endowed with an initial wealth w_i and, assuming that he is also a price taker, will maximize his utility subject to his budget constraints. As is typical in such problems, the utility function is assumed to be either Cobb–Douglas

$$u^i(d) \prod_c d_c^{\lambda_c}, \quad \lambda_c \geqslant 0, \quad \sum_c \lambda_c = 1,$$

or constant elasticity of substitution (CES)

$$u^i(d) \left(\sum_{c=1}^{\ell} \lambda_c x_c^{\rho} \right)^{1/\rho}, \qquad \lambda_c \geqslant 0, \quad c = 1, \ldots, \ell,$$

where λ_c are given parameters dependent on the consumer i. The consumer solves

$$\max\{u^i(d^i) : p^T d^i \leqslant w_i, \quad d^i \geqslant 0\}, \tag{2}$$

which in the case of Cobb–Douglas utility gives

$$d_c^i(p, w_i) = \frac{\lambda_c w_i}{p_c}, \tag{3}$$

whereas for CES utility it gives,

$$d_c^i(p, w_i) = \frac{(p_c/\lambda_c)^{r-1} w_i}{\sum_{j=1}^{\ell} p_j^r / \lambda_j^{r-1}}, \quad r = \rho/(\rho - 1). \tag{4}$$

Typically, wealth is determined from an initial endowment of commodities e^i, so that $w_i = p^T e^i$.

While the prices were assumed above to be given to each of the agents in the economy, the model actually uses Walras's law to determine the prices so that markets clear. This clearance condition is stated succinctly as follows:

$$p \geqslant 0, \quad Az - \sum_{i=1}^{m} [d^i(p, p^T e^i) - e^i] \geqslant 0, \quad p^T \left(Az - \sum_{i=1}^{m} [d^i(p, p^T e^i) - e^i] \right) = 0$$

If the price p is positive then supply $Az + \sum_{i=1}^{m} e^i$ must equal demand $\sum_{i=1}^{m} d^i(p, w_i)$. Otherwise, when excess supply is available, the price must be zero.

Summing all the information regarding production and consumption results in the following complementarity problem:

$$z \geqslant 0, \quad -A^T p \geqslant 0, \quad p^T Az = 0,$$

$$p \geqslant 0, \quad Az - \sum_{i=1}^{m} (d^i(p, p^T e^i) - e^i) \geqslant 0, \quad p^T \left(Az - \sum_{i=1}^{m} (d^i(p, p^T e^i) - e^i) \right) = 0.$$

Note that if (p, z) solves the above problem, then so does $(\lambda p, z)$ for any $\lambda > 0$. This frequently causes algorithms difficulty in determining a solution. Typically this singularity is removed by fixing a numeraire, that is, fixing a price and dropping the corresponding complementary relationship.

The data for a particular instance of this model is available in MCPLIB (Dirkse and Ferris, 1995a). Many extensions of this basic framework are possible (Ferris and Pang, 1997b; Rutherford, 1995, 1999; examples using specialized software are described in Ferris et al. (1999a), Ferris and Munson (2000), and Rutherford (1995, 1999).

Another example of the use of complementarity problems is in the field of structural analysis. One of the aims of structural analysis is to find displacement, stress, and strain fields in a given structure under given conditions. Typically, loads F are specified and the problem is to find displacements u, stresses Q, and strains q. Compatibility or kinematic relationships relate u and q, equilibrium or statics relationships relate F and Q, and the constitutive laws relate Q and q. There are many models in use by engineers due to different assumptions for displacement and constitutive laws. For example, small or large displacements can be assumed, different discretization or finite element models can be applied, and time independent plasticity relationships can be used or not.

Starting in the 1970s, Maier and colleagues (Maier, 1970, 1971; Maier and Novati, 1990; Tin-Loi and Misa, 1996; Tin-Loi and Pang, 1993) proposed the use of complementarity in mathematical models for structural mechanics. The general form of the model is as follows. The compatibility relationships are determined from the geometry of the structure using $q = q(u)$. The equilibrium relationships are normally given by

$$C^T Q = F, \quad C = \frac{\partial q}{\partial u},$$

so that in a small displacement model $q = Cu$. The constitutive laws can be written

$$Q = Se, \qquad p = N\lambda, \qquad q = e + p,$$

where e and p represent elastic and plastic strains, S corresponds to elastic stiffnesses for each member of the structure, N contains unit outward normals, and λ are the plastic multipliers. The yield function satisfies the following complementarity relationship

$$\lambda \geqslant 0, \qquad -N^T Q + Q_* + h(\lambda) \geqslant 0, \qquad \lambda^T(-N^T Q + Q_* + h(\lambda)) = 0.$$

The elastic limit Q_*, and h (representing softening behavior) are material properties and are typically given. Most of the previous work in this area has been concerned with linear models of this format; typically, many of the above variables are eliminated to construct a standard form linear complementarity model. A more reasonable modeling approach maintains the sparsity and the mixed complementarity nature of the above formulation (see Section 5.1 and Tin-Loi and Ferris, 1997). Nonlinear extensions of the models are detailed in Bolzon et al. (1995), Ferris and Pang (1997b), and Tin-Loi and Pang (1993).

3. Existence

In this section, we deal with the following questions:

Does the NCP have at least one solution?

When is the solution unique?

What is the structure of the solution set?

In general, it is impossible to give a satisfactory answer to any of these questions without posing certain conditions on the function *F*. In order to give an idea of what can happen, consider the following simple examples.

Example 1. *In all examples given here, the function* $F : \mathbb{R}^n \to \mathbb{R}^n$ *is defined for* $n = 1$

(a) *If* $F(x) = -x - 1$, *then NCP has no solution since* $F(x) < 0$ *for all* $x \geqslant 0$.

(b) *If* $F(x) = x - 1$, *then NCP has a unique solution* $x^* = 1$.

(c) *If* $F(x) \equiv 0$, *then NCP has infinitely many solutions, namely the half line* $[0, +\infty)$.

(d) *If* $F(x) = (x-1)^2 - 1$, *then NCP has two isolated solutions, namely* $x^* = 0$ *and* $x^{**} = 2$.

In order to state some existence and, possibly, uniqueness results, we now introduce some classes of functions that play the role of corresponding convexity properties in optimization problems.

Definition 1. *Let* $X \subseteq \mathbb{R}^n$ *be a nonempty set and* $F : X \to \mathbb{R}^n$ *be a given function. Then* F *is said to be*

(i) *monotone (on* X*) if* $(x-y)^T(F(x) - F(y)) \geqslant 0$ *for all* $x, y \in X$;

(ii) *strictly monotone (on* X*) if* $(x-y)^T[F(x) - F(y)] > 0$ *for all* $x, y \in X$ *with* $x \neq y$;

(iii) *strongly monotone (on* X*) if there exists a constant* $\mu > 0$ *(called modulus) such that* $(x-y)^T[F(x) - F(y)] \geqslant \mu \| x - y \|^2$ *for all* $x, y \in X$.

Using these definitions, we can state the following result.

Theorem 1. *Let* $F : X \to \mathbb{R}^n$ *be a continuous function on* $X = \mathbb{R}^n_+$. *Then the following statements hold:*

(a) *If* F *is monotone on* X, *then NCP has a convex (possibly empty) solution set.*

(b) *If* F *is strictly monotone on* X, *then NCP has at most one solution.*

(c) *If* F *is strongly monotone on* X, *then NCP has a unique solution.*

Unfortunately, the assumption in Theorem (1c) is rather strong and usually not satisfied in many practical applications. On the other hand, the monotonicity assumption in Theorem (1a) is a much weaker condition, which, in fact, can be guaranteed to hold in quite a few situations. So it would be desirable to also have an existence result for monotone complementarity problems. Fortunately, such a result exists if, in addition, we assume that there is a strictly feasible point.

Definition 2. *A vector* $\hat{x} \in \mathbb{R}^n$ *is called strictly feasible for NCP if* $\hat{x} > 0$ *and* $F(\hat{x}) > 0$.

The strict feasibility assumption is closely related to the well-known Slater condition in convex optimization problems and allows us to state the following result.

Theorem 2. *Let* $F : X \to \mathbb{R}^n$ *be a continuous function on* $X = \mathbb{R}^n_+$. *Assume that* F *is monotone on* X *and there is a strictly feasible point* $\hat{x}$ *for NCP. Then NCP has a nonempty and compact solution set.*

Moré and Rheinboldt (1973) provide proofs of most of the results stated in this section (see also Harker and Pang, 1990 and Cottle et al., 1992, for some generalizations and specializations).

4. Algorithms

Here we want to describe the basic ideas of some solution methods for the nonlinear complementarity problem. Suitable algorithms should have the following desirable properties:

1. Globally convergent.
2. Locally fast convergent.

The first requirement says that, if possible, we want to have a method that is guaranteed to find a solution for the NCP, independent of how good or bad our starting point is. This can usually be guaranteed only if F has certain properties like being monotone or strongly monotone.

For the second point, a method is viewed as being locally fast convergent if it is locally superlinearly or quadratically convergent. Fast convergence can often be established under a condition called strong regularity (Robinson, 1980).

Definition 3. *Let* x^* *be a solution of NCP and define the index sets*

$$\alpha := \{i \mid x_i^* > 0 = F_i(x^*)\} \quad \text{and} \quad \beta := \{i \mid x_i^* = 0 = F_i(x^*)\}.$$

Then x^* *is called a strongly regular solution if the submatrix* $F'(x^*)_{\alpha\alpha}$ *is nonsingular and the Schur-complement*

$$F'(x^*)_{\alpha\cup\beta,\alpha\cup\beta}/F'(x^*)_{\alpha\alpha} := F'(x^*)_{\beta\beta} - F'(x^*)_{\beta\alpha}F'(x^*)_{\alpha\alpha}^{-1}F'(x^*)_{\alpha\beta} \in \mathbb{R}^{|\beta|\times|\beta|}$$

has positive principal minors.

The following subsections discuss several different algorithms. Most of these algorithms are based on a suitable reformulation of the complementarity problem as a nonsmooth equation.

4.1. Projection-Type Methods

Projection-type methods are based on the following simple observation.

Proposition 1. *A vector* x^* *is a solution of NCP if and only if* x^* *satisfies the fixed-point equation*

$$x = (x - \gamma F(x))_+,$$

where $\gamma > 0$ *is an arbitrary constant and* z_+ *denotes the Euclidean projection of a vector* $z \in \mathbb{R}^n$ *onto the nonnegative orthant.*

Projection-type methods typically attempt a fixed-point iteration. The simplest consists of the iteration

$$x^{k+1} := (x^k - \gamma F(x^k))_+, \quad k = 0, 1, 2, \ldots, \tag{5}$$

where $x^0 \in \mathbb{R}^n_+$ is any given starting point (see Auslender, 1976; Bertsekas and Tsitsiklis, 1989; Sibony, 1970). Using Banach's fixed-point theorem, it follows easily that this method is globally convergent if F is strongly monotone and globally Lipschitzian, provided the constant $\gamma > 0$ is sufficiently small. Furthermore, the rate of convergence one can expect is linear.

Modifications of this projection method have better global convergence properties. For example, the extragradient method (Khobotov, 1987; Korpelevich, 1976; Marcotte, 1991; Nagurney, 1993) generates iterates using the formula

$$x^{k+1} := (x^k - \gamma F((x^k - \gamma F(x^k))_+))_+, \quad k = 0, 1, 2, \ldots, \tag{6}$$

that is, the difference with (5) is that the function F is not evaluated at the previous iterate x^k, but at a point that would already be the next iterate for (5).

The extragradient method can be shown to be globally convergent if F is monotone (rather than strongly monotone) and globally Lipschitzian provided that the constant $\gamma > 0$ is sufficiently small (see Nagurney, 1993).

In fact, it is even possible to remove the assumptions that F is globally Lipschitzian and $\gamma > 0$ has to be sufficiently small (in practice, it is usually not known how small γ has to be). This is usually done by choosing γ dynamically in a suitable line search within the extragradient method (see Solodov and Tseng, 1996; Sun, 1996; Iusem and Svaiter, 1997; Solodov and Svaiter, 1999).

These modified extragradient methods have nice global convergence properties for monotone complementarity problems, but they are at best linearly convergent. Due to their slow convergence, we recommend their use only if either the complementarity problem is huge or if it is impractical to form F'.

4.2. Merit Functions

We first give the definition of a merit function as it will be used within this survey.

Definition 4. *A function* $\Psi : \mathbb{R}^n \to \mathbb{R}$ *is called a merit function for the NCP if it has the following properties:*

(a) $\Psi(x) \geqslant 0$ *for all* $x \in \mathbb{R}^n$

(b) $\Psi(x) = 0$ *if and only if* x *is a solution of NCP.*

This definition suggests replacing the NCP by the unconstrained minimization problem

$$\min \Psi(x), \quad x \in \mathbb{R}^n.$$

In order to apply standard optimization tools to this minimization problem, the merit function should satisfy certain smoothness conditions. In fact, the merit function Ψ should be continuously differentiable at least once, if possible even twice. It turns out, however, that most known merit functions are only continuously differentiable once. For example, the implicit Lagrangian (Mangasarian and Solodov, 1993)

$$\Psi(x) := \sum_{i=1}^{n} \{x_i F_i(x) + \frac{1}{2\alpha}(\max^2\{0\ x_i - \alpha F_i(x)\} - x_i^2 + \max^2\{0, F_i(x) - \alpha x_i\} - F_i(x)^2)\} \tag{7}$$

(where $\alpha > 1$ is a fixed parameter); the Fischer–Burmeister merit function (Facchinei and Soares, 1997; Geiger and Kanzow, 1996)

$$\Psi(x) := \frac{1}{2}\sum_{i=1}^{n}\left(\sqrt{x_i^2 + F_i(x)^2} - x_i - F_i(x)\right)^2, \tag{8}$$

or the penalized Fischer–Burmeister merit function (Chen et al., 1997)

$$\Psi(x) := \frac{1}{2}\sum_{i=1}^{n}\left\{\lambda\left(\sqrt{x_i^2 + F_i(x)^2} - x_i - F_i(x)\right) - (1-\lambda)\max\{0, x_i\}\max\{0, F_i(x)\}\right\}^2 \tag{9}$$

(where $\lambda \in (0,1)$ is a fixed parameter) are all known to be merit functions that are once but not twice continuously differentiable.

This is one of the reasons why we do not suggest minimizing a merit function directly using existing unconstrained optimization software. Another reason is that most merit functions can be derived from corresponding equation reformulations (see the following subsection), and it is the interplay between such an equation reformulation and its merit function that makes these methods successful.

Apart from this, merit functions are often used in order to monitor global convergence of different algorithms. We believe this is actually the major contribution of a merit function

We close this subsection by stating two types of results for the three merit functions mentioned above. The first result indicates sufficient conditions to ensure that a stationary point of the merit function Ψ is a global minimum and hence a solution of the NCP.

Theorem 3. *Assume that* $F : \mathbb{R}^n \to \mathbb{R}^n$ *is continuously differentiable. If* F *is strongly monotone, then every stationary point of the implicit Lagrangian (7) is a solution of the NCP. The same statement holds for the Fischer–Burmeister merit function (8) and the penalized Fischer–Burmeister merit function (9) under the weaker assumption that* F *is monotone.*

Most algorithms guarantee that any accumulation point of the iterates is at least a stationary point of Ψ and therefore, under the conditions of Theorem 3, a solution of the underlying complementarity problem. The next result concerns the compactness of the level sets

$$\mathcal{L}(c) := \{x \in \mathbb{R}^n \mid \Psi(x) \leqslant c\}$$

for a given constant c. This is a critical result, since a compact level set implies that the sequence of iterates will have an accumulation point.

Theorem 4. *Assume that* $F : \mathbb{R}^n \to \mathbb{R}^n$ *is continuous and let* $c \in \mathbb{R}$ *be any given constant. If* F *is strongly monotone, then the level sets* $\mathcal{L}(c)$ *for any of the three merit functions (7)–(9) are compact (possibly empty). The same statement holds for the penalized Fischer–Burmeister merit function (9) under the weaker assumption that* F *is monotone and the NCP has a strictly feasible point.*

There have been many other merit functions proposed in the literature (see Luo and Tseng, 1997; Kanzow et al., 1997; Kanzow and Kleinmichel, 1998; Sun and Qi, 1999). Currently, the penalized Fischer–Burmeister merit function and those in Kanzow et al. (1997) and Sun and Qi (1999) have the strongest properties.

4.3. Equation-Based Methods

It is possible to reformulate the NCP as a nonlinear system of equations using so-called NCP functions.

Definition 5. *A function* $\phi : \mathbb{R}^2 \to \mathbb{R}$ *is called an NCP function if it satisfies the condition*

$$\phi(a,b) = 0 \Longleftrightarrow a \geqslant 0, b \geqslant 0, ab = 0.$$

The following functions are NCP functions:

(a) $\phi(a,b) = \min\{a,b\}$;

(b) $\phi(a,b) = \sqrt{a^2+b^2} - a - b$;

(c) $\phi(a,b) = \lambda\left(\sqrt{a^2+b^2} - a - b\right) - (1-\lambda)\max\{0,a\}\max\{0,b\}$ with $\lambda \in (0,1)$ fixed;

(d) $\phi(a,b) = -ab + \frac{1}{2}\min^2\{0, a+b\}$.

Normally (a) is called the minimum function, and (b) is called the Fischer–Burmeister function (it first appeared in Fischer, 1992, where it is attributed to Burmeister). Example (c) is termed the penalized Fischer–Burmeister function (Chen et al., 1997). Example (d) is the only NCP function listed that is differentiable on the whole space $\mathbb{R}^2$ (Evtushenko and Purtov, 1984). Many other NCP functions are known in the literature (see Mangasarian, 1976; Sun and Qi, 1999).

For any given NCP function ϕ, define $\Phi : \mathbb{R}^n \to \mathbb{R}^n$ by

$$\Phi(x) := \begin{pmatrix} \phi(x_1, F_1(x)) \\ \vdots \\ \phi(x_n, F_n(x)) \end{pmatrix}. \tag{10}$$

We call Φ the equation operator corresponding to ϕ. As an immediate consequence of the definition of Φ, we obtain the following characterization of the complementarity problem.

Proposition 2. *Let* ϕ *be an NCP function and* Φ *be the corresponding equation operator defined by (10). Then* x^* *is a solution of NCP if and only if* x^* *is a solution of the nonlinear system of equations* $\Phi(x) = 0$.

If F and the NCP function ϕ are continuously differentiable, then the same holds for the equation operator Φ. In this case, we can apply the classical Newton method to $\Phi(x) = 0$ to solve the underlying NCP. This leads to the iteration

$$x^{k+1} := x^k - \Phi'(x^k)^{-1}\Phi(x^k), \quad k = 0,1,2,\ldots$$

for a given starting point $x^0 \in \mathbb{R}^n$. We can expect local fast convergence if the standard assumptions for the Newton method are satisfied; in particular, the Jacobian matrix $\Phi'(x^*)$ has to be nonsingular at a solution x^* of NCP. Unfortunately, the following result from Kanzow and Kleinmichel (1995) shows that this matrix is singular at any degenerate solution. (Here, a solution x^* is called *degenerate* if the index set β from Definition 3 is nonempty.)

Proposition 3. *Let x^* be a degenerate solution of NCP. Assume that F and the NCP function ϕ are differentiable, and let Φ denote the corresponding equation operator. Then the Jacobian $\Phi'(x^*)$ is singular.*

One way to overcome the singularity problem is to use a nondifferentiable NCP function. This, in general, leads to a nondifferentiable equation operator Φ. However, Φ is usually at least locally Lipschitzian, so Φ is differentiable almost everywhere by a result due to Rademacher. Let us denote by D_Φ the set of differentiable points of Φ. Then we define the so-called B-subdifferential (Robinson, 1987) as

$$\partial_B\Phi(x) := \{H \in \mathbb{R}^{n\times n} \mid \exists\{x^k\} \subseteq D_\Phi : x^k \to x \text{ and } \Phi'(x^k) \to H\}$$

as well as its convex hull

$$\partial\Phi(x) := \text{conv}\{\partial_B\Phi(x)\},$$

which is called the generalized Jacobian (Clarke, 1983). Elements from $\partial_B\Phi(x)$ or $\partial\Phi(x)$ can be used instead of the classical (and in our situation sometimes not existing) Jacobian $\Phi'(x)$, and a typical nonsmooth Newton method then leads to the iteration

$$x^{k+1} := x^k - H_k^{-1}\Phi(x^k), \quad k = 0, 1, 2, ..., \tag{11}$$

where H_k is an arbitrary element from $\partial_B\Phi(x^k)$.

Although, in general, it is difficult to compute elements from the B-subdifferential, we are interested only in particular operators Φ, and for those it is usually not difficult to compute an element $H_k \in \partial_B\Phi(x)$ (see De Luca et al., 1996; Chen et al., 1997).

If Φ is the equation operator corresponding to either the minimum function, the Fischer–Burmeister function or the penalized Fischer–Burmeister function, and x^* is a strongly regular solution of NCP, then it is possible to show that

(a) all elements $H \in \partial_B\Phi(x^*)$ are nonsingular, and
(b) Φ satisfies further smoothness assumptions (called semismoothness/strong semismoothness in Qi and Sun, 1993).

These two requirements are exactly those that are needed in order to prove local superlinear or quadratic convergence of the iterates generated by (11) (see Qi and Sun, 1993; Qi, 1993).

Furthermore, it is often relatively easy to make these methods globally convergent at least for monotone problems. To this end, one typically uses the vector

$$d^k := -H_k^{-1}\Phi(x^k)$$

as a search direction and performs a line search along this direction to get sufficient decrease in the corresponding merit function

$$\Psi(x) := \tfrac{1}{2}\Phi(x)^T\Phi(x). \tag{12}$$

Such a decrease is possible if the merit function is continuously differentiable (De Luca et al., 1996). This condition is satisfied for the Fischer–Burmeister function and the penalized Fischer–Burmeister function discussed in the previous subsection. In general, these methods work well even for nonmonotone complementarity problems.

Unfortunately, the merit function Ψ corresponding to the minimum function is not differentiable everywhere, and therefore the nonsmooth Newton iteration (11) based on the minimum function is more difficult to globalize. The references (Pang, 1990, 1991; Pang and Gabriel, 1993; Gabriel and Pang, 1992; De Luca et al., 2000) give possible strategies.

4.4. Smoothing and Interior-Point Methods

Given an NCP function ϕ and the corresponding equation operator Φ defined by (10), Proposition 3 suggests using nondifferentiable NCP functions and nonsmooth algorithms. Another possibility to deal with nondifferentiable functions is the smoothing technique, the basis of which will be described in this section.

Smoothing methods for the NCP typically approximate a nonsmooth NCP function ϕ by a smooth function ϕ_μ that depends on a smoothing parameter $\mu > 0$. For example, it is possible to approximate the Fischer–Burmeister function by

$$\phi(a,b) = \sqrt{a^2+b^2} - a - b \approx \sqrt{a^2+b^2+\mu} - a - b =: \phi_\mu(a,b). \tag{13}$$

Using similar techniques, it is possible to smooth most known NCP functions.

As soon as we have such a smooth approximation ϕ_μ, we define the corresponding equation operator

$$\Phi_\mu(x) := \begin{pmatrix} \phi_\mu(x_1, F_1(x)) \\ \vdots \\ \phi_\mu(x_n, F_n(x)) \end{pmatrix},$$

which is now a smooth approximation of Φ if F itself is smooth.

The main idea of any smoothing method is to apply a standard Newton method to the nonlinear system of equations $\Phi_\mu(x) = 0$ and to drive the smoothing parameter μ down to zero. If we denote the value of μ at the kth iteration by μ_k, a typical iteration of a smoothing method is given by

$$x^{k+1} = x^k - \Phi'_{\mu_k}(x^k)^{-1}\Phi_{\mu_k}(x^k), \quad k = 0, 1, 2, \ldots.$$

A critical point of all smoothing methods is the way the smoothing parameter μ_k gets updated at each iteration. If this is done in an appropriate way, it is possible to prove global convergence for monotone problems, and even local fast convergence if a solution x^* of the NCP satisfies the strong regularity assumption (see Burke and Xu 1998; Chen and Chen, 1999; Chen et al., 1998; Chen and Ye, 1999; and Xu and Burke, 1999), for a couple of different smoothing methods).

Smoothing methods have an interesting connection to interior point methods as pointed out in Kanzow (1996). For example, let ϕ_μ denote the function defined in (13). Then it is easy to see that

$$\phi_\mu(a,b) = 0 \Longleftrightarrow a > 0, b > 0, ab = \mu.$$

Hence solving the nonlinear system of equations

$$\Phi_\mu(x) = 0 \tag{14}$$

is equivalent to solving the central path conditions

$$x_i > 0, \quad F_i(x) > 0, \quad x_i F_i(x) = \mu \quad \forall i = 1, \ldots, n, \tag{15}$$

which are typically used as the basis of path-following interior-point methods. However, the way smoothing methods and interior-point methods solve their (theoretically equivalent) subproblems (14) and (15), respectively, is completely different. Furthermore, smooth approximations ϕ_μ of NCP functions ϕ exist that are not related to the central path conditions (see Chen and Mangasarian, 1996).

So far, the theory developed for interior-point methods (see the books by Wright, 1997; Ye, 1997; and Roos et al., 1997) theory is much stronger than the one developed for smoothing methods. For example, interior-point methods can often be shown to have polynomial complexity and local fast convergence even in the absense of the strong regularity condition. On the other hand, smoothing-type methods seem to be quite promising from a numerical point of view (see Chen and Mangasarian, 1996; Billups et al., 1997; Kanzow and Pieper, 1999). We therefore expect further theoretical improvements in the field of smoothing methods in the near future.

4.5. Linearization Methods

Here, the fundamental idea is to linearize F about the current iterate x^k and solve the corresponding linear complementarity problem

$$x \geqslant 0, \quad F(x^k) + F'(x^k)(x - x^k) \geqslant 0, \quad x^T(F(x^k) + F'(x^k)(x - x^k)) = 0 \tag{16}$$

to generate x^{k+1}. In the context of generalized equations, this method, which has been commonly termed SLCP (for sequential LCP), was shown to be locally well-defined and fast convergent under the strong regularity condition (Josephy, 1979). Mathiesen (1987, 1985) successfully used this scheme for solving equilibrium models such as those outlined in Section 2. While this method works extremely well in the neighborhood of a solution, there are substantial difficulties in globalizing the approach. Two key difficulties are subproblems not having solutions and choosing appropriate merit functions that do not destroy fast local convergence. It is no mere coincidence that these problems plague sequential quadratic programming methods too—their theory can be treated analogously

For strongly monotone variational inequalities, convergence of line search schemes based on (16) are shown in Fukushima (1992) and Taji et al. (1993). A theoretically more general approach is via reformulation of the problem as a nonsmooth system of equations. The normal map (Robinson, 1992) defined as

$$F_+(y) := F(y_+) + y - y_+ \tag{17}$$

is equivalent to solving the NCP in the following sense. If y is a zero of (17), then y_+ solves the NCP. Furthermore, if x is a solution of the NCP, then $x - F(x)$ is a zero of the normal map. Note two points that contrast the normal map from previously mentioned nonsmooth equivalents of the NCP. First, it is defined for all points in $\mathbb{R}$ even if F is only defined on $\mathbb{R}^n_+$. It is known to be a piecewise smooth map, the pieces correspond to the orthants of $\mathbb{R}^n$, a subdivided manifold (Eaves, 1976), typically called the normal manifold. Secondly, the nonsmoothness is inside the function F as opposed to around the function F, so that F is only evaluated on the positive orthant.

The Newton method proposed in Robinson (1994), in the context of the normal map, approximates $F_+(y)$ with a piecewise linear normal map

$$A^k(y) := F(y^k_+) + F'(y^k_+)(y_+ - y^k_+) + y - y_+.$$

The Newton point is a zero of A^k. It is well known that this is equivalent to SLCP by taking $x^k = y^k_+$. However, Ralph (1994) proposed a globalization scheme that extends damped Newton methods for smooth equations to the above situation. Instead of using Lemke's method (Cottle and Dantzig, 1968; Lemke and Howson, 1964) to solve (16), a path $p^k(t)$ is constructed (Dirkse and Ferris, 1995b) from y^k to y^{k+1} with the property that

$$A^k(p^k(t)) = (1 - t)F_+(y^k).$$

This path has exactly the same properties that are required for standard line search methods, namely that small steps guarantee sufficient decrease. That theory can be extended to this case, provided that $\|F_+(y)\|$ is used as the merit function, to prove standard global and local convergence results (Dirkse and Ferris, 1995b; Ralph, 1994).

Considerable success has been reported for an implementation of this approach called the PATH solver (Dirkse and Ferris, 1995b; Ferris and Munson, 1999). This implementation uses standard large-scale linear programming (simplex) technology (Nazareth, 1987) to generate the required path. SLCP methods typically use similar technology, except the subproblems use ray starts in Lemke's method as opposed to the required regular point starts in PATH. Some advanced techniques for choosing the initial piece of the normal manifold can be found in Anstreicher et al. (1992) and Dirkse and Ferris (1997).

Most of the above implementations rely on heuristics to deal with the case where the subproblems have no solution. Recent work in this area has extended global convergence theory along different lines. Ferris et al. (1999b) proposed a theoretical framework for globalization of feasible descent methods that uses (8) as a merit function. The main features of this framework are: (a) it is well defined for an arbitrary complementarity problem, (b) it generates only feasible iterates, (c) it has a strong global convergence theory, and (d) it is locally fast convergent under standard regularity assumptions. The key observations of such a globalization are that if the subproblems are unsolvable, then a projected gradient step can be taken for the merit function, guaranteeing progress and that eventually the full Newton step will be accepted. Numerical results in the cited paper indicate this framework leads to substantial computational improvements.

5. Extensions

5.1. *Mixed Complementarity Problems*

While most of this paper has been written in the context of nonlinear complementarity problems, many practical applications generate problems for which some of the variables are nonnegative, others are bounded, and others are completely free. To accommodate these problems, many researchers have investigated the mixed complementarity problem (MCP) of finding a vector $x \in [l, u]$ such that exactly one of the following holds:

$$x_i = l_i \quad \text{and} \quad F_i(x) > 0,$$

$$x_i = u_i \quad \text{and} \quad F_i(x) < 0,$$

$$x_i \in [l_i, u_i] \quad \text{and} \quad F_i(x) = 0.$$

Note that l_i and u_i can be infinite. Two special cases are of importance. When $l_i = -\infty$ and $u_i = \infty$, then the MCP is exactly the problem of finding a zero of F. This shows why many of the techniques for solving complementarity problems are closely related to nonlinear equation theory, as opposed to standard optimization techniques. Furthermore, when $l_i = 0$ and $u_i = \infty$ the NCP is recovered, and thus it is clear that MCP is a generalization of NCP. Although there are many computational issues related to exploiting this framework, it is well known that appropriate changes of variables can reduce the MCP to a (larger dimensioned) NCP. Therefore, for theoretical purposes, it is entirely adequate to deal exclusively with NCPs. In practice, however, exploiting the MCP structure is of paramount importance (Dirkse and Ferris, 1995a; Ferris and Pang, 1997b).

5.2. *Variational Inequalities*

Given a nonempty, closed, and convex subset $X \subseteq \mathbb{R}^n$ and a function $F : X \rightarrow \mathbb{R}^n$, the variational inequality problem (VIP) is to find a vector $x^* \in \mathbb{R}^n$ such that

$$F(x^*)^T (x - x^*) \geqslant 0 \quad \text{for all} \quad x \in X.$$

It is not difficult to see that the variational inequality problem VIP reduces to NCP if the feasible set X is equal to the nonnegative orthant $\mathbb{R}^n_+$, and becomes the MCP if $X = [l, u]$.

Many results and algorithms described above for the complementarity problem can actually be extended to the more general variational inequality problem. For example, Theorem 1 still holds if NCP is replaced everywhere by VIP. Also Theorem 2 has an analogous result for variational inequalities (Harker and Pang, 1990).

For solution techniques, all the projection-type methods outlined in Section 4.1 can be applied to VIP and have the same global convergence properties if we interpret the symbol z_+ as the Euclidean projection of a vector $z \in \mathbb{R}^n$ onto the feasible set X (rather than the projection onto $\mathbb{R}^n_+$).

Also the linearization methods have a direct translation for variational inequalities. For example, the SLCP method for VIP generates a sequence $\{x^k\}$ in such a way that x^{k+1} is a solution of the linearized variational inequality

$$(F(x^k) + F'(x^k)(x - x^k))^T (y - x) \geqslant 0 \quad \text{for all} \quad y \in X,$$

with similar convergence to that described in Section 4.5.

Furthermore, a couple of gap function approaches exist for the solution of VIP, which may be viewed as a generalization of the merit function approaches used for complementarity problems although the situation is much more complicated for VIP.

Auslender (1976) introduced the gap function, defined as

$$g(x) := \sup_{y \in X} \{F(x)^T (x - y)\}.$$

It is not difficult to see that g is nonnegative on the feasible set X and that $g(x) = 0$ and $x \in X$ if, and only if, x solves VIP. Hence VIP can be rewritten as the constrained minimization problem

$$\min g(x) \quad \text{subject to} \quad x \in X.$$

Some algorithms based on this are given in Marcotte (1985) and Marcotte and Dussault (1987). However, the gap function g is not differentiable in general. Moreover, when X is unbounded, g is in general not finite valued.

These drawbacks can be avoided by using the *regularized gap function* (Auchmuty, 1989; Fukushima, 1992)

$$f_\alpha(x) := \max_{y \in X} \left\{ F(x)^T (x - y) - \frac{\alpha}{2} \|x - y\|^2 \right\},$$

where α is any positive parameter. It is also nonnegative on the set X and x solves VIP if, and only if, $f_\alpha(x) = 0$ and $x \in X$. Hence, VIP is equivalent to the constrained optimization problem

$$\min f_\alpha(x) \quad \text{subject to} \quad x \in X. \tag{18}$$

In addition, f_α can be shown to be continuously differentiable and finite valued everywhere.

However, the regularized gap function has a serious drawback: If x^* denotes a constrained stationary point of (18), then F has to be strongly monotone in order to guarantee that x^* is a solution of VIP. Another extension, the D-gap function (Peng, 1997; Yamashita et al., 1997) also suffers from the same drawback.

5.3. Mathematical Programs with Equilibrium Constraints

A mathematical program with equilibrium constraints (MPEC) is a constrained optimization problem with some of the constraints representing certain equilibrium conditions. For example, if these equilibrium conditions are described by a complementarity problem, then an MPEC can be formulated as

$$\min f(x, y) \quad \text{subject to} \quad x \in X, y \geqslant 0, F(x, y) \geqslant 0, y^T F(x, y) = 0; \tag{19}$$

here, $f : \mathbb{R}^n \times \mathbb{R}^m \to \mathbb{R}$ and $F : \mathbb{R}^n \times \mathbb{R}^m \to \mathbb{R}^m$ are given functions and $X \subseteq \mathbb{R}^n$ is an abstract constraint, which can usually be represented by standard equality/inequality constraints.

Note that the MPEC (19) has two types of variables, the design variables x and the state variables y. The complementarity constraints in (19) are parameterized in the design variables. While extensions or particular instantiations of (19) are possible, we will restrict ourselves to this form in order to convey the main algorithmic ideas.

The first point is that (19) should not be treated as a standard constrained optimization problem, since it is well known that it does not even satisfy the Mangasarian–Fromovitz constraint qualification (Chen and Florian, 1995). Therefore, standard methods from constrained optimization are prone to failure.

Several researchers have proposed more specialized solution methods for MPEC problems. Given that there is much current research in this area, we will only summarize a few central ideas.

One way to overcome the previously mentioned difficulty in (19) is to add the constraint $y^T F(x, y) = 0$ to the objective function and then to solve a sequence of penalized problems

$$\min f(x, y) + \rho y^T F(x, y) \quad \text{subject to} \quad x \in X, y \geqslant 0, F(x, y) \geqslant 0$$

with penalty parameter $\rho > 0$. In general, this gives only an inexact penalty reformulation unless some strong conditions hold (Luo et al., 1996; Mangasarian and Pang, 1997; Scheel and Scholtes, 1998; Scholtes and Stöhr, 1999). Some researchers also proposed exact penalty functions for the MPEC (19), which, however, are severely nondifferentiable and therefore difficult to deal with from a numerical point of view.

Another technique for the solution of the MPEC (19) is the implicit programming approach. These methods assume that for any given $x \in X$, one can find a solution $y(x)$ of the complementarity constraints in (19). They replace the variable y by the function $y(x)$ in the objective function f and try to solve the optimization problem

$$\min f(x, y(x)) \quad \text{subject to} \quad x \in X, \tag{20}$$

which is a problem in the design variables only. The function $y(x)$ is usually nondifferentiable, so one has to use nondifferentiable optimization methods to solve (20) (see Outrata, 1994; Outrata and Zowe, 1995).

Another possible way to deal with the MPEC (19) is to reformulate the complementarity constraints using an NCP function ϕ. One obtains a constrained optimization problem of the form

$$\min f(x, y) \quad \text{subject to} \quad x \in X, \phi(y_i, F_i(x, y)) = 0 \quad (i = 1, ..., m).$$

In order to avoid the nonsmooth constraints $\phi[y_i, F_i(x, y)] = 0$, we can use the smoothing idea and approximate this problem by the smooth optimization problem

$$\min f(x, y) \quad \text{subject to} \quad x \in X, \phi_\mu[y_i, F_i(x, y)] = 0 \quad (i = 1, ..., m),$$

where, of course, ϕ_μ is a smooth approximation for ϕ (see (13)). This idea has been exploited in Facchinei et al. (1999), Fukushima et al. (1998), and Jiang and Ralph (1997). Alternatively, and closely related to the smoothing idea, we can apply interior-point techniques to the perturbed problem

$$\min f(x, y) \quad \text{subject to} \quad x \in X, y_i > 0, F_i(x, y) > 0, y_i F_i(x, y) = \mu \quad (i = 1, ..., m).$$

For further techniques and several theoretical results about MPECs, we refer the reader to Luo et al. (1996) and Outrata et al. (1998).

ACKNOWLEDGMENTS
Michael C. Ferris's research was partially supported by National Science Foundation Grants CCR-9619765 and CDA-9726385 and Air Force Office of Scientific Research Grant F49620-98-1-0417. Christian Kanzow's research was supported by the DFG (Deutsche Forschungsgemeinschaft).

References

Anstreicher, K. M., J. Lee, and T. F. Rutherford. 1992. "Crashing a maximum-weight complementarity basis." *Math. Programming* **54**: 281–294.

Arrow, K. and G. Debreu. 1954. "Existence of equilibrium for a competitive economy." *Econometrica* **22**: 265–290.

Auchmuty, G. 1989. "Variational principles for variational inequalities." *Numerical Funct. Anal. Optim.* **10**: 863–874.

Auslender, A. 1976. *Optimization Méthodes Numériques*. Paris: Masson.

Bertsekas, D. P. and J. N. Tsitsiklis. 1989.*Parallel and Distributed Computation: Numerical Methods*. Englewood Cliffs, NJ: Prentice-Hall Inc.

Billups, S. C., S. P. Dirkse, and M. C. Ferris. 1997. "A comparison of large scale mixed complementarity problem solvers." *Comput. Optim. Appl.* **7**: 3–25.

Bolzon, G., G. Maier, and F. Tin-Loi. 1995. "Holonomic and nonholonomic simulations of quasi-brittle fracture: a comparative study of mathematical programming approaches." In *Fracture Mechanics of Concrete Structures,* Second International Conference on Fracture Mechanics of Concrete Structures (FRAMCOS 2), edited by F. H. Wittmann, pp. 885–898. Freiburg: Aedificatio Publishers.

Burke, J. V. and S. Xu. 1998. "The global linear convergence of a non-interior path-following algorithm for linear complementarity problems." *Math. Oper. Res.* **23**: 719–735.

Chen, B. and X. Chen. 1999. "A global and local superlinear continuation-smoothing method for $P_0 - R_0$ NCP or monotone NCP." *SIAM J. Optim.* **9**: 624–645.

Chen, B., X. Chen, and C. Kanzow. 1997. "A penalized Fischer–Burmeister NCP-function: theoretical investigation and numerical results." Preprint 126. Hamburg: Institute of Applied Mathematics, University of Hamburg.

Chen, C. and O. L. Mangasarian. 1996. "A class of smoothing functions for nonlinear and mixed complementarity problems." *Comput. Optim. Appl.* **5**: 97–138.

Chen, X. and Y. Ye. 1999. "On homotopy-smoothing methods for variational inequalities." *SIAM J. Control Optim.* **37**: 589–616.

Chen, X., L. Qi, and D. Sun. 1998. "Global and superlinear convergence of the smoothing Newton method and its application to general box constrained variational inequalities." *Math. Comput.* **67**: 519–540.

Chen, Y. and M. Florian. 1995. "The nonlinear bilevel programming problem formulations, regularity, and optimality conditions." *Optimization* **32**: 193–209.

Clarke, F. H. 1983. *Optimization and Nonsmooth Analysis.* New York: John Wiley & Sons.

Cottle, R. W. and G. B. Dantzig. 1968. "Complementary pivot theory of mathematical programming." *Linear Algebra Appl.* **1**: 103–125.

Cottle, R. W., J. S. Pang, and R. E. Stone. 1992. In *The Linear Complementarity Problem.* Boston: Academic Press.

De Luca, T., F. Facchinei, and C. Kanzow. 1996. "A semismooth equation approach to the solution of nonlinear complementarity problems.' *Math. Programming* **75**: 407–439.

De Luca, T., F. Facchinei, and C. Kanzow. 2000. "A theoretical and numerical comparison of some semismooth algorithms for complementarity problems." *Comput. Optim. Appl.* **16**: 173–205.

Dirkse, S. P. and M. C. Ferris. 1995a. "MCPLIB: a collection of nonlinear mixed complementarity problems." *Optim. Methods Softw.* **5**: 319–345.

Dirkse, S. P. and M. C. Ferris. 1995b. "The PATH solver: a non-monotone stabilization scheme for mixed complementarity problems." *Optim. Methods Softw.* **5**: 123–156.

Dirkse, S. P. and M. C. Ferris. 1997. 'Crash techniques for large-scale complementarity problems." In *Complementarity and Variational Problems: State of the Art*, edited by M. C. Ferris and J. S Pang, pp. 40–61. Philadelphia: SIAM Publications.

Eaves, B. C. 1976. "A short course in solving equations with PL homotopies" In *Nonlinear Programming*, edited by R. W. Cottle and C. E. Lemke, pp. 73–143. Providence: American Mathematical Society, SIAM-AMS Proceedings.

Evtushenko, Y. G. and V. A. Purtov. 1984. "Sufficient conditions for a minimum for nonlinear programming problems." *Sov. Math. Doklady* **30**: 313–316.

Facchinei, F., H. Jiang, and L. Qi. 1999. "A smoothing method for mathematical programs with equilibrium constraints." *Math. Programming* **85**: 107–134.

Facchinei, F. and J. Soares. 1997. "A new merit function for nonlinear complementarity problems and a related algorithm." *SIAM J. Optim.* **7**: 225–247.

Ferris, M. C. and T. S. Munson. 2000. "Complementarity problems in GAMS and the PATH solver." *J. Econ. Dynamics Control* **24**: 165–188.

Ferris, M. C. and T. S. Munson. 1999. "Interfaces to PATH 3.0: Design, implementation and usage." *Comput. Optim. Appl.* **12**: 207–227.

Ferris, M. C. and J. S. Pang (eds). 1997a. In *Complementarity and Variational Problems: State of the Art.* Philadelphia: SIAM Publications.

Ferris, M. C. and J. S. Pang. 1997b. "Engineering and economic applications of complementarity problems." *SIAM Rev.* **39**: 669–713.

Ferris, M. C., R. Fourer, and D. M. Gay. 1999a. "Expressing complementarity problems and communicating them to solvers." *SIAM J. Optim.* **9**: 991–1009.

Ferris, M. C., C. Kanzow, and T. S Munson. 1999b. "Feasible descent algorithms for mixed complementarity problems." *Math. Programming* **86**: 475–497.

Fischer, A. 1992. "A special Newton-type optimization method." *Optimization* **24**: 269–284.

Fukushima, M. 1992. "Equivalent differentiable optimization problems and descent methods for asymmetric variational inequality problems." *Math. Programming* **53**: 99–110.

Fukushima, M., Z.-Q. Luo, and J. S. Pang. 1998. "A globally convergent sequential quadratic programming algorithm for mathematical programs with linear complementarity constraints." *Comput. Optim. Appl.* **10**: 5–34.

Gabriel, S. A. and J. S. Pang. 1992. "An inexact NE/SQP method for solving the nonlinear complementarity problem." *Comput. Optim. Appl.* **1**: 67–91.

Geiger, C. and C. Kanzow. 1996. "On the resolution of monotone complementarity problems." *Comput. Optim. Appl.* **5**: 155–173.

Harker, P. T. 1993. *Lectures on Computation of Equilibria with Equation-Based Methods.* CORE Lecture Series. Université Catholique de Louvain, Louvain-la-Neuve: CORE Foundation.

Harker, P. T. and J. S. Pang. 1990. "Finite-dimensional variational inequality and nonlinear complementarity problems: a survey of theory, algorithms and applications." *Math. Programming* **48**: 161–220.

Harrison, G. W., T. F. Rutherford, and D. Tarr. 1997. "Quantifying the Uruguary round." *Econ. J.* **107**: 1405–1430.

Iusem, A. N. and B. F. Svaiter. 1997. "A variant of Korpelevich's method for variational inequalities with a new search strategy." *Optimization* **42**: 309–321.

Jiang, H. and D. Ralph. 1997. "Smooth SQP methods for mathematical programs with nonlinear complementarity constraints." Technical report. Melbourne, Australia: Department of Mathematics and Statistics, The University of Melbourne.

Josephy, N. H. 1979. "Newton's method for generalized equations." Technical Summary Report 1965. Madison, Wisconsin: Mathematics Research Center, University of Wisconsin.

Kanzow, C. 1996. "Some noninterior continuation methods for linear complementarity problems." *SIAM J. Matrix Anal. Appl.* **17**: 851–868.

Kanzow, C. and H. Kleinmichel. 1995. "A class of Newton-type methods for equality and inequality constrained optimization." *Optim. Methods Softw.* **5**: 173–198.

Kanzow, C. and H. Kleinmichel. 1998. "A new class of semismooth Newton-type methods for nonlinear complementarity problems." *Comput. Optim. Appl.* **11**: 227–251.

Kanzow, C. and H. Pieper. 1999. "Jacobian smoothing methods for nonlinear complementarity problems." *SIAM J. Optim.* **9**: 342–373.

Kanzow, C., N. Yamashita, and M. Fukushima. 1997. "New NCP-functions and their properties." *J. Optim. Theory Appl.* **94**: 115–135.

Khobotov, E. N. 1987. "A modification of the extragradient method for the solution of variational inequalities and some optimization problems." *Zh. Vych. Matematik. i Matematich. Fiz.* **27**: 1462–1473.

Korpelevich, G. M. 1976. "The extragradient method for finding saddle points and other problems." *Matecon* **12**: 747–756.

Lemke, C. E. and J. T. Howson. 1964. "Equilibrium points of bimatrix games." *SIAM J. Appl. Math.* **12**: 413–423.

Luo, Z.-Q., J. S. Pang, and D. Ralph. 1996. In *Mathematical Programs with Equilibrium Constraints.* Cambridge, England: Cambridge University Press.

Luo, Z.-Q. and P. Tseng. 1997. "A new class of merit functions for the nonlinear complementarity problem." In *Complementarity and Variational Problems: State of the Art*, pp. 204–225. Philadelphia: SIAM Publishers.

Maier, G. 1970. "A matrix structural theory of piecewise linear elastoplasticity with interacting yield planes." *Meccanica* **5**: 54–66.

Maier, G. 1971. "Incremental plastic analysis in the presence of large displacement and physical instabilizing effects." *Int. J. Solids Struct.* **7**: 345–372.

Maier, G. and G. Novati. 1990. "A shakedown and bounding theory allowing for nonlinear hardening and second order geometric effects with reference to discrete structural models." In *Inelastic Solids and Structures, A. Sawczuk Memorial Volume*, pp. 451–471. Swansea, Wales: Pineridge Press.

Mangasarian, O. L. 1976. "Equivalence of the complementarity problem to a system of nonlinear equations." *SIAM J. Appl. Math.* **31**: 89–92.

Mangasarian, O. L. and J. S. Pang. 1997. "Exact penalty functions for mathematical programs with linear complementarity constraints." *Optimization* **42**: 1–8.

Mangasarian, O. L. and M. V. Solodov. 1993. "Nonlinear complementarity as unconstrained and constrained minimization." *Math. Programming* **62**: 277–298.

Manne, A. S. and T. F. Rutherford. 1993. "International trade in oil, gas and carbon emission rights: an intertemporal general equilibrium model." *Energy J.* **14**: 1–20.

Marcotte, P. 1985. "A new algorithm for solving variational inequalities with application to the traffic assignment problem." *Math. Programming* **33**: 339–351.

Marcotte, P. 1991. "Application of Khobotov's algorithm to variational inequalities and network equilibrium problems." *Inf. Syst. Oper. Res.* **29**: 258–270.

Marcotte, P. and J.-P. Dussault. 1987. "A note on a globally convergent Newton method for solving monotone variational inequalities." *Oper. Res. Lett.* **6**: 35–42.

Mathiesen, L. 1985. "Computation of economic equilibria by a sequence of linear complementarity problems." *Math. Programming Study* **23**: 144–162.

Mathiesen, L. 1987. "An algorithm based on a sequence of linear complementarity problems applied to a Walrasian equilibrium model: an example." *Math. Programming* **37**: 1–18.

Moré, J. J. and W. C. Rheinboldt. 1973. "On *P*- and *S*-functions and related classes of *n*-dimensional nonlinear mappings." *Linear Algebra Appl.* **6**: 45–68.

Nagurney, A 1993. In *Network Economics—A Variational Inequality Approach*. Dordrecht, The Netherlands: Kluwer Academic Publishers.

Nash, J. F. 1950. "Equilibrium points in N-person games." *Proc. Nat. Acad. Sci.* **36**: 48–49.

Nazareth, J. L. 1987. In *Computer Solution of Linear Programs*. Oxford: Oxford University Press.

Outrata, J. V. 1994. "On optimization problems with variational inequality constraints." *SIAM J. Optim.* **4**: 340–357.

Outrata, J. V. and J. Zowe. 1995. "A numerical approach to optimization problems with variational inequality constraints." *Math. Programming* **68**: 105–130.

Outrata, J. V., M. Kočvara, and J. Zowe. 1998. In *Nonsmooth Approach to Optimization Problems with Equilibrium Constraints*. Dordrecht, The Netherlands: Kluwer Academic Publishers.

Pang, J. S. 1990. "Newton's method for B-differentiable equations." *Math. Oper. Res.* **15**: 311–341.

Pang, J. S. 1991. "A B-differentiable equation based, globally and locally quadratically convergent algorithm for nonlinear programs, complementarity and variational inequality problems." *Math. Programming* **51**: 101–132.

Pang, J. S. 1994. "Complementarity problems." In *Handbook in Global Optimization*, edited by R. Horst, and P. Pardalos. Boston: Kluwer Academic Publishers.

Pang, J. S. and S. A. Gabriel. 1993. "NE/SQP: a robust algorithm for the nonlinear complementarity problem." *Math. Programming* **60** 295–338.

Peng, J. M. 1997. "Equivalence of variational inequality problems to unconstrained optimization." *Math. Programming* **78**: 347–356.

Qi, L. 1993. "Convergence analysis of some algorithms for solving nonsmooth equations." *Math. Oper. Res.* **18**: 227–244.

Qi, L. and J. Sun. 1993. "A nonsmooth version of Newton's method." *Math. Programming* **58**: 353–368.

Ralph, D. 1994. "Global convergence of damped Newton's method for nonsmooth equations, via the path search." *Math. Oper. Res.* **19**: 352–389.

Robinson, S. M. 1980. "Strongly regular generalized equations." *Math. Oper. Res.* **5**: 43–62.

Robinson, S. M. 1987. "Local structure of feasible sets in nonlinear programming, Part III: stability and sensitivity." *Math. Programming Study* **30**: 45–66.

Robinson, S. M. 1992. "Normal maps induced by linear transformations." *Math. Oper. Res.* **17**: 691–714.

Robinson, S. M. 1994. "Newton's method for a class of nonsmooth functions." *Set Valued Anal.* **2**: 291–305.

Roos, C., T. Terlaky, and J.-Ph. Vial. 1997. In *Theory and Algorithms for Linear Optimization: An Interior-Point Approach*. Chichester, England: John Wiley & Sons.

Rutherford, T. F. 1995. "Extensions of GAMS for complementarity problems arising in applied economic analysis." *J. Econ. Dynamics Control* **19**: 1299–1324.

Rutherford, T. F. 1999. "Applied general equilibrium modeling with MPSGE as a GAMS subsystem: an overview of the modeling framework and syntax." *Comput. Econ* **14**: 1–46.

Scarf, H. E. 1973. In *The Computation of Economic Equilibria*. New Haven: Yale University Press.

Scheel, H. and Scholtes, S. 1998. "Mathematical programs with complementarity constraints: stationarity, optimality, and sensitivity." Technical report. Cambridge, England: Judge Institute of Management Studies, University of Cambridge.

Scholtes, S. and M. Stöhr. 1999. "Exact penalization of mathematical programs with equilibrium constraints." *SIAM J. Control Optim.* **37**: 617–652.

Sibony, M. 1970. "Methodes iteratives pur les equations et inequations aux derives partielles nonlineaires de type monotone." *Calcolo* **7**: 65–183.

Smeers, Y. 1997. "Computable equilibrium models and the restructuring of the European electricity and gas markets." *Energy J.*

Solodov, M. V. and B. F. Svaiter. 1999. "A new projection method for variational inequality problems." *SIAM J. Control Optim.* **37**: 765–776

Solodov, M. V. and P. Tseng. 1996. "Modified projection-type methods for monotone variational inequalities." *SIAM J. Control Optim* **34**: 1814–1830.

Sun, D. 1996. "A class of iterative methods for nonlinear projection equations." *J. Optim. Theory Appl.* **91**: 123–140.

Sun, D. and L. Qi. 1999. "On NCP-functions." *Comput. Optim. Appl.* **13**: 201–220.

Taji, K., M. Fukushima, and T. Ibaraki. 1993. "A globally convergent Newton method for solving strongly monotone variational inequalities." *Math. Programming* **58**: 369–383.

Tin-Loi, F. and M. C. Ferris. 1997. "Holonomic analysis of quasibrittle fracture with nonlinear softening." In *Advances in Fracture Research*, volume 2, edited by B. L. Karihaloo, Y. W Mai, M. I. Ripley, and R. O. Ritchie, pp. 2183–2190. Oxford: Pergamon Press.

Tin-Loi, F. and J. S. Misa. 1996. "Large displacement elastoplastic analysis of semirigid steel frames." *Int. J. Numer. Methods Engng* **39**: 741–762.

Tin-Loi, F. and J. S. Pang. 1993. "Elastoplastic analysis of structures with nonlinear hardening." *Comput. Methods Appl. Mech. Engng* **107**: 299–312.

Wright, S. J. 1997. In *Primal–Dual Interior-Point Methods*. Philadelphia: SIAM.

Xu, S. and J. V. Burke. 1999. "A polynomial time interior-point path-following algorithm for LCP based on Chen–Harker–Kanzow smoothing techniques." *Math. Programming* **86**: 91–104.

Yamashita, N., K. Taji, and M. Fukushima. 1997. "Unconstrained optimization reformulations of variational inequality problems." *J. Optim. Theory Appl.* **92**: 439–456.

Ye, Y. 1997. *Interior-Point Algorithms: Theory and Analysis*. New York: John Wiley & Sons.

14 DATA ENVELOPMENT ANALYSIS

José H. Dulá

ABSTRACT Data Envelopment Analysis (DEA) is a nonparametric frontier estimation methodology based on linear programming for measuring relative efficiencies of a collection of firms or entities in transforming their inputs into outputs. DEA is a standard tool in OR/MS whose number and domain of applications is rapidly growing. This chapter introduces DEA, focusing on the algorithmic and computational aspects of the methodology. As more analysts apply the methodology and as studies grow in scale and sophistication, data sets become larger, analyses become more frequent, even dynamic, and more effort is spent on extracting information efficiently. These issues place algorithmic and computational aspects of DEA at the forefront.

1. Introduction

Data Envelopment Analysis (DEA) is a standard, widely used, and powerful quantitative tool for performing efficiency studies. DEA evaluates the relative efficiency of a collection of functionally similar firms or entities (called decisionmaking units or DMUs) in transforming inputs into outputs. The comparison is strictly in relation to the members of the subject group. Evaluating and comparing efficiency and performance of many similar units is an important part of managing a complex organization. DEA provides decisionmakers with a measure of how well subordinate units transform the resources they manage locally into the outputs that are necessary to achieve the operation's mission.

DEA is a recent addition to the toolbox of the management scientist. DEA was formally introduced in the article by Charnes, Cooper, and Rhodes in 1978 (Charnes et al., 1978). At the time of writing there are well over one thousand published works on DEA dealing with theory, modeling issues, applications, and computations (see Seiford, 1996, for a historical account and an extensive bibliography). The success of DEA as a quantitative tool is due, in part, to the fact that it is a nonparametric methodology; that is, the evaluation of efficiency is not based on theoretical limits requiring assumptions about the functional form of a frontier and estimation of its parameters. Rather, DEA measures the efficiency of DMUs relative to each other. The classification of a DMU as efficient or inefficient depends only on its performance in transforming inputs into outputs as compared to other DMUs in the data domain.

Another reason for DEA's vitality, and a compelling argument for its validity, is that the mechanism involved in the classification of DMUs is amenable to acceptance from the part of the subjects of the study. The classification essentially grades each DMU using standard theoretical definitions for efficiency by applying a scale specific to each unit and such that its grade or score is maximized. With DEA, it is assured that no DMU can select a grading scale for itself that would improve its efficiency position with respect to the rest of the DMUs.

DEA has several models and orientations. A model in DEA is defined by the economic assumptions about the returns to scale of the transformation process. The four standard models in DEA are: (i) the constant returns model also known as the CCR model, which was the first model and was introduced by Charnes et al. (1978); (ii) the variable returns, BCC, model of Banker et al. (1984); and (iii) and (iv) the increasing and decreasing returns models of Banker et al. (1984) and Seiford and Thrall (1990). The constant returns model is used when the uniform positive scaling of inputs and outputs of any of the DMUs does not affect its efficiency classification. The variable returns DEA model is used when uniform scaling above and below maximum observed values for any of the inputs and outputs cannot be assumed to be possible. The increasing and decreasing returns extensions for DEA models are used when the assumption that scaling down or scaling up is not valid.

An efficiency study involving DEA is computationally intensive. DEA is based on the repeated application of linear programming. As many linear programs as they are DMUs may need to be solved to perform one DEA efficiency analysis and several analyses may be involved in an elaborate study. Compare this with, say, least-squares regression where, for similar data structures, only a single system of linear equations needs to be solved. In DEA this would be multiplied by the number of iterations to solve an LP and the total number of LPs in the complete study.

The evolution of DEA as an analytical tool has been and will remain closely linked to computational and algorithmic advances. Computational issues become increasingly relevant as studies involve larger numbers of DMUs, as analysts ask more questions about different economic assumptions governing the transformation process, and as the studies become more dynamic requiring more frequent evaluations. This chapter is about the algorithmic and computational aspects of DEA; its history, its background, and the state of the art.

2. Notation and Modeling Issues

The data domain for a DEA study is the set $\mathcal{A}$ of n data points, $a^1, \ldots, a^n$; one for each DMU. Each data point is composed of two types of components, those pertaining to the m_1 inputs, $0 \neq x^j \geqslant 0$, and those corresponding to the m_2 outputs, $0 \neq y^j \geqslant 0$. We organize the data in the following way:

$$\mathcal{A} = [a^1, \ldots, a^n], \quad \text{where} \quad a^j = \begin{bmatrix} -x^j \\ y^j \end{bmatrix}; \quad j = 1, \ldots, n,$$

and $\mathcal{A}$ is the $m_1 + m_2$ by n matrix, the columns of which are the data points.

The assumption that no data point is such that all its input or output components are zeroes is standard in DEA. This assumption excludes the absurdity of a DMU that transforms nothing into something or the uninteresting case of a unit with no outputs.

Another standard assumption is that the data be reduced. This has different meanings depending on the DEA model. In all cases, a reduced data set contains no duplicate data points. This assumption is necessary since in all DEA algorithms, duplicate data points confound the procedure by not permitting certain sufficient conditions needed for conclusive classifications to emerge. For the case of the constant returns model, excluding duplication is not enough to avoid ambiguities since here two points that are proportional to each other are effectively equivalent. In this case, a reduced data set also excludes proportionality among points.

Each data point, a^j, collects a list of deterministic values that summarizes the essence of the input/output transformation process. All n DMUs will be evaluated using the same input and output attributes. The choice of these attributes is a modeling decision. The OR/MS analyst needs to interact closely with the decisionmaker to arrive at a final list of attributes that will adequately reflect the activity of the units and provide meaningful measurements of their relative efficiency.

Another modeling decision regards the type of DEA model that will be used. As stated earlier, there are four standard DEA models. The choice of a model depends on what assumptions about returns to scale apply in the specific process under study. When this is not known precisely, the analyst may perform different analyses using several models and from that make inferences about the actual returns to scale.

An important decision in the modeling process regards what is referred to as the orientation of the study. If the analysis is to provide recommendations for benchmark references for inefficient DMUs, these can be based on considerations about decreasing inputs or about increasing outputs. This defines the orientation of the model as either input or output oriented. However, a study may be orientation free, in which case the analysis, results, and benchmarking recommendations are based on the optimization of a weighted sum of slacks and surpluses. This is the case in the additive models and their generalizations as described by Thrall (1996b). Recent work by Cooper et al. (1998) offers yet more variations on orientation-free analyses.

3. Derivation of the Basic Models

Given data for n decisionmaking units (DMUs) composed of a single input and output, the following standard formula can be used to compare their efficiencies:

$$\text{efficiency} = \frac{\text{output of DMU } j}{\text{input of DMU } j}, \quad j = 1, \ldots, n.$$

Given multiple incommensurate inputs and outputs the extension of the efficiency formula above is

$$\text{efficiency} = \frac{\text{weighted sum of outputs of DMU } j}{\text{weighted sum of inputs of DMU } j}, \quad j = 1, \ldots, n.$$

The problem with this formula is the assignment of weights. A fixed set of input and output weights applied across the DMUs could be considered arbitrary and unfair from the point of view of some of them; namely, those who score low in such a scheme. One solution to this problem is the one proposed by Charnes et al. in 1978; it is, in a sense, to allow each DMU to select its own weights.

There must be rules if each DMU gets to select its own weights. The first rule is that, after a DMU picks its weights, these will be applied to the rest of the DMUs. Of special interest will be DMUs that manage to score higher using the same weights selected by the DMU being evaluated. The presence of such DMUs is the indication used in DEA that the DMU being scored is not operating efficiently. To be certain that a DMU is inefficient by this standard, the weights must maximize the efficiency score of the DMU under evaluation. Another rule is that weights will be restricted so that the total efficiency score any DMU can attain cannot exceed some positive constant; this is usually 1. Besides normalizing scores and preventing trivialities, this rule offers an immediate advantage; it provides a readily recognized sufficient condition for a DMU to be efficient, namely, that it attain a score of unity for some set of weights.

Another condition imposed on the weighting scheme for evaluating a particular DMU's comparative efficiency is that none of the weights can be zero. If zero weights were allowed, it would be possible for two DMUs identical in input and output values except for, say, output $\hat{i}$, to have the same score simply by choosing zero for the weight used for that output. This would be a particularly pernicious problem for an evaluation scheme if the weights were such that the final score was unity since it would classify a clearly inferior DMU (the one with smaller value for $\hat{i}$) in the same category as the one with a greater value for that output.

The mathematical program that models the efficiency measuring scheme described above is given below (Charnes et al., 1978):

$$\max_{\pi > 0, \sigma > 0} \frac{\langle \sigma, y^{j^*} \rangle}{\langle \pi, x^{j^*} \rangle}$$

$$\text{subject to} \quad \frac{\langle \sigma, y^j \rangle}{\langle \pi, x^j \rangle} \leqslant 1; \quad j = 1, \ldots, n,$$

where $x^j, y^j, j = 1, \ldots, n$ are the vectors containing the m_1 input and m_2 output data, respectively, for the n DMUs in the model. Also, $\pi = (\pi_1, \ldots, \pi_{m_1})$ and $\sigma = (\sigma_1, \ldots, \sigma_{m_2})$ are the weights, or multipliers, of the inputs and outputs. Finally, j^* is the index of the DMU chosen to be scored.

It would appear that we are up against a problematic nonlinear program since the objective function is a ratio and the variables are restricted to be strictly positive. The form of the constraints is not a concern since these can be made linear after a simple algebraic manipulation. The nonlinearity of the objective function is actually easily dispatched as shown by Charnes and Cooper in 1962. The issue of strict positivity of the variables, however, requires special treatment. The approach is to introduce a non-Archimedean constant, ϵ, defined to be simply a sufficiently small positive constant. After this, the fractional program has the following LP equivalent:

$$\max_{\pi \geqslant \epsilon \mathbf{e}, \sigma \geqslant \epsilon \mathbf{e}} \quad \langle \sigma, y^{j^*} \rangle$$

$$\text{subject to} \quad -\langle \pi, x^j \rangle + \langle \sigma, y^j \rangle \leqslant 0, \quad j = 1, \ldots, n;$$

$$\langle \pi, x^{j^*} \rangle \qquad = 1,$$

where $\mathbf{e}$ is an appropriately dimensional vector of 1s. This LP is the multiplier version of the input-oriented constant-returns DEA model. It is input oriented because the derivation is based on fixing the denominator of the original fractional program by making it a constraint that must equal unity. The assumption of constant returns was made initially by defining efficiency as the simple ratio of weighted outputs and inputs. By this definition, a DMU's efficiency score is unaffected by the uniform positive scaling of all inputs and outputs.

If the objective function value of this multiplier LP is strictly less than 1, DMU j^* is inefficient since even after selecting specific multipliers to maximize its score, other DMUs obtain a better score (recall that an optimal extreme point solution to this LP will have $m_1 + m_2$ tight constraints; therefore, one less than this number of DMUs will attain unity score). If the objective function value is unity then the DMU being scored is classified as efficient.

The dual to the input-oriented constant-returns multiplier form DEA LP is:

$$\min_{\lambda \geqslant 0, \mathscr{S}^+ \geqslant 0, \mathscr{S}^- \geqslant 0, \theta} \quad -\epsilon\langle \mathbf{e}, \mathscr{S}^+ \rangle - \epsilon\langle \mathbf{e}, \mathscr{S}^- \rangle + \theta$$

$$\text{subject to} \quad \sum_j x^j \lambda_j + I_{m_1} \mathscr{S}^+ \qquad - x^{j^*} \theta = 0,$$

$$\sum_j y^j \lambda_j \qquad - I_{m_2} \mathscr{S}^- \qquad = y^{j^*}.$$

Here $\mathscr{S}^+$ and $\mathscr{S}^-$ are m_1 and m_2 dimensional vectors of slacks associated with the input and output components, respectively, and I_{m_1} and I_{m_2} unit matrices with dimensions given by their subscripts. This dual is known as the envelopment version of the input oriented constant return DEA model. This LP formulation answers the question: Is it possible to find a positively scaled combination of the input and output data of DMUs in the data domain to define a virtual DMU with inputs and outputs such that the inputs are uniformly less than the inputs of the DMU being scored with at least as many outputs? The answer is affirmative if the value for θ is strictly less than one. This situation is the indication used to conclude that the DMU being scored is inefficient relative to other specific DMUs in the data set; namely, those used to define the virtual DMU. The collection of DMUs used to define the virtual DMU is known as the reference set. If the optimal objective function value is unity, there is no better virtual DMU and DMU j^* is efficient.

The envelopment version provides a more intuitive explanation as to why this is an input orientation. If DMU j^* is inefficient, then the optimal solution to the LP will provide values for the variables, λ, that can be used to construct a virtual DMU, which, if it existed, would produce at least as much as DMU j^* with strictly less amounts of inputs. The virtual DMU can serve as a benchmark for the inefficient DMU, one where improvement is achieved through the reduction of inputs.

Either LP in the primal-dual pair can be solved in a DEA study. Note that a different LP must be solved for each DMU. The envelopment version has become standard in DEA studies and computations because of its intuitive appeal and because, as taught in LP courses, fewer rows than columns is preferable to the converse for the purposes of the simplex algorithm.

The DEA LPs can be derived using an output orientation. This starts from a symmetric definition of efficiency as the ratio of inputs and outputs. In this case, a smaller value is better. This approach generates the following primal-dual pair of LPs:

$$\underbrace{\begin{aligned} &\min_{\pi \geqslant \epsilon\mathbf{e}, \sigma \geqslant \epsilon\mathbf{e}} \quad \langle \pi, x^{j^*} \rangle \\ &\text{subject to} \quad -\langle \pi, x^j \rangle + \langle \sigma, y^j \rangle \leqslant 0, \quad j = 1, \ldots, n; \\ &\qquad\qquad \langle \sigma, y^{j^*} \rangle = 1. \end{aligned}}_{\text{Output-oriented, constant returns, non-Archimedean, multiplier version}}$$

$$\underbrace{\begin{aligned} &\max_{\lambda_j \geqslant 0, \mathscr{S}^+ \geqslant 0, \mathscr{S}^- 0, \phi} \quad \epsilon\langle \mathbf{e}, \mathscr{S}^+ \rangle - \epsilon\langle \mathbf{e}, \mathscr{S}^- \rangle + \phi \\ &\text{subject to} \quad \sum_j x^j \lambda_j + I_{m_1} \mathscr{S}^+ = x^{j^*}, \\ &\qquad\qquad \sum_j y^j \lambda_j - I_{m_2} \mathscr{S}^- - y^{j^*} \phi = 0. \end{aligned}}_{\text{Output-oriented, constant returns, non-Archimedean, envelopment version}}$$

In an output orientation, the objective function values of the LPs are greater than or equal to 1. DMU j^* is classified as efficient if the optimal objective function value is unity. We can use the envelopment form to explain this classification scheme since if the optimal solution is greater than unity, it is possible to scale and combine DMUs in the data domain to define a virtual DMU, which transforms at most as many inputs as DMU j^* into strictly more outputs; this satisfies DEA's inefficiency standard. It is important to note that the efficiency classification of a DMU in a constant returns environment is independent of the form or orientation; that is, DMU j^* is constant returns efficient if, and only if, the optimal solution to any and all of the constant returns LPs is unity.

The efficiency classification of a DMU, however, is affected by the type of DEA model. The oriented LPs for the other three DEA models corresponding to the variable, increasing, and decreasing returns to scale assumptions are as follows. (We present here only the LPs in input orientation since, with this, it is clear what the form should be for the output orientation.) First the multiplier form

$$\underbrace{\begin{aligned} &\max_{\pi \geqslant \epsilon\mathbf{e}, \sigma \geqslant \epsilon\mathbf{e}, \beta} \quad \langle \sigma, y^{j^*} \rangle + \beta \\ &\text{subject to} \quad -\langle \pi, x^j \rangle + \langle \sigma, y^j \rangle + \beta \leqslant 0, \quad j = 1, \ldots, n \\ &\qquad\qquad \langle \pi, x^{j^*} \rangle = 1; \end{aligned}}_{\text{Input-oriented, variable returns, non-Archimedean, multiplier version}}$$

where β unrestricted, $\beta \leqslant 0$, and $\beta \geqslant 0$ for the variable, increasing, and decreasing returns models, respectively. The corresponding envelopment forms have an extra constraint. They are

$$\underbrace{\begin{aligned} &\min_{\lambda \geqslant 0, \mathscr{S}^+ \geqslant 0, \mathscr{S}^- \geqslant 0, \theta} \quad -\epsilon\langle \mathbf{e}, \mathscr{S}^+ \rangle - \epsilon\langle \mathbf{e}, \mathscr{S}^- \rangle + \theta \\ &\text{subject to} \quad \sum_j x^j \lambda_j + I_{m_1} \mathscr{S}^+ - x^{j^*} \theta = 0, \\ &\qquad\qquad \sum_j y^j \lambda_j - I_{m_2} \mathscr{S}^- = y^{j^*}, \\ &\qquad\qquad \sum_j \lambda_j \begin{pmatrix} = \\ \leqslant \\ \geqslant \end{pmatrix} 1, \end{aligned}}_{\text{Input-oriented, variable returns, non-Archimedean, envelopment versions}}$$

where = in the last constraint applies to the variable returns model; $\leqslant$ to the increasing returns model; and $\geqslant$ to the decreasing returns model.

A DEA model can be made more realistic and its results more discriminating by the addition of specific conditions on the multipliers. As long as these can be expressed as linear relations, a multiplier LP can be augmented with information about upper or lower bounds of the values of individual multipliers, pairwise relations between multipliers, including ratios, and more complex interrelations among multipliers. Note that the impact of additional linear constraints on a multiplier LP is a potential reduction of the number of DMUs that will be classified as efficient.

We close this section with a discussion of orientation-free LPs in DEA. These LPs are presented in the envelopment version and are given as follows:

$$\underbrace{\begin{aligned} &\max_{\lambda_j \geqslant 0, \mathcal{S}^+ \geqslant 0, \mathcal{S}^- \geqslant 0} \quad \langle \mathbf{e}, \mathcal{S}^+ \rangle + \langle \mathbf{e}, \mathcal{S}^- \rangle \\ &\text{subject to} \quad \sum_j x^j \lambda_j + I_{m_1} \mathcal{S}^+ = x^{j^*}, \\ &\qquad \sum_j y^j \lambda_j - I_{m_2} \mathcal{S}^- = y^{j^*}, \\ &\qquad \sum_j \lambda_j \begin{pmatrix} \cdot \\ = \\ \leqslant \\ \geqslant \end{pmatrix} 1; \end{aligned}}_{\text{"Additive," envelopment versions}}$$

where the last constraint is omitted in the constant returns model and present in the other three models according to our conventions. This is the additive form of Charnes et al. (1985). It is characterized by the objective function, which maximizes the sum of the slacks and surpluses and the absence of non-Archimedean constants. Extensions and generalization by Thrall (1996b), Cooper et al. (1998), and Tone (1998) explore the impact of different values for the coefficients of the slacks in the objective function. The original role of the additive model serves two purposes. The first is as an alternative to the oriented forms. Additive LPs yield scores that are different from oriented forms without affecting final efficiency classifications. Efficient reference sets and resultant recommendations and benchmarks, however, may be different. Another feature of additive LPs is that they are not subject to the complications connected with the non-Archimedean constants. We will see in the next section that this represents an important feature of these formulations.

4. Traditional DEA Computational Procedures

An algorithm for performing DEA is implicit in its definition and derivation. That is, evaluate the efficiency of each DMU, one at a time, by modifying the appropriate linear program relative to the efficiency of the other DMUs. This is what we refer to as the traditional approach:

Traditional DEA Procedure

For $j = 1$ to n **do:**

Step 1. $j^* \leftarrow j$.

Step 2. Solve appropriate LP.

Step 3. Classify DMU j^*.

The simple description for the traditional DEA procedure can mask an onerous computational task. Every iteration in the procedure requires the complete solution of a dense LP. The modifications required for each LP depend on its form. Multiplier forms require that the objective function and the last constraint be updated with the values for the DMU being evaluated. In the envelopment this translates dually to changes in the right-hand side and, with oriented forms, in the last column of the LP.

One of the first computational issues that needed to be addressed with the traditional procedure dealt with the non-Archimedean constant in the oriented forms. It was understood early on that the presence of such a constant in the LPs was problematic. The specification for the value of ϵ is that it be large enough to eliminate the possibility of spurious classification of DMUs as efficient by essentially ignoring an inconvenient input or output attribute, but not so large that it became a significant condition on the values of the multipliers and truly efficient DMUs would be misclassified. This type of specification eventually leads to problems in actual implementations.

Ali (1990) describes a solution to this dilemma in the form of a two-stage approach. In the first stage the appropriate oriented LP is solved with $\epsilon = 0$. At this stage, an objective function value different from unity is still sufficient to conclude that the DMU being evaluated is inefficient. A score of unity, however, is necessary but insufficient to conclude that the DMU is efficient. When a score of unity is attained by a DMU in an oriented DEA LP where $\epsilon = 0$, it could be such that one of the slacks is strictly positive in the envelopment version or a multiplier is zero in the multiplier version. In either case this means that it is possible to construct a DMU by combining the data (respecting the returns to scale assumption) such that it is identical to the DMU being scored except for a better value (less input or more output) for the component where the positive slack or zero multiplier occurred. Therefore, the DMU being scored would have to be classified as inefficient although the score is still unity. Such DMUs are denoted as weakly efficient although a more appropriate name would be spuriously efficient.

A positive slack in the optimal solution to an oriented envelopment LP for a DMU that scores unity is a sufficient condition for that DMU to be weakly efficient. Conversely, a set of strictly positive multipliers in the optimal solution of a multiplier form when the score is unity is sufficient to conclude that the DMU being scored is efficient. The problem arises when a DMU scores unity and the slacks are zero in the envelopment LP or when the multipliers are not all strictly positive in the multiplier LP. In either case, the situation is inconclusive as to the status of the DMU since there exist alternate optima. Therefore, a second test is needed if the score is unity for an oriented DEA LP with $\epsilon = 0$ and none of the sufficient conditions have been satisfied. This is done in a second stage. The modification needed to permit using $\varepsilon = 0$ in DEA LPs while arriving to conclusive results appears in the next procedure.

Two Stage Traditional DEA Procedure (Oriented Envelopment Forms)
For $j = 1$ to n **do:**
Step 1. $j^* \leftarrow j$.
Step 2. Solve appropriate oriented LP with $\epsilon = 0$.
Step 3. If optimal objective function value is not unity:
Then.
Classify DMU j^* as inefficient.
Else.
If a slack is positive at optimality
Then
Classify DMU as weakly coefficient.
Else
Solve "additive" form and classify accordingly.

A newcomer to DEA may wonder: why solve oriented forms at all when an uncomplicated conclusive classification is possible with an additive form? Oriented forms are the original DEA formulations directly linked to standard efficiency definitions used in theory, analyses, and recommendations. They are also intuitively appealing to both the analyst and the decisionmaker. Unoriented forms are not indicated for benchmarking. The vast majority of applications to date apply oriented forms. Although new, interesting, and useful additive forms are being studied and proposed, oriented forms will remain an important tool in the toolbox of the OR/MS analyst.

The description of the traditional DEA procedures above are sufficiently straight forward to allow OR/MS practitioners to program their own routines to perform DEA studies; and surely, many have done just this. Until recently, the major decision that the programmer had to make was the choice for the LP solver. If the program was in a language such as FORTRAN, the programmer had several choices, including Marsten's (1981) XMP subroutines; Saunders and Wright's (1986) commercially available LP solver, LSSOL; or standard subroutines such as IMSL's DLPRS. An important criterion in making the choice for the LP solver is whether it is optimized for sparsity as in the case of XMP. This is an issue because DEA LPs are typically very dense, making XMP less indicated than say, LSSOL and DLPRS.

These days OR/MS analysts who wish to program traditional DEA procedures directly are likely to turn to a commercial spreadsheet. An excellent candidate for this would be Microsoft's Excel since it offers as part of the package an LP solver that is adequate for the task. It is a simple matter to program the spreadsheet to perform the single-stage traditional procedure. A complete macro using no more than fifteen to twenty instructions is possible and quite effective for small problems ($\leqslant 100$ DMUs). More sophisticated programming would be needed to implement the traditional two-stage approach. Almost certainly this has already been done and macros may be available through contacts or commercially. It is reasonable to expect that, one day, a macro for DEA will be bundled with the spreadsheet package in the same way as for, say, least-squares regression.

Besides the potential for problems with times and memory during execution using large data sets ($\geqslant 2500$ DMUs), a naïve implementation based on the procedures above must contend with the possibility of other complications. A particularly serious one is the problem of degeneracy and cycling. Barr and Duchholz (1997) state, "Our experience indicates that cycling in DEA codes is not only prevalent but likely, in the absence of anticycling procedures." This concern has lead to works such as Charnes et al. (1993) where an antidegeneracy/cycling linear programming method especially for data envelopment analysis is introduced. The primary factor inducing degeneracy in DEA codes is the inclusion of the column of the DMU being scored in the constraint matrix of envelopment LPs. If DMU j^* is efficient, one solution to the original LP formulation is $\lambda_{j^*} = 1$ and $\lambda_j = 0 \;\; \forall j \neq j^*$. In fact, this is the only solution for a certain category of efficient DMUs. Clearly this solution is highly degenerate.

In the next section we will present some of the ideas and techniques put forth to make the traditional procedures more efficient.

5. Enhancements to Traditional DEA Procedures

The heavy computational burden of efficiency studies with DEA makes ideas for computational savings important contributions to the field.

One idea originally proposed by Charnes et al. (1985) and formalized by Ali (1994) is early identification of efficient DMUs (EIE). It follows from the following result.

Result 1. *DMUs corresponding to basic structural variables in an optimal and dual feasible basis of an envelopment LP are themselves efficient and, conversely, no inefficient DMUs variable can be basic at optimality for these LPs.*

The consequence of this result is that the optimal solution of an envelopment DEA LP, arrived at by applying an extreme point LP method, establishes conclusively the status of as many as all DMUs corresponding to the columns in the optimal basis. The status of these DMUs is that they are DEA efficient. This additional information is attained with a minimum of additional computations requiring only a simple check to identify the basic variables. One fewer LP will have to be solved for any previously unknown efficient DMU, which is identified this way.

Another consequence of the result above is that any variable corresponding to a known inefficient DMU may be blocked from entering a basis in the course of solving any subsequent envelopment LPs using the simplex method since every optimal and dual feasible basis is composed only of vectors from columns corresponding to efficient DMUs. This leads to the restricted basis entry, or RBE, scheme appearing in Ali (1994). In this scheme, as knowledge of the inefficient DMUs is successively acquired, the effective size of the LPs in the analysis is reduced.

In the same spirit as the opportunistic schemes above, another idea takes advantage of access to an optimal (dual feasible) basis of the envelopment LP for the given DMU being scored. The idea is formalized in the next result (Dulá and Hickman, 1995).

Result 2. *Consider an optimal and dual feasible basis to an envelopment LP when scoring DMU* j^*. *A dual pivot minimum ratio for any leaving basic variable (except* θ *or* ϕ *in oriented forms), when attained at the nonbasic structural variable* $\hat{j}$, *implies that DMU* $\hat{j}$ *is efficient.*

The results above need to be adjusted for implementations that avoid the non-Archimedean constant. RBE remains valid either way; but the distinction between true and weak efficiency is lost in EIE and in the idea based on a dual pivot. All DMUs that emerge as not conclusively inefficient will need to be processed again in a second additive LP. However, this second LP is potentially much smaller since it need not include any of the known inefficient DMUs.

An effective idea that helps resolve many problems with traditional DEA computations is the exclusion of the DMU being scored from the coefficient matrix of envelopment LPs. This is known as the deleted domain technique for DEA. This idea has been around since early in DEA computations and was probably practiced by many. It is mentioned in Banker and Gifford (1991) and Charnes et al. (1992). It was not formally presented and discussed as an alternative to the traditional approach until the papers by Andersen and Petersen (1993) and Bogetoft (1994). A rigorous study of the impact of domain deletion for the case of input oriented constant returns model appears in Thrall (1996a). Dulá and Hickman (1997) provide the definitive theoretical study of this technique. Seiford and Zhu (1999) also studied the deleted domain technique and presented necessary and sufficient conditions for infeasibility. The technique essentially provides all the same information as the traditional approach plus more insights about efficient DMUs including sufficient conditions to identify extreme efficient DMUs, an important category of efficiency (we will see more about extreme efficient DMUs and their role in the design of algorithms in the next section). Extreme efficient DMUs are by far the most common type of efficient DMUs and are never weakly efficient; this is useful information when the LP is not non-Archimedean. The fact that LPs can be infeasible with this technique actually adds to its value since this is sufficient to conclude that the DMU being scored is extremely efficient. Another obvious advantage of this technique is the fact that it eliminates the conditions for induced degeneracy.

A recent contribution to speed up DEA computations is based on partitioning the data domain. The idea is by Barr and Durchholz (1997) and applies the basic principle that if a DMU is inefficient with respect to a subset of the DMUs, it will be inefficient with respect to any superset. The idea was implemented in a procedure using hierarchical decomposition. The idea is to partition the DMUs into similarly sized blocks and apply the DEA analysis to these blocks independently identifying inefficient DMUs within blocks. The procedure proceeds to join processed blocks of DMUs with unknown status and running these through the analysis. This process is repeated until a final block is created. By then it is expected that the bulk of inefficient DMUs have been culled. All DMUs are scored in a second phase using LPs composed of the DMUs that survived the culling; it is hoped that a much smaller LP than would otherwise be used in the traditional approach. Barr and Durchholz observed that the performance of the scheme is affected by the size of the initial and intermediary blocks. This poses a problem since this decision has to be made explicitly by the analyst. Their experiments suggest that performance is quasi convex in that times are monotonically nondecreasing as the block size deviates from an optimal value. Their tests indicate substantial time reductions in ideal implementations.

All schemes and procedures for DEA computations discussed up to now are based on the close relationship between the definition of DEA efficiency and the solution to LPs. This relationship, though, is actually more complex. It involves polyhedral sets and regions within them that define a data point's DEA status. Recent contributions to DEA computations result from the study of these sets.

6. Recent Trends in DEA Computations

An important realization about DEA is that DEA efficiency is determined by the location of the data point of the DMU being evaluated in a special polyhedral set known as the production possibility set. This is the set of all points in the space where the points $a^j = (-x^j, y^j) \in \mathcal{R}^m$ reside, such that the relation between inputs and outputs is theoretically admissible in the context of the economic assumptions. The same set of n DMUs defines four different production possibility sets, one for each of the DEA models. A DMU is efficient for a given model if the point it represents in $\mathcal{R}^m$ belongs to boundary regions known as the efficient frontier. The four polyhedral sets, or hulls, are explicitly defined in what follows (Banker et al., 1984; Seiford and Thrall, 1990; Dulá and Venugopal, 1995):

$$\mathcal{P}^\ell = \{z \in \mathcal{R}^m \mid z \leqslant A\lambda; \quad \text{for } \lambda \in \Lambda^\ell\}, \quad \ell = 1, 2, 3, 4;$$

where Λ^ℓ is the subset of $\Lambda = \{\lambda \mid \lambda_j \geqslant 0, j = 1, \ldots, n\}$ defined by

Constant returns $\Lambda^1 = \Lambda;$

Variable returns $\Lambda^2 = \left\{\lambda \in \Lambda \mid \sum_{j=1}^n \lambda_j = 1\right\};$

Increasing returns $\Lambda^3 = \left\{\lambda \in \Lambda \mid \sum_{j=1}^n \lambda_j \leqslant 1\right\};$

Decreasing returns $\Lambda^4 = \left\{\lambda \in \Lambda \mid \sum_{j=1}^n \lambda_j \geqslant 1\right\}.$

It is important to stress here that the production possibility sets are independent of orientation considerations. That is, orientation is not a factor in the production possibility set's geometry nor does it affect the location of its efficient frontier.

Production possibility sets are finitely generated convex polyhedral sets. This has several important implications relevant to DEA. First, they are defined by their extreme elements; that is, extreme rays and extreme points. Second, all extreme elements (except those associated with the unit directions of recession defined by the slacks) correspond to the data points, that is, actual DMUs in the data set. Finally, the production possibility set's extreme elements are sufficient to describe the set. Therefore, any DEA inference about a DMU can be obtained using only data points that correspond to extreme elements of the production possibility set. These special DMUs are necessarily efficient to their corresponding model and are known as extreme efficient DMUs.

The set of extreme efficient DMUs of production possibility set $\mathcal{P}^\ell$ is its frame and it is denoted by $\mathcal{F}^\ell$. An important result is that a frame is unique in DEA given our assumptions about the data (Dulá and Thrall, 2001). Since the production possibility sets generated by the frame and by the entire data domain are one and the same, all questions concerning the efficiency status of a DMU can be answered using only the frame. This suggests the following two stage approach to DEA:

DEA Procedure Using Frames

Stage 1. Find frame $\mathcal{F}^\ell$ for specified DEA model $\ell = 1, 2, 3$ or 4.

Stage 2. Score DMUs in $\mathscr{A}\setminus\mathscr{F}^{\ell}$ using LPs composed of the frame elements.

In the first stage of the procedure, the frame is identified for the specific DEA model; in the second stage all DMUs are scored using LPs composed of the frame elements. Therefore, when the frame is a small subset of the data set, the DMUs can be processed using much smaller LPs. Another interesting advantage of using this two-stage method is that the first stage is orientation-free. This means that when an orientation decision is made in the second stage, the bulk of the calculations are behind. This extends flexibility to the process since many differently oriented analyses can be performed with relatively little additional work.

Finding a frame for Stage 1 of the procedure could be done using the deleted domain scheme of Dulá and Hickman (1997). However, the problem of finding extreme elements of finitely generated polyhedral sets has a long tradition in convex analysis (see, e.g., Wets and Witzgall, 1967). New contributions for finding these essential subsets have been made by Dulá and Helgason (1996), Dulá et al. (1998) have substantially decreased the computational times for these calculations. The new algorithms erect the frame, one element at a time. It turns out that these procedures can be adapted to work in DEA (Dulá, 1996). Implementations based on these new frame algorithms substantially reduce computational times.

Recent work by Dulá and Thrall (2001) has focused on the interrelations among the four frames, $\mathscr{F}^{\ell}; \ell = 1, 2, 3, 4$. Given standard reduction assumptions on the data, the relations are as follows:

Result 3.

(i) $\mathscr{F}^2 = (\mathscr{F}^3 \cup \mathscr{F}^4)$.

(ii) $\mathscr{F}^1 = (\mathscr{F}^3 \cap \mathscr{F}^4)$.

(iii) (Corollary)$\mathscr{F}^1 \subset \mathscr{F}^3 \subset \mathscr{F}^2$.

(iv) (Corollary)$\mathscr{F}^1 \subset \mathscr{F}^4 \subset \mathscr{F}^2$.

These results have immediate computational implications since, after calculating any two frames from the list, $\{\mathscr{F}^3, \mathscr{F}^4, \mathscr{F}^1\}$, the third may be obtained through simple set operations. One of these ideas is as follows: consider three routines VRFRAME, IRFRAME, and DRFRAME inputs of which are a DEA data set such as $\mathscr{A}$ and outputs will be frames as follows:

$\mathscr{F}^2 \leftarrow$VRFRAME($\mathscr{A}$);

$\mathscr{F}^3 \leftarrow$IRFRAME($\mathscr{A}$);

$\mathscr{F}^4 \leftarrow$DRFRAME($\mathscr{A}$).

An important realization is that routines IRFRAME and DRFRAME can be used as follows:

$\mathscr{F}^3 \leftarrow$IRFRAME($\mathscr{F}^2$);

$\mathscr{F}^4 \leftarrow$DRFRAME($\mathscr{F}^2$).

With this we have enough to design a procedure for DEA analysis based on the identification of all four frames (Dulá and Thrall, 2001):

DEA Procedure for all Models

Stage 1. Identify all four frames of the DEA data set $\mathscr{A}$.

Step 1. $\mathscr{F}^2 \leftarrow$VRFRAME($\mathscr{A}$).

Step 2. $\mathscr{F}^3 \leftarrow$IRFRAME($\mathscr{F}^2$)

Step 3. $\mathscr{F}^4 \leftarrow$DRFRAME($\mathscr{F}^2$).

Step 4. $\mathscr{F}^1 \leftarrow \mathscr{F}^3 \cap \mathscr{F}^4$.

Stage 2. Select models and orientation and score DMUs using appropriate frames.

Let us analyze some scenarios. In the worst case, the cardinalities of the frame and the entire data set are almost the same. In this case, the effort to find the frames of the data for the four models will require roughly three times the time to find the frame of the variable returns model, since Step 4 is computationally inexpensive. So, in a sense, we obtain four frames for the price of three. In real applications, however, such dense data domains do not seem to be the norm. A more realistic scenario is that the set of efficient DMUs is relatively small compared to the full DEA data set. In actual testing with this type of data, the four steps in Phase 1 are completed in little more than the time needed to execute Step 1 alone (Dulá and Thrall, 1999). This, combined with the fact that decisions about orientation are left to the second stage where they can be performed using only frames, translates to significant increases in flexibility and speed especially when DEA analyses are over several models and multiple orientations.

7. Conclusions

Research on algorithmic and computational aspects of DEA is resulting in faster and more flexible procedures. It is not unrealistic to predict large scale dynamic DEA studies where DEA scoring is updated virtually instantaneously in data sets with tens of thousands of DMUs. Applications for such procedures are not hard to conceive especially when we look around us and find already complex, financial, social, and technical systems and processes steadily growing and becoming more sophisticated.

References

Ali, I. 1990. "IDEAS: Integrated Data Envelopment Analysis System." Technical Report. Amherst, MA: Department of General Business and Finance, University of Massachusetts at Amherst.

Ali, A. I. 1994. "Computational aspects of Data Envelopment Analysis." In *DEA: Theory, Methodology and Applications*, edited by A. Charnes, W. W. Cooper, A. Y. Lewin, and L. M. Seiford. Boston: Kluwer Academic Publishers.

Andersen, P. and N. C. Petersen. 1993. "A procedure for ranking efficient units in data envelopment analysis." *Mgmnt Sci.* **39(10)**: 1261–1264.

Banker, R. D., A. Charnes, and W. W. Cooper. 1984. "Some models for estimating technological and scale inefficiencies in data envelopment analysis." *Mgmnt Sci.* **30(9)**: 1078–1092.

Banker, R. D. and J. L. Gifford. 1991. "Relative efficiency analysis." Unpublished manuscript.

Barr, R. S. and M. L. Durchholz. 1997. "Parallel and hierarchical decomposition approaches for solving large-scale Data Envelopment Analysis models." *Ann. Oper. Res.* **73**: 339–372.

Bogetoft, P. 1994. "Incentive efficient production frontiers: an agency perspective on DEA." *Mgmnt Sci.* **40(8)**: 959–968.

Charnes, A. and W. W. Cooper. 1962. "Programming with linear fractional functionals." *Naval Res. Logistics Q.* **9**: 181–185.

Charnes, A., W. W. Cooper, and E. Rhodes. 1978. "Measuring the efficiency of decision making units." *Eur. J. Oper. Res.* **2(6)**: 429–444.

Charnes, A., W. W. Cooper, B. Golany, L. Seiford, and J. Stutz. 1985. "Foundations of data envelopment analysis for Pareto–Koopmans efficient empirical production functions." *J. Econometrics* **30**: 91–107.

Charnes, A., S. Haag, P. Jaska, and J. Semple. 1992. "Sensitivity of efficiency classifications in the additive model of data envelopment analysis." *Int. J. Syst. Sci.* **23(5)**: 789–798.

Charnes, A., J. Rousseau, and J. Semple. 1993. "An effective non-Archimedean anti-degeneracy/cycling linear programming method especially for data envelopment analysis and like methods." *Ann. Oper. Res.* **47**: 271–278.

Cooper, W. W., K. S. Park, and J. T. Pastor. 1999. "RAM: a range adjusted measure of inefficiency for use with additive models and relations to other models and measures in DEA." *J. Productivity Anal.* **11**: 5–42.

Cooper, A. 1995. In *About Face: The Essentials of User Interface Design*. Foster City, CA: IDG Books Worldwide.

Dulá, J. H. 1996. "An algorithm for Data Envelopment Analysis (DEA)." MS 38677. School of Business, University of Mississippi.

Dulá, J. H. and R. V. Helgason. 1996. "A new procedure for identifying the frame of the convex hull of a finite collection of points in multidimensional space." *Eur. J. Oper. Res.* **92**: 352–367.

Dulá, J. H. and B. L. Hickman. 1995. "An opportunistic scheme for identifying efficient DMUs in DEA." Technical Report 95-4. Omaha, NE 68182: Computer Science Department, The University of Nebraska at Omaha.

Dulá, J. H. and B. L. Hickman. 1997. "Effects of excluding the column being scored from the DEA envelopment LP technology matrix." *J. Oper. Res. Soc.* **48**: 1001–1012.

Dulá, J. H. and R. M. Thrall. 2001. "A computational framework for accelerating DEA." *J. Productivity Anal.* **16**: 63–78.

Dulá, J. H. and N. Venugopal. 1995. "On characterizing the production possibility set for the CCR ratio model in DEA." *Int. J. Syst. Sci.* **26(12)**: 2319–2325.

Dulá, J. H., R. V. Helgason, and N. Venugopal. 1998. "An algorithm for identifying the frame of a pointed finite conical hull." *INFORMS J. Comput.* **10(3)**: 323–330.

Marsten, R. 1981. "The design of the XMP linear programming library." *ACM Trans. Math. Softw* **7**: 481–497.

Saunders, M. A. and M. H. Wright. 1986. "A FORTRAN package for constrained linear, least-squares, and convex quadratic programming." Technical Report SOL 86-1. Stanford, CA: Systems Optimization Laboratory, Operations Research Department, Stanford University.

Seiford, L. 1996. "Data envelopment analysis: the evolution of the state of the art (1978–1995)." *J. Productivity Anal.* **7**: 99–137.

Seiford, L. M. and R. M. Thrall. 1990. "Recent developments in DEA: the mathematical programming approach to frontier analysis." *Econometrics* **46**: 7–38.

Seiford, L. M. and J. Zhu. 1999. "Infeasibility of super-efficiency data envelopment analysis models." *INFOR* **37**: 174–187.

Thrall, R. M. 1996a. "Duality, classification and slacks in DEA." *Ann. Oper. Res* **66**: 109–138.

Thrall, R. M. 1996b. "Goal vectors for DEA efficiency and inefficiency." Working Paper No. 128. Houston, TX 77005-1892: J. H. Jones School of Administration, Rice University.

Tone, K. 1998. "A slacks-based measure of efficiency in Data Envelopment Analysis." In *GRIPS Research Report Series I-98-0001*. Saitama, Japan: National Graduate Institute for Policy Studies, Urawa.

Wets, R. J. B. and C. Witzgall. 1967. "Algorithms for frames and lineality spaces of cones." *J. Res. Nat. Bureau Standards-B* **71B(1)**: 1–7.

15 PARALLEL ALGORITHMS IN OPTIMIZATION

Yair Censor and Stavros A. Zenios

ABSTRACT Parallel computing technology is extending the size and complexity of mathematical programming models that can be represented and solved on a computer by huge amounts. This article discusses mathematical algorithms for structured optimization problems that efficiently utilize parallel architectures. Parallel algorithms are presented that exploit parallelism either due to the mathematical structure of the algorithm's operations, or due to the structure of the optimization problem at hand.

1. Introduction

As we enter the twenty-first century we are witnessing an unprecedented growth in the power of computer systems. This growth has been brought about, to a large extent, by parallel computer architectures: multiple, semiautonomous processors coordinating to obtain the solution of a single problem. The late 1980s witnessed the advent of massive parallelism whereby the number of processors brought to bear on a single problem could be several thousands. The 1990s witnessed the proliferation of high-performance workstations, high-speed communications, and the rapid growth of the Internet. Within this technological landscape the multicomputer—a computer system composed of heterogeneous, autonomous processors, linked through a network and with an architecture that is mostly transparent to the user—gained widespread acceptance and this trend is expected to continue. Parallel computers are not found solely in research laboratories and academia but are emerging, in self-organizing forms through networking, in industry and commerce.

Parallelism is improving by quantum leaps the size and complexity of models that can be represented on and solved by a computer. But these improvements are not sustained by the architecture of the machines alone. Equally important is the development of appropriate mathematical algorithms and the proper decomposition of the problem at hand, in order to effectively utilize parallel architectures. In this article we introduce methods of parallel computing for large-scale constrained optimization and structured linear programs, drawing heavily from the recent book on this topic by Censor and Zenios (1997). We take foremost an algorithmic approach presenting specific implementable algorithms from three broad families of algorithms and we discuss their suitability for parallel computations. The treatment in this article is unavoidably not exhaustive of the vast area of parallel computing methods for optimization. Additional material can be found in Censor and Zenios (1997), parallel methods for global optimization are presented in Pardalos (1992), and numerical methods for parallel and distributed computing, including extensive coverage of optimization problems, are given in Bertsekas and Tsitsiklis (1989). Parallel methods for unconstrained optimization and for the solution of nonlinear equations are found in Schnabel (1995). A snap-shot of current research in parallel optimization can be found in Butnariu, Censor and Reich (2001).

Several of the algorithms and their underlying theories have a long history that dates back to the 1920s. At that time the issue of computations—that is, implementation of the algorithm as a computer program—was irrelevant. (The calculations involved in an algorithm were in those days carried out by humans, or by humans operating mechanical calculators. The issues underlying the efficient execution of these calculations are different from the issues of computation, as we understand them today.) Here we present the algorithms not just with a view toward computation, but more importantly, toward parallel computations. It is when viewed within the context of parallel computing that these algorithms are understood to possess some important features. They are well-suited for implementation on a wide range of parallel architectures, with as few as two processors or as many as millions. It is our aim to show that parallel optimization methods lead to implementable algorithms whose efficiency is independent of the specific parallel machine. The theory underlying parallel optimization algorithms and the algorithms themselves (and, of course, the applications) are independent of the rapidly changing landscape of parallel computer architectures.

This article is organized as follows: Section 2 gives a classification of parallel algorithms, Section 3 discusses model decomposition algorithms in more detail and highlights their parallel nature. Section 4 discusses the parallelization of interior point algorithms.

2. A Classification of Parallel Algorithms

A parallel algorithm is any algorithm that can be implemented on a parallel machine in a way that results in clock-time savings when compared to its implementation, for the solution of the same problem, on a serial computer. This property of an algorithm, that is, of being parallel or not, stems from either its mathematical structure or from the features of the problem to which it is applied, or from both. Even an algorithm that is mathematically sequential can become, for certain decomposable problems, a parallel algorithm. We look first at several mathematical structures of algorithms and discuss their parallelism, and then turn to algorithms that become parallel due to problem structure.

We deal with algorithms drawn from three broad algorithmic classes: (i) iterative projection algorithms, (ii) model decomposition algorithms, and (iii) interior-point algorithms. These classes of algorithms exploit parallelism in fundamentally different ways. Algorithms in the first class do so mostly due to their mathematical structure, while algorithms in the second and third classes do so for problems with special structure. Iterative projection methods are the most flexible in exploiting parallelism since they do so by suitable reorganization of their operations. The reorganization of operations can be accomplished in many different ways, and a classification of the alternative approaches is given in this section. Model decomposition and interior point algorithms can exploit parallelism by taking advantage of the structure of the model of the problem that is being solved. A general preview of the potential parallelism of these two classes of algorithms is also given in this section.

2.1. Parallelism Due to Algorithm Structure: Iterative Projection Algorithms

We consider applications that can be modeled by a system of linear or nonlinear equations or inequalities, that is, a system of the form

$$f_i(x) * 0, \quad i = 1, 2, ..., m, \tag{1}$$

where $x \in \mathbb{R}^n$ ($\mathbb{R}^n$ is the n-dimensional Euclidean space), $f_i : \mathbb{R}^n \to \mathbb{R}$, for all i, are real-valued functions, and $*$ stands for equality signs, inequality signs, or a mixture of such. This generic description allows us to classify a broad set of iterative procedures and discuss their suitability for parallel computations. The mathematical classification of algorithms we give next depends on how each algorithm makes use of row-action or block-action iterations. This classification is purely logical and independent of the architecture of the machine on which the algorithm will be implemented or of the structure of the problem to be solved.

Solving a *feasibility problem* (1) may be attempted if the system is *feasible*, that is, if the set $\{x \in \mathbb{R}^n \mid f_i(x) * 0, i = 1, 2, ..., m\}$ is not empty, or even if the system is infeasible. The mathematical model may also be set up as an optimization problem with an objective function $f_0 : \mathbb{R}^n \to \mathbb{R}$ imposed and (1) serving as constraints. The optimization problem can be written as:

$$\text{Optimize } f_0(x) \tag{2}$$

$$\text{subject to } f_i(x) * 0, \quad i = 1, 2, ..., m. \tag{3}$$

Here optimize stands for either minimize or maximize. Additional constraints of the form $x \in Q$ can also be imposed, where $Q \subseteq \mathbb{R}^n$ is a given subset describing, for example, box constraints.

Special-purpose iterative algorithms designed to solve these problems may employ iterations that use a single row of the constraints system (1) or a group of rows in each iterative step. A row-action iteration has the functional form

$$x^{\nu+1} \doteq R_{i(\nu)}(x^\nu, f_{i(\nu)}), \tag{4}$$

where ν is the iteration index, and $i(\nu)$ is the control index, $1 \leqslant i(\nu) \leqslant m$, specifying the row that is acted upon by the algorithmic operator $R_{i(\nu)}$. The algorithmic operator generates, in some specified manner, the new iterate $x^{\nu+1}$ from the current iterate x^{ν} and from the information contained in $f_{i(\nu)}$. $R_{i(\nu)}$ may depend on additional parameters that vary from iteration to iteration, such as relaxation parameters, weights, tolerances, and so on.

The system (1) may be decomposed into M groups of constraints (called blocks) by choosing integers $\{m_t\}_{t=0}^{M}$ such that

$$0 = m_0 < m_1 < \cdots < m_{M-1} < m_M = m, \tag{5}$$

and defining for each t, $1 \leqslant t \leqslant M$, the subset

$$I_t \doteq \{m_{t-1}+1, \quad m_{t-1}+2, \ldots, m_t\}. \tag{6}$$

This yields a partition of the set:

$$I \doteq \{1, 2, \ldots, m\} = I_1 \cup I_2 \cup \cdots \cup I_M. \tag{7}$$

A block-action iteration then has the functional form

$$x^{\nu+1} \doteq B_{t(\nu)}(x^{\nu}, \{f_i\}_{i \in I_{t(\nu)}}), \tag{8}$$

where $t(\nu)$ is the control index, $1 \leqslant t(\nu) \leqslant M$, specifying the block that is used when the algorithmic operator $B_{t(\nu)}$ generates $x^{\nu+1}$ from x^{ν} and from the information contained in all rows of (1) whose indices belong to $I_{t(\nu)}$. Again, additional parameters may be included in each $B_{t(\nu)}$, such as those specified for the row-action operator $R_{i(\nu)}$. The special-purpose iterative algorithms that we consider in our classification scheme may address any problem of the form (1) or (2)–(3) and may be classified as having one of the following four basic structures.

1. *Sequential algorithms.* For this class of algorithms we define a control sequence $\{i(\nu)\}_{\nu=0}^{\infty}$ and the algorithm performs, in a strictly sequential manner, row-action iterations according to (4), from an appropriate initial point until a stopping rule applies.

2. *Simultaneous algorithms.* Algorithms in this class first simultaneously execute row-action iterations on all rows

$$x^{\nu+1,i} \doteq R_i(x^{\nu}, f_i), \quad i = 1, 2, \ldots, m, \tag{9}$$

using the same current iterate x^{ν}. The next iterate $x^{\nu+1}$ is then generated from all intermediate ones $x^{\nu+1,i}$ by an additional operation

$$x^{\nu+1} \doteq S(\{x^{\nu+1,i}\}_{i=1}^{m}). \tag{10}$$

Here S and R_i are algorithmic operators, the R_i's all being of the row-action type.

3. *Sequential block-iterative algorithms.* Here the system (1) is decomposed into fixed blocks according to (7), and a control sequence $\{t(\nu)\}_{\nu=0}^{\infty}$ over the set $\{1, 2, \ldots, M\}$ is defined. The algorithm sequentially performs, according to the control sequence, block iterations of the form (8).

4. *Simultaneous block-iterative algorithms.* In this class, block iterations are first performed using the same current iterate x^{ν}, on all blocks simultaneously

$$x^{\nu+1,t} \doteq B_t(x^{\nu}, \{f_i\}_{i \in I_t}), \quad t = 1, 2, \ldots, M. \tag{11}$$

The next iterate $x^{\nu+1}$ is then generated from the intermediate ones $x^{\nu+1,t}$ by

$$x^{\nu+1} \doteq S(\{x^{\nu+1,t}\}_{t=1}^{M}). \tag{12}$$

Here S and B_t are algorithmic operators, the B_t's being block-iterative (8).

An important generalization of block-iterative algorithms refers to variable block-iterative algorithms, wherein block iterations are performed according to formula (8). The blocks $I_{t(\nu)}$ however, are not determined *a priori* according to (7) and kept fixed, but the algorithm employs an infinite sequence $\{I_{t(\nu)}\}_{\nu=0}^{\infty}$ of nonempty blocks $I_{t(\nu)} \subseteq I$. This means that as the iterations proceed, both the sizes of the blocks and the assignment of constraints to blocks may change in a much more general manner than the scheme with fixed blocks. Of course, convergence theory of such an algorithm could still impose certain restrictions on how the sets $I_{t(\nu)}$ may be constructed.

PARALLEL COMPUTING WITH ITERATIVE ALGORITHMS

We examine two approaches for introducing parallelism in both the sequential and the simultaneous algorithms described above. We address in particular the problem of task partitioning, that is, identifying operations of the algorithm that can be executed concurrently.

A sequential row-action algorithm uses information contained in a single row $f_{i(\nu)}(x)$ *0 and the current iterate x^ν in order to generate the next iterate $x^{\nu+1}$. The control sequence $\{i(\nu)\}_{\nu=0}^{\infty}$ specifies the order in which rows are selected. Is it possible to select rows in such a way that two (or more) rows can be operated upon simultaneously without altering the mathematical structure of the algorithm? This can be achieved if at every iteration only different components of the updated vector are changed. If the algorithmic operators $R_{i(\nu)}$ and $R_{i(\nu+1)}$, which are vector-to-vector mappings, use different components of x^ν and $x^{\nu+1}$ to operate upon and leave all other components unchanged, then the operations can proceed concurrently. Identifying row-indices $i(\nu)$ so that the operators can be applied concurrently on such rows depends on the structure of the family $\{f_i(x)\}_{i=1}^m$. Several important problems have a structure that allows us to identify such row-indices. The parallel algorithm is, with this approach, mathematically identical to the serial algorithm.

The parallelism in the simultaneous algorithms is obvious. The iterations (9) can be executed concurrently using up to m parallel processors. The step (10) is a synchronization step where the processors must cooperate, and exchange information contained in the m vectors $\{x^{\nu+1,i}\}_{i=1}^m$ in order to compute the next iterate $x^{\nu+1}$.

The block-iterative algorithms—sequential or simultaneous—lead to parallel computing identical to the two ways outlined above. However, the introduction of blocks permits more flexibility in handling the task scheduling problem. The blocks can be chosen so as to ensure that all processors receive tasks of the same computational difficulty. Hence, all processors complete their work in roughly the same amount of time and delays while processors wait for each other are minimized. The specification of block-sizes will typically depend on the computer architecture. For fine-grain, massively parallel systems it is preferable to create a very large number of small blocks. Fewer, but larger, blocks are needed for coarse-grain parallel computers.

2.2. *Parallelism Due to Problem Structure: Model Decomposition and Interior-Point Algorithms*

We now consider the solution of structured optimization problems. We introduce first the following definition.

Definition 1 (Block Separability). *A function* F *is block-separable if it can be written as the sum of functions of subsets of its variables. Each subset is called a block.*

Consider, for example, the vector $\mathbf{x} \in \mathbb{R}^{nK}$, which is the concatenation of K subvectors, each of dimension n,

$$\mathbf{x} = ((x^1)^T, (x^2)^T, ..., (x^K)^T)^T,$$

where $x^k \in \mathbb{R}^n$, for all $k = 1, 2, ..., K$. (Boldface letters denote vectors in the product space $\mathbb{R}^{nK}$.) Then $F : \mathbb{R}^{nK} \rightarrow \mathbb{R}$ is block-separable if it can be written as

$$F(\mathbf{x}) \doteq \sum_{k=1}^{K} f_k(x^k),$$

where $f_k : \mathbb{R}^n \rightarrow \mathbb{R}$.

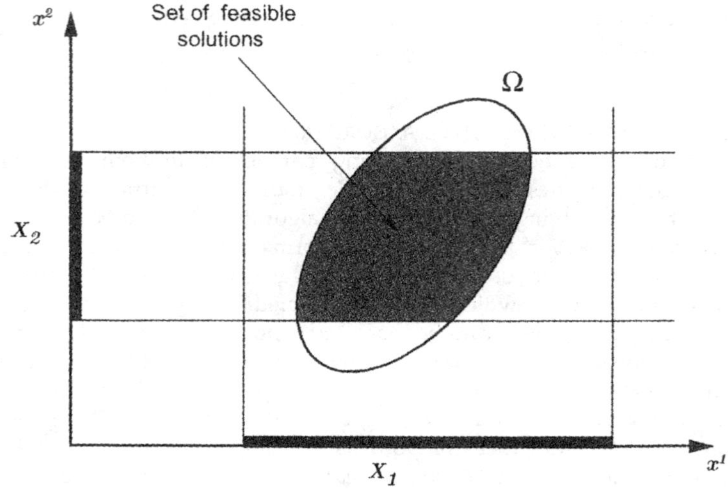

Figure 1 Constraint sets and set of feasible solutions of problem [$\mathscr{P}$] in $\mathbb{R}^2$ with $k = 2$

We now consider the following constrained optimization problem:

$$\text{Optimize } F(x) \doteq \sum_{k=1}^{K} f_k(x^k) \tag{13}$$

$$[P] \text{ subject to } x^k \in X_k \subseteq \mathbb{R}^n, \quad \text{for all } k = 1, 2, \ldots, K, \tag{14}$$

$$x \in \Omega \subseteq \mathbb{R}^{nK}. \tag{15}$$

The sets X_k, $k = 1, 2, \ldots, K$, and Ω are assumed to be closed and convex; Figure 2 illustrates the structure of this problem in two dimensions.

The block-separability of the objective function and the structure of the constraint sets of this problem can be exploited for parallel computing. Two general algorithms that exploit this structure are described below.

Model Decomposition Algorithms

If the product set $X_1 \times X_2 \times \cdots \times X_K$ of problem [P] is a subset of Ω, then the problem can be solved by simply ignoring the constraints $x \in \Omega$. A solution can be obtained by solving K independent subproblems in each of the x^k vector variables. These subproblems can be solved in parallel.

In general, however, the complicating constraints $x \in \Omega$ cannot be ignored. A model decomposition applies a modifier to problem [P] to obtain a problem [P'] in which the complicating constraints are not explicitly present. It then employs a suitable algorithm to solve [P']. Since the modified problem does not have complicating constraints it can be decomposed into K problems, each one involving only the x^k vector and the constraint $x^k \in X_k$, for $k = 1, 2, \ldots, K$. These problems are not only smaller and simpler than the original problem [P], but they are also independent of each other, and can be solved in parallel. If the solution to the modified problem [P'] is sufficiently close to a solution of the original problem [P] then the process terminates. Otherwise the current solution is used to construct a new, modified problem, and the process repeats until some stopping criterion is satisfied (see Figure 2). A wide variety of algorithms fall under this broad framework, and specific instances are discussed in Section 3.

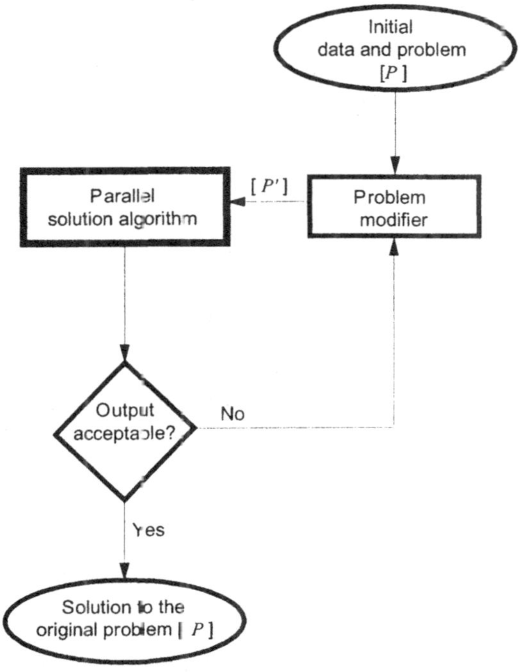

Figure 2 Structure of model decomposition algorithms: Modifications [P'] of the original problem [P] are solved repeatedly using a parallel solution algorithm. The solution of each modified problem is used to construct the next modification until an acceptable solution to the original problem is reached

Of particular interest is the application of model decomposition algorithms to problems with linear constraints, with a constraint matrix that is block-angular or dual block-angular. For example, the constraint matrix of the classic multicommodity network flow problem (Rockafellar, 1984; Censor and Zenios, 1997) is of the following, block-angular form

$$A = \begin{pmatrix} A_1 & & & \\ & A_2 & & \\ \vdots & & \ddots & \\ & & & A_K \\ I & I & \cdots & I \end{pmatrix}. \qquad (16)$$

Here $\{A_k\}$, $k = 1, 2, \ldots, K$, are $m \times n$ network flow constraints matrices, and I denotes the $n \times n$ identity matrix. The constraint set of this problem can be put in the form of problem [P] by defining X as the product set

$$X \doteq \prod_{k=1}^{K} X_k,$$

where $X_k = \{x^k \mid A_k x^k = b^k,\ 0 \leqslant x^k \leqslant u^k\}$, where b^k denotes the right-hand side of the equality constraints and u^k are vectors of upper bounds on the variables, and by defining $\Omega \doteq \{x \mid x \in X,\ Ex \leqslant U\}$, where $E = [I \mid I \mid \ldots \mid I]$. A model decomposition algorithm can now be applied to solve the multicommodity flow problem (Censor and Zenios, 1997).

The constraint matrix of linear programming formulations of two-stage stochastic programming problems (Birge and Louveaux, 1997; Censor and Zenios, 1997; Kall and Wallace, 1994) is of the dual block-angular form

$$A = \begin{pmatrix} A_0 & & & \\ T_1 & W_1 & & \\ \vdots & & \ddots & \\ T_N & & & W_N \end{pmatrix}. \tag{17}$$

The matrix A_0 is of dimension $m_0 \times n_0$. The matrices W_l are of dimensions $m_l \times n_l$, for $l = 1, 2, ..., N$. The matrices T_l are of dimensions $m_l \times n_0$. It is assumed that A_0 and W_l have full row rank, with $m_l \leqslant n_l$ for all $l = 1, 2, ..., N$, and that $n_0 \leqslant \sum_{l=1}^{N} n_l$.

The dual formulation of this problem, which is formulated using the transpose A^{T} of the constraint matrix, can be put in the form of [P] using definitions for X and Ω similar to those introduced above for the multicommodity flow problem.

INTERIOR-POINT ALGORITHMS

Consider now special cases of problem [P], when the objective function is linear or separable quadratic, and the constraints sets X and Ω are defined by linear equalities and inequalities. Then we can solve the problem directly using a general-purpose interior point algorithm for linear or quadratic programs (Kojima et al., 1991; Lustig et al., 1991; Censor and Zenios, 1997).

The major computational effort of these algorithms appears in the factorization and solution of a system of equations, for the unknown vector Δy, of the form

$$(A\Theta A^{\mathrm{T}})\Delta y = \psi, \tag{18}$$

where A is the constraint matrix of the problem, ψ is a given vector, Θ is a given diagonal matrix, and Δy denotes a dual step direction of the interior point algorithm.

When the constraint matrix A is of the structure shown in (16) it is possible to design special-purpose procedures that solve the system of equations (18) by solving subsystems involving the smaller matrices $A_1, ..., A_K$, or A_0, $W_1, ..., W_N$, $T_1, ..., T_N$ (see Section 4). These subsystems are independent of each other and can be solved in parallel. Hence, while interior-point algorithms are not parallel algorithms *per se* they can exploit parallel architectures when executing the numerical linear algebra calculations for problems with special structures.

3. Algorithms for Problem Modifiers and Solutions

It is the judicious combination of a problem modifier and a suitable algorithm for the solution of the modified problem [P'] that leads to a decomposition of the original problem [P] suitable for parallel computations. We discuss in this section modifiers and algorithms suitable for solving the modified problems.

3.1. Problem Modifiers

We now present two modifiers for problem [P], drawing on the theory of penalty and barrier methods, and on augmented Lagrangian methods (Fiacco and McCormick, 1990; Bertsekas, 1982; Censor and Zenios, 1997).

Definition 2 (Penalty Functions). *Given a nonempty set* $\Omega \subseteq \mathbb{R}^n$, *a function* $p : \mathbb{R}^n \to \mathbb{R}$ *is called an (exterior) penalty function with respect to* Ω *if the following hold: (i)* $p(x)$ *is continuous, (ii)* $p(x) \geqslant 0$, *for all* $x \in \mathbb{R}^n$, *and (iii)* $p(x) = 0$ *if and only if* $x \in \Omega$.

Definition 3 (Barrier Functions). *Given a robust set* $\Omega \subseteq \mathbb{R}^n$, *a function* $q : \mathbb{R}^n \to \mathbb{R}$ *is called a barrier function with respect to* Ω *(also called an interior penalty function) if the following hold: (i)* $q(x)$ *is continuous on int* Ω, *(ii)* $q(x) \geqslant 0$, *for all* $x \in \text{int}\,\Omega$, *and (iii)* $\lim q(x) = +\infty$, *as* x *tends to any boundary point of* Ω.

MODIFIER I: PENALTY OR BARRIER FUNCTIONS

The first modifier eliminates the complicating constraints $x \in \Omega$ by using a penalty or a barrier function. Using a penalty function $p : \mathbb{R}^{nK} \rightarrow \mathbb{R}$ with respect to the set Ω defining the complicating contraints the modified problem can be written as:

$$\text{Minimize } F(x) + cp(x) \tag{19}$$

$$[P'] \qquad \text{subject to } x^k \in X_k, \quad \text{for all } k = 1, 2, ..., K. \tag{20}$$

It is possible to construct penalty functions that are "exact", that is, for which there exists some constant $\bar{c} > 0$, such that for $c > \bar{c}$ any solution of $[P']$ is also a solution to $[P]$. Hence, a solution to $[P]$ can be obtained by solving $[P']$. Note that $[P']$ has a simpler constraint set than $[P]$ because the complicating constraints $x \in \Omega$ have been removed. However, problem $[P']$ still cannot be solved by solving K independent subproblems, since the function p is not necessarily block-separable; see Definition 1. The next subsection explores algorithms that induce separability of this function.

Consider now situations when the set Ω has a nonempty interior. Such sets arise when inequality constraints are used to define them, for example,

$$\Omega = \{x \mid g_l(x) \leqslant 0, \quad \text{for all } l = 1, 2, ..., L\}, \tag{21}$$

where $g_l : \mathbb{R}^{nK} \rightarrow \mathbb{R}$. In this case we can use a barrier function (see Definition 3) to establish a barrier on the boundary of Ω so that the iterates of an algorithm that starts with an interior point remain in the interior of the set, therefore satisfying the constraints $x \in \Omega$. For example, a barrier function for the set Ω defined by (21) can be constructed with the aid of Burg's entropy as:

$$q(x) = -\sum_{l=1}^{L} \log g_l(x),$$

where g_l for $l = 1, 2, ..., L$ are the functions used in the definition of Ω. With the use of such a barrier function the modified problem is written as:

$$\text{Minimize } F(x) + cq(x) \tag{22}$$

$$[P'] \qquad \text{subject to } x^k \in X_k, \quad \text{for all } k = 1, 2, ..., K. \tag{23}$$

A solution to the problem $[P]$ can be approximated by solving the modified problem with the barrier function for a sufficiently small value of the parameter c. It is also possible to solve the barrier-modified problem repeatedly for a sequence of barrier parameters $\{c_\nu\}$, such that $c_\nu > c_{\nu+1} > 0$. If $\{x^\nu\}$ denotes the sequence of solutions of these barrier problems, then it is known that $\{x^\nu\}$ converges to a solution of $[P]$ as $c_\nu \rightarrow 0$.

Like the penalty-modified problem $[P']$, the barrier-modified problem has a simpler constraint structure than problem $[P]$. However, it still cannot be decomposed into independent components since the barrier function is not necessarily block-separable. The algorithms of the next subsection can be used to induce separability of penalty and barrier functions.

MODIFIER II: VARIABLE SPLITTING AND AUGMENTED LAGRANGIAN

The second modifier first replicates (or splits) the components x^k of the vector x into two copies, one of which is constrained to belong to the set X_k and the other constrained to satisfy the complicating constraints. Let $z \in \mathbb{R}^{nK}$ denote the replication of x, where $z = ((z^1)^T, (z^2)^T, ..., (z^K)^T)^T$ and the vector $z^k \in \mathbb{R}^n$ for all $k = 1, 2, ..., K$. Consider now the equivalent split-variable formulation of $[P]$:

$$\text{Minimize } \sum_{k=1}^{K} f_k(x^k) \tag{24}$$

$$[\text{Split} - P] \qquad \text{subject to } x^k \in X_k, \quad \text{for all } k = 1, 2, \ldots, K, \tag{25}$$

$$z \in \Omega, \tag{26}$$

$$z^k = x^k, \quad \text{for all } k = 1, 2, \ldots, K. \tag{27}$$

The constraints $z^k = x^k$ link the variables that appear in the constraint sets X_k with the variables that appear in the set Ω. An augmented Lagrangian formulation is now used to eliminate these complicating constraints. We let $\pi \doteq ((\pi^1)^T, (\pi^2)^T, \ldots, (\pi^K)^T)^T$ where $\pi^k \in \mathbb{R}^n$ denotes the Lagrange multiplier vector for the complicating constraints $z^k = x^k$, and let $c > 0$ be a constant. Then a partial augmented Lagrangian for (24)–(27) can be written as:

$$\mathscr{L}_c(\boldsymbol{x}, \boldsymbol{z}, \boldsymbol{\pi}) = \sum_{k=1}^{K} f_k(x^k) + \sum_{k=1}^{K} \langle \pi^k, z^k - x^k \rangle + \frac{c}{2} \sum_{k=1}^{K} \| z^k - x^k \|^2. \tag{28}$$

A solution to problem [Split-P] can be obtained by solving the dual problem:

$$[P''] \ \underset{\boldsymbol{\pi} \in \mathbb{R}^{nK}}{\text{Maximize}} \ \varphi_c(\boldsymbol{\pi}), \tag{29}$$

where $\varphi_c(\boldsymbol{\pi}) \doteq \min_{x^k \in X_k, z \in \Omega} \mathscr{L}_c(\boldsymbol{x}, \boldsymbol{z}, \boldsymbol{\pi})$. This is the modified problem whose solution yields a solution of $[P]$.

An algorithm for solving convex optimization problems using augmented Lagrangians is the method of multipliers, which is an instance of the augmented Lagrangian algorithmic scheme (Bertsekas, 1982; Censor and Zenios, 1997). It proceeds by minimizing the augmented Lagrangian for a fixed value of the Lagrange multiplier vector, followed by a simple update of this vector. Using the method of multipliers to solve the dual problem $[P'']$ we obtain the following algorithm:

Algorithm 1. Method of Multipliers for Solving the Modified Problem $[P'']$.

Step 0: (Initialization.) Set $\nu = 0$. Let $\boldsymbol{\pi}^0$ be an arbitrary Lagrange multiplier vector.

Step 1: (Minimizing the augmented Lagrangian.)

$$(\boldsymbol{x}^{\nu+1}, \boldsymbol{z}^{\nu+1}) = \underset{x^k \in X_k, z \in \Omega}{\text{argmin}} \, \mathscr{L}_c(\boldsymbol{x}, \boldsymbol{z}, \boldsymbol{\pi}^\nu) \tag{30}$$

Step 2: (Updating the Lagrange multiplier vector.) For $k = 1, 2, \ldots, K$, update:

$$(\pi^k)^{\nu+1} = (\pi^k)^\nu + c((z^k)^{\nu+1} - (x^k)^{\nu+1}). \tag{31}$$

Step 3: Replace $\nu \leftarrow \nu + 1$ and return to Step 1.

The minimization problem in Step 1 has a block-decomposable constraint set. The problem, however, still cannot be decomposed into K independent subproblems since the augmented Lagrangian is not block-separable due to the cross-products $\langle z^k, x^k \rangle$ in the quadratic term of (28). The next subsection explores algorithms that induce separability of this term. Step 2 consists of simple vector operations that can be executed very efficiently on parallel architectures.

3.2. Solution Algorithms

Both modified problems $[P']$ and $[P'']$ have a block-decomposable constraint set, but the objective function is not block-separable. These problems can be written in the general form:

$$\text{Minimize } \Phi(x) \doteq \Phi(x^1, x^2, \ldots, x^K) \tag{32}$$

$$\text{subject to } x \in X \doteq X_1 \times X_2 \times \cdots \times X_K. \tag{33}$$

In this subsection we consider two solution algorithms that—when applied to problems of this form—give rise to block-separable functions and thus decompose the problem into K subproblems, which can then be solved in parallel. The first algorithm uses linear approximations to the nonlinear function Φ. These linear approximations are block-separable. The second algorithm uses a diagonal approximation for cases where Φ is a nonseparable quadratic function, such as the last summand appearing in the augmented Lagrangian (28).

Solution Algorithms Based on Linearization

One of the earliest algorithms suggested for the solution of nonlinear optimization problems using linear approximations is the Frank–Wolfe algorithm. It uses Taylor's expansion formula to obtain a first-order approximation $\hat{\Phi}$ to Φ around the current iterate x^{ν}, that is,

$$\hat{\Phi}(x) = \Phi(x^{\nu}) + \langle \nabla\Phi(x^{\nu}), x - x^{\nu} \rangle,$$

ignoring second- and higher-order terms. The Frank–Wolfe algorithm now minimizes this linear function, subject to the original constraints. The solution of this linear program, $\bar{y}$, is a vertex of the constraint set, which determines a direction of descent for the original nonlinear function, given by $p = \bar{y} - x^{\nu}$. The algorithm then performs a one-dimensional search along this direction to determine the step length where the nonlinear function attains its minimum. Figure 3 illustrates the algorithm.

Solution Algorithms Based on Diagonalization

Consider now a special structure of the objective function (32) arising from the modified problem $[P'']$. In particular, we assume that $\Phi(\cdot)$ can be written as

$$\Phi(x, z) \doteq \sum_{k=1}^{K} \|x^k - z^k\|^2. \tag{34}$$

This is the structure of the quadratic terms of the augmented Lagrangian (28). (We consider only the quadratic terms of the augmented Lagrangian, since these are the nonseparable terms that prevent us from decomposing the minimization of Step 1 of Algorithm 1 into K independent subproblems.) We will approximate this nonseparable function using a separable quadratic function. The term *diagonal quadratic approximation* is also used, which indicates that the Hessian matrix of $\Phi(x, z)$ is approximated by a diagonal matrix.

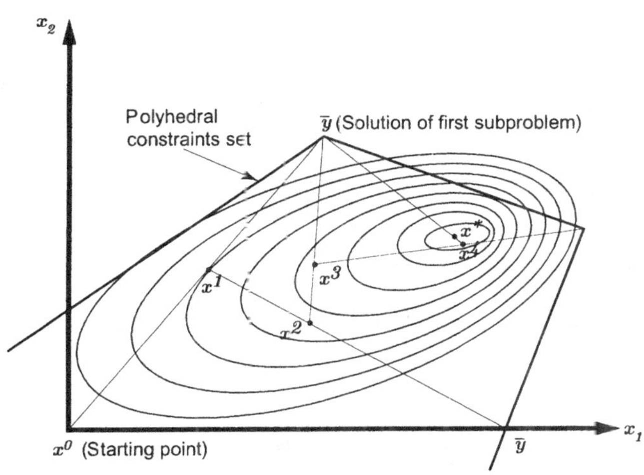

Figure 3 Illustration of the Frank–Wolfe linearization algorithm for a function $\Phi(x_1, x_2)$. The curves denote level sets

The terms in (34) can be expanded as

$$\|x^k - z^k\|^2 = \|x^k\|^2 + \|z^k\|^2 - 2\langle x^k, z^k\rangle,$$

and we only discuss the cross-product terms $\langle x^k, z^k\rangle$ for $k = 1, 2, ..., K$. Using Taylor's expansion formula we obtain a first-order approximation of the cross-product terms around the current iterate (x^ν, z^ν) as:

$$\langle x^k, z^k\rangle \approx \langle x^k, (z^k)^\nu\rangle - \langle (x^k)^\nu, (z^k)^\nu\rangle + \langle (x^k)^\nu, z^k\rangle,$$

for $k = 1, 2, ..., K$. With this approximation of its cross-product terms, the function (37) is approximated by the expression:

$$\sum_{k=1}^{K} \|x^k - (z^k)^\nu\|^2 - \sum_{k=1}^{K} \|(x^k)^\nu - (z^k)^\nu\|^2 + \sum_{k=1}^{K} \|(x^k)^\nu - z^k\|^2.$$

A solution algorithm, based on diagonalization solves problem $[P'']$ using the method of multipliers (Algorithm 1) but instead of minimizing the augmented Lagrangian in Step 1 it minimizes the diagonal quadratic approximation. This approximation is block-separable into the variable blocks x^k, $k = 1, 2, ..., K$, and the minimization is decomposed into K independent problems. These problems can be solved in parallel.

4. Parallel Matrix Factorization Procedures for the Interior-Point Algorithm

We turn now to the parallelization of interior point algorithms. In particular, we consider the solution of the linear systems (18) for structured problems. We assume that the constraint matrix A has the block-angular structure (17), which arises in the deterministic equivalent formulation of two-stage stochastic programs with recourse and in the dual formulation of multicommodity network flow problems. The stochastic programming problems and multicommodity network flow problems encountered in practical applications are of extremely large size, and matrices of this form occasionally have millions of columns and hundreds of thousands of rows. Furthermore, while these matrices are sparse the product matrix $A\Theta A^T$ could be completely dense due to the presence of the matrices T_l, $l = 1, 2, ..., N$.

This section develops a matrix factorization procedure for solving the linear systems of equations (18) for the calculation of Δy, that exploits the special structure of the constraint matrix A. The vectors Δy and ψ in equation (18) are written as concatenations of subvectors, that is, as $[(\Delta y^0)^T, (\Delta y^1)^T, ..., (\Delta y^N)^T]^T$ and $[(\psi^0)^T, (\psi^1)^T, ..., (\psi^N)^T]^T$, respectively, with Δy^l, $\psi^l \in \mathbb{R}^{m_l}$ for $l = 0, 1, ..., N$. Θ is assumed to be diagonal, as is the case for linear programs and for separable quadratic programs.

4.1. The Matrix Factorization Procedure for the Dual Step Direction Calculation

The procedure for solving (18) is based on the use of a generalized version of the Sherman–Morrison–Woodbury formula. This formula is stated in the following lemma, where I denotes the identity matrix.

Lemma 1. *For matrices* $A \in \mathbb{R}^{n\times n}$, $U \in \mathbb{R}^{n\times K}$, *and* $V \in \mathbb{R}^{n\times K}$ *such that both* A *and* $(I + V^T A^{-1} U)$ *are invertible,*

$$(A + UV^T)^{-1} = A^{-1} - A^{-1}U(I + V^T A^{-1} U)^{-1} V^T A^{-1}.$$

Proof. *See, for example, Householder (1975) or Censor and Zenios (1997, p. 228).*

The following theorem provides the foundation for the matrix factorization routine that solves system (18) for Δy.

Theorem 1. *Let* $M \doteq A\Theta A^T$, *where* Θ *is a diagonal matrix, and let* $S \doteq diag(S_0, S_1, S_2, ..., S_N)$, $S_l \doteq W_l \Theta_l W_l^T \in \mathbb{R}^{m_l\times m_l}$, $l = 1, 2, ..., N$, *where the matrix* $S_0 \doteq I$ *is an* $m_0 \times m_0$ *identity matrix, and for* $l = 0, 1, 2, ..., N$, $\Theta_l \in \mathbb{R}^{n_l \times n_l}$ *is the diagonal submatrix of* Θ *corresponding to the lth block. Also, denoting* $(\Theta_0^{-1})^2$ *by* Θ_0^{-2}, *let*

$$G_1 \doteq \Theta_0^{-2} + A_0^{\mathrm{T}} A_0 + \sum_{l=1}^{N} T_l^{\mathrm{T}} S_l^{-1} T_l, \tag{35}$$

$$G \doteq \begin{pmatrix} G_1 & A_0^{\mathrm{T}} \\ -A_0 & 0 \end{pmatrix}, \quad U \doteq \begin{pmatrix} A_0 & I \\ T_1 & 0 \\ \vdots & \vdots \\ T_N & 0 \end{pmatrix}, \quad V \doteq \begin{pmatrix} A_0 & -I \\ T_1 & 0 \\ \vdots & \vdots \\ T_N & 0 \end{pmatrix}.$$

If A_0 and $W_l, l = 1, 2, \ldots, N$, have full row rank then the matrices M and $G_2 \doteq -A_0 G_1^{-1} A_0^{\mathrm{T}}$ are invertible, and

$$M^{-1} = S^{-1} - S^{-1} U G^{-1} V^{\mathrm{T}} S^{-1}. \tag{36}$$

Proof. *See Birge and Qi (1988) or Censor and Zenios (1997, p. 229).*

It is easy to verify, using equation (36), that the solution of the linear system $(A\Theta A^{\mathrm{T}})\Delta y = \psi$ is given by $\Delta y = p - r$ where p solves the system $Sp = \psi$, and r is obtained from the system

$$Gq = V^{\mathrm{T}} p \text{ and } Sr = Uq. \tag{37}$$

The vector p can be computed componentwise by solving $S_l p^l = \psi^l$, for $l = 1, 2, \ldots, N$. In order to solve for q we exploit the block structure of G and write:

$$Gq = \begin{pmatrix} G_1 & A_0^{\mathrm{T}} \\ -A_0 & 0 \end{pmatrix} \begin{pmatrix} q^1 \\ q^2 \end{pmatrix} = \begin{pmatrix} \hat{p}^1 \\ \hat{p}^2 \end{pmatrix}, \quad \text{where} \quad \begin{pmatrix} \hat{p}^1 \\ \hat{p}^2 \end{pmatrix} \doteq V^{\mathrm{T}} p. \tag{38}$$

Hence, we get

$$q^2 = -G_2^{-1}(\hat{p}^2 + A_0 G_1^{-1} \hat{p}^1), \tag{39}$$

$$q^1 = G_1^{-1}(\hat{p}^1 - A_0^{\mathrm{T}} q^2). \tag{40}$$

Once q is known, r can be computed componentwise from (40). The procedure for calculating Δy can be summarized as follows. (We adopt here the notation $A_{i\cdot}$ for the ith row of A, and $A_{\cdot j}$ for the jth column of A.)

Procedure 1. Matrix factorization for dual step direction calculation.

Step 1: Solve $Sp = \psi$.

Step 2: (Solve $Gq = V^{\mathrm{T}} p$.)

a. For all $l = 1, 2, \ldots, N$, solve $S_l (u^l)^i = (T_l)_{\cdot i}$, for $(u^l)^i$, $i = 1, 2, \ldots, n_0$ thus computing the columns of the matrix $S_l^{-1} T_l$.
b. For all $l = 1, 2, \ldots, N$ multiply $T_l^{\mathrm{T}} (u^l)^i$, for $i = 1, 2, \ldots, n_0$ to form $T_l^{\mathrm{T}} S_l^{-1} T_l$. Form G_1 (eq. (35)). Compute $\hat{p}^1$, $\hat{p}^2$ (eq. (38)).
c. Solve $G_1 u = \hat{p}^1$ for u and set $v = \hat{p}^2 + A_0 u$ (eq. (39)).
d. Form G_2 by solving $G_1 w^i = (A_0^{\mathrm{T}})_{\cdot i}$ for w^i, for $i = 1, 2, \ldots, m_0$, and setting $G_2 = -A_0 [w^1 \; w^2 \; \cdots \; w^{m_0}]$.
e. Solve $G_2 q^2 = -v$ for q^2, and solve $G_1 q^1 = \hat{p}^1 - A_0^{\mathrm{T}} q^2$ for q^1, (eq. (40)).

Step 3: (Solve $Sr = Uq$.) Set $r^0 = A_0 q^1 + q^2$ and for all indices $l = 1, 2, \ldots, N$ solve $S_l r^l = T_l q^1$ for r^l.

Step 4: (Form Δy.) For $l = 1, 2, \ldots, N$, set $\Delta y^l = p^l - r^l$.

Decompositions for Parallel Computing

Procedure 1 is well-suited for parallel implementation because it relies largely on matrix (sub)block computations that can be performed independently of one another. The computation begins with the submatrices T_l, W_l and Θ_l and the vector segment ψ^l located at the lth processor. Processor l can compute S_l and proceed independently with all computations involving only local data.

Interprocessor data communication is necessary at only three instances in the parallel algorithm. After forming the terms $T_l^T S_l^{-1} T_l$ in Steps 2a–2b of Procedure 1 in parallel, the processors communicate to form the matrix G_1 and the vectors $\hat{p}^1$ and $\hat{p}^2$ in Step 2b. The results can be accumulated at a single processor, designated as the master. The computations involving the dense matrix G_1 can be done serially on the master processor. Steps 2d–2e, which involve the dense matrix G_2, are also done serially on the master processor.

The master processor must then broadcast the computed vector q to all other nodes. The remaining Steps 3 and 4 require only the distributed data S_l, T_l, and p^l on the lth processor and may be carried out with full parallelism. A final communication step accumulates the partial vectors Δy^l at the master processor. This vector can then be made available to all processors for use in the calculation of the primal step and the slack step directions, respectively, Δx and Δz of an interior point algorithm, see, e.g. Censor and Zenios (1997, Algorithm 8.2.1).

An alternative parallel implementation of Procedure 1 would distribute the matrices G_1 and G_2 to all processors and let the processors proceed locally (and redundantly) with all calculations involving these matrices. The approach based on a master processor described above uses an all-to-one communication step to accumulate the dense matrices at the master, followed by a one-to-all communication step to distribute the results to the processors. The alternative approach suggested here combines these two communication steps into a single all-to-all communication step that distributes the dense matrices to all processors. Both of these alternatives are very efficient on present day distributed memory machines with high-bandwidth communication networks.

Yet another alternative approach is to distribute the dense matrices across processors and use parallel dense linear algebra techniques for all calculations that involve these matrices. This approach is more suitable to shared-memory, tightly coupled multiprocessors, or when the dense matrices G_1 and G_2 are large.

5. Notes and References

The material of this chapter is based on the book by Censor and Zenios (1997). The development of programs for parallel machines is discussed by Carriero and Gelernter (1992). For alternative viewpoints on parallel computing in general, see Buzbee and Sharp (1985), Deng et al. (1992), and the special issue of the journal *Daedalus* (1992). A collection of references on parallel computing has been compiled by the Association for Computing Machinery (1991). Parallel computer architectures are treated by Hwang and Briggs (1984) and Hwang (1987). Kumar et al. (1994) also contains material on hardware aspects. A historical note on the development of parallel architectures is given by Desrochers (1987). A taxonomy of parallel architectures based on the interaction of instructions and data streams is due to Flynn (1972).

For perspectives on parallel computing for optimization problems see Kindervater et al. (1989), and Zenios (1994). Advanced treatment of numerical methods for parallel and distributed computing, including extensive coverage of optimization problems, is given in Bertsekas and Tsitsiklis (1989). Special issues of journals focusing on parallel optimization were edited by Meyer and Zenios (1988), Mangasarian and Meyer (1988, 1991, 1994), Rosen (1990). Pardalos et al. (1992) treat some aspects of mathematical programming with respect to their potential for parallel implementations. Extensive lists of references on parallel computing for optimization are collected in Schnabel (1985) and Zenios (1989).

Methods of parallel computing for global optimization problems are collected in the volume by Pardalos (1992), see also Pardalos (1986). A discussion of parallel methods for unconstrained optimization and for the solution of nonlinear equations is found in Schnabel (1995).

The use of array processors for linear programming was first discussed by Pfefferkorn and Tomlin (1976), and the use of vector computing for the solution of large-scale nonlinear optimization problems by Zenios and Mulvey (1986, 1988).

Extensive discussion on the use of block-separability for the solution of large-scale optimization problems can be found in Griewank and Toint (1982). Block-separability was exploited, in the context of parallel computing, by Lescrenier (1988), Lescrenier and Toint (1988), and Zenios and Pinar (1992). Problems with block- or dual-block angular structure appear in many application domains.

For a general introduction to interior-point algorithms see Goldfarb and Todd (1989), Kojima et al. (1991), and Marsten et al. (1990). For a general discussion of direct matrix inversion methods see Householder (1975). Parallel implementation of interior point algorithms for dense problems is discussed in Qi and Zenios (1994) and Eckstein et al. (1992). The exploitation of matrix structure in computing the dual step direction for interior point algorithms has been addressed in Choi and Goldfarb (1993) and Czyzyk et al. (1995). The use of the Sherman–Morrison–Woodbury updating formula for computing dual steps for stochastic programs was suggested by Birge and Qi (1988), and was further extended by Birge and Holmes (1992) who also performed extensive numerical experiments with this method. The extension to multistage stochastic programming problems is discussed in Holmes (1993). Its use for solving multicommodity network flow problems has been suggested by Choi and Goldfarb (1990, 1993). The parallel matrix factorization procedure and its implementation on hypercubes and other parallel machines was developed by Jessup et al. (1994). Yang and Zenios (1996) discuss the application of the parallel matrix factorization procedure within an interior point algorithm for the solution of large-scale stochastic programs.

References

Association for Computing Machinery. 1991. *Resources in Parallel and Concurrent Systems*. ACM Press.

Bertsekas, D. P. 1982. *Constrained Optimization and Lagrange Multipliers Methods*. New York: Academic Press.

Bertsekas, D. P. and J. N. Tsitsiklis. 1989. In *Parallel and Distributed Computation: Numerical Methods*. Englewood Cliffs, NJ: Prentice-Hall.

Birge, J. R. and D. F. Holmes. 1992. "Efficient solution of two-stage stochastic linear programs using interior point methods." *Comput. Optim. Appl.* **1**: 245–276.

Birge, J. R. and F. Louveaux. 1997. *Introduction to Stochastic Programming*. Heidelberg: Springer-Verlag.

Birge, J. R. and L. Qi. 1988. "Computing block-angular Karmarkar projections with applications to stochastic programming." *Mgmnt Sci.* **34**: 1472–1479.

Buzbee, B. L. and D. H. Sharp. 1985. "Perspectives on supercomputing." *Science* **227**: 591–597.

Carriero, N. and D. Gelernter. 1992. *How to Write Parallel Programs: a First Course*. Cambridge, MA: The MIT Press.

Censor, Y. and S. A. Zenios. 1997. *Parallel Optimization: Theory, Algorithms, and Applications*. New York, NY: Oxford University Press.

Choi, I. and D. Goldfarb. 1990. "Solving multicommodity network flow problems by an interior point method." In *Large Scale Numerical Optimization*, edited by T. Coleman and Y. Li, pp. 58–69. Philadelphia: SIAM.

Choi, I. C. and D. Goldfarb. 1993. "Exploiting special structure in a primal–dual path-following algorithm." *Math. Programming* **58**: 33–52.

Czyzyk, J., R. Fourer, and S. Mehrotra. 1995. "A study of the augmented system and the column-splitting approaches for solving two-stage stochastic linear programs by interior point methods." *ORSA J. Comput.* **7**: 474–490.

Daedalus. 1992. Journal of the American Academy of Arts and Sciences.

Deng, Y., J. Glimm, and D. H. Sharp. 1992. "Perspectives on parallel computing." *Daedalus* **121**: 31–52.

Desrochers, G. R. 1987. *Principles of Parallel and Multi-Processing*. New York: McGraw-Hill.

Eckstein, J., R. Qi, V. I. Ragulin, and S. A. Zenios. 1992. "Data-parallel implementations of dense linear programming algorithms." Report 92-05-06. Decision Sciences Department, The Wharton School, University of Pennsylvania, Philadelphia, PA 19104.

Fiacco, A. V. and G. P. McCormick. 1990. *Nonlinear programming: sequential unconstrained minimization techniques. Classics in Applied Mathematics,* Volume 4. Philadelphia: Society for Industrial and Applied Mathematics (Republication of the work first published by Wiley, New York, 1968).

Flynn, M. J. 1972. "Some computer organizations and their effectiveness." *IEEE Trans. Comput.* **C-21**: 948–960.

Goldfarb, D. and M. J. Todd. 1989. "Linear programming." In *Handbooks in Operations Research and Management Science,* volume 1, edited by G. L. Nemhauser, A. H. G. Rinnooy Kan, and M. J. Todd, pp. 73–170. Amsterdam: North-Holland.

Griewank, A. and Ph. L. Toint. 1982. "On the unconstrained optimization of partially separable functions." In *Nonlinear Optimization 1981,* edited by M. J. D. Powell, pp. 301–312. New York: Academic Press.

Holmes, D. 1993. "An explicit factorization for solving multistage stochastic linear programs using interior point methods." Report 93-18. Department of Industrial and Operations Engineering, The University of Michigan, Ann Arbor, MI.

Householder, J. S. (1975). The Theory of Matrices in Numerical Analysis, New York: Dover Publications.

Hwang, K. 1987. *Advanced Computer Architecture: Parallelism, Scalability, Programmability*. New York: McGraw-Hill.

Hwang, K. and F. A. Briggs. 1984. *Computer Architecture and Parallel Processing*. New York: McGraw-Hill.

Jessup, E. R., D. Yang, and S. A. Zenios. 1994. "Parallel factorization of structured matrices arising in stochastic programming." *SIAM J. Optim.* **4**: 833–846.

Kall, P. and S. W. Wallace. 1994. *Stochastic Programming*. New York: John Wiley & Sons.

Kindervater, G. A. P., J. K. Lenstra, and A. H. G. Rinnooy Kan. 1989. "Perspectives on parallel computing." *Oper. Res.* **37**: 984–990.

Kojima, M., N. Megiddo, T. Noma, and A. Yoshise. 1991. *A unified approach to interior point algorithms for linear complementarity problems. Lecture Notes in Computer Science*. New York: Springer-Verlag.

Kumar, V., A. Grama, A. Gupta, and G. Karypis. 1994. *Introduction to Parallel Computing*. Redwood City, CA: The Benjamin/Cummins Publishing Company, Inc.

Lescrenier, M. 1988. "Partially separable optimization and parallel computing." *Ann. Oper. Res.* **14**: 213–224.

Lescrenier, M. and Ph. L. Toint. 1988. "Large scale nonlinear optimization on the FPS164 and CRAY X-MP vector processors." *Int. J. Supercomputer Appl.* **2**: 66–81.

Lustig, I. J., R. E. Marsten, and D. F. Shannon. 1991. "Computational experience with a primal–dual interior point method for linear programming." *Linear Algebra Appl.* **152**: 191–222.

Mangasarian, O. L. and R. R. Meyer. (eds). 1988. "Parallel optimization." *Math. Programming, Series B.* **42**, Issue 2.

Mangasarian, O. L. and R. R. Meyer. (eds). 1991. "Parallel optimization II." *SIAM J. Optim.* **1**, Issue 4.

Mangasarian, O. L. and R. R. Meyer. (eds). 1994. "Parallel optimization III." *SIAM J. Optim.* **4**, Issue 4.

Marsten, R., R. Subramanian, M. Saltzman, and D. Shanno. 1990. "Interior point methods for linear programming: Just call Newton, Lagrange, Fiacco and McCormick!" *Interfaces* **20(4)**: 105–116.

Meyer, R. R. and S. A. Zenios. (eds). 1988. *Parallel optimization on novel computer architectures. Ann. Oper. Res.* **14**. Basel, Switzerland: J.C. Baltzer Scientific Publishing Co.

Pardalos, P. M. 1992. 1986. *Proccedings of the 24th Annual Allerton Conference on Communication, Control and Computing,* pp. 812–821. Urbana: University of Illinois.

Pardalos, P. M. (ed). 1992. *Advances in Optimization and Parallel Computing*. Amsterdam: North-Holland.

Pardalos, P. M., A. T. Phillips, and J. B. Rosen. 1992. *Topics in Parallel Computing in Mathematical Programming*. New York: Science Press.

Pfefferkorn, C. E. and J. A. Tomlin. 1976. "Design of a linear programming system for the Illiac IV." Technical report. Department of Operations Research, Stanford University, Stanford, CA.

Qi, R-J. and S. A. Zenios. 1994. "On the scalability of data-parallel decomposition algorithms for stochastic programs." *J. Parallel Distributed Comput.* **22**: 565–570.

Rockafellar, R. T. 1984. *Network Flows and Monotropic Programming*. New York: John Wiley and Sons.

Rosen, J. B. (ed). *Supercomputers and Large-Scale Optimization: Algorithms, Software and Applications,* Volume 22, Annals of Operations Research. Switzerland: J.C. Baltzer Scientific Publishing Co.

Schnabel, R. B. 1985. "Parallel computing in optimization." *Computational Mathematical Programming,* edited by K. Schittkowski, pp. 357–382. Berlin: Springer-Verlag.

Schnabel, R. B. 1995. "A view of the limitations, opportunities and challenges in parallel nonlinear optimization." *Parallel Comput.* **21**: 875–905.

Yang, D. and S. A. Zenios. 1997. "A scalable parallel interior point algorithm for stochastic linear programming and robust optimization." *Computational Optimization and Applications* **7**: 143–158.

Zenios, S. A. and J. M. Mulvey. 1986. "Nonlinear network programming on vector supercomputers: a study on the CRAY X-MP." *Oper. Res.* **34**: 667–682.

Zenios, S. A. and J. M. Mulvey. 1988. "Vectorization and multitasking of nonlinear network programming algorithms." *Math. Programming* **42**: 449–470.

Zenios, S. A. 1989. "Parallel numerical optimization: Current status and an annotated bibliography." *ORSA J. Comput.* **1(1)**: 20–43.

Zenios, S. A. 1994. "Parallel and super-computing in the practice of management science." *Interfaces* **24(5)**: 122–140.

Zenios, S. A. and M. C. Pinar. 1992. "Parallel block-partitioning of truncated newton for nonlinear network optimization." *SIAM J. Sci. Stat. Comput.* **13(5)**: 1173–1193.

16 RANDOMIZATION IN DISCRETE OPTIMIZATION: ANNEALING ALGORITHMS

Sanguthevar Rajasekaran

ABSTRACT Annealing algorithms have been employed extensively in the past decade to solve myriads of optimization problems. Several intractable problems such as the traveling salesman problem, graph partitioning, and circuit layout have been solved to get satisfactory results. In this article we survey convergence results known for annealing algorithms. In particular we deal with simulated annealing and nested annealing.

Simulated Annealing (SA) is a randomized heuristic that can be used to solve any combinatorial optimization problem. SA is typically used to produce quasi-optimal results. In practice SA has been applied to solve some presumably hard (e.g., NP-hard) problems. The level of performance obtained has been promising (Golden and Skiscim, 1986; ElGamal and Shperling, 1984; Johnson et al., 1989; Vecchi and Kirkpatrick, 1985). The success of this heuristic technique has motivated the study of convergence of this technique. One of the early results in this direction is due to Mitra et al. (1986) who proved that SA converges in the limit to a globally optimal solution with probability 1. Later results proved certain time bounds within which SA is guaranteed to converge with high probability. In this chapter we provide one such proof (due to Rajasekaran, 2000).

Nested Annealing (NA) (Rajasekaran and Reif, 1988) is a variation of SA that has been proven to perform better for optimization problems for which the cost function has some special properties (Rajasekaran and Reif, 1988). Herein we provide a summary of NA and its convergence properties.

1. Introduction

SA is a heuristic technique proposed by Kirkpatrick et al. (1983) for solving combinatorial optimization problems such as the traveling salesman problem, graph partitioning, and so on. This randomized algorithm was obtained in analogy with a physical system, for example, a fluid. A fluid can be brought to a low energy state using annealing. SA uses a procedure similar to annealing to compute the minimum value of a given objective function. In particular, SA employs the Metropolis algorithm for the computer simulation of annealing.

Annealing a substance involves melting the substance at a very high temperature and then cooling it slowly. At each temperature the system should reach a steady state and hence enough time should be given. The lower the temperature, the higher the time given should be. SA imitates this process using the Metropolis' algorithm.

Each basic step of Metropolis' simulation perturbs the value of one of the variables by a small amount. The resultant configuration will be accepted if it has a lower cost. Even if the new configuration has a higher cost, it will be accepted with certain probability. This enables the system to hill-climb from a locally optimal state.

SA is the repeated application of the above basic step until no more improvement in the cost function is possible.

SA can be modeled as a time inhomogeneous Markov chain. Mitra et al. (1986) have employed such a model to prove the convergence of SA. This convergence proof involved proving that the probability state vector of the Markov chain converges in the limit to an optimal stationary probability state vector with probability 1. An optimal stationary probability state vector is one in which the only nonzero entries correspond to globally optimal states of the Markov chain. There are also results that compute the rate of convergence of the probability state vector to the optimal vector. Such results are extremely important from a theoretical point of view since they provide an explanation for why SA works in practice.

The result of Rajasekaran (2000) shows that SA converges in time $M(n, d, D)$ with probability $\geqslant (1 - n^{-\Omega(1)})$ for some appropriate function $M(.)$. Here n is the number of states, D is the diameter, and d is the degree of the underlying Markov chain. Convergence in this context means that the Markov chain has been in a globally optimal state at least once. Perhaps this notion of convergence is more relevant than the earlier notions, from a computing point of view. By keeping track of the state with the minimum cost visited so far by the Markov chain the globally optimal solution can be found.

It can be shown that the performance of SA can be improved when the cost functions have some special properties. For cost functions that are small separable, faster convergence has already been proved by Rajasekaran and Reif (1988). They call their algorithm nested annealing (abbreviated as NA). A simpler convergence proof for NA has been given in Rajasekaran (2000). We describe this proof in this chapter.

NA employs the following idea. For any optimization problem, we can define a graph. If this graph is known to be small separable then NA leads to faster convergence. Several other algorithms have also been designed that exploit the small separability of the underlying graphs (see Lipton and Tarjan, 1980). For a given arbitrary problem, the separability of the underlying graph may not be known. In this chapter we provide tight bounds on the separability of a random graph (as given in Rajasekaran, 2000). These separability bounds are then used to analyze the expected behavior of NA on an arbitrary optimization problem.

2. Preliminaries

SA is a family of randomized algorithms. Classical approaches to introducing randomness in algorithms are based on certain assumptions on the input distribution. Based on these assumptions the average run times of algorithms are computed. When the assumptions made do not hold, the analysis fails. Good performance can still be achieved without making assumptions on input distributions, by introducing randomness in the algorithm itself (see Rabin, 1976; Solovay and Strassen, 1977). An algorithm employing this technique is called a randomized algorithm.

A randomized algorithm is one which makes coin flips to make certain decisions. Such an algorithm is typically shown to have a certain performance measure with high probability (this probability will be over the space of all possible outcomes for coin flips made in the algorithm and not over the space of all possible inputs).

We say a randomized algorithm has a resource (like time, space, etc.) bound of $\tilde{O}(f(n))$ if there exists a constant c such that the resource used by the algorithm is no more than $c\alpha f(n)$ on any input of size n, with probability $\geqslant(1-n^{-\alpha})$.

By an optimization problem we mean the problem of computing the minimum value of a given objective function or cost function $C(.)$ of n parameters $p_1, p_2, \ldots, p_n$, subject to some given constraints.

For any nonnegative random variable X with mean μ, Markov's inequality (see Feller, 1966) states that Prob. $[X \geqslant k\mu] \leqslant 1/k$, for any $k>0$.

3. Modeling Simulated Annealing

Here we state the mathematical model to be employed for SA.

Given a cost function $C(p_1, p_2, \ldots, p_n)$, a configuration or state of the optimization problem (OP) is defined to be an assignment of values to its n parameters. Neighbors of a given state are all those states that differ from the given state only in the value of one parameter by a small amount. Temperature in the case of an OP is simply a control parameter that has the units as that of the cost function.

```
procedure Anneal (start_state, start_temperature);
    while (not frozen) do
        while (not steadystate) do
            Generate a random neighbor of the current state;
            Let Δ = cost of old state—cost of the new state;
            Accept the new state with probability min{1,exp(−ΔT)}, T
            being the current temperature;
```

```
Update the temperature; steady state = false;
```

In the above algorithm, frozen is a boolean variable that takes the value *true* when no improvement in the given cost function has been observed in a long time (e.g., during the past few temperatures). The boolean variable steady state becomes *true* when the system attains steady state at the given temperature.

Several schemes (also known as annealing schedules) have been employed to derive the sequence of temperatures the system goes through. For example, the temperature can be decreased by a constant factor each time. Other annealing schedules can be found, for example, in Mitra et al. (1986).

For any given OP, we can define a directed graph $G(V, E)$ as follows: Nodes in G are simply the states of OP and the edges going out of any node (or state) are to the neighbors of this state. SA can be thought of as a random walk on this graph. State transition made at any step is dependent only on the current state. For a given temperature, the transition probabilities from any node are fixed. These facts suggest modeling SA as a Markov chain. SA can be modeled as a time homogeneous Markov chain at any given temperature. However, the transition probabilities might change with temperature. Thus the whole of SA can be modeled as a time inhomogeneous Markov chain (Mitra et al., 1986). We assume G is strongly connected (i.e., there is a directed path from i to j for any two nodes i and j in V).

For any node i in G, let $N(i)$ stand for the set of neighbors of i. Let $C(i)$ be the cost of state i. Under the assumption that, when the Markov chain is in state i, each one of its $|N(i)|$ neighbors is equally likely to be generated next, the transition probability from state i to state j at temperature T, $P_{ij}(T)$, is given by

$$P_{ij}(T) = \begin{cases} 0 & \text{if } j \notin N(i) \,\&\, j \neq i \\ \dfrac{1}{|N(i)|} \min(1, \exp\{(C(j) - C(i))/T\}) & \text{if } j \in N(i) \end{cases}$$

and

$$P_{ii}(T) = 1 - \sum_{j \in N(i)} P_{ij}(T)$$

We make use of this model to prove the convergence time of SA (Rajasekaran, 2000).

4. Convergence of Simulated Annealing

SA is said to have converged if the underlying Markov chain has been in a globally optimal state at least once. Another way to define convergence is to require that the probability state vector reaches an optimal vector (see Mitra et al., 1986). Since our goal is to find an optimal solution, our definition of convergence is more appropriate.

Now we describe the convergence proof of (Rajasekaran, 2000). Given a cost function $C(p_1, p_2, \ldots, p_n)$, let $G(V, E)$ be its state graph. Let T denote the minimum temperature that SA was ever in. Also let $\Delta = \max_{i \in v, j \in N(i)} \{C(i) - C(j)\}$. Denote the degree and diameter of $G(V, E)$ by d and D, respectively.

For any $i \in V, P_{ij}$ will be at least $\exp(-\Delta/T)$ if $j \in N(i)$ (for any temperature). The following facts are needed.

Fact 1. *Let* X *be the state of a Markov chain at time* $t = t_0$. *The probability that a global minimum state is visited during the next* q *steps (for any* q*) is dependent only on* X *and* q *and not on the states visited before.*

Lemma 1. *Let* X *be any state in* V. *The expected number of steps before a global optimal state is visited starting from* X *is* $\leqslant \left(\frac{1}{d}\exp(-\Delta/T)\right)^{-D}$.

Proof. *If* g *is any globally optimal state, then there exists a directed path from* X *to* g *in* G(V, E) *of length* $q \leqslant D$. *Let* $e_1, e_2, \ldots, e_q$ *be the sequence of edges in the path. The* *probability that g is visited starting from X is at least the probability that each one of the edges* e_i, $1 \leqslant i \leqslant q$ *is traversed in succession. The later probability is at least*

$$\left[\left(\frac{1}{d}\right)\exp(-\Delta/T)\right]^q \geqslant \left[\left(\frac{1}{d}\right)\exp(-\Delta/T)\right]^D,$$

(under the assumption that each neighbor of a state is equally likely to be generated next).

Thus the probability that g will ever be visited starting from X is $\geqslant [(1/d)\exp(-\Delta/T)]^D$. *This implies that the expected number of steps before g is visited is* $\leqslant [d\exp(\Delta/T)]^D$.

Theorem 1. *SA converges in time* $\leqslant 2k[d\exp(\Delta/T)]^D$, *with probability* $\geqslant (1-2^{-k})$, *independent of the start state.*

Proof. *Let* $E = 2[d\exp(\Delta/T)]^D$. *We can prove that the probability of a global optimal state g not being visited in kE steps is* $\leqslant 2^{-k}$, *by induction on k.*

Induction Hypothesis. *Irrespective of the start state, probability that g is not visited in kE steps is* $\leqslant 2^{-k}$.

Base case. *When* $k = 1$, *for any start state X, the expected number of steps before g is visited is* $\leqslant E/2$ *(using Lemma 1). An application of Markov's inequality implies that the probability of g not being visited starting from X in E steps is* $\leqslant 1/2$.

Induction step. *Assume the hypothesis for all* $k \leqslant (r-1)$. *We'll prove the hypothesis for* $k = r$.

Let $X_E, X_{2E}, \ldots, X_{(r-1)E}$ *be the states of the Markov chain during time steps* $E, 2E, \ldots, (r-1)E$, *respectively. Let A denote the event that g is not visited during the first E steps, and B denote the event that g is not visited during the next* $(r-1)E$ *steps.*

Probability that g is not visited in rE steps, P, is given by

$$P = Prob.[B/A] \times Prob.[A].$$

Using Fact 1, Prob.[B/A] depends only on what state the Markov chain is in at time step E and the time duration $(r-1)E$. *And hence,*

$$P = Prob.[A]\sum_{i\in V} Prob.[B/X_E = i] \times Prob.[X_E = i].$$

But, Prob.[A] is $\leqslant 1/2$ *and* $Prob.[B/X_E = i]$ *is* $\leqslant 2^{-(r-1)}$ *for each* $i \in V$ *(using the induction hypothesis). Therefore, we have,*

$$P \leqslant \tfrac{1}{2}2^{-(r-1)} = 2^{-r}.$$

Corollary 1. *If* Δ, *T, and d are assumed to be constants and* $D = \theta(\log|V|)$, *then, SA converges in polynomial time in* $|V|$ *with probability* $\geqslant (1-2^{-\Omega(|V|)})$.

Note that the above analysis is oblivious to the annealing schedule used.

SA is typically used to obtain quasi-optimal solutions by running it for only a small amount of time. The solutions obtained have been very satisfactory.

The above proof assumes that each neighbor of a state is equally likely to be generated next. This proof can be extended to the case where this assumption is invalid as follows.

In some applications, it may be necessary to generate certain neighbors with higher probability. Mitra et al. (1986) assume that the probability of generating state j from state i is $\frac{g(i)}{g(j)}$, where $g(i, j)$ is the weight of j as a neighbor of i and $g(i)$ is a normalizing function such that $\sum_{j\in N(i)} g(i,j) = g(i)$. Under this assumption the state transition probabilities become:

$$P_{ij}(T) = \begin{cases} 0 & \text{if } j \notin N(i) \,\&\, j \neq i \\ \dfrac{g(i,j)}{g(i)}\min(1, \exp\{(C(j)-C(i))/T\}) & \text{if } j \in N(i) \end{cases}$$

and

$$P_{ii}(T) = 1 - \sum_{j \in N(i)} P_{ij}(T).$$

If $p = \min_{i \in V, j \in N(i)} \frac{g(i,j)}{g(i)}$, for the above general model we can prove a convergence result similar to the one given by Theorem 1. We can prove the following

Theorem 2. *The SA algorithm converges in time* $\leqslant 2k\left[\frac{1}{p}\exp(\Delta/T)\right]^{D}$ *with probability* $\geqslant (1-2^{-k})$, *no matter what the start state is.*

5. Nested Annealing

Rajasekaran and Reif (1988) have shown that if the cost function being optimized is small separable, then SA can be modified to get faster convergence. The modified algorithm is called Nested Annealing (NA). A simple proof of convergence for NA (Rajasekaran, 2000) is described below.

5.1. *Some Definitions*

A graph $G(V, E)$ with n nodes is said to be $s(n)$–separable if there exist constants $\alpha < 1$, $\beta > 0$ such that V can be partitioned into three subsets V_1, S, V_2 such that no vertex in V_1 is adjacent to any vertex in V_2, both $|V_1|$ and $|V_2|$ are less than αn, and $|S|$ is less than $\beta s(n)$. Also, the induced subgraphs of G on V_1, and on V_2 are $s(|V_1|)$–separable and $s(V_2)$–separable, respectively. S will be referred to as the separator set or simply the separator. Intuitively, by eliminating (the nodes in) S from G we end up with two roughly equal disjoint subgraphs.

If C is a cost function on n parameters $p_1, p_2, ...p_n$, we can define the separability of C as follows. Write Cas $C = C_1 + C_2 + \cdots + C_k$, where each C_i is a product of functions of the parameters. Call each $C_i, 1 \leqslant i \leqslant k$ as a clause. Define a bipartite graph $G_C(V, E)$ whose nodes are the parameters and the clauses. There is an edge between a clause node and a parameter node if that parameter occurs in that clause. G_C is called the graph of C. We say C is $s(n)$-separable if G_C is.

As an example, consider the problem of CNF-satisfiability where the input is a boolean formula in conjunctive normal form, F, on n variables. The problem is to decide if there is an assignment to the variables that makes F *true*. The graph corresponding to this problem will consist of nodes, one for each variable and each clause. There is an edge between a clause node and a variable node if, and only if, that variable occurs in that clause.

Small separability (i.e., small $s(n)$) of a cost function implies that by assigning values to a small number of parameters we can obtain two independent subproblems such that the parameters involved in one subproblem are disjoint from the parameters of the other subproblem.

5.2. *The Algorithm*

Given a cost function C on the n parameters $p_1, p_2, ..., p_n$, construct the graph of C, $G_C(V, E)$. If G_C is $s(n)$–separable, we can partition V into V_1, S, and V_2 as explained above.

Assume that each parameter is binary. The minimum value of C can be found as follows. For each possible assignment of values to parameters in S, find the minimum value of C, and pick the minimum of these minima. Finding the minimum of C under a particular assignment for S, is now easy. We need to find the minimum of two functions C_1 and C_2, where C_1 involves only parameters from V_1, and C_2 involves only parameters from V_2.

Let $G_1(V_1, E_1)$ and $G_2(V_2, E_2)$ be the restrictions of G on V_1 and V_2, respectively. Finding the minimum of C_1 and C_2 can be done recursively by finding separators for G_1 and G_2, respectively.

At the top level, we are given a set S for which we need to find an optimum assignment (an assignment that corresponds to a global minimum for C). We can think of this as an OP on $|S|$ parameters. There are thus $2^{|S|} \leqslant 2^{\beta s(n)}$ states of the OP. The cost of each state is the minimum of C under that particular assignment to S. Instead of considering each possible state of this OP, and computing the cost of each state, we can run a simulated annealing algorithm on this OP with $\leqslant \beta s(n)$ parameters.

If we employ SA on the above OP with $|S|$ parameters, the number of states visited will be much less than $2^{\beta s(n)}$ and hence the run time of the overall recursive algorithm will be much less. This is the whole idea behind nested annealing.

NA can be specified as follows:

```
procedure Nested_Annealing (G_C(V, E));
  Find a separator set S for G_C. Let V_1, S, and V_2 be the partition of V.
  Also let G_1 and G_2 be restrictions of G on V_1 and V_2, respectively.
  Find an optimal assignment for S by running an SA algorithm on these
  parameters.
  For each state of the corresponding Markov chain visited by SA we need to
  compute the cost.
  To compute this cost, we need to find minimum of two other functions C_1 and C_2
  (see the discussion above). Each of C_1 and C_2 involves ≤αn parameters.
  Find these two minima recursively by finding separators for G_1 and G_2,
  respectively.
```

An analysis of the above follows. Let $T(n)$ be the expected run time of NA for finding a globally optimal solution to a given OP with n parameters. To obtain an upper bound on $T(n)$, we need to know how many of the $2^{|S|}$ states at the top level will be visited (including repeated visits) by the corresponding SA algorithm, and on each state visited the time needed to compute the cost of the state. Let the worst case run time of SA on an OP with N parameters be $2^{M(n)}$. The number of states visited by NA at the top level will be no more than $2^{M(\beta s(n))}$.

Computing the cost of each state involves computing the minimum of two other functions involving no more than αn parameters each, accounting for a total expected cost of $\leqslant 2\ T(\alpha n)$.

As a result we have,

$$T(n) \leqslant 2^{M(\beta s(n))} 2T(\alpha n),$$

which solves to

$$T(n) \leqslant 2 \sum_{i=1}^{\log n} M(\beta s(\alpha^i n)).$$

If $s(n)$ is assumed to be $O(n^\sigma)$ for some $\sigma < 1$, we have

$$T(n) \leqslant 2^{\gamma M(\beta s(n))} = 2^{O(M(\beta s(n)))}.$$

Here $\gamma \leqslant \frac{1}{1-\sqrt{\alpha}}$. Throughout we have used the fact that the expected value of the sum of any random variables is the sum of the expected values of the individual random variables.

Let $L = 2^{\gamma M(\beta s(n))}$. The probability that the run time of NA exceeds kL is less than $1/k$, using Markov's inequality. Thus we have the following

Theorem 3. *Nested Annealing converges in time* $\leqslant n^\alpha L$ *with probability* $\geqslant (1 - n^{-\alpha})$.

For problems with $M(n) = O(n)$, we have the following

Corollary 2. *If* $M(n)$ *is* $O(n)$, *Nested_Annealing converges in time* $\leqslant n^\alpha 2^{O(s(n))}$ *with probability* $\geqslant (1 - n^{-\alpha})$.

The above corollary has already been proven in Rajasekaran and Reif (1988).

We can strengthen Theorem 2 in the following way. Probability that the convergence time of NA exceeds $2L$ is $\leqslant 1/2$. Make $k \log n$ independent runs of the procedure and terminate each procedure after exactly $2L$ steps. Pick the minimum of the minima found in the log n runs. The probability that none of the runs finds the global minimum is $\leqslant 2^{-k}$. Thus we have the following

Theorem 4. *Nested_Annealing converges in time* $\leqslant 2\mathrm{kL}$ *with probability* $\geqslant (1-2^{-\mathrm{k}})$ *(for any* $\mathrm{k}>0$*).*

A number of important problems like planar traveling salesman, planar satisfiability, and so on (which have been proven to be NP-complete) have $s(n) = \sqrt{n}$. For all these problems NA converges in time $2^{O[M(\beta\sqrt{n})]}$, whereas SA has a convergence time of $2^{O(M(n))}$.

6. Separability of Random Graphs

Algorithms that exploit the separability of the underlying graphs presuppose that a separator is known for the given graph. Algorithms have been devised for finding separators of restricted classes of graphs. For instance efficient algorithms exist for computing the separator set of planar graphs (Lipton and Tarjan, 1980). No such algorithms are known for general graphs. In fact, deciding if a given graph is $s(n)$-separable is NP-hard (Garey and Johnson, 1979). In this section we state tight bounds that are known for the separability of random graphs. These bounds can be used to study the expected behavior of NA on arbitrary OPs.

6.1. *A Modification of Nested Annealing*

The separability results of Rajasekaran (2000) indicate that random graphs are not small separable. Let $G_C(V, E)$ be the graph of a given OP. If G_C is $s(n)$-separable, then V can be partitioned into V_1, S, and V_2 such that $|S| = s(n)$ and there is no $V_1 - V_2$ edge. Moreover, $|V_1|$ and $|V_2|$ are less than αn for some $\alpha < 1$. If the cardinality of the separator set, S, is itself a constant fraction (or more) of $|V_1|$ and $|V_2|$, then there is no gain in running all the levels of recursion of the NA algorithm (given in Section 5.2). It may suffice to stop the procedure at the top level. We can modify NA in the following way: replace the instructions that call for computing the minimum of C_1 and C_2 recursively, with instructions to compute these two minima using SA.

This modified NA will have an expected convergence time of $2^{M(|S|)}[2^{M(|V_1|)} + 2^{M(|V_2|)}] = O(2^{M(|S|+\max[|V_1|,|V_2|])})$. Here $2^{M(n)}$ stands for the convergence time of SA on an OP with n parameters.

If the separability of the given OP is not known this modified NA together with the procedure for finding a separator in a random graph (given in Rajasekaran, 2000) can be used to solve the OP. In this section we state expected bounds on the convergence time of this modified NA on an arbitrary OP. The expected convergence of the original NA will be nearly the same since a random graph is not small separable.

6.2. *The Separator Theorems*

There are two popular models of random graphs (Bollobás, 1985). The first consists of all graphs with n vertices and $M(n)$ edges (for some specified $M(n)$), each such graph having equal probability. A member of this model is denoted as $G_{M(n)}$ or simply G_M. The second model consists of all graphs with n nodes in which each of the $n^2/2$ possible edges is chosen independently with probability p. A member of this model is denoted as G_p.

We assume the second model. A random bipartite graph in the second model is denoted as $G_p(A, B, E)$ where both A and B have n nodes each and each possible edge is chosen independently with probability p. For any set of nodes X, we let $\Gamma(X)$ stand for the neighbors of X.

The following separability results are proven in Rajasekaran (2000).

Theorem 5. *Let* p *be* $> \frac{100}{\mathrm{n}}$. *Then, almost every* $\mathrm{G_p(V,E)}$ *has the following property. If* X *is any subset of nodes of* $\mathrm{G_p}$ *with* $\frac{1}{\mathrm{p}}$ *nodes, then, the set* $\mathrm{T_X} = \{\mathrm{y} \in \mathrm{V}-\mathrm{X} : \Gamma(\mathrm{y}) \cap \mathrm{X} = \Phi\}$ *has at least* $\frac{1}{\mathrm{p}}$ *nodes.*

Theorem 6. *Modified NA converges in an expected* $O(2^{M(n-(1/p))})$ *time on an arbitrary OP with* n *parameters, given that the graph of the OP is a member of* G_p.

Theorem 7. *Let* $\frac{1}{p} \rightarrow \infty$. *For almost every* $G_p(V, E)$, *the following holds. If* X *is any set of* δn *nodes (for any* δ*), then the cardinality of* $T_X = \{y \in V - X : \Gamma(y) \cap X = \Phi\}$ *is at the most* $\left[\left(\frac{2-\delta}{\delta \log e}\right) + \varepsilon\right]\frac{1}{p}$, *for any constant* $\varepsilon > 0$.

Corollary 3. *If* X *is any set of* δn *nodes (where* δ *is a constant), then,* T_X *has cardinality O(1/p).*

7. Conclusions

In this chapter we have described SA and NA which are randomized algorithms that can be used to solve any combinatorial optimization problem. NA is a variation of SA that exploits the separability of the underlying graphs. We have summarized convergence results known for SA and NA. Separability results for random graphs have been stated and applied to study the expected behavior of SA and NA.

References

Bollabás, B. 1985. *Random Graphs*. Academic Press.

ElGamal, A. and I. Shperling. 1984. "Design of good codes via simulated annealing." In *Workshop on Statistical Physics in Engineering and Biology*.

Feller, W. 1966. *An introduction to Probability Theory and its Applications*, volume I and II. John Wiley and Sons.

Garey, M. and D. S. Johnson. 1979. *Computers and Intractability: a Guide to the Theory of NP-Completeness*. W.H. Freeman & Co.

Golden, B. L. and C. C. Skiscim. 1986. "Using simulated annealing to solve routing and location problems." *Naval Res. Logistics Q.* **33**: 261–279.

Johnson, D. S., C. R. Aragon, L. A. McGeoch, and C. Schevon. 1989. "Optimization by simulated annealing: an experimental evaluation (Part I)." *Oper. Res.* **37**: 865–892.

Kirkpatrick, S., C. D. Gelatt, and M. P. Vecchi. 1983. "Optimization by simulated annealing." *Science* **220**: 671–680.

Lipton, R. J. and R. E. Tarjan. 1980. "Applications of a planar separator theorem." *SIAM J. Comput.* **9**: 615–627.

Mitra, D., F. Romeo, and A. Sangiovanni-Vincentelli. 1986. "Convergence and finite-time behavior of simulated annealing." *Adv. Appl. Probability* **18**: 747–771.

Rabin, M. O. 1976. "Probabilistic algorithms." In *Algorithms and Complexity*, edited by J. F. Traub, pp. 21–36. Academic Press.

Rajasekaran, S. 2000. "Simulated annealing and nested annealing." *J. Global Optim.* **16**: 43–56.

Rajasekaran, S. and J. H. Reif. 1988. "Nested annealing: a provable improvement to simulated annealing." *Lecture Notes Comput. Sci.* **317**: 455–472.

Solovay, R. and V. Strassen. 1977. "A fast Monte-Carlo test for primality." *SIAM J Comput.* **6**: 84–85.

Vecchi, M. P. and S. Kirkpatrick. 1985. Global wiring by simulated annealing. Technical report IBM Thomas J. Watson Research Center.

PART TWO

APPLICATIONS

17 PROBLEM TYPES

17.1 Optimization and Heuristics in Scheduling

Chung-Yee Lee and Michael Pinedo

ABSTRACT This chapter gives an overview of the optimization techniques and heuristics that are commonly used in scheduling. The first section describes a framework and introduces the notation for the scheduling models considered. The following section discusses priority rules, also known as dispatching rules. The third, fourth, and fifth sections describe how optimization techniques are applied to scheduling, in particular dynamic programming and integer programming with branch and bound. The next two sections give an overview of the best-known heuristic procedures, namely decomposition techniques and local search methods. The final section contains a number of remarks with regard to areas of applications and comparative studies that have been done.

1. Introduction

One of the more important application areas of applied optimization is concerned with the scheduling of jobs. The importance of scheduling is based on its wide applicability in manufacturing as well as in services. The field of scheduling itself can be subdivided into various different subfields, for example, project scheduling, job shop scheduling, workforce scheduling, and so on.

In this chapter we will focus mainly on job shop scheduling, which at times is also referred to as machine scheduling. In a job shop there are typically m machines and n jobs. The processing time of job j on machine i is p_{ij}. If job j needs processing on only one machine, and its processing time is independent of the machine on which it is processed, then its processing time is denoted by p_j. If job j has to undergo processing on various machines, we say that job j consists of a number of operations and operation (i, j) then refers to the processing of job j on machine i. Job j may have a priority or weight w_j. The weight w_j of a job could be viewed as the cost incurred for keeping the job in the system for each unit of time. Job j may also have a release date r_j and a due date d_j. The processing of a job cannot start before its release date and should be completed preferably before its due date. It is possible to complete a job after its due date, but then a penalty is incurred.

A job shop scheduling problem is usually completely characterized by the machine environment, the objective function, and the processing restrictions and constraints. In the remaining part of this section we elaborate on these characteristics.

There are various machine environments that are of interest. The most basic machine environment is the single machine. All jobs have to be processed once and only once on this machine. A more general machine environment is a setting with m identical machines in parallel. In this setting each job again has to be processed exactly once and any machine will do. A third machine environment is the so-called flow shop environment. In this environment each job has to undergo processing on each one of the m machines. Each job has to be processed first on machine 1; after it has completed its processing on machine 1, it has to proceed to machine 2; after completion of its processing there it has to go to machine 3; and so on. Since the routes of all the jobs in a flow shop are identical, one can set the machines up in series. A fourth machine environment is a generalization of the flow shop, which is called the job shop. Again each job has to be processed on each one of the machines, but now each job has its own route and the routes may be different.

In deterministic scheduling there are numerous performance measures of interest. In this chapter we consider the following objective functions.

1. The makespan. The makespan is denoted by $C_{\max}$ and refers to the time it takes to complete all n jobs, that is, $C_{\max} = \max(C_1, ..., C_n)$.
2. The total weighted completion time, which is denoted by $\sum w_j C_j$.

3. The maximum lateness, which is denoted by L_{max}. The maximum lateness is formally defined as $L_{max} = \max(L_1, ..., L_n)$, where $L_j = C_j - d_j$.
4. The total weighted tardiness, which is denoted by $\sum w_j T_j$, where the tardiness T_j of job j is defined as $T_j = \max(L_j, 0) = \max(C_j - d_j, 0)$.

The processing of the jobs may be subject to several types of restrictions and constraints. In this chapter we only consider the restrictions and constraints that are the most common. In some environments preemptions of jobs being processed are not allowed, whereas in other environments preemptions are allowed. When a job that is being processed is preempted by another job, then the preempted job is put back in the queue of jobs that are waiting for that machine. In this chapter we assume that the amount of work already done on a preempted job is not lost; when the preempted job is put back on the machine its processing resumes where it had left off when it was preempted.

In a single or a parallel machine environment there may be precedence constraints. These constraints may specify that a job only may start with its processing after a given set of other jobs have been completed. The precedence constraints may be depicted by a precedence graph.

In job shops there are typically routing constraints that specify the route each job has to follow through the shop.

2. Priority Rules

A dispatching rule prioritizes all the jobs that are waiting for processing on a machine. The prioritization scheme may take into account the jobs' attributes, the machines' attributes as well as the current time. Whenever a machine has been freed, a dispatching rule selects, among the jobs waiting, the job with the highest priority. Research in dispatching rules has been active for several decades and many different rules have been developed and studied in the literature. There are many basic dispatching rules (see Pinedo, 1995). In what follows we list a few of the most important ones.

The earliest release date first (ERD) rule. This rule is equivalent to the well-known first-come-first-served rule. This rule minimizes the variation in the waiting times of the jobs at a machine.

The earliest due date first (EDD) rule. Whenever a machine is freed the job with the earliest due date is selected to be processed next. This rule tends to minimize the maximum lateness among the jobs waiting for processing. Actually, in a single machine setting, with n jobs available at time zero, the EDD rule does minimize the maximum lateness.

The minimum slack first (MS) rule is a variation of the EDD rule. If a machine is freed at time t the remaining slack of each job at that time, defined as $\max(d_f - p_j - t, 0)$, is computed. The job with the minimum slack is scheduled next. This rule tends to minimize due date related objectives.

The weighted shortest processing time first (WSPT) rule. Whenever a machine is freed, the job with the highest ratio of weight w_j over processing time p_j is scheduled next, that is, the jobs are ordered in decreasing order of w_j/p_j. This rule tends to minimize the weighted sum of the completion times, that is, $\sum w_j C_j$. In a single machine setting, with n jobs available at time zero, the WSPT rule does minimize the $\sum w_j C_j$. When all the weights are equal the WSPT rule reduces to the shortest processing time first (SPT) rule.

The critical path (CP) rule is used with operations subject to precedence constraints. It selects as the next job the one at the head of the longest string of processing times in the precedence constraint graph.

Table 1 gives a summary of these dispatching rules. A number of these rules result in optimal schedules in certain machine environments and are reasonable heuristics in others. All of these rules have variations that can be applied in more complicated settings.

The basic dispatching rules described above all run in $O(n \log n)$ time, that is, they are fast. However, they are of limited use. When a complex objective has to be minimized it is not likely that any of these dispatching rules will perform satisfactorily.

Table 1 A Summary of the Dispatching Rules

	Rule	Data	Objectives
Rules dependent on release dates and due dates	ERD	r_j	Variance in throughput times
	EDD	d_j	Maximum lateness
	MS	d_j	Maximum lateness
Rules dependent on processing times	SPT	p_j	Sum of completion times, WIP
	WSPT	p_j, w_j	Weighted sum of completion times, WIP
	CP	p_j, *prec*	Makespan

3. Dynamic Programming

Scheduling problems can often be solved optimally by various types of mathematical programming approaches, such as dynamic programming, integer programming, and branch and bound. Among them, dynamic programming is probably the most popular one, especially for small- and medium-size problems. The dynamic programming approach has often been applied to problems that are pseudo-polynomially solvable.

Basically, dynamic programming is a total enumeration method. It uses a divide-and-conquer approach and first finds optimal solutions to subproblems. Then, based on the principle of optimality, it solves the problem recursively. An algorithm is typically characterized by the initial condition(s), recursive equation(s), and an optimal solution. A time complexity analysis typically yields a measure of efficiency of the dynamic programming. The justification is usually based on the principle of optimality.

In this section, we give an example of one polynomial and two pseudopolynomial dynamic programming algorithms.

Consider the following batch processing problem that occurs in semiconductor burn-in operations (Lee et al., 1992; Chandru et al., 1993). There are n jobs to be processed on a single machine and the machine can process up to B jobs simultaneously. The time required for processing those jobs simultaneously is the maximum of the processing requirements of the jobs. Assume that the job sequence is already fixed. Our objective is to partition jobs into batches so that the sum of completion times is minimized. Note that jobs in the same batch will have the same completion time. We use a backward dynamic programming approach to solve the problem. Assume that we index jobs from the end of the sequence. Let $f(j)$ be the minimum objective value, given that we have already sequenced jobs $1, ..., j$.

Initial conditions:

$f(0) = 0$

Recursive equation:

For $j = 1, ..., n$,

$$f(j) = \min_{1 \leqslant i \leqslant \max\{B,j\}} \{f(j-i) + j \max\{p_k : k = j-i+1, ..., j\}\}$$

Optimal value:

$f(n)$

The time complexity of this algorithm is $O(nB)$, where $B \leqslant n$, and hence the complexity is not worse than $O(n^2)$. So the algorithm solves the problem optimally in polynomial time. The justification of this algorithm is based on the fact that the last batch can contain i jobs where $1 \leqslant i \leqslant \max\{B, j\}$. The term $j \max\{p_k : k = j-i+1, ... j\}$ in the recursive equation is the cost increase due to the last batch which contains the last i jobs, that is, job $j-i+1$ up to job j.

Our second and third examples of dynamic programming algorithms are pseudopolynomial algorithms. Consider a classical scheduling problem with m machines in parallel and the objective of minimizing total weighed completion time. We demonstrate two different dynamic programming approaches: forward and backward. Note that in most combinatorial problems, solving a problem by forward dynamic programming results in the same time complexity as by the backward approach. Interestingly, in this particular problem, these two approaches result in different time complexities. The forward approach by Rothkopf (1966) and Lawler and Moore (1969) has time complexity $O(nmT^{m-1})$, where T is the sum of all processing times. The backward approach, by Lee and Uzsoy (1992), has time complexity $O(nmW^{m-1})$, where W is the sum of all weights.

Note that in an optimal solution, jobs must be processed in weighted shortest processing time (WSPT) order on each machine.

3.1. Forward Approach

Re-index the jobs in increasing order of p_j/w_j. Let $T = \sum_{j=1}^{n} p_j$ and define $f(j, t_1, ..., t_m)$ as the minimum total weighted completion time after assigning jobs $1, ..., j$ with the total processing time on machine i being t_i, $i = 1, ..., m$.

Initial conditions:

$f(0, 0, ..., 0) = 0,$

$f(j, t_1, ..., t_m) = \infty$ if any $t_i < 0$ or if $\sum t_i > T$.

Recursive equation:

$$f(j, t_1, ..., t_m) = \min_{1 \leqslant i \leqslant m} \{f(j-1, t_1, ..., t_i - p_j, ..., p_m) + w_j t_i\} \quad \text{for } j = 1, ..., n; \quad t_i = 0, ..., T,$$

where

$i = 1, ..., m.$

Optimal value:

$\min\{f(n, t_1, ..., t_m) | \sum t_i = T\}.$

The justification for this algorithm is based on the fact that each job j can be assigned to any one of the m machines and the increase in the objective value is $w_j t_i$ if it is assigned to machine i. The time complexity of this algorithm is $O(nmT^m)$. Hence it is a function of T and the algorithm can solve the problem optimally in pseudopolynomial time. Note that the time complexity can be improved to $O(nmT^{m-1})$ because the time assigned to the mth machine can be calculated easily once the total time assigned to the first $m-1$ machines is specified.

3.2. Backward Approach

Re-index the jobs in decreasing order of p_j/w_j. We assign jobs from right to left. Let $W = \sum_{j=1}^{n} w_j$ and define $f(j, W_1, ..., W_m)$ as the minimum total weighted completion time after assigning jobs $1, ..., j$ with the total weight on machine i being W_i, $i = 1, ..., m$.

Initial conditions:

$f(0, 0, ..., 0) = 0,$

$f(j, W_1, ..., W_m) = \infty$ if any $W_i < 0$ or if $\sum W_i > W$.

Recursive equation:

$$f(j, W_1, ..., W_m) = \min_{1 \leqslant i \leqslant m} \{f(j-1, W_1, ..., W_i - w_j, ..., W_m) + W_i p_j\}$$

for $j = 1, ..., n;$ $W_i = 0, ..., W$ where $i = 1, ..., m.$

Optimal value:

$\min\{f(n, W_1, ..., W_m) | \sum W_i = W\}.$

The justification of this algorithm is based on the fact that each job j can be assigned to any one of m machines. Since we assign jobs from right to left, the increase in the objective value is $W_i p_j$ if job j is assigned to machine i. The time complexity of the algorithm is $O(nmW^m)$. Similar to the argument in the forward approach, the time complexity can be improved to $O(nmW^{m-1})$.

Remark. *In many industrial applications, weights are assigned based on the relative priorities of the jobs. Namely, a manager may prioritize jobs based on their relative importance and assign each one a weight between 1 and n, with n being the highest priority. In such a case,* $W = n(n+1)/2$, *which is polynomial in n and is smaller than* T *in most cases. Thus the complexity* $O(nmW^{m-1})$ *is usually smaller than* $O(nmT^{m-1})$. *For example, for the two machine problem* $nmW^{m-1} = 2nW$. *Hence, if* $W = n(n+1)/2$, *the time complexity is* $O(n^3)$.

4. Integer Programming

Linear programming (without integrality constraints) is not widely applicable in the scheduling field. It can be applied to some problems with preemptions and some problems with all jobs having identical processing requirements (see Pinedo, 1995, p. 72). Integer programming is much more widely applicable and has become more and more popular recently for solving large-scale and strongly NP-hard problems. See Blazewicz et al. (1991) and Pinedo (1995) for a detailed survey. Here we provide two integer programming formulations for a parallel machine scheduling problem with the jobs subject to precedence constraints.

Formulation 1

Let

$x_{ij}^t = 1$ if job j is processed on machine i with starting time t, and zero otherwise,

p_{ij} = processing time of job j if it is processed on machine i.

T = planning horizon

Min $f(C_1, C_2, \ldots, C_n)$

$$\text{Subject to} \quad \sum_{t=0}^{T-1} \sum_{i=0}^{m} x_{ij}^t = 1 \quad j = 1, \ldots, n, \tag{1}$$

$$\sum_{j=1}^{n} \sum_{s=\max\{t-p_{ij},0\}}^{t-1} x_{ij}^s \leqslant 1 \quad i = 1, \ldots, m; \quad t = 1, \ldots, T, \tag{2}$$

$$\sum_{t=0}^{T-1} \sum_{i=1}^{m} (t + p_{ij}) x_{ij}^t = C_j \quad j = 1, \ldots, n, \tag{3}$$

$$x_{lk}^t \leqslant \sum_{i=1}^{m} \sum_{s=0}^{t-p_{ij}} x_{ij}^s, \quad \text{all } j \text{ and } k \text{ such that } j \rightarrow k, \quad l = 1, \ldots, m, \tag{4}$$

$$x_{ij}^t = 0 \text{ or } 1 \quad \text{all } i, j \text{ and } t \tag{5}$$

Equation (1) guarantees that any given job, starting at any particular time t, can be processed on only one machine. Equation (2) ensures that for each machine at any particular time, there is at most one job that will be processed by that machine. Equation (3) enforces the completion time C_j. Equation (4) ensures the job precedence constraints are satisfied. The objective function $f(C_1, C_2, \ldots, C_n)$ can be makespan, total weighted completion time, or a due date related performance measure.

In the formulation above, the binary decision variables are a function of t, which can be very large. In the following, we provide another integer programming formulation where the binary variables are a function of n.

Formulation 2

Let

$x_{ij} = 1$ if job j is processed by machine i, and zero otherwise,

$y_{jk} = 1$ if job j is assigned before job k, and zero otherwise, for those j and k $(j, k = 1, \ldots, n)$ without given precedence constraints,

$M =$ a sufficiently large number.

Min $f(C_1, C_2, \ldots, C_n)$

subject to

$$\sum_{i=1}^{m} x_{ij} = 1 \qquad j = 1, \ldots, n \tag{6}$$

$$C_k - \sum_{i=1}^{m} x_{ik} p_{ik} \geqslant C_j \qquad j,\ k \text{ with } j \rightarrow k \tag{7}$$

$$y_{jk} + y_{kj} = 1 \quad \text{no precedence constraint between job } j \text{ and job } k \tag{8}$$

$$C_k - x_{ik} p_{ik} \geqslant C_j - (2 - x_{ik} - x_{ij} + y_{kj})M \quad i = 1, \ldots, m; \quad \text{no precedence constraint between job } j \text{ and job } k \tag{9}$$

$$C_j \geqslant \sum_{i=1}^{m} x_{ij} p_{ij} \qquad j = 1, \ldots, n \tag{10}$$

$$x_{ji} = 0 \text{ or } 1 \qquad j = 1, \ldots, n; \quad i = 1, \ldots, m \tag{11}$$

$$y_{jk} = 0 \text{ or } 1 \qquad j, k = 1, \ldots, n. \tag{12}$$

Equation (6) guarantees that there is exactly one machine that will process job j. Equation (7) ensures that if it is given that job $j \rightarrow$ job k, job k cannot start before job j is completed. Now suppose that there is no precedence constraint between job j and job k. In such a case, we may have $y_{lk} = 1$ or $y_{kl} = 1$ (but not both) by equation (8). Since there is no precedence constraint between job j and job k, the only chance that these two jobs will interact with each other is if they are both assigned to the same machine, say machine i. In such a case, the job assigned later, say job k, cannot start before the completion of the previously assigned job, say job j. In this situation $y_{kj} = 0$ (and $y_{jk} = 1$) and $x_{ik} = x_{ij} = 1$ for some i in equation (9). In turn, this implies that $C_k - x_{ik} p_{ik} \geqslant C_j$, and hence the starting time of job k cannot be earlier than the completion time of job j. In all other cases, the right-hand side of equation (9) will be a large negative number and equation (9) will automatically be satisfied. Equation (10) provides a lower bound on the completion time for each job j.

Although Formulation 2 looks much more compact than Formulation 1, it can be shown that the linear programming relaxation of Formulation 1 provides a good approximation. The linear programming relaxation of Formulation 1 usually generates stronger lower bounds and hence solutions generated from it are more useful in constructing good feasible solutions to the scheduling problem.

The use of large problem formulations in general has a long history (see Gilmore and Gomory, 1961, 1963; Nemhauser and Wolsey, 1988; Pochet and Wolsey, 1988; Wolsey, 1989; Boyd, 1992). However, the use of large problem formulations to solve machine scheduling problems is relatively new (Luh and Hoitomt, 1993). Recently, a column generation approach, to be discussed next, has been shown to be an effective tool for solving certain large-scale scheduling problems.

5. Branch and Bound: Set Partitioning and Column Generation

In this section, we describe branch-and-bound algorithms based on column generation, which have been developed recently for certain scheduling problems. Column generation methods have been used in the literature to solve large-scale combinatorial optimization problems, including vehicle routing (Desrochers et al., 1992), cutting stock (Vance et al., 1994), capacitated lot sizing (Cattrysse et al., 1993), air crew scheduling (Lavoie et al., 1988), graph coloring (Mehrotra and Trick, 1996), and the generalized assignment problem (Savelsbergh, 1997). Recently, it has been successfully applied to large-scale scheduling problems; see, for example, Van den Akker et al. (1999), Chen and Powell (1998, 1999a, 1999b), Lee and Chen (2000), and Chen and Lee (1999). Chan et al. (1997) provide an error bound analysis of column generation applied to parallel-machine scheduling problems. In the column generation approach the problem can be solved as follows:

- Formulate the scheduling problem as a set partition problem.
- Solve the linear relaxation problem by using column generation: first generate a partial solution set and then solve one machine subproblem to minimize reduced costs. Then add new columns to the partial solution sets.
- Apply a branch-and-bound approach to find an integer solution.

In what follows we demonstrate the column generation approach for solving an m-parallel machine scheduling problem with the total weighted completion time objective. We describe here the approach used by Chen and Powell (1999a) and Van den Akker et al (1999).

5.1. Set Partitioning Type Formulations

Let S denote the set of all feasible partial schedules on a single machine. For any $s \in S$, define

f_s = the total weighted completion time of the jobs in schedule s,

a_{js} = 1 if schedule s covers job j, and 0 otherwise,

x_s = 1 if schedule s is selected in the optimal solution and 0 otherwise,

C_{js} = the completion of job j in s, and is defined only if $a_{js}=1$.

Note that for any given partial schedule $s \in S$, the parameters f_s and a_{js} are known. The problem can be formulated as the following set partitioning program.

$$\text{Min} \sum_{s \in S} f_s x_s \tag{13}$$

subject to

$$\sum_{s \in S} a_{js} x_s = 1, \quad j \in N \tag{14}$$

$$\sum_{s \in S} x_s = m \tag{15}$$

$$x_s = 0 \text{ or } 1, \quad s \in S. \tag{16}$$

The objective (13) is to minimize the total weighted completion time of all the selected partial schedules. The constraint (14) guarantees that each job is covered exactly once. The constraint (15) ensures that exactly m partial schedules are selected since we have m machines.

The integer program above is solved by first relaxing the integer constraints and then solving the linear relaxation problem. Since the number of all possible partial schedules in most practical problems is extremely large, it is impossible to list all of them. We thus use column generation to solve the linear relaxation problem. We first use a limited number of columns (a restricted version of the problem), and then add more columns later based on the results of a single-machine problem and the selection of the column with the most negative reduced cost. We continue adding columns until no negative reduced cost can be found. The result of the linear relaxation problem will be used in a branch-and-bound algorithm.

5.2. *Column Generation Approach*

Each column of the above program represents a feasible partial schedule for a single machine. In order to learn whether we need to add more columns to the program, we need to find the minimum reduced cost and see whether it is negative. Let π_j denote the value of the dual variable corresponding to job j in constraint (14), $j = 1, \ldots, n$, and σ denote the corresponding value to the constraint (15). Then the reduced cost r_s of column s is

$$r_s = f_s - \sum_{j=1}^{n} a_{js}\pi_j - \sigma.$$

Since σ is constant and

$$f_s = \sum_{j=1}^{n} w_j C_{js} = \sum_{j=1}^{n} \left[w_j \left(\sum_{k=1}^{j} a_{ks} p_k \right) a_{js} \right],$$

minimizing r_s is equivalent to minimizing

$$f_s - \sum_{j=1}^{n} a_{js}\pi_j = \sum_{j=1}^{n} \left[w_j \left(\sum_{k=1}^{j} a_{ks} p_k \right) - \pi_j \right] a_{js}. \tag{17}$$

If any possible partial schedule s has $r_s \geqslant 0$, then we know that we have solved the linear relaxation problem optimally. Hence the next problem is to find an algorithm to test whether the value in (17) is nonnegative for all possible s, which is a single-machine problem. We call it a single-machine subproblem.

5.3. *Dynamic Programming Algorithm for the Single-Machine Subproblem*

Re-index the jobs such that $p_1/w_1 \leqslant p_2/w_{12} \leqslant \ldots \leqslant p_n/w_n$. Let $T = \sum_{j=1}^{n} p_j$. Also, let $F(j, t)$ denote the minimum objective value (total weighted completion times minus total dual variable value) if we have scheduled jobs from the subset $\{1, \ldots, j\}$ and job j is the last job and is completed at time t. Define that $B_j = \{i \mid i \text{ can precede } j \text{ in a feasible partial schedule on a single machine}\}$.

Initial values:

$F(j, t) = \infty \quad \text{for } t < 0, \; j = 0, \ldots, n$

$F(0, t) = 0 \quad \text{for } t \geqslant 0.$

Recursive equations:

$$F(j, t) = \min\{F(k, t - p_j) + tw_j - \pi_j \mid k \in B_j\}; \quad \text{for } j = 1, \ldots, n; \quad t = 0, \ldots, T.$$

Optimal solution:

$\min\{F(j, t) : 1 \leqslant j \leqslant n, 0 \leqslant t \leqslant T\}.$

The time complexity of this algorithm is $O(nT)$. Hence the algorithm solves the problem in pseudo-polynomial time.

Remark. *Chen and Powell (1999a) show that the single-machine subproblem is NP-hard.*

5.4. *Branch and Bound Scheme*

Note that in the traditional branching strategy for integer programming, we set a variable x_s as either zero or one, that is, we either disregard a specific column or we add it to the main program. However, this scheme cannot be implemented effectively as it is difficult to avoid the same column being picked up by the pricing algorithm again even though we have fixed it at zero. To avoid this difficulty, there are two approaches in the literature, one by Chen and

Powell (1999a) and the other by Van den Akker et al. (1999). Chen and Powell use a relative precedence relationship to branch from a node, while Van den Akker et al. use upper and lower bound constraints on the completion time to branch from a node. Here we describe the Chen and Powell approach.

Recall that B_j is the set of jobs that can be processed before job j. At the beginning, because of the WSPT property, B_j contains all jobs i with $p_i/w_i \leqslant p_j/w_j$. However, B_j will be updated in the branch-and-bound process. Given the solution of a branch-and-bound node, x_s for $s \in S$, define for each pair of jobs j and $k \in B_j$ the value of ordering relation λ_{kj} as follows:

$$\lambda_{kj} = \sum_{s \in S} b_{skj} x_s \tag{18}$$

where b_{skj} is 1 if both jobs j and k are contained in schedule s, and job k is scheduled immediately before job j, and 0 otherwise. In general, the bigger the value λ_{kj}, the stronger the ordering relation of the job pair (k, j). Chen and Powell (1999a) prove that the x-variable solution is integra if, and only if, the corresponding λ-solution given by (18) is an integer.

Given the solution of a node in the branching tree, if the solution is integral, that is, each value x_s is either 0 or 1, then each λ_{kj} is 0 or 1 as well. We do not branch out of this node and this node is pruned. If this solution is better than the best integer solution generated so far, then this solution replaces the previous best integer solution. If the solution is fractional and the integer part of its objective value is equal to or greater than that of the best integer solution generated so far, we do not need to branch out from this node and this node is also pruned. Now suppose that the solution is fractional and the integer part of its objective value is smaller than that of the best integer solution generated so far. In order to be effective, we select a pair of jobs that violates the integrality conditions the most, that is, we select a pair of jobs (s, t) such that

$$\lambda_{s,t} = \arg \min_{ij}\{| \lambda_{i,j} - 0.5 |\}.$$

Two child nodes are then created; one branch fixes $\lambda_{s,t} = 0$, while the other branch fixes $\lambda_{s,t} = 1$. For the branch with $\lambda_{s,t} = 0$, if $s \neq 0$ then the initial node will consider the problem that consists of all the columns of its parent node excluding those in which job t is scheduled immediately after job s. If $s = 0$ then it will consider the problem with all columns of its parent node except the ones in which job t is scheduled first on a machine. We also update the job ordering restriction by setting $B_t := B_t \backslash \{s\}$. For the branch with $\lambda_{s,t} = 1$, the initial node will consider the problem that consists of all the columns of its parent node except the ones in which t is scheduled immediately after a job other than s and the ones in which a job other than t is scheduled immediately after job s. We will also update job ordering restriction by setting $B_t := \{s\}$.

5.5. Computational Results

The advantage of the column generation approach is based on the fact that the solution of the linear relaxation problem is usually very close to an integer solution. Hence very few branch-and-bound steps are needed to bring the solution back to an integer. For the scheduling problem with m machines in parallel, most empirical tests result in an integer solution. The average gap between the integer programming solution and the linear relaxation solution is less than 0.1 percent (Van den Akker et al., 1999; Chen and Powell, 1999a). For the total weighted number of late jobs objective, the gap is less than 0.8 percent (Chen and Powell, 1999a). For the parallel machine scheduling problem with maintenance constraints (Lee and Chen, 2000), the average gap is less than 0.2 percent.

As we have shown above, the parallel-machine scheduling problem with total weighted completion time objective can be solved by dynamic programming in $O(nmT^{m-1})$ or $O(nmW^{m-1})$. The problem is pseudo-polynomially solvable. However, in practice, it is not practical to solve a large-scale problem using this type of approach due to the high time complexity. Belouadah and Potts (1995) use a branch-and-bound algorithm to solve the problem for up to 20 jobs and 5 machines and 30 jobs and 4 machines. Applying the column generation approach to solve the same problem, Chen and Powell (1999a) solve problems with 20 machines and 100 jobs, while Van den Akker (1999) is able to solve problems with 10 machines and 100 jobs. They all demonstrate that column generation is a very effective approach for solving parallel-machine scheduling problems.

6. Heuristics: Decomposition Methods

In this section we describe a class of heuristic procedures that are typically referred to as decomposition methods. One of the most successful heuristic procedures developed for the minimization of the makespan in a job shop is a decomposition procedure that is called the shifting bottleneck heuristic (see Adams et al., 1998; Morton and Pentico, 1993; Pinedo, 1995; Pinedo and Chao, 1999).

The problem of minimizing the makespan in a job shop without recirculation can be represented by a so-called disjunctive graph. Consider a directed graph G with a set of nodes N and two sets of arcs A and B. The nodes N correspond to all of the operations (i, j) that must be performed on the n jobs. The conjunctive (solid) arcs A represent the routes of the jobs. If arc $(i,j) \rightarrow (k,j)$ is part of A, then job j has to be processed on machine i before it is processed on machine k, that is, operation (i, j) precedes operation (k, j). Two operations that belong to two different jobs and that have to be processed on the same machine are connected to one another by two disjunctive (broken) arcs going in opposite directions. The disjunctive arcs B form m cliques of double arcs, one clique for each machine. (A clique is a term in graph theory that refers to a graph in which any two nodes are connected to one another; in this case a connection between two nodes within a clique consists of a pair of disjunctive arcs.) All operations (nodes) in the same clique have to be done on the same machine. All arcs emanating from a node, conjunctive as well as disjunctive, have as their length the processing time of the operation that is represented by that node. In addition there is a source U and a sink V, which are dummy nodes. The source node U has n conjunctive arcs emanating to the first operations of the n jobs. The sink node V has conjunctive arcs coming in from all the last operations. The arcs emanating from the source have length zero (see Figure 1). Such a graph is typically denoted by $G = (N, A, B)$.

A feasible schedule corresponds to a selection of one disjunctive arc from each pair such that the resulting directed graph is acyclic. This implies that each selection from arcs within a clique must be acyclic. Such a selection determines the sequence in which the operations are to be performed on that machine. That a selection from a clique has to be acyclic can be argued

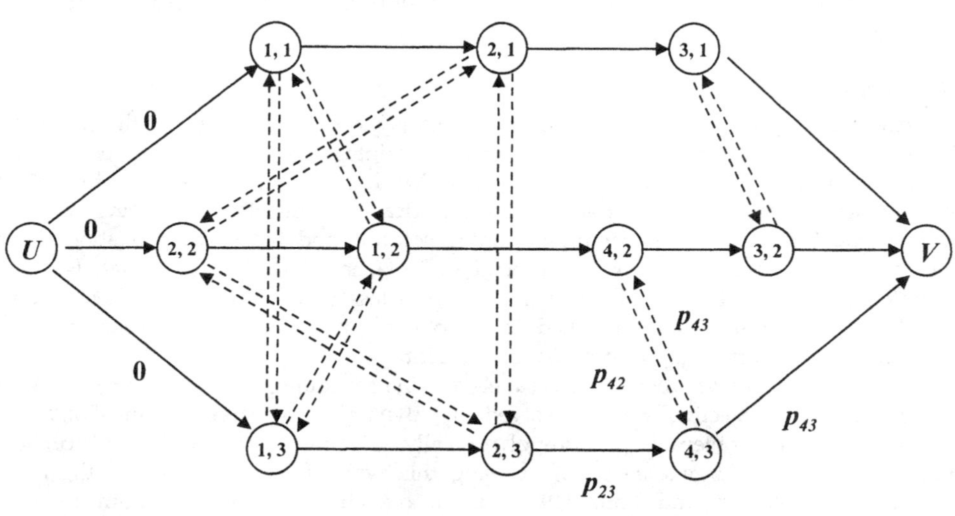

Figure 1 Disjunctive Graph of an Instance of the Job Shop Problem

as follows. If there were a cycle within a clique, a feasible sequence of the operations on the corresponding machine would not have been possible. So, if D denotes the subset of the selected disjunctive arcs and the graph $G(D)$ is defined by the set of conjunctive arcs and the subset D, then D corresponds to a feasible schedule if, and only if, $G(D)$ contains no directed cycles.

The makespan of a feasible schedule is determined by the longest path in $G(D)$ from the source U to the sink V. This longest path consists of a set of operations of which the first starts at time 0 and the last finishes at the time of the makespan. Each operation on this path is immediately followed either by the next operation on the same machine or by the next operation of the same job on another machine. The problem of minimizing the makespan is therefore reduced to finding a selection of disjunctive arcs that minimizes the length of the longest path (i.e., the critical path).

The shifting bottleneck heuristic schedules the m machines in a job shop one at a time. Each iteration of the procedure selects one machine among the machines remaining to be scheduled and sequences all the operations on the chosen machine. So, at the start of an iteration there is a set of machines that has already been scheduled in the previous iterations and a set of machines that still needs to be scheduled. The sequence (or the order) of the n operations on a machine that has already been scheduled is completely determined. That is, a set of disjunctive arcs has been selected, or, equivalently, a set of precedence constraints has been specified for those n operations. On the other hand, the sequence of the operations on a machine that has not yet been scheduled is still completely open, that is, the disjunctive arcs have not yet been selected. Because of that, a machine that has not been scheduled yet is basically, for the time being, equivalent to a workstation with a number of identical machines in parallel, implying that the operations destined for the original machine can be processed, at this point in the procedure, if necessary, simultaneously and in parallel. Given the routing constraints with respect to the operations of the jobs, that is, the conjunctive arcs, and given the precedence constraints enforced by the sequencing of the operations on the machines already scheduled, that is, the disjunctive arcs that have been selected, one can come up with a graph G, of which the longest path from the source to the sink is a lower bound for the makespan.

In order to select a machine from the set of unscheduled machines, each machine has to be evaluated separately. The evaluation of a machine can be done as follows. The current precedence constraints graph provides a lower bound on the makespan (i.e., the longest path from the source to the sink), and the objective is to sequence the operations on the machine in such a way that the lower bound does not go up, that is by selecting the disjunctive arcs we do not create a new path from the source to the sink that is, longer than any existing one. Or, if that cannot be avoided, we would like to find a sequence for the operations on the machine in such a way that the longest path goes up by a minimum amount. The process of finding the sequence of operations on that machine with a minimum increase in the length of the longest path is typically a hard combinatorial problem. However, heuristics have been developed that yield good solutions in a reasonably fast time. In each iteration of the shifting bottleneck heuristic all the unscheduled machines have to be evaluated this way. The machine that causes the largest increase in the longest path is in a sense the most critical and the hardest to schedule; it has the greatest impact on the makespan and may be regarded, among the remaining machines, as the bottleneck. Therefore, it is this machine for which the schedule must be fixed in the current iteration.

Before the heuristic proceeds to the next iteration, it looks back at all the machines sequenced in previous iterations and attempts to resequence them in order to see if a better schedule can be obtained. After this step has been completed, the heuristic starts with the next iteration.

The shifting bottleneck heuristic can be summarized as follows.

Shifting Bottleneck Heuristic

Step 1. Initial conditions

Step 2. Analysis of machines still to be scheduled

Step 3. Bottleneck selection and sequencing

Step 4. Resequencing of all machines scheduled earlier

Step 5. Stopping criterion

So the shifting bottleneck heuristic is an algorithm of the constructive type. It starts without a schedule and gradually constructs a schedule by sequencing the machines one machine at a time. Shifting bottleneck heuristics have also been developed for other objectives, for example, $L_{\max}$, $\sum w_j T_j$ (see Ovacik and Uzsoy, 1997; Pinedo and Singer, 1997; Yang et al., 1997).

7. Heuristics: Local Search Methods

We now describe a heuristic of the improvement type. Heuristics of the improvement type are conceptually completely different from those of the constructive type. They start out with a complete schedule, which may be selected arbitrarily, and then try to obtain a better schedule by making changes in the current schedule. An important class of improvement type algorithms are the local search procedures. This class includes simulated annealing, tabu search, and genetic algorithms (see Andersen et al., 1995; Aarts et al., 1994).

A local search procedure does not guarantee an optimal solution. It usually attempts to find a better schedule than the current one in the neighborhood of the current one. Two schedules are neighbors, if one can be obtained through a well-defined modification of the other. At each iteration, a local search procedure performs a search within the neighborhood and evaluates the various neighboring solutions. The procedure either accepts or rejects a candidate solution as the next schedule to move to, based on a given acceptance-rejection criterion.

One can compare the various local search procedures on the following four design criteria:

1. The schedule representation needed for the procedure.
2. The neighborhood design.
3. The search process within the neighborhood.
4. The acceptance-rejection criterion.

The representation of a schedule may at times be nontrivial. A nonpreemptive single machine schedule can be specified by a simple permutation of the n jobs. A nonpreemptive job shop schedule can be specified by m consecutive strings, each one representing a permutation of n operations on a specific machine; based on this information, the starting and completion times of all operations can be computed. However, when preemptions are allowed, the format of the schedule representation becomes significantly more complicated.

The design of the neighborhood is a very important aspect of a local search procedure. For a single machine a neighborhood of a particular schedule may simply be defined as all schedules that can be obtained by performing a single adjacent pairwise interchange. This implies that there are $n-1$ schedules in the neighborhood of the original schedule. A larger neighborhood for a single machine schedule may be defined by taking an arbitrary job in the schedule and inserting it in another position in the schedule. Clearly, each job can be inserted in $n-1$ other positions. The entire neighborhood contains less than $n(n-1)$ neighbors as some of these neighbors are identical. The neighborhood of a schedule in a more complicated machine environment is usually more complex.

An example is a neighborhood designed for the job shop problem with the makespan as objective. In order to describe this neighborhood, we have to use the concept of a critical path. A critical path in a job shop schedule consists of a set of operations of which the first one starts out at time $t = 0$ and the last one finishes at time $t = C_{\max}$. The completion time of each operation on a critical path is equal to the starting time of the next operation on that path; two successive operations either belong to the same job or are processed on the same machine. A schedule may have multiple critical paths that may overlap. Finding the critical path(s) in a given schedule for a job shop problem with the makespan as objective is relatively straightforward. It is clear that in order to reduce the makespan, changes have to be made in the sequence(s) of the operations on the critical path(s). A simple neighborhood of an existing schedule can be designed as follows: the set of schedules whose corresponding sequences of operations on the machines can be obtained by interchanging a pair of adjacent operations on the critical path of the current schedule. Note that in order to interchange a pair of operations on the critical path, the operations must be on the same machine and belong to different jobs. If there is a single critical path, then the number of neighbors within the neighborhood is at most the number of operations on the critical path minus 1. Experiments have shown that this type of neighborhood for the job shop problem is too simple to be effective. The number of neighboring schedules that are better than the existing schedule tends to be very limited. More sophisticated neighborhoods have been designed that perform better (see Pinedo, 1995)

The search process within a neighborhood can be performed in a number of ways. A simple way is to select schedules in the neighborhood at random, evaluate these schedules, and decide which one to accept. However, it may pay to do a more organized search and first select schedules that appear promising. One may want to consider swapping those jobs that affect the objective the most. For example, when the total weighted tardiness has to be minimized, one may want to move very tardy jobs closer to the beginning of the schedule.

The acceptance-rejection criterion is usually the design aspect that distinguishes a local search procedure the most. For example, the difference between two of the best-known local search procedures, simulated annealing and tabu search, lies mainly in their acceptance-rejection criteria. In simulated annealing the acceptance–rejection criterion is based on a probabilistic process while in tabu search it is based on a deterministic process.

In what follows we discuss simulated annealing in some detail (see Van Laarhoven et al., 1992). This procedure is a search process that has its origin in the fields of material science and physics. It was first developed as a simulation model for describing the physical annealing process of condensed matter.

The simulated annealing procedure goes through a number of iterations. At iteration k of the procedure, there is a current schedule S_k as well as a best schedule found so far, S_0. For a single-machine problem these schedules are sequences (permutations) of the jobs. Let $G(S_k)$ and $G(S_0)$ denote the corresponding values of the objective function. Note that $G(S_k) \geqslant G(S_0)$. The value of the best schedule obtained so far, $G(S_0)$, is often referred to as the aspiration criterion. The algorithm, in its search for an optimal schedule, moves from one schedule to another. At iteration k, a search for a new schedule is conducted within the neighborhood of S_k. First, a so-called candidate schedule, for example, S_c, is selected from the neighborhood. This selection of a candidate schedule can be done at random or in an organized, possibly sequential way. If $G(S_c) < G(S_k)$, a move is made, setting $S_{k+1} = S_c$. If $G(S_c) < G(S_0)$ then S_0 is set equal to S_c. However, if $G(S_c) \geqslant G(S_k)$, a move is made to S_c with probability

$$P(S_k, S_c) = \exp\left(\frac{G(S_k) - G(S_c)}{\beta_k}\right),$$

with probability $1 - P(S_k, S_c)$ schedule S_c is rejected in favor of the current schedule, setting $S_{k+1} = S_k$. Schedule S_0 does not change when it is better than schedule S_c. The $\beta_1 \geqslant \beta_2 \geqslant \beta_3 \geqslant \ldots \geqslant 0$ are control parameters referred to as cooling parameters or temperatures (in analogy with the annealing process mentioned above). Often β_k is chosen to be a^k for some a between 0 and 1.

From the above description of the simulated annealing procedure it is clear that moves to worse solutions are allowed. This is the major difference from regular neighborhood searches. The reason for allowing these moves is to give the procedure the opportunity to move away from a local minimum and find a better solution later on. Since β_k decreases with k, the acceptance probability for a nonimproving move decreases as the search process goes on. The definition of the acceptance probability also ensures that if a neighbor is significantly worse, its acceptance probability is very low and the move is unlikely to be made.

Several stopping criteria are used for this procedure. One way is to let the procedure run for a prespecified number of iterations. Another is to let the procedure run until no improvement has been obtained for a given number of iterations.

The method can be summarized as follows:

Simulated Annealing Algorithm

Step 1.

Set $k = 1$ and select β_1.

Select an initial sequence S_1 using some heuristic.

Set $S_0 = S_1$.

Step 2.

Select a candidate schedule S_c from the neighborhood of S_k.

If $G(S_0) < G(S_c) < G(S_k)$, set $S_{k+1} = S_c$ and go to Step 3.

If $G(S_c) < G(S_0)$, set $S_0 = S_{k+1} = S_c$ and go to Step 3.

If $G(S_c) > G(S_k)$ generate a random number U_k from a Uniform (0, 1) distribution;

If $U_k \leqslant P(S_k, S_c)$ set $S_{k+1} = S_c$, otherwise set $S_{k+1} = S_k$ and go to Step 3.

Step 3.

Select $\beta_{k+1} \leqslant \beta_k$.

Increment k by 1.

If $k = N$ then STOP, otherwise go to Step 2.

The effectiveness of simulated annealing depends on the design of the neighborhood as well as how the search is conducted within this neighborhood. If the neighborhood is designed in a way that facilitates moves to better solutions and moves out of local minima, then the procedure will perform well. The search within a neighborhood can be done randomly or in a more organized way. For example, the contribution of each job to the objective function can be computed and the job with the highest impact on the objective can be selected as a candidate for an interchange. Over the last two decades simulated annealing has been applied to many scheduling problems, in academia as well as in industry, with considerable success.

Other local search procedures, for example, tabu search and genetic algorithms, have also received a considerable amount of attention. Tabu search has been implemented and used by Barnes et al. (1995) and Dell'Amico and Trubian (1993). Genetic algorithms have been studied and experimented with by Della Croce et al. (1995), Herrmann et al. (1995), and Lee et al. (1997).

8. Discussion

Optimization models for scheduling problems have gained popularity recently due to the progress in computing power. Dynamic programming is popular for solving pseudo-polynomially solvable problems. Branch and bound and integer programming have been used to solve problems that are categorized as strongly NP-hard. However, their practical applications have been limited to small size problems. Recently, branch and bound using column generation has been shown to solve large-scale problems quite effectively. However, applications have been limited to parallel machine scheduling problems and other problems that are related, such as the single-machine earliness-tardiness problem, which is similar to a two machines in parallel problem. Using column generation to solve other scheduling problems is an interesting and challenging task.

A number of comparative studies have been conducted to compare heuristic methods with one another and with other more classical mathematical programming approaches. There are several important parameters that can be compared. The most important one is the quality of the solution, that is, which approach leads to a better answer. Another important aspect is the amount of CPU time that is required to arrive at an acceptable solution. Also, a ratio of these two factors could be of interest. A third characteristic on which an approach can be judged is the development time or the implementation time. Neighborhood search approaches usually need less development time than the more classical mathematical programming approaches.

It has to be kept in mind that there are a number of issues that affect the outcome of each comparative study, namely the initial solution, the setting of parameters, the language and manner in which the procedure is coded, and the platform on which the study is conducted.

ACKNOWLEDGMENT
This research was supported in part by the National Science Foundation under Grant No. DMI-9610229.

References

Aarts, E. H. L., P. J. M. Van Laarhoven, J. K. Lenstra, and N. L. J. Ulder. 1994. "A computational study of local search algorithms for job shop scheduling." *ORSA J. Comput.* 6: 118–125.

Adams, J., E. Balas, and D. Zawack. 1988. "The shifting bottleneck procedure for job shop scheduling." *Mgmnt Sci.* **34**: 391–401.

Andersen, E. J., C. A. Glass, and C. N. Potts. 1995. "Local search in combinatorial optimization: applications in machine scheduling." Preprint Series No. OR56. Southampton, England: Faculty of Mathematical Studies, University of Southampton

Barnes, J., M. Laguna, and F. Glover. 1995. "An overview of tabu-search approaches to production scheduling problems." In *Intelligent Scheduling Systems*, edited by D. E. Brown and W. T. Scherer. Boston: Kluwer Academic Publisher, pp. 101–127.

Belouadah, H. and C. N. Potts. 1995. "Scheduling identical parallel machine to minimize total weighted completion time." *Discrete Appl. Math.* **48**: 201–218.

Blazewicz, J., M. Dror, and J. Weglarz. 1991. "Mathematical programming formulations for machine scheduling: a survey." *Eur. J. Oper. Res.* **51**: 283–300.

Boyd, E. A. 1992. "A pseudopolynomial network flow formulation for exact knapsack separation." *Networks* **22**: 503–514.

Cattrysse, D., M. Salomon, R. Kuik, and L. N. Van Wassenhove. 1993. "A dual ascent and column generation heuristic for the discrete lot sizing and scheduling problem with setup times." *Mgmnt Sci.* **39**: 477–486.

Chan, L. M. A., P. Kaminsky, A. Muriel, and D. Simchi-Levi. 1997. "Machine scheduling, linear programming and list scheduling heuristic." Technical Report. Evanston, IL: Northwestern University.

Chandru, V., C.-Y. Lee, and R. Uzsoy. 1993. "Minimizing total completion time on a batch processing machine with job families." *Oper. Res. Lett.* **13**: 61–67.

Chen, Z.-L. and C.-Y. Lee. 1999. "A column generation algorithm for parallel machine common due window scheduling." To appear in *Eur. J. Oper. Res.*

Chen, Z.-L. and W. B. Powell. 1998. "Network representation, column generation and branch and bound for parallel machine scheduling with multiple job families." Working Paper 98-02. Philadelphia, PA 19104-6315: Department of Systems Engineering, University of Pennsylvania.

Chen, Z.-L. and W. B. Powell. 1999a. "Solving parallel machine scheduling problems by column generation." *INFORMS J. Comput.* **11**: 78–94.

Chen, Z.-L. and W. B. Powell. 1999b. "A column generation based decomposition algorithm for a parallel machine scheduling Just-In-Time scheduling problem." *Eur. J. Oper. Res.* **116**: 221–233.

Della Croce, F., R. Tadei, and G. Volta. 1995. "A genetic algorithm for the job shop problem." *Comput. Oper. Res.* **22**: 15–24.

Dell'Amico, M. and M. Trubian. 1993. "Applying tabu-search to the job shop scheduling problem." *Ann. Oper. Res.* **41**: 231–252.

Desrochers, M., M. Desrochers, J. Desrosiers, and M. Solomon. 1992. "A new optimization algorithm for the vehicle routing problem with time windows." *Oper. Res.* **40**: 332–342.

Gilmore, P. C. and R. E. Gomory. 1961. "A linear programming approach to the cutting stock problem—Part 1." *Oper. Res.* **9**: 849–859.

Gilmore, P. C. and R. E. Gomory. 1963. "A linear programming approach to the cutting stock problem—Part 2." *Oper. Res.* **11**: 863–887.

Herrmann, J. W., C.-Y. Lee, and J. Hinchman. 1995. "Global job shop scheduling with a genetic algorithm." *Prod. Oper. Mgmnt* **4**: 30–45.

Lavoie, S., M. Minoux, and E. Odier. 1988. "A new approach for crew pairing problems by column generation with an application to air transport." *Eur. J. Oper. Res.* **35**: 45–58.

Lawler, E. L. and J. M. Morore. 1969. "A functional equation and its application to resource allocation and scheduling problems." *Mgmnt Sci.* **16**: 77–84.

Lee, C.-Y. and Z. L. Chen. 2000. "Scheduling of jobs and maintenance activities on parallel machines." *Naval Res. Logistics* **47**: 145–165.

Lee, C.-Y. and R. Uzsoy. 1992. "A new dynamic programming algorithm for the parallel machine total weighted completion time problem." *Oper. Res. Lett.* **11**: 73–75.

Lee, C.-Y., R. Uzsoy, and L. A. Martin-Vega. 1992. "Efficient algorithms for scheduling semiconductor burn-in operations." *Oper. Res.* **40**: 764–775.

Lee, C.-Y., S. Piramuthu, and Y.-K. Tsai. 1997. "Job shop scheduling with a genetic algorithm and machine learning." *Int. J. Prod. Res.* **35**: 1171–1191.

Luh, B. P. and D. J. Hoitomt. 1993. "Scheduling of manufacturing systems using the Lagrangian relaxation technique." *IEEE Trans. Aut. Control* **38**: 1066–1079.

Morton, T. E. and D. W. Pentico. 1993. In *Heuristic Scheduling Systems*. New York: John Wiley.

Mehrotra, A. and M. A. Trick. 1996. "A column generation approach to graph coloring." *INFORMS J. Comput.* **8**: 344–354.

Nemhauser, G. L. and L. A. Wolsey. 1988. In *Integer and Combinatorial Optimization*. New York: Wiley and Sons.

Ovacik, I. and R. Uzsoy. 1997. In *Decomposition Methods in Large Scale Job Shops*. Boston: Kluwer Academic Press.

Pinedo, M. 1995. *Scheduling: Theory, Algorithms and Systems*. Englewood Cliffs, NJ: Prentice-Hall.

Pinedo, M. and X. Chao. 1999. In *Operations Scheduling with Applications in Manufacturing and Services*. Boston: Irwin McGraw-Hill.

Pinedo, M. and M. Singer. 1997. "A shifting bottleneck heuristic for minimizing the total weighted tardiness in a job shop." *Naval Res. Logistics* **46**: 1–17.

Pochet, Y. and L. A. Wolsey. 1988. "Lot-sizing models with backlogging: strong formulations and cutting planes." *Math. Programming* **40**: 317–335.

Rothkopf, M. H. 1966. "Scheduling independent tasks on parallel processors." *Mgmnt Sci.* **12**: 437–447.

Savelsbergh, M. 1997. "A branch-and-price algorithm for the generalized assignment problem." *Oper. Res.* **45**: 831–841.

Van den Akker, J. M., J. A. Hoogeveen, and S. L. van de Velde. 1999. "Parallel machine scheduling by column generation." *Oper. Res* **47**: 862–872.

Van Laarhoven, P. J. M., E. H. L. Aarts, and J. K. Lenstra. 1992. "Job shop scheduling by simulated annealing." *Oper. Res.* **40**: 113–125.

Vance, P. H., C. Barnhart, E. J. Johnson, and G. L. Nemhauser. 1994. "Solving binary cutting stock problems by column generation and branch-and-bound." *Comput. Optim. Appl.* **3**: 111–130.

Wolsey, L. A. 1989. "Strong formulations for mixed integer programming: a survey." *Math. Programming* **45**: 173–191.

Yang, Y., S. Kreipl, and M. Pinedo. 1997. "Heuristics for minimizing the total weighted tardiness in flexible flow shops." Technical Report. New York: Stern School of Business, New York University.

17.2 The Vehicle Routing Problem

John E. Beasley, Abilio Lucena, and Marcus Poggi de Aragão

ABSTRACT This article is concerned with mathematical models of the vehicle routing problem. Initially, three models are discussed for the simplest form of the problem. These models can be categorized, respectively, in the following classes: a polynomial number of variables and an exponential number of constraints; a polynomial number of variables and constraints; and an exponential number of variables and a polynomial number of constraints. The first model is a direct extension of the Dantzig, Fulkerson, and Johnson formulation of the traveling salesman problem. The second one is a two-commodity flow-based model, while the third model is based on a set partitioning reformulation of the problem. The models are then adapted to address a number of relevant extensions of the basic problem.

1. Introduction

The vehicle routing problem (VRP) is the generic name given to a large class of problems involving the distribution of goods, services, information, or personnel. As the name implies, the problem involves a fixed number of vehicles, which are stationed at a central depot (to which they must return at the end of the operation), and which have identical, fixed capacities. These vehicles are used to visit customers, typically at a visit supplying each customer with a desired fixed quantity of a certain commodity. The routing decisions involve allocating customers to vehicles without exceeding vehicle capacities, and determining the sequence in which customers allocated to a vehicle should be visited. These routing decisions must be made so as to optimize an appropriate objective function, such as minimizing the total distance (time) travelled by the vehicles.

The problem (or minor variations of it) has been known in the literature by a variety of names: the VRP (Christofides et al., 1981a, 1981b), the vehicle scheduling problem (Clarke and Wright, 1964; Gaskell, 1967), the truck dispatching problem (Dantzig and Ramser, 1959; Christofides and Eilon, 1969; Krolak et al., 1972), or the classical VRP (Laporte, 1997). In this article the problem will be referred to as the VRP.

The VRP is clearly NP-hard. Indeed it generalizes the traveling salesman problem (TSP; see the book by Lawler et al., 1985 or the survey by Jünger et al., 1995) in that a TSP can be viewed as a VRP with a single vehicle of infinite capacity. The VRP (and generalizations of it) has been the object of a considerable amount of research for the past four decades. For surveys on exact and approximate algorithms, and bibliographic sources, the reader is referred to Christofides (1985), Laporte and Norbert (1987), Fisher (1995), Laporte and Osman (1995), and Laporte (1997).

A number of extensions to the VRP are frequently mentioned in the literature. They can be viewed, to a great extent, as involving the addition of extra constraints to the problem. The three most commonly encountered extensions are:

1. Limits on the length or duration of each vehicle route.
2. Time windows for the arrival of vehicles at customers (waiting times are allowed).
3. Precedence constraints relating to the order in which given sets of customers may be visited.

Further extensions to the VRP allow the possibility of either pickup (collection) or delivery at customers, the use of a nonhomogeneous fleet, and the presence of multiple depots.

This article is organized as follows. In Section 2 three different formulations for the VRP are described. These formulations are then generalized, in Section 3, to model the extensions mentioned above.

2. Formulations for the Basic VRP

2.1. Notation

The basic VRP can be defined in graph theoretical terms. Let $G=(V_0, E)$ denote an undirected graph with a vertex set V_0 with $n-1$ vertices indexed $\{0, 1, \ldots, n\}$ and an edge set E. For simplicity, we assume G to be a complete graph. Furthermore, assume that $V \subset V_0$ is the subset of vertices indexed by $\{1, \ldots, n\}$. Vertices V represent the customers while the vertex indexed by 0 represents the depot. Costs $\{c_e: e \in E\}$ that may represent distances travelled or time taken are associated with the edges of G. For ease of presentation we will be somewhat lax and use i to denote, simultaneously, vertex $i \in V_0$ and that vertex in V_0 indexed by i, assuming, as a result, that those two vertices are the same. The same convention will also apply to an edge $e \in E$ and that edge of E indexed by e.

The demand (quantity to be delivered) at customer (vertex) $i \in V$ is given by d_i and the time required to process the delivery at i is p_i (by definition, $d_0 = 0$ and $p_0 = 0$). The capacity for each one of the k vehicles stationed at the depot is given by b.

For a given set of vertices $S \subseteq V_0$ we use $d(S)$ to denote the sum of the demands for the vertices in S. $E(S) \subseteq E, S \subseteq V_0$, is used for the set of edges of G with both extremities in S. The complement in V_0 for a set of vertices $S \subset V_0$ is denoted $\bar{S} = V_0 \backslash S$. Those edges with one extremity in $S \subseteq V$ and the other in $\bar{S}$ form the cut-set $E(S, \bar{S})$. The set of edges incident on vertex $i \in V_0$ is denoted $\delta(i)$. Finally, we use $r(S)$ to denote $\lceil d(S)/b \rceil, S \subseteq V_0$, where $\lceil y \rceil$ is the smallest integer larger than, or equal to, the scalar y.

It may be found advantageous, in terms of the design of solution algorithms, to model the problem based on an underlying directed graph $D = (V_0, A)$ (derived from the undirected graph $G = (V_0, E)$). To that effect, every edge $e = (i, j), i, j \in V_0, e \in E$, should give rise to two arcs $[i, j]$ and $[j, i]$ in D with costs $c'_{ij} = c'_{ji} = c_e$. Following the notation used for the undirected graph case, let $\delta^+(i), i \in V_0$ denote the set of arcs of D that are directed toward i (i.e., $[j, i] \in A, j \in V_0$). Let $\delta^-(i)$ play a similar role for those arcs that are directed away from i (i.e., $[i, j] \in A, j \in V_0$). Finally, for a given subset of the vertices in V_0, say S, let $A(S)$ denote the set of arcs in A with both end points in S.

2.2. *A Formulation with a Polynomial Number of Variables and an Exponential Number of Constraints*

Let variables x_e, $e \in E$, control the inclusion ($x_e > 0$), or not ($x_e = 0$), of an edge in a VRP route and consider the polyhedral region R_1 defined by

$$\sum_{e \in \delta(i)} x_e = 2, \quad i \in V, \tag{1}$$

$$\sum_{e \in \delta(0)} x_e = 2k, \tag{2}$$

$$\sum_{e \in E(S, \bar{S})} x_e \geqslant 2r(S), \quad S \subseteq V, |S| \geqslant 2, \tag{3}$$

$$0 \leqslant x_e \leqslant 1, \quad e \in E \backslash \delta(0), \tag{4}$$

$$0 \leqslant x_e \leqslant 2, \quad e \in \delta(0). \tag{5}$$

Constraint (1) ensures that a vehicle visits each customer while (2) enforces the condition that there must be exactly k routes that include the depot. Inequalities (3), while breaking subtours, also impose a lower bound on the number of vehicles that are necessary to feasibly service the customers in S. Finally, notice that (5) allows for the *duplication* of edges incident on 0 (i.e., the depot), thus allowing routes that contain only a single customer. Note, however, that all vehicles must be used, that is, each vehicle must visit at least one customer. The VRP can then be formulated as

$$\min\left\{\sum_{e \in E} c_e x_e : x \in R_1 \cap \mathbb{Z}^{|E|}\right\}, \tag{6}$$

where the use of $\mathbb{Z}^{|E|}$ ensures that x must be integer valued.

Formulation (6) has been used by Cornuejols and Harche (1993) and, with minor modifications, by Laporte et al. (1985). Inequalities (3) are known as generalized subtour elimination constraints (GSECs) and were introduced by Laporte and Norbert (1983) in an equivalent description. These inequalities generalize the subtour elimination constraints introduced by Dantzig et al. (1954) for the TSP. Constraints (3) can be made stronger by replacing $r(S)$ with the optimal solution to the bin packing problem associated with bins of size b and a set of $|S|$ elements with weights $\{d_i: i \in S\}$. This strengthening has been credited by Laporte and Norbert (1987) to Gavish. Other possibilities for strengthening also exist (e.g., see Fisher, 1994). Formulation (6) can be easily adapted to become a purely 0–1 model (see, Campos et al., 1991).

2.3. *A Formulation with a Polynomial Number of Variables and Constraints*

In association with the directed graph $D = (V_0, A)$, consider two sets of real-valued variables: $\{y_{ij}: [i, j] \in A\}$ and $\{z_{ij}: [i, j] \in A\}$. Each of these sets will be used, in turn, to model two single commodity network flow subproblems that are to be linked by additional constraints. These subproblems will be complementary in the sense that, for each vertex $i \in V$, d_i will act as a vertex demand for one subproblem and a vertex supply for the other. Define the following polyhedral region, R_2,

$$\sum_{[i,j]\in\delta^-(i)} y_{ij} - \sum_{[j,i]\in\delta^+(i)} y_{ji} = d_i, \quad i \in V, \tag{7}$$

$$\sum_{[i,j]\in\delta^-(i)} z_{ij} - \sum_{[j,i]\in\delta^+(i)} z_{ji} = -d_i, \quad i \in V, \tag{8}$$

$$\sum_{[i,j]\in\delta^-(i)} (y_{ij} + z_{ij}) = b, \quad i \in V, \tag{9}$$

$$\sum_{[0,j]\in\delta^-(0)} (y_{0j} + z_{0j}) = kb, \tag{10}$$

$$y_{ij}, z_{ij} \geqslant 0, \quad [i,j] \in A. \tag{11}$$

The basic VRP can then be formulated as

$$\min\left\{ \sum_{[i,j]\in A} (c'_{ij}/b)(y_{ij} + z_{ij}) : (y,z) \in R_2 \cap \{(y_{ij} + z_{ij}) \in \{0,b\}, [i,j] \in A\} \right\}. \tag{12}$$

This formulation is due to Lucena (1986) and generalizes the TSP formulation of Finke et al. (1984).

Constraints (7), (8), and (11) define feasible flow paths for each one of the two different commodities involved. Constraint (10), together with $\{(y_{ij} + z_{ij}) \in \{0, b\}, [i,j] \in A\}$ in (12), ensures that there are exactly k arcs leaving vertex 0 (i.e., the depot). Each of these arcs will be carrying a total amount of b units of flow. Analogously, one can show that exactly one arc, with a total combined amount of b units of flow, must be leaving any vertex in V. Taking (7) and (8) together one can show that there must be exactly one arc (with a total combined amount of b units of flow) entering any vertex in V. It then follows that there must be exactly k arcs entering vertex 0 (each one with a total combined amount of b units of flow). Therefore, as far as the vertex degrees are concerned, a feasible solution to (12) is feasible for the basic VRP. Additionally, notice that there must be a path from vertex 0 to any vertex $i \in V$ and also a path back from i to 0 (the supply/demand constraints (7)–(8) in the formulation guarantee that). A feasible solution to (12) is thus connected and induces k routes centered on 0. One is then just left to show that vehicle capacities cannot be exceeded. That is guaranteed by noticing that, along any of these routes, the flows modeled by $\{z_{ij} : [i,j] \in A\}$ can only decrease from a maximum of b units (when leaving vertex 0) to a minimum of zero units (when entering vertex 0). It is immediately clear that a similar conclusion may be reached by focusing the analysis on $\{y_{ij} : [i,j] \in A\}$.

2.4. *A Formulation with an Exponential Number of Variables and a Polynomial Number of Constraints*

Let $I = \{1, \ldots, n\}$ be a finite set and $\{I_j : j \in J\}$, $J = \{1, \ldots, m\}$ be a collection of subsets of I. A partition of I is any set $P, P \subseteq J$, of disjoint subsets $\{I_j: j \in P\}$ where $\cup_{j\in P} I_j = I$. Assuming that a cost h_j is associated with each I_j, the set partitioning problem (SPP) is to find a partition of I with minimum cost. This results in the cardinality constrained SPP (CCSPP) when one restricts partitions to have exactly a given, fixed number of subsets, for example, $k \geqslant 1$. Consider a polyhedral region, R_3,

$$\sum_{j\in J} a_{ij}\lambda_j = 1, \quad i \in I, \tag{13}$$

$$\sum_{j \in J} \lambda_j = k, \tag{14}$$

$$0 \leqslant \lambda_j \leqslant 1, \quad j \in J, \tag{15}$$

where a_{ij} is 1 if $i \in I_j$, 0 otherwise, and where λ_j represents the inclusion ($\lambda_j > 0$), or not ($\lambda_j = 0$), of a subset j in a partition. A formulation for the CCSPP is thus given by

$$\min\left\{ \sum_{j \in J} h_j \lambda_j : \lambda \in R_3 \cap \mathbb{Z}^{|J|} \right\}. \tag{16}$$

It is straightforward to reformulate the VRP as a CCSPP. First, assume that a one-to-one correspondence between elements of I and VRP customers exists. Then, let J represent the set of all nonempty subsets of customers that can feasibly be serviced by a single vehicle (i.e., J represents possible feasible routes). Finally, let the cost h_j, for $j \in J$, be equal to the optimal solution value for the TSP associated with the subgraph of G induced by the depot vertex and those vertices representing customers in I_j. Subject to those conditions, (16) may be understood as a VRP formulation.

The cardinality of J in (16) may be enormous as the number of possible feasible routes grows exponentially with the number of VRP customers. Nevertheless, strictly speaking, one need not have generated all feasible routes in order to solve the problem. This is because the set partitioning model for the VRP is very amenable to column generation (see Wolsey, 1988 for a discussion of column generation algorithms). To that effect, initially consider a (small) subset J' of J and define a polyhedral region R_4, equivalent to R_3, but for variables $\{\lambda_j : j \in J'\}$. For an optimal solution to the linear programming (LP) problem

$$\min\left\{ \sum_{j \in J'} h_j \lambda_j : \lambda \in R_4 \right\}, \tag{17}$$

let $\{u_i : i \in I\}$ and v denote the corresponding dual values (for $\sum_{j \in J'} a_{ij}\lambda_j = 1, i \in I$, and $\sum_{j \in J'} \lambda_j = k$, respectively). In order to check whether an optimal solution for (17) is also optimal for (16), note that $I = V$ and define a polyhedral region R_5 as

$$\sum_{i \in V} d_i \alpha_i \leqslant b, \tag{18}$$

$$\sum_{e \in \delta(i)} \gamma_e = 2\alpha_i, \quad i \in V, \tag{19}$$

$$\sum_{e \in \delta(0)} \gamma_e = 2, \tag{20}$$

$$\sum_{e \in E(S)} \gamma_e \leqslant \sum_{i \in S \setminus \{j\}} \alpha_i, \quad S \subseteq V, \ j \in S, \tag{21}$$

$$0 \leqslant \alpha_i \leqslant 1, \quad i \in V, \tag{22}$$

$$0 \leqslant \gamma_e \leqslant 1, \quad e \in E, \tag{23}$$

and solve the following single route (column generation) problem:

$$\min\left\{ \sum_{e \in E} c_e \gamma_e - \sum_{i \in V} u_i \alpha_i - v : (\alpha, \gamma) \in R_5 \cap (\mathbb{Z}^{|V|}, \mathbb{Z}^{|E|}) \right\}. \tag{24}$$

In (18)–(24), α_i represents the inclusion (or not) of a customer in the solution and γ_i represents the inclusion (or not) of an edge in the solution. This model (24) of a prize-collecting TSP, with minor variations, is due to Balas (1989). Constraint (18) is the vehicle capacity constraint and (19) ensures that there are two edges associated with each customer that is included in the tour. Constraint (20) ensures that there are two edges associated with the depot. Constraint

(21) eliminates subtours by ensuring that the number of edges used in any subset S is related to the number of customers that are visited in S.

Notice that a nonempty optimal solution (α^*, γ^*) to (24) corresponds to a feasible VRP route for those elements of I (i.e., customers) for which $\alpha_i^* = 1$. Denote the corresponding set I_{j^*}. The cost for this route is $\sum_{e\in E} c_e\gamma_e^*$. Furthermore, notice that a reduced cost of $(\sum_{e\in E} c_e\gamma_e^* - \sum_{i\in V} u_i\alpha_i^* - v)$ would result if a variable λ_{j^*} of cost $h_{j^*} = \sum_{e\in E} c_e\gamma_e^*$ were to be included in R_5 (i.e., if j^* were to be introduced into J'). Assume this reduced cost to be negative. One may thus conclude that there is at least one $j \in J\setminus J'$ whose addition to J' would result in a decrease in the optimal solution value of (17). In particular, any feasible solution to (24) with a negative value would serve that purpose.

At this point, three remarks should be made. First, observe that single customer routes (i.e., routes such as $0-i-0$ for some customer i) are not allowed in (24) [because of (20)]. Hence we require their prior enumeration with the corresponding indices be introduced into J' initially. That, by itself, would guarantee nonnegative reduced costs for all variables associated with single customer routes in (17). The second remark accounts for a possible modification of the objective function in (24). That objective function associates dual variable values in (17) to vertex variables in R_5. It is possible, nevertheless, to redistribute dual values exclusively among the edge variables. This is accomplished through an objective function such as

$$\min\left\{ \sum_{e\in E} \bar{c}_e\gamma_e - v : \gamma \in R_5 \cap \mathbb{Z}^{|E|} \right\}, \tag{25}$$

where $\bar{c}_{ij} = c_{ij} - u_i/2 - u_j/2$. Finally, a third and most important remark is that one has no guarantee that a partition of I into k subsets is contained in J' (or even in J for that matter). This limitation may be partially overcome with the introduction of the n single route variables $(0-i-0 : i \in V)$ mentioned above. One should nevertheless notice that even if (17) is feasible with respect to some initial subset J' of J, this does not of itself ensure the feasibility of (16). In this situation, infeasibility may only be detected during an implicit enumeration of the problem.

An optimal VRP solution for (16) is achieved whenever the solution to (17) is integral and the optimal solution to the column generation subproblem (24) is nonnegative. Otherwise one would have to resort to implicit enumeration (branch and bound). In this case, an additional difficulty would become apparent: that of dealing with the set of columns implicitly. An exact solution procedure, such as the one just sketched out above, is commonly referred to as a branch-and-price algorithm. A more detailed description of branch-and-price algorithms can be found in Barnhart et al. (1998).

A critical step in a branch-and-price algorithm is that of modifying a column generation subproblem, such as (24) above, as a result of branching. Ideally, one wishes to keep the basic subproblem structure intact throughout implicit enumeration. A commonly used branching rule that achieves this, for set partitioning problems with an exponential number of variables, is the following: at each enumeration node, decide whether two given elements of I should belong to the same subset or not. Let $i, j \in I$ be the elements involved. Each such branching would modify the column generation subproblem, (24), with the addition of either a constraint $\alpha_i = \alpha_j$ or a constraint $\alpha_i + \alpha_j = 1$.

The set partitioning reformulation of the VRP was introduced by Balinski and Quandt (1964). A column generation algorithm based on this reformulation was suggested by Agarwal et al. (1989). The column generation proposed there differs from the one suggested in this article. Column generation schemes for generalizations of the VRP have been proposed as early as Desrosiers et al. (1984).

3. Extensions to the Basic VRP

The three VRP formulations given in Section 2 above can be adapted to deal with the three common extensions to the basic VRP discussed in Section 1 above (namely, limits on the length of each vehicle route, time windows, and precedence constraints). In this section we indicate how this can be achieved.

3.1. *Limits on the Length of Each Vehicle Route*

The length or duration of a route is frequently limited for operational or legal reasons. Without loss of generality, we let ℓ_e denote the length of edge e, $e \in E$ (where length can be taken to mean duration if we so wish). Additionally, let L denote the maximum allowed length of a route.

The first VRP formulation we discussed above cannot be easily adapted to deal with this length restriction. The difficulty lies in ensuring that the number of vehicles serving a subset of customers, $S \subseteq V$, are capable of satisfying the length constraints. For a single vehicle serving S, for example, this would involve the optimal solution to the TSP defined on $S \cup \{0\}$. Laporte et al. (1985) adopt an approach whereby $r(S)$ must be chosen as the maximum between $\lceil \sum_{i \in S} d_i / b \rceil$ and $\lceil \sum_{e \in E(S) \cup E(S,\{0\})} \ell_e x_e / L \rceil$ and any violated inequalities, inequality (3), identified as they occur.

For the second formulation (the flow-based model) one may choose to enforce the length restriction through an additional network flow problem that is embedded into the original formulation.

Let f_{ij} be a real-valued flow variable associated with the length commodity. Let us restrict R_2 by introducing into it the following inequalities

$$\sum_{[i,j] \in \delta^-(i)} f_{ij} - \sum_{[j,i] \in \delta^+(i)} f_{ji} \geqslant \sum_{[j,i] \in \delta^+(i)} \ell_{ji}(y_{ji} + z_{ji})/b, \quad i \in V, \tag{26}$$

$$f_{i0} + \ell_{i0}(y_{i0} + z_{i0})/b \leqslant L, \quad i \in V, \tag{27}$$

$$f_{ij} \leqslant (y_{ij} + z_{ij})L/b, \quad [i,j] \in A, \tag{28}$$

$$f_{ij} \geqslant 0, \quad [i,j] \in A. \tag{29}$$

Constraint (26) increases a flow variable through a customer by at least the length of the arc used into the customer and constraint (27) ensures that the flow back to the depot cannot exceed the vehicle length limit. Constraint (28) limits the flow in an arc appropriately. Denote the resulting polyhedral region as R_6. Then restrictions on the length of a route can be imposed by solving

$$\min\left\{ \sum_{[i,j] \in A} (c_{ij}'/b)(y_{ij} + z_{ij}) : (y, z, f) \in R_6 \cap \{(y_{ij} + z_{ij}) \in \{0, b\}, [i,j] \in A\} \right\}. \tag{30}$$

Note here that if we wished to enlarge the first VRP formulation significantly, then through the introduction of flow variables, that formulation could be amended to incorporate restrictions on the length of a vehicle route in a similar manner to that presented above.

For the set partitioning–based model, the length restriction can be imposed with the introduction of a single constraint,

$$\sum_{e \in E} \ell_e \gamma_e \leqslant L, \tag{31}$$

into region R_5 (assuming that the subset J' of initial feasible routes used in (17) satisfy the length constraint). Let R_7 denote the resulting polyhedral region. Then, feasible routes are obtained by solving the column generation subproblem

$$\min\left\{ \sum_{e \in E} c_e \gamma_e - \sum_{i \in V} u_i \alpha_i - v : (\alpha, \gamma) \in R_7 \cap (\mathbb{Z}^{|V|}, \mathbb{Z}^{|E|}) \right\}. \tag{32}$$

Notice that if a single customer route $(0-i-0)$ has length greater than L then the problem becomes infeasible (assuming distances satisfy the triangle inequality). Notice further that whenever L denotes time, the processing times at customers, that is, $\{p_i : i \in V\}$, must be taken into account. In that respect, one could replace $\ell_e, e = (i,j) \in E$, by $\ell_e + (p_i + p_j)/2$ for the third formulation. Similarly, one could replace $\ell_{ij}, [i,j] \in A$, by $\ell_{ij} + p_i$, for the second formulation.

The modeling of the length of a route as in (26)–(29) is due to Lucena and Wolsey (1987) and Langevin et al. (1993).

3.2. *Time Windows*

When a customer visit must fall in a given time interval (i.e., a time window), one should notice that a directed graph, such as $D = (V_0, A)$, must be brought into effect. Assuming this is the case, a fairly straightforward way of modeling time windows is to rely on an auxiliary network flow problem, akin to (26)–(29). Let $[T_i^-, T_i^+]$, $T_i^- \leqslant T_i^+$, denote the time interval in which the visiting of vertex $i \in V$ is allowed. Also let $w_i, i \in V$, be a real-valued variable for the time spent by the vehicle while it sits, idle, at vertex i, prior to actually servicing that vertex.

For the first VRP formulation directly incorporating time window constraints is difficult and (if we wished) we would need to amend the formulation to incorporate flow variables.

For the second VRP formulation the imposition of time windows can be modeled by introducing into R_2 the following inequalities

$$\sum_{[i,j]\in\delta^-(i)} f_{ij} - \sum_{[j,i]\in\delta^+(i)} f_{ji} \geqslant \sum_{[j,i]\in\delta^+(i)} \ell_{ji}(y_{ji} + z_{ji})/b + w_i, \quad i \in V, \tag{33}$$

$$T_i^- \leqslant \sum_{[i,j]\in\delta^-(i)} f_{ij} \leqslant T_i^+, \quad i \in V, \tag{34}$$

$$f_{ij} \leqslant (y_{ij} + z_{ij})L/b, \quad [i,j] \in A, \tag{35}$$

$$f_{ij} \geqslant 0, \quad [i,j] \in A, \tag{36}$$

where L is any upper bound on the total time consumed by a single vehicle route and ℓ_{ij} are interpreted as the times to traverse an arc. Constraint (33) increases a flow variable out of i by at least the time consumed in traversing an arc to arrive at i plus the waiting time w_i. Constraint (34) is the time window constraint and constraint (35) limits the flow in any arc appropriately. Let the resulting polyhedral region be denoted by R_8. Routes satisfying time window constraints are obtained by solving

$$\min\left\{\sum_{[i,j]\in A} (c'_{ij}/b)(y_{ij} + z_{ij}) : (y, z, f) \in R_8 \cap \{(y_{ij} + z_{ij}) \in \{0, b\}, [i,j] \in A\}\right\}. \tag{37}$$

Adopting a similar approach for the set partitioning model, then, in order to incorporate the time window restriction one needs to introduce into R_5

$$\sum_{[i,j]\in\delta^-(i)} f_{ij} - \sum_{[j,i]\in\delta^+(i)} f_{ji} \geqslant \sum_{[j,i]\in\delta^+(i)} \ell_{ji}(y_{ji} + z_{ji})/b + w_i, \quad i \in V, \tag{38}$$

$$T_i^- \alpha_i \leqslant \sum_{[i,j]\in\delta^-(i)} f_{ij} \leqslant T_i^+ \alpha_i, \quad i \in V, \tag{39}$$

$$f_{ij} \leqslant (y_{ij} + z_{ij})L/b, \quad [i,j] \in A, \tag{40}$$

$$\gamma_e = (y_{ij} + z_{ij})/b, \quad [i,j] \in A, e \in E, e = (i,j), \tag{41}$$

$$f_{ij} \geqslant 0, \quad [i,j] \in A. \tag{42}$$

Let R_9 denote the polyhedral region that results. Time windows can then be introduced by solving the column generation subproblem

$$\min\left\{\sum_{e\in E} c_e\gamma_e - \sum_{i\in V} u_i\alpha_i - v : (\alpha, \gamma, f) \in R_9 \cap (\mathbb{Z}^{|V|}, \mathbb{Z}^{|E|}, \mathbb{R}^{|A|}) \cap \{(y_{ij} + z_{ij}) \in \{0, b\}, [i,j] \in A\}\right\}. \tag{43}$$

3.3. *Precedence Constraints*

There does not seem to be a clear consensus in the literature as to the exact meaning of precedence constraints for the VRP. Let $P_i \subseteq V \backslash \{i\}$ denote a set of precedence vertices for vertex $i \in V$. Then there are two possible interpretations as to the meaning of precedence constraints:

1. The first possibility is that all vertices in P_i must be visited in the same route as i (prior to i being visited). This implies that all vertices in $\{i\} \cup P_i$ need to be visited by the same vehicle.
2. The second possibility is that only those vertices in P_i that are in the same route as i must be visited before i is visited (i.e., vertices in $\{i\} \cup P_i$ need not necessarily be visited by the same vehicle).

In either case the only VRP model that appears to be capable of being easily adapted to cope with such precedence constraints is the set partitioning model. Therefore, this subsection will focus uniquely on that model.

Once again, we adopt the approach of incorporating an additional routing restriction though the introduction into the column generation subproblem of an associated network flow problem. As one may appreciate, the options here are various but, possibly, the simplest option is to consider the introduction into R_5 of constraints

$$\sum_{[0,i] \in \delta^-(0)} f_{0i} = 1, \tag{44}$$

$$\sum_{[i,j] \in \delta^-(i)} f_{ij} - \sum_{[j,i] \in \delta^+(i)} f_{ji} = \alpha_i, \quad i \in V, \tag{45}$$

$$\sum_{[j,k] \in \delta^+(k)} f_{jk} - r_k(1 - \alpha_i) \leqslant \sum_{[j,i] \in \delta^+(i)} f_{ji}, \quad k \in P_i, \quad i \in V, \tag{46}$$

$$f_{ij} + f_{ji} \leqslant (\min\{r_i, r_j\})\gamma_e, \quad e \in E, e = (i,j), \tag{47}$$

$$f_{ij} \geqslant 0, \quad [i,j] \in A, \tag{48}$$

where r_i gives an upper bound on the maximum number of edges that can be feasibly used on any route containing vertex i. Constraint (44) initializes the flow and constraint (45) increases a flow variable by one each time a vertex is visited. Constraint (46) ensures that all visited vertices in P_i are visited before i is visited (if i is visited) and constraint (47) ensures that there can be no flow if an edge is not used.

Equations (44)–(48) above deals with the second interpretation of precedence constraints. To deal with the first interpretation we merely need to add to (44)–(48) the constraint

$$\alpha_i = \alpha_k, \quad k \in P_i, \quad i \in V. \tag{49}$$

Let R_{10} be the corresponding polyhedral region in either case. Then precedence constraints are accommodated by solving the (single route) column generation subproblem

$$\min\left\{\sum_{e \in E} c_e \gamma_e - \sum_{i \in V} u_i \alpha_i - v : (\alpha, \gamma, f) \in R_{10} \cap (\mathbb{Z}^{|V|}, \mathbb{Z}^{|E|}, \mathbb{R}^{|A|})\right\}. \tag{50}$$

3.4. *Further Extensions*

The VRP formulations presented so far model situations where all vehicles are being used, simultaneously, in just one of two possible modes of operation, that is, either delivery or pickup (collection). However, in real-world applications, some vehicles may be allowed to operate in delivery mode while others are operating in pickup mode. A vehicle may even be allowed to switch, during a route, from one mode of operation to the other. When switching modes is permitted, it appears reasonable to allow pickups only after vehicles have completed their assigned deliveries (barring the case where a route involves pickups only). This is known in the literature as the VRP with backhauls (VRPB) problem.

Among the VRP models described here, the one that can most naturally be extended to the VRPB is the set partitioning model. Again, that can be achieved by introducing an auxiliary network flow problem into the column generation subproblem (24). In the process, the vertex set V can be split into two mutually disjoint subsets according to the type of operation required, say V_d and V_p, respectively, for deliveries and pickups. Proceeding in this way makes it simpler to impose the constraint that allows pickups only after vehicles have completed their assigned deliveries.

Another common extension involves the use of a nonhomogeneous (capacity-wise) fleet. The set partitioning–based model is, once again, the most straightforward model to amend in order to impose this constraint. First, one should introduce into (17) as many constraints as there are different types of vehicle capacity. Each such constraint is used to control the routes (and their number) that may be allocated to vehicles of a given capacity. Notice that, in the design of these constraints, a route that can be allocated to a particular vehicle of a certain capacity can also be allocated to vehicles of greater capacities.

In more detail, for a master problem solution, one different column generation subproblem will have to be solved for each vehicle capacity range. Say the different vehicle capacities are integral values $b_1 < ... < b_t$, where t is the number of different vehicle types. Capacity ranges should therefore be $[(b_0 + 1) = 1, b_1], [b_1 + 1, b_2], ... [b_{t-1} + 1, b_t]$. Let $v_1, ... v_t$ be the dual variables associated with the vehicle capacity constraints. Two changes to subproblem (24) are now necessary. One change is to replace $\sum_{i \in V} d_i \alpha_i \leqslant b$ (Equation (18)) by $b_{s-1} + 1 \leqslant \sum_{i \in V} d_i \alpha_i \leqslant b_s$ where s indicates the type of vehicle considered. The other change is to replace the constant term v in the objective function of (24) by the sum of $v_s, v_{s+1}, ..., v_t$.

For the other formulations, the constraint relating to a nonhomogeneous fleet can be imposed through an auxiliary network flow problem. In this case, the original depot vertex should be replaced by $k+1$ artificial vertices: one for the original depot and the others for each of the vehicles involved. Arcs directed away from the depot are the only ones that may reach vehicle vertices. Accordingly, only arcs directed away from customer vertices may reach the depot vertex. An upper limit equal to the corresponding vehicle capacity should be imposed for the flow on arcs between the depot vertex and each vehicle vertex. One should notice that all VRP extensions treated before can be further extended to the nonhomogeneous fleet case.

Finally, instead of a single depot, one may be confronted with multiple depots. In this case, it is normal practice to insist that a vehicle must return to its origin depot at the end of its route. A number of possibilities for modeling this constraint do exist. They basically evoke, in a more laborious way, the same techniques used above.

4. Conclusions

In this article we have considered the vehicle routing problem. We presented three different formulations of the basic problem and indicated how these formulations can be modified to deal with some of the more common extensions found in the literature. The VRP is a problem that, because of its practical importance and theoretical difficulty, has attracted much attention in previous years. We expect that this attention will continue into the future.

References

Agarwal, Y., K. Mathur, and H. M. Salkin. 1989. "A set-partitioning based exact algorithm for the vehicle routing problem." *Networks* **19**: 731–749.

Balas, E. 1989. "The prize collecting traveling salesman problem." *Networks* **19**: 621–636.

Balinski, M. L. and R. E. Quandt. 1964. "On an integer program for a delivery problem." *Oper. Res.* **12** 300–304.

Barnhart, C., E. L. Johnson, G. L. Nemhauser, M. W. P. Savelsbergh, and P. H Vance. 1998. "Branch-and-price: column generation for huge integer programs." *Oper. Res.* **46**: 316–329.

Campos, V., A. Corberan, and E. Mota. 1991. "Polyhedral results for a vehicle routing problem." *Eur. J. Oper. Res.* **52**: 75–85.

Christofides, N. 1985. "Vehicle routing." *The Traveling Salesman Problem*, edited by E. L. Lawler, J. K. Lenstra, A. H. G. Rinnooy Kan, and D. B. Shmoys. Chichester, England: J. Wiley & Sons.

Christofides, N. and S. Eilon. 1969. "An algorithm for the vehicle-dispatching problem." *Oper. Res. Q.* **20**: 309–318.

Christofides, N., A. Mingozzi, and P. Toth. 1981a. "Exact algorithms for the vehicle routing problem, based on spanning tree and shortest path relaxations." *Math. Programming* **20**: 255–282.

Christofides, N., A. Mingozzi, and P. Toth. 1981b. "State space relaxation procedures for the computation of bounds to routing problems." *Networks* **11**: 145–164.

Clarke, G. and J. W. Wright. 1964. "Scheduling of vehicles from a central depot to a number of delivery points." *Oper. Res.* **12**: 568–581.

Cornuejols, G. and F. Harche. 1993. "Polyhedral study of the capacitated vehicle routing problem." *Math. Programming* **60**: 21–52.

Dantzig, G. B., D. R. Fulkerson, and S. M. Johnson. 1954. "Solution of a large scale traveling salesman problem." *Oper. Res.* **2**: 393–410.

Dantzig, G. B. and J. H. Ramser. 1959. "The truck dispatching problem." *Mgmnt Sci.* **6**: 80–91.

Desrosiers, J., F. Soumis, and M. Desrochers. 1984. "Routing with time windows by column generation." *Networks* **14**: 545–565.

Finke, G., A. Clauss, and E. A. Gunn. 1984. "A two-commodity network flow approach to the traveling salesman problem." *Congressus Numerantium* **41**: 167–178.

Fisher, M. L. 1994. "Optimal solution of vehicle routing problems using minimum k-trees." *Oper. Res.* **42**: 626–642.

Fisher, M. L. 1995. "Vehicle routing problem." In *Networks Models, Handbooks in Operations Research and Management Science*, edited by T. L. Magnanti, C. L. Monma, and G. L. Nemhauser. Amsterdam: Elsevier Publisher B.V.

Gaskell, T. J. 1967. "Bases for vehicle fleet scheduling." *Oper. Res. Q.* **18**: 281–295.

Jünger, M., G. Reinelt, and G. Rinaldi. 1995. "The traveling salesman problem." In *Networks Models, Handbooks in Operations Research and Management Science*, edited by T. L. Magnanti, C. L. Monma, and G. L. Nemhauser. Amsterdam: Elsevier Publisher B.V.

Krolak, P. D., W. Felts, and J. H. Nelson. 1972. "A man-machine approach toward solving the generalized truck-dispatching problem." *Transport. Sci.* **6**: 149–170.

Langevin, A., M. Desrochers, J. Desrosiers, S. Gélinas, and F. Soumis. 1993. "A two-commodity flow formulation for the traveling salesman and the makespan problems with time windows." *Networks* **23**: 631–640.

Laporte, G. 1997. "The vehicle routing problem." In *Annotated Bibliographies in Combinatorial Optimization*, edited by M. Dell'Amico, F. Maffioli, and S. Martello. Chichester, England: J. Wiley & Sons.

Laporte, G. and Y. Norbert. 1983. "A branch and bound algorithm for the capacitated vehicle routing problem." *Oper. Res. Spektr.* **5**: 77–85.

Laporte, G. and Y. Norbert. 1987. "Exact algorithms for the vehicle routing problem." In *Surveys in Combinatorial Optimization, Annals of Discrete Mathematics, 31*, edited by S. Martello, G. Laporte, M. Minoux, and C. Ribeiro. Amsterdam: North-Holland.

Laporte, G., Y. Norbert, and M. Desrochers. 1985. "Optimal routing under capacity and distance restrictions." *Oper. Res.* **33**: 1050–1073.

Laporte, G. and I. H. Osman. 1995. "Routing problems: a bibliography." In *Freight Transportation, Annals of Discrete Mathematics, 61*, edited by G. Laporte and M. Gendreau. Amsterdam: North-Holland.

Lawler, E. L., J. K. Lenstra, A. H. G. Rinnooy Kan, and D. B. Shmoys (eds). 1985. *The Traveling Salesman Problem*. Chichester, England: J. Wiley & Sons.

Lucena, A., 1986. "Exact solution approaches for the vehicle routing problem." Ph.D. diss. London, England: Imperial College, University of London.

Lucena A. and L. A. Wolsey. 1987. "Two-commodity flow formulations for routing and scheduling problems." Unpublished manuscript.

Wolsey, L. A. 1998. "Column generation algorithms." In *Integer Programming*, chapter 11. Chichester, England: John Wiley & Sons.

17.3 Network Designs: Approximations for Steiner Minimum Trees

Ding-Zhu Du

ABSTRACT The Steiner tree is a typic optimization problem in network designs. Study of Steiner tree problems received great attention in the 1990s since many important open problems, including the Gilbert–Pollak conjecture on the Euclidean Steiner ratio, the existence of better approximation, and the existence of the polynomial-time approximation scheme, were solved, which influenced the general theory of designs and analysis of approximation algorithms for combinatorial optimizations. In this survey, we follow those recent developments to review results regarding the approximation of Steiner minimum trees and related problems. Throughout this review, we summarize and propose open problems to encourage research work in this direction.

1. Introduction

Designing optimal networks for certain problems, such as railroad design, telecommunication, and computer networks, has been an important research topic in applied mathematics and theoretical computer science for a long time. Many such problems are very difficult. Thus, finding polynomial-time approximation solutions plays an important role.

For example, given a set of points in the Euclidean plane, the shortest network interconnecting the points in the set is called the Steiner minimum tree. The Steiner minimum tree may contain some vertices that are not the given points. Those vertices are called Steiner points while the given points are called regular points. The Steiner minimum tree for three points was first studied by Fermat (1601–1665). Recent research on mathematical history showed that the Steiner minimum tree for more than three points was first studied by Gauss. In a letter to a friend, he discussed all possible topologies of Steiner minimum trees for four points. It is well known that the Steiner minimum tree is a NP-hard problem (Garey et al., 1977), that is, the optimal solution is unlikely to be computable in polynomial-time. Therefore, efforts have to be made to find good polynomial-time approximation solutions.

Shor and Smith (1992) proposed a greedy construction as follows: Start with all n regular points, regarded as a forest of n 1-node trees. At any stage, add the shortest possible line segment to the current forest, which causes two trees to merge. Continue until the forest is completely merged into a single tree. They conjectured that this greedy heuristic has a performance ratio of $(3 + 2\sqrt{3})$, which is smaller than $2/\sqrt{3}$. (Here, the performance ratio for an approximation algorithm is defined as the least upper bound for the ratio of lengths between the approximation solution and the Steiner minimum tree for the same set of given points.) Du (1995) disproved their conjecture by proving that this greedy heuristic has a performance ratio of $2/\sqrt{3}$. This terminated one of the efforts to find better polynomial-time approximations for Steiner minimum trees in the Euclidean plane.

In fact, it was a long-standing open problem as to whether there was a polynomial-time approximation for Steiner minimum trees in the Euclidean plane with a performance ratio smaller than $2/\sqrt{3}$. The number $\sqrt{3}/2$ is the value of the Steiner ratio in the Euclidean plane. The Steiner ratio in a metric space is the largest lower bound for the ratio of lengths between the Steiner minimum tree and the minimum spanning tree for the same set of points in the space. In other words, it is the inverse of the performance ratio of the minimum spanning tree as an approximation of the Steiner minimum tree. Gilbert and Pollak (1968) conjectured that the Steiner ratio in the Euclidean plane equals $\sqrt{3}/2$. This conjecture received a lot of attention and efforts (Graham and Hwang, 1976; Chung and Hwang, 1978; Pollak, 1978; Du and Hwang, 1983; Du et al., 1985; Chung and Graham, 1985; Friedel and Widmayer, 1989; Rubinstein and Thomas, 1989; Booth, 1991) and was finally proved by Du and Hwang (1990a). Their work received a lot of publicity (Kolata, 1990; Peterson, 1990; Cipra, 1991; Stewart, 1991), as it was the first breakthrough on the study of Steiner trees in the 1990s.

In general, the problem as to whether a polynomial-time approximation existed for Steiner minimum trees in each metric space with a performance ratio smaller than the inverse of the Steiner ratio was called the better approximation problem.

Zelikovsky (1993a) made the second breakthrough. He found a polynomial-time 11/6-approximation for Steiner minimum trees in graphs which beats the inverse of the Steiner ratio in graphs, $\rho_2^{-1} = 2$. Not long after this, Berman and Ramaiyer (1994) gave a polynomial-time 92/72-approximation for the Steiner minimum tree in the rectilinear plane and Du et al. (1991) gave a general solution for the open problem. They showed that in any metric space, there is a polynomial-time approximation with a performance ratio better than the inverse of the Steiner ratio provided that for any set of a fixed number of points, the Steiner minimum tree is polynomial-time computable. A major part of these works is the establishment of the lower bound for the k-Steiner ratio. Let us explain it as follows.

A tree interconnecting a regular point set is called a Steiner tree if every leaf is a regular point. However, a regular point in a Steiner tree may not be a leaf. A Steiner tree is full if every regular point is a leaf. When a regular point is not a leaf, the tree can be decomposed into several small trees at this point. In this way, every Steiner tree can be decomposed into smaller trees in each of which every regular point is a leaf. These smaller trees are called full components of the tree. The size of a full component is the number of regular points in the full component.

A k-size Steiner tree is a Steiner tree with all full components of size at most k. The k-size Steiner minimum tree is the shortest one among all k-size Steiner trees. The 2-size Steiner minimum tree is also called the minimum spanning tree. The k-Steiner ratio in a metric space is the least ratio of lengths between the Steiner minimum tree and the k-Steiner minimum tree for the same set of points in the metric space. The 2-Steiner ratio is exactly the Steiner ratio. A better lower bound for the k-Steiner ratio will give a better performance ratio for approximations of Zelikovsky's type. Zelikovsky (1993a) showed that the 3-Steiner ratio in graphs is at least 3/5. Du et al. (1991) showed that the k-Steiner ratio in graphs is at least $\lfloor \log_2 k \rfloor / (1 + \lfloor \log_2 k \rfloor)$. Recently, Borchers and Du (1995) completely determined the exact value of the k-Steiner ratio in graphs for all k and Borchers et al. (1998) determined the exact value of the k-Steiner ratio in the rectilinear plane for all k. These papers contain several new techniques for studying this type of problem.

Arora (1995) made the third breakthrough. He found a polynomial-time approximation scheme for Euclidean and rectilinear Steiner minimum trees. This surprising work is certainly a great achievement in the study of geometric optimization problems.

In this survey, we would like to review existing ideas in the analysis and construction of approximations for Steiner minimum trees and related problems, including the above developments. More importantly, we want, through this review, to find out how to make further progress from existing work and how to improve the existing techniques by summarizing and raising open problems.

2. The Steiner Ratio

A minimum spanning tree on a set of points is the shortest network interconnecting the points in the set with all edges between the points. While the Steiner tree problem is intractable, the minimum spanning tree can be computed pretty fast. As we mentioned in the introduction, the Steiner ratio is a measure of performance of the minimum spanning tree as a polynomial-time approximation of the Steiner minimum tree. Determining the Steiner ratio in each metric space is a traditional problem on Steiner trees. Hwang (1976) determined that the Steiner ratio in the rectilinear plane is 2/3. However, it took twentytwo years to complete the story of determining the Steiner ratio in the Euclidean plane. In 1968, Gilbert and Pollak conjectured that the Steiner ratio in the Euclidean plane is $\sqrt{3}/2$. Through efforts made by Graham and Hwang (1976), Chung and Hwang (1978), Pollak (1978), Du and Hwang (1983), Du et al. (1985), Chung and Graham (1985), Friedel and Widmayer (1989), Rubinstein and Thomas (1989), and Booth (1991), the conjecture was finally proved by Du and Hwang (1990a,b). The significance of their proof also stems from the potential applications of the new approach included in the proof.

In their approach, the central part is a new minimax theorem about minimizing the maximum value of several concave functions over a polytope as follows:

Minimax Theorem. *Let* $f(x) = \max_{i \in I} g_i(x)$ *where* I *is a finite set and* $g_i(x)$ *is a continuous, concave function in a polytope* X. *Then the minimum value of* f(x) *over the polytope* X *is achieved at some critical point, namely, a point satisfying the following property:*
(*) *There exists an extreme subset* Y *of* X *such that* $x \in Y$ *and the index set* $M(x)(= \{i \mid f(x) = g_i(x)\})$ *is maximal over* Y.

The Steiner ratio problem is first transformed to such a minimax problem [$g_i(x)$ = (the length of a Steiner tree) − (the Steiner ratio) × (the length of a spanning tree with graph structure i) where x is a vector whose components are edge-lengths of the Steiner tree] and the minimax theorem reduces the minimax problem to the problem of finding the minimax value of the concave functions at critical points. Then each critical point is transformed back to an input set of points with special geometric structure; it is a subset of a lattice formed by equilateral triangles. This special structure enables us to verify the conjecture corresponding to the nonnegativeness of minimax value of the concave functions.

Using the same approach, Gao et al. (1995) proved that in any Minkowski plane (or two-dimensional Banach space), the Steiner ratio is at least 2/3. They also proved an upper bound 0.8686 for the Steiner ratio in the Minkowski plane. This upper bound is very close to the conjectured upper bound $\sqrt{3}/2 = 0.8666\ldots$ that was made independently by Cieslik (1990) and Du et al. (1993). Actually, many interesting open problems have a possible solution with this approach. For example, the upper bound problem above can be transformed to the following max−min problem:

$$\max_{\|\cdot\|} \min_{(A,B)} MST_{\|\cdot\|}(A, B, C) \leqslant \sqrt{3}$$

where the maximization is over all norm $\|\cdot\|$ in the plane, the minimization is over all possible directions of edge (A, B) in the equilateral triangle ABC with unit edges, and $MST_{\|\cdot\|}(A, B, C)$ is the length of the minimum Steiner tree for three points A, B, C in the plane with norm $\|\cdot\|$. Clearly, the Euclidean norm is the most symmetric one. However, the above minimax theorem cannot be used. One of the reasons is that possible directions of edge (A, B) form an interval that is not a finite discrete set. Du and Pardalos (1994) proved a continuous version of the minimax theorem, which could be useful for this upper bound problem.

The research on this minimax approach has covered the following three aspects:

- Develop new minimax theorems of the above type.
- Find new methods to determine critical structures.
- Generate new proof techniques for each open problem on critical structures.

Any research development on this approach would help to solve the following open problems about the Steiner ratio.

2.1. *Chung–Gilbert's Conjecture on the Steiner Ratio in Euclidean Space*

Steiner trees in Euclidean spaces have applications in constructing phylogenetic trees (Cavalli-Sforza and Edwards, 1967) and the Internet (Smith, 1995). A conjecture was also suggested by Gilbert and Pollak (1968) that in any Euclidean space the Steiner ratio is achieved by the vertex set of a regular simplex. Chung and Gilbert (1976) constructed a sequence of Steiner trees on regular simplices. The lengths of constructed Steiner trees decreases to $\sqrt{3}/(4-\sqrt{2})$. Although the trees constructed are not known to be Steiner minimum trees, Chung and Gilbert conjectured that $\sqrt{3}/(4-\sqrt{2})$ is the best lower bound for Steiner ratios in Euclidean spaces. Clearly, if $\sqrt{3}/(4-\sqrt{2})$ is the limiting Steiner ratio in d-dimensional Euclidean space as d goes to infinity, then Chung–Gilbert's conjecture is a corollary of Gilbert and Pollak's general conjecture. However, recently, Smith (1992) and Du and Smith (1996) disproved Gilbert–Pollak's general conjecture for dimension more than two. This leaves a question mark over

Chung and Gilbert's conjecture. It is quite interesting that Chung–Gilbert's conjecture could still be true. In fact, we could get rid of Gilbert–Pollak's general conjecture and use another method to reach the conclusion that the limiting Steiner ratio for regular simplex is the best lower bound for Steiner ratios in Euclidean spaces. Consider n points in $(n-1)$-dimensional Euclidean space. Then all of $n(n-1)/2$ distances between the n points are independent. Suppose that we could do a similar transformation and the minimax theorem could apply to these n points to obtain a similar result in the proof of Gilbert–Pollak's conjecture for the Euclidean plane, that is, a point set with critical geometric structure has the property that the union of all minimum spanning trees contains as many equilateral triangles as possible. Then such a critical structure must be a regular simplex. Following this direction, Smith and Smith (1995) proved that the regular simplex of four points gives the smallest Steiner ratio over all sets of four points in a Euclidean space.

2.2. *Graham–Hwang's Conjecture on the Steiner Ratio in Rectilinear Space*

A Steiner tree with rectilinear distance is called a rectilinear Steiner tree. While rectilinear Steiner trees in plane have many applications on CAT and VLSI, rectilinear Steiner trees in high-dimensional space can be found in biology (Cavalli-Sforza and Edwards, 1967; Foulds and Graham, 1982; Thompson, 1973) and optimal traffic multicasting for some communication networks (Bharath-Kumar and Jaffe, 1983; Chang, 1972b). Although the Steiner ratio in the rectilinear plane was determined by Hwang (1976) many years ago, there has still been no further progress on the Steiner ratio in rectilinear spaces up to now. The Steiner ratio in a d-dimensional rectilinear space was conjectured to be $d/(2d-1)$ by Graham and Hwang (1976). The difficulty in extending Hwang's approach to proving Graham–Hwang's conjecture is due to the lack of knowledge on the full rectilinear Steiner trees in high dimensional spaces. In fact, for a full rectilinear Steiner tree in plane, all Steiner points lie on a path. However, it is not known whether a similar result holds for full rectilinear Steiner trees in a space of dimension more than two. Graham–Hwang's conjecture can easily be transformed to a minimax problem by our minimax approach. For example, choose lengths of all straight segments of a Steiner tree. When the connection pattern of the Steiner tree is fixed, the set of original points can be determined by such segment-lengths, the length of the Steiner tree is a linear function and the length of a spanning tree is a convex function of such segment-lengths, so that g_i is a concave function of such segment-lengths. However, for this transformation, it is hard to determine the critical structure.

2.3. *The Steiner Ratio in Banach Spaces*

A Minkowski space is a Banach space of finite dimension. In the other words, it is a finite dimensional linear space with a norm. Some applications of Steiner minimum trees with L_p norm can be found in Smith and Liebman (1979). Examining the proof of Gilbert–Pollak's conjecture in the Euclidean plane, we observe that the proof has nothing to do with special properties of the Euclidean norm except the last part, verification of the conjecture on point sets with critical structure. This means that using the minimax approach to determine the Steiner ratio in Minkowski planes, we would have no problem in finding a transformation and determining critical structures. We need to work only on point sets with critical structure. The following are three interesting conjectures on the Steiner ratios in Minkowski planes and in Banach spaces.

- In any Minkowski plane, the Steiner ratio is between $2/3$ and $\sqrt{3}/2$. (The first half is proved in Gao et al. (1995).)
- In any infinite dimensional Banach space, the Steiner ratio is between $1/2$ and $\sqrt{3}/(2-\sqrt{2})$.
- The Steiner ratio in a Banach space equals that in its dual space.

Note that each Banach space is a linear space with a norm. The first and the second conjectures mean that if the linear space is fixed, the Steiner ratio will achieve the minimum for the rectilinear norm and the maximum for the Euclidean norm. More conjectures and open problems on the Steiner ratio in L_p normed plane can be found in Liu and Du (1991).

2.4. *Smiths' Sausage Conjecture on the Steiner Ratio in Three-Dimensional Euclidean Space*

Smith and Smith (1995) studied sausages consisting of a sequence of regular simplices. They found that in a Euclidean space of dimension more than two, the Steiner ratio decreases as the length of the sausage increases. In three-dimensional Euclidean space, the limiting ratio is

$$\rho_2(E^3) = \sqrt{\frac{283}{700} - \frac{3\sqrt{21}}{700} + \frac{9\sqrt{11-\sqrt{21}}\sqrt{2}}{140}}.$$

They conjectured that this value is the Steiner ratio in three-dimensional Euclidean space. This conjecture is quite interesting. In fact, so far a Banach space has not been found in which the Steiner ratio cannot be achieved by some sausage (including infinite sausage like that in Smiths' conjecture). Thus, one might conjecture that in any Banach space, the Steiner ratio equals the least upper bound for the Steiner ratios on finite sausages.

Pratt (1991) proposed a problem of determining the Steiner ratio on a sphere. He observed that the telephone network is distributed in a large area so that it should be considered on a sphere rather than an Euclidean plane because the earth is a sphere. He also claimed that $\sqrt{3}/2$ should also be the Steiner ratio on a sphere and proved that his claim is true for three points. There is a difficulty in using the minimax approach directly for Pratt's problem because the concavity of g_i would be lost when a similar transformation is used. Recently, Rubinstein and Weng (1997) showed that Pratt's claim is true. They transform the Steiner minimum tree on a sphere to a Steiner tree in the Euclidean plane and showed that each minimum spanning tree becomes shorter after this transformation. This result indicates that it is possible to do a nonlinear transformation from a metric space to another metric space such that the Steiner ratio keeps the same value and it is possible to use minimax approach.

3. Better Approximations

Starting from a minimum spanning tree, improve it by adding Steiner points. This is a natural idea to obtain an approximation solution for the Steiner minimum tree. Every approximation solution obtained in this way would have a performance ratio at most the inverse of the Steiner ratio. The problem is how much better than the inverse of the Steiner ratio one can get.

Before 1990, numerous heuristics (Karp, 1972; Chang, 1972a; El-Arbi, 1978; Kou et al., 1981; Smith et al., 1981; Rayward-Smith, 1983; Bharath-Kumar and Jaffe, 1983; Wu et al., 1986; Suzuki and Iri, 1986; Kou and Makki, 1987; Waxman and Imase, 1988; Beasley, 1989) for Steiner minimum trees have been proposed for points in various metric spaces. Their superiority over minimum spanning trees were often claimed by computation experiments. But no theoretical proof of superiority was ever given. It was a long-standing problem whether a polynomial-time approximation existed with a performance ratio better than the inverse of the Steiner ratio. For simplicity, a polynomial-time approximation with a performance ratio smaller than the inverse of the Steiner ratio will be called a better approximation. The first significant work on better approximations was made by Bern (1986). He proved that for the rectilinear metric and Poisson distributed regular points, a greedy approximation, obtained by a very simple improvement over a minimum spanning tree, has a shorter average length. Later, Hwang and Yao (1990) extended this result to the usual case when the number of regular points is fixed.

In the following, we study some approximation algorithms possibly with or without worst-case better performance ratio.

3.1. *Chang's Idea*

Chang (1972a,b) proposed the following approximation algorithm for Steiner minimum trees in the Euclidean plane: Start from a minimum spanning tree and at each iteration choose a Steiner point such that using this Steiner point to connect three vertices in the current tree could replace two edges in the minimum spanning tree and this replacement achieves the maximum saving among such possible replacements.

Smith et al. (1981) also use the idea of the greedy improvement. But, they start with Delaunay triangulation instead of a minimum spanning tree. Since every minimum spanning tree is contained in Delaunay triangulation, the performance ratio of their approximation algorithm can also be bounded by the inverse of the Steiner ratio. The advantage of the Smith–Lee–Liebman algorithm is in the running time. While Chang's algorithm runs in $O(n^3)$ time, the Smith–Lee–Liebman algorithm runs only in $O(n \log n)$ time.

Motivated by the work of Georgakopoulos and Papadimitriou (1987), Kahng and Robins (1990) proposed an approximation algorithm for Steiner minimum trees in the rectilinear plane by using the same idea as that of Chang. For these three algorithms, it can be proved that for any particular set of points, the ratio of lengths of the approximation solution and the Steiner minimum tree is smaller than the inverse of the Steiner ratio. Some experimental results also show that the approximation solution obtained by these algorithms are very good. However, no proof has been found to show any one of them being a better approximation.

3.2. *Zelikovsky's Idea*

As we mentioned in the introduction, Zelikovsky's idea (1993a) is based on the study of k-size Steiner trees. Clearly, for any $k \geqslant 3$, a k-size Steiner minimum tree usually has a shorter length compared with a minimum spanning tree. It is natural to think about using a minimum k-size Steiner tree to approximate the Steiner minimum tree. However, this does not work because computing a k-size Steiner minimum tree is still an intractable problem. Zelikovsky's idea is to approximate the Steiner minimum tree by a 3-size Steiner tree generated by a polynomial-time greedy algorithm. The key fact is that the length of such a heuristic is smaller than the arithmetic mean of lengths of a minimum spanning tree and a 3-size Steiner minimum tree; that is, the performance ratio of his approximation satisfies

$$PR \leqslant \frac{\rho_2^{-1} + \rho_3^{-1}}{2}$$

where ρ_k is the k-Steiner ratio. Thus, if the 3-Steiner ratio ρ_3 is bigger than the Steiner ratio ρ_2, then this greedy algorithm is a better approximation for the Steiner minimum tree. Zelikovsky was able to prove that the 3-Steiner ratio in graphs is at least 3/5, which is bigger than 1/2, the Steiner ratio in graphs (Karp, 1972). So, he solved the better approximation problem in graphs. Zelikovsky's idea has been extensively studied in the literature.

Du et al. (1991) generalized Zelikovsky's idea to the k-size Steiner tree. They showed that a generalized Zelikovsky's algorithm has performance ratio

$$PR \leqslant \frac{(k-2)\rho_2^{-1} + \rho_k^{-1}}{k-1}.$$

Berman and Ramaiyer (1994) employed a different idea to generalize Zelikovsky's result. They obtained an algorithm with the performance ratio satisfying

$$PR \leqslant \rho_2^{-1} - \sum_{i=3}^{k} \frac{\rho_{i-1}^{-1} - \rho_i^{-1}}{i-1}.$$

They also showed that in the rectilinear plane, the 3-Steiner ratio is at least 72/94, which is bigger than 2/3 (Hwang, 1976), the Steiner ratio in the rectilinear plane. So, they solved the better approximation problem in the rectilinear plane. Some improvements on this solution were made by Foßmeier et al. (1993), Berman et al. (1994), and Zelikovsky (1992).

Du et al. (1991) proved a lower bound for the k-Steiner ratio in any metric space. This lower bound goes toward one as k goes to infinity. So, in any metric space with a Steiner ratio less than one, there exists a k-Steiner ratio bigger than the Steiner ratio. Thus, they proved that the better heuristic exists in any metric space satisfying the following conditions:

1. The Steiner ratio is smaller than one.

2. The Steiner minimum tree on any fixed number of points can be computed in polynomial-time.

These metric spaces include Euclidean plane and Euclidean spaces.

Zelikovsky (1995) used a different potential function in his greedy approximation and obtained an approximation with a performance ratio satisfying

$$PR \leqslant \rho_k^{-1}(1 - \ln \rho_2).$$

3.3. *Comparison of the Two Ideas*

Although Zelikovsky's idea starts from a point different from Chang's, the two approximations are actually similar. To see this, let us describe Zelikovsky's algorithm as follows: Start from a minimum spanning tree and at each iteration choose a Steiner point such that using this Steiner point to connect three regular points could replace two edges in the minimum spanning tree, achieving the maximum saving among such possible replacements.

Clearly, they both start from a minimum spanning tree and improve it step by step by using a greedy principal to choose a Steiner point to connect a triplet of vertices. The difference is only that this triplet in Chang's algorithm may contain some Steiner points while it contains only regular points in Zelikovsky's algorithm. This difference makes Chang's approximation hard to analyze. Which one will give a better approximation solution? This is an interesting open problem.

3.4. *The k-Steiner Ratio ρ_k*

As we saw in Section 3.2, the determination of the k-Steiner ratio plays an important role in estimation of the performance ratio of several recent better approximations. Recently, Borchers and Du (1995) have completely determined the k-Steiner ratio in graphs that for $k = 2^r + h \geqslant 2$,

$$\rho_k = \frac{r2^r + h}{(r+1)2^r + h}$$

and Borchers et al. (1998) have completely determined that for the k-Steiner ratio in the rectilinear plane $\rho_2 = 2/3$, $\rho_3 = 4/5$, and for $k \geqslant 4$, $\rho_k = (2k-1)/(2k)$.

3.5. *Variable Metric Method*

Berman and Ramaiyer (1994) introduced an interesting approach to generalize Zelikovsky's greedy approximation. Let us call the Steiner minimum tree for a subset of k regular points a k-tree. Their approach consists of two steps. The first step processes all i-trees, $3 \leqslant i \leqslant k$, sequentially in the following way: For each i-tree T with positive saving in the current graph, put T in a stack and if two leaves x and y of T are connected by a path p in a minimum spanning tree without passing any other leaf of T, then put an edge between x and y with weight equal to the length of the longest edge in p minus the saving of T. In the second step, it repeatedly pops i-trees from the stack remodifying the original minimum spanning tree for all regular points and keeping only i-trees with the current positive saving. Adding weighted edges to a point set would change the metric on the points set. Let E be an arbitrary set of weighted edges such that adding them to the input metric space makes all i-trees for $3 \leqslant i \leqslant k$ have a nonpositive saving in the resulting metric space M_E. Denote by $t_i(P)$ a supremum of the length of a minimum spanning tree for the point set P in metric space M_E over all such Es. Then Berman–Ramaiyer's algorithm produces a k-size Steiner tree with total length at most

$$t_2(P) - \sum_{i=3}^{k} \frac{t_{i-1}(P) - t_i(P)}{i-1} = \frac{t_2(P)}{2} + \sum_{i=3}^{k-1} \frac{t_i(P)}{(i-1)i} + \frac{t_k(P)}{k-1}.$$

The bound for the performance ratio of Berman–Ramaiyer's approximation in Section 3.2 is obtained from this bound and the fact that $t_k(P) \leqslant \rho_k^{-1} SMT(P)$, where $SMT(P)$ is the length of the Steiner minimum tree for point set P.

Based on the above observation, we may have the following questions. Could we find another way to vary metric for a better bound? Could we forget the greedy idea and design a better approximation with only the variable metric idea? Answering these questions requires deeper understanding of the variable metric method. We attempt to obtain new algorithms from this study.

3.6. *Preprocessing*

Recently, Karpinski and Zelikovsky (1997) proposed a preprocessing procedure to improve the variable metric method. First, they use this procedure to choose some Steiner points and then run a variable metric method on the union of the set of regular points and the set of chosen Steiner points. This preprocessing improves the performance ratio for the variable metric method.

The preprocessing procedure is similar to the algorithm of Berman and Ramaiyer, but it uses a related gain instead of the saving of the greedy function. One possible idea is to modify Chang's algorithm in the following way: At each iteration, if a Steiner point is introduced, then compute its related gain, and later consider only triplets of regular points and Steiner points with positive related gain. Would this approximation perform better? It is an open problem.

A more accurate analysis for the performance ratios of Karpinski–Zelikovsky's preprocessing requires bounds for t_k and a similar number t^k. The techniques in Borchers and Du (1995) and Borchers et al. (1998) for determining the k-Steiner ratio seem very promising for establishing tight upper bounds for t_k and t_k.

3.7. *Decompositions*

Jiang et al. (1994) and Jiang and Wang (1994) brought a quite different idea from previous ones to approximate Steiner minimum trees. They decompose the set of regular points based on the lengths of edges in a minimum spanning tree. By an interesting analysis, they proved that if the ratio of lengths between the longest edge and the shortest edge in a minimum spanning tree is bounded by a constant, then there is a polynomial-time approximation scheme for Steiner minimum trees in the rectilinear plane and in the Euclidean plane.

Arora (1995) found a revolutionary way to do the decomposition. His method is quite powerful; it gives not only a polynomial-time approximation scheme to Euclidean and rectilinear Steiner minimum trees, but also polynomial-time approximation schemes to other geometric optimization problems, including the geometric traveling salesman problem.

3.8. *On the Performance Ratio*

Arora's work (1995) almost completed the story of the theoretical study for approximations of Euclidean and rectilinear Steiner minimum trees. However, his approach cannot apply to Steiner trees in graphs. For this problem, none of the existing polynomial-time approximations has a performance ratio smaller than the inverse of the 3-Steiner ratio. The inverse of the 3-Steiner ratio seems to be the barrier.

Arora et al. (1994) conjectured that their backtrack greedy technique gives a polynomial-time approximation scheme to the 3-size Steiner minimum trees. If their conjecture is true, then their algorithms also give approximations for Steiner minimum trees in graphs with the performance ratio approach to the inverse of the 3-Steiner ratio. This is possibly the best possible performance ratio. Thus, the conjecture of Arora et al. is an attractive problem for further research. It is worth mentioning that recently, Martins et al. (1999) presented a heuristic with the best experimental result for Steiner minimum trees in graphs.

The knowledge for the lower bound of the performance ratio is wide open. It is only known that for Steiner minimum trees in graphs, if $NP \neq P$ then a lower bound larger than one exists, because the problem in this case is MAX SNP-complete (Bern and Plassmann, 1989).

4. Related Problems

Based on various applications, many different Steiner tree problems have been proposed in the literature. They are often NP-hard. Therefore, approximation is also an important topic to them. In the following, we mention a few of them.

Steiner Minimum Trees in Directed Graphs. Given a weighted directed graph G, a vertex r, and a subset P of n vertices, find a shortest tree with root r such that for each $x \in P$ a path exists from r to x. If $NP \neq P$, then the best performance ratio of polynomial-time approximation for this problem is $O(n \log n)$. But, for a version of this problem in the rectilinear plane, there is a polynomial-time approximation with performance ratio 2 (Rao et al., 1992). Recently, Lu and Ruan (1999) presented a polynomial-time approximation scheme for this problem.

On-Line Steiner Trees. When a new customer is out of the original telephone network, the company has to build a new line to connect the customer to the network. This situation brings us an on-line Steiner tree problem as follows: Assume that a sequence of points in a metric space are given step by step. In the ith step, only locations of the first n_i points in the sequence are known. The problem is to construct a shorter network at each step based on the network constructed in previous steps. The study of on-line problems was initiated by Sleator and Tarjan (1985) and Manase et al. (1988). A criterion for the performance of an on-line algorithm is to compare the solution generated by the on-line algorithm with the solution of the corresponding off-line problem. In the Euclidean plane, it has been known that the worst-case ratio of lengths between the on-line solution and off-line solution is between $O(n \log n / \log \log n)$ and $O(n \log n)$.

Steiner Trees in λ-Geometry Plane. A λ-geometry plane is a Minkowski plane with a center-symmetric convex polygon as the unit disc. Steiner trees in the λ-geometry plane have applications in VLSI designs. Many interesting open problems in this research direction can be found in Du and Hwang (1992) and Lee and Shen (1996).

Intersection of High Ways (Du et al., 1999). Given n straight segments in the Euclidean plane, find a shortest tree interconnecting the n segments. This is wide open for many issues.

Multiphase Spanning Network (Du and Miller, 1988). Given an edge-weighed complete graph with vertex set X ($|X| = n$) and subsets $X_1, ..., X_m$ of vertices, the problem is to find a minimum weighed subgraph G such that for every $i = 1, ..., m$, G contains a spanning tree for X_i. If $NP \neq P$, then the best performance ratio of polynomial-time approximation for this problem is $O(n \log n)$. But, in the unit weight case (this case has more applications), it is unknown whether a constant-bounded polynomial-time approximation exists.

Multiphase Steiner Network (Du and Miller, 1988). Given an edge-weighed graph B with vertex set X and subsets $X_1, Y_1, ..., X_m, Y_m$ of X with $X_i \cap Y_i = \emptyset$, the problem is to find a minimum weighted subgraph G such that for every $i = 1, ..., m$, G contains a Steiner tree for X_i without using vertices not in Y_i.

The last two problems arose in communication network design (Prisner, 1992) and vacuum system design (Du and Miller, 1988). For the former, when the solution is a forest, the system $(X_1, ..., X_m)$ is called a subtree hypergraph. Such a system has various applications in computer database schemes (Beeri et al., 1983) and statistics. It is also related to chordal graphs (Duchet, 1978; Flament, 1978). Tarjan and Yannakakis (1984) gave an $O(m + n)$-time algorithm to tell whether a set system is a subtree hypergraph.

Constructing phylogenetic trees is an important topic in computer biology. One of the formulations is as follows: For a fixed alphabet A, let d denote the Hamming distance on A^n, that is, $d[(a_1, ..., a_n), (b_1, ..., b_n)]$ equals the number of indices i such that $a_i \neq b_i$. Given a set P of points in the metric space (A^n, d), find a Steiner minimum tree for P. This problem is known to be NP-hard (see Foulds and Graham, 1982). So, one looks for various heuristics (Hwang and Richards, 1992). Comparing the phylogenetic tree problem with the multiphase Steiner network problem, we find some similarities between them if we look at each coordinate as a phase. For the multiphase Steiner tree problem, if the solution is a tree, then we have either a good heuristic or a polynomial-time computable exact solution (Du and Miller, 1988). This suggests to us to study the relationship between the two problems. It is hoped that a new construction of phylogenetic trees may be found from this study.

Finally, we would like to point out that several books (Hwang et al., 1992; Ivanov and Tuzhilin, 1994; Cieslik, 1998) on Steiner trees have been published in the 1990s. The reader may find more information from them.

ACKNOWLEDGMENTS

This research was supported in part by the National Science Foundation under grant CCR-9530306 and in part by the National 973 Information Technology and High-Performance Software Program of China.

References

Arora, S. 1995. "Polynomial time approximation schemes for Euclidean TSP and other geometric problems." In *Proceedings of 37th FOCS*.

Arora, V., V. Santosh, H. Saran, and V. Vazirani. 1994. "A limited-backtrack greedy schema for approximation algorithms." Manuscript.

Beasley, J. E. 1989. "A heuristic for the Euclidean and rectilinear Steiner problems." Technical Report. London, England: Management School, Imperial College.

Beeri, C., R. Fagin, D. Maier, and M. Yannakakis. 1983. "On the desirability of acyclic database schemes." *J. ACM* **30**: 479–513.

Berman, P. and V. Ramaiyer. 1994. "Improved approximations for the Steiner tree problem." *J. Algorithm* **17**: 381–408.

Berman, P., U. Foßmeier, M. Karpinski, M. Kaufmann, and A. Zelikovsky. 1994. "Approaching the 5/4-approximations for rectilinear Steiner trees." *LNCS* **855**: 60–71.

Bern, M. 1986. "Two probabilistic results on rectilinear Steiner trees." In *Proceedings of the 18th STOC*, pp. 433–441.

Bern, M. and D. Eppstein. 1997. "Approximation algorithms for geometric problems." In *Approximation Algorithms for NP-hard Problems*, edited by D. S. Hochbaum, pp. 296–345: PWS Publishing Company.

Bern, M. and P. Plassmann. 1989. "The Steiner problem with edge lengths 1 and 2." *Inf. Processing Lett.* **32**: 171–176.

Bharath-Kumar, K. and J. M. Jaffe. 1983. "Routing to multiple destinations in computer networks." *IEEE Trans. Commun.* **31**: 343–351.

Borchers, A. and D.-Z. Du. 1995. "The *k*-Steiner ratio in graphs." In *Proceedings of the 27th STOC*, pp. 641–649.

Borchers, A., D.-Z. Du, B. Gao, and P.-J. Wan. 1998. "The *k*-Steiner ratio in the rectilinear plane." *J. Algorithms* **29**: 1–17.

Booth, R. S. 1991. "The Steiner ratio for five points." *Discrete Comput. Geometry*.

Cavalli-Sforza, L. L. and A. W. Edwards. 1967. "Phylogenetic analysis: models and estimation procedures." *Am. J. Human Genetics* **19**: 233–257.

Chang, S.-K. 1972a. "The generation of minimal trees with a Steiner topology." *J. ACM* **19**: 699–711.

Chang, S.-K. 1972b. The design of network configurations with linear or piecewise linear cost functions. In *Symposium on Computer-Communications, Networks, and Teletraffic*, pp. 363–369.

Chung, F. R. K. and E. N. Gilbert. 1976. "Steiner trees for the regular simplex." *Bull. Inst. Math. Acad. Sinica* **4**: 313–325.

Chung, F. R. K. and R. L. Graham. 1985. "A new bound for Euclidean Steiner minimum trees." *Ann. N.Y. Acad. Sci.* **440**: 328–346.

Chung, F. R. K. and F. K. Hwang. 1978. "A lower bound for the Steiner tree problem." *SIAM J. Appl. Math.* **34**: 27–36.

Cieslik, D. 1990. "The Steiner ratio of Banach–Minkowski planes." In *Contemporary Methods in Graph Theory*, edited by R. Bodendiek, pp. 231–248. Mannheim, Germany: BI-Wissenschatteverlag.

Cieslik, D. 1998. In *Steiner Minimal Trees*: Kluwer Academic Publishers.

Cipra, B. A. 1991. "Euclidean geometry alive and well in the computer age." *SIAM News* **24(1)**.

Cole, T. 1995. "A property of T_3-trees." Manuscript.

Du, D.-Z. 1995. "On greedy heuristics for Steiner minimum trees." *Algorithmica* **13**: 381–386.

Du, D.-Z. and F. K. Hwang. 1983. "A new bound for the Steiner ratio." *Trans. Am. Math. Soc.* **278**: 137–148.

Du, D.-Z. and F. K. Hwang. 1987. "Steiner minimal trees for bar waves." *Acta Math. Appl. Sinica* **3**: 246–256.

Du, D.-Z. and F. K. Hwang. 1990a. "The Steiner ratio conjecture of Gilbert–Pollak is true." *Proc. Natl Acad. Sci.* **87**: 9464–9466.

Du, D.-Z. and F. K. Hwang. 1990b. "An approach for proving lower bounds: solution of Gilbert–Pollak's conjecture on Steiner ratio." In *Proceedings of the 31th FOCS*, pp. 76–85.

Du, D.-Z. and F. K. Hwang. 1992. "Reducing the Steiner problem in a normed space with a *d*-dimensional polytope as its unit sphere." *SIAM J. Comput.* **21**: 1001–1007.

Du, D.-Z. and Z. Miller. 1988. "Matroids and subset interconnection design." *SIAM J. Discrete Math.* **1**: 416–424.

Du, D.-Z. and P. M. Pardalos. 1994. "A continuous version of a result of Du and Hwang." *J. Global Optim.* **5**: 127–129.

Du, D.-Z. and W. D. Smith. 1996. "Three disproofs for Gilbert–Pollak conjecture in high dimensional spaces." *J. Combinat. Theory* **74**: 115–130.

Du, D.-Z., E. N. Yao, and F. K. Hwang. 1982. "A short proof of a result of Pollak on Steiner minimal trees." *J. Combinat. Theory, Series A* **32**: 396–400.

Du, D.-Z., F. K. Hwang, and E. Y. Yao. 1985. "The Steiner ratio conjecture is true for five points." *J. Combinat. Theory, Series A* **38**: 230–240.

Du, D.-Z., Y. Zhang, and Q. Feng. 1991. "On better heuristic for Euclidean Steiner minimum trees." In *Proceedings of the 32nd FOCS.*

Du, D.-Z., B. Gao, R. L. Graham, Z.-C. Liu, and P.-J. Wan. 1993. "Minimum Steiner trees in normed planes." *Discrete Comput. Geometry* **9**: 351–370

Du, D.-Z., F. K. Hwang, and G. Xue. 1999. "Interconnecting highways." *SIAM Discrete Math.* **12**: 252–261.

Duchet, P. 1978. "Propriete de Helly et problemes de representation." *Colloque Int. Paris–Orsay* **260**: 117–118.

El-Arbi, C. 1978. "Une heuristique pour le probleme de l'Arbre de Steiner." *RAIRO* **12**: 207–212.

Flament, C. 1978. "Hypergraphes arbores." *Discrete Math.* **21**: 223–227.

Friedel, J. and P. Widmayer. 1989. "A simple proof of the Steiner ratio conjecture for five points." *SIAM J. Appl. Math.* **49**: 960–967.

Foßmeier, U., M. Kaufman, and A. Zelikovsky. 1993. "Fast approximation algorithms for the rectilinear Steiner tree problem." *LNCS* **762**: 533–542.

Foulds, L. R. and R. L. Graham. 1982. "The Steiner problem in phylogeny is NP-complete." *Adv. Appl. Math.* **3**: 43–49.

Gao, B., D.-Z. Du, and R. L. Graham. 1995. "A tight lower bound for the Steiner ratio in Minkowski planes." *Discrete Math.* **142**: 49–63.

Garey, M. R., R. L. Graham, and D. S. Johnson. 1977. "The complexity of computing Steiner minimal trees." *SIAM J. Appl. Math.* **32**: 835–859.

Georgakopoulos, G. and C. H. Papadimitriou. 1987. "The 1-Steiner tree problem." *J. Algorithms* **8**: 122–130.

Gilbert, E. N. and H. O. Pollak. 1968. "Steiner minimal trees." *SIAM J. Appl. Math.* **16**: 1–29.

Graham, R. L. and F. K. Hwang. 1976. "Remarks on Steiner minimal trees." *Bull. Inst. Math. Acad. Sinica* **4**: 177–182.

Hwang, F. K. 1976. "On Steiner minimal trees with rectilinear distance." *SIAM J Appl. Math.* **30**: 104–114.

Hwang, F. K. and D. S. Richards, 1992. "Steiner tree problems." *Networks* **22**: 55–89.

Hwang, F. K. and Y. C. Yao. 1990. "Comments on Bern's probabilistic results on rectilinear Steiner trees." *Algorithmica* **5**: 591–598.

Hwang, F. K., D. S. Richards, and P. Winter. 1992. *Steiner Tree Problems.* Amsterdam: North-Holland.

Ivanov, A. O. and A. A. Tuzhilin. 1994. *Minimal Networks: The Steiner Problem and Its Generalizations.* CRC Press.

Jamik, V. and O. Kossler. 1934. "O minimalnich gratech obsahujicich n danych bodu (in Czech)." *Casopis Pesk. Mat. Fyr.* **63**: 223–235.

Jiang, T. and L. Wang. 1994. "An approximation scheme for some Steiner tree problems in the plane." *LNCS* **834**: 414–422.

Jiang, T., E. L. Lawler, and L. Wang. 1994. "Aligning sequences via an evolutionary tree: complexity and approximation." In *Proceedings of the 26th STOC.*

Kahng, A., and G. Robins. 1990. "A new family of Steiner tree heuristics with good performance: the iterated 1-Steiner approach." In *Proceedings of the IEEE International Conference on Computer-Aided Design,* Santa Clara, CA, pp. 428–431.

Karp, R. M. 1972. "Reducibility among combinatorial problems." In *Complexity of Computer Computation,* edited by R. E. Miller and J. W. Tatcher. pp. 85–103. New York: Plenum Press.

Karpinski, M. and A. Z. Zelikovsky. 1997. "New approximation algorithms for Steiner tree problems." *J. Combinat. Optim.* **1**: 47–65.

Kolata, G. 1990. "Solution to old puzzle: how short a shortcut?" *New York Times* 30 October 1990.

Korthonen, P. 1979. "An algorithm for transforming a spanning tree into a Steiner tree." In *Survey of Mathematical Programming (2),* pp. 349–357. Amsterdam: North-Holland.

Kou, L. and K. Makki. 1987. "An even faster approximation algorithm for the Steiner tree problem in graph." *Congressus Numerantium* **59**: 147–154.

Kou, L., G. Markowsky, and L. Berman. 1981. "A fast algorithm for Steiner trees." *Acta Informatica* **15**: 141–145.

Lee, D. T. and C. F. Shen. 1996. "The Steiner minimal tree problem in the λ-geometry plane." *Lecture Notes Comput. Sci.* **1178**: 247–255.

Levin, A. J. 1971. "Algorithm for the shortest connection of a group of vertices." *Sov. Math. Doklady* **12**: 1477–1481.

Liu, Z. C. and D. Z. Du. 1991. "On Steiner minimal trees with L_p distance." *Algorithmica* **7**: 179–191.

Lu, B. and L. Ruan. 1999. "Polynomial-time approximation scheme for the rectilinear Steiner arborescence problem." Manuscript.

Lundy, M. 1985. "Applications of the annealing algorithm to combinatorial problems in statistics." *Biometrika* **72**: 191–198.

Lundy, M. and A. Mees. 1986. "Convergence of an annealing algorithm." *Math. Programming* **34**: 111–124.

Manase, M. S., L. A. McGeoch, and D. D. Sleator. 1988. "Competitive algorithms for on-line problems." In *Proceedings of the 20th STOC*, Chicago.

Martins, S. L., M. G. C. Resende, C. C. Ribeiro, and P. M. Pardalos. 1999. "A parallel grasp for the Steiner tree problem in graphs using a hybrid local search strategy." *J. Global Optim.*

Pollak, H. O. 1978. "Some remarks on the Steiner problem." *J. Combinat. Theory, Ser. A* **24**: 278–295.

Pratt, V. 1991. Personal communication.

Peterson, I. 1990. "Proven path for limiting shortest shortcut." *Science News*, 22–29 December 1990, p. 389.

Prisner, E. 1992. "Two algorithms for the subset interconnection design." *Networks* **22**: 385–395.

Prisner, E. 1988. "Familien zusammenhangender Teilgraphen eines Graphen und ihre Durchschnittsgraphen." Dissertation. Hamburg, Germany: Universitat Hamburg.

Provan, I. S. 1988. "Convexity and the Steiner tree problem." *Networks* **18.**

Rayward-Smith, V. J. 1983. "The computation of nearly minimal Steiner trees in graphs." *Int. J. Math. Ed. Sci. Tech.* **14**: 15–23.

Rao, S. K., P. Sadayappan, F. K. Hwang, and P. W. Shor. 1992. "The rectilinear Steiner arborescence problem." *Algorithmica* **7**: 277–288.

Richards, D. 1989. "Fast heuristic algorithms for rectilinear Steiner trees." *Algorithmica* **4**: 191–207.

Rubinstein, J. H. and D. A. Thomas. 1989. "A variational approach to the Steiner network problem." In *Proceedings of NATO Workshop on Topological Networks*, Copenhagen.

Rubinstein, J. H. and D. A. Thomas. 1991. "The Steiner ratio conjecture for six points." *J. Combinat. Theory, Ser. A* **58**: 54–77.

Rubinstein, J. H. and J. F. Weng. 1997. "Compression theorems and Steiner ratios on spheres." *J. Combinat. Optimization* **1**: 67–78.

Sleator, D. D. and R. E. Tarjan. 1985. "Amortized efficiency of list update and paging rules." *Commun. ACM* **28**: 202–208.

Smith, J. M. 1995. Private Communication.

Smith, J. M. and J. S. Liebman. 1979. "Steiner trees, Steiner circuits and interference problem in building design." *Engng Optim.* **4**: 15–36.

Smith, J. M., D. T. Lee, and J. S. Liebman. 1981. "An $O(N \log N)$ heuristic for Steiner minimal tree problems in the Euclidean metric." *Networks* **11**: 23–39.

Smith, W. D. 1992. "How to find Steiner minimal trees in Euclidean d-space." *Algorithmica* **7.**

Smith, W. D. and P. W. Shor. 1992. "Steiner tree problems." *Algorithmica* **7**: 329–332.

Smith, W. D. and J. M. Smith. 1995. "On the Steiner ratio in 3-space." *J. Combinat. Theory, Series A.*

Soukup, J. 1975. "On minimum cost networks with nonlinear costs." *SIAM J. Appl. Math.* **29**: 571–581.

Soukup, J. 1977. "Minimum Steiner trees, roots of a polynomial and other magic." *ACM/SIGMAP Newsletter* **22**: 37–51.

Stewart, I. 1991. "Trees, telephones and tiles." *New Scientist* **16**: 26–29.

Suzuki, A. and M. Iri. 1986. "A heuristic method for Euclidean Steiner problem as a geographical optimization problem." *Asia–Pacific J. Oper. Res.* **3**: 106–122.

Tarjan, R. E. and M. Yannakakis. 1984. "Simple linear-time algorithms to test chordality of graphs, test acyclicity of hypergraphs, and selectively reduce acyclic hypergraphs." *SIAM J. Comput.* **13**: 566–579.

Thompson, E. A. 1973. "The method of minimum evolution." *Ann. Human Genet.* **36**: 333–340.

Waxman, B. M. and M. Imase. 1988. "Worst-case performance of Rayward-Smith's Steiner tree heuristic." *Inf. Processing Lett.* **29**: 283–287.

Wu, Y. F., P. Widmayer, and C. K. Wong. 1986. "A faster approximation algorithm for the Steiner problem in graphs." *Acta Informatica* **23**: 223–229.

Zelikovsky, A. Z. 1992. "An 11/8-approximation algorithm for the Steiner problem on networks with rectilinear distance." *Coll. Math. Soc. J. Bolyai* **60**: 733–745.

Zelikovsky, A. Z. 1993a. "The 11/6-approximation algorithm for the Steiner problem on networks." *Algorithmica* **9**: 463–470.

Zelikovsky, A. Z. 1993b. "A fast approximation algorithm for the Steiner tree problem in graphs." *Inf. Processing Lett.* **46**: 79–83.

Zelikovsky, A. Z. 1995. "Better approximation bounds for the network and Euclidean Steiner tree problem." Manuscript.

17.4 Approximate Solutions to Bin Packing Problems

Edward G. Coffman, Jr., János Csirik, and Gerhard J. Woeginger

1. Introduction

The following abstract packing problem arises in a wide variety of contexts in the real world. A list $L = \langle a_1, a_2, ..., a_n \rangle$ of items must be packed into, that is, partitioned among, a minimum-cardinality set of bins $B_1, B_2, \ldots$ subject to the constraint that the set of items in any bin fits within that bin's capacity. To give just a few of the applications modeled by bin packing, we mention storage allocation for computer networks, assigning advertisements to newspaper columns, packing fruit slices into tins, assigning commercials to station breaks on television, copying a collection of files to several floppy disks, packing trucks with a given weight limit, and the cutting-stock problems of various industries like those producing lumber and cable.

Let A denote an arbitrary packing algorithm and let OPT denote an algorithm that produces optimal packings. Let $A(L)$ and $\mathrm{OPT}(L)$ denote the numbers of bins used by algorithms A and OPT. To measure worst-case behavior, the asymptotic worst-case ratio R_A^∞ of algorithm A is defined by

$$R_A^\infty := \limsup_{m \to \infty} R_A^m,$$

where

$$R_A^m := \sup_L \{A(L)/\mathrm{OPT}(L) | \mathrm{OPT}(L) = m\}.$$

This ratio, which we shall just call the asymptotic ratio, is the performance metric of choice for bin packing algorithms; the standard objectives of bin packing problems are algorithms with small asymptotic ratios. Although we focus on these ratios, comments on absolute ratios, $R_A := \sup_L\{A(L)/\mathrm{OPT}(L)\}$, will sometimes be appropriate. Certain classes of algorithms have the undesirable property that a small asymptotic ratio is obtained at the expense of a large absolute ratio.

In some applications, an *a priori* upper bound α $(0 < \alpha \leqslant 1)$ on the size of all items is known. To gauge the performance of approximation algorithms on lists of such items, the parametric asymptotic ratio $R_A^\infty(\alpha)$ is introduced as

$$R_A^\infty(\alpha) := \limsup_{m \to \infty} R_A^m(\alpha),$$

where

$$R_A^m(\alpha) := \sup_L \{A(L)/\mathrm{OPT}(L) | \mathrm{OPT}(L) = m \text{ and all } a_i \in L \text{ are } \leqslant \alpha\}.$$

The problem defined so far is quite general. Implicitly, we have allowed the bins B_i to have varying capacities and both bins and items to be multidimensional. The bulk of this paper, not to mention bin packing literature, is confined to the one-dimensional problem with equal bin capacities, which we normalize to 1 for convenience; in this case, any set of items whose total size is at most 1 fits into a bin. Correspondingly, items are always assumed to be numbers drawn from an interval $[0, \alpha]$ for some $\alpha \leqslant 1$. We keep with these classical assumptions throughout the next section; the third and last section will cover more general versions of bin packing. We note once and for all that the versions of bin packing of interest to us here are NP-hard, except where stated otherwise.

Many extensions and variants have been instrumental in maintaining interest in the general area of bin packing. We mention a few basic illustrations here; others can be found in the more comprehensive treatment by Coffman et al. (1998). Three principal parameters of bin packing problems are the number of items, the bin capacity, and the number of bins. Choosing the number of bins as the objective function is the classical problem, but choosing one of the two others as the objective function also generates interesting problems. Multiprocessor scheduling actually predates bin packing and is the problem with a given number m of bins and the objective of finding the minimum common capacity, such that all items on the given list can be packed into m bins. The problem falls more properly within scheduling theory and is surveyed in that context in the article by Lawler et al. (1993). The second problem is that of finding a maximum cardinality subset of the given list of items that can be packed into a given number of bins with a given capacity (see e.g., Coffman et al., 1977 and Coffman and Leung, 1979). The dual of packing is covering, where the item sizes in a bin must add up to at least 1. As a final variant, we mention dual bin packing, the aim of which is to maximize the number of unit capacity bins covered by a given list of items (Assmann et al., 1984).

2. The Classical Problem

A bin packing algorithm is called on-line if it packs every item a_i solely on the basis of the sizes of the items a_j, $1 \leqslant j \leqslant i$, that is, without any information on subsequent items. The decisions of an on-line algorithm are irrevocable; packed items cannot be repacked at later times. A bin packing algorithm that can use full knowledge of all items in packing L is called off-line. Below, we cover on-line algorithms first and then discuss a number of off-line algorithms.

2.1. *On-Line Algorithms*

Simple Algorithms

The simplest approximation algorithm for bin packing is probably the next-fit (NF) algorithm: NF makes a single scan through the list L and shortsightedly packs the items one after the other into a unique, active bin. When an item does not fit into the active bin, the bin is closed (never to be used again), and an empty bin is opened and becomes the new active bin. The running time of NF is linear for the number of items packed. It is easy to see that NF has an asymptotic ratio $R^{\infty}_{NF} = 2$: The list $L = \langle \frac{1}{2}, \varepsilon, \frac{1}{2}, \varepsilon, \frac{1}{2}, \varepsilon, \ldots \rangle$ that consists of n pairs $\frac{1}{2}$, ε where $0 < \varepsilon < \frac{1}{n}$ shows that the asymptotic ratio cannot be better than 2. Moreover, the total contents assigned to two consecutive active bins is always at least 1, and this yields an upper bound of 2 on the asymptotic ratio. Johnson (1973, 1974) gave a complete analysis of the parametric asymptotic ratio of NF; he showed that for $1/2 \leqslant \alpha \leqslant 1$, $R^{\infty}_{NF}(\alpha) = 2$ and for $\alpha < 1/2$, $R^{\infty}_{NF}(\alpha) = 1/(1-\alpha)$.

An obvious drawback of NF is that it never uses the empty space in closed bins. A simple modification gives the first-fit (FF) algorithm: FF also makes a single scan through the list L, but it never closes an active bin. When packing a new item, FF puts it into the lowest indexed bin into which it will fit. A new bin is started only if the item does not fit into any nonempty bin. The running time of FF is $O(n \log n)$ and thus greater than the $O(n)$ running time of NF. However, FF has a much better asymptotic ratio. Johnson et al. (1974) proved the following: Let m be a positive integer with $1/(m+1) < \alpha \leqslant 1/m$. Then for $m = 1$, $R^{\infty}_{FF}(\alpha) = 17/10$ holds and for $m \geqslant 2$, $R^{\infty}_{FF}(\alpha) = (m+1)/m$ holds.

Other classical on-line algorithms are best fit (BF), worst fit (WF), and almost worst fit (AWF). BF behaves like FF, except that it puts the next item into the bin into which it will fit with the smallest gap left over; ties are broken arbitrarily. WF puts the next item into a nonempty bin with the largest gap, starting a new bin only if this largest gap is not big enough. AWF tries first to put the next item into a nonempty bin with the second largest gap; if the item does not fit there then AWF behaves like WF. All three variants belong to the class of so-called any-fit (AF) algorithms: An AF algorithm scans once through the list L while packing the items. It never puts an item a_i into an empty bin, unless the item does not fit into any partially filled bin. Similarly, an almost-any-fit (AAF) algorithm is an AF algorithm that never puts an item into a partially filled bin with the lowest level, unless there is more than one bin having this level or unless the bin of lowest level is the only one that has enough room. Johnson (1974) proved a beautiful and surprising result: an AF algorithm can never have an asymptotic ratio better than FF, and all AAF algorithms have the same asymptotic ratio as FF; these statements hold even for the parametric ratios. As a consequence we can easily calculate that the AAF algorithms BF and AWF have asymptotic ratios of 17/10. The asymptotic ratio of the AF algorithm WF, however, is 2, the same as NF.

Bounded Space On-line Algorithms

An on-line bin packing algorithm is said to use k-bounded-space if for each new item, the choice of bins into which it may be packed is restricted to a set of k or fewer active bins. A bin becomes active when it receives its first item; once a bin is closed, it can never become active again. NF uses 1-bounded-space, whereas FF, WF, and AWF all use unbounded space. There are four very natural bounded-space bin packing algorithms that are defined via simple packing rules for items and simple closing rules for bins: a new item can always be packed into the lowest indexed bin (as in FF) or into the bin with the smallest remaining gap (as in BF). If the new item does not fit into any active bin, some active bin has to be closed; in this case one can always choose the lowest indexed bin (the first bin) or the fullest bin (the best bin). The corresponding four algorithms are called AFF_k, AFB_k, ABB_k, and ABF_k: here A stands for algorithm, the second letter denotes the packing rule (best fit or first fit), the third letter denotes the closing rule (best bin or first bin), and k is the upper bound on the number of active bins.

- *Algorithm AFF_k.* This algorithm was defined under the name next-k fit by Johnson (1974). Provably tight bounds were not in hand until almost twenty years later. Csirik and Imreh (1989) constructed a sequence of worst-case examples that show $R^{\infty}_{AFF_k} \geqslant \frac{17}{10} + \frac{3}{10(k-1)}$, and Mao (1993b) proved a matching upper bound on $R^{\infty}_{AFF_k}$.
- *Algorithm ABF_k.* Mao (1993a) proved that $R^{\infty}_{ABF_k} = \frac{17}{10} + \frac{3}{10k}$.
- *Algorithm AFB_k.* Zhang (1994) adapted the analysis of Mao (1993b) to this algorithm and proved that $R^{\infty}_{AFB_k} = \frac{17}{10} + \frac{3}{10(k-1)}$.
- *Algorithm ABB_k.* This algorithm was investigated by Csirik and Johnson (1991). In comparison to the other three algorithms, ABB_k uses better packing and closing rules and it has a better asymptotic ratio; $R^{\infty}_{ABB_k} = 17/10$ holds for any $k \geqslant 2$.

So, except for small k, the asymptotic ratios of all four algorithms are around 17/10; since they are of the AF type, they cannot of course outperform FF.

Lee and Lee (1985) introduced a new class of bounded space algorithms using bin reservation techniques. Their algorithm harmonic is based on a partition of the interval (0, 1] into k subintervals, where the partitioning points are $1/2, 1/3, \ldots, 1/k$; to each of these subintervals there corresponds a single active bin, and only items belonging to this subinterval are packed into this bin. If a new item arrives that does not fit into its corresponding active bin, the bin is closed and a new bin is activated. In Lee and Lee (1985) it is proved that $R^{\infty}_{H_k}$ tends to the number $h_{\infty} \approx 1.69103$. This number is defined by $h_{\infty} = \sum_{i=1}^{\infty} \frac{1}{t_i - 1}$, where $t_1 = 2$ and $t_{i+1} = t_i(t_i - 1) + 1$ for $i \geqslant 1$; it is a prominent number in bin packing. Woeginger (1993) introduced the simplified harmonic (SH_k) algorithm, a modification of harmonic with a different interval structure and with a slightly better asymptotic ratio for small values of k.

A summary of the asymptotic ratios of the bounded-space algorithms for some small values of k is given in Table 1. The asymptotic ratios of all five algorithms always remain above h_{∞}. In fact, Lee and Lee (1985) showed that a bounded-space algorithm cannot have an asymptotic ratio better than h_{∞}.

Better Algorithms

The first on-line algorithm for bin packing with $R^{\infty}_A < h_{\infty}$ was Yao's (1980) revised FF (RFF). Both the definition and the analysis of RFF are fairly involved; it is something of a cross between FF and harmonic with the asymptotic ratio $R^{\infty}_{RFF} = 5/3$. All other known on-line algorithms that beat this h_{∞} bound are variants of harmonic with a special treatment of large items $\geqslant 1/3$. Lee and Lee (1985) described the refined harmonic (RH) algorithm, which was based on H_{20}, and proved the asymptotic ratio $R^{\infty}_{RH} = 373/228 \approx 1.639$. Ramanan et al. (1989) introduced the modified harmonic (MH) with $R^{\infty}_{MH} \approx 1.616$. Finally, Richey (1991) introduced the harmonic + 1 algorithm, currently the best algorithm. The definition of harmonic + 1 uses a

Table 1 Asymptotic Ratios for Bounded-space Bin Packing Algorithms, Rounded to Five Decimal Places

k	AFF_k	ABF_k	ABB_k	H_k	SH_k	*Champion*
2	2.00000	1.85000	1.70000	2.00000	2.00000	ABB
3	1.85000	1.80000	1.70000	1.75000	1.75000	ABB
4	1.80000	1.77500	1.70000	1.71429	1.72222	ABB
5	1.77500	1.76000	1.70000	1.70000	1.70000	ABB, H, SH
6	1.76000	1.75000	1.70000	1.70000	1.69444	SH
7	1.75000	1.74286	1.70000	1.69444	1.69388	SH
8	1.74286	1.73750	1.70000	1.69388	1.69106	SH
9	1.73750	1.73333	1.70000	1.69345	1.69104	SH
∞	1.70000	1.70000	1.70000	1.69103	1.69103	H, SH

partition of [0, 1] into more than seventy intervals. Its design and analysis were performed with the help of linear programming. Its asymptotic ratio is less than 1.5888.

Finally, we discuss lower bounds on the performance of any on-line bin packing algorithm. Roughly speaking, a lower bound argument can proceed as follows. Suppose an on-line algorithm A is confronted with a huge set of tiny items. If A packs these tiny items very tightly, it will not be able to find an efficient way of packing for the larger items that might arrive later; if such items actually do arrive, A is going to lose. On the other hand, if A leaves lots of room for large items while packing the tiny items, the large items might not arrive; in that case, A is again going to lose. Yao (1980) formulated this idea mathematically; he used $1/7+\varepsilon$ as the size of the tiny items, and $1/3+\varepsilon$ and $1/2+\varepsilon$ as sizes of large items. He proved that, with certain lists of such items, the asymptotic ratio of every on-line bin packing algorithm A must satisfy $R_A^\infty \geqslant 1.5$. Brown (1979) and Liang (1980) independently generalized this lower bound to 1.53635. Ten years later, Van Vliet (1992, 1995) found an elegant linear programming formulation for the Brown–Liang construction. Van Vliet gave an exact analysis and increased the lower bound to $R_A^\infty \geqslant 1.5401$. This is currently the best lower bound known for on-line bin packing.

2.2. *Off-Line Algorithms*

SIMPLE ALGORITHMS

The FF and BF algorithms do not work well on lists ordered by increasing item size. Intuitively, the large items at the end of the list should be combined with the small items at the beginning of the list, whereas FF and BF produce very dense packings for the small items and very poor packings for the large items. Hence, it is natural to design off-line algorithms that first sort the list by decreasing size, and afterward behave like one of the simple on-line algorithms. This yields first-fit decreasing (FFD), best-fit decreasing (BFD), and next-fit decreasing (NFD). Not surprisingly, the performance of NFD is not very strong: Baker and Coffman (1981) proved that $R_{NFD}^\infty = h_\infty$ holds. On the other hand, the improvement of FFD and BFD over the on-line packing algorithms is dramatic: Johnson (1973) showed that $R_{FFD}^\infty = R_{BFD}^\infty = 11/9 \approx 1.22$. The proof of this result is notorious: more than seventy pages of tedious case analysis. In fact, Johnson proved the inequality $FFD(L) \leqslant (11/9)\mathrm{OPT}(L)+4$. The additive constant of 4 in this inequality was later reduced to 3 by Baker (1985) and then to 1 by Yue (1991); the proofs in Baker (1985) and Yue (1991) are simpler than Johnson's but are based on similar ideas.

BETTER ALGORITHMS

After Johnson's 11/9 theorem in the early 1970s, quite a few years passed before a polynomial-time algorithm beating the 11/9 bound was published. Yao (1980) showed that the bound was beaten by his revised FFD (RFFD) algorithm that behaves as FFD but has a special treatment for items of size >1/3. The running time deteriorates to $O(n^{10} \log n)$, and the asymptotic ratio is improved only very slightly to $11/9-10^{-7}$; but the result shows that there is nothing magic about the 11/9 bound.

Fernandez de la Vega and Lueker (1981) constructed a polynomial-time approximation scheme (PTAS) for bin packing. In other words, they proved that for $\varepsilon > 0$ an approximation algorithm A_ε exists that has a running time polynomial in the size of the input list L (but exponential in $1/\varepsilon$), and computes a packing with $A_\varepsilon(L) \leqslant (1+\varepsilon)\text{OPT}(L) + \text{const}(\varepsilon)$, where $\text{const}(\varepsilon) \to \infty$ as $\varepsilon \to 0$. This result was considerably strengthened by Karmarkar and Karp (1984), who designed another PTAS that guarantees $A_\varepsilon(L) \leqslant \text{OPT}(L) + \log^2(\text{OPT}(L))$. The algorithm in Karmarkar and Karp (1984) uses many tricks: elimination of small items, rounding techniques, linear programming formulations, the ellipsoid method, and so on. The running time of the algorithm is roughly $O(n^9)$. In this area, development stops with Karmarkar and Karp, although many questions remain open. Is it possible to guarantee that $A_\varepsilon(L) \leqslant \text{OPT}(L) + \log(\text{OPT}(L))$ in polynomial time (i.e., replace the $\log^2$ in Karmarkar and Karp, 1984, by a simple log)? We feel that the answer to this question should be yes. Is it possible to guarantee that $A_\varepsilon(L) \leqslant \text{OPT}(L) + 1$ in polynomial time? The answer to this question might indeed be yes! (And it seems to be completely out of reach of our current techniques.)

3. Generalizations

We now relax certain of the assumptions of the classical problem.

3.1. Variable-Sized Bin Packing

In variable-sized bin packing, the items are packed into bins of several different types $B_1, \ldots, B_r$ with sizes $1 = s(B_1) > s(B_2) > \ldots > s(B_r)$. There is an infinite supply of bins of each type. The goal is to pack the items $a_i \in [0,1]$ into a set of bins with *smallest total size* (observe that the special case of a single bin type B_1 of size $s(B_1) = 1$ is the classical one-dimensional bin packing problem). For a list L of items and an approximation algorithm A, denote by $s(A, L)$ the total size of bins used by algorithm A. Denote by $s(\text{OPT}, L)$ the total size of bins used in an optimal packing. The quality of algorithm A is measured by

$$R_A^{var} = \lim_{k \to \infty} \sup\{s(A,L)/s(\text{OPT},L) | s(\text{OPT},L) \geqslant k\}.$$

In the on-line version of variable-sized bin packing, every time a new bin is opened, the on-line algorithm decides which bin type to use next.

Friesen and Langston (1986) give three simple approximation algorithms for variable-sized bin packing with the asymptotic ratios 2, 3/2, and 4/3. Only the first of these three algorithms is on-line. This algorithm always chooses the largest bin size when a new bin is opened, and otherwise behaves just like NF. Not surprisingly, it has the same asymptotic ratio as NF. It can be shown that any on-line algorithm that always chooses the largest possible bin size has an asymptotic ratio of at least 2, and also any on-line algorithm that always chooses the smallest available bin size has an asymptotic ratio of at least 2. Kinnersley and Langston (1988) design a hybrid strategy called FFf. FFf uses the packing strategy of FF. The bin type selection is based on a so-called filling factor f with $1/2 \leqslant f \leqslant 1$. Suppose that FF$f$ must start a new bin as item a_i arrives. If $a_i \leqslant 1/2$, then FFf starts a new bin of size 1. If $a_i > 1/2$, then FFf chooses the smallest bin size in the interval $[a_i, a_i/f]$; in case no such bin size exists, FFf chooses size 1. The asymptotic ratio of FFf is at most $1.5 + f/2$, and values of f (e.g., $f = 3/5$) exist for which this bound is tight. Zhang (1996) shows that the asymptotic ratio of FF$\frac{1}{2}$ equals 17/10, thus matching the guarantee of FF.

Csirik (1989) designed an on-line algorithm for variable-sized bin packing that is based on the harmonic algorithm. For every item size a_i, the algorithm computes a corresponding bin type B_j. All items that are assigned to the same bin type B_j are then packed by a harmonic algorithm into bins of size $s(B_j)$. For any collection of bin types, the asymptotic ratio of this algorithm can be made arbitrarily close to $h_\infty \approx 1.691$. The algorithm performs even better for special collections of bin types; for example, for two bin types with sizes $s(B_1) = 1$ and $s(B_2) =$

0.7, the asymptotic ratio is 1.4. These constitute the strongest results currently known for the on-line version of variable-sized bin packing.

For the off-line version of variable-sized bin packing, Murgolo (1987) constructs a polynomial-time approximation scheme. This scheme is based on and extends the techniques of Karmarkar and Karp (1984).

3.2. *Packings in Higher Dimensions*

There are several generalizations of bin packing to higher dimensions; we will first discuss packings in which items are vectors and then discuss packings of rectangles in two dimensional arenas.

VECTOR PACKING

In vector packing, instead of an item being a single number, it is a d-dimensional vector $a_i = [v_1(a_i), \ldots, v_d(a_i)]$, where $0 \leqslant v_j(a_i) \leqslant 1$ holds for $1 \leqslant j \leqslant d$. The goal is to pack all items into the minimum number of bins in such a way that, in every bin, the sum of all vectors is simultaneously at most one in every coordinate. The vector packing problem arises as a crucial subproblem in scheduling with resource constraints (cf. Garey et al., 1976).

Kou and Markowsky (1977) call an approximation algorithm A for d-dimensional vector packing reasonable, if A never yields packings in which the contents of two nonempty bins could be combined into a single bin (in one dimension, this obviously corresponds to AF algorithms). Kou and Markowsky show that any reasonable vector packing algorithm A obeys the bound $R_A^\infty \leqslant d+1$. Since the obvious generalization of FF to d-dimensional vector packing is reasonable, d-dimensional FF has an asymptotic ratio of at most $d+1$. Garey et al. (1976) performed an exact analysis of d-dimensional FF and showed that it has a slightly better asymptotic ratio of $d+7/10$. This currently provides the best asymptotic ratio known for d-dimensional on-line vector packing. Galambos et al. (1994) and Blitz et al. (1996) discuss lower bounds for on-line vector packing. Their lower bounds are rather weak and tend to 2 as d goes to infinity. For $d = 2$ dimensions, the best-known lower bound is 1.6712.

For off-line algorithms, the method of Fernandez de la Vega and Lueker (1981) yields asymptotic ratios of $d+\varepsilon$ in d dimensions, where ε can be made arbitarily close to 0. Yao (1980) proved that in a reasonable (but somewhat restricted) model of computation, any off-line approximation algorithm with $O(n \log n)$ running time must have an asymptotic ratio $R_A^\infty \geqslant d$. Chekuri and Khanna (1999) give polynomial time approximation algorithms for d-dimensional vector packing whose asymptotic ratios grow like $O(\ln d)$. Their approach is based on a linear programming relaxation of this problem. Clearly, their algorithm does not fall into Yao's (1980) restricted model of computation. It is an open problem as to whether $R_A^\infty <$ const is possible for some polynomial-time approximation algorithm A in any dimension $d \geqslant 2$ where the value of const is independent of d. In fact, it would even be interesting to obtain asymptotic ratios growing like $O(\ln \ln d)$. Woeginger (1997; see also Chekuri and Khanna, 1999) observed that unless $P = NP$, a polynomial-time approximation scheme cannot exist for two-dimensional vector packing.

PACKING IN STRIPS AND TWO-DIMENSIONAL BINS

Strip packing originated in the work of Baker et al. (1980). In this extension to two dimensions, the items are rectangles and the goal is to pack them into a unit-width semi-infinite strip so as to minimize the total length of the strip spanned by the packing. Packed rectangles cannot overlap each other or the boundaries of the strip. We consider the strip to be vertical and require that rectangles be packed orthogonally in their given orientations, that is, their edges must be parallel to the bottom or sides of the strip and no rotations are allowed. If we place boundaries at the integers and require that no rectangle overlap any such boundary, then we have the two-dimensional packing setup of Chung et al. (1982), where the objective is to minimize the number of bins (unit squares) needed, that is, the nearest integer no smaller than the height of the corresponding strip packing.

The bottom-left algorithms of Baker et al. (1980) and the level-oriented (or shelf) algorithms by Coffman et al. (1980a) were fundamental to packing in two dimensions. In a bottom-left packing, each successive item is placed as near the bottom of the strip as possible and then as far left at that height as possible. If the rectangles are packed in decreasing-width order, an asymptotic ratio of 2 is achieved. As the name suggests, in level-oriented packings, each rectangle is placed on one of a sequence of levels. The bottom-most level is the bottom of the strip; a higher level is a horizontal boundary running through the tallest rectangle in the next lowest level. Thus, levels correspond to (horizontal) bins in the one-dimensional problem. The algorithms NF, FF, and BF can be adapted in an obvious way to level packings. Prior to applying any of these algorithms, the list may be ordered by decreasing height or width. To cite a typical result (see Coffman et al., 1980a), we mention that FF with rectangles in decreasing height order has the 17/10 asymptotic ratio of FF in the one-dimensional case. Baker et al. (1981) extended the basic ideas and designed an algorithm with an asymptotic ratio of 5/4. As a final note, we mention that a polynomial-time approximation scheme for a (somewhat restricted) special case of strip packing was devised by Fernandez de la Vega and Zissimopoulos (1995). Finally, Kenyon and Remila (1998) succeeded in constructing an approximation scheme for the general strip packing problem.

Shelves are preset levels and were introduced as the basis of on-line level-oriented algorithms by Baker and Schwarz (1983). As an example, assume that 1 is an *a priori* bound on rectangle heights and suppose that the only shelf heights possible are r^k, $k \geqslant 0$ for some given $r \in (0, 1)$. Rectangles with heights in $(r^{k-1}, r^k]$ must be placed on shelves of height r^k, starting a new shelf on top of the current packing whenever necessary according to the given packing algorithm. A principal result of Baker and Schwarz (1983) is that the FF shelf algorithm has an asymptotic ratio of $1.7/r$, but an absolute ratio that grows like $1/(1-r)$. In an analogous way, Csirik and Woeginger (1997) transform the harmonic algorithm into the corresponding harmonic shelf algorithm. The asymptotic ratio of the harmonic shelf algorithm can be made arbitrarily close to 1.691. Achieving an asymptotic ratio better than 1.691 for on-line strip packing is probably difficult; in Csirik and Woeginger (1997) it is argued that no shelf algorithm can do this, in which case a completely different approach will be necessary.

The algorithmic techniques of one-dimensional and strip packing are easily applied to two-dimensional bin packing. Combining FFD bin packing with FFDH strip packing yields an asymptotic ratio known to be in the interval [2.022, 2.125]; cf. Chung et al. (1982).

Shelf algorithms were adapted for on-line two-dimensional bin packing by Coppersmith and Raghavan (1989). Variants and extensions of their algorithms were introduced by Csirik et al. (1990) and by Li and Cheng (1990a); in the latter paper, an algorithm was presented with the provably tight asymptotic ratio of $2\frac{39}{40}$. The harmonic algorithm (see Section on Bounded Space On-line Algorithms) was extended to the two-dimensional problem by Li and Cheng (1990b); their algorithm has an asymptotic ratio of $(1.691\ldots)^2$, the square of the corresponding result in one dimension. Csirik and Van Vliet (1993) subsequently described an improvement that substantially reduces the absolute ratio of the Li–Cheng algorithm while leaving the asymptotic ratio unchanged.

Lower bounds for on-line two-dimensional bin packing were taken up by Galambos (1991). He proved a lower bound of 1.6. A later collaboration with Van Vliet (Galambos and Van Vliet, 1994) yielded 1.808. The Ph.D. dissertation of Van Vliet (1995) contains 1.851 The best bound currently known is $R_A^\infty \geqslant 1.907$ (Blitz et al., 1996).

4. Final Remarks

While worst-case analysis of bin packing remains quite active, it must be noted that in the last fifteen years or so, average-case analysis has received at least as much attention. The monograph by Coffman and Lueker (1991) describes basic results and analytic techniques. A more recent survey of average-case results for one-dimensional bin packing can be found in Coffman et al. (1996). To illustrate typical average-case estimates, suppose the n items of L are independent, uniform random draws from [0, 1], and consider the wasted space W in the classical NF, FF, and BF packings (i.e., the difference in the number of bins used and the total size of the items). Then we have $E[W_n^{NF}] \sim n/6$, $(n \to \infty)$ (Coffman et al., 1980b), $E[W_n^{FF}] = \Theta(n^{2/3})$ (Shor, 1986; Coffman et al., 1997), and $E[W_n^{BF}] = \Theta(\sqrt{n} \log^{3/4} n)$ (Leighton and Shor, 1989; Shor, 1986; Rhee and Talagrand, 1988). These results can be compared to $E[W_n^{OPT}] = \sqrt{n}$ (Lueker, 1982).

Our discussion of generalizations can be continued to problems of three or more dimensions. In general, the algorithmic extensions from one to two dimensions can essentially be repeated as extensions from d to $d+1$ dimensions for $d \geqslant 2$; as one might expect, asymptotic ratios tend to worsen as d increases. We refer the reader to Coppersmith and Raghavan (1989), Galambos and van Vliet (1994), and Li and Cheng (1990a, 1990c, 1992).

ACKNOWLEDGMENTS

János Csirik's research was supported by the Project 20u2 of the Austro-Hungarian Action Fund. Gerhard J. Woeginger's research was supported by the START program Y43-MAT of the Austrian Ministry of Science and by the Project 20u2 of the Austro-Hungarian Action Fund.

References

Assmann, S. B., D. S. Johnson, D. J. Kleitman, and J. Y. -T. Leung. 1984. "On a dual version of the one-dimensional bin packing problem." *J. Algorithms* **5(4)**: 502–525.

Baker, B. S. 1985. "A new proof for the first-fit-decreasing bin-packing algorithm." *J. Algorithms* **6**: 49–70.

Baker, B. S. and E. G. Coffman. 1981. "A tight asymptotic bound for next-fit-decreasing bin-packing." *SIAM J. Algorithms Discrete Meth.* **2**: 147–152.

Baker, B. S. and J. S. Schwarz. 1983. "Shelf algorithms for two-dimensional packing problems." *SIAM J. Comput.* **12**: 508–525.

Baker, B. S., E. G. Coffman, Jr., and R. L. Rivest. 1980. "Orthogonal packings in two dimensions." *SIAM J. Comput.* **9**: 846–855.

Baker, B. S., D. J. Brown, and H. P. Katseff. 1981. "A 5/4 algorithm for two-dimensional packing." *J. Algorithms* **2**: 348–368.

Blitz, D., A. Van Vliet, and G. J. Woeginger. 1996. "Lower bounds on the asymptotic worst-case ratio of on-line bin packing algorithms." Unpublished manuscript.

Brown, D. J. 1979. "A lower bound for on-line one-dimensional bin packing algorithms." Technical Report R-864. Urbana, Illinois: Coordinated Science Laboratory.

Chekuri C. and S. Khanna. 1999. "On multi-dimensional packing problems." In *Proceedings of the Tenth Annual ACM–SIAM Symposium on Discrete Algorithms*, pp. 185–194.

Chung, F. R. K., M. R. Garey, and D. S. Johnson. 1982. "On packing two-dimensional bins." *SIAM J. Algorithmm, Discrete Meth.* **3**: 66–76.

Coffman, Jr., E. G. and J. Y. Leung. 1979. "Combinatorial analysis of an efficient algorithm for processor and storage allocation." *SIAM J. Comput.* **8**: 202–217.

Coffman, Jr., E. G. and G. S. Lueker. 1991. In *Probabilistic Analysis of Packing and Partitioning Algorithms.* New York: Wiley.

Coffman, Jr., E. G., J. Y. Leung, and D. W. Ting. 1977. "Bin packing: maximizing the number of pieces packed." *Acta Informatica* **9**: 263–271.

Coffman, Jr., E. G., M. R. Garey, D. S. Johnson, and R. E. Tarjan. 1980a. "Performance bounds for level-oriented two-dimensional packing algorithms." *SIAM J. Comput.* **9**: 808–826.

Coffman, Jr., E. G., K. So, M. Hofri, and A. C. Yao. 1980b. "A stochastic model of bin-packing." *Inf. Cont.* **44**: 105–115.

Coffman, Jr., E. G., M. R. Garey, and D. S. Johnson. 1996. "Bin packing approximation algorithms: a survey." In *Approximation Algorithms for NP-Hard Problems*, edited by D. Hochbaum. Boston: PWS Publishing Company.

Coffman, Jr., E. G., D. S. Johnson, P. W. Shor, and R. R. Weber. 1997. "First-Fit bin packing: a tight bound on expected wasted space." *Probab. Theory Related Fields* **10**: 69–101.

Coffman, Jr., E. G., G. Galambos, S. Martello, and D. Vigo. 1998. "Bin packing approximation algorithms: combinatorial analysis." In *Handbook of Combinatorial Optimization*, edited by D.-Z. Du and P. M. Pardalos: Kluwer Academic Publishers.

Coppersmith, D. and P. Raghavan. 1989. "Multidimensional on-line bin packing: algorithms and worst-case analysis." *Oper. Res. Lett.* **8**: 17–20.

Csirik, J. 1989. "An on-line algorithm for variable-sized bin packing." *Acta Informatica* **26**: 697–709.

Csirik, J. and B. Imreh. 1989. "On the worst-case performance of the NkF bin packing heuristic." *Acta Cybernetica* **9**: 89–105.

Csirik, J. and D. S. Johnson. 1991. "Bounded space on-line bin packing: best is better than first." In *Proceedings of the 2nd ACM–SIAM Symposium on Discrete Algorithms*, pp. 309–319.

Csirik, J. and A. Van Vliet. 1993. "An on-line algorithm for multidimensional bin packing." *Oper. Res. Lett.* **13**: 149–158.

Csirik, J. and G. J. Woeginger. 1997. 'Shelf algorithms for on-line strip packing." *Inf. Process. Lett.* **63**: 171–175.

Csirik, J., J. B. G. Frenk, and M. Labbé. 1990. "Two dimensional rectangle packing: on line methods and results." ARIDAM V, Rutgers University.

Fernandez de la Vega, W. and G. S. Lueker. 1981. "Bin packing can be solved within $1+\varepsilon$ in linear time." *Combinatorica* **1**: 349–355.

Fernandez de la Vega, W. and V. Zissimopoulos. 1995. "An approximation scheme for strip-packing of rectangles with bounded dimensions." Technical Report. Paris, France: LRI, University of Paris XI.

Friesen, D. K. and M. A. Langston. 1986. 'Variable sized bin packing." *SIAM J. Comput.* **15**: 222–230.

Galambos, G. 1991. "A 1.6 lower-bound for the two-dimensional on-line rectangle bin-packing." *Acta Cybernetica* **10**: 21–24.

Galambos, G. and A. Van Vliet. 1994. "Lower bounds for 1-, 2- and 3-dimensional on-line bin packing algorithms." *Computing* **52**: 281–297.

Galambos, G., H. Kellerer, and G. J. Woeginger. 1994. "A lower bound for on-line vector packing algorithms." *Acta Cybernetica* **11**: 23–34.

Garey, M. R., R. L. Graham, D. S. Johnson, and A. C. C. Yao. 1976. "Resource constrained scheduling as generalized bin packing." *J. Combinat. Theory Series A.* **21**: 257–298.

Johnson, D. S. 1973. *Near-optimal bin packing algorithms*. Ph.D. diss. Cambridge, MA: MIT.

Johnson, D. S. 1974. "Fast algorithms for bin packing." *J. Comput. Sys. Sci.* **3**: 272–314.

Johnson, D. S., A. Demers, J. D. Ullman, M. R. Garey, and R. L. Graham. 1974. "Worst-case performance bounds for simple one-dimensional packing algorithms." *SIAM J. Comput.* **3**: 256–278.

Karmarkar, N. and R. M. Karp. 1984. "An efficient approximation scheme for the one-dimensional bin packing problem." In *Proceedings of the 23rd Annual Symposium on Foundations of Computer Science*, pp. 289–298.

Kenyon, C. and E. Remila. 1998. "Approximate strip-packing." To appear in *Math. Oper. Res.*

Kinnersley, N. G. and M. A. Langston. 1988. "On-line variable-sized bin packing." *Discrete Appl. Math.* **22**: 143–148.

Kou, L. T. and G. Markowsky. 1977. 'Multidimensional bin packing algorithms." *IBM J. Res. Dev.* **21**: 443–448.

Lawler, E. L., J. K. Lenstra, A. H. G. Rinnooy Kan, and D. B. Shmoys. 1993. "Sequencing and scheduling: algorithms and complexity." In *Handbook of Operations Research and Management Science, Logistics of Production and Inventory*, edited by S. C. Graves, A. H. G. Rinnooy Kan, and P. H. Zipkin, Vol. 4, pp. 445–522. Amsterdam: North-Holland.

Lee, C. C. and D. T. Lee. 1985. "A simple on-line bin packing algorithm." *J. Assoc. Comput. Mach.* **32**: 562–572.

Leighton, T. and P. Shor. 1989. "Tight bounds for minimax grid matching with applications to the average case analysis of algorithms." *Combinatorica* **9(2)**: 161–187.

Li, K. and K.-H. Cheng. 1990a. "Generalized first-fit algorithms in two and three dimensions." *Int. J. Foundations Comput. Sci.* **1(2)**: 131–150.

Li, K. and K.-H. Cheng. 1990b. "A generalized harmonic algorithm for on-line multidimensional bin packing." Technical Report TR UH-Cs-90-2. University of Houston.

Li, K. and K.-H. Cheng. 1990c. "On three dimensional packing." *SIAM J. Comput.* **19**: 847–867.

Li, K. and K.-H. Cheng. 1992. "Heuristic algorithms for on-line packing in three dimensions." *J. Algorithms* pp. 589–605.

Liang, F. M. 1980. "A lower bound for on-line bin packing." *Inf. Processing Lett.* **10**: 76–79.

Lueker, G. S. 1982. "An average-case analysis of bin packing with uniformly distributed item sizes." Technical Report 181. Irvine: Department of Information and Computer Science, University of California at Irvine.

Mao, W. 1993a. "Best-k-fit bin packing." *Computing* **50**: 265–270.

Mao, W. 1993b. "Tight worst-case performance bounds for Next-*k*-Fit bin packing." *SIAM J. Comput.* **22**: 46–56.

Murgolo, F. D. 1987. "An efficient approximation scheme for variable-sized bin packing." *SIAM J. Comput.* **16**: 149–161.

Ramanan, P., D. J. Brown, C. C. Lee, and D. T. Lee. 1989. "On-line bin packing in linear time." *J. Algorithms* **10**: 305–326.

Rhee, W. T. and M. Talagrand. 1988. "Exact bounds for the stochastic upward matching problem." *Trans. Am. Math. Soc.* **307(1)**: 109–125.

Richey, M. B. 1991. "Improved bounds for harmonic-based bin packing algorithms." *Discrete Appl. Math.* **34**: 203–227.

Shor, P. W. 1986. "The average-case analysis of some on-line algorithms for bin packing." *Combinatorica* **6(2)**: 179–200.

Van Vliet, A. 1992. "An improved lower bound for on-line bin packing algorithms." *Inf. Processing Lett.* **43**: 277–284.

Van Vliet, A. 1995. "Lower and upper bounds for on-line bin packing and scheduling heuristics." Ph.D. diss. Rotterdam, The Netherlands: Erasmus University.

Woeginger, G. J. 1993. "Improved space for bounded-space on-line bin packing." *SIAM J. Discrete Math.* **6**: 575–581.

Woeginger, G. J. 1997. "There is no asymptotic PTAS for two-dimensional vector packing." *Inf. Processing Lett.* **64**: 293–297.

Yao, A. C. C. 1980. "New algorithms for bin packing." *J. Assoc. Comput. Mach.* **27**: 207–227.

Yue, M. 1991. "A simple proof of the inequality $FFD(L) \leqslant \frac{11}{9}\mathrm{OPT}(L) + 1 \ \forall L$, for the FFD bin-packing algorithm." *Acta Math. App. Sinica* **7**: 321–331.

Zhang, G. 1994. "A tight worst-case performance bound for AFB-k." Technical Report 015. Beijing, China: Institute of Applied Mathematics, Academia Sinica.

Zhang, G. 1996. "Worst-case analysis of the FFH algorithm for on-line variable-sized bin packing." *Computing* **56**: 165–172.

17.5 The Traveling Salesman Problem

Rainer E. Burkard

ABSTRACT This article surveys traveling salesman problems (TSPs). After stating different formulations for the traveling salesman problem, some applications are outlined. Then exact solution procedures like branch and bound, as well as branch and cut, are sketched and heuristics for the TSP are surveyed. Finally, special cases of the TSP that can be solved in polynomial time are addressed.

1. Introduction

One of the most well-known hard combinatorial optimization problems is the traveling salesman problem (TSP). Consider a salesperson who has to visit n cities. The TSP asks for the shortest tour through all the cities such that no city is visited twice and the salesperson returns at the end of the tour back to the starting place. Mathematically, this problem can be modeled by a permutation φ on the ground set $N = \{1, 2, ..., n\}$, where $\varphi(i)$ describes the city that is visited immediately after city i. The permutation φ is required to be cyclic (tour). Given the distances c_{ij} between city i and city j, the length of a tour is defined as

$$z(\varphi) := \sum_{i=1}^{n} c_{i\varphi(i)}.$$

The TSP can be written as

$$\min_{\varphi \text{ cyclic}} \sum_{i=1}^{n} c_{i\varphi(i)}.$$

Sometimes instead of a sum objective, a bottleneck objective function is considered. The bottleneck TSP has the form

$$\min_{\varphi \text{ cyclic}} \max_{1 \leqslant i \leqslant n} c_{i\varphi(i)}.$$

We speak of a symmetric TSP, if for all pairs i, j the distance c_{ij} is equal to the distance c_{ji}. If one-way roads are incurred, this will not be the case. In this case one speaks of an asymmetric TSP. If the cities can be represented as points in the plane such that c_{ij} is the Euclidean distance between point i and point j, then the corresponding TSP is called the Euclidean TSP. Euclidean TSPs obey in particular the triangle inequality $c_{ij} \leqslant c_{ik} + c_{kj}$ for all i, j, k.

Often the cities are not given as locations in space, but just as vertices of a finite graph $G = (V, E)$ with vertex set V and edge set E. Every edge e is supposed to have a length $c(e)$ In this model a tour corresponds to a so-called Hamiltonian cycle in the graph G. Not every graph contains a Hamiltonian cycle, and whether or not a given graph contains a Hamiltonian cycle (is Hamiltonian) is already an NP-complete question. Provided a graph is Hamiltonian, the TSP on a graph G asks for a Hamiltonian cycle C of shortest length $z(C) := \sum_{e \in C} c(e)$.

A related question is the Hamiltonian path problem: Given n cities (vertices of a graph), find the shortest path from city (vertex) 1 to city (vertex) n that visits all cities. Thus it is not required to return to the start from the last city. As for the TSP, the shortest Hamiltonian path problem is NP-hard.

The (asymmetric) TSP can be formulated as an integer linear program in the following way. Let an $n \times n$ distance matrix $C = (c_{ij})$ be given. We introduce binary variables x_{ij} by

$$x_{ij} := \begin{cases} 1 & \text{if } j \text{ is visited immediately after } i, \\ 0 & \text{otherwise.} \end{cases}$$

A permutation is described by

$$\sum_{i=1}^{n} x_{ij} = 1 \quad \text{for all} \quad j = 1, 2, \ldots, n \tag{1}$$

$$\sum_{j=1}^{n} x_{ij} = 1 \quad \text{for all} \quad i = 1, 2, \ldots, n \tag{2}$$

$$x_{ij} \in \{0, 1\} \quad \text{for all} \quad i, j = 1, 2, \ldots, n. \tag{3}$$

The property that our permutation has to be cyclic, is described by the so-called subtour elimination constraints: Let W be an arbitrary subset of $\{1, 2, \ldots, n\}$ and denote the number of elements of W by $|W|$. Assume $1 \leqslant |W| \leqslant n-1$. Every tour has at most $|W| - 1$ edges with both endpoints in the set W. Therefore a solution x_{ij} has to obey the following inequalities:

$$\sum_{i \in W} \sum_{j \in W} x_{ij} \leqslant |W| - 1 \quad \text{for all} \quad W \subset \{1, 2, \ldots, n\}, \quad 1 \leqslant |W| \leqslant n-1.$$

Thus, it can be shown that a TSP can be written as the following binary linear program

$$\min \sum_{i=1}^{n} \sum_{j=1}^{n} c_{ij} x_{ij} \quad \text{subject to} \tag{4}$$

$$\sum_{i=1}^{n} x_{ij} = 1 \quad \text{for all} \quad j = 1, 2, \ldots, n \tag{5}$$

$$\sum_{j=1}^{n} x_{ij} = 1 \quad \text{for all} \quad i = 1, 2, \ldots, n \tag{6}$$

$$\sum_{i \in W} \sum_{j \in W} x_{ij} \leqslant |W| - 1 \quad \text{for all} \quad W \subset \{1, 2, \ldots, n\}, \quad 1 \leqslant |W| \leqslant n-1 \tag{7}$$

$$x_{ij} \in \{0, 1\} \quad \text{for all} \quad i, j = 1, 2, \ldots, n. \tag{8}$$

Note that the subtour elimination constraints (7) can also be written as

$$\sum_{i \in W} \sum_{j \notin W} x_{ij} \geqslant 1 \quad \text{for all} \quad W \subset \{1, 2, ..., n\}, \quad 1 \leqslant |W| \leqslant n-1.$$

The symmetric TSP can be defined on a complete, undirected weighted graph $G = (V, E)$ with vertex set V, edge set E, and weights c_{ij}, $(i, j) \in E$. The vertices correspond to the cities. A tour is a cycle in this graph that passes through all vertices just once. Thus, a tour can be defined as a set of edges such that every vertex is incident with exactly two edges (see equation (10)) and without cycles of length k, $k < n$, (see the subtour elimination constraints (11)). Therefore we get the following description of a symmetric TSP as a binary linear program:

$$\min \sum_{i=1}^{n-1} \sum_{j=i+1}^{n} c_{ij} x_{ij} \quad \text{subject to} \tag{9}$$

$$\sum_{j \neq i} x_{ij} = 2 \quad \text{for all} \quad i = 1, 2, ..., n \tag{10}$$

$$\sum_{i \in W} \sum_{\substack{j \in W \\ j > i}} x_{ij} \leqslant |W| - 1 \quad \text{for all} \quad W \subset \{1, 2, ..., n\}, \quad 2 \leqslant |W| \leqslant n-1 \tag{11}$$

$$x_{ij} \in \{0, 1\} \quad \text{for all} \quad i, j = 1, 2, ..., n. \tag{12}$$

The TSP belongs to the class of NP-hard optimization problems. This means that no polynomial-time algorithm is known for its solution. Thus, clever enumeration procedures are necessary to find an optimal solution. Today it would be impossible to solve TSPs of size $n =$ 25 by complete enumeration, even if supercomputers were used. Therefore it is very remarkable that very large TSP instances can be solved by elaborate mathematical methods such as branch and cut. The largest traveling salesman instance ever solved is an optimal tour through all the 13,509 cities in the United States that have more than five hundred inhabitants (Applegate et al., 1998).

Since it is so hard to find an optimal solution for TSPs, there is a special need for good heuristics that find a good suboptimal solution (see Section 4). Moreover the question arises whether special structures of the underlying graph or of the cost matrix C lead to an efficiently solvable problem (see Section 5). In the following section we address some applications of the TSP. In Section 3 we outline branch-and-bound as well as branch-and-cut procedures as examples for exact solution routines.

2. Applications

A classical application of TSPs concerns vehicle routing problems. A delivery vehicle has to serve several customers. Thus the shortest tour starting from the depot and serving all customers has to be found for this vehicle. In many applications there are several, say k, vehicles available for serving n customers. This situation can also be modeled by a TSP: Instead of one depot we generate k vertices $n+1, n+2, ..., n+k$ and we choose the distances between these vertices as very large numbers. Moreover, the distance from any vertex $n + i$ $(1 \leqslant i \leqslant k)$ to a vertex j $(1 \leqslant j \leqslant n)$ is the distance between customer j and the depot in the original model. Solving such a TSP yields a Hamiltonian cycle that can be broken up in up to k single cycles, all starting in the depot, such that all customers are served. More about vehicle routing problems can be found in Fischer (1995) and Laporte (1997).

A mathematically similar problem concerns the economical manufacturing of printed circuit boards (PCBs); see Grötschel et al. (1991). To position pins of integrated circuits, holes of different diameter have to be drilled through the board. To drill two holes of different diameters consecutively, the head of the machine has to move to the tool box and has to change the drilling equipment. Since this is time consuming, all holes of the same diameter are drilled in one sequence. If the distance between two holes of the same diameter is given by the time needed to move the drilling head from one position to the other, drilling all holes such that the travel time for the head of the machine is minimized amounts to solving a series of TSPs.

Very large TSPs also arise in the production process of VLSI chips. Programmable logic arrays (PLAs) are VLSI chips initially fabricated as a regular array of interconnected components, which can be converted to PLAs by using a laser to vaporize specific interconnections. Finding the order in which the interconnections are to be vaporized leads to a large TSP.

A classical application of shortest Hamiltonian path problems is for machine scheduling problems with job-dependent setup times. Let n jobs be given that are to be processed on one machine. Let c_{ij} be the machine setup time between job i and job j. Such setup times arise, for example, if the machine has to be cleaned after processing job i and before processing job j. The cleaning time depends on the job pair i, j. Finding a sequence for processing the jobs such that the total setup time is the minimum leads to a shortest Hamiltonian path problem.

A similar problem was reported in X-ray crystallography (see Bland and Shallcross, 1989). A four-circle diffractometer is used to obtain information about the crystalline structure of some material. A small sample is mounted in the apparatus, which has computer-driven motors that can accurately position the crystal. When the crystal is positioned appropriately, the detector measures the intensity of monochromatic X-rays scattered by the crystal. Typically, five thousand to thirty thousand different positions are used for analyzing the crystal. The measurements can be taken very quickly, however the time to reposition the apparatus between two readings is significant and nowadays takes about fifteen to twentyfour hours. Sequencing the readings in order to minimize the total time needed for repositioning leads to a TSP. Solving this large (non-Euclidean) TSP suboptimally speeds up the time for performing an experiment significantly.

Another example of TSPs arising in a scheduling context are job shop problems with no wait conditions (see Lenstra and Rinnooy Kan, 1975). Some polynomially solvable TSPs arising in practice are described in Section 5 (see also Burkard et al., 1998). Reinelt (1991) provides a library of TSP instances and their solutions on the Web. It can be found at the following Web address

`http://www.iwr.uni-heidelberg.de/iwr/comopt/soft/TSPLIB95/TSPLIB.html`

3. Exact Solution of TSPs

Since the TSP is NP-hard, only enumeration procedures are known for finding optimal solutions. The two most important methods in this respect are branch and bound and branch and cut.

3.1. *Branch-and-Bound Methods*

In branch-and-bound methods the following four concepts play a crucial role: relaxation, branching, bounding, and a selection rule that chooses the next problem to be investigated. Relaxing a problem P means that we replace it by a more easily solvable problem Q such that all feasible solutions of P are also feasible in Q. During the branch-and-bound procedure we keep a list of subproblems that still have to be investigated. At the beginning, this list contains only our original problem. We determine a lower bound for the optimal solution of this problem by means of a relaxation of this problem. If it happens that the optimal solution of the relaxed problem is feasible for the TSP, we are done and this solution is also the optimal solution of the TSP. Otherwise we apply the branching rule and generate new subproblems with the property that all feasible tours of the original problem are also feasible in one of the subproblems, but on the other hand the current nonfeasible solution of the relaxed problem is not feasible for any of the subproblems. Then we apply the selection rule and choose a subproblem to be investigated next. Now a lower bound on the length of all tours belonging to this subproblem is determined. If this lower bound is larger than or equal to the length of an already known feasible solution of the TSP (the reference value, which can easily be obtained by heuristics in advance), then this subproblem cannot contain a better solution. Therefore, we can cancel this subproblem. We say it is fathomed. If, during the lower bound computation, a feasible solution of the TSP with a smaller objective function value than the reference value is obtained, we record this solution and update our reference value. Otherwise we branch again on this subproblem. The optimal solution of the TSP has been found if all subproblems have been investigated.

A natural relaxation of the asymmetric TSP is the corresponding assignment problem (4)–(6) and (8), where the condition that the permutation φ is a cyclic permutation is relaxed to φ being an arbitrary permutation. A lower bound is obtained by the optimal value of this assignment problem. If the optimal solution of the assignment problem is cyclic, then this solution is also feasible for the TSP and no branching is required. Otherwise the optimal solution of the assignment problem contains subtours. If the corresponding optimal solution value does not allow this subproblem to be fathomed, we can branch on it in the following way. Let $S = (i_1, i_2, \dots, i_t)$ be a subtour of minimum cardinality. (The notation above means that we are traveling from city i_1 to city i_2, and from there to i_3, and so on, until we return from city i_t to the start i_1.) Using the subtour elimination constraints we can define t new subproblems as follows:

- In subproblem (P_1) the arc (i_1, i_2) is forbidden (i.e., set $x_{i_1,i_2} = 0$).
- Subproblem (P_2) contains the arc (i_1, i_2), but forbids the arc (i_2, i_3) (i.e., set $x_{i_1,i_2} = 1, x_{i_2,i_3} = 0$).
- Subproblem (P_3) contains the arcs (i_1, i_2) and (i_2, i_3), but forbids the arc (i_3, i_4).
- ...
- Subproblem (P_t) contains the arcs $(i_1, i_2), \dots, (i_{t-1}, i_t)$, but forbids the arc (i_t, i_1).

As selection rule, usually the LIFO rule (last in first out), is used. This means that the last subproblem added is investigated next. An alternative is selecting a subproblem with the lowest lower bound. But this rule requires considerable storage space.

For symmetric TSPs it turned out that the 2-matching relaxation that corresponds to the assignment relaxation in the asymmetric case is not very strong. Therefore often minimum 1-trees are used as the relaxation of symmetric TSPs. A minimum 1-tree in the underlying undirected graph G can be obtained as follows: Delete vertex 1 and find a minimum spanning tree in the remaining graph on the vertices $2, 3, \dots, n$. Then add those two edges incident with vertex 1 to the tree that have the smallest lengths. It is easy to see that every tour in G is a 1-tree. Thus 1-trees are really a relaxation of the symmetric TSP. The bound obtained by this relaxation can be strengthened considerably by taking the constraints (10) via Lagrangian parameters in the objective function

$$L(\lambda) = \min_x \left\{ \sum_{i=1}^{n-1} \sum_{j=i}^{n} c_{ij} x_{ij} + \sum_{i=1}^{n} \lambda_i \left(\sum_{j \neq i} x_{ij} - 2 \right) \right\}. \tag{13}$$

Then this piecewise linear function $L(\lambda)$ is maximized with respect to λ. Since this function is concave, but not differentiable in λ, so-called subgradient methods have to be used for the maximization of $L(\lambda)$.

According to Volgenant and Jonker (1982) the following branching rule can be used for the symmetric TSP: Choose a vertex i whose degree in the current 1-tree exceeds 2, and which is incident to two free edges (i, j_1), (i, j_2). (This means that the variables x_{ij_1} and x_{ij_2} are not fixed in this node of the branching tree.) Generate the following three new subproblems:

- Subproblem (P1). Set $x_{ij_1} = x_{ij_2} = 1$ and set all other variables $x_{ij} = 0$.
- Subproblem (P2). Set $x_{ij_1} = 1$ and set $x_{ij_2} = 0$.
- Subproblem (P3). Set $x_{ij_1} = 0$.

More about branch-and-bound methods can be found in the TSP book of Lawler et al. (1985).

3.2. Branch-and-Cut Methods

Branch-and-cut methods rely on the polyhedral structure of TSPs. We define the asymmetric TSP polytope $ATSP(n)$ as the convex hull of all vectors x fulfilling constraints (5)–(8). Similarly, the symmetric TSP polytope $STSP(n)$ is defined as the convex hull of all vectors x fulfilling constraints (10)–(12). Solving a TSP would be easy if the facial structure of the polytopes $ATSP(n)$ and $STSP(n)$ could be explicitly described by a (not too large) system of inequalities. Then an optimal solution could be found simply by solving a linear program. Unfortunately, the facial structure of these polytopes is only partly known. Nevertheless, the knowledge of some facets enables us to solve very large TSPs today. The first step in solving a large TSP consists of solving its subtour relaxation, which results by replacing $x_{ij} \in \{0, 1\}$ by $0 \leqslant x_{ij} \leqslant 1$ in the integer programming formulations (4)–(8) and (9)–(12), respectively. In principle this is a very large linear program (LP) with exponentially many constraints. But this large LP can be solved in polynomial time by a cutting-plane approach. The solution yields either an optimal tour (in which case the TSP is solved) or at least a very good lower bound c_L on the optimum length of a tour. If c_{opt} denotes the length of an optimal tour, it can be shown that c_L/c_{opt} is at least two-thirds, for any problem instance fulfilling the triangle inequality.

Let us suppose that the optimal solution of the subtour relaxation is not binary. In this case one has two options: either the problem is branched by setting some x_{ij} to 0 and 1, respectively, or one searches for a violated inequality, which defines a facet of the underlying polytope. Classes of such inequalities include comb inequalities, clique tree inequalities, ladder inequalities, crown inequalities, and others. To find a cutting plane one has to solve the so-called separation problem: given a point outside the polytope, find an inequality within one of the above classes that separates this point from the polytope. No exact polynomial separation procedure is known for the above-mentioned inequalities. (Polynomial separation methods are only known for the subtour elimination constraints and the two-matching constraints.) Thus heuristics are used to find suitable cutting planes. Finally, a new LP has to be solved that takes the additional cutting planes into account. Its solution offers two possibilities: either a tour is found, which allows this subproblem to be fathomed in the branch-and-cut tree, or at least a new lower bound is found, which may also be used to fathom the subproblem. If this is not possible, one continues either with branching or with finding new cutting planes. The skill in solving very large problems consists of finding the right cuts, handling large LPs, selecting the right variables for branching, and of solving the problem in parallel.

A detailed discussion of branch-and-cut methods can be found in Jünger et al. (1995), where many classes of facet-defining inequalities are also explicitly described. A detailed description on how to solve large TSP instances by branch and cut is also given in Applegate et al. (1998). The computer code (written in C programming language) of Applegate et al. is available for research purposes on the following internet page:

`http://www.caam.rice.edu/~keck/concorde.html`

4. Heuristics

Since it is so hard to find optimal solutions for TSPs, there is a great need for powerful heuristics that find good suboptimal solutions. We can distinguish between three classes of heuristics: construction methods, which build up a tour step by step; improvement methods, which try to improve a given tour; and metaheuristics such as simulated annealing, tabu search, and genetic algorithms. Construction methods are usually fast, but in general do not yield a very good solution. Solutions found by construction methods can be further improved by some elementary exchange operations; see Section 4.2. Metaheuristics find quite good solutions, but have the disadvantage of long running times.

4.1. Construction Methods

Construction methods build up a tour step by step. One of the simplest methods is the nearest neighbor rule: Start from an arbitrary city and go to its nearest neighbor. Then proceed from there in the same manner. The nearest neighbor rule is very myopic. So it is no surprise that examples can be constructed where it behaves arbitrarily badly. On the other hand it makes, usually, only a few wrong decisions. Therefore, a tour found by the nearest neighbor rule is well suited to be processed afterward by an improvement method (see Section 4.2).

A second class of construction methods consists of insertion heuristics: They start with an arbitrary cycle (which could also be a single vertex) and insert the next city according to a certain criterion. Such criteria are:

- Nearest insertion: Insert a vertex which has the shortest distance to a vertex of the cycle.
- Farthest insertion: Insert a vertex whose minimum distance to a vertex on the cycle is maximum.
- Cheapest insertion: Insert a vertex that can be added at the lowest cost to the cycle.

Rosenkrantz et al. (1977) have shown that for Euclidean TSPs tours, which are computed by the nearest or cheapest insertion heuristic, are at most twice as long as an optimal tour.

Another class of heuristics for symmetric TSPs that satisfy the triangle inequality is based on spanning trees in the underlying graph. They use the fact that any cycle through all vertices of the graph (where the cycle can meet some vertices more than once) can be transformed into a Hamiltonian tour without increasing the length of the tour: if an edge goes from vertex i to vertex j and another one from vertex j to vertex k, one can shortcut the tour by a direct edge from i to k, which due to the triangle inequality is not longer than the original path from i to k. In the double tree heuristic, determine a minimum spanning tree in the underlying graph, create a cycle by doubling all its edges, and transform this cycle by shortcuts in a Hamiltonian tour. In the Christofides' heuristic, also determine a minimum spanning tree. Then solve a perfect matching problem for all odd degree vertices of the tree. By adding the matching edges to the spanning tree, a cycle is created, which can be a shortcut due to the triangle inequality. It can be shown that the Christofides' heuristic yields a tour that is at most 1.5 times as long as an optimal tour.

The last class consists of saving heuristics, which were originally motivated by vehicle routing problems. Start with a vertex (called depot) in which $n-1$ subcycles to all other vertices meet. For every pair of subcycles compute the savings by deleting an edge leading to the depot in each cycle and by connecting the two open ends. Merge those two subcycles that provide the largest savings and continue until a tour is obtained.

4.2. Improvement Methods

Improvement methods start with a tour and try to transform it into a shorter tour. In the 2-opt transformation two edges are deleted and the open ends are connected in a different way as to obtain another tour. A 2-opt algorithm performs this transformation, if it yields a smaller length, until no further improvement of the objective function can be reached.

A 3-opt transformation is defined in a similar way. Three edges are deleted. The remaining three paths can now be rearranged to obtain a tour in eight different ways. Thus a 3-opt procedure offers more possibilities, but takes considerably more time.

An even more complicated, yet powerful, transformation has been described by Lin and Kernighan (1973). If carefully implemented, Lin–Kernighan heuristics yield the best results of all heuristics that require a reasonably small amount of computing time.

4.3. Metaheuristics

A breakthrough was obtained with the introduction of the so-called metaheuristics, such as simulated annealing, tabu search, and genetic algorithms. The first two of these are similar to improvement methods, but have the possibility to find their way out of local optima. In simulated annealing this is achieved by allowing the length of the tour even to increase with a certain probability after a transformation like 2-opt or 3-opt. Gradually the probability allowing the objective function value to increase is lowered until no more transformations are possible: the system is then frozen. It can be shown that, by a careful setting of the parameters, an optimal solution of the TSP can be reached with probability 1. These good settings from a theoretical point of view are, however, too slow for practical implementation. Nevertheless, very good suboptimal solutions can be reached with simulated annealing.

Tabu search uses another technique to get out of local optima. The algorithm keeps a list of forbidden transformations. In this way it may be necessary to use a transformation that deteriorates the objective function value in the next step. Computational results show that by a careful implementation of a tabu search method similarly good solutions can be obtained as for simulated annealing.

Genetic algorithms mimic the evolution process in nature. Their basic operation is the mating of two tours in order to form a new tour. Moreover, they use algorithmic analogs to mutation and selection. Combined with other improvement techniques genetic algorithms may find very good tours considerably faster than simulated annealing or tabu search, but they are not yet competitive with the best implementations of Lin–Kernighan type algorithms, if computer time is limited.

There are also some papers applying neural network algorithms to the TSP and training the network in order to find good tours. But, up to now, this approach is not competitive to the other heuristics described above. More about heuristics can be found in the case study of Johnson and McGeoch (1997).

5. Efficiently Solvable Special Cases

Since in general the TSP is NP-hard, the question arises under which conditions it can be solved in polynomial time. For an extensive survey of this field, see Burkard et al. (1998). In particular, a TSP is solvable in linear time if its cost coefficients c_{ij} fulfill the so-called Monge property:

$$c_{ij} + c_{kl} \leqslant c_{il} + c_{kj} \quad \text{for all} \quad 1 \leqslant i < k \leqslant n, 1 \leqslant j < l \leqslant n.$$

We call a TSP a product TSP if its cost coefficients fulfill $c_{ij} = a_i b_j$. In general, the product TSP is NP-hard. But if $a_1 \leqslant a_2 \leqslant \ldots \leqslant a_n$ and $b_1 \geqslant b_2 \geqslant \ldots b_n$, then this special product TSP has the Monge property. Such special product TSPs occur in many applications, like in statistics (minimization of the autocorrelation coefficient), in assembling rotors in a gas turbine, and in balancing the rest periods of bus drivers. For closer details and references, see the survey mentioned above.

Other classes of efficiently solvable TSPs can be described in geometric terms. For example, if all cities lie on k parallel (or at least nearly parallel) lines in the plane, a dynamic programming algorithm can find an optimal TSP solution in $O(n^k)$ time. Another case is the so-called necklace TSP: Let n cities in the plane be given. If it is possible to circumscribe circles around the cities such that each circle contains just one city (as center point) and any circle intersects just two neighboring circles (i.e., the circles form a necklace), then the TSP is solvable in polynomial time. Note in particular that these conditions can be tested in polynomial time.

References

Applegate, D., R. Bixby, V. Chvátal, and W. Cook. 1998. "On the solution of travelling salesman problems." *Documenta Mathematica*, pp. 645–656. Also available at: http://www.mathematik.uni-bielefeld.de/documenta/xvol-icm/ICM.html.

Bland, R. G. and D. F. Shallcross. 1989. 'Large traveling salesman problems arising from experiments in X-ray cristallography: a preliminary report on computation." *Oper. Res. Lett.* **8**: 125–128.

Burkard, R. E., V. G. Deĭneko, R. van Dal, J. A. A. van der Veen, and G. Woeginger. 1998. "Well-solvable special cases of the TSP: a survey." *SIAM Rev.* **40**: 496–546.

Fischer, M. L. 1995. "Vehicle routing." In *Network Routing (Handbooks in Operations Research and Management Science)*, edited by M. O. Ball, T. L. Magnanti, C. L. Monma, and G. L. Nemhauser, Vol. 3 1–33. Amsterdam: North-Holland.

Grötschel, M., M. Jünger, and G. Reinelt. 1991. "Optimal control of plotting and drilling machines: a case study." *Z. Oper. Res.* **35**: 61–84.

Johnson, D. S. and L. A. McGeoch. 1997. "The traveling salesman problem: a case study." In *Local Search in Combinatorial Optimization*, edited by E. Aarts and J. K. Lenstra,, pp. 215–310. Chichester, England: J. Wiley and Sons.

Jünger, M., G. Reinelt, and G. Rinaldi. 1995. "The traveling salesman problem." In *Network Routing (Handbooks in Operations Research and Management Science)*, edited by M. O. Ball, T. L. Magnanti, C. L. Monma, and G. L. Nemhauser, Vol. 8: pp. 225–330. North-Holland.

Jünger, M., G. Reinelt, and G. Rinaldi. 1997. "The Travelling Salesman Problem." In *Annotated Bibliographies in Combinatorial Optimization*, edited by M. dell'Amico*et al.*,, pp. 199–221. Chichester, England: J. Wiley and Sons.

Laporte, G. 1997. "Vehicle routing." In *Annotated Bibliographies in Combinatorial Optimization*, edited by M. dell'Amico*et al.*,, pp. 223–240. Chichester, England: J. Wiley and Sons.

Lawler, E. L., J. K. Lenstra, A. H. G. Rinnooy Kan, and D. B. Shmoys. 1985. *The Travelling Salesman Problem*. Chichester, England: J. Wiley and Sons.

Lenstra, J. K. and A. H. G. Rinnooy Kan. 1975. "Some simple applications of the travelling salesman problem." *Oper. Res. Q.* **26**: 717–733.

Lin, S. and B. W. Kernighan. 1973. "An effective heuristic algorithm for the traveling salesman problem." *Oper. Res.* **21**: 498–516.

Reinelt, G. 1991. "TSPLIB – a traveling salesman library." *ORSA J. Comput.* **3**: 376–384.

Rosenkrantz, D. J., R. E. Stearns, and P. M. Lewis. 1977. "An analysis of several heuristics for the traveling salesman problem." *SIAM J. Comput.* **6**: 563–581.

Volgenant, T. and R. Jonker. 1982. "A branch and bound algorithm for the symmetric traveling salesman problem based on the 1-tree relaxation." *Eur. J. Oper. Res.* **9**: 83–89.

17.6 Inventory Management

Dukwon Kim and Boghos D. Sivazlian

ABSTRACT Inventory management has been one of the most important areas in optimization theory and in practical applications to various industries. Optimizing inventories becomes more complex and sophisticated as the number of parameters to be controlled increases. For example, in supply-chain management, integrating the inventory-production systems as well as transportation and information systems is a very challenging problem in the area of applied optimization. This is due not only to the size of the problem but also to the problem structure (typically minimizing concave cost functions). In order to understand this attractive application, we study some fundamental inventory models as well as inventory models associated with the integrated supply-chain system. In addition, an interesting solution approach for this application is briefly presented.

1. Introduction

Inventory management of goods or other products is an integral part of logistic systems common to all economic sectors such as business, industry, agriculture, and defense. In a perfectly predictable economic situation, inventory may be needed to take advantage of the economic features of a particular technology, or to synchronize human tasks, or to regulate the production process to meet the changing trends in demand. However, when uncertainty is present, inventories may be used as a protection against risk of shortage in stock.

The fundamental problem of inventory management can be described by two questions: (1) when should an order be placed? and (2) how much should be ordered? The complexity of the resulting inventory model depends upon the assumptions one makes about the various parameters of the system. The major distinction is between models that assume known demand (deterministic) and those that assume random demand (stochastic). In addition, the structure of the cost functions and the assumptions one makes about the physical characteristics of the system play an important role in determining the complexity of the resulting models.

In general, the models that we will discuss can be used interchangeably to describe either replenishment from outside vendors or internal production. This means that from the point of view of the model, inventory control and production planning are often synonymous. The main issue here is not the label placed on a technique, but whether it is being effectively applied to the problem under consideration. In the following sections, two fundamental inventory models and an application in supply-chain management are presented, and a solution approach is discussed.

2. Inventory Models

Inventory models can be classified as deterministic models and stochastic models based on the fact that all input parameters are known (deterministic) or some parameters are unknown (stochastic). Due to the variety of such models depending on different assumptions, we only discuss the basic model for each category. Basic descriptions for these models can be given as follows.

2.1. Deterministic Inventory Models

Deterministic inventory models are characterized by the certainty of all input parameters. With the absence of uncertainty, the only motive to carry inventory would be of the transaction type. We only investigate situations in which inventory is used to synchronize the outflow of goods (demand) with their inflow.

Deterministic inventory models have been investigated for many situations. In particular, solution methods have been developed for continuous review systems involving continuous demand and for periodic review systems involving discrete demand. Most of these models are applicable to many practical systems. Here we discuss the most well-known and fundamental inventory model, the so-called EOQ (Economic Order Quantity) model. It describes the important trade-off between fixed order costs and holding costs, and is the basis for the analysis of more complex systems.

The Basic EOQ Model

Assumptions for the model are as follows:

1. Immediate delivery of orders occurs and the replenishment rate is infinite.
2. The demand rate is constant D.
3. The system operates over an infinite horizon period.
4. No shortages are allowed.
5. Replenishment of stock is accomplished based on the (s, S) policy; that is, whenever the inventory level reaches s, a quantity $Q = S - s$ is ordered to bring its level up to S.
6. The initial inventory level is S.
7. There are two competing cost elements, which are the cost of procurement associated with placing an order and the cost of carrying stock.
8. The objective function is to minimize the long-run total cost per unit time or the average cost per unit time.

Referring to Figure 1, let t be the time elapsed between two successive orders; t is known as the length of a cycle. Consider an arbitrary cycle, and assume the time origin is measured at the beginning of a cycle coinciding with the receipt of an order or procurement. Then

$$\frac{dI(\theta)}{d\theta} = -D, \quad 0 < \theta < t \tag{1}$$

with

$$I(0) = S = s + Q. \tag{2}$$

Integrating (1) and using (2) as initial conditions we get

$$I(\theta) = s + Q - D\theta, \quad 0 \leqslant \theta < t. \tag{3}$$

Now, at the end of the cycle the inventory level is s; that is,

$$I(t) = s. \tag{4}$$

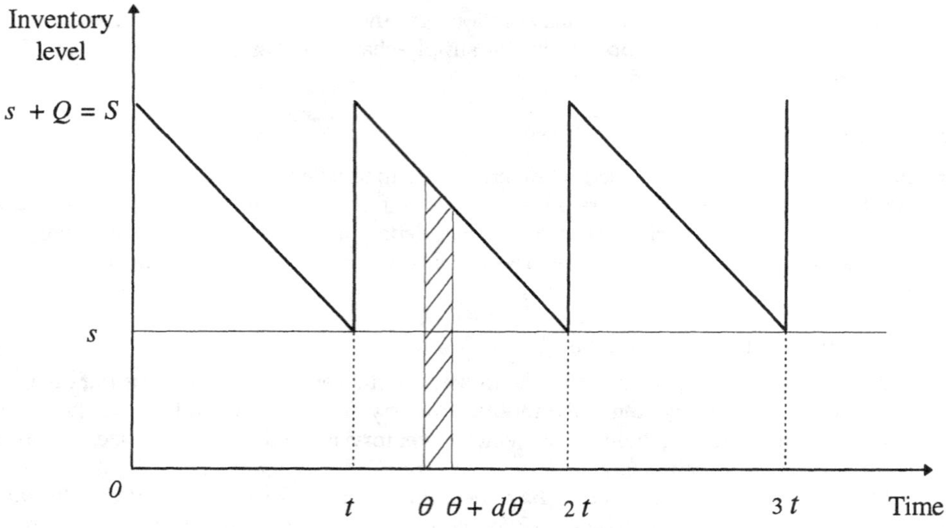

Figure 1 Inventory movement in the basic model

But from equation (3)

$$I(t) = s + Q - Dt. \tag{5}$$

From equations (4) and (5), it follows that

$$Q = Dt. \tag{6}$$

Since exactly one procurement of the amount of Q is made over one cycle, the total procurement cost per cycle is $K + cQ$. Therefore,

$$\text{the average procurement cost/unit time} = \frac{K + cQ}{t}. \tag{7}$$

Since no shortage is allowed, the inventory level $I(\theta)$ is equal to the stock level $H(\theta)$; thus the average stock level over one cycle is

$$\frac{1}{t}\int_0^t H(\theta)\, d\theta = \frac{1}{t}\int_0^t I(\theta)\, d\theta \tag{8}$$

and

$$\text{the average holding cost/unit time} = \frac{h}{t}\int_0^t I(\theta)\, d\theta. \tag{9}$$

Therefore, the average total cost per unit time $\tilde{F}(Q, s, t)$ is the sum of the average procurement cost per unit time and the average holding cost per unit time, which is given by

$$\tilde{F}(Q, s, t) = \frac{K + cQ}{t} + \frac{h}{t}\int_0^t I(\theta)\, d\theta. \tag{10}$$

Using relation (3), we obtain

$$\tilde{F}(Q, s, t) = \frac{K + cQ}{t} + \frac{h}{t}\int_0^t (s + Q - D\theta)\, d\theta = \frac{K + cQ}{t} + h(s + Q) - h\frac{Dt}{2}.$$

The problem now is to determine the optimum values of two decision variables s and Q so as to minimize the average total cost per unit time, taking into account the operating constraints on the system. Mathematically, this problem can be formulated as

$$\min_{Q,s,t}\left\{\tilde{F}(Q, s, t) = \frac{K + cQ}{t} + h(s + Q) - h\frac{Dt}{2}\right\}.$$

subject to

$$Q > 0, \quad s \geqslant 0, \quad Q = Dt.$$

Moreover, using the constraint $Q = Dt$, we can eliminate the variable t in the model. The problem is now

$$\min_{Q,s}\left\{F(Q,s) = cD + \frac{KD}{Q} + hs + h\frac{Q}{2}\right\}$$

subject to

$$Q > 0, \quad s \geqslant 0.$$

Since for $s \geqslant 0$ the minimum value of the linear function $cD + hs$ occurs at $s = 0$, it follows that the optimal value s^* of s is $s^* = 0$. This implies that the most economic reorder level is selected when the stock on hand reaches a zero level. To determine the optimal value Q^* of Q, we consider the problem

$$\min_{Q>0}\left\{\hat{F}(Q) = \frac{KD}{Q} + h\frac{Q}{2}\right\}.$$

Now, since $\hat{F}(Q)$ is a strictly convex function of Q, we can obtain the optimal value Q^* by solving the first order condition

$$\frac{d\hat{F}(Q)}{dQ} = -\frac{KD}{Q^2} + \frac{h}{2} = 0,$$

which yields

$$Q^2 - \frac{2KD}{h} = 0, \quad Q > 0.$$

Therefore, the optimal values of s and Q are

$$s^* = 0$$

$$Q^* = \sqrt{\frac{2KD}{h}}. \qquad (11)$$

These values minimize the objective function $F(Q, s)$, while simultaneously satisfying the constraints. The corresponding value of the minimum cost function is

$$F(Q^*, s^*) = cD + KD\sqrt{\frac{h}{2KD}} + \frac{h}{2}\sqrt{\frac{2KD}{h}} \qquad (12)$$

$$= cD + \sqrt{2KDh}. \qquad (13)$$

The optimum value of S, S^* is given by

$$S^* = s^* + Q^* = \sqrt{\frac{2KD}{h}}.$$

The optimum value of Q^* involving the square root expression in (11) is often referred to as the Wilson formula or the economic order quantity (EOQ) formula. It is also called economic lot size in production systems.

Many variants based on the basic EOQ model as described above can be obtained by relaxing the initial assumptios, for example, on lead time, discounted cost, shortage, production rate, and review system. The detail derivations of such variants can be found in Sivazlian (1975).

2.2. *Stochastic Inventory Models*

In an inventory system, an element of randomness may be present in one or more of the input parameters, mainly due to the inability to provide exact estimation or prediction on the nature of these parameters. Thus, in general, to account for the uncertainty over time in the values taken by these parameters, it is necessary to describe them as random variables.

Of the set of input parameters that normally characterize an inventory system, demand is perhaps the most important element whose predictability could significantly affect the structure of the inventory problem. In this section, we restrict our analysis to inventory situations in which demand is the only element incorporating randomness.

Although, in general, the fluctuation in the demand pattern may incorporate trends and seasonalities—that is, elements which are functions of time—we shall only be concerned with instances in which the stochastic process generating demand is time invariant, or stationary.

The mathematical formulation of stochastic inventory models will depend very much on whether a continuous or periodic review system is analyzed and on the stochastic process describing the demand. The stochastic process generating demand will, in turn, structure the movement of inventory level at various times. Thus depletion of inventory occurs in a discrete fashion, and the resultant stochastic process for the inventory level is a discrete-valued continuous-parameter process. Detailed derivations of the stochastic process can be found in Sivazlian (1975).

We consider the case when demand is a homogeneous Poisson process where units occur one at a time. Now, let us assume that the expected demand per unit time $\bar{D}$ is given from the stochastic process. Then we can introduce the following basic stochastic inventory model. We select the steady-state total expected cost as our objective function; the decision variables s and Q are to be selected so as to minimize the object function. The total expected cost per unit time, expressed as a function of s and Q, is

$$F(Q, s) = K\frac{\bar{D}}{Q} + c\bar{D} + h\left[s + \frac{Q+1}{2}\right] \quad \text{for } s \geqslant 0, \quad Q \geqslant 1. \tag{14}$$

It is clear that $F(Q, s)$ is a separable function of s and Q and that the optimum value of s is $s^* = 0$. The optimum value of Q is obtained by minimizing the function

$$\tilde{F}(Q) = K\frac{\bar{D}}{Q} + h\frac{Q}{2}$$

over the set of positive integers Q.

It can be shown that an optimum value of Q, Q^*, exists and that, for such value(s), the following conditions hold:

$$\tilde{F}(Q^*) - \tilde{F}(Q^* + 1) \leqslant 0 \tag{15}$$

$$\tilde{F}(Q^*) - \tilde{F}(Q^* - 1) \leqslant 0. \tag{16}$$

From the first inequality (15), we obtain

$$K\frac{\bar{D}}{Q^*} + h\frac{Q^*}{2} - \left(K\frac{\bar{D}}{Q^*+1} + h\frac{Q^*+1}{2}\right) \leqslant 0.$$

After simplifying the inequality, we have

$$\frac{2K\bar{D}}{h} \leqslant Q^*(Q^*+1).$$

Similarly, the second inequality (16) implies

$$\frac{2K\bar{D}}{h} \geqslant Q^*(Q^*-1).$$

Combining these two conditions, we obtain

$$Q^*(Q^*-1) \leqslant \frac{2K\bar{D}}{h} \leqslant Q^*(Q^*+1).$$

Using this condition, the optimum Q^* may be selected for a given value of $\frac{2K\bar{D}}{h}$. For example, if $\frac{2K\bar{D}}{h} = 10$, $Q^* = 3$, while if $\frac{2K\bar{D}}{h} = 42$, then $Q^* = 6$ or 7.

As a special case, the present model covers the deterministic situation when the time interval between demand occurrence is a constant. The model then corresponds to the discrete-demand version of the basic model. In fact, for large values of Q, the optimal value of Q, Q^*, can be approximated by the square root formula

$$Q^* \approx \sqrt{\frac{2K\bar{D}}{h}}.$$

3. Inventory Management in Supply Chain

Management of the supply chain is becoming an increasingly important issue in virtually all industries. As firms seek greater market share, reduced costs and improved efficiency become critical. Logistics and supply-chain management is one area in which cost reductions and strategic advantages are being sought.

A supply chain is the entire product delivery structure including the plant, storage warehouses, distributors, and retailers. In this global structure, there are costs associated with moving the material (or products) from one location to another, with building facilities for storage, and for transactions at each stage of the chain. However, the costs of holding inventories are perhaps the most significant. It has been observed that a serial supply-chain structure has the tendency to amplify demand variation, resulting in much-larger-than-anticipated inventories (bullwhip effect; Burns and Sivazlian, 1978; Lee et al., 1997).

In order to see the importance of inventory management, we need to discuss logistics as an integrated supply-chain system. Logistics is the set of functions associated with the flow of goods, information, and payments among suppliers and consumers from the origination of raw material to final recycling or disposal of finished goods. Companies increasingly view logistics as a crucial element in their corporate strategy. If modern enterprises are to meet customer demands for responsiveness, customization, quality, and value, they must effectively and efficiently link their entire global supply chain.

Logistics supply chains are the physical components including the manufacturing plants, warehouses, vehicles, and transportation infrastructure. Logistics information and decision technologies are the computer-related technologies utilized in the design, planning, and operation of supply-chains including monitoring status within the supply-chain and communicating among supply-chain elements. A totally integrated logistics value chain is seamless across all supply-chain elements and optimized with regard to both service and cost.

4. A Network Approach

Unfortunately, due to the complexity of supply-chain structures including production and inventory systems, it is very difficult to develop a global solution technique for supply-chain management problems. Recently, many companies have tried to locally optimize their supply-chain systems (e.g., component by component), and then integrate the components and implement the local solutions to their global systems. It is clear that the integrated local solutions may not provide a global solution to the entire supply-chain management problem.

There is an interesting approach to handling these problems globally in various network structures. Although the approach may have some difficulties in representing very specific and complicated systems, we still have two advantages in using this approach. The first advantage is that once the problem is formulated in a network structure, we can obtain a global solution by using an appropriate network optimizing technique that can handle concave costs (fixed charge and/or price discounts). This network solution is also very useful for viewing the entire flow diagram of goods, so that one can monitor the current status of each component in the supply chain. The second advantage is more realistic: once a network structure for the problem is obtained it can be utilized in designing a data communication network for supply-chain management.

Historically, a network approach for production-inventory problems was originally introduced by Zangwill (1969), as a concave-cost network flow problem. Consider the following formulation of the classical Wagner–Whitin model (Whitin, 1953; Wagner and Whitin, 1959). Let x_j be the amount produced at the beginning of period j and I_j be the inventory at the end of period j. Assuming the market demand in period j is given (or estimated) by d_j, we obtain the following formulation.

$$\min\sum_{j=1}^{n}P_j(x_j)+\sum_{j=1}^{n}H_j(I_j) \tag{17}$$

subject to

$$\sum_{j=1}^{n}x_j=\sum_{j=1}^{n}d_j \tag{18}$$

$$x_j+I_{j-1}-I_j=d_j,\quad j=1,2,\dots,n \tag{19}$$

$$x_j\geqslant 0,\quad I_j\geqslant 0,\quad j=1,2,\dots,n \tag{20}$$

$$I_0=I_n=0, \tag{21}$$

where $P_j(x_j)$ is a production cost function and $H_j(I_j)$ is an inventory cost function. These costs are assumed concave, especially fixed-charge and/or concave piecewise linear, in most real-world situations. Note that, clearly, Equation (18) is a redundant constraint, and the above system describes a network structure. An example of the network representation of the model is shown in Figure 2.

Now let us consider a typical supply chain with four levels, from a manufacturing system to customers (see Figure 3). Clearly, it has a network structure. In most cases, this network problem will involve transportation and/or vehicle routing problem(s), which can be solved efficiently by an appropriate network optimization algorithm.

The main idea here is that the network model described in (17)–(21) can be embedded into the main network structure of a given supply chain. Then we have an extended network structure for the supply chain, which may have fixed-charges and/or concave piecewise linear costs. As a result, the problem becomes a fixed-charge or concave piecewise linear network flow problem, which can be transformed to a 0–1 mixed-integer programming problem and solved by the branch-and-bound method (Guisewite and Pardalos, 1990; Horst and Pardalos, 1995).

Due to the limitations (computing time and memory) of the branch-and-bound method, a number of approximation methods have been developed in this field over the last few decades. Useful information can be found in Guisewite and Pardalos (1995). Recently, Kim and Pardalos (1999) developed an efficient heuristic approach for the fixed-charge and concave piecewise linear network flow problems. Their computational results show that the heuristic algorithm, the so-called dynamic slope scaling procedure (DSSP), provides quite efficient and effective solutions exploiting the separability of network structures.

5. Concluding Remarks

Based on the brief descriptions of fundamental inventory models, the importance of inventory management incorporating other systems, such as production systems, logistics, and the supply chain, has been discussed in this article. In particular, the supply chain, viewed as an integrated system of production, inventory, and distribution modules, can be an important example in which inventory management is a key to success.

The network approach introduced in Section 4 can provide useful mathematical and strategic information for supply-chain management. In this global approach, although it seems to be difficult to consider detailed variations in such parameters as lead time, variable demand rate, and decay, such variations can be easily optimized in local structures.

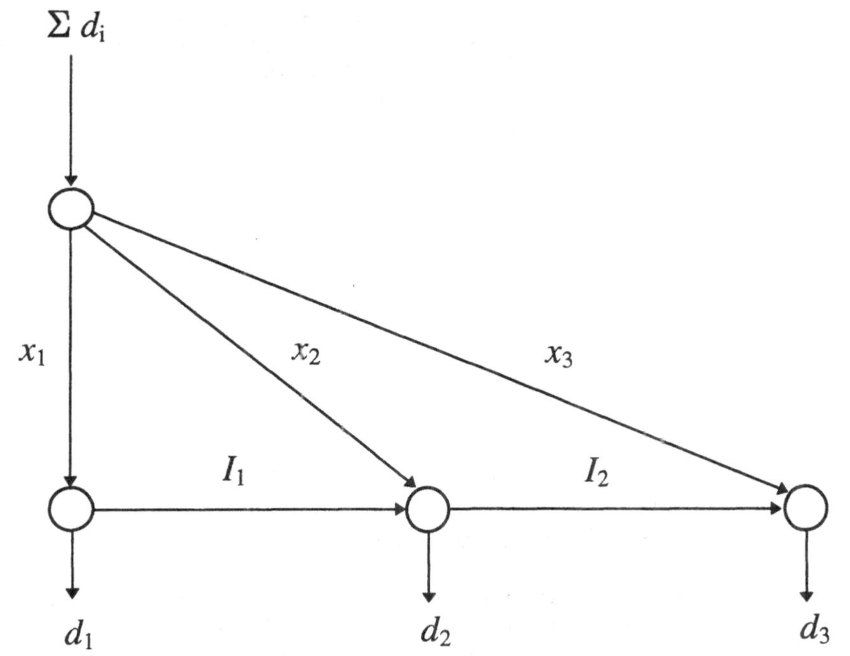

Figure 2 The network representation of the Wagner–Whitin model

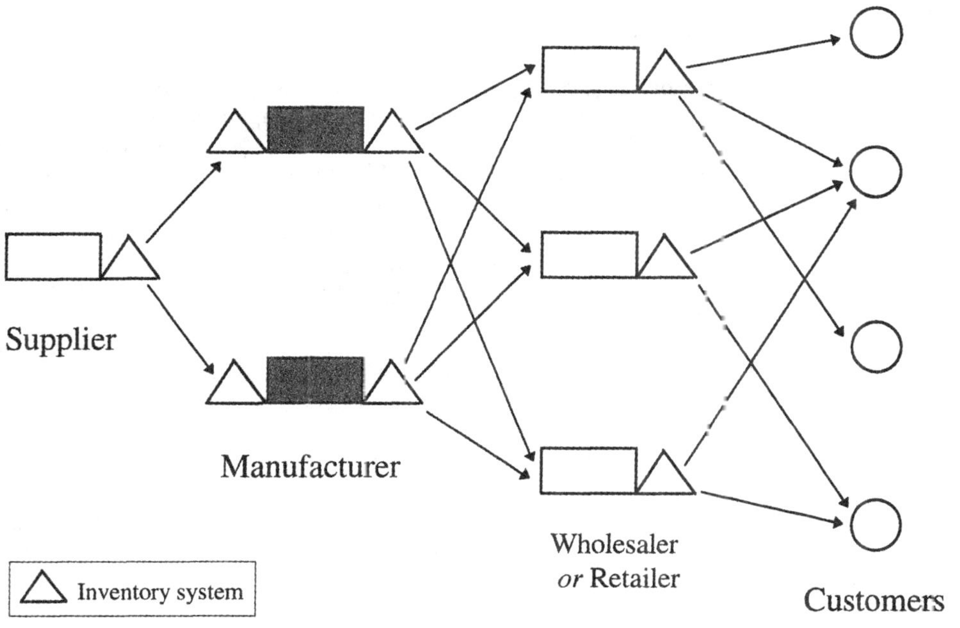

Figure 3 A typical supply-chain model

References

Burns, J. F. and B. D. Sivazlian. 1978. "Dynamic analysis of multi-echelon supply systems." *Comput. Ind. Engng* **2**: 181–193.

Guisewite, G. M. and P. M. Pardalos. 1990. "Minimum concave-cost network flow problems." *Ann. Oper. Res.* **25**: 75–100.

Horst, R. and P. M. Pardalos, (eds). 1995. *Handbook of Global Optimization*. Kluwer Academic Publishers.
Kim, D. and P. M. Pardalos. 1999. "A solution approach to the fixed charge network flow problem using a dynamic slope scaling procedure." *Oper. Res. Lett.* **24**: 195–203.
Lee, H. L., V. Padmanabhan, and S. Whang. 1997. "Information distortion in a supply chain: the bullwhip effect." *Mgmnt Sci.* **43(4)**: 546–558.
Sivazlian, B. D. 1975. In *Analysis of Systems in Operations Research*. Prentice-Hall, Inc.
Wagner, H. M. and T. M. Whitin. 1959. "Dynamic version of the economic lot size model." *Mgmnt Sci.* **5(1)**: 89–96.
Whitin, T. M. 1953. In *The Theory of Inventory Management*. Princeton University Press.
Zangwill, W. I. 1969. "A backlogging model and a multi-echelon model of a dynamic economic lot size production system—a network approach." *Mgmnt Sci.* **15(9)**: 506–527.

17.7 Location

Zvi Drezner

ABSTRACT Location problems deal with finding the best location for one or more facilities that optimize some objective. Such facilities may provide service to a given set of customers, may be a nuisance to a set of points, or may compete with other facilities to attract customers. Location models differ by the nature and the number of the facilities, the way the distance is measured, the existence of constraints on the locations, the nature of the parameters, and other considerations. Solution techniques vary from combinatorial procedures to analytical techniques. Some problems may require unfeasible computer resources for solving reasonable size problems. Heuristic approaches are developed for such problems.

1. Introduction

An important aspect of optimizing company operations is locating its facilities in their optimal location. This is part of analyzing the logistics aspects of the company's operations. The area of operations research has benefited significantly from the introduction of digital computers. Optimization techniques have become more accessible, more powerful, and therefore more useful. The last twenty to thirty years have witnessed a vast increase in interest in operations research problems in general, and location problems in particular, among researchers from a variety of disciplines including economics, marketing, engineering, urban planning, geography, and management. This diverse interest has yielded a multitude of research approaches and solutions to such problems.

A location problem is a spatial resource allocation problem. In the general location paradigm, one or more service facilities serve a spatially distributed set of demands (customers). The objective is to locate facilities to optimize a spatially dependent objective. Location decisions include the addition of new facilities to an existing infrastructure, rearranging an existing layout, or planning a completely new system.

The application of location theory to industrial problems was first introduced by Alfred Weber (1909), who considered locating a single warehouse so as to minimize the total travel distance between the warehouse and a set of spatially distributed customers. Another early location problem was formulated by Hotelling (1929), an economist who considered locating two competing retail facilities on a line such as main street.

The classic application of locational analysis is Weber's location of a warehouse (or a distribution center) servicing a set of retail facilities. Other service type location examples include the location of schools, day care facilities, bank branches, telephone switchboards; the location of emergency facilities such as hospitals, ambulance depots, or fire stations; the location of military and naval installations, international couriers, and international airline hubs. These location decisions represent different company objectives. The location objectives range from minimizing total travel time as is the case of school location, to the minimization of travel time for the farthest customer as is the case of emergency facility location, to location on a globe for airline routing or military installations. Other location models are used in competitive environments, where companies compete for customers, such as grocery stores, department stores, shopping malls, restaurants, or gas stations. There are situations where facilities are considered obnoxious. Such obnoxious facility location models are used for locating airports, recycling centers, nuclear power plants, prisons, waste disposal sites, pollution generating factories, or refineries. In this case the location represents a balance between two conflicting objectives: protecting residents but still being within commuter range.

This section is based on the book chapter by Drezner and Drezner (1998) and the books Drezner (1993) and Drezner and Hamacher (2001). For a more detailed review the reader is referred to this book chapter. The various dimensions of locational analysis are summarized below.

1.1. Facility Location Models and Facility Layout

The distinction between facility location and facility layout consists of the presence or absence of demand. While facility location models deal with locating facilities among customers who require service, facility layout models are concerned with the optimal layout of facilities in a given area in relation to each other. In facility layout problems, the facilities are interrelated and there are no demand points. Examples of facility layout include the layout of a firm's offices, the layout of an instrument control panel in an airplane, and the layout of a building complex. The facility layout problem is sometimes referred to in the literature as the quadratic assignment problem. For a review see the chapter by Burkard (1990) in Mirchandani and Francis (1990).

Facilities can be either desirable or obnoxious. Desirable facility location models assume that the closer the facility is to demand, the better the objective is met. There are two main objectives for desirable facilities models: minisum and minimax. The first and most common is the classical model of minimizing the sum of weighted distances (sometimes called the Weber problem, named after Alfred Weber, 1909). The second, the minimax model, is concerned with minimizing the maximal distance between facilities and demand points. This is the appropriate model for locating emergency facilities where the level of service to the farthest customer is the relevant criterion. For reviews of these two models, see Love et al. (1988) and Francis et al. (1992).

Another type of desirable facilities are competitive facilities such as retail outlets. One or more facilities are to be located in an area in which competition already exists. Competitive facility location models are applied mainly by marketers. Facilities compete for customers and their objective is to maximize the market share they capture. Assuming that profit increases when market share increases, maximizing profit is equivalent to maximizing market share. It follows that the location objective is to locate the retail outlet at the location that maximizes its market share. One unique feature of competitive facility models is facility attractiveness (its appeal to customers). Facility attractiveness level, therefore, needs to be incorporated in the model. Other location models do not consider differences in attractiveness among facilities and therefore allocate customers to facilities by proximity. Also unique to competitive facility location is the modeling of demand in terms of buying power. Income levels and discretionary spending become a measure of demand. For a review of competitive models, see Drezner (1995).

Obnoxious facility location models seek to locate facilities as far as possible from customers. Commonly, a secondary consideration is that the facilities are still within reach. The airport location is a good example of this type. An airport is a source of noise pollution and traffic congestion, yet it needs to be easily accessible. Both objectives need to be accommodated. For a review of obnoxious facility location models, see Erkut and Neuman (1989) and Drezner (1995).

It becomes clear that facility location analysis is useful for both public and private sector facilities. A common objective in the public sector is equity such as in service provision, therefore the minimax objective is used. A common objective in the private sector is minimization of cost (e.g., transportation cost) thus the minisum objective is appropriate.

1.2. *The Number of Facilities*

A distinction exists in the analysis between locating one facility and locating multiple (two or more) facilities. Multiple-facility location is conceptually and mathematically more complex. Applications for single-facility location problems usually become multiple-facility problems when the area covered is large and one facility cannot service the whole area. The area is then divided between several facilities and the problem becomes a combination of simultaneously finding both the proper allocation of customers to facilities and the best location for the facilities. In location literature this is referred to as the location-allocation model (Ghosh and Rushton, 1987). It is quite common that several new facilities need to be located in an area in which some existing facilities are already located. Such problems are called conditional location problems.

1.3. *Location Environment*

Three main environments are considered in location analysis. Customers and facilities can be located on a network, in the plane, or on the surface of the globe. Each environment entails its own formulation. The nature of the environment determines the measure of distance between customers and facilities.

Network problems deal with facility location on a network such as a road or a rail system, a communication network, or a computer chip. A graph (or a network) is defined by nodes and arcs of given lengths. Movement from one point on the graph to another is restricted to arcs of the network. The location of facilities is usually restricted to arcs and nodes. The shortest distance between two points on the graph needs to be calculated as part of each network modeling. For a review of network problems, see Daskin (1995).

Planar problems, where demand points and facilities are located in the plane, have received the most attention by researchers. There are three main distance measures on the plane: Euclidean distance, or the distance along a straight line between two points; rectangular distance (sometimes called rectilinear or Manhattan distance), where movement between two points is allowed only along North–South or East–West axes roads; and the general ℓ_p distance, which approximates actual road distances most accurately. For a review of planar models, see Love et al. (1988) and Francis et al. (1992).

Global facility location deals with location on a sphere for air travel or naval operations. The facilities and the demand are located on the surface of the earth. Facility location on a sphere is an extension of facility location on the plane. Whereas planar problems deal with demand points and facilities on a small scale (where they are relatively close and can be assumed to be on a plane), spherical models deal with long-range distances where location on the plane can no longer be assumed. In this case, distance is measured along great circles. For a review of spherical algorithms, see Wesolowsky (1983).

1.4. *Solution Domain*

The locations of the new facilities can be either found in a discrete space, continuous space, or constrained space. Discrete models seek the optimal location from among a prespecified set of possible sites. Continuous models seek the best location anywhere in the problem space (planar, spherical, or network). Some models restrict the location of the new facilities by imposing constraints on their location.

1.5. *The Nature of the Parameters*

In most location models it is assumed that problem parameters (such as demand level or its distribution) are deterministic (known). Other models deal with uncertainty about the parameters, assuming that some parameters are drawn from some statistical distribution and are not deterministic. These are referred to as stochastic models. In dynamic models it is assumed that the parameters change over time in a predictable way. The problem becomes finding the best location that will minimize cost over a time horizon.

These models constitute the bulk of the location literature. There are other problems that lend themselves to formulation as a location model. For a review, see Drezner and Drezner (1998).

An important issue in all location models is the modeling of demand. For the distribution center problem, demand is generated as a finite set of users called demand points. For many other applications, demand is generated over an area since it is prohibitive to use each household as a separate demand point. For that purpose the area is partitioned into small subareas and each is assigned one demand point at its center. This demand point represents the total demand generated by all households in that subarea. In competitive location models, demand is modeled in terms of aggregated buying power. For many practical applications, census data are available for census tracts, zip codes, or cities. In such cases the demand points are defined by the type of available data.

For further and more in-depth discussion of the various location issues, we recommend Daskin (1995) for network problems, Love et al. (1988) and Francis et al. (1992) for planar problems, and Wesolowsky (1983) for spherical problems. The book by Drezner (1995) provides a state-of-the-art discussion on location, analysis models and provides an excellent bibliography of over one thousand two hundred references. For a review of location models we recommend the chapters by Plastria (1995) and Eiselt and Laporte (1995).

2. Selected Special Algorithms

Many location problems can be solved by standard mathematical programming software such as the solver in Excel. However, when the number of demand points or facilities is large, the problem contains too many variables or constraints that exceed Excel's capacity. In such cases a special program is needed for the solution of these problems In this section we review selected special, but simple, algorithms for solving various location problems. These methods are independent of Excel and require coding a special program. They can be coded in any programming language such as Basic, FORTRAN, Pascal, or others If coded in Basic (or visual basic) the program can be part of the spreadsheet. The special methods are also useful as part of complicated algorithms that require many solutions to simple problems. For example, solving multiple-facility location problems often requires repeated solutions to single-facility location problems. Therefore, a simple approach to solving single facility location problems is essential for employing such algorithms for the solution of multiple-facility location problems.

2.1. *Notation*

n is the number of demand points;

(a_i, b_i) is the location coordinates of demand point i, for $i = 1, .., n$.

w_i is the weight associated with demand point i, for $i = 1, ..., n$

(x, y) is the unknown location of the facility.

$d_i(x, y)$ is the distance between demand point i and the facility location (x, y).

Euclidean distance: $d_i(x,y) = \sqrt{(x-a_i)^2 + (y-b_i)^2}$

Rectilinear distance: $d_i(x,y) = | x-a_i | + | y-b_i |$

Chebishev distance: $d_i(x,y) = \max\{| x-a_i |, | y-b_i |\}$

ℓ_p distance: $d_i(x,y) = \sqrt[p]{| x-a_i |^p + | y-b_i |^p}$

Before turning to specific algorithms we elaborate on the issue of distance. Euclidean distance (or straight-line distance) is the most common distance measure. The rectilinear distance is used for square grid distances. Another useful distance is the max (or Chebishev) distance. These distance measures are special cases of the ℓ_p distance. Euclidean distances use $p = 2$, rectilinear distances use $p = 1$, and the Chebishev distance uses $p = \infty$. The general ℓ_p distance was found useful in approximating road distances.

2.2. *Single-Facility Location Using Euclidean Distances*

The single-facility location problem using Euclidean distances is the most basic facility location problem. The effective procedure commonly used for the single-facility problem is the Weiszfeld procedure first proposed by Weiszfeld (1936). The single-facility location problem consists of minimizing: $\sum_{i=1}^{n} w_i\sqrt{(x-a_i)^2 + (y-b_i)^2}$.

The derivatives of this objective function by x and y are equated to zero, yielding after simple algebraic manipulation:

$$x = \frac{\sum_{i=1}^{n} \frac{w_i a_i}{\sqrt{(x-a_i)^2 + (y-b_i)^2}}}{\sum_{i=1}^{n} \frac{w_i}{\sqrt{(x-a_i)^2 + (y-b_i)^2}}}; \quad y = \frac{\sum_{i=1}^{n} \frac{w_i b_i}{\sqrt{(x-a_i)^2 + (y-b_i)^2}}}{\sum_{i=1}^{n} \frac{w_i}{\sqrt{(x-a_i)^2 + (y-b_i)^2}}}.$$

This equation is an implicit solution for x and y meaning that these variables appear in both sides of the equation. The Weiszfeld algorithm starts from a starting site (x, y) and calculates a new x and y using this equation by substituting the previous x and y in the right-hand side of the equation. The Weiszfeld procedure is proven to converge to the solution point, and is very effective. Acceleration methods were proposed for this algorithm (for the latest, see Drezner, 1996). This algorithm is very easy to code and can serve as a subroutine for more complicated algorithms.

2.3. *The Generalized Weiszfeld Procedure*

In many instances the minimization of transportation cost involves a function of the distance rather than the distance itself. One such example is competitive facility location based on gravity models. We show how to design a Weiszfeld-type procedure for general functions of the Euclidean distance.

The objective is to minimize: $\sum_{i=1}^{n} f_i[d_i(x, y)]$ where $d_i(x, y)$ is the Euclidean distance. By equating the derivatives of the function by x and y to zero and using the chain rule we get:

$$\sum_{i=1}^{n} \frac{\partial f_i}{\partial d_i} \frac{x-a_i}{d_i} = 0; \quad \sum_{i=1}^{n} \frac{\partial f_i}{\partial d_i} \frac{y-b_i}{d_i} = 0,$$

which leads to:

$$x = \frac{\sum_{i=1}^{n} \frac{1}{d_i} \frac{\partial f_i}{\partial d_i} a_i}{\sum_{i=1}^{n} \frac{1}{d_i} \frac{\partial f_i}{\partial d_i}}; \quad y = \frac{\sum_{i=1}^{n} \frac{1}{d_i} \frac{\partial f_i}{\partial d_i} b_i}{\sum_{i=1}^{n} \frac{1}{d_i} \frac{\partial f_i}{\partial d_i}}.$$

These formulas can be applied in an iterative procedure similar to the Weiszfeld procedure. Note that for the minisum Euclidean $f_i(d_i) = w_i d_i$ and the formula reduces to the Weiszfeld procedure. For squared Euclidean distances the formula is $f_i(d_i) = w_i d_i^2$ and the formula reduces to the center of gravity formula.

2.4. *Single-Facility Location Using Rectilinear Distances*

The single-facility minisum objective using rectilinear distances is the minimization of: $w_i[|x-a_i| + |y-b_i|]$. This problem is separable into two independent optimization problems: one for x and one for y. The problem is solved twice, once for each dimension. Consider the solution of the x dimension. The best solution is the median point of the sequence $a_1, \ldots, a_n$, each coordinate with its associated weight. The median point is either at one a_i or the segment connecting two consecutive coordinates a_i and a_{i+1}. In any case, one of the optimal solutions must be on a point a_i for some i. In order to find the median point, all coordinates a_i are sorted, points with the same coordinate are collapsed, and the median found.

2.5. *The Single-Facility Minimax Problem Using Euclidean Distances*

The unweighted Euclidean Distance minimax problem is equivalent to finding the smallest circle that surrounds all demand points. Such a circle is either defined by three points, or its diameter is defined by two points. The center of such a circle is the minimax solution. Elzinga and Hearn (1972) proposed an exchange algorithm for the solution of this problem. The algorithm proceeds as follows:

1. A starting point (center) is selected.
2. The three farthest demand points from the selected center are found.
3. The circle passing through these three points is calculated. (If the triangle is obtuse, the center of the largest side of the triangle is the center of the circle.)
4. If the circle encloses all demand points, then the center of this circle is the optimal solution.
5. If there are points outside this circle (these points are farther from the center than the radius of the circle), select the farthest point from the center as a fourth point.
6. Find the smallest circle enclosing these four demand points. This circle passes through three of these four demand points. There are four possible choices of three points out of four but the original three demand points cannot be the group enclosing all four points because the newly selected point is outside the circle enclosing the original three points. Therefore, there are only three possible selections of three points out of four to check. Each time, one of the original three points is dropped from the group and the circle enclosing the remaining three points is found. The group yielding the largest radius must enclose all four points. This set of three points is the new set of three points and defines a new circle with a new center. Go back to Step 4.

This algorithm is empirically fast because few iterations are required for most problems. Even a problem with ten thousand points, which is beyond Excel's capability, is solved in a fraction of a second. The weighted version of the problem is solved in a similar manner (Hearn and Vijay, 1982).

2.6. *The Single-Facility Rectilinear Distance Minimax*

The rectilinear distance has the following property: If the coordinates are rotated 45 degrees, the rectilinear distance becomes a Chebishev distance. Therefore, to find the minimax rectilinear distance location, one rotates the coordinates 45 degrees and finds the smallest enclosing rectangle. The center of this rectangle is one of the optimal solutions. In Drezner (1987b) a simple formula is given for such a solution.

It is interesting that the two-center solution is also very simple to find even though the objective function is not convex. See Drezner (1987b) for the explicit formula for the two-center problem.

2.7. *Multiple-Facility Location Problems*

There are two types of multifacility problems. In one type, each customer is serviced by all facilities. The more useful type is the location of several identical facilities such that customers are serviced by the facility closest to them. We discuss problems of the second type. These problems are the location-allocation problems. Another common name (especially on networks) is the p-median problem for the minisum objective (because the solution to the single-facility problem is the median) and the p-center problem for the minimax multifacility problem (because the solution to the one facility minimax problem is the center of a circle). These problems are not convex and may have many local optima. It is therefore necessary to check all possible local minima in order to identify the global minimum. For these problems heuristic algorithms are proposed.

Cooper (1963, 1964) was the first to propose a heuristic approach to solving the location-allocation problem for any distance measure. His idea is as follows:

1. Random locations for p facilities are selected.
2. Each demand point is assigned to its closest facility (this is the allocation phase).
3. Each facility defines a set of demand points closest to it.
4. p single-facility problems are solved by finding the best location for each subset of demand points (this is the location phase).

5. If the calculated locations of all p facilities are the same as in the previous iteration, stop with this as a proposed solution. Otherwise go to Step 2.

This is a simple and effective heuristic procedure. In order to operationalize this procedure, one needs a procedure for solving the single-facility problem. Note that a solution to many single-facility problems (whether minisum or minimax) is required. Since it is a heuristic algorithm, it is recommended that the procedure is repeated many times from randomly generated starting locations, and the best solution among these experiments is selected as the solution to be used. Metaheuristic algorithms such as simulated annealing, tabu search, and genetic algorithms can be applied to this problem utilizing the above approach as the descent base for these procedures.

2.8. *Voronoi Diagrams*

A useful tool in constructing special algorithms for various location problems is the Voronoi diagram (Okabe et al., 1992). To explain the concept, consider a set of n points (demand points or facilities) in the plane. Each point defines a polygon of all the points in the plane that are closest to it. Some of these polygons are finite and some are open to infinity. Constructing such a diagram is done very efficiently by special construction algorithms. The diagram consists of a set of generating points (the original points) and the sides and vertices of all polygons. In location-allocation models, the Voronoi diagram generated by the facilities defines the allocation of customers to facilities. An example of the use of a Voronoi diagram is the solution to the obnoxious facility location problem maximizing the shortest distance to all demand points. The solution point for this maximin problem is a vertex of the Voronoi diagram.

One can construct a Voronoi diagram where each generator point defines a polygon whose points are farthest from the generator point. The solution for the single-facility minimax location problem is on a vertex of such a Voronoi diagram (Okabe et al., 1992).

2.9. *Network Problems*

Location problems on a network are usually formulated as integer linear programming models. The Solver option in Excel can solve such problems, but the number of integer variables is large even for small problems and would normally exceed Excel's capacity. Location problems on a network were first introduced by Hakimi (1964, 1965). Hakimi proved the main property of these problems: "the solution for the one facility minisum problem on a network must be on a node of the network." (Note that it is possible to have many equally valued optimal solutions but one of these must be on a node.) This restricts the search for the optimal solution to the finite set of nodes.

Special cases of networks are much easier to analyze. A commonly researched network is a tree network. A tree network does not contain cycles and there is only one way to travel from any node to any other node. Such a network lends itself to simpler algorithms.

Solution algorithms on a network are combinatorial in nature. This means that there are many combinations to be tested and algorithms perform these tests in an efficient way trying to exclude as many trials as possible. For a review of network location problems, see the textbook by Daskin (1995).

2.10. *Facility Layout*

The facility layout problem is different from other location models in that there are only facilities to be located in relation to each other and no demand points. An area is available for the location of facilities, and each facility has a given size (and sometimes given shape). A weight describing the relationship between each pair of facilities is given. This weight quantifies the attraction between pairs of facilities. A large weight means that the two facilities should be located close to one another, while a zero weight (we may have negative weights as well) indicates indifference to the relative location of the two facilities. For example, in the design of an office building complex, the weight may indicate the amount of traffic between buildings. In planning an instrument panel for an airplane the weight may indicate the frequency of consecutive or simultaneous watch of a pair of displays.

The first heuristic procedure for this problem was craft (Armour and Buffa, 1963). A random arrangement is selected. All pairwise exchanges of facilities are checked and the exchange that most reduces the objective function is performed. The iterations are repeated until no improvement is possible by pairwise exchanges.

DISCON (Drezner, 1980) is a different approach to formulating and solving the facility layout problem. The locations for the facilities are not preset to a certain pattern. A configuration in the plane is obtained by a simulation of the big bang, which is believed to have created the universe. All facilities are placed at one point and are allowed to explode (the dispersion phase). The configuration is then concentrated back to the center (the concentration phase) yielding the final configuration. This configuration is finished by the user to fit his or her needs. The configuration is good in the sense that facilities that need to be close to each other usually are in the final configuration. DISCON was simplified in Drezner (1987a) where the locations for the facilities are determined by the eigenvectors of a certain matrix. This approach was used in Marcoulides and Drezner (1993) to map n-dimensional data in two dimensions. The weights are calculated as reciprocals of the distances in the n-dimensional space raised to some power and the layout problem is solved. The result is a two-dimensional map where points that are close to one another in the n-dimensional space tend to be close to one another in that two-dimensional solution and vice versa.

Recently, metaheuristics such as tabu search, simulated annealing, and genetic algorithms were applied to the layout problem. For a review, see the book edited by Pardalos and Wolkowicz (1994).

References

Armour, G. C. and E. S. Buffa. 1963. "A heuristic algorithm and simulation approach to relative location of facilities." *Mgmnt Sci.* **9**: 294–309.

Burkard, R. E. 1990. "Locations with spatial interactions: the quadratic assignment problem." In *Discrete Location Theory,* edited by P. B. Mirchandani and R. L. Francis, pp. 337–437. New York: John Wiley and Sons.

Cooper, L. 1963. "Location-allocation problems." *Oper. Res.* **11**: 331–343.

Cooper, L. 1964. "Heuristic methods for location-allocation problems." *SIAM Rev.* **6**: 37–53.

Daskin, M. S. 1995. In *Network and Discrete Location: Models, Algorithms, and Applications.* New York: John Wiley & Sons.

Drezner, Z. 1980. "DISCON—a new method for the layout problem." *Oper Res.* **28**: 1375–1384.

Drezner, Z. 1987a. "A heuristic procedure for the layout of a large number of facilities." *Mgmnt Sci.* **33**: 909–915.

Drezner, Z. 1987b. "On the rectangular p-center problem." *Naval Res. Logist.* **34**: 229–234.

Drezner, Z. (ed). 1995. *Facility Location: A Survey of Applications and Methods.* New York: Springer-Verlag.

Drezner, Z. 1996. "A note on accelerating the Weiszfeld procedure." *Location Sci.* **3**: 275–279.

Drezner, Z. and T. Drezner. 1998. "Applied location theory models." In *Modern Methods for Business Research,* edited by G. A. Marcoulides, pp. 79–120. Mahwa, NJ: Lawrence Erlbaum Associates.

Drezner, Z. and H. Hamacher (eds). 2001. In *Facility Location: Applications and Theory.* Berlin: Springer-Verlag.

Eiselt, H. A. and G. Laporte. 1995. "Objectives in location problems." In *Facility Location: a Survey of Applications and Methods,* edited by Z. Drezner, pp. 151–180. New York: Springer-Verlag.

Elzinga, D. J. and D. W. Hearn. 1972. 'Geometrical solutions for some minimax location problems." *Transport. Sci.* **6**: 379–394.

Erkut, E. and S. Neuman. 1989. "Analytical models for locating undesirable facilities." *Eur. J. Oper. Res.* **40**: 275–291.

Francis, R. L., L. F. McGinnis, Jr., and J. A. White. 1992. 2d ed. *Facility Layout and Location: an Analytical Approach.* International Series in Industrial and Systems Engineering, Englewood Cliffs, NJ: Prentice Hall.

Ghosh, A. and G. Rushton. 1987. In *Spatial Analysis and Location-Allocation Models.* NY: Van Nostrand Reinhold Company.

Hakimi, S. L. 1964. "Optimal location of switching centers and the absolute centers and medians of a graph." *Oper. Res.* **12**: 450–459.

Hakimi, S. L. 1965. "Optimum distribution of switching centers in a communication network and some related graph theoretic problems." *Oper. Res.* **13**: 462–475.

Hearn, D. W. and J. Vijay. 1982. "Efficient algorithms for the (weighted) minimum circle problem." *Oper. Res.* **30**: 777–795.

Hotelling, H. 1929. "Stability in competition." *Econ. J.* **39**: 41–57.

Love, R. F., J. G. Morris, and G. O. Wesolowsky. 1988. In *Facilities Location: Models and Methods*. NY: North-Holland.

Marcoulides, G. and Z. Drezner. 1993. "A procedure for transforming points in multi-dimensional space to two-dimensional." *Educ. Psych. Measurement* **53**: 933–940.

Mirchandani, P. B. and R. L. Francis (eds). 1990. *Discrete Location Theory*. New York: John Wiley and Sons.

Okabe, A., B. Boots, and K. Sugihara. 1992. In *Spatial Tessellations. Concepts and Applications of Voronoi Diagrams*. Chichester, England: John Wiley & Sons.

Pardalos, P. M. and D.-Z. Du (eds). 1998. *Network Design: Connectivity and Facilities Location*, DIMACS Series vol. 40. New York: American Mathematical Society.

Pardalos, P. M. and H. Wolkowicz (eds). *Quadratic Assignments and Related Problems*. New York: American Mathematical Society.

Plastria, F. 1995. "Continuous location problems." In *Facility Location: A Survey of Applications and Methods*, edited by Z. Drezner, pp. 225–262. New York: Springer-Verlag.

Weber, A. 1909. *Uber Den Standort Der Industrien, 1. Teil: Reine Theorie Des Standortes*, Tubingen. English translation (1929): *On the location of industries*. University of Chicago Press, Chicago. (English Translation by C. J. Friedeich, 1957. *Theory of the Location of Industries*. Chicago: Chicago University Press.)

Weiszfeld, E. 1936. "Sur Le Point Pour Lequel La Somme Des Distances De N Points Donnes Est Minimum." *Tohoku Math. J.* **43**: 355–386.

Wesolowsky, G. O. 1983. "Location problems on a sphere." *Regional Sci. Urban Econ.* **12**: 495–508.

17.8 Algorithms for the Satisfiability (SAT) Problem

Jun Gu, Paul W. Purdom, John Franco, and Benjamin W. Wah

ABSTRACT The satisfiability (SAT) problem is a core problem in mathematical logic and computing theory. In practice, SAT is fundamental in solving many problems in automated reasoning, computer-aided design, computer-aided manufacturing, machine vision, database, robotics, integrated circuit design, computer architecture design, and computer network design. In this article, we describe four major classes of algorithms: variable setting and resolution procedures, local search algorithms, nonlinear unconstrained methods, and nonlinear constrained techniques, for the SAT problem. A number of advanced techniques for SAT problems, such as general Boolean representations, the binary decision diagram (BDD), multispace search method, parallel SAT algorithms and architectures, and the multiSAT algorithm, are described in more comprehensive discussions (Gu et al., 1997a,b, 2000).

1. Introduction

An instance of the satisfiability (SAT) problem is a Boolean formula that has three components (Cook, 1971; Garey and Johnson, 1979):

- A set of n variables: $x_1, x_2, ..., x_n$.
- A set of literals. A literal is a variable ($Q = x$) or a negation of a variable ($Q = \bar{x}$).
- A set of m distinct clauses: $C_1, C_2, ..., C_m$. Each clause consists of only literals combined by just logical or ($\vee$) connectives.

The goal of the satisfiability problem is to determine whether there is an assignment of truth values to variables that makes the following conjunctive normal form (CNF) formula satisfiable:

$$C_1 \wedge C_2 \wedge \ldots \wedge C_m, \tag{1}$$

Where $\wedge$ is a logical and connective.

The SAT problem is the core of a large family of computationally intractable NP-complete problems (Cook, 1971; Garey and Johnson, 1979). Such NP-complete problems have been identified as central to a number of areas in computing theory and engineering. Since SAT is NP-complete, it is unlikely that any SAT algorithm has a fast worst-case time behavior. However, clever algorithms can rapidly solve many SAT formulas of practical interest. There has been great interest in designing efficient algorithms to solve most SAT formulas.

In practice, SAT is fundamental in solving many problems in automated reasoning, computer-aided design, computer-aided manufacturing, machine vision, database design, robotics, integrated circuit design automation, computer architecture design, and computer network design (see Gu et al., 1997b, 2000). Therefore, methods to solve SAT formulas play an important role in the development of efficient computing systems.

In this article, we briefly describe four major classes of algorithms; variable setting and resolution procedures, local search algorithms, nonlinear unconstrained methods, and nonlinear constrained techniques, for the SAT problem. A number of advanced techniques for the SAT problems, such as general Boolean representations, binary decision diagram (BDD), multispace search method, parallel SAT algorithms and architectures, and the MultiSAT algorithm, are described in detail in Gu et al. (1997b, 2000).

1.1. Basic SAT Algorithm Classes

SAT algorithms can be grouped into the following classes. Most existing SAT algorithms can be grouped into these categories (see Table 1).

- *Discrete, constrained algorithms.* Algorithms in this category treat a SAT formula as an instance of a constrained decision problem, applying discrete search and inference procedures to determine a solution. One straightforward way to solve an instance of SAT is to enumerate all possible truth assignments and check to see if one satisfies the formula. Many improved techniques, such as consistency algorithms (Gu et al., 1987; Gu and Wang, 1992; Mackworth, 1977), backtracking algorithms (Bitner and Reingold, 1975; Brown and Purdom, 1981; Bugrara and Purdom, 1993; Larrabee, 1992; Purdom, 1983), term-rewriting (Dershowitz et al., 1983; Hsiang, 1985), production systems (Siegel, 1987), multivalued logic (Shensa, 1989), binary decision diagrams (Bryant, 1986; Ashar et al., 1992), chip-and-conquer (Gelder, 1988), resolution and regular resolution (Genesereth and Nilsson, 1987; Loveland, 1978; Nilsson, 1980; Robinson, 1965; Tseitin, 1968; Urquhart, 1995; Winston, 1984), independent set algorithm (Iwama, 1989), matrix inequality system (Truemper, 1994), and the discrete Lagrangian algorithm (Wah and Shang, 1998), have been proposed.

  Other specific algorithms using these principles include simplified DP algorithms (Galil, 1977; Goldberg et al., 1982; Purdom and Brown, 1985), and a simplified DP algorithm with strict ordering of variables (Hu et al., 1992). The DP algorithm improved certain aspects compared to Gilmore's proof method (Gilmore, 1960). Analyses of SAT algorithms often concentrate on algorithms that are simple because it is difficult to do a correct analysis of the best algorithms. Under those conditions where simple algorithms are fast, related practical algorithms are also fast. (It is difficult to tell whether a practical algorithm is slow under conditions that make the corresponding simplified algorithm slow.)

  A number of special SAT problems, such as 2-satisfiability and Horn clauses, are solvable in polynomial time (Aho et al., 1974; Cook, 1971; Nilsson, 1980). Several linear time (Aspvall et al., 1979; Even et al., 1976) and polynomial time algorithms (Papadimitriou, 1991; Schaefer, 1978) exist.

- *Discrete, unconstrained algorithms.* In this approach, the number of unsatisfiable CNF (or satisfiable DNF) clauses is formulated as the value of the objective function, transforming the SAT formula into a discrete, unconstrained minimization problem to the objective

Table 1 Some Typical Algorithms for the SAT Problem

Continuous		Discrete	
Unconstrained	Constrained	Unconstrained	Constrained
1987: UniSAT models (Gu, 1998, 1992b, 1994)	1986: Branch-and-bound (APEX) (Blair et al., 1986)	1987: Randomized local search (SAT1) (Gu, 1988, 1993)	1960: Davis–Putnam (DP) algorithm (Davis and Putnam, 1960)
1987: Nonlinear optimization (SAT6) (Gu, 1988, 1994)	1988: Programming models (Blair et al., 1986; Jeroslow, 1988)	1987: Parallel local search (SAT1.6) (Gu, 1988, 1993)	1965: Resolution (Robinson, 1965)
1989: Neural net models (Johnson, 1989; Chakradhar et al., 1990)	1988: Cutting plane (Hooker, 1988a,c)	1988: Local search for n-queen (Gu, 1989; Sosič and Gu, 1988a,b)	1971: Consistency algorithms (Clowes, 1971; Gu, 1989; Huffman, 1971; Waltz, 1972)
1990: Nonlinear optimization & backtracking (Gu, 1994)	1989: Branch-and-cut (Hooker and Fedjki, 1989)	1990: Unison algorithm and hardware (Sosič et al., 1996, 1992)	1978: Loveland's Davis–Putnam (DPL) (Davis et al., 1962; Loveland, 1978)
1991: SAT14 algorithms (Gu, 1994)	1989: Interior point method (Kamath et al., 1989, 1990)	1991: Local search complexity (Gu and Gu, 1994; Papadimitriou, 1992)	1986: Parallel consistency chips (Gu, 1989; Gu et al., 1987; Gu and Wang, 1992)
		1991: Local search for 2-SAT (Papadimitriou, 1991)	1986: Binary decision diagrams (BDD) (Ashar et al., 1992; Bryant, 1986)
		1992: Local search with traps (SAT1.5) (Gu, 1992a, 1993)	1988: Chip-and-conquer (Gelder, 1988)
		1992: Greedy local search—GSAT (Selman et al., 1992)	1990: DPL plus heuristic (DPLH) (Jeroslow and Wang, 1990)
			1989: Local search & backtracking (Gu, 1993)
			1993: Backtracking and probing (Purdom and Haven, 1993)
			1994: Parallel DP algorithm (Bohm and Speckenmeyer, 1994)
			1994: Matrix inequality system (Truemper, 1994)
			1996: CSAT (Dubois et al., 1996)
			1996: Lagrangian algorithm (Wah and Shang, 1998)

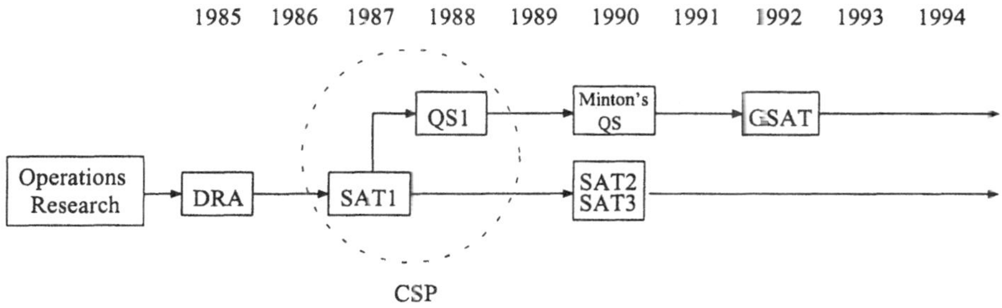

Figure 1 Early development of local search algorithms for the SAT problem. There were two major approaches: randomized local search (SAT1) and greedy local search (GSAT). SAT1 was the first local search algorithm developed for VLSI engineering and scheduling applications. The GSAT algorithm was derived from early local search algorithms for the n-queen problem

function. Local search is a major class of discrete, unconstrained search methods (Gu, 1992a, 1993; Gu and Gu, 1994; Papadimitriou, 1992; Selman et al., 1992). It can be used to solve the transformed formula.

Early work in constraint satisfaction problem (CSP) and complexity study contributed to the development of local search algorithms for the SAT problem (see Gu et al. (1997b, 2000) and Figure 1). There were two major approaches in this area: randomized local search (SAT1) and greedy local search (GSAT). The SAT1 algorithm was the first local search algorithm developed from VLSI engineering and scheduling applications. The GSAT algorithm was derived from the early local search algorithms for the n-queen problem.

- *Constrained programming algorithms.* Methods in this class were developed based on the fact that CNF or DNF formulas can be transformed to instances of integer programming, and possibly solved using linear programming relaxations (Blair et al., 1986; Hooker, 1988b,c; Jeroslow, 1988; Kamath et al., 1989, 1990; Papadimitriou and Steiglitz, 1982; Williams, 1987). Many approaches, including branch-and-bound (Blair et al., 1986), cutting plane (Hooker, 1988a,c), branch-and-cut (Hooker and Fedjki, 1989), interior point (Kamath *et al.* 1989, 1990), and improved interior point (Shi et al., 1992), have been proposed to solve the integer program representing the inference problem. Researchers found integer programming methods faster than resolution for certain classes of problems, although these methods do not possess a robust convergence property and often fail to solve hard instances of satisfiability (Blair et al., 1986; Hooker, 1988b,c; Jeroslow, 1988; Kamath et al., 1989, 1990).
- *Unconstrained, nonlinear optimization algorithms.* Special models have been formulated to transform a discrete formula on Boolean space $\{0, 1\}^n$ (a decision problem) into an unconstrained UniSAT problem on real space E^n (an unconstrained nonlinear optimization problem). The transformed formulas can be solved by many existing nonlinear optimization methods (Gu, 1988, 1992a,b, 1994) (see Section 5).

1.2. Parallel SAT Algorithms

In practice, most sequential SAT algorithms can be mapped onto parallel computer systems, resulting in parallel SAT algorithms (Gu, 1995). Accordingly, as given in Table 2, there are four classes of parallel algorithms for solving SAT.

- *Parallel, discrete, constrained algorithms.* Many discrete, constrained SAT and CSP algorithms have been implemented in parallel algorithms or put on special-purpose, hardware VLSI architectures. These include parallel consistent labeling algorithms (Tyler, 1983; McCall et al., 1985), parallel discrete relaxation (DRA) chips (Gu et al., 1987; Gu, 1989; Gu and Wang, 1992), parallel arc consistency (PAC) algorithms (Samal and Henderson, 1987), parallel constrained search architectures (Gu, 1989, Gu and Wang,

Table 2 Some Parallel SAT/CSP Algorithms

Continuous		Discrete	
Unconstrained	Constrained	Unconstrained	Constrained
1987: UniSAT models (Gu, 1988, 1994)	1989: Interior point method (Kamath et al., 1989, 1990)	1987: CNF local search (Gu, 1988, 1993)	1983: Parallel CLP algorithms (Tyler, 1983; McCall et al., 1985)
1987: Nonlinear optimization (SAT6) (Gu, 1988, 1994)		1987: DNF local search (Gu, 1988, 1994)	1986: parallel DRA chips (Gu, 1989; Gu et al., 1987; Gu and Wang, 1992)
1991: SAT14 algorithms (Gu, 1994)		1987: Parallel local search (Gu, 1988, 1993)	1987: Parallel DP algorithm (Chen and Liu, 1987)
1991: Parallel nonlinear optimization (Gu, 1992b, 1994)		1991: Discrete $\alpha\beta$ relaxation (Gu et al., 1997b, 2000)	1988: Parallel AC algorithms (Samal and Henderson, 1987)
1992: Neurocomputing (Gu, 1995)		1993: Multiprocessor local search (Sosič and Gu, 1993a,b)	1988: Parallel CSP architectures (Gu, 1989; Gu and Wang, 1992)
			1990: Unison algorithm and hardware (Sosič et al., 1996, 1992)
			1992: Vectorized DP algorithm (Fang and Chen, 1992)
			1994: MIMD DP algorithm (Bohm and Speekenmeyer, 1994)

1992), parallel Unison algorithms (Sosič et al., 1996), parallel Unison architectures (Sosič et al., 1992), parallel DP algorithms (Bohm and Speckenmeyer, 1994; Chen and Liu, 1987; Fang and Chen, 1992), and parallel logical programming languages (Conery, 1987; Lin, 1988; Wah and Li, 1986; Wah, 1987; Wah et al., 1989)

- *Parallel, discrete, unconstrained algorithms.* A number of discrete local optimization algorithms were implemented on parallel computing machines. These include CNF local search (Gu, 1988, 1993), DNF local search (Gu, 1988, 1994), parallel local search (Gu, 1988, 1993), and multiprocessor local search (Sosič and Gu, 1993a,b).
- *Parallel, constrained programming algorithms.* Kamath et al. implemented an interior point zero–one integer programming algorithm on a KORBX(R) parallel/vector computer (Kamath et al., 1989, 1990).
- *Parallel, unconstrained, nonlinear optimization algorithms.* Several of these algorithms have been implemented: UniSAT models (Gu, 1988, 1994), and parallel nonlinear optimization algorithms (Gu, 1992b, 1994).

1.3. Algorithm Categories

Some SAT algorithms are complete (they definitely determine whether an input has a solution or not) (Bryant, 1986; Davis and Putnam, 1960; Davis et al., 1962; Loveland, 1978; Robinson, 1965), while others are incomplete (they sometimes determine whether or not the input has a solution, but in other cases they cannot find one) (Gu, 1993, 1994; Kamath et al., 1990; Papadimitriou, 1991).

Most incomplete algorithms find one solution (or perhaps several solutions) in favorable cases, but give up or do not terminate in other cases. In such cases one does not know whether the input has no solution or the algorithm did not search hard enough. Some incomplete algorithms can verify that a formula has no solution but cannot find one if at least one solution exists. Such is the case for incomplete algorithms that check for patterns that imply unsatisfiability. Incomplete algorithms, however, are of particular interest for inputs that are so difficult that a complete algorithm cannot solve them in reasonable time.

Complete algorithms can determine whether or not a solution exists, give the variable settings for one solution, find all solutions or an optimal solution, or prove that there is no solution. The techniques used in complete SAT algorithms can usually be adapted to provide exact solutions to optimization problems. The techniques used in incomplete SAT algorithms can usually be adapted to provide approximate solutions to optimizations problems. They normally lead to algorithms that produce low (but not necessarily the lowest) value of the function.

1.4. Performance Evaluation

The performance of an algorithm can be determined experimentally or analytically. It is feasible to do experimental studies with typical or random formulas, but not with worst-case formulas (there are too many formulas of a given size to experimentally determine which one leads to the worst-case time). It is feasible to do analytical studies with random or worst-case formulas but not with typical formulas (typical sets of formulas seldom have a mathematical structure suitable for analysis).

Experimental studies are sometimes inconclusive because they consider a relatively small number of input possibilities. Such restrictions are often forced because the space of likely input formulas, and even the size of such formulas, is so large. Analytical studies are intended to determine performance over broad families of inputs where each family typically represents a class of formulas of a particular size. However, such studies have the drawback that only the simplest of algorithms can be analyzed.

The analytical studies of SAT algorithms involve the following.

WORST-CASE STUDIES

Unless $P = NP$, all SAT algorithms have a worst-case time that is superpolynomial in input size (Cook, 1971). A number of studies have concentrated on the worst-case analysis of variable setting algorithms for solving SAT (Gelder, 1988; Monien and Speckenmeyer, 1983; Kullmann, 1996).

Probabilistic Studies

Since the typical performance of many satisfiability algorithms is much better than any proven worst-case results, there is considerable interest in evaluating the probabilistic performance of these algorithms. Such studies use some model for generating random formulas and then calculate the performance of algorithms on these formulas. The two most widely used measures of performance are average time and probabilistic time.

Average time is a weighted average of the time (or some related measure, such as the number of nodes) to solve a given sample of formulas. An algorithm must solve each formula for the average to be defined. In probabilistic time studies, an algorithm is given a deadline (usually specified as a polynomial in the length of input formulas), and one studies the fraction of formulas that are solved within the deadline. Probabilistic time studies can be performed on algorithms that give up on some fraction of the formulas so long as that fraction is less than the goal fraction.

For incomplete algorithms, the average time is not defined so only the fraction of inputs solved can be studied. One can also use various hybrid measures, such as the average time used to solve the easiest 90 percent of the inputs.

The literature contains a number of studies of the average time and probabilistic time performance of certain SAT algorithms (Brown and Purdom, 1981; Franco and Paull, 1983; Franco and Ho, 1988; Goldberg, 1979a,b; Goldberg et al., 1982; Hu et al., 1992; Purdom, 1978, 1990; Purdom et al., 1981). Despite the worst-case complexity of SAT, algorithms and heuristics with polynomial average time complexities have been reported (Chao and Franco, 1986, 1990; Franco, 1986; Purdom, 1983; Purdom and Brown, 1983, 1985, 1987; Wu and Tang, 1992).

2. SAT Input Models

The running time of a SAT algorithm depends on the type of input being solved. The following SAT input models are often used to generate a variety of input types.

2.1. *Random Input Models*

- *Hardest formulas.* Generate that formula which is the most difficult for the algorithm being measured. This approach is often used for analytical studies. There are too many possible formulas to use this approach in experimental studies.
- Experimental studies sometimes include results for the hardest formulas from the set of formulas tested, but such results are quite different from what the results would be if the entire set of possible formulas had been tested.

Most analytical studies use the following two basic models to generate random CNF formulas. Each model has several variations depending on whether identical clauses are permitted, whether a variable and its negation can occur in a clause, and so on.

- *The l-SAT model.* A randomly generated CNF formula consists of m independently generated random clauses. Each clause is chosen uniformly from the set of all possible clauses of exactly l literals that can be composed from a variable set $X = \{x_1, ..., x_n\}$ such that no two literals are equal or complementary. The number of possible clauses is $2^l\binom{n}{l}$. This model is sometimes called the fixed-clause-length model. Similar models were used in Brown and Purdom (1981), Chao and Franco (1986, 1990), Franco and Paull (1983), Gu (1993, 1994), Mitchell et al. (1992), Méjean et al. (1995), and Purdom and Brown (1983).
- *The average l-SAT model.* A randomly generated CNF formula consists of m independently generated random clauses. In each clause, each of n variables occurs positively with probability $p(1-p)$, negatively with probability $p(1-p)$, both positively and negatively with probability p^2, and not at all with probability $(1-p)^2$, where p can be a function of m and n. The average number of literals in a clause is $l = 2pn$. This model is also called

the random-clause-length model. This model and variations were used in Franco (1986, 1991), Goldberg et al. (1982), Gu (1992a, 1993, 1994), Hooker (1988c), Hooker and Fedjki (1989), Kamath et al. (1989, 1990), Shi et al. (1992).

2.2. *Hardness*

Various SAT algorithms differ greatly in the amount of time they need to solve particular inputs. For example, Iwama's algorithm (Iwama, 1989) is fast for random formulas with lots of solutions and slow for random formulas with few solutions, while simple backtracking (Purdom and Brown, 1987) is fast on formulas with few solutions and slow on formulas with many solutions. Therefore, the hard-and-easy distributions of SAT formulas depend not only on the inherent property of the SAT input models but also on the algorithms used to solve the formulas. Any particular SAT formula is easy for some algorithm (for example a table lookup algorithm with that formula in its lookup table). Thus, hardness is a property of large sets of formulas rather than individual formulas.

For sets of formulas generated by random models with parameters, the probability of finding a solution varies with the parameter settings. Those sets generated with parameters set in regions where solutions are going from unlikely to common are particularly difficult for all algorithms that have been studied.

For random l-SAT formulas fewer literals and larger number of clauses reduce the possibility of making all clauses jointly satisfiable. Therefore the computing time for random l-SAT formulas increases, up to a point, when m/n increases or the number of literals l ($l > 3$) in each clause decreases.

2.3. *Practical Input Models*

Random input models generate formulas that possess a symmetry that can be exploited for analysis. Actual formulas often have a different structure. They are essential to evaluate the performance of SAT algorithms. Examples of these are the following:

- *Regular SAT models.* Models derived from problems such as graph coloring and n-queens, are used to assess the performance of SAT algorithms (Gu, 1992b, 1994).
- *Practical application problems.* Models derived from practical application domains, such as integrated circuit design, mobile communications, computer architecture and network design, computer-aided manufacturing, and real-time scheduling, have a variety of special characteristics (see Gu et al., 1997b, 2000).

Some experiments strongly suggest that there is little correlation between the performance of a SAT algorithm tested through random input models and the performance of the same algorithm tested through practical input models. Local search is faster for some random inputs but can be slower than a complete SAT algorithm for some problems raised from practical applications. The boundary phenomenon discussed in random models is an artifact of some probabilistic models It has not yet been observed in practical input models.

3. Splitting and Resolution

Recursive replacement of a formula by one or more other formulas, the solution of which implies the solution of the original formula, is an effective paradigm for solving CNF formulas. Recursion continues until one or more primitive formulas have been generated and solved to determine the satisfiability of the original. One way to achieve this is through splitting.

In splitting, a variable v is selected from a formula, and the formula is replaced by one subformula for each of two possible truth assignments to v. Each subformula has all the clauses of the original except those satisfied by the assignment to v and otherwise all the literals of the original formula except those falsified by the assignment. Neither subformula contains v, and the original formula has a satisfying truth assignment if, and only if, either subformula has one. Splitting insures that a search for a solution terminates with a result

Another effective paradigm is based on resolution (Robinson, 1965). In resolution, a variable v is selected and a resolvent (see below) obtained using v is added to the original formula. The process is repeated to exhaustion or until an empty clause is generated. The original formula is not satisfiable if, and only if, an empty clause is a resolvent. Although there is only one new formula on each step, the total number of steps (or resolvents) can be extremely large compared to the number of clauses in the original formula. Many algorithms that use resolution form all possible resolvents using a particular variable at one time. When this is done, the original clauses that contain the variable and its negation may be dropped. An algorithm may use both splitting and resolution.

Early examples of these approaches are the two forms of the Davis–Putnam procedure. The original DP procedure used resolution (Davis and Putnam, 1960) while the revised version, that is, the Davis–Putnam–Loveland (DPL) procedure, used splitting (Davis et al., 1962; Loveland, 1978). Combining splitting with depth-first search in the DPL procedure avoids memory explosion that occurs on many inputs when they are solved by the original DP procedure.

3.1. Resolution

Given two clauses $C_1 = (v \vee x_1 \vee ... \vee x_{l1})$ and $C_2 = (\bar{v} \vee y_1 \vee ... \vee y_{l2})$, where all x_i and y_j are distinct, the resolvent of C_1 and C_2 is the clause $(x_1 \vee ... \vee x_{l1} \vee y_1 \vee ... \vee y_{l2})$, that is, the disjunction of C_1 and C_2 without v or $\bar{v}$. The resolvent is a logical consequence of the logical and of the pair of clauses. Resolutions is the process of repeatedly generating resolvents from original clauses and previously generated resolvents until either the null clause is derived or until no more resolvents can be created (Robinson, 1965). In the former case (a refutation) the formula is unsatisfiable and in the latter case it is satisfiable.

For some formulas the order in which clauses are resolved can have a big effect on how much effort is needed to solve it. The worst-case associated with the best possible order (the order is selected after the formula is given) has received considerable study (Galil, 1977; Tseitin, 1968; Haken, 1985; Urquhart, 1987). These studies all used formulas that have no solution, but where this is not obvious to the resolution algorithm. Eventually a much stronger result was shown: nearly all random l-SAT formulas need exponential time when the ratio of clauses to variables is above a constant (whose value depends on l) (Chvátal and Szemerédi, 1988). The constant is such that nearly all of the formulas in this set have no solution.

A number of restrictions and at least one extension to resolution have been proposed and applied to CNF formulas. Restrictions aim to shorten the amount of time needed to compute a resolution derivation by limiting the number of possible resolvents to choose from at each resolution step. The extension aims to provide shorter derivations than possible for resolution alone by adding equivalences which offer more clauses on which to resolve. A nice treatment of these refinements can be found in Büning and Lettmann (1996, Chapter 4). We mention here a few of these.

- *Set of Support (Wos et al., 1965).* Split a given formula into two sets of clauses $\mathcal{F}_1$ and $\mathcal{F}_s$ such that $\mathcal{F}_1$ is satisfiable. Permit only resolutions involving one clause either in $\mathcal{F}_s$ or an appropriate previous resolvent. Set $\mathcal{F}_s$ is called the support set. This restriction can be useful if a large portion of the given formula is easily determined to be satisfiable.
- *P- and N-resolution.* If one of the two clauses being resolved has all positive literals (respectively negative literals), then the resolution step can be called a P-resolution (respectively N-resolution) step. In P-resolution (respectively N-resolution) only P-resolution (respectively N-resolution) steps are used. Clearly there is great potential gain in this restriction due to the usually low number of possible resolvents to consider at each step. However, it has been shown that some formulas solved in polynomial time with general resolution require exponential time with N-resolution.
- *Linear resolution.* We have linear resolution if every resolution step except the first involves the most recently generated resolvent (the other clause can be a previous resolvent or a clause in the given formula). Depending on the choice of initial clause and previous resolvents it is possible not to complete a refutation.
- *Regular resolution (Tseitin, 1968).* In every path of a resolution tree no variable is eliminated more than once.

- *Davis–Putnam resolution.* Once all the resolvents with respect to a particular variable have been formed, the clauses of the original formula containing that variable can be dropped. Doing this does not change the satisfiability of the given formula, but it does change the set of solutions to the extent that the value of that variable is no longer relevant. When dropping clauses, it is natural to first form all the resolvents for one variable, then all the resolvents for a second variable, and so on. When doing resolution in this way, it is easy to find one satisfying assignment if the formula is satisfiable. At the next to last step the formula has just one variable, so each value can be tested to see which one satisfies the formula (perhaps both will). Pick a satisfying value and plug it into the formula for the next step, converting it into a one variable formula. Solve that formula and proceed in this manner until an assignment for all variables is found.
- *Extended resolution (Tseitin, 1968).* For any pair of variables, a, b, in a given formula, $\mathscr{F}$ create a variable z not in $\mathscr{F}$ and append the following expression to $\mathscr{F}$: $(z \vee a) \wedge (z \vee b) \wedge (\bar{z} \vee \bar{a} \vee \bar{b})$. Judicious use of such extensions can result in polynomial size refutations for problems that have no polynomial size refutations without extension.

3.2. *Backtracking*

Backtracking algorithms are based on splitting. During each iteration, the procedure selects a variable and generates two subformulas by assigning the two values, *true* and *false*, to the selected variable. In each subformula, those clauses containing the literal that is *true* for the variable assignment are erased from the formula, and those clauses that contain the literal that is *false* have that literal removed. Backtrack algorithms differ in the way they select which variable to set at each iteration. The unit clause rule, the pure literal rule, and the smallest clause rule, are the three most common ones. We state each algorithm informally.

The flow of control in splitting-based algorithms is often represented by a search tree. The root of the tree corresponds to the initial formula. The internal nodes of the tree correspond to subformulas that cannot be solved directly, whereas the leaf nodes correspond to subformulas that can be solved directly. The nodes are connected with arcs that can be labeled with variable assignments.

- *Simple backtracking (Brown and Purdom, 1981).* If the formula has an empty clause (a clause that always has value *false*) then exit and report that the formula has no solution If the formula has no variables, then exit and report that the formula has a solution. (The current assignment of values to variables is a solution to the original formula.) Otherwise, select the first variable that does not yet have a value. Generate two subformulas by assigning each possible value to the selected variable. Solve the subformulas recursively. Report a solution if any subformula has a solution, otherwise report no solution.
- *Unit clause backtracking (Purdom, 1983).* If some clause contains only one of the unset variables then select that variable and assign it a value that satisfies the clause containing it; otherwise, select the first unset variable.
- *Clause order backtracking (Bugrara and Purdom, 1993).* If this setting does not solve the formula, then select the first clause that can evaluate to both *true* and *false* depending on the setting of the unset variables. Select variables from this clause until its value is determined.

  By setting only those variables that affect the value of clauses, this algorithm sometimes avoids the need to assign values to all the variables. The algorithm as stated finds all the solutions, but in a compressed form. The solutions come in cylinders, where some variables have the value "don't care." Thus, a single solution with unset variables represents the set of solutions obtained by making each possible assignment to the unset variables.
- *Probe order backtracking (Purdom and Haven, 1993).* Temporarily set all the unset variables to some predetermined value. Select the first clause that evaluates to *false* with this setting. Return previously unset variables back to unset and continue as in clause order backtracking.

For practical formulas one should consider adding the following five refinements to probe order backtracking: stop the search as soon as one solution is found; carefully choose the probing sequence instead of just setting all variables to a fixed value (Lieberherr and Specker, 1981; Sosič and Gu, 1991, 1994); probe with several sequences at one time (Buro and Bunning, 1992, 1993); carefully select which variable to set (Buro and Buning, 1992, 1993); and use resolution when it does not increase the input size (Franco, 1991).

Franco (1986) noticed that a random assignment solves a nonzero fraction of the formulas in the average l-SAT model when pn is large compared to $\ln m$. Simple uses of that idea do not lead to good average time (Purdom and Haven, 1993), but combining the idea with clause order backtracking leads to probe order backtracking, which is fast when pn is above $\ln m$.

- *Shortest-clause backtracking.* This algorithm is the same as clause order backtracking except for the clause selected. In this case, select the shortest clause.

  The corresponding idea for constraint satisfaction is to first set a variable in the most constraining relation. This idea is quite important in practice (Bitner and Reingold, 1975).

3.3. Backtracking and Resolution

Some algorithms have adapted ideas inspired by resolution to splitting algorithms. From the resolution view point, pure literals are interesting in that they lead to a single subformula that is no more complex than the original formula, while from the perspective of splitting, pure literals lead to two subformulas, but the solutions to the subformula where the literal has the value *false* are a subset of the one where the literal has the value *true*. Therefore, the original formula has a solution if, and only if, the formula associated with the *true* literal does.

- *The pure literal rule algorithm (Goldberg, 1979a).* Select the first variable that does not have a value. (If all variables have values, then the current setting is a solution if it satisfies all the clauses.) If some value of the selected variable results in all clauses that depend on that variable having the value *true*, then generate one subformula by assigning the selected variable the value that makes its literals *true*. Otherwise, generate a subformula for both values of the selected variable. Solve the one or two subformulas.

4. Local Search

Local search is a major class of discrete, unconstrained optimization procedures that can be applied to a discrete search space. Such procedures can be used to solve SAT by introducing an objective function that counts the number of unsatisfiable (CNF) clauses and solving to minimize the value of this function (Gu, 1988, 1992a, 1993; Gu and Gu, 1994; Papadimitriou, 1992; Selman et al., 1992).

4.1. Randomized Local Search

Most discrete local search procedures were developed based on a discrete, unconstrained optimization model, the SAT1 model (Gu, 1988, 1992a, 1993; Gu and Gu, 1994). In this model, the truth values assigned to the variables are defined as:

$$x_i = \begin{cases} 1 & \text{if the variable has value } \textit{true} \\ -1 & \text{if the variable has value } \textit{false} \end{cases} \tag{2}$$

The objective function in the SAT1 model counts the number of unsatisfied clauses as its objective value. A CNF is true if, and only if, the objective function takes the global minimum value 0 on the corresponding x.

- *Basic local search.* A basic local search algorithm, SAT1 algorithm, is shown in Figure 2. It consists of an initialization stage and a search stage. At the beginning of search, a SAT formula is generated. An initial random solution is chosen. The number of unsatisfiable clauses is computed and is assigned as the value of the objective function. During each

```
procedure SAT1 ()
begin
    /* initialization */
    get_a_SAT_instance();
    x_0 := select_a_random_initial_point();
    F(x_0) := evaluate_objective_function(x_0);

    /* search */
    k := 0;
    while F(x_k) ≠ 0 do
    begin
        for each variable i := 1 to n do
            /* if flip(x_i, x̄_i) does not increase F */
            if test_flip(x_i, x̄_i) then
            begin
                x_{k+1} := perform_flip(x_i, x̄_i);
                F(x_{k+1}) := evaluate_objective_function(x_{k+1});
            end;
        /* random flips */
        if local then local_handler();
        k := k + 1;
    end;
end;
```

Figure 2 SAT1: A Randomized Local Search Procedure for the SAT Problem (Gu 1992a, 1993; Gu and Gu, 1994). Random flips are introduced (1) to disorder the sequence with which the variables are selected for location optimization, and (2) to perturb local search with randomized downhill, tunneling, or uphill moves (Gu and Gu, 1994; Gu, 1993)

iteration, function test_flip() performs a test to see if the objective function would not increase after a flip. If test_flip() returns *true*, a flip operation is performed by procedure perform_flip(). Then function evaluate_objective_function() updates the objective function.

The procedure terminates when the objective function is reduced to zero, that is, a solution to the given SAT instance is found. In practice, before the objective function reduces to zero, the procedure may become stuck at local minima.

In the SAT1 algorithm (Gu, 1992a; Gu and Gu, 1994), a simple randomized local handler performing random flips was used (Figure 3). It combines local descent (reducing objective function) with the random uphill moves (increasing objective function), effectively improving SAT1's convergence performance. In the SAT1 algorithm family, one or more local handlers were implemented (Gu and Gu, 1994, Gu, 1992a, 1993). If the algorithms have difficulty in proceeding, it will call the local handlers and use other heuristics (see Gu et al., 1997b, 2000) to improve the algorithms' convergence performance.

```
procedure Local_handler ()
begin
    random select some variable x_t's;
    x_{k+1} := perform_flip(x_t's, x̄_t's);
    F(x_{k+1}) := evaluate_objective_function(x_{k+1});
end;
```

Figure 3 A Simple Randomized Local Handler Used in the SAT1 Algorithms (Gu 1992a, 1993; Gu and Gu, 1994). Random flips or a new random solution were applied to the algorithm if (1) $F \neq 0$ (SAT1 algorithm), (2) $F > 0$ (SAT1.1 algorithm), (3) $F(x_{k+1}) = F(x_k)$ (SAT1.2 algorithm), and (4) $F > 0$ and $F(x_{k+1}) = F(x_k)$ (SAT1.3 algorithm)

The random flips used in SAT1 algorithms make the order of selecting which variable for local examination (i.e., the *for* loop) trivial (Gu and Gu, 1994; Gu, 1992a, 1993). One can essentially select any variable randomly for examination.

- *Tunneling heuristic.* A local handler and its activating condition(s) have a significant effect on the performance (e.g., running time and average running time) of a local search algorithm for the SAT problem. The conditions for activating local handlers differ from algorithm to algorithm (see Figure 3). In the SAT1.1 algorithm, the local handler is called if the objective function is not zero (an aggressive strategy) (Gu and Gu, 1994; Gu, 1993). In the SAT1.2 algorithm, the local handler is called if the objective function does not increase (Gu and Gu, 1994; Gu, 1993). In the SAT1.3 algorithm, the local handler is called if the objective function does not increase or the objective function is greater than zero for some iterations (Gu and Gu, 1994; Gu, 1993). In the last two algorithms, the condition "objective function does not increase" means that the objective value is either reduced (local descent) or remains unchanged (tunneling heuristic). The behavior of local search with tunneling illustrates the fact that seemingly innocuous changes in an optimization routine can have a surprisingly large effect on its performance.
- *Random flips (noise).* If the local search procedure becomes stuck at a local minimum, it may proceed by using a noise perturbation to change its current location in the search space. The effectiveness with which local minima are handled significantly affects the performance of a local search algorithm. Researchers have proposed a number of techniques, such as jumping, climbing, annealing, and indexing, to handle local minima (Gu, 1990). In simulated annealing, a search process occasionally moves up rather than down in the search space, with large uphill moves being less likely than small ones. The probability of large uphill moves is gradually reduced as the search progresses.

  A variety of local handlers have been designed for use in local search algorithms. SAT1 uses a local handler that may randomly negate the truth values of one or up to n variables (Gu, 1988, 1992a, 1993, Gu and Gu, 1994). The basic idea is to generate random exchanges in some current solution points when the search is stuck at a local minimum. The search accepts a modified point as a new current solution not only when the value of the objective function is better but also when it is worse (Gu and Gu, 1994; Gu, 1988, 1992a, 1993). This differs from the traditional local search such as GSAT that used greedy local descent (Selman et al., 1992).

  Past lessons showed that four components, that is, min-conflict heuristics, best-neighbor heuristics, random value/variable selection heuristics, and trap handling heuristics (Gu et al., 1997b, 2000), are crucial to the development of an efficient local search algorithm for satisfiability and NP-hard problems.

```
procedure GSAT ()
begin
  for i := 1 to MAX-TRIES
    T := a randomly generated truth assignment
      for j := 1 to MAX-FLIPS
        if T satisfies α then return T
        p := a propositional variable such that a change
              in its truth assignment gives the largest
              increase in the total number of clauses
              of α that are satisfied by T
        T := T with the truth assignment of p reversed
      end for
  end for
  return "no satisfying assignment found"
end
```

Figure 4 GSAT: A Greedy Local Search Procedure for the SAT Problem (Selman et al., 1992). MAX-FLIPS, MAX-TRIES are constants, and α is a set of clauses. During each search step, GSAT takes the best neighbor that gives the maximum descent to the objective function

4.2. Randomized Local Search with Trap Handling

For the SAT problem, with high probability, a greedy local search will fall into a trap much more easily. In this case some variables are updated very quickly. The related clauses oscillate between the *sat* and *unsat* states. The search is limited to these trapping states. Without any help, there is little chance of getting out to explore other solution states.

Based on early observation of trap phenomenon and the development of a three-level search space model, Gu et al. (1997b, 2000) developed a SAT1.5 algorithm with trap handling ability (Gu, 1992a; Gu et al., 1995). The SAT1.5 algorithm can monitor and break local minimum loops and can handle multiple traps during the search. The current version of the SAT1.5 algorithm contains advanced data structures and complicated trap detection/handling methods (Gu, 1992a, 1993; Gu et al., 1995). The algorithm can solve large size random CNF formulas effectively (Gu et al., 1997b, 2000).

4.3. Greedy Local Search

Selman et al. proposed a greedy local search procedure, that is, GSAT, for the SAT problem (Selman et al., 1992) (see Figure 4). During each search step, the algorithm evaluates all the moves and selects the best move that gives the greatest decrease in the total number of unsatisfied clauses. If the algorithm becomes stuck at a local minimum, GSAT uses side-walk (a form of tunneling heuristic) to move aside. In GSAT procedure, two parameters, MAX-TRIES and MAX-FLIPS, were used to control the algorithm's maximum running state. Based on greedy local search, GSAT pursues only paths where every step leads to a largest improvement; this leads to a procedure that becomes stuck much more often than the randomized local search.

To overcome the weakness of the greedy GSAT algorithm Selman et al. recently combined several early random search techniques (Sosič and Gu, 1988a, 1991, 1994) in the

GSAT algorithm (Selman et al., 1994, 1996). They found that these random heuristics improved the performance of the GSAT algorithm significantly (Selman et al., 1994).

5. Nonlinear Unconstrained Methods

Using UniSAT (Universal SAT Problem Models), we can transform an SAT formula represented as an instance of a discrete constrained decision problem in Boolean {0, 1} space into a nonlinear unconstrained optimization problem (Gu, 1988, 1992b, 1994).

The UniSAT models (Gu, 1988, 1992a,b) suggest:

$$\min_{y \in E^n} f(y) = \sum_{i=1}^{m} c_i(y), \tag{3}$$

where

$$c_i(y) = \prod_{j=1}^{n} q_{i,j}(y_j) \tag{4}$$

$$q_{i,j}(y_j) = \begin{cases} |\, y_j - T \,| & \text{if } x_j \text{ in } C_i \\ |\, y_j + F \,| & \text{if } \bar{x}_j \text{ in } C_i \\ 1 & \text{otherwise,} \end{cases} \tag{5}$$

where T and F are positive constants.

Two special formulations to (5) exist. In the UniSAT5 model (Gu, 1992b, 1994)

$$q_{i,j}(y_j) = \begin{cases} |\, y_j - 1 \,| & \text{if } x_j \text{ is in } C_i \\ |\, y_j + 1 \,| & \text{if } \bar{x}_j \text{ is in } C_i \\ 1 & \text{otherwise} \end{cases} \tag{6}$$

and in the UniSAT7 model (Gu, 1992b, 1994):

$$q_{i,j}(y_j) = \begin{cases} (y_j - 1)^2 & \text{if } x_j \text{ is in } C_i \\ (y_j + 1)^2 & \text{if } \bar{x}_j \text{ is in } C_i \\ 1 & \text{otherwise.} \end{cases} \tag{7}$$

Values of y that make $f(y) = 0$ are solutions to the original formula in (1). UniSAT5 can be solved with efficient, discrete, local search algorithms (Gu, 1994). UniSAT7 requires computationally expensive continuous optimization algorithms, rendering them applicable to small size formulas.

A good property of the transformation is that UniSAT models establish a correspondence between the global minimum points of the objective function and the solutions of the original SAT formula. A CNF $\mathcal{F}$ has value *true* if, and only if, $f(y)$ takes the global minimum value 0 on the corresponding solution y^*.

6. Nonlinear Constrained Method

The SAT problem can be formulated as a constrained nonlinear optimization problem either in the continuous space or in the discrete space.

6.1. Continuous Lagrangian Search Algorithms

The SAT problem can first be transformed into a continuous constrained optimization problem.

$$\min_{y \in E^n} F(y) = \sum_{i=1}^{m} c_i(y) \tag{8}$$

$$\text{subject to } c_i(y) = 0 \quad \forall i \in \{1, 2, \ldots, m\},$$

where $y = (y_1, y_2, ..., y_n)$, and $c_i(y)$ is defined in (4) and (5) and repeated as follows.

$$c_i(y) = \prod_{j=1}^{n} q_{i,j}(y_j)$$

$$q_{i,j}(y_j) = \begin{cases} (1-y_j)^2 & \text{if } x_j \text{ in } C_i \\ y_j^2 & \text{if } \bar{x}_j \text{ in } C_i \\ 1 & \text{otherwise.} \end{cases}$$

Here, $F(y)$ is a scalar differentiable function that takes the norm of its argument so that $F(y) = 0$ iff $c_i(y) = 0$ for all i.

In Lagrangian methods, Lagrange multipliers are introduced to gradually resolve constraints through iterative updates. They are exact methods that optimize the objective using Lagrange multipliers to meet the Kuhn–Tucker conditions (Luenberger, 1984). Equation (8) can be reformulated using Lagrange multipliers into the following unconstrained problem:

$$L(y, \lambda) = F(y) + \lambda^T c(y) \quad \text{(Lagrangian function)} \tag{9}$$

$$\mathscr{L}(y, \lambda) = F(y) + \|c(y)\|_2^2 + \lambda^T c(y) \quad \text{(Augmented Lagrangian)}, \tag{10}$$

where $c = (c_1(y), c_2(y), ..., c_m(y))$, and λ^T is the transpose of the set of Lagrange multipliers. The augmented Lagrangian formulation is often preferred because it provides better numerical stability.

According to classical optimization theory (Luenberger, 1984), all the extrema of (10), whether local or global, are roots of the following sets of equations.

$$\nabla_y \mathscr{L}(y, \lambda) = 0 \quad \text{and} \quad \nabla_\lambda \mathscr{L}(y, \lambda) = 0. \tag{11}$$

These conditions are necessary to guarantee the (local) optimality to the solution of (8).

Search methods for solving (10) can be classified into local and global algorithms. Local minimization algorithms, such as gradient-descent and Newton's methods, find local minima efficiently and work best in uni-modal problems. Global methods, in contrast, employ heuristic strategies to look for global minima and do not stop after finding a local minimum (Pardalos, 1993; Törn and Žilinskas, 1989; Luenberger, 1984). Note that gradients and Hessians can be used in both local and global methods (Törn and Žilinskas, 1989).

Local search methods can be used to solve (11) by forming a Lagrangian dynamic system that includes a set of dynamic equations to seek equilibrium points along a gradient path. These equilibrium points correspond to the constrained minima of (8). The Lagrangian dynamic system of equations are as follows.

$$\frac{dy(t)}{dt} = -\nabla_y \mathscr{L}[y(t), \lambda(t)] \quad \text{and} \quad \frac{d\lambda(t)}{dt} = \nabla_\lambda \mathscr{L}[y(t), \lambda(t)] \tag{12}$$

Locally optimal solutions to the continuous formulation are governed by the Saddle Point Theorem, which states that y^* is a local minimum to the original problem defined in (8) if there exists λ^* such that (y^*, λ^*) constitutes a saddle point of the associated Lagrangian function $F(y, \lambda)$. Note that the converse of the above condition is not always true. Here, a saddle-point (y^*, λ^*) of Lagrangian function $F(y, \lambda)$ is defined as one that satisfies the following condition.

$$F(y^*, \lambda) \leqslant F(y^*, \lambda^*) \leqslant F(y, \lambda^*) \tag{13}$$

for all (y^*, λ) and all (y, λ^*) sufficiently close to (y^*, λ^*).

It is important to note out that a Lagrangian search modeled by (12) is incomplete: if it does not find a solution in a finite amount of time, it does not prove whether the original SAT problem is satisfiable or not. Hence, infinite time will be required to prove unsatisfiability.

Unfortunately, continuous gradient-based local search methods for solving (12) are very time consuming. Our experience indicates that continuous descent methods are several orders of magnitude more complex that discrete descent methods. For instance, it takes over one hour of CPU time on a Sun SS10/51 workstation to solve a problem with two hundred variables and sixty constraints.

6.2. *Discrete Lagrangian Search Algorithms*

A discrete constrained formulation of an SAT problem and its solution using a discrete Lagrangian method is much more efficient.

Recall the following discrete constrained formulation of an SAT problem (Gu et al., 1997b, 2000).

$$\min_{y \in \{0,1\}^n} N(y) = \sum_{i=1}^{m} U_i(y) \tag{14}$$

$$\text{subject to } U_i(y) = 0 \quad \forall i \in \{1, 2, ..., m\}.$$

Without showing all the details (Wah and Shang, 1998), the continuous Lagrangian method can be extended to work on discrete problems. The discrete Lagrangian function for (14) is defined as follows.

$$L(y, \lambda) = N(y) + \lambda^T U(y), \tag{15}$$

where $y \in \{0,1\}^n$, $U(y) = [U_1(y), ..., U_m(y)] \in \{0,1\}^m$, and λ^T is the transpose of $\lambda = (\lambda_1, \lambda_2, ..., \lambda_m)$ that denotes the Lagrange multipliers. (Note that λ_i can be continuous variables.)

In a definition similar to that in (13), a saddle points (y^*, λ^*) of $L(y, \lambda)$ in (15) is defined as one that satisfies the following condition.

$$L(y^*, \lambda) \leqslant L(y^*, \lambda^*) \leqslant L(y, \lambda^*) \tag{16}$$

for all λ sufficiently close to λ^* and for all y whose Hamming distance between y^* and y is 1.

It can be proved that, in discrete space, the set of all saddle points is equal to the set of all solutions that satisfy the following two first-order necessary conditions for discrete problems:

$$\Delta_y L(y, \lambda) = 0 \tag{17}$$

$$\nabla_\lambda L(y, \lambda) = 0, \tag{18}$$

where $\Delta_y L(y, \lambda)$ is the discrete gradient of $L(y, \lambda)$ at point y for fixed λ. Informally, $\Delta_y L(y, \lambda)$ is a vector that points from y to a neighborhood point of $y \in \mathcal{N}(y) \cup \{y\}$ with the minimum L value, for a user-defined set $\mathcal{N}(y)$ of neighborhood points.

Similar to (12), the discrete Lagrangian method (DLM) for solving SAT problems can be defined as a set of difference and vector equations,

$$y^{k+1} = y^k \oplus \Delta_y L(y^k, \lambda^k) \tag{19}$$

$$\lambda^{k+1} = \lambda^k + U(y^k), \tag{20}$$

where $\oplus$ is the vector-add operator $(x \oplus y = (x_1 + y_1, ..., x_n + y_n))$.

7. Conclusions

The SAT problem is at the core of the class of NP-complete problems and has many practical applications. In recent years, many optimization methods, parallel algorithms, and practical techniques have been developed for solving the SAT problem. The past two decades have seen the proliferation of many SAT algorithms: resolution, local search, nonlinear optimization, BDD SAT solver, and multispace search, among others. Existing methods complement rather than exclude each other by being effective for particular instances of SAT. In this article, we describe several major classes of SAT algorithms. The performance evaluation of existing SAT algorithms and a set of practical applications of SAT can be found in our recent work (Gu et al., 1997b, 2000). The area of SAT research is rich in well-developed theory and methods. To apply theoretical/algorithmic results to practical problems seems the ultimate way to test and benchmark SAT algorithms. Not only will the end results of such an endeavor have a major scientific and industrial impact, but in the process it will push optimization technology to its limit.

References

Aho, A. V., J. E. Hopcroft, and J. D. Ullman. 1974. *The Design and Analysis of Computer Algorithms*: Addison-Wesley.

Ashar, P., A. Ghosh, and S. Devadas. 1992. "Boolean satisfiability and equivalence checking using general binary decision diagrams." *Integration* **13**: 1–16.

Aspvall, B., M. F. Plass, and R. E. Tarjan. 1979. "A linear-time algorithm for testing the truth of certain quantified Boolean formulas." *Inform. Process. Lett.* **8**: 121–132.

Chvátal, V. and E. Szemerédi. 1988. "Many hard examples for resolution." *J. ACM* **35**: 759–770.

Bitner, J. R. and E. M. Reingold. 1975. "Backtrack programming techniques." *Commun. ACM* **18**: 651–656A.

Blair, C. E., R. G. Jeroslow, and J. K. Lowe. 1986. "Some results and experiments in programming techniques for propositional logic." *Comput. Oper. Res.* **5**: 633–645.

Bohm, M. and E. Speckenmeyer. 1994. 'A fast parallel SAT-solver—efficient workload balancing." Presented at the 3rd International Symposium on AI & Mathematics

Brown, C. A. and P. W. Purdom. 1981. "An average time analysis of backtracking." *SIAM J. Comput.* **10**: 583–593.

Bryant, R. E. 1986. "Graph-based algorithms for Boolean function manipulation." *IEEE Trans. Comput.* **C-35**: 677–691.

Bugrara, K. and P. Purdom. 1993. "Average time analysis of clause order backtracking." *SIAM J. Comput.* **22**: 303–317.

Büning, H. K. and T. Lettmann. 1996. In *Aussagenlogik: Deduktion und Algorithmen*: B.G. Teubner.

Buro, M. and Buning, H. K. 1992. "Report on a SAT competition." Technical Report FB-17. Paderborn, Germany: Mathematik/Informatik, Universitat Paderborn

Buro, M. and H. K. Buning. 1993. "Report on a SAT competition." *Bull. Eur. Assoc. Theoretical Comput. Sci.* **49**: 143–151.

Chakradhar, S., V. Agrawal, and M. Bushnell. 1990. "Neural net and Boolean satisfiability model of logic circuits." *IEEE Design Test Comput.* **7**: 54–57.

Chao, M. T. and J. Franco. 1986. "Probabilistic analysis of two heuristics for the 3-satisfiability problem.' *SIAM J. Comput.* **15**: 1106–1118.

Chao, M. T. and J. Franco. 1990. "Probabilistic analysis of a generalization of the unit-clause literal selection heuristics for the *k* satisfiable problem." *Inform. Sci.* **51**: 289–314.

Chen, W. T. and L. L. Liu. 1987. "Parallel approach for theorem proving in propositional logic." *Inform. Sci.* **41**: 61–76.

Clowes, M. B. 1971. "On seeing things." *Artificial Intelligence* **2**: 79–116.

Conery, J. S. 1987. *Parallel Execution of Logic Programs*: Kluwer Academic Publishers.

Cook, S. A. 1971. "The complexity of theorem-proving procedures." In *Proceedings of the Third ACM Symposium on Theory of Computing*, pp. 151–158.

Davis, M., G. Logemann, and D. Loveland. 1962. "A machine program for theorem proving." *Commun. ACM* **5**: 394–397.

Davis, M. and H. Putnam. 1960. "A computing procedure for quantification theory." *J. ACM* **7**: 201–215.

Dershowitz, N., J. Hsiang, N. Josephson, and D. Plaisted. 1983. "Associative-commutative rewriting." In *Proceedings of IJCAI*, pp. 940–944.

Dubois, O., P. Andre, Y. Boufkhad, and J. Carlier. 1996. "SAT versus UNSAT." In *DIMACS Series Volume: Clique, Graph Coloring, and Satisfiability—Second DIMACS Implementation Challenge*, Volume 26, edited by D. S. Johnson and M. A. Trick, pp. 415–433: American Mathematical Society.

Even, S., A. Itai, and A. Shamir. 1976. 'On the complexity of timetable and multicommodity flow problems." *SIAM J. Comput.* **5**: 691–703.

Fang, M. Y. and W. T. Chen. 1992. "Vectorization of a generalized procedure for theorem proving in propositional logic on vector computers." *IEEE Trans. Knowledge Data Engng* **4**: 475–486.

Franco, J. 1986. "On the probabilistic performance of algorithms for the satisfiability problem." *Inform. Process. Lett.* **23**: 103–106.

Franco, J. 1991. "Elimination of infrequent variables improves average case performance of satisfiability algorithms." *SIAM J. Comput.* **20**: 1119–1127.

Franco, J. and Y. C. Ho. 1988. "Probabilistic performance of heuristic for the satisfiability problem." *Discrete Appl. Math.* **22**: 35–51.

Franco, J. and M. Paull. 1983. "Probabilistic analysis of the Davis–Putnam procedure for solving the satisfiability problem." *Discrete Appl. Math.* **5**: 77–87.

Galil, Z. 1997. "On the complexity of regular resolution and the Davis–Putnam procedure." *Theoretical Comput. Sci.* **4**: 23–46.

Garey, M. R. and D. S. Johnson. 1979. In *Computers and Intractability: a Guide to the Theory of NP-completeness*: W.H. Freeman and Company.

Gelder, A. V. 1988. "A satisfiability tester for non-clausal propositional calculus." *Inform. Comput.* **79**: 1–21.

Genesereth, M. R. and N. J. Nilsson. 1987. In *Logical Foundations of Artificial Intelligence*: Morgan Kaufmann Publishers.

Gilmore, P. C. 1960. "A proof method for quantification theory." *IBM J. Res. Dev.* **4**: 28–35.

Goldberg, A. 1979a. "Average case complexity of the satisfiability problem." In *Proceedings of the Fourth Workshop on Automated Deduction*, pp. 1–6.

Goldberg, A. 1979b. "On the complexity of the satisfiability problem." Technical Report 16. New York: Courant Computer Science, New York University.

Goldberg, A., P. W. Purdom, and C. A. Brown. 1982. "Average time analysis of simplified Davis–Putnam procedures." *Inform. Process. Lett.* **15**: 72–75.

Gu, J. 1988. "How to solve very large-scale satisfiability problems." Technical Report UUCS-TR-88-032. Utah: Department of Computer Science, University of Utah, 1988. See also Report UCECE-TR-90-002, Department of Electrical and Computer Engineering, University of Calgary, 1990

Gu, J. 1989. "Parallel algorithms and architectures for very fast search." Ph.D. diss. Utah: Department of Computer Science, University of Utah.

Gu, J. 1990. "Optimization by multispace search." Technical Report UCECE-TR-90-001. Department of Electrical and Computer Engineering, University of Calgary.

Gu, J. 1992a. "Efficient local search for very large-scale satisfiability problem." *SIGART Bull.* **3**: 8–12.

Gu, J. 1992b. "The UniSAT problem models (appendix)." *IEEE Trans. Pattern Anal. Machine Intell.* **14**: 865.

Gu, J. 1993. "Local search for satisfiability (SAT) problem." *IEEE Trans. Sys. Man, Cybernet* **23**: 1108–1129 Correction in vol. 24, pp. 709, 1994.

Gu, J. 1994. "Global optimization for Satisfiability (SAT) problem." *IEEE Trans. Knowledge Data Engng* **6**: 361–381.

Gu, J. 1995. "Parallel algorithms for satisfiability (SAT) problem." In *Parallel Processing of Discrete Optimization Problems*. Volume 22, edited by P. M. Pardalos, M. G. C. Resende, and K. G. Ramakrishnan, DIMACS Series on Discrete Mathematics and Theoretical Computer Science, pp. 105–161. American Mathematical Society.

Gu, J. and Q. P. Gu. 1994. "Average time complexities of several local search algorithms for the Satisfiability problem." *Lecture Notes Comput. Sci.* **834**: 146–154.

Gu, J. and W. Wang. 1992. "A novel discrete relaxation architecture." *IEEE Trans. Pattern Anal. Machine Intell.* **14**: 857–865.

Gu, J., W. Wang, and T. C. Henderson. 1987. "A parallel architecture for discrete relaxation algorithm." *IEEE Trans. Pattern Anal. Machine Intell.* **9**: 816–831.

Gu, J., H. Li, Z. Zhou and B. Du. 1995. "An efficient implementation of SAT 1.5 algorithm." Technical report. Department of Computer Science, USTC.

Gu, J., D. Z. Du, and P. M. Pardalos. 1997a. In *Satisfiability Problem: Theory and Applications*, Volume 35, DIMACS Series on Discrete Mathematics and Theoretical Computer Science: American Mathematical Society.

Gu, J., P. W. Purdom, J. Franco, and B. W. Wah. 1997b. "Algorithms for satisfiability (SAT) problem: a survey." In *Satisfiability Problem: Theory and Applications*, Volume 35, edited by J. Gu, D. Z. Du, and P. M. Pardalos, DIMACS Series on Discrete Mathematics and Theoretical Computer Science, pp. 19–151. American Mathematical Society.

Gu, J., P. W. Purdom, J. Franco, and B. Wah. 2000. In *Algorithms for the Satisfiability Problem*: Cambridge University Press.

Haken, A. 1985. "The intractability of resolution." *Theoretical Comput. Sci.* **39**: 297–308.

Hooker, J. N. 1988a. "Generalized resolution and cutting planes." *Ann. Oper. Res.* **12**: 217–239.

Hooker, J. N. 1988b. "A quantitative approach to logical inference." *Decision Support Sys.* **4**: 45–69.

Hooker, J. N. 1988c. "Resolution vs. cutting plane solution of inference problems: some computational experience." *Oper. Res. Lett.* **7**: 1–7.

Hooker, J. N. and Fedjki, C. 1989. "Branch-and-cut solution of inference problems in propositional logic." Technical Report 77-88-89. GSIA, Carnegie Mellon University.

Hsiang, J. 1985. "Refutational theorem proving using term-rewriting systems." *Artificial Intell.* **25**: 255–300.

Hu, T. H., C. Y. Tang, and R. C. T. Lee. 1992. "An average case analysis of a resolution principle algorithm in mechanical theorem proving." *Ann. Math, Artificial Intell.* **6**: 235–252.

Huffman, D. A. 1971. "Impossible objects as nonsense sentences." In *Machine Intelligence*, edited by B. Meltzer and D. Michie, pp. 295–323. Edinburgh, Scotland: Edinburgh University Press.

Iwama, K. 1989. "CNF satisfiability test by counting and polynomial average time." *SIAM J. Comput.* **18**: 385–391.

Jeroslow, R. E. and J. Wang. 1990. "Solving propositional satisfiability problems." *Ann. Math. Artificial Intell.* **1**: 167–187.

Jeroslow, R. G. 1988. "Computation-oriented reductions of predicate to propositional logic." *Decision Support Sys.* **4**: 183–197.

Johnson, J. L. 1989. "A neural network approach to the 3-satisfiability problem." *J. Parallel Distributed Comput.* **6**: 435–449.

Kamath, A. P., Karmarkar, N. K., Ramakrishnan, K. G. and Resende, M. G. C. 1989. "Computational experience with an interior point algorithm on the satisfiability problem." Technical report. Mathematical Sciences Research Center, AT & T Bell Laboratories.

Kamath, A. P., N. K. Karmarkar, K. G. Ramakrishnan, and M. G. C. Resende. 1990. "Computational experience with an interior point algorithm on the satisfiability problem." *Ann. Oper. Res.* **25**: 43–58.

Kullmann, O. 1996. "A systematic approach to 3-SAT-decision, yielding 3-SAT-decision in less than 1.5045^n steps." Manuscript.

Larrabee, T. 1992. "Test pattern generation using Boolean satisfiability." *IEEE Trans. Comput. Aided Design* **11**: 4–15.

Lieberherr, K. J. and E. Specker. 1981. "Complexity of partial satisfaction." *J. ACM* **28**: 411–421.

Lin, Y.-J. 1988. "A parallel implementation of logic programs." Ph.D. diss. TX: Department of Computer Science, The University of Texas at Austin.

Loveland, D. W. 1978. *Automated Theorem Proving: a Logical Basis.* North-Holland.

Luenberger, D. G. 1984. *Linear and Nonlinear Programming.* Addison-Wesley.

Mackworth, A. K. 1977. "Consistency in networks of relations." *Artificial Intell.* **8**: 99–119.

McCall, J. T., J. G. Tront, F. G. Gray, R. M. Haralick, and W. M. McCormack. 1985. "Parallel computer architectures and problem solving strategies for the consistent labeling problem." *IEEE Trans. Comput.* **C-34**: 973–980.

Méjean, H.-M., H. Morel, and G. Reynaud. 1995. "A variational method for analysing unit clause search." *SIAM J. Comput.* **24**: 621–649.

Mitchell, D., B. Selman, and H. Levesque. 1992. "Hard and easy distributions of SAT problems." In *Proceedings of AAAI'92*, pp. 459–465.

Monien, B. and E. Speckenmeyer. 1983. "Solving satisfiability in less than 2^n steps." *Discrete Appl. Math.* **10**: 117–133.

Nilsson, N. J. 1980. In *Principles of Artificial Intelligence.* Tioga Publishing Company.

Papadimitriou, C. H. 1991. "On selecting a satisfying truth assignment." In *Proceedings of the 32nd Annual Symposium of the Foundations of Computer Science*, pp. 163–169.

Papadimitriou, C. H. 1992. Private communications.

Papadimitriou, C. H. and K. Steiglitz. 1982. In *Combinatorial Optimization: Algorithms and Complexity* Prentice Hall.

Pardalos, P. M. 1993. In *Complexity in Numerical Optimization.* World Scientific.

Purdom, P. W. 1978. "Tree size by partial backtracking." *SIAM J. Comput.* **7**: 481–491.

Purdom, P. W. 1983. "Search rearrangement backtracking and polynomial average time." *Artificial Intell.* **21**: 117–133.

Purdom, P. W. 1990. "A survey of average time analyses of satisfiability algorithms." *J. Inform. Process.* **13**: 449–455.

Purdom, P. W. and C. A. Brown. 1983. "An analysis of backtracking with search rearrangement." *SIAM J. Comput.* **12**: 717–733.

Purdom, P. W. and C. A. Brown. 1985. "The pure literal rule and polynomial average time." *SIAM J. Comput.* **14**: 943–953.

Purdom, P. W. and C. A. Brown. 1987. "Polynomial-average-time satisfiability problems." *Inform. Sci.* **41**: 23–42.

Purdom, P. W. and Haven, G. N. 1993. "Backtracking and probing." Technical Report 387. USA Department of Computer Science, University of Indiana.

Purdom, P. W., C. A. Brown, and E. L. Robertson. 1981. "Backtracking with multi-level dynamic search rearrangement." *Acta Inform.* **15**: 99–113.

Robinson, J. A. 1965. "A machine-oriented logic based on the resolution principle." *J. ACM* **12**: 23–41.

Samal, A. and T. C. Henderson. 1987. "Parallel consistent labeling algorithms." *Int. J. Parallel Programming* **16**: 341–364.

Schaefer, T. J. 1978. "The complexity of satisfiability problems." In *Proceedings of the 10th ACM Symposium on Theory of Computing*, pp. 216–226.

Selman, B., H. Levesque, and D. Mitchell. 1992. "A new method for solving hard satisfiability problems." In *Proceedings of AAAI'92*, pp. 440–446.

Selman, B., H. Kautz, and B. Cohen. 1994. "Noise strategies for improving local search." In *Proceedings of the 12th National Conference on Artificial Intelligence*, pp. 337–343.

Selman, B., H. A. Kautz, and B. Cohen. 1996. "Local search strategies for satisfiability testing." In *DIMACS Series Volume: Clique, Graph Coloring, and Satisfiability—Second DIAMACS Implementation Challenge*, Volume 26, edited by D. S. Johnson and M. A. Trick, pp. 521–532. American Mathematical Society.

Shensa, M. J. 1989. "A computational structure for the propositional calculus." In *Proceedings of IJCAI*, pp. 384–388.

Shi, R. C.-J., A. Vannelli, and J. Vlach. 1992. "An improvement on Karmarkar's algorithm for integer programming." In *Special issue on Computational Aspects of Combinatorial Optimization*, Volume 21, COAL Bulletin, pp. 23–28. Mathematical Programming Society.

Siegel, P. 1987. "Representation et Utilization de la Connaissances en Calcul Propositionel." Ph.D. diss. France: University Aix-Marseille II.

Sosič, R. and J. Gu, 1988a. "How to search for million queens." Technical Report UUCS-TR-88-088. Utah: Department of Computer Science, University of Utah.

Sosič, R. and J. Gu, 1988b. "Quick *n*-queen search on VAX and Bobcat machines." CS 547 AI Class Project. Utah: University of Utah.

Sosič, R. and J. Gu. 1991. "Fast search algorithms for the *n*-queens problem." *IEEE Trans. Sys., Man, Cybernet.* **21**: 1572–1576.

Sosič, R. and J. Gu. 1993a. "A parallel local search algorithm for Satisfiability (SAT) problem." Submitted for publication.

Sosič, R. and J. Gu. 1993b. "A parallel local search algorithm for the *n*-queen problem." Submitted for publication.

Sosič, R. and J. Gu. 1994. "Efficient local search with conflict minimization." *IEEE Trans. Knowledge Data Engng* **6**: 661–668.

Sosič, R., J. Gu, and R. Johnson. 1992. "A universal Boolean evaluator, 1992." To appear in *IEEE Trans. Comput.*

Sosič, R., J. Gu, and R. Johnson. 1996. "The Unison algorithm: fast evaluation of Boolean expressions." *ACM Trans. Design Automation Electr. Sys.* **1**: 456–477.

Törn, A. and A. Žilinskas. 1989. *Global Optimization*. Berlin: Springer-Verlag.

Truemper, K. 1994. "Polynomial algorithms for problems over *d*-systems." Presented at the 3rd International Symposium on AI & Mathematics.

Tseitin, G. S. 1968. "On the complexity of derivations in propositional calculus." In *Structures in Constructive Mathematics and Mathematical Logic, Part II*, edited by A. O. Slisenko, pp. 115–125. Consultants Bureau.

Tyler, J. E. 1983. "Parallel computer architectures and problem solving strategies for the consistent labeling problem." Master's thesis. Virginia: Department of Electrical Engineering, Virginia Polytechnic Institute State University.

Urquhart, A. 1987. "Hard examples for resolution." *J. ACM* **34**: 209–219.

Urquhart, A. 1995. "The complexity of propositional proofs." *Bull. Symbolic Logic* **1**: 425–467.

Wah, B. W. (ed). In *New Computers for Artificial Intelligence Processing*, Volume 20. IEEE Computer Society Press.

Wah, B. and G. J. Li. 1986. *Computers for Artificial Intelligence Applications*: IEEE Computer Society Press.

Wah, B. W. and Y. Shang. 1998. "A discrete Lagrangian-based global-search method for solving satisfiability problems." *J. Global Optim.* **12**: 61–100.

Wah, B. W., M. B. Lowrie, and G.-J. Li. 1989. "Computers for symbolic processing." In *Proceedings of the IEEE* **77**: 509–540.

Waltz, D. 1972. "Generating semantic descriptions from drawings of scenes with shadows." Technical Report AI271. MIT.

Williams, H. P. 1987. "Linear and integer programming applied to the propositional calculus." *Sys. Res. Inform. Sci.* **2**: 81–100.

Winston, P. H. 1984. *Artificial Intelligence*. Addison-Wesley.

Wos, L., G. Ã. Robinson, and D. Carson. 1965. "Efficiency and completeness of the set of support strategy." *J. ACM* **12**: 536–541.

Wu, L. C. and C. Y. Tang. 1992. "Solving the satisfiability problem by using randomized approach." *Inform. Process. Lett.* **41**: 187–190.

17.9 Assignment Problems

Eranda Çela

ABSTRACT Assignment problems arise in different situations where we have to find an optimal way to assign *n* objects to *m* other objects in an injective fashion. Depending on the objective we want to optimize, we obtain different problems ranging from linear assignment problems to quadratic and higher-dimensional assignment problems. Assignment problems are a well-studied topic in combinatorial optimization. These problems find numerous applications in production planning, telecommunication, VLSI design, economics, and so on. We introduce the basic problems classified into three groups: linear assignment problems, three- and higher-dimensional assignment problems, and quadratic assignment problems and problems related to it. For each group of problems we mention applications, show basic properties, and describe briefly some of the most successful algorithms used to solve these problems.

1. Introduction

Assignment problems deal with the question of how to assign n objects to m other objects in an injective fashion in the best possible way. An assignment problem is completely specified by its two components: the assignments, which represent the underlying combinatorial structure, and the objective function to be optimized, which models "the best possible way."

In the classical assignment problem, one has $m = n$ and most of the problems with $m > n$ can be transformed or are strongly related to analogous problems with $m = n$. Therefore, we will consider $m = n$ throughout the rest of this article, unless otherwise specified.

From the mathematical point of view an assignment is a bijective mapping of a finite set $N = \{1, 2, ..., n\}$ into itself, that is, permutation ϕ assigning some $j = \phi(i)$ to each $i \in N$. The set of all permutations (assignments) of n items will be denoted by $\mathcal{S}_n$ and has $n!$ elements. Every permutation ϕ of the set $N = \{1, ..., n\}$ corresponds uniquely to a permutation matrix $X_\phi = (x_{ij})$ with $x_{ij} = 1$ for $j = \phi(i)$ and $x_{ij} = 0$ for $j \neq \phi(i)$. Thus, a permutation matrix $X = (x_{ij})$ can be defined as a matrix that fulfills the following conditions, so-called assignment constraints

$$
\begin{aligned}
&\sum_{i=1}^{n} x_{ij} = 1 \quad \text{for all} \quad j = 1, ..., n \\
&\sum_{j=1}^{n} x_{ij} = 1 \quad \text{for all} \quad i = 1, ..., n \qquad (1) \\
&x_{ij} \in \{0, 1\} \quad \text{for all} \quad i, j = 1, ..., n.
\end{aligned}
$$

By replacing the conditions $x_{ij} \in \{0, 1\}$ by $x_{ij} \geqslant 0$ in (1), we get a doubly stochastic matrix. The set of all doubly stochastic matrices forms the assignment polytope P_A. Due to the famous result of Birkhoff (1946), the assignment polytope P_A is the convex hull of all assignments, or equivalently, every doubly stochastic matrix can be written as a convex combination of permutation matrices.

The concept of an assignment is strongly related to another well-known concept in graph theory and in combinatorial optimization, matching in bipartite graphs. A bipartite graph G is a triple $(V, W; E)$, where the vertex sets V and W have no vertices in common and the edge set E is a set of pairs (i, j) where $i \in V$ and $j \in W$. A subset M of E is called "matching" if every vertex of G is incident with at most one edge from M. The cardinality of M is called "cardinality of the matching." The maximum matching problem asks for a matching with as many edges as possible. A matching M is called a "perfect matching," if every vertex of G is incident with exactly one edge from M. Evidently, every perfect matching is a "maximum matching." A perfect matching in a bipartite graph $G = (V, W; E)$ with $V = \{v_1, v_2, ..., v_n\}$, $W =$

$\{w_1, w_2, ..., w_n\}$ can be represented by a permutation ϕ_M of $\{1, 2, ..., n\}$ such that $\phi_M(i) = j$ if, and only if, $(v_i, w_j) \in M$. Hence a perfect matching in a bipartite graph is an "assignment."

Hopcroft and Karp (1973) gave an $O(|E|\sqrt{|V|})$ algorithm, which constructs a perfect matching if it exists. Even and Tarjan (1975) gave an $O(\sqrt{|V|}\,|E|)$ algorithm for the maximum flow problem on unit capacity simple networks, which can also be applied to find a matching of maximum cardinality in a bipartite graph. Alt et al. (1991) gave an $O(|V|^{1.5}\sqrt{|E|/\log|V|})$ implementation for the Hopcroft–Karp algorithm. Based on ideas similar to those in Hopcroft and Karp (1973), a fast randomized Monte Carlo algorithm is given by Mulmuley et al. (1987). This algorithm finds a perfect matching at costs of a single matrix inversion. The reader is referred to the bibliography in Burkard and Çela (1999) for further reference pointers related to algorithms for cardinality matching problems.

2. Linear Assignment Problems

The linear assignment problem (LAP) is one of the oldest and most-studied problems in combinatorial optimization. Many algorithms have been developed to solve this problem. Also, other aspects of the problem such as asymptotic behavior or special cases have been thoroughly investigated. The reader is referred to Dell'Amico and Martello (1997) for a comprehensive annotated bibliography and to Burkard and Çela (1999) for a recent review on assignment problems.

2.1. Problem Definition and Applications

Recall the original model, where n items are to be assigned to n other objects in the best possible way. Let c_{ij} be the cost incurred by the assignment of object i to object j. We are looking for an assignment ϕ, that minimizes the overall cost $\sum_{i=1}^{n} c_{i\phi(i)}$. Thus, the linear assignment problem (LAP) is given as follows

$$\min_{\phi \in \mathcal{S}_n} \sum_{i=1}^{n} c_{i\phi(i)}, \tag{2}$$

where $\mathcal{S}_n$ is the set of permutations of $\{1, 2, ..., n\}$. Based on the description (1) of the set of all assignments (see Section 1), the LAP can also be formulated as follows:

$$\min \sum_{ij} c_{ij}x_{ij} \text{ over all matrices } X = (x_{ij}), \text{ which fulfill (1).}$$

Due to Birkhoff's result we can relax the conditions $x_{ij} \in \{0, 1\}$ to $x_{ij} \geqslant 0$ and obtain the linear programming formulation of the LAP. Any basic solution of this linear program corresponds to a permutation matrix.

$$\begin{aligned} \min \quad & \sum_{i=1}^{n} c_{ij}x_{ij} \\ & \sum_{i=1}^{n} x_{ij} = 1 \quad j = 1, ..., n \\ & \sum_{j=1}^{n} x_{ij} = 1 \quad i = 1, ..., n \\ & x_{ij} \geqslant 0 \quad i, j = 1, ..., n. \end{aligned} \tag{LP}$$

As we will mention in the next section, many algorithms for the LAP are based on linear programming techniques and often deal with the dual linear program:

$$\begin{aligned} \max \quad & \sum_{i=1}^{n} u_i + \sum_{j=1}^{n} v_j \\ & u_i + v_j \leqslant c_{ij} \quad i,j = 1,..,n \\ & u_i, v_j \in \mathbb{R} \quad i,j = 1,...,n, \end{aligned} \tag{3}$$

where u_i and v_j, $1 \leqslant i, j \leqslant n$, are dual variables.

Among the numerous applications of LAP the so-called personnel assignments are the most typical. In the personnel assignment we want to assign people to objects, that is, jobs, machines, rooms; to other people; and so on. Each assignment has a cost and we want to make an assignment so as to minimize the overall sum of the costs. For example, one company might want to assign graduates to vacant jobs. In this case the cost c_{ij} is given by $c_{ij} = -p_{ij}$ where p_{ij} is the proficiency index for placing candidate i to job j, and the goal is to assign each candidate i to some vacancy $\phi(i)$ such that the overall cost $\sum_i c_{i\phi(i)}$ is minimized, or equivalently, the overall proficiency $\sum_i p_{i\phi(i)}$ is maximized.

There are many other applications of the LAP, for example, in locating and tracing objects in space, scheduling on parallel machines, inventory planning, vehicle and crew scheduling, wiring of typewriters, and so on. The reader is referred to Ahuja et al. (1995) and Burkard and Çela (1999) for a detailed description of some applications of the LAP and literature pointers to other applications.

2.2. *Algorithms for the LAP*

The LAP can be solved efficiently, and the design of efficient solution methods for this problem has been an object of research for many years. There are an amazing number of algorithms, sequential and parallel, for the LAP, ranging from primal-dual combinatorial algorithms, to simplex-like methods. The worst-case complexity of the best sequential algorithms for the LAP is $O(n^3)$, where n is the size of the problem. From the computational point of view very large-scale dense assignment problems with about 10^6 nodes can be solved within a couple of minutes by sequential algorithms (see Lee and Orlin, 1994).

There are a number of survey papers and books on algorithms, among others Derigs (1985), Dell'Amico and Toth (1999), and the book on the first DIMACS challenge edited by Johnson and McGeoch (1993). Among papers reporting on computational experience we mention Carpaneto et al. (1988), Lee and Orlin (1994), and Dell'Amico and Toth (1999), and some of the papers in Johnson and McGeoch (1993).

Most sequential algorithms for the LAP can be classified into primal-dual algorithms and simplex-based algorithms. Primal-dual algorithms work with a pair consisting of an infeasible solution x_{ij}, $1 \leqslant i, j \leqslant n$, of LP (called primal solution) and a feasible solution u_i, v_j, $1 \leqslant i, j \leqslant n$ of the dual (3) (called dual solution). These solutions fulfil the complementarity slackness conditions:

$$x_{ij}(c_{ij} - u_i - v_j) = 0, \quad \text{for } 1 \leqslant i,j \leqslant n. \tag{4}$$

These solutions are updated iteratively until the primal solution becomes feasible, while keeping the complementary slackness conditions fulfilled and the dual solution feasible. At this point the primal solution would be optimal, according to duality theory.

Different primal-dual algorithms differ on (i) the way they obtain a starting pair of a primal and a dual solution fulfilling the conditions described above, and (ii) the way the solutions are updated. A starting dual solution can be obtained as in the Hungarian method by setting $u_i := \min\{c_{ij} : 1 \leqslant j \leqslant n\}$, for $1 \leqslant i \leqslant n$, and then $v_j := \min\{c_{ij} - u_i : 1 \leqslant i \leqslant n\}$, for $1 \leqslant j \leqslant n$. An infeasible primal starting solution could be given by a matching of maximal cardinality in the bipartite graph $\bar{G} = (V, W; \bar{E})$, where $V = W = \{1, 2, ..., n\}$, and $\bar{E} = \{(i,j) : \bar{c}_{ij} := c_{ij} - u_i - v_j = 0\}$. Then, set $x_{ij} := 1$ if (i, j) is an edge of the matching and $x_{ij} = 0$, otherwise. Clearly, the pair of solutions obtained in this way fulfil the complementarity slackness conditions. One way of updating the pair of solutions is the shortest augmenting path method. This method gives rise to a whole class of algorithms that meet the best-known time complexity bound for the LAP, namely $O(n^3)$. For a given pair of solutions, as above, construct a weighted directed bipartite graph $\tilde{G} = (V, W; \tilde{E})$ with arc set $E = D \cup R$ with set of forward arcs $D = \{(i,j) : (i,j) \in E, x_{ij} = 0\}$ and set of backward arcs $R = \{(j,i) : (i,j) \in \bar{E}, x_{ij} = 1\}$. The weights of the backward arcs are set equal to 0, whereas the weights of the forward arcs are set equal to the corresponding reduced costs $\bar{c}_{ij}$. Then select a node r in V that has not been assigned yet, and solve the single-source shortest-path problem, that is, compute the shortest paths from r to all nodes of $\tilde{G}$. The shortest among all paths from r to some free (i.e., not yet assigned) node in W is used to augment the current primal solution by swapping the free and matched edges (i.e., edges belonging to the matching). The dual solution and the reduced costs are then accordingly updated. It can be shown that after n augmentations an optimal primal solution results (see Derigs, 1985).

There are various shortest-augmenting-path algorithms for the LAP. Basically they differ in the way they determine a starting pair of primal and dual solutions, and by the subroutine they use for computing the shortest paths. Most of the existing algorithms use the Dijkstra algorithm for the shortest-path computations.

Simplex-based algorithms are special implementations of the primal or the dual simplex algorithm for linear programming applied to LP. Simplex-based algorithms for the LAP are specific implementations of the network simplex algorithm. The latter is a specialization of the simplex method for linear programming to network problems. The specialization relies on exploiting the combinatorial structure of network problems to perform efficient pivots acting on trees rather than on the coefficient matrix.

It is well known that there is a one-to-one correspondence between primal (integer) basic solutions of the LAP and spanning trees of the bipartite graph G related to assignment problems as described in Section 1. Moreover, given a spanning tree, one can uniquely determine the values of the corresponding dual variables so as to fulfill the complementarity slackness conditions, as soon as the value of one of those variables is fixed (arbitrarily). Every integer primal feasible basic solution is highly degenerate because it contains $2n-1$ variables and $n-1$ of them are equal to 0. Hence degeneracy poses a problem, and the first simplex-based algorithms for the LAP were exponential. The first steps toward the design of polynomial-time simplex-based algorithms were made by introducing the concept of so-called strongly feasible trees, introduced by Cunningham (1976). There are implementations of the simplex-based algorithm for the LAP that match the best-known time complexity bound of $O(n^3)$. The reader is referred to Burkard and Çela (1999) for references and further details.

More recently Ramakrishnan et al. (1993) applied an interior point algorithm to the LAP and got promising results, in particular for large-size instances (see Johnson and McGeoch, 1993).

Since the late 1980s a number of parallel algorithms for the LAP have been proposed. The speedup achieved by such algorithms is limited by the sparsity of the cost matrices and/or by the decreasing load across the iterations. For a good review on parallel algorithms for the LAP and network flow problems in general the reader is referred to Bertsekas et al. (1995).

2.3. *Asymptotic Behavior and Probabilistic Analysis*

When dealing with the asymptotic behavior of the LAP, it is always assumed that the cost coefficients c_{ij} are independent random variables (I.R.V.) with a common prespecified distribution. The main question concerns the behavior of the expected optimal value of the problem as its size tends toward infinity.

For cost coefficients c_{ij} being I.R.V. with a uniform distribution on [0, 1], it has been shown that the optimal value of the LAP remains within constant bounds as the size n of the problem tends to infinity. The best upper bound equals 2 and is due to Karp (1987). The best lower bound equals 1.51 and is due to Olin (1992). Although the gap between the current lower and upper bounds on the expected optimal value of the LAP is large, it is believed that the expected value is close to 1.6 or more exactly $\frac{\pi^2}{6}$ in the case of independent cost coefficients c_{ij} uniformly distributed on [0, 1]. For a discussion in some detail and for more references, see Burkard and Çela (1999).

A more general scenario where the coefficients c_{ij} are I.R.V. with a common arbitrary general distribution has been investigated by Frenk et al. (1987) and Olin (1992). Under mild assumption on the probability distribution of the coefficients, the authors derive constant upper and lower bounds on the expected optimal value of the LAP.

In the case of coefficients c_{ij} being I.R.V. uniformly distributed on [0, 1], the LAP can be solved in expected $O(n^2 \log n)$ time by a randomized algorithm proposed by Karp (1980). Faster randomized algorithms that produce a solution whose objective function value is within a constant factor of the optimal objective function value have been proposed by several authors. There is, for instance, an algorithm by Karp et al. (1994), that runs in expected linear time $[O(n)]$ and provides a solution within a factor of $3 + O(n^{-a})$ of the optimum with probability $1 - O(n^{-a})$, where a is some fixed positive number.

2.4. *Linear Assignment Problems with Other Objective Functions*

The best-known linear assignment problem with an objective function different from that of the LAP is the bottleneck linear assignment problem (BLAP)

$$\min_{\phi} \max_{1 \leqslant i \leqslant n} c_{i\phi(i)}. \tag{5}$$

considered originally by Fulkerson et al. (1953). This problem occurs, for example, in the assignment of jobs to parallel machines so as to minimize the latest completion time. One of the first algorithms proposed for the BLAP is the so-called threshold algorithm. The threshold algorithm chooses a cost element c_{ij}^* [not defined] K—the threshold value—and constructs a matrix $\bar{C}$ defined as follows

$$\bar{c}_{ij} := \begin{cases} 1 & \text{if } c_{ij} > K \\ 0 & \text{if } c_{ij} \leqslant K. \end{cases}$$

Then the algorithm checks whether the bipartite graph with adjacency matrix $\bar{C}$ contains a perfect matching or not. The algorithm repeats this procedure for a new (and eventually smaller) value of the threshold K. The smallest value K for which the corresponding bipartite graph contains a perfect matching is the optimum value of the BLAP. The best time complexity known today amounts to $O(n\sqrt{nm})$ and is due to Punnen and Nair (1994). (Here m is the number of finite elements in the coefficient matrix (c_{ij}), which would correspond to the number of edges of the bipartite graph in the graph theoretical setting.)

A randomized algorithm with quadratic expected running time $[O(n^2)]$ has been given by Pferschy (1996). A computational study on the comparison of different deterministic algorithms for the BLAP has been given by Pferschy (1997).

Another linear assignment problem introduced by Burkard and Rendl (1991) is the so-called lexicographic bottleneck assignment problem (LexBAP). In the LexBAP we want to find a permutation (assignment) ϕ^*, that lexicographically minimizes c_ϕ over all permutations ϕ, where c_ϕ is the vector of costs $c_{i\phi(i)}$, $1 \leqslant i \leqslant n$, sorted nonincreasingly.

Martello et al. (1984) have considered the so-called balanced assignment problem (BalAP). Given a real $n \times n$ matrix $C = (c_{ij})$, the balanced assignment problem can be formulated as

$$\min_{\phi} \left[\max_{i} c_{i\phi(i)} - \min_{i} c_{i\phi(i)} \right].$$

The problem can be solved efficiently in $O(n^4)$ time.

A more general LAP, which includes as special cases the LAP and the BLAP, is the algebraic assignment problem (AAP) introduced by Burkard et al. (1977). In the AAP the coefficients c_{ij} are elements of a totally ordered semigroup $(H, *, \prec)$ with composition $*$ and order relation $\prec$. The AAP can then be formulated as follows:

$$\min_{\phi \in \mathcal{S}_n} c_{1\phi(1)} * c_{2\phi(2)} * \cdots * c_{n\phi(n)}. \tag{6}$$

The AAP can be solved efficiently if the order relation and the composition fulfill some natural algebraic properties. For further results and reference pointers consult the survey on algebraic optimization by Burkard and Zimmermann (1982).

2.5. *Available Computer Codes and Test Instances*

FORTRAN listings of codes for the LAP and the BLAP can be found in the book by Burkard and Derigs (1980). The code for the LAP is a primal-dual algorithm based on shortest-path computations done by a version of Dijkstra's algorithm.

Source codes of another primal-dual algorithm for the LAP and the BLAP can be downloaded from `http://207.158.230.188/assignment.html`. One can choose among C++, Pascal, and FORTRAN implementations of an algorithm by Jonker and Volgenant (1986).

A (compressed) FORTRAN source file—called 548.Z—of an implementation of the Hungarian algorithm, due to Carpaneto and Toth (1980), can be downloaded from `ftp://netlib.att.com` in `/netlib/toms`. Other listings of FORTRAN codes for the LAP can be found in Carpaneto et al. (1988). The codes are available from the floppy disk included in the book.

The C code of an efficient implementation of the scaling push-relabel algorithm of Goldberg and Kennedy (1995) for the LAP can be downloaded from Goldberg's network optimization library at `http://www.neci.nj.nec.com/homepages/avg/soft.html`. Finally, listings of five FORTRAN codes of auction algorithms for the LAP can be found in Bertsekas' homepage at `http://web.mit.edu/dimitrib/www/auction.txt`. Test instances of the LAP can be downloaded as ASCII files from the homepage of the OR-Library maintained by J. Beasley at `http://mscmga.ms.ic.ac.uk/pub`. Other test instances can be obtained from the ELIB library at `ftp://ftp.zib.de/pub/Packages/mp-testdata/assign/index.html`.

Clearly, since the LAP can be formulated as a minimum-cost flow problem, algorithms developed for the latter can also be applied to the LAP. However, such algorithms are not supposed to exploit the specific features of the LAP, and hence may not be competitive with algorithms developed especially for the LAP. Besides Goldberg's network optimization library and Bertsekas' homepage, other codes for network optimization can be found in `Netlib` at `http://www.OpsResearch.com/OR-Links/index.html`. Furthermore, C codes of implementations of the primal and the dual network simplex algorithm, due to Löbel, can be obtained through `http://www.zib.de/Optimization/index.de.html`.

3. Multidimensional Assignment Problems

3.1. *General Remarks and Applications*

Multidimensional (sometimes referred as multi-index) assignment problems (MAP) are natural extensions of the LAP. They were considered for the first time by Pierskalla (1967). The most prominent representatives of this class are axial and planar three-dimensional assignment problems to be considered in the next section. The MAP asks for $d-1$ permutations $\phi_1, \phi_2, ..., \phi_{d-1}$, which minimize the following objective function:

$$\min_{\phi_1,\phi_2,\ldots,\phi_{d-1}} \sum_{i=1}^{n} c_{i\phi_1(i)\phi_2(i)\ldots\phi_{d-1}(i)}.$$

In terms of graphs, a multidimensional assignment problem can be described as follows: Let a complete d-partite graph $G = (V_1, V_2, ..., V_d; E)$ with vertex sets V_i, $|V_i| = n$, $i = 1, 2, ..., d$, and edge set E be given. A subset X of $V = \bigcup_{i=1}^{d} V_i$ is a clique if it meets every set V_i in exactly one vertex. A d-dimensional assignment is a partition of V into n pairwise disjoint cliques. If c is a real-valued cost function defined on the set of cliques of $G = (V_1, V_2, ..., V_d; E)$, the d-dimensional assignment problem asks for a d-dimensional assignment of minimum cost. Special cases, where the costs c of a clique are not arbitrary, but are n as a function of elementary costs attached to the edges of the complete d-partite graph (e.g., sum costs, star costs, tour costs or tree costs), have been investigated and the performance of simple heuristics has been analyzed in these cases. For more information, see Burkard and Çela (1999) and references therein.

MAPs in their general form have found some applications as a means to solve data association problems in multi-target tracking and multisensor surveillance. The data association problem consists of partitioning the observations into tracks and false alarms in real time. General classes of these problems can be formulated as MAPs. Other applications of MAPs are related to track initiation, track maintenance, and multisensor tracking. Another interesting MAP arises in the context of tracking elementary particles. For more information on these two applications, see Burkard and Çela (1999) and references therein.

3.2. *Axial Three-Dimensional Assignment Problems*

Consider n^3 cost coefficients c_{ijk}. The axial three-dimensional assignment problem (3-DAP) can then be stated as

$$\begin{aligned} \min \quad & \sum_{i=1}^{n}\sum_{j=1}^{n}\sum_{k=1}^{n} c_{ijk}x_{ijk} \\ \text{subject to} \quad & \sum_{j=1}^{n}\sum_{k=1}^{n} x_{ijk} = 1, \quad i = 1,2,\ldots,n, \\ & \sum_{i=1}^{n}\sum_{k=1}^{n} x_{ijk} = 1, \quad j = 1,2,\ldots,n, \\ & \sum_{i=1}^{n}\sum_{j=1}^{n} x_{ijk} = 1, \quad k = 1,2,\ldots,n, \\ & x_{ijk} \in \{0,1\} \quad \forall\ 1 \leqslant i,j,k \leqslant n. \end{aligned} \tag{7}$$

We can think of c_{ijk} as the cost of assigning job j to be performed by worker i on machine k. It follows that $x_{ijk} = 1$, if job j is assigned to worker i on machine k, and $x_{ijk} = 0$, otherwise

Equivalently, a 3-DAP can be described with the help of two permutations ϕ and ψ

$$\min_{\phi,\psi \in \mathcal{S}_n} \sum_{i=1}^{n} c_{i\phi(i)\psi(i)}. \tag{8}$$

Thus this problem has $(n!)^2$ feasible solutions. It has been shown that the dimension of the axial three-index assignment polytope (i.e., the convex hull of feasible solutions to Equation (7)) is $n^3 - 3n + 2$. Furthermore, several classes of facet-defining inequalities have been identified and efficient algorithms for the separation of these facets have been developed. For more information on these topics, see the description in Burkard and Çela (1999) and consult the references cited therein.

In contrast to the LAP, the 3-DAP cannot be solved efficiently and branch-and-bound algorithms are the most-used algorithms to solve this problem. The lower bounds are usually computed by solving some Lagrangian relaxation of the 3-DAP by subgradient optimization approaches (see Balas and Saltzman, 1991). The authors in Balas and Saltzman (1991) also introduce a nontrivial branching strategy, that exploits the structure of the problem and allows several variables to be fixed at each branching node.

A heuristic for solving the 3-DAP has been proposed by Pierskalla (1967).

Finally, there are a number of efficiently solvable special cases of the 3-DAP, for example, if the cost coefficients are taken from a three-dimensional Monge array or if the cost coefficients are decomposable, that is, $c_{ijk} = u_i v_j w_k$, and u_i, v_j, and w_k are nonnegative The reader is referred to Burkard and Çela (1999) and the references therein for more information on this topic.

3.3. *Planar Three-Dimensional Assignment Problems*

Let c_{ijk}, $1 \leqslant i,j,k \leqslant n$, be n^3 cost coefficients. The planar three-dimensional assignment problem (3-PAP) is stated as follows:

$$\min \quad \sum_{i=1}^{n}\sum_{j=1}^{n}\sum_{k=1}^{n} c_{ijk}x_{ijk}$$

subject to

$$\begin{aligned}
&\sum_{i=1}^{n} x_{ijk} = 1, \quad j,k = 1,2,\ldots,n,\\
&\sum_{j=1}^{n} x_{ijk} = 1, \quad i,k = 1,2,\ldots,n,\\
&\sum_{k=1}^{n} x_{ijk} = 1, \quad i,j = 1,2,\ldots,n,\\
&x_{ijk} \in \{0,1\} \quad i,j,k = 1,2,\ldots,n.
\end{aligned} \tag{9}$$

The 3-PAP has interesting applications in timetabling problems. See Euler and Le Verge (1996) for a recent study on timetables and related polyhedra.

It is easily seen that the feasible solutions of the 3-PAP correspond to Latin squares. Thus, the number of feasible solutions of a 3-PAP of size n equals the number of Latin squares of order n, and hence increases very fast. Similarly to the 3-DAP, the 3-PAP is also a hard problem and cannot be solved efficiently. There are not many algorithms known for the 3-PAP. Besides two branch-and-bound algorithms due to Vlach (1967) and Magos and Miliotis (1994) there is also a tabu search algorithm for the 3-PAP due to Magos (1996). It uses a neighborhood structure based on the relationship between the 3-PAP and the Latin squares. See Burkard and Çela (1999) for more information on algorithms for the 3-PAP.

4. The Quadratic Assignment Problem

The quadratic assignment problem (QAP) was introduced in 1957 by Koopmans and Beckmann as a model for a plant location problem. Since then the QAP has been the object of intensive investigations concerning different aspects of the problem ranging from algorithms to asymptotic behavior and special cases. In contrast to its linear counterpart the QAP cannot be solved efficiently. However, there are some restricted cases of the QAP that can be solved in polynomial time. Such special cases are QAPs whose coefficient matrices show special combinatorial properties, for example, Monge and Monge-like properties. The interested reader is referred to Çela (1998) for a detailed discussion on efficiently solvable special cases of the QAP.

Also, from the practical point of view, the QAP is widely considered as one of the hardest problems in combinatorial optimization and there are instances of size less than thirty that have not yet been solved to optimality, for example, the Nugent instances (see QAPLIB Burkard et al., 1977).

The reader is referred to Burkard and Çela (1997) for a comprehensive annotated bibliography, to Çela (1998) for a recent monograph; and to Burkard et al. (1998) for a recent review on QAPs and for reference pointers.

4.1. Problem Definition and Applications

Among the most common and best-known applications of the QAP are those that arise in a facility location context. This is one reason for choosing the facility location terminology to introduce the problem.

Consider the problem of allocating n facilities to n locations, with costs depending on the distance between the locations and the flow between the facilities, plus costs associated with a facility being placed at a certain location. The objective is to assign each facility to a location such that the total cost is minimized. More specifically, we are given three $n \times n$

real matrices $A=(a_{ij})$, $B=(b_{kl})$ and $C=(c_{ik})$, where a_{ij} is the flow between the facility i and facility j, b_{kl} is the distance between the location k and location l, and c_{ik} is the cost of placing facility i at location k. The Koopmans–Beckmann version of the QAP can be then formulated as follows:

$$\min_{\phi\in\mathcal{S}_n}\left(\sum_{i=1}^{n}\sum_{j=1}^{n}a_{ij}b_{\phi(i)\phi(j)}+\sum_{i=1}^{n}a_{i\phi(i)}\right). \tag{10}$$

where $\mathcal{S}_n$ is the set of all permutations of N. A product $a_{ij}b_{\phi(i)\phi(j)}$ is the cost of assigning facility i to location $\phi(i)$ and facility j to location $\phi(j)$.

A more general version of the QAP was introduced by Lawler in 1963. In this version we are given a four-dimensional array $D=(d_{ijkl})$ of coefficients instead of the two matrices A and B, and the problem can be stated as

$$\min_{\phi\in\mathcal{S}_n}\left(\sum_{i=1}^{n}\sum_{j=1}^{n}d_{ij\phi(i)\phi(j)}+\sum_{i=1}^{n}c_{i\phi(i)}\right). \tag{11}$$

Besides applications in facility location, the QAP finds applications in VLSI design, computer manufacturing, scheduling, process communications, backboard wiring, and typewriter keyboard design. We describe here a lesser known application on turbine balancing. Hydraulic turbine runners as used in electricity generation consist of a cylinder around which a number of blades are welded at regular spacings. Due to inaccuracies in the manufacturing process, the weights of these blades differ slightly, and it is desirable to locate the blades around the cylinder in such a way that the distance between the center of mass of the blades and the axis of the cylinder is minimized. This problem was introduced by Mosevich (1986) and has been formulated as a QAP by Laporte and Mercure (1988). The places at regular spacings on the cylinder are modeled by the vertices $v_1,\ldots,v_n$ of a regular n-gon on the unit circle in the Euclidean plane. Thus

$$v_i=\left(\sin\left(\frac{2i\pi}{n}\right),\cos\left(\frac{2i\pi}{n}\right)\right),\quad 1\leqslant i\leqslant n.$$

The masses of the n blades are given by the positive reals $0<m_1\leqslant m_2\leqslant\cdots\leqslant m_n$. The goal is to assign each mass m_i to some vertex $v_{\phi(i)}$ in such a way that the center of gravity,

$$\sum_{i=1}^{n}m_{\phi(i)}\begin{pmatrix}\sin\left(\frac{2i\pi}{n}\right)\\ \cos\left(\frac{2i\pi}{n}\right)\end{pmatrix},$$

of the resulting mass system is as close to the origin as possible. Minimizing the Euclidean norm of the above vector is equivalent to minimizing the expression

$$\sum_{i=1}^{n}\sum_{j=1}^{n}m_{\phi(i)}m_{\phi(j)}\cos\left(\frac{2(i-j)\pi}{n}\right). \tag{12}$$

Thus we have a QAP, where the coefficient matrices $A=(a_{ij})$ and $B=(b_{ij})$ have a very special structure: $a_{ij}=\cos\left(\frac{2(i-j)\pi}{n}\right)$ and $b_{ij}=m_i\cdot m_j$.

Finally, notice that there are a number of other well-known combinatorial optimization problems, that can be formulated as QAPs, for example, graph partitioning, maximum clique, the traveling salesman problem, the linear arrangement problem, and the minimum-weight feedback arc set problem.

To conclude this section notice that the QAP is a very hard problem from the theoretical point of view. Not only can the QAP not be solved efficiently, but it cannot even be approximated efficiently within some constant approximation ratio. Furthermore, finding local optima is also, in general, a nontrivial task. For information and references concerning the computational complexity of the QAP, see Çela (1998).

4.2. *Alternative Formulations of the QAP*

There are several equivalent formulations of the QAP. Different formulations stress different characteristics of the problem and lead to different solution approaches. Recall that there is a one-to-one correspondence between the permutations of $N = \{1, 2, ..., n\}$ and the $n \times n$ permutation matrices defined in Section 1. Let $\mathbf{X}_n$ be the set of $n \times n$ permutation matrices. In terms of permutation matrices QAP(A, B, C) can be formulated as the following quadratic integer program:

$$\min \sum_{i=1}^{n}\sum_{j=1}^{n}\sum_{k=1}^{n}\sum_{l=1}^{n} a_{ij}b_{kl}x_{ik}x_{jl} + \sum_{i,j=1}^{n} c_{ij}x_{ij} \tag{13}$$

$$\text{subject to } (x_{ij}) \in \mathbf{X}_n.$$

Let us define an inner product between $n \times n$ matrices as follows

$$\langle A, B\rangle := \sum_{i=1}^{n}\sum_{j=1}^{n} a_{ij}b_{ij}.$$

Clearly, we have $XAX^T = (a_{\phi(i)\phi(j)})$, for some $n \times n$ matrix A, a permutation $\phi \in \mathcal{S}_n$ and the associated permutation matrix $X \in \mathbf{X}_n$. Thus the QAP (13) can be formulated alternatively as

$$\min \quad \langle A, XBX^T\rangle + \langle C, X\rangle \tag{14}$$

$$\text{subject to } X \in \mathbf{X}_n.$$

Finally consider the trace formulation of the QAP. The trace tr(A) of an $n \times n$ matrix $A = (a_{ij})$ is defined as the sum of its diagonal elements: $\text{tr}(A) = \sum_{i=1}^{n} a_{ii}$. Let us denote $B^{\phi} := XB^tX^t$, where X is the permutation matrix corresponding to ϕ. We get

$$\text{tr}(AB^{\phi}) = \sum_{i,j=1}^{n} a_{ij}b_{ji}^{\phi} = \sum_{i,j=1}^{n} a_{ij}b_{\phi(i)\phi(j)},$$

since $b_{ij}^{\phi} = b_{\phi(i)\phi(j)}$, for $i, j \in \{1, 2, ..., n\}$. Since $\text{tr}(CX^t) = \sum_{i=1}^{n} c_{i\phi(i)}$, the QAP in (14) can be formulated as

$$\min \quad \text{tr}(AXB^T + C)X^T \tag{15}$$

$$\text{subject to } X \in \mathbf{X}_n.$$

4.3. *Linearizations*

A first attempt to solve the QAP would be to eliminate the quadratic term in the objective function in (13), in order to transform the problem into a (mixed) 0–1 linear program (MILP). The linearization of the objective function is usually achieved by introducing new variables and new linear (and binary) constraints. The very large number of new variables and constraints, however, poses an obstacle for efficiently solving the resulting linear integer programs. The optimal value of an LP relaxation of some MILP formulation is a lower bound for the QAP. In this context the tightness of the continuous relaxation of the linear integer program mentioned above is a desirable property. Several linearizations of the QAP have been proposed in the literature, for example, by Lawler (1963), Kaufmann and Broeckx (1978; this linearization has the smallest number of variables and constraints), Frieze and Yadegar (1983), and by Adams and Johnson (1994). The linearization of Adams and Johnson unifies most of the previous linearizations and is important for getting lower bounds. The authors prove that the QAP with an array of coefficients $C = (d_{ijkl})$ is equivalent to the following MILP:

$$\min \sum_{i,j=1}^{n}\sum_{k,l=1}^{n} d_{ijkl}y_{ijkl} \tag{16}$$

subject to $(x_{ik}) \in \mathbf{X}_n$,

$$\sum_{i=1}^{n} y_{ijkl} = x_{jl}, \quad j,k,l = 1,...,n,$$

$$\sum_{k=1}^{n} y_{ijkl} = x_{jl}, \quad i,j,l = 1,2,...,n,$$

$$y_{ijkl} = y_{jilk}, \quad i,j,k,l = 1,...,n.$$

$$y_{ijkl} \geqslant 0, \quad i,j,k,l = 1,2,...,n,$$

where each y_{ijkl} represents the product $x_{ik}x_{jl}$. Although, as noted by Adams and Johnson (1994), a significant smaller formulation in terms of both variables and constraints could be obtained, the structure of the continuous relaxation of the above formulation is favorable for solving it approximately by Lagrangian dual methods. The theoretical strength of the linearization (16) relies on the fact that the constraints of the continuous relaxations of previous linearizations can be expressed as linear combinations of the constraints of the continuous relaxation of (16). Moreover, many of the previously published lower-bounding techniques can be explained based on the Lagrangian dual of this relaxation. For more details on this topic, refer to the section on linear programming relaxations, below.

4.4. *Lower Bounds*

Lower-bounding techniques are used within implicit enumeration algorithms, such as branch and bound, to perform a limited search of the set of feasible solutions, until an optimal solution is found. Numerous bounding techniques have been developed for the QAP so far. The most successful bounding techniques for the QAP can be classified in three groups: Gilmore–Lawler-type lower bounds, bounds based on LP relaxation, and eigenvalue related bounds.

GILMORE–LAWLER-TYPE LOWER BOUNDS

The basic idea of this type of lower bounds goes back to the early 1960s when Gilmore (1962) and Lawler (1963) developed the so-called Gilmore–Lawler bound (GLB) for the QAP. Nowadays, Gilmore–Lawler-type lower bounds, and especially the GLB, are the most frequently-used bounds within branch-and-bound algorithms for the QAP. The most advantageous property of these bounds is that they can be computed efficiently. The main drawback is the fast deterioration of their quality with increasing problem size.

To compute the GLB for a given QAP of size n, one has to solve $n^2 + 1$ LAPs, n^2 of them of size $n - 1$ and the last one of size n, as described below.

Consider an instance of the Lawler QAP (11) with coefficients $D = (d_{ijkl})$. For each ordered pair of indices (i, k), $1 \leqslant i, k \leqslant n$, solve an LAP with coefficient matrix $D^{(i,k)} = (d_{ijk\cdot})$ under the additional constraint that $\phi(i) = k$. Let us denote by l_{ik} the optimal solution of the above-mentioned LAP. Finally, solve an LAP with coefficient matrix $(l_{ij} + c_{ij})$; its optimal value is the GLB. Since the LAP can be solved efficiently, the GLB can also be computed efficiently.

Several Gilmore–Lawler-type lower bounds have been developed aiming at improving the quality of the GLB. One of the ideas on which such bounds are based relies on so-called reduction methods. These methods decompose each quadratic cost coefficient into several terms, some of these terms being linear coefficients. The latter are then moved to the linear term of the objective function. This would yield a bound that is tighter that the original GLB because the LAP can be solved exactly.

More recently another bounding procedure, that shares the basic idea of the GLB, was proposed by Hahn and Grant (1998). This procedure combines GLB ideas with reduction steps in a general framework. The resulting bound (HGB) shows a good trade-off between computation time and bound quality when tested in instances from QAPLIB.

BOUNDS BASED ON LINEAR PROGRAMMING RELAXATIONS

Consider a mixed-integer linear programming (MILP) formulation of the QAP. Clearly, the optimal solution of the continuous relaxation of the MILP formulation is a lower bound for the optimal value of the corresponding QAP. Moreover, each feasible solution of the dual of this relaxation is also a lower bound. Unfortunately, it is practically impossible to solve to optimality the LP relaxations of any MILP formulation, even for QAPs of moderate size. Generally, such formulations are highly degenerated and the memory requirements already become prohibitive for QAPs of moderate size, as shown by Resende et al. (1995). The LP relaxations of the MILPs are therefore approximately solved by applying subgradient optimization techniques or Lagrangian relaxations. Adams and Johnson (1994) have shown that a particular Lagrangian relaxation of the MILP (16) can be solved efficiently for each fixed set of Lagrangian multipliers. The Lagrangian multipliers are then updated iteratively in the fashion of a dual ascent procedure. The strength of the method of Adams and Johnson relies on the fact that it can produce all Gilmore–Lawler-like bounds described in the previous section on Gilmore–Lawler-type Lower Bounds and others, except the HGB and the bound CMB of Carraresi and Malucelli (1994), for different settings of the Lagrangian multipliers.

Recently, Karisch et al. (1999) have shown that although HGB cannot be obtained by applying the algorithm of Adams and Johnson, both AJB, CMB, and HGB can be obtained as feasible solutions of the dual of the continuous relaxation of the MILP formulation (16). Karisch et al. propose an iterative algorithm to approximately solve this dual, and show that AJB, HGB, and all other Gilmore–Lawler-like bounds can be obtained by applying this algorithm with specific settings for the control parameters.

EIGENVALUE-BASED LOWER BOUNDS

These bounds were introduced by Finke et al. (1987) and can be applied to the Koopmans–Beckmann QAP in (10). They are based on the relationship between the objective function of the QAP in the trace formulation (15) and the eigenvalues of its coefficient matrices. When designed and implemented carefully, these techniques produce bounds of good quality in comparison with Gilmore–Lawler-like bounds or, more generally, with bounds based on linear relaxations. However, these bounds are quite expensive in terms of computation time requirements and are, therefore, not appropriate for use within branch-and-bound algorithms.

4.5. *Exact Solution Methods*

Since QAP is a hard problem from the theoretical (and also the practical) point of view, no efficient algorithms are known for this problem. All existing exact algorithms are in principle enumeration methods like branch and bound, cutting-plane algorithms, and branch-and-cut algorithms.

BRANCH-AND-BOUND ALGORITHMS

Nowadays branch-and-bound algorithms appear to be the most efficient exact algorithms for solving the QAP.

The most efficient branch-and-bound algorithms for the QAP employ the Gilmore–Lawler bound (GLB). The reason is that other bounds that outperform GLB in terms of bound quality are simply too expensive in terms of computation time. More recently some efforts have been made to employ other Gilmore–Lawler-like bounds. The bound of HGB has been used in a branch-and-bound algorithm by Hahn et al. (1998) and the results are promising.

Three types of branching strategies are generally used for the QAP: single assignment branching, pair assignment branching, and branching based on relative positioning. The most efficient strategy is single assignment branching. It assigns a facility to a location in each branching step, that is, each problem is divided into subproblems by fixing the location of one of the facilities that is not assigned yet. Several rules for the choice of the facility location pairs, that determine the subproblems of a new level of the search tree, have been proposed by different authors. The appropriate rule usually depends on the bounding technique. If the GLB is employed, the branching rule is frequently formulated in terms of the reduced costs of the last assignment problem solved to bound the subproblem, that is currently being branched.

As for the selection rule there seems to be no clear winner among different strategies tested for the QAP, ranging from problem independent depth or breadth-first search, to instance-dependent criteria related to the maximization of lower bounds or reduced costs.

More recently a number of parallel branch-and-bound algorithms have been developed for the QAP, for example, by Pardalos and Crouse (1989), Bruengger et al. (1997), and Clausen and Perregaard (1997).

CUTTING-PLANE METHODS

Cutting-plane methods in general can be classified into traditional cutting-plane methods and polyhedral cutting-plane methods. Traditional cutting-plane algorithms for the QAP have been developed by different authors, for example, Bazaraa and Sherali (1982), Balas and Mazzola (1984a,b), and Kaufmann and Broeckx (1978). These algorithms make use of MILP formulations for the QAP, that are suitable for Benders' decomposition. Generally, the time needed for these methods to converge is too large, hence these methods may solve to optimality only very small QAPs. However, heuristics derived from cutting-plane approaches produce good suboptimal solutions in early stages of the search.

Also polyhedral cutting planes or branch-and-cut algorithms make use of MILP formulations of the QAP. Additionally, polyhedral cutting-plane methods make use of a class of (nontrivial) valid or facet-defining inequalities known to be fulfilled by all feasible solutions of the original problem. Some properties and a few facet-defining inequalities of the QAP polytope are already known, but polyhedral cutting-plane methods for the QAP are still not backed by a strong theory. Some efforts to design branch-and-cut algorithms for the QAP have been made by Padberg and Rijal (1996) and Kaibel (1997). The numerical results are encouraging, although the developed software is of a preliminary nature, as claimed by the authors.

4.6. Heuristics

The QAP is a very difficult problem from the practical point of view, instances of dimension $n > 20$ still not being practical to solve because of very high computation time requirements. This is probably the reason why the literature abounds in heuristics, that are the only available algorithms to provide good quality solutions for the QAP in a reasonable computational time. The reader is referred to Burkard et al. (1998) for numerous literature pointers to numerous heuristics applied to the QAP.

The heuristic approaches developed for the QAP can be classified as construction methods, limited enumeration methods, local search algorithms including improvement methods, tabu search and simulated annealing algorithms, genetic algorithms, greedy randomized adaptive search procedures (GRASP), and ant systems. In the case of the QAP there is no widely-accepted winner among these strategies. The construction methods, limited enumeration methods, and improvement methods seem, however, to be outperformed by the other heuristic approaches.

We refer the reader to other chapters of this handbook for a detailed description of the general metaheuristic approaches mentioned above. In the following we briefly mention the most frequently-used neighborhood structures included in local search algorithms for the QAP.

Frequently-used neighborhoods for the QAP are the pair-exchange neighborhood and the cyclic triple-exchange neighborhood. In the case of pair exchanges, the neighborhood of a given solution (permutation) consists of all permutations that can be obtained from the given one by applying a transposition to it. The size of this neighborhood is

$$O\left(\binom{n}{2}\right).$$

In the case of cyclic triple exchanges, the neighborhood of a solution (permutation) π consists of all permutations obtained from π by a cyclic exchange of some triple of indices. The size of this neighborhood is

$$O\left(\binom{n}{3}\right).$$

In general, cyclic triple exchanges do not lead to better results when compared with pair exchanges.

4.7. Available Computer Codes for the QAP

Burkard et al. (1997) have compiled a library of QAP instances (QAPLIB), that is widely used to test bounds, exact algorithms, and heuristics for the QAP. Many of these instances have not been solved to optimality yet, the most celebrated among them being the instances of Nugent et al. (1969) of size larger than 25. QAPLIB can be found at `http://www.opt.math.tu-graz.ac.at/~karisch/qaplib`.

Two codes for computing lower bounds are also available from the QAPLIB Web page: a FORTRAN code due to Burkard and Derigs (1980) to compute the GLB for instances of size up to 256, and another FORTRAN code to compute the elimination bound (ELI) for symmetric QAP instances of size up to 256. Finally, a FORTRAN code of the branch-and-bound algorithm developed by Burkard and Derigs (1980) can also be downloaded from the QAPLIB Web page.

Recently, Espersen et al. (1999) have developed QAPpack which is a JAVA package containing a branch-and-bound algorithm to solve the QAP. In QAPpack a number of bounds based on linearization are implemented: the Gilmore–Lawler bound, the bound of Carraresi and Malucelli (1994), the bound of Adams and Johnson (1994), the bound of Hahn and Grant (1998), and the bound of Karisch et al. (1999). The implementation is based on the dual framework provided by Karisch et al. (1999). QAPpack can be found at `http://www.imm.dtu.dk/~te/QAPpack`.

There are also some codes of heuristics available. The (compressed) FORTRAN source file—608.Z—of a heuristic due to West (1983) can be downloaded at `ftp://netlib.att.com` in `/netlib/toms`.

The source files (compressed tar files) of two FORTRAN implementations of GRASP for dense QAPs by Resende et al. (1996) and sparse QAPs by Pardalos et al. (1997) can be downloaded from Resende's Web page at `http://www.research.att.com/~mgcr/src/index.html`.

The source file of a FORTRAN implementation of the simulated annealing algorithm of Burkard and Rendl (1984) can be downloaded from the QAPLIB Web page.

The source file of a C++ implementation of the simulated annealing algorithm of Connolly (1990), due to Taillard, can be downloaded from Taillard's Web page at `http://www.idsia.ch/~eric/codes.dir/sa_qap.c`. A source file of a Pascal implementation of the robust tabu search algorithm by Taillard (1991) can also be found at Taillard's Web page.

4.8. Asymptotic Behavior

While being a very difficult problem both from the theoretical and the practical point of view, the QAP shows an interesting asymptotic behavior, which suggests that, under certain probabilistic conditions on the problem data, QAPs that are large enough are trivial to solve. Namely, it can be shown that the ratio between the best and worst values of the objective function approaches 1, as the size of the QAP approaches infinity. Thus the relative error of every heuristic method vanishes as the size of the problem tends to infinity, that is, every heuristic almost always finds an almost optimal solution when applied to QAP instances that are large enough.

A number of authors have investigated the asymptotic behavior of the QAP. Burkard and Fincke (1983, 1985) and Frenk et al. (1985) have shown the convergence of the above mentioned ratio to 1 in probability. Later Szpankowski (1995) improved the convergence almost to certainty. Summarizing we get the following theorem.

Theorem. *Consider a sequence* $P_n, n \in \mathbb{N}$, *of QAPs, where* P_n *has size* n, *and coefficient matrices* $A^{(n)} = \left(a_{ij}^{(n)}\right)$ *and* $B = \left(b_{ij}^{(n)}\right)$. *Assume that* $a_{ij}^{(n)}$ *and* $b_{ij}^{(n)}, n \in \mathbb{N}, 1 \leqslant i, j \leqslant n$, *are independently distributed random variables on* $[0, M]$, *where* M *is a positive constant.*

Moreover, assume that entries $a_{ij}^{(n)}$, $n \in \mathbb{N}$, $1 \leqslant i, j \leqslant n$, *have a common distribution, and entries* $b_{ij}^{(n)}$, $n \in \mathbb{N}$, $1 \leqslant i, j \leqslant n$, *also have a common distribution (which does not necessarily coincide with that of* $a_{ij}^{(n)}$). *Let* $\pi_{opt}^{(n)}$ *and* $\pi_{wor}^{(n)}$ *denote an optimal and a worst solution of* P_n *with objective function values* $z(\pi_{opt}^{(n)})$ *and* $z(\pi_{wor}^{(n)})$, *respectively. Then the following equality almost currently holds:* $\lim_{n\to\infty} z(\pi_{opt}^{(n)})/z(\pi_{wor}^{(n)}) = 1$.

The asymptotic behavior of the QAP has been exploited by Dyer et al. (1968) to analyze the performance of branch-and-bound algorithms for QAPs with coefficients generated randomly as described above. Dyer et al. have shown that any branch-and-bound algorithm that uses single assignment branching and employs a bound obtained by solving the continuous relaxation of the linearization of Frieze and Yadegar (1983) would branch on at least $n^{(1-o(1))n/4}$ nodes with probability tending to 1 as the size n of the problem tends to infinity.

4.9. *The Biquadratic Assignment Problem*

A generalization of the QAP arises if we consider objective functions of a higher degree and obtain in this way cubic, biquadratic, and generally N-adic assignment problems as introduced by Lawler (1963). In this way we also get among others the biquadratic assignment problem, denoted by BiQAP and stated as follows:

$$\min \sum_{i,j=1}^{n} \sum_{k,l=1}^{n} \sum_{m,p=1}^{n} \sum_{s,t=1}^{n} a_{ijkl} b_{mpst} x_{im} x_{jp} x_{ks} x_{lt}$$

$$\text{subject to} \quad X = (x_{ij}) \in \mathbf{X}_n,$$

where $A = (a_{ijkl})$ and $B = (b_{mpst})$ are two $n^4 \times n^4$ arrays.

An application of the BiQAP arises in very large scale integrated (VLSI) circuit design. A detailed description of the mathematical modeling of the VLSI problem as a BiQAP is given by Burkard et al. (1994). Similarly to the QAP, the BiQAP is also a hard problem and cannot be solved efficiently. Gilmore–Lawler-like lower bounds, branch-and-bound algorithms, and some local search heuristics, as well as a GRASP implementation for the BiQAP, can be found in the literature. See Burkard and Çela (1997) or Çela (1998) for more information and reference pointers.

4.10. *The Bottleneck QAP*

Another problem related to the QAP is the bottleneck quadratic assignment problem (BQAP), obtained by substituting the sum by a max operation in the objective function of the QAP:

$$\min_{\phi \in \mathcal{S}_n} \max\{a_{ij} b_{\phi(i)\phi(j)} : 1 \leqslant i, j \leqslant n\}.$$

The first occurrence of the BQAP in the literature is due to Steinberg (1961) and arises as an application in backboard wiring while trying to minimize the maximum length of the involved wires. A well-studied problem in graph theory that can be modeled as a BQAP is the bandwidth problem. In the bandwidth problem we are given an undirected graph $G = (V, E)$ with vertex set V and edge set E, and seek a labeling of the vertices of G by the numbers $1, 2, ..., n$, where $|V| = n$, such that the maximum distance of 1-entries of the resulting adjacency matrix from the diagonal is minimized, that is, the bandwidth of the adjacency matrix is minimized.

Analogously to the QAP, the BQAP is also a hard problem that cannot be solved efficiently. Some enumeration algorithms to solve BQAP to optimality have been proposed by Burkard (1974). Those algorithms employ a Gilmore–Lawler-like bound for the BQAP.

Burkard and Fincke (1982) investigated the asymptotic behavior of the BQAP and proved results analogous to those obtained for the QAP: Under certain probabilistic constraints on the problem data, the relative difference between the worst and the best value of the objective function approaches 0 with probability tending to 0 as the size of the problem approaches infinity.

ACKNOWLEDGMENT

Research for this article was supported by the Spezialforschungsbereich F003 "Optimierung und Kontrolle," Projektberich Diskrete Optimierung.

References

Adams, W. P. and T. A. Johnson. 1994. "Improved linear programming-based lower bounds for the quadratic assignment problem." In *Quadratic Assignment and Related Problems*, Volume 6, edited by P. M. Pardalos and H. Wolkowicz. DIMACS Series on Discrete Mathematics and Theoretical Computer Science, pp. 43–75. Providence: AMS.

Ahuja, R. K., T. L. Magnanti, J. B. Orlin, and M. R. Reddy. 1995. "Applications of network optimization." In *Network Models—Handbooks of Operations Research and Management Science*, edited by M. O. Ball, T. L. Magnanti, C. L. Monma, and G. L. Nemhauser **7**: 1–83. Amsterdam: Elsevier.

Alt, H., N. Blum, K. Mehlhorn, and M. Paul. 1991. "Computing maximum cardinality matching in time $o(n^{1.5}\sqrt{m/\log n})$." *Inf. Processing Lett.* **37**: 237–240.

Balas, E. and J. B. Mazzola. 1984a. "Nonlinear programming: I Linearization techniques." *Math. Programming* **30**: 1–21.

Balas, E. and J. B. Mazzola. 1984b. "Nonlinear programming: II Dominance relations and algorithms." *Math. Programming* **30**: 22–45.

Balas, E. and M. J. Saltzman. 1991. "An algorithm for the three-index assignment problem." *Oper. Res.* **39**: 150–161.

Bazaraa, M. S. and H. D. Sherali. 1982. "On the use of exact and heuristic cutting plane methods for the quadratic assignment problem." *J. Oper. Res. Soc.* **33**: 991–1003.

Bertsekas, D. P., D. A. Castañon, J. Eckstein, and S. Zenios. 1995. "Parallel computing in network optimization." In *Network Models—Handbooks in Operations Research and Management Science 7*, edited by M. O. Ball, T. L. Magnanti, C. L. Monma, and G. L. Nemhauser, pp. 330–399. Amsterdam: Elsevier.

Birkhoff, G. 1946. "Tres observaciones sobre el algebra lineal." *Rev. Univ. Nac. Tucumán (A)* **5**: 147–151.

Bruengger, A., J. Clausen, A. Marzetta, and M. Perregaard. 1997. "Joining forces in solving large-scale quadratic assignment problems in parallel." In *Proceedings of the 11th IEEE International Parallel Processing Symposium (IPPS)*, pp. 418–427.

Burkard, R. E. 1974. "Quadratische bottleneckprobleme." *Oper. Res. Verfahren* **18**: 26–41.

Burkard, R. E. and E. Çela. 1997. "Quadratic and three-dimensional assignments." In *Annotated Bibliographies in Combinatorial Optimization*, edited by M. Dell'Amico, F. Maffioli, and S. Martello, pp. 373–392. Chichester, England: Wiley.

Burkard, R. E. and E. Çela. 1999. "Linear assignment problems and extensions." In *Handbook of Combinatorial Optimization 4*, edited by P. Pardalos and D.-Z. Du: Kluwer Academic Publishers.

Burkard, R. E. and U. Derigs. 1980. In *Assignment and Matching Problems: Solution Methods with FORTRAN Programs.* Berlin: Springer.

Burkard, R. E. and U. Fincke. 1982. "On random quadratic bottleneck assignment problems." *Math. Programming* **23**: 227–232.

Burkard, R. E. and U. Fincke. 1983. "The asymptotic probabilistic behavior of the quadratic sum assignment problem." *Z. Oper. Res.* **27**: 73–81.

Burkard, R. E. and U. Fincke. 1985. "Probabilistic asymptotic properties of some combinatorial optimization problems." *Discrete Appl. Math.* **12**: 21–29.

Burkard, R. E. and F. Rendl. 1984. "A thermodynamically motivated simulation procedure for combinatorial optimization problems." *Eur. J. Oper. Res.* **17**: 169–174.

Burkard, R. E. and F. Rendl. 1991. "Lexicographic bottleneck problems." *Oper. Res. Lett.* **10**: 303–308.

Burkard, R. E. and U. Zimmermann. 1982. "Combinatorial optimization in linearly ordered semimodules: a survey." In *Modern Applied Mathematics*, edited by B. Korte, pp. 392–436. Amsterdam: North Holland.

Burkard, R. E., W. Hahn, and U. Zimmermann. 1977. "An algebraic approach to assignment problems." *Math. Programming* **12**: 318–327.

Burkard, R. E., E. Çela, and B. Klinz. 1994. "On the biquadratic assignment problem." In *Quadratic Assignment and Related Problems*, Volume 16, edited by P. M. Pardalos and H. Wolkowicz. DIMACS Series on Discrete Mathematics and Theoretical Computer Science, pp. 117–147. Providence: AMS.

Burkard, R. E., S. E. Karisch, and F. Rendl. 1997. "QAPLIB—a quadratic assignment problem library." *J. Global Optim.* **10**: 391–403 An on-line version is available via World Wide Web at the following URL: http://www.opt.math.tu-graz.ac.at/~karisch/qaplib/.

Burkard, R. E., E. Çela, P. Pardalos, and L. Pitsoulis. 1998. "The quadratic assignment problem." In *Handbook of Combinatorial Optimization 3*, edited by P. Pardalos and D.-Z. Du, pp. 241–338: Kluwer Academic Publishers.

Carpaneto, G., S. Martello, and P. Toth. 1988. "Algorithms and codes for the assignment problem." *Ann. Oper. Res.* **13**: 193–223.

Carpaneto, G. and P. Toth. 1980. "Solution of the assignment problem." *ACM Trans. Math. Softw.* **6**: 104–111.

Carraresi, P. and F. Malucelli. 1994. "A reformulation scheme and new lower bounds for the qap." In *Quadratic Assignment and Related Problems*, Volume 16, edited by P. Pardalos and H. Wolkowicz. DIMACS Series in Discrete Mathematics and Theoretical Computer Science, pp. 147–160. Providence: AMS.

Çela, E. 1998. In *The Quadratic Assignment Problem: Theory and Algorithms*. Dordrecht, The Netherlands: Kluwer Academic Publishers.

Clausen, J. and M. Perregaard. 1997. "Solving large quadratic assignment problems in parallel." *Comput. Optim. Appl.* **8**: 111–127.

Connolly, D. T. 1990. "An improved annealing scheme for the QAP." *Eur. J. Oper. Res.* **46**: 93–100.

Cunningham, W. H. 1976. "A network simplex method." *Math. Programming* **11**: 105–116.

Dell'Amico, M. and S. Martello. 1997. "Linear assignment." In *Annotated Bibliographies in Combinatorial Optimization*, edited by M. Dell'Amico, F. Maffioli, and S. Martello, pp. 355–372. Chichester, England: Wiley.

Dell'Amico, M. and P. Toth. 1999. "Algorithms and codes for dense assignment problems." *Discrete Appl. Math.* **100**: 17–48.

Derigs, U. 1985. "The shortest augmenting path for solving assignment problems—motivation and computational experience." In *Algorithms and Software for Optimization—Part I*, Volume 4, edited by C. L. Monma. Annals of Operations Research, pp. 57–102. Basel: Baltzer.

Dyer, M. E., A. M. Frieze, and C. J. H. McDiarmid. 1986. "On linear programs with random costs." *Math. Programming* **35**: 3–16.

Espersen, T., S. E. Karisch, E. Çela, and J. Clausen. 1999. "QAPPACK—a java package for solving quadratic assignment problems." Working Paper. Denmark: Department of Mathematical Modelling, Technical University of Denmark, and Austria: Institute of Mathematics, Technical University Graz.

Euler, R. and H. Le Verge. 1996. "Time-tables, polyhedra and the greedy algorithm." *Discrete Appl. Math.* **65**: 207–221.

Even, S. and R. E. Tarjan. 1975. "Network flow and testing graph connectivity." *SIAM J. Comput.* **4**: 507–518.

Finke, G., R. E. Burkard, and F. Rendl. 1987. "Quadratic assignment problems." *Ann. Discrete Math.* **31**: 61–82.

Frenk, J. B. G., M. van Houweninge, and A. H. G. R. Kan. 1985. "Asymptotic properties of the quadratic assignment problem." *Math. Oper. Res.* **10**: 100–116.

Frenk, J. B. G., M. van Houweninge, and A. H. G. Rinnooy Kan. 1987. "Order statistics and the linear assignment problem." *Computing* **39**: 165–174.

Frieze, A. M. and J. Yadegar. 1983. "On the quadratic assignment problem." *Discrete Appl. Math.* **5**: 89–98.

Fulkerson, R., I. Glicksberg, and O. Gross. 1953. "A production line assignment problem." Technical Report RM-1102. Santa Monica, CA: The Rand Corporation.

Gilmore, P. C. 1962. "Optimal and suboptimal algorithms for the quadratic assignment problem." *SIAM J. Appl. Math.* **10**: 305–313.

Goldberg, A. V. and R. Kennedy. 1995. "An efficient cost scaling algorithm for the assignment problem." *Math. Programming* **75**: 153–177.

Hahn, P. and T. Grant. 1998. "Lower bounds for the quadratic assignment problem based upon a dual formulation." *Oper. Res.* **46**: 912–922.

Hahn, P., T. Grant, and N. Hall. 1998. "Solution of the quadratic assignment problem using the hungarian method." *Eur. J. Oper. Res.* **108**: 629–640.

Hopcroft, J. E. and R. M. Karp. 1973. "An $n^{5/2}$ algorithm for maximum matchings in bipartite graphs." *SIAM J. Comput.* **2**: 225–231.

Johnson, D. S. and C. C. McGeoch. 1993. In *Network Flows and Matching—First DIMACS Implementation Challenge*, Volume 12. DIMACS Series in Discrete Mathematics and Theoretical Computer Science. Providence: AMS.

Jonker, R. and A. Volgenant. 1986. "Improving the hungarian assignment algorithm." *Oper. Res. Lett.* **5**: 171–175.

Kaibel, V. 1997. "Polyhedral combinatorics of the quadratic assignment problem." Ph.D. diss. Cologne: Universität zu Köln.

Karisch, S. E., E. Çela, J. Clausen, and T. Espersen. 1999. "A dual framework for lower bounds of the quadratic assignment problem based on linearization." *Computing.* **63**: 351–403.

Karp, R. M. 1980. "An algorithm to solve the $m \times n$ assignment problem in expected time $o(mn \log n)$." *Networks* **10**: 143–152.

Karp, R. M. 1987. "An upper bound on the expected cost of an optimal assignment." In *Discrete Algorithms and Complexity,* Volume 15, edited by D. Johnson, T. Nishizeki, A. Nozaki, and H. Wilf. Perspectives in Computing, pp. 1–4. Boston: Academic Press.

Karp, R. M., A. H. G. R. Kan, and R. V. Vohra. 1994. "Average case analysis of a heuristic for the assignment problem." *Math. Oper. Res.* **19**: 513–522.

Kaufmann, L. and F. Broeckx. 1978. "An algorithm for the quadratic assignment problem using benders' decomposition." *Eur. J. Oper. Res.* **2**: 204–211.

Laporte, G. and H. Mercure. 1988. "Balancing hydraulic turbine runners: a quadratic assignment problem." *Eur. J. Oper. Res.* **35**: 378–382.

Lawler, E. L. 1963. "The quadratic assignment problem." *Mgmnt Sci.* **9**: 586–599.

Lee, Y. and J. B. Orlin. 1994. "On very large scale assignment problems." In *Large Scale Optimization, State of the Art,* edited by W. W. Hager, D. W. Hearn, and P. M. Pardalos, pp. 206–244. Dordrecht, The Netherlands: Kluwer Academic Publishers.

Magos, D. 1996. "Tabu search for the planar three-index assignment problem." *J. Global Optim.* **8**: 35–48.

Magos, D. and P. Miliotis. 1994. "An algorithm for the planar three-index assignment problem." *Eur. J. Oper. Res.* **77**: 141–153.

Martello, S., W. R. Pulleyblank, P. Toth, and D. de Werra. 1984. "Balanced optimization problems." *Oper. Res. Lett.* **3**: 275–278.

Mosevich, J. 1986. "Balancing hydraulic turbine runners—a discrete combinatorial optimization problems." *Eur. J. Oper. Res.* **26**: 202–204.

Mulmuley, K., U. V. Vazirani, and V. V. Vazirani. 1987. "Matching is as easy as matrix inversion." *Combinatorica* **7**: 105–113.

Nugent, C. E., T. E. Vollmann, and J. Ruml. 1969. "An experimental comparison of techniques for the assignment of facilities to locations." *J. Oper. Res.* **16**: 150–173.

Olin, B. 1992. "Asymptotic properties of random assignment problems." Ph.D. diss. Stockholm: Division of Optimization and Systems Theory, Department of Mathematics, Royal Institute of Technology.

Padberg, M. W. and M. P. Rijal. 1996. In *Location, Scheduling, Design and Integer Programming.* Boston: Kluwer Academic Publishers.

Pardalos, P. and J. Crouse. 1989. "A parallel algorithm for the quadratic assignment problem." In *Proceedings of the Supercomputing Conference 1989,* pp. 351–360. ACM Press.

Pardalos, P. M., L. S. Pitsoulis, and M. G. C. Resende. 1997. "Fortran subroutines for approximate solution of sparse quadratic assignment problems using grasp." *ACM Trans. Math. Softw.* **23**: 196–208.

Pferschy, U. 1996. "The random linear bottleneck assignment problem." *RAIRO Oper. Res.* **30**: 127–142.

Pferschy, U. 1997. "Solution methods and computational investigations for the linear bottleneck assignment problem." *Computing* **59**: 237–258.

Pierskalla, W. P. 1967. "The tri-substitution method for the three-multidimensional assignment problem." *Canadian ORS J.* **5**: 71–81.

Punnen, A. P. and K. P. K. Nair. 1994. "Improved complexity bound for the maximum cardinality bottleneck bipartite matching problem." *Discrete Appl. Math.* **55**: 91–93.

Ramakrishnan, K. G., N. K. Karmarkar, and A. P. Kamath. 1993. "An approximate dual projective algorithm for solving assignment problems." In *Network Flows and Matching—First DIMACS Implementation Challenge,* Volume 12, edited by D. S. Johnson and C. S. McGeoch. DIMACS Series in Discrete Mathematics and Theoretical Computer Science, pp. 431–449. Providence: AMS.

Resende, M. G. C., P. M. Pardalos, and Y. Li. 1996. "Fortran subroutines for approximate solution of dense quadratic assignment problems using grasp." *ACM Trans. Math. Softw.* **22**: 104–118.

Resende, M. G. C., K. G. Ramakrishnan, and Z. Drezner. 1995. "Computing lower bounds for the quadratic assignment problem with an interior point algorithm for linear programming." *Oper. Res.* **43**: 781–791.

Steinberg, L. 1961. "The backboard wiring problem: a placement algorithm." *SIAM Rev.* **3**: 37–50.

Szpankowski, W. 1995. "Combinatorial optimization problems for which almost every algorithm is asymptotically optimal!" *Optimization* **33**: 359–367.

Taillard, E. 1991. "Robust tabu search for the quadratic assignment problem." *Parallel Comput.* **17**: 443–455.

Vlach, M. 1967. "Branch and bound method for the three-index assignment problem." *Ekonomicko-Matematicky Obzor* **12**: 181–191.

West, D. H. 1983. "Algorithm 608: approximate solution of the quadratic assignment problem." *ACM Trans. Math. Softw.* **9**: 461–466.

18 APPLICATION AREAS

18.1 Transportation and Logistics

Warren B. Powell

ABSTRACT Transportation and logistics involves the planning of the movements of product, equipment and people over space and time at the lowest cost, providing the highest possible service to customers and employees. Static versions of these problems have been studied for decades, creating a variety of specialized subproblems such as vehicle routing, crew scheduling, network design, and facility location. This article focuses on the emerging field of dynamic problems in transportation and logistics. With this emphasis, the article reviews the major problem classes, and highlights important issues that need to be addressed. Modeling frameworks are given for both static and dynamic information processes (often referred to as deterministic and stochastic models). The article then closes with discussions of specific problem classes of interest.

1. Introduction

The modeling of problems in transportation and logistics has grown into a rich and exciting area, where new modeling approaches and powerful new algorithms are combining with modern communication technologies to create powerful new information tools. There is a rapidly growing demand for systems that help with strategic, tactical, and operational (real-time, or near-real-time) planning. The purpose of this article is to provide an overview of the types of problems that are being addressed, and an introduction to how these can be modeled and solved.

Our discussion focuses on problems that involve the movement of goods. Although there is considerable overlap between modeling issues for passengers and freight, we confine our discussion to problems that arise in freight transportation. We also focus most of our attention on problems where the flow of information is an important dimension. There is a vast array of models and algorithms for problems where all the information is assumed known (commonly referred to as deterministic problems). We do not ignore this important literature, since this work is relevant even in a stochastic setting, but our presentation is biased toward models and solution methods in which information is assumed to evolve over time.

Problems in the movement of goods tend to be divided into two broad classes. The first, popularly but imprecisely referred to as supply chain management covers the manufacture of new products and their delivery to the customer. These problems generally involve the coordination of activities among different suppliers, manufactures, distribution channels, and retail outlets. The second is the movement of goods from the perspective of a transportation company. The company may be dedicated to a single shipper or it might be a common carrier, serving the needs of many shippers. Models that address the shipper perspective usually focus on the assembly and distribution of product, with relatively simple models of the transportation process (typically represented as little more than a cost function). By contrast, models of transportation operations focus on the flows of drivers, tractors and trailers (trucking), locomotives, crews and boxcars (rail), pilots, aircraft, and freight (air) or ships and containers (ocean).

Each industry tends to create unique modeling issues. For example, the relatively high expense of aircraft and ships means that the management of these resources dominates the planning of operations in air and shipping. By contrast, drivers, tractors, and trailers are all important resources in trucking, and need to be modeled simultaneously. Despite this, we are not going to attempt in our brief presentation to describe specific models for individual industries. Rather, we are going to develop a general framework that will provide an introduction to modeling frameworks and solution approaches.

There are several themes that emerge in the modeling of complex operations. These include:

1. *Managing the flows of different resource classes.* Many models focus on the management of a vehicle (truck, locomotive or aircraft) or a human operator (drivers, pilots, crews). In practice, it is often necessary to simultaneously manage the flows of different resource classes.
2. *Modeling the flow of information.* It is easiest to assume that we have all the information available to us, and then have to find the best set of decisions. In reality, information evolves over time, and some information never becomes known to the system.
3. *Handling multiagent control structures.* Classical models assume a single decisionmaker. Real problems are characterized by multiple decisionmakers, possibly distributed spatially (or with spatial areas of control); alternatively, decisionmakers may control different parts of the problem (for example, leasing new trailers versus moving them) or may work for different companies or organizations. Coordinating different agents is at the heart of supply chain management.
4. *Designing robust systems.* Often, it is too late to take certain actions once the information becomes known. As a result, we have to consider different outcomes, and plan accordingly.

Ultimately, our ability to solve realistic models must reflect the data that we are able to collect and store, and the tools that are available to solve these models. Clearly, our field has benefited from dramatic improvements in data collection, transmission, and storage technologies, and from faster computers to manipulate the data. In parallel, researchers in the field of operations research have contributed their own dramatic advances, without which even today's powerful computers would make little headway in the complex problems arising in large-scale operations. Specific algorithmic advances that have directly affected our ability to solve realistic problems include:

1. *Large-scale linear programming.* One of our most important foundational tools, algorithms for linear programs have advanced dramatically during the 1990s, partly reflecting breakthroughs such as interior point algorithms, but mostly as a result of extensive engineering of our old standby, the simplex algorithm. The dramatic improvement in speed of these tools, combined with the generality of linear programming, has to a large extent replaced the specialized network algorithms that were very popular in the 1980s.
2. *Set partitioning.* Advances in our ability to solve integer set partitioning problems has produced a powerful new tool for large-scale practical problems such as airline crew scheduling. These tools work well when the problem of finding optimal schedules for a single crew are not too hard.
3. *Tabu search.* Tabu search has emerged as an extremely flexible tool for intractably difficult combinatoric problems. It appears to pick up where set partitioning begins to fail (large, but relatively loosely constrained problems). The best vehicle routing algorithms today are variations on tabu search.
4. *Dynamic programming.* Long relegated to toy problems under the burden of the curse of dimensionality, dynamic programming is making a resurgence, combining the simulation characteristics of forward dynamic programming with structural results that capture the properties of specific problems.

The remainder of this article is structured as follows. We start by highlighting the major elements of dynamic resource management problems that arise in transportation and logistics in Section 2. Then, Section 3 Offers a general modeling framework that formalizes the major elements of our problem class. Finally, Section 4 discusses specific problems that are of general interest, and the algorithms that are bested suited to solving them.

2. Elements of Dynamic Resource Management Problems

We suggest three broad dimensions for identifying major problems classes: (1) the types of resources being managed, (2) the nature of the information processes, and (3) the control structure.

2.1. Resources

There are three important dimensions of resources.

1. *The heterogeneity of the resources.* How many different types of resources are there? Are there one or two attributes of a resource, or more?
2. *Resource layering.* Resources can be combined to form composite resources (a driver in a truck, a truck pulling a load). What are the rules that govern the coupling resources? Can different resources of one type (e.g., a driver) be coupled with a given resource of another type (such as a load)? Is the coupling one to one (one driver, one tractor), several to one (several locomotives pulling one train), or many to one (many packages in one trailer)?
3. *How discrete is the problem?* Is there typically one resource of a particular type (a particular pilot or driver), or are there many (pools of trailers or containers)? When there is typically one resource of a given type, the problem is usually modeled as discrete and approached using specialized algorithms for integer programs. When there are many resources of a given type, then the problem is often accurately modeled as continuous.

2.2. Information Processes

Information processes cover the arrival of information to the system. There are three classes of information processes:

1. *Static.* There are no changes to the initial set of information. Static information processes give rise to deterministic problems.
2. *Dynamic with immediate action* Information about activities in the future can be acted on the instant the information arrives. Simulation models often make this assumption.
3. *Dynamic with delayed action.* Information arrives about an activity before the activity takes place. For example, a customer may call in an order a day in advance; we know about the information, but cannot act on it right away. Such information is useful only if the model plans into the future.

2.3. Control Structure

Problems may be organized based on:

1. *Single agent.* The problem is modeled as if there is a single decisionmaker controlling the system.
2. *Multiagent.* There is more than one decisionmaker. These problems can be further divided into:
   - *Single organization.* All agents are within a single organization.
   - *Multiple organization.* Decision agents may exist in different organizations, but with a need to coordinate.

3. A Modeling Framework

Once we have settled on a problem, we usually want to express it mathematically and then search for tractable algorithms. We suggest two modeling paradigms: one for deterministic problems (static information processes) and the second for stochastic problems (dynamic information processes).

The diversity of problems listed above can actually be captured within a single modeling framework developed by Powell and Shapiro (1999). The problem class is called a dynamic resource transformation problem (DRiP), and consists of three primary dimensions: Information || Processes || Controls. In a typical deterministic problem, this representation offers dimensions that are not really of interest. Just the same, we prefer to adopt a consistent notational system, and feel that the structure of a DRiP helps even with the expression of deterministic problems.

3.1. Static Information Processes

When the information process is static, we have a classical deterministic optimization problem. In this setting, it is more natural to replace the first dimension of our problem, information with resources since information about parameters other than resources is static. We may express our problem as follows:

Resources

C^R = Set of resource classes, each representing a different type of commodity. For example, we might have C^R = {drivers, tractors, trailers}.

$\mathscr{R}^{(c)}$ = Set of resources in class c.

$\mathscr{A}^{(c)}$ = State space (set of potential attributes) of a resource in class $c \in \mathscr{C}^R$.

r_{at} = Number of resources with attribute vector a.

$r_t = \{..., r_{at}, ...\}$ = Vector of resources available at time t.

r^0_{at} = Number of resources that first enter the system into state a at time t.

Processes

$\tau_{a,a'}$ = Time to convert a resource from state a to state a'. This is called the transfer time.

$\tau^{\max}$ = Largest transfer time.

$A_{t',t}$ = Matrix where element $A_{i,j,t',t} = 1$ if row i corresponds to attribute a', and j corresponds to the decision to move resources from attribute a' to attribute a [thus, j corresponds to the pair (a', a)] and where $\tau_{a',a} = t - t'$.

B_t = Matrix where element $B_{i,j,t} = 1$ if row i corresponds to attribute a, and j corresponds to the decision to move resources from attribute a to attribute a' [thus, j corresponds to the pair (a, a')].

$u^c_{a,a',t}$ = Restriction on the total flow of resources from state a to state a' in time t of resources of class c.

$u^c_t = \{..., u^c_{a,a',t}, ...,\}$

Controls

$x^r_{a,a',t}$ = Flow of resource r from state a to state a', where $a, a' \in \mathscr{A}$.

$x^r_t = \{...,, x^r_{a,a',t}, ...,\}$

$\bar{x}^c_t = \sum_{r \in \mathscr{R}^{(c)}} x^r_t$

$c^r_t(x^r_t)$ = The cost of the vector of flows x^r_t.

$C_t(x_t) = \sum_{c \in C^R} > \sum_{c \in \mathscr{R}^{(c)}} c^r_t(x^r_t).$ = Total costs in time t.

We now wish to solve:

$$\min_x \sum_{t \in \tau} C_t(x_t) \tag{1}$$

subject to:

$$\sum_{\tau=0}^{\tau^{\max}} A^c_{t-\tau,t}, \bar{x}^c_{t-\tau} - B^c_t \bar{x}^c_t = r^{(0,c)}_t \quad \forall r \in \mathscr{R}^{(c)}, c \in \mathscr{C}^R \tag{2}$$

$$\bar{x}_t^c \leqslant u_t^c \quad \forall c \in \mathscr{C}^R \tag{3}$$

$$x_t^r \geqslant 0 \quad \forall r \in \mathscr{R}^{(c)}, c \in \mathscr{C}^R. \tag{4}$$

Many problems require the inclusion of the integrality constraint:

$$x_t^r \in (0, 1) \tag{5}$$

Finally, it is not uncommon to have discrete design variables that govern whether flow can move between two points. For this, we add the decision variable:

$z_{a,a',t}^c = 1$ if flow for resources in class c is allowed from a to a' at time t, and 0 otherwise.

With this variable, we replace constraint (3) with:

$$\bar{x}_{a,a',t}^c \leqslant z_{a,a',t,}^c u_{a,a',t}^c \quad \forall a, a' \in \mathscr{A}, c \in \mathscr{C}^R. \tag{3a}$$

Our formulation is quite general, although not completely so. However, it captures many problems arising in transportation and logistics. This way of presenting dynamic resource management problems is fairly new. It is more common to formulate what are called multicommodity flow problems, where we might let $x_{i,j}^k$ be the flow of commodity k from location i to location j. The problem with this notational style is that it does not reflect different commodity classes. Each pilot might be a commodity, and different types of aircraft might be commodities. Also, the multicommodity flow literature is oriented toward deterministic models, which do not use a state vector. Stochastic models require a state vector in order to model the sequencing of exogenous information and decisions.

3.2. *Dynamic Information Processes*

Our model for dynamic information processes must address several issues that are generally ignored in the context of deterministic problems. We start with the notation that we introduced above for deterministic problems. Some elements (such as the matrices $A_{t,t'}$ and B_t) will be ignored, and some will be changed (such as the decision variable $z_{a,a'}$).

INFORMATION

A resource is simply a form of information. Since information evolves over time, we have to represent all forms of information. We describe information using:

$\mathscr{C}^I$ = Set of information classes, where $\mathscr{C}^R \subseteq \mathscr{C}^I$ is a subset of information classes.

$\mathscr{E}_t^{(c)}$ = Set of information elements in class c at time t.

a_e = The vector of data attributes for element $e \in \mathscr{E}^{(c)}$, $c \in \mathscr{C}^I$, and where $a_e \in \mathscr{A}^{(c)}$. The vector a_e contains all the data that describes element e. One important element of a_e is the time at which all the elements of a_e becomes effective, meaning that it can be used to change the system.

$K_t = \{a_e \mid e \in \mathscr{E}_t^{(c)}, c \in C^I\}$ = The knowledge base of our system at time t.

An information class may be static (the data does not evolve over time), or dynamic. If it is dynamic, the data may change purely as a result of exogenous forces, but some classes may change due to endogenous controls. A resource is an endogenously controllable information class that constrains the system over time.

PROCESSES

There are two types of information that enter our system. The first is exogenous information, that we represent using:

K_{et} = Information arriving at time t updating the attributes of element $e \in \mathscr{E}^{(c)}$ at time t.

$\Omega_t^{(c)}$ = The set of potential information arriving for elements in class c over time periods $t \in \mathscr{T}_t^{ph}$, where $\mathscr{T}_t^{ph}$ are the time periods in the planning horizon at time t.

New information k_{et} is assumed to update the knowledge base of our system using the function:

$$K_{t+1} = U^K(K_t, k_t)$$

The second type of information is the decisions we make endogenously. Thus, a decision represents a form of endogenous information. Decisions are represented using:

$\mathscr{C}^D$ = Set of decision classes.

$\mathscr{D}^{(c)}$ = Decision within class $c \in \mathscr{D}^D$.

$x_{a,d,t}$ = The number of times we act on resources with attribute a with type d at time t.

In contrast with our deterministic model, which used the notation $x_{a,a',t}$, our decision notation $x_{a,d,t}$ does not explicitly contain the outcome of the decision (which may be uncertain). The impact of a decision d on our information set is captured through the modify function:

$$M_t(a, d, K_t) \rightarrow (a', c, \tau), \qquad (6)$$

where a is the attributes of the resource being acted on with decision d at time t with knowledge base K_t. The result is a modified resource with attribute a', generating a cost (or contribution) c, and τ is the time required to complete the decision (the transfer time).

It is useful to point out special types of decisions. There are only three basic types of decisions: couple, uncouple, and modify. Couple decisions put two or more resources together (a driver in a truck, packages in a trailer); uncouple decisions take two (or more) coupled resource classes and separate them; and modify decisions take a particular resource (that may be a coupled set of several resource classes) and modify its attributes. In transportation, the most important modify decision is move (change location), but many others are also important (refuel the truck, clean the trailer, repair the locomotive, put the driver to bed). Especially important is the hold decision, which is a way of expressing doing nothing. For a further discussion of these issues, see Powell and Shapiro (1999).

CONTROLS

Our control structure is captured using:

$\mathscr{Q}$ = Set of decision agents.

$x_q^\pi(I_q)$ = The decision function for agent q, where x^π, $\pi \in \Pi$ is the space of potential functions, and where: I_q = is the information set that may be used by the function x_q^π.

$\overrightarrow{\mathscr{M}}_q$ = The set of subproblems where decisions made in subproblem q may directly impact the resources available in subproblems in $\overrightarrow{\mathscr{M}}_q$. We call $\overrightarrow{\mathscr{M}}_q$ the forward reachable set.

$\overleftarrow{\mathscr{M}}_q$ = The set of subproblems where decisions impact subproblem q. We call $\overleftarrow{\mathscr{M}}_q$ the backward reachable set.

$C_q(x_q^\pi(I_q))$ = Total costs in subproblem q using decision function $x_q^\pi(I_q)$ and information set I_q.

A subproblem $q \in \mathscr{Q}$ consists of resources in a subset of the attribute space, $\mathscr{A}_q$, acted on using a subset of decision classes $\mathscr{C}_q$, over a subset of time periods $\mathscr{T}_q$.

Decision functions can be divided into three broad classes, based on the nature of the information set I_q:

1. $I_q = (K_q)$, meaning that the information set contains only the information in the knowledge base.
2. $I_q = (K_q, \Omega_q)$, meaning that the information set contains both the knowledge base, and forecasts of future events. Ω_q is the set of outcomes that may arrive between now and when subproblem q needs to be implemented.

3. $I_q = (K_q, \Omega_q, V_{\vec{\mathcal{M}}_q})$, where $V_{\vec{\mathcal{M}}_q}$ is a class of functions that captures the impact of a decision made in subproblem q on subproblems in $\mathcal{M}_q$, which are the subproblems that are impacted by decisions made in q.

If $I_q = K_q$, then we have a classical myopic decision rule. If $I_q = (K_q, \Omega_q)$, then we have a rolling horizon procedure. Finally, using $I_q = (K_q, \Omega_q, V_{\vec{\mathcal{M}}_q})$ puts us in the category of dynamic programming or stochastic linear programming. In this setting, $V_{\vec{\mathcal{M}}_q}$ might represent a value function approximation (see Bertsekas and Tsitsiklis, 1996, or Sutton and Barto, 1998), or Bender's cuts from stochastic linear programming (for example, Higle and Sen, 1991; Van Slyke and Wets, 1969 and Birge, 1985).

Recall that x_{adt} is the number of times decision d is applied to resources with attribute a. The outcome of this decision may be uncertain. It is useful to write the conditional decision $x_{aa't}(\omega_q)$, where a' is the outcome of decision d given ω_q.

It is possible to write out our problem in a relatively compact way, taking advantage of the decomposition $\mathcal{Q}$. We start by writing out the local subproblem denoted LSP_q. The cost function is given by:

$$C_q(x_q, \omega_q) = \sum_{a \in \mathcal{A}_q} \sum_{a' \in \mathcal{A}_q} c_{aa'}(x_{aa'} \omega_q)$$

The optimization problem is then:

$$\min_{x_q(\omega_q)} C_q(x_q, \omega_q) \tag{7}$$

subject to:

$$\sum_{a' \in \mathcal{A}_q} x_{aa'}(\omega_q) - \sum_{a' \in \mathcal{A}_q} x_{a'a}(\omega_q) = r_a + \hat{r}_a(\omega_q) \quad \forall_a \in \mathcal{A}_q \tag{8}$$

$$x_{aa'}(\omega_q) \leqslant u_{aa'}(\omega_q) \tag{9}$$

$$x_{aa'}(\omega_q) \geqslant 0 \quad \forall a, a' \in \mathcal{A}_q, \tag{10}$$

As stated, the local subproblem ignores the impact of decisions in q on other subproblems. In a stochastic environment, this is a common strategy (keep in mind that LSP_q can incorporate future events). A general formulation that captures the impact of decisions in subproblem q on other subproblems is given by the global subproblem (GSP_q).

$$\tilde{V}_q(r_q) = E\left\{ \max_{x_q} C_q(x_q, \omega_q) + \sum_{q' \in \vec{\mathcal{M}}_q} \hat{V}_{qq'}(x_{qq'} \mid \omega_q) \right\}. \tag{11}$$

Subject to the local constraints (8)–(10) and equations that handle flow conservation between subproblems:

$$r_a = \sum_{q' \in \overleftarrow{\mathcal{M}}_q} x_{q'a} \quad a \in \mathcal{A}_q. \tag{12}$$

It has been traditionally assumed that stochastic problems are much harder than deterministic problems. Thus, it may come as a surprise that our stochastic formulation is actually relatively easy to solve (but only approximately), since the subproblems $x_q^{\pi}(I_q)$ are typically quite small. The challenge is designing the function $x_q^{\pi}(I_q)$. Myopic functions can be optimal in certain steady state problems [consider, for example, the (s, S) decision rule of inventory theory], but are generally very suboptimal. It is common to solve rolling horizon procedures where $I_q = (K_q, \Omega_q)$ and where $|\Omega_q| = 1$ (a deterministic forecast), updating the model as new information arrives. This is the easiest and most popular method of solving dynamic problems, but it needs to be emphasized that these methods can be suboptimal, possibly even by significant margins. Furthermore, if the planning horizon $\mathcal{T}_q^{ph}$ is large, then these subproblems can be quite hard to solve.

An emerging line of research is the construction of approximate value functions $V_{\bar{\mathscr{U}}_q}$. Linear approximations have been used for single commodity (Powell and Carvalho, 1998a; Carvalho and Powell, 2000) and multicommodity flow problems (Powell and Carvalho, 1997), which have been applied to a complex operational problem involving the optimization of trailers and containers for intermodal rail operations (Powell and Carvalho, 1998b). Powell et al. (1998) adapted the method to heterogeneous resources, producing the first algorithm for optimizing the massive linehaul scheduling problem faced by the national less-than-truckload motor carriers.

4. Special Classes

In this section, we discuss some of the important problem classes that arise frequently in transportation and logistics.

4.1. Dynamic Assignment: The Load Matching Problem

The dynamic assignment problem is the foundational problem for many problems in operations. One important application is the assignment of drivers to loads in the truckload trucking industry, where one driver pulls one load. Loads are called in randomly over the day, and the challenge faced by the company is to match the right driver to the right load.

Using our vocabulary, this problem involves two resources layers, such as drivers and loads. We refer to loads as a passive resource class, because a load can only be coupled to a driver, or it can do nothing. By contrast, a driver is an active resource class, because we can couple it to a load, or we can move the driver without a load (called "moving empty").

Let $\mathscr{R}_t^{(1)}$ be the set of drivers available at time t, and $\mathscr{R}_t^{(2)}$ be the set of loads available at time t. It is sometimes possible to aggregate drivers into buckets. Let a_r be the attributes of driver r. Following our earlier notation. we would let r_{at}^1 be the number of drivers with attribute a at time t. The decision classes would be $\mathscr{C}^D = \{\text{move\_load}, \text{move\_empty}\}$, and $\mathscr{D}^{(\text{move\_load})} = \{\mathscr{R}_t^{(2)}\}$, meaning that an instance of a decision for a driver to move load is a particular load to be moved.

The set $\mathscr{D}^{(\text{move\_empty})}$ would be the set of locations a driver can move empty to. We assume that the outcome of a move load decision is that the load is removed from the system, revenue is generated, and the attributes of the driver are updated to reflect what is required to move the load (this logic is buried in the modify function). Finally, let $\mathscr{D} = \mathscr{D}^{\{\text{move\_load}\}} \cup \mathscr{D}^{\{\text{move\_empty}\}}$ be the set of all decisions.

Assume that a subproblem q corresponds to the entire system, and that $\mathscr{T}_t^{ph} = t$. Further assume that $\hat{V}_t(r_t)$ approximates the value of having a vector r_t of resources (drivers and loads) at time t. Then we wish to solve, at time t:

$$\min \; c_t x_t + \hat{V}_{t+1}[r_{t+1}(x_t)] \tag{13}$$

subject to:

$$\sum_{d \in \mathscr{D}} x_{a,d,t} = r_{a,t}^1 \quad a \in \mathscr{A}^1 \tag{14}$$

$$\sum_{a \in \mathscr{A}^1} x_{a,d,t} \leqslant 1 \quad \forall d \in \mathscr{D}^{\text{move\_load}} \tag{15}$$

$$x_{a,d,t} \geqslant 0. \tag{16}$$

Equation (15) captures the coupling constraints between drivers and loads. This notation allows for very general substitution rules between drivers handling a particular load. Note that our vector r_t may include drivers who are due to arrive at some point in the future, as well as loads that have been called in, but cannot be picked up until some time in the future. If $\hat{V}_{t+1}(r_{t+1}) = 0$, then we have a myopic model.

If $\hat{V}_t(r_t)$ is linear (or zero) then the optimization problem (13)–(16) is an assignment problem that can be solved using a fast network algorithm.

4.2. *Dynamic Resource Management*

Section 3.2 provides a very general model for dynamic resource management. For this reason, we do not attempt to repeat the model here. Instead, we review important special cases and discuss how they would be modeled.

1. *Inventory distribution.* Resources are different types of products to be delivered to warehouses and retail outlets. The product may go from plant to intermediate storage and distribution centers before eventual delivery to retail outlets or end customers. There are two resource classes for this problem. One class is all the different types of products that are being delivered, and the second class is the customer orders themselves. If the product is computers, and if a customer order requires different components such as the CPU, monitor, keyboard and other accessories, then each type of product could be a different class (where we couple different classes together to satisfy the customer).
2. *Fleet management.* The management of trucks, containers, boxcars, and locomotives are very similar to the management of distribution problems, with one exception: fleets of containers represent a reusable resource; after satisfying a customer order, the resource remains in the system. The basic model, however, is almost identical to that for distribution of product (see Powell et al., 1997; Crainic et al., 1993; Dejax and Crainic, 1987).
3. *Crew scheduling.* A driver, pilot, operator, or crew is simply another instance of a reusable resource. The difference is that the attribute vector tends to be very complicated, which means the vector r_t tends to be full of zeroes and ones. As a result, the discrete nature of the problem is much more important. The crew scheduling problem tends to be very similar to the assignment problem, with the significant difference that we are assigning an active resource (the person) to a sequence of tasks, rather than just one. As a result, the subproblem q is no longer a simple network problem, but rather is usually a set partitioning problem (for a good reference on this topic, see Desrosiers et al., 1997).

All the problems in this class are fundamentally linear in nature. They can either be solved as network problems (yielding integer solutions automatically) or linear programs (typically yielding solutions that are nearly integer). If the information process is dynamic, then it is necessary to solve sequences of linear programs (either as rolling horizon procedures or using value function approximations). The crew scheduling problems may require the use of set partitioning problems, but these have yielded readily to integer programming techniques that, as a rule, only work well when the linear relaxation is tight.

4.3. *Dynamic Vehicle Routing*

Vehicle routing generally involves routing a fleet of vehicles through a set of customers to perform pickups or deliveries (good references include Bodin et al., 1983; Fisher, 1997, and Laporte, 1992). Our goal is to find the tours and involve the fewest miles and/or the fewest vehicles, without violating the capacity of the truck. Important problem variations include serving a customer within a specified time window, and producing tours that do not violate restrictions on the total time that a driver can drive.

The classical view of a vehicle routing problem is that it involves what we would call two resource classes: the vehicle, and the customer order. The vehicle is the only active resource, so the movement of vehicles is the decision variable. Each vehicle is typically represented as a different commodity.

In some problems, it is necessary to separate the physical vehicle from the driver. In the chemical distribution industry, there are five resource classes: drivers, tractors, trailers, product to be delivered, and the customer tanks. In the chemical industry, trailers may be very expensive and have to be modeled explicitly. In this case, it is popular to assume that each vehicle has a generic driver who starts the day with the ability to drive, say, ten hours (see, for example, Bell et al., 1983 and Bausch et al., 1995). However, this assumption is not always valid. Federal law (in the United States) limits not only the number of hours a driver can work in a day, but also the number of hours a driver can work over eight days. As a result of this second rule, it is not uncommon to have a set of drivers who all have a different limit on how much they can work on a given day.

Another twist offered by some industries is that the customer has an inventory of product. Delivering to the customer one day does not eliminate the order from the system; instead, it simply refills his inventory. As a result, the driver, tractor, trailer and customer tank are all examples of reusable resources.

Vehicle routing problems have long challenged the operations research community from an algorithmic perspective. They are highly discrete problems that are not well approximated by linear programming relaxations. In theory these could be solved using the column generation/set partitioning techniques that have been so successful with crew scheduling, but the subproblem that involves finding an optimal tour for a single vehicle is very difficult if there are wide time windows on the tasks (pickups or deliveries). For this reason, researchers have focused on heuristics. Among these, tabu search has emerged as the leading contender (Gendreau et al., 1994). Simply stated, tabu search will make a local change, and allow the change to make the solution temporarily worse (for a good review by its inventory, see Glover, 1989, 1990). Parameters control the ability of the algorithm to explore solutions that are worse in order to eventually find a better solution. As with any local search procedure, the algorithm is extremely flexible and easily handles complex operational rules. In contrast with column generation/set partitioning techniques that work best on highly constrained problems, tabu search excels on relatively unconstrained problems.

There has been a flurry of recent research looking into the use of tabu search in the context of dynamic vehicle routing problems. In this setting, it is assumed that orders to serve customers arrive randomly. At any one point in time, then, there is a set of vehicles and a partial set of orders. The standard assumption is to then try to find the best tour given this set of orders (equivalent to using $I_t = K_t$). The challenge is that a new order may arrive before the algorithm has satisfied its stopping criteria. Research in dynamic vehicle routing that explicity uses forecasts of future demands is in its infancy.

4.4. Service Network Design

Service network design problems are perhaps the hardest problems in logistics (harder even than vehicle routing problems). In these problems, there is usually the flow of goods that are constrained by the movement of vehicles, or the flow of vehicles that are constrained by a published schedule. These are captured by a variable $z_{a,a',t}$ in equation (3a). Static versions of these problems are exceptionally difficult (an excellent review is contained in Crainic and Rousseau, 1988; see also Farvolden and Powell, 1994; Crainic and Roy, 1992) because the number of integer variables is large, and the linear programming relaxations, needed within branch-and-bound algorithms, are not very tight. Dynamic versions of these problems can be solved using heuristic rules for when to send a vehicle. These are essentially simulation models, often used in practice but rarely documented, with unproven solution quality.

References

Bausch, D., G. Brown, and D. Ronen. 1995. "Consolidating and dispatching truck shipments of mobil heavy petroleum products." *Interfaces* **25(2)**: 1–17.

Bell, W., L. Dalberto, M. Fisher, A. Greenfield, R. Jaikumar, P. Kedia, R. Mack, and P. Prutzman. 1983. "Improving the distribution of industrial gases with an online computerized routing and scheduling optimizer." *Interfaces* **13**: 4–23.

Bertsekas, D. and J. Tsitsiklis. 1996. In *Neuro-dynamic Programming*. Belmont, MA: Athena Scientific.

Birge, J. 1985. "Decomposition and partitioning techniques for multistage stochastic linear programs." *Oper. Res.* **33(5)**: 989–1007.

Bodin, L., B. Golden, A. Assad, and M. Ball. 1983. "Routing and scheduling of vehicles and crews." *Comput. Oper. Res.* 63–211.

Carvalho, T. A. and W. Powell. 2000. "A multiplier adjustment method for dynamic resource allocation problems." *Transport. Sci.* (Forthcoming).

Crainic, T. and J.-M. Rousseau. 1986. "Multicommodity, multimode freight transportation: a general modeling and algorithmic framework for the service network design problem." *Transport Res. B* **20B**: 290–297.

Crainic, T. and J. Roy. 1992. "Design of regular intercity driver routes for the LTL motor carrier industry." *Transport. Sci.* **26**: 280–295.

Crainic, T., M. Gendreau, and P. Dejax. 1993. "Dynamic stochastic models for the allocation of empty containers." *Oper. Res.* **41**: 102–126.

Dejax, P. and T. Crainic. 1987. "A review of empty flows and fleet management models in freight transportation." *Transport. Sci.* **21**: 227–247.

Desrosiers, J., M. Solomon, and F. Soumis. 1997. "Time constrained routing and scheduling." In *Network Routing, Handbooks in Operations Research and Management Science, Volume 8*, edited by M. O. Ball, T. L. Magnanti, C. L. Monma, and G. L. Nemhauser, pp. 35–139. Amsterdam: North Holland.

Farvolden, J. and W. Powell. 1994. "Subgradient optimization for the service network design problem." *Transport. Sci.* **28**: 256–272.

Fisher, M. "Vehicle routing." 1997. In *Network Routing, Handbooks in Operations Research and Management Science, Volume 8*, edited by M. O. Ball, T. L. Magnanti, C. L. Monma, and G. L. Nemhauser, pp. 1–33. Amsterdam: North Holland.

Gendreau, M., A. Hertz, and G. Laporte. 1994. "A tabu search heuristic for the vehicle routing problem." *Mgmnt Sci.* **40**: 1276–1290.

Glover, F. 1989. "Tabu search, part I." *ORSA J. Comput.* **1**: 190–206.

Glover, F. 1990. "Tabu search, part II." *ORSA J. Comput.* **2**: 4–32.

Higle, J. and S. Sen. 1991. "Stochastic decomposition: an algorithm for two stage linear programs with recourse." *Math. Oper. Res.* **16(3)**: 650–669.

Laporte, G. 1992. "The vehicle routing problem: an overview of exact and approximate algorithms." *Eur. J. Oper. Res.* **59**: 345–358.

Powell, W. and T. A. Carvalho. 1997. "Dynamic control of multicommodity fleet management problems." *Eur. J. Oper. Res.* **98**: 522–541.

Powell, W. and T. A. Carvalho. 1998a. "Dynamic control of logistics queueing network for large-scale fleet management." *Transport. Sci.* **32(2)**: 90–109.

Powell, W. and T. A. Carvalho. 1998b. "Real-time optimization of containers and flatcars for intermodal operations." *Transport. Sci.* **32**: 110–126.

Powell, W. B. and J. A. Shapiro. 1999. "A representational paradigm for dynamic resource transformation problems." In *Annals of Operations Research*, edited by R. Fourer, C. Coullard, and J. Owens. J.C. Baltzer AG.

Powell, W. B., P. Jaillet, and A. Odoni. 1997. "Stochastic and dynamic networks and routing." In *Network Routing, Handbooks in Operations Research and Management Science, Volume 8*, edited by M. O. Ball, T. L. Magnanti, C. L. Monma, and G. L. Nemhauser, pp. 141–295. Amsterdam: North Holland.

Powell, W. B., J. A. Shapiro, and H. P. Simao. 1998. Dynamic management of heterogeneous resources. Technical Report SOR-98-06. Department of Operations Research and Financial Engineering, Princeton University.

Sutton, R. and A. Barto. 1998. In *Reinforcement Learning*. Cambridge, MA: The MIT Press.

Van Slyke, R. and R. Wets. 1969. "L-shaped linear programs with applications to optimal control and stochastic programming." *SIAM J. Appl. Math.* **17**: 638–663.

18.2 Airline Optimization

Gang Yu and Benjamin G. Thengvall

ABSTRACT The airline industry provides an application-rich environment for the field of operations research. This article will introduce many problems faced by the airline industry that have been modeled and solved using operations research techniques. Along with a general description of each problem, the models and algorithms that have been applied to solve them are provided. The models developed and solution approaches utilized to solve these airline problems vary widely, from complex mathematical programming techniques to simple heuristic approaches. The problems addressed include network design and schedule construction, fleet assignment, aircraft routing, crew pairing, crew rostering, revenue management, irregular operations, air traffic control and collaborative decisionmaking, and ground delay programs.

1. Introduction

Many large-scale practical applications of operations research methodologies and techniques can be found in the airline industry. As early as the 1950s operations researchers were beginning to study how the developing fields of mathematical programming could be used to address a number of very difficult airline-related problems. Since then, many problems in the airline industry have been topics of active research (see Richter, 1989, for a general survey).

Optimization of an airline's entire operations in a single closed-form optimization model is impractical for a number of reasons. Even if an optimal flight schedule were known *a priori*, such a model would have to include the concurrent scheduling of pilots, flight attendants and aircraft, as well as aircraft maintenance and gate assignments. In addition, due to the dynamic nature of the airline business, many aspects of airline services vary over time. For example, flight schedules are constructed in accordance with anticipated demand, but as the operating environment changes it may be best to change flight offerings or swap fleet assignments. In addition, day-to-day problems ranging from unforeseen aircraft maintenance to extreme weather conditions force airlines to readjust their planned service in real time. Finally, the very large scale of airline problems makes the modeling and solution process too complex to solve in a single framework. For these reasons, the many planning and control problems faced by airlines are handled individually.

Most optimization-related research in the airline industry can be placed in one of the following areas:

- Network design and schedule construction
- Fleet assignment
- Aircraft routing
- Crew pairing
- Crew rostering
- Revenue management
- Irregular operations
- Air traffic control and collaborative decision making
- Ground delay programs

In the following, each of these problem areas will be defined along with a brief discussion of some of the operations research techniques that have been applied to solve them. Generally, each of these application areas is handled separately, but in some instances models have been developed to handle two or more of the above listed areas in a single framework. In addition to these major areas, a section on miscellaneous topics is included at the end of the article. This section will introduce the topics of gate assignment, fuel management, and human resource planning. The models developed and solution approaches utilized to solve airline problems vary widely, covering the spectrum from complex mathematical programming techniques to simple heuristic approaches.

2. Network Design and Schedule Construction

Construction of the flight schedule is the starting point for all other airline optimization problems. Therefore, network design and flight schedule construction are critical planning tasks for an airline. The flight schedule defines a set of flight segments that an airline will service along with corresponding origin and destination points and departure and arrival times for each flight segment. Given a flight schedule, a large portion of an airline's costs and revenues are fixed (Etschamaier and Mathaisel, 1985). An airline's decision to offer certain flights will depend to a large extent on market demand forecasts, available aircraft operating characteristics, available human resources, and the behavior of competing airlines (Dobson and Lederer, 1993).

Of course, prior to the construction of flight schedules, an airline must decide which markets it will serve. This is termed the airline network design problem. Before the 1978 Airline Deregulation Act, airlines had to fly routes as assigned by the Civil Aeronautics Board regardless of the demand for service. During this period, most airlines emphasized long point-to-point routes. Since deregulation, airlines have gained the freedom to choose which markets to serve and how often to serve them. This change has led to a fundamental shift in most airlines' routing strategies from point-to-point flight networks to hub-and-spoke oriented networks. Lederer and Nambimadom (1998) explore some of the trade-offs between point-to-point and hub-and-spoke networks using a model from which they derive analytic closed-form expressions for airline and passenger costs.

Research activities have explored not only optimal network design, but also related subproblems, including hub location, maintenance base location, and flight frequencies, once a flight network is determined. Solution methodologies for determining optimal hub locations include a branch-and-bound algorithm by Aykin (1994) and integer linear programming models solved using heuristic schemes based on mathematical programming (Jaillet et al., 1997). In one example of research focused on optimal maintenance base planning, Feo and Bard (1989) use an integer minimum-cost multicommodity network flow formulation. Solutions are provided using a two-phase approach where a set covering problem is solved and followed by a local improvement heuristic. A model to design the airline network and determine flight frequencies using fuzzy set theory is proposed by Teodorovic et al. (1994). This approach also consists of two phases. In the first, fuzzy logic is used to determine a set of candidate routes to fly. The second phase, based on fuzzy linear programming, determines flight frequencies on the previously selected candidate routes.

3. Fleet Assignment

Following network design and schedule construction, an aircraft type must be assigned to each flight segment in the schedule. This is called the fleet assignment problem. Airlines generally operate a number of different fleet types, each having different characteristics and costs such as seating capacity, landing weights, and crew and fuel costs. The most profitable assignment of fleets is sought, which incorporates not only these physical constraints, but also a variety of marketing and operational requirements.

The majority of fleet assignment methods represent the flight schedule via some variant of a space-time network with flight arcs between stations and inventory arcs at each station. Nodes at each station represent the arrival or departure of a flight leg. A multicommodity network flow problem can then be formulated with arcs and nodes duplicated as appropriate for all fleets that can take a particular flight. Side constraints must be implemented to ensure each flight segment is assigned to only one fleet. Other side constraints are included to account for other operational constraints such as a limit on the number of aircraft of each fleet type. In domestic fleet assignment problems, a common simplifying assumption is that every flight is flown every day of the week. Under this assumption, the network model need only account for one day's flights and a looping arc connects the end of one day with the beginning of the next. The resulting models are mixed-integer programs (Rushmeier and Kontogiorgis, 1997).

Many research papers have been published dealing with this basic model for fleet assignment and its extensions. Examples of applications of this type of model for major commercial airlines can be found in Abara (1989) and Subramanian et al. (1994) for American Airlines and Delta Airlines, respectively. Special modeling devices for including maintenance and crew considerations are presented in Clarke et al. (1996). A number of specialized solution techniques including model aggregation, branching on set-partitioning constraints, and prioritizing the order of branching can be found in Hane et al. (1995). Results on the complexity of these problems can be found in Gu et al. (1994).

Work by Desaulniers et al. (1997) solves the fleet assignment problem while incorporating a portion of the schedule construction problem. By allowing flight legs with known departure time windows, solutions specify not only the fleet type, but also the departure time for each operational flight leg. Two models are presented: a set partitioning type formulation and a time-constrained multicommodity network flow formulation. Linear relaxation followed by branch-and-bound are applied to both problems to generate solutions. To solve the linear relaxation problems, a column generation technique is applied to the first model and a Dantzig–Wolfe decomposition approach to the second. Another solution technique proposed for the fleet assignment problem is a heuristic adaptation of the Frank–Wolfe algorithm that utilizes the underlying network structure of the problem (Soumis et al., 1980).

Fleet assignment is initially done months in advance of the actual day of operations. However, once an initial assignment is made, it is common to reevaluate and make adjustments as demand projections are updated or market conditions change. In addition, as stated above, fleet assignment is often done under the assumption that the flight schedule is the same for every day of the week. Most airlines, though, have the same schedule for the five weekdays with a subset of those flights on the weekend. For these reasons, changes to the initial fleet assignments are often made. A network flow model has been developed by Klincewicz and Rosenwein (1995) to make these short term adjustments to the initial fleet assignments. To make such adjustments Yan and Young (1996) present a multicommodity network with side constraints solved using a Lagrangian heuristic and a modified subgradient method. Another approach to solving this problem is a simple algorithm described by Talluri (1996).

4. Aircraft Routing

Aircraft routing is the fleet-by-fleet process of assigning individual aircraft to fly each flight segment assigned to a particular fleet. There are three primary considerations at this stage: (1) flight coverage—each flight assigned to the fleet can be flown only once; (2) flow balance in aircraft routings—a flight arrival at station i must be followed by a flight departure from station i; and (3) maintenance requirements—all aircraft must be maintained at required times and bases.

There are four different types of maintenance activities mandated by the Federal Aviation Administration (FAA). The most frequent and most important during the aircraft routing process is called an A check and must be performed every 60 flight hours (Gopolan and Talluri, 1998). Industry maintenance practices, however, are generally more stringent and it is common for A checks to be performed every three or four days. These maintenance checks are performed overnight, but not all stations are equipped with proper maintenance facilities for all fleets. During the aircraft routing process, individual aircraft assignments must be made in a manner that provides maintenance opportunities for all aircraft at appropriate stations within the required time intervals.

The aircraft routing problem has been formulated and solved in a number of ways. One example is the general integer programming problem presented by Daskin and Panayotopoulos (1989), which they solve using Lagrangian relaxation. They provide heuristics for converting the Lagrangian solutions into primal feasible solutions and for improving these feasible solutions. A number of algorithmic approaches have also been proposed including the work of Wollmer (1990), which minimizes the number of aircraft used to meet the flight schedule. For the special case in which routings during the day are fixed, Gopalan and Talluri (1998) give a simple polynomial time algorithm for determining routings that satisfy maintenance requirements. In another special case, the problem can be solved as a bipartite matching problem (Talluri, 1998).

As described above, the problems of fleet assignment and aircraft routing have historically been solved in a sequential manner. Recently, work has been done solving both problems simultaneously using a string-based model and a branch-and-price solution approach (Barnhart et al., 1998). A string is defined as a sequence of connected flights that begins and ends at a maintenance station, satisfies flow balance and is maintenance feasible. This model can capture costs associated with aircraft connections and complicating constraints such as maintenance requirements. One drawback of this approach is the potentially enormous number of possible strings. This relegates the solution space to a subset of all possible strings; however, methods are proposed for intelligent string selection.

5. Crew Pairing

Crew scheduling, like aircraft routing, is done following fleet assignment. In crew scheduling two separate problems are solved sequentially. The first is the crew pairing problem. A crew pairing is a sequence of flight legs beginning and ending at a crew base that satisfies all governmental and contractual restrictions (sometimes called legalities). Restrictions include limits on the number of hours worked per day and minimum times for overnight rest enforced by the FAA. The sequence of flight legs must be connected, and is restricted in terms of number of flight legs and total time away from the crew base. The length in days of crew pairings varies at different airlines, but generally covers a period of two to five days. The problem is to find a minimum cost set of crew pairings such that all flight legs are covered.

Besides enforced legality constraints, additional considerations go into constructing crew pairings. Crew assignments must consider the number of crews available at various bases. This number translates into a minimum and maximum number of flying hours available at each crew base. Ideally, crews would stay with one aircraft through each pairing. This would eliminate many crew-related problems when a plane is delayed or the flight schedule is changed. Crews need more rest, however, and cannot fly as many hours as a plane. In addition, the constraint that crews must begin and end at a crew base is not shared by aircraft. For these reasons crews generally change planes during a pairing. However, it is desirable to limit the number of times these switches occur.

The crew pairing problem, in a manner similar to that used for the fleet assignment problem, is solved sequentially for most large airlines. The final crew schedule must specify dates and times, because fully dated pairings must be assigned to crew members. Since the fully dated problem is too difficult to solve in one step, the problem is often first simplified by assuming the flight schedule repeats daily. While this is frequently true Monday through Friday, a subset of these flights is typically flown on weekends. After solving the daily problem, a weekly problem is solved, which allows adjustments to be made to incorporate these weekend differences. Information from the daily problem solution is used to fix certain problem variables or as an advanced starting point. Finally, the fully-dated problem is solved allowing for any other special periods such as holidays, one-time events or for changes in the flight schedule over time.

Finding an optimal set of crew pairings can easily be modeled as a set partitioning problem in which each row represents a flight to be covered and the columns consist of all potential crew pairings. Each column contains a 1 in each row that the pairing covers, and a 0 in all other rows. However, this problem is known to be NP-complete (Housos and Elmroth, 1997), and thus for large airlines it is impossible to enumerate exhaustively all possible crew pairings and then solve this problem optimally. The enormous number of possible crew pairings rules out a complete enumeration scheme for building the columns in the set partitioning problem. Therefore, in practice a subset of possible crew pairings are generated from which the best combination of pairings is selected to cover the given flight schedule and fleet assignment. Because solutions are only obtained from a subset of the possible crew pairings, solution quality depends heavily on the generation of good pairings.

When using a set partitioning approach to model the crew pairing problem, solution procedures must encompass both the generation of feasible crew pairings and the optimization of the set partitioning problem. The papers of Graves et al. (1993) and Hoffman and Padberg (1993) are some representative, recent examples of such an approach. A very influential early reference is that of Rubin (1973), in which a method of repeatedly solving set covering problems is presented. Each of the sequence of set covering problems contains a subset of all possible crew pairings. Additional columns are generated at each subsequent iteration. Housos and Elmroth (1997) have developed a series of subproblems that reduces the size of the general set partitioning problem to be solved and improves the quality of solutions. Their techniques have been implemented at a number of international airlines. Solving many sets of small subproblems rather than looking for a global solution to the set partitioning problem is also advocated by Gershkoff (1989). In this paper reporting work done at American Airlines, it is estimated that the switch to an integer linear programming approach from earlier enumeration-based techniques will save $18 million per year. Further improvement in crew pairing at American Airlines is reported by Anbil et al. (1992) with a move to a more global approach. After initially generating millions of possible pairings, the very large set partitioning problem is solved using specialized pricing schemes and branching rules.

For two completely different approaches to the crew pairing problem see Lagerholm et al. (1997) and Levine (1996). In the former, a neural network is employed with Potts neurons to solve the crew pairing problem. The latter discusses the development of a hybrid genetic algorithm for the problem and then compares it with results obtained using branch-and-cut and branch-and-bound algorithms.

A practice called "deadheading" is often used to reposition flight crews in which a crew will fly a flight leg as passengers. When deadheading is allowed the traditional set partitioning problem becomes a set covering problem. In solving the crew pairing problem with deadheading, all flight segments must be covered, but they may be covered by more than one crew. Lavoie et al. (1988) provides an example of such a set covering problem solved using a combination of a linear relaxation of the integer program and column generation. Ball and Roberts (1985) use a graph partitioning approach to provide an initial solution to the set partitioning problem. By partitioning on a graph, they decompose the problem into a series of subproblems that can be solved as matching problems. This initial solution is improved using heuristic procedures that include the addition of deadheads.

6. Crew Rostering

The second problem to be solved relating to crew scheduling is the monthly crew rostering problem. Once crew pairings have been generated to cover all flights in the schedule, individual crew members must be assigned to crew pairings to create their monthly schedules. These schedules must also incorporate employee time off and training, as well as union regulations and other contractual obligations. While this section has the general title crew rostering, the problem described has been approached in practice in two different ways. In what is traditionally called rostering, each employee selects a number of desired activities, such as specific pairings or rest periods on certain dates. Personalized schedules are then constructed in which an effort is made to respect, when possible, the wishes of each employee. In the second approach, called a bidline procedure, schedules are constructed first, independent of employee desires. After employees indicate their preferences for several of these schedules, airlines assign one to each employee according to a predefined priority system. Bidline procedures are more common among large North American carriers, while rostering is found more frequently in Europe.

Whether building personalized schedules or developing general schedules to be bid on by individuals, crew schedulers seek assignments that will provide maximum benefit to the company. Some important considerations include maximizing the utilization of each crew member, evenly distributing flying time for each crew member over the scheduling period, and evenly distributing the number of working days and off days for each crew member.

Solutions to crew rostering problems are obtained separately for each homogeneous group of crew members. Pilots require specialized training for different aircraft types and are usually only scheduled to fly one type of aircraft in a month. Therefore, monthly schedules are made for pilots of each aircraft type sequentially. Flight attendants, on the other hand, are generally licensed to work on a number of different aircraft types.

Gamache et al. (1998) details the rostering approach used at Air Canada where schedules for each employee are constructed sequentially, from the most senior to the most junior. For each employee an integer programming problem is solved that determines an employee's maximum-score schedule while taking into account all remaining employees. This preferential bidding system is used to make assignments in which each personalized schedule includes an employee's preassigned activities and takes into account weighted bids representing an employee's preferences. Nicoletti (1975) provides a capacitated network model to solve the rostering problem in which crew seniority is not taken into account. Assignments are made day-by-day, up to the end of the month. Set partitioning problems are presented by Ryan (1992) and Gamache and Soumis (1998) as another option for solving the crew rostering problem. As part of his solution methodology, Ryan (1992) includes a right-hand-side perturbation scheme to avoid degeneracy and specialized branching strategies. His methods have been implemented at Air New Zealand to solve problems with more than 200,000 binary variables. Gamache and Soumis (1998) solve their generalized set partitioning problem using a branch-and-bound procedure that includes an embedded column generation algorithm. New columns are generated by solving constrained shortest path problems for each employee. A two-step approach for crew rostering is given by Lucic and Teodorovic (1999). In the first step, a feasible crew assignment is made by assigning each employee sequentially. Then a simulated annealing technique is used to improve the initial feasible solution. A bidline approach is demonstrated by Marsten et al. (1979) in which a set partitioning problem is used to build schedules for employees. The development of a system for automated generation of bidlines at American Airlines is presented by Jones (1989). This heuristic approach includes many techniques used in expert system design.

7. Revenue Management

Given a flight schedule, revenue management can be described as the control and management of the reservations inventory in a way that maximizes the profitability of an airline. Two essential components in revenue management are determining fare classes for each flight in the flight schedule as well as the allocation of available seats to each fare class. Not only are seats on an airplane partitioned physically into sections such as first class and coach, but also, seats in the same section are priced at many different levels and are sold with various restrictions. Determining how may seats to allocate to the different fare classes is termed the seat allocation problem. Effective revenue management can save airlines hundreds of millions of dollars each year. Smith et al. (1992) gives an overview of revenue management at American Airlines that includes a number of operations research models and an estimation of their benefits to the airline.

The need for effective revenue management can be motivated with the following explanation. Compared with the relatively high fixed costs of operating a flight, the marginal cost of carrying additional passengers is very low. Therefore, increasing the load factor on flights can have a significant impact on an airline's expected revenues. The primary issue is to maximize the expected monetary gain from a particular flight segment by finding the proper balance between gaining additional revenue by selling less expensive seats and losing revenue by turning away higher fare customers.

There have been two basic approaches to the seat allocation problem: nested and nonnested. Early airline reservation systems followed a simple nonnested approach in which distinct and separate inventories were established for each fare class. In this type of approach, seats are partitioned with a certain number of seats in each fare class such that the sum of the seats in each fare class equals the total seating capacity of the aircraft. Today's airline reservation systems employ a nested approach in which each fare class is assigned a booking limit that is the total number of bookings allowed for that class and all classes with lower fare prices. The booking limit is an increasing function of the fare price. The booking level for the class with the highest fare price is the total number of seats on the aircraft. In this manner, a higher fare class customer will not be refused a seat as long as any seats remain available in any lower fare class.

In modeling approaches found in the literature, a standard assumption is that fare classes are filled sequentially from the lowest to the highest. This is often the case where discounted fares are offered in advance, while last minute tickets are sold at a premium. Under this assumption, a calculus-based policy was developed by Brumelle and McGill (1993) to establish booking limits. A dynamic programming formulation that does not require any assumptions about the arrival patterns for various fare classes is presented by Lee and Hersh (1993). Another common simplification is to consider the booking policy for each flight leg separately. One exception to this is the work of Curry (1990), who examines booking limits when fare classes are nested for a network of flights. Other representative work under the two assumptions mentioned can be found in Wollmer (1992) and Robinson (1995). Work fundamental to many airlines' current pricing models is the probabilistic decision model of Belobaba (1989). His expected marginal seat revenue (EMSR) decision model takes into account the uncertainty associated with estimates of future demand as well as the nested structure of booking limits in airline reservation systems.

Overbooking is another area of study in revenue management. Airlines commonly overbook flights to deal with the propensity of airline customers to cancel flight reservations or fail to show up for a flight. In doing so, an airline must balance the extra revenue generated by ensuring a maximum load with the cost of not having sufficient capacity to accommodate all passengers. Overbooking that results in insufficient capacity costs an airline both in financial penalties and in loss of customer goodwill because some passengers will be denied boarding and must be compensated. This problem was recently addressed employing a dynamic programming approach by Chatwin (1998). Under certain assumptions, an optimal booking limit policy is presented, and it is shown that these optimal booking limits are easy to calculate and are nondecreasing as a function of the time to flight departure. Another overbooking model is presented by Alstrup et al. (1989) for nonstop flights with two passenger types. Assuming Poisson arrivals for reservations in a given time period, a two-dimensional stochastic dynamic programming model is developed. The model demonstrated potential savings of $2 million yearly at Scandinavian Airlines.

The basic input data for all of the above revenue management problems are the demand distributions for seats in particular fare classes. When setting overbooking policies it is also desirable to know the distribution of no-shows, that is, people who have bought tickets but do not show up for a flight. Forecasting these demand and no-show distributions is very difficult because large airlines service hundreds or thousands of origin-destination pairs, each having their own seasonal and weekly effects. In addition, forecasting is further complicated by competitor's actions and special one-time events.

The majority of forecasting approaches used in industry are simple time-series models such as moving averages or exponentially smoothed averages (Sun et al., 1998). The history of past flights is weight-averaged and then adjusted for seasonal effects. Of course, when forecasting future market behavior using time-series models, the more accurate the booking histories of past flights, the better the forecast will be. To overcome incomplete histories, Popovic and Teodorovic (1997) provide a Bayesian approach for estimating demand distributions in cases where historical data may be missing or limited. This is often the case when an airline opens a new route or when a competitor enters a market.

Multivariate regression is another technique sometimes employed to forecast demand. In multivariate regression the quantitative relationship between demand and any number of independent factors that may influence it is modeled. Factors thought to be important in determining demand include income, population, distance between cities, proximity of competing airports, competition from other airlines on a particular route, frequency of flights, level of customer service, aircraft size, and fare price (Jorge-Calderon, 1997). These models can adjust to changes in the marketplace without the time delay involved when analyzing historical data. However, as many different factors may affect demand in different origin-destination pairs, specialized models must be built for each unique market. This requires expert knowledge and is time intensive. Also, as time passes influential factors may change, requiring constant monitoring and frequent updating of such regression models. A neural network approach for forecasting demand and no-shows is given by Sun et al. (1998).

8. Irregular Operations

When faced with a lack of resources, airlines often are not able to fly their published flight schedule. This is frequently the result of aircraft mechanical difficulties, inclement weather, or crew shortages. As situations like these arise, decisions must be made to deal with the shortage of resources in a manner that returns the airline to the originally planned flight schedule in a timely fashion while attempting to minimize operational costs and keep passengers satisfied. To deal with these unforeseen disturbances airlines most commonly use flight delays, flight cancellations, equipment substitutions, ferry flights, and aircraft diversions. This general situation is called the airline irregular operations problem and involves aircraft routing, crew scheduling, gate assignment, and passenger recovery.

Aircraft routing and crew scheduling during irregular operations are generally handled in an iterative fashion. After an aircraft routing is found, an attempt is made to pair it with a feasible crew schedule. If no feasible crew schedule can be found, the aircraft routing is adjusted until a crew assignment can be made.

Among irregular operations topics found in the literature, the aircraft schedule recovery problem has received the majority of the attention. The aircraft schedule recovery problem is defined as finding an assignment of flights to all available aircraft during the period of irregular operation such that once normal operations resume, aircraft are positioned to execute all flights as originally scheduled. Teodorovic and Guberinic (1984) developed one of the first models for aircraft schedule recovery. They suggest a solution procedure based on branch-and-bound to minimize total customer delay. Their work was extended by Teodorovic and Stojkovic (1990) to include cancellations and station curfews. Solutions were obtained using a hierarchical objective and a sequential dynamic programming-based heuristic with the primary objective of minimizing flight cancellations. Jarrah et al. (1993) present two models based on a space-time network: one for delays and the other for cancellations. While delays and cancellations are not addressed simultaneously, the approach was considered practical enough to be implemented by United Airlines (Rakshit et al., 1996).

More recently, Yan and Yang (1996) were the first to incorporate flight cancellations, delays, and ferry flights in a single model. Their most comprehensive models were solved using Lagrangian relaxation. The framework introduced was extended by Yan and Lin (1997) to handle station closures and by Yan and Tu (1997) to handle multiple fleets. Argüello et al. (1997) present a greedy randomized adaptive search procedure (GRASP) to reconstruct aircraft routings. Argüello et al. (1998) provide a time-band optimization model to solve the same problem. Their formulation is an integral minimum-cost network flow problem with side constraints. Cao and Kanafani (1997a,b) present a 0–1 quadratic programming (QP) model based on the delay model of Jarrah et al. (1993). An algorithm similar to successive linear programming is used to solve the QP, though global optimality is not guaranteed since the problem is nonconvex. Thengvall et al. (2000) extended the work of Yan and Young (1996) and Yan and Tu (1977) to build recovery schedules for aircraft by including an incentive in the formulation to minimize deviation from the original aircraft routings. In practice, most airlines currently use heuristic methods or a combination of simple models and expert knowledge to solve these problems, as their real-time nature does not allow for lengthy optimization run times.

While decisions regarding aircraft are the most important when considering impact on revenues, crew management is often the bottleneck during irregular operations. The biggest difficulty in crew management during irregular operations is abiding by complicated and restrictive limitations imposed by federal regulatory rules, union contracts, and company work rules. These rules include duty time and flight time limitations for safety, minimum rest times, scheduled time off, and seniority rules. A few papers dealing with crew scheduling during irregular operations have been published recently. A multicommodity integer network flow model and a heuristic search algorithm are presented by Wei et al. (1997). The problem has also been formulated as a set partitioning type problem by Stojkovic et al. (1998). To solve it they employ a column generation method embedded in a branch-and-bound search tree.

9. Air Traffic Control and Collaborative Decisionmaking

The United States Federal Aviation Administration (FAA) is charged with managing the nation's air traffic, which encompasses over sixty thousand daily flights of commercial, military, and general aviation aircraft. The national air space has been partitioned into twenty Air Route Traffic Control Centers and further divided into seven hundred airspace sectors (Helme, 1998). The daily flight schedules of commercial airlines are designed assuming there will be good weather and little congestion at stations. However, problems such as severe weather or equipment failure at a station are not unusual. In such instances, congestion is often the result, with more flights scheduled than can be flown safely. The FAA is responsible for generating strategies to solve air traffic flow problems in these situations in a manner that reduces congestion and delay. To accomplish this, they employ methods including ground delays, miles-in-trail restrictions, rerouting, and airborne delays. Ground delays will be described and examined in detail later. Mile-in-trail restrictions mandate spacing between planes en route to prevent sector overloading. Rerouting is used to maneuver aircraft around inclement weather. Airborne delays are used when an airport lacks arrival capacity as aircraft approach.

Current approaches to address the traffic flow management problem make decisions based primarily on capacity forecasts for affected destination airports and are made independently for each affected station (Lindsay et al., 1998). Some effort in recent years in the operations research community has focused on using optimization to discover better solutions for the traffic flow management problem faced by the FAA. Dembo et al. (1989) model and solve an air traffic problem for a single airspace sector over multiple time periods as a large-scale nonlinear network. The model determines optimal altitudes for flights as well as possible delays to avoid congestion. A more advanced model that represents ground delays and air delays simultaneously at multiple locations is presented by Helme (1998). The model is formulated as a multicommodity network and is solved using an improvement heuristic to update an initial feasible solution. Lindsay et al. (1998) present several optimization models to address the traffic flow management problem, including linear, integer, mixed-integer, and network programming models.

Air traffic control in the United States is currently in a state of transition. The collaborative decisionmaking (CDM) program, officially formed in 1995, is a joint government/ industry initiative charged with improving air traffic flow management in situations where an airport's arrival capacity is reduced. CDM replaced the current ground delay program and was expected to be fully implemented in the year 2000. Key elements implemented in this new system are data exchange, ground delay program (GDP) advisories, ration by schedule, and compression (`http://www.metsci.com/cdm`).

As the sharing of information is critical to the success of CDM, data exchange is the backbone of the program. National airspace users send in their operational schedules and schedule changes on a continuous basis, including but not limited to, delays, cancellations and newly created flights. This demand information is used to determine whether a capacity/ demand imbalance exists and warrants some type of ground hold strategy. Users receive information in an aggregate demand list that allows them to see where they fit in the traffic flow and to plan operations accordingly.

The GDP advisory represents the strongest truly collaborative component of the program. Presently, advisories are sent out stating the time and airport acceptance rate at which a GDP will be run. In the new system, a CDM advisory will be sent in advance of a GDP and will include a cut-off time (the time when the program will actually be run). Users will have until the cut-off time to re-schedule, cancel, or delay flights and send in their new operational schedule. In this manner, demand could potentially be reduced sufficiently to either delay the start of the GDP, or eliminate the need for the program altogether.

Currently, if an airline reports flight cancellations before a GDP is implemented, those flights are dropped from the FAA database and the airline is not able to use the assigned arrival slots for substitution. If an airline reports a mechanical delay on a flight, the system will re-project its arrival time. If a GDP were then run, that flight would likely receive an additional delay on top of its mechanical delay. These effects have become known as the double penalty issue and represent a barrier to implementing open data exchange between airlines and the FAA. Ration by schedule removes this disincentive in the following manner.

When arrival capacity is reduced, the limited arrival resources will be rationed based upon the original schedule, and not the current projections of demand.

Compression is a process whereby unusable arrival slots are shifted in time. For example, consider an airline with two flights scheduled to arrive at Chicago; Flight 1 at 1300 and Flight 2 at 1500. After a GDP is run, Flight 1 is assigned a 1400 arrival slot and Flight 2 receives a 1700 arrival slot. If Flight 1 is canceled, Flight 2 cannot make use of the 1400 arrival slot because it occurs before its scheduled arrival time of 1500. Compression will allow the vacated slot to move down to where Flight 2 can make use of it.

Implementation of these different elements of CDM will change the environment under which airlines operate. A set of integer programming models is presented in Carlson (1999) to help airlines make real-time decisions when this new system is in place. These models make decisions regarding the assignment of inbound flights to the limited arrival slots allocated by the FAA in situations of reduced station capacity. Implicit in these decisions is the cancellation or delay of those flights without a slot. Exploiting the structure of the problem, it is shown that under the assumption of integer arrival slot allocations, a much simpler transportation problem can be solved.

10. Ground Delay Programs

The main strategy the FAA currently uses to deal with air traffic flow problems when congestion is anticipated at one or more destination stations is the implementation of a GDP. During a ground delay, aircraft departing for a congested station are held on the ground prior to departure in an effort to slow the pace of arrivals and alleviate congestion. The rationale for this behavior is that delays of flights bound for a heavily congested station are inevitable and ground delays are less expensive and safer than airborne delays. Currently, the FAA estimates the arrival capacity of the congested station and assigns landing times to scheduled flights on a first-come-first-served basis. Delays are calculated based on these assignments and aircraft are held at their origin stations instead of being delayed in the air.

Several optimization models have been formulated in the last decade with the intent of improving current FAA practice during a GDP. A common objective is to decrease the total minutes of delay experienced throughout the system during GDP. The earliest and simplest versions deal with the problem locally, determining ground holds for aircraft departing for a single destination airport. Terrab and Odoni (1993) present models for the single airport case with deterministic and stochastic runway capacities and single and multiple time periods. A minimum-cost flow algorithm is used to solve the deterministic version. An exact dynamic programming formulation is proposed for the stochastic case as well as some heuristic solution approaches. Richetta and Odoni (1993) introduce a solution technique that includes dynamic updates of ground-hold assignments as the day progresses using stochastic linear programming with recourse. Luo and Yu (1997) propose a single airport integer programming model to minimize delays among flights out of the airport where the ground delay program is initiated. To solve their model, they derive valid inequalities to strengthen the LP relaxation bound and then implement a heuristic procedure to find good feasible solutions. This work is continued in Luo and Yu (1998a,b) where they examine how to handle schedule perturbations due to the ground delay program under the assumptions of splitable and nonsplitable resources. The assumption of nonsplitable resources implies that an aircraft and the crew associated with it arrive and depart together from a congested station. For both cases mathematical models and heuristic approaches are developed.

Work for a single station has been extended to cover multiple stations. Including multiple airports adds realism to the problem by allowing delays to propagate through the system by aircraft performing consecutive flights. Vranas et al. (1994) formulate several integer programming models and provide a heuristic to find feasible solutions by rounding optimal solutions to the LP relaxation. Other integer programming formulations for the multiple airport case have been recently proposed by Bertsimas and Stock Patterson (1998) and Navazio and Romanin-Jacur (1998).

11. Other Topics

Optimization based methods have also been applied to a myriad of other airline related topics such as gate assignment, fuel management, and human resource planning.

Assigning arriving flights to airport gates is a critical issue in airport station operations. The large volumes of baggage and passengers transferring between flights at large connecting airports make this a very complex task. Considerations when assigning aircraft to gates include aircraft size, passenger walking distance, baggage transfer, and aircraft service requirements. Due to the real-time nature of the problem, multiple possible objectives, and the uncertainty associated with the arrival of flights, many airlines use expert systems to schedule gate activity. Some examples of expert system methodology for gate assignment include Gosling (1990) and Srihari and Muthukrishnan (1991). Brazile et al. (1994) present a comparison of integer programming, simulation, and expert systems for the gate assignment problem. An integer programming formulation is provided by Haghani and Chen (1998) along with a heuristic solution procedure.

Fuel costs are the largest operating cost faced by an airline. Substantial cost reductions may result from tankering or fuel ferrying if prices differ significantly from station to station. A minimum cost fuel management policy for an airline flight schedule has been developed by Stroup and Wollmer (1992). Based on fuel prices, bounds on the amount of fuel that can be purchased at a particular station, and bounds on the amount of fuel that can be purchased from a particular supplier, a linear program is formulated. By a series of transformations on constraints and variables, they show that the linear programming formulation can often be converted to a generalized network and sometimes to a pure network, resulting in significant savings in computational effort and time.

Manpower planning is a broad subject that includes the hiring, training, and scheduling of human resources. Two topics covered here are the ground crew scheduling and the pilot training problems. The airline industry faces a complex tour-scheduling environment for their ground station personnel. Scheduling these customer service representatives and ramp service workers is a difficult problem that must take into account scheduled aircraft arrivals and departures as well as planned passenger, baggage, and cargo volumes. Personnel tour scheduling also involves the determination of work and nonwork days during the week, along with associated daily shift starting and ending times. Brusco et al. (1995) report on the development and implementation of a two-part approach that enhanced the tour-scheduling process at United Airlines. Initially, a column generation approach is used to solve an integer programming model. Then a local search heuristic based on simulated annealing is employed to improve this initial feasible schedule.

Due to the dynamic nature of the industry, airlines are constantly analyzing which markets to serve, the number of fleets to operate, the number of aircraft of each fleet to operate, and the utilization of each fleet. The number of pilots needed to execute the flight schedule in this dynamic environment is constantly changing. This makes planning for new pilot hires and transitional training of current pilots a difficult task. Yu et al. (1998) present an optimization model for the pilot training assignment problem. The solution to this integer program accounts for new hires and transitional training for current pilots while covering the current flight schedule with qualified pilots. An iterative heuristic to schedule pilots for training is also presented. Both the mathematical programming model and the heuristic approach discretize the time horizon into one week segments. Another approach to the pilot training assignment problem is given by Verbeek (1991). He develops a framework for the design of a decision support system for pilot training. A set of integer subproblems is solved to assist a decisionmaker. The system described has been implemented at KLM Airlines.

References

Abara, J. 1989. "Applying integer linear programming to the fleet assignment problem." *Interfaces* **19(4)**: 20–28.

Alstrup, J., S.-E. Andersson, S. Boas, O. B. G. Madsen, and R. V. V. Vidal. 1989. "Booking control increases profit at Scandinavian Airlines." *Interfaces* **19(4)**: 10–19.

Anbil, R., R. Tanga, and E. L. Johnson. 1992. "A global approach to crew-pairing optimization." *IBM Sys. J.* **31(1)**: 71–78.

Argüello, M. F., J. F. Bard, and G. Yu. 1997. "A GRASP for aircraft routing in response to groundings and delays." *J. Combinat. Optim.* **5**: 211–228.

Argüello, M. F., J. F. Bard, and G. Yu. 1998 "Models and methods for managing airline irregular operations aircraft routing." In *Operations Research in Airline Industry*, edited by G. Yu,, pp. 1–45. Boston: Kluwer Academic Publishers.

Aykin, T. 1994. "Lagrangian relaxation based approaches to capacitated hub-and-spoke network design problem." *Eur. J. Oper. Res.* **79(3)**: 501–523.

Ball, M. and A. Roberts. 1985. "A graph partitioning approach to airline crew scheduling." *Transport. Sci.* **19(2)**: 107–126.

Barnhart, C., N. L. Boland, L. W. Clarke, E. L. Johnson, G. Nemhauser, and R. G. Shenoi. 1998. "Flight string models for aircraft fleeting and routing." *Transport. Sci.* **32(3)**: 208–220.

Belobaba, P. P. 1989. "Application of a probabilistic decision model to airline seat inventory control." *Oper. Res.* **37(2)**: 183–197.

Bertsimas, D. and S. Stock Patterson. 1998. "The air traffic flow problem with en route capacities." *Oper. Res.* **46(3)**: 406–422.

Brazile, R. P., K. M. Swigger, and D. L. Wyatt. 1994. "Selecting a modelling technique for the gate assignment problems: integer programming, simulation, or expert system." *Int. J. Modelling Simulation* **14(1)**: 1–5.

Brumelle, S. L. and J. I. McGill. 1993. "Airline seat allocation with multiple nested fare classes." *Oper. Res.* **41(1)**: 127–137.

Brusco, M. J., L. W. Jacobs, R. J. Bongiorno, D. V. Lyons, and B. Tang 1995. "Improving personnel scheduling at airline stations." *Oper. Res.* **43(5)**: 741–751.

Cao, J.-M. and A. Kanafani. 1997a. "Real-time decision support for integration of airline flight cancellations and delays. Part I: mathematical formulations." *Transport. Planning Technol.* **20**: 183–199.

Cao, J.-M. and A. Kanafani. 1997b. "Real-time decision support for integration of airline flight cancellations and delays. Part II: algorithms and computational experiments." *Transport. Planning Technol.* **20**: 201–217.

Carlson, P. 1999. "Exploiting the opportunities of collaborative decision making in aviation transportation." To appear in *Transport. Sci.* **33.**

Chatwin, R. E. 1998. "Multiperiod airline overbooking with a single fare class." *Oper. Res.* **46(6)**: 805–819.

Clarke, L. W., C. A. Hane, E. L. Johnson, and G. L. Nemhauser. 1996. "Maintenance and crew considerations in fleet assignment." *Transport. Sci.* **30(3)**: 249–260.

Curry, R. E. 1990. "Optimal airline seat allocations with fare classes nested by origins and destinations." *Transport. Sci.* **24(3)**: 193–204.

Daskin, M. S. and N. D. Panayotopoulos. 1989. "A Lagrangian relaxation approach to assigning aircraft to routes in hub and spoke networks." *Transport. Sci.* **23(2)**: 91–99.

Dembo, R. S., J. M. Mulvey, and S. A. Zenios. 1989. "Large-scale nonlinear network models and their application." *Oper. Res.* **37(3)**: 353–372.

Desaulniers, G., J. Desrosiers, Y. Dumas, M. M. Solomon, and F. Soumis. 1997. "Daily aircraft routing and scheduling." *Mgmnt Sci.* **43(6)**: 841–855.

Dobson, G. and P. J. Lederer. 1993. "Airline scheduling and routing in a hub-and-spoke system." *Transport. Sci.* **27(3)**: 281–297.

Etschamaier, M. M. and D. F. X. Mathaisel. 1985. "Airline scheduling: an overview." *Transport. Sci.* **9(2)**: 127–138.

Feo, T. A. and J. F. Bard. 1989. "Flight scheduling and maintenance base planning." *Mgmnt Sci* **35(12)**: 1415–1432.

Gamache, M. and F. Soumis. 1998. "A method for optimally solving the rostering problem." In *Operations Research in Airline Industry*, edited by G. Yu,, pp. 124–157. Boston: Kluwer Academic Publishers.

Gamache, M., F. Soumis, D. Villeneuve, and J. Desrosiers. 1998. "The preferential bidding system at Air Canada." *Transport. Sci.* **32(3)**: 246–255.

Gershkoff, I. 1989. "Optimizing flight crew schedules." *Interfaces* **19(4)**: 29–43.

Gopalan, R. and K. T. Talluri. 1998. "The aircraft maintenance routing problem." *Oper. Res.* **46(2)**: 260–271.

Gosling, G. D. 1990. "Design of an expert system for aircraft gate assignment." *Transport. Res. Part A* **24(1)**: 59–69.

Graves, G. W., R. D. McBride, I. Gershkoff, D. Anderson, and D. Mahidhara. 1993. "Flight crew scheduling." *Mgmnt Sci.* **39(6)**: 736–745.

Gu, Z., E. L. Johnson, G. Nemhauser, and Y. Wang. 1994. "Some properties of the fleet assignment problem." *Oper. Res. Lett.* **15(2)**: 59–71.

Haghani, A. and M.-C. Chen. 1998. "Optimizing gate assignments at airport terminals." *Transport. Res. Part A* **32(6)**: 437–454.

Hane, C. A., C. Barnhart, E. L. Johnson, R. E. Marsten, G. L. Nemhauser, and G. Sigismondi. 1995. "The fleet assignment problem: solving a large-scale integer program." *Math. Programming* **70(2)**: 211–232.

Helme, M. P. 1998. "A selective multicommodity network flow algorithm in air traffic control." In *Operations Research in Airline Industry*, edited by G. Yu,, pp. 101–123. Boston: Kluwer Academic Publishers.

Hoffman, K. L. and M. Padberg. 1993. "Solving airline crew scheduling problems by branch-and-cut." *Mgmnt Sci.* **39(6)**: 657–680.

Housos, E. and T. Elmroth. 1997. "Automatic optimization of subproblems in scheduling airline crews." *Interfaces* **27(5)**: 68–77.

Jaillet, P., G. Song, and G. Yu. 1997. "Airline network design and hub location problems." *Location Sci.* **4(3)**: 195–212.

Jarrah, A. I. Z., G. Yu, N. Krishnamurthy, and A. Rakshit. 1993. "A decision support framework for airline flight cancellations and delays." *Transport. Sci.* **27(3)**: 266–280.

Jones, R. D. 1989. "Development of an automated airline crew bid generation system." *Interfaces* **19(4)**: 44–51.

Jorge-Calderon, J. D. 1997. "A demand model for scheduled airline services on international European routes." *J. Air Transport Mgmnt* **3(1)**: 23–35.

Klincewicz, J. G. and M. B. Rosenwein. 1995. "The airline exception scheduling problem." *Transport. Sci.* **29(1)**: 4–16.

Lagerholm, M., C. Peterson, and B. Soderberg. 1997. "Airline crew scheduling with Potts neurons." *Neural Comput.* **9(7)**: 1589–1599.

Lavoie, S., M. Minoux, and E. Odier. 1988. "A new approach for crew pairing problems by column generation with an application to air transportation." *Eur. J. Oper. Res.* **35**: 45–58.

Lederer, P. J. and R. S. Nambimadom. 1998. "Airline network design." *Oper. Res.* **46(6)**: 785–804.

Lee, T. C. and M. Hersh. 1993. "A model for dynamic airline seat inventory control with multiple seat bookings." *Transport. Sci.* **27(3)**: 252–265.

Levine, D. 1996. "Application of a hybrid genetic algorithm to airline crew scheduling." *Comput. Oper. Res.* **23(6)**: 547–558.

Lindsay, K., E. A. Boyd, G. Booth, and C. Harvey. 1998. "The use of optimization to perform air traffic flow management." In *Operations Research in Airline Industry*, edited by G. Yu,, pp. 287–311. Boston: Kluwer Academic Publishers.

Lucic, P. and D. Teodorovic. 1999. "Simulated annealing for the multi-objective aircrew rostering problem." *Transport. Res. Part A* **33(1)**: 19–45.

Luo, S. and G. Yu. 1997. "On the airline schedule perturbation problem caused by the ground delay program." *Transport. Sci.* **31(4)**: 298–311.

Luo, S. and G. Yu. 1998a. "Airline schedule perturbation problem: landing and takeoff with nonsplitable resource for the ground delay program." In *Operations Research in Airline Industry*, edited by G. Yu, 404–432. Boston: Kluwer Academic Publishers.

Luo, S. and G. Yu. 1998b. "Airline schedule perturbation problem: ground delay program with splitable resources." In *Operations Research in Airline Industry*, edited by G Yu,, pp. 433–460. Boston: Kluwer Academic Publishers.

Marsten, R. E., M. R. Muller, and C. L. Killion. 1979. "Crew planning at Flying Tiger: a successful application of integer programming." *Mgmnt Sci.* **25(12)**: 1175–1183.

Navazio, L. and G. Romanin-Jacur. 1998. "The multiple connections multi-airport ground holding problem: models and algorithms." *Transport. Sci.* **32(3)**: 268–276.

Nicoletti, B. 1975. "Automatic crew rostering." *Transport. Sci.* **9(1)**: 33–48.

Popovic, J. and D. Teodorovic. 1997. "An adaptive method for generating demand inputs to airline seat inventory control models." *Transport. Res. Part B* **31(2)**: 159–175.

Rakshit, A., N. Krishnamurthy, and G. Yu. 1996. "Systems operations advisor: a real-time decision support system for managing airline operations at United Airlines." *Interfaces* **26(2)**: 50–58.

Richetta, O. and A. R. Odoni. 1993. "Solving optimally the static ground-holding policy problem in air traffic control." *Transport. Sci.* **27(3)**: 228–238.

Richter, H. 1989. "Thirty years of airline operations research." *Interfaces* **19(4)**: 3–9.

Robinson, L. W. 1995. "Optimal and approximate control policies for airline booking with sequential nonmonotonic fare classes." *Oper. Res.* **43(2)**: 252–263.

Rubin, J. 1973. "A technique for the solution of massive set covering problems, with application to airline crew scheduling." *Transport Sci.* **7(1)**: 34–48.

Rushmeier, R. A. and S. A. Kontogiorgis. 1997. "Advances in the optimization of airline fleet assignment." *Transport. Sci.* **31(2)**: 159–169.

Ryan, D. M. 1992. "The solution of massive generalized set partitioning problems in air crew rostering." *J. Oper. Res. Soc.* **43**: 459–467.

Smith, B. C., J. F. Leimkuhler, and R. M. Darrow. 1992. "Yield management at American Airlines." *Interfaces* **22(1)**: 8–31.

Soumis, F., J. A. Ferland, and J.-M. Rousseau. 1980. "A model for large-scale aircraft routing and scheduling problems." *Transport. Res. Part B* **14B**: 191–201.

Srihari, K. and R. Muthukrishnan. 1991. "An expert system methodology for aircraft-gate assignment." *Comput. Ind. Engng* **21(1)**: 101–105.

Stojkovic, M., F. Soumis, and J. Desrosiers. 1998. "The operational airline crew scheduling problem." *Transport. Sci.* **32(3)**: 232–245.

Stroup, J. S. and R. D. Wollmer. 1992. "A fuel management model for the airline industry." *Oper Res.* **40(2)**: 229–237.

Subramanian, R., Scheff , Jr. R. P., J. D. Quillinan, D. S. Wiper, and R. E. Marsten. 1994. "Coldstart: fleet assignment at Delta Air Lines." *Interfaces* **24(1)**: 104–120.

Sun, X., E. Brauner, and S. Hormby. 1998. "A large-scale neural network for airline forecasting in revenue management." In *Operations Research in Airline Industry,* edited by G. Yu,, pp. 46–67. Boston: Kluwer Academic Publishers.

Talluri, K. T. 1998. "The four-day aircraft maintenance routing problem." *Transport. Sci.* **32(1)**: 43–53.

Talluri, K. T. 1996. "Swapping applications in a daily airline fleet assignment." *Transport. Sci.* **30(3)**: 237–248.

Teodorovic, D. and S. Guberinic. 1984. "Optimal dispatching strategy on an airline network after a schedule perturbation." *Eur. J. Oper. Res.* **15**: 178–182.

Teodorovic, D. and G. Stojkovic. 1990. "Model for operational daily airline scheduling." *Transport. Planning Technol.* **14**: 273–285.

Teodorovic, D., M. Kalic, and G. Pavkovic. 1994. "The potential for using fuzzy set theory in airline network design." *Transport. Res. Part B* **28(2)**: 103–121.

Terrab, M. and A. R. Odoni. 1993. "Strategic flow management for air traffic control." *Oper. Res.* **41(1)**: 138–152.

Thengvall, B. G., J. F. Bard, and G. Yu. 2000. "Balancing user preferences for aircraft schedule recovery during irregular operations." *IIE Transactions on Operations Engineering.* **32(3)**: 181–193.

Verbeek, P. J. 1991. "Decision support systems—an application in strategic manpower planning of airline pilots." *Eur. J. Oper. Res.* **55(3)**: 368–381.

Vranas, P. B., D. J. Bertsemas, and A. R. Odoni. 1994. "The multi-airport ground-holding problem in air traffic control." *Oper. Res.* **42(2)**: 249–261.

Wei, G., G. Song, and G. Yu. 1997. "Model and algorithm for crew management during airline irregular operations." *J. Combinat. Optim.* **1(3)**: 80–97.

Wollmer, R. D. 1990. "An airline tail routing algorithm for periodic schedules." *Networks* **20(1)**: 49–54.

Wollmer, R. D. 1992. "An airline seat management model for a single leg route when lower fare classes book first." *Oper. Res.* **40(1)**: 26–37.

Yan, S. and C.-G. Lin. 1997. "Airline scheduling for the temporary closure of airports." *Transport. Sci.* **31(1)**: 72–82.

Yan, S. and D.-H. Yang. 1996. "A decision support framework for handling schedule perturbations." *Transport. Res. Part B* **30(6)**: 405–419.

Yan, S. and H.-F. Young. 1996. "A decision support framework for multi-fleet routing and multi-stop flight scheduling." *Transport. Res. Part A* **30(5)**: 379–398.

Yan, S. and Y.-P. Tu. 1997. "Multifleet routing and multistop flight scheduling for schedule perturbation." *Eur. J. Oper. Res.* **103(1)**: 155–169.

Yu, G., S. Dugan, and M. Argüello. 1998. "Moving toward an integrated decision support system for manpower planning at continental airlines: optimization of pilot training assignments." In *Industrial Applications of Combinatorial Optimization,* edited by G. Yu,, pp. 1–24. Boston: Kluwer Academic Publishers.

18.3 Optimization in the Rail Industry

Alexandra M. Newman, Linda K. Nozick, and Candace Arai Yano

ABSTRACT This article describes optimization problems that arise in the rail industry related to:

- infrastructure (track and siding) planning and track maintenance,
- sizing of fleets of locomotives and railcars,
- locomotive, railcar, and container repositioning,
- train scheduling, including service frequency planning, day-of-week train scheduling, freight routing, meet-pass planning, and timetable construction.

We focus on problems that are specific to the rail industry because of technological or cost considerations. We survey models that provide a broad perspective of rail decision problems and describe optimization methods employed to address them. We also suggest directions for future research.

1. Introduction

Deregulation from the Staggers Rail Act along with improved technology and productivity has led to a 50 percent increase in rail freight transportation since 1980. Railroads transport over 40 percent of the ton-miles of intercity freight, including 70 percent of new automobiles, 40 percent of farm products, and 65 percent of coal in the United States of America (Association of American Railroads, 1999). With increased profit incentives and better availability of computer systems, railroads have begun to adopt optimization-based decision support with some success.

We focus on problems that are specific to the rail industry because of special features of infrastructure, operations, or cost structures, and that are amenable to solution by optimization methods. We first describe these industry-specific characteristics and then discuss problem areas and commonly-used modeling constructs.

1.1. Special Characteristics of the Rail Industry

The nature of travel and freight handling are the most distinctive characteristics of rail transportation. Trains operate on limited-capacity tracks over long distances, and their hauling capabilities depend on the assigned set of locomotives. Moreover, railcars, containers (traveling on *stack cars*), and truck trailers (traveling on *flatcars*) can only be transferred between trains at railyards.

Due to the high cost of installing and maintaining track ($9 billion is spent annually in the United States of America on maintenance alone), many railway networks consist of single tracks with periodic sidings, where trains may wait while others pass. This necessitates detailed *meet-pass plans* that specify when and where trains meet and pass or overtake one another, and related train *timetables*.

Train crew costs exhibit strong economies of scale; they depend on the number of crew-shifts and are fairly insensitive to the number of transported railcars. Locomotive operating costs include a fixed-charge component for their use, and variable (per ton-mile) costs for the transported freight. Different combinations of locomotives, or *consists*, may be assigned to a train. These assignments must satisfy constraints on minimum tractive power or number of locomotives, that depend on the topography of the rail segment. Train scheduling problems are complicated by fixed charges for each train and for each additional locomotive. Moreover, the possibility of assigning multiple locomotives to a train results in motive planning problems quite different from those encountered in truck or air transportation. The long travel distances involved in rail transport also increase the importance of minimizing equipment deadheading and empty car movement because of both direct costs and opportunity costs of idling expensive equipment (e.g., $1 million locomotives or $65,000 railcars) for long periods.

Both space for storage and capacity to handle railcars and containers are limited at railyards, which have various track configurations and use different mechanisms for sorting or *classifying* railcars. Typically, each track segment is allocated to a *block*, which consists of a group of cars traveling together for one or more portions of the journey. *Train makeup* refers to the assignment of these blocks to trains. The layout of the yard and the capacity of the handling equipment constrain the number of blocks, and the workload imposed on the yard affects in-transit delays. Containerized freight may be handled differently because of its fragility. The containers, rather than the stack cars, may be rearranged, but doing so imposes additional workload on expensive, specialized cranes, which are often the bottlenecks at intermodal rail terminals.

Figure 1 depicts a simple rail network. Trains may carry grain and coal to urban areas and manufacturing sites. Container ships may generate intermodal traffic bound for domestic sites, while an automobile factory may provide intermodal exports. The trains may visit railyard(s) where cars are reclassified. Sidings allow trains to meet and overtake along single-tracked lines. Signals at sidings and intersections govern traffic flow.

1.2. Rail Decision Problems

We review several major categories of rail optimization models. Section 2 covers infrastructure planning models concerned with the design of a railway network, locations of track sidings, and track maintenance and improvement. These are strategic decisions that require a long-term, system-wide view. Problems of sizing fleets of locomotives and railcars, which are covered in Section 3, also are strategic decisions. However, they commonly are addressed using more detailed data on demand, train schedules, and motive power constraints because these factors strongly influence fleet requirements.

Section 3 also surveys models for short- to medium-term locomotive, railcar, and container repositioning. Some are tactical models that rely on steady-state analysis and suggest long-run rebalancing strategies, but most are designed to consider the state of the system and detailed demand forecasts, train schedules, and so on, to determine when and how locomotives, and empty railcars and containers, should be repositioned.

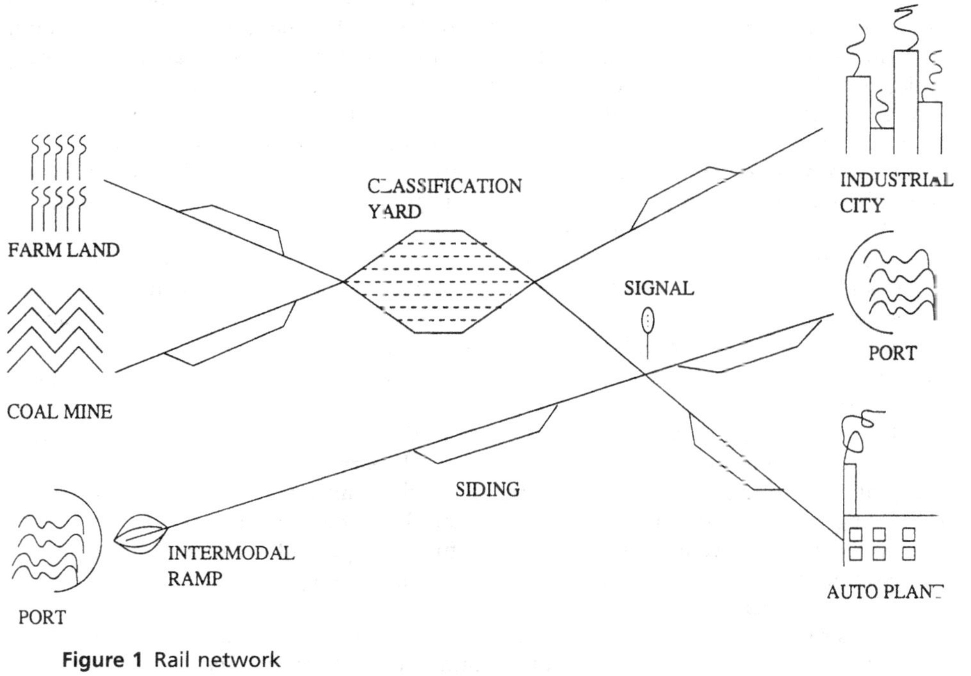

Figure 1 Rail network

Section 4 discusses train scheduling and freight routing. Tactical models in this section are concerned with the service frequency of direct and indirect trains, blocking patterns (i.e., paths specifying intermediate transfer locations) to be utilized, and allocation of freight to the blocking patterns. As in other tactical models, these decisions are based on average costs and demand rates. Operational models focus on constructing more detailed train schedules and/or freight routing to satisfy demands while considering other factors (such as due dates, blocking plans, and locomotive availability). The level of precision may extend to minute-to-minute timetables, which are often determined concurrently with meet-pass plans.

Space limitations preclude us from covering models of railyard operations; most are descriptive models based on queuing concepts. Related discussion appears in Petersen (1977a,b) and Turnquist and Daskin (1982). We also omit real-time decisions for train scheduling, freight routing, and equipment repositioning. Most work in these areas relies on short-term planning approaches for decision support, and does not address the real-time problem directly. More commonly, real-time decisions are based on judgment and experience. We omit railcar loading and unloading. Interesting applications of optimization to these areas appear in Bard (1997) and Vasko et al. (1994). Finally, we do not discuss crew scheduling. Models for rail crew management and relevant references can be found in Caprara et al. (1997).

Several survey papers cover topics that we discuss. Assad (1980) provides excellent background on institutional aspects of, and operating policies for, rail transportation. We cite more focused surveys within the corresponding section of this chapter. We have attempted to describe representative models that collectively provide the reader with both a broad perspective of decision problems and an understanding of how researchers and practitioners have addressed them using optimization models.

1.3. *Modeling Constructs*

Several modeling constructs are commonly used for rail planning and scheduling problems. One important construct is a *time-space network*; a simple example with instantaneous travel times appears in Figure 2. Each node represents a time-location pair. Arcs connect nodes if it is feasible for a railcar or locomotive to move from one time-location pair to the other. Arcs from one period to the next at the same location represent holding inventory, and reverse arcs from one period to a previous period represent back-ordering of demand for applicable equipment.

Time–space networks lead naturally to problem representations as single or multicommodity networks. When train capacity and timing are given, the network contains one capacitated arc per train and the flows represent railcars, locomotives, and/or containers. The commodities in the network may be distinguished by their origin, destination, priority, and their physical attributes (e.g., coal versus automobiles). The latter distinction is important due to differences in the railcars used for various goods. When the train service and/or timing are to be decided, the resulting models become *network design problems*. In some cases, arc capacities may also be decided (e.g., by selecting the number of locomotives). Some fleet sizing problems may be modeled by adding arcs for external supply of capital equipment to the basic time–space network. Research on single and multicommodity network flow and design problems has advanced rapidly in the past decade; many algorithms applicable to rail planning and scheduling problems can be found in Ahuja et al. (1993) and Magnanti and Wong (1984).

The rail industry commonly uses *string diagrams* to represent detailed scheduling problems in continuous time and space domains. In string diagrams, the locations of trains, usually along a single rail line, are plotted, with the x-axis representing time and the y-axis representing the distance from one end of the line. Thus, the slope of line segment corresponds to the velocity of the train, and horizontal segments represent waiting at a siding or a railyard. Note that string diagrams are designed for representing trains, not railcar, container, or locomotive movements.

The string diagram in Figure 3 illustrates a meet-pass plan and timetable for two southbound trains (I and II) and a northbound train (III) over three hours. There are three

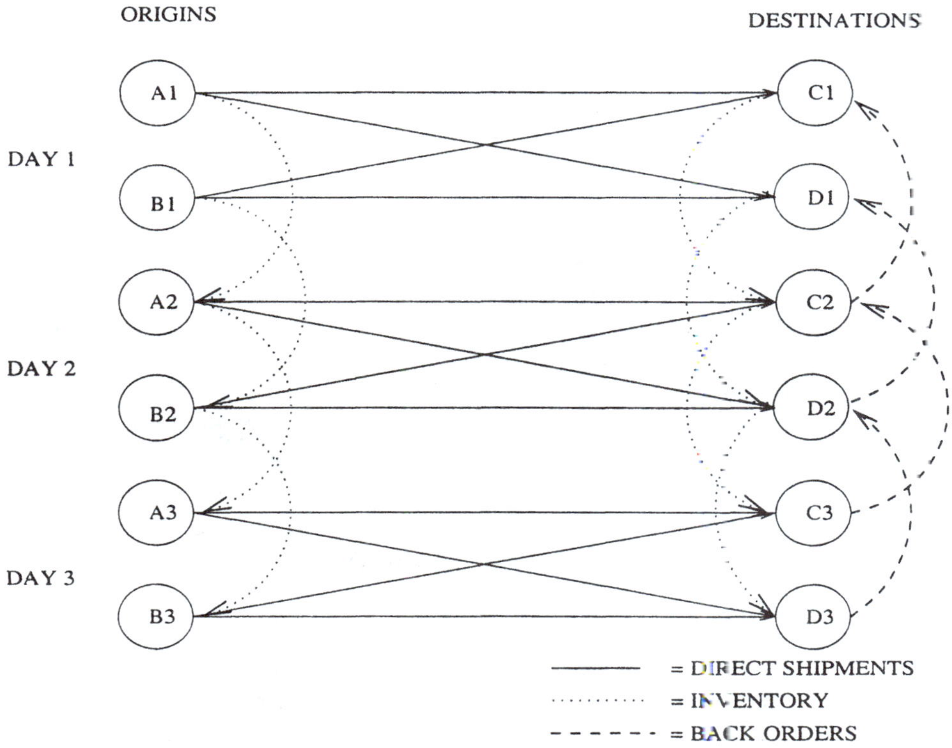

Figure 2 Time–space diagram

sidings, *A*, *B*, and *C*. The pass that occurs between trains II and III at siding *C* is feasible. However, a collision would then ensue between trains I and III, which could be avoided if train I waits at siding *B*.

2. Infrastructure Design and Maintenance

U.S. railroads are making significant investments in both track and railyards. Burlington Northern Santa Fe Railway (BNSF) has improved and added track along its mainline in five states in 1998. Union Pacific has expanded its lines in Iowa and recently upgraded a bridge crossing in Louisiana. Norfolk Southern opened new intermodal terminals in North Carolina and, jointly with Kansas City Southern Railway, in Texas. Despite this growth, few optimization models have been developed for rail infrastructure planning.

Higgins et al. (1995) address the problem of determining the location of sidings on a single rail line to minimize total tardiness for a planned train schedule. Construction costs are not considered. The solution method involves iteratively optimizing the train schedule (meet-pass plan, departure and arrival times of trains) for revised siding locations, and optimizing the locations of sidings for a revised train schedule. They report that modifying siding locations leads to a substantial reduction in tardiness.

The ambitious task of planning expansions of the Brazilian Railway is addressed by Crainic et al. (1990). Much of their work involves parameter estimation. The optimization model itself seeks to minimize the average cost per unit time of loaded and empty freight flows, expressed in tons, for a given network configuration. The model permits multiple transportation modes (e.g., different track gauges) and freight transfers between modes. The authors present an application of the model for proposed rail lines.

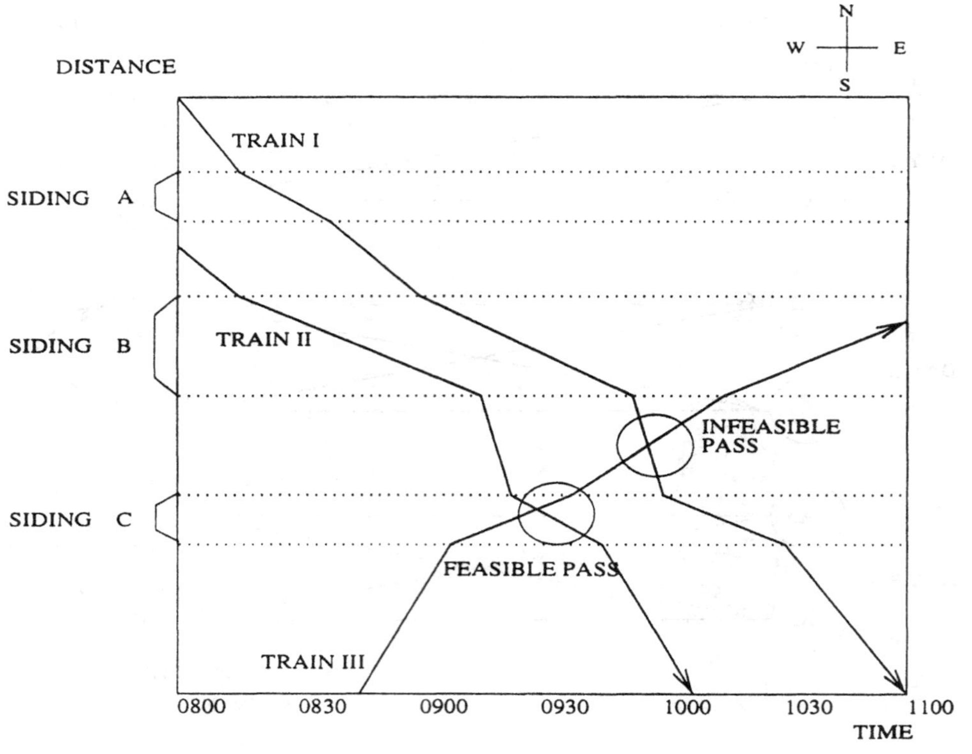

Figure 3 String diagram

Improved track maintenance reduces travel times, which increases labor productivity and customer service, and reduces the probability of derailment. LeBlanc (1976) develops a nonlinear optimization model in which investments in track improvements have decreasing marginal returns for improving train speed and decreasing operating costs. The problem is to simultaneously select investment levels (including implied abandonments) for each track segment and freight routing to minimize overall costs. For a particular relation between the investment cost and the variable transportation cost on each segment, he shows that the problem can be solved optimally.

Re-laying track involves moving used rail to a lower-traffic or lower-speed location where it can still be used effectively. Acharya et al. (1989) address the problem of determining when stretches of track should be re-laid after an expert system has identified candidates for replacement. There are economies of scale from re-laying nearby stretches of track, even if immediate replacement is not required. Operating costs, future maintenance and derailment costs, the salvage value of the re-laid rail, and the total value of the rail are considered. Constraints ensure that track is replaced before its wear limits have been reached. Labor, operating budgets, and equipment use may be limited. This constrained shortest-path problem is solved via Lagrangian relaxation of the capacity constraints.

A discussion of issues and models for rail maintenance planning, including the scheduling of rail repair facilities, appears in Genser (1982). Ferreira and Murray (1997) highlight the need for methods to maximize net financial benefit considering design standards, maintenance expenditures, and the capability of a rail segment to handle increased loads or train speeds.

3. Fleet Sizing and Repositioning

Minimizing empty car and locomotive repositioning contributes to reducing the ownership and operating costs for locomotives and railcars. In this section we address these issues for a fixed train schedule. The integrated problem of scheduling trains and routing loaded and empty railcars is discussed in Section 4. We first concentrate on repositioning and then discuss fleet sizing.

3.1. *Fleet Repositioning*

For a fixed train schedule, fleet management entails the assignment of loaded and empty railcars and locomotives to trains to satisfy shipment requirements and empty car demands. Most research on fleet management, however, takes the freight movements as given and optimizes empty car and/or locomotive repositioning.

Turnquist and Markowicz (1989) address the problem of minimizing the total cost of moving, holding inventory, and back-ordering empty cars. Their model permits multiple car types and limited substitutability among them. The problem is formulated as a single-commodity minimum-cost network flow problem. A decision support tool based on this model was used regularly by CSX from 1990 to 1996.

An interesting repositioning problem pertains to railroad autoracks, which are multilevel railcars that carry automobiles. An empty autorack generally had been returned to its last loading location, a practice that resulted in many empty car miles. In 1982, all railroads that move automobiles entered into a pooling agreement under which an unloaded autorack is repositioned to an economically efficient location, possibly another auto manufacturer's site.

Sherali and Suharko (1998) develop a decision support system, which is used daily by RELOAD, a central management group that controls autorack repositioning. It incorporates two discrete-time optimization models. The first model seeks, in effect, to minimize the maximum tardiness in filling demands for each autorack type at each location. Constraints ensure that each automaker's cumulative car-day utilization is consistent with its contribution to the pool. Each automaker's prioritization of its plants is incorporated via priority weights. Travel time uncertainty is considered by using chance constraints to exclude repositioning options that are unlikely to meet the target transit time. This problem can be decomposed by autorack type and solved by standard network flow algorithms. For the second model, which also considers blocking, the authors evaluate several heuristic procedures based on priority schemes and approximate dual prices. They also present a method to modify the solution to the first model to account for blocking.

The need to select among available locomotive-consists for assignment to trains is a major distinction between railcar and locomotive management problems. Ziarati et al. (1997) seek to optimize the schedule for locomotive movements between *power change points* (where locomotives may be transferred between trains) to support a one-week train schedule, requirements of *outposts* for locomotives handling local pickup and delivery, and scheduled maintenance. The subnetwork for each locomotive type is a single-commodity time-space network, with costs for hauling and deadheading. Bundle constraints limit the total locomotives of each type, and additional constraints ensure that assigned consists satisfy power requirements. The authors develop a branch-and-bound procedure with lower bounds derived using Dantzig–Wolfe decomposition. There is a subproblem for each locomotive type and the master problem links the types. The authors test the procedure on a data set from Canadian National North America for one week in 1994 (about 2000 trains) and report the potential for substantial savings.

Fleet management problems are typically formulated as large, integer programs that become extremely difficult to solve if there are complex constraints. Some researchers have developed approximate decomposition approaches as a step toward surmounting these challenges.

Powell and Carvalho (1998a) consider the problem of routing vehicles and assigning full or empty loads to them to maximize profit over a finite horizon. Each full load earns revenue only if delivered within a specified time window, and costs are incurred for both loaded and empty vehicle movements. They develop an approximate decomposition method for their dynamic program. The state of the system is defined by the vector of empty-vehicle inventories by location and the set of uncovered tasks. They assume that the value-to-go for additional vehicles at each time and location is a linear function, with the estimated incremental value as its coefficient. With this, the optimal solution for each state is determined easily using a greedy procedure: available vehicles should be dispatched or held in inventory in descending order of the incremental benefit of the available assignments. The coefficients are updated and the process repeats until the objective value converges.

Powell and Carvalho (1998b) apply the aforementioned approach to container and flatcar management. They also implement it within a decision support system for locomotive fleet management to be launched at the Norfolk Southern in 1999. A forecasting module predicts loads on outbound trains for the next 14 days. Using these forecasts, a scheduling system determines the assignments of locomotives to trains and estimates the incremental value of each locomotive type at each yard in each period. A real-time heuristic is employed to select a consist for each train during the upcoming 24 hours, considering factors such as repositioning costs and the value of each consist at the train's destination.

In reality, supplies and demands of railcars and travel times are stochastic. Jordan and Turnquist (1983), Crainic et al. (1993), and Beaujon and Turnquist (1991; see also Section 3.2), among others, model the stochastic aspects of this problem. For surveys on fleet management and empty flows, see Dejax and Crainic (1987), Haghani (1987), and Cordeau et al. (1998). Recent references also appear in Holmberg et al. (1998).

3.2. *Fleet Sizing*

Beaujon and Turnquist (1991) develop a multiperiod stochastic program to simultaneously determine fleet size and routing decisions to maximize expected revenues from moving loads less costs of empty and loaded car movement, ownership, lease, and shortages. Demands for loaded movements are stochastic and nonstationary. Travel times are represented by probability mass functions on the integers. The authors approximate the inventory and back-order costs as deterministic, nonlinear functions of the mean and variance of on-hand inventory or back orders at the corresponding time and location. They propose a heuristic that iterates between solving the underlying network flow problem for assumed variances (from the last iteration, if available) and computing updated variances resulting from the network flow solution. In computational experiments, the procedure produces solutions superior to those obtained from heuristics based on deterministic travel times.

Sherali and Tuncbilek (1997) address the problem of minimizing the number of additional autoracks required to satisfy demand in all time periods and locations. They propose a network flow model with time-varying demands, and with recirculation of flows from sink to source to account for end-of-horizon effects. The problem is decomposed heuristically into overlapping time intervals. This model is used annually by RELOAD to develop recommendations for autorack purchases and the apportionment of costs among the automakers.

Little research has been done on locomotive fleet sizing. Gertsbach and Gurevich (1977) address the problem of minimizing the size of a fleet of homogeneous locomotives to cover a fixed schedule by assigning a chain of transport segments to each locomotive. Their procedure relies on properties of good chains in constructing an optimal solution and is applicable to both finite-horizon and periodic, repeating schedules. Florian et al. (1976) appears to be the only paper that permits nonhomogeneous locomotive consists. The objective is to minimize the total cost of capital investment and maintenance for the locomotives. The model is an aggregate one that ensures a balance of locomotive flows among locations, but does not account for the effects of the train schedule and connections on fleet requirements.

Table 1 characterizes the articles reviewed in this section, including the decisions, major assumptions, objective function, and important constraints. Note that self-evident constraints, such as conservation of flow, are omitted. For reviews of fleet management and sizing, including empty car repositioning, see Dejax and Crainic (1987) and Haghani (1987).

Table 1 Literature Classification: Fleet Management and Fleet Sizing

Paper/Year	*Decisions*	*Major Assumptions*	*Objective Function*	*Constraints*
Turnquist and Markowicz (1989)	Empty car routing	• Back orders allowed • Fixed train schedule • Partial substitutability of car types	Minimize railcar transportation, holding, and back-order costs	• Supply and demand of cars by time and location
Sherali and Suharko (1998)	Empty car repositioning	• Back orders allowed • Fixed train schedule • Travel time uncertainty considered	Minimize maximum weighted tardiness	• Consistent cumulative car-day use vs. pool contribution
Ziarati et al. (1997)	Locomotive scheduling	• Known train makeup • Fixed train schedule	Minimize locomotive transportation costs	• Power requirements • Scheduled maintenance • Shop capacities • Locomotive demands at outposts
Powell and Carvalho (1998a)	Scheduling of empty and loaded movements	• Fixed train schedule • Single equipment type	Maximize revenue less vehicle movement costs	• Only loaded movements within time window earn revenue
Powell and Carvalho (1998b)	Flatcar/container or locomotive routing	• Fixed train schedule • Multiple equipment types	Maximize profit (as above)	• Container assignments to flatcars
Jordan and Turnquist (1983)	Empty car distribution	• Stochastic empty car supplies, demands, and travel times	Maximize expected revenue less car trans., holding, and shortage costs	• Car availability
Crainic et al. (1993)	Empty container distribution	• One container type • Demands, supplies partially dynamic and uncertain	Minimize sum of operating and expected holding and leasing costs	• Demand satisfaction • Limits on container movements between depots
Beaujon and Turnquist (1991)	Fleet sizing and vehicle routing	• Dynamic and stochastic demands • Travel times uncertain	Maximize revenue less expected car movement, use and shortage costs	—
Sherali and Tuncbilek (1997)	Fleet sizing	• Fixed train schedule • Time-varying demands	Minimize fleet size	• Demand satisfaction
Gertsbach and Gurevich (1977)	Fleet sizing, locomotive scheduling	• Fixed train schedule • Homogeneous fleet	Minimize fleet size	• Arrival and departure time
Florian et al. (1976)	Fleet sizing, locomotive scheduling	• Fixed train schedule • Multiple locomotive types	Minimize capital and maintenance costs	• Power requirements

4. Scheduling and Routing

In this section, we discuss train scheduling and freight routing, including blocking and train makeup decisions. These problems are addressed, both in research articles and in practice, at different levels of demand aggregation across time. The most aggregate models use average demand rates over an appropriate time horizon. At an intermediate level, customer shipments may be aggregated within each origin-destination pair, but not across time, leading to time-varying demands. At the most detailed level, the current system status is considered for real-time decisionmaking (not discussed here).

We first discuss meet–pass planning and closely related timetabling problems, followed by routing problems for conventional railcars, where blocking decisions are integral, and for intermodal traffic where blocking is not required. We then continue with joint train scheduling and freight routing problems, and finally discuss more comprehensive models.

4.1. Meet-Pass Planning and Timetabling

Most timetabling models involve single-track paths shared by multiple trains, often with bi-directional traffic. Brännlund et al. (1998) address a timetabling problem on a single rail line, assuming a constant velocity for each train. The track is divided into segments, each with a capacity of one train. The goal is to maximize profits from the selected train itineraries, where profits depend on the departure time and en route delay. The problem is solved via a Lagrangian heuristic in which the segment capacities are relaxed. See Carey and Lockwood (1995) for related work.

Sauder and Westerman (1983) develop a meet-pass planning system for a set of intersecting tracks at Southern Railway (a predecessor of Norfolk Southern Railway). Their model seeks to minimize total weighted delay, where tardiness is weighted according to the train's priority, and delays within the scheduled arrival time have unit weight. The optimal solution is obtained by enumerating all meets and passes. The meet-pass plans are integrated with a real-time simulator to support on-line decisions. In one and a half years of monitored performance, delay declined by 15 percent with steady or increasing volume. Savings in fuel, crew, and cost of capital were estimated at over $300,000 annually during the early 1980s.

Jovanovic and Harker (1991) develop the SCAN-I model to construct timetables and meet–pass plans over a set of interlinked traffic lanes, with a focus on robustness against travel time randomness. To assess whether a timetable is feasible under deterministic assumptions, they employ a branch-and-bound variant called a process-interaction simulation, which proceeds forward in time, resolving conflicts between trains sequentially, while avoiding deadlock. Backtracking occurs when a feasible continuation cannot be constructed. Whenever a schedule is identified that is feasible for the deterministic travel times, simulation is used to estimate the probability that the schedule is achievable for random travel times. The SCAN-I model has been used by BNSF for special studies.

See Kraay et al. (1991) for a timetabling approach allowing variable train velocity, which permits more flexibility in scheduling and fuel cost reductions.

4.2. Routing

The need to make blocking decisions complicates the routing of conventional railcars. Virtually all blocking models are based on aggregate, deterministic, steady-state assumptions. The first successful solution of a blocking optimization model appears to be that of Bodin et al. (1980) who utilize commercial software to solve their mixed-integer program. Van Dyke (1986) provides an excellent discussion of practical issues related to blocking.

Newton et al. (1998) seek to minimize total mileage, handling and delay costs subject to on-time delivery constraints that vary with priority class, and constraints on blocks and railcar handling at the yards. For each origin-destination pair, a set of blocking patterns and the railcar flow on each must be decided. They devise a branch-and-price procedure in which column generation is used to construct and price candidate paths. The binary blocking decisions are handled via branch-and-bound. The model is being extended and tested with a view toward implementation at CSX.

Kwon et al. (1998) formulate a multicommodity flow problem to determine car routes assuming the train schedule, blocking plan, and block-to-train assignments are given. The goal is to minimize late delivery penalties while ensuring demand is met, cars are appropriately assigned to blocks, blocks are appropriately assigned to trains, and train capacity restrictions are enforced. They use a column generation approach to solve realistic problem instances. This model has proved useful for modifying train schedules when the initial train schedule does not provide adequate customer service.

All major North American railroads currently use MultiRail software, which, among other things, assigns railcars to existing blocks. The objective is to minimize variable transportation and handling costs incurred for all transit segments, but costs may be modified to reflect routing preferences. Block and yard capacities are not explicitly modeled, allowing car routing to be determined using a shortest-path algorithm (Van Dyke, 1999).

Blocking decisions are not essential for intermodal freight because less reclassification occurs en route. Furthermore, intermodal goods are promised more rapid delivery, so due dates are important and steady-state models are no longer adequate. Nozick and Morlok (1997) address a finite-horizon, discrete-time problem of minimizing the total variable cost of moving loaded and empty trailers and flatcars given a fixed train schedule while satisfying due dates. They develop a procedure that involves iteratively solving a linear programming relaxation and rounding some of the resulting fractional values until a feasible integral solution is found. The heuristic is shown to provide good results.

4.3. Combined Scheduling and Routing Models

Morlok and Peterson (1970) introduce one of the first models for concurrent routing and scheduling decisions. The objective is to minimize the sum of fixed costs for trains, variable costs for transportation, handling and storage of freight, and opportunity costs of using rail equipment, while providing on-time deliveries of time-sensitive goods. Each potential train has a departure time, routing, set of stops, and an upper limit on cars. Decisions are which trains to operate and which freight to assign to each train. The authors apply branch-and-bound to solve a small instance of the resulting multicommodity network design problem.

Gorman (1998) treats the discrete-time problem of simultaneously deciding train service on all possible nonstop links and freight allocation. The model seeks to minimize the cost of labor, locomotive utilization, fuel, freight handling, and the opportunity cost for equipment use subject to on-time delivery, train capacity, railyard handling, and aggregate track capacity constraints. The proposed tabu-enhanced genetic search procedure is tested on small problems. Because of difficulties making widespread changes, a heavily constrained version of this model is used by BNSF for evaluating new routing and scheduling plans. A similar problem is addressed by Newman and Yano (2000), who present both centralized and decentralized approaches for solving the problem.

Huntley et al. (1995) devise a procedure for scheduling trains and routing cars to transport the one billion bushels of grain handled by CSX Transportation each fall. Each potential train is defined by a nonstop route between two locations and a departure time. Routing decisions are made for batches of cars with the same origin, time of availability at origin, and destination. The objective is to minimize a complex nonlinear cost function, including labor, fuel, freight car rental, and locomotive capital expenditures, while ensuring that no cars miss train connections, and no trains exceed car-carrying capacity. Starting from a simple initial solution, simulated annealing is used to search for improved solutions by considering addition or deletion of stops or changing train departure time. The company reports substantial savings from using this procedure.

Marín and Salmerón (1996) address an aggregate, steady-state freight planning model in which train routes (including stops), their frequency, and the number of cars using each service are determined. Costs include a fixed charge for each train, handling and delay costs, and costs of investments in additional trains. Constraints are imposed on the number of cars transported on each track segment, the number of cars using each yard, and the number of trains. They suggest heuristics in which service frequency decisions are handled by simulated annealing or tabu search and freight routing is addressed using a network flow model.

For surveys of rail scheduling and routing for freight, see Cordeau et al. (1998) and Assad (1980).

4.4. Integrated Models

More comprehensive models can provide better-coordinated decisions. The resulting size of such models, however, dictates that detailed decisions (e.g., blocking or train timing) and local constraints (e.g., yard capacity) must be ignored or modeled approximately.

Keaton (1989) examines the problem of simultaneously deciding which pairs of terminals are provided with direct train service and its frequency, car routing, and allocation of blocks to trains, to satisfy constant demand rates among the terminals. The objective is to minimize the average cost per unit time incurred from a fixed cost for each unlimited-capacity train, variable costs for the use of cars, and delays for classification and train assembly. Constraints are imposed on the number of blocks formed at each yard but delays due to block formation are not modeled. Keaton develops a Lagrangian procedure, relaxing constraints on the number of blocks, as well as a simpler heuristic in which, starting from a feasible solution, train frequencies are reduced and some connections are eliminated.

Haghani (1989) considers the discrete-time problem of simultaneously determining train service, train makeup, and loaded and empty car movements. The objective is to minimize the sum of fixed charges for trains, variable car movement, classification, delay and empty car back-order costs, and penalties for undelivered goods. Constraints ensure that the allocation of locomotives to each route is adequate to handle the flow of loaded and empty cars. The solution procedure involves heuristically rounding the train variables. An example illustrates that the integrated approach provides better customer service at a lower cost than a sequential solution approach.

Abel et al. (1981) develop a model for scheduling sugar cane harvesting times at growers and scheduling trains to transport it to mills in Queensland, Australia. Only one route exists between each grower and its associated mill, but a train need not visit all growers on its route. The goals, in ranked order, are to minimize total operating cost, minimize staling of sugar cane, minimize locomotive and railcar capital costs, and minimize the number of shifts required at the mills. The authors develop a heuristic to find the lowest-cost set of locomotive trips while satisfying demand. The timing of the trips is then determined considering the mill operating schedules. Mill labor cost considerations are handled by imposing constraints to ensure that sugar cane arrivals at the mills allow them to operate continuously while open. Equipment is added only if needed. The authors report that the solution provides improvements in operating cost, cane age, and capital requirements.

Tables 2 and 3 characterize the major articles reviewed in this section.

5. Conclusions and Directions for Further Research

Our review of many dozens of papers revealed areas where opportunities exist for development of new frameworks and paradigms. An important opportunity lies in the development of formal methods for linking decisions at different levels of the hierarchy. For example, aggregate, steady-state scheduling and routing models specify train service frequency and average freight flow patterns, but no systematic procedures exist for making short-term train scheduling and freight routing decisions as demand evolves using the results from the aggregate model as guidelines or constraints. Scheduling and routing models do exist for short-term, discrete-time problems with time-varying demands, but they do not account for any longer-term considerations.

We similarly found few papers that incorporate the effects of short-term and local dynamics upon longer-term, aggregate decision models. Turnquist and Daskin (1982) and Powell and Carvalho (1998a) take important steps, but more research is needed.

Another opportunity lies in extending models to handle system-wide circulation of locomotives and railcars. Nearly all discrete-time, finite-horizon scheduling and routing models

Table 2 Literature Classification: Scheduling and Routing, I

Paper/Year	*Decisions*	*Major Assumptions*	*Objective Function*	*Constraints*
Timetabling				
Brännlund et al. (1998)	Detailed train itineraries	• Single track • Fixed train velocity	Maximize value of selected itineraries	• Train capacity on each segment
Carey and Lockwood (1995)	Train paths and schedules	• One-way traffic • Fixed velocity	Minimize deviations from preferred departure time	• Arrival and departure time windows
Sauder and Westerman (1983)	Train timetables, meet-pass plans	—	Minimize total weighted delay	• Maximum train velocity • Siding length
Jovanovic and Harker (1991)	Detailed train timetables, meet-pass plans	• Fixed velocity • Travel times uncertain	(High probability of feasibility)	• Deadlock avoidance • Siding capacity
Kraay et al. (1991)	Train velocity profile and meet-pass plan	• Single track • Variable velocity	Minimize fuel consumption, deviation from schedule	• Maximum train velocity
Blocking				
Bodin et al. (1980)	Car classification strategy for all yards	• No freight priorities • Constant demand	Minimize transportation, classification and delay costs	• Demand satisfaction • Yard and block capacity • Block formation and strategy
Van Dyke (1999)	Assignment of cars to blocks	• Existing blocking plan • Train schedules fixed • Constant demand	Minimize transportation and handling costs	• Demand satisfaction
Newton et al. (1998)	Blocking patterns, assignment of cars to blocks	• Blocks with different priority classes • Constant demand	Minimize mileage, handling and delay costs	• Demand satisfaction • Number of containers classified and blocks built at each rail yard
Kwon et al. (1998)	Car routing	• Train schedules, block definitions, block-to-train assignments fixed • Time-varying demand	Minimize late delivery penalties	• Train capacity • Correct car-to-block, block-to-train assignments

Table 3 Literature Classification: Scheduling and Routing, II

Paper/Year	*Decisions*	*Major Assumptions*	*Objective Function*	*Constraints*
Intermodal Routing				
Nozick and Morlok (1997)	Assign equipment and loads to trains	• Fixed train schedule • Known demands	Minimize costs of repositioning and satisfying demand	• Demand satisfaction • Fleet size • Terminal capacity
Scheduling				
Morlok and Peterson (1970)	Train schedules and freight routes	• Deterministic demand and transit times	Minimize fixed and variable operating, storage, and opportunity costs	• Delivery windows • Power requirements • Train capacity
Gorman (1998)	Train schedules and freight routes	• Deterministic transit times • Time-varying demand • Limited feasible routes	Minimize fixed and variable operating and opportunity costs	• Freight due dates • Train, yard, and line capacity
Newman and Yano (2000)	Train schedules and freight routes	• Deterministic transit times • Time-varying demand	Minimize fixed and variable operating and storage costs	• Freight due dates • Train and line capacity
Huntley et al. (1995)	Train schedules and routes for car batches	• Deterministic transit times • Constant demand • No handling costs	Minimize fixed and variable transit, capital, and car rental costs	• Demand satisfaction • Train capacity
Marín and Salmerón (1996)	Train routes and car routing frequencies	• Deterministic transit times • Constant demand	Minimize fixed and variable transit, holding, handling, and investment costs	• Demand satisfaction • Yard, line, and train capacity • Limited cars on track
Integrated Models				
Keaton (1989)	Frequency of direct train service, block allocation, car routes	• Constant demand • Block formation delays ignored	Minimize fixed and variable car use and delay costs	• Demand satisfaction • Yard and train capacity
Haghani (1989)	Locomotive and loaded car routing and empty car distribution	• Constant demand	Minimize transit, congestion, classification, undelivered goods, and back-order costs	• Tractive power
Abel et al. (1981)	Train and harvesting schedule, fleet size	• Perishable freight	Minimize operating costs, staling loss, and capital costs	• Mill operation • Equipment availability • Train capacity

terminate in states of the system that may be undesirable in the longer term. Optimization methods can handle minimum cost circulation problems for periodic, repeating plans, but the problems become much more difficult when demands are not periodic. Likewise, considerable opportunity exists to account for randomness. Most of the models described above are deterministic, and because optimization methods identify extreme point solutions, they are often not robust to uncertainty.

There is opportunity to develop models with greater realism and solution methods that facilitate more effective implementation. Martland and Sussman (1995) point out that even fairly realistic optimization models must be customized to suit a railroad's operating policies and user capabilities.

Research areas that have become popular in other transportation industries are beginning to touch the rail industry, including demand management, yield management, and crew scheduling. Although there are many dozens of articles on optimization in the rail industry, the use of optimization in the industry is young, leaving many doors open for further contributions.

ACKNOWLEDGMENTS

We gratefully acknowledge information, comments, and suggestions provided by Carl Van Dyke, Multimodal; Ajith Wijeratne and Erick Wickum, Norfolk Southern; Todd McClain, Burlington Northern Santa Fe; Michael Bodden and David Hunt, ALK Associates; Harry Newton, U.S. Air Force; and Carl Martland, Joseph Sussman and Cynthia Barnhart, Massachusetts Institute of Technology.

References

Abel, D. J., K. P. Stark, C. R. Murry, and Y.-M. Demoulin. 1981. "A routing and scheduling problem for a rail system: a case study." *J. Oper Res. Soc.* **32**: 767–774.

Acharya, D., C. D. Martland, and J. M. Sussman. 1989. Using expert systems and optimization techniques for rail relay scheduling. In *Proceedings of the ASCE First International Conference on Advanced Technologies in Transportation Engineering*, San Diego, CA, pp. 378–383.

Ahuja, R. K., T. L. Magnanti, and J. B. Orlin. 1993. *Network Flows: Theory Algorithms, and Applications.* Englewood Cliffs, NJ: Prentice Hall.

Association of American Railroads. 1999. Web site: http://www.aar.com.

Assad, A. A. 1980. "Models for rail transportation." *Transport. Res. Part A* **14A**: 205–220.

Bard, J. F. 1997. "An analysis of a rail car unloading area for a consumer products manufacturer." *J. Oper. Res. Soc.* **48**: 873–883.

Beaujon, G. J. and M. A. Turnquist. 1991. "A model for fleet sizing and vehicle allocation." *Transport. Sci.* **25**: 19–45.

Bodin, L. D., B. L. Golden, A. D. Schuster, and W. Romig. 1980. "A model for the blocking of trains." *Transport. Res. Part B* **14B**: 115–120.

Brännlund, U., P. O. Lindberg, A. Nou, and J.-E. Nilsson. 1998. "Railway timetabling using Lagrangian relaxation." *Transport. Sci.* **32**: 358–369.

Caprara, A., M. Fischetti, P. Toth, D. Vigo, and P. L. Guida. 1997. "Algorithms for railway crew management." *Math. Programming* **79**: 125–141.

Carey, M. and D. Lockwood. 1995. "A model, algorithms and strategy for train pathing." *J. Oper. Res. Soc.* **46**: 988–1005.

Cordeau, J.-F., P. Toth, and D. Vigo. 1998. "A survey of optimization models for train routing and scheduling." *Transport. Sci.* **32**: 380–404.

Crainic, T. G., M. Florian, and J.-E. Leal. 1990. "A model for the strategic planning of national freight transportation by rail." *Transport. Sci.* **24**: 1–24.

Crainic, T. G., M. Gendreau, and P. Dejax. 1993. "Dynamic and stochastic models for the allocation of empty containers." *Oper. Res.* **41**: 102–126.

Dejax, P. J. and T. G. Crainic. 1987. "A review of empty flows and fleet management in freight transportation." *Transport. Sci.* **21**: 227–247.

Ferreira, L. and M. H. Murray. 1997. "Modelling rail track deterioration and maintenance: current practices and future needs." *Transport Rev.* **17**: 207–221.

Florian, M., G. Bushell, J. Ferland, G. Guerin, and L. Nastansky. 1976. "The engine scheduling problem in a railway network." *INFOR* **14**: 121–138.

Genser, R. 1982. "Maintenance planning in rail transportation." *Eur. J. Oper. Res.* **11**: 35–41.

Gertsbach, I. and Y. Gurevich. 1977. "Constructing an optimal fleet for a transportation schedule." *Transport. Sci.* **11**: 20–36.

Gorman, M. F. 1998. "Operating plan model improves service design at Santa Fe Railway." *Interfaces* **28(4)**: 1–12.

Haghani, A. E. 1987. "Rail freight transportation: a review of recent optimization models for train routing and empty car distribution." *J. Adv. Transport.* **21**: 147–172.

Haghani, A. E. 1989. "Formulation and solution of a combined train routing and makeup, and empty car distribution model." *Transport. Res. Part B* **23B**: 433–452.

Higgins, A., L. Ferreira, and E. Kozan. 1995. "Modeling single-line train operations." *Transport. Res. Record* **1489**: 9–16.

Holmberg, K., M. Joborn, and J. T. Lundgren. 1998. "Improved empty freight car distribution." *Transport. Sci.* **32**: 163–173.

Huntley, C. L., D. E. Brown, D. E. Sappington, and B. P. Markowicz. 1995. "Freight routing and scheduling at CSX transportation." *Interfaces* **25(3)**: 58–71.

Jordan, W. C. and M. A. Turnquist. 1983. "A stochastic, dynamic network model for railroad car distribution." *Transport. Sci.* **17**: 123–145.

Jovanovic, D. and P. T. Harker. 1991. "Tactical scheduling of rail operations: the SCAN-I system." *Transport. Sci.* **25**: 46–64.

Keaton, M. H. 1989. "Designing optimal railroad operating plans: Lagrangian relaxation and heuristic approaches." *Transport. Res. Part B* **23B**: 415–431.

Kraay, D., P. T. Harker, and B. T. Chen. 1991. "Optimal pacing of trains in freight railroads: model formulation and solution." *Oper. Res.* **39**: 82–99.

Kwon, O. K., C. D. Martland, and J. M. Sussman. 1998. "Routing and scheduling temporal and heterogeneous freight car traffic on rail networks." *Transport. Res. Part E* **34E**: 101–115.

LeBlanc, L. J. 1976. "Global solutions for a nonconvex nonconcave rail network model." *Mgmnt Sci.* **23**: 131–139.

Magnanti, T. L. and R. T. Wong. 1984. "Network design and transportation planning: models and algorithms." *Transport. Sci.* **18**: 1–55.

Marín, A. and J. Salmerón. 1996. "Tactical design of rail freight networks—Part I: Exact and heuristic methods." *Eur. J. Oper. Res.* **90**: 26–44.

Martland, C. D. and J. M. Sussman. 1995. Perspectives on rail systems modeling: what works and what doesn't. Presented at the Spring 1995 Meeting of the Institute for Operations Research and the Management Sciences, April 1995

Morlok, E. K. and R. B. Peterson. 1970. "Final report on a development of a geographic transportation network generation and evaluation model." *J. Transport. Res. Forum* **11**: 71–105.

Newman, A. and C. A. Yano. 2000. "Centralized and decentralized scheduling for intermodal operations." *IIE Transactions* **32:** 743–754.

Newton, H. N., C. Barnhart, and P. H. Vance. 1998. "Constructing railroad blocking plans to minimize handling costs." *Transport. Sci.* **32**: 330–345.

Nozick, L. K. and E. K. Morlok. 1997. "A model for medium-term operations planning in an intermodal rail-truck service." *Transport. Res. Part A* **31A**: 91–107.

Petersen, E. R. 1977a. "Railyard modeling: Part I. Prediction of put-through time." *Transport. Sci.* **11**: 37–49.

Petersen, E. R. 1977b. "Railyard modeling: Part II. Effect of yard facilities on congestion." *Transport. Sci.* **11**: 50–59.

Powell, W. B. and T. A. Carvalho. 1998a. "Dynamic control of logistics queuing networks for large-scale fleet management." *Transport. Sci.* **32**: 90–109.

Powell, W. B. and T. A. Carvalho. 1998b. "Real-time optimization of containers and flatcars for intermodal operations." *Transport. Sci.* **32**: 110–126.

Sauder, R. L. and W. M. Westerman. 1983. "Computer aided train dispatching: decision support through optimization." *Interfaces* **13(6)**: 24–37.

Sherali, H. D. and A. B. Suharko. 1998. "A tactical decision support system for empty railcar management." *Transport. Sci.* **32**: 306–329.

Sherali, H. D. and C. H. Tuncbilek. 1997. "Static and dynamic time–space strategic models and algorithms for multilevel rail-car fleet management." *Mgmnt Sci.* **43**: 235–250.

Turnquist, M. A. and M. S. Daskin. 1982. "Queueing models of classification and connection delay in railyards." *Transport. Sci.* **16**: 207–230.

Turnquist, M. A. and B. P. Markowicz. 1989. An interactive microcomputer-based model for railroad car distribution. Working Paper. Ithaca, NY: Cornell University.

Van Dyke, C. D. 1986. "Dynamic management of railroad blocking plans." *Transport. Res. Forum* **27**: 116–121.

Van Dyke, C. D. 1999. Algorithm-based blocking plan analysis. Working Paper. MulitModal Applied Systems.

Vasko, F. J., J. A. McNamara, D. D. Newhart, and F. E. Wolf. 1994. "A practical solution to a cargo loading problem at Bethlehem Steel." *J. Oper. Res. Soc.* **45**: 1285–1292.

Ziarati, K., F. Soumis, J. Desrosiers, S. Gelnis, and A. Saintonge. 1997. "Locomotive assignment with heterogeneous consists at CN North America." *Eur. J. Oper. Res.* **97**: 281–292.

18.4 Forestry Industry

Andrés Weintraub Pohorille and John Hof

ABSTRACT This article describes the methods and applications of operations research to forestry problems at the strategic, tactical, and operational levels. The impacts of these models on both public and private enterprises is also discussed. Special sections are devoted to environmental concerns and to dealing with risk and uncertainty.

1. Introduction

The field of forestry requires decisions at different levels. Forests produce timber and provide other important goods and services, such as range, recreation, and preservation of wildlife species and vegetation. The way forests are managed plays an important role in preserving the environment, through avoiding land erosion and preserving the water quality of rivers that flow through them.

The use of operations research (OR) tools in forestry is widespread and can be considered one of the successful areas in OR implementations. Linear programming (LP) was first used in the 1960s, mainly for strategic, long-term planning of forest harvesting. This initiative was pioneered mainly by the USDA Forest Service. Timber RAM (Navon, 1971) can be considered the first LP model with widespread use in forest planning. Developments followed different lines. As the concept of multiple use of forest and environmental concerns increased in the late 1970s and became very prevalent in the 1990s, models incorporated these issues and required more sophistication.

In parallel, the use of models to handle artificial plantations, whose output is primarily timber (for use at sawmills and pulp plants), posed different challenges. In these cases, no multiple use is expected, so environmental concerns focus on avoiding harmful consequences. Models in this setting concentrate more on efficient harvesting operations.

This article will consider problems in forestry at strategic, tactical, and operational levels, and will analyze models, solution processes, and their impact on decisonmaking in public and private enterprises. A special section will be devoted to spatial aspects derived from environmental concerns. We will also cover problems dealing with risk and uncertainty.

2. Strategic Planning: Land Allocation

Strategic planning looks at long-term horizons, typically involving more than one rotation (the time it takes for a full growth cycle). Depending on the species, this horizon can vary from 40 years for fast-growing pine plantations to 200 years for some native species. Decisions at this level are very aggregated, both in terms of land characterization and actions to be taken. When dealing with plantations, timber planting and harvesting is the main issue, whereas in native forests, other multiple uses and environmental issues are also incorporated. Models dealing with strategic decisions have been in use successfully for two or three decades. They are typically LP models.

If the only product included is timber, then the typical approach (referred to as Model I by Johnson and Scheurman, 1977) defines discrete time periods and limits to the age of harvest, so as to create a finite number of scheduling options on specified units of land. Assuming that the objective function is discounted net revenue maximization, the problem is typically formulated as:

Maximize:

$$\sum_{t=1}^{p} R/(1+r)^{T_t} \cdot H_t - \sum_{t=1}^{p} L/(1+r)^{T_t} \cdot H_t - \sum_{i=1}^{m} \sum_{j=1}^{n_i} \sum_{t=1}^{p} C_{ijt}/(1+r)^{T_t} \cdot X_{ij}$$

Subject to:

$$\sum_{j=1}^{n_i} X_{ij} \leqslant A_i \quad \forall i$$

$$\sum_{i=1}^{m} \sum_{j=1}^{n_i} V_{ijt} X_{ij} = H_t \quad \forall t$$

Where:

p = number of time periods in the planning horizon

m = number of initial age and productivity classes (analysis areas)

n_i = number of harvesting regimes for the ith analysis area

H_t = total harvest in the tth time period

X_{ij} = number of acres in the ith analysis area allocated to the jth management regime

A_i = number of acres in the ith analysis area

V_{ijt} = the yield volume per acre in the ith analysis area, in the jth harvesting regime, and in the tth time period

T_t = the time (in years) at the midpoint of time period t

r = the discount rate

R = the nominal revenue per unit of volume harvested

L = the nominal cost per unit of volume harvested

C_{ijt} = the nominal per-acre cost in time period t for the ith analysis area and in the jth harvesting regime.

Notice that discrete harvesting regimes with a limited number of harvesting ages are defined with discrete time periods, and yields are approximated accordingly.

It should be noted that the linear programming (discrete timing variable) approach is quite conducive to constraints such as nondeclining yield, formulated as:

$$H_t \leqslant H_{t+1} \quad t = 1, \ldots, p-1.$$

Nondeclining yield is a typical way to express the need for timber supply continuity (see also, Hof et al., 1986, for an alternative to these constraints). Another example of this type of constraint would be to define constraints on areas with different tree ages at the end of the planning horizon. This insures that the forest is well-regulated in terms of having a desirable distribution of different tree ages. Another common model structure for timber harvest scheduling is the one that Johnson and Scheurman (1977) refer to as Model II, originally discussed by Nautiyal and Pearse (1967) and Hallanger (1973). Instead of defining activities on specific units of land, as in Model I, Model II allows units of land to be regrouped if they have similar characteristics and if a harvest or other activity is applied to them in the same time period.

Well-known models that have been widely used in strategic planning are FORPLAN (Kent et al., 1991) and Spectrum (USDA Forest Service, 1995), both developed by the USDA Forest Service. Other strategic models have been developed for private plantations (Weintraub and Bare, 1996).

3. Tactical Planning

Decisions at this level are of horizons smaller than a rotation. Typically, for native forests with rotations of about 80 years, 20 to 30 years are considered, while for plantations, such as pines with rotations of 25 years, typical tactical horizons are 3 to 5 years. The main decisions usually involve the sequencing of the harvesting process, for example, what unit will be harvested what year or season so as to satisfy expected timber demands at different destinations over time. Timber is considered here in an aggregate way, as a single total volume or divided into a few categories, such as saw timber and pulp timber. Spatial definition of activities is considered at this level, in particular through the requirements of roads to access the areas to be harvested.

These models are mixed-integer LPs, with the integer variables representing road building or upgrading (improving the standard of the road from dirt to gravel, for example, to be able to use the road in rainy seasons). One of the first implementations of these models for native forests was carried out by the USDA Forest Service (Kirby et al., 1986) and a similar version was later implemented in FORPLAN, integrated with additional spatial issues related to nontimber aspects. Sessions and Sessions (1991) developed a widely used system, SNAP 3, which considers not only harvesting and road building but other factors such as wildlife habitat and water quality.

For private plantations the decision process is similar but simpler, as the environmental questions are less prominent. Solution procedures for these problems include heuristic approaches (Sessions and Sessions, 1991), which incorporate a sequence of partial decision processes to reach a good feasible solution. Other solution approaches are local search methods. In many cases, however, exact model representations have been used for native forests as well as plantations.

The models for both cases are similar, and next we describe a basic version (Kirby et al., 1986; Andalaft et al., 1996).

Let:

X_s^t = Number of hectares of unit s harvested in period t.

Y_{ik}^t = Volume of timber type k harvested in origin i in period t.

$F_{ij}^{k,t}$ = Flow of timber type k on arc (ij) in period t.

W_{ij}^t = 1 if road (ij) is built in period t.

0 otherwise

Constraints:

1. *Production by origin.*

$$\sum_{s\in[i]} a_{st}^k X_{st} = Y_{it}^t$$

for all origins i, products k, periods t, where a_{st}^k is the timber volume of type k produced if unit s is harvested in period t, and $[i]$ is the set of units that load at origin i.

2. *Harvesting by area.*

$\sum_t X_{st} \leqslant A_s$ for all units s.

The total area of a unit s harvested in all periods is bounded by the total area of the unit, A_s.

3. *Flow conservation.*

$\sum_p F_{rp}^{k,t} - \sum_q F_{qr}^{k,t} = -Y_{rk}^t$ if r is an origin node

$= 0$ if r is an intersection node

$= d_k^t$ if r is a demand node, d_k^t being the demand for product k in period t.

There are cases where a given demand can be satisfied by different products. For example, higher-quality timber can be used for lower-quality needs.

4. *Road building.*

$$\sum_k F_{ij,r}^{k,t} \leqslant U_{ij,r}^t \cdot \sum_{\theta \leqslant t} W_{ij}^{\theta} \quad \forall ij, t.$$

If (ij) is a potential road, there can be a flow on it up to the road capacity U_{ij}^t in period t only if it has been previously built.

Typical models also consider the use of multiple road standards, with the possibility of upgrading roads, storage of timber between seasons, and environmental concerns. Because this model is a mixed-integer LP model, it is often difficult to solve using commercial packages in a straightforward form. Improved solution approaches have included heuristics combined with LP models (Weintraub et al., 1994), adding logical inequalities, lifting and careful use of priorities (Guignard et al., 1998), as well as Lagrangian relaxation (Andalaft et al., 1996).

4. Operational Planning

Operational decisions typically have horizons of one day to several months and deal with harvesting standing timber and transporting it to destinations. These decisions must follow the guidelines provided by higher-level, longer-term decisions. This means that operational decisions must function within an already-established framework of total timber volumes to be harvested, budgets, demands, use of machinery, and existing road networks. There are different ways in which the operational decisions are considered. Often in public natural forests, independent firms carry out the operational tasks, usually after winning bids for the right to harvest. On the other hand, in large private plantations, it is typically the firm that owns the land that carries out the operational activities (often integrated with plants such as sawmills or pulp plants that are supplied by the harvested timber). This has meant that the more intense development of operational tools has been carried out by large private firms to manage their holdings, or by public land owners, such as the USDA Forest Service, in support of contractors.

The operational decisions involve the following aspects:

1. *Stands or cutting units harvested in each period (week, month) from among those available for harvesting.* A stand is available if it has mature trees and is accessible through existing main roads.

2. *How to satisfy demand from existing trees.* Trees have variable forms and, given their dimensions, can satisfy different needs. The lowest part of the tree of higher dimension is more costly and has uses in different sawmill options. The higher, thinner part of the tree, of lower value, is used primarily for pulp. In this form, there are different products for different uses, which carry varying prices, depending on the length, diameter, and grade (quality) of each piece. Defects such as knots limit the use of a log and considerably reduce the value that can be obtained from it. To match a determined product demand with existing standing trees is a hard combinatorial problem. If not well programmed, it is possible that there will be a poor fit between the characteristics of the products demanded and the standing timber supply, and losses will be incurred such as using higher-valued timber as pulp. This leads to the third operational decision, related to the first two.

3. *How to buck (cut into pieces) the trees in order to satisfy demands.* Bucking occurs in different forms. In some cases instructions are given to loggers in the field, to obtain specific products. Then the different products are transported to their destinations. At the other extreme, logs are cut up at the plant, controlled by a computerized system that considers the demand and price for alternative products and the characteristics of each tree (often determined through the use of X-rays to detect defects).

4. *Use of harvesting machinery*. Typically, once logs are cut they need to be transported to a landing to be loaded onto trucks. For flat terrain, skidders or tractors are used. For steeper slopes, cable logging or towers are used, where logs are pulled by cables in the air to a landing. Cable logging causes less soil damage in these circumstances. In addition to the machinery, secondary short roads must be built to link the harvesting operations with main roads

5. *Transportation of logs*. Once on a road or landing, logs are transported to their destination: mills, stocking yards, ports for exports. Typically trucks are used for this purpose. In some instances, trains can be a complementary alternative. In a few cases, helicopters (which are very expensive) and barges over rivers are also options.

The first three decisions can be viewed in an integrated way, since they interact, but there have also been applications that solve a subset of these three problems. The first two problems relate to which areas to harvest and the corresponding log allocation to satisfy demands. These problems have typically been addressed through LP, where decision variables indicate which units are to be harvested in a given period (weeks or months) and the distribution of the logs harvested to the different demand centers. There are a number of systems in use in Canada, the United States, Chile, New Zealand, and other countries. Examples include the FOLPI system used in New Zealand (Garcia, 1990) and LOGPLAN II (Newnham, 1991) in Canada. Papps and Manley (1992) link short- and long-term harvesting decisions.

A basic model for short-term harvesting and log allocation could be as follows.

Let:

$X_t^i = 1$ if unit i is harvested in period t.

Y_{ijk}^t = volume of product type k sent from unit i to destination j in period t.

a_i^k = volume of product of type k obtained in unit k.

r_{jk}^t = unit revenue obtained in period t from product type k in destination j.

C_i^t = cost of harvesting unit i in period t.

Then maximize:

$$V = \sum_t \left(\sum_j \sum_k r_{jk}^t \cdot \sum_i Y_{ijk}^t - \sum_i C_i^t X_i^t \right)$$

Subject to:

1. $\sum_t X_i^t = 1 \quad \forall i$

   All units are harvested.

2. $\sum_j Y_{ijk}^t = a_i^k \cdot X_i^t \quad \forall i, k, t$

Timber produced in each unit for each product type is sent to destinations. Additional constraints can relate to demand restrictions and machine availability.

In some cases, the log bucking decisions are also integrated with the decisions about which units to harvest and how to satisfy demands for specific products. OPTICORT (Epstein et al., 1999) is such a system, that integrates the first three operational decisions described. It is in use in several Chilean firms, and is based on an LP model with column generation, which supports decisions for units to be harvested, bucking patterns to be used, and allocation of products to demands. The column generation scheme is needed to generate satisfactory additional bucking patterns and is tied to a stand simulation model that indicates, for any given bucking pattern, the volume to be obtained for each product. For this integrated problem, Eng et al. (1986) proposed a Dantzig–Wolfe approach where the log-bucking patterns were produced in the subproblems. The difficulty with this approach was that too many patterns were created to be handled by operators in the field. Another column generation scheme was proposed by Mendoza and Bare (1986). Sessions et al. (1989) proposed a combined tree bucking and allocation approach based on heuristics. A similar scheme combining LP and tabu search was proposed by Laroze (1999) to support decisions that determine bucking patterns for a set of forest units, to satisfy given demands. CONDOR, a New Zealand system developed recently, supports decisions on: volume of timber to obtain, which harvest units to cut, and what bucking patterns should be allocated to individual logging crews.

A simplified form of the OPTICORT model, assuming one destination, is as follows (Epstein et al. 1999).

Let:

Y_{ikt} = Volume of product k (m^3) transported from stand i, in period t to meet the demand.

K_{ijt} = Volume of timber (m^3) produced in stand i using bucking pattern j in period t.

Constraints:

1. $\sum_{j,t} K_{ijt} \leqslant \mathrm{Vol}_i \quad \forall i$

In each stand i, the total timber harvested is bounded by the existing volume, Vol_i.

2. $\sum_j \mathrm{REN}_{ijk} \cdot K_{ijt} = Y_{ikt} \quad \forall i, k, t$

The amount of product k produced in stand i considering all bucking patterns used is transported. REN_{ijk} is the volume obtained of product k when pattern j is used in stand i.

3. $\sum_i Y_{ikt} = \mathrm{DD}_{k,t} \quad \forall k, t$

The demand DD_{kt} for each product k in each period t must be satisfied. Additional constraints relate to harvesting limitations due to machinery capacity, the existence of different destinations, and the need to satisfy quality (diameter) constraints for some products.

In many cases, both in the literature and in practice, bucking has been considered as an independent problem. Dynamic programming has been the traditional way of tackling this problem (Briggs, 1989). In the last few years, local search algorithms such as tabu search have also been proposed. A highly successful system was developed by Lemberski and Chi (1986), which combined an interactive visual system with dynamic programming and was used by Weyerhaeuser for personnel training and bucking decision processes.

The fourth operational decision corresponds to machine harvesting use. At present these decisions are based on digital terrain information provided by a GIS, which yields information on topography (height level curves), land quality, terrain accidents, water courses, existing roads, and standing timber characteristics. Several systems have been used. PLANS was developed by the USDA Forest Service (Twito et al., 1987) and PLANZ is used in New Zealand (Cossens et al., 1992). They are basically simulation tools that, for a given location of machinery indicated by the user, determine (in a visual interactive scheme) the area harvested by each machine and the design of access roads. PLANEX (Epstein et al., 1999) integrates a visual interactive interface with a decision process that simultaneously decides on optimal machine location and access roads to be built, based on a heuristic algorithm. This system is being used by the main forest firms in Chile. There are also systems that deal with the cable engineering part of the process; for example, the maximum payload a cable will be able to handle (Brown and Sessions, 1996).

Finally, the transportation issue is usually of high importance, as this item constitutes a significant proportion of overall operational costs. In most cases, the scheduling of transportation is done without the support of mathematical models, using radio or similar means for communication. Instructions are given by one or more operators who base their decisions on experience and predetermined criteria. At times, they are supported by visual aids such as a viewing screen showing the location of trucks. Decision support models have occasionally been implemented. These include ASICAM (Weintraub et al., 1996), which is a system that is used by about eight firms in Chile, as well as in South Africa and Brasil. The system is based on a simulation model with heuristic rules to support the simulation assignments. The system runs on a PC and takes a few minutes to schedule fleets of two hundred trucks. Another system based on an approximation of a real-time assignment model was developed by Ronnqvist and Ryan (1995) and was used in New Zealand.

5. Environmental/Spatial Concerns

At all levels of decisionmaking, environmental considerations are receiving ever-increasing emphasis in most resource management settings. In the simplest of cases, this can be handled analytically by including environmental variables as additional outputs, just as we included the timber outputs in Section 2. The primary reason that this does not always work is that environmental concerns often include a strong spatial element. In the forestry literature, the predominant perception of spatial considerations has been that of adjacency relationships. Concerns regarding adjacent timber harvests emerged from legal and regulatory restrictions on the effective size of harvests—if two seemingly legal-size clearcuts, for example, occur within a short time of each other in adjacent areas, the effective size of the clearcut may not be legal. Because the original motivations for the size limits included concerns for aesthetics, wildlife habitat, water quality, and so forth, addressing adjacency restrictions has often been interpreted (or even advertised) as addressing those underlying concerns. Hof and Bevers (1998) attempt to capture the specific spatial relationships or processes involved, and argue that adjacency constraints are only adequate when the problem is undesired adjacency of management actions, *per se*.

The basic formulation for avoiding adjacent harvests is:

$$X_1 + X_2 \leqslant 1$$

$$X_1, X_2 \in \{0, 1\},$$

where X_1 and X_2 are management prescriptions for areas that are adjacent. If "n-tuples" (triplets, quadruplets, etc.) of mutually adjacent areas can be defined for $i = 1, \ldots, I$, then:

$$\sum_i X_i \leqslant 1$$

$$X_i \in \{0, 1\} \quad \forall i$$

can cover the set. Many other sophistications are possible. These approaches require that the management variables be integer, and considerable effort has been expended toward solution of these problems—typically with heuristic methods such as tabu search or simulated annealing (Martell et al., 1998).

In many cases, it may be easier and more effective to directly include the spatial relationships of concern. For example, wildlife habitat fragmentation (spatial division into disaggregated units) is a common concern with regard to placement of timber harvests. Adjacency constraints may actually create fragmentation by spatially scattering harvests and the resultant habitats associated with different timber age classes over time. The underlying biological process that makes habitat fragmentation important is wildlife population dispersal (the process by which animals spread out into new areas). Whereas dispersal processes are nonlinear, they can be piecewise approximated for discrete time periods, much like we capture nonlinear growth and yield in linear programs for timber harvest scheduling. The basic approach is to include constraints of the form:

$$S_{it} \leqslant \sum_{j=1}^{J} g_{ij}(1 + r_j)S_{j(t-1)},$$

which relates population in habitat area i and in time period t, S_{it}, to the populations in areas $j = 1, \ldots, J$ in time period $t - 1$. The populations in $t - 1$ are expanded by the r-value growth rate $(1 + r_j)$ and dispersed by the g_{ij} coefficients, which account for the immigration from each area j to area i between time period $t - 1$ and t. When $i = j$, the g_{ij} coefficient indicates the portion of population in area i that remains in place during that time span. Population is determined by the most limiting factor: dispersal modeled in this fashion or carrying capacity as affected by management-variable solution values. Many other spatial problems involving water, exotics, spatial autocorrelation, and recreation can be addressed with optimization models using this basic approach (Hof and Bevers, 1998). This is an emerging area of study that has not yet been widely used.

6. Risk and Uncertainty

Uncertainty comes about both from biological/ecological systems and from market conditions. Market uncertainty is reflected primarily through future timber prices, and has generally been dealt with through stochastic dynamic programming and Monte Carlo simulation (Martell et al., 1998).

We have come to expect managed ecosystems to exhibit considerable random (some also say chaotic) behavior. The most common (and probably the most powerful) way to account for randomness in natural resource optimization models is the chance constraint. Charnes and Cooper (1963) provide what has become a classic analysis of chance constraints in linear programming. Their analysis focuses on problems where some or all of the right-hand sides are random variables. Charnes and Cooper's approach to this problem is to derive linear, deterministic formulations that are equivalent to the probabilistic problems. Their approach applies to the case where one wishes to constrain the model such that each right-hand side (input or output) is obtained with a prespecified probability. An example of an application of this approach to natural resources is Hunter et al. (1976).

Van de Panne and Popp (1963) extended the Charnes and Cooper analysis to the case of a random A-matrix (the matrix of decision variable coefficients for all constraints in the model). They start with a constraint of the form:

$$\sum_{j=1}^{n} a_{ij}x_j \geqslant b_i \quad \text{for a given } i.$$

Van de Panne and Popp assume that the a_{ij} are stochastically independent, but this assumption can be relaxed by applying the covariance matrix, as in Miller and Wagner (1965). Assuming that the a_{ij} have distributions with means α_{ij} and variances/covariances σ^2_{ijh}, a chance constraint that requires a certain probability of meeting the right-hand-side b_i can be written as:

$$\sum_{j=1}^{n} \alpha_{ij}x_j + \delta_i\left(\sum_{j=1}^{n}\sum_{h=1}^{n} x_j x_h \sigma^2_{ijh}\right)^{1/2} \geqslant b_i,$$

where δ_i is the standard deviation (sometimes called the tabular value, such as the z value for normally distributed random variables) corresponding to the required probability. This mathematical program is nonlinear, but Van de Panne and Popp (1963) show that (for normally distributed a_{ij}) it is convex if all required probabilities are $\geqslant \frac{1}{2}$. In this case, only a single, global optimum exists. With a 0.5 confidence level, $\delta_i = 0$ and only the mean levels of the a_{ij} (the α_{ij}) are needed. Another way of interpreting this point is that if means of the a_{ij} are inserted into the original linear program and variances/covariances are ignored, the solution should be interpreted as having a 0.5 probability of meeting each binding constraint individually. Between-row independence is necessary and relaxation of this assumption is, at this time, analytically intractable. In renewable resource optimization problems with multiple time periods, coefficients in different rows may be interrelated because one time period's status or yield is indirectly correlated with another time period's status or yield. The impact of ignoring between-row covariances when they are, in fact, nonzero is difficult to predict. Miller and Wagner (1965) extend the Van de Panne and Popp (1963) approach to a case where the aim is to constrain the joint probability of meeting a set of right-hand sides to be at least some prespecified constant, rather than to limit each constraint to meet its right-hand side with a certain probability. Examples of applications of this approach to natural resources include Hof et al. (1996) and Weintraub and Vera (1991). Given the difficulties in model implementation and in developing reliable probabilistic data, these modeling approaches have rarely been used in actual decisionmaking.

7. Conclusion

There are a number of applications of optimization to forestry problems that have not been discussed, such as multiobjective programming, multilevel or hierarchical optimization, fire and fuels management, and budget management. Obviously, a wide variety of problems in forestry are amenable to optimization analysis. Perhaps the greatest problem in the field is to integrate all of the different levels and types of decision analysis into a cohesive management strategy. In the future, an even wider variety of optimization applications to forestry is anticipated, as we embrace the challenge of creating a sustainable world society.

References

Andalaft, N., P. Andalaft, M. Guignard, A. Magendzo, A. Wainer, and A. Weintraub. 1996. "A problem of forest harvesting and road building." In *Proceedings VII CLAIO, Latin-Ibero American Congress on Operation Research and System Engineering, Brazil, August 1996*, pp. 1132–1136. Edited by OR Society of Brasil, Rio de Janeiro.

Briggs, D. G. 1989. "Tree value system: Description and assumptions." USDA Forest Service. General Technical Report PNW-239, Portland, Oregon.

Brown, C. and J. Sessions. 1996. "The standing skyline: a maximum log load solution process." *Forest Sci.* **42(2)**: 220–227.

Charnes, A. and W. W. Cooper. 1963. "Deterministic equivalents for optimizing and satisfying under chance constraints." *Oper. Res.* **11**: 18–39.

Cossens, P. 1992. "Planning and control of shortterm log allocation in New Zealand." *Integrated Decision-Making in Planning and Control of Forest Operations, Proceedings of IUFRO Conference*, January 27–31, 1992, Christchurch, New Zealand, New Zealand School of Forestry, University of Canterbury, Christchurch, New Zealand, pp. 46–56.

Eng, G., H. G. Daellenbach, and A. G. D. Whyte. 1986. "Bucking tree-length stems optimally." *Canadian J. Forest Res.* **16**: 1030–1035.

Epstein R., R. Morales, J. Serón, and A. Weintraub. 1999. "Use of OR systems in the Chilean Forest Industries." *Interfaces* **29(1)**: 7–29.

Epstein, R., E. Nieto, A. Weintraub, P. Chevalier, and J. Gabarró. 1999. "A system for short term harvesting." *Eur. J. Oper. Res* **119**: 427–439.

García, O. 1990. "Linear programming and related approaches in forest planning." *New Zealand J. Forest Sci.* **20(3)**: 307–331.

Guignard, M., C. Ryn, and K. Spielberg. 1998. "Model tightening for integrating timber harvest and transportation planning." *Eur. J. Oper. Res.* **111(3)**: 448–460.

Hallanger, W. 1973. "The linear programming structures of the problem." In *Sustainable Harvest Analysis 1971 and 1972* ed. C. J. Chambers and R. N. Pierson, pp. 67–71. Washington State Department of Natural Resources, Harvest Regulation Report 5.

Hof, J. and M. Bevers. 1998. *Spatial Optimization for Managed Ecosystems.* New York: Columbia University Press.

Hof, J. G., J. B. Pickens, and E. T. Bartlett. 1986. "A MAXMIN approach to nondeclining yield timber harvest scheduling problems." *Forest Sci.* **32(3)**: 653–666.

Hof, J., M. Bevers, and J. Pickens. 1996. "Chance-constrained optimization with spatially autocorrelated forest yields." *Forest Sci.* **42(1)**: 118–123.

Hunter, D. H., E. T. Bartlett, and D. A. Jameson. 1976. "Optimum forage allocation through chance-constrained programming." *Ecol. Modell.* **2**: 91–99.

Johnson, K. N. and H. L. Scheurman. 1977. "Techniques for prescribing optimal timber harvest and investment under different objectives: discussion and synthesis." *Forest Sci. Monograph* 18.

Kent, B., B. B. Bare, R. C. Field, and G. A. Bradley. 1991. "Natural resource land management planning using large-scale linear programs: the USDA forest service experience with FORPLAN." *Oper. Res.* **39(1)**: 13–27.

Kirby, M. W., W. Hager, and P. Wong. 1986. "Simultaneous planning of wildland and transportation alternatives." In *TIMS Studies in The Management Science Vol. 21*, edited by Kallio et al., pp. 371–387. New York: Elsevier Science Publishers.

Laroze, A. 1999. "A linear programming, tabu search method for solving forest-level bucking optimization problems." Forest Sci. **45(1)**: 108–116.

Lembersky, M. R. and K. N. Johnson. 1975. "Optimal policies for managed stands: An infinate horizon Markov decision process approach." Forest Sciente, 21(1): 109–122.

Martell, D., E. Gunn, and A. Weintraub. 1998. "Forest management challenges for operational researchers." Eur. J. Oper. Res. **104(1)**: 1–17.

Mendoza, G. A. and B. B. Bare. 1986. "A two-stage decision model for bucking and allocation." Forest Products J. **36(10)**: 70–74.

Miller, B. L. and H. M. Wagner. 1965. "Chance constrained programming with joint constraints." Oper. Res. **13**: 930–945.

Nautiyal, J. C. and P. H. Pearse. 1967. "Optimizing the conversion to sustained yield: a programming solution." Forest Sci. **13**: 131–139.

Navon, D. 1971. Timber RAM: a long term planning method for commercial timber lands under multiple use management. USDA Forest Service Pacific Southwest Research Paper 70, Berkeley.

Newnham, R. M. 1991. "LOGPLAN II: a model for planning logging and regeneration activities." Petawawa National Forestry Institute Information Report PI-X-102, p. 39. Forestry Canada, Chalk River, Ont.

Papps, S. R. and B. R. Manley. 1992. "Integrating short-term planning with long-term forest estate modeling using FOLPI." In Integrating Forest Information over Space and Time: Proceedings of a Conference, Canberra, Australia, 13–17 January 1992, pp. 188–198. ANUTECH Pty. P.O. Box 4, Canberra, ACT 2601, Australia.

Ronnqvist, M. and D. Ryan. 1995. "Solving truck dispatch problems in real time." In Proceedings on the 31th Annual Conference of the Operational Research Society of New Zealand, 31 August–1 September 1995,
pp. 165–172. Wellington. Edited by OR Society of New Zealand, Auckland.

Sessions J. and B. Sessions. 1991. "Tactical harvest planning." In Proceedings of 1991 Society of American Foresters National Convention, Bethesda, MD, 4–7 August 1991, Publication 91-05, pp. 362–368.

Sessions, J., E. Olsen, and J. Garland. 1989. "Tree bucking for optimal stand value with log allocation constraints." Forest Sci. **35**: 271–276.

Twito, R. H., S. E. Reutebuch, R. J. McGaughey, and C. N. Mann. 1987. "Preliminary logging analysis system (PLANS): Overview." USDA forest Service, general technical report PNW-199, Portland, Oregon.

USDA Forest Service, 1995. Spectrum User's Guide. Ecosystem Management Analysis Center, Fort Collins, CO.

Van de Panne, C. and W. Popp. 1963. "Minimum-cost cattle feed under probabilistic protein constraints." Mgmnt Sci. **9**: 405–430.

Weintraub, A. and B. Bare. 1996. "New issues in forest land management from an operations research perspective." Interfaces **26(5)**: 9–25.

Weintraub, A. and J. Vera. 1991. "A cutting plane approach for chance constrained linear programs." Oper. Res. **39**: 776–785.

Weintraub, A., G. Jones, A. Magendzo, M. L. Meacham, and M. W. Kirby. 1994. "A heuristic system to solve mixed-integer forest planning models." Oper. Res. **42**: 1010–1024.

Weintraub, A., R. Epstein, R. Morales, J. Serón, and P. Traverso. 1996. "A truck scheduling system improves efficiency in the forest industries." Interfaces, 26(4): 1–12.

18.5 Manufacturing Planning and Control

Stephen C. Graves

ABSTRACT Manufacturing planning and control entails the acquisition and allocation of limited resources to production activities so as to satisfy customer demand over a specified time horizon. As such, planning and control problems are inherently optimization problems, where the objective is to develop a plan that meets demand at minimum cost or that fills the demand that maximizes profit. The underlying optimization problem will vary due to differences in the manufacturing and market context. This article provides a framework for discrete-parts manufacturing planning and control and provides an overview of applicable model formulations.

1. Introduction

Manufacturing planning and control addresses decisions on the acquisition, utilization, and allocation of production resources to satisfy customer requirements in the most efficient and effective way. Typical decisions include work-force level, production lot sizes, assignment of overtime, and sequencing of production runs. Optimization models are widely applicable for providing decision support in this context.

In this article we focus on optimization models for production planning for discrete-parts, batch manufacturing environments. We do not cover detailed scheduling or sequencing models (e.g., Graves, 1981), nor do we address production planning for continuous processes (e.g., Shapiro, 1993). We consider only discrete-time models and do not include continuous-time models such as that developed by Hackman and Leachman (1989).

Our intent is to provide an overview of applicable optimization models; we present the most generic formulations and briefly describe how these models are solved. There is an enormous range of problem contexts and model formulations, as well as solution methods. We make no effort to be exhaustive in the treatment herein. Rather, we have made choices of what to include based on personal judgment and preferences.

We have organized the article into four major sections. In the first section we present a framework for the decisions, issues, and trade-offs involved in implementing an optimization model for discrete-part production planning. The remaining three sections present and discuss three distinct types of models. In the second section we discuss linear programming models for production planning, in which we have linear costs. This category is of great practical interest, as many important problem features can be captured with these models and powerful solution methods for linear programs are readily available. In the third section, we present a production-planning model for a single aggregate product with quadratic costs; this model is of historical significance as it represents one of the earliest applications of optimization to manufacturing planning. In the final section we introduce the multi-item capacitated lot-size problem, which is modeled as a mixed-integer linear program. This is an important model as it introduces economies of scale in production, due to the presence of production setups.

2. Framework

There are a variety of considerations that go into the development and implementation of an optimization model for manufacturing planning and control. In this section we highlight and comment upon a number of key issues and questions that should be addressed. Excellent general references on production planning are Thomas and McClain (1993), Shapiro (1993), and Silver et al. (1998).

Any planning problem starts with a specification of customer demand that is to be met by the production plan. In most contexts, future demand is at best only partially known, and often is not known at all. Consequently, one relies on a forecast for the future demand. To the extent that any forecast is inevitably inaccurate, one must decide how to account for or react to this demand uncertainty. The optimization models described in this article treat demand as being known; as such they must be periodically revised and rerun to account for forecast updates.

A key choice is what planning decisions to include in the model. By definition, production-planning models include decisions on production and inventory quantities. But in addition, there might be resource acquisition and allocation decisions, such as adding to the work force and upgrading the training of the current work force.

In many planning contexts, an important construct is to set a procedural hierarchy. Namely, one orders the decisions according to their relative importance. Hax and Meal (1975) introduced the notion of hierarchical production planning and provide a specific framework for this, whereby there is an optimization model with each level of the hierarchy. Each optimization model imposes a constraint on the model at the next level. Bitran and Tirupati (1993) provide a comprehensive survey of hierarchical planning methods and models.

The identification of the relevant costs is also an important issue. For production planning, one typically needs to determine the variable production costs, including setup, inventory holding, and resource acquisition. There might also be costs associated with imperfect customer service, such as when demand is back-ordered.

A planning problem exists when there are limited production resources that cannot be stored from period to period. Choices must be made as to which resources to include and how to model their capacity and behavior, and their costs. Also, there may be uncertainty associated with the production function, such as with yields or lead times. One might only include the most critical or limiting resource in the planning problem, for example, a bottleneck. Alternatively, when there is not a dominant resource, one must model the resources that could limit production. In this article we describe two types of production functions. The first assumes a linear relationship between production quantity and resource consumption. The second assumes that a fixed charge or setup is required to initiate production and then a linear relationship between the production quantity and resource usage.

Related to these choices is the selection of the time period and planning horizon. The literature distinguishes between big-bucket and small bucket time periods. A time period is big bucket if multiple items are typically produced within a certain time period; a small bucket is such that at most one item would be produced in the time period. For big-bucket models, one has to worry about how to schedule or sequence the production runs assigned to any time period. The choice of planning horizon is dictated by the lead times to enact production and resource-related decisions, as well as the quality of knowledge about future demand.

Planning is typically done in a rolling horizon fashion. A plan is created for the planning horizon, but only the decisions in the first few periods are implemented before a revised plan is issued. Indeed, as noted above, periodic revision is necessary due to the uncertainties in demand forecasts and production. For instance a firm might envisage their strategy the next twenty six weeks, but revise once a month to incorporate new information on demand and production.

Production planning is usually done at an aggregate level, for both products and resources. Distinct but similar products are combined into product families consolidated so as to reduce planning complexity. Similarly, production resources, such as distinct machines or labor pools, are consolidated into a machine or labor resource. Care is required when specifying these aggregates to assure that the result can be reasonably separated into feasible production schedules.

Finally, for complex products, one must decide the level and extent of the product structure to prepare for. For instance, in some contexts it is sufficient to just arrange the production of end items; components and subassemblies are subservient to the master schedule for end items. In other contexts, planning just the end items is suboptimal, as there are critical resource constraints applicable to multiple levels of the product structure. In this instance, a multistage model allows for simultaneously mapping out end items and components or subassemblies. Of course, this produces a much larger model.

3. Production Planning: Linear Programming Models

In this section we develop and state the most basic optimization model for production planning for the following context:

- multiple items with independent demand,
- multiple shared resources,
- big-bucket time periods,
- linear costs.

We define the following notation

Decision variables

p_{it} production of item i during time period t

q_{it} inventory of item i at end of time period t

Parameters

T, I, K number of time periods, items, resources, respectively
a_{ik} amount of resource k required per unit of production of item i
b_{kt} amount of resource k available in period t
d_{it} demand for item i in period t
cp_{it} unit variable cost of production for item i in time period t
cq_{it} unit inventory holding cost for item i in time period t.

We now formulate the linear program P1:

$$P1: \quad \text{Min} \sum_{t=1}^{T} \sum_{i=1}^{I} cp_{it}p_{it} + cq_{it}q_{it} \tag{1}$$

subject to

$$q_{i,t-1} + p_{it} - q_{it} = d_{it} \quad \forall i, t \tag{2}$$

$$\sum_{i=1}^{I} a_{ik}p_{it} \leqslant b_{kt} \quad \forall k, t \tag{3}$$

$$p_{it}, q_{it} \geqslant 0 \quad \forall i, t.$$

The objective function (1) minimizes the variable production costs plus the inventory holding costs for all items over the planning horizon of T periods.

Equation (2) is a set of inventory balance constraints that equate the supply of an item in a period with its demand or usage. In any period, the supply for an item is the inventory from the prior phase $q_{i,t-1}$, plus the production in the period p_{it}. This supply can be used to meet demand in the period d_{it} or held in inventory as q_{it}. As we require the inventory to be non-negative, these constraints assure that demand is satisfied for each item at each stage. We are given as input the initial inventory for each item, namely q_{i0}.

Equation (3) is a set of resource constraints. Production in each period is limited by the availability of a set of shared resources, where the making of one unit of item i requires a_{ik} units of resource k, for $k = 1, 2, \ldots K$. Typical resources are various types of labor, process and material handling equipment, and transportation modes.

The number of decision variables is $2IT$, and the number of constraints is $IT + KT$. For any realistic problem size, we can solve P1 by any good linear programming algorithm, such as the simplex method.

We briefly describe next a number of important extensions to this basic model. We introduce these as if they were independent; however, we note that many contexts require a combination of these extensions.

3.1. Demand Planning: Lost Sales

For some problems we have the option of not meeting all demand in each time period. Indeed, there might not be sufficient resources to meet all demand. In effect, the demand parameters represent the demand potential, and the optimization problem is to decide what demand to meet and how. We assume that demand that cannot be met in a period is lost, thus reducing revenue. In addition, a firm might incur a loss of customer goodwill that would manifest itself in terms of reduced future sales. This lost-sales cost is very difficult to quantify as it represents the future unknown impact from poor service today.

We pose a new planning problem to maximize revenues net of the production, inventory, and lost-sales costs. We introduce additional notation and then state the model:

Decision variables

u_{it} unmet demand of item i during time period t

Parameters

r_{it} unit revenue for item i in period t

cu_{it} unit cost of not meeting demand for item i in time period t

$$P2: \quad \text{Max} \sum_{t=1}^{T} \sum_{i=1}^{I} [r_{it}(d_{it} - u_{it}) - cp_{it}p_{it} - cq_{it}q_{it} - cu_{it}u_{it}]$$

subject to (3)

$$q_{i,t-1} + p_{it} - q_{it} + u_{it} = d_{it} \quad \forall i, t$$

$$p_{it}, q_{it}, u_{it} \geqslant 0 \quad \forall i, t.$$

The objective function has been modified to include revenue as well as the cost of lost sales. The potential revenue, $\Sigma\Sigma r_{it}d_{it}$, is a constant and could be dropped in the objective function. In this case, we can restate the problem as a cost-minimization problem, where the cost of lost sales includes the lost revenue.

Also, in P2, the inventory balance constraint has been modified to permit the option of not meeting demand; thus demand in a period can be met from production or inventory, or not satisfied at all. The resource constraint (3) remains unchanged.

3.2. *Demand Planning: Back Orders*

A related problem variation is when it is possible to reschedule or back-order demand. That is, we might defer current demand until a later period, when it can be served from production. Of course there is a cost for doing this, which we term the back-order cost. Like the lost-sales cost, the back-order cost includes hard-to-quantify costs due to loss of customer goodwill, as well as reduced revenue and additional processing or expediting costs due to the deferral of the demand fulfillment. We assume that this cost is linear in the number of back orders in each period.

We define additional notation and then state the model.

Decision variables

v_{it} back-order level for item i at end of time period t

Parameters

cv_{it} unit cost of back order for item i in time period t

$$P3: \quad \text{Min} \sum_{t=1}^{T} \sum_{i=1}^{I} [cp_{it}p_{it} + cq_{it}q_{it} + cv_{it}v_{it}]$$

subject to (3)

$$q_{i,t-1} - v_{i,t-1} + p_{it} - q_{it} + v_{it} = d_{it} \quad \forall i, t$$

$$p_{it}, q_{it}, v_{it} \geqslant 0 \quad \forall i, t$$

In comparison with P1, we now include a back-order cost on the objective function for P3. The inventory balance equation is modified to account for the back orders, which in effect behave like negative inventory. We would typically add a terminal constraint on the back orders at the end of the planning horizon; for instance, we might require $v_{iT} = 0$, so that over the T-period planning horizon all demand is eventually met by the production plan. Any initial back orders, namely v_{i0}, can be dropped by adding them to the first-period demand; that is, we restate the demand as $d_{i1} := d_{i1} + v_{i0}$, and then drop v_{i0} from the formulation.

In this formulation, when demand is back-ordered, the cost of this event is linear for the size and duration of the back order. That is, if it takes n time periods to fill the back-order, the cost is proportional to n. In contrast, in some cases, the back order cost might not depend on the duration but only on the occurrence and size of the back-order. We can modify the formulation for this case by defining a variable to represent new back orders in period t, given by max $[0, v_{it} - v_{i,t-1}]$; then we would apply the back order cost to this variable in the objective function. This modification assumes that we fill back orders in a last-in, last-out fashion, as there is no incentive to do otherwise for this cost assumption.

3.3. Piecewise Linear Production Cost Functions

In P1 the relevant production cost is linear in the production quantity. In many contexts the actual cost function is nonlinear. In this section, we consider a convex-cost function, and assume that it is well modeled as a piecewise linear function. We model concave-cost functions that exhibit economies of scale in later sections.

Let Cost(p_{it}) denote the cost function for item i in period t; we present the case where this cost function is the same in each period, and introduce the following notation:

Decision variables

p_{ist} production of item i during time period t, that falls in cost segment s

Parameters

S number of segments in cost function

cp_{ist} unit variable cost of production for item i in time period t in cost segment s

P_{is} upper bound on cost segment s for item i.

Thus, we assume that Cost(p_{it}) is given by:

$$Cost(p_{it}) = \sum_{s=1}^{S} cp_{is}p_{ist}$$

where

$$p_{it} = \sum_{s=1}^{S} p_{ist}$$

$$0 \leqslant p_{ist} \leqslant P_{is} \quad \forall s.$$

In order for Cost(p_{it}) to be convex, we require that the unit variable costs be strictly increasing from one segment to the next:

$$0 < cp_{i1t} < cp_{i2t} < \ldots < cp_{iSt}.$$

This cost function applies when there are multiple options or sources for production, and these options can be ranked by their variable costs. A common example is when one models regular time and overtime production. We have two cost segments ($S = 2$), where the first segment corresponds to production during regular time, and the second is overtime production. The variable production cost is usually more during overtime, as workers earn a rate premium. Another example is when the firm works multiple shifts and the variable costs differ between these shifts. A third example is when there are subcontracting or outsourcing options; there are multiple-costs segments, one to represent in-house production and the others to represent the outsourcing options ranked by cost.

We model the planning problem with convex, piecewise linear production cost functions by replacing the production cost in P1 with the above formulation for Cost(p_{it}):

$$P4: \quad \text{Min} \sum_{t=1}^{T} \sum_{i=1}^{I} \left(cq_{it}q_{it} + \sum_{s=1}^{S} cp_{is}p_{ist} \right)$$

subject to (2)

$$\sum_{i=1}^{I} \sum_{s=1}^{S} a_{iks}p_{ist} \leqslant b_{kt} \quad \forall k, t$$

$$p_{it} - \sum_{s=1}^{S} p_{ist} = 0 \quad \forall i, t$$

$$0 \leqslant p_{ist} \leqslant P_{is} \quad \forall i, s, t$$

$$p_{it}, p_{ist}, q_{it} \geqslant 0 \quad \forall i, s, t.$$

In P4 we have modified the resource constraints (3) to accommodate the possibility that the usage of the shared resources depends on the production quantity by source or cost segment, that is, p_{ist} rather than just on p_{it}. In this case, a_{iks} denotes the amount of resource k required per unit of item i produced at source or cost segment s. This form permits great flexibility in modeling production costs as well as resource constraints.

In one of the first papers on production planning, Bowman (1956) formulates this problem as a transportation problem, when there are multiple time periods and multiple production options, but only one item and one resource type.

3.4. *Resource Planning*

Up to now we have assumed that the resource levels are fixed and given. In some cases, an important element of the planning problem is to decide how to adjust the resource levels over the planning horizon. For instance, one might be able to change the work force level, by means of hiring and firing decisions. Hansmann and Hess (1960) provide an early example of this type of model.

Suppose for ease of notation that we have just one type of resource, namely the work force. We introduce additional notation and then state the model:

Decision variables

w_t work-force level in time period t

h_t change to work-force level by hiring in time period t

f_t change to work-force level by firing in time period t

Parameters

a_i amount of work force (labor) required per unit of production of item i

cw_t variable unit cost of work force in time period t

ch_t variable hiring cost in time period t

cf_t variable firing cost in time period t

$$P5: \quad \text{Min} \sum_{t=1}^{T} [cw_t w_t + ch_t h_t + cf_t f_t] + \sum_{t=1}^{T} \sum_{i=1}^{I} [cp_{it}p_{it} + cq_{it}q_{it}]$$

subject to (2)

$$\sum_{i=1}^{I} a_i p_{it} - w_t \leqslant 0 \quad \forall t$$

$$w_{t-1} + h_t - f_t - w_t = 0 \quad \forall t$$

$$p_{it}, q_{it}, w_t, h_t, f_t \geqslant 0 \quad \forall i, t.$$

We add the variable cost for the work force to the objective function, along with expenses for hiring and firing workers. Hiring costs include finding and training applicants. Firing costs include outplacement and retraining of displaced workers, as well as severance costs; there might also be a cost of lower productivity due to lower work-force morale, when firings or layoffs occur.

The inventory balance constraint (2) remains the same as for P1, and we restate the resource constraint, reflecting the work force as a decision variable and as the sole resource. We then add a new set of balance constraints for planning the work force: the work force in period t is that from the prior period plus new hires minus the number fired.

We have stated P5 for a single resource, representing the work force. The model extends immediately to include other resources that might be managed in a similar fashion over the planning horizon. In addition, there might be other considerations to model such as time lags when adjusting a resource level. There might be limits on how quickly new workers can be added due to training requirements. If there were limited training resources, then this imposes a constraint on h_t. Alternatively, new hires might be less productive until they have acquired some experience. In this case, we modify the formulation to model different categories of workers, depending on their tenure and experience level.

Another common variation of this model is when there are two labor classes, for example, permanent employees and temporary employees. These classes differ in terms of their cost coefficients and possibly their efficiency factors. Permanent employees have higher hiring and firing cost, as they receive more training and have more rights and protection from layoffs. But their variable production cost, normalized by their productivity, should be lower than that for temporary workers. The planning problem then entails the management and planning of both work classes over the planning horizon. For completeness, we revise P5 for two work-force classes:

Decision variables

w_{jt} work-force level in time period t, for labor class j

h_{jt} change to work-force level by hiring in time period t, for labor class j

f_{jt} change to work-force level by firing in time period t, for labor class j

p_{ijt} production of item i during time period t, using labor class j

Parameters

a_{ij} amount of labor required per unit of production of item i, using labor class j

cw_{jt} variable unit cost of labor class j in time period t

ch_{jt} variable hiring cost for labor class j in time period t

cf_{jt} variable firing cost for labor class j in time period t

$$P6: \quad \text{Min} \sum_{t=1}^{T} \sum_{j=1}^{2} [cw_{jt}w_{jt} + ch_{jt}h_{jt} + cf_{jt}f_{jt}] + \sum_{t=1}^{T} \sum_{i=1}^{I} [cp_{it}p_{it} + cq_{it}q_{it}]$$

subject to (2)

$$p_{it} - \sum_{j=1}^{2} p_{ijt} = 0 \quad \forall i, t$$

$$\sum_{i=1}^{I} a_{ij}p_{ijt} - w_{jt} \leqslant 0 \quad \forall j, t$$

$$w_{j,t-1} + h_{jt} - f_{jt} - w_{jt} = 0 \quad \forall j, t$$

$$p_{it}, q_{it}, w_{jt}, h_{jt}, f_{jt} \geqslant 0 \quad \forall i, j, t.$$

In comparison to P5, we have decision variables for both labor classes, as well as for their hiring and firing decisions, in order to model the cost differences. We also introduce production decision variables, by labor class, to capture the differences in productivity.

3.5. Multiple Locations

In P1, there is a single supply location or production facility that serves demand for all items. Often there are multiple production facilities that are geographically dispersed and that supply multiple distribution channels. The planning problem is to determine the production, inventory, and distribution plans for each facility to meet demand, which is now modeled by geographic region.

There are many ways to formulate this type of problem. We provide an example, and then comment on some variants. We define the notation and then state the model.

Decision variables

p_{ijt} production of item i at facility j during time period t

q_{ijt} inventory of item i at facility j at end of time period t

x_{ijmt} shipment of item i from facility j to demand location m in time period t

Parameters

T, I, K number of time periods, items, resources, respectively

J, M number of facility locations, demand locations, respectively

a_{ijk} amount of resource k required per unit of production of item i at facility j

b_{jkt} amount of resource k available at facility j in period t

d_{imt} demand for item i at location m in period t

cp_{ijt} unit variable cost of production for item i at facility j in time period t

cq_{ijt} unit inventory holding cost for item i at facility j in time period t

cx_{ijmt} unit transportation cost to ship item i from facility j to demand location m in time period t

$$P7: \quad \text{Min} \sum_{t=1}^{T} \sum_{i=1}^{I} \sum_{j=1}^{J} \left[cp_{ijt}p_{ijt} + cq_{ijt}q_{ijt} + \sum_{m=1}^{M} cx_{ijmt}x_{ijmt} \right]$$

subject to

$$q_{ij,t-1} + p_{ijt} - q_{ijt} - \sum_{m=1}^{M} x_{ijmt} = 0 \quad \forall i, j, t$$

$$\sum_{j=1}^{J} x_{ijmt} = d_{imt} \quad \forall i, m, t$$

$$\sum_{i=1}^{I} a_{ijk}p_{ijt} \leqslant b_{jkt} \quad \forall j, k, t$$

$$p_{ijt}, q_{ijt}, x_{ijmt} \geq 0 \quad \forall i, j, m, t.$$

The decision variables for production and inventory are now specified by location, where inventory is held at the production locations. In addition, we have aggregated demand into a set of demand regions or locations, and introduce a new set of decision variables to denote transportation from the production facilities to the demand locations. The objective function captures production and inventory holding cost, which depends on the facility, plus transportation or distribution costs for moving the product from a facility to the demand location. The inventory balance constraints assure that the supply of an item at each facility is either held in inventory or shipped to a demand location to meet demand. The second set of constraints assures that the shipments satisfy the demand for each period at each location. The resource constraints are structurally the same as in P1, but for multiple production locations.

A key variant of this model occurs when there are additional stocking locations, such as a network of warehouses or distribution centers. These stocking locations not only provide additional space to store inventory close to the demand locations, but also permit economies of scale in transportation from the production sites to these stocking locations. In this case, one defines shipments to and from each stocking location, and has inventory balance constraints for each stocking location. The shipment costs would capture any differences in transportation modes that might be employed.

The size of P7 creates a challenge for implementing and maintaining such a model. There are now $IJT(2+M)$ decision variables and $(IJT+IMT+JKT)$ constraints. A typical problem might have 20–100 aggregate item families, 5–10 facilities, 50–100 demand locations, 12–20 time periods, and 1–5 resource types. Thus, the model might have in the order of 100,000 to 1 million decision variables, and in the order of 100,000 constraints. As the problem is still a linear program, such problems are readily solved by commercial optimization packages. However, the real difficulty in such implementations is the development and maintenance of the parameters. There can be in the order of 1 million demand forecasts, 100,000 to 1 million cost coefficients, and 10,000 resource coefficients. The success of many applications often rests on whether this data can be obtained and kept accurate.

3.6. Dependent-Demand Items

So far we have considered production plans for end items or finished goods, which see independent demand. But these end items are usually comprised of many fabricated components and subassemblies, for which there also needs to be a production plan. These items differ from the end items, in that their demand depends on the end-item production plan. Material requirements planning (MRP) systems are designed to characterize this dependent demand and to facilitate the planning for these dependent-demand items in a coordinated and systematic way (Vollman et al. 1992). Nevertheless, it is quite easy to incorporate dependent-demand items into the optimization models presented here.

We first need to define what goes into matrix $A = \{\alpha_{ij}\}$, where α_{ij} is the number of units of item i required per unit of production of item j. Then for the model P1 we replace the inventory balance constraints (2) with the following:

$$q_{i,t-1} + p_{it} - q_{it} - \sum_{j=1}^{I} \alpha_{ij} p_{jt} = d_{it} \quad \forall i, t.$$

In this general form, the demand for item i has two parts: exogenous demand given by d_{it}, and endogenous or induced demand given by $\Sigma \alpha_{ij} p_{jt}$. For end items, we expect there to be no induced demand, whereas for components, there will typically be no or limited exogenous demand.

One variation to this model is when there are manufacturing lead times, whereby production of item j in time period t requires that component i be available at time $t-L$ where L is the lead time for producing j. The inventory balance constraints can easily be modified to accommodate this, given that the lead times are known and deterministic.

Billington et al. (1983) provide a comprehensive treatment of this model and problem, and present methods for reducing the size of the problem so as to facilitate its solution. In the literature, this problem is referred to as a multiple-stage problem, where end items, subassemblies, and components might represent distinct stages in a manufacturing process.

4. Quadratic Cost Models and Linear Decision Rule

One of the earliest production-planning modeling efforts was that of Holt et al. (1960), who developed a model for the Pittsburgh Paint Company. They assume a single aggregate product, and then define three decision variables:

p_t production of the aggregate item during time period t

q_t inventory of the aggregate item at end of time period t

w_t work-force level in time period t.

Holt et al. (1960) assume that the cost function in each period has four components. The first component is the regular payroll costs that are a linear function of the work-force level. The second component is the hiring and layoff costs, which were assumed to be a quadratic function in the change in work force from one period to the next.

The next cost component is for overtime and idle-time costs. Holt et al. assume there is an ideal production target that is a linear function of the work-force level. If production is greater than this target there is an overtime cost, while if production is less than this target there is an idle-time cost. Again, Holt et al. assume that the cost is quadratic about the variance between the actual production and the production target for the work-force level.

The final component is inventory and back-order costs. Similar to the overtime and idle-time costs, there is an inventory target for each period, which is a linear function of the demand in the period. The inventory and back-order cost is a quadratic function of the deviation between the inventory and the inventory target.

The Holt et al. (1960) optimization is to minimize the sum of the expected costs over a fixed horizon, subject to an inventory balance constraint. The expectation is over the demand random variables, where we are given an unbiased forecast of demand over the planning horizon. The analysis of this optimization yields two key results.

First, the optimal solution can be characterized as a linear decision rule, whereby the aggregate production rate in each period is a linear function of the future demand forecasts, as well as the work force and inventory level in the prior period. There is a similar linear function for specifying the work-force level in each period.

Second, the optimal decision rule is derived for the case of stochastic demand, but only depends on the mean of the demand random variables. That is, we only need to know (or assume) that the demand forecasts are unbiased in order to apply the linear decision rule.

5. Production Planning: Lot-Size Models

In this section we consider production-planning problems for which there are economies of scale associated with the production activity or function. The most common example occurs when there is a required setup to initiate the production of an item. For instance, this initiation might require a change in tooling, dies, or raw material; the setup might also require a change in the production control settings, as well as an initial run to assure that the production output meets quality specifications. There might be a set-up cost, corresponding to labor costs, plus direct expenditures for materials and tools. There might also be resource requirements, usually referred to as the set-up time. The production resource cannot produce until the setup is completed; thus the set up consumes production capacity, equal to its duration, namely the setup time.

Once an item is set up to produce, we may want to manufacture a large batch or lot size so as to cover demand over a number of future periods and hence defer the next time when the item will be set up and produced. Whereas producing large batches will reduce the set-up costs, this also increases inventory as more demand is produced earlier in time. The lot-sizing problem, as described here, is to determine the relative frequency of setups so as to minimize set-up and inventory costs, within the resource and service constraints of the production-planning problem.

We start with the simplest model and then briefly discuss variants to it. We develop and state this model for the following context:

- multiple items with independent demand,
- a single shared resource,
- big-bucket time periods,
- linear costs, except for set-up costs.

For ease of presentation we assume a single resource; the extension to consider multiple resources is straightforward.

We define the following notation

Decision variables

p_{it} production of item i during time period t

q_{it} inventory of item i at end of time period t

y_{it} binary decision variable to denote setup of item i in time period t

Parameters

T, I number of time periods, items, respectively

a_{i1} amount of resource required per unit of production of item i

a_{i2} amount of resource required for setup of production of item i

b_t amount of resource available in period t

d_{it} demand for item i in period t

B a large constant

cp_{it} unit variable cost of production for item i in time period t

cq_{it} unit inventory holding cost for item i in time period t

cy_{it} set-up cost for production for item i in time period t.

We now formulate the mixed-integer linear program P8:

$$P8: \quad \text{Min} \sum_{t=1}^{T} \sum_{i=1}^{I} cp_{it}p_{it} + cq_{it}q_{it} + cy_{it}y_{it} \tag{4}$$

subject to

$$q_{i,t-1} + p_{it} - q_{it} = d_{it} \quad \forall i, t \tag{5}$$

$$\sum_{i=1}^{I} a_{i1}p_{it} + a_{i2}y_{it} \leqslant b_t \quad \forall t \tag{6}$$

$$p_{it} \leqslant By_{it} \quad \forall i, t \tag{7}$$

$$p_{it}, q_{it} \geqslant 0; \quad y_{it} = 0, 1 \quad \forall i, t.$$

The objective function (4) minimizes the sum of variable production costs, the inventory holding costs, and the set-up costs for all items over the planning horizon of T periods.

Equation (5) is the same as (2), the inventory balance constraints that equate the supply of an item in a period with its demand or usage.

The resource constraints (6) reflect the resource consumption both due to the production quantity for each item, and due to the setup. Production of one unit of item i consumes a_{i1} units of the shared resource, while the set-up requires a_{i2} units.

The constraint set (7) is for the so-called forcing constraints. These constraints relate the production variables to the set-up variables. For each item and time period, if there is no setup ($y_{it} = 0$), then this constraint assures that there can be no production ($p_{it} = 0$). Conversely, if there is production in a period ($p_{it} > 0$), then there must also be a setup ($y_{it} = 1$). In (7), B is any large positive constant that exceeds the maximum possible value for p_{it}; for instance, one might set B equal to the sum of all demand.

This problem is now a mixed-integer linear program, with IT binary decision variables. For modest-size problems with, for example, a few hundred binary decision variables, this problem can be reliably solved by commercial optimization packages. But specialized approaches are warranted for increasing problem size and complexity. We discuss one of these approaches next.

5.1. Dual-Based Solutions

In this section we develop a dual problem for P8. We identify a generalized linear program for solving this dual problem, which is equivalent to a convexification of P8. We can solve this problem by column generation to obtain a lower bound on P8; we also discuss how this solution can be used to identify near-optimal solutions to P8.

The approach is based on the original work of Manne (1958), who suggested the generalized linear program given below. Dzielinski and Gomory (1965) extended the model of Manne to include resource planning decisions (i.e., hiring and firing labor) and applied the Dantzig–Wolfe decomposition method to introduce column generation. Lasdon and Terjung (1971) reformulated the linear program so as to provide a more efficient and effective column generation approach for solving the linear program; they also address a variant of P8, where there are small-time buckets for planning production and multiple resources corresponding to scarce machines and dies.

To develop the dual problem, we first define a Lagrangian function $L(\pi)$ by dualizing the resource constraints (6):

$$L(\pi) = \text{Min} - \sum_{t=1}^{T} b_t \pi_t + \sum_{t=1}^{T} \sum_{i=1}^{I} (cp_{it} + a_{i1}\pi_t)p_{it} + cq_{it}q_{it} + (cy_{it} + a_{i2}\pi_t)y_{it}$$

$$\text{subject to } (5), (7)$$

$$p_{it}, q_{it} \geqslant 0; \quad y_{it} = 0, 1 \quad \forall i, t$$

where $\pi = (\pi_1, \pi_2, \ldots \pi_T) \geqslant 0$ is a vector of dual variables.

We state the following observations:

- The Lagrangian separates by item, where we have a single-item dynamic lot-size problem with no capacity constraints for each item. This is the so-called Wagner–Whitin (1958) problem and can be solved by dynamic programming.
- If $q_{i0} = 0$, the extreme points to $L(\pi)$, namely the single-item dynamic lot-size problem, have the property whereby $p_{it}q_{it-1} = 0$. As a consequence the optimal production quantities are a sum of consecutive demands. That is, if $p_{it} > 0$, then $p_{it} = d_{it} + \ldots + d_{is}$ for $t \leqslant s \leqslant T$. We will refer to this as a Wagner–Whitin schedule.
- Without loss of generality, we will assume that $q_{i0} = 0$, and the above property applies to optimal solutions. If $q_{i0} > 0$, then we use this initial inventory to reduce the item's demand. That is, we restate the demand as follows:

$$d_{it} := 0 \quad \text{for } t = 1, 2, \ldots s - 1$$

$$d_{is} := d_{i1} + d_{i2} + \dots + d_{i,s-1} - q_{i0},$$

where s is such that $d_{i1} + d_{i2} + \dots + d_{i,s-1} \leqslant q_{i0} < d_{i1} + d_{i2} + \dots + d_{is}$.

We can use the Lagrangian function to define a dual problem to P8:

$$D8: \quad \text{Max } L(\pi)$$

subject to $\pi \geqslant 0$.

The dual solution need not, and usually will not, identify a primal feasible solution to P8. In such instances, a duality gap exists and the dual solution provides a lower bound to P8. One might consider two procedures for resolving the duality gap. First, one could use the dual problem for generating bounds in a branch-and-bound procedure; the effectiveness of this depends on the tightness of the bounds and the number of integer variables in P8. The second approach incorporates the solution of the dual problem into a heuristic procedure. For each iteration in the solution of the dual, we might generate, somehow, a corresponding feasible solution to P8. The best such feasible solution can be compared with the solution of the dual to assess its near optimality; the procedure stops once the best feasible solution is sufficiently close to the optimum or after a predetermined number of iterations.

We might solve the dual D8 directly by means of a method such as subgradient optimization, or a dual-ascent procedure. Alternatively, we can follow the general derivation provided by Magnanti et al. (1976) to reformulate the dual problem as a generalized linear program. We denote the extreme points of the convex hull defined by constraint sets (5) and (7), the non-negativity constraints, and the binary constraints by

$$u^j = (u^j_1, u^j_2, \dots u^j_I) = [(p^j_{1t}, q^j_{1t}, y^j_{1t}), \dots (p^j_{I,t}, q^j_{I,t}, y^j_{I,t})],$$

where for each item i, u^j_i is a Wagner–Whitin schedule.

We define the cost for the jth extreme point and resource usage for the jth extreme point in time period t as

$$cu^j = \sum_{i=1}^{I} \sum_{t=1}^{T} cp_{it}p^j_{it} + cq_{it}q^j_{it} + cy_{it}y^j_{it}$$

$$a^j_t = \sum_{i=1}^{I} a_{i1}p^j_{it} + a_{i2}y^j_{it}.$$

We can rewrite D8 in terms of the extreme points as the following equivalent linear program:

Max z

subject to

$$z + \sum_{t=1}^{T} (b_t - a^j_t)\pi_t \leqslant cu^j \quad \forall j$$

$$\pi_t \geqslant 0, \quad \forall t,$$

where z is unconstrained in sign. The dual of this problem is:

$$P9: \quad \text{Min} \sum_{j=1}^{J} cu^j x_j$$

subject to

$$\sum_{j=1}^{J} x_j = 1$$

$$\sum_{j=1}^{J} a_t^j x_j \leqslant b_t \quad \forall t$$

$$x_j \geqslant 0, \quad \forall j,$$

where J is the number of extreme points; the decision variable x_j indicates the fraction of the schedule given by the jth extreme point. Problem P9 is a convexification of the primal problem P8 in which we replace the feasible region, defined by constraints (5) and (7), the nonnegativity and binary constraints, with the convex hull of this region. The solution of P9 provides a lower bound for P8. However, P9 is not all that useful due to the large number of variables, in the order of 2^{IT}; and the solution to P9 will typically be fractional and not suggestive of good, near-optimal feasible solutions.

We can reformulate P9 by noting that we can express the set of extreme points $U = \{u^j\}$ in terms of extreme points for the individual items. That is,

$$U = U_1 \times U_2 \times \ldots \times U_I,$$

where

$$U_i = \{u_i^k\}$$

is the set of extreme points or Wagner–Whitin schedules for item i. For the kth extreme point for item i, we define its cost and resource usage parameters:

$$cu_i^k = \sum_{t=1}^{T} cp_{it} p_{it}^k + cq_{it} q_{it}^k + cy_{it} y_{it}^k$$

$$a_{it}^k = a_{i1} p_{it}^k + a_{i2} y_{it}^k.$$

We can now rewrite P9 in terms of the Wagner–Whitin schedules for the items:

$$P10: \quad \text{Min} \sum_{i=1}^{I} \sum_{k=1}^{K_i} cu_i^k x_{ik}$$

subject to

$$\sum_{k=1}^{K_i} x_{ik} = 1 \quad \forall i$$

$$\sum_{i=1}^{I} \sum_{k=1}^{K_i} a_{it}^k x_{ik} \leqslant b_t \quad \forall t$$

$$x_{ik} \geqslant 0, \quad \forall i, k,$$

where K_i denotes the number of Wagner–Whitin schedules for item i and the decision variable x_{ik} is the fraction of the kth Wagner–Whitin schedule for item i in the solution. The first constraint assures that these fractions sum to one, so that the demand for each item is met. The second constraint enforces the resource constraint.

The optimal solution to P10 solves the dual D8 and provides a lower bound to P8. If the optimal solution to P10 is all integer ($x_{ik} = 0$ or 1 for all i, k), then this solution is optimal to P8. When the solution to P10 is not integral, it provides the basis for finding near-optimal solutions. We describe this solution strategy in two parts: first, how we solve P10, and second, how we use the solutions to P10 to obtain good feasible solutions to P8.

Manne (1958) first proposed solving P10 as an approximation to solving P8. However, as with P9, there can be an enormous number of decision variables, in the order of $I2^T$ in this case. Rather than generate all of these decision variables and their parameters, we solve P10 by means of column generation (Dzielinski and Gomory, 1965; Lasdon and Terjung, 1971). At each iteration, we solve a master problem, namely, a reduced version of P10 with a subset of columns. Then, using the dual values from the master problem, we solve a Wagner–Whitin problem for each item to find a new candidate schedule to enter into the master problem. The procedure iterates between the master problem and the item subproblems until the solutions to the Wagner–Whitin subproblems yield no new item schedules. One would typically terminate this procedure when the gap between the lower bound provided by the master problem and a known feasible solution to P8 is suitably small.

Manne (1958) noted that although solutions to P10 are usually fractional, there is an integer solution for most items. An optimal solution to P10, as well as the master problem in the column generation procedure, has $I + T$ basic variables, as there are $I + T$ constraints. Thus, there are at most $I + T$ fractional variables in the solution. Each item must have at least one basic variable, in order to satisfy the convexity constraint in P10. As a consequence, no more than T items can have two or more basic variables. Thus, at least $I - T$ items have a single basic variable, which must be integral due to the convexity constraint.

For most planning problems, the number of items would be much greater than the number of time periods. For instance, we might be planning for 50–100 items, with 12 or 13 time periods. For $I = 100$ and $T = 13$, a solution to P10 provides a single Wagner–Whitin schedule for at least 87 items. For the remaining items, the solution suggests a convex combination of two or more Wagner–Whitin schedules. These schedules satisfy the demand constraints in P8, but require more set-up time and cost than assumed by P10. (P10 assumes that we only incur a fraction of the set-up cost and time, for a schedule at a fractional level of activity.) Thus, the set of schedules from P10 might violate the resource constraints in P8. Manne notes that in some contexts, the resource constraints are soft constraints and minor violations can be ignored. In other cases, one might apply a heuristic to modify the solution from P10 so as to make it satisfy the resource constraints. One expects that it is relatively easy to find a feasible solution to P8 from the solution to P10, as most of the items have a single schedule. Furthermore, one expects that the feasible solutions are near optimal, again due to the fact that the solution to P10 is nearly integral. Computational experience in Manne (1958), Dzielinski and Gomory (1965), and Lasdon and Terjung (1971) supports these claims. Also, Trigeiro et al. (1989) describe and test a heuristic smoothing procedure for constructing feasible solutions.

5.2. *Variable Redefinition*

Eppen and Martin (1987) report on an alternative solution approach to P8, based on variable redefinition. They reformulate P8 so that its LP relaxation provides a very tight bound, comparable to that from the dual-based approaches of Lagrangian relaxation or column generation. They then solve the mixed-integer program using general-purpose optimization codes to obtain optimal or near-optimal solutions for problems with up to 200 products and 10 time periods.

Their approach is based on the optimal property of Wagner–Whitin schedules: when a production activity occurs, the production quantity covers the demand in an integer number of consecutive periods beginning with the period of production. They then define decision variables for each such production opportunity for each item. With these new variables, the schedule for each item is a shortest-path problem through a network of T nodes. These shortest-path problems are coupled by resource constraints that cut across the individual items.

For completeness we present the Eppen–Martin model (Eppen and Martin, 1987) for multi-item capacitated lot-sizing.

Decision variables

z_{itk} binary decision variable to denote production of item i during time period t, where the production quantity is to satisfy demand for periods t through k

y_{it} binary decision variable to denote setup of item i in time period t

Parameters

T, I number of time periods and items, respectively

a_{itk} amount of resource required in period t, if $z_{itk} = 1$

b_t amount of resource available in period t

cz_{itk} variable production and inventory holding cost for item i, for production in period t to satisfy demand from period t to k for item i.

cy_{it} set-up cost for production for item i in time period t.

We now formulate the mixed-integer linear program P11:

$$P11: \quad \text{Min} \sum_{i=1}^{I} \sum_{t=1}^{T} \sum_{k=t}^{T} cz_{itk} z_{itk} + \sum_{i=1}^{I} \sum_{t=1}^{T} cy_{it} y_{it}$$

subject to

$$\sum_{i=1}^{I} \sum_{k=t}^{T} a_{itk} z_{itk} \leqslant b_t \quad \forall t$$

$$\sum_{k=1}^{T} z_{i1k} = 1 \quad \forall i$$

$$\sum_{k=t}^{T} z_{itk} - \sum_{k=1}^{t-1} z_{ik,t-1} = 0 \quad \forall i, t$$

$$\sum_{k=t}^{T} z_{itk} \leqslant y_{it} \quad \forall i, t$$

$$z_{itk} \geqslant 0, \quad y_{it} = 0, 1 \quad \forall i, t, k.$$

The first set of constraints corresponds to the resource constraint (6) in P8. The next two sets model the flow balance for the underlying shortest-path problem for each item on nodes $1, 2, ..., T$, where z_{itk} corresponds to the flow on the arc from node t to node $k+1$. The last set is the forcing constraint, equivalent to (7) in P8. Note that the demand parameters d_{it} do not appear in P11. Rather they are embedded in the definition of the cost parameters cz_{itk} and the resource coefficients a_{itk}. Also, whereas we define z_{itk} to be a binary variable, we do not make this an explicit requirement when solving P11. Eppen and Martin (1987) find that P11 provides a tight lower bound to P8 and also identifies near-optimal feasible solutions.

5.3. *Extensions to the Lot-Size Model*

We can extend formulation P8 to incorporate all of the problem variations that were introduced for the linear programming production-planning model. In addition we mention here three other common extensions.

In P8 we assume that any quantity could be produced, subject to the resource constraint. Furthermore, the resource usage for a production quantity consists of a fixed set-up time and a variable amount linear in the production quantity. In some contexts, production occurs as a batch process, for example, a heat treat or diffusion process. Each batch produces a fixed amount of product and consumes a fixed amount of the limited resource. We can model this by introducing an integer decision variable for the number of batches produced of an item in a time period.

A second variation is when the setups are sequence dependent; that is, the setup for an item will depend upon what was just previously processed. There is not an easy way to modify P8 to accommodate this feature. Indeed, in general, the standard representation of sequence-dependent setups is to map this into a traveling salesman problem, which results in a new level of complexity.

The third variation is when setups can be carried over from one period to the next. In P8 we assume that this is not possible; that is, every period that we produce an item we incur a setup. In some contexts, though, we might be able to preserve the last setup in the period. Thus, we would incur only one setup if we produce item i last in one period and first in the next period. Karmarkar et al. (1987) examine a single-product version of this problem. Another variation of this problem is when there are multiple products and we assume small-time buckets, so that at most one item is produced in a period. Lasdon and Terjung (1971) formulate and address this problem by means of column generation, and Eppen and Martin (1987) show how to solve both of these problems efficiently by variable redefinition.

References

Billington, P. J., J. O. McClain, and L. J. Thomas. 1983. "Mathematical approaches to capacity-constrained MRP systems: review, formulation and problem reduction." *Mgmnt Sci.* **29(10)**: 1126–1141.

Bitran, G. R. and D. Tirupati. 1993. "Hierarchical production planning." In *Handbooks in Operations Research and Management Science, Volume 4, Logistics of Production and Inventory*, edited by S. C. Graves, A. H. G. Rinnooy Kan, and P. H. Zipkin, pp. 523–568. Amsterdam: Elsevier Science Publishers B.V.

Bowman, Edward H. 1956. "Production scheduling by the transportation method of linear programming." *Oper. Res.* **4(1)**: 100–103.

Dzielinski, B. P. and R. E. Gomory. 1965. "Optimal programming of lot sizes, inventory and labor allocations." *Mgmnt Sci.* **11(9)**: 874–890.

Eppen, G. D. and R. K. Martin. 1987. "Solving multi-item capacitated lot-sizing problems using variable redefinition." *Oper. Res.* **35(6)**: 832–848.

Graves, S. C. 1981. "A review of production scheduling." *Oper. Res.* **29(4)**: 646–675.

Hackman, S. T. and R. C. Leachman. 1989. "A general framework for modeling production." *Mgmnt Sci.* **35(4)**: 478–495.

Hansmann, F. and S. W. Hess. 1960. "A linear programming approach to production and employment scheduling." *Mgmnt Technol.* **1(1)**: 46–51.

Hax, A. C. and H. C. Meal. 1975. "Hierarchical integration of production planning and scheduling." In *Studies in Management Sciences, Volume 1, Logistics*, edited by M. A. Geisler, pp. 53–69. New York: Elsevier.

Holt, C. C., F. Modigliani, J. F. Muth, and H. A. Simon. 1960. *Planning Production, Inventories and Work Force*. Englewood Cliffs, NJ: Prentice-Hall.

Karmarkar, U. S., S. Kekre, and S. Kekre. 1987. "The dynamic lotsizing problem with startup and reservation costs." *Oper. Res.* **35(3)**: 389–398.

Lasdon, L. S. and R. C. Terjung. 1971. "An efficient algorithm for multi-item scheduling." *Oper. Res.* **19(4)** 946–969.

Magnanti, T. L., J. F. Shapiro, and M. H. Wagner. 1976. "Generalized linear programming solves the dual." *Mgmnt Sci.* **22(11)**: 1195–1203.

Manne, A. S. 1958. "Programming of economic lot sizes." *Mgmnt Sci.* **4(2)**: 115–135.

Shapiro, J. F. 1993. "Mathematical programming models and methods for production planning and scheduling." In *Handbooks in Operations Research and Management Science, Volume 4, Logistics of Production and Inventory*, edited by S. C. Graves, A. H. G. Rinnooy Kan, and P. H. Zipkin, pp. 371–443. Amsterdam: Elsevier Science Publishers B.V.

Silver, E. A., D. F. Pyke, and R. Peterson. 1998. *Inventory Management and Production Planning and Scheduling*. 3d ed. New York: John Wiley.

Thomas, L. J. and J. O. McClain. 1993. "An overview of production planning." In *Handbooks in Operations Research and Management Science, Volume 4, Logistics of Production and Inventory*, edited by S. C. Graves, A. H. G. Rinnooy Kan, and P. H. Zipkin, pp. 333–370. Amsterdam: Elsevier Science Publishers B.V.

Trigeiro, W. W., L. J. Thomas, and J. O. McClain. 1989. "Capacitated lot sizing with setup times." *Mgmnt Sci.* **35(3)**: 353–366.

Vollman, T. E., W. L. Berry, and D. C. Whybark. 1992. *Manufacturing Planning and Control Systems.* 3d ed. Burr Ridge, IL: Richard D. Irwin Inc.

Wagner, H. M. and T. Whitin. 1958. "Dynamic version of the economic lot size model." *Mgmnt Sci.* **5(1)**: 89–96.

18.6 Semiconductor Production Planning

Robert C. Leachman

ABSTRACT The planning of semiconductor manufacturing presents formidable challenges for the successful application of mathematical optimization. Formulation techniques are described that precisely model noninteger flow times, time-phased capacity consumption by reentrant process flows, proper horizon behavior, multiple priority classes of demands and multiple objectives, binning and substitution in product structures, and arbitrary distributions of initial work in process. We also formulate exact capacity analyses of nonhomogeneous resources, machine arrangement constraints, and alternative combinations of resources necessary to execute manufacturing operations. Brief histories are provided of key modeling contributions and of optimization applications to production planning in the industry.

1. Introduction

Semiconductors form the basis of electronics, the world's largest industry except for agriculture. The manufacture of integrated circuits is one of the most capital-intensive and complex manufacturing processes extant. A state-of-the-art wafer fabrication facility processing twentyfive thousand twelve-inch silicon wafers per month, each printed with integrated circuits of electrical switches as small as 180 nanometers in length, represents an investment of several billion U.S. dollars. Most of the processing equipment installed in this factory will need to be replaced within a single-digit number of years in order to keep up with the advances in process technology.

While most other capital-intensive process industries have long been fruitful areas for the application of mathematical optimization to production planning, regular application in semiconductor production planning was unknown until the second half of the 1980s, and did not become widespread until the second half of the 1990s. This lag compared to other process industries is due to the extraordinary challenges posed by semiconductor manufacturing to the formulation of a computationally tractable yet sufficiently accurate mathematical programming model.

The semiconductor manufacturing process consists of hundreds of process steps performed by tens of machine types on discrete manufacturing lots. The process flow is re-entrant in the sense that a given type of processing machine performs a number of steps in the process flow, interspersed with process steps performed by other machine types. Process flow times between different steps performed by the same machine type range from as little as hours to as long as many weeks. The overall flow time through a wafer fabrication process ranges from twenty to sixty days; the subsequent wafer testing, device assembly, burn-in, final testing, and packing processes add another five to fifteen days of flow time. Complexities for planning arising from the manufacturing process include steep learning curves on process yields, process times, flow times, and equipment efficiencies; uncertainties in yields, flow times, and capacities; nonhomogeneous machines and tooling; and the integrality of production lots.

Another aspect of complexity arises from the semiconductor product structure. Testing of production lots generates a distribution of quality-graded products known as bin splits. Products that are sold and bins of quality have a many-to-many relationship. There is an embedded optimization problem concerning how many (pretested) products should be fabricated and assembled, and how the consequent quality bins should be allocated, in order to best service demands. There are also complexities arising from the nature of the demands: there may be significant uncertainties in market demands; and there may be a great variety of demand types for each finished good, ranging from firm orders to flexible customer contracts to reliable forecasts to risky forecasts.

The sheer size of the planning problem in most semiconductor manufacturing companies is daunting: even a single wafer fabrication plant performs hundreds of process steps on hundreds of wafer types using dozens of machine and tool types. A typical company has thousands or even tens of thousands of finished goods types manufactured in dozens of manufacturing areas.

Planning in the industry is performed both incrementally and in a regenerative fashion. Incremental planning involves adding production to an existing plan in order to meet new demands, without other adjustment of the existing plan; regenerative planning involves a complete reassessment of the plan in light of revised demands or other changes in the input data. Incremental planning is of course a much smaller problem with a much shorter turnaround time, but regenerative planning offers the ability to more fully optimize production. We shall treat the more difficult regenerative planning problem in this article.

We remark that production planning is not the only potential application of optimization within the domain of semiconductor manufacturing management. There have also been efforts to apply it to establish short-term goals for factory floor operations (e.g., Fordyce et al., 1992; Kang, 1996), and, more recently, to apply it to the problem of capacity planning (e.g., Liu, 1998; Stray et al., 2000). However, we shall restrict our attention here to the production planning problem, in which production is scheduled subject to process flow times and machine capacities that are prespecified.

2. Formulation of Semiconductor Planning Models

We now discuss the key techniques for formulating a practical linear programming model for planning semiconductor production.

2.1. *A Brief History of Key Contributions*

Pioneering work on the formulation of practical optimization models for semiconductor production planning was carried out beginning in 1985 at the University of California at Berkeley, under the direction of Prof. R. C. Leachman. This effort continues to the present day. The Berkeley researchers have helped various semiconductor manufacturers sponsoring their research to implement software incorporating their optimization formulations, and these implementations have sometimes been referred to as the Berkeley Planning System (BPS). Working independently, researchers at IBM also made useful contributions beginning in the early 1990s, particularly to the capacity analysis of nonhomogeneous machines. Software incorporating these contributions has been utilized in IBM's microelectronics manufacturing management.

Key contributions are summarized as follows:

1. Variables of the planning model represent constant production rates in given planning periods, expressed either as rates of release into process flows or rates of output from process flows (Leachman, 1986, 1993).
2. Noninteger time lags are used to map, through continuous working time, the release rates into process output rates; inventory balance constraints are formulated at noninteger time points in order to insure mass conservation through continuous time (Hackman and Leachman, 1989).
3. Noninteger time lags also are used to map release rates to workload rates for process steps, which are summarized into total workload rates on capacitated resources (Leachman, 1986, 1993).

4. A steady-state horizon condition is formulated to mitigate undesirable horizon effects induced by the presence of reentrant flows in a finite horizon model (Leachman, 1993).
5. In the case of time-varying bottlenecks, an iterative simulation-optimization scheme has been developed to jointly establish flow times and production quantities (Hung and Leachman, 1996).
6. Binning and substitution at product structure junctions are exactly formulated (Leachman, 1986, 1993).
7. Iterative optimization of multiple demand priority classes is used to treat varying types of demands and market uncertainties. Multiobjective programming is used, with bounds on variables placed to ensure there is no switching between alternative optimal solutions (Leachman, 1993).
8. Capacity analysis of nonhomogeneous machines, of machine assignment constraints, and of alternative sets of machines and tools has been progressively refined (Leachman and Carmon, 1992; Fordyce and Sullivan, 1995; Lin, 1999).
9. Scheduling of the initial work in process is explicitly incorporated into the planning model while maintaining computational tractability (Lin, 1999).

2.2. *Preliminary Model*

We first consider an unrealistic, textbook-type LP formulation of semiconductor production planning. Subsequently, we introduce modifications that make the model more suitable for industrial application. For simplicity of exposition, we first consider a single-stage system, that is, demand items and items released to production are identical, hereafter simply referred to as products.

Wafer fabrication areas can often be modeled as a single-stage system with re-entrant flows. The process flows may include upward of six hundred process steps, but the product identity at the first and last step in each flow may be identical. When there are many products to be processed, it is impractical to formulate a planning model incorporating scheduling variables and balance constraints for each step. Instead, we first formulate a model including only variables representing the release of raw material for processing each product. Flow time parameters are employed to estimate subsequent workloads associated with the release of raw material at each downstream process step. The model includes capacity constraints for multiple resource types and inventory balance constraints on completed products.

Subscripts

t = time period $(t-1, t]$, $t = 1, 2, 3, ..., T$, where T is the planning horizon.

i = product, $i = 1, 2, 3, ..., N_I$.

j = processing step, $j = 1, 2, 3, ..., N_i$.

k = machine type or other scarce resource, $k = 1, 2, 3, ..., K$.

Parameters

d_{it} = demand for product i in period t.

a_{kij} = processing time per unit on machine type k when performing step j on product i.

L_i = estimated flow time from release of raw material until completion of product i.

L_{ij} = estimated flow time from release of raw material until step j is initiated on product i.

b_i = estimated yield from release of raw material until completion of product i.

b_{ij} = estimated yield from release of raw material until step j is initiated on product i.

c_{kt} = capacity (machine-hours) of machine type k in period t.

h_i = holding cost per unit per period of completed product i.

s_i = backorder cost per unit per period of completed product i

r_i = avoidable cost (e.g., raw materials) per unit of product i released into production.

Variables

$x_{i,t}$ = quantity of raw material for the manufacture of product i to be released during period t. $x_{i,0}, x_{i,-1}, x_{i,-2}, \ldots, x_{i,-Li+1}$ must be given.

$I_{i,t}$ = inventory of completed product i at time t. $I_{i,0}$ must be given.

$BO_{i,t}$ = shortage of completed product i at time t. $BO_{i,0}$ must be given.

Formulation

$$\text{Minimize } \sum_{i=1}^{N}\sum_{t=1}^{T}(r_i x_{i,t} + h_i I_{i,t} + s_i BO_{i,t})$$

Subject to

$$\sum_{\tau=1}^{t} b_i x_{i,\tau-L_i} - I_{i,t} + BO_{i,t} = \sum_{\tau=1}^{t} d_{i,\tau} - I_{i,0} + BO_{i,0}, \quad i = 1, \ldots, N_I, \quad t = 1, \ldots, T.$$

$$\sum_{i=1}^{N_I}\sum_{j=1}^{N_i} a_{kij} b_{ij} x_{i,t-L_{ij}} \leqslant c_{kt} \quad k = 1, \ldots, K, \quad t = 1, \ldots, T.$$

$$x_{it} \geqslant 0, \ I_{it} \geqslant 0, \ BO_{it} \geqslant 0, \quad i = 1, \ldots, N_I, \quad t = 1, \ldots, T.$$

We have expressed a classical production planning formulation allowing inventory and backorders with respect to given demands, and making production subject to multiple resource capacity constraints. Integer time lags between raw material release and resource consumption and between raw material release and availability of finished product are incorporated.

It is straightforward to expand the formulation to model multistage systems, that is, to accommodate a product structure using inventory balance constraints that express mass conservation of intermediate products. Alternative process flows for the products also are easily incorporated.

2.3. *Needed Modifications*

We now point out weaknesses in the classical formulation that cause it to generate inaccurate or undesirable production plans for semiconductor manufacturing, and then introduce modifications to overcome these weaknesses.

Accommodating Real-Valued Time Lags

The first undesirable characteristic of the formulation that we shall deal with is the integrality of the time lags. The technique for formulation of an exact LP model incorporating noninteger time lags, first reported by Hackman and Leachman (1989), is summarized as follows. Let us view the release of raw material to manufacture product i as a continuous time function $x_i(t)$. Assuming the rate of release is uniform within each time period, we can relate the LP variables to this time function by observing that the value of $x_i(t)$ over the interval $(t-1, t]$ is x_{it}. That is, $x_i(t)$ is restricted to be a step function by the LP formulation. Similarly, we may view the demand for product i as a continuous time function $d_i(t)$, again assuming uniform rates within each given period. Using these continuous functions, we may rewrite the inventory balance constraint for product i as

$$\int_{-L_i}^{t-L_i} x_i(\tau)d\tau - I_i(t) + BO_i(t) = \int_0^t d_i(\tau)d\tau - I_i(0) + BO_i(0), \quad \text{all } t \geqslant 0, \quad i = 1, \ldots, N_I. \qquad (1)$$

Here, $I_i(t)$ denotes the inventory level of product i at an arbitrary point t in continuous time, and $BO_i(t)$ denotes the shortage of product i at time t.

These integrals reduce to linear combinations of the x_{it} variables and the d_{it} parameters when the constraint is evaluated at any arbitrary point t, even when t is noninteger. For example, suppose $L_i = 0.7$ and $t = 2.5$. Then the first integral of (1) reduces to

$$0.7x_{i,0} + x_{i,1} + 0.8x_{i,2}.$$

Thus, constraints suitable for the application of linear programming are obtained even when the L_i parameters are fractional. It remains to determine at what values of t do the constraints (1) need to be enforced in order to insure inventory balance throughout continuous time. Hackman and Leachman (1989) show that if the constraints are enforced at all points at which either integrand may change rates, then inventory balance is assured throughout continuous time. In the case of (1), these are precisely the positive integers (i.e., the epochs when the rate of outflow from inventory, $d_i(t)$, may change), and the points of the form $(t-L_i)$ where t is a positive integer greater than or equal to L_i (i.e., the epochs when the rate of inflow to inventory may change). Note that the implementation of the constraints at the latter kind of points induces the formulation of inventory variables $I_i(t-L_i)$ in addition to the usual inventory variables $I_i(t)$ defined for the integer points.

Using the continuous time functions, we may also rewrite the capacity constraint for machine type k as

$$\int_{t-1}^{t} \sum_{i=1}^{N_I} \sum_{j=1}^{N_i} a_{kij} x_i(\tau - L_{ij}) d\tau \leqslant c_{kt}, \quad k = 1, \ldots, K, \quad t = 1, \ldots, T. \tag{2}$$

This constraint also reduces to a linear constraint in the $x_{i,t}$ variables when evaluated at a given value of t, even if the L_{ij} parameters are noninteger.

Given a large number of steps performed on a given machine type for each product, and given a large number of products, there are potentially many time points per period where the integrand may change rate. Strictly speaking, each interval of constant capacity consumption needs to be individually constrained by the resource capacity in order to ensure resource conservation throughout continuous time. While it is possible to formulate linear constraints for every interval between epochs at which the integrand may change rate, it may not be practical to do so. Moreover, as a practical matter, one cannot expect the epochs t at which the input rate $x_i(t)$ is adjusted to precisely map to epochs $t + L_{ij}$ at which the workload at step j, $x_i(t + L_{ij})$, changes rate, given the uncertainties in the production process and the level of detail at which the model is formulated.

Fortunately, the wide variety of values for the L_{ij} parameters in a practical case tends to result in level loads in each given planning period, especially when capacity is fully utilized and when flows are very re-entrant. As a practical approximation, it is suggested to simply enforce constraints (2) only for the intervals defined by the given planning periods $(t-1, t]$, $t = 1, \ldots, T$. Using this approach, Hung and Leachman (1996) demonstrate very close agreement between machine utilizations and product flow times predicted by such an LP planning model and utilizations and flow times predicted by detailed, discrete-event simulation on actual industrial data.

Accommodating Factory Working Calendars and Variable-Length Planning Periods

Time lags as above are measured in terms of working time. The lengths of planning periods (in terms of working time) in the planning model may need to be different from unity. This is because of working holidays occurring in the planning periods, and/or because of a desire for increased granularity in early periods of the plan, for example, weekly periods over the first several months of the planning horizon, and monthly periods after that. A generalization of the approach in the previous section is required. Basically, the limits of integration appearing in equations (1) and (2) must be properly established in terms of epochs of continuous working time measured from the start of the planning horizon. See Hackman and Leachman (1989) for details. Moreover, shortage and holding cost parameters also may need to vary by period, appropriately scaled to reflect the lengths of the periods (Hung, 1991; Lin, 1999).

CONTROLLING HORIZON EFFECTS

In the basic finite-horizon LP formulation above, optimal values of the variables for the release rates of raw material into capacitated reentrant process flows exhibit a peculiar, undesirable behavior. Variables for product release in periods within one flow time of the horizon will be set equal to zero, since they do not contribute to demands included in the formulation. Variables for the release rates in periods just before this will typically have surprisingly large values. Such unreasonably large values are feasible since they do not have to compete for capacity with releases in subsequent periods. The tacit assumption that production is permanently terminated at the horizon T is the root cause of this undesirable behavior.

To overcome this problem, Leachman (1993) incorporates into the formulation a steady-state horizon condition. For each product i, the interval $[T-L_i, T]$ is termed the frozen interval for product i. Releases of product i in all time periods that intersect the frozen interval are constrained to be equal, that is, a single variable is utilized to represent releases in each of these periods (multiplied by an appropriate scaling constant if there are differences in period lengths). For the purposes of computing inventory balance in constraint (1) and resource capacity consumption in constraint (2), an additional time period is appended on to the planning horizon, with length equal to L_i working days. Demand for product i in this period is set to be the same rate as the average rate in the frozen period. Inventory and backorders of completed product i are measured in the objective function at both the start and the end of the extra period.

Formulations including this steady-state condition generate optimal plans that exhibit smooth production.

INCORPORATING MARKETING CONSIDERATIONS

The total demand for each product is an aggregate of components with varying priorities. One component corresponds to firm customer orders with promised delivery dates. Another may correspond to replenishment of inventory stocks supporting long-term customer contracts. Even within the component of product demand corresponding to market forecasts unrealized by on-hand customer orders or contracts, priorities are varying. One portion of the forecast may be fairly certain to be realized; the final portion may be more risky, that is, there is a significant probability that not that much demand will materialize.

Treating the entire demand for products with uniform, linear tasks for backorders and inventory is an oversimplification of the planning problem, leading to optimal solutions that may be undesirable to marketing management. That is, in a capacity-constrained situation, an optimal solution may deny capacity to confirmed customer orders for one product and award the capacity to risky forecasts for another.

Leachman (1993) proposes a structure of demand classes and iterative optimization calculations to cope with this situation. In this structure, the overall market demands for each product are categorized into R priority classes, with the understanding that demands in class $n+1$ must not inhibit the servicing of demands in class n as on-time as is feasible. Linear programs are formulated and solved for each class $r = 1, 2, \ldots, R$. Let $d_i^r(t)$ denote the rate at time t of demands in class r for product i. The demand rate used in Equation (1), the inventory balance constraint for the rth formulation, is

$$\sum_{s=1}^{r} d_i^s(t),$$

that is, demands are cumulative over classes. To this formulation one appends an upper bound on the back-order variables, specifically,

$$BO_{i,t} \leqslant BO_{i,t}^{r-1} + \int_0^t \sum_{s=1}^{r} d_i^s(\tau)d\tau - \int_0^t \sum_{s=1}^{r-1} d_i^s(\tau)d\tau, \quad \text{all } i, \quad \text{all } t,$$

where

$$BO_{i,t}^{r-1}$$

denotes the optimal value of the back orders of product i at time t from the $r-1$th formulation. This bound ensures that the maximum allowed back orders for the demand classes 1 through r equals the optimal back orders in classes 1 through $r-1$ plus the increment in cumulative demand between classes $r-1$ and r.

Note that the only difference between formulations for class $r-1$ and class r are the right-hand-side demand parameters, the bounds on the back-order variables, and possibly some objective function coefficients, if relative product priorities vary from class to class. Also note that the optimal solution to class $r-1$ is a feasible solution to class r, provided the values of the backorder variables are adjusted to reflect the increment in cumulative demands. Thus solving the series of R linear programs is not too much more computationally difficult than solving a single linear program. Leachman and Raar (1994) report that in industrial practice with very large planning problems (in excess of 100,000 rows), the time to solve four demand classes is about 2 to 2.5 times the time required to solve the first class.

Another marketing consideration concerns a policy choice between build-to-order and build-to-forecast. Where forecast accuracy is high and raw material costs low, it may be desirable to build to forecast in order to compress lead times offered to customers. But for products with significant raw material costs and unreliable demand forecasts, it may be more prudent to delay production until receipt of a customer order. In the case of semiconductors, back end production (assembly and test) is typically built to order, except for high-volume standard or commodity products. On the other hand, wafers are usually fabricated to forecast.

We may handle this consideration as follows. For each product in the overall product structure we suppose an input parameter is defined indicating whether the product is build-to-order or build-to-forecast. (In a rational case, if a product is declared build-to-order, then all child products are also build-to-order.) If the parameter indicates the product is build-to-order, then all production in the first planning period must be for order-based demands only. Planned production in subsequent periods can be for both orders and forecasts, assuming the plan is recalculated on a rolling horizon basis.

Suppose class $r-1$ is the last demand class corresponding to protection of customer commitments, and class r is the first class corresponding to forecasted demand. When the parameter indicates the product is build-to-order, upper bounds on the release variables x_{it} are introduced into the formulations for classes $r, r+1, \ldots, R$, as follows:

$$x_{i,t} \leqslant x_{i,t}^{r-1},$$

where

$$x_{i,t}^{r-1}$$

is the optimal value of the release variable for the formulation of class $r-1$.

In summary, marketing considerations can be conveniently incorporated into the planning model through the use of upper bounds on production and back-order variables in multiple optimization calculations over prioritized demand classes. This approach does not expand the formulation size, yet it enables the generation of much more desirable production plans.

Handling Product Structures Featuring Binning and Substitution

Product structures in most manufacturing environments consist almost exclusively of one-to-many relationships, that is, end-item demands may be uniquely sourced to intermediate products and raw materials. These environments are suitable for the application of materials requirements planning (MRP) logic to translate end-item demands into demands for intermediate products and raw materials. Semiconductor product structures may feature a considerable number of many-to-many relationships. These many-to-many relationships arise because of frequent design revisions and because of quality grading of tested products. In the semiconductor environment, optimization is required to correctly translate demands back through the product structure.

Design revisions lead to many-to-many relationships as follows. Suppose a new revision is offered of an original product. The revision offers higher performance, and it corrects some shortcomings in the design of the original product. Depending on the customer application, the revision may require some customers to revise their system in order to utilize the revision, while others may utilize the original product or the revision indiscriminately. New customers may choose to design their systems to solely utilize the revision. The semiconductor manufacturer then faces three kinds of demands: customers who insist on purchasing only the original product, customers who insist on purchasing only the revision, and customers willing to accept either original product or the revision. The last type of demand cannot be uniquely sourced.

Many semiconductor products undergo binning, whereby a distribution of quality-graded products emerges from testing a single manufacturing lot of source product. Each bin defines a specific range or performance for one or several electrical attributes of performance, for example, a bin definition might be of the form 'speed between 200 and 300 MHz and power consumption less than 100 mA'. The fractions of the source product lot falling into each bin of quality are known as bin splits. The bin splits are characteristic of the manufacturing process and must be regarded for planning purposes as prespecified (but probabilistic).

Each finished goods type has specific requirements for electrical performance. A particular bin is generally suitable for a number of different finished goods types. Conversely, there may be a number of bins whose attributes fulfill the requirements of a particular finished goods type. When a bin of higher quality is used to satisfy demands, this is known as substitution.

It is often the case that there are alternative testing procedures generating different bin distributions. For example, the same basic packaged device can be put through an elaborate testing procedure that generates a large set of bins, or through a simpler testing procedure that generates only lower-grade bins. The simpler testing procedure offers a reduction in test capacity consumption that could be attractive, depending on the mix of demands.

We now modify the formulation of Section 2.2 to admit a two-stage product structure including constraints modeling binning and substitution, as developed in Leachman (1993). New variables are introduced to allocate bins to releases of the second-stage products. To admit the case of testing and processing alternatives, product shall be taken to mean a particular device passed through a particular processing and testing alternative.

Subscripts

t = time period $(t-1, t]$, $t = 1, 2, 3, ..., T$, where T is the planning horizon.

i = first-stage product, $i = 1, 2, 3, ..., N_I$.

m = second-stage product, $m = 1, 2, 3, ..., N_M$.

n = quality bin, $n = 1, 2, 3, ..., N_B$. $m \varepsilon n$ denotes that demands for product m may be sourced using bin n. $n \varepsilon m$ denotes that bin n is suitable for meeting demands for product m.

j = processing step, $j = 1, 2, 3, ..., N_i$ for first-stage process flows, and $j = 1, 2, 3, ..., N_m$ for second-stage flows.

k = machine type or other scarce resource, $k = 1, 2, 3, ..., K$.

Parameters

d_{mt} = demand for product m in period t.

a_{kij} = processing time per unit on machine type k when performing step j on product i. (a_{kmj} is similar.)

f_{ni} = bin split to bin n from completed product i.

L_i = estimated lead time from release of raw material until binning of product i. (L_m is similar.)

L_{ij} = estimated lead time from release of raw material until step j is initiated on product i. (L_{mj} is similar.)

b_i = estimated yield from release of raw material until binning of product i. (b_m is similar.)

b_{ij} = estimated yield from release of raw material until step j is initiated on product i. (b_{mj} is similar.)

c_{kt} = capacity (machine-hours) of machine type k in period t.

h_m = holding cost per unit per period of completed product m.

s_m = shortage cost per unit per period of completed product m.

r_i = avoidable cost (e.g., raw materials) per unit of product i released into production. (r_m is similar.)

Variables

$x_{i,t}$ = quantity of raw material for the manufacture of product i to be released during period t. $x_{i,0}, x_{i,-1}, x_{i,-2}, \ldots, x_{i,Li+1}$ must be given.

$y_{n,m,t}$ = quantity of bin n allocated to release of product m in period t.

$z_{m,t}$ = quantity of raw material for the manufacture of product m to be released during period t. $z_{m,0}, z_{m,-1}, z_{m,-2}, \ldots, z_{m,-Lm+1}$ must be given.

$I_{n,t}$ = inventory of bin n at time t. $I_{n,0}$ must be given.

$I_{m,t}$ = inventory of completed product m at time t. $I_{m,0}$ must be given.

$BO_{m,t}$ = shortage of completed product m at time t. $BO_{m,0}$ must be given.

Formulation

$$\text{Minimize} \sum_{i=1}^{N_I}\sum_{t=1}^{T} r_i x_{i,t} + \sum_{m=1}^{N_M}\sum_{t=1}^{T}(r_m z_{m,t} + h_m I_{m,t} + s_m BO_{m,t})$$

Subject to

$$\sum_{\tau=1}^{t} b_m z_{m,\tau-L_i} - I_{m,t} + BO_{m,t} = \sum_{\tau=1}^{t} d_{m,\tau} - I_{m,0} + B_{m,0}, \quad m = 1,\ldots,N_M, \quad t = 1,\ldots,T.$$

$$\sum_{\tau=1}^{t} f_{ni} b_i x_{i,\tau-L_i} + I_{n,0} - \sum_{m\in n}\sum_{\tau=1}^{t} y_{n,m,\tau} = I_{n,t}, \quad n = 1,\ldots,N_B, \quad t = 1,\ldots,T.$$

$$\sum_{n\in m} y_{n,m,t} = z_{m,t}, \quad m = 1,\ldots,N_M, \quad t = 1,\ldots,T.$$

$$\sum_{i=1}^{N_I}\sum_{j=1}^{N_i} a_{kij} b_{ij} x_{i,t-L_{ij}} + \sum_{m=1}^{N_M}\sum_{j=1}^{N_m} a_{kmj} b_{mj} z_{m,t-L_{mj}} \leqslant c_{kt}, \quad k = 1,\ldots,K, \quad t = 1,\ldots,T.$$

$$x_{i,t} \geqslant 0, \; z_{m,t} \geqslant 0, \; y_{n,m,t} \geqslant 0, \; I_{n,t} \geqslant 0, \; I_{m,t} \geqslant 0, \; BO_{m,t} \geqslant 0,$$

$$i = 1,\ldots,N_I, \quad n = 1,\ldots,N_B, \quad m = 1,\ldots,N_m, \quad t = 1,\ldots,T.$$

The modifications introduced in the sections on pages 4–7 may also be incorporated into a multistage formulation modeling binning and substitution such as the above.

Scheduling Initial WIP

The reentrant nature of semiconductor manufacturing flows means that new releases must compete for capacity with work in process (WIP). The basic formulation of Section 2.2 captures this competition by making use of parameters specifying past production releases $x_{i,0}$, $x_{i,-1}$, $x_{i,-2}$, and so on. However, the formulation implicitly assumes the following:

1. the initial WIP is distributed through the manufacturing process precisely according to the given time lags (i.e., flow times) and the given history of product releases,
2. the initial WIP must keep moving through the process according to the given flow times, and
3. the initial WIP has priority over new releases for allocation of scarce capacity.

In stochastic environments, assumption (1) may be unrealistic. Because of a severe process disruption or because of a downgrading of priority, it is possible that the initial WIP of a product seriously lags the target distribution calculated based on its release dates and the planned flow times. In such a case, the initial WIP will consume more resource capacity to exit the factory than the amount implied by assumption (1), and hence the planning model may overestimate the amount of new releases that may be made without extending flow times.

To circumvent this problem, one could precalculate the capacity required to flush the initial WIP according to the given flow times from each step to product completion. To do this, one can compute a product completion schedule equivalent to the initial WIP (calculated in terms of the given yields and flow times), and then apply a backward version of the capacity consumption formula introduced in Section 2.2. If we let

$$\bar{x}_i(t)$$

denote the output rate at time t calculated from the initial WIP, then the consumption of resource k capacity by initial WIP in the planning period $(t-1,\ t]$ is computed as

$$\sum_{i=1}^{N_I}\sum_{j=1}^{N_i}\int_{t-1}^{t} a_{kij}\bar{x}_i(\tau + L_i - L_{ij})d\tau.$$

Instead of incorporating capacity consumption estimated from the history of past releases, one could subtract the capacity consumption precalculated as above from the right-hand sides c_{kt} of capacity constraints placed on new releases. This was the approach taken in the Berkeley Planning System (Leachman et al., 1996), which served as the planning engine of the IMPReSS planning system at Harris Corporation.

This precalculation approach overcomes problems arising from assumption (1), but it continues to make assumptions (2) and (3). Because of changed product forecasts, it may be desired in some applications to give higher priority to new product releases of one product over the initial WIP of another.

To handle this consideration requires an expansion of the planning model to schedule not only releases of raw material at the first manufacturing step, but to also schedule releases of initial WIP from manufacturing steps throughout the process. When process flows consist of upward of six hundred process steps, a model formulated with release variables defined for every step is likely to be computationally impractical. A more practical strategy is to formulate WIP release variables only for major process steps, that is, steps utilizing scarce resources included in the capacity analysis. In typical applications in the author's experience, there are in the order of fifty or less major steps per process flow. At the other minor steps, initial WIP may be assumed to flow according to planned time lags up to the next major step, where its release downstream will be scheduled by the model. We shall therefore assume in the following that the index j applies only to major process steps.

This strategy is implemented by introducing new parameters, new variables, and both new and modified constraints into the formulation of Section 2.2 as follows:

Parameters

w_{ijt} = initial WIP arriving at major step j during period t.

$L_{i\alpha j}$ = estimated flow time from initiation of step α to initiation of step j on product i, equal to $L_{ij} - L_{i\alpha}$.

$b_{i\alpha j}$ = estimated yield from initiation of step α to initiation of step j on product i, equal to $b_{ij}/b_{i\alpha}$.

b_{ij}^{d} = estimated yield from initiation of step j on product i to completion of product i, equal to b_i/b_{ij}.

Variables

$x_{i,j,t}$ = planned release of initial WIP at major step j of the process flow for product i during period t

Constraints

New and modified constraints of the formulation are as follows:

$$\sum_{\tau=1}^{t} x_{i,j,\tau} \leqslant \sum_{\tau=1}^{t} w_{ij\tau}, \quad \text{all } i, \text{ all } j, \text{ all } t$$

$$\sum_{j=1}^{N_i} \sum_{\tau=L_i-L_{ij}+1}^{t} b_{ij}^{d} x_{i,j,\tau-L+L_{ij_i}} + \sum_{\tau=1}^{t} b_i x_{i,\tau-L_i} - I_{it} + BO_{it} = \sum_{\tau=1}^{t} d_{i,\tau} - I_{i,0} + B_{i,0},$$

$$i = 1, \ldots, N_I, \quad t = 1, \ldots, T.$$

$$\sum_{i=1}^{N_I} \sum_{j=1}^{N_i} a_{kij}(b_{ij} x_{i,t-L_{ij}} + x_{i,j,t} + \sum_{\alpha=1}^{j-1} b_{i\alpha j} x_{i,\alpha,t-L_{i\alpha j}}) \leqslant c_{kt}, \quad k = 1, \ldots, K, \quad t = 1, \ldots, T.$$

$$x_{i,j,t} \geqslant 0, \ x_{i,t}0, \ I_{i,t} \geqslant 0, \ BO_{i,t} \geqslant 0, \quad j = 1, \ldots, N_i, \quad i = 1, \ldots, N_I, \quad t = 1, \ldots, T.$$

We remark that, in order to implement the steady-state horizon condition proposed in the section on Controlling Horizon Effects, one cannot allow independent WIP release variables to be formulated for periods intersecting the frozen interval. It is proposed to define WIP release variables only for time periods preceding the frozen interval, that is, all initial WIP that is to be released must be released in time to be realized as completed product before the planning horizon. We also remark that enforcing exact inventory balance of finished goods is likely to be impractical when there are many major process steps with noninteger time lags. A practical approach is to formulate the balance constraints over a sufficiently fine time grid.

ACCOMMODATING MACHINE ARRANGEMENT CONSTRAINTS

Nonhomogeneous sets of machines that perform a basic kind of semiconductor fabrication step such as photolithography exposure are increasingly prevalent in semiconductor factories. To accurately represent capacity relationships in the planning model, one must expand the model. Depending on the nature of the alternatives, there are various formulation strategies that minimize the model complexity necessary to accurately model capacity.

The simplest case is when process times are independent of the resource alternative selected, and the alternative machine types are nested. For example, suppose there are two types of exposure machines, type A and type B. Type B is a newer model that can perform any exposure step; type A is an older model that can perform only noncritical exposure steps. For this type of case, Leachman and Carmon (1992) show that ordinary capacity constraints defined for appropriate groups of machine types constitute an exact capacity model. For this particular example, two capacity constraints per period constitute an exact model. One constraint limits the total workload of critical steps by the available processing time of machine type B, and the other limits the total workload of all exposure steps to the sum of available processing times of both machine types. This approach also provides an exact model when process times of the nested machine types are proportional, since available processing times of alternative machine types may be appropriately scaled according to the process times of one machine type chosen as a standard.

A more difficult case arises when machine usage patterns are not nested. For example, suppose now there are three exposure machine types. Suppose some process steps must be performed on either machine types A or B, other process steps must be performed on either machines types B or C, and still others must be performed on machine types A or C. These more general patterns of allowed allocation of machines arise when engineering effort is expended to qualify machines one by one for critical process steps and certain machines are found to perform better than others. The restrictions placed on machine allocation are thus an avenue for securing better process control and higher yields, albeit at the potential expense of reduced capacity and longer flow times.

When alternative machine types exhibit this more general pattern of allowed assignments to process steps, Leachman and Carmon (1992) show that the most compact exact model requires introduction into the model of new variables that allocate step workloads to the resource types. Fordyce and Sullivan (1995) formulated a planning model with static demand rates incorporating this approach. Allocation variables can also be incorporated into a dynamic planning model as we shall now illustrate.

We shall term the allowed pattern of assignments of machine types to process steps the machine arrangement table. For the case we consider, machine arrangements at one step are independent of arrangements made at other steps; we term this case one of static machine arrangement constraints. We introduce the following new notation and variables, and we modify the constraints of the formulation of Section on Scheduling Initial WIP as follows.

Notation

$k \in (i, j)$ denotes that machine type k is one that is qualified to perform step j on product i.

$(i, j) \in k$ denotes that step j on product i is one of the steps that machine type k is qualified to perform.

Variables

$A_{i,j,k,t}$ = workload of step j on product i in period t allocated to machine type k.

Constraints

Modified capacity constraints are as follows. Inventory balance constraints are unchanged.

$$b_{ij}x_{i,t-L_{ij}} + x_{i,j,t} + \sum_{\alpha=1}^{j-1} b_{i\alpha j}x_{i,\alpha,t-L_{i\alpha j}} = \sum_{k\in(i,j)} A_{i,j,k,t}, \quad j = 1,\ldots,N_i, \quad i = 1,\ldots,N_I, \quad t = 1,\ldots,T.$$

$$\sum_{(i,j)\in k} a_{kij}A_{i,j,k,t} \leqslant c_{kt}, \quad k = 1,\ldots K, \quad t = 1,\ldots,T.$$

$$A_{i,j,k,t} \geqslant 0, \; x_{i,j,t} \geqslant 0, \; x_{i,t} \geqslant 0, \; I_{it} \geqslant 0, \; BO_{i,t} \geqslant 0,$$
$$k = 1,\ldots,K, \quad j = 1,\ldots,N_i, \quad i = 1,\ldots,N_I, \quad t = 1,\ldots,T.$$

Dynamic Machine Arrangement Constraints

The most difficult case of machine arrangement involves dynamic machine arrangement constraints, whereby the set of qualified process machines at one step is dependent on the machine type assigned at some previous process step. Efforts to achieve process control on the most advanced digital process technologies in the industry sometimes include dynamic machine allocation constraints between critical photolithography exposure steps, or between the lithography exposure step and the following etching step. (In such cases, most of the machine types are individual machines.) To illustrate dynamic arrangement constraints, suppose the qualified machines for the first critical exposure step are machines *A*, *B*, and *C*. If machine *A* is selected, then the qualified machines for the second critical exposure step are machines *A* or *C*. If machine *B* is selected at the first step, then the qualified machines for the second critical exposure step are machines *B* or *D*. Thus the qualified machines for performing the second critical step vary according to which machine was utilized at the first critical step.

To properly model capacity constraints in this case, the allocation of workloads to machines at different steps must be constrained so as to be consistent, if planned flow times are to be observed. To illustrate, we assume the case where one general type of machine, such as the photolithography exposure machines, is subject to dynamic machine arrangement constraints. We shall designate process steps whose machine assignment influences or is influenced by the machine assignment at other steps as critical steps. We assume the case where the machine assignment decision at the first critical step constrains the assignment at the second critical step, the assignment at the second critical step constrains the assignment at the third critical step, and so on.

Lin (1999) shows that this case is most efficiently modeled using routing variables that schedule the release of WIP for movement through particular machine types at downstream critical steps. In this approach, machines utilized at each critical step are categorized into machine groups. A machine group may consist of a single machine or several machines. Machine groups at the first critical step are defined with the property that product processed by any machine in the group has identical machine assignment possibilities at the next downstream critical step. Machine groups at subsequent critical steps are defined with the property that product processed by any machine in the group has identical machine assignment possibilities at both the immediately preceding upstream and the next downstream critical step.

Initial WIP at each process step is categorized into WIP groups. Each WIP group is distinguished based on which machine group was utilized at the preceding critical step. (If there are no preceding critical steps, then all WIP belongs to the same group. Also, if two WIP groups have identical machine group possibilities at the next downstream critical step, they may be consolidated into a single WIP group.) At the next critical step downstream, each WIP group will have one or more machine groups qualified to process that WIP.

In Lin's approach, routing variables are defined that route the initial WIP in each WIP group into all possible machine groups at all downstream critical steps. Constraints insuring consistency of these assignments are incorporated into the formulation.

New notation, parameters, variables, and revised constraints to revise the model of Section on Scheduling Initial WIP to precisely model dynamic arrangement constraints are as follows:

Notation

j = major process step; $j \in C_i$ denotes that major process step j is a critical process step for product i.

j_u = the first critical process step upstream from major process step j.

j_d = the first critical process step downstream from major process step j.

j_i = the first critical process step for product i.

g = machine group; $g \in k$ denotes that machine group g is one that includes machine type k.

a = WIP group; $a \in j$ denotes that WIP group a is one that applies to WIP at major process step j.

(a, j') = WIP group a at process step j'; $(g, j)\ \varepsilon\ a$ denotes that machine group g at downstream critical step j is suitable for processing WIP in WIP group a.

Parameters

$w_{i,(a,j),t}$ = initial WIP in the WIP group that is a resident at steps between major steps $j-1$ and j arriving at major step j during period t.

Variables

$x_{i,t}^{(g,j)}$ = quantity of raw material for the manufacture of product i to be released during t and to be processed by maching group g at critical step j.

$x_{i,(a,j),t}^{(g,j')}$ = release of initial WIP from WIP group a at major step j to be processed by machine group g at downstream critical step j'.

$A_{i,k,t}^{(g,j)}$ = workload for machine group g at critical step j on product i allocated to machine type k in period t.

$A_{i,k,t}^{j}$ = workload at noncritical major step j allocated to machine type k in period t.

Constraints

$$x_{i,t} = \sum_{(g,j_i)} x_{i,t}^{(g,j_i)}, \quad t=1,\dots,T, \quad i=1,\dots,N_I.$$

$$\sum_{(g',j_u)} x_{i,t}^{(g',j_u)} = \sum_{(g,j)} x_{i,t}^{(g,j)}, \quad \forall j \in C_i, \quad t=1,\dots,T, \quad i=1,\dots,N_I.$$

$$x_{i,j,t} = \sum_{a\in j} \sum_{(g,j_d)\in a} x_{i,(a,j),t}^{(g,j_d)}, \quad j=1,\dots,N_i, \quad t=1,\dots,T, \quad i=1,\dots,N_I.$$

$$\sum_{(g',j_u)\in a} x_{i,(a,j'),t}^{(g',j_u)} = \sum_{(g,j)\in a} x_{i,(a,j'),t}^{(g,j)} \quad \forall j \in C_i \text{ such that } j>j', \quad t=1,\dots,T, \quad i=1,\dots,N_I.$$

$$\sum_{\tau=1}^{t} \sum_{(g,j_d)\in a} x_{i,(a,j),\tau}^{(g,j_d)} \leqslant \sum_{\tau=1}^{t} w_{i,(a,j),\tau}, \quad j=1,\dots,N_i, \quad i=1,\dots,N_I, \quad t=1,\dots,T.$$

$$A_{i,k,t}^{(g,j)} = b_{ij} x_{i,t-L_{ij}}^{(g,j)} + \sum_{j'=1}^{j} \sum_{\{a\in j' \mid (g,j)\in a\}} b_{ij'j} x_{i,(a,j'),t-L_{i,j',j}}^{(g,j)}, \quad j \in C_i, \quad i=1,\dots,N_I, \quad t=1,\dots,T.$$

$$A_{i,k,t}^{j} = b_{ij} x_{i,t-L_{ij}} + \sum_{j'=1}^{j} \sum_{a\in j'} \sum_{(g,j'_d)\in a} b_{ij'j} x_{i,(a,j'),t-L_{i,j',j}}^{(g,j)}, \quad j \notin C_i, \quad i=1,\dots,N_I, \quad t=1,\dots,T.$$

$$\sum_{i=1}^{N_I} \sum_{j\in C_i} \sum_{g\in k} a_{kij} A_{ikt}^{(g,j)} + \sum_{i=1}^{N_I} \sum_{j\notin C_i} a_{kij} A_{ikt}^{j} \leqslant c_{kt}, \quad k=1,\dots,K, \quad t=1,\dots,T.$$

All variables are constrained to be nonnegative. Inventory balance constraints are unchanged.

MODELING ALTERNATIVE RESOURCE COMBINATIONS

There are instances of scarce tooling in semiconductor manufacturing whereby capacity may be a function not only of the available machines, but also of the available tools, such as photomasks, burn-in boards, test handlers, and fixtures, and so on. An added dimension of complexity arises when alternative machine types for a process step each require distinct tooling. That is, there are alternative combinations of resources that can be assigned to carry out a process step. Modeling the capacities of machines and tools independently may lead to production plans impossible to carry out, because the plans may implicitly require joint application of incompatible tools and machines. The model must explicitly represent the assignment of a valid combination of resources to carry out the process step.

For example, in semiconductor test operations, several different types of tools are required, in addition to the testing CPU. These may include handlers, interface boards, software programs, jigs, and fixtures. There may be several alternative tools for any one of these tool types, for example, handlers capable of only room temperature tests, handlers capable of both room temperature and high temperature tests, and handlers capable of cold temperature, room temperature and high temperature tests. A similar phenomenon occurs in burn-in operations, where the set of resources to be applied includes the burn-in oven chambers, the burn-in board, and the load/unload equipment. There can be two or more alternative sizes of burn-in boards that may be assigned to burn in a given device, each compatible only with similar-sized oven chambers and load/unload machines. Each instance of a valid set of resources to carry out the process is what we term a valid resource combination.

We may extend the model of Section on Accommodating Machine Arrangement Constraints to handle such cases by introducing variables that assign resource combinations to process steps. New notation, parameters and variables, as well as revised capacity constraints, are introduced as follows.

Notation

c denotes a resource combination, that is, a set of resource types required to perform one or more process steps. $k \in c$ denotes that resource type k is a member of resource combination c.

$c \in (i, j)$ denotes that resource combination c is one that is qualified to perform step j on product i.

$(i, j, c) \in k$ denotes that step j on product i is one of the steps that machine type k performs as part of resource combination c.

Variables

$A_{i,j,c,t}$ = workload of step j on product i in period t allocated to resource combination c.

Constraints

Modified capacity constraints are as follows. Inventory balance constraints are unchanged.

$$b_{ij}x_{i,t-L_{ij}} + x_{i,j,t} + \sum_{\alpha=1}^{j-1} b_{i\alpha j}x_{i,\alpha,t-L_{i\alpha j}} = \sum_{c\in(i,j)} A_{i,j,c,t}, \quad j = 1, \ldots, N_i, \quad i = 1, \ldots, N_I, \quad t = 1, \ldots, T.$$

$$\sum_{(i,j,c)\in k} a_{kij}A_{i,j,c,t} \leqslant c_{kt}, \quad k = 1, \ldots, K, \quad t = 1, \ldots, T.$$

$$A_{i,j,c,t} \geqslant 0, \quad x_{i,j,t} \geqslant 0, \quad x_{i,t} \geqslant 0, \quad I_{i,t} \geqslant 0, \quad BO_{i,t} \geqslant 0,$$

$$k = 1, \ldots, K, \quad j = 1, \ldots, N_i, \quad i = 1, \ldots, N_I, \quad t = 1, \ldots, T.$$

2.4. *Handling Integrality of Production Lots*

Work in process in wafer fabrication plants is typically released into production and moves from step to step in lots of twentyfive or fifty wafers, less yield losses. In device assembly and testing areas production lots typically consist of thousands of packaged chips, with specific upper limits based on the capacity of lot carriers. The foregoing linear programming formulations make no allowance for lot integrality when calculating release schedules. It is computationally unattractive to enforce integer constraints on release schedules, so heuristic techniques are typically used to alter input data and/or solution results into a workable schedule.

For example, cumulative demands for each product may be rounded up into lot-sized step functions, where the lot size is chosen according to the desired starting lot size and the expected process yield. If the total demand over the horizon is less than one lot, the input demand can be rounded up to one lot, or alternatively, the model may be allowed to schedule the release of a reduced-size lot, if overproduction is not desired. If the demand is satisfied on time in the LP solution, the resulting release schedule will be lot-sized as desired. If the release schedule is fractional, it nevertheless can be interpreted as a target rate (remember the variables were defined as production rates) to which the actual discrete lot releases should adhere as closely as possible. Of course, the integrality problem is most acute for low-volume products and becomes insignificant for high-volume products.

2.5. *Multiobjective Programming*

It is sometimes difficult practically or institutionally to capture all the desirable qualities of a production plan in a single economic objective such as formulated in Section 2.2. Successive LP runs, possibly with different objectives for different priority classes of demands, were suggested in Section 2.4 as a means of preventing the LP from making trade-offs between servicing prior customer commitments and responses to other market opportunities. But there may be other objectives of interest as well.

For example, it may be desired to control the distribution of bin inventories remaining at the end of the planning horizon, encouraging the planning model to use the lowest-quality bins and the least capacity-consuming test flows necessary to meet the demands. If costs on inventories and costs on test flows were incorporated into the same objective function that minimizes shortage costs, it may admit solutions that trade-off product shortage costs with production and inventory costs, yet no such trade-off is desired. Handling bin inventory costs in a second objective function, to be optimized subject to the optimal value of the first, is a more desirable approach.

It is straightforward to add follow-on, multiple objectives to the basic formulation discussed herein.

3. Brief History of Optimization Applications to Semiconductor Production Planning

The first regularly-used optimization-based planning system in the industry known to the author was the A-Plus system at Intel Corporation, based on the Berkeley Planning System software, installed in 1986 to plan across the fabrication lines in that company. BPS was implemented to plan three fabrication plants at Harris Corporation in 1987, and beginning in 1989 an effort was undertaken to enhance and expand BPS for fully automated, company-wide production planning. This culminated in the IMPReSS planning system, fully implemented in 1992. The Franz Edelman Award from the Institute for Operations Research and the Management Sciences (INFORMS) was awarded to Harris in 1995 in recognition of the achievements of IMPReSS. Subsequently, enhanced versions of BPS were implemented at Advanced Micro Devices and at Samsung Electronics, Co., Ltd.

Commercial supply chain management and production planning software products incorporating linear programming calculations began to be marketed to the semiconductor industry following the success of IMPReSS. Major products currently available include MIMI from Chesapeake Decision Sciences (later Aspen Technologies), Rhythm from i2 Technologies, and the Supply Chain Planner from Paragon Management Systems (later Adexa Management Systems). Leachman and Associates LLC offers customized optimization-based planning systems to the industry.

At the start of the millennium, a substantial and growing fraction of the global semiconductor manufacturing industry employ optimization calculations to assist in the generation of production plans. This remarkable growth has been paced not only by the development of modeling software as above, but also by the rapid development of computer hardware capabilities and of optimization software. At the time of implementation of the A-Plus system at Intel, the leading commercial LP software could solve problems with up to

10,000 rows in a few hours using the simplex method operating on a top-of-the-line IBM mainframe computer. Six years later, in 1992, the IMPReSS system at Harris routinely optimized formulations with upward of 130,000 rows in a few hours using interior point methods operating on a UNIX workstation computer.

Despite this progress, it must be acknowledged that optimization still plays a limited role in the generation of production plans within most commercial planning systems. For example, Rhythm uses optimization calculations only to generate guidelines such as routings through the product structure and allocations of capacity to products. Heuristics and artificial intelligence rules subsequently act on the guidelines to actually compute schedules. With the exception of products from Leachman and Associates LLC, the commercial products utilize optimization models that incorporate only a subset of the capabilities described in Section 2.3. For example, formulations in most commercial systems require time lags to be integers. Some offer only crude capabilities for capacity analysis, yet others model alternative resource combinations. Some do not accommodate binning and substitution product structures, yet others readily handle such complex product structures.

While optimization modeling has made considerable progress over the last fifteen years in serving the semiconductor industry, it is the author's opinion that the opportunity remains for much more progress to be made.

References

Fordyce, K. and G. Sullivan. 1995. "A dynamically generated rapid response fast capacity planning model for semiconductor fabrication facilities." In *The Impact of Emerging Technologies on Computer Science and Operations Research*, edited by S. Nash and A. Sofer. Boston: Kluwer Academic Publishers.

Fordyce, K., D. Dalton, B. Gerard, R. Jesse, and G. Sullivan. 1992. "Daily output planning: integrating operations research, artificial intelligence, and real-time decision support with APL2." *Expert Sys. Appl.* **5**: 245–256.

Hackman, S. T. and R. C. Leachman. 1989. "A general framework for modeling production." *Mgmnt Sci.* **35(4)**: 478–495.

Hung, Y.-F. 1991. *Corporate-level production planning with simulation feedback of parameters*. Ph.D. diss. Berkeley: College of Engineering, University of California at Berkeley.

Hung, Y.-F. and R. C. Leachman. 1996. "A production planning methodology for semiconductor manufacturing based on iterative simulation and linear programming calculations." *IEEE Trans. Semiconductor Manufacturing* **9(2)**: 257–269.

Kang, J. 1996. *A method for target scheduling of semiconductor wafer fabrication based on event-based optimization modeling and discrete event simulation*. Ph.D. diss. Berkeley: College of Engineering, University of California at Berkeley.

Leachman, R. C. 1986. Preliminary design and development of a corporate-level production planning system for the semiconductor industry. ORC Report 86-11. Berkeley: Operations Research Center, University of California at Berkeley.

Leachman, R. C. 1993. "Modeling techniques for automated production planning in the semiconductor industry." In *Optimization in Industry*, edited by T. A. Ciriani and R. C. Leachman, pp. 1–30. Chichester, England: John Wiley and Sons, Ltd.

Leachman, R. C. and T. F. Carmon. 1992. "On capacity modeling for production planning with alternative machine types." *IIE Trans.* **24(4)**: 62–72.

Leachman, R. C. and D. J. Raar. 1994. "Optimized production planning and delivery quotation for the semiconductor industry." In *Optimization in Industry 2*, edited by T. A. Ciriani and R. C. Leachman, pp. 63–72. Chichester, England: John Wiley and Sons, Ltd.

Leachman, R. C., R. F. Benson, C. Liu, and D. J. Raar. 1996. "IMPReSS: an automated production-planning and delivery-quotation system at Harris Corporation—semiconductor sector." *Interfaces* **26(1)**: 6–37.

Lin, V. 1999. *Advanced Semiconductor Production Planning*. Ph.D. diss. Berkeley: College of Engineering, University of California at Berkeley.

Liu, T.-Y. 1998. *Equipment acquisition planning in the semiconductor industry considering learning effects in equipment efficiency*. PhD diss. Berkeley: College of Engineering, University of California at Berkeley.

Stray, J., J. W. Fowler, and W. M. Carlyle. 2000. "Enterprise wide strategic and logistics planning for semiconductor manufacturing." In *Proceedings of the International Conference on Modeling and Analysis of Semiconductor Manufacturing (MASM 2000)*, edited by W. Cochran, John W. Fowler, and Steven Brown, pp. 353–356. San Diego: The Society for Computer Simulation International.

18.7 Optimization in the Aerospace Industry

Matthew E. Berge, John T. Betts, Sharon K. Filipowski, William F. Huffman, and David P. Young

ABSTRACT The aerospace industry is a source of a large number of challenging problems, and optimization plays a significant role in the solution of many of these. In this article, we discuss how optimization is used in aerodynamic design of aircraft wings, data association for multitarget tracking, and trajectory optimization. These three examples give a sample of application areas and uses of optimization in the aerospace industry.

1. Introduction

The aerospace industry gives rise to a large number of challenging problems. In this article, we discuss three application areas where the use of optimization enables quality solutions to difficult underlying problems. In the first example, we discuss the role of optimization in the aerodynamic design of the external shape of the wings of an aircraft. The optimization methods developed have resulted in significant savings in both the time and cost of designing airplanes. In the second example, we describe data association in multitarget tracking. In this case, optimization-based approximation algorithms enable real-time solutions of these problems so that they can be used in onboard systems. Finally, we describe the role of optimization in determining trajectories for a wide range of applications, including defining the path for a spacecraft in orbit and determining the reentry path of the space shuttle.

2. Aerodynamic Design Optimization

2.1. *Introduction*

The idea of optimizing aerodynamic configurations by minimizing drag is classical and has been applied in many special cases such as drag minimization in supersonic linear flow. There are also classical optimization results for induced drag in linear flow and for area ruling (Ashley and Landahl, 1965). These classical methods are based mostly on analytic solutions and calculus of variations.

However, the application of numerical optimization to aerodynamic design had to await sufficient maturity of the fundamental analysis tools in the field of computational fluid dynamics (CFD) with the ability to solve a complex enough model of the flow to include all the effects crucial to performance with reasonable accuracy and reliability. Progress was fairly slow until the 1990s, due to unresolved difficulties with compressible flow solvers and the inadequacy of available scientific computers.

The desire in the aerospace industry was originally to use optimization as a tool to improve airplane performance by finding the best wing shape to minimize drag at the airplane's cruising lift condition. At first wind tunnel testing was used with a cut-and-try methodology to improve the design. When CFD tools first became available, it was natural to use them in this same methodology but without formal optimization. This methodology relied on engineering understanding and intuition to achieve better designs. As wind tunnel testing became sophisticated enough to allow measurements of detailed surface pressure distributions and as experience with CFD tools accumulated, a body of knowledge about desirable surface pressure distributions developed. In most CFD codes, it is straightforward to build a capability to do inverse design in which a geometry yielding a specified target pressure distribution could be constructed. However, the approximations involved in relating surface curvature or slope to local pressure make the process iterative. In addition, these inverse design methods cannot in general directly minimize measures of performance such as drag, nor can they directly satisfy equality or inequality constraints.

By the time CFD codes became robust and accurate enough to predict drag levels near the resolution levels of the wind tunnel, the goals for CFD optimization had changed substantially. The desire was to cut cost and cycle time in designing new airplanes from years to months. These new goals required consideration of constraints from other disciplines (e.g., structures, manufacturing, controls), as well as the performance issue. Use of drag as the design objective led to the well-known occurrence of a point design, (a shape that substantially reduces drag at the flow condition where the design is being performed, but has considerably worse drag than the original at other flow conditions). In fact, single point airfoil analysis and design, and therefore wing design, are mathematically poorly posed as discussed in Young et al. (1996). In addition, because an airplane will fly at a number of speeds and lift conditions during a typical flight, it became clear that multipoint design was necessary. The multipoint design problem can be considered a simplification of the more general multicriteria optimization problem, where all the criteria are the same, namely drag. Many of the complexities introduced by multipoint design into an already computationally intensive process (Pareto optimality, etc.) are yet to be fully addressed. Typically an assumption is made that the weights between the different flow conditions are proportional to the amount of time spent at the flow condition during a standard flight.

2.2. *Solution Approaches*

At least three major methodologies have been developed to attack the aerodynamic design optimization problem. All have been driven by the need to keep the required number of full CFD analysis solutions to a small number, because of CPU time/expense. The first method is the easiest to implement because it treats the CFD analysis tool as a black box. A set of design variables is selected that parameterize potential geometry changes and a strategy is chosen for sampling the design space when these variables are changed. CFD analysis solutions are then requested on this sample space (a perfectly parallel process) and cataloged. A model of the design objective and constraints is constructed through response-surface techniques. Optimization can easily be applied to the resulting models of the objective and constraints (Herling et al., 1998). Strengths of this methodology include the ability to detect multiple extrema (a global optimization process), easily explore changes in objective/constraint definition, detect variables with little effect, and the ability to use any CFD solver. Weaknesses include the rapid growth in number of CFD analyses needed as the number of design variables is increased and the often slow convergence to an optimum.

The second major approach uses the fact that the CFD design problem can be considered as an optimal control problem for a boundary value problem. To solve such a problem one constructs the adjoint equations for the original set of flow equations. The same solution technology used to solve the flow system of PDEs can often be used to solve the adjoint equations. The gradient of the objective can be computed inexpensively from the adjoint variables. Thus, the cost per gradient evaluation is at most the cost of a single extra analysis (Jameson, 1998). A gradient based optimizer can now be applied to solve the problem because the gradients are no longer tremendously expensive. The strengths of the method are more rapid local convergence to an optimum, the ability to have a large number of design variables (the cost of a single gradient evaluation is virtually independent of the number of design variables), and the ability to have a one-pass process to a final design. Drawbacks are due to being a local optimization process (if multiple extrema are possible), and the difficulty (or expense) in adding flow constraints to the optimization problem (these would require extra adjoint equation solves).

The third major approach embeds an optimization step inside the CFD flow solver by using a local sensitivity model of the flow to define an approximate optimization subproblem to which standard nonlinear programming algorithms can be applied. In the algorithm used in Jou et al. (1995), the sensitivity model is constructed by solving multiple right-hand sides using the flow Jacobian matrix already available in that particular CFD solver. This again attacks the gradient expense problem, because multiple right-hand-sides of the linear system can be solved by two orders of magnitude more rapidly than a single flow solution. The strengths of the method are more rapid local convergence to an optimum, full availability of any flow quantity as a constraint/objective, and the sensitivity modeling of the flow, which allows multipoint design. Drawbacks include being a local optimization process, and being dependent on a CFD solver that forms a flow Jacobian as part of its solution process. A significant innovation that is feasible using this approach is the use of design optimization in the context of a CFD solver that uses solution adaptive grid refinement.

There is continuing controversy about the merits of adjoint based methods versus sensitivity based methods. The usual statement is that on a per iteration basis, the adjoint method will be preferred if the number of state equation constraints is less than or equal to the number of design variables. However, this comparison ignores global convergence issues. There is some reason to believe that the model of the objective that can be constructed easily from sensitivities will accelerate the generation of second order information and hence converge to the optimum more rapidly than an adjoint based method. Furthermore, an adjoint based method must rely on BFGS and other orthogonality based methods for acceleration beyond the usually unacceptable expected convergence rate of steepest descent. These accelerations work well only if the gradient calculation and the function values used in the line search are discretely consistent. This consistency can be difficult to achieve in practice. On the other hand, with sensitivities no such consistency problems arise because the theory of inexact Newton methods justifies any approximations that are required. Furthermore, adaptive solution methods such as those based on solution adaptive grid refinement or adaptive higher order methods pose no special difficulties.

2.3. *Summary*

The success of the methods just described in improving the performance of aircraft during their design phase has made the optimization approach to aerodynamic design an accepted tool in engineering practice. In addition, the success has helped to reduce the number of wind tunnel models built and tested for new designs (giving considerable savings in both money and flow time). Some of the major issues currently being addressed are ease of use (graphical interface, process control), parallel processing in both the CFD solvers and the optimization hook-ups, the ability to extract trade study information from the interdisciplinary constraints imposed, and insight into the potential and practicality of solving the tougher, multidisciplinary optimization (MDO) problem for aircraft design.

3. Data Association for Multitarget Tracking

3.1. *Introduction and Problem Definition*

An aerospace problem that has received considerable attention is one of utilizing sensor measurement data to track multiple targets in a surveillance volume (Bar-Shalom, 1990; Blackman, 1990). An example involves Airborne Early Warning aircraft such as the AWACS. An important aspect of this problem is data association that is characterized by the existence of multiple measurements, either from the same sensor or from multiple sensors, and a desire to determine which of these measurements are associated in the sense that they are generated by the same tracked object.

Consider the case in which there is a single sensor that obtains measurement data on each scan of the surveillance volume. The simplest of association problems is one in which a set of maintained tracks is to be extended into a single scan of measurements. The problem is to associate the tracks with the new measurements in a way that best explains the data, thus producing the best track extensions. Characteristics of the tracks and measurements are used to quantify the likelihood of each track-measurement combination. After transformation, this problem can be formulated and easily solved as a bipartite (2-D) network assignment problem (Bertsekas, 1991). This simple approach can work well for easy tracking problems but can be inadequate for more difficult cases (e.g., problems with maneuvering, crossing, or low observable targets). For these problems it is desirable to formulate the association problem as extending a set of existing tracks into multiple scans of measurement data. The idea is to generate a large set of reasonable associations (candidate tracks), compute a cost (i.e., negative log-likelihood) for each, and then solve the data association problem by selecting a subset of associations of minimum total cost while accounting for each maintained track and each measurement. This problem can be formulated as multidimensional assignment problem (MAP), as shown in the next section. The advantages are that data is fused across multiple scans and there is the possibility of revisiting decisions concerning individual measurements. The difficulty is that for MAP dimensionality larger than two the problem is known to be NP-hard; yet, the interest is in algorithms that solve, or approximately solve, these problems fast enough to support real-time applications.

3.2. *Multidimensional Assignment Problem (MAP) Formulation*

Consider the case in which a set of N_0 maintained tracks is to be associated with S scans of sensor data where scan i contains N_i measurements. Also, consider the corresponding index sets $T_0, M_1, \ldots, M_S$ containing $N_0+1, N_1+1, \ldots,$ and N_S+1 elements, respectively. The additional element in each set refers to a 0th index with a special use described below. These sets are summarized as follows:

$$T_0 \equiv \{0, 1, \ldots, i_0, \ldots, N_0\},$$

$$M_1 \equiv \{0, 1, \ldots, i_1, \ldots, N_1\},$$

$$\vdots$$

$$M_S \equiv \{0, 1, \ldots, i_S, \ldots, N_S\}$$

Furthermore, there is a set of associations (and costs), each of which is an $(S+1)$-tuple defined in terms of these index sets. For example, when existing tracks are to be extended into two scans of data (i.e., $S=2$), then (3, 2, 5) asserts that the object associated with track 3 generates measurement 2 from the first scan and measurement 5 from the second scan. The 0-index enables the representation of false alarms and missed detections. For example, (0, 0, 4) asserts that measurement 4 from the second scan is false while (2, 0, 4) asserts that the object associated with track 2 generated measurement 4 from the second scan but experienced a missed detection on the first scan. Finally, the objective in the data association problem is to select a set of associations of minimum total cost subject to the constraint that each nonzero index in each set is used exactly once. The 0-index may be used any number of times. A mathematical programming formulation for the sparse representation of this $S+1$-dimensional MAP is given below.

Association set $A \subseteq T_0 \times M_1 \times \ldots \times M_S$

Define $A(j,k) \equiv \{(i_0, i_1, \ldots, i_j, \ldots, i_s) \in A \mid i_j = k\}$

$$(MAP) \quad \min \sum_{A} c_{i_0 i_1 \ldots i_S} x_{i_0 i_1 \ldots i_S}$$

$$\sum_{A(0,k_0)} x_{i_0 i_1 \ldots i_S} = 1 \qquad k_0 = 1, 2, \ldots, N_0$$

$$\sum_{A(1,k_1)} x_{i_0 i_1 \ldots i_S} = 1 \qquad k_1 = 1, 2, \ldots, N_1$$

$$\vdots$$

$$\sum_{A(S,k_S)} x_{i_0 i_1 \ldots i_S} = 1 \quad k_S = 1, 2, \ldots, N_S \quad x_{i_0 i_1 \ldots i_S} \in \{0, 1\}$$

3.3. *Solution Approaches*

MAPs arising in a tracking context often have the property that they are decomposable into independent subproblems. This property may be caused, for example, by spatial separation between clusters of objects and can be detected by graph theoretic methods. The advantages of decomposition are that it enables a divide and conquer approach and a parallel processing strategy. Furthermore, individual subproblems may be identified for solution by different approaches depending on their characteristics. Small problems, for example, can usually be solved quickly and exactly by branch-and-bound while larger problems may require approximation techniques. Several methods have been explored for the solution of MAPs besides branch-and-bound including GRASP (Murphey et al., 1998) and Lagrangian relaxation (LR) (Deb et al., 1997; Poore and Rijavec, 1993; Robertson and Poore, 1998). The remainder of this discussion will be focused on LR-based solution approaches.

Lagrangian relaxation is a widely used method for addressing difficult combinatorial optimization problems (Beasley, 1993). A subset of (difficult) constraints is identified for relaxation, that is, multiplied by a value (Lagrange multiplier) and moved into the objective function. For fixed multipliers the relaxed problem, with its new objective and remaining (easy) constraints, is typically much easier to solve and provides a lower bound on the optimal objective function for the original combinatorial minimization problem. Thus, it is useful to find the values of the multipliers that maximize the objective of the relaxed problem. This new problem is called the Lagrangian dual, and is typically a problem in nonsmooth optimization. A complication in LR is that an optimal solution to the dual may not be a feasible solution to the original combinatorial problem. Consequently, a Lagrangian heuristic is often needed to adjust the dual solution to one that is both feasible and, hopefully, close to an optimal solution for the original problem.

In applying LR to solving the MAP one must first decide on a relaxation strategy. One approach is to relax all of the index set constraints. An alternate, and more common, approach is to relax the index set constraints for all but two sets. This approach results in a relaxed problem that can be transformed and efficiently solved as a bipartite assignment problem. This problem is similar from iteration to iteration and so there is an opportunity for network reoptimization. Once the dual is solved then a Lagrangian heuristic is needed to find a good feasible solution. One approach is based on the sequential solution of bipartite assignment problems, where at each stage, the dimensionality of the problem is reduced by one.

The solution of the dual requires the use of nonsmooth optimization (Lemarechal, 1989). The subgradient method has been used extensively in LR settings; however, because of drawbacks in the method, more sophisticated approaches, such as bundle algorithms, may be attractive alternatives. Finally, in this tracking environment, a sequence of related MAPs needs to be solved creating the opportunity for nonsmooth reoptimization.

To summarize, an effective overall approach for solving the MAP combines a graph theoretic approach for decomposition, branch-and-bound for solving small problems, and a Lagrangian relaxation approach that makes use of nonsmooth optimization to update the multipliers and network optimization to solve the subproblems. Opportunities for reoptimization exist for both nonsmooth and network optimization.

4. Trajectory Optimization

4.1. Introduction

A problem that occurs frequently in the aerospace industry is trajectory optimization. In simple terms this application requires finding the best way to steer a vehicle as it travels from one point to another. For a high performance fighter aircraft it may be desirable to go from one altitude to another as quickly as possible (i.e., minimize the flight time). An optimal trajectory for a spacecraft in orbit may require definition of a path that minimizes the fuel consumed. When computing the reentry path for a vehicle such as the space shuttle it is often useful to construct the steering to minimize the aerodynamic heating. A wide variety of applications such as these are encountered by performance analysts, and they can all be formulated mathematically as optimal control problems.

4.2. *Optimal Control Problem*

Although more general definitions are possible, for the sake of discussion let us introduce the essential elements of an optimal control problem. Typically it is necessary to satisfy a set of dynamic equations of the form

$$\dot{\mathbf{y}} = \mathbf{f}[\mathbf{y}(t), \mathbf{u}(t), t] \tag{1}$$

$$\mathbf{0} \leqslant \mathbf{g}[\mathbf{y}(t), \mathbf{u}(t), t] \tag{2}$$

for $t_0 \leqslant t \leqslant t_f$. In addition, boundary conditions of the form

$$\boldsymbol{\psi}[\mathbf{y}(t_f), \mathbf{u}(t_f), t_f] = \mathbf{0} \tag{3}$$

are usually imposed. The variables $\mathbf{y}$ are called state variables and $\mathbf{u}$ are called control variables. The goal is to find the dynamic variables $[\mathbf{y}(t), \mathbf{u}(t)]$ that minimize the objective:

$$J = \phi[\mathbf{y}(t_f), \mathbf{u}(t_f), t_f]. \tag{4}$$

In terms of general purpose software it is reasonable to require the user to define the items in this description. In fact, these items are probably the minimal user requirement.

A typical example of an optimal control problem is illustrated in Figure 1. In this application it is necessary to compute the optimal orientation of a spacecraft during a single thrusting arc or burn, such that the minimum fuel is consumed (Betts, 1993). The optimal maneuver requires eight revolutions of the earth to complete the transfer from the initial (park) orbit to the final (mission) orbit.

4.3. *Direct Transcription Method*

Historically the approach used to solve an optimal control problem can be derived using the calculus of variations and is called the indirect method. This technique leads to the solution of a multipoint boundary value problem. Unfortunately, in practice the indirect method is often

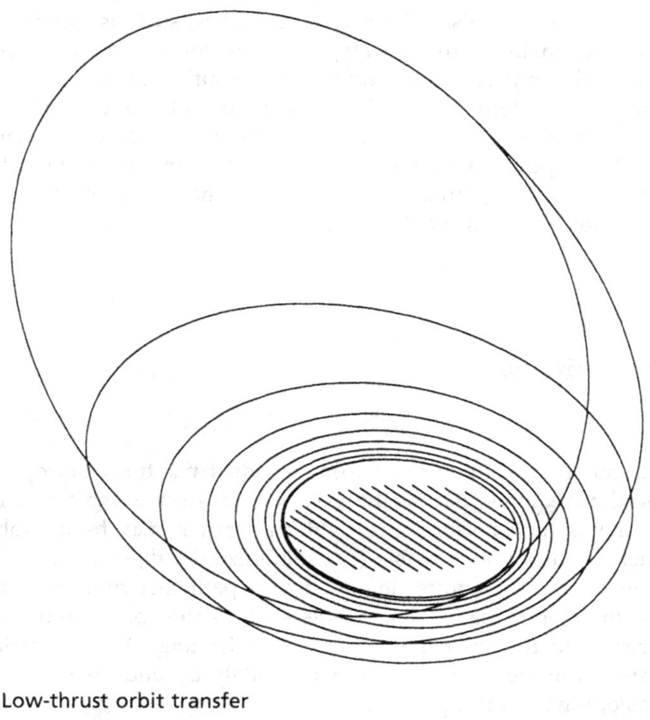

Figure 1 Low-thrust orbit transfer

not robust, lacks flexibility, and requires considerable expertise in control theory to obtain the correct formulation.

In an attempt to eliminate the deficiencies of the indirect method an approach called direct transcription has been investigated (Betts, 1994; Betts and Huffman, 1993, 1998). Briefly, the method can be summarized as follows:

Transcription. Transcribe the optimal control problem into a nonlinear programming (NLP) problem by discretization;

Sparse Nonlinear Program. Solve the sparse NLP using, for example, sequential quadratic programming, in other words

1. Solve a sparse quadratic program (QP) to estimate the NLP solution
2. If the solution is acceptable, terminate; otherwise, update the NLP solution estimate and solve a new QP subproblem;

Mesh Refinement. Assess the accuracy of the approximation (i.e., the finite-dimensional problem), and if necessary refine the discretization, and then repeat the optimization steps.

The fundamental idea is to convert the original optimal control problem into a constrained optimization problem (i.e., a nonlinear program) involving a finite number of variables and constraints. All approaches divide the trajectory into n_s segments or intervals

$$t_0 < t_1 < t_2 < \ldots < t_f = t_{n_s}, \tag{5}$$

where the points are referred to as node, mesh or grid points. Define the number of mesh points as $M \equiv n_s + 1$ and let $\mathbf{y}_k \equiv \mathbf{y}(t_k)$ denote the value of the state variable at a grid point. Using similar notation for the control variable, one can define a set of NLP variables

$$\mathbf{x} = (\mathbf{y}_1, \mathbf{u}_1, \ldots, \mathbf{y}_M, \mathbf{u}_M). \tag{6}$$

Furthermore, let $\mathbf{f}_k \equiv \mathbf{f}[\mathbf{y}_k, \mathbf{u}_k, t_k]$ and $\mathbf{g}_k \equiv \mathbf{g}[\mathbf{y}_k, \mathbf{u}_k, t_k]$. Then, using a trapezoidal discretization, the differential equation (1) can be approximated by the set of nonlinear constraints

$$\zeta_k = \mathbf{y}_{k+1} - \mathbf{y}_k - \frac{h_k}{2}(\mathbf{f}_k + \mathbf{f}_{k+1}) = 0 \tag{7}$$

where the stepsize $h_k = t_k - t_{k-1}$. Thus, the original optimal control problem can be posed as an NLP with objective function

$$F(\mathbf{x}) = \phi[\mathbf{y}_M, \mathbf{u}_M, t_{n_s}] \tag{8}$$

and the following constraints

$$\mathbf{c}_L \leqslant \mathbf{c}(\mathbf{x}) \leqslant \mathbf{c}_U, \tag{9}$$

where

$$\mathbf{c}(\mathbf{x}) = [\zeta_1, \zeta_2, \ldots, \zeta_f, \boldsymbol{\psi}_f, \mathbf{g}_0, \mathbf{g}_1, \ldots, \mathbf{g}_f]^{\mathrm{T}}. \tag{10}$$

Because the Jacobian and Hessian matrices for the resulting NLP are sparse, very efficient optimization algorithms for solving NLPs can be exploited to solve these applications.

The algorithm described above has been implemented in the SOCS software system (Betts and Huffman, 1997) and has been successfully applied to hundreds of applications not only in trajectory optimization but also in chemical process control and machine tool path design. In practice this technique has become very attractive because it eliminates most of the drawbacks found with the classical indirect techniques. In particular, the direct transcription method derives significant robustness from the underlying nonlinear programming software. Because finite difference derivatives can be used quite effectively, the software implementations are very flexible. And finally, it is attractive for the naive user because much of the underlying optimal control theory can be incorporated into the software.

References

Ashley, H. and M. Landahl. 1965. In *Aerodynamics of Wings and Bodies*. Addison-Wesley.

Bar-Shalom, Y. (ed.). 1990. In *Multitarget-Multisensor Tracking: Advanced Applications*. Artech House.

Beasley, J. E. 1993. "Lagrangian relaxation." In *Modern Heuristic Techniques for Combinatorial Problems*, edited by C. R. Reeves, pp. 243–303. John Wiley & Sons, Inc.

Bertsekas, D. P. 1991. *Linear Network Optimization—Algorithms and Codes*. The MIT Press.

Betts, J. T. 1993. "Using sparse nonlinear programming to compute low thrust orbit transfers." *J. Astronautical Sci.* **41**: 349–371.

Betts, J. T. 1994. "Issues in the direct transcription of optimal control problems to sparse nonlinear programs." In *Computational Optimal Control*, Volume 115, edited by R. Bulirsch and D. Kraft, pp. 3–18. Birkhäuser Verlag.

Betts, J. T. and W. P. Huffman. 1993. "Path constrained trajectory optimization using sparse sequential quadratic programming." *AIAA J. Guidance, Control, Dynamics* **61**: 59–68.

Betts, J. T. and W. P. Huffman. 1997. Sparse optimal control software package SOCS. Technical report. Mathematics and Engineering Technical Document MEA-LR-085. The Boeing Company, PO Box 3707, Seattle, WA 98124-2207. More information on software availability can be found at `http://www.boeing.com/assocproducts/socs/`.

Betts, J. T. and W. P. Huffman. 1998. "Mesh refinement in direct transcription methods for optimal control." *Optimal Control Appl. Methods* **19**: 1–21.

Blackman, S. S. 1990. *Multiple Target Tracking with Radar Applications*. Artech House.

Deb, S., M. Yeddanapudi, K. Pattipati, and Y. Bar-Shalom. 1997. "A generalized S-D assignment algorithm for multisensor-multitarget state estimation." *IEEE Trans. Aerospace Electr. Syst.* **33**: 523–536.

Herling, W., S. T. LeDoux, R. R. Ratcliff, D. A. Treiber, and M. J. Warfield. 1998. 3DOPT—an integrated system for aerodynamic design optimization. Technical Report 98-2514. AIAA.

Jameson, A. 1998. "Aerodynamic design via control theory." *J. Sci. Comput.* **3**: 233–260.

Jou, H., W. P. Huffman, D. P. Young, R. G. Melvin, M. B. Bieterman, C. L. Hilmes, and F. T. Johnson. 1995. Practical considerations in aerodynamic design optimization. Technical Report 95-1730. AIAA.

Lemarechal, C. 1989. "Nondifferentiable optimization." In *Handbooks in Operations Research and Management Science*, edited by G. L. Nemhauser et al. **1**: 529–572. Elsevier Science Publishers B.V.

Murphey, R. A., P. M. Pardalos, and L. Pitsoulis. 1998. "A greedy randomized adaptive search procedure for the multitarget multisensor tracking problem." In Network Design: Connectivity and Facilities Location, volume 40 of DIMACS Series in Discrete Mathematics and Theoretical Computer Science, edited by P. M. Pardalos and D.-Z. Du, pp. 131–142. American Mathematical Society.

Poore, A. and N. Rijavec. 1993. "A Lagrangian relaxation algorithm for multidimensional assignment problems arising from multitarget tracking." SIAM J. Optim. **3**: 544–563.

Robertson, A. and A. Poore. 1998. A Lagrangian relaxation algorithm for the multidimensional assignment problem. Working paper available from alexander.robertson@lmco.com.

Young, D. P., W. P. Huffman, R. G. Melvin, F. T. Johnson, C. L. Hilmes, and M. B. Bieterman. 1996. "Requirements for optimization in engineering." In Linear and Nonlinear Conjugate Gradient-Related Methods, edited by L. Adams and L. Nazareth, pp. 37–49. SIAM.

18.8 Energy

18.8.1 Optimization in Electrical Power Systems

Gerson Couto de Oliveira, Sergio Granville, and Mario Pereira

ABSTRACT We describe some major applications of optimization techniques for power systems planning and operation. Power systems optimization models are widely used, ranging from long-term planning studies to real-time operation of generation and transmission facilities, involving the representation of electrical laws governing power plants and transmission networks, the behavior of private agents in a market structure, and different sources of uncertainty. The large scale nature of nationwide transmission networks that provide access to producers and consumers requires the use of specialized decomposition methods that also have to deal with stochasticity in supply and demand. Game theory is extensively applied in conjunction with operation and planning models.

1. Introduction

Power systems are complex due to several factors:

- Nonlinear equations relate power production and transmission over large meshed networks;
- There are diverse technologies for power plants (nuclear, hydro, fossil fueled, etc.), each one having specific operating characteristics (e.g., nonlinearities, ramping constraints, etc.) and environmental aspects;
- A large spectrum of studies ranging from long-term planning to real-time scheduling, requiring consideration of multiple sources of uncertainty;
- Multiple private agents interacting in a market structure, competing among themselves.

We present an overview of some major problems in this area and of optimization models currently used in the industry. In Sections 2 and 3 we describe the hydrothermal system scheduling and planning problems, the main modeling issues, and basic decomposition solution approaches. In Section 5 we present the classic optimal power flow problem, and a state-of-the-art nonlinear interior point method capable of solving large scale problem instances. Sections 4 and 6 discuss new problems related to the recent deregulation of the power industry, implemented in a market structure. One problem is related to optimal bidding of generators in a wholesale market, and the other deals with cost allocation schemes among players which share transmission resources. Each section provides a detailed description of the application and the optimization techniques used; the applications are illustrated with real world examples.

2. Long-Term Scheduling of Hydrothermal Systems

2.1. Hydrothermal System Characteristics

In purely thermal systems, the operating cost of each plant depends basically on its fuel cost. In this case, the thermal dispatch problem is to determine the plant combination that minimizes the total fuel cost required to meet the system load, subject to plant capacity constraints. This problem is solved by inspection: generators are dispatched by increasing operating cost until demand is met. Although the actual scheduling problem is more complex due to factors such as losses, transmission network constraints, start-up costs, ramping rates, et cetera the thermal problem retains some basic characteristics:

- it is decoupled in time: an operating decision in a given stage does not affect the next stage's operating decision;
- generating units have a direct operating cost that does not depend on the output of the other system plants; besides that, plant operation does not affect the generation capacity or availability of other plants: this provides a natural coordination mechanism for energy purchase and sale.

A hydro plant, in turn, can use its reservoir storage to turbine free water for energy production, thus avoiding fuel expenses with thermal units. If reservoirs are in a cascade in a river basin, the outflow release of upstream plants contributes to the inflow of downstream hydro plants. Figure 1 shows the schematic diagram of a hydro plant with reservoir.

The availability of this hydro energy is limited by reservoir storage capacity. Therefore, there is a relationship between the operative decision in a given stage and the future consequences of this decision. For example, if the stored hydroelectric energy is used today, and a drought occurs, it may be necessary to use expensive thermal generation in the future, or even interrupt the energy supply. If, on the other hand, reservoir levels are kept high through a more intensive use of thermal generation today, and high inflows occur in the future, reservoirs may spill and waste energy, resulting in increased operation costs. Figure 2 illustrates this decision tree.

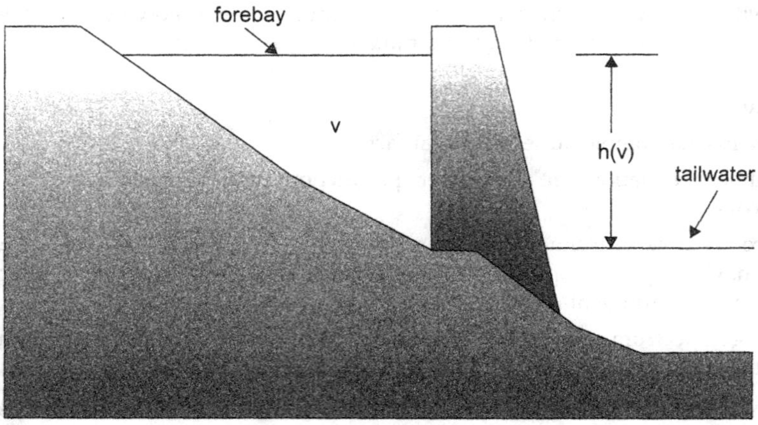

Figure 1 Schematic diagram of a hydro plant with reservoir

In contrast with thermal systems, whose long-term operation is decoupled in time, hydro system operation is coupled in time, so a decision today affects operating costs in the future. Also, since future inflows are unknown and difficult to forecast, the scheduling of hydrothermal systems is essentially stochastic. The tradeoff between immediate and future operating costs is illustrated in Figure 3.

The immediate cost function (ICF) is related to thermal generation costs in the present stage. The more the stored water is used for energy production, the cheaper the ICF will be today, since less thermal generation is needed to meet the load. However, using more water today leaves less storage for future use. So, in terms of the final storage, the ICF increases for higher final storage values. In turn, the future cost function (FCF) is

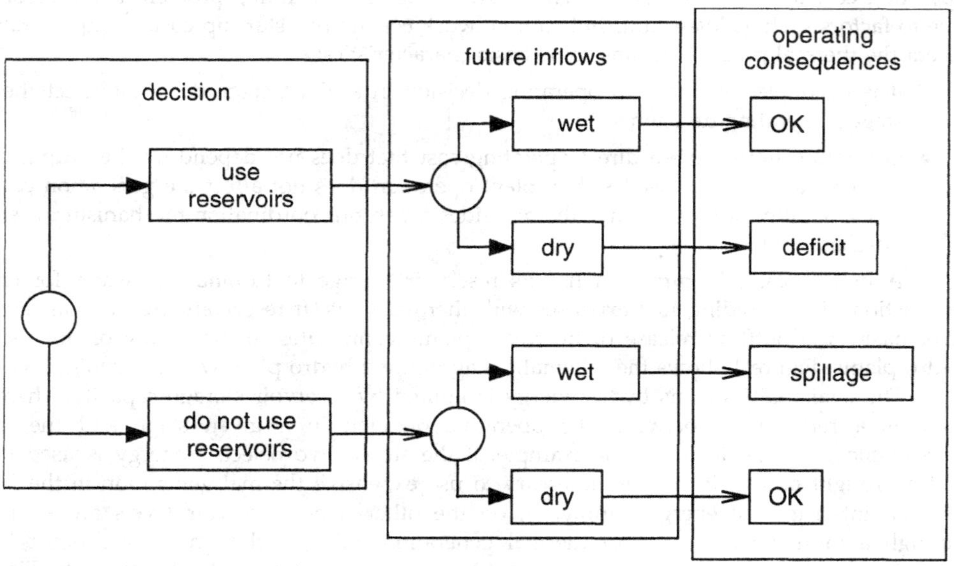

Figure 2 Decision tree for hydrothermal systems

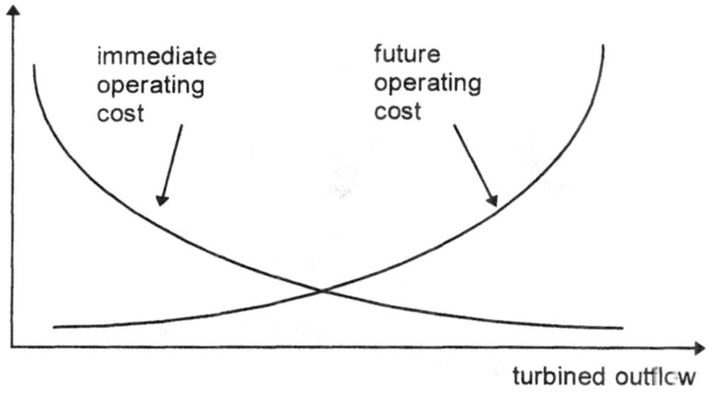

Figure 3 Immediate and future operation costs as a function of final storage

associated with the expected thermal generation expenses from the next stage to the end of the study period. We see that the FCF decreases with final storage, as more water becomes available for future use.

Conceptually, the FCF can be obtained by simulating the system operation in the future for different starting values of initial storage and calculating the operation costs. If the capacity is relatively small, as in the Spanish or Norwegian system, the impact of a decision is diluted in several months. If the capacity is substantial, as in Brazil, the simulation horizon may reach five years. The simulation must take into account the variability of inflows to reservoirs, which fluctuate seasonally, regionally, and from year to year. Inflow forecasts are generally inaccurate, in particular when it comes from rainfall, not snowmelt. Therefore, inflows are usually modeled as a multivariate stochastic process which preserves relevant serial and spatial dependencies observed in the past. As a consequence, FCF calculation has to be carried out on a probabilistic basis, using a large number of hydrological scenarios.

The optimal use of stored water corresponds to the point that minimizes the sum of immediate and future costs. As shown in Figure 4, this is also where the derivatives of ICF and FCF (in absolute value) with respect to storage are equal. These derivatives are known as water values. The optimal hydro dispatch is at the point that equalizes immediate and future water values.

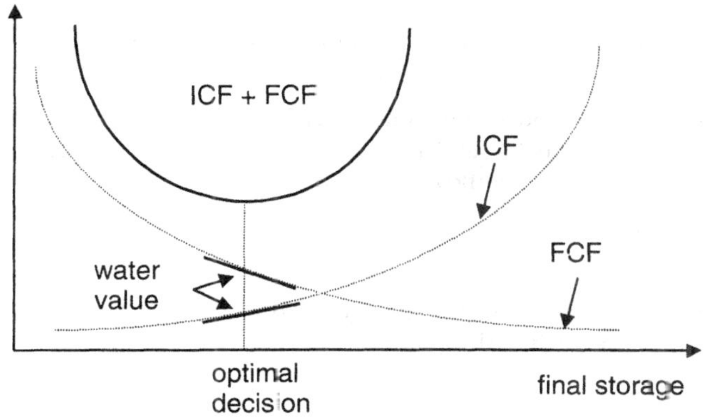

Figure 4 Optimal hydro scheduling

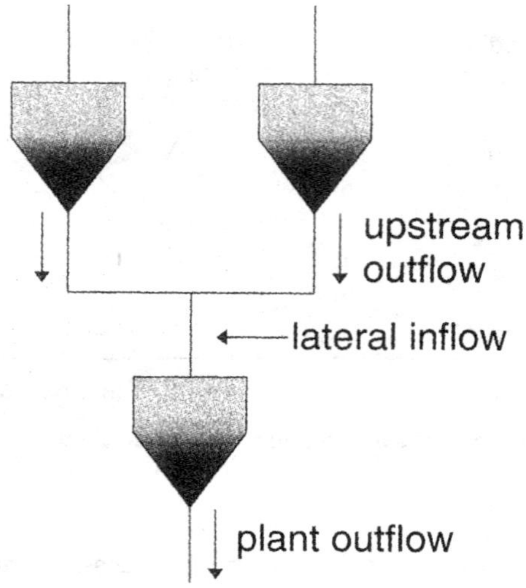

Figure 5 Reservoir water balance

2.2. *Formulation of One-Stage Hydrothermal Dispatch*

The immediate cost is given by the thermal operating costs in stage t, $\sum_{j\in J} c(j)g_t(j)$, where J denotes the set of thermal plants, c is the vector of thermal unit operating costs and variable g_t (MWh) is the vector of thermal generations in stage t. In turn, the future cost is represented by the function $\alpha_{t+1}(v_{t+1})$, where variable v_{t+1} is the vector of reservoir levels in stage $t+1$. Initially, let us consider independent inflow scenarios. Given the initial storage vector v_t, the objective is to minimize the sum of immediate and future discounted operating costs (β is the discount factor):

$$z_t(v_t) = \min \sum_{j\in J} c(j)g_t(j) + \beta\alpha_{t+1}(v_{t+1}). \tag{1}$$

Plant operation is modeled through the following constraints. The water balance equation represents the coupling between successive stages, as illustrated in Figure 5: the reservoir storage v_{t+1} at stage $t+1$ is equal to the initial storage v_t minus outflow volumes (turbined variable u_t and spilled variable s_t) plus inflow volumes (lateral inflow a_t plus releases from immediately upstream plants belonging to set U), all in stage t, for all hydro plants in set I:

$$v_{t+1}(i) = v_t(i) - u_t(i) - s_t(i) + a_t(i) + \sum_{m\in U(i)} [u_t(m) + s_t(m)], \quad i \in I. \tag{2}$$

The load supply equation relates total thermal and hydro generation to system load d_t (MWh). The hydro generation for unit i is given by the product of its production coefficient $\rho(i)$ (MWh/m^3) and its turbined outflow $u_t(i)$, resulting in

$$\sum_{i\in I} \rho(i)u_t(i) + \sum_{j\in J} g_t(j) = d_t. \tag{3}$$

Finally, there are bounds on thermal generation ($g_{\max}$), maximum storage ($v_{\max}$), and turbine capacity ($u_{\max}$) for each hydro plant:

$$g_t(j) \leqslant g_{\max}, \quad j \in J. \tag{4}$$

$$v_{t+1}(i) \leqslant v_{\max}(i), \quad i \in I \tag{5}$$

$$u_t(i) \leqslant u_{\max}(i), \quad i \in I. \tag{6}$$

For simplicity, network constraints are not represented in the above formulation. These constraints are not coupled in time, and are expressed as linearized power flow equations with transmission limits.

2.3. *Calculation of Future Cost Function*

In principle, the FCF could be obtained through a stochastic dynamic programming (SDP) recursion. Reservoir levels are discretized, and starting from the last stage T, problem (1–6) is solved assuming the first storage level for each reservoir. Since we are at the last stage, the FCF is zero. Because of inflow uncertainty, the hydro scheduling problem is successively solved for N different inflow scenarios and the expected operation cost is calculated as the mean of the costs over the N scenarios. For each remaining storage state in stage T, repeat the calculation of expected operation costs and interpolate in order to produce the FCF $\alpha_T(v_T)$ for stage $T-1$. This process is then repeated for all states in stages $T-1$, $T-2$ et cetera. Note that the objective in those stages is to minimize immediate operation plus expected future cost, given by previously calculated FCF. The final result of the SDP scheme outlined above is the set of future cost functions $\alpha_{t+1}(v_{t+1})$ for each stage t.

However, due to the discretization, the SDP computational effort increases exponentially with the number of reservoirs, the well-known curse of dimensionality of DP. Therefore, it is not practical for systems with many reservoirs. In the next section we present an alternative approach that overcomes this drawback.

2.4. *The Dual Dynamic Programming Scheme*

The stochastic dual dynamic programming scheme (SDDP) proposed independently by Pereira and Pinto (1991) and Read and George (1990) is based on the observation that the FCF can be represented as a piecewise linear function, so there is no need to create an interpolated table. Furthermore, the slope of the FCF around a given point can be analytically obtained from the one-stage dispatch problem (1–6). The last-stage dispatch problem is shown below (note that the FCF in this stage is zero).

$$z_T = \min \sum_{j \in J} c(j) g_T(j) \tag{7}$$

$$v_{T+1}(i) = v_T(i) - u_T(i) - s_T(i) + a_T(i) + \sum_{m \in U(i)} [u_T(m) + s_T(m)], \quad i \in I \tag{8}$$

$$\sum_{i \in I} \rho(i) u_T(i) + \sum_{j \in J} g_T(j) = d_T \tag{9}$$

$$g_T(j) \leqslant g_{\max}, \quad j \in J \tag{10}$$

$$v_{T+1}(i) \leqslant v_{\max}(i), \quad i \in I \tag{11}$$

$$u_T(i) \leqslant u_{\max}(i), \quad i \in I. \tag{12}$$

The Lagrange multiplier vector π_h associated to the water balance equation (3), also known as the water value, represents the derivative of z_T with respect to a variation in the initial storage v_T, and corresponds to the slope of FCF for stage $T-1$. Figure 6 shows the calculation of the operation cost and FCF slopes for each state in stage T. It can be seen that the FCF for stage $T-1$ corresponds to the piecewise cost surface produced by taking the linear segment with the highest cost value in each state (the convex hull).

The dispatch problem for stage $T-1$ is now

$$z_{T-1}(v_{T-1}) = \min \sum_{j \in J} c(j) g_T(j) + \beta \alpha_T \tag{13}$$

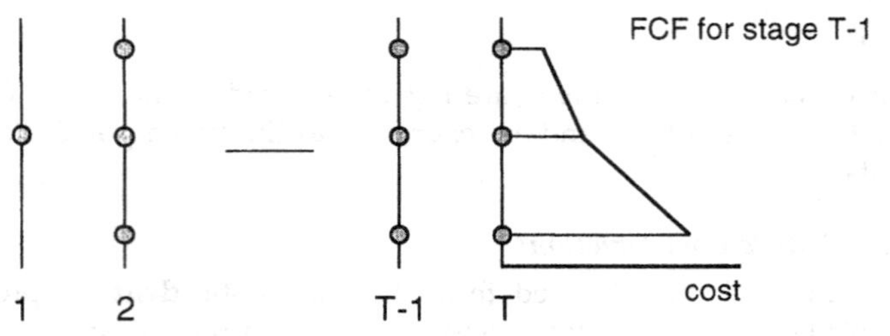

Figure 6 Calculation of piecewise FCF for stage $T-1$

$$v_T(i) = v_{T-1}(i) - u_{T-1}(i) - s_{T-1}(i) + a_{T-1}(i) + \sum_{m \in U(i)} [u_{T-1}(m) + s_{T-1}(m)], \quad i \in I \tag{14}$$

$$\sum_{i \in I} \rho(i) u_{T-1}(i) + \sum_{j \in J} g_{T-1}(j) = d_{T-1} \tag{15}$$

$$g_{T-1}(j) \leqslant g_{\max}, \quad j \in J \tag{16}$$

$$u_{T-1}(i) \leqslant u_{\max}(i), \quad i \in I \tag{17}$$

$$\alpha_T \geqslant \varphi_T^n v_T + \delta_T^n, \quad n = 1, \ldots, N. \tag{18}$$

The future cost function is represented by the scalar variable α_T and the N linear constraints $\{\alpha_T \geqslant \varphi_T^n v_T + \delta_T^n\}_{n=1,N}$ where N is the number of linear segments. As shown in Figure 7, these inequalities represent the piecewise characteristic of this function.

Therefore in general, for each stage t, the FCF is represented by the scalar variable α_t and N linear constraints.

2.5. Backward Recursion and Lower Bound Calculation

The recursive calculation of the piecewise linear FCF is very similar to the standard SDP scheme. In order to take into account that future inflows are unknown consider K inflow scenarios. The backward recursion scheme is shown below.

Set N equal to M, the number of initial storage values;

initialize the FCF for stage T as zero: δ_{T+1}^n and φ_{T+1}^n are null, for $n = 1, \ldots, N$;

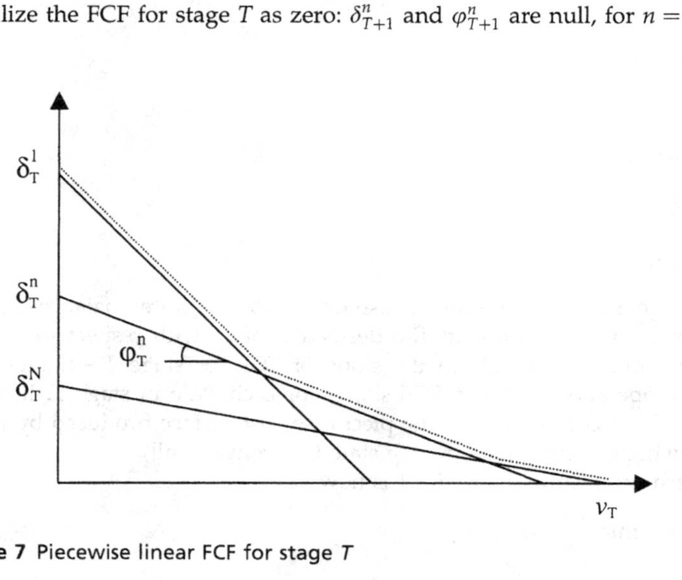

Figure 7 Piecewise linear FCF for stage T

for $t = T, T-1, \ldots, 1$

for each storage value v_t^m, $m = 1, \ldots, M$

for each inflow scenario a_t^k, $k = 1, \ldots, K$

solve the one-stage scheduling problem for initial storage v_t^m and inflow a_t^k :

$$z_t^k(v_t^m) = \min \sum_{j \in J} c(j) g_t(j) + \beta \ \alpha_{t+1} \tag{19}$$

$$v_{t+1}(i) = v_t^m(i) - u_t(i) - s_t(i) + a_t^k(i) + \sum_{m \in U(i)} [u_t(m) + s_t(m)], \quad i \in I \tag{20}$$

$$\sum_{i \in I} \rho(i) u_t(i) + \sum_{j \in J} g_t(j) = d_t \tag{21}$$

$$g_t(j) \leqslant g_{\max}, \quad j \in J \tag{22}$$

$$u_t(i) \leqslant u_{\max}(i), \quad i \in I \tag{23}$$

$$\alpha_{t+1} \geqslant \varphi_{t+1}^n v_{t+1} + \delta_{t+1}^n, \quad n = 1, \ldots, N \tag{24}$$

end;

calculate the volume coefficients and constant term for the mth linear segment of FCF in the previous stage by taking averages over all scenarios (scenario k at stage t has conditional probability p_{kt}):

$$\varphi_t^m = \sum_k p_{kt} \pi_{ht}^k \tag{25}$$

$$\delta_t^m = \sum_k p_{kt} \alpha_t^k(v_t^m) - \varphi_t^m v_t^m \tag{26}$$

end;

end.

The main difference of the SDDP scheme with respect to DP is that the piecewise linear segments can be used to extrapolate the FCF values, that is it is not necessary to use all combinations of points to obtain a complete FCF. Moreover, if a smaller number of initial storage values is used, a smaller number of linear segments will result. The resulting FCF, which is based on the maximum value over all segments, will then be a lower bound to the true FCF. As a consequence, the FCF for the first stage is a lower bound z^L to the optimal solution of the hydrothermal scheduling problem:

$$z^L = z_1(v_1). \tag{27}$$

2.6. *Forward Simulation and Upper Bound Calculation*

If we use the FCF produced by the backward recursion scheme, an upper bound to the optimal solution of the hydrothermal scheduling problem can be obtained by Monte Carlo simulation of system operation [Read and George (1990) use a complete representation of the piecewise linear function; their approach was limited to two reservoirs]. This is due to the fact that the only FCF that can result in the optimal operation cost is the optimal function itself; all others, by definition, will have higher operation costs. The simulation scheme is shown below.

Define inflow scenarios a_t^m, $m = 1, \ldots, M$ for all stages $t = 1, \ldots, T$:

for each inflow scenario a_t^m, $m = 1, \ldots, M$

initialize storage value for stage 1 as $v_t^m = v_1$;

for $t = 1, ..., T$

solve the one-stage scheduling problem (19–24);

calculate the total operating cost z^m for scenario m as the sum of all immediate thermal costs along the study period;

end;

end.

An upper bound for the expected operation cost is estimated as the mean total cost over all scenarios:

$$z^U = M^{-1} \sum_{m=1,M} z^m. \tag{28}$$

This estimator is unbiased, converging to the population value. Due to the sampling variation, there is an uncertainty around the true expected value. A 95 percent confidence interval $[z^U - 1.96\sigma,\ z^U + 1.96\sigma]$ can be derived by estimating the variance of the estimator as

$$\sigma^2 = M^{-1} \sum_{m=1,M} (z^m - z^U)^2. \tag{29}$$

2.7. *Optimality Check and New Iteration*

Optimality is achieved when the lower bound z^L is within the confidence interval of the upper bound. Note that because of sampling variation, the lower bound may exceed the upper bound mean estimate z^U.

If the lower bound is outside the confidence interval, the backward recursion step described in Section 2.5 is repeated with an additional set of storage values. These values are produced by the forward simulation step described in Section 2.6. Note that all linear constraints produced along the iterative process are retained, since the piecewise FCF is given by the convex hull. Therefore, the representation of the FCF is gradually improved along the process until convergence is achieved.

2.8. *Analytical Representation of Inflow Uncertainties in Dual DP*

Inflows to reservoirs present serial (seasonal and yearly) and spatial dependencies. Long years of droughts observed in many historical records must also be represented, by means of a multivariate inflow model. One such model is the multivariate periodic autoregressive (MPAR), which has the following basic characteristics. Each hydro plant inflow for stage t is a periodic autoregressive (PAR) process of lag k_t. We illustrate with a simple PAR lag 1 model. Let μ_t denote the inflow mean, σ_t denote the inflow standard deviation, ξ_t is an independent random variable, and ϕ_1, ϕ_2 are model parameters. The model is

$$(a_t - \mu_t)/\sigma_t = \phi_1(a_{t-1} - \mu_{t-1})/\sigma_{t-1} + \phi_2\xi_t. \tag{30}$$

Parameters ϕ_j, $j = 1, ..., k_t$ are estimated so as to preserve historical serial correlations up to lag k_t, which varies from season (stage) to season. Wet seasons usually present lower lags than dry seasons. Historical seasonal inflow mean and standard variation are also preserved. Historical cross correlations among inflows to different hydro plants are represented by the correlation matrix of variables ξ_t, which are spatially correlated. These variables are the result of a linear transformation applied to an independent random vector of the same dimension.

Now, when solving the one-stage dispatch problem (19–24), we express the inflow for stage t, a_t^k, as a function of previous stage inflow using the following equation derived from (30):

$$a_t^k = \sigma_t\phi_1(a_{t-1}^m - \mu_{t-1})/\sigma_{t-1} + \phi_2\xi_t^k + \mu_t, \tag{31}$$

where ξ_t^k is sampled from the corresponding probability distribution.

The linear approximation of the FCF for the previous stage will now have two coefficients, based on the derivative of z_t^k with respect to v_t^m and a_{t-1}^m at the optimal solution. The first coefficient is calculated in the same way as the independent case:

$$\partial z_t^k / \partial v_t = \pi_{ht}^k. \tag{32}$$

The second is calculated through the chain rule:

$$\partial z_t^k / \partial a_{t-1} = (\partial z_t^k / \partial a_t)(\partial a_t / \partial a_{t-1}). \tag{33}$$

The term $\partial z_t^k / \partial a_t$ also corresponds to π_{ht}^k because both v_t^m and a_t^k are in the same water balance equation. In turn, the term $\partial a_t / \partial a_{t-1}$ is calculated from the inflow model equation (30):

$$\partial a_t / \partial a_{t-1} = \sigma_t \phi_1 / \sigma_{t-1}. \tag{34}$$

The linear constraint (24) now becomes

$$\alpha_{t+1} \geqslant \varphi_{t+1}^n v_{t+1} + \gamma_{t+1}^n a_t^k + \delta_{t+1}^n, \quad n = 1, ..., N, \tag{35}$$

where the coefficients and constant term for the nth linear segment of FCF are calculated by taking averages over all scenarios:

$$\varphi_{t+1}^n = \sum_k p_{kt} \pi_{ht}^k \tag{36}$$

$$\gamma_{t+1}^n = \sum_k p_{kt} \partial z_t^k / \partial a_{t-1} \tag{37}$$

$$\delta_{t+1}^n = \sum_k p_{kt} \alpha_t^k (v_t^m) - \varphi_{t+1}^n v_t^m - \gamma_{t+1}^n a_t^k. \tag{38}$$

Figure 8 illustrates the backward recursion considering the previous inflow vector as a state variable.

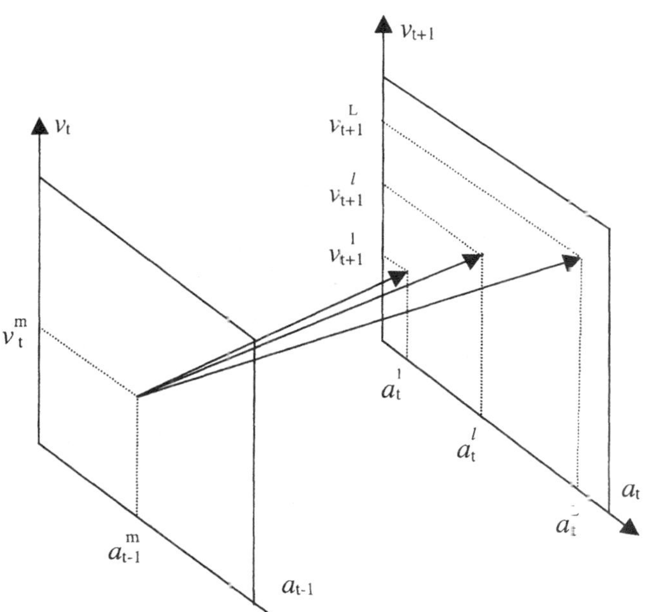

Figure 8 Illustration of backward recursion with a_{t-1} as a state variable

Table 1 Monthly System Loads During the Period (MW)

Jan	*Feb*	*Mar*	*Apr*	*May*	*Jun*	*Jul*	*Aug*	*Sep*	*Oct*	*Nov*	*Dec*
3602	3527	3775	3653	3736	3655	3711	3812	3708	3863	3724	3857

Table 2 Convergence of the Dual SDP Method

Iter	Z^L	Z^U	*Gap*	2σ
1	0.20375E + 06	0.34835E + 06	0.14460E + 06	0.11470E + 06
2	0.20701E + 06	0.32717E + 06	0.12015E + 06	0.11538E + 06
3	0.20734E + 06	0.31228E + 06	0.10494E + 06	0.10403E + 06
4	0.20915E + 06	0.31775E + 06	0.10860E + 06	0.10719E + 06
5	0.21009E + 06	0.30799E + 06	97 900	0.10102E + 06

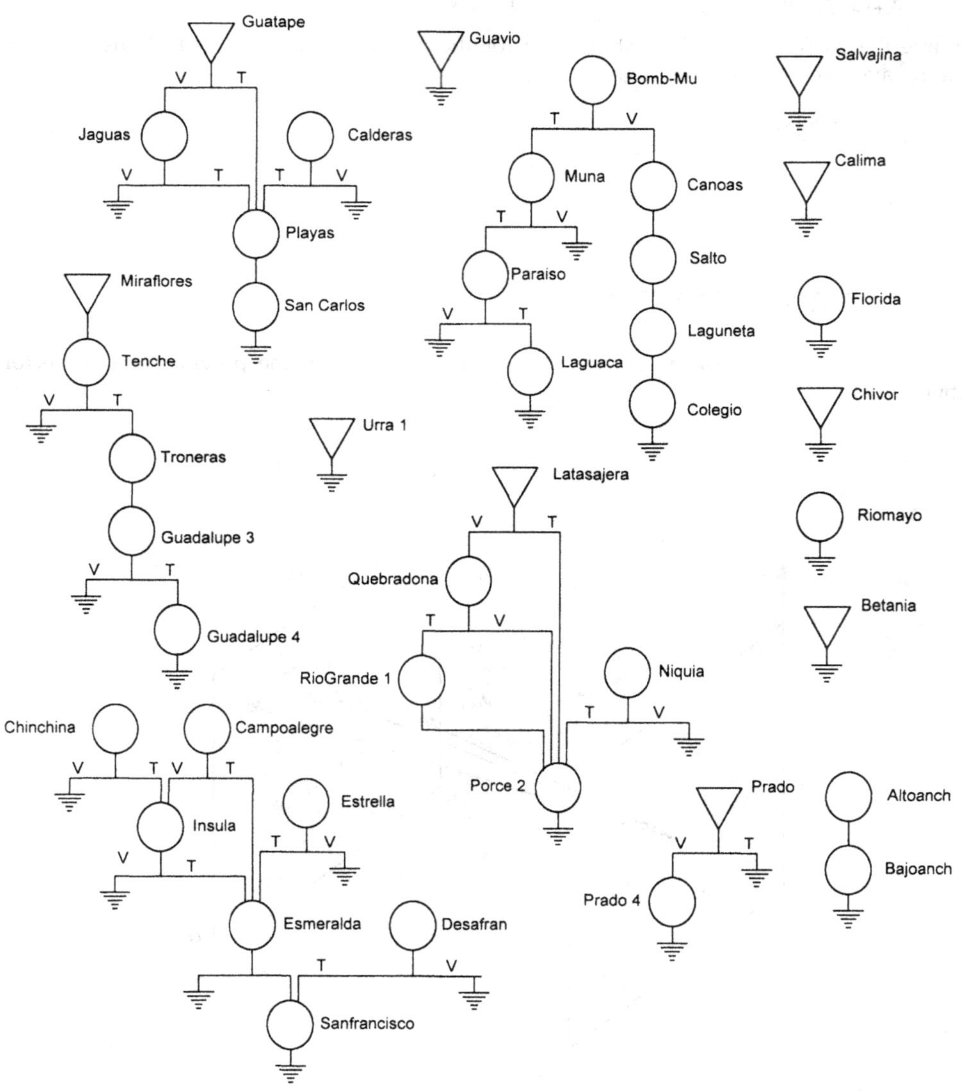

Figure 9 Topology of hydro plants: circles are run-of-the river, triangles are reservoirs

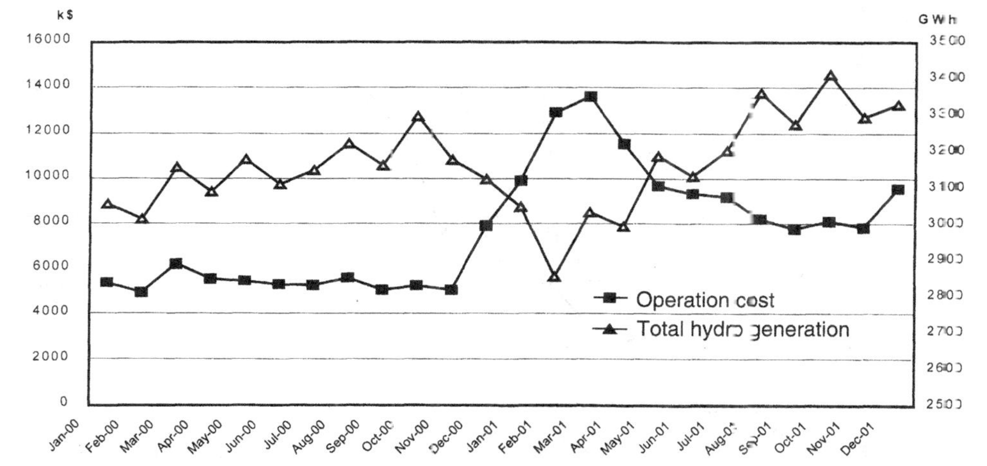

Figure 10 Operating cost and total hydro generation along the period

Note that a_t^k is a known value when the one-stage dispatch problem (19–24) is solved. Therefore, $\gamma_{t+1}^n a_t^k$ is included in the constant term δ_{t+1}^n. In other words, the computational effort of the Dual SDP scheme does not change with the increase in the state space, since the derivatives are calculated analytically.

2.9. Example with Colombia System

The Colombia system consists of thirty six hydro plants and forty five thermal plants. A two-year study period starting in 2000 on a monthly basis was used, with an additional year being considered in order to take into account tail-end effects. The total hydro capacity is 8910 MW and total thermal capacity is 5187 MW. A \$338/MWh cost for the first 1.5 percent of rationing, and \$613/MW thereon was considered for a fictitious rationing thermal unit of unlimited capacity. The discount rate is 12 percent per year. The monthly system loads along the period are presented in Table 1. For each monthly stage, the load profile was represented by five load levels. Figure 9 presents the topology of hydro plants.

When calculating the optimal scheduling policy, fifteen inflow scenarios were used in forward simulations, all of them starting from the same initial inflow. During backward recursions, ten inflow scenarios were sampled. These scenarios were represented by means of a MPAR model adjusted to the historical monthly inflows to the hydro plants.

Table 2 presents the convergence of the dual SDP method applied to this problem. The last column refers to two times the standard variation of the upper bound. Convergence was reached after five iterations using a 95 percent confidence interval criterion.

The total computing time of the backward steps was 734 seconds, and the forward steps took 257 seconds on a Pentium II 400 MHz microcomputer. A total of $5940 \times 5 = 29{,}700$ linear programs were solved.

The average operating cost for the period is \$7.7 million. Figure 10 presents the system operating costs and total hydro energy throughout the study period.

The total monthly available energy is the sum of all the hydro plants monthly available energy, each one being the product of the plant lateral volume multiplied by the sum of the production coefficients of upstream hydro plants, including the hydro plant itself. As expected, we can observe a negative correlation between total hydro generation and operating cost: when the hydro condition is favorable, thermal generation goes down, and when the hydro condition is bad, the opposite occurs. Note that there is no rationing during the period.

Figure 11 shows the monthly average marginal costs. Note that marginal costs variation follows total operation cost and also present seasonal variation.

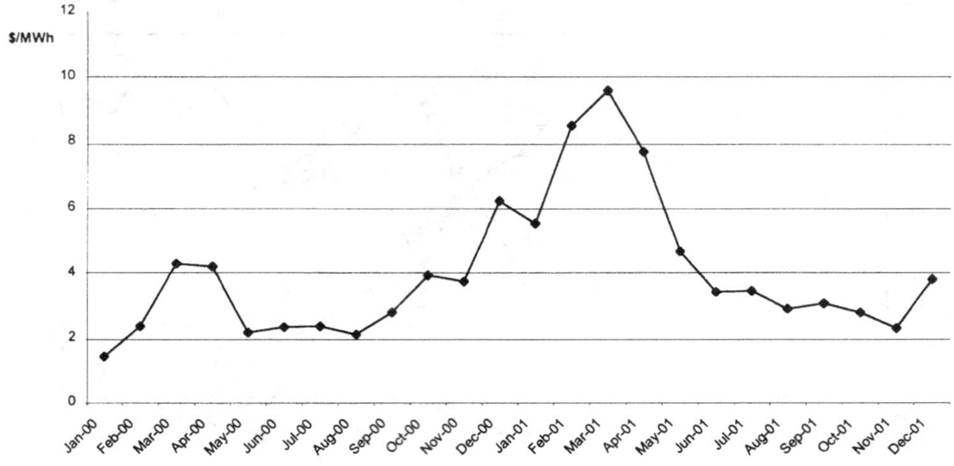

Figure 11 Monthly average marginal costs along the period ($/MWh)

3. Expansion Planning Under Uncertainty

Expansion planning aims at determining the optimal sequence of new generation plants and interconnection links so as to ensure an economic and reliable supply of the predicted load along the study period. The total cost is the sum of investment and operation costs, the latter corresponding to fuel costs of thermal units and transmission losses.

Long range planning studies must consider all sorts of uncertainty, such as load variation, equipment availability, inflows to hydro plants, load growth rates, fuel costs, interest rates, environmental constraints, and so on. These factors are difficult to predict, many are strongly dependent on economics, and therefore in order to take them into account we need a wider scope of methodological tools and new ways to present results. It is also necessary to produce expansion strategies that take into account the tree of possible future scenarios and the dynamics of the decision making process as the uncertainties get resolved in time. Finally, the traditional expected value objective function used in planning studies must be reformulated, since it does not capture the exposure of a plan to unfavorable but plausible scenarios.

We present a methodology for expansion planning under uncertainty that is based on stochastic optimization (allowing the implicit representation of alternative investment alternatives), decision analysis (representing the dynamic aspects of decisionmaking as uncertainties are resolved in time), and hedging objectives from trade-off analysis to help select flexible and resilient expansion strategies.

3.1. General Minimax Formulation

The use of expected cost as the decision criterion in stochastic optimization problems is adequate for high frequency phenonema, in which a representative sample of all possible factors is expected to occur along the study period. One typical example is the thermal stochastic scheduling problem, where the most likely combinations of generator outages (the scenarios) are expected to occur.

The same is not true for low frequency phenonema, such as total load in expanding systems. Due to the uncertainty in load growth rates, there will be a great variability in investment and operation costs for load growth rate scenarios, and we can question the meaning of investment decisions that are on average optimal.

An alternative approach is to calculate the loss, or regret, associated with each combination of scenario and decision. The regret is the difference between the actual cost and the cost that would have been incurred if there were prior knowledge that a given scenario would take place. The minimax criteria minimizes the maximum regret associated with each scenario. To illustrate this concept, consider a two-stage stochastic linear decision process, where at the second stage m scenarios may occur. The deterministic linear program associated to scenario $i = 1, ..., m$ is formulated as:

$$z_i = \min c_1 x_1 + c_2 x_{2i} + d_1 y_1 + d_2 y_{2i} \tag{39}$$

$$A_1 x_1 \geqslant b_1 \tag{40}$$

$$A_2 x_{2i} \geqslant b_{2i} \tag{41}$$

$$G_1 x_1 + F_1 y_1 \geqslant h_1 \tag{42}$$

$$G_1 x_1 + G_{2i} y_1 + F_{2i} y_{2i} \geqslant h_{2i}, \tag{43}$$

where:

x_1, y_1 are the investment and operation vector variables for stage 1;

x_{2i}, y_{2i} are the investment and operation vector variables for stage 2 and scenario i;

c_1, c_2 are investment cost vectors for stages 1 and 2;

d_1, d_2 are operation cost vectors for stages 1 and 2;

z_i is the total (investment plus operation) reference cost for scenario i.

Constraints (40) and (41) are investment constraints for stages 1 and 2, (42) is a coupling constraint for stage 1, and (43) is a coupling constraint for stage 2. Corresponding matrices and vectors are of appropriate dimensions. Note that only first-stage variables are unique, and second-stage investment and operation variables (x_{2i} and y_{2i}) are defined for each scenario $i = 1, ..., m$. Therefore, non anticipativity first-stage constraints are implicit.

We can now formulate the two-stage minimax linear program as:

$$z = \min \alpha \tag{44}$$

$$\alpha \geqslant c_1 x_1 + c_2 x_{2i} + d_1 y_1 + d_2 y_{2i} - z_i, \quad i = 1, ..., m \tag{45}$$

$$A_1 x_1 \geqslant b_1 \tag{46}$$

$$A_2 x_{2i} \geqslant b_{2i}, \quad i = 1, ..., m \tag{47}$$

$$G_1 x_1 + F_1 y_1 \geqslant h_1 \tag{48}$$

$$G_1 x_1 + G_{2i} y_1 + F_{2i} y_{2i} \geqslant h_{2i}, \quad i = 1, ..., m. \tag{49}$$

Because α is greater than or equal to each regret, it will be greater than or equal to the maximum regret, and since we are minimizing α, it will be equal to the maximum regret. Note that we need not attribute probabilities to scenarios using this Savage criteria. If we do attribute probabilities to scenarios, the minimization of the expected regret is known as the Laplace criteria. Also note that the decision process is a strategy, not a plan, meaning that the decision in a given stage depends on the observed parameters on previous stages.

This formulation is easily generalized to any number of stages It consists of a two-phase approach: in phase 1, a reference cost for each scenario is obtained by solving the deterministic linear program (39–43), followed by phase 2 where the minimax linear program (44–49) is solved.

3.2. Problem Characteristics and Solution Approach

Expansion planning decisions present a delay between the stage where they are made and when the new facilities (thermal and hydro plants, transmission reinforcements) are built. Investment decisions are discrete and stage dependent, and one must take into account these delay times for each type of facility, by means of linear investment constraints (financial constraints can also be represented). Also, as seen in Section 2, hydrothermal system scheduling (that calculates the operation cost for a given plan) is coupled in time and stochastic. Investment decisions affect the scheduling problem, and these relations along with investment and operation variables are represented by linear coupling constraints. The uncertainties described in this section must also be taken into account. Therefore, we are faced with a multistage stochastic mixed integer program of huge dimension.

We model this problem using the minimax criterion, in order to take into account low-frequency uncertainties such as load growth rates, and use a stochastic Benders decomposition approach, where expected operation costs are minimized when solving the hydrothermal stochastic scheduling subproblem. The investment subproblem is a mixed-integer linear program using the minimax regret criterion. Therefore, one must initially solve (also by Benders decomposition) a stochastic expansion plan for each (e.g., load) scenario. This is phase 1 of the solution approach. Due to space limitations, we will not present the complete mathematical formulation. As a reference, consider the generic model presented in the previous section, assuming now an arbitrary number of stages and scenarios.

For each scenario, the investment subproblem expansion problem is a multistage mixed-integer program, solved by branch-and-bound. It produces a trial investment solution for the whole study period, which is a lower bound of the optimal solution. For this investment plan, the operation subproblem is a stochastic hydrothermal scheduling problem, solved by the SDDP method. The investment cost plus the operation cost resultant from the scheduling problem is an upper bound for the optimal solution, since it is feasible (artificial thermal plants with high costs are included in the scheduling problem to guarantee feasibility). The consequences of the investment plan on the expected operation cost are translated to the investment subproblem as Benders cuts, which are incorporated into the investment subproblem, and, if the upper bound is not within the confidence interval of the lower bound, a new iteration starts.

Once these reference costs are obtained for all scenarios, we begin phase two, which corresponds to the minimax formulation. Also due to space limitations, we will not present the complete mathematical formulation (see the generic model presented in the previous section).

The corresponding optimization model is solved using the same stochastic Benders decomposition method. The investment subproblem now has the continuous α regret variable,

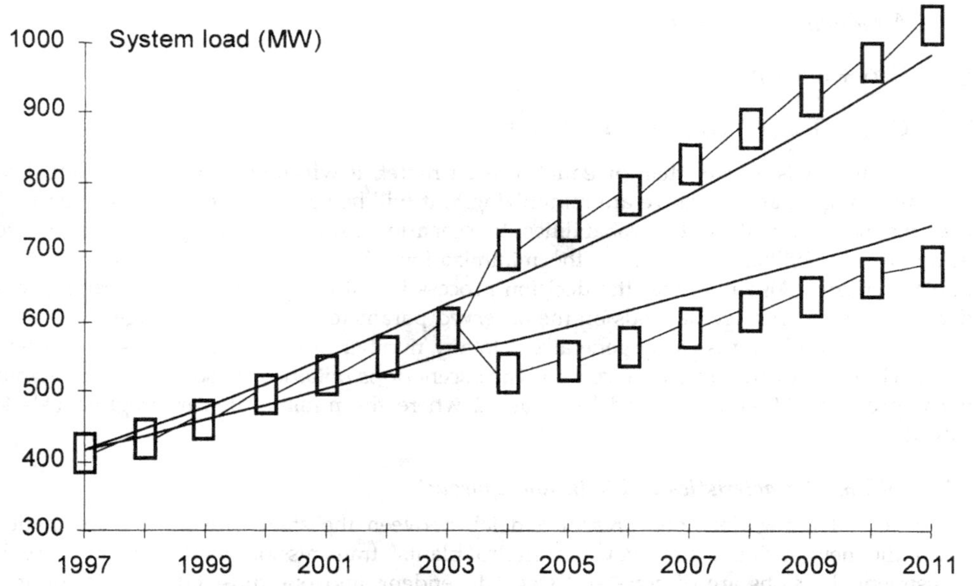

Figure 12 Load scenarios and decision tree for Costa Rica system

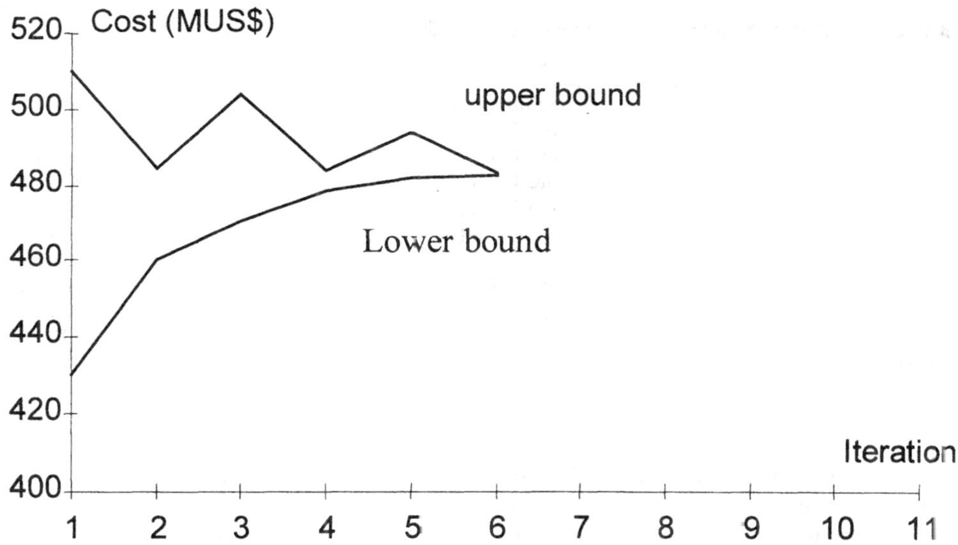

Figure 13 Convergence of decompcsition method—phase 1, scenario 1

and additional linear constrains relating α to x and y variables (one for each scenario). The operation subproblem is the same hydrothermal stochastic scheduling problem. Details can be found in Gorenstin et al. (1993), and Pereira et al. (1995).

3.3. Example with the Costa Rica System

The generation system is composed of six hydro and six thermal plants. There are nine hydro projects and eight thermal projects under consideration during the study period, from 1997 to 2011. Two load scenarios are considered, shown in Figure 12, where the decision tree is also presented.

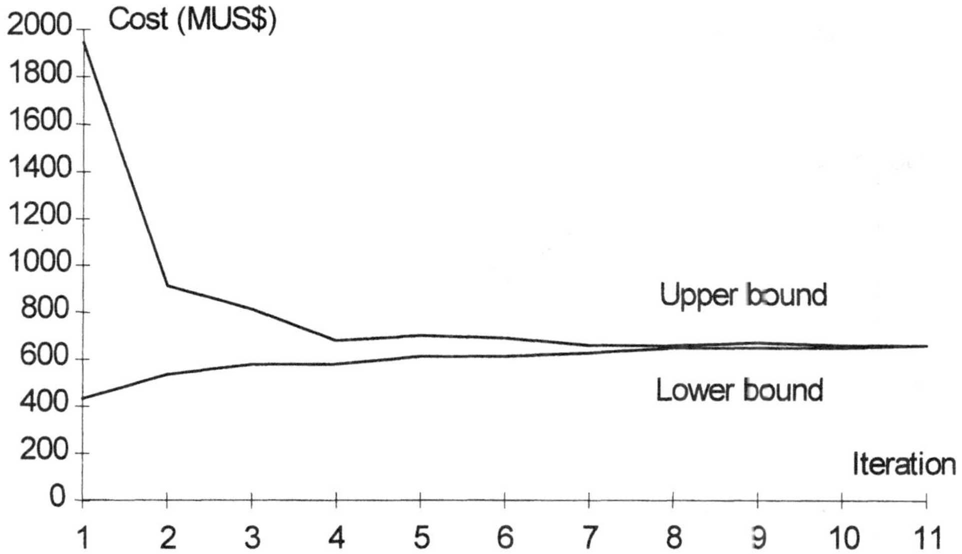

Figure 14 Convergence of decomposition method—phase 1, scenario 2

Table 3 Phase 1 Expansion Plans and Reference Costs for Scenarios 1 and 2

Year	*Scenario 1*	*Scenario 2*
1998	Sandillal (H)	Sandillal (H)
1999	Mira I (T)	Mira I (T)
2000	Toro I (T)	Toro I (T)
	Toro II (T)	Toro II (T)
	Mira II (T)	Mira II (T)
2004	Angostura (H)	Angostura (H)
2006		Guayabo (H)
2008	Pirris (H)	Pirris (H)
		Mira III (T)
2010		San Fernando (H)
Ref Cost (M$)	457	623

Note that uncertainty in load is relevant only after year 2003. Thereafter the scenarios are distinguishable, and therefore different decisions are expected.

First we present the convergence of the Benders decomposition method used to calculate the reference cost for each load scenario (phase 1). The investment subproblem is a mixed-integer program with 255 binary variables, solved by branch-and-bound. The operation subproblem is solved by the dual SD method. For scenario 1 (the one with lower loads), global convergence was reached after six iterations, as shown in Figure 13.

For scenario 2 (the one with higher loads), global convergence was also reached after eleven iterations, as shown in Figure 14.

Table 3 presents the expansion plan and reference cost for each scenario, obtained in phase 1. The plans differ in the construction of plants Guayabo, Mira III, and San Fernando in scenario 2, due to its the higher load. The reference costs will be used in phase 2.

Phase 2 consists of solving the minimax regret problem by Benders decomposition. The investment subproblem is a mixed-integer program with 391 binary variables solved by branch-and-bound. The operation subproblem is solved by the dual SD method. Figure 15 presents the convergence of the decomposition method.

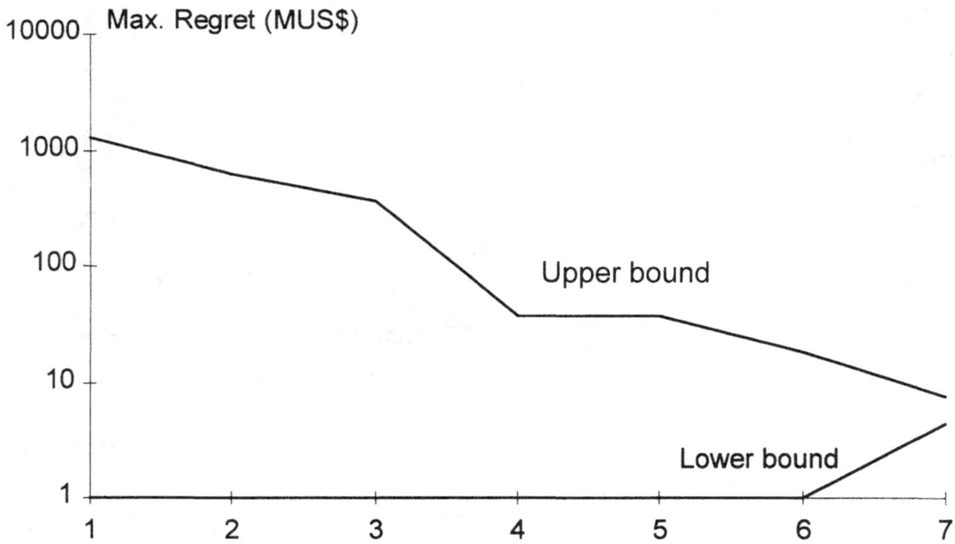

Figure 15 Convergence of decomposition method—phase 2

Table 4 Scenario 1 Strategy and Reference Plan

Year	*Strategy*	*Reference Plan*
1998	Sandillal (H)	Sandillal (H)
1999	Mira I (T)	Mira I (T)
2000	Toro I (T)	Toro I (T)
	Toro II (T)	Toro II (T)
	Mira II (T)	Mira II (T)
2004	Angostura (H)	Angostura (H)
2006	Pirris (H)	
2009		Pirris (H)

Table 5 Scenario 2 Strategy and Reference Plan

Year	*Strategy*	*Reference Plan*
1998	Sandillal (H)	Sandillal (H)
1999	Mira I (T)	Mira I (T)
2000	Toro I (T)	Toro I (T)
	Toro II (T)	Toro II (T)
	Mira II (T)	Mira II (T)
2004	Angostura (H)	Angostura (H)
2006	Pirris (H)	Guayabo (H)
2008	Guayabo (H)	Pirris (H)
	Mira III (T)	Mira III (T)
2010	San Fernando (H)	San Fernando (H)

Tables 4 and 5 present the expansion strategy for scenarios 1 and 2, along with the corresponding expansion plans obtained in phase 1.

Note that the strategies are chosen without knowing the outcomes of the load scenarios, while each reference plan is customized for its load scenario. The main differences between the strategies and the plans are:

- a delay in constructing hydro plant Guayabo from year 2006 to year 2008
- anticipation of constructing hydro plant Pirris from year 2008 to year 2006

The delay is due to the five-year lead time for constructing Guayabo hydro plant. Since in 2001 this plant is not chosen in both scenario reference plans, the expansion strategy delay of this plant allows waiting until 2003 in order to start building it, since by then the load scenarios differ. This delay is compensated by means of anticipating the construction of Pirris hydro plant. Note that due to its four-year lead time, the construction of this plant starts now in 2002, and although the reference plans have different construction dates for this plant, the plant is built in both reference plans.

Figure 16 presents the regrets in terms of investment and operation costs for each scenario. It can be seen that the minimax strategy results in an over-investment of $6.9 million in scenario 1 and an under-investment of $15.8 million in scenario 2, compared to the corresponding reference plans. The resulting operation costs reflect these investment cost differences: the operating cost for scenario 1 is $1.5 million less than its reference cost, and the operating cost for scenario 2 is $23.1 million greater than its reference cost. The total regret for scenario 1 and 2 is $5.4 and $7.3 million, respectively.

Enamorado et al. (1999) used this approach for interconnection planning under uncertainty.

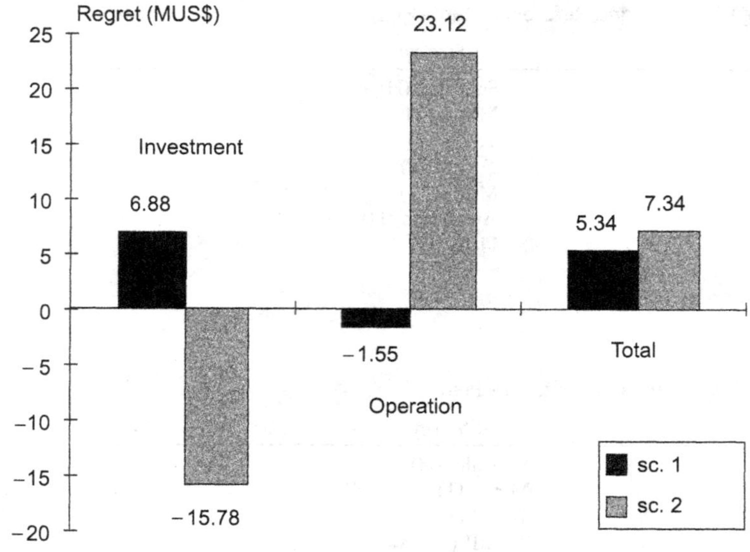

Figure 16 Total regrets and regrets with respect to investment and operation costs

4. Optimal Bidding of Generators in a Wholesale Energy Market (WEM)

The modeling of the behavior of generators (agents) competing among themselves in a wholesale market requires using game theory. The short term problem of generation bidding is presented, based on a Bertrand model. A heuristic algorithm is used, considering stochasticities in demand and equipment availability. We also formulate the long-term problem of market behavior of thermal generator agents, and the Nash–Cournot equilibrium solution of the game is obtained. We then extend the analysis for hydrothermal systems. We formulate the problem for the multistage case and solve it using an extended stochastic dynamic programming scheme.

4.1. Bertrand Model for Short Term Bidding of Generators

In a market oriented environment, thermal generators bid hourly prices for their energy production in a WEM (Mielczarski et al., 1999; Zhang et al., 1999). Units are then loaded by increasing price until demand is met. Dispatched generators are remunerated on the basis of the system spot price, which corresponds to the offer of the most expensive loaded unit. Under perfect competition, it has been argued that the bidding scheme induces an efficient use of system resources in system dispatch.

The Bertrand model is based on prices. We will consider the daily price based thermal dispatch with hourly steps. Each day, generators offer hourly prices and capacity. Given these offers, and forecasted hourly demands, the following least cost dispatch problem is solved:

$$z = \min \sum_{t=1,T} \sum_{j \in J} \lambda_t(j) g_t(j) \tag{50}$$

$$\sum_{j \in J} g_t(j) = d_t, \quad t = 1, \ldots, T \tag{51}$$

$$g_t(j) \leqslant g_{\max}(j), j \in J, \quad t = 1, \ldots, T, \tag{52}$$

where $\lambda_t(j)$ is the energy price of plant j at hour t. Note that this price based dispatch is similar to the thermal scheduling problem.

Now suppose there are K utilities, each utility k owning U_k plants. The hourly net revenue of utility k is the sum of the hourly net revenues of all its plants:

$$r_t^k = \sum_{j \in Uk} r_t(j), \tag{53}$$

where the hourly net revenue of plant j is given by the difference between its total revenue and its operating cost:

$$r_t(j) = (\lambda_t(j) - c(j))g_t(j), \quad j \in J. \tag{54}$$

Each utility k determines the set of its plant hourly prices λ_t^k, which maximize, its hourly net revenue:

$$r_t^k(\lambda_t^k) = \max \sum_{j \in Uk} (\pi_t(\lambda_t(j)) - c(j))g_{tj}(\lambda_t(j)), \tag{55}$$

where $\pi_t(\lambda_t(j))$ is the system spot price [Lagrange multiplier associated to the load supply constraint (51)] and $g_{tj}(\lambda_t(j))$ is the plant generation resulting from the solution of (50–52), when the hourly energy prices for plants j owned by utility k are $\lambda_t(j)$.

Note that primal and dual variables of problems (50–52), π_t and g_{tj}, respectively, appear in (55). Using the primal and dual feasibility, and linear programming primal-dual optimality conditions, problem (55) for stage t can be rewritten as the following nonlinear program:

$$r_t^k = \max \sum_{j \in Uk} (\pi_t - c(j))g_{tj} \tag{56}$$

$$\sum_{j \in J} g_{tj} = d_t \tag{57}$$

$$g_{tj} \leqslant g_{\max}(j), \quad j \in J \tag{58}$$

$$\pi_t + \mu_{tj} - \lambda_{tj} \leqslant 0, \quad j \in U_k \tag{59}$$

$$\pi_t + \mu_{tj} \leqslant \lambda_{tj}^*, \quad j \notin U_k \tag{60}$$

$$\sum_{j \in Uk} \lambda_{tj} g_{tj} + \sum_{j \notin Uk} \lambda_{tj}^* g_{tj} - \pi_t d_t - \sum_{j \in J} \mu_{tj} g_{\max}(j) = 0, \tag{61}$$

where variable μ_{tj} is the Lagrange multiplier with respect to (52), λ_{tj}^* are the price offers of the plants not owned by utility k, and λ_{tj} are the price variables of plants owned by the utility. Constraints (57–58) express primal feasibility, constraints (59–60) express dual feasibility, and constraint (61) is the primal-dual equality. Note that quadratic terms appear in the objective function ($\pi_t\ g_{tj}$) and the primal-dual equality constraint ($\lambda_{tj}\ g_{tj}$).

Besides being nonlinear, another complexity of the static (one-stage) problem (56–61) is due to the fact that variables π_t and g_{tj} for utility k depend on price offers of all utilities, which are not known. One way to deal with this is to consider scenarios of price offers and also hourly demands, and maximize the expected net revenue over all scenarios. We now are faced with a nonlinear stochastic program, and to solve it we will apply a heuristic method using an iterative scheme, which we describe next.

The initialization step sets all price offers equal to the corresponding plant operating costs. At each iteration, each agent (utility) maximizes its price offers given the offers of the other agents. This is done for each plant of the utility, in order of decreasing operating costs. The reasoning for this ordering is that since the net revenue is the difference between the spot price and the operating cost, this difference will be smaller for the more expensive plants and greater for the cheaper ones, therefore it pays to sacrifice the more expensive plants of the utility first since if these plants are not dispatched the utility will lose less than it would lose with a cheaper plant. The offer price of the plant is obtained by a linear search, keeping the prices of the remaining utility plants fixed and increasing its price until convergence is

Table 6 Generator Costs and Capacities

Generator	1	2	3	4	5	6	7	8
Cost($/MWh)	8	10	14	7	15	16	12	21
Capacity(MW)	15	10	15	10	15	10	20	20

obtained. For each trial price of the plant, a Monte Carlo simulation is made to estimate the expected net revenue of the utility, where for each sampled scenario a static dispatch problem is solved to find scenario dispatch and spot prices which, when summed up, compose the estimator of the utility net revenue.

The optimal prices obtained for the utility are used to update the scenario spot prices, and the process of setting plant offer prices is repeated for the next utility. When all utilities are examined, a global convergence test is made by comparing the price offers obtained with the ones in the previous iteration, and the whole procedure is repeated until convergence is obtained.

The resulting offer prices correspond to a situation where no utility has any advantage in changing its offers, and is therefore a Nash equilibrium point $(\lambda_1^*, \lambda_2^*, \ldots, \lambda_K^*)$ such that:

$$\lambda_k^* = \text{argmax} \;\; E[r^k(\lambda_k \mid \lambda_j^*, j \neq k)], \quad k = 1, \ldots, K. \tag{62}$$

We illustrate this approach with a system whose demand in a given hour is normally distributed with mean 57 MW and standard deviation 5.7 MW. There are three utilities: the first owns generators 1, 2, and 3, the second owns generators 4, 5, and 6, and the third owns generators 7 and 8. Table 6 shows operating costs and capacities of the plants. The plant outage affect was considered using a 10 percent unavailability factor.

The heuristic method applied to this case is illustrated in Figure 17. Convergence was obtained after four iterations, where for each iteration we plot the net revenue evolution (left vertical axis) and the demand payment (right vertical axis) for utilities 1, 2, and 3.

Comparing the initial and final values of the total payment by consumers, we can see that due to the fact that we have few utilities, the utilities were able to double the electricity

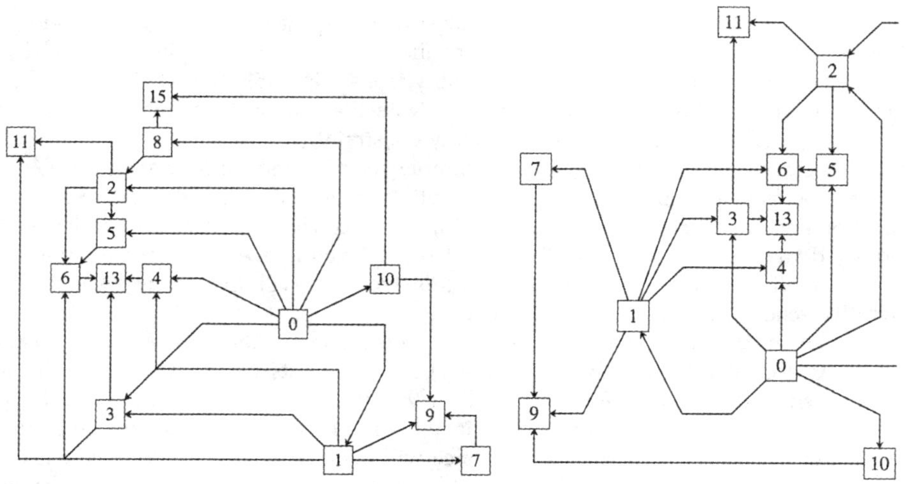

Figure 17 Convergence of the heuristic method

cost by market manipulation. This behavior was also observed by means of the spot price evolution, which started at about \$15/MWh and at equilibrium it reached about \$30/MWh.

4.2. *Cournot Model for Long-Term Behavior of Market Power Agents*

The strategic behavior of power agents can be modeled by noncooperative game theory. In particular, in the Cournot model, agents decide on quantities offered and market price is defined through an inverse demand function. We first consider the static (one-stage) thermal system case.

Alongside the market makers, consider price takers, which do not have the power to influence the system spot price. Assume that they have the quadratic operating cost $\delta(O) = O^2/(2\alpha)$, where O is the total price taker production and α is a scalar parameter. It follows that the price-taker production is given by $O(p) = \alpha p$, where p is the spot price. The price makers will meet the residual demand $d - \alpha p$, where d is the system demand. The total price market production q is the sum of all price maker energy offers, and therefore the spot price is given by $p(q) = (d - q)/\alpha$.

Consider N price makers, with variable operating cost vector c and offered energy production vector g. Given the production of the remaining agents, the net revenue of agent j is the difference between its total revenue and its operating cost:

$$r(j) = (p(q) - c(j))g(j), \quad j = 1, \ldots, N. \tag{63}$$

The energy production $g_j^*(j)$ that maximizes net revenue $r(j)$ is obtained by the zero of the derivative of $r(j)$ with respect to $g(j)$. Substituting the expressions for q and $p(q)$, results in the condition

$$2g_j^*(j) = d - \alpha c(j) - \sum_{k \neq j} g(k), \quad j = 1, \ldots, N. \tag{64}$$

The above expression allows the determination of the profit maximization production of an agent, assuming that other agents' productions are known. Nevertheless, all agents are simultaneously trying to maximize their own net revenue without knowing the remaining agents' decisions. The Nash–Cournot equilibrium provides a solution to this problem: agents will reach a standoff situation where no agent can unilaterally increase its net revenue by changing its production level. The solution of the linear system (64) for unknown vector g is given by

$$g_j^*(j) = (d - N\alpha c(j) + \alpha \sum_{k \neq j} c(j))/(N + 1), \quad j = 1, \ldots, N. \tag{65}$$

Total price makers production q, system spot price $p(q)$, and agents net revenue r can now be obtained by substitution. In a case where all price makers have equal cost c, expression (65) reduces to

$$g^* = (d - \alpha c)/(N + 1). \tag{66}$$

The total equilibrium production in this case is

$$q^e = (N/(N + 1))(d - \alpha c). \tag{67}$$

This expression will be used to illustrate the difference between the Nash–Cournot equilibrium solution and the ideal least-cost solution. The least-cost dispatch problem for the same static case can be formulated as

$$z = \min \delta(d - q) + cq, \tag{68}$$

where $\delta(d-q)$ is the quadratic operating cost of price takers when they supply the residual load $(d-q)$. The price taker production q that maximizes overall production cost is the zero of the derivative of z with respect to q, that is

Table 7 Hydro Plant Characteristics

Name	*Inst. cap. (MW)*	*max. Storage (Hm^3)*	*Prod. Coeff. ($MWh/m^3/s$)*
A	1312	22 950	0.745
B	4082	34 432	0.3832

Table 8 Thermal Plant Characteristics

Plant #	*Inst. cap. (MW)*	*Oper. Cost ($/MWh)*
1	480	2.2
2	300	2.9
3	320	7.2
4	480	7.2
5	450	7.2
6	500	7.2
7	657	8.3
8	1309	8.4
9	1309	8.5
10	450	10.8
11	256	15.4
12	390	15.4
13	131	20.1
14	84	29.7
15	84	29.7
16	220	29.7
17	220	29.7
18	32	29.7
18	200	37.6
20	270	37.6
21	36	41.9
22	16	43.2
23	16	47.8
Total	8210	

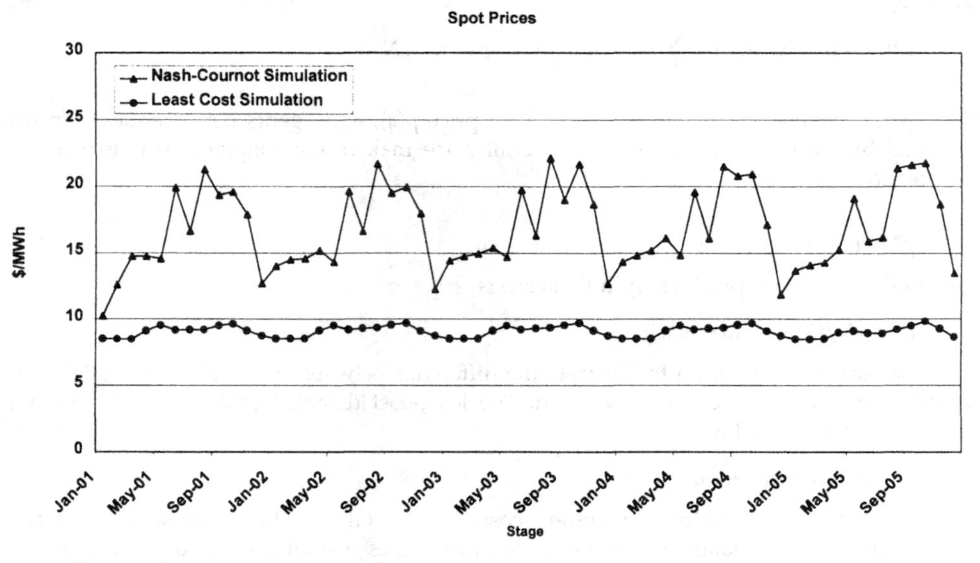

Figure 18 Nash–Cournot and least-cost spot prices

$$q^c = (d - \alpha c). \tag{69}$$

Comparing (67) and (69), we see that $q^e < q^c$. This means that price makers tend to decrease their production to force an increase in spot price. Also note that these solutions converge as the number of agents increase (more competition). The Nash–Cournot equilibrium and least-cost spot prices also converge to the same value as competition increases.

Now consider the multi-stage problem for hydrothermal systems. As seen in Section 2, the operating decisions are coupled in time due to the hydro plants' capability of storing water in reservoirs. The FCF captures the opportunity cost of hydro plants and translates it in terms of the expected thermal cost from each stage to the end of the study period. It can be calculated by a SDP scheme, as described in Section 2.3. In order to simulate the Nash–Cournot equilibrium in a market environment, instead of obtaining the decision that minimizes the sum of immediate and future operating costs for each stage and state, we calculate the Nash–Cournot equilibrium where each agent tries to maximize the sum of immediate and future operating net revenues. In this case, the operating cost $c(i)$ of each hydro agent i represents the expected net revenue opportunity of saving water for future use.

The equilibrium solution is obtained by a heuristic method: each agent chooses its offer so as to maximize its net revenue, assuming other agent offers are fixed. The first agent sets its offer under the assumption that the other agents will not offer, the second sets its offer assuming the first will maintain its offer at the level that was calculated for it in the previous iteration, and so on. This process is repeated for all agents, until no agent can take advantage from changing its offer given the offer of other agents. Thus, at the Cournot equilibrium, each agent is offering its maximized net revenue given the offers of all other agents in the market.

To illustrate, consider two hydro plant price makers A and B, whose characteristics are shown in Table 7.

The price takers are twenty three thermal plants, whose characteristics are shown in Table 8.

The Nash–Cournot SDP scheme was applied considering a five year study period with monthly steps, and allowing five additional years as buffers to avoid emptying the reservoirs. After calculating the Nash–Cournot equilibrium, system operation was simulated for one thousand inflow scenarios. A least-cost operating policy and simulation was also carried out using the same data and inflow scenarios, in order to measure the effects of market power. Figure 18 shows the expected monthly spot prices for the Nash–Cournot and least-cost solutions along the study period.

It can be seen that the power makers were able to substantially increase the system spot prices. Figure 19 shows the total production of both hydro plants in both cases. Note that hydro plants not only reduce their production but also reduce water transfers from wet to dry seasons, thus amplifying the spot price differences.

5. Optimal Power Flow

The optimal power flow (OPF) consists of optimizing an objective function subject to a set of physical and operating constraints associated to an electric power system in steady state. Besides electrical power flow equations, the OPF is composed of an objective function, a list of controls that can be optimized, and a set of operating constraints that should be obeyed at its optimal solution. The objective functions specified in OPF applications are generation cost, load shedding cost, load maximization, maximum active power transfer throughout the network, and so on. The controls are generator output, taps of transformers, phase-shifter angles, DC links controls, and so on.

The first set of constraints is related to the representation of the AC network in steady state. In this model each network bus (node) is associated to two nonlinear equations (active and reactive power balance equations):

$$\sum_{j \in \Omega_i} p_{ij}(v, \theta, a) + PG_i = PL_i \quad i = 1, \ldots, Nbus \tag{70}$$

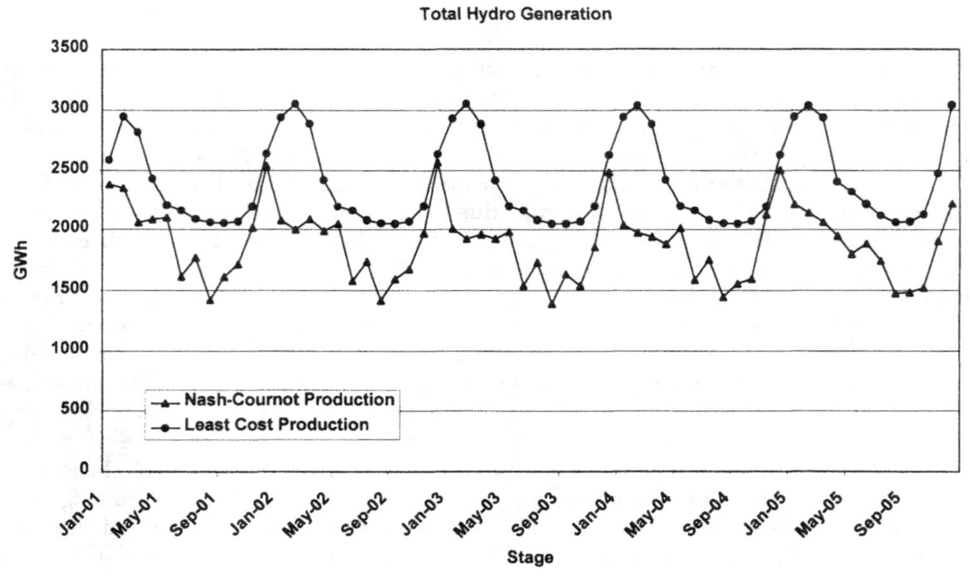

Figure 19 Nash–Cournot and least-cost total hydro generation

$$\sum_{j\in\Omega_i} Q_{ij}(v,\theta,a) + V_i^2 b_{shi} + QG_i = QL_i \quad i = 1, ..., N\text{bus}, \tag{71}$$

where:

PL_i, QL_i are the active and reactive loads at bus i;

PG_i, QG_i, are the active and reactive generations at bus i;

$P_{ij}(v,\theta,a) = (a_{ij}v_i)^2 g_{ij} - (a_{ij}v_i)v_j \; g_{ij}\cos(\theta_{ij}+\varphi_{ij}) - (a_{ij}v_i)v_j \; b_{ij}\sin(\theta_{ij}+\varphi_{ij})$ is the active flow on circuit (i, j);

$Q_{ij}(v,\theta,a) = (a_{ij}v_i)^2(b_{ij}+b_{ij}^{sh}) - (a_{ij}v_i)v_j \; b_{ij}\cos(\theta_{ij}+\varphi_{ij}) - (a_{ij}v_i)v_j \; g_{ij}\sin(\theta_{ij}+\varphi_{ij})$ is the reactive flow on circuit (i, j);

$\theta_{ij} = (\theta_i - \theta_j)$ is the difference of terminal bus voltage angles of circuit (i, j);

v is the vector of bus voltage magnitudes;

a is the vector of circuit tap variables;

Ω_i is the set of buses connected to bus i;

Nbus is the number of buses (nodes) of the network.

In the general case the vectors v, θ, PG, QG, a, φ correspond to variables and g, b, PL, QL to system parameters in these equations. The remaining constraints are circuit power flow limits, constraint on power factor at area interconnections, constraints on the sum of active power flow at a subset of circuits specified by the user, security constraints, and bounds on control and state variables.

OPF can be viewed as a stand-alone problem or as a tool for on-line applications, reliability evaluation of power systems, optimal expansion of reactive power sources, voltage collapse analysis, and so on.

A key aspect of a code which can deal with all the aspects mentioned above and a wide variety of power system networks is the optimization algorithm employed in its development. Computationally, the OPF is a large-scale nonlinear programming problem with thousands of variables and nonlinear constraints. Since its initial formulation in the sixties, several optimization methods have been proposed to solve it. Among them, we mention the differential injection method of Carpentier (1962), the reduced gradient method of Dommel and

Tinney (1968), the GRG method of Abadie and Carpentier (1969), the projected Lagrangian method of Murtagh and Saunders (1982), sequential quadratic programming methods (Sun et al., 1984), specific algorithms based on the resolution of a sequence of linear problems (Alsac et al., 1990), and interior-point methods (Granville, 1994; Vanderbei and Shanno, 1999).

In this section we describe the basic steps of implementation of an interior-point method based on the primal-dual algorithm to solve the OPF problem (Granville, 1994). Extensive numerical experiences have shown that direct interior-point methods are flexible, very effective in dealing with large-scale, ill-conditioned, and voltage problem networks, and efficient in optimizing a wide variety of real and reactive related objective functions.

5.1. The Interior-Point Algorithm

The OPF can be mathematically stated as:

$$\text{Min } g(z) \tag{72}$$

$$h(z) = 0 \tag{73}$$

$$a \leqslant z \leqslant b. \tag{74}$$

The first step in the application of the primal-dual algorithm to problem (72–74) is to incorporate constraint (74) to a logarithmic barrier function, resulting in the equality constrained problem

$$\text{Min } \left\{ g(z) - \mu \sum_j \log(z_j - a_j) - \mu \sum_j \log(b_j - z_j) \right\} \tag{75}$$

$$h(z) = 0, \tag{76}$$

where μ is the barrier parameter.

The first order optimality conditions associated to problem (75–76) are:

$$\nabla g(z) - J(z)^t \lambda - \pi_1 - \pi_2 = 0 \tag{77}$$

$$h(z) = 0 \tag{78}$$

$$\mu e - S_1 \pi_1 = 0 \tag{79}$$

$$\mu e + S_2 \pi_2 = 0, \tag{80}$$

where:

$\nabla g(z)$ is the gradient of $g(z)$;

$J(z)$ is the Jacobian of $h(z)$;

λ corresponds to the Lagrange multipliers (dual variables) associated with constraint (78);

e is the unitary vector, $e = (1, ..., 1)^t$;

S_1, S_2 are diagonal matrices whose diagonal elements are $z_j - a_j$ and $b_j - z_j$, respectively.

At the beginning, the primal and dual variables are chosen in such a way that:

$$\pi_1 > 0, \quad \pi_2 < 0, a < z < b. \tag{81}$$

Increments for the primal Δz and dual ($\Delta\lambda$, $\Delta\pi_1$, $\Delta\pi_2$) variables are obtained through the application of one iteration of the Newton–Raphson method to system (77–80). Let (Δz, $\Delta\lambda$) be the solution of the following system:

$$\begin{bmatrix} W & -J^t \\ -J & 0 \end{bmatrix} \begin{bmatrix} \Delta z \\ \Delta\lambda \end{bmatrix} = \begin{bmatrix} t \\ h(z) \end{bmatrix}, \tag{82}$$

where:

$$W = \nabla^2 g(z) - \sum_i \lambda_i \nabla^2 h_i(z) + S_1^{-1}\Pi_1 - S_2^{-1}\Pi_2; \tag{83}$$

$$t = -\nabla g(z) + J(z)^t \lambda + \mu S_1^{-1} - \mu S_2^{-1}; \tag{84}$$

$\nabla^2 g(z)$, $\nabla^2 h_i(z)$ are Hessian matrices for $f(z)$ and $h_i(z)$;
J is the Jacobian of h;
Π_1, Π_2 are diagonal matrices whose diagonal elements are π_{1j} and π_{2i}, respectively.
The remaining variables are computed through the following:

$$S_1 \Delta\pi_1 = \mu e - S_1\pi_1 - \Pi_1 \Delta z \tag{85}$$

$$S_2 \Delta\pi_2 = -\mu e - S_2\pi_2 + \Pi_2 \Delta z. \tag{86}$$

Next, the step length in the primal and dual is computed as:

$$\alpha_p = \min\left[\min_{\Delta z_j < -\delta} \frac{(z_j - a_j)}{|\Delta z_j|}, \min_{\Delta z_j > \delta} \frac{(b_j - z_j)}{|\Delta z_j|}, 1.0 \right] \tag{87}$$

$$\alpha_d = \min\left[\min_{\Delta\pi_{1j} < -\delta} \frac{\pi_{1j}}{|\Delta\pi_{1j}|}, \min_{\Delta\pi_{2j} > \delta} \frac{-\pi_{2j}}{|\Delta\pi_{2j}|}, 1.0 \right] \tag{88}$$

where δ is a tolerance. A new approximation to the optimal solution is then:

$$z := z + \sigma\alpha_p \Delta z \tag{89}$$

$$\lambda := \lambda + \sigma\alpha_d \Delta\lambda \tag{90}$$

$$\pi_1 := \pi_1 + \sigma\alpha_d \Delta\pi_1 \tag{91}$$

$$\pi_2 := \pi_2 + \sigma\alpha_d \Delta\pi_2, \tag{92}$$

where $\sigma = 0.9995$. A critical aspect in the primal-dual algorithm is the choice of the barrier parameter μ. In our implementation the initial value for μ is established in the program and at intervening iterations it is determined by:

$$\mu = \frac{s_1^t \pi_1 - s_2^t \pi_2}{2n\beta}, \tag{93}$$

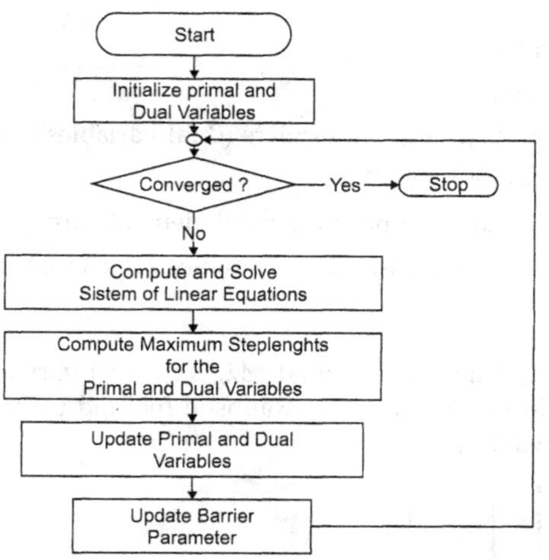

Figure 20 Primal-dual direct IP method

where β is a parameter and n is the number of variables in the optimization problem. Figure 20 shows the basic steps of the interior-point method applied to OPF.

Most of the work in the algorithm is in the resolution of system (82). Its matrix is symmetric and indefinite and there are several proposed ways to deal with it in the context of interior-point methods. The approach adopted here (based on Sun et al., 1984) consisted of making a symmetric permutation of the rows and columns so that the matrix is firstly partitioned into two blocks—light border and heavy border as follows:

$$\begin{bmatrix} W & -J \\ -J^t & \end{bmatrix} = \begin{bmatrix} D & L \\ L^t & H \end{bmatrix}, \tag{94}$$

where D is a diagonal matrix associated to primal variables that are not linked to any constraints like tap values, active and reactive generation, and so on. The heavy border matrix H corresponds to primal variables, which are coupled to constraints. For instance, phase-angle with active power balance equation in the same bus, voltage level with the corresponding reactive power balance equation, slack associated to a circuit flow or Benders cut constraints with the corresponding constraint, and so on. Based on these associations, matrix H is further partitioned into 2 × 2 matrix blocks. Each diagonal block has the following form:

$$\begin{bmatrix} w & -j_1 \\ -J_1 & \end{bmatrix}. \tag{95}$$

There is some flexibility in defining the association variables × constraints, but proper care should be taken to avoid the singularity of some diagonal block in the factorization. The original system is solved by first eliminating the variables associated to the light border and then the heavy border matrix H is factorized by block. This procedure proved to be quite appropriate and robust in dealing with the indefinite matrix (82).

With respect to the application of line search procedures to guarantee and/or improve convergence of the iterative procedure, many numerical experiments have been done during recent years. Despite the superior performance of the basic algorithm to solve large-scale network OPF problems, no significant improvements were observed when line-search procedures were considered. As a final word, the IP algorithm described above belongs to the class of infeasible interior-point methods and one of its important aspects is the fact that feasibility is attained together with optimality in the iterative process, that is, at intermediate iterates only the bound constraints are enforced by the logarithm barrier function. This means that the optimization does not require a starting point that satisfies the power balance equations. This aspect is important for voltage collapse analysis where we have to deal with unsolvable power flows.

5.2. *Numerical Results*

To give a flavor of the behavior of the algorithm it will be illustrated using a network derived from a North American generation/transmission system with 6313 buses, 13,129 circuits, 4408 LTC controls, 1,159 voltage controls, and a system load of 410,690 MW

The OPF consists of an optimization problem with 18,193 variables and 25,755 nonlinear equality and inequality constraints. Two kinds of objective functions will be considered: minimization of reactive power injection to keep voltage levels throughout the network within bounds, and active power loss minimization. Bounds for voltage levels were defined based on system operating procedures.

The initial value and minimum value for the barrier parameter were 5.0 and 5.0×10^{-4} respectively and the value of the β parameter (see equation (93)) was 10. The computational work performed below was carried out on a Pentium II 400 MHz microcomputer using single precision arithmetic. Using the first objective function, the algorithm converged in fifty three iterations with a total reactive allocation of 0 MVAr. The total amount of active power loss at the optimal solution was 7175 MW. Figure 21 shows the evolution of the maximum error (mismatch) in attaining the active (MW) and reactive (MVAr) power balance equations.

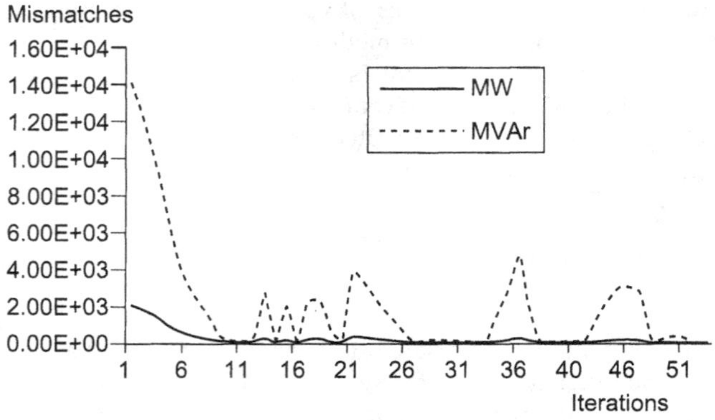

Figure 21 Active and reactive power mismatches—minimization of reactive injection

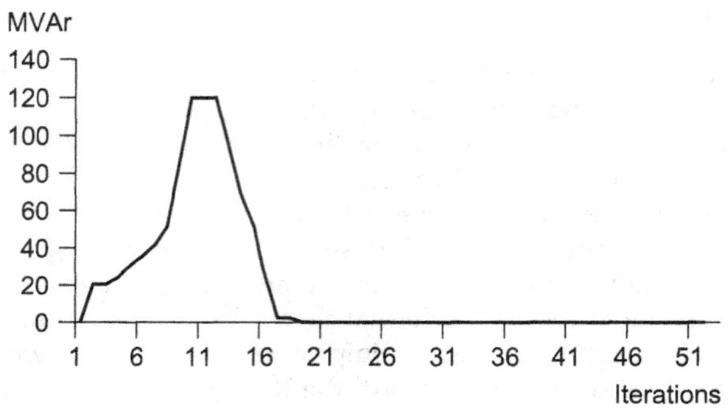

Figure 22 Objective function—minimization of reactive injection

Observe that the initial mismatches are high, $-2.0E+3$ MW and $1.6E+4$ MVAr, and that the reactive power mismatches drop at a lower rate than active power mismatches. This reflects the nonlinear behavior of reactive power in power systems. Figure 22 shows the evolution of the objective function. Initially it is zero and increases to a maximum of 120 MVAr at the tenth iteration and then it starts to decrease toward zero. This fluctuation reflects the trade-off of satisfying the constraints against minimizing the objective function during the iterations of the infeasible interior-point method.

Figure 23 displays the evolution of the barrier parameter. As can be seen, it does not decrease monotonically but goes to zero quite fast, reaching its minimum value (5.0×10^{-4}) at iteration 26.

In the second run for this system the algorithm converged in twelve iterations and active power losses at the optimal solution were 6196 MW. This represents a 13.5 percent reduction in active power with respect to the previous case (7175 MW). Figure 24 displays the evolution of the objective function showing the fluctuation behavior mentioned before. CPU time for this run was sixteen seconds.

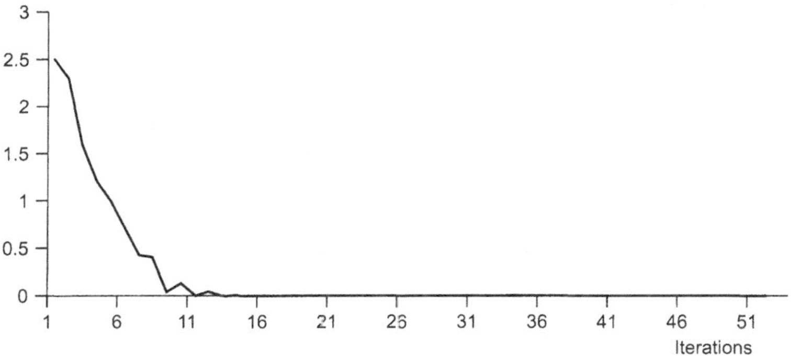

Figure 23 Barrier parameter—minimization of reactive injection

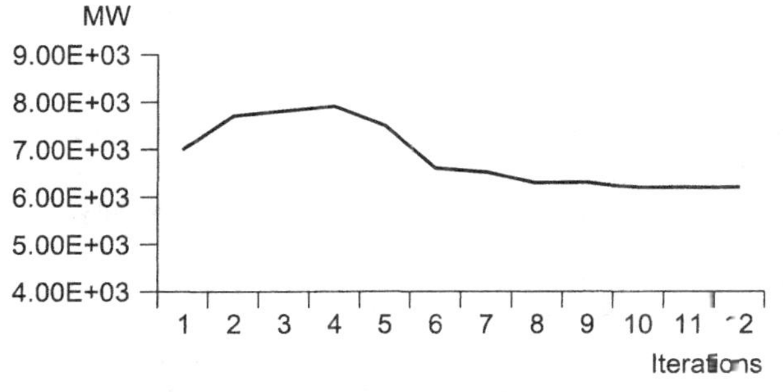

Figure 24 Objective function—loss reduction

6. Cost Allocation

In an open trading environment, considering the trends on competition in generation and a regulated transmission system, a set of services called ancillary services are needed to ensure that the power system will operate adequately and efficiently. These services, which were normally provided by the vertically integrated electric power utilities, will be unbundled. The number of ancillary services identified by the power industry is high and there are several ways of classifying them. Examples of such services are reactive support, provision for active power losses, several types of reserve, load following service, and frequency control, among many others.

Some ancillary services, such as load following, can be negotiated directly by the transmission consumer agent with any qualified provider, while some others are more effectively handled by the transmission entity. Taking, for instance, reactive support, it is likely that the transmission entity will be primarily responsible for ensuring adequate system-wide reactive support. To accomplish this task it should buy reactive power from the reactive providers, and the cost incurred in acquiring this ancillary service should be somehow shared by loads as illustrated in Figure 25. The same would be true for provision for active power losses, reserve, and so on.

Several cost allocation rules may be devised in the framework above, but to promote economic efficiency in resource utilization the one based on marginal cost is the most desirable because it is compatible with a competitive economic environment. However, in the application of any allocation rule, two issues have to be examined: fairness when more than a transmission access agent is considered, and revenue adequacy. With respect to the latter issue there are many papers in the literature dedicated to the problem of revenue reconciliation in transmission service pricing because it has been observed that short-run marginal cost price

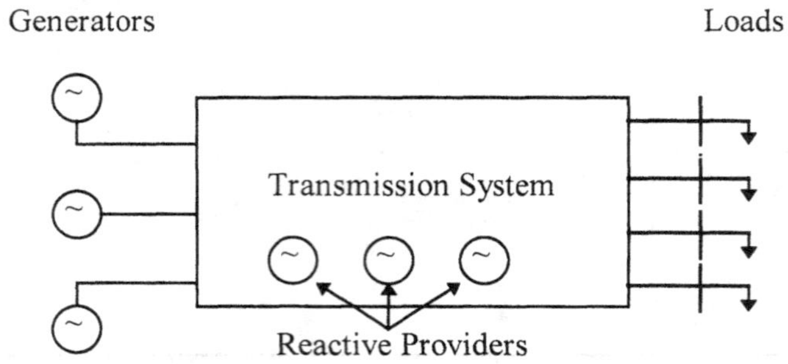

Figure 25 Reactive support transactions

does not always recover the cost of the transmission system. In the following sections we will illustrate cost allocation problems for the case of reactive power.

6.1. The Cost Allocation Problem

As it will be seen in the example, system-wide cost of reactive support is marginally increasing. This implies that if marginal cost is applied to reactive power the revenue collected from the transmission agents is greater than the total cost of the service. In this section the cost allocation problem is analyzed using a simple example.

Suppose three agents *A*, *B*, and *C* are all going to use a resource. Let P_A, P_B, and P_C be the amount of their transactions, and $c(P_A, P_B, P_C)$ be the total cost in providing the resource as a function of P_A, P_B, P_C, respectively. The cost function $c(P_A, P_B, P_C)$ is given by $P_A + (P_B + P_C)^3$. The problem is how to split the cost between agents *A*, *B*, and *C*.

Considering that $P_A = 1$, $P_B = 2$, and $P_C = 1$ are the transaction amounts of agents *A*, *B*, and *C*, respectively, the service cost to be allocated among the three agents will be equal to $28.

Direct application of marginal cost would lead to:

$$x_i = \frac{\partial c(P_A, P_B, P_C)}{\partial P_t} \cdot P_i = 1, \quad i = A, B, C,$$

resulting in $x_A = 1$, $x_B = 54$ and $x_C = 27$. We see that the total amount recovered ($82) exceeds the service cost, which is $28. This over-recovery is due to the marginally increasing behavior of the service cost function.

Another approach would be incremental cost allocation. The idea is to charge each agent the amount of cost he/she adds to total system cost. In this approach there is no revenue conciliation problem but there is a difficulty related to the order in which the transactions of agents are taken. For instance, Table 9 shows the incremental service costs for the entrance order *A*-to-*B*-to-*C*:

Table 10 shows the incremental service costs if the entrance order is changed to *A*-to-*C*-to-*B*.

We see that x_C was reduced from $19 to $1, whereas x_B increased from $8 to $26. In summary, the incremental cost allocation recovers cost, but is very sensitive to the entrance order.

Table 9 Incremental Service Costs: Sequence *A*, *B*, *C*

Agents	$C(P_A, P_B, P_C)$	$X_i = \Delta C(P_A, P_B, P_C)$
A	1	$X_A = 1$
A, *B*	$1 + 2^3 = 9$	$X_B = 8$
A, *B*, *C*	$1 + 3^3 = 28$	$X_C = 19$

Table 10 Incremental Service Costs: Sequence *A, C, B*

Agents	$C(P_A, P_B, P_C)$	$X_i = \Delta C(P_A, P_B, P_C)$
A	1	$X_A = 1$
A, C	$1 + 1^3 = 2$	$X_C = 1$
A, C, B	$1 + 3^3 = 28$	$X_B = 26$

Table 11 Shapley Allocation

Sequence	X_A	X_B	X_C
A, B, C	1	8	19
A, C, B	1	26	1
B, A, C	1	8	19
B, C, A	1	8	19
C, A, B	1	26	1
C, B, A	1	26	1
Mean	1	17	10

To correct this problem we can consider permutations of entrance orders. This allocation rule, which corresponds to the Shapley value (Young, 1994), is ordering independent. The charge for each transaction is the mean of the incremental costs when all transaction permutations are taken into account. The results are shown in Table 11.

This scheme is intuitively fair, as it allows all agents to be first, second, and so on, but it is not yet completely satisfactory because, given that agents *B* and *C* have a similar effect on the service cost function, we would expect them to have the same per-unit charges. However, we see in Table 11 that the allocation for those agents is, respectively, \$8.5/unit and \$10/unit. The reason is that agents with higher transaction amounts are less sensitive to the entrance order than those with lower values.

In order to correct the observed limitation in the Shapley allocation scheme, we allow smaller agents to enter after just a fraction of the larger agent has been served. In terms of the example in Table 11, we split agent *B* into two subagents, *B*1 and *B*2, both with unit size. Table 12 summarizes the allocation results (for simplicity, only results for agents *B* and *C* are shown).

We observe now that, as desired, the unit charge is the same for both agents. Note, however, that in real cases the total number of agents and subagents may be very large. In this case the number of combinations can increase exponentially, resulting in a high computational effort. In an extreme case, the calculation can even be infeasible.

The next step in defining the allocation rule is to make a partition of each transaction amount, P_A, P_B, P_C, into a large number of small subtransactions of equal size. Let Δ be the size of the small transactions, $\Delta \rightarrow 0$. It can be shown (Vieira et al., 1997) that as a result of this

Table 12 Shapley Allocation with Agent Splitting

Sequence	X_{B1}	X_{B2}	X_C
*B*1, *B*2, *C*	1	7	19
*B*1, *C*, *B*2	1	19	7
C, *B*1, *B*2	7	19	1
C, *B*2, *B*1	19	7	1
*B*2, *B*1, *C*	7	1	19
*B*2, *C*, *B*1	19	1	7
Mean	9	9	9

limiting process we obtain the Aumann–Shapley allocation rule (Aumann and Shapley, 1974), which is given by the following expression:

$$x_i = P_i \cdot \tilde{\pi}_j \quad i = A, B, C, \tag{96}$$

where $\tilde{\pi}_j$ is called the Aumann–Shapley unit cost for agent i, and is given by the expression below:

$$\tilde{\pi}_i = \int_{t=0}^{1} \frac{\partial c(tP_A, tP_B, tP_C)}{\partial P_i} dt \quad i = A, B, C. \tag{97}$$

Applying the Aumann–Shapley methodology in the previous cost function example, we obtain $x_A = 1$, $x_B = 18$, and $x_C = 9$.

Therefore, beside fairness, this new allocation rule leads to revenue reconciliation. Other aspects of its interesting features are: additivity, the cost allocation for the sum of more than one cost function is equal to the sum of the cost allocation of each one taken separately; monotonicity, if $c(P_A, P_B, P_C)$ is nondecreasing then $\tilde{\pi}_A \geqslant 0$, $\tilde{\pi}_B i \geqslant 0$; and aggregation invariant. Besides, it is the only cost allocation procedure that satisfies all of the properties above (Billera and Heath, 1982).

In this example, the fixed cost is null $[c(0,0,0) = 0]$. When $c(0,0,0) \neq 0$, the Aumann–Shapley allocation rule only recovers the additional cost to be shared among the agents $[c(P_A, P_B, P_C) - c(0,0,0)]$. In this case the fixed cost could be shared using a pro rata-based method, such as a postage stamp methodology, or other cost allocation scheme based in cooperative game theory. Besides this, the Aumann–Shapley allocation rule does not depend on any particular shape of the cost function.

6.2. *Case Study 1: Bilateral Contracts*

Consider a network where besides the native loads there are bilateral transaction contracts: generator and load. Each transaction k is associated to a generation at bus i where active power is supplied, a load bus j where active power is delivered, and the amount of contracted active power transaction P_k and reactive load associated to the transaction Q_k. The reactive load associated to the transaction is not part of the contract generator-load and is supplied by the transmission system.

Once the transactions are implemented, besides their reactive load there is an increase in active power flow throughout the network implying an additional need for reactive support. The first step in calculation of reactive power pricing is to determine the total cost of reactive power support supplied by the reactive providers. This amounts to solving an optimal power flow problem.

The loads and generation in the OPF problem are computed from the native loads, bilateral transactions, and the associated reactive power. The total cost to be shared among the users is equal to $CT(P_1, ..., P_K) - CT(0, ..., 0)$, where $CT(0, ..., 0)$ is the reactive power cost associated to the native loads and $CT(P_1, ..., P_K)$ is the total cost of supplying reactive power as a function of the transactions.

As we have seen in the example, the main point in the Aumann–Shapley application is the calculation of the marginal cost associated with each transaction. In this computation it will be assumed that the associated reactive load will be increased in the same proportion to the active power transaction.

The reactive power marginal cost for each transaction k is:

$$\pi_k = \frac{\partial CT(P_1, ..., P_K)}{\partial P_k} = \lambda_{pi} - \frac{Q_k}{P_k}\lambda_{qj} - \lambda_{pj}, \tag{98}$$

where:

λ_{pi}, λ_{pj} are Lagrange multipliers associated with the active power balance equations of the OPF problem in the transaction supplying and delivering buses, respectively;

Table 13 Bilateral Transactions

Trans. No.	*Load Bus*	*Gen. Bus*	*Trans. V. (MWh)*	*Reactive power (MVArh)*
1	1	1	60.52	12.33
.	.	.	.	.
11	8	16	14.01	2.85
12	8	21	81.82	16.77
.	.	.	.	.
25	19	23	17.37	3.55
26	20	23	71.73	14.57

λ_{qj} is the Lagrange multiplier associated with the reactive power balance equations in the transaction delivering the bus, and P_k, Q_k are the values for transaction k and its associated reactive load.

Using the Aumann–Shapley expression (97), the reactive power allocation rule of each transaction k is based on:

$$\tilde{\pi}_k = \int_{t=0}^{1} \frac{\partial CT}{\partial P_k}(tP_1, \ldots, tP_K)dt, \quad k = 1, \ldots, K. \tag{99}$$

This integral is computed numerically taking a discretization of the variable t in the interval [0, 1]. For each value of t, the OPF problem is solved and formula (98) is used to compute $\partial CT/\partial P_k$.

The ideas presented above will be illustrated in a twenty four bus system, derived from the IEEE Reliability Test System (RTS). Part of the original network load, (1709 + 348j MVAh) and generation data was specified as native load, and twenty six bilateral transactions were also specified. The hourly transactions correspond to new generations and loads that are added to the native ones. Table 13 shows some of the transactions. Note that some of them involve generators and loads at the same bus while others involve generators and loads at different buses. For instance, transaction 11 involves a generator and load quite far from each other. The total hourly active load associated with the transactions is 1596 MWh.

Voltage limits of (0.95, 1.05) were specified for all buses and a set of twelve reactive providers was selected. Figure 26 shows the reactive cost, as a function of the transaction parameter t, $0 \leqslant t \leqslant 1$. This curve was obtained from a discretization of the unit interval into a mesh of 500 points and for each point an OPF problem, which corresponds to an optimal reactive dispatch, was solved by an interior-point method. At $t = 0$ (no transactions) the reactive cost is zero, so no capacitive reactive support is necessary for native loads. But, as the transaction amounts increase, additional capacitive reactive support is needed in the system. At $t = 1$ (full transactions) the reactive cost required to make transactions feasible amounts to

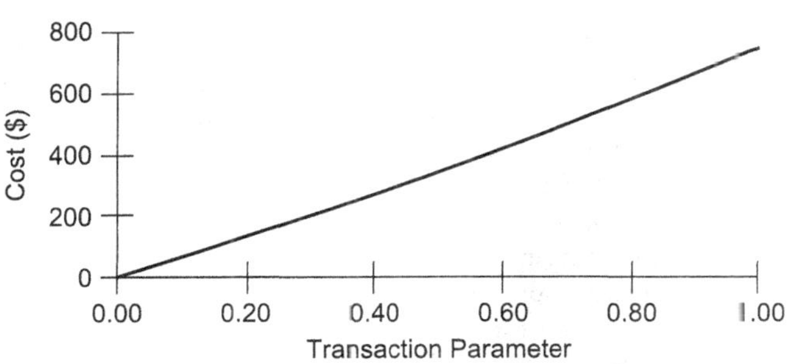

Figure 26 Reactive cost—twenty four bus system

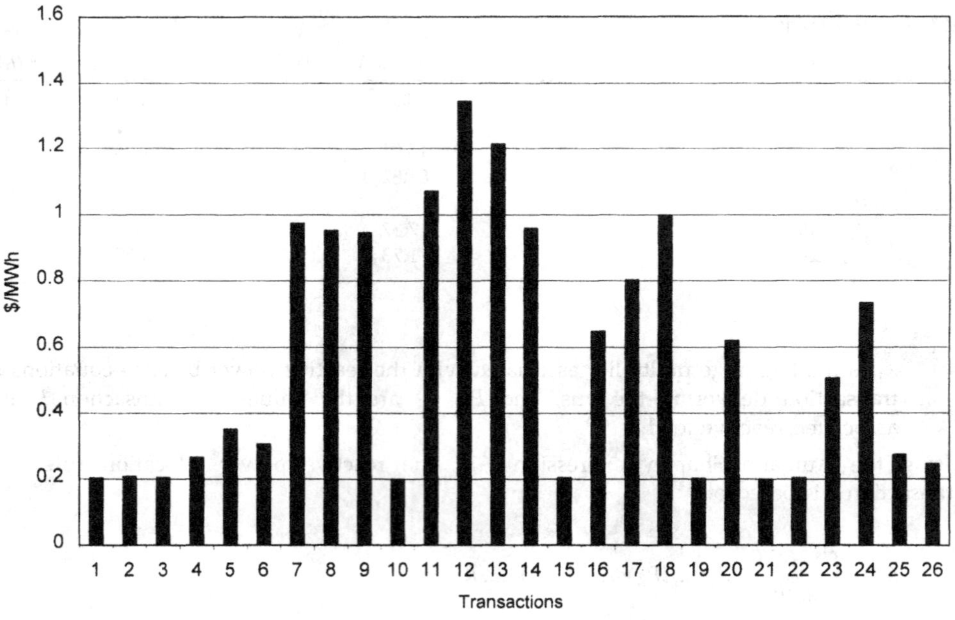

Figure 27 Transaction marginal cost

\$747.35. The additional reactive power cost to be shared among the transaction agents is \$747.35.

Figure 27 shows reactive marginal costs associated to the transactions. Note how they vary across them, for instance, transaction 1 involves a generator and load at the same bus. An increase of 1 MWh in its amount implies an increase of $12.33/60.52 = 0.204$ in its reactive load. As the reactive price is \$1/MVArh and the increase in transaction amount does not imply any change of circuit flow, the marginal cost associated with it is \$0.204/MWh as shown in the figure. Transaction 12, on the other hand, involves a generator and load far from each other and its marginal cost is much higher—\$1.343/MWh. This is due to the fact that, besides the reactive load associated with the transaction, reactive consumption by the transmission lines grows at an increasing rate as this transaction amount increases. Total hourly revenue collected from the transmission access agents based on this scheme is equal to \$879.54, which is much greater than the hourly cost of acquiring reactive power by the transmission system (\$747.35)—a difference of 18 percent.

One way to solve this problem is to consider an adjustment factor to be applied to all charges so that the original cost is recovered. But by doing so the charge for transaction 1 would be less than 0.204, which means that there would be a cross-subsidy from other transactions to transaction 1.

Next the Aumann–Shapley unit charge is computed using an approximation of formula (99) based on the discretization of interval [0, 1] into a mesh of five hundred points. At each point an OPF problem is solved to obtain the corresponding marginal cost. Figure 28 shows the adjusted marginal cost and the Aumann–Shapley unit cost.

Observe that the Aumann–Shapley unit charge for some transactions is greater than the adjusted marginal cost and for others it is lower. This is because, when the transaction amounts are scaled down, the marginal cost associated with transactions involving a generator and load close to each other are approximately constant while for those involving a generator and load far apart it is smaller. The total hourly revenue collected with the Aumann–Shapley unit charge (\$740.73) is approximately equal to the hourly additional cost for the transmission system in acquiring reactive power (\$747.35)—a difference of less than 1 percent.

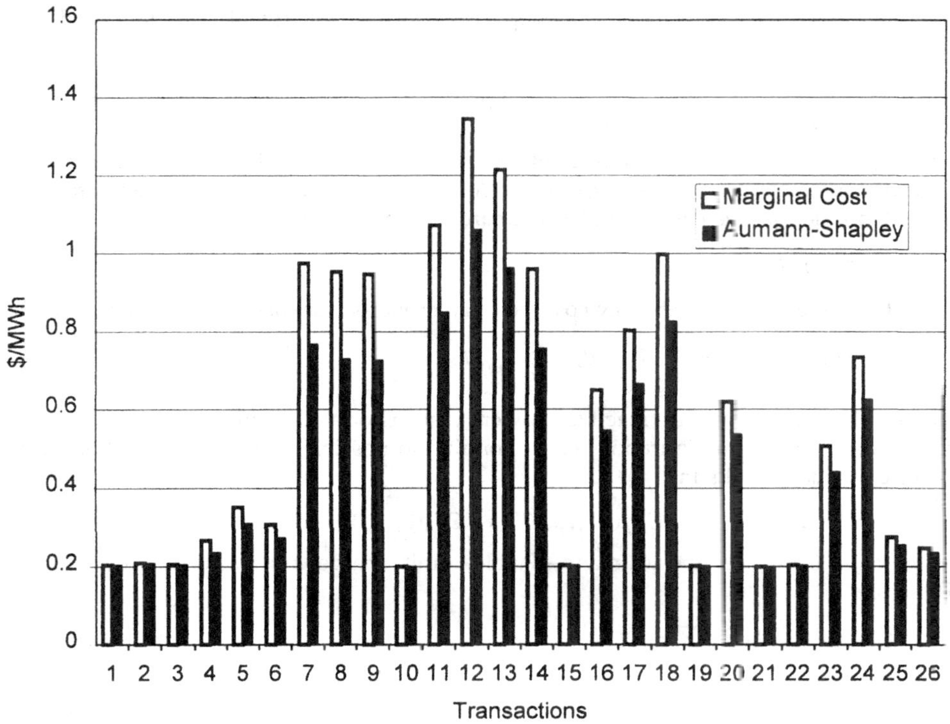

Figure 28 Transaction marginal cost and Aumann–Shapley unit charge

6.3. Case Study 2: Nodal Pricing

In this second scheme the transactions are between the loads and the remainder of the generation–transmission system. Note that once the transactions are implemented there is a need for additional reactive power in order for the generators to supply their loads. This reactive power is bought by the transmission system from the reactive providers and the cost incurred will be shared by the transaction agents. In this case equipment failure rates will also be taken into account in the computation of reactive power pricing.

In the analysis of each network configuration derived from a scenario of one or more generation and/or transmission equipment outages, the following steps are considered:

1. Verify if any active power related violation exists. If yes, go to (2), otherwise freeze active power dispatch and go to (3);
2. Consider the minimum generation/load shedding cost OPF problem, taking into account only active power related network constraints. In the load shedding process power factors are kept constant. After optimization, implement load shedding if any, freeze active power generation dispatch, and go to (3);
3. Compute the minimum cost of reactive power supplied by the reactive power providers taking into account all network constraints. At this stage active power dispatch is considered fixed.

The purpose of step 2 is to eliminate any system-active power-related violations. It takes into account active power balance equations, circuit flow limits, generation redispatch capabilities, and load shedding.

Let $CT_\mu(PL_1, ..., PL_K)$ be the solution value of the OPF problem solved in step 3, the minimum cost of reactive power allocation at configuration μ as a function of the new active load amounts $PL_1, ..., PL_N$ and let Γ_μ be the vector of fractions of the load kept after solving the minimum load shedding problem in step 2. The reactive marginal cost associated to a given load PL_k is equal to:

$$\frac{\partial CT_\mu(PL_1, ..., PL_N)}{\partial PL_K} = -\Gamma_{\mu k}\left(\beta_{Pk} + \frac{QL_k}{PL_k}\beta_{Qk}\right), \tag{100}$$

where β_{Pk}, β_{Qk} are the Lagrange multipliers associated with the active and reactive power balance equations for bus k. Let $\pi_{\mu k}$ be the Aumann–Shapley unit charge associated to agent k in configuration μ. The corresponding Aumann–Shapley charge is equal to

$$x_{\mu k} = \pi_{\mu k} PL_k. \tag{101}$$

From the Aumann–Shapley cost recovery property, for each configuration μ we have

$$CT_\mu(PL_1, ..., PL_K) - CT_\mu(0, ..., 0) = \sum_k \pi_{\mu k} PL_k, \tag{102}$$

where $CT_\mu(0, ..., 0)$ is the reactive power cost associated with the native loads. Now, considering a set of configurations corresponding to possible system configurations and taking expected values, results in

$$CT(PL_1, ..., PL_K) = E\mu[CT_\mu(PL_1, ..., PL_K) - CT_\mu(0, ..., 0)], \tag{103}$$

where $\pi_k = E_\mu[\pi_{\mu k}]$ is the Aumann–Shapley unit charge associated to bus k, $k = 1, ..., N$. Therefore,

$$CT(PL_1, ..., PL_K) = \sum_k \pi_{\mu k} PL_k. \tag{104}$$

A Monte Carlo simulation is used to estimate the Aumann–Shapley unit charges. Besides sampling a system configuration for each draw, the transaction parameter is also drawn based on the interpretation of the Aumann–Shapley allocation scheme as a limiting process of expected values.

To illustrate this scheme, an application was made to a Brazilian system with 1,653 buses, 2,378 circuits, and 175 generators. The total hourly system load is 41,098 MWh but only 1,523 MWh were associated with transactions, the remainder were considered to be native loads. Costs charged by reactive providers were equal to $1/MVArh. The main results were:

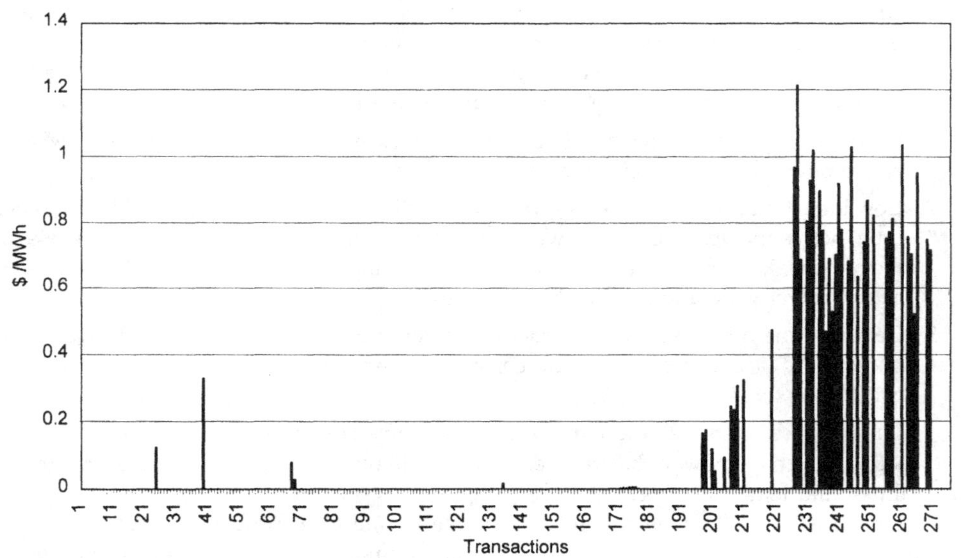

Figure 29 Transaction Aumann–Shapley unit charge

- the hourly expected need for additional reactive power due to the transactions was 180 MVArh at a cost of $180, to be shared among transaction agents;
- for the expected Aumann–Shapley charge estimate, the Monte Carlo sample size was three thousand. The expected hourly revenue based on marginal cost was $348, and the hourly expected revenue based on Aumann–Shapley was $183.

It can be seen that the revenue collected based on marginal pricing is much higher than the service cost, whereas the Aumann–Shapley pricing scheme recovered the service cost. Figure 29 shows the Aumann–Shapley unit charge associated with each transaction. Note that it can be very different across transactions, depending on the load bus location in the network.

References

Abadie, J. and J. Carpentier. 1969. "Generalization of the Wolfe gradient method to the case of nonlinear constraints." In *Optimization*, edited by R. Fletcher. London and New York: Academic Press, pp. 37–47.

Alsac, O., J. Bright, M. Prais, and B. Stott. 1990. "Further developments in LP-based optimal power flow." *IEEE Trans. PAS* **5**: 697–711.

Aumann, R. and L. Shapley. 1974. In *Values of Non-Atomic Games*. Princeton, NJ: Princeton University Press.

Billera, L. and D. Heath. 1982. "Allocation of shared costs: a set of axioms yielding a unique procedure." *Math. Oper. Res.* **7(1)**: 32–39.

Carpentier, J. 1962. "Contribution a l'étude du Dispatching Economique." *Bulletin Soc. Française Electriciens* **3**: **Ser. 8:** 431–447.

Dommel, H. W. and W. F. Tinney. 1968. "Optimal power flow solutions." *IEEE Trans. PAS* **87**: 1866–1876.

Enamorado, J. C., T. Gomes, and A. Ramos. 1999. "Multi-area regional interconnection planning under uncertainty." In *Proceedings of the 13th Power Systems Computing Conference, 1999.*

Gorenstin, B., N. Campodonico, J. Costa, and M. Pereira. 1993. "Power system planning under uncertainty." *IEEE Trans. Power Syst.* **8(1)**: 129–136.

Granville, S. 1994. "Optimal reactive dispatch through interior point methods." *IEEE Trans. Power Syst.* **9(1)**: 136–146.

Mielczarski, W., G. Michalik, and M. Widjaja. 1999. "Bidding stategies in electricity markets." In *Proceedings of the 21st International Conference on Power Industry Computer Applications*, pp. 71–76.

Murtagh, B. and M. Saunders. 1982. "A projected Lagrangian algorithm and its implementation for sparse nonlinear constraints." *Math. Programming Study* **16.**

Pereira, M. and L. Pinto. 1991. "Multi-stage stochastic optimization applied to energy planning." *Math. Programming.*

Pereira, M., N. Campodonico, B. Gorenstin, and J. Costa. 1995. "Application of stochastic optimization to power system planning and operation." In *Proceedings of the IEEE Stockholm Power Tech*, pp. 234–239.

Pereira, M., N. Campodonico, and R. Kelman. 1998. "Long-term hydro scheduling based on stochastic models." In *Proceedings of EPSOM Conference.* Zurich.

Read, E. and J. George. 1990. "Dual dynamic programming for linear production/inventory systems." *Comput. Math. Appl.* **19(11)**: 29–42.

Scott, T. and E. Read. 1996. "Modeling hydro reservoir operation in a deregulated electricity market." *Int. Trans. Oper. Res.* **3(3–4)**: 243–253.

Sun, D., B. Ashley, H. Brewer, and W. Tinney. 1984. "Optimal power flow by Newton approach." *IEEE Trans. PAS* **103.**

Vanderbei, R. J. and D. F. Shanno. 1999. "An interior-point algorithm for nonconvex nonlinear programming." *Comput. Optim. Appl.* 3: 231–252.

Vieira, X., S. Granville, M. Pereira, B. Gorenstin, J. Mello, and A. Melo. 1997. "Ancillary services: how to allocate costs in an optimization-based framework." In *Proceedings of the CIGRÉ Symposium* Neptun.

Young, H. 1994. "Cost allocation." In *Handbook of game theory with economic applications*, edited by Robert Aumann and Sergiu Hart. N.H. Elsevier.

Zhang, D., Y. Wang, and P. Luh. 1999. "Optimization based bidding strategies in the deregulated market." In *Proceedings of the 21st International Conference on Power Industry Computer Applications*, pp. 63–69.

18.8.2 Optimization Applications in Oil and Gas Recovery

Roland N. Horne

ABSTRACT Optimization in upstream oil and gas recovery has generally been focused in three areas: (a) production scheduling, (b) well placement, and (c) production facilities design. The problems are often complex, and usually highly nonlinear. A variety of different optimization techniques have been applied successfully, including gradient-base methods such as Newton's method and the polytope approach, and nongradient methods such as the genetic algorithm and simulated annealing.

1. Introduction

Petroleum engineers face a wide variety of optimization problems. Every time a production strategy is designed or a location for a well is chosen, a decision problem has been solved. Yet it is surprising how rarely a formal optimization technique is used to solve these problems. In fact the word optimization in the oil recovery industry is misused widely in the sense of analyzing a few cases and choosing the best one. Perhaps the explanation for this is that petroleum optimization problems are extremely difficult. Principle characteristics of oil and gas recovery are that: (a) expenditures are enormous, (b) risks are large, (c) important parameters that affect the production are subject to large uncertainty, and (d) the cost of modeling is huge and may not allow multiple evaluations. However, with the improvement of computer modeling over time, it is now becoming feasible to apply optimization techniques to address several petroleum recovery issues. This paper describes the application of formal optimization techniques to common petroleum engineering problems.

2. Oilfield Optimization Problems

Although there are many places in the field of petroleum engineering in which optimization methods can be applied, only a selection of the major ones will be described here. In general, the development of an oil or gas reservoir requires the management of three principal components: (a) the reservoir, (b) the wells, and (c) the production equipment. These components are illustrated schematically in Figure 1, which shows a reservoir producing a mixture of oil and gas into a single well, which conducts the fluids to the surface through a tubing string with the assistance of a gas lift system that reduces the density of the fluid. At the surface, the fluids are controlled by a choke, and separated into oil and gas streams for separate sale. Part of the gas stream may be injected back into the well in the gas lift system. The reservoir will be tapped by multiple wells, and the surface equipment will be interconnected at multiple wellheads.

The interaction between these many components is complex and interdependent, and is therefore an ideal candidate for optimization. There are three major classes of problems that have been addressed.

1. *Production scheduling*. The reservoir is heterogeneous in flow properties and is penetrated by multiple wells, including both injectors and producers. Injecting water can improve recovery by raising pressure and sweeping oil through the reservoir, but once injected water starts to be produced at the production wells, the recovery efficiency drops dramatically. The optimal approach is to inject at such a rate that oil is swept uniformly throughout the reservoir so that some production wells do not water out before others. The problem is complicated by the fact that the producers may need to meet variable production schedules, for example, if they are producing gas and are required to meet high winter demands and smaller summer demands. For these reasons, deciding on the rates at which to inject and produce fluids in a given set of wells is an optimization problem.

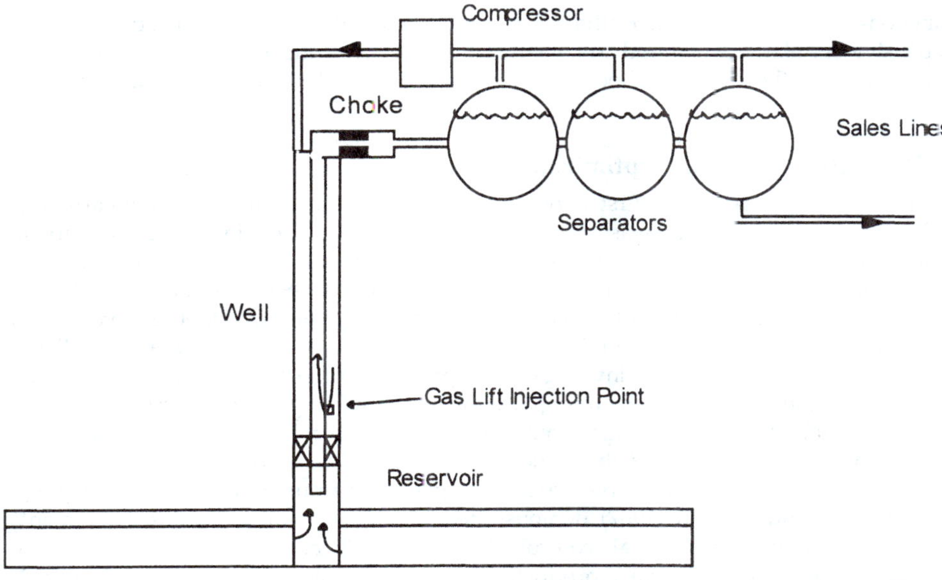

Figure 1 Example production scenario in which reservoir fluids are separated at surface. Part of the gas flow is used for gas lifting the well. From Palke and Horne (1997b), copyright Society of Petroleum Engineers, Dallas, Texas

2. *Well placement.* When a new well is drilled, it needs to be located in the reservoir at a location that will drain as much oil or gas as possible, yet not interfere with production from other wells. Due to heterogeneity in the reservoir properties, there may be places where wells need to be placed closer together or further apart The problem is also time dependent, in that some areas of the reservoir may be well swept by earlier recovery operations while other regions of the reservoir may remain untapped. To complicate the issue further, the placement of the well must also consider the decision as to the type of completion (for example, horizontal well or vertical well) that would produce the best recovery.
3. *Well and production facilities.* Once the wells have been placed, the production facilities have to be optimized to maximize recovery and minimize cost. Due to the very nonlinear phase behavior of hydrocarbon mixtures, the quantity and even the quality of the oil and gas produced are a function of the sequence of separations to which the fluids are subjected in the wellbore and surface equipment. Wells that are interconnected via the same surface facilities may influence each other, for example, if a high producing pressure in one well causes the constraint of flow in another. Deciding on the values of production system variables therefore becomes an optimization problem. Since the production characteristics of the reservoir change with time, the optimization must also take account of historical and expected future variations.

Each of these types of problems depends on determination of the likely future performance of the reservoir. This future performance can be estimated using reservoir simulation. However, petroleum reservoir simulation is an enormously challenging task in that it requires large-scale computations and is subject to significant uncertainty as to the values of reservoir properties. Due to the cost of simulation, it is never feasible to enumerate all possible scenarios, and optimization in the oil and gas industries has focused mainly on either simplifying the models to speed up their computation or refining the optimization techniques to reduce the number of function evaluations or model runs that need to be made.

Because of the large computational cost of evaluating the performance functions, most petroleum optimization problems focus on one of three approaches: (a) linearizing the objective functions so that simpler algorithms can be applied, (b) developing cheaper proxies for the expensive simulations required to construct the objective functions, (c) improving the rates of convergence of the search methods to reduce the number of function evaluations needed.

3. Production Schedule Optimization

Optimization methods were first adopted by the petroleum industry in the early 1950s. The bulk of the early work was on linear programming techniques applied to reservoir management on a macrolevel. For a detailed treatment of linear programming methods in petroleum engineering, both past and recent, see Aronofsky (1983) and Lo et al. (1995).

Aronofsky and Lee (1958) developed a linear programming model to maximize profit by scheduling production from multiple single-well reservoirs. Aronofsky and Williams (1962) extended the same model to investigate the problems of scheduling production for a fixed drilling program and also scheduling drilling for a fixed production schedule. Attra et al. (1961) developed a linear programming model to maximize flow rate subject to several production constraints. All of these models used linear reservoir models based on material balance considerations, and the reservoirs were generally assumed uniform and single phase.

Rowan and Warren (1967) demonstrated how to formulate the reservoir management problem in terms of optimal control theory. O'Dell et al. (1973) developed a linear programming model to optimize production scheduling from a multireservoir system. Huppler (1974) developed a dynamic programming model to optimize well and facility design given the delivery schedule and using a material balance reservoir model.

Starting in the 1970s, investigators began coupling numerical reservoir simulation with linear programming models. The idea has been to use the reservoir simulator to generate a linearized unit response matrix that could be used with linear programming models. Wattenbarger (1970) developed a linear programming model to schedule production from a gas storage reservoir using this technique. Murray and Edgar (1979) used influence functions in a mixed-integer linear programming model that optimized well placement and production scheduling. See and Horne (1983), extending on work performed by Coats (1969) and Crichlow (1977), demonstrated how to refine the unit response matrix using nonlinear regression techniques. This work was expanded by Lang and Horne (1983) to consider dynamic programming techniques. McFarland et al. (1984) used nonlinear optimization to optimize reservoir production scheduling.

In field applications, Asheim (1978) studied petroleum developments in the North Sea by coupling reservoir simulation and optimization. A similar approach was used by Nesvold et al. (1996) to design the redevelopment of the Ekofisk field, also in the North Sea. Notice that virtually all of this research attempted to model reservoir performance by linearizing the reservoir performance and feeding this to some variation of linear programming.

A somewhat different approach has been to substitute for the expensive reservoir simulation with some kind of proxy. Pan (1995) and Pan and Horne (1998) experimented with the replacement of the simulations by using a limited number of simulation runs to develop either a multivariate polynomial approximation or a kriged surface (Delhomme, 1978, Doctor, 1979). The initial search space was populated as fully as possible with a reduced number of function evaluations, using the algorithm of Fang (1980). In the groundwater field, Rogers (1992) and Rogers et al. (1995) used the artificial neural network approach to develop the proxy for the reservoir simulation, and then performed the optimization using the genetic algorithm (Goldberg, 1989).

4. Optimal Well Placement

A typical petroleum development involves many variables that affect the operational schedule involved in its management. These variables are usually used as input to a reservoir simulator that generates a forecast of the production profile. Using this forecast, the production engineer has to consider several hypotheses to achieve the best strategy for development of the field.

One of the most important decisions to be made is where to place the wells, both for production as well as for injection.

The difficulty with well placement is that the number of variables is large, and the interaction between them is complex and nonlinear. In general, it is difficult to perform a sufficient number of reservoir simulations for function evaluation, since each simulation is computationally expensive. Most approaches have addressed this difficulty by reducing the number of simulations by careful tuning of the optimization methods, often by using knowledge of reservoir physics to good effect.

Aanonsen et al. (1995) used kriging to replace the search space with a smooth function obtained by matching a reduced number of simulations chosen by experimental design (Damsleth et al., 1992). A similar approach was also used by Pan and Horne (1998) and for history matching by Watkins (1993).

Well placement problems have also been addressed using randomized search methods such as simulated annealing and the genetic algorithm (Holland, 1975; Goldberg, 1989). Beckner and Song (1995) used simulated annealing for both well placement and scheduling problems. Bittencourt and Horne (1997) used the genetic algorithm for well placement in an offshore development, but showed that it was possible to greatly improve the performance of the algorithm by introducing a hill-climbing step such as polytope (Nelder and Mead, 1965), as had been described by Ackley (1987). In the field problem investigated by Bittencourt and Horne (1997), the hybrid genetic algorithm approach proposed a well placement that resulted in a predicted overall project profit that was $68 million greater than with the well placement scheme proposed initially by the engineering design team.

5. Well and Production Facility Optimization

The optimization of a petroleum development depends not only on the condition of the reservoir, but also on the configuration of the well and surface equipment. Design of the well was performed historically by a graphical technique known as total system analysis; however, this did not usually consider the complex and time-varying nature of reservoir production and generally performed optimization only in the crude sense of trial and error. Carroll and Horne (1992) applied multivariate optimization techniques to a model of a field produced by a single well, including reservoir performance and the full influence of the hydrocarbon composition in the separator, although no wellbore parameters were allowed to vary with time. Carroll and Horne (1992) used two types of optimization routines, gradient methods and nongradient methods, and found that the nongradient polytope method (Nelder and Mead, 1965) worked best since the objective function surface is not smooth (see Figure 2 for example). In 1992, Ravindran followed a similar approach and allowed engineering parameters to be varied with time. Fujii and Horne (1995) continued this line of study by including a network of wells connected at the surface. Fujii and Horne (1995) determined that the polytope method used by earlier authors was no longer suitable for interconnected network problems because of the large number of variables, and instead used the genetic algorithm approach (Goldberg, 1989). Palke and Horne (1997a,b) extended the approach to problems with physical feedback loops, such as occur when the wells are produced using gas lift (Figure 1).

6. Conclusions

Oil and gas recovery can be improved significantly by the application of optimization techniques. Maximizing recovery and minimizing cost can be achieved by the use of common optimization algorithms such as linear and dynamic programming, polytope, simulated annealing, and the genetic algorithm. Although these methods are applied commonly in many other fields, the most successful applications in the petroleum industry have been those that made modifications to the algorithms to take account of reservoir physics (for example, by choosing a combination of variables that are related to each other physically in an approximately linear way, instead of by imposing a naive mathematical linearization).

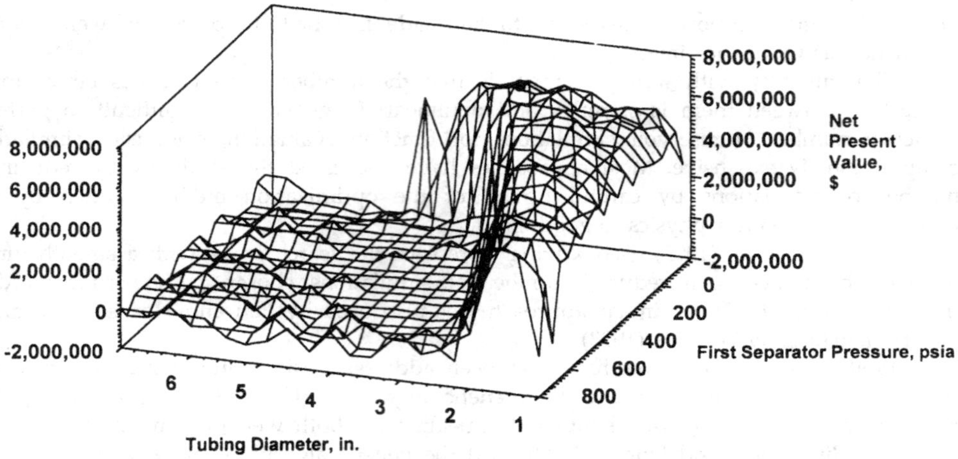

Figure 2 A well and separator optimization problem. Decision variables are tubing diameter and first separator pressure. From Palke and Horne (1997b), copyright Society of Petroleum Engineers, Dallas, Texas

Finally, it should be noted that maximizing profit is not the only type of optimization problem addressed in the petroleum industry. Another class of problems is the minimization of uncertainty. This class of problems has not been addressed here (although an example can be found in Palke and Horne, 1997a). Copyright Society of Petroleum Engineers, Dallas, Texas.

References

Aanonsen, S. I., A. L. Eide, L. Holden, and J. O. Aasen. 1995. "Optimizing reservoir performance under uncertainty with application to well location." Paper SPE 30710 presented at the 1995 SPE Annual Technology Conference and Exhibition, Dallas, TX, October 22–25.

Ackley, D. H. 1987. *A Connectionist Machine for Genetic Hillclimbing*. Boston: Kluwer Academic Publishers.

Aronofsky, J. S. 1983. Optimization methods in oil and gas development. Paper SPE 12295. Available from the Society of Petroleum Engineers, Richardson, TX.

Aronofsky, J. S. and A. S. Lee. 1958. "A linear programming model for scheduling crude oil production." *J. Pet. Technol.* **10**: 51–54.

Aronofsky, J. S. and A. C. Williams. 1962. "The use of linear programming and mathematical models in underground oil production." *Mgmnt Sci.* **8**: 394–407.

Asheim, H. A. 1978. "Offshore petroleum exploration planning by numerical simulation and optimization." Ph.D. diss. TX: Department of Petroleum Engineering, University of Texas at Austin.

Attra, H. D., W. B. Wise, and W. M. Black. 1961. "Application of optimizing techniques for studying field producing." *J. Pet. Technol.* **13**: 82–86.

Beckner, B. L. and X. Song. 1995. "Field development planning using simulated annealing—optimal economic well scheduling and placement." Paper SPE 30650 presented at the 1995 SPE Annual Technical Conference and Exhibition, Dallas, TX, October 22–25.

Bittencourt, A. C. and R. N. Horne. 1997. "Reservoir development and design optimization." Paper SPE 38895 presented at the 72nd Annual Technical Conference & Exhibition, San Antonio, TX, October 5–8.

Carroll III, J. and R. N. Horne. 1992. "Multivariate production system optimization." *J. Pet. Technol.* **44**: 782–789.

Coats, K. H. 1969. "An approach to locating new Wells in heterogeneous, gas producing fields." *J. Pet. Technol.* **21**: 549–558.

Crichlow, H. B. 1977. In *Modern Reservoir Engineering—A Simulation Approach*. Englewood Cliffs, NJ: Prentice-Hall.

Damsleth, E., A. Hage, and R. Volden. 1992. "Maximum information at minimum cost: a North Sea development study with an experimental design." *J. Pet. Technol.* **44**: 1350–1356.

Delhomme, J. P. 1978. "Kriging in the hydrosciences." *Adv. Water Resources* **1**: 251–266.

Doctor, P. G. 1979. "An evaluation of kriging techniques for high level radioactive waste repository site characterization." Department of Energy, Pacific Northwest Laboratory.

Fang, K. 1980. "Application of number theory in test design." *ACTA Math. Appl. Sinica* (In Chinese).

Fujii, H. and R. N. Horne. 1995. "Multivariate optimization of networked production systems." *SPE Prod. Facilities* **10**: 165–171.

Goldberg, D. E. 1989. In *Genetic Algorithms in Search, Optimization, and Machine Learning*. Reading: Addison-Wesley.

Holland, J. 1975. In *Adaptation in Natural and Artificial Systems*. Cambridge: MIT Press.

Huppler, J. D. 1974. "Scheduling gas field production for maximum profit." *SPE J.* **16**: 274–279.

Lang, Z. X. and R. N. Horne. 1983. "Optimum production scheduling using reservoir simulators: a comparison of linear programming and dynamic programming techniques." Paper SPE 12159 presented at the 1983 SPE Annual Technical Conference and Exhibition, San Francisco, October 5–8.

Lo, K. K., G. P. Starley, and C. W. Holden. 1995. "Application of linear programming to reservoir development evaluations." *SPE Reservoir Engng.* **10**: 52–58.

McFarland, J. W., L. Lasdon, and V. Loose. 1984. "Development planning and management of petroleum reservoirs using tank models and nonlinear programming." *Oper. Res.* **32**: 270–289.

Murray III, J. E. and T. F. Edgar. 1979. "Optimal scheduling of production and compression in gas fields." *J. Pet. Technol.* **31**: 109–116.

Nelder, J. A. and R. Mead. 1965. "A simplex method for function minimization." *Comput. J.* **7**: 308–313.

Nesvold, R. L., T. R. Herring, and J. C. Currie. 1996. "Field development optimization using linear programming coupled with reservoir simulation—Ekofisk field." Paper SPE 36874 presented at the 1996 SPE European Petroleum Conference, Milan, Italy, October 22–24.

O'Dell, P. M., N. W. Steubing, and J. W. Gray. 1973. "Optimization of gas field operation." *J. Pet. Technol.* **35**: 419–425.

Palke, M. R. and R. N. Horne. 1997a. "Determining the value of reservoir data by using nonlinear production optimization techniques." Paper SPE 38047 presented at the Society of Petroleum Engineers Asia Pacific Oil and Gas Conference Kuala Lumpur, April 14–17.

Palke, M. R. and R. N. Horne. 1997b. "Nonlinear optimization of well production considering gas lift and phase behavior." Paper SPE 37428 presented at the 1997 SPE Production Operations Symposium, Oklahoma City, OK, March 9–11.

Pan, Y. 1995. "Application of least squares and kriging in multivariate optimizations of field development scheduling and well placement design." MS report. Stanford, CA: Stanford University.

Pan, Y. and R. N. Horne. 1998. "Multivariate optimization of field development scheduling and well placement design." *J. Pet. Technol.* **50**: 83–85.

Ravindran, N. 1992. "Multivariate optimization of production systems—The time dimension." MS report. Stanford, CA: Stanford University.

Rogers, L. L. 1992. "Optimal groundwater remediation using artificial neural networks and the genetic algorithm." Ph.D. diss. Stanford, CA: Stanford University.

Rogers, L. L., F. U. Dowla, and V. M. Johnson. 1995. "Optimal field-scale groundwater remediation using neural networks and the genetic algorithm." *Environ. Sci. Technol.* **29**: 1145–1155.

Rowan, G. and J. E. Warren. 1967. "A systems approach to reservoir engineering, optimum development planning." *J. Can. Pet. Technol.* 84–94.

See, B. A. and R. N. Horne. 1983. "Improved application of linear programming optimization to nonlinear reservoir simulations using a least-squares determination of the influence functions." *SPE J.* **35**: 717–726.

Watkins, A. J. 1993. "Linking universal kriging and reservoir history matching." Paper SPE 19464 presented at the 1993 SPE Annual Technical Conference and Exhibition, Houston, October 3–6.

Wattenbarger, R. A. 1970. "Maximizing seasonal withdrawals from gas storage reservoirs." *J. Pet. Technol.* **22**: 994–998.

18.8.3 Natural Gas Pipeline Optimization

Roger Z. Ríos-Mercado

ABSTRACT In this article we introduce the reader to the problem of minimizing the fuel cost incurred by compressor stations driving the gas in steady-state pipeline networks. This is one of the problems that has received most attention from both practitioners and researchers by virtue of its tremendous economical impact. After presenting the basic mathematical framework and discussing the most important modeling assumptions, we establish the fundamental network topologies, discuss their importance in the selection of the solution technique, and present current solution methodologies based on dynamic programming. We conclude by highlighting the research potential of this exciting application area of operations research.

1. Introduction

A typical pipeline network for delivering natural gas requires a tremendous amount of fuel per day to operate the compressor stations driving the gas. Efficient design and operation of these complex networks can substantially reduce airborne emissions, increase safety, and decrease the often multimillion dollar daily operating costs.

Models describing gas flows in networks are in a reasonably well-developed descriptive stage. However, algorithms for actually finding good solutions relative to some cost criterion are not nearly as well developed up to now. From an optimization standpoint, the domain of feasible solutions prescribed by the models is mathematically complicated by virtue of the existence of both nonlinearity and nonconvexity. In addition, the related cost functions are typically nonlinear and nonconvex, and, in many cases, discontinuous, depending on the modeling assumptions. While classical nonlinear programming techniques (Chapter 5) can handle these complications on a practical basis for small pipeline networks, they are wholly inadequate for many of the large and growing pipeline networks in operation today.

We would also like to emphasize that despite the great amount of research on network flows by discrete and nonlinear optimizers within the operations research community, little work has been devoted by this community to optimization problems for compressor driven network flows.

In this article we introduce the reader to the problem of minimizing the fuel cost incurred by the compressor stations driving the gas in steady-state pipeline networks. This is one of the problems that has received most attention from both practitioners and researchers by virtue of its tremendous economical impact. Other application areas in gas networks where optimization techniques have been used include pipeline design (Edgar et al., 1978; Mariani et al., 1997; De Wolf and Smeers, 1996) and operations planning (Tomlin, 1988; Creegan and Monforte, 1989) to name a few.

In Section 2 we present a basic mathematical framework for the steady-state case, and discuss the most important modeling assumptions. In Section 3 we establish the fundamental network topologies and discuss their importance in the selection of the solution technique. In Section 4 we overview the current solution methodologies, and present a basic solution technique within a dynamic programming framework. We conclude in Section 5 with a summary of the direction of current research trends, putting in perspective the research potential of this exciting application area of operations research.

2. Model Formulation

2.1. *Gas Transmission Elements*

In modeling gas transmission networks, we consider three basic types of entities: pipelines and compressor stations, which are represented by arcs; and interconnection points, represented by nodes. Let $\mathcal{A}$ denote the set of arcs and $\mathcal{V}$ the set of nodes. We call $\mathcal{A}_p$ and $\mathcal{A}_c$ the set of pipeline and compressor (station) arcs, respectively. Note that $(\mathcal{A}_p, \mathcal{A}_c)$ is a partition of $\mathcal{A}$, that is, $\mathcal{A} = \mathcal{A}_p \cup \mathcal{A}_c$ and $\mathcal{A}_p \cap \mathcal{A}_c = \emptyset$. A node $i \in \mathcal{V}$ is used to represent the physical interconnection among the arcs. We also have different types of nodes. A receipt or supply node represents a gas production/storage facility. A delivery node represents a place where gas is to be taken out of the system. Let $\mathcal{V}_s$ and $\mathcal{V}_d$ be the set of supply and delivery nodes, respectively.

Likewise, each compressor arc $e = (i,j) \in \mathcal{A}_c$ defines another two types of nodes technically referred to as suction and discharge nodes (i and j, respectively). In this case, variable x_e represents the gas mass flow rate through arc $e \in \mathcal{A}$. In addition to the flow variables, there are pressure variables p_i that must be determined at every node $i \in \mathcal{V}$.

For a given compressor station $(i,j) \in \mathcal{A}_c$, there are three relevant decision variables: the mass flow rate through the station, the inlet (suction) pressure p_i at node i, and the outlet (discharge) pressure p_j at node j. In the compressor, these variables are further restricted by a set of constraints that depend on the operating attributes of the compressor. Let $\mathcal{D}_{ij}$ denote the set of feasible operating values for x_{ij}, p_i, and p_j within compressor $(i,j) \in \mathcal{A}_c$.

At each compressor station (i, j), the amount of fuel used g_{ij} is a function of x_{ij}, p_i, and p_j, that is $g_{ij} : \mathcal{D}_{ij} \rightarrow R$. In general, the nature of this function g_{ij} is very complex and depends also on considerations such as the number of compressor stations running within the compressor station, how the compressor units are configured (i.e., in series, in parallel, combination of both, etc.), physical properties of the compressor units themselves, and type of compressor unit. One of the most common configurations encountered in practice is that of compressor stations consisting of identical centrifugal compressor units operating in parallel. Even with this assumption, g_{ij} is in general nonlinear and nonconvex. The overall gas consumption is obtained by adding up all the g_e for each $e \in \mathcal{A}_c$.

For pipeline $(i,j) \in \mathcal{A}_p$, there is an equation that describes the relationship between the mass flow rate through the pipe and the pressure at the end points of the pipe. Let $\mathcal{P}_{ij}$ be the set of feasible values for mass flow rate and pressures in pipe (i, j). As will be illustrated later, $\mathcal{P}_{ij}$ is a nonconvex set.

The problem is to determine $(x, p) \in R^{|\mathcal{A}|+|\mathcal{V}|}$ (flow rates and pressures) that minimizes the total fuel consumption $g(x, p) = \sum_{(i,j)\in\mathcal{A}_c} g_{ij}(x_{ij}, p_i, p_j)$ subject to (a) mass balance constraints at each node, (b) compressor operating limits, (c) pipeline constraints, and (d) single lower, upper bounds for all the variables.

2.2. *The Basic Model*

In the development we make use of the following notation.

Indices and sets

i, j, k receipt, delivery, interconnection points; $i, j, k \in \mathcal{V} = \{1, 2, \ldots, n\}$

$\mathcal{A}_p$ set of pipeline arcs

$\mathcal{A}_c$ set of compressor station arcs

$\mathcal{A}$ set of all arcs; $\mathcal{A} = \mathcal{A}_p \cup \mathcal{A}_c$

$\delta^+(j)$ set of outgoing arcs from point j; $\delta^+(j) \subset \mathcal{A}$; $j \in \mathcal{V}$

$\delta^-(j)$ set of ingoing arcs into point j; $\delta^-(j) \subset \mathcal{A}$; $j \in \mathcal{V}$

$\mathcal{V}_S$ set of supply points; $\mathcal{V}_S \subset \mathcal{V}$

$\mathcal{V}_D$ set of delivery points; $\mathcal{V}_D \subset \mathcal{V}$

Input data

S_i supply limit at point i; $i \in \mathcal{V}_S$

D_i forecasted demand at point i; $i \in \mathcal{V}_D$

p_i^L, p_i^U lower, upper pressure bounds at node i; $i \in \mathcal{V}$

C_{ij} capacity of pipeline (i, j); $(i, j) \in \mathcal{A}$

Decision variables

x_{ij} flow rate on arc (i, j); $(i, j) \in \mathcal{A}$

p_i pressure at node i; $i \in \mathcal{V}$

The problem (P_0) has the general mathematical framework:

$$\text{Minimize } f(x,p) = \sum_{(i,j)\in\mathscr{A}_c} g_{ij}(x_{ij}, p_i, p_j) \tag{1a}$$

$$\text{subject to } \sum_{k\in\delta^+(j)} x_{jk} - \sum_{i\in\delta^-(j)} x_{ij} = b(j) \quad j\in\mathscr{V} \tag{1b}$$

$$x_{ij} \in [0, C_{ij}] \quad (i,j)\in\mathscr{A} \tag{1c}$$

$$(x_{ij}, p_i, p_j) \in \mathscr{D}_{ij} \quad (i,j)\in\mathscr{A}_c \tag{1d}$$

$$(x_{ij}, p_i, p_j) \in \mathscr{P}_{ij} \quad (i,j)\in\mathscr{A}_p \tag{1e}$$

$$p_i \in [p_i^L, p_i^U] \quad i\in\mathscr{V}, \tag{1f}$$

where

$$b(j) = \begin{cases} S_j & \text{if } j\in\mathscr{V}_S \\ -D_j & \text{if } j\in\mathscr{V}_D \\ 0 & \text{otherwise} \end{cases}$$

Constraints (1b)–(1c) are the classical network flow constraints. Note that the overall objective function is an aggregate of the fuel consumption at every compressor station. The difficulty of this problem is due to the nonconvexity of the objective function (1a), and the nonconvexity of both $\mathscr{D}_{ij}$ and $\mathscr{P}_{ij}$ [compressor station operating domain (1d) and pipeline domain (1e)]. In the following sections, a more detailed discussion of these formulation components will be given.

2.3. *Compressor Stations*

In general, compressor stations in gas pipeline networks are very complex entities because they may consist of several compressor units (typically fifteen to twenty) connected in a number of different configurations (e.g., in series, in parallel, combinations of both). In addition, there are different types of compressor units, two of the most important being centrifugal and reciprocating units. Of these two, it is the centrifugal type that is most commonly used in today's industry. Thus, it is no surprise that most of the research done up to date has focused on centrifugal units. For a detailed description of these and other types of compressor units, the reader is referred to Osiadacz (1987). In this article we present the reader with a modeling description based on centrifugal compressor units.

We recall there are three modeling variables involved in a compressor station (i, j), namely the mass flow rate, and the inlet and outlet pressures, x_{ij}, p_i, and p_j, respectively. However, the variables that are actually manipulated in the compressor are (i) compressor speed s, (ii) volumetric flow rate q, and (iii) adiabatic head h. These variables are related by the following equation

$$h/s^2 = \phi(q, s),$$

where ϕ is a polynomial function of the ratio q/s, typically of degree 3. There are also physical operational limits set on both on s and q, given by

$$s \in [s^L, s^U]$$

$$q \in [q^L, q^U].$$

A consequence of these is an operational limit on the ratio q/s technically referred as surge and stonewall for the lower and upper bound limit, respectively.

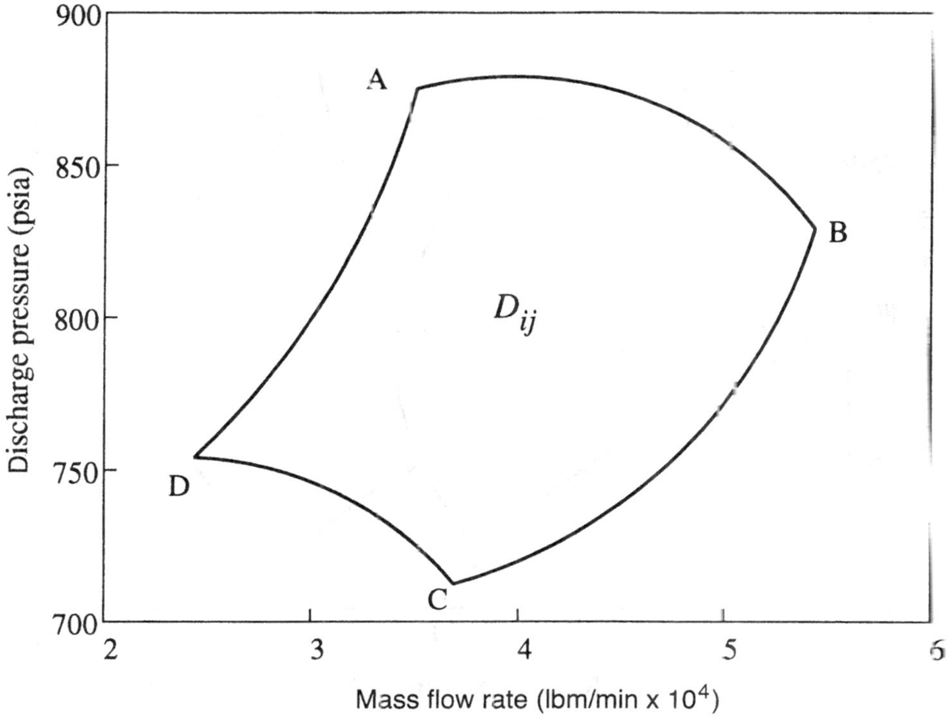

Figure 1 Feasible domain $\mathscr{D}_{ij}$ for a single unit compressor station

The (x_{ij}, p_i, p_j) and (s, q, h) spaces are related by the following transformation.

$$h = \beta[(p_j/p_i)^m - 1]$$

$$q = \beta(x_{ij}/p_i),$$

where β and m are parameters that depend on gas properties, with $m \in (0, 1)$. Hence, the feasible operating domain $\mathscr{D}^{\text{unit}}$ for a single centrifugal compressor unit in the (x_{ij}, p_i, p_j) space is given by $\mathscr{D}^{\text{unit}} = \{(x, p_i, p_j) : \text{constraints (2a)–(2c) below are satisfied}\}$.

$$(s^L)^2\phi(s^L, \beta x_{ij}/p_i) \leqslant \frac{\beta}{m}\left[\left(\frac{p_j}{p_i}\right)^m - 1\right] \leqslant (s^U)^2\phi(s^U, \beta x_{ij}/p_i) \tag{2a}$$

$$\left(\frac{s^U}{q^U}\right)^2 \phi(s^U, q^U)(\beta x_{ij}/p_i)^2 \leqslant \frac{\beta}{m}\left[\left(\frac{p_j}{p_i}\right)^m - 1\right] \leqslant \left(\frac{s^L}{q^L}\right)^2 \phi(s^L, q^L)(\beta x_{ij}/p_i)^2 \tag{2b}$$

$$p_i \in [p_i^L, p_i^U]. \tag{2c}$$

$\mathscr{D}^{\text{unit}}$ is nonconvex, as can be seen in Figure 1, which shows the feasible domain as a function of x_{ij} and p_j when p_i is fixed in a centrifugal compressor unit. In this figure, the function joining points A and D (and C and B) is convex, whereas the function joining points A and B (and C and D) is nonconvex.

Now, let us consider a compressor station with N identical units running in parallel. In this case, the suction and discharge pressures of the station are the same as those of the individual units in the station. If the units are identical, it is fair to assume that the mass flow rate x_{ij} through the station is equally divided among the individual units. Hence, if only one unit is running, then the feasible domain, denoted as $\mathscr{D}^1$, is the same as the feasible domain

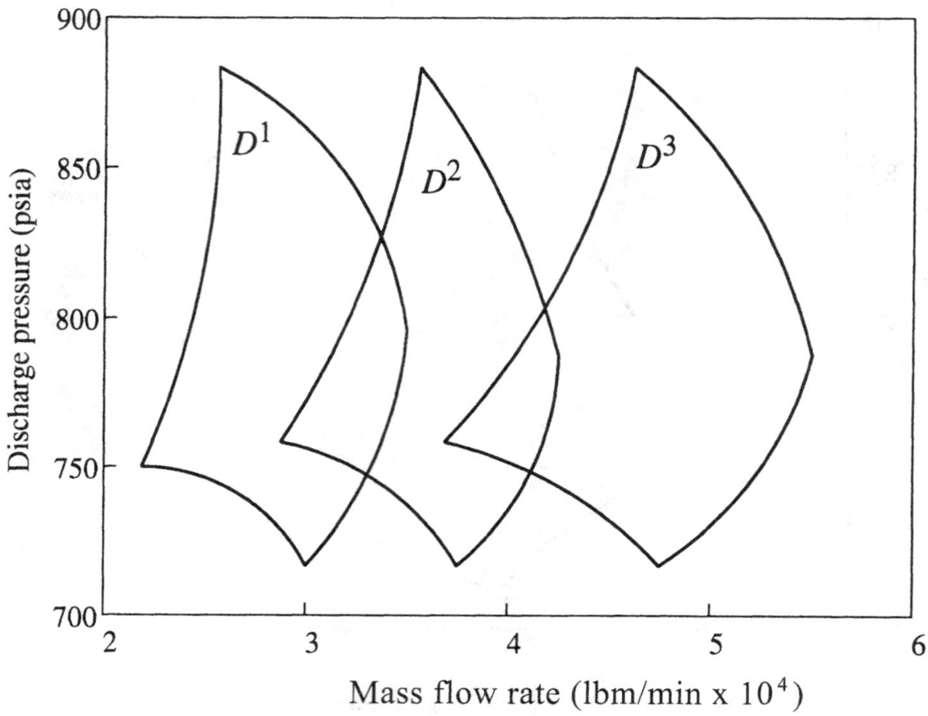

Figure 2 Profile of domain $\mathscr{D}$ for a three-unit station

$\mathscr{D}^{\text{unit}}$ for a single unit described above. When r units are selected to run, $1 \leqslant r \leqslant N$, then the feasible domain, denoted as $\mathscr{D}^r$, is given by

$$\mathscr{D}^r = \{(v, p_s, p_d) : (v/r, p_s, p_d) \in \mathscr{D}^1\}. \tag{3}$$

As we can see, if r is selected beforehand, $\mathscr{D}^r$ can be seen as extending domain $\mathscr{D}^1$ r times in the direction of x_{ij} and is essentially the same type of domain. However, r can be another decision variable. If this is the case, and we assume that at least one unit must be running, the feasible domain $\mathscr{D}$ is thus given by

$$\mathscr{D} = \bigcup_{r=1}^{N} \mathscr{D}^r. \tag{4}$$

The (x_{ij}, p_j) profile of domain $\mathscr{D}$ when p_i is fixed in a station with three identical units is shown in Figure 2.

2.4. *Pipeline Equations*

The steady-state relationship between pressure and mass flow rate along the pipe can be described by different equations depending on the working pressure of the network. Basically, three types of equations are considered (for low-, medium-, and high-pressure networks). For isothermal models (e.g., gas temperature assumed constant throughout the pipe), the general formula takes the following form

$$\phi(\Delta p) = K_{ij} x_{ij}^2, \tag{5}$$

where ϕ is the pressure function, and K_{ij} is the pipe parameter which depends on the physical properties of the pipeline. For low-pressure networks, we have $\phi(\Delta p) = p_i - p_j$, $K_{ij} = 11.7 \times 10^3 L/D^5$ (Lacey's equation). For medium-pressure networks we have $\phi(\Delta p) = p_i^2 - p_j^2$,

$K_{ij} = 27.24L/E^2D^5$ (Polyflo equation). For high-pressure networks we have $\phi(\Delta p) = p_i^2 - p_j^2$, $K_{ij} = 18.43L/E^2D^5$ (Panhandle equation). Here, the physical properties L (length of pipe), D (inside pipe diameter), and E (efficiency factor) are assumed to be known with certainty.

Note that in some instances, the direction of the flow through a given pipeline may not be known in advance and could go either way [this is represented by an undirected edge (i, j)]. If this is the case, the undirected edge e is replaced by two directed arcs e^+ and e^- and equation (5) becomes:

$$\phi(\Delta p) = K_{ij}(x_{ij}^+ + x_{ij}^-)(x_{ij}^+ - x_{ij}^-). \tag{6}$$

When doing this, a constraint $(x_{ij}^+)(x_{ij}^-) = 0$ must be added since the objective function is nonconvex. Note that there is only one equation (6) per pipeline. For nonisothermal models, it is still possible to derive a pipe-leg equation describing the relationship between mass flow rate and pressure drops. This equation, though, is more complex and includes gas temperature as a variable.

2.5. *Objective Function*

The fuel consumption in a single centrifugal compressor unit $(i, j) \in \mathscr{A}_c$ is governed by the following equation:

$$g_{ij}^{\text{unit}} = \gamma \frac{x_{ij}}{\eta(x_{ij}, p_i, p_j)} \frac{\alpha}{m} \left[\left(\frac{p_j}{p_i} \right)^m - 1 \right],$$

where γ is a parameter, and η is the adiabatic efficiency function. In general, η is a complex expression, not implicitly defined. To make one function evaluation of η, one must solve a linear system of equations. In practice, though, one employs polynomial approximation functions that fit the function relatively well and are simpler to evaluate. In other cases, when the fluctuations of η are small enough, η can be assumed to be a constant, leaving us with the function:

$$g_{ij}^{\text{unit}} = \alpha' x_{ij} \left[\left(\frac{p_j}{p_i} \right)^m - 1 \right], \tag{7}$$

which is not polynomial, but easy to evaluate ($\alpha' = (\eta\alpha)/m$ is a constant). This function is nonconvex.

For the case of r identical units connected in parallel, the flow entering the compressor station is equally divided among the units, and its fuel consumption can be described with the following mixed-integer nonlinear function:

$$g_{ij}(x_{ij}, p_i, p_j, r) = r g_{ij}^{\text{unit}}(x_{ij}/r, p_i, p_j).$$

3. Network Topologies

There are two fundamental classes of gas network topologies: nonlooped (or tree) networks and looped networks. For this classification, rather than considering the original topology, an associated network is formed in the following form:

1. By temporarily removing all compressor arcs from the original network, we are left with a set of connected components or subnetworks each consisting of nodes and pipeline arcs.
2. Each of these subnetworks is merged into one single node.
3. The compressor arcs are restored and the result is the associated reduced network.

This is illustrated in Figure 3. The solid lines represent original arcs (pipelines and compressor stations). In the reduced network, each subnetwork shown within the dotted line is viewed as a node. Here, the reduced network is a tree with 4 nodes and 3 arcs. We call a network a nonlooped network if, and only if, the undirected version of its corresponding reduced graph has no cycles. Otherwise, we say the network is a looped network. A special

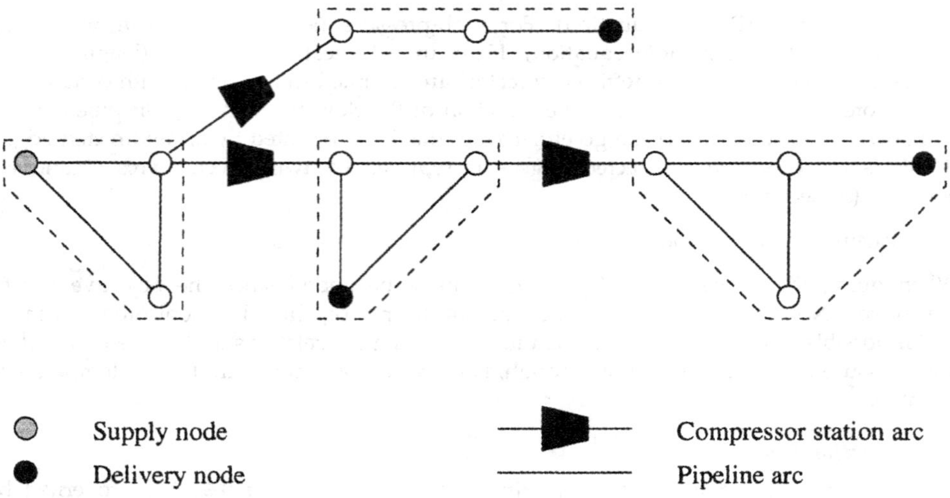

Figure 3 A pipeline network and its reduced graph

case of a nonlooped network where the associated reduced network is a simple path is called a serial (or gun-barrel) system.

It is important to realize that looped networks pose a higher degree of difficulty than nonlooped networks. First, for nonlooped networks it can be shown that if the parameter $b(i)$ in the mass balance constraints (1b) is specified for every node i in the network (which is usually the case), then the flow variables can indeed be uniquely determined for each arc in the network, even if some of the subnetworks have cycles. To prove this result one uses the mass balance constraints in combination with the pipeline constraints to derive a system of equations with a unique solution for the flow variables. The proof can be found in Ríos-Mercado et al. (2000). Note that this result is unique for this type of gas network and in general does not apply to other network flow problems.

Hence, for nonlooped networks, it is possible to fix the flow variables and remove them from the problem, reducing its size significantly. This has a direct impact on the selection of the solution procedure. For instance, for nonlooped networks, if the continuous pressure variables are discretized, dynamic programming can be applied to find the optimal pressures. However, when considering looped networks, the flow variables cannot be uniquely determined, and the use of dynamic programming now has to deal with the issue of multi-dimensionality. More will be said about optimization algorithms in Section 4.

4. Optimization Algorithms

Dynamic programming (DP) has been by far the most popular technique for solving many classes of natural gas pipeline networks since the late 1960s. One of the main reasons is that, in a DP framework, it is relatively straightforward to satisfy the pipeline constraints and to handle the nonconvexity of the feasible domain. Today, DP can be successfully applied to optimally solve very large nonlooped networks, and some subclasses of looped systems.

DP approaches include that of Wong and Larson (1968), applied to gun-barrel networks; Lall and Percell (1990), applied to tree networks; and, more recently, Carter (1998), applied to looped networks. Other approaches based on local optima techniques (Percell and Ryan, 1987) and hierarchical methods (Anglard and David, 1988; Luongo et al., 1989; Osiadacz, 1994) have been applied with modest success. These methods deliver local optimum solutions, which makes them unattractive for this particular type of problem given the typical nonconvex behavior of the objective function and its large number of local optima. We must point out, though, that many of the methodologies in the operations research repertoire have yet to be explored. See Carter (1998) for a survey.

4.1. Dynamic Programming on Nonlooped Networks

We assume the reader is familiar with the basic concepts of DP (Chapter 3.4). As we have discussed, in nonlooped systems it is possible to determine unique values for mass flow rate variables so they can be eliminated from further consideration, leaving us with the problem of determining the optimal set of pressures. Let $\bar{x}$ denote this unique set of mass flow rates. Let us also assume throughout this section that we are dealing with high-pressure networks, that is pipeline constraints (5) are given by

$$p_i^2 - p_j^2 = K_{ij} x_{ij}^2 \quad (i,j) \in \mathcal{A}_p. \tag{8}$$

Furthermore, we assume without loss of generality that compressor station arcs are joined by a single pipeline arc. Note that this is possible when the mass flow rates are fixed due to the presence of the pipeline constraints (8). To see this, consider compressor arcs (i_s, i_d) and (j_s, j_d) (subindex s and d stand for suction and discharge nodes, respectively) and suppose that between these two there is a subnetwork $G' = (\mathcal{V}', \mathcal{A}')$. Let P be a path (of arcs) in $\mathcal{A}'$ from node i_d to j_s. Then, by adding up each pipeline equation corresponding to arcs in P, we obtain $p_{i_d}^2 - p_{j_s}^2 = K'_{i_d, j_s}$, where $K'_{i_d, j_s} = \sum_{(k,l) \in P} K_{kl} \bar{x}_{kl}$ is a constant.

SERIAL SYSTEMS

We now define the DP formulation for a serial system consisting of T compressor stations. We identify each compressor station (i, j) with a stage t, $t = 1, 2, ..., T$. Boundary stages 0 and $T+1$ are associated with single nodes 1 and n, respectively. For each stage (see Figure 4), we define:

y_t state variable; $y_t \equiv p_i$ (suction pressure), $t = 1, ..., T+1$

u_t decision (control) variable; $u_t \equiv p_j$ (discharge pressure), $t = 0, ..., T$

$U_t(y_t)$ feasible domain; $U_t(y_t) = \{u:(\bar{x}_{ij}, y_t, u)$ satisfies (1d)$\}$, $t = 1, ..., T$, and $U_0(y_0) = \{u: p_1^L \leqslant u \leqslant p_1^U\}$

$f_t(u_t)$ state transition function; from equation (8), $y_{t+1} = f_t(u_t) = (u_t^2 - K_{jk} \bar{x}_{jk}^2)^{1/2}$, $t = 0, ..., T$, where (j, k) is the pipeline arc succeeding compressor (i, j)

$g_t(y_t, u_t)$ cost function; from equation (7), (for a single unit) given by $g_t(y_t, u_t) = g_{ij}^{\text{unit}}(\bar{x}_{ij}, y_t, u_t)$

The backward recurrence relationship is given by

$$J_t(y_t) = \min_{u_t \in U_t(y_t)} \{g_t(y_t, u_t) - J_{t+1}(y_{t+1})\}, \quad t = 0, ..., T,$$

where $J_t(y_t)$ is the minimum fuel cost from station t to T when the suction pressure at stage t is y_t. The boundary condition is given by $J_{T+1}(y_{T+1}) = 0$, and the solution by $J_0(y_0)$, where

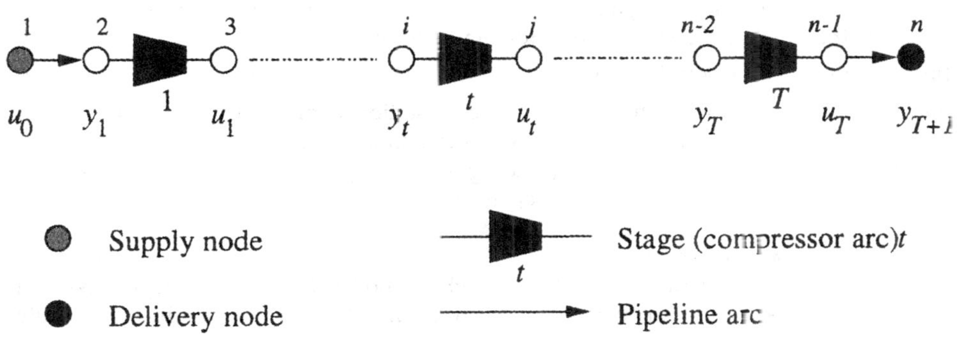

Figure 4 DP formulation for a serial system

$g(y_0, u_0) = 0$ for all (y_0, u_0). Note that for given y_t, the feasible domain $U_t(y_t)$ is merely a closed interval for u, so the nonconvexity issue is easily handled in DP.

The application of the recurrence relationship is straightforward. A discharge pressure grid is constructed at each stage by discretizing over its feasible range. The recurrence relationship is thus applied to a finite number of discharge pressures. The optimal discharge pressure at each stage is obtained by applying the recurrence relationship backwards, that is, by first analyzing stage T, then $T-1$, and so on, until all stages are included.

It is also possible to derive a forward formulation for a serial system. To this end, we define the following entities at each compressor (i, j).

- z_t — state variable; $z_t \equiv p_j$ (discharge pressure), $t = 0, ..., T$
- v_t — decision (control) variable; $v_t \equiv p_i$ (suction pressure), $t = 1, ..., T+1$
- $V_t(z_t)$ — feasible domain; $V_t(z_t) = \{v : (\bar{x}_{ij}, v, z_t) \text{ satisfies (1d)}\}$, $t = 1, ..., T$, and $V_{T+1}(z_{T+1}) = \{v : p_n^L \leqslant v \leqslant p_n^U\}$
- $f_t(u_t)$ — state transition function; from equation (8), given by $z_{t-1} = f_t(v_t) = (v_t^2 + K_{ki}\bar{x}_{ki}^2)^{1/2}$, $t = T+1, ..., 1$, where (k, i) is the pipeline arc preceding compressor (i, j)
- $h_t(v_t, z_t)$ — cost function; from equation (7), (for a single unit) given by $h_t(v_t, z_t) = g_{ij}^{\text{unit}}(\bar{x}_{ij}, v_t, z_t)$

The forward recurrence relationship is thus given by

$$J_t(z_t) = \min_{v_t \in V_t(z_t)} \{h_t(v_t, z_t) + J_{t-1}(z_{t-1})\}, \quad t = 1, ..., T+1,$$

where $J_t(z_t)$ is the minimum fuel cost from station 1 to t when the discharge pressure at stage t is z_t. The boundary condition is given by $J_0(z_0) = 0$, and the solution by $J_{T+1}(z_{T+1})$, where $h(v_{T+1}, z_{T+1}) = 0$ for all (v_{T+1}, z_{T+1}).

Diverging Branch Systems

Diverging branch systems (i.e., networks whose nodes have all indegree equal to one, except for a single node (root) that has indegree equal to zero) can be solved with relatively the same amount of effort as the one required for serial systems. Let node 1 denote the root node and r the corresponding root station. The DP formulation is modified as follows:

(i) Let $J_t(y_t)$ denote the minimum fuel cost of the subnetwork rooted at station t when the suction pressure is y_t; $t = 1, ..., T$.

(ii) The (backward) recurrence relationship is given by

$$J_t(y_t) = \min_{u_t \in U_t(y_t)} \left\{ g_t(y_t, u_t) + \sum_{w \in \mathscr{C}_t} J_w(y_w) \right\}, \quad t = 1, ..., T$$

where $\mathscr{C}_t$ is the set of compressor stations succeeding t. Note that for each $w \in \mathscr{C}_t$, the transition function $y_w = f_t(u_t)$ is accordingly defined by $f_t(u_t) = (u_t^2 - K_{jk}\bar{x}_{jk}^2)^{1/2}$, where (j, k) is the pipeline arc joining compressors t and w.

(iii) For given y_r, the solution is given by $J_r(y_r)$.

Note that if t is a station in the boundary (i.e., a station with no succeeding stations) then we have $\mathscr{C}_t = \varnothing$. The stages must be carefully labeled so as to ensure that a stage with diverging branches is computed after its corresponding succeeding stages. Figure 5 shows a diverging stage with three branches, where $C_t = \{w_1, w_2, w_3\}$.

Converging Branch Systems

Converging branch systems (i.e., networks whose nodes have all outdegree equal to one, except for a single node (root) that has outdegree equal to zero) are handled essentially in the same way as diverging branch systems. In this case, a forward recurrence formulation is employed.

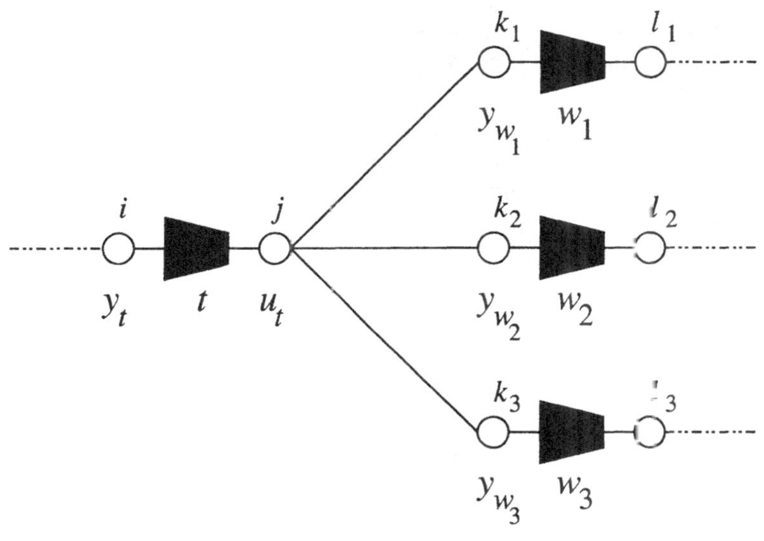

Figure 5 DP formulation for a diverging system

General Tree Systems

In the case of general tree structures with multiple converging and diverging branches, the DP formulation makes use of both backward and forward recursion to deal with diverging and converging branches, respectively. To help understand the formulation, let us introduce some notation. Let a be any (source) node with zero indegree and let b be any node with zero outdegree (reachable from a). The path from a to b is called the main path or main stream of the network. The path merely acts as a reference and its selection is arbitrary. We then proceed to label all paths in the remaining branches as follows. Starting with a leave node (i.e., either indegree or outdegree equal to one), which does not belong to the main path, we form a path by iteratively adding adjacent nodes whose indegree and outdegree are both equal to one (i.e., right before we reach a branching node). We call this path (along with its respective nodes and stations contained in it) diverging if the leave node where we started has indegree equal to one; otherwise, we call it converging. We remove the path from the network (just for labeling purposes) and repeat all over again until all nodes in the network have been labeled as either diverging or converging. As a convention, the main path is referred to as diverging.

We now introduce the supported subnetwork concept for each stage t. Let (i, j) be the compressor arc associated with compressor t and let $G = (V, E)$ be the undirected version of the network. The subnetwork supported by t, G_t, is the subnetwork (containing t) that remains after the edges in E in the a–i or a–j (whichever is shorter) path are removed. This is illustrated in Figure 6, where each of the supported subnetworks is shown within dotted frames. Consider $t = 4$, for instance. A path from a to 4 is formed by arcs 1 and C. By removing these arcs, G_4 is the connected component containing station 4.

We now give a DP formulation for the entire network based on backward recursion for the main and diverging paths and forward recursion for converging paths. These are the fundamental components:

(i) Let $J_t(\alpha_t)$ denote the minimum fuel cost of the network supported by t when the state variable at stage t is α_t; $t = 1, \ldots, T$. If t is a diverging stage, $\alpha_t = y_t$ refers to the suction pressure at t. Otherwise, $\alpha_t = z_t$ refers to the discharge pressure at t.

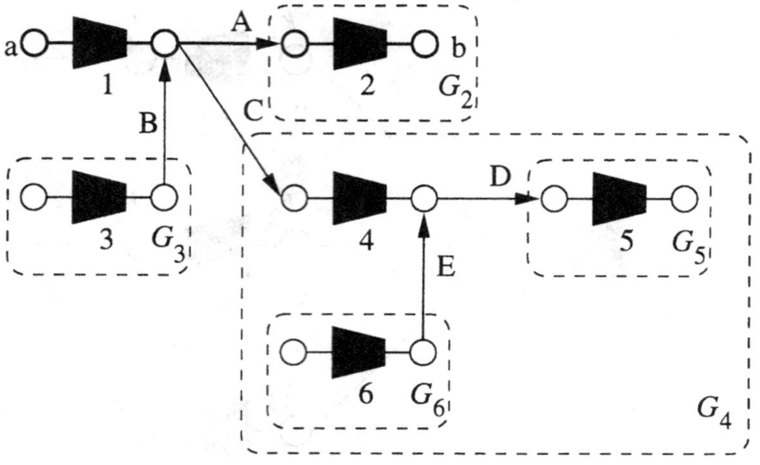

Figure 6 A network and its supported subnetworks

(ii) The recurrence relationship is thus given by

$$J_t(y_t) = \min_{u_t \in U_t(y_t)} \left\{ g_t(y_t, u_t) + \sum_{w \in \mathscr{C}_t^{\mathrm{div}}} J_w(y_w) + \sum_{w \in \mathscr{C}_t^{\mathrm{conv}}} J_w(z_w) \right\}, \text{if } t \text{ is diverging,}$$

and

$$J_t(z_t) = \min_{v_t \in V_t(z_t)} \left\{ h_t(v_t, z_t) + \sum_{w \in \mathscr{C}_t^{\mathrm{div}}} J_w(y_w) + \sum_{w \in \mathscr{C}_t^{\mathrm{conv}}} J_w(z_w) \right\}, \text{if } t \text{ is converging,}$$

where $\mathscr{C}_t^{\mathrm{div}}$ ($\mathscr{C}_t^{\mathrm{conv}}$) is the set of all compressor stations in G_t succeeding (preceding) t. Note that for $w \in \mathscr{C}_t^{\mathrm{div}}$, the transition function $y_w = f_t(u_t)$ is defined by $f_t(u_t) = (u_t^2 - K_{jk}\bar{x}_{jk}^2)^{1/2}$ [pipeline arc (j, k) succeeding compressor (i, j)], whereas for $w \in \mathscr{C}_t^{\mathrm{conv}}$, the transition function is $z_w = f_t(v_t) = (v_t^2 + K_{ki}\bar{x}_{ki}^2)^{1/2}$ [pipeline arc (k, i) preceding (i, j)].

(iii) For given y_a, the solution is given by $J_a(y_a)$, where a is the preselected source node.

Again, the stages are carefully labeled so as to ensure that, a stage with branches is computed after its corresponding stages in the subnetwork supported by it.

4.2. Dynamic Programming on Looped Networks

As mentioned in Section 3, looped networks are more difficult to handle as we must deal with both pressure and flow variables. One way to simplify this task is to assume the flow variables are known. This is a very common assumption from the practical standpoint. By doing this, we avoid the multidimensionality issue in DP; however, DP is applied in a slightly different way. Rather than attempting to formulate DP recursively, we proceed to combine simple elements of the system into an equivalent composite element. This process is applied iteratively until there is exactly one element left, which completely characterizes the behavior of the entire system. At a given iteration, the elements can be selected from any position in the system, so this is referred as nonserial dynamic programming (NDP) (Bertelè and Brioschi, 1972). This technique has been successfully applied to solve large instances with complex looped network topologies in reasonable times. The details of the implementation of NDP for looped networks can be found in Carter (1998).

5. Conclusion

The research potential in the natural gas industry from the optimization perspective is tremendous. In this article, we have presented only one of the problems in the field that has received a significant amount of attention over the past thirty years. We have seen how DP can be used to solve (within a given pressure discretization) nonlooped network structures and how NDP is an attractive technique for addressing looped systems (for the particular modeling assumptions). However, we feel many of the important operations research techniques have not yet been used at full capacity on this type of problem. For instance, in addition to DP, there are global optimization methods (Chapter 6) that can be used to attempt to solve the problem optimally. In that direction, we have seen some preliminary work consisting of development of lower bounding schemes (Wu et al., 2000) that can be used within these methodologies.

There is also an area of opportunity for discrete optimization methods, such as metaheuristics (Chapter 3.6), to be applied to this continuous optimization problem. One can devise discrete neighborhoods in such a way that it is computationally efficient to generate feasible neighbors and then use local search techniques such as tabu search (Chapter 3.6.7), for instance, to try to overcome local optimality.

Likewise, it is an entire open area of research to investigate other variations of this problem.

- Other compressor station configurations such as reciprocating units, nonidentical compressor units, nonserial units.
- Nonisothermal models, which would involve considering gas temperature as a system variable.
- Medium- and low-pressure networks, which would lead to different pipeline equations (5).
- Nonsteady models, which implies considering analyzing the system as a function of time and undoubtedly stands as one of the most important challenges.

It is expected that within the next few years, more researchers with optimization background will become involved in helping address these issues.

References

Anglard, P. and P. David. 1988. "Hierarchical steady state optimization of very large gas pipelines." In *Proceedings of the 20th PSIG Annual Meeting*. Toronto. Available from the Pipeline Simulation Interest Group Web site: `http://www.psig.org/`.

Bertelè, U. and F. Brioschi. 1972. In *Nonserial Dynamic Programming*. New York: Academic Press. This book is out of print, but available at many university libraries.

Carter, R. G. 1998. "Pipeline optimization: dynamic programming after 30 years." In *Proceedings of the 30th PSIG Annual Meeting*. Denver. Available from the Pipeline Simulation Interest Group Web site: `http://www.psig.org/`.

Creegan, J. B. and F. A. Monforte. 1989. "Modeling nonconvexities arising in a gas distribution planning problem with special ordered sets." In *Proceedings of the 21st PSIG Annual Meeting*. El Paso. Available from the Pipeline Simulation Interest Group Web site: `http://www.psig.org/`.

De Wolf, D. and Y. Smeers. 1996. "Optimal dimensioning of pipe networks with application to gas transmission networks." *Oper. Res.* **44(4)**: 596–608.

Edgar, T. F., D. M. Himmelblau, and T. C. Bickel. 1978. "Optimal design of gas transmission networks." *SPE J.* **18**: 96–104.

Lall, H. S. and P. B. Percell. 1990. "A dynamic programming based gas pipeline optimizer." In *Analysis and Optimization of Systems*, edited by A. Bensoussan and J. L. Lions. Lecture Notes in Control and Information Sciences **144**: 123–132. Berlin: Springer-Verlag.

Luongo, C. A., B. J. Gilmour, and D. W. Schroeder. 1989. "Optimization in natural gas transmission networks: a tool to improve operational efficiency." Presented at the 3rd SIAM Conference in Optimization, Boston.

Mariani, O., F. Ancillai, and E. Donati. 1997. "Design of a pipeline: optimal configuration." In *Proceedings of the 29th PSIG Annual Meeting*. Tucson. Available from the Pipeline Simulation Interest Group Web site: `http://www.psig.org/`.

Osiadacz, A. J. 1987. In *Simulation and Analysis of Gas Networks*. Houston, TX: Gulf Publishing Company. This book is out of print, but available at many university libraries.

Osiadacz, A. J. 1994. "Dynamic optimization of high pressure gas networks using hierarchical systems theory." In *Proceedings of the 26th PSIG Annual Meeting*. San Diego. Available from the Pipeline Simulation Interest Group Web site: `http://www.psig.org/`.

Percell, P. B. and M. J. Ryan. 1987. "Steady-state optimization of gas pipeline network operation." In *Proceedings of the 19th PSIG Annual Meeting*. Tulsa. Available from the Pipeline Simulation Interest Group Web site: `http://www.psig.org/`.

Ríos-Mercado, R. Z., S. Wu, L. R. Scott, and E. A. Boyd. 2000. Preprocessing on natural gas transmission networks. Technical Report PISIS-2000-01. San Nicolás de los Garza, México: Graduate Program in Systems Engineering, Universidad Autónoma de Nuevo León. Available at author's Web site: `http://yalma.fime.uanl.mx/~roger/`.

Tomlin, J. A. 1988. "Special ordered sets and an application to gas supply operations planning." *Math. Programming* **42(1)**: 69–84.

Wong, P. J. and R. E. Larson. 1968. "Optimization of natural-gas pipeline systems via dynamic programming." *IEEE Trans. Automat. Control* **13(5)**: 475–481.

Wu, S., R. Z. Ríos-Mercado, E. A. Boyd, and L. R. Scott. 2000. "Model relaxations for the fuel cost minimization of steady-state gas pipeline networks." *Mathematical and Computer Modelling* **31(2–3)**: 197–220. Available at author's Web site: `http://yalma.fime.uanl.mx/~roger/`.

18.9 Optimization of Telecommunications Networks

G. Anandalingam

ABSTRACT We provide an exposition of the application of optimization for solving telecommunication network design problems. The design of large-scale telecommunications networks involves making decisions on the type of network, communication network architecture, switching, routing, network control and monitoring, as well as the design of the network. We provide the mathematical programming formulation of the optimal design problem for access, backbone, and overall telecommunications networks. We then give a sense of the literature in each of the subproblems and also describe some heuristics for solving these NP-hard problems.

1. Introduction

In this article, we will provide an exposition of the application of optimization for solving telecommunication network design problems. Telecommunications networks have become ubiquitions in public and private organizations all over the world. Organizations are comprised of entities that need to communicate with each other. These entities (or stations) can be computer workstations, databases, facsimile machines, local area networks, or other entities. Stations will communicate over links, which can be various types of cables suspended in the air or buried underground, or the links can be microwave or satellite transmission.

The design of large-scale telecommunications networks involves making decisions on the following subjects:

- type of network: centralized or distributed;
- type of communication network architecture;
- type of switching: circuit switching or packet switching;

Given	— node locations
	— peak traffic demand
Obtain	— topology
	— link capacities
	— routing policy
To minimize	— total network cost
Subject to	— link capacity constraint
	— node capacity constraint
	— delay constraint
	— survivability constraint

Figure 1 General Network Formulation

- type of routing: static or dynamic;
- type of network control and monitoring: centralized or distributed;
- design of the network: network resources and capacity.

We will provide the mathematical programming formulation of the optimal design problem for access, backbone, and overall telecommunications networks. We will then give a sense of the literature in each of the subproblems and also describe some heuristics for solving these NP hard problems.

For the general telecommunications network design problem, there are quite a few issues to consider. The design problem can be formulated in many different ways depending on the given scenario. One general formulation is shown in Figure 1, which is quite complex and large. The problem is frequently split into two or more pieces. One problem can be topology design with no other considerations. Assigning link capacities could be a different problem. Still another problem is the routing of traffic on the network. All three problems are dependent on each other. A truly optimal design may be achieved if all three problems could be optimized at one time.

The topological design consists of two subproblems: the backbone network design, and the local access network design.

The backbone network consists of switches (backbone nodes) interconnected by high-speed trunks. The backbone network design problem is stated as follows.

Given:

- potential sites for backbone node location,

- traffic data, and
- available link types and their cost,

obtain a least-cost local access network configuration; that is,

- optimum number of backbone nodes and their locations,
- optimum set of links interconnecting the backbone nodes and their capacities,
- routing paths,

subject to constraints on:

- delay,
- throughput,
- reliability, and
- link capacity.

The local access network is the network that connects the user terminals to the backbone network nodes either directly or through local access switches. The local access network design problem is stated as follows:

Given:

- user (terminal) locations,
- potential sites for concentrator or access switch location,
- traffic data, and
- available link types and their cost,

obtain a least local access network configuration; that is,

- optimum number of local access switches and their locations,
- assignment of access switches to backbone network nodes,
- assignment of each terminal to a local access switch,
- type and capacity of switches and communication links in the access network,

subject to constraints on

- link capacity, and
- concentrator capacity.

In the exposition that follows we will provide specific formulation of optimization models used for telecommunications network design. We do not claim that these are the only useful models, or even that these models account for every nuance of optimizing telecommunications networks. We also do not claim that the literature review is complete and exhaustive. We apologize in advance for those fine papers that have been left out. Finally, it should be noted that the optimization heuristics described are for illustrative purposes only, and not meant to suggest that they are exclusive in terms of performance.

2. Designing Telecommunication Networks

2.1. Overview

In this section, we will provide a formulation of the overall network design problem, which integrates access and backbone network topologies and also provisions capacity for the entire network. Large-scale networks are interconnected by appropriate internetworking devices. The need for devices arises in two areas, one that requires complex routing and another that provides simple connectivity at fast speeds. Specifically, a backbone network requires a device that can provide all the necessary processing and routing for internetwork traffic from the networks it connects. Here the different protocols and different speeds are relevant issues in providing efficient internetworks. However, in an access network, simple routing is usually sufficient, since the internetworking device only needs to filter intratraffic and provide the simplest connection.

2.2. *Model Formulation*

In order to formulate the problem of designing large-scale private telecommunications networks, we need to use the following definitions:

Let, $G(X) = [S, A(X)]$ model the overall network, where S is the set of all nodes, $A(X)$ is the set of all arcs, and X is the set of links available; $GR(X) = [SR, AR(X)]$ model the access network, where SR is the set of all access nodes, and $AR(X)$ is the set of all access arcs; and $GB(X) = [SB, AB(X)]$ model the backbone network, where SB is the set of all backbone nodes, and $AB(X)$ is the set of all backbone arcs. Note that $(SR \cup SB) = S : SR, SB \subset S$, and $AB(X) + AR(X) = A(X)$. od are the origin (o) and destination (d) nodes; od_r is the od pair within the access network (internetwork traffic), and od_i is the od pair, where origin and destination nodes are not in the same network and must cross the backbone network (internetwork traffic). There are no od pairs between nodes in the backbone network.

The following assumptions are made:

- There are only two types of internetworking devices, remote devices (bridges or remote routers) in the access network and backbone devices (routers) in the backbone network.
- The buffer size at all device sites are large enough so that packets will not be dropped.
- All nodes have the potential to be device node locations. A device node is either a remote device or a backbone device location. Therefore, we provide flexibility in the backbone location.
- No overhead traffic is created by address tables. Traffic on the network is only generated by traffic flow from each node (for an od pair).
- Packets are served on a first-come-first-served basis. There are no priority packets.
- There is no propagation delay.

The model formulation is a mixed-integer program.

Let,

$u, v \in S,$

$i, j, p, q \in SR,$

$k, l \in SB,$

$c1_{ij}$ = cost of having a link from node i to node j in the access network,

$c2_{kl}$ = cost of having a link from node k to node l in the backbone network,

$$X_{ij} = \begin{cases} 1 & \text{if there is a link from access node } i \text{ to remote node } j \\ 0 & \text{otherwise} \end{cases}$$

$$X_{ik} = \begin{cases} 1 & \text{if there is a link from remote node } i \text{ to backbone node } k \\ 0 & \text{otherwise} \end{cases}$$

$$X_{kl} = \begin{cases} 1 & \text{if there is a link from backbone node } k \text{ to backbone node } l \\ 0 & \text{otherwise} \end{cases}$$

$$Y_i = \begin{cases} 1 & \text{if there is a remote device at node } i \\ 0 & \text{otherwise} \end{cases}$$

$$Z_k = \begin{cases} 1 & \text{if there is a backbone device at node } k \\ 0 & \text{otherwise} \end{cases}$$

CR_i = cost of having a remote device at node i which is known

CB_k = cost of having a backbone device at node k which is known

where

IR is the maximum number of input ports the remote device can support and is equal to the number of output ports,

IB is the maximum number of input ports the backbone device can support and is equal to the number of output ports,

FR is the filter rate in *pps* of remote device,

TR is the transfer (forwarding) rate in *pps* of the remote device,

FB is the filter rate in *pps* of backbone device,

TB is the transfer (forwarding) rate in *pps* of the backbone device,

Let

Z denote the total cost that we are trying to minimize,

F_{ij}^{pq} be flows on links from node i to j created by origin p for destination q,

T^{pq} denote the traffic flow generated by origin p to destination q, which is known,

LM_{uv} denote the maximum link capacity on link from node u to node v, which is known,

$$(P1) \quad \min Z = \sum_{\substack{i,j \in SR \\ i \neq j}} c1_{ij} X_{ij} + \sum_{\substack{k,l \in SB \\ k<l}} c2_{kl} X_{kl} + \sum_{i \in SR} CR_i * Y_i + \sum_{k \in SB} CB_k * Z_k \tag{1}$$

Subject to

$$\sum_{\substack{j \in SR \\ i \neq j}} X_{ij} + Y_i = 1 \quad \forall i \in SR \tag{2}$$

$$X_{ij} - Y_j \leqslant 0 \quad \forall i,j \in SR, i \neq j \tag{3}$$

$$\sum_{\substack{i \in SR \\ i \neq j}} X_{ij} - Y_j \geqslant 0 \quad \forall j \in SR \tag{4}$$

$$\sum_{\substack{i \in SR \\ i \neq j}} X_{ij} - IR * Y_j \leqslant 0 \quad \forall j \in SR \tag{5}$$

$$\sum_{k \in SB} X_{ik} - 2Y_i \geqslant 0 \quad \forall i \in SR \tag{6}$$

$$\sum_{\substack{l \in SB \\ k \neq l}} X_{kl} \geqslant 2 \quad \forall k \in SB \tag{7}$$

$$\sum_{\substack{l \in SB \\ k \neq l}} X_{kl} \leqslant IB \quad \forall k \in SB \tag{8}$$

$$X_{kl} - X_{lk} = 0 \quad \forall k,l \in SB, k < l \tag{9}$$

$$\sum_{\substack{l \in SB \\ k \neq l}} F_{kl}^{ij} - F_{lk}^{ij} = 0 \quad \forall k \in SB, i,j \in SR, i < j \tag{10}$$

$$\sum_{\substack{p,q \in SR \\ p \neq q}} F_{uv}^{pq} \leqslant LM_{uv} X_{uv} \quad \forall u, v \in S, u < v \tag{11}$$

$$\sum_{j \in SR} F_{ik}^{ji} \leqslant FR * Y_i \quad \forall i \in SR, k \in SB \tag{12}$$

$$\sum_{ij \in SR} F_{ik}^{ji} \leqslant FB * Z_k \quad \forall k \in SB \tag{13}$$

$$\sum_{j \in SR} F_{ik}^{ji} \leqslant TR * Y_i \quad \forall i \in SR, k \in SB \tag{14}$$

$$\sum_{i,j \in SR} F_{ik}^{ji} \leqslant TB * Z_k \quad \forall k \in SB \tag{15}$$

$$\sum_{\substack{p,q \in SR \\ p \neq q}} F_{uv}^{pq} \geqslant 0 \quad \text{and} \quad X_{u,v}, Y_u \in (0,1) \quad \forall u, v \in S \tag{16}$$

This model minimizes costs as depicted by (1). The first portion of the cost equation denotes link costs in the access network and in the backbone network. The costs of placing remote and backbone devices are represented by the last two terms, respectively. The constraints are made up of three parts. The first part, given by (2)–(5), defines the characteristics of the access network. The second part, given by (7)–(9), represents the backbone network. The third part, given by (10)–(15), represents capacity considerations. The lone equation (6) defines the linkage between the access and backbone networks.

Constraint (2) implies that a node i is either a remote device or that it has exactly one outgoing link (denoting the location of an end user). Constraint (3) ensures that these nonremote device nodes (i.e., user locations) are not connected to each other directly. Constraint (4) indicates that for user nodes, the one outgoing link must go to a device node, whereas for remote device nodes, many links can enter the nodes. Constraint (5) limits the number of links that the remote device can support. Constraint (7) ensures survivability in the backbone network. Constraint (8) limits the capacity of the input (output) ports of the backbone device. Duplexity in the backbone network is given by (9), and the flow constraint (10) forces connectivity in the backbone network. Note that flow constraints are not necessary to ensure connectivity in the access networks, since all nodes are connected to a remote device and each remote device has at least two links to the backbone network. Thus, constraint (10) is strictly to invoke backbone connectivity. Capacity constraint (11) refers to link capacity of all nodes in the network. Internetworking devices have a limit in the amount of packets they can filter and forward. These are depicted by constraints (12) and (14) for remote devices, and (13) and (15) for backbone devices, respectively. Constraints (16) are the nonnegative and integral constraints.

The network design problem given by ($P1$) can be solved by addressing three different subnetwork problems separately. The first problem is to design the access network. The problem here is twofold: first, all nodes must be divided into subgroups, and second, for each subgroup, a remote device location must be chosen and all nodes linked to it. The second problem is the backbone network design problem. For the backbone network, a survivable topology is needed. The connectivity of the access and backbone network must also be considered. The objective is to locate a backbone node for each subgroup and then connect them at minimum cost in a way to make them survivable. The final problem is the capacity provisioning problem. Once the access and backbone topologies are determined, link and node capacities can be determined. Solving model $P1$ by decomposing into three subproblems obviously does not guarantee an optimal solution. The main issue then is how close one can get to the optima of the overall problem, and how long this would take.

2.3. Selected Literature on Large-Scale Network Design

This section provides a review of the literature of large-scale network design, as well as a review of algorithms for these design problems. A number of vintage survey papers (Boesch, 1976; Boorstyn and Frank, 1977; Gerla and Kleinrock, 1977) cover a variety of formulations and solution techniques for distributed networks. Textbooks on data networks consider optimal design to be extremely important (Bertsekas and Gallager, 1992; Schwartz, 1987; Sharma, 1990; Tanenbaum, 1988). Clearly, optimization models based on mathematical programming were the design method of choice in all these studies.

There are many star-star network formulations and solution techniques. The star-star hierarchical network problem was solved by a two-phase algorithm in Lo and Kershenbaum (1988) and by three different Lagrangian relaxations in Mirzaian (1985). The star-star single echelon, single commodity problem for a planning horizon rather than a single time period was studied in Klincewicz et al. (1988). Gavish and Pirkul (1986) discussed the hierarchical problem of computer resource and database location in wide area networks. Vernekar et al. (1990) solved the optimization of resource location in hierarchical computer networks using a modified simulated annealing technique.

Lee et al. (1992) introduced the design of a hierarchical two-level network that consists of a ring in the upper level and a star in the lower level. This ring-star network was formulated into a mixed-integer program and its structures were exploited and solved by a dual based lower bounding procedure. Altenkemer (1994) introduced an enhanced ring architecture that consists of a two-level ring-ring network connecting the local rings with a size k limit by a backbone ring. The ring-ring design is also examined by Laguna (1994). This ring design is preferred by the standardized SONET (Synchronous Optical Network) and the network design consists of many ring levels. The rings are connected by two links to provide survivability. The authors used heuristics based on tabu search to assign nodes to rings while meeting total demand, minimizing costs, and satisfying the capacity constraint.

A related set of papers examines the problem of locating two facility types when two types of nodes, primary and secondary nodes, are given in an undirected network. Balakrishnan et al. (1994b) described the worst-case performance of any heuristics used to solve the model, while in Balakrishnan et al. (1994a) and Pirkul and Narasimhan (1988) the model was solved by the dual-based algorithm and a Lagrangian relaxation with a branch-and-bound technique, respectively.

There have been various papers discussing LAN-LAN network design with one internetworking device. Marucheck and Sulek (1987) discussed the problem of LAN technology choice and formulated a dynamic integer program. Their model provided an optimal design over a planning horizon and included network topology, transmission media, and channel capacity. Fetterolf and Fetterolf and Anandalingam (1991; 1992b) provided an MIP model for the optimal design of LAN-LAN networks with transparent bridges. They used simulated annealing (Fetterolf and Anandalingam, 1991), and later used a Lagrangian relaxation (Fetterolf and Anandalingam, 1992a) to solve the problem. The authors also introduced a similar model for optimal design of LAN-WAN networks (Fetterolf and Anandalingam, 1992b).

Leblanc et al. (1994) expanded the idea and presented a topology for networks connecting a number of token rings connected by source routing bridges rather than transparent bridges. They employed a drop algorithm using the simplex method and simulated annealing using flow deviation for the same subproblem.

Liang and Yee (1989) formulated a nonlinear program that minimized the joint problem of gateway location and routing delay. The trade-off involved was interconnection of networks and the costs of the gateways. They also formulated an MIP to determine which bridges and which LAN-bridge connections to establish to interconnect a set of Ethernets (Liang and Yee, 1990).

3. Access Network Design

3.1. Problem Formulation

In all of the studies cited above, the final network design was hierarchical with a backbone at the core of the network with several access networks at the edge or periphery. The main purpose of an access network is to provide a communication link to the backbone network for users (nodes) who are close to each other geographically. The access network problem only solves part of the overall problem defined by (*P*1). In particular, we consider constraints (2)–(5), which are reproduced below in problem (*P*2) as (18)–(21) The integral constraints are also included in (22). The objective function is modified to relate to the cost of linking users to remote internetworking devices and the cost of the devices themselves, see (17) below.

Thus, the access network design model is:

$$(P2)\ \min Z_{P2} = \sum_{\substack{i,j \in S \\ i \neq j}} c1_{ij} X_{ij} + \sum_{i} CR_i * Y_{ii} \tag{17}$$

subject to

$$\sum_{\substack{j \in SR \\ i \neq j}} X_{ij} + Y_{ii} = 1 \quad \forall i \in SR \tag{18}$$

$$X_{ij} - Y_j \leqslant 0 \quad \forall i, j \in SR, i \neq j \tag{19}$$

$$\sum_{\substack{i \in SR \\ i \neq j}} X_{ij} - Y_j \geqslant 0 \quad \forall j \in SR \tag{20}$$

$$\sum_{\substack{i \in SR \\ i \neq j}} X_{ij} - IR * Y_j \leqslant 0 \quad \forall j \in SR \tag{21}$$

$$X_{ij}, Y_i \in (0,1) \quad \forall i, j \in SR \tag{22}$$

It should be noted that problem (*P*2) is similar to the concentrator location problem, also known as the warehouse location problem that is well known in operations research. The most popular methods to solve this problem are the ADD (Kuehn and Hamburger, 1963) and DROP (Feldman et al., 1966) heuristics. A related problem is the facility location problem. The dual-descent algorithm initiated by Erlenkotter (1978) and Bilde and Krarup (1977) is commonly used to solve this problem.

3.2. Selected Literature on Optimizing Access Networks

The literature on access network design is quite extensive. Gersht and Weihmayer (1987) solved a network design problem jointly considering capacity, flow, performance, and reliability issues. There is vast literature regarding the concentrator (or access switch) location problem. In addition to the ADD (Kuehn and Hamburger, 1963) and DROP (Feldman et al., 1966) algorithms, there have been other variations and solutions. Bahl and Tang (1972) presented a 0–1 integer programming model, and used a combination of Frazer's algorithm (Frazer, 1967) and the DROP algorithm to solve the problem. Lo and Kershenbaum (1988) examined the star-star network and the star-star concentrator location problem (SSCLP) They used a two-phase algorithm to solve the binary integer problem, where in the first phase, a Lagrangian relaxation and a heuristic yielded lower and upper bounds; a tighter lower bound was achieved by a branch-and-bound technique in the second phase. Narasimhan (1989) introduced variable coverage concentrator location problems where nodes have a varying number of links to concentrators. They solved the MIP by exploiting the relaxation of the problem.

There are still other methods and emphasis on the concentrator location problem. Kleinrock and Karmoun (1980), Krolak-Schwerdt and Eckes (1992), and Smith (1993) all dealt with finding the optimal threshold for the number of clusters for datasets. Anandalingam and Lee (2000) used a clustering heuristic based on statistical analysis to solve the access network problem. McGregor and Shen (1977) utilized the center of mass (COM) concepts, which allowed for a multilevel hierarchy where the capacity, cost, and capabilities of the concentrator varied as a function of placement.

3.3. *An Illustrative Method: The DROP Algorithm*

A simple way to solve the access network problem is by using the DROP algorithm, which is very popular with facility (and switch) location problems. If one boils down problem (*P*2) to its basic components, there is a set of nodes, *i*, and a set of potential switch locations, *j*. A cost, cl_{ij}, is associated to connection nodes *i* to *j*. An additional cost occurs for the switches themselves, CR_j. The DROP algorithm starts by assigning all nodes as switches (making each node an access cluster). In this case, the only cost is the cost of switches; there are no costs of connecting users to switches. The other extreme would be to have just one switch at a node, and all users connected to this switch. In this case, the cost of switches will be very small, but interconnect costs will be high. Clearly there is a trade-off between having less switches versus more interconnections.

In the DROP algorithm, access clusters are merged by the greedy method, combining with the nearest-neighbor cluster. Clusters are merged in the order that results in the most cost-effective assignment. Thus, the first switches to be dropped are the ones that provide the most savings when they are dropped. This is continued until dropping switches no longer reduces the total cost of the network.

The DROP heuristic is as follows:

1. Initialize each node as its own cluster and as a switch location. Let $K = N$, the total number of nodes.
2. For cluster $k(k = 1, ..., K)$, locate the nearest-neighbor cluster (other than itself) and its switch location. Find savings of dropping one switch and merging all nodes connected to that switch to the nearest-neighbor switch. That is, savings $\Delta_k =$ cost of dropping one switch less the cost of reconnecting all nodes from that switch to the nearest-neighbor *cluster switch*, as well as connecting the dropped switch to the new switch. Let $k_{\max} = \max\{\Delta_k\}$.
3. If any $\Delta_k > 0$, merge links from cluster $k_{\max}$ to the *nearest neighbor switch*. $K = K - 1$.
4. If $K = 1$ or all $\Delta_k \leqslant 0$ (i.e., there were no merges), then Stop. Otherwise go to Step 2.

A detailed DROP algorithm can be found in Feldman et al. (1966). Another popular method of solving the concentrator/switch location problem is the dual descent algorithm proposed by Erlenkotter (1978) and Bilde and Krarup (1977).

In cases where there are a small (e.g., less than twenty) number of user locations, and there are no capacity constraints (i.e., equation (21) is dropped from problem (*P*2)), both the DROP and variations of dual descent perform extremely well. In recent experimentations, we have shown that these methods, along with a statistical clustering heuristic obtain solutions within 5 percent of a simple lower bound based on linear relaxation (Anandalingam and Lee, 2000).

In the case where the actual access network has various complicating features such as link capacity, equipment type and capacity, limitations on the number of clusters (in access networks and/or concentrators) and so on, the DROP heuristic would form part of a hybrid algorithm. In either the simple or the complex case, the solution strategy will be based on using a cuisine (or combination) of well-established optimization techniques for solving the access network design in telecommunications.

4. Backbone Network Design

4.1. Problem Formulation

The main purpose of the backbone network is to provide the long-haul movement of communications traffic with little delay and high reliability. As in the case of the access network, only part of the overall problem (*P*1) is solved where one optimizes the backbone network.

We reproduce below that part of problem *P*1 that pertains to topology of the backbone network:

$$(P3) \quad \min Z_{p4} = \sum_{\substack{k,l \in SB \\ k<l}} c2_{kl} X_{kl}$$

subject to:

$$\sum_{\substack{l \in SB \\ k \neq l}} X_{kl} \geqslant 2 \quad \forall k \in SB$$

$$\sum_{\substack{l \in SB \\ k \neq l}} X_{kl} \leqslant IB \quad \forall k \in SB$$

$$X_{kl} - X_{lk} = 0 \quad \forall k, l \in SB, k < l$$

Clearly, that portion of the backbone capacity constraints (10)–(15) need to be included in problem *P*3 when we consider the optimization of minimum cost capacitated survivable backbone network design.

Designing minimum cost backbone networks that consider node capacity and routing delay, and are survivable, that is, have at least two disjoint paths for every demand pair, is an NP-hard problem. We cannot use the usual set of integer programming methods for this problem. In this paper, we will describe a system of heuristics that can be used to design survivable backbone networks in telecommunications.

4.2. Selected Literature on Optimizing Backbone Networks

Over the past twenty years, a number of papers have attempted to address the issue of designing backbone network topologies that are survivable in the sense that there are at least *k* pairwise vertex-disjoint paths for every demand pair. (See survey by Robertson and Seymour, 1985). One of the earliest works is by Steiglitz et al. (1969), in which they design heuristics using the relationship between the minimum cost survivable network (MCSN) design and the traveling salesman problem. This idea has subsequently been used by a number of researchers including Monma and Shallcross (1989). Suurballe has been quite active in examining disjoint paths in network algorithms that use variants of Dijkstra's shortest-path algorithm (Suurballe, 1974; Suurballe and Tarjan, 1984). Another basic design approach is to start with a tree and augment it to a *k* edge-connected graph (Ueno et al., 1988) or a *k* edge-connected directed path (Kajitani and Ueno, 1986). In a recent paper, we introduced the bootstrap heuristic for designing minimum-cost survivable networks (Clarke and Anandalingam, 1994). The bootstrap heuristic starts with a lower bounding procedure, and obtains a feasible solution by bootstrapping onto the best lower bound solution.

There is related literature on designing survivable networks with capacity constraints. Agarwal (1989) developed a mathematical programming approach for an edge-disjoint network with capacity constraints. He used a heuristic Lagrangian relaxation technique with a partial branch-and-bound approach. Another interesting mathematical programming formulation is by Gavish et al. (1989). This formulation is for link and node failures with capacity constraints. It requires specifying operational states of the network and allowable routes for each operational state. Their heuristic approach solves repeated LPs fixing the largest fractional value to one at each iteration. Schumacher (1984) developed an algorithm to minimize the number of edges and the diameter of a graph with k node-disjoint paths. Newport (1990) developed a quantitative measure of how survivable a network is and incorporated it into a cut-saturation algorithm. LeBlanc and Reddoch (1990) developed a mathematical programming formulation that requires a set of disjoint routes as input and solves the problem as an LP.

4.3. *Capacitated Survivable Backbone Network Design*

In this section, we provide a general formulation for the problem of designing survivable telecommunications networks with capacity constraints. We do not use the problem described by the formulation (*P*3) directly, but the relationship between that and problem (*P*4) below is obvious, and is lift as an exercise to the reader.

Let the network be represented by a graph $G = (N, E)$, where N is a set of nodes or locations that need to be connected, and E is a set whose elements are pairs of nodes, called edges or links. These edges are the set of allowable direct node to node connections in the network design. A cockle of a path, $\omega(S)$, is a set of edges with one end point being any of the nodes in the set $S \subseteq N$ and the other endpoint in the set $N \backslash S$.

We denote x_{ij} as the 0/1 decision variable that has a value of 1 when the solution indicates building the direct link (i, j) between node i and j and has a value of 0 if nothing is to be done. The cost of building such a link is c_{ij}. Let g_{ij} be the capacity of link (i, j). and f_{ij} is the flow on link (i, j). We denote b^{od} as the flow requirement for a particular origin-destination (hereafter $o-d$) pair. The formulation for a node-disjoint network with flow constraints is presented in (23)–(29) below.

$$(P4) \quad \min_{x,f} \sum_{\forall (i,j) \in E} c_{ij} x_{ij} \tag{23}$$

$$\text{subject to} \sum_{\forall (i,j) \in \omega(S)} x_{ij} \geqslant 2 \quad \forall S \subseteq N, \varnothing \neq S \neq N \tag{24}$$

$$\sum_{\forall (i,j) \in \omega_{G-z}(S)} x_{ij} \geqslant 1 \quad \forall z \in N, \forall S \subseteq N, \backslash z, \varnothing \neq S \neq N \backslash z \tag{25}$$

$$f_{ij} \leftarrow \text{routing algorithm } (b^{od}) \tag{26}$$

$$f_{ij} \leqslant g_{ij} x_{ij} \quad \forall (i,j) \in E \tag{27}$$

$$f_{ij} \geqslant 0 \quad \forall (i,j) \in E \tag{28}$$

$$x_{ij} \in (0,1) \quad \forall (i,j) \in E. \tag{29}$$

The objective (23) is to minimize the sum of all link costs used in the design. Constraint (24) states that for each possible subset of nodes, there must be two edges with one endpoint in the subset and the other endpoint outside the subset (i.e., in the complement of the subset). This constraint ensures link survivability. Constraint (25) requires that two edges leaving a subset of nodes must not both be incident on the same node. This constraint ensures node survivability.

Equation (26) says that there is some algorithm that takes the demand matrix b^{od} and routes the demands on the existing network and returns the resulting flow, f_{ij}, on each link. This algorithm can be a multicommodity flow constraint. A more appropriate interpretation would be an application-specific shortest-path approach. Constraint (27) says that the sum of the flows for all demands that traverse a particular edge must be less than the capacity of that edge if it exists or zero if no arc exists. If one ignores capacity constraints and only considers the design of feasible topologies for the backbone network, one gets the minimum-cost survivable network (MCSN).

4.4. Heuristic for Obtaining Feasible Topologies

As defined earlier, survivable networks have two disjoint paths for each pair of nodes. This leads to the simple statement that each node must have at least two edges. A mathematical program to minimize cost with the constraint that each node has a degree of two is written as:

$$\min_{x,f} \sum_{\forall (i,j) \in E} c_{ij} x_{ij} \tag{30}$$

$$\text{subject to} \sum_{\forall (i,j) \in \omega(i)} x_{ij} \geqslant 2 \quad \forall i \in N \tag{31}$$

$$x_{ij} \in (0,1) \quad \forall (i,j) \in E. \tag{32}$$

The above integer program is a lower bound of the MCSN. The continuous relaxation, which is generated by replacing (32) with $0 \leqslant x_{ij} \leqslant 1$, $\forall (i,j) \in E$, is a weak bound for the MCSN. Since it is easy to solve, we will use the LP version (30)–(32), and call it the Degree 2 problem. We should note here that the network solutions resulting from the Degree 2 constraints are typically disconnected and thus are infeasible from the point of view of the MCSN problem.

We can strengthen the Degree 2 solution by adding the constraint that the network is connected. This will be achieved by a cutting-plane method. The constraints that we add are simple subtour elimination constraints. They are in the form of

$$\sum_{\forall (i,j) \in \omega(S)} x_{ij} \geqslant 2 \quad \forall S \in \text{disjoint subgraphs} \tag{33}$$

Initially the LP relaxation of (30)–(32) is solved. Some variables may have fractional values. Upon examination of the solution in search of connected subgraphs, it is assumed that if the variable corresponding to an edge has a nonzero value, then the edge exists. Constraints are added for each connected subgraph and another linear program is solved. This process continues until the solution (with fractions rounded up) results in one connected subgraph. We call this procedure, LP Bound 1. The final objective value for LP Bound 1 does not use the rounded solution, but rather it uses the fractional solutions.

LP Bound 1 results in a connected subgraph. Continuing from there, an algorithm is used to find biconnected subgraphs. We call this LP Bound 2. Additional cuts are added of the form:

$$\sum_{\forall (i,j) \in \omega(S)} x_{ij} \geqslant 2 \quad \forall S \in \text{biconnected subgraphs} \tag{34}$$

For each biconnected subgraph connected by a single edge, constraints are added that require two edges from each of the subgraphs. Constraints are continually added until the rounded up variables result in one biconnected (i.e., survivable) subgraph.

LP Bound 2 is an efficient method for generating good lower bounds for the MCSN. Clearly the solution from the lower bounding procedure LP Bound 2 will not be feasible—some of the variables will have fractional values. The bootstrap heuristic is straightforward from this point onward. We take the fractional solutions from LP Bound 2 and round them all up to a value of one. This will yield a feasible network.

In a previous paper (Clarke and Anandalingam, 1994) we showed that the bootstrap heuristic performed exceptionally well and gave solutions that were 0.06 to 0.3 percent from the optimal for twenty-five node problems.

4.5. Heuristic for Provisioning Capacity

We need to extend the feasibility (i.e., survivability) heuristics defined in the previous section to incorporate link capacity and routing delay considerations. A traffic matrix for all o—d pairs of users on the network will be given as input. When we consider capacitated telecommunications networks, two features must be incorporated. The first is to route demand via the minimum hop path. The other is to ensure that link flows do not exceed capacity when this routing scheme is used, that is, we need to provision capacity in order to accommodate the routed traffic flows. We introduce a heuristic that will take a network with a survivable topology and ensure that flow and capacity constraints are also satisfied.

The heuristic uses the concept that very large demands should traverse as few links as possible (one direct link is ideal) to keep from creating congestion problems on several links in the network. The heuristic adds links for high demand pairs, therefore reducing flow on other potentially overused links.

For each node (i.e., user station) it finds the largest demand entering or leaving that node from another node (or user station). If that demand pair is directly connected, the heuristic moves on to the next highest demand. Otherwise, a direct link is added and the flows are tested. If the flows are satisfied, the heuristic stops. If they are not satisfied, the procedure continues.

4.6. Improvement Heuristics

Once the feasibility heuristic and the capacity provisioning heuristic have been exercised, we would have a feasible telecommunications network. The goal is now to reduce the cost of designing this network while maintaining feasibility. Presented below is the brief description of branch exchange heuristics that improve upon the feasible network (for details, see Clarke, 1992)).

1. *2for2 heuristic.* The 2for2 heuristic consists of examining all pairs of edges. If the triangle inequality is in effect, the edges will be checked to see if they cross before continuing. The two edges in question are replaced with two different edges not in the network that cause the overall cost to be reduced. If feasibility (i.e., survivability and within capacity) is maintained, the exchange is kept, otherwise the previous network structure returned. This procedure is repeated for all pairs of edges. After the exchange is made, the network is tested. If the network is feasible, the change is kept. If it is not feasible the structure prior to attempting this 2for2 exchange is returned.
2. *2for1 heuristic.* The 2for1 heuristic is similar to the Pretzel heuristic of Monma and Shallcross (1989). The 2for1 heuristic consists of examining all pairs of edges, and replacing two edges (links) with one. Note that each link connects two nodes, and thus, when we consider removing two links (edges), we are also implicitly considering four nodes that will be affected. In the 2for1 heuristic, two edges are replaced with one edge not currently in the network, but which connects two of the four nodes associated with the two edges that were removed. This exchange should cause a reduction in cost. If feasibility (i.e., survivability and within capacity) is maintained, the exchange is kept, otherwise the previous structure is returned. This procedure is repeated for all pairs of edges.
3. *1for1 heuristic.* The 1for1 (or swing) heuristic is similar to the one-optimal heuristic of Monma and Shallcross (1989). The 1for1 exchange replaces one edge with a different edge. There are many different ways of removing an edge at random. This exchange will only consider a limited number of possible replacement edges. This algorithm is called swing because each edge considered has a fixed end and a free end. The free end is removed and swings toward a closer node. The fixed end of the edge will be the node of lesser degree. The free end of the edge must have a degree greater than two, or it cannot be a candidate to be moved. The 1for1 heuristic proceeds by selecting an edge that has one endpoint with a degree of three or greater. The free end swings into the node that creates the least-cost edge. The changes remain if an edge does not already exist in this position, and the network is still feasible. If the change is not valid, nodes are tried that create edges of increasing cost until a valid change is reached or the original node is reached. This procedure is repeated for all edges.
4. *Delete step.* The delete step is self explanatory. It removes edges. Each time an edge is removed the remaining network is checked for feasibility. There are a few practicalities to make the heuristic more efficient. An edge cannot be deleted if either endpoint has a

degree of two. Also, it is best to first remove edges at higher degree nodes. The delete heuristic begins with the most costly edge. It checks to see if the network would be feasible without the edge in question. If the network is feasible without this edge, it is removed. The next edge in the list is examined. The heuristic continues until all edges have been examined.

Thus the complete backbone network with a survivable topology and feasible capacity will be fully designed at near-optimal cost using the three heuristics (feasible topology heuristic, capacity provisioning heuristic, and improvement heuristic) described above. Although the backbone design problem can be formulated as a mixed-integer program, we need these heuristics because the problem is NP-hard.

5. Concluding Remarks

Optimization techniques as practiced by operations research scholars have played a vital role in the design of telecommunications networks. Whether one is considering the overall network, or its components (access and backbone networks), the design problem is formulated as mixed-integer programming problem and solved using intelligent heuristics.

Telecommunications networks have isomorphic relationships to transportation and logistics networks. Sometimes, as in the design of simple access networks, one could and does use established techniques, such as DROP and dual ascent, to solve the problems. Most often, however, one needs to use heuristics based on established optimization strategies such as Lagrangian relaxation, cutting planes, and valid inequalities to solve the problems. In this paper, we have provided a flavor of the types of telecommunications network design problems that are of concern to industry and the methodologies used to solve them.

References

Agarwal, Y. K. 1989. "An algorithm for designing survivable networks." *AT&T Technical J.*, pp. 64–76.

Altenkemer, K. 1994. "Topological design of ring network." *Comput. Oper. Res.* **21(4)**: 421–431.

Anandalingam, G. and F. Lee. 2000. "A statistical heuristic for designing access telecommunications networks." Working paper. Pennsylvania: Department of Systems Engineering, University of Pennsylvania.

Bahl, L. R. and D. Tang. 1972. "Optimization of concentrator locations in teleprocessing networks." Technical report. Presented at the Symposium on Computer-Communications Networks and Teletraffic Polytechnic Institute of Brooklyn.

Balakrishnan, A., T. Magnanti, and P. Mirchandani. 1994a. "A dual-based algorithm for multi-level network design." *Mgmnt Sci.* **40(5)**: 567–581.

Balakrishnan, A., T. Magnanti, and P. Mirchandani. 1994b. "Modeling and heuristic worst-case performance analysis of the two-level network design problem." *Mgmnt Sci.* **40(7)**: 846–867.

Bertsekas, D. and R. Gallager. 1992. In *Data Networks*. Englewood Cliffs, NJ: Prentice-Hall.

Bilde, O. and J. Krarup. 1977. "Sharp lower bounds and efficient algorithms for the simple plant location problem." *Ann. Discrete Math.* **1**: 79–97.

Boesch, F. T. (ed). 1976. In *Large-scale Networks: Theory and Design*. New York: IEEE Press.

Boorstyn, R. R. and H. Frank. 1977. "Large-scale network topological optimization." *IEEE Trans. Commun.* **25(1)**: 29–47.

Clarke, L. W. 1992. "The Design of Survivable Telecommunication Networks." Ph.D. diss. University of Pennsylvania, Philadelphia.

Clarke, L. W. and G. Anandalingam. 1994. "A bootstrap heuristic for designing minimum cost survivable networks." *Comput. Oper. Res.* **22**: 921–934.

Erlenkotter, D. 1978. "A dual-based procedure for uncapacitated facility location." *Oper. Res.* **26(6)**: 992–1009.

Feldman, E., F. Lehrer, and T. Ray. 1966. "Warehouse location under continuous economies of scale." *Mgmnt Sci.* **12**: 670–684.

Fetterolf, P. and G. Anandalingam. 1991. "Optimizing interconnections of LAN: an approach using simulated annealing." *ORSA J. Comput.* **3(4)**: 275–287.

Fetterolf, P. and G. Anandalingam. 1992a. "A Lagrangian relaxation technique for optimizing interconnection of local area network." *Oper. Res.* **40(4)**: 678–688.

Fetterolf, P. and G. Anandalingam. 1992b. "Optimal design of LAN–WAN internetworks: an approach using simulated annealing." *Ann. Oper. Res.* **36**: 275–298.

Frazer, W. 1967. "An approximate algorithm for plant location under piecewise linear concave costs." Technical Report RC-1875. IBM Research Center.

Gavish, B. and H. Pirkul. 1986. "Computer and database locating in distributed computer systems." *IEEE Trans. Comput.* **35(7)**: 583–590.

Gavish, B., Trudeau, Dror, Gendreau, and Mason. 1989. "Fiberoptic circuit network design under reliability constraints." *IEEE J. Selected Areas in Commun.* **7(8)**: 1181–1187.

Gerla, M. and L. Kleinrock. 1977. "On the topological design of distributed computer networks." *IEEE Trans. Commun.* **25(1)**: 48–60.

Gersht, A. and R. Weihmayer. 1987. "A descrete optimization model for joint communications network design and facility engineering." Technical Report. GTE Laboratories, Telecommunications Research Laboratory, 40 Sylvan Road, Waltham, MA 02254.

Kajitani, Y. and S. Ueno. 1986. "The minimum augmentation of a directed tree to a *k*-edge-connected directed graph." *Networks* **16**: 181–197.

Kleinrock, L. and F. Karmoun. 1980. "Optimal clustering structures for hierarchical topological design of large computer networks." *Networks* **10**: 221–248.

Klincewicz, J., H. Luss, and C.-S. Yu. 1988. "A large-scale multilocation capacity planning model." *Eur. J. Oper. Res.* **34**: 178–190.

Krolak-Schwerdt, S. and T. Eckes. 1992. "A graph theoretic criterion for determining the number of clusters in a data set." *Multivariate Behav. Res.* **27(4)**: 541–565.

Kuehn, A. A. and M. J. Hamburger. 1963. "A heuristic program for locating warehouses." *Mgmnt Sci.* **9**: 643–666.

Laguna, M. 1994. "Clustering for the design of sonet rings in interoffice telecommunications." *Mgmnt Sci.* **40(11)**: 1533–1541.

Leblanc, L., J. Park, V. Sridhar, and J. Kalvene. 1994. "Topological design and bridge capacity assignment for interconnecting token rings." Working Paper. Nashville, TN: Owen Graduate School of Management, Vanderbilt.

LeBlanc, L. J. and R. Reddoch. 1990. "Reliable link topology/capacity design and routing in backbone telecommunication networks." Working Paper 90-08. Nashville, TN 37203: Owen Graduate School of Management, Vanderbilt University.

Lee, C. H., H.-B. Ro, and D.-W. Tcha. 1992. "Topological design of a two-level network with a ring-star configuration." Working Paper. Incheon, Korea: Department of Industrial Engineering, Inha University.

Liang, S. and J. Yee. 1989. "A gateway allocation algorithm for interconnecting existing data networks." In *Proceedings IEEE INFOCOMM*, pp. 468–473.

Liang, S. and J. Yee. 1990. "Algorithms for interconnecting ethernets with multiport bridges." Technical Report. CA: Communication Science Institute, University of Southern California.

Lo, C.-C. and A. Kershenbaum. 1988. "A two-phase algorithm for the star-star concentrator location problem." *IEEE Trans. Commun.* **6**: 946–955.

Marucheck, A. and J. Sulek. 1987. "An algorithm for technology choice in local area network design." *Mgmnt Sci.* **33(1)**: 58–72.

McGregor, P. and D. Shen. 1977. "Network design: an algorithm for the access facility location problem." *IEEE Trans. Commun.* **25(1)**.

Mirzaian, A. 1985. "Lagrangian relaxation for the star–star concentrator location problem: approximation algorithm and bounds." *Networks* **15**: 1–20.

Monma, C. L. and D. F. Shallcross. 1989. "Methods of designing survivable communication networks." *Oper. Res.* **37**.

Narasimhan, S. 1989. "The concentrator location problem with variable coverage." Working Paper. Atlanta, GA: College of Management, Georgia Institute of Technology.

Newport, K. T. 1990. "Incorporating survivability considerations directly into the network design process." In *IEEE INFOCOM '90*, pp. 215–220.

Pirkul, H. and S. Narasimhan. 1988. "Primary and secondary route selection in backbone computer networks." Working Paper. Columbus, OH: College of Business, Ohio State University.

Robertson, N. and P. D. Seymour. 1985. "Disjoint paths—a survey." *SIAM J. Algebraic Discrete Meth.* **6(2)**.

Schumacher, U. 1984. "An algorithm for construction of a *k*-connected graph with minimum number of edges and quasiminimal diameter." *Networks* **14**: 63–74.

Schwartz, M. 1987. *Telecommunication Networks: Protocols, Modeling and Analysis*. Reading, MA: Addison-Wesley.

Sharma, R. L. 1990. *Network Topology Optimization: The Art and Science of Network Design*. New York: Van Nostrand Reinhold.

Smith, S. P. 1993. "Threshold validity for mutual neighborhood clustering." *IEEE Trans. Pattern Anal. Machine Intell.* **15(1)**: 89–92.

Steiglitz, K., P. Weiner, and D. J. Kleitman. 1969. "The design of minimum-cost survivable networks." *IEEE Trans. Circuit Theory* **16(4)**: 455–460.

Suurballe, J. W. 1974. "Disjoint paths in a network." *Networks* **4**: 125–145.

Suurballe, J. W. and R. E. Tarjan. 1984. "A quick method for finding shortest pairs of disjoint paths." *Networks* **14**: 325–336.

Tanenbaum, A. S. 1988. *Computer Network*. Englewood Cliffs, NJ: Prentice-Hall.

Ueno, S., Y. Kajitani, and H. Wada. 1988. "Minimum augmentation of a tree to a *k*-edge-connected graph." *Networks* **18**: 19–25.

Vernekar, A., G. Anandalingam, and C. N. Dorny. 1990. "Optimization of resource location in hierarchical computer networks." *Comput. Oper. Res.* **17(4)**: 375–388.

18.10 Optimization of Test Intervals in Nuclear Engineering

Stanislav Uryasev

ABSTRACT This paper considers modeling and optimization approaches for scheduling of test intervals of unreliable components and systems. From a formal point of view, optimization of test intervals is a challenging multiextremum nonsmooth optimization problem. In practical applications, such as scheduling of testing of safety systems of nuclear power plants, this problem is solved using engineering experience and the brute force (trial-and-error) approach. Although in general form, the problem is intractable (when many systems and components are involved), under simplifying assumptions, such as the randomness of the beginnings of tests, it can be reduced to a relatively simple smooth nonlinear programming problem. Optimization techniques for finding optimal test intervals are discussed and explained with an example.

1. Introduction

Probabilistic safety analysis (PSA) is generally used to find the weak points of a plant and to make recommendations for possible changes in design. The same technique can also be used to assess technical specifications and to develop guidance for operator training and accident management. The calculation of the optimal test strategies for standby systems is one of the problems that can be solved with this approach. The test intervals affect the global safety of the plant. The optimization of operation schedules is a complex problem due to the sophisticated dynamic interrelations of many variables and the large dimension.

Many useful equations for components with periodic testing (as well as for monitored components) were developed by Vesely and Goldberg (1977a,b), and the effects of aging and degradation are incorporated in other papers (Vesely, 1991; Kim et al., 1992). Some analytical formulas have been proposed to calculate the optimal test intervals for one isolated component (Apostolakis and Chu, 1980; Lofgren et al., 1981; Mankamo and Pulkkinen, 1988; Sim and Endrenyi, 1988; Vaurio, 1979, 1982). More advanced analytical expressions also exist for cases where the problem can be decomposed and test intervals for different groups of components can be treated independently. For example, this is the case when a group of components may be treated as one super component (Apostolakis and Chu, 1980, 1981). Analytical expressions for series-parallel redundancy systems are available for staggered testing (Vaurio, 1980, 1981) and with consecutive schemes (Vaurio, 1982). Dialynas and Michos (1993) developed recurrent formulas for the time-dependent analysis of unavailability of standby components incorporating arbitrary failure distributions and three inspection/maintenance policies.

There are several tools (computer codes) to assess the global unavailability of a plant that take into account dependencies between test intervals of the components; see, for example, FRANTIC (Vesely and Goldberg, 1977a; Ginsburg and Powers, 1986), SOCRATES (Wagner et al., 1987), and MARELA (Papazoglou, 1988). These codes can be used to compare different variants, but not to optimize model parameters. The trial-and-error method works poorly for systems with many interrelated control parameters.

We consider a general system with subsystems (groups of components) that are tested at the same time. The system could be very large and include, for example, the whole safety system of the plant. There are some technological constraints on the test periods (technical specifications, manufacturer's recommendations, and other engineering factors, e.g., dry-out of oil films, or build up of boron crystals with long standby times) and some management constraints (the availability of technical personnel could be limited). We estimate the time-dependent unavailability of entire system. The unavailabilities of components are presented as periodic piecewise linear curves. The approach is based on a simple unavailability model of standby components with a constant standby failure rate. This model neglects imperfectness of tests and degradation caused by testing, which may overestimate the benefits of shortening test intervals (Mankamo and Pulkinnen, 1988; Pulkkinen et al., 1989; Laakso, 1990; Kim et al., 1992). Also, we assume that common cause failures are included directly in the fault trees (and, consequently, minimal cut sets); different approaches to common cause failures can be found, for example, in the Special Issue on Dependent Failure Analysis (1991). The fault tree analysis (minimal cut set representation) was used to link all components into one system. The paper by Uryasev and Vallerga (1993) considers three cases where the average unavailability of the system is presented as a function of the average unavailabilities of the system's components:

1. System with independent components (in each cut set, at the most one tested component);
2. System with independent series groups (tested components can be united in super components, in each cut set, at the most one super component);
3. System with random starts of the tests (the first regular inspection for each component starts at random time, then inspections are made with the same test interval).

For these three cases, the problem of optimization of test intervals can reduced to a nonlinear optimization problem. The system with random starts of tests can be considered as a reasonable approximation of more realistic (but much more complicated) systems. Here, we consider the case with random starts of tests. We provide explicit formulas for the gradient of the performance function (with respect to the test intervals). The gradient can be calculated during one run through the cut sets. We apply numerical nonlinear programming techniques to optimize test intervals of the system with random starts of the tests.

2. General Description of the Model

We consider a system failure in the time interval $0 \leqslant t \leqslant T$. The system consists of a set of components which we denote by A. To assure the low unavailability of the system, some groups of components are tested periodically. The set A is divided into $L+1$ subsets U_l $(0 \leqslant l \leqslant L)$. Here, U_l $(1 \leqslant l \leqslant L)$ is a subset of components with the same interval between tests, and U_0 is a subset of components that should not be tested during the interval $[0, T]$. The subset U_l can be divided into redundant groups of components. We suppose that during testing, components have some probability of being unavailable (unavailability). These redundant groups should not be tested simultaneously because this could lead to a high unavailability of the system. Let us denote by V_l the number of redundant groups of components in U_l, and by v a group in U_l $(1 \leqslant v \leqslant V_l)$. We also define M_{lv} as the number of components in the group v from the set l. Let m be the component number in the group. We denote by $a_{lvm} \in A$ the component m in the group v from the set l.

As an example, let us assume that for the subset U_l the number V_l is equal to 4 (see Figure 1). The time schedule for groups of components is staggered. The interval between the start of successive periodic tests of the same group of components is equal to Δ_l. The shortest

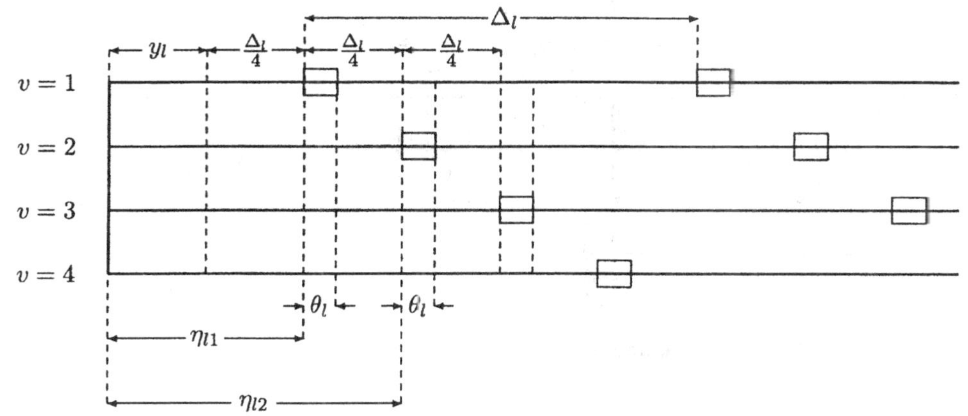

Figure 1 The time schedule for the set l

interval between the beginning of tests of different groups is equal to Δ_l/V_l. We denote a shift in the schedule by y_l, and the first periodic inspection interval of group v from the set l by η_{lv}. The test intervals in the set l are staggered and η_{lv} satisfies the equation (see Figure 1)

$$\eta_{lv} = y_l + \frac{\Delta_l}{V_l} v, \quad l = 1, ..., L, \quad v = 1, ... V_l. \tag{1}$$

We consider that fault trees were constructed for the plant in the standard way (Robert et al., 1981). The fault trees are presented by Boolean equations, which are used to determine minimal cut sets.

Any fault tree has a finite number of minimal cut sets that are unique for the top event. The minimal cut set expression for the top event can be written in the general form

$$M = M_1 \vee M_2 \vee \cdots \vee M_W, \tag{2}$$

where M is the top event, and M_w $(w = 1, ..., W)$ are the minimal cut sets. Each minimal cut set consists of a combination of component failures; therefore the general minimal cut set can be expressed as

$$M_w = \bigwedge_{(lvm) \in C_w} \alpha_{lvm},$$

where α_{lvm} is an event of failure of the component a_{lvm}, and C_w is a set of components in the cut set w $(1 \leqslant w \leqslant W)$. Thus, equation (2) can be presented as

$$M = \bigvee_{w=1}^{W} \bigwedge_{(lvm) \in C_w} \alpha_{lvm}. \tag{3}$$

We assume that all events α_{lvm} are mutually stochastically independent and the probability of each cut set M_w is considerably less than 1, that is

$$P(M_w) \ll 1, \quad w = 1, ..., W.$$

If we take into account only linear terms, then

$$P(M) = P\left(\sum_{w=1}^{W} \prod_{(lvm) \in C_w} \alpha_{lvm}\right) \approx \sum_{w=1}^{W} \prod_{(lvm) \in C_w} P(\alpha_{lvm}). \tag{4}$$

Let us denote by $p_{lvm}(t)$ the unavailability of the component a_{lvm} at time t. In this case, the time dependent unavailability $p(t)$ of the system can be approximated as

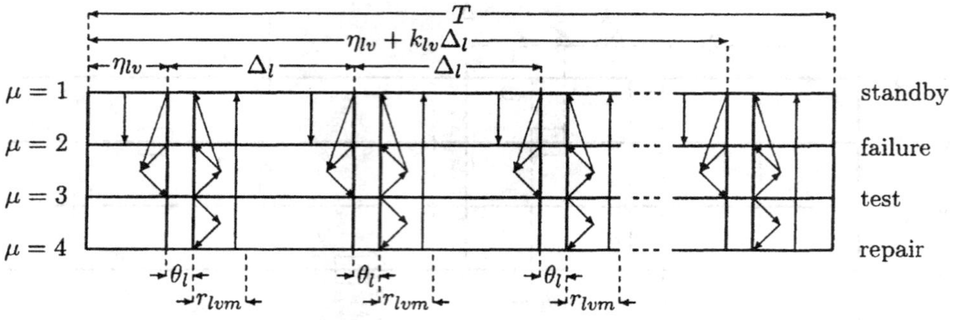

Figure 2 States of the component

$$p(t) = \sum_{w=1}^{W} \prod_{(lvm)\in C_w} p_{lvm}(t). \tag{5}$$

Further, we will use the value $p(t)$ to calculate (approximately) the failure probability of the system.

3. One Component

In this section, we describe the instantaneous unavailability of tested components. We consider that each component $a_{lvm}, l = 1, ..., L$ at time t can be in one of the following four states (see Figure 2):

- $\mu = 1$ standby state (component is available);
- $\mu = 2$ failure state (state with a latent failure);
- $\mu = 3$ test (or maintenance) state;
- $\mu = 4$ repair state.

A component is available in a standby state and is unavailable in the failure and repair states. After a failure, the component enters the failure state. To identify the failure, the component is periodically tested (test state). If it is found that the component has failed, it is repaired (repair state).

We assume that in the test state ($\mu = 3$), the component a_{lvm} is unavailable with probability ρ_{lvm}. After the test, the component is unavailable with some probability (for instance, human errors). This part of unavailability of the component is included in the time independent unavailability part h_{lvm} of the component a_{lvm}. The time independent unavailability part h_{lvm}, also, includes the probability of failure due to a true demand (e.g., an event exceeding the design-basis criteria of a component) and the probability that a system cannot operate during a mission period.

During their lifetime, components periodically pass through three different phases. Each phase defines an interval; we divide the time interval $0 \leqslant t \leqslant T$ into three types:

- Test (or maintenance) duration intervals are intervals in which the component is tested (or serviced) and is in the test state ($\mu = 3$). These intervals are denoted by θ_l. Here, θ_l are deterministic variables chosen by the plant operator or the designer.
- Repair intervals, following the test duration intervals, are scheduled for potential repairs. During these intervals the component could be in the standby state ($\mu = 1$), failure state ($\mu = 2$), or repair state ($\mu = 4$). The component changes the repair state to the standby state during the repair interval of length r_{lvm}. Also, the component can go from a standby state to a failure state during repair intervals.

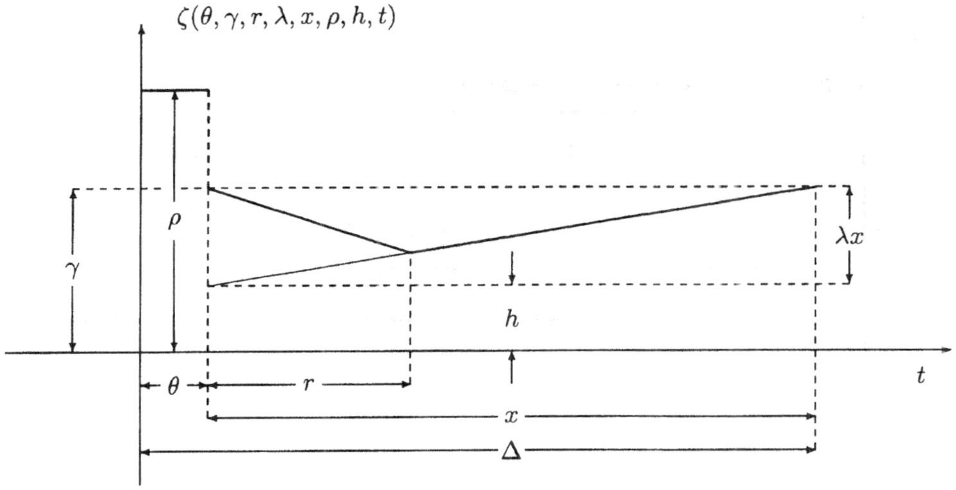

Figure 3 The function ζ

- Standby intervals are intervals in which components are not being tested or repaired. During these intervals, a component can be in a standby or failure state. After a failure, it goes from a standby to a failure state.

To describe one period of the component unavailability $v_{lvm}(t)$, let us introduce a function $\zeta(\theta, \gamma, r, \lambda, x, \rho, h, t)$ on the interval $[0, \Delta]$ (see Figure 3),

$$\zeta(\theta,\gamma,r,\lambda,x,\rho,h,t)=\begin{cases} 1, & \text{if } 0 \leqslant t < \theta, \\ (t-\theta)\left(\lambda+\dfrac{h-\gamma}{r}\right)+\gamma, & \text{if } \theta \leqslant t < \theta + r, \\ (t-\theta)\lambda + h, & \text{if } \theta + r \leqslant t \leqslant \theta + x \end{cases} \tag{6}$$

With the function ζ, the unavailability $p_{lvm}(t)$ of the component a_{lvm} is given by the formula (see Figure 4):

$$p_{lvm}(t)=\begin{cases} \lambda_{lvm}t + h_{lvm}, & \text{if } 0 \leqslant t < \eta_{lv}; \\ \zeta(\theta_l, \lambda_{lvm}\eta_{lv} + h_{lvm}, r_{lvm}, \lambda_{lvm}, x_l, \rho_{lvm}, h_{lvm}, t-\eta_{lv}), & \text{if } \eta_{lv} \leqslant t < \eta_{lv} + \Delta_l; \\ \zeta(\theta_l, \lambda_{lvm}x_l + h_{lvm}, r_{lvm}, \lambda_{lvm}, x_l, \rho_{lvm}, h_{lvm}, t-(k-1)\Delta_l-\eta_{lv}), & \text{if } \eta_{lv} + (k-1)\Delta_l \leqslant t, \\ & t < \eta_{lv} + k\Delta_l, \\ & k = 2, \dots, k_{lv}; \\ \zeta(\theta_l, \lambda_{lvm}x_l + h_{lvm}, r_{lvm}, \lambda_{lvm}, & \\ T-\eta_{lv}-k_{lv}\Delta_l-\theta_l, \rho_{lvm}, h_{lvm}, t-\eta_{lv}-k_{lv}\Delta_l), & \text{if } \eta_{lv} + k_{lv}\Delta_l \leqslant t, \\ & t \leqslant T. \end{cases}$$

The last formula and formula (5) are used to calculate the point-wise unavailability of the whole system.

4. Calculation of the Average Unavailability

The function $p(t)$ is integrated with respect to t to calculate the average unavailability p_u of the whole system

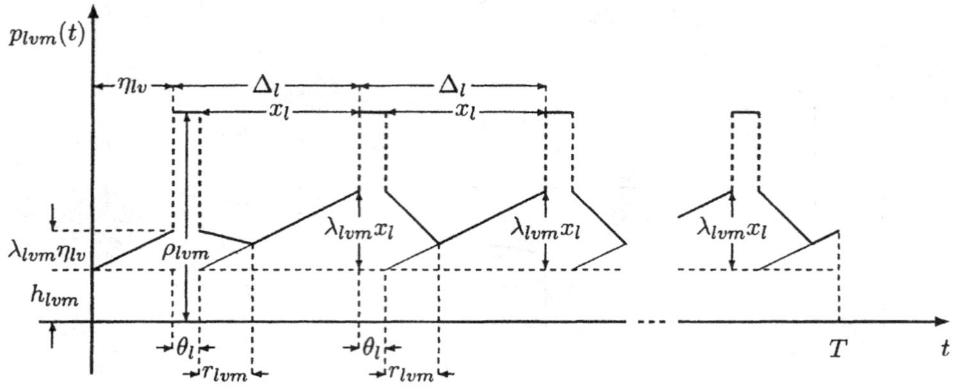

Figure 4 Unavailability $p_{lvm}(t)$

$$p_u = \frac{1}{T}\int_0^T p(t)\,\mathrm{d}t = \frac{1}{T}\int_0^T \sum_{w=1}^{W} \prod_{(lvm)\in C_w} p_{lvm}(t)\,\mathrm{d}t. \tag{7}$$

Here, we consider that T is very large in comparison with $\Delta_l, l = 1, \ldots, L$. Therefore, we can omit the beginning and the end of the operation interval and consider only the periodical part of the schedule for each component. The average unavailability p_u can be approximated as

$$p_u \approx \lim_{T\to\infty} \frac{1}{T}\int_0^T \sum_{w=1}^{W} \prod_{(lvm)\in C_w} p_{lvm}(t)\,\mathrm{d}t \stackrel{\text{def}}{=} \tilde{p}_u. \tag{8}$$

The paper by Uryasev and Vallerga (1993) considers various cases when the approximate unavailability of the function $\tilde{p}_u$ can be expressed in terms of the limit of average unavailability $\lim_{T\to\infty}\frac{1}{T}\int_0^T p_{lvm}(t)\,\mathrm{d}t$ of the components a_{lvm}. In this paper, we discuss only the case with the random starts of the tests.

Since one period of the function $p_{lvm}(t)$ for $l > 0$ can be calculated with the function

$$\zeta(\theta_l, \lambda_{lvm}x_l + h_{lvm}, r_{lvm}, \lambda_{lvm}, x_l, \rho_{lvm}, h_{lvm}, t-(k-1)\Delta_l - \eta_{lv}),$$

we have (see Figure 4)

$$\frac{1}{T}\int_0^T p_{lvm}(t)\,\mathrm{d}t \approx \lim_{T\to\infty}\frac{1}{T}\int_0^T p_{lvm}(t)\,\mathrm{d}t = \frac{1}{\Delta_l}\int_0^{\Delta_l} \zeta(\theta_l, \lambda_{lvm}x_l + h_{lvm}, r_{lvm}, \lambda_{lvm}, x_l, \rho_{lvm}, h_{lvm}, t)\,\mathrm{d}t.$$

The integral of the function ζ is denoted by ψ,

$$\psi(\theta, r, \lambda, x, \rho, h) \stackrel{\text{def}}{=} \frac{1}{\theta + x}\int_0^{\theta+x} \zeta(\theta, \lambda x + h, r, \lambda, x, \rho, h, t)\,\mathrm{d}t = \frac{\theta\rho + 0.5\lambda x(r+x) + hx}{\theta + x}. \tag{9}$$

With the previous notation, the average unavailability of the component a_{lvm} is approximated by the following function ψ_{lvm},

$$\psi_{lvm} = \begin{cases} p_{lvm} & \text{if } l = 0, \\ \psi(\theta_l, r_{lvm}, \lambda_{lvm}, x_l, \rho_{lvm}, h_{lvm}), & \text{otherwise.} \end{cases} \tag{10}$$

4.1. Average Unavailability of a System with Random Starts of Tests

Now, we consider the case when each subset $U_l, l = 1, \ldots, L$ has only one component (subset U_0 can have several components). Further, this restriction is relaxed with additional constraints in the optimization problem. We imposed constraints that assure that the appropriate test intervals are equal (see (14)). We supposed that for each $l = 1, \ldots, L$, the shift of schedule y_l is a random variable uniformly distributed on the interval $[0, \Delta_l]$. All shifts are independent. With this condition, the expected unavailability $\mathbb{E}\tilde{p}_u$ can be calculated as:

$$\mathbb{E}\tilde{p}_u = \mathbb{E} \lim_{T\to\infty} \frac{1}{T}\int_0^T \sum_{w=1}^{W} \prod_{(l1m)\in C_w} p_{l1m}(t)\, dt = \lim_{T\to\infty} \frac{1}{T}\int_0^T \sum_{w=1}^{W} \prod_{(l1m)\in C_w} \mathbb{E}p_{l1m}(t)\, dt.$$

If $l>0$ and $\Delta_l/4+2\Delta_l \leqslant t \leqslant \Delta_l/4+k_{lv}\Delta_l$, then the mathematical expectation $\mathbb{E}p_{l1m}(t)$ is equal to the integral of the function $p_{l1m}(t)$ over one period, that is

$$\mathbb{E}p_{l1m}(t) = \psi_{l1m}.$$

Consequently,

$$\mathbb{E}\tilde{p}_u = \lim_{T\to\infty} \frac{1}{T}\int_0^T \sum_{w=1}^{W} \prod_{(l1m)\in C_w} \psi_{l1m}\, dt = \sum_{w=1}^{W} \prod_{(l1m)\in C_w} \psi_{l1m}. \tag{11}$$

The last formula is a good approximation of the unavailability (7) for the general case. Most probabilistic risk assessments of nuclear systems are based on similar formulas.

5. Optimization of Test Intervals

Let us consider the minimization problem for the unavailability estimate $\tilde{p}_u$ with respect to the test intervals $x_1, \dots, x_L$

$$\tilde{p}_u(x) \longrightarrow \min_{x\in X}, \tag{12}$$

subject to the constraints

$$f_i(x) \leqslant 0, \quad i = 1, \dots, I, \tag{13}$$

where $x = (x_1, \dots, x_L) \in R^L$ and $X = \{x \in R^L : \underline{x}_l \leqslant x_l \leqslant \bar{x}_l, l = 1, \dots, L\}$; and the constants $\underline{x}_l$ and $\bar{x}_l$ are the lower and upper bounds of the test interval x_l. Constraints (13) (possibly nonlinear) can take into account relations between test intervals of different groups of components. For example, we can impose the following constraints:

$$x_i = x_j = \cdots = x_k, \tag{14}$$

where $i, j, \dots, k$ are some numbers from the set $l = 1, \dots, L$. In this case, we can introduce a new variable $\hat{x}$

$$\hat{x} = x_i = x_j = \cdots = x_k$$

and reduce the dimension of the problem.

5.1. System with Random Starts of Tests

Let us consider the following optimization problem for a system with random starts of the tests:

$$\mathbb{E}\tilde{p}_u = \sum_{w=1}^{W} \prod_{(l1m)\in C_w} \psi_{l1m} \longrightarrow \min_{x\in X\subset R^L} \tag{15}$$

$$X = \{x \in R^L : \underline{x}_l \leqslant x_l \leqslant \bar{x}_l, \ \text{for}\ l = 1, \dots, L\},$$

subject to

$$f_i(x) \leqslant 0, \quad i = 1, \dots, I. \tag{16}$$

We used gradient methods to solve this problem. The gradient of the function $\mathbb{E}\tilde{p}_u$ is calculated as follows:

$$\nabla_x \mathbb{E}\tilde{p}_u(x) = \sum_{w=1}^{W} \sum_{(l1m)\in C_w} \psi_{l1m}^{-1} \nabla_{x_l} \psi_{l1m} \prod_{(l1m)\in C_w} \psi_{l1m}. \tag{17}$$

The gradient $\nabla_{x_l}\psi_{l1m}$ in this equation is given by the following formula:

$$\nabla_{x_l}\psi_{lvm} = \begin{cases} 0, & \text{if } l = 0, \\ \dfrac{\theta_l(h_{lvm} + 0.5\lambda_{lvm}r_{lvm} - \rho_{lvm}) + \lambda_{lvm}x_l(\theta_l + 0.5x_l)}{(\theta_l + x_l)^2}, & \text{otherwise.} \end{cases}$$

The objective function $\mathbb{E}\tilde{p}_u(x)$ and the gradient $\nabla_x\mathbb{E}\tilde{p}_u(x)$ can be calculated during one run through the cut sets.

6. An Example

Here, we consider a relatively small system that exemplifies the suggested methodology. To estimate the unavailability of the system we assumed that this is a system with random starts of the tests. Although this assumption introduces an error in the estimate, it allows construction of efficient algorithms for estimation and optimization of the unavailability. We optimized the unavailability function (15) using the gradient variable metric algorithm (Uryasev, 1991). The gradient for all test intervals was calculated during one run through the cut sets (see (17)). The calculation of the gradient with respect to all variables took approximately the same time as the calculation of the unavailability function.

The nuclear power plant safety system considered includes several systems: a reactor protection system (RP), a high pressure injection system (HPI), a low pressure injection system (LPI), and one support (electric) system. We consider only one initiating event, loss of coolant (LOCA). The initiating event LOCA is presented in Figure 5, where sequence 1 is the success sequence, and sequences 2 to 5 lead to different degrees of core damage; we choose only the minimal cut sets of sequence 4. The system flow chart is drawn in Figure 6 and the support system in Figure 7.

The general framework for the system with random starts of the tests allows any nonlinear constraints for the test intervals. We impose two types of constraints:

1. Test intervals are the same for the following systems:
   - HPI that includes tanks and valves (TA, VD; TB, VE; TC, VF);
   - LPI that includes pumps and valves (PA, VA; PB, VB; PC, VC);
   - Electric power supply (BOH, BOL; BEH, BEL);
2. Each test interval has a lower and upper bound that reflect technical specifications and management constraints.

The first set of constraints can be eliminated by the proper change of variables (equal test intervals can be denoted by the same variable). After the change of variables, we have a relatively simple nonlinear programming problem with three variables and box constraints

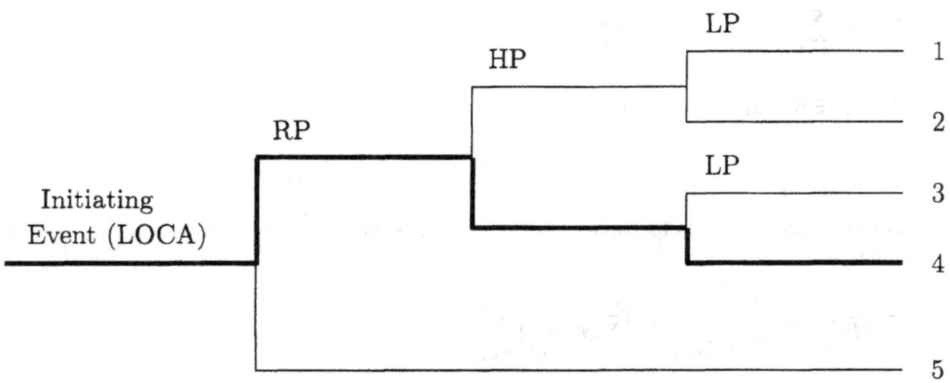

Figure 5 Event tree

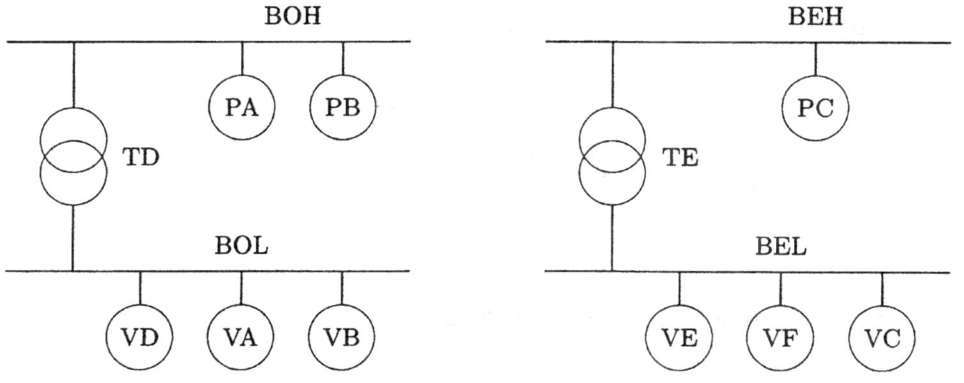

Figure 6 System flow chart

(upper and lower bounds on variables). The input data are shown in Table 1. It was considered that the test unavailabilities equal 1 and the time independent parts of unavailabilities equal 0 (the data were taken from the standard example available to the author). These assumptions lead to an increase in the optimal test intervals.

The system was solved (the minimal cut sets were generated) by the PSAPACK fault event/tree computer code (see, Advisory Group Meeting Report on the Development of Computer Software for Using PSA in NPP Operational Safety Management, 1988; and Bojadiev et al., 1989) and the result was 81 minimal cut sets; the average unavailability was 0.003. Table 2 shows the box constraints and the initial and optimal values for the test intervals. At the optimal point, the average unavailability is 0.00054. The optimization improved the average unavailability about five times.

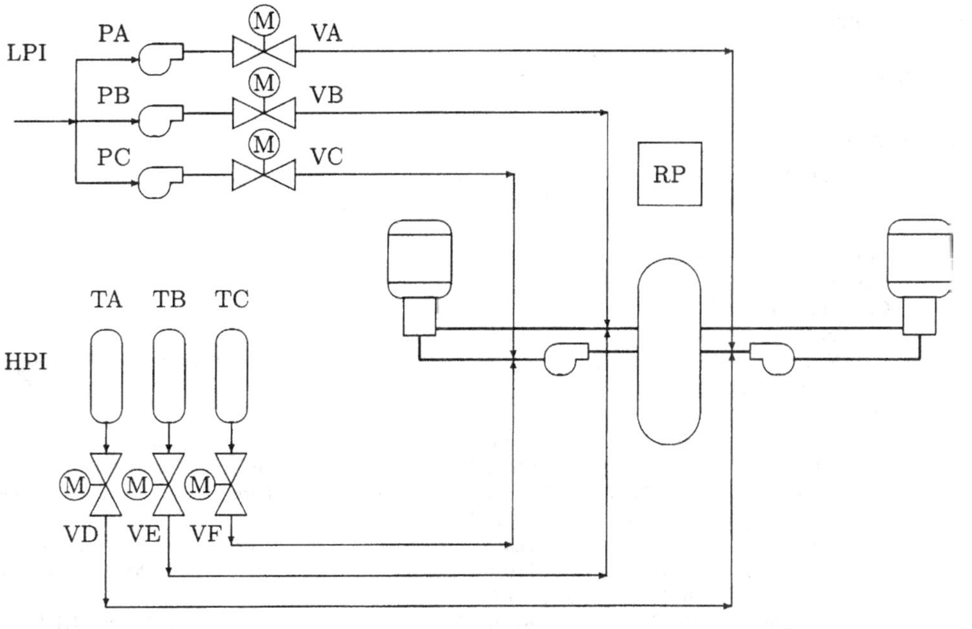

Figure 7 Electric power supply

Table 1 Input Data

Component	*Failure rate* (hr^{-1})	*Repair time* (hr)	*Test duration* (hr)	*Initial test interval* (Days)
PUMP	2×10^{-5}	18	2	120
VALVE I	7×10^{-7}	3	2	120
VALVE II	7×10^{-7}	3	2	360
TANK	1×10^{-8}	24	4	360
BUS I	6×10^{-7}	10	4	360
BUS II	2×10^{-7}	8	4	360

Table 2 Bounds, and Initial and Optimal Values of the Test Intervals

Subsystem	*Lower bound*	*Upper bound*	*Initial value*	*Optimal value*
HPI	7	360	360	360
LPI	7	360	120	360
EP	7	8000	360	4121

7. Conclusion

The paper considers the modeling approach for estimating the average unavailability of a system with periodically tested components. The components, which are tested at the same time, are united in groups. To simplify the estimates, we considered that the system has random starts of the tests. Although this assumption introduces error in the unavailability expression, it allows for developing efficient algorithms for evaluating and optimizing unavailability. Similar formulas (obtained under different assumptions) are presently used in most practical studies of nuclear safety systems. For a system with random starts of the tests, the problem of finding optimal test intervals was reduced to a nonlinear programming problem. For this problem, we developed formulas for the calculation of the gradients of the performance function and constraints. A gradient algorithm with variable metric was used to optimize the average unavailability of the system. The approach was tested with an example including several nuclear power plant safety systems: a reactor protection system (RP), a high pressure injection system (HPI), a low pressure injection system (LPI), and one support (electric) system. The approach demonstrated that considerable reduction of unavailability can be achieved by optimizing the test intervals.

References

Advisory Group Meeting Report. 1988. "Development of Computer Software for Using PSA in NPP Operational Safety Management." IAEA Internal Report. Vienna: IAEA.

Apostolakis, G. and T. Chu. 1980. "The unavailability of systems under periodic tests and maintenance." *Nuclear Technol.* **50**: 5–15.

Bojadiev, A., L. Lederman, and H. Vallerga. 1989. "PSAPACK: an event/fault tree package for PSA using PC." In *PSA 89 ANS ENS Topical Meeting*. Pittsburgh.

"Dependent Failure Analysis." 1991. *Reliability Engng Sys. Safety*, Special Issue **34(3)**.

Dialynas, E. N. and D. G. Michos. 1993. "Time dependent unavailability analysis of standby components incorporating arbitrary failure distributions and three inspection/maintenance policies." *Reliability Engng Sys. Safety* **39**: 25–35.

Ginsburg, T. and J. Powers. T. 1986. *FRANTIC-III—a computer code for the time-dependent reliability analysis.* Technical Report 3230, 8-20-86. Long Island, NY: Brookhaven National Laboratory (BNL).

Kim, I. S., Martorell, S., Vesely, W. E., and Samanta, P. K. 1992. *Quantitative evaluation of surveillance test intervals including test-caused risks.* NUREG/CR-5775.

Laakso, K. (Ed.) 1990. "Optimization of technical specifications by use of probabilistic methods—a nordic perspective." Final Report of the NKA Project RAS-450. Prepared by K. Laakso, M. Knochenhauer, T. Mankamo, and K. Pörn. Nord Series 1990:33.

Lofgren, E., F. Varcolik, and W. E. Vesely. 1981. In *Optimum Test Intervals for Online Testing*. NUREG/CR - 2158 SAI 528951-S.

Mankamo, T. and U. Pulkkinen. 1988. In *Test Interval Optimization of Standby Equipment*. Research Notes 892. Espoo, Finland: Technical Research Centre of Finland.

Papazoglou, I. A. 1988. In *A Code for Markovian Reliability Analysis of Systems. User's Guide. Demokritos*. The Greek Atomic Energy Commission.

Pulkkinen, U., Huovinen, T., Mankamo, T., Norros, L. and Vanhala, J. 1989. "Reliability of diesel generators in the Finnish and Swedish nuclear power plants." Report SÄH 7/82, June (1982). Technical Research Centre of Finland, Espoo, Finland. Enhanced version published as VTT Research Notes 1070, 1989.

Robert, N. H., W. E. Vesely, D. F. Haasl, and F. F. Goldberg. 1981. In *Fault Tree Handbook*: System and Reliability Research Office of Nuclear Regulatory Research, US Nuclear Regulatory Commission.

Sim, S. H. and J. Endrenyi. 1988. "Optimal preventive maintenance with repair." *IEEE Trans. Reliability* **37**: 92–96.

Uryasev, S. 1991. "Adaptive variable metric algorithms for nonsmooth optimization problems." *J. Optim. Theory and Appl.* **71(2)**: 359–388.

Uryasev, S. and H. Vallerga. 1993. "Optimization of test strategies: a general approach." *Reliability Engng Sys. Safety* **41**: 155–165.

Vaurio, J. K. 1979. "Unavailability of components with inspection and repair." *Nucl. Engng Des.* **54**: 309.

Vaurio, J. K. 1981. "Availability analysis of standby safety systems." In *Proceedings of the International ANS/ENS Meeting on Probabilistic Risk Assessment, Port Chester, New York, September 20–24.*

Vaurio, J. K. 1982. "Practical Availability Analysis of Standby Systems." In *Proceedings of Annual Reliability and Maintainability Symposium*, pp. 125–131.

Vaurio, J. K. and D. Sciadone. 1980. "Unavailability analysis of redundant safety systems." In *Proceedings of 1980 Reliability Conference for the Electric Power Industry*. Madison, WI: American Society for Quality Control.

Vesely, W. E. 1991. "Incorporating aging effects into probabilistic risk analysis using a Taylor expansion approach." *Reliability Engng* **32**: 315–337.

Vesely, W. E. and F. F. Goldberg. 1977a. "Time dependent unavailability analysis of nuclear safety systems." *IEEE Trans. Reliability* **26(3)**: 220.

Vesely, W. E. and F. F. Goldberg. 1977b. In *FRANTIC—a Computer Code for Time-dependent Unavailability Analysis, NUREG-0193*.

Wagner, D. P., L. A. Minton, S. E. Rose, and D. J. Hesse. 1987. In *PC-SOCRATES Version 1.02 User's Guide*. Electric Power Research Institute. CA.

18.11 Optimization in VLSI Design: Target Distance Models for Cell Placement

Hussein A. Y. Etawil and Anthony Vannelli

ABSTRACT Analytical VLSI placement has traditionally used linear or quadratic models to reduce total circuit wirelength. The methodology iterates between two basic steps: solving an optimization problem and repartitioning. In this work, we extend the classical quadratic wirelength objective in such a way that overlap between connected cells is diminished and no partitioning is used. The new formulation is a convex optimization problem. The new placer is tested using five industrial benchmarks (MCNC, 1992). The quality of the placements obtained is comparable to, and for some benchmarks is better than, that obtained using a quadratic placer combined with full partitioning. Moreover, the new placer outperforms a simulated annealing based placement package (TimberWolf4.2; Sechen et al., 1988) in terms of quality and efficiency.

1. Introduction

From its humble beginning in the early 1950s to the manufacture of circuits with millions of components today, VLSI (Very Large Scale Integration of electronic circuits) design has brought the power of a mainframe computer to a laptop. This tremendous growth in the area of VLSI design has been made possible by the development of sophisticated design tools and software. To deal with the complexity of millions of components, VLSI design tools must be computationally fast and generate layouts close to optimality. The future growth of VLSI systems depends critically on the research and development of circuit layout (physical design) automation tools. Before we describe the design cycle of VLSI systems, we would like to point out that most of the background described in this section is well established. For those who are interested in more information, the following references are recommended: Ohtsuki et al. (1979), Preas and Lorenzetti (1988), Sherwani (1993), and Zobrist (1994).

The VLSI design cycle starts with a formal specification of a VLSI chip, follows a series of steps, and eventually produces a packaged chip. The steps of the VLSI design cycle can be briefly outlined as follows (Sherwani, 1993; Ullman, 1984):

System specification. As in any design process, the first step is to lay down the specifications of the system to be designed. This necessitates creating a high-level representation of the system. The factors to be considered in this process include: performance, functionality, and the physical dimensions. The choice of fabrication technology and design techniques are also considered. The end results are specifications for the size, speed, power, and functionality of the VLSI system to be designed.

Functional design. In this step, the behavioral aspects of the system are considered. The outcome is usually a timing diagram or other relationships between subunits. This information is used to improve the overall design process and to reduce the complexity of the subsequent phases.

Logic design. In this step, the logic structure that represents the functional design is derived and tested. The design achieved is represented by a textual, schematic, or graphic description. The logic design is usually represented by Boolean expressions. These expressions are minimized to achieve the smallest logic design that conforms to the functional design. Logic design of the system is simulated and tested to verify its correctness.

Circuit design. The purpose of circuit design is to develop a circuit representation based on the logic design. The Boolean expressions are converted into circuit representation by taking into consideration the speed and power requirements of the original design. The electrical behavior of the various components are also considered in this phase. The circuit design is usually expressed in a detailed circuit diagram.

Circuit layout. In this step, the circuit representation of each component is converted into geometric representation. This representation is in fact a set of geometric patterns that perform the intended logic function of the corresponding component. Connections between different components are also expressed as geometric patterns. As stated earlier, this geometric representation of a circuit is called a layout. The exact details of the layout also depend on design rules, which are guidelines based on the limitations of the fabrication process and the electrical properties of the fabrication materials. Circuit layout is a very complex process, therefore, it is usually broken down into various substeps in order to handle the complexity of the problem. In fact, circuit layout is arguably the most time consuming step in the VLSI design cycle.

Design verification. The layout is verified in this step to ensure that the layout meets the system specifications and fabrication requirements. Design verification consists of design rule checking and circuit extraction. Design rule checking is a process which verifies that all geometric patterns meet the design rules imposed by the fabrication process. After checking the layout for design rule violations and removing the design violations, the functionality of the layout is verified by circuit extraction. This is a reverse engineering process and generates the circuit representation from the layout. This reverse engineering circuit representation can then be compared with the original circuit representation to verify the correctness of the layout.

Fabrication. After verification, the layout is ready for fabrication. The fabrication process consists of several steps: preparation of wafer, deposition and diffusion of various materials on the wafer according to the layout description. Before the chip is mass produced, a prototype is made and tested.

Packaging, testing, and debugging. Finally, the wafer is fabricated and diced in a fabrication facility. Each chip is then packaged and properly tested.

The VLSI design cycle involves several iterations, both within each step and between different steps. The entire design cycle may be viewed as transformations of representation in various steps. In each step, a new representation of the system is created and analyzed. The representation is iteratively improved to meet system specifications For instance, a layout is iteratively improved so that it meets the timing specifications of the system.

In this work, we focus on the circuit layout (physical implementation) task. As we indicated above, the physical implementation of VLSI circuits is a complex process that attempts to perform a layout of an electronic circuit through several independent steps. Due to its tremendous complexity, the layout task has always been performed in two phases, placement phase and routing phase. In the placement phase, the circuit cells are assigned to geometric locations on the layout grid. In the routing phase, the connections among the cells are realized. In this report, we are going to confine ourselves to the placement phase. We start by reviewing the most popular existing placement techniques.

Min-cut placement. The procedure is based on the recursive application of the bipartitioning algorithm (Fiduccia and Mattheyses, 1982). Basically, the algorithm partitions the set of cells into two subsets by either a horizontal or vertical line such that the number of cut-nets between the two subsets is minimized and a certain area-based criterion is satisfied. This procedure is recursively applied to each subset until the subset contains only one cell.

Constructive placement. In this method, a cell is first selected as a seed cell. Recursively, a new cell is selected and placed adjacent to the ensemble by applying a best fit heuristic scheme. There are many variations to this method regarding the selection of the seed cell and the selection of the new cell (Preas and Lorenzetti, 1988; Zobrist, 1994).

Search heuristics based placement. In a search heuristic based placement algorithm, the solution space is usually the set of all possible assignments of cells to the locations. Specifically, a solution space is specified by the locations of all the cells on the chip as well as their rotations and reflections (in the case of macro-cells design style) from standard orientations. Search heuristics begin with an initial placement and attempt to obtain a better placement by perturbing the current placement such that the new arrangement of cells has a lower cost. A neighboring solution can be generated via cell displacement, where a cell is moved to a new location, or via cell interchange, where two cells are selected and their centers are exchanged (Sechen et al., 1988). To increase the flexibility of the solution space, overlap of cells is usually allowed. The cost functions usually account for an estimation of the total wirelength together with a penalty term for module overlap.

Analytic methods. In this approach, the placement problem is formulated as a continuous linear or nonlinear mathematical program. Mathematical programming techniques are then applied. The result is a relative placement in which a cell is expected to be placed in a region that would most likely contain the geometric location where the cell should eventually reside. In the case of row-based placement, cells are snapped into the rows using some heuristic. Local improvement is then applied to the relative placement to obtain the final placement. The force-directed method and electrical network analogy based method (Kang et al., 1994; Gao et al., 1992) represent typical examples of analytic methods.

Currently the best up to date placers are based on the iterative improvement method, that is, simulated annealing (Sechen et al., 1988) and analytic methods (Kleinhans et al., 1991).

Traditionally, analytical placement has used a linear or a quadratic wirelength objective function. Placement algorithms optimize such objectives by approximately solving a series of sparse linear systems. Usually cells are placed on top of each other as a result of the attraction force and the tendency of the optimizer to make the wirelength as short as possible. Typically, partitioning is applied to recursively create subproblems in which new constraints are introduced to restrict the movement of the cells. As a result, partitions are refined in each iterate and overlap between the cells is reduced. Our method is the analytic approach based method. It still minimizes wirelength, but it deals with cell overlap in a completely different manner. To be more specific, partitioning of the placement area becomes irrelevant, and if needed, the number of partitioning steps required is drastically reduced.

Our main contributions can be summarized as follows:

- We introduced the concept of target distance based placement in which, upon minimization, a prespecified distance is maintained between each pair of connected cells. As a result, reducing overlap using partitioning becomes less important.
- We discovered a class of convex functions that can easily be adapted to the target distance (TD) problem.
- The fact that our models are convex suggests that any solution methodology for convex optimization can be applied.
- Besides being convex, our new models are easy to tune and do not impose any restrictions on the value of the desired distance between the cells. For instance, the target distance can be adaptively computed during the optimization process based on the information in the problem-domain.
- The method allows one to investigate the feasibility of integrating the task of assigning cells to rows (in the case of row-based design style) during the computation of the relative placement.

The remainder of this report is organized as follows. Section 2 presents the problem formulation and describes the quadratic measure of wirelength. In Section 3, the new objective functions are described and proof of their convexity is presented. Section 4 outlines the details of applying the new models to the TD problem. In Section 5, a description of how to integrate relative placement and placement legalization is presented. In Section 6, analysis of the new formulation is introduced. Section 7 presents the results obtained using the TDP (target distance placer) and compares them to other known placers. Section 8 concludes the report.

2. Problem Formulation and the Quadratic Measure

A circuit is represented by a hypergraph $G(V, E)$, where the vertex set $V = \{v_1, v_2, ..., v_n\}$ represents the nodes of the hypergraph, that is, set of cells to be placed, and $E = \{e_1, e_2, ..., e_m\}$ represents the set of edges of the hypergraph, that is, the set of nets connecting the cells. The two-dimensional placement region is represented as an array of legal placement locations. The hypergraph is transformed into a graph (a hypergraph with all hyperedge sizes equal to two) via a clique model for each net. Each edge e_j is an unorder pair of vertices with a nonnegative weight w_j assigned to it. The placement task seeks to assign all cells of the circuit to legal locations so that cells do not overlap. Each cell i is assigned a location (x_i, y_i) on the XY-plane. The cost of an edge connecting two cells i and j with locations (x_i, y_i) and (x_j, y_j) is computed as the product of the squared l_2 norm of the difference vector $(x_i - x_j, y_i - y_j)$ and the weight of the connecting edge w_{ij}. In other words, the cost contributed by each two connected cells i and j is the Euclidean distance between their geometric locations. The total cost, denoted $\phi(x, y)$, can then be given as the sum of the cost over all edges, that is,

$$\phi(x, y) = \sum_{1 \leqslant i<j \leqslant N} w_{ij}[(x_i - x_j)^2 + (y_i - y_j)^2]. \tag{1}$$

Formulation (1) can be written in matrix form:

$$\phi(x,y) = \tfrac{1}{2}\mathbf{x}^T\mathbf{C}\mathbf{x} + \mathbf{d}_x^T\mathbf{x} + \tfrac{1}{2}\mathbf{y}^T\mathbf{C}\mathbf{y} + \mathbf{d}_y^T\mathbf{y} + t. \tag{2}$$

Vectors **x** and **y** denote the coordinates of the N movable cells; matrix **C** is the Hessian matrix; vectors $\mathbf{d}_x^T$ and $\mathbf{d}_y^T$ and the constant term t result from the contributions of the fixed cells. Normally the first moment constraints are added to force the distribution of the cells to be uniform around the center of the placement area. It follows that the quadratic placement model is given as:

$$\begin{aligned} \text{Min} \quad & \phi(x,y) \\ \text{subject to} \quad & A_x x = b_x \\ & A_y y = b_y \\ & l_x \leqslant x_i \leqslant u_x \\ & l_y \leqslant y_i \leqslant u_y, \end{aligned}$$

where A_x and A_y are $q \times n$ matrices; q is the number of regions into which the placement area has been partitioned. The $q \times 1$ vectors b_x and b_y represent the centers of the q regions. The parameters l_x, u_x, l_y, and u_y are lower and upper bounds on the x and y coordinates of the cells. Clearly, the above optimization problem can be split into two one-dimensional subproblems and each subproblem can then be solved independently.

As pointed out previously, minimizing the quadratic model for wirelength as given by (1) yields a placement where cells overlap (Figures 6, 8, and 10). Consequently, extra efforts need to be made to remove the overlap. In all previous attempts, the overlap problem was handled by partitioning the placement area and changing the constraints on cell locations before solving another optimization problem. An intuitive idea to avoid partitioning is to maintain a target distance between each connected pair of cells. In other words, place the cells such that a lower bound on the distance between their respective locations is maintained. For example, the following formulation accomplishes this aim (Du and Vannelli, 1998):

$$\psi(x,y) = \sum_{1 \leqslant i<j \leqslant N} w_{ij}[(x_i - x_j)^2 + (y_i - y_j)^2 - d_{ij}]^2. \tag{3}$$

Minimizing (3) yields a placement with no overlap between connected cells. However, this model is not convex and, therefore, convergence to a global optimal answer is not guaranteed. This difficulty motivated us to seek a different model for the target distance-based placement. Such a model should be convex in the first place, and flexible enough to incorporate other performance issues (i.e., delays, clock, power dissipation, etc.).

3. New Convex Models

In an endeavor to reduce overlap among cells through maintaining a target distance between their geometric locations, we discovered a class of functions that achieves this aim. The following theorem summarizes the main results.

Theorem 1. *Let* $\zeta : \mathscr{R}^m \to \mathscr{R}$ *and be given by* $\zeta = \|\mathbf{v}\|_2^2$, *where* $\mathbf{v} = (v_1, v_2, \ldots, v_m)$ *is an m-dimensional vector, then the function* $\eta(\zeta) = \sigma(\zeta) + \rho(\zeta)$ *is convex for* $\zeta \in [1, \infty)$, *and the ordered pair of functions* $(\sigma(\zeta), \rho(\zeta)) \in \pi$ *and* $\pi = \{(\zeta, -\ln(\zeta)), (\zeta, e^{1-\zeta}), (0, 1/\zeta)\}$.

To prove this theorem, we use the fact that if a function $f(x) \in C^2$, then $f(x)$ is convex over a convex set Θ if the Hessian matrix of $f(x)$ is positive semidefinite through Θ (Luenberger, 1984). First, we consider the case when the elements of **v** are independent, then we consider the case when the elements of **v** are dependent; specifically when $\mathbf{v} = \mathbf{u} - \mathbf{r}$ and **u** and **r** are also m-dimensional vectors.

3.1. *Elements of v are Independent*

We start by defining

$$\dot{\rho}(\zeta) = \partial\rho(\zeta)/\partial\zeta$$

$\ddot{\rho}(\zeta) = \partial^2 \rho(\zeta)/\partial \zeta^2.$

The gradient of $\eta(\zeta)$ is given by

$$\nabla \eta(\zeta) = \begin{bmatrix} 2v_1(1+\dot{\rho}(\zeta)) \\ 2v_2(1+\dot{\rho}(\zeta)) \\ \vdots \\ 2v_m(1+\dot{\rho}(\zeta)) \end{bmatrix}.$$

Subsequently, the Hessian matrix $\nabla^2 \eta(\zeta)$ can be computed as follows:

$$\nabla^2 \eta(\zeta) = B + C,$$

where B and C are given as follows:

$$B = 4\ddot{\rho}(\zeta)\mathbf{v}^T\mathbf{v}$$

$$C = \mathrm{diag}\{2(1+\dot{\rho}(\zeta))\}. \tag{4}$$

Clearly when $\ddot{\rho}(\zeta) > 0$, matrix B is a rank one matrix with $m-1$ zero eigenvalues and only one positive eigenvalue. Hence, matrix B is a positive semidefinite matrix. At $\dot{\rho}(\zeta) = -1$, the diagonal matrix C vanishes, that is, $C = 0$. Thus, the Hessian matrix $\nabla^2 \eta(\zeta)$ is a positive semidefinite matrix when $\dot{\rho}(\zeta) = -1$. On the other hand, when $\dot{\rho}(\zeta) > -1$, C is a positive matrix because $C > 0$. In this case C adds to the positiveness of the Hessian matrix $\nabla^2 \eta(\zeta)$. As a result, the Hessian matrix $\nabla^2 \eta(\zeta)$ is a positive definite matrix when $\dot{\rho}(\zeta) > -1$.

We can now examine if $\rho(\zeta) \in \pi$ satisfies Theorem 1. In fact we only need to examine $\dot{\rho}(\zeta)$.

Case 1. $\rho(\zeta) = -\ln(\zeta)$. In this case $\dot{\rho}(\zeta) = -1/\zeta$. Matrix C is 0 when $\dot{\rho}(\zeta) = -1$ or $\zeta = 1$, thus in this case $\nabla^2 \eta(\zeta) = B$ is positive semidefinite. When $\dot{\rho}(\zeta) > -1$ or $\zeta > 1$, matrix $C > 0$ and as a consequence $\nabla^2 \eta(\zeta)$ is positive definite. Thus Theorem 1 is satisfied and accordingly $\eta(\zeta)$ is convex for $\zeta \geqslant 1$.

Case 2. $\rho(\zeta) = \mathbf{e}^{1-\zeta}$. Function $\dot{\rho}(\zeta) = -\mathbf{e}^{1-\zeta} \geqslant -1$ implies that $1-\zeta \leqslant 0$ or $\zeta \geqslant 1$. Again when $\dot{\rho}(\zeta) = -1$ or $\zeta = 1$, matrix C is 0 and $\nabla^2 \eta(\zeta) = B$ is positive semidefinite. When $\dot{\rho}(\zeta) > -1$ or $\zeta > 1$, C is positive and $\nabla^2 \eta(\zeta)$ is positive definite. Thus, Theorem 1 is satisfied and $\eta(\zeta)$ is convex for $\zeta \geqslant 1$.

Case 3. $\rho(\zeta) = 1/\zeta$. Function $\dot{\rho}(\zeta) = -1/\zeta^2$ and $\dot{\rho}(\zeta) \geqslant -1$ when $\zeta \geqslant 1$ (and $\zeta \geqslant -1$). But negative ζ means that $\ddot{\rho}(\zeta) = 2/\zeta^3$ is negative, which in turn leads to negative B and negative $\nabla^2 \eta(\zeta)$. Thus $\zeta \geqslant 1$ is essential for $\nabla^2 \eta(\zeta)$ to be positive semidefinite/definite and consequently for $\eta(\zeta)$ to be convex.

The following corollary demonstrates the effect on $\nabla^2 \eta(\zeta)$ as $\dot{\rho}(\zeta) \to 0$ or equivalently as $\zeta \to \infty$.

Corollary 1. *As $\dot{\rho}(\zeta) \to 0$, matrix $B \to 0$, matrix $C \to diag\{2\}$, and, consequently, the Hessian matrix $\nabla^2 \eta(\zeta) \to diag\{2\}$. That is, $\nabla^2 \eta(\zeta) \approx 2I$, where I is the identity matrix.*

Corollary 1 implies that as $\dot{\rho}(\zeta) \to 0$ (which implies that ζ goes to ∞), the Hessian matrix $\nabla^2 \eta(\zeta)$ become less dependent on the variables $v_1, v_2, \ldots, v_m$. In other words, the Hessian matrix approaches the Hessian of the quadratic function ζ. Since the quadratic function is a perfectly convex function, we conclude that as $\zeta \to \infty$, the curvature of $\eta(\zeta)$ resembles that of a quadratic function and it is fairly reasonable to assume $\eta(\zeta)$ is a perfectly convex function.

3.2. *Elements of v are Dependent*

We now turn to investigate the case where the elements of vector $\mathbf{v}$ are not independent variables. In particular, we are interested in the case where $\mathbf{v}$ is given as the difference between m-dimensional vectors p_i and p_j. We limit our analysis to the two-dimensional case as distances

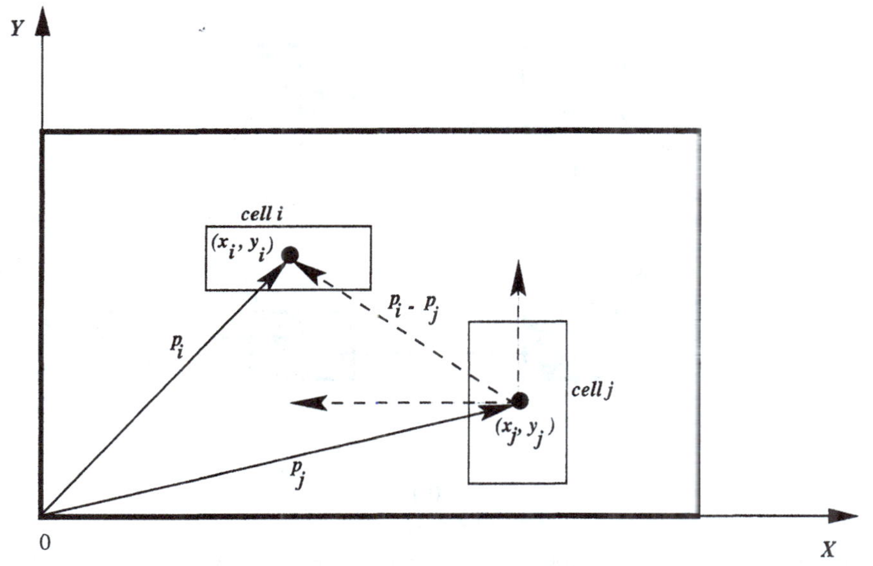

Figure 1 Coordinates of a Two Cell Net

in two-dimensional space constitute the objective function in VLSI placement (see Figure (1)), and also to simplify the analysis.

It is well known that the standard quadratic formulation given by (2) is positive semidefinite, and is positive definite if one (or more) variables is (are) fixed (Kleinhans et al., 1991; Kennings, 1997). Now the question to be answered is: is the function $\eta(\|p_i - p_j\|_2^2)$ convex? We will answer this question and use VLSI placement as our framework. In the context of VLSI placement p_i and p_j represent the geometric locations (x_i, y_i) and (x_j, y_j) of cells i and j. We can then be more specific and ask: is $\eta(\|p_i - p_j\|_2^2)$ convex if cell i and cell j are movable (free) cells, and what would be the effect of fixing one cell (or adding a fixed cell to the group) on the convexity of $\eta(\|p_i - p_j\|_2^2)$? Each interconnection between a pair of cells contributes $\eta(\|p_i - p_j\|_2^2)$ units to the overall objective function. In theory, the number of $\eta(\|p_i - p_j\|_2^2)$ terms constituting the objective function can be ∞. Practically, the number of terms may be very large. In both cases it is not feasible to examine such instances. However, a reasonable alternative is to limit our investigation to small instances and draw a general conclusion about larger ones. Without loss of generality and to keep the analysis simple, we consider a two cell netlist (the simplest netlist that can ever exist) and look at different scenarios with regard to whether all cells are movable or some of them are fixed. In the first scenario, we consider the case where the two cells are movable. In the second scenario, we consider the case where a fixed cell is connected to one of the two movable cells.

Scenario 1. We consider the case of two interconnected movable cells i and j (see Figure 2(a)). The quadratic formulation of wirelength is given by

$$\zeta_{ij} = \|p_i - p_j\|_2^2 = (x_i - x_j)^2 + (y_i - y_j)^2 \tag{5}$$

and the objective function to be minimized $\eta(\zeta_{ij}) = \zeta_{ij} + \rho(\zeta_{ij})$. It follows that the Hessian matrix $\nabla^2\eta(\zeta_{ij})$ is given as

$$\nabla^2\eta(\zeta_{ij}) = M_1 + M_2,$$

where

$$M_1 = \ddot{\rho}(\zeta_{ij})\theta^T\theta$$

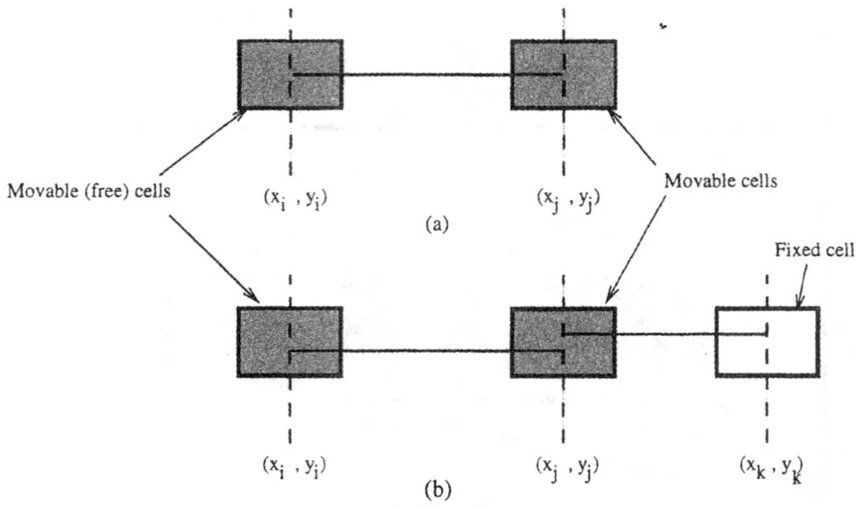

Figure 2 Simple Two Cell Netlist. (a) Both cells are movable. (b) A fixed cell is added to the netlist

$$\theta = \left[\frac{\partial \zeta_{ij}}{\partial x_i} \quad \frac{\partial \zeta_{ij}}{\partial x_j} \quad \frac{\partial \zeta_{ij}}{\partial y_i} \quad \frac{\partial \zeta_{ij}}{\partial y_j} \right]$$

and

$$M_2 = \begin{bmatrix} 1+\dot{\rho}(\zeta_{ij}) & -(1+\dot{\rho}(\zeta_{ij})) & 0 & 0 \\ -(1+\dot{\rho}(\zeta_{ij})) & 1+\dot{\rho}(\zeta_{ij}) & 0 & 0 \\ 0 & 0 & 1+\dot{\rho}(\zeta_{ij}) & -(1+\dot{\rho}(\zeta_{ij})) \\ 0 & 0 & -(1+\dot{\rho}(\zeta_{ij})) & 1+\dot{\rho}(\zeta_{ij}) \end{bmatrix}.$$

Clearly matrix M_1 is rank one matrix (with all zeros but one positive eigenvalue). Matrix M_1 is, therefore, positive semidefinite matrix. Matrix M_2 is positive semidefinite if $\dot{\rho}(\zeta_{ij}) \geqslant -1$. As a consequence, the Hessian matrix $\nabla^2 \eta(\zeta_{ij})$ is positive semidefinite for $\dot{\rho}(\zeta_{ij}) \geqslant -1$. For $\dot{\rho}(\zeta_{ij}) < -1$, M_2 is indefinite matrix and consequently $\nabla^2 \eta(\zeta_{ij})$ can be indefinite matrix. Note that when $\rho(\zeta_{ij}) \in \pi$, $\dot{\rho}(\zeta_{ij}) \geqslant -1$ implies that $\zeta \geqslant 1$.

Scenario 2. We consider the case where one of the cells (cell j for instance) is connected to cell k with a fixed geometric location $p_k = (x_k, y_k)$ (see Figure 2(b)). For the sake of simplicity, let $(x_k, y_k) = (2, 2)$. The objective function f is then given as

$$f = \eta(\zeta_{ij}) + \eta(\zeta_{jk}),$$

where ζ_{ij} is given by Equation (5) and

$$\zeta_{jk} = \|\mathbf{p}_j - \mathbf{p}_k\|_2^2 = (x_j - 2)^2 + (y_j - 2)^2.$$

The Hessian matrix $\nabla^2 f$ is given by

$$\nabla^2 f = M_1 + M_2 + M_3,$$

where

$$M_1 = \ddot{\rho}(\zeta_{ij})\theta^T \theta$$

$$M_2 = \ddot{\rho}(\zeta_{jk})\lambda^T \lambda$$

$$\ddot{\rho}(\zeta_{ij}) = \frac{\partial^2 \rho(\zeta_{ij})}{\partial \zeta_{ij}^2}$$

$$\ddot{\rho}(\zeta_{jk}) = \frac{\partial^2 \rho(\zeta_{jk})}{\partial \zeta_{jk}^2}$$

$$\lambda = \begin{bmatrix} 0 & \frac{\partial \zeta_{jk}}{\partial x_j} & 0 & \frac{\partial \zeta_{jk}}{\partial y_j} \end{bmatrix}$$

and

$$M_3 = \begin{bmatrix} 1+\dot{\rho}(\zeta_{ij}) & -(1+\dot{\rho}(\zeta_{ij})) & 0 & 0 \\ -(1+\dot{\rho}(\zeta_{ij})) & 2+\dot{\rho}(\zeta_{ij})+\dot{\rho}(\zeta_{jk}) & 0 & 0 \\ 0 & 0 & 1+\dot{\rho}(\zeta_{ij}) & -(1+\dot{\rho}(\zeta_{ij})) \\ 0 & 0 & -(1+\dot{\rho}(\zeta_{ij})) & 2+\dot{\rho}(\zeta_{ij})+\dot{\rho}(\zeta_{jk}) \end{bmatrix},$$

where

$$\dot{\rho}(\zeta_{ij}) = \frac{\partial \rho(\zeta_{ij})}{\partial \zeta_{ij}}$$

and

$$\dot{\rho}(\zeta_{jk}) = \frac{\partial \rho(\zeta_{jk})}{\partial \zeta_{jk}}.$$

Evidently for $\ddot{\rho}(\zeta_{ij}) \geqslant 0$ and $\ddot{\rho}(\zeta_{jk}) \geqslant 0$, M_1 and M_2 are rank one matrices and therefore they are positive semidefinite matrices. Matrix M_3 is positive semidefinite when $\dot{\rho}(\zeta_{ij}) = -1$ and $\dot{\rho}(\zeta_{jk}) = -1$. Also it is unmistakable that the determinant of every principal submatrix of M_3 is positive if $\dot{\rho}(\zeta_{ij}) > -1$ and $\dot{\rho}(\zeta_{jk}) > -1$. Matrix M_3 is therefore positive definite when $\dot{\rho}(\zeta_{ij}) > -1$ and $\dot{\rho}(\zeta_{jk}) > -1$. It follows that $\nabla^2 f$ is also positive definite when $\dot{\rho}(\zeta_{ij}) > -1$ and $\dot{\rho}(\zeta_{jk}) > -1$. Again, $\dot{\rho}(\zeta_{ij}) \geqslant -1$ and $\dot{\rho}(\zeta_{jk}) \geqslant -1$ imply that $\zeta_{ij} \geqslant 1$ and $\zeta_{jk} \geqslant 1$.

Explanation: Theorem 1 and Corollary 1 manifest that adding the nonlinear function $\rho(\zeta)$, where $v_1, v_2, ..., v_n$ are independent variables, causes the Hessian matrix to be dependent on the problem variables $v_1, v_2, ..., v_n$. Specifically, the function $\dot{\rho}(\zeta)$ rectifies the positive definiteness of matrix C (given by Equation (4)), which in turn regulates the positive definiteness of the Hessian matrix $\nabla^2 \eta(\zeta)$. Based on what we have seen from Scenarios 1 and 2, this is applicable to the case where $v_1, v_2, ..., v_n$ are not independent. In Scenario 1, the positive semidefiniteness of matrix M_2 is regulated by the term $\dot{\rho}(\zeta_{ij})$, which in turn regulates the semidefiniteness of the associated Hessian matrix $\nabla^2 \eta(\zeta_{ij})$. Similarly, in Scenario 2 the positive definiteness of M_3 is regulated by the terms $\dot{\rho}(\zeta_{ij})$ and $\dot{\rho}(\zeta_{jk})$, which in turn regulates the positive definiteness of the associated Hessian matrix $\nabla^2 f$. These results ascertain the following points:

- As in the case of traditional quadratic formulation, the existence of fixed cells adds positiveness to the Hessian matrix. Such augmentation forces the Hessian matrix to be positive definite. The existence of at least one fixed cell attached to at least one free cell is sufficient to provide adequate positiveness to the Hessian matrix. This can be explained in light of the fact that the netlist hypergraph G (presented in Section 2) is a connected graph. Consequently, cells are globally interconnected and the effect of a fixed cell propagates and accordingly sustains the positive definiteness of the associated Hessian matrix. In general, the amount of positiveness added to the associated Hessian matrix is proportional to the number of fixed cells in the netlist.
- The positive definiteness of the Hessian matrix is controlled by the terms $\dot{\rho}(\zeta_{ij})$, $i, j \in \{1, 2, ..., N\}$, where N is the total number of cells in the netlist.

We now state the following theorem, which is, in essence, a generalization of Theorem 1

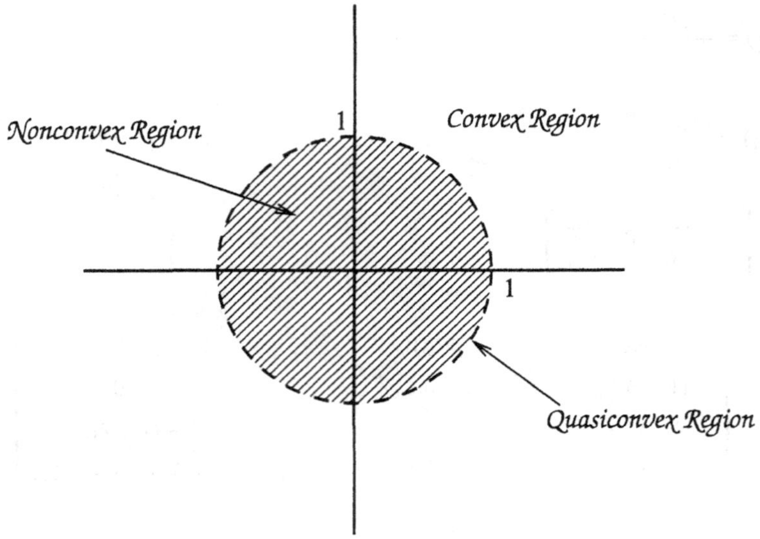

Figure 3 Representation of the regions in which $\sum_{i,j}\eta(\zeta_{ij})$ is convex, quasiconvex, and nonconvex

Theorem 2. *Let $\zeta_{ij} : \mathscr{R}^m \rightarrow \mathscr{R}$ and be given by $\zeta_{ij} = \|p_i - p_j\|_2^2$; $p_i \in \mathscr{R}^m$ and $p_j \in \mathscr{R}^m$; $i \in \mathscr{S}_1$ and $j \in \mathscr{S}_2$; $\mathscr{S}_1 \subseteq \mathscr{S}$ and $\mathscr{S}_2 \subseteq \mathscr{S}$; $\mathscr{S}_1 \cap \mathscr{S}_2 \neq \phi$; $\mathscr{S} = \{1, 2, ..., N\}$, then $\sum_{i,j}\eta(\zeta_{ij})$ is a convex function provided that $\zeta_{ij} \geqslant 1$ $\forall i$ and $\forall j$, and at least one vector $p_h, h \in \mathscr{S}$, is constant.*

Geometrically, each ζ_{ij} corresponds in essence to a circle of radius $r_{ij} \in \mathscr{R}$. According to Theorem 2, function $\sum_{i,j}\eta(\zeta_{ij})$ is convex if the loci of each ζ_{ij} occur outside the unit circle. It is quasiconvex if these loci fall on the contour of the unit circle and nonconvex if some of these loci fall inside the unit circle. Figure 3 depicts the different regions in $\mathscr{R}^2$ for which $\sum_{i,j}\eta(\zeta_{ij})$ is convex, quasiconvex, and nonconvex with respect to the unit circle. Clearly, the nonconvex region is insignificant compared to the convex region. This implies that the model can be applied to real-world problems without expecting any major difficulties.

Besides Theorem 2, the following corollary (analogous to Corollary 1) emphasizes the relationship between the Hessian matrix and ζ_{ij} as $\zeta_{ij} \rightarrow \infty$.

Corollary 2. *As $\zeta_{ij} \rightarrow \infty$, $\forall i$ and $\forall j$, the Hessian matrix of $\sum_{i,j}\eta(\zeta_{ij})$, become less dependent on the problem variables, and eventually approaches the Hessian matrix of quadratic function.*

Generally the elements of the Hessian matrix of $\sum_{i,j}\eta(\zeta_{ij})$ are combinations of variable and fixed terms. Each variable term is comprised in part by $\dot{\rho}(\zeta_{ij})$. Corollary 2 demonstrates that as each ζ_{ij} becomes large, the variable terms vanish, and the Hessian matrix become independent on the problem variables.

Before we conclude this section, we would like to stress again that in real circuits, the number of fixed cells is normally adequate to add sufficient positiveness to the Hessian matrix.

4. Adaptation To VLSI Placement

Since $\eta(\zeta_{ij})$ is convex in the interval $[1,\infty)$, it is quite obvious to choose $[1, \infty)$ as our working interval. Function $\eta(\zeta_{ij})$ is a convex and monotonically increasing function for each $\zeta_{ij} \in [1, \infty)$.

If we let

$$z_{ij} = \frac{\zeta_{ij}}{d} = \frac{(x_i - x_j)^2 + (y_i - y_j)^2}{d}, \tag{6}$$

where $d > 0$ is a constant, we can then present the convex repeller model for the pair of

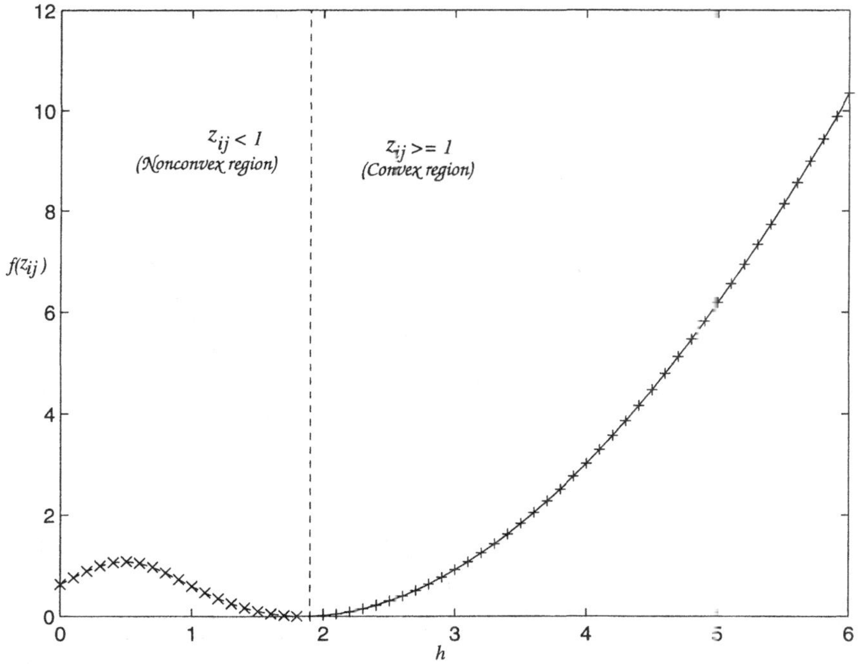

Figure 4 $f(z_{ij}) = z_{ij} - \ln(z_{ij}) - 1$. The portion of the curve marked with "x's" represents $f(z_{ij})$ for $z_{ij} < 1$ and the one marked with "+'s" represents $f(z_{ij})$ when $z_{ij} \geqslant 1$. $f(z_{ij})$ is not convex for $z_{ij} < 1$. Note that we subtracted 1 so that at convergence, $z_{ij} = 1$, $f(z_{ij}) = 0$

connected cells i and j as:

$$f(z_{ij}) = \begin{cases} \eta(z_{ij}) & \text{if } z_{ij} \geqslant 1 \\ 0 & \text{otherwise} \end{cases}, \tag{7}$$

where $\eta(z_{ij}) = z_{ij} + \rho(z_{ij})$ and $\rho(z_{ij}) \in \{-\ln(z_{ij}) - 1, e^{1-z_{ij}} - 2\}$.

Figures 4 and 5 illustrate, in one dimension, examples of $f(z_{ij})$ for $\rho(z_{ij}) = -\ln(z_{ij}) - 1$ and $\rho(z_{ij}) = e^{1-z_{ij}} - 2$, respectively. (Note that the nonconvex part of $f(z_{ij})$ is also shown.) In these examples, $z_{ij} = \|\bar{x} + h\Delta\bar{x} - \bar{y} + h\Delta\bar{y}\|_2^2$, $h \in \{0, 0.1, 0.2, \ldots, 6\}$, and vectors $\bar{x}$, $\Delta\bar{x}$, $\bar{y}$, and $\Delta\bar{y}$ are randomly generated.

The fact that $f(z_{ij})$ is flat in the interval $z_{ij} \in [0, 1]$ implies that $f(z_{ij})$ has multiple solutions in this region. However, line search methods conclude the search when the first optimal answer is encountered. Thus, if the initial solution is outside the flat region, the search will be concluded when $z_{ij} = 1$. It follows that at convergence ($z_{ij} = 1$) the square of the distance equals a target estimate of d units. The cell repeller based placement model can then be given as

$$\begin{aligned} \text{Min} \quad & \mathcal{F} = \sum_{1 \leqslant i < j \leqslant N} w_{ij} f(z_{ij}) \\ \text{subject to} \quad & A_x x = b_x \\ & A_y y = b_y \\ & l_x \leqslant x_i \leqslant u_x \\ & l_y \leqslant y_i \leqslant u_y, \end{aligned} \tag{3}$$

where w_{ij} is the connectivity weight between cells i and j.

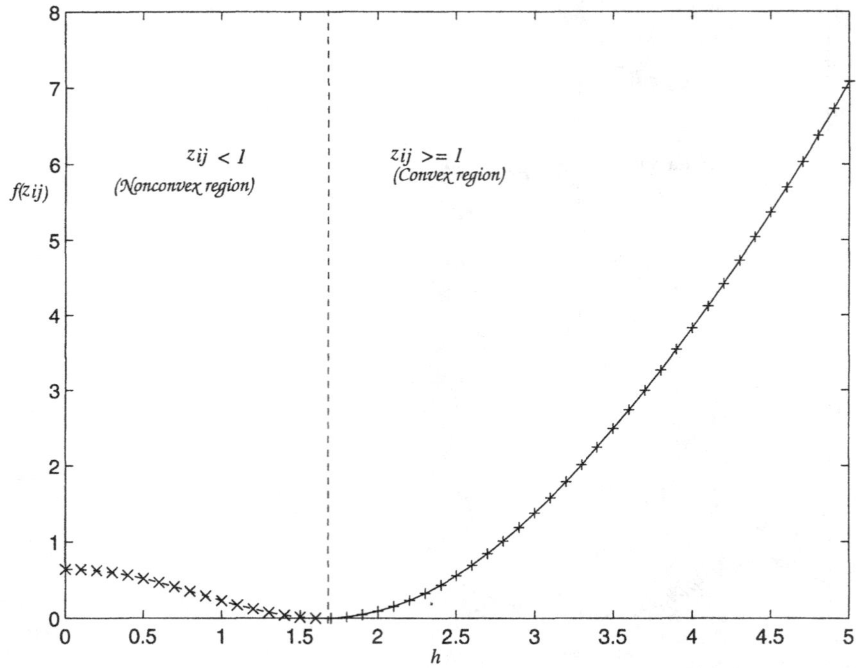

Figure 5 $f(z_{ij}) = z_{ij} + e^{1-z_{ij}} - 2$. The portion of the curve marked with "x's" represents $f(z_{ij})$ for $z_{ij} < 1$ and the one marked with "+'s" represents $f(z_{ij})$ when $z_{ij} \geqslant 1$. Again, $f(z_{ij})$ is not convex for $z_{ij} < 1$. Again, note that we subtracted 2 from $f(z)$ so that at convergence, $z_{ij} = 1$, $f(z_{ij}) = 0$

The result of the global optimization is a relative placement where each cell is placed at or near its final destination. The next step is to adapt the relative placement to the specific design style (standard cell, macro-cell, etc.). In standard cells or row-based design style, cells are snapped to rows. In macro-cell design style, an optimization of the area utilization is needed and cells need to be packed in a compact slicing structure.

Up to now, we have only experimented with standard cell circuits. In this design style, cells are of the same height but not necessarily the same width. The chip area is determined by the width of the routing channels between the rows and the lengths of the rows. The objective is to accomplish a smaller chip area by reducing channel widths and not exceeding the maximum row length. The process of snapping cells to rows is called legalizing the relative placement, or in short, legalization. A widely used technique for legalization was proposed by Dunlop and Kernighan (1985). In this technique, cells are sorted based on their y-coordinates and divided into k subsets (k is the desired number of rows). Each subset is then assigned to a row and the sequence of the cells within the rows is determined based on their x-coordinates. Despite its popularity and despite the fact that the quality of answers obtained is generally acceptable, we believe it is not a robust technique and it is best not to totally rely on it. Perhaps one of the reasons is its greedy nature, which we suspect is responsible for deteriorating the total wirelength as spreading of cells increases. The latter reason motivated us to seek a cumbersome legalization technique that is part of cell relative placement. The idea will be presented in the next section.

5. Relative Placement and Legalization

While working with the TD model, we realized that it is quite feasible to drive the cells to cluster at the row locations during the computation of the relative placement. Specifically, the TD model can be extended by adding another working force that will push each cell to a row rather than any other location across the y-axis. As we pointed out in previous sections, our method is a force-directed method. The x and y locations of cell i are computed based on the attraction force with other cells sharing its nets. Thus, nets act like springs and the optimal placement is analogous to the state of equilibrium in a mechanical system. In light of this analogy, the new working force should shift the equilibrium state to a new one in which cells are distributed among the rows and the length of each row is maintained below the maximum row length.

Define $r_1, r_2, ..., r_k$ to be the row locations across the y-axis. We add another force ω_i working on cell i and given by

$$\omega_i = \prod_{j=1}^{k} (y_i - r_j)^2.$$

The function ω_i is characterized with a minimum value of zero and it is obtained only if the y coordinate of cell i, (i.e., y_i), matches a row location. Thus ω_i seems to satisfy our requirement. However, practically the implementation of ω_i is not feasible. The first reason is the fact that ω_i is the product of other functions and consequently it is not necessarily a convex function (even though it is the product of a convex function). The second reason is the difficulty of differentiating the product of more than 2 or 3 functions. To overcome these difficulties, we suggest taking the minimum of the terms comprising ω_i as our additional force. We define Ω_i to be the additional force as follows

$$\Omega_i = \min\{(y_i - r_1)^2, (y_i - r_2)^2, ..., (y_i - r_k)^2\}.$$

We are now in a position to give the complete TD placement formulation that also accounts for assigning cells to rows while computing the relative placement

$$\begin{aligned} &\text{Min} \quad \mathscr{F} + \sum_{1 \leqslant i \leqslant N} \Omega_i \\ &\qquad l_x \leqslant x_i \leqslant u_x \\ &\qquad l_y \leqslant y_i \leqslant u_y \end{aligned} \tag{9}$$

Where $\mathscr{F}$ is given by (8). We delete the moment constraints since they are irrelevant (the new term would distribute the cells among the rows). Constraints to account for the finite length of each row need to be added. However, at this point, we have not yet figured out the best way to formulate them, but we hope to propose a formulation in the near future. Hence, we would like to make it clear that the results presented in this report are based on the model given by (8).

6. Analysis of the Formulation

In Kennings (1997), a thorough analysis of the number of variable requirements for the quadratic programming formulation is presented. Since our formulation is an extended quadratic formulation, the analysis quoted holds for our formulation.

Let $\mathscr{N}$ represent the number of nets, $\mathscr{M}$ represent the number of cells, and $\mathscr{R}$ be the number of regions into which the placement area has been partitioned. When investigating the sizes and storage requirements of the optimization problem, the quantities of interest, besides the number of variables, are the number of constraints and the number of nonzeros in any matrices required in the formulation (Kennings, 1997).

For the quadratic formulation (as well as the TD-based formulation), the number of variables equals $\mathscr{M}$ and the number of constraints equals $2\mathscr{R}$. The number of nozeros in A_x and A_y is also $\mathscr{M}$.

The density of the Hessian matrix depends on the sparsity of the circuit. For instance, if a net connecting q cells exists, a $q \times q$ dense block of nonzero entities appears in the Hessian matrix. It follows that worst-case scenario takes place if a net connecting every cell exists. In such a scenario, the Hessian matrix is totally dense and requires the storage of $\mathscr{M}^2$ nonzero entries. The storage requirements are $\mathcal{M}^2 + 2\mathcal{M}$ if the Hessian matrix, A_x, and A_y are considered. However, in a typical circuit, the number of nonzero entries in the Hessian matrix is lower than $\mathcal{M}^2$ due to the sparsity of the circuit. Nonetheless, for large circuits, the existence of even one dense block in the Hessian matrix may cause the solver to take a large amount of time to converge, besides the large amount of storage required.

In our implementation of the TD formulation, we used a limited-memory quasi-Newton based solution methodology in which explicit evaluation of the Hessian matrix is not needed (an estimate of the Hessian matrix is computed based on the first order information supplied by the user), namely, L_BFGS_B (Zhu et al., 1997). Also, since our method does not require partitioning of the placement area (and in the worst-case scenario, the number of partitions required is minimal), the number of hard constraints is expected to be the minimum.

7. Experimental Results

The proposed TD formulation is implemented in C language on a SunSparc 5 workstation. Total wirelength is measured as the sum of the half perimeter of the enclosing rectangle for each net. Computation times are measured in units of CPU seconds. The target distance parameter d is computed as the square root of the average cell width. Starting placements are randomly generated such that cells are not on top of each other to satisfy the constraint $z_{ij} \geqslant 1$. The number of partitions required is minimum, and the number of hard constraints is expected to be minimum.

Table 1 summarizes the statistics of each test case (benchmark). Each column represents cells, nets, I/O pads, and number of rows. The test cases are taken from the MCNC benchmark set. The circuits referred to as *ckt1* and *ckt2* are modified versions of the MCNC benchmarks *primary1* and *primary2*.

TDP performance is compared to a quadratic placer combined with full partitioning (referred to as QPP), and to a simulated annealing placer, namely, TimberWolf 4.2 (Sechen et al., 1988; Kennings, 1997). Comparisons are made based on total wirelength (in meters) and execution time (in CPU seconds) for relative and final placements. We tested TDP when $\rho(z_{ij}) = -\ln(z_{ij})$ and $\rho(z_{ij}) = e^{1-z_{ij}}$, and we refer to their associated placers as LTDP and ETDP, respectively.

To draw a general conclusion about the likeness of the different placers, we compare statistics drawn from the results obtained using each placer. The idea is that if similar statistics are identified for each placer, then it is very reasonable to assume that they are equivalent. An intuitive and adequate statistic is the average wirelength and the average computation time computed over all test cases. If the ratios of average wirelength and average computation time are around 1, it is quite logical to assume that the placers are similar, or more precisely, equivalent (see Figures 6 and 7).

The last row of Tables 2 and 3 list the average wirelength and computation time for relative and final placements for each TDP. Examining the numerical data presented there, we find that the ratio between the initial (legalized) placement average wirelength of the LTDP to that of the ETDP is 1.023, and their final placement average wirelength ratio is 1.01. Moreover, the initial placement average computation time ratio is 0.998, and their final placement average computation time ratio is 0.994. This observation implies that the relative placements obtained

Table 1 Test Circuit Statistics

Circuit	*Cells*	*Nets*	*Pads*	*Rows*
ckt1	833	904	81	21
ckt2	3014	3136	107	21
biomed	6417	5742	97	27
industry1	2271	2478	580	15
industry2	12 142	13 419	495	70
industry3	15 059	21 938	375	50

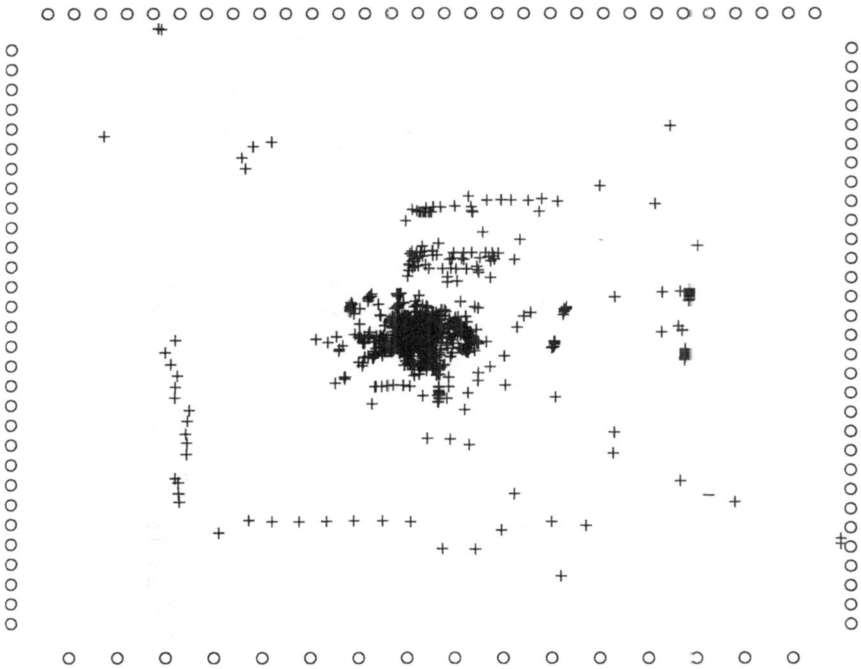

Figure 6 QPP solution (one iteration with no partitioning) for *ckt2* test case

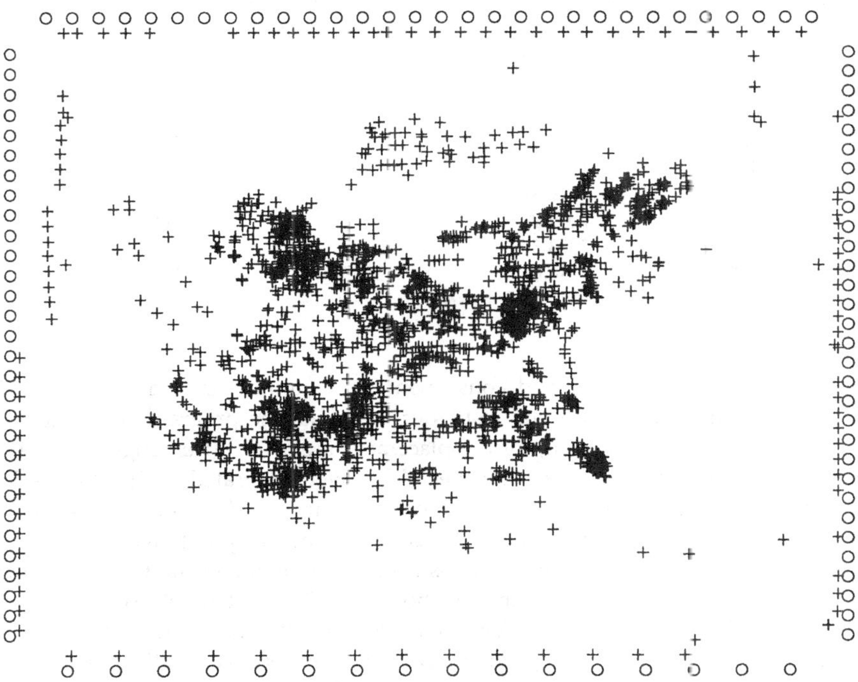

Figure 7 TDP solution for *ckt2* test case

Table 2 LTDP Initial and Final Placements

Circuit	*Initial placement*		*Final placement*		
	WL	*Time*	*WL*	*Time*	*Total time*
ckt1	1.65	13.1	1.09	238.1	251.2
ckt2	6.89	92.3	4.402	670	762.2
biomed	3.91	256.1	2.74	2640	2896.0
industry1	2.61	66.11	1.85	1120	1186.1
industry2	33.4	502.2	21.3	3100	3602.2
industry3	92.2	654.1	56.98	4713	5367.1
Average	23.44	263.985	14.727	2080.2	2344.15

Table 3 ETDP Initial and Final Placements

Circuit	*Initial Placement*		*Final Placement*		
	WL	*Time*	*WL*	*Time*	*Total time*
ckt1	1.67	13.29	1.09	244.1	257.39
ckt2	7.1	85.2	4.56	663	748.2
biomed	3.85	253.1	2.73	2647	2900.1
industry1	2.57	68.11	1.85	1109	1177.11
industry2	35.4	505.2	22.01	3051	3556.2
industry3	87.2	661.1	55.2	4844.4	5505.2
Average	22.9	264.33	14.57	2093.1	2357.42

Table 4 Initial and Final Placements of Quadratic Placer Combined with Full Partitioning (QPP)

Circuit	*Initial Placement*		*Final Placement*		
	WL	*Time*	*WL*	*Time*	*Total time*
ckt1	1.66	29.48	1.19	275	304.48
ckt2	6.66	308.2	4.8	964	1272
biomed	4.01	1230.1	2.8	3034	4264.1
industry1	2.23	119.87	1.78	1221	1340.87
industry2	29.0	2097.0	19.7	3611.47	5708.47
industry3	76.6	2303.3	55.4	4865.36	7168.66
Average	20.02	1014.6	14.278	2328.5	3343.1

from each TDP are quite similar and consequently it is quite fair to assume that LTDP and ETDP are equivalent. Table 4 lists the results of the quadratic programming placer combined with exhaustive slicing (partitioning) of the placement area, QPP (Kennings, 1997). We ran the QPP package on the same machine, however, the solver integrated within the package is an interior-point-based algorithm. An iterative method is employed to solve the system of linear equations needed to obtain the search directions for the interior point algorithm. The efficiency of the integrated solver can, therefore, be considered comparable to the L_BFGS_B. Comparing the quality of the QPP initial placements to those of LTDP and ETDP, we observe that QPP produced noticeably better results for some test cases and worse results for others. Interestingly, the final placements, generated by the iterative method using the initial placements of LTDP and the ETDP are the same (or better) than those obtained using initial placements generated by QPP (see Figures 8 and 9). In particular, QPP obtained a significantly

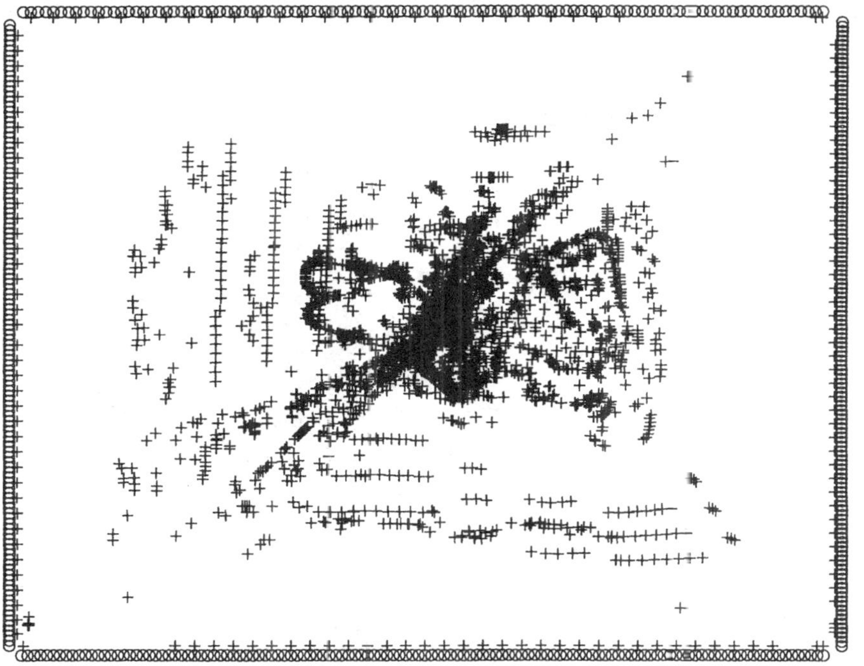

Figure 8 QPP solution (one iteration with no partitioning) for *industry2* test case

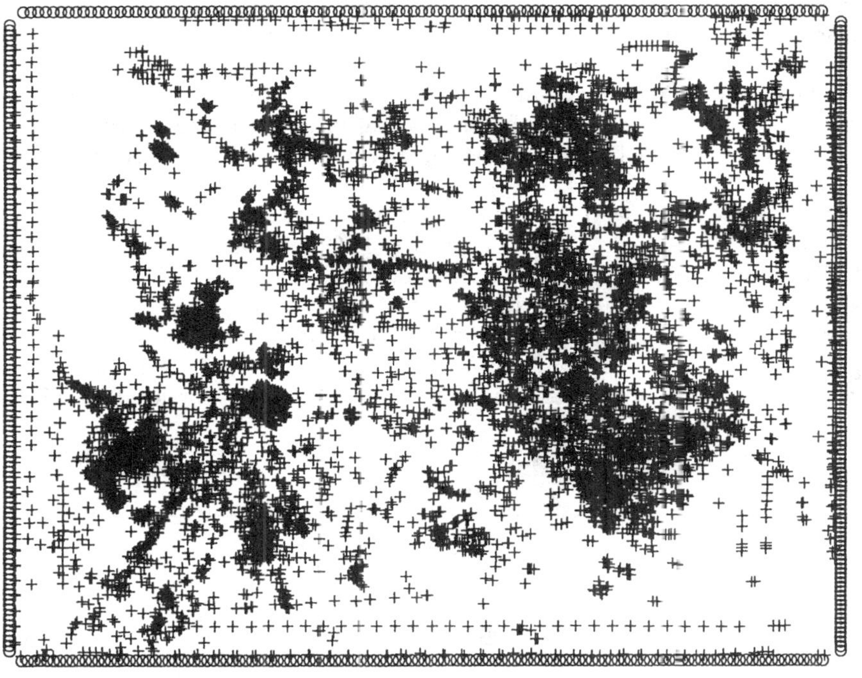

Figure 9 TDP solution for *industry2* test case

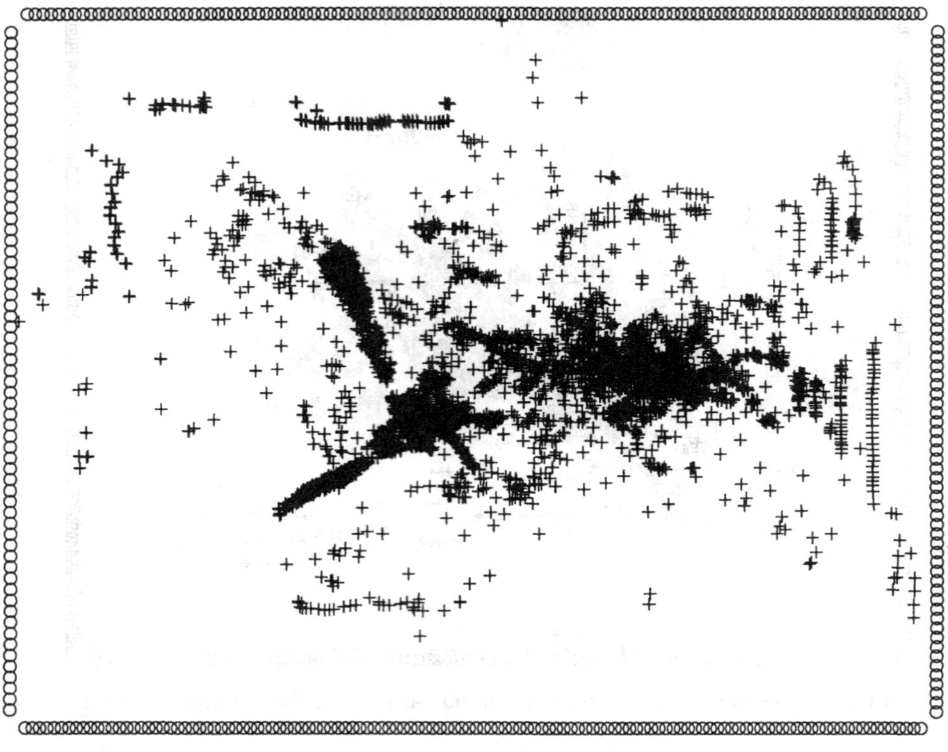

Figure 10 QPP solution (one iteration with no partitioning) for *industry3* test case

better initial placement for *industry3* compared to the TDPs, but the quality of the final answer is the same as that of the ETDP (see Figures 10 and 11). On the other hand, the efficiency of the TDPs is far better than that of the QPP. Specifically, the average computation times (in CPU seconds) of the initial placements of LTDP and ETDP are 263.98 and 264.33, where as that of the QPP is 1014.6. Numerically, this means that on average, the LTDP and ETDP are 3.8 times faster than the QPP. This proves the superiority of the TDP approach over the classic QP approach.

Finally, Table 5 lists the results obtained using a simulated annealing placer, TimberWolf v4.2. The average wirelength over all circuits is 16.83 and the average computation time is 13,204. The ratio of the average wirelength of the LTDP and ETDP to that of simulated annealing is approximately 0.87. This implies that using initial placements produced by either TDP, the iterative improvement algorithm produces final placements with wirelength (on average) roughly 13 percent shorter than by simulated annealing. Equivalently we can say that on average, simulated annealing produces placements with wirelength 1.15 times those

Table 5 Wirelengths and Computation Times of TimberWolf4.2

Circuit	*WL*	*Time (CPU secs)*
ckt1	1.18	656
ckt2	4.82	3961
biomed	2.84	10 309
industry1	1.93	2629
industry2	21.71	25 405
industry3	68.5	36 267
Average	16.83	13 204

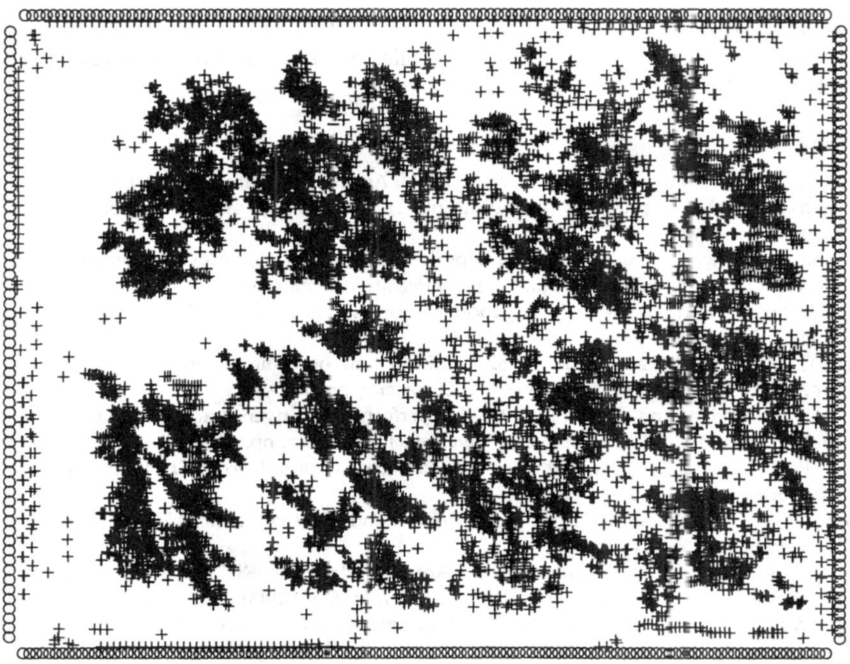

Figure 11 TDP solution for *industry3* test case

produced by the iterative improvement algorithm when initial placements are supplied by LTDP or ETDP. Regarding efficiency, the ratio of the maximum average computation time (ETDP case) to that of simulated annealing is 0.178. This implies that on average, either TDP needs only 17.8 percent of the time required by simulated annealing to achieve the same answer. Clearly, TDPs achieve better results in terms of quality and efficiency compared to simulated annealing. Again, this demonstrates the superiority of the TD approach.

8. Conclusions

The fact that future placement tasks are much more complicated implies that faster placement tools should be developed to handle such immense complexity. Future placement tools must be capable of placing larger circuits in a reasonable time, besides they must be flexible enough to handle any modifications in the design style of VLSI circuits.

In light of these facts, we proposed a new approach for performing placement tasks. Our approach adds new forces to the classical quadratic wirelength approach to avoid the need for partitioning the placement area in order to remove cell overlap. We demonstrated the feasibility of the new approach and showed its superiority to the partitioning based approach; that is, the classical quadratic programming approach. We have also demonstrated the superiority of final placement qualities obtained iteratively using initial placements produced by the TD approach compared to simulated annealing.

Our approach is general in the sense that it can be applied to other target distance problems in different fields of science and engineering. For instance, determining geometric structures of molecules of chemical compounds based on the distances between their atoms, (Alfakih et al., 1997).

Finally, we would like to point out that the efficiency of the new models can be further improved through the use of efficient data structure techniques.

ACKNOWLEDGMENT

Special thanks are due to our friend and colleague Dr Andrew Kennings for his help and constant support.

References

Alfakih, A., A. Khandani, and H. Wolkowicz. 1997. "An interior point method for the Euclidean distance matrix completion problem." Technical report. Department of Combinatorics and Optimization, University of Waterloo.

Du, Y. and A. Vannelli. 1998. "A nonlinear programming and local improvement method for standard cell placement." In *Proceedings of IEEE Custom Integrated Circuit Conference.*

Dunlop, A. and B. Kernighan. 1985. "A procedure for placement of standard cell VLSI placement." *IEEE Trans. CAD* **4(4)**: 92–98.

Fiduccia, C. M. and R. M. Mattheyses. 1982. "A linear time heuristic for improving network partitions." In *Proceedings of the 19th Design Automation Conference*, pp. 175–181.

Gao, T., P. M. Vaidya, and C. L. Liu. 1992. "A performance driven macro-cell placement algorithm." In *Proceedings of ACM/IEEE 29th Design Automation Conference*, pp. 147–152.

Kang, E., R. B. Lin, and E. Shragowitz. 1994. "Fuzzy logic approach to VLSI placement." *IEEE Trans. Very Large Scale Integ. Sys.* **2**: 489–501.

Kennings, A. 1997. "Cell placement using constructive and iterative methods." Ph.D. diss. Ontario, Canada: University of Waterloo.

Kleinhans, J. M., G. Sigl, F. Johannes, and K. Antreich. 1991. "GORDIAN: VLSI placement by quadratic programming and slicing optimisation." *IEEE. Trans. CAD* **10(3)**: 356–365.

Luenberger, D. G. 1984. In *Linear and Nonlinear Programming.* Addison-Wesley Publishing Company.

MCNC. 1992. Available at www.cbl.ncsu.edu/benchmarks/layoutsynth92/.

Ohtsuki, T., H. Mori, E. Kuh, T. Kashiwabara, and T. Fujisawa. 1979. "One-dimensional logic gate assignment and interval graphs." *IEEE Trans. Circ. Sys.* 26.

Preas, B. and M. Lorenzetti. 1988. In *Physical Design Automation of VLSI Systems.* Benjamin Cummings.

Sechen, C., D. Braun, and A. Sangiovanni-Vincentelli. 1988. "ThunderBird: a complete standard cell layout package." *IEEE J. Solid State Circuits* **23(2)**: 410–420.

Sherwani, N. 1993. In *Algorithms for VLSI Physical Design Automation.* Kluwer Academic publishers.

Ullman, J. D. 1984. In *Computational Aspects of VLSI.* Computer Sceince Press Inc.

Zhu, C., R. H. Byrd, P. Lu, and J. Nocedal. 1997. "Algorithm 778: L-BFGS-B: Fortran subroutines for large scale boundary constrained optimization." *ACM Trans. Math. Softw.* **23(4)**: 550–560.

Zobrist, G. 1994. In *Routing, Placement, and Partitioning.* Ablex Publishing Co.

18.12 Optimization Models in Transportation Planning

Michael Florian and Donald W. Hearn

ABSTRACT In this article we present several optimization models that are used frequently in the practice of transportation planning. These include the spatial interaction model, several variants of the network equilibrium model, and a transit route choice model. Solution algorithms are presented for each class of models.

1. Introduction

Quantitative approaches to transportation planning construct models that predict the demand for travel for an urban area or region based on the socioeconomic characteristics of the resident population and the levels of service from origins to destinations provided by the existing transport infrastructure. The aim of the descriptive travel demand models is to predict when trips originate, their destinations, which mode they use, and which routes are chosen. The theory and practice of travel demand models are very rich and a wide variety of econometric and optimization models and methods are used to formulate and calibrate such models by using survey data. The transportation planning process uses such descriptive models in order to evaluate future scenarios against a base or current scenario and obtain indications for the best alternative to be adopted.

Optimization models have played a very important part since the 1970s in developing appropriate models for destination, mode, and route choice models and efficient algorithms for their numerical solution. In this article three classes of such optimization models are presented. These are: spatial interaction, or trip distribution models; network equilibrium, or trip assignment models; and transit route choice models. These models are static network optimization or variational inequality problems, in spite of the fact that all traffic phenomena are temporal. One considers a given period of time, for which the demand for travel has been quantified. In order to derive a prediction of the flows between origins and destinations and on the links (paths) of the network, a network model is used to represent the supply of transportation infrastructure and services. Spatial interaction models are employed to determine demand origin-destination matrices. Network equilibrium models serve to model route choice on congested road networks, while transit route choice models consider the frequency of service provided by transit lines.

The spatial interaction models are presented next.

2. Spatial Interaction Models

Consider a transportation network that permits the flow of one type of traffic (vehicles or passengers) on its links. The nodes $n \in N$ represent origins, destinations, and intersection of links. The links $a \in A \subset N \times N$ represent the transportation infrastructure. If the number of trips that start from origins $r \in R \subset N$ is O_r and the number that are destined to destinations $s \in S \subset N$ is D_s, then the issue of interest is to determine t_{rs} [or t_p where $p = (r, s)$] given the time or cost of travel u_{rs}.

The classical model that is used to determine the origin-destination matrix t_{rs} is known as the entropy model. Conservation of flow at origins and destinations imply that

$$\sum_{s \in S} t_{rs} = O_r, \quad r \in R \tag{1}$$

and

$$\sum_{r \in R} t_{rs} = D_s, \quad s \in S \tag{2}$$

and, evidently the demand for travel is nonnegative

$$t_{rs} \geqslant 0, \quad r \in R, \quad s \in S. \tag{3}$$

In the absence of other information, it is postulated that the origin-destination matrix most likely to occur is that which leads to the objective function

$$\max - \sum_r \sum_s t_{rs} \ln t_{rs} \tag{4}$$

subject to (1)–(3). The objective has the interpretation of entropy maximization and the formalism originates from information theory (see Jaynes, 1957a,b). The model was introduced to transportation and regional analysis by Wilson (1967, 1970) when some *a priori* information is known about the matrix, say t^0_{rs}, $\forall (r, s)$. Kullback (1959) and Snickars and Weibull (1977) have suggested the use of the objective

$$\max - \sum_r \sum_s t_{rs} \ln \left(\frac{t_{rs}}{t^0_{rs}} \right). \tag{5}$$

In order to characterize the dispersion of trips a constraint is added to the total travel time, where C is an observed total travel time

$$\sum_r \sum_s t_{rs} u_{rs} = C, \tag{6}$$

which leads to the objective function

$$\min \sum_r \sum_s t_{rs}(\ln t_{rs} + \theta u_{rs}). \tag{7}$$

θ has the interpretation of the dual variable associated with constraint (6). In applications θ is calibrated from observed data so that (6) holds. It is trivial to verify that, by applying the Karush–Kuhn–Tucker conditions, any solution of (7) subject to (1)–(3) has the general form

$$t_{rs} = \exp(\alpha_r + \beta_s - 1)\exp(-\theta u_{rs}), \quad \forall (r, s) \tag{8}$$

$$= A_r B_s \exp(-\theta u_{rs}),$$

where α_r and β_s are the dual variables associated with the conservation of flow constraints. With the convention $t_{rs} \ln t_{rs} = 0$ when $t_{rs} = 0$, it is possible to obtain solutions to this class of problems by using any primal convex programming algorithm. However, this problem has the property that the primal variables may be expressed in terms of the dual variables as will be shown next. The Lagrangian dual problem of (7) subject to (1)–(2) is

$$\max_{\alpha,\beta}\left\{\min_t \sum_r \sum_s t_{rs}(\ln t_{rs} + \theta u_{rs}) + \sum_r \alpha_r\left(O_r - \sum_s t_{rs}\right) + \sum_s \beta_s\left(D_s - \sum_r t_{rs}\right)\right\}. \tag{9}$$

By using (8) to replace the primal variables t_{rs} one obtains, after some simplifications

$$\max_{\alpha,\beta}\left\{\sum_r \alpha_r O_r + \sum_s \beta_s D_s - \sum_r \sum_s \exp(\alpha_r + \beta_s - 1 - \theta u_{rs})\right\}. \tag{10}$$

This property led to the development of an efficient solution procedure known as the balancing method, which is a dual ascent method for one variable at a time. Let t_{rs}^l, α_r^l, β_s^l be a solution at an iteration l. Then β_s^{l+1}, is obtained by annuling the partial derivative of the objective function with respect to β_s

$$D_s - \sum_r \exp(-\theta u_{rs} + \beta_s^{l+1} + \alpha_r^l - 1) = 0, \tag{11}$$

which has the solution

$$\exp(\beta_s^{l+1}) = D_s \Big/ \sum_r \exp(-\theta u_{rs} + \alpha_r^l - 1). \tag{12}$$

By multiplying and dividing the right-hand side of (12) by $\exp(\beta_s^l)$ one obtains

$$\exp(\beta_s^{l+1}) = \frac{D_s}{\sum_r t_{rs}^l}\exp(\beta_s^l). \tag{13}$$

The ratio $(D_s/\sum_r t_{rs}^l)$ is referred to as the balancing factor at iteration l. Since the same one-dimensional maximization may be used for each α_r and β_s one obtains the primal-dual method, which is known as the balancing method. It is stated next.

Balancing method

Step 1. (Initialization) $l = 0$, $t_{rs}^0 = \exp(-\theta u_{rs})\ \forall (r, s)$

$$A_r^0 = 1, \quad \forall r; \quad B_s^0 = 1, \quad \forall s$$

Step 2. ($l = 1, 2, \ldots$)

(a)

$$A_r^{l+1} = \frac{O_r}{\sum_s t_{rs}^l} A_r^l \quad \forall r; \quad t_{rs}^{l+1/2} = \frac{O_r}{\sum_s t_{rs}^l} t_{rs}^l, \quad \forall r, s$$

(b)

$$B_s^{l+1} = \frac{D_s}{\sum_r t_{rs}^{l+1/2}} B_s^l \quad \forall s; \quad t_{rs}^{l+1} = \frac{D_s}{\sum_r t_{rs}^{l+1/2}} t_{rs}^{l+1/2}, \quad \forall r, s$$

The algorithm is terminated when $\|A_r^{l+1} - A_r^l\| \leqslant \epsilon$ and $\|B_s^{l+1} - B_s^l\| \leqslant \epsilon$. It is interesting to note that $A_r = \ln \alpha_r$ and the difference between two successive values of the dual variables $(\alpha_r^{l+1} - \alpha_r^l) = \ln(O_r / \sum_s t_{rs}^l)$.

This method dates back to at least 1937 when Kruithof used it for the prediction of telephone traffic distribution (Kruithof, 1937). Deming and Stephan (1940) independently rediscovered this method and applied it to a cross-classification problem in statistics as a simplification of least squares fitting.

This model has other interesting properties. If $u_{rs} < \infty$ and $\theta > 0$ then $t_{rs} > 0$, $\forall (r, s)$ and the nonnegativity constraints (3) are not required. To clarify this, if t_{rs} were equal to zero, then, from (8) either α_r or $\beta_s = -\infty$, which would imply that $\sum_s t_{rs} = 0$ or $\sum_r t_{rs} = 0$, which in turn would violate (1) or (2).

Another interesting variant of this model is when the double constraints (1) and (2) are replaced by inequalities. For instance, when

$$\sum_s t_{rs} \leqslant O_r, \quad \forall r \tag{14}$$

replaces (1), the solution may be found by modifying the balancing method by replacing the balancing factor in Step 1(a) to be

$$\min\left(1, \frac{O_r}{\sum_s t_{rs}^l}\right)$$

instead of

$$\left(\frac{O_r}{\sum_s t_{rs}^l}\right).$$

This variant of the model was studied by Jefferson and Scott (1979). Lamond and Stewart (1981) showed that the balancing method and its variants may be viewed as a special case of Bregman's (1967) nonorthogonal projection method to solve certain problems of convex programming. Variants of the balancing method were studies by Robillard and Stewart (1974), Evans and Kirby (1974), and Andersson (1981). The text of Erlander et al. (1979) is dedicated to a complete exposition of the entropy model and its variants.

3. Traffic Assignment Models and Methods

Traffic assignment models are designed to describe the traffic patterns formed by users of a transportation network such as an urban street system. They also adapt to serve as models for travel on rail or airline networks. It is assumed that certain attributes of the network are known and that the travel demand is defined by an origin-destination demand matrix, as described in the preceding section, or it is defined by demand functions.

This section summarizes the most important traffic assignment models and a few basic algorithms for solving them. Proofs of the results stated, convergence analysis of the algorithms, and more detailed development of alternative models may be found in the literature (see the texts by Sheffi, 1985, the monograph by Patricksson, 1994, and the survey article by Florian and Hearn, 1995 and their references).

For notational simplicity, assume a transportation network model with one type of vehicular flow on the directed links of the network. The nodes i, $i \in N$, represent origins, destinations, and intersections; the arcs a, $a \in A$, are the transportation links. Origin-destination

(O-D) demands give rise to link flows v_a, $a \in A$ and the cost of traveling on a link is given by a user cost function $s_a(v)$, where v is the vector $(v_a)_{a \in A}$ of link flows over the entire network. Cost functions model time delay on a link or more general costs such as tolls or fuel consumption, and are assumed to be nonnegative. Let P be the set of O-D pairs, K_p, $p \in P$ be a set of directed paths connecting O-D pair p, and K be the set of all paths. The demand between O-D pair p, t_p, uses directed paths and the path flows h_k obey conservation of flow and nonnegativity

$$\sum_{k \in K_p} h_k = t_p \quad \forall p \in P \tag{15}$$

$$h_k \geqslant 0 \quad \forall k \in K. \tag{16}$$

Link flows are given by

$$v_a = \sum_{p \in P} \sum_{k \in K_p} \delta_{ak} h_k \quad \forall a \in A, \tag{17}$$

where $\delta_{ak} = 1$ if link a belongs to path k and is zero otherwise.

Define Δ as the $|A| \times |K|$ arc-path incidence matrix (δ_{ak}) so that $v = \Delta h$, where h is the vector $(h_k)_{k \in K}$ of path flows for all O-D pairs. The cost $s_k(= s_k(h))$ of each path k is then defined by

$$s_k = \sum_{a \in A} \delta_{ak} s_a(v) = \sum_{a \in A} \delta_{ak} s_a(\Delta h) \quad \forall k \in K_p, p \in P, \tag{18}$$

and $u_p(= u_p(h))$ is by definition the cost of the least cost path for O-D pair p:

$$u_p = \min_{k \in K_p} s_k \quad \forall p \in P. \tag{19}$$

For each $p \in P$, the travel demand t_p, may be obtained from a fixed O-D demand matrix, in which case we write $t_p = \bar{t}_p$, or it is given by a specific demand function $t_p(u)$, where u is the vector of least cost travel values, $(u_p)_{p \in P}$, for all the O-D pairs of the network:

$$t_p = t_p(u) \quad \forall p \in P. \tag{20}$$

3.1. *System and User Optimization Models*

System optimal traffic assignment models assume that travel on the network follows paths such that network utilization is for the common good. If the demands t_p are fixed, then the objective is to satisfy a normative principle that states the average travel cost (or time) is to be minimized. Since total demand is a constant, it is equivalent to minimize total system cost, and the fixed demand system optimization model is

SOPT-FD

$$\min \sum_{a \in A} s_a(v) v_a \tag{21}$$

subject to (15), (16), and (17) with $t_p = \bar{t}_p$.

If, however, travel demand is elastic, that is, dictated by demand functions (20), the system optimization model aims to maximize the net economic benefit to the network users. From standard economic principles, the benefit to travelers between any O-D pair $p \in P$ is measured by the area under their demand curve $t_p(u)$. It is assumed that this function has an inverse $w_p(t_p) = u_p$. Hence the economic benefit can be expressed as $\int_0^{t_p} w_p(y) dy$. Therefore, in this case, the system optimization model becomes

SOPT-ED

$$\max \sum_{p} \int_{0}^{t_p} w_p(y)dy - \sum_{a \in A} s_a(v)v_a \tag{22}$$

subject to (15), (16), (17) and (20).

These system problems have counterparts in user optimization models that aim to more accurately describe the situation where travelers on the network distribute themselves so that no single user can unilaterally improve their travel costs. The descriptive models of traffic flow, therefore, assume the users are in a Wardrop equilibrium (Wardrop, 1952), a special case of Nash equilibrium. The mathematical statement is:

Determine h^* and u^* such that the following conditions are satisfied:

$$(s_k(h^*) - u_p^*)h_k^* = 0 \quad \forall k \in K_p, \quad p \in P \tag{23}$$

$$s_k(h^*) - u_p^* \geqslant 0 \quad \forall k \in K_p, \quad p \in P \tag{24}$$

$$\sum_{k \in K_p} h_k^* - t_p = 0 \quad \forall p \in P \tag{25}$$

$$h^* \geqslant 0, \quad u^* \geqslant 0, \tag{26}$$

where $t_p = \bar{t}_p$ for fixed demand and $t = t(u^*)$ when demand is elastic. The equilibrium link flows, v^*, are calculated from the path flows h^* using (17).

The first two conditions ensure that, for all $p \in P$, only minimum-cost paths are used, and the third equates the total path flows to the total demand, given the minimum path costs. This general version of the problem is known as the network equilibrium model (NEM), which specializes for many other applications, including electrical networks, water pipe networks, and spatial price equilibrium problems. Florian and Hearn (1995) provide numerical examples of these applications. They also summarize research results of many authors, showing that mathematical analysis of the NEM can be facilitated by reformulation as a nonlinear complementarity problem, as a fixed-point problem or as a variational-inequality problem.

The basic NEM reformulations used in transportation planning, however, are optimization problems. The primary assumptions are that the cost and demand functions are separable, that is, they have the form $s_a(v) = s_a(v_a)$ and $t_p(u) = t_p(u_p)$, respectively. In other words, the cost on a link depends only on the flow on that link, and the demand for O-D pair p depends only on the minimum travel time for that O-D pair. It is further assumed that the cost functions are convex and the demand functions are strictly monotone. Under these conditions, the elastic demand user optimization problem may be stated as the convex program

UOPT-ED

$$\min \sum_{a \in A} \int_{0}^{v_a} s_a(x)dx - \sum_{p \in P} \int_{0}^{t_p} w_p(y)dy \tag{27}$$

subject to (15), (16), (17) and (20).

In the fixed demand case, the user problem becomes

UOPT-FD

$$\min \sum_{a \in A} \int_{0}^{v_a} s_a(x)dx \tag{28}$$

subject to (15), (16) and (17) with $t_p = \bar{t}_p$.

That the solutions of these problems are equivalent to (23)–(26), in the elastic and fixed demand cases, respectively, follows directly from the Karush–Kuhn–Tucker conditions for the two problems. The connection between the NEM conditions and the SOPT models is also revealed by their Karush–Kuhn–Tucker conditions. It is straightforward to verify that they have the same form, with the terms $s_k(h^*)$ calculated from the marginal link costs, $s_a(v_a^*) + s_a'(v_a^*)v_a^*$. Thus, the interesting and important connection is that the solution to a SOPT model is in equilibrium with respect to marginal costs, while the solution of an UOPT model is in equilibrium with respect to average costs.

3.2. *Algorithms for UOPT Traffic Assignment*

Three basic methods, two for UOPT-FD and one for UOPT-ED, are given in this section. Since the two UOPT and SOPT models differ only in the objective function terms that contain the link cost functions, algorithms designed for UOPT-FD and UOPT-ED will apply to SOPT-FD and SOPT-ED, respectively. The primary adjustment that must be made is in the calculation of these objective function derivatives, which, in turn, affect the costs of shortest paths. In the UOPT models these are simply the given $s_a(v_a)$, but in the SOPT models these take the form $s_a(v_a) + s_a'(v_a)v_a$. This adjustment is consistent with the observation above relating the Karush–Kuhn–Tucker conditions of the models.

Cyclic O-D Decomposition with Path Equilibration

The UOPT models emphasize path flows and path costs for each O-D pair, therefore the most intuitive solution approach works in the space of path flows with an outer decomposition by O-D pair. This general approach, which is equivalent to a Gauss–Seidel decomposition (or relaxation) is also referred to as a cyclic decomposition, since in one cycle of the algorithm a single UOPT problem is solved for each O-D pair, while keeping the flows for all other O-D pairs fixed. The overall cyclic strategy is stated as follows:

Cyclic O-D Decomposition

Step 0. Given an initial solution set $p = 0$, $p' = 0$.

Step 1. If $p' = |P|$, the total number of O-D pairs, stop; otherwise set $p = p \mod |P| + 1$ and continue.

Step 2. If the current solution is optimal for the pth subproblem (see (29)–(33) below), set $p' = p' + 1$ and return to step 1; otherwise, solve the pth subproblem, update the current flows, set $p' = 0$, and return to Step 1.

The following is a statement of the subproblem when applied to the UOPT-FD model.

$$\min \sum_{a \in A} \int_0^{v_a^p + \bar{v}_a} s_a(x)dx \tag{29}$$

subject to

$$\sum_{k \in K_p} h_k = \bar{t}_p \tag{30}$$

$$h_k \geq 0, \quad \forall k \in K_p, \tag{31}$$

where

$$\bar{v}_a = \sum_{p \neq p} \sum_{k \in K_p} \delta_{ak} h_k \tag{32}$$

and

$$v_a^p = \sum_{k \in K_p} \delta_{ak} h_k, \quad \forall a \in A. \tag{33}$$

Path equilibration algorithms used to solve (29)–(33) operate in the space of path flows and obtain a solution where all used paths are of equal cost. The simplest such algorithm finds the shortest path and the longest path and then transfers flow between these in order to equalize their cost. Letting $K_p^+ = \{k \in K_p \mid h_k > 0\}$ be the set of paths that have positive flow, the algorithm is:

Path Equilibration Algorithm

Step 0. Given an initial solution v_a^p; $s_a = s_a(v_a^p + \bar{v}_a)$ and the initial K_p^+.

Step 1. Compute the costs of the currently used paths s_k, $k \in K_p^+$. Find k_1 such that $s_{k_1} = \min_{k \in K_p^+} s_k$ and k_2 such that $s_{k_2} = \max_{k \in K_p^+} s_k$. If $(s_{k_2} - s_{k_1}) \leqslant \epsilon$, go to Step 4. Otherwise continue to Step 2.

Step 2. Redistribute the path k_2 flow to path k_1 until it is reduced to zero or the two paths costs are equal.

Step 3. With the new path flows h_{k_1} and h_{k_2} update the link flows, v_a^p, the link costs s_a, and the path costs s_k.

Step 4. Compute the shortest path $\tilde{k}$ with cost $\tilde{s}_k = \min_{k \in K_p} s_k$; if $\tilde{s}_k < \min_{k \in K_p^+} s_k$, then the path $\tilde{k}$ is added to the set of kept paths, $K_p^+ = K_p^+ \cup \tilde{k}$ and return to Step 1; otherwise stop.

The algorithm presented above is just one of many possible path equilibration schemes. For instance, the adaptation of the reduced gradient or the projected gradient algorithms for the subproblem results in an equilibration step on all the paths k, $k \in K_p^+$. Also, each subproblem need not be solved to optimality; only one or two equilibration steps may be performed for each O-D pair.

This algorithm is suitable for UOPT-FD when the number of paths used for an O-D pair is relatively small, such as in an airline or railway network, and when the number of O-D pairs is relatively small as well. For many models of urban street networks, the storage requirements of the paths and their flows become prohibitively large very quickly. For this reason, the following methods are more prevalent since they work in the space of link flows.

LINEAR APPROXIMATION METHODS

One of the simplest convergent algorithms for minimizing a convex function subject to linear constraints is the linear approximation method. Its adaptation for the solution of the two UOPT models yields methods that primarily require only the computation of shortest paths and one-dimensional minimization of a convex function.

We first consider adaptation to the UOPT-FD model. Starting from an initial feasible solution, the method generates a feasible descent direction by solving a subproblem, which is obtained by linearizing the objective function. Then, an improved solution is found on the line segment between the current solution and the solution of the subproblem. Letting the objective function be $S(v)$, the linearized approximation at an intermediate iteration l, when the current solution is v^l, is

$$S(v^l) + \nabla S(v^l)(y - v^l). \tag{34}$$

Since $S(v^l)$ and $\nabla S(v^l)v^l$ are constants, the linearized subproblem to be solved reduces to

$$\min \sum_{p \in P} \sum_{k \in K_p} \sum_{a \in A} s_a(v_a^l)\delta_{ak} h_k \tag{35}$$

subject to

$$\sum_{k \in K_p} h_k = \bar{t}_p \quad \forall p \in P \tag{36}$$

$$h_k \geqslant 0 \quad \forall k \in K. \tag{37}$$

By changing the order of summation in (35) and by using (18) the objective becomes

$$\min \sum_{p \in P} \sum_{k \in K_p} s_k^l h_k. \tag{38}$$

As the terms of the objective function (38) may be separated now by O-D pair p, the solution of the linearized subproblem may be obtained by computing shortest paths for each O-D pair p and allocating the demand $\bar{t}_p$ to the links of these paths. Such a demand allocation or assignment is referred to as an all-or-nothing assignment. This yields the arc flow vector

$$y_a^l = \sum_{k\in K} \delta_{ak} h_k^l \quad \forall a \in A, \tag{39}$$

and the direction of descent is

$$d_a^l = (y_a^l - v_a^l) \quad \forall a \in A. \tag{40}$$

An iteration of the linear approximation algorithm is completed by doing a line search between v^l and y^l. This is the one variable problem

$$\min_{0\leqslant\lambda\leqslant 1} S(v^l + \lambda d^l). \tag{41}$$

The adaptation of the linear approximation method results in the following algorithm:

Linear Approximation Method

Step 0. Find an initial solution v^l; $s^l = s(v^l)$; $l = 1$.

Step 1. Perform an all-or-nothing assignment based on the current arc costs $s(v^l)$ to obtain the arc flow vector y^l. Let $d^l = (y^l - v^l)$.

Step 2. Verify if a predetermined stopping criterion is satisfied. If it is, stop; otherwise continue to

Step 3. Find the optimal step size λ^l by solving (41).

Step 4. Update arc flows $v^{l+1} = v^l + \lambda^l d^l$ and arc costs $s^{l+1} = s(v^{l+1})$; set $l = l + 1$, and return to Step 1.

The algorithm has several advantages for the solution of the fixed demand symmetric NEM. The paths that are used to compute the descent direction at a given iteration are generated as required and need not be kept in successive iterations. Thus the storage requirements are modest and do not increase with the number of iterations. At each iteration only the flows v^l and the link costs s^l need to be kept, in addition to the network link data.

Since $S(v)$ is a convex function and $\nabla S(v) = s(v)$, at any iteration,

$$S(v^*) \geqslant S(v^l) + s(v^l)(y^l - v^l). \tag{42}$$

Thus the best lower bound (BLB) obtained up to a current iteration l is

$$\text{BLB} = \max_{i=1\ldots l} S(v^i) + s(v^i)(y^i - v^i). \tag{43}$$

Hence a natural stopping criteria is when the relative objective error,

$$\frac{S(v^l) - \text{BLB}}{S(v^l)}, \tag{44}$$

is smaller that some predetermined parameter.

It is worthwhile to note that this algorithm has the intuitive interpretation of the travelers adjusting their route choice from congested routes to less congested routes. This explains its close relation to many heuristic algorithms that have been suggested and used for solving this problem. On the other hand, the principal drawback of the linear approximation method is that it often exhibits slow convergence in the vicinity of the optimal solution due to the fact that its asymptotic rate of convergence is arithmetic. This fact has motivated the development of variants of this algorithm which attempt to improve its convergence rate. Two of the more successful ones are based on the PARTAN method and on simplicial decomposition; both are covered in the survey by Florian and Hearn (1995) and computational results are in the referenced literature.

The UOPT-ED model may be solved by a partial linear approximation method, which proceeds by linearizing only some of the variables of the objective function. Given a solution (v^l, t^l) at iteration l, the aim of the subproblem is to determine (y^l, z^l) so that a line search can be performed in the combined link flow and O–D demand space. For this case, it is natural to linearize only the arc cost functions. Letting $W(z) = \sum_p \int_0^{z_p} w_p(x)dx$, the resulting subproblem at iteration l is

$$\min \sum_{p\in P} \sum_{k\in K_p} \sum_{a\in A} s_a(v_a^l)\delta_{ak}h_k - W(z) \tag{45}$$

subject to

$$\sum_{k\in K_p} h_k - z_p = 0 \quad \forall p \in P \tag{46}$$

$$h_k \geqslant 0 \quad \forall k \in K, \quad z_p \geqslant 0 \quad \forall p \in P. \tag{47}$$

This subproblem is solved by first determining u_p^l, $p \in P$, to be the costs of the shortest paths based on the current link costs $s(v^l)$, and then simplifying (45) by using (46) and (47) to obtain

$$\min \sum_{p\in P} (u_p^l z_p - W_p(z_p)) \tag{48}$$

subject to

$$z_p \geqslant 0 \quad \forall p \in P. \tag{49}$$

From the subproblem optimality conditions, the z_p^l are determined analytically as follows:

$$z_p^l = \begin{cases} t_p(u_p^l) & \text{if } t_p(u_p^l) \geqslant 0 \\ 0 & \text{otherwise.} \end{cases} \tag{50}$$

The line search between (v^l, t^l) and (y^l, z^l) then follows.

The convergence criteria for the partial linear approximation method may be based on the best lower bound or the maximum number of iterations. In particular, the lower bound obtained from the partial linear approximation for UOPT-ED is

$$\text{BLB} = \max_l S(v^l) + s(v^l)(y^l - v^l) - W(t^l) - w(t^l)(z^l - t^l). \tag{51}$$

3.3. The User Problem as a Variational Inequality

The success of optimization algorithms in solving applications of the NEM with separable demand and cost functions has motivated researchers to seek formulations for the more general NEM where the separability assumptions are not made. As previously mentioned, fixed point and nonlinear complementarity formulations have also been developed. However, most recent interest in more general models has centered on the variational inequality (VI) formulation of the problem. This section will give a summary of some of the results obtained for the fixed demand user problem, extensions to the elastic demand problem are in the references such as Florian and Hearn (1995).

The basic theoretical result is that v^* solves the fixed demand NEM problem if, and only if, it solves the following VI problem:

$$\sum_{a\in A} s_a(v^*)(v_a - v_a^*) \geqslant 0 \tag{52}$$

for all (v, t) such that

$$\sum_{k\in K_p} h_k = \bar{t}_p, \quad \forall p \in P \tag{53}$$

$$h_k \geqslant 0, \quad \forall k \in K \tag{54}$$

$$v_a = \sum_{k\in K} \delta_{ak} h_k, \quad \forall a \in A. \tag{55}$$

We refer to this as the user VI model, and for convenience in describing two algorithms below, we write it more compactly by defining Θ as the set of all feasible flows in terms of the arc flow vector v:

UVI-FD
Find v^* in Θ so that

$$s(v^*)(v - v^*) \geqslant 0, \quad \forall v \in \Theta, \tag{56}$$

where

$$\Theta = \left\{ v : v_a = \sum_{k \in K} \delta_{ak} h_k, \sum_{k \in K_p} h_k = \bar{t}_p, \quad \forall p \in P, a \in A \right\}.$$

A Projection Method for UVI-FD

A popular method for solving this model is based on projecting successive iterates into the space of link flows Θ. It should be noted that it can also be solved by cyclic O-D decomposition, as described earlier, with a projection method used to solve the resulting subproblem for each O-D pair in the space of path flows.

Let Q be a symmetric positive definite matrix and let $\bar{v}$, $\bar{v} \in \Theta$ be a feasible flow. A new cost function $\hat{s}(v)$ is defined over Θ as follows:

$$\hat{s}(v) = Qv + c, \tag{57}$$

where

$$c = \rho s(\bar{v}) - Q\bar{v}, \tag{58}$$

and ρ is a positive constant. The choice of appropriate values of ρ is discussed below. Since $\hat{s}(v)$ is linear and has a symmetric Jacobian the solution $\hat{v}$ of

$$\hat{s}(\hat{v})(v - \hat{v}) \geqslant 0 \quad \forall v \in \Theta \tag{59}$$

may be obtained as the solution of a symmetric NEM by using one of the algorithms of the preceding section. Inequality (59) is equivalent to

$$(Q\hat{v} + \rho s(\bar{v}) - Q\bar{v})(v - \hat{v}) \geqslant 0 \quad \forall v \in \Theta \tag{60}$$

and this defines the mapping M_ρ on Θ, which takes $\bar{v}$ into $\hat{v}$, the unique solution of (60). This implies that every fixed point of M_ρ is a solution of (56). To see this let $\bar{v}$ be such a fixed point, that is, $M_\rho(\bar{v}) = \bar{v}$. Then setting $\hat{v} = \bar{v}$ in (60) gives $\rho s(\bar{v})(v - \bar{v}) \geqslant 0 \;\; \forall v \in \Theta$ and dividing by $\rho > 0$ shows that, indeed, $\bar{v}$ is a solution of (56).

The resulting algorithm is a projection method since $\hat{v}$ is the projection of the point $\bar{v} - \rho Q^{-1} s(\bar{v})$ onto the set Θ, where the projection is defined with respect to the Q norm, that is, $\hat{v} = \mathrm{Proj}_{Q,\Theta}(\bar{v} - \rho Q^{-1} s(\bar{v}))$. The algorithm may be stated as follows:

Projection Algorithm

Step 0. Find an initial solution v^l. Let $\hat{s}(v^l) = Qv^l + c$ and set $l = 1$.

Step 1. Determine v^{l+1} as $\mathrm{PROJ}_{Q,\Theta}(v^l - \rho Q^{-1} s(v^l))$ or equivalently solve

$$\min (v - v^l)\hat{s}(v^l) + \frac{1}{2\rho}(v - v^l)Q(v - v^l)$$

$$\text{subject to} : v \in \Theta.$$

Step 2. If $\|v^{l+1} - v^l\| \leqslant \epsilon$, stop. Otherwise, let $l = l + 1$; $\hat{s}(v^{l+1}) = Qv^{l+1} + c$, and go to Step 1.

The convergence of the algorithm is ensured if s is required to be continuously differentiable and strongly monotone, that is,

$$[s(v_1) - s(v_2)](v_1 - v_2) \geqslant \alpha \mid v_1 - v_2 \mid^2 \quad \forall v_1, \;\; v_2 \in \Theta, \tag{61}$$

where $\alpha > 0$ is the modulus of strong monotonicity, provided that ρ is sufficiently small. This technical condition may be difficult to verify in practice because ρ depends on both α and the maximum eigenvalues (over Θ) of the matrix $\nabla s(v)Q^{-1}\nabla s(v)$.

SIMPLICIAL DECOMPOSITION FOR UIV-FD

The success of the linear approximation method for large-scale separable fixed demand models stems from the fact that it works in the space of link flows and that it alternates between shortest-path computations and line searches. Trying to generalize that method for the UIV-FD model leads to significant modifications. First, for UIV-FD, an objective function is lacking and it is necessary to utilize an auxiliary function to generate costs for the shortest-path computations and to monitor the overall progress of the method. Simple line searches also have to be replaced with solving VI subproblems over simplices.

The auxiliary function G is known as the gap function and is defined as follows. For any $\tilde{v}$ in Θ, let

$$G(\tilde{v}) = \max_{v \in \Theta} s(\tilde{v})(\tilde{v} - v).$$

Note that $G(\tilde{v}) \geqslant 0$ for all $\tilde{v} \in \Theta$. Further, unless $\tilde{v}$ solves (56), $G(\tilde{v}) > 0$ and $G(\tilde{v}) = 0$ if, and only if, $\tilde{v} = v^*$, the solution. Thus solving (56) is equivalent to finding a v^*, which solves

$$\min_{\tilde{v} \in \Theta} G(\tilde{v}), \tag{62}$$

which may also be written as

$$\min_{\tilde{v} \in \Theta}\left\{\max_{v \in \Theta}\{s(\tilde{v})(\tilde{v} - v)\}\right\}, \tag{63}$$

where for each $\tilde{v}$, the inner maximization problem is equivalent to the linear programming (shortest-path) subproblem of the linear approximation method.

In the context of equilibrium traffic assignment, $G(\tilde{v})$ has an important interpretation, which makes (62) intuitively clear. To see this, note that

$$G(\tilde{v}) = s(\tilde{v})\tilde{v} + \max_{v \in \Theta} - s(\tilde{v})v \tag{64}$$

$$= s(\tilde{v})\tilde{v} - \min_{v \in \Theta} s(\tilde{v})v \tag{65}$$

$$= \sum_{a \in A} s_a(\tilde{v}_a)\tilde{v}_a - \sum_{p \in P} \tilde{u}_p \bar{t}_p, \tag{66}$$

where $\tilde{u}_p = u_p(\tilde{v})$ is, by definition (19), the least cost path, given $\tilde{v}$, for O-D pair p. The first term in (66) is the total system cost and the second term is the total system cost if all demand is routed along minimum paths with respect to $\tilde{v}$. Clearly these two terms are equal if, and only if, $\tilde{v} = v^*$, the user optimal flow vector.

Simplicial Decomposition

Step 0. Let Θ^1 be a set initialized to contain one extreme point of Θ. Let $\delta > 0$ be chosen. Set $\bar{G}^1 = \infty$ and $l = 1$. Let $\{\epsilon_l\}$ be a strictly monotonically decreasing positive sequence converging to zero.

Step 1. Find $v^l \in$ convex hull (Θ^l) such that $s(v^l)(v - v^l) \geqslant -\epsilon_l$, for every $v \in$ convex hull (Θ^l). Let D^l denote the elements of Θ^l with zero weight in the expression of v^l as a convex combination of the extreme points of θ^l.

Step 2. Solve:

$$\min s(v^l)y \ \text{ subject to: } \ y \in \Theta.$$

Denote the solution of the linear program by y^l.

If $G(v^l) = s(v^l)(v^l - y^l) = 0$, stop. Otherwise,

(i) if $G(v^l) \geqslant \bar{G}^l - \delta$, let $\Theta^{l+1} = \Theta^l \cup \{y^l\}$

(ii) if $G(v^l) < \bar{G}^l - \delta$, let $\Theta^{l+1} = (\Theta^l - D^l) \cup \{y^l\}$.

Let $\bar{G}^{l+1} = \min\{\bar{G}^l, G(v^l)\}$. Set $l = l + 1$. Go to Step 1.

Note that Step 1 involves a relaxed variational inequality, implying that it only need be approximately solved. Any of the methods available can be used but a projection method is often employed because the domain is a simplex. Step 2 is just the familiar calculation of shortest paths as in the linear approximation method. Since $\{G(v^l)\}$ is not monotonic, $\bar{G}^l$ is introduced to obtain a monotonically decreasing sequence and this makes it possible to prove convergence of the method.

4. Transit Route Choice Models

Transit route choice models or transit assignment models aim to describe traffic flows on a network of transit lines, which operate at known frequencies. The main difference with the traffic assignment models and methods of Section 3 is the waiting phenomenon: transit travelers experience a waiting time for the first vehicle (bus) of the line that they have chosen. In addition, the access to the transit line stop implies an access time, which is usually a walk time, transfers between lines if more then one line is taken and the in-vehicle time. The following exposition is based on the work of Spiess and Florian (1989). Consider a transit network that consists of a set of nodes, a set of transit lines, each defined by an ordered list of nodes at which boarding and alightings are permitted, and a set of walk links, each defined by two nodes. The times associated with each walk link and each transit line segment are constant. At each node that is on the itinerary of a transit line, the distribution of the interarrival times of the vehicles is known for each line that serves the node. As a consequence one can compute the combined expected time for the arrival of the first vehicle, for any subset of lines incident at a node, as well as the probability that each line arrives first.

In order to state the mathematical model that corresponds to the transit route choice selection, it is noted that each walk link may be replaced (conceptually) by a transit line of one link with a zero waiting time (infinite frequency). Also, it is assumed that the underlying network is strongly connected. The objective is taken to be the minimization of expected waiting and travel time, or the expected total generalized cost if waiting times and travel times may have different weights (e.g., waiting is more onerous than in-vehicle time).

The network is composed of four types of arcs: wait arcs (no travel time), in vehicle (no waiting), alighting (no travel time, no waiting), and walk arcs (travel time, no waiting). Thus, the segment of a transit line is an arc that is served by a vehicle at given intervals and the transit traveler waits for the link to be served by a vehicle. The arcs that will be included in a solution of the model are denoted by $\bar{A} \in A$, where A is the set of arcs and N is the set of nodes. Thus the solution for a single destination s is a subgraph $b_s = (N, \bar{A})$. The demand for travel from nodes i, $i \in N$ to the destination s is denoted $\bar{t}_i$. Among the links included in a solution $\bar{A}$, at each node i, $i \in N$, a traveler boards the first vehicle that serves any of the lines in the $\bar{A}_i^+$ ($\bar{A} = \cup_i \bar{A}_i^+$). The set $\bar{A}_i^+$ corresponds to the lines that will be chosen by the traveler to yield one or more routes from i to s in a solution of the model. At each stop i, it is convenient to refer to the set $\bar{A}_i^+$ as the set of attractive lines.

Let $W(\bar{A}_i^+)$ be the expected waiting time for the arrival of the first vehicle serving any of the links $a \in \bar{A}_i^+$, which is denoted as the combined waiting time of links $a \in \bar{A}_i^+$. Let $P_a(\bar{A}_i^+)$ be the probability that link a is the first line to be served among the links $\bar{A}_i^+$. If an exponential distribution of interarrival time is assumed then

$$W(\bar{A}_i^+) = \frac{1}{\sum_{a \in \bar{A}_i^+} f_a} \tag{67}$$

and

$$P(\bar{A}_i^+) = \frac{f_a}{\sum_{a' \in \bar{A}_i^+} f_a}, \quad a \in \bar{A}_i^+, \tag{68}$$

where f_a is the frequency of link (line) a.

Since $\bar{A}$ is not known, *a priori* the single destination model is formulated by using binary variables x_a

$$x_a = \begin{cases} 0 & \text{if } a \neq \bar{A} \\ 1 & \text{if } a \in \bar{A}. \end{cases}$$

The optimization model may be stated as follows by letting A_i^+ and A_i^- denote forward and backward stars of node i:

$$\min \sum_{a \in A} s_a v_a + \sum_{i \in N} \frac{V_i}{\sum_{a \in A_i^+} f_a x_a} \tag{69}$$

subject to

$$v_a = \frac{x_a f_a}{\sum_{a' \in A_i^+} f_{a'} x_{a'}}, \quad a \in A_i^+, \quad i \in N \tag{70}$$

$$V_i = \sum_{a \in A_i^-} v_a + \bar{t}_i, \quad i \in N \tag{71}$$

$$V_i \geqslant 0, \quad i \in N \tag{72}$$

$$x_a = 0 \text{ or } 1, \quad a \in A, \tag{73}$$

where s_a is the travel cost on link a and V_i is the total volume at node i. At first sight, the problem (69)–(73) is a mixed-integer nonlinear optimization problem. Fortunately, the problem may be reduced to a much simpler linear programming problem by making the following observations. (72) may be replaced by the nonnegativity constraints of the link volumes $v_a \geqslant 0$, $a \in A$ since $\sum_{a \in A_i^+} v_a = V_i$, $i \in N$. Then by introducing new variables w_i, which denote the total waiting time of all trips at node i,

$$w_i = \frac{V_i}{\sum_{a \in A_i^+} f_a x_a}, \quad i \in N,$$

one obtains the equivalent problem

$$\min \sum_{a \in A} s_a v_a + \sum_{i \in N} w_i \tag{74}$$

subject to

$$v_a = x_a f_a w_i, \quad a \in A_i^+, \quad i \in I \tag{75}$$

$$\sum_{a \in A_i^+} v_a - \sum_{a \in A_i^-} v_a = \bar{t}_i, \quad i \in N \tag{76}$$

$$v_a \geqslant 0, \quad a \in A. \tag{77}$$

The objective function (74) is now linear and the 0–1 variables are only used in constraints (75), which are the only nonlinear constraints. These constraints may be relaxed by replacing (75) with

$$v_a \leqslant f_a w_i, \quad a \in A_i^+, \quad i \in N, \tag{78}$$

which yields the linear programming problem (74), (78), (76), (77). It may be shown, by using the extreme point properties of the solutions of a linear programming model, that this problem is equivalent to (74)–(77). The dual problem of this last linear program is

$$\max \sum_{i \in N} \bar{t}_i u_i \tag{79}$$

subject to

$$u_j + s_a + \mu_a \geqslant u_i, \quad a \in A \tag{80}$$

$$\sum_{a \in A_i^+} f_a \mu_a = 1, \quad i \in N \tag{81}$$

$$\mu_a \geqslant 0, \quad a \in A, \tag{82}$$

where u_i, u_j are the dual variables corresponding to (76) and μ_a are the dual variables corresponding to (75).

Let (v^*, w^*) and (u^*, μ^*) denote the optimal solutions of the primal and dual problems. The weak complementary slackness conditions are

$$(v_a^* - f_a w_i^*)\mu_a^* = 0, \quad a \in A_i^+, \quad i \in N \tag{83}$$

and

$$(u_j^* + s_a + \mu_a^* - u_i^*)v_a^* = 0, \quad a \in A. \tag{84}$$

In both the primal and dual formulations, the transit route choice model has a close resemblance to the shortest-path route choice model. The latter corresponds to the shortest-path problem when none of the links of the network involves waiting; thus $f_a \to \infty$ and $w_i \to 0$. The solution algorithm, which solves the transit route choice model bears a strong resemblance to the label-setting algorithm for computing shortest paths.

The solution algorithm is composed of two parts. In a first pass, from the destination nodes to all origins, the arcs which carry flow, $\bar{A}^*$, and the expected travel times u_i^* from each node i, $i \in N$, to the destination nodes are computed. In a second pass, from all origins to the destination, the demand is assigned to the arcs a, $a \in \bar{A}^*$. The algorithm is stated below (the convention $0\ \infty = 1$ is used):

Transit Route Choice Algorithm

Part 1: Find $\bar{A}^*$

Step 1.1 (Initialization): $u_i \leftarrow \infty$, $i \in N - \{s\}$; $u_s \leftarrow 0$

$$f_i \leftarrow 0, \quad i \in N$$

$$S \leftarrow A; \quad \bar{A} \leftarrow \phi$$

Step 1.2. (Choose next link): If $S = \phi$ STOP; otherwise find $a = (i,j) \in S$ such that $u_j + s_a \leqslant u_{j'} + s_{a'}$; for any $a' = (i',j') \in S$. $S \leftarrow S - \{a\}$.

Step 1.3. (Update node labels): If $u_i \geqslant u_j + s_a$

$$u_i \leftarrow \frac{f_i u_i + f_a(u_j + s_a)}{f_i + f_a}$$

$$f_i \leftarrow f_i + f_a; \quad \bar{A} = \bar{A} + \{a\}$$

Go to Step 1.2;

Part 2: Assign demand to $\bar{A}^*$

Step 2.1 (Initialization): $V_i \leftarrow \bar{t}_i$, $i \in N$

Step 2.2 (Loading): For every link $a \in A$ in decreasing order of $(u_j + s_a)$ do:

if $a \in \bar{A}$ then $v_a \leftarrow \frac{f_a}{f_i} V_i$,

$V_j \leftarrow V_j + v_a$;

otherwise $v_a \leftarrow 0$.

The auxiliary variables f_i, $i \in N$, contain the combined frequencies of all selected links at node i.

By using the primal and the dual formulations of the transit route choice model, one can prove that the algorithm indeed find the solution of (74), (78), (76), (77). The algorithm is applied for each destination in turn.

It is possible to extend the algorithm to a nonlinear version of this problem where the link travel times are no longer constants, but are continuous functions $s_a(v_a)$, $a \in A$ of the arc flows v_a. The resulting model may be solved by an adaptation of the linear approximation algorithm. Further details may be found in Spiess (1984) and Spiess and Florian (1989).

5. Conclusions

Some of the principal optimization models used frequently in transportation planning applications have been presented. These include the spatial interaction or trip distribution model, several variants of the network equilibrium models and corresponding solution algorithms, and a transit route choice model.

References

Andersson, P. A. 1981. "On the convergence of iterative methods for the distribution balancing problem." *Transport. Res.* **15B**: 173–201.

Bregman, L. M. 1967. "The relaxation method for finding the common point of convex sets and its application to the solution of problems in convex programming." *USSR Comput. Math. Mathemat. Phys.* **7**: 200–217.

Deming, W. E. and F. F. Stephan. 1940. "On a least squares adjustment of a sampled frequency table when the expected marginal totals are known." *Ann. Math. Stat.* **11**: 427–444.

Erlander, S., S. Nguyen, and N. Stewart. 1979. "On the calibration of the combined distribution-assignment model." *Transport. Res.* **13B**: 259–267.

Evans, S. P. and H. R. Kirby. 1974. "A three-dimensional Furness procedure for calibrating gravity models." *Transport. Res.* **8**: 105–122.

Florian, M. and D. W. Hearn. 1995. "Network equilibrium models and algorithms." In *Handbooks in OR & MS*, edited by M. O. Ball *et al.*, pp. 485–550. Amsterdam, The Netherlands: Elsevier Science.

Jaynes, E. 1957a. "Information theory and statistical mechanics." *Physical Rev.* **106**: 171–190.

Jaynes, E. 1957b. "Information theory and statistical mechanics." *Physical Rev.* **108**: 620–630.

Jefferson, T. R. and C. H. Scott. 1979. "The analysis of entropy models with equality and inequality constraints." *Transport. Res.* **13B**: 123–132.

Kruithof, J. 1937. "Calculation of telephone traffic." *De Ingenieur* **52**: E-15–E-25 [In Flemish.].

Kullback, S. 1959. *Information Theory and Statistics*. New York: Wiley.

Lamond, B. and N. F. Stewart. 1981. "Bregman's balancing method." *Transport. Res.* **15B**: 239–248.

Patricksson, M. 1994. *The Traffic Assignment Problem—Models and Methods*. Utrecht: VSP.

Robillard, P. and N. F. Stewart. 1974. "Iterative numerical methods for trip distribution problems" *Transport. Res.* **8**: 575–582.

Sheffi, Y. 1985. *Urban Transportation Networks. Equilibrium Analysis with Mathematical Programming Methods*. Englewood Cliffs, NJ: Prentice-Hall.

Snickars, F. and J. W. Weibull. 1977. "A minimum information principle, theory and practice." *Regional Sci. Urban Econ.* **7**: 137–168.

Spiess, H. 1984. "Contributions à la Théorie et aux outils de planification des réseaux de transport urbains." Ph.D. diss. Montréal: Département d'informatique et de recherche opérationnelle, Universite de Montréal.

Spiess, H. and M. Florian. 1989. "Optimal strategies: a new assignment model for transit networks." *Transport. Res.* **23B**: 83–102.

Wardrop, J. 1952. "Some theoretical aspects of road traffic research." In *Proceedings of the Institute of Civil Engineers, Part II*, **1**: 325–378.

Wilson, A. G. 1967. "A statistical theory of spatial distribution models." *Transport. Res.* **1**: 253–269.

Wilson, A. G. 1970. *Entropy in Urban and Regional Modelling*. London: Pion.

18.13 Optimization in Computational Molecular Biology

Guoliang Xue

ABSTRACT Computational molecular biology has been a subject of extensive study by researchers in biology, physics, chemistry, mathematics, and computer science. There are many optimization problems in this area that are both interesting and challenging. The goal of this paper is to describe some of the optimization problems and current challenges in computational molecular biology. These problems include molecular conformation, fast computation of the force field of a molecular cluster, minimum interparticle distance at stable states of a molecular cluster, and a simplified protein folding model.

1. Introduction

For many years, computational molecular biology has been a field of biologists, physicists, and chemists. In recent years, however, many researchers from mathematics and computer science have paid close attention to this area. These researchers bring many advanced techniques in numerical optimization and in data structures and computer algorithms to this interdisciplinary research field. As a result of these combined efforts, many important results have been obtained. For example, polynomial time algorithms have been designed that transform the genome of cabbage into the genome of turnip (Hannenhalli and Pevzner, 1999). As another example, with the help of advanced algorithms and high performance computers, many lower energy states for the Lennard–Jones clusters have been discovered by Xue (1994a,b).

In this article, we will concentrate on optimization problems in the minimization of potential energy functions arising from molecular conformation or protein folding problems and related problems. In Section 2, we discuss the molecular conformation problem. In Section 3, we discuss fast computation of the force field of a molecular cluster. In Section 4, we discuss the minimum inter-particle distance at stable states of a molecular cluster. In Section 5, we present a simplified protein folding model. In each section, we will define the optimization problem, describe the methods used in solving the problem, and present some current challenges. We conclude this paper in Section 6.

2. Conformation of Lennard–Jones Clusters

The molecular conformation problem is to find a global minimizer of a predefined potential energy function of a cluster of particles. Global minimizers of properly defined potential energy functions are of great importance because the minimizers often correspond to stable states of the cluster. The global minimizers are often called native states or ground states of the given cluster. In this section, we will concentrate on the Lennard–Jones potential (to be formally defined in the following) which has applications in metal clusters (Honeycutt and Andersen, 1987; Kostrowicki et al., 1991; Kostrowicki and Scheraga, 1992; Maranas and Floudas, 1992; Northby, 1987; Phillips and Rosen, 1994; Xue et al., 1992). In the next three subsections, we will define the optimization problem, review the algorithms developed for solving this problem, and discuss current challenges.

2.1. The Optimization Problem

Of the many cases of interest, the simplest potential energy function is that of two-body central forces between the component particles. Although this potential energy function does not actually model the real system in many cases—ignoring angle-dependent conditions, bond constraints, and dihedral-angle conditions—it provides a useful prototype that enormously simplifies the design of computer programs. This simplest potential energy function is also very important because it has been the most expensive part of the more general potential energy functions.

Given a cluster of n particles in three-dimensional space, the potential energy function of the cluster is defined as the summation (over all of the pairs) of the two-body interparticle pair potentials. Let the center of the n particles be $a_1, ..., a_n$. The potential energy function of the cluster is defined as

$$V_n(a_1, ..., a_n) = \sum_{1 \leqslant i<j \leqslant n} v_{\text{pair}}(\|a_i - a_j\|), \tag{1}$$

where $\|a_i - a_j\|$ is the Euclidean distance between a_i and a_j, and $v_{\text{pair}}(r)$ is an interparticle pair potential. In this proposal, n will be used to represent the number of particles or beads in the given system if not stated otherwise.

Although many types of function $v_{\text{pair}}(r)$ may be used in physical models, it is necessary to apply some restrictions if we want $v_{\text{pair}}(r)$ to have satisfactory behavior. In general, we would like $v_{\text{pair}}(r)$ to be continuous at least down to a hard core $r_{\min}$ and to possess derivatives up to the second order over the interval $(r_{\min}, \infty)$. More specifically, we are only interested in cases where $v_{\text{pair}}(r)$ is a well potential satisfying the following conditions:

(C_1) $v_{\text{pair}}(r) \to 0^-$ as $r \to \infty$;

(C_2) $v_{\text{pair}}(r) \to \infty$ as $r \to r_{\min}$ and $r_{\min} \geqslant 0$;

(C_3) $v'_{\text{pair}}(r_0) = 0$ for a unique r_0 with $r_{\min} < r_0 < \infty$;

(C_4) $v''_{\text{pair}}(r_0) > 0$ and $v(r_0) < 0$.

There are a number of pair potential energy functions that satisfy the above conditions (Pardalos et al., 1994). A well-known example is the Lennard–Jones pair potential function

$$v_{LJ}(r) = \left(\frac{\sigma}{r}\right)^{12} - 2\left(\frac{\sigma}{r}\right)^{6}, \tag{2}$$

where σ is a positive constant. For the Lennard–Jones potential function, the global minimum of -1 is achieved at a unique point $r = \sigma$. A plot of this function with $\sigma = 1$ is given in Figure 1. Although finding the unique global minimizer of one Lennard–Jones pair potential $v_{LJ}(r)$ is easy, as one can see from Figure 1, finding a global minimizer of $V_n(\bullet)$ is extremely difficult except for very small cluster sizes. The difficulty is that while it is always possible with a local minimization algorithm to compute a local minimizer from any reasonable initial configuration, unless the starting configuration is in the catchment basin of the global minimizer, the minimizer found may not be the global minimizer. Hoare (1979) has claimed that the number of local minimizers in the potential energy surface of an n-particle Lennard–Jones cluster is about $O(e^{n^2})$. Since cluster sizes of interest are as large as one thousand or even ten thousand, molecular conformation problems with Lennard–Jones (and many other) potential energy functions have long been a challenge to researchers in molecular physics and computer science.

2.2. *Optimization Methods*

Computational methods proposed for molecular conformation problems can be classified into three categories. We will briefly discuss all three of them with no prejudice.

The first class of methods is called spatial averaging methods (Pardalos et al., 1994), which use a spatially-averaged objective function to approximately locate the global minimizer, and then iteratively increase accuracy by sequential minimizations with smoothed objective functions that have been averaged over smaller and smaller regions of space. Some of the representative methods in this class are the diffusion equation method (Piela et al., 1989) and the effective energy method (Coleman et al., 1993, 1994; Shalloway, 1992). The success of spatial averaging approaches depends on the validity of what is called the strong spatial-scaling hypothesis: that the position of the global minimizer varies in a continuous manner as the energy (objective) function is progressively smoothed by averaging over increasingly large regions of space. Even when the strong hypothesis does not hold, the weak spatial-scaling hypothesis—that the global minimizer varies in a continuous manner except for a bounded number of discontinuities of bounded extent—may be valid. The weak condition can also be

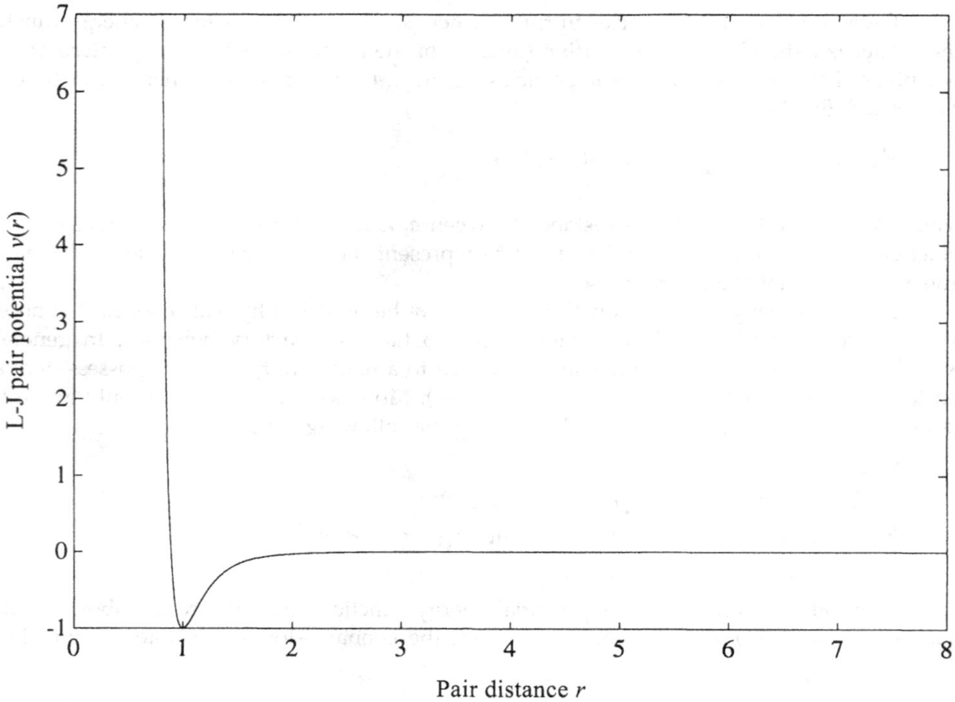

Figure 1 The Lennard-Jones pair potential function

exploited to organize searches although more complicated algorithms will be needed. The validity of these hypotheses for specific problems is not known and is an interesting topic for future examination.

The second class of methods is called lattice-based search methods (Pardalos et al., 1994). Chemical physicists have learned from previous research that the ground states of Lennard–Jones clusters exhibit certain kinds of lattice structures. Therefore, methods are proposed to find the best lattice configurations and then relax these lattice configurations by applying continuous minimization from the lattice configurations and let the particles move off the lattice points. A critical assumption for lattice search-based algorithms is that a well-defined set of lattice structures contains at least one initial cluster configuration that relaxes to the ground state. As described and supported by computational results in Hoare (1979), Northby (1987), Maier et al. (1992), and Xue (1994a,b), the *IC* and *FC* lattices (see Xue, 1994a,b for definitions and descriptions of these lattices) are well-defined lattice structures for pure Lennard–Jones clusters.

So far, the most successful algorithms for computing ground states of Lennard–Jones clusters are based on lattice search followed by local minimization from the lattice minimizers, represented by the milestone paper of Hoare (Hoare, 1979) and the Northby algorithm (Northby, 1987). Extending the work of Hoare (1979), Northby (1987) proposed to search the best lattice configuration by moving a particle at a time (this is called a pivot) on the surface of the *IC* and *FC* lattices. After finding the lattice configurations, which have the lowest potential energy function among all lattice configurations, Northby then applies continuous minimization methods starting from the best lattice configurations (allowing the particles to move off the lattice sites). Northby (1987) reported putative global minima of Lennard–Jones clusters for $n \leqslant 153$. For a long time, no other algorithm could find solutions as good as the Northby algorithm for all cluster sizes in this range.

Since Northby only relaxes the lattice configurations with the best possible potential function value, and his algorithm has been successful, it is generally believed that in most cases the relaxation of a global lattice minimizer will result in a configuration with a lower energy than the relaxation of a nonglobal lattice local minimizer.

Xue (1994) noticed that the lattice search problem in the Northby algorithm is equivalent to a set partitioning problem, that is, partition the surface sites into empty and filled sites. Xue introduced simple data structures into the problem and improved the Northby algorithm from $O(n^{\frac{5}{3}})$ operations per pivot to $O(n^{\frac{2}{3}})$ operations per pivot for an n-particle Lennard–Jones cluster. With a powerful parallel two-level simulated annealing algorithm, Xue was able to produce putative global minima for the lattice search problem for cluster sizes as large as one hundred thousand on the CM-5 (Xue, 1994a). Xue also made another improvement on the Northby algorithm by relaxing every lattice local minimizer found instead of relaxing only those lattice local minimizers with the best-known potential function value. In this case, the algorithm requires more time. However, lower energy configurations were found with this improved algorithm for $n = 65, 66, 75, 76, 77$, and 134. These findings also show that, in some cases, the relaxation of a lattice local minimizer with a worse potential function value may lead to a local minimizer with a better potential function value. The results reported by Xue (1994a,b) are currently the best-known results for Lennard–Jones clusters in the range $n \leqslant 153$.

Lattice-based search methods are very successful for the kind of clusters for which a lattice structure is known, such as the Lennard–Jones cluster. However, for many other clusters, such a lattice structure is not known. Therefore we need to apply other methods for smaller clusters and study the configurations in order to find a suitable lattice structure and then apply lattice-based search methods for larger clusters. Xue's findings also indicate that even for the Lennard–Jones cluster where the lattice structure is known, relaxing the best lattice configuration may not result in a global minimizer in all cases.

The third class of methods is general global minimization methods. These are often mutistart local minimization methods or simulated annealing or genetic algorithms. These methods are not as successful as the lattice-based search methods for clusters where knowledge about the underlying lattice structure is known. However, they have the advantage that no knowledge about the physical problem is assumed and therefore can be applied to essentially any molecular conformation problem. Some of the most efficient algorithms in this class are the ones of Maranas and Floudas (1992, 1993, 1994). Instead of applying the global minimization methods to the potential function without using any knowledge of the physical problem at all, these algorithms use some knowledge of the physical problem. For example, when trying to compute the global minimizer of a cluster of n particles, they first compute the global minimizer of a cluster of $n-1$ particles and then try to find an optimal position of the nth particle while holding the positions of the first $n-1$ particles at positions which form the global minimizer of the cluster with $n-1$ particles. After this is done, a few particles are allowed to move in an attempt to reduce the potential while others are fixed. These methods were able to produce most of the putative global minima reported by Northby (1987).

However, for some of the cluster sizes, these methods could not find the lower potential configurations reported by Northby (1987). In any case, these methods use less knowledge about the physical problem and therefore apply to a wider range of problems. Improving these methods to make them more effective is one focus of proposed future research.

2.3. *Challenges*

Although many algorithms have been designed to minimize nonconvex potential energy functions, there is still very much to learn in this area. It is of great importance to enlarge or improve the table of putative global minima for Lennard–Jones clusters because this will allow scientists to study various properties of such clusters. On the other hand, these known minima can serve as a yardstick for other global optimization algorithms. The best-known putative global minima for Lennard–Jones clusters can be found in Gu and Du (1995), Northby (1987), and Xue (1994a,b). It is the experience of the author that specially designed algorithms that take advantage of the problem structures are usually more successful than general purpose global optimization algorithms.

3. Fast Computation of Force Field

Fast algorithms for force field evaluation have important applications in molecular conformation, molecular dynamics, and astrophysical simulations. Given a cluster of n particles in three-dimensional space, we need to compute the force exerted on each particle by the other particles. Since there are $n(n-1)/2$ pairs, the direct method requires $\Theta(n^2)$ time for force-evaluation, which is very expensive for astronomical simulations.

In astrophysical simulations, the force exerted by one particle on another is given by the gravitational force. In molecular dynamics and molecular conformation, the Lennard–Jones potential is widely used. In both cases, the force exerted on one particle by another particle vanishes as the distance between them approaches infinity. This observation leads to several fast approximation algorithms. Two famous $O(n \log n)$ time hierarchical tree algorithms were published by Appel (1985) and by Barnes and Hut (1986). In 1987, Greengard and Rokhlin (1987) published the fast multipole algorithm, which computes the force field in $O(n)$ time. These algorithms have made great impacts on the computational study of molecular conformation/dynamics and astronomical simulations. Due to the big constant in the fast multipole algorithm and the simplicity and efficiency of the tree algorithms, hierarchical tree algorithms have received more attention in computational studies (Anderson, 1996).

3.1. Overview

A central idea behind Appel's algorithm and the Barnes–Hut algorithm is the monopole approximation (Appel, 1985).

Appel showed that when a group of n_1 particles and a group of n_2 particles are well separated from each other, we can approximate the $n_1 \times n_2$ pair interactions by a single pair interaction between two big particles (one at the gravitational center of the first group and one at the gravitational center of the other). The tree algorithms consist of two phases. In the first phase, an oct-tree is constructed, which hierarchically partitions the particles into many smaller clusters. In the second phase, the oct-tree is used to compute an approximation to the force field. In most simulations, the particles are almost homogeneously distributed. In this case, the oct-tree for an n-particle cluster has a height of $\Theta(\log n)$. The oct-tree was built using the following top-down approach. The root node corresponds to a computation box (a cube) big enough to contain all the particles in the given cluster. The n particles are inserted in the root of the tree one by one. Whenever a node in the tree has two or more particles, the corresponding computation box is subdivided into eight smaller computation boxes, which correspond to the eight children of the current node. The particles in the current node are then inserted in the children nodes according to their spatial positions. It is clear that $O(n \log n)$ time is required to build the oct-tree this way if the height of the tree is bounded by $O(\log n)$. Both Appel's algorithm and the Barnes–Hut algorithm require $O(n \log n)$ time to build the tree for homogeneously-distributed clusters. In a recent paper (Xue, 1998), Xue showed that the oct-tree can be constructed bottom-up in $O(n)$ time. He then showed that the force field can be computed in $O(n)$ time as well. We will describe Xue's algorithm in this section.

Throughout this, we make the following assumption on the distribution of the particles:

Assumption 1. *Two positive constants* c_1 *and* c_2 *exist such that the minimum interparticle distance is at least* c_1 *and the maximum interparticle distance is smaller than* $c_2 n^{1/3}$.

Assumption 1 is strongly believed to be true for most applications and is supported by many computer simulations. For the Lennard–Jones cluster it has been proved that the minimum interparticle distance has a positive lower bound, which is independent of the number of particles in the cluster (Xue, 1997).

3.2. Building the Oct-Tree Bottom-Up in O(n) Time

The top-down construction of the oct-tree described in the previous section is so simple that most researchers assume the existence of the oct-tree in the study of force field evaluation algorithms. However, it is possible to build the oct-tree faster if we are looking for an asymptoticly faster algorithm. Such an algorithm is presented in Xue (1998) through a bottom-up approach.

A computation box is defined by a point base in three-dimensional space and a positive number size. Let *baseX, baseY, baseZ* be the coordinates of point base. Then the computation box defined by base and size is

$$[baseX, baseX + size) \times [baseY, baseY + size) \times [baseZ, baseZ + size). \tag{3}$$

Note that each interval in the Cartesian product (3) is closed on the left but open on the right. This convention will make the partition of a computation box much easier as we will see later. A computation box is illustrated in Figure 2(a). When a computation box is partitioned, we obtain 8 nonintersecting computation boxes of equal size whose union is the original computation box. An example is illustrated in Figure 2(b).

Xue uses the following data structure in the description of his algorithm:

```
typedef struct _node{
struct _node *parent; struct _node *child[8];
int    isLeaf; int    weight; int    pindex;
double coordX; double coordY; double coordZ;
double forceX; double forceY; double forceZ;
double baseX;  double baseY;  double baseZ; double size;
}NODE;
```

Each node in the oct-tree is of type NODE. For every node in the tree, weight is the number of particles contained in the corresponding computation box. If weight is 0 or 1, we have a leaf node and the field *isLeaf* is 1. If weight is 2 or larger, we have an interior node and the field isLeaf is 0. For an interior node, the fields *coordX, coordY, coordZ* represent the coordinates of the gravitational center of the particles contained in the computation box corresponding to this node. The fields *baseX, baseY, baseZ* and size define the computation box corresponding to the node in the tree. The fields *forceX, forceY* and *forceZ* are used in the evaluation of force field whose use will be discussed later. For a leaf node whose weight is 1, pindex is the index of the unique particle that is contained in the computation box corresponding to the leaf node. The field parent contains a pointer to the parent node in the tree. For the root node, parent is NULL. For any interior node, *child*[*j*] is a pointer to the *j*th child of the current node ($j = 0, 1, ..., 7$). We make the assumption that every interior node has exactly eight children whose computation boxes have the same size, which is half the size of

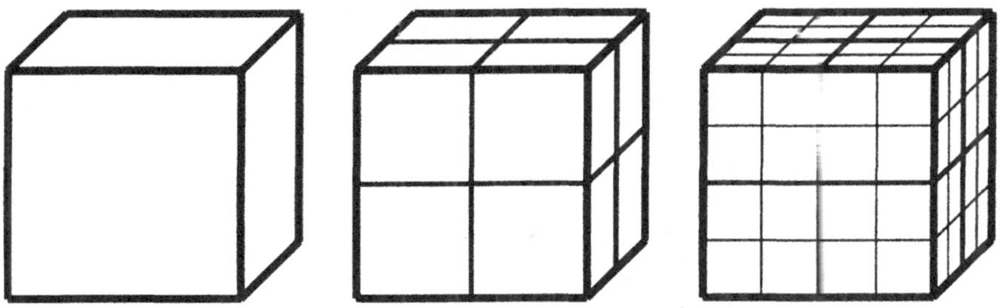

(a) level 0 (b) level 1 (c) level 2

Figure 2 Computation boxes associated with the first three levels of the oct-tree

the computation box of the current node. If the computation box of the current node is as given in (3), then the computation boxes for *child*[0], *child*[1], ..., *child*[7] are defined as follows.

$$[baseX, baseX + \tfrac{1}{2}size) \times [baseY, baseY + \tfrac{1}{2}size) \times [baseZ, baseZ + \tfrac{1}{2}size),$$
$$[baseX, baseX + \tfrac{1}{2}size) \times [baseY, baseY + \tfrac{1}{2}size) \times [baseZ + \tfrac{1}{2}size, baseZ + size),$$
$$[baseX, baseX + \tfrac{1}{2}size) \times [baseY + \tfrac{1}{2}size, baseY + size) \times [baseZ, baseZ + \tfrac{1}{2}size),$$
$$[baseX, baseX + \tfrac{1}{2}size) \times [baseY + \tfrac{1}{2}size, baseY + size) \times [baseZ + \tfrac{1}{2}size, baseZ + size),$$
$$[baseX + \tfrac{1}{2}size, baseX + size) \times [baseY, baseY + \tfrac{1}{2}size) \times [baseZ, baseZ + \tfrac{1}{2}size),$$
$$[baseX + \tfrac{1}{2}size, baseX + size) \times [baseY, baseY + \tfrac{1}{2}size) \times [baseZ + \tfrac{1}{2}size, baseZ + size),$$
$$[baseX + \tfrac{1}{2}size, baseX + size) \times [baseY + \tfrac{1}{2}size, baseY + size) \times [baseZ, baseZ + \tfrac{1}{2}size),$$
$$[baseX + \tfrac{1}{2}size, baseX + size) \times [baseY + \tfrac{1}{2}size, baseY + size) \times [baseZ + \tfrac{1}{2}size, baseZ + size),$$

Note that the partition of a computation box takes constant time. The computation box in Figure 2(a) is partitioned into eight smaller computation boxes in Figure 2(b), which in turn are partitioned into a total of sixtyfour even smaller computation boxes in Figure 2(c).

3.3. Building the Tree Bottom-Up

We assume that the n particles are given in an array of points so that $part[i].x$, $part[i].y$, $part[i].z$ represent the coordinates of particle i ($i = 0, 1, 2, ..., n-1$).

In $O(n)$ time, we can compute the base point of the computation box for the root node by computing the minimum of the coordinates of the n particles for each of the three dimensions. Similarly, we can compute the maximum of the coordinates of the n particles for each of the three dimensions in $O(n)$ time. By Assumption 1, we can now decide the size of the smallest computation box as well as the size of the largest computation box, in constant time. By then we should know an $O(\log n)$ upper bound on the height of the oct-tree. Therefore we can dynamically allocate space for every possible tree node. We assume that all the fields of a tree node are initialized to zero at the time the memory is allocated. The total space allocated is $O(n)$ as will be analyzed after the presentation of the algorithm.

Instead of inserting the particles to the tree from the root node, we insert the particles directly to the nodes corresponding to the smallest computation boxes. We then pass information from one layer of the tree to the layer above, starting from the bottom layer. Although there are $\Theta(\log n)$ layers of the tree, the amount of time required is decreased by a factor of eight every time we move up one layer. This is the key to achieving the $O(n)$ time complexity. The complete algorithm is presented below.

Algorithm 1. Building the oct-tree bottom-up.

Step_1 {Compute the root box}

Let $Xmin := \min_{i=1,n} part[i-1].x$; $Ymin := \min_{i=1,n} part[i-1].y$; $Zmin := \min_{i=1,n} part[i-1].z$. Let $Xmax := \max_{i=1,n} part[i-1].x$; $Ymax := \max_{i=1,n} part[i-1].y$; $Zmax := \max_{i=1,n} part[i-1].z$. Let $\delta := \sqrt{\tfrac{3}{3}}c_1$. Let *maxlevel* be the smallest positive integer such that $\delta 2^{maxlevel} > \max\{Xmax - Xmin, Ymax - Ymin, Zmax - Zmin\}$. Let $\Delta := \delta \times 2^{maxlevel}$.

Step_2 {Allocate space}

We will use a three-dimensional array of NODE for the nodes on each level of the oct-tree. Let *tree*[*l*] be a pointer to the three-dimensional array of NODE with $2^l \times 2^l \times 2^l$ elements ($l = 0, 1, \ldots, maxlevel$). It is clear that we require to allocate $\Theta(n)$ space because there are 8^l tree nodes on level-l of the tree. These arrays are dynamically allocated at this time.

Step_3 {Construct the leaf nodes}

for $p = 1$ to n do

```
    Let i = ⌊(part[p−1].x − Xmin)/δ⌋;  j = ⌊(part[p−1].y − Ymin)/δ⌋;  k = ⌊(part[p−1].z − Zmin)/δ⌋.
    tree[maxlevel]→node[i][j][k].size := δ;
    tree[maxlevel]→node[i][j][k].baseX := Xmin + iδ;
    tree[maxlevel]→node[i][j][k].baseY := Ymin + jδ;
    tree[maxlevel]→node[i][j][k].baseZ := Zmin + kδ;
    tree[maxlevel]→node[i][j][k].weight := 1;
    tree[maxlevel]→node[i][j][k].pindex := p;
    tree[maxlevel]→node[i][j][k].coordX := part[p−1].x;
    tree[maxlevel]→node[i][j][k].coordY := part[p−1].y;
    tree[maxlevel]→node[i][j][k].coordZ := part[p−1].z.
  endfor
Step_4 {Building the tree bottom-up}
 for l := maxlevel−1 downto 0 do
  for i := 0 to 2^l−1 do for j := 0 to 2^l−1 do for k := 0 to 2^l−1 do
    tree[l]→node[i][j][k].child[0] := tree[l+1]→node[2i+0][2j+0][2k+0];
    tree[l]→node[i][j][k].child[1] := tree[l+1]→node[2i+0][2j+0][2k+1];
    tree[l]→node[i][j][k].child[2] := tree[l+1]→node[2i+0][2j+1][2k+0];
    tree[l]→node[i][j][k].child[3] := tree[l+1]→node[2i+0][2j+1][2k+1];
    tree[l]→node[i][j][k].child[4] := tree[l+1]→node[2i+1][2j+0][2k+0];
    tree[l]→node[i][j][k].child[5] := tree[l+1]→node[2i+1][2j+0][2k+1];
    tree[l]→node[i][j][k].child[6] := tree[l+1]→node[2i+1][2j+1][2k+0];
    tree[l]→node[i][j][k].child[7] := tree[l+1]→node[2i+1][2j+1][2k+1];
    Also set the parent field for each of the eight children nodes;
    tree[l]→node[i][j][k].baseX := Xmin + i2^(maxlevel−l)δ;
    tree[l]→node[i][j][k].size := 2^(maxlevel−l)δ;
    tree[l]→node[i][j][k].baseY := Ymin + j2^(maxlevel−l)δ;
    tree[l]→node[i][j][k].baseZ := Zmin + k2^(maxlevel−l)δ;
    Let tree[l]→node[i][j][k].weight be the sum of the weights of its
    children;
    Let tree[l]→node[i][j][k].coordX, tree[l]→node[i][j][k].coordY, and
    tree[l]→node[i][j][k].coordZ be the coordinates of the weighted center
    of
    the particles contained in the computation box of the current node;
    if tree[l]→node[i][j][k].weight == 0 then
     tree[l]→node[i][j][k].isLeaf := 1;
    elseif tree[l]→node[i][j][k].weight == 1 then
     tree[l]→node[i][j][k].isLeaf := 1;
     Let tree[l]→node[i][j][k].pindex be the index of the only particle
     contained in the current computation box;
    endif
  endfor endfor endfor
 endfor
```

With some careful analysis, Xue proved the following theorem.

Theorem 1. *Algorithm 1 builds the oct-tree for* n *particles using* Θ(n) *time and* Θ(n) *space, provided that the particles satisfies Assumption 1. The constant behind the asymptotic notation is proportional to* $\left(\frac{c_2}{c_1}\right)^3$.

Note that although we have allocated space for a full oct-tree, the actual oct-tree may not be a full oct-tree in most cases. As a result, the leaf nodes of the oct-tree may be at different levels of the tree. Note also that under the same assumption about the distribution of the particles, the top-down construction of the oct-tree requires $O(n \log n)$ time (Appel, 1985; Barnes and Hutt, 1986). In our bottom-up construction, we are allocating space for some tree nodes that will never be used (i.e., the descendants of a node whose weight is 1). We should note that the asymptotic memory requirements for the top-down algorithm and the bottom-up algorithm are both $O(n)$. However, the time complexity of the bottom-up algorithm is a $\Theta(\log n)$ factor lower than that of the top-down algorithm. In the next section, we will show that Appel's algorithm for computing the force field of a cluster of n particles can be implemented in $O(n)$ time after the oct-tree is constructed. Therefore the improved time complexity of the construction of the oct-tree has a great impact on the simulation of large clusters.

3.4. Computing Force Fields Top-Down in O(n) Time

Given a cluster of n particles, we need to compute the potential energy function and the force exerted on each particle by the other particles. In many applications, the potential energy function of a cluster is the sum of the pair-wise potential functions.

Let p_1 and p_2 be the positions of two particles of unit charge each, the Lennard–Jones potential function between this pair of particles is defined by

$$f_{LJ}(p_1, p_2) = \frac{\sigma_1}{\|p_2 - p_1\|^{12}} - \frac{\sigma_2}{\|p_2 - p_1\|^6}, \tag{4}$$

where $\|\bullet\|$ stands for the Euclidean norm and σ_1 and σ_2 are given positive constants. This potential energy function is widely used in molecular conformation, molecular dynamics and protein folding. In other applications, the following gravitational potential

$$f_G(p_1, p_2) = \frac{\sigma_3}{\|p_2 - p_1\|} \tag{5}$$

is often used, where σ_3 is a given positive constant. In both cases, the potential energy function approaches zero as the distance between the two particles approaches infinity. In both cases, the force exerted on p_1 by p_2 is computed as the negative of the gradient of the potential energy function with respect to p_1. These are given by

$$\left(\frac{-12\sigma_1}{\|p_2 - p_1\|^{14}} - \frac{-6\sigma_2}{\|p_2 - p_1\|^8}\right)(p_2 - p_1), \tag{6}$$

and

$$\frac{-\sigma_3}{\|p_2 - p_1\|^3}(p_2 - p_1) \tag{7}$$

for the Lennard–Jones pair potential and the gravitational pair potential, respectively.

Since there are $n(n-1)/2$ pairs for a cluster of n particles, the conventional algorithm for computing the potential energy function and force field requires $O(n^2)$ time.

In 1985, Appel (1985) proposed a divide-and-conquer algorithm for computing the force field of a gravitational cluster. His algorithm has a proved time complexity of $O(n \log n)$. Appel's algorithm is based on the following idea: given two clusters of particles consisting of n_1 and n_2 unit weight particles each, there are $n_1 \times n_2$ particle pairs with one particle from the first cluster and the other particle from the other cluster. If the two clusters are well separated (i.e., the ratio of the maximum diameter of the clusters over the distance between the clusters is small), we may consider the first cluster as a big particle located at the gravitational center P_1 of the first cluster with a mass of n_1 and consider the second cluster as another big particle located at the gravitational center P_2 of the second cluster with a mass of n_2. We may then approximate the total interactions between particles from the first cluster and particles from the second cluster by the following weighted pair potential:

$$F_G(P_1, n_1; P_2, n_2) = n_1 n_2 \frac{\sigma_3}{\|P_2 - P_1\|}. \tag{8}$$

The negative of the gradient of the above function with respect to P_1 is

$$n_1 n_2 \frac{-\sigma_3}{\|P_2 - P_1\|^3}(P_2 - P_1). \tag{9}$$

Therefore, the force exerted on each particle in the first cluster by all the particles in the second cluster can be approximated by

$$n_2 \frac{-\sigma_3}{\|P_2 - P_1\|^3}(P_2 - P_1), \tag{10}$$

since there are n_1 particles in the first cluster. Similarly, the force exerted on each particle in the second cluster by all the particles in the first cluster can be approximated by

$$n_1 \frac{-\sigma_3}{\|P_2 - P_1\|^3}(P_1 - P_2). \tag{11}$$

In the case where Lennard–Jones potential energy function is used, the force exerted on each particle in the first cluster by all the particles in the second cluster can be approximated by

$$n_2\left(\frac{-12\sigma_1}{\|P_2 - P_1\|^{14}} - \frac{-6\sigma_2}{\|P_2 - P_1\|^8}\right)(P_2 - P_1), \tag{12}$$

and the force exerted on each particle in the second cluster by all the particles in the first cluster can be approximated by

$$n_1\left(\frac{-12\sigma_1}{\|P_2 - P_1\|^{14}} - \frac{-6\sigma_2}{\|P_2 - P_1\|^8}\right)(P_1 - P_2). \tag{13}$$

In this way, we can spend constant time to compute an approximation to the potential that requires $\Theta(n_1 \times n_2)$ time using the conventional method. If the ratio of the maximum of the radii of the clusters over the distance between the clusters is δ, the relative error in this approximation is $O(\delta^2)$ (Appel, 1985).

The performance of Appel's algorithm depends on the parameter that defines well separateness. If this parameter is close to 0, we have more accuracy but need more computing time. If this parameter is close to 1, we have less accuracy but need less computing time. A generic description of Appel's algorithm is given in the following. We can prove that Appel's algorithm actually requires only $O(n)$ time when the oct-tree is given.

After the oct-tree is constructed, Appel's algorithm can be implemented using the procedures oneNODE, twoNODES, and pushDOWN. We assume that there is a global variable FUNC, which is initialized to 0 and is used to accumulate the potential energy function of the cluster. We also assume that the fields *forceX*, *forceY*, and *forceZ* at every tree node are all initialized to 0 before the computation. These fields are used to hold partial values of the force field during the computation.

```
void oneNODE(NODE *A, double delta)
{ int i, j;
  if (A->isLeaf == 0) {
    for (i = 0; i < 8; i++) oneNODE(A->child[i], delta);
    for (i = 0; i < 7; i++)
      for (j = i + 1; j < 8; j++)
        twoNODES(A->child[i], A->child[j], delta);
  }
}
void twoNODES(NODE *A, NODE *B, double delta)
{ double d; int i, j;
  if(A->weight == 0||B->weight == 0) return;
  d = distance(A,B);
  if ((A->size/d) <= delta) && (B->size/d) <= delta))
    compGRAD(A, B);
  else
    for (i = 0; i < 8; i++)
      for (j = 0; j < 8; j++)
        twoNODES(A->child[i], B->child[j], delta);
}
```

```
void pushDOWN(NODE *A)
{ int i;
  while (A->weight >= 2) {
    for (i = 0; i < 8; i++) {
      A->child[i].forceX += A->forceX/A->weight;
      A->child[i].forceY+= A->forceY/A->weight;
      A->child[i].forceZ+= A->forceZ/A->weight;
      pushDOWN(A->child[i]);
    }
  }
}
```

The function *distance(A, B)* measures the distance between the centers of the two computation boxes associated with the two tree nodes pointed to by A and B. It is convenient to use the L_∞ norm, that is, the largest difference in the three dimensions. The procedure *compGRAD(A, B)* treats the particles contained in the computation box of A as a big particle at the gravitational center of these particles with weight equal to $A\rightarrow$weight and treats the particles contained in the computation box of B as a big particle at the gravitational center of these particles with weight equal to $B\rightarrow$weight. It then approximates the $A\rightarrow$weight* $B\rightarrow$weight pair potential energy by a single weighted pair potential. It also computes the negative of the gradient of this weighted pair potential with respect to the gravitational center of A and stores this information in the fields $A\rightarrow forceX$, $A\rightarrow forceY$, and $A\rightarrow forceZ$. The opposite of this force is stored in the fields $B\rightarrow forceX$, $B\rightarrow forceY$, and $B\rightarrow forceZ$.

The computation of force field starts with the procedure call oneNODE(root, delta) where root is a pointer to the root of the oct-tree and delta is the well separateness parameter. This procedure call is followed by the procedure called pushDOWN(root). When the second call returns, the total potential energy function can be found in the global variable FUNC and the force field exerted on each particle can be found in the fields *forceX, forceY*, and *forceZ* at the corresponding leaf node of the oct-tree. The memory can now be freed using the array of pointers level[] so that there is no memory leek during the computation.

Let

$$\delta = \frac{1}{2d+2} \quad \text{or} \quad d = \frac{1}{2\delta} - 1.$$

Xue proved the following theorem on the time complexity of his sequential algorithm.

Theorem 2. *Given a cluster of n particles satisfying Assumption 1, the force field can be computed using Appel's algorithm in* $O\left(\left(\frac{c_2}{c_1} \times \frac{1}{\delta}\right)^3 n\right)$ *time, where* c_1 *and* c_2 *are the distribution parameters and* δ *is the well separateness parameter.*

3.5. *Challenges*

In this section, we have presented an $O(n)$ time variant of Appel's algorithm for force field evaluation in N-body simulations. A key to this improved complexity is an $O(n)$ time bottom-up construction of the oct-tree that was constructed top-down using $O(n \log n)$ time in previous studies. We have also studied the dependency of the constant behind the asymptotic notation on the distribution parameters c_1 and c_2 and on the *well separateness parameter* δ. In a more recent paper (Xue, 1999), Xue presented a cost-optimal CREW PRAM algorithm for force field computation. When the particles are not uniformly distributed we may use the fair-split tree algorithm due to Callahan and Kosaraju (1995). Computational results using the fair-split tree algorithm are presented in Xue and Lasher (1998). Parallel implementations of the Barnes–Hut algorithm are presented in Warren and Salmon (1993). More computational studies are required.

4. Minimum Interparticle Distance

Computer simulation has become a very important tool in the study of molecular conformation and the prediction of the three-dimensional structure of a protein, given the sequence of amino acids (Richards, 1991). It is generally believed that the optimal structure of a protein or a cluster of particles corresponds to a global minimizer of a potential energy function associated with the cluster (Hoare, 1979; Pardalos et al., 1994).

While these computer simulations require weeks or months of CPU time on a modern supercomputer, a significant part of the computing time is spent in the evaluation of potential energy functions and force fields (Pardalos et al., 1994). Therefore, many algorithms for fast evaluation of potential energy functions and force fields are proposed in the literature (Appel, 1985; Barnes and Hut, 1986; Esselink, 1992; Greengard and Rokhlin, 1987). Xue et al. (1995) noticed that the most expensive part of the commonly used potential energy functions is the pair-wise Lennard–Jones potential energy function and therefore concentrated on fast evaluation of Lennard–Jones potential energy functions of a cluster of n particles. However, most of these algorithms assume that the particles are uniformly distributed in order to guarantee good performance.

In a 1997 paper, Xue (1997) proved that the minimum interparticle distance at any global minimizer of a Lennard–Jones cluster is bounded away from zero by a positive constant that is independent of the number of particles in the cluster. Although it seems simple from a pure physics point of view, that is the first mathematical proof for the problem model. Based on previous computational results, it is believed that the maximum interparticle distance at any global minimizer of an n particle Lennard–Jones cluster is bounded from above by $O(n^{\frac{1}{3}})$. This essentially says that the particles are almost uniformly distributed near the global minimizer and therefore the algorithms (Appel, 1985; Barnes and Hut, 1986; Esselink, 1992; Greengard, 1987; Xue, 1998; Xue et al., 1995) will all have good performance near the global minimizer. Therefore, we can concentrate on the design of force field evaluation algorithms that are efficient for almost uniformly distributed clusters. As a by-product, we also prove that the global minimum of an n particle Lennard–Jones cluster is bounded between two linear functions. This is an improvement over a result in Xue et al. (1992). In this section, we present the main results of the 1997 paper of Xue.

4.1. *The Lennard–Jones Potential Energy Function*

Definition 1. *The Lennard–Jones pair potential function is defined as*

$$v(r) = \frac{1}{r^{12}} - \frac{2}{r^6}, \quad r > 0. \tag{15}$$

This function defines the pair-wise reaction between two particles at a distance of r. It can be seen from Figure 1 that the two particles push away from one another when the distance is smaller than 1 and attract one another when the distance is greater than 1. The reaction becomes extremely strong when r approaches 0 and vanishes when r approaches ∞. A Lennard–Jones cluster of n particles is associated with the following potential energy function.

Definition 2. *Let* $\mathrm{x}_1, \mathrm{x}_2, \ldots, \mathrm{x}_n$ *be a cluster of* n *particles in the three-dimensional Euclidean space. The Lennard–Jones potential energy function for this cluster is defined as*

$$f(x) = \sum_{1 \leqslant i<j \leqslant n} v(\|x_i - x_j\|), \tag{16}$$

where $\| \bullet \|$ *represents the Euclidean norm.*

Definition 3. *Let* $\mathrm{x}_1, \mathrm{x}_2, \ldots, \mathrm{x}_n$ *be a cluster of* n *particles in the three-dimensional Euclidean space. The potential energy function due to* x_i *is*

$$f_i(x) = \sum_{j \neq i} v(\|x_i - x_j\|). \tag{17}$$

Clearly, the Lennard–Jones potential energy function for a cluster of particles is half of the sum of potentials due to each of the particles. In molecular conformation and protein folding, people are interested in the global minimizers of the potential energy function. Here we are interested in the minimum interparticle distance at global minimizers of $f(x)$.

Theorem 3. *The following are true for the Lennard–Jones pair potential function.*

- $v(r) \geqslant -1 \quad \text{for } 0 < r < \infty.$
- $v(r) > 0 \quad \text{for } 0 < r < \left(\frac{1}{2}\right)^{1/6}.$
- $v(r) = 0 \quad \text{for } r = \left(\frac{1}{2}\right)^{1/6}.$
- $v(r) < 0 \quad \text{for } r > \left(\frac{1}{2}\right)^{1/6}.$
- $v'(r) < 0 \quad \text{for } 0 < r < 1.$
- $v'(r) = 0 \quad \text{for } r = 1.$
- $v'(r) > 0 \quad \text{for } r > 1.$
- $v''(r) > 0 \quad \text{for } 1 < r < \left(\frac{13}{7}\right)^{1/6}.$
- $v''(r) = 0 \quad \text{for } r = \left(\frac{13}{7}\right)^{1/6}.$
- $v''(r) < 0 \quad \text{for } r > \left(\frac{13}{7}\right)^{1/6}.$

Theorem 4. *Let* $f_{min}(n)$ *be the global minimum of the Lennard–Jones function for a cluster of n particles. Then*

$$f_{\min}(n) \leqslant -(n-1). \tag{18}$$

Theorem 5. *Let* y *be a global minimizer of* f(x). *Then* $f_i(y) < 0, (i = 1, 2, \ldots, n)$.

4.2. *Minimum Interparticle Distance*

Theorem 6. *The minimum inter-particle distance at any global minimizer of* f(x) *is greater than or equal to 0.5.*

In molecular conformation, people are interested in finding the global minimizer of the potential energy function (Hoare, 1979). Since the Lennard–Jones potential function has many local minimizers, it is very hard to verify that a local minimizer is a global one. If we know a tight lower bound on the optimal objective function value, we know how good a local minimizer is.

For the Lennard–Jones cluster, $-O(n^2)$ is a trivial lower bound. In Xue et al. (1992), the first nontrivial lower bound of $-O(n^{\frac{19}{13}})$ on the optimal objective function value was proven. They also proved that the optimal objective function value is bounded between two linear functions under the assumption that the minimum inter-particle distance is bounded away from zero. The following theorem is a corollary of Theorem 5 in this article and Theorem 2.2 of Xue et al. (1992).

Theorem 7. *Positive constants* α *and* β *exist such that*

$$-\alpha n \geqslant f_{\min}(n) \geqslant -\beta n, \quad n = 2, 3, \ldots; \tag{19}$$

where $f_{min}(n)$ *is the global minimum function value of the Lennard–Jones potential energy function of an n particle cluster.*

4.3. *Challenges*

In this section, we have shown that the minimum interparticle distance at any global minimizer of a Lennard–Jones cluster is bounded away from zero by a positive constant, which is independent of the number of particles in the cluster As a by-product, we also established the first linear lower bound on the optimal objective function value. While these theoretical results are important further studies are necessary to make them more useful in applications. It is very important to obtain a linear lower bound for the optimal objective function for the Lennard–Jones cluster that is tight for many values of n. Based on previous computational results (Hoare, 1979), we make the following conjecture.

Conjecture. *The maximum interparticle distance at any global minimizer of Lennard–Jones clusters is bounded from above by* $O(n^{\frac{1}{3}})$, *where n is the number of particles in the cluster.*

Provided that it is true, Conjecture and Theorem essentially says that the particles are almost uniformly distributed near a global minimizer. Therefore, most of the algorithms for fast evaluation of potential energy functions and force fields (Appel, 1985; Barnes and Hut, 1986; Greengard, 1987) would have good performance near the global minimizer.

5. Conclusions

In this paper, we have presented several optimization problems in computational molecular biology, together with current methods and remaining challenges More references can be found in Neumaier (1997); Pardalos et al. (1994, 1995); Pardalos and Xue (1994, 1997), and Wille (1999). It is not our intention to provide a complete reference on this subject here. It is our hope that this article will provide a general picture of the optimization problems in computational molecular biology to the readers and lead the interested researcher to some of the challenging problems in this field.

ACKNOWLEDGMENTS

This research was supported in part by ARO grant DAAD 19-00-1-0377 (Xue) and DOE grant DE-FG02-00ER45828 (Xue).

References

Appel, A. W. 1985. "An efficient program for many-body simulation." *SIAM J. Sci. Stat. Comput.* **6**: 85–103.

Anderson, R. J. 1996. "Tree data structures for N-body simulation." In *37th Annual Symposium of Foundations of Computer Science*, pp. 224–233. IEEE.

Barnes, J. and P. Hut. 1986. "A hierarchical O(*n* log *n*) force-calculation algorithm." *Nature* **324**: 446–449.

Callahan, P. B. and S. R. Kosaraju. 1995. "A decomposition of multidimensional point sets with applications to *k*-nearest-neighbors and *n*-body potential fields." *J. ACM* **42**: 67–90.

Coleman, T., D. Shalloway, and Z. Wu. 1993. "Isotropic effective energy simulated annealing searches for low energy molecular cluster states." *Comput. Optim. and Appl.* **2**: 145–170.

Coleman, T., D. Shalloway, and Z. Wu. 1994. "A parallel build-up algorithm for global energy minimizations of molecular clusters using effective energy simulated annealing." *J. Global Optim.* **4**: 171–185.

Esselink, K. 1992. "The order of Appel's algorithm." *Inf. Process. Lett.* **41**: 141–147.

Greengard, L. and V. Rokhlin. 1987. "A fast algorithm for particle simulations." *J. Comput. Phys.* **73**: 325–348.

Gu, J. and B. Du. 1995. "Bin A multispace search algorithm for molecular energy minimization." In *Global Minimization of Nonconvex Energy Functions: Molecular Conformation and Protein Folding, DIMACS Series in Discrete Mathematics and Theoretical Computer Science*, edited by P. M. Pardalos and G. L. Xue, **23**: 65–87.

Hannenhalli, S. and P. Pevzner. 1999. "Transforming cabbage into turnip: polynomial algorithm for sorting signed permutations by reversals." *J. ACM* **46**: 1–27.

Danna Honeycutt, J. and Hans C. Andersen. 1987. "Molecular dynamics study of melting and freezing of small Lennard–Jones clusters." *J. Phys. Chem.* **91**: 4950–4963.

Hoare, M. R. 1979. "Structure and dynamics of simple microclusters." *Adv. Chem. Phys.* **40**: 49–135.

Kostrowicki, J. and H. A. Scheraga. 1992. "Application of the diffusion equation method for global optimization to oligopeptides." *J. Phys. Chem.* **96**: 7442–7449.

Kostrowicki, J., L. Piela, B. J. Cherayil, and H. A. Scheraga. 1991. "Performance of the diffusion equation method in searches for optimum structures of clusters of Lennard–Jones atoms." *J. Phys. Chem.* **95**: 4113–4119.

Maier, R. S., J. B. Rosen, and G. L. Xue. 1992. "A discrete-continuous algorithm for molecular energy minimization." In *IEEE/ACM Supercomputing'92*, pp. 778–786. IEEE Computer Society Press.

Maranas, C. D. and C. A. Floudas. 1992. "Global Optimization of Lennard–Jones Microclusters." *J. Chem. Phys.* **97**: 7667–7678.

Maranas, C. D. and C. A. Floudas. 1993. "Global optimization for molecular conformation problems." *Ann. Oper. Res.* **42**: 85–117.

Maranas, C. D. and C. A. Floudas. 1994. "Global minimum potential energy conformations of small molecules." *J. Global Optim.* **4**: 135–170.

Neumaier, A. 1998. "Molecular modeling of proteins and mathematical prediction of protein structure." *SIAM Review* **39**: 407–460.

Northby, J. A. 1987. "Structure and binding of Lennard–Jones clusters: $13 \leqslant n \leqslant 147$." *J. Chem. Phys.* **87**: 6166–6178.

Pardalos, P. M. and G. L. Xue. (eds). 1994. "Advances in computational chemistry and molecular biology." Special issue of *J. Global Optim.*, **4**(2).

Pardalos, P. M. and G. L. Xue. (eds). 1997. "Computer simulations in molecular and protein conformations." Special issue of *J. Global Optim*, **11**(1).

Pardalos, P. M., D. Shalloway, and G. L. Xue. 1994. "Optimization methods for computing global minima of nonconvex potential energy functions." *J. Global Optim.* **4**: 117–134.

Pardalos, P. M., D. Shalloway, and G. L. Xue. (eds). 1995. "Global minimization of non-convex energy functions: molecular conformation and protein folding." In *DIMACS Series in Discrete Mathematics and Theoretical Computer Science*, Volume 23. American Mathematical Society.

Phillips, A. T. and J. B. Rosen. 1994. "A quadratic assignment formulation of the molecular conformation problem." *J. Global Optim.* **4**: 229–242.

Piela, L., J. Kostrowicki, and H. A. Scheraga. 1989. "The multiple-minima problem in the conformational analysis of molecules." *Deformation of the potential energy hypersurface by the diffusion equation method. J. Phys. Chem.* **93**: 3339–3346.

Richards, F. M. 1991. "The protein folding problem." In *Scientific American*, January 1991, pp. 54–63.

Shalloway, D. 1992. "Application of the renormalization group to deterministic global minimization of molecular conformation energy functions." *J. Global Optim.* **2**: 281–311.

Shalloway, D. 1992. "Packet annealing: a deterministic method for global minimization. Application to molecular conformation." In *Recent Advances in Global Optimization*, edited by C. Floudas and P. Pardalos, pp. 433–477. Princeton: Princeton University Press.

Warren, M. S. and J. K. Salmon. 1993. "A parallel hashed oct-tree N-body algorithm." In *IEEE/ACM Supercomputing'93*, pp. 54–65. IEEE Computer Society Press.

Wille, L. T. 1999. "Lennard–Jones clusters and the multiple-minima problem." In *Annual Reviews of Computational Physics*, edited by D. Stauffer.

Xue, G. L. 1994a. "Molecular conformation on the CM-5 by parallel two-level simulated annealing." *J. Global Optim.* **4**: 187–208.

Xue, G. L. 1994b. "Improvement on the Northby algorithm for molecular conformation: better solutions." *J. Global Optim.* **4**: 425–440.

Xue, G. L. 1997. "Minimum inter-particle distance at global minimizers of Lennard–Jones clusters." *J. Global Optim.* **11**: 83–90.

Xue, G. L. 1998. "An $O(n)$ time algorithm for force field calculation in N-body simulations." *Theor. Comput. Sci.* **197**: 157–169.

Xue, G. L. 1999. "A cost optimal parallel algorithm for computing force field in N-body simulations on a CREW PRAM." *Theor. Comput. Sci.* **255(2001)**: 555–568.

Xue, G. L. and Mark Lasher. 1998. "Fast evaluation of potential and force field in particle systems using a fair-split tree spatial structure." *Optim. Methods Softw.* **14(2000)**: 245–266.

Xue, G. L., R. S. Maier, and J. B. Rosen. 1992. "Minimizing the Lennard–Jones potential function on a massively parallel computer." In *proceedings of 1992 ACM International Conference on Supercomputing*, pp. 409–416. ACM Press.

Xue, G. L., A. Zall, and P. M. Pardalos. 1995. "Rapid evaluation of potential energy functions in molecular and protein conformations." In *DIMACS Series in Discrete Mathematics and Theoretical Computer Science* **23**: 237–249.

18.14 Optimization in the Financial Services Industry

Anna Nagurney

ABSTRACT In this article we overview several applications of optimization in the financial services industry, specifically: portfolio selection, market equilibrium analysis, risk management, asset/liability management, and option pricing. Our primary goal here is to describe, within the financial applications setting, some fundamental mathematical models and the appropriate optimization framework. A bibliography is provided for further in-depth reading.

1. Introduction

Finance is concerned with the study of capital flows over space and time in the presence of risk. The field over the past three decades has witnessed extraordinary innovations, which have occurred during a time of economic growth in response to an environment characterized by uncertainty and volatility associated with interest rates and other asset prices, and foreign exchange rates, as well as the desire of investors and institutions to respond appropriately to government-imposed regulation. Indeed, since the collapse of the Bretton Woods regime of fixed exchange rates (Shapiro, 1992), along with the oil crisis of the same time, and the United States Federal Reserve Board's efforts to control inflation through money supply targeting, as opposed to interest rate targeting, the environment has become more volatile with financial business operations becoming necessarily those of risk taking.

Parallel to the dynamic global economic and financial environment and marketplace a vast increase in computing technology has occurred, which has revolutionized the financial services industry (*Economist*, 1996). The technological advances in computing power have been congruent with theoretical advances in both financial economics and mathematical programming, yielding a wealth of tools and opportunities for financial and quantitative analyses. Indeed, the recognition of the existence of distinct sectors in economies, the proliferation of new financial instruments, coupled with the increasing diversification of portfolios internationally, along with governmental policy interventions, as well as the presence of transaction costs, have all brought added complexities and a greater degree of realism to the mathematical modeling and computation of solutions to financial phenomena.

The primary objective of this article is to present applications of optimization in the financial services industry. Applications of optimization to financial services can contribute to the efficiency, profitability, and competitiveness of the financial entity in the current market scenario. The applications have been selected based on their timely relevance as well as historical importance. In particular, we focus on the following: portfolio selection, market equilibrium analysis, risk management, asset/liability management, and option pricing.

2. Portfolio Selection

Portfolio selection is a classical application of optimization to finance. In its most basic form, the problem is to select a portfolio of assets that yields the highest expected return for a given level of risk; alternatively, the problem can be viewed as one of minimizing the level of risk for a given expected rate of return. A fundamental concept, rigorously formulated by Markowitz (1952, 1959, 1989), is the recognition that risk reduction comes with effective diversification.

Markowitz claimed that investors select assets in such a way that the risk of their portfolio matches their risk preferences. In other words, individuals who cannot bear risk will invest in assets with low risk, whereas people more comfortable with risk will accept investments of higher risk. His work also suggested that the trade-off between risk and return is different for each investor, but the preferences of all people lie upon a fictitious curve that is usually called the frontier of efficient portfolios. Along this curve lie all the diversified portfolios that have the highest return for a given risk, or the lowest risk for a given return.

2.1. *Mean-Variance Portfolio Selection*

Markowitz's model was based on mean-variance portfolio selection, where the average and the variability of portfolio returns were determined in terms of the mean and covariance of the corresponding investments. The mean is a measure of an average return and the variance is a measure of the distribution of the returns around the mean return. Markowitz formulated the portfolio optimization problem as:

$$\text{Minimize} \quad V = X^T QX \quad \text{(Minimize risk)}$$

subject to:

$$R = \sum_{i=1}^{n} X_i r_i \quad \text{(attainment of given expected rate of return)}$$

$$\sum_{i=1}^{n} X_i = 1 \quad \text{(budget constraint)}$$

$$X_i \geqslant 0, \quad i = 1, \ldots, n, \quad \text{(no short sales allowed),}$$

where n is the total number of securities available in the economy; X_i represents the relative amount of capital invested in security i, with the securities being grouped into the column vector X. Q denotes the $n \times n$ variance-covariance matrix on the return of the portfolio; r_i denotes the expected value of the return of security i; and R denotes the expected rate of return on the portfolio.

In his work, Markowitz suggested that for a fixed set of expected values r_i and covariances of the returns of all assets i and j, all investors can find an (R, V) combination that best fits their taste, solely limited by the constraints of the specific problem. Hence, according to the original work of Markowitz (1952), the efficient frontier had to be identified and then every investor had to select a portfolio through a mean-variance analysis that fit his or her preferences.

A related mathematical optimization model (Markowitz, 1959) to the one above, which can be interpreted as the investor seeking to maximize his returns while minimizing his risk can be expressed by the quadratic programming model:

$$\text{Maximize} \quad \alpha R - (1 - \alpha) V$$

subject to:

$$\sum_{i=1}^{n} X_i = 1$$

$$X_i \geqslant 0, \quad i = 1, \ldots, n,$$

where α denotes an indicator of how risk-averse a specific investor is.

Many versions and extensions of Markowitz's model have appeared in the literature, a collection of which can be found in Francis and Archer (1979), with $\alpha = 1/2$ being a frequently accepted value. A recent interpretation of the model as a multicriteria decisionmaking model along with theoretical extensions to multiple sectors can be found in Dong and Nagurney (1998).

The first important simplification of Markowitz's model was by Sharpe (1963), through a model known as the diagonal model, where "the individual covariances between all securities are assumed to be zero." According to Sharpe's model, the variance-covariance matrix Q has zeros in all positions other than the diagonal. Clearly, in this model the number of covariances required for portfolio analysis was smaller than in the original formulations and, consequently, the solution process was simplified, requiring less data and less computation time.

In addition, the mean-variance portfolio analysis that was introduced and mathematically formulated by Markowitz (1952, 1959) and later simplified by the diagonal model of Sharpe (1963), was further extended by Pogue (1970) and Francis (1978), with the introduction of variance–covariance matrices for both assets and liabilities, applied to the asset-liability management of banks. We introduce such matrices in the next section and focus on the asset and liability management application in Section 5.

In practice, there may be more constraints on the portfolio manager than the ones revealed in the above models. For example, other constraints may include: the attainment of certain cash yields, guaranteeing specific asset holdings, following particular index trackings, and so on. Such constraints can readily be incorporated into the basic quadratic programming formulation. In addition, one may modify the objective function to include a linear term to represent transaction costs.

Furthermore, in present-day practice, the portfolio selection problem may be formulated as an alternative mathematical programming problem depending on the selection of the objective function, the definition of the decision variables, and the particular constraints underlying the specific situation. Hence, the solution of the portfolio selection problem may involve one or more, or a combination thereof, of the following optimization techniques:

- If the risk of the portfolio can be measured as a ranking of assets or by the linear distance from the target, then the portfolio selection problem can be formulated as a linear programming problem.
- Quadratic programming is applied when the model is a mean-variance one as in the above Markowitz-type models.
- Nonlinear programming is applied when the portfolio selection model is characterized by an objective function that seeks to maximize utility as a function of the portfolio composition with the utility function being a nonlinear function.
- Mixed-integer and nonlinear programming is applied when some of the variables must take on integer values as would be the case when the financial products must be bought or sold in batches.
- Network programming takes advantage of the underlying network structure of portfolio optimization and other financial problems that may yield enhanced computational speed.
- Stochastic programming allows for the incorporation of stochastic variables.
- Dynamic programming may be utilized when the portfolio selection problem involves more than a single time period in which case the problem is no longer static but a multiperiod one.

In order to give an illustration of the size of problems that are being or need to be solved, we note that in 1984 (see Perold, 1984) portfolio optimization problems with about five hundred securities could be solved whereas it is not unusual to have more than three thousand securities to select from. Recent research has focused on the development of enhanced computational methods, which can increase the efficiency of determining optimal portfolios by exploiting, for example, some special underlying problem structure, including network structure (see Pardalos et al., 1994; Nagurney and Siokos, 1997, and the references therein) as well as on the simplification of the data that is needed and data reduction.

2.2. Capital Asset Pricing Model

The most significant extension of the models by Markowitz (1952) and Sharpe (1963) was the capital asset pricing model (CAPM), which was based on the work of Sharpe (1964), Lintner (1965), and Mossin (1966) and which explicitly recognizes that one set of assets may be related to another set. Hence, the CAPM is an equilibrium model, a topic we will elaborate on further in the next section where we describe a model in which there are multiple sectors in the economy that are no longer assumed to be homogeneous. In the CAPM, the concept of a risk-free asset and market portfolio were introduced. A risk-free asset is an asset with a positive expected rate of return and a zero standard deviation, with the U.S. Treasury bill being a good example. A market portfolio, on the other hand, is a portfolio on the efficient frontier of the Markowitz model, which is considered to be desirable by all investors.

The CAPM assumes that all investors will select a portfolio that will be a linear combination of the risk-free asset and the market portfolio and, hence, the equilibrium prices of all assets can be expressed as a linear combination of the risk-free price and the price of the market portfolio. The assumptions of the CAPM concern the investors, the firms, and the structure of the existing capital market. Investors are assumed to be risk-averse, trying to maximize utility functions that can be expressed in terms of the mean and the variance of the returns. Moreover, returns are assumed to be normally distributed, and investors to be price takers with homogeneous expectations and without any tax obligations. All the firms are equity financed and the market is assumed to be perfect, that is, there are no transaction costs and the investors cannot influence the price of any product in the economy. Although some of the assumptions governing the CAPM are not realistic, and subsequent modeling efforts have strived to relax one or more of them, the model is considered to be one of the major breakthroughs in modern economic and finance theory and forms the basis for most of the financial models that exist today. See, for example, the extension to a multiperiod economy by Merton (1973), the consumption CAPM by Breeden (1978; which, however, failed empirically due to the difficulty in observing and computing consumption) and, finally, the inclusion of transaction costs by Milne and Smith (1980).

2.3. *Arbitrage Pricing Theory*

The dissatisfaction with the empirical tests of the CAPM led to more advanced models such as the arbitrage pricing theory (APT) by Ross (1976). APT's biggest contribution was the inclusion of multiple risk factors and the generalization of the CAPM, which was considered to be a special case of APT with only a single risk factor. According to APT, there is a linear relation among the expected returns of all assets in the economy. In particular, Ross assumed that the rate of return of every security can be expressed as a linear combination of some basic risk factors. The factors used in APT are nondiversifiable and not pre-identified by the model and, consequently, empirical results were not easily obtainable. The APT of Ross was subsequently refined by Chamberlain (1983) and Connor (1984).

3. Market Equilibrium Analysis

Market equilibrium analysis has its roots in the last half of the nineteenth century. The work of Gossen (1854), Jevons (1871), and Walras (1874) initiated the analysis of equilibrium theory. In the 1930s, the study of market equilibrium became more formalized, with the work of Wald (1935, 1936) and Hicks (1939) providing proofs of different qualitative properties of the equilibrium, along with a detailed study of the conditions under which an equilibrium could be modeled and derived. Subsequently, Arrow (1951) and Debreu (1959) initiated a new era in equilibrium analysis by bringing uncertainty into equilibrium theory, which led to the current status of market equilibrium theory. A notable example of the evolution of equilibrium theory was the contributions of Tobin (1963, 1969, 1971), who developed a partial equilibrium model for the financial sector, as opposed to a general equilibrium model in which all sectors are included, based on the classical Walrasian (1874) model.

The basic assumption that governs most of the existing models that address the theory of market equilibrium is that of perfect competition. As mentioned in Section 2, perfect competition forbids any participant in the economy (buyer or seller) from having control over the prices of different products or over the actions of other participants. The price, hence, of a product is considered to be a variable, the value of which is determined by the combined actions of all the buyers and sellers. (Note that in the classical Markowitz-type models, in contrast, the prices were assumed to be known by the investors.) Buyers are, hence, "price takers," in that they modify their holdings of a product according to the price, ignoring the effects that their behavior may have on that price. In addition, under perfect competition it is assumed that all participants in the economy have perfect information about the products available, the current price, as well as the bids of a specific product. Furthermore, the number of the participants in the economy is assumed to be large enough so that the market activity regarding a specific product will be small compared to the transactions in the overall market.

The mechanism that derives the prices of specific products and the amount that will be bought or sold by a particular investor is governed by the aggregated supply and demand functions of every product, which in finance is usually termed a financial instrument. Examples of financial instruments include: savings, corporate equities, mutual funds, money market funds, mortgages, and so on.

For definiteness, we now present a basic market equilibrium model.

3.1. *A Market Equilibrium Model*

Assume that there are n financial instruments with a typical instrument denoted by i, and m investors with a typical investor denoted by j. Let p_i denote the price of instrument i, which is assumed to be nonnegative, and group the instrument prices into the column vector $p \in R^n_+$. Here we consider the general case where the demand for an instrument may depend on the price of not only the instrument but also of the other instruments. The aggregated demand $D_i = D_i(p)$ of an instrument i at a vector of instrument prices p is the summation of the total demands for that instrument by all investors. The total demand for instrument i is, hence, given by:

$$D_i(p) = \sum_{j=1}^{m} d_i^j(p),$$

where $d_i^j(p)$ denotes the demand for instrument i by investor j at p and which is assumed to be derived from utility maximization. Similarly, one can define the total supply $S_i = S_i(p)$ of instrument i, as:

$$S_i(p) = \sum_{j=1}^{m} s_i^j(p),$$

with $s_i^j(p)$ denoting the supply of instrument i by investor j at p.

3.2. *Equilibrium Conditions*

In the equilibrium state, the demand for an instrument is equal to its supply (assuming a positive price), and no one can benefit from buying or selling that instrument. Under the assumption of perfect competition and of nonnegative instrument prices, we have the following economic equilibrium conditions: for each instrument i; $i = 1, \ldots, n$:

$$S_i(p^*) - D_i(p^*) \begin{cases} = 0, & \text{if } p_i^* > 0 \\ \geqslant 0, & \text{if } p_i^* = 0. \end{cases}$$

Specifically, this system of equalities and inequalities states that if the price of a financial instrument is positive in equilibrium, then the market must clear for that instrument and if the price is zero, then either there is an excess supply of that instrument in the economy or the market clears.

These equilibrium conditions can be formulated as a finite-dimensional variational inequality problem: determine $p^* \in R^n_+$, satisfying:

$$\langle S(p^*) - D(p^*)^T, p - p^* \rangle \geqslant 0, \quad \forall p \in R^n_+,$$

where S and D denote, respectively, the n-dimensional vectors of supplies and demands and $\langle \cdot\,, \cdot \rangle$ denotes the inner product in R^n. Moreover, this variational inequality can be put in standard form as: determine $x^* \in \mathscr{K}$, satisfying:

$$\langle F(x^*)^T, x - x^* \rangle \geqslant 0, \quad \forall x \in \mathscr{K},$$

if we let $x \equiv p$, and $F(x)$ be the n-dimensional vector with component $F_i(x) = S_i(p) - D_i(p)$, and $\mathscr{K} \equiv R^+_n$. For additional background on variational inequalities, qualitative theory, computational methods, applications (ranging from traffic network equilibrium to oligopolistic market equilibrium), and references, see the book by Nagurney (1999).

In the special symmetric case, where the Jacobian matrix $\nabla F(x)$ is symmetric (and positive definite) the solution to the variational inequality problem can be obtained as the solution to the convex optimization problem:

$$\text{Minimize}_{\mathscr{K}} \quad \int F(x),$$

where $\int$ denotes the line integral, which in the context of the market equilibrium application becomes:

$$\text{Minimize}_{R^n_+} \int S(p) - D(p).$$

In particular, in the case of separable supply and demand functions, where $S_i = S_i(p_i)$, and $D_i = D_i(p_i)$, for all investors i; $i = 1, ..., n$, we obtain the optimization problem:

$$\text{Minimize} \quad \sum_{i=1}^{n} \int_0^{p_i} S_i(x)dx - \sum_{i=1}^{n} \int_0^{p_i} D_i(y)dy$$

subject to:

$$p_i \geqslant 0, \quad i = 1, ..., n.$$

The objective function then has the economic interpretation of minimizing producer's surplus.

Financial equilibrium modeling and analysis provide a useful tool for financial analysts and practitioners since the role of financial equilibrium is of increasing importance in many applications in the real world. Specifically, one of the major contributions that exposed the practicality of the study and application of financial equilibrium is the work of Black and Litterman (1992), who suggested that investors could increase the efficiency of their capital allocation processes by combining their personal views on financial products with the information contained in an equilibrium pattern. In particular, they implied that an equilibrium pattern provided a "neutral view" of the market, and that personal views of each investor could then be used to make adjustments to their capital allocation decisions. Hence, equilibrium models can provide an unbiased estimate of price movements for every financial instrument.

For definiteness, we now present a financial equilibrium model due to Nagurney et al. (1992; see also Nagurney and Siokos, 1997, and the references therein). The model relaxes the CAPM assumptions of homogeneous expectations (Sharpe, 1964; Lintner, 1965; Mossin, 1966) without imposing restrictions as to the nature of different sectors (e.g., Gonedes, 1976). The mathematical framework that is utilized to develop the multisector, multi-instrument financial equilibrium model is finite-dimensional variational inequality theory, already noted above. The methodology of finite-dimensional variational inequalities was first suggested for the modeling, analysis, and computation of multisector, multi-instrument financial equilibrium problems by Nagurney et al. (1992) and was further explored by Nagurney (1994). For complete references, qualitative results, as well as a plethora of financial equilibrium models and computational approaches, see Nagurney and Siokos (1997).

3.3. *A Multisector, Multi-Instrument Financial Equilibrium Model*

Recall the classical mean-variance model presented in Section 2, which is based on the pioneering work of Markowitz (1959). Now, however, we assume that there are m sectors, each of which seeks to maximize his return and, at the same time, to minimize the risk of his portfolio, subject to the balance accounting and nonnegativity constraints. Examples of sectors are: households, businesses, state and local governments, banks, and so on. We denote a typical sector by j and assume that there are liabilities in addition to assets held by each sector. Denote the volume of instrument i that sector j holds as an asset, by X^j_i, and group the (nonnegative) assets in the portfolio of sector j into the column vector $X^j \in R^n_+$. Further, group the assets of all sectors in the economy into the column vector $X \in R^{mn}_+$. Similarly, denote the volume of instrument i that sector j holds as a liability, by Y^j_i, and group the (nonnegative) liabilities in the portfolio of sector j into the column vector $Y^j \in R_+^n$. Finally, group the liabilities of all sectors in the economy into the column vector $Y \in R_+^{mn}$. Let r_i denote the nonnegative price of instrument i and group the prices of all the instruments into the column vector $r \in R_-^r$.

It is assumed here that the total volume of each balance sheet side of each sector is exogenous. Recall that a balance sheet is a financial report that demonstrates the status of a company's assets, liabilities, and the owner's equity at a specific point of time. The left-hand side of a balance sheet contains the assets that a sector holds at a particular point of time, whereas the right-hand side accommodates the liabilities and owner's equity held by that sector at the same point of time. According to accounting principles, the sum of all assets is equal to the sum of all the liabilities and the owner's equity. Here, the term "liabilities" is used in its general form and, hence, also includes the owner's equity. We let S^j denote the financial volume held by sector j. Finally, assume that the sectors under consideration act in a perfectly competitive environment.

3.4. A Sector's Portfolio Optimization Problem

Recall that in the mean-variance approach for portfolio optimization, the minimization of a portfolio's risk is performed through the use of the variance-covariance matrix. Hence, the portfolio optimization problem for each sector j is the following:

$$\text{Minimize} \quad \begin{pmatrix} X^j \\ Y^j \end{pmatrix}^T Q^j \begin{pmatrix} X^j \\ Y^j \end{pmatrix} - \sum_{i=1}^{n} r_i (X_i^j - Y_i^j)$$

subject to:

$$\sum_{i=1}^{n} X_i^j = S^j$$

$$\sum_{i=1}^{n} Y_i^j = S^j$$

$$X_i^j \geqslant 0, \quad Y_i^j \geqslant 0, \quad i = 1, 2, \ldots, n$$

where Q^j is a symmetric $2n \times 2n$ variance-covariance matrix associated with the assets and liabilities of sector j. Moreover, since Q^j is a variance-covariance matrix, we can also assume that it is positive definite and, as a result, the objective function of each sector's portfolio optimization problem, given by the above, is strictly convex.

We partition the symmetric matrix Q^j, as

$$Q^j = \begin{pmatrix} Q_{11}^j & Q_{12}^j \\ Q_{21}^j & Q_{22}^j \end{pmatrix},$$

where Q_{11}^j and Q_{22}^j are the variance-covariance matrices for only the assets and only the liabilities, respectively, of sector j. These submatrices are each of dimension $n \times n$. The submatrices Q_{12}^j and Q_{21}^j, in turn, are identical since Q^j is symmetric. They are also of dimension $n \times n$. These submatrices are, in fact, the symmetric variance-covariance matrices between the asset and the liabilities of sector j. We denote the ith column of matrix $Q_{(\alpha\beta)}^j$, by $Q_{(\alpha\beta)i}^j$, where α and β can take on the values of 1 and/or 2.

3.5. Optimality Conditions

The necessary and sufficient conditions for an optimal portfolio for sector j are that the vector of assets and liabilities, $(X^{j^*}, Y^{j^*}) \in K_j$, where K_j denotes the feasible set for sector j, given above, satisfies the following system of equalities and inequalities: for each instrument i; $i = 1, \ldots, n$, we must have:

$$2(Q_{(11)i}^j)^T \cdot X^{j^*} + 2(Q_{(21)i}^j)^T \cdot Y^{j^*} \cdot r_i^* - \mu_j^1 \geqslant 0,$$

$$2(Q^j_{(22)i})^T \cdot Y^{j^*} + 2(Q^j_{(12)i})^T \cdot X^{j^*} + r^*_i - \mu^2_j \geqslant 0,$$

$$X^{j^*}_i [2(Q^j_{(11)i})^T \cdot X^{j^*} + 2(Q^j_{(21)i})^T \cdot Y^{j^*} - r^*_i - \mu^1_j] = 0,$$

$$Y^{j^*}_i [2(Q^j_{(22)i})^T \cdot Y^{j^*} + 2(Q^j_{(12)i})^T \cdot X^{j^*} + r^*_i - \mu^j_2] = 0,$$

where μ^1_j and μ^2_j are the Lagrange multipliers associated with the accounting constraints, respectively.

Let $\mathscr{K}$ denote the feasible set for all the asset and liability holdings of all the sectors and all the prices of the instruments, where $\mathscr{K} \equiv \{K \times R^n_+\}$ and $K \equiv \prod_{i=1}^m K_j$.

3.6. Economic System Conditions

Moreover, the economic system conditions, which relate the supply and demand of each financial instrument and the instrument prices, are given by: for each instrument i; $i = 1, ..., n$, an equilibrium asset, liability, and price pattern, $(X^*, Y^*, r^*) \in \mathscr{K}$, must satisfy:

$$\sum_{j=1}^{J} (X^{j^*}_i - Y^{j^*}_i) \begin{cases} = 0, & \text{if } r^*_i > 0 \\ \geqslant 0, & \text{if } r^*_i = 0. \end{cases}$$

We now state the definition of financial equilibrium and then give the variational inequality formulation. Combining the optimality conditions for each sector, along with conditions along with the economic system conditions for each instrument, we obtain the following definition of equilibrium.

Definition (Multisector, Multi-Instrument Financial Equilibrium). *A vector* $(X^*, Y^*, r^*) \in \mathscr{K}$ *is an equilibrium of the multisector, multi-instrument financial model if, and only if, it satisfies the optimality conditions and the economic system conditions, for all sectors* j; j = 1, ..., m, *and for all instruments* i; i = 1, ..., n, *simultaneously.*

The variational inequality formulation of the equilibrium conditions, due to Nagurney et al. (1992) is given by:

Theorem (Variational Inequality Formulation for the Quadratic Model). *A vector of assets and liabilities of the sectors, and instrument prices,* $(X^*, Y^*, r^*) \in \mathscr{K}$, *is a financial equilibrium if, and only if, it satisfies the variational inequality problem:*

$$\sum_{j=1}^{m} \sum_{i=1}^{n} [2(Q^j_{(11)i})^T \cdot X^{j^*} + 2(Q^j_{(21)i})^T \cdot Y^{j^*} - r^*_i] \times [X^j_i - X^{j^*}_i] + \sum_{j=1}^{m} \sum_{i=1}^{n} [2(Q^j_{(22)i})^T \cdot Y^{j^*}$$

$$+ 2(Q^j_{(12)i})^T \cdot X^{j^*} + r^*_i] \times [Y^j_i - Y^{j^*}_i] + \sum_{i=1}^{n} \sum_{j=1}^{m} [X^{j^*}_i - Y^{j^*}_i] \times [r_i - r^*_i] \geqslant 0,$$

$$\forall (X, Y, r) \in \mathscr{K}.$$

For completeness, we present the standard form of variational inequality. We define the N-dimensional column vector $x \equiv (X, Y, r) \in \mathscr{K}$, and the N-dimensional column vector $F(x)$ such that:

$$F(x) \equiv D \begin{pmatrix} X \\ Y \\ r \end{pmatrix} \quad \text{where } D = \begin{pmatrix} 2Q & B \\ -B^T & 0 \end{pmatrix},$$

$$Q = \begin{pmatrix} Q_{11}^1 & & & Q_{21}^1 & & \\ & \ddots & & & \ddots & \\ & & Q_{11}^J & & & Q_{21}^J \\ Q_{12}^1 & & & Q_{22}^1 & & \\ & \ddots & & & \ddots & \\ & & Q_{12}^J & & & Q_{22}^J \end{pmatrix}_{2mn \times 2mn},$$

and $B^T = (-\mathcal{I} \quad \dots \quad -\mathcal{I} \quad \mathcal{I} \quad \dots \quad \mathcal{I})_{n \times mn},$

where $\mathcal{I}$ is the $n \times n$-dimensional identity matrix.

Therefore, it is clear that the above variational inequality problem can be put into standard variational inequality form:

Determine $x^* \in \mathcal{K}$, satisfying:

$$\langle F(x^*)^T, x - x^* \rangle \geqslant 0, \quad \forall x \in \mathcal{K}.$$

Note that the above model is a special case of the financial equilibrium model due to Nagurney (1994), in which each sector j seeks to maximize his utility function,

$$U^j(X^j, Y^j, r) = u^j(X^j, Y^j) + r^T \cdot (X^j - Y^j),$$

which, in turn, is a special case of the model with a sector j's utility function given by the general form: $U^j(X^j, Y^j, r)$. Interestingly, it has been shown (Nagurney and Siokos, 1997) that, in the case of utility functions of the form $U^j(X^j, Y^j, r) = u^j(X^j, Y^j) + r^T \quad (X^j - Y^j)$, of which the above described quadratic model is an example, one can obtain the solution to the above variational inequality problem by solving the optimization problem:

$$\text{Maximize} \quad \sum_{j=1}^{J} u^j(X^j, Y^j)$$

subject to:

$$\sum_{j=1}^{J} (X_i^j - Y_i^j) = 0, \quad i = 1, \dots, n$$

$$(X^j, Y^j) \in K_j, \quad j = 1, \dots, m,$$

with Lagrange multiplier r_i^* associated with the ith market clearing constraint. Moreover, this optimization problem is actually a network optimization problem as revealed in Nagurney and Siokos (1997).

Other financial equilibrium models, including models with transaction costs, hedging instruments such as futures and options, and international financial equilibrium models, can be found in Nagurney and Siokos (1997) and the references therein.

4. Risk Management

Risk management represents a broad application area of financial optimization. Risk management models are applied in order to choose portfolios with specified exposure to different risks. First, however, one must select which risks one is to be exposed to and which risks one is to be immunized against. Well-known risks include, among others: interest rate risk, which denotes the risk associated with the movement of market rates for assets and liabilities; liquidity risk, which is the risk associated with the possibility that the bid-ask spread on security transactions may change; exchange rate risk, which is the risk associated with exchange rate fluctuations; credit risk, which is the risk associated with a borrower's credit worthiness; volatility risk, which is commonly associated with options (see Section 6), since these financial instruments are characterized by very asymmetric returns; and residual risk, which denotes nonsystematic and security-specific risk.

In this section, we present several optimization models in risk management applications. The *Economist* (1996) notes that risk management is an area of finance in which the tension between the theoretical foundations and the daily practice of finance has been particularly fruitful.

We begin with a bond portfolio immunization model that seeks to manage interest rate risk. Portfolio immunization refers to the problem of finding a set of bonds whose present value matches that of a prespecified set of liabilities. Hence, in the case of fund management, specifically pension fund management, a fund would have similar values for its assets and liabilities, regardless of the changes in the interest rate. References to portfolio immunization include the work by Granito (1984), Platt (1986), Bierwag (1987), and Fabozzi et al. (1991).

4.1. *A Linear Programming Model for Bond Portfolio Immunization*

As in Section 2, we let X_i denote security i and assume that there are n securities and a single investor. Now, however, we consider time explicitly and let t denote the discrete time index, where $t = 1, ..., T$. Let r_i denote the interest rate for security i, let C_{it} denote the cashflow from security i at time t, and let P_i denote the present value of security i, where

$$P_i = \sum_{t=1}^{T} C_{it}(1 + r_i)^{-t}.$$

Also, let the dollar duration for each security i, denoted by d_i, be the partial derivative of P_i with respect to r_i, that is,

$$d_i = -\sum_{t=1}^{T} tC_{it}(1 + r_i)^{-t-1},$$

with the dollar duration for the entire portfolio, denoted by d, given by:

$$d = \sum_{i=1}^{n} d_i X_i.$$

The most common objective in this specific risk management application is to maximize the asset portfolio yield, since the portfolio return is considered to be risk free. A first order approximation to the true portfolio yield (cf. Dahl et al., 1993) is given by the term: $\sum_{i=1}^{n} d_i r_i X_i$, which yields the following optimization problem for bond immunization:

$$\text{Maximize} \quad \sum_{i=1}^{n} d_i r_i X_i$$

subject to:

$$\sum_{i=1}^{n} P_i X_i = P_l$$

$$\sum_{i=1}^{n} d_i X_i = d_l$$

$$X_i \geqslant 0, \quad \forall i = 1, ..., n,$$

where P_l denotes the present value of the liabilities and d_l denotes the dollar duration of the liabilities.

The first constraint in the optimization model guarantees that the present values on asset and liabilities in the immunized portfolio are equal, whereas the second constraint ensures that the dollar durations on the assets and liabilities are equal.

Various extensions of this basic model can be developed to handle other types of risk, including liquidity risks, which may then yield a mixed-integer optimization problem. For further details, see Dahl et al. (1993) and the references therein. The data for this model, however, may vary over time since the present value depends on the discount rates, which may, in turn, be affected by the interest rates. The problem, hence, may be rephrased as a stochastic system and an approach termed scenario optimization (see Dembo, 1993 and the references therein) may be used in order to handle the resulting optimization problem under uncertainty in which the uncertain component may be represented by discrete scenarios.

Immunization models match some function of cash flows. Dedication models, on the other hand, attempt to track the cash flows directly. Dembo (1993) presents a dedication model that allows for stochastic cash flow matching.

4.2. A Quadratic Programming Model for Risk Management

Mean-variance models, such as those described in Section 2, diversify residual risk, if it is assumed that residual risk can be represented as a function of the mean and variance of the returns of the securities only. In order to reduce or eliminate other types of risk, one needs to add additional constraints to the basic model as, for example,

$$\text{Maximize} \quad \alpha R - (1 - \alpha)V$$

subject to:

$$AX = b$$

$$X_i \geqslant 0, \quad i = 1, \ldots, n,$$

where A is a $k \times n$-dimensional matrix denoting the coefficients associated with the constraints, and b is an n-dimensional column vector representing the right-hand-side values of the constraints, and where, as before, α denotes an indicator of how risk-averse a specific investor is. The k constraints would then include the budget constraint as well as any linear constraints associated with risk exposure, including portfolio target constraints. In addition, one might include, as in the preceding model, liquidity constraints, in which case the model becomes then a mixed-integer nonlinear programming model.

A common technique for risk management is the approach termed value at risk (VAR), since it offers a measure of a firm's overall exposure to market risk, which in the fixed income market is synonymous with interest rate risk and which in the stock market is associated with the movement of the market index of portfolio returns. Nevertheless (see *Economist*, 1996), it may have certain drawbacks in that:

1. It relies principally on historical volatility as a guide to future volatility and
2. It is of limited use strategically since it does not permit for meaningful comparisons across financial markets.

An alternative approach (see discussion in *Economist*, 1996) has been termed the benchmark value at risk (B-VAR) and measures traders' performance against different benchmarks appropriate to those markets, in particular, the risk of underperforming the benchmark.

5. Asset and Liability Management

Asset and liability management is one of the most challenging and demanding applications in quantitative finance in general and in financial optimization in particular. Succinctly stated, the principal objectives of asset and liability management (ALM) are to manage risks, meet regulatory requirements, and handle growth. ALM, which is sometimes referred to as balance-sheet management, can be interpreted as a process whereby a financial institution's, typically a bank's, total assets and liabilities are managed and controlled in an integrated manner. The major risks that need to be managed (see also Section 4) are the interest-rate risk, the credit risk, the foreign exchange risk, as well as the liquidity risk. Recall that the interest-rate risk influences the return on both assets and liabilities and, hence, the profitability of a financial institution. The credit risk, or default risk, is sometimes referred to as the "risk of ruin" by actuaries with the primary source of default risk, in the case of banks, being the loans that the bank has issued. The foreign exchange risk is the risk associated with the movement of foreign exchange rates. The liquidity risk concerns the risk that the financial institution will not have sufficient funds to honor its commitments.

In order to fix ideas, it is necessary to underline the basic economic activity of the specific institution involved in ALM. For example, banks serve as intermediaries between deposit-makers (with the deposits then serving as the supply) and the loan takers (with the loans reflecting the demand) (see also Section 3). Hence, in order for a bank to perform the intermediary function in an effective and efficient manner it has to consider both the demand for products (and services) as well as the supply and, hence, to avoid a mismatch between assets and liabilities. In theory, a financial institution such as a bank or insurance company would wish to match perfectly its cash flows on the asset and liability sides of the balance sheet, in which case there would be no exposure to such risks as the interest-rate risk, liquidity risks, and/or credit risks. However, in practice, this may not be realistic or even desirable since institutions make their profits by taking precisely such risks. Indeed, banks and insurance companies do take such risks in order to achieve a return but, unlike their customers, they have the advantage that they may pool the risks.

ALM is usually considered a tool for the long-term view and action and consists of both asset management, with a focus on asset allocation and, hence, includes portfolio selection and liability management, with an emphasis on liability choice, including deposit modeling. Heretofore, ALM has been utilized principally for risk management, but there is an increasing interest on the part of banks to also use ALM for return management. Moreover, ALM is concerned not only with the measurement of risks, but also with the development and application of optimization strategies in order to attain the risk objectives. Hence, the management of an institution, be it a bank or an insurance company, must clearly identify its ALM objectives, which may vary according to the environment or management itself. In addition, in order for any optimization to be able to take place, be it in a tactical or strategic setting, the decision variables need to be identified and should include the range of financial products, including those for hedging and pricing. Finally, in order to be able to develop ALM strategies, an optimization framework would also need to include a variety of constraints, with some principal constraints being those concerning:

- asset and liability flows, which would express the relationship between the flow of deposits (in the case of banks) as well as the evolution of loans. These would also handle growth or decrease in the demand for the financial products;
- shareholders, in order to determine the distribution of the dividends;
- the budget (both initial and final);
- liquidity (relevant for both banks and insurance companies);
- any relevant political and legal constraints, which come mainly in the form of regulatory constraints);
- taxes and other relevant pricing policies.

It is worth noting that banks may also be interested in objectives besides the management of risk as a part of their ALM strategy. For example, such objectives may be to expand the balance sheet, maintain the reserve ratio, and to match risk and return, among others. Hence, the optimization problem involved in ALM may very well be a multicriteria one.

Kusy and Ziemba (1986) present a bank asset and liability management model, which is a stochastic linear programming model that includes the basic institutional, legal, financial, and policy constraints. They also present an application of the model that has been operationalized. The article contains references to deterministic as well as stochastic asset and liability management models to that date.

6. Option Pricing

The liberalization of interest rate policies in the United States in the 1970s led to the development of new financial products that could provide some risk protection. The majority of these financial products were simple adaptations of existing products from the markets for raw materials, where hedging strategies had existed for many years. Recall that hedging refers to any type of strategy, the primary goal of which is to offset an investment risk. The goal of a hedger is to reduce or to eliminate the risk associated with the price level of a financial instrument at the time of delivery, so that he can rely on profitability and price calculations with a greater degree of certainty. In times of high price volatility, a greater financial exposure can have significant effects on the financial performance of a firm or of an individual investor and, hence, hedging strategies are needed. The most important and commonly used financial products for hedging financial risk are financial futures and options.

A futures contract is a firm commitment to deliver or receive a specific quantity of a financial instrument, referred to as an underlying instrument, at a predetermined price and point in time. An option, on the other hand, is the right, but not the obligation, to purchase or sell an instrument, at a specified price, before or at a specific point in time.

The first interest rate futures contract appeared at the Chicago Board of Trade (CBOT) in 1975, and it was a futures certificate by the Government National Mortgage Association (GNMA). In subsequent years, new futures markets, such as the London International Financial Futures Exchange (LIFFE), established in 1982, followed by financial markets in Frankfurt (DTB), Singapore (SIMEX), and Tokyo (TSE), along with a variety of products, were introduced (see Andersen, 1993).

Investors are interested in financial futures because they can eliminate or reduce the uncertainty as to the price of an instrument at the time of delivery. The price expectations of investors are directly related to their actions on the futures market. More precisely, investors can hedge against increasing prices by selling futures contracts, whereas investors that expect decreasing prices and want to get a fixed price will have to buy a futures contract.

Options, on the other hand, had existed for many years, but it was not until 1973 that the first organized market, the Chicago Board of Exchanges (CBOE) opened. An option contract is characterized by its expiration date, that is, the date before which the option can be exercised. If an option is not exercised by the time of its expiration date it has no value. There are two types of options available: the put option, which gives the holder the right to sell an instrument at the exercise price before the expiration date, and the call option, which gives the holder the right to buy an instrument at the exercise price before the expiration date.

We now, for completeness, outline several terms that are relevant to the understanding of options. The relationship between the instrument's spot market price and the option's strike price determines whether or not an option has an actual value, which is referred to as its intrinsic value. The life of the option (its time value) is reflected in the current price of an option as the amount in excess of the option's intrinsic value. Time value plays more of a role in American options, which can be exercised at any point in time prior to expiration, than in European options, where the options can be exercised only at the expiration date. Since an option represents a right and not an obligation its value can never be negative.

Investors in options can either be long or short on their positions. An investor who is buying an option is opening an options transaction and is said to hold a long position, whereas an investor that writes an option is considered to hold a short position. Obviously, the decision of an investor to be short or long mirrors his expectations of the market. Those who expect to profit with an increase of the underlying instrument price (buy calls or sell puts) are said to be bullish, whereas investors who anticipate a decline as to the price of the underlying instrument in order to profit (sell calls or buy puts) are said to be bearish.

Moreover, an option can be in, out of, or at the money. Specifically, an option is said to be at the money if the current market price of the underlying instrument is equal to the exercise price of the contract, and it is out of the money if the exercise price is higher (lower) than the current market price of the underlying instrument for a call (put) option. Finally, a call (put) option is in the money if the current market price of the underlying instrument is below (above) the contract exercise price.

The pricing of options has been extensively studied and many models have been presented and applied (see Black and Scholes, 1973; Brennan and Schwartz, 1977; Karatzas, 1988). The most well-known and widely used model, however, is still the one by Black and Scholes (1973), which turns out to be accurate enough in the pricing of options, even though some of its assumptions cannot be completely fulfilled (that is, interest rates are assumed to be unchanged, price developments are assumed to follow a lognormal distribution, and there are no taxes and transaction costs).

We now present a brief overview of the Black and Scholes option pricing model, which is considered to be the basis of option pricing. For more detailed discussions of this model, as well as a detailed analysis of other existing option pricing models, see the original paper by Black and Scholes (1973), and the books by Kolb (1991), McMillan (1992), Hull (1993), and Wilmott et al. (1993).

Note that models for price contracts such as options and futures and other more exotic financial products may have closed form solutions as provided by the Black and Scholes formula for a class of options contracts. However, if this is not the case, the pricing of such contracts may require numerical techniques. Such techniques may include binomial or trinomial methods in which the state space is discretized and the option price is determined by using a dynamic programming method (Hull and White, 1992); Monte Carlo or other simulation methods (Boyle, 1977), and by finite difference or other methods used to solve partial differential equations (Courtadon, 1982; Hull and White, 1990).

6.1. The Black and Scholes Option Pricing Model

The price of an option, be it a put or a call option, depends on a series of different financial factors, which are directly related to the underlying security, its time to expiration, and the existing financial environment. Black and Scholes suggested a relatively simple formula that takes into consideration different factors and manages to derive a relatively accurate price for every option contract.

Specifically, the Black and Scholes option pricing model is based on the following assumptions about these factors: the existence of a constant risk-free interest rate; the prices are distributed lognormally with constant volatility; no taxes or transaction costs; the trading of the securities is continuous; there are no dividends during the life of the underlying security; the market participants can sell short; and the options are exercisable only at expiration (European).

Although almost all the assumptions of the Black and Scholes model are violated in practice, the model can, nevertheless, be adjusted to accommodate situations that violate some of its assumptions (see Hull, 1993; Wilmott et al., 1993). For example, even though security trading is assumed to be continuous, discrete filters are acceptable on a portfolio basis. Moreover, adjustments can be made to the basic framework to handle transaction costs, taxes, and derivatives, as well as the case of American options. However, the theory will not behave well in markets in which participants cannot sell short as is the case in Japanese equity warrants, or in cases that the price distributions of the underlying assets are not lognormally distributed.

The following expression gives the Black and Scholes option pricing model for a call option:

$$C = [sN(\beta_1)] - [EN(\beta_2)e^{-rt}]$$

with

$$\beta_1 = \frac{\left[\ln\left(\frac{s}{E}\right) + (r + 0.5\sigma^2)t\right]}{\sqrt{\sigma^2 t}}$$

and

$$\beta_2 = \beta_1 - \sqrt{\sigma^2 t},$$

where $N(\cdot)$ denotes the cumulative normal distribution of a normally distributed variable, with a mean value equal to zero and a standard deviation equal to one. In addition, S denotes the spot price of the underlying stock, E denotes the exercise price of the option contract, r and t denote, respectively, the risk-free interest rate of the market and the time to expiration of the specific option contract. Finally, σ^2 corresponds to the variance of the underlying stock.

A major part of the subsequent literature focused on different approaches to, simplifications of, and variations of the Black and Scholes model (BSM). A significant simplification of the BSM was done by Cox et al. (1979), and an important application was demonstrated by Merton (1974) as to the pricing of corporate debt. The Black and Scholes (1973) model managed to offer a simple but reliable tool for option pricing, where the number of variables that determine an option price is small and the derivation of their value is simple. The model of Black and Scholes was further explored and significantly generalized (see Duffie and Huang, 1985; Cox, et al., 1985; Duffie, 1986).

Nagurney and Siokos (1997) develop hedging models, including international ones, which focus on the determination of equilibrium prices as well as asset and liability holdings of the sectors. They also identify the network structure of such problems and provide computational procedures for effective computation based on the theory of variational inequalities and network theory.

ACKNOWLEDGMENTS

The preparation of this article was supported by the John F. Smith Memorial Fund at the Isenberg School of Management at the University of Massachusetts at Amherst while the author was on sabbatical in Sweden. Specifically, acknowledgment is given to the Stockholm School of Economics for the use of its library resources. The support is gratefully acknowledged.

References

Andersen, T. J. 1993. *Currency and Interest Hedging*. New York: New York Institute of Finance.

Arrow, K. J. 1951. "An extension of the basic theorems of classical welfare economics." *Econometrica* **51** 1305–1323.

Bierwag, G. 1987. *Duration Analysis: Managing Interest Rate Risk*. Cambridge, MA: Ballinger Publishing Co.

Black, F. and R. Litterman. 1992. "Global portfolio optimization." *Financial Analysts J.* **48**: 28–43.

Black, F. and M. Scholes. 1973. "The pricing of options and corporate liabilities." *J. Political Econ.* **3**: 637–654.

Boyle, P. 1977. "Options: A Monte Carlo approach." *J. Financial Econ.* **4**: 323–338.

Breeden, D. T. 1978. "An intertemporal asset pricing model with stochastic consumption and investment opportunities." *J. Financial Econ.* **7**: 265–296.

Brennan, M. J. and E. S. Schwartz. 1977. "The valuation american put option." *J. Finance* **32**: 449–462

Chamberlain, G. 1983. "Funds, factors and diversification in arbitrage pricing models." *Econometrica* **51** 1305–1323.

Connor, G. 1984. "A unified beta pricing theory." *J. Econ. Theor.* **34**: 13–31.

Courtadon, G. 1982. "A more accurate finite difference approximation for the valuation of options." *J. Financial Quantit. Anal.* **17**: 697–703.

Cox, J. C. and C. Huang. 1987. "Option pricing theory and its applications." In *Frontiers of Financial Theory*, edited by G. Constantinides and S. Bhattacharya. Totowa, NJ: Rowman and Littlefield.

Cox, J. C., J. Ingersoll, and S. A. Ross. 1985. "An intertemporal general equilibrium model of asset prices." *Econometrica* **53**: 363–384.

Cox, J. C., S. A. Ross, and M. Rubinstein. 1979. "Option pricing: a simplified approach." *J. Financial Econ.* **7**: 229–263.

Dahl, H., A. Meeraus, and S. A. Zenios. 1993. "Some financial optimization models: I Risk management." In *Financial Optimization*, edited by S. A. Zenios. Cambridge, UK: Cambridge University Press.

Debreu, G. 1959. In *Theory of Value*. New Haven: Yale University Press.

Dembo, R. 1993. "Scenario immunization." In *Financial Optimization*, edited by S. A. Zenios. Cambridge, UK: Cambridge University Press.

Dong, J. and A. Nagurney. 1998. "Bicriteria decision making and financial equilibrium: a variational inequality perspective." To appear in *Comput. Econ.*

Duffie, D. 1986. "Stochastic equilibria: existence, spanning number and the 'no expected financial gain from trade' hypothesis." *Econometrica* **54**: 1161–1184.

Duffie, D. 1988. *Security Markets. Stochastic Models*. San Diego: Academic Press, Inc.

Duffie, D. and C. Huang. 1985. "Implementing Arrow–Debreu equilibria by continuous trading of few long-lived securities." *Econometrica* **53**: 1337–1356.

Economist, 1996. A survey of technology in finance. *The Economist*, 26 October 1996.

Elton, E. J. and M. J. Gruber. 1987. *Modern Portfolio Theory and Investment Analysis*. 3d ed. New York: John Wiley & Sons.

Fabozzi, F. J., T. D. Fabozzi, and I. M. Pollack (eds). 1991. In *The Handbook of Fixed Income Securities*, third edition, New York: John Wiley & Sons.

Francis, J. C. 1978. "Portfolio analysis of asset and liability management in small-, medium-, and large-sized banks." *J. Monetary Econ.* **3**: 112–134.

Francis, J. C. and S. H. Archer. 1979. *Portfolio Analysis*. Englewood Cliffs, NJ: Prentice-Hall Inc.

Gonedes, N. J. 1976. "Capital market equilibrium for a class of heterogeneous expectations in a two-parameter world." *J. Finance* **31**: 1–15.

Gossen, H. 1854. *Entwicklung der Gesetze des Menschlichen Verkehrs*. Berlin: Prager.

Granito, M. R. 1984. *Bond Portfolio Immunization*. Lexington, MA: Lexington Books.

Hicks, J. R. 1939. *Value and Capital*. Oxford, UK: Clarendon Press.

Ho, D. and T. Schneeweis (eds). 1999. In *Applications in Finance, Investments, and Banking*. Dordrecht, The Netherlands: Kluwer Academic Publishers.

Huang, C. and R. H. Litzenberger. 1988. In *Foundations for Financial Economics*. Englewood Cliffs, NJ: Prentice-Hall Inc.

Hull, J. 1993. *Options, Futures, and Other Derivatives*. Englewood Cliffs, NJ: Prentice-Hall.

Hull, J. and A. White. 1990. "Valuing securities using the explicit finite difference method." *J. Financial Quantit. Anal.* **25**: 87–100.

Hull J. and A. White. 1992. "A note on the valuation of mortgage-backed securities." Presented at the Recent Developments in Pricing and Hedging Options Seminar, organized by the University of Toronto, June 1992

Jevons, W. 1871. *The Theory of Political Economy*. London: Macmillan Press Ltd.

Karatzas, I. 1988. "On the pricing of american options." *Appl. Math. Optim.* **17**: 37–60.

Kolb, R. 1991. *Options An Introduction*. Miami, FL: Kolb Publishing Company.

Kusy, M. I. and W. T. Ziemba. 1986. "A bank asset and liability management model." *Oper. Res.* **34**: 356–376.

Lintner, J. 1965. "The valuation of risk assets and the selection of risky investments in stock portfolios and capital budgets." *Rev. Econ. Studies* **47**: 13–37.

Markowitz, H. M. 1952. "Portfolio selection." *J. Finance* **7**: 77–91.

Markowitz, H. M. 1959. *Portfolio Selection: Efficient Diversification of Investments*. New York: John Wiley & Sons.

Markowitz, H. M. 1989. *Mean Variance Analysis in Portfolio Choice and Capital Markets*. Oxford: Basil Blackwell.

McMillan, L. G. 1992. *Options as a Strategic Investment*. New York: New York Institute of Finance.

Merton, R. C. 1973. "An intertemporal capital asset pricing model." *Econometrica* **41**: 867–887.

Merton, R. C. 1974. "On the pricing of corporate debt: the risk structure of interest rates." *J. Finance* **29**: 449–470.

Milne, F. 1995. *Finance Theory and Asset Pricing*. New York: Oxford University Press Inc.

Milne, F. and C. Smith. 1980. "Capital asset pricing with proportional transaction costs." *J. Financial and Qualitat. Anal.* **15**: 253–265.

Modigliani, F. and M. Miller. 1958. "The cost of capital, corporate finance and the theory of corporation finance." *Am. Econ. Rev.* **48**: 261–297.

Mossin, J. 1966. "Equilibrium in a capital asset market." *Econometrica* **34**: 768–783.

Nagurney, A. 1994. "Variational inequalities in the analysis and computation of multi-sector, multi-instrument financial equilibria." *J. Econ. Dynam. Control* **18**: 161–184.

Nagurney, A. 1999. *Network Economics: A Variational Inequality Approach*. 2d ed rev. Dordrecht, The Netherlands: Kluwer Academic Publishers.

Nagurney, A. and J. Dong. 1998. "Financial networks and optimally sized portfolios." To appear in *Comput. Econ.*

Nagurney, A., J. Dong, and M. Hughes. 1992. "Formulation and computation of general financial equilibrium." *Optimization* **26**: 339–354.

Nagurney, A. and S. Siokos. 1997. *Financial Networks: Statics and Dynamics.* Heidelberg: Springer-Verlag.

Pardalos, P. M., M. Sandstrom, and C. Zopounidis. 1994. "On the use of optimization models for portfolio selection: a review and some computational results." *Comput. Econ.* **7**: 227–244.

Perold, A. F. 1984. "Large-scale portfolio Optimization." *Mgmnt Sci.* **30**: 1143–1160.

Platt, R. B. (ed.). 1986. *Controlling Interest Rate Risk: New Techniques and Applications for Money Management.* New York: John Wiley & Sons.

Pogue, G. A. 1970. "An extension of the Markowitz portfolio selection model to include transaction costs, short sales, leverage policies, and taxes." *J. Finance* **25**: 1005–1027.

Roll, R. 1977. "A critique of the asset pricing theory's tests. Part 1: On past and potential testability of the theory." *J. Financial Econ.* **4**: 129–176.

Ross, S. A. 1976. "Arbitrage theory of capital asset pricing." *J. Econ. Theor.* **13**: 341–360.

Shapiro, A. 1992. *Multinational Financial Management.* Needham Heights, MA: Allyn and Bacon.

Sharpe, W. F. 1963. "A simplified model for portfolio analysis." *Mgmnt Sci.* **9**: 277–293.

Sharpe, W. F. 1964. "Capital asset prices: a theory of market equilibrium under conditions of risk." *J. Finance* **19**: 425–443.

Tobin, J. 1963. "Commercial banks as creators of money." In *Banking and Monetary Studies,* edited by D. Carson. Homewood, IL: Irwin.

Tobin, J. 1969. "A general equilibrium approach to monetary theory." *J. Money, Credit, Banking* **1**: 15–29.

Tobin, J. 1971. *Essays in Macroeconomics.* Amsterdam: North-Holland Inc.

Wald, A. 1935. "Über die Eindeutige Positive Losbarkeit der Neuen Productionsgleichungen." *Ergebnisse eines Mathematischen Kolloquiums* **6**: 12–20.

Wald, A. 1936. "Über die Productionsgleichungen der Ökonomischen Wertlehre." *Ergebnisse eines Mathematischen Kolloquiums* **7**: 1–6.

Walras, L. 1874. *The Eléments d'Economie Politique Pure.* Paris: Guillaumin.

Wilmott, O., J. Dewynne, and S. Howison. 1993. In *Option Pricing: Mathematical Models and Computation.* Oxford: Oxford Financial Press.

Further Reading

This list contains suggestions for in-depth reading in addition to those references cited in the text.

Cox, J. C. and C. Huang. 1987. "Option pricing theory and its applications." In *Frontiers of Financial Theory,* G. Constantinides and S. Bhattacharya, editors, Totowa, NJ: Rowman and Littlefield.

Duffie, D. 1988. Security Markets. Stochastic Models. San Diego, CA: Academic Press, Inc.

Elton, E. J. and M. J. Gruber. 1987. *Modern Portfolio Theory and Investment Analysis.* 3d ed, New York: John Wiley & Sons.

Ho, D. and T. Schneeweis. (eds). 1999. *Applications in Finance, Investments, and Banking.* Dordrecht, The Netherlands: Kluwer Academic Publishers.

Huang, C. and R. H. Litzenberger. 1988. *Foundations for Financial Economics.* Englewood Cliffs, NJ: Prentice-Hall, Inc.

Milne, F. 1995. *Finance Theory and Asset Pricing.* New York: Oxford University Press Inc.

Modigliani, F. and M. Miller. 1958. "The cost of capital, corporate finance and the theory of corporation finance." *Am. Econ. Rev.* **48**: 261–297.

Nagurney, A. and J. Dong. 1998. "Financial networks and optimally sized portfolios." To appear in *Comput. Econ.*

Roll, R. 1977. "A critique of the asset pricing theory's tests. Part 1: On past and potential testability of the theory." *J. Financial Econ.* **4**: 129–176.

18.15 Applied Large-Scale Nonlinear Optimization for Optimal Control of Partial Differential Equations and Differential Algebraic Equations

J. B. Rosen, John H. Glick, and E. Michael Gertz

ABSTRACT The computation of optimal controls for many important systems in science and engineering is now possible. These systems are typically described by either time-dependent partial differential equations (PDEs) or differential algebraic equations (DAEs). In the former case, the PDEs are first converted to DAEs in order to obtain a finite system. The optimal controls for a DAE system are then computed by solving a single large-scale, nonlinear optimization problem. Methods for doing this are summarized, and illustrated with several important applications.

1. Introduction

Many important problems in science and engineering can be formulated as the optimal control of a system of time-dependent partial differential equations (PDEs). These include boundary control of nonlinear heat conduction (Petzold et al., 1997; Banks, 1982; Butkovskiy, 1991; Chen et al., 1991; Leibfritz and Sachs, 1994; Elden, 1995), reservoir simulation model parametric estimation (Ewing, 1991; Ewing and George, 1985), and shape optimization of turbine blades (Schulz, 1998). In addition to the system of partial differential equations describing the process, an objective function to be minimized (or maximized), is specified. The desired solution is typically obtained in terms of time-dependent state variables and control variables. The aim is to determine the control variables so as to minimize the objective function subject to satisfaction of the PDEs. Both the control variables and the state variables may be subject to constraints, either equality or inequality.

The theory of this type of optimal control problem has been extensively investigated, giving conditions for which optimal solutions exist and properties of these solutions (Banks, 1982; Banks and Kunisch, 1982; Bulirsch et al., 1993; Butkovskiy, 1991; Chen et al., 1991; Lions, 1988; Markus, 1991; Sakawa, 1974; Hager and Pardalos, 1998). However, except in the simplest cases, it is necessary to use numerical methods to solve problems of this type. In this survey, we will limit consideration to some of the more recent computational methods for large-scale systems. Such methods usually begin by approximating the original PDE system by a system of differential algebraic equations (DAEs). This is accomplished by either the use of finite elements or finite differences in the spatial domain. The accuracy of the DAE approximation will improve as the size of the DAE system increases, so that typically to get reasonable accuracy, the DAE system will be large-scale. For example, with a uniform grid in two spatial dimensions, and n grid points per dimension, there will be n^2 ordinary differential equations, plus any required algebraic equations.

There are also many engineering processes that can be formulated directly as the optimal control of a system of DAEs (Kumar and Daoutidis, 1999; Pantelides et al., 1994; Cuthrell and Biegler, 1989; Bulirsch and Kraft, 1994; Sargent, 1998). These include multistage batch reactor control (Kumar and Daoutidis, 1999; Pantelides et al., 1994), power plant design and operation (Bausa and Tsatsaronis, 1994; Beato et al., 1992), optimal paths for satellite mounted robots (Schulz, 1998), and optimal flight trajectory determination (Gill et al., 1994; Ringertz, 1999; Hargraves and Paris, 1987). Software for the solution of DAE systems is described in Ascher and Spiteri (1994).

The reduction of a PDE system to a DAE system is very problem dependent (see Section 5). For the purpose of this review article, we assume that this reduction has been done, and that the corresponding DAE system is known. Therefore we formulate the optimal control problem to be solved as a large-scale DAE system, with both equality and inequality constraints. This formulation therefore includes both time-dependent PDE systems and systems that can be directly stated as DAEs.

Another important problem that can be formulated as the optimal control of a PDE or DAE is that of parameter estimation. For this type of problem the control variables are parameters that do not vary with time. The optimum value of these parameters is then usually determined by minimizing the error in some suitable norm, between the model predicted value of some state variable (as a function of time), and the experimentally measured value of the same variable. This is described in more detail in Section 6.

In order to actually compute the optimal control solution to the DAE system, it is discretized in time to give a large, sparse system of linear and nonlinear equality and inequality constraints in continuous variables. These are to be determined so as to satisfy the constraints, and minimize an objective function. This discrete approximation to the DAE system is described in Section 3.

This large-scale sparse continuous optimization problem can be solved by any one of several nonlinear optimization software codes (Gill et al., 1996, 1986; Gould et al., 1992). These codes take advantage of the sparse structure of the constraints, and can solve problems with thousands of variables and constraints. In this survey, we will briefly describe only one of these codes, the general-purpose sparse nonlinear optimizer, SNOPT. This description is given in Section 4.

The size of this large-scale optimization problem is determined by the size of the DAE system and the number, N, of grid points in the time interval $[0, t_f]$ being considered. Thus, if the DAE system is obtained by a spatial discretization of a PDE in 2 spatial dimensions, there will be n^2 DAEs, so that the number of variables in the optimization problem will be proportional to n^2N. Since the error in the approximate solution decreases as n and N increase, the size of the problem may need to be very large to obtain the desired accuracy.

To illustrate more clearly how an approximate numerical solution is obtained, we present four specific examples of different typical problems from the literature. These consist of a boundary temperature control problem (Section 5), a reservoir parameter estimation problem (Section 6), the dynamic optimization of a chemical reactor system (Section 7), and a minimum time to climb aircraft trajectory problem (Section 8). These typical examples are summarized to show how some difficult nonlinear optimal control problems can now be solved efficiently with existing computational methods.

2. Optimal Control Problem Formulation

We will consider the computational solution of the following problem:

$$\min_{u,v,w} \int_0^{t_f} \psi(v(t), w(t), u(t), t)dt \tag{1}$$

subject to

$$\begin{aligned} f(v(t), \dot{v}(t), w(t), u(t), t) = 0, \qquad & t \in [0, t_f] \\ g_l(t) \leqslant g(v(t), \dot{v}(t) w(t), u(t), t) \leqslant g_u(t), \quad & t \in [0, t_f] \\ v(0) = \bar{v}_0, \qquad & \end{aligned} \tag{2}$$

where $v(t) \in \mathcal{R}^{n_1}$ and $w(t) \in \mathcal{R}^{n_2}$ represents the state variables, $u(t) \in \mathcal{R}^{n_3}$ represents the control variables, and $f \in \mathcal{R}^{n_1}$ defines an implicit system of ordinary differential equations for $v(t)$, with $\dot{v}$ representing the time-derivative of v. In the important case where (2) is an explicit system of differential equations, it takes the form

$$\dot{v} = f(v(t), w(t), u(t), t). \tag{3}$$

The algebraic inequalities on the vectors v, w, and u given by $g \in \mathcal{R}^{n_4}$ in (2) are additional conditions that must be satisfied. These may include lower and upper bounds on $v(t)$, $\dot{v}(t)$, $w(t)$, and $u(t)$, and they may also include equalities, by setting the corresponding elements of the bounds $g_l(t)$ and $g_u(t)$ to equal. Also, final conditions may be imposed by specifying the values of $g_l(t_f)$ and $g_u(t_f)$.

It is assumed that given the initial condition v_0 and the control function $u = u(t)$, $t \in [0, t_f]$, the state vector function $v = v(t)$ is uniquely determined by the DAE system (2). Conditions on f that ensure this are discussed, for example, in Campbell and Petzold (1995). Specifically, it is assumed that the $n_1 \times n_1$ Jacobian matrix $\partial f / \partial \dot{v}$ is nonsingular. We also assume that the control $u(t)$ satisfies some standard conditions needed for the existence of an optimal control (see, e.g., Leitmann, 1981).

For simplicity of presentation, we assume that v_0 is given and fixed. However, there is no difficulty in treating v_0 as a vector of parameters determined by the optimization. It may also be desired to determine the control $u(t)$ so that the final value of the state vector $v(t)$ attains a specified target value v_f in minimum time. That is, we want $v(t_f) = v_f$, with t_f as small as possible. This is easily accomplished by introducing the new time variable $\tau \in [0, 1]$ and making the substitutions:

$$t_f \tau \longrightarrow t$$

$$t_f^{-1} \frac{dv}{d\tau} \longrightarrow \dot{v} \tag{4}$$

and replacing (1) by

$$\min_{u,v} t_f. \tag{5}$$

To illustrate how system (1) and (2) is modified to solve a parameter estimation problem we first replace $u(t)$ by the parameter vector p in f and g. Also assume that one of the state variables, say $v_1(t)$ is measured, as $\hat{v}_1(t)$, $t \in [0, t_f]$. Now the computed model value of $v_1(t)$ will depend on the parameter vector p, so that $v_1(t) = v_1(t, p)$. To determine the optimal value of p, we solve

$$\min_p \int_0^{t_f} [v_1(t, p) - \hat{v}_1(t)]^2 dt,$$

subject to (2), with $u(t) = p$. For a more detailed examples see Section 6.

3. Discrete Approximation to the DAE System

The computational solution of the optimal control problem (1) and (2) requires that the time interval [0, t_f] be discretized by a finite grid. For simplicity, we consider a uniform grid, $t_i = i\Delta t$, $i = 1, \ldots, N$, with $\Delta t = t_f / N$. The values of u, v, $\dot{v}$, and w at t_i are now given by u_i, v_i, $\dot{v}_i$, and w_i for $i = 0, \ldots, N$. The relationship between the state vectors v_i and their derivatives $\dot{v}_i$ is given by a simple collocation scheme:

$$v_{i+1} - v_i = \frac{\Delta t}{2}(\dot{v}_i + \dot{v}_{i+1}), \quad i = 0, \ldots, N-1.$$

Other collocation schemes can also be used, but will not be described here.

The objective function (1) is replaced by the equivalent differential equation, to give

$$\min \nu(t_f) \tag{6}$$

subject to

$$\dot{\nu} = \psi(v, w, u, t), \quad \nu(0) = 0, \quad t \in [0, t_f]. \tag{7}$$

The problem (1) and (2) is now approximated by the large, but finite, optimization problem:

$$\min_{u, v, \dot{v}, w} \nu_N \tag{8}$$

subject to

$$\left.\begin{aligned} v_0 &= \bar{v}_0 \\ f(v_i, \dot{v}_i, w_i, u_i) &= 0 \\ g_l(t_i) \leqslant g(v_i, \dot{v}_i, w_i, u_i) &\leqslant g_u(t_i) \end{aligned}\right\} i = 0, ..., N$$

$$\left.\begin{aligned} v_{i+1} - v_i - \frac{\Delta t}{2}(\dot{v}_i + \dot{v}_{i+1}) &= 0 \\ \mathrm{v}_{i+1} - \mathrm{v}_i - \frac{\Delta t}{2}(\psi_i + \psi_{i+1}) &= 0 \end{aligned}\right\} i = 0, ..., N-1 \tag{9}$$

There are $(N + 1)(2n_1 + n_2 + n_3 + 1)$ variables in this problem, and $(N + 1)(2n_1 + 2n_4 + 1)$ constraints, including both equalities and inequalities.

The control vector $u(t)$, in typical applications, will be at least a continuous function of time, and often will be differentiable. The simplest continuous, piecewise linear case is given by:

$$u(t) = u_i + \mu_{i1}(t - t_i), \quad t \in [t_i, t_{i+1}], \quad i = 1, ..., N-1, \tag{10}$$

where the u_i must satisfy $u_{i+1} = u_i + \mu_{i1}\Delta t$. Bounds on the μ_{i1} can be imposed to prevent the controls from changing too rapidly. If smoother controls are desired a piecewise quadratic can be used, so that:

$$u(t) = u_i + \mu_{i1}(t - t_i) + \mu_{i2}(t - t_i)^2, \quad t \in [t_i, t_{i+1}], \quad i = 1, ..., N-1. \tag{11}$$

Now, in addition to continuity, we can require a continuous first derivative at each t_i, so that for $i = 1, ..., N-1$:

$$\begin{aligned} u_{i+1} &= u_i + \mu_{i1}\Delta t + \mu_{i2}(\Delta t)^2 \\ \mu_{i+1,1} &= \mu_{i1} + 2\mu_{i2}\Delta t \end{aligned} \tag{12}$$

For this case, we can bound both the first and second derivatives of $u(\cdot)$, by imposing bounds on the vectors μ_{i1} and μ_{i2}, $i = 1, ..., N-1$. Note that the requirements (12) represent additional linear equations (imposed on the control variables), which must be satisfied. Higher order piecewise polynomials, or splines, can also be used in a similar manner.

4. A Sparse Nonlinear Optimizer

In this section we discuss the application of the general-purpose sparse nonlinear optimizer (SNOPT) to solve the discretized optimal control problem. The discretized problem of (8) and (9) has several important characteristics: (a) many variables and constraints; (b) sparse constraint and objective derivatives; (c) objective and constraint functions (and their first derivatives) that are expensive to evaluate; and (d) many constraints binding at the solution. Sequential quadratic programming (SQP) methods are particularly well suited to problems with these characteristics.

To simplify the presentation in this section we let the vector p denote the variables (v_i, $\dot{v}_i$, w_i, u_i), $i = 1, ..., N$, and $r(p)$ the constraints, which occur in the optimization problem (9). The objective function (8), will be denoted by $\theta(p)$. At a constrained minimizer p^*, the objective gradient $\nabla\theta$ can be written as a linear combination of the constraint gradients. The multipliers in this linear combination are known as the Lagrange multipliers. The Lagrange multipliers for an upper bound constraint are nonpositive, the multipliers for a lower bound constraint are nonnegative. The vector of Lagrange multipliers associated with the nonlinear constraints of (8) is denoted by π^*.

As their name suggests, SQP methods are a class of optimization methods that solve a quadratic programming subproblem at each iteration. Each QP subproblem minimizes a quadratic model of a certain modified Lagrangian function subject to linearized constraints. A merit function is reduced along each search direction to ensure convergence from any starting point. The basic structure of an SQP method involves major and minor iterations. The major iterations generate a sequence of iterates (p_k, π_k) that converge to (p^*, π^*). At each iterate a QP subproblem is used to generate a search direction towards the next iterate (p_{k+1}, π_{k+1}). Solving such a subproblem is itself an iterative procedure, with the minor iterations of an SQP method being the iterations of the QP method. (For an overview of SQP methods, see, for example, Gill et al., 1981).

Each QP subproblem minimizes a quadratic model of the modified Lagrangian

$$\mathcal{L}(p, p_k, \pi_k) = \theta(p) - \pi_k^T d_L(p, p_k), \tag{13}$$

which is defined in terms of the constraint linearization,

$$r_L(p, p_k) = r(p_k) + J(p_k)(p - p_k), \tag{14}$$

and the departure from linearity, $d_L(p, p_k) = r(p) - r_L(p, p_k)$, where $J(p)$ is the Jacobian matrix of $r(p)$.

Given estimates (p_k, π_k) of (p^*, π^*), an improved estimate is found from $(\hat{p}_k, \hat{\pi}_k)$, the solution of the following QP subproblem:

$$\min_{p \in \mathcal{R}^n} \quad \theta(p_k) + \nabla\theta(p_k)^T(p - p_k) + \tfrac{1}{2}(p - p_k)^T H_k (p - p_k)$$

$$\text{subject to} \quad b_l \leqslant \begin{Bmatrix} p \\ Ap \\ r(p_k) + J(p_k)(p - p_k) \end{Bmatrix} \leqslant b_u,$$

where H_k is a positive-definite approximation to $\nabla_p^2 \mathcal{L}(p_k, p_k, \pi_k)$.

Once the QP solution $(\hat{p}_k, \hat{\pi}_k)$ has been determined, the major iteration proceeds by determining new variables (p_{k+1}, π_{k+1}) as

$$\begin{pmatrix} p_{k+1} \\ \pi_{k+1} \end{pmatrix} = \begin{pmatrix} p_k \\ \pi_k \end{pmatrix} + \alpha_k \begin{pmatrix} \hat{p}_k - p_k \\ \hat{\pi}_k - \pi_k \end{pmatrix},$$

where α is found from a line search that enforces a sufficient decrease in an augmented Lagrangian merit function (Gill et al., 1996).

In this SQP formulation, the problem derivatives $\nabla\theta$ and J are required once for each major iteration. The calculation of $\nabla\theta(p)$ and $J(p)$ is avoided in the line search by using a safeguarded quadratic interpolation scheme that requires only $\theta(p)$ and $r(p)$. The objective and constraint derivatives are needed, however, to define the objective and constraints of the QP subproblem. The structure of the derivatives for the inequality constraints $g \geqslant 0$ will depend upon the particular application.

The QP algorithm is of reduced-gradient type, with the QP reduced Hessian being computed at the first feasible minor iterate. The QP solver must repeatedly solve linear systems formed from rows and columns of the structured derivatives. In the current version of SNOPT, these sparse systems are solved using the general-purpose sparse LU package LUSOL.

The definition of the QP Hessian H_k is crucial to the success of an SQP method. In SNOPT, H_k is a positive-definite approximation to $G = \nabla_p^2 \mathcal{L}(p_k, p_k, \pi_k)$, the Hessian of the modified Lagrangian. The exact Hessian is highly structured. In SNOPT, H_k is a limited-memory quasi-Newton approximate Hessian. On completion of the line search, let the change in p and the gradient of the modified Lagrangian be

$$\delta_k = p_{k+1} - p_k \quad \text{and} \quad y_k = \nabla\mathcal{L}(p_{k+1}, p_k, \pi_{k+1}) - \nabla\mathcal{L}(p_k, p_k, \pi_{k+1}). \tag{15}$$

The approximate Hessian is updated using BFGS quasi-Newton update,

$$H_{k+1} = H_k - \rho_k q_k q_k^T + \theta_k y_k y_k^T, \tag{16}$$

where $q_k = H_k \delta_k$, $\rho_k = 1/q_k^T \delta_k$ and $\theta_k = 1/y_k^T \delta_k$.

The limited-memory scheme used in SNOPT is based on the observation that the SQP computation can be arranged so that the approximate Hessian H_k is only required to perform matrix-vector products of the form $H_k u$. This implies that H_k need not be stored explicitly, but may be regarded as an operator involving an initial diagonal matrix H_r and a sum of rank-two matrices held implicitly in outer-product form. With this approach, a preassigned fixed number (say ℓ) of these outer products are stored and products $H_k u$ are computed using $O(\ell)$ inner-products.

Currently, SNOPT uses a simple limited-memory implementation of the BFGS quasi-Newton method. As the iterations proceed, the two vectors (q_k, y_k) defining the current update are added to an expanding list of most recent updates. When ℓ updates have been accumulated, the storage is reset by discarding all information accumulated so far. For a related description of SQP methods see Gill et al. (1998).

In the next four sections we summarize typical examples of optimal control of PDEs and DAEs, which will illustrate the methods described in the previous sections. In Section 5, boundary control of a two-dimensional heating problem is formulated (Petzold et al., 1997). This is a two-dimensional nonlinear parabolic PDE, where the temperature of an interior domain is required to vary according to a specified function of time. In Section 6, a parameter estimation problem is summarized, the problem being to determine a good set of parameters in an underground permeable reservoir simulation model. This is done by minimizing the difference between the measured, time-dependent output pressures at a set of production wells and the corresponding pressures predicted by the simulation model (Ewing, 1991).

The next example, in Section 7, is the dynamic optimization of a chemical batch reactor operation. The amount of product is determined by the rate at which one of the reactants is added. The optimum addition rate is determined so as to maximize the total production, subject to constraints on the maximum allowed temperature (Pantelides et al., 1994). This problem is formulated directly as a DAE system. Finally, in Section 8, the optimum trajectory for an aircraft is to be determined, so as to minimize the time required to climb to 65,000 feet, starting at sea level (Gill et al., 1994). This problem is also formulated directly as a DAE system.

5. Boundary Control Heating Problem

A boundary control heating problem in two spatial dimensions will now be described to show the effectiveness of the large-scale optimizer SNOPT. This model problem is described by a nonlinear parabolic PDE. It is a two-dimensional generalization of the model problem described in Leibfritz and Sachs (1994). A rectangular domain in space is heated by controlling the temperature on its boundaries. It is desired that the transient temperature in a specified interior subdomain follows a prescribed temperature-time trajectory as closely as possible. The domain Ω is given by

$$\Omega = \{(x, y) \mid 0 \leqslant x \leqslant x_{\max}, \quad 0 \leqslant y \leqslant y_{\max}\},$$

and the control boundaries are given by

$$\partial\Omega_1 = \{(x, y) \mid y = 0\}, \quad \text{and} \quad \partial\Omega_2 = \{(x, y) \mid x = 0\}.$$

The temperature distribution in Ω, as a function of time, is controlled by the energy input across the boundaries $\partial\Omega_1$ and $\partial\Omega_2$, as discussed below. The other two boundaries ($x = x_{\max}$ and $y = y_{\max}$) are assumed to be insulated, so that no energy flows into or out of Ω along the normals to these boundaries. The temperature must be controlled in the subdomain

$$\Omega_c = \{(x, y) \mid x_c \leqslant x \leqslant x_{\max}, \quad y_c \leqslant y \leqslant y_{\max}\}.$$

This is illustrated in Figure 1.

The control problem is to be solved for the time interval $t \in [0, t_{\max}]$. The temperature $T = T(x, y, t)$ is then determined by the nonlinear parabolic PDE given below, for $(x, y, t) \in \Omega \times [0, t_{\max}]$.

The temperature T is controlled by heat sources located on the boundaries $\partial\Omega_1$ and $\partial\Omega_2$. These heat sources are represented by control functions $u_1(x, t)$ on $\partial\Omega_1$, and $u_2(y, t)$ on $\partial\Omega_2$. The control functions are to be determined. The objective is to control the temperature-time trajectory on the subdomain Ω_c. A target trajectory $\tau(t)$, $t \in [0, t_{\max}]$, is specified. The actual temperature in Ω_c should approximate $\tau(t)$ as closely as possible.

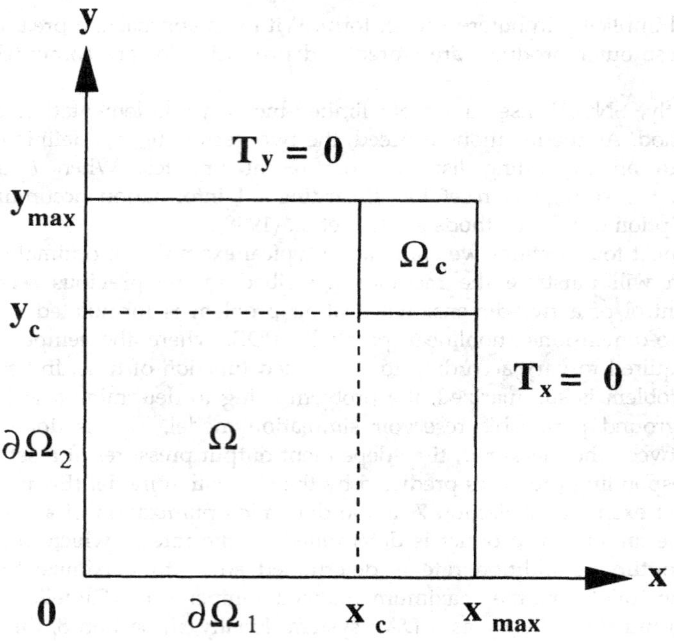

Figure 1 Two-dimensional spatial domain for the parabolic control problem

We measure the difference between $T(x, y, t)$ and $\tau(t)$ on Ω_c by the function

$$\phi(u) = \int_0^{t_{\max}} \int_{y_c}^{y_{\max}} \int_{x_c}^{x_{\max}} w(x, y, t)[T(x, y, t) - \tau(t)]^2 dxdydt, \quad (17)$$

where $w(x, y, t) \geqslant 0$ is a specified weighting function. The control functions u_1 and u_2 are determined so as to

$$\underset{u}{\text{minimize}} \ \phi(u), \quad (18)$$

subject to $T(x, y, t)$ satisfying the PDE and other constraints.

The temperature $T(x, y, t)$ must satisfy the following PDE, boundary conditions, and bounds

$$\begin{aligned}
&\alpha(T)[T_{xx} + T_{yy}] + S(T) = T_t, \qquad (x, y, t) \in \Omega \times [0, t_{\max}] \\
&\qquad T(x, 0, t) - \lambda T_y = u_1(x, t), \qquad x \in \partial\Omega_1 \\
&\qquad T(0, y, t) - \lambda T_x = u_2(y, t), \qquad y \in \partial\Omega_2 \\
&\qquad T_x(x_{\max}, y, t) = 0, \\
&\qquad T_y(x, y_{\max}, t) = 0, \\
&\qquad 0 \leqslant T(x, y, t) \leqslant T_{\max}
\end{aligned} \quad (19)$$

The controls u_1 and u_2 are also required to satisfy the bounds

$$0 \leqslant u_1, u_2 \leqslant u_{\max}.$$

The initial temperature distribution $T(x, y, 0)$ is a specified function. The coefficient $\alpha(T) = \lambda/c(T)$, where λ is the heat conduction coefficient and $c(T)$ is the heat capacity. The source term $S(T)$ represents internal heat generation, and is given by

$$S(T) = S_{\max} e^{-\beta_1/(\beta_2+T)},$$

where $S_{\max}$, β_1, $\beta_2 \geqslant 0$ are specified nonnegative constants.

A uniform rectangular grid is constructed on the domain Ω

$$x_i = i\Delta x, \quad i = 0, 1, ..., m, \quad \Delta x = x_{\max}/m$$

$$y_j = j\Delta y, \quad j = 0, 1, ..., n, \quad \Delta y = y_{\max}/n.$$

Then let

$$T_{ij}(t) = T(x_i, y_j, t), \quad u_{1i}(t) = u_1(x_i, t), \quad \alpha_{ij}(t) = \alpha(T_{ij}(t)),$$

$$S_{ij}(t) = S(T_{ij}(t)), \quad u_{2j}(t) = u_2(y_j, t).$$

The PDE is then approximated in the interior of Ω by the following system of $(m-1)\times(n-1)$ ODEs

$$\frac{dT_{ij}}{dt} = \frac{\alpha_{ij}}{\Delta x^2}[T_{i-1,j} - 2T_{ij} + T_{i+1,j}] + \frac{\alpha_{ij}}{\Delta y^2}[T_{i,j-1} - 2T_{ij} + T_{i,j+1}] + S_{ij}, \tag{20}$$

for $i = 1, 2, ..., m-1$, $j = 1, 2, ..., n-1$. Each of the $2(m+n)$ boundary points also satisfies a differential equation similar to (20). These will include values outside Ω, which are eliminated by using the boundary conditions. Specifically, we use

$$T_{i,n+1} = T_{i,n-1}, \quad i = 0, 1, ..., m$$

$$T_{m+1,j} = T_{m-1,j} \quad j = 0, 1, ..., n$$

to approximate the conditions $T_y = 0$ and $T_x = 0$.

The finite-difference approximations to the boundary conditions on $\partial\Omega_1$ and $\partial\Omega_2$ are given by

$$\begin{aligned} T_{i0} - \frac{\lambda}{2\Delta y}(T_{i1} - T_{i,-1}) &= u_{1i}, \quad i = 0, 1, ..., m \\ T_{0j} - \frac{\lambda}{2\Delta x}(T_{1j} - T_{-1,j}) &= u_{2j}, \quad j = 0, 1, ..., n \end{aligned} \tag{21}$$

These relations are used to eliminate the values $T_{i,-1}$ and $T_{-1,j}$ from the differential equations [as in (20)], for the functions T_{ij} on $\partial\Omega_1$ and $\partial\Omega_2$. As a result, the control functions u_{1i} and u_{2j} are explicitly included in these differential equations, giving $2(m+n)$ additional differential equations. Together with the $(m-1)(n-1)$ ODEs given by (20), this gives a total of $(m+1)\times(n+1)$ ODEs for the same number of unknown functions $T_{ij}(t)$. This system of $(m+1)(n-1)$ ODEs can now be represented by

$$\frac{dv(t)}{dt} = f(v, u(t), t), \quad v(0) = v_0, \tag{22}$$

where v_0 represents the initial value of $v(t)$, and $u = u(t)$ the control functions. The vector function $u(t)$ has elements $u_{1i}(t)$, $i = 0, 1, ..., m$, and $u_{2j}(t)$, $j = 0, 1, ..., n$.

We have shown how the original nonlinear parabolic PDE given by (19) is reduced to the system of ODEs given by (22). The system of ODEs was then discretized as described in Section 3, and then solved by SNOPT as described in Section 4. Computational results were obtained for ten cases, including cases with bounds on the interior temperatures. Results for two cases are shown in Figures 2 and 3.

Both figures show the desired temperature-time trajectory $\tau(t)$ (solid line) and the computed trajectory $T_{ij}(t)$ for $(x,y) \in \Omega_c$ (dash-dot lines). The boundary control, $u(t)$ (dotted line), and the temperature at the origin, $T_{00}(t)$ (dashed line), are also shown. The grid size used

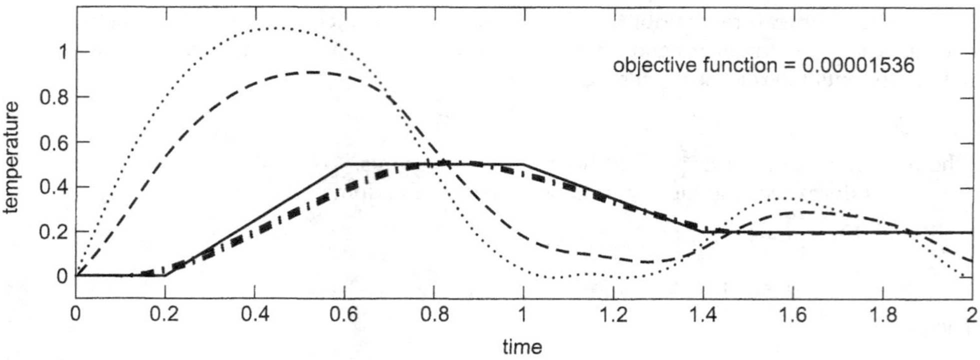

Figure 2 Optimal solution computed by SNOPT on a 9 × 17 grid. Solid line: $\tau(t)$. Dotted line: $u(t)$. Dashed line: $T_{00}(t)$. Dash-dot lines: $T_{ij}(t)$ in Ω_c

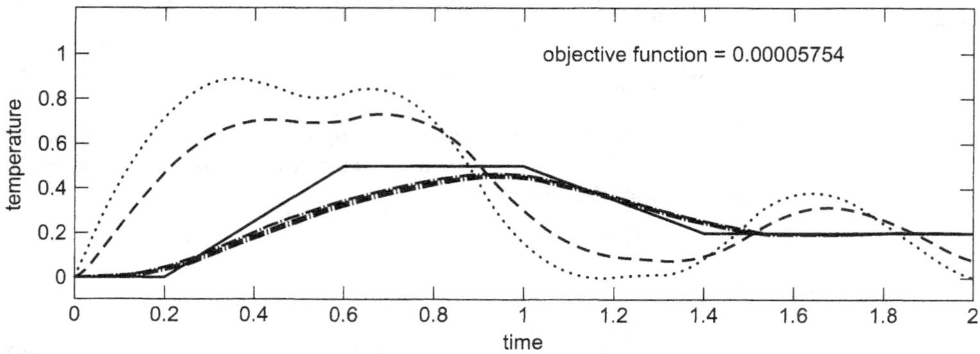

Figure 3 Optimal solution computed by SNOPT with temperature upper bound. Solid line: $\tau(t)$. Dotted line: $u(t)$. Dashed line: $T_{00}(t)$. Dash-dot lines: $T_{ij}(t)$ in Ω_c

was 9 × 17, giving a total of 180 ODEs. No state constraints were imposed for the case shown in Figure 2, and it can be seen that the desired trajectory $\tau(t)$ is matched very well. For the case shown in Figure 3, a bound of $T_{ij}(t) \leqslant 0.7$ was imposed at every grid point. It can be seen that this requires a decrease in the maximum value of $u(t)$, and results in a significant decrease in $T_{00}(t)$, so that it satisfies the imposed state variable bound. As a result the temperature in Ω_c cannot match $\tau(t)$ as closely, for $0.4 \leqslant t \leqslant 1.0$. For the 10 cases, it typically required approximately 5 minutes of Cray C90 CPU time to compute the optimal solution. For a complete description of the problem and its solution see Petzold et al. (1997).

6. Parameter Estimation in Reservoir Simulation

The purpose of mathematical reservoir simulation models is to be able to optimize the recovery of valuable minerals and hydrocarbons from underground permeable reservoirs. To do this, one must be able to predict the performance of the reservoir under various exploitation and production schemes. A series of models is constructed to yield information about, and understanding of, the complex chemical, physical, and fluid flow processes that accompany the different recovery mechanisms. Reservoir simulators are also used to monitor the fluids and ensure the control and containment of any possible contamination agents see Ewing and George (1985), and Ewing (1991).

The mathematical techniques that are used to model multiphase or multicomponent flow are representative of those needed for many other applications, such as chemically reactive or thermally driven flows, and will be used to illustrate the source of ill-posed and inverse problems in the modeling process. Modeling of large-scale physical processes involves four major interrelated stages. First, a physical model of the processes must be developed incorporating as much physics as is deemed necessary to describe the essential phenomena. A careful list of the assumptions made in establishing this physical model should be compiled together with expected properties of the process, such as bifurcation or physical instabilities, that might be expected and should be modeled. Second, a mathematical formulation of the physical model is developed, usually involving coupled systems of nonlinear partial differential equations. The properties of this mathematical model, such as existence, uniqueness, and regularity of the solution, are then obtained and related to the physical process in order to check the model. This part of the modeling process becomes exceedingly difficult when large, coupled systems of nonlinear partial differential equations are involved. These systems of differential equations have function parameters that describe properties of the fluids or the media and may be highly nonlinear. Since these parameters often cannot be measured directly, they must be determined via inverse problems. Third, a discretized numerical model of the mathematical equations is produced. This numerical model must have the properties of accuracy and stability and produce solutions that represent the basic physical features as well as possible without introducing spurious phenomena associated with the specific numerical schemes. Fourth, a computer program capable of efficiently performing the necessary computations for the numerical model is sought. Properties of the computer architecture to be used in the computation must be considered strongly in the development of efficient computational algorithms. Although the total modeling process encompasses aspects of each of these four intermediate stages, the process is not complete with one pass through the steps. Usually, several iterations of this modeling loop are necessary to obtain reasonable models for the highly complex physical phenomena involved in many applications.

The model equations used in hydrology for a slightly compressible, single phase fluid will be used as an example. For this application, the model equations are

$$\phi c_t \frac{\partial p}{\partial t} = \nabla \cdot \left(\frac{k}{\mu} \nabla p\right) + q, \quad x \in \Omega, \quad t \in J, \tag{23}$$

$$p(x, 0) = p_0(x), \quad x \in \Omega, \tag{24}$$

where p is the pressure, c_t is the compressibility of the rock/fluid system and p_0 is a given initial pressure or head distribution.

The process of determining unknown parameter values that are used in a mathematical or computer model to give the best fit to measured output variables, such as production history from wells, is commonly called "history matching." We would like to have an automatic routine for history matching that can determine a set of parameters giving a good history match with a reasonable amount of computational time and effort. Then, we would like to have some estimate of the accuracy of the process, taking into account, as much as possible, the various sources of error in the modeling process.

Although an inhomogeneous reservoir physically requires an infinite number of parameters for complete property description, a computational reservoir simulator can allow specification of only a finite number of parameters. By using a discrete computer model for the simulation of a continuous process, we are clearly making a type of discretization error. If we use a computer model with constant media properties on each grid block and apply this model to a large-scale simulation problem, we may have as many as 50,000 unknowns to determine via a nonlinear output least-square minimization problem. Therefore, the feature of the history-matching problem in reservoir simulation that distinguishes it from parameter identification problems in other fields of science and engineering is the large dimensionality of both the system state space and the unknown parameter space. In order to reduce both the statistical uncertainty and the computational complexity, one must decrease the number of unknowns and, if possible, utilize any additional available information. To reduce the large number of variables and obtain a tractable problem, several effective schemes are in current use. The most common process, called zonation, divides the reservoir into a smaller set of zones, each of which is assumed to have reservoir properties that can be described in a functional manner with very few parameters in each zone. As the number of zones is decreased for computational efficiency, the modeling error increases. On the other hand, if the number of unknowns is not reduced by some technique, accurate solution of the nonlinear optimization problem will be impossible. This is an example of the need to balance errors from different aspects of the modeling process.

We consider equations 23 and 24 and try to determine the unknown functions ϕ and k by matching the output variable pressure at production wells. The history-matching problem can then be viewed as a nonlinear, constrained optimization (least-squares) problem of minimizing the sum of the squares of differences between observed and calculated values. For example, if pressures at N_w wells are the observable at times $t_1, t_2, \ldots, t_{N_0}$, and N unknown parameters α_i are to be determined, the following problem is formulated:

Problem: Find a vector $\alpha = (\alpha_1, \ldots, \alpha_N)$ that minimizes

$$J = \sum_{i=1}^{N_w} \sum_{k=1}^{N_0} w_{ik}[p^*(x_i, y_i, t_k) - P^*(x_i, y_i, t_k; \alpha)]^2, \tag{25}$$

where $p^*(x_i, y_i, t_k)$ is the measured pressure at well i and time t_k, $P^*(x_i, y_i, t_k; \alpha)$ is the computed pressure from the simulator at well i and time t_k using the vector α as parameters, and w_{ik} are specified weights. The weights w_{ik} are used to give a bias to certain data that may be believed to be more or less accurate than other data. If certain *a priori* knowledge is assumed about the mean values of correlations of the unknown parameters, the objective functional J in (25) could be modified to include a Bayesian term penalizing the weighted deviations of the parameters from their mean values.

The problem of minimizing J from (25) over some admissible finite-dimensional parameter set A has several difficulties:

1. The number of unknowns may be quite large.
2. The least-squares minimization problem is nonlinear.
3. Many applications yield quite low sensitivity to the unknowns.
4. There are usually many local minima and relatively flat surfaces of the minimization functional.

These features lead to a very large, highly ill-conditioned, nonlinear minimization problem. Typical techniques for obtaining a better-conditioned problem are various forms of regularization that involve modifying the objective functional by adding a stabilizing term that incorporates prior information about the measurement error or prior statistics.

7. Optimization of Batch Reactor Operation

Here we consider the optimal operation of a batch reactor used to carry out the chemical reaction

$$A + B \rightarrow C$$

in the presence of the side reaction

$$B + C \rightarrow D.$$

Both reactions are strongly exothermic, and therefore direct mixing of the entire necessary amounts of the reactants A and B must be avoided. Instead, the operation is performed by initially charging an amount of A in the reactor vessel, with B being added slowly during the reaction. The reactor vessel is fitted with a cooling water jacket, which is used to remove the heat generated by the reaction.

Under a suitable set of assumptions, the transient behavior of the above system is described by the following mathematical model:

$$\max_{v,w,u,t_f} \quad v_3(t_f)$$

subject to

$$\begin{aligned} v(0) &= v_0 \\ \dot{v} &= f(w,u) \\ 0 &= g(v,w) \\ 0 \leqslant w(t) &\leqslant \bar{w}, \quad t \in [0, t_f] \\ 0 \leqslant u(t) &\leqslant \bar{u}, \quad t \in [0, t_f] \\ 0 < t_f &\leqslant \bar{t} \end{aligned}$$

The size of this system of DAEs is given by $v \in \mathscr{R}^5$, $w \in \mathscr{R}^8$, $u \in \mathscr{R}^2$, and $g \in \mathscr{R}^8$. The vector v represents the state variables (concentrations of A, B, C, D, and total energy holdup). v_3 is the concentration of product C. The algebraic variables are represented by the vector w. These include temperature, volume of liquid in system, and two reaction rates. The controls are represented by the vector $u(t)$ with $u_1(t)$ = rate of addition of B, and $u_2(t)$ = external cooling load. The optimization determines $u_1(t)$, $u_2(t)$, and t_f, so as to maximize $v_3(t_f)$.

In addition to the upper bounds on the two control variables, the final temperature must lie in a narrow range [295°K, 300°K] to ensure that the reactions are sufficiently quenched without excessive subcooling. This problem was solved both with and without path constraints, to limit the maximum allowed temperature. Typical solution times were less than three minutes on a SUN SPARCStation 2 workstation. For complete details see Pantelides et al. (1994).

8. Aircraft Trajectory Optimization

The flight dynamics of a spacecraft or aircraft are completely determined by initial and terminal conditions on the flight trajectory and by (differential) equations of motion. The equations of motion include control functions that can be chosen to optimize some performance criterion, such as the amount of fuel burned, or the elapsed time of the mission. Such optimal trajectory problems may be formulated in the following general form. Associated with each problem is the definition of an N-stage dynamical system, whose ith stage is

$$\dot{x}^i(t) = \psi^i(x(t), u(t), d, t), \qquad t \in [t_i, t_{i+1}], \tag{26}$$

where $x(t)$ and $u(t)$ are vector-valued functions of state and control variables, and d is a vector of design variables (e.g., platform area, shape parameters, rocket nozzle diameter, etc.). At the start of each stage, a boundary condition of the form

$$x(t_i) = b^i(v_i, t_i, d) \tag{27}$$

is applied, where v_i is a vector of variables that determine the position of the trajectory at t_i. The initial and final values of t_i are denoted by t_0 and t_f. In general, t_0 is given, but other t_is may be variables to be optimized.

The optimal trajectory problem is to find a local minimizer of a nonlinear function $\psi^0[t_f, x(t_f)]$, subject to the differential constraints (26) and (27), and the so-called path constraints

$$b_l(t) \leqslant \phi(x(t), u(t), d, t) \leqslant b_u(t), \quad t \in [t_0, t_f], \tag{28}$$

where $b_l(t)$ and $b_u(t)$ are prescribed vector functions of upper and lower bounds.

Recently, an approach based on Hermite collocation and the SQP algorithm NPSOL has been employed within the optimal trajectory package OTIS (Hargraves and Paris, 1987). The resulting system has had a significant impact on space-vehicle design, and is being used in the calculation of trajectories for the National Aerospace Plane, the Mars Lander, and the single-stage-to-orbit test vehicle.

Here we give a highly simplified description of the discretization scheme used in OTIS. The idea is to obtain a discrete set $x_1, x_2, x_3, \ldots,$ whose elements are the values of $x(t)$ at a set of so-called *nodes* $\tau_1, \tau_2, \tau_3, \ldots$. The x-values are adjusted by the optimization routine, and they must satisfy a set of functional equalities and inequalities that represent discretized versions of the continuous constraints.

The state equations (26) and the path constraints (28) are discretized, giving the finite-dimensional problem

$$\min_{x \in \mathscr{R}^n} \quad f(x)$$
$$\text{subject to} \quad l \leqslant \begin{pmatrix} x \\ Ax \\ c(x) \end{pmatrix} \leqslant u, \tag{29}$$

where x is a set of variables, $f(x)$ is a nonlinear function, A is an $m_L \times n$ matrix, and $c(x)$ is an m_N-vector of nonlinear functions.

SQP methods require the specification of the derivatives of the nonlinear functions $f(x)$ and $c(x)$. When derivatives are not available a common approach is to approximate the derivatives by finite differences. In general, derivatives of ψ and ϕ (and hence of $c(x)$ and $f(x)$) are not provided. Fortunately, the problem derivatives are sparse, since each variable is associated with a particular node that is linked only to nodes at nearby points. In OTIS the sparsity pattern and the structure implied by the discretization is exploited to reduce the number of function evaluations required to approximate the derivatives. Nevertheless, the derivatives are very expensive to approximate.

This formulation treats both $x(t)$ and $u(t)$ as independent variables, that is, there is no attempt to follow the traditional approach of using the differential constraints (26) to eliminate the state variables from the problem. An important consequence of this approach is that the differential constraints are satisfied only as the discretized variables converge to their optimal values. This increases the importance of the reliability of the optimizer, since termination at a point other than a solution will not provide a valid trajectory.

SQP methods solve a nonlinearly constrained optimization problem using a sequence of quadratic programming subproblems. SQP methods have a two-level structure of major and minor iterations. The major iterations generate a sequence of iterates $\{x_k\}$ such that

$$x_{k+1} = x_k + \alpha_k p_k, \tag{30}$$

where x_k is the current iterate, the nonnegative scalar α_k is the steplength, and p_k is the search direction. The defining feature of an SQP method is that the search direction p_k is the solution of a quadratic programming (QP) subproblem. The minor iterations are then those of the method used to solve the quadratic program. The steplength α_k is usually chosen to yield a sufficient decrease in a merit function (a suitable combination of the objective and constraint functions).

NPSOL is an SQP method that utilizes an augmented Lagrangian merit function. Under suitable conditions, the iterates will converge from any starting point to a point that satisfies the first-order conditions necessary for a solution of (29). Furthermore, the augmented Lagrangian merit function does not impede superlinear convergence, in the sense that when the iterates converge superlinearly with $\alpha_k = 1$ the unit step gives a sufficient decrease in the merit function.

In Table 1 we give the results of applying the SQP method NZOPT to a standard optimal trajectory problem: the F4 minimum time to climb. In this problem, a pilot cruising in an F-4 at Mach 0.34 at sea level wishes to ascend to sixtyfive thousand feet and be at Mach 1.0 in the minimum time. The problem has two path constraints: a constraint on the maximum altitude and a constraint on the maximum dynamic pressure. The table gives the details of six runs with increasing numbers of nodes. The columns n and m give the number of variables and constraints (excluding the simple upper and lower bounds). The columns Majors and n_f give the total number of major iterations and function evaluations in each case. Note that at each major iteration, OTIS approximates the gradients of the problem functions. The gradient

Table 1 NZOPT Applied to the F4 Minimum Time to Climb

Nodes	*n*	*m*	n_Z	*Majors*	n_f	*CPU secs*
20	184	232	7	19	32	60.5
40	364	472	22	31	35	562.9
60	544	712	40	39	43	3002.9
80	724	934	55	45	49	9292.7
100	904	1192	74	46	50	18843.1

approximations dominate the computational effort for small n. The column n_Z gives the number of degrees of freedom associated with the solution of each problem. (n_Z is the dimension of the underlying unconstrained problem, i.e., the number of variables n less the number of binding constraints at the solution.) All runs were performed on a Sun SPARCStation 10/41.

The results in Table 1 indicate that the dense SQP algorithm used by NZOPT is extremely effective for problems with a small to moderate number of nodes. These results are typical of those obtained on practical problems, that is, NZOPT converges rapidly and with great reliability. The problem functions, which are very expensive to evaluate, are only computed during the major iterations and so the remarkably few major iterations needed makes NZOPT very well suited to trajectory calculations provided n is not too large. Note that only fifty evaluations of the problem functions were necessary to solve a problem with almost one thousand variables. For a more complete description of this problem, see Gill et al. (1994), and for a more recent related application, see Ringertz (1999).

References

Ascher, U. M. and R. J. Spiteri. 1994. "Collocation software for boundary value differential-algebraic equations." *SIAM J. Sci. Comput.* **15(4)**: 938–952.

Banks, H. T. 1982. *A Survey of Some Problems and Recent Results for Parameter Estimation and Control in Delay and Distributed Parameter Systems*, pp. 3–24. Marcel Dekker.

Banks, H. T. and K. Kunisch. 1982. "An approximation theory for nonlinear partial differential equations with applications to identification and control." *SIAM J. Control Optim.* **20**: 815–819.

Bausa, J. and G. Tsatsaronis. 1994. "Dynamic optimization of start-up and load-increasing processes in power plants—methodology and applications." Berlin: Institute for Energy Engineering, Technical University of Berlin.

Beato, A., V. Bolis, P. Colombo, A. De Marco, S. Quatela, A. Zizzo, and F. Marziali. 1992. "Study of the start-up control of a drum boiler by means of a mathematical model." In *IFAC Control of Power Plants and Power Systems.*

Bulirsch, R. and D. Kraft. (eds). 1994. In *Computational Optimal Control*, International Series of Numerical Mathematics, Volume 115. Birkhauser.

Bulirsch, R., A. Miele, J. Stoer, and K. H. Wells. (eds). 1993. In *Optimal control: calculus of variations, optimal control theory and numerical methods*. International Series of Numerical Mathematics, Volume 111. Birkhauser.

Butkovskiy, A. G. 1991. "Recent research on control of distributed and lumped systems: a brief summary." In *Distributed Parameter Control Systems: New Trends and Applications*, pp. 205–232. Marcel Dekker.

Brenan, K. E., S. L. Campbell, and L. R. Petzold. 1995. In *Numerical Solution of Initial-Value Problems Control Theory and Numerical Methods.* 2d ed. Philadelphia: SIAM Publications.

Chen, G., E. B. Lee, W. Littman, and L. Markus. (eds). 1991. In *Introduction to the Theory of Distributed Control Systems.* Marcel Dekker.

Cuthrell, J. E. and L. T. Biegler. 1989. "Simultaneous optimization and solution methods for batch reactor control profiles." *Comput. Chem. Engng* **13(1/2)**: 49–62.

Elden, L. 1995. "Numerical solution of the sideways heat equation." In *Inverse Problems in Diffusion Processes*, edited by H. Engl and W. Rundell, pp. 130–150. SIAM Publishers.

Ewing, R. E. 1991. "Large-scale scientific computation for distributed systems." In *Distributed Parameter Control Systems: New Trends and Applications*, pp. 139–170. Marcel Dekker.

Ewing, R. E. and J. H. George. 1985. "Identification and control of distributed parameters in porous media flow." In *Distributed Parameter Systems*, edited by M. Thomas. Lecture Notes in Control and Information Sciences **70**: 145–161. Berlin: Springer-Verlag.

Gill, P. E., W. Murray, and M. H. Wright. 1981. *Practical Optimization*. Academic Press.

Gill, P. E., W. Murray, M. A. Saunders, and M. H. Wright. 1986. "User's guide for NPSOL (version 4.0): a fortran package for nonlinear programming." Report SOL 86-2. Stanford, CA: Department of Operations Research, Stanford University.

Gill, P. E., W. Murray, and M. A. Saunders. 1994. *Large-scale SQP Methods and Their Application in Trajectory Optimization*. International Series of Numerical Mathematics, **115**: 29–42. Birkhauser.

Gill, P. E., W. Murray, and M. A. Saunders. 1996. "SNOPT: an SQP algorithm for large-scale constrained optimization." Numerical Analysis Report 96-2. La Jolla, CA: Department of Mathematics, University of California, San Diego.

Gill, P. E., A. Barclay, and J. B. Rosen. 1998. "SQP methods in numerical optimal control." In *Variational Calculus, Optimal Control and Applications*, edited by R. Bulirsch, pp. 207–222. Birkhauser.

Gould, N. I. M., A. R. Conn, and Ph. L. Toint. 1992. *LANCELOT: a FORTRAN package for large-scale nonlinear optimization (Release A)*. Springer Series in Computational Mathematics, Volume 17. Berlin: Springer-Verlag: Software available online at `http://www.rl.ac.uk/departments/ccd/numerical/lancelot`.

Hager, W. W. and P. M. Pardalos. (eds). 1998. In *Optimal Control: Theory, Algorithms and Applications*. Kluwer Academic Publishers.

Hargraves, C. R. and S. W. Paris. 1987. "Direct trajectory optimization using nonlinear programming and collocation." *AIAA J. Guidance, Control Dynamics* **10(4)**: 338–342.

Kumar, A. and P. Daoutidis. 1999. *Control of Nonlinear Differential Algebraic Equation Systems*. Chapman and Hall.

Leibfritz, F. and E. W. Sachs. 1994. "Numerical solution of parabolic state constrained control problems using SQP and interior point methods." In *Large Scale Optimization: the State of the Art*, edited by W. W. Hager, D. W. Hearn, and P. M. Pardalos, pp. 245–258. Kluwer Academic Publishers.

Leitmann, G. 1981. *The Calculus of Variations and Optimal Control*. Mathematical Concepts and Methods in Science and Engineering, Volume 24. New York: Plenum Press.

Lions, J. L. 1988. "Exact controllability, stabilization and perturbations for distributed systems." *SIAM Rev.* **30**: 1–68.

Markus, L. 1991. "Introduction to the theory of distributed control systems." In *Distributed Parameter Control Systems: New Trends and Applications*, edited by G. Chen, E. B. Lee, W. Littman, and L. Markus, pp. 1–60. Marcel Dekker.

Pantelides, C. C., R. W. H. Sargent, and V. S. Vassiliadis. 1994. "Optimal control of multistage systems described by high-index differential algebraic equations." In *Computational Optimal Control*, edited by R. Bulirsch and D. Kraft, pp. 171–191. Birkhauser.

Petzold, L., J. B. Rosen, P. E. Gill, L. O. Jay, and K. Park. 1997. "Numerical optimal control of parabolic pdes using dasopt." In *Large Scale Optimization with Applications, Part II: Optimal Design and Control*, edited by L. T. Biegler, T. F. Coleman, A. R. Conn, and F. N. Santosa. IMA Volumes in Mathematics and its Applications, **93**: 271–300. Berlin: Springer-Verlag.

Ringertz, U. 1999. "Multistate trajectory optimization using large scale nonlinear programming." Technical Report 99-25. Department of Aeronautics, Royal Institute of Technology, Stockholm.

Sakawa, Y. 1974. "Controllability for partial differential equations of parabolic type." *SIAM J. Control* **12**: 389–400.

Sargent, R. W. H. 1998. "Necessary conditions for optimal control of inequality-constrained DAE systems." Centre Reports Series B98.17. Centre for Process Systems Engineering: Imperial College of Science and Technology and Medicine.

Schulz, V. H. 1998. *Solving Discretized Optimization Problems by Partially Reduced SQP Methods*. Computing and Visualization in Science, **1**: 83–96. Berlin: Springer-Verlag.

18.16 Optimization in Water Reservoir Systems

Kumaraswamy Ponnambalam

ABSTRACT Optimization in water reservoir systems is challenging due to the following factors: multiperiods, multi-reservoirs, multiobjectives, uncertainty, and nonlinearity. Additionally, the need for public participation in a multiregional and multinational framework makes it necessary to convey the results of optimization in an easily understandable manner both for the general public and also for the many operators of such systems. Because of the technical challenges, it is fair to say that researchers have tried almost all known optimization techniques in order to solve these problems. While such research is good, not many real-world applications have been found. This may be partly due to operators of water reservoir systems having no easy access to these techniques due to the lack of availability of suitable software. In this article, a review of the state of the art is made for those readers who are well-motivated but require some guidance on the suitability of a technique for a given problem.

1. Introduction

Water reservoirs (both surface reservoirs and underground reservoirs called aquifers) augment water supply either in space (where there are wet and dry regions) or in time (when supply is nonuniform in time, for example, in countries with heavy monsoon rainfall in three months of a year and almost none in the remaining months). Generally, surface water reservoirs provide high flows and high potential heads and hence are used for producing hydroelectric energy that supplies up to 50 percent of the energy needs of the world. On the other hand, to get water from a groundwater reservoir, pumping is needed, which consumes energy. In this article, the focus will be on surface water reservoirs. The other uses of surface water reservoirs include water-based recreation including wildlife enhancement, navigation, fishing, irrigation, water quality control, and flood control. Excepting for flood control, all the uses require nearly full reservoirs. Flood control requires empty reservoirs to catch the high flows coming from upstream and reduce floods downstream. The time scales of reservoir studies may vary from many years in the planning and design stage to only hours in real-time operations. The other important aspect of reservoir optimization is that both supply (called inflows) and demand (releases) are highly varying and uncertain. Nonlinearity arises due to both nonlinear objective functions (Niedda and Sechi, 1996) and nonlinear system equations (Niewiadomsk-Szynkiewicz et al., 1996; Fletcher and Ponnambalam, 1998; Sadjadi and Ponnambalam, 1999) as described later. Optimal power flow problems are a special kind of problem where hydro and thermal power scheduling, requiring real-time reservoir operations and unit operations, represented by integer variables, with a duration of an hour or less, need to be considered (Momoh et al., 1999a,b).

1.1. *Mathematical Model of Reservoirs*

A typical multireservoir system has many reservoirs connected either in serial or in parallel or both and is a complex network of rivers and reservoirs, human made channels, dams, hydroelectric power plants, and pumps sprawled over a large distance spanning tens to thousands of kilometers.

In Figure 1 a schematic of a five-reservoir system (based on Ponnambalam and Adams, 1996) is shown. The mathematical model of this system can be written as:

$$s_i^{t+1} = s_i^t + i_i^t - e_i^t - \sum_i (d_i^t + u_i^t - l_i^t), \quad i = 1, ..., N; \quad t = 1, ..., T, \tag{1}$$

where

i, t reservoir index, time period index, respectively

N, T Total reservoirs and total periods, respectively

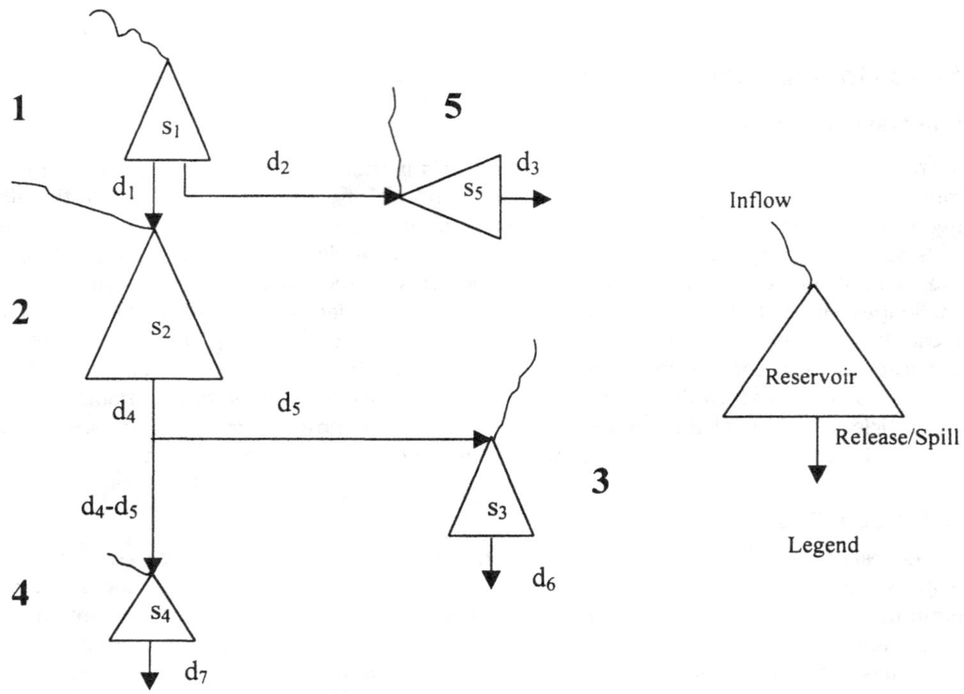

Figure 1 A schematic diagram of a multireservoir system

s_i^t available water in reservoir i at the beginning of period t

The following variables are defined, corresponding to reservoir i during period t

i_i^t natural inflow (stochastic)

e_i^t evaporation and seepage losses (nonlinear and stochastic)

The variables below are subtracted when leaving a reservoir, added when entering a reservoir, and equal zero when not connected to the reservoir. This is represented by the algebraic sum.

d_i^t deterministic release decision (deterministic), $\geqslant 0$

u_i^t surplus releases over the demand d_i^t (stochastic), $\geqslant 0$

l_i^t shortage in releases corresponding to d_i^t (stochastic), $\geqslant 0$

If the reservoir system can be modeled just with equation (1) and if linearity is assumed in all variables, then the resulting system is a linear system. In the limit (time interval $\rightarrow 0$) equation (1) is a stochastic differential equation. However, real-world reservoirs are doubly bounded, that is the state variable s_i^t is bounded by the minimum and maximum storage levels of a physical reservoir $s\ \text{min}_i^t$ and $s\ \text{max}_i^t$, respectively. Adding these we get

$$s\ \text{min}_i^t \leqslant s_i^t \leqslant s\ \text{max}_i^t. \tag{2}$$

Because of (2), both the superposition and scaling principle necessary for linearity is lost and hence a reservoir system is inherently nonlinear (for example, if you double a storage volume, the resulting volume may not be in the feasible space and hence the space of the storage variable is nonlinear). This is not a major problem in deterministic optimization but in the stochastic case this has caused much hardship, as will be explained later (Fletcher and Ponnambalam, 1996). See Georgakakos and Yao (1993) for a method using bounded sets. Also, release d_i^t can be doubly bounded. The surplus and shortage variables cannot both be simultaneously nonzero for any i and t and can be omitted when deterministic release decisions are feasible. Lastly, the losses due to evaporation are typically nonlinear functions of the available water and hence can also make the reservoir dynamic equation (1) nonlinear (Loucks, et al., 1981). The large-scale nature of the problem is due to potentially large values for N and T. The objective function is usually nonlinear and depends on the problem and will be explained later. An optimization model produces a solution for s, d, u, and l such that they are feasible and optimal (at least locally) for a given objective function. A linear programming formulation is straightforward; in addition to equations of form (1) and (2) a linear objective function in decision variables s, d, u, and l is used. We will describe a dynamic programming formulation next where application of equation (1) requires one state variable for each reservoir resulting in an N-dimensional recursive equation.

$$f^t(s_1^t, s_2^t, \ldots, s_N^t) = \max_{d_1^t, d_2^t, \ldots, d_N^t} \{C^t(d_1^t, d_2^t, \ldots, d_N^t, s_1^t, s_2^t, \ldots, s_N^t) + f^{t+1}(s_1^{t+1}, s_2^{t+1}, \ldots, s_N^{t+1})\}, \quad t = T, \ldots, 1 \tag{3}$$

Equation (3) is the well-known dynamic programming backward recursive equation where t represents the stage and the state variable transition from s_i^t to s_i^{t+1} is given by equation (1). The equation is solved backward starting from known boundary values for $f^{T+1}(\cdot)$. The function $f^t(\cdot)$ represents the total optimal cost from period t to the end of horizon T given that the system state is represented by s_i^t, $i = 1, \ldots, N$. The release decisions are functions of the current state in the current stage. The immediate cost from the decision d_i^t in period t is given by the function $C^t(\cdot)$ in (3), and $f^{t+1}(\cdot)$ represents the cost-to-go value due to the transition. We are assuming one decision variable d per reservoir that can be easily extended to multiple decision variables for each reservoir as in Figure 1. When the state variable is discretized, as is common in water reservoir problems, then equation (3) is referred to as discrete dynamic programming backward recursion. However, it is clear that if N is large, then the number of possible states at each stage can be exponentially large, resulting in the curse of dimensionality. When inflow is stochastic, the state transition from s_i^t to s_i^{t+1} is probabilistic given by the transition probability. Therefore the value within the brackets $\{\ldots\}$ is calculated as the expected value using transition probabilities. Such a formulation is called stochastic dynamic programming (SDP), which is even more computationally expensive for large N. In a later section we will describe how approximations of state vector in equation (3) are made to reduce the dimensionality from N to a smaller value. In the above model, the inflows in each period are assumed probabilistically independent, that is serial correlation is ignored. This is not a serious problem in long-term planning models but may not be acceptable for short-term (such as weekly) or real-time models. In order to consider serial correlation one needs to add more state variables that reflect the state of the inflow in previous periods, which makes the DP problem computationally expensive to solve.

1.2. Uncertainty Issues and Decomposition

In reservoir modeling, inflows and losses are considered uncertain. The case of demand uncertainty is usually considered only in the planning models (Ng and Kuczera, 1993). This, however, is changing due to deregulation of the power industry that now requires the consideration of variance in prices (dual multipliers) caused by demand variability. Deterministic models and some linear quadratic Gaussian (LQG) models are sufficient when the system is certainty equivalent, for example, when the system has quadratic cost functions, the system dynamic is linear, there are no inequality constraints, and uncertain inputs are independent and normally distributed. See Philbrick and Kitanidis (1999) for a study that describes the benefits of using an appropriate stochastic optimization model. The reservoir storage volume depends on these uncertain variables and hence is itself uncertain. Let us consider only the case of inflow uncertainty because the magnitude and variation of uncertainty in inflows are relatively much larger than in losses. Hydrologic models of inflows are used to understand the statistical nature of inflows (in many new areas where such data do not exist, data from similar areas are used). After a time-series analysis, any trend, periodic, and probabilistic distributions of inflows (including spatial and temporal correlation) as a function of the season in the year are determined. It is desirable to consider this information explicitly in all optimization models. One common assumption is that, in the absence of any trend, stochastic inflows are considered stationary (at least in the lower-order moments, such as mean and covariance—called weakly stationary). So yearly statistics remain the same throughout all time horizons. This also means that the reservoir storage volumes, which depend on this stationary process, are also assumed stationary, bringing further difficulties in stochastic optimization models such as the LQG based models (Fletcher and Ponnambalam, 1998). The advantage of an implicit stochastic optimization model is that deterministic models are solved many times for each sample of the random variable and the results are aggregated to get the optimal solution. The disadvantages are: (i) it requires a large amount of computing time; and (ii) determining a single suitable solution from the many solutions corresponding to random samples is nontrivial.

For practical reasons reservoir optimization models are divided into planning and design models, and operations models, although each of these require interactions with the other (see Figure 2).

Figure 2 presents the three-level decompositions generally based on horizons and time intervals of study, the input and output variables at each level, and the interactions that take place between the levels. A planning model is used to decide the demand over the planning horizon, the capacity required, and their sequencing. This model also requires the uncertain project costs and benefits over the planning horizon. The design model is used to estimate system yields where yield is defined as the probability of meeting certain demands in the long run. Planning and design models are also used to decide the network topology of reservoirs and their actual sizes, which depend on geography and hydrology. Typically, calculating yields requires the results of reservoir operations, which are available only from operating models. In order to make this process solvable, operating policies called rule curves are determined at the planning and design stages and verified and fine-tuned by operating models later.

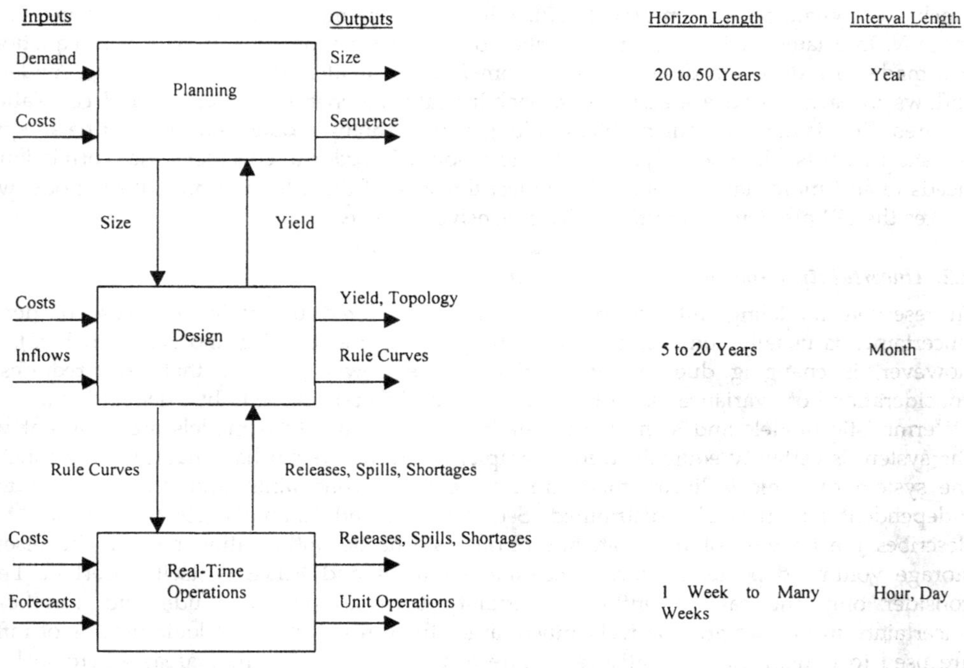

Figure 2 Overview of reservoir optimization problems

2. Planning and Design (Long Term)

In planning problems, the size of the reservoir(s) and flow or spillway capacities are the decision variables; the costs, the demand for water supply, or storage availability in the case of flood control, are considered stochastic. An implicit decision variable is network topology if there is more than one reservoir. In general, the problem is (i) what capacities to assign to each reservoir and (ii) what should the scheduling or timing of the construction of various reservoirs be such that total costs are minimized. In general a nonlinear integer programming problem must be solved. For practical reasons, the sizes of the various reservoirs are considered discrete and the feasible projects are divided into sets of projects such that a dynamic programming model can be used to solve the problem (Loucks et al., 1981). If there are three possible sizes for each project including the zero size and for example, five projects, then the number of feasible sets equal $3^5 = 243$, which is the number of discrete states in each stage of the dynamic programming problem. The objective function of the dynamic program minimizes the total discounted cost and the sequencing possibilities are given by the demand structure. It is clear that if either the number of reservoirs is large or the possible number of sizes for each reservoir is large, then the problem becomes very hard due to the curse of dimensionality. Global optimization algorithms, such as genetic algorithms, are one of the modern choices for solving this problem especially in the presence of integer variables (Orero and Irving, 1998).

2.1. *Calculation of the Reservoir Yield*

The typical design question is: given the demand and uncertain inflows, what would be the yield of a reservoir of a given size. Yield is defined as the probability of meeting or exceeding a given demand. In multireservoir systems both the topology of the network and the sizes of the various reservoirs and their interconnections are given. The earliest models used discrete Markov chains where the reservoir storage volume was divided into many discrete states. Transition probability matrices were determined based on the reservoir size and inflow distributions and releases. Using the transition probability matrix, it was straightforward to derive the stationary (or steady-state) probabilities of each discrete state where one of the states might define the full storage and one the empty storage. The yield of the reservoir is equal to 1 − probability of empty reservoir. While this method was quite adequate for a single reservoir system, for multireservoir systems one has to deal with the problem of dimensionality. Serial and cross correlations would make this problem even harder and are generally considered only at the verification stage using Monte Carlo simulations.

Alternatively, chance-constrained programming was used because it does not suffer from the curse of dimensionality. However, the reservoir sizes determined are usually found to be overestimates (Loucks et al., 1981). Recently, Fletcher and Ponnambalam (1996) used a new model to estimate yields and were able to offer a reasonably accurate estimate. This model did not suffer from dimensionality. The model is based on either a first- or second-order Taylor series approximation of combinations of equations (1) and (2) and the resulting equations are not only approximate but are highly nonlinear. Models based on LQG controllers do not provide accurate estimates for the probability distributions of storage, as LQG moment equations do not consider the reservoir bounds explicitly in the reservoir dynamic equation (Fletcher and Ponnambalam, 1998). It is possible to use stochastic linear programming (see Loucks et al., 1981) but it has a similar dimensionality problem as in stochastic dynamic programming.

2.2. *Approximate Methods*

When considering uncertainty explicitly (that is, using statistical information and not samples of inflows), in any of these methods, the only choice to reduce the dimensionality is to use some sort of approximation that leads to suboptimal solutions. The approximation could be in the modeling of the multireservoir system (Valdes et al., 1992; Saad et al., 1996) or in the solution process as in Bender's decomposition method (Pereira et al., 1985), to be explained later.

2.3. *Multilevel Approximate Model*

This model approximates the multireservoir system and is based on the aggregation-decomposition principle. It is usually solved using stochastic dynamic programming. Instead of the N-dimensional state vector in equation (3), a K-dimensional vector is used where K is of dimension 2 to 4 (Ponnambalam and Adams, 1996). The $K-1$ state variables are used to model the reservoir storage as in equation (3) and the Kth variable is used to represent the rest of the system as an aggregated storage reservoir. The aggregation function and the transition of the aggregated storage reservoir depend on the objective function of the original problem. For example, if it is maximizing hydropower energy, the aggregation coefficients are based on a nominal operation of hydropower generation. The size of K depends on how big a stochastic dynamic programming problem one can solve with the available computing capability; $K=2$ is easy and $K=4$ is possible even with the personal computers of today. Another modification of this method depends on deciding which of the variables (representing storages and inflows) should be chosen so that the dimensionality is reduced, or in what order the reservoirs should be chosen for aggregation so that the suboptimal solution is nearly as good as the optimal solution. In both cases, methods such as principal component analysis and neural networks are used, in addition to physically based models. In practice this method seems to be the most promising and is amenable to parallel or distributed computing for speedup.

2.4. *Benders Decomposition (Pereira and Pinto, 1985)*

Consider a two-stage linear programming problem

$$\begin{aligned} &\min_{x,y} c^T x + d^T y \\ &\text{subject to:} \\ &\quad Ax \geqslant b \\ &Ex + Fy \geqslant g, \end{aligned} \tag{4}$$

where T is a transpose operator, x and y are decision variables, c, d, b, and g are vectors of coefficients, and A, E, and F are appropriate matrices. If the dimensions of x and y are very large with g as the vector of stochastic coefficients, then the optimization problem is potentially large. Benders method decomposes equation (4) into two one-stage optimization problems where the second stage is parametrized iteratively with the first-stage solution x as the parameter:

$$\begin{aligned} &\text{First Stage:} \\ &\min_{x} c^T x + \alpha(x) \\ &\text{Subject to:} \\ &\quad Ax \geqslant b \\ &\pi^j(g - Ex) - \alpha \leqslant 0, \quad \forall j \\ &\text{Second Stage:} \\ &\alpha(x) = \min_{y} d^T y \\ &\text{Subject to:} \\ &\quad Fy \geqslant g - Ex \end{aligned} \tag{5}$$

where π^j is the dual multiplier of the second stage solution at jth iteration. In equation (5) the term $c^T x$ represents the immediate cost and the term $\alpha(x)$ represents the cost-to-go. The correspondence to the dynamic programming in equation (3) is immediate. So is the extension of (5) to the stochastic case when g is stochastic. This method is also called dual dynamic programming. The disadvantage of the method is that the number of solutions required to construct a good approximation is not known *a priori* and is generally large, especially in multiperiod problems. In the stochastic case, where second-stage solutions are required for many samples of g, use of parallel or distributed computing to find the solution corresponding to each sample can reduce the CPU time considerably. The extension of this method to nonlinear problems is still in its infancy.

2.5. *Hierarchical Decomposition (Haimes, 1977)*

Easily applicable hierarchical decomposition algorithms are based on the nonfeasible decomposition method where an upper-level optimization problem determines the prices of shared resources while the lower-level (one for each reservoir) optimization determines an optimal release based on the given price. The upper level objective function as a function of the upper level price is now constructed using the information obtained from the lower levels. A new direction for price is determined and sent to the lower levels. This iterative process is continued until convergence. The lower-level problem is the primal problem and the upper-level problem is the dual problem. A major advantage, in addition to the breaking of a large problem into many smaller subproblems, is the economic interpretation of the price used to coordinate the two levels. In essence, the price is used to make the demand equal the supply. In the stochastic case it will make the expected demand equal to the expected supply. A disadvantage of the method is the possibility of a duality gap, that is, the maximization of the dual solution does not produce the objective function value obtained from the minimization of the primal problems.

2.6. *Multiobjective Optimization*

There are many multiobjective optimization methods such as constraint-based, weighted objective functions, surrogate-tradeoff method, and so on. The weighted penalty function method is the most commonly used in practice. Multiple users are asked to weight from 0 to 1 the deviation from a target (for storages and releases) in which the penalty function is a linear piecewise function. For easy understanding, only a maximum of five segments is allowed. A convex supporting function (or a similar quadratic function) is generated that approximates the cumulative penalty function using weights for the various users' penalty functions. Weights may be decided depending upon their order of priority defined by a working group representing the users. After determining the optimal policy, Monte Carlo results are generated for the various alternative weights. Usually about five different choices are used, each of which may weigh a particular user's penalty function more than the others. A forum is arranged where the results are displayed for all the user groups. The results include simulated levels of storages and releases, total penalty for each group, total power generated, and variance of penalties. Each group is asked to rank the various results according to their own preferences. Based on these discussions new weights are formulated and corresponding optimal policies and simulated results are generated and brought back to the forum. Usually, two to three meetings are sufficient to determine a set of results acceptable to all. The weights and the optimal policy corresponding to these results are chosen for further operations. Because each group has different utilities for storage and releases this approach is preferable to converting benefits to purely economic values (or monetary values). For example, in the past, use of monetary values had made the optimal policies heavily biased toward hydropower generation. Recently, in addition to the maximization of total benefits, one of the multi-objectives, namely the minimization of variation of the benefits, has also been considered using stochastic dynamic programming (Liang et al., 1996; Ponnambalam et al., 1999).

3. Operations (Real-Time)

Long-term optimization (planning and design) is used to determine targets for reservoir storages and releases called the rule curves. Long-term optimization is also done using past data and assuming stationarity. In real-time optimization, however, deterministic or forecasted data is used to reflect current conditions (Yang et al., 1995; Brdys and Ulanicki, 1994; see

Tejada-Guibert et al., 1995, regarding the value of information). In addition, real-time cost coefficients may be given as an alternative to the penalty functions especially if the system is predominantly operated for energy generation. For example, if the system is part of a hydrothermal power system then a possible objective function may be to minimize the total cost of the thermal component of the power system by adjusting reservoir releases and storages. Of course, real-time optimization must be done to determine policies that are optimal in real-time but also optimal in the long term. For this reason, a horizon length is taken that is as long as possible (but a longer horizon makes the problem larger) and the end of the horizon function values for storages are taken from the long-term optimization. Another major technical problem in real-time optimization is that the time lag of flow from one reservoir to another becomes an important consideration in a multireservoir problem. This usually excludes the dynamic programming type methods that require additional state variables. Lastly, real-time optimization may include not only nonlinear objective functions and equations but also unit operations requiring integer variables, thus making the problem very time consuming to solve. We will provide a typical real-time optimization model assuming a hydrothermal system; if the system is not hydrothermal then long term penalty functions described above can be also used.

3.1. Real-Time Operations Model

We will consider a hydrothermal system with the real-time objective of minimization of thermal generation costs for a given horizon-end storage target in reservoirs to account for the long-term objective. In addition, the cumulative releases from the reservoirs in the real-time horizon (for example, a week) should meet the weekly target releases determined by the long-term optimization. There are hydro generation plants and thermal generation plants where the cost of hydro generation is assumed to be zero while the cost of thermal generation is given by a quadratic function. The total generation should satisfy the demand (if not, the surplus demand will be satisfied by buying power from external sources) and hydraulic flow constraints should also be satisfied. The cost of thermal generation depends on whether the unit is on or off and similarly the production of thermal and hydro generation depends on the respective units having been turned on. An added complexity is the ramping rates of some thermal units.

Additional variables to be defined are:

G_h Set of hydro generation plants

G_c Set of thermal generation plants

$S_{iu,i}$ Set of iu upstream reservoirs whose release flow into reservoir i

DH_i^t Power generation of ith hydro plant in period t (a nonlinear function of average reservoir volume)

DC_i^t Power generation of ith thermal plant in period t

DP^t Total power demand including transmission losses in period t

$\tau_{iu,i}$ Water transport delay from upstream reservoir iu to reservoir i

$\mathrm{F}_i(DC_i^t)$ Cost function of ith thermal unit

The objective function is:

$$\underset{d_i^t \in G_h, DC_i^t \in G_c}{\mathrm{Min}} \sum_{i \in G_c} \sum_t F_i(DC_i^t) \tag{6}$$

where, for convenience, we assume that each reservoir release d is used to generate power (if not, simply use a zero power generation function for such reservoirs). If there is a need to buy power from external sources, new variables and their corresponding cost coefficients should be added to equation (6). Also, in case of possible spills and shortages, appropriate variables as in equation (1) should be added with penalties for their violations. If the thermal unit i is not on, then the cost is zero; this feature introduces integer variables in the above problem. Some important constraints to add are

subject to:

- $\sum_{i \in G_h} DH_i^t + \sum_{i \in G_c} DC_i^t = DP^t, \quad t = 1, T$ (Power demand constraint)

- $s_i^{t+1} = s_i^t + i_i^t - e_i^t - d_i^t + \sum_{i \in S_{iu,i}} (d_i^{t - \tau_{iu,i}}), \quad i = 1, N; \quad t = 1, T,$

  (Reservoir volume continuity equation)

- $s_i^T = s_i^{\text{Target}}$

- $\sum_{t=1}^{T} d_i^t = d_i^{\text{Target}}$

- Other constraints include:
- Hydro- and thermal power generation lower and upper limits
- Reservoir storage volume limits
- Release limits
- Hydropower generation function
- Forecasting models

When all power units are on and continuously variable (eliminating integer variables) then many nonlinear programming algorithms are available. When N and T are large, the above is a large sparse nonlinear programming problem. Recent successful methods include the trust-region-based Lagrangian function methods (Sadjadi and Ponnambalam, 1999), although all of them can only guarantee a local optimal solution. New methods include fuzzy logic-based algorithms (Russell and Campbell, 1996) and genetic algorithm methods that are especially suitable when integer variables need to be considered (Orero and Irving, 1998). However, these methods are much slower than trust-region-based methods. In real-time optimization, speed is essential, as the above problem may have to be solved at every decision period using what is considered a rolling horizon. That is, after solving, only the first period decision is implemented, the horizon is shifted and the problem is solved again. This is essential if continuously forecasted data are to be used.

Table 1 Reservoir Optimization Problems and Suitable Mathematical Programming Methods

Problem	*Suitable methods*
Capacity expansion (planning)	Dynamic programming, integer programming, nonlinear programming, global optimization
Reservoir yield study (design)	Markov models, yield models, chance-constrained programming (linear and nonlinear)
Single reservoir operations (long-term)	Dynamic programming, network linear programming, stochastic linear programming
Single reservoir operations (real-time)	Linear programming, nonlinear programming
Multireservoir operations (long-term)	Aggregation-based dynamic programming, Bender's decomposition (Dual DP), neural networks
Multireservoir operations (real-time)	Linear, integer and nonlinear programming

4. Conclusions

Table 1 presents a summary of various reservoir optimization problems and some preferable solution methods. The challenges remain the same. That is, how to rapidly solve some of the largest nonlinear integer optimization problems faced. New research includes developing modern computing methods based on distributed computing. Promising directions also include hybrid algorithms that take advantage of the fast convergence to local solutions of continuous methods of optimization and global optimality of slow but more general algorithms. In addition, the stochastic nature of the problem may require good approximate models of real-world systems to reduce dimensionality.

ACKNOWLEDGMENTS

The author is grateful to Professor Jiri Vlach, Henry David Venema, and Seyed Jamshid Mousavi for their constructive suggestions.

References

Brdys, M. A. and B. Ulanicki. 1994. In *Operational Control of Water Systems: Structures, Algorithms, and Applications*. New York: Prentice Hall.

Fletcher, S. G. and K. Ponnambalam. 1998. "A constrained state formulation for the stochastic control of multi-reservoir systems." *Water Resource Res.* **34(2)**: 257–270.

Fletcher, S. G. and K. Ponnambalam. 1996. "Estimation of reservoir yield and storage distribution using moments analysis." *J. Hydrology* **182**: 259–275.

Georgakakos, A. P. and H. Yao. 1993. "New control concepts for uncertain water resources systems, 1. Theory." *Water Resource Res.* **29(6)**: 1505–1516.

Haimes, Y. 1977. In *Hierarchical Analyses of Water Resources Systems*. New York: McGraw-Hill Inc.

Liang, Q., L. E. Johnson, and Y-S. Yu. 1996. "A comparison of two methods for multiobjective optimization for reservoir operation." *Water Resource Bull.* **32(2)**: 333–339.

Loucks, D. P., J. R. Stedinger, and D. A. Haith. 1981. In *Water Resources Systems Planning and Analysis*. NJ: Prentice-Hall Inc.

Momoh, J. A., M. E. El-Hawary, and R. Adapa. 1999a. "A review of selected optimal power flow literature to 1993, Part I: Nonlinear and quadratic programming approaches." *IEEE Trans. Power Syst.* **14(1)**: 96–104.

Momoh, J. A., M. E. El-Hawary, and R. Adapa. 1999b. "A review of selected optimal power flow literature to 1993, Part II: Linear and interior point methods." *IEEE Trans. Power Syst.* **14(1)**: 105–111.

Ng, W. S. and G. Kuczera. 1993. "Incorporating demand uncertainty in water supply headwork simulation." *Water Resources Res.* **29(2)**: 469–477.

Niedda, M. and G. M. Sechi. 1996. "Mixed optimization technique for large-scale water-resources system." *ASCE J. Water Resource Plan. Mgmnt* **122(6)**: 387–393.

Niewiadomska-Szynkiewicz, E., K. Malinowski, and A. Karbowski. 1996. "Predictive methods for real-time control of flood operation of a multireservoir system: methodology and comparative study." *Water Resources Res.* **32(9)**: 2885–2895.

Orero, S. O. and M. R. Irving. 1998. "A genetic algorithm modelling framework and solution technique for short term optimal hydrothermal scheduling." *IEEE Trans. Power Syst.* **13(2)**: 501–518.

Pereira, M. V. P. and L. M. V. G. Pinto. 1985. "Stochastic optimization of a multireservoir hydroelectric system: a decomposition approach." *Water Resource Res.* **21(6)**: 779–792.

Philbrick, C. R. and P. K. Kitanidis. 1999. "Limitation of deterministic optimization applied to reservoir operations." *ASCE J. Water Resource Plan. Mgmnt* **125(3)**: 135–142.

Ponnambalam, K. and B. J. Adams. 1996. "Stochastic optimization of multireservoir systems using a heuristic algorithm: case study from India." *Water Resource Res.* **32(3)**: 733–741.

Ponnambalam, K., S. G. Fletcher, and A. W. Heemink. 1999. "Design and management of water resources systems: a unified approach with case studies." Manuscript. Ontario, Canada: Department of Systems Design Engineering, University of Waterloo.

Russell, S. O. and P. F. Campbell. 1996. "Reservoir operating rules with fuzzy programming, real-time operations." *ASCE J. Water Resource Plan. Mgmnt* **122(3)**: 165–170.

Saad, M., P. Birgas, A. Turgeon, and R. Duquette. 1996. "Fuzzy learning decomposition for the scheduling of hydroelectric power systems." *Water Resource Res.* **32(1)**: 179–186.

Sadjadi, S. J. and K. Ponnambalam. 1999. "Advances in trust region algorithms for constrained optimization." *Appl. Numer. Math.* **29**: 423–443.

Tejada-Guibert, J. A., A. J. Sharon, and J. R. Stedinger. 1995. "The value of hydrologic information in stochastic dynamic programming models of a multireservoir system." *Water Resources Res* **31(10)**: 2571–2579.

Valdes, J. B., J. M-D. Filippo, K. M. Strzepek, and P. J. Restrepo. 1992. "Aggregation-disaggregation approach to multireservoir operation." *ASCE J. Water Resource Plan. Mgmnt* **118(4)**: 423–444.

Yang, X., E. Parent, C. Michael, and P.-A. Roche. 1995. "Comparison of real-time reservoir-operation techniques." *ASCE J. Water Resource Plan. Mgmnt* **121(5)**: 345–351.

18.17 Optimization Problems in Air-Pollution Modeling

Ivan Dimov and Zahari Zlatev

ABSTRACT Environmental problems are becoming more and more important for modern society, and their importance will certainly be increased in the next century. High pollution levels (high concentrations and/or depositions of certain chemical species) may cause damage to plants, animals, and humans. Moreover, some ecosystems can also be damaged (or even destroyed) when pollution levels are very high. This is why pollution levels must be carefully studied in order to be able to predict the appearance of high pollution levels and/or to decide what can be done to prevent exceeding prescribed critical levels. Mathematical models can successfully be used to resolve these problems. Very often optimization problems have to be applied. The appearance of optimization problems in the field of air-pollution modeling and their importance will be discussed in this paper. Some applications of adjoint equations in the treatment of optimization problems arising in air-pollution modeling will be considered. We shall present a review of some approaches that are based on the adjoint equations formulation.

1. Need for Optimal Environmental Solutions

The control of the pollution levels in different highly developed and densely populated regions of Europe and North America is an important task for modern society. Its importance has been steadily increasing during the last two decades. The need to establish reliable control strategies for air-pollution levels will become even more important in the next century. Large-scale air-pollution models can successfully be used to design reliable control strategies.

The nature of the environmental problems is such that it is highly desirable (and sometimes necessary) to find optimal solutions. This fact can be illustrated by the following simple example. It is very easy to suggest a reduction of all emissions by some large amount (e.g., by 50 percent everywhere in Europe). Most of the environmental problems would certainly be resolved if such a decision was carried out in practice. However, there is a serious danger of creating new and even greater problems because the reduction of the emissions is an expensive process and could cause a great economic crisis in the society if unjustified large reductions were carelessly enforced. This is why optimal solutions are needed in this situation. One has (i) to justify the need for reductions, and (ii) to find out where to reduce the emissions and by how much to reduce them. These tasks can be solved successfully by developing and using reliable mathematical models for studying different pollution phenomena (Zlatev, 1995; Marchuk, 1985; McRae et al., 1984). These models must satisfy several important requirements (Zlatev et al., 1996):

1. The mathematical models must be defined on large space domains, because the long-range transport of air pollution is an important environmental phenomenon and high pollution levels are not limited to the areas where the high emission sources are located.
2. All relevant physical and chemical processes must be adequately described in the models used.

3. Enormous files of input data (both meteorological data and emission data) are needed.
4. The output files are also very big, and fast visualization tools must be used in order to represent the trends and tendencies, hidden behind many megabytes (or even many gigabytes) of digital information, so that even nonspecialists can easily understand them.

2. Mathematical Formulation of an Air Pollution Model

All important physical and chemical processes must be taken into account when an air-pollution model is to be developed. Systems of partial differential equations (PDEs) are often used to mathematically describe an air-pollution model. Consider a three-dimensional space domain Ω and assume that $c(r,t) \equiv (c_1(r,t), c_2(r,t), \dots c_q(r,t))^T$, where $r \equiv (x, y, z) \in \Omega$. Then the PDE systems are of the following type:

$$\frac{dc(r,t)}{dt} - A(r,t)c(r,t) = f(r,t), \tag{1}$$

where

$$A(r,t) \equiv (A_1(r,t), A_2(r,t), \dots, A_q(r,t))^T, \tag{2}$$

$$A_s(r,t)c_s = -\frac{\partial(uc_s)}{\partial x} - \frac{\partial(vc_s)}{\partial y} - \frac{\partial(wc_s)}{\partial z} + \frac{\partial}{\partial x}\left(K_x \frac{\partial c_s}{\partial x}\right) + \frac{\partial}{\partial y}\left(K_y \frac{\partial c_s}{\partial y}\right) + \frac{\partial}{\partial z}\left(K_z \frac{\partial c_s}{\partial z}\right)$$
$$-(\kappa_{1s} + \kappa_{2s})c_s, \tag{3}$$

$$f \equiv (f_1, \dots, f_q)^T, \tag{4}$$

and

$$f_s = E_s + Q_s(c_1, c_2, \dots, c_q), \quad s = 1, 2, \dots, q. \tag{5}$$

The different quantities that are involved in the mathematical model have the following meaning:

- the concentrations are denoted by c_s;
- u, v, and w are wind velocities;
- K_x, K_y, and K_z are diffusion coefficients;
- the emission sources in the space domain are described by the functions E_s;
- κ_{1s} and κ_{2s} are deposition coefficients;
- the chemical reactions are described by the nonlinear functions $Q_s(c_1, c_2, \dots, c_q)$.

The nonlinear functions Q_s, representing the chemical reactions in which the sth pollutant is involved, are of the form:

$$Q_s(c_1, c_2, \dots, c_q) = -\sum_{i=1}^{q} \alpha_{si} c_i + \sum_{i=1}^{q}\sum_{j=1}^{q} \beta_{sij} c_i c_j, \qquad s = 1, 2, \dots, q.$$

This is a special kind of nonlinearity (it is seen that the chemical terms are described by quadratic functions), but it is not clear if this property can efficiently be exploited. To the authors' knowledge, it is not exploited in the existing large-scale air-pollution models.

The models defined by (1)–(5) are traditionally used to calculate some concentration fields by using both meteorological and emission data as input (Zlatev et al., 1996). This gives an answer to the question: what are the concentration levels and/or the deposition levels caused by the existing emissions under the particular meteorological conditions that take place in the time period under consideration? However, it is much more important to study the question: how can the concentrations be kept under certain critical levels? Different types of optimization problems must be solved when this and similar questions are to be answered.

The formulation of some of the optimization problems arising in air-pollution modeling will be studied in the rest of this paper.

3. Data Assimilation in Air-Pollution Models

Initial concentration fields are needed when the model defined by (1)–(5) is to be treated numerically (and this is, of course, true for any other air-pollution model). The problem of finding good initial concentration fields is very difficult. Normally, one starts with some background concentrations and runs the model over a certain sufficiently large period (e.g., five days) to generate initial concentration fields. It is expected that the initial fields found in this way are good, but there is no guarantee that this is so. Moreover, this is an unnatural solution in the case where an air-pollution forecast for the next two to three days is to be produced (because the time to start up the model, five days, is much larger than the period of two to three days that is needed for the forecast).

An alternative approach is based on the use of time-series measurements to analyze the initial data that is to be used in the air-pollution model under consideration. The experiments indicate the remarkable fact that time series of only a few key chemical species are sufficient to produce initial fields (also for species that are not measured, by exploiting their coupling with the measured species).

It must be emphasized here that data assimilation is a very powerful approach. Not only can it be used to analyze the initial data (as sketched above), but also to obtain optimal estimations of (i) emission fields, (ii) deposition rates, and (iii) other model parameters. The use of data assimilation can be considered as applying a sequence of corrections performed at successive times in the treatment of the model. Each correction combines the appropriate background field of results produced by the model with new observations.

Data assimilation has been used successfully in meteorology. The basic ideas can be explained by sketching the application of data assimilation in this field. Before the development of data assimilation techniques in meteorology, the most successful device used was optimal analysis or optimal interpolation. Such a technique is presented in the works of McPherson (1979), Lorenc (1981), and Hollingsworth et al. (1985). The advantage of optimal analysis is that it provides a simple and internally consistent procedure for treating a large number of observations with different distributions, nature, and accuracy. However, optimal analysis, as it has been used, only takes into account the dynamics of meteorological processes very indirectly. It is because the temporal evolution of the statistical covariances of the predicted error is represented by a very simple low. This low is taken independently or almost independently of the current state of the wind field, which determines the real evolution of the predicted error. In particular, it ignores all effects due to advection. In the paper by Talagrand and Courtier (1987) and Courtier and Talagrand, 1987 a variational data assimilation technique is considered. This technique uses the adjoint equations of the model. This approach is based on the Lions' theory of optimal control (see Lions, 1971). Generally speaking, the theory of optimal control deals with the problem of how the output parameters of a given numerical model can be controlled by acting on the input parameters of the model. Among different approaches to control theory the approach of the adjoint equations is one of the most efficient. For meteorological problems this idea was first suggested by Marchuk (1974) to compute the local gradient of a multivariable function. The computed gradient is then used for performing a descent step in the space of initial conditions and the process is continued until some satisfactory approximation of the initial conditions minimizing the distance function in the corresponding functional space has been fulfilled. This specific application of adjoint equations was implemented in a model by Penenko and Obraztsov (1976), but the feasibility and usefulness of this approach was demonstrated in the works of Derber (1985) and Lewis and Derber (1985). In the works of Le Dimet and Talagrand (1986) and Agoshkov (1994) one can find a rather general presentation of the use of the adjoint equation technique in data assimilation analysis.

There are several different ways of implementing the data assimilation approach in the field of air-pollution modeling. Since the mid-1980s an approach based on the principle of optimal control theory has become more and more popular. In this approach the initial values are viewed as control parameters. Then a distance function is defined. This function provides weighted and accumulated distances between the available measurements and the values of the corresponding state variables calculated by the model during a predefined assimilation window. An optimization procedure is applied to minimize the distance function. This requires some knowledge of local gradients with respect to the initial state. To obtain these gradients one has to use the adjoint equation of (1). The procedure is normally called a four-dimensional variational data assimilation.

3.1. *Introduction of the Distance Function*

Consider a state variable x that is an element of a Hilbert space $\mathscr{H}$. The evolution in time of the state variable x can be expressed by the differential equation:

$$\frac{dx}{dt} = M(x), \tag{6}$$

where M is a nonlinear operator. Denote the measurements by $\hat{x}$ and let

$$\mathcal{O}(x(t)) \stackrel{\text{def}}{=} \tfrac{1}{2}(\hat{x}(t) - x(t))O^{-1}(\hat{x}(t) - x(t)),$$

where O is the covariance matrix of the observation and representativeness errors. By using this notation and Lagrangian multipliers $\lambda(t)$, the distance function can be defined by

$$\mathscr{J}(x(t)) \stackrel{\text{def}}{=} \int_{t_0}^{t_N} \left[\mathcal{O}(x(t)) + \left\langle \lambda(t), \frac{dx(t)}{dt} - Mx(t) \right\rangle \right] dt, \tag{7}$$

where the inner product $\int_\Omega f(x)g(x)dx$ of two functions f and g in the Hilbert space $\mathscr{H}$ is denoted by $\langle f, g \rangle$. It should be mentioned here that the background concentrations can also be used in the definition of the distance function by adding an extra term

$$\tfrac{1}{2}(x_b - x(t_0))^T B^{-1}(x_b - x(t_0))$$

to the right-hand side of (7).

3.2. *Calculating the Gradient of the Distance Function*

A perturbation equation (or a variational equation), corresponding to (6), can be defined by

$$\frac{d(\delta x)}{dt} = M'(\delta x), \tag{8}$$

where δx is some small deviation of x and M' is the tangential linear operator of M. Denote by g_1 and g_2 the inner products:

$$g_1 = \left\langle \delta\lambda(t), \frac{dx(t)}{dt} - Mx(t) \right\rangle,$$

$$g_2 = \left\langle \frac{\delta\lambda(t)}{dt}, \frac{d(\delta x(t))}{dt} - M'\delta x(t) \right\rangle.$$

In this notation, the gradient (the variation) of the distance function introduced in the previous paragraph can be defined as follows:

$$\delta\mathscr{J} \stackrel{\text{def}}{=} \int_{t_0}^{t_N} (\langle \delta\mathcal{O}, (\delta x(t)) \rangle + g_1 + g_2) dt.$$

After some modifications of this equation (the most important of which is the traditionally used integration by parts), the application of the extremal principle $\delta\mathscr{J} = 0$ results in the adjoint equation:

$$-\frac{d\lambda(t)}{dt} - M'^{*}\lambda(t) = O^{-1}(\hat{x}(t) - x(t)),$$

where the adjoint operator M'^{*} is defined by $\langle y, M'z \rangle = \langle M'^{*}y, z \rangle$.

In order to estimate the evolution in time of the given initial perturbation of the data at time t_0 we have to integrate equation (8):

$$\delta x(t_n) = R(t_n, t_0)\delta x(t_0), \quad n = 1, ..., N, \tag{9}$$

where $R(t_n, t_0)$ is the resolvent operator of M'. In general, we have to use some numerical integration scheme in order to calculate the resolvent operator $R(t_n, t_0)$. If this is done, then the resolvent operator is expressed as

$$R(t_n, t_0) = \prod_{i=0}^{n-1} R(t_{i+1}, t_i), \quad n = 1, ..., N,$$

where $R(t_{i+1}, t_i)$ is a sufficiently accurate numerical operator for stepwise integration.

Let us now consider the operator S as the resolvent of M'^*. From the duality principle it follows that

$$\langle \lambda(t_{t_N}), R(t_N, t_0)\delta x(t_0)\rangle = \langle S(t_0, t_N)\lambda(t_N), \delta x(t_0)\rangle$$

and

$$\nabla_{x(t_0)}\mathcal{J} = \sum_{m=1}^{N} \prod_{i=0}^{m-1} S(t_i, t_{i+1})\nabla_x \mathcal{O}(t).$$

It remains to be shown that $\nabla_{x(t_0)}\mathcal{J} = \lambda(0)$. After defining the backward initial condition $\lambda(t_N) = 0$ and applying the duality principle, one can get

$$-\frac{\partial}{\partial t} S(t, t')\nabla_{x(t')}\mathcal{O}(t') = -\frac{d\lambda(t)}{dt} = -M'^* S(t, t')\nabla_{x(t')}\mathcal{O}(t'). \tag{10}$$

After the integration of (10) one can get

$$\lambda(t) = \int_t^{t_N} S(t, t')\nabla_{x(t')}\mathcal{O}(t')dt'.$$

The computations concerning the calculation of the values of the distance function $\mathcal{J}(x(t_0))$ and of its gradient $\delta\mathcal{J}(x(t_0))$ can be carried out in two steps.

- *Forward step.* Some approximations of the state variable x and δx are calculated, from (6) and (8), respectively, for $t = t_1, t_2, ..., t_N$. The values of x are used to calculate the values of the distance function $\mathcal{J}[x(t_0)]$ in the same points.
- *Backward step.* The adjoint equation (or some transformed form of this equation) is integrated backward for $t = t_{N-1}, t_{N-2}, ..., t_0$ to obtain values of the perturbed state variable $\lambda(t)$ starting with $\lambda(t_N) = 0$. The values of $\lambda(t)$ and $\delta x(t)$ are used to calculate values of $\delta\mathcal{J}$.

The values obtained for $\mathcal{J}$ and $\delta\mathcal{J}$ are needed in the minimization procedure, which will be outlined in the next section.

3.3. *Minimizing the Distance Function and Beneficial Impact from the Assimilation of the Selected Species*

Different standard optimization algorithms can be applied to minimize the distance function. The algorithm proposed by Broyden–Fletcher–Goldfarb–Shanno (BFGS) is often used in this field in its limited-memory version. It should be stressed, however, that nonnegative solutions are required. Therefore, it is necessary to insert some constraints that ensure nonnegative solutions. A procedure of this kind is proposed by Bertsekas (1982).

Let us describe briefly a practical minimization procedure of the distance function in order to demonstrate to what extent measurements of a limited set of species can reduce the manifold of possible initial perturbations to fit to the observations. Let the observations of species $1, ..., p$ out of $p \leqslant q$ be available at N time steps. Denote by $\delta_p \hat{x}(t_n)$ the following vector:

$$\delta_p \hat{x}(t_n) = (\delta\hat{x}_1(t_n), ..., \delta\hat{x}_p(t_n))^T, \quad n = 1, ..., N, \tag{11}$$

where p is the number of the components of the vector and $\delta\hat{x}_i(t_n) = \hat{x}_i(t_n) - x_i(t_n)$, $i = 1, 2, ..., p$. The system (9) can be rewritten as

$$\delta_p\hat{x}(t_n) = [\mathbf{I}, \mathbf{0}]R(t_n, t_0), \quad n = 1, ..., N, \tag{12}$$

where $[\mathbf{I}, \mathbf{0}] \in \mathbb{R}^{p \times q}$ is the truncation operator composed of the identity matrix $\mathbf{I} \in \mathbb{R}^{p \times p}$ and the zero matrix $\mathbf{0} \in \mathbb{R}^{p \times (q-p)}$ (for the case $q > p$). In system (12) the rows of unobserved species are truncated. It is convenient to comprise the matrices of the resolvent operator and the truncation operator in a matrix $\mathcal{R}(t_n, t_0)$

$$\delta_{Np}\hat{x}(t) = \mathcal{R}(t_n, t_0)\delta_{Np}x(t_0), \quad n = 1, ..., N, \tag{13}$$

where $\mathcal{R}(t_n, t_0) \in \mathbb{R}^{Np \times q}$, $\delta_{Np}x(t_0) \in \mathbb{R}^{Np}$, and

$$\delta_{Np}\hat{x}(t) = (\{\delta_p\hat{x}(t_1)\}^T, ..., \{\delta_p\hat{x}(t_N)\}^T)^T \in \mathbb{R}^{Np}$$

(the row vectors $\delta_p\hat{x}(t_n)$, $n = 1, ..., N$ are defined by (11)).

It is easy to see that for $Np < q$ the system (13) is underdetermined. In the case where $Np > q$ the system (13) is overdetermined. In this case one can estimate the condition numbers of the matrix $\mathcal{R}'(t_n, t_0) \in \mathbb{R}^{q \times q}$, part of $\mathcal{R}(t_n, t_0)$, which is a square matrix of size $q \times q$, that is $cond(\mathcal{R}'(t_n, t_0))$ with any value of $mh_i = q$, where h_i is any combination of measured species and m is the number of the time steps. The condition numbers $cond(\mathcal{R}'(t_n, t_0))$ for different combinations give us an estimate of what beneficial impact may be expected from the assimilation of the selected species i and observation times m. These condition numbers can provide some evidence of reasonable weighting of the observations, which is an important task. For more details see Elbern et al. (1997).

4. Keeping the Concentrations in a Given Sensitive Area Under a Prescribed Level

The problem of environmental protection of a given sensitive area is an optimization problem. This problem can be formulated as follows. Let a sensitive point $r_1 \in \Omega$ be given. Assume that the concentrations at point $r_1 \in \Omega$ in time $t_1 \in [0, T]$ should not exceed some critical (for point r_1) level C. This means that the requirement

$$c(r_1, t_1) \leqslant C \tag{14}$$

must be satisfied. Assume also that some source of a prescribed power $E(t)$ is to be located somewhere in Ω. The problem is to find a subdomain ω of Ω where the source is located, so that condition (14) is satisfied. This problem could also be considered as a problem of optimal planning in an effort to achieve sustainable development.

It is convenient to simplify the problem in order to facilitate the understanding of the main principles that can be used to find the fundamental solution of the adjoint problem. This can be done as follows. Consider the solution of $1D$-transport diffusion problem:

$$\left(\frac{\partial}{\partial t} - A\right)c = \frac{\partial c}{\partial t} - \frac{\partial}{\partial x}(\alpha x c) - D\frac{\partial^2 c}{\partial x^2} = E(t)\delta(x - x_0), \tag{15}$$

$$c = c_0(x), \quad \text{for} \quad t = 0. \tag{16}$$

Assume that $c(x, t) \in W_1^2$, where W_1^2 is a Sobolev space, $c(x, t) > 0$, $D > 0$, $\alpha = \text{const}$, $-\infty < x < \infty$, $t \in [0, T]$, and the conditions providing the existence of the unique solution of the problem (15)–(16) in a weak formulation are fulfilled.

It is not possible to directly obtain an analytical solution to the problem defined by (15)–(16), due to the presence of the term $E(t)\delta(x - x_0)$. In addition, standard numerical methods used to solve the equation (15) are not efficient because there is a δ-function in (15).

The condition $c(x, t) \in W_1^2$ implies that $\int_0^T dt \int_{-\infty}^{\infty} c^{(2)}(x, t)dx < \infty$, which is a natural condition.

In fact, the operator used in (15) contains a rather artificial wind field defined by $u = \alpha x$. Nevertheless, we demonstrate in this way how the approach under consideration can be used for problems with a nonconstant advection.

The problem of finding the subdomain ω is not so complicated if the solution of the adjoint formulation of the original problem (15)–(16) is used. The conditions formulated here will be relaxed in Section 4.3.

4.1. Adjoint Formulation

Multiply equation (15) by a function c^*, which will be defined later (see the paragraph after equations (25)–(26)). Integrate the resulting equation in time and space:

$$\int_0^T dt \int_{-\infty}^{\infty} c^* \left(\frac{\partial c}{\partial t} - \frac{\partial}{\partial x}(\alpha x c) - D \frac{\partial^2 c}{\partial x^2} \right) dx = \int_0^T E(t) dt \int_{-\infty}^{\infty} c^* \delta(x - x_0) dx. \tag{17}$$

Assume that $c = c^* = 0$ when $x \to \infty$ or $x \to -\infty$. Apply integration part by part to (17). The following results will be obtained in this way:

$$\int_0^T dt \int_{-\infty}^{\infty} c^* \frac{\partial c}{\partial t} dx = \int_{-\infty}^{\infty} c c^* dx \Big|_{t=0}^{t=T} - \int_0^T dt \int_{-\infty}^{\infty} c \frac{\partial c^*}{\partial t} dx$$

$$= - \int_{-\infty}^{\infty} c_0(x) c_0^*(x) dx - \int_0^T dt \int_{-\infty}^{\infty} c \frac{\partial c^*}{\partial t} dx, \tag{18}$$

$$\int_0^T dt \int_{-\infty}^{\infty} c^* \frac{\partial^2 c}{\partial x^2} dx = \int_0^T \left(c^* \frac{\partial c}{\partial x} - c \frac{\partial c^*}{\partial x} \right) dx \Big|_{x=-\infty}^{\infty} + \int_0^T dt \int_{-\infty}^{\infty} c \frac{\partial^2 c^*}{\partial x^2} dx, \tag{19}$$

and

$$\int_0^T dt \int_{-\infty}^{\infty} c^* \frac{\partial (xc)}{\partial x} dx = \int_{-\infty}^{\infty} x c c^* dx \Big|_{t=0}^{t=T} - \int_0^T dt \int_{-\infty}^{\infty} x c \frac{\partial c^*}{\partial x} dx$$

$$= - \int_{-\infty}^{\infty} x c_0(x) c_0^*(x) dx - \int_0^T dt \int_{-\infty}^{\infty} x c \frac{\partial c^*}{\partial x} dx. \tag{20}$$

From (18), (19), and (20), one can obtain:

$$\int_0^T dt \int_{-\infty}^{\infty} c \left(- \frac{\partial c^*}{\partial t} + \alpha x \frac{\partial c^*}{\partial x} - D \frac{\partial^2 c^*}{\partial x^2} \right) dx - \int_{-\infty}^{\infty} (1 - x) c_0(x) c_0^*(x) dx = \int_0^T E(t) c^*(x_0, t) dt. \tag{21}$$

Assume that function c^* satisfies the following adjoint equation:

$$- \frac{\partial c^*}{\partial t} + \alpha x \frac{\partial c^*}{\partial x^*} - D \frac{\partial^2 c}{\partial x^2} = \psi(x, t),$$

with an initial condition $c^*(x, T) = 0$, where the function $\psi(x, t)$ will be defined later.

It is possible to rewrite (21) in the following form:

$$\int_0^T dt \int_{-\infty}^{\infty} \psi(x, t) c(x, t) dx = \int_0^T E(t) c^*(x_0, t) dt + \int_{-\infty}^{\infty} (1 - x) c_0(x) c_0^*(x) dx = J. \tag{22}$$

In fact, functions $c(x, t)$ and $c^*(x, t)$, as well as functional J, depend on parameters x_0, x_1, and t_1, so that $c(x, t) = c(x, t, x_0, x_1, t_1)$, $c^*(x, t) = c^*(x, t, x_0, x_1, t_1)$, and $J = J(x_0, x_1, t_1)$. When we do not use these parametric dependences, they will be omitted.

It is necessary to find an efficient numerical method to compute J. One can choose the function $\psi(x, t)$ in the form of a product of two δ-functions: $\psi(x, t) = \delta(x - x_1)\delta(t - t_1)$.

Let us also assume that $E(t)$ can be represented as

$$E(t) = e(t) H(t), \tag{23}$$

where $H(t)$ is a Heaviside function ($H(t) = 1$ for $t > 0$ and $H(t) = 0$ for $t \leqslant 0$). The second term of the right-hand side of (22) vanishes when such an assumption is made.

Let us now consider the following linear functional of the solution of the original Equations (15)–(16):

$$J = \int_0^T dt \int_{-\infty}^{\infty} \psi(x,t)c(x,t)dx. \tag{24}$$

Under the above assumptions, we have $J = \int_0^T dt \int_{-\infty}^{\infty} c(x,t)\delta(x-x_1)\delta(t-t_1)dx = c(x_1,t_1)$.

Obviously, the value of the functional (24) is equal to the value of the concentration in point $x = x_1$ at moment in time $t = t_1$. From the representation (22) it follows that instead of solving the original problem (Equations (15)–(16)) one can solve the adjoint problem for c^*:

$$-\frac{\partial c^*}{\partial t} + \alpha x \frac{\partial c^*}{\partial x} - D\frac{\partial^2 c^*}{\partial x^2} = \delta(x-x_1)\delta(t-t_1) \tag{25}$$

$$c^* = 0, \quad \text{for} \quad t = T. \tag{26}$$

From the solution of the problem formulated by (25)–(26) one can estimate the functional (22) as a function depending of a parameter x_0 and solve the problem under consideration. In fact, if the function $c^*(x, t, x_0, x_1, t_1)$ is the solution of the adjoint problem (25)–(26), then the presentation (22) defines a function of $x_0 - J(x_0, x_1, t_1)$. Now, the set ω of points x_0 can be determined by solving the inequality $J(x_0, x_1, t_1) \leqslant C$. That is the solution of the problem under consideration.

4.2. Evaluating the Functionals

One needs an efficient method for evaluating the linear functionals

$$\int_0^{t_i} e(t)c^*(x_0, t)dt \tag{27}$$

of the solution to the adjoint problem (25)–(26) ($e(t)$ is defined by (23)).

It is efficient to apply a statistical numerical method in the evaluation of (27). This is because the complexity of the statistical numerical methods for computing a linear functional of the solution is very low, of the same order as the complexity of the solution at only one point of the domain (see Sobol, 1973; Dimov and Tonev, 1993a). The statistical methods give statistical estimates for the functional of the solution by performing samples of a certain random variable, the mathematical expectation of which is the desired functional. These methods have proved to be very efficient in solving multidimensional problems in composite domains (Curtiss, 1954, 1956; Hammersley and Handscomb, 1964; Sobol, 1973; Ermakov et al., 1984). Moreover, it has been shown that for some problems (including one-dimensional ones) the statistical methods have a better convergence rate in the corresponding functional spaces than the optimal deterministic methods in the same functional spaces (Bahvalov, 1964; Dimov and Tonev, 1989, 1993b; Ermakov, 1982). It is also very important that the statistical methods are very efficient when parallel or vector processors or computers are available, because the above-mentioned methods are inherently parallel and have loose dependencies. They are also well vectorizable. By using powerful parallel computers, it is possible to apply a Monte-Carlo method or a particle-tracking method for evaluating large-scale nonregular problems, which are sometimes difficult to solve by traditional numerical methods.

The difficulties with finite difference and finite element methods are caused by the nonregularity of the right-hand side of Equation (25) and by artificial diffusion (or diffusion caused by the numerical scheme). This explains why the functional (27) can be evaluated by using a statistical numerical method when A is a general linear operator. In our case, the fundamental solution of the adjoint equation can be used.

In fact, introduce a new variable $t^* = T - t$, where $t^* \in [0, T]$, and present the adjoint problem in the following form:

$$\frac{\partial c^*}{\partial t^*} + \alpha x \frac{\partial c^*}{\partial x} - D\frac{\partial^2 c^*}{\partial x^2} = \delta(x-x_1)\delta(T-t^*-t_1) \tag{28}$$

$$c^* = 0, \quad \text{for} \quad t^* = 0, \quad \forall x.$$

The fundamental solution of Equation (28) can be found by using the techniques discussed by Bitzadze (1982) and Tikhonov and Samarski (1977):

$$\hat{c}^*(x,t) = \frac{1}{[\pi a(t)]^{1/2}} \exp\left\{ - \frac{[x - b(t)]^2}{a(t)} \right\}, \tag{29}$$

where $a(t) = \frac{2D}{\alpha}(1 - e^{-2\alpha t})$ and $b(t) = x_0 e^{-\alpha t}$.

The solution of the adjoint problem can be expressed in the following form:

$$c^*(x,t^*) = \int_0^T dt^* \int_{-\infty}^{\infty} \hat{c}^*(x-\xi, t^* - \tau)\delta(\xi - x_1)\delta(T - \tau - t_1)d\xi d\tau. \tag{30}$$

The result of integration (30) is the function defined by (29) with new arguments:

$$c^*(x - x_1, t_1 - t) = \frac{1}{[\pi a(t_1 - t)]^{1/2}} \exp\left\{ - \frac{[x - x_1 - b(t_1 - t)]^2}{a(t_1 - t)} \right\}. \tag{31}$$

Substituting (31) into the functional (27) one can get:

$$\int_0^{t_1} e(t) \frac{1}{[\pi a(t_1 - t)]^{1/2}} \exp\left\{ - \frac{[x_0 - x_1 - b(t_1 - t)]^2}{a(t_1 - t)} \right\} dt = J(x_0, x_1, t_1).$$

Under the assumptions formulated in this section the problem has been reduced to a problem of numerical integration. We have to evaluate a 1-D integral in our particular case. For this purpose we do not need to use a special statistical numerical method since the problem is simplified. The proposed numerical method is based on a numerical integration scheme for evaluating the last parameterized functional. Using a Newton–Cotes quadrature one can get

$$J = \sum_{i=0}^{k-1} e_i \frac{1}{[\pi a(t_1 - \tau_i)]^{1/2}} \exp\left\{ - \frac{[x_0 - x_1 - b(t_1 - \tau_i)]^2}{a(t_1 - \tau_i)} \right\} (\tau_{i+1} - \tau_i) + O(\Delta\tau), \tag{32}$$

where $\Delta\tau = \max_i(\tau_{i+1} - \tau_i)$ is the maximal time step and $e_i = e(\tau_i)$. The numerical scheme (32) is of the order $O(\Delta\tau)$. It is possible to apply more complicated schemes of high-order accuracy (say, $O((\Delta\tau)^2)$, or $O((\Delta\tau)^4)$). But from a computational point of view, when the integrand is a sufficiently smooth function, which corresponds to the case under consideration, it is better to use an adaptive numerical scheme. In our practical computations an adaptive numerical scheme is used. The integration starts with a regular coarse grid with a low number k of grid points. After evaluating the first coarse approximation to the functional (32), the total and local *a posteriori* variances are estimated. The information obtained is used for refinement of the mesh—in the subregions in which the local variance is too large the mesh is refined, such that after each step of refinement the number of nodes (mesh points) increases two times. The refinement procedure is continued until the required accuracy of the integration is reached or until the maximum number of evaluations is executed.

4.3. Application to Other Problems

The approach studied can be easily applied to transport problems with another linear operator A. One can, for example, include deposition terms kc in the operator considered in the previous subsections:

$$\left(\frac{\partial}{\partial t} - A\right)c = \frac{\partial c}{\partial t} - u\frac{\partial c}{\partial x} + \kappa c - D\frac{\partial^2 c}{\partial x^2} = E(t)\delta(x - x_0), \tag{33}$$

$$c = c_0(x), \quad \text{for} \quad t = 0, c = 0, \quad \text{for} \quad x \to \infty, \quad \text{and} \quad x \to -\infty.$$

Using the technique of the adjoint equations described above, one can reconstruct the location and power of a steady-state point source from two observations $c(x_1, t_1)$ and $c(x_2, t_2)$. A possible application would be the identification and characterization of a pollutant source in an inaccessible area. This formulation leads to a well-conditioned problem and can be generalized to a large class of more complicated cases (see Dimov et al., 1996).

Here we consider the following

Problem. *Find the place of location* x_0 *and the power e of the source* $E(t) = E = eH(t)$ ($H(t)$ *is a Heaviside function) if two measurement data* $c(x_1, t_1)$ *and* $c(x_2, t_2)$ *are available (it is assumed, that* $x_1 \neq x_2$).

Let the power of source e as well as its location x_0 be unknown. Two measurement data points $c(x_1, t_1)$ and $c(x_2, t_2)$ are assumed to be available. One can then define two functions of x_0, $F_1(x_0)$ and $F_2(x_0)$, using the following formal presentations:

$$F_i(x_0) = \frac{\hat{J}(x_0, x_i, t_i)}{c(x_i, t_i)}, \quad i = 1, 2,$$

where

$$\hat{J}(x_0, x_i, t_i) = \frac{J(x_0, x_i, t_i)}{e}, \quad i = 1, 2.$$

(It is possible to do this because e is strongly positive.)

One can see that x_0 is the solution of equation

$$F_1(x) = F_2(x),$$

being the point of the location of the source of pollutants. The value of the power of the source e can be determined using the expression:

$$e = \frac{1}{F_i(x_0)}.$$

The problem consists of using an efficient numerical method for evaluating the functional

$$\int_0^{t_i} ec^*(x_0, t)dt.$$

Such a method for finding the place of location and the power of the sources for problem (33) is formulated and experimented with in Dimov et al. (1996, 1997). Computational results are given in Dimov et al. (1996) for different values of D/u, where D is the diffusion coefficient and u is the wind velocity (see (33)).

5. Optimization Problems in the Theory of Measurements and Planning the Experiments

Suppose that n functionals J_{p_i}, $i = 1, \ldots, n$, are available. For every functional J_{p_i}, we consider the corresponding weight function $u^*_{p_i}$, $i = 1, 2, \ldots, n$ for a nonperturbed problem. [It should be emphasized that in this section the functionals J_{p_i} are measurements, u and v are weight functions, not wind velocities, and p_i are also functions.] We consider a model in which the operator L, as well as the domain of definition are supposed to be known. Consider the following n problems:

$$L^* u^*_{p_i} = p_i, \quad i = 1, 2, \ldots, n.$$

Let us find n weight functions $u^*_{p_i}$, $i = 1, 2, \ldots, n$ and solve the following problem:

$$Lu = E. \tag{34}$$

The nonperturbed operator L is adjoint to the operator L^*. In our consideration, we assume that $u \in U$ and $u^* \in U^*$, where U and U^* are the domains of definition of the operators L and L^*, respectively.

The goal is to reconstruct the coefficients α'_k and β'_k in the expression, by which the perturbed operator L' is defined:

$$L' = \sum_{k=1}^{m} \{\alpha_k' A_k + B_k(\beta_k' C_k)\}, \tag{35}$$

where A_k, B_k, and C_k are simple linear operators (for example, operators of differentiation, integration, and so on), and $\alpha_k'(x)$ and $\beta_k'(x)$ are perturbation coefficients.

Use the theory of small perturbations to construct n formulas:

$$(u_{p_i}^*, \delta L u) = -\delta J_{p_i}, \quad i = 1, 2, ..., n, \tag{36}$$

where δL is the difference between the perturbed operator L' and the nonperturbed operator L.

Assume that the operator L can be presented as

$$L = \sum_{k=1}^{m} \{\alpha_k A_k + B_k(\beta_k C_k)\}, \tag{37}$$

where $\alpha_k(x)$ and $\beta_k(x)$ are unknown coefficients that are usually approximately known for nonperturbed (model) problem.

Using expressions (37) and (35) one can get

$$\delta L = \sum_{k=1}^{m} \{\delta\alpha_k A_k + B_k(\delta\beta_k C_k)\}, \tag{38}$$

where

$$\delta\alpha_k = \alpha_k' - \alpha_k \quad \text{and} \quad \delta\beta_k = \beta_k' - \beta_k.$$

Substituting (38) into (36) one can get the following system of equations:

$$\sum_{k=1}^{m} [(u_{p_i}^*, \delta\alpha_k A_k u) + (B_k^* u_{p_i}^*, \delta\beta_k C_k u)] = -\delta J_{p_i}, \tag{39}$$

$$i = 1, 2, ..., n.$$

Now the problem consists of parameterization of the variances $\delta\alpha_k$ and $\delta\beta_k$. It has been shown (Marchuk, 1980) that if $\delta\beta_k = 0$ and $\delta\alpha_k$ are constants, then (39) is reduced to the linear algebra problem

$$\Lambda y = F, \tag{40}$$

where Λ is a matrix with elements $\lambda_{ik} = (u_{p_i}^*, A_k u)$, F is a vector with components $-\delta J_{p_i}$ and y is a vector with components $\delta\alpha_k$.

If the number of functionals (measurement data) n is equal to the number of varied coefficients α_k, then the system allows us, in principle, to determine $\delta\alpha_k$. If n is greater than m, the solution could be found by minimizing the following quadratic functional:

$$\|\Lambda y - F\|_2^2 = \min, \tag{41}$$

where $\| \cdot \|_2^2$ is the Euclidean norm.

This optimization problem can be efficiently solved using a gradient optimization procedure like the procedure described in Section 3.2.

This approach could be extended for the case when $\delta\alpha_k(x)$ and $\delta\beta_k(x)$ are functions. In this case, the solution of the inverse problem could be found using the parameterization approach. In this approach we suppose that as a result of the statistical and correlation analysis, orthogonal systems of functions $u_{k,l}(x)$ and $v_{k,l}(x)$ could be found. Using such functions, one can obtain good approximations to functions α_k and β_k for a sufficiently small $n(k)$ (n depends on k), so that

$$\delta\alpha_k(x) = \sum_{l=1}^{n(k)} a_{k,l} u_{k,l}(x), \tag{42}$$

$$\delta\beta_k(x) = \sum_{l=1}^{n(k)} b_{k,l} v_{k,l}(x), \tag{43}$$

where $a_{k,l}$ and $b_{k,l}$ are unknown coefficients. Substituting (42) into (39), one can get

$$\sum_{k=1}^{m} \sum_{l=1}^{n(k)} [a_{k,l}[u^*_{p_i}, u_{k,l} A_k u) + b_{k,l}(B^*_k u^*_{p_i}, v_{k,l} C_k u)] = -\delta J_{p_i}, \quad i = 1, 2, ..., n. \tag{44}$$

It is easy to see that system (44) is equivalent to system (40). Then one could again consider the minimization problem (41).

Consider the problem of planning a complicated experiment in environmental science. Let us formulate the following

Problem. *Consider a set of all possible (practically realizable) experiments and choose the subset that is best, from the point of view of solving the given inverse problem of reconstruction of the parameters of the environment (that is, coefficients of the equations).*

Such a problem is very difficult and does not have a solution in the general case. But there are some special approaches to this problem. Suppose that one considers the model of nonperturbed problem (34) before doing the experiment and that one describes linear functionals of the solution. One can find the necessary accuracy for measuring the functionals by taking into account the information about measurement accuracy. Assume that some prescribed requirements to the accuracy of the measurements are fulfilled, that is, $\|\delta J_{p_i}\| < \varepsilon$, where $\varepsilon > 0$ is a given small value. Then one could consider different sets of experiments and choose the set that leads to the matrix Λ defined by (40) with the best condition number (i.e., the Λ with the smallest $\|\Lambda\|_2 \|\Lambda^\dagger\|_2$, where $\|\Lambda\|_2$ is any Euclidean norm of the matrices Λ and $\Lambda^\dagger$ is the pseudoinverse of the matrix Λ defined by $\Lambda^\dagger = (\Lambda^T \Lambda)^{-1} \Lambda^T$). In fact, by all this we are assuming that the system of linear algebraic equations can be solved and that the solution found is sufficiently accurate. Such a plan for the experiments (the subset of the experiments leading to matrix Λ with the smallest condition number) can be considered as optimal among the set of subsets considered.

More details about this approach, as well as some applications, can be found in the works of Marchuk (1994) and Marchuk and Agoshkov (1993).

6. Optimization of Emissions from Enterprises by Using Some Economical Criteria

Here we consider the problem of determining the quantity of enterprise emissions, such that the concentrations of any pollutant of a given region will be less than or equal to the pollution level, which ensures that the ecosystem will not be destroyed. Another problem that is very close to this one is the problem of optimization of the location of new enterprises so that the pollution of a given region will not exceed prescribed critical levels that are damaging for plants, animals, and humans. We are trying to determine emission levels that do not damage the ecosystem, but are not too low, in order to achieve the maximum economical efficiency for the given restrictions. It is important to consider such kinds of optimization problems especially for Central and East European countries, which must quickly achieve sustainable development of their industry and agriculture.

Let us consider the region Ω and its boundary $\partial\Omega$. Let n enterprises (or sources of pollutants) A_i be located in the points r_i, $i = 1, 2, ..., n$. Assume that all sources can be considered as point sources E_i ($i = 1, 2, ..., n$). Introduce m ecological domains ω_k, $k = 1, 2, ..., m$. Suppose that the limits of the pollution levels for the integrated interval of time $[0, T]$ are defined for every ecological domain.

In this case the model of air pollution transport for a fixed sth pollutant (1) can be presented in the following form

$$\frac{dc(r,t)}{dt} - A(r,t)c(r,t) = \sum_{i=1}^{n} E_i \delta(r - r_i). \tag{45}$$

Consider the functional induced by the domain ω_k.

$$Y_{sk} = \int_0^T dt \int_{\omega_k} p_c c_s d\omega, \quad k = 1, ..., m. \tag{46}$$

Thus, functional (46) describes the admissible pollution levels in the ecodomain ω_k. The problem is to find the set of planned emissions E_i ensuring

$$Y_{sk} \leqslant (C_s)_k, \quad k = 1, 2, ..., m, \tag{47}$$

for minimal economical losses for technical reconstruction of the enterprises, where $(C_s)_k$ is the critical level of the considered sth pollutant for the domain ω_k.

It is clear that in this case we have to consider the minimizing functional

$$I = \int_{i=1}^{n} \xi_i (\hat{E}_i - E_i) \tag{48}$$

together with constraints (47), where $\hat{E}_i$ and E_i are the initial and planning power of the emissions, ξ_i describes the economical investments in technologies needed to achieve the same level of production with the reduced emissions (calculated per unit of emission power). Then the functional (48) describes the losses needed to improve the technologies of all A_i enterprises in order to switch from emission powers $\hat{E}_i$ to the planned emission powers E_i. This means that we can formulate the following minimization problem: find the power of the sources E_i in model (45) such that

$$I = \int_{i=1}^{n} \xi_i (\hat{E}_i - E_i) = \min, \tag{49}$$

$$Y_{sk} \leqslant (C_s)_k, \quad k = 1, 2, ..., m. \tag{50}$$

The problem (45), (49), (50) can be considered as a linear programming problem (Gill et al. 1982). There are two approaches to solving this problem: the first one uses the original equations, the second one is based on the theory of the adjoint equations. The approach that is based on the theory of the adjoint equations can be derived by using a technique that is similar to that presented in Section 4.

ACKNOWLEDGMENTS

This study was partially supported by grants from NATO (OUTR. CGR.960312), the European Union (WEPTEL Esprit 22727 and EUROAIR Esprit 24618), and the Bulgarian Ministry of Education, Science and Technology (I-811/98). A grant from the Danish Natural Sciences Research Council gave us access to all Danish supercomputers.

References

Agoshkov, V. 1994. "Control theory approaches in: data assimilation processes, inverse problems and hydrodynamics." In *HERMIS'94*, edited by E. A. Lipitakis Vol. 1: pp. 21–39. Helenic Mathematical Society, Printed in Hellas.

Bahvalov, N. S. 1964. "On the optimal estimations of convergence of the quadrature processes and integration methods." In *Numerical methods for solving differential and integral equations*, pp. 5–63. Moscow, Russia: Nauka.

Bertsekas, D. P. 1982. In *Constrained Optimization and Lagrange Multiplier Methods*. New York: Academic Press.

Bitzadze, A. V. 1982. *Equations of the Mathematical Physics*. Moscow, Russia: Nauka.

Courtier, P. and O. Talagrand. 1987. "Variational assimilation of meteorological observations with the adjoint vorticity equation. II: Numerical results." *Q. J. R. Meteorol. Soc.* **113**: 1329–1347.

Curtiss, J. H. 1954. "Monte Carlo methods for the iteration of linear operators." *J. Math Phys.* **32(4)**: 209–232.

Curtiss, J. H. 1956. "A theoretical comparison of the efficiencies of two classical methods and a Monte Carlo method for computing one component of the solution of a set of linear algebraic equations." In *Proceedings of Symposium on Monte Carlo Methods*, pp. 191–233. New York: John Wiley and Sons.

Derber, J. C. 1985. "The variational four-dimensional assimilation of analyses using filtered models as constraints." Ph.D. diss. Madison WI: University of Wisconsin.

Dimov, I., U. Jaekel, and H. Vereecken. 1996. "A numerical approach for determination of sources in transport equations." *J. Comput. Math. Appl.* **32(5)**: 31–42.

Dimov, I., U. Jaekel, H. Vereecken, and D. Wendt. 1997. "A numerical approach for determination of sources in reactive transport equations, Second International Conference on Monte Carlo and Quasi-Monte Carlo methods in scientific computing, University of Salzburg, 8–12 July, 1996." In *Proceedings of MC & QMC 96, Springer Notes in Statistics*, edited by H. Niederreiter, P. Hellekalek, G. Larcher, and P. Zinterhof, pp. 189–204. Berlin: Springer-Verlag.

Dimov, I. and O. Tonev. 1989. "Monte Carlo methods with overconvergent probable error." In *Numerical Methods and Applications, Proceedings of International Conference on Numerical Methods and Applications, Sofia*, pp. 116–120. Sofia, Bulgaria: Publishing House of Bulgarian Academy of Science.

Dimov, I. and O. Tonev. 1993a. "Random walk on distant mesh points Monte Carlo methods." *J. Stat. Phys.* **70(5/6)**: 1333–1342.

Dimov, I. and O. Tonev. 1993b. "Monte Carlo algorithms: performance analysis for some computer architectures." *J. Comput. Appl. Math.* **48**: 253–277.

Elbern, H., H. Schmidt, and A. Ebel. 1997. *Variational data assimilation for tropospheric chemistry modeling*. Cologne: Institute of Geophysics and Meteorology, University of Cologne. January 3, No. 97.0030.

Ermakov, S. M. and G. A. Mikhailov. 1982. In *Statistical Modeling*. Moscow, Russia: Nauka.

Ermakov, S. M., V. V. Nekrutkin, and A. S. Sipin. 1984. In *Random Processes for Solving Classical Equations of Mathematical Physics*. Moscow, Russia: Nauka.

Gill, P. E., W. Murray, and M. H. Wright. 1982. In *Practical Optimization*. London: Academic press.

Hammersley, J. M. and D. C. Handscomb. 1964. In *Monte Carlo Methods*. New York: John Wiley & Sons, Inc.

Hollingsworth, A., A. C. Lorenc, M. S. Tracton, K. Arpe, G. Cats, S. Uppala, and P. Kallberg. 1985. "The response of numerical weather prediction systems to FGGE level IIb data. Part I: Analysis." *Q.J.R. Meteorol. Soc.* **111**: 1–66.

Le Dimet, F. X. and O. Talagrand. 1986. "Variational algorithms for analysis and assimilation of meteorological observations: theoretical aspects." *Tellus* **38A**: 97–110.

Lewis, J. M. and J. C. Derber. 1985. "The use of adjoint equations to solve a variational adjustment problem with advective constraints." *Tellus* **37A**: 309–322.

Lions, J. L. 1971. In *Optimal Control of Systems Governed by Partial Differential Equations*. Berlin: Springer-Verlag.

Lorenc, A. C. 1981. "A global three-dimensional multivariate statistical interpolation scheme." *Monthly Weather Rev.* **109**: 701–721.

Marchuk, G. I. 1974. In *Numerical Solution of the Problem of Dynamics of the Atmosphere and the Ocean*. Leningrad, Russia: Gidrometeoizdat.

Marchuk, G. I. 1980. *Methods of Computational Mathematics*. Moscow, Russia: Nauka.

Marchuk, G. I. 1985. *Mathematical Modeling for the Problem of the Environment, Studies in Mathematics and Applications, No. 16*. Amsterdam: North-Holland.

Marchuk, G. I. 1994. "Applications of adjoint equations to problems of global change and environmental protection." In *HERMIS'94*, edited by E. A. Lipitakis, Vol. 1: pp. 1–20. Helenic Mathematical Society, Printed in Hellas.

Marchuk, G. I. and V. I. Agoshkov. 1993. "On solvability and numerical solution of data assimilation problems." *Russ. J. Numer. Math. Math. Model.* **8**: 1–16.

McPherson, R. D., K. H. Bergman, R. E. Kistler, G. E. Rasch, and D. S. Gordon. 1979. "The NMC operational global data assimilation scheme." *Monthly Weather Rev.* **107**: 1445–1461.

McRae, G. J., W. R. Goodin, and J. H. Seinfeld. 1984. "Numerical solution of the atmospheric diffusion equations for chemically reacting flows." *J. Comput. Phys.* **45**: 1–42.

Penenko, V. V. and N. N. Obraztsov. 1976. "A variational initialization model for the fields of meteorological elements." *Sov. Meteorol. Hydrol.* **11**: 1–11.

Skelboe, S. and Z. Zlatev. 1997. "Exploiting the natural partitioning in the numerical solution of ODE systems arising in atmospheric chemistry." In *Numerical Analysis and Its Applications*, edited by L. Vulkov, J. Wasniewski, and P. Yalamov, pp. 458–465. Berlin: Springer.

Sobol, I. M. 1973. *Monte Carlo Numerical Methods*. Moscow, Russia: Nauka.

Talagrand, O. and P. Courtier. 1987. "Variational assimilation of meteorological observations with the adjoint vorticity equation. I: Theory." *Q.J.R. Meteorol. Soc.* **113**: 1311–1328.

Tikhonov, A. N. and A. A. Samarski. 1977. *Equations of the Mathematical Physics.* Moscow, Russia: Nauka.
Zlatev, Z. 1995. *Computer Treatment of Large Air Pollution Models.* Dordrecht, The Netherlands: Kluwer Academic Publishers.
Zlatev, Z., I. Dimov, and K. Georgiev. 1996. "Three-dimensional version of the Danish Eulerian Model." *Z. Angewandte Math. Mech.* **76(S4)**: 473–476.

18.18 Applied Optimization in Agriculture

Charles B. Moss

ABSTRACT Most of the applied research in agriculture emphasizes the formulation of current research questions into mathematical specifications that can be solved using well-tested algorithms. In addition, most of the applied optimization occurs in the economic or management dimensions of agriculture. This article focuses on such applications in nine categories: food and diet, farm and agribusiness management, farm firm development, production response, interregional trade and spatial economics, natural resources, econometric and statistical applications, policy analysis, and nonparametric efficiency analysis. The article then presents the dominant algorithms used in the field followed by emerging issues both in the area of formulations and algorithms.

1. Introduction

Applied optimization is a broad topic in agricultural research. As a general field, agriculture encompasses several disciplines, from biological chemistry to agricultural economics and rural sociology. Furthermore, many disciplines in agriculture maintain closer theoretical and epistemological ties to parent disciplines than to other fields within agriculture. Another defining characteristic of agricultural research involves its applied nature rather than basic science nature. This distinction can be attributed to research enabling legislation and funding from government sources. For the purpose at hand, the applied nature of the disciplines typically pushes agricultural applications of optimization toward the use of previously developed and well-tested techniques, and away from the development of new algorithms. Hence, innovations in agriculture tend to focus on new formulations of objective functions or constraints and not in the direct development of new procedures.

A related dichotomy that exerts influence on the application of applied optimization in agriculture involves the contrast between descriptive and prescriptive models. While descriptive models are often developed to test hypotheses about the reaction of farms to changes in agricultural and trade policy, technology, and so on, prescriptive modeling efforts in agriculture are far more common. This prescriptive focus can be attributed to the triple mission of research, education, and extension in land grant universities in the United States. The research and education functions are typical to institutions of higher education, but the extension mission is unique to agriculture. Land-grant universities are specifically directed to provide support to agriculture and rural communities. Therefore, research conducted on agricultural topics typically involves prescriptive models for use by farmers in the management of agricultural enterprises.

This article provides an overview of applied optimization in agriculture in three phases. First, an overview of a variety of agricultural applications of applied optimization techniques is provided. Next, algorithms used for agricultural applications are discussed. Finally, emerging issues in applied optimization within agricultural sciences, both from an algorithmic and epistemological perspective, are addressed.

2. Applications

Because of its applied nature, agricultural research has been historically well disposed to applied optimization. A search through agricultural literature for applications of applied optimization results in numerous example from agricultural economics, forestry, dairy, poultry science, food science and human nutrition, agronomy, and biological chemistry. The dominant use of applied optimization occurs in the form of mathematical programming.

Day and Sparling (1977) propose classification of linear programming in agricultural economics into six categories: (1) food and diet, (2) farm and agribusiness management, (3) farm firm development, (4) production response, (5) interregional trade and spatial economics, and (6) natural resources. The advancements in algorithms since their review have made nonlinear programming feasible for a broader spectrum of problems. Partially as a result of the increased feasibility of nonlinear programming, three additional categories may be added to this list: (7) econometric and statistical application, (8) policy analysis, and (9) nonparametric efficiency analysis.

2.1. *Food and Diet*

Agricultural applications in the area of food and diet include both human consumption studies, such as Stigler (1945), and models minimizing the cost of animal feed, such as Heady (1951). In general, these models solve the problem of minimizing the cost of the combination of feed components subject to nutrient requirements such as minimum protein requirements, carbohydrate levels, and so on.

The earliest human food and diet problem is attributed to Stigler, who solved the problem for "...cost of the most economical diets in August 1939 and 1944 for an active economist (weighing 70 kilograms) who lives in a large city." Stigler produced a cost-minimizing diet, without a linear programming algorithm, costing \$39.93 for August 1939, and \$59.88 for August 1944, using a two-stage procedure. In the first stage, he generated an efficient set of foods that dominated all other foods in the cost of key nutrients. This procedure reduced the number of foods from seventy seven to fifteen. The second stage was to directly enumerate the combinations of the remaining fifteen foods subject to nine nutritional constraints. Stigler recognized that his procedure may not contain the precise cost-minimizing solution, but concluded that the cost function was fairly flat around his final diet. While the least-cost human diet was limited to Stigler, Christensen and Mighell (1951) justify their analysis of minimum-cost hog rations by looking at the lost caloric opportunity due to the misallocation between corn and protein supplements.

The development of Becker's household production theory gave rise to an alternative specification of the food and diet problem. The household production model links the traditional consumption decision to a household level production decision.

Within this model, households not only must decide what food groups to consume but also how to produce them. In a development context, this framework has been used to analyze the production of food crops versus market crops in Korean households (Ahn et al., 1981) and the effectiveness of food aid (Bezuneh et al., 1988). In these studies, the household is assumed to choose the level of inputs, x, and outputs, y, that maximize utility, $U(.)$, subject to a budget constraint

$$
\begin{aligned}
\max_{x,y} \quad & U(y) \\
\text{subject to} \quad & y = F(x) \\
& p'x \leqslant I,
\end{aligned}
\tag{1}
$$

where $F(.)$ denotes some production relationship between purchased foodstuffs and consumable goods, p is a vector of prices for the purchased food stuffs, and I is the level of income. The first complication in the optimization problem in Equation (1) typically involves the inclusion of labor income and a constraint that incorporates the labor requirement in the production process. The development scenarios evaluate the production of marketable goods versus goods grown for household consumption.

Development of optimal-feed rations follow a similar line. Initial studies used linear programming to minimize the cost of a particular diet subject to nutritional constraints. The first application of linear programming to this problem is attributed to Waugh (1951). Like the human diet problem, the simple linear formulation has given way to more complicated formulations. Hertzler (1988) made the feed ration problem dynamic in recognition of the fact that the nutrient requirements of animals change over time. Cravener et al. (1994) brought in the stochastic nature of feed stuffs. Henry et al. (1995) analyzed the dairy ration problem considering the effect of animal waste on the environment. Specifically, they formulate the ration problem to include both purchased inputs and those inputs that can be grown by the dairy. In addition, they consider the use of wastewater from the animal production facilities on the crops grown. The crops are then selected to minimize the cost of feeding subject to the constraint that the crops grown by the dairy utilize sufficient percentages of nutrients found in animal waste.

2.2. *Farm and Agribusiness Management*

The largest share of applied optimization techniques in agriculture falls under the heading of farm and agribusiness management. This general category also represents a great diversity within the general category. Early applications in farm and agribusiness management grew out of the crop selection problem. Specifically, at planting time, farmers were hypothesized to choose the combination of crops to be grown subject to various resource constraints such as land, labor, and capital. Early applications typically assumed a linear objective function and linear constraints:

$$\begin{aligned} &\max_x \quad c'x \\ &\text{subject to} \quad Ax \leqslant b, \end{aligned} \tag{2}$$

where c is a vector of returns for each crop enterprise, x is a conformable vector of crop enterprise levels, A is a matrix of constraint coefficients, and b is a conformable vector of resource constraints. Early reformulations of the basic farm management problem involved more elaborate specification of the constraints. One of the key factors in agricultural production is often the number of growing days available to a crop. This led to the expansion of constraints to depict equipment and operator labor timing in the selection of crops. Shadow values from these expanded models could then be used to determine the value of machinery.

A second major area of expansion for farm management models came through the incorporation of risk into the farm-planning problem. The dispersion of agricultural production decisions over time increases the risk of the farm enterprise. Numerous mathematical programming approaches have been developed in agriculture to analyze the effect of these risks. Freund (1958) developed a mean-variance specification of the crop selection problem that was consistent with the von Neumann–Morgenstern expected utility model under certain utility and distribution functions. (Specifically, Freund assumes that returns are normally distributed and the utility function is negative exponential. The equivalence between the mean-variance programming formulation and the risk-efficient decision set is actually less restrictive.) Freund's formulation involves the respecification of the farm-planning model in Equation (2) using a nonlinear objective function in place of the linear cost function

$$\begin{aligned} &\max_x \quad f(x, \rho) \\ &\text{subject to} \quad Ax \leqslant b, \end{aligned} \tag{3}$$

where $f(x, \rho)$ is a quadratic objective function of the mean and variance of the profit for a given vector of production activities x, and ρ is the absolute risk aversion coefficient.

The quadratic specification of Equation (3) made this nonlinear programming problem fairly tractable, but alternative specifications such as Hazell's (1971) minimization of total absolute deviation (MOTAD) were developed that yielded linear specifications for decision-making under risk. As developed by Mapp and Helmers (1984), MOTAD reformulates the risk problem by appending artifical constraints to the original linear programming problem in Equation (2):

$$\begin{aligned} \max_{x} \quad & c'x - \lambda 1'(d_1 + d_2) \\ \text{subject to} \quad & Ax \leqslant b \\ & Bx = d_1 - d_2, \end{aligned} \tag{4}$$

where λ is a risk aversion coefficient along the lines of ρ in the mean-variance specification, 1 is a conformable vector of ones, d_1 is a vector of positive deviations from the mean, d_2 is a vector of negative deviations from the mean, and B is a matrix of observed deviations from the sample mean.

While the linearity of MOTAD increased its adoption during the 1970s and early 1980s, increased availability of nonlinear algorithms in the 1980s and 1990s led to a return of the mean-variance approach and the growth of direct utility maximization. Following the original Freund specification in Equation (2), Kaylen et al. (1987) defined the direct utility maximization problem as

$$\begin{aligned} \max_{x} \quad & E[U(x)] \\ \text{subject to} \quad & Ax \leqslant b, \end{aligned} \tag{5}$$

where $E[.]$ is the expectation operator and $U(x)$ is a utility function defined on the choice variables x. In addition, to generalize the decision problem away from well-known conjugate priors, Kaylen et al. proposed the use of numerical quadrature for the expectation.

2.3. *Farm Firm Development*

The farm firm development literature can be viewed as an extension to the farm and agribusiness management literature that incorporates the time dimension into the planning problem. Specifically, within the single period framework of the farm management problem, it is unlikely that a firm will purchase land or equipment. The purchase of land can occasionally be incorporated through a terminal value, but the initial depreciation on equipment typically precludes expansion. Most farm firm development models address these firm size issues by using a multiperiod programming framework.

Expanding the problem formulation in Equation (2):

$$\begin{aligned} \max_{x_1, x_2, \Lambda x_n} \quad & \begin{bmatrix} c_1 & c_2 & \cdots & c_n \end{bmatrix} \begin{bmatrix} x_1 \\ x_2 \\ \vdots \\ x_n \end{bmatrix} \\ \text{subject to} \quad & \begin{bmatrix} A_{11} & 0 & \cdots & 0 \\ T_{12} & A_{22} & \cdots & 0 \\ \vdots & \vdots & \text{O} & \vdots \\ 0 & 0 & \cdots & A_{44} \end{bmatrix} \begin{bmatrix} x_1 \\ x_2 \\ \vdots \\ x_4 \end{bmatrix} = \begin{bmatrix} b_1 \\ b_2 \\ \vdots \\ b_4 \end{bmatrix}, \end{aligned} \tag{6}$$

where A_{ii} are production constraints within period i, T_{ij} is the coefficient transferring resources between period i and period j, x_i is the vector of activities in period i, and b_i is the original resource available (without purchases or sales) in period i. This formulation has been used to model production response (Swanson, 1956; Loftsgard and Heady, 1959), financial constraints

(Irwin and Baker, 1962; Martin and Plaxico, 1967), and tax effects on growth (Moss, 1984). Like the original approaches in the farm and agribusiness management area, these approaches were deterministic. Johnson et al. (1967) extended the formulation by allowing crop yields to be stochastic. Their optimization problem used stochastic dynamic programming. An alternative approach utilizing discrete stochastic programming was developed by Cocks (1968), and was applied by Featherstone et al. (1990). This approach expands the multiperiod constraint matrix in Equation (6) by adding the possibility of multiple states of nature in each year. For example, assume that for each year there are two possible states of nature (high crop prices and low crop prices). The transition matrix for the first year becomes

$$\begin{bmatrix} A_{11} & 0 & 0 \\ T_{12}^{h} & A_{22}^{h} & 0 \\ T_{12}^{l} & 0 & A_{12}^{l} \end{bmatrix} \begin{bmatrix} x_1 \\ x_2^{h} \\ x_2^{l} \end{bmatrix} \leqslant \begin{bmatrix} b_1 \\ b_2^{h} \\ b_2^{l} \end{bmatrix}, \tag{7}$$

where T_{12}^{h} is the transfer to the second period under the high crop price outcome, T_{12}^{l} is the transfer to the second period under the low crop price outcome, A_{22}^{h} are the production constraints following the high crop price, A_{22}^{l} are the production constraints following the low price outcome, x_1 is the vector of crop activities in the intitial period, x_2^{h} and x_2^{l} are vectors of crop activities under high and low crop prices, respectively, and the bs are the respective resource constraints. The primary difficulty with this approach is that the size of the constraint matrix expands rapidly with the richness of the probability distribution and the number of periods.

2.4. *Production Response*

In a similar vein, the production response models build on the farm and agribusiness planning models by aggregating over farms to analyze sector response to policy changes. The most famous agricultural model in this group was the CARD (Center for Agricultural and Rural Development) model produced by Earl O. Heady and associates at Iowa State University in the 1960s and 1970s. This model was used to evaluate the regional impact of such policies as the change in agricultural price supports. However, these models are typically one-sided or do not incorporate output price responses. As such, computable general equilibrium models have replaced the production response models.

2.5. *Interregional Trade and Spatial Economics*

The first steps to price endogeniety are found in the interregional trade models first proposed by Takayama and Judge (1964). In these models, a supply is allocated between various demand regions considering transportation cost. Mathematically, these models can be expressed as

$$\begin{aligned} \max_{d_1, d_2, s} \quad & \int_0^{d_1} p_1^{d}(x_1)dx_1 + \int_0^{d_2} p_2^{d}(x_2)dx_2 - \int_0^{s_1} p_1^{s}(y_1)dy_1 - t_1 d_2 \\ \text{subject to} \quad & d_1 + d_2 \leqslant s_1, \end{aligned} \tag{8}$$

where d_1 and d_2 are the levels of the goods consumed in markets 1 and 2, respectively, and s_1 is the amount of the good produced in market 1. In this formulation, $p_1^{d}(.)$ and $p_2^{d}(.)$ are the inverse demand functions for the good in markets 1 and 2, $p_1^{s}(.)$ is the inverse supply function for good 1, and t_1 is the cost of shipping the good to market 2. Differentiation of Equation (8) with respect to each market quantity results in the first order condition that the price in market 1 equals the marginal cost of producing the total level of output, while the price in market 2 less the transportation cost equals the price in market 1. Algorithmically, this problem is typically solved as a quadratic programming problem by assuming that the inverse demand functions are linear in the quantity consumed in each market, and the supply function is linear in the quantity produced.

2.6. Natural Resources

Applied optimization in natural resources involves several different formulations. At one extreme, these formulations simply append environmental constraints, such as maximum nitrogen runoff, onto existing farm management problems. In these applications the shadow value for the environmental constraint can be used to value environmental restrictions. At the other extreme, Henry et al. (1995) build on the multiperiod firm growth framework to develop a multiperiod model to examine the effect of nitrogen retention and uptake requirements on the operation of a dairy in northern Florida. Similar multiperiod formulations have been used to examine aquifer and reservoir management issues.

2.7. Econometric and Statistical Applications

The emergence of econometric and statistical applications of applied optimization techniques can be traced to the emergence of nonlinear statistical models. Historically, econometrics relied on ordinary or generalized least squares that yielded closed-form solutions of the parameter estimates. However, the emergence of more complicated formulations such as nonlinear demand and cost systems (Theil et al., 1989; Featherstone and Moss, 1994), autoregressive conditional heterscedasticity formulations (Engle, 1982; Moss et al., 1990), and models of nonnormality of crop yields (Nelson and Preckel, 1989; Moss and Shonkwiler, 1993; Ramirez et al., 1994) have resulted in models without closed-form solutions. Featherstone and Moss (1994) estimate a system of cost equations including the cost function and three factor demand equations. The residuals for this system can be expressed as:

$$e_i(\theta) = \begin{bmatrix} c_i - (\alpha_0 + \alpha w_i + w_i' A w_i + \beta y_i + y_i' B y_i + y_i' \Pi w_i) \\ x_{1i} - (\alpha_1 + A_{11} w_{1i} + A_{12} w_{2i} + A_{13} w_{3i} + \Pi_{11} y_{1i} + \Pi_{21} y_{2i} + \Pi_{31} y_{3i} + \Pi_{41} y_{4i} + \Pi_{51} y_{5i}) \\ x_{2i} - (\alpha_2 + A_{21} w_{1i} + A_{22} w_{2i} + A_{32} w_{3i} + \Pi_{12} y_{1i} + \Pi_{22} y_{2i} + \Pi_{32} y_{3i} + \Pi_{42} y_{4i} + \Pi_{52} y_{5i}) \\ x_{3i} - (\alpha_3 + A_{31} w_{1i} + A_{32} w_{2i} + A_{33} w_{3i} + \Pi_{13} y_{1i} + \Pi_{23} y_{2i} + \Pi_{33} y_{3i} + \Pi_{43} y_{4i} + \Pi_{53} y_{5i}) \end{bmatrix}, \tag{9}$$

where $e_i(\theta)$ is the residual vector for observation i that is a function of the parameter vector $\theta = \{\alpha_0, \alpha_j, A_{jk}, \Pi_{kl}\}$ $(j,k = 1,2,3$ and $l = 1, \ldots 5), c_i$ is the observed cost for observation i, x_{ji} are the observed input levels for observation i, w_{ji} are the observed prices for each input for observation i, and y_{ki} are the observed output levels for observation i. This system is then estimated by searching for the set of parameters that maximizes the concentrated system likelihood function

$$\begin{aligned} &\max_{\theta} - \frac{T}{2} \ln | \Omega(\theta) | \\ &\Omega(\theta) = \frac{1}{T} \sum e_i(\theta) e_i'(\theta), \end{aligned} \tag{10}$$

where $\Omega(\theta)$ is the sample variance matrix conditioned on the parameter estimates. Direct estimation of this formulation can be reduced to a generalized least squares form with a closed form optimum. However, unrestricted estimates of the parameters using the generalized least squares approach typically fail the theoretical concavity restrictions for a cost function. Thus, Featherstone and Moss (1994) impose concavity by estimating a Cholesky decomposition of the coefficients in the A matrix

$$A = - \begin{bmatrix} a_{11} & 0 & 0 \\ a_{12} & a_{22} & 0 \\ a_{13} & a_{23} & a_{33} \end{bmatrix} \begin{bmatrix} a_{11} & a_{12} & a_{13} \\ 0 & a_{22} & a_{23} \\ 0 & 0 & a_{33} \end{bmatrix} \tag{11}$$

that results in a concave cost function with respect to input prices. Substituting this formulation into Equations (9) and (10) yields a nonlinear optimization problem that can only be solved using iterative techniques. (Shumway et al., 1990, solve the same problem by restricting the eigenvalues of the original A matrix to be nonpositive yielding an optimization problem with a nonlinear constraint.)

2.8. *Policy Analysis*

Another recent development is the emergence of policy models to analyze the effect of agricultural and resource policies on the entire agricultural economy. These models, referred to as computable or doable general equilibrium models, build on the alternative production opportunities in the production response models and the price endogeniety of the interregional trade models. Collapsing Equation (8) to a single consumption region but expanding social benefit to include multiple goods yields the societal optimization problem.

$$\max_{d_i, s_i} \int_{d_1}\int_{d_2}\cdots\int_{d_n} p^d(x_1, x_2, \cdots x_n)dx_1 dx_2 \Lambda dx_n - \int_{s_1}\int_{s_2}\cdots\int_{s_n} p^s(y_1, y_2, \Lambda y_n)dy_1 dy_2 \Lambda dy_n - \sum_{i=1}^{n} t_i x_i \tag{12}$$

$$\text{subject to} \quad d_i = s_i \quad i = 1, \Lambda n,$$

where, as in the interregional trade formulation, p^d is the demand curve for each commodity, p^s is the supply curve for each commodity, d_i is the level of each good demanded, and s_i is the level of each good supplied. However, the vector t is now a vector of taxes or subsidies on each output. The first order conditions for this model then become

$$p_i^d(d_1, d_2, \Lambda d_n) - p_i^s(s_1, s_2, \Lambda s_n) - t_i = 0 \ \text{ for } \ i = 1, \Lambda n \tag{13}$$

with one redundant equation because of the adding-up restriction. Within this framework, the relative effects of agricultural price supports or tariff policies can be analyzed by varying t_i for the commodity in question. Espinosa and Smith (1995) use this framework to analyze the environmental effect of trade policy, Kilkenny (1993) examines the rural/urban effect of agricultural programs, and Sherony et al. (1991) examine the economic impact of crop losses. Methodologically, since the social objective function itself is often of little interest, this applied optimization problem can be solved by applying Gauss–Siedel to the first-order conditions in Equations (13).

2.9. *Nonparametric Efficiency Analysis*

A final agricultural specification that has been growing in popularity is the nonparametric analysis of production efficiency. Traditionally, the economic efficiency of farms and agribusinesses have been analyzed by first estimating a parametric cost or profit function as developed in Featherstone and Moss (1994). The estimated parameters could then be used to decompose economic inefficiencies into allocative and technical inefficiencies based on the estimated parameters (Akridge, 1989). Unfortunately, the efficiency results in these studies are conditioned on the choice of structural form used to estimate the parametric production function as well as the distributional assumptions used in estimation. An alternative approach is to allow the most efficient firms to form an efficient technological envelope. Assuming that a firm produces a vector of m output from n inputs, a vector of outputs, y^*, could be produced using some combination of outputs from firms in the sample. From the cost perspective, the most efficient of such combinations could be found by choosing the vector z that

$$\min_z c^* = \sum_{i=1}^{n} c_i z_i$$

$$\text{subject to} \quad \sum_{i=1}^{n} y_{ij} z_i \geqslant y_j^* \quad j = 1, \Lambda m \tag{14}$$

$$\sum_{i=1}^{n} z_i = 1,$$

where c_i is the observed level of cost for firm i. The ratio c_i/c^* then provides a measure of economic efficiency that approaches one as the firm becomes more efficient. Recent applications of nonparametric efficiency analysis include Preckel et al. (1997), who analyze the economic efficiency of retail fertilizer firms, and Tauer (1993) and Weersink et al. (1990), who analyze the economic efficiency of dairy farms.

3. Algorithms

As previously indicated, applied optimization in agriculture has been dependent on the availability of commercial code. Early efforts such as the diet and farm management models made extensive use of the simplex algorithm. The interaction between problem specification and availability of commercial code can be seen in the development of farm management specifications under risk. Specifically, Freund solved the mean-variance optimization problem using the Kuhn–Tucker–Hildreth algorithm (Hildreth, 1954) in 1956. This was followed by Hazell's reformulation of the risk-programming problem as a linear programming problem in 1971. The widespread adoption of the mean-variance framework evolved with the availability of commercial code in the mid and late 1980s.

After the advent of nonlinear programming techniques, most agricultural applications have focused on the optimization of smooth nonlinear objective functions with linear inequality constraints. As such, the primary optimization algorithm has been a Newton–Raphson or quasi-Newton conjugate gradient algorithm. Notable exceptions to this generalization include the imposition of concavity on profit function by Shumway et al. (1990) who constrained the eigenvalues of the profit function.

With the development of MINOS by the System Optimization Laboratory at Stanford University (Murtagh and Saunders, 1995), the solution of mathematical programming problems with nonlinear objective functions and constraints became fairly tractable. Using these routines for nonlinear programming involves writing subroutines in Fortran 77 to generate the nonlinear objective function or constraint, and the derivatives of the nonlinear functions with respect to the search variables. These subroutines are compiled with the original MINOS code to yield an executable program.

The reliance on Fortran increases the overall flexibility of problems that can be solved with MINOS. This flexibility can be seen in Kaylen et al.'s (1987) use of a numerical integration technique to define their objective function. However, this flexibility is purchased at the price of specialization. One alternative is the use of GAMS to provide a user-friendly front-end to MINOS. Another alternative frequently used in statistical and econometric applications is Gauss. Gauss contains two subroutines (OPTMUM and MAXLIK) that minimize a function specified as a subroutine using a variety of conjugate gradient routines.

4. Emerging Issues

Emerging issues in applied optimization in agriculture can be divided along the lines of issues involving problem formulations, and issues involving the availability of algorithms.

4.1. Formulations

Two contradictory pressures will affect applied optimization formulations in agriculture over the next decade. These pressures are the expansionary effect of precision, or site specific, agriculture, which will give rise to larger optimization problems as management decisions extend to smaller tracts of land and a growing distrust of overly large or complex model specifications. The basic premise of precision agriculture is that the productivity of agricultural assets such as farmland or animals varies across units. In addition, increases in technology currently allow for more disaggregated management. Thus, instead of balancing the ration for a pen of dairy cattle or a field, new technologies allow for management at the cow or hectare level. Advocates of precision agriculture argue that such disaggregated management will lead to increases in agricultural profitability. However, design and implementation of these disaggregated management strategies will undoubtedly lead to increased size and complexity in optimization formulations.

Balancing the expansionary effect of precision agriculture is a growing mistrust of large or numerically complex formulations. Intuitively, as the number of inputs and outputs in an agricultural production problem grows, the number of coefficients necessary to parameterize the mathematical programming problems grows. Concomitantly, the statistical information used to determine each parameter falls as the number of required coefficients grows. As a result, the numerical stability and decisionmaking relevance of the resulting optimization problem becomes questionable. Paris and Howitt (1998) characterize this problem as an ill-posed problem in production economics. They also propose a maximum entropy approach to address this problem. A similar argument can be constructed around the complexity of function forms.

4.2. *Algorithms*

From an algorithmic perspective, the emerging issue for applied optimization in agriculture involves past reliance on smooth functions with well-defined derivatives. As previously discussed, applied optimization in agriculture has traditionally relied on conjugate gradient algorithms. If the researcher did not explicitly define the gradients, the algorithm approximated smooth gradients using some form of central differencing. However, the emergence of process models, such as ceres-maize and soygrow, are inconsistent with traditional notions of smooth, well-defined functions. Two possible solutions to this problem are the use of approximating functions to transform the results of the process models into traditional smooth functions, or the use of new algorithms such as simulated annealing.

References

Ahn, C. Y., I. Singh, and L. Squire. 1981. "A model of an agricultural household in a multi-crop economy: the case of Korea." *Rev. Econ. Stat.* **61**: 520–525.

Akridge, J. T. 1989. "Measuring productive efficiency in multiple product agribusiness firms: a dual approach." *Am. J. Agric. Econ.* **71**: 116–125.

Bezunch, M., B. J. Deaton, and G. W. Norton. 1988. "Food aid impacts in rural Kenya." *Am. J. Agric. Econ.* **70**: 181–191.

Christensen, R. P. and R. L. Mighell. 1951. "Food production strategy and the protein–feed balance." *J. Farm Econ.* **33**: 183–203.

Cocks, C. W. 1968. "Discrete stochastic programming." *Mgmnt Sci.* **15**: 72–79.

Cravener, T. L., W. B. Roush, and T. H. D'Alfonso. 1994. "Laying hen production response to least cost rations formulated with stochastic programming." *Poultry Sci.* **73**: 1290–1295.

Day, R. H. and E. Sparling. 1977. "Optimization models in agricultural and resource economics." In *A Survey of Agricultural Economics Literature, Volume 2,* edited by G. G. Judge, R. H. Day, S. R. Johnson, G. C. Rausser, and L. R. Martin, pp. 93–156. Minneapolis: University of Minnesota Press.

Engle, R. F. 1982. "Autoregressive conditional hetersocedasticity with estimates of the variance of United Kingdom inflation." *Econometrica* **50**: 987–1007.

Espinosa, J. A. and K. V. Smith. 1995. "Measuring the environmental consequences of trade policy: a nonmarket CGE analysis." *Am. J. Agric. Econ.* **77**: 772–777.

Featherstone, A. M. and C. B. Moss. 1994. "Measuring economies of scale and scope in agricultural banking." *Am. J. Agric. Econ.* **76**: 669–674.

Featherstone, A. M., P. V. Preckel, and T. G. Baker. 1990. "Modeling farm financial decisions in dynamic and stochastic environment." *Agric. Finance Rev.* **50**: 80–99.

Freund, R. J. 1956. "The introduction of risk into a programming model." *Econometrica* **24(3)**: 253–263.

Hazell, P. B. R. 1971. "A linear alternative to quadratic and semivariance programming for farm planning under uncertainty." *Am. J. Agric. Econ.* **53**: 53–62.

Heady, E. O. 1951. "A production function and marginal rates of substitution in the utilization of feed resources by dairy cows." *J. Farm Econ.* **33**: 485–498.

Henry, G. M., M. A. DeLorenzo, D. K. Beebe, C. B. Moss, and W. G. Boggess. 1995. "Determining optimal nutrient management strategies for dairy farms." *J. Dairy Sci.* **78**: 693–703.

Hertzler, G. 1988. "Dynamically optimal and approximately optimal beef cattle diets formulated by nonlinear programming." *Western J. Agric. Econ.* **13**: 7–17.

Hildreth, C. G. 1954. "Point estimates of ordinates of concave functions." *J. Am. Stat. Assoc.* **49**: 598–619.

Irwin, G. D. and C. B. Baker. 1962. *Effects of lender decisions on farm financial planning.* Illinois Agricultural Experiment Station Bulletin 688.

Johnson, S. R., K. R. Tefertiller, and D. S. Moore. 1967. "Stochastic linear programming and feasibility problems in farm growth analysis." *J. Farm Econ.* **49**: 908–919.

Kaylen, M. S., P. V. Preckel, and E. T. Loehman. 1987. "Risk modeling via direct utility maximization using numerical quadrature." *Am. J. Agric. Econ.* **69**: 701–706.

Kilkenny, M. 1993. "Rural/urban effects of terminating farm subsidies." *Am. J. Agric. Econ.* **75**: 968–980.

Loftsgard, L. D. and E. O. Heady. 1959. "Application of dynamic programming models for optimum farm and home plans." *J. Farm Econ.* **41**: 51–67.

Mapp, H. P and G. A. Helmers. 1984. "Methods of risk analysis for farm firms." In *Risk Management in Agriculture*, edited by P. J. Barry, pp. 116–128. Ames, IA: Iowa State University Press.

Martin, J. R. and J. S. Plaxico. 1967. "Polyperiod analysis of growth and capital accumulation of farms in the rolling plains of Oklahoma and Texas." USDA Technical Bulletin 1381.

Moss, C. B. 1984. "A polyperiod analysis of financial performance with alternative resource endowments, credit constraints, and organizational forms." (Unpublished Masters thesis, Oklahoma State University, Stillwater, Oklahoma).

Moss, C. B. and J. S. Shonkwiler. 1993. "Estimating yield distributions with a stochastic trend and nonnormal errors." *Am. J. Agric. Econ.* **75**: 1056–1062.

Moss, C. B., J. S. Shonkwiler, and S. A. Ford. 1990. "A risk endogenous model of aggregate agricultural debt." *Agric. Finance Rev.* **50**: 73–79.

Murtagh, B. A. and M. A. Saunders. 1995. "Minos 5.4 User's Guide." Technical Report SOL83-20R. Department of Operations Research, Stanford University.

Nelson, C. H. and P. V. Preckel. 1989. "The conditional beta distribution as a stochastic production function." *Am. J. Agric. Econ.* **71**: 370–378.

Paris, Q. and R. E. Howitt. 1998. "An analysis of ill-posed production problems using maximum entropy." *Am. J. Agric. Econ.* **80**: 124–138.

Preckel, P. V., J. T. Akridge, and M. A. Boland. 1997. "Efficiency measures for retail fertilizer dealers." *Agribusiness: Int. J.* **13**: 497–509.

Ramirez, O., C. B. Moss, and W. G. Boggess. 1994. "Estimation and use of the inverse hyperbolic sine transformation to model nonnormal correlated random variables." *J. Appl. Stat.* **21**: 289–304.

Sherony, K. R., G. J. Knowles, and R. Boyd. 1991. "The economic impact of crop losses: a computable general equilibrium approach." *Western J. Agric. Econ.* **16**: 144–155.

Shumway, C. R., W. P. Alexander, and H. Talpaz. 1990. "Texas field crops: estimation with curvature." *Western J. Agric. Econ.* **15**: 45–54.

Stigler, G. J. 1945. "The cost of subsistence." *J. Farm Econ.* **27**: 303–314.

Swanson, E. R. 1956. "Application of programming analysis to corn belt farms." *J. Farm Econ.* **38**: 408–419.

Takayama, T. and G. G. Judge. 1964. "Spatial equilibrium and quadratic programming." *J. Farm Econ.* **46**: 67–93.

Tauer, L. W. 1993. "Short-run and long-run efficiencies of New York dairy farms." *Agric. Resource Econ. Rev.* **22**: 1–9.

Theil, H., C. F. Chung, and Seale, J. L. Jr. 1989. In *International Evidence on Consumption Patterns.* Greenwich, CT: JAI Press.

Waugh, F. V. 1951. "The minimum-cost dairy feed." *J. Farm Econ.* **33**: 299–305.

Weersink, A., C. G. Turvey, and A. Godah. 1990. "Decomposition measures of technical efficiency for Ontario dairy farms." *Can. J. Agric. Econ.* **38**: 439–456.

18.19 Optimization in Graph Drawing

Petra Mutzel

ABSTRACT Graph drawing is a relatively new and growing area. Recently, it has attracted many researchers from different fields such as computer science, engineering, and mathematics. In graph drawing the task is to visualize discrete structures of objects and their relations between each other in such a way that it is easy to read and to understand. Naturally, in graph drawing many optimization problems arise. This article gives recent advances on the algorithmic and computational aspects of applied optimization in the field of graph drawing.

1. Introduction

The diagram in Figure 1 shows the relations in a database (Rauh and Stickel, 1997). Diagrams like this appear more and more in newspapers, magazines, journals, books, reports, and presentations.

The discrete structure here consists of the objects in the database and their relationships to each other. A discrete structure can be modeled as a graph $G = (V, E)$ where the set of vertices V is associated with the set of objects and the set of edges $E \subseteq V \times V$ with the relations.

Many of the published diagrams are drawn using simple editing tools in which the objects and the connecting lines are generated via mouse clicks. The user needs to decide exactly where to place each object and line. This task is time-consuming and can only be done for small diagrams.

If the diagrams get bigger, or the layout task needs to be repeated over and over again, we would like to have a tool that automatically places the objects and lines in the diagram. This is what graph drawing is about. The aim of graph drawing is to generate nice layouts (in two or three dimensions) so that the displayed structures can be read and understood easily.

Graph drawing has many applications. For example, recently UML (unified modeling language) was established as a standard for modeling processes and systems in many application domains. Another example is computational biology in which visualization helps to study the interaction of proteins. Other applications include business process modeling, information systems, printed circuit layout, and petri-nets.

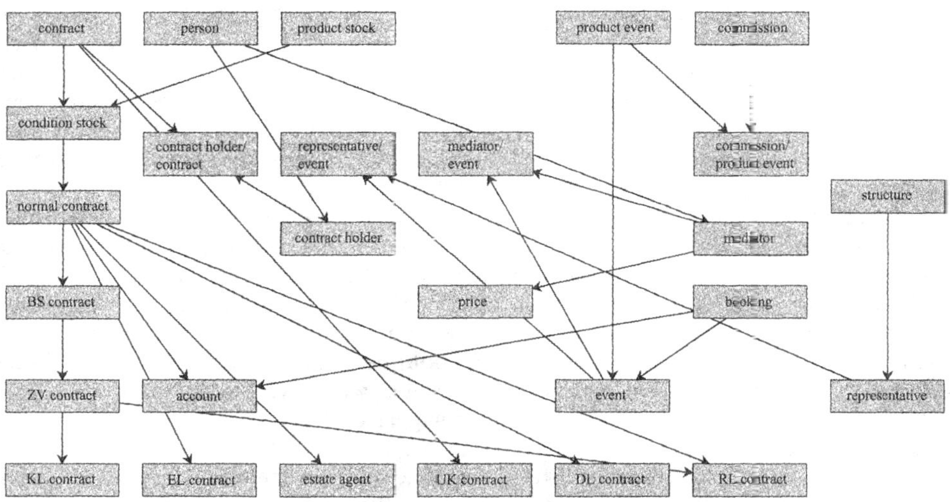

Figure 1 A hand drawing of a database graph

It is difficult to model niceness of a layout, since this often depends on the particular application. However, there are some criteria that lead to readable drawings in many applications. The objects representing the vertices should not overlap each other. Moreover, lines representing the edges should not traverse any object. Objects should be distributed evenly within the drawing area, the length of the edges should be kept short.

One of the most important optimization criteria is to keep the number of edge crossings in a drawing small. The vertex resolution of a drawing is the minimum distance between a pair of vertices. For a given drawing size, we like to have a drawing with maximum resolution. Reasonable vertex resolution is often obtained using grid drawings, in which the vertices are placed on an integer grid, while keeping the size of the drawing small.

Users often prefer orthogonal drawings in which the edges are represented as a path of orthogonal line segments that are parallel to the coordinate axes. In orthogonal drawings, it is advantageous to keep the number of bends small.

A classical drawing style is the straight-line drawing, in which each vertex is represented as a circle or box and each edge as a straight line between the coordinates of its endpoints. If the data is symmetric or hierarchic, this should be respected in the drawing.

Many optimization problems arise when trying to get a drawing satisfying one or more of the above criteria. Many of these problems are NP-hard. In the following, we will introduce the main drawing methods used in practice, the optimization problems that arise, and the recent advances of algorithmic and combinatorial optimization in theory and practice.

2. Force-Directed Methods

Force-directed methods are useful for displaying tree-like networks such as computer networks or social networks. The idea of force-directed methods is to model the graph as a physical system, vertices are modeled as mass points (or rings) that repel each other and the edges as springs. Moreover, it is possible to integrate additional forces between the mass points and springs. The aim is to get an even distribution of the vertices and edges by finding the state of minimum potential energy of the modeled physical system.

This idea, originally used in the layout of printed circuit boards, has been introduced to graph drawing by Eades (1984). He has suggested a spring embedder algorithm in which the forces of the springs obey Hooke's law.

Force-directed methods proceed in two steps. First, the physical force model needs to be defined. In the second step, the state of minimum total potential energy needs to be determined via computing adequate coordinates for the vertices. For both steps many approaches have been suggested in the literature.

The additional forces can depend on vertex distribution, edge length, the number of edge crossings, the vertex-edge distances or the drawing area. Moreover, using forces it is easy to model user constraints such as, for example, vertex u must be placed to the right of vertex v.

The state of minimum energy can be approximated using iterative local search heuristics like simulated annealing, genetic programming or tabu search methods (see Chapter 3.5). Sometimes special methods can be used for solving certain systems, for example, the Newton–Raphson method for a system of partial differential equations.

Force-directed methods are easy to understand and easy to implement. Moreover, they work in two and three dimensions. Figure 2 shows the database diagram of Figure 1 drawn using the spring embedder method suggested by Eades (1984). The vertices are distributed evenly in the drawing area. Edges are always drawn as straight lines and are generally kept short. Often symmetric drawings appear for symmetric graphs. One drawback of this method is that the layouts often contain crossings that could be eliminated easily by visual inspection. Furthermore, straightline drawing methods do not work well when the node sizes are relatively large. Recently, various authors have suggested ideas to get around these problems (Gansner and North, 1998; Bertault, 1999).

For nontree like structures, the result of force-directed methods is, in many cases, not competitive to other methods. An overview of force-directed methods is given in Brandenburg et al. (1996) and Di Battista et al. (1999).

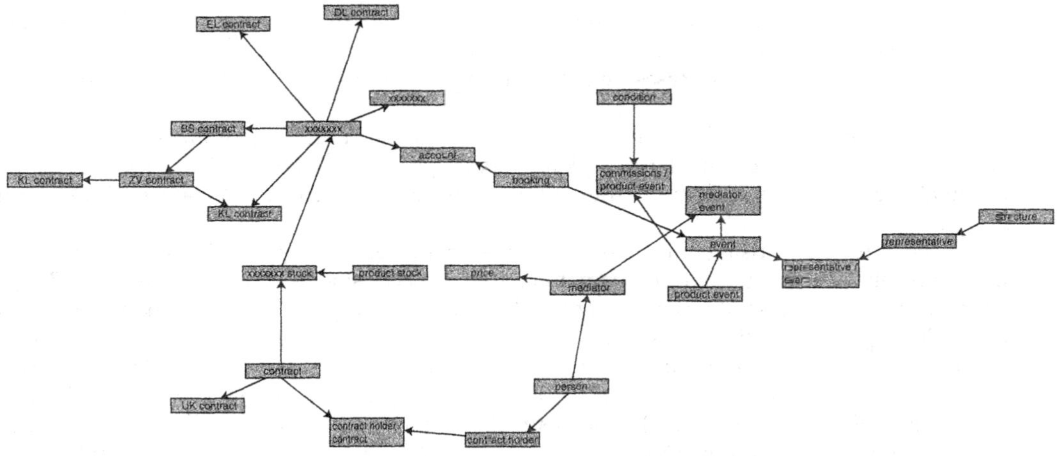

Figure 2 A force-directed drawing of the database graph

Many software tools implement force-directed graph drawing methods. For example, Graphlet, a toolkit for graph editors and graph algorithms (Brandenburg and Himsolt, 2000), contains many different force-directed drawing methods. Figure 2 was produced using AGD, a library of algorithms for graph drawing (AGD, 2000).

3. Crossing Minimization

The number of edge crossings is among the most important aesthetic criteria for nice two-dimensional drawings. However, to find a drawing of a given graph G with the minimum number $\nu(G)$ of edge crossings is NP-hard. Although there is a lot of research in this field, no algorithm for solving practical instances to optimality is known. Moreover, so far no tight lower bound for the number of crossings in a given graph is known.

Bhatt and Leighton (1984) and Chung and Yau (1994) have developed a polynomial time algorithm that produces a drawing of a given graph with n vertices and bounded degree with at most $O((\nu(G)+n)\log^2 n)$ crossings. The algorithm is based on successively bisecting the graph into smaller graphs.

The best practical method to generate drawings with a small number of crossings is based on planarization. A graph G is planar if it can be drawn in the plane without any edge crossings. A graph drawn in this way is also called a plane graph. Planarity can be tested in linear time. A planar drawing divides the drawing space into regions that are called the faces of the drawing.

In general, the graphs in graph drawing applications are not planar. However, many of them are almost planar, in the sense that only a few edges need to be deleted in order to get a planar graph. This observation can be used for crossing minimization.

The idea is to first delete the minimum number of edges of the given nonplanar graph so that the resulting graph P becomes planar. Then, in the second step, a (planar) combinatorial embedding Π of P is computed. Formally, a (planar) combinatorial embedding Π of a planar graph P is a set of circular lists of the neighbors for each vertex that corresponds to a planar drawing of P. A combinatorial embedding essentially fixes the faces of a planar drawing. Using a planar combinatorial embedding makes it easier to insert the deleted edges in the following step (Step 3) while keeping the number of crossings small.

The first problem is the (NP-hard) maximum planar subgraph problem. Jünger and Mutzel (1996) have shown that this problem can be solved to provable optimality for many practical instances, in particular, if the number of edges to be deleted from G is small. Their algorithm is based on an integer linear programming formulation excluding the forbidden

Kuratowski subgraphs. Kuratowski subgraphs are the subdivisions of the complete graph on five vertices K_5 and the complete bipartite graph on six vertices $K_{3,3}$. This integer linear program can in many cases be solved using a methodology called branch-and-cut (see Chapter 3.3).

Calinescu et al. (1998) have given a polynomial time algorithm that is able to approximate the size of the largest planar subgraph by a factor of 2/5. Their approach essentially computes a maximum triangular structure called triangular cactus in the input graph.

The problem arising in Step 3 of the planarization approach is also NP-hard. Unfortunately, an integer linear programming approach suggested in Mutzel and Ziegler (1999) works for very small graphs only and hence, so far no practically efficient algorithm for producing optimal solutions is known. In practice, various heuristics are used based on the observation that the problem of inserting only one edge into a plane graph $P(V, E, \Pi)$ with fixed combinatorial embedding Π can be solved optimally by computing a shortest-path in the dual graph of P (see Chapter 8). The heuristics proceed by iteratively inserting the edges $e_1, ..., e_l$. After each insertion step i, the crossings generated by edge e_i are substituted by artificial vertices so that the resulting graph $G \cup \{e_1, ..., e_i\}$ gets planar again ($i = 1, ..., l$).

Ziegler (2000) has investigated the influence of the combinatorial embedding computed in Step 2. A planar combinatorial embedding can be computed in linear time. Using SPQR-trees (Di Battista and Tamassia, 1990), it is possible to encode the set of all combinatorial embeddings of a planar graph in linear time and space, although the set of all combinatorial embeddings may be exponential. Recently, Gutwenger et al. (2000) have suggested a simple linear time algorithm using SPQR-trees for inserting one edge $e \in V \times V$ into a planar graph $P = (V, E)$ so that the number of crossings in $P \cup \{e\}$ over the set of all possible planar combinatorial embeddings of P is minimized. Computational experiments have shown that iterative use of the new algorithm for the set of all deleted edges $e_1, ..., e_l$ leads to a smaller number of crossings.

At the end of the planarization phase, the given graph G has been transformed into a planar graph. We also say that this is the planarized graph G_P. The number of crossings in the final drawing of G is the number of artificial vertices of G_P. It is now possible to use any planar drawing algorithm (see Section 4) in order to draw the planar graph G_P. However, the result is not always satisfying, since the crossings are not always drawn in the correct way. Consider the graph shown in Figure 3(a). It has been drawn using a planar algorithm based on canonical orderings (Gutwenger and Mutzel, 1998). The crossing between the edges (1,6) and

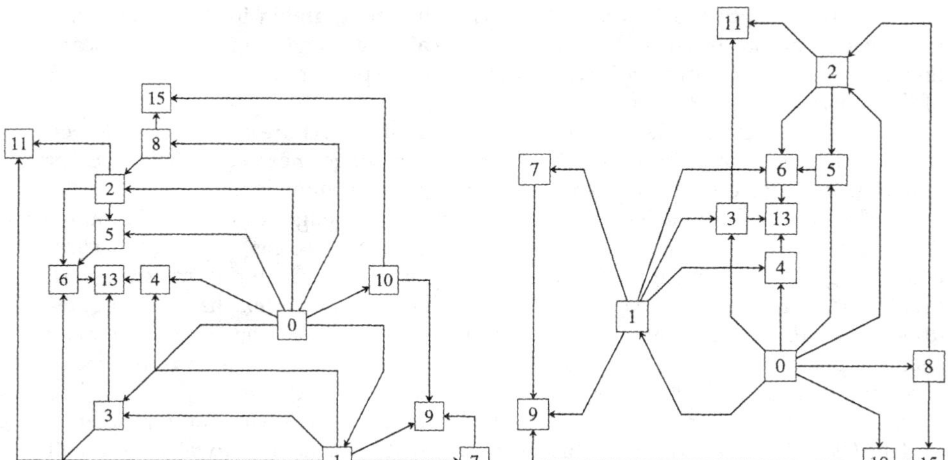

Figure 3 Two drawings of a graph produced using planarization

(3,11) is very difficult to read. Like this algorithm, many planar drawing algorithms still need to be adapted so that they are able to deal correctly with planarized graphs. However, the orthogonal bend minimization algorithm based on network flow and many of its variants (see Section 4) automatically draw crossings in the correct way. Figure 3(b) shows the graph drawn using the quasi-orthogonal algorithm based on network flow (Klau and Mutzel, 1998; Tamassia, 1987). For further reading on crossing minimization we refer the reader to (Ziegler, 2000).

Since it is very time consuming to implement a drawing method based on planarization, only a few implementations exist; these are GDToolkit (GDToolkit, 2000), an object-oriented library for handling and drawing graphs, Graph Drawing Server (Graph Drawing Server, 2000), a graph drawing and translation service; and AGD (AGD, 2000; Mutzel et al., 1998), a library of algorithms for graph drawing.

4. Planar Drawing Methods

We will focus on the two main classes of planar drawing methods used in practice. Both classes generate drawings with integer coordinates in the plane (grid drawings) and keep the drawing size polynomially bounded.

The first class runs in linear time, is based on a data structure called canonical ordering, and works in two phases. In the first phase, a canonical ordering of the vertices is computed. In the second phase, the vertices are placed iteratively in the plane. Whenever one vertex or a chain of vertices is placed, some of the vertices already placed are shifted to some extent to the right.

This method was first proposed by De Fraysseix et al. (1990). They settled the long-standing conjecture that there is always a planar straight line graph drawing of a planar graph with n vertices of a size that is polynomially bounded in n. Their algorithm produced such a drawing of size $(2n-4) \times (n-2)$.

Figure 4 shows the big drawbacks of this theoretically perfect algorithm for practical usage. Since the algorithm needs as input a triangulated graph, many artificial edges need to be introduced that will be deleted afterward. Since vertices are always placed above vertices which have already been placed, the shape of the drawings produced is always triangular. Although the size of $(2n-4) \times (n-2)$ appears small, in practice, this is far too big. Since the grid size is so big, in order to fit on a screen, the whole drawing needs to be shrunk so that it is hard to read the text in the labels.

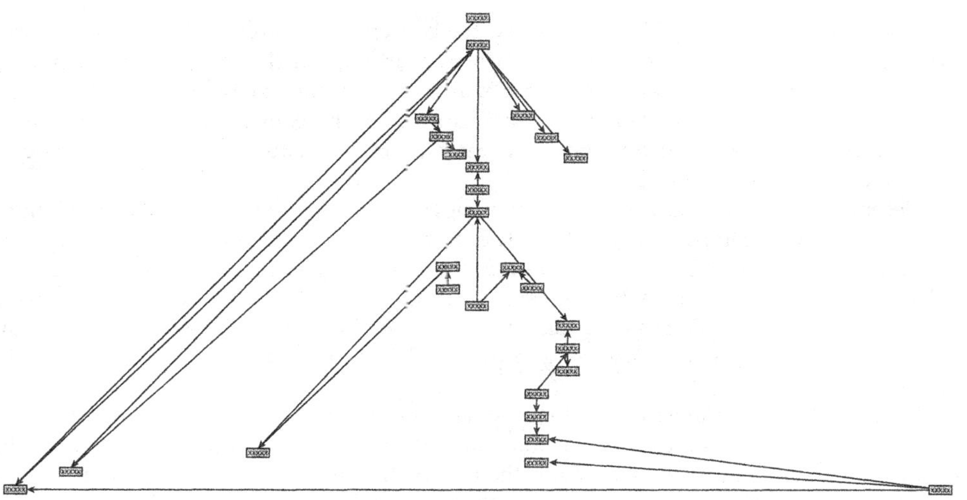

Figure 4 A drawing of the database graph using the algorithm in De Fraysseix et al. (1990)

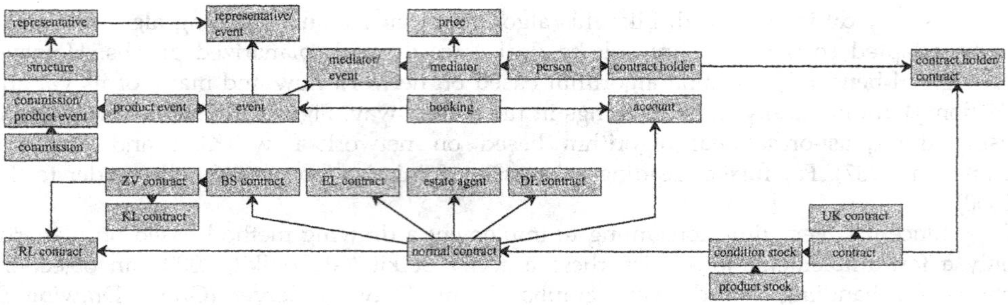

Figure 5 A drawing in mixed-model style using AGD (AGD, 2000)

Hence, a lot of effort went into improving the algorithm. The area bound has been improved to $(n-2) \times (n-2)$, the angular resolution was improved by allowing up to three bends in the drawings, and the applicability was improved by generalizing the canonical ordering from triangulated to biconnected graphs. Eventually, this lead to a set of practically useful algorithms, one of which is called the mixed-model algorithm (Gutwenger and Mutzel, 1998). Figure 5 shows a drawing of our database graph using the mixed-model implementation of AGD.

The second class of planar drawing algorithms is based on network flow algorithms. It originated from Tamassia (1987), in which the author suggested a polynomial time algorithm for computing a bend-minimal orthogonal drawing for a graph G with vertex degree bounded by four when a fixed planar combinatorial embedding of G is given. The algorithm proceeds in two steps.

First, the graph is transformed into a network in which a minimum-cost flow (see Chapter 8.4) is computed. From the flow, the angles between adjacent edges and the bends can be determined thus fixing the shape of the drawing (orthogonal representation).

In the second step, the task is to determine the lengths of the edge segments that are consistent with the given orthogonal representation. The optimization goal is to keep the total edge length, the maximal edge length, or the total drawing area small. Many compaction algorithms are known, however, they only work in one dimension. Since two-dimensional compaction is NP-hard, the heuristics first iteratively use the one-dimensional algorithms in a horizontal direction, and then in a vertical direction, and repeat this until no improvement is made.

Bridgeman et al. (1999) have introduced the concept of turn-regularity and show that for a turn-regular orthogonal representation H a planar orthogonal drawing of H with minimum area can be computed in linear time, and a planar orthogonal drawing of H with minimum area and minimum total edge length within that area can be computed in $O(n^{7/4} \log n)$ time. Although, in general, a given orthogonal representation is not turn-regular, the ideas can be used to obtain a new heuristic method.

Recently, Klau and Mutzel (1999) have suggested a new combinatorial formulation for the two-dimensional compaction problem. The new formulation introduces the concept of so-called constraint graphs. The set of feasible solutions to the two-dimensional compaction problem can be characterized by the existence of certain paths in the set of extended constraint graphs. The new characterization leads directly to an integer linear program that can be attacked via branch-and-cut thus yielding a practically efficient algorithm for two-dimensional compaction.

The drawback of Tamassia's (1987) original bend minimization algorithm is that it only works for graphs with vertex degree bounded by four. Many efforts have been made to generalize it. The bound four comes from the observation that at a vertex v in a grid there is only space for one incident edge leaving v at the top, at the bottom, to the left and to the right. In order to get around this problem, the following ideas have been proposed.

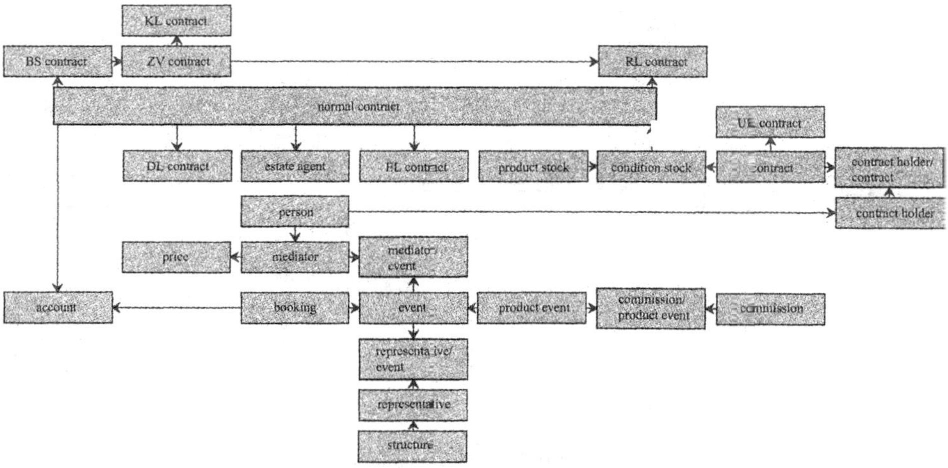

Figure 6 A drawing of the database graph with increased vertex sizes

- The size of the vertices of high degree can be increased (e.g., Batini et al., 1984; Fößmeier and Kaufmann, 1996). The drawback of this method is that in some applications the sizes of the given vertices are not allowed to be changed. Moreover, some vertices get much bigger than others, although their degree is not really that much larger. Figure 6 shows the database graph drawn using this idea (implementation in AGD).
- Vertices of degree $k>4$ can be transformed into a set of vertices of degree smaller or equal to four (Klau and Mutzel, 1998). The drawback of this approach is that the number of bends is not the minimum anymore. However, the drawings look pleasant in many cases. A drawing based on this idea is shown in Figure 7 (implementation in AGD).
- Two grids are used, a coarse grid for the vertices and a fine grid for the edges (Fößmeier and Kaufmann, 1996). This allows more incident edges to leave a vertex at the same side. The drawback of this method is that the model gets much more complex and the

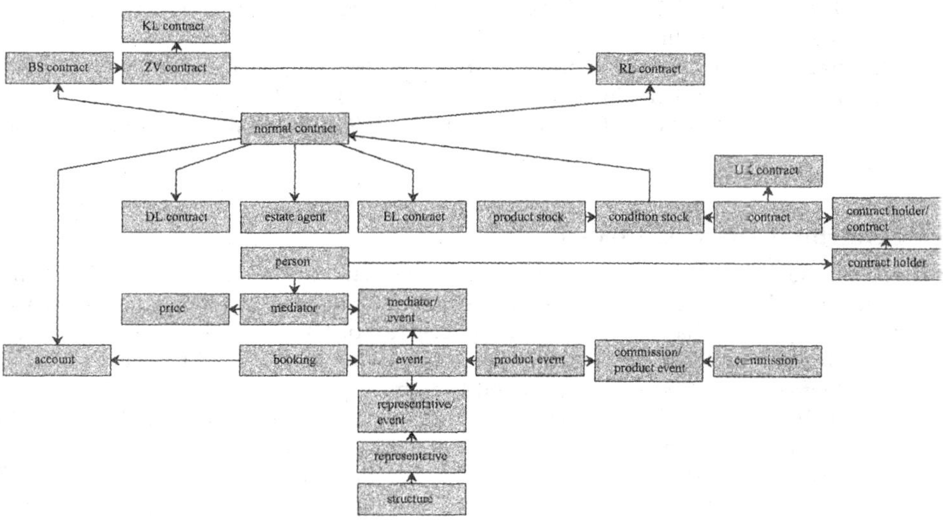

Figure 7 A drawing in quasi-orthogonal style (Klau and Mutzel, 1998)

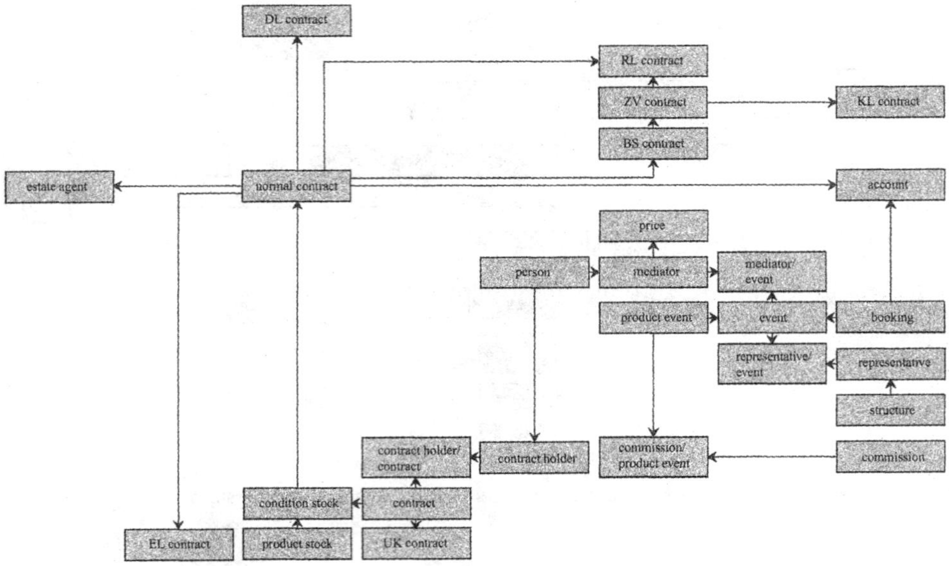

Figure 8 A drawing in Kandinsky style (AGD, 2000)

problem of assigning consistent lengths to the edge segments has not been solved satisfactorily. Figure 8 shows a drawing using the implementation in AGD.

Practical experiments show that, indeed, drawings with fewer bends look nicer and get smaller. Also theoretically one can show that the drawing size is bounded by the number of bends. For a graph $G = (V, E)$ with bounded vertex degree four there is always an orthogonal planar drawing that needs at most $2 \mid V \mid - \mid E \mid + B$ grid lines, where B is the number of bends.

However, the number of bends in a drawing depends on the combinatorial embedding chosen. It is NP-hard to compute a planar orthogonal drawing of a planar graph of vertex degree at most four with the minimum number of bends over the set of all combinatorial embeddings. Recently, Bertolazzi et al. (1998a) have presented an algorithm based on branch-and-bound (see Chapter 3.2) and SPQR-trees, which is able to solve moderately sized instances of the problem to optimality.

Several software tools exist for planar graph drawing. Many different kinds of planar drawing algorithms are contained in the system PIGALE (PIGALE, 2000), a graph drawing toolkit, and in AGD (AGD, 2000, Mutzel et al., 1998), a library of algorithms for graph drawing.

5. Hierarchical Drawing Methods

Sometimes, the given data contains a hierarchical structure. This structure should be displayed in the layout. Hierarchical structures appear, for example, in dataflow diagrams, workflow diagrams, or activity charts. Formally, a directed graph G is called hierarchic if it does not contain a directed cycle. In hierarchical diagrams, all the edges should point in one direction, for example, from top to bottom or from left to right.

By far the most practical software used for laying out hierarchical data is based on the layering method suggested by Sugiyama et al. in 1981 (Sugiyama et al., 1981).

Their algorithm proceeds in three steps. In the first step, the vertex set V of the directed graph $G = (V, E)$ is partitioned into k sets $L_1, \ldots, L_k$, called the layers of G, so that for each edge $(u, v) \in E$, $u \in V_i$, $v \in V_j$ with $i, j \in \{1, \ldots, k\}$ the condition $i < j$ holds. All vertices in the same set L_i, $i = 1, \ldots, k$, finally get the same y-coordinate, thus ensuring that all edges point into the same

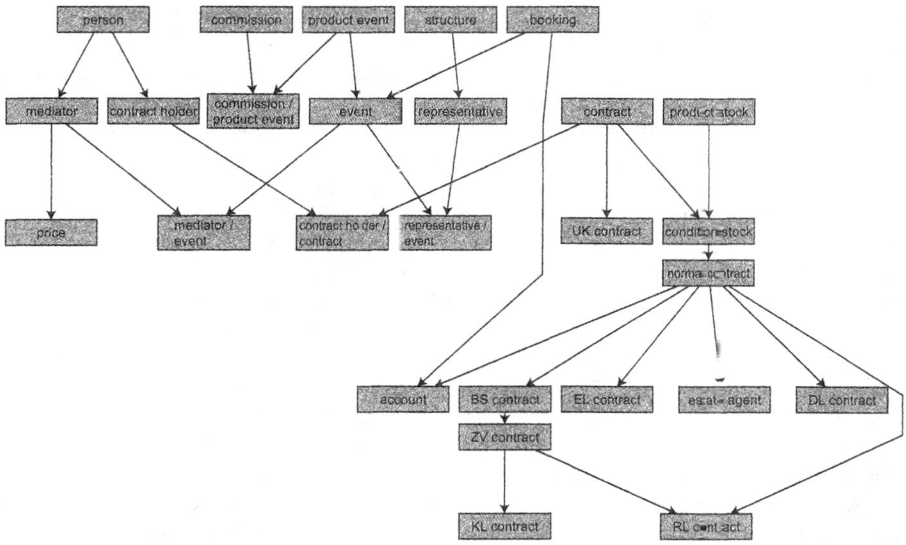

Figure 9 A Sugiyama drawing of the database graph (using AGD)

direction, for example, downward. In the second step, the vertices within each layer are ordered so that the number of edge crossings gets small. In the final step, the absolute y-coordinates of each layer and the x-coordinates of each vertex are computed. Figure 9 shows a Sugiyama-style drawing of our database graph (AGD, 2000).

Some details of the algorithm still need to be discussed. In some cases, the given graph is almost hierarchical. Here, the optimization problem that arises is to delete as few edges as possible in order to obtain a hierarchical graph. This problem is called the minimum feedback arcset problem and has already been investigated in Grötschel et al. (1990). Recent experiments using AGD (AGD, 2000) show that their algorithm based on polyhedral combinatorics and branch-and-cut is able to solve large instances of the problem to provable optimality in a very short computation time.

For Step 1 of the Sugiyama approach various methods have been suggested. Topological sorting computes a layering with minimal height. The Coffman–Graham algorithm finds a layering of width at most w and height $h \leqslant (2-2/w)h_{\min}$, where $h_{\min}$ is the minimum height of a layering of width w. Gansner et al. (1993) have suggested an integer linear program that can be solved in polynomial time for finding a layering with the minimum total edge length in y-direction. Recently, Healy and Nikolov (2000) have suggested an algorithm based on integer linear programming, which attacks the problem of computing a layering subject to any dimension constraints concerning the width and the height of the drawing.

The crucial step of the Sugiyama approach is the crossing minimization step. Crossing minimization for a layered graph is NP-hard even for two layers. Jünger et al. (1997) have suggested an integer linear programming formulation for the layered crossing minimization problem. Further polyhedral studies in Jünger et al. (1997) and Healy and Kuusik (1999) turned out to be useful in order to get a practically efficient algorithm for solving instances of moderate size to provable optimality (Healy and Kuusik, 1999). Since the branch-and-cut code is quite involved, so far only heuristics are used in practice.

Before the heuristics can be applied, the graph $G=(V,E\ L_1,\dots,L_k)$ needs to be transformed into a proper hierarchy $G'=(V',E',L'_1,\dots,L'_k)$ in which each edge (u, v) with $u \in L_i$, $v \in L_j$, and $i<j-1$ is substituted by a path of edges $(u,w_1),(w_1,w_2),\dots,(w_{j-i-1},v)$, $w_l \in L_{i+l}$ for $l=1,\dots,j-i-1$. The heuristics are all based on the idea of successively considering two adjacent layers in which the ordering of the vertices in the first layer L_1 is already fixed

whereas an ordering in the second layer L_2 is searched so that the number of crossings between the set of edges $\{(u,v) \in E' \mid u \in L_1', v \in L_2'\}$ is minimized. The edges are drawn as a straight line between two adjacent layers. Since this fixed two-layer crossing minimization problem is also NP-hard, many heuristics for solving the problem have been suggested in the literature (see, e.g., Di Battista et al., 1999). Jünger and Mutzel (1997) have transformed the problem into the linear ordering problem, which can be solved efficiently in practice using polyhedral combinatorics and branch-and-cut (Grötschel et al., 1990).

Mutzel (1995) has suggested an alternative approach for Step 2, namely, counting the number of edges that are crossed in the drawing instead of counting the crossings. The idea is based on the planarization idea for general graph drawing. The problem is to find the minimum number of edges that need to be deleted so that the resulting layered graph contains no crossings. In Mutzel (1995) a branch-and-cut algorithm is suggested for solving the two-layer planarization problem to optimality. Recently, Healy and Kuusik (1999) have generalized this algorithm to $k > 2$ layers. Their approach is based on the concept of the vertex-exchange graph. Planar layered graphs $G = (V, E, L_1, \ldots, L_k)$ can be characterized via simple forbidden substructures in the associated vertex-exchange graph. This directly yields a simple planarity testing algorithm for layered graphs. However, the running time is $O(n^2)$ for a graph with n vertices, whereas Jünger et al. (1998) have developed a linear algorithm. The above characterization can also be used to get minimal nonplanar substructures in a layered graph and thus yields an important basis for the integer linear programming approach. Experiments in Healy and Kuusik (1999) show that their approach leads to a practically efficient algorithm that can drastically improve hierachical drawings.

An alternative to the Sugiyama approach is to combine Steps 1 and 2 in one step. Here, the question arises if, for a given hierarchical graph, a layering and a permutation of the vertices within the layers exists so that the associated layered drawing does not contain edge crossings. This problem is called the upward planarity testing problem and is, in general, NP-hard. However, if the hierarchical graph contains only one source or sink, upward planarity can be tested in linear time (Bertolazzi et al., 1998b). This may be the starting point for a practically efficient alternative algorithm in hierarchical graph drawing. Once an upward planar combinatorial embedding is computed, an upward planar drawing algorithm can be used to draw the graph.

An overview of hierarchical drawing methods is given, for example, in Di Battista et al. (1999). The idea of the Sugiyama method can also be applied to three dimensions (cone trees).

There is a vast amount of software tools using the Sugiyama approach. A classical tool is the program dag (Gansner et al., 1993), a program that draws directed graphs, and a relatively diverse tool is GDToolkit (GDToolkit, 2000), an object-oriented library for handling and drawing graphs.

6. Miscellaneous

Graph drawing has become very popular in recent years. This article can only cover a small part of it. Many more results in algorithms and optimization have been obtained. We will only mention a few others briefly in the following.

Recently, many researchers have investigated three-dimensional straightline and orthogonal drawing. The questions studied so far concern bounds on the volume and the number of bends in these drawings.

Another line of research is to draw large (or huge) graphs that cannot be displayed on the screen. The idea is to successively contract vertices to clusters and to display the resulting graph. This yields two new problems: the problem of finding good clusters and of displaying clustered graphs.

Further popular graph drawing problems appear in dynamic or interactive graph drawing. Here, the users are allowed to add or delete some vertices or edges of a displayed graph. The question arises as to how to redraw the layout without changing the so-called mental map of the drawing.

Symmetry detection (e.g., axial, central or rotational symmetries) is still a challenging task, although many results have recently been obtained in this field.

For further reading, we suggest the book on graph drawing (Di Battista et al., 1999) and the yearly proceedings of the International Symposium on graph drawing (e.g., Kratochvíl, 1999).

References

AGD. 2000 *A library of algorithms for graph drawing*. Max-Planck-Institut Saarbrücken, TU Wien, Universität Halle, Universität Köln. Available via http://www.ads.tuwien.ac.at/AGD/.

Batini, C., M. Talamo, and R. Tamassia. 1984. "Computer aided layout of entity relationship diagrams." *Journal of Systems and Software* **4**: 163–173.

Bertault, F. 1999. "A force-directed algorithm that preserves edge crossing properties." In *Graph Drawing (Proc. GD '99)*, Volume 1731, edited by J. Kratochvíl. Lecture Notes in Computer Science, pp. 351–358. Berlin: Springer-Verlag.

Bertolazzi, P., G. Di Battista, and W. Didimo. 1998a. "Computing orthogonal drawings with the minimum number of bends." In *WADS '97*, Volume 1272, Lecture Notes in Computer Science, pp. 331–344. Berlin: Springer-Verlag.

Bertolazzi, P., G. Di Battista, C. Mannino, and R. Tamassia. 1998b. "Optimal upward planarity testing of single-source digraphs." *SIAM J. Comput.* **27(1)**: 132–169.

Bhatt, S. N. and F. T. Leighton. 1984. "A framework for solving VLSI layout problems." *J. Comput. Sys. Sci.* **28**: 300–343.

Brandenburg F. and M. Himsolt. 2000. *Graphlet*. Available via http://infosun.fmi.uni-passau.de/Graphlet/.

Brandenburg, F. J., M. Himsolt, and C. Rohrer. 1996. "An experimental comparison of force-directed and randomized graph drawing algorithms." In *Graph Drawing (Proc. GD '95)*, Volume 1072, edited by J. Brandenburg, Lecture Notes in Computer Science, pp. 76–87. Berlin: Springer-Verlag.

Bridgeman, S. S., G. Di Battista, W. Didimo, G. Liotta, R. Tamassia, and L. Vismara. 1999. "Turn-regularity and planar orthogonal drawings." In *Graph Drawing (Proc. GD '99)*, Volume 1731, Lecture Notes in Computer Science, pp. 8–26 Kratochvíl (ed.). Springer-Verlag.

Calinescu, G., C. G. Fernandes, U. Finkler, and H. J. Karloff. 1998. "A better approximation algorithm for finding planar subgraphs." *J. Algorithms* **27(2)**: 269–302.

Chung, F. R. K. and S. T. Yau. 1994. "A near optimal algorithm for edge separators." In *ACM Symposium on the Theory of Computation (Proc.)*, pp. 1–8. New York: ACM.

De Fraysseix, H., J. Pach, and R. Pollack. 1990. "How to draw a planar graph on a grid." *Combinatorica* **10(1)**: 41–51.

Di Battista, G. and R. Tamassia. 1990. "On-line graph algorithms with SPQR-trees." In *Automata, Languages and Programming (Proc. 17th ICALP)*, Volume 442, Lecture Notes in Computer Science, pp. 598–611. Berlin: Springer-Verlag.

Di Battista, G., P. Eades, R. Tamassia, and I. G. Tollis. 1999. In *Graph Drawing*. Upper Saddle River, New Jersey 07548: Prentice Hall.

Eades, P. 1984. "A heuristic for graph drawing." *Congr. Numer.* **42**: 149–160.

Fößmeier, U. and M. Kaufmann. 1996. "Drawing high degree graphs with low bend numbers." In *Graph Drawing (Proc. GD '95)*, Volume 1027, edited by F. J. Brandenburg, Lecture Notes in Computer Science, pp. 254–266. Berlin: Springer-Verlag.

Gansner, E. R. and S. C. North. 1998. "Improved force-directed layouts." In *Graph Drawing (Proc. GD '99)*, Volume 1547, edited by S. H. Whitesides, Lecture Notes in Computer Science, pp. 364–373. Berlin: Springer-Verlag.

Gansner, E., E. Koutsofios, S. North, and K. Vo. 1993. "A technique for drawing directed graphs." In *IEEE Transactions on Software Engineering* **19(3)**: 214–229. Piscataway, NJ: IEEE.

GDToolkit. 2000. Available at http://www.dia.uniroma3.it/~gdt.

Graph Drawing Server. 2000. Available at http://www.cs.brown.edu/cgc/graphserver.

Grötschel, M., M. Jünger, and G. Reinelt. 1990. "Facets of the linear ordering polytope." *Math. Programming* **33**: 43–60.

Gutwenger, C. and P. Mutzel. 1998. "Planar polyline drawings with good angular resolution." In *Graph Drawing, 6th International Symposium, GD '98*, Volume 1547, edited by S. H. Whitesides, Lecture Notes in Computer Science, pp. 167–182. Berlin: Springer-Verlag.

Gutwenger, C., P. Mutzel, and R. Weiskircher. 2000. "Inserting an edge into a planar graph." In ACM-SIAM Symposium on Discrete Algorithms (Proc.), pp. 246–255. New York: ACM.

Healy, P. and A. Kuusik. 1999. "The vertex-exchange graph: a new concept for multi-level crossing minimization." In *Graph Drawing (Proc. GD '99)*, Volume 1731, edited by J. Kratochvíl, Lecture Notes in Computer Science, pp. 205–216. Berlin: Springer-Verlag.

Healy, P. and N. S. Nikolov. 2000. "An optimal solution to dimension-constrained graph layering." Technical Report. Limerick, Ireland: Department of Computer Science, University of Limerick.

Jünger, M., E. Lee, P. Mutzel, and T. Odenthal. 1997. "A polyhedral approach to the multi-layer crossing number problem." In *Graph Drawing (Proc. GD '97)*, Volume 1353, edited by G. Di Battista, Lecture Notes in Computer Science, pp. 13–24. Berlin: Springer-Verlag.

Jünger, M., S. Leipert, and P. Mutzel. 1998. "Level planarity testing in linear time." In *Graph Drawing, 6th International Symposium, GD '98*, Volume 1547, edited by S. H. Whitesides, Lecture Notes in Computer Science, pp. 224–237. Berlin: Springer-Verlag.

Jünger, M. and P. Mutzel. 1996. "Maximum planar subgraphs and nice embeddings: practical layout tools." *Algorithmica* **16(1)**: 33–59.

Jünger, M. and P. Mutzel. 1997. "2-layer straightline crossing minimization: performance of exact and heuristic algorithms." *J. Graph Algorithms Appl.* **1(1)**: 1–25.

Klau, G. W. and P. Mutzel. 1998. "Quasi-orthogonal drawing of planar graphs." Technical Report MPI-I-98-1-0130. Saarbrücken, Germany: Max-Planck-Institut für Informatik.

Klau, G. W. and P. Mutzel. 1999. "Optimal compaction of orthogonal grid drawings." In *Integer Programming and Combinatorial Optimization (IPCO '99), Proceedings of the Seventh Conference*, Volume 1610, edited by G. P. Cornuejols, Lecture Notes in Computer Science, pp. 304–319. Berlin: Springer-Verlag.

Kratochvíl, J., (ed). 1999. In *Graph Drawing (Proc. GD '99)* Volume 1731, Lecture Notes in Computer Science. Berlin: Springer-Verlag.

Mutzel, P. 1995. "A polyhedral approach to planar augmentation and related problems." In *Algorithms—ESA '95, Third Annual European Symposium*, Volume 979, edited by P. Spirakis, Lecture Notes in Computer Science, pp. 494–507. Berlin: Springer-Verlag.

Mutzel, P., C. Gutwenger, R. Brockenauer, S. Fialko, G. Klau, M. Krüger, T. Ziegler, S. Näher, D. Alberts, D. Ambras, G. Koch, M. Jünger, C. Buchheim, and S. Leipert. 1998. "AGD: a library of algorithms for graph drawing (poster-abstract)." In *Graph Drawing, 6th International Symposium, GD '98*, Volume 1547, edited by S. Whitesides, Lecture Notes in Computer Science, pp. 456–457. Berlin, Germany: Springer-Verlag.

Mutzel, P. and T. Ziegler. 1999. "The constrained crossing minimization problem." In *Graph Drawing, 7th International Symposium, GD '99* Volume 1731, edited by J. Kratochvíl, Lecture Notes in Computer Science, pp. 175–185. Berlin, Germany: Springer-Verlag.

PIGALE. 2000. Available at ftp://www.pr.cams.ehess.fr/pub/pigale.tgz.

Rauh, O. and E. Stickel. 1997. In *Fallstudien zum Datenbankentwurf*. Th. Gabler, Wiesbaden.

Sugiyama, K., S. Tagawa, and M. Toda. 1981. "Methods for visual understanding of hierarchical systems." *IEEE Trans. Sys. Man Cybernet.* **11(2)**: 109–125.

Tamassia, R. 1987. "On embedding a graph in the grid with the minimum number of bends." *SIAM J. Comput.* **16(3)**: 421–444.

Ziegler, T. 2000. "Crossing minimization in automatic graph drawing." Ph.D. diss. Saarbrücken: Max-Planck-Institut für Informatik.

18.20 Optimization for Modeling of Nonlinear Interactions in Mechanics

G. E. Stavroulakis

ABSTRACT General nonlinear interaction laws in mechanics can be expressed by means of monotone or nonmonotone, possibly multivalued, relations. Mathematical optimization provides a systematic and rigorous framework for modeling and numerical calculation of coupled field problems (for example, in fluid-structure or structure-foundation cases) with the above mentioned nonlinear interaction effects. The link is established through convex and nonconvex, in general nonsmooth, potential energy optimization. The optimality (respectively, critical point) conditions of the energy minimization problem are the governing relations of the mechanical model. A number of typical applications, including unilateral and adhesive contact modeling, are discussed in this article. Simple examples demonstrate the proposed approach. Hints for further applications, including the modeling of shape memory alloys cf smart structures, and for efficient, parallel solution techniques are also included with appropriate references.

1. Introduction

Highly nonlinear interaction effects can be described by using the theory of variational inequalities and of its extensions (for example, implicit variational inequalities or hemivariational inequalities). These variational or weak formulations are tightly connected to the critical point conditions for appropriately defined, convex or nonconvex and generally nonsmooth energy potentials. In this framework the kinematical restriction of nonpenetration, which arises in a unilateral contact problem, leads to a variational inequality problem, which should be seen as a generalization of the classical variational equality description of the nonlinear interaction problem. Furthermore, again on this unilateral contact example, when contact does not take place between two parts of a structure, a nonnegative gap arises. Clearly, contact tractions (which are only compressive) and gaps are complementary variables, since one of them must be equal to zero. One of the main difficulties of these problems is the appearance of a variable structure, which depends on the external action (loading) and the corresponding difficulty in determining the free boundaries. In this example one has to determine the areas of contact and separation. Optimization and complementarity help us model the problem in a rigorous way and use suitable algorithms for its numerical solution. Analogously one treats more complicated nonlinear interaction effects, which include problems with free boundaries or a varying mechanical behavior (for example, plasticity-like interfaces, multiphase modeling in fluids, etc.).

Extension to nonconvex problems leads to systems of variational inequalities or implicit (mixed) variational inequalities or hemivariational inequalities. Global energy optimization problems or, less ambitious, critical-point problems arise in this framework. This class of problems includes delamination modeling in composites, adhesive interactions, and pseudoelastic modeling of shape memory layers.

Variational inequalities and complementarity problems arise in various scientific disciplines, as is discussed in several places in this handbook. Thus, the formulation of highly nonlinear structural analysis and optimization problems within this framework is meaningful. By doing so one is able to use results and established algorithms for the solution of industrial applications. The same problems are usually very difficult if they are modeled with more classical methods and if black-box-type, iterative algorithms are used for their solution. Thus, from the engineering point of view, mathematical optimization gives us more reliable iterative methods and, what is more important, a useful view and understanding of the mechanical behavior of complicated structures (for example, multiplicity of solutions coming from multiple local minima of a potential energy, stable and unstable solutions due to critical points, etc.)

In this paper, monotone and nonmonotone interaction laws with complete vertical branches (the so-called multivalued or complete laws) are considered. They are derived from convex and nonconvex, in general, nondifferentiable (nonsmooth), potential functions. Virtual work relations, which are related to critical-point conditions involving directional derivatives, lead to the above-mentioned variational and hemivariational inequalities. Elastostatic analysis problems in mechanics are used as model examples, where several parts of an elastic structure are connected through nonlinear interaction laws. The problems that arise are outlined and appropriate references are given for further details. A few useful application details, which include finite element and boundary element formulations, and further mechanical models are discussed in the last section of this paper. Hints for effective numerical methods are also given. It is mentioned that the nature of the mechanical application considered can be exploited by parallel numerical solution algorithms, but this discussion exceeds the scope of this review article. Extension to other coupled-field models in computational science (e.g., fluid-structure interactions) are not discussed in detail here. It is hoped that the interested reader will be able to develop the corresponding models without difficulty.

2. Monotone Laws and Nonconvex Critical Points

Elements of nonlinear, monotone, and possibly multivalued interaction laws, of the corresponding potentials, and of the resulting variational forms as they arise in the modeling of mechanical systems are given in this section. For smooth potentials, critical points are characterized as solutions of nonlinear equations (i.e., the gradient of the function vanishes) or of variational equalities (i.e., the directional derivative is equal to zero for every admissible direction). For convex nonsmooth potentials one has, respectively, a convex set-valued equation (i.e., inclusion) and a variational inequality problem. Details of more complicated problems can be found in Panagiotopoulos (1985) and Mistakidis and Stavroulakis (1998).

2.1. *One-Dimensional Spring*

Let us assume a simple one-dimensional elastic spring with the spring stiffness c. The linear elastic constitutive law of the spring, which connects stress s and deformation e, that is,

$$s = ce, \tag{1}$$

can be derived by differentiating the following spring elastic potential:

$$\pi(e) = \tfrac{1}{2}\, ce^2. \tag{2}$$

2.2. *Locking Spring*

For the nonlinear, monotone constitutive law of Figure 1(a), that is,

$$s = \begin{cases} s_2 + c_1 e, & \text{for} \quad e \geqslant 0, \\ [-s_1, s_2], & \text{for} \quad e = 0, \\ -s_1 + c_2 e, & \text{for} \quad e \leqslant 0, \end{cases} \tag{3}$$

one needs the introduction of a nondifferentiable, convex potential (the so-called superpotential):

$$\pi(e) = \max\{\pi_1(e), \pi_2(e)\} = \max\left\{s_2 e + \tfrac{1}{2}\, c_1 e^2,\ -s_1 e + \tfrac{1}{2}\, c_2 e^2\right\}. \tag{4}$$

At the point of nondifferentiability, $e = 0$, the vertical branch of relation (3), that is, the interval $[-s_1, s_2]$, is produced by the convex analysis subdifferential ∂ of the nonsmooth function $\pi(e)$,

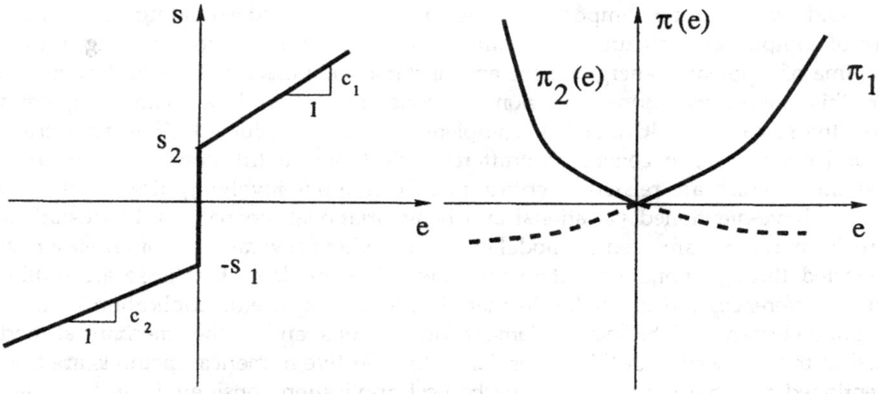

Figure 1 A locking spring, of a plastic-hardening type. (a) The law and (b) the corresponding nonsmooth potential

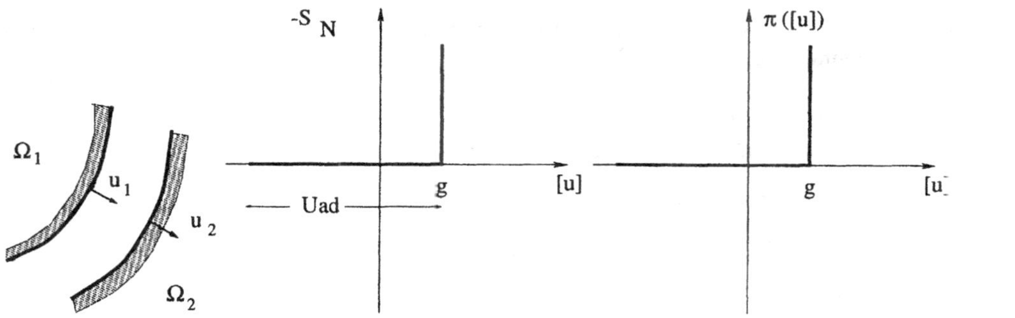

Figure 2 A unilateral contact element. (a) Mechanical interface, (b) multivalued law, and (c) corresponding nonsmooth potential

which is constructed by means of the convex hull of the classical gradients of the two smooth constituents, that is, $\partial\pi(e=0)=\text{co}\{\nabla\pi_1(\epsilon),\nabla\pi_2(e)\}$.

2.3. *Unilateral Contact*

Let us assume further the unilateral contact relation between two parts Ω_1 and Ω_2 of a deformable structure. See for example the two-dimensional example of Figure 2(a). Using the notion of relative normal displacement $[u]=u_2-u_1$, and the initial distance g, the unilateral contact relations read:

$$[u]-g\leqslant 0,\ -S_N\geqslant 0,\ -S_N([u]-g)=0. \tag{5}$$

Here S_N denotes the contact traction. The multivalued, monotone contact law reads:

$$-S_N=\begin{cases} 0, & \text{for}\quad [u]\leqslant g,\\ [0,+\infty], & \text{for}\quad [u]=g. \end{cases} \tag{6}$$

The latter relation can be produced by subdifferentiating the following potential function, which has the form of an indicator function $I_{U_{\text{ad}}}$ for the kinematically admissible relative displacements set: $U_{\text{ad}}=\{[u]\in R^1,[u]-g\leqslant 0\}$,

$$\pi([u])=I_{U_{\text{ad}}}([u])=\begin{cases} 0, & \text{for}\quad [u]\leqslant g,\\ +\infty, & \text{for}\quad [u]=g. \end{cases} \tag{7}$$

Let us consider further a finite element discretized structure, with a number of unilateral contact interfaces whose nonpenetration relation can be written in matricial form as:

$$\mathbf{Nu}-\mathbf{g}\leqslant\mathbf{0}. \tag{8}$$

The elastostatic analysis problem, in view of restrictions (8), can be derived from the minimization of the potential energy:

$$\min_{\mathbf{u}\in\mathbf{R}^n\cap\{\mathbf{u}:\mathbf{Nu}-\mathbf{g}\leqslant\mathbf{0}\}}\left\{\Pi(\mathbf{u})=\tfrac{1}{2}\,\mathbf{u}^T\mathbf{K}\mathbf{u}-\mathbf{p}^T\mathbf{u}\right\}. \tag{9}$$

Here $\mathbf{K}$ is the stiffness matrix of the discretized structure and $\mathbf{p}$ is the loading vector. For the quadratic minimization problem (9) the Karush–Kuhn–Tucker (KKT) optimality conditions read:

$$\mathbf{Ku}=\mathbf{p}-\mathbf{N}^T\lambda, \tag{10}$$

$$\mathbf{Nu}-\mathbf{g}\leqslant\mathbf{0},\quad \lambda\geqslant\mathbf{0},\quad (\mathbf{Nu}-\mathbf{g})^T\lambda=\mathbf{0}.$$

One recognizes in the previous relations the equilibrium of the structure, which includes the contribution of the contact stresses represented by the Lagrange multipliers λ, and the complete set of inequalities and complementarity relations of (5), (8). Problem (10) is a linear complementarity problem (LCP).

2.4. *Variational Inequalities*

Let us assume the min-type nonsmooth potential:

$$\pi(e) = \max_{e}\{c_1 e, -c_2 e\}, \tag{11}$$

and derive from it the following subdifferential law:

$$s \in \partial\pi(e). \tag{12}$$

For the definition of the convex analysis subdifferential the reader is referred, for example, to Ekeland and Temam (1976) or to Sewell (1987; see also relation (17)). It is clear that the law (12) includes complete vertical branches, for example for $e = 0$ it reads $s \in [-c_2, c_1]$. Relation (12) is by definition equivalent to the variational inequality:

$$s(e)\delta e \leqslant \pi(e + \delta e) - \pi(e), \quad \forall e, \delta e \in R^1. \tag{13}$$

On the other hand the directional derivative of function $\pi(e)$ at point e in the direction δe reads:

$$\pi'(e, \delta e) = \max_{s \in \partial\pi(e)} \langle s, \delta e \rangle. \tag{14}$$

In this case, $\langle ., . \rangle$ denotes the scalar product, but in the general case with more than one dimensions, it denotes an inner product.

Furthermore, from relation (14) the following piecewise linear approximation of the function $\pi(e)$ can be constructed:

$$\pi(e + \delta e) = \pi(e) + \pi'(e, \delta e) = \pi(e) + \max_{s \in \partial\pi(e)} \langle s, \delta e \rangle. \tag{15}$$

It is clear that, since (15) holds, the inequality (13) can be written for every $s \in \partial\pi(e)$. Function $\pi(e)$ attains a minimum at point e if the following inclusion holds true: $0 \in \partial\pi(e)$.

In mechanics, the directional derivative plays the role of the virtual work (for a small virtual displacement or deformation δe). In this framework, the unilateral contact problem of relations (5)–(7) is expressed by:

$$-S_N \in \partial I_{U_{\mathrm{ad}}}([u]), \tag{16}$$

$$-S_N \delta[u] \leqslant I_{U_{\mathrm{ad}}}([u] + \delta[u]) - I_{U_{\mathrm{ad}}}([u]), \quad \forall \delta[u] \in R^1, \tag{17}$$

or, equivalently, by

$$-S_N \delta[u] \leqslant 0, \quad \forall \delta[u] \in U_{\mathrm{ad}}. \tag{18}$$

The above variational inequalities describe local nonlinear effects which are expressed by means of monotone, possibly multivalued, subdifferential laws (cf. (5), (12)). They can be introduced into the principle of virtual work of mechanics and lead to variational formulations of structural analysis problems which have the form of variational inequalities (see, among others, Ekeland and Temam, 1976; Panagiotopoulos, 1985; Antes and Panagiotopoulos, 1992; and Mistakidis and Stavroulakis, 1998 for more details).

2.5. *Examples*

ONE-DIMENSIONAL LAWS

The following one-dimensional law (see Figure 3(a)):

$$s = \begin{cases} [s_1, +\infty) & \text{for} \quad e = e_1 \\ c_1 e & \text{for} \quad e \leqslant e_1, \end{cases}$$

with $c_1 = \frac{s_1}{e_1}$, can be derived from the convex and nonsmooth potential

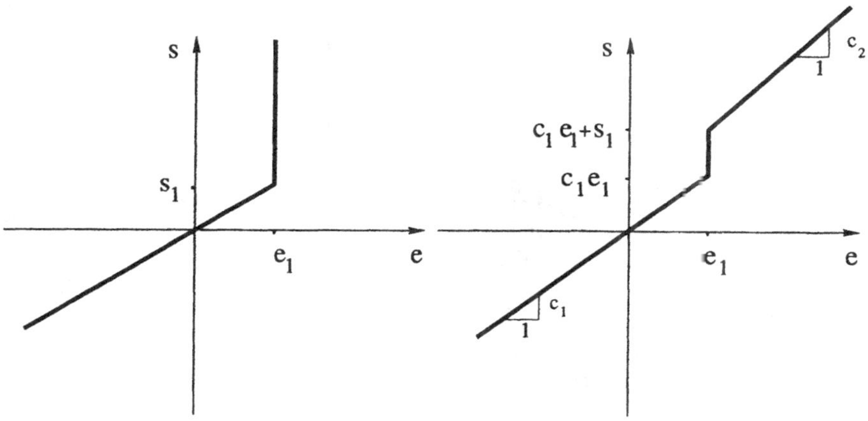

Figure 3 Two typical monotone multivalued laws

$$\pi(e) = \max\left\{\tfrac{1}{2}\, c_1 e^2, I_{(-\infty, e_1]}(e)\right\} = \begin{cases} +\infty & \text{for} \quad e > e_1, \\ \tfrac{1}{2}\, c_1 e^2 & \text{for} \quad e \leqslant e_1, \end{cases}$$

Analogously, for the law of Figure 3(b):

$$s = \begin{cases} (c_1 e_1 + s_1) + c_2 (e - e_1)_+ & \text{for} \quad e \geqslant e_1 \\ [c_1 e_1, c_1 e_1 + s_1] & \text{for} \quad e = e_1 \\ c_1 e & \text{for} \quad e \leqslant e_1, \end{cases}$$

with $x_+ = \max\{0, x\}$, one may use the following potential

$$\pi(e) = \max\left\{\tfrac{1}{2}\, c_1 e^2, \tfrac{1}{2}\, c_1 e_1^2 + (c_1 e_1 + s_1)(e - e_1) + \tfrac{1}{2}\, c_2 (e - e_1)^2\right\}.$$

A Simple Unilateral Contact Problem

Let us consider the point-mass of Figure 4 coming in frictionless contact with a rigid support. The corresponding equilibrium configuration will be calculated by means of the quadratic programming problem (9) or the linear complementarity problem (10) with:

$$\mathbf{K} = \begin{bmatrix} k_x & 0 \\ 0 & k_y \end{bmatrix}, \quad \mathbf{p} = \begin{bmatrix} P_x \\ P_y \end{bmatrix}, \quad \mathbf{u} = \begin{bmatrix} u_x \\ u_y \end{bmatrix}.$$

For the concrete application we will consider the following spring stiffness data: $k_x = k_y = 1.0$ and various loading vectors, as well as various positions of the rigid support (i.e., different contact inequality restrictions (8)).

1. Consider the following loading $P_x = 10, P_y = 10$ without the rigid support. The equilibrium position of the spring-supported mass can be found by the solution of the equilibrium conditions $\mathbf{Ku} = \mathbf{p}$, which is a system of linear equations and corresponds to the optimality (KKT) condition of the optimization problem (9). The solution is trivial: $u_x = u_y = 10$. For further reference note, that the loading vector $P_x = 10$, $P_y = -10$ without unilateral restriction leads to the solution: $u_x = 10$, $u_y = -10$.
2. Consider now the loading $P_x = 10$, $P_y = -10$ and a unilateral support that can be described by means of the inequality $u_x \leqslant 5$. One can easily see that the system (10) admits the following solution: $u_x = 5, u_y = -10$, and $\lambda = 5$. Thus, contact is realized, the corresponding nonpenetration inequality holds as an equality and the force equilibrium equation (10) is satisfied with the contribution of the contact force $\lambda = 5$.

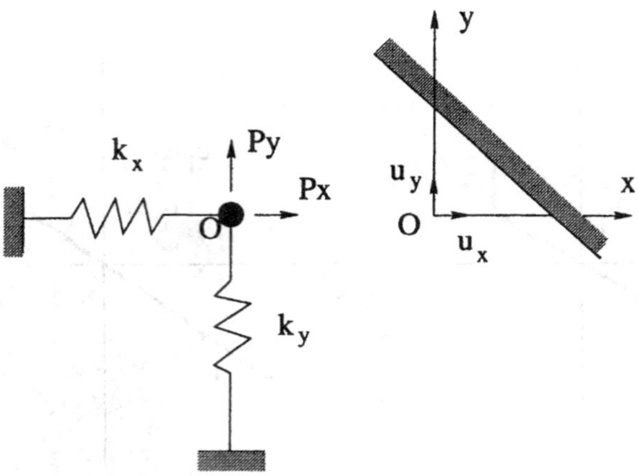

Figure 4 A small-scale unilateral contact example

3. Consider now the loading $P_x = 10$, $P_y = 10$, and unilateral supports that are expressed by the inequalities $\frac{u_x}{2} + \frac{u_y}{1} \leqslant g$. Working analogously one finds, for example, for $g = 1$ the solution: $u_x = 4.4, u_y = -1.2$, and $\lambda = 11.2$; and for $g = 0$ the solution: $u_x = 4, u_y = -2$, and $\lambda = 12$.

Of course, the solution of the previously considered small scale LCPs (or equivalently, the solution of the optimization problem (9)) was done by hand using a trial-and-error strategy. For real-life applications the use of mathematical programming and of numerical optimization algorithms is indispensable.

A MAX-TYPE POTENTIAL

Figure 5 shows a max-type function of two variables e_1 and e_2:

$$\pi(\mathbf{e}) = \max\{e_1, e_2, -e_1, -e_2\}.$$

It is clear that the point of nondifferentiability $e_1 = e_2 = 0$ is a local maximum of this function, since at this point the following set-valued inclusion holds true:

$$\begin{bmatrix} 0 \\ 0 \end{bmatrix} \subset \partial\pi(\mathbf{s} = \mathbf{0}) = \mathrm{co}\left\{ \begin{bmatrix} 1 \\ 0 \end{bmatrix}, \begin{bmatrix} 0 \\ 1 \end{bmatrix}, \begin{bmatrix} -1 \\ 0 \end{bmatrix}, \begin{bmatrix} 0 \\ -1 \end{bmatrix} \right\}.$$

Here co denotes the convex hull of the indicated set. The last function and its generalizations can be used for the construction of multidimensional nonlinear, monotone interaction laws in mechanics.

3. Nonmonotone Laws and Nonconvex Critical Points

Nonmonotone relations, possibly with complete vertical branches, can be derived from nonconvex and possibly nonsmooth potential energy functions. Relations of this kind arise, for instance, in the modeling of delamination and, in general, strength degradation effects, in the modeling of nonmonotone friction laws (cf., the so-called Stribeck curve: Pfeiffer and Glocker, 1996) and in the pseudoelastic modeling of shape memory alloys (see, for instance, Frémond, 1996; Seelecke, 1996; Pagano and Alart, 1999; Polyakova and Stavroulakis, 2000). The reader may see the schematic laws of Figure 6, the examples given at the end of this section and, for more details, the following publications: Panagiotopoulos (1993), Demyanov et al. (1996), Mistakidis and Stavroulakis (1998), and Haslinger et al. (1999).

Let us consider a nonmonotone relation and assume that the corresponding potential energy $\pi(e)$ can be expressed as a difference of convex functions (d.c. function):

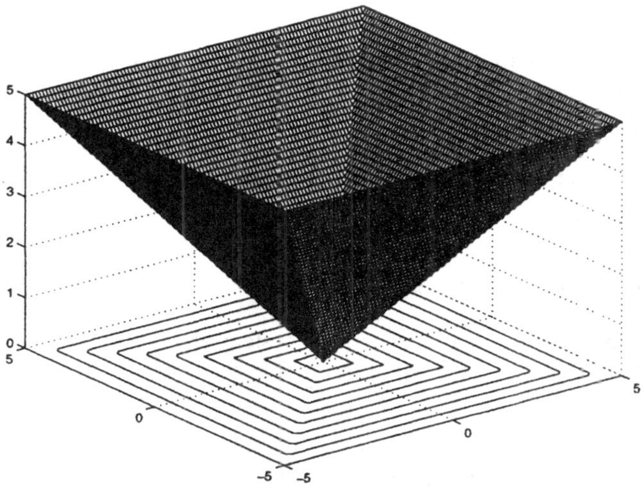

Figure 5 A convex nondifferentiable function

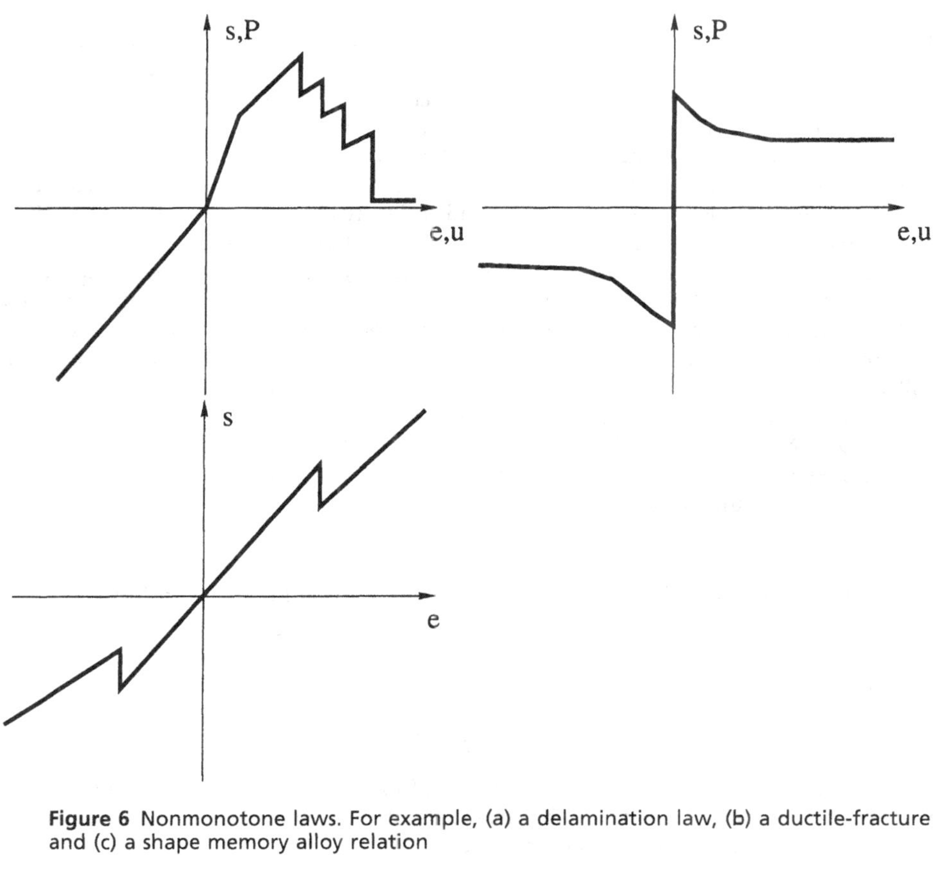

Figure 6 Nonmonotone laws. For example, (a) a delamination law, (b) a ductile-fracture law, and (c) a shape memory alloy relation

$$\pi(e) = \pi_1(e) - \pi_2(e). \tag{19}$$

Using results from d.c. optimization one may express the law by means of the subdifferentials of the convex constituents, that is,

$$s = s_1 - s_2, s_1 \in \partial\pi_1(e), s_2 \in \partial\pi_2(e). \tag{20}$$

The directional derivative of the d.c. potential $\pi(e)$ reads:

$$\pi'(e, \delta e) = \max_{s_1 \in \partial\pi_1(e)} \langle s_1, \delta e\rangle - \max_{s_2 \in \partial\pi_2(e)} \langle s_2, \delta e\rangle$$

$$= \max_{s_1 \in \partial\pi_1(e)} \langle s_1, \delta e\rangle + \min_{s_2 \in -\partial\pi_2(e)} \langle s_2, \delta e\rangle. \tag{21}$$

Exploiting the latter relation one may write a system of variational inequalities, or an implicit variational inequality, and the following optimality condition for the d.c. potential:

$$\partial\pi_2(e) \subset \partial\pi_1(e). \tag{22}$$

Let us explain this point starting from the primitive optimality condition concerning the directional derivative and the linear approximation of function $\pi(e)$:

$$\pi'(e, \delta e) = \langle s, \delta e\rangle \geqslant 0, \quad \forall \delta e. \tag{23}$$

Taking into account relations (21) and (23) one gets the following system of variational inequalities find a position such that:

$$\forall s_2 \in \partial\pi_2(e), \quad \text{that is, } \langle s_2, \delta e\rangle \leqslant \pi_2(e + \delta e) - \pi_2(e), \quad \forall \delta e, \tag{24}$$

the following inequality holds true

$$\langle s, \delta e\rangle + \langle s_2, \delta e\rangle = \max_{s_1 \in \partial\pi_1(e)} \langle s_1, \delta e\rangle \leqslant \pi_1(e + \delta e) - \pi_1(e), \quad \forall \delta e. \tag{25}$$

Details can be found in Panagiotopoulos and Stavroulakis (1992) or in Demyanov et al. (1996), as well as in the examples that follow. In particular, one notes that the d.c. decomposition is not unique and that at least every finite-dimensional function can be decomposed into a difference of convex constituents. A more general way of separating convex and concave parts of a nonconvex function and of following this decomposition through nonlinear function transformations (calculus rules) is provided by the theory of quasidifferentiability, due to Demyanov and Rubinov. Details can be found in Demyanov et al. (1996) (see also, Demyanov and Rubinov). Another point of view for analyzing systems (24) and (25), seen as a set-valued resolvent equation or as a mixed variational inequality, has recently been proposed (see, for example, Noor et al., 1998). Finally, the notion of a generalized subdifferential in the sense of F. H. Clarke can also be used for the characterization of certain critical points of nonsmooth and nonconvex potentials and, in this sense, for the description of nonmonotone, possibly multivalued, interaction laws. This notion gives rise to the theory of hemi-variational inequalities, after P. D. Panagiotopoulos, which have been studied in detail in Panagiotopoulos (1985, 1993), Naniewicz and Panagiotopoulos (1995), Motreanu and Panagiotopoulos (1998), and Haslinger et al. (1999). Some comments are given in the discussion of numerical algorithms later in this paper.

3.1. Examples

Let us consider the following one-dimensional nonconvex and nonsmooth function:

$$\pi(e) = \min_e\{-e, 2e\}.$$

A possible d.c. decomposition reads:

$$\pi(e) = \pi_1(e) - \pi_2(e) = 0 - (-\pi(e)) = 0 - \max_e\{-e, 2e\}.$$

The complete directional derivative information of function $\pi(e)$ at a given point can be derived by means of the convex analysis subdifferentials of the involved convex constituents. Clearly the nondifferentiability point $e = 0$ is of interest:

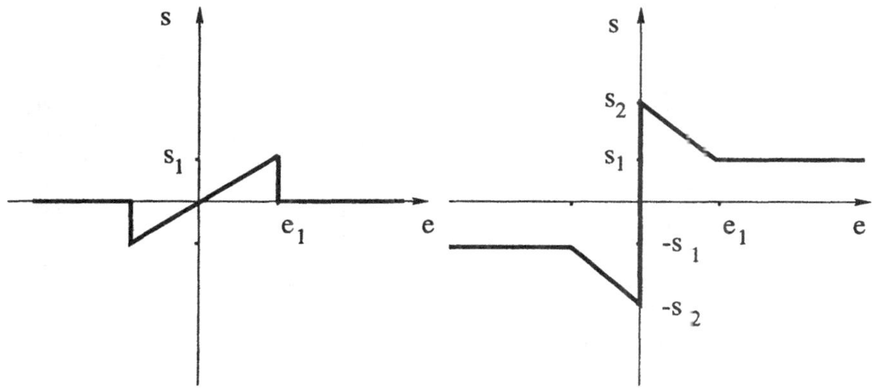

Figure 7 One-dimensional nonmonotone laws

$$\partial\pi_2(e=0) = \partial(-\pi(e)) = (-2, 1].$$

In fact, the directional derivative can be calculated as:

$$\pi'(e=0,\delta e) = 0 - \max_{e\in\partial\pi_2(e=0)} \langle e, \delta e\rangle = \begin{cases} -1, & \text{for } \delta e = +1, \\ -2, & \text{for } \delta e = -1. \end{cases}$$

One-Dimensional Nonmonotone Laws

Let us consider the following one-dimensional law with complete falling branches (see Figure 7(a)):

$$s = \begin{cases} 0 & \text{for} \quad e < -e_1, \\ [-s_1, 0] & \text{for} \quad e = -e_1, \\ \dfrac{s_1}{e_1}e & \text{for} \quad -e_1 < e < e_1, \\ [0, s_1] & \text{for} \quad e = e_1, \\ 0 & \text{for} \quad e > e_1. \end{cases}$$

This law can be derived by differentiating the following nonsmooth and nonconvex potential (which is also a d.c. function):

$$\pi(e) = \min\left\{\tfrac{1}{2}c_1e^2,\ \tfrac{1}{2}c_1e_1^2\right\}$$

$$= \tfrac{1}{2}c_1e^2 - \left\{\begin{matrix} 0, & \text{for } e < e_1, \\ s_1e + \tfrac{1}{2}c_1(e-e_1)^2, & \text{for } e > e_1 \end{matrix}\right\}$$

$$= \pi_1(e) - \pi_2(e), \qquad (26)$$

with $c_1 = \frac{s_1}{e_1}$. Analogously one treats the relation of Figure 7(b), which can be used for the modeling of slip-dependent frictional effects in mechanics (cf. the so-called Stribeck curve).

A Simple Structure with an Element Obeying a Nonmonotone Law

For the simple system of Figure 8 the equilibrium condition for several loadings will be found by using nonconvex optimization for the corresponding potential energy. The right-hand-side spring element is linearly elastic (cf. (1), (2), with $e = u$) with a spring constant equal to $c = 4$. The left-hand-side spring obeys the law shown in Figure 8, that is, it has an initial stiffness $c = \frac{1.0}{0.1} = 10.0$ and breaks after a displacement equal to $u = 0.1$. For the four loading cases, $p =$

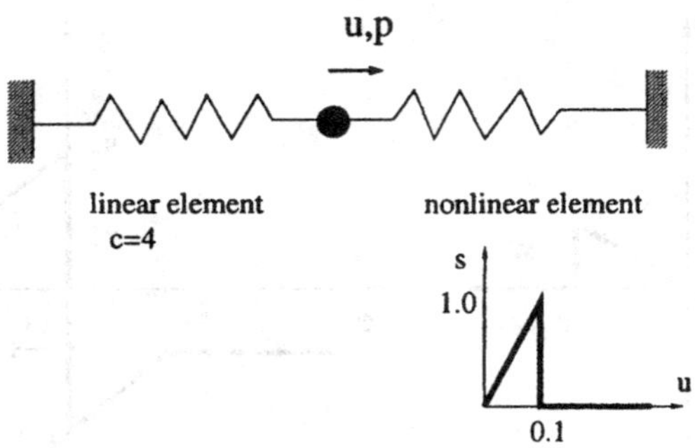

Figure 8 A simple structure with a linear and a nonlinear spring

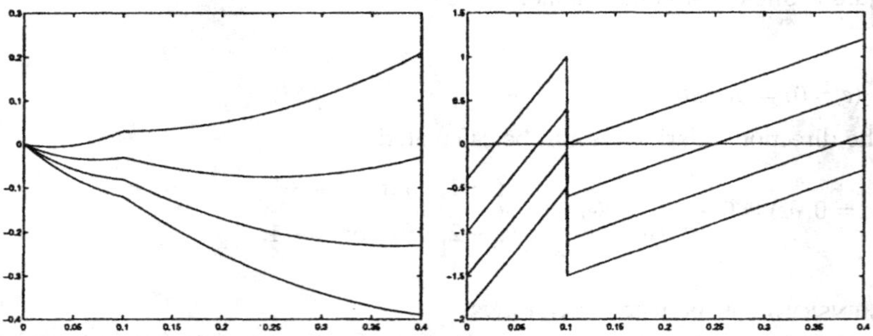

Figure 9 A simple structure with a linear and a nonlinear spring

0.4, 1.0, 1.5, and 1.9, the corresponding potential and its first derivative are plotted in Figure 9 (from the top to the bottom). From this figure it is obvious that local minima of the potential, as well as critical points (in this case, a local maximum) provide equilibrium configurations for the mechanical system considered.

A MIN–MAX-TYPE POTENTIAL

Figure 10 shows a min-max-type function of two variables e_1 and e_2:

$$\pi(\mathbf{e}) = \max\{e_1, -e_1,\} + \min\{e_2, -e_2\}$$

$$= \max\{e_1, -e_1,\} - \max\{e_2, -e_2\}$$

$$= \pi_1(e) - \pi_2(e). \tag{27}$$

It is clear that the point of nondifferentiability $e_1 = e_2 = 0$ is a critical point of this function, but it is clearly not a minimum point. One may check it by observing that the set-valued inclusion (22) does not hold true, while the following less restrictive relation holds:

$$\partial\pi_2(e) \cap \partial\pi_1(e) \neq \emptyset.$$

Analogously to the convex case, the last function can be used for the construction of multidimensional nonlinear, nonmonotone interaction laws in mechanics (for more details, see Demyanov et al., 1996, where the d.c. optimization framework used above is extended by using the notion of quasi-differentiability).

4. Details of Modeling and Numerical Realization

Mathematical programming and convex optimization techniques have provided a number of theoretical insights and effective algorithms for the solution of several complicated modeling

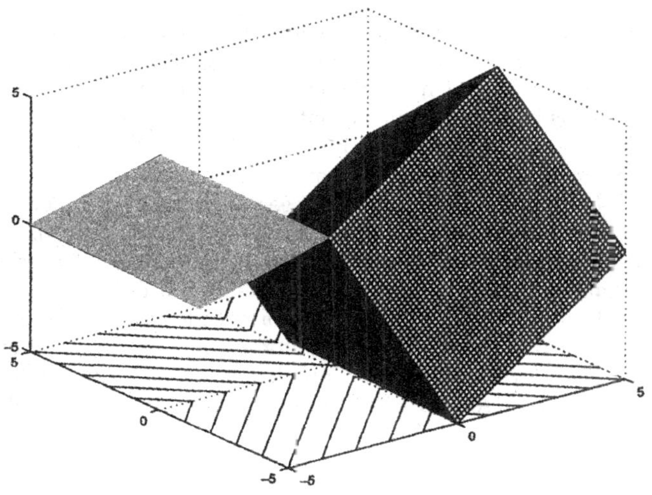

Figure 10 A nonconvex, nondifferentiable function

problems. The unilateral problem, stick-slip friction problems, and elastoplasticity problems previously outlined are representative examples. In particular, one mentions the class of linear and nonlinear complementarity problems, which have provided efficient modeling tools for a number of industrial applications (see, e.g., Pfeiffer and Glocker, 1996, and Brogliato, 1999).

Effective algorithms for convex optimization problems and for complementarity problems are always needed for the solution of large-scale problems in mechanics, that is, in the framework of finite element or boundary element methods and nonlinear interaction or material laws. The engineering references given in this paper actually explain what has already been tested. A large number of new algorithms from mathematical optimization remain more or less unknown to the engineering circles.

The interested reader will find in the references previously mentioned technical details on finite element or boundary element applications (e.g., Panagiotopoulos, 1985, 1993; Antes and Panagiotopoulos, 1992; Demyanov et al., 1996; Mistakidis and Stavroulakis, 1998, and Haslinger et al., 1999). Since the author is more familiar with problems of structural analysis, this class of problems has been discussed in more detail here. In other areas of engineering modeling similar problems arise, for instance, in multiphase flows, in semipermeability interfaces for flow mechanics, in thermal conductivity, in production modeling, and so on.

5. Further Extensions

A number of open problems exist in the area of nonconvex optimization and critical point theory in mechanics. Besides the insight and efficient computation tools for highly nonlinear computational mechanics, one mentions, along the lines of Panagiotopoulos (1993) or Mistakidis and Stavroulakis (1998), the very important engineering applications that can be solved in this way, among others, delaminations in composites and modeling of shape memory alloys.

The exploitation of solvability conditions for the study of important engineering problems is also an area of active research. One mentions here solvability conditions for convex optimization problems or complementarity problems, which gave a quantitative measure for the robotic grasping problem (see Stavroulakis et al., 1991). The corresponding studies for nonconvex problems are still in their infancy (see Goeleven (1996) for a first attempt in this direction). Several results on nonconvex duality, including engineering applications, are given in Gao (2000).

The importance of effective computer algorithms needs no explanation here. Let us only underlie the growing significance of innovative algorithms based on the modeling of dynamical systems and of neural networks (see, among others, Nagurney and Zhang, 1996). The great promise is the hardware implementation of such algorithms, which will make online, real-time, complicated, and detailed modeling feasible with obvious benefits (e.g., in robotics and control). Numerical effectiveness of various algorithms and the development of, for example, parallel solution schemes have not been discussed here. For convex problems the reader may consult, among others, Bertsekas and Tsitsiklis (1989), Ladevèze (1999), or Champaney et al. (1997). For nonconvex problems several questions are still open.

Finally, a number of optimal design and identification problems are naturally connected with optimization. In this framework it is of importance to formulate the mechanical model in the language of optimization as well. This way, for example, a structural optimization problem is recognized to be a bilevel optimization problem for which more information or effective algorithms may be available (see, for instance, Stavroulakis, 1995; Luo et al., 1996; Stavroulakis, 2000).

References

Antes, H. and P. D. Panagiotopoulos. 1992. *An Integral Equation Approach to the Static and Dynamic Contact Problems. Equality and Inequality Methods.* Stuttgart: Birkhäuser Verlag.

Bertsekas, D. P. and J. N. Tsitsiklis. 1989. In *Parallel and Distributed Computation: Numerical Methods.* Englewood Cliffs, NJ: Prentice-Hall.

Brogliato, B. 1999. In *Nonsmooth Mechanics: Models, Dynamics and Control.* London: Springer.

Champaney, L., J. Y. Cognard, D. Dureisseix, and P. Ladevèze. 1997. "Large scale applications on parallel computers of a mixed domain decomposition method." *Comput. Mech.* **19(4)**: 253–263.

Demyanov, V. F. and A. Rubinov. (eds). 2000. In *Quasidifferential Calculus and Related Topics.* Dordrecht, The Netherlands: Kluwer Academic Publishers.

Demyanov, V. F., G. E. Stavroulakis, L. N. Polyakova, and P. D. Panagiotopoulos. 1996. *Quasidifferentiability and Nonsmooth Modelling in Mechanics, Engineering and Economics.* Dordrecht, The Netherlands: Kluwer Academic Publishers.

Ekeland, I. and R. Temam. 1976. *Convex Analysis and Variational Problems.* Amsterdam: North-Holland.

Frémond, M. 1996. "Shape memory alloy. A thermomechanical macroscopic theory." In *CISM Lecture Notes* **351**: 2–68. Wien, Austria: Springer.

Gao, D. Y. 2000. *Duality Principles in Nonconvex Nonsmooth Systems: Problems, Methods and Solutions.* Boston: Kluwer Academic Publishers.

Goeleven, D. 1996. *Noncoercive Variational Problems and Related Results.* Pitman Research Notes in Mathematics Series, Volume 357. New York: Addison-Wesley Longman.

Haslinger, J., M. Miettinen, and P. Panagiotopoulos. 1999. *Finite Element Approximation of Hemivariational Inequalities: Theory, Numerical Methods and Applications.* Dordrecht, The Netherlands: Kluwer Academic Publishers.

Ladevèze, P. 1999. *Nonlinear Computational Structural Mechanics: New Approaches and Non-incremental Methods of Calculation.* New York: Springer (French original published by Paris: Hermes, 1996).

Luo, Z. Q., J. S. Pang, and D. Ralph. 1996. In *Mathematical Programs with Equilibrium Constraints.* Cambridge, UK: Cambridge University Press.

Mistakidis, E. S. and G. E. Stavroulakis. 1998. *Nonconvex Optimization in Mechanics. Algorithms, Heuristics and Engineering Applications by the F.E.M.* Dordrecht, The Netherlands: Kluwer Academic Publishers.

Motreanu, D. and P. D. Panagiotopoulos. 1998. *Minimax Theorems and Qualitative Properties of the Solutions of Hemivariational Inequalities.* Dordrecht, The Netherlands: Kluwer Academic Publishers.

Nagurney, A. and D. Zhang. 1996. *Projected Dynamical Systems and Variational Inequalities with Applications.* Dordrecht, The Netherlands: Kluwer Academic Publishers.

Naniewicz, Z. and P. D. Panagiotopoulos. 1995. *Mathematical Theory of Hemivariational Inequalities and Applications.* New York: Marcel Dekker.

Noor, M. A., K. I. Noor, and T. M. Rassias. 1998. "Set-valued resolvent equations and mixed variational inequalities." *J. Math. Anal. Appl.* **220(2)**: 741–759.

Panagiotopoulos, P. D. 1985. *Inequality Problems in Mechanics and Applications. Convex and Nonconvex Energy Functions.* Stuttgart: Birkhäuser.

Panagiotopoulos, P. D. 1993. *Hemivariational Inequalities. Applications in Mechanics and Engineering.* New York: Springer.

Panagiotopoulos, P. D. and G. E. Stavroulakis. 1992. "New type of variational principles based on the notion of quasidifferentiability." *Acta Mech.* **94**: 171–194.

Pagano, S. and P. Alart. 1999. "Solid–solid phase transition modelling: relaxation procedures, configurational energies and thermomechanical behaviours." *Int. J. Engng Sci.* **37(4)**: 1821–1840.

Pfeiffer, F. and Ch. Glocker. 1996. *Multibody Dynamics with Unilateral Contacts*. New York: John Wiley.

Polyakova, L. N. and G. E. Stavroulakis. 2000. "QD and DC optimization for pseudoelastic modeling of shape memory alloys." In *Quasidifferential Calculus and Related Topics*, edited by V. F. Demyanov and A. Rubinov. Dordrecht, The Netherlands: Kluwer Academic Publishers.

Seelecke, S. 1996. "Equilibrium thermodynamics of pseudoelasticity and quasiplasticity." *Continuum Mech. Thermodynamics* **8**: 309–322.

Sewell, N. J. 1987. *Maximum and Minimum Principles. A Unified Approach with Applications*. Cambridge, UK: Cambridge University Press.

Stavroulakis, G. E. 1995. "Optimal prestress of cracked unilateral structures: finite element analysis of an optimal control problem for variational inequalities." *Comput. Meth. Appl. Mech. Engng* **123**: 231–246.

Stavroulakis, G. E. 2000. *Inverse and crack identification problems in engineering mechanics*. Dordrecht, The Netherlands: Kluwer Academic Publishers.

Stavroulakis, G. E., P. D. Panagiotopoulos, and A. M. Al-Fahed. 1991. "On the rigid body displacements and rotations in unilateral contact problems and applications." *Comput. Structures* **40(3)**: 599–614.

PART THREE

SOFTWARE

19 OPTIMIZATION MODELING LANGUAGES

Emmanuel Fragnière and Jacek Gondzio

ABSTRACT Access to advanced optimization software needs more and more sophisticated modeling tools. Optimization modeling languages are tools that facilitate the decisionmaking process based on the optimization paradigm. Among the variety of modeling tools, algebraic modeling languages seem to be the leaders. Optimization modeling tools evolve together with progress in optimization techniques. Links with solvers are crucial, especially in nonlinear optimization. New classes of problems such as complementarity, stochastic programming, combinatorial optimization or global optimization problems can be modeled with these tools. In a wider context, interesting developments of modeling tools also emerge from other fields such as chemical engineering and computer science. Indeed, approaches based on object-oriented modeling or constraint programming also influence the evolution of optimization modeling languages. Optimization modeling tools evolve toward fully integrated modeling management systems opening access to databases, spreadsheets, and graphical user interfaces.

1. Introduction

This handbook describes the rich area of optimization including algorithms that prove efficient in practice and powerful in solving real-world problems. Numerous problems are solved routinely by optimization tools. Today's state of the art in optimization techniques enables the use of optimization modeling in decision making. Access to advanced optimization software needs equally sophisticated optimization modeling tools. We discuss these issues in this chapter. Many decision problems can benefit from an optimization-type approach. Indeed, the paradigm of optimization seems to be well adapted to needs in different domains such as economics, engineering, management, and physics. In many applications, it is quite natural to formulate the problem as a choice of one solution from a set of possible solutions. Mathematically this can be written as

$$\min z = f(x) \quad \text{subject to} \quad x \in X, \tag{1}$$

where f is an objective function defined over a domain X of feasible solutions. Typically, models involve relations defined over many different dimensions that inevitably lead to an increase in the complexity of Equation (1). Optimization modeling languages bridge the gap between model formulation and the appropriate solution technique. Optimization modeling languages are computer programs that help to develop and maintain optimization models.

Modeling is a highly iterative process. It includes such steps as problem definition, collection of data, formulating and solving mathematical problems, and the analysis of results. Optimization modeling languages support this process. We develop these issues at the beginning of Section 2. To introduce the reader to this subject we first discuss one particular type of modeling tool—the algebraic modeling languages. We illustrate their use with an example from economics that needs nonlinear optimization techniques

Optimization modeling started from the need for interfacing to the simplex method for linear programming. Modeling tools followed the progress in optimization techniques giving access, for instance, to mixed integer and nonlinear programming solvers. In these cases, more complicated data had to be sent to the solvers. For instance, the nonlinear optimization solver may require the derivatives of the objective and constraints.

Independently, new needs arose on the modeling side, such as the need to include uncertainty in planning models or to handle complementarity constraints in economics. They have become a driving force in the development of optimization modeling tools. Several examples of such needs that led to techniques that are currently being implemented are discussed in Section 4. Summing up, the evolution of optimization modeling languages follows both the progress in optimization techniques and the needs of modelers.

Note that important contributions to optimization tools come from other fields such as chemical engineering and computer science. Specific optimization requirements in these fields led to independent developments in modeling. Typically, chemical engineers build their models of chemical processes from reusable modules, which naturally favors object-oriented modeling. These issues are discussed in Section 5.

Finally, there are other tools such as databases, model viewers, debuggers, and graphical user interfaces that may facilitate the modeling process. Today's tendency is to integrate them into modeling management systems. These issues are discussed in Section 6.

2. Optimization Modeling Languages

In this section, we explain the general process used to model real-life problems and justify the use of modeling languages. (We focus our attention on modeling that by default involves optimization.) We then present the category of modeling languages that are used most, that is, algebraic modeling languages.

2.1. *The Need of Optimization Modeling Tools*

Modeling is a highly iterative process as is illustrated in Figure 1. We distinguish five steps in this process. Usually a modeler will go through them repeating certain steps many times.

Recognition and definition of the real problem is often the most difficult step. For such a problem, the modeler can formulate the objective function and the set of constraints so producing an optimization model. Then, the third step, the collection of data is, according to many researchers and consultants, the most demanding task. Typical difficulties at this stage are the lack of data (or its confidentiality) or just the opposite, the oversupply of information and the need for selecting the appropriate data from it (e.g., data-mining techniques). The data collected must then be calibrated and validated if the results are to gain credibility. These first three steps of modeling end up with a mathematical model of the real process. The model now has to be translated into a form that is understandable to an appropriate solver. This step needs a computer-based tool, an optimization modeling language. Indeed, the purpose of the

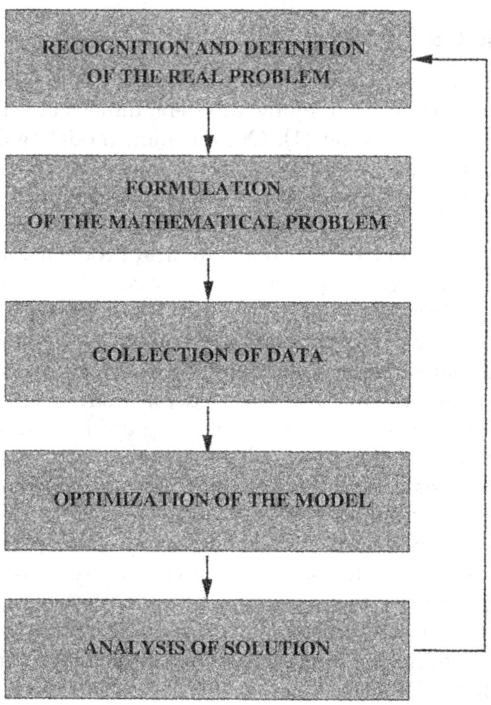

Figure 1 Modeling process

modeling language is to bridge the gap between the natural formulation of the problem proposed by the modeler and the input data scheme for the solver. Finally, when the solver finds an optimal solution to the optimization problem, the modeling support system provides the modeler with solutions that can be validated and interpreted. This interpretation may indicate the need to revise the original problem definition. We should point out that in practice the modeling process does not have to follow these five steps in the order indicated in Figure 1. Also the learning process about the situation being modeled is often as important as the results obtained from the solver. Indeed, knowledge gathered during this process helps to develop a pertinent expertise.

Optimization modeling languages are computer-based mathematical modeling tools. Their main role is to help the modeler to formulate the model as a mathematical programming problem and to translate it into the form in which optimization code can handle it. Real-life problems may involve thousands of constraints and variables and it is impossible to generate each constraint and each variable individually by hand. Instead, the constraints are partitioned into classes, each class having its generic representative. The model builder just provides the mathematical formulation for the constraint or variable representative. This step mimics the mathematical way model builders describe their problem in a clear, concise, and efficient way using indices to specify variables and constraints with a common structure.

There is a great variety of modeling languages. In the next section, we address one particular class, algebraic modeling languages.

2.2. *Algebraic Modeling Languages*

An algebraic modeling languages (AML) enables decision models to be formulated with an algebraic notation. They use a generic model description in the form of a data file. The models developed with AMLs can easily be modified. AMLs are employed within the scheme displayed in Figure 2. The user builds the model and provides the AML with the appropriate data. The AML translates the model into a form that is understandable to a solver and evokes the appropriate solver. In this setting, the solver is seen as a black box. The optimization code may query the AML about additional information on the problem. For example, nonlinear optimization code may ask for the function values as well as the first and the second derivatives at a given point. Once the solution of the mathematical program is found, it is returned to the AML and the results are reported to the user.

AML enables a modeler to express the problem in an index-based mathematical form with abstract entities: sets, indices, parameters, variables, and constraints. The key notion in the AML is the ability to group conceptually similar entities into a set. Once the entities are grouped in a given set, they can be referenced by indices to the elements of this set. That way the groups of entities (such as variables or constraints) can be represented in a compact way with one algebraic expression. For instance, the mathematical formulation $\sum_{i \in I} X_i$ is translated into the expression SUM(I, X(I)) in the GAMS modeling language. This leads to a problem formulation that is very close to the formulation using algebraic notations. The role of the

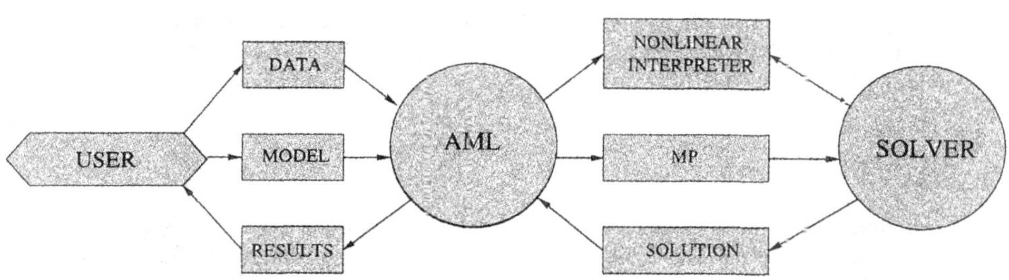

Figure 2 Links between an AML and a solver

AML is to expand this compact problem formulation (problem structure and data) into the problem instantiation, the one ready to be solved by an appropriate optimization code. This operation is realized within the AML by replicating every entity over the different elements of the set. This is often referred to as a set-indexing ability of the AML. The user of an AML can define generic expressions that are indexed over several sets. Set indexing in such cases involves compound sets.

As opposed to programming languages, in which the use of procedural statements dominates over the use of declarative ones, the AMLs essentially use declarative statements. The few procedural statements used in AMLs are read/write data and solve commands. The need for dealing with more complicated models encouraged AML developers to add more statements typical for programming languages, like for example, if-then-else or commands creating loops. Such commands even enable the modeler to write certain solution algorithms directly in the modeling language. Seminal papers concerning AMLs are Bisschop and Meeraus (1982) and Fourer (1983). For a survey of recent conceptual development in AMLs, we refer the reader to Huerlimann (1999), and the references therein. The book by Williams (1993) is a standard reference to modeling.

An important feature of some algebraic modeling languages is the capability of dealing with nonlinear models. To illustrate this point we recall briefly an intertemporal aggregate growth model stemming from work of Ramsey (1928). This model involves three decision variables in each time period $t \in T$: consumption, investment, and capital stock (C_t, I_t, K_t). Consider an economy with a single agent acting as producer, consumer, investor, and saver. Given initial levels of our decision variables (C_0, I_0, K_0), and a Cobb–Douglas production function that is a function of capital and labor [$f(K_t, L_t) = A * K_t^{\beta} L_t^{(1-\beta)}$], and given exogenous labor supplies, L_t, we want to find an optimal level of consumption, investment, and capital stock. Labor L_t has a fixed growth rate, that is, $L_t = L_0 * (1+g)^t$. To find this optimal level, we maximize a discounted logarithmic utility function under a capital stock constraint and a production constraint. With a fixed growth rate, g, and a utility discount factor, udf, the model is written as:

$$
\begin{aligned}
\max \quad & \sum_t (udf)^t * \log(C_t) \\
\text{subject to} \quad & K_{t+1} = K_t + I_t, && t \in T \\
& a_t * K_t^{\beta} = C_t + I_t, && t \in T \cdot \\
& a_t = A * (L_t)^{1-\beta}, && t \in T \\
& L_t = L_0 * (1+g)^t, && t \in T \\
& C_t, K_t, I_t \geqslant 0, && t \in T
\end{aligned}
\tag{2}
$$

An extract of the corresponding model written in GAMS algebraic modeling language is displayed in Figure 3.

In this model `C(T)`, `I(T)` and `K(T)` correspond to consumption, investment, and capital stock (C_t, I_t, K_t) at the time t, respectively. `T` is an index (time period) that varies from one to twenty. The line `CC(T)...AL(T) * K(T) * * B = E = C(T) + I(T);` corresponds to a capacity constraint

$$a_t * K_t^{\beta} = C_t + I_t.$$

We notice that the notation in the AML is close to the mathematical formulation of problem (2). This example demonstrates well the capabilities of today's algebraic modeling languages. These capabilities are the result of many years of development of optimization modeling languages.

```
SET T          TIME PERIODS   / 1*20/
    TFIRST(T) FIRST PERIOD
    TLAST(T)  LAST PERIOD

PARAMETERS BETA(T)   DISCOUNT FACTOR
           ALPHA(T) PRODUCTION FUNCTION CONSTANT.
 . . .
TFIRST(T)= YES$(ORD(T) EQ 1);
TLAST(T)= YES$(ORD(T) EQ CARD(T);

BETA(T)= BET**ORD(T);
BETA(TLAST)= BETA(TLAST)/(1-BET);
A= (C0+I0)/K0**B;
AL(T)= A*(1+G)**((1-B)*(ORD(T)-1));

VARIABLES C(T) CONSUMPTION (TRILLION RUPEES PER YEAR)
          I(T) INVESTMENT (TRILLION RUPEES PER YEAR)
          K(T) CAPITAL STOCK (TRILLION RUPEES)
          UTILITY

EQUATIONS CC(T) CAPACITY CONSTRAINT (TRILLION RUPEES PER YEAR)
          KK(T) CAPITAL BALANCE (TRILLION RUPEES)
          TC(T) TERMINAL CONDITION (PROVIDES FOR POST-TERMINAL GROWTH)
          UTIL DISCOUNTED LOG OF CONSUMPTION: OBJECTIVE FUNCTION;

CC(T).. AL(T)*K(T)**B= E= C(T)+I(T);
KK(T+1).. K(T+1)= E= K(T)+I(T);
TC(TLAST)..G*K(TLAST)= L= I(TLAST);
UTIL.. UTILITY= E= SUM(T, BETA(T)*LOG(C(T)));

MODEL RAMSEY /CC, KK, TC, UTIL/;

K.LO(T)= K0; C.LO(T)= CC; I.LO(T)= I0;
K.FX(TFIRST)= K.LO(TFIRST);
I.UP(T)= I0*((1+AC)**(ORD(T)-1));

SOLVE RAMSEY MAXIMIZING UTILITY USING NLP;
```

Figure 3 Extract of the Ramsey model written by Alan Manne

3. Links with Solvers

The growth of modeling languages was motivated by the need to get easy access to modern optimization technology. The word modern referred to different solvers at the beginning of the 1960s (the only mature optimization software at that time was the simplex method for linear programming) and at the end of the nineties when the whole variety of optimization algorithms became available. The interface between the modeling language and the solvers evolved together with the development of new modeling features and new optimization techniques.

At the very beginning of the practical use of optimization algorithms the needs were limited to formulating and maintaining linear programs. Linear programs have a compact formulation involving only the constraint matrix and a few vectors associated with it: hence their generation and storing was fairly obvious. In the 1960s IBM's linear optimization code, MPSX, specified a file format for the matrices and vectors in a linear programming model and

$$\begin{array}{ll} \max & 120x_1 + 80x_2 \\ \text{subject to} & 2x_1 + x_2 \leqslant 6 \\ & 7x_1 + 8x_2 \leqslant 28 \\ & x_1 \geqslant 0 \quad x_2 \geqslant 0 \end{array}$$

```
NAME Production Problem
ROWS
N PROFIT
L MATERIAL
L WORKTIME
COLUMNS
  PRODUCT1  PROFIT     120.0
  PRODUCT1  MATERIAL     2.0  WORKTIME 7.0
  PRODUCT2  PROFIT      80.0
  PRODUCT2  MATERIAL     1.0  WORKTIME 8.0
RHS
  RHS       MATERIAL     6.0
  RHS       WORKTIME    28.0
ENDATA
```

Figure 4 Production problem: LP formulation and MPS file

this quickly became the standard. It is called the MPS format and is illustrated in Figure 4. It became common to write programs called matrix generators to produce the models in the MPS format.

MPS format allowed the standardization of the formulation of linear programs. Its design was clearly influenced by the methodology used to solve linear programs at that time. A linear program is represented in MPS format as a sparse matrix accessed columnwise with the vectors providing constraint right-hand sides and ranges, and variable bounds. MPS has several extensions that followed the progress in optimization. In particular, MPS format can be used today to formulate mixed-integer, quadratic, or stochastic optimization problems. In Figure 4 we give an example of a simple linear program and the corresponding MPS file. The linear program is a production model in which we maximize the profit from the sales of two products. The products x_1 and x_2 are subject to material and work-time constraints.

Matrix generators were data-driven tools that could run on mainframes of the 1960s and 1970s. They were in fact specialized languages designed to write the linear program and to generate the solution report. We mention three such languages: PDS, MAGEN, and OMNI. The last one is still used today, for example, in the modeling of refineries. Data transfer between the matrix generator and the solver involved writing the problem in the form of a text file onto a disk (or tape) and reading it from there. Clearly, such a connection was not efficient. In the next step, the matrix generator and the solver were integrated to enable direct passing of columnwise linear program data from the generator to the solver. Until the 1970s such integrated systems were tailored to each specific application. Naturally, the maintenance of such programs and models could not be done by a nonspecialist. At the end of 1970s matrix generators gave way to algebraic modeling languages (Bisschop and Meeraus, 1982; Fourer, 1983). The first modeling language, GAMS (Brooke et al., 1992) was developed at the World Bank at the end of 1970s.

Independently of the progress in modeling tools, nonlinear optimization software became available. The need for solving more complicated (nonlinear) optimization problems led Brooke et al. (1985) to include in GAMS the feature to model nonlinear problems and to interface GAMS to MINOS—a nonlinear optimization code of Murtagh and Saunders (1983). Nonlinear solvers need to be able to evaluate problem functions and possibly derivatives at points that only become known when the optimization proceeds. This is quite different from the case with linear problems where all the data of the problem can be transferred to the solver before the solution starts. Nonlinear optimization codes may differ significantly between one another in the type of information they need to solve the problem. Thus the standardization of the interface between algebraic modeling languages and the nonlinear solvers is a complex issue.

As shown in Figure 2, AMLs integrate nonlinear interpreters. These interpreters rely on symbolic and/or automatic differentiation tools to provide the solvers with the required information. Most interpreters included in today's AMLs are based on automatic differentiation. Automatic differentiation (Griewank and Corliss, 1991) exploits the chain rule for computing derivatives. This technique is based on the acyclic expression graph, where nodes of the graph correspond to the computation of partial derivatives. The graph can be explored with the forward method of automatic differentiation, meaning that the nodes are analyzed sequentially. A more sophisticated backward method identifies common factors that may be exploited while computing different derivatives. Although this complicates the expression graph, it enables more efficient differentiation especially in the case of larger dimensions. The backward method first applies a forward exploration of the intermediate nodes and then a backward exploration, where derivatives are computed from a triangular set of adjoint equations.

The last few years have seen a fast development of new optimization techniques and led to the creation of codes able to solve very large stochastic optimization problems, linear complementarity problems, and global optimization problems. To make these algorithms as efficient as possible, one has to pass fairly complex information to the solver and the solver has to query the modeling language while solving the problem. For example, in integer optimization the modeler can provide the solver with the branching priorities; in stochastic optimization an event tree has to be passed to the solver, and the solver may use sampling techniques to examine only selected branches of the event tree. Global optimization algorithms often rely on a good starting point or a hint of how to branch to examine the most interesting local optima. Summing up: to be used in the most efficient way, today's optimization software often needs to access nonstandard information. This creates a need for an open interface between the AMLs and the solver. The open interface between AMLs and the solver must be sufficiently flexible to pass the type of information required by recently developed optimization algorithms (e.g., in global optimization) and by the need to model new classes of problems (e.g., complementarity problems). Yet these extensions cannot continue indefinitely because they inevitably complicate the interface. We shall present some of these extensions in Section 4.

Rather than complicating the interface, one can change the modeling process, leaving more freedom in building models from smaller elements, each communicating on its own with the necessary optimization tool. This gives rise to modeling management systems that we shall discuss in Section 6.

4. Optimization Modeling Extensions

In this section, we discuss four extensions of optimization modeling languages that are currently included or are being considered for inclusion in algebraic modeling languages: complementarity problems, combinatorial optimization, stochastic programming, and global optimization.

4.1. Complementarity Problems

An important class of problems are those containing complementarity constraints. These constraints imply that in two inequalities that must be satisfied at least one should hold as an equality. This means for instance that either $x_j = 0$ and $g_j(x) \geqslant 0$, or $x_j \geqslant 0$ and $g_j(x) = 0$. This particular constraint can be represented naturally by a logical constraint such as $x_j = 0$ or $g_j(x) = 0$ or as a nonlinear constraint such as $x_j g_j(x) = 0$.

Complementarity problems typically arise when one deals with variational inequalities. Such applications appear in computational economics (e.g., computational equilibrium models) as well as in other areas. The possibility of writing complementarity problems directly in modeling languages greatly reduces the work on model formulation; additionally, it simplifies passing the appropriate problem to a suitable solver such as MILES (Rutherford, 1995) and PATH (Dirkse and Ferris, 1995). Syntax extensions to deal with complementarity problems have been added to most AMLs.

4.2. *Combinatorial Optimization*

Many combinatorial optimization problems may be reformulated as integer optimization ones. Since AMLs can invoke mixed-integer programming solvers, this offers a possibility of modeling combinatorial optimization problems. However, this way of modeling is certainly not the most efficient one. Combinatorial optimization problems arise, for instance, in scheduling, sequencing, location, and assignment. The modeling of these problems needs logical expressions (and, or, not) as well as procedural operators such as do-while, if-then-else (Fourer and Gay, 1995). In this sense, it highlights the need for procedural statements in the modeling environment as noted in Section 2. Some production models, for example, involve fixed costs. It is natural to express the overall production cost through the following procedural construct

$$\text{if } \mathit{Prod}_i \geqslant \mathit{MinProd} \quad \text{then} \quad \mathit{FixedCost}_i + \mathit{VarCost}_i \times \mathit{Prod}_i \quad \text{else } 0.$$

Another example justifies the use of logical operators in a scheduling problem. We assume that a set of different jobs is to be executed on a machine and there are set-up times to be taken into account. The corresponding constraint can be expressed with the following procedural construct

$$\{\forall i, j \in \mathit{JOBS} : i < j\} \quad \mathit{Start}_j \geqslant \mathit{Start}_i + \mathit{SetTime}_{ij} \quad \texttt{or} \quad \mathit{Start}_i \geqslant \mathit{Start}_j + \mathit{SetTime}_{ji}.$$

Both these representations are much more natural than the representation with integer variables. For more details, see Chapter 3 of this handbook.

4.3. *Stochastic Programming*

In classical optimization problems, exogenous parameters are deterministic. Typically, in planning applications, these parameters can be replaced with random variables. The corresponding branch of mathematical programming is known as stochastic programming (see Chapter 11 of this handbook for a comprehensive introduction to the topic).

Unfortunately, AMLs are not yet well adapted to handle stochastic programming modeling. Although there is a lot of research going on that aims to allow the modeler to write stochastic programming models directly in the AML, it is still not clear which standard will be adopted. The complexity and variety of formulations (see Gassmann and Ireland, 1995, for a comprehensive taxonomy of stochastic programming problems) are slowing down the integration of specific language items. We should emphasize that the possibility of modeling stochastic programming problems directly in AMLs is not the only issue in this field. Indeed, the size of these problems tend to explode since it grows exponentially with the number of periods and the number of outcomes at each period. Another important point is that these huge problems are structured and they can only be solved by specialized optimization techniques if their structure is exploited. Several powerful codes have been developed such as MSLiP, DECIS, and SP/OSL, but they still need to be linked with modeling languages. Recent research in this area shows the need to model multistage stochastic programming problems directly in AMLs. Gassmann and Ireland (1995) note that stochastic programming modeling could benefit largely from specific contributions to AMLs. Scenarios, for instance, can be seen like a collection of data. New features must thus be provided, like the implicit declaration of scenarios through the declaration of random parameters. Among other interesting directions, we could mention the recursive definition of stochastic programming problems to express their dynamic character within AMLs (Buchanan et al., 1999). This brief discussion shows the versatility of approaches proposed. It is thus understandable that AMLs developers hesitate and wait before imposing a language syntax that should engage them in the long run. Consequently, these concepts are still only incorporated into prototype AMLs.

4.4. *Global Optimization*

To simplify the optimization problem and to render it tractable, the mathematical programming community used to require the objective function and the constraint functions to be convex. However, many real-life problems are inherently nonconvex. Just to simplify our explanations we focus solely on continuous optimization problems (many combinatorial optimization problems can be viewed as global optimization ones, but they were discussed in the earlier section). Nonconvex optimization problems may have multiple local optima. The objective of global optimization is to find a global optima (or, more realistically, a good one satisfying the decisionmaker). The general global optimization problems are usually very difficult and this is certainly one of the reasons why AMLs do not offer global optimization as an option. However, for many classes of the global optimization problem (see Chapter 6), there exist quite efficient solution approaches that work well in practice. They often require some additional knowledge of the problem. For instance, the modeler should be able to provide his own heuristics and search strategies to explore the search spaces or to provide a good starting point. Such additional knowledge can hardly be quantified in a general way, which is certainly the second obstacle in including global optimization into modeling tools. On the other hand, the progress in global optimization has encouraged the search for a specialized modeling environment dedicated to global optimization such as NUMERICA (Van Hentenryck et al., 1997) and LGO (Pintér, 1997).

5. Nonalgebraic Modeling Languages

In the previous sections, particular attention was given to AMLs. However, we should not forget the progress in optimization modeling languages motivated by developments in other areas. We mention two classes of such modeling languages: object-oriented languages that arose from the field of chemical engineering, and constrained programming languages that arose from the field of computer science. We begin this section with the concept of structured modeling. This research trend has influenced developers of modeling languages, in particular non-algebraic modeling languages.

5.1. *Structured Modeling*

The birth of the idea of structured modeling (Geoffrion, 1987) in the 1980s was prompted by the limitations of most languages available at this time. Structured modeling proposed a list of criteria intended to lead to good modeling. The main concept underlying structured modeling is that every model can be viewed as a set of smaller models or distinct elements. As stated in structured modeling literature, elements are categorized into five types: primitive entity, compound entity, attribute, function, and test. Moreover, these elements are organized into a hierarchical structure. A directed acyclic graph is often used to represent the dependencies that exist among them. The development of certain languages has been inspired by ideas found in structured modeling. It is especially the case for those giving the possibility of building models from smaller objects (object-oriented modeling features) as can be seen in the next section.

5.2. *Optimization Modeling in Chemical Engineering*

The modeling of chemical engineering processes is so particular that optimization modeling tools were developed to specifically handle these kinds of problems. To understand the nature of these problems, we present in Figure 5 a flow diagram containing two turbines, a cooler, two pumps, and a boiler that are linked together in a two-turbine power generation cycle.

The first modeling tools developed in this field had some peculiarities inherent to chemical processes. These problems involve certain reusable units. Typically, for simulating the flows that pass in the unit, a set of input variables is specified (energy, temperature, pressure, entropy) that describe the chemical process. Those variables are linked through possibly very complicated physical or chemical laws (equations) and determine the output of the unit. The blocks are linked with each other via specialized syntax. The links use conservation laws to reproduce the flow diagram. From the beginning, chemical engineering modeling languages

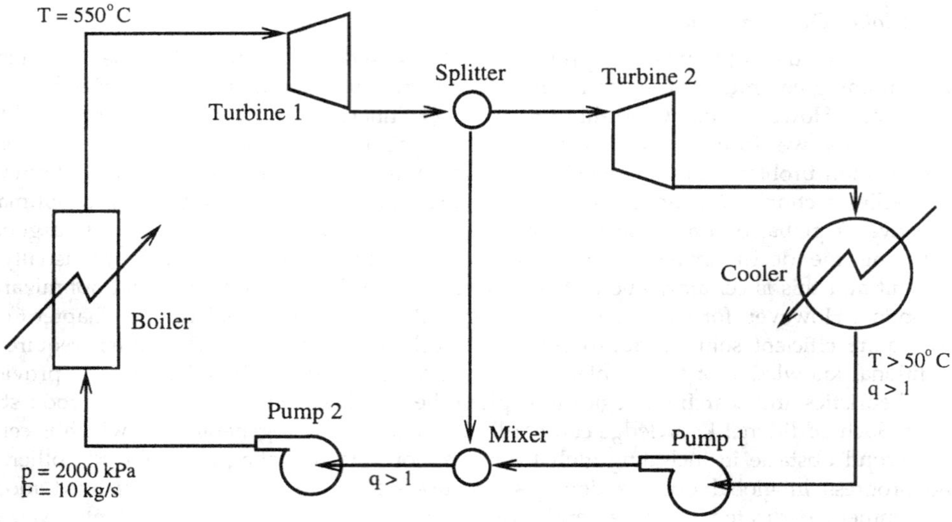

Figure 5 Chemical model

were linked to simulation software (a Monte Carlo–type simulation). In the 1970s, a new trend emerged called equation-based modeling that employed optimization to help decisionmaking. The problem was viewed as designing a flow process (capacity variables of each unit had to be determined) in conjunction with the optimal use of the system. In economic terms, this means investing in capacity and minimizing the overall operating costs.

The modeling environments developed by chemical engineers were flexible enough to preserve their object-type modeling approach while incorporating optimization. Many modeling languages exist that are specifically devoted to chemical engineering. They are built of blocks that can use simulation or complex mathematical relations describing the given process (e.g., ordinary and partial differential equations). Two such modeling languages are certainly worth a mention at this point, namely gPROMS (Barton et al., 1991) and ASCEND (Piela et al., 1991). The model is built of separate objects linked through constraints, usually being conservation laws (energy, mass, et cetera), and this creates a problem instance that is readable for a solver. Typically, these models lead to nonlinear optimization; moreover, the convexity of constraints can rarely be satisfied, hence the models often give rise to global optimization problems.

Chemical engineering optimization modeling languages are not the only ones to build models from reusable blocks. In modeling manufacturing processes, simulation tools such as EXTEND use a modeling interface to create a discrete event production process from building blocks. Such an interface known as drag-n-flop flowsheeting enables modelers without sophisticated skills to build a complicated model. We could imagine the evolution of optimization modeling languages in a similar direction. Summing up, different research areas come up with different modeling styles, sometimes including the optimization paradigm. Let us discuss another modeling approach called constraint programming.

5.3. Constraint Programming

Consider the following problem. A message "send more money" is to be coded as the addition:

code (SEND)
+ code (MORE)
\-\-\-\-\-\-\-\-\-\-
= code (MONEY).

The question is to choose distinct digits associated with all eight letters used in the message in such a way that the arithmetical operation gives the correct result. This problem can be modeled as an integer programming problem. There are different formulations of it, but each of them involves about a hundred integer variables, which render the associated integer program quite difficult. Strictly speaking, we should say that these integer programming formulations are difficult for classical integer optimization approaches, for example, for branch-and-bound solvers.

The logical analysis of this problem makes it possible to guess a feasible solution. For example, we note that neither S nor M can be zero. Next, we find that the only possible code for M is equal to 1. (We could never produce $M = 2$ or larger on a fifth position from adding digits of a thousand $S + M$.) Next, we realize that as M is 1, S can only be 9. We also realize that O has to be coded with 0. By analyzing the addition of digits corresponding to hundreds we find that since O is coded with zero and E and N have to be different, they must satisfy $N = E + 1$. By continuing such a logical analysis of constraints, we eventually come to a feasible solution of the coding problem: $E = 5$, $N = 6$, $D = 7$, $R = 8$ and $Y = 2$.

This example shows the obvious limits of the classical integer programming approach, as well as highlights the need to incorporate logical information into the optimization algorithm. The main difficulty in modeling this problem consists of a constraint that requires all variables (corresponding to distinct letters) to be different. This problem can be modeled in a compact form if a logical predicate All_different is used. Unfortunately, the classical modeling languages do not offer such predicates.

Two main techniques used in constraint logic programming are consistency techniques and local propagation. The idea behind consistency techniques is to reduce the assignable domains of variables. Local propagation means solving an equation in one variable and propagating the value of that variable to all other equations, which may then become solvable. In constraint logic, programming indices can be variables and logical predicates can be used. Moreover, every constraint is treated as a separate object. The objects communicate with each other trying to restrict their feasible domains and eventually coming up with a feasible solution (or at least a limited number of possible solutions). Clearly, each object can solve auxiliary integer optimization problems; for this an access to a fast integer optimization solver is crucial. The most natural implementation of constraint logic programming takes full advantage of the object-oriented design. Constraint logic programming is a new, successful area of optimization. The applications of this technique have increased very quickly during the last few years. The technique was applied to solve a number of problems that could not be solved with the classical integer optimization approach.

Constraint logic programming (see Jaffar and Maher, 1994) is implemented in ILOG software. It has been applied, for instance, to solve the complicated logistics problem of scheduling airplanes to parking slots. This optimization problem relies on a sophisticated information system and combines modeling, difficult online planning, and advanced optimization technology. The problem has been solved and a fully automated decision support system combining a graphical user interface and real-time access to a database has been implemented. ILOG developed this system primarily for Orly, one of Paris's international airports. Today, the same solution is used by a number of airports through the world, which is certainly a spectacular success of constraint logic programming. This approach shows a lot of promise in solving different kinds of scheduling problems.

6. Other Modeling Aid Tools

There is a crucial need for integrated systems that include optimization modeling languages. As shown in Figure 1, the analysis of results is an important element of modeling process.

Most modeling languages include or make use of different kinds of tools such as the following ones: debuggers to check the model formulation syntax, high-level data checking procedures, model analyzers, text editors for model or report writing, database systems, and graphical user interface (GUI).

In this section, we focus on two such tools: model analyzers and database systems with a particular emphasis on geographical information systems (GIS). We conclude this section by commenting on the recent wide dissemination of spreadsheet modeling, which, in a sense, is a modeling language that emerged from a database environment.

6.1. Model Analyzers

Programming language compilers include powerful user-friendly debuggers. The needs in modeling languages are similar. However, syntax errors in the model are not the only difficulty the user may encounter. Another modeling concern is checking the feasibility of the model and detecting the source of possible infeasibility. Tools such as ANALYZE (Greenberg, 1983) and MPROBE exist that were especially designed to analyze mathematical programs written with optimization modeling languages. ANALYZE was designed to analyze linear programs and to be used with matrix generators. MPROBE is a more recent tool inspired by ANALYZE that is devoted to nonlinear modeling. For instance, it offers useful features such as function shape analysis, constraint effectiveness checking, objective effect analysis, and variables and constraints sorting.

The ability to view a mathematical programming model has also attracted a lot of attention. To be able to represent different dependencies among elements of linear programs one can use different types of graphs, for example, an activity-constraint graph or a value-dependency graph, or an index-dependency graph. The visualization techniques can help the modeler to build the model and to correct mistakes. Graph representation may sometimes be a more efficient way to assist the modeler than using uniquely algebraic expressions (see Jones, 1996). Certain integrated modeling environments (e.g., MIMI) offer such visual supports in their modeling environment.

6.2. Links with Databases and GIS Extensions

A current trend is to combine optimization modeling languages and databases systems. An AML-style model file is an ASCII file containing definitions of sets and parameters in the form of scalars, vectors, and tables. Each line is either a comment, a set definition, a parameter definition or a continuation line. An import-export filter can translate the data into a database format. The interface between a user, without advanced modeling skills, and the language can thus be facilitated by the development of links between database management systems and the model and data files. These capabilities are particularly well suited to handle data. Because data appears to be more volatile than the problem structure, most modeling language designers insist on data and model structure being separated (Huerlimann, 1999); hence data management should benefit from appropriate tools. Data storage of the optimization modeling language should respect the structure query language standards thus becoming an element of a wider database environment. This enables an easy exchange of information between the modeling language database and other (external) databases. AMPL PLUS is an example of a modeling language that enables data queries to be made with MS-Windows databases or spreadsheets, thus allowing users to produce customized modeling products.

The technology of geographical information systems (GIS), has shown a very rapid development in recent years. A GIS is a database in which the information is associated with a spatial location indicator. Most GIS environments provide standard tools for creation and visualization of geographic data. Data management is handled using the relational database paradigm with its standard structured query language and its extension to incorporate spatial operators. These tools can be particularly useful for handling network modeling (e.g., routing, shortest-path) or environmental analysis. The possibility of displaying the solutions resulting from the use of the model together with maps that could be further analyzed with GIS tools, give very appealing insights to decisionmakers. Moreover, results obtained from optimization may become an input to some other modeling tools (e.g., simulation) also linked with the GIS environment. An example of merging integer programming, network optimization, and GIS is given in Camm et al. (1997).

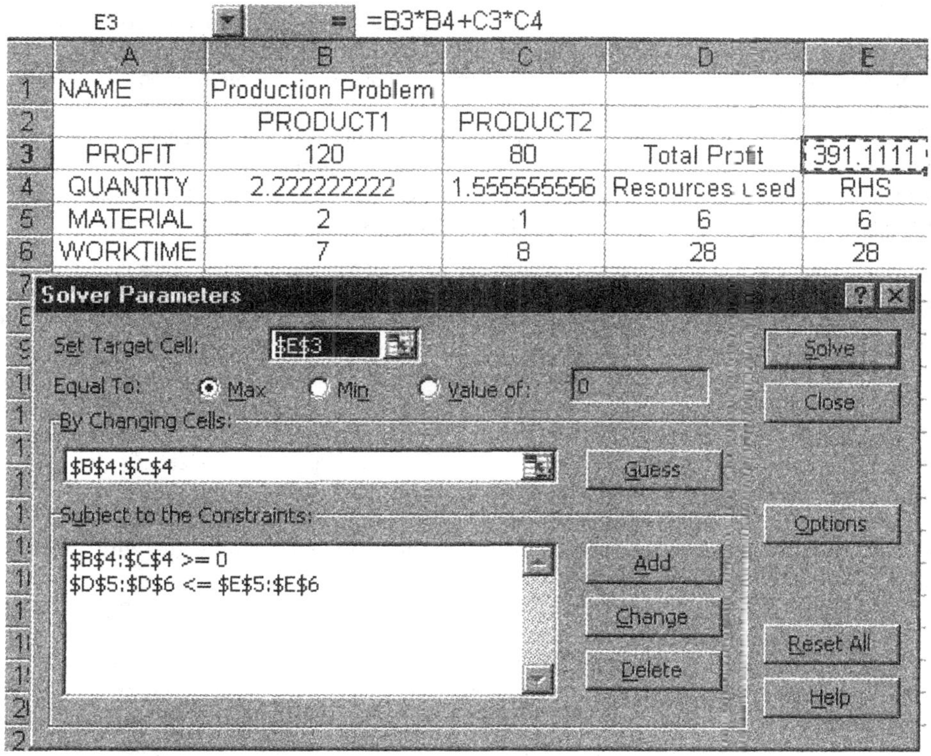

Figure 6 Excel modeling

6.3. Spreadsheet Modeling

Compared with what was presented in previous sections, spreadsheet modeling could be considered as a very simple modeling approach. Figure 6 shows an extract of an optimization problem written in EXCEL. Spreadsheets provide a set of simple functions such as the scalar product and copy and paste operations, which makes modeling possible. Using copying and pasting to replicate equations and variables does not ensure that the model formulation will remain compact, and this limits the size of problems that can be modeled. It is the responsibility of the modeler to organize the cells of the spreadsheet in the most convenient way. Models can, for example, be made of data spread across the spreadsheet. This may lead to creating so-called spaghetti models with a danger that modeling errors can be difficult to detect. Indeed the main disadvantage of spreadsheet modeling is that the mathematical structure of the model cannot be visualized.

The solver interface of the spreadsheet is quite simple to use. The objective cell is the variable that will be optimized and changing cells correspond to the decision variables. Concerning the definition of constraints, here, the logic is to define a column of cells where the use of resources is computed for a given solution (value entered in the changing cells). These values are then put into relation with the right-hand-side vector and the user needs to choose the type of the constraint (the sense of inequality) for each of them. Cells may also compute nonlinear function values; this opens the possibility to model nonlinear problems. As the simple example in Figure 6 shows, spreadsheet modeling is not the most convenient modeling tool. However, it is sufficient to develop a small prototype. It is certainly a controversial statement but we wonder if due to millions of licenses sold across the world, spreadsheet modeling of optimization problems is not the most widely used modeling tool today. EXCEL contains an optimization solver (Fylstra et al., 1998) that provides most PC owners with access to optimization techniques.

On this point, we should mention that many management science courses are based on spreadsheet modeling, enabling the teaching of optimization within a user-friendly environment. As stated in Erkut (1998), optimization courses have often been devoted to large-scale modeling. In many companies there is a need to solve small optimization problems that can be handled directly by managers and probably do not require the presence of specialized consultants. This can give meaningful business insights. Students seem to take readily to this way of learning optimization modeling. Also they can incorporate this knowledge into simple programs in visual basic and develop quite impressive decision support systems.

7. Conclusions

The ultimate aim of optimization-based modeling is to help decisionmakers to find the best solution to their problem. It is often said that decision modeling systems have to be designed in a way that allows modelers to formulate their models in the most natural way. However, depending on the background of the modeler, natural can mean very different things. For instance, chemical engineers often use graphical representation to model continuous and batch processes. Economists are used to expressing their models with algebraic notations. This observation on different modeling styles in different professional groups extends to students; new generation of modelers usually follow the modeling style of their teachers. The modern way of teaching operations research by using specific modeling languages will certainly give rise to another natural way of writing models.

This chapter gives an overview of tools that facilitate the decisionmaking process based on the optimization paradigm. It seems to us that the evolution of optimization modeling languages is driven by two factors: the need to formulate particular mathematical problems and the availability of new solution techniques. We have paid special attention in this chapter to the historical development and the current state of algebraic modeling languages, because these tools seem to be widely accepted by the optimization research community.

New developments in optimization modeling languages do not come uniquely from the research of the optimization community. There are important contributions from different research fields (e.g., chemical engineering and computer science).

Finally, we note the increasing influence of market needs on the developments in optimization modeling languages. Optimization-based modeling has to respond to the new requirements imposed by the fast development of information technology, especially in the field of databases.

ACKNOWLEDGMENTS

We are grateful to our colleagues: Christian Condevaux-Lanloy, Tony Huerlimann, Ken McKinnon, and Bill Morton for many helpful remarks.

The research of the first author was supported by the Fonds National de la Recherche Scientifique Suisse, grant number 1214-049696.96/1. The research of the second author was supported by the Engineering and Physical Sciences Research Council of the UK, EPSRC grant GR/M68169.

References

Barton, P. I., E. M. B. Smith, and C. C. Pantelides. 1991. "Combined discrete/continous process modelling using gPROMS." Centre for Process Systems Engineering, Imperial College of Science, Technology and Medicine, London SW7 2AZ, UK. Prepared for presentation at the 1991 AIChE Annual Meeting.

Bisschop, J. and A. Meeraus. 1982. "On the development of the general algebraic modeling system in a strategic planning environment." *Math. Programming Study* **20**: 1–29.

Brooke, A., A. Drud, and A. Meeraus. 1985. "High level modeling systems and nonlinear programming." In *Numerical Optimization 1984*, edited by P. T. Boggs, R. H. Byrd, and R. B. Schnabel. Philadelphia: SIAM.

Brooke, A., D. Kendrick, and A. Meeraus. 1992. *GAMS: A User's Guide*. Redwood City, CA: The Scientific Press.

Buchanan, C. S., K. I. M. McKinnon, and G. K. Skondras. 1999. "The recursive definition of stochastic linear programming problems within an algebraic modeling language." To appear in *Ann. Oper. Res.*

Camm, J. D., T. E. Chorman, F. A. Dill, J. R. Evans, D. J. Sweeney, and G. W. Wegryn. 1997. "Blending OR/MS, judgment, and GIS: restructuring P&G's supply chain." *Interfaces* **27(1)**: 128–142.

Dirkse, S. P. and M. C. Ferris. 1995. "The PATH solver: a non-monotone stabilization scheme for mixed complementarity problems." *Optim. Meth. Softw.* **5**: 123–156.

Erkut, E. 1998. "How to 'EXCEL' in teaching management science." *OR/MS Today*, October 1998.

Fourer, R. 1983. "Modeling languages versus matrix generators for linear programming." *ACM Trans. Math. Softw.* **9**: 143–183.

Fourer, R. and D. Gay. 1995. "Expressing special structures in an algebraic modeling language for mathematical programming." *ORSA J. Comput.* **7**: 166–190.

Fylstra, D., L. Lasdon, A. Warren, and J. Watson. 1998. "Design and use of the Microsoft Excel Solvers." *Interfaces* **28(5)**: 29–55.

Gassmann, H. I. and A. M. Ireland. 1995. "Scenario formulation in an algebraic modelling language." *Ann. Oper. Res.* **59**: 45–75.

Geoffrion, A. M. 1987. "An introduction to structured modeling." *Mgmnt Sci.* **33**: 547–588.

Greenberg, H. J. 1983. "A functional description of ANALYZE: a computer-assisted analysis system for linear programming models." *ACM Trans. Math. Softw.* **9**: 18–56.

Griewank, A. and G. F. Corliss. 1991. *Automatic Differentiation of Algorithms: Theory, Implementation, and Application.* Philadelphia: SIAM.

Hürlimann, T. 1999. Mathematical Modelling and Optimization: An Essay for the Design of Computer-Based Modeling Tools. Dordrecht: Kluwer Academic Publishers.

Jaffar, J. and M. J. Maher. 1994. "Constraint logic programming: a survey." *J. Logic Programming* **19/20**: 503–581.

Jones, C. V. 1996. *Visualization and Optimization.* Boston: Kluwer Academic Publishers.

Murtagh, B. A. and M. A. Saunders. 1983. MINOS 5.0 User's guide. Report SOL 83-20. Stanford: Department of Operations Research, Stanford University.

Piela, P., T. G. Epperly, and K. M. Westerberg. 1991. "Ascend: an object-oriented computer environment for modeling and analysis: the modeling language." *Comput. Chem. Engng* **15**: 53–72.

Pintér, J. D. 1997. *LGO—A Model Development System for Continuous Global Optimization, User's Guide.* Canada: Halifax.

Ramsey, F. P. December 1928. "A mathematical theory of saving." *Economics Journal.*

Rutherford, T. F. 1995. "Extensions of GAMS for complementarity problems arising in applied economics." *J. Econ. Dynam. Control* **19(8)**: 1299–1324.

Van Hentenryck, P., L. Michel, and Y. Deville. 1997. *Numerica A Modeling Language for Global Optimization.* Cambridge, MA: The MIT Press.

Williams, H. P. 1993. *Model Building in Mathematical Programming.* Chichester, England: John Wiley & Sons.

A. Web Sites of Optimization Modeling Languages

AIMMS: http://www.aimms.com/

AMPL: http://www.ampl.com/ampl/

ASCEND: http://www.cs.cmu.edu/afs/cs.cmu.edu/project/ascend/home/

GAMS: http://www.gams.com/

gPROMS: http://www.psenterprise.com/gPROMS/index.html

ILOG: http://www.ilog.com/

LGO: http://www.tuns.ca/~pinter/

LINGO: http://www.lindo.com/lingo.html

LPL: ftp://www-iiuf.unifr.ch/pub/lpl

MAGIC: http://www.magic-MP.com/

MPL: http://www.maximal-usa.com/mplmain.html

MPROBE: http://www.sce.carleton.ca/faculty/chinneck/mprobe.html

OMNI: http://www.haverly.com/omni.htm

XPRESS-MP: http://www.dash.co.uk/

20 OPTIMIZATION SOFTWARE PACKAGES

Stephen J. Wright

ABSTRACT We discuss software packages for solving optimization problems, focusing on fundamental software that assumes that the problem has already been formulated in mathematical terms. Such packages can be used directly in many applications, linked to modeling languages or to graphical user interfaces, or embedded in complex software systems such as logistics and supply chain management systems.

1. Introduction

Optimization software is used in a wide variety of applications, including the fundamental sciences, engineering, operations research, finance, and economics. This plethora of applications has led to the development of a large number of fundamental, general-purpose software packages that can be used efficiently in a variety of contexts.

This chapter discusses software packages that solve optimization problems that have been specified precisely in a mathematical sense. The specification may take the form of a data file, user-supplied code that fills out given data structures or evaluates functions at given values of the argument, or a model defined via a modeling language such as AMPL, AIMMS, GAMS, or MPL. We do not discuss optimization software that forms a part of larger applications, such as spreadsheets or supply chain management systems, unless the optimization code is also available separately.

Optimization software development has gone hand in hand with the theoretical development and analysis of algorithms during the past thirty years. In many instances, the software authors themselves contributed heavily to the theoretical analysis. This fact is hardly surprising, since the experience gained in applying algorithms to practical problems reveals many features of their performance that are not obvious from the theory, thereby driving the modification and invention of new algorithms.

The growing commercial market, particularly in linear and integer programming, has yielded rapid improvements in the quality of the software in some areas and sometimes in the underlying algorithms as well. In other areas, improvements have been more gradual, due in some cases to the lack of demand and in other cases (such as nonlinear programming) to the inherent difficulty of the problem class.

The state of the art in optimization software, circa 1993, is discussed in Moré and Wright (1993). Information from this book was used as the initial basis of the *NEOS Guide to Optimization Software*, which can be found on the Web at `www.mcs.anl.gov/otc/Guide/SoftwareGuide/`. This site, which is updated continuously, contains more detailed descriptions of most of the packages mentioned in this chapter. We refrain from giving bibliographic references or URLs in the text below in cases where this information can be found easily in the NEOS Guide. Other notable online resources also can be accessed through the NEOS Guide; go to `www.mcs.anl.gov/otc/Guide/OtherSites/`. We mention in particular the decision tree for optimization Software at `plato.la.asu.edu/guide.html`, and Michael Trick's Operations Research Page at mat.gsia.cmu.edu. Surveys of optimization software are carried out regularly in *OR/MS Today*, the magazine published by INFORMS, the leading professional society in operations research.

Some of the codes mentioned in this article can be executed remotely on the NEOS Server at `www.mcs.anl.gov/neos/Server/`. Users submit data to the server and receive results through an e-mail or Web interface to the Internet. The server is a convenient way for users to benchmark solvers and to test their suitability for a specific application.

In the following sections, we discuss the availability of software tools in several important areas, including linear programming, integer programming, and nonlinear programming. We mention a few of the major packages in each area, discuss the underlying algorithms, and outline recent developments and current research that is likely to have an impact on future software. In the final section, we present some impressions of the possible future directions of the field of optimization software.

2. Linear Programming

The birth of modern optimization dates back to George Dantzig's invention of the simplex algorithm for linear programming in the late 1940s. This extremely effective method is still the basis of most linear programming codes today. Linear programming codes consume far more computer cycles than codes for any other class of optimization problems. The reasons are partly historical and partly practical. Generations of modelers in economics and finance have been trained to build linear models. Even when the underlying application is actually nonlinear, as is often the case, a linear model suffices to yield useful results, and the unavailability of much of the data associated with the problem makes the use of a nonlinear model unnecessarily sophisticated in any case. Finally, the global solutions of linear programs are comparatively easy to find, while this is not the case with general nonlinear programs.

The linear programming problem can be formulated mathematically as follows:

$$\min_{x} c^T x \quad \text{subject to} \quad Ax = b, \quad x \geqslant 0, \tag{1}$$

where $x \in \mathbf{R}^n$, $c \in \mathbf{R}^n$, $b \in \mathbf{R}^m$, and $A \in \mathbf{R}^{n \times m}$, for some integers m and n with $m \leqslant n$. The set of feasible points for (1) is a polyhedral subset of $\mathbf{R}^n$.

The simplex method generates a sequence of iterates $x^k, k = 0, 1, 2, \ldots$ that are vertices (extreme points) of the feasible polyhedron. At most m components of each x^k are nonzero, and each iterate is obtained by moving along an edge of the polyhedron from the preceding iterate. The objective value decreases as the iterations proceed, that is, $c^T x^{k+1} \leqslant c^T x^k$ for all k. Codes based on the simplex method were available from the 1950s onward, but their efficiency increased steadily through the 1970s due mainly to advances in the linear algebra techniques and in pricing, the process of deciding the edge along which to move away from the current iterate x^k. By the late 1980s, commercial simplex codes appeared to have stabilized, but a new spurt of development was motivated by the appearance of codes based on a rival class of algorithms: interior-point methods (see Wright, 1997).

Today, interior-point codes coexist with simplex codes in the commercial domain. Commercial simplex codes include CPLEX Simplex, MINOS, and those distributed by XPRESS-MP, OSL, and LINDO. Commercial interior-point codes include CPLEX Barrier and OSL. All these codes are undergoing continual development, and significant improvements are still being reported on many large practical problems. When coupled with the improvements in computer hardware, the resulting solvers are vastly more powerful than was the case in the mid-1980s.

Some codes are inexpensive or freely available to researchers and nonprofit users. The outstanding simplex code in this category is SOPLEX, which is available at the following URL: `www.zib.de/Optimization/Software/Soplex/`. Academic licensing of the OSL package for Windows platforms is also currently free of charge. Free or inexpensive interior-point codes include BPMPD, PCx, HOPDM, LIPSOL, and LOQO. Interior-point codes are more widely available for the simple reason that they are less complicated to program than simplex codes. However, they frequently outperform even the best simplex codes on some problems, particularly large problems. Simplex codes have the important advantage that they can exploit prior information, such as a good estimate of the solution of (1), whereas interior-point codes do not derive much benefit from such information. Warm starts are available in one of the most significant uses of linear programming codes, namely, their use in solving continuous relaxations of integer programming problems.

Computational comparisons of many of the codes mentioned above have been performed by Hans Mittelmann, and the results can be viewed at the following URL: `http //plato.la.asu.edu/bench.html`.

The traditional format for entering data into linear programming codes is the MPS file, a format dating to the 1950s and based on 80-character-wide punch cards. (See the NEOS Guide for an explanation of the format.) A much more natural way to specify a linear programming problem is via an algebraic modeling language such as AMPL, AIMMS, GAMS, or MPL, which allows variables, constraints, and objectives to be written in much the same way as on the modeler's note pad. Most linear programming codes can be accessed through at least one of these languages. In other cases (for instance LINDO) the linear programming solver and the modeling language are sold as an integrated package. Some packages, notably XPRESS-MP and CPLEX, come with their own algebraic input format.

Another way to interface to the software is through subroutine calls, which are used to build up the model, modify it, solve the problem, and extract the results. This model is useful when the software must be embedded into a larger system or an existing application, and most commercial codes come with the required subroutine library interface.

3. Integer Programming

Integer linear programming problems have the form (1), except that some variables are required to take on integer values. (In the important special case of binary variables, we require $x_i \in \{0, 1\}$.) In many problems, there is a subset of variables that is not required to be integral. Such problems are known as mixed-integer (linear) programs.

Problems with integer constraints are much more difficult to solve than continuous linear programs, due to the fact that the set of feasible points is no longer convex or even connected. The software is considerably more complicated.

Most algorithms proceed by solving a sequence of continuous linear programs of the form

$$\min_x c^T x \quad \text{subject to} \quad Ax = b, \quad x \geqslant 0, Cx \leqslant d,$$

where the additional constraints $Cx \leqslant d$ are changed as the algorithm proceeds. In these so-called continuous relaxations, the integrality requirements are not imposed on x. The purpose of the additional constraints $Cx \leqslant d$ is to either fix or bound some of the components of x at integer values or to shrink the feasible set $\{x | Ax = b, x \geqslant 0\}$ by cutting off regions of the continuous feasible set that do not contain integer-feasible points. Constraints of the latter type are known as cutting planes, and were the basis of the first methods developed in the 1950s. Constraints that fix or bound some of the components at integer values arise in the branch-and-bound approach, which has formed the basis of almost all commercial codes. Recent research has focused on combining the cutting plane and branch-and-bound approaches, and on identifying powerful new classes of cuts. Surprising advances are still being made in this area, as algorithms and heuristics are improved and combined in imaginative ways, and the interaction between the integer programming code and its simplex method substrate is made more efficient. For a brief introduction to methods for integer programming, see Wolsey (1998).

High-quality codes for integer programming are available only in the commercial sector, partly because of the need for a highly efficient simplex code to solve the continuous relaxations and partly because of the intrinsic complexity of the heuristics used in the best methods to generate cutting planes and to perform the branching. The CPLEX, XPRESS-MP, and OSL codes are possibly the best known, and all are undergoing continual development. More information on these and other solvers can be found on the NEOS Guide.

Integer nonlinear programs also arise in many applications, but these are very difficult to solve, combining as they do the vagaries of nonconvex nonlinear programming and integer programming. There are a number of research codes for this problem class. The code MIQP for mixed-integer quadratic programming is available on the NEOS Server.

4. Quadratic Programming

Quadratic programming problems have the general form

$$\min_x c^T x + \tfrac{1}{2} x^T Q x \quad \text{subject to} \quad Ax = b, \quad x \geqslant 0, \tag{2}$$

where Q is a symmetric matrix whose properties and structure are highly application-dependent. When Q is positive semidefinite, the problem (2) is a convex quadratic program, and is much easier to solve than in the complementary case in which Q has some negative eigenvalues. Another important special case arises when all the constraints are bounds, that is, the general linear equalities $Ax = b$ are not present.

The simplex method for linear programming, described in Section 2, can be extended to problem (2). In this setting it is known by the more general term active-set method, since each step can be characterized by a subset $A \subset \{1, 2, ..., n\}$ of indices of x, which are fixed at their bound of 0. As in linear programming, this algorithm generates a sequence of feasible iterates, but it no longer confines its search to the vertices of the feasible polyhedron because the solution may lie in the interior of an edge or face, or even strictly inside the polyhedron. If the component x_i reaches its bound on the current step, the index i is added to A. An index j may be deleted from A on some iteration if the Lagrange multiplier for the constraint $x_j \geqslant 0$ indicates that the step generated by relaxing this constraint is a descent direction for the objective function in (2).

Codes based on the active-set approach appear in OSL and LINDO, while the packages BQPD, QPOPT, and SNOPT also implement this approach.

The interior-point approach can be used as an alternative to the active-set approach in the convex case, that is, Q positive semindefinite. Interior-point solvers are well suited to large-scale problems, and they can often be tailored to exploit special structure in the matrix Q to improve the efficiency of the calculations. The basic approach is a straightforward extension of Mehrotra's primal–dual algorithm for linear programming. Packages that implement this approach include CPLEX, LOQO, OSL, and BPMPD.

For problems in which only bound constraints are present, algorithms based on the projected gradient approach are useful, particularly when the problem is large. This approach has been investigated in many applications, but no codes are generally available for this specific problem. However, this approach is available in more general codes such as L-BFGS-B or LANCELOT, which are designed for problems with a nonlinear objective.

5. Unconstrained Optimization

In unconstrained optimization, we seek to find the minimizer of a function $f(x)$. Techniques differ according to whether the function is smooth (at least twice differentiable) or nonsmooth, and to whether we seek a local or a global minimizer.

Algorithms related to Newton's method have proved to be particularly effective when we seek a local minimizer of a smooth function. In general, we form a model of the function f around the current iterate x^k—for instance, the quadratic model

$$q^k(d) \stackrel{\text{def}}{=} f(x^k) + \nabla f(x^k)^T d + \tfrac{1}{2} d^T H_k d, \quad (3)$$

where H_k is either the Hessian $\nabla^2 f(x^k)$ or some approximation to it, obtained either from a quasi-Newton update formula or a finite-difference formula. The problem (3) is solved, possibly approximately and possibly with the inclusion of a bound on the size of d. Line-search methods search along the resulting step d to find a new iterate $x^{k+1} = x^k + td$ (for some $t > 0$) with a significantly smaller value of f. Trust-region methods adjust the bound on the size of d until the point $x^{k+1} = x^k + d$ gives an improved f value.

The code LANCELOT, though written for more general classes of problems, reduces to a Newton-like method (with approximate Hessians H^k) when applied to unconstrained minimization problems. L-BFGS and VE08 are designed for large-scale problems; they use a line-search approach in which H^k is a quasi-Newton approximation that can be stored in $O(n)$ locations (where n is the dimension of the unknown vector x), rather than the $n^2/2$ memory locations required by a dense matrix. TN/TNBC and TNPACK, which are also designed for large problems, take H^k to be the true Hessian, and use an iterative method to find an approximate minimizer of (3). UNCMIN and NMTR (a code from the MINPACK-2 collection) implement a variety of trust-region approaches. The code TENMIN augments the second-order model (3) with third-order terms from the Taylor series for f. It tends to perform more reliably than Newton-type methods when the Hessian $\nabla^2 f$ is singular at the solution. The implicit filtering approach described in Kelley (1999, chapter 7) uses finite-difference approximations to obtain an approximation to the gradient $\nabla^2 f(x^k)$, while a quasi-Newton approximation is used for H^k. This approach is appealing when the function is noisy, and when derivatives are not available.

Nonlinear conjugate gradient methods are popular for large-scale problems in which gradients $\nabla^2 f$ can be supplied by the user. They do not use a quadratic model and require only $O(n)$ storage. Several such methods are implemented in CGplus, which is available through the NEOS Server.

Direct search methods such as Nelder–Mead simplex, Hooke–Jeeves, and multi-directional search are popular because they require only function values to be supplied. These methods, which are designed for small n, evaluate f on a regular pattern of points. This pattern is modified to investigate regions of the space that appear to be promising, and f is reevaluated at any new points so generated. For details on these methods and information on software, see Kelley (1999, chapter 8). The simulated annealing approach can also be used when only function values are available. For an implementation, see the code ASA.

Often we wish to confine the search to a rectangular box defined by $l \leqslant x \leqslant u$, where l and u are vectors of upper and lower bounds, respectively. The most popular algorithms in this class are active-set and gradient-projection methods. Active-set methods hold a subset of the components of x at one or other of their bounds on each iteration, and take steps like those of unconstrained optimization methods in the other components. The active set, which consists of the fixed components of x, generally changes from one iteration to the next. Codes of this type include TN/TNBC. In the gradient projection approach, a search direction d is calculated in the space of all variables, and candidates for the new iterate are obtained by projecting points of the form $x^k + td$ onto the feasible box, where $t > 0$ is the line search parameter. The new point obtained in this way may be improved by a sequence of Newton-like steps. Codes that use this approach include LANCELOT, L-BFGS-B, IFFCO, and VE08.

6. Nonlinear Programming

In nonlinear programming problems, both the objective function and constraints can be general nonlinear functions. By adding slack variables to the inequalities, we obtain the following general formulation:

$$\min_x f(x) \quad \text{subject to} \quad c(x) = 0, \quad l \leqslant x \leqslant u, \tag{4}$$

where l and u are vectors of upper and lower bounds, respectively. (Some of the components of l may be $-\infty$, indicating that there is no lower bound on the corresponding component of x, while similarly some of the components of u may be $+\infty$.)

Algorithms that are currently available in software include sequential quadratic programming, the projected gradient method, and the augmented Lagrangian method, otherwise known as the method of multipliers. Interior-point methods, both primal (log-barrier) and primal-dual, are the subject of much research but are not currently available in production software. Other penalty and barrier techniques, while still investigated by theoreticians from time to time, do not appear in general-purpose codes.

In sequential quadratic programming (SQP), a model problem for (4) is formed by taking Taylor-series approximations about the current iterate x^k. In the case of SQP, this model problem typically has the form

$$\min_d \nabla f(x^k)^T d + \tfrac{1}{2} d^T H_k d \quad \text{subject to} \quad c(x^k) + \nabla c(x^k)^T d = 0, \quad l \leqslant x^k + d \leqslant u, \tag{5}$$

where H^k represents the Hessian of the Lagrangian function for problem (4) and is constructed either from exact second derivatives or from a quasi-Newton update formula. If this step d makes progress toward solving (4), as measured by some merit function, we define the new iterate to be $x^{k+1} = x^{k+d}$. Otherwise, a trust-region or line-search strategy is used to obtain a new candidate step d.

The codes NLPSPR, SPRNLP, and SNOPT use the SQP approach. All of these packages exploit sparsity in the linear algebra calculations, while the codes DONLP2, NLPQL, and NPSOL use dense linear algebra and are therefore suited to problems with fewer variables and constraints. The package FSQP uses a variant of SQP in which all iterates are confined to the feasible region. Such a restriction is useful when the objective function is not defined outside the feasible region, or when it is essential to the user that the code return a feasible point, even when it is not able to converge to full accuracy to a solution of the problem (4).

The method of multipliers makes use of the augmented Lagrangian function for (4), which is defined as

$$L_a(x, \lambda; \gamma) \overset{\text{def}}{=} f(x) + \lambda^T c(x) + \frac{1}{2\gamma} c(x)^T c(x),$$

where $\gamma > 0$ is a positive parameter and λ is a vector of Lagrange multiplier estimates. At iteration k, a new value x^{k+1} is obtained by approximately solving the problem

$$\min_x L_a(x, \lambda_k; \gamma_k) \quad \text{subject to} \quad l \leqslant x \leqslant u, \tag{6}$$

and the multiplier estimates are updated by setting

$$\lambda^{k+1} \leftarrow \lambda^k + \frac{1}{\gamma_k} c(x^{k+1}).$$

The parameter γ_k may either be held constant or decreased prior to the next iteration. The subproblem (6) is a bound-constrained optimization problem that can be solved by using the techniques discussed in the previous section.

LANCELOT implements the method of multipliers for general constrained problems. This code also makes use of trust-region techniques, gradient projection techniques for solving (6), efficient iterative linear algebra methods, quasi-Newton updates, and sophisticated heuristics for updating the parameter γ_k, which is actually weighted differently for different components of $c(\cdot)$.

The reduced gradient approach for solving (4) is an extension of the simplex method for linear programming, in which nonbasic variables (those at their upper or lower bound) and basic variables (those whose values are determined by the constraint $c(x) = 0$) take their place alongside superbasic variables, which are allowed to move freely. Techniques from unconstrained minimization are applied to find steps for the superbasic variables. Variables may be reclassified among the three categories when superbasic or basic variables encounter bounds, or when Lagrange multiplier estimates indicate that a nonbasic variable should move off its bound. Codes based on this approach include CONOPT and LSGRG2, both of which exploit sparsity in the Jacobian matrix $\nabla c(x)$.

The sequential linearly constrained algorithm used by the popular code MINOS combines the augmented Lagrangian approach with projected gradient and penalty techniques.

It is not currently possible to make general statements about the relative efficiency of the techniques discussed above. The paradigm (4) encompasses a wide variety of problems with a wide variety of peculiarities, and a technique that performs relatively well on one type of problem might perform quite poorly on another type.

7. Other Problems

We mention here packages that are aimed at other widely solved classes of problems.

Nonlinear least-squares problems often arise in the context of fitting models to data. In these problems the objective function has the special form

$$f(x) = \tfrac{1}{2} \sum_{i=1}^{m} r_i^2(x),$$

so that the Hessian $\nabla^2 f(x)$ has the special property that it can be approximated by the term

$$\sum_{i=1}^{m} \nabla r_i(x) \nabla r_i(x)^T, \tag{7}$$

which consists of only first-order information about the functions r_i. Hence, in a sense, we get some information about the second-order quantity $\nabla^2 f(x)$ for free, and algorithms usually exploit this fact by setting H_k in the quadratic model (3) to the approximation (7). MINPACK contains an implementation of a trust-region algorithm in which the approximation (7) is used for the Hessian of f—essentially the Levenberg–Marquardt algorithm. TENSOLVE implements both line-search and trust-region strategies that include second-order information in the Hessian. Like its unconstrained minimization counterpart TENMIN, this code performs well on degenerate problems. The PORT3 subroutine library contains an implementation of the algorithm NL2SOL, which combines the approximation (7) with a quasi-Newton estimate of the remaining terms of $\nabla^2 f(x)$. VE10 uses a line-search method in which the conjugate gradient method is used to find an approximate minimizer of the quadratic model (3). Other codes that exploit the least-squares structure include DFNLP, MODFIT, PROCNLP, and NLSSOL. The package ODRPACK also implements Levenberg–Marquardt, and additionally addresses an alternative to the least-squares paradigm known as orthogonal distance regression (ODR).

Global optimization problems, in which we seek the global minimizer of a nonconvex function f, are becoming increasingly important in applications, and the increasing power of computers has made it possible to solve them in many interesting cases. Usually it is essential for the algorithm to exploit the special properties of f. While general algorithmic strategies such as branch-and-bound, smoothing, multistart, and so on, are useful in designing algorithms for specific applications, general-purpose codes are less useful. Nevertheless, several such codes exist. We mention ASA, the adaptive simulated annealing code discussed in Section 5, LGO, and OPTECH. The new code MCS is implemented in Matlab and is freely available from the site `http://solon.cma.univie.ac.at/~neum/software/mcs/`. This Web site contains a link to a Web page with much useful information about global optimization techniques.

Optimization problems involving networks arise in many applications, and a great deal of research has gone into devising algorithms that exploit the structure of such problems. Linear network problems (such as linear minimum-cost flow, or maximum flow) are often solved by general simplex solvers, which sometimes make an effort to detect and exploit the network structure in the way they perform their linear algebra calculations. Other solvers such as NETFLOW and LNOS exploit the structure explicitly, often applying special-purpose algorithms. Nonlinear network problems are addressed by LSNNO.

8. Future Directions

Though the current generation of optimization software has been extremely useful in a wide variety of applications, there is plenty of room for improvement in the next generation. Much of the current software does not make use of advanced software concepts, such as object-oriented design and data-structure independence. Much of it is programmed in Fortran and is invoked through subroutine-call interfaces whose flexibility is limited (though the increasing use of modeling languages is improving the ease of use of many codes). Moreover, most of the software is tied to the standard taxonomy of optimization problems that is reflected in the section titles in this article. While the structure of certain problems can often be exploited by using sparse linear algebra routines at each iteration of the optimization algorithm, there are many other problems that do not fit neatly into the standard taxonomy, and that would benefit from specialized linear algebra routines or new algorithms.

We believe that object-oriented design and component software ideas will become key features of the next generation of optimization software. Components that implement successful algorithms such as sequential quadratic programming, augmented Lagrangian methods, and interior-point methods can be programmed in a way that is independent of the special features of the underlying problem and the computational platform on which it is to be executed. These components can be reused efficiently in a wide variety of applications when coupled with modules that implement the linear algebra operations specific to the application type and the architecture of the machine. Powerful toolkits for application areas such as statistics, process design and control, scheduling, and inverse problems can be built up from components that implement optimization algorithms, specialized and general linear algebra operations, mesh generation algorithms, discretization algorithms for differential equations, visualization techniques, and graphical user interfaces.

Given the continuing improvements in such long-studied areas as linear programming, it seems certain that algorithmic advances will continue to play a role in improvement of the software tools. Moreover, the scope of optimization software will be widened to handle more difficult and computationally challenging problem classes, particularly those involving uncertainty, discrete decision variables, and matrix variables. Areas such as stochastic programming, stochastic integer programming, semidefinite programming, and second-order cone programming have been widely studied, but are not well represented in the current generation of production software. This is because of the complexity of the algorithms, the sophisticated demands they place on modelers, and their high computational requirements. This situation will change in future generations of optimization software as the benefits of these algorithms are demonstrated in applications.

ACKNOWLEDGMENT

Research supported by the Mathematical, Information, and Computational Sciences Division subprogram of the Office of Advanced Scientific Computing Research, U.S. Department of Energy, under Contract W-31-109-Eng-38.

References

Kelley, C. T. 1999. In *Iterative Methods for Optimization. Frontiers in Applied Mathematics.* Philadelphia: SIAM Publications.

Moré, J. J. and S. J. Wright. 1993. *Optimization Software Guide. Frontiers in Applied Mathematics*, No. 15. Philadelphia: SIAM Publications.

Wolsey, L. A. 1998. In *Integer Programming.* Series in Discrete Mathematics and Optimization. New York: John Wiley & Sons.

Wright, S. J. 1997. In *Primal–dual Interior-point Methods.* Philadelphia: SIAM Publications.

21 OPTIMIZATION SOFTWARE LIBRARIES

Andreas Fink, Stefan Voß, and David L. Woodruff

ABSTRACT Optimization software libraries provide building blocks for application-specific software systems. After making a distinction between callable packages, numerical libraries, and component libraries, we give an overview of available libraries, which provide reusable functionality with respect to different optimization methodologies.

1. Introduction

Optimization problems in practice are diverse and evolve over time, so one requires both ready-to-use optimization software packages, which are discussed in Chapter 20 of this handbook, and optimization software libraries, which provide more or less adaptable building blocks for application-specific software systems. A general classification provides a distinction between callable packages, numerical libraries, and component libraries. In this chapter, we discuss some general aspects of corresponding libraries. To allow for wider applicability, we restrict the exposition to general approaches and mostly neglect problem-specific software such as for, for example, vehicle routing problems. Our survey must necessarily be cursory and incomplete; it is not intended to be judgmental and in some cases one has to rely on descriptions provided by software vendors. Therefore, we include several references (literature and WWW) that provide further information.

In order to apply optimization methods to a new type of problem, corresponding models and algorithms have to be coded so that they are accessible to a computer. One way to achieve this step is the use of a modeling language. Such modeling systems provide an excellent interface between models and solvers, but only for a limited range of model types (in some cases, for example, linear) due, in part, to limitations imposed by the solvers. Furthermore, while modeling systems for heuristic search is an active research topic, it is still an open question as to whether such an approach may be generally successful (Michel and van Hentenryck, 1999). Modeling languages treat the solvers as a black box with numerous controls. Due to variations, for example, with respect to the pursued objective or specific problem properties, addressing real-world problems often requires special-purpose methods. Thus, we are faced with the difficulty of efficiently adapting and applying appropriate methods to these problems. Optimization software libraries are intended to make it relatively easy and cost effective to incorporate advanced planning methods in application-specific software systems.

1. Callable packages allow users to embed optimization functionality in applications, and are designed primarily to allow the user's software to prepare the model and feed it to the package. Such systems also typically include routines that allow manipulation of the model and access to the solver's parameters. As with the modeling language approach, the solver is treated essentially as an opaque object, which provides a classical functional interface, using procedural programming languages such as C. While there are only restricted means to adapt the corresponding coarse-grained functionality, the packages do often offer callbacks that facilitate execution of user code during the solution process.
2. Numerical libraries provide similar functionality, except that the model data is treated using lower levels of abstraction. For example, while modeling languages and callable packages may allow the user to provide names for sets of variables and indexes into the sets, numerical libraries only facilitate the manipulation of vectors and matrices as numerical entities. Well-known solution techniques can be called as subroutines, or can be built from primitive operations on vectors and matrices. These libraries provide support for linear algebra, numerical computation of gradients, and support for other operations of value, particularly for continuous optimization.
3. Component libraries provide useful abstractions for manipulating algorithm and problem concepts. Object-oriented software technology is generally used to build and apply

corresponding components; see Chapter 25 of this handbook. To enable adaptation these components are often provided at source-code level. Corresponding class libraries support the development of application-specific software systems by providing a collection of adaptable classes intended to be reused. However, the reuse of algorithms may still be regarded as "a challenge to object-oriented programming" (Weihe, 1997).

2. Callable Packages

2.1. *Linear Programming*

Linear programming problems can be efficiently solved to optimality by the techniques introduced in Chapter 1 of this handbook. Respective algorithms are based on a declarative specification of problems, which can be implemented using well-defined low-level schemes such as vectors and matrices. This facilitates the use of highly advanced methods that are, in general, independent from specific types of problems. At present, large-scale linear programs having hundreds of thousands of continuous variables are regularly solved by the usage of advanced techniques such as sparse matrices. There are several corresponding commercial software systems; for example, CPLEX (1999), LINDO (1999), OSL (1999), XPRESS-MP (1999).

While these products are applicable as ready-to-use packages, they generally include subroutine libraries (such as shared/dynamic link libraries), which can be integrated in application-specific software systems. Corresponding libraries include optimizing functions that may be called with arguments defining problem instances and/or solution vectors. Moreover, such libraries generally include routines to set parameters, set, get and modify problem and/or solution instances, and to store information to or retrieve information from a file system. While these libraries commonly include some type of callback mechanism, which allows user-specific additions to the solver process, the libraries are generally designed to be utilized as a single entity to solve problems rather than as a set of components to be mixed and matched to solve problems. For additional information, see for instance Fourer (1999).

Linear programming algorithms are also available as spreadsheet add-ons, which afford users the opportunity to manipulate their data using the spreadsheet interface. Such packages are available from, for example, LINDO or Frontline; see Savage (1997)

While minimum-cost network flow problems can be modeled as linear programs, specialized algorithms generally solve such problems more efficiently. Corresponding methods are often included in linear programming packages. There are also specialized packages such as NETFLOW (1999).

2.2. *Integer Linear Programming*

Integer and mixed-integer programming problems, which are in general NP-hard, can be addressed by well-known combinatorial optimization algorithms such as those discussed in Chapter 3 of this handbook. While these problems may in theory be intractable due to high computational requirements, recent advances in both corresponding algorithms and computing technology have led to a broad applicability of a variety of methods. All libraries mentioned in the preceding section also include methods for solving integer programming problems by branch-and-bound algorithms exploiting linear programming relaxations. In the following, we discuss some additional specialized libraries, which follow different approaches.

MINTO

MINTO (Mixed INTeger Optimizer) is a software system that solves mixed-integer linear programs by a branch-and-bound algorithm with linear programming relaxations (using respective callable libraries for linear programming). It also provides automatic constraint classification, preprocessing, primal heuristics, and constraint generation. Moreover, the user can enrich the basic algorithm by providing a variety of specialized application routines that can customize MINTO to a problem class. Further information can be found at MINTO (1999).

EMOSL

EMOSL (Entity Modelling and Optimisation Subroutine Library) of DASH Associates is a combined modeler and optimizer library, which is based on the linear and integer programming capabilities of XPRESS-MP. EMOSL aims at techniques that provide mechanisms for exploiting model structure. The EMOSL toolkit enables developers to access entities expressed in the modeler's notation from a high level programming language. It allows developers to refer to generic model entities such as data tables, parameters, subscripted variables and constraint names, special ordered sets, and so on. Further information can be found at XPRESS-MP (1999).

ABACUS

ABACUS (A Branch-And-CUt System) is a framework, implemented in C++, that incorporates cutting plane and column generation methods for combinatorial optimization problems. A framework defines a general template, which has to be completed by application-specific functionality. ABACUS provides support for the implementation of branch-and-bound algorithms using linear programming relaxations that can be complemented with the dynamic generation of cutting planes or columns (branch-and-cut, branch-and-price, branch-and-cut-and-price). ABACUS provides a variety of general algorithmic concepts, for example, enumeration and branching strategies, from which the user of the system can choose the best alternative for his application. For further information, see ABACUS (1999) and Thienel (1997).

2.3. *Nonlinear and Global Optimization*

Once the assumption of linearity is dropped, the problems can become much more difficult to solve. This is particularly true when convexity is not present. Problems that require a search over a large space for the global minimizer to a nonlinear function are often referred to as global optimization problems.

BARON

BARON (Branch And Reduce Optimization Navigator) is a FORTRAN package/library for solving nonconvex optimization problems to global optimality. It derives its name from its combining of interval analysis and duality in its reduce arsenal with enhanced branch-and-bound concepts as it searches the solution space. The code consists of a core module for global optimization of arbitrary problems as long as problem-specific lower and upper bounding routines are provided by the user. A variety of specialized modules for which no coding from the user is required are also provided. The list of specialized modules currently includes: separable concave quadratic minimization, separable concave minimization, programming with economies of scale (Cobb–Douglas functions), fixed-charge programming, fractional programming, univariate polynomial programming, linear multiplicative programming, general linear multiplicative programming, mixed-integer convex quadratic programming, mixed-integer linear programming, indefinite quadratic programming, and factorable nonlinear programming. With the exception of the factorable nonlinear programming module, all other modules solve linearly constrained problems. The factorable nonlinear programming module can handle most types of typical optimization problems. All modules can solve problems where some or all of the variables are required to have integral values. Further information can be found at BARON (1999).

LGO

LGO (Lipschitz Global Optimizer) is a package for continuous global optimization, which also provides corresponding callable modules. The LGO program system solves (Lipschitz)-continuous global optimization problems on finite *n*-interval ('box') regions, in the possible presence of additional (Lipschitz)- continuous constraints, (cf. Pinter, 1996).

2.4. *Stochastic Programming*

Stochastic programming is a methodology for bringing uncertainty into the decisionmaking process, especially with respect to models based on linear programming. For example, IBM's *Optimization Library Stochastic Extensions* enhance the OSL package by multistage stochastic optimization algorithms, which may be applied as callable modules (OSL, 1999).

3. Numerical Libraries

Numerical libraries are typically implemented as classic function libraries, which provide related functionality by the definition of a set of routines that transform data at lower levels of abstraction (for example, vectors and matrices). Function libraries are relatively inflexible, as such routines commonly perform well-defined functionality without mechanisms that can be varied or extended. However, function libraries have been proven useful as they provide numerical algorithms that are often needed as building-blocks of optimization software.

"Numerical recipes" provide a collection of numerical algorithms in the form of a "cookbook," including both mathematical descriptions and actual implementations (available in FORTRAN, Pascal, and C; see Press et al., 1993). The functionality includes algorithms of numerical analysis and linear algebra, as well as simple optimization methods.

The NAG and IMSL libraries are collections of routines for the solution of numerical and statistical problems (both available for different languages such as FORTRAN and C; cf. NAG, 1999; IMSL, 1999).

LAPACK (Linear Algebra PACKage) is a library of FORTRAN subroutines for solving the most commonly occurring problems in numerical linear algebra (Anderson et al., 1995).

MATLAB is an integrated technical computing environment that combines numeric computation, visualization, and a high-level programming language. MATLAB thus provides a more abstract and easy-to-use interface to the functionality of the libraries discussed before. Special toolboxes extend the functionality of MATLAB. The optimization toolbox provides methods for the optimization of different types of nonlinear problems; cf. MATLAB (1999)

4. Component Libraries

4.1. *Constraint Programming*

Constraint logic programming may be regarded as a backtracking search that is based on the assumption that it may be advantageous to solve combinatorial optimization problems by investing more effort in a reduction of the solution domain by applying constraint propagation rules; see Marriott and Stuckey (1998). Problems can often be specified more naturally as constraint programs than as integer programs, which is particularly supportive in connection with respective modeling languages (Fourer, 1998; van Hentenryck, 1999). Corresponding approaches have been quite successfully applied to problems with a significant number of logical constraints (for example, special scheduling and assignment problems). In the following, we discuss several such libraries.

ILOG SOLVER

ILOG Solver is a C++ library, which embodies constraint logic programming concepts such as logical variables, incremental constraint satisfaction, and backtracking (Puget and Leconte, 1995). Application-specific add-ons such as ILOG Scheduler or ILOG Dispatcher provide functionality for special types of problems (constraint-based scheduling and vehicle routing/technician dispatching, respectively). For example, ILOG Scheduler provides respective classes that support modeling of concepts such as resources, activities, and scheduling constraints. Further information can be found at ILOG (1999).

ECLIPSE

ECLiPSe is a development environment for constraint logic programming applications. It contains several constraint solver libraries and provides a high-level modeling language to facilitate the development of programs to solve combinatorial problems in planning, scheduling, resource allocation, timetabling, transport, and so on. ECLiPSe has an open architecture for constraint extensions based on attributed variables and its constraint solvers allow the users to define and use new constraints at different conceptual levels. Further information can be found at ECLiPSe (1999).

CLAIRE

CLAIRE is a high-level functional and object-oriented language with rule processing capabilities. It is intended to allow the programmer to express complex algorithms in a concise and natural manner. To achieve this goal, CLAIRE provides means such as a rich type system, parametric classes and methods, an object-oriented logic with set extensions, and dynamic versioning. CLAIRE is a complete programming system with an interpreter, a compiler, and a set of tools (tracer, debugger, object inspector) that works both on compiled and interpreted code. CLAIRE can also be used as a C++ preprocessor because it generates human-readable C++ code. On top of CLAIRE, CLAIRE SCHEDULE is a library of constraint propagation algorithms for preemptive, elastic, and mixed scheduling; that is, scheduling with both interruptible and noninterruptible activities. It consists of a set of functions allowing the definition of preemptive and mixed scheduling problems, and the propagation of the basic types of decisions that can be made in the process of solving such problems. For further information see Caseau and Laburthe (1996).

4.2. *Libraries for Heuristic Optimization*

Most discrete optimization problems are NP-hard. Moreover, as the modeling of the problem is often an approximate one, and the data are often imprecise, heuristics are a primary way to tackle these problems. The use of appropriate metaheuristics generally meets the needs of decisionmakers to efficiently generate satisfactory solutions; see Osman and Kelly (1996), Reeves (1993), Voß et al. (1999). The common incorporation of advanced heuristics in application systems requires the user to find a way to reuse much of such software and to redo as little as possible each time. However, in comparison to the exact optimization field discussed above, corresponding software libraries that meet practical demands offer less support with respect to, for example, robustness and ease of use.

The difficulties in developing reusable and adaptable software components for heuristic search are as follows. Compared to the field of mathematical programming, which relies on well-defined, problem-independent representation schemes for problems and solutions on which algorithms may operate, metaheuristics are based on abstract definitions of solution spaces and neighborhood structures. Moreover, for example, memory-based tabu search approaches are generally based on abstract problem-specific concepts such as solution and move attributes. With this, different heuristics can be described on a high level.

The crucial problem of local search based metaheuristics libraries is a generic implementation of heuristic approaches as reusable software components, which must operate on arbitrary solution spaces and neighborhood structures. The drawback is that the user must, in general, provide some kind of a problem/solution definition and a neighborhood structure, which is usually done using sophisticated computer languages such as C++.

An early C++ class library for heuristic optimization by Woodruff (1997) included both local search-based methods and genetic algorithms. This library raised issues that illustrate both the promise and the drawbacks to the adaptable component approach. From a research perspective such libraries can be thought of as providing a concrete taxonomy for heuristic search. So concrete, in fact, that they can be compiled into machine code. This taxonomy sheds some light on the relationships between heuristic search methods for optimization and on ways in which they can be combined. Furthermore, the library facilitates such combinations as the classes in the library can be extended and/or combined to produce new search strategies.

From a practical and empirical perspective, these types of libraries provide a vehicle for using and testing heuristic search optimization. A user of the library must provide the definition of the problem-specific abstractions and may systematically vary and exchange heuristic strategies and corresponding components.

In the following, we discuss several heuristic optimization libraries from the research field, which differ, for example, in design concept, the chosen balance between ease of use and flexibility and efficiency, and overall scope. All of these approaches are based on the concepts of object-oriented programming.

HotFrame

HotFrame, a Heuristic OpTimization FRAMEwork implemented in C++, provides both adaptable components that incorporate different metaheuristics and an architectural description of the collaboration among these components and problem-specific complements. All typical application-specific concepts are treated as objects or classes: problems, solutions, neighbors, and solution and move attributes. On the other hand, metaheuristics' concepts such as different methods and their building blocks such as tabu criteria and diversification strategies are also treated as objects. HotFrame uses genericity as the primary mechanism to make these objects adaptable. That is, common behavior of metaheuristics is factored out and grouped into generic classes, applying static type variation. Metaheuristics template classes are parameterized by aspects such as solution spaces and neighborhood structures.

HotFrame defines an architecture for the interplay between heuristic classes and application-specific classes, and provides several such classes, which implement classic methods that are applicable to arbitrary problem types, solution spaces, and neighborhood structures:

- A simple local search frame, which can be instantiated by, for example, different neighbor selection rules resulting in methods such as steepest descent or iterated local search.
- Several simulated annealing and threshold accepting methods, which can be customized by different acceptance criteria, cooling schedules, and reheating schemes. This includes an implementation of the robust simulated annealing algorithm of Johnson et al. (1989).
- A classic tabu search frame, which is based on treating both the tabu criterion and the diversification strategy as parameters. The tabu criterion, which may be used to dynamically vary the neighborhood by prohibiting or penalizing certain moves, is constituted taking into account information about the search history (performed moves and traversed solutions). Move attributes, trajectory information, and tabu criteria can be defined flexibly. Static, strict, and reactive tabu criteria are built in. Other tabu search methods can be implemented exploiting basic building blocks included in the framework.
- Candidate list templates can be used as filters in connection with meaningful yet arbitrary neighborhood structures and heuristics. On the other hand, variable neighborhood depth methods may be used to enlarge the neighborhood.
- A general evolutionary algorithm template, which is customizable with respect to arbitrary solution representations, selection and replacement strategies, recombination (not necessarily restricted to two parents), and mutation operators. Common implementations of these modules are included.

All heuristics are implemented in a consistent way, which facilitates the easy embedding of arbitrary methods into application systems or as parts of more advanced/hybrid methods. Both new metaheuristics and new applications can be added to the framework. HotFrame includes built-in support for solution spaces representable by binary vectors or permutations, in connection with corresponding standard neighborhood structures, solution and move attributes, and recombination operators. Otherwise, the user may derive specialized classes from suitable built-in classes or implement corresponding classes from scratch according to a defined interface. For further information about HotFrame see Fink and Voß (1999) and HotFrame (1999).

TEMPLAR

The Templar framework, implemented in C++, provides a method, and software components, for constructing systems to solve optimization problems. The Templar framework is based on problem classes and engine classes. The engine's source code is representation and problem independent. The problem class is an abstract base class, which is used to derive specific problem classes. These specific problem classes embody the characteristics of the problem type, and are capable of loading problem instances. Problems make operators, such as neighborhood operators or genetic operators, available for use by an engine. The representation used by the problem (for example, permutation) also provides its own operators. Engines can then use these operators to produce good solutions to a given problem instance. The base engine abstract base class is also used for derivation of specific engine classes. Examples include simulated annealing engines, tabu search engines, and genetic algorithm engines. Although a single instance of an engine is connected to a particular problem, each specific engine class is problem, and representation, independent. As well as being easy to control, engines are designed so that a driving process can make a group of engines work cooperatively. For further information, see Templar (1999).

NEIGHBORSEARCHER

Andreatta et al. (1998) describe NeighborSearcher, an object-oriented framework for local search heuristics. Their main goal is to provide an architectural basis for the implementation and comparison of different local search heuristics. Accordingly, they define a coarse-grained modularization of the domain in abstract classes that encapsulate concepts such as the construction of the initial solution, the local search algorithm, the solution, or the movement model. With this, one may implement a specific search strategy by selecting derived classes that provide the respective functionality. This provides an adaptation mechanism to the higher-level client code.

GENETIC ALGORITHM LIBRARIES

Several libraries exist for genetic algorithms; surveys, as well as software, can be found at Heitkötter and Beasley (1999). In principle, an advantage of using classic genetic algorithm libraries such as Genitor (1999) or GAlib (1999) is that no neighborhood needs to be specified. If the built-in genomes of a genetic algorithm library adequately represent one's problem, a user-specified objective function may be the only problem-specific code that must be written. Unfortunately, genetic algorithms without a local search component have not generally proven to be very effective.

Next, we briefly discuss the functionality of the GAlib library, a C++ library, which provides the application programmer with a set of genetic algorithms objects (cf. GAlib, 1999). GAlib is flexible with respect to arbitrary data representations and standard or custom selection, crossover, mutation, scaling, replacement, and termination methods. Overlapping (steady-state GA) and nonoverlapping (simple GA) populations are supported. Built-in selection methods include rank, roulette wheel, tournament, stochastic remainder sampling, stochastic uniform sampling, and deterministic sampling. One can use chromosome types built in to the library (bit-string, array, list, tree) or derive a chromosome based on user-defined objects. All chromosome initialization, mutation, crossover, and comparison methods can be customized. Built-in mutation operators include random flip, random swap, Gaussian, destructive, swap subtree, and swap node. Built-in crossover operators include arithmetic, blend, partial match, ordered, cycle, single point, two point, even, odd, uniform, and node- and subtree-single point.

5. Conclusions and Outlook

Optimization libraries enable the broad application of sophisticated methods by facilitating their inclusion in applications as well as supporting extension and adaptation of the methods. We should note that, at present, the success of respective applications often critically depends on appropriate adaptations of library components to the specific problem at hand. Clearly, research will continue along various dimensions (such as an increased exploitation of the techniques provided by object-oriented programming to build fine-grained components), which should result in more effective optimization libraries.

References

ABACUS. 1999. http://www.informatik.uni-koeln.de/ls_juenger/projects/abacus.html. A Branch-And-CUt System.

Anderson, E., Z. Bai, C. Bischof, J. Demmel, and J. J. Dongarra. 1995. In *LAPACK Users Guide*. 2d ed. Philadelphia: SIAM.

Andreatta, A. A., S. E. R. Carvalho, and C. C. Ribeiro. 1998. "An object-oriented framework for local search heuristics." In *Proceedings of the 26th Conference on Technology of Object-Oriented Languages and Systems (TOOLS USA'98)*, pp. 33–45. Piscataway: IEEE.

BARON. 1999. http://archimedes.scs.uiuc.edu/baron.html. The Branch-And-Reduce Optimization Navigator.

Caseau, Y. and F. Laburthe. 1996. "CLAIRE: combining objects and rules for problem solving." In *Multiparadigm Logic Programming: Proceedings of the JICSLP'96 Post-Conference Workshop*, Report No. 96-28, edited by M. M. T. Chakravarty, Y. Guo, and T. Ida, pp. 105-114. Berlin: Technische Universität Berlin, Fachbereich Informatik.

CPLEX. 1999. http://www.cplex.com. Mathematical programming software.

ECLiPSe. 1999. http://www.icparc.ic.ac.uk/eclipse/. Constraint logic programming system.

Fink, A. and S. Voß. 1999. "Generic metaheuristics application to industrial engineering problems." *Comput. Ind. Eng.* **37**: 281–284.

Fourer, R. 1998. "Extending a general-purpose algebraic modeling language to combinatorial optimization: a logic programming approach." In *Advances in Computational and Stochastic Optimization Logic Programming, and Heuristic Search*, edited by D. L. Woodruff, pp. 31–74. Boston: Kluwer.

Fourer, R. 1999. "Software survey: linear programming." *OR/MS Today* **26(4)**: 64–65 Software survey available in abridged form online at http://lionhrtpub.com/orms/surveys/LP/LP-survey.html.

GAlib. 1999. http://lancet.mit.edu/ga/. A C++ library of genetic algorithm components.

Genitor, 1999. http://www.cs.colostate.edu/~genitor/. A genetic algorithm library.

Heitkötter, J. and D. Beasley. 1999. "The hitch-hiker's guide to evolutionary computation (FAQ for comp.ai.genetic)." http://surf.de.uu.net/encore/www/. Issue 7.3, 20 September 1999.

van Hentenryck, P. 1999. *The OPL Optimization Programming Language*. Cambridge, MA: MIT Press With contributions by I. Lustig, L. Michel, and J.-P. Puget.

HotFrame. 1999. http://www.winforms.phil.tu-bs.de/research/hotframe.htm. Heuristic OpTimization FRAMEwork.

ILOG. 1999. http://www.ilog.com. Constraint logic programming libraries.

IMSL. 1999. http://www.imsl.com. Mathematics and statistics libraries.

Johnson, D. S., C. R. Aragon, L. A. McGeoch, and C. Schevon. 1989. "Optimization by simulated annealing: an experimental evaluation: Part I. graph partitioning." *Oper. Res.* **37**: 865–892.

LINDO. 1999. http://www.lindo.com. Mathematical programming software.

Marriott, K. and P. J. Stuckey. 1998. *Programming with Constraints: An Introduction*. Cambridge, MA: MIT Press.

MATLAB. 1999. http://www.mathworks.com/products/matLab/. An integrated technical computing environment.

Michel, L. and P. van Hentenryck. 1999. "Localizer: a modeling language for local search." *INFORMS J. Comput.* **11**: 1–14.

MINTO. 1999. http://akula.isye.gatech.edu/~mwps/projects/minto.html. A Mixed INTeger Optimizer.

NAG. 1999. http://www.nag.com. Numerical libraries.

NETFLOW. 1999. ftp://dimacs.rutgers.edu/pub/netflow. Network flow packages.

OSL. 1999. http://www6.software.ibm.com/es/oslv2/features/welcome.htm The IBM Optimization Subroutine Library.

Osman, I. H. and J. P. Kelly. (eds). 1996. *Meta-heuristics: Theory and Applications*. Boston: Kluwer.

Pinter, J. D. 1996. *Global Optimization in Action: Continuous and Lipschitz Optimization—Algorithms, Implementations and Applications*. Boston: Kluwer.

Press, W. H., S. A. Teukolsky, W. T. Vetterling, and B. P. Flannery. 1993. *Numerical Recipes in C: the Art of Scientific Computing*. 2d ed. Cambridge, England: Cambridge University Press.

Puget, J.-F. and M. Leconte. 1995. "Beyond the glass box: constraints as objects." In *Proceedings of the International Symposium on Logic Programming*, edited by J. Lloyd, pp. 513–527. Cambridge MA: MIT Press.

Reeves, C. (ed). 1993. *Modern Heuristic Techniques for Combinatorial Problems*. Oxford, England: Blackwell.

Savage, S. 1997. "Weighing the pros and cons of decision technology in spreadsheets." *OR/MS Today* **24(1)**: 42–45. Software survey available in abridged form online at http://lionhrtpub.com/orms/surveys/SSA/SSA.html.

Templar. 1999. http://www.sys.uea.ac.uk/~msj/templar/. The Templar framework.

Thienel, S. 1997. "ABACUS—A Branch-And-CUt System, version 2.0, user's guide and reference manual." Technical report No. 97.298, Köln, Germany: Universität Köln.

Voß, S., S. Martello, I. H. Osman, and C. Roucairol. (eds). 1999. *Meta-heuristics: Advances and Trends in Local Search Paradigms for Optimization*. Boston: Kluwer.

Weihe, K. 1997. "Reuse of algorithms: still a challenge to object-oriented programming." *ACM SIGPLAN Notices* **32(10)**: 34–48. Proceedings of the 1997 ACM SIGPLAN Conference on Object-Oriented Programming, Systems, Languages & Applications (OOPSLA'97).

Woodruff, D. L. 1997. "A class library for heuristic search optimization." *INFORMS Comput. Sci. Tech. Sect. Newsletter* **18(2)**: 1–5.

XPRESS-MP. 1999. http://www.dash.co.uk. Mathematical programming software.

22 OPTIMIZATION TEST PROBLEM LIBRARIES

John E. Beasley

ABSTRACT In this chapter we detail a number of sources (libraries) of test problems for optimization problems commonly dealt with by the operations research community. All of these test problem libraries are accessible via the Internet.

1. Introduction

One difficulty faced by the developers of algorithms is that of obtaining test problems that can be used to compare their algorithms with those of other researchers. In recent years this difficulty has been much alleviated by the increasing availability of test problems accessible via the Internet. Frequently these test problems are concentrated at a number of sources, and these sources are typically known as test problem libraries.

We shall assume throughout this chapter some familiarity on the part of the reader with the Internet and the World Wide Web, and with basic functions such as e-mail, ftp (file transfer protocol), newsgroups, and Web browsing (e.g., using Netscape Navigator or Internet Explorer). Readers without such knowledge can easily gain it from any one of the many books written to introduce the beginner to the Internet/Web.

2. Why Use Standard Test Problems?

Before detailing the libraries available it is helpful to review the scientific reasons for making use in one's own work of standard test problems that are publicly available.

The primary reason for using standard test problems is that it enables a clear scientific comparison to be made between one's own work and the work of others (who will also have used the same test problems).

A secondary reason relates to getting one's work published in refereed journals. Increasingly authors will find that journal referees will expect to see such test problems used in new work.

In summary, our advice to others would be: if standard publicly available test problems exist, then use them.

3. History

Readers who are younger than the author may not have considered how things were done in the days before libraries of test problems existed. In those (olden) times, different authors typically generated and used their own test problems. This led to the situation where making a relative comparison between the merits of different algorithms for the same mathematical problem was difficult. Naturally, different authors used different computers, and so with the added complication of them using different test problems as well, comparison between algorithms was typically judgmental, rather than having a more scientific base.

For some mathematical problems (e.g., vehicle routing, plant location) standard test problems did exist but these were typically small test problems that could be conveniently written out in full on a few sheets of paper. As time, and algorithms, advanced larger test problems were needed.

While a few brave souls made their test problems available via magnetic tape, this tended to be a somewhat time-consuming, cumbersome, and expensive process.

Early test-problem libraries (dating from the mid-1980s) made use of e-mail to distribute problems. Typically one sent an e-mail containing the names of the files one wished to obtain and these were returned via e-mail. A number of test-problem libraries still retain such a distribution mode.

As the Internet developed, ftp (file transfer protocol) became a common access and distribution mode. Using ftp, one can obtain access to another computer but with limited privileges, typically only sufficient privileges to download certain public files.

Finally, as the World Wide Web (WWW) has exploded, accessing test problems via a Web browser has become the dominant access and distribution mode.

4. Libraries

We will now go on to detail a number of the test-problem libraries that are available. Note here that the libraries that do exist can be conveniently categorized as single problem or multiple problem:

(a) by whether they contain test instances for a single mathematical (optimization) problem; or

(b) by whether they contain test instances for a number of different mathematical (optimization) problems.

It is common nowadays for one library to refer to another so that while one library may contain no test instances for a particular problem it will contain a hypertext link to a library that does. Probably the library that best makes use of such linkages is the OR-Library, maintained by the author. As a consequence of this linkage we will refer in this paper to:

(a) all relevant libraries of which we are aware that can be categorized as multiple-problem libraries; and

(b) a selection of the libraries of which we are aware that can be categorized as single-problem libraries.

For each library we give details of how to access it (using e-mail, ftp, and WWW as appropriate) and also outline the test problems that it contains. Naturally, readers will appreciate that over time the test problems available will change, as may access details.

Note here that for all the ftp addresses given below, log in as anonymous.

4.1. DIMACS

`ftp: address: dimacs.rutgers.edu`
`directories: pub/netflow` and `pub/challenge/graph`

`WWW: http://dimacs.rutgers.edu/Challenges`

DIMACS has test problems available for graph coloring, matching, maximum clique, and network flow.

4.2. ELIB

`ftp: address: elib.zib.de`
`directory: pub/Packages/mp-testdata`

`WWW: http://elib.zib.de/`

ELIB has test problems available for: assignment, clustering, linear programming, integer and mixed-integer programming, matrix decomposition, matching, maximum flow, minimum-cost flow and transportation, set partitioning, Steiner tree, traveling salesman, and vehicle routing. It also has available a wide selection of computer codes for various problems.

4.3. Handbook of Test Problems in Local and Global Optimization

`WWW: http://titan.princeton.edu/TestProblems/`

This site has test problems available for: quadratic programming (integer programming, quadratic assignment, maximum clique); quadratically constrained problems (separable quadratic constraints, complementarity-type constraints, integer-type constraints); univariate polynomial; bilinear (pooling and blending, separation sequencing, heat exchanger network design, multicommodity network flow); biconvex and difference of convex functions (phase and chemical equilibrium); generalized geometric (process design, stability analysis); twice continuously differentiable nonlinear programs (process design, phase and chemical equilibrium, computational chemistry, VLSI chip design, portfolio optimization); bilevel programming (various economics, civil engineering, and chemical engineering problems); complementarity (Nash equilibrium, Walrasian equilibrium, traffic assignment); semidefinite programming (combinatorial optimization, control theory); mixed-integer nonlinear programming (process and network synthesis, molecular design); combinatorial optimization (satisfiability, traveling salesman, assignment, graph coloring, maximum clique, Steiner problem in networks); nonlinear systems of equations (multiple steady-state identification, locating all azeotropes); and dynamic optimization (optimal control, parameter estimation for dynamic models, reactor network synthesis).

4.4. *LOLIB*

`WWW: http://www.iwr.uni-heidelberg.de/iwr/comopt/software/LOLIB`

LOLIB contains test problems for the linear ordering problem.

4.5. *MIPLIB*

`ftp: address: softlib.rice.edu`
`directory: pub/miplib`

`WWW: http://www.crpc.rice.edu/softlib/catalog/miplib.html`

MIPLIB has test problems available for both pure and mixed-integer programming.

4.6. *NETLIB*

`email: message: send index`
`address: netlib@ornl.gov`

`ftp: address: ftp.netlib.org`

`WWW: http://www.netlib.org/index.html`

NETLIB has test problems available for linear programming. It also has available a wide selection of computer codes for various problems.

4.7. *OR-Library*

`email: message: info`
`address: o.rlibrary@ic.ac.uk`

`ftp: address: mscmga.ms.ic.ac.uk`
`directory: pub`

`WWW: http://mscmga.ms.ic.ac.uk/info.html`

OR-Library has details of test problems available for a large number of optimization problems. Specifically it has details of test problems for: assignment, bin packing, biquadratic assignment, capacitated minimal spanning tree, crew scheduling, complementarity, data envelopment analysis, equitable partitioning, fixed-charge transportation, frequency assignment, generalized assignment, graph coloring, graph planarization, ground holding, index tracking, integer programming, linear ordering, linear programming, location (capacitated, *p*-hub, *p*-median: uncapacitated and capacitated, uncapacitated), matching, matrix decomposition, maximum clique, min-cut clustering, multidimensional knapsack, network flow, nonlinear programming, portfolio optimization, quadratic assignment, resource-constrained project scheduling, resource-constrained shortest-path, satisfiability, scheduling (aircraft landing, common due date, flow shop, job shop, open shop, weighted tardiness), semidefinite programming, set covering, set partitioning, Steiner (Euclidean, rectilinear, in graphs), stochastic programming, three-dimensional packing, time-series forecasting, timetabling, traveling salesman (period salesman, single period), two-dimensional cutting (assortment, constrained guillotine, constrained nonguillotine, unconstrained guillotine), unconstrained binary quadratic programming, and vehicle routing (fixed areas, fixed routes, multidepot, period routing, single period, single period with time windows, sparse feasibility graph).

OR-Library is maintained by the author of this paper and researchers with publicly available test problems are invited to contact him (`j.beasley@ic.ac.uk`) so that a link to their test problems can be included in OR-Library.

4.8. PSPLIB

`WWW: http://www.bwl.uni-kiel.de/Prod/psplib/library.html`

PSPLIB contains test problems for resource constrained project scheduling.

4.9. QAPLIB

`WWW: http://www.imm.dtu.dk/~sk/qaplib/`

QAPLIB contains test problems for the quadratic assignment problem.

4.10. TSPLIB

`WWW: http://www.iwr.uni-heidelberg.de/iwr/comopt/software/TSPLIB95`

TSPLIB contains traveling salesman and vehicle routing test problems.

5. Finding test problems

It may be that the reader is concerned with a problem for which the libraries listed above do not contain appropriate test problems. Before embarking on generating their own problems we would advise checking to see if appropriate test problems exist elsewhere. This can be done:

(a) by using the standard Web search engines (e.g., using Alta Vista `http://www.altavista.com`)

(b) by searching the Internet newsgroups (e.g., using Deja.com `http://www.deja.com`)

(c) by checking out Michael Trick's compilation of problem instances `http://mat.gsia.cmu.edu/resource.html`)

(d) by appealing to the readers of relevant Internet newsgroups for information. Relevant newsgroups for readers of this paper might include `comp.ai.genetic`, `sci.math`, `sci.math.num-analysis`, `sci.math.research`, `sci.nonlinear`, and `sci.op-research`.

23 PARALLEL COMPUTING ENVIRONMENTS

Simone de L. Martins, Celso C. Ribeiro, and Noemi Rodriguez

ABSTRACT This article presents a survey of parallel computing environments available for the implementation of parallel algorithms for optimization problems. These parallel environments are composed by programming tools, performance evaluation tools, debuggers, and optimization libraries. Programming tools, such as PVM, MPI, and Linda, and parallel optimization libraries, such as CPLEX parallel solvers and OSLp, are described. Performance evaluation tools and debuggers are also presented. References to parallel implementations of optimization algorithms for solving combinatorial problems are given.

1. Introduction

The demand for faster and more efficient computer systems for both scientific and commercial applications has grown considerably in recent times. Physical limitations and high costs make it impossible to increase processor speed beyond certain limits. To overcome these difficulties, new architectures emerged introducing parallelism in computing systems. Currently, different types of high-performance machines based on the integration of many processors are available. Among these, some of the most important are (Lau, 1996): vector multiprocessors (a small number of high-performance vector processors); massively parallel processors (hundreds to millions of processors connected together with shared or distributed memory); and networked computers (workstations connected by a medium- or high-speed network, working as a high-performance virtual parallel machine).

Parallelism leads to the need for new algorithms, specifically designed to exploit concurrency, because many times the best parallel solution strategy cannot be achieved by just adapting a sequential algorithm (Foster, 1995; Kumar et al., 1994). In contrast to sequential programs, parallel algorithms are strongly dependent on the computer architecture for which they are designed. Programs implementing these algorithms are developed and executed in parallel computing environments, composed by a parallel computer and the associated software (Chapman et al., 1999; Wolf and Kramer-Fuhrmann, 1996). The software consists of parallel programming tools, performance tools and debuggers associated to them, and some libraries developed to help in solving specific classes of problems (Blackford et al., 1997; IBM, 1999b; Merlin et al., 1999; Nieplocha et al., 1994; Plastino et al., 1999). Programming tools typically provide support for developing programs composed by several processes that exchange information during execution. The performance of systems developed with a programming tool and the available support for communication and synchronization among the various components of a parallel program are some important issues. The programming tool should also match the programming model underlying the parallel algorithm to be implemented.

Ease of programming and the time spent by a programmer to develop a program may sometimes be fundamental. Performance tools can help a programmer to understand and improve parallel programs, by monitoring the execution of a parallel program and producing performance data that can be analyzed to locate and understand areas of poor performance. Debugging parallel programs can be a hard task, due to their inherent nondeterminism. Parallel debugging tools developed to specifically debug parallel programs can be very helpful in finding bugs in a program.

This text is organized as follows. In Section 2 we introduce some basic parallel programming concepts related to memory organization, communication among processors, and parallel programming models. Some programming tools available for the development of parallel programs are discussed in Section 3. Next, we present in Section 4 some performance tools associated with these programming tools. The use of debuggers is discussed in Section 5. Two parallel optimization libraries are presented in Section 6. Some concluding remarks are drawn in the last section.

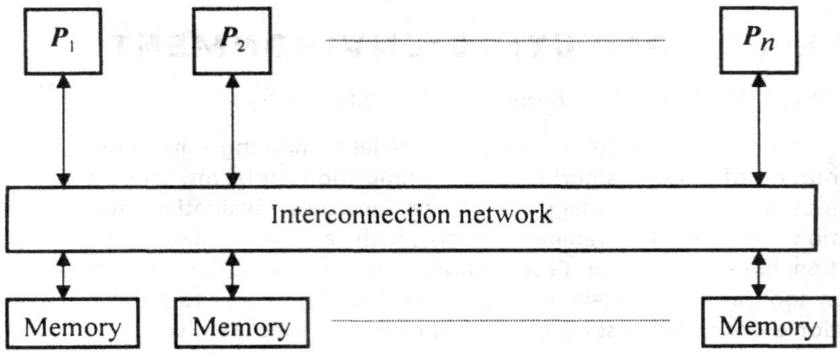

Figure 1 Shared memory architecture

2. Parallel Programming Concepts

Parallel programs consist of a number of processes working together. These processes are executed in parallel machines, based on different memory organization methods: shared memory or distributed memory (DCTC, 2001; Kumar et al., 1994). In shared-memory machines (Protic et al., 1998), all processors are able to address the whole memory space, as shown in Figure 1. Task communication is achieved through read and write operations performed by the parallel tasks on the shared memory. The main advantage of this approach is that access to data is very fast and processor communication is done through the common memory. However, system scalability is limited by the number of paths between the memory and processors, which can lead to poor performance due to access contention. In order to eliminate this disadvantage, some local memory can be added to each processor. This memory stores the code to be executed and data structures that are not shared by the processors. The global data structures are stored in the shared memory, as shown in Figure 2. One extension to this type of architecture consists of removing all physical memory sharing, as shown in Figure 3. For both models, references made by one processor to the memory of another processor are mapped by the hardware that interconnects memory and processors. Since access time to local memory is much smaller than to remote memory, this model can improve memory access time. A drawback of these architectures is that the control of memory access, necessary to maintain data consistency, has to be done by the programmer.

In the context of distributed-memory machines, memory is physically distributed among the processors. Each processor can only address its own memory, as shown in Figure 4. Communication among processes executing on different processors is performed by messages passed through the communication network. Networks of workstations are important examples of this architecture, whose importance has been growing in the last years. Besides allowing for scalability and fault tolerance, networks of computers present good cost/

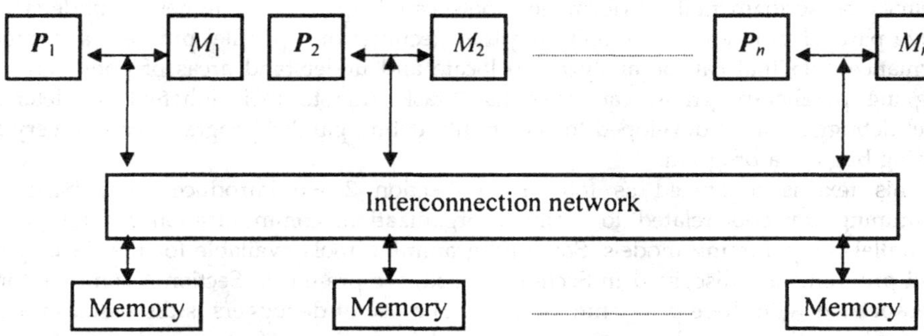

Figure 2 Shared memory architecture with local memory for processors

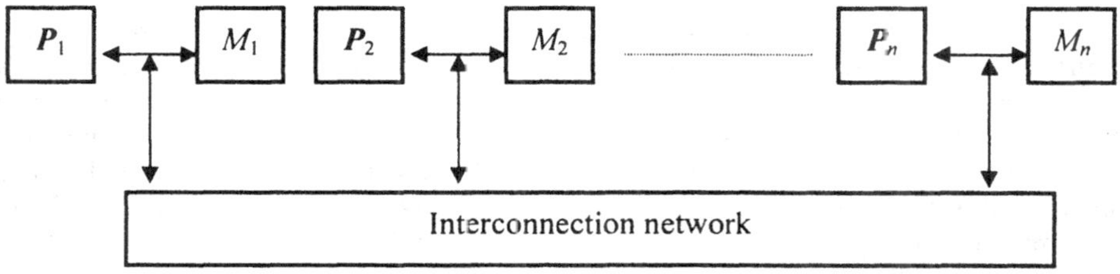

Figure 3 Shared memory architecture with only local memory for processors

performance ratios when compared to other parallel machines. This has led to a recent explosion in the use of clusters of computers, which are basically high-speed networks connecting commercial off-the-shelf hardware components.

Two basic communication modes are used for the exchange of information among processes executing on a parallel machine. Tasks may communicate by accessing a common memory (shared-memory paradigm), or by exchanging messages (message-passing paradigm) (Andrews and Schneider, 1983). In the shared-memory communication model, tasks share a common address space, which they can asynchronously access. Mechanisms such as locks and semaphores (Andrews and Schneider, 1983; Dijkstra, 1968; Vahalia, 1996) are used to control access to the shared-memory. This communication model can be directly used on different shared memory parallel machines, but can also be simulated on distributed-memory machines. In the latter, the user works with a shared memory physically distributed among the processors. One way to implement this mechanism is to use a managing process that (i) translates each address of the shared memory to its actual physical location, and (ii) is responsible for sending messages to the processes in order to read or write data requested by a task.

In the message passing model, processors communicate by sending and receiving messages in blocking or nonblocking mode. In the blocking sending mode, a task sending a message stops its processing and waits until the destination task receives it, while in the nonblocking mode it sends the message and continues its execution. In the blocking receiving mode, a task stops its processing until a specific message arrives, while in the nonblocking mode it just checks whether there is a message for it and carries on its processing if not. The message-passing model can be directly used for distributed-memory machines. It can also be very easily simulated for shared-memory machines (Tanenbaum, 1992).

Many parallel programming tools using shared-memory or message-passing models are essentially sequential languages augmented by a set of special system calls. These calls provide low-level primitives for message passing, process synchronization, process creation, mutual exclusion, and other functions. Some of these tools are presented in the next section.

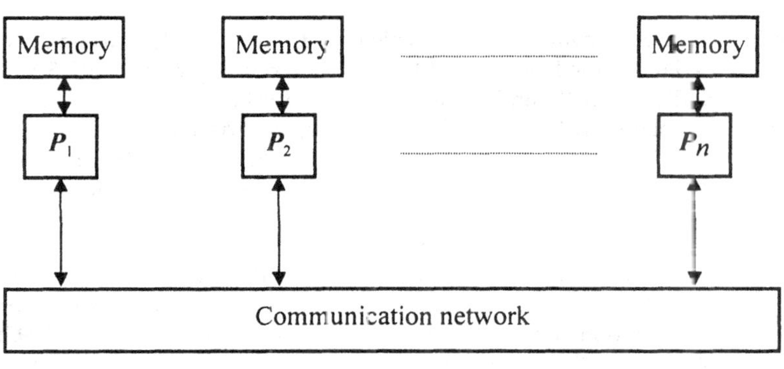

Figure 4 Distributed memory architecture

There are two basic types of parallelism that can be explored in the development of parallel programs: data parallelism and functional parallelism (Foster, 1995; Kumar et al., 1994; Morse, 1994). In the case of data parallelism, the same instruction set is applied to multiple items of a data structure. This programming model can be easily implemented for shared memory machines. Data locality is a major issue to be considered, regarding the efficiency of implementations in distributed memory machines. Several languages called data parallel programming languages have been developed to help with the use of data parallelism. One of these languages, HPF, is presented in the next section.

In the case of functional parallelism, the program is partitioned into cooperative tasks. Each task can execute a different set of functions/code and all tasks can run asynchronously. Data locality and the amount of processing within each task are important concerns for efficient implementations.

The most suitable approach varies from application to application. As discussed in Foster (1995), functional and data parallelism are complementary techniques that may sometimes be applied together to different parts of the same problem or may be applied independently to obtain alternative solutions.

3. Parallel Programming Tools

Many programming tools are available for the implementation of parallel programs, each of them being more suitable for some specific problem type. The choice of the parallel programming tool to be used depends on the characteristics of each problem to be solved. Some tools are better, for example for numerical algorithms based on regular domain decomposition, while others are more appropriate, for example for applications that need dynamic spawning of tasks and irregular data structures. Many parallel programming tools have been proposed, and it would be impossible to survey all of them in this chapter. We have chosen a few tools that are currently in a stable state of development and have been used in a number of applications, being representative of different programming paradigms.

3.1. Parallel Virtual Machine

PVM (Parallel Virtual Machine) (Geist et al., 1994) is a widely used message-passing library, created to support the development of distributed and parallel programs executed on a set of interconnected heterogeneous machines.

A PVM program consists of a set of tasks that communicate within a PVM by exchanging messages. A configuration file created by the user defines the physical machines that comprise the virtual machine. A managing process executes on each of these machines and controls the sending and receiving of messages among them. There are subroutines for process initialization and termination, for message sending and receiving, for group creation, for coordinating communication among tasks of a group, for task synchronization, and for querying and dynamically changing the configuration of the parallel virtual machine. The application programmer writes a parallel program by embedding these routines into a C, C++, or FORTRAN code.

A public domain software can be downloaded from `http://www.netlib.org/pvm3/`. PVM has been used on all the following systems: (i) PCs running Win 95, NT 3.5.1, NT 4.0, Linux, Solaris, SCO, NetBSD, BSDI, and FreeBSD, (ii) workstations and shared-memory servers executing Sun OS, Solaris, AIX, Hpux, OSF, NT-Alpha, and Iris, and (iii) parallel computers, such as CRAY YMP, T3D, T3E, Cray2, Convex Exemplar, IBM SP-2, 3090, NEC SX-3, Fujitsu, Amdahl, TMC CM5, Intel Paragon, Sequent Symmetry, and Balance. The developers provide efficient support through e-mail.

PVM has been used in several implementations of exact algorithms to optimization problems. A parallel implementation of the nested Benders solution algorithm for the large-scale certainty equivalent LP of a dynamic stochastic linear program was developed and tested on networks of workstations and on IBM SP-2 (Dempster and Thompson, 1997). Linderoth et al. (1999) describe a parallel linear programming-based heuristic to solve set partitioning problems producing good solutions in a short amount of time. Zakarian (1995) presents an efficient method for large-scale convex cost multicommodity network flow problems and its parallelization using a cluster of workstations. A unified platform for implementing parallel branch-and-bound algorithms was developed at the Prism Laboratory (Benaïchouche et al., 1996, Le Cun et al., 1995).

Several parallel metaheuristics for combinatorial optimization problems were also implemented using PVM. Taillard et al. (1997) describe a parallel implementation of the tabu search metaheuristic (Glover, 1989, 1990; Glover and Laguna, 1997) for the vehicle routing problem with soft time windows. A synchronous parallel tabu search algorithm for task scheduling in heterogeneous multiprocessor systems under precedence constraints was developed and implemented by Porto and Ribeiro (1995a,b, 1996) and Porto et al. (1999). Aiex et al. (1998) present an implementation of a cooperative multithread parallel tabu search for circuit partitioning (Andreatta and Ribeiro, 1994). These last two implementations were executed on an IBM SP-2. The implementation of master-slave and multiple-population parallel genetic algorithms can be found (Cantú-Paz, 1998).

3.2. *Message Passing Interface*

MPI (Message Passing Interface) (Foster, 1995; MPI Forum, 1994, 1995; Gropp et al., 1994, 1998) is a proposal for the standardization of a MPI for distributed-memory parallel machines. The aim of this standard is to enable program portability among different parallel machines. It just defines a message-passing programming interface, not a complete parallel programming environment. For this reason, it does not handle issues such as parallel program structuring and debugging. It does not provide any definition for fault tolerance support and assumes that the computing environment is reliable.

The core of MPI is formed by routines for point-to-point communication between pairs of tasks. These routines can be activated in two basic modes: blocking or nonblocking. Three communication modes are available: ready (a process can only send a message if there is a corresponding reception operation initiated); standard (a message can be sent even if there is no reception operation for it), and synchronous (similar to the standard mode, but with a sending operation considered as completed only after its destination processor initiates a reception operation). The selection of the mode to be used depends on the type of communication needs of the application and can have a great influence on the efficiency of the parallel program. Group communication routines are defined to coordinate the communication among tasks belonging to a predefined group of processors. These routines allow group data transmission providing the following operations (see Figure 5): sending the same data from one member of the group to all others (broadcast); sending different data from one member to all others (scatter); sending data from all members to one specific member (gather); sending data from all members to all members (allgather); and sending different data from each member to all others (alltoall).

There is also a reduction operation for computing functions over data belonging to members of a group of processors. The reduction operation applies a predefined or a user-defined function to data placed on the different participating processors. The result of this function can be sent to all group members, or just to one of them. One example is the determination of the maximum of a set of numbers, in which each number belongs to the address space of a different group task.

Many implementations of the MPI standard are available. All of them are based on a parallel virtual machine, composed by several connected heterogeneous computers (workstations, PCs, multiprocessors) and each of which executes a process used to control the exchange of messages among these machines. Among these implementations, mpich was developed at the Argonne National Laboratory based on the MPI 1.1 standard. It is free and can be downloaded from `http://www-unix.mcs.anl.gov/mpi/mpich`. This tool currently runs on the following parallel machines: IBM SP-2, Intel Paragon, Cray T3D, Meiko CS-2, TMC CM-5, NCUBE NCUBE-2, Convex Exemplar, SGI Challenge, Sun Multiprocessors, CRAY YMP, and CRAY C-90. It also runs on PCs (running Linux, FreeBSD, or Windows), and Sun, HPO, SGI, IBM RS/6000, and DEC workstations. LAM is another free package that can

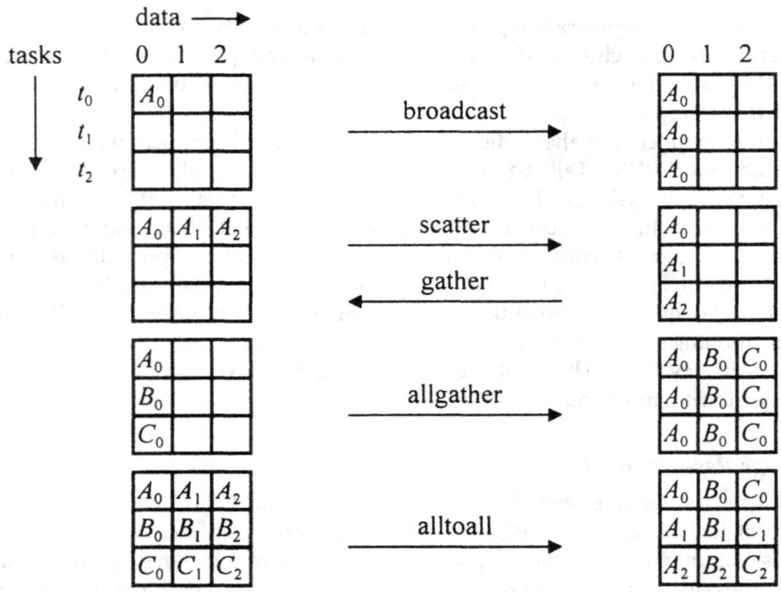

Figure 5 MPI routines for global communication

be downloaded from `http:/www.mpi.nd.edu/lam`, based on the MPI 2.0 standard and developed at the Laboratory for Scientific Computing of the University of Notre Dame. This implementation provides some monitoring functions, which are very helpful for managing processes running on the physical machines. It has already been tested and verified on Solaris, IRIX, AIX, Linux, and HP-UX. Both tools can be quite easily installed and used.

Some companies have also developed specific implementations for their parallel machines, such as Alpha, Cray, IBM, SGI, DEC, and HP. The Cornell Theory Center provides many helpful tutorials and FAQ sections about MPI at `http://www.tc.cornell.edu/`.

We can find some parallel implementations of exact algorithms for solving optimization problems using MPI. A portable C++ class library was developed by Eckstein et al. (1997) for portable, modular, parallel implementation of branch-and-bound and related algorithms. ZRAM (Brüngger et al., 1999) is a portable parallel library of exhaustive search algorithms and has been modeled on a subset of the standard MPI. Brüngger et al. (1997) show the use of this library for solving some test-instances of the quadratic assignment problem. Homer (1997) describes the development and implementation of a parallel algorithm to approximate the maximum weight cut in a weighted undirected graph. A generic framework for coarse grained parallel branch-and-bound algorithms supporting dynamic load balancing can be found in Batoukov and Serevik (1999).

Several parallel algorithms using metaheuristics were implemented using MPI. Martins et al. (1998, 1999) developed parallel implementations of greedy adaptive randomized search procedures (GRASP, see e.g., Feo and Resende, 1995) to solve the Steiner problem in graphs. A general object-oriented library based on C++ and MPI, developed by Kliewer and Tschöke (1998), provides a framework for a simulated annealing optimization system, which can be applied to many different optimization problems. Fachat and Hoffmann (1997) describe an implementation of ensemble-based simulated annealing with dynamic load balancing to get maximum use of the available processing power. Salhi (1998) reports on a parallel implementation of a tool for symbolic regression, the algorithmic mechanism of which is based on genetic programming. Rivera–Gallego (1998) presents a fast genetic algorithm to determine three-dimensional configurations of points that generate circulant Euclidean distance matrices and its parallelization.

3.3. Linda

Linda (Carriero et al., 1989, 1994) is a language that offers a collection of primitives for process creation and communication, based on the concept of an associative shared memory. The shared memory is referred to as the tuplespace and consists of a collection of data registers (or tuples). Primitives offered by Linda are such that processes may write into the tuplespace, as well as read and delete tuples from the tuplespace according to some specified pattern. In the case of associative access to the memory, information is accessed by its contents, not by its address. Although the tuplespace is logically shared by the processes, it can be implemented in distributed memory environments, with different memory partitions distributed among the processors.

Whenever a process wants to communicate with another one, it generates a new data object (a tuple) and places it in the tuplespace. The receiver process may access this tuple because it is written in the shared-memory space. To create a new concurrently executing process, the generator process performs an eval operation that creates a process tuple consisting of the fields specified as its arguments and then returns. This process tuple implicitly creates a process to evaluate each argument. While any of these processes runs, the process tuple is referred to as a live tuple. When all processes complete their execution, the values to be returned are placed in the corresponding field and the resulting data tuple is placed into the tuplespace. The live tuple carries out some specified computation on its own, independently of the process that generated it. The operations to manage tuples are extensions to C, C++, and FORTRAN. The Linda compiler generates executable programs. The programmer develops a parallel program implementation by writing a C, C++, or FORTRAN code using the basic Linda operations to access the tuplespace: read a tuple, write a tuple, place a new tuple in the tuplespace, take a tuple from the tuplespace, and create a new process.

Linda is a commercial system implemented and distributed by Scientific Computing Associates for shared memory and distributed memory machines. It runs on Cray T3D/T3E, DEC Alpha, Hitachi SR2201, and HP900/7xx machines. Network Linda is another product developed by Scientific Computing Associates (Scientific Computing Associates, 1995) to implement Linda in distributed memory environments. This version runs on IBM RS/6000, IBM PowerPC, IBM SP-2, SGI Iris, Indy, Onyx, Power Indy and Power Challenge, and Sun-Sparc machines.

An implementation in Linda of a genetic algorithm can be found in Davis and Elias (1994). Aiex et al. (1998) describe an implementation in Linda, running on an IBM SP-2, of a cooperative multithread parallel tabu search for the circuit partitioning problem, comparing the results with those obtained with the use of PVM.

3.4. High-Performance FORTRAN

FORTRAN 90 provides constructs for specifying concurrent execution based on data parallelism. HPF (High Performance FORTRAN Forum, 1997; Knies et al., 1994; Koelbel et al., 1994; Merlin and Hey, 1995) extends FORTRAN 90 with additional parallel constructs and data placement directives. A data parallel program developed using both languages is a sequence of operations that are applied to some or all elements of an array.

There are implicit and explicit constructs. In FORTRAN 90, the explicit construct A = B * C can be used to specify that each element of an array A is to be assigned the product of the corresponding elements of arrays B and C. A do-loop is an implicit parallel construct: the compiler determines whether the iterations are independent from one another, and, if so, executes them concurrently. A data parallel program is a sequence of explicit and implicit parallel statements. Typically, a compiler translates these statements into a program that controls the distribution and the execution of these operations on the data at each processor. Thus, the responsibility for low-level programming details is transferred from the programmer to the compiler.

There are six HPF compiler directives that specifically allow the programmer to control processor mapping and data distribution onto the physical processors: ALIGN, DISTRIBUTE,

PROCESSORS, TEMPLATE, REDISTRIBUTE, REALIGN, and DYNAMIC. Initially, the programmer has a group of arrays or subsets of arrays that are related by the ALIGN directive. These aligned objects are then distributed to an abstract set of processors via the DISTRIBUTE directive. Finally, the resulting abstract set of processors is mapped to the physical processors via a compiler dependent procedure, which is not specified in HPF and can vary according to the implementation. REALIGN and REDISTRIBUTE are the dynamic forms of the DISTRIBUTE and ALIGN directives, allowing data mapping to change during program execution. The PROCESSORS directive is used to specify the shape and the size of a set of abstract processors. The TEMPLATE directive defines a conceptual object to be used for alignment and distribution operations.

There are some commercial implementations of HPF provided by Applied Parallel Research (Applied Parallel Research, 1999), IBM (Gupta et al., 1995), Digital (Harris et al., 1995), and Portland Group (Schuster, 1994). Adaptor (SCAI-GMD, 1999) is a public domain HPF compilation system, which offers the use of HPF for parallel computing on distributed memory machines with explicit message-passing routines. FORGE (Applied Parallel Research, 1999) is an integrated collection of interactive tools to support the parallelization of sequential FORTRAN programs. It consists of a FORTRAN program browser for code analysis, a distributed memory parallelizer, and a parallel performance profiler and simulator. It also provides FORTRAN 90 and HPF support.

The parallelization of a commodity trade model using FORGE is described in Bergmark (1995) and Bergmark and Pottle (1994). The resulting program was executed on a SP1 machine.

3.5. *Threads*

Threads are a general operating systems concept, not specifically related to parallel programming. However, due to their importance in providing support for concurrent programming, and to their extensive use in many of the tools discussed in other sections, their understanding is essential to a parallel programmer.

Many operating systems and libraries have used the term thread in recent years to describe the execution of a sequence of statements, maintaining the process as the resource allocation unit. In this approach, several threads may be executed in the context of a single process. These threads (sometimes called lightweight processes) can then communicate using the global memory allocated to the associated process (global variables). Programming with threads is specially useful in shared-memory machines, since in this case the global variables model the shared memory to which all processors have access. However, threads can be useful for parallel programming even in distributed memory environments, since it is often useful to use concurrency within each machine, for instance to allow overlapping communication and computation.

In some systems, the operating system is aware of the existence of threads, while in others they are implemented as user-level libraries. Switching between threads is much faster when thread management is done at user-level. Since in this case the scheduler is not aware of the existence of separate threads, care must be taken to avoid invoking any blocking system call, for this would imply blocking all threads in the process. Both schemes are currently in use.

In 1995, IEEE defined a standard application program interface (API) for programming with threads, known as POSIX threads or Pthreads (IEEE, 1995). Several operating systems currently adopt this standard, which is also implemented by a public library (Provenzano, 1998, Butenhof, 1997). The Pthreads interface basically defines routines for control, synchronization, and scheduling. The control group includes functions such as thread creation, elimination, and suspension. Synchronization is provided by mutual exclusion locks and condition variables. Finally, scheduling routines allow the programmer to define relative priorities for the several threads within a process.

A parallel-cut generation procedure for service cost allocation in transmission systems (Aiex et al., 1999) was implemented using PThreads on a Sun Starfire ENT10000 and results showed a significant reduction in terms of the overall elapsed times, with respect to the sequential version.

Although the Pthreads API has been used in a number of parallel implementations, it is often considered a low-level standard, more targeted at concurrent and distributed applications than at high-performance computing (Dagum, 1997, 1998). Besides, Pthreads is defined for C, and there is currently no standard binding between Pthreads and FORTRAN, making it difficult to use threads to parallelize existing FORTRAN codes on multi-processor architectures. These reasons led to the proposal of the OpenMP industry standard. OpenMP was jointly defined by Digital Equipment Corp., IBM, Intel Corporation, Kuck & Associates Inc. (KAI), and Silicon Graphics/Cray Research. Information about OpenMP can be obtained from `http://www.openmp.org`. OpenMP consists basically of a set of compiler directives and API that extend FORTRAN (OpenMP Architecture Review Board, 1997) (or alternatively, C and C++; OpenMP Architecture Review Board, 1998).

Kuhn et al. (1999) compare the use of Pthreads and OpenMP in a genetic application, a linkage analysis program. Hybrid approaches, using threads for fine-grained parallelism on multiprocessor machines and libraries such as PVM and MPI for coarse-grained distributed-memory programming, are discussed in Bova et al. (1999).

4. Performance Evaluation Tools

It is very important to use empirical data in order to evaluate the performance of parallel implementations. The first step toward performance improvement consists of collecting suitable and reliable execution data (Foster, 1995). There are three basic categories of data collectors: profilers, counters, and event traces.

Profilers record the amount of time spent in different parts of a program, usually in subroutines. This information is often obtained by sampling the program counter and generating a histogram of the execution frequencies. These frequencies are combined with compiler symbol table information to estimate the amount of time spent in different program components. Most profiling tools are vendor supplied and machine specific. Usually, the user specifies an option for the compiler to generate data for profiling during the execution of the program. A parallel profiler usually gathers and displays information from each task of the parallel program. Profilers can be very helpful in identifying the most time-consuming parts of a program, allowing program optimization to be more focused. They can also identify program components that do not scale well with the input size or with the number of processors.

Counters record each time a specific event occurs and may require some additional coding by the programmer. They can be used to record global events of the program, such as the number of procedure calls, the total number of messages, and the number of messages exchanged between pairs of processors, which cannot be obtained by profilers.

Event traces record each occurrence of specific events (such as procedure calls or messages exchanged) and generate log files containing these records. They can be used to investigate relationships between communications and determine sources of idle time and communication bottlenecks. This approach generates a huge volume of data, which may be difficult to handle and store. The logging of this type of data tends to change program characteristics, masking the original sources of problems (probe-effect). Post-processing is difficult to perform, due to the large amount of data, and the programmer may have to spend some effort tuning the data collection process in order to record only the relevant events. This technique usually requires some extra coding by the programmer.

Raw data produced by such data collectors is usually transformed to reduce the total data volume, so as to be more easily interpreted and exhibited by visualization tools. Histograms are often used to display data collected by profilers, counters, and event traces. They can display information such as average message size, total computation time, or communication volume, as a function of time. For event traces, some forms of display can provide detailed views of temporal relations between different processors. For example, charts where each bar represents the status of a processor (computing, communicating, or idle) as a function of time. In the remainder of this section, we present some public domain and commercial packages available for performance analysis of the software tools presented in Section 3.

Gprof is a commercial tool provided by IBM to profile parallel programs developed using the IBM MPI library. It shows execution times of the subroutines executed by each processor. Except gprof, all other tools described in this section are based on event traces.

Nupshot is a performance tool distributed with the Argonne National Laboratory mpich implementation of MPI. The user inserts calls to logging routines in the source code, compiles and links it with the profiling library called MPE MPI. When the code executes, a log file of time stamped events is generated. This log file can be visualized with nupshot, displaying the state of the processes during the execution of the program.

AIMS (automated instrumentation and monitoring system) (Yan et al., 1995; Yan and Sarukkai, 1996) is a public domain package that can be downloaded from `http://science.nas.nasa.gov/Software/AIMS/` for measurement and analysis of FORTRAN 77 and C message-passing programs written using communication libraries such as MPI and PVM. It can also be used to analyze programs developed with the Portland Group HPF. The source code should be instrumented by a graphical user interface that allows the programmer to specify what constructs should be traced. The source files are compiled and linked with the AIMS monitor library and can be run in the usual manner to produce trace or statistics files. These files can be analyzed by AIMS tools to show the execution of the program based on the trace file.

Paradyn (Miller et al., 1995) is a tool for interactive run-time analysis for parallel programs, using message passing libraries such as PVM and MPI. It can be downloaded from `http://www.cs.wisc.edu/paradyn/`. Before running the program, the user defines performance visualizations, in terms of metric–focus pairs. Metrics can be, for example, processing time, blocking time, or message rates. Focus is a list of program components, such as procedures, processors, or message channels. For example, the user may specify a time visualization with a metric of processing time and focus on a particular subset of processes.

The Pablo performance analysis environment (DeRose et al., 1998; Reed et al., 1993) available from `http://vibes.cs.uiuc.edu/` is a public domain package that consists of several components for instrumenting and tracing parallel MPI programs, as well as for analyzing the trace files produced by the instrumented executables.

VAMPIR (visualization and analysis of MPI resource) (Galarowicz and Mohr, 1998; Nagel et al., 1996) is a commercial trace visualization tool from Pallas GmbH available for many different platforms. The programmer instruments the code and links it with the VAMPIR trace library. The trace files generated by the MPI processes along program execution are automatically collected and merged into a single trace file. The latter can be viewed using VAMPIR with visualization displays configured by the user. It can also be used to analyze programs implemented with HPF translated into an MPI program. Information about this tool is available at `http://www.pallas.de/pages/vampir.htm`.

VT (visualization tool) can be used for post-mortem trace visualization or for on-line performance monitoring of parallel programs developed with the IBM implementation of MPI. It is a commercial tool provided by IBM and can only be used on the IBM SP parallel environment.

XPVM is a public domain software available from `http://www.netlib.org/pvm3/xpvm` that provides a graphical console and monitor for PVM. The user starts a PVM program using XPVM and every spawned task returns trace event information for real time analysis or for post-mortem playback from saved trace files.

5. Debugging a Parallel Program

In parallel programming, issues related to concurrency and distribution lead to nondeterminism: repeated executions of one same program, over the same input data, may exhibit different behaviors. This can make the task of debugging a parallel program very hard. Some authors even suggest that the programmer should avoid nondeterminism whenever possible (for instance, by restricting message origins when executing receive statements)

(Foster, 1995). This, however, may lead to the subexploitation of the potential parallelism of the application.

One common, straightforward way to debug a parallel program is by inserting instructions in the code to print relevant information. Although helpful, the output data lacks temporal information, since the information is not necessarily printed in the same order as generated by the parallel program, due to data buffering in the processors.

PVM and the LAM public domain implementation of MPI allow the programmer to launch debuggers on remote nodes. Tasks can then be executed under the debugger control, making it easier to follow program execution. However, this kind of technique may interfere with the execution of the parallel program, making it behave differently from the normal way it runs. XPVM also provides some simple debugging assistance to PVM programmers, by listing textual details from trace events and remote task output.

Total View is a commercial debugger developed by Etnus Inc. (Etnus, 1999) that can be used to debug MPI, PVM, HPF, and multithreaded programs. It allows the programmer to set breakpoints and action points within parallel programs to control their execution. It also provides data visualization for variables associated with threads or processes, as well as to distributed HPF arrays. Although helpful for debugging parallel programs, it can demand some learning effort from the programmer before its use.

Tuplescope (Scientific Computing Associates, 1995) is a very helpful tool for debugging Linda programs. This tool emulates the execution of the parallel program within one single machine. The program can execute in normal or step mode, as with a sequential debugger. Visual indications of process interaction during the whole duration of the program run and data in specified tuples can be displayed using a graphical interface.

6. Parallel Optimization Libraries

Optimization libraries consist of a set of subroutines that can be used by application programmers to solve large optimization problems. They include functions to solve linear and mixed-integer programming problems and provide some features, such as control variables (routines to allow users to change a variety of default parameter settings controlling the algorithms), debugging capabilities (routines to provide statistics about element values, such as objective and constraint senses, bounds and coefficients in the constraint matrix, and storage used during execution), and file reading and writing (routines to read various input formats and to generate output solution files). Two widely known commercial optimization systems, CPLEX and OSL, provide parallel implementations of their serial optimization libraries.

CPLEX (CPLEX, 1999) is a registered trademark from ILOG. It provides an optimization library for Unix systems and for PCs running DOS, Windows 3.1, Windows NT, Windows 95, and OS/2. The CPLEX parallel solvers work on shared memory multiprocessor parallel architectures, such as DEC, Hewlett-Packard, Silicon Graphics, Sun, and Intel-based computer versions for which CPLEX parallel support has been established. Parallel versions of the barrier solver and mixed-integer solver are available for all parallel platforms. Parallel simplex availability varies by platform. The user chooses the degree of parallelism to be exploited before starting an optimization and CPLEX automatically implements the parallelization.

OSLp, IBM's parallel optimization subroutine library (IBM, 1999a), includes all functions of the serial library OSL. It runs on SP IBM RS/6000 Scalable POWERparallel systems and on clusters of RS/6000s workstations. OSLp runs under the AIX parallel environment. OSLp provides a set of over 60 subroutines callable from application programs written in XL FORTRAN or C, to solve linear programming, mixed integer programming, and quadratic programming optimization problems. Some of these routines use serial algorithms, while others use algorithms that exploit the capabilities of the parallel environment, so that multiple processors may concurrently perform computations on subtasks of the problem to be solved. Only minor changes to serial OSL applications are required to generate parallel application programs and no explicit parallel coding is required. More information about this library can be found at `http://www.rs6000.ibm.com/software/sp_products/oslp.html`.

7. Concluding Remarks

The development of portable, efficient, and easy to use software tools, together with the availability of a wide range of parallel machines with large processing capacities, increasing reliability, and decreasing costs, has brought parallel processing into reality as an effective strategy for the implementation of computationally efficient techniques for the solution of large, difficult optimization problems. In this Chapter, we have given descriptions of some parallel processing environments, together with examples of their application in the exact or approximate solution of several optimization problems and applications.

Judicious choice of a programming environment is essential to the development of a parallel program. Implementation issues may lead to programs that perform very poorly, independently of the quality of the underlying parallel algorithm. One possible source of problem is a mismatch between the proposed algorithm and the programming model supported by the development tool. It is also common for the program, once implemented, to have a different behavior from what was expected due to unforeseen communication and processing bottlenecks. Performance evaluation tools have a fundamental role in solving this kind of problem, since they can help the programmer in identifying the origin of such bottlenecks.

We surveyed in this Chapter only a small fraction of existing work on parallel computing environments. The amount of ongoing research activity in this field is witnessed by the number of workshops and conference sessions devoted to it, such as the "Support Tools and Environments" series in the Euro-Par conferences (Lengauer et al., 1997; Pritchard et al., 1998; Amestoy et al., 1999).

Among the current trends in parallel processing, the concept of cluster computing is one of the leading tendencies. Basically, a cluster is a collection of PCs or workstations connected via a network and using off-the-shelf components. The significant improvements in performance and reliability of PCs, workstations, and high-speed networks strongly contributed to turn the use of clusters into an appealing, low-cost alternative for parallel applications. Clusters can be put together very fast, using operating systems such as public domain Linux, networks such as Ethernet, SCI, or Myrinet, and software environments such as MPI or PVM. Some cluster systems are currently among those with a better cost/performance ratio, ranking among the fastest and most powerful machines. Clusters have been used for solving challenging applications in combinatorial optimization, stochastic programming, power systems, weather modeling, simulations, life sciences, computational fluid dynamics, image processing, data mining, aerodynamics, and astrophysics.

References

Aiex, R. M., S. L. Martins, C. C. Ribeiro, and N. R. Rodriguez. 1998. "Cooperative multi-thread parallel tabu search with an application to circuit partitioning." *Lecture Notes Comput. Sci.* **1457**: 310–331.

Aiex, R. M., C. C. Ribeiro, and M. V. Poggi de Aragão. 1999. "Parallel cut generation for service cost allocation in transmission systems." In *Extended Abstracts of the Third Metaheuristics International Conference*, pp. 1–7. Brazil: Catholic University of Rio de Janeiro.

Amestoy, P., P. Berger, M. J. Daydé, I. S. Duff, V. Frayssé, L. Giraud, and D. Ruiz. (eds). 1999. In *Proceedings of the 5th International Euro-Par Conference (LNCS 1685)*, pp. 89–162. Berlin, Heidelberg, New York: Springer-Verlag.

Andreatta, A. and C. C. Ribeiro. 1994. "A graph partitioning heuristic for the parallel pseudo-exhaustive logical test of VLSI combinational circuits." *Ann. Oper. Res.* **50**: 1–36.

Andrews, G. and F. Schneider. 1983. "Concepts and notations for concurrent programming." *Comput. Surveys* **15**: 3–43.

Applied Parallel Research. 1999. *Applied Parallel Research Home Page.* http://www.apri.com.

Batoukov, R. and T. Serevik. 1999. "A generic parallel branch and bound environment on a network of workstations." In *Proceedings of High Performance Computing on Hewlett-Packard Systems.* Norway: University of Tromso, pp. 474–483. Also available at http://www.ii.uib.no/~tors/publications/.

Benaïchouche, M., V.-D. Cung, S. Dowaji, B. Le Cun, T. Mautor, and C. Roucairol. 1996. "Building a parallel branch-and-bound library." *Lecture Notes Comput. Sci.* **1054**: 201–231.

Bergmark, D. 1995. "Optimization and parallelization of a commodity trade model for the IBM SP1/2 using parallel programming tools." In *Proceedings of 1995 International Conference on Supercomputing*, pp. 227–236. Barcelona: ACM Press.

Bergmark, D. and M. Pottle. 1994. "The optimization and parallelization of a commodity trade model for the IBM SP1, using parallel programming tools." Technical Report CTC94TR181. Cornell Theory Center. Available at http://cs-tr.cs.cornell.edu:80/Dienst/UI/1.0/Download/ncstrl.cornell.tc/94-181.

Blackford, L. S., J. Choi, A. Cleary, E. Dázevedo, J. Demmel, I. Dhillon, J. Dongarra, S. Hammarling, G. Henry, A. Petitet, K. Stanley, D. Walker, and R. C. Whaley. 1997. In *ScaLAPACK User's Guide*. Philadelphia: Society for Industrial and Applied Mathematics.

Bova, S., C. Breshears, R. Eingenmann, H. Gabb, G. Gaertner, B. Kuhn, B. Magro, S. Salvini, and V. Vatsa. 1999. "Combining message-passing and directives in parallel applications." *SIAM News* **32(9)**: 110–111.

Brüngger, A., A. Marzetta, J. Clausen, and M. Perregaard. 1997. "Joining forces in solving large-scale quadratic assignment problems in parallel." In *Proceedings of the 11th International Parallel Processing Symposium*. Geneva, Switzerland: IEEE Computer Society Press, pp. 418–427. Also available at http://nobi.ethz.ch/group/sppiuk/progress98.html.

Brüngger, A., A. Marzetta, K. Fukuda, and J. Nievergelt. 1999. "The parallel search bench ZRAM and its applications." *Ann. Oper. Res.* **90**: 45–63. Also available at http://www.baltzer.nl/anor/contents/1999/90.html.

Butenhof, D. 1997. In *Programming with POSIX Threads*. Reading, MA: Addison-Wesley.

Cantú-Paz, E. 1999. "Implementing fast and flexible parallel genetic algorithms." In *Practical Handbook of Genetic Algorithms*, edited by L. D. Chambers **III**: 65–84. Boca Raton, FL: CRC Press.

Carriero, N., D. Gelernter, and T. Mattson. 1989. "Linda in context." *Commun. ACM* **32**: 444–458.

Carriero, N., D. Gelernter, T. Mattson, and A. H. Sherman. 1994. "The Linda alternative to message-passing systems." *Parallel Comput.* **20**: 633–655.

Chapman, B., F. Bodin, L. Hill, J. Merlin, G. Viland, and F. Wollenweber. 1999. "FITS—a light-weight integrated programming environment." *Lecture Notes Comput. Sci.* **1685**: 125–134.

CPLEX. 1999. *Home page of CPLEX, a division of ILOG.* http://www.cplex.com.

CTC-Cornell Theory Center. 2001. Parallel Programming Concepts. Virtual Workshop Module. Available at http://www.tc.cornell.edu/Services/Edu/Topics/ParProgCons.

Dagum, L. 1997. *OpenMP: a proposed industry standard API for shared memory programming*. Available at http://www.openmp.org/specs/index.cgi?mp-documents/paper/paper.html.

Dagum, L. and R. Menon. 1998. "OpenMP: an industry standard API for shared memory programming." *IEEE Comput. Sci. Engng.* **5**: 46–55.

Davis, M., L. Liu, and J. G. Elias. 1994. "VLSI circuit synthesis using a parallel genetic algorithm." In *Proceedings of the IEEE'94 Conference on Evolutionary Computing*, pp. 104–109. New York: IEEE Computer Society Press.

Dempster, M. A. H. and R. T. Thompson. 1997. "Parallelization and aggregation of nested Benders decomposition." *Ann. Oper. Res.* **81**: 163–187.

DeRose, L., Y. Zhang, and D. A. Reed. 1998. "SvPablo: A multi-language performance analysis system." In *10th International Conference on Computer Performance Evaluation—Modeling techniques and tools*, pp. 352–355. Palma de Mallorca, Spain: Springer-Verlag.

Dijkstra, E. W. 1968. "Cooperating sequential processes." In *Programming Languages*, edited by F. Genvys, pp. 43–112. New York: Academic Press.

Eckstein, J., W. E. Hart, and C. A. Philips. 1997. "An adaptable parallel toolbox for branching algorithms." In *XVI International Symposium on Mathematical Programming* 82. Lausanne. Also available at http://dmawww.epfl.ch/roso.mosaic/ismp97.

Etnus Com. 1999. *Home page for Etnus Com.* http:www.etnus.com.

Fachat, A. and K. H. Hoffman. 1997. "Implementation of ensemble-based simulated annealing with dynamic load balancing under MPI." *Comput. Phys. Commun.* **107**: 49–53.

Feo, T. A. and M. G. C. Resende. 1995. "Greedy randomized adaptive search procedures." *J. Global Optim.* **6**: 109–133.

Foster, I. 1995. In *Designing and Building Parallel Programs: Concepts and Tools for Parallel Software Engineering*. Reading, MA: Addison-Wesley.

Galarowicz, J. and B. Mohr. 1998. "Analyzing message passing programs on the Cray T3E with PAT and VAMPIR." In *Proceedings of Fourth European CRAY-SGI MPP Workshop, IPP-Report des MPI für Plasmaphysik, IPP R/46*, edited by H. Lederer and F. Hertwick, pp. 29–49. Also available at http://www.kfa-juelich.de/zam/docs/autoren98/galarowicz.html. Only by the web at http://www.rzg.mpg.de/mpp-workshop/papers/ipp-report.html.

Geist, A., A. Beguelin, J. Dongarra, W. Jiang, R. Manchek, and V. Sunderman. 1994. In *PVM: Parallel Virtual Machine—A User's Guide and Tutorial for Networked Parallel Computing*. Cambridge, Mass: MIT Press. Also available at http://www.netlib.org/pvm3/book/pvm-book.html.

Glover, F. 1989. "Tabu Search—Part I." *ORSA J. Comput.* **1**: 190–206.

Glover, F. 1990. "Tabu Search—Part II." *ORSA J. Comput.* **2**: 4–32.

Glover, F. and M. Laguna. 1997. In *Tabu Search*. Boston, MA: Kluwer Academic Publishers.

Gropp, W., E. Lusk, and A. Skjellum. 1994. In *Using MPI: Portable Parallel Programming with the Message Passing-Interface*. Cambridge, MA: MIT Press.

Gropp, W., S. Huss-Lederman, A. Lumsdaine, E. Lusk, B. Nitzberg, W. Saphir, and M. Snir. 1998. In *MPI: The Complete Reference, Volume 2—The MPI Extensions*. Cambridge, MA: MIT Press.

Gupta, M., S. Midkiff, E. Schonberg, V. Seshadri, D. Shields, K.-Y. Wang, W.-M. Ching, and T. Ngo. 1995. "An HPF compiler for the IBM SP2." In *Proceedings of the 1995 ACM/IEEE Supercomputing Conference*: ACM Press. Also available at http://www.supercomp.org/sc95/proceedings/417_SAMM/SC95.HTM.

Harris, J., J. Bircsak, M. Bolduc, J. Diewald, I. Gale, N. Johnson, S. Lee, C. A. Nelson, and C. Offner. 1995. "Compiling high performance FORTRAN for distributed-memory systems." *Digital Technical J.* **7**: 5–38.

High Performance FORTRAN Forum. 1997. "High Performance FORTRAN language specification version 2.0." Also available at http://www.crpc.rice.edu/HPFF/home.html.

Homer, S. 1997. "Design and performance of parallel and distributed approximation algorithms for maxcut." *J. Parallel Distributed Comput.* **41**: 48–61.

IEEE (Institute of Electric and Electronic Engineering). 1995. *Information Technology—Portable Operating Systems Interface (POSIX)—Part 1—Amendment 2: Threads Extension.*

IBM. 1999a. *Parallel ESSL Version 2 Release 1.2 Guide and Reference (SA22-7273)*. Also available at http:/www.rs6000.ibm.com/resource/aix_resource/sp_books/essl.

IBM, 1999b. *Users Guide to Parallel OSL*, SC23-3824.

Kliewer, G. and S. Tschöke. 1998. "A general parallel simulated annealing library (parSA) and its applications in industry." Presented at *PAREO'98: First Meeting of the PAREO Working Group on Parallel Processing in Operations Research.* Also available at http://www.uni-paderborn.de/~parsa/.

Knies, A., M. O'Keefe, and T. MacDonald. 1994. "High performance FORTRAN: a practical analysis." *Sci. Programming* **3**: 187–199.

Koelbel, C., D. Loveman, R. Schreiber, G. Steele, Jr., and M. Zosel. 1994. In *The High Performance FORTRAN Handbook*. Cambridge, MA: The MIT Press.

Kuhn, B., P. Petersen, E. O'Toole, and M. Daly. 1999. "Using SMP parallelism with OpenMP," *CCP11 Newsletter*, Issue 9. Available at http://www.hgmp.mrc.ac.uk/CCP11/newsletter.

Kumar, V., A. Grama, A. Gupta, and G. Karypis. 1994. In *Introduction to Parallel Computing Design and Analysis of Parallel Algorithms*. Redwood City, CA: Benjaming/Cummings.

Lau, L. 1996. "Implementation of scientific applications in a heterogeneous distributed network." PhD. Thesis. Australia: University of Queensland, Department of Mathematics. Available at http://www.acmc.uq.edu.au/~ll/thesis.

Le Cun, B., C. Roucairol, M. Benaïchouche, V.-D. Cung, S. Dowaji, and T. Mautor. 1995. "BOB: a unified platform for implementing branch-and-bound like algorithms." Technical Report RR-95/16. France: PRiSM Laboratory, Université de Versailles. Available at http://www.prism.uvsq.fr/Rapports/1995.

Lengauer, C., M. Griebl, and S. Gorlatch, (eds.). 1997. *Proceedings of the 3th International Euro-Par Conference. Lecture Notes Comput. Sci.* **1300**: 89–165.

Linderoth, J., E. K. Lee, and M. W. P. Savelsbergh. 1999. "A parallel linear programming based heuristic for large scale set partitioning problems." Industrial & Systems Engineering Technical Report. Available at http://www.isye.gatech.edu/faculty/Eva_K_Lee.

Martins, S. L., C. C. Ribeiro, and M. C. Souza. 1998. "A parallel GRASP for the Steiner problem in graphs." *Lecture Notes Comput. Sci.* **1457**: 285–297.

Martins, S. L., M. G. C. Resende, C. C. Ribeiro, and P. Pardalos. 2000. "A parallel GRASP for the Steiner tree problem in graphs using a hybrid local search strategy." *J. Global Optim.* **17**: 267–283.

Merlin, J., S. Baden, S. Fink, and B. Chapman. 1999. "Multiple data parallelism with HPF and KeLP." *J. Future Generation Comput. Syst.* **15**: 393–405.

Merlin, J. and A. Hey. 1995. "An introduction to high performance FORTRAN." *Sci. Programming* **4**: 87–113.

Miller, B. P., M. D. Callaghan, J. M. Cargille, J. K. Hollingsworth, R. B. Irvin, K. L. Karavanic, K. Kunchithapadam, and T. Newhall. 1995. "The Paradyn parallel performance tools." *IEEE Comput.* **28**: 37–46.

Morse, H. S. 1994. In *Practical Parallel Computing*. Boston, MA: AP Professional.

MPI Forum. 1994. "A message-passing interface standard." *Int. J. Supercomput Appl. High Perform. Comput.* **8**: 138–161.

MPI Forum, 1995. "A message-passing interface standard." Available at http://www.mpi-forum.org/docs/docs.html.

Nagel, W., A. Arnold, M. Weber, H. Hoppe, and K. Solchenbach. 1996. "VAMPIR: visualization and analysis of MPI resources." *Supercomputer* **63**: 69–80.

Nieplocha, J., R. J. Harrison, and R. J. Littlefield. 1994. "Global arrays: a portable "shared-memory" programming model for distributed memory computers." In *Proceedings of Supercomputing 94*: 340–349. Also available at http://citeseer.nec.com/nieplocha94global.html/.

OpenMP Architecture Review Board. 1997 *OpenMP FORTRAN Application Program Interface.*

OpenMP Architecture Review Board. 1998 *OpenMP C and C++ Application Program Interface.*

Plastino, A., C. C. Ribeiro, and N. Rodriguez. 1999. "A tool for SPMD application development with support for load balancing." In *Proceedings of the International Conference ParCo99, Parallel Computing. Fundamentals and Applications* pp. 45–71. Also available at http://www.inf.puc-rio.br/~celso.

Porto, S. C. and C. C. Ribeiro. 1995a. "Parallel tabu search message-passing synchronous strategies for task scheduling under precedence constraints." *J. Heuristics* **1**: 207–223.

Porto, S. C. and C. C. Ribeiro. 1995b. "A tabu search approach to task scheduling on heterogeneous processors under precedence constraints." *Int. J. High Speed Comput.* **7**: 45–71.

Porto, S. C. and C. C. Ribeiro. 1996. "A case study on parallel synchronous implementations of tabu search based on neighborhood decomposition." *Investigación Operativa* **5**: 233–259.

Porto, S. C., J. P. Kitajima, and C. C. Ribeiro. 2000. "Performance evaluation of a parallel tabu search task scheduling algorithm." *Parallel Comput* **26**: 73–90.

Pritchard, D. and J. Reeve, 1998. eds. *Proceedings of the 4th International Euro-Par Conference. Lecture Notes Comput. Sci.* **1470**: 80–189.

Protic, J., M. Tomasevic, and V. Milutinovic. 1998. In *Distributed Shared Memory*: IEEE Computer Society.

Provenzano, C. 2000. Pthreads. Available at http://ub.mit.edu/people/proven/IAP_2000.

Reed, D. A., R. A. Aydt, R. J. Noe, P. C. Roth, K. A. Shields, B. Schwartz, and L. F. Tavera. 1993. "Scalable performance analysis: the Pablo performance analysis environment." In *Proceedings of the Scalable Parallel Libraries Conference*, pp. 104–113. IEEE Computer Society.

Rivera-Gallego, W. 1998. "A genetic algorithm for circulant euclidean distance matrices." *J. Appl. Math Comput.* **97**: 197–208.

Salhi, A. 1998. "Parallel implementation of a genetic-programming based tool for symbolic regression." *Inf Process. Lett.* **66**: 299–307.

SCAI-GMD. 1999. Home page for Adaptor. http://www.gmd.de/SCAI/lab/adaptor.

Scientific Computing Associates. 1995. *Linda User's Guide & Reference Manual*, version 3.0.

Schuster, V. 1994. "PGHPF from the Portland Group." *IEEE Parallel Distrib. Technol.* **2**: 72.

Taillard, E., M. Gendreau, F. Guertin, J.-Y. Potvin, and P. Badeau. 1997. "A parallel tabu search heuristic for the vehicle routing problem with time windows." *Transport. Res. C* **5**: 109–122.

Tanenbaum, A. 1992. In *Modern Operating Systems*. Engelwood Cliffs: Prentice-Hall.

Vahalia, U. 1996. In *Unix Internals—The New Frontiers*. Upper Saddle River, NJ: Prentice-Hall.

Wolf, K. and O. Kraemer-Furhmann. 1996. "An integrated environment to design parallel object-oriented applications." *Lecture Notes Comput. Sci.* **1123**: 120–127.

Yan, J., S. Sarukhai, and P. Mehra. 1995. "Performance measurement, visualization and modeling of parallel and distributed programs using the AIMS toolkit." *Softw. Practice Experience* **25**: 429–461.

Yan, J. and S. Sarukhai. 1996. "Analyzing parallel program performance using normalized performance indices and trace transformation techniques." *Parallel Comput.* **22**: 1215–1237.

Zakarian, A. A. 1995. "Nonlinear Jacobi and ε-relaxation methods for parallel network optimization." Ph.D. Thesis. Mathematical Programming Technical Report 95-17. Madison: University of Wisconsin. Also available at http://www.cs.wisc.edu/math-prog/tech-reports/.

24 EXPERIMENTAL ANALYSIS OF OPTIMIZATION ALGORITHMS

Catherine C. McGeoch

ABSTRACT This chapter surveys methodological issues that arise in experimental research on optimization algorithms. Guidelines are presented for selecting research problems, input classes, and program metrics; for ensuring correctness and reproducibility of the results; for applying appropriate data analysis techniques; and for presenting informative and convincing conclusions. The chapter contains many references to related work on these topics.

1. Introduction

There is a long tradition of experimental research on algorithms and heuristics for optimization problems. The motivation for doing experiments is clear: our questions about program performance and behavior far outstrip our ability to obtain answers by purely analytical means. Worst-case input models are too pessimistic; realistic input models are too complicated; standard models of computation are too simple; and many important algorithms are too complex to allow us to develop precise predictions of performance, or to build better algorithms and implementations, without the insights gained by experimentation and observation.

Other sections of this handbook describe experimental research on specific algorithms, heuristics, and metaheuristics for problems in optimization. To complement these case studies, this chapter surveys work that explicitly addresses the methodology and practice of computational research on optimization algorithms. Because the topic is broad, the treatment here is necessarily superficial: references to works containing more detailed and rigorous presentations appear throughout the chapter. The discussion is organized around three topics.

- *Goals and strategies.* Experimental research on optimization algorithms can be undertaken for many reasons; Section 2 identifies some goals of experimental research and shows how goals influence strategy, for example in the selection of performance measure and of input classes to study.
- *Doing experiments.* The power of an experiment, informally, is a measure of the quality of insight obtained per unit of effort. Strategies for ensuring correctness of results, selecting input parameters wisely, and choosing effective data analysis tools, are discussed in Section 3.
- *Presentation of results.* Section 4 briefly surveys guidelines for presenting results in published work. A final section of this chapter presents concluding remarks.

Several writers on experimental methodology have suggested that there are two flavors of computational research on algorithms. Various authors have used different terms to draw more or less the same distinction. For example, Ahuja et al. (1993) comment on the difference between the "typical experimentation study" and one emphasizing "asymptotic empirical analysis." Barr et al. (1995) describe two general goals of computational experimentation: the first is to compare algorithms, and the second is to describe algorithm performance. Hooker, (1994, 1996) mentions two strategies for empirical research: collecting a set of benchmark problems for comparison of algorithms; and investigating how algorithm performance depends on problem characteristics. And in McGeoch (1996) a contrast is drawn between experiments that measure programs to identify the best ones for particular scenarios, and experiments that simulate algorithms to produce models of their behavior.

Most authors who draw this distinction, between comparison-based and model-building styles of research, do so while suggesting that more attention be paid to the latter. The main difference in the two may be the level of ambition involved: a well-done comparison study can reveal the best algorithms, best implementation strategies, and best choices of parameters for certain sets and classes of problem instances; a well-done model-building study can do all this, and produce new insights that suggest improved algorithms, and permit extrapolation of results well beyond the scope of the experiments. Issues arising in both approaches are discussed in this chapter, although more emphasis is placed on the model-building style.

Some good general references are available that address methodological issues in experimental research on optimization algorithms with greater detail and rigor than is possible here; see also the extensive biblographies they contain.

Golden et al. (1986) demonstrate that even within the seemingly straightforward comparison model of research, layers of complexity arise in the design of good experiments for optimization algorithms. They discuss problems of comparing performance of algorithms that sometimes fail, sometimes fail to converge, and sometimes require human interaction. Difficulties can also arise when algorithmic templates (such as simulated annealing) present myriad implementation decisions and when there are several dimensions to the performance criteria. The authors give examples and guidelines for handling problems of these types, and they illustrate their ideas with case studies drawn from projects in nonlinear optimization, simulated annealing, and application-specific testing. Barr et al. (1995) and McGeoch (1996) describe hazards and pitfalls of experimental research on algorithms, and present guidelines and standards for setting goals, choosing appropriate performance measures and parameters, performing experiments, and analyzing and presenting results.

2. Goals and Strategies

This section identifies some goals for experimental research on optimization algorithms, and shows how goals affect strategy in the selection of performance measures and input classes.

2.1. Choosing a Research Problem

Certainly for most researchers, one goal is publishability. Johnson (1996) presents principles, guidelines, and criteria for publishable experimental research on algorithms. Anderson (1997) describes an ideal symbiotic relationship between empirical and theoretical research, and gives examples of good and bad research projects in this context. Hooker (1996) argues forcefully that the model-building approach should replace competitive testing in operations research culture. Crowder et al. (1978), Greenberg (1990), and Jackson et al. (1991) address the question of how much experimentation and analysis is enough in the context of a given research topic.

One problem that arises in research on algorithm and program efficiency is deciding what kind of properties deserve scrutiny. Let us define environmental factors as those properties that depend significantly on the choice of computing environment; implementation factors as those parts that contribute constant factors to the overall running time of a program; and algorithmic factors as those components and operations that change in frequency according to input size or other parameters. (Bentley, 1982, describes a systematic methodology for obtaining significant constant-factor improvements by tuning code and exploiting environmental properties.)

Some anecdotal evidence exists (see for example Bentley, 1982, 1986, Chapter 5; McGeoch, 1996; Johnson and McGeoch, 1991; and Johnson and Trick, 1993) to suggest, roughly, that changing environmental factors by porting code to different machines typically gives speedups by factors of two to fifty; that constant-factor improvements in the implementation can produce speedups by factors of five to one hundred; and that algorithmic improvements by orders greater than $\Theta(n)$ give speedups by factors of one hundred to one thousand or more.

Thus the likely most important source of performance improvements depends upon the maturity of the research area. If an algorithm or heuristic is not understood to within a $\Theta(n)$ term, then experiments developed to study algorithmic behaviors and to suggest big-on improvements might be most beneficial; on the other hand, if the algorithm is well understood, then experiments to study implementation and environmental factors can be useful for suggesting finer-grained improvements.

2.2. *Measuring Time Performance*

The performance of an optimization algorithm can be measured along several different dimensions, including time efficiency, quality of solution, range of solutions produced, space usage, and generality of application. In general the choice of performance measures is straightforward and suggested by the research questions motivating the experimental study. However, it is notoriously difficult to find a measure of time efficiency that is both generalizable and precise.

One common strategy is to measure CPU times for the program and/or its major subroutines. Such measurements are of course quite precise indicators of the real time required by a given program; but in cases where time is dominated by the cost of I/O and access to the memory hierarchy, even CPU times can be inadequate for predicting performance in practice. Knuth (1993) has proposed a mems metric based on counts of certain types of memory access performed by a program, arguing that this metric is both more predictive of real computation time and more stable across machine platforms and over time. Some promising work on modeling costs due to caching and hierarchical memory access has been presented by LaMarca and Ladner (1996) and Sanders (1999).

Another problem with CPU times is the difficulty of extending results to other platforms and environments. It is clear that no simple scale exists for making such translations. For example, Fredman et al. (1995) tested identical TSP programs and identical inputs on one CISC and two RISC architectures. They observed very different performance profiles between the two machine classes; indeed results between the two RISC machines cannot be calibrated by a constant scaling term. A similar observation arises from papers contributed to the first DIMACS Challenge (Johnson and McGeoch, 1991), where participants at several research sites ran identical tests of network flow programs in their local environments. One might hope that a program run on big and small inputs would show a ratio Time(Big)/Time(Small) that is fairly stable across platforms. But in a test of C programs in about twenty five different environments, these ratios varied by as much as 20 percent of their means. (Another test produced less variation, but problem sizes were too small to be reliable.)

The usual alternative to CPU measurements is to insert code in the program to report the number of key operations executed. This kind of combinatorial measurement is both general and portable, but it may be too imprecise for some predictive purposes; and in some cases may not be possible to find a common set of operations with which to compare different implementations. Ahuja et al. (1993) address the problem of selecting combinatorial measures of performance that are stable across machine environments as well as highly predictive of performance on individual machines. They give excellent guidelines and examples for choosing such measures, and illustrate the effectiveness of their approach on a study of the network simplex algorithm.

Common sense would suggest that combinatorial measures are most appropriate in studies of algorithmic performance factors; and that runtimes are most suitable when implementation and environmental factors are of interest. In many cases the best option is to gather both kinds of information about performance; this must be done with care, though, as the insertion of code to measure a program can of course alter its running time.

2.3. *Selecting Input Classes for Study*

The problem of finding a good set of input classes for testing is widely acknowledged. Real-world instances give highly specific information about performance in practical scenarios, but are extremely difficult to generalize; artificially generated inputs can give results that hold over broad classes of instances, but their relevance to real performance can be hard to justify.

Standard benchmark testsets produce a degree of comparability among algorithms tested over different times and places, and they often provide a broad scope in the variety of instance types. But benchmarks also have drawbacks, especially if used exclusively in experimental research: there are obvious limits of generalizability; programmers tend to tune for the benchmark set; and benchmark problems tend to become obsolete over time. Most well-known benchmark sets contain both real-world and generated instances.

Barr et al. (1995), Golden et al. (1986), Jackson et al. (1991), and Johnson (1996) give guidelines for choosing input classes to meet particular goals of experimental research on optimization algorithms.

3. The Experimental Process

This section considers issues that arise during the process of experimental research, and surveys techniques that can enhance the quality of the experimental study.

3.1. *Ensuring Replicability*

Many articles addressing issues of presentation in the publication of computational research on algorithms (see Section 4) point out that any respectable experimental research paper must provide enough information to allow readers to replicate the experiments.

Even before publication, the experimenter should take steps to replicate his or her own work. McGeoch (1996) describes several reported examples where backup experiments undertaken to check the results of a study in progress, revealed system-dependent artifacts and other problems that would have produced misleading and erroneous conclusions, if they had not been caught in time.

One such experience is enough to convince most researchers that replication of key experiments is an important principle of experimental integrity. Such replication takes many forms. Porting a program to a different computing environment can reveal dependencies on machine precision and caching policies. Running identical tests with different classes of random number generators can expose problems due to patterns in the pseudorandom number stream. Comparing different implementations of the same algorithm can reveal variations in interpretation that produce significant performance differences and/or incorrect code.

3.2. *Designed Experiments*

The statistics literature contains many strategies and techniques for identifying the most important input and program parameters in an experimental research project; choosing parameter settings to maximize the information gained; finding the number of different parameter settings to use; and selecting good sample sizes. A small pilot study, undertaken before the main project begins, can provide the information a researcher needs to make good choices about parameters, their settings, and the overall scope of the experiments.

Any introductory textbook on statistics presents a basic discussion of these issues. Cohen's (1995) text on empirical methods for artificial intelligence presents statistical techniques for computational research on all kinds of algorithms and heuristics, not just those in AI. Cohen brings together information that is otherwise widely dispersed in the statistical literature, and presents illustrations that are immediately accessible to researchers studying algorithmic problems. Nance et al. (1987) show how to apply analysis of variance (ANOVA) to the problem of selecting important parameters in studies of heuristic algorithms, with examples drawn from a capacity assignment problem.

3.3. *Plan for Data Analysis*

It is often possible to plan ahead and select experimental parameters and performance measures that generate better results in the data analysis stage of the research.

For example, experts in exploratory data analysis, a subfield of statistical data analysis, recognize that data sets fall into distinct categories, such as: counts, which usually have a fixed lower bound and a skewed distribution; ratios, which have fixed lower and upper bounds; and counted fractions, which are ratios formed from integers in a small range. The quality of information gained from analysis of a data set depends to some extent on the match between the data category and the analysis method.

Experimental research on optimization algorithms provides ample opportunities for adjusting the data category to fit the analysis technique. For example, the researcher usually has a choice about whether to report solution quality as a count (e.g., the size of the max clique found) or as a ratio (perhaps max clique size/number of nodes, or max clique size/optimal clique size). The (hypothetical) fact that one quantity grows with problem size, and the other converges to a constant from above, in combination with variance properties, can make a difference in the suitability of these data sets for curve-fitting analysis—as a general rule, functions that grow are easier to analyze than functions that converge. Tukey (1977) is credited with creating the field of exploratory data analysis.

The selection of sample points can also make a difference in the quality of data analysis. For example, some graphical analysis techniques begin by taking logarithms of the $\langle x, y \rangle$ data points. When x represents problem size n, the data analysis will be easier if problem sizes increase by a constant factor (e.g., multiples of two) rather than by a constant increment, so that the log data will appear evenly spaced on the graph.

In experimental studies on randomized optimization algorithms, or those using randomized input instances, where the goal is to estimate or compare average performance, large variance in the data set can present difficulties. Two basic strategies exist for overcoming the variance problem. The first is to obtain more data points by running more experiments at the sample point—roughly, variance is inversely proportional to the square root of the number of points in the sample. Fishman and Rubin (1992) describe a method for estimating bounds on variance in order to determine good sample sizes.

The second is to apply variance reduction techniques (VRTs). In the context of experimental research on algorithms, a VRT is a strategy for modifying the experimentation program so that the data points generated are ranged more tightly around their means. Note that these techniques should only be applied if the mean is the only measure of interest—it is not appropriate to evaluate the variance or deviation in the data set after a VRT has been applied. But when applicable, VRTs can produce quite dramatic improvements in the quality of information gained from an experiment.

Indeed some VRTs are known informally in the folklore of experimentation on algorithms. For example, the paired differences method is a formalization of the notion that two algorithms are best compared by running them on identical input sets and looking at the differences in matched tests. The batch means method can be applied to an iterative algorithm where the goal is to estimate mean cost at time t: the idea is to form that estimate by averaging costs at times $t - m$ through $t + m$, where $m/2$ is the batch size.

The control variates technique involves subtracting a source of random noise from a performance measure, in order to obtain a new measure that has less variance. This technique has been successfully applied in studies of algorithms for bin packing described in Bentley et al. (1983) [although the technique is not mentioned there—see McGeoch, 1986, 1992 for a discussion]. Johnson et al. (1996) also use control variates in a study of lower bounds on traveling salesman problems.

Technical discussion and examples of variance reduction techniques may be found in most statistical textbooks, such as Bratley et al. (1983). McGeoch (1992) presents a tutorial discussion of VRTs and their application to experimental research on algorithms. Higle (1997) also describes several VRTs and applies them to the evaluation of objective functions in stochastic linear programs. Fishman and Yarberry (1997) present guidelines for application of the batch means technique to optimization problems, and Cheng (1982) does the same for the antithetic variates method.

3.4. *Choosing Data Analysis Tools*

Certainly the statistical literature provides an abundance of tools and techniques for analyzing data—the real problem is knowing which ones to use. Results of experiments on optimization algorithms tend to have certain general properties, such as: extremely large data sets; multi-dimensional point sets relating several parameters; large sample sizes at each sample point (sometimes); data that is known to be causally related by (complex and unknown) mathematical functions; and parameters that are controlled and often evenly spaced in their range. These properties, and the kinds of questions researchers in optimization ask, suggest broad categories of applicable techniques. For example, graphical methods are most appropriate for large and multidimensional data sets. Methods of parametric statistics begin with an assumption about the functional model underlying the data set: when the correct model is known, these methods can be used to generate powerful inferences. Even when the model is not known to be correct, parametric methods can provide useful descriptive characterizations of relationships in data. Methods of exploratory data analysis and nonparametric statistics, which do not start with an assumed functional relationship, are most appropriate to the model-building style of experimentation, where the primary goal is to find the right model.

Fitting a correct model to a set of data is a very difficult and error-prone task, even in simple scenarios. Figure 1, for example, shows data from a hypothetical algorithm, plotting cost c against problem size n. Suppose it is known (it is) that this data comes from a random function that belongs in either $\Theta(n \log n)$ or $\Theta(n^2)$: how would you decide which one applies?

Two regression curves are superimposed on the graph, showing the best least squares fits found for the two simple models, corresponding to $c = 0.1165251n \ln n - 48.5838$, and $c = 0.001136n^2 + 23.274$. Neither fit is particularly great, although the n^2 model looks a little better; a comparison of the curve of the data to the curves in the fits suggests that both functional forms underestimate the growth rate of the data.

Other data analysis strategies might be applied. For example, an application of the power rule appears in Figure 2. A log-log plot of the data is shown with a linear least squares fit to the higher-value data points. The slope of this line suggests that c grows approximately as $n^{1.85}$. Other analysis strategies (such as plotting n vs $c/n \ln n$ and n vs c/n^2 and looking for convergence) are similarly inconclusive.

In fact, the function is $c = n \ln n - 6n + 200 + \epsilon$, where ϵ is a random uniform variate. Here the negative second order term was sufficiently large to subvert these data analysis techniques, causing them to produce inconclusive and even misleading results. The obvious solution in this case is to increase the maximum problem size so that the leading term is able to dominate; unfortunately that solution is not always available to a researcher studying optimization algorithms. McGeoch et al. (1997) compare several strategies for finding bounds on functional growth in data, with data sets drawn in part from typical algorithmic problems.

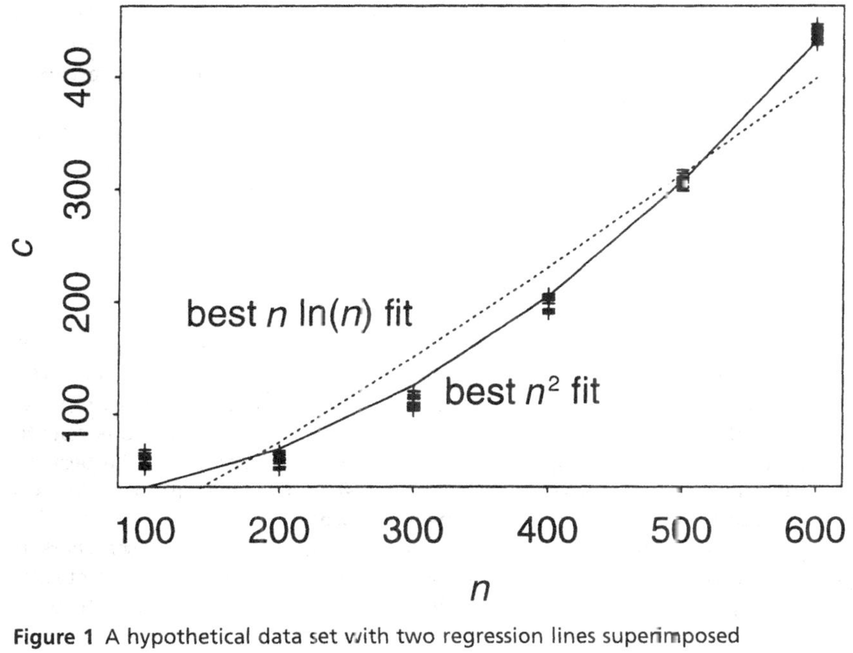

Figure 1 A hypothetical data set with two regression lines superimposed

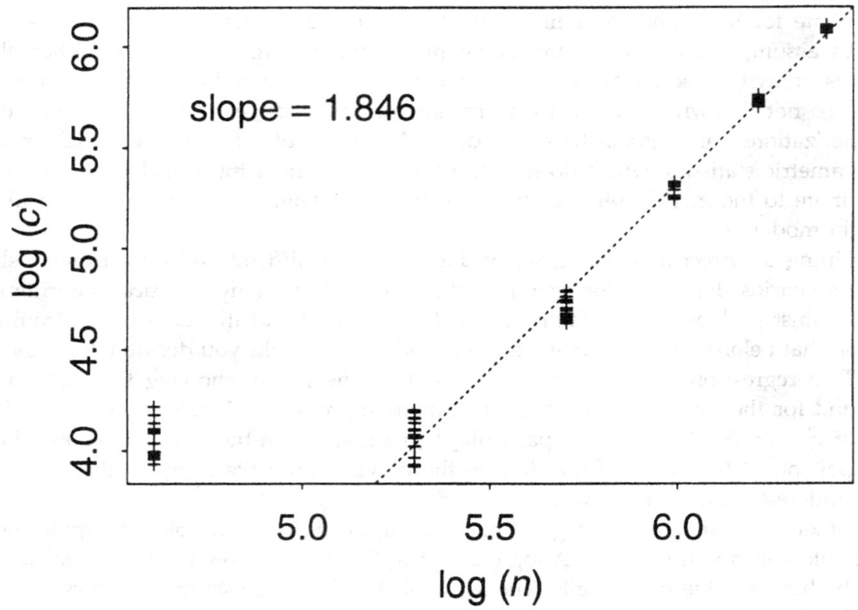

Figure 2 The power law applied to the data set

For excellent and readable textbook presentations of data analysis techniques that are most suitable to research on optimization algorithms, see Box et al. (1978), Bratley et al. (1983), Chambers et al. (1983), Cleveland (1993), Cohen (1995), and Tukey (1977).

4. Presenting Experimental Results

Crowder et al. (1978) present a discussion of standards for reporting results of computational experiments in mathematical programming that stand the tests of time. Their checklist for referees of computational papers is especially useful. A later report by Jackson et al. (1991) reinforces their conclusions and provides new information about issues arising in modern computational research, such as reporting work on distributed systems and advanced architectures, and dealing with proprietary software. Barr et al. (1995) also provide excellent guidelines for reporting results of experiments on heuristic algorithms. Johnson (1996) discusses guidelines for presenting sufficient explanations, measurements, and information to justify the conclusions in the text.

5. Concluding Remarks

Research on the methodology of computational experimentation is in a state of rapid development. Whether the goals are to compare algorithms and implementations, or to build models of underlying mechanisms, we desire experiments that are rigorous, replicable, insight-giving, newsworthy, and correct. The foundation has been laid for rigorous scientific research in this area, but many methodological questions are not yet answered. The best way to move forward is to continue to carry out experimental research on algorithms of all kinds, and to promote thoughtful reflection on the experimental process.

To that end, this chapter surveys general references of use to researchers undertaking computational research on optimization algorithms; highlights some interesting research directions; and provides some basic guidelines for high-quality research on algorithmic problems.

References

Ahuja, R. K., T. L. Magnanti, and J. B. Orlin 1993. In *Network Flows: Theory, Algorithms and Applications*. Chapter 18. Prentice Hall.

Anderson, R. J. 1997. "The role of experiment in the theory of algorithms." Position paper. To appear in *Proceedings of the Fifth DIMACS Challenge*. Available at www.cs.washington.edu/homes/anderson/papers/method.ps.

Barr, R. S., B. L. Golden, J. P. Kelly, M. G. C. Resende, and W. R. Stewart. 1995. "Designing and reporting on computational experiments with heuristic methods." *J. Heuristics* **1(1)**: 9–32.

Bentley, J. L. 1982. In *Writing Efficient Programs*. Englewood Cliffs, NJ: Prentice-Hall.

Bentley, J. L. 1986. In *Programming Pearls*. Reading, MA: Addison-Wesley.

Bentley, J. L. January 1990. "Experiments on traveling salesman heuristics." In *Proceedings of the 1st Annual ACM–SIAM Symposium on Discrete Algorithms*, pp. 91–99.

Bentley, J. L. 1991. "Tools for experiments on algorithms." In *CMU Computer Science: A 25th Anniversary Commemorative*, edited by R. F. Rashid, pp. 99–123. ACM Press.

Bentley, J. L., D. S. Johnson, T. Leighton, and C. C. McGeoch. October 1983. "An experimental study of bin packing." In *Proceedings of 21st Annual Allerton Conf. on Communication, Control, and Computing*, pp 51–60.

Box, G. E. P., W. G. Hunter, and J. S. Hunter. 1978. In *Statistics for Experimenters* John Wiley and Sons.

Bratley, P., B. L. Fox, and L. E. Schrage. 1983. In *A Guide to Simulation*. Berlin: Springer-Verlag.

Chambers, J. M., W. S. Cleveland, B. Kleiner, and P. A. Tukey. 1983. In *Graphical Methods for Data Analysis* Boston: Duxbury Press.

Cheng, R. C. H. 1982. "The use of antithetic variates in computer simulation." *J. Oper. Res. Soc.* **33**: 229–237.

Cleveland, W. S. 1993. In *Visualizing Data*. Summit, NJ: Hobart Press.

Cohen, P. R. 1995. In *Empirical Methods for Artificial Intelligence*. Cambridge, MA: The MIT Press.

Crowder, H. P., R. S. Dembo, and J. M. Mulvey. 1978. "Reporting computational experiments in mathematical programming." *Math. Programming* **15**: 361–329.

Fishman, G. S. and D. S. Rubin. 1992. "Bounding the variance in Monte Carlo experiments." *Oper. Res. Lett* **11**: 243–248.

Fishman, G. S. and L. S. Yarberry. 1997. "An implementation of the batch means method." *INFORMS J Comput.* **9(3)**: 296–310.

Fredman, M. L., D. S. Johnson, L. A. McGeoch, and G. Ostheihmer. 1995. "Data structures for travelling salesmen." *J. Algorith.* **18(3)**: 432–479.

Gent, I. P. and T. Walsh. 1994. "How not to do it." Manuscript.

Golden, B. L., A. A. Assad, E. A. Wasil, and E. Baker. 1986. "Experimentation in optimization." *Eur. J Oper. Res.* **27**: 1–16.

Greenberg, H. J. 1990. "Computational testing: why, how, and how much." *ORSA J. Comput.* **2(1)**: 94–97.

Higle, J. L. 1997. "Variance reduction and objective function evaluation in stochastic linear programs." *INFORMS J. Comput.* **10(2)**: 236–247.

Hoaglin, D. C. and D. F. Andrews. 1980. "The reporting of computation-based results in statistics." *Am. Stat.* **29**: 122–126.

Hooker, J. N. 1994. "Needed: an empirical science of algorithms." *Oper. Res.* **42**: 201–212.

Hooker, J. N. 1996. "Testing heuristics: we have it all wrong." *J. Heuristics* **1**: 33–42.

Jackson, R. H. F., P. T. Boggs, S. G. Nash, and S. Powell. 1991. "Guidelines for reporting results of computational experiments. Report of the ad hoc committee." *Math. Programming* **49**: 413–425.

Johnson, D. S. 1996. "A theoretician's guide to the experimental analysis of algorithms." To appear in *Proceedings of the Fifth DIMACS Challenge*. Available at http://www.research.att.com/dsj/papers/exper.ps.

Johnson, D. S. and C. C. McGeoch. (eds). 1991. In *Network Flows and Matching: Proceedings of the First DIMACS Implementation Challenge*, Volume 12, DIMACS Series in Discrete Mathematics and Theoretical Computer Science. American Mathematical Society.

Johnson, D. S. and M. A. Trick. (eds). 1993. In *Cliques, Coloring and Satisfiability: Proceedings of the Second DIMACS Implementation Challenge*, Volume 26, DIMACS Series in Discrete Mathematics and Theoretical Computer Science. American Mathematical Society.

Johnson, D. S., L. A. McGeoch, and E. E. Rothberg. 1996. "Asymptotic experimental analysis for the Held-Karp traveling salesman bound." In *Proceedings of the 7th Annual ACM–SIAM Symposium on Discrete Algorithms*, pp. 341–350.

Knuth, D. E. 1993. In *The Stanford GraphBase*. ACM Press.

LaMarca, A. and R. E. Ladner. 1996. "The influence of caches on the performance of heaps." *ACM J. Exp. Algorith., 1, Article 4*. Also available at www. jea.acm.org/1996/LaMarcaInfluence.

Lin, B. W. and R. L. Rardin. 1980. "Controlled experimental design for statistical comparison of integer programming algorithms." *Mgmnt Sci.*, pp. 1255–1271.

McGeoch, C. C. 1986. "Experimental analysis of algorithms." Ph.D. diss. Technical Report CMU-CS-87-124, Carnegie Mellon University, Department of Computer Science.

McGeoch, C. C. 1992. "Analyzing algorithms by simulation: variance reduction techniques and simulation speedups." *ACM Comput. Surveys* **245(2)**: 195–212.

McGeoch, C. C. 1996. "Toward an experimental method for algorithm simulation." *INFORMS J. Comput.* **1(1)**: 1–15.

McGeoch, C. C., P. Cohen, and D. Precup. August 1997. "How to find big-oh in your data set (and how not to)." Presented at Second International Symposium on Intelligent Data Analysis (IDA-97), London.

Nance, R. E., R. L. Moose, Jr., and R. V. Foutz. 1987. "A statistical technique for comparing heuristics: an example from capacity assignment strategies in computer network design." *Commun. ACM* **30(5)**: 430–442.

Sanders, P. 1999. "Fast priority queues for cached memory." To appear in *Proceedings of the Workshop on Algorithm Engineering and Experimentation (ALENEX)*, edited by M. J. Goodrich and C. C. McGeoch, pp. 321–327. Berlin: Springer-Verlag.

Tukey, J. W. 1977. In *Exploratory Data Analysis*. Addison-Wesley.

25 OBJECT-ORIENTED PROGRAMMING

Andreas Fink, Stefan Voß, and David L. Woodruff

ABSTRACT The fundamental idea of object-oriented program design is to embody relevant abstractions of the application domain as encapsulated, yet adaptable, classes. Classes categorize objects of the same type with respect to common data structure and behavior. Variability of type can be represented by means of inheritance, where specialized classes are derived from general classes representing commonality. The object-oriented software construction method provides the ability to flexibly produce modularized large-scale programs that are correct and employ reusable components. Prominent practically oriented object-oriented programming languages are C++, Eiffel, Java, and Smalltalk.

1. Introduction

We use the words computer programming to refer to the task of creating software systems aimed at supporting computer users. Programming languages that support the object-oriented style of programming allow the programmer to represent concepts in a more human-oriented way. This is accomplished by creating software components called objects that correspond to entities of interest to the user. When compared with traditional programming languages, object-oriented programming enables improvements in the ability to produce large-scale programs that are correct and employ reusable components.

Ideally, software systems might be built by selecting, possibly adapting, and combining appropriate components from a large set of existing components (McIlroy, 1976). The need for reusability comes from the observation that software systems often follow similar patterns. It should be possible to exploit this commonality and avoid reinventing solutions to problems that have been encountered before (Coplien et al., 1998). By capturing such patterns, reusable and adaptable software components can serve for the construction of different applications. Object-oriented software technology provides a way to build reusable software, but does not guarantee reusability *per se*. Especially, the reuse of algorithms may still be regarded as "a challenge to object-oriented programming" (Weihe, 1997).

Building large-scale software systems generally requires an incremental system construction by means of respective analysis and design models, and eventually code in a programming language. The object-oriented approach promises a consistent system design and implementation through common basic mechanisms. Meyer (1997) regards modularization as the primary condition for reusable and adaptable systems. Modularization should allow decomposition of a complex problem domain into less complex subdomains, which may be independently addressed (divide-and-conquer). Respective modules should be both self-contained as much as possible and adaptable by appropriate robust techniques. The structured programming concept, which underlies classic imperative (procedural) programming languages such as FORTRAN, ALGOL, Pascal, and C, is connected with a top-down functional decomposition of the problem. This approach generally lacks robust techniques to build autonomous and adaptable modules, from which extensible large-scale software systems may easily be composed. On the other hand, an object-based approach, which combines classic top-down refinement with bottom-up elements, offers a promising way to manage the complexity in building extensible large-scale software systems.

In the following, we describe the fundamental concepts of object-oriented programming, followed by a brief discussion of the most prominent object-oriented programming languages. The object-oriented paradigm focuses on abstractions of objects or types of objects. An object is any distinguishable thing that may be comprehended intellectually, whether of material form or a pure cognitive artifact. Every object-oriented programming language provides some kind of mechanism to handle abstractions of objects. This includes constructs to classify objects of the same type in classes, and to factor out commonalities between objects of similar type by means of generalized and specialized classes. Objects of corresponding class hierarchies represent polymorphic abstractions of the application domain. There is some variety in the object-oriented terminology (and in the actual definition of object-orientation as well); we primarily use the mainstream conventions of C++ and Java. In-depth discussions of object-oriented software construction are given by Meyer (1997) and Booch (1994); for a more theoretical treatment see Cardelli and Wegner (1985).

2. Classes and Objects

The fundamental idea of object-oriented program design is to embody relevant abstractions of the application domain as classes (object types). The class concept is based on the key idea of abstract data types. An abstract data type is a formal description of a set of objects by a specification of common functions that operate on these objects. The only way to access an object is by use of the interface functions provided. This enforces the principle of information hiding. The fact that an abstract data type encapsulates the state (data) of an object facilitates building correct and adaptable software modules (cf. Hoare, 1972; Parnas, 1972a,b).

A class is a definition of an object type. Objects (instances of a class) have

- a state, which is defined by its data,
- behavior, defined by its methods (synonyms: functions, operations, procedures, routines) that operate on that data, and
- an identity (e.g., a variable name or a position in a named list).

That is, classes define objects of the same semantics. Programming is primarily done by specifying the construction of objects, which can be regarded as variables, that send messages to one another (call member functions of other objects) to cooperatively accomplish tasks.

As a classic example, consider a simple graphics software system designed to draw circles on screens. There are two fundamental concepts of the application domain, screens and circles, which lead to two corresponding classes. A circle is characterized by its center position, color, and radius (state/data). The functionality of a circle class may include methods to move it, to set a new color, to set a new radius, and to draw it. A screen includes a vector of references to circles positioned on that screen. There are methods to add a new circle to the screen and to redraw the screen; the latter method essentially delegates the actual drawing to the corresponding circle objects. This simple object-oriented model is depicted in Figure 1, using the notation of the unified modeling language (UML), a graphical language for visualizing, specifying, constructing, and documenting the artifacts of software-intensive systems (Booch et al., 1999). Classes are represented by rectangles showing its name and, possibly, its data and method interfaces. The line connecting both classes depicts a structural relationship: an object of the screen class is associated with objects of the circle class that are positioned on the screen. Two more classes, point and color, represent other user-defined types used in the model.

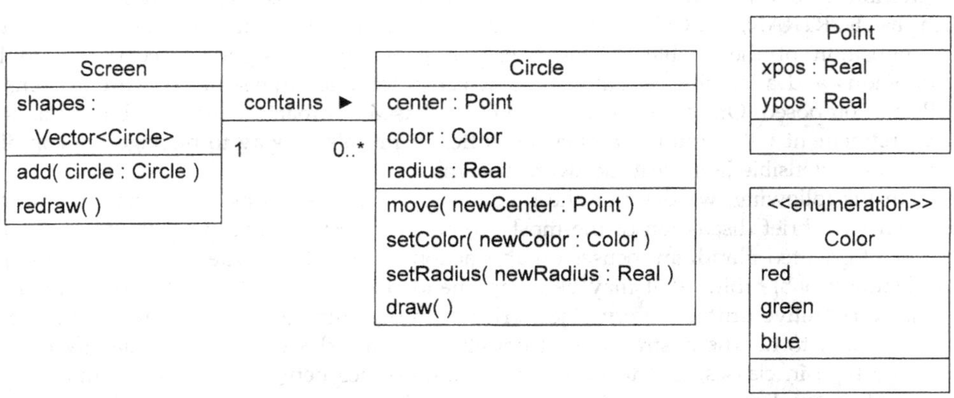

Figure 1 A simple class diagram

Object-based programming languages allow the definition of user-defined types, which enables a natural way of specifying problem-solving instructions to the computer. With a pure object-oriented system design the system dynamics are representable by the interaction of objects via method calls. A corresponding specification of the possibilities of this interaction requires an agreement between classes regarding their rights and obligations such as pre- and post-conditions of a method call and class invariants, which express semantic properties of instances of a class. Such a design-by-contract facilitates the development of reliable software (Helm et al., 1990, Meyer, 1997).

3. Inheritance and Polymorphism

Obviously, there are objects with substantial commonalities but some differences of type. That is, these objects have a similar but not an identical type. The distinguishing characteristic of object-oriented design is the concept of inheritance: grouping of common structure and behavior in general classes, from which special subclasses can be derived, which add or modify specific structure and behavior. In general, inheritance refers to a relation between classes. Inheriting classes automatically embrace the data and methods of the general class (superclass), which provides an extension mechanism for classes.

The traditional object-oriented style of programming is based on applying inheritance. Reconsidering the example introduced above, one may wish to draw rectangles on the screen as well. Circles and rectangles share attributes such as position and color, yet differ in the attributes radius versus length and width. The method to set a new color may be identical for both shape types, the method to draw it may exist for both but must be implemented differently, while there are type-specific methods such as setting a new radius versus setting a new width or height. Such semantics may be embodied in only one class by the use of type fields and extended methods that embody respective algorithms in a complicated if-then-else way. Such a design precludes adaptations and extensions without modifying its definition and implementation, which leads to severe inflexibility. The inheritance concept enables a way out of this problem by explicitly defining commonalities in general classes from which specializing classes, which capture the variability of the domain, may inherit. In general, a class may inherit from more than one superclass (multiple inheritance).

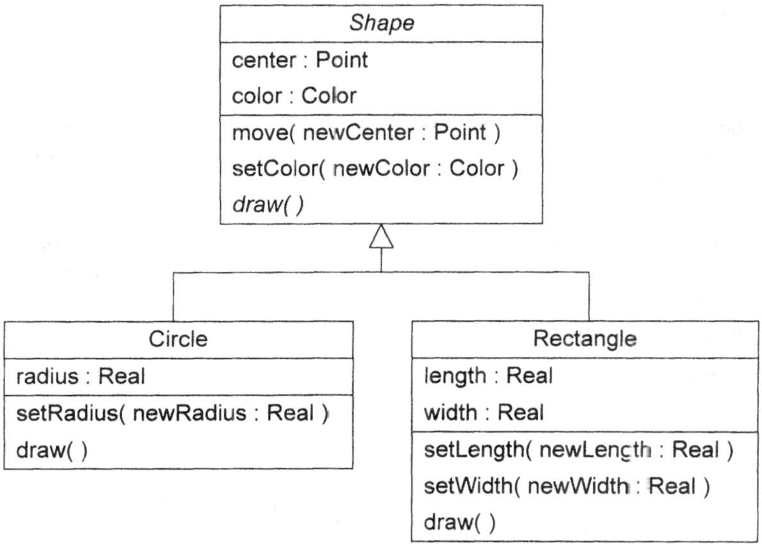

Figure 2 Inheritance

Considering the example, it seems reasonable to define a general shape class that includes the data and methods that all shape types have in common. Specific data and behavior of circles and rectangles is represented by two classes, which inherit from the general shape class. Figure 2 shows a corresponding design model.

Classes that are incomplete in some sense, so that no objects of such classes can be instantiated, are called abstract. For example, the shape class shown in Figure 2 is abstract, indicated by the italicized name. In particular, its draw method is abstract, which means that the interface (signature) is specified while an implementation of the method is deferred to concrete inheriting classes. With that, one can use objects of classes derived from some base class by relying only on the behavioral semantics specified in an abstract base class. This concept enables an important kind of extensibility as one may add new application-specific behavior when corresponding general abstractions already exist. In a properly designed object-oriented system, every object of some class A must be substitutable by an object of every class B that inherits from A (Liskov principle). The corresponding ability of objects to appear at run-time in many concrete forms from an inheritance hierarchy is called polymorphism. That is, the concrete behavior (implementation of methods) of an object is determined at run-time dependent on the type of the object. In this sense, inheritance leads to the view of classes as polymorphic implementations of abstract data types. This type of polymorphism, which is generally considered to be a requirement of any true object-oriented programming language, is sometimes called dynamic polymorphism. On the other hand, static polymorphism refers to compile-time concepts such as overloading and generic classes (parameterized types).

While the inheritance hierarchy shown in the example seems rather straightforward, in general, the design of reasonable class hierarchies is nontrivial. This can be illustrated by considering the design of a class hierarchy to represent the concept of constrained optimization problems. One might begin by observing that problems with a linear objective function and linear constraints (LP) are a special case of those with potentially nonlinear attributes (NLP). We would then define a class for NLP and derive a class for LP from it. The fact that the data representations for LPs and NLPs are typically very different is not a source of difficulty since the structure of the data is typically not part of the public interface of a class. A bigger problem is the public interface itself. A canonical form of an NLP is

$$\text{(NLP)} \quad \text{minimize } f_0(x)$$

subject to

$$f_i(x) = 0 \qquad i = 1, \ldots, s$$

$$f_i(x) \geqslant 0 \quad i = s+1, \ldots, s+t,$$

where $f_i(\cdot)$ are functions from $\mathcal{R}^n$ to $\mathcal{R}$ for $i = 0, 1, \ldots, s+t$, and the decision variable, x is intended to be a vector of length n. One can design a class that stores information about the functions f_i or else pointers to the functions. A reasonable public interface might allow access to the functions. This might work well for NLPs, but this is not the way one typically thinks of an LP and not the way one would design the interface to objects that represent an LP. Although LPs are a special case of NLPs in a mathematical sense, that is not really the way most people think of them. The objective function is specified and referenced as a vector of n coefficients and the constraints are given as a matrix with m rows and n columns and a vector of length m as in

$$\text{(LP)} \quad \text{minimize } cx$$

subject to

$$Ax \leqslant b.$$

Although the rows of the matrix can be considered to be linear functions, specifying and accessing them as a function is not common.

Perhaps a better approach is to define a class for constrained optimization with a very simple interface as shown in Figure 3. The classes NLP and LP are derived from this class and therefore inherit its interface. Programmers who wish to write code that interacts with a constrained optimization program of either type can write their programs so that they make use only of the interface provided by the ConstrainedOptimizationProblem class and be assured that both LP and NLP objects will provide implementations of the methods named objectiveFunction and isFeasible. Of course, an actual design would have more access functions specified. This design has pros and cons, which supports our contention that the design of object hierarchies can be challenging.

4. Object-Oriented Software Construction

While the fundamental techniques of object-oriented design and implementation are easily obtainable, a reasonable application of these techniques cannot be reduced to some simple rules. Identifying objects and classes, describing their relationships, finding commonality and understanding variability in an applications domain, and transferring these in a good object-oriented design is a demanding task. Nevertheless, at least there are some object-oriented design heuristics which guide the developer in a mostly informal way to good design (Riel, 1996). Moreover, design patterns capture the experience of expert software developers in proven general solutions to recurring problems (Gamma et al., 1995; Buschmann et al., 1996). Corresponding knowledge is an indispensable complement to methods that structure the process of object-oriented software construction.

With respect to extensibility and adaptability, Meyer (1997) stresses two principles. First, a simple design will always be easier to adapt to changes than a complex one. This may cause a restriction of the functionality and the generality of the software. Second, the more autonomous the classes, the higher the likelihood that a simple change will affect just one class, or a small number of classes, rather than triggering a chain reaction of changes over the whole system. That is, decentralization simplifies extension and adaptations of such software systems. In this respect, the object-oriented paradigm may be regarded as a system architecture method that helps designers to produce large systems whose structure remains both simple (even for large systems) and decentralized.

Class libraries support the development of application-specific software systems by providing a collection of (possibly semifinished) classes intended to be reused. The approach of building software by using class libraries corresponds to the basic idea of object-oriented

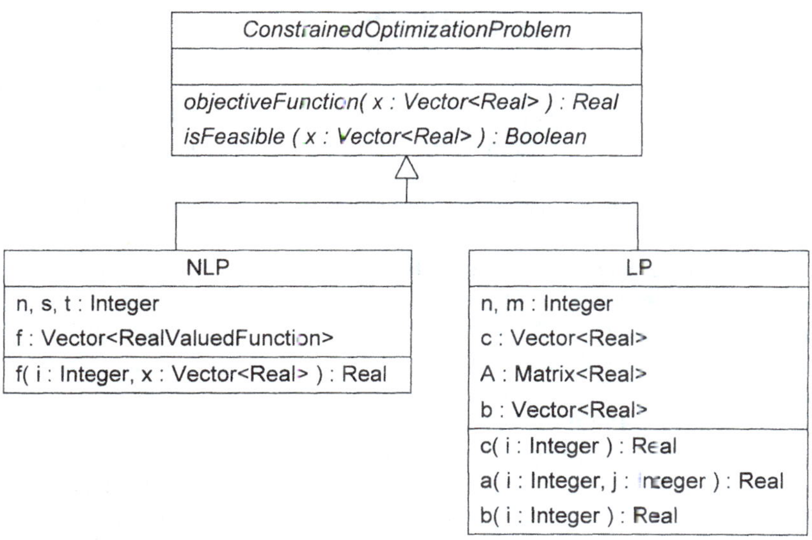

Figure 3 One hierarchy for constrained optimization

software construction, which may be defined as building software systems as "structured collections of possibly partial abstract data type implementations" (Meyer, 1997). Class libraries are mostly based on dynamic polymorphism by factoring out common behavior in general classes and providing the specialized functionality needed by subclassing (inheritance). Genericity, which enables one to leave certain types and values unspecified until the code is actually instantiated and used (compiled) is another way—applicable orthogonal to inheritance—to define polymorphic classes. For a discussion of some class libraries for optimization, see Chapter 22.

One approach primarily devoted to the goal of achieving a higher degree of reuse is the framework approach (Fayad et al., 1999). Taking into account that for the development of application systems for given domains quite similar software is needed, it is reasonable to implement such common aspects by a generic design and embedded reusable software components. Here, one assumes that reuse on a large scale cannot only be based on individual components, but there has, to a certain extent, to be a reuse of comprehensible design. Thus, the components have to be embedded in a corresponding architecture, which defines the collaboration between the components. A framework may be defined as a set of classes that embody an abstract design for solutions to a family of related problems; in this sense, frameworks may be described as reference application architectures (skeletons) or software families. A framework provides abstract applications in a particular domain, that may be tailored for individual applications. That is, frameworks provide not only reusable software elements but also some type of reuse of architecture and design patterns, which may simplify software development considerably. Frameworks represent implementation-oriented generic models for specific domains. There is no clear dividing line between class libraries and frameworks. Whereas class libraries may be more flexible, frameworks often impose a broader structure on the whole system. Frameworks, sometimes termed as component libraries, may be gradually differentiated from class libraries by the activeness of components, that is, components of the framework define application logic and call application-specific code. This generally results in a bidirectional flow of control.

A variety of protocols have been established that carry this one step further. Entire programs are treated as objects by the industry-wide CORBA standard (Szyperski, 1997), which specifies the mechanisms for named objects to access and alter the states of other objects via previously defined methods. Microsoft standards such as COM+ and .NET have a similar function. Components may be developed in different programming languages and even run in a distributed, heterogeneous environment. A possible definition of a (software) component, as given by Szyperski (1997, p. 34), reads as follows: "A software component is a unit of composition with contractually specified interfaces and explicit context dependencies only. A software component can be deployed independently and is subject to composition by third parties." That is, a component-oriented software construction in this sense differs from object-oriented program design especially because of its focus on directly executable (binary) software modules.

A successful construction of complex large-scale software systems requires a structured process that ties together the concepts discussed so far. The development process is commonly divided in an analysis stage (with focus on understanding the problem to be solved), design stage (creating a structured abstract solution), and implementation stage (actually writing and testing code). There are several concrete process models that focus on the object-oriented paradigm. For example, UML-based methodologies are described by Jacobson et al. (1999) and D'Souza and Wills (1999). In general these methodologies describe an iterative and incremental developmental process, which is commonly regarded as favorable to a one-time sequential (waterfall-like) process through the steps of system analysis, system design, and system implementation. The application of a methodology should be supported by the use of CASE (computer aided software engineering) tools.

5. Object-Oriented Programming Languages

The general way of thinking recognizes a programming language as object-oriented if it provides direct support for some kind of objects and classes in connection with inheritance mechanisms that enable some kind of dynamic (run-time) polymorphism. While the pioneering object-oriented language, Simula (Nygaard and Dahl, 1981), was already designed in the 1960s, widespread practical usage of object-oriented programming did not occur until the 1990s. The most prominent practically oriented object-oriented programming languages are C++ (Stroustrup, 1997), Eiffel (Meyer, 1997), Java (Arnold and Gosling, 1997), and Smalltalk (Goldberg and Robson, 1989).

In the following, we illustrate object-oriented programming using C++ (Stroustrup, 1997). While C++ is a hybrid language, which supports different programming styles, it was, as a successor of C, especially designed to efficiently support the object-oriented paradigm by offering the fundamental constructs of a class and inheritance. Moreover, C++ has been shown to efficiently support a multiparadigm design, especially including generic programming by exploiting the template construct, which favors its use in the field of applied optimization (cf. Barton and Nackman, 1994; Coplien, 1999).

The class and inheritance mechanism of C++ may be used as follows to implement some parts of the graphics system example discussed above.

```
class Shape
{
  public:
    Shape(Point firstCenter, Color firstColor)
      : center(firstCenter), color(firstColor) { }
    void move(Point newCenter) {center = newCenter;}
    void setColor(Color newColor) {color = newColor;}
    virtual void draw( ) = 0;
  protected:
    Point center;
    Color color;
};
```

An instantiation of an object of class shape is done using a constructor method, which initializes the center and color to corresponding values. Access authorizations to member data and member functions of a class are defined using the keywords private, protected, and public. Private definitions meet the highest level of encapsulation as only member functions of the corresponding class can access these members. The member data of the shape class (center, color) is defined as protected, which means that only member functions of this and derived classes have direct access to it. Public definitions (member functions move, setColor, and draw) result in no access restrictions at all. Sometimes, two classes are so strongly coupled and designed to work together that granting access to internal structure may be reasonable. Classes declared as friends of a class *A* are granted access to all members of class *A*. However, extensive use of friendship should be avoided as it breaks encapsulation.

The declaration of a member function as virtual, as shown for the draw function, enables polymorphic behavior in this respect. That is, virtual functions define an interface, which can be (re-)implemented in a derived class. The compiler makes sure that the replacement is always called whenever the object in question is actually of the derived class, even if the object is accessed by a base pointer rather than a derived pointer. This allows algorithms of the base class to be replaced in derived classes. A derived class can either fully reimplement the base class member function, or the derived class member function can augment the base class member function by also calling the base class member function. The shape class is an abstract class as it includes a pure virtual function (marked by "=0"), which means that the shape class does not implement this member method. Correspondingly, objects of the shape class cannot be constructed. The inheritance mechanism is illustrated in the following code excerpt.

```
class Circle
: public Shape
{
  public:
    Circle(Point firstCenter, Color firstColor, float firstRadius)
      : Shape(firstCenter, firstColor), radius(firstRadius) { }
    void setRadius(float newRadius) {radius = newRadius; }
    virtual void draw( ) {/* ... code to draw a circle ... */}
  protected:
    float radius;
};
```

The circle class is publicly derived from the shape class, which makes the derived class a subtype of its base class. (Protected and private derivation restrict access to members of the base class and is mostly used to represent implementation details; private inheritance can be regarded as a syntactic variant of a composition relationship.) The circle class includes all members of the shape class and the additionally defined member data radius and the corresponding member function to set the radius. The member function draw defines the circle-specific implementation of the draw interface.

In C++, polymorphic behavior of objects is connected with virtual member functions and pointers (or references). A pointer to an object declared as a pointer to some base class may actually point to a derived object. When some client calls a virtual method through such a pointer, the specific member function to be used is determined at run-time based on the dynamic type of the object (dynamic/late binding). This may be illustrated by the following definition of a screen class, which includes some kind of a vector (from the standard C++-library) of pointers to shape objects.

```
class Screen
{
  public:
    void add(Shape* newShape) {shapes.push_back(newShape);}
    void draw( )
      {
        for(int i = 0; i < shapes.size(); ++i)
          shapes[i] → draw();
      }
  protected:
    vector⟨Shape*⟩ shapes;
};
```

While some other object-oriented programming languages such as Small-talk, a pure object-oriented language that treats all entities as objects, use a more flexible general dynamic member function invocation, C++ requires the programmer to decide about what functions to be defined as virtual. This enables the compiler to efficiently statically type-check and resolve member function invocation for nonvirtual functions. In Smalltalk, the determination of whether a method can be applied to an object is done dynamically at run-time. Because Smalltalk was one of the first object-oriented languages that possessed, from the beginning, generally attractive supporting programming environments, it received considerable attention in the 1980s and early 1990s. For example, the Smalltalk approach makes it relatively easy to prototype and experiment with software systems with visual interfaces. However, its lack of static typing in connection with some deficiencies with respect to run-time efficiency makes it a questionable choice for applied optimization software systems.

Due to its upward-compatibility to C, its multiparadigm supporting constructs, and its strict adjustment to run-time efficiency, C++ is widely regarded as a rather complex language with many subtleties. This view is supported by the long-lasting problems in constructing C++-compilers with ANSI-C++ conformance. Thus, it seemed reasonable to throw away some of this ballast of C++, designing a new programming language taking into account some of the lessons learned, as was done by Java. However, there is some criticism of Java with respect to missing constructs directly supporting genericity, multiple inheritance, and assertions, which might be added in an extended language. The principal contribution of Java is the definition of a virtual universal machine, through its low-level portable bytecode format, on which the execution of Java programs rests. This enables portable software development, however, paying the price of some efficiency problems.

6. Conclusions and Outlook

Object-oriented programming is in no way a panacea (or a "silver bullet," as expressed by Brooks,, 1995), that eventually solves productivity problems in software construction. However, object-oriented concepts are widely regarded as supportive of effective construction of flexible software systems—when embedded in a sensible organization and process environment.

At present, although a great deal of effort is going into improving the languages and programming tools, the more important thrust is toward development of methodologies for object-oriented analysis and design. The successful use of object technologies requires that the lines between analysis, design, and programming are blurred. This is because the object classes and their interfaces must correspond to entities in the problem domain. New analysis and design techniques as well as new ways of viewing the division of labor are evolving in response to the opportunities presented by object-oriented programming.

References

Arnold, K. and J. Gosling. 1997. *The Java Programming Language.* 2d ed. Reading, MA: Addison-Wesley.

Barton, J. J. and L. R. Nackman. 1994. *Scientific and Engineering C++.* Reading, MA: Addison-Wesley.

Booch, G. 1994. *Object-Oriented Analysis and Design: With Applications.* 2d ed. Reading, MA: Addison-Wesley.

Booch, G., J. Rumbaugh, and I. Jacobson. 1999. *The Unified Modeling Language User Guide.* Reading, MA: Addison-Wesley.

Brooks, F. P. 1995. *The Mythical Man-Month: Essays on Software Engineering.* 20th Anniversary ed. Reading, MA: Addison-Wesley.

Buschmann, F., R. Meunier, H. Rohnert, P. Sommerlad, and M. Stal. 1996. *Pattern-oriented Software Architecture.* Chichester, England: Wiley.

Cardelli, L. and P. Wegner. 1985. "On understanding types, data abstraction, and polymorphism." *ACM Comput. Surveys* **17(4)**: 471–522.

Coplien, J. O. 1999. *Multi-Paradigm Design for C++.* Reading, MA: Addison-Wesley.

Coplien, J. O., D. Hoffman, and D. Weiss. 1998. "Commonality and variability in software engineering." *IEEE Softw.* **15(6)**: 37–45.

D'Souza, D. F. and A. C. Wills. 1999. *Objects, Components, and Frameworks with UML.* Reading, MA: Addison-Wesley.

Fayad, M. E., D. C. Schmidt, and R. E. Johnson. 1999. *Building Application Frameworks: Object-oriented Foundations of Framework Design.* Chichester, England: Wiley.

Gamma, E., R. Helm, R. E. Johnson, and J. Vlissides. 1995. *Design Patterns. Elements of Reusable Object-oriented Software.* Reading, MA: Addison-Wesley.

Goldberg, A. and D. Robson. 1989. *Smalltalk 80: The Language.* Reading, MA: Addison-Wesley.

Helm, R., I. M. Holland, and D. Gangopadhyay. 1990. "Contracts: specifying behavioral compositions in object-oriented systems." *ACM SIGPLAN Notices* **25(10)**: 169–180.

Hoare, C. A. R. 1972. "Proof of correctness of data representations." *Acta Inf.* **1**: 271–281.

Jacobson, I., G. Booch, and J. Rumbaugh. 1999. *The Unified Software Development Process*. Reading, MA: Addison-Wesley.

McIlroy, M. D. 1976. "Mass-produced software components." In *Software Engineering Concepts and Techniques (Proceedings of 1968 NATO Conference on Software Engineering)*, edited by J. M. Buxton, P. Naur, and B. Randell, pp. 88–98. New York: Van Nostrand Reinhold.

Meyer, B. 1997. *Object-oriented Software Construction*. 2d ed. Englewood-Cliffs, NJ: Prentice Hall.

Nygaard, K. and O.-J. Dahl. 1981. "The development of the SIMULA languages." In *History of Programming Languages*, edited by R. L. Wexelblat, pp. 439–493. New York: Academic Press.

Parnas, D. L. 1972a. "A technique for software module specification with examples." *Commun. ACM* **15(5)**: 330–336.

Parnas, D. L. 1972b. "On the criteria to be used in decomposing systems into modules." *Commun. ACM* **15(12)**: 1053–1058.

Riel, A. J. 1996. *Object-oriented Design Heuristics*. Reading, MA: Addison-Wesley.

Stroustrup, B. 1997. *The C++ Programming Language*. 3d ed. Reading, England: Addison-Wesley.

Szyperski, C. 1997. *Component Software, Beyond Object-oriented Programming*. Reading, MA: Addison-Wesley.

Weihe, K. 1997. "Reuse of algorithms: still a challenge to object-oriented programming." *ACM SIGPLAN Notices* **32(10)**: 34–48.

26 OPTIMIZATION AND THE INTERNET

Michael A. Trick

ABSTRACT The Internet has created enormous opportunities for optimization researchers. These opportunities can be described in three major areas: increased information transfer, alternative software distribution channels, and the creation of an online optimization community. While past academic and professional organization efforts are now being supplemented by extensive commercial operations, there are still tremendous opportunities for further efforts, particularly in the creation of online, specialized communities.

1. Introduction

There is little argument that the most significant development in computing over the last decade has been the tremendous growth and development of the Internet. In 1989, the Internet was a collection of approximately one hundred thousand machines linking researchers through e-mail, online discussions (USENET), and simple file transfers (like FTP). Ten years later, the Internet consists of millions of machines, and a wide array of communication and transfer programs, with the World Wide Web (WWW or Web) the most ubiquitous. The purpose of this chapter is to give an overview of the world of optimization on the Internet. The Internet has been used in many ways by the optimization community. I will focus on three major uses: the Internet as an information source, the Internet as a distribution channel for optimization implementations, and the Internet as a community. An alternative approach to this topic is offered by Bhargava and Krishnan (1998), who looked at particular technologies and commented on how the technology affects researchers in operations research and management science, with much of interest to the optimization community.

Such an overview is fraught with difficulties, however. Foremost is the sheer mass of information available. As we will see, there are sites that attempt to keep track of the available online resources. The largest such sites list more than one thousand sites relevant to optimization. Clearly it is impossible to survey more than a few selected sites in this chapter. There has been no attempt to be exhaustive in this coverage. Second, Internet sites can be somewhat ephemeral. While some of the sites we will discuss have existed almost since the beginnings of the Web, many others have existed only for a year or two. Some sites that were important three years ago no longer exist, or have not changed in that time.

To combat these difficulties, I have attempted to extract the key types of optimization-oriented Web sites and give some illustrative examples. While the examples may change, the type of information being transmitted will continue.

I will begin by outlining some of the uses of the Internet to provide information (Section 2). These efforts range from extensive efforts to classify available optimization software to optimization tutorials to general guides to the optimization world. Section 3 then examines the use of the Internet as a distribution channel. There are three ways optimization software is made available on the Internet: as downloadable code, as a Java program within a Web page, and as an online optimizer, where the code remains resident on a host computer. Section 4 looks at the growing area of online communities and identifies a number of attempts to create an online optimization community. The final section looks at some future possibilities and opportunities for optimization on the Internet.

2. Information Resources

A common use of the Internet is to provide information about optimization and its algorithms. These sites come in a number of different flavors. Some sites provide information about a particular software implementation or individual's research program. We will see some of these sites in Section 3. These sites provide the raw material for a number of online guides to available software, whose purpose is to collect and classify what is available on the Web. Other sites synthesize research by providing a tutorial or glossary on an optimization topic.

This section describes some of the important guide and tutorial sites on the web and then closes with some other uses of the Web to provide information.

2.1. Guides to the Available Software

There are literally thousands of software packages available in optimization. One key activity is providing guides to the available resources in a usable and intuitive form.

- *Decision tree for optimization software.* (`http://plato.la.asu.edu/guide.html`) by H. D. Mittelman and P. Spellucci organizes available software, with an emphasis on noncommercial software. The software is organized in a tree form, where users can find the appropriate software by working through the type of problem they wish to solve. For instance, the uppermost level of the tree contains global optimization, linear/nonlinear optimization (which emphasizes local optimization), zero (finding zeros of a function), and four other areas. Within linear/nonlinear optimization, the next level is unconstrained, constrained, and special codes for least squares. Constrained leads to linear, quadratic, semidefinite and five other types of constrained optimization. Finally, the linear part gives twenty four pages relevant to linear constrained optimization. This approach makes it very easy to find the software appropriate for any particular problem.
- *Optimization software guide.* (`http://www-fp.mcs.anl.gov/otc/Guide/SoftwareGuide/index.html`) from NEOS (Network Enabled Optimization System). While this listing began with the book by Moré and Wright (1993), it has been updated consistently since then and provides guides to hundreds of software packages. All packages are placed into one of fourteen categories (linear programming, integer programming, etc.) and the authors provide a full page of information on each package giving a description of the applicability and availability of each package. This guide also includes a large number of commercial packages.
- *Guide to available mathematical software.* (`http://gams.nist.gov`) from the National Institute of Standards and Technology provides a guide (through both a Java interface and regular Web interface) to more than ten thousand software modules, almost all of which are FORTRAN subroutines. This system also uses a tree structure to guide users to the correct modules, though it lacks the explanatory notes of the Mittelman and Spellucci tree. This system, while more limited than the other two listed in the breadth of software included, goes much deeper in identifying specific modules for use in FORTRAN programs.

2.2. Optimization Tutorials

Another critical use of the Internet is to provide tutorials and information on optimization. This information is aimed at nonspecialists who may need access to optimization information for the first time and need pointers on where and how to look.

- *Linear and nonlinear frequently asked questions.* (`http://www-unix.mcs.anl.gov/otc/Guide/faq`) established by John Gregory and currently maintained by Robert Fourer and the Optimization Technology Center of Argonne National Laboratory and Northwestern University. These pages represent a long-running (since 1993) effort to provide up-to-date and clearly organized answers to the most common questions about linear and nonlinear optimization. The questions range from the most basic "What is linear programming?" to more specialized questions on decomposition of problems, stochastic optimization and so on. The FAQs (as they are known) include pointers to software, textbooks, journals, and other Web systems and are an invaluable first resource in this area.
- *Mathematical programming glossary.* (`http://www.cudenver.edu/~hgreenbe/glossary/glossary.html`) by Harvey Greenberg. Greenberg's glossary addresses perhaps the most common question of nonspecialists and specialists alike: what does this word mean? From abstract program through zigzag phenomenon, this glossary defines

hundreds of terms and, more importantly, cross-references the definitions to other terms that occur in the glossary (making the glossary much easier to use than a printed version would be). Greenberg also provides a short course in linear programming to help people learn the basics (http://www.cudenver.edu/~hgreenbe/courseware/LPshort/intro.html).

- *NEOS guide.* (http://www-fp.mcs.anl.gov/otc/Guide) from NEOS (Network Enabled Optimization System) at the Optimization Technology Center, Argonne National Laboratory and Northwestern University and is described in Czyzyk et al. (1997). This site (which also contains the Software Guide and FAQs mentioned above) contains thumbnail sketches of optimization algorithms through its optimization tree. The site also contains a number of case studies which further illustrate important optimization concepts.
- *OR-notes.* (http://mscmga.ms.ic.ac.uk/jeb/or/contents.html) by J. E. Beasley. These are a series of introductory notes on topics generally in operations research and including many aspects of optimization, including linear and nonlinear optimization, stochastic optimization, and more.

2.3. *Other Uses*

- *Test instances and benchmarks.* One hallmark of science is replicability and continuous advancement. This cannot be done without some consistent repository of benchmark instances. The most extensive collection is OR-Library (http://graph.ms.ic.ac.uk/info.html) by J. E. Beasley (Beasley, 1990). This collection includes instances for many combinatorial optimization problems. Another famous example is TSPLIB, the collection of traveling salesman problems maintained by Gerhard Reinelt (http://www.iwr.uni-heidelberg.de/iwr/comopt/software/TSPLIB95). The pages of Mittelman and Spellucci has an extensive list of test instances (http://plato.la.asu.edu/topics/testcases.html) and benchmarks (http://plato.la.asu.edu/topics/benchm.html).
- *Specialized software listings.* As the number of software packages increases, so does the need for increasingly specialized listings. For instance, SAL (Scientific Application on Linux) has an optimization page (http://SAL.kachinatech.com/B/3/index.shtml) devoted to optimization packages available for the linux operating system.
- *Online journals.* In the past few years, practically every journal published in optimization has created an online version. While the journal contents themselves are typically only available to subscribers, most provide the table of contents and abstracts to any user. Greenberg (http://www.cudenver.edu/~hgreenbe/sites/journals.html) has a particularly extensive listing of optimization journals.

3. Software on the Web

Perhaps the greatest boon of the Internet is the ease in which code and optimization systems can be made available to the broader community. Researchers have used three ways to make their work available. These are through downloadable programs, Java applets, and online optimizers.

3.1. *Downloadable Programs*

The most obvious way to make computer programs available though the Internet is to make the code downloadable. This involves placing either the source code (the program written in C or FORTRAN or other language) or an executable code (created for a particular machine type) in an area of a Web server that is accessible to the public. There is a huge amount of software that is available this way: the software guides given previously have extensive listings of available codes.

The advantage of this form of distribution is the ease with which the developer can make work available and the flexibility the user has in the use of the resulting software. A disadvantage of this is the resulting lack of control the developer has once the software has been made available. Very little commercial code is distributed as source code. In addition, for complicated optimization systems, it is difficult to write programs that will work consistently from machine to machine. This is particularly true when there is a graphical user interface, which typically are nonstandard from system to system. The alternative of providing just an executable version for a particular system leads to a limited audience for the software and clearly makes support of the product (handling upgrades to improved operating systems, for instance) more difficult.

3.2. Java Optimizers

Java is a general computer language that is very well supported on the Web. A program written in Java designed for the web is called an applet. An applet will run in a Web page independent of which operating system the user is using. This allows the developer to write one program that can be run on a wide variety of machines. Furthermore, the developer can choose only to provide an intermediate semicompiled code rather than the full source code, retaining control of the source code of the work. Also, the code can simply be updated by the developer and the user will automatically get the new program the next time the Web page is accessed. Finally, Java comes with a large standard library of graphics routines that make it easy to provide simple graphical user interfaces that will work from system to system.

The primary disadvantage of Java as a distribution method is speed: Java programs are typically interpreted rather than fully compiled. Furthermore, the machine used to run the optimization is the client machine (of the user), which may be a much slower machine than the server machine (of the developer). The result is a program that is much slower than a compiled program. Because of this, Java programs are typically used to illustrate concepts or provide a graphical view of an algorithm rather than provide high-speed stand-alone optimization systems.

Some examples of optimization-based Java applets follow.

- *RIOT (Remote Interactive Optimization Testbed).* (`http://riot.ieor.berkeley.edu/riot`) by Ilan Adler, Ken Goldberg, and Dorit Hochbaum. This provides an extensive list of interesting optimization problems with a graphical interface to problem entry and solution display. Examples include facility layout, engineering design, sports elimination, and many other problems.
- *Pivot tools.* (`http://www.princeton.edu/~rvdb/LPbook/index.html`) by Robert Vanderbei. Created as supplement to the book by Vanderbei (1999), this set of applets is one of a number of linear programming based applets available. Due to the speed of Java applets, these tools tend to concentrate on illustrating algorithms rather than providing serious computational tools. As educational tools, however, these types of applets take much of the drudgery out of learning algorithms.

3.3. Online Optimizers

The final approach to distributing software is not to distribute it at all, but rather make it available on the developer's machine. In this case, a user fills out a Web form describing the problem to be solved and the optimization gets done on the server machine. The optimization can be coded in any language due to the wide availability of programs to take information from the Web and feed it to server-side programs. The advantage of this approach is the absolute control it gives the developer on the use of the software: the developer can even withdraw the program completely knowing there are no copies of the software elsewhere. Software updates are simply a matter of replacing the single copy of the software on the server machine. Furthermore, if the developer has a machine significantly faster than the user, the resulting service may be an appealing way to attract users to the program. Drawbacks of this approach include the possibility of excessive load on the server machine as more users try to access the system and user frustration over being unable to include the program in other systems.

Examples of this approach span the optimization world.

- *NEOS Server.* (`http://www-neos.mcs.anl.gov`) from the Optimization Technology Center of Argonne National Laboratory and Northwestern University. The group that provides the NEOS Guide and the Optimization Software Guide also offers a large selection of optimization codes to be run on the NEOS servers. These include linear and integer programming solvers, a large variety of nonlinear solvers, and specialized solvers for stochastic optimization and specific problem classes. The system currently has more than twenty five solvers available.
- *InterTools.* (`http://rtm.science.unitn.it/intertools/`) from the University of Trento. A collection of discrete optimization problems including satisfiability, maximum clique, and graph partitioning. Results can be received either by e-mail or through the Web.

4. Community

The preceding sites concentrate on what exists: either surveys or computer programs or tutorials. Another group of resources exist in order to bring together a community of those interested in optimization. These sites are typified by constantly changing content and the opportunity for participants to add to the site. The first three sites I list are more operations research sites rather than optimization sites, but they have a strong optimization component.

- *USENET: sci.op-research.* (available through USENET browsers; archives at `http://www.deja.com`) USENET consists of thousands of discussion areas on various topics The optimization community mainly uses the operations research discussion at `sci.op-research`. This group was formed due to the actions of Mohan Sodhi in 1993. This newsgroup gets five–ten messages per day from those interested in all aspects of operations research, with a very strong optimization contingent (the FAQs listed previously are posted on this newsgroup and grew out of common questions sent to the group). Unlike previous sites, there is no administrator for USENET or for `sci.op-research`: any interested person can post any message they like to the newsgroup. This makes the group ideal for quick answers to questions and rapid announcement of the availability of research and software. Currently, the newsgroup has a daily readership in excess of seventeen thousand people, which makes it larger than INFORMS and the British Operational Research Society combined.
- *Michael Trick's Operations Research Page.* (`http://mat.gsia.cmu.edu`) by Michael Trick. This site, while ostensibly about operations research, shows a strong optimization bias. It provides pointers to all aspects of the field and allows people to add their own pointers to the pages. It also includes a discussion area for optimization and other questions.
- *INFORMS Online.* (`http://www.informs.org`) by the Institute for Operations Research and the Management Sciences. This professional organization has a site that creates a community through its discussion area, working paper archive, presentation database, index of member's home pages, and other offerings. Again, while the site is about operations research in general, there is much of interest to the optimization community.
- *Opt-Net.* (`http://optnet.itwm.uni-kl.de/opt-net`) organized by the Special Interest Group in Optimization of the Deutsche Mathematiker-Vereinigung (German Mathematical Society). An electronic community for people interested in optimization that revolves around a weekly mailing list. Members of the community have an entry giving their addresses and other details.
- *Semidefinite programming.* (`http://www.zib.de/helmberg/semidef.html`) by Christoph Helmberg. This site is aimed at the semidefinite programming community and

provides links to working papers, lists of researchers in the area, a calendar of upcoming conferences, and much more. While not interactive at this point, the information makes this a gathering point for people in the semidefinite community.

5. Challenges and Opportunities

There are a tremendous number of other sites of interest to the optimization community. These include journal sites, test instances, conferences, and so on. Most of the major indexing sites provide pointers to the available resources. This chapter can only provide pointers to a representative cross section.

The Internet has given the optimization community a tremendous opportunity. Research and computer code distribution has never been easier. It has also made a huge audience aware of the community's contributions. Through general search engines like those at Lycos (http://www.lycos.com), AltaVista (http://www.altavista.com) and many others, nonspecialists from around the world are finding sites in optimization. The number of such people far outnumbers those normally thought of as the optimization community. For instance, approximately seventeen thousand different people per week visit INFORMS Online, many looking for information on optimization techniques. Over a six-month period, this rises to more than two hundred and fifty thousand different people. This contrasts with the twelve thousand members of INFORMS. These people greatly need the surveys and guides to the Internet that many sites provide. They also need guidance on the use of appropriate optimization codes: these codes are not going to be used just by specialists. This clearly will lead to issues such as interface design, robustness, and reliability, in addition to such traditional algorithmic worries as speed and applicability.

If there is one service that has not yet arisen in the optimization community it is the creation of a true online community. A true community allows participants to identify themselves and participate fully in online activities, which might include discussions, research postings, referral services, and other activities. While some sites have some aspects, no site is currently the meeting point for all in the optimization community.

One interesting possibility is being developed by the software company ILOG. They are creating a site called e-optimization.com(munity) (http://www.e-optimization.com), a vendor-neutral site to bring optimization researchers and companies together with business people who wish to learn more about optimization. This site is too preliminary at this point for review, but might fill this need for a central meeting point.

Despite this (current) lack of a central point, there is a tremendous amount of information available about optimization on the Web, and the amount is increasing at a rapid pace.

List of Relevant Web Sites

Providing full references to Web sites is difficult. Often, it is unclear who is responsible for which aspects of large Web systems. In the following, I attempt to give credit for the sites referred to in the article.

Adler, I., K. Goldberg, and D. Hochbaum. RIOT (Remote Interactive Optimization Testbed). http://riot.ieor.berkeley.edu/riot

Beasley, J. E. OR-Library. http://graph.ms.ic.ac.uk/info.html

Beasley, J. E. OR-Notes. http://mscmga.ms.ic.ac.uk/jeb/or/contents.html

Fourer, R. Linear and Nonlinear Frequently Asked Questions. http://www-unix.mcs.anl.gov/otc/Guide/faq

Greenberg, H. Mathematical Programming Glossary. http://www.cudenver.edu/~hgreenbe/glossary/glossary.html

Greenberg, H. Journals in Optimization. http://www.cudenver.edu/~hgreenbe/sites/journals.html

Helmberg, C. Semidefinite Programming. http://www.zib.de/helmberg/semidef.html

ILOG. e-optimization.com(murity). http://www.e-optimization.com

INFORMS. INFORMS Online. http://www.informs.org

Mittelman, H. D. and P. Spellucci. Benchmarks for Optimization Software. http://plato.la.asu.edu/topics/benchm.html

Mittelman, H. D. and P. Spellucci. Decision Tree for Optimization Software. http://plato.la.asu.edu/guide.html

Mittelman, H. D. and P. Spellucci. Test Cases for Optimization Software. http://plato.la.asu.edu/topics/testcases.html

NEOS (Network Enabled Optimization System). NEOS Guide. http://www-fp.mcs.anl.gov/otc/Guide

NEOS (Network Enabled Optimization System). Optimization Software Guide. http://www-fp.mcs.anl.gov/otc/Guide/SoftwareGuide/index.html

NEOS (Network Enabled Optimization System). NEOS Server. http://www-neos.mcs.anl.gov/

NIST (National Institute of Standards and Technology). Guide to Available Mathematical Software (GAMS) http://gams.nist.gov

Optimization SIG. DMV Opt-Net. http://optnet.itwm.uni-kl.de/opt-net

Reinelt, G. TSPLIB. http://www.iwr.uni-heidelberg.de/iwr/comopt/software/TSPLIB95

SAL (Scientific Application on Linux). Optimization with Linux. http://SAL.kachinatech.com/B/3/index.shtml

Trick, M. A. Michael Trick's Operations Research Page. http://mat.gsia.cmu.edu.

University of Trento. Intertools. http://rtm.science.unitn.it/intertools/.

Vanderbei, R. Pivot Tools. http://www.princeton.edu/~rvdb/LPbook/index.html

References

Beasley, J. 1990. "OR-Library: distributing test problems by electronic mail." *J. Oper. Res. Soc.* **41**: 1069–1072.

Bhargava, H. and R. Krishnan. 1998. "The World Wide Web: opportunities for operations research and management science." *INFORMS J. Comput.* **10(4)**: 359–383.

Czyzyk, J., J. H. Owen, and S. H. Wright. 1997. "Optimization on the Internet." *OR/MS Today*, October.

Moré, J. J. and S. J. Wright. 1993. In *Optimization Software Guide*. Philadelphia SIAM Publications.

Vanderbei, R. 1999. In *Linear Programming: Foundations and Extensions*. New York: Kluwer Academic Publishers.

CONTRIBUTORS

E. H. L. Aarts
Department Head, Philips Research Laboratories; and Professor of Computing Science, Eindhoven University of Technology, Netherlands

Ravindra K. Ahuja
Professor of Industrial and Systems Engineering, University of Florida, Gainesville

G. Anandalingam
Professor of Management Science, R. H. Smith School of Business and Institute for Systems Research, University of Maryland, College Park

John E. Beasley
Senior Lecturer in Operations Research, The Management School, Imperial College, London, United Kingdom

Matthew E. Berge
Technical Fellow, The Boeing Company, Seattle, Washington

John T. Betts
Technical Fellow, The Boeing Company, Seattle, Washington

John R. Birge
Dean, McCormick School of Engineering and Applied Science, Northwestern University, Evanston, Illinois

Rainer E. Burkard
Professor of Mathematics, Technische Universität Graz, Austria

Eranda Çela
Assistant Professor of Applied Mathematics, Institut für Mathematik, Technische Universität Graz, Austria

Yair Censor
Professor of Mathematics, University of Haifa, Israel

Pierre Chardaire
Reader in Computer Science, School of Information Systems, University of East Anglia, United Kingdom

Edward G. Coffman, Jr.
Professor of Electrical Engineering, Columbia University, New York

Edith Cohen
Technology Consultant, AT&T Labs Research, Florham Park, New Jersey

Gerson Couto de Oliveira
Consultant of Power Systems Research, Inc., and Professor of Electrical Engineering, Catholic University of Rio de Janeiro, Brazil

Yves Crama
Professor of Operations Research and Production Management, Ecole d'Administration des Affaires, University of Liège, Belgium

János Csirik
Professor of Computer Science, József Attila University of Szeged, Hungary

Ivan Dimov
Professor, and Director of the Central Laboratory for Parallel Processing, Bulgarian Academy of Sciences, Sofia

Gianni Di Pillo
Professor in Operations Research, Università di Roma "La Sapienza," Italy

Zvi Drezner
Professor of Management Science, California State University, Fullerton

Ding-Zhu Du
Professor of Computer Science, University of Minnesota, Minneapolis; and Research Professor, Institute of Applied Mathematics, Chinese Academy of Sciences, Beijing

José H. Dulá
Associate Professor of Production and Operations Management, School of Business Administration, University of Mississippi

Augustine O. Esogbue
Professor, School of Industrial and Systems Engineering, Georgia Institute of Technology, Atlanta

Hussein A. Y. Etawil
Senior Member of the Technical Staff, Sequence Design Incorporated, Santa Clara, California

Michael C. Ferris
Professor of Computer Sciences and Industrial Engineering, University of Wisconsin, Madison

Sharon K. Filipowski
Mathematics and Modeling Analyst, The Boeing Company, Seattle, Washington

Andreas Fink
Staff Member, Department of Business Administration, Information Systems, and Information Management, Technische Universität Braunschweig, Germany

Michael Florian
Professor of Operations Research, Center for Research on Transportation, University of Montreal, Quebec, Canada

Christodoulos A. Floudas
Professor of Chemical Engineering, Princeton University, New Jersey

Emmanuel Fragnière
Lecturer of Management Science, University of Bath, United Kingdom

John Franco
Geier Professor of Computer Science, University of Cincinnati, Ohio

Manlio Gaudioso
Professor of Operations Research, Università della Calabria, Rende, Italy

E. Michael Gertz
Postdoctoral Researcher, Northwestern University and Argonne National Laboratory, Illinois

John H. Glick
Associate Professor of Mathematics and Computer Science, University of San Diego, California

Fred Glover
MediaOne Professor of Systems Science, School of Business, University of Colorado, Boulder

Bruce L. Golden
Professor and France-Merrick Chair in Management Science, University of Maryland, College Park

Jacek Gondzio
Reader in Computational Optimization, University of Edinburgh, United Kingdom

Sergio Granville
Technical Director of Power Systems Research, Inc., Rio de Janeiro, Brazil

Stephen C. Graves
Abraham J. Siegel Professor of Management Science and Engineering Systems, Massachusetts Institute of Technology, Cambridge

Jun Gu
Professor of Computer Science, Hong Kong University of Science and Technology, Kowloon

Monique Guignard
Professor of Operations and Information Management, The Wharton School; and Professor of Systems Engineering, University of Pennsylvania, Philadelphia

Peter L. Hammer
Professor, Rutcor, Rutgers University, New Brunswick, New Jersey

Pierre Hansen
Professor of Operations Research, École des Hautes Études Commerciales; and Groupe d'études et de recherches en analyse des decisions (GERAD), Montréal, Québec, Canada

Donald W. Hearn
Professor and Chair, Department of Industrial and Systems Engineering, and Co-Director Center for Applied Optimization, University of Florida, Gainesville

John Hof
Chief Economist, Rocky Mountain Research Station, USDA Forest Service, Fort Collins, Colorado

Roland N. Horne
Professor and Chairman, Petroleum Engineering, Stanford University, California

William P. Huffman
Mathematics and Modeling Analyst, The Boeing Company, Seattle, Washington

Toshihide Ibaraki
Professor of Informatics, Kyoto University, Japan

Christian Kanzow
Assistant Professor, Department of Mathematics, University of Hamburg, Germany

Suat Kasap
Graduate Assistant, School of Industrial Engineering, University of Oklahoma, Norman

Dukwon Kim
Member of the Center for Applied Optimization, University of Florida, Gainesville

Manuel Laguna
Associate Professor of Operations Management, Graduate School of Business Administration, University of Colorado, Boulder

Robert C. Leachman
Professor of Industrial Engineering and Operations Research, University of California, Berkeley

Chung-Yee Lee
Professor of Industrial Engineering, Texas A&M University, College Station

Eva K. Lee
Assistant Professor of Radiation Oncology, Emory University School of Medicine; and Assistant Professor of Industrial and Systems Engineering, Georgia Institute of Technology, Atlanta

Abdel Lisser
Head of Research and Development Group, France Telecom, Issy-Moulineaux

Arne Løkketangen
Professor of Informatics, Molde College, Norway

Abilio Lucena
Full Professor, Departamento de Administração, Universidade Federal do Rio de Janeiro, Brazil

Nelson Maculan
Professor of Systems Engineering and Computer Science, Federal University of Rio de Janeiro, Brazil

Thomas L. Magnanti
Dean of Engineering and Institute Professor, Massachusetts Institute of Technology, Cambridge

Philippe Mahey
Professor, Institut Supérieur d'Informatique de Modélisation et de leurs Applications (ISIMA), Université de Clermont-Ferrand II, and LIMOS, Centre National de la Recherche Scientifique (CNRS), Aubière, France

Simone de L. Martins
Assistant Researcher, Bioinformatics Laboratory, Department of Applied Mathematics, The National Laboratory of Scientific Computation, LNCC, Rio de Janeiro, Brazil

S. Thomas McCormick
W. J. Van Dusen Professor of Management, University of British Columbia, Vancouver, Canada

Catherine C. McGeoch
Professor of Computer Science, Amherst College, Massachusetts

John E. Mitchell
Associate Professor of Mathematical Sciences, Rensselaer Polytechnic Institute, Troy, New York

Nenad Mladenović
Higher Scientific Associate, Mathematical Institute, Serbian Academy of Sciences and Arts, Belgrade, Yugoslavia

Pablo Moscato
Research Scientist, Meme Pool Project, Universidade Estadual de Campinas, Brazil

Charles B. Moss
Professor of Food and Resource Economics, University of Florida, Gainesville

Petra Mutzel
Professor for Algorithms and Data Structures, Vienna University of Technology, Austria

Anna Nagurney
John F. Smith Memorial Professor, Department of Finance and Operations Management, Isenberg School of Management, University of Massachusetts, Amherst

Alexandra M. Newman
Assistant Professor of Operations Research, Division of Economics and Business, Colorado School of Mines, Golden

Linda K. Nozick
Associate Professor of Systems Engineering, School of Civil and Environmental Engineering, Cornell University, Ithaca, New York

James B. Orlin
Edward Pennell Brooks Professor of Operations Research, Massachusetts Institute of Technology, Cambridge

Laura Palagi
Researcher in Operations Research, Università di Roma "La Sapienza," Italy

Panos M. Pardalos
Professor and Co-Director, Center for Applied Optimization, Industrial and Systems Engineering Department, University of Florida, Gainesville

Mario Pereira
President of Power Systems Research, Inc., Rio de Janeiro, Brazil

Michael Pinedo
Professor of Operations Management, Stern School of Business, New York University, New York

Leonidas S. Pitsoulis
Associate Professor, Department of Statistics, University of the Aegean, Greece

Marcus Poggi de Aragão
Professor of Computer Science, Catholic University of Rio de Janeiro, Brazil

Kumaraswamy Ponnambalam
Associate Professor of Systems Design Engineering, University of Waterloo, Ontario, Canada

Warren B. Powell
Professor of Operations Research and Financial Engineering, Princeton University, New Jersey

Paul W. Purdom
Professor of Computer Science, Indiana University at Bloomington

Sanguthevar Rajasekaran
Professor of Computer and Information Science and Engineering, University of Florida, Gainesville

Mauricio G. C. Resende
Technology Consultant, Algorithms and Optimization Research Department, AT&T Labs Research, Florham Park, New Jersey

Celso C. Ribeiro
Professor of Computer Science, Catholic University of Rio de Janeiro, Brazil

Roger Z. Ríos-Mercado
Assistant Professor, Systems Engineering Program, Universidad Autónoma de Nuevo Leon San Nicolás de los Garza, Mexico

Noemi Rodriguez
Associate Professor, Computer Science Department, Catholic University of Rio de Janeiro, Brazil

Kees Roos
Professor of Applied Mathematics, Department of Information Technology and Systems, Delft University of Technology, Netherlands

J. B. Rosen
Adjunct Research Professor of Computer Science, University of California, San Diego

Boghos D. Sivazlian
Professor of Industrial and Systems Engineering, University of Florida, Gainesville

G. E. Stavroulakis
Associate Professor of Mechanics, University of Ioannina, Greece; and Privatdozent for Mechanics, Technical University of Braunschweig, Germany

Éric D. Taillard
Director of the Institute for Applied Computer Sciences, École d'Ingénieurs du Canton de Vaud (EIVD), University of Applied Sciences of Western Switzerland, Yverdon

H. M. M. Ten Eikelder
Assistant Professor of Biomedical Engineering, Eindhoven University of Technology, Netherlands

Tamás Terlaky
Professor of Optimization, Department of Computing and Software, Information Technology Center, McMaster University, Hamilton, Ontario, Canada

Benjamin G. Thengvall
Operations Research Scientist, CALEB Technologies Corporation, Austin, Texas

Theodore B. Trafalis
Associate Professor of Industrial Engineering, University of Oklahoma, Norman

Michael A. Trick
Associate Professor of Operations Research, Graduate School of Industrial Administration, Carnegie Mellon University, Pittsburgh, Pennsylvania

Hoang Tuy
Professor of Mathematics, Institute of Mathematics, National Center for Science and Technology, Hanoi, Vietnam

Stanislav Uryasev
Associate Professor, Department of Industrial and Systems Engineering, University of Florida, Gainesville

Andrés Weintraub Pohorille
Professor of Industrial Engineering, University of Chile, Santiago

Anthony Vannelli
Professor and Chair, Department of Electrical and Computer Engineering, University of Waterloo, Ontario, Canada

Stefan Voß
Professor of Business Administration, Information Systems, and Information Management, Technische Universität Braunschweig, Germany

Benjamin W. Wah
R. T. Chien Professor of Electrical and Computer Engineering, University of Illinois at Urbana-Champaign

Edward A. Wasil
Professor of Management Science, Kogod School of Business, American University, Washington, D.C.

Gerhard J. Woeginger
Professor of Mathematics, Technische Universität Graz, Austria

Henry Wolkowicz
Professor of Mathematics, University of Waterloo, Ontario, Canada

David L. Woodruff
Professor of Management, University of California, Davis

Stephen J. Wright
Professor of Computer Sciences, University of Wisconsin-Madison

Guoliang Xue
Professor, Department of Computer Science and Engineering, Arizona State University, Tempe

Mutsunori Yagiura
Lecturer, Graduate School of Informatics, Kyoto University, Japan

Candace Arai Yano
Professor of Industrial Engineering and Operations Research, University of California, Berkeley

Yinyu Ye
Henry B. Tippie Research Professor, Department of Management Sciences, University of Iowa, Iowa City

David P. Young
Technical Fellow, The Boeing Company, Seattle, Washington

Gang Yu
Professor of Management Science and Information Systems, University of Texas at Austin

Stavros A. Zenios
Professor of Management Science, and Dean of the School of Economics and Management, University of Cyprus, Nicosia

Zahari Zlatev
Senior Researcher, Department for Atmospheric Environment, National Environmental Research Institute, Roskilde, Denmark

INDEX

The manufacturer's authorised representative in the EU for product safety is Oxford University Press España S.A. of El Parque Empresarial San Fernando de Henares, Avenida de Castilla, 2 – 28830 Madrid (www.oup.es/en or product.safety@oup.com). OUP España S.A. also acts as importer into Spain of products made by the manufacturer.

Printed in the USA/Agawam, MA
February 13, 2025

882720.013